The Amphibians and Reptiles of Costa Rica

The Amphibians and Reptiles of Costa Rica

A Herpetofauna
between
Two Continents,
between
Two Seas

Jay M. Savage

With Photographs by Michael Fogden and Patricia Fogden

The University of Chicago Press
Chicago and London

JAY M. SAVAGE is emeritus professor of biology at the University
of Miami and adjunct professor of biology at San Diego State
University. He is the author of *Evolution*, coauthor of *Introduction
to the Herpetofauna of Costa Rica*, and editor of *Ecological Aspects
of Development in the Humid Tropics*.

The University of Chicago Press, Chicago 60637
The University of Chicago Press, Ltd., London
© 2002 by The University of Chicago
All rights reserved. Published 2002
Printed in China

11 10 09 08 07 06 05 04 03 02   1 2 3 4 5

ISBN: 0-226-73537-0 (cloth)

The University of Chicago Press gratefully acknowledges the
American Society of Ichythyologists and Herpetologists (ASIH),
the Herpetologists' League, the Instituto Nacional de Bioversidad
(INBio), the Organization for Tropical Studies (OTS), and the Society
for the Study of Amphibians and Reptiles (SSAR) for their generous
contributions toward publication of this work.

Library of Congress Cataloging-in-Publication Data
Savage, Jay Mathers.
   The amphibians and reptiles of Costa Rica : a herpetofauna
between two continents, between two seas / Jay M. Savage ; with
photographs by Michael Fogden and Patricia Fogden.
       p.   cm.
   Includes bibliographical references (p.).
   ISBN 0-226-73537-0 (alk. paper)
   1. Reptiles—Costa Rica.   2. Amphibians—Costa Rica.
I. Title.
QL656.C783 S27 2002
597.9′097286—dc21

                                        2001043717

⊗ The paper used in this publication meets the minimum
requirements of the American National Standard for Information
Sciences—Permanence of Paper for Printed Library Materials,
ANSI Z39.48-1992.

To my friends, the people of Costa Rica

Quedar bien

Contents

List of Keys   ix

Preface   xi

Acknowledgments   xiii

Photographers' Acknowledgments   xvii

Illustration Credits   xix

PART 1 THE BASICS

1   Discovering a Tropical Herpetofauna   3

2   The Costa Rican Environment   60

3   Organization of the Systematic Accounts
(How to Use This Book)   91

PART 2 LIVING AMPHIBIANS

4   Amphibians (Class Amphibia)   105

5   Caecilians (Order Gmynophiona)   113

6   Salamanders (Order Caudata)   121

7   Frogs and Toads (Order Anura)   158

PART 3 LIVING REPTILES

8   Reptiles (Class Reptilia)   409

9   Squamates (Order Squamata)   412

10   Lizards (Suborder Sauria)   414

11   Snakes (Suborder Serpentes)   535

12   Turtles (Order Testudinata)   738

13   Crocodilians (Order Crocodilia)   772

PART 4 BIOGEOGRAPHY AND EVOLUTION

14   Ecological Distribution of the Herpetofauna   783

15   Geographic Distribution: Historical Units, Faunal Areas,
Endemism, and General Patterns   794

16   Development of the Herpetofauna   815

Addendum   839

Glossary   841

Literature Cited   859

Systematic Index   911

Subject Index   927

Keys

Key to the Major Groups within the Costa Rican Herpetofauna   102

Key to the Principal Groups of Costa Rican Anamniote Eggs and Developing Embryos   107

Key to the Genera of Costa Rican Caecilians   116

Key to the Caecilians of Costa Rica   117

Key to the Genera of Costa Rican Salamanders   126

Key to Costa Rican Salamanders of the Genus *Bolitoglossa*   128

Key to Costa Rican Salamanders of the Genus *Nototriton*   140

Key to the Costa Rican Salamanders of the Genus *Oedipina*   147

Key to the Genera of Costa Rican Frogs and Toads   173

Key to the Tadpoles of Costa Rican Anurans   178

Key to the Costa Rican Harlequin Frogs of the Genus *Atelopus*   186

Key to the Costa Rican Toads of the Genera *Bufo* and *Crepidophryne*   193

Key to the Costa Rican Species of the Genus *Leptodactylus*   214

Key to the Costa Rican Rainfrogs of the Genus *Eleutherodactylus*   230

Key to the Costa Rican Subfamilies of the Treefrog Family Hylidae   277

Key to Costa Rican Phyllomedusine Treefrogs (Family Hylidae)   277

Key to Costa Rican Hemiphractine and Hyline Treefrogs (Family Hylidae)   288

Key to the Costa Rican Treefrogs of the Genus *Duellmanohyla*   297

Key to the Costa Rican Treefrogs of the Genus *Scinax*   343

Key to the Costa Rican Treefrogs of the Genus *Smilisca*   349

Key to the Costa Rican Glassfrogs of the Genera *Centrolenella* and *Cochranella*   359

Key to the Costa Rican Glassfrogs of the Genus *Hyalinobatrachium*   367

Key to the Costa Rican Species of the Genus *Colostethus*   377

Key to the Costa Rican Poison Frogs of the Genus *Dendrobates*   382

Key to the Costa Rican Poison Frogs of the Genus *Phyllobates*   389

Key to the Costa Rican Species of the Genus *Rana*   398

Key to the Genera of Costa Rican Lizards   422

Key to the Costa Rican Species of the Genus *Basiliscus*   427

Key to the Costa Rican Species of the Genus *Ctenosaura*   434

Key to the Costa Rican Species of the Genus *Sceloporus*   440

Key to Costa Rican Anoles, Genera *Ctenonotus*, *Dactyloa*, and *Norops*   449

Key to the Costa Rican Geckos of the Genus *Hemidactylus*   483

Key to the Costa Rican Geckos of the Genus *Sphaerodactylus*   492

Key to the Costa Rican Species of the Genus *Lepidophyma*   498

Key to the Costa Rican Species of *Ameiva* and *Cnemidophorus*   507

Key to the Costa Rican Lizards of the genus *Celestus*   527

Key to the Costa Rican Species of the Genus *Diploglossus*   530

Key to the Genera of Costa Rican Snakes   544

Key to New World Genera of Blind Burrowing Snakes   553

Key to Costa Rican Snakes of the Families Loxocemidae, Boidae, and Ungaliophiidae   560

Key to the Costa Rican Species of the Genus *Corallus*   564

Key to the Costa Rican Species of the Genus *Clelia*   573

Key to the Costa Rican Species of the Genus *Erythrolamprus*   578

Key to the Costa Rican Species of the Genus *Enulius*   588

Key to the Costa Rican Species of the Genus *Coniophanes*   592

Key to the Costa Rican Species of the Genus *Dipsas*   596

Key to the Costa Rican Species of the Genus *Geophis*   599

Key to the Costa Rican Species of the Genus *Imantodes*   606

Key to the Costa Rican Species of the Genus *Leptodeira*   610

Key to the Costa Rican Species of the Genus *Ninia*   616

Key to the Costa Rican Species of the Genus *Rhadinaea*   622

Key to the Costa Rican Species of the Genus *Sibon*   628

Key to the Costa Rican Species of the Genus *Trimetopon*   637

Key to the Costa Rican Species of the Genus *Urotheca*   641

Key to the Costa Rican Snakes of the Genus *Chironius*   648

Key to the Costa Rican Species of the Genera *Dendrophidion* and *Drymobius*   653

Key to the Costa Rican Species of the Genus *Leptophis*   668

Key to the Costa Rican Species of the Genus *Oxybelis*   675

Key to the Costa Rican Species of the Genus *Stenorrhina*   688

Key to the Costa Rican Species of the Genus *Tantilla*   690

Key to the Costa Rican Garter Snakes of the Genus *Thamnophis*   700

Key to the American Genera of Elapidae   703

Key to the Coral Snakes of Costa Rica   706

Key to the Costa Rican Species of Venomous Coral Snakes, Genus *Micrurus*   708

Key to the Genera of American Pitvipers (Family Viperidae: Subfamily Crotalinae)   717

Key to the Costa Rican Pitvipers of the Genus *Atropoides*   720

Key to the Costa Rican Pitvipers of the Genus *Bothriechis*   723

Key to the Costa Rican Pitvipers of the Genus *Lachesis*   730

Key to the Costa Rican Pitvipers of the Genus *Porthidium*   732

Key to the Genera of Costa Rican Turtles   742

Key to the Costa Rican Species of Mud Turtles, Genus *Kinosternon*   745

Key to Costa Rican Species of Green Turtles, Genus *Chelonia*   756

Key to the Costa Rican Species of the Genus *Rhinoclemmys*   764

Key to Costa Rican Crocodilians   775

Preface

How did it all begin, my fascination with amphibians and reptiles? I don't know, but the signs were there early on. One of my family's great stories shows that my *destino* was to be a herpetologist. When I was eight we lived in South Africa for a year. As we were about to end our stay, my father decided we should visit Victoria Falls before returning to the United States. En route we stopped over in Bulawayo, Rhodesia (now Zimbabwe).

Among other places of interest my father wanted to visit was the tomb of Cecil Rhodes, set among giant boulders in the Matopo Hills outside the town. The moment we arrived at the site I began chasing bright-colored rainbow lizards *(Agama)* across the boulders, and as they tried to escape I followed them right across the large bronze plaque covering the great British imperialist's grave. Two guards, visibly upset at my transgression, came running after me, but my mother interceded and I was released to her care. This incident was a clear indication of my future career and also as close as I ever got to being a Rhodes scholar.

I do know when my love affair with Costa Rica began. In 1958 Andy and Holly Starrett, who had previously collected in Costa Rica, and I developed a proposal to study the ecological distribution of the country's herpetofauna. The National Science Foundation (NSF) awarded us a grant for fieldwork the following year. However, as tenure-seeking assistant professors at the University of Southern California (USC), we initially could not spend long periods in the field. Consequently we arranged for three graduate students— Arden H. Brame Jr. II, Arnold G. Kluge, and Robert J. Lavenberg *(el trio)*—to undertake field collections. After many adventures driving overland to Costa Rica from Los Angeles, they commenced fieldwork in March 1959. They were later joined by the Starretts and continued collecting until the end of August. Kluge had decided to label and catalog the specimens with the initials CRE, standing for Costa Rica Expeditions, although at the time none of us had any reason to expect there would be more than one expedition. After forty years the CRE collections have now grown to about thirty thousand specimens and are deposited in the Natural History Museum of Los Angeles County.

El trio, besides amassing the significant herpetological collections that are the wellspring of this book, made many valuable contacts in the biological community, especially at the University of Costa Rica (UCR), then on the outskirts of the capital city of San José. The potential for cooperative programs suggested through these contacts led John L. Mohr, chairman of Biological Sciences at USC, fellow faculty member Thomas R. Pray, and me to fly to Costa Rica in the spring of 1960.

It was the dry season on the Meseta Central, and the sky was that clear tropical blue that seems to let you see every tree in the almost black forest covering the mountains surrounding the city. A few puffy white clouds contrasted with the pinks, purples, yellows, and reds of the many flowering trees that radiated brilliance under the tropical sun. Each afternoon the prevailing northeast trade winds pushed Atlantic moisture in the form of gray fingers of clouds through the passes between the volcanoes north of the city, and by late afternoon the clouds completely obscured the peaks. In the rainy season these fingers cross the divide about noon, rapidly extend over the valley, and by afternoon they produce tropical downpours.

During our two-week stay we were welcomed by our Costa Rican colleagues at the university, Rafael Lucas Rodríguez, director of the Escuela de Biología, José Joaquín Trejos, dean of Sciences and Letters (later president of Costa Rica), and the vice dean, John de Abate. Together we developed a joint UCR-USC plan for instructing United States faculty members in tropical biology in the Tropics. Between meetings we drove around like excited teenagers collecting material on the volcanoes and passes and the Cerro de la Muerte, where a dozen salamanders could be found under nearly every rock or log.

As you will see below, I have been back to Costa Rica many times. Perhaps I will never again experience the same excitement, but each time the plane crosses the Pass of Desengaño through the Cordillera Central, a wave of feeling envelops me. My heart speeds up, I instantly become alert, and I find myself once more filled with anticipation of the new tropical places, people, and creatures I will encounter on this journey.

In 1961 the first UCR-USC course program, supported by NSF, was held during the rainy season. Andy Starrett co-coordinated this effort, which was centered at UCR but included many field trips throughout the country. The program was expanded in the summers of 1962 and 1963 and became the basis for the course program of the Organization for Tropical Studies (OTS). John de Abate, Rafael Rodríguez, and I were among the founders of OTS in 1963 (see Stone 1988 for a more detailed history). Now internationally renowned as a center for education, research, and conservation in tropical biology, OTS is operated by a consortium of sixty-four universities and research organizations from the United States, Canada, Costa Rica, Mexico, Peru, South Africa, and Australia. Fortunately, I had received a John Simon Guggenheim Fellowship for 1963 to 1964, which allowed me to carry on fieldwork and assist for eighteen months during the first years of OTS's operations. I was also fortunate to have my graduate students William A. Bussing, Roy W. McDiarmid, Norman J. Scott Jr., and James L. Vial and my dear friend the late Charles F. Walker work in the field with me for long periods during this time, covering all the major life zones in the country and most of the accessible areas except around Tortuguero and Santa Rosa, both sites of future national parks. I can never fully express my gratitude for their contributions, then and later, to our understanding of the Costa Rican herpetofauna and for our lifelong friendship.

Since 1964 I have returned more than sixty times to this beautiful country, with its incredible biodiversity and its kind and friendly citizens. Each visit was its own special adventure, and I went home to the United States each time rejuvenated, filled with energy and consciousness. Only the most important of the visits in terms of herpetology are highlighted below.

During the 1970s much of my fieldwork centered on understanding the composition and ecological factors regulating distribution in lowland forests. Comparative study of Atlantic (La Selva) and Pacific (La Pacifica) lowland sites was undertaken in 1972–74. My friend Ian R. Straughan and I designed a project in which several graduate students, now colleagues, took part (see "A History of Costa Rican Herpetology" in chapter 1 for details). I had first visited La Selva in 1961 on the invitation of its owner, the tropical forester Leslie R. Holdridge. Holdridge later sold the property to OTS, and it now forms the major portion of the expanded, world-famous La Selva Biological Station of OTS. I have subsequently taught courses, carried on research, and attended conferences at La Selva numerous times.

I had also begun fieldwork on the Península de Osa in 1962, with other forays in 1964 (with Norman Scott), 1968,

and March 1973. Major collections were made around Rincón de Osa by parties led by Roy McDiarmid and me (see chapter 1 for details) in the summer of 1973.

My first work in the area of what is now the OTS Las Cruces Biological Station was in 1964, with additional field time in the region in 1968, 1974, 1976, 1979, 1987, 1993, 1994, and 1998. Norm Scott and I first collected in the Monteverde area in 1963 and did so again in 1964. I returned in 1980, 1983, 1986, 1988, 1994, and 1999 but did no collecting on the last three trips.

I carried out fieldwork at the OTS Palo Verde Biological Station in 1974 and 1983 and made a subsequent visit in 1999 to review the new facilities.

From 1974 to 1980 I was honored to serve as the president of OTS, which meant that on many trips to Costa Rica I had responsibilities that restricted fieldwork to only a few hours or days at a time. Subsequent service with OTS also curtailed herpetological fieldwork, which was now carried out primarily by my graduate students and others, including an emerging cadre of Costa Rican herpetologists. However, in 1986 I was able to participate in a major expedition to explore a transect from La Selva southward up the slopes of Volcán Barba to its summit (see chapter 1). From then until the present I have made a number of trips to the Monteverde Forest Preserve and the La Selva Biological Station and visited Tortuguero National Park.

Aside from the fieldwork, most of my trips to Costa Rica in the 1990s were to examine specimens in the Museo de Zoología, UCR, and to collect the information necessary to complete this book. A memorable trip to the Santa Rosa National Park and the Guancaste Conservation Area in 1995, made at the instigation of Daniel Janzen and sponsored by the Instituto Nacional de Biodiversidad, Costa Rica (INBio), provided much background on this region that I had not previously explored.

Forty years is a long time to devote oneself to the biodiversity of a single small country. But even as I write, the blue skies and billowing white clouds, the dense and sometimes forbidding forests, the rushing streams, the exuberant plant and animal life, and the thrill of discovery call me back. I know that I must return soon to that rich coast, still searching for golden frogs after all these years.

Acknowledgments

In a project of this magnitude, which has developed and matured over a forty-year period, I have been dependent on many people for help in innumerable ways. I acknowledge with pleasure all the individuals mentioned below and thank them effusively for their assistance and support, especially when given at some inconvenience to themselves. If I have overlooked anyone, I apologize: it wasn't intended; my memory just isn't as good as it used to be.

First I want to thank the many students, associates, and graduates of the University of Southern California and the University of Miami who have contributed specimens, notes, and field companionship as well as testing me and the keys and criticizing various parts of the manuscript. I especially thank David P. Bickford, Arden H. Brame Jr. II, Shyh-Hwang Chen, Brian I. Crother, James E. DeWeese, Charles F. Dock, Maureen A. Donnelly, Sharon B. Emerson, Craig Guyer, Ronald T. Harris, Marc P. Hayes, W. Ron and Miriam Heyer, David M. Hillis, Jason D. Johnson, John T. Kitasako, Arnold G. Kluge, Däna M. Krempels, Peter N. Lahanas, Robert J. Lavenberg, Carl S. Lieb, Susan S. Lieberman, Sandra Limerick, Karen R. Lips, Roy W. McDiarmid, John R. Meyer, Michael M. Miyamoto, David J. Morafka, Kirsten E. Nicholson, David T. Nicolson, William F. Presch, Allan A. Schoenherr, Norman J. Scott Jr., Joseph B. Slowinski, Andrew and P. H. "Holly" Starrett, Ian R. Straughan, James J. Talbot, James L. Vial, David B. and Marvalee H. Wake, and Steven D. Werman.

At the University of Costa Rica, William A. and Myrna Bussing, Douglas C. Robinson, and Federico Bolaños were extremely helpful during many visits. At the University of Southern California John L. Mohr provided support for my first trip to Costa Rica, and Leslie A. Chambers and Bernard C. Abbott, successive directors of the Allan Hancock Foundation, continuously supported my work during their tenures. At the University of Miami my colleagues Michael S. Gaines, Casimer T. Grabowski, Carol G. Horvitz, and David P. Janos have provided much encouragement in the preparation of this book.

Personnel of the Organization for Tropical Studies (OTS), whose efforts expedited fieldwork essential for completing the book, include Ana Lorena Bolaños, Jorge R. Campabadal, David B. and Deborah A. Clark, Luis Diego Goméz, Gary S. Hartshorn, Gail Hewson, Jorge Jiménez, Pedro León, Steven B. Preston, Charles E. Schnell, Donald R. and Beverly Stone, and Flor M. Torres. Neither rain, hurricanes, earthquakes, nor flood kept them from providing good advice and strong logistics for many trips into the field.

Numerous other people have contributed in various ways. Particularly appreciated and acknowledged with the deepest thanks are the many people who aided my work in Costa Rica. Rafael Rodríguez, John de Abate, and Dean José J. Trejos at the University of Costa Rica and Leslie R. Holdridge and J. Robert Hunter made the prospect for working in the republic most inviting and provided many kindnesses during the initial period of my studies there. Norman J. and Joan Scott and James L. and Lynda Vial, Roy W. McDiarmid, W. Ron and Miriam Heyer, and Ruth L. Savage all contributed to field successes during this period. Others who were especially helpful or who joined me in the field in Costa Rica were Rafael Acuña, Juan Diego Alfaro, Róger Bolañas, William A. and Myrna Bussing, David C. Cannatella, Richard S. Casebeer, Anny Chaves, Gerardo "Cachí" Chaves, Darrell Cole, Michael J. Corn, Martha L. Crump, Michael Franzen, Darrel Frost, David A. Good, George C. Gorman, Harry W. Greene, José María Gutiérrez, David D. Hardy Jr., Norman E. Hartweg, David G. Huckaby, Jerry James, Daniel H. Janzen, César Jarmillo, Alfonso Jiménez, José M. Jiménez, Salvador Jiménez, Jerry D. Johnson, James Kezer, Laurence M. Klauber, Gunther Köhler, Brian Kubicki, Thomas C. LaDuke, John M. Legler, Ronald B. Linsky, J. R. "Randy" McCranie, Charles A. McLaughlin, Eberhard Meyer, Manuel M. Murillo, George S. Myers, Dennis R. Paulson, James A. Peters, Luis Porras, J. Alan Pounds, Mason Ryan, Mahmood Sasa, Charles E. Shaw, Peter Sigfried, Paul Slud, Hobart M. Smith, Alejandro Solórzano, Ian R. and Isdale Straughan, Laurence C. Stuart, Fred G. Thompson, Walter T. and Karen Timmerman, Fred S. Truxal, R. Wayne Van Devender, Charles F. Walker, Larry David Wilson, Allen M. Young, George R. Zug, and the many students in various OTS courses who carried out herpetological field problems and stimulated me to look anew at my assumptions and conclusions regarding the biology of amphibians and reptiles in the Tropics.

In preparing this book I have been dependent on the curators of all the museums and other collections housing Costa Rican amphibians and reptiles. They have been uniformly helpful in lending critical specimens and extending many other courtesies. Without their generous aid this book could never have been finished. *Mil gracias* to each of you, listed below by institution: James E. Böhlke and John E. Cadle, Academy of Natural Sciences, Philadelphia; Charles M. Bogart, George Foley, Linda S. Ford, Charles W. Myers, Janis Roze, and Richard G. Zweifel, American Museum of Natural History; E. N. Arnold, Barry T. Clarke, Alice G. C. Grandison, H. W. Parker, and Andrew Stimson, Natural History Museum, London; Robert C. Drewes, Alan E. Leviton, and Jens V. Vindum, California Academy of Sciences; C. J. "Jack" McCoy and Ellen J. Censky, Carnegie Museum of Natural History; Robert F. Inger, Hymen Marx, Alan Resetar, and Harold Voris, Field Museum of Natural History; Walter Auffenberg and David L. Auth, Florida Museum of Natural History; Hobert M. Smith, University of Illinois Museum of Natural History and University of Colorado Museum; Henryk Szarski, Krakow University; William E. Duellman, John E. Simmons, and Linda Trueb, Museum of Natural History, University of Kansas; Edward H. Taylor, University of Kansas; Kent Beaman, Robert L. Bezy, Arden H. Brame Jr. II, James R. Dixon, David A. Kizirian, and John W. Wright, Natural History Museum of Los Angeles County; Douglas A. Rossman, Louisiana State University; John E. Cadle, Ernest E. Williams, and José P. Rosado, Museum of Comparative Zoology, Harvard University; Günther Peters and Rainer Günther, Museum für Naturkunde, Humboldt-Universität zur Berlin; Harry W. Greene, David B. and Marvalee H. Wake, and David A. Good, Museum of Vertebrate Zoology, University of California (Berkeley); Josef Eislet, Naturhistorischens Museums, Vienna; James R. Dixon, Texas Cooperative Wildlife Collection, Texas A&M University; Federico Bolaños and Douglas C. Robinson, Museo de Zoología, Universidad de Costa Rica; Norman Hartweg, Arnold G. Kluge, and Charles F. Walker, Museum of Zoology, University of Michigan; Harold A. Dundee, Tulane University; Doris M. Cochran, W. Ron Heyer, Roy W. McDiarmid, James A. Peters, and George Zug, United States National Museum of Natural History.

The Amphibians and Reptiles of Costa Rica is a descendant of earlier works. The first, *A Handlist with Preliminary Keys to the Herpetofauna of Costa Rica* (Savage 1980b) was prepared at the urging of A. Stanley Rand of the Smithsonian Tropical Research Institute. For that encouragement and many other courtesies, I am most grateful. *¡Muchissimas gracias!*

The second is the *Introduction to the Herpetofauna of Costa Rica/Introducción a la herpetofauna de Costa Rica* (Savage and Villa 1986). That book was developed through the good offices of Kraig Adler, Cornell University, and the text was translated into Spanish by Jaime Villa. Both have my great thanks for the boost that publication gave to completion of the present work.

My studies in Costa Rica have been supported by the National Science Foundation (G6089, BMS 7301610, and DEB 9200081); the American Philosophical Society; the Organization for Tropical Studies; the Allan Hancock Foundation, University of Southern California; the College of Arts and Sciences of the University of Miami; and the Bache Fund of the National Academy of Sciences. A John Simon Guggenheim Fellowship allowed me to carry out research in Costa Rica during an eighteen-month period in 1963–64. The Giles W. and Elise G. Mead Foundation provided funding specifically for fieldwork leading to completion of the book. This support is gratefully acknowledged; without it I would have never begun this book, let alone completed it.

I also especially wish to thank Cristina Ugarte, who prepared most of the text figures, and Cathy Gibbs Thornton, graphic artist, who rendered the figures into final form. Others who contributed one or more line drawings or base maps are Patricia Brame, Russell Cangialosi, Maureen A. Donnelly, Sharon B. Emerson, Michael P. Fahay, Robert Kanter, Däna M. Krempels, Peter N. Lahanas, Karen R. Lips, Charles Messing, Theodore S. Pietsch, Norman J. Scott Jr., Joseph B. Slowinski, and Steven D. Werman. Figure 1.1 was produced by Marco Castro (INBio) and figure 3.3 by James A. Polisini. Ronald T. Harris recreated the Amerind herpetological motifs decorating the book.

Steven F. Oberbauer, Kirsten E. Nicholson, David T. Nicolson, and Paul Richards provided computer expertise. Dave designed the system for making all distribution maps, which I prepared on Aldus Freehand with assistance from Paul. I also thank Maureen Grover, who prepared several drafts of all parts of the manuscript. Additional word processing was done by Barbara Merritt González, Beretha Howard, Von Reid, Vivian Roe, Marcy Thorne, Sandra Quarles, and Susan Milne.

Funding to make it possible to include the many colored plates and still maintain a low price for the finished product was received from the American Society of Ichthyologists and Herpetologists, Instituto Nacional de Biodiversidad, Costa Rica, the Herpetologists' League, the Organization for Tropical Studies, and the Society for the Study of Amphibians and Reptiles.

The Direcciones de Fauna Silvestre and Parques Nacionales de Costa Rica encouraged fieldwork at the various parks, reserves, and wildlife refuges under their control. They graciously expedited the work of my associates and me by providing various kinds of logistical support and issuing all appropriate permits for sampling the country's herpetofauna.

The debt of gratitude I owe to Michael and Patricia Fogden for lending their artistry to illustrate this book is immense. Along many muddy trails and through many rainy nights they patiently sought and photographed the rarest forest gems, the subjects of this book. I suspect most readers will spend more time appreciating the Fogdens' pictures than reading my verbiage! Thanks Mike and Patricia, again and again!

Alejandro Solórzano and R. Wayne Van Devender were especially helpful in providing color photographs of many taxa. Unless otherwise credited, all photographs in the book are by the Fogdens.

Greatly appreciated were the comments and criticisms by the three anonymous reviewers selected by the University of Chicago Press to evaluate the manuscript. The reviewers, Jonathan A. Campbell, Martha L. Crump, and Harry W. Greene, later identified their role in the editorial process. These reviews aided me a great deal in improving the text and were most positive, even though the reviewers did not have the opportunity to see the Fogdens' photographs during their review.

At the University of Chicago Press the Fogdens and I es-pecially wish to thank editors Susan Abrams for initial en-couragement to prepare the book, Christie Henry for guid-ing it to completion, and Alice Bennett for suggesting many improvements to the original manuscript.

Maureen A. Donnelly and Bruce E. Grayson also are pro-fusely acknowledged for helping me to maintain my sanity during the final phases of finishing the manuscript.

I can never thank my beloved wife, Rebecca Papendick, enough for her emotional support during the long gestation of this book. Without her encouragement and determina-tion that I complete the project, I might have faltered in overcoming the many challenges posed in wrapping up a project of this size. She also prepared and edited the final draft of the manuscript.

The constant aid of the five greatest brains ever assembled as a group is also much appreciated.

Photographers' Acknowledgments

MANY OF Costa Rica's reptiles and amphibians are rare or secretive or both. The success we have had in finding and photographing many of the more difficult species owes much to the generous help of numerous friends and colleagues. Special thanks are due to Alejandro Solórzano, whose contribution has been outstanding. To him we owe the photographs of more than thirty species, mainly snakes. For help and information we are also grateful to Rodrigo Aymerich and the Instituto Clodomiro Picado, Róger Blanco, Federico Bolaños, Marlene Brenes, David and Deborah Clark, Eladio Cruz, Maureen Donnelly, Randy Figeroa-Sandí, Luis Diego Gómez, Harry Greene, Wolf Guindon, Craig Guyer, David Hardy, Marc Hayes, Dana Krempels, William Lamar, Richard LaVal, Michael Linley, Karen Lips, Mario Méndez, David Norman, Norman Obando, Alan Pounds, Manuel Santana, Jay Savage, Martin Schlaepfer, Miguel Solano, Walter and Karen Timmerman, Orlando Vargas, David and Marvalee Wake, and Doug Wechsler. Our apologies to anyone we have unintentionally left out.

We also gratefully acknowledge those photographers, herpetologists, and institutions who have generously provided photographs of species we failed to encounter: Richard Bartlett, Jonathan A. Campbell, Dave Cannatella, Pete Carmichael, Tom Devitt, William E. Duellman, Scott Eckert, James Hanken, Allan Jaslow, Frank Joyce, Karl-Heinz Jungfer, Günther Köhler, Twan Leenders, Pedro León, Karen Lips, John H. Malone, Roy W. McDiarmid, Jack Musick, Charles W. Myers, Ken Neumuras, David Norman, Louis Porras, Peter Pritchard, Carlos de la Rosa, Martin Schlaepfer, Gustavo Serrano, Kevin Shafer, the Smithsonian Tropical Research Institute, and photographers Marcos Guerra and Carl C. Hansen, Alejandro Solórzano, R. Wayne Van Devender, Andrés Vega, Doug Wechsler, Kentwood Wells, and Steve Wilson. As always, we thank Sue Fogden for help in a multitude of ways.

Finally, we thank Jay Savage for giving us the opportunity to participate in his monumental book.

Illustration Credits

THE FOLLOWING individuals and institutions are acknowledged for permission to use their photographs or copyrighted materials in illustrating *The Amphibians and Reptiles of Costa Rica.*

Institutions
American Society of Ichthyologists and Herpetologists: *Copeia*, figs. 7.44, 7.45, 7.46, 7.52, 11.36, 11.37
Field Museum of Natural History, figs. 7.47, 7.49a–e, 7.51
Geological Society of America, fig. 16.1
Herpetologists' League: *Herpetologica*, figs. 7.17, 7.19, 7.115, 11.3c, 11.34; *Herpetological Monographs*, fig. 6.6
Missouri Botanical Garden, figs. 15.2, 16.18
Museum of Vertebrate Zoology, pls. 29, 30, 35
Natural History Museum, University of Kansas, figs. 7.11d–g, 7.12c–f, 7.36b,c, 7.48a,c,e, 7.55a,b, 7.60, 7.64, 7.70, 7.72, 7.85, 7.94, 10.29, 10.30, 10.31b, 10.32a–e, 10.34b,e–f
Smithsonian Tropical Research Institute, pls. 109, 122, 126, 152, 201, 202, 259, 301, 331
Society for the Study of Amphibians and Reptiles: *Intro-duction to the Herpetofauna of Costa Rica* (Savage and Villa 1986), figs. 3.3, 5.2, 6.2, 7.6a, 7.7, 10.3, 11.4, 11.5, 11.6a–b, 11.8, 12.9, 13.3c–d; *Journal of Her-petology*, figs. 6.7, 6.8, 6.9, 6.10
John Wiley and Sons Ltd., figs. 2.3, 2.6, *Tropical Clima-tology*, McGregor and Nieuwolt (1998).

Individuals
R. Bartlett, pls. 69, 156, 310, 382
J. Campbell, pls. 248, 294
D. Cannatella, pls. 49, 52
J. Carmichael, pl. 179
T. Devitt, pl. 434
W. Duellman, pl. 166
S. Eckert, pl. 503
L. D. Gómez, pls. 9, 11, 15, 508
J. Hanken, pl. 28
A. Jaslow, pls. 42, 116, 279
F. Joyce, pl. 80
K.-H. Jungfer, pls. 162, 163, 177, 212, 226
G. Köhler, pl. 295
T. Leenders, pl. 21
P. León, pl. 33

K. Lips, pl. 97
J. Malone, pl. 329
R. McDiarmid, pls. 54, 150, 205, 231
C. Messing, fig. 12.14
J. Musick, pl. 499
C. Myers, pls. 107, 145, 391
K. Nemuras, pl. 64
D. Norman, pls. 85, 228
L. Porras, pl. 143
P. Pritchard, pl. 501
J. Renjifo, pl. 103
C. de la Rosa, pl. 82
M. Schlaepfer, pls. 123, 297
G. Serrano, pls. 83, 90
K. Shafer, pls. 498, 502, 504
A. Solórzano, pls. 24, 112, 142, 181, 194, 254, 282, 283, 309, 314, 326, 358, 362, 376, 385, 395, 413, 423, 433, 505, 509
R. W. Van Devender, pls. 31, 32, 40, 43, 44, 71, 99, 132, 149, 175, 176, 192, 253, 255, 286, 298, 316, 328, 353, 401, 415, 426, 445, 446, 448, 458, 512
D. Wechsler, pl. 291, 313
K. Wells, pl. 213
S. Wilson, pls. 25, 292, 315

Additional figures were redrawn from the following sources:

fig. 2.10, D. Janzen, *Costa Rican Natural History*, Univer-sity of Chicago Press (1983)
figs. 4.4 (part), 4.5b–c, G. Orton, *Turtox News* 28 (1950), 29 (1951)
fig. 6.4, D. Wake, *Mem. South. Calif. Acad. Sci.* 4 (1966)
fig. 6.9b, A. Brame and W. Duellman/artist P. Brame, *Contrib. Sci. Nat. Hist. Mus. Los Angeles County* 201 (1970)
figs. 7.2, 10.14c, 10.33, J. Lee, *The Amphibians and Rep-tiles of the Yucatán Peninsula*, Cornell University Press (1996)
fig. 7.13, K. Lips and J. Savage, *Stud. Neotrop. Fauna and Environ.* 31 (1996)
figs. 7.48b,d J. Lynch and C. Myers, *Bull. Amer. Mus. Nat. Hist.* 175 (1983)
figs. 7.71, 7.73, J. Savage and W. Heyer, *Rev. Biol. Trop.* 16 (1969)

fig. 7.115, P. Starrett and J. Savage, *Bull. South. Calif. Acad. Sci.* 72 (1973)

fig. 10.2, C. Myers, *Amer. Mus. Nat. Hist. Novitates* 2523 (1973)

fig. 10.5, E. Williams and W. Duellman, *Breviora* 256 (1967)

figs. 10.8, 10.31a, 10.25e, 10.34c, T. Avila-Pires, *Zool. Verh. Natl. Mus. Leiden* 299 (1995)

fig. 10.10, R. Stebbins, *Amphibians and Reptiles of Western North America,* McGraw-Hill Book Co. (1954)

fig. 10.11, W. Burt, *Bull. U.S. Nat. Mus.* 154 (1931)

fig. 10.12a, C. Myers and M. Donnelly, *Amer. Mus. Nat. Hist. Novitates* 3027 (1991)

fig. 10.12c, M. Hoogmoed, *Biogeographica* 4 (1973)

fig. 10.15, G. Köhler and B. Streit, *Sencken. Biol.* 75 (1996)

fig. 10.22, E. Williams, *Breviora* 502 (1995)

fig. 10.23, C. Myers, *Amer. Mus. Nat. Hist. Novitates* 2471 (1971)

fig. 10.36, Harris and Kluge, *Occ. Pap. Mus. Zool. Univ. Michigan* 706 (1984)

fig. 10.34a, Oftedal, *Arq. Zool. Mus. Zool. Univ. São Paulo* 25 (1974)

fig. 10.34d, Uzzell, *Bull. Amer. Mus. Nat. Hist.* 132 (1966)

fig. 10.36, J. Savage and K. Lips, *Rev. Biol. Trop.* 41 (1993)

fig. 10.37a, C. Meszoely, *Bull. Mus. Comp. Zool.* 139 (1970)

fig. 10.41, Oliver, *Bull. Amer. Mus. Nat. Hist.* 92 (1948)

figs. 11.2, 11.3a–b, H. Dowling and J. Savage, *Zoologica* 45 (1960)

fig. 11.7b–d, J. Peters and B. Orejas, *Bull. U.S. Nat. Mus.* 297 (1970)

figs. 11.8, 11.20, 11.49a, J. Campbell and W. Lamar/artist C. Garrett, *The Venomous Reptiles of Latin America,* Cornell University Press (1989)

figs. 11.9, 11.11–13, 11.16, 11.43, 11.45, G. Jan and F. Sordelli, *Iconographie Générale des Ophidiens,* J. Cramer (1961 reprint)

fig. 11.12e, J. Chippaux, *Faune Tropicale* 27 (1986)

fig. 11.30, F. Downs, *Misc. Pub. Mus. Zool. Univ. Michigan* 131 (1967)

figs. 11.35b–c, 11.36a, J. Peters, *Misc. Pub. Mus. Zool. Univ. Michigan* 114 (1960)

fig. 12.5, L. Stejneger, *Bull. U.S. Nat. Mus.* 58 (1907)

figs. 12.6–7c–d, 12.15–16, 12.18, L. Brongersma, *Zool. Verhand. Natl. Natuurhist. Mus. Leiden* 121 (1972)

fig. 13.2, C. Behler, *Zoologica* 58 (1971)

fig. 16.3, E. Barron, C. Harrison, J. Sloan II, and W. Hay, *Eclog. Geol. Helvetiae* 74 (1981)

figs. 16.7–8, J. Jackson, A. Budd, and A. Coates, *Evolution and Environment in Tropical America,* University of Chicago Press (1996)

1

The Basics

1 | Discovering a Tropical Herpetofauna

"And on that rich coast teemed life. Being near death from our long voyage, it was most wonderous to behold."

Anonymous eighteenth-century voyager

INTRODUCTION TO THE HERPETOFAUNA: A PRELIMINARY RECONNAISSANCE

I AM SLOWLY walking along a trail at the La Selva Biological Station on the Atlantic lowlands of Costa Rica. It is early morning, and lingering patches of mist from yesterday's rains shroud some of the emergent trees. Soon the mist will be gone as the tropical sun rises higher in the sky. It is shady within the forest even at high noon because of the dense canopy formed by the contiguous giant treetops. Scattered patches of light on the forest floor bring out a kaleidoscope of colors—greens, reds, browns, yellows, and blacks—from the litter of dying leaves in various stages of returning their nutrients to the soil.

As I walk a small anole lizard, facing head downward on the stem of a small shrub, fans out its orange dewlap at my intrusion. A few steps later a rainfrog leaps out from underfoot, bounds down the trail, then disappears at a right angle into the forest litter. As the morning progresses I hear the rattle of dry leaves on the side of the trail and spy a large bluish green lizard with a broad middorsal yellow field nervously probing the leaf litter and searching the soil for prey. When it spots me it runs pell-mell to a refuge under a fallen tree.

Later still I am in the territory of competing male strawberry dart-poison frogs. The little bright red males, with dark purple legs, give a series of harsh "chirp-chirp-chirp" calls to advertise their presence to other males. Today an intruding male has come too close, and he and the resident are locked in a wrestling contest, standing upright on their hind legs and trying to force one another to the ground. Finally the resident pins his opponent, who when released quickly hops away. On my return toward the main laboratory buildings of the station a large frog comes bursting across the trail, chased by a glossy black snake two meters long.

I pause and reflect on the previous night's exploration of the forest. A heavy rain had fallen in the afternoon, and lighter showers continued during the evening. By dark I began to hear the voices of many kinds of nocturnal frogs. I followed the loudest cacophony to one of the temporary ponds that fill with water at the beginning of the rainy season. Five or six frog species were breeding at this site, males calling loudly and females arriving from their daytime retreats or already locked in the embrace of a lusty male. Many bright green, red-eyed frogs, their purple flanks barred with yellow, were laying green eggs on leaves above the pond. Other species were calling or laying eggs at various sites within the pond itself. As the evening wore on, several marauding pop-eyed arboreal snakes appeared uninvited to gorge themselves on the energy-rich green eggs. Returning from the pond through the forest, I heard the constant "dink-dink-dink" of small rainfrogs almost impossible to find hiding in the low herbaceous vegetation and bromeliads. I stopped once on the trail to observe a beautiful red-, black-, and yellow-ringed coral snake that, disturbed by my presence, uncoiled and slipped away.

Continuing my morning's foray, I rest briefly on the bridge leading across the river from the forest research area to station headquarters. On a log below, several black river turtles are basking while a large male green iguana follows suit on a branch of a huge tree high above the river. This morning is especially unusual because as I turn around to look back on the forest I spot a rarely seen stray crocodile two and a half meters long stretched out just above the water's edge on the opposite shore.

After lunch heavy rain begins to fall, typical of each afternoon during the rainy season, confining me for a time to my *cabina*. My visit has stimulated many memories of other times at both lowland and highland sites in Costa Rica during my searches for and study of the country's amphibians and reptiles.

I vividly recall nights on mountain slopes, with clouds and rainy winds boiling over the ridges as I followed the calls of unknown frogs emitted from bromeliads on giant oaks or from tree holes far above the forest floor. There was the excitement of discovering these rarities among the jumbled pile of branches and epiphytes one dark night when one

3

of the giants came crashing down from a lightning strike. There was also the shock on another night when the vine I had reached for to prevent a fall on the muddy trail turned out to be a treeviper indistinguishable in that light from the green creeper it perched on.

Other memories abound of working my way up the beds of torrential mountain brooks, the rushing waters glittering like silver in my headlamp's glow, to ferret out incredibly beautiful treefrogs and glassfrogs whose calls were barely discernible above the water's roar. Of the time I first plucked tank bromeliads from their lofty seat on the branches of a mountain oak to find a treasure of wiggling, mostly red salamanders, a delicate green arboreal long-tailed lizard, and two brightly colored snail-eating snakes. Of searching through a moss bank for salamanders but to my surprise finding instead bright little frogs guarding their egg masses and a small yellow-bellied earthworm-eating snake. Although the objectives were always scientific, the creatures of study constantly have led me to experience adventure and the thrill of discovery.

One aim of this book is to encourage others to share in the adventure by introducing tropical amphibians and reptiles and showing where to look for them. The organisms include some of the most interesting and beautiful of tropical animals, many likely to be seen almost every day or night at an appropriate site. Others are so secretive or rare as to be discovered once in a lifetime, if ever. The process of finding, observing, and understanding the lives of tropical amphibians and reptiles is ongoing. I hope that this book conveys the excitement of this quest and that you will join the search for a deeper understanding of these fascinating and often maligned creatures. Perhaps, like me, you will also come to a deeper understanding of yourself.

SCOPE AND PLAN OF TREATMENT

This book attempts to summarize what is known about the herpetofauna of the small Central American country of Costa Rica (fig. 1.1). By herpetofauna I mean the living species of amphibians and reptiles (exclusive of birds). The scientific field concerned with study of the world's herpetofauna is herpetology. The term "herpetology" is also often used to denote what is known about amphibians and reptiles (e.g., the herpetology of Costa Rica) or even as a synonym of herpetofauna (e.g., the herpetology of La Selva). As much as possible. I have avoided the third usage.

The subject of this book, the known herpetofauna of Costa Rica, consists of 396 species of amphibians and reptiles (including introduced species). These are distributed among 37 families and 140 genera, including 174 species of

Figure 1.1. Map of Costa Rica showing provinces and principal cities and towns.

amphibians and 222 species of reptiles. Of these species, 16% (44 amphibians and 17 reptiles) are endemic to Costa Rica. The composition of the herpetofauna by major taxonomic groups is given in table 3.1.

Of the 366 names of amphibians and reptiles listed from Costa Rica by Savage and Villa (1986), 41 are no longer valid owing to systematic changes at the generic or specific level or synonymization. Other changes include 17 new species described from Costa Rica, 23 resurrected from synonymy, and 11 previously unknown in the country and added to the faunal list.

The Amphibians and Reptiles of Costa Rica is written for a variety of audiences. First is the general reader with some interest in the Costa Rican flora and fauna but with relatively little background in biology or without much knowledge of amphibians and reptiles. Second is the person with some biological training, an amateur or part-time naturalist or ecotourist, who wants to expand his or her horizons by undertaking a greater challenge in the field than is offered by bird or butterfly watching or snorkeling. Third is the biology student, especially one majoring in that subject at a Costa Rican institution or attending one of the several environmentally oriented field courses offered in Costa Rica by groups like the Organization for Tropical Studies (OTS). Such a student first wants to be able to identify common species but later may undertake research projects, including those leading to an advanced degree based on studies of amphibians and reptiles. Fourth is the professional biologist or environmental scientist who needs to recognize the amphibians and reptiles encountered in the field and have a ready source of data about these animals' biology. I hope this book will encourage use of the herpetofaunal component of the biota in behavioral, ecological, physiological, and related research, including that centered on conserving the republic's biodiversity. Finally, this book is written for the serious amateur or professional herpetologist who is intensely interested in all aspects of the biology of one or more species or higher taxa of amphibians and reptiles and who will use this book as a foundation to further advance our knowledge of these remarkable animals.

The Amphibians and Reptiles of Costa Rica consists of four main parts. The first (part 1) presents introductory materials basic to using the rest of the book. Important sections relate to the principles of biological systematics, scientific names and measurements, the location, capture, and care of amphibians and reptiles in captivity, preservation for scientific study, snakebite, the history of Costa Rica and of herpetological discovery, sources of materials, conservation, and future research opportunities.

This section also includes a major chapter on Costa Rica's natural environments. Part 1 concludes with a general description of the materials contained in the systematic accounts in parts 2 and 3, suggestions on how to use them to best advantage, and a key to major groups represented in the herpetofauna.

Parts 2 and 3 provide species accounts for all Costa Rican amphibians and reptiles, respectively. These chapters also include descriptions of the definitive characteristics of the major groups, families, and genera and the principal known features of their biology and distribution.

The final section of the text (part 4) is an analysis of the current patterns of distribution of Costa Rican taxa and a synthesis of data to explain the history producing these biogeographic patterns.

An extensive glossary and a list of publications cited in the text provide support for those unfamiliar with herpetological terminology and an entrance into the voluminous literature on Costa Rica's herpetofauna.

The book covers about 55 to 60% of the total Central American (Guatemala to Panama) herpetofauna. In this regard 89% of the herpetofauna of Nicaragua and 87% of the herpetofauna of western Panama and the Canal Zone area are covered.

A Note on Word Usage and Spanish Personal Names

This seems a logical place to explain some conventions regarding word usage. Throughout the text the terms "the country" and "the republic" refer the Republic of Costa Rica. Costa Ricans are properly called *Costarricense* but call themselves *ticos;* both will be used. Tico is the nickname given to the Costarricense by other Central Americans because of their frequent use of -*tico* or -*tica* as a diminutive. Where other Spanish speakers use an ending such as -*ito(a)* or -*ico(a)* to mean little (as *ranita*, little frog), Costa Ricans combine the two *(ranitica)* or use double diminutives like *chiquitico* or *chiquitica* for a little boy or a little girl. When I first visited Costa Rica it was all right for one tico to call another "tico" but bad form for a foreigner to use the term. This no longer applies. Foreigners may be called *macho* or *macha*, words also used colloquially for fair-haired people, and Americans may be called *gringos* with no animosity intended.

Spanish personal names may confuse the uninitiated. Because there is matrilineal inheritance, a person's legal name, such as José María Castro Madriz, consists of one or more personal names (José María) and two surnames *(apellidos)*, one's father's family name *(apellido primero)*, Castro, and one's mother's family name *(apellido segundo)*, Madriz. Often this is abbreviated to José María Castro M. If a woman married señor Castro she would retain her father's family name and add her husband's; for example, Pacífica Fernández de Castro. My tico friends are much amused by my legal Costa Rican name, which is Jay Savage Bird. In this book I will normally use only the *apellido primero,* but in some cases of well-known individuals both *apellidos* will be given the first time they are mentioned.

In most Spanish-speaking countries the titles *don* (gentleman, esquire) and *doña* (lady), originally applied to feudal

Box 1.1
Systematic Terminology

Taxon (pl. taxa): Any named or unnamed group of
   organisms, such as a species or a genus.

Natural taxon: A species or monophyletic group.

Monophyletic group: An ancestral species and all its
   descendants; a clade.

Unnatural group: A group comprising a common
   ancestor but not all its descendants (paraphyletic)
   or comprising two or more sets of taxa but not
   their most recent common ancestor (polyphyletic).

Systematic character: Any feature that differen-
   tiates one natural taxon from another. These include primitive features, derived features, and
   homoplasies.

Homoplasy: A resemblance between taxa independently derived from different ancestral species or a
   reversal of a derived feature to the ancestral state
   during the evolution of a group. For example, the
   convergence in digital disk structure in different
   families of frogs, the parallel development of green
   bones in different species within the same genus
   of glassfrogs, and the loss of limbs in snakes (a
   reversal).

Phylogeny: The evolutionary history of a group or
   lineage of organisms.

Clade. *See* Monophyletic group (above)

Cladistics: The science of phylogenetic reconstruction based on shared derived characteristics.

Cladogram: A branching diagram representing phylogenetic relationships among taxa derived from a
   cladistic analysis.

knights and their ladies, are of restricted application and
usually used with full names: Don José María Castro. The
democratic ticos call nearly everyone a gentleman or lady
but usually use the titles with the personal name in an af-
fectionate or sometimes teasing manner, as in my case Don
Jota (from the Spanish for the letter *J*), a course followed in
some places in this book.

SYSTEMATICS, TAXONOMY, AND PHYLOGENY

The science of discovering the diversity and evolutionary
relationships of organisms (species and natural groups) is
called systematics. In the context of the general evolutionary theory of descent with modification, natural groups are
defined as clusters of species that have descended from a
common ancestor and that form a hierarchical pattern of
relationships from less to more inclusive relatedness. Species

and natural groups are collectively called taxa (sing. taxon).
The ultimate aim of systematics is to elucidate the natural
hierarchy and its causes.

To attain this goal, systematics involves four interrelated
activities:

Taxonomy: recognition, identification and differentiation,
   description, classification, and naming of taxa

Phylogenetics: determination of evolutionary relationships
   (phylogeny) through reconstruction or estimation of the
   pattern of ancestry and descent (usually represented as a
   branching diagram)

Systematization: representation of the pattern of evolutionary relationships in a unitary system

Taxagenetics: explanation of the processes responsible for
   the origin, adaptation, and extinction of taxa based on
   patterns of phylogeny

The species is the basic unit in systematics, and all higher
or more inclusive taxa are composed of an ancestral species
and all of its descendants. Species living today are the end
products of a long and complex history of evolutionary diversification. Their unique pattern of common ancestry and
genetic relatedness is embodied in their history and provides
the basis for constructing phylogenies. These in turn form
the basis for a phylogenetic classification of organisms that
maximizes information content and allows systematists to
make predictions about species and their properties.

The cladistic method of phylogenetic inference is a technique that clusters species based on shared derived characters. In this system, if two related species (A and B) share a
derived feature not shared by a third (C), they are regarded
as more closely related to one another (more recently derived from a common ancestor) than either is to C (fig. 1.2).
Other kinds of characters, such as those that are primitive,
convergences, and parallelisms or reversals, are uninforma-
tive for this purpose. A clade consists of an ancestral species
and all its descendants and is characterized by the ancestral and descendant species' sharing one or more derived
features. The smallest clade includes a common ancestor
and two derived species (e.g., species A + B + their ancestor; fig. 1.2).

Phylogenetic relationships among clades are determined
in a similar manner based on all members of a group's sharing one or more derived characteristics. For example, all
frogs and toads have two ankle bones that are greatly elongated to form a fourth segment in the leg (all other limbed
amphibians and reptiles have three leg segments). This condition provides increased leverage for jumping and distinguishes this group of animals, the order Anura, to which all
frogs and toads belong. Actually, an evaluation of many
characters is used to produce a hierarchical or nested set of
taxa (fig. 1.3).

The process of systematization simply involves converting
the cladogram into a natural sequence of more inclusive to
less inclusive clades that reflects the hierarchical nature of

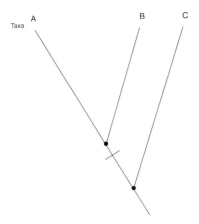

Figure 1.2. Minimal cladogram or three-taxon statement showing the phylogenetic relationships of taxa A, B, and C. The bar represents the derived feature shared by A + B and their ancestor.

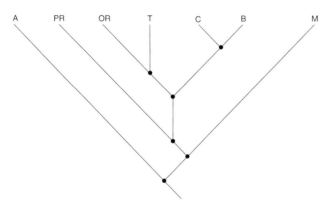

Figure 1.3. Cladogram of phylogenetic relationships among taxa: amphibians (A), primitive reptiles (PR), tuatara, lizards, and snakes (OR), turtles (T), crocodilians (C), birds (B), and mammals (M).

their evolutionary relationships. The most inclusive taxon in this example contains all seven terminal groups (A, PR, OR, T, C, B, M), and each in turn contains two subgroups: (A) and (PR, OR, T, C, B, M), and so on as follows:

Group: A, PR, OR, T, C, B, M
  Subgroup 1a: A
  Subgroup 1b: PR, OR, C, T, B, M
   Subgroup 2a: PR, OR, T, C, B
    Subgroup 3a: PR
    Subgroup 3b: OR, T, C, B
     Subgroup 4a: OR, T
      Subgroup 5a: OR
      Subgroup 5b: T
     Subgroup 4b: C, B
      Subgroup 6a: C
      Subgroup 6b: B
  Subgroup 2b: M

At present it is usual to represent the discovered phylogenetic relationships using the hierarchical system of notation originally developed by the great Swedish biologist Carolus

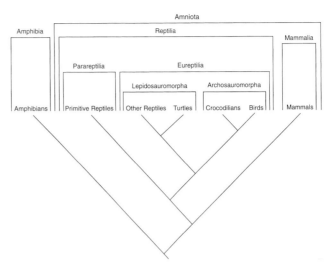

Figure 1.4. Conversion of cladogram in figure 1.3 to a natural classification.

Linnaeus, later Carl von Linné (1707–78), and formalized in the various international codes of botanical, zoological, and bacterial nomenclature (see below). The example (fig. 1.3) just discussed may be translated into the following natural classification (fig. 1.4):

Group: Infraphylum Tetrapoda
  Subgroup 1a: Superclass Batrachomorpha
   Class Amphibia
  Subgroup 1b: Superclass Amniota
   Subgroup 2a: Class Reptilia
    Subgroup 3a: Subclass Parareptilia
    Subgroup 3b: Subclass Eureptilia
     Subgroup 4a: Infraclass Lepidosauromorpha
      Subgroup 5a: Superorder Lepidosauria (several orders)
      Subgroup 5b: Superorder Testudines (two orders)
     Subgroup 4b: Infraclass Archosauromorpha
      Subgroup 6a: Order Crocodilia
      Subgroup 6b: Order Aves
  Subgroup 2b: Class Mammalia

Just as the comparative study of characteristics is used to discover evolutionary relationships, these relationships are expressed by the unique names provided to the various levels in the hierarchy. A list of species and other taxa arranged to reflect the hierarchy in nature is far more predictive of their characteristics than an alphabetical list or artificial classification. At the same time, it provides an efficient basis for storing and retrieving information about any group of organisms.

NAMES, TYPES, CLASSIFICATION, AND RELATIONSHIPS

Scientists assign Latinized names to groups of organisms to provide universality and stability and to avoid translating the names into various languages. No matter what one's native language, the Latinized name is immediately associated with the same group of organisms. The *International Code of Zoological Nomenclature* (International Commission on Zoological Nomenclature 1999) has been established to ensure the application of these principles and to avoid duplication of names.

It should be clear from what I have said above that the species is the fundamental category in classification. The name of a species consists of two words (is binomial): the first word is the name of the genus (the generic name) and indicates its closest relatives; the second name (the specific epithet) is uniquely used for that species within a genus. The generic name is always capitalized, the specific epithet is always written in lowercase, and both are always printed in italics. For example, the name of the yellow-bellied sea snake is *Pelamis platurus* (Linné, 1758). Specific epithets frequently are based on a distinctive feature (in this case the flat tail), refer to the species' place of origin, or honor a person by Latinizing that person's name. In this book the name of the person who first described and named the species and the date of the original description immediately follow the specific name. If the author's name and the date are in parentheses, the species has subsequently been transferred to a genus different from the one in which it was first described.

Scientific nomenclature requires that each species name be associated with a particular specimen—the type—as the point of reference for determining if material collected later is of the same species. Sometimes a species is described by two authors at different times and in different places, and the name-bearing type specimens must be compared to clarify the situation. In such a case the name and description that appeared first chronologically have priority and the second name is a junior synonym.

The following terminology is used when discussing type specimens:

Holotype: the single specimen on which the original description was based

Syntype: one of several specimens on which the original description was based if no single specimen was designated as the holotype

Lectotype: one syntype subsequently selected as the name bearer

Neotype: a specimen designated as the type where no holotype, lectotype, or syntype is believed to exist

Paratype: specimens other than the holotype or lectotype used to prepare the original description.

Holotypes, lectotypes, and neotypes are unequivocal name-bearing types. Sound nomenclatural practice recommends that a lectotype be selected from among the syntypes, whenever possible, to avoid ambiguity, especially if specimens of more than one valid species were inadvertently included among the syntypes. Paratypes and lectoparatypes, examples from the syntypic series other than the lectotype,

are non-name-bearing types that may be consulted to ascertain the original describer's concept of a nominal species' limits. The *International Code of Zoological Nomenclature* (article 16.4) now requires that a holotype or syntypes be designated in the original description of any newly discovered species to make the species name available to compete for priority.

Scientists use a standardized set of terms (categories) for each level in the hierarchical arrangement of taxa, and each of these has a single Latin name (is uninomial), is always capitalized, and should not be italicized. Amphibians and reptiles belong to the kingdom Animalia. This taxon contains all organisms that are multicellular, obtain their energy by ingesting food from their surroundings, and usually develop from the coming together of two cells—a large egg and a small sperm. The kingdom Animalia in turn belongs to the all-inclusive group that would include all living organisms, that is, all the descendants of the single original ancestral organism.

The kingdom Animalia is composed of about thirty-five phyla, each phylum representing a distinctive major biological organization and lifestyle. Amphibians and reptiles belong to the phylum Chordata, which consists of animals that have a single dorsal nerve cord, a notochord (at least embryonically) that lies below the nerve cord and extends the length of the body, and gill arches and gill pouches (present only embryonically in reptiles, birds, and mammals). In other chordates, including larval amphibians, the gill arches support functional gills and the gill pouches in the throat are open.

Three subphyla are recognized within the Chordata; amphibians and reptiles are placed in the subphylum Craniata (or Vertebrata). Animals belonging to this group have chambered hearts, semicircular canals in the inner ear, skeletal gill supports (integrated into the skull in terrestrial vertebrates), and a protective cartilaginous or bony cranium around the brain. Most members of this group also have segmentally arranged skeletal elements (vertebrae) surrounding and protecting the spinal cord (fig. 1.5).

The next division in the hierarchy is the class. Amphibians and reptiles are placed in different classes, Amphibia and Reptilia. Classes are further divided into orders; there are three living orders of amphibians and five of reptiles. Orders are divided into families, families are divided into genera, and each genus contains one or more species. The classification adopted in this book may also use additional categories in some cases for intermediate groupings of taxa, including subclass, infraclass, superorder, suborder, and infraorder in descending order of inclusiveness; subdivisions between the family and genus used for some groups are subfamily and tribe.

These hierarchical categories form a set of taxa diagnosed by shared derived similarities. This means that species within a genus are more closely related to one another than to species in other genera within the same family; genera in the

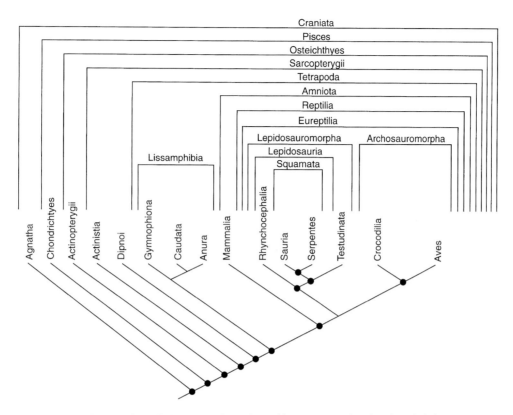

Figure 1.5. Cladogram of the phylogenetic relationships of living groups placed in the subphylum Craniata.

same family are more closely related to one another than to genera in other families; and so on. It also means that, progressing up the hierarchy, fewer and fewer features are shared by the species included in an order, a class, a phylum, or a kingdom. The hierarchy thus reflects in reverse the branching pattern of the evolution of life, because the decrease in shared derived similarity correlates with time of divergence: the greater the similarity, the more recent the divergence.

An example of the hierarchical classification for the yellow-bellied sea snake is:

Kingdom: Animalia
  Phylum: Chordata
    Subphylum: Craniata
      Class: Reptilia
        Subclass: Eureptilia
          Superorder: Lepidosauria
            Order: Squamata
              Suborder: Serpentes
                Family: Elapidae
                  Subfamily: Hydrophiinae
                    Tribe: Hydrophiini
                      Genus: *Pelamis*
                        Species: *Pelamis platurus* (Linné, 1758)

Throughout the systematic accounts (chapters 4–13), the name of each higher taxon represented in Costa Rica, when first cited or at the beginning of an account, is followed by

a citation of the author and date of the publication where the name was first used. The author and date are not formally a part of the name but provide entry into the literature. When the date on the publication differs from the actual date when it appeared, the former is included in quotation marks followed by the actual date of publication (e.g., "1858," 1859). A few family names are followed by two dates, neither in quotation marks. The first date is that of the original publication of the currently accepted family name. The second date, in parentheses, is the date used in establishing priority under article 40 of the *International Code of Zoological Nomenclature.*

Unlike the species, the genus name is based not on a specimen but on a type species. Family names in turn are based on a type genus, and the name is formed by adding *-idae* to the stem of the generic name (e.g., *Rana*, Ranidae); the suffix *-inae* is used for subfamily names, and *-ini* is used for tribal names. Names at the ordinal level and above are Latinized but have no standardized endings.

The sine qua non of systematization is to accurately represent the evolutionary history of organisms in a taxonomy that reflects their relatedness based on the patterns of ancestry and descent. As a result, classifications are modified to reflect newly understood evolutionary relationships. In this way the information content of the system increases. Classifications therefore evolve, and they must be understood as statements of current knowledge. Scientists also may disagree regarding parts of a classification because of different interpretations of available information. The classification of amphibians and reptiles used in this book is eclectic, if not idiosyncratic. I have generally followed the latest published work on any particular taxon but have not hesitated to use an alternative scheme where I seriously disagree with the published analysis or its interpretation.

Species

Recently there has been a resurgence of interest in, and controversy about, what constitutes a species (see Cracraft 1989; Templeton 1989; Frost and Hillis 1990; Mayr and Ashlock 1991; Graybeal 1995). Most biologists would agree that species are genetically and ecogeographically cohesive populations that constitute independent evolutionary entities. Often these entities are reproductively isolated from their close relatives (are closed genetic systems). In other cases, among sexually reproducing forms, some or considerable hybridization may occur.

Different emphases on reproductive isolation, cohesive or ecological mechanisms, and recovered history (phylogeny) have led to different concepts of species. For the purposes of this book, I have attempted to follow the evolutionary species concept espoused by Frost and Hillis (1990), which views species as the largest cohesive populations (lineages) that have become independent evolutionary entities. These entities are recognized by diagnostic features that separate them from other such entities.

From an ecogeographic point of view the geographic ranges of these entities are sympatric or allopatric. The former term is applied to species whose ranges overlap, the latter to those whose ranges are geographically separate. Sympatric species that share the same habitat in the area of overlap are termed syntopic, and those that do not are allotopic. There is usually little question as to the status of similar, related entities that occur together and maintain their distinctive features (that have genetic cohesion and are diagnosable). More difficult to interpret is the situation where similar and presumably related populations have separate, nonoverlapping geographic ranges. Ideally the evolutionary independence of any of these entities, regardless of geographic or ecological characteristics, would be confirmed through analysis of phylogenetic relationships, but few Costa Rican genera have been subjected to this kind of rigorous analysis.

Consequently the following operational definition of species has been applied throughout: Species are the most inclusive populations of organisms that differ from one another in one or more discrete characteristics (are diagnosable) and do not intergrade in nature, although they may occasionally hybridize with one another. Intergradation in this context means that although certain portions of a population appear to be diagnosable, numerous intermediate individuals confirm that they are part of the same entity. This may apply to

an intermediate geographic area between superficially different sets of populations or to situations where there is overlap in features among otherwise allopatric populations (e.g., among several mountaintops), as in the snake *Ninia psephota*.

Most species are diagnosable based on coloration and morphology, which reflect genetic cohesion. However, recent advances in chromosomal and molecular techniques have given scientists the ability to more directly assay the genetic characteristics of populations. These studies have led to the discovery of separate, previously unsuspected genetically and ecogeographically cohesive units within what appeared to be a single taxon based on external features. These cryptic entities are diagnosable on chromosomal and molecular features and clearly represent distinct species. Fortunately, only a few such cryptic species are known among Costa Rican taxa (e.g., the salamanders of the *Oedipina uniformis* complex), although others are sure to be discovered.

As I indicated above, sometimes hybridization may occur between species with syntopic distributions. Frequently such mismatings produce nonviable or sterile offspring, but this is not necessarily the case. Hybridization among reptiles in nature is rare because of their morphological and behavioral requirements for internal fertilization. Closely related species, however, may be forced to breed in captivity. It is unlikely that Costa Rican caecilians or salamanders will mismate, for this same reason. Because most frogs and toads practice external fertilization, there is a greater chance for the occasional production of hybrids. I am unaware of any definite hybrid individuals from the republic. If a presumptive hybrid is collected, it will be difficult to identify because it may share a mixture of diagnostic features. It is best to have someone knowledgeable in herpetology confirm the status of such a specimen.

On Geographic Variation and "Subspecies"

The characteristics of many entities in the species category differ in different parts of their geographic range. When coloration or scalation differences appear substantial, the variations have often been recognized as subspecies or geographic races and given a trinomial name (e.g., *Coluber mentovarius striolatus*).

Unfortunately, the notion of subspecies has been used to denote several very different situations. The first of these involves recognition of arbitrary convenience classes. For example, the populations of the snake *C. mentovarius* from Central America and eastern Mexico differ from those in northwestern Mexico and have been dubbed *C. m. mentovarius* and *C. m. striolatus*, respectively. These two population systems are connected along the west coast of Mexico by a series of populations that gradually change in color from one extreme to the other along a north-south gradient. As Wilson and Brown (1953) pointed out long ago, full study of such situations usually shows that features other than the one emphasized have a different and discordant geographic pattern of variation. Other authors have recognized as subspecies any two populations whose members may be separated on some variable feature that is more common in one than the other. Often in these cases intermediate geographic areas are populated by individuals intermediate in the emphasized subspecific feature. Frequently the intermediates have a larger geographic range than either of the recognized subspecies.

The recognized "races" in either case are neither diagnosable nor historical entities. They remain pattern classes arbitrarily delimited from within a geographic continuum. Some argue that they are a useful convenience for describing geographic variation, but recognizing them implies a biological reality that does not exist. Actually, using the trinomial usually obfuscates other elements of variation within the so-called race.

Another use of subspecies relates to similar and presumably related allopatric populations such as those on the mainland and an island or on a series of mountaintops. Even when diagnosable, these entities often have been incorporated into a single species but called different subspecies in a tradition established by and advocated by Mayr (1942) and discussed in Mayr and Ashlock (1991). Whether such isolates are reproductively isolated or not, they currently represent independent historical lineages by reason of their allopatry and should be recognized as distinct species.

The subspecies originally was invented to preserve familiar names that would otherwise have been synonymized and to describe a limited aspect of geographic variation. It is not recognized as a taxonomic category in this book. Where appropriate, geographic variation is described in the species accounts by geographic area, as recommended by Wilson and Brown (1953) and endorsed by Savage and Heyer (1967) for herpetological analysis. The diagnosable allopatric subspecies simply evaporates as a concept, since it always was a species in disguise.

SCIENTIFIC MEASUREMENT

All measurements used in this book are in the metric system. In this system, unlike the American and British ones, all multiples and submultiples of the units are formed by standard prefixes no matter what kind of measurement is involved. For those uncomfortable with the metric system, the following *approximate* conversion factors may be used for the most frequently mentioned units:

Linear: millimeters (mm) to inches, divide by 25; centimeters (cm) to inches, divide by 2.5; mm to feet, divide by 320; cm to feet, divide by 32; meters (m) to feet, divide by 3.3; kilometers (km) to miles, divide by 1.6

Area: m² to square feet, multiply by 11; hectares (ha) to acres, multiply by 2.5; km² to square miles, divide by 2.5

Liquid measure (U.S.): milliliters (ml) or cubic centimeters (cc) to ounces, divide by 30; liters to gallons, divide by 3.8

Weight: grams (g) to ounces, divide by 28; kilograms (kg) to pounds, multiply by 2.2.

Temperature: Celsius (centigrade) to Fahrenheit:

   35°C = 95°F
   30°C = 86°F
   20°C = 68°F
   10°C = 50°F
    5°C = 41°F
    0°C = 32°F

LOCATING, OBSERVING, AND PHOTOGRAPHING AMPHIBIANS AND REPTILES IN THE WILD

Locating and Observing Wildlife

The best tools for locating wild amphibians and reptiles are awareness of and concentration on the sights and sounds of the environment. When you are walking trails, frequent stops to visually survey the ground ahead, look up into the overhanging vegetation, and scan vines, tree trunks and branches, and the foliage of understory plants will increase your chances of seeing immobile or fleeing individuals. As you walk, small frogs or lizards or an occasional snake may be startled and try to escape from right under your feet. If you remain motionless, often they will stop after a short flight and sit on the surface. Listen carefully as some diurnal frogs call during the day from the leaf litter or from fallen logs or other elevated spots. The sudden rustling in the leaves just off the trail usually means a lizard is trying to escape the perceived threat of your movements.

A pair of binoculars is particularly useful for observing many of these animals at a distance, since most will hide or flee when closely approached. Binoculars with magnifications between 6× and 8× are best for field study. Adequate near-point focusing and light-gathering power are essential; the binoculars must be focusable at short as well as long range. Light-gathering power can be estimated by dividing the diameter of the front (object) lens by the magnification. Lens diameter is usually marked on field glasses right after the magnification (e.g., 7 × 35). In this case the power would be 5, but any value between about 4 and 7 is satisfactory.

Some forms are less skittish than others and may be approached if you move slowly at an angle to them, watching them out of the corner of your eye. Once you are in a position for further observation, stop and remain quiet and motionless. Many lizards perched on tree trunks or fallen logs or fences escape by scampering to the side of the trunk away from an observer. If there are two people, one can slowly move to the other side of the tree, and the lizard will re-escape by returning to near its original perch. The stationary observer can then identify it or watch its activities.

Nocturnal or secretive forms may often be found under debris on the ground, in piles of trash or leaves, or under fallen logs, rocks, and such. You can carefully lift, turn, or move these objects to discover who may be hiding below. If you are using this method, always turn the log or other object toward you so it is between you and any venomous snake that might be underneath. Do not put your hands on the far side of a large object that you mean to turn without first looking to see what may be there. If you find any animals, return them to the original spot of capture once you identify them. As much as possible, replace the logs and rocks and other disturbed materials as well. In more open, drier, or human-modified situations, many lizard species may be found on fence posts, stone walls, or walls or around human habitations.

Freshwater turtles and crocodilians are best observed along large streams during the day. They are frequently seen from bridges, but a leisurely canoe trip that mixes drifting with the current and paddling will usually be more productive. You are also likely to see other large reptiles during such an outing along the stream banks, in the trees overhanging the stream course, or even in the water. These include iguanas, basilisk lizards, boas, and other large snakes. If you have the gear, snorkeling is an excellent way to observe turtles in clear water.

Visual surveys at night will sample another subset of species. When walking trails you should follow the same procedures as during the day. It is particularly important to scan leaf surfaces and branches of the understory plants for salamanders, frogs, and sleeping lizards and to look up into the overhanging vegetation for arboreal frogs and snakes active at night. Watch where you are standing before you look around, not so much because you are likely to step on a venomous snake as because you may be standing on a stinging ants' nest! Again, a canoe trip at night is an exciting experience and will usually turn up some species not seen during the day.

Most frogs and toads are nocturnal and are readily located by their calls. During the breeding seasons large choruses may be heard from a considerable distance calling from marshes, ponds, or the shallows of slow-moving streams. A visit to a chorus will introduce you to many new forms. Beside breeding amphibians, you are likely to observe several species of frog-eating snakes around the breeding site and in the overhanging vegetation. In montane situations, the hardy or daring observer will find that walking up small torrential streams usually leads to the discovery of another subset of frogs and toads. In both cases, take advantage of the opportunity to learn the calls of the different anurans. Since calls and calling sites are quite specific, you will soon learn where to look for the calling males of particular species.

In the subhumid northwestern region of Costa Rica, you may discover terrestrial turtles by cruising roads at slow speeds (under 40 km/h), especially early in the morning.

Similarly, cruising paved roads at night is an excellent way to look for snakes. On nights of heavy rain many frogs and toads may be found in ditches and temporary ponds along roadsides. Occasionally freshwater turtles are found crossing the road from one pond to another in the rain. Mixed breeding choruses of frogs and toads may also be located by driving slowly along the road, stopping from time to time, then turning off the vehicle's engine and listening for the cacophony of sounds.

Sea turtles are best observed on the nesting beaches at Santa Rosa (August to December) and Tortuguero (July to October) National Parks on the Pacific and Atlantic coasts, respectively. Leatherback sea turtle nesting, unlike that of other species, peaks in November and December on the Pacific coast and in April and May on the Atlantic.

Photographing Wildlife

Wildlife photography is now a major hobby for many people. Herpetologists also take photographs to record life colors, behaviors, and habitats. Since most amphibians are small, macrolenses and flash or strobe lights are essential for good shots. A telephoto lens is a must for photographing shy or small species from a distance. In most cases a 35 mm single-lens reflex camera with or without extension rings is adequate for species that may be approached closely. For close-ups of small animals a macrolens is useful, as is a 135 mm or 200 mm telephoto lens for photographing animals at a long distance. The camera should have a through the lens (TTL) light meter with TTL flash and flash synchronization at speeds up to 1/250 for on-site or nighttime photography. The increasing availability of moderate-priced digital cameras and rapid improvements in their technology may make them the cameras of choice for fieldworkers in the future. With small or nervous individuals, often the best results are obtained by bringing them back to camp, field station, or laboratory and posing them using natural backgrounds under good lighting conditions. After the photographic session they should be released where they were captured, unless there is a scientific reason to keep them and you have the necessary collecting permits.

Amphibians, anole lizards, and the background materials need to be kept moist during these sessions. Considerable patience is usually required to pose the specimen, since most animals will try to escape. If repeatedly posed and recaptured as they try to flee, most small specimens become exhausted and are then quiescent and easy to pose. Some photographers keep the subject in the dark under an upside-down box or other container. When the animal is suddenly exposed to light by removing the container, it usually freezes, and a quick shot often produces excellent results. Of course, for this to work the photographer needs a second person to handle the animal and container. I have found it useful to cool down especially frisky individuals before photographing them by placing them in a container in a refrigerator or ice chest for a short time. Many lowland species are very cold sensitive, so if you use this trick they should be checked frequently to prevent hypothermia.

Salamanders and frogs are best posed facing the camera at a three-quarter angle. Setting lights at a distance or using a reflector will reduce the light reflected from the mucus-coated skin. Remember, many frogs and toads have toxic skin secretions. If you are photographing several species, wash your hands before changing specimens, or these toxins may harm or kill animals you handle later. Terrestrial and freshwater turtles are particularly difficult to photograph, since they frequently withdraw the head and limbs inside the shell and refuse to extend them while the photographer is near. The best shots are obtained when the shell is clean and moist so that the color pattern is evident and when the head, neck, and front limbs are extended. You may have to wait quietly for a considerable time.

Ashton and Ashton (1985) recommend photographing turtles underwater in a small (38 l) clean and unscratched aquarium filled with clear water and with clean sand or gravel on the bottom. Natural props of aquatic vegetation or waterlogged leaves may be added to create a natural look. Dual flash or photo lights placed above and on the sides will eliminate shadows.

Lizards and young crocodilians should be posed looking toward the camera or slightly to one side. The eyes should be open, the head erect, the digits extended, and the tail parallel to the body but in full view. With calm or cooled-down individuals the appendages can be slowly and gently moved with a pencil point or twig. Be careful in handling lizards because many have very fragile tails. Even small crocodilians and some lizards can inflict painful bites, so handle them with due respect. Caecilians and snakes are best photographed when partially coiled with both head and tail included. Side views or those taken from directly above produce the best results. Snakes often bite, so take reasonable care in handling them. Three genera of harmless snakes, *Enulius*, *Scaphiodontophis*, and *Urotheca*, have very fragile tails, which you should avoid touching.

Sea turtles and large crocodilians, on land or in the water, and large lizards and snakes offer good photo opportunities in the field for cameras equipped with telephoto lenses. Nocturnal photographs, especially of calling frogs and toads, are the specialty of some photographers. Flash or strobe lights are obviously a must for these shots, and it is best to use a headlamp for locating the subject and focusing the camera.

Aquatic eggs and tadpoles present special problems. They should be photographed underwater in an aquarium, with lighting from above. The floor of the aquarium should be black. This can be achieved by placing the tank on a black background (e.g., tarpaulin or poster board). A more permanent setup could include lining the dry aquarium with plaster of Paris dyed with india ink or some other nontoxic,

waterproof coloring material. Sticks, leaves, rocks, and such may be added before the plaster dries. Once dry, water may be added and the plaster plus other materials will look like a natural background in the photographs. It is also best to place blue or light green paper on the outside of the back of the aquarium so extraneous objects will not be included in the photograph.

The recently published *How to Photograph Reptiles and Amphibians* (West and Leonard 1997) provides many useful suggestions on photographing these animals in nature. It is an excellent resource for herpetologically inclined photographers.

Field Notes

You will probably find that your pleasure in discovering and photographing Costa Rican amphibians and reptiles is enhanced by keeping a field diary or notebook. In it you may record sightings of and observations on animals you see and details regarding your photographs. Most people keep their notes by date, but you may want to keep separate species sections as well. Durable, bound data books about 12 × 20 cm are the most convenient, since they may be slipped into a pocket. The amount of detail recorded will vary greatly depending on the depth of each individual's interest. The detailed kind of field book kept by the professional herpetologist, as described below, is at one extreme, and a simple life list of species observed is at the other. Whatever format you select, I am sure you will greatly enjoy reading this record in later years, since it will remind you of the excitement of your adventures in tropical herpetology.

COLLECTING AND TRANSPORTING LIVING SPECIMENS

Costa Rica has recently enacted stringent laws regarding the collection of specimens of native animals for any purpose. Both amphibians and reptiles are covered by these laws. Consequently the following sections on collecting, transport, and preservation are for the information of students, herpetologists, or other environmental scientists who are properly permitted by the authorities to collect specimens for research purposes.

The section on care and feeding of captive herpetofauna also is primarily for the serious student holding appropriate permits. However, since it is not illegal under present Costa Rican law to keep captive-bred and some nonnative species in captivity, that section may also be of value to the herpetoculturalist.

Knowing the specific habitat and general habits of particular species is the first step in trying to locate them. What information is known is included in the species accounts in this book. Observing individual animals as described in the previous section is the next step in finding out more about them. However, the serious student who needs to confirm identifications, obtain additional biological data, or take voucher specimens for a scientific collection will find the following sections useful. They outline general procedures, but in a few cases more specialized techniques will be mentioned in the individual species accounts. These are no means exhaustive; most field collectors have favorite methods that work well for them.

The tropical herpetologist requires, as a minimum, four items of equipment (fig. 1.6: a three-prong potato rake, a headlamp, "tico" boots, and collecting bags. The rake may be used as a snake stick, to turn over logs or rocks, to pull bromeliads and moss down from trees, and as a staff when the mud is deep and the going gets tough. It should have a handle about 1.3 m long, long enough to fend off an irate terciopela *(Bothrops asper)* but short enough to be maneuvered readily. A headlamp is essential for night collecting and leaves both hands free to catch the prey. The best are rechargeable miners' or hunters' lamps. Most give powerful illumination for about eight hours on one charge. The light from these does not gradually fade as the battery discharges but goes out suddenly, and it is best to take along a less

powerful reserve light so you can find your way home if need be. Rubber boots or, as the British say, gum boots that come up to the knee are sold everywhere in Costa Rica and are the best general wear for soggy or wet conditions. Muslin bags or pillowcases are best for general collecting. Many collectors use plastic bags when catching amphibians or small lizards, but these warm up very quickly during the day, and the animals need to be transferred to other containers or else hyperthermia and death may result. However, plastic bags may be used to restrain specimens while they are being identified and then released.

Three other essential items, not specifically related to collecting, are (fig. 1.7) insect repellent, a Swiss army knife, and duct tape. The first has an obvious use, the other two form a repair kit that will allow you to fix almost any other equipment used in collecting.

Mano a Mano

Many kinds of amphibians and reptiles may be simply picked up or grabbed using only your hands. Quick grabs are required for specimens sitting on leaf surfaces, twigs, or

a

b

Figure 1.6. Basic collecting gear for the tropical herpetologist: (a) three-prong potato rake, headlamp, boots, and collecting bags; (b) using the rake.

Figure 1.7. Essential items of field gear for the tropical herpetologist: insect repellent, Swiss army knife, and duct tape.

small branches. A slapping-down technique is effective for animals that are motionless on the ground. Don't be too vigorous, since you may physically harm the more delicate species and some may become overheated if held too long. Most small amphibians and lizards are best restrained by holding on to a limb or limbs. Large frogs and toads, once captured, should be held firmly around the waist; they are very slippery and may kick themselves free if held by a limb.

Caecilians are extremely slimy and hard to hold. I recommend a three-finger viselike "eel hold," preferably using both hands, where the middle finger encircles the body and presses the caecilian against the tops of fingers one and three.

All but the smallest lizards and small crocodilians should be firmly grasped about the body and the neck to avoid a sudden lunge to escape and to prevent your being bitten. This technique will also apply for handling the largest Costa Rican salamanders. You will need both hands for large lizards, one at the neck and the other around the body. Never grab or hold a salamander or lizard by the tail, which in most species is easily shed.

Small snakes may be grabbed by hand. A moderate-sized, especially swift, or even large snake may be initially immobilized by gently stepping on it. Use the other foot or a branch, rake, or snake stick to pin down the head. Then you can pick up the snake by grasping it firmly behind the jaws with the thumb and middle finger, with the index finger on top of the head. Hold on to the body with your other hand or else the snake will start thrashing around and pull free of the neck hold. The best way to secure small animals is to gently toss them into a container or collecting bag. The container is immediately closed with a cap that has been held in the free hand. A similar strategy works with bags, but the bag is quickly twisted to close it. Shaking the bag will force the specimen to the bottom before you knot the top of the bag. Larger specimens are usually held in one hand, placed inside the bag, and grasped through the bag with the other hand. The now free hand is withdrawn from the bag and the

upper portion of the bag twisted shut. The bagged animal is then released and the bag is knotted closed. Large snakes are often not very accommodating about being put in a jar or bag, and this is usually a two-person job. Bagging a venomous snake is not recommended without prior training.

Snapping turtles *(Chelydra)* are easily held by their long tails, but keep the head aimed away from your legs while holding or carrying them. Catching any but small juvenile crocodilians (under 500 mm in total length) by hand is foolhardy, and few people have a legitimate reason for capturing large ones, since they are fully protected. Turtles and very large lizards and snakes are usually placed in heavy canvas sacks or wooden boxes with lids that can be closed and secured.

Rolling Logs, Raking Leaves, and Turning Stones

The potato rake is invaluable for these purposes. You can also use it to take apart rotten logs and remove loose bark or shag from downed trees or snags. All this can be done with your hands, a pinch bar or crowbar, a hay hook, or a geologist's pick, but the rake ensures a safe distance from any surprises nipping at your fingers. Even so, wearing gloves is a good idea, because you are likely to end up sorting through litter and other debris by hand after you use the rake. An especially good use for the rake is rummaging through the leaf litter between the root buttresses at the base of large trees.

Many people prefer a snake stick to a rake. These come in several varieties (fig. 1.8). The simplest is an L-shaped metal angle iron attached to a wooden handle 1.2 to 1.5 m long (e.g., a broom handle or a dowel) or welded to a metal handle. An old golf putter with the head filed down also makes a good snake stick. The free edge of the short arm (bar) of the stick should be beveled so it will slide under a snake on a hard surface. In catching a snake the bar may be used to hook or pull the snake from a resting or hiding place or to pin it down until you can grasp it securely by the neck. Captive snakes usually become habituated to being picked up on the stick's bar, and they will balance themselves there when being carried short distances. Don't expect naive wild snakes to follow the same pattern or you will surely lose valuable specimens.

The stick may also be used much the same way as the potato rake in looking for animals. However, the simple angle iron on a wooden handle is unsatisfactory for this because it tends to break or bend with heavy work. The classic forked stick is virtually useless for catching snakes. If the fork is too large small snakes slip right through, if it is too small it will harm a large snake. It also makes it almost impossible to grasp the snake around the neck so it can't bite you. Canvas or leather gloves are useful, too, when dealing with large snakes so as to avoid nasty bites.

Some herpetologists use metal snake tongs (fig. 1.8) to capture or manipulate captive snakes. These tongs are available commercially. They consist of a long handle (0.9 to

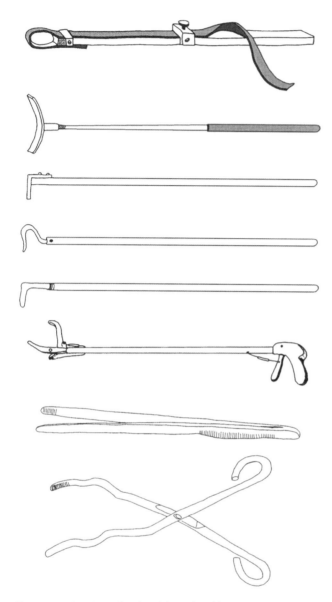

Figure 1.8. A variety of snake sticks and grabbers.

1.2 m long), connected to a pair of jaws that operate by spring action. Pressure on the hand grip closes the jaws around the snake's body. Relaxing the grip releases the snake. The tongs are not as versatile for all aspects of collecting as well-constructed snake sticks or the potato rake. Also, the spring action may jam when it picks up small rocks, sticks, or other debris while grasping a snake. Many collectors use long (about 300 mm) forceps of various styles to pick up small snakes and medium-sized lizards.

Noosing and Hooking

Many lizards and snakes are too wary or too fast to catch by hand. A slip noose of heavy thread, nylon fishing line, or copper wire attached to the end of a 1 m or longer pole is an excellent way to catch such animals without harming them

(fig. 1.9a). Use number 50 thread for small lizards (100 to 120 mm in standard length) and number 8 thread or fishing line for larger specimens. The noose should be tied to the notched end of a flexible stick, around the tip of an old car radio antenna, or through the last eye of a telescopic fishing rod, so it won't slip off. The shank of the noose should be about 100 to 150 mm long when the noose is open. To make the loop, first make a small loop about 6 mm in diameter at the end of the thread or line and tie it with a square knot so it won't close. Then thread the shank through the small loop and tie it to the pole. If cotton or nylon nooses are used, the experienced collector rubs the loop with paraffin, other wax, or saliva from time to time to help keep it open. Copper wire nooses have the advantage of not being blown about by any slight breeze or closing on themselves, but they often break and need to be replaced. They can be made of a single copper strand from an old electrical cord. Twist a small loop 4 to 5 mm at the end of the shank and then pass the shank through the loop and attach to the pole. The best way to do this is by wrapping the wire several times, then winding the free end around the base of the shank. After each use reshape the noose with the fingers to eliminate kinks and straighten the shank to make it the desired diameter.

Catching animals with these nooses is all in the wrist. Slip the noose slowly and quietly over the animal's head until it reaches the neck. Then jerk the pole upward and slightly backward and you've made your capture. Because the lizards weigh very little, they suffer no damage from the noose. Most lizards are rarely bothered by the approaching noose; some even grab it, apparently confusing it with a flying insect. Some need to be approached more slowly with an occasional pause so that the noose does not frighten them away. You can distract wary individuals from watching or avoiding the noose by slowly moving your laterally extended free hand or fingers. For lizards without distinct necks (e.g., skinks) one needs to open the noose wide and wait until a forelimb passes through before flipping it closed.

Larger and stronger nooses are required for capturing crocodilians, large lizards, or snakes. Care must be taken in handling the first two so you aren't seriously damaged by claws and tails while avoiding a bite. Professional collectors of live crocodilians usually immobilize the jaws by wrapping duct tape around the animal's snout, with its mouth tightly closed, before handling it further.

Some species of freshwater turtles may be caught from near shore with a hook baited with meat or fish or with a series of hooks on a trotline. The snapping turtle *(Chelydra)* is the only Costa Rican species reported to be collected in this fashion, but mud turtles of the genus *Kinosternon* are caught by this method elsewhere in their range.

Netting

Many a frog and lizard has been captured with an insect net (fig. 1.9b). Animals may be scooped up from the ground or

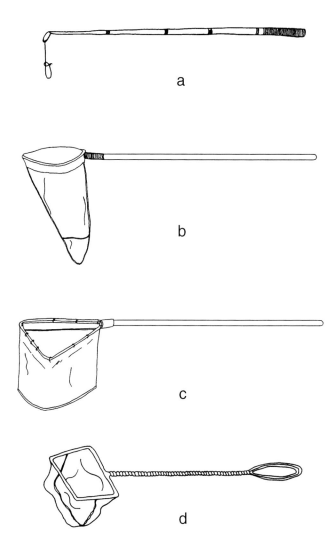

Figure 1.9. Additional collecting devices: (a) lizard noose on a fishing pole; (b) insect net; (c) aquatic net; (d) aquarium net for collecting eggs and tadpoles.

swept out of vegetation. Other times you can quickly slam the net over them from above. In most cases you should then rapidly rotate the pole to ensnare the frog or lizard in the netting and simultaneously close the net by wrapping the bag around the hoop. These nets are made of fine-mesh nylon, and the bag should be at least 900 mm deep and have an opening of 550 mm. They are attached to a spring trap hoop mounted on a lightweight pole about 1.5 to 2 m long.

Dip nets may be used to capture frogs calling from aquatic vegetation, small aquatic turtles, and crocodilians. The best sizes have 250 mm or 300 mm openings with depths of 250 mm and 450 mm, respectively. They usually need long handles, so get one with a ferrule and locking device so it can be taken apart for transport. With a bag made of fine mesh, you can catch small animals or even tadpoles. For serious sampling of aquatic forms hiding on the bottom or in marginal aquatic vegetation, a triangular double-bag net is very useful (fig. 1.9c). The inner net of fine mesh is protected

by an outer canvas bag. This net is used like a small dredge except that you push it along the bottom by the handle to stir up specimens, then scoop them up. Near shore, you slip the net under the vegetation and vigorously move it up and down to shake animals into it. A small fine-mesh aquarium net with a 125 mm handle is particularly useful for collecting tadpoles (fig. 1.9d). These nets are easy to carry and may be purchased at any aquarium store.

To capture turtles the serious student may also use larger nets, particularly trammel nets of various lengths by 2.5 m deep. These nets are composed of three nets hung from a single float and attached to a single lead line. The two outer nets have relatively large meshes, and the inner net is loosely hung and of smaller mesh. Turtles trap themselves in pockets of the fine webbing pushed through the larger openings in the outer nets on either side. Trammel nets must be checked frequently to prevent the submerged and trapped turtles from drowning. There may be some restrictions on their use, and you need to obtain clearance from the appropriate fisheries office before starting a project including them in its design.

Trapping

Traps are most commonly used to collect freshwater turtles and in long-term field studies of other taxa. The two most often used for turtles are funnel traps (fig. 1.10a) and barrel traps (fig. 1.10b). The former may be a box constructed of 5.5 mm hardware cloth or a hoop trap made of 25 mm mesh cord netting. Both have a funnel entrance at one end that a turtle can enter easily but that is reduced to a small slit entering the main trap space. The main portion of the trap should contain a bait can suspended from the upper frame and filled with chopped raw fish, chicken entrails, or carrion. The bait can, about the size of a soup can, should have a tight fitting lid with holes punched in it and be completely submerged. The trap needs to be kept rigidly in place by long stakes driven into the substrate at the four corners. The upper part must be above water so the turtles will not drown. Sizes vary, but a typical hoop trap would be 1.2 to 1.5 m long and 750 mm in diameter.

A barrel or basking trap consists of a smooth-sided empty barrel or drum. It is used to catch herbivorous turtles that will not be enticed into a funnel trap. The trap is set with its rim just above the waterline near the turtles' basking log and weighted down with cement blocks or rocks. A two-by-four with a counterbalance is placed over the trap opening so that a turtle's weight will drop it into the trap. The two-by-four then springs back into position to await the next turtle to try out the "new" log. The size of the trap varies with the species of interest.

Pitfall traps (fig. 1.11a) are simply containers placed so the top is flush with the ground. The most commonly used sizes are 19 l plastic buckets or 8 l cans. In the Tropics pitfall traps need to have a cover that is slightly elevated above

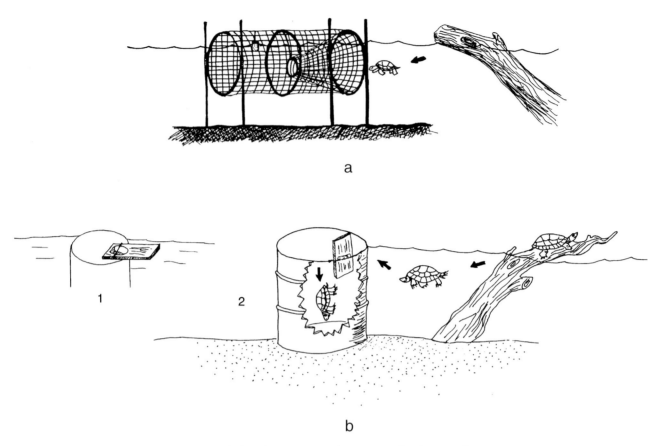

Figure 1.10. Turtle traps: (a) funnel hoop trap; (b) barrel trap: (1) set with counterweight; (2) sprung by turtle's weight.

the opening to keep out water and allow animals to crawl under it and fall into the trap. Straight-line drift fences are short barriers (5 to 15 m) that direct amphibians and reptiles moving across the ground into traps at the ends or beside the barrier (fig. 1.11c). The traps may be pitfalls but usually are small wire funnel traps (fig. 1.11b). Drift fences may also be designed to encircle a breeding pond or other study site. Drift fences are best used for long-term studies, since the material and labor costs for construction are high and monitoring the traps is time consuming.

Sampling Leaf Litter, Moss Banks, and Bromeliads

Leaf litter organisms are frequently caught in pitfall and funnel traps. If a group such as a class is available, a roundup of litter amphibians and reptiles also can produce excellent results. In this process an 8 × 8 m plot is laid out, and individuals are spaced around the periphery. They then begin working their way through the litter, all moving at about the same rate toward the center of the plot and capturing any specimens they uncover. If only a few people are available, the same basic procedure may be followed while putting all the litter into large plastic bags to be sorted through after catching all specimens scared up during the process. The litter can be scooped up with small plastic wastebaskets to expedite sampling. Large mats of mosses from the ground or from large branches of trees may similarly be searched and bagged for further examination.

Tank bromeliads serve as a hiding place for some salamanders, frogs, lizards, and snakes and as breeding sites for some of the amphibians. They may be carefully brought down from the trees, placed in plastic bags, and later carefully searched for eggs, tadpoles, or adult specimens. It is important to keep the plants upright while transporting them so as not to spill the water where the early amphibian stages are located.

Road Running

On any long trip you will occasionally see animals on or near the road, as described in the section on field observation. The most effective and widely used method of collecting them is road running—driving slowly along paved roads at night. Some snakes and geckos tend to rest on the blacktop because it retains heat longer than the adjacent air and soil. Speeds of 40 km/h or less allow you to see most specimens except the very smallest. Many objects on the road—sticks, banana peels, pebbles, parts of tires, fan belts, and

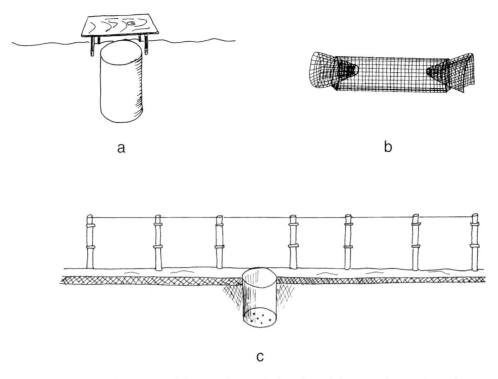

Figure 1.11. Terrestrial traps: (a) pitfall trap and cover; (b) funnel trap (a larger version may be used as an aquatic turtle trap); (c) drift fence with pitfall trap.

such—may resemble your prey, but always check to make sure they are not dead animals, also called DORs ("dead on road"). A DOR may represent a significant scientific record and usually is in good enough shape to be preserved.

It is helpful to mark the spot where you see a potential specimen. A collecting bag with a rock tied in it will serve this purpose and may be tossed out the window as you pass. Then you can back up until the marker is in the headlights. Be careful not to run over the suspected specimen, by backing up on the side of the road opposite to where you saw your prey. Watch the other traffic on the road at all times, especially before leaping out of the vehicle. Relatively little-traveled roads in northwestern Costa Rica are the most favorable for this technique. You can cruise gravel or dirt roads in a similar fashion, but they usually do not yield as many specimens.

Road running at night is also a way to locate many kinds of frogs and toads. This works best when it has rained in the late afternoon and during or immediately after evening rains. At such times you can capture many frogs and toads crossing the road from roadside ditches or temporary ponds. You can also cruise to locate breeding choruses of anurans as previously described.

Shooting

Many herpetologists collect animals for preservation by using firearms. Lizards in particular are hunted with a smoothbore .22-caliber pistol, preferably a revolver with a long barrel. These use shells containing fine "dust shot," since the shot stuns or kills the lizard without much damage. For large lizards and crocodilians you need a rifle firing solid shells. Generally, Costa Rican authorities frown on using firearms for collecting even if you have permission to carry out a project in a particular area.

In open areas lizards and larger frogs may be captured using slingshots, rubber-band guns, or large rubber bands. These normally stun the prey, so you must be quick to catch it while it is still dazed. Fine gravel, lead shot, or dry beans can be used as ammunition for a slingshot. A typical rubber-band gun is handheld, with a barrel about 450 mm long and a clothespin trigger attached to the handle with its jaws opening upward. A 12.5 mm rubber band is stretched from the end of the barrel to the trigger and fired by opening the clothespin. Two large rubber bands, about 400 to 500 mm long when tied together, may be used without a gun. One end is held between the thumb and index finger and the other is stretched backward by the thumb and index finger of the other hand as if drawing a bow. Opening the drawn-back hand will send the band flying free, just as when we fired small rubber bands at one another in grammar school. All these shooting methods take some practice before using them in the field.

Eyeshine, Voice, and Triangulation

At night several kinds of animals may be located by light reflected from their eyes. This is most effective when wearing a headlamp, since the light source needs to be near your own eyes so you can see the reflection. The color of the eyeshine may be white, pale yellow, bluish green, pink, or red depending on the species. Usually only those forms with large eyes can be spotted by this method, and one is often fooled by the conspicuous shines of spiders (silver, blue, or green) and moths (yellow or red). Eyeshine is most obvious in many frogs and toads and in crocodilians.

Although geckos squeak and crocodilians grunt, bellow, and roar, frogs and toads are the main animals in the herpetofauna that may be located by sound. Often you can find individual calling males in the forest or along a stream margin without much difficulty. In large choruses some species are also easily located, but others call from hiding or appear to be ventriloquists. In these circumstances two or more people may use a triangulation technique to locate a particular male. They approach from different directions, and both aim headlights at the spot where they think the call originates. The point where the beams from the lights intersect will be where the caller is. Particularly shy or hidden animals may need to be retriangulated one or more times, and even then they may be hard to find and capture. Triangulation may be used when working alone by taking bearings from two or more directions. Many anurans are unfazed by the headlamp's illumination. Others will stop calling when approached but will usually start up again if you turn the light off for a while.

Trying to imitate the call and writing down a description are useful ways to learn it. Calls may be recorded as grunts, whistles, tongue clicks, snorts, or other sounds to approximate your perception of their qualities and, if present, the sequence of different sounds. Serious studies of anuran vocalizations require a portable field tape recorder with or without a separate microphone. A relatively inexpensive system should suffice for most people, but the higher-quality recordings needed for audiospectral analysis mandate more sophisticated recorders.

Local Help

Local youngsters often have a surprising knowledge of and interest in the animals from their immediate locale. You can usually enlist them to look for amphibians and reptiles for a small fee per animal. They are often the source of apparently rare or particularly secretive species that you haven't found. It is best to set a low payment initially and, as animals are brought in over a few days, let the collectors know you will not pay for any more examples of common species, while raising the price for ones of particular interest. Sometimes the adults in a community will join in by bringing you some specimens, usually dead snakes they believe are venomous. Pay them the same prices set for the children, but let them know you don't need many venomous snakes.

Transporting

Basic containers needed in the field include several cloth bags, sturdy plastic bags of various sizes, and a few plastic bottles or plastic refrigerator dishes (fig. 1.12). Unbleached muslin is preferred to heavier cloth bags. Be sure the seams are well sewn, and wash new bags before use to remove any sizing. Heavier canvas bags or wooden boxes with securable tops will be needed for turtles and large lizards and snakes.

These bags may be used for all amphibians and reptiles but should be kept moist. Anurans and small lizards should have some moss, leaves, or paper toweling inside to retain moisture and give them a place to hide. Once a specimen is placed in the bag, close it by tying a tight overhand knot in the top. Be sure the catch is in the bottom of the bag so it isn't caught in the knot. Plastic bags are good temporary containers but tend to heat up and become contaminated with repeated use even if washed out each time. Salamander collectors prefer to place their catch in plastic bottles with a little vegetation. Both plastic and glass bottles, if used, should have small holes in the lid for ventilation. Drill the holes from the inside surface out so the rough edges don't hurt the specimens (fig. 1.12a). Do not overcrowd any container, and be careful not to mix frogs that produce skin toxins with other species or carnivorous frogs, lizards, or snakes with other individuals. Most of us will never need to catch and transport a really large live Costa Rican turtle, lizard, or crocodilian. These would all require special means of confinement and transport and will not be discussed here.

You must take care in transporting live material within Costa Rica because of the great elevational differences at various sites and the concomitant differences in temperature. Lowland amphibians and anole lizards are especially sensitive to cold, and many montane species suffer heat shock if suddenly transported to the lowlands. Styrofoam chests are useful for insulating specimens from temperature extremes while en route to the home base or new sites.

Other important considerations when transporting live specimens:

Never leave the animals or their containers in direct sunlight or in a vehicle sitting in the sun, especially one closed up and locked. They may be placed under the vehicle in the shade when you are stopped in a secure spot.

Be sure every container is labeled. Water-resistant paper labels written on in soft pencil or india ink or temporary numbered plastic tags go inside each bag. The advantage of the plastic tags is that they can be washed and reused after the day's catch has been preserved and labeled. Outside labels written on masking tape with ballpoint or felt-tip pens are an easy choice for bottles.

For eggs and tadpoles that you are rearing, use small plastic refrigerator dishes with a few holes drilled in the lid (fig. 1.12c). These should be placed upright in small terrycloth bags that are soaked in water from time to time. Fill

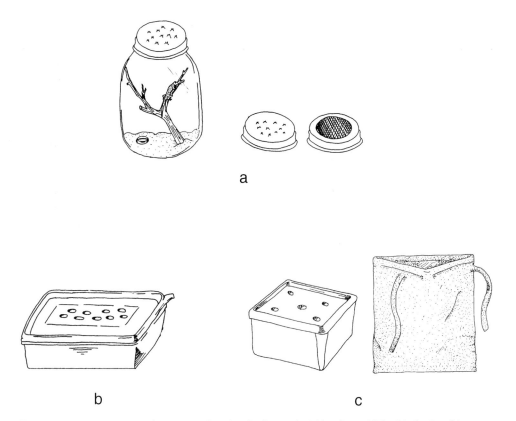

Figure 1.12. Transporting and temporary housing for livestock: (a) bottles and lids; (b) plastic refrigerator dish; (c) special transporter for eggs and larvae.

the containers to the brim to reduce sloshing and the resulting damage to eggs or larvae.

THE SNAKEBITE PROBLEM

I cannot overemphasize that *you should never try to catch a venomous snake unless you have received training in how to do so*. I strongly recommend that you read the sections on venomous snakes (families Elapidae and Viperidae) in chapter 11 before going into the field and that you learn to recognize these dangerous animals.

In Costa Rica about five hundred bites by venomous snakes are reported annually. Most are on the extremities of *campesinos* clearing undergrowth, grass, or weeds. Probably many bites are never reported because no medical aid is sought. The death rate of those receiving treatment is about 2%, but another 8 to 10% have serious permanent damage (Bolaños 1984; Hardy 1994a; Russell et al. 1997).

The chances of being bitten are slim if you take reasonable precautions such as wearing boots, watching where you place your hands and feet, and not handling venomous snakes. It is always best to do fieldwork in pairs so that if any emergency arises there is aid nearby. Even with these precautions, accidents do happen from time to time, and the following paragraphs offer advice on what to do in case of a venomous snake bite. They are derived from the recommendations of the Instituto Clodomiro Picado (Chaves et al. 1993) and Hardy (1994a, 1994b):

1. Get yourself or other victim to a hospital or other medical facility as soon as possible and try and alert the facility that you are on your way.
2. Most people are very frightened, so it is important to reassure them that death from a snakebite is rare and calmly arrange matters to expedite treatment; try to remain calm if you are the victim.
3. If the bite is on an extremity, try to immobilize the arm or leg; if necessary, cut away clothing to reach the wound. If the bite is by a coral snake, an elastic bandage may be applied over the site of the bite and wrapped around the limb as one would do for a sprain. Do not do this for a viper bite.
4. Extract as much venom as possible from the bite, using a rubber bulb or other device (Aspivenin or Sawyer Extractor). *Do not* make any incisions at or near the bite—they do no good and may cause serious loss of blood. Washing around the bite shortly after envenomation is also recommended.
5. If possible, kill the offending snake and take it with you

to the medical facility; be careful, snakes may reflexively bite for some time even when apparently dead.

6. Remove any tight-fitting clothing, rings, watches, and such before swelling begins.

7. Keep the victim warm to offset the possible effects of shock; you may give coffee as a stimulant, but never any alcoholic drink.

The accepted treatment for snakebite in Costa Rica is the use of antivenins produced by the Instituto Clodomiro Picado. Two types are available. One is polyvalent (labeled POLIVALENTE), for all the pitvipers native to the republic. The second (labeled ANTI-CORAL) is used for most coral snake bites but is not effective for the bite of *Micrurus mipartitus,* the gargantilla. All public hospitals, most field stations, and ecotourism facilities routinely stock these antivenins. Many field biologists carry several ampules of antivenin when working at remote sites.

Hardy (1994a) recommends field administration of POLIVALENTE antivenin only when the bite is inflicted by a pitviper over 1 m long, there are symptoms of envenomation, the time to the nearest medical facility is over four hours, and there is no known allergy to horse dander or serum.

The Instituto recommends that the antivenin be administered as soon as possible after the bite has been determined to be by a venomous snake. Sometimes venomous snakes bite but do not inject any venom (a "dry" bite). In the case of pitvipers, wait a few minutes to see if symptoms of envenomation appear. These will include intense pain, swelling, hemorrhaging, and darkening of the area around the bite. Confirmation need not wait until all of these symptoms are present. Coral snake venoms do not show these local effects, so begin treatment as soon as you are confident that the bite is from a coral snake, not a mimic.

There is some danger in using the antivenins for people who may have a severe or even fatal allergic reaction to the horse serum that is their basis. Avoid injecting antivenin if no envenomation has occurred. To test for an allergic reaction, put a small drop of the antivenin diluted 1:10 or 1:100 in one of the victim's eyes. If after fifteen minutes the eye becomes red, swollen, or teary, the victim is allergic and there is urgency in getting to the medical facility before using antivenin. Russell et al. (1997) prefer to use a skin test in which a dilute saline solution of the antivenin (1:100 or 1:10) is injected under the skin on a smooth surface of an arm to produce a small welt. Sensitivity is indicated if the welt becomes red or gets bigger within fifteen to twenty minutes. If no allergy is indicated, take the following steps:

1. Slowly inject one ampule of the antivenin intramuscularly into the anterior thigh, shoulder (deltoid), or buttock (a little bit above and lateral to the center of the muscle mass). Repeat on the opposite thigh, shoulder, or buttock (two vials total); this usually is sufficient for coral snake bites. Do *not* inject the serum rapidly.

2. Recommended dosages are two to four ampules of the appropriate antivenin depending on the severity of the bite. Use whole ampules, not fractions of their contents; do *not* repeat treatment in the field if the time to the medical facility is under two hours.

3. For severe pitviper bites use six ampules; in all cases use adult dosages for children.

The Instituto further recommends keeping on hand no fewer than six doses of each antivenin and immediately replacing them when used. Antivenins need not be refrigerated but should be kept cool. Never place them in a freezer.

BASIC PRESERVATION TECHNIQUES FOR SCIENTIFIC STUDY

Scientific collections of amphibians and reptiles comprise specimens stored in a solution of ethyl (or isopropyl) alcohol or prepared as dry or cleared and stained skeletons, photographs, slides of karyotypes, and tissues frozen in liquid nitrogen or fixed in alcohol. Emphasis here will be on the most accepted methods of preserving whole specimens in the field as vouchers or for future study. Interested readers will find greater detail, alternative methods, and rationales in McDiarmid (1994) and Simmons (1987).

There are four stages in preserving complete specimens: euthanasia, fixation, labeling, and transfer to final preservative. The various chemicals required may normally be purchased in Costa Rica. Several, however, are controlled substances in some countries, and hypodermic syringes may also have restrictions.

Euthanasia

Formerly herpetologists simply killed animals they collected, but now we "euthanize" them, with the same result. Amphibians may be painlessly put to death by immersion in a solution of chlorobutanol (chloretone) in water. It is best to make up in advance a small container of stock solution, a saturated solution of chlorobutanol in 95% ethyl alcohol. A killing jar can then be set up in the field by adding a few milliliters of the stock solution to a 750 ml wide-mouthed jar filled with water. This solution may be used for several weeks, but it gradually loses its potency and will need to be replaced. Amphibians gradually relax in the solution and die within ten to twenty-five minutes. They must be removed from the solution shortly after death or they stiffen and are difficult to position properly.

If chlorobutanol is not available, salamanders may be killed by placing them in warm water at about 36 to 37°C (about human body temperature); 40 to 43°C is used for frogs and toads. Immersion in weak alcohol solutions (15 to 25%) or coating the head with benzocaine-containing gel sold in most pharmacies as a toothache medication is also effective.

The most humane way to euthanize reptiles is by injecting a small amount of aqueous sodium pentobarbital (Nembutal) near the heart. This kills them quickly and leaves them relaxed. A dose of 1 ml should suffice for an average lizard, or 5 ml for a 1 m snake. Larger amounts will be needed for larger animals. The Nembutal may be diluted in water for use with small specimens. Coating the upper head surface with benzocaine gel also seems effective for small lizards. For further information on acceptable, conditionally acceptable, and unacceptable methods of euthanasia see American Society of Ichthyologists and Herpetologists (1987) and American Veterinary Medical Association (2000).

Fixation

Once the specimen is dead and relaxed it should be fixed in a preserving tray lined on the bottom with white paper toweling saturated with 10% formalin. Formalin is a 37 to 40% solution of formaldehyde gas dissolved in water (100% formalin). The fixing solution is made by diluting one part formalin with nine parts water. Formaldehyde is very irritating to the mucous membranes, and formalin is highly toxic if taken internally. Using formalin for long without latex gloves will damage the skin. Some people develop an allergy to it and may even break out in a rash in response to the fumes. It is best to do specimen preservation in well-ventilated areas and avoid direct contact with the preservative as much as possible.

Before positioning, large amphibians and all reptiles should be injected at multiple sites, particularly the abdominal cavity, tail, and limbs, with the 10% formalin fixing solution (fig. 1.13). If syringes are not available, you must make a long incision in the venter to open the body cavity and shorter ventral incisions in the limbs and base of the tail to ensure that the formalin penetrates and fixes the internal organs and large muscle masses. Salamanders, frogs, toads, and lizards should be positioned in natural resting postures. The tails of salamanders are usually laid out straight, and those of large lizards are brought forward to parallel the body. Turtles need to be fixed with the limbs and neck extended from the shell. Crocodilians are positioned like lizards, but to save space caecilians and snakes are usually coiled, one loop around another. Usually the mouths of caecilians, salamanders, and turtles are propped open with a stick or other object to allow for later examination of tooth or beak characteristics. The male genitalia of caecilians and reptiles may be everted at this time by injecting preservative a short distance anterior to the vent in caecilians or at some distance posterior to the vent in reptiles. It is very important not to crowd specimens in the fixing trays; otherwise they will remain soft.

After labeling (see below), place another paper towel saturated with 10% formalin over the specimens, add more fixing solution, and cover the tray. In a few hours for small

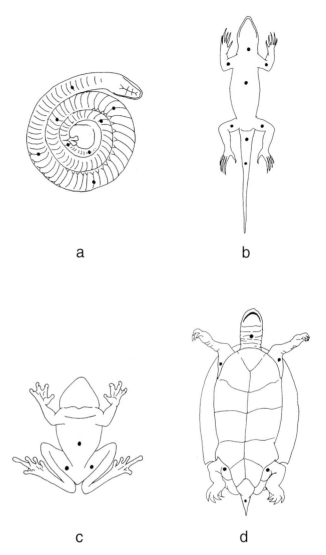

Figure 1.13. Injection points and positioning of amphibians and reptiles for final fixation: (a) caecilians and snakes; (b) salamanders, lizards, and crocodilians; (c) frogs and toads; (d) turtles.

specimens or a day or so for larger ones, the specimens will have hardened enough to be transferred to a container filled with 10% formalin. They should remain there at least four to five days for further hardening.

Animals found dead (e.g., DORs) are generally salvageable if not totally decayed. Open the abdominal cavity and thoroughly wash it out, and if necessary remove the digestive tract. Place the animals in a strong formalin solution (50% of full strength) for an hour or more depending on size and condition, then transfer them to the fixing container and position them as above.

The eggs and larvae of amphibians are rather delicate and may be placed in vials or larger containers of buffered 10% formalin on capture or after euthanasia as for adults, respectively. This solution can be made in the field by adding about two teaspoons of powdered magnesium carbonate per liter of 10% formalin. After a few hours in fixative, the specimens may be placed in smaller containers.

Labeling

Each individual specimen should have a numbered tag tied on with linen thread: to the right leg at the knee in large salamanders, anurans, turtles, lizards, and crocodilians and a short distance behind the neck in caecilians, small salamanders, and snakes. This tag must be of formalin- and alcohol-resistant heavy rag paper. Write the number on the tag in permanent ink or use prenumbered tags purchased commercially. Each number will be documented in the field notes and cross-referenced to photographs, tissue, tape recordings, and so on. Basic data include date, time of collection, locality, collector, sampling method, and habitat. Especially important is a record of the altitude at the collecting site. Some workers attach larger labels to each specimen that contain the essential information, but when large series are involved this is not feasible.

Eggs and larvae are usually given a lot number, which is floated in the container containing the specimens. If rough sorting is done in the field, clusters of larvae of similar morphotype may be given separate numbers and placed in separate vials. Do not mix egg clutches or larvae taken on different days.

Transfer to Final Preservative

Scientific collections of preserved specimens are usually stored permanently in 65% (for amphibians) and 75% (for reptiles) solutions of ethyl alcohol. Remove the formalin by soaking the specimens in fresh water, changed every day, for several days (small specimens) or a week or two (large specimens). Transfer the specimens to alcohol when the formalin odor is no longer present. Some herpetologists prefer to soak amphibians in water briefly and after five to ten minutes transfer them to the first of several alcohol solutions beginning at 15% and finally reaching 65% after about a day. Others briefly soak amphibians in water and then soak them in 65 to 75% ethyl alcohol over a period of a month or two, changing the solution several times.

Amphibian eggs and larvae are stored permanently in buffered formalin.

Special Preparations

Osteological

Specimens of small species that will have the tissues cleared (made transparent) and the bones stained are fixed in formalin as described above. If the animals are moderate in size but are going to be made into dry skeletons later, they should be preserved in 75% alcohol. Both these preparations are made after return from the field as described for the former by

Hanken and Wassersug (1981) and for the latter by Simmons (1986, 1987). For a preliminary preparation of larger specimens as dry skeletons, the animal must be eviscerated and skinned, with as much muscle as possible removed from the bones. The carcass is then wrapped loosely in string to form a compact ball and hung to dry in a sheltered place safe from predators. A screened cage or cheesecloth bag may be used to protect against flies. Drying may be expedited by immersion in alcohol and redrying. As with the alcohol-preserved specimens, final preparation will be made on return from the field.

Karyological

Preparation of chromosomes for karyotypic analysis may be done in the field using a series of special techniques. You need to acquire specific chemicals and master the techniques involved before going into the field. Readers interested in this approach should see Sessions (1996) for more information.

Molecular

Molecular analyses of several kinds require fresh tissues from field-collected animals. Most workers use two principal methods of preparation. The first involves removing heart, liver, kidney, and muscle tissue before fixation. These are placed in cryogenic tubes and frozen temporarily in dry ice or put directly into liquid nitrogen, the permanent storage medium. The second method requires only that tissue be chopped into small pieces and preserved in 95% ethyl alcohol, although lower strengths (down to 75%) may serve this purpose, as may isopropyl and propanol alcohols (see Dessauer, Cole, and Hafner 1996). To make good preparations, you need some instruction in these procedures before doing fieldwork.

Packing and Shipping

Packing preserved material for in-country transport by vehicle from a particular site or at the end of a field season is relatively simple. Small, tagged specimens may be consolidated into containers, and larger ones may be loosely wrapped in paper towels or cheesecloth and placed in larger containers or fixing trays with some formalin to keep them moist. Make sure all lids are tightly closed. Plastic trays can be sealed with duct tape but must still be kept flat to prevent leakage.

Specimens to be shipped are a different matter. Preserved specimens should be wrapped in paper towels or cheesecloth and put in a plastic bag with just enough fluid to keep them moist. The bag should be sealed or tied shut and placed in a second bag after checking for leaks. The sealed double bag may then be put with others in plastic jars and or trays and taped shut. These containers are then packed into sturdy boxes or fiberboard drums designed for shipping wet materials. These should be clearly marked with address labels both inside and on the container.

Dry skeletal materials can be packed in sturdy cardboard boxes. Frozen tissues will require special metal Dewar flasks filled with liquid nitrogen or, if shipping only a short distance, Styrofoam containers of dry ice.

To export materials from Costa Rica via a shipper or as luggage, you need to obtain an export permit, and all appropriate collecting permits must first be in order. It is also a good idea to make prior arrangements with the carrier for shipping liquid nitrogen in "dry shippers," or dry ice packages, if you are leaving the country by air (see Dessauer, Cole, and Hafner 1996, for details).

FIELD STUDY

There is much to be learned about Costa Rican amphibians and reptiles through field observation and collecting specimens. However, many scientists will be interested in more detailed and long-term research on particular species or communities or at individual sites. Heyer et al. (1994) provide an outstanding review of many of the kinds of studies that may be undertaken on amphibians. Similar studies may be made on reptiles, although aspects of technique may differ. Refer to that volume for ideas.

Essential to the success of any project is accumulating data. Basic data for most field biologists are recorded in a field notebook. Usually this is a three-ring loose-leaf notebook (180 × 250 mm), although some workers use a pocket-sized book to temporarily record data before transcribing the notes to a permanent record. The notes should be written on 100% cotton rag paper with permanent ink.

A standard field notebook is organized by date, like a diary. On the date you first work at a particular site you record the following minimum data: as specific a locality as possible, including latitude and longitude, if available; elevation; date; habitat type; and sampling method(s). Under each locality, list all specimens collected on the same date by field number and give preliminary species identification with the following data: method and time of collection; collector(s); microhabitat; environmental temperature; specimen's temperature; weather conditions; activity or behavior; color notes; associated separate materials: photographs, tape recordings, karyotype preparations; tissue samples; and so on. The separate materials should be cross-referenced by the field number so they can be associated with a specific specimen.

Some workers prefer a bipartite field book with a journal section and a section organized by species. In this system the journal entries include the basic data by locality and other general information or observations, including route taken, changes in landscape or vegetation, and such. Data on particular species are grouped, and each entry is by specimen field number and includes the data for that specimen, as in the standard notebook style. Locality data for the specimen are referenced to the journal account, which does not list specimens but usually records the specimen numbers used at a particular site.

CAPTIVE HUSBANDRY

Amphibians and reptiles are kept captive for a number of reasons. Captivity may be temporary while you await the best time to make notes or take photographs. In other cases you may want to retain your catch longer to record behavior or for other reasons related to a research project. Most professional herpetologists started out keeping amphibians or reptiles—usually snakes—as pets, and those experiences were crucial to their maturation into scientists.

There is much to be learned about behavior, nutrition, reproduction, and even physiology from captive individuals properly housed and maintained. Over the years a wealth of information was obtained from hobbyists, especially in Europe (Zimmerman 1986), and from zoological parks. Then about twenty years ago herpetoculture—the keeping and breeding of captive amphibians and reptiles—became immensely popular and came to combine the interests of a mixture of hobbyists, serious amateurs, and zoo herpetologists. Today thousands of individuals are involved in these pursuits, and supplying the needs of herpetoculturalists and their charges is a multimillion-dollar business (De Vosjoli 1994). The result has been an exponential growth in information on many species and their captive husbandry and a huge literature on herpetoculture.

It is not possible to do more than introduce the basic requirements of captive care in this section, since each species may have its own requirements. For further information consult the many publications of Krieger Publishing Company (Krieger Drive, Malabar, Florida 32950, USA), and Tropical Fish Hobbyist (TFH; One TFH Plaza, Neptune, New Jersey 07753, USA). For entry into the literature I also recommend the semipopular journals *Reptilia* (Muntaner 88, 08011 Barcelona, Spain), now published in English, German, and Spanish editions, and *Reptiles* (P.O. Box 58700, Boulder, Colorado 80322-8700, USA). The emphasis here will be on captive care under the tropical field and laboratory conditions that prevail in Costa Rica.

Remember that the care and feeding of captive individuals requires time and effort, even for species relatively easy to maintain. There should be a serious purpose in keeping them in captivity for a long time, one that will enhance our knowledge of herpetology.

Housing

Large wide-mouth glass jars with perforated lids may be used as temporary housing for most small to moderate-sized species (fig. 1.12a). Place moist crumpled paper towels in jars

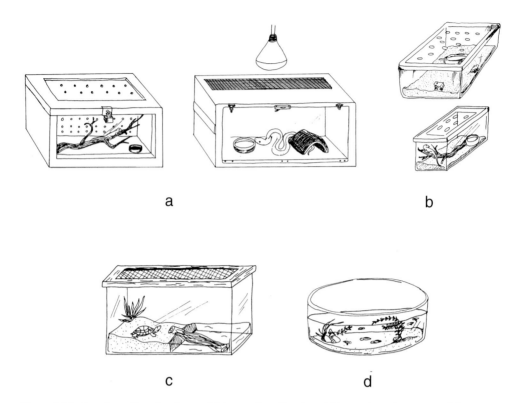

Figure 1.14. A variety of housing for amphibians and reptiles: (a and b) terrariums for terrestrial forms; (c) semiaquatic terrarium; (d) finger bowl aquarium for raising tadpoles.

for amphibians and anoles to hide in and dry ones for most other lizards and small snakes. The best all-round permanent cages are clear-sided glass or Plexiglas aquariums with tight-fitting tops of Peg-Board or fine-mesh plastic screen in a wooden frame (fig. 1.14a). Wooden cages are harder to clean and have low resistance to high temperature and humidity. Avoid wire screening, since animals usually develop abrasions from rubbing against it. You can construct a more elaborate lid with hinged doors. The lids must be completely closed or weighted down, since amphibians and reptiles are escape artists and will slip through even a very narrow gap. Care must also be taken in opening the lids or doors: frogs and many lizards will sit and wait for an opportunity to leap to freedom.

There is an infinite variety of sizes and interior arrangements for these cages, depending on the species being kept and on individual preferences. A good size for general use is 370 × 750 × 450 mm. Three general arrangements are most common: terrariums for terrestrial forms (fig. 1.13a,b), semiaquatic terrariums for turtles and some frogs and snakes (fig. 1.13c), and aquariums for larval frogs and toads (fig. 1.14d). Substrates for terrariums vary with species requirements; sand for reptiles from drier habitats, and leaf litter, humus, or moss on a drainage layer of gravel for species from humid areas. Add rocks and pieces of bark or a box as hiding places, or branches for arboreal forms.

Semiaquatic terrariums need to be set up so one end provides a terrestrial situation, sloping down to a small pond (fig. 1.14c). The terrestrial substrate should be similar to that of a humid-habitat terrarium, and the pond needs a fine gravel bottom. A water filtration system may be added for cleaning and recycling the water.

Aquariums for raising frog and toad larvae need not be as large as for many other purposes (fig. 1.14d). Essentials are providing clean, oxygenated water and avoiding high temperatures and overcrowding. These containers may be set up like aquariums for tropical fish, with aeration and filtration systems. A heater is probably not needed for lowland species raised on the Meseta Central. Near the time of metamorphosis, place a floating object in the aquarium so the froglets or toadlets can crawl out. The water in these aquariums should be completely changed every two or three weeks.

Small to moderate or even large cages of the types described are usually kept indoors but could be set up in a yard or shade house outside to take advantage of natural light and light cycles. You can also build large outdoor enclosures to study large lizards, turtles, or small crocodilians. Turtles are the easiest to accommodate; they may be released in any area well fenced with wood, metal, or masonry (but not wire), including backyard gardens. For aquatic or semiaquatic species, a children's plastic swimming pool sunk in the ground will serve if there is no fishpond. Lizards need to be caged or have special fencing to keep them from climbing out of an enclosure. Most people will prefer not to have much to do with confined crocodilians, but if they are kept they must have adequate terrestrial and aquatic habitat. Protection against predators such as rats, cats, and birds is a problem if the outdoor enclosures are not covered by wire mesh.

Light and Temperature

Amphibians and reptiles are ectothermic and depend on external sources of heat in order to carry out the basic functions of feeding, digestion, metabolism, and reproduction. The principal sources of heat under natural conditions are direct solar radiation and reradiation from the substrate. Nocturnal reptiles and most amphibians depend on the substrate as the external heat source, and their body temperatures fluctuate with it.

Unfiltered sunlight is the best source of illumination and heat for reptiles (mostly diurnal lizards and turtles) that prefer direct solar radiation (are heliotherms). Overhead incandescent lighting is usually used as a substitute primary heat source for terrariums (fig. 1.14a). Do not use heat lamps (infrared); they are too hot and will injure the animals. Since reptiles thermoregulate behaviorally, a reflector light is ideal for establishing a thermal gradient in the cage, with a basking spot at one end. Other areas of the cage must be cooler than 24 to 27°C so the animal can retreat and not overheat. Fluorescent tubes give off light close to the color spectrum of visible light and are routinely used to illuminate cages but not to heat them. None of these lights provide enough ultraviolet (UV) radiation to ensure adequate synthesis of vitamin D. Consequently vitamin D and calcium supplements must be added to the diets of heliotherms not regularly exposed to enough sunlight to prevent bone diseases.

In the lowlands temperatures are high enough all year so that animals may be maintained without supplemental heat sources. On the Meseta Central few homes or other facilities have any kind of heating. If the captives are housed there, supplemental heating of the cages may be required for some lowland species during the coldest times of the year.

Water

Most terrariums need a water dish for drinking and soaking. Some species of lizards prefer to take water from a drip bottle or to lick drops off vegetation. Many frogs, some lizards, and a few snakes need to be misted with water from a spray bottle every few days to stay in good condition. For animals in large outdoor enclosures a constant water supply is a necessity.

Food and Nutrition

Check the species accounts to see what kind of food your captives prefer. Caecilians, salamanders, and some small snakes do well on earthworms taken from the soil and slugs taken from vegetation. Salamanders, frogs and toads, many lizards, and some snakes feed on insects or other arthropods. Many insects can be collected around lights at night or shaken out of leaf litter. Some pet stores in San José sell mealworms and crickets. Herbivores and omnivores feed mostly on vegetable matter, especially flowers, fruits, and leaves. Omnivores and carnivores eat a range of animals, including invertebrates, but most snakes eat frogs, lizards, birds, or mammals. If snakes are being fed birds and mammals, these should be presented to them dead. Crocodilians prefer raw meat and dead or live fish.

Larval frogs and toads are mostly herbivorous and do well on boiled lettuce (only the leafy part) and dry dog or cat food placed in the water. Tadpoles of some dendrobatids thrive on the yolks of birds' eggs. Recent metamorphs have a high mortality rate but may be fed vestigial-wing fruitflies (Drosophila) or other small insects. In all cases avoid toxic plants as dietary items, and do not feed your animals any food that has been sprayed with herbicides or pesticides.

Eggs

Anuran eggs may be raised in aquariums or large finger bowls. If placed in aquariums they should be at approximately the same depth of water as under natural conditions. Reptile eggs may be raised in a plastic bag or plastic tray. Make a substrate of damp earth, peat moss, sand, or vermiculite and smooth the surface, then make well-spaced slight depressions in the surface for each egg. If using a plastic bag, blow it full of air and seal the top tightly with a rubber band or duct tape. The plastic tray incubator should have air holes for ventilation in the lid. In both cases the eggs should be maintained at 25 to 33°C or slightly lower for montane forms, under lighted conditions, to inhibit the growth of molds. Keep the substrate slightly damp. Unfertilized eggs and those containing dead embryos tend to collapse and become covered with fungus. Remove them immediately to reduce the chance of infecting the remaining eggs.

General

Cages and aquariums must be checked frequently to be sure that the captives are healthy and that the water supply is clean and uncontaminated and to remove dead food items. Any sick or dead specimens should be removed quickly for treatment and to prevent contamination or infection of other animals. Always check that a new or refurbished cage or enclosure is secure so that neither the captive nor its prey can escape.

Careful note taking and record keeping are a must if you are to profit fully from keeping and studying captive animals. Notes may be kept separately or in a section of your field notebook. In addition to the basic collection data recorded in the field book, you will want to note especially (by date, time, and usually temperature) what was offered and eaten, defecation, reproductive or other behaviors, number and characteristics of eggs or neonates (size and, for reptiles, weight), duration of incubation or gestation, length of larval life, size at metamorphosis (anurans), changes in size and weight, and duration of juvenile period (time to sexual maturity).

Parasites and Disease

Amphibians and reptiles, just like all other organisms, suffer from a variety of diseases induced by viruses, bacteria, fungi, parasites, and even algae as well as from those caused by vitamin and mineral deficiencies and other factors. Two outstanding recently published references, *Amphibian Medicine and Captive Husbandry* (Wright and Whitaker 2001) and *Reptile Medicine and Surgery* (Mader 1996), provide in-depth reviews of these subjects. Make every effort to have these books available or acquired by a library to which you have access for reference. Zimmermann (1986) gives a short list of the most common diseases encountered in captive amphibians and reptiles and recommends treatment. The subject is too vast to be summarized in this book. Those seriously interested in herpetoculture should be familiar with these publications and their coverage of the most prevalent aliments of captive amphibians and reptiles before undertaking a project requiring care of captive species.

A BRIEF HISTORY OF COSTA RICA

Pre-Columbian Times

When the first humans moved from the north into what is now Costa Rica 10,000 to 15,000 years ago, it was indeed a "rich coast." Almost the entire country was densely forested with an incredible variety of landscapes, inspiring vistas, and multitudinous flora and fauna whose remnants excite visitors' imaginations even today. By the time the Spaniards arrived, three main cultural groups were present, with a total population estimated at 27,000 for what is now the republic.

A group of tribes, often collectively called "Chorotegas," inhabited the seasonally deciduous forests of the Pacific northwest sector of the country. They spoke an Uto-Aztecan language, lived in houses in large stockaded towns centered on a plaza, and practiced an agriculture based on corn and beans but grew other crops as well. These peoples were part of the general Mesoamerican cultural sphere, whose complex social organization, calendar, writing, polychrome ceramics, and occasional human sacrifice link them to more northern cultures.

A second group of many local tribes, usually referred to as "Huetars" or "Guyamies," were found throughout the Atlantic slope rainforests and across central Costa Rica to the Pacific coast. They spoke Chibchan languages, lived in small, dispersed villages composed of circular houses built of poles and roofed with thatch, and grew root crops (particularly yucca) using swidden (slash-and-burn) agriculture. The fruits of the pejibaye palm *(Bactris gasipaes)* were also a major dietary component. Their social organization, lack of a calendar and writing, and head-hunting warfare allied them to similar cultures in northwestern South America. Nevertheless, they constructed complex ceremonial centers such as those found at Guayabo de Turrialba, Cartago Province, Las Mercedes between Juntas and Guápiles, Limón Province, and other sites.

The third cultural group, often called "Diquís," "Coto," or "Brunca," inhabited the rainforests of the southern Pacific region. They shared most of the features of the Atlantic group but excelled in the production of stone statuary, including huge stone balls, and of gold jewelry using the lost-wax casting method. Much of the jewelry depicts animals, including the famous *aquila* (eagle) pendants. These actually represent vultures, which were highly regarded because they were believed to extract the spirit of a dead person as they consumed the flesh and to carry it to heaven.

European Discovery and Conquest

The first Europeans to visit Costa Rica were Columbus and crew on his fourth voyage to the New World. He landed at Cariarí, now Puerto Limón, on the Caribbean coast on September 18, 1502 (September 25 by the Julian calendar). Not much impressed with the beauty of his surroundings but much impressed by the gold jewelry the Indians gave him, he called the area *costa rica*. He little realized that the name was so appropriate for this beautiful land with its riches of biological diversity.

The Spaniards paid little attention to the region for many years. Early on (1519), during a voyage of discovery sent out from Panama, Gaspar de Espinoza entered the Gulf of Nicoya and subsequently sailed north along the northwest coast of what would become Costa Rica (Oviedo 1535). Along that coast he discovered a bay teeming with sea snakes *(Pelamis platurus)* and named it Bahía de las Culebras, a name it retains today (Taylor 1953). This constitutes the first known record for an amphibian or reptile from the future Costa Rica. Subsequently Gil Gonzáles explored the Península de Nicoya on the Pacific versant in 1522 and received much gold, so that after a time the area between Nicaragua and Panama was officially called Costa Rica. However, it was not until 1561 that a group from Nicaragua established the first settlement, Castillo de Garcimuñoz, near present-day Santa Ana, San José Province, on the Meseta Central Occidental.

In 1562 Juan Vásquez de Coronado, the younger brother of Francisco Vásquez de Coronado of North American fame, undertook an official *entrada* to secure Costa Rica for king and God. His expeditions included exploration of the Pacific lowlands, crossing the Cordillera de Talamanca to the Caribbean slope, exploring the Atlantic lowlands, and returning to the Pacific coast via the Valle Central. He arranged for the settlers at Garcimuñoz to move to a much more favorable site on the Atlantic portion of the Meseta, which he named Cartago. It became the first permanent settlement.

The Colonial Period

Costa Rica shortly thereafter (1568) became a province of the captaincy general of Guatemala. It remained a neglected, poor backwater of the Spanish New World empire for the next 260 years. The imagined rich sources of gold never materialized, and the Indians refused to be serfs, so they were killed off, were isolated, or became working partners in the subsistence agriculture required of all to survive. Many Costa Ricans believe that this time, when everyone suffered more or less equally just to get by, is the basis for their commitment to equality, expressed in freedom, universal education, and democratic institutions.

By 1575 the Valle Central was pacified, and during this period the Valle Occidental was gradually settled with the future San José, Heredia, and Alajuela, established in 1737, 1763, and 1781, respectively.

Almost all the population of the country lived on the Valle Central, having little communication with the outside. The few goods Costa Ricans could afford came to the uplands primarily from the Pacific coast through the settlements of Esparza and Suerre and were brought overland following the course of the Río Grande de Tárcoles.

Emancipation and Nationhood

The sleepy little backwater province of 40,000 inhabitants was taken completely by surprise in October 1821 when its people learned that, without a bit of effort on their part, all of Central America had been proclaimed free from Spain on September 15, 1821. This had followed Mexico's successful revolution and its invitation for the Central American provinces to join in the new independence. Costa Rica immediately accepted this gift, and for a brief period it joined other Central American provinces as part of the Mexican empire.

In Costa Rica as elsewhere, conservatives favored the empire while republicans wanted a separate Central American state. The leaders in Cartago and Heredia were mostly of the former persuasion; those from Alajuela and San José were of the latter. The issue was decided on April 5, 1823, in a brisk fight between the two groups on the Ochomogo ridge separating the Valle Oriental from the Valle Occidental. The republicans won, and to the *josefinos*' delight San José became the capital, replacing the colonial capital of Cartago. Costa Rica and the other Central American provinces shortly thereafter joined in a federation that declared independence from Mexico. The federation in turn fell apart by 1839, with each of the provinces becoming a sovereign state.

Costa Rica had instituted an elected government as early as 1824, which had regularly chosen prominent individuals to be chief of state. One of these, Braulio Carrillo, became a virtual dictator *(caudillo)* and established government sta-

bility. But more important, he promoted the cultivation of coffee.

The agriculturists of the Meseta Central had for many years tried to find export crops. At first wheat had been grown on the rich volcanic soils and then tea, but foreign competition—wheat from California and Chile and tea from India—made these experiments unpopular. As luck would have it, in 1843, after Carrillo had been forcibly evicted from office, William Le Lacheur, an English merchant, docked at the Pacific port of Caldera. On speculation he rode a mule to San José looking for cargo for his homeward trip. He arranged with Santiago Fernández, a coffee dealer, to carry 5,500 hundredweight of coffee to England on credit. Le Lacheur returned the next year with several ships, paid Fernández for the earlier shipment, and purchased more coffee with pounds sterling. A coffee boom that followed underpinned the Costa Rican economy well into the twentieth century.

In 1848 Costa Rica proclaimed its absolute independence from the long moribund federation after electing its first president, José María Castro, in 1847. During his career he established one of the early newspapers, inaugurated the Universidad de Santo Tomás, and as president supported freedom of the press and founded a school for young women. He was too liberal for the coffee barons, who had risen to power in a mere five years, and he was forced to resign. His elected successor, Juan Rafael Mora Porras, a coffee baron himself, served from 1849 to 1859. It was during his administration that Costa Ricans had to fight for the liberty handed to them on a platter thirty-five years earlier.

This crisis arose from the ambitions of an American adventurer, William Walker, and the machinations of several Southern slaveholders who dreamed of adding the Central American countries to the United States as slaveholding states. Walker invaded Nicaragua in 1855 with a force of fifty-eight men. He soon defeated organized opposition and had himself proclaimed president. He next invaded Guanacaste Province, which in a plebiscite in 1824 had chosen to be annexed to Costa Rica rather than Nicaragua.

President Mora, with the authorization of Congress, called up an army composed of Costa Ricans from every walk of life to repel the invasion and remove Walker from power. With Mora in command, the army routed Walker's force in a battle at what is now the headquarters of Santa Rosa National Park and pursued it into Nicaragua. There at the battle of Rivas (1857) the Costa Rican army prevailed. It was in this battle that Private Juan Santamaría sacrificed his life to become the nation's hero while making a sure throw of a bomb that blew up many of the enemy and a munitions magazine. The Guerra de Filibustros stands as a watershed experience for the country. The fear of invasion and the enthusiasm of war in a common cause united Costa Ricans as never before. Many even today refer to this crisis as their war for independence, and it surely led to the country's emerging sense of nationalism.

Ironically, President Mora was later removed from office in a coup d'état. Thereafter the coffee barons called the tune, making or breaking presidents, controlling the military or taking office themselves. Mora ultimately was executed by a firing squad when he attempted to regain power in 1869.

In 1871 General Tomás Guardia became president in a successful coup. A dictator but a man of great vision, he brought the coffee elite under some control, raised their taxes, and began to modernize the country. In his administration major investments were made in education, public health, and especially transportation.

Transporting coffee to the clamoring Europeans and processed goods back to the clamoring ticos had long been a problem. Numerous attempts had been initiated to connect the Atlantic coast to the Meseta Central, especially to reduce the costs associated with the long voyage around Cape Horn, but none were satisfactory. Most commerce passed through the Pacific ports of Caldera and Puntarenas. To handle the increasing commerce the Carretera Nacional, an oxcart road, was built in 1843 to join San José with the Pacific ports.

Guardia was determined to have the Atlantic ports connected to the Meseta Central by railroad. This vital connection was completed by the legendary entrepreneur Minor C. Keith. Guardia originally contracted with Keith's uncle, Henry Meiggs, who had won fame in building railroads up the face of the Andes in Chile and Peru. In 1872 Meiggs undertook what must have seemed to him the infinitely easier task of building the Costa Rican railroad. Minor's older brother was placed in charge but died, and the younger man took over. As a way to provide freight for the railroad as it was a-building, Keith began to grow bananas and started shipping them in 1874. Out of this grew the combine of Keith and Andrew W. Preston that formed the United Fruit Company and its Costa Rican subsidiary, La Companía Bananera de Costa Rica. Without doubt, during the late nineteenth and early twentieth centuries Minor C. Keith was the most powerful economic and political force in Central America, and the United Fruit Company continued its dominance until these countries became more self-determining in the mid-twentieth century.

The immense effort involved in constructing the railroad saw continuous recruitment of workers for the road because there were hundreds of deaths, mostly from malaria. Finally hundreds of Jamaican Negroes were brought in because of their higher resistance to disease, and they remained to work in the banana plantations. The original plan to connect Puerto Limón and the Meseta Central called for the line to run from the coast to Guápiles and then across the La Palma pass to San José. By 1882 the railroad was completed between Limón and Carrillo. The physical difficulties of frequent washouts, changes in the stream beds, landslides, and floods in the heavy rainfall zone between Guápiles and La Palma forced abandonment of this attempt. The remnant of this line now runs from La Junta to Guápiles (La Linea Vieja).

Keith then regrouped and replanned his project, deciding to follow the old colonial trail up the Río Reventazón through Peralta and Turrialba to Cartago (the new line). The Meseta Central was finally connected to the Caribbean by rail in 1890, after nearly twenty years of effort and four thousand deaths from disease among its construction workers.

Advances toward Popular Democracy

The new constitution of 1871 emphasized presidential power as the means of providing a better life for Costa Ricans, reflecting the past dependence on a few gifted leaders. After Guardia's death in 1881 liberal ideas began to predominate. Among them were the separation of church and state, the social responsibility of government, a free press, and most especially the power of formal education as a basis for solving the problems of individuals and the republic.

In the 1880s Costa Ricans, now numbering 180,000 people, saw many changes as halting steps were taken to translate liberalism into reality. The activities of religious orders were curtailed, the Universidad de Santo Tomás was closed because of its supposed ties to the Catholic Church, and education was secularized. But even more important, under President Bernardo Soto's minister of education, Mauro Fernández, the entire educational system was revised. Public education became totally funded by the government and compulsory through the seventh grade. Fernández established the first school for teacher training (Instituto Alajuela), public high schools for boys (Liceo de Costa Rica) and girls (Colegio Superior de Señoritas), the Museo Nacional, and the Biblioteca Nacional (national library). In addition, in a successful move to raise education to a new level he imported a cadre of European teachers to staff the new institutions. By these actions Fernández contributed mightily to a society that became accustomed to having educated people carry on reasoned dialogue on public issues.

The first unrigged election for president was held in 1889 when President Soto tried to impose his candidate on the electorate. At the last moment, angry crowds marched on San José demanding the right to vote and forced an honest election. There followed a period of material progress fueled by the twin engines of coffee and bananas and sparked by a series of progressive presidents. Electricity, telephones, and streetcars were introduced to the capital; the gold standard was established for the colón; the Teatro Nacional opened in 1897; great improvements were made in infrastructure, especially in the port facilities of Limón; the plan for the interoceanic railroad was revived, though the line was not completed to Puntarenas until 1910.

By 1900 the population had increased to 313,000, most

of whom lived in the country and were very poor independent farmers working at a subsistence level. The economic growth of the previous decades began to stall as the price of coffee on the world market dropped. Two of the most beloved and revered presidents, "Don Cleto" González and "Don Ricardo" Jiménez, held office for much of the economically difficult time before World War I.

However, when Jiménez proposed an income tax as a way to solve the country's financial woes, further exacerbated by the war, his minister of war, Federico Tinoco, staged a successful coup in 1917.

Tinoco and his brother Joaquín reflected the fears of the coffee elite and business community over the tax, which would have benefited less well-to-do Costa Ricans. Nevertheless when the Tinocos behaved like caudillos even their original supporters turned against them, and President Tinoco went into exile in 1918.

The following decades were difficult as coffee prices continued to drop and banana production was established outside Central America. The years of the Great Depression also took their toll. However, whereas previously Costa Rica had looked to England and France for markets and ideas, now the focus shifted more and more toward the United States. During this period Don Cleto served a second term as president and Don Ricardo two more. Their style was personal government, although committed to democracy, direct elections, and stability, but as old-style liberals they represented the status quo. The liberal philosophy had run out of ideas to solve the terrible poverty of most ticos or meet their needs for employment, housing, health care, transportation, or even an adequate diet. New power groups and the spread of socialist, communist, and fascist ideas led to demands for change and conflict with the conservative position of the coffee-business complex.

Sturm und Drang: The Decade of the 1940s

In 1940 the population had reached 656,000, and Rafael Ángel Calderón Guardia won an overwhelming victory in the presidential race over Manuel Mora, the chief of the Communist Party. Calderón immediately instituted a series of then revolutionary social reforms including universal health insurance, a minimum wage, mandatory collective bargaining, and the right of the landless to obtain title to land not in cultivation. Of course members of the coffee-business complex strongly opposed these measures as being at their expense, but workers and the emerging middle class of small businessmen and professionals enthusiastically supported them. The strength of the opposition caused Calderón to ally himself more and more with Manuel Mora's communists.

The Costa Rican constitution forbade more than one successive term as president, so Calderón designated a puppet candidate to run for his party in 1944. Although the opposition candidate apparently won, the Calderón government

declared its man the winner. The next four years saw the coalescence of many diverse elements that wished to defeat Calderón, who could run again for president in 1948. In that election the opposition candidate, Otilio Ulate, appeared to be winning when the Calderón-controlled congress annulled the election with the intent that the incoming congress would make Calderón president.

Within days the country was in a state of civil war, with the opposition led by the charismatic José Figueres, whom Calderón had exiled for political reasons. The anti-Calderón forces won several bloody engagements between March 10 and April 13, 1948, and a peace treaty was signed shortly thereafter. Although ostensibly the war was to ensure Ulate's presidency, the phenomenon that was "Don Pepe" Figueres had other plans.

Don Pepe convinced Ulate that an interim revolutionary junta, with Figueres at its head, should act as the government for eighteen months, after which time Ulate would assume the nation's highest office. Under Figueres' control the junta completely transformed Costa Rica. Among the most important changes were abolishing the army; establishing a civil service; expanding the powers of the congress; granting full citizen rights to anyone born in Costa Rica, since previously most of the descendants of Keith's railroad workers were denied this right; extending the right to vote to all citizens, including women and illiterates; eliminating private banks; taxing the wealthy 10% of their current wealth; accepting all Calderón's reforms (modest by comparison); developing a series of semiautonomous institutions outside direct control of the executive and congress to provide public services such as insurance, water, and electricity; and allocating a minimum of 10% of the budget to public education, including the new University of Costa Rica, which had been established in 1940. Finally, the Communist Party was outlawed, and laws were passed that prohibited a president from running for reelection until he or she had been out of office for two terms (eight years).

As promised, Ulate assumed the presidency in late 1949, and no subsequent government has been installed by force. The independent Supreme Electoral Tribunal established in the new constitution has successfully guaranteed honest elections during the rest of the century. Figueres' role as the shaper of the new Costa Rica was recognized by his overwhelming presidential election victory in 1953 and his reelection in 1970.

Don Pepe had the opportunity to make his program work during his first term. His National Liberation Party (PLN) and its program dominated political, economic, and social life in Costa Rica for the rest of the century. Costa Ricans traditionally vote out the party in power at each national election, but every other president from 1958 on was a Liberacionista, with the party winning the presidency for consecutive terms in 1970 and 1974 and in 1982 and 1986. The opposition to the PLN usually consists of a coalition of conservative parties. The 1950s ended on a conciliatory note when the conservative president Mario Echandi allowed Calderón to return from exile in Mexico.

The 1960s and 1970s

The 1960s brought new challenges to Figueres' "Second Republic." In 1964 the population had swelled to 1.3 million, and throughout the decade Costa Rica had the world's highest birthrate—4.2%. The growing demands for a better life, fueled now by television and the ready availability of consumer goods, and the baby boom had far-reaching repercussions. The government made every effort to encourage industry and foreign investment and rushed plans to open new lands on the Meseta to the burgeoning masses. Extensive tracts of primary forest thus came under the chain saw for new settlements and to satisfy the demands for building material. Vast areas, especially on the Atlantic lowlands, were cleared to establish large cattle ranches to provide for home consumption and export to the United States. Banana production, which had been reestablished on the Atlantic coast in the 1950s, boomed. Keith's original operations had been abandoned in 1942 and transferred to the humid Pacific lowlands because of the sigatoka disease that infected the plants.

The economic benefits had their downside. During this period deforestation accelerated, so that by the 1970s only 35% of the country's land that was originally in forest remained so. Increases in erosion, floods, and adverse effects on watersheds were dramatic.

The 1970s were not a pleasant time for Costa Ricans. Figueres' second term was a mix of small successes and controversy. During his term Costa Rica recognized the Soviet Union and legalized the Communist Party, still led by Manuel Mora. The first highway linking the Meseta and Puerto Limón was completed, and the Atlantic (Northern) railroad was nationalized. The hint of corruption swirled around Figueres' administration because he protected the notorious United States investment swindler Robert Vesco.

The country's economy was weak during this period even though industry now contributed twice as much (24%) to the gross national product as in 1950. The oil crises of 1973 and 1978 were terrible blows that set off an inflationary spiral, plunged the country into a deep depression, and geometrically increased its negative balance of payments and foreign debt.

It was during this difficult period, however, that Costa Ricans developed an emerging awareness of the values of their natural environment and biological diversity. Catalyzed by two young leaders in the environmental movement, Mario Boza and Alvaro Ugalde, President Daniel Oduber (1974–78) and President Rodrigo Carazo (1978–82) expanded and enhanced the nation's system of national parks and natural reserves, which now constitute 15% of the country's surface area.

War and Peace

The revolution of 1948 laid the groundwork for a new kind of state, one that emphasized modernization, social justice, a high level of government services, and a government run by well-educated citizens. By the mid-1970s the population had increased to 2.5 million and the welfare state was showing signs of stress, with 45% of the basic payroll being fringe benefits (insurance, retirement, etc.), at least 20% of the labor force being government employees, and the semiautonomous institutions accounting for 50% of the gross national product.

Agriculture, coffee, and bananas were still the mainstay of the economy, but mass production industry had substantially increased and cattle production, which boomed in the late 1960s and early 1970s, had declined. A new industry, ecotourism, was beginning to expand, taking advantage of the enlightened vision of Costa Ricans in preserving in national parks and reserves many areas of natural beauty and maximum biological diversity.

Even during this difficult period and on into the 1980s and 1990s, the attention of Costa Ricans was focused on the drama taking place just to north of their border in the neighboring country of Nicaragua. Nicaragua was having problems similar to those of all the Central American republics that depend on agriculture. However, the traditions were very different from those in Costa Rica. A few families owned most of the land and national wealth. Most of the populace was illiterate and even poorer than the Costa Rican *campesinos*. There was no history of even patriarchal democracy, and the country had been ruled since independence by a series of brutal dictators.

Among the most brutal was Anastasio Somoza García, who was trained by United States marines during the United States occupation of Nicaragua from 1926 to 1934. After the Americans left, Somoza seized power in 1936 and imposed his personal dictatorship on the country for thirty years. Costa Ricans had no love of the Somoza regime, which exemplified all the repressive forces they had overcome to establish a democracy. Furthermore, Somoza had invaded Costa Rica in 1948 in support of Calderón and again in 1955.

Unexpectedly, Somoza was assassinated in 1956, but his son Luis Somoza Debayle immediately became "president" (dictator), ruling from 1956 to 1963, followed by a Somoza puppet (1963–67). The younger son, Anastasio Somoza Debayle, took over on his brother's death. This Somoza was if anything more brutal, avaricious, and cruel than his father.

Somoza's tight grasp on all things Nicaraguan began to come under attack by a variety of groups in the 1970s. The most daring were the radical Marxist Sandinistas (FSLN). They took their name from Augusto Sandino, who had led the resistance to the American occupation and was assassinated on the elder Somoza's order after the American withdrawal. By 1978 widespread public support led the Sandinista leaders, Daniel and Humberto Ortega, to armed insurrection against the Somoza regime. Costa Rican sympathies were always behind any group resisting the undemocratic Somozas, so there was dancing in the streets in all the major cities in Costa Rica on July 19, 1979, when Somoza was ousted from power.

Joy gradually turned to dismay as the generally anticommunist ticos saw their neighbor transformed into a repressive Marxist state that under the Ortegas aligned itself with Castro's Cuba. President Ronald Reagan of the United States also became disenchanted with the Sandinista government. In 1981 clandestine operations by the United States Central Intelligence Agency began to aid counterrevolutionary Nicaraguan forces (contras) attempting to oust the Sandinistas. As the Contra War expanded, Costa Rica reluctantly became involved.

The Contra War had both good and bad effects on all things Costa Rican. President Reagan's obsession with the Sandinistas led to massive grants and loans to Costa Rica. United States military personnel aided in the construction of new highways and infrastructure, and the economy boomed. Negatives included fears that the United States would use Costa Rica to stage contra attacks across the border or launch an invasion itself from Costa Rican soil. Actual negatives were the increased armaments brought into the country, converting the rural police into a virtual army, the use of Costa Rica as a refuge for contras, the presence of hundreds of refugee *nicas* fleeing the Sandinista regime and the fighting, and an increasingly difficult relationship with the United States.

"Democracy and peace" might well be the motto for Costa Rica, and the 1986 presidential election came to center on the latter. The PLN candidate was Oscar Arias Sanchéz and the opposition candidate was Rafael Angel Calderón Fournier, the son of former president Calderón, who is usually called "Junior." Calderón was heavily favored to win, but Arias became the peace candidate and easily beat Calderón's anti-Sandinista appeal.

Once Arias became president he moved rapidly to reduce the contras' presence in the country and eliminate United States clandestine operations to support them. After months of effort and negotiations Arias was able to bring together in Guatemala all the presidents of Central American countries, including Daniel Ortega. After two days of dealing with hard issues, on August 7, 1987, all signed the Arias peace plan for Central America, which included provisions for amnesty, cease-fires, freedom of the press, radio, and television, the cutoff of aid to any rebel or irregular force, and refusal to allow national territory to be used to destabilize other governments. Later, in 1989, Ortega agreed to schedule free elections in Nicaragua in 1990.

On October 13, 1987, "Don Oscar" was awarded the Nobel Prize for Peace. In February 1988 the United States Congress cut off aid to the contras, and the Contra War ended. On February 25, 1990, Violeta Chamorro, running on an anti-Sandinista and pro-democracy platform, was elected president of Nicaragua in a stunning defeat of her opponent, Daniel Ortega, and Costa Ricans were dancing in the streets again.

La Lucha sin Fin

"Junior" Calderón was elected Costa Rica's president in 1990. He promised to concentrate on internal affairs, streamline the bureaucracy, and encourage the business sector. Once in office he faced enormous economic problems that his immediate predecessors had been able to ignore because of the artificial boost to the economy of United States dollars during the Contra War. President Carlos Monge during his administration (1982–86) had attempted to enforce an austerity program after the disastrous recession of 1982–84. It might have succeeded but for the sudden inflow of funds the war stimulated and the runaway spending this inspired in the next several years.

The end of the war brought some economic benefits. Ecotourism blossomed as a major economic force, rivaling coffee and bananas as a source of foreign currency. No longer frightened by the war, nature lovers from all over the world flocked to Costa Rica to visit the many national parks and reserves (Boza 1978; Boza and Cevo 1998) established beginning in the Oduber administration. For a while, if one wandered about San José it seemed that ecotourists outnumbered the ticos. This benefit was offset by a depression in coffee prices and reduction in banana exports to Europe and a new round of inflation.

Calderón soon had to raise taxes and fees, and there was talk of reducing government services and some welfare programs, under pressure from the International Monetary Fund (IMF). A series of general strikes and protests by labor unions, followed but "Junior" was able to weather the storm. Few, however, believed that any of his measures addressed the underlying problems long term.

In 1994 José María Figueres, son of Don Pepe, ran for president as the Liberación candidate. He surprised even members of his own party by espousing not fewer, but more government programs and encouraging involvement in the economy. After his election, President Figueres presented an ambitious plan to make Costa Rica a model of sustainable development focused on environmentally sound principles. The plan was very "green," and among other components it emphasized ecotourism, reforestation, and revised land-use policies.

By 1995 reality set in. Costa Rica now had a population of nearly 3.6 million people, estimated to double in thirty years. Although the birthrate had been lowered to 2.7% since the high in the 1960s, life expectancy at birth is now seventy-five years. The labor force was divided into 27% agriculture, 35% industry and commerce, and 33% service, mostly government, little changed from the early 1980s. The bonanza of ecotourism that had seen continuous expansion and had increased foreign exchange in the first half of the decade faltered and, contrary to rosy expectations, remained constant at 1992 levels. In addition, Costa Ricans and their governments, long bolstered by foreign aid, had been living far beyond their means, with a negative balance of payments every year since 1936, and spending close to 50% of the gross national product on government.

When the World Bank under pressure from the IMF held back a $100 million grant because of fears of the country's bankruptcy, Figueres was forced to abandon many of his plans. After a series of conferences with leaders of the opposition and the business community, he agreed to take drastic steps to resolve the situation. Among these were liberalizing the banking laws to allow private banks, supporting privatization of many government-run programs, reducing the bureaucracy by eight thousand positions, and ending state monopolies in insurance, telecommunications, water and power, and petroleum.

These measures brought increasing resistance to Figueres from within his own party and widespread public opposition to privatization. The continuing low economic growth and high inflation (11.5%), combined with the opposition to Figueres' policies, saw little substantial change during his administration. He was able to trim the federal deficit somewhat, and he increased spending on education to 6% of the republic's gross domestic product by extending the school term from 160 to 200 days per year.

The presidential election in early 1998 saw general disillusionment by the voters with both the PLN and opposition party candidates. As expected, the candidate of the opposition, businessman Manuel Ángel Rodríguez, won. Both he and the PLN candidate bowed to the antiprivatization block (unions, government employees, etc.) and promised not to further dismantle the welfare state. This meant retaining the government monopolies for health care, telecommunications, electricity, petroleum, the national distillery, and so on. The apathy of the electorate was manifest by a 70% turnout at the polls, where voting is mandatory and presidential elections usually see 90% of registered voters balloting.

Both Figueres in his final days in office and Rodríguez immediately after taking over tried to emphasize the strengths of Costa Rica's literate populace and the potential for foreign investment in assembly plants for high-tech electronic and telecommunications equipment. In 1998 the economy remained strained, however, by an unemployment rate of 13%, foreign debt of $3.5 billion, and among the lowest income per capita ($2,882) in Latin America. Without strong measures to attack the fundamental problems of government subsidies, tariffs, state monopolies, and welfare entitlements, there seemed little prospect of escape from the current morass. Most observers believed that President Rodríguez had little stomach for the political fights that would be necessary to solve the basic economic problems.

The Rodríguez administration had an early success in signing a contract with computer chip giant Intel to establish a plant in Costa Rica with a workforce of two thousand. The ripple effect from the late 1990s economic boom in the United States, Costa Rica's main trading partner, saw unemployment drop to 5.6% and per capita income rise to $3,700 by 2000. Fundamental problems remained, with inflation at above 10% per year, the population hitting 4 million and increasing by 1.64% each year (projected to be nearly 6 million in 2025 and 8 million by 2036), and the government still deficit spending.

As expected, attempts to privatize the telecommunications and electricity agency (ICE) led to general strikes. This was followed by a Supreme Court decision against selling ICE to private companies. An ongoing conflict with Nicaragua over use of the Río San Juan reduced tourism and distracted the government from other matters.

Overshadowing everything is the government's inability to upgrade and repair the country's infrastructure because of the huge debt service on internal loans to cover deficits and a foreign debt of nearly $4 billion. The high-tech meltdown in the United States also had a delayed but negative effect on tourism and manufacturing areas. The illusion of sustainable development in the Tropics under the economic models espoused both by Rodríguez's party (Unidad Social Cristiano), which also controls the unicameral congress, and by the PLN has recently been countered in a detailed critique (Hall, Léon, and Leclerc 2000). Nevertheless, President Rodríguez deserves credit for trying to push for reforms that, because of his initiatives, the next president may find easier to implement. Observers wonder what magic the 2002 presidential candidates hope to use to solve the country's economic problems and to confront the vested interests in making the needed reforms.

As a result, there are deep concerns whether any government initiatives will succeed in ensuring the continuance of a Costa Rica based on the hard-fought-for principles of democracy, peace, universal education, and universal suffrage. Can the economic stresses be overcome without undermining the foundations of the Costa Rican state that has evolved as a model for other societies to emulate? At this time the issue remains in doubt. Nevertheless, those of us that know the Costa Ricans well have faith that the national characteristics that allowed them to successfully meet the challenges to their democracy in the past will reassert themselves to meet these new challenges. For those dedicated to these principles, the struggle never ends *(la lucha sin fin)*.

A HISTORY OF COSTA RICAN HERPETOLOGY

In this section I trace the progress of the scientific discovery of the Costa Rican herpetofauna. The emphasis will be on those who advanced our knowledge of the country's amphibians and reptiles by collecting or describing Costa Rican taxa, and I also show how these individuals and events fit into the broader picture of intellectual and historical trends in the overall development of herpetology. Adler's

(1989) marvelous, though incomplete, "Herpetologists of the Past," in *Contributions to the History of Herpetology*, gives additional details and an overview of the progress of herpetological discovery worldwide.

For this purpose I have referenced only major publications of historical significance. Particularly when dealing with events occurring in my lifetime, I have attempted to identify the most influential contributors, but most of their publications are cited in appropriate species accounts and in the literature cited rather than redundantly when they appear on the historical stage. Although the history is generally chronological and for the past half-century organized by decades, I have not hesitated to follow the work of individuals when their research spans two or more periods.

Pre-Colombian Times

Long before the first Europeans arrived in Central America, the Amerind peoples in what is now Costa Rica had accumulated much information on the flora and fauna they met in their daily lives, including amphibians and reptiles. Unfortunately they did not leave a written record, so we are deprived of that knowledge. We do know that the subjects of this book formed an important part of their culture, as evidenced by the many replicas of amphibians and reptiles most commonly found as funerary adornments or offerings.

Among the organisms represented are crocodilians in most media (ceramics, jade, stone, and gold); turtles and toads in ceramic ocarinas; basilisk lizards (often misidentified as birds by archaeologists and collectors) on tripod vases; and frogs and snakes in ceramic dishes and urns. Probably the most commonly seen objects are gold pendants representing a variety of charming little frogs and elegant caimans and crocodiles (Ferrero 1987). In terms of religious rites crocodilians were especially important, since mastery over these huge beasts was a symbol of manhood and of humans' superiority to their animal nature (as in the bullfight). Frogs represented the coming of rain and the growth of plants, including maize, that it ensured. By association, they stood as symbols of human fertility.

Pre-Linnaean Period

In reading this section, bear in mind that even into the early nineteenth century, though most scientific works on natural history were still written in Latin, there was no uniform system of nomenclature. Binomial nomenclature was first consistently applied by Carl von Linné in his *Species Plantarum* (1753) and *Systema Naturae*, tenth edition (1758), but it was not universally accepted until the nineteenth century.

The principle of priority as a means of selecting among two or more names for the same species was established only in 1843. The basis of priority differed among several codes of zoological nomenclature, in which different editions of the *Systema Naturae* were used as the starting date

until 1905, when the tenth edition was adopted by the zoological community (Melville 1995). Under the present *International Code of Zoological Nomenclature* (International Commission on Zoological Nomenclature 1999), no names published before 1758 have any nomenclatural status, even if the taxa are clearly identifiable.

After Colombus's discovery and during the conquests and early settlement of the Americas, many strange and new creatures were brought to Europe. There is little likelihood that any of them were from the future Costa Rica, but specimens of some widespread species of amphibians and reptiles that occur in the republic were doubtless among those taken to Europe early on. In 1554 the great Swiss naturalist Conrad Gessner already included two large species, *Chelonia mydas* and *Caiman crocodilus*, in his monumental work on zoology, albeit not by these names.

The first certifiable field herpetologists to collect in tropical America were Francisco Hernández and his son, who with a large retinue undertook to prepare a natural history and geography of Mexico. Their fieldwork covered 1570 to 1577. Long after his death in 1587, a much modified version of Hernández's manuscript was printed in 1648. Smith and Smith (1973) have detailed the history of this work. Smith (1969) also partially succeeded in identifying many of the amphibians and reptiles described in Latin by Hernández, using Náhuatl names for them. These included thirteen large species commonly found today in Costa Rica: *Bufo marinus, Basiliscus vittatus, Iguana iguana, Agkistrodon bilineatus, Boa constrictor, Bothrops asper, Crotalus durissus, Drymarchon corais, Oxybelis aeneus, O. fulgidus, Pseustes poecilonotus, Spilotes pullatus,* and *Crocodylus acutus.*

Two other early field naturalists, Georg Marcgraf (1611–64), who collected in Brazil from 1638 to 1644, and Mark Catesby (1683–1749), who spent time in Florida, the Carolinas, and the Caribbean region in 1722–26, also described species, later named by Linné, that occur in Costa Rica (Marcgraf 1648; Catesby 1743, 1748). Both recognized *Iguana iguana*, Marcgraf described *Chelonia mydas*, and Catesby included *Caretta caretta, Sibon nebulatus,* and *Caiman crocodilus* in his books.

During the early 1700s two residents of the Netherlands, Albertus Seba (1655–1736) and Lorenz Theodore Gronovius (1730–77), amassed large natural history collections. Much of the tropical American material probably came from Suriname, a Dutch colony seized from the British in 1667 and awarded to the Netherlands at the Treaty of Breda. These animals were often labeled simply "America" or "Indiis," leading to some confusion in later times. The first two of Seba's monumental folio volumes (1734–65) included the amphibians and reptiles (1734, 1735) and contained figures that were the partial basis for names of fourteen species of amphibians and reptiles found in Costa Rica as recognized by Linné (1758, 1766). Gronovius's two-volume work (1756, 1763) contained figures of species now found in Costa Rica, and these were the partial basis for ten Linnaean names (1758, 1766). It is almost certain that none of Seba's or Gronovius's specimens came from Middle America.

Carl von Linné and the Late 1700s

The great Swedish systematist Carolus Linnaeus (1707–78), gradually developed the system of binomial nomenclature over a period of years in early editions of the *Systema Naturae*. Originally the names were shorthand for the longer Latin description, but by the tenth edition the species name came to represent a concept reduced from the Latin description to the genus and species names (a binomen or binomial). For his many accomplishments and contributions to science, Linnaeus was late in life raised to the nobility, with a name change to Carl von Linné.

Linné earned his medical degree in the Netherlands early in his career, knew Seba and Gronovius, and saw their collections. Although many of his descriptions in the *Systema* cite the figures of Marcgraf, Catesby, Seba, and Gronovius, contrary to some authors Linné based all his herpetological names on actual specimens now mostly housed in the Zoological Museum at the University of Uppsala or the Royal Museum in Stockholm (Lönnberg 1896; Andersson 1899, 1900).

Linné (1758) is responsible for the names of one amphibian and nineteen reptiles now known to occur in Costa Rica. Four additional reptile names date from Linné (1766). Only a few other valid species now known to occur in Costa Rica were named outside the tenth and twelfth editions of the *Systema*. Domenico Vandelli (1761) described the leatherback turtle *(Dermochelys coriacea)*; Joseph Nicholas Laurenti (1768), two frogs, *Leptodactylus pentadactylus* and *Phrynohyas venulosa*; Martin Houttuyn (1782), the lizard *Thecadactylus rapicauda*; and the Comte de Lacépède (1788), the snake *Lampropeltis triangulum*. None of these names were based on specimens from Costa Rica.

Early Nineteenth Century

The period from 1800 to 1830 saw a steady increase in the discovery of the herpetofaunas of the Americas. François-Marie Daudin (1801–3) described three snakes *(Clelia clelia, Oxybelis fulgidus*, and *Tripanurgos compressus)*, and Georges Cuvier (1807) described the crocodile *(Crocodylus acutus)*, species now known to occur in Costa Rica. Similarly, several German scientists, Friedrich Boie (1827), Johann F. von Eschscholtz (1829), Heinrich Kuhl (1820), Blasius Merrem (1821), Johann G. Wagler (1824), Maximilian, Prinz von Wied-Neuwied (1824), and Arned Wiegmann (1828) described a sea turtle, three lizards, and three snakes now found in Costa Rica in a variety of works. Thomas Say (1823) was the first American to recognize and name a valid species that occurs in Costa Rica, the ribbon snake *Thamnophis proximus*.

In Europe the 1830s and 1840s were a time of enthusiastic interest in natural history and formed an important period in the founding or enhancement of the great natural history museums. The most famous scientist of his era, Alexander von Humboldt (1769–1859), arranged for Ferdinand Deppe and Count Graf von Sack to collect material in Mexico (1824–27) for the Berlin and Vienna museums. Subsequently Deppe returned to Mexico with Christian Schiede (1829–30) for further collecting. These materials became the basis of Arned Wiegmann's papers on Mexican amphibians and reptiles, the most important being the *Herpetologia Mexicana* (1834a), which introduced to science the valid Costa Rican species *Ameiva undulata, Cnemidophorus deppii, Norops biporcatus*, and *Sceloporus variabilis*. He also described *Phyllodactylus tuberculosus* Wiegmann, 1834b.

John E. Gray (1800–1875), whose career spanned forty-three years at the British Museum of Natural History (now the Natural History Museum), was to make its zoological collections the largest in the world at the time, foreshadowing the museum's coming dominance as a center for herpetology. He published many books and papers in the 1820s and 1830s that included descriptions of the large lizard *Ctenosaura similis* (1831a) and the turtle *Chrysemys ornata* (1831b), both prominent members of the Costa Rican herpetofauna.

This period also saw the development of the Rijksmuseum van Natuurlijke Historie in Leiden (now the Nationaal Naturhistorisch Museum) as a force in herpetology. The German scientist Hermann Schlegel (1804–84) was brought to the museum from Vienna in 1825 and devoted himself initially to the study of reptiles. Although he was later best remembered as the leading ornithologist of his day, his 1837 *Essai su le physionomie des serpens* was groundbreaking in the study of snakes. Two species of snakes now known to occur in Costa Rica were described in this work: *Drymobius margaritiferus* and *Scolocophis atrocinctus*.

However, the most important center of herpetological study in this era was the Muséum National d'Histoire Naturelle in Paris, and the most significant work was the *Erpétologie générale, ou Histoire naturelle complète des reptiles* (a ten-volume series plus an atlas of 120 plates). This comprehensive account of the world's amphibians and reptiles was written by Constant Duméril (1774–1860) with the assistance of Gabriel Bibron and later Duméril's son, Auguste. The *Erpétologie* appeared over several years (1834–54) and contains the original descriptions of three frogs and twelve reptiles now recognized as valid Costa Rican species.

In the 1840s Gray initiated the catalog series describing the extensive collections in the British Museum that would see their apogee in the volumes published later in the century by Albert C. L. G. Günther and George A. Boulenger. In the first of these on lizards Gray (1845) described *Anadia ocellata*, and earlier he had described *Ctenosaura quinquecarinatus* (1842).

During this decade Eduard Rüppell of Senckenberg recognized *Atropoides nummifer* (1845), and Arnold Berthold (1846) of the Göttingen Museum described *Polychrus gutturosus*, *Stenorrhina degenhardtii*, and *Bothriechis schlegelii* from Colombia. The five species mentioned above all are now known from Costa Rica.

Mid-Nineteenth Century

The 1850s were a period of accelerating knowledge of the herpetofauna of tropical America that added many valid species now known from Costa Rica. Auguste Duméril described the night lizard *Lepidophyma flavimaculatum*, and Constant Duméril and Bibron described the turtle *Kinosternon leucostomum* (in Duméril and Duméril 1851). The final volume of the *Erpétologie* appeared in 1854 and included the first descriptions of eight snakes now included in the Costa Rican fauna. Gray (1855) recognized *Rhinoclemmys pulcherrima*, the first species of land turtle now included in the biota of the republic, based on a specimen of unknown provenance.

Charles Girard (1854) and Edward Hallowell (1856) were the first herpetologists associated with United States institutions to describe amphibians and reptiles from lower Central America. The former named the frog *Dendrobates auratus* and the coral snake *Micrurus nigrocinctus* from Panama. The latter named the anole *Norops cupreus* from Nicaragua. Girard, of course, was an associate of Spencer F. Baird, then assistant secretary of the Smithsonian Institution and builder of the National Museum of Natural History's great collections. Hallowell was a Philadelphia physician who was associated with the oldest natural history museum in the United States, the Academy of Natural Sciences of Philadelphia. Both of these institutions were to have a prominent role in documenting the diverse herpetofauna of Costa Rica.

During this same period Martin Lichtenstein and Carl von Martens (1856), at the Berlin Museum, described *Atelopus varius* and *Ameiva festiva* from material almost certainly collected by Josef Warszewicz in western Panama (Savage 1970, 1972a). Wilhelm Peters (1857), who would become Lichtenstein's successor as director of the museum, named another species, the snake *Liotyphlops albirostris*, which Warszewicz collected in Panama. Warszewicz material from Panama sent to the Vienna museum formed the basis for the descriptions of several valid species of frogs, *Eleutherodactylus fitzingeri*, *Dendrobates pumilio*, *Phyllobates lugubris*, and *Rana warszewitschii*, named by Oskar Schmidt (1857). Not surprisingly, all ten of the taxa mentioned above were later found in Costa Rica.

In 1857 Gray hired Günther, who ultimately succeeded Gray as keeper of zoology for the museum and wrote over 220 books and papers devoted wholly or in part to amphibians and reptiles. The first of his major works was the *Catalogue of Colubrine Snakes in the Collection of the British Museum (Natural History)*, based on 3,100 specimens and completed within four months of his appointment by Gray. The catalog (1858a) contained the original descriptions of three snakes, *Coniophanes fissidens*, *Rhadinaea decorata*, and *Pseustes poecilonotus*, now part of the Costa Rican herpetofauna.

With the appearance of Günther's (1858a) book, thirteen anurans and seventy-seven reptiles that occur in Costa Rica and are today regarded as valid had been described. Not one of these forms had been based on Costa Rican specimens, nor had any examples from Costa Rica been collected or associated with these names.

All of that was about to change.

The Beginnings of a Herpetology of Costa Rica

In 1853 Alexander von Humboldt wrote to Juan Rafael Mora, president of Costa Rica, asking him to aid two young Germans who were medical doctors and scientists, Alexander von Frantzius (1821–77) and Carl Hoffmann (1823–59), in their studies of the biology and geology of the country. The two arrived in Costa Rica early in 1854, accompanied by the horticulturist Julian Carimol. The letter from Humboldt (Guillén 1989) assured them of a fine welcome from President Mora, who encouraged and supported their scientific studies. Among their activities Frantzius and Hoffmann began sending collections of plants and animals back to Berlin, which was now Humboldt's home base. Both men made a living from their skill as physicians while dedicating their spare time to scientific work. Carimol and his son became well-known bird collectors while continuing their horticultural interests, and they sent extensive ornithological materials to the Smithsonian Institution in Washington, D.C.

Hoffmann seems to have been the more restless and energetic of the two German doctors. He collected numerous plants and animals that he sent to the museums in Berlin. He also explored the volcanoes of the Cordillera Central (Alvarado 1989), climbing Volcanes Irazú (Hoffmann 1856) and Barva (Hoffmann 1857). He served as surgeon-general under Juan Rafael Mora in the Costa Rican army (1855–57) that fought against the filibuster William Walker, and he was at the decisive battle of Rivas. He died in 1859 and is buried in Esparza. His name is commemorated in the names of a dozen or so Costa Rican plant and animal species.

Frantzius, though working at a less hectic pace than his short-lived companion, became the preeminent resident naturalist in Costa Rica during this period, remaining in the country for fifteen years. He made many general collections that were at first deposited in Berlin but later went to the Smithsonian Institution. His botanical explorations called attention to the diversity of the Costa Rican biota, and he prepared the first annotated lists of birds (1869a) and mammals (1869b) for the republic. He explored the craters of Volcán Irazú in 1859 and Volcán Poás in 1860 (1861). He also published on the geology and geography of Costa Rica (1862, 1869c,d, 1873).

Frantzius's importance goes beyond these contributions. In his role of medical doctor, he decided to open a drugstore in San José, where his young apprentice, José Castulo Zeledón (1846–1923), became interested in natural history. Zeledón later became a prominent ornithologist of international reputation and a major figure at the Museo Nacional de Costa Rica. The Frantzius and Zeledón drugstore soon became a gathering place for those interested in natural history and served as an incubator for the development of the first series of native Costa Rican naturalists. In addition to Zeledón, Juan J. Cooper and later J. Fidél Tristán and Anastasio Alfaro, among others, became part of the "drugstore gang." One can well imagine the excitement of those days when Frantzius and his boys would take off on weekend collecting trips loaded down with plant presses, butterfly nets, collecting jars, pillboxes, and other outlandish paraphernalia. Or when one of their number returned from a longer foray to areas far distant from San José, with new wonders of plant and animal life and tales of the trail.

The drugstore, the Botica de Frantzius, quickly became known *al a tico* as the Botica Francés, and that name changed in time to the Botica Francesa as the business was passed on through several generations. Those interested in this bit of history can visit the Botica now located in San Francisco de Guadalupe antes del 31 de Marzo Marque!

For Costa Rican herpetology 1859 was a banner year. Wilhelm Peters (1815–83), the distinguished herpetologist at the Berlin Museum (Museum für Naturkunde zu Berlin) and professor of anatomy at the University of Berlin, where the museum was then located, described three new species of snakes from the collections of Carl Hoffmann. These were the first reptile names ever based on Costa Rican specimens. All three were placed in new genera as *Colobognathus* (= *Geophis*) *hoffmanni*, *Hydromorphus concolor*, and *Bothriechis nigroviridis*.

The 1860s saw the greatest explosion in knowledge of the diversity of amphibians and reptiles of Costa Rica. Sixty-three valid species now represented in the country's herpetofauna were described in that decade, although only ten were based on material from Costa Rica. Peters continued to describe material from the Panama collections of Warszewicz and various German naturalists (seven species later found in Costa Rica and six forms from the republic, five of which definitely were collected by Hoffmann or Frantzius; see Bauer, Günther, and Klipfel 1995 for a detailed biography of Peters). Although he described additional taxa from lower Central America into the 1970s, Peters's leading role was soon eclipsed by the incredible outpouring of papers on the region by the dynamic American genius Edward Drinker Cope (1840–97).

Cope began his career as an eighteen-year-old at the Academy of Natural Sciences in Philadelphia and studied anatomy with the famous paleontologist Joseph Leidy (1823–91) at the University of Pennsylvania. He became a protégé of Spencer F. Baird at the Smithsonian Institution, spending time there in the winters of 1861–62 and 1862–63. At age twenty-two he visited all the major European collections, meeting Gray and Günther in London, encountering Schlegel in Leiden, and staying with Georges Jan in Italy as well as visiting the museums at Berlin and Paris. From 1863 to 1883 he was closely associated with the Academy of Natural Sciences, and throughout his life he had ties with the Smithsonian Institution. Although he was perhaps best known for his works on vertebrate paleontology, herpetology was his first love, and he published about 170 papers on living amphibians and reptiles out of a total output of 1,395 books, articles, and scientific papers. In later life he was a professor at the University of Pennsylvania, became a member of the United States National Academy of Sciences, and was elected first president of the American Association for the Advancement of Science. You can learn much more about this most illustrious of herpetologists from the brief biographies by his friend Henry F. Osgood (1930) and by Adler (1989) and in the detailed one by Osgood (1931).

Cope's interests were broad, and he published many papers in ichthyology, geology, ornithology, and mammalogy. His early papers were all herpetological. His later ones revolutionized herpetology by emphasizing internal anatomy, both soft parts and skeleton, as a basis for classification. The 1860s saw him describing numerous new taxa from the Americas, including thirty valid species found in Costa Rica. The first of these, *Sceloporus malachiticus* (Cope 1864a), had been sent to the Smithsonian Institution by Charles N. Riotte and was said to be from "near Arriba, Costa Rica." The second two, *Bufo coccifer* and *Hypopachus variolosus* (Cope 1866a), were also sent to the National Museum by Riotte and were labeled simply "Arriba, Costa Rica." This ambiguous locality has usually been interpreted to mean upland Costa Rica, probably the Meseta Central, since all three species are found in the cities of San José and Cartago and environs. The late Douglas C. Robinson of the University of Costa Rica thought the locality might be San Juan Arriba, Heredia Province, and I suggested San Rafael Arriba, San José Province, because Riotte's descendants live near there.

During 1864 and 1865 the geologist Karl von Seebach (1865a,b) explored the country and discovered the first species of salamander from Costa Rica. The specimen was sent to the Museum at Göttingen and named as a new genus and species, *Oedipina uniformis*, by Wilhelm Keferstein (1868a). Keferstein (1867, 1868b) also documented six frogs that Seebach collected in Costa Rica, three of which he described as new. These are now known by other names, but Keferstein's generic name *Hypopachus* is valid.

Other active field collectors were busy elsewhere in Central America during this period. Marie-Fermin Bocourt (1819–1904) was assigned leadership of a scientific commission to study the fauna of Mexico. This project was sponsored by the government of Napoleon III as part of the effort to establish Maximilian von Hapsburg, the younger brother of Emperor Franz Josef of Austria-Hungary, as emperor of Mexico. When Benito Juárez and the Mexicans vigorously rejected this notion, in 1864 Bocourt was sent to Guatemala instead. Bocourt had worked in the section of herpetology and ichthyology with Constant Duméril and Bibron since 1834 and had recently returned from an expedition to Thailand. For the *Mission* work he visited Belize, but he spent most of the time from 1864 to 1866 in Guatemala, with brief stops in El Salvador, Costa Rica, and Panama on his return home.

Bocourt's collections of amphibians and reptiles formed the basis of the *Mission scientifique au Mexique et dans l'Amérique Centrale*. The reptile section was variously written by Auguste Duméril, Marie-Firmin Bocourt, François Mocquard, Léon-Louis Vaillant, and Fernand Angel between 1870 and 1909, although usually credited to the first three. The amphibians were treated by Paul Brocchi (1839–98) in three *livraisons*, or parts, in 1881, 1882, and 1883. Interestingly, only five valid species described by Bocourt and one from the amphibian volume form part of the Costa Rican fauna.

The 1860s closed with Frantzius taking José Zeledón with him to visit the Smithsonian Institution in Washington, whence Frantzius returned to Germany. This trip cemented the relationship between the National Museum of Natural History and Costa Rican science that would be expanded on during the rest of Zeledón's life.

Foundations of Costa Rican Herpetology

The single most important herpetological publication on Costa Rica of the nineteenth century was Cope's 1875 work *On the Batrachia and Reptilia of Costa Rica*. The Costa Rican components of the studied collections had been obtained mostly by William More Gabb in his survey of the natural resources of the Talamanca region of southeastern Costa Rica (Savage 1970). These explorations were supported by President Tomás Guardia, urged on no doubt by railroad builder Minor C. Keith, and took advantage of the Smithsonian Institution connections of Keith and of José Zeledón.

Gabb carried out his studies of the region for seventeen months in 1873 and 1874 (Gabb 1875). Zeledón and Juan Cooper accompanied Gabb on most of his forays, including scaling Cerro Utyum (3,084 m). The vertebrate collections were made primarily by the two Costarricense, and the bulk of the material was sent to the United States National Museum of Natural History. Cope's paper is based on Gabb's material and specimens from the neighborhood of the city of San José collected by Dr. Charles Van Patten and previously listed, and on new species described in Cope (1871). Van Patten (1814–89) was a dentist and physician originally from New York who received his degrees from Harvard. He became a member of the Frantzius-Zeledón group after settling in San José in 1865. Many of his descendants were and are prominent in Costa Rican public affairs.

Between the two collections, twenty-six valid species from Costa Rica were described for the first time. Cope's Costa Rican paper and those associated with it have caused some confusion. The original letterpress separates of the second series, volume 8, of the *Journal of the Academy of Natural Sciences of Philadelphia* consist of articles 4, 5, and 6. They were published together on November 26, 1875. The same articles were published as part of complete volume 8 when it was issued in early 1876. Unfortunately, several authors perpetuate the erroneous notion that the names date from 1876, not 1875 (e.g., Henderson 1997; McDiarmid, Campbell, and Toure 1999). Some of the articles were also sent out together with a cover sheet titled "On the Batrachia and Reptilia of Costa Rica with Notes on the Herpetology and Ichthyology of Nicaragua and Peru. Extracted from the Journal of the Academy of Natural Sciences Philadelphia, 1875." Article 4 is the paper on the herpetology of Costa Rica. Article 5 is titled "On the Batrachia and Reptilia Collected by Dr. John M. Bransford during the Nicaraguan Canal Survey of 1874." As I have pointed out elsewhere, the cover sheet and title page are misleading, since all of Bransford's 1874 specimens were collected in Panama (Savage 1973a). Article 6 is on Peruvian animals and causes no problems. (For more information on Bransford's activities in Central America, see Bransford 1881, 1884.)

By the end of the 1870s there appeared the first issue of the monumental sixty-seven volumes of the *Biologia Centrali-Americana*, based on the extensive private collections of Osbert Salvin (1835–98) and Frederick Du Cane Godman (1834–1919). Material forming the basis for these comprehensive treatments of the biota of Mexico and Central America had been accumulated by the active fieldwork of Salvin and Godman themselves and a series of associates. Salvin and Godman (to 1898) and Godman and George C. Champion from 1898 to 1915 edited the entire series.

Collection of Costa Rican amphibians and reptiles for the *Biologia* originated in the mid-1870s, and H. Rodgers, who resided in Cachí, Cartago Province, obtained most of the early specimens. Although a few new species now known from Costa Rica were named by Günther from the *Biologia* collections in the 1860s and 1870s, none were based on specimens from the republic.

During the 1880s Cope continued to describe new forms from Middle America, eight of them recognized as valid Costa Rican species, but none were based on examples from Costa Rica. It was at the beginning of this decade (1881) that Günther hired a promising young Belgian, George Albert Boulenger (1858–1937), as assistant in charge of lower vertebrates. Since Günther was himself at the time most interested in fish systematics, he assigned Boulenger the task of preparing a new edition of the catalogs of amphibians and reptiles in the British Museum.

Boulenger produced nine volumes between 1882 and 1896, covering all then known species (8,469) of amphibians and reptiles. He also published hundreds of other papers on herpetology and freshwater fishes and doubtless knew the world's herpetofauna better than anyone before or since. He was honored by being elected to the Royal Society of London and received honorary degrees from several universities. He certainly ranks with Constant Duméril and Cope among the outstanding systematic herpetologists of all time (Adler 1989).

As material for the *Biologia* was gradually donated to the British Museum over time, Boulenger described an increasing number of Central American forms. Günther, who had originally agreed to write the herpetological volume for Salvin and Godman, obviously encouraged him in this regard. During his career Boulenger named to our faunal list seventeen valid species that occur in Costa Rica, seven based on examples from the country.

In the 1880s the Costa Rican educational system underwent a major reformation, as discussed above in the section on the history of Costa Rica. As part of the effort to upgrade the quality of teaching and modernize its methods, the government recruited a number of educators, principally from Switzerland. Among these were Paul (Pablo) Biolley (1863–1908) and Henri Pittier (1857–1950), who arrived in 1886 and 1887, respectively. Biolley was for many years the professor of natural sciences at the Liceo de Costa Rica and the Colegio de las Señoritas. He was a broadly interested field naturalist with a special interest in entomology.

Pittier came to the country as a professor of mathematics for the Liceo, but within a year he was appointed director of the Instituto Físico y Geográfico. From that base this human dynamo stimulated and dominated every aspect of scientific discovery in the country, published copiously, and encouraged scientists from elsewhere to study Costa Rica's rich natural environment and resources (Conejo 1975; Gómez and Savage 1983, 1991).

Under Pittier's influence the Museo Nacional de Costa Rica was established in 1888 (San Román et al. 1987) with José Zeledón's protégé Anastasio Alfaro (1865–1951) as its twenty-two-year-old managing director. Pittier, Biolley, and Zeledón were three of the six members of the overseeing board of directors. Although none of them were herpetologists per se, each collected Costa Rican material that Günther used in his volume of the *Biologia,* as did Alfaro.

Shortly after the Museo was organized, George K. Cherrie (1865–1946) joined the staff as taxidermist and also sent herpetological material to Günther. In 1889 Cecil F. Underwood (1867–1943), encouraged by Zeledón, came to Costa Rica from Britain, making his living by collecting natural history specimens, mostly birds, sold to museums. He too sent herpetological materials to the British Museum and Smithsonian Institution.

Günther's Reptilia and Batrachia contributions to the *Biologia* began to appear in 1885 and were not completely published until 1902. Coverage was much more comprehensive than the *Mission* volumes and integrated into Günther's accounts most of Cope and Boulenger's innovations in classification and the many new forms discovered since the 1870s. This volume became the foundation for all future work on the herpetofauna of Costa Rica.

But Cope wasn't through yet! In 1893 he published *An Addition to the Knowledge of the Batrachia and Reptilia of Costa Rica* and in 1894 another. The former was based on material Cherrie collected in southwestern Costa Rica; it lists a few other forms and includes descriptions of the valid species *Eleutherodactylus stejnegerianus, Phyllobates vittatus,* and *Dendrophidion percarinatum.* The enigmatic salamander *Haptoglossa pressicauda,* never again collected or definitely synonymized with another taxon, was also described. Unfortunately, all this material has been lost.

Cherrie remained in Costa Rica for several years but thereafter was associated with a number of museums in the United States. He later joined the staff at the American Museum of Natural History in New York and accompanied Theodore Roosevelt on the museum's expeditions to the Amazon and central Asia.

Cope's 1894 paper contains the descriptions of the valid species *Bolitoglossa robusta, Rana vibicaria, Celestus cyanochloris, Trimetopon pliolepis, Dendrophidion paucicarinatum,* and *Tantilla ruficeps.* Although said by Cope to be on loan from the Museo Nacional and variously collected by Alfaro, Biolley, Cherrie, Cooper, and Zeledón, these specimens are by some odd chance housed at the American Museum of Natural History. One wonders if Cherrie might have had something to do with this mystery.

A new name to appear on the roster of herpetologists contributing to the knowledge of the fauna of Costa Rica in the last decade of the nineteenth century was Oskar Boettger (1844–1910). He was the curator at the Senckenberg Museum in Frankfurt am Main, Germany, and he described two of the commonest glassfrogs, *Centrolenella prosoblepon* (1892a) and *Hyalinobatrachium fleischmanni* (1893b), plus *Eleutherodactylus fleischmanni* (1892a), all collected by Carl Fleischmann in 1892. Fleischmann also sent material to the British Museum from Costa Rica.

Pittier's forceful personality made him many enemies, and in 1901 the Costa Rican government without his knowledge transferred his contract to the United Fruit Company. After two years of virtual exile in Puerto Limón, he emigrated to the United States. For eighteen years he lived there, employed by various United States government agencies. In 1918 he moved to Venezuela to establish an agricultural college. During the next thirty-two years he more or less repeated for science in Venezuela what he had done for Costa Rica, with a much happier personal result.

The Early Twentieth Century

By the end of the nineteenth century Cope was dead, Pittier was about to be forced out of Costa Rica, Günther's herpetology volume for the *Biologia* was almost complete, and new faces were appearing in herpetology internationally. One of these was Leonhard H. Stejneger (1851–1943), a Norwegian ornithologist hired to assist Spencer Baird at the Smithsonian in 1881 and later pressed into duty to be in charge of amphibians and reptiles. Stejneger was one of the most distinguished systematists of the period and served as head curator of biology at the National Museum of Natural History from 1911 to 1943.

In herpetology he established the practices of detailed description of a single, identifiable individual as the basis of both the description of new or previously named species and of definite designations of type specimens and type localities. This approach was typological rather than populational in philosophy and is still followed by many herpetologists today, at least in descriptions of new taxa. Stejneger's descriptions contrasted markedly with the very short and sometimes ambiguous ones of Cope and the somewhat longer ones of Boulenger, who did not clearly indicate specific specimens as types or as the source of his redescriptions.

Stejneger's role in our story was brief, since he described only *Oedipina collaris* (1907) and *Sibon longifrenis* from Panama (1909) and *Nototriton picadoi* (1911) and *Hyla phlebodes* (1906) from mainland Costa Rica. The first two are now also regarded as members of our fauna. He further named *Norops townsendi* (1900) and *Sphaerodactylus pacificus* (1903) from Isla del Coco. The former was collected by Charles H. Townsend of the USS *Albatross* and the latter by Biolley. Stejneger's most important contribution to Costa Rican herpetology, however, was his encouraging the young Emmett Reid Dunn to study salamanders, a project that led inevitably to Dunn's interest in the herpetofauna of lower Central America.

The sinking coffee prices of the 1900s and World War I and its aftermath left Costa Rican field biology represented essentially by the Museo Nacional. Alfaro soon became director, and thus the lineal intellectual descendant of Alexander von Humboldt out of Frantzius and Zeledón's "drugstore gang" became the first Costarricense officially recognized as a scientific leader. Through Zeledón, doubtless encouraged by Keith, Alfaro had good relations with the Smithsonian Institution and especially the distinguished curator of ornithology Robert Ridgeway (1850–1929).

Both Zeledón and Alfaro always welcomed visiting scientists to Costa Rica, a habit instilled through Zeledón's association with Frantzius. They responded especially well to a long string of North Americans sent to them over the years by their Washington associates. These included Ridgeway himself, whose semipopular account of his 1904 visit (1922) recalls the beauty and serenity of Costa Rica and its unspoiled beauty in another era. It also reminds us of a slower-paced existence and of the clublike atmosphere engendered by gentlemen scientists working together, which has long since passed.

Amelia and Philip Calvert's classic 1917 account *A Year of Costa Rican Natural History* (1909–10) reflects a similar ambiance. Although the book contains little of herpetological interest, its bibliography, together with those of Biolley (1899, 1902), provides a snapshot of the knowledge of Costa Rica's biota up to that date.

The Calverts' principal associate in Costa Rica was J. Fidel Tristán (1874–1932), then director of the Colegio de las Señoritas. Tristán was a Costarricense who was of the new generation of high-school teachers educated in Chile. Primarily interested in insects, he was a generalist like most other naturalists living in Costa Rica at that time and collected a number of herpetological specimens sent to various foreign museums.

Alfaro was also of necessity a general naturalist writing on a wide variety of subjects, including botany, entomology, mammalogy, geography, geology, archaeology, and especially ornithology. He published a number of short papers in herpetology, the most important of which are Alfaro 1906a,b,c,d, 1907a,b, 1912. These indicated that 204 species of amphibians and reptiles were known from Costa Rica by 1912. Many of the cited names are synonyms of one another or of older ones, and a few are based on misidentifications, but the lists contain about 159 of the species that are known to occur in Costa Rica today.

Although Alfaro and company were clearly most strongly allied with United States science, before World War I many of the Costa Rican intelligentsia looked for inspiration to Europe, especially France. For that reason Clodomiro Picado (1887–1944) was sent to Paris (Sorbonne, Institute of Colonial Medicine or Institut Pasteur) for advanced education. His doctor of science dissertation was a groundbreaking study of the epiphytic bromeliads as a special environment for a variety of animals (Picado 1913). This project had been encouraged by Alfaro and Tristán following some of the Calvert's initial observations, but Picado's work was much more comprehensive than any undertaken before and may be considered the first scientific study of canopy ecology.

Picado found one species of salamander and five of frogs in tank bromeliads up to 30 m above the ground. We now know that two of these frogs, *Anotheca spinosa* and *Hyla zeteki,* habitually lay their eggs in the water trapped in tree holes or bromeliads and that their tadpoles complete development there. Alfaro sent some of Picado's specimens to Stejneger (1911), who described them as new species,

Nototriton picadoi and *Anotheca spinosa,* a synonym of *Gastrotheca coronata (= Anotheca coronata).*

On Picado's return to Costa Rica in 1914 he established the first clinical laboratory in the country at the Hospital San Juan de Dios. He published many papers on the biology of Costa Rica, and I will return to those on venomous snakes in a later section.

The Dunn Era: 1920 to 1945

Emmett Reid Dunn (1894–1956), who received his doctorate from Harvard (1921), became the world's foremost authority on salamanders and the higher classification of colubrine snakes. His pursuit of the former led him to carry out much fieldwork in tropical America. He first collected in Costa Rica in 1920. Using the influence of Thomas Barbour, his mentor at Harvard, with the United Fruit Company, by then headquartered in Boston, he was able to travel about the country. On that trip he met Alfaro and established a long-term relationship with the Museo Nacional. He received his degree the following year based on his monograph on the plethodontid salamanders, published in 1926.

Subsequent expeditions to Panama over the next decade and two return trips to Costa Rica (1929 and 1936) solidified his plan to prepare a comprehensive systematic account of the herpetofauna of Costa Rica and Panama. By the 1930s a more or less formal agreement among the leading United States herpetologists had divided Mesoamerica into spheres of influence. Edward Harrison Taylor and Hobart Muir Smith (University of Kansas) concentrated on Mexico, Laurence Cooper Stuart (University of Michigan) on Guatemala, and Karl Patterson Schmidt (Field Museum of Natural History) on Honduras, British Honduras (now Belize), and El Salvador, with Dunn by default adding Nicaragua to his planned monograph.

In 1928 Dunn was awarded a John Simon Guggenheim Fellowship, which during 1928 and 1929 allowed him to visit all the major European museums to examine their herpetological holdings, especially the types of Central American forms, and to carry out fieldwork in Panama and Costa Rica. He returned to a position on the faculty at Haverford College, where Cope had also served briefly. Among his most important publications of this period was his biogeographic synthesis of the herpetofaunas of the Americas (1931b). In 1937 Dunn was named honorary curator of herpetology at the Academy of Natural Sciences in nearby Philadelphia, a collection with which he was already fully familiar.

He served as longtime editor of the journal *Copeia* (1923–36) and was responsible in large part for the publication of a Cope centenary number (1940, no. 2). In that issue Dunn and his wife Merle published a listing of all genera of recent amphibians and reptiles proposed by Cope, with bibliographic citations and type species. He also maintained a card file listing the location, type locality, and specimen number(s) for all the species names of recent amphibians and reptiles first described by Cope. Unfortunately this list was never published and could not be found after Dunn's death.

Dunn knew and corresponded with the leading Costa Rican zoologists of the time, including Alfaro, Picado, Manuel Valerio (director of the Museo after Alfaro), and Carlos Viquez, and named one or more species from the republic in their honor.

Dunn was brilliant, incisive, and often impatient with others less intellectually gifted. His species descriptions were summary because he believed that only the examination of types could definitely establish the identity of species names. He was conservative in naming new forms, trying always to match them up with those previously proposed. Unfortunately his trip to European museums occurred very early in his career, before he was familiar with many lower Central American forms in life. His notes on many types were consequently inadequate for comparison with newly acquired specimens, leading to many misuses of older names, especially for anurans.

By the late 1930s Dunn had developed considerable antipathy toward his contemporary Taylor, the other great contributor to our knowledge of the herpetofauna of Costa Rica. That his feelings were reciprocated is evident from the published exchanges over the status of the worm snakes of the genus *Anomalepis* (Taylor 1939a; Dunn 1941b; Dunn and Tihen 1944; Taylor 1951b). The rivalry was no doubt exacerbated by Taylor's (1944) work on plethodontid salamanders, where he thoroughly revised much of the generic classification proposed by Dunn (1926) in his classic *The Salamanders of the Family Plethodontidae.*

By this time Dunn was well advanced on his projected herpetofauna of lower Central America. He had prepared synonymies, listings of specimens examined and localities, notes on type species, and some keys. As the work progressed he had published a number of papers preliminary to completion of the larger work (e.g., Dunn 1935, 1937a, 1994b, 1938, 1942a). The proposed monograph was delayed by his acceptance of an Interamerican Cultural Exchange Fellowship to Colombia for 1943–44, sponsored by the Nelson Rockefeller Committee.

Dunn spent a year collecting and writing papers about the Colombian herpetofauna, with additional publications appearing through 1946. Four of the principal ones prepared at that time (Dunn 1944 a,b,c, 1945, all reprinted in a single volume 1957) form the foundation for all subsequent herpetological studies for Colombia. It was apparently during Dunn's absence that Taylor was able to copy Dunn's outline and notes for Taylor's subsequent forays into Costa Rican herpetology.

During the period of Dunn's active fieldwork in Costa Rica and Panama he described twenty-two valid forms that are now known to occur in Costa Rica. Ten are based on Costa Rican specimens, and he was the first herpetologist to have named taxa from the republic that he had collected or seen in life. Barbour and Dunn (1921) described one valid form from Panama that also occurs in Costa Rica.

Barbour (1928) named several new frogs from Panama collected by Dunn, of which two are valid species that are found in Costa Rica. He also described two from Costa Rica whose type specimens were collected by Charles H. Lankester (1879–1969). One of these, with the incorrect patronym *Eleutherodactylus lancasteri*, is valid. As an aside, Lankester was an Englishman, nephew of the famous zoologist E. Ray Lankester (1847–1929), who emigrated to Costa Rica in 1897 and became a very successful coffee baron and a devoted naturalist. Charles Lankester's private collection of plants became the University of Costa Rica's Lankester Gardens (established in 1973) at Las Concavas near the city of Cartago. He gave the herpetological specimens in question to Dunn during Dunn's 1920 visit to Costa Rica.

Another important contribution in this period was the report of Otto Wettstein (1934) of the Vienna Museum on collections made in 1930 by O. Koller and R. Zimara. This paper contained additions to the Costa Rican herpetofauna but described no new taxa. It was the first to record examples from the Península de Osa.

The study by Schmidt (1936) of Central American *Micrurus* was particularly important in establishing the distinctiveness of the three tricolor venomous coral snakes that occur in Costa Rica. Previously it had been thought that only one taxon of this group occurred in Costa Rica and western Panama (Picado 1931; Viquez 1935).

The years of the Great Depression adversely affected all aspects of Costa Rican society. The Museo was no exception, and its role shifted from research to public education. Alfaro retired in 1930, and his last publication was a book of essays in 1935. He lived for another sixteen years, showered with honors from throughout Latin America for his contributions. Leadership at the Museo passed first to Fidel Tristán, who had retired from his directorship at the women's high school, then to Manuel Valerio (1932–34), then to Juvenal Valerio (1935–43), and finally to the latter's brother, Romulo Valerio (1943–49). Stressed by the financial situation, the museum turned more and more toward collecting and exhibiting archaeological materials, and the systematic collections were neglected.

Costa Rican contributions to herpetology in this period related primarily to the increasing concern about the dangers of bites by venomous snakes. Picado, who by this time had established a reputation as a microbiologist and an immunologist, published a series of papers on venoms and their treatment beginning in 1926. His major book on the subject, *Serpientes venenosas de Costa Rica* (1931, 1976), describes the snakes, the venoms, treatment of bites by antivenin, and the manufacture of antivenin.

Carlos Viquez (1890–1957) also published *Animales venenosos de Costa Rica* (1935), in which he listed 113 snakes known from the republic. If misidentifications and synonyms are taken into account, about 93 represent valid taxa. Later works by Viquez (1941, 1942) contained some additional information but were essentially based on the 1935 book.

In 1940–41 the University of Costa Rica was reestablished, and shortly thereafter responsibility for the Museo Nacional was transferred to the university. During the period of university control (1944–52) little attention was given to the zoological collections. This is attested to by the listing of many specimens in the museum's collections in Dunn's notes, practically none of which could be found by the 1960s, since most had been allowed to dry out or had been discarded. As the Dunn era drew to a close, Costa Rican herpetology was at its nadir.

Edward Harrison Taylor and the Transformation of Costa Rican Herpetology

Edward Harrison Taylor (1889–1978), the last of the truly great explorer-herpetologists, had a long and varied career in the Far East before returning to his home state and to the faculty of the University of Kansas in 1926. He had made major collections and published numerous papers on the Philippines, where he had been a schoolteacher before World War I and a member of the Bureau of Fisheries after the war. He received his Ph.D. from the University of Kansas the following year (1927) for his monograph on Philippine mammals. It was shortly thereafter that he and then graduate student Hobart Muir Smith (1912–; Ph.D., University of Kansas, 1936) decided to concentrate their field and laboratory efforts on elucidating the Mexican herpetofauna.

Over his career Taylor became known as the authority on skinks of the genus *Eumeces* (1935), the caecilians (1968), and the herpetofauna of the Philippines, Mexico, Thailand, Ceylon, and finally of Costa Rica. He was without question the greatest field collector of amphibians and reptiles that ever lived, and he collected more species in their native habitats than any other herpetologist. He named over five hundred new species, mostly based on examples he collected himself, and published 202 papers, monographs, and other books.

By the mid-1940s Smith and Taylor had completed the first phases of their Mexican fieldwork and systematic studies as summarized in their checklists and keys (1945, 1948, 1950). Smith continued to focus his efforts on Mexico for the rest of his career. Taylor was looking for new fields to conquer. Why not Costa Rica?

Because of a developing relationship between the University of Kansas and several Costa Rican government institutions, Taylor arranged to receive an invitation in 1947 to begin fieldwork in the republic. He and his son, Richard C. Taylor, collected at a wide range of localities during June to September of that year. Taylor returned for further fieldwork in the rainy seasons (June to September) in 1951, 1952, and 1954, and these collections were the basis for several short papers and his monographs on the snakes (1951b, 1954b), frogs (1952a), salamanders and caecilians (1952b), and lizards (1956). Taylor could not have missed the irony that he, not his archrival Dunn, should publish the basic works on the Costa Rican herpetofauna.

Taylor was a man of enormous energy and focus in his fieldwork. He was a veritable whirlwind, leaving a broad swath of destruction behind him as he turned every log or rock, searched every nook and cranny, and knocked down and searched all but the largest trees and shrubs to capture his prey. However, he clearly knew his animals well, using every device at his command to find them in their hiding places (Smith 1975). His data recording in the field was somewhat cavalier, and he often depended on his exceptional memory to label specimens several days after their capture and preservation. Note taking was not for him—there was collecting to be done!

In his science Taylor was essentially a Boulengerian alpha taxonomist. He followed Stejneger's topological procedures in describing new forms, often emphasizing extremes of variation or anomalies to justify their description. He was interested in novelty and was little concerned with attempts to allocate older names, preferring to introduce new ones. His forceful personality sometimes was expressed as great impatience or even irascibility. He sometimes described himself as an outsider snubbed by the eastern establishment of the Harvard-Washington axis, with Michigan and Chicago thrown in for good measure. Nevertheless he could be the most charming host or companion in a social situation.

During the period of Taylor's Anschluss (1947–58) of Costa Rican herpetology he published sixteen papers containing 1,306 pages and describing thirty-five valid species from the republic. In addition, his collections added to the herpetofaunal list many other species not previously collected in Costa Rica. Taylor collected in Costa Rica briefly in 1952, 1964, and 1966, but no publications were forthcoming from these visits. I recall especially the 1964 visit when I showed him living specimens of the as yet undescribed Monteverde golden toad *(Bufo periglenes)*. In typical Taylorian fashion he opined, "The good Lord has been busy since I last collected in Costa Rica."

The Early Post-Taylor Period

Archibald Fairly Carr Jr. (1909–87), known to everyone as Archie, was at midcentury the United States authority on the systematics and ecology of turtles. His Ph.D. was from the University of Florida (1937), and he was a protégé of Thomas Barbour at Harvard, where Archie spent his summers until 1945. Although Carr collected few specimens and named no new taxa from the republic, he was to be involved in two of the most important events shaping Costa Rican herpetology in the coming decades.

By the 1950s Carr had focused most of his attention on the biology of sea turtles. In 1954 he surveyed a number of sites as locales for long-term studies on the reproduction and migration of the green turtle *(Chelonia mydas)*. As a result he selected the turtle nesting beaches of northern Limón Province for his studies and established a field camp at the mouth of the Río Tortuguero near the little village of the same name.

Fieldwork at the site began in 1955, and since that time the primitive camp has been replaced by modern facilities. Many herpetologists, hundreds of graduate students, thousands of volunteers, and quite a few ecotourists have been involved in this, the longest continuous study of the population of any amphibian or reptile. Numerous papers by Archie, his students, and others on the behavioral ecology and other aspects of the life of turtles that nest at Tortuguero have appeared and continue to appear. These are summarized in the appropriate accounts later in this book.

Archie also became deeply concerned about the need for conservation of the Tortuguero turtles, and he and several others formed the Caribbean Conservation Corporation to further this end. His activities led President Francisco J. Orlich to establish the Tortuguero beaches as a protected area in 1963, and a greatly expanded national park was created in 1970. Many believe Carr's efforts were the start of the conservation movement in Costa Rica that has resulted today in the country's model system of national parks and natural reserves.

Carr's other major contribution is less well known. By 1954 it had become apparent that the reestablished University of Costa Rica was not meeting the country's needs in higher education. A complete reorganization of the institution and a major revision of its curricula thus were undertaken in 1955. Carr was asked to serve as an outside consultant to advise on how to transform the Escuela de Biología into a strong scientific component of the university in education and research and to modernize its curriculum. Archie worked closely with the new director of the school, Rafael Lucas Rodríguez (1915–81), who had recently returned to his native land after receiving his Ph.D. (1954) in botany at the University of California, Berkeley. Their joint efforts produced an organization and curriculum along United States university lines that served as the foundation for all future developments in biology and conservation in Costa Rica.

Rodríguez and his faculty were especially open to the participation of foreign scientists at the university and in biological studies throughout the country. Their enthusiasm and their encouragement of these activities by their own students and visiting scientists led, as we will see below, to an explosion of studies on the biodiversity of Costa Rica. Because so many individuals are involved from the late 1950s on, I will chronicle the work of only a selection of herpetologists who made particularly significant contributions during the rest of the century. Those others not mentioned here are documented in the various species accounts and in the literature cited.

One other noteworthy event should be mentioned. In

1953, before the Escuela de Biología was created, the various professional schools at the university began publication of the journal *Revista de Biología Tropical.* Rodríguez was one of the founders and more or less continuously on the editorial committee until his death. Early issues emphasized parasitology and microbiology, but a broader array of topics and an increasing emphasis on systematics and ecology characterize the growth and continuance of the journal (now in its forty-eighth year). The first herpetological paper to appear within its covers was by Carr (1957) on sea turtles, and many others on amphibians and reptiles have been published through the years.

Institutional Bases and Founders of the Modern Era: The 1960s

This decade was one of great excitement in the advancement of knowledge of the herpetofauna of the republic. Earlier in this book I traced my own involvement in this process from 1959 on, so in this section and those following I will highlight the contributions of others. Archie Carr's sea turtle work, of course, continued throughout the next three decades, as described above.

William E. Duellman (1930–) was active in Costa Rica from February to June 1961, overlapping slightly with the summer course program of the University of Costa Rica and the University of Southern California, which I was codirecting and in which James L. Vial was a participant (see below).

Duellman, one of the most prolific modern herpetologists, had been appointed to the staff at the University of Kansas shortly after receiving his Ph.D. (1956) from the University of Michigan. His principal research at the time was the preparation of his classic monograph *The Hylid Frogs of Middle America* (1970). However, in addition to collecting hylids and their larvae and recording their calls, he made general collections. These materials and those from later fieldwork in Costa Rica in 1964 and 1966 for the hylid project with Linda Trueb (1942–; Ph.D., University of Kansas, 1968) added greatly to the holdings at Kansas originally obtained by Taylor, making it the largest collection of Costa Rican amphibians and reptiles.

Of great importance to this period, Rafael Rodríguez and the University of Costa Rica welcomed the University of Southern California for a joint program of education in tropical biology for United States university and college professors. The first such course was given in 1961. One of the participants, James L. Vial (1924–), returned to carry out research on the population biology of the high-altitude salamander *Bolitoglossa pesrubra* on the Cerro de la Muerte (1962–64). This classic study (Vial 1967), the first of its kind on a tropical salamander, served as Vial's 1965 Ph.D. dissertation at the University of Southern California.

Also active on the Costa Rican front was John M. Legler (1930–; Ph.D., University of Kansas, 1960) of the University of Utah, who collected significant series of freshwater turtles in Costa Rica, primarily in the summer of 1961 but also in July 1960 and 1962. Most of this material was never reported on but is the partial basis for a classic study of *Chrysemys ornata* (Moll and Legler 1971) and Legler's systematic review (1990). A mud turtle, the type of which had been collected by Taylor in 1952, was described as *Kinosternon angustipons* with the collection of additional material by Legler (1965).

Arden H. Brame Jr. II (1934–; M.S., University of Southern California, 1967) prepared a model study of the salamander genus *Oedipina* for his thesis (published in 1968) based principally on material that he collected in Central America on the first CRE (Costa Rica Expeditions) trip to Costa Rica (see preface for details).

In the years 1963 and 1964 Vial was a professor at the University of Costa Rica and began a small herpetological research collection there. During Vial's tenure a precocious young Nicaraguan named Jaime Villa (1944–; Ph.D., Cornell University, 1978), who had already written a book on venomous snakes (Villa 1962), entered the university as a first-year student. He came under Vial's influence, and during the rest of his years in Costa Rica he contributed to the growing herpetological collection.

A welcome visitor to Vial's laboratory during his Costa Rican stay was James Kezer (1908–; Ph.D., Cornell University, 1949) of the University of Oregon. Kezer is a cytogeneticist interested in salamander and genome evolution. It was during his two trips to Costa Rica that Kezer made one of the most important discoveries in salamander cytogenetics, the presence of XY (sex) chromosomes in *Oedipina* (Sessions and Kezer 1991). Unfortunately, though these results were included in Morescalchi (1975), Kezer did not publish them until much later (Kezer, Sessions, and Leon 1989).

W. Ronald Heyer (1941–; Ph.D., University of Southern California, 1969) and his wife Miriam carried out the first detailed ecogeographic study of herpetofaunal distribution for Costa Rica during the summer of 1964. Their material was used, together with previous collections, to establish patterns of distribution across the northern end of the Cordillera de Tilarán, from Pacific slope lowland dry forest to Atlantic lowland rainforest (Heyer 1967). Heyer (1968) also published an important paper reviewing the Costa Rican frogs of the genus *Leptodactylus.*

During this same period the Organization for Tropical Studies (OTS) was established in 1963, based in part on the success of the University of Costa Rica–University of Southern California course program. Since that time OTS has grown from a consortium of six United States universities and the University of Costa Rica to sixty-four members, including five Costa Rican ones. Since its inception OTS has provided educational and research opportunities to thousands of graduate students and faculty, many of them herpetologists. Stone (1988) provided a detailed history of OTS's first twenty-five years, and it need not be repeated here. OTS, however, did provide the opportunity and support for many of the herpetological studies reported in the body of this book.

Norman J. Scott Jr. (1934–), who carried out research on the distribution of Costa Rican snakes for his Ph.D. (University of Southern California, 1967), succeeded Vial as a professor at the University of Costa Rica during 1965 and 1966. He continued to encourage Villa in his herpetological studies, and together they added Costa Rican and Nicaraguan material to the Escuela de Biología's herpetological collections. Scott worked for OTS from 1966 to 1968, primarily as a course coordinator, but he retained an association with the University of Costa Rica collections. His unpublished dissertation (1969) provides much of the ecogeographic data on snakes incorporated in this book.

Scott was in turn succeeded as a professor at the University of Costa Rica by another *norteamericano*, Douglas C. Robinson (1936–91), who received his Ph.D. from Texas A&M University in 1968. Robinson had at one time taken courses from Dunn at Haverford College, and his own principal interest was salamanders.

Robinson was an ardent field biologist, and between teaching a wide variety of courses he began an extensive effort to add to the university's collections of amphibians and reptiles. During the next twenty-five years he recruited many Costa Rican students into these efforts, taking every opportunity to go into the field. Déjà vu the days of Frantzius: the somewhat chaotic putting together of field gear, loading it into ancient jeeps or their clones, and taking off for a weekend or a week to uncollected parts of the country.

Although Robinson published little, his contributions were enormous. Besides collecting extensive materials from areas never previously explored and inspiring a generation of Costa Rican biologists, he instituted long-term studies on the massive nesting migrations of the marine turtle *Lepidochelys olivacea*. Because the Museo Nacional zoological collections were moribund or in danger of becoming so, he worked to establish the Museo de Zoología at the university (1968) to house and administer the extensive, fish, herpetological, and bird collections developed there.

In 1974–75 the director of the Museo de Zoología, Myrna López, obtained funding to construct a modern two-story structure to house the museum's collections; it was completed in 1976. At Robinson's death the herpetological collections numbered over twelve thousand specimens, and they are unrivaled both for the number of species and for material from previously uncollected Costa Rican localities.

During this period Jaime Villa prepared his *Anfibios de Nicaragua* (1972a). This work is based in considerable part on the Museo collections, compilations from the literature, and various unpublished materials from course syllabi and laboratory manuals and manuscripts by Robinson or me.

Ed Taylor may have believed he had collected examples of all the Lord's herpetological creations in Costa Rica by the 1960s. If that were so the Lord was indeed busy, for a multitude of new salamanders and frogs (sixteen species), a liz-

ard, one turtle, and a snake were described during the decade. Several other forms first described in the 1960s from Panama or Colombia are now also known from Costa Rica.

Changing Themes in Herpetological Research: The 1970s

This period began with two propitious events. First, the Costa Rican government formally established a national park system in recognition of the immense value of natural environments to the country and their need for protection. The emerging environmental consciousness of Costa Ricans and the successive support of each government has led to development of over sixty national parks, biological and other reserves, and wildlife refuges, today comprising 15% of the country's land area. These provide protection for nearly 75% of the known herpetofauna.

The second event was the founding of the Instituto Clodomiro Picado in 1972. Since 1960 a group of Costa Rican scientists led by microbiologist Róger Bolaños (1931–; Ph.D., Louisiana State University, 1964) had been urging the development of a national program for the production of snake antivenins. Shortly after production of sera began in 1967, responding to the need for a freestanding laboratory and production center, the University of Costa Rica and the Ministry of Public Health set up the Instituto at Finca Dulce Nombre, near San Isidro de Coronado, San José Province. Bolaños became the first director (1970–80), followed by Luis Cerdas (1942–88; Lic., immunology, University of Costa Rica, 1979), and in 1988 by José Maria Gutiérrez (1954–; Ph.D., physiological sciences, Oklahoma State University, 1979). During the interim members of the Instituto staff have published over 180 papers on various aspects of the biology and distribution of venomous snakes, the karyology of Costa Rican species, and the biochemistry and effects of their venoms (Gutiérrez 1996).

The current director (1996–) is Gustavo Rojas (1960–; Lic., microbiology, University of Costa Rica, 1982). Gutiérrez and Bruno Lomonte (1958–; Ph.D., clinical immunology, University of Götenberg, 1994) continue on the Instituto's staff, carrying out research on biochemical and immunological aspects of snake venoms.

Between August 1969 and June 1970, Michael J. Corn (1944–; Ph.D., University of Florida, 1981) spent eight months collecting amphibians and reptiles on the Atlantic slope, where well-developed forest was being felled by the Standard Fruit Company for a banana plantation. His extensive samples were taken near the town of Río Frío (100 m), primarily between the Río Sucio and Río Chirripó, and include many rare forest canopy species represented poorly or not at all in earlier collections. This material is now at the University of Florida. Unfortunately, Corn's dissertation (1981) on the ecology of anole lizards in this rainforest was never published.

Probably the most important contributor to Costa Rican

herpetology during this decade was the ecologist Henry S. Fitch (1909–). Fitch received his Ph.D. from the University of California, Berkeley, in 1937. From 1948 on he has been at the University of Kansas Natural History Reservation, where his detailed long-term studies, primarily on reptiles, are among the most meticulous and data-rich autecologies ever published. Doubtless stimulated by the University of Kansas's active association with OTS and looking for new fields to conquer, Fitch accepted Norm Scott's invitation to teach in an OTS course. This was followed by stints of fieldwork from 1968 to 1970 and 1973 to 1974. The many publications based on these studies principally on lizard biology did not begin to appear until 1971 (see literature cited) and form the starting point for all subsequent research on lizard ecology in the republic.

As part of the development of its educational and research objectives, in 1968 OTS purchased the core property of what is now the world famous La Selva Biological Station from Leslie R. Holdridge. In 1972 Ian R. Straughan (1938–; Ph.D., University of Queensland, 1967) and I, stimulated by Scott's fieldwork in 1970 and 1971 that was the basis for his seminal paper (Scott 1976), initiated a comparative study of the leaf litter herpetofauna at La Selva and La Pacifica, a dry forest locality in Guanacaste Province. Quantitative samples were taken for thirteen months, from December 1972 to December 1973, by a field party composed of Charles F. Dock, Carl S. Lieb, James J. Talbot, and R. Wayne Van Devender. All subsequently attained Ph.D. degrees and published papers related to the project.

The doctoral dissertation of Susan S. Lieberman (1951–; Ph.D., University of Southern California, 1982) was based on the La Selva component of this material. She analyzed differences in species composition and abundance between primary forest and an abandoned cacao grove according to a set of environmental variables to determine how these variables interact to influence seasonality. Her dissertation and published paper (Lieberman 1986) form the basis for much of the ecological information contained in the species accounts in this book for taxa that occur at La Selva.

At the invitation of Doug Robinson a flood of herpetologists invaded Costa Rica for the fifty-third annual meeting of the American Society of Ichthyologists and Herpetologists (ASIH) in 1973, the first meeting of the society outside the United States or Canada. Numerous papers, later published, on Costa Rican amphibians and reptiles, many by Robinson's students and Róger Bolaños's associates, highlighted the program. For this event I prepared a handlist of the currently recognized amphibians and reptiles of the republic with an indication of general distributions (Savage 1973b). Postconference field trips also added considerable material to many a herpetological collection.

The meeting also provided an opportunity for a major effort to sample the herpetofauna of the relatively undisturbed and isolated rainforests of the Península de Osa. Roy W. McDiarmid (1940–; Ph.D., University of Southern Califor-

nia, 1968), who first worked in Costa Rica in 1964 with Scott and me, had been to the peninsula several times while teaching OTS courses (1966, 1967, 1969, 1971). His fascination with the biotic diversity in this exceptionally tall and mature forest led him to undertake extensive fieldwork on the Osa before and after the ASIH meetings.

McDiarmid had previously carried out a number of other studies at various sites both on the Península de Osa and elsewhere in conjunction with the soon to be well known ornithologist Mercedes S. Foster (1942–; Ph.D., University of South Florida, 1974). During the summers of 1967, 1969, and 1971, his fieldwork concentrated on the ecology and reproductive biology of glassfrogs at numerous sites throughout Costa Rica. Foster and McDiarmid also spent considerable time studying the ecology of dry forest amphibians, reptiles, and birds in 1971, 1972, 1974, and 1977 at Taboga (Finca Jiménez) in Guanacaste Province. Collections made during these trips are now housed at the Natural History Museum of Los Angeles County and the United States National Museum of Natural History.

In June and July 1973 I led a group of notable herpetologists on an intensive two-week inventory of the Osa herpetofauna. This expedition involved at various times J. D. DeWeese, W. Ron Heyer, Carl S. Lieb, Ian R. Straughan, Priscilla H. Starrett, James J. Talbot, R. Wayne Van Devender, David B. and Marvalee H. Wake, and George R. Zug. We overlapped with McDiarmid and his graduate students during part of this period. These several collections, mostly from near Rincón, Puntarenas Province, established basic knowledge of the Osa herpetofauna and are deposited in collections at the University of California, University of Costa Rica, Natural History Museum of Los Angeles County, and National Museum of Natural History.

In 1975, at the urging of many Costa Rican and international scientists, President Daniel Oduber created the Corcovado National Park to preserve intact 41,800 ha of the Península de Osa forest. The area around Rincón was not included in the park, and efforts to elucidate the herpetofauna of the peninsula shifted to within the park boundaries. Collections by Robinson and his "gang" and others, principally near Sirena and Llorona on the outer slope of the peninsula but also near Rincón, have fleshed out knowledge of the herpetofauna, now known to be composed of forty-nine species of amphibians and eighty-five species of nonmarine reptiles.

The study of Costa Rican salamanders also received new impetus during the 1970s, partially because of Robinson's own interest in them. David B. Wake (1936–; Ph.D., University of Southern California, 1964), the world's leading authority on Caudata and for twenty-seven years (1971–98) director of the Museum of Vertebrate Zoology at the University of California, Berkeley, first visited Costa Rica in 1961 (see preface). In 1971 he and his wife, Marvalee H. Wake (1939–; Ph.D., University of Southern California, 1967), equally well known as an authority on caecilians, joined Foster and McDiarmid at Taboga and later carried on research at Tapantí, Cartago Province, and other sites. They were also participants in the Península de Osa expedition of 1973 described above. Dave returned again in 1984 as an instructor in an OTS course and renewed his relationship with Doug Robinson and Pedro León (see below).

A series of short trips followed in each year from 1985 to 1990 on which Dave and his students and associates collected significant series of salamanders and their tissues for both morphological and molecular analyses. Marvalee accompanied him on several of these trips, most notably in 1986 and 1988, to collect caecilian material. Her M.S. thesis (1964) was on the ecogeographic distribution of Costa Rican lizards; that never-published opus is the basis for much of the data on lizard distributions incorporated in this book.

At first active in Robinson's laboratory was Pedro León (1944–; Ph.D., University of Oregon, 1974), who carried out his doctoral work with Jim Kezer. Many of León's papers are on salamander karyology and genetics, although his research currently encompasses a wide range of studies in cytogenetics and molecular biology. He and the Wakes have a longtime working relationship, and he spent his 1988 John Simon Guggenheim Fellowship at the Berkeley campus of the University of California. León is among the most distinguished of Costa Rican scientists and, besides contributing to understanding salamander evolution, has been a leader in the conservation movement. He was a founder and served as president of the Fundación de Parques Nacionales (1979–91) and is currently chairman of the Board of Directors of the Organization for Tropical Studies.

In 1976 Róger Bolaños hosted the fifth International Symposium on Animal, Plant, and Microbial Toxins, which particularly emphasized snake venoms and antivenin production. Again Costa Rican contributions and herpetologists were given visibility, as at the ASIH meetings in 1973. A revised version of the earlier handlist of the herpetofauna was published for participants at this meeting (Savage 1976).

By the end of the decade opportunities for herpetological field research had increased dramatically. The national park and reserve system, consisting at that time of twenty protected natural areas, offered unique opportunities for study. OTS now owned or managed the La Selva and Las Cruces Biological Stations and had a field station in the Palo Verde National Park. The study of Atlantic sea turtles continued in the Tortuguero Park, and Pacific sea turtles were receiving increased attention in the Santa Rosa Park and at Playa Ostional. The famous Monteverde Forest Reserve in the Cordillera de Tilarán also was becoming a favored research site.

The trend in herpetology during the 1970s was away from systematically oriented studies and toward ecological research. This trend would continue through the next two decades. As evidence of this shift, only seven new species were described from Costa Rica during the 1970s, and five in the 1980s.

Expanding Vistas: The 1980s

In 1980 I published a third edition of the handlist of Costa Rican amphibians and reptiles (Savage 1980b). Simultaneously, at the insistence of A. Stanley Rand (1932–), longtime researcher at the Smithsonian Tropical Research Institute in Panama, who had used a set of my field keys while teaching in an OTS course, I published a series of keys to the then known Cost Rican species (1980b). This seems to have marked a close to the 1960s and 1970s period of exploration and the search for herpetological novelty.

By this time Robinson's effort to train Costa Rican herpetologists was paying off. Among his more notable and still active protégés were Pedro León, Julián Monge (1960–; M.Sc., University of Costa Rica, 1987), José María Mora, Rafael Acuña, Anny Chaves, Federico Bolaños, and Alejandro Solórzano, who was also involved at the Instituto Picado. During the mid-1970s through the 1980s Robinson's principal interest was the immense nesting migrations (arribadas) of the olive ridley sea turtle, Lepidochelys olivacea, at Playa Ostional, on the Pacific side of the Península de Nicoya. All his students cut their herpetological teeth by working on this project in one form or another. Both Acuña and Chaves continue to study turtles, though Acuña now focuses on nonmarine forms.

An indication of the level of knowledge of Costa Rica's biodiversity that had accumulated since the founding of OTS in 1963 was the book edited by Daniel H. Janzen (1983), Costa Rican Natural History. The herpetological section was divided into three parts. The Introduction by Scott and Limerick (1983) provides an overview and synthesis of what was known about many aspects of the biology of Costa Rican species. Scott, Savage, and Robinson (1983) presented a checklist of the amphibians and reptiles for the principal sites used in field studies by OTS courses. Finally, the treatment included descriptions by twenty-two authors of basic natural history for forty-two commonly encountered species. In total these sections, together with Savage (1980b), established the state of our knowledge of the systematics and ecology of the herpetofauna at that time.

Significant advances in that knowledge were being made even as the Janzen volume was in press. In addition to the ongoing sea turtle studies on the Atlantic and Pacific coasts, beginning in the early 1980s herpetological studies involving long-term research of population and reproductive biology and behavioral ecology were carried out primarily at the La Selva Biological Station and the Monteverde Forest Reserve. At La Selva Maureen A. Donnelly (1954–), Harry W. Greene (1945–), and Craig Guyer (1952–) initiated research on Dendrobates pumilio, snake predation, and anole ecology, respectively. In addition, Donnelly and Guyer collaborated on a thirteen-month study of a pond frog assemblage and the ecological correlates of length-mass relationships in the La Selva snake fauna (Guyer and Donnelly 1990).

Greene (Ph.D., University of Tennessee, 1977), in the 1980s a faculty member at the University of California, Berkeley, and now at Cornell University, used radiotelemetry to track and observe the feeding behaviors of the larger snake species. Many of his observations are contained in his wonderful semipopular book *Snakes* (1997). Harry was joined in his venomous snake research in 1984 by David L. Hardy Sr. (1933–; M.D., University of Kansas, 1959) of the University of Arizona. Wendy E. Roberts (1956–; Ph.D., University of California, Berkeley, 1994) carried out her doctoral dissertation under Greene's supervision on arboreal egg-laying frogs at La Selva (1989–92).

Both Donnelly and Guyer received their Ph.D. degrees from the University of Miami (1987 and 1986, respectively) based on work at La Selva. They are now faculty members at Florida International University and Auburn University, respectively, and continued to carry out additional studies at La Selva into the 1990s (Donnelly 1994a,b; Guyer 1990, 1994a,b).

Martha "Marty" Crump (1946–; Ph.D., University of Kansas, 1974), in the 1980s a faculty member of the University of Florida, began long-term study of the behavioral ecology of frogs at the Monteverde Forest Preserve in 1980. She had previously carried out very original studies on tropical amphibian communities in Amazonian Ecuador (Crump 1974). Marty spent considerable time at Monteverde from 1980 to 1989, including a full year from August 1982 to August 1983. Her studies focused on the population biology and behavioral ecology of the frogs *Atelopus varius* and *Hyla pseudopuma*. Her then graduate student, J. Alan Pounds (1953–; Ph.D., University of Florida, 1987), worked with her while carrying out his doctoral dissertation study on the ecology of the anole lizard assemblage. Also resident in Monteverde at that time (1980, 1982) was Marc P. Hayes (1950–; Ph.D., University of Miami, 1991), who studied the biology of glassfrogs (Centrolenidae) and especially the role of male brooding behavior in *Hyalinobatrachium fleischmanni*. Unfortunately, his dissertation research has never been published.

In 1984 Róger Bolaños, the founder of the Instituto Clodomiro Picado, published *Serpientes venenosos y ofidismo en Centromérica*. This book provides an overview of distribution, characteristics, karyotypes, venoms, and bites of Costa Rican venomous snakes, and treatment of their bites, with emphasis on epidemiological data. This account is invaluable to any herpetologist working in the region and nicely supplements the broad coverage in *Snake Venom Poisoning* (Russell 1983).

The year 1986 was an especially significant one for Costa Rican herpetology. Early in the year Stephen C. Cornelius published a noteworthy review of the biology of sea turtles on the Pacific coast of the republic, especially at Santa Rosa National Park. Anny Chaves (1955–; Lic., biology, University of Costa Rica, 1986) won the Florida Audubon Society's prize for conservation in 1985 for her contributions to the preservation of the *Lepidochelys* nesting beaches. The year 1986 also marked the appearance of the first in a continuing series of papers on the life histories of Costa Rican snakes by Alejandro Solórzano, often in collaboration with colleagues at the Instituto Picado.

In this same year a major biological survey was undertaken of the extension of the Braulio Carrillo National Park running down the north side of Volcán Barva to the La Selva Biological Station. This tract had been purchased through the efforts of OTS and given to the government to annex to the park. A preliminary exploration of this corridor, originally the Zona Protectora de la Selva, was carried out in 1983 (Almeda and Pringle 1988) with Harry Greene surveying the herpetofauna.

The 1986 effort involved sampling by teams from a variety of biological specialties at six sites between 300 and 2,600 m over a two-month period (Timm et al. 1990). The herpetological unit represented a cooperative effort between University of Costa Rica and United States scientists and included Doug Robinson, Federico Bolaños, Federico Muñoz, Manuel Santana, David A. Good, Craig Guyer, Christopher d'Orgeix, Jay Savage, and David Wake. Material collected is now at the Museo de Zoología, the University of California, Berkeley, and the Natural History Museum of Los Angeles County. This was the first substantial collection of amphibians and reptiles from the windward slopes of the volcanoes, and it greatly increased knowledge of their zonal distribution.

At the end of the year the *Introduction to the Herpetofauna of Costa Rica/Introducción a la Herpetofauna de Costa Rica* appeared (Savage and Villa 1986). This book was based on Savage (1980b) and contained a table of distribution, revised keys, illustrations of characteristics used in the keys, and an annotated bibliography. The entire work is in both English and Spanish and stands as the precursor to the present book.

If 1986 was a banner year in Costa Rican herpetology, 1988 was the reverse, for that was when the marked declines in amphibian populations became apparent. The most striking of these cases involved the Monteverde golden toad *(Bufo periglenes)*, originally discovered in the 1960s. This gaudy-colored species had a very restricted range along the continental divide near Monteverde, and the Monteverde Reserve had been created to protect it. Suddenly these toads failed to breed for the first time since their discovery twenty-five years before. In 1988 only ten adult toads were found during the normal reproductive period, with no sign of eggs or larvae; in 1989 one adult was found, and none have been seen since that date.

Nevertheless, the decade finished on an upbeat note with the appearance of *An Annotated List and Guide to the Amphibians and Reptiles of Monteverde, Costa Rica* (Hayes, Pounds, and Timmerman 1989), which replaced an earlier, less complete edition by Timmerman and Hayes (1981); the review of all Costa Rican venomous snakes in Campbell and Lamar's (1989) monumental *The Venomous Reptiles of Latin America;* and the opening of the Serpentario Nacional in the same year. The Serpentario exhibits live snakes from Costa Rica and was established and is operated by Alejandro Solórzano (1958–; Bach., biology, University of Costa Rica, 1982). Also appearing near the end of this period was a useful but error-plagued checklist and bibliography of the Middle American herpetofauna compiled from the literature (Villa, Wilson, and Johnson 1988).

Toward a new Millennium: The 1990s

By the end of the 1980s the knowledge of the composition and distribution of the herpetofauna seemed to be on firm ground. Distributional gaps had been substantially filled in by the work of Robinson and his associates. Robinson and especially his student Federico Bolaños (1961–; M.Sc., University of Costa Rica, 1991) had carried out inventories of almost all the natural preserves within the country. Bolaños, now a professor at the university, is the son of Róger Bolaños, noted previously for his work in snake venoms. The son, however, is primarily interested in amphibian biology and continued into the 1990s the ongoing inventory of new areas being opened up as reserves or newly established field stations at remote sites.

The principal geographic areas needing study at the start of the decade remained the mostly inaccessible southern portion of the Cordillera de Talamanca, especially the upland areas near the Panama frontier and the isolated peaks in the Cordillera de Guanacaste. Preliminary fieldwork in the uplands in the vicinity of Las Alturas and Las Tablas and on the slopes of Cerro Pando on the Panama border was carried out in the late 1980s by Bolaños and Robinson. A detailed inventory of the region's herpetofauna was initiated by Karen R. Lips (1966–; Ph.D., University of Miami, 1995) during her research in the field for several periods from 1990 to 1994. Bolaños and a field party from the University of California, Berkeley, led by David Wake also added to the herpetofaunal list during this time.

The construction of a field station on the slopes of Cerro Cacao led to preliminary collecting there by several groups beginning in 1987, among them David A. Good (1956–; Ph.D., University of California, Berkeley, 1985). Michael Franzen, who had earlier spent time at Santa Rosa National Park (1986), undertook an inventory of the west slopes of Volcán Orosi and Cerro Cacao in 1993 as a basis for his diploma in biology, Universität Bonn (1988, 1994).

In 1991 and again in 1997, Michael Schmid of the Universität Würzburg, now the leading student of anuran karyology, collected extensive material, particularly of *Eleutherodactylus,* for cytogenetic study in the republic. To date only two short papers (Schmid et al. 1995; Kaiser et al. 1996) have appeared on this ongoing research, which prom-

ises to clarify relationships among species of this notoriously difficult genus.

Throughout the 1980s and 1990s a number of German and Dutch scientists and enthusiasts published a series of semipopular articles on the amphibians and reptiles of the republic. Many of these contain important observations on the biology of Costa Rican species. The most informative of these papers are Köhler and Ziegler (1997), Koller (1996a,b), Pröhl (1997a), and Weimer et al. (1993a,b,c, 1994).

A number of books containing detailed accounts of the herpetofauna appeared in the 1990s. Rafael Acuña (1953–; M.Sc., biology, University of Costa Rica) published *Las tortugas continentales de Costa Rica* (1993, 1998), the first account of a major component of the herpetofauna by a Costa Rican. A comprehensive volume on the biota of the La Selva Biological Station (McDade et al. 1994) contains reviews of amphibians' diversity and ecology by Donnelly and those of reptiles by Guyer. Similarly, Pounds (2000) treats the herpetofauna of the Monteverde region as part of an in-depth treatment on the biodiversity of that area (Nadkarni and Wheelwright 2000). In addition, a Spanish edition of Janzen's (1983) *Costa Rican Natural History* was issued in 1991, making the sections on amphibians and reptiles accessible to a wider audience. Also of considerable significance was Duellman's (1990) comparison of the composition of the herpetofauna of the La Selva Biological Station with that at other major field sites in Panama, Peru, and Brazil as part of a large volume on Neotropical rainforests edited by Gentry (1990).

Other notable contributions during this period include a summary of the herpetofauna of Santa Rosa National Park by Mahmood Sasa (1969–; Ph.D., University of Texas, Arlington, 2000) and Alejandro Solórzano (Sasa and Solórzano 1995), and the several papers by Edmund D. Brodie III (1963–; Ph.D., University of Chicago, 1991) on the experimental field study of coral snake mimicry. Of substantial significance were the papers by Good and Wake (1993, 1997) revising the systematics of the Costa Rican members of the salamander genera *Nototriton* and *Oedipina,* respectively, based on morphological and biochemical features.

Another significant herpetological event was the decision by the organizers of the twenty-first International Herpetological Symposium to hold their meeting in Costa Rica in 1997. The unique venue for the meeting, the first of these symposia held outside the United States, was the Hotel Boyeros in Liberia, Guanacaste Province, far from the madding crowd on the Meseta Central. A broad spectrum of papers on Neotropical herpetology formed the program, with emphasis on ongoing studies in Costa Rica in presentations by Daniel H. Janzen, David L. Hardy Sr., Alejandro Solórzano, J. Alan Pounds, José Maria Gutiérrez, Gustavo Rojas, and Mahmood Sasa.

During the 1990s new taxa continued to be found or named, primarily through fieldwork in previously un-

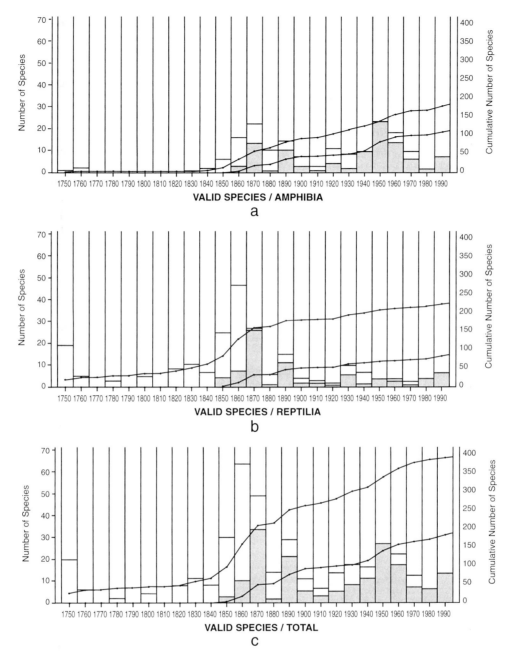

Figure 1.15. Rates of discovery of the Costa Rican herpetofauna by decade (1758–1999): (a) discovery of valid species of amphibians; (b) discovery of valid species of reptiles; (c) discovery of valid species of total herpetofauna. Open bars indicate the number of valid species; shaded bars represent the number of valid species based on type material from Costa Rica; the upper curve is the cumulative number of valid species; the lower curve is the cumulative number based on type material from the republic.

sampled localities. A new caecilian, three new salamanders, three new frogs, three new lizards, and three new snakes, all based on Costa Rican material, were described during the decade. Figure 1.15 summarizes the history of the discovery of the Costa Rican herpetofauna through 1999.

Through much of the 1990s major attention was focused on declining amphibian populations. Catastrophic disappearances of over 50% of the amphibian species at essentially pristine or protected sites began in 1988 at the upland

sites of Monteverde, Cerro Chompipe, and Tapantí. Subsequent investigation indicated similar declines at a somewhat later date for the northern Talamancas and between 1992 and 1993 for the upland area near the Panama boundary. Various possible explanations have been put forward, especially global climate change (Pounds and Crump 1994) or climate change leading to the spread of a pathogen (Lips 1998; Berger et al. 1998). That the declines are real has been demonstrated by Crump, Hensley, and Clark (1992), Pounds et al. (1997), Pounds, Fogden, and Campbell (1999), and Lips (1998, 1999). Study of this phenomenon is ongoing at this time.

The declines and extinctions in lower Central America are one of the better-known parts of a worldwide decline in amphibians. The significance of declines merited a full chapter in the Worldwatch Institute's *State of the World 2001* (Mattoon 2001). That report emphasizes the role of the Declining Amphibian Populations Task Force (DAPTF) in catalyzing monitoring programs and documenting population changes throughout the world. Monitoring of crash sites in Costa Rica continues, with no sign of population rebounds. Although the disease model has received wide acceptance as a proximate cause of declines in Australia, Costa Rica, and western North America, most biologists suspect that a complex of interacting factors is responsible. A recent study (Kiesecker, Blaustein, and Belden 2001) implicates global warming as a precursor of pathogen-mediated declines. The resulting increase in exposure of developmental stages to UV-B radiation apparently sets up conditions for a fungal pathogen outbreak. However, this may be only one of several chains of events that produce a decline or extinction at a particular site (Pounds 2001). For ongoing updates on the occurrence and causes of amphibian declines connect to http://elib.cs.berkeley.edu/aw/.

CONSERVATION AND PROTECTION

Before the 1970s Costa Ricans gave little thought to protecting, let alone conserving, their seemingly endless supply of natural resources. The population explosion, immigration into new lands, and demands for a higher quality of life all contributed to an ever increasing rate of deforestation. Growing concern with this issue that clearly threatened the national patrimony was articulated by leaders in the biological community who had organized the Colegio de Biólogos in 1968.

The developing awareness of environmental problems and the general demand by Costarricense for a park system dedicated to preserving natural areas and their biodiversity led President José Figueres to form the Servicio de Parques Nacionales (now Dirección de Parques Nacionales) in 1970. Under the leadership of Mario A. Boza (1942–, M.S., Centro Agronómico Tropical de Investigación y Enseñanza [CATIE], 1969) and Alvaro F. Ugalde (1949–; B.S., University of Costa Rica, 1971), with the support of Presidents Daniel Oduber and Rodrigo Carazo, by 1980 twenty major parks, reserves, and biological and wildlife refuges had been established. Additional natural areas (plate 1) have been added to this system in the interim so that today there are twenty-six national parks, four other biological reserves, and twenty-three refuges, and 15% of Costa Rica's land surface is devoted to protected natural areas. About 75% of the known herpetofauna of the republic occurs within the parks and reserves. The wildlife refuges have been administrated by a separate directorate (Dirección de Fauna Silvestre) and contain a few additional species not protected by the parks and reserves. Currently all parks, reserves, and refuges are being administratively grouped into eleven regional conservation units (table 1.1; plate 1). In addition there are twenty-one reserves for indigenous peoples where some features of their original cultures are preserved and these offer some protection for the environment (Bozzolli de Wille 1975).

Other important preserves under nongovernment control are the Monteverde Forest Reserve, the adjacent Children's Rainforest, and the La Selva and Las Cruces Biological Stations of the Organization for Tropical Studies. The expanding ecotourism produced by the availability of parks and reserves also has stimulated many private landowners to establish small field stations near or in remaining forest tracts, especially for bird-watchers. All of these contribute to conserving herpetofaunal diversity against the continuing onslaught of habitat destruction and pollution. Nevertheless, marked declines in populations of amphibians within essentially pristine preserves occurred during the 1980s and 1990s.

Simply establishing a series of natural protected areas is clearly not enough to conserve and maintain the health of ecosystems. With this in mind the Instituto Nacional de Biodiversidad (INBio) was established in 1989 based on the idea that the country's biodiversity could be protected in perpetuity only through its responsible and nondestructive use for the benefit of the citizens of Costa Rica. Under the leadership of Rodrigo Gámez (1936–; Ph.D., virology, University of Illinois, 1967), a distinguished virologist and conservationist, INBio is carrying out an ambitions program of conservation (saving biodiversity), biological inventory (knowing biodiversity), and biodiversity prospecting and environmental education (using biodiversity).

INBio now has major responsibility for providing scientific advice on the management of Costa Rica's system of conservation areas and has pushed forward comprehensive inventories of higher plants, fungi, arthropods, and mollusks. Particularly important have been studies on the Guanacaste, La Amistad, and Osa Conservation Areas, which have substantially increased knowledge of the *biota costarricense.*

Costa Rica has major legislation that protects all plants and animals against unauthorized killing or capture. In addition, the Convention on International Trade in Endangered Species of Wild Flora and Fauna (CITES) lists the following Costa Rican taxa as endangered, and these may not be traded among party nations for primarily commercial purposes (Appendix I): all sea turtles, members of the genera *Caretta, Chelonia, Dermochelys, Eretmochelys,* and *Lepidochelys; Crocodylus acutus;* and the toad *Bufo periglenes.* Costa Rican taxa that may become endangered if trade is not regulated (Appendix II) are *Caiman crocodilus, Iguana iguana,* all boas, *Clelia clelia,* and all species of *Dendrobates* and *Phyllobates.* The following Costa Rican

Table 1.1. Protected Areas of Costa Rica

Área de Conservación Guanacaste
1 Parque Nacional Santa Rosa
2 Parque Nacional Guanacaste
3 Parque Nacional Rincón de la Vieja
4 Refugio de Vida Silvestre Junquillal
5 Estación Experimental Forestal Horizontes
6 Refugio de Vida Silvestre Iguanita
7 Refugio de Vida Silvestre Corredor Fronterizo Costa Rica–Nicaragua (Guanacaste section)

Área de Conservación Tempisque
1 Reserva Biologíca Lomas Barbudal
2 Parque Nacional Palo Verde
3 Parque Nacional Marino Las Baulas
4 Parque Nacional Barra Honda
5 Refugio de Vida Silvestre Ostional
6 Reserva Biologícas Islas Negritos, Guayabo y de los Pájaros
7 Refugio de Vida Silvestre Curú
8 Reserva Natural Absoluta Cabo Blanco
9 Humedal Riberino Zapandí
10 Reserva Forestal Taboga
11 Humedal Laguna Madrigal
12 Humedal Palustrino Corral de Piedra
13 Refugio de Vida Silvestre Laguna Mata Redonda
14 Refugio de Vida Silvestre Bosque Nacional Diriá
15 Humedal Río Cañas
16 Refugio de Vida Silvestre Camaronal
17 Zona Protectora Cerro La Cruz
18 Zona Protectora Nosara
19 Zona Protectora Península de Nicoya
20 Zona Protectora Abangares
21 Reserva Natural Absoluta Nicolás Wessberg

Área de Conservación Cordillera Volcánica Central
1 Parque Nacional Braulio Carrillo
2 Monumento Nacional Guayabo
3 Parque Nacional Volcá Irazú
4 Zona Protectora La Selva
5 Parque Nacional Volcán Poás
6 Refugio de Vida Silvestre Bosque Alegre
7 Zona Protectora Río Toro
8 Zona Protectora El Chayote
9 Reserva Forestal Grecia
10 Reserva Forestal Cordillera Volcánica Central
11 Zona Protectora Río Grande
12 Zona Protectora Cerros de La Carpintera
13 Zona Protectora Tiribí
14 Parque Nacional Volcán Turrialba
15 Zona Protectora Cerro Atenas
16 Refugio de Vida Silvestre Corredor Fronterizo Costa Rica–Nicaragua (Cordillera Central section)
17 Zona Protectora Cuenca del Río Tuis

Área de Conservación Tortuguero
1 Parque Nacional Tortuguero
2 Refugio de Vida Silvestre Barra del Colorado
3 Zona Protectora Tortuguero
4 Zona Protectora Acuíferos de Guácimo y Pococí
5 Refugio de Vida Silvestre Archie Carr
6 Reserva Forestal Matina
7 Humedal Nacional Cariari
8 Refugio de Vida Silvestre Corredor Fronterizo Costa Rica–Nicaragua (Tortuguero section)

Área de Conservación la Amistad Caribe
1 Parque Internacional La Amistad (Caribe section)
2 Parque Nacional Cahuita
3 Refugio de Vida Silvestre Gandoca-Manzanillo
4 Zona Protectora Cuenca del Río Banano
5 Reserva Forestal Río Pacuare
6 Parque Nacional Barbilla
7 Refugio de Vida Silvestre Limoncito
8 Humedal Lacustrino Bonilla-Bonilllita
9 Reserva Biológíca Hitoy Cerere

Área de Conservación La Amistad Pacífico
1 Zona Protectora Río Navarro y Río Sombrero
2 Reserva Forestal Río Macho
3 Zona Protectora Las Tablas
4 Humedal de San Vito
5 Humedal Palustrino Laguna del Paraguas
6 Parque Nacional Chirripó
7 Parque Nacional Tapantí-Macizo de La Muerte
8 Parque Internacional La Amistad (Pacifico section)

Área de Conservación Osa
1 Parque Nacional Marino Ballena
2 Reserva Biológica Isla del Caño
3 Parque Nacional Corcovado
4 Refugio de Vida Silvestre Golfito
5 Humedal Nacional Térraba-Sierpe
6 Reserva Forestal Golfo Dulce
7 Parque Nacional Piedras Blancas
8 Humedal Lacustrino Pejeperro-Pejeperrito

Área de Conservación Pacífico Central
1 Parque Nacional Carara
2 Parque Nacional Manuel Antonio
3 Zona Protectora Tivives
4 Zona Protectora El Rodeo
5 Zona Protectora Cerros de Escazú
6 Zona Protectora Caraigres
7 Zona Protectora Cerros de Turrubares
8 Refugio de Vida Silvestre Fernando Castro Cervantes
9 Zona Protectora Cerros de La Cangreja
10 Zona Protectora Cerro Nara
11 Reserva Forestal Los Santos
12 Zona Protectora Montes de Oro
13 Refugio de Vida Silvestre Finca Barú del Pacífico
14 Refugio de Vida Silvestre Portalón
15 Reserva Biológica Cerro Vueltas
16 Refugio de Vida Silvestre Peñas Blancas

Área de Conservación Arenal
1 Parque Nacional Volcán Arenal
2 Parque Nacional Volcán Tenorio
3 Zona Protectora Arenal-Monteverde
4 Reserva Biológica Alberto Manuel Brenes
5 Zona Protectora Miravalles
6 Zona Protectora Tenorio

Área de Conservación Huetar Norte
1 Refugio de Vida Silvestre Caño Negro
2 Parque Nacional Juan Castro Blanco
3 Refugio de Vida Silvestre Corredor Fronterizo Costa Rica-Nicaragua (Huetar section)
4 Refugio de Vida Silvestre Laguna Las Camelias
5 Reserva Forestal Cerro El Jardín
6 Reserva Forestal Cureña-Cureñita
7 Humedal Palustrino Laguna Maquenque
8 Humedal Lacustrino de Tamborcito

Área de Conservación Marina Isla del Coco
1 Parque Nacional Isla del Coco

species are protected under the United States Endangered Species Act as well: *Chelonia mydas, Dermochelys coriacea, Eretmochelys imbricata,* and *Crocodylus acutus.* Collecting these animals, their eggs, or any of their parts, even if related to scientific study, requires special permits from Costa Rican agencies, as does importing them into the United States.

Unfortunately, even with legal safeguards and expanded educational programs by INBio and others, prejudice against amphibians and reptiles, especially snakes, remains high among Costa Ricans and visitors to the country, so that random and senseless slaughter of harmless and, indeed, beneficial animals occurs all too frequently. I hope this book will serve as a vehicle to support conservation efforts by calling attention to the beauty and essential ecological value of these often maligned and underappreciated creatures.

SOURCES OF INFORMATION

In the course of preparing this book I have examined most of a total of approximately eighty thousand specimens of Costa Rican origin in major scientific collections. The courtesies extended to me by these institutions and their curators are gratefully acknowledged earlier in the book as well as here. In this process I was able to examine the type specimens of all but a few species names, including synonyms. In addition I have seen most of the species in the field or in life, except for a few rare forms known from single specimens. Reference to particular examples in the text is by collection number, usually using the institutional abbreviations listed in Leviton et al. (1985); CRE refers to the Costa Rica Expeditions collection housed since 1998 at the Natural History Museum of Los Angeles County.

The scientific and popular literature on Costa Rican amphibians and reptiles is immense. Although I have reviewed most of it, the literature cited at the end of the book includes only works cited in the text and is not exhaustive. All references in the text are cited by the date appearing on the publication. When the actual date differs from the published one, the former is indicated in the literature cited within quotation marks followed by the actual publication date: "1872," 1873. Citations following taxon names in the various accounts include both dates in the same style as above when they differ. While writing, I continued to update the manuscript through literature published in 2000. A few especially significant subsequent publications are cited and were added during editing. I have not attempted to develop partial synonymies or complete ones (i.e., chresonymies). Interested readers should see the works of Taylor (1951b, 1952a,b, 1954b, 1956, 1958b) for synonymies of names used up to those dates. Changes in names or synonyms discovered after these reviews appeared are discussed in the remarks sections in the species accounts.

The Spanish and Latin American custom of using a double surname, the first element of which is the father's surname, has been discussed earlier in this chapter. In the text and literature cited I have referenced publications by Spanish authors only by the father's surname, with two exceptions. The first is the case of a married woman who consistently uses her father's surname coupled with that of her husband (e.g., María E. Bozzolli de Wille). She would be cited as Bozzolli de Wille. The second is where an author is well known and himself consistently uses both *apellidos* (e.g., Mario García-París). He would be cited as García-París.

Four key works essential to a student of Central American herpetology cause some confusion as to dates of publication and authorship. The first of these is the *Erpétologie générale, ou Histoire complète des reptiles.* The following outline provides the essential information:

Authors	Date	Volume
CD + B	1834	1
CD + B	1835	2
CD + B	1836	3
CD + B	1837	4
CD + B	1839	5
CD + B	1844	6
CD + B + AD	1854	7 (1)
CD + B + AD	1854	7 (2)
CD + B	1841	8
CD + B + AD	1854	9
CD + B + AD	1854	Atlas

AD = Auguste H. A. Duméril; B = Gabriel Bibron; CD = A. M. C. D. Constant Duméril.

Similar data are provided for the *Études sur les reptiles: Mission scientifique au Mexique et dans l'Amérique Centrale. Recherches zoologiques,* pt. 3, sec. 1, by M. F. Boucourt.

Livraison	Date	Author	Pages	Plates
1	1870	D + B	1–32	1–7, 9, 11–12
2	1873	B	33–120	8, 10, 13–15
3	1874	B	121–216	16–18B
4	1874	B	217–80	19–20C, 23
5	1878	B	281–360	20, 20D–G, 21A–C
6	1879	B	361–440	21–22D
7	1881	B	441–88	22E–J
8	1882	B	489–528	27–30
9	1883	B	529–92	31–35
10	1886	B	593–656	36–41
11	1888	B	657–96	42–47
12	1890	B	697–732	25–26,48–51
13	1893	B	733–80	52–57
14	1895	B	781–830	24, 58–62
15	1897	B	831–60	63–68
16	1908	B + M	861–932	69–74
17	1909	B + M	933–89	75–77

D = Auguste H. A. Duméril; B = Marie-Firmin Bocourt; M = François Mocquard.

Similar data are provided for the *Études des batraciens de l'Amérique Centrale: Mission scientifique au Mexique et dans l'Amérique Centrale. Recherches zoologiques*, pt. 3, sec. 2, by Paul Brocchi.

Livraison	Date	Pages	Plates
1	1881	1–56	1–5, 9–10

Contents: *Rana, Leptodactylus, Leiuperus, Scaphiopus, Pyxicephalus, Hyla, Opisthodelphis, Nototrema, Exerodonta, Hylodes* (pt.)

2	1882	57–96	6, 8, 11, 13–15

Contents: *Hylodes* (pt.), *Ixalus, Cauphias, Hylarana, Phyllobates, Phyllomedusa, Bufo, Hylaplesia, Hypopachus, Engystoma, Crepidius, Atelopus*

3	1883	97–122	7, 12, 16–17, 17B, 18–18B, 21

Contents: *Phryniscus, Rhinophrynus, Urodela*

Data for *Biologia-Centrali Americana: Reptilia and Batrachia*, by Albert C. L. G. Günther, include:

Signature	Date	Pages
1	April 1885	1–8
2	April 1885	9–16
3	May 1885	17–24
4	August 1885	25–32
5	October 1885	33–40
6	October 1885	41–48
7	October 1885	49–56
8	February 1890	57–64
9	February 1890	65–72
10	February 1890	73–80
11	April 1893	81–88
12	May 1893	89–96
13	August 1893	97–104
14	October 1893	105–12
15	February 1894	113–20
16	February 1894	121–28
17	July 1894	129–36
18	October 1894	137–44
19	January 1895	145–52
20	January 1895	153–60
21	March 1895	161–68
22	May 1895	169–76
23	July 1895	177–84
24	August 1895	185–92
25	October 1895	193–96
26	February 1900	197–204
27	February 1900	205–12
28	April 1900	213–20
29	June 1900	221–28
30	August 1900	229–36
31	February 1901	237–44
32	February 1901	245–52
33	May 1901	253–60
34	June 1901	261–68
35	September 1901	269–76
36	September 1901	277–84
37	September 1901	285–92
38	December 1901	293–300
39	January 1902	301–8
40	May 1902	309–16
41	May 1902	317–24
42	May 1902	325–26

Date	Plates
April 1885	1–15
May 1885	16–18
August 1885	19–21
October 1885	22–25
January 1890	26–28
February 1890	29–30
April 1893	31–32
May 1893	33
August 1893	34–35
October 1893	36–37
November 1893	38–40
February 1894	41
April 1894	42–43
July 1894	44–45
October 1894	46–48
January 1895	49
March 1895	50
May 1895	51–52
July 1895	53–55
August 1895	56–57
October 1895	58–59
February 1900	60–61
March 1900	62–63
April 1900	64–66
August 1900	67–68
April 1901	69
May 1901	70
June 1901	71–73
September 1901	74
January 1902	75–76

For those interested in more information on amphibians and reptiles in general, I recommend the following works.

General

Cogger, H. G., and R. G. Zwiefel, eds. 1998. *Amphibians and reptiles*. 2d ed. San Diego: Academic Press.

Halliday, T. R., and K. Adler, eds. 1986. *The encyclopedia of amphibians and reptiles*. New York: Facts on File.

Pough, F. H., R. M. Andrews, J. E. Cadle, M. L. Crump, A.

H. Savitzky, and K. D. Wells. 1997. *Herpetology.* Upper Saddle River, N.J.: Prentice Hall.

Amphibians

Duellman, W. E., and L. Trueb. 1994. *Biology of amphibians.* Baltimore: Johns Hopkins University Press.

Heatwole, H., senior ed. 1994 et seq. *Amphibian biology.* Chipping Norton, NSW, Australia: Surrey Beatty. Vol. 1 (1994); vol. 2 (1995); vol. 3 (1998); vol. 4 (2000) (continuing series).

Hofrichter, R., ed. 2000. *Amphibians: The world of frogs, toads, salamanders and newts.* Buffalo, N.Y.: Firefly Books.

Stebbins, R. C., and N. W. Cohen. 1995. *A natural history of amphibians.* Princeton: Princeton University Press.

For an up-to-date listing of the amphibians of the world, check Amphibian Species of the World: http://research .amnh.org/herpetology/amphibia/.

Frogs and Toads

Mattison, C. 1987. *Frogs and toads of the world.* New York: Facts on File.

Vial, J. L., ed. 1973. *Evolutionary biology of the anurans: Contemporary research on major problems.* Columbia: University of Missouri Press.

Reptiles

Bellairs, A. 1969. *The life of reptiles.* Vols. 1–2. London: Weidenfeld and Nicolson.

Gans, C., senior ed. 1969 et seq. *Biology of the reptilia.* London: Academic Press, vol. 1 (1969) and vol. 13 (1982); New York: John Wiley, vol. 14 (1985) and vol. 15 (1985); New York: A. Liss, vol. 16 (1988); Chicago: University of Chicago Press, vol. 17 (1992) and vol. 18 (1992); Oxford, Ohio: Society for the Study of Amphibians and Reptiles, vol. 19 (1998) (continuing series).

For an up-to-date listing of reptiles of the world check the EMBL Reptile Database: *www.embl-heidelberg.de/~uetz/ livingreptiles.html.*

Lizards

Huey, R. B., E. R. Pianka, and T. W. Schoener, eds. 1983. *Lizard ecology: Studies of a model organism.* Cambridge: Harvard University Press.

Mattison, C. 1989. *Lizards of the world.* New York: Facts on File.

Vitt, L. J., and E. R. Pianka, eds. 1994. *Lizard ecology: Historical and experimental perspectives.* Princeton: Princeton University Press.

Snakes

Campbell, J. A., and W. W. Lamar. 1989. *The venomous reptiles of Latin America.* Ithaca: Cornell University Press.

Greene, H. W. 1997. *Snakes: The evolution of mystery in nature.* Berkeley: University of California Press.

McDiarmid, R. W., J. A. Campbell, and T. Touré. 1999. *Snake species of the world: A taxonomic and geographic reference.* Vol. 1. Washington, D.C.: Herpetologists' League.

Mattison, C. 1987. *Snakes of the world.* New York: Facts on File.

Seigel, R. A., and J. T. Collins. 1993. *Snakes: Ecology and behavior.* New York: McGraw-Hill.

Seigel, R. A., J. T. Collins, and S. S. Novak, eds. 1987. *Snakes: Ecology and evolutionary biology.* New York: Macmillan.

Turtles

Bjorndal, K. A., ed. 1995. *Biology and conservation of sea turtles.* Rev. ed. Washington, D.C.: Smithsonian Institution Press.

Ernst, C. H., and R. W. Barbour. 1989. *Turtles of the world.* Washington, D.C.: Smithsonian Institution Press.

Lutz, P. L., and J. A. Musick. 1997. *The biology of sea turtles.* Boca Raton, Fla.: CRC Press.

Pritchard, P. C. H. 1979. *Encyclopedia of turtles.* Neptune City, N.J.: Tropical Fish Hobbyist.

Crocodilians

Ross, C. A. 1989. *Crocodiles and Alligators.* New York: Facts on File.

Webb, G. J. W., S. C. Manolis, and P. J. Whitehead, eds. 1987. *Wildlife management: Crocodiles and alligators.* Chipping Norton, NSW: Surrey Beatty.

An excellent Web site for information of crocodilians is www.crocodilian.com.

For reasonably up-to-date summaries of the composition and general distribution of the amphibians and reptiles found in Mesoamerica, the following are most useful.

Middle America

Campbell, J. A. 1999. Distributional patterns of amphibians in Middle America. In *Patterns of distribution of amphibians: A global perspective,* ed. W. E. Duellman, 111–210. Baltimore: Johns Hopkins University Press.

Panama

Young, B. E., G. Sedaghatkish, E. Roca, and Q. D. Fuenmayor. 1999. *El estatus de la conservación de la herpetofauna de Panamá.* Arlington, Va.: Nature Conservancy.

Nicaragua

Köhler, G. 1999. The amphibians and reptiles of Nicaragua: A distributional checklist with keys. *Cour. Forsch. Senckenberg* 213:1–121.

Guatemala

Campbell, J. A., and J. P. Vannini. 1989. Distribution of amphibians and reptiles in Guatemala and Belize. *Proc. West. Found. Vert. Biol.* 4 (1): 1–21.

Mexico

Flores, O. 1993. Herpetofauna Mexicana: Annotated list of the species of amphibians and reptiles of Mexico, recent taxonomic changes, and new species. *Spec. Pub. Carnegie Mus. Nat. Hist.* 17:1–72.

Honduras

McCranie, J. R., and L. D. Wilson. 2002. *The amphibians of Honduras.* Ithaca: Society for the Study of Amphibians and Reptiles.

Myer, J. R., and L. D. Wilson. 1973. A distributional checklist and key to the turtles, crocodilians, and lizards of Honduras. *Contr. Sci. Nat. Hist. Mus. Los Angeles County* 244:1–39.

Wilson, L. D., and J. R. Meyer. 1985. *The snakes of Honduras.* 2d ed. Special Publications in Biology and Geology. Milwaukee: Milwaukee Public Museum.

Wilson, L. D., and J. R. McCranie. 1994. Second update on the list of amphibians and reptiles known from Honduras. *Herp. Rev.* 25 (4): 146–50.

Belize

Meyer, J. R., and C. F. Foster. 1996. *A guide to frogs and toads of Belize.* Malabar, Fla.: Krieger.

Stafford, P. J., and J. R. Meyer. 2000. *A guide to the reptiles of Belize.* San Diego: Academic Press.

Yucatán Peninsula

Campbell, J. A. 1998. *Amphibians and reptiles of northern Guatemala, the Yucatán, and Belize.* Norman: University of Oklahoma Press.

Lee, J. C. 1996. *The amphibians and reptiles of the Yucatán Peninsula.* Ithaca: Cornell University Press.

Journals and Organizations

As interest continues to grow in the herpetofauna of Latin America, including Costa Rica, knowledge of taxa and significant biological study of those described in this book will appear in the scientific literature. Two important Costa Rican journals that contain articles of herpetological import are *Brenesia* and the *Revista de Biología Tropical.* These should be routinely consulted by anyone interested in the *herpetofauna costarricense.* Information on *Brenesia* may be obtained from the Museo Nacional de Costa Rica, apartado 749-1000, San José, Costa Rica. The *Revista* is published by the University of Costa Rica, and copies or further information are available from *Revista Biología Tropical,* Biología 15, Universidad de Costa Rica, 2060 San José, Costa Rica.

From time to time the antivenin institute, the Instituto Clodomiro Picado, University of Costa Rica, which is physically located in San Isidro de Coronado, San José Province, publishes information on venomous snakes. Occasional popular booklets also are issued by the privately owned Serpentario Nacional, La Sabana, San José, Costa Rica. Live examples of mostly venomous snake species are exhibited free of charge every Friday afternoon at the Instituto and daily (for an entrance fee) at the Serpentario (Monday to Friday, 9:00 A.M. to 6:00 P.M., Saturday and Sunday 10:00 A.M. to 5:00 P.M.).

Five major Northern Hemisphere herpetological societies publish journals that contain many articles on Latin American herpetology:

American Society of Ichthyologists and Herpetologists
c/o Dr. Maureen A. Donnelly, College of Arts and Sciences
Florida International University
North Miami, FL 33181
USA
(Publisher of the quarterly *Copeia*)
www.utexas.edu/depts/asih

The Herpetologists' League
c/o Mac F. Given
Department of Biology
Neumann College
One Neumann Drive
Aston, PA 19014-1298
USA
(publisher of the annual *Herpetological Monographs* and the quarterly *Herpetologica*)
www.inhs.uiuc.edu/cbd/HL/HL.html

La Sierra University
Department of Biology
Riverside, California 92515-8247
USA
(sponsor of the biannual *Herpetological Natural History*)
hkaiser@lasierra.edu

Societas Europaea Herpetologica
c/o Dr. Michael R. K. Lambert
Natural Resources Institute
Central Avenue, Chatham Maritime
Kent ME44TB
UK
(publisher of the quarterly *Amphibia-Reptilia*)
www.gli.cas.cz/SEH

Society for the Study of Amphibians and Reptiles
c/o John N. Matter
Department of Biology
Juniata College
Huntington, PA 16652
USA
(publisher of the quarterlies *Journal of Herpetology* and
Herpetological Reviews and of *Facsimile Reprints in
Herpetology* and *Contributions in Herpetology*, among
other publications)
www.ukans.edu/~ssar/SSAR.html

Inquiries regarding memberships and publications may be
addressed to the individuals named.

SPECIES OF POSSIBLE OCCURRENCE

One species, the lizard *Cnemidophorus lemniscatus*, is
known both from Honduras to the north of Costa Rica and
Panama to the south. It favors relatively open lowland situations and might be found on beaches on the Atlantic versant near the Nicaragua frontier. It is not known to occur in
western Panama. I have included this lizard in the keys and
species accounts.

A considerable number of species known from western
Panama but not yet recorded from Costa Rica actually may
range into the republic. These include the following upland forms from the slopes of Cerro Pando and Volcán Barú:
the anurans *Bufo peripetates, Dendrobates speciosus, Epipedobates maculatus, Hemiphractus fasciatus, Hyla fuscata, Eleutherodactylus monnichorum,* and *Eleutherodactylus museosus;* the lizard *Norops kemptoni;* and the snakes
Geophis championi and *Hydromorphus dunni.* These species possibly range along both Pacific and Atlantic slopes
onto the southern Cordillera de Talamanca of Costa Rica.
The skink *Sphenomorphus rarus* (Myers and Donnelly,
1991) is known from a single locality in Bocas del Toro
Province, about 40 km southeast of the Costa Rican boundary. The introduced gecko *Sphaerodactylus argus* occurs on
several islands in the Bocas del Toro archipelago. These two
forms may ultimately be discovered in extreme southeastern
Costa Rica.

Less likely to be discovered in Costa Rica are a number of
species known from farther to the east near the continental divide that forms the boundary between Bocas del Toro
and Chiriquí Provinces. These include the caecilian *Caecilia
volcani,* the frogs *Dendrobates arboreus, Hyla graceae,
Eleutherodactylus emcelae,* and *Eleutherodactylus jota;* the
lizards *Dactyloa casildae, Norops exsul,* and *Norops fortunensis;* and the snake *Pseudoboa neuwiedii.* These taxa
are most likely to be found along the Atlantic slope farther
to the west, and some may range into extreme southeastern
Costa Rica.

ERRONEOUS RECORDS

Most cases where non–Costa Rican species have been listed
for the country are based on misidentifications or subsequent improved taxonomic discrimination that gave a different name to Costa Rican forms. These sources of erroneous records reflect increases in taxonomic knowledge as
documented in the synonymies in Taylor (1951b, 1952ab,
1954b, 1956, 1958b) or discussed in the remarks sections
of the species accounts.

Two species purportedly from Costa Rica clearly based
on non–Costa Rican taxa are the snakes described as a new
form, *Helicops wettsteini,* by Amaral (1929a) and those
identified as *Tropidonotus sipedon* by Günther (1894).

Rossman and Scott (1968) discovered that the known
examples of *H. wettsteini* (NHMW 18726:1, holotype;
18726:2, paratype), supposedly from Juan Viñas, Cartago
Province, were representatives of the common Asian homo-
lapsine snake *Enhydris plumbea.* The data associated with
these specimens clearly are in error.

I have recently examined the snakes (BMNH 71.11.1–4)
that were the basis for Günther's record of *T. sipedon* from
Cartago, Cartago Province. These specimens are representatives of a common water snake, *Nerodia sipedon,* which is
not otherwise known to occur south of extreme northeastern Texas. According to the records of the Natural History
Museum (BMNH), they were purchased from a Mr. Janson.
Since no other examples of this species have ever been collected anywhere in Mexico or Central America, there is no
support for regarding these snakes as from Costa Rica.

CHALLENGES FOR READERS OF THIS BOOK

In the 147 years since Frantzius and Hoffmann initiated the
discovery of the Costa Rican herpetofauna, it has become
the most studied of any in tropical America, save perhaps
that of Mexico. Nevertheless, species new to science or new
to the republic continue to be discovered. Additional fieldwork, especially in the unexplored reaches of the Cordillera
de Talamanca between the Cerro de la Muerte region and the
Panama border and the upper slopes of the volcanoes comprising the Cordillera de Guanacaste, doubtless will produce
many novelties. Exploration of the intact forests on these
relatively inaccessible, cold, and wet mountain slopes and
peaks will provide unusual physical challenges but exceptional rewards for those committed to completing the catalog of riches that constitutes the Costa Rican herpetofauna.

Modern methods of phylogenetic analysis and the application of refined cytogenetic and molecular genetic techniques promise to clarify the status of many taxa and doubtless will uncover numerous cryptic species. It seems likely
that these approaches will lead to the validation of many allopatric forms, now regarded as single species, as distinct
evolutionary entities. The most obvious challenges in this
area of study will involve comparisons of populations from

the rainforests of the Atlantic lowlands with those of the Golfo Dulce region and among populations from the various mountain massifs and isolated volcanic peaks. I would predict that the author of the next treatment of the Costa Rican herpetofauna will need to account for about 450 species. Sadly, another 25 or so will probably become extinct through habitat destruction before they are even discovered.

Although the basic biology of a number of species is well known, the ecology and life history of most forms remain partially known at best. There are numerous opportunities for detailed studies of population dynamics, feeding and foraging ecology, antipredator defenses, reproductive biology, and community ecology for those willing to accept the challenge. Similarly, these tropical amphibians and reptiles offer a rich but thus far relatively untouched source for studies in comparative physiology, including all aspects of gaseous exchange, temperature and water relations, energetics, and performance.

From a purely descriptive standpoint two areas where the present work is incomplete call out for organized and systematic study. The first is anuran vocalizations. I have included data on advertisement calls and male call sites for as many species as possible, but the calls of many others are unknown or have never been recorded. Courtship calls by males and reciprocation calls by females where these occur, as well as possible interspecies differences in release calls, have been little documented. These are serious deficiencies for any attempt to understand the role of communication in the social life of individual species and the structuring of anuran species assemblages.

I had intended to include a graphic representation of the vocalizations of the known recorded calls of all anurans as part of the species accounts in this book. Such figures usually include a sonogram (a plot of frequencies against time), a representation of the wave form that plots amplitude against time, and a spectrogram that plots amplitude against frequency (fig. 7.2). Unfortunately, most call recordings are on old tapes that have degenerated over time, making them unusable for this purpose.

New recordings using digital technology electronic equipment are required before satisfactory graphic representations for Costa Rican taxa will be produced. This project, alas, must be left for the next generation of tropical anuran biologists. Reference to any published description of calls is made in the species accounts.

The second area of incompleteness is inadequate description of the eggs of Costa Rican amphibians. Oviposition site, characteristics of individual eggs, differences in numbers of eggs in a clutch, and the nature of egg masses are unknown for most species. Someone needs to make a concerted effort to fill in this major gap in our knowledge to ensure a base for future studies on development, survivorship, and kin and sexual selection, among other matters.

The study of behavior using Costa Rican amphibians and reptiles is in its infancy. Many of the country's salamanders, anurans, and lizards are ideal but unexploited subjects for analysis of interesting and biologically significant behaviors. Great opportunities exist for both field and laboratory analyses of ethology, behavioral ecology, and the evolution of behavior.

Although I spend a considerable portion of this book on the biogeography of the herpetofauna, it is far from the last word on the subject. More detailed study of the ecological determinants of current distributions and the insight into historical determinants provided by phylogenetic analysis should refine our understanding of the evolution of not only the Costa Rican herpetofauna but that of all of tropical Mesoamerica. We need more phylogenies correlated with a better understanding of geologic history for the region! I hope someone reading this book will take up the challenge of providing the next step in approximating the biogeographic process.

Another challenge relates to declining amphibian populations. The declines in lower Central America are now well documented and clearly are not the result of natural population fluctuations (Pounds et al. 1997). Proximate and ultimate causes for the declines remain controversial. Are they the result of an epidemic (Lips 1998), or are the influences of climate change at work? How do these declines relate to declines elsewhere (Laurance, McDonald, and Speare 1996)? Surely these are issues that require immediate attention and additional research.

Although many of the challenges reviewed above can be made only by scientific study, the final challenge is to all readers of this book. Succinctly, by what means do we protect the Costa Rican habitats and their amphibian and reptile inhabitants from destruction and extinction so that future generations may learn from and enjoy the intrinsic natural values they provide? Costa Rica has an unrivaled system of conservation areas, but this system is under constant negative pressure, both demographic and economic. Conservation education and citizen involvement are cornerstones of any program at the national and international level. In these areas Costa Rica continues to make great progress, and non–Costa Ricans can contribute to this progress by supporting in-country and international conservation agencies and organizations. However, high human population growth rates in Costa Rica, as elsewhere in the Tropics, constantly threaten conservation efforts. Boza (1993) has compared tropical national parks to Gothic cathedrals, Renaissance castles, the Greek Acropolis, and the Egyptian pyramids. In these forests each species is a unique creation of the evolutionary process, richer in content than any painting by van Gogh or concerto by Beethoven. If we allow these masterpieces of nature to be destroyed, how long afterward will we ourselves endure?

2 | The Costa Rican Environment

THE REPUBLIC of Costa Rica is the third smallest of the Central American countries, with an area of 50,900 km², and shares a northern border with Nicaragua and an eastern (often called southern) one with Panama. The country lies along an axis running generally from southeast to northwest and is bordered on the west by the Pacific Ocean and on the east by the Caribbean Sea (fig. 2.1). From east to west Costa Rica is about 259 km at its widest point and 119 km at it narrowest. The extent of the country along its longest axis is approximately 464 km. The western coastline is relatively long and complex at 1,248 km, while the 212 km eastern coast forms a relatively smooth curve from southeast to northeast. In addition to the mainland territory and adjacent islands, Costa Rica includes the Isla del Coco, 550 km to the southwest at 5°30–33′ south latitude. The republic is formed of seven political units called *provincias* or provinces, and these in turn are divided into *cantones* or counties.

PHYSIOGRAPHY AND HYDROGRAPHY

The geography of Costa Rica is exceedingly complex for an area approximately the same size as Denmark, West Virginia, or San Bernardino County, California (19,575 square miles in the English system). Much of this complexity is derived from the geologically recent uplift of the main mountain systems and the continuing activity reflected at present by volcanic eruptions and frequent earth tremors. These dynamic events have produced a very uneven land surface with numerous undulating ridges, hills, and small valleys as well as the sharp relief of the steep and tall mountain ranges. Along the main southeast-to-northwest geographic axis of the country is a tectonic backbone of high mountains that effectively separates the Atlantic and Pacific coastal regions. The most extensive mountain mass is the Cordillera de Talamanca, which runs northwestward from the southeast to approximately 10° north latitude. This cordillera is continuous with the Chiriquí (Barú) massif of western Panama. Three fingerlike projections protrude westward from the north end of the Talamanca range (in order from south to north)—the Cerros de Dota, the Cerros de Bustamante, and the lower Cerros de Candeleria (sometimes called the Fila or Macizo Cedral). The continental divide along the Cordillera de Talamanca is uniformly above 2,000 m, with a host of mountains rising over 3,000 m. The highest peaks *(cerros)* are in the Chirripó massif, with five over 3,600 m, the highest being Chirripó Grande at 3,820 m; Kamuk, near the Panama border (3,563 m); and Buenavista (3,491 m), just south of the Interamerican Highway where it crosses the continental divide (along the ridge called the Cerro de la Muerte) at 3,100 m. Jutting out northeastward from this range are a series of high and precipitous ridges separated by deep river gorges. The most important of these (in order from southeast northward) are the Filia Dúrika, the Filia de Matama, which terminates at its northeast tip in Cerro Matama (2,251 m), and the ridge that separates the Río Chirripó Atlántico from the Río Pacuare. A shorter, narrower, and lower ridge projects northward to the Río Reventazón between two major tributaries of that river, the Río Pejibaje on the east and Río Grande de Orosi on the west. This ridge ends as a somewhat isolated cluster of cerros (maximum elevation 1,865 m) lying in the arms of the three rivers between Tapantí, Cachí, and Tucurrique.

To the north of the Talamancas and their extensions lie the series of active volcanoes that form the Cordillera Central or Volcánica. The main axis of this range is slightly north of an east-west baseline. The principal masses center on (in order from east to west) Volcán Turrialba (3,328 m), Volcán Irazú (3,433 m), Volcán Barva (2,820 m), Volcán Poás (2,704 m), and Volcán Viejo (2,050 m). Passes from the Pacific to Atlantic slopes of this range are at 1,500 m between Irazú and Barva, 2,100 m between Barva and Poás, and 2,000 m between Poás and Viejo.

Extending northwestward from the main mass of the Cordillera Central is the Cordillera de Tilarán, which begins at about 84°30′ W, to the west and north of San Ramón, and terminates at the Laguna de Arenal. The highest peaks in this mass are at a maximum of about 1,600 m. The currently active Volcán Arenal (1,638 m) is a northeastern outlier of this range. The pass between this mountain chain and that to the north is about 700 m at the headwaters of the Río Sábolo, a tributary of Laguna de Arenal.

The Cordillera de Guanacaste is a series of isolated peaks, mostly volcanic, rising from an upland block of low relief that runs northwest from Laguna de Arenal to 11° N. The ridge is generally between 500 and 1,000 m in altitude, with the following distinctive peaks and volcanoes (arranged from south to north): Cerro Jilguero (1,221 m), Volcán

Tenorio (1,916 m), Cerro la Giganta (1,490 m), Volcán Miravalles (2,028 m), Volcán Santa María (1,916 m), Volcán Rincón de la Vieja (1,806 m), Volcán Cacao (1,659 m), and Volcán Orosi (1,487 m). Low passes between the peaks connect the Pacific and Atlantic lowlands.

Nestled between the cordilleras are several large intermountain valleys. The most important of these are:

Pacific

Valle de Dota: between the Cerros de Dota and Bustamante in the Río (Pirrís) Parrita drainage

Valle de Candelaria: Between the Cerros de Bustamente and Candelaria in the Río (Pirrís) Parrita drainage

Valle Central Occidental: between the Candelaria range and the Cordillera Central, forming the headwaters basin for the Río Grande de Tárcoles

Atlantic

Valle Central Oriental: between the Cordillera de Talamanca and the Cordillera Central, forming the headwaters basin of the Río Reventazón

The eastern and western central valleys are separated by a relatively low ridge (1,550 m) that runs northeastward from the Candelaria range between the cities of San José on the west and Cartago to the east to the southern slope of Volcán Irazú. Costa Ricans usually refer to the two basins together as the Valle Central or Meseta Central.

The lowlands of the Atlantic versant form a series of rolling hills separating broad plains or *llanuras* lying northeast of the central mountain axis. The lowlands are rather extensive north of Puerto Limón but narrow toward the Panamanian border. Three prominent headlands are the only interruptions in the monotonous sandy beaches forming the Atlantic coastline: Punta Mona near the mouth of the Río Sixaola (marking the Costa Rica–Panama boundary) and Punta Cahuita and Punta Limón, two uplifted coral reefs. Living coral reefs lie offshore at Cahuita and on the north shore at Limón. North of Limón to the mouth of the Río San Juan the sandy beaches form barrier islands separated from the mainland by saltwater lagoons with occasional connections to the sea. In the northern zone the lowlands are continuous with those of eastern Nicaragua, and to the south they extend into western Panama for a short distance east of the Laguna de Chiriquí.

The northern sector is crossed by a series of rivers that

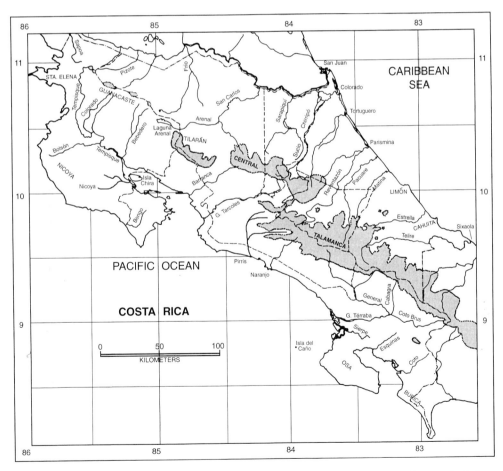

Figure 2.1. Map of Costa Rica showing principal physiographic features (in capitals) and rivers. Shaded areas are above 1,500 m in elevation.

have their sources on the rain-drenched windward Caribbean slopes of the Cordillera de Guanacaste, Cordillera de Tilarán, and Cordillera Central and the extreme northeastern Cordillera de Talamanca. These streams may be grouped into two divisions: those flowing generally northward into Lake Nicaragua or its outlet the Río San Juan and those flowing generally eastward to empty into the Caribbean. The rivers draining into Lake Nicaragua originate on the eastern and north slopes of the Guanacaste range and are mostly very short (30–40 km long); the largest, longest, and most eastern is the Río Frío. The major tributaries of the Río San Juan have their headwaters in the Cordillera de Tilarán and Cordillera Central, and their stream courses cross the extensive northeastern plains for 100–150 km. The most important streams in the latter group are the Río San Carlos, Río Sarapiquí, and Río Chirripó. A major tributary of the former, the Río Arenal, is the outlet for the lake of the same name. These rivers have a markedly seasonal regime and carry large amounts of sediments eroded from soft volcanic materials. In consequence the Río San Juan in its lower reaches has a sandy, shifting bed that is hardly navigable at certain times of the year. The San Juan has an extensive delta interlaced with connecting waterways, swamps, and coastal lagoons. It disgorges its waters to the sea through two main mouths, one at San Juan del Norte (Graytown) in Nicaragua and the other via the Río Colorado at Boca del Colorado in Costa Rica.

Numerous voluminous, short rivers drain directly into the Caribbean Sea along the eastern coast, as do several large ones that arise from the interior. The gradients of these latter are extremely steep, and the flow is torrential in their upper and middle sections, with a continuous steady flow throughout the year. The principal streams in this category (in order from north to south) are the Río Tortuguero, Río Parismina, Río Matina, and Río Sixaola. A major tributary of the Parismina, the 155 m Río Reventazón, drains the Meseta Central Oriental and courses downward through a deep gorge to finally discharge its waters on the coastal plain near Siquirres. The Río Sixaola forms the southeastern boundary with Panama and has its sources from the extreme northern portion of the Chiriquí range of western Panama and from the streams in the Talamanca massif of Costa Rica that flow southeastward across a relatively level alluvial plain, the Valle de Talamanca. The upper portion of this valley is Alta Talamanca, and the lower reaches (from a line drawn between Uatsi and the mouth of the Río Yorkín to the coast) form Baja Talamanca. In the latter region the Sixaola is a broad, deep, sinuous lowland river surrounded by extensive swamp forests toward the coast.

The physiography of the Pacific versant lowlands of Costa Rica is considerably more complex than that of the Caribbean versant because of a longer and more complicated coastline and greater topographic variation in other features. The Pacific littoral is marked by many rocky headlands and four large peninsulas, in contrast to the Caribbean shore. The northernmost of these, the Península de Santa Elena, projects westward into the Pacific Ocean a short distance south of the Nicaragua frontier between the Golfo de Santa Elena and the Golfo de Papagayo. The Península de Nicoya is connected to the mainland on the north and parallels it to form the southern boundary of the large, shallow Golfo de Nicoya. The peninsula is divided by a series of low ridges that run down its center from northwest to southeast, ranging from 500 m to 900 m in elevation. The Golfo de Nicoya contains a number of low-lying islands, the largest of which, Isla Chira, is near the head of the gulf. Most of these islands, except off the tip of the peninsula, and the shoreline of the gulf itself, are bordered by extensive mudflats and mangrove swamps. The similarly shaped and oriented Península de Osa forms the southwestern border to another shallow embayment, the Golfo Dulce, along the southern coast. The Península de Osa is divided along a northwest-to-southeast axis by a series of low hills that reach a maximum elevation of 745 m but are usually under 500 m. Another important geographic feature in this sector is the rocky Isla del Caño (90 m in elevation) that lies to the northwest of the Península de Osa. It is surrounded by a sublittoral coral reef, one of the few such reefs in the eastern Pacific Ocean. Finally, the elongate Península Burica projects directly south from the mainland at the Panama–Costa Rica boundary.

Two principal areas may be recognized within the Pacific lowlands. The first of these areas, the southwest Pacific lowlands, extends from the Panama frontier northwestward to just south of the mouth of the Río Grande de Tárcoles and includes the Península de Osa, the coastal plains, and the interior valleys of the Río General and the Río Coto Brus. These valleys lie southwest of the main Cordillera de Talamanca and are separated from the coastal plain by a pair of extensions of that range, one running from the northwest and the other from the southwest. These two ridges collectively form the Cordillera Costeña, two hilly areas that are generally below 1,000 m in elevation, both with high spots reaching 1,700 m toward their southeast sectors.

In this region many short rivers empty directly into the Pacific Ocean or the Golfo Dulce. The major northern stream is the Río Grande de Pirrís (or Parrita), which drains the leeward side of the Candelaria and northern Talamanca ranges. Its main branch, the Río Grande de Candelaria, originally flows westward through the Valle de Candelaria, then turns abruptly around the western end of the Cerros de Bustamante and continues directly south to the ocean.

The other principal river system in the region consists of two major branches: a southeastward-flowing one, the Río General, and the northwestward-flowing Río Coto Brus. These streams drain the southern slopes of the Cordillera de Talamanca and the northern slopes of the Costeña range. At their confluence near Potrero Grande the combined rivers become the Río Grande de Térraba (or Diquís), which plunges due west through a deep, narrow gorge in the Cordillera Costeña to reach the narrow coastal plain near Palmar Norte and continue to the sea. The Río Térraba and Río Sierpe, the next rivers to the south, are involved in an extensive and complex delta that forms a vast swampy area over much of the coastal plain directly north of the Península de Osa. The Térraba system has high stream flow most of the year, but this is greatly reduced, especially in the tributaries of the General and Coto Brus, during the moderately long dry season (January through April).

Because of differences in rainfall patterns at their headwaters and in the length and intensity of rainfall in the lowlands, the two major stream systems show considerable differences. To the north the Río Grande de Pirrís (or Parrita) closely resembles the Río Grande de Tárcoles in cyclic seasonal fluctuations. Farther to the south, around the Golfo Dulce, extremely high rainfall (between 4,000 mm and 6,000 mm annually), combined with or occurring without a short but definite dry season produce incredible differences in stream volume at different times of the year.

The second area includes all of the Pacific lowlands in the relatively dry region of northwestern Costa Rica. This area extends from the mouth of the Río Grande de Tárcoles to the Nicaragua border and includes much of mainland northwestern Puntarenas Province, almost all of Guanacaste Province, and most of the Península de Nicoya. The central portion of this zone (from the Nicaragua border to about 10°5′ N) is composed of a series of broad, undulating plains. Farther to the south these areas are restricted to a narrow coastal belt around the margins of the Golfo de Nicoya. The Pacific coastal belt shows considerable relief, and rocky headlands and ridges are typically found intermixed with sandy beaches along the outer coast of the Península de Nicoya.

Many short streams drain this northwestern area along the Pacific coast and both margins of the Golfo de Nicoya. The major river system of the Río Tempisque (including its large tributary the Río Bebedero) drains the western slopes of the Guanacaste range and the Guanacaste Plain, flowing southward in a deep meandering channel into the head of the Golfo de Nicoya. Because of the long dry season in this region, all these streams exhibit marked differences in volume and silt load during dry and rainy seasons.

The only other important river on this portion of the Pacific slope is the Río Grande de Tárcoles, which lies near the boundary between the northern and southern lowlands. The Tárcoles arises from streams on the leeward side of the Cordillera Central and northern Talamanca range and drains the Meseta Central through a deep and narrow caldera before discharging its substantial flow and silt load near the mouth of the Golfo de Nicoya. As would be expected, seasonal differences in flow and load also are substantial between dry and rainy seasons for this river.

Costa Rica's uninhabited, isolated Isla del Coco is the steep-sided, rugged tip (634 m above sea level) of an isolated mountain that rises some 3,000 m directly from the submarine Cocos Ridge. Its position on the ridge is about halfway between the Galápagos Islands (600 km to the southwest) and the mainland of Central America. The sheer walls of this island plunge directly into the sea and are intersected by numerous streams and waterfalls. The island consists mostly of consolidated tufa and lava and is surrounded by submarine coral formations, lava flows, and complex rocky formations.

CLIMATE

General

Mainland Costa Rica lies between 11°13′ N and 8°2′ N, completely within the Tropics or tropical latitudinal region. This climatic region corresponds approximately to the belt between the tropics of Cancer and Capricorn, where relatively constant and large amounts of solar radiation (range 720 to 1,000 langleys, mean 860 per day) reach the earth's atmosphere throughout the year and where there is an annual surplus of incoming solar radiation over outgoing radiation lost to space. As a result, seasonal fluctuations are minimal at any given locality, regardless of altitude, and there is no distinct cold (winter) season. Because of the unequal distribution of land and water on the earth's surface, effects of physiography, the impact of prevailing winds, cold marine currents, and the action of local atmospheric conditions, the outer boundaries of the tropical region are best defined by temperature characteristics as a reflection of the radiation balance. Although a number of approaches have been used for delimiting the zone by temperature (McGregor and Nieuwolt 1998), a scheme that reduces temperature data to sea level equivalents best defines the areas that by consensus are regarded as tropical and support vegetation, soil, animal life, agriculture, and patterns of human utilization of natural resources that are characteristically tropical.

In this approach (Holdridge 1967, 1982) the tropical region includes all areas where the mean annual biotemperature (BT°) in the lowlands (or for upland temperatures reduced to sea level) is greater than 24°C. Biotemperature is calculated as follows:

$$BT° = \frac{\text{the sum of unit period temperatures °C}}{\text{number of unit periods (hours, days, weeks, etc.)}}.$$

As an example, in the calculation of BT° for Puerto Limón, Costa Rica (3 m above sea level), the mean monthly Celsius temperatures for January through December are 25.0°, 25.4°, 25.6°, 26.0°, 26.2°, 26.4°, 26.2°, 26.0°, 26.4°, 26.3°, 25.4°, 25.3°. Thus

$$BT° = \frac{310.2}{12} = 25.85°C.$$

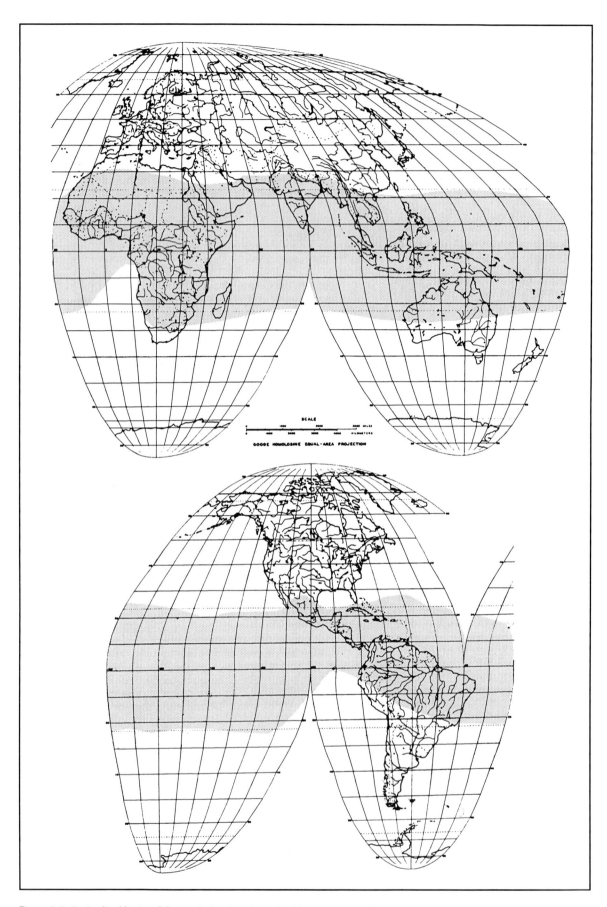

Figure 2.2. Latitudinal limits of the tropical region determined by mean annual biotemperature (BT°) on land and mean sea surface temperature of coldest month more than 21°C.

In this formula all values below 0°C and above 30°C are recorded as 0°C, since it is assumed that plant growth does not occur below 0°C and is substantially or completely inhibited where mean daily or monthly temperatures are above 30°C; hence the name biotemperature. In most cases the BT° for tropical localities corresponds to the mean annual temperature calculated from climatic data. At some lowland sites and in latitudinally peripheral deserts the mean annual BT° will be lower than the mean annual temperature. At extremely high altitudes in the Tropics the mean annual BT° will be higher than the mean annual temperature. The mean annual biotemperature isotherm of 24°C for air temperatures over land corresponds to the 20°C isotherm for ocean surface temperatures for the coldest month of the year. The latitudinal limits of the tropical region defined by these parameters is shown on the accompanying map (fig. 2.2).

Note that the actual heating of the atmosphere is indirect. During the day insolation heats the earth's surface, which in turn warms the atmosphere from below to increase the temperature. At night and on cloudy days the energy received from the sun is reradiated in the form of long-wave (infrared) or terrestrial radiation. When clouds are present this radiation rapidly heats up the lower atmosphere and raises the temperature of the earth, and the adjacent air comes to equilibrium. When there are no clouds, terrestrial radiation heats a much thicker layer of the atmosphere and consequently produces a rapid decrease in temperature at night. Ultimately all of the energy received from the sun is converted into far infrared radiation and is lost to outer space. Most of this outgoing radiation comes from the top of the cloud cover; only a little comes directly from the earth's surface.

As I mentioned above, because of the surplus of incoming to outgoing radiation in the Tropics, temperature characteristics at any site are remarkably constant. All mean daily temperatures are within a few degrees of one another, and mean monthly temperatures rarely vary by more than 1°C. As might be expected, there is little annual variation among mean daily maxima and minima (about 2 to 5°C). The difference between the daily maximum and minimum, however, may be considerable (10 to 20°C) during times of the year when cloud cover is reduced and substantial reradiation occurs at night, especially at higher elevations.

Extratropical regions of the earth, by contrast, show marked seasonal temperature regimens, and the difference between the highest summer temperatures and lowest winter temperatures increases with increasing latitude. The amount of solar radiation (the source of this difference) reaching the earth's atmosphere is consequently much more variable (0 to 1,200 langleys, mean 600/day) than in the Tropics, and extratropical regions have a net radiation deficit of incoming solar radiation to outgoing radiation lost to outer space. The polar regions are always in this deficit zone, and zones between the polar and tropical regions show seasonal fluctuations between a net surplus and a net deficit. Although values for the insolation of the outer layers of

the atmosphere have been used here to emphasize the difference in radiation regimens, losses to the atmosphere reduce the amount reaching the earth's surface (fig. 2.3). The amount of the loss depends mostly on differences in the sun's angle with latitude and the amount of cloud cover and averages about 15% for the whole earth. Solar radiation received at the earth's surface ranges from 0 to 500 langleys per day for the extratropics as opposed to 350 to 450 for the tropical region. Highest values at the earth's surface are at about 30° north or south latitude in the summer, but tropical values are high and relatively constant throughout the year.

The main reason for the relative thermal uniformity of the tropical region is the small amount of difference in the net amount of solar radiation received at the earth's surface at various localities. The total amount of radiation received at any place on earth depends on two factors; the duration of insolation and its intensity. Both are controlled by the astronomical movements of the earth and its relation to the sun. Because the earth's axis of rotation is tilted at 66°33′ to the plane of the earth's orbit around the sun, to an observer on the earth's surface the position of the sun relative to the horizon appears to change during the year. In addition, changes in day length are correlated with the apparent movement of the sun overhead during the annual orbit. Thus in the Northern Hemisphere at middle and high latitudes the sun appears to move higher in the sky (farther north) beginning in December and continuing until June as day length increases, while the opposite occurs from June to January as day length decreases. The opposite occurs in the Southern Hemisphere. The apparent "march" of the sun and its changing position (elevation) are usually expressed in degrees above the horizon at noon local time (table 2.1). At the equinoxes (September 22 and March 22) the sun is directly over the equator at noon. At the June 21 solstice the sun is directly over the tropic of Cancer, its farthest progress northward. At the December 21 solstice the sun is directly overhead at the tropic of Capricorn, its farthest progress southward.

Day length is directly correlated with the time of year. At the equator all days of the year have equal lengths of twelve hours and seven minutes (table 2.2). In low latitudes the difference between the shortest and longest day is about seven minutes per degree, but at higher latitudes (50° to 60°) the difference is larger—about twenty-eight minutes per degree. At the poles six months of continuous daylight are followed by six months of darkness. In the Tropics seasonal variation in day length is insignificant.

Differences in the intensity of solar radiation of greatest significance for climate are caused by the sun's elevation. The higher the sun, the greater the intensity of insolation. There are three reasons for this relationship: first, rays coming from a sun in a high position are spread over a smaller surface than oblique ones from a low sun; second, rays from a high sun have a shorter passage through the atmosphere,

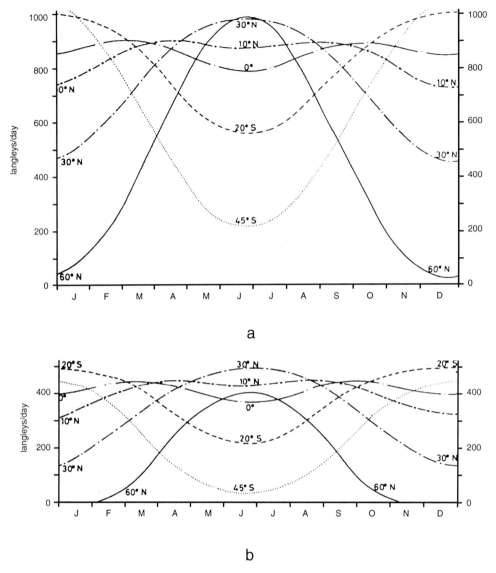

Figure 2.3. Radiation: (a) received at outer limits of the atmosphere; (b) received at earth's surface assuming 60 percent loss in transmission through the atmosphere.

and there is less loss from the scattering effects of dust particles; third, under standard conditions more insolation is reflected with lower angles of incidence. In terms of insolation the Tropics have slight seasonal variation and continuous high intensities. Extratropical areas show large seasonal differences caused by low intensities during the winter, but because of the combined effects of duration and intensity the higher latitudes (with their long summer days) have more insolation than the Tropics during midsummer.

A secondary component in maintaining the thermal uniformity of the Tropics is the heat storage capacity of the ocean. In this region the ocean surface is proportionately much greater than the land surface, and as a heat reservoir the sea prevents the development of extremely cold air masses. During the year the earth's thermal equator (the line

that connects the warmest temperatures or their sea level equivalents) remains close to the geographic equator over the ocean. Over land there are seasonal movements in the thermal equator of up to 40° in latitude associated with the changing overhead position of the sun (fig. 2.4).

Within the tropical region, the primary differences in temperature are related to altitude, with an average mean annual biotemperature decrease of 5 to 6.5°C per 1,000 m increase in altitude, subject to some variability as determined by local atmospheric conditions. Upland and mountain sites in the region, although having lower temperatures than those typical of lowland areas, are characteristically tropical in having a surplus of radiation and minimal seasonal temperature fluctuations throughout the year. For this reason a locality such as Villa Mills (mean annual BT° =

Table 2.1. Altitude of the Sun at Noon in Degrees of Arc above the Horizon (Assuming No Atmosphere) at Different Latitudes at the Equinoxes and on the Summer and Winter Solstices

North and South Latitudes	Equinoxes	Solstice		Solstice	
		Winter	Summer	Summer	Winter
		South	North	South	North
90	0	−23.5	23.5	23.5	−23.5
85	5	−18.5	28.5	28.5	−18.5
80	10	−13.5	33.5	33.5	−13.5
75	15	−8.5	38.5	38.5	−8.5
70	20	−3.5	43.5	43.5	−3.5
65	25	1.5	48.5	48.5	1.5
60	30	6.5	53.5	53.5	6.5
55	35	11.5	58.5	58.5	11.5
50	40	16.5	63.5	63.5	16.5
45	45	21.5	68.5	68.5	21.5
40	50	26.5	73.5	73.5	26.5
35	55	31.5	78.5	78.5	31.5
30	60	36.5	83.5	83.5	36.5
25	65	41.5	88.5	88.5	41.5
20	70	46.5	86.5	86.5	46.5
15	75	51.5	81.5	81.5	51.5
10	80	56.5	76.5	76.5	56.5
5	85	61.5	71.5	71.5	61.5
0	90	66.5	66.5	66.5	66.5

Table 2.2. Day Length at Various Latitudes on Different Dates of the Year

North Latitudes	January 21	February 21	March 21	April 21	May 21	June 21	July 21	August 21	September 21	October 21	November 21	December 21
0	12−7	12−7	12−7	12−7	12−7	12−7	12−7	12−7	12−7	12−7	12−7	12−7
10	11−38	11−52	12−7	12−24	12−37	12−43	12−37	12−24	12−8	11−52	11−38	11−32
20	11−7	11−36	12−8	12−43	13−9	13−20	13−10	12−43	12−9	11−36	11−7	10−55
30	10−32	11−19	12−9	13−3	13−47	14−5	13−48	13−4	12−11	11−18	10−31	10−13
40	9−48	10−57	12−11	13−30	14−34	15−1	14−36	13−32	12−13	10−56	9−48	9−19
50	8−47	10−29	12−13	14−7	15−58	16−23	15−44	14−9	12−17	10−26	8−47	8−4
60	7−7	9−44	12−17	15−6	17−35	18−52	17−42	15−10	12−23	9−41	7−6	5−52
65	5−38	9−10	12−21	15−52	19−26	22−3	19−36	15−57	12−28	9−6	5−37	3−34
70	2−28	8−18	12−25	17−7	24−0	24−0	24−0	17−14	12−35	8−13	2−24	
75		6−43	12−34	19−39	24−0	24−0	24−0	19−53	12−47	6−33		
80		1−53	12−50	24−0	24−0	24−0	24−0	24−0	13−11	1−15		
85			13−19	24−0	24−0	24−0	24−0	24−0	14−36			
90			24−0	24−0	24−0	24−0	24−0	24−0	24−0			

Note: This is an example; because of changes in the earth's axis of rotation, poles "wobble" around their mean position so there are slight differences in day lengths at a particular latitude between years.

10.6°C) on the south-facing slope of the Cordillera de Talamanca of Costa Rica, at 3,096 m in altitude and mean monthly biotemperatures (in order from January through December) of 9.7°, 10.4°, 10.9°, 11.2°, 11.4°, 11.2°, 11.0°, 10.8°, 10.7°, 10.4°, 10.2°, and 9.7°, has every bit as much a tropical climate as Puerto Limón on the Caribbean coast.

In the tropical region seven major temperature-defined zones may be recognized. The boundaries between these zones are at mean annual BT°s of approximately 24°, 18°, 12°, 6°, 3°, and 1.5°. Ranges of mean annual BT°s for the various zones are as follows: lowland, greater than 24°C; premontane between 18 and 24°C; lower montane, between 12 and 18°C; montane, between 6 and 12°C; subalpine, between 3 and 6°C; and alpine, between 1.5 and 3°C. Areas having mean annual BT°s less than 1.5°C are covered by per-

manent snow. The mean annual isotherm of 24°C usually lies at about 500 m but may extend locally to elevations as high as 800 m. For this reason the accompanying diagram (fig. 2.5) uses the former elevation as the upper limit for the lowland zone. Because of the relation to temperature, each altitudinal belt above the lowland sector extends over approximately the same vertical span regardless of geographic location: premontane, 1,000 m; lower montane, 1,000 m; montane, 1,000 m; subalpine, 500 m; and alpine, 500 m. Contrary to the statements of many authors, there is no gradual lowering of the BT°s and upper altitudinal limits of each zone with increasing latitude. Even near the boundaries of the tropical region the 24°C mean annual BT° isotherm and the upper limit of the lowland zone are near the 500 m contour because of the influence of the radiation

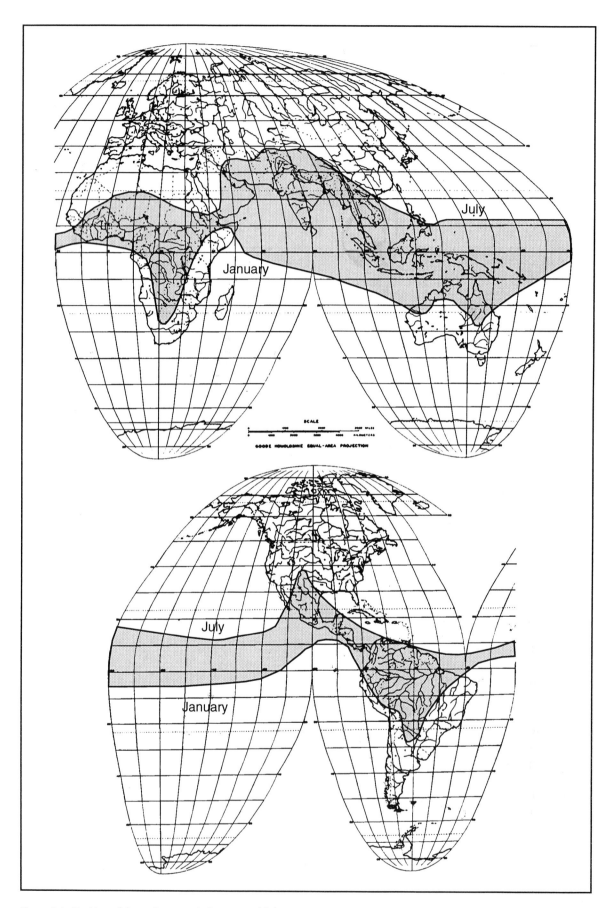

Figure 2.4. Position of thermal equator in January and July.

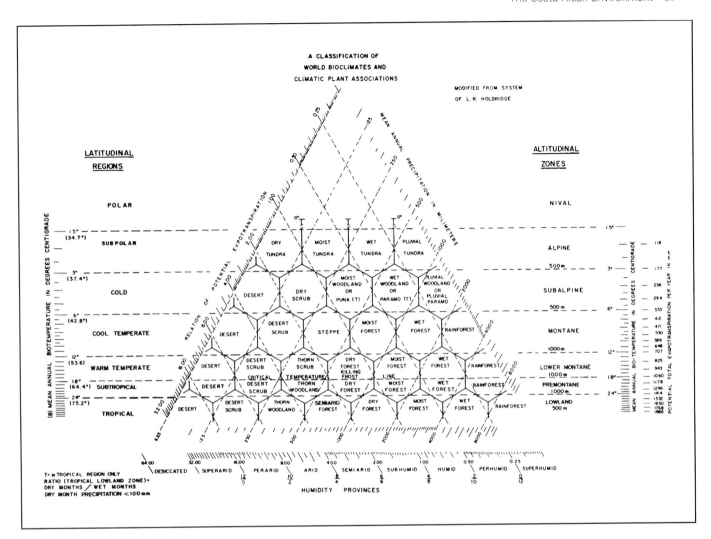

Figure 2.5. A classification of world bioclimates and plant formations.

surplus. A rather abrupt change thus occurs at the boundary between tropical and subtropical regions, since the latter has a surplus of radiation during only part of the year and the upper and lower thermal limits for the altitudinal zones are rapidly depressed over a short latitudinal distance.

Since insolation and temperature are relatively uniform in the Tropics, differences in the amount and temporal distribution of available water, principally in the form of precipitation, account for regional differences in climate. Annual precipitation in the Tropics varies from zero (e.g., lowland deserts of northern Peru) to 11,600 mm (e.g., premontane forest region of Assam), and many areas have marked seasonal differences in rainfall. Although other factors influence rainfall patterns, the principal element regulating seasonal variation is the effect of solar radiation on the earth's heat budget during the apparent annual north and south movement of the sun.

For this discussion, bear in mind that all atmospheric pre-

cipitation is caused by upward movements of air that result in cooling by expansion beyond the level of condensation. In the Tropics the main cause of rising air is convection, best indicated by the high and almost constant presence of cumulus clouds and thunderstorms. In addition, tropical air masses are generally warmer and more humid than extratropical ones. These masses tend to be unstable, since warm, humid air reaches the condensation point at relatively high temperatures and forms clouds consisting mainly of water droplets and rarely of ice crystals. Consequently, copious rainfall is often produced by tropical clouds at elevations where temperatures are well above freezing. By contrast, at higher latitudes uplifting of the air tends to take place along clearly defined fronts, and cloud formation involves an ice phase.

Because of these differences in origin, tropical rains are generally more intense, of shorter duration, and more localized than those in other areas. In addition, any feature that causes ascent of the air, including heating of air over land, airflow from over the ocean onto land (with subsequent heating), or airflow up the slope of mountains, is likely to trigger precipitation.

As pointed out by Riehl (1979), Nieuwolt (1977), McGregor and Nieuwolt (1998), and Jackson (1977), there is no completely satisfactory explanation of the general circulation of the tropical atmosphere and the origins and constraints on precipitation patterns. The following section attempts to identify the principal features believed to produce seasonal differences, but it is clear that some characteristics of circulation that are relatively constant in some areas are weakly developed or nonexistent in others or vary temporally. Particularly confounding are a wide variety of small-scale (thunderstorms) to large-scale (monsoon depressions, waves, hurricanes, etc.) atmospheric disturbances that contribute significantly to rainfall production, but to different degrees in different areas.

The major elements of tropical circulation are a series of subtropical high-pressure cells (STHs), the trade winds, and the equatorial trough or intertropical convergence zone (ITCZ). In concert they act as a heat pump that dissipates surplus heat energy accumulated from the intensity of solar radiation in low latitudes to other areas of the globe. In addition, the constantly maintained difference in temperature between extratropical regions and the Tropics is the main driving force of the surface air currents. In an idealized model, heated, moisture-laden air rises from near the equator (fig. 2.6) and is carried away from this zone at high altitudes to slowly sink back to the earth's surface near the tropics of Cancer and Capricorn. This produces an area of relatively high surface pressure, the STHs. The system continues as massive, constant airflows (the trade winds) moving at low altitudes back toward the equator. Because of the effects of the earth's rotation (the Coriolis force), the trade winds of the Northern Hemisphere are deflected clockwise (the northeast trade winds) and those in the Southern Hemisphere counterclockwise (the southeast trade winds).

The system is completed when the two trade winds systems converge on an area of low pressure, the intertropical trough or ITCZ. As usually described, this low-pressure system forms a relatively broad belt (several degrees in latitudinal extent) characterized by rising air movements and convergence of air masses. Since the converging masses have similar temperature and humidity characteristics, neither will override the other, and a complicated and dynamic interaction produces ascent of first one component and then the other to produce precipitation. The ITCZ is a zone of precipitation because of this feature and because disturbances, particularly thunderstorms, are readily triggered by any factor that initiates convectional lifting of the very humid and warm air masses to altitudes where cooling causes condensation of the water vapor.

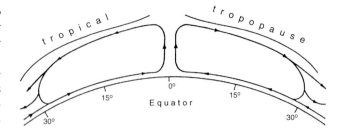

Figure 2.6. Model of heat transport by the atmosphere in the tropical region in cross section; vertical scale exaggerated.

This model clearly oversimplifies the complexity of tropical circulation, since it is now recognized that there is not a single heat source near the equator but a number of centers of varying dimensions—that there is significant seasonal variation in intensity and location of the STHs and the tropical trough and in the directionality of the trade winds. Consequently the ITCZ is not a continuous feature but is ephemeral or nonexistent in certain areas during most of the year. Seasonality of rainfall in the Tropics nevertheless seems to be associated with the annual march of the sun across the tropical sky, which correlates in a broad way with seasonal changes in location and degree of development of the ITCZ.

During the course of the annual cycle, as the sun moves northward and crosses the equator at the March equinox, the high intensity of insolation heats the ocean and atmosphere. However, because of the effects of the unequal proportion of land and water in the Southern Hemisphere and northern tropical belt and of cool water marine currents flowing far northward on the western margins of the southern continents, the thermal equator trails behind the overhead position of the sun by a month or six weeks over the continents. In addition, because of these factors the mean annual position of the thermal equator lies mostly north of the equator (fig. 2.4). Since the thermal equator lies in the area of highest temperature, the northward movement of the ITCZ follows a similar pattern (fig. 2.7). As this system moves, the location and intensity of the northern STHs change, as does the directionality of the southeast trade winds, which now cross the equator and may become westerly winds. The interaction of moisture-laden trade winds, the ITCZ, and disturbances produces rains in a broad belt in this dynamic system. When the sun reaches its northernmost progression at the June solstice, the process is reversed and the thermal equator and ITCZ follow a delayed return southward that continues across the equator and duplicates and completes the cycle in the Southern Hemisphere at the December solstice. The mean annual position of the ITCZ, or meteorological equator, lies at 5° north latitude and, as expected, corresponds to the latitude with the greatest amount of annual rainfall (mean of 2,000 mm worldwide).

Generally speaking, the system just described produces

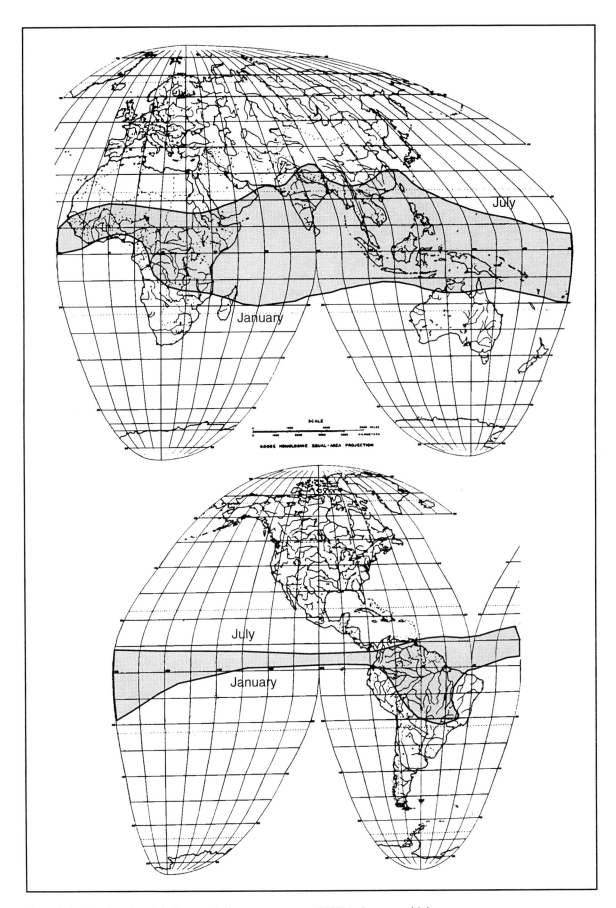

Figure 2.7. Mean location of the intertropical convergence zone (ITCZ) in January and July.

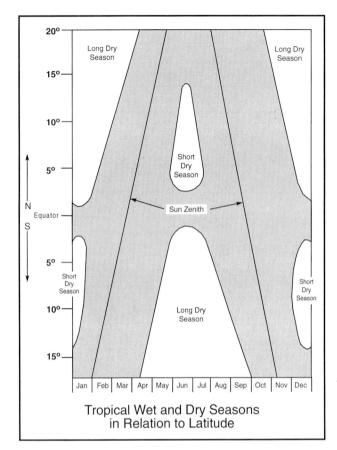

**Tropical Wet and Dry Seasons
in Relation to Latitude**

Figure 2.8. Generalized pattern of seasonal rainfall over tropical land masses.

the following seasonal patterns of rainfall at comparable latitudes in both the Northern and Southern Hemispheres (fig. 2.8).

Near the equator, a double rainfall maximum but no real dry season

From 3° to 10° N or S, two rainy and two dry seasons; the two rainy seasons are about equal in length and separated by a long dry season during the low-sun period and a short dry season during the high-sun period after the ITCZ has passed overhead briefly on its progress toward the tropic

From 10° to 20° N or S, one long wet season and one long dry season; the former during the high-sun period, the latter during the low-sun period

Peripherally in Africa, Australia, and the Americas (20° to 25° N or S), a short rainy season in the high-sun period and a long dry season, grading into a area where rainfall is rare and unpredictable

All the many complexities previously mentioned as affecting tropical circulation have modifying effects on this simple scheme, so its application is not universal. Let me briefly note several obvious exceptions. Equatorward from the STHs, the areas with a single short rainy season and deserts are ex-

panded along the western margins of the continents. Desert conditions are particularly well developed on the western margins of the southern continents because of northward-flowing cold marine currents offshore. These currents also account for the northward displacement of the tropical boundary along the eastern margins of the southern oceans.

In the region of the Asian tropics and northern Australia the ITCZ is very broad and moves much farther northward during the high-sun period than over other continents. It often obscures or eliminates the pattern of two rainy seasons plus one long and one short dry season. In addition, the strong periodic winds or monsoons of Asia and Australia disturb the system between 5° and 20° N or S by concentrating seasonal rainfall into one relatively short period (four months) of exceptionally heavy precipitation followed by a long dry season.

Determinants of Costa Rican Climates

Important reviews of the factors influencing the climates of the republic include Coen (1953, 1983), Gómez (1986b), Herrera (1986), and Herrera and Gómez (1993). The works of Vivó Escoto (1964) and Portig (1965) provide background on patterns for the tropics of Mexico and Central America. Much of what appears below is derived from these sources, especially Herrera's comprehensive work.

Costa Rica has a mean latitude of 10° N, and consequently day length varies only slightly during the year. The difference between the shortest day of the year (December 21), eleven hours, thirty-two minutes, and the longest (June 21), twelve hours, forth-three minutes, is one hour and eleven minutes (seventy-one minutes). At the equinoxes the days are twelve hours, seven minutes. The greatest difference between time of sunrises and sunsets during the year is thirty-three minutes, the earliest sunrise being at 5:42 and the latest sunset at 6:15. As is true for every locality on earth, the mean day length per year is twelve hours. Nevertheless, because of cloud cover the number of hours of sunshine varies with the season and locality and has an annual mean of about 5.5 hours, with very low values in the mountains during most of the year. For example, at Santa Rosa in the Pacific northwest lowlands the annual mean is high at 7.1 hours, with a lowest mean monthly value of 5 hours (September) and a highest of 9.7 (February); at Coliblanco (2,200 m) on the Atlantic slope of Volcán Irazú, the mean is 3.4 hours, with the high at 4.7 hours (January) and the low at 2.1 (July).

The sun has a high elevation throughout the year and is directly overhead at noon on April 16 and August 28. At the December solstice the elevation at noon is 56°6′, at the equinoxes 80°, and at the June solstice 76°6′. As is to be expected from these relationships and the slight variation in day length, all localities are typically tropical, with uniform temperatures and daily patterns of oscillation throughout the year. However, because of the location of the republic in the Northern Hemisphere, the south-facing (Pacific) slopes of its major mountain chain receive slightly more annual insolation than the north-facing Atlantic ones. This factor, combined with the effect of the longer dry season, raises land temperatures on the Pacific slope disproportionately. In consequence the mean annual BT° at sea level on the Atlantic versant is 26°C versus 27.6°C on the Pacific, with lapse rates of 0.57°/100 m and 0.52°/100 m, respectively. Altitudinal temperature zones thus are displaced upward by several hundred meters on the Pacific slope of the Talamanca range compared with the Atlantic slope. Atlantic winds and cloud cover mitigate this thermal effect in the cordilleras farther to the north. These low lapse values for both versants undoubtedly reflect the buffering effect on air temperatures provided by the isthmian location of Costa Rica. The isthmian factor is also important in several other ways. Because no point in Costa Rica is more than 130 km from the ocean, the country is mostly under maritime influences. Consequently climatic disturbances that originate in the Caribbean frequently affect the Pacific slope, and to a lesser extent disturbances that develop over the Pacific have an impact on the Atlantic. Most important, extensive cloud formations originating on the Atlantic versant flow through the low mountain passes of the Cordillera Central and Cordillera de Guanacaste and over the low-lying Cordillera de Tilarán to affect the climates of the Valle Central Occidental and northwestern Costa Rica.

Another aspect of the isthmian influence is produced by the marine currents that flow along each coast. The Caribbean shore is bathed by an eddy system formed from the very warm (surface temperatures of over 25°C), highly saline branch of the North Equatorial Current that flows north and westward through the Lesser Antilles to pass through the Yucatán Channel into the Gulf of Mexico. The very warm (surface temperature over 28°C) Costa Rican Coastal Current flows northward along the western margin of Central America. The northward extent of this current is related to the location of the ITCZ. In June it continues to Cabo Corrientes, Mexico, but during the low-sun period it reaches only to Costa Rica, where it turns westward to form the North Equatorial Current. The very high temperatures of these waters provide warm, moist air that is carried inland by prevailing winds and contributes substantial amounts of precipitation on uplift and cooling.

The most important atmospheric systems affecting Costa Rican climates are those associated with seasonal variation in rainfall. These include the primary influences of the ITCZ and the trade winds, major atmospheric disturbances produced by modified northern cold fronts, easterly waves, and hurricanes and local (onshore and offshore) winds. Also important in this regard are orographic effects that modify prevailing conditions locally by blocking or fragmenting air masses and marine incursions of landward flowing air.

The general seasonal pattern of precipitation for Costa Rica is described in this section with reference to the trade winds and the ITCZ, which is a particularly well delineated feature in the Caribbean and Central American region. In most of Central America the Atlantic and Pacific slopes have markedly different seasonal rainfall patterns. The Atlantic versant has the heaviest rain in two peaks, one in November to January and a second in July and August. On the Pacific slope, except in southwestern Costa Rica and adjacent Panama, there is hardly any rain from December through April. At inland locations of intermediate altitudes the rainfall pattern will be a mix of Atlantic and Pacific types determined primarily by topography and seasonal airflow patterns. Ridges and windward-facing slopes tend to be moist, while lee sides of hills, ridges, and mountains and the floors of valleys will be dry.

In Costa Rica rainfall patterns are controlled primarily by Atlantic air masses and movements during the low-sun period (November to March) in which the strong northeast trade winds predominate. During the rest of the year the ITCZ is the major influence, and Pacific air masses and air movements dominate on the Pacific slope. Throughout the year the northeast trade winds blow across the Caribbean Sea and the Atlantic lowlands of Costa Rica to be intercepted by the cordilleras or continue over onto the Pacific slope. Their strength, however, is reduced relatively while the ITCZ is ascendant. During this high-sun period the southeast trade winds of the Southern Hemisphere follow the movement of the ITCZ northward across the equator. Here they are deflected eastward to form strong southwesterly winds that affect Pacific versant Costa Rica.

At the beginning of the year the thermal equator and ITCZ lie well to the south near the geographic equator. The climate of the country at this time is predominantly under the influence of the northeast trade winds, which bring abundant rainfall to the Atlantic slope directly or through interactions with modified polar air masses. The latter masses have their origins far to the north. They are part of the general world air circulation, which at midlatitudes is driven by westerly winds that carry a series of well-organized disturbances, warm and cold fronts, across the continents from west to east. When cold fronts are very intense during the North American winter, a front and the polar air behind it are occasionally driven far southward along the western margin of the Gulf of Mexico and Caribbean by deflected westerly winds. These invasions of cold air, usually accompanied by strong winds, are called *nortes* in Texas and Mexico, where they often produce sudden snowstorms or ice storms. A similar phenomenon occurs in Central America, where the polar air mass has its cold, dry air rapidly modified as it passes over the Gulf of Mexico. When the edge of the modified cold front meets the humid, warm tropical air over land, heavy precipitation and strong north winds are produced. The resulting storm is called a *norte* in most of Central America but is referred to as a *temporal* or more specifically a *temporal del Atlántico* in Costa Rica. These storms produce heavy, continuous rainfall for one to several days. At this time of year the strong northeast trade winds flow across the country's mountain backbone to create an inversion layer over the Pacific slope. The inversion inhibits the development of precipitation and is in effect from November until April or May (the dry season). Farther south the dry season is shorter (December or January to March).

Conditions remain much the same during February and March except that there is a perceptible decrease in rainfall on the Atlantic versant and in other areas influenced by the northeast trade winds. As the ITCZ moves northward during the next months (April through May), the influence of the northeast trades is reduced on the Pacific versant and the land becomes so heated that the air rises, cools, and drops its water. On this slope the rains first appear in the southern area and follow the ITCZ as it moves northward. By the end of May the entire country is under the influence of the ITCZ and into the definitive rainy season. As the ITCZ continues to move northward, southwesterly winds (modified southeast trade winds) bring additional rain to the southern portions of the Pacific slope. June conditions are essentially a continuation of the situation in late May.

July and August are times of very high rainfall along the Atlantic versant. It is during this period that low-pressure disturbances produced by elongation of tropical cyclones (easterly waves) are most frequent. Easterly waves may appear at any time from June to November and move from east to west across the Caribbean at approximately right angles to the trade winds. These disturbances may come as often as every three to five days during July and August and bring copious rainfall to the Caribbean slope. They may also trigger severe storms (*temporales del Pacífico*) brought on by westward-moving air masses coming off the Pacific in the ITCZ zone.

Although less frequent than easterly waves, the powerful cyclonic storms or hurricanes of the Caribbean region may contribute increased rainfall to both Atlantic and Pacific areas of Costa Rica. Peripheral effects from these storms include high winds and torrential rains. Hurricanes are most frequent during June to November, but generally those that affect lower Central America form in September through November. Before 1988 only one of these disturbances (Hurricane Martha, November 1969) is known to have hit the Costa Rican coast in one hundred years of record keeping. Then in October 1988 Hurricane Juana swept across Nicaragua and Costa Rica. Although the eye of the hurricane made landfall at Bluefields in Nicaragua, the storm ravaged Costa Rica as it passed across the Isthmus and dumped enormous amounts of rain on the Pacific slope. Especially hard hit was southwestern Costa Rica, where torrential rains produced major floods and landslides on the mountain slopes. The hurricane then continued out over the Pacific Ocean and moved northward as it became much reduced in intensity and finally dissipated. This was followed eight years later by Hurricane César, which tracked a similar course and again caused widespread and even greater damage on the Atlantic coast and Pacific slopes.

In late July to early August when the ITCZ reaches its farthest point north, a brief period of little or no rainfall occurs for a week or two on the Pacific versant. This brief dry season is followed by the southward return of the ITCZ and a season of heavy rainfall that ends with the start of the dry season in November in the northwest and somewhat later (December) farther to the south on the Pacific versant. At the same time the northeast trade winds strengthen, and Atlantic air masses once again become the predominant climate control on both Atlantic and Pacific slopes. A period of especially heavy rains on the Atlantic begins at this time and continues into the next year.

Four distinctive rainfall regimens may be recognized for Costa Rica based on these patterns:

Northern Pacific regimen: characterized by a long dry season (November through April) and a long wet season (May to November) interrupted briefly by a short period of reduced rainfall in July or August or both; the months of heaviest rain are September and October (the Pacific versant including mountain slopes, except for the Golfo Dulce area, and the Atlantic slope Llanuras de Guatusos)

Golfo Dulce regimen: defined by no dry season or a very short one (January through March) and a long wet season; heaviest rains are in October and November (the area around the Golfo Dulce, including the Peninsulas de Osa and Burica)

Atlantic coastal regimen: characterized by having no definite dry season (all months with more than 100 mm of rainfall); heaviest rains are in November to January and July through mid-August (the Atlantic lowlands except Llanuras de Guatusos)

Atlantic slope regimen: typified by rains throughout the year; heaviest rains in July and August (the Atlantic slopes of the mountains and higher elevations of the Pacific slopes of the Cordillera de Talamanca)

Although the Pacific regimen is usually described as a seasonal one and the Atlantic coastal and slope regimens are considered aseasonal, all three exhibit a definite reduction in amount of rainfall during all or part of the low-sun period and during July or August or both (Pacific regimen) or during September (Atlantic regimens). The Golfo Dulce regimen also shows this pattern (reduced rainfall in July or August), even at localities where the short dry season is eliminated by local rains. The overall pattern seems to be the lower Central American equivalent of the two rainy seasons and one long and one short dry season climate expected at 10° N latitude (fig. 2.8), as modified by isthmian and orographic effects.

Seasonal differences in temperature (although very slight) and rainfall are recognized by Costa Ricans and are the basis for the legal calendar, school schedules, and vacation periods. Although the country is in the Northern Hemisphere, December to May, which is the hottest and driest time of the year on the Pacific side, is called summer, or *verano*. The cooler rainy season, May through November, is referred to as winter, or *invierno*, and these terms are used throughout the country. The short period of low rainfall in the midst of the rainy season is the *veranillo* or "little summer."

Classification of Tropical Climates

The system used in this book to describe the major tropical climates is modified from the scheme developed by L. R. Holdridge (1967) as a classification of the world's vegetation based on simple climatic data. Hartshorn (1983) gives a succinct account of the rationale and applicability of this system. In my own experience I have found the Holdridge classification to be especially useful in discussing animal distributions on a broad scale. It is of particular value because its author was a distinguished tropical biologist with unparalleled experience in evaluating tropical vegetation and the system has been widely applied to the ecological mapping of vegetation, especially in Latin America but in other areas as well.

The Holdridge system utilizes three parameters—mean annual biotemperature (BT°), mean annual rainfall (in mm), and the ratio of mean annual potential evapotranspiration (PET) to mean annual precipitation—to define a series of approximately 115 major climates on earth. Holdridge refers to each of these units as a "life zone," but that term has usually been used by biologists for another concept developed by C. Hart Merriam (1890) based on altitudinal zonation related to temperature. Consequently I prefer to call the units "bioclimates" for clarity, since one of their defining parameters is biotemperature and since each unit is given a descriptor derived from the most characteristic vegetation type (climatic association or formation) that occurs under the stated weather conditions (fig. 2.5). It is, however, appropriate to refer to the altitudinal zones recognized in this arrangement as life zones. The underlying concept of this classification is that each bioclimate supports a distinctive vegetation adapted in physiognomy and structure to the bioclimate's unique temperature, precipitation, and water availability regimens.

The parameters of biotemperature and rainfall have already been defined. Evapotranspiration is a term that describes the origin of atmospheric moisture from the earth's surface, where water and ice are transformed into water vapor and may be carried upward by air movements. It combines two processes. The first involves evaporation of water vapor from water and ice surfaces and from various wet terrestrial surfaces. The second process is produced by transpiration, the uptake of water from the soil by living plants and its evaporation through leaf stomata. Over water surfaces evapotranspiration is not limited, and the values are termed potential evapotranspiration. Over land surfaces water is frequently a limiting factor, and the actual evapotranspiration is often lower than the potential. This is reflected by the fact that the oceans have a mean actual evapotranspiration rate of about 930 mm per year, while continents average 500 mm per year. However, actual evapotranspiration over land exceeds the potential over adjacent tropical oceans in some areas because of high precipitation values and the development of luxuriant broadleaf rainforests with their multiple tree strata.

On figure 2.5 the three parameters are scaled logarithmically and arranged isogonally. The two independent variables are mean annual biotemperature (BT°) and precipitation. The third parameter, the ratio of annual potential evapotranspiration (PET) to annual precipitation, is a dependent variable, since PET is calculated by multiplying the BT° by the constant 58.93 to give a value in millimeters (Holdridge et al. 1971). The PET ratio is an attempt to derive a biologically meaningful index of moisture availability; for example, a PET ratio of 1.00 indicates that mean annual precipitation equals PET. Where the PET ratio is less than 1.00 an excess of moisture is present, and where it is greater than 1.00 a water deficit is predicted. Where the PET ratio is low, plant and animal life is able to develop in its most exuberant forms. Where it is high, structural and physiological adaptations to conserve water (including periods of dormancy) are most fully developed by the vegetation and its animal associates.

In the bioclimate diagram the intersection of any two logarithmic isograms defines a boundary between bioclimates. Transitional situations are those formed by the six small triangles bordering the core of the bioclimate hexagon. Each bioclimate is denoted by a combination of words indicating its latitudinal region, altitudinal zone, and climatic plant association. The latitudinal regions are determined by mean annual biotemperatures, reduced to sea level (e.g., Rome, Italy, with a mean annual BT° = 15.5°C, lies in the warm temperate latitudinal region). Altitudinal zones are also temperature determined. Within any latitudinal region the altitudinal zone that extends upward from sea level to the next temperature boundary is called the lowland zone. Rome lies at 100 m in elevation and is in the lowland zone. Annual precipitation at this locale is 700 mm. Consequently Rome is in the Warm Temperate Lowland Dry Forest bioclimate. Helena, Montana, which may serve as another example, has a mean annual BT° = 8.8°C, has mean annual rainfall of 322 mm, and is at an altitude of 1,252 m. From the temperature and rainfall data the vegetation may be denoted as steppe. If the temperature data are compensated to sea level (5.6° per 1,000 m), we see that this site is in the Warm Temperate Latitudinal region (compensated BT° = 16°C). Because of its actual BT° and elevation it cannot be associated with the lowland altitudinal zone, which at this latitude has a maximum extent of 1,000 m. Consequently Helena has a Warm Temperate Montane Steppe bioclimate.

Within the Tropical Latitudinal region thirty-eight bioclimates may be recognized: eight lowland, seven premontane, seven lower montane, six montane, five subalpine, four alpine, and one of permanent snow. Each of these bioclimates, since they all occur in the Tropical Latitudinal region, properly should have "Tropical" in its name (e.g., Tropical Subalpine Paramo). However, since all the bioclimates represented in Costa Rica belong in this grouping, any bioclimate mentioned in this book, unless otherwise noted, will be a tropical one, and to save space and eliminate redundancy each will be denoted by its altitudinal and formation descriptors (in the cited case simply as Subalpine Paramo).

In the lowland life zone, where BT°s are uniformly high, seasonal patterns of precipitation correlate closely with total annual precipitation in most areas, and the limits of each bioclimate may be defined by a ratio of dry months (those receiving less than 100 mm of precipitation) to wet months as noted along the base of the bioclimate diagram (fig. 2.5). As an example, lowland dry forest bioclimates typically will have four to six dry months per year.

Other aspects of the temperature relations for the altitudinal belts need to be mentioned. The boundary between the premontane and lower montane zones seems to be defined by the frequency of low nocturnal temperatures (3 to 6°C) that injure cold-sensitive plants rather than by the occurrence of rare or occasional killing frosts in the lower montane life zone. Although drawn at 18°C on the diagram, this critical temperature line lies between 12 and 18°C locally, depending on air drainage, slope orientation, vegetative cover, and cloud cover in the dry season. Nights with killing frosts occur every year at locations having a mean annual BT° of 12°C or less. Obviously the number of nights with frosts increases with increasing altitude in the montane and subalpine zones.

In extratropical regions trees cannot grow in areas having mean annual BT°s of 3°C or lower latitudinally or altitudinally. These are areas where the soil is frozen permanently (permafrost) except for the superficial layer that thaws during the brief summer. The Boreal Latitudinal region and the Alpine Altitudinal zone have permafrost that prevents rooting and growth of trees and supports only a tundra vegetation of lichens, mosses, and small herbaceous annual plants. In the Tropics permafrost occurs altitudinally at the 3°C level, but the tree line lies at the 6°C level. Whereas the 3 to 6°C zone supports tree growth in the extratropics, this zone supports a series of special scrub, grasslands, and herbaceous formations in the Tropics. On a tropical mountain near the equator the tree line will be at about 3,500 m, and permafrost and tundra vegetation appear at 4,000 m elevation.

The Bioclimates of Costa Rica

In this section the principal climates of Costa Rica are characterized using the bioclimate system previously described. Within this system each bioclimate is named for the charac-

teristic vegetation to be expected under the defining climatic conditions and growing on soils typical for the climate (the zonal soil). This formation is the climatic association for the bioclimate (fig. 2.5). Under similar climatic conditions differences in soils, water relations (flooding, high water table, etc.), and atmospheric conditions (air drainage, prevailing winds, etc.) may produce other kinds of vegetation that constitute edaphic, hydric, and atmospheric associations differing from the climatic vegetation. Where such associations are of significance to amphibian and reptile distributions, they will be discussed in the section on vegetation and habitats below.

Herrera (1986) has presented a detailed description of the local climates of Costa Rica based on temperature, precipitation, and PET data and derived indexes of aridity, humidity, and the relation of precipitation to PET using the system of Thornthwaite (Mather 1974). Herrera defines ninety-five climates in this fashion grouped by eight humidity provinces —subhumid dry (A); subhumid humid (B); humid (C–F); perhumid (G); and superhumid (H)—that correspond to those in the Holdridge system. Note, however, that Herrera's method for calculating PET provides substantially higher values than Holdridge's formula. Consequently Herrera believes that perhumid (rainforest) areas as shown on Tosi's (1969) ecological map based on the Holdridge scheme are overestimated. Nevertheless the two approaches are comparable.

In the Holdridge system, as previously discussed, PET is calculated by multiplying the BT° by 58.93. In the Herrera calculation PET at 26°C is 1,710 mm (or 65.7 mm per °C) and decreases by approximately 41 mm/degree to the 12°C level; then it decreases by 48 mm to the 3°C level. The Holdridge calculation estimates a PET at 26°C as 1,532 mm. Therefore, in the range of BT°s from 12 to 26°C, Herrera's values will be 178 mm plus 17.93 mm/degree higher than Holdridge's; from 3 to 12°C, Herrera's values will be 178 mm plus 10.93 mm/degree higher.

The Herrera formulation appears to offer an excellent framework for intensive local analysis of the relations of climate and biota, but it is too fine-grained for the purposes of this book. For that reason I regard the climatic units recognized by Herrera to represent microclimates that reflect the diversity of geographic conditions in Costa Rica. Each of these microclimates forms a subdivision of a bioclimate.

Recently Herrera and Gómez (1993) have mapped the biotic units of Costa Rica based on a modification of the Holdridge (1967) system as influenced by their previous work (Gómez 1986b; Herrera 1986). In this revision they recognized fifty-five units based on a combination of temperature, rainfall, actual and potential evaportranspiration, and distribution of rainfall. Their classification may prove very useful for small-scale analysis, but it is too detailed to be useful at the level of coverage of the present account.

According to the strict application of the Holdridge

Table 2.3. Elevational Limits for Altitudinal
(Life) Zones in Costa Rica

Bioclimate	Atlantic and Northwest Pacific Area	BT°C	Southwest Pacific Area
Lowlands	0–500 m	>24	0–650 m
Premontane	500–1,500 m	18–24	650–1,800 m
Lower Montane	1,500–2,500 m	12–18	1,800–2,700 m
Montane	2,500–3,500 m	6–12	2,700–3,600 m
Subalpine	>3,500 m	3–6	>3,600 m

system, twelve bioclimates are recognizable in Costa Rica (fig. 2.5) based on temperature, precipitation, and relation of PET to precipitation. However, the marked difference between Pacific and Atlantic rainfall patterns requires that we recognize two additional bioclimates for the seasonally dry but high-rainfall climates of Pacific lowland Costa Rica. Principal differences among localities are related to altitude, and the vertical zones listed in table 2.3 are recognized and used in subsequent discussions and statements of distribution.

The altitudinal limits indicated are approximate, since they are temperature related and vary with slope exposure, cloud cover, wind direction, and air drainage. The lowland zone is the banana belt of many authors; the premontane is the coffee belt of most authors. Herrera's (1986) scheme delimits more or less equivalent zones, designated *muy caliente* (torrid) for the lowlands, *caliente* (hot) for the premontane, and *templado* (temperate) and *frio* (cold) for the upper two zones, montane and subalpine, respectively. Herrera and Gómez (1993) modify this system to recognize tropical, subtropical, temperate, cool temperate, and boreal zones.

The bioclimates, their principal climatic parameters, and their general distribution are briefly diagnosed below. See Herrera (1986) and Herrera and Gómez (1993) for a detailed characterization of Costa Rican microclimates. The following abbreviations indicate the rainfall regimens, where appropriate, that primarily control the bioclimates within an altitudinal zone: northern Pacific regimen (NP); Golfo Dulce regimen (GD); Atlantic coastal regimen (AC); and Atlantic slope regimen (AS). Climatic data for select sites within major bioclimatic zones are provided in figure 2.9.

Lowland Bioclimates

Pacific Region

Dry Forest: 1,100 to 2,000 mm annual precipitation; a long dry season (November through May), with five to six months having less than 100 mm of rainfall. Lowlands of northwestern Costa Rica, except for the central portion of the Península de Nicoya (NP).

Moist Forest: 2,000 to 4,000 mm annual precipitation; a definite dry season (December or January to March) of three to four months. Discontinuous in the lowlands of southwestern Costa Rica and the interior valleys of the Río Grande de Térraba (GD), and the central area of the Península de Nicoya, Llanuras de Guatusos (Atlantic drainage), and lower Pacific slopes (300–600 m) of the Cordillera de Guanacaste and Cordillera de Tilarán (NP).

Wet Forest: 4,000 to 8,000 mm annual precipitation; no dry season or only a short one (one to two months) (in January to March). Lowlands of southwestern Costa Rica (GD).

Atlantic Region

Moist Forest: 2,000 to 4,000 mm annual precipitation; reduced rainfall (less than 200 mm/month) during February and March and in September. Most of the Atlantic plains (AC).

Wet Forest: 4,000 to 8,000 mm annual precipitation; reduced rainfall (less than 200 mm/month) in February or March or both at some localities. Northeastern plains and the foothills all along the Atlantic slope (AC).

Premontane Bioclimates

Moist Forest: 1,000 to 2,000 mm annual precipitation; a long dry season of four to five months (December through April). Meseta Central, San Ramón valley, Valle de Candelaria, and higher ridges of the Península de Nicoya (NP).

Wet Forest: 2,000 to 4,000 mm annual precipitation; a definite three- to five-month dry season (during December to April). Pacific versant: slopes of the Cordillera de Tilarán and Cordillera de Guanacaste and those above the Valle de General and intermediate slopes around the Meseta Central and the slopes of the southern Cordillera de Talamanca and the tops of the Cordillera Costeña (NP); Atlantic slope: from Cartago southeastward along the northern margin of the Talamanca range (AS, but in rain shadow).

Rainforest: 4,000 to 8,000 mm annual precipitation; no dry season. Both slopes of the Cordillera de Guanacaste and Cordillera de Talamanca and windward Caribbean slopes of the Cordillera Central and Cordillera de Tilarán (AS).

Lower Montane Bioclimates

Moist Forest: 1,000 to 2,000 mm annual precipitation; a definite four- to five-month dry season (December to April). Three small areas in rain shadow near Zarcero, Alajuela Province, southwest of Cartago, Cartago Province, and on the south-facing slope of Volcán Irazú (NP).

Wet Forest: 2,000 to 4,000 mm annual precipitation; a definite four- to five-month dry season (December to April). Pacific slope of the Cordillera de Tilarán and Cordillera Central and the Dota valley (NP).

Rainforest: 4,000 to 8,000 mm annual precipitation; no

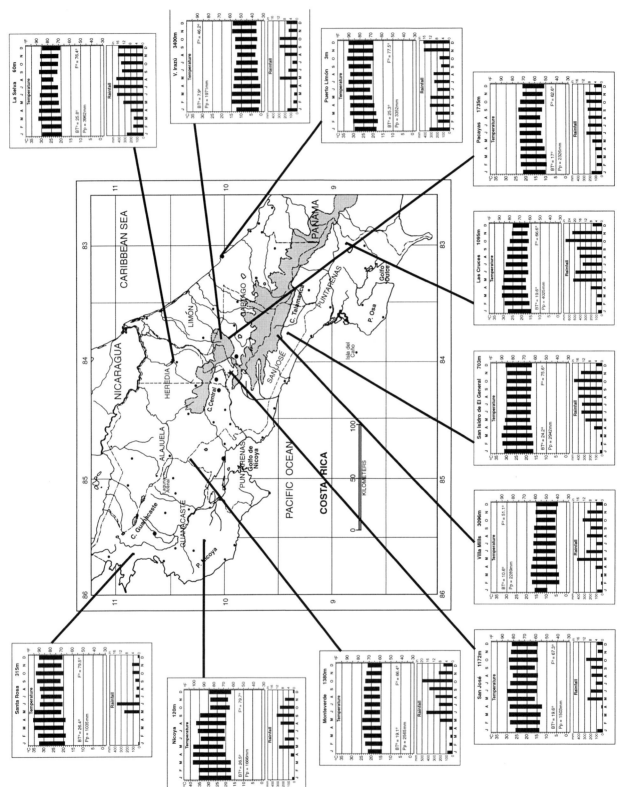

Figure 2.9. Climatographs for selected localities in Costa Rica. Temperature ranges are delimited by mean monthly maxima and minima; precipitation is based on mean monthly amounts. Dots represent other population centers as shown on figure 1.1 for reference.

dry season or only a short one (one to two months). Both slopes of the Cordillera de Guanacaste and Cordillera de Talamanca, windward slopes of the Cordillera Central, and the top of the Cordillera de Tilarán (AS).

Montane Bioclimates

Wet Forest: 1,000 to 2,000 mm annual precipitation; a three- to four-month dry season (December to April). Top of Volcán Irazú (NP).

Rainforest: 2,000 to 6,300 mm annual precipitation; no dry season or only a short one (one to two months). Upper slopes of the Cordillera de Talamanca and around the summits of Volcanes Poás, Barva, Irazú, and Turrialba (AS).

Subalpine Bioclimate

Pluvial Paramo: 2,000 to 7,400 mm annual precipitation; no dry season or only a short one (one to two months). Top of Cerro Chirripó Grande and Cerro Kamuk.

The bioclimates for Isla del Coco are of the lowland wet type (5,000 to 6,000 mm annual precipitation) from zero to 375 m in altitude, and premontane rainforest (6,000 to 7,000 mm annual precipitation) between 375 m and the island's peak at 634 m. There is a definite short period of low rainfall during February and March, when precipitation averages between 175 and 275 mm/month.

SOILS

Soils are produced by a complex of interactions over time among bioclimate, parent material, and the biota. Generally the occurrence of major kinds of soils broadly correlates with bioclimates, but variations in the composition of the parental material, age, and slope among other factors lead to many local differences. Soils are characterized based on three main features:

The soil profile: the composition and thickness of the various vertical layers (horizons)
Color
Physical properties
    Texture: clay, silt, sand, gravel, or loam mixture
    Structure: different kinds of aggregates of particles
    Acidity
Moisture content

Soils are classified based on the features above in a hierarchical system from more to less inclusive as orders, suborders, great groups, subgroups, families, and series (U.S. Soil Conservation Service 1975). Only the first two categories are of interest here, and the distribution of the most important of these in Costa Rica has been mapped in Hartshorn and Hartshorn (1982). Although seven soil orders contain-

ing sixteen suborders are represented in the republic, several occur rather locally and need not be discussed here. The most widespread soil orders in the republic are Inceptisols, Ultisols, and Entisols (fig. 2.10).

Inceptisols are soils of humid regions that have horizons lacking bases of iron or aluminum but retain some silicates. Two important suborders occur in Costa Rica. Tropepts are moderately fertile brownish to reddish soils, typical of the premontane zone but also found along lowland rivers. Andepts are the dark-colored fertile volcanic soils containing substantial amounts of silica found on the slopes of all the cordilleras.

Utisols are yellow, reddish, to dark red very acidic soils containing high concentrations of iron or aluminum, or both, with most of the silica leached out. These soils predominate throughout the humid lowlands and on the slopes of the Cordillera de Talamanca. They are essentially absent from the dry forest region of the northwestern Costa Rica, where most of the soils are Tropepts.

Entisols are soils with little or no development of horizons that may be light or dark in color. The most important suborder in Costa Rica is Aquents. These are soils of marshes, swamps, and other poorly drained areas that are most extensive in the most humid lowland coastal and riverine areas.

VEGETATION AND HABITATS

The following paragraphs describe the principal habitats where amphibians and reptiles are found in Costa Rica. For convenience the habitats have been divided into two categories, aquatic and terrestrial, with emphasis on differences in vegetation. In drawing up this review I have relied heavily on the works of Holdridge et al. (1971) and Hartshorn (1983) but have found Gómez's publications (1984, 1986a) of special value in organizing my own observations and in providing an overview of habitat diversity.

The delineation of aquatic from terrestrial habitats is somewhat arbitrary, since there is a gradation and interrelation between them. In addition, a number of amphibians and reptiles may continue to use both types of habitat, while others use one and not the other seasonally for feeding or reproduction. Also bear in mind that the transition between brackish and fresh water is gradual, and some kinds of vegetation range across the salinity gradient.

Gómez (1984, 1986a) recognized nineteen aquatic plant associations in Costa Rica, ten in saline situations (marine, littoral, or in brackish water) and nine in fresh water. Several of these are essentially similar in physiognomy and species composition, differing primarily in whether they are inundated permanently or seasonally. Not all of these associations are hospitable to representatives of the herpetofauna, and only those where amphibians and reptiles occur are described below. Where appropriate I have also given the

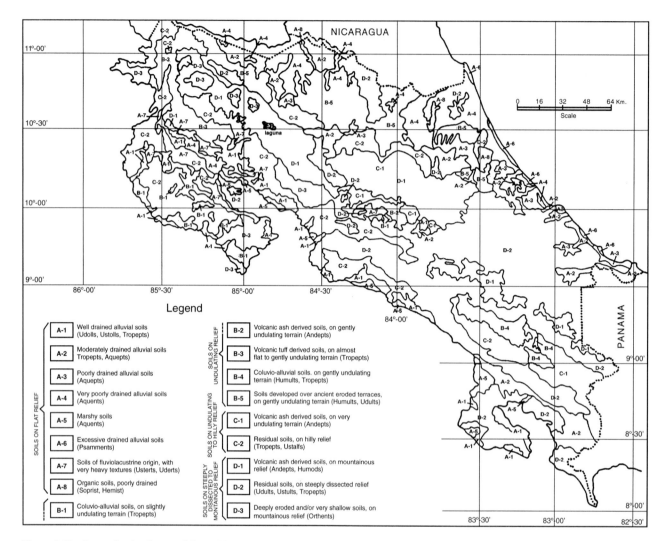

Figure 2.10. Generalized soil map of Costa Rica.

Spanish name for particular vegetation and habitat types to aid readers seeking the detail provided by Gómez (1984, 1986b) and to help in locating particular situations in the field. An illustration of one of the important aquatic habitats (plate 3) is also included. Additional aquatic habitats not included in Gómez's review that are important for our purposes are marine, temporary ponds and ditches, permanent ponds and lakes, small streams, and creeks and rivers.

Aquatic Habitats

Marine

The warm seas bathing both coasts of Costa Rica differ substantially in characteristics on the opposite coasts. The Caribbean is very saline, shows a very small tidal range (typically 0.6 m), and has little suspended particulate matter, which makes for very clear water with great visibility during most of the year. The coast consists mostly of long sand beaches, but coral formations occur north of Puerto Limón and at Cahuita. The Pacific is less saline and warmer than the Caribbean. Its tidal range is greater (typically 3.6 m) on the average, particularly in the enclosed Golfo de Nicoya and Golfo Dulce. Rocky reefs and headlands dominate the outer coasts but are intermixed with sandy beaches. Reptiles found in this habitat on both coasts are the several species of sea turtles that range far out to sea but return to lay their eggs on sandy beaches, and the crocodile, which usually remains close to shore. The completely pelagic yellow-bellied sea snake *(Pelamis)* is restricted to Pacific waters.

Salt Marshes

This kind of habitat is found along both coasts in areas of low relief where freshwater rivers and streams flow relatively slowly and gradually mix with ocean water. Because of tidal effects and seasonal flooding with fresh water this environment has a range of salinities, as is typical for estuarine situations. Salt marshes typically grow in areas where the substrate consists of several feet of decaying vegetation and mud.

Some portions are regularly flooded by high tides, others are only seasonally flooded by the highest tides and may also be affected by seasonal flushing with fresh water. Two kinds of vegetation occur under these conditions. The first *(marisma)* is dominated by salt-tolerant plants including *Acrostichum, Nymphaea,* and *Montichardia.* The second *(carrizal litoral)* is composed principally of reeds, sedges, and rushes and tends to occur where the water is brackish and there is only seasonal flooding. Very often the salt marsh is gradually replaced inland by freshwater marshes of similar structure. In many areas salt marshes occur adjacent to mangrove forests. This environment is not favored by many reptiles, and of course the high salinities prevent amphibians from breeding or living for long under these conditions. The crocodile is to be expected in such estuarine situations where it has not been eliminated by hunters.

Mangrove Swamp *(Manglar)*

This habitat has an extensive distribution on both coasts of Costa Rica and is very well represented on the Pacific coast (plate 3). The swamp is actually a tidal forest, since it is subject to the daily effects of the tides as well as to seasonal flooding by fresh water. In its original form this forest was spectacular, with individual trees more than 18 m tall. The red mangrove *(Rhizophora mangle)* grows on the seaward side of the forest. Farther inland where the substrate is more often exposed by the tides, the red mangrove is replaced by black mangroves *(Avicennia)* followed by the white mangrove *(Laguncularia racemose)* and finally by buttonwoods *(Conocarpus erectus)* in the highest high-tide zone. Because of the mangrove's commercial value, most of the large trees were cut many years ago, especially in the Golfo de Nicoya, and present stands often consist only of small or shrubby plants. Environmental conditions for amphibians and reptiles in the mangrove forest are similar to those described for salt marshes, except that the substrate is more shaded. The woody vegetation, however, provides conditions that allow more reptiles to use this habitat, and lizards of the genus *Norops* and arboreal snakes, *Leptodeira* and *Oxybelis,* are regularly found here. Individuals of several species of treefrogs apparently use the forest as a feeding site at night during the rainy season, but they return inland during the day to avoid further exposure to saline conditions.

Freshwater Marshes

Substantial areas of marshlands occur in Costa Rica and are particularly important features of the environment in the lowlands. A wide variety of plant species may occur in different marshes because of degree of permanency, soil drainage conditions, or fluctuations in water level. Many marshes have areas of open water and often have free-floating vegetation (e.g., water lilies or water hyacinths). Some marshlands in Costa Rica cover large areas, but many are only marginal zones along streams or larger bodies of fresh water.

Two major kinds of freshwater marshes may be recognized. The tule swamps *(carrizal),* are characterized by a homogeneous tall (1 m high) vegetation dominated by bullrushes *(Typha)* and reeds. The large Palo Verde swamp near the mouth of the Ríos Bebedero and Tempisque, at the head of the Golfo de Nicoya, is the best-developed example of this vegetation. The inland portions of the Palo Verde swamp dry up during the low-sun period but refill seasonally. The second common type is dominated by grasses, sedges, and various herbaceous plants and is called *pantano herbaceo* by Gómez (1986a). Marshes of this kind are found throughout the lowlands but are particularly well developed along the northeast Atlantic coast and on the Río Sierpe floodplain of southwestern Costa Rica.

In most of Costa Rica, except for the high, cold, and sterile volcanic lakes, there are no extensive areas of permanent open water (permanent lakes and ponds). The availability of water at such sites combined with the yearlong growing season leads to rapid modification by plant invaders. In the typical situation, once aquatic vegetation becomes established around the margins of a body of water, the lake or pond fills in rather rapidly. Vegetation expands and spreads out over the deeper areas of the system by using as a substrate an accumulation of dead and decaying plant material that forms floating mats on the water surface. After a time the entire surface is virtually covered by a nearly continuous raft of living plants rooted in the underlying dead material. Laguna Arenal (shown as a large, open lake on all Costa Rican maps) was formerly such a vegetation-covered lake, with practically no open water. The damming of the lake's outlet in the late 1970s as part of a major hydroelectric project and the subsequent rise in water level has formed a much enlarged open water impoundment that is as yet free of substantial surface vegetation. At a later stage of succession the accumulating organic matter gradually fills in the lake and reduces it to a shallow swamp, which will in turn be replaced by terrestrial conditions and vegetation. The Laguna de Corcovado is a swamp of this type (Boza 1978, 1986; Boza and Cevo 1998).

Freshwater swamps are especially used by a wide variety of seasonally breeding frogs, and many frog-eating snakes are concentrated at these sites during the same period. In permanent swamps caimans and a variety of frogs, turtles, and snakes are present throughout the year.

Temporary Ponds and Roadside Ditches

During the rainy season, especially in areas under the Pacific rainfall regimen, low-lying areas, poorly drained depressions, roadside ditches, and even tire tracks in the soil fill with water temporarily. When filled these areas become centers of frog breeding activity and concentrations of their snake predators. Turtles of the genera *Kinosternon* and *Rhinoclemmys* and small caimans are frequently found in these temporary situations.

Permanent Ponds and Lakes

These open-water habitats are relatively uncommon features of the Costa Rican environment. In the uplands there are a number of clear, sterile volcanic lakes, including Lago de Cote, Lagos Río Cuarto, Congo, and Hule, and the Lago del Poás. These lakes are acidic and do not appear to serve as habitats for amphibians or reptiles. Lowland lakes and ponds tend to be encroached on by aquatic vegetation (see discussion in freshwater marshes above), but where they occur, as at Lago Caño Negro in northeastern Costa Rica, a diversity of aquatic and semiaquatic amphibians and reptiles inhabit them and many frogs use their margins seasonally for breeding.

Small Streams and Creeks

These watercourses and their adjacent banks are important habitats and feeding and breeding sites for many frogs. In lowland areas many of these streams flow too rapidly or fluctuate too much in depth during the wet seasons to allow frog eggs and tadpoles to survive (except for some glassfrogs), but they are used for reproduction in the dry season. Very small stream courses within the lowland forests that are often reduced to trickles at some times of the year are favored sites for breeding by toads *(Bufo)*. Lizards of the genus *Basiliscus* are commonly found along stream margins, and the semiaquatic freshwater snake *(Hydromorphus)* occurs in small streams in the lowlands and on the Meseta Central. In the highlands a number of frogs breed in small, torrential streams and have evolved specialized egg-laying modes or tadpoles to allow survival in this environment. An important factor in montane stream habitats is the absence of native freshwater fishes at sites above 1,500 m.

Rivers

Costa Rica has a plenitude of rivers that originate for the most part from its tectonic backbone. In the uplands the gradient and flow of most large rivers keep amphibians and reptiles from using them as habitats, but stream banks often provide feeding grounds, nesting sites, and shelter for a variety of species. In the lowlands, rivers are the principal habitat for turtles, caimans, crocodiles, and the snake *Tretanorhinus*. Backwater areas are often frog breeding sites as well.

Swamp Forests

These are lowland forests that occur on alluvial soils or near river mouths and are inundated throughout the year or during the rainy season. The most widespread and distinctive of this forest type are associations dominated by palms *(Rhaphia, Manicaria, Astrocaryum,* and/or *Scheela* and *Bactris)* that occur along the Caribbean coast and the southwest Pacific area inland to the mangrove forests. Other swamp forests may have as dominants dicotyledonous trees of the genera *Mora, Pterocarpus, Erythrina, Prioria, Brosimum,* and *Pachira.* In the dry northwest a distinctive swamp forest *(mezquital)* grows on soils with a superficial saline cap, just inland from the mangrove, with *Acacia, Caesalpina coriaria, Prosopis juliflora,* and *Opuntia* as dominants. Aquatic and semiaquatic reptiles and seasonally breeding frogs from the adjacent terrestrial forests are commonly found in swamp forests that are in fresh water or minimally affected by the highest tides.

Terrestrial Habitats

The description of terrestrial habitats will center on the climatic associations for the fourteen bioclimates found in Costa Rica (plate 2), with some reference to other distinctive associations found under similar conditions, following Gómez (1986a). I have tried to list a few representative plants for each vegetation type depending heavily on Gómez (1986a) and on Holdridge et al. (1971) and emphasizing tree species. Obviously, when one considers the diversity in many of these forests (e.g., over 450 tree species now are known at the La Selva Biological Station on the Atlantic lowlands), these short lists provide an inadequate sense of the magnificence and structural complexity of the habitat. I hope the accompanying illustrations will help users of this book recognize some of the major forest types. In addition, I include descriptions of several habitats—coastal beaches and dunes, pastures, and secondary growth—that do not readily fit within a single bioclimate.

It is well to recall at the outset that Costa Rica, until rather recently, was covered by dense forests and that these forests formed the preeminent habitats for most amphibians and reptiles and indeed for most terrestrial animals. Although the dry forest area of the northwest and the relatively dry Meseta Central were extensively modified by human activity by the early nineteenth century, most other areas in the country remained heavily forested until the 1960s. Since that time deforestation has dramatically reduced forest cover everywhere except in the national parks and reserves and in the higher mountain areas. From a land that originally had more that 90% of its area in forest, Costa Rica has become a land of farms, coffee fincas, and pasture, with only about 20% of its area now supporting primary or secondary forests. Nevertheless, this section will be emphasize the characteristics of undisturbed or relatively undisturbed forest communities.

In general terms the forests of Costa Rica may be classified as broadleaf deciduous, semideciduous, or evergreen. In deciduous forests nearly all the trees lose their leaves during the dry season. Semideciduous forests have no more than 50% of the species losing most of their leaves during the dry season. Evergreen forests have most of the species of trees (at least 75%) evergreen. Costa Rica has no native stands of needle-leaf forests. Pines and cypresses do not occur naturally south of Nicaragua, although species of an allied gymnosperm family (Podocarpaceae: *Podocarpus*) are found in the cordilleras. In this book the term evergreen forest, unless otherwise qualified, refers to broadleaf evergreen associations.

Some or all tropical evergreen forests are often referred to as rainforests. This somewhat ambiguous term is usually applied to forest vegetation that characteristically is found where there is high annual rainfall and no long dry season. In the Holdridge (1967) system a rainforest is precisely defined as one that occurs where mean annual rainfall is four or more times as great as the potential evapotranspiration (the PET ratio is less than 0.25). Most tropical biologists and laypersons use the term rainforest less exclusively for all broadleaf evergreen tropical forests.

In this book rainforest (lowercase) is applied collectively to vegetation growing under high-rainfall conditions where there is no definite dry season. Rainforest in this sense includes the vegetation described below as Lowland Atlantic Moist and Wet Forest, Lowland Pacific Wet Forest and Premontane, Lower Montane and Montane Wet, and Rainforest. That portion of the evergreen seasonal Pacific-Atlantic Moist Forest found from around the Río Grande de Tárcoles northward is not considered to belong here. In cases where the more exclusive Holdridge terminology is employed Rainforest will be capitalized.

Two kinds of vegetation that are often recognized in tropical areas, riparian or gallery forest and cloud forest, are not confined to a particular bioclimate. Riparian forest is a general term used to describe vegetation that is established along the banks of streams. Members of this type of association require more moisture than species away from the stream in the same bioclimate. Usually the riparian species are a subset of plants derived from upstream sources in bioclimates with higher rainfall. Gallery forests form distinct and continuous corridors along stream margins. The term is usually applied to lowland situations and is most obvious in northwestern Costa Rica, where evergreen bands of forest follow stream courses in the midst of deciduous or semideciduous vegetation. In the Holdridge terminology such riparian vegetation would form a hydric association within a bioclimate. Some reptiles and amphibians are restricted to gallery forests in the dry forest zone, although they are abundant in other more humid forests.

The term cloud forest is often applied to forests that develop at an altitude where the temperature (6 to 10°C) causes water condensation that produces clouds, fog, and rain. This zone may be at any elevation, and its degree of development is related to the amount of water vapor in the air. Cloud forests usually occur where there are prevailing onshore winds that have their air masses uplifted along ocean-facing mountains. A forest of this kind that is familiar to many is the famous redwood forest of northern California and southern Oregon, which occurs from near sea level to about 800 m. During much of the year prevailing westerlies bring very moist air onto this mountainous area. During the dry summers when the amount of water vapor is lower, the forest is bathed with dense, dripping fog.

In Central America cloud forests develop principally on the windward slopes affected by the northeast trade winds. In Costa Rica these forests are maintained primarily by moisture that evaporates from the surface of the adjacent lowlands during the daily cycle and condenses into clouds at 1,500 to 2,000 m in elevation that are then pushed against the mountain slopes by prevailing winds. Although some authors (Myers 1969a; Campbell 1983) regard cloud forest as representing a coherent and distinctive biogeographic unit, Gómez (1986a) has demonstrated the local nature of these associations. In the Holdridge (1967) system cloud forests are regarded as atmospheric associations within bioclimates that, in Central America, usually develop in the lower portion of the lower montane life zone under the influence of strong prevailing winds. During much of the year these forests receive precipitation in the form of light mists. In the drier seasons, much of the time they are enveloped in dense, dripping fog.

Coastal Beaches and Dunes

These sandy areas frequently separate the adjacent forests from direct contact with the sea. Back from the bare sand a number of salt-resistant low-lying herbaceous plants and grasses *(Ipomoea pes-caprae, Canavalis maratima, Hibiscus tiliaceous, Andropogon littoralis)* occur, and coconut palms *(Cocos)* are usually present. Slightly inland, shrubby species (e.g., *Chrysobalanus* and *Cocolobis*) often form a dense bordering thicket of vegetation. Farther inland typical secondary growth species (e.g., *Heliconia* and *Gynerium*) also occur. This habitat is affected by intense insolation, and relatively few species of reptiles are found there. Most conspicuous are several lizards, *Ameiva quadrilineata, Basiliscus vittatus* (on the Caribbean), and *Cnemidophorus deppei* and *Sceloporus variabilis* (on the Pacific).

Pastures

These human-manipulated habitats are relatively permanent areas where exotic grasses have been planted for forage. Usually, scattered individual trees have been left standing and there are logs that are remnants of the destroyed forest. During the wet season temporary ponds often form in depressions. Frequently small streams wind through the pasture and are used for watering stock. Surprisingly, a number of lizards may be found on the trees or fallen logs and along fencerows, and the streams and ponds serve as breeding sites for some amphibians. Upland pasturelands generally are modified oak forests, and several kinds of salamanders and snakes may be found under rocks or logs in these fields.

Lowland Dry Forest

This forest is the southernmost extension of a more or less continuous lowland belt running southward along the Pacific coast from southern Sonora, Mexico (plate 4). It is also the formation that has been under heaviest human impact in Costa Rica and elsewhere in Mesoamerica (Janzen 1986) for over four hundred years, and it has been so converted and modified that little of its original vegetation persists. Dry forest habitat is relatively easy to clear by cutting and burning, and grain crops and cotton grow well in this environment. It is also an area of major cattle production, and the pastures, which consist mostly of introduced grasses, are maintained by seasonal burning. In addition, because of the subhumid climate, different soil conditions of drainage and nutrient content have an obvious effect on vegetation, producing a number of distinctive plant associations. These factors (human disturbance and soils) make the dry forest region one of considerable original complexity simplified by human management practices. The most typical and widespread forest in the region is described by Hartshorn (1983) as follows (plates 5–6):

> The Tropical Dry Forest is a low semideciduous forest with only two strata of trees. Canopy trees are usually 20–30 m tall, with short, stout trunks and large, spreading, flat-topped crowns, usually not in lateral contact with each other. Many canopy trees have thin, often compound leaves that are dry-season deciduous. Bipinnately leaved mimosoid and caesalpinioid leguminous trees are the most conspicuous canopy component. Understory trees are 10–20 m tall, with slender crooked or leaning trunks and small open crowns with more evergreen species than in the canopy. Rubiaceae is a prominent understory family. The shrub layer is 2–5 m tall, dense in openings, often multiple stemmed, and armed with thorns or spines. The ground layer is sparse except in openings. Woody vines are common, but herbaceous vines are uncommon. Epiphytes are occasional, with bromeliads the most conspicuous.

Characteristic species include *Achras zapota, Anacardium excelsum, Andira inermis, Annona holosericea, Apeiba tibourbou, Bombacopsis quinatum, Bursera simaruba, B. tomentosa, Byrsonima crassifolia, Calycophyllum candidissimum, Casearia aculeata, Cecropia peltata, Ceiba pentandra, Cochlospermum vitifolium, Dalbergia retusa, Dipterodendron coastaricense, Enterolobium cyclocarpum, Erythroxylon havanense, Ficus glabrata, Godmania aesculifolia, Guazuma ulmifolia, Hippocratea obovata, Hymenaea courbaril, Karwinski calderoni, Lonchocarpus costaricense, Luehea speciosa, Lysiloma seemannii, Machaerium biovalatum, Pseudobombax septenatum, Sapranthus palanga, Simaruba glauca, Swietenia humilis, Tabebuia* spp., and *Xylosma flexuosum.*

On well-drained soils a lower, two-strata forest is found. The canopy is 15–25 m tall, with a lower layer 8–15 m high. Otherwise similar to the semideciduous forest, this association is composed of fewer tree species, a subset of deciduous forms from the semideciduous association. Another association that develops on pumice-ash soils is a species-poor evergreen oak forest that comprises primarily *Quercus oleoides.* This association has been seriously affected by human activity and replaced by pasture, but scattered stands occur as far south as Bagaces in Guanacaste Province. Other associations found within the dry forest area (Gómez 1986a) include shrubland types: *sonzocuital,* dominated by *Acacia* and *Crescentia alata* on Vertisols, and thorn woodland *(matorral espinoso),* found on saline soils and composed of a mixture of genera from the adjacent shrublands along with cacti and other thorny plants. Gómez (1986a) also describes two savanna associations, a *Trachypogon*-dominated short bunchgrass savanna and one found on Entisols where open areas covered with native grasses are interspersed with concentrations of shrubs and trees within the dry forest zone. Hartshorn (1983) and Janzen (1986) believe that these open grassy areas are artifacts of annual burning that reduce the natural cover to fire-resistant species (e.g., *Byrsonima crassifolia* and *Curatella americana*). In either case the savannas are now very characteristic features of the dry forest region.

As may be seen from this brief review, the seasonal dry lowlands of Costa Rica have a variety of habitats for amphibians and reptiles, and a substantial number of species are restricted to this zone. An important aspect of future research will be ascertaining how closely individual species are tied ecologically to particular plant associations.

Lowland Pacific-Atlantic Moist Forest

This is a tall multistratal, semideciduous forest in which most emergent trees are evergreen (plate 7). Where it is best developed the canopy trees are 40–50 m tall, with slender, smooth boles and light-colored bark. The boles are unbranched for 25–35 m, and the tree crowns are mostly wide. Most large trees have high, thin buttresses. Subcanopy trees usually have narrow crowns and are up to 30 m tall. The understory is rich in palms. Understory trees are mostly 8–20 m high with round to conical crowns; leaves often have long drip tips. The shrub layer consists of dwarf palms and giant broadleaf herbs; the ground layer is relatively bare. Herbaceous vines, woody lianas, and epiphytes are abundant. Characteristic species include *Alchornea costaricensis, A. latifolia, Aspidosperma megalocarpum, Bellucia costaricensis, Brosimum utile, B. alicastrum, Cassia spectabilis, Cecropia* spp., *Ceiba pentandra, Compsoneura sprucei, Cordia alliodora, Didymopanax morototoni, Dipterodendron costaricense, Geothalsia meiantha, Hampea appendiculata, Hieronyma alchorneoides, Inga oerstediana, Jacaranda lasiogyne, J. copaia, Lacistema aggregatum, Ladenbergia brenesii, Leandra lasiopetala, Luehea seemannii, Pithecolobium arboreum, Pourouma aspera, Saccoglottis amazonica, Sloanea laurifolia, Sterculia recordiana,* and *Virola sebifera.*

Lowland forests of this type are affected by the seasonality of rainfall on the Pacific versant and along the southern borderlands of Lake Nicaragua. In the areas of occurrence in northwestern Costa Rica (and on the Atlantic slope Llanuras de Guatusos), the long dry season produces a forest type that grades into the semideciduous dry forest. Farther south in the Valle de General this vegetation develops in the rain shadow provided by the Cordillera Costeña. This formation supports a great diversity of amphibian and reptile life, with specialists for existence at every vertical level in the forest (fossorial, leaf litter, shrub, understory, and canopy). However, it must be noted that the Península de Nicoya forests of this type closely resemble the adjacent dry forests in herpetofaunal composition rather than lowland moist forests found elsewhere in the country.

Within this bioclimate where there is poor drainage and a high water table, a moist savanna association is also found, principally at a number of sites in the Valle de General region. This savanna resembles the llanos of northern South America in physiognomy, with widely scattered trees, especially the palm *Scheelea rostrata*, as well as *Byrsonima*, *Cochlospermum*, and *Ceiba*, on a grassy, almost marshy plain. These probably represent natural savannas, although much modified by human activity, most recently in the form of extensive pineapple plantations.

Lowland Pacific Wet Forest

By far the richest habitat in terms of species diversity and biotic interactions are the lowland wet forests of southwestern Costa Rica (plate 8). Hartshorn (1983) characterizes this type of forest as follows:

> Tropical Wet Forest is a tall, multistratal, evergreen forest. A few canopy species are briefly deciduous, but this does not change the overall evergreen aspect of the forest. Canopy trees are 45–55 m tall, with round to umbrella-shaped crowns, and have clear boles to 30 m and attaining 100–200 cm dbh [diameter at breast height]. Smooth, thin, light-colored bark and high buttresses are common. Subcanopy trees are 30–40 m tall, with round crowns and slender trunks, generally lacking buttresses. Understory trees are 10–25 m tall, with narrow conical crowns and slender boles, often twisted or crooked, usually with smooth, dark bark, occasionally cauliflorous. Stilt-rooted palms are often abundant. Shrub layer is 1.5–2.5 m tall with abundant dwarf palms; unbranched treelets and giant broad-leaved herbs are occasional. The ground layer is sparse, with a few ferns and *Selaginella*. Woody lianas are not common, and epiphytic shrubs and strangling trees are rare.

Plants typical for this formation include *Achras zapota*, *Alchornea costaricensis*, *Anacardium excelsum*, *Andira inermis*, *Apeiba aspera*, *A. tibourbou*, *Aspidosperma megalocarpum*, *Billia colombiana*, *Brosimum alicastrum*, *B. terrabanum*, *B. utile*, *Bursera simaruba*, *Calatola costaricensis*, *Calophyllum brasiliense*, *Carapa guianensis*, *Cariniana pyriformis*, *Castilla elastica*, *Ceiba pentandra*, *Chimarris latifolia*, *Coccoloba standleyana*, *Compsoneura sprucei*, *Couratari panamensis*, *Cymbopetalum costaricense*, *Dialium*

guianense, *Dialanthera otoba*, *Dipterodendron costaricense*, *Dussia cuscatlanica*, *Ficus nymphaefolia*, *Grias fendleri*, *Guarea* spp., *Hieronyma alchorneoides*, *Hernandia sonora*, *Hymenaea courbaril*, *Inga coruscans*, *Lacistema aggregatum*, *Lacmellea panamensis*, *Lonchocarpus* spp., *Minquartia guianensis*, *Pachira aquatica*, *Parkia pendula*, *Peltogyne purpurea*, *Pithecolobium arboreum*, *Platymiscium pinnatum*, *Pourouma aspera*, *Pouteria neglecta*, *Protium* spp., *Qualea paraensis*, *Rheedia madruno*, *Saccoglottis amazonica*, *Sapium* spp., *Schizolobium parahybum*, *Sloanea laurifolia*, *Sterculia recordiana*, *Swartzia simplex*, *Symphonia globulifera*, *Tachygalia versicolor*, *Talisia nervosa*, *Terminalia amazonia*, *T. bucidolies*, *Tococa grandifolia*, *Vantanea barbouri*, *Virola* spp., *Vochysia ferruginea*, *V. hondurensis*, and *Welfia georgii*.

A similar kind of forest is found in the lowlands on Isla del Coco, but with a reduced plant species diversity and lower canopy. Lowland Pacific Wet Forest occurs over most of the Pacific lowlands south of the Río Grande de Tárcoles and is most fully developed around the Golfo Dulce. As might be expected, in correlation with the overall diversity in this forest a very rich herpetofauna is represented, with a minimum of 125 species of amphibians and reptiles occurring at relatively well collected sites.

Lowland Atlantic Moist Forest

The Atlantic versant version of this forest occurs under more constant humid conditions than those affected by the Pacific rainfall regimen. Although rainfall is reduced during January to March, this does not constitute a significant dry season. The major features for this environment have been succinctly summarized by Hartshorn (1983):

> Tropical Moist Forest is a tall, multistratal, evergreen forest. Canopy trees are 40–50 m tall, mostly with wide crowns and tall, slender boles unbranched for 25–35 m, mostly less than 100 cm dbh, often with high, thin buttresses and smooth, light-colored bark. Subcanopy trees are to 30 m tall, mostly with narrow crowns. Palms, especially *Scheelea rostrata*, are usually abundant, except in cool transitional areas. Understory trees are mostly 8–20 m tall, with round to conical crowns; leaves often have long drip tips. The shrub layer consists of dwarf palms and giant broad-leaved herbs. The ground layer is generally bare except for occasional ferns. Herbaceous vines and woody lianas are abundant, as are epiphytes.

Characteristic plants include *Acalypha* sp., *Alfaroa costaricensis*, *Allophylus psilospermus*, *Anacardium excelsum*, *Ardisia* sp., *Brosimum costaricanum*, *Bursera simaruba*, *Calophyllum brasiliense*, *Carapa guianensis*, *Casilla* sp., *Cephaelis* sp., *Chamaedorea* sp., *Chrysophyllum panamense*, *Compsoneura sprucei*, *Coussapoa* sp., *Croton glabellus*, *C. panamensis*, *Cymbopetalum costaricense*, *Cyphomandra* sp., *Dendropanax gonatopodus*, *Dussia macrprophyllata*,

Eugenia spp., *Ficus tonduzii*, *Genipa caruto*, *Goethalsia meiantha*, *Guarea aligera*, *Hasseltia floribunda*, *Inga marginata*, *I. sapinodoides*, *Inga* sp., *Lacistema aggregatum*, *Lafoensia punicifolia*, *Lonchocarpus rugosus*, *Luehea seemannii*, *Myriocarpa* sp., *Ouratea tuerckheimii*, *Palicourea* sp., *Persea pallida*, *Piper candelarianum*, *Pithecolobium* sp., *Protium* sp., *Pseudolmedia* sp., *Psidium friedrichsthalianum*, *Psychotria* sp., *Rheedia edulis*, *Rheinhardtia* sp., *Simaruba amara*, *Sorocea affinis*, *Spondias mombin*, *Terminalia lucida*, *Trema micrantha*, *Trichilia* sp., *Trophis* sp., *Virola sebifera*, *Xylosma excelsum*, and *Zexmenia frutescens*.

Species diversity is very high at most sites in this forest, with about forty-eight species of amphibians and eighty-seven species of reptiles present.

Lowland Atlantic Wet Forest

Hartshorn (1983) characterizes this kind of vegetation as follows (plate 9):

> Tropical Wet Forest is a tall, multistratal, evergreen forest. A few canopy species are briefly deciduous, but this does not change the overall evergreen aspect of the forest. Canopy trees are 45–55 m tall, with round to umbrella-shaped crowns, and have clear boles to 30 m and attaining 100–200 cm dbh. Smooth, thin light-colored bark and high buttresses are common. Subcanopy trees are 30–40 m tall, with round crowns and slender trunks, generally lacking buttresses. Understory trees are 10–25 m tall, with narrow conical crowns and slender boles, often twisted or crooked, usually with smooth, dark bark, occasionally cauliflorous. Stilt-rooted palms are often abundant. Shrub layer is 1.5–2.5 m tall with abundant dwarf palms; unbranched treelets and giant broad-leaved herbs are occasional. The ground layer is sparse, with a few ferns and *Selaginella*. Woody lianas are not common, and epiphytic shrubs and strangling trees are rare.

Most of the plants listed for the Atlantic Moist Forest occur in the Wet Forest, and other typical ones are *Alchornea costaricensis*, *Allophylus psilospermus*, *Anaxagorea costaricensis*, *Astrocaryum alatum*, *Bravaisia integerrima*, *Brosimum panamense*, *Capparis pittieri*, *Carpotroche platypetra*, *Casearia* spp., *Cespedezia macrophylla*, *Cordia alliodora*, *Cynometra retusa*, *Dendropanax arboreus*, *Dipteryx panamensis*, *Erythrina cochleata*, *Gloeospermum diversipetalum*, *Hedyosmum calloso-serratum*, *Hernandia* spp., *Hura crepitans*, *Jacaratia* spp., *Laetia procera*, *Lecythis costaricensis*, *Mortoniodendron membranaceum*, *Pentaclethra macroloba*, *Sloanea medusula*, *Sterculia apetala*, *Stryphnodendron excelsum*, *Swartzia simplex*, *Terminalia oblonga*, *Tovomita nicaraguensis*, *Venconcibea pleiostemona*, *Warscewiczia coccinea*, and *Zanthozylum panamense*.

This is the richest habitat for amphibians and reptiles on the Atlantic versant and is comparable in species diversity to the Golfo Dulce forests (plate 10).

Premontane Moist Forest

This forest grows at intermediate elevations in central Costa Rica and is transitional between the lowland dry and moist habitats of northern Costa Rica and the humid vegetation of the uplands. Hartshorn (1983) describes it as follows:

> Tropical Premontane Moist Forest is a two-layered, semideciduous, seasonal forest of medium height. Canopy trees are mostly dry-season deciduous, about 25 m tall, with characteristically broad, flat, or umbrella-shaped crowns and relatively short, stout trunks, often with thick, fissured, or flaky bark. Compound leaves are very common. Understory trees are 10–20 m tall, evergreen, with round to conical crowns and short, twisted, or crooked boles with smooth or moderately rough bark. The shrub layer is dense, 2–3 m tall, of single- or multiple-stemmed woody plants, some armed with spines. The ground layer is sparse. Epiphytes are rare. Tough, supple, thin-stemmed woody vines are abundant.

This forest has been eliminated for all practical purposes by human activity, since it was the native vegetation of the prime coffee-producing areas of Costa Rica and is now the area of greatest human population density, the Meseta Central. For this reason it is difficult to list characteristic plants, although species of the following genera are still represented in scattered areas, mostly in steep gullies: *Anacardium*, *Aracia*, *Bombacopsis*, *Bursera*, *Byrsonima*, *Cecropia*, *Cedrela*, *Ceiba*, *Cordia*, *Enterolobium*, *Eugenia*, *Gauzuma*, *Godmania*, *Luehea*, *Lysiloma*, *Persea*, *Pseudobombax*, *Randia*, *Tabebuia*, *Trichilia*, *Xylosoma*, and *Zanthoxylum*.

Although urbanization has taken its toll, several species of frogs, lizards, and snakes (e.g., *Hyalinobatrachium fleischmanni*, *Agalychnis annae*, *Norops intermedius*, *Sceloporus malachiticus*, *Ninia maculata*, and *Micrurus nigrocinctus*) are frequently found in gardens and vacant lots even within the city limits of San José. Many other species still persist in scattered habitats intermixed with coffee fincas and suburban sprawl.

Premontane Wet Forest

This kind of forest is found along the lower slopes of the mountains where there is a long dry season because of the northern Pacific regimen of rainfall or a rain shadow. The vegetation has been described by Hartshorn (1983) as follows (plate 11):

> Tropical Premontane Wet Forest is medium to tall, semi-evergreen forest with two or three strata, with a few canopy species dry-season deciduous. Canopy trees are mostly 30–40 m tall, with mostly round to spreading crowns and relatively short clear boles. Buttresses are

common but small. Bark is mostly brown or gray, moderately thick and flaky or fissured. Leaves are often clustered at the twig ends. Understory trees are 10–20 m tall with deep crowns and smooth, often dark bark. Stilt roots and long, strap-shaped leaves are common. Tree ferns are occasional. The shrub layer is 2–3 m tall and often dense. The ground layer is generally bare except for ferns. Epiphytes are present but not conspicuous. Climbing herbaceous vines are abundant. Most trees are covered by a thick layer of moss.

Trees commonly found in the forest are *Alchornea latifolia, Brosimum panamense, B. utile, Calophyllum brasiliense, Casearia arborea, C. javitensis, Cassipourea guinanensis, Cecropia obtusifolia, Cephaelis elata, C. tomentosa, Clusia pithecobia, Cordia toqueve, Croton glabellus, Dendropanax arboreus, Enterolobium guatemalense, Eugenia* sp., *Euterpe panamensis, Ficus* sp., *Guarea* sp., *Heisteria densifrons, Hirtella racemosa, Inga* sp., *Ladenbergia brenesii, Laplacea semiserrata, Miconia* spp., *Mosquitoxylum jamaicense, Mouriri cyphocarpa, Nectandra* sp., *Ocotea cernua, Ouratea lucens, Pourouma aspera, Protium copal, Psychotria* spp., *Rapanea pellucido-punctata, Roupala complicata, Simaruba amara, Symphonia globulifera, Terminalia amazonia, Tococa guyanensis, Virola sebifera, Vochysia ferruginea, V. hondurensis,* and *Zinowiewia integerrima.*

The premontane forests of this type and the following support a wide variety of amphibians and reptiles. Because this life zone lies in the foothills region, many lowland species range upward into the premontane zone, and a number of lower montane forms range downward into the zone as well. A diverse group of stream-breeding frogs and members of the genus *Eleutherodactylus* are prominent components of the herpetofauna. Many anoles and other lizards are represented, and both lizards and frogs are among the animals preyed on by a substantial array of snakes.

Premontane Rainforest

This kind of forest has a wide distribution in Costa Rica on all mountain slopes (plates 12–13), with the major exceptions of the Pacific versant slopes of the Cordillera de Tilarán and the south-facing slopes of the Cordillera Central. It is in a zone of constant high precipitation, and Coen (1983) reports a site in this rainforest where rain falls 320 to 359 days each year. Hartshorn (1983) provided the following characterization:

Tropical Premontane Rain Forest is an evergreen forest, intermediate in height, with two or three strata. Canopy trees are mostly 30–40 m tall, with round or umbrella-shaped crowns and straight branches. Buttresses are common but small. Bark is brown, black, or gray, moderately thick, mostly flaking or fissured. The subcanopy is very dense, with trees 15–25 m tall, having slender trunks often unbranched for most of their length; nar-

row, round to conical crowns; and thin light- or dark-colored bark. Palms are common in well-drained situations. Understory is also very dense and may be difficult to distinguish from the subcanopy stratum. Understory trees are 8–15 m tall, often with leaning, crooked, or twisted trunks and relatively long crowns with horizontal branches; many trees have stilt roots. Tree ferns are common in the understory. The shrub layer is 2–3 m tall and very dense. Dwarf palms are uncommon in the shrub layer. The ground layer consists of a nearly complete cover of ferns, *Selaginella* and broad-leaved herbs, often with bluish leaves. Epiphytes, woody vines, and herbaceous climbers are very abundant. Moss and epiphytes cover practically all surfaces.

Typical trees and shrubs include *Alchornea costaricensis, Alnus jorullensis, Billia colombiana, Brosimum* sp., *Brunellia costaricensis, Calocarpum mammosum, Casearia* sp., *Cassia* sp., *Cecropia polyphlebia, Cedrela tonduzii, Cestrum panamense, Clethra gelida, Clusia alata, C. pithecobia, Cornus disciflora, Cornutia grandifolia, Croton glabellus, Cupania* sp., *Cyathea divergens, C. mexicana, Dendropanax* sp., *Engelhardtia mexicana, Eugenia* spp., *Ficus tonduzii, F. torresiana, Guarea* sp., *Hampea* sp., *Heisteria* sp., *Heliocarpus appendiculatus, Inga punctata, Mauria* sp. *Melisoma vernicosa, Miconia* sp., *Myrica arborescens, Myriocarpa* sp., *Nectandra panamensis, Ochroma lagopus, Olmedia falcifolia, Piper* sp., *Pothomorphe peltata, Protium* sp., *Quercus* spp., *Randia* sp., *Rapanea ferruginea, Sapium jamaicense, Sloanea* spp., *Sorocea trophoides, Tabernaemontana aphlebia, Tovomitopsis psychotriaefolia, Trichilia pittieri, Ulmus mexicana, Viburnum* sp. *Virola* sp., *Vitex* sp., *Weinmannia* sp., and *Zanthoxylum* sp.

The vegetation on the upper slopes of Isla del Coco is best regarded as a Premontane Rainforest.

Lower Montane Moist Forest

This formation is restricted to small areas in rain shadows adjacent to the Meseta Central. Hartshorn's (1983) résumé of its features is as follows:

Tropical Lower Montane Moist Forest is an open evergreen forest of intermediate height with two tree strata. Canopy trees are mostly *Quercus*, 30–35 m tall, with heavy, gnarled branches and thick, twisted boles and thick, scaling, or rough bark. The understory is fairly open, with evergreen trees to 20 m in height, having slender trunks and round to conical crowns. The shrub layer is 2–5 m tall, fairly dense, with soft-wooded plants and often with large leaves. The ground layer is mostly open, with scattered broad-leaved herbs and grasses. Although a few epiphytic trees occur, epiphytic herbs and mosses are inconspicuous.

Characteristic plants in this much disturbed formation are *Alnus jorullensis, Ardisia* sp., *Bocconia fratescens, Cestrum*

Box 2.1
Paso de Desengaño

The Paso de Desengaño is on the continental divide between Volcán Poás and Volcán Barva and provides a passage at 1,080 m from the Meseta Central Occidental to the Atlantic lowlands along the Río Sarapiquí. Its name is from the Spanish word *desengañar,* which means to be de-illusioned—to lose one's illusions or be disabused of error. Appropriately, the Río Desengaño flows from the pass to the Pacific, and the Quebrada Desengaño is a tributary of the eastward-flowing Río Sarapiquí (plate 13).

In earlier times the road across the pass followed the route of an even more ancient oxcart road, and until rather recently it was the only way to travel to the world-famous La Selva Biological Station of the Organization for Tropical Studies (OTS) near Puerto Viejo de Sarapiquí. In its traverse of the Cordillera Central this route passes through five major bioclimatic zones. Formerly the area of the Meseta Central at the foot of the pass was covered by Premontane Moist Forest, but over a century ago this area was pretty much converted to coffee production. The road quickly climbs out of the coffee zone into Premontane Wet Forest with its luxuriance of ferns, climbing vines, and moss-covered trees, and then on through Lower Montane Wet Forest. At the top of the pass and down onto the Atlantic slope the road follows the rim of the Río Sarapiquí gorge through a tall closed canopy of Lower Montane Rainforest, still intact on slopes away from the road. Farther down the slope the road follows the rim of the Río Sarapiquí gorge, surrounded by an extraordinarily rich Premontane Rainforest with its closed canopy, numerous tree ferns, dense understory, and overwhelming diversity of epiphytes. In this zone one passes by a magnificent waterfall (El Salto de La Paz Grande) and then continues down to the Atlanta plains.

I crossed the Pass of Desengaño on my first visit to Costa Rica in 1960. Since then I have crossed it many times. Each passage was different, often enveloped by clouds driven into the pass by the northeast trade winds, sometimes in driving rain and occasionally in bright sunlight on a clear day when the Atlantic plains and Caribbean Sea were visible from the top of the pass. For me the pass has come to symbolize my Costa Rican experience: The ever-changing interaction of light and shadow on the forest, the sudden showers, the unexpected fog or landslide, the earthquakes and volcanic eruptions, and the discovery of plants and animals new to my experience that, in the words of Walt Whitman, "bring me tokens of myself."

aurantiacum, Citharexylum lankesteri, Cleyra theaeoides, Clusia sp., *Cornus disciflora, Fuchsia arborescens, Garrya laurifolia, Ilex pallida, Meliosma irazuensis, Ocotea seibertii, Oreopanax capitatum, O. xalapense, Phoebe mollicella, Quercus aaata, Q. eugeniaefolia, Rapanea guinanensis, Rhacoma tonduzii, Rhamnus capreaefolia, Senecio copeyensis, Solanum* sp., *Styrax polyneurus, Urera caracasana, Viburnum stelato-tomentosum, Xylosma intermedium,* and *Zanthoxylum chiriquinu.*

Lower Montane Wet Forest

This forest is found under conditions of relatively low rainfall on Pacific slopes. It lies under the northern Pacific regimen of precipitation in the Cordillera de Tilarán and Cordillera Central but is at least partially an effect of rain shadow where it occurs in the northern Cordillera de Talamanca. Hartshorn's (1983) characterization is as follows:

> Tropical Lower Montane Wet Forest is an evergreen forest of intermediate height with two tree strata. Canopy trees are mostly 20–25 m tall, but some *Quercus* are taller, with short, stout trunks dividing into numerous long, heavy, twisting, ascending branches, producing wide, umbrella-shaped, billowing crowns. Buttresses are uncommon. Bark is thick, mostly flaking or fissured. The understory is fairly open, with trees 5–10 m tall, having spreading crowns. The shrub layer is relatively dense, 2–3 m tall, and palms are uncommon. The ground layer is well covered with ferns, begonias, aroid vines, and a thick layer of moist, rotting leaves. Small orchids, bromeliads, and ferns are common epiphytes. A thin layer of moss grows on tree trunks. Herbaceous vines, especially Araceae, are common at and near ground level. Large, coiled lianas are occasional to common.

Common plants found in this kind of forest include *Aiouea costaricensis, Ardisia* sp., *Casearia* sp., *Cedrela tonduzii, Cestrum aurantiacum, Chiococca phaenostemon, Citharexylum donnell-smithii, Citronella costaricensis, Conostegia oerstediana, Cornus disciflora, Ehretia austen-smithii, Eugenia storkii, Ficus* sp., *Guarea* spp., *Hasseltia floribunda, Hieronyma poasana, Leandra costaricensis, Mastichodendron capiri, Mauria biringo, M. glauca, Meliosma glabrata, M. irazuensis, Microtropis occidentalis, Ocotea seibertii, Oreopanax xalapense, Palicourea angustifolia, Persea americana, Phoebe mollicella, P. valeriana, Picramnia carpinterae, Pithecolobium costaricense, Prunus annularis, Prunus* sp., *Quercus oocarpa, Randia karstenii, Rhacoma tonduzii, Rhamnus capreaefolia, Roupala complicata, Sapium sulciferum, Sloanea megaphylla, Sorocea trophoides, Symplocos costaricana, Trema micrantha, Trichilia havanensis, Turpinia occidentalis, Viburnum costaricanum, Zanthoxylum limoncello,* and *Zinowiewia integerrima.*

This kind of forest and the one described next have the following conspicuous representatives: a rich complement of salamanders, frogs, including many that breed in torrential streams, several anole lizards, a number of small snakes (*Geophis, Ninia,* and *Rhadinaea*) and the arboreal pitviper *Bothreichis nigroviridis.*

Lower Montane Rainforest

Forests of this kind have often been called cloud forests, as discussed earlier, because they lie at an elevation where clouds form and may be driven against mountain slopes or ridges by prevailing winds (plate 14). Gómez (1986a) has described the elfin cloud forest of this type that occurs on windswept ridges. Premontane rainforest areas, or at least their upper elevations, have also been denoted by this not well defined term. Hartshorn (1983) provides the following summary of this vegetation type:

> Tropical Lower Montane Rain Forest is an evergreen forest of low to intermediate height, with two tree strata. Canopy trees are mostly 25–30 m tall, but *Quercus* may reach 50 m, having short, stout, often twisted trunks with rough, dark bark. Branches are thick, sinuous, and relatively short. Crowns are relatively small and compact. Buttresses are uncommon. Understory stratum is often dense, with trees 10–20 m tall. Trunks are slender, straight or sinuous, with small, brushlike crowns of twisted branches. Bark is smooth, thin, and mostly dark. Suckers are common at the base of the trunk. The shrub layer is very dense, 1.5–3 m tall, often with flat sprays of small leaves. The ground layer is very well covered with ferns, sedges, delicate trailing herbs, and patches of moss. Epiphytes (orchids, bromeliads, gesneriads, and aroids) are common in the moss covering trunks and branches. Ericaceae and Melastomataceae are abundant shrubby epiphytes. Large-leaved vines are occasional, but large lianas are uncommon.

Characteristic species of plants include *Alnus jorullensis, Ardisia* sp., *Brunellia costaricensis, Clusia* sp., *Cornus* sp., *Didymopanax pittieri, Drimys winteri, Escallonia poasana, Eugenia* sp., *Fuchsia arborescens, Gaiadendron poasense, Garrya laurifolia, Hedyosmum mexicanum, Hesperomeles heterophylla, Hieronyma poasana, Holodiscus fissus, Ilex* sp., *Magnolia poasana, Miconia* spp., *Microtropis occidentalis, Myrica arborescens, Nectandra* sp., *Oreopanax nubigenum, Palicourea* sp., *Persea* sp., *Phoebe* spp., *Podocarpus oleifolius, Prunus annularis, Quercus seemanni, Q. copeyensis, Q. oocarpa, Rapanea ferruginea, Rhamnus humboldtiana, Sapium* sp., *Solanum* sp., *Styrax polyneurus, Ternstroemia seemannii, Trichilia* sp., *Vaccinium consanguineum, Weinamannia pinnata, W. wercklei,* and *Zanthoxylum melanostictum.*

On poor soils the oaks *Quercus copeyensis* and *Q. oocarpa* may occur as virtually monospecific stands. At one time these giants lined the route of the Carretera Interamericana across the Talamanca range, and the spectacular drive rivaled a trip along the Redwood Highway in California.

Box 2.2
Cerro de la Muerte

The name Cerro de la Muerte (mountain of death) was originally applied to the area of the pass across the northern Cordillera de Talamanca at 3,300 m between Cerro Asunción and Cerro Buenavista (plate 17). This was normally an overnight stop for those traveling by horseback between Cartago to the north and the El General valley to the south. Travelers from these warmer regions were ill prepared for the sudden and extensive temperature drops at the altitude of the pass, and over the years several died of hypothermia—hence the name.

After the Carretera Interamericana (Route 2) was completed to carry motor cars and trucks across the cordillera in the 1940s, the term Cerro de la Muerte gradually came into use, at least by field biologists, for the entire ridge (continental divide) that the highway follows from Cangreja (1,900 m) on the north to División (2,286 m) or even farther south on the Pacific slope. Because the highway runs along the boundary between Cartago and San José Provinces, it is not always clear which side of the road is the source of a particular collection. Although some recent maps identify Cerro Buenavista as Cerro La Muerte, informally many biologists simply call this area "the Cerro." In this book Cerro de la Muerte follows the broader usage and refers to the entire area on both sides of the highway along the continental divide as described above.

Unfortunately, hardly anything now remains of this forest near the highway, but large tracts of this kind of habitat are found to the southeast throughout this cordillera (plate 15).

Montane Wet Forest

The small area on the upper southwest summit of Volcán Irazú that formerly supported this type of vegetation was heavily affected by the eruptions of 1963 to 1965. The heavy ash load and direct physical damage seem to have eliminated this forest type from the vicinity of the crater where it used to be found.

Montane Rainforest

Forests of this kind are confined to the highest areas in the Cordillera Central and occur over most of the highlands in the Talamanca range (plates 16 and 18). Hartshorn (1983) described the vegetation as follows:

> Tropical Montane Rain Forest is an evergreen forest of low to intermediate height with two tree strata. Canopy trees are 25–30 m tall, having short, stout, unbuttressed

trunks with rough bark. Crowns are small, compact, and rounded, with many thick, short, twisting branches. Leaves are often clustered at the twig tip. The understory is fairly open, with trees mostly 5–15 m tall, having slender, crooked trunks and compact, much-branched, brushlike crowns. Tree ferns are common in the understory. The shrub layer is dense, with dwarf bamboos up to 5 m tall. The ground layer is open under the bamboo. Trunks and branches of trees are thickly covered with moss and small herbaceous epiphytes; orchids and ferns are common in the moss. Large epiphytes are restricted to a few species of bromeliads. Woody vines with thick, fleshy leaves are common as canopy epiphytes.

Typical plants found in this forest are *Alsophilia* sp., *Arctostaphylos rubescens, Byrsonima densa, Buddleia alpina, Clethra gelida, Clusia alata, Cornus disciflora, Cyathea maxoni, Didymopanax pittieri, Drimys winteri, Escallonia poasana, Fuchsia microphylla, Gaiadendron poasense, Garrya laurifolia, Hesperomeles obovata, Hypericum silenoides, H. strictum, Ilex* sp., *Miconia biperulifera, Monnia* sp., *Oreopanax xalapense, Pernettya coriacea, Quercus costaricensis, Rapanea pittieri, Rhamnus* sp., *Solanum storkii, Solanum* sp., *Styrax polyneurus, Vaccinium consanguineum, Viburnum* sp., *Weinamannia pinnata,* and *Zanthoxylum* sp.

At a number of sites within this forest sphagnum bogs occupy open areas, often of considerable extent. These highly acidic sites are dominated by the bamboo *Chusquea subtesselata, Rapanea ferruginea, Vaccinium* spp., tree ferns (*Blechum* spp.), and dense stands of the terrestrial bromeliad *Puya dasylirioides.* Gómez (1986a) pointed out that the cooccurrence of *Escallonia poasana, Vaccinium* spp., *Fuchsia arborescens, Blechum loxense,* and *Puya* indicates that a site has regular nocturnal frosts during the year.

Subalpine Pluvial Paramo

This type of vegetation originally was found only on the highest peaks in the Cordillera de Talamanca (Weber 1958) and is the northernmost outlier of this essentially Andean formation. Human disturbance along the Carretera Interamericana on the Cerro de la Muerte has led to a downward expansion of paramo into the upper portions of the montane zone where the trees have been cut down, and it is maintained by occasional fires (usually accidentally set).

The paramo is a cold, wet area above timberline and is dominated by bamboos of the genus *Chusquea,* bunchgrasses *(Cortaderia* and *Calamagrostis),* thickets of *Diplostephium* and *Senecio,* low shrubs (e.g., *Myrrhidendron donnell-smithii),* and a variety of low-growing umbels, composites, and other herbaceous plants (plate 17). Gómez (1986a) notes that undisturbed paramo (plate 18) is characterized by the presence of the grasses *Aciachne pulvinata* and *Lorenzochloa rectifolia.* Where drainage is poor sphagnum bogs similar to those described for the Montane Rainforest environment also occur. Only a salamander *(Bolitoglossa pesrubra)* and two lizards *(Sceloporus malachiticus* and *Mesaspis monticola)* are found in the paramo of Costa Rica.

One area within the subalpine zone, La Sabana de los Leones, on the southwestern slope of Chirripó Grande, is a grassland without the shrubs found in typical paramo. It appears to represent a possible atmospheric association that Gómez (1986a) calls moist puna, since it grows under higher rainfall conditions than the puna vegetation of South America.

Secondary Growth

In all forested areas that have been disturbed either by natural events (treefalls, landslides, flooding, etc.) or human intervention, successional stages replace the original vegetation if there is no further perturbation. The replacement vegetation gives the impression that it is of a uniform age, since it has a low upper tree layer, and it usually has a high diversity of species and density at all levels. At lower elevations it includes many sun-tolerant trees and shrubs (e.g., *Cecropia* and *Heliconia),* especially around its margins. Many species of reptiles, especially diurnal lizards and snakes, and some amphibians are more abundant in these disturbed situations than they were in the original forest.

3 | Organization of the Systematic Accounts
(How to Use This Book)

THIS BOOK is designed for use by a variety of people having different degrees of interest in and knowledge of amphibians and reptiles. Many will use the book primarily to identify species encountered in the field or brought into the camp or laboratory to confirm an identification or for further study. Although it is not exclusively an identification manual or a field guide, the illustrations, keys, and diagnostics sections in each species account, and the distribution maps, will in combination serve those functions.

IDENTIFICATION

For the general reader with only a budding interest in these animals, the first step will be to thumb through the illustrations and become familiar with the general appearance of members of the major groups: caecilians, salamanders, frogs and toads, lizards, snakes, turtles, and crocodilians. Even those unfamiliar with the tropical fauna are unlikely to confuse representatives of these groups with one another, although you may have some initial difficulty in distinguishing between salamanders and lizards, lizards and small crocodilians, and snakes, elongate lizards with reduced limbs, and caecilians. In this process you will probably learn to recognize and subsequently remember particularly unusual or brightly colored species as well. Individuals with some biology training likely will spend more time perusing the illustrations to familiarize themselves with similarities among representatives of the principal subgroups (families and genera).

Both populations of readers should next turn to the introductory section for each major group, where the characters used to distinguish among the families, genera, and species are described and illustrated. Although the colored illustrations of individual species will often allow for immediate identification, only comparison with the diagnostics and description sections in each species account will yield a positive determination.

Students of biology at all levels, including professionals who are not herpetologists, usually will have some knowledge of the kinds of features used to characterize amphibians and reptile taxa and some experience in using dichotomous keys in identification. However, they too may find it useful to follow the steps recommended above for the non-professionals before using this book for identification.

Users with a considerable depth of knowledge of amphibians and reptiles, including professional herpetologists, will doubtless be able to refer many specimens immediately to family or genus. For identification they may prefer to go directly to the appropriate descriptions for these categories and depend on the keys and the species diagnostics and description sections for confirmation.

Remember that, unlike birds and butterflies, except for some large or very conspicuous diurnal species amphibians and reptiles tend to be secretive and furtive. You are unlikely to see them unless you make a careful search. Most are small, shy creatures whose distinguishing features are not readily observed from a distance. Consequently, attempts at identification usually require that the specimen in question be in hand. In some cases it is easier to identify small forms with a hand lens, magnifying glass, or microscope.

Most amphibians and reptiles can be handled without injury to them or the handler, but you must take care with turtles, larger lizards and snakes, and crocodilians of any size to avoid painful to serious bites. Since a considerable number of Costa Rican snakes are venomous, take extreme care not to handle or molest an unfamiliar snake. The venomous snakes are illustrated in plates 461–95; learn to recognize them. In no circumstances should you handle them unless you have been given professional training in how to manipulate them with a snake stick, tongs, or noose. Crocodilians of any size should be similarly treated with respect.

Identification may proceed in several ways. Once the specimen is identified to major group, simply compare it with the illustrations until you find one that matches or is very similar. Then turn to the account of the family or genus the illustrated species belongs to and compare your specimen's characteristics with the description of that group. If the description and specimen agree, you may then refer to the subsequent descriptions of included genera or species until you find one that matches the example in question.

You may shortcut this method by going directly to the account of the illustrated species most similar to the animal at hand. If the diagnostics or description section for that form is inconsistent with the characteristics of your specimen, compare it with accounts of other species in the same genus and those listed as similar species.

Another way to achieve rapid identification, once you are sure of the generic allocation for the unknown form, is to look over the distribution maps to see which species are known from the source locale of the specimen. You can then determine which species may be involved by comparing illustrations and diagnostics of species of that genus that occur in the subset of this particular area.

Whatever the outcome, once you are convinced you have identified your specimen to a particular species, it is well to confirm your assignment by comparing it with the appropriate illustration(s).

Although illustrations may greatly expedite identification, most professional biologists rely on systematic keys. A key is a simple device that presents a series of alternatives of characters, only one of which should match the features of the specimen being identified. In this book all the keys are dichotomous; that is, they consist of couplets (e.g., 1a and 1b) of contrasting characters. As you work through the key you choose one of the two alternatives, which may lead to another couplet (e.g., 2a and 2b) or to the name of a taxon. Each alternative in the couplet is a clue, so logically following the chain of evidence leads you to a correct identification of a specimen.

The principal advantages of keys over the illustrations of live whole animals are that they usually use structural features that are not always visible in a photograph and that they may be used with preserved material, which usually loses bright colors or pattern a short time after preservation. In addition, the features included in the key frequently are employed in determining evolutionary relationships among taxa and are ones professional herpetologists need to be familiar with.

When using a key, you sometimes reach a couplet where it seems that neither alternative or both will fit the specimen at hand. This may be because you have misinterpreted characters listed in an earlier couplet, and some backtracking may be necessary to confirm the choices up to that point. Where both alternatives appear equally plausible, you will need to follow the two alternative paths through the rest of the key until a determination can be made.

These complications, if persistent, have two possible sources. First, the key may be flawed in that some presumed difference(s) are not consistent or there is greater variation in a taxon than has been discovered from available material. This problem is most likely to appear where a genus or species is known from only one or a few specimens. Since the keys in this book have been tested and refined in the field over the past forty years, such effects are unlikely.

Another, perhaps even less likely, cause for problems with the keys may be that the example being examined represents an undescribed species or one described after this book went to press. New species of reptiles and amphibians continue to be discovered throughout Central America as previously unexplored areas are investigated, and Costa Rica is no exception. Although it often turns out that suspected new forms have been previously described or are unusual variants of recognized species, if the identification is ambiguous the specimen or a photograph and description should be referred to a professional herpetologist to determine its status.

When trying to identify any specimen, remember that most species exhibit considerable variation in many features of morphology and coloration. Intraspecific variation is usually of one of the following types.

Individual

Temporal: Some anurans and lizards can rapidly change color depending on various internal (neurohormonal) and external stimuli (time of day, social communication, etc.).

Continuous: A species may have a characteristic range of variation in certain features (e.g., ventral scales, coloration) so that the extremes are quite different from one another but are bridged by intermediate individuals.

Polymorphic and polychromatic: Other species have discontinuous variation in morphological features (e.g., smooth dorsum versus longitudinal ridges) or coloration (e.g., middorsal light stripe versus dark dorsal chevrons) so that individuals may differ markedly in appearance and initially be thought to represent different species. Rare individual genetic variants including albinistic (white or whitish) or melanistic (all or mostly black) ones are also occasionally found, as are individuals with reddish, yellowish, or bluish colors unusually emphasized.

Populational

Ontogenetic: In some species the juveniles differ markedly in morphology (e.g., head crest absent versus present in adults) or coloration (e.g., bright green versus gray in adults) so that the two might be thought to represent different forms.

Sexual dimorphism or dichromatism: In many cases females are much larger than males or vice versa; males may differ morphologically (e.g., large dewlap versus small or no dewlap in females), and males are often more brightly colored than females.

Geographic variation: Individuals from particular geographic areas tend to differ from those from other areas in features such as scalation or coloration.

Whenever possible these kinds of variations are briefly described in the species accounts, especially in the case of sexual dimorphism or dichromatism and ontogenetic changes. Unusual color variants (e.g., melanistic and albinistic individuals) are best identified by close attention to the morphological features they share with normally colored conspecifics.

In frogs and toads male vocalizations usually are another means of species identification. These calls are as distinctive and species specific to the trained ear as the songs of birds.

Descriptions of the calls, where known, are included in the species accounts. A detailed discussion of sound production and call characteristics in anurans is found in chapter 7.

In addition to the features reviewed above I have included descriptions of eggs and larvae of Costa Rican amphibians as far as they are known. Regrettably, few eggs have been described in detail, and this remains a rich area for future study. Most anuran larvae have been identified, and a key to known Costa Rican tadpoles is included in chapter 7.

OTHER USES

Although identification is an important process that is fundamental to any knowledge of a particular organism, this book is not simply an aid to identification. It has the additional function of providing an overview of what is known about the ecology, biology, and biogeography of the Costa Rican herpetofauna. Each chapter on the major components of amphibian and reptile diversity contains a review of the most significant features of morphology, systematics, distribution ecology, and biology. Each family and genus account summarizes salient aspects of these features for that taxon. Finally, the species accounts include extensive information and references to the literature on ecological, biological, and distributional data.

This information is provided so that generally interested readers may learn more about the habitat and habits of these important components of the tropical biota. In addition, I hope these accounts will stimulate biology students, professional biologists, and environmental scientists to use this database in designing their own research projects or conservation efforts. Finally, tropical herpetologists may use these accounts to ascertain what is known or not known about the herpetofauna so as to integrate their new observations and experiments with present knowledge, to determine what

areas are in greatest need of further study, and to identify particularly intriguing research possibilities.

DESCRIPTION OF THE MAJOR SYSTEMATIC GROUPS

Part 2 (Amphibians) and part 3 (Reptiles) are devoted to the detailed description of the herpetofauna of Costa Rica (table 3.1). Each part begins with a chapter (chapter 4 for amphibians and chapter 8 for reptiles) that summarizes the distinct morphological and biological features shared by members of each of these groups. In addition, an outline classification (also see fig. 1.5) and a review of their worldwide geographic distribution are provided. These chapters are comparative, including both obvious and more technical characteristics distinguishing amphibians from reptiles.

Part 2 continues with a chapter on each of the major groups of living amphibians: chapter 5 on the caecilians (order Gymnophiona), chapter 6 for salamanders (order Caudata), and chapter 7 for frogs and toads (order Anura). Each of these chapters begins with a general description of the group that includes a listing of significant features distinctive to each order, a systematic classification of family taxa placed in the order, and their geographic distributions. Additionally, there is a summary of the principal biological features shared by members of a particular order, such as basic bodily functions, the senses, locomotion, food and feeding, sexual dimorphism, courtship, mating and reproduction, and special structural or behavioral adaptations. This section ends with a description of the kinds of features most often used to characterize and identify taxa within the order, followed by a dichotomous key to Costa Rican genera belonging to that order.

The chapters in part 3 are similarly organized to cover the principal groups of living reptiles: after the introductory

Table 3.1.   Composition of the Costa Rican Herpetofauna

	Families		Genera		Species	
	N	%	N	%	N	%
Caecilians	1	2.7	3	2	4	1
Salamanders	1	2.7	3	2	37	9
Frogs and toads	8	22	26 + 1*	19 (19)	131 + 2*	34 (34)
Total amphibians	10	27	32 + 1*	23 (24)	172 + 2*	44 (44)
Lizards	11	30	30 + 3*	22 (24)	69 + 4*	18 (18)
Snakes	9	24	63	46	133	34
Subtotal squamates	20	54	93 + 3*	68 (69)	202 + 4*	52 (52)
Turtles	5	14	9	7	14	4
Crocodilians	2	5.4	2	1.5	2	0.5
Total reptiles	27	73	104 + 3*	76 (76)	218 + 4	56 (56)
Total herpetofauna	37	100	136 + 4*	100 (100)	390 + 6*	(100) (100)

*Introduced taxa.

Percentage in parentheses includes introduced taxa.

chapter come chapter 9 (order Squamata) for the two subdivisions of squamate reptiles; chapter 10 for lizards (suborder Sauria); chapter 11 for snakes (suborder Serpentes); chapter 12 for turtles (order Testudinata); and chapter 13 for crocodilians (order Crocodilia). Since lizards and snakes are closely related and similar in many features, chapter 9 treats characteristics of structure and biology shared by both groups rather than repeating them in chapters 11 and 12.

In all chapters in parts 2 and 3 of the book, except for chapters 4, 8, and 9, the key to genera is followed by individual accounts of the families, genera, and species of each major group known to occur in Costa Rica. Family accounts include a summary of characteristic and definitive internal and external features that will distinguish them from all other families placed in a particular order. In larger families the genera may be grouped into a number of subfamilial units that will also be diagnosed, and the geographic distribution of families and subfamilies will be summarized. Numbers of genera and species given here and elsewhere in this book are as current as possible, but since new taxa are being constantly described, they often are only approximate. The family accounts are arranged in phylogenetic order from the most ancient to the most recently derived clades. There is some controversy about the branching patterns of several groups, so not all herpetologists will agree with the sequences adopted here, but this should not bother other readers, since it has little effect on using the keys or descriptive accounts.

DESCRIPTION OF THE GENERA

Within all families or subfamilies recognized in this treatment each genus is characterized by its most trenchant features. These are usually external, but in some groups differences in diagnostic internal structure are often used for herpetological classification, so they may also be described. Because some users of this book are studying taxa from other areas in tropical America, I have tried to present diagnostic descriptions that will separate the genera included from all others in a particular family or subfamily. I remember my own frustration in my early days in Costa Rica trying to identify an unfamiliar species without many library sources and wishing the literature on the herpetofauna of other countries clearly stated the diagnostic features of a genus as a whole rather than those distinguishing it only from other confamilial genera found in the same country. I have, however, pointed out the ways each genus differs from other Costa Rican genera it might be confused with. Significant biological features shared by all members of a genus are also listed, especially in cases where little is known about the habits and ecology of its individual member species. In addition, each account gives the approximate number of species currently recognized worldwide as belonging to a genus and its geographic distribution. Most

generic accounts conclude with a dichotomous key to the Costa Rican species referred to the genus if more than one occurs in the republic. In a few cases the generic keys are to all species in two or more related genera that differ primarily in internal features but may be separated at the species level by external characteristics.

MORE ABOUT THE DICHOTOMOUS KEYS

I have already discussed the rationale and process of using keys. Four principal kinds of keys appear in this book. First is a key (at the end of this chapter) to the seven major groups of amphibians and reptiles found in Costa Rica. If you are uncertain which of these groups the specimen you wish to identify belongs to, begin with that key (p. 102). It will guide you to the next key in the sequence by page number. The next key will be one of those to all Costa Rican genera of a particular major group (e.g., Anura).

The reason for including all genera of a particular order or suborder in a single key rather than having keys to families and then to their component genera is utilitarian. The definitions of most families of amphibians and reptiles are based to a considerable extent on detailed knowledge of internal, especially skeletal features. Often rather distantly related genera resemble one another closely in superficial external appearances. The prime purpose of the keys is to aid in rapid and authoritative species identification. Keys to families, which are nearly impossible to construct without reference to internal anatomical features, would add an additional and at times confusing step to the identification process and consequently have been omitted.

The keys to genera will guide you to a specific generic account, again by reference to a page number. For each genus having more than one Costa Rican representative, a third kind of key is provided—a species key. Using the species key will lead to a tentative identification and a page reference to a particular species account.

In addition, a fourth kind of key, one for all the known larvae (tadpoles) of Costa Rican frogs and toads, is also presented (p. 178). Again, including all species of tadpoles in a single key is done for convenience, since larvae of distantly related species often superficially resemble one another because of adaptive convergences. This leads to unnecessarily complicated keys if one attempts to construct them at the generic level. The initial identification of a particular tadpole may be confirmed by the description in the page-referenced species account and illustrations.

One additional key, for identifying venomous coral snakes and their mildly toxic or harmless mimics, is included in chapter 11 (p. 706).

The keys presented here have had a long gestation and are much refined and expanded to include additional taxa not in the keys published earlier (Savage 1980b; Savage and Villa 1986) as well as the synopsis of tadpoles (Savage 1981a), recently revised into a dichotomous format by Lips and Savage (1996a). Several words of caution are in order, however. In any couplet listing several characters, the feature listed first is the most reliable. The remaining features are usually listed in order from most to least reliable if used singly, except that colors in life are often decisive. In keys to species within large or systematically complex genera I have often added supplemental characters of size, scale counts, ontogenetic change, and sexual dimorphism or dichromatism for each form to warn of a possible misidentification. Where variation in measurements or discrete variables is indicated, the values are for Costa Rican populations. The term *usually* means that a feature is present in 80% or more of the specimens examined; *occasionally* means that fewer than 15% of the specimens exhibit the feature; and *rarely* means that no more than 5% of available examples show the character. The genera *Bolitoglossa, Eleutherodactylus,* and *Norops* are among the most speciose and difficult in Costa Rica. For this reason it is especially important to verify any identification made by using the keys to species in these taxa through comparison with the illustrations and descriptions in the species accounts.

SPECIES ACCOUNTS

Probably the most important parts of this book are the accounts of individual species. Each account, where appropriate, includes the following in the order indicated: scientific and common names; diagnostics; descriptions of adults and juveniles, larvae; voice; comparison with similar species; habitat; biology; remarks; distribution. The contents of these sections and any pertinent terminology used are explained more fully in the paragraphs below.

Names

Scientific Name, Author, and Date

I have adopted the name currently recognized as valid for each taxon. Where differences of opinion exist as to the correct name, alternative interpretations are presented in the remarks section. The species name is followed by the name of the person who first described the species and the date of publication. Sometimes the date of publication that appears on a particular work is known to differ from the actual date of publication. In such a case the date stated on the publication is cited in quotation marks, followed by the date when the work actually appeared (e.g., "1860," 1861).

Common Name

Most genera and species of amphibians lack standardized vernacular English or Spanish names. Often very different taxa are called by the same name, as are all members of some genera or even families. Where long-standing, well-known

common names have been consistently applied, I have listed them. I abhor the pernicious trend of coining common names for genera and species in either English or Spanish and then substituting them for the scientific ones. Learning the correct generic name seems no more difficult than memorizing a counterfeit one. Remembering Latin names may take a little more effort, but this should be preferred to using a common name made up by someone who has never even seen the species in life. It certainly is no more difficult to learn and remember *Oedipina alfaroi* than "Alfaro's worm salamander." There can be no doubt as to the species identified as *Agalychnis callidryas,* but there is some doubt about which genus and species is meant by the "red-eyed leaf frog."

Diagnostics

This section enumerates the most obvious distinguishing characteristics for the species. Emphasis is on life colors and readily determined morphological features including adult size. Where pertinent, major differences in the appearance of juveniles versus adults and males versus females are also included. Most readers will use this section, in combination with the illustrations, for field identifications.

Description

This section provides a description of the size, morphology, coloration in life, and variation in these features for adults and juveniles. The description is not exhaustive but includes only those features that will expedite comparison with related or similar species. The order of character presentation is uniform for all species within a particular order or suborder. Within each genus the same set of features is described in parallel for each species.

Size and Measurements

Primary measurements include total length and standard length. Other measurements, including tail length, are usually given as percentages of standard length. The definition of standard length varies among orders but usually applies to the length of the head plus body or, in turtles, to the length of the upper shell (carapace). Details on size and mode of measurement are given in each account of orders and suborders.

Amphibians and reptiles, unlike mammals, are characterized by continuous or indeterminate growth. That is, they may continue to increase in size throughout life so that unusually large individuals of a particular species may occasionally be encountered. The maximum total length known to be authentic is given for each species. This may be for an extralimital animal, so the usual range in adult sizes in Costa Rica may also be given. Where there is marked sexual dimorphism in size, the ranges for adult males and females are listed separately. Measurements larger than 30 mm usually are rounded to the nearest 0.5 mm. Measurements less than 30 mm are rounded to the nearest 0.1 mm. Total and standard lengths for very large species or examples are given in meters, rounded to the nearest 0.1 m. Weights for large turtles and crocodilians are in kilograms, rounded to the nearest 0.1 kg.

Remember that living caecilians and snakes are rather flexible, so length measurements are affected by how much one stretches the specimen at hand. All preserved specimens tend to shrink over time. Changes in amphibians are unpredictable (Lee 1982); some examples even of the same species, sex, and time of preservation increase in length, and others shrink.

Morphology

The principal morphological features used in distinguishing among species of amphibians and reptiles are illustrated and described in the appropriate chapters. Additional line drawings illustrating differences between particular genera or among allied species are included elsewhere in the text. Sexual dimorphism in size and morphology are also described, as well as ontogenetic changes in morphology. Juveniles usually have proportionately larger heads, limbs, and eyes than adults, and their tails may be relatively shorter.

In the species accounts on snakes a description of the male genitalia (hemipenes) is included if it is known and differs markedly among species within a genus. Otherwise the hemipenes description is in the generic account.

For measurements and discrete variables the values listed are for Costa Rican specimens, followed in parentheses by the range of variation for the species as a whole if that exceeds in-country variation. Where quantitative features are expressed by a three-number notation (26–28.2–34), the values are for the lower limit of variation, the mean, and the upper limit. The same notation is used in the biology sections for variable features.

Coloration

Life colors and color pattern are of great importance in distinguishing among many species of amphibians and reptiles. Although colors in life are emphasized throughout this book, most preserved specimens rapidly lose bright colors (yellow, orange, red, green). Usually yellow, orange, or red fades to a pale yellow or white. In other cases the color changes in preservative (e.g., green to blue). Dark markings (black or brown) tend to intensify in preservative so that the contrasting light and dark pattern in the living animal is retained in death. Where there is a characteristic and distinguishing color in preserved material of a taxon, this is noted in addition to the colors in life. The principal types of dorsal color pattern (fig. 3.1) may be defined by the presence or absence of distinct marking as follows:

Uniform: essentially unicolor without contrasting dark or light markings

Blotches: large regular to irregular light or dark markings that contrast with the ground color

Spots: small regular light or dark markings that contrast with the ground color

Speckled: small light or dark spots on a contrasting ground color

Ocelli: large light spots outlined by a dark border

Reticulum: a dark or light network of lines overlying a ground color of a contrasting color

Bands: transverse contrasting bands that run across the body and extend onto the venter only slightly or not at all

Dorsal light or dark field(s): entire dorsum contrasting markedly in color to flanks (dorsal band *auctorum*); contrasting light or dark areas between longitudinal dorsal and dorsolateral stripes

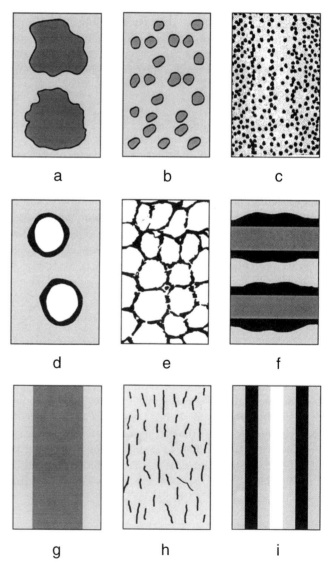

Figure 3.1. Principal color patterns in amphibians and reptiles: (a) blotches; (b) spots; (c) speckles; (d) ocelli; (e) reticulum; (f) bands; (g) dark field; (h) lines; (i) stripes.

Rings: dark or light transverse markings that completely encircle the body across both the dorsum and the venter

Lines: short lineate light or dark markings

Stripes: continuous longitudinal dark or light lines of various widths that extend over long distances on the body; the ground color between the stripes is a field, as described above

Two-tone: entire body marked by a broad dark or light middorsal stripe contrasting with broad lateral stripes of a different color

The shape, size, and other particulars of the markings show considerable variation and are described in greater detail at appropriate places in the descriptive accounts. Of course some species exhibit a combination of pattern elements that often are different on different regions of the body. The patterns on the undersurfaces of the head, venter, and tail frequently contrast with those of the upper surfaces, and there may be regional differences along the body axis.

The description of coloration in each species account is based on Costa Rican examples, and if marked extralimital variation in coloration occurs, it is discussed in the remarks section. If known, the colors in life of the iris and tongue are noted. When appropriate, dichromatism is described, as are marked ontogenetic differences in coloration.

Larvae

The tadpole of each species of Costa Rican frog and toad, if known, is briefly described. Characteristics used for identifying tadpoles are described and illustrated in chapter 7 (p. 177).

Voice

The call, if known, is described for the anurans of the republic. Features that differ among calls are defined in chapter 7 (p. 162).

Similar Species

Comparisons are made in this section with other species that might be confused with the subject of the specific account. Frequently these include congenerics, so there may be some redundancy with the keys and diagnostics section. However, when dealing with a puzzling specimen or one difficult to identify, you should consult all three as well as the descriptions.

Habitat

This section is designed to answer the question, Where am I likely to find this species? First the occurrence of each species in the major habitats as described in chapter 2 is documented, including, for terrestrial environments, a specific plant formation or formations. A species is usually not found uniformly throughout a formation, so further detail is provided (e.g., in riparian situations, under logs, etc.).

Biology

This section summarizes what is known about the life of the species. It contains information from the literature and general and specific observations and data from my own fieldwork. Where no citation is given, the content is from the latter source. For many forms little or nothing is known, whereas others have been extensively studied. By reading this section the serious student or professional biologist will discover many opportunities for research to fill in gaps in our knowledge or undertake more sophisticated studies than carried out up to now. When information is available, the following main topics are covered in the following order: abundance, general habits, behaviors, and folklore.

Abundance

Reference is usually made to the relative abundance of each species at some point in the species account (diagnostics, habitat, biology, or remarks), based on my experience in the field. These are subjective impressions but will convey the likelihood of seeing a particular species in its usual habitat during the diel and seasonal periods when it is active. The terms used for this purpose are *rare* (occasionally seen; known from one specimen or a few); *uncommon* (unlikely to be seen more than once every few months); *common* (likely to be seen every week); and *very common* (likely to be seen every day).

There are no very common snakes using these definitions, so in their case *very common* equals likely to be seen every week; *common*, likely to be seen every few weeks. Conversely, many pond-breeding frogs are encountered in great numbers at reproductive sites but may rarely be seen during the nonreproductive portions of the year. Reference to their relative abundance is based on the situation in the breeding season. Explosive breeders in particular are not predictably present, so some of them are scored as uncommon although they might be present in great numbers at a breeding site a few nights each year.

Diel and Seasonal Activity Cycles

The following terms describe the diel period of usual activity: *diurnal* (active during the day); *nocturnal* (active at night); and *crepuscular* (active near dawn or dusk or both).

Habits

The following descriptors are used to denote ecological valence:

Fossorial: residing in the soil, in burrows created either by the organism itself or by other animals

Semifossorial: living in the leaf litter or under debris on the soil surface

Terrestrial: carrying out activities at ground level

Semiarboreal: occurring in the interface between ground level and the herbaceous shrubby stratum; some time spent both on the ground and in bushes and small shrubs

Arboreal: living in the three-dimensional space of large shrubs and trees; rarely found on the ground

Saxicolous: hiding or living in crevices or under rocks

Semiaquatic: living along the margins of streams or other bodies of water but entering water readily to feed or escape

Aquatic: spending all or most of adult life in the water (freshwater, estuarine, or marine areas); larvae of anurans are also aquatic

Population Characteristics

Population characteristics include size, density, age structure, and cyclic or other fluctuations.

Food and Feeding

Food and feeding covers foraging and prey location, diet, and prey size. Animals are usually classified as generalist or specialist feeders. Generalists eat a wide range of food items during their lifetimes. Specialists consume a much more restricted number of foods. Species feeding on only one kind of prey are usually regarded as specialists. If the predator eats many kinds of prey within a particular taxonomic group—for example, insects—it is referred to as an insect generalist. Another predator may specialize on only one or two types of insect prey—for example, ants and termites—and be regarded as an insect specialist.

The following additional terminology is also used to describe the food habits of the species treated in this book:

Cannibalistic: feeding on conspecifics

Carcinivorous: feeding on crustaceans

Carnivorous: feeding on animals; used here for predators eating terrestrial vertebrates

Herbivorous: feeding on plants

Insectivorous: feeding on insects, but usually used for predators that eat other arthropods as well

Molluscivorous: feeding on mollusks

Omnivorous: feeding on both plants and animals

Oophagus: feeding on eggs

Ophiophagus: feeding on snakes

Piscivorous: feeding on fish

Vermivorous: feeding on earthworms

Social Interactions

Social interactions include agonistic behavior, aggregations, social hierarchy, and sexual behavior (periodicity, courtship, and mating).

Reproduction

Reproduction covers oviparity or viviparity, oviposition, eggs, incubation, gestation, and parturition.

Life History

Life history includes biology of eggs and larvae; hatching, metamorphosis, or birth; parental care; growth, sexual maturity, and longevity.

Antipredator Defenses

Defenses against predators include crypsis, flight behaviors, bluffing, threatening and attacking behaviors, urotomy, secretion of noxious or toxic chemicals, warning coloration and behaviors, and mimicry.

Other Special Behaviors

Other special behaviors include locomotion, movements, migration, homing, and sound production.

Folklore

Folklore covers traditional beliefs and tall tales, mostly false, about Costa Rican amphibians and reptiles preserved orally by country folk.

Remarks

The remarks section will be of primary interest only to herpetologists. In it I will attempt to clarify nomenclatural problems, questionable records, and misidentifications. Changes in usage of species names from those in Taylor's classic reviews (1951b, 1952a,b, 1954b, 1956) of the Costa Rican herpetofauna also will be noted and explained. In addition, I will review and evaluate extralimital variation in the species when the variation substantially exceeds that found in Costa Rican samples. Miscellaneous comments on other matters may also be presented.

I also include a brief description of the chromosome complement (karyotype) of the species, if known. This description, at a minimum, will consist of the number of chromosomes in the diploid (2N) complement. Depending on the information available in published or manuscript descriptions, I will give a more detailed summary of karyotypic features.

Chromosomes are classified by the location of the centromere, a structure important in nuclear division. The two parts of the chromosome on either side of the centromere are referred to as arms. The following kinds of chromosomes (fig. 3.2) are recognized after Green and Sessions (1991b):

Metacentric: chromosomes having the centromere near the center of the chromosome (M); these chromosomes are biarmed

Submetacentric: chromosomes having the centromere away from the center of the chromosome and the short arm no less than 25% of the length of the chromosome (SM); these chromosomes are biarmed

Subtelocentric: chromosomes with the centromere much closer to the end of the chromosome than to its center and

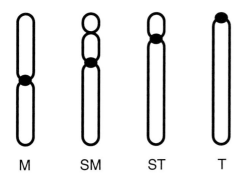

Figure 3.2. Morphology of chromosomes: M = metacentric, SM = submetacentric, ST = subtelocentric, T = telocentric; black dots represent centromeres. The submetacentric chromosome shows a secondary constriction on the upper arm.

the distance from the centromere to the nearest tip less than 25% of the length of the chromosome (ST); these chromosomes are regarded as uniarmed in reptile karyotypes (Gorman 1973) but as biarmed in amphibian karyotypes (King 1990)

Telocentric: chromosomes having the centromere near one end of the chromosome (T); these chromosomes are uniarmed

These four kinds of chromosomes are often called macrochromosomes, in contrast to very short microchromosomes, which are uniarmed and are present or not in different taxa.

The NF (*nombre fondamental*) of the karyotype is the total number of major arms in the diploid number (2N) of chromosomes. For amphibians this is calculated by multiplying the number of telocentrics by one and summing that figure with the number of other chromosomes multiplied by two (e.g., 2 telocentrics × 1 + 12 others × 2 = 26) (King 1990). However, for reptiles only the metacentrics and submetacentrics are considered biarmed; they are multiplied by two, and all other chromosomes are multiplied by one (e.g., 12 biarmed × 2 + 10 other chromosome × 1 = 34) (see Gorman 1973). The 2N chromosome counts in these two examples are 14 and 22, respectively.

Distribution

The statements of ranges for species included in this book contain a general description of the extralimital geographic range and a more detailed summary of the distribution in Costa Rica, including altitude limits. Because of the shape of the republic, with its main axis running southeast to northwest, it is sometimes difficult to communicate direction to those unfamiliar with lower Central America. Costa Ricans orient themselves primarily with reference to the Meseta Central or the capital city, San José, at 9°56′ north latitude and 84°5′ west longitude. Although it goes against local usage, I have found that dividing the country into four

quadrants provides a consistent and biotically useful system of reference (fig. 3.3). I have divided the country into northern and southern parts roughly along a line running from Cabo Blanco and Jacó, Puntarenas Province, on the west to Puerto Limón, Limón Province, that bows slightly northward along the divide of the Cordillera Central (fig. 3.3). Eastern and western regions are separated rather arbitrarily along the main axis of the Cordillera de Talamanca, Cordillera Central, Cordillera de Tilarán, and Cordillera de Guanacaste (fig. 3.3). According to this system four major geographic lowland and foothill regions may be defined as follows:

Northeast: Atlantic slope Costa Rica north or east of the continental divide from 10° N, the level of Puerto Limón, northward to the border with eastern Nicaragua

Southeast: Atlantic slope Costa Rica northeast of the continental divide of the Cordillera de Talamanca and south of 10° N latitude to the border with northwestern Panama

Northwest: Pacific slope Costa Rica from the continental divide westward, north of 9°40′ N to the level of Cabo Blanco and Jacó, Puntarenas Province, on the south and to the border with western Nicaragua on the north, including the Península de Nicoya

Southwest: Pacific slope Costa Rica southwest of the continental divide of the Cordillera de Talamanca and south of 9°40′ N southeastward to the boundary with southwestern Panama

The northeast sector includes the areas usually called the Zona Norte by Costa Ricans (Heredia Province north and east of the continental divide) and Atlántico Norte (Heredia Province north of the Cordillera Central and Limón Province from the Río Moín northward). The northwest area includes Guanacaste Province, the Península de Nicoya, northern Puntarenas Province, and contiguous low-elevation areas and the Meseta Central Occidental. The southeast sector comprises southern Limón Province (Atlántico Sur) and the Meseta Central Oriental. The southwest region includes central Puntarenas Province and adjacent San José Province and the Zona Sur, the area south of the Cordillera de Talamanca from the level of Domnical southeastward.

The highland areas of Costa Rica (about 1,500 m and higher) form two units sharing many species but with considerable endemism:

The Cordillera Central, Cordillera de Tilarán, and Cordillera de Guanacaste: the first two form a continuous mass, but remember that the Guanacaste range consists of a series of separate volcanoes of relatively low elevation

The Cordillera de Talamanca

Unless otherwise stated, reference to distributions in the various cordilleras implies that a particular species occurs on both Atlantic and Pacific slopes.

In some cases species distributions may not conform to

the general areas defined above and a more specific delimination is necessary. For that purpose the term central Costa Rica is applied to the region south of the Cordillera Central and north of the Talamanca range and includes the upland valleys from Moravia de Chirripó on the east through Turrialba to Cartago and Orosi on the west, all in Cartago Province, on the Atlantic versant, as well as the drainage basins of the Río Barranca and Río Grande de Tárcoles from the San José area to the Pacific coast.

For describing extralimital distributions of Costa Rican species and for the analysis and synthesis of biogeography in chapters 14 and 15, I use the following terminology for major geographic areas frequently referred to in these sections:

North America: the continental mass from the Isthmus of Tehuantepec in southern Mexico northward

Central America: the land east and south of the Isthmus of Tehuantepec to the border between Panama and Colombia

South America: the continent south of the border between Colombia and Panama

Mesoamerica or Middle America: Mexico and Central America

Nuclear Central America: the northern portion of Central America (Mayan and Chortis blocks) from the Isthmus of Tehuantepec to Nicaragua

Isthmian Link: Costa Rica and Panama (Chorotega and Chocó blocks)

Upper Central America: Mexico east of the Isthmus of Tehuantepec to the Honduras-Nicaragua border

Lower Central America: Nicaragua, Costa Rica, and Panama

Areas of Nicaragua and Panama adjacent to Costa Rica often cited in the distributional summaries are consistently referred to using these geographic units:

Western Nicaragua: the narrow Pacific coastal slope and the drainages of the great Nicaraguan lakes including the western portions of the Departments of Estilí, Matagalpa,

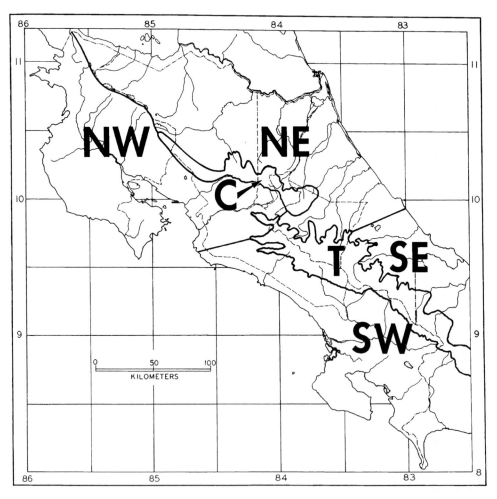

Figure 3.3. Distributional areas for the Costa Rican herpetofauna as described in text: NE = northeast, SE = southeast, NW = northwest, SW = southwest, C = Cordillera de Tilarán and Cordillera Central, T = Cordillera de Talamanca.

Boaca, Chontales, and Río San Juan; contiguous to north-western Costa Rica

Eastern Nicaragua: the slopes and lowlands of the Atlantic drainage, exclusive of the areas draining into the great lakes; continuous with northeastern Costa Rica

Northwestern Panama: the Atlantic coast and slopes of western and central Bocas del Toro Province from Punta Valiente to the Costa Rican boundary; adjacent to southeastern Costa Rica

Southwestern Panama: essentially the Pacific coast and slopes of Chiriquí Province; adjacent to southwestern Costa Rica

In Central America the general distribution of most amphibians and reptiles in the lowlands tends to be confined to either the Atlantic or the Pacific slope. This pattern is related to the differences in climate and vegetation on the two versants (Savage 1966a, 1982). Except for the northern portions of the Yucatán Peninsula, the prevailing northeast trade winds and the moisture they bring create an environment suitable for broadleaf evergreen forests (rainforests) more or less continuously from Veracruz, Mexico, to eastern Panama.

On the Pacific slope rainfall is more limited, and subhumid to semiarid conditions, supporting at best seasonally deciduous forests, range from Sinaloa, Mexico, to central Costa Rica. Southwestern lowland Costa Rica and the immediately adjacent area of Panama, however, were covered originally by humid forests similar to those on the Atlantic slope and contain isolated populations of many Atlantic rainforest forms. Farther east in Panama to the Colombian border, forest having marked seasonal differences in rainfall predominates on the Pacific versant.

In a number of other areas throughout their range, species typical of the Atlantic forests extend onto the Pacific slope. Some of these are disjunct from Atlantic versant humid forest habitats, but others are continuous with them across low points in the continental divide. The former disjunctions are best explained as fragmentation of formerly continuous ranges by fairly recent mountain uplift events. The others occur where the influence of the northeast trade winds extends across low spots in the continental divide to affect climate and vegetation on the Pacific versant so that humid conditions continue to prevail. As uplift continues these areas may also become discontinuous, disjunct Pacific slope fragments of the Atlantic slope biota.

The most important of these areas as reflected in herpetofaunal distributions are the foothills of the Sierra Madre de Chiapas (Mexico) and the lower slopes of the Guatemalan highlands (disjunct); slopes of and gaps in the Cordillera de Guanacaste of northwestern Costa Rica (continuous); slopes around the Prensa Fortuna, Chiriquí Province, El Valle de Antón, Coclé Province, and La Campana, Panama Province, Panama (continuous); and the central Panama depression (continuous).

Similarly, a number of amphibians and reptiles characteristic of Pacific lowland subhumid to semiarid habitats are found disjunctly on the Atlantic slope, primarily in the rain shadow valleys of Guatemala and Honduras, but some on the dry outer portions of the Yucatán Peninsula (Lee 1996). In Costa Rica several typical lowland dry forest taxa occur as isolated populations within the Atlantic lowland forests or on the Meseta Central Oriental (e.g., the lizard *Norops cupreus*) or at a few Atlantic lowland sites (e.g., the snake *Imantodes gemmistratus*).

Also bear in mind that the pattern of vegetation distribution described above for Mexico and Central America switches in northwestern Colombia near the Panama border. Here rainforest conditions are continuous from both Atlantic and Pacific southeastern Panama through western Colombia to central Ecuador (Pacific drainage). Consequently many species found in the Atlantic slope forests of Central America also range southward along the Pacific coast of northwestern South America. Conversely, northern Colombia and northwestern Venezuela (Caribbean drainage) support deciduous forest and other open habitats, and a number of taxa found disjunctly in similar situations in central and western Panama are shared between the two areas (Dunn 1940a; Rand and Myers 1990).

Vertical limits for distribution are described in general terms by the five altitudinal zones recognized in the republic (lowland, premontane, lower montane, montane, and subalpine) as defined in chapter 2 (fig. 2.5) and by elevational ranges. These ranges are for the species in Costa Rica unless otherwise stated. Many species are known from only a few localities, so the altitudinal zone information will often eliminate some species when making an identification.

Also note that political units mentioned in the text or distribution sections usually are cited by country and by province, department, or territory (e.g., Panama, Chiriquí Province). Exceptions are Mexico, where country and state are simply listed (e.g., Oaxaca, Mexico), and the United States, where only the state is given (e.g., Texas). In addition, when referring to Costa Rican localities I cite only the province (e.g., Heredia Province). When citing a specific locality in the text I use the following convention: country: department, etc.: locality (for example, Nicaragua: Río San Juan: El Castillo). Appropriate diacritical marks are used for all localities in Spanish, except that I use the English Mexico (not México) and Panama (not Panamá) throughout.

The accompanying maps show the geographic distribution of each species in Costa Rica. The principal mountain masses are delineated by the dotted 1,500 m contour on each map. At this scale the extent of the Cordillera de Tilarán, which is generally below 1,500 m in elevation with a few scattered ridges and peaks with higher altitudes, is exaggerated. The continental divide in this region lies on a line roughly extending in a gentle curve from the southwestern corner of the Cordillera Central to just southwest of the Laguna de Arenal near 10°28′ N. On these maps each dot indicates a locality, or several localities where they are so close together as not to be separated on the scale of the maps. The maps have been compiled from the examination of Costa Rican specimens in most of the world's scientific collections and from the latest revisionary studies in the literature and well-documented photographs or authoritative sight records. The last two sources are especially important in delimiting the range of large species (e.g., iguanas, turtles, boas, large colubrid snakes, and crocodilians), since adults of these animals are rarely preserved for museum collections. One snake species, *Liotyphlops albirostris*, is not known from any definite locality in Costa Rica, as indicated by its distribution map.

Maps of the distribution of the several species of sea turtles indicate nesting beaches and a few oceanic sightings. The map for the sea snake *Pelamis* indicates beachings where specimens have been collected. This form may be expected to be seen at sea at any point along the Pacific coast of tropical Mesoamerica.

ABBREVIATIONS

The following abbreviations are used for museums and other collections in referring to specimen identification numbers in the text and in the species accounts.

AMNH	American Museum of Natural History
ANSP	Academy of Natural Sciences, Philadelphia
BMNH	Natural History Museum (British Museum of Natural History)
CAS	California Academy of Sciences
CHP	Circulo Herpetológico de Panamá
CMNH	Carnegie Museum of Natural History
CRE	Costa Rica Expeditions (now at LACM)
FMNH	Field Museum of Natural History
ICP	Instituto Clodomiro Picado (to be deposited at UCR)
KU	Museum of Natural History, University of Kansas
LACM	Natural History Museum of Los Angeles County
LSUMZ	Louisiana State University, Museum of Natural Science
MCZ	Museum of Comparative Zoology, Harvard University
MNCR	Museo Nacional de Costa Rica
MNHNP	Museum National d'Histoire Naturelle, Paris
MSNG	Museo Civico di Storia Naturale di Genova "Giacomo Doria"
MVZ	Museum of Vertebrate Zoology, University of California, Berkeley
NHMG	Naturchistorisches Museet, Göteborg
NHMW	Naturhistorisches Museum in Wien
SMF	Forschungsinstitut und Naturmuseum Senckenberg
UCR	Museo de Zoología, University of Costa Rica
UF	Florida Museum of Natural History
UMMZ	University of Michigan, Museum of Zoology
USC	Field series to be deposited at LACM
USNM	United States National Museum of Natural History
ZMB	Museum für Naturkunde, Humboldt-Universität zur Berlin
ZMH	Zoologisches Museum Hamburg

KEY TO THE MAJOR GROUPS WITHIN THE COSTA RICAN HERPETOFAUNA

1a. Body covered with smooth to strongly tuberculate skin, moist or slimy to the touch in life; never any externally visible horny or bony scales or plates; digits always lacking claws . . . . Amphibia 2

1b. Body, head, and limbs covered with horny scales or plates and/or encased in a bony, often scale-covered shell; skin dry to touch; digits, if present, almost always with claws . . . . Reptilia 4

2a. Adults and juveniles with four limbs and movable eyelids; no larval stage, *or* larvae aquatic, fishlike tadpoles with strongly compressed tails and dorsal and ventral caudal fins 3

2b. Adults and juveniles wormlike, lacking limbs; eyes covered by bone and/or skin; no larval stage . . . . Gymnophiona (p. 116)

3a. Postcloacal tail in adults and juveniles; hind limbs with three segments (femoral, tibial, and foot); no larval stage . . . . Caudata (p. 126)

3b. No tail in adults or juveniles; hind limbs with four segments (femoral, tibial, tarsal, and foot); usually an aquatic fishlike larva (tadpole) with strongly compressed tail and dorsal and ventral caudal fins, sometimes no larval stage . . . . Anura (p. 173)

4a. Body not encased in a bony shell; teeth on jaws 5

4b. Body encased in a bony shell; no teeth on jaws . . . . Testudinata (p. 738)

5a. Transverse cloacal opening; if limbs present, toes rarely webbed; teeth attached to lingual side of jaw (pleurodont) 6

5b. Longitudinal cloacal opening; toes webbed; teeth set in sockets (thecodont) . . . . Crocodilia (p. 772)

6a. Both forelimbs and hind limbs clearly evident; usually with an ear opening and movable eyelids . . . . Sauria (p. 422)

6b. No forelimbs; hind limbs at most reduced to vestigial clawlike structures, usually entirely absent; never with an ear opening or movable eyelids . . Serpentes (p. 535)

2

Living Amphibians

4 | Amphibians
Class Amphibia Linné, 1758

THE LIVING amphibians (caecilians, salamanders, frogs, and toads) are highly specialized descendants of the first terrestrial vertebrates that made their debut about 370 million years ago in the Devonian period. These earliest tetrapods gave rise to both the living amphibians and, through the reptiles, all other groups of terrestrial vertebrates. They differed, as do their descendants, from their closest relatives among the bony fishes (class Osteichthyes: subclass Sarcopterygii), most importantly as follows:

1. Paired forelimbs and hind limbs; pectoral girdle not attached to skull; pelvic girdle dorsally expanded to form sacral attachment to vertebral column (limbs and girdles variously reduced or lost secondarily in some forms)
2. Respiration by gills restricted to larvae (or secondarily retained in aquatic adults); adult respiration by lungs or cutaneous respiration or both
3. Three-chambered heart (right and left atria and single ventricle) and double circulation of blood; pulmonary arteries provide deoxygenated blood to lungs, systemic arteries provide oxygenated blood to body (mammals and birds are further specialized in having four-chambered hearts)
4. Hyomandibular bone no longer provides support for jaws but becomes middle ear ossicle (the columella) for sound transmission to inner ear
5. Remaining gill arches form hyobranchial apparatus to support tongue

Unlike other tetrapods, amphibians have eggs and development much as in fishlike vertebrates and lack the extraembryonic membranes (amnion, serosa or chorion, and allantois) characteristic of reptiles (including birds) and mammals. Primitively and in most living species, the eggs develop into an aquatic larva that has functional gills and develops further in a body of water. Subsequently the larva typically undergoes a period of rather rapid morphological and physiological change and metamorphoses into a terrestrial organism lacking gills. This diphasic life history gives the class its name (from the Greek *amphi*, two, plus *bios*, mode of life). So far as we know, based on fossilized remains of gilled larvae, this pattern of life history was characteristic of all ancient amphibians as well. However, as we will see, some living species have evolved different life history modes that circumvent the need for an aquatic larval period.

After their initial invasion of the land amphibians had a spectacular evolutionary radiation in the Paleozoic and early Mesozoic eras (360 to 213 million years ago), but none of the many major groups living in that time frame have survived to the present.

Modern amphibians all belong to a later evolutionary pulse, and the exact relation of the three extant orders (approximately 461 genera and 5,000 species) to the earlier, now extinct lineages has not been established. Although subject to some continuing controversy, the consensus at present (Milner 1988; Trueb and Cloutier, 1991) is that the living orders together forming the Lissamphibia are more closely related to one another than to any other amphibian stock.

Although one may encounter some difficulty when attempting to separate certain fossil forms of amphibians from primitive reptiles because of the great diversity among the extinct members of both classes, the Lissamphibia may be consistently distinguished from all living reptiles as follows:

1. Body lacking a covering of horny or bony scales or plates
2. Skin moist or slimy to touch in life
3. No claws on digits (although horny tips are present is some forms extralimital to Costa Rica)
4. Two occipital condyles
5. A single sacral vertebra in limbed forms
6. Teeth pedicellate
7. Ribs straight, short (not extending into flank muscles), or absent
8. A specialized sensory area, the *papilla amphibiorum*, in the wall of the sacculus in the inner ear that receives sensory vibrations below 1,000 Hz
9. A levator bulbi muscle consisting of a thin sheet under the orbit that elevates the eye to assist in the intake of air during respiration in salamanders and anurans (reduced but present in caecilians)

The following classification of living amphibians (fig. 4.1) is used in this book. In the older literature the group is often referred to as the class Batrachia.

Class Amphibia Linné, 1758
  Subclass Lissamphibia Haeckel, 1866
    Order Gymnophiona: caecilians
    Order Caudata: salamanders
    Order Anura: frogs and toads

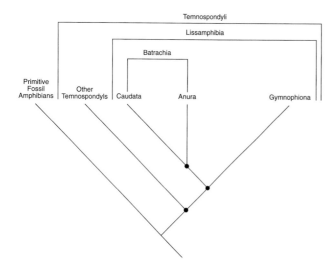

Figure 4.1. Phylogenetic relationships of living orders of amphibians.

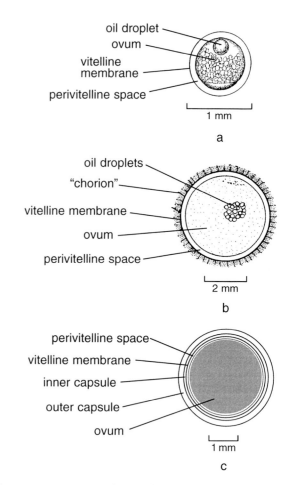

Figure 4.2. Comparison of typical fish and amphibian eggs: (a) pelagic nonadherent fish egg; (b) demersal adherent fish egg; (c) amphibian egg.

As I emphasized above, although amphibians and reptiles may be technically distinguished based on a combination of internal morphological features, they differ primarily in characters associated with life history. Amphibian eggs, like those of other anamniotes, must be deposited in water or very moist situations to protect them from desiccation. The cleidoic eggs of reptiles, with their protective shells and extraembryonic membranes, do not face the same hazards, nor do reptiles have a vulnerable aquatic larval stage.

The eggs of most amphibians closely resemble those of oviparous freshwater fishes (fig. 4.2). Typically, amphibian eggs are surrounded by a series of gelatinous capsules and are laid in standing water (marsh, swamp, pond, or lake) or in a stream. After a period of development the embryo hatches from the egg to become a free-swimming, feeding larva. Subsequently the larva undergoes a rapid transformation and metamorphoses into a small terrestrial organism. There are, however, a substantial number of amphibians that exhibit various modifications of this pattern (fig. 4.3). In Costa Rica these modifications include:

1. Eggs deposited out of the water on vegetation, with the larvae subsequently falling or being washed into a pond or stream: several treefrogs of the family Hylidae and all glassfrogs, family Centrolenidae
2. Eggs laid in a foam nest in aquatic situations: several species of *Leptodactylus* and *Physalaemus pustulosus*
3. Eggs laid in an incubation chamber and larvae washed into a nearby body of water: the frogs *Leptodactylus labialis* and *L. poecilochilus*
4. Eggs laid in axils of aerial plants or tree holes: some treefrogs, family Hylidae
5. Eggs deposited on land in moist situations; no larval stage (direct development): salamanders of the family Plethodontidae and frogs of the genus *Eleutherodactylus* (family Leptodactylidae)

6. Eggs carried in the dorsal pouch of the female; no larval stage (direct development): the frog *Gastrotheca cornuta*
7. Eggs retained in the mother's oviduct, where fetal nutrition is provided; no larval stage (viviparous): the caecilians of the family Caeciliidae

In contrast, a few Costa Rican freshwater fishes lay a considerable number of small, buoyant eggs that are suspended and float freely and singly in the water column (i.e., are pelagic). These eggs are bounded by a single thin, nonadhesive membrane, are translucent, and lack the gelatinous capsules typical of aquatic amphibian eggs, so they are unlikely to be mistaken for them. Most Costa Rican freshwater fishes have demersal eggs (they are heavy and sink to the bottom or may be stuck to vegetation or rocks). Although these eggs are larger than the pelagic ones, they are bounded only by a vitelline membrane that is adhesive to other eggs or to vegetation or the substrate or that becomes much thickened at the moment of fertilization to form a very tough, smooth, nonadhesive protective layer. In a few forms gelatinous capsules surround the eggs, in which case they may be distinguished by one or more of the special features listed in the

ning at the head and may be sloughed as a single unit or in patches, in most cases assisted by the hands and feet. Generally the shed skin is then eaten.

The sense organs that provide amphibians with information regarding the external environment are similar to those found in other amniotes. These include chemoreceptors (taste and olfaction); mechanoreceptors (tactile, equilibrium, gravity source, acceleration, and hearing); thermoreceptors; and photoreceptors. I should point out, however, that amphibians and most reptiles have a dual olfactory system—the main olfactory sac and the vomeronasal organs (Jacobson's organs). The sensory cells of the paired olfactory sacs sample chemical particles that are inhaled through the nares, while the vomeronasal organs usually pick up sensory information from food in the mouth. Each vomeronasal organ lies lateral to an olfactory sac but is not completely separated from it as in most reptiles. In caecilians and squamate reptiles the vomeronasal organs are especially well developed and obtain chemical clues from the external environment.

In addition, as in fishlike vertebrates, the larval amphibians and adults of many aquatic forms have a series of lateral-line organs distributed along the sides of the body and especially over the surfaces of the head. Typically these organs resemble small papillae (in larvae) or are porelike. Each organ contains a number of ciliated sensory hair cells that are stimulated by movements in the surrounding water. A second, somewhat simpler porelike series of lateral-line organs, also found on the heads of larval caecilians and aquatic salamanders, are electroreceptors, apparently stimulated by the low-level electric impulses generated by some fishes. Because of the significant diversity among the major groups of amphibians in chemoreception, hearing, and photoreception, descriptions of these features will be included as appropriate under the individual major groups. Similarly distinctive aspects of locomotion, food and feeding, and reproductive biology will be treated in the same fashion.

DECLINING AMPHIBIAN POPULATIONS: THE PHENOMENON AND POSSIBLE CAUSES

Some 150 million years ago today's two dominant groups of amphibians—salamanders (order Caudata) and frogs and toads (order Anura)—were already well represented in the earth's biota. In the interim these groups and the other surviving order of amphibians (the Gymnophiona or caecilians), which appeared in the fossil record somewhat earlier, have survived the many vicissitudes in the planet's environments. These include most significantly the mass extinctions at the Cretaceous-Tertiary boundary, the continentalization of climate, glacial expansions and contractions, and the general cooling and drying trends characteristic of the Quaternary period. In the twentieth century the biodiversity of these stocks is equal to that of the class Mammalia.

The survival and success of these animals over this vast stretch of time attests to their resilience to environmental perturbations. It was thus a great shock to scientists studying these organisms to note a dramatic and alarming decline in some amphibian populations and the apparent extinction of others at many sites worldwide during the mid-1980s.

At first this decline was attributed to possible local causes (e.g., habitat destruction, pesticide pollution, etc.), since most observers were aware of changes only in their own geographic areas. Obviously, some of the changes in amphibian populations were due to such influences. However, a number of individuals recorded very striking declines in population size or complete disappearance of formerly abundant forms at a number of sites, including protected reserves and preserves in North, Central, and South America, tropical Asia, Europe, and Australia between the 1987 and 1988 breeding seasons. As word of these events filtered out, a growing international concern began to develop that some broader, even worldwide, environmental effect was responsible.

The apparent pervasiveness of the decline in amphibian populations became convincingly evident at the first World Congress of Herpetology, held at the University of Kent, United Kingdom, in September 1989. Both formal and informal reports emphasized that the perplexing degree to which declines and extinctions were both widespread internationally and increasing in areas seemingly protected from human influence was of particular concern. The concern was heightened by the appreciation that because of their biphasic life history (aquatic larvae and usually terrestrial adult stages), their low and high status in food chains, and their permeable skin, amphibians are particularly sensitive to environmental disturbance and may be affected by environmental deterioration well in advance of other components of the biota.

As a response to these discussions, a workshop addressing the proposition that worldwide amphibian populations are in decline was developed and sponsored by the Board of Biology of the National Research Council (Blaustein and Wake 1990). The panel participating in the workshop concluded that several declines or extinctions were not readily explainable by proximate causes. A number of anthropogenic effects all seemed plausible as factors that might be implicated. In addition, it was concluded that the unexplained declines might well be harbingers of major degrading factors in the global environment.

Costa Rica, unfortunately, was one of the first areas in the American tropics to show such declines, with the apparent extinction of the famous golden toad, *Bufo periglenes*, and several other less spectacular species. Marked declines in salamander populations (e.g., *Bolitoglossa pesrubra*), harlequin frogs *(Atelopus),* and several other anurans also occurred between 1987 and 1988 and continue to this time. As anticipated in a review of the situation in lower Central America in 1991 (Savage 1991), the number and the extent of declines have increased, even in areas far removed from human population centers or protected from habitat modification. The recent papers by Pounds et al. (1997), Pounds, Fogden, and Campbell (1999), and Lips (1998) clearly demonstrate the reality of the declines in the republic. In addition, these authors concluded that the number of synchronous declines cannot be attributed to natural population fluctuations (contra Pechmann and Wilbur 1994).

Current emphasis among scientists has shifted from an initial phase of documenting the worldwide extent of the decline (Blaustein 1994; Blaustein, Wake, and Sousa 1994) to evaluating causative factors. Particularly targeted are soil and water acidification, increased ultraviolet radiation, disease or parasitism, and immunosuppression in response to environmental stress or global climate change. It seems likely that one or more of these factors may be operating to produce declines in a particular instance, but there seems little doubt that in many places the declines and extinctions exceed expected natural fluctuations.

Declines in upland Costa Rica and western Panama (Lips 1998, 1999) mirroring those in eastern, upland Australia (Laurance, McDonald, and Speare 1995) led workers in both areas to postulate that the proximate cause of the declines was an epidemic, rapidly spreading disease. This possibility was further supported by the implication of a motile chytrid fungus as the proximate cause of death in frogs from both Australia and Central America (Berger et al. 1998). It is unclear at this time whether other factors may so stress the animals that they are then susceptible to infection.

It remains open whether anthropogenic stresses are involved and what serious consequences to the health of the global environment may be signaled by declining amphibian populations (see p. 52).

In the accounts that follow, I will identify Costa Rican amphibian populations especially affected by declines and discuss what is known about the chronology, extent, and putative causes of these declines. A few populations appear to be rebounding from near extinction, and the factors involved in these cases will also be reviewed. Since amphibians are the most sensitive vertebrate bioindicators of environmental health, the Costa Rican examples will illuminate issues in evaluating and conserving tropical biodiversity overall.

5 | Caecilians
Order Gymnophiona Rafinesque, 1814 (Apoda)

THIS STRICTLY tropical group of elongate, limbless, wormlike amphibians ranks among the most bizarre and specialized vertebrates. Most naturalists, including many professional biologists, are unaware that these amphibians exist, and many experienced tropical biologists and herpetologists have never seen a caecilian in the field. A recently discovered fossil from the Lower Jurassic (about 200 million years ago) has limbs but is clearly assignable to this order (Jenkins and Walsh 1993) and testifies to the long independent history of caecilians. Two other ancient fossil species are known, from the Paleocene epoch of Bolivia (about 65 million years ago) and Brazil (about 60 million years ago), both presumed to lack limbs. Most caecilians are terrestrial burrowing forms, but one extralimital family (the Typhlonectidae of South America) is completely aquatic and restricted to fresh water.

When first seen many terrestrial caecilians might be mistaken for large earthworms, since the body is elongate and segmented by annular grooves and the eyes are obscure or hidden. Caecilians most obviously differ from these worms in having a well-developed mouth, jaws, teeth, and nostrils. To the uninitiated, gymnophionans might be confused with snakes, amphisbaenians, or lizards with the limbs reduced or absent. However, these reptiles have the body covered with epidermal scales and have long postcloacal tails (caecilians have no epidermal scales and no tails or extremely short ones). Similarly, salamanders, even very attenuate ones, have limbs (pectoral, pelvic, or both) and a long postcloacal tail.

The principal distinctive features of the living members of the order include:

1. Body greatly elongate (67 to 286 trunk vertebrae), tail short or absent
2. No limbs, girdles, or sternal elements
3. Eyes vestigial, buried under skin or skin and skull bones
4. No tympanum, middle ear, and Eustachian tube or associated muscles
5. Annular grooves and folds that form rings running transversely around body
6. A protrusible chemosensory tentacle on side of head between eye and nostril
7. Larvae with true teeth (enamel and dentine present); elongate, lacking limbs, and having a single gill slit
8. Males with an eversible median copulatory organ (a phallodeum) formed from a portion of the cloacal wall
9. Palatoquadrate articulating to skull
10. Atlas articulating to skull only by atlantal cotyles
11. Teeth present on upper and lower jaws, vomer, and palatine portion of maxillopalatine
12. Frontal and parietal bones distinct

Many of the most distinctive features of caecilians (elongation, loss of limbs, reduction of eyes and auditory apparatus, etc.) seem to be obvious correlates of their fossorial habits. In addition, the skull is extremely compact and well ossified, with several elements fused to increase its rigidity for effectively pushing through the soil. The posterior trunk vertebrae are also modified from those in other amphibians by having an additional articulation between them to create greater strength in the specialized burrowing locomotion described below.

The characteristic annulation of these animals is the result of the development of a series of primary folds that have a one-to-one correspondence with the vertebrae except in the cervical and tail regions. Each fold is demarcated anteriorly and posteriorly by an annular groove. Several to many primary folds are divided in some cases by a secondary groove and in a few cases (in forms extralimital to Costa Rica) by a tertiary groove. In the former there are two folds per affected body segment and in the latter four folds. Within the annular grooves, embedded in the skin, there is often a series of one to ten rows of tiny bony scales that form a series of corselets around the body.

Most gymnophionans are slate gray, bluish purple, or black and usually somewhat lighter below, but in some forms the annular grooves are marked with a contrasting light or dark color. Others, however, have a uniformly colored bright yellow body (*Schistometopum thomense*) or one marked with bold yellow stripes.

The classification of caecilians (fig. 5.1) is based for the most part on differences in internal anatomy, primarily of the skeleton (especially the skull), the musculature, and the kinds of annulation. Approximately 168 species of gymnophionans have been described, but doubtless many more remain undiscovered because of their cryptic habits. They are placed into thirty-six genera and six families as follows:

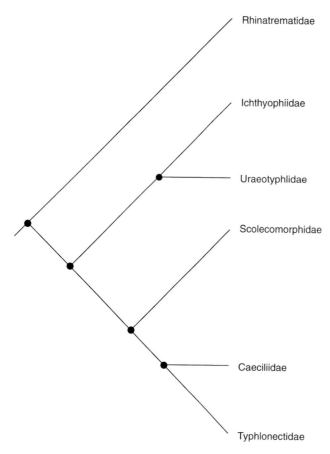

Figure 5.1. Phylogenetic relationships among living families of caecilians.

Family Rhinatrematidae
Family Ichthyophiidae
Family Uraeotyphlidae
Family Scolecomorphidae
Family Caeciliidae
Family Typhlonectidae

The caecilians have a circumtropical distribution but are absent from Madagascar, the eastern East Indies, New Guinea, Australia, and the Pacific islands. Only the Caeciliidae (twenty-four genera) are represented in Mexico (to near 20° N) and Central America, and they are also found throughout tropical South America, in East and West Africa (between 10° N and S), and on the Seychelles islands in the Indian Ocean. The distribution of the other families is as follows: the Rhinatrematidae (two genera) occur in tropical northern South America from Ecuador and Peru to Venezuela and the Guianas; the Ichthyophiidae (two genera) are found in southern India and Sri Lanka and from northern India (near 27° N) south and east to western Indonesia (10° S) and the Philippine Islands; the Uraeotyphlidae (one genus) are restricted to southern India; the Scolecomorphidae (two genera) are found disjunctly in tropical East and West Africa;

and the Typhlonectidae (five genera) occur disjunctly in the Magdalena, Orinoco, and Amazon drainages of northern South America and from extreme southern Brazil to near 35° S in Argentina.

Certainly the most unusual of the many unusual features characteristics of caecilians is the unique chemosensory tentacle on the side of the head between the eye and the nostril. Superficially the tentacle looks like a small papilla when protruded, but this is only the free tip that lies in the tentacle aperture of the elongate organ. The rest of the tentacular apparatus consists of the tentacle sheath, which is continuous with the free tip and encloses a long retractor muscle (modified from the retractor muscle of the eye found in other amphibians) attached to the base of the tentacle, and the large Harderian (orbital) gland that, in other vertebrates, lubricates the eye.

When the tentacle is protruded by pressure of the secretion from the orbital gland, it picks up chemical particles from the environment. When the muscle is retracted the tentacle is pulled posteriorly and the tentacular aperture closes over. Simultaneously the tentacle sheath elongates and narrows, forcing the orbital gland secretion across the tentacle tip to carry the fluid and chemical particles through small ducts mediad to the tentacle aperture and thence to the large vomeronasal or Jacobson's organ (Billo 1986). This latter structure lies lateral to the olfactory sac and is a separate chemoreception center. It appears that the tentacular apparatus functions for chemosensory perception primarily while caecilians are burrowing, since their nostrils are presumed to be closed during such times and olfaction consequently nonoperational.

In line with their burrowing existence, late-stage larval and adult caecilians have the vestigial eyes covered by bone or skin or both. Although thus modified and reduced in size, the eyes retain a lens and retina and are apparently sensitive to differences in light intensity. The reduced auditory apparatus also seems to be a correlate of the burrowing habit. However, a very large single ear ossicle (the fused columella plus operculum or columella only) articulates with the quadrate distally. This arrangement is ineffective in receiving airborne sounds, but vibrations from the substrate or surrounding water are transmitted through the lower jaw via the quadrate to the columella and thence to the inner ear. Not surprisingly, sensitivities to these vibrations are for frequencies below 1,500 Hz, and highest sensitivities are between 200 and 400 Hz.

When on the ground, surface caecilians use curvilinear (serpentine) locomotion in which lateral undulations throw the body into a series of more or less S-shaped curves that push against irregularities (bumps, rocks, twigs, etc.) on the substrate and propel the animal forward. Aquatic caecilians use a similar behavior in swimming, with the loops of the body pushing against the surrounding medium. When burrowing, caecilians use the compact, solid, spatulate skull to push through the soil. Propulsion is provided by lateral undulations of the vertebrae and surrounding muscles that thicken and then thin the body locally. The thickened part of body comes in contact with the walls of the burrow, driving the caecilian forward as the waves of thickening pass down the body (Gans 1974). O'Reilly, Ritter, and Carrier (1997) concluded that this pattern of locomotion is powered hydrostatically by force applied by a crossed-helical array of tendons that surrounds the body cavity.

Interestingly, the structure of the terminal region of the body shows considerable diversity within the group (Nussbaum and Wilkinson 1989). In some forms there is a definite tail (families Ichthyophiidae, Rhinatrematidae, and Uraeotyphlidae) that is completely annulated and supported by several postcloacal vertebrae and associated segmental musculature. In most genera (families Caeciliidae and Typhlonectidae) there are no postcloacal vertebrae and associated segmentation. In these groups annulation is incomplete, consisting of a few annuli that intercept the cloaca, or the cloaca is included in a nonannulated terminal shield. A number of species, however, have a pseudotail consisting of a few postcloacal vertebrae, but annulation is incomplete and a terminal shield is consistently present. A pseudotail is characteristic of the families Scolecomorphidae and Typhlonectidae and shows interspecific variation in several genera of the Caeciliidae.

Terrestrial caecilians feed primarily on elongate prey, particularly earthworms, which the tentacular organs probably find during burrowing. Well-developed recurved teeth on the jaws and the roof of the mouth grasp the slippery prey and move it posteriorly before it is swallowed. Unlike terrestrial salamanders and most anurans, the tongue in caecilians is not protrusible and is not used to capture prey. Caecilians have a unique dual jaw-closing mechanism consisting of the usual vertebrate adductor muscle that pulls the lower jaw upward and a second muscle that attaches to the elongate posterior part of the lower jaw and pulls back and downward on it. During prey capture the lower jaw is pressed against the substrate and the skull is raised. Closing the mouth involves the simultaneous contraction of both jaw-closing muscles, and a single bite takes about half a second (Bemis, Wake, and Schwenk 1983). This system seems to represent an adaptation for securing prey in a subterranean tunnel.

Little is known regarding predation on caecilians, although we may assume that, as with other amphibians, a wide variety of carnivorous animals feed on them. Based on limited data from stomach contents, various semifossorial and fossorial snakes appear to be important predators.

Fertilization is internal in all known cases and presumably is universal in the order. Oviparous forms lay large (up to 35 × 42 mm) and heavily yolked terrestrial eggs that may hatch into free-living aquatic larvae or undergo direct development without a larval stage. The females of at least some species guard the eggs within a burrow or incubation chamber. Members of the aquatic family Typhlonectidae are viviparous, and the large fetal gills form a placental connection with the enlarged oviduct wall of the mother. The young in this case are born as aquatic larvae.

Many terrestrial caecilians, including most Costa Rican species, are also viviparous. In these amphibians the small eggs are retained in the oviduct, and the developing fetuses quickly exhaust the yolk supply. They subsequently obtain their nutrition by feeding on rich secretions (the hydrotroph) from the thickened oviduct wall and on cells of the oviduct lining. The fetal caecilians are characterized by a series of specialized multicuspidate, deciduous teeth (lost at birth) that are used to scrape material from the oviduct wall (Wake 1977). This life history mode is known to occur among genera found in the Americas and Africa that are representatives of three of the six families of caecilians.

CAECILIAN IDENTIFICATION: KEY FEATURES

Species identification of caecilians is based primarily on the location and visibility of the eye, the placement of the tentacle in relation to the eye and external nostril, the presence or absence of a terminal shield, and the number of primary, secondary, and total folds (fig. 5.2). You should encounter no difficulty in discriminating between the various expressions of these features as used in the keys and the diagnostics sections. Bear in mind that the first and often quite a few additional secondaries are formed by grooves that do not completely encircle the body. These folds, however, are to be included in the secondary count. Because there is no tail

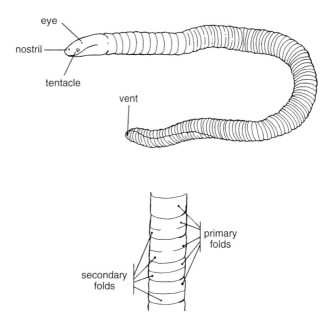

Figure 5.2. External features of caecilians.

in Costa Rica caecilians, all measurements of size are of total length.

Most caecilians are small to moderate-sized forms under a meter in total length. *Caecilia thompsoni* of Colombia, however, attains a total length of 1,520 mm. *Idiocranium russeli* of Nigeria and *Grandsonia brevis* of the Seychelles have a maximum size of a little over 100 mm. The largest Costa Rican species, *Gymnopis multiplicata,* has a maximum total length of 480 mm.

KEY TO THE GENERA OF COSTA RICAN CAECILIANS

1a. Tentacle lying well posterior to nostril (fig. 5.3b,c); head and body not contrasting in color. 2
1b. Tentacle lying directly below nostril (fig. 5.3a); head lighter in color than body . . . .*Oscaecilia* (p. 120)
2a. Eye not visible externally, orbit roofed by squamosal bone; tentacle at lower anterior tip of an oblique, posteriorly directed light spot (fig. 5.3b); tentacle just anterior to eye, tentacular foramen at or near center of

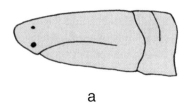

a

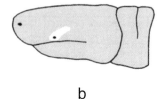

b

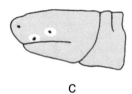

c

Figure 5.3. Lateral view of heads of Costa Rican caecilians showing differences in placement of nostrils and tentacles and visibility of eyes: (a) *Oscaecilia,* tentacle under nostril; (b) *Gymnopis,* eye not visible, tentacle posterior to nostril; (c) *Dermophis,* eye visible, tentacle posterior to nostril.

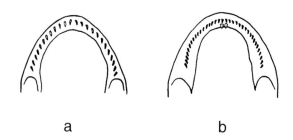

a   b

Figure 5.4. Lower jaw and floor of mouth in Costa Rican caecilians: (a) no splenial teeth; (b) splenial teeth present.

maxillary bone; a single splenial tooth on each ramus of lower jaw so that there are two tooth series (dentary and splenial) (fig. 5.4b) . . . . *Gymnopis* (p. 118)
2b. Eye visible externally (fig. 5.3c), orbit not roofed by bone; tentacle in a small oval light spot; tentacle about halfway between eye and nostril, tentacular foramen in anterior margin of maxillary bone; no splenial teeth, so only one tooth row (dentary) on each lower jaw (fig. 5.4a) . . . .*Dermophis* (p. 117)

Family Caeciliidae Rafinesque, 1814

Caecilians, Solda con Soldas, Suelo con Suelos, Dos Cabezas

Representatives of this the largest family of caecilians (twenty-four genera and eighty-nine species) are all fossorial forms typically found in humid tropical forests, although several species occur in seasonal (dry) forest habitats. Members of this family have a recessed mouth, have the tentacle anterior to the orbit, and lack a true tail, though a short pseudotail is present in some species. Because of the vermiform habitus, most features distinguishing caecilian families are based on internal morphology, and the Caeciliidae are defined by the following combination of characters: skull with temporal fossae narrow or usually roofed over by bone; nasal and premaxillary fused; septomaxillary, prefrontal, and postfrontal lost or fused to other bones; maxillopalatine widely separated from quadrate, which is firmly articulated to the columella; M-shaped ceratohyal; right lung well developed and elongate, left lung rudimentary; no tail; some, no, or all primary annuli subdivided by secondary grooves; never any tertiary grooves; external gills of larvae and embryos in three rami.

This family is disjunctly circumtropical in distribution with ten Neotropical genera *(Brasilotyphlus, Caecilia, Dermophis, Gymnopis, Lutkenotyphlus, Microcaecilia, Mimosiphonops, Oscaecilia, Parvicaecilia,* and *Siphonops);* two East *(Boulengerula* and *Sylvacaecilia)* and three West African *(Geotrypetes, Idiocranium,* and *Herpele)* endemics; one *(Schistometopum)* on the Gulf of Guinea islands and in East Africa; three *(Grandisonia, Hypogeophis,* and *Praslinia)* on the Seychelles; and two from India *(Gegenophis* and *Indotyphlus).*

KEY TO THE CAECILIANS OF COSTA RICA

1a. Tentacle lying well posterior to nostril (fig. 5.3b,c); head and body usually not markedly different in color. . . .2

1b. Tentacle lying directly below nostril (fig. 5.3a); head much lighter in color than body; eye not visible externally; splenial teeth present (fig. 5.4b); 175 primaries, no secondaries. . . . . *Oscaecilia osae* (p. 120)

2a. Eye visible externally; tentacle in a small oval light spot (fig. 5.3c); no splenial teeth (fig. 5.4a) 3

2b. Eye not visible externally; tentacle in a dorsally directed oblique light spot (fig. 5.3b); a single splenial tooth on each lower jaw; primaries 112–33, secondaries 84–117; total folds 193–250 . . . .
. . . . *Gymnopis multiplicata* (p. 119)

3a. Head and body similar in color; secondaries 65–96. *Dermophis gracilior* (p. 117)

3b. Head lighter than body; secondaries 11–26 . . . . *Dermophis parviceps* (p. 118)

Genus *Dermophis* Peters "1879," 1880

Members of this genus (five species) are moderate-sized (to 600 mm in total length), relatively robust (total length to body width ratio 15 to 34) caecilians having the eye usually visible (visible in all Costa Rican species) and the tentacle about halfway between the eye and nostril, showing secondary grooves, and lacking an unsegmented terminal shield and terminal vertical keel. Other diagnostic but internal features are eye in a socket, not covered by bone; tentacular foramen in anterior margin of maxillopalatine bone; no temporal fossae; no splenial teeth; no narial plugs; no diastema between vomerine and palatine teeth; scales present; karyotype 2N = 26, with five pairs of microchromosomes in one species, *Dermophis mexicanus* (Wake and Case 1975).

In Costa Rica the species of *Dermophis* are likely to be confused with *Gymnopis multiplicata*, but in that form the eyes are covered by bone and are not visible externally.

Dermophis ranges from Veracruz, Mexico, to western Honduras and from Jalisco and Michoacán, Mexico, to northwestern Colombia. Although there are no records for the Atlantic versant from Nicaragua, the genus may occur in humid forest situations in that area. The apparent absence of the genus from dry forest areas on Pacific versant lowlands of northwestern Costa Rica and Panama probably is real. One member of the genus *(Dermophis mexicanus)* has been reported recently from a Pacific coastal Quaternary site (1200 to 1350 B.C.) in Chiapas, Mexico (Wake and Wake 1999).

So far as is known, all species in this genus are viviparous.

Dermophis gracilior (Günther, 1902)

(map 5.1)

DIAGNOSTICS: A relatively robust species, plumbeous in dorsal color, having the eye visible (fig. 5.3c) and the head not contrasting in color with the body.

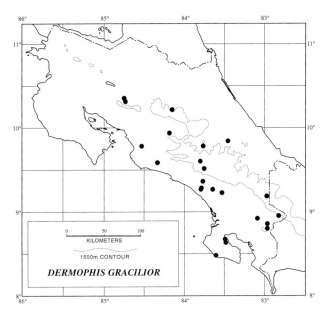

Map 5.1. Distribution of *Dermophis gracilior*.

DESCRIPTION: A moderate-sized (to 387 mm in total length), relatively stout (total length to body width ratio 23 to 34 in adults) species; primary annuli 91 to 117, secondaries 65 to 96, total annuli 159 to 208. Upper surface plumbeous, venter uniform cream mottled with dark or most of venter gray to dull black; annular grooves not lighter or darker than adjacent areas.

SIMILAR SPECIES: (1) The eye of *Gymnopis multiplicata* is hidden under bone (fig. 5.3b). (2) *Oscaecilia osae* is extremely attenuate, has a pink head, and lacks secondary annuli. (3) The several blind snakes and worm snakes *(Anomalepis, Helminthophis, Lepotyphlops, Liotyphlops,* and *Typhlops)* are covered by epidermal scales. (4) *Bachia blairi* has paired limbs and is covered by epidermal scales.

HABITAT: Lowland Moist Forest, Premontane Wet Forest and Rainforest, and Lower Montane Rainforest, where it is usually found under logs and surface debris.

BIOLOGY: An uncommon to rare fossorial species infrequently encountered on the surface, although it may be found under rotting logs or other surface litter. Food is mostly earthworms and termites, but it also eats larvae and instars of other insects. In the related species *Dermophis mexicanus* (Wake, 1980), development takes place in the maternal oviduct, and the embryo uses the yolk for nutrition before emerging from the egg membrane at a total length of 10 to 15 mm. Subsequent nutrition is provided by material (hydrotroph) secreted by the oviduct. The specialized fetal dentition is used to scrape the oviduct to stimulate hydrotroph production. These teeth are spoon shaped with a double (early in development) or single median spike. Litter sizes are two to sixteen, and the neonates are 110 to 150 mm in total length at birth. Sexually mature females are 300 mm or more in total length.

REMARKS: More than one species may be represented by the several taxa recognized by Taylor (1968), including *Dermophis balboai*, *D. costaricense*, *D. glandulosus*, and *D. gracilior*. Savage and Wake (1972) placed *D. costaricense* and *D. gracilior* in the synonymy of *Dermophis mexicanus* (Mexico to Nicaragua), but that form differs most strikingly from *D. gracilior* in having the ventral portion of the grooves black, in contrast to the lighter ventral ground color. In *D. mexicanus* there are 94 to 112 primary annuli, 35 to 85 secondaries, and a total annulus count of 152 to 196. Individuals here placed in *D. gracilior* form two geographic segments that may prove to be distinct. Examples from Atlantic slope premontane Costa Rica have 107 to 117 primaries, 74 to 96 secondaries, and 186 to 208 total folds. Those from the Pacific versant of Costa Rica and western Panama have 91 to 102 primaries, 65 to 78 secondaries, and 159 to 176 total folds. If two species were recognized, the name *D. costaricense* would apply to the Atlantic slope form, with *D. gracilior* restricted to the Pacific versant population. Pacific lowland specimens here tentatively referred to *D. gracilior* have low numbers of secondaries (29 to 60) and may not be conspecific with the upland populations. See addendum for update (p. 839).

DISTRIBUTION: Humid Pacific lowlands and premontane slopes of Costa Rica and western Panama; premontane zone of Atlantic slope Costa Rica and southern lower montane belt of the Cordillera de Talamanca (404–2,000 m).

Dermophis parviceps (Dunn, 1924c)

(plate 21; map 5.2)

DIAGNOSTICS: A small, moderately slender caecilian with a pinkish head that contrasts with the purplish slate dorsum; the eye is visible through the skin (fig. 5.3c).

DESCRIPTION: A small, slender species, to 217 mm in total length and total length to body width ratio 22 to 28; primary annuli 85 to 102, secondaries 11 to 26, total annuli 97 to 126. Upper surfaces purplish slate, head pinkish; venter grayish; annular grooves not lighter or darker than adjacent areas.

SIMILAR SPECIES: (1) *Oscaecilia osae* has no secondary annuli, and the eye is not visible (fig. 5.3a). (2) The several blind snakes and worm snakes *(Anomalepis, Helminthophis, Leptotyphlops, Liotyphlops,* and *Typhlops)* have the body covered with epidermal scales. (3) *Bachia blairi* has tiny limbs and is covered by epidermal scales.

HABITAT: Lowland Moist and Wet Forests and Premontane Moist and Wet Forests and Rainforest. Usually found under logs or in leaf litter.

BIOLOGY: A rare fossorial species. Most often collected in undisturbed forest under debris or by digging to depths of 400 to 600 mm through the leaf litter and upper soil in the angles between tree buttresses.

REMARKS: Savage and Wake (1972) synonymized Taylor's *Dermophis balboai, Dermophis occidentalis,* and *Der-*

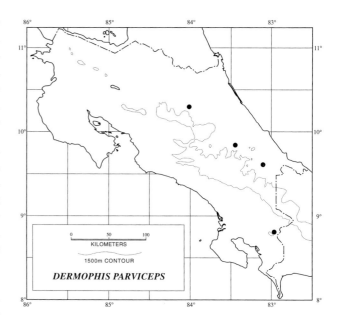

Map 5.2. Distribution of *Dermophis parviceps*.

mophis glandulosus with this form. Recently collected material suggests that this may be an oversimplification. Although one Pacific slope specimen is tentatively placed with *D. parviceps* for the purposes of this book (map 5.2), further study is likely to change the status of the other taxa. See addendum for update (p. 839).

DISTRIBUTION: Definitely known from scattered localities on the lowlands and premontane slopes of the Atlantic versant from southeastern Costa Rica to one premontane central Panama locality (40–1,220 m). Other examples from lowland and premontane southwestern Costa Rica are questionably conspecific.

Genus *Gymnopis* Peters, 1874

Representatives of this genus (two species) are moderate-sized (to 480 mm in total length), relatively robust (body length to body width ratio 23 to 37) to slender (ratio 37 to 52) forms having the eye covered by bone and not visible externally, the tentacle just anterior to the eye, not the nostril, with secondary grooves present, and lacking a terminal shield and terminal vertical keel. Other diagnostic internal features include tentacular foramen at or near the center of the maxillopalatine bone; no temporal fossae; splenial teeth present; no narial plugs; no diastema between vomerine and palatine teeth; scales present; karyotype 2N = 24 or 26, with four or five pairs of microchromosomes (Wake and Case 1975).

The single Costa Rican species differs most obviously from the somewhat similar appearing species of the allied genus *Dermophis* by having the eyes hidden under bone; in *Dermophis* the eyes are visible externally, though covered by the integument and associated tissues.

Members of this genus are viviparous.

The genus occurs from Guatemala to western Panama on the Atlantic versant and on the Pacific slope in Costa Rica and western Panama.

Gymnopis multiplicata Peters, 1874

(plates 22–23; map 5.3)

DIAGNOSTICS: A relatively robust caecilian, purplish slate in dorsal color, with numerous secondary annuli and the eyes not visible externally (fig. 5.3b).

DESCRIPTION: A moderate-sized (to 480 mm in total length), relatively stout (total length to body width ratio 23 to 37) species; primary annuli 112 to 133, secondaries 84 to 117, total annuli 199 to 250: Upper surfaces purplish slate, venter lighter lavender to cream; annular grooves not lighter or darker than adjacent areas.

SIMILAR SPECIES: (1) The species of *Dermophis* have the eye visible externally (fig. 5.3c). (2) *Oscaecilia osae* is extremely attenuate, has a pink head, and lacks secondary annuli. (3) The several blind snakes and worm snakes *(Anomalepis, Helminthophis, Leptotyphlops, Liotyphlops,* and *Typhlops)* are covered with epidermal scales. (4) *Bachia blairi* has paired limbs and is covered by epidermal scales.

HABITAT: Forest floor, meadows, pastures, and other areas with surface debris in gallery forest in the Lowland Dry Forest zone, in Lowland Moist and Wet Forests and Premontane Moist and Wet Forests, and marginally in Lower Montane Wet Forest.

BIOLOGY: An uncommon fossorial species that is rarely encountered on the surface during the day except under leaf litter, debris, or fallen logs. The animals often emerge from burrows in the soil or other hiding places at night during rains to forage on the surface. Their food is mostly earthworms and termites, but they also eat soft-bodied larvae or instars of insects (beetles, hemipterans, earwigs, and orthopterans) (Wake 1983). Development takes place in the mother's oviduct, and when the embryos have exhausted the yolk supply at a total length of 10 to 15 mm they emerge from the egg membrane. As described earlier in the general account on caecilians, they then feed on the maternal nutritive secretions (hydrotroph) by using the specialized fetal dentition. In this species the fetal teeth are biconvex and topped with a row of small enamel spikes. Litter sizes are 2 to 10, and the young are 100 to 120 mm at birth. Females are sexually mature between 300 and 400 mm in total length (Wake 1980).

REMARKS: Taylor (1952b) recognized two geographic races in Costa Rica, *G. m. multiplicata* (Atlantic slope) and *G. m. proxima* (Pacific versant). He subsequently elevated them to specific status (Taylor 1968), but Savage and Wake (1972) demonstrated intergradation between Atlantic and Pacific populations in the area of Laguna Arenal, Tilarán, Guanacaste Province. Atlantic versant samples have lower annular counts (112 to 124 primaries and 84 to 107 secondaries) than Pacific slope specimens (119 to 133 primar-

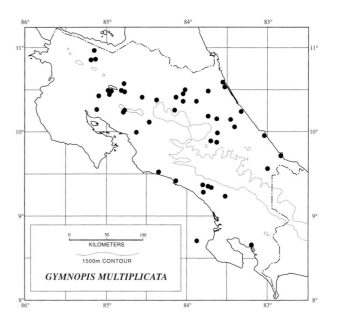

Map 5.3. Distribution of *Gymnopis multiplicata.*

ies and 97 to 117 secondaries), but the series from Laguna Arenal to Tilarán are intermediate. Wake (1988) provided a comprehensive review of *G. multiplicata.* Savage and Wake (1972) included the morphologically distinctive nominal form *Gymnopis oligozona,* known only from Atlantic versant Guatemala, within their overall concept of *G. multiplicata* with some reservations. Subsequently Nussbaum (1988) conclusively demonstrated the validity of the Guatemalan form, which he believed to be conspecific with Cope's (1866a) enigmatic *Siphonops syntremus.* The type and only known specimen of *S. syntremus* had apparently disappeared long ago, but Nussbaum proposed, following up on the suggestion of Dunn (1942a, 470), that Cope actually described the same specimen twice under different names, the second being *S.* (presumably for *Siphonops*) *oligozonus* (Cope 1877). If this interpretation is correct, the holotype of *S. oligozonus* (USNM 25187) is also the holotype of *S. syntremus,* they are objective synonyms, and the species should be called *Gymnopis syntrema.* The paper containing the original description was based on material collected by a Dr. Parsons from Belize and "the neighboring region of Honduras," and the holotype was stated to be from the latter locality. Duellman (1993) incorrectly cites Belize as the type locality. Taylor (1968) recognized *G. oligozona* as valid and placed *syntrema* in a new genus, *Copeotyphlinus* (although no specimens were then known to exist in any collection). Wake and Campbell (1983), misled by the striking differences between *G. syntrema (oligozona)* and *G. multiplicata,* redescribed the former as a new genus and species *(Minascaecilia sartoria)* based on additional material from Guatemala. Nussbaum (1988) established the identity of *G. oligozona* and *M. sartoria* in the course of concluding that *G. syntrema* should be applied to the species involved. *G. syn-*

trema differs most obviously from *G. multiplicata* (features for this species in parentheses) in being smaller, total length to 307 mm (480 mm); slender, total length to body width ratio 37 to 52 (23 to 37); and having the annular grooves cream to white (body coloration uniform). *G. syntrema* occurs on the Atlantic slope of Guatemala and in neighboring Belize (400 to 900 m).

DISTRIBUTION: Atlantic lowlands from Guatemala, Belize, and western Honduras to northwestern Panama (1–900 m) and on the Pacific versant on premontane slopes and in gallery forest in the lowlands in northwestern Costa Rica and on the lowlands of southwestern Costa Rica and adjacent Panama (1–1,400 m).

Genus *Oscaecilia* Taylor, 1968

The members of this genus are moderate-sized to large (maximum adult total lengths 334 to 750 mm), slender (total length to body width ratio 47 to 58) to extremely gracile (ratio 77 to 107) caecilians characterized as follows: eye covered by bone and not visible externally, the tentacle just below the nostril, secondary grooves present to absent, terminal shield present or absent, and no terminal keel. Additional internal diagnostic features are tentacular foramen near suture between nasopremaxilla and maxillopalatine bones, no temporal fossae, splenial teeth present; narial plugs present; no diastema between vomerine and palatine teeth.

The species of this genus form two distinctive groups in terms of habitus: the extremely attenuate forms *Oscaecilia bassleri* (Amazonian Ecuador and Peru), *O. elongata* (provenance unknown, possibly eastern Panama), *O. equatorialis** (western Ecuador), *O. hypereumeces** (southern Brazil), and *O. osae** (Costa Rica) and the stouter *O. koepckeorum* (Amazonian Peru), *O. ochrocephala* (Panama), *O. polyzona* (north-central Colombia)*, and *O. zweifeli* (Guyana). Within these two groups species having an unsegmented terminal shield are indicated by an asterisk (*).

All members of this genus are probably viviparous.

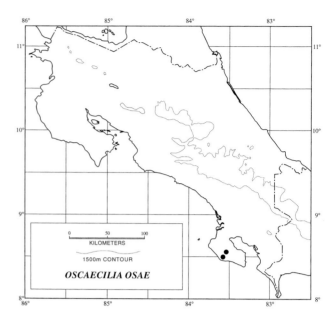

Map 5.4. Distribution of *Oscaecilia osae*.

Oscaecilia osae Lahanas and Savage, 1992

(plate 24; map 5.4)

DIAGNOSTICS: This very slender caecilian with a pink head and the eye not visible externally (fig. 5.3a) is unlikely to be confused with any other species.

DESCRIPTION: A moderate-sized (to 382 mm in total length), extremely attenuate form (total length to body width ratio 91); 232 primary annuli and no secondaries. Body uniform lavender, becoming lighter anterior and ventrally; head pinkish.

SIMILAR SPECIES: (1) Species of *Dermophis* have secondary folds and the eye visible. (2) *Gymnopis multiplicata* has the head and body uniform in coloration and has secondary folds. (3) The burrowing snakes *Anomalepis*, *Helminthophis*, *Leptotyphlops*, *Liotyphlops*, and *Typhlops* are covered by epidermal scales. (4) *Bachia blairi* has paired limbs and is covered by epidermal scales.

HABITAT: Lowland Wet Forest.

DISTRIBUTION: Known only from the Península de Osa of southwestern Costa Rica (3–40 m).

6 | Salamanders
Order Caudata Oppel, 1811 (Urodela)

O F LIVING groups of vertebrates, terrestrial salamanders most closely approach the earliest tetrapods in general habitus in having a body of moderate length (13 to 25 presacral vertebrae), short limbs, and long tails. However, several highly specialized fossorial genera, including *Oedipina* in Costa Rica, have reduced limbs, an attenuate body, and extremely long tails, and a number of strictly aquatic salamanders families (Sirenidae, Amphiumidae, and the genus *Proteus* of the Proteidae) are eeliform, with reduced limbs.

The order first appears in the fossil record in the Upper Jurassic (approximately 160 million years ago), and most of the modern families probably date back at least to late Cretaceous times (70 million years ago). Most members of this lineage are secretive semifossorial or fossorial forms, but some are saxicolous, a number are completely aquatic in fresh water, and many tropical forms (family Plethodontidae) are semiarboreal, scansorial, or arboreal.

Laypeople often confuse salamanders with lizards, but lizards are reptiles that are covered by cornified epidermal scales and that usually have five fingers and toes and a transverse cloacal opening (vent). The skin of salamanders is smooth and protected by moist mucous secretions; no more than four fingers are ever present, and the vent is longitudinal. A number of elongate fossorial forms (e.g., *Oedipina*) are sometimes mistaken for earthworms, which of course differ from salamanders in lacking eyes, teeth, and limbs (all present in caudate amphibians). The wormlike salamanders also differ from caecilians in having limbs, exposed functional eyes, and a long tail (the latter group has no limbs, eyes concealed under bone or skin and vestigial, and no tail or an extremely short one).

Among the primary characteristics (including many other features of internal morphology) that define the order are the following:

1. Body moderate or somewhat elongate (13 to 61 presacral vertebrae), tail long (20 to 100 or more caudal vertebrae)
2. Pectoral limbs, girdle, and sternal elements present; pelvic limbs and girdle present except in family Sirenidae
3. Pelvic limbs comprising three segments (femur, tibia-fibula, and foot)
4. Eyes present, exposed, and functional except in a few subterranean species
5. No tympanum, middle ear, or Eustachian tube
6. Larvae with external gills and four limbs (forelimbs only in Sirenidae), true teeth (enamel, dentine, or both present)
7. Body not annulate (although costal grooves may be present); no specialized cephalic chemosensory tentacles; no male intromittent organ
8. Palatoquadrate fused by processes to cranium
9. Atlas articulating to skull by atlantal cotyles and medioventral forward-directed process that meets the walls of foramen magnum on either side
10. Teeth present on upper and lower jaws and on vomer (often in two segments, an anterior series and a posterior patch underlying the parasphenoid)
11. Frontal and parietal bones distinct

The classification of salamanders is based to a considerable extent on differences in internal anatomy, especially the vertebrae and associated spinal nerve pathways, skull bones, trunk musculature, and ear structure. Salamanders are unusual among vertebrates in that many species are neotenic, reaching sexual maturity while retaining a larval morphology. All the members of several salamander families are obligatory neotenics (**), and a number of species in five other families (*) exhibit this feature (as indicated below). There are approximately 454 living species of salamanders placed in sixty-four genera and ten families as follows (fig. 6.1):

Suborder Sirenoidea Goodrich, 1930 (Meantes)
**Family Sirenidae (sirens)
Suborder Cryptobranchoidea Dunn, 1922
  *Family Hynobiidae (hynobiids)
**Family Cryptobranchidae (giant salamanders and hellbenders)
Suborder Salamandroidea Noble, 1931
  *Family Plethodontidae (lungless salamanders)
**Family Amphiumidae (amphiumas)
  *Family Rhyacotritonidae (torrent salamanders)
**Family Proteidae (mudpuppies and the olm)
  *Family Salamandridae (salamanders and newts)
  *Family Ambystomatidae (mole salamanders)
  *Family Dicamptodontidae (dicamptodontids)

The greatest diversity of salamander major groups occurs in the north temperate region, with centers in eastern and

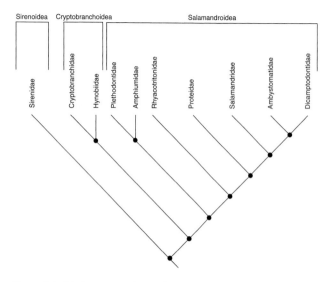

Figure 6.1. Phylogenetic relationships among living families of salamanders.

western Eurasia and eastern and western North America. A major radiation of genera and species of one family (Plethodontidae) has taken place on the tropical American mainland, but otherwise salamanders are virtually absent from the Tropics.

The Hynobiidae (eight genera) are found disjunctly from the Caspian Sea in Iran eastward to eastern China and from the Ural Mountains to Kamchatka and Japan and north to the Arctic Circle. The Cryptobranchidae (two genera) are found in eastern and central North America and western China and Japan. The Proteidae (two genera) are confined to eastern North America and southeastern Europe. Families endemic to North America include the Sirenidae (two genera), Amphiumidae (one genus), Rhyacotritonidae (one genus), Dicamptodontidae (one genus), and Ambystomatidae (two genera). The first two are restricted to the southeastern United States and the second two to the Far West and Northwest, and the last family is found over most of the continent (to 50° N and 18° S). The Salamandridae (sixteen genera) occur disjunctly in eastern and western North America, western Eurasia, and southeastern China and Vietnam (to 20° N). Members of the family Plethodontidae, the only salamanders found in Costa Rica, occur disjunctly in eastern and western North America (to 50° N) and from northern Mexico to Bolivia (15° S) and in southern Europe.

Although salamanders lack external and middle ear structures, the auditory system is highly variable and complex. Typically two sound-conducting ossicles are present: an anterior columella with a short stylus that articulates with the squamosal bone and a footplate that sits in the fenestra ovalis and abuts the wall of the inner ear, and an operculum that also lies in the fenestra ovalis. In aquatic larvae and fossorial forms sound is transmitted from the surrounding medium or substrate through the head tissue and jaws to the squa-

mosal, then via the columella to the inner ear (there is no operculum or associated muscle). In terrestrial and arboreal species an operculum is usually present and is variously connected to the pectoral girdle (suprascapula) by an "opercularis" muscle or muscles.

Traditionally (Kingsbury and Reed 1909) it has been thought that in these forms vibrations are transmitted from the substrate through the forelimbs and girdle via the "opercularis" muscle(s) and operculum to the inner ear. Wever (1985), however, discounts this interpretation and proposes that the opercularis muscle, when contracted, protects the inner ear against excessive stimulation by reducing movement of the operculum. He further concludes that the salamander ear is designed to function effectively in an aquatic situation and in most terrestrial forms during their return to water for breeding. Wever's view seems to have some validity, since both aquatic and terrestrial species appear to have good sensitivity only to low-frequency vibrations, approximately 100 to 600 Hz and 100 to 800 Hz, respectively.

The eyes in terrestrial and arboreal species have protective eyelids, as well as glands that are lacking in larval and neotenic, aquatic salamanders. Only the lower eyelid moves to cover the eye when it is closed, and there is no special elaboration of the lid as in frogs. Accommodation is accomplished by muscles that move the lens forward toward the cornea or the reverse when they are relaxed. Many plethodontid salamanders have some degree of binocular vision. The retinas of salamanders and anurans are made up of four kinds of receptor cells: single cones, double cones, red rods, and green rods. The red rods are similar to the rods in other vertebrates, but the green rods are unique to amphibians, although absent in caecilians, neotenic salamanders, and some subterranean salamanders. The visual cells all lack oil droplets. Salamanders lack color vision, and the several types of retinal cells seem to function differentially depending on the quality and intensity of light.

Aquatic salamanders use lateral undulations of the body and tail to propel themselves through the water, much like fishes. Terrestrial locomotion usually involves the sprawling walk (each foot on the substrate more than half the time) typical of nonmammalian tetrapods. In this mode of locomotion only one foot is off the ground at any one time, and the forefoot strikes the ground next after the hind foot on the same side (lateral sequence gait) followed by the hind foot striking the ground on the opposite side. Some terrestrial salamanders use sudden flipping movements based on maximum contraction of the trunk or tail muscles as a means of rapid locomotion to escape potential predators. These include rapidly coiling and uncoiling the body, flipping the tail to propel themselves through the air for short distances, and writhing laterally so that the coils push against irregularities in the substrate.

Primitively, salamanders have four fingers and five toes, with phalangeal formulas of 2-2-3-3 and 2-2-3-3-2. The first finger is often reduced in length, with or without the loss of a phalanx. Among the aquatic sirenids, one *(Pseudobranchus)* has only three digits on the hand. The toes may also be shortened, with or without the loss of a phalanx on toe IV. Toe V (the outermost) has been completely lost in several genera in diverse families. The aquatic amphiumas have three, two, or one fingers and toes through successive loss of I, IV, and II. The reductions in terrestrial forms are associated with burrowing habits.

Salamanders are primarily carnivorous as larvae and adults. Most feed on arthropods and earthworms, although crustaceans are important for some large aquatic species. Large species may also capture relatively large vertebrates. Prey are located primarily by vision and olfaction. In larvae and in strictly aquatic, neotenic salamanders the hyobranchial apparatus supports and moves the gills for respiration and expands and contracts the mouth cavity during feeding, and the tongue is rudimentary. Typically these salamanders lunge at the prey in a rapid strike motion and suck water and prey into the mouth by movements of the mouth floor.

Terrestrial and arboreal salamanders all have well-developed tongues that are protrusible, and in species with lungs the hyoid apparatus functions as a respiratory air pump and in tongue protrusion. In these forms the tongue is attached anteriorly, and the posterior half is protruded a short distance to strike the prey, which is trapped on its sticky surface and drawn into the mouth. In all plethodontid salamanders and some salamandrids the hyoid apparatus is no longer used in respiration. In a number of these forms the modified and elongate hyoid apparatus supports a projectile tongue composed of a sticky tongue pad and a sheathed stalk that contains folded muscle and skeletal elements and nerves and blood vessels. The tongue has lost its anterior attachment or has a very elastic one, and it can be projected by extension of the stalk structure to hit a prey item, then quickly retracted to bring the prey into the mouth. Some of these salamanders can project the tongue distances equal to 45 to 80% of their body length in ten milliseconds. All species of Costa Rican salamanders have highly projectile tongues without anterior attachments. Binocular vision is associated with the projectile tongue mechanism.

Salamanders, like anurans, are preyed on by an wide variety of invertebrates and of usually small vertebrate animals. In terrestrial and arboreal species, including all Costa Rican forms, the eggs are commonly eaten by insects, especially beetles, by other salamanders, and probably by small snakes. Snakes are doubtless the most important predators on adults. Since salamanders are eaten by almost any opportunistic feeding carnivore and there are very few accounts of predation on Costa Rican salamanders, this statement can be taken to apply generally for all species.

Hynobiids and cryptobranchids have relatively simple courtship patterns, and breeding takes place in an aquatic situation. Fertilization is external; in most cases males of these groups are excited by the sight of the eggs and fertilize them directly. Sirenids are presumed to have a similar mating pattern. In all other salamanders, whether aquatic or terrestrial breeders, fertilization is internal, accomplished by an unusual method of sperm transfer. In these forms the spermatophore, a special structure elaborated in the cloaca, is deposited by the male on objects (leaves, twigs, grass, etc.) floating in the water or on the substrate. Spermatophores are roughly conical, with a gelatinous basal portion capped by a spherical packet of sperm. Females pick up the sperm packet with the lips of their cloaca and store the sperm in a series of tubules, known collectively as spermatotheca, that open through a duct or ducts into the anterodorsal region of the cloaca. Sperm may be stored in this fashion from a few hours to two and a half years and are released when the female ovulates.

Courtship in these species involves complex behaviors that are variously distinctive at the family, generic, and specific levels. All involve males' approaching, pursuing, and stimulating females in various ways. Courtship usually culminates in the female's following the male, which then deposits a spermatophore that she picks up. In aquatic species eggs may be deposited singly or in clumps or strings (clutch sizes vary from 8 eggs to 5,000, usually 200 to 300) and range in size (including the protective layers) from about 3 to 18 mm. Eggs laid in terrestrial situations are large and richly provided with yolk (capsular diameter to 9.5 mm) and are usually stalked and laid singly or in small clumps (clutch sizes 9 to 37).

Salamanders exhibit a great variety of life history patterns, as briefly summarized below (after Duellman and Trueb 1986, 1994)

I. Fertilization external
   a. Eggs and larvae aquatic (Sirenidae, Hynobiidae, Cryptobranchidae)
II. Fertilization internal
   a. Eggs and larvae aquatic (Proteidae, Dicamptodontidae, Rhyacotritonidae, Amphiumidae, most Salamandridae and Ambystomatidae, and many Plethodontidae)
   b. Eggs terrestrial, larvae aquatic (Ambystomatidae: two species of *Ambystoma;* Plethodontidae: *Hemidactylium* and *Stereochilus*)
   c. Eggs terrestrial, larvae nonfeeding, terrestrial (Plethodontidae: *Desmognathus aeneus*)
   d. Eggs terrestrial, direct development, no larval stage (most Plethodontidae)
   e. Eggs retained in oviduct
      i. Ovoviviparous (Salamandridae: montane *Salamandra salamandra,* some *Mertensiella caucasica*)
      ii. Viviparous (Salamandridae: *Salamandra atra,* and some *Mertensiella luschani*)

The general features of aquatic salamander larvae were described and compared with those of other amphibians in chapter 4. Differences in the length of the gills and development of the tail fins characterize larvae developing in ponds or lakes (well developed), streams (moderately developed), and fast-moving mountain brooks (greatly reduced) as correlates of water flow. In forms having direct development, the enlarged gills function in aerial respiration. In viviparous salamanders, gills are associated with the thickened oviduct wall to form a placental connection for gaseous exchange between the embryo and the mother's circulatory system. Nutrition is also provided by a hydrotroph of mucoid secretions mixed with blood cells, which the embryo consumes. All Neotropical salamanders, including all Costa Rican species, have direct development.

Females of a number of aquatic breeders (*Dicamptodon, Amphiuma,* and several plethodontids) and species that lay eggs on land but have free-living larvae (*Ambystoma opacum* and two plethodontids) attend their egg clutches and usually are found near them, as is *Desmognathus aeneus.* So far as is known, the females of many species of lungless salamanders that have direct development also attend the eggs, but males have been observed to do so in a few species. Attendance probably functions primarily to reduce egg predation by arthropods and by other salamanders, including conspecifics.

The time of development from egg to metamorphosis is relatively long in species having aquatic eggs and larvae (106 to 1,826 days) but shorter in those having direct development (42 to 169 days).

Many salamanders, particularly of the family Salamandridae, secrete noxious to very toxic skin secretions and exhibit one or more behaviors that may discourage a predator or warn of toxicity (Brodie 1983). These may include one or more of the following: curling the body back to expose the belly; lashing, wagging, or undulating the tail; elevating or coiling the body; butting with the head; biting; and vocalizing. Many of these forms are brightly colored, and the behavior may emphasize this coloration. So far as is known, no Costa Rican species produces highly toxic secretions, but several brightly colored forms exude noxious ones that may discourage small predators, especially snakes.

Several modes of urotomy (tail breakage) besides that produced by trauma reflect different levels of specialized antipredator defenses in the family Plethodontidae and the genus *Chioglossa* (family Salamandridae). Tail breakage in all cases involves pseudautotomy, since the tail must be grasped or requires mechanical resistance to separate from the body (Wake and Dresner 1967). Whatever the mode of urotomy, the tail regenerates and develops vertebrae (unlike autotomy in lizards) and all other tissue except the notochord, in contrast to pseudautotomy in lizards and snakes (no regeneration).

SALAMANDER IDENTIFICATION: KEY FEATURES

The external morphology of salamanders provides a relatively limited number of characteristics that are useful for species identification. The following paragraphs describe the character conditions used in the keys and the diagnostics sections and clarify how to determine them where it would not be obvious to the general reader.

Costal grooves (fig. 6.2): these structures mark the boundaries of the body segments between the forelimbs and hind limbs. They are counted along one side of the body, with the first one recorded for the axilla and the last at the groin. The number of costal grooves between the adpressed limbs is counted between the longest finger and toe when the arm is pressed against the flank posteriorly and the leg anteriorly.

Head shape and proportions: the shape of the snout and the relative proportions of nostril diameter, head length (snout to gular fold), and width to standard length are often distinctive.

Hand and foot structure: the degree of hand and foot webbing, presence or absence of subdigital pads, and whether the digits are fused into a single element (are syndactylous) are variable within Costa Rican forms as follows (fig. 6.3):

SW: digits slightly webbed, well over two phalanges of longest digit free of web, digital tips free of web, rounded, and with a well-defined subterminal pad

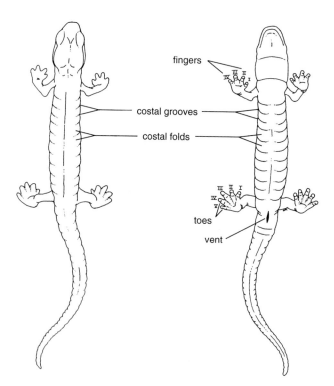

Figure 6.2. External characteristics of salamanders.

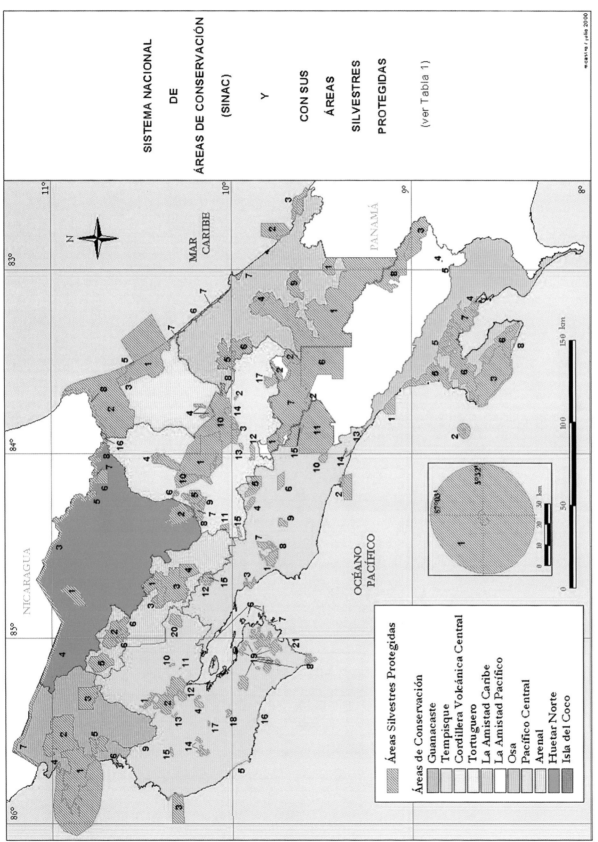

SISTEMA NACIONAL

DE

ÁREAS DE CONSERVACIÓN

(SINAC)

Y

CON SUS

ÁREAS

SILVESTRES

PROTEGIDAS

(ver Tabla 1)

Áreas Silvestres Protegidas

Áreas de Conservación

- Guanacaste
- Tempisque
- Cordillera Volcánica Central
- Tortuguero
- La Amistad Caribe
- La Amistad Pacífico
- Osa
- Pacífico Central
- Arenal
- Huetar Norte
- Isla del Coco

NICARAGUA

MAR CARIBE

PANAMÁ

OCÉANO PACÍFICO

N

150 km

100

50

0

30 km

20

10

0

Plate 1. National parks, biological reserves, wildlife refuges, and other protected areas in Costa Rica. See table 1.1 for key to numbered areas. Prepared by Instituto Nacional de Biodiversidad (INBio), Costa Rica.

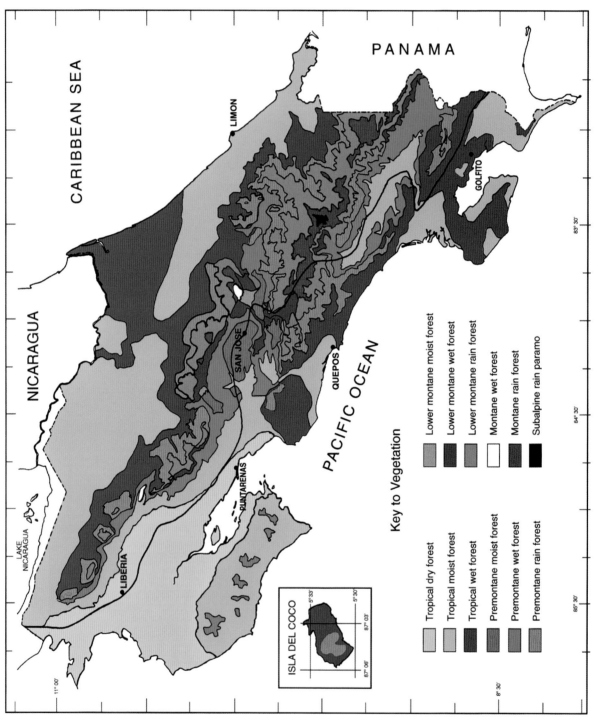

Key to Vegetation

Tropical dry forest		Lower montane moist forest
Tropical moist forest		Lower montane wet forest
Tropical wet forest		Lower montane rain forest
Premontane moist forest		Montane wet forest
Premontane wet forest		Montane rain forest
Premontane rain forest		Subalpine rain paramo

Plate 2. Generalized distribution of Costa Rican vegetation based on the Holdridge (1967) system as modified from Tosi (1969). Note that the distributions of the distinctive Pacific Lowland Humid Forests (Moist and Wet) are indicated by the same color code as their Atlantic Lowland counterparts; Tropical = Lowland.

Plate 3. Mangrove swamp (*manglar*). Costa Rica: Puntarenas: mouth of Río Grande de Tárcoles.

Plate 4. Lowland Dry Forest. Costa Rica: Guanacaste: Estero Real and Playa Naranjo, Santa Rosa National Park.

Plate 5. Lowland Dry Forest, in dry season, with cortez tree (*Tabebuia ochracea*) in flower. Costa Rica: Guanacaste: Lomas Barbudal Biological Reserve, about 30 m.

Plate 6. Lowland Dry Forest in dry season; *indio desnudo* tree (*Bursera simaruba*) in foreground and stand of terrestrial bromeliads (*Aechmea mariaereginae*).

Plate 7. Lowland Pacific Moist Forest. Costa Rica: Puntarenas: Carara National Park, about 40 m.

Plate 8. Lowland Pacific Wet Forest. Costa Rica: Puntarenas: Corcovado National Park, about 20 m.

Plate 9. Lowland Atlantic Wet Forest; canopy. Costa Rica, about 100 m.

Plate 10. Lowland Atlantic Wet Forest (interior). Costa Rica: Heredia: La Selva Biological Station, about 60 m.

Plate 11. Pacific slope Premontane Wet Forest. Costa Rica: San José: Río Chirripó Pacífico, about 1,200 m.

Plate 12. Atlantic slope Premontane Rainforest. Costa Rica: Alajuela: Monteverde Cloud Forest Preserve, 750 m.

Plate 13. Atlantic slope Premontane Rainforest. Costa Rica: Alajuela-Heredia: gorge of the Río Sarapiquí, 1,100 down to 600 m.

Plate 14. Lower Montane Rainforest (interior). Costa Rica: Alajuela-Puntarenas: Monteverde Cloud Forest Preserve, 1,550 m.

Plate 15. Lower Montane Rainforest; oak stand in background, deforestation in foreground. Costa Rica: Cartago: near El Empalme, about 2,200 m.

Plate 16. Montane Rainforest. Costa Rica: San José: Cordillera de Talamanca: Cerro de la Muerte: valley of the Río Savegre, about 2,800 m.

Plate 17. Subalpine Pluvial Paramo. Costa Rica: San José: Cordillera de Talamanca: Cerro de la Muerte, about 3,600 m.

Plate 18. Montane Rainforest. Costa Rica: San José: Cordillera de Talamanca: Cerro de la Muerte: oak stand in valley of the Río Savegre, about 3,300 m.

Plate 19. *Dermophis costaricense*. Costa Rica: Cartago: Tapantí National Park, about 1,500 m; see addendum p. 839.

Plate 20. *Dermophis costaricense*; lateral view of head. Costa Rica: Cartago: Tapantí National Park, about 1,500 m; see addendum p. 839.

Plate 21. *Dermophis parviceps*. Costa Rica: Heredia: Rara Avis, about 700 m.

Plate 22. *Gymnopis multiplicata*. Costa Rica: Puntarenas: Monteverde, 1,400 m.

Plate 23. *Gymnopis multiplicata*, view of head. Costa Rica: Puntarenas: Monteverde, 1,400 m.

Plate 24. *Oscaecilia osae*, dead on trail. Costa Rica: Puntarenas: Península de Osa: Los Patos, 40 m.

Plate 25. *Bolitoglossa alvaradoi*. Costa Rica: Alajuela: Fortuna District, about 250 m.

Plate 26. *Bolitoglossa cerroensis*. Costa Rica: Cordillera de Talamanca: Cerro de la Muerte, about 2,600 m.

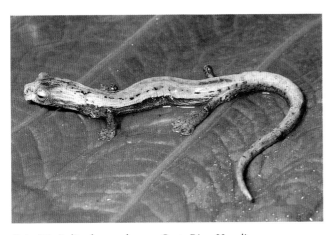

Plate 27. *Bolitoglossa colonnea*. Costa Rica: Heredia: La Selva Biological Station, about 60 m.

Plate 28. *Bolitoglossa compacta*. Panama: Chiriquí: Cerro Respingo, about 2,300 m.

Plate 29. *Bolitoglossa diminuta*. Costa Rica: Cartago: Tapantí National Park, about 1,500 m.

Plate 30. *Bolitoglossa epimela*. Costa Rica: Cartago: Río Chitaría, about 900 m.

Plate 31. *Bolitoglossa gracilis*. Costa Rica: Cartago: Tapantí
National Park, about 1,500 m.

Plate 32. *Bolitoglossa gracilis,* lateral view of head. Costa Rica:
Cartago: Tapantí National Park, about 1,500 m.

Plate 33. *Bolitoglossa lignicolor*. Costa Rica: Puntarenas:
Las Pilas, 245 m.

Plate 34. *Bolitoglossa lignicolor,* view of head. Costa Rica:
Puntarenas: Península de Osa, about 40 m.

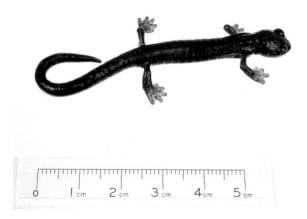

Plate 35. *Bolitoglossa sooyorum*. Costa Rica: Cordillera de
Talamanca: Cerro de la Muerte, about 2,700 m.

Plate 36. *Bolitoglossa minutula*. Costa Rica: Puntarenas: Las Tablas,
about 1,900 m.

Plate 37. *Bolitoglossa* sp., *B. nigrescens* complex. Costa Rica: Puntarenas: Las Tablas, about 1,800 m.

Plate 38. *Bolitoglossa pesrubra*. Costa Rica: Cordillera de Talamanca: Cerro de la Muerte, about 2,500 m.

Plate 39. *Bolitoglossa pesrubra*. Costa Rica: Cerro de la Muerte, *"B. torresi"* color phase, about 3,000 m.

Plate 40. *Bolitoglossa pesrubra*, female with eggs. Costa Rica: Cerro de la Muerte, 3,000 m.

Plate 41. *Bolitoglossa robusta*. Costa Rica: Puntarenas: Monteverde, about 1,400 m.

Plate 42. *Bolitoglossa schizodactyla*. Panama: Coclé: El Copé, about 1,100 m.

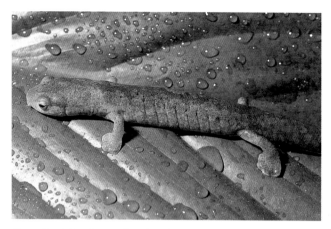

Plate 43. *Bolitoglossa schizodactyla*. Panama: Coclé: El Copé, about 1,200 m.

Plate 44. *Bolitoglossa striatula*. Costa Rica: Alajuela: Bijagua, 425 m.

Plate 45. *Bolitoglossa subpalmata*. Costa Rica: Heredia: Volcán Barva: Braulio Carrillo National Park, 1,500 m.

Plate 46. *Bolitoglossa subpalmata*. Costa Rica: Heredia: Volcán Barva: Braulio Carrillo National Park, 2,600 m.

Plate 47. *Nototriton abscondens*. Costa Rica: Heredia: Volcán Barva: Braulio Carrillo National Park, about 1,800 m.

Plate 48. *Nototriton gamezi*. Costa Rica: Alajuela: Monteverde Cloud Forest Preserve, about 1,600 m.

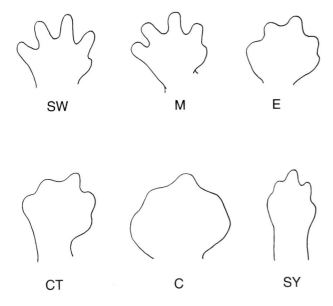

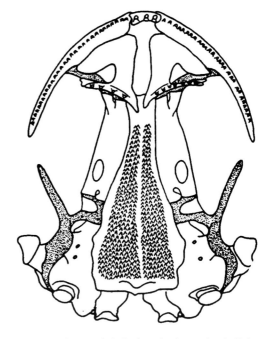

Figure 6.3. Foot structure in Central American salamanders: SW = slightly webbed, M = moderately webbed, E = extensively webbed, CT = completely webbed, some digits projecting from web, C = completely webbed, some digits projecting slightly from web, SY = syndactylous, no web and digits bound tightly together.

Figure 6.4. Ventral view of plethodontid salamander skull showing dentition. Teeth are present on the premaxillary and maxillary and on the anterior and posterior portions of the vomer.

M: digits moderately webbed, about two phalanges on longest digit free of web, digital tips rounded to truncate, subterminal pads on two or three longest digits

E: digits extensively webbed, digital tips rounded or slightly pointed and separated by definite indentations, with or without subterminal pads

CT: digits completely webbed, only longest digits projecting from web, little or no indentation between digits, and subterminal pads present

C: digits completely webbed, only longest digits projecting slightly from web, little or no indentation between digits, and no subterminal pads

SY: no webs, but digits inseparable for most of length (syndactylous), usually no indentations between digits, but sometimes slight ones between some digits, digital tips rounded to pointed, no subterminal pads

Additional mensural features: the relative proportions of trunk length (distance between axilla and groin), greatest trunk width, limb length, width of feet, and tail length to standard length are often used to differentiate among adult salamanders that are otherwise morphologically very similar in appearance.

Differences in coloration, especially in life, are among the most diagnostic features within the genus *Bolitoglossa* and among some species of *Oedipina*. Most members of the genus *Nototriton* and the *Oedipina uniformis* group, however, are dark brown to black above (lighter below) and lack a distinctive pattern.

Finally, three internal features are important when working with preserved material:

Maxillary teeth: the total number of teeth on the two maxillary bones (including missing teeth indicated by gaps) (fig. 6.4)

Vomerine teeth: total number of ankylosed teeth on the two anterior patches immediately posterior to each choana (fig. 6.4)

Sublingual fold: a fleshy fold running transversely across the mouth below the tongue

It is well to bear in mind that teeth (both maxillary and vomerine) generally increase in number with increasing age and size.

Size alone will distinguish some forms. The standard length measurement for salamanders is the distance from the tip of the snout to the posterior margin of the vent. In the following sections and elsewhere in this book, salamander size refers to total length as follows:

Tiny: under 50 mm
Small: 50 to 100 mm
Moderate: 100 to 200 mm
Large: 200 to 500 mm
Very large: 500 to 1,200 mm
Giant: greater than 1,200 mm

The approximate maximum total length is given in each species account. Since salamanders frequently lose their tails and regenerate ones that approach the originals very closely in appearance and length, it is difficult to be sure that an original or complete tail is present. Consequently the maximum standard length is also indicated. In Costa Rica the largest species of salamander is 260 mm in total length and the smallest is 66 mm.

Sexual dimorphism in size is not marked in Costa Rican salamanders. Adult males, however, have the terminus of the nasolabial groove swollen or elongated into a cirrus, have the premaxillary teeth elongate and piercing the lip, and develop a distinctive large mental gland under the chin during the breeding season.

The largest salamanders are the aquatic species *Andrias davidianus* (total length 1,520 mm) of China, *A. japonicus* (1,440 mm) of Japan, and *Amphiuma tridactylum* (1,015 mm) of the southeastern United States. The largest terrestrial species is *Dicamptodon ensatus* (to 325 mm) of the Pacific Northwest region of North America. The smallest salamanders belong to the terrestrial Mexican genus *Thorius* (to about 27 mm in total length). The largest Costa Rican salamander is *Bolitoglossa robusta,* with a total length of 260 mm.

KEY TO THE GENERA OF COSTA RICAN SALAMANDERS

1a. Costal grooves not more than 14; body not markedly elongate; limbs moderate to well developed; fewer than 7 costal folds exposed when limbs adpressed to sides of body 2
1b. Costal grooves 17–20; body extremely elongate; limbs very short; 6–15 costal folds exposed when limbs adpressed to sides of body; well-developed sublingual fold . . . . *Oedipina* p. 145)
2a. Hands and feet narrow, width of hand less than length (fig. 6.3); well-developed sublingual fold . . . . *Nototriton* (p. 139)
2b. Hands and feet broad, width of hand equal to or exceeding length (fig. 6.3); no sublingual fold . . . . *Bolitoglossa* (p. 127)

Family Plethodontidae Gray, 1850

Lungless Salamanders, Salamandras, Escorpiones

Salamanders of this, the largest family (twenty-nine genera, 299 species) in the order are tiny (to a maximum total length of 27 mm) to moderately large (325 mm) forms including aquatic, semiaquatic, terrestrial, fossorial, scansorial, saxicolous, and arboreal species. Members of this family are unique in having a distinctive nasolabial groove on the side of the head, running from the lower margin of the nostril to the edge of the upper lip. This structure picks up chemical cues from the substrate and carries them to the olfactory

sac; it is used in prey and mate location and inter- and intraspecific communication. The groove, however, is not always easy to see with the unaided eye in small specimens, but this is not a problem for users of this book, since all Costa Rican salamanders are plethodontids.

Another characteristic feature of plethodontids is the absence of lungs and the ventrally located ypsiloid cartilage and associated muscles that in other salamanders control the shape of the lungs. Consequently respiration is primarily cutaneous, supplemented by gaseous exchange across the buccopharyngeal membranes. A number of neotenic North American species are completely aquatic and use external gills as the principal respiratory organs. Lunglessness is known elsewhere among salamanders only in the two species of the East Asian genus *Onychodactylus* (family Hynobiidae), although the lungs are greatly reduced in several other genera.

Other features (internal) distinguishing the family from other salamanders are exoccipital, prootic, and opisthotic bones fused, internal carotid foramen absent; no angular, lacrimal, or pterygoid bones in adults (pterygoid in larvae); vertebrae opisthocoelous, all but first three spinal nerves exiting intravertebrally; vomerine teeth in anterior and posterior patches or two patches continuous; karyotype 2N = 26 or 28.

The family consists of two subfamilies distinguished as follows (Wake 1966):

Desmognathinae: Four larval gill slits and four epibranchials in larvae; posterior atlantal area of atlas raised for muscle attachment involved in mouth opening, in which mandibles are held rigid and skull is raised; hypapophysial keels on anterior vertebrae: *Desmognathus* (fourteen species), *Leurognathus* (one species), and *Phaeognathus* (one species); eastern North America.

Plethodontinae: one to three larval or embryonic gill slits and one to three epibranchials; no enlargement of atlantal area of atlas, mouth opened by lowering mandible; no hypapophysial keels on vertebrae. This taxon contains four distinctive subgroups of genera differing principally in life history, modifications of the tongue, and supportive hyobranchial skeleton and musculature for different feeding modes (Lombard and Wake 1987; Wake and Larson 1987; Deban, Wake, and Roth 1997):

1. Aquatic larvae; protrusible tongue attached to mandible by a moderately elongate and stretchable genioglossus muscle; no anterior extensions on first basibranchial; second basibranchial ossified: eastern North America: *Hemidactylium* (one species)
2. Aquatic larvae; freely projectile tongue, no muscle attachment to mandible; no anterior extensions on first basibranchial; second basibranchial ossified: eastern North America to Texas: *Eurycea* (twelve species), *Gyrinophilus* (three species), *Haideotriton* (one species), *Pseudo-*

triton (two species), *Stereochilus* (one species), *Typhlo-molge* (two species), and *Typhlotriton* (one species)

3. Direct development, no larvae; protrusible tongue attached to mandible by a stout genioglossus muscle (as in Desmognathinae) or attached projectile tongue with moderately elongate and stretchable muscle; anterior extensions present on first basibranchials; second basibranchials ossified: North America: *Aneides* (five species) and *Plethodon* (forty-two species) have protrusible tongues, *Ensatina* (two species) has a projectile tongue

4. Direct development, no larvae; attached projectile tongue with greatly elongate and stretchable genioglossus muscle or freely projectile tongue without that muscle; no anterior extensions on first basibranchials; no second basibranchials: western North America: *Batrachoseps* (eleven species) has an attached tongue; all other genera have a freely projectile tongue as follows: Europe and western North America: *Hydromantes* (ten species); Mexico: *Chiropterotriton* (nine species), *Ixalotriton* (one species), *Lineatriton* (one species), *Parvimolge* (one species), *Thorius* (twenty-one species); Mexico and Central America: *Dendrotriton* (six species), *Nyctanolis* (one species), *Pseudoeurycea* (thirty-two species); Mexico to northern South America: *Oedipina* (twenty-three species); Central America: *Bradytriton* (one species), *Cryptotriton* (six species), *Nototriton* (twelve species); Mexico to Bolivia: *Bolitoglossa* (seventy-six species)

There is substantial sexual dimorphism in all genera within the family. Females attain larger sizes that males. Adult males of most species have a circular mental gland under the chin. During courtship the male rubs the gland on various parts of the female's body or introduces the gland's secretions into her superficial circulation by biting her or abrading the skin. Breeding males of most genera have the premaxillary teeth elongate and protruding through the lip. The margin of the lip surrounding the terminus of the nasolabial groove becomes swollen and may be elongated into a cirrus in temperate zone plethodontines. The swollen protuberances are present year round in all tropical salamanders studied to date, and most of these species are apparently reproductive throughout the year. In addition, breeding males have papillate vents.

Unspecialized pseudautotomy, in which the tail is twisted off (as in the snake *Thamnophis*), is characteristic of many plethodontids (e.g., *Desmognathus*, *Leurognathus*, and *Phaeognathus*). In these forms the tail has a thick base, breakage may occur anywhere along the length of the tail, and there is no specialization for wound healing. One kind of specialized pseudautotomy is found in other forms (e.g., *Batrachoseps* and *Plethodon*) having the base of the tail slender and the fracture occurring anywhere along its length. Additionally, the skin and muscle break at different points (posterior portion of last retained vertebra and anterior por-

tion of first lost vertebra, respectively), so that a free cylinder of skin closes over the wound to initiate healing and regeneration (Wake and Dresner 1967).

An even more specialized form of pseudautotomy is found in species of the genera *Ensatina*, *Chiropterotriton*, many *Bolitoglossa*, *Hemidactylium*, and some *Pseudoeurycea*. These salamanders have a conspicuous constriction at the base of the tail immediately behind the vent, and urotomy most frequently occurs at that site. The differential breakage of skin and muscle in this case occurs at the posterior and anterior portions, respectively, of the first caudal segment. The free cylinder of skin then closes over the terminal wound, and healing and tail regeneration commence. Although they lack well-developed tail constrictions, several other genera of plethodontids (e.g., *Lineatriton*, *Oedipina*, and *Thorius*) and perhaps all other Neotropical salamanders have the same caudal morphology and breakage pattern involving the first caudal vertebra as described above and also may be considered representatives of this type of specialized pseudautotomy.

As this summary shows, the family as a whole is distributed in three major and two minor disjunct centers. Areas of high diversity at both the generic and specific levels are eastern North America, Pacific costal North America, and the Neotropical region; one genus (*Hydromantes*) occurs in southern Europe, and two species of the genus *Plethodon* and one of *Aneides* occur in the Rocky Mountain region of western North America.

Genus *Bolitoglossa* C. Duméril, Bibron, and A. Duméril, 1854

This is the largest genus of salamanders (seventy-six species) and is composed of small (to 66 mm in total length, 31 mm in standard length) to moderately large (to 260 mm in total length, 131 mm in standard length), somewhat robust forms having well-developed limbs, hands, and feet, the digits partially or completely webbed, relatively stout and moderate tails (44 to 58% of total length) and no sublingual fold. Other features (all osteological) that in combination distinguish the genus from other Neotropical salamanders are premaxillaries fused; prefrontals present; parietals approaching one another closely on midline; tibial spur well developed or reduced to a ridge; arrangement of wrist skeleton unspecialized (ulnare and intermedium discrete) and that of ankle specialized (distal tarsals four and five fused); and 14 trunk vertebrae.

Species of *Bolitoglossa* are unlikely to be confused with any other salamanders from lower Central America, since all members of the genera *Nototriton* and *Oedipina* have reduced limbs and a sublingual fold. *Bolitoglossa* are much more robust and have a shorter tail than members of the other two genera, even though one *Oedipina* (*O. collaris*) has a total length approaching that of the largest *Bolitoglossa*.

The members of this genus occupy a variety of habitats and include terrestrial, log dwelling, semifossorial, saxicolous, scansorial, epiphyllous, and arboreal forms. The degree of digital webbing generally correlates with ecology on a spectrum from terrestrial to arboreal. Forms with slight to moderate webbing, rounded digital tips, and subterminal pads (SW) are mostly terrestrial. Species with extensive (E) to complete webbing (C), are scansorial to arboreal. In the most specialized arboreal forms the digital tips are often pointed, and no subterminal pads are present (fig. 6.3).

There is considerable sexual dimorphism in species of this genus. As in other plethodontids, females attain greater sizes than males. Breeding males have elongate premaxillary teeth that protrude through the lip, nasolabial protuberances, a mental gland, and papillate vents. In addition, males tend to have broader heads, longer legs, and larger hands and feet than females.

Taylor (1944, 1952b) recognized those species having discrete digits and slight webbing as forming a genus, *Magnadigita* (type *Bolitoglossa rufescens* Taylor, 1944, by original designation), distinct from the extensively to fully webbed *Bolitoglossa*, but since these features may appear in otherwise closely related species Wake and Brame (1963a,b) retained all forms in a single genus. However, should the beta section (see below) be recognized as a distinct genus, this name is applicable but would have a very different composition than proposed by Taylor (1944, 1952b).

Wake and Lynch (1976) divided the genus into two major series based on the degree of specialization for tail breakage and reductions in size of several bones:

Alpha series: caudosacral and first caudal vertebrae not specialized, tail not markedly constricted at base; tibial spur and frontal bones reduced (forty-nine species): San Luis Potosí, Mexico, to southern Brazil and northern Bolivia, exclusive of the major part of the Yucatán Peninsula, on the Atlantic slope and from Oaxaca, Mexico, to Ecuador on the Pacific versant, exclusive of the dry lowlands from Guatemala to Costa Rica

Beta series: caudosacral and first caudal vertebrae with a specialized arrangement of transverse processes, tail markedly constricted at base; tibial spur and prefrontal bones well developed (twenty-seven species): Guerrero and Veracruz, Mexico, to Guatemala and Honduras and Costa Rica

Wake and Lynch (1976) recognized ten species groups within the alpha series, most of which are monotypic and based on phenetics. Five of these occur in our area of interest, with the number of Costa Rican species indicated (*adspersa* group, one species; *epimela* group, two species; *lignicolor* group, two species; *striatula* group, two species; *subpalmata* group, eight species). Only a single species of the beta division in the monotypic *alvaradoi* species group is found in Costa Rica.

The general distribution of *Bolitoglossa* includes lowland to montane zones (to 3,610 m in elevation) under humid climatic conditions, from northeastern and southern Mexico to Amazonian southern Brazil, Amazonian Bolivia, and Pacific slope Ecuador.

KEY TO COSTA RICAN SALAMANDERS OF THE GENUS *BOLITOGLOSSA*

1a. No transverse fleshy fold or ridge across frontal region. 2

1b. A distinct transverse fleshy fold or ridge across frontal region; hands and feet completely webbed . . . . *Bolitoglossa colonnea* (p. 131)

2a. No light ring around base of tail 3

2b. A cream to reddish orange ring around base of tail; hands and feet moderately webbed; large, adult males 59–114 mm in standard length, adult females 65–134 mm . . . . *Bolitoglossa robusta* (p. 136)

3a. Hands and feet flattened, palmate; webbing complete, only longest digits projecting slightly, at most, from webbed pad (fig. 6.3C), usually no subterminal pads; no definite indentations between digits 4

3b. Hands and feet not palmate; webbing incomplete; tips of most digits free of webs; definite indentations present between most digits; digits rounded (fig. 6.3M,E); subterminal pads present 8

4a. Venter cream or white, immaculate or with a broad median dark stripe or several to many longitudinal tan stripes 5

4b. Venter mostly black or brown, uniform or with light streaks, spots, or mottling 6

5a. Paired dorsal and lateral longitudinal dark stripes on a light background; venter cream to yellow, usually with several to many narrow tan streaks running length of body; underside of tail cream, usually with a narrow median tan stripe; adult males 37–54 mm in standard length, adult females 40–62 mm . . . . *Bolitoglossa striatula* (p. 138)

5b. Upper surfaces black or brown and no longitudinal stripes; venter immaculate yellow or with a broad median dark brown stripe; underside of tail clear yellow; adult males 38–61 mm in standard length, adult females 46–62 mm . . . . *Bolitoglossa schizodactyla* (p. 137)

6a. No subterminal pads; large species, adults more than 42 mm in standard length 7

6b. Subterminal pads weak but present; small species, adult males 28–36 mm in standard length, adult females 30–37 mm; dorsum uniform black brown or median field uniform dull orange or bordered by narrow dark stripe . . . . *Bolitoglossa minutula* (p. 134)

7a. Total maxillary teeth 27–48 in adults; digital webs fleshy, thickened; dorsum usually creamy tan with many short darker streaks; venter dark brown with or without lighter mottling or streaks; adult males 47–68 mm

in standard length, adult females 46–81 mm. . . . . . .
. . . . *Bolitoglossa lignicolor* (p. 133)

7b. Total maxillary teeth 47–89 in adults; dorsal ground color olive green to dark brown, nearly uniform or with some black markings or broad, irregular fawn band down middle of back; adult males 57–65 mm in standard length, adult females 68–79 mm . . . .
. . . .*Bolitoglossa alvaradoi* (p. 129)

8a. Digits moderately webbed; one or two phalanges free of webs on fingers II–III and toes II–V (fig. 6.3M) . 9

8b. Digits extensively webbed; webs including all phalanges on finger IV and toe V (fig. 6.3E); dark above and below; total maxillary teeth 32–43; adult male 43 mm in standard length, adult females 38–46 mm.
. . . . *Bolitoglossa epimela* (p. 132)

9a. Dorsum and venter almost uniform black 10

9b. Dorsum with light flecks, spots, blotches, or stripes; venter not black, or dark with light markings 11

10a. Total maxillary teeth 45–78; 0–3 costal folds between adpressed limbs; adult females 38–67 mm in standard length. 16

10b. Total maxillary teeth 17–37; 2½–4 costal folds between adpressed limbs; adult females 58–94 mm in standard length. . . . .*Bolitoglossa nigrescens* (p. 135)

11a. Dorsum and venter lavender to purplish brown (sometimes brown to black in preservative), with a pair of irregular dorsolateral light stripes and/or numerous light flecks and/or small spots 12

11b. Dorsum and venter light or dark, uniform or with small spots or blotches, but never purplish brown with light dorsal and ventral stripes or dorsal and ventral flecks and/or spots. 13

12a. Dorsum and venter with yellowish light spots and a pair of irregular dorsolateral light stripes; total maxillary teeth 19–45 in adults; costal folds between adpressed limbs ½–1 in adult males, 1–2 in adult females; adult males 54–72 mm in standard length, adult females 49–76 mm . . . .
. . . .*Bolitoglossa cerroensis* (p. 130)

12b. Dorsum and venter with numerous yellow flecks and small spots; no irregular dorsal light stripes; total maxillary teeth 43–79 in adults; costal folds between adpressed limbs 0–½ barely overlapping in adult males, 0–1 in adult females; adult males 58–65 mm in standard length, adult females 34–72 mm . . . .
. . . . *Bolitoglossa sooyorum* (p. 137)

13a. Variously colored, but never with a broad middorsal light field bordered by a scalloped dark brown lateral stripe on each flank; adults more than 35 mm in standard length. 14

13b. A broad middorsal light field bordered by a scalloped dark brown lateral stripe on each flank; adult female 31 mm in standard length. . . . .
. . . . *Bolitoglossa diminuta* (p. 132)

14a. Ground color various but no midventral dark stripe;

adult males 44–57 mm in standard length, adult females 53–67 mm. 15

14b. Ground color yellowish with sparse and irregular dark dorsal markings and a midventral dark stripe; adult males 38 mm in standard length, adult females 38–42 mm . . . . *Bolitoglossa gracilis* (p. 133)

15a. Venter light, variously marked with dark spots, blotches, or marbling, or dark with some white, red, or yellow spots or punctations; dorsum never with stripes; adult males 53–57 mm in standard length, adult females 51–67 mm. 16

15b. Venter uniform brown (black in preservative); dorsum dull to reddish brown (black in preservative) with light (reddish orange to yellow in preservative) blotches that may fuse into a pair of irregular, indistinct dorsolateral light stripes; adult males 44–53 mm in standard length, adult females 68–74 mm . . . .
. . . . *Bolitoglossa compacta* (p. 131)

16a. Total maxillary teeth in adults 58–78; tail relatively long, 51–58% of total length; found in the Cordillera de Guanacaste, Cordillera de Tilarán, and Cordillera Central . . . . *Bolitoglossa subpalmata* (p. 139)

16b. Total maxillary teeth in adults 45–65; tail moderate, 49–51% of total length; found in the Cordillera de Talamanca . . . . *Bolitoglossa pesrubra* (p. 135)

Bolitoglossa alvaradoi Taylor, 1954a

(plate 25; map 6.1)

DIAGNOSTICS: A moderate-sized member of the genus having completely webbed, flattened, and palmate hands and feet (fig. 6.3C), a uniform black to dark brown venter, and the dorsum olive green to dark brown, marked with black spots or blotches or most of middorsal area light in color.

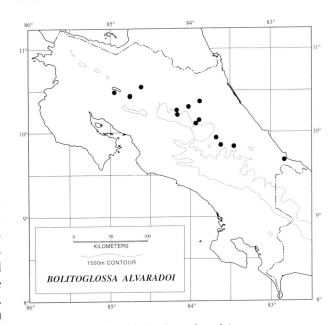

Map 6.1. Distribution of *Bolitoglossa alvaradoi.*

DESCRIPTION: Adults 106 to 155 mm in total length; adult males 57 to 65 mm in standard length, adult females 68 to 79 mm; tail moderate, 46 to 49% of total length; eyes large and protuberant; 42 to 89 maxillary teeth in adults extending nearly to back of eye; 26 to 30 vomerine teeth; 1½ to 2 costal folds between adpressed limbs in males, 3 or 4 in females; hands and feet palmate without subterminal pads; proportions as percentages of standard length: head width 14 to 17%; leg length in adults 23 to 27%. Upper surface olive to dark brown marked with small black spots or black blotches on upper head and caudal surfaces and flanks or most of dorsum and tail covered by pinkish to fawn frosting; venter uniform black.

SIMILAR SPECIES: *Bolitoglossa lignicolor* usually has a broad light-colored dorsal field streaked or washed with darker brown and a venter marked with light spots or streaks and fewer maxillary teeth (27 to 48) in adults.

HABITAT: Lowland Moist and Wet Forests and Premontane Rainforest.

BIOLOGY: A nocturnal arboreal form rarely seen but probably living in the canopy. It has been collected on several occasions from tank bromeliads.

REMARKS: Because of the great variation in coloration, several names have been applied to this species. Taylor (1952b) listed the only specimen (BMNH 96.10.8.77) collected from Costa Rica (Alajuela: San Carlos: Boca de Arenal) up to that time as *Bolitoglossa platydactyla* but questioned that allocation, which had been made by Dunn (1926). This is the specimen cited by Günther (1902) as part of his *Spelerpes variegatus*. In Dunn's unpublished notes he later referred this example to *Oedipus mexicanus*. None of these names are applicable to Costa Rican material. Later Taylor (1954a) described as new both *Bolitoglossa arborescandens* and *B. alvaradoi*, from Moravia de Chirripó, Cartago Province. The former is based on a single male (holotype: KU 34925) having only a few small black dorsal spots and the latter on three males and the female holotype (KU 30484) having extensive middorsal and caudal frosting. Comparison of these examples and more recently collected specimens shows that the differences Taylor emphasized are extremes in the color variation for this species and are bridged by intermediate conditions in other individuals. As first reviser I select the name *B. alvaradoi* over *B. arborescandens* for this species under article 24.2, *International Code of Zoological Nomenclature*. This is the only Costa Rican salamander of the *Bolitoglossa* beta division.

DISTRIBUTION: Humid lowland and premontane areas on the Atlantic versant of Costa Rica (15–1,116 m).

Bolitoglossa cerroensis (Taylor, 1952b)

(plate 26; map 6.2)

DIAGNOSTICS: A moderate-sized *Bolitoglossa* with moderate digital webs (fig. 6.3M) and a purplish brown ground color marked laterally and ventrally with yellow

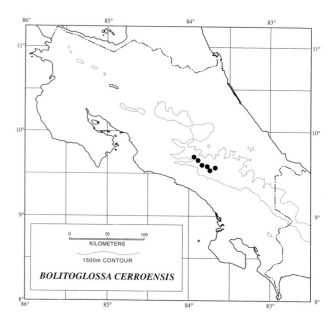

Map 6.2. Distribution of *Bolitoglossa cerroensis*.

green stippling and streaking. The light markings usually form a pair of irregular dorsolateral light stripes or vertical light bands along the costal grooves.

DESCRIPTION: Adults 111 to 156 mm in total length; adult males 54 to 72 mm in standard length, adult females 49 to 47 mm; tail moderate, 46 to 52% of total length; eyes large, protuberant; 19 to 45 maxillary teeth in adults, extending to middle of eye; 19 to 25 vomerine teeth in adults; ½ to 1 costal folds between adpressed limbs in adult males, 1 to 2 in adult females; hands and feet moderately webbed, about two phalanges on longest digit free of web; with subterminal pads on two or three longest digits; proportions as percentages of standard length: head width 13 to 15%; leg length in adults 22 to 29%. All surfaces purplish brown marked with yellow green streaking and stippling, especially laterally and ventrally; usually yellow green irregular, paired dorsolateral longitudinal stripes or narrow vertical bands along costal grooves. Preserved specimens purplish with cream markings, rarely appearing uniform.

SIMILAR SPECIES: (1) *Bolitoglossa sooyorum* is lavender brown with distinct yellow flecking, longer limbs (one costal fold or none between adpressed limbs), and more maxillary teeth (43 to 79 in adults). (2) *Bolitoglossa compacta* has reddish markings, a broader head, and shorter legs.

HABITAT: Montane Rainforest in talus on steep slopes.

BIOLOGY: This uncommon salamander co-occurs with *B. sooyorum* at several localities. The smallest available juvenile is 40 mm in standard length.

REMARKS: A member of the *Bolitoglossa adspersa* species group.

DISTRIBUTION: Humid montane zone of the Cordillera de Talamanca of Costa Rica (2,530–2,990 m).

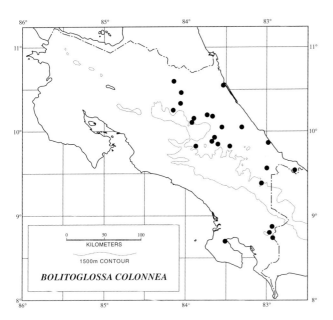

Map 6.3. Distribution of *Bolitoglossa colonnea*.

Bolitoglossa colonnea (Dunn, 1924a)

(plate 27; map 6.3)

DIAGNOSTICS: This uncommonly seen small to moderate-sized brownish gray species having fully webbed digits (fig. 6.3C) is immediately recognized by the unique fleshy fold that extends across the head at the level of the anterior margin of the eyes.

DESCRIPTION: Adults 56 to 113 mm in total length; adult males 32 to 54 mm in standard length, adult females 38 to 53 mm; tail moderate, 44 to 49% of total length; eyes large and protuberant; 0 to 6 maxillary teeth in adults, rows not extending to eye; 17 to 36 vomerine teeth in adults; 1½ to 2½ costal folds between adpressed limbs; hands and feet completely webbed, flattened, and palmate; no subterminal pads; proportions as percentages of standard length: head width 15 to 20%; leg length in adults 19 to 24%. Gray brown above, lighter brown below, often some dark mottling or short streaks on dorsum; interorbital ridge usually cream, demarcated anteriorly and posteriorly by narrow dark transverse stripes; limbs darker than body; hands and feet cream.

SIMILAR SPECIES: There are none.

HABITAT: Lowland Moist and Wet Forests and Premontane Rainforest.

BIOLOGY: A nocturnal, epiphyllous salamander characteristically found on herbaceous vegetation a meter or so above the ground. Bruce (1997) found twelve to sixteen large, yolked ovarian follicles in a sample of twelve gravid females with no seasonality indicated. Males appear to produce sperm throughout the year as well. The smallest known juvenile is 17 mm in standard length.

REMARKS: *B. colonnea* is a member of the *Bolitoglossa*

striatula species group. This species is listed from Barro Colorado Island, Panama Province, Panama, by Frost (1985) but not by Rand and Myers (1990).

DISTRIBUTION: Humid lowlands and premontane slopes of the Atlantic versant of Costa Rica and western Panama and Pacific versant southwestern Costa Rica and Cerro Campana, Panama (40–1,245 m).

Bolitoglossa compacta Wake, Brame, and Duellman, 1973

(plate 28; map 6.4)

DIAGNOSTICS: An uncommon moderate-sized, rather robust species that has moderately webbed fingers and toes (fig. 6.3M) and is brown with some reddish to orange markings forming blotches, a middorsal field, or indistinct paired dorsolateral stripes.

DESCRIPTION: Adults 100 to 143 in total length; adult males 44 to 53 mm in standard length, adult females 68 to 74 mm; tail 44 to 50% of total length; eyes large, extending somewhat beyond midpoint of eye, protuberant; 11 to 50 maxillary teeth in adults, extending somewhat beyond midpoint of eye; 19 to 33 vomerine teeth in adults; 1½ to 2½ costal folds between adpressed limbs; hands and feet moderately webbed, at least two phalanges free of web on two or three longest digits; subterminal pads present; proportions as percentages of standard length: head width 14 to 16%; leg length 22 to 24% in adult males, 25 to 28% in adult females. Dorsum and venter brown (black in preservative) with irregular reddish (orange to yellow in preservative) to orange spots and blotches, sometimes forming indistinct dorsolateral stripes or broad reddish dorsal field; venter uniform; iris pale brown.

SIMILAR SPECIES: (1) *Bolitoglossa cerroensis* has yellowish green markings and a narrower head and longer legs.

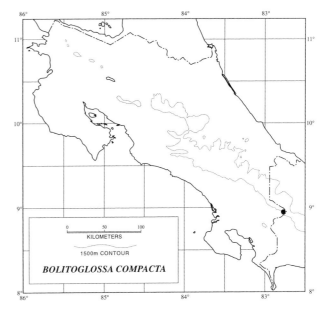

Map 6.4. Distribution of *Bolitoglossa compacta*.

(2) *Bolitoglossa pesrubra* and *B. subpalmata* are smaller (30 to 61 mm in standard length) and have more maxillary teeth (58 to 78 in adults).

HABITAT: Lower Montane Rainforest.

BIOLOGY: This nocturnal, scansorial salamander is found on the ground or usually on low vegetation.

REMARKS: This species is a member of the *Bolitoglossa subpalmata* group. It was first reported from Costa Rica by Lips (1993a).

DISTRIBUTION: Humid lower montane areas on or near the Costa Rica–Panama border; on Pacific versant Costa Rica (1,650–1,980) and Atlantic and Pacific slopes in western Panama (1,810–2,780 m).

Bolitoglossa diminuta Robinson, 1976

(plate 29; map 6.5)

DIAGNOSTICS: This rare form is the smallest species within the genus and is characterized by moderate digital webbing (fig. 6.3M), a broad middorsal light field with scalloped lateral margins, and a light-colored venter.

DESCRIPTION: Total length 66 mm; standard length 31 in adult female; tail moderately long, 53% of total length; eyes large, protuberant; 34 maxillary teeth, extending posterior to midpoint of eye; 12 vomerine teeth; 2 costal folds between adpressed limbs; digits moderately webbed; subterminal pads weak; proportions as percentages of standard length: head width 15%; leg length 22%. Dorsum brown with darker flank color producing a laterally scalloped middorsal field; venter light with scattered melanophores (all colors in preservative).

SIMILAR SPECIES: *Bolitoglossa gracilis* has a yellowish ground color marked with irregular dark brown lines and a dark midventral longitudinal stripe.

HABITAT: Premontane Rainforest.

BIOLOGY: The female holotype (UCR 5217) was collected from a mat of liverworts on a small branch that had presumably fallen from a tree very recently (Robinson 1976). The same vegetation mat contained seven eggs, five of which hatched into salamanders (about 8 mm in standard length), apparently of *B. diminuta*. Robinson takes this as evidence that the associated female was a sexually mature individual and that the species is a canopy dweller.

REMARKS: *B. diminuta* seems to be part of the variable and systematically difficult *Bolitoglossa subpalmata* complex that also includes *B. gracilis*, *B. pesrubra*, and several other putative but undescribed species from Costa Rica and western Panama. See the remarks section under *B. subpalmata* for further comment.

DISTRIBUTION: Known from a single locality, near Quebrada Valverde, near Tapantí, Cartago Province, Atlantic slope Costa Rica (1,555 m).

Bolitoglossa epimela Wake and Brame, 1963c

(plate 30; map 6.6)

DIAGNOSTICS: A small, rather slender, mostly dark brown or black species with extensive digital webs, rounded digital tips (fig. 6.3E), and definite subterminal pads.

DESCRIPTION: Adults 76 to 97 mm in total length; adult males 43 mm in standard length, adult females 38 to 46 mm; tail long, 52 to 56% of total length; eyes large, protuberant; 32 to 43 maxillary teeth in adults, extending to posterior one-fourth of eye; 19 to 24 vomerine teeth in adults; 1 costal fold between adpressed limbs in male, 2 to 3 in females; hands and feet extensively webbed, fewer than two phalanges free of web on all digits, digital tips rounded; subterminal pads present; proportions as percentages of standard

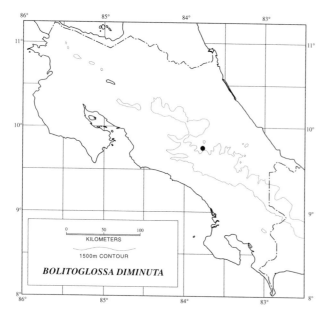

Map 6.5. Distribution of *Bolitoglossa diminuta*.

Map 6.6. Distribution of *Bolitoglossa epimela*.

length: head width 14 to 15%; leg length in adults 23 to 26%. Dark brown to black upper surfaces, nearly uniform or with some tan, orange, pinkish, and/or whitish spots or irregular blotches; venter dark, nearly uniform or with a few purplish to pinkish blotches.

SIMILAR SPECIES: (1) *Bolitoglossa pesrubra* and *B. subpalmata* have less digital webbing (one or two phalanges free of web on fingers II–IV, toes II–V) (fig. 6.3M) and more maxillary teeth (58 to 78). (2) *Bolitoglossa minutula* is smaller in adult size (males 28 to 36 mm, females 30 to 37 mm in standard length) and has completely webbed digits (fig. 6.3C).

HABITAT: Premontane Wet Forest and Rainforest.

BIOLOGY: This rare nocturnal, scansorial salamander forages on elevated surfaces including fallen logs and leaf surfaces within a meter or so of the ground. The extensively webbed digits allow it to walk upside down on the underside of leaves (Wake and Brame 1963c).

REMARKS: This salamander belongs to the *Bolitoglossa epimela* species group.

DISTRIBUTION: Humid premontane areas of Atlantic slope, central Costa Rica (775–1,550 m).

Bolitoglossa gracilis Bolaños, Robinson, and Wake, 1987

(plates 31–32; map 6.7)

DIAGNOSTICS: A rare small, rather slender yellowish salamander with moderately webbed digits (fig. 6.3M) and a dark midventral longitudinal stripe.

DESCRIPTION: Adults 83 to 91 mm in total length; adult male 38 mm in standard length, adult females 37 to 42 mm; tail long, 55 to 58% of total length; eyes moderately large, protuberant; 31 to 58 maxillary teeth in adults, extending about two-thirds through eye; 3 to 3½ costal folds between adpressed limbs; hands and feet moderately webbed, with two phalanges free of web on longest digits and all digits somewhat free of web; all but first digit with subterminal pads; proportions as percentages of standard length: head width 14 to 16%; leg length in adults 22 to 25%. Dorsal ground color bright yellowish (juveniles) to golden tan; dorsal pattern of irregular, scattered black to brown streaks and spots; venter silver to cream with a reddish brown midventral stripe; iris gold.

SIMILAR SPECIES: (1) *Bolitoglossa diminuta* has a laterally scalloped pattern to the middorsal field and lacks a midventral dark stripe. (2) *Bolitoglossa pesrubra* and *B. subpalmata* are larger forms with many more maxillary teeth (58 to 78) in adults.

HABITAT: Premontane Rainforest.

BIOLOGY: One example was collected from a moss mat on a tree, 100 mm above the ground. The smallest juvenile is 22.9 mm in standard length.

REMARKS: This species (Bolaños, Robinson, and Wake 1987) is closely allied to the widespread and variable *Bolitoglossa subpalmata* complex. It differs in size from *B. pesru-*

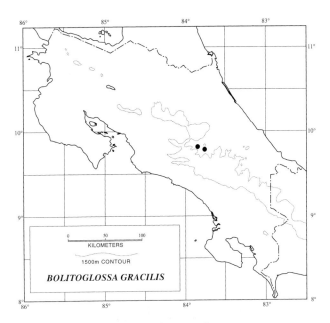

Map 6.7. Distribution of *Bolitoglossa gracilis*.

bra of the Cordillera de Talamanca, which is restricted to areas above 2,000 m, and correlated with size is a reduced number of maxillary teeth. *B. gracilis* is essentially sympatric with the enigmatic *B. diminuta*, also a member of the *subpalmata* complex. The latter form is known from a single presumed adult female (UCR 5217) and five juveniles hatched from eggs associated with her at the time of capture. Bolaños, Robinson, and Wake (1987) state that *B. diminuta* differs from *B. gracilis* by having more digital webbing and also differs subtly in other respects besides coloration. These differences may be within the range of variation for one species; only future sampling can resolve this question.

DISTRIBUTION: Humid premontane zone in the vicinity of Tapantí, Cartago Province, Atlantic slope, central Costa Rica (1,225–1,280 m).

Bolitoglossa lignicolor (W. Peters, 1873a)

(plates 33–34; map 6.8)

DIAGNOSTICS: A moderate-sized, robust mostly brown salamander having fully webbed digits (fig. 6.3C). It usually has a broad light-colored dorsal field that is streaked or washed with darker brown and a venter marked with large light spots. Some individuals are uniform brown.

DESCRIPTION: Adults 76 to 160 mm in total length; adult males 47 to 68 mm in standard length, adult females 46 to 81 mm; tail moderate, 45 to 51% of total length; eyes moderate, slightly protuberant; 27 to 48 maxillary teeth in adults, extending to center of eye; 22 to 40 vomerine teeth in adults; 2 to 3½ costal folds between adpressed limbs in males, 2½ to 4 in females; hands and feet completely webbed; webs thick and digits not greatly flattened; no subterminal pads; proportions as percentages of standard length: head width 14 to 18%; leg length 22 to 26% in adult males, 21

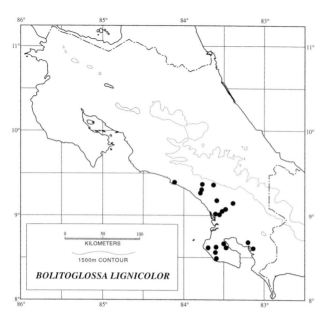

Map 6.8. Distribution of *Bolitoglossa lignicolor*.

to 24% in adult females. Color variable, flanks and venter dark chocolate and usually sharply demarcated from broad dorsal cream to tan field; dorsal field unmarked in juveniles but with dark brown streaks or wash in adults; other adults uniform dark brown dorsally or with only a few streaks of light color; venter uniform dark brown in juveniles and a few adults, usually with many large light blotches or spots.

SIMILAR SPECIES: *Bolitoglossa alvaradoi* has a dorsal pattern of irregular light frosting or black spots or blotches, a uniformly dark venter, and many more maxillary teeth (42 to 89 in adults).

HABITAT: Lowland Moist and Wet Forests and marginally into Premontane Rainforest.

BIOLOGY: This uncommon nocturnal salamander is arboreal and epiphyllous. During the day it hides in bromeliads or other plants' axils or in or under logs on the ground. The smallest juvenile is 23.4 mm in standard length; a hatchling is 11 mm in standard length.

REMARKS: This salamander is placed in the *B. lignicolor* species group. The nominal species *Bolitoglossa palustris* Taylor, 1949c has been shown by Brame and Wake (1963) to be a synonym of *B. lignicolor*. Cope's (1893) record of *Oedipus variegatus* (specimen now lost; Puntarenas: Buenos Aires) is probably based on this species.

DISTRIBUTION: Humid lowlands and lower portion of the premontane zone of southwestern Costa Rica and adjacent western Panama and the Azuero Peninsula, Panama (2–884 m).

Bolitoglossa minutula Wake, Brame, and Duellman, 1973

(plate 36; map 6.9)

DIAGNOSTICS: A small salamander having the digits completely webbed (fig. 6.3C) and varying in coloration from nearly uniform black above to dull orange on dorsum and upper surface of tail.

DESCRIPTION: Adults 54 to 76 mm in total length; adult males 28 to 36 mm in standard length; adult females 30 to 37; tail moderately long, 47 to 53% of standard length; eyes moderately large, slightly protuberant; 34 to 55 maxillary teeth in adults, extending four-fifths the length of eye; 15 to 40 vomerine teeth in adults; ½ to 2 costal folds between adpressed limbs in adult males, 2 to 4 in adult females; hands and feet completely webbed, subterminal pads present; proportions as percentages of standard length: head width 15 to 16%; leg length 21 to 24% in adult males, 25 to 27% in adult females. Upper surfaces nearly uniform dark brown to black, or with a light brown median field bordered by dark brown or black longitudinal stripes on each side or dorsum dull orange; venter dark brown or black or mottled yellow and dark; iris pale brown.

SIMILAR SPECIES: *Bolitoglossa epimela* has less digital webbing (fig. 6.3E) and larger adult size (male 43 mm, females 38 to 46 mm in standard length).

HABITAT: Lower Montane Rainforest.

BIOLOGY: An uncommon nocturnal, scansorial salamander that forages on vegetation up to 1 m above the ground and hides during the day, sometimes in bromeliads.

REMARKS: This form is a member of the *Bolitoglossa epimela* species group and was first reported from Costa Rica by Lips (1993b). However, one of her specimens (CRE 5324) from Las Cruces, Puntarenas Province, is an example of *Bolitoglossa colonnea*.

DISTRIBUTION: Humid lower montane zone on both slopes of the southern Cordillera de Talamanca of Costa Rica (1,670–2,100 m) and its extension into western Panama (1,810–2,660 m).

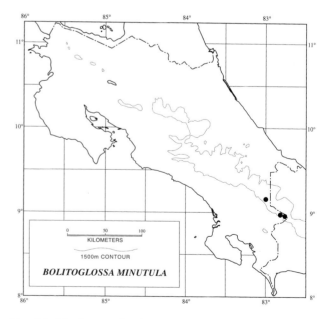

Map 6.9. Distribution of *Bolitoglossa minutula*.

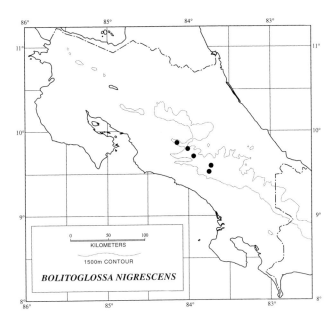

Map 6.10. Distribution of *Bolitoglossa nigrescens*.

Bolitoglossa nigrescens (Taylor, 1949c)

(map 6.10)

DIAGNOSTICS: A moderate-sized black species having moderate digital webs (fig. 6.3M) and lacking a light ring around the base of the tail.

DESCRIPTION: Adult females 89 to 169 mm in total length; adult females 58 to 94 in standard length; tail moderate, 42 to 45% of total length; eyes large, protuberant; 18 to 37 maxillary teeth in adults; 12 to 31 vomerine teeth in adults; 2½ to 4 costal folds between adpressed limbs in adults; hands and feet moderately webbed, two phalanges free of web on longest two or three digits, subterminal pads present; proportions as percentages of standard length: head width 15 to 17%; leg length 19 to 25%. Dorsum, venter, tail, and most areas of limbs black.

SIMILAR SPECIES: (1) *Bolitoglossa robusta* has a cream to red orange ring around the base of the tail. (2) Black *Bolitoglossa pesrubra* and *subpalmata* have more maxillary teeth (45 to 78) and longer limbs (0 to 3 costal grooves between adpressed limbs in adult males, 1 to 3 in adult females).

HABITAT: Lower Montane and Montane Rainforests.

REMARKS: The holotype (KU 23816; Costa Rica: San José: Boquete) is a small juvenile 44 mm in standard length. Subsequent to Taylor (1952b) it usually has been regarded as a juvenile of *Bolitoglossa robusta*, but recently collected material confirms the validity of *B. nigrescens*. The name has also been misapplied to two similar as yet undescribed large black species from western Panama (Hanken and Wake 1985) and adjacent Costa Rica (Lips 1993c). The latter form (plate 37) also lacks a caudal light ring but has a much narrower head than *B. nigrescens*.

BIOLOGY: A rare nocturnal terrestrial species.

DISTRIBUTION: Humid lower montane and montane zones of the northern Cordillera de Talamanca of Costa Rica (1,900–3,000 m).

Bolitoglossa pesrubra (Taylor, 1952b)

(plates 38–40; map 6.11)

DIAGNOSTICS: A very common small to moderate-sized species having moderate digital webbing (fig. 6.3M), 45 to 57 maxillary teeth in adults, a moderately long tail, and extreme color variation.

DESCRIPTION: Adults 60 to 131 mm in total length; adult males 26 to 57 mm in standard length, adult females 38 to 67 mm; tail moderate, 49 to 51% of total length; eyes large, protuberant; 45 to 65 maxillary teeth in adults, extending three-fourths the length of eye; 21 to 33 vomerine teeth in adults; 0 to 2½ costal folds between adpressed limbs in adult males, 1 to 3 in adult females; hands and feet moderately webbed, at least two phalanges free of web on longest two or three digits; subterminal pads present; proportions as percentages of standard length: head width 14 to 17%; leg length 22 to 25%. Dorsal ground color varies from glossy brown through uniform light gray to black, uniform or mottled light and dark brown to black, or black with light spotting; light areas often red or pinkish; sometimes with a broad brick red to yellow middorsal field; limbs and throat often pink or red; venter brown with darker markings to nearly uniform black with a few white spots or blotches.

SIMILAR SPECIES: (1) *Bolitoglossa subpalmata* has more maxillary teeth (58 to 78) and a longer tail (51 to 58% of total length). (2) *Bolitoglossa nigrescens* is always nearly uniform black and has fewer maxillary teeth (17 to 37 in

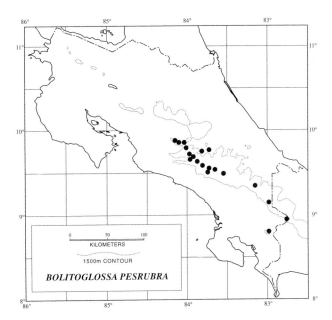

Map 6.11. Distribution of *Bolitoglossa pesrubra*.

adults) and shorter limbs (2½ to 4 costal folds between adpressed limbs in adult females). (3) *Bolitoglossa gracilis* is smaller (38 to 42 mm in standard length), with fewer maxillary teeth (31 to 58), overall yellowish coloration, and a midventral dark stripe. (4) *Bolitoglossa compacta* is larger (44 to 74 mm in standard length) and has fewer maxillary teeth (11 to 50), a pattern of reddish orange blotches, and a middorsal field or paired dorsolateral stripes.

HABITAT: Lower Montane Wet Forest and Rainforest, Montane Wet Forest and Rainforest, and Subalpine Pluvial Paramo.

BIOLOGY: Vial (1967) carried out a detailed pioneering study on this tropical species, called *B. subpalmata* by all authors cited in this section. Much of what follows is from Vial's account and field notes. This is a nocturnal generalist that is active throughout the year. Activity temperatures are near the substrate temperature and range from 0 to 20°C, but examples have been observed to be active at air temperatures between 0 and 4°C (Scott 1983a). At lower elevations it is semiarboreal and is often found hiding in bromeliads during the day. At higher elevations it is mostly terrestrial and forages on the ground or on low vegetation, moss mats, and such at night and hides under surface debris during the day. These salamanders are rather sedentary, and movements away from original capture sites during six months of study ranged from 0.1 to 50 m. Average distances moved were 2.8 m for juveniles, 3.7 m for females, and 5.4 m for males. Average home ranges are 44 m². Food consists of a variety of small arthropods. Jiménez (1994) reported this species climbing *Puya* plants to find insects and other prey. Egg clutches consist of 13 to 38 eggs about 5 mm in diameter surrounded by a vitelline membrane and two gelatinous layers. The clutches are deposited in cavities under rocks and logs, at least over most of the area studied by Vial on the continental divide of the northern Cordillera de Talamanca. At lower elevations bromeliads or other plants with large leaf axils may also be oviposition sites. The eggs are attended by an adult, usually a female (plate 40), but in four of thirty-one cases a male was in attendance. The attending adult coils tightly around the eggs with the head and throat resting on top of the mass. Adults rotate the eggs every few days with their arms and tails and rarely leave the clutch. All embryos in unattended clutches fail to develop. Vial estimated incubation times as four to five months. Hatchlings were observed to rupture the egg membranes, but time of emergence from the capsule varied from five seconds to seventeen minutes. Hatchlings are about 9 to 11 mm in standard length. Growth under natural conditions is slow, averaging about 3 mm per year for an estimated twelve years to sexual maturity for females and six years for males. Houck (1982) found that laboratory-reared members of this species grow five times as fast and reach sexual maturity in three years for females and one and a half years for males. Presumably food availability may account for the major differ-

ence implied by the two studies. Vial estimated longevity as about eighteen years. Extremely high populations densities (9,297/ha) are known for this species at high elevations (3,200 m). At the lowest studied elevation (2,438 m) densities are much lower (756/ha). Population sizes fluctuate seasonally, with the highest densities in the rainy season (May to November) and the lowest in the driest month (March). Brodie, Ducey, and Baness (1991) found that snakes of the genus *Thamnophis* suffered severe discomfort or were incapacitated when fed salamanders of this species. Further experiments (Ducey and Brodie 1991; Ducey, Brodie, and Baness 1993) showed populational variation in tail displays as an antipredator defense, with displays less likely for *B. pesrubra* from higher elevations where there are no snakes. In experiments the salamanders rarely shed their tails; only one out of eighty-one did so when seized by a snake.

REMARKS: This species has usually been regarded as conspecific with *Bolitoglossa subpalmata* of the more northern Costa Rican cordilleras. Recent biochemical evidence (García-París et al. 2000) indicates genetic distinctness of the two forms. As least two other Talamancan populations of the *B. subpalmatus* complex appear to represent other valid species but have not been formally described. One of these, called species B by García-París et al. (2000), occurs at lower elevations in the northern Cordillera de Talamanca and in the Fila Cedral south of San José (1,870–2,600 m). Vial (1966) has demonstrated that the nominal species *Bolitoglossa torresi* Taylor, 1952b (plate 39) is based on individuals conspecific with *B. pesrubra* as confirmed by García-París et al. (2000). Variation in morphological and coloration characters in this species and *B. subpalmata* makes definitive identification difficult, especially for smaller individuals. So far as can be determined, the two taxa are completely allopatric, so geographic origin may serve most needs for identification. The karyotype is 2N = 26, NF = 52, since all chromosomes are biarmed. There is no sex chromosome heteromorphism (Kezer 1964; León and Kezer 1978).

DISTRIBUTION: Humid lower montane, montane, and subalpine zones of the Cordillera de Talamanca, including the Fila Cedral (1,870–3,620 m).

Bolitoglossa robusta (Cope, 1894a)

(plate 41; map 6.12)

DIAGNOSTICS: A relatively uncommon large black salamander with a cream to red orange ring around the base of the tail.

DESCRIPTION: The largest Costa Rican salamander, adults 122 to 260 mm in total length; adult males 59 to 114 mm in standard length, adult females 65 to 134 mm; tail relatively long, 48 to 54% of total length; eyes large, protuberant; 46 to 105 maxillary teeth in adults, extending one-half the length of eye; 18 to 41 vomerine teeth in adults; 2 to 4 costal folds between adpressed limbs; digits moderately webbed, two phalanges free of web on longest two or three

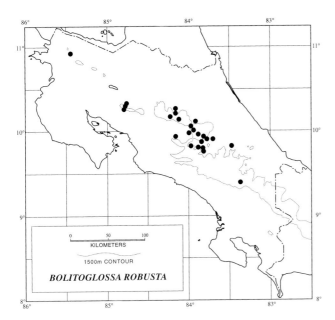

Map 6.12. Distribution of *Bolitoglossa robusta*.

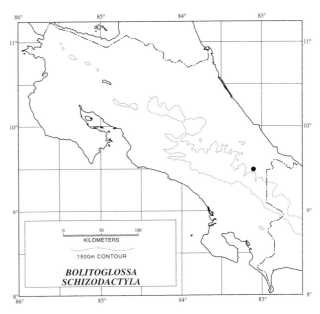

Map 6.13. Distribution of *Bolitoglossa schizodactyla*.

digits (fig. 6.3M), subterminal pads present; proportions as percentages of standard length: head width 16 to 20%; leg length 21 to 26%. Mostly black dorsally and ventrally with a cream, pink, or reddish orange ring around base of tail.

SIMILAR SPECIES: *Bolitoglossa nigrescens* does not have a light-colored tail ring.

HABITAT: Premontane Wet Forest and Rainforest and marginally in Lower Montane Wet Forest and Rainforest.

BIOLOGY: This nocturnal terrestrial species is usually found under fallen logs and other debris while hiding during the day. The smallest juvenile is 47 mm in standard length.

REMARKS: *B. robusta* is a distinctive member of the *Bolitoglossa subpalmata* species group.

DISTRIBUTION: Humid premontane and lower montane areas in the cordilleras of Costa Rica (500–2,048 m) and adjacent western Panama (830–2,050 m).

Bolitoglossa schizodactyla Wake and Brame, 1966

(plates 42–43; map 6.13)

DIAGNOSTICS: A rare moderate-sized species with completely webbed digits (fig. 6.3C) and a yellow venter and subcaudal region, the former often split by an irregular dark median stripe.

DESCRIPTION: Adults 96 to 147 mm in total length; adult males 38 to 61 mm in standard length, adult females 46 to 62 mm; tail long, 50 to 58% of total length; eyes moderate, slightly protuberant; 43 to 104 maxillary teeth in adults, extending three-fourths the length of eye; 27 to 61 vomerine teeth in adults; 1½ to 3 costal folds between adpressed limbs; hands and feet completely webbed, no subterminal pads; proportions as percentages of standard length: head width 15 to 18%; leg length 20 to 28%. Most

upper surfaces dark brown to black; venter and subcaudal area yellow; venter split by irregular dark brown to black central stripe or immaculate.

SIMILAR SPECIES: *Bolitoglossa striatula* has a cream to light yellow ground color striped and streaked with dark brown above and below.

HABITAT: Premontane Rainforest.

BIOLOGY: Probably a nocturnal arboreal species.

REMARKS: This form belongs to the *Bolitoglossa lignicolor* species group. The specimen mentioned by Wake and Brame (1966) from Barro Colorado Island, Panama Province, Panama, was an escapee accidentally released on the island.

DISTRIBUTION: Humid lowlands and premontane slopes of the Atlantic versant from southeastern Costa Rica (780 m) to Central Panama (1–750 m); also from the Pacific slope at El Valle de Antón, Coclé Province, and La Campana, Panama Province, Panama (300–850 m).

Bolitoglossa sooyorum Vial, 1953

(plate 35; map 6.14)

DIAGNOSTICS: A moderate-sized long-legged and big-footed purplish brown salamander that is heavily flecked and spotted with yellow and has moderately webbed digits (fig. 6.3M).

DESCRIPTION: Adults 62 to 42 mm in total length; adult males 58 to 65 mm in standard length, adult females 34 to 72 mm; tail moderately long, 49 to 52% of total length; eyes large, protuberant; 43 to 79 maxillary teeth in adults, extending to mideye level; 22 to 33 vomerine teeth; 0 to ½ costal folds separating adpressed limbs in adult males, 0 to 1 in females; hands and feet moderately webbed, two phalanges

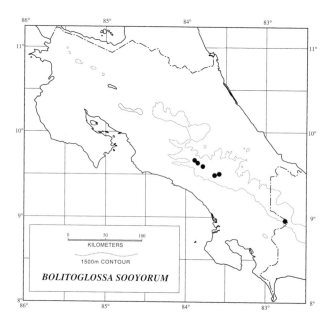

Map 6.14. Distribution of *Bolitoglossa sooyorum*.

free of web on longest digits; subterminal pads present; proportions as percentages of standard length: head width 13 to 15% in adult males, 14 to 17% in adult females; leg length 28 to 29% in adult males, 25 to 31% in adult females. Dorsum, flanks, and venter purplish brown heavily flecked and spotted with yellow.

SIMILAR SPECIES: *Bolitoglossa cerroensis* has yellowish green stippling and streaking that tends to form irregular paired dorsolateral light stripes or vertical bands along the costal grooves, shorter legs (1½ to 2 costal folds between adpressed limbs), smaller hands and feet, and fewer maxillary teeth (19 to 45 in adults).

HABITAT: Lower Montane and Montane Rainforests.

BIOLOGY: This uncommon nocturnal species is found hiding in talus on steep slopes in the northern Talamanca range.

REMARKS: Another form belonging to the *Bolitoglossa subpalmata* species group. David Wake (pers. comm.) informs me that *Bolitoglossa sooyorum* differs markedly in molecular features from *Bolitoglossa marmorea* (Tanner and Brame, 1961) of western Panama, which it resembles morphologically. See addendum (p. 839).

DISTRIBUTION: Humid montane and upper portions of the lower montane zones on both Atlantic and Pacific slopes in the Cordillera de Talamanca of Costa Rica (2,355–3,000 m) and possibly its extension into western Panama.

Bolitoglossa striatula (Noble, 1918)

(plate 44; map 6.15)

DIAGNOSTICS: A moderately small light-colored salamander having completely webbed digits (fig. 6.3C), paired dorsal and lateral dark longitudinal stripes, and the venter usually marked with several dark longitudinal streaks or stripes.

DESCRIPTION: Adults 81 to 130 mm in total length; adult males 37 to 54 mm in standard length, adult females 40 to 65 mm; tail long, 49 to 55% of total length; eyes moderate, not protuberant; 38 to 46 maxillary teeth in adults, extending to middle of eye; 22 to 24 vomerine teeth in adults; 2½ to 3 costal folds between adpressed limbs; hands and feet completely webbed; no subterminal pads; proportions as percentages of standard length: head width 14 to 16%; leg length 20 to 23%. Upper and ventral surfaces cream to yellow with paired brown dorsal stripes; paired brown lateral stripes from neck onto base of tail; narrow lateral, cream longitudinal stripe bordering dark lateral stripe and edged below by black line; middorsal light field and lateral light area marked with small brown spots; venter with two indistinct to distinct ventrolateral stripes and several indistinct dark streaks or stripes, rarely immaculate.

SIMILAR SPECIES: *Bolitoglossa schizodactyla* has a brown to black dorsal ground color without longitudinal dark stripes.

HABITAT: Lowland Moist and Wet Forests and marginally into Premontane Wet Forest and Rainforest.

BIOLOGY: This common species is nocturnal and forages on the leaves of herbaceous vegetation or even tall grass or reeds near ponds. Its color pattern blends in well with dead banana leaves, and during the day it hides among them and in other debris. A hatchling is 20.2 mm in standard length.

REMARKS: This salamander is a member of the *B. striatula* species group. A specimen (USNM 37772) of this species from Hacienda Miravalles, Guanacaste Province, is the basis for Dunn's (1926) record of *Oedipus* (= *Bolitoglossa*)

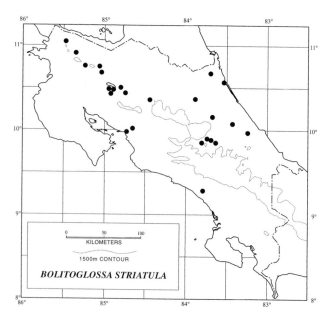

Map 6.15. Distribution of *Bolitoglossa striatula*.

dard length). (2) *Nototriton major* is a slender form (trunk width 9 to 10% of standard length), with less prominent parotoid glands and a short and narrow head (length 17.7% and width 11% of standard length).

HABITAT: In moss mats in Lower Montane Rainforest.

BIOLOGY: This uncommon species is most usually found in moss growing on the trunks and branches of the pygmy forest on the northernmost volcanoes in the Cordillera de Guanacaste. Temperature depression around these low peaks produced by the prevailing northeast trade winds allows typical stunted Lower Montane Rainforest to develop well below its usual elevational limits. Good and Wake (1993) reported discovering two clutches of five and seven eggs (plate 49) collected in or under moss on August 22. The first was in an early stage of development. The embryos in the second clutch were at an advanced stage, and one embryo from this clutch hatched on August 26. Franzen (1999) notes that he collected specimens on Cerro Cacao from moss mats 10 to 20 cm thick and 0.5 to 5 m above the ground. These animals used rapid body coiling-uncoiling and tail flipping as defensive behaviors to propel themselves 30 to 50 cm. He also reported collecting two eggs and the presumptive parent from an air root 2 m above the ground on September 7. The female later laid seven more eggs on the same night. He collected another clutch of five eggs from a moss mat growing on a vertical tree trunk on the same date. There is no indication that the mothers attend the eggs. These data suggest that this species is arboreal in undisturbed forest.

DISTRIBUTION: Humid lower montane areas on the summits of Volcán Orosí and Cerro Cacao, in the Cordillera de Guanacaste of northern Costa Rica (1,400–1,580 m).

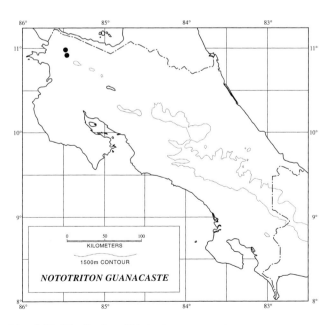

Map 6.19. Distribution of *Nototriton guanacaste*.

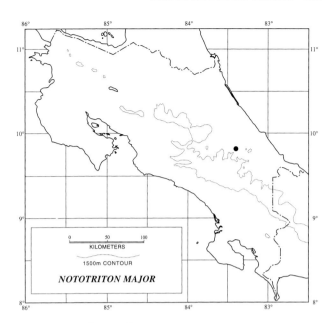

Map 6.20. Distribution of *Nototriton major.*

Nototriton major Good and Wake, 1993

(fig. 6.6f; map 6.20)

DIAGNOSTICS: This small, rather slender uniform dark brown salamander is the largest Costa Rican *Nototriton*. The head is short and narrow, the nostrils are tiny (fig. 6.6f), and the legs are long compared with those of other members of the genus.

DESCRIPTION: Adult male 93 mm in total length, 38 mm in standard length, adult females probably larger; tail moderately long, 59% of total length; eyes moderate; parotoid glands present, not prominent; hands and feet with fully differentiated, rounded digits and subterminal pads; proportions as percentages of standard length: head length 17.7%; head width 11%; nostril diameter 0.3%; trunk width 9 to 10%; leg length 19%. Dorsum and flanks essentially uniform dark brown with some very obscure light chevrons on dorsum and a few obscure light spots on flanks; parotoid glands slightly lighter brown than dorsum; venter lighter than dorsum, gray brown with many small white spots.

SIMILAR SPECIES: (1) *Nototriton abscondens* is smaller (adults 21 to 33 mm in standard length), with a longer and broader head (length 18 to 21% and width 12 to 14% of standard length) and a shorter tail (101 to 136% of standard length). (2) *Nototriton guanacaste* is smaller (to 34 mm in standard length) and robust (trunk width 11 to 15% of standard length), with prominent parotoid glands and a long and broad head(length 18 to 22% and width 13 to 16% of standard length).

HABITAT: Premontane Rainforest.

BIOLOGY: A rare species that probably lives in moss mats.

DISTRIBUTION: Known only from one locality on the north-facing premontane slope of the Cordillera de Talamanca, near Moravia de Chirripó, Cartago Province, Costa Rica (1,100–1,200 m).

Nototriton picadoi (Stejneger, 1911)

(fig. 6.6e; plate 50; map 6.21)

DIAGNOSTICS: Another small, rather robust brown salamander with irregular light stripes from the occipital region and groin down the back and flanks. The head is moderate in length and width, the nostrils are large (fig. 6.6e), and the legs are relatively short compared with those of other *Nototriton*.

DESCRIPTION: Adults to 66 mm in total length, 20 to 32 mm in standard length; tail moderate, 54 to 59% of total length; eyes moderate; parotoid glands present; hands and feet with fully differentiated, rounded digits and subterminal pads; proportions as percentages of standard length: head length 18 to 22%; head width 12 to 14%; nostril diameter 1 to 1.6%; trunk width 10 to 11%; leg length 18 to 19%. Dorsum brown with irregular yellow dorsolateral and lateral light stripes; parotoid glands usually light; venter light brown.

SIMILAR SPECIES: *Nototriton abscondens* is a more slender form (trunk width 8 to 10% of standard length) and has small nostrils (diameter 0.3 to 0.8% of standard length) (fig. 6.6d).

HABITAT: Found in moss mats and bromeliads in Premontane Rainforest and Lower Montane Wet Forest.

BIOLOGY: The holotype was originally collected from a bromeliad, as noted by Picado (1913). Dunn (1937a) reported a second example found in a bromeliad (see remarks below). All subsequent finds have been in mosses hanging from tree trunks or branches or in moss balls up to 8 m above the ground (Good and Wake 1993; Bruce 1999). Good and Wake suggest that *N. picadoi* is arboreal in relatively undisturbed situations. These authors report on a clutch of ten eggs very near hatching collected from under moss in August. The embryos had triramous gills and were about 7 mm in standard length. No female was guarding this egg clutch. Bruce (1998, 1999) found eighteen egg clutches of this form in moss mats from ground level to 2 m above the surface variously on trees, shrubs, lianas, or stumps. Clutch size was 1–4.3–11, and the eggs had a diameter of about 5 mm including the capsules. Females were not seen in attendance or brooding the eggs. Sixteen clutches raised in the laboratory indicated that hatching takes place six weeks to two months after oviposition. The smallest hatchling was 6.7 mm in total length. Based on the dates of collection of these clutches and observations on clusters of empty egg capsules, Bruce suggests that the species has a seasonal reproductive period (May to December).

REMARKS: *Nototriton abscondens* as originally described by Taylor (1948a) and included in his studies (1952b, 1954a) of Costa Rican amphibians was synonymized with this form by most other authors (e.g., Wake and Lynch 1976). As a matter of fact, most examples referred to *N. picadoi* (as *Oedipus* by Dunn 1926 or *Chiropterotriton picadoi* by Taylor 1952b) were actually *N. abscondens*. Good and Wake (1993) rediscovered this uncommon species, previously known from only two specimens (one now lost), and demonstrated its distinctness from *N. abscondens*. They mention the lost specimen listed by Dunn (1937a) as taken by Manuel Valerio "at 2200 m on Escazú." Reference to Dunn's notes shows that this example actually came from Piedra Blanca, usually called Pico Blanco (2,271 m), in the mountains to the south of the city of Escazú, San José Province (1,100 m). This peak is in that portion of the Mazico de Cedral or Candeleria often called the Cerros de Escazú. There seems little reason to doubt the identification of this specimen as *N. picadoi*.

DISTRIBUTION: Humid premontane and lower montane forests of the northern Cordillera de Talamanca of Costa Rica (1,200–2,200 m).

Nototriton richardi Group

Hands and feet syndactylous (fig. 6.3SY), forming a single pad, digital tips pointed and lacking webbing and subterminal pads; maximum standard length 24 mm.

Nototriton richardi (Taylor, 1949c)

(fig. 6.6b; plate 51; map 6.22)

DIAGNOSTICS: a rare diminutive, rather slender brown salamander having some small lighter spots or other markings and syndactylous hands and feet with toes I and V only outlined by grooves externally. The head is moderate in length and width, the nostrils are very large (fig. 6.6b), and the legs are shorter than those of most other members of the genus.

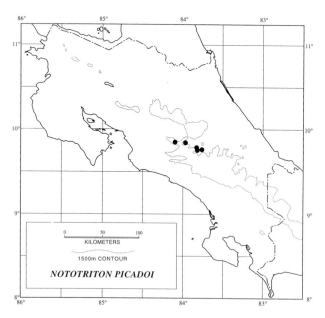

Map 6.21. Distribution of *Nototriton picadoi*.

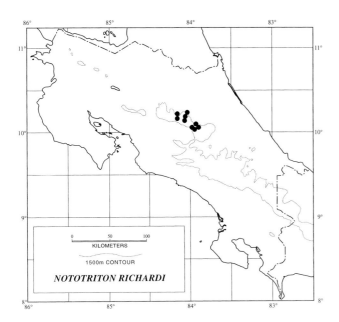

Map 6.22. Distribution of *Nototriton richardi*.

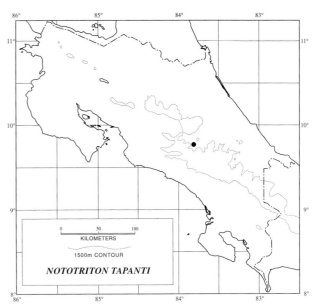

Map 6.23. Distribution of *Nototriton tapanti*.

DESCRIPTION: Tiny, adults to 50 mm in total length, 21 to 24 mm in standard length; tail moderate, 52 to 60% of standard length; eyes moderate; parotoid glands present, prominent; hands and feet syndactylous, with pointed tips that lack subterminal pads; toes I and V only outlined by grooves externally; proportions as percentage of standard length: head length 19 to 21%; head width 13 to 15%; nostril diameter 1.2 to 1.6%; trunk width 8 to 9.6%; leg 17 to 19%. Dorsum brown with some small light spots and/or costal grooves light; venter light lavender brown with numerous light flecks.

SIMILAR SPECIES: (1) *Nototriton tapanti* has greatly reduced hands and feet, toe V only outlined by a groove externally, and 41 maxillary teeth. (2) *Nototriton abscondens* has nonsyndactylous digits with rounded tips and subterminal pads (fig. 6.5). (3) *Oedipina gracilis, O. pacificensis,* and *O. uniformis* have 19 costal grooves.

HABITAT: Found in leaf litter and moss banks in Lower Montane Rainforest and marginally into the upper portions of Premontane Rainforest.

BIOLOGY: This species has been collected in moss on stumps, tree trunks, and moss mats on road cuts and typically in leaf litter at undisturbed sites. A small juvenile is 13.5 mm in standard length.

DISTRIBUTION: Humid lower montane and upper premontane zones on the Atlantic slopes of the Cordillera Central of Costa Rica (1,370–1,800 m).

Nototriton tapanti Good and Wake, 1993

(fig. 6.6a; plate 52; map 6.23)

DIAGNOSTICS: Another rare tiny to small, rather slender brown salamander having syndactylous hands and feet with only toe V outlined by a groove externally. The head is moderate in length and width, the nostrils are very large (fig. 6.6a), and the legs are shorter than those of other *Nototriton* except *N. richardi*.

DESCRIPTION: Adult female 51 mm in total length, 24 mm in standard length, adult males probably slightly smaller; tail moderate, 53% of total length; eyes moderate; parotoid glands rather prominent; hands and feet very small and digits barely demarcated from one another; tips of longest digits pointed, all digits without subterminal pad; toe V only outlined by groove externally; proportions as percentage of standard length: head length 19%; head width 13%; nostril diameter 1.3%; trunk width 9.4%; leg length 17%. Dorsum light brown; flanks darker brown; obscure light spots on either side of midline; paired reddish brown spots on back of head anterior to light parotoid glands; venter dark brown.

SIMILAR SPECIES: (1) *Nototriton richardi* has both toes I and V outlined by grooves externally and no more than 24 maxillary teeth. (2) *Nototriton abscondens* has nonsyndactylous digits with rounded tips and subterminal pads (fig. 6.5). (3). *Oedipina gracilis, O. pacificensis,* and *O. uniformis* have 19 costal grooves.

HABITAT: In leaf litter in Premontane Rainforest.

DISTRIBUTION: Known only from a humid premontane Atlantic slope forest near Tapantí, Cartago Province, Costa Rica (1,300 m).

Genus *Oedipina* Keferstein, 1868a

Tropical Worm Salamanders

Salamanders of this genus (twenty-three species) are mostly extremely slender and elongate (to 254 mm in total length, to 77 mm in standard length), with short to tiny limbs and long to very long tails (50 to 75% of total length), and have a sublingual fold. Internal features that in combination diagnose the genus are premaxillary bones fused; no prefrontals; parietals approaching one another on the midline; no tibial spur present; wrist and ankle skeletons specialized (ulnare and intermedium fused and tarsals four and five fused); and 18 to 22 trunk vertebrae. These usually attenuate, wormlike salamanders should be instantly recognizable because of their elongate bodies, long tails, short to tiny limbs, and small hands and feet that are webbed (fig. 6.3E, CT,C) to syndactylous (fig. 6.3SY). One species, *Oedipina carablanca*, does not fit this general mold and superficially resembles some members of the genus *Bolitoglossa*. *Bolitoglossa* and *Nototriton* both have less specialized body and tail morphology and no more than 14 costal grooves, and all *Oedipina* have 18 to 22 costal grooves and a sublingual fold (lacking in *Bolitoglossa*).

All *Oedipina* that are not in the process of tail regeneration have obviously long to very long tails compared with those of most species of *Bolitoglossa* and *Nototriton*. The ratio of tail length to total length for *Oedipina* does not always adequately reflect this difference because of the great concurrent elongation of the body associated with more trunk vertebrate. Consequently reference to tail lengths in the species accounts below will emphasize relative tail lengths within the genus. One species, *Oedipina collaris* (255 mm), approaches the maximum total length of the largest plethodontid salamander, *Bolitoglossa robusta* (260 mm), but they have maximum standard lengths of 77 mm and 125 mm, respectively.

Most species in this genus are semifossorial or fossorial in habits, but at least two forms live under slabs of bark on fallen trees or in insect burrows within rotten logs. Several species of *Oedipina* typically occur near the margins of watercourses, partially buried in sand or gravel or hidden under rocks.

When disturbed these salamanders frequently remain motionless or form a coil as a first line of antipredator defense. Coiling is a form of crypsis that probably increases apparent size (too big a mouthful for some predators). Body flipping is also used if the disturbance is violent or the salamander is touched (see above, p. 122).

So far as is known *Oedipina*, like *Nototriton*, is characterized by the absence of parental care of the eggs (Good and Wake 1993). The usual plethodontid suite of male sexually dimorphic features are found in this genus: smaller size, mental gland, elongate premaxillary teeth, nasolabial protuberances, and papillate vent.

Brame (1968) divided the genus into two major subdivisions, the *Oedipina parvipes* species group and the *Oedipina uniformis* species group. These were elevated to subgeneric status as *Oedopinola* and *Oedipina*, respectively, by García-París and Wake (2000).

Oedipina ranges from Chiapas, Mexico, over most of Central America, except the driest parts, to northwestern Colombia and northwestern Ecuador.

Closely related species within this genus are notoriously difficult to identify without comparative material. Although some forms are immediately recognizable, others are subtly distinguishable primarily based on head, limb, and foot proportions. When taken in combination, adults of each taxon share a distinctive Gestalt that is not fully reflected in comparisons of relative proportions, which frequently overlap to some degree.

Consequently, as the key to identification I have included the outstanding illustrations of typical examples of all species originally published in Arden H. Brame Jr. II's (1968) classic review of the genus and in Brame and Duellman (1970). These figures (figs. 6.7 to 6.10) marvelously capture

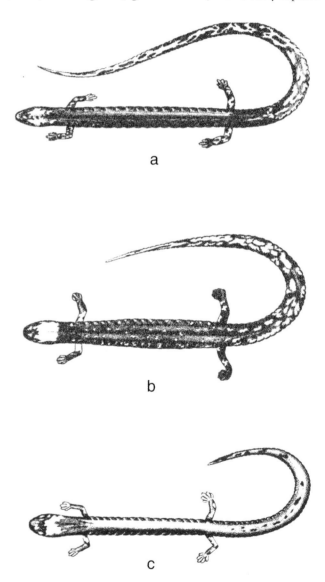

Figure 6.7. Dorsal views of salamanders of the genus *Oedipina* showing differences in head and limb proportions and coloration: (a) *O. alleni;* (b) *O. carablanca;* (c) *O. complex* (a Panama species).

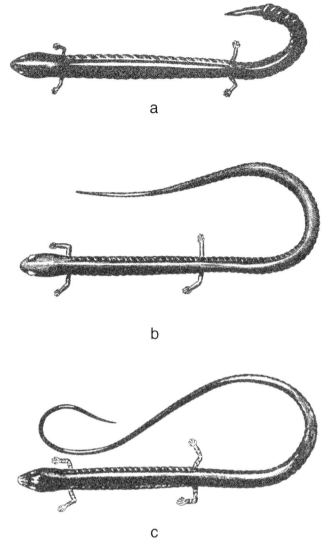

a

b

c

Figure 6.8. Dorsal views of Costa Rican salamanders of the genus *Oedipina* showing differences in head and limb proportions and coloration: (a) *O. alfaroi*; (b) *O. cyclocauda*; (c) *O. collaris*.

the essential features and their nuances for each species in a way that no written description can adequately express. For this reason the descriptions of head, limb, and foot proportions in the diagnostics sections are presented in general terms (e.g., broad head, short legs, etc.). Proportions expressed as percentages of standard lengths are included in the key or in detailed descriptions.

Some workers might suspect the status of several species Brame recognized, in light of the rather slight morphological differences between them and related forms. However, in a study of protein variation in the genus Good and Wake (1997) discovered significant genetic differences between those taxa most difficult to separate on morphological grounds. In addition, they conclude that the nominal species *Oedipina uniformis* is composed of three genetically differentiated cryptic forms, each meriting species status. I

follow their conclusions here, although no morphological feature may be used to separate the members of the O. *uniformis* complex.

KEY TO THE COSTA RICAN SALAMANDERS OF THE GENUS *OEDIPINA*

1a. 17–18 costal grooves; hands and feet extensively to completely webbed (fig. 6.3CT,C,E) 2
1b. 19–22 costal grooves; hands and feet syndactylous (fig. 6.3SY). 4
2a. Hands and feet not flattened or broadly palmate, distal portions of finger and toes free of webbing (fig. 6.3E); 6½–9½ costal folds between adpressed limbs; total maxillary teeth 0–45. 3

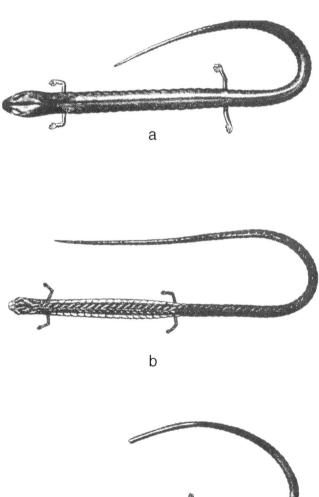

a

b

c

Figure 6.9. Dorsal views of Costa Rican salamanders of the genus *Oedipina* showing differences in head and limb proportions and coloration: (a) *O. altura*; (b) *O. grandis*; (c) *O. paucidentata*.

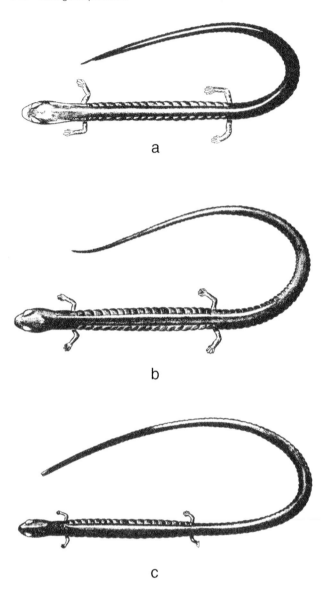

Figure 6.10. Dorsal views of Costa Rican salamanders of the genus *Oedipina* showing differences in head and limb proportions and coloration: (a) *O. poelzi;* (b) *O. pseudouniformis;* (c) *O. uniformis.*

2b. Hands and feet flattened and broadly palmate, digits completely included in webbing (fig. 6.3C); 8 costal folds between adpressed limbs; no maxillary teeth; adults 47–54 mm in standard length . . . . *Oedipina carablanca* (p. 149)

3a. Snout long and bluntly pointed; eyes small; dorsum uniform light brown; eyes small; maxillary teeth 0–4, toe III bluntly rounded; adults 40–58 mm in standard length. . . . . *Oedipina alleni* (p. 149)

3b. Snout short and rounded; eyes moderate; dorsum with a light field; maxillary teeth 7–18; toe III pointed; adults 34–39 mm in standard length . . . . *Oedipina savagei* (p. 150)

4a. Snout long and narrow, tip truncate or pointed . . . 5
4b. Snout short, rounded. 7
5a. Snout truncate; dorsum gray black; total maxillary teeth 19–90. 6
5b. Snout pointed (fig. 6.8a); dorsum brown; no maxillary teeth; adults 53–60 mm in standard length . . . . *Oedipina alfaroi* (p. 150)
6a. 80–98 maxillary teeth; 9–11 costal folds between adpressed limbs; adults 58–77 mm in standard length . . . . *Oedipina collaris* (p. 151)
6b. 19–27 maxillary teeth; 11–13 costal folds between adpressed limbs; adult male 58 mm in standard length . . . . *Oedipina altura* (p. 151)
7a. Dorsum dark to medium brown, separated from deep black venter by thin light lateral stripe 8
7b. Dorsum uniform gray black, venter slightly lighter 9
8a. Head narrow, width 8.5–10% of standard length; feet small, width 2.1–2.8% of standard length; adults 55–72 mm in standard length. .*Oedipina grandis* (p. 153)
8b. Head broad, width 10–11.8% of standard length; feet broad, width 3.5–4% of standard length; adults 41–64 mm in standard length. . .*Oedipina poelzi* (p. 155)
9a. Legs moderate, 11–15% of standard length; feet larger, width 2.7–3.7% of standard length; head broader, width 8.8–11.3% of standard length . . . 10
9b. Legs tiny, 7.8–11.7% of standard length; feet tiny, width 1.3–2.9% of standard length; head narrower, 7.5–9.7% of standard length 11
10a. Head narrow, 8.8–10% of standard length; legs shorter, 11–14% of standard length; feet smaller, width 2.7–3.5% of standard length; adults 36–46 mm in standard length. . . . *Oedipina cyclocauda* (p. 152)
10b. Head broader, 10–11.3% of standard length; legs longer, 12.3–15% of standard length; feet larger, width 2.9–3.7% of standard length; adults 39–59 mm in standard length. . . . . *Oedipina pseudouniformis* (p. 156)
11a. Head short and relatively broad, width 8.2–10% of standard length; total maxillary teeth 29–55; adults 36–57 mm in standard length. 12
11b. Head long and narrower, width 7.5–9.5% of standard length; total maxillary teeth 14–43; adults 44–62 mm in standard length. . . . . *Oedipina paucidentata* (p. 154)
12a. Found in lowlands to 730 m 13
12b. Head usually same color as dorsum; found in uplands (750–2,200 m) . . . . *Oedipina uniformis* (p. 156)
13a. Head usually same color as dorsum; found on Atlantic lowlands (3–710 m). . . *Oedipina gracilis* (p. 153)
13b. Head often with light markings; found on southwestern Pacific lowlands (15–730 m). . . . . *Oedipina pacificensis* (p. 154)

Oedipina parvipes Group

The group is characterized by having 17 to 18 costal grooves (18 or 19 trunk vertebrae) and relatively large, broad webbed feet. In addition to the three species found in Costa Rica, other members of the group are *Oedipina complex* (Panama), *O. elongata* (Mexico, Guatemala, Belize), *O. maritima* (Panama), *O. gephyra* (Honduras), and *O. parvipes* (western Colombia and Ecuador).

Oedipina alleni Taylor, 1954a

(fig. 6.7a; plate 53; map 6.24)

DIAGNOSTICS: A relatively uncommon moderate-sized light brown long-limbed salamander having 17 costal grooves, a long, bluntly pointed snout, fingers and toes with the distal portions free of the web (fig. 6.3CT), and usually considerable light markings on the head and tail.

DESCRIPTION: Adults 114 to 151 mm in total length, adult males 40 to 48 mm in standard length, adult females 42 to 58 mm; tail robust, moderately short, 56 to 62% of total length; eyes small; 0 to 4 maxillary teeth not extending posteriorly beyond choanae; 11 to 19 vomerine teeth: 6½ to 8½ costal grooves between adpressed limbs; hands and feet not flattened or broadly palmate, webbed; toe III bluntly rounded; proportions as percentages of standard length: head width 8.5 to 10.6%; leg length 17 to 22%. Upper surfaces light brown with cream markings on tail and usually on head and body, including an occipital light blotch; venter reticulated with brown around cream spots.

SIMILAR SPECIES: (1) *Oedipina savagei* has a broader head (11.3 to 11.5% of standard length), broadly rounded snout, and 7 to 18 maxillary teeth that continue posterior to the level of the choanae; (2) *Oedipina collaris* has a truncate snout and 19 to 22 costal grooves.

HABITAT: In leaf litter in Lowland Moist and Wet Forests and Premontane Wet Forest and Rainforest.

REMARKS: Taylor (1952b) included *Oedipina parvipes* (W. Peters, 1879) in his key to Costa Rican forms but noted its distribution as from Colombia to Panama. Taylor (1954a) described *O. alleni* from Costa Rica. Brame (1968) synonymized *O. alleni* with *O. parvipes*. Good and Wake (1997) determined that salamanders of this complex from Costa Rica and western Panama are not conspecific with those from central Panama to Colombia and recommended that *O. alleni* be applied to the taxon described here.

DISTRIBUTION: Humid lowlands and lower premontane slopes on the Pacific versant of southwestern Costa Rica and extreme western Panama (2–880 m).

Oedipina carablanca Brame, 1968

(fig. 6.7b; map 6.25)

DIAGNOSTICS: A moderate-sized, robust, vividly marked dark brown to black broad-headed, long-legged salamander with a large occipital light blotch on top of the head, many light spots on the body and tail, a broadly rounded, short snout, 17 to 18 costal grooves, and palmate hands and feet (fig. 6.4C).

DESCRIPTION: Adult males 107 to 128 mm in total length; 47 to 54 mm in standard length, adult females probably larger; tail relatively short and thick, 56 to 58% of total length; eyes moderate; no maxillary teeth; 10 to 12 vomerine teeth; 8 costal grooves between adpressed limbs; hands and feet greatly flattened, palmate; proportions as

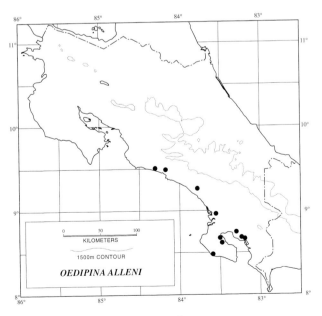

Map 6.24. Distribution of *Oedipina alleni*.

Map 6.25. Distribution of *Oedipina carablanca*.

percentages of standard length: head width 11%; leg length 19%. Upper surfaces very dark brown or black; large occipital light blotch and light spots on body and especially on tail; venter gray with large white patches; iris dirty yellow.

SIMILAR SPECIES: Not likely to be confused with any other Costa Rican species.

HABITAT: Inside rotten logs or under bark on fallen trees in Lowland Moist Forest.

DISTRIBUTION: Humid Atlantic lowlands of Costa Rica (60–260 m).

Oedipina savagei García-París and Wake, 2000

(plate 54; map 6.26)

DIAGNOSTICS: An uncommon small to moderate-sized brown, long-limbed species with a light-colored dorsal field, usually with a large white occipital blotch. Other significant features are 17 to 18 costal grooves; a short, broadly rounded snout; hands and feet with the distal portions free from the web: and toe III definitely pointed (fig. 6.3CT).

DESCRIPTION: Adults 78 to 98 mm in total length; adult males 36 to 39 mm in standard length, adult females 34 to 39 mm; tail relatively short, 52 to 65% of total length; eyes moderate; 7 to 18 maxillary teeth extending posteriorly behind choanae; 11 to 22 vomerine teeth; 6 to 7 costal grooves between adpressed limbs; hands and feet webbed, not flattened or broadly palmate; proportions as percentages of standard length: head width 11.3 to 11.5%; leg length 15 to 16%. Upper surfaces brown with a lighter middorsal field, often with small light spots or patches on the limbs and tail and usually a large occipital light spot; venter light brown with small light spots.

SIMILAR SPECIES: (1) *Oedipina alleni* has a narrower

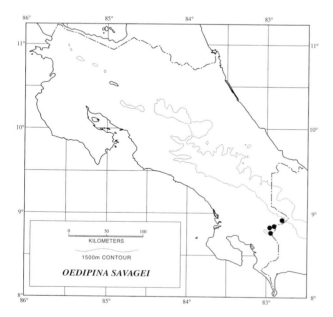

Map 6.26. Distribution of *Oedipina savagei.*

head (8.5 to 10.6% of standard length), a narrow, bluntly pointed snout, 0 to 4 maxillary teeth, and toe III bluntly rounded. (2) *Oedipina carablanca* is much more robust, with much more robust limbs, and the hands and feet are palmate. In addition, it lacks maxillary teeth.

HABITAT: Found in leaf litter in Premontane Rainforest.

REMARKS: This recently described form was referred to as *Oedipina complex* (fig. 6.7c) by Savage and Villa (1986). Biochemical differences strongly support the morphological ones that separate this form from the allied *Oedipina alleni* (García-París and Wake 2000).

DISTRIBUTION: Humid premontane slopes of southwestern Costa Rica (1,200–1,400 m).

Oedipina uniformis Group

Members of this group have 19 to 20 costal grooves (20 to 23 trunk vertebrae), slender limbs, and small to tiny syndactylous hands and feet (fig. 6.3SY). Brame's (1968) dendrogram indicates that two subgroups could be recognized within the *uniformis* species group:

An *Oedipina collaris* subgroup containing the more generalized semifossorial species having relative broad heads and moderately short limbs: *O. altura* and *O. poelzi* (Costa Rica and Panama), *O. cyclocauda* (Honduras to Panama), *O. grandis* (Costa Rica and Panama), *O. collaris* (Nicaragua to Panama), and *O. pseudouniformis* (Nicaragua and Costa Rica).

An *Oedipina uniformis* subgroup including the following highly specialized, primarily fossorial forms having relative narrow heads, very short limbs, and tiny hands and feet: *O. alfaroi*, *O. gracilis*, *O. paucidentata*, and *O. uniformis* (Costa Rica), *O. pacificensis* (Costa Rica and Panama); *O. ignea* and *O. stenopodia* (Guatemala), *O. taylori* (Guatemala and El Salvador), and *O. stuarti* (Honduras).

Good and Wake's (1997) analysis of protein characteristics generally supports this arrangement. However, their data indicate that *O. poelzi* might best be included in the *uniformis* subgroup, a course not followed here.

Oedipina alfaroi Dunn, 1921

(fig. 6.8a; map 6.27)

DIAGNOSTICS: A moderate-sized dark brown salamander having a narrow, pointed snout, 20 costal grooves, the upper surfaces of the head a lighter brown than the dorsum, very short limbs and tiny hands and feet, and no maxillary teeth.

DESCRIPTION: Adults 131 to 155 mm in total length; adult females 53 to 60 mm, males probably somewhat shorter; tail moderately long, 56 to 66% of total length; head bullet shaped; eyes tiny; no maxillary teeth; 4 to 13 vomerine teeth; 12 to 14 costal grooves between adpressed limbs; hands and feet syndactylous; proportions as percentages of standard length: head width 7.1 to 7.3%; leg length

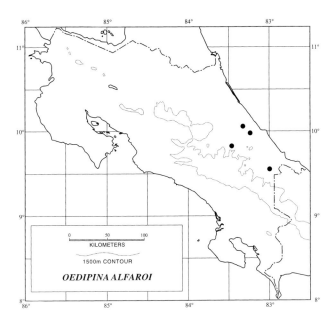

Map 6.27. Distribution of *Oedipina alfaroi.*

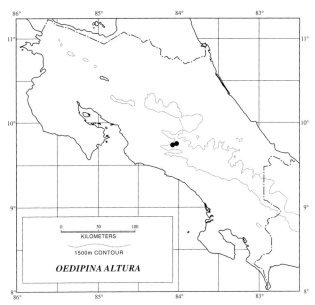

Map 6.28. Distribution of *Oedipina altura.*

9.4 to 10.5%; foot width 27%. Upper surfaces dark brown, except light brown head; venter pale brown.

SIMILAR SPECIES: (1) *Oedipina pseudouniformis* has many maxillary teeth, has large eyes, and a short, rounded snout (fig. 6.10b). (2) *Oedipina collaris* has many maxillary teeth, a narrow, long, truncate snout (fig. 6.8c), and 9 to 11 costal grooves between the adpressed limbs.

HABITAT: Lowland Moist and Premontane Rainforests.

BIOLOGY: The holotype (MCZ 6944) and paratype (MCZ 9826) were collected from fallen logs, and this uncommon species is syntopic with *O. pseudouniformis* at some localities. The smallest available specimen has a standard length of 28 mm.

DISTRIBUTION: Humid lowland and premontane areas on the Atlantic versant of Costa Rica and in extreme northwestern Panama (19–850 m).

Oedipina altura Brame, 1968

(fig. 6.9a; map 6.28)

DIAGNOSTICS: A moderate-sized, very slender gray black salamander having a long, truncate snout, 19 to 20 costal grooves, 19 to 27 maxillary teeth, and moderately short limbs and small feet.

DESCRIPTION: Adult males 131 mm in total length, 58 mm in standard length, females probably somewhat larger; tail moderate, 56% of total length; eyes moderate; vomerine teeth 9 to 17; 11 to 13 costal grooves between adpressed limbs; hands and feet syndactylous; proportions as percentages of standard length of adult male: head width 9.4%; leg length 12%; foot width 1.9%. Adult gray black above with some small white spots on head, neck, shoulder, and upper limb surfaces; tail black; venter black. Juvenile

bronze above, dorsum separated from black venter by narrow silver lateral stripe.

SIMILAR SPECIES: (1) *Oedipina cyclocauda* is smaller (36 to 44 mm in standard length) and has a bluntly rounded snout (fig. 6.8b) and more maxillary teeth (28 to 47). (2) *Oedipina pseudouniformis* has a much broader head, large eyes (fig. 6.10b), and more maxillary teeth (36 to 57). (3) *Oedipina gracilis, O. pacificensis,* and *O. uniformis* have shorter legs, tiny hands and feet (fig. 6.10c), and many more maxillary teeth (29 to 55).

HABITAT: Lower Montane Rainforest.

BIOLOGY: The holotype (LACM 1739) was collected from under moss. The small paratype (LACM 1740) is 25.5 mm in standard length.

REMARKS: This upland species is known from only three specimens, all collected near El Empalme, San José Province. This species was collected originally by James L. Vial in 1961 and 1964 at the type locality of *O. paucidentata.* Taylor collected the types of the latter in 1951 and 1952, but it has never been taken again.

DISTRIBUTION: Humid lower montane area of the extreme northern Cordillera de Talamanca of Costa Rica (2,286–2,320 m).

Oedipina collaris (Stejneger, 1907)

(fig. 6.8c; map 6.29)

DIAGNOSTICS: A rare large gray black salamander having a long, narrow truncate snout, 19 to 20 costal grooves, long, moderately robust legs, and 80 to 98 maxillary teeth. This is the largest member of the genus and has a very attenuate and long tail.

DESCRIPTION: Adults 200 to 255 mm in total length;

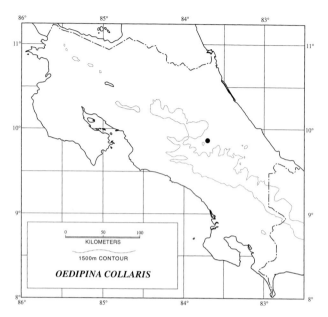

Map 6.29. Distribution of *Oedipina collaris*.

adult males 58 to 72 mm in standard length, adult females to 77 mm; tail very long, 65 to 73% of total length; small eyes; 22 to 33 vomerine teeth; 9 to 11 costal grooves between adpressed limbs; hands and feet syndactylous; proportions as percentages of standard length: head width 10 to 11%; leg length 15 to 19%; foot width 4.6 to 5%. Upper and lower surfaces gray black, sometimes a narrow silver to cream lateral light stripe (extralimital); usually some white marks on head, limbs, and at limb insertions.

SIMILAR SPECIES: *Oedipina alleni* has 17 costal grooves and a short, bluntly pointed snout (fig. 6.7a).

HABITAT: Lowland Wet Forest and Premontane Rainforest

BIOLOGY: Taylor (1952a) collected a large adult from under a log at the edge of a stream bank.

REMARKS: Taylor (1954a) described *Oedipina serpens* from near Turrialba, Cartago Province. Brame (1963) provided evidence that the holotype of *O. serpens* (BMNH 1868.8.17.5; Nicaragua: Atlántico Sur: Topaz Mine) was conspecific with *O. collaris* and synonymized the former with the latter.

DISTRIBUTION: Humid Atlantic lowlands and adjacent premontane slopes from eastern Nicaragua to central Panama (600 m).

Oedipina cyclocauda Taylor, 1952b

(fig. 6.8b; plates 55–56; map 6.30)

DIAGNOSTICS: A small to moderate-sized gray black salamander having 19 to 20 costal grooves, 28 to 47 maxillary teeth, moderately broad head, short, rounded snout, short legs, and small hands and feet. This is the smallest Costa Rican *Oedipina*.

DESCRIPTION: Adults 91 to 140 mm in total length; adult males 36 to 44 mm in standard length, adult females 37 to 46 mm; moderately robust and with moderate to long tail, 51 to 70% of standard length; large eyes; 14 to 23 vomerine teeth; 11 to 12½ costal grooves between adpressed limbs; hands and feet syndactylous; proportions as percentages of standard length: head width 8.8 to 10%; leg length 11 to 14%; foot width 2.7 to 3.5%. Gray black above, grayish below.

SIMILAR SPECIES: (1) *Oedipina pseudouniformis* has a much broader head, large eyes, and larger legs and feet (fig. 6.10b). (2) *Oedipina altura* has a truncate snout (fig. 6.9a), is much larger (58 mm in standard length), and has 19 to 27 maxillary teeth. (3) *Oedipina gracilis*, *O. pacificensis*, and *O. uniformis* have a narrower head, shorter legs, tiny feet and a more robust body (fig. 6.10c). (4) *Oedipina paucidentata* is extremely slender, has a very long, narrow head, very short legs, and tiny feet (fig. 6.9c), and is larger (41 to 62 mm in standard length).

HABITAT: Lowland Moist and Wet Forests.

BIOLOGY: A relatively common salamander, found in rotting logs, under moss on tree trunks, in piles of rotting weeds, or around tree stumps. Syntopic with *O. gracilis* at some localities. The smallest specimen of this species has a standard length of 29 mm.

REMARKS: Good and Wake (1997) comment on the difficulty of distinguishing some examples of *O. cyclocauda* from *O. pseudouniformis* on morphological grounds. However, the two species are markedly distinct in protein (genetic) characters and do not occur in sympatry, confirming the validity of their species status. Putative *O. cyclocauda* from extreme southeastern Costa Rica and adjacent north-

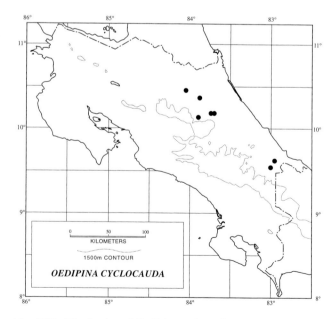

Map 6.30. Distribution of *Oedipina cyclocauda*.

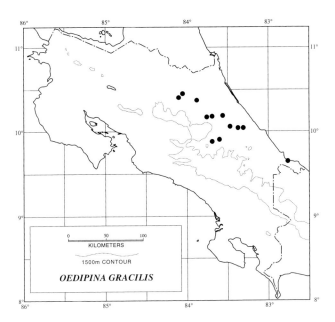

Map 6.31. Distribution of *Oedipina gracilis*.

western Panama may not be conspecific with those studied biochemically by Good and Wake.

DISTRIBUTION: Humid Atlantic lowlands from western Honduras to central Panama (60–500 m).

Oedipina gracilis Taylor, 1952b

(plates 57–58; map 6.31)

DIAGNOSTICS: A moderate-sized, very long-tailed gray black salamander having a moderately broad head, bluntly rounded snout, very short legs, tiny hands and feet, 32 to 53 maxillary teeth, and 19 to 20 costal grooves. Indistinguishable morphologically from *O. pacificensis* and *O. uniformis*.

DESCRIPTION: Adults 98 to 150 mm in total length; adult males 41 to 44 mm in standard length, adult females 32 to 47 mm; tail very long, 64 to 72% of total length; eyes tiny; 14 to 23 vomerine teeth; 12½ to 14½ costal grooves between adpressed limbs; hands and feet syndactylous; proportions as percentages of standard length: head width 8.9 to 9.7%; leg length 9.2 to 10%; foot width 1.3 to 2.9%. Gray black above and below; head sometimes lighter.

SIMILAR SPECIES: (1) *Oedipina cyclocauda* is smaller (adults 36–41.5–46 mm in standard length), with a broader head, longer legs, and larger feet (fig. 6.8b). (2) *Oedipina altura* has a truncate snout, longer legs, larger hands and feet (fig. 6.9a), and fewer maxillary teeth (19 to 27). (3) *Oedipina pseudouniformis* has a broader head, large eyes, longer legs, and larger hands and feet (fig. 6.10b). (4) *Oedipina paucidentata* has a longer, narrower head and fewer maxillary teeth (33–36.4–43 in adults). (5) *Nototriton richardi* and *N. tapanti* have 13 to 14 costal grooves.

HABITAT: Lowland Moist and Wet Forests and Premontane Rainforest.

BIOLOGY: Moderately common in leaf litter and under rotten logs, moss on tree trunks near the ground, and rocks. Taken syntopically with *O. cyclocauda*. The smallest juvenile has a standard length of 17.5 mm.

REMARKS: Taylor (1952a) described this species as new, but Brame (1968) synonymized it with *O. uniformis*. Good and Wake (1997) showed that there were major genetic differences among populations of *O. uniformis* (sensu lato). Consequently they recognized this as a cryptic species with a lowland Atlantic versant geographic range. *O. gracilis* appears to be smaller than *O. uniformis*.

DISTRIBUTION: Humid Atlantic lowlands of Costa Rica and extreme northwestern Panama (3–710 m).

Oedipina grandis Brame and Duellman, 1970

(fig. 6.9b; plate 59; map 6.32)

DIAGNOSTICS: A relatively common moderately large salamander, among the largest in the genus, having a silvery to creamy narrow dorsolateral light stripe separating the brown dorsum from the black venter, 19 to 20 costal grooves, 38 to 60 maxillary teeth, the upper surface of the tail black, a relatively narrow head, short, rounded snout, short limbs, and small feet.

DESCRIPTION: Adults 154 to 211 mm in total length; adult males 55 to 68 mm in standard length, adult females 61 to 72 mm; tail long, 64 to 69% of total length; eye small; 36 to 72 maxillary teeth; 16 to 23 vomerine teeth: hands and feet syndactylous; proportions as percentages of standard length: head width 8.5 to 10%; leg length 9.9 to 12.3%;

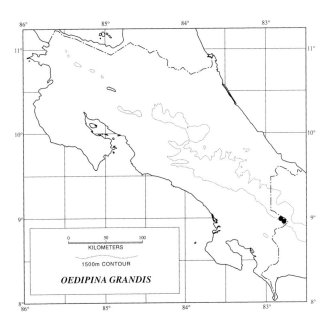

Map 6.32. Distribution of *Oedipina grandis*.

foot width 2.1 to 2.8%. Dorsum brown; venter and tail black; narrow silver to cream lateral stripe; iris dark brown.

SIMILAR SPECIES: *Oedipina poelzi* is smaller (41-56-64 mm in standard length) and has a broader head, longer legs, and broader feet.

HABITAT: Lower Montane Rainforest.

BIOLOGY: Found under decaying logs or in leaf litter around tree buttresses. The smallest juvenile is 32.5 mm in standard length.

REMARKS: This large species was described (Brame and Duellman 1970) after the appearance of Brame's (1968) monograph. Lips (1993f) reported the first record of the species from Costa Rica.

DISTRIBUTION: Humid lower montane areas of the Cordillera de Talamanca in extreme southern Costa Rica and immediately adjacent western Panama (1,810–1,950 m).

Oedipina pacificensis Taylor, 1952b

(map 6.33)

DIAGNOSTICS: A moderate-sized, very long-tailed gray black salamander having a moderately broad head, bluntly rounded snout, very short legs, tiny hands and feet, 32 to 43 maxillary teeth, and 19 to 20 costal grooves. It cannot be distinguished morphologically from *O. gracilis* and *O. uniformis*.

DESCRIPTION: Adults 108 to 175 mm in total length; adult males 26 to 48 mm in standard length, females 39 to 51 mm; tail very long, 64 to 72% of total length; eyes tiny; 20 to 26 vomerine teeth; 12½ to 13½ costal grooves between adpressed limbs; hands and feet syndactylous; proportions as percentages of standard lengths: head width 9.2 to 10%; leg length 9.9 to 10%; foot width 1.3 to 2.9%. Upper and

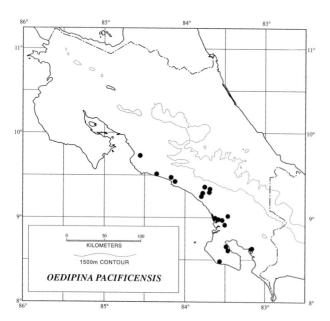

Map 6.33. Distribution of *Oedipina pacificensis*.

lower surfaces gray black, often with a whitish area in parietal region and/or near limb insertions or joints.

SIMILAR SPECIES: (1) *Oedipina cyclocauda* has a broader head, longer legs, and larger feet (fig. 6.9a) and is smaller (adults 36–41.5–46 mm in standard length). (2) *Oedipina altura* has a truncate snout, longer legs, larger hands and feet (fig. 6.8b), and fewer maxillary teeth (19 to 27). (3) *Oedipina pseudouniformis* has a broader head, large eyes, longer legs, and larger hands and feet (fig. 6.10b). (4) *Oedipina paucidentata* has a longer, narrower head (fig. 6.9c) and fewer maxillary teeth (33–36.4–43 in adults). (5) *Nototriton richardi* and *N. tapanti* have 13 to 14 costal grooves.

HABITAT: Lowland Moist and Wet Forests and Premontane Rainforest.

BIOLOGY: Commonly found in leaf litter, usually around tree buttresses. The smallest juvenile is 18.2 mm in standard length.

REMARKS: Taylor (1952a) described this form, but Brame (1968) included it in his concept of *O. uniformis*. Good and Wake (1997) documented major genetic differences among populations of *O. uniformis* (sensu lato) and recognized this as a cryptic species. In addition to the genetic differences, this is a smaller species than *O. uniformis*. Taylor emphasized the light-colored parietal spots as a distinguishing feature. In available material most examples have this character, and in some the entire head appears whitish. However, many individuals have the light areas obscured by dark pigment, or the head is entirely dark. The light head markings also occur occasionally in *O. gracilis* and *O. uniformis*. The genetic characters and the geographically restricted Pacific lowland distribution seem to justify recognizing *O. pacificensis* as a valid species. Taylor (1952a) regarded *Haptoglossa pressicauda* Cope, 1893 (type lost; Puntarenas: Palmar Norte) as a valid genus and species that differed from *Oedipina* solely in having an adherent tongue and unossified carpals and tarsals. Brame (1968) presented evidence that it was probably based on a small *Oedipina uniformis* (= *O. pacificensis*). If the unusual morphological features reported by Cope and the geographic provenance are accurate, *Haptoglossa* may be a valid taxon waiting to be rediscovered.

DISTRIBUTION: Humid lowlands and premontane slopes of southwestern Costa Rica and adjacent southwestern Panama (5–730 m).

Oedipina paucidentata Brame, 1968

(fig. 6.9c; map 6.34)

DIAGNOSTICS: a rare very slender, very short-legged, moderate-sized gray black salamander with a very narrow head, short, rounded snout, tiny hands and feet, and 14 to 43 maxillary teeth.

DESCRIPTION: Adults 100 to 139 mm in total length; adult males 44 to 52 mm in standard length, adult females

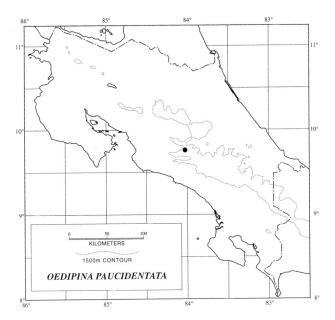

Map 6.34. Distribution of *Oedipina paucidentata*.

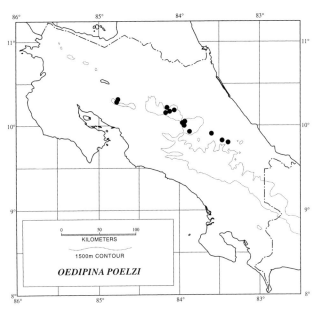

Map 6.35. Distribution of *Oedipina poelzi*.

41 to 62 mm; tail slender, moderate, 59 to 66% of total length; eyes tiny; 10 to 23 vomerine teeth; 12½ to 15 costal grooves between adpressed limbs; hands and feet syndactylous; proportions as percentages of standard length: head width 7.5 to 9.5%; leg length 7.8 to 10.1%; foot width 1.8 to 2.4%. Dorsal surfaces gray black; venter gray.

SIMILAR SPECIES: (1) *Oedipina cyclocauda* is more robust, has a broader head, longer legs, and larger hands and feet (fig. 6.8b), and is smaller (adults 36–41.5–46 mm in standard length). (2) *Oedipina gracilis, O. pacificensis,* and *O. uniformis* have broader heads (fig. 6.10c) and more maxillary teeth (29–44.5–55).

HABITAT: Lower Montane Rainforest.

BIOLOGY: Secretive and probably semifossorial. The smallest juvenile is 28 mm in standard length.

REMARKS: This species is known from the same general area as the type locality of *O. altura.* No *O. paucidentata* have been obtained since their discovery and collection by E. H. Taylor in 1951 and 1952.

DISTRIBUTION: Humid lower montane zone of the extreme northern Cordillera de Talamanca (2,286 m).

Oedipina poelzi Brame, 1963

(fig. 6.10a; plate 60; map 6.35)

DIAGNOSTICS: A rather strikingly colored moderate-sized, very broad-headed brown-backed, black-tailed salamander having a creamy or silvery lateral stripe separating the dorsal color from the black venter. Other distinguishing features are 19 to 20 costal grooves, 42 to 70 maxillary teeth, a short blunt snout, moderately long and robust legs, and small feet.

DESCRIPTION: Adults 107 to 181 mm in standard length; adult males 41 to 63 mm in standard length, adult females

42 to 64 mm; tail moderate, 57 to 68% of total length; eyes large; 14 to 34 vomerine teeth; 9½ to 11½ costal grooves between adpressed limbs; hands and feet syndactylous; proportions as percentages of standard length: head width 10 to 11.8%; leg length 10 to 10.9%; foot width 3.5 to 4%. Brown dorsum; venter and tail black; narrow creamy to silver lateral stripe separates dorsal color from ventral color; iris black.

SIMILAR SPECIES: *Oedipina grandis* is larger (adults 55-65-72 mm in standard length) and has a narrower head, shorter legs, and smaller feet (fig. 6.9b).

HABITAT: Premontane Wet Forest and Rainforest and Lower Montane Rainforest.

BIOLOGY: Moderately common in moss mats near streams and under rocks and logs. *Oedipina poelzi* seems to prefer very wet situations where a layer of surface water is present. The smallest juvenile has a standard length of 24.5 mm.

REMARKS: Good and Wake (1997) found considerable genetic divergence between populations of this species from the Cordillera Central and Cordillera de Talamanca, in the isolated valley of Moravia de Chirripó, Cartago Province. The differences are similar in magnitude to those between *O. alfaroi* and *O. uniformis* (sensu stricto). However, because there was little morphological differentiation they refer the Moravia sample to *O. poelzi.* The karyotype is 2N = 26, with definite XY heteromorphism; females are XX. All chromosomes are biarmed, and NF = 52 (Sessions 1984; Sessions and Kezer 1991). Up to two supernumerary chromosomes are characteristic of the genotype (Green 1991).

DISTRIBUTION: In the humid lower montane and premontane zones of the Cordillera de Tilarán, Cordillera

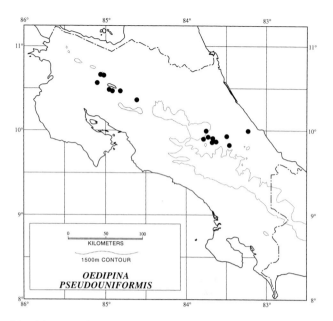

Map 6.36. Distribution of *Oedipina pseudouniformis*.

Central, and Cordillera de Talamanca of Costa Rica (775–2,050 m).

Oedipina pseudouniformis Brame, 1968

(fig. 6.10b; map 6.36)

DIAGNOSTICS: Another small to moderate-sized gray black salamander lacking lateral light stripes and having a very broad head, short, rounded snout, short, moderately robust legs, small feet, 19 to 20 costal grooves, and 36 to 57 maxillary teeth.

DESCRIPTION: Adults 75 to 165 mm in total length; adult males and females 39 to 53 mm in standard length; tail moderately long, robust, 59 to 67% of total length; eyes large; 17 to 34 vomerine teeth; 9 to 12 costal grooves between adpressed limbs; hands and feet syndactylous; proportions as percentages of standard length: head width 10 to 11.3%; leg length 12.3 to 15.1%; foot width 2.9 to 3.7%. Gray black above, gray below, often with some light dorsal blotches.

SIMILAR SPECIES: (1) *Oedipina cyclocauda* is smaller (adults 36–41.5–46 mm in standard length) and has smaller eyes, a narrower head, shorter legs, and smaller feet (fig. 6.9a). (2) *O. altura* has a truncate snout, smaller eyes, a narrower head (fig. 6.8b), and fewer maxillary teeth (19 to 27). (3) *Oedipina alfaroi* has a bullet-shaped head, tiny eyes, and a narrow, pointed snout (fig. 6.8a).

HABITAT: Lowland Wet Forest and Premontane Wet Forest and Rainforest.

BIOLOGY: Moderately common in moss banks and under logs. Syntopic with *O. uniformis* at some localities and *O. alfaroi* at others. The smallest juvenile is 20.1 mm in standard length.

REMARKS: Specimens of this form were the basis for Taylor's (1952b) description of *O. uniformis*. Without comparative material, and sometimes even with it, *C. pseudouniformis* is especially difficult to separate from *C. cyclocauda* on morphological features. The genetic differences between the two populations convinced Good and Wake (1997) to regard both as valid. On the Atlantic slope their distributions are complementary, and the two forms are not known to occur in sympatry. Nevertheless the geographic relationships are unusual. As understood at present, *O. pseudouniformis* occurs on the central Atlantic versant of Costa Rica and on both slopes in northern Costa Rica. *O. cyclocauda* is found in both northeastern and extreme southeastern Costa Rica and adjacent northwestern Panama but not in between. It does not overlap the range of *O. pseudouniformis* in northern Costa Rica. Good and Wake's biochemical samples are from the northern segment of the range of *O. cyclocauda*, and it is possible that the more southern populations are different, related to *O. pseudouniformis* or conspecific with it.

DISTRIBUTION: Humid lowland and premontane areas of Atlantic slope central Costa Rica and on both slopes in northern Costa Rica (19–1,253 m).

Oedipina uniformis Keferstein, 1868

(fig. 6.10c; plate 61; map 6.37)

DIAGNOSTICS: A common moderate-sized, very long-tailed gray black salamander having a moderately broad head, bluntly rounded snout, very short legs, tiny hands and feet, 29 to 55 maxillary teeth, and 19 to 20 costal grooves. This species cannot be separated from *O. gracilis* and *O. pacificensis* morphologically.

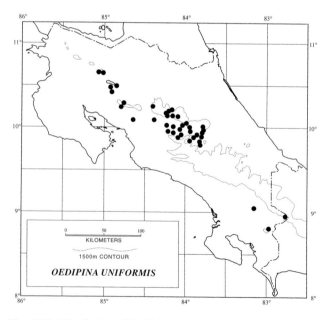

Map 6.37. Distribution of *Oedipina uniformis*.

DESCRIPTION: Adults 101 to 215 in total length; adult males 37 to 57 mm in standard length, adult females 38 to 55 mm; tail very long, 61 to 74% of total length; eyes tiny; 12 to 25 vomerine teeth; 12½ to 14½ costal grooves between adpressed limbs; hands and feet almost always syndactylous; proportions as percentages of standard length: head width 8.3 to 9.7%; leg length 7.7 to 11.7%; foot width 1.3 to 2.9%. Dorsal surfaces gray black, venter less intense gray black.

SIMILAR SPECIES: (1) *Oedipina cyclocauda* has a broader head and longer legs (fig. 6.9a) and is smaller (adults 36–41.5–46 mm in standard length). (2) *Oedipina altura* has a truncate snout, longer legs, larger hands and feet (fig. 6.8b), and fewer maxillary teeth (19 to 27). (3) *Oedipina pseudouniformis* has a broader head, larger eyes, longer legs, and larger hands and feet (fig. 6.10b). (4) *Oedipina paucidentata* has a longer, narrower head and fewer maxillary teeth (33–36.4–43 in adults). (5) *Nototriton richardi* and *N. tapanti* have 13 to 14 costal grooves.

HABITAT: Premontane Moist Forest, Wet Forest, and Rainforest and Lower Montane Wet Forest and Rainforest.

BIOLOGY: Found in and under decaying logs, in moss banks, and in leaf litter. Syntopic with *O. pseudouniformis* at some localities. The smallest juvenile is 21 mm in standard length.

REMARKS: Taylor's (1952a) concept of *O. uniformis* was based on material now referred to *O. pseudouniformis*. In addition to *O. gracilis* and *O. pacificensis* (Taylor 1952a)

Taylor also described as new taxa *O. syndactyla* (Taylor 1948a), as well as *O. bonitaensis*, *O. inusitata*, and *O. longissima* (Taylor 1952a) from the uplands of Costa Rica, and he regarded *O. vermicularis* Gray, 1868 as valid. Brame (1968), after examining all the type specimens involved, placed these seven names in the synonymy of *O. uniformis* Keferstein, 1868a. Note that Keferstein provided a second, fuller description of this species (1868b). Good and Wake (1997) found marked genetic differences among upland, Atlantic lowland, and Pacific lowland populations of *O. uniformis* (sensu Brame 1968) and resurrected the names *O. gracilis* and *O. pacificensis*, respectively, for the two lowland populations. These authors placed all highland material in *O. uniformis* (sensu stricto), and regarded the other five names as synonyms of that form. Taylor (1952a), through a *lapsus calami*, had created another available name, *O. longicauda*, on a map meant to show the type locality of *O. longissima*. This is also a synonym of *O. uniformis*. The karyotype is 2N = 26, with definite XY heteromorphism; females are XX. All chromosomes are bi-armed, and NF = 52 (Kezer, Sessions, and León 1989; Sessions and Kezer 1991). Up to eight supernumerary chromosomes occur in the genome of this species (Green 1991).

DISTRIBUTION: Humid premontane and lower montane slopes from Volcán Tenorio south through the cordilleras and Meseta Central to the Panama frontier (750–2,150 m).

7 | Frogs and Toads
Order Anura Rafinesque, 1815 (Salientia)

THE THIRD ORDER of living amphibians (Anura) is one of the most distinctive groups of vertebrates. Adult frogs and toads are unmistakable in appearance, having the head not distinct from the short robust body, elongate hind limbs that are much longer than the forelimbs, and no postcloacal tail. Frogs and toads are highly modified from any presumed primitive amphibian ancestor and specialized for jumping or hopping, as reflected in the shortened body, elongate hind limbs, modified pelvic and caudal regions, reduction of skull elements, and simplified external morphology.

When at rest or between jumps, adult anurans sit with the posteroventral portion of the body and the ventral surface of the laterally folded hind limbs pressed against the substrate. The rest of the body and the head are elevated at an oblique angle and supported by the extended forelimbs to provide optimal conditions for receiving olfactory, photic, and auditory stimuli. The anuran larva (tadpole or pollywog) is also distinct from the larvae of fishes or other amphibians in having the head and body fused into a single structural unit, no pectoral fins, no external gills (except shortly after hatching), no bony rays in the caudal fins, and no teeth on the jaws.

The first definitive frogs made their appearance in the fossil record about 200 million years ago (early Jurassic) (Shubin and Jenkins 1996). A presumed ancestral stock (*Triadobatrachus* of Madagascar) is known from the early Triassic (approximately 240 million years ago.)

Modern frogs and toads are primarily terrestrial or arboreal as adults, but a number are strictly aquatic, semiaquatic, semifossorial, or fossorial. Most species lay their eggs in water, but many have evolved life histories that partially or completely circumvent the need for development in an aquatic environment. Most anurans are nocturnal, and in the temperate zones, arid regions, and tropical situations with wet and dry seasons they exhibit seasonal activity patterns positively correlated with rainfall. Others, especially those from tropical humid forests, are diurnal, and many tropical frogs are active and breed throughout the year.

The order Anura is characterized by the following principal features, among a number of others (primarily anatomical) that will not be detailed here:

1. Body short, relatively robust (only five to nine, usually eight, presacral vertebrae); no tail in adults; postsacral vertebrae fused to form rodlike coccyx (urostyle) that supports the pelvic girdle
2. Pectoral and pelvic limbs and girdle and sternal elements present
3. Pelvic limbs composed of four segments (femur, tibia and fibula, elongate tibiale and fibulare, and foot)
4. Eyes present, exposed, and functional
5. Usually a well-developed tympanum, middle ear, and Eustachian tube
6. Larvae lacking true teeth, although keratinized beaks and denticles are usually present; gills covered by an operculum (except in very early stages), as are the forelimbs until just before metamorphosis; opercular chambers open to the outside through one or two spiracles
7. Body not annulate or with costal grooves; no specialized cephalic chemosensory tentacles; no phallodeum
8. Palatoquadrate fused by processes to cranium
9. Atlas articulates to skull by atlantal cotyles
10. No teeth on lower jaw (except in the hylid *Gastrotheca guentheri*); upper jaw and vomerine teeth variably present
11. Frontal and parietal bones on each side fused into a single element (a frontoparietal)

Frogs and toads are classified into major groups primarily based on internal features, since there is great similarity in external appearance among species from similar habitats and widely distant continents. Differences in vertebrae, pectoral girdles, hyoid apparatus, and hand and foot skeletons as well as limb musculature are among the most informative features and will be detailed subsequently for Costa Rican families. There are approximately 4,400 recognized species of living frogs that are placed in 361 genera and twenty-eight families as given below (fig. 7.1). Because the higher classification of anurans is in flux, I have clustered the families usually thought to be related to one another into informal categories (sections) below:

Section A
Family Ascaphidae ("tailed" frogs)
Family Leiopelmatidae (New Zealand frogs)

Family Bombinatoridae (fire-bellied frogs and allies)
Family Discoglossidae (discoglossids)
Section B
Family Megophryidae (megophryids)
Family Pelobatidae (spadefoots)
Family Pelodytidae (parsley frogs)
Family Rhinophrynidae (cone-nosed frog)
Family Pipidae (pipas and "clawed" frogs)
Section C (Bufonoidea Gray, 1825)
Family Allophrynidae (the allophrynid frog)
Family Brachycephalidae (saddleback frogs)
Family Bufonidae (toads)
Family Heleophrynidae (ghost frogs)
Family Leptodactylidae (Neotropical frogs)
Family Myobatrachidae (Australopapuan frogs)
Family Sooglossidae (Seychelles frogs)
Family Rhinodermatidae (Darwin's frogs)
Family Hylidae (New World treefrogs)
Family Pelodryadidae (Australopapuan treefrogs)
Family Pseudidae (natator frogs)
Family Centrolenidae (glassfrogs)
Family Dendrobatidae (dart-poison frogs)
Section D Ranoidea Gray, 1825
Family Microhylidae (microhylids)
Family Hemisotidae (shovel-nosed frogs)
Family Arthroleptidae (squeakers)
Family Ranidae (Holarctic and Paleotropic frogs)
Family Hyperoliidae (reed and lily frogs)
Family Rhacophoridae (Old World treefrogs)

Anuran diversity is greatest in tropical areas, with many species also occurring in subtropical and warm temperate

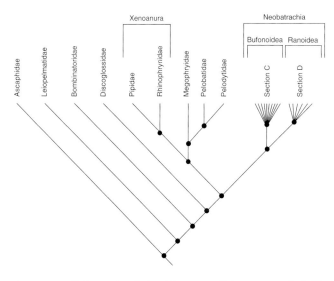

Figure 7.1. Phylogenetic relationships among living anurans; individual families in sections C and D are not indicated because of their unresolved relationships. Several well-supported superfamily groupings are indicated.

latitudinal zones. Anurans range to 71° N in Eurasia and 60° N in North America (species of *Rana* in both cases), to the Straits of Magellan in South America (51° S), to land's end in Africa (35° S) and Tasmania (42° S), and to 41° S in New Zealand. Eight families of frogs and toads are represented in Costa Rica, and their distributions are given in the appropriate accounts later in this book. Extralimital families exhibit a variety of distribution patterns. Those in section A are primarily temperate in occurrence: Ascaphidae (one genus) in northwestern North America; Leiopelmatidae (one genus) in New Zealand; and Discoglossidae (two genera) in western Eurasia and North Africa. But the Bombinatoridae (two genera) are found in Borneo and the Philippine Islands in addition to eastern and western Eurasia, and the Megophryidae (eight genera) range from Pakistan and western China through tropical Southeast Asia to the Philippines.

Among section B families the Pelodytidae (one genus) and Pelobatidae (three genera) are north temperate in distribution and found in Europe and southwestern Asia (the former) and Europe, North Africa, western Asia, and North America (the latter). The Pipidae (five genera) are found in tropical areas in Africa and South America.

Frogs included in section C have two centers of high diversity, the Neotropics and the Australopapuan region. Families found exclusively in South America include Allophrynidae (one genus), Brachycephalidae (two genera), Pseudidae (two genera), and Rhinodermatidae (one genus). Those restricted to the New Guinea–Australian region are Myobatrachidae (twenty-three genera) and Pelodryadidae (four genera). The Heleophrynidae (one genus) are restricted to extreme southern Africa and the Sooglossidae (two genera) to the Seychelles Islands in the Indian Ocean. Within section D the Arthroleptidae (three genera) and Hemisotidae (one genus) are restricted to sub-Saharan Africa. The Hyperoliidae (nineteen genera) occur in sub-Saharan Africa, Madagascar, and the Seychelles. The Rhacophoridae (ten genera) range widely in tropical Africa, Madagascar, southern India, Japan, and southern China to the Philippine Islands.

At this point it seems well to answer one of the questions most commonly asked about anurans: What is the difference between a frog and a toad? Of course these words are English of Anglo-Saxon origin and are based on the kinds of amphibians found in Great Britain. The British Isles has a depauperate amphibian fauna consisting of three species of the genus *Rana* (Ranidae) and two of the genus *Bufo* (Bufonidae). The British forms of *Rana* have smooth skins and are relative slim and long legged, whereas the species of *Bufo* have warty skins, have large paired parotoid (poison) glands, and are stocky and short legged. Consequently English-speaking people wherever they explored or settled applied the term frog to any smooth-skinned anurans resembling the *Rana* of Europe and used the term toad for the warty ones. Unfortunately this classification broke down as more was learned about anuran relationships. It was soon discovered that different species within the same families may be froglike or toadlike in general appearance. Consequently most herpetologists now use the term frog for all anurans. Within this inclusive usage only the toadlike frogs of the family Bufonidae properly are called toads.

It is also interesting that the word tadpole, the name commonly applied to anuran larvae, is also Anglo-Saxon in origin and means "toadhead," in reference to the fusion of the head and body into a single headlike unit. The corresponding words in Spanish are *rana* (frog), *sapo* (toad), and *renacuajo* (tadpole). When one is nervous or afraid in Costa Rica one has "renacuajos en mi estómago" instead of butterflies.

The skin of anurans has several unusual properties. Whereas in caecilians and salamanders the integument is connected to the underlying bones and muscles by a more or less continuous sheath of connective tissue, in frogs the skin is loose and attached to the body wall at discrete intervals. In addition the granulate ventral pelvic and femoral area (the seat patch) of anurans is highly vascularized and serves as the major organ used in water uptake. Frogs thus "drink" by sitting on a moist substrate or in a puddle or other free water source.

The auditory sense is an extremely important feature of anuran biology. Sound reception is especially significant in social interactions (e.g., territorial and reproductive behavior) and predator avoidance. The evolution of the elongate hind limbs and the associated characteristic anuran posture that elevates the head above the substrate appear to have enhanced dependence on the reception of airborne pressure (sound) waves. Most frogs have a well-developed complete auditory apparatus: a large external tympanum; a middle ear cavity that contains the columella (= stapes), which is attached to the tympanum by a cartilaginous extracolumella and has a footplate lying in the fenestra ovalis; a second middle ear ossicle, the small cartilaginous operculum lying posterior to the stapes in the fenestra ovalis; and a Eustachian tube with an opening into the throat (ostium pharyngeum) and middle ear. Each of the middle ear skeletal elements also has a muscular attachment to the pectoral girdle (suprascapula).

In some species the thin tympanum is covered by a layer of skin (internal tympanum), and in many semifossorial and fossorial forms the tympanum, middle ear, columella, Eustachian tubes, and ostia are variously lost. In these species the operculum is retained, and sound reception was traditionally thought to involve transmission from the substrate through the shoulder girdle via the opercularis muscle and operculum to the inner ear as described for salamanders (chapter 6). Recent studies indicate that the opercularis muscle has another function in anurans and that transmission of sound waves to the operculum is generally through the tissues of the head, especially the skull (Wever 1985).

This system is also functional in those species with a fully developed auditory apparatus, so they pick up low-frequency sounds (less than 1,000 Hz) from the substrate and also high-frequency sounds (above 1,000 Hz) from the air. Apparently the opercularis and columella muscles act primarily to protect the inner ear against overstimulation and to reduce mobility in one or the other of the ossicles to select for reception of low-frequency (columella dampened) or high-frequency (operculum dampened) sounds.

In association with the elevated posture of anurans, vision is also an extremely important sense, and the eyes are well developed except in the strictly aquatic Pipidae and a number of secretive or burrowing forms. Upper and lower eyelids are absent in larvae and in adult pipids but are present in all adult terrestrial and arboreal frogs. Only the upper thinned, translucent, and retractable portion of the lower eyelid is used to close the eye except in *Rhinophrynus*, in which the upper eyelid is involved.

The eyeball as a whole is capable of considerable movement. As you will recall, it may be protruded dorsally to aid in respiration (chapter 4). It may also be pulled downward into the head by a special muscle that simultaneously closes the eye. This action creates a bulge in the roof of the mouth that aids in swallowing food. However, the coordinated retraction of the eyeball and eye closure appear to be primarily a protective behavior, although the eye may be closed without complete retraction. Accommodation is accomplished by the action of two muscles that move the lens forward toward the cornea or the reverse when they relax (as in salamanders). Frogs and toads generally have binocular vision with a wider field of view than plethodontid salamanders. The retinas of frogs contain the same four kinds of photosensitive cells found in salamanders (single and double cones, red and green rods). The cones of at least some diurnal frogs (e.g., ranids) contain oil droplets (some of them yellow), although they are absent in most anurans. Current evidence indicates that although diurnal anurans may have some ability to discriminate among different wavelengths, anurans must be regarded as functionally color-blind.

All adult anurans are capable of hopping or jumping by straightening the folded hind limbs simultaneously to push the soles of the feet against the substrate and propel the body forward. The animal moves forward in an upward arc with the hind limbs extended and the forelimbs folded along the sides. At the end of the leap the frog descends headfirst in a downward arc and lands on its hands. The short forelimbs are used to orient the anterior part of the body before jumping and to absorb the shock of landing. All anurans can walk using the same sequence of footfalls typical of most tetrapods other than mammals (the lateral sequence gait as described for salamanders in chapter 6), but only a few short-limbed species use this as a major form of locomotion. Saltation is extremely effective for movements about the environment, and especially for predator avoidance. Most anurans can leap many times the length of their bodies; the record for one South African frog is an average of 3.28 m on three consecutive jumps.

Within these capacities the various species of frogs and toads tend to use predominantly one or two locomotor modes (after Emerson 1988)—jumping, hopping, walking, or swimming, with a number combining walking and hopping or jumping and walking. Strictly aquatic frogs (e.g., Pipidae) have long legs and short arms; terrestrial jumpers (those that can travel ten or more standard lengths in a single bound) have long legs and moderately long arms (e.g., many leptodactylids and ranids), and arboreal jumpers or jumper-walkers have long arms and legs (e.g., most treefrogs, representing several families).

All hoppers, walkers, and walker-hoppers (e.g., Bufonidae) have relatively short legs, but there is considerable variation in arm development. This situation is confounded by the burrowing habits of many species, and short arms characterize these forms. Differences in ligamentous connections of the pelvic girdle to the sacrum are associated with locomotor mode, and specializations for burrowing tend to enhance walking ability. These specializations, however, permit only backward burrowing using the hind limbs to displace the soil.

Facultative backward burrowing is characteristic of a wide variety of anurans and also occurs in several taxa that are composed of primarily subterranean forms (e.g., *Rhinophrynus*, Pelobatidae). Many forms ("sliding" backward burrowers) move directly backward as they excavate the soil (e.g., *Hypopachus*), but others (e.g., *Rhinophrynus*) rotate corkscrew fashion during burrow construction (circular backward burrowers). Headfirst burrowing is much less common and, though used by a few facultative burrowers, is best developed in a number of subterranean species (several microhylids and *Hemisus*). These forms use the head like a shovel to construct underground tunnels.

Several treefrogs of the families Hylidae and Rhacophoridae have been called "flying frogs" because they can reduce the speed of descent after leaping from a tree. Oliver (1951) has designated this behavior parachuting locomotion, since the angle of descent deviates no more than 45° from the vertical. Anurans that do this have greatly enlarged, extensively webbed hands and feet that they extend from the flattened body, which is oriented downward and parallel to the ground. These in combination expose the greatest possible surface and slow the descent.

Several Costa Rican hylids (some *Agalychnis* and the *Hyla miliaria* species group) exhibit this behavior. No anuran is known to use gliding locomotion, as occurs in tree squirrels or lizards of the genus *Draco*, where the angle of descent is more than 45° from the vertical, aided by the development of a fold of skin between the arms and legs or connecting them.

Typically anurans have four fingers and five toes with phalangeal formulas (in order, from inner to outer digits) of 2-2-3-3 and 2-2-3-4-3. All frogs and toads retain at least rudiments of the four fingers, although some species have a reduced number of phalanges. Several taxa have lost one or more phalanges from the toes, and two Neotropical genera outside Costa Rica have only four toes (loss of digit I). The reductions are generally found in terrestrial forms that are hoppers or walkers or show miniaturization.

A number of families and one genus of microhylids are characterized by an intercalary cartilage or bone between the outer two phalanges of each digit. The intercalary cartilages appear to be an adaptation that increases the efficiency of locomotion on vertical surfaces and are a typical feature of the several treefrog families, although found in other families as well. The intercalary bones occur in the strictly aquatic Pseudidae and, by extending the length of the digits, provide greater leverage for swimming.

Larval anurans swim by lateral undulations of the muscular tails. Aquatic and semiaquatic frogs use essentially the same leg thrust typical of terrestrial saltation. However, "tailed" frogs, genus *Ascaphus*, and the New Zealand *Leiopelma* use alternate leg kicks. Arboreal species also leap, but representatives of several families have enlarged adhesive disks on the fingers and toes that aid in climbing (see account for the Hylidae for further detail). Anurans are principally carnivores as adults, but most larvae feed on green algae and other aquatic plants. Adult or larval arthropods and earthworms are the principal food sources for adults of most species. Several frogs are crustacean specialists, at least one species feeds on gastropods, and several large forms eat large prey such as snakes, small turtles, birds, and small mammals (including bats). Many anurans are sit-and-wait predators, but others forage widely. Emerson (1976, 1988) has suggested that walker-hoppers tend to be wide foragers that have small mouths and feed on small, slow-moving prey. In contrast, jumpers are usually sit-and-wait predators with large gapes that help them capture large, fast-moving prey. Both kinds of predators depend primarily on visual detection of movement to locate prey, but olfaction probably plays a substantial role in fossorial forms.

Feeding behavior in most adult anurans consists of one of two principal modes. In group A, most of group B, and many group C frogs and in some genera of the Microhylidae, the round, sticky tongue is attached to the floor of the mouth for most of its length and is barely free posteriorly. During feeding the head is bent downward, the lower jaw is depressed as the mouth is opened, and a forward lunge of the body brings the slightly protrusible tongue in contact with the prey. This combination of movements rotates the tongue pad around the tips of the lower jaws so that it comes to face ventrally at maximum protraction. During this process the entire body is launched toward the prey, and the jaws grasp the food.

In several group B and C genera, all members of the family Bufonidae, and all group D frogs (except for some microhylids), the tongue is highly protrusible and is the principal organ of capture. Anurans using this feeding mode have the tongue attached anteriorly to the floor of the mouth with the posterior one-third to one-half free and extensible. They usually obtain food by flipping the retracted organ out of the mouth for some distance so that what was the postero-dorsal surface becomes the anteroventral surface. The latter sticky area strikes and captures the prey and is then folded back into the mouth.

The aquatic pipid frogs lack tongues and feed principally by sucking in small prey and water, but they sometimes use their fingers to push larger food items into their mouths. The unusual tongue and prey capture technique of the species *Rhinophrynus dorsalis* will be described in the account for that form.

Tadpoles exhibit a number of feeding adaptations. Many are filter feeders that pump water through the mouth and gills and extract food particles before expelling it through the spiracle(s). Many, perhaps most, graze on green algae or decaying vegetation, using their horny beaks and denticles to rasp or chop it into small particles that they then ingest. Others feed on small aquatic organisms (primarily crustaceans and arthropods and their larvae), and a few catch larger prey, including other tadpoles, again using the often enlarged and serrated beaks and denticles to grasp whole prey items.

At all stages of their life history, anurans are preyed on by nearly all kinds of carnivorous invertebrates (especially arthropods) and vertebrates that co-occur with them. Aquatic stages particularly are eaten by arthropods, fishes, snakes, and wading birds.

The same suite of predators also are the most important ones feeding on breeding congregations of adult frogs, along with turtles, crocodilians, and some aquatic mammals (e.g., otters). Spiders and land crabs are frequent predators on terrestrial anurans, and many snakes are toad or frog specialists. Large carnivorous frogs, passerine and raptorial birds, and several bats are also important. Because of this wide range, references in the species accounts are restricted to unusual predators or prey-predator interactions, and I have made no attempt to fully document this aspect of each species' biology.

Among the most interesting and complex attributes of frogs are those related to reproductive biology, life history, and larvae. Before we can survey these important features, we must consider anuran vocalization as the primary mediator of social interactions. Males of almost all frogs and toads are vocal and produce relatively loud and characteristic calls. The most frequently heard vocalization is the advertisement call that serves a territorial function in spacing males and attracts females to males at the breeding site during mating season. These calls function in long-distance social communication and may be combined in a monophasic call (Wells 1977b). Other vocalizations are also produced, with different functions, though not in all species:

Close-range
Courtship call by males when females approach
Encounter call by males during agonistic interactions when a rival male vocalizes or approaches closely
Release call by males or unreceptive females
Reciprocation call by females in response to courtship call
Long-range
Distress call by either sex

Except for distress calls, which are usually given with the mouth open, the various calls are produced in the same fashion and depend on the mechanism of buccopulmonary respiration described in chapter 4. The fundamental frequency of the call is produced by air pumped out of the lungs over the vocal cords, which are contained within the now open larynx, and into the mouth. None of this air escapes, since the mouth and nostrils are kept closed. Males of most frogs have two openings in the floor of the mouth (vocal slits) that lead to a more or less inflatable vocal sac or sacs. These structures act as a resonator to amplify the sound and are primarily responsible for the dominant frequency (always a multiple of the fundamental frequency) of the call. Most of the air in the vocal sac is then shunted back to the lungs and used repeatedly to produce more calls.

At the end of the calling cycle the nostrils are opened and air is pumped out of the mouth cavity. Females can also produce limited sounds but never have vocal sacs. The structure of the inflated vocal sac is of some systematic significance. Three major configurations may be recognized: median subgular (a single ventral sac), paired subgular (bilobed, with or without a complete median septum), and paired lateral (separate sacs below the angle of the jaw on either side of the head). The sac(s) may be internal (covered with unmodified skin and bulging slightly when inflated) or external (skin thin and sacs prominent when inflated).

Advertisement calls are species specific and differ markedly between related forms, especially among those using a common breeding site. Acoustical features of frog calls are usually described by analyzing tape recordings of vocalizations using one of several kinds of electronic equipment (usually an oscilloscope or sonograph) that produce a pictorial representation of the sound (an oscillogram and an audiospectrogram, respectively). The oscillogram shows the waveform of the sound—how call amplitude (vertical axis) changes with time (horizontal axis). The audiospectrogram demonstrates the change in frequency (vertical axis) with time (horizontal axis). In audiospectrograms the darkness of the pattern reflects the amplitude of the sound components. Also included is a display plotting amplitude against frequency (fig. 7.2).

A particular call or call group consists of a single note ("wonk"), a series of similar notes ("wonk-wonk-wonk"), or groups of different notes ("co-qui"). The simplest vocalizations are single-note calls consisting of a single pulse perceived as a click or those produced as a series of clicks.

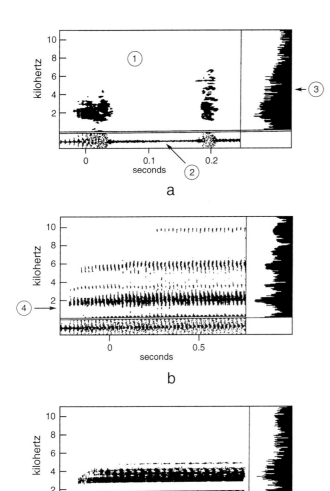

Figure 7.2. Graphic displays of typical anuran calls. (a) *Agalychnis callidryas;* (b) *Bufo valliceps;* (c) *Hypopachus variolosus.* 1 = audiospectrogram, 2 = waveform, 3 = amplitude against frequency, 4 = dominant frequency. Note that the trilled *Bufo marinus* vocalization shows definite pulses and a series of harmonics (see text for further discussion).

Many calls are pulsed, that is, divided into successive bursts (pulses) of sound that may or may not be distinguished by the human ear. Those calls, which are long and composed of distinct pulses that are discernible by the human ear, are usually described as trilled or pulsatile. Although a number of other components of anuran vocalizations may be recognized on an audiospectrogram, the most significant given the limitations of the frog's auditory apparatus are the dominant frequency (pitch), duration, and pulse rate. These are measured in cycles per seconds, or hertz (Hz), time in seconds, and number per second, respectively. The dominant frequency is analogous to the fixed frequency band of a radio transmitter, and the female or rival male's ear (the receiver) is tuned to a particular frequency channel. Different species calling at the same time differentially use the available channels much as do radio stations broadcasting at the same time. In large multispecies breeding choruses the number of channels may not be sufficient to separate all species calls, and species-specific differences within the same acoustical channel are the result of call complexity, single, multiple, or heterogeneous call notes, and their duration and pulse rate.

Other features of sound production that need to be mentioned are frequency modulation and tuning, or the distribution of sound through the frequency spectrum. Modulation is a change in dominant frequency over the course of a note or call. Tuning refers to the distribution of energy in respect to frequency. Finely tuned calls sound musical or melodious to the human ear, and energy is concentrated in several narrow bands of frequencies (harmonics). In poorly tuned calls, the sound produced is harsh or noisy because the energy is spread over the spectrum. No evidence suggests that female anurans can discern these differences, but they are striking to the human listener and useful in field identification of calling frogs and toads.

Two other factors also affect components of the advertisement call. Generally, larger individuals of a species produce louder and lower-frequency calls than smaller ones. Differences in temperature also affect several features of the call. At higher temperatures call and pulse rates and spectral frequencies increase and duration decreases. Detailing these and other complexities of frog and toad vocalizations is beyond the scope of this book, and I focus on the characteristic advertisement call usually heard in the field. The descriptions of voice presented in the species accounts are of the typical advertisement call of single males recorded at a temperature characteristic of the species' habitat during the breeding season.

Vocalization plays a major role in courtship and mating in most anurans. Typically, at the time of the breeding season (spring in the temperate zone or the onset of the rainy season in the Tropics) males congregate near available watercourses or standing water and begin to call. Somewhat later females, apparently attracted by the male chorus, arrive in the area and locate and approach calling males of the same species. Mating takes places as the male climbs onto the female's back and grasps her (amplexus), characteristically around the inguinal or pectoral region.

Male frogs in breeding condition are almost totally indiscriminate in selecting objects for amplexus. Around a breeding pond males may be seen grasping branches, rocks, empty beer cans, old boots, and such, trying to stimulate them to lay eggs. If another male (of the same or another species) of approximately the right size approaches they will also amplex with him. In many species the misidentified male will then emit a release call consisting of a series of short, rapid croaks or peeps, accompanied by a series of vibrations of the body wall (and sometimes the hind limbs), and the amplexing male will release. Females of many species that are not gravid or have already laid their ages exhibit the release behavior and call if amplexed, with a similar result.

In addition to having specific advertisement calls, males have a species-specific call site from which they vocalize during the mating season. Around a single pond the call sites may include the shoreline, shrubs or overhanging tree branches, emergent reeds, or other vegetation in the pond at the water's edge or floating on the surface. Other sites include terrestrial burrows, tree holes, rocks in the middle of streams, under submerged or partially exposed rocks and terrestrial vegetation (logs, leaves, epiphytes, etc.).

Each species also has a characteristic breeding season ranging from a few nights to year-round reproduction. For most tropical species reproduction is concentrated during the rainy season, but a number breed only during the dry season. The reproductive behavior of frogs and toads generally follows one of two patterns at the local population level (Wells 1977a). Species exhibiting prolonged reproduction have breeding periods of more than a month. Those that are explosive breeders have breeding periods of a few days to a few weeks. The breeding season over the entire geographic range of an explosive breeder may last several months, but each local population within the species has a short, explosive breeding period.

Females of prolonged breeders arrive at the reproductive site irregularly (asynchronously) over the breeding period and in some species return to mate several different times. Females of explosive breeders tend to arrive at the reproductive site more or less at the same time (synchronously) and are less likely to lay multiple clutches.

Males of most prolonged breeders use vocalizations to attract females to the calling site. In these forms the males engage in vocal competition and may defend territories of one kind or another (e.g., calling, courtship, or oviposition sites). Males of most explosive breeders, when present in high densities, actively search for females, and male-male physical interactions are typical (scramble competition). In a few species males may remain in about the same place for most of the night and attempt to clasp only frogs that move into the immediate area (limited-area searching).

At low densities, when individuals first arrive at the reproductive site male-female and male-male interactions are often similar to those of prolonged breeders. At intermediate densities a mix of active searching and vocal attraction may take place. In this case silent (satellite or parasitic) males may locate themselves near a calling male and attempt to intercept females attracted to his call. Similar satellite behavior occurs in some prolonged breeders. The problem of acoustic interference in large choruses is met in most species by antiphonal alternation or synchrony of calls. In alternation the calls are spaced so as not to overlap those of neighboring males, and this pattern is characteristic of species that emphasize vocal competition among males and female choice of a mate. Precise alternation leads to the forma-

tion of duets, trios, quartets, or other groups within the chorus, and such choruses often have a syncopated quality. Synchronous calls involve males' rapidly answering one another's advertisement calls so that the individual calls overlap to produce rhythmic bursts of sound by the chorus as a whole or simply a cacophonous din. Males of a few synchronous calling species (e.g., *Hyla microcephala*) are unusual in that they frequently alternate individual notes of their calls with those of other males. Synchronous chorusing probably attracts females to the chorus as a whole and is typical of species exhibiting scramble competition. It also has been suggested that synchronous calling may confuse frog-eating predators that locate prey acoustically. Nocturnal multispecies frog choruses are often encountered in the lowlands of Costa Rica during the wet season, especially after the first heavy rains. The presence of calling males, however, does not necessarily indicate reproductive activity. Males of many species call from time to time throughout the wet season from sites at or near places used for mating and oviposition. In mesic areas males of several forms vocalize throughout the year, including the dry season. In most cases there is considerable temporal overlap in reproductive periods among species at any particular locale, but not all of them breed at the same time.

Among frogs laying their eggs near or in lentic situations (slow-moving or standing water), assemblages may include up to fifteen species. At a typical Atlantic versant, lowland rainforest site (e.g., the La Selva Field Station, Heredia Province) the following species use the same ponds. Hylidae: *Agalychnis callidryas*, *A. saltator*, *Hyla ebraccata*, *H. loquax*, *H. phlebodes*, *Scinax boulengeri*, *S. elaeochroa*, and *Smilisca baudinii*; Leptodactylidae: *Leptodactylus melanonotus*, *L. pentadactylus*; Microhylidae: *Gastrophryne pictiventris*; Ranidae: *Rana vaillanti*. As Donnelly and Guyer (1994) demonstrated for the hylids in these assemblages, none of these species breed throughout the year. Although there is slight temporal overlap between adults and juveniles, the peak in numbers for each age-group is temporally separated, further confirming the seasonality of reproduction.

Similarly, the pond-breeding frog assemblage of a Pacific lowland rainforest (e.g., Rincón de Osa, Puntarenas Province) consists of Hylidae (*Agalychnis callidryas*, *A. spurrelli*, *Hyla ebraccata*, *Phrynohyas venulosa*, *Scinax boulengeri*, *S. elaeochroa*, and *Smilisca phaeota*) and Leptodactylidae (*Leptodactylus bolivianus*, *L. pentadactylus*, and *Physalaemus pustulosus*). The reproductive period is also restricted to the rainy season, although males of some species may call off and on all year.

In the lowland dry forest regimen of northwestern Costa Rica (e.g., Taboga area, Guanacaste Province) the pond-breeding frog assemblage usually includes Bufonidae (*Bufo coccifer* and *B. luetkenii*); Hylidae (*Hyla microcephala*, *Phrynohyas venulosa*, *Scinax boulengeri*, *S. staufferi*, and *Smilisca baudinii*); Leptodactylidae (*Leptodactylus bolivianus*, *L. labialis*, *L. melanonotus*, *L. pentadactylus*, and

Physalaemus pustulosus); Microhylidae *(Hypopachus variolosus);* Rhinophrynidae *(Rhinophrynus dorsalis);* and Ranidae *(Rana forreri* and *R. vaillanti).* These frogs reproduce only during the wet season (May to November).

In upland areas the principal pond-breeding frogs are Hylidae *(Agalychnis annae, Hyla pseudopuma,* and *Phyllomedusa lemur)* and Ranidae *(Rana vibicaria* and *R. taylori).* During the reproductive period these species may co-occur in disturbed situations with other higher elevational forms that usually lay their eggs in shallow puddles within the forest (e.g., *Bufo holdridgei* and *Hyla angustilineata).*

Stream-breeding frogs are less likely to form multispecies choruses than pond breeders, especially since some lack calls or have very weak calls. A typical dry season (January to February) assemblage in the humid lowland forest zone includes Bufonidae *(Bufo marinus* and *B. melanochlorus)* and Hylidae *(Smilisca sordida).*

During the wet season, particularly along smaller streams, several anurans that do not lay their eggs directly in the water form mixed breeding choruses. These include species of *Colosthethus* and *Phyllobates* (Dendrobatidae; reproductive pattern 18, table 7.1) that call during the day plus two or more glassfrogs (Centrolenidae; pattern 22) and Leptodactylidae, *Eleutherodactylus* (pattern 21) that call at night.

At higher elevations a typical assemblage of calling males contains an *Atelopus* (Bufonidae), one or two species of the *Hyla pictipes* group (Hylidae), one or two kinds of glassfrogs (Centrolenidae), and an *Eleutherodactylus* of the *rugulosus* species group (Leptodactylidae). The calls of the last-named group may be too weak to be heard above the stream background noise. Although frogs of this assemblage tend to call throughout the year, reproduction may occur at different times depending on the species.

Primitive frogs (all section A and B families and some

Table 7.1. Reproductive Patterns in Frogs and Toads

I. Eggs laid in bodies of water (aquatic)
   A. Eggs deposited in water
      *1. Eggs and feeding tadpoles in still (lentic) water
      *2. Eggs and feeding tadpoles in moving (lotic) water
      *3. Eggs and early larval stages in natural or constructed basins; flooding carries feeding tadpoles to ponds or streams
      *4. Eggs and feeding tadpoles in water in tree holes or aerial plants
      *5. Eggs and nonfeeding tadpoles in water-filled depressions
      6. Eggs and nonfeeding tadpoles in water in tree holes or aerial plants
      7. Eggs deposited in stream and swallowed by female; eggs and tadpoles complete development in stomach (egg site unknown, may be terrestrial)
   B. Eggs in a nest
      *8. Foam nest on puddle or pond; feeding tadpoles in puddle or pond
      9. Foam nest on pool; feeding tadpoles in stream
      *10. Foam nest in flooded burrow or excavated nest; feeding tadpoles develop in situ, or flooding releases them into larger pond
      11. Folded-leaf nest on or in pond; feeding tadpoles in pond
   C. Eggs embedded in dorsum of aquatic female
      12. Eggs hatch into feeding tadpoles in ponds
      13. Eggs hatch into froglets (direct development)
II. Eggs laid out of water (terrestrial or arboreal)
   D. Eggs on ground or in burrows
      *14. Eggs and early tadpoles in excavated nest; after flooding, feeding tadpoles in ponds or streams
      15. Eggs on ground or rock above water, in leaf litter, or in excavated nest; upon hatching, feeding tadpoles move to water
      16. Foam nest in burrow; following flooding, feeding tadpoles in ponds or streams
      17. Foam nest in burrow; nonfeeding tadpoles complete development in nest
      *18. Eggs hatch into feeding tadpoles that are carried to water by adult
      19. Eggs hatch into nonfeeding tadpoles that complete their development in nest
      20. Eggs hatch into nonfeeding tadpoles that complete their development on dorsum or in pouches of adult
      *21. Eggs hatch into froglets (direct development)
   E. Eggs arboreal
      *22. Eggs hatch into tadpoles that drop into puddles, ponds, or streams
      *23. Eggs hatch into tadpoles that drop into water-filled depressions or cavities in fallen trees
      *24. Eggs hatch into tadpoles and slide into water in tree holes or aerial plants
      25. Arboreal foam nest; hatchling tadpoles drop into ponds or streams
      26. Eggs hatch into feeding tadpoles that are carried to water by adult
      *27. Eggs hatch into froglets (direct development)
   F. Eggs carried by adult
      28. Eggs carried on legs of male; feeding tadpoles in ponds
      29. Eggs carried in vocal sac of male; feeding tadpoles in streams
      30. Eggs carried in vocal sac of male; nonfeeding tadpoles complete development in vocal sac
      31. Eggs carried in dorsal pouch of female; feeding tadpoles in ponds
      32. Eggs carried on dorsum or in dorsal pouch of female; nonfeeding tadpoles in bromeliads
      *33. Eggs carried on dorsum or in dorsal pouch of female; direct development into froglets
III. Eggs retained in oviducts
      34. Ovoviviparous
      35. Viviparous

*Occurs in a Costa Rican species.

section C families) use inguinal amplexus. Most other frogs exhibit axillary (pectoral) amplexus, but cephalic, head straddle, and other modifications are known for a few species. Fertilization is usually external, and the male releases sperm onto the eggs as they are laid. Internal fertilization is documented for a few species, including the viviparous forms, but it may be more widespread, especially in taxa having direct development.

All anurans reproduce sexually, but a number are unusual in being bisexual and having entire sets of their chromosomes added to the normal diploid number (2N). Most of these polyploid species are tetraploids (4N complement of chromosomes), but one is a hexaploid (6N), and several are octoploids (8N). Polyploidy is found in the Pipidae (basic karyotype 20 or 36, polyploids of 40, 72, and 108); Bufonidae (basic complement 20 or 22, polyploids 40 and 44); Myobatrachidae (basic karyotype 24, polyploids 48); Leptodactylidae (Ceratophryinae: basic number 26, polyploids 104; Leptodactylinae and Telmatobiinae: basic complement 22, polyploids 44); Hylidae (basic number 24 or 26, polyploids 48 and 52); Microhylidae (basic number 24 or 26, polyploids 48 and 52); Ranidae (basic complement 26, polyploids 39 and 52). Rhacophoridae (basic number 26, polyploid 52). These polyploid taxa are thought to originate through hybridization between two or more diploid species (allopolyploidy).

The European hybridogenetic complex of "*Rana esculenta*" consists of diploid and triploid individuals produced originally by hybridization between two other species. During egg production the genetic complement of one parental species is eliminated and there is differential mating with that species to produce a new generation of triploid or diploid hybrids.

Primitively and most frequently, frogs lay eggs in the water, where they hatch as tadpoles that after a period of aquatic existence metamorphose into a terrestrial or an arboreal form (exotrophy). There are numerous exceptions to this scenario, and in many species the eggs undergo some development or complete development out of the water (on land or vegetation), and an aquatic larval stage may be present or eliminated entirely.

Forms depending on maternal investment of nutrition (egg stores or via an oviductal placenta) and lacking feeding tadpoles are called endotrophic and include all lecithotrophic (oviparous and ovoviviparous) and viviparous species (table 7.3). In anuran species with aquatic eggs, the ova may be laid singly, in small packets, in single or double long strings, as masses of various shapes (cylinders, balls, lumps, spheres, plinths), or as a surface film. Single eggs are usually attached to vegetation or rocks, and strings or masses are often attached to aquatic plants or debris. Generally eggs are submerged, but in addition to those deposited as a film, some species may construct foam nests floating on the surface.

The oviducts produce gelatinous secretions around the ova that swell by taking up water when the eggs are laid. They form translucent capsules around single eggs, or the successive envelopes of many eggs coalesce to form a mass, string, or film.

The kind, amount, and distribution of pigment in the eggs are distinctive. Most eggs contain melanin, which is concentrated at the animal (upper) pole to produce an egg that is gray, brown, or black above and cream white to yellow below. The two colors grade into each other in the equatorial region. It has been suggested that this countershaded pattern camouflages the egg from predators. From above the egg is similar in color to the background substrate of a pond or stream, while from below it blends in with the bright background of the sky. Its melanin concentration may also shield the egg and developing embryo from ultraviolet radiation.

Some frogs (e.g., *Centrolenella prosoblepon*) have a heavy density of melanin throughout the eggs, which are uniform black. Others (e.g., *Hyalinobatrachium fleischmanni*) have entirely green eggs. Still others lack egg pigment (e.g., *Atelopus*), and the eggs appear white, cream, or yellow. Significantly, in Costa Rica the uniform black eggs are laid on the upper surface of leaves out of the water, green eggs usually on lower leaf surfaces, and unpigmented eggs usually under rocks or surface debris or in burrows. These features are discussed further in the individual species accounts.

Most frogs and toads are oviparous and lay their eggs in aquatic situations. These eggs range in size (including protective capsules) from 1.5 to 15 mm, with clutch sizes between 27 and 15,000. Eggs laid in arboreal or terrestrial situations are usually deposited in clumps or short, adherent strings, are richly provided with yolk, and are generally about twice as large as aquatic ones but do not exceed 14 mm in diameter (including capsules). Clutch sizes have been reported to vary between 3 and 540 in these latter forms.

Relatively few anurans are ovoviviparous or viviparous. As used in this context the term ovoviviparity (table 7.1, pattern 34; table 7.3) applies only to forms in which the eggs are retained in the oviduct and the embryos complete their development using only the yolk supply for nutrition. Viviparity (pattern 35) applies to species in which the eggs are retained in the oviduct and the embryos complete their development with nutrition provided by the mother. Nutrition is in the form of oviductal secretions (hydrotroph), possibly supplemented by a tail placenta that seems to be involved in gaseous exchange. In this case the vascularized tail of the two viviparous frogs (*Nimbaphrynoides liberiensis* and *N. occidentalis*) forms the embryonic portion of the placenta and the oviductal lining the maternal portion.

Placentas involved in gaseous exchange between the maternal circulation and the embryo's external gills are characteristic of the genus *Gastrotheca* (family Hylidae; patterns 31 and 33), and tail placentas are present in *Pipa* spp. (family Pipidae; patterns 12 and 13). Gill placentas are also found among viviparous caecilians and salamanders.

If we take into account the site of egg deposition or retention in the oviducts and the location and degree of larval development or its elimination, some thirty-five or so different life histories may be recognized among anurans (table 7.1), after Duellman and Trueb (1986, 1994) as modified from Livezey and Wright (1947). Most of these involve a feeding larval stage, but a number have nonfeeding tadpoles that complete their development from the food stored in the yolk. The situation in which the eggs are laid out of water and hatch into nonfeeding tadpoles (table 7.1, patterns 17, 19, and 20) is referred to as semidirect development. Direct development (no larval stage) also occurs in a wide variety of forms (table 7.1, patterns 21, 27, and 33). Fifteen of these primary lifestyles are exemplified by various Costa Rican frogs, as will be detailed in the family and species accounts.

There is also considerable diversity of parental care in anurans involving the following principal behaviors, all represented in one form or another in Costa Rican anurans as indicated:

Attendance of aquatic eggs: male *Hyla rosenbergi* (family Hylidae)

Attendance of terrestrial eggs and nonfeeding larvae or directly developing embryos: probably all species of the genus *Eleutherodactylus* (family Leptodactylidae), since all probably have direct development except the Puerto Rican *E. jasperi*, which is ovoviviparous

Attendance of arboreal eggs: males of all members of the genus *Hyalinobatrachium* (family Centrolenidae)

Larvae carried by adult to body of water from a terrestrial or arboreal site: all members of the family Dendrobatidae

Eggs and larvae carried on or within adult: female *Gastrotheca cornuta* (family Hylidae)

Either parent or both parents may be involved in parental care depending on the species and its life history.

Fish predation on eggs and larvae in particular is thought to be an important historical influence molding reproductive patterns in tropical anurans (Heyer, McDiarmid, and Weigmann 1975). Among lowland species breeding in lentic situations, most use temporary ponds or pools where fish are absent, or the frogs or toads have unpalatable eggs, or the larvae produce noxious skin secretions that make them distasteful to fishes. Stream breeders also reduce egg and larval mortality by reproducing in the dry season and laying eggs in isolated pools or backwaters that tend to lack piscine predators, or they have nonpalatability defenses. Since no native Costa Rican fish species occurs above 1,500 m in altitude, it is not surprising that upland anurans form a radiation predominantly of stream breeders (especially hylids and centrolenids).

Roberts (1994a) emphasized the importance of conspecific and heterospecific predation on eggs by tadpoles as a possible force in the evolution of nonaquatic patterns of egg deposition in tropical assemblages. She also pointed out that tadpoles appear to be one of the major predators on developing or recently hatched embryos of their own or other species.

The general features of anuran larvae were discussed and compared with those of caecilians and salamanders in chapter 4. Although other amphibian larvae are similar to adults in morphology and feeding habits, tadpoles bear little resemblance to adult anurans in these features and have become astonishingly diversified in structure and ecology.

Four major kinds of free-living tadpoles have been generally recognized based on details of internal and external morphology that seem to have phylogenetic significance (fig. 7.3):

Type L: a single midbranchial spiracle; an oral disk and horny beaks and denticles; external nares present (section A families)

Type A: a single lateral, sinistral spiracle; an oral disk and usually horny beaks and denticles; external nares present (most section B and D families and all section C families)

Type S: a single posterior ventral spiracle; no oral disk and usually no horny beaks (present in two genera); external nares not opening until near time of metamorphosis (family Microhylidae)

Type X: a pair of ventral spiracles; no oral disk and no horny beaks or denticles; external nares present (families Rhinophrynidae and Pipidae)

Larval adaptations that reflect ecological differences relate primarily to whether the tadpoles occur in still (lentic) or flowing (lotic) water, where they feed in the water column, and their prey items and feeding mode. Convergence in morphological characteristics in species having similar ecologies, regardless of taxonomic position, led Altig and

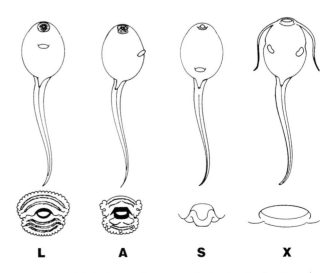

L   A   S   X

Figure 7.3. Principal types of anuran tadpoles showing location of spiracle(s), vent tube, and generalized mouthparts (see text for additional information).

Table 7.2. Ecomorphological Tadpole Guilds, with Costa Rican Examples in Parentheses

I. Lentic larvae found in still waters (puddles, ponds, lakes, backwaters, eddies, tree holes, and bromeliads)
   A. Feeding on small particles
      1. Benthic: generalized tadpole feeding near bottom *(Bufo marinus)*
      2. Nektonic: pelagic tadpole, with rasping beaks and denticles, feeding in open water *(Scinax staufferi)*
      3. Neustonic: surface-feeding tadpole with umbelliform oral disk that may be directed dorsally *(Colostethus nubicola)*
      4. Suspension feeder: type S and X tadpoles (without beaks or denticles) feeding in the water column *(Rhinophrynus dorsalis)*
      5. Suspension-rasper: type A (beaks and denticles present), feeding in the water column and on the bottom *(Agalychnis species)*
      6. Fossorial: vermiform tadpoles living in stream bottoms *(Centrolenella species)*
   B. Feeding on relatively large organisms or particles
      7. Macrophagous: tadpoles with beaks but reduced oral disk and denticles feeding on zooplankton *(Hyla microcephala)*
      8. Carnivorous: tadpoles with strong jaw muscles and beaks feeding on large prey, often cannibalistic *(Leptodactylus pentadactylus)*
      9. Oophagous: tadpoles feeding on frog eggs *(Anotheca spinosa)*
II. Lotic larvae found in flowing waters (creeks, streams, rivers)
   A. Living in slow-moving waters
      10. Clasping: generalized tadpole with large ventral oral disk *(Smilisca sordida)*
   B. Living in fast-moving streams
      11. Mountain brook: elongate tadpole using oral disk to maintain position *(Hyla rufioculis)*
      12. Gastromyzophorous: tadpole with a ventral suctorial disk and together with oral disk used to maintain position *(Atelopus varius)*

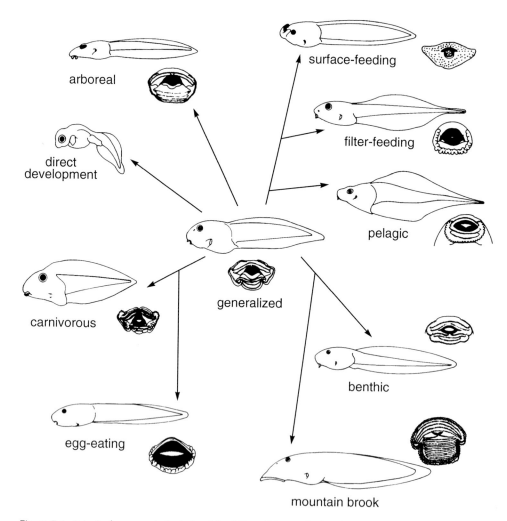

Figure 7.4. Principal ecomorphological guilds of Costa Rican tadpoles, showing conformation and mouthparts. Also see table 7.2.

Johnston (1989) to recognize some thirty-two major eco-morphological guilds among tadpoles. I have simplified their system for this book, as briefly characterized and illustrated in table 7.2 and figure 7.4. Other highly derived tadpoles are also shown to illustrate other extremely specialized morphologies and habits (fig. 7.5).

Descriptions of tadpoles in the species accounts are based on stages having the mouthparts fully developed, which usually coincides with the time when the gills are enclosed by the fleshy operculum and the spiracle(s) and hind limb buds appear. As metamorphosis approaches, first the forelimbs erupt, the gills are reabsorbed, and the larval mouthparts degenerate. The head and mouth are then restructured, and finally the tail is reabsorbed to complete the transformation.

The time of development from egg to metamorphosis varies considerably for species with aquatic eggs, from as little as 20 days (2 days to hatching) in seasonal, temporary pond breeders to 1,110 days (30 days to hatching and involving overwintering in some temperate species). In most tropical forms 30 to 60 days is usual. In taxa that lay the eggs out of water and have terrestrial or arboreal eggs but aquatic larvae, variation is approximately 20 (2 days to hatching) to 390 days (30 days to hatching). In direct developers known development times range between 15 and 49 days.

Johnson and Altig (1989) and McDiarmid and Altig (1999b) proposed a classification of anuran development patterns based on the source(s) of nutrition for embryos and larvae. A modified version of this scheme is outlined (table 7.3). Note that they used the term "nidicolous" for semidirect development. In this book nidicolous refers to organisms that develop for some time in a nest regardless of the source of nutrition.

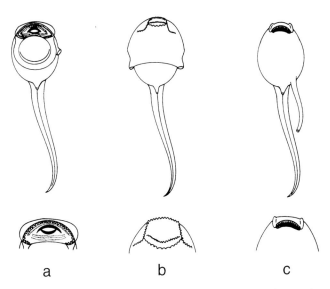

Figure 7.5. Highly derived anuran tadpoles. (a) *Atelopus* of tropical America showing abdominal sucker used to maintain position in torrential streams; (b) *Lepidobatrachus*, a voracious carnivorous tadpole of the South American Chaco, having paired spiracles and lacking horny beaks and denticles; (c) *Otophryne* of Colombian Amazonia to the Guianan region, with fanglike denticles and elongate sinistral spiracle; this tadpole hides under leaves in tiny streams flowing over sandy soil, and the denticle structure suggests it may be predaceous.

Many frogs have a brightly colored posterior thigh surface, groin, and anterior thigh surface. These areas are concealed when the animal is in its usual resting position with the hind limbs folded up against the body. When the anuran is disturbed and jumps to avoid a potential predator, the colors are suddenly displayed or "flashed." It has been proposed that the predator sees the color flash and continues to

Table 7.3. Developmental Patterns in Frogs and Toads Based on Nutritional Sources

I. Exotrophs: having free-living and feeding larvae
  *A. Orthotrophs: eggs hatch where laid, larvae finish development in aquatic situations (most frogs and toads)
  B. Neotrophs: fertilized eggs develop in mother's pouch, hatch as tadpoles that finish development in pond (some *Gastrotheca*, family Hylidae)
  C. Peponotrophs: terrestrial eggs; after some development, father introduces eggs into his vocal sac; tadpoles later released into stream (*Rhinoderma rufum*, family Rhinodermatidae)
II. Endotrophs: nonfeeding larvae or embryos, maternal investment source of nutrition
  A. Lecithotrophic: deutoplasm (yolk) of eggs sole source of nutrition
    1. Entotrophic: aquatic larvae (*Platyhyla grandis*, family Microhylidae)
    2. Semidirect developers: eggs laid out of water, larvae complete development in nests (e.g., *Pleophryne* spp., family Bufonidae)
    *3. Direct developers: no larval stage, froglet hatches from egg laid in terrestrial or arboreal situation (e.g., *Eleutherodactylus*, family Leptodactylidae)
    4. Exoviviparous: terrestrial eggs, father caregiver
      a. After some development, father introduces eggs into his vocal sac and they develop into froglets before release (*Rhinoderma darwinii*, family Rhinodermatidae)
      b. After hatching, tadpoles migrate onto father's back and metamorphose there (two *Leiopelma* spp., family Leiopelmatidae)
      c. After hatching, tadpoles invade father's inguinal pouches and are carried there until metamorphosis (all *Assa*, family Myobatrachidae)
    *5. Paraviviparous: yolk-rich eggs develop into froglets on back, in dorsal pouch(es), or in stomach of mother (e.g., most *Gastrotheca*, family Hylidae)
    6. Ovoviviparous: eggs retained and hatched in mother's oviduct with birth of small froglets (e.g., *Eleutherodactylus jasperi*, family Leptodactylidae)
  B. Viviparous: eggs retained and hatched in mother's oviduct, nutrition provided by oviductal secretions (hydrotroph) after exhaustion of deutoplasm; placenta present (all *Nimbaphrynoides* spp., family Bufonidae).

*Occurs in a Costa Rican species.

concentrate momentarily on the residual image on the retina, even after the frog has come to rest with the colors again concealed. Often the underside of the hind limb and upper and lower surface of the foot may also be brightly colored as part of the flash system.

Among many sympatric and similar species, the best identification characters are the different patterns on the posterior thigh surface. At least some observations indicate that females use the differences to recognize conspecific males. In these species (e.g., some *Eleutherodactylus*) the advertisement call is used primarily to space males in the environment (to establish territoriality), and calls may be so similar among related forms that female call discrimination is reduced. Clasping discrimination by females in these cases apparently depends on visual recognition of the exposed thigh, and amplexus follows if the thigh is right.

As I said earlier, anurans, like other amphibians, produce or contain a variety of noxious and toxic skin secretions (see Erspamer 1994 for a review). These may be elaborated by granular glands located throughout the integument or concentrated in warts, parotoid, femoral, and inguinal glands, or in dorsolateral and dorsal ridges. Among Costa Rican anurans, members of the genera *Atelopus* and *Bufo* secrete a variety of steroid cardiotoxins, and species of *Atelopus* produce tetrodotoxin-containing secretions. Frogs of the genera *Dendrobates* and *Phyllobates* produce a variety of alkaloids. Although it was long thought that the frogs synthesized these alkaloids, recent research implicates dietary uptake and subsequent concentration in the skin glands (Daly, Garraffo, et al. 1994; Daly, Secunda, et al. 1994) as the proximate source of at least some of the toxins. Ultimate sources (the actual synthesizers) could be the frogs' leaf litter arthropod prey or the plant food from which the arthropods are able to accumulate the alkaloids.

Whatever the source, the most toxic of these are several batrachotoxins found in *Phyllobates*, which are among the most powerful natural low-molecular-weight toxins known. These are the substances some South American Amerindians use on the darts for their blowguns. If injected into a bird or small mammal, the batrachotoxins cause heart failure in a fairly short time. So far as is known, the original inhabitants of Costa Rica did not use anuran toxins in this way. The skin of the Colombian species *Phyllobates terribilis* contains enough toxin to kill about 20,000 white mice weighing 20 g, estimated as equivalent to several human beings. Unfortunately these compounds are not destroyed by gastric juices, and they are very irritating to the eyes and mucous membranes. I recommend that you not kiss, lick, or swallow any frogs or toads!

In Costa Rica the species of *Atelopus*, *Bufo periglenes*, *Dendrobates*, and *Phyllobates* are diurnal and brightly and presumably aposematically colored (to warn off predators). The nontoxic *Eleutherodactylus gaigeae* is unusual among members of the genus in being uniform dark above with two lateral yellow to orange stripes. It is presumed to be a mimic of the much smaller and toxic *Phyllobates lugubris*, which has the same coloration and is sympatric with *E. gaigeae* in Costa Rica and Panama.

Frogs and toads are relatively short-lived, and most small to medium-sized species probably survive no more than two to five years. Large species have been kept in captivity for a maximum of thirty-six years, but natural longevity is known for only a few forms.

FROG AND TOAD IDENTIFICATION: KEY FEATURES

Although the body is greatly shortened and the overall morphology simplified as an adaptation to saltation, anurans have many more external characteristics that may be used for species identification than do other amphibians. Most of these are illustrated (fig. 7.6), and the following paragraphs should clarify differences among other features that might be ambiguous to the general reader.

Head form: There are differences in the dorsal outline of the head and snout when viewed from above, the snout profile (fig. 7.6b,c), and relative eye and tympanum sizes. The last are usually measured relative to one another or to the size of the largest finger disk.

Pupil shape: The pupil in bright light and preservative is round or forms either a horizontal slit (fig. 7.11d) or a vertical one (fig. 7.11e), and the condition is consistent at the family or generic level.

Tympanum: Most anurans have an external thin, transparent auditory membrane supported by the raised tympanic annulus that prominently rings the tympanum except above, where it is frequently obscured by a supratympanic fold of skin. This condition is referred to as an external, distinct tympanum. In other species the tympanum may be internal (covered by a layer of skin). If the skin is thin the tympanum and the annulus are partially visible, and this is termed an indistinct tympanum. If the skin is thicker so that the tympanum is not visible, it is called an internal hidden tympanum. Finally, in some species the tympanum is completely absent.

Cranial crests: These structures are bony ridges on the skull, usually overlain by thickened or co-ossified integument, that are found in most species of bufonid toads and one group of *Eleutherodactylus*. The crests develop after metamorphosis and are not evident in recently transformed and young juvenile specimens. Terminology applied to the crest system is discussed in greater detail and illustrated (fig. 7.20) in the section on the family Bufonidae.

Parotoid and other glands: These concentrations of many granular glands form discrete structures above the tympanum and behind the eye in bufonids (fig. 7.20a). Similar gland concentrations may be present on the area between the eye and the snout (canthis rostralis) and dorsal regions

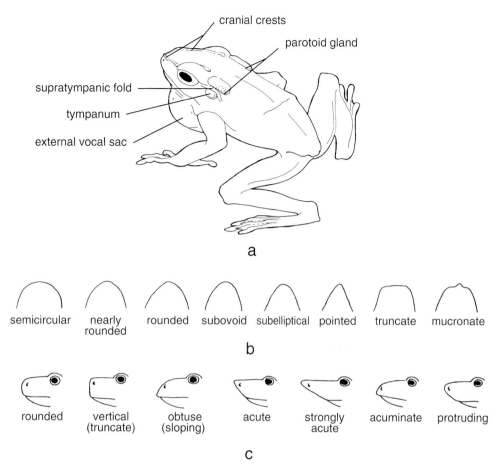

cranial crests

parotoid gland

supratympanic fold

tympanum

external vocal sac

a

semicircular | nearly rounded | rounded | subovoid | subelliptical | pointed | truncate | mucronate

b

rounded | vertical (truncate) | obtuse (sloping) | acute | strongly acute | acuminate | protruding

c

Figure 7.6. Morphological features of anurans: (a) external morphology; (b) dorsal outlines of head; (c) snout profiles.

or on the elbow, the heel, or the dorsal region of the hind limbs (fig. 7.17b,c)

Vocal sacs: Vocal sacs may be internal, with the skin covering the sac unmodified, or external with the skin of the throat modified (thinned or folded) to allow full extension of the sac when inflated. Internal vocal sacs vary by species, bulging slightly to fully when inflated. The sacs may be single (median) subgular, bilobate subgular, paired subgular, or paired lateral structures (fig. 7.12). Paired vocal slits under the posterolateral margins of the tongue are present in most sound-producing species, but they may be absent, or only one slit may be present. When present the slits usually extend from the midlateral base of the tongue nearly to the angle of the jaw. A few species have the slits lying along the medial margin of the lower jaw, and others have small round or slitlike openings located posteriorly, near the angle of the jaw.

Axillary web or membrane: A web of skin extends from the axilla to the posterior margin of the upper forearm in a few species (fig. 7.70d).

Ventral disk: In some genera, the ventral integument is somewhat thickened and clearly demarcated by a bounding groove that passes behind the chest along the ventrolateral region and across the venter just anterior to the groin area or extends to the femurs (fig. 7.10e)

Nuptial pads: In many species whitish, brown, or black glandular or keratinized excrescences develop on the thumbs in sexually mature males.

Hand and arm structures: The configuration, number, and presence or absence of features on the anterior limbs involve an externally visible prepollical spine, various tubercles (fig. 7.7a), and a fold or series of warty tubercles along the outer ventral margin of the lower arm or a definite fleshy fold. The relative length of fingers I and II when both are adpressed is often a significant feature.

Foot and leg structures: The form, number, and presence or absence of tubercles and fold(s) are illustrated (fig. 7.7b). In some species one or more enlarged tubercles may be present on the heel, or a fleshy flap may occur along the outer margin of the metatarsal segment. In some genera the relative length of toes III and V is a useful character.

Disks: Many species have the fingers or toes variously expanded and often have a pad on the tip of the digit; the digital disks are more complex structures consisting of an upper

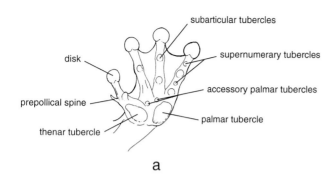

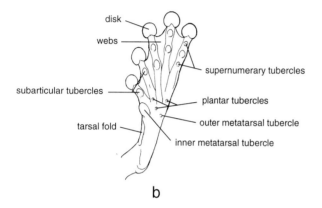

Figure 7.7. Morphological features of anurans: (a) palmar view of hand; (b) plantar view of foot.

disk cover and a ventral disk pad, separated by a terminal transverse groove; the finger and toe disks are of various sizes and shapes, and difference in shape between the disk cover and disk pad may be characteristic (fig. 7.7; 7.51).

Webbing: Differences in the presence and extent of finger and toe webs may distinguish similar species. The degree of webbing is expressed in terms of how many segments of the digits (phalanges), the distal metacarpal on the hand, and the distal metatarsal on the foot are free of webbing, as proposed by Savage and Heyer (1967) and modified by Myers and Duellman (1982) and Savage and Heyer (1997). In this system each finger and toe is represented by a roman numeral and the webbing between the digits by an arabic numeral. If the web extends to the tip of the digit or disk it is recorded as 0. If it extends to the base of the terminal phalanx the notation would be 1, if two phalanges are free of the web it would be 2, and so forth. Since the division of the fingers and toes continues between the metacarpals and metatarsals, the webbing formula refers to segments free from the web. A typical webbing formula for a hand would be I 2–2 II 2–3 III 2–2 IV; for a foot, I 2–2 II 1–2 III 1–1½ IV 2–1 V.

Superscript plus or minus signs indicate that the web reaches proximal or distal to the subarticular tubercle or terminal digit joint (e.g., 3⁺ means three segments plus are free of web). When the web in disked forms extends beyond the joint between the outer two phalanges it is scored in fractions as ¾, ⅔, ½, ⅓, or ¼. If the web reaches the tip of the digit the notation is 0. In those taxa having an intercalary cartilage or bone between the outer two phalanges, the convention is to consider the intercalary element as part of the distal phalanx. In this case a notation of 1 means the web reaches the joint between the penultimate phalanx and the intercalary cartilage, and fractions indicate the position of the web that extends distally beyond that point. In forms having deeply incised webs or extensive digital fringes, the web position is estimated by drawing a straight line between the points where the web connects to the two bordering digits. In the descriptions, where appropriate, the modal webbing formulas are given; remember that there will be some individual variation around the modes.

Teeth and odontoids: Teeth are present or absent on the upper jaws and vomerine bones (fig. 7.36b,c), depending on the taxon. The tooth-bearing portion of the vomer is called an odontophore, and its shape and position are characteristic. Vomerine odontophores and teeth appear during ontogeny, and small juveniles of species having these teeth as adults may lack them. Paired pointed bony projections (odontoids) are found at the symphysis of the lower jaws in one species *(Hyla picadoi)* (fig. 7.71b)

Coloration: Among the best ways to identify species in the field is by differences in color and pattern. Although many bright colors are lost in preservative, patterns are usually retained and are useful for identification. Overall coloration of the upper surfaces may be uniform or variously marked with dark or light spots, blotches, bands, or stripes (fig. 3.1). The markings on the lips (uniform, barred, or striped), the presence of a dark eye mask (fig. 7.47a) and its posterior limit, and supra- or posttympanic dark markings (fig. 7.46d,e) are often diagnostic. Differences also exist in the color of the groin and posterior thigh surface, and these are particularly useful in separating otherwise morphologically similar species. Ventral color and markings may also be diagnostic.

The size of anurans is a useful distinguishing feature in many instances. The standard length measurement for these animals is the distance from the tip of the snout to the posterior margin of the vent. In the species accounts and throughout the rest of this book reference to size in adult (sexually mature) frogs and toads refers to length as follows:

Tiny: less than 20 mm
Small: 20 to 30 mm
Moderate: 30 to 60 mm
Large: 60 to 200 mm
Giant: greater than 200 mm

References to other periods in the life history use the following terminology, which generally corresponds to the developmental stages in the Gosner (1960) staging system:

Embryos: developing individuals within the egg capsules; stage 1 to hatching, which may occur from stage 17 to stage 21

Hatchlings: embryos recently hatched from the egg, usually retaining external gills and lacking mouthparts; stages from hatching to stage 25, or froglets hatching directly from an egg

Larvae: free-swimming individuals with mouthparts, internal gills, a tail, and hind limb buds to definite hind limbs; stages 26–42

Metamorphs: recently transformed individuals lacking gills and larval mouthparts in which the tail has been nearly or completely reabsorbed; individuals at stage 46 and older but usually not sufficiently resembling adults to permit positive identification

Juveniles: sexually immature froglets and toadlets, identifiable to species

Animals at stages 41–45 are in transition between larval and terrestrial life, and during this period the tail is reabsorbed, the larval mouthparts are lost, the head becomes distinct, and the forelimbs erupt.

Larval sizes are defined in this book by total length (the distance from the tip of the snout to the tip of the tail) of advanced tadpoles at stages 35–40, using the following categories:

Tiny: less than 10 mm
Small: 10 to 25 mm
Moderate: 25 to 40 mm
Large: 40 to 60 mm
Giant: greater than 60 mm

There is sexual dimorphism in size in most frogs and toads, and males usually are obviously smaller than females. Secondary sexual characters of adult males that make sex determination relatively easy for Costa Rican species include:

The presence of vocal sac, vocal slits, or both (many species)
Glandular nuptial thumb pads (many species)
Prepollical or thumb spines (several species)
Chest spines (*Leptodactylus pentadactylus*)
Humeral hooks (*Centrolenella*)
Hypertrophied forelimbs (some *Leptodactylus*)
Swollen third finger (some *Colostethus*)
Brownish ventrolateral glands (*Leptodactylus melanonotus*)
Small spicules on chin (breeding *Gastrophryne, Hypopachus,* and *Nelsonophryne*)
Abdominal nuptial gland (*Gastrophryne, Hypopachus,* and *Nelsonophryne*)

The tympanum of most anurans is about the same size in both sexes or slightly larger in males. In the genus *Eleutherodactylus* however, many males have a tympanum that is nearly twice as large as in females (fig. 7.47).

Most anurans are relatively small. *Conraua goliath*, a West African semiaquatic ranid, is the largest frog and attains a length of 300 mm and a weight of 3.3 kg. The toads *Bufo blombergi* (Ecuador) and *B. marinus* (widespread in the Neotropics and where introduced) reach maximum lengths of 250 and 241 mm, respectively. *Psyllophryne didactyla* (Brachycephalidae) of Brazil is the smallest known tetrapod at a maximum length of 10.2 mm. *Bufo marinus* and *Leptodactylus pentadactylus* (to 185 mm) are the largest Costa Rica forms.

Leptodactylus pentadactylus has the largest tadpoles (total length 83 mm) of any Costa Rican anuran. These are dwarfed by the South American larva of the aquatic *Pseudis paradoxa*, with a length of 230 mm.

KEY TO THE GENERA OF COSTA RICAN FROGS AND TOADS

1a. No conspicuous tarsal tubercle 2
1b. A conical tarsal tubercle at proximate end of inner tarsal fold (fig. 7.8); no teeth on vomer or maxilla; toad-like, with numerous white gland-tipped warts and ridges . . . . *Physalaemus* (p. 223)
2a. A conspicuous enlarged parotoid gland, posterior to orbit or just anterior to level of axilla (fig. 7.6) 3
2b. Usually no enlarged parotoid gland; *if* gland is indicated, no cranial crests or tarsal fold present. 4
3a. Fingers and toes not fully webbed; inner digits well developed . . . . *Bufo* (p. 191)

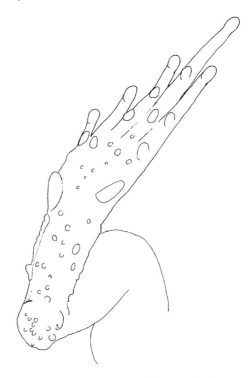

Figure 7.8. Underside of left tarsus and foot in *Physalaemus pustulosus* showing inner tarsal fold and tarsal tubercle.

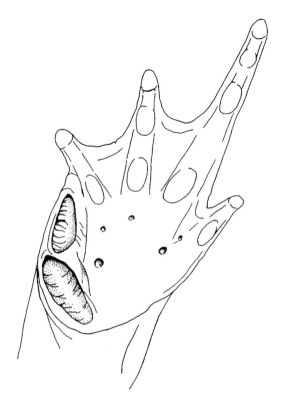

Figure 7.9. Plantar view of left foot of *Rhinophrynus dorsalis* showing two enlarged spadelike metatarsal tubercles and four well-developed toes.

3b. Fingers (except III) and toes (except IV) nearly fully webbed; inner digits rudimentary, enclosed by fleshy web (fig. 7.22a) . . . . *Crepidophryne* (p. 211)

4a. Never two elongate spadelike inner metatarsal tubercles, with free edge; five well-developed toes 5

4b. Two elongate spadelike inner metatarsal tubercles, with a free edge; four well-developed toes (fig. 7.9). . . . . *Rhinophrynus* (p. 183)

5a. A pair of dorsal scutelike fleshy flaps on tips of digits (fig. 7.10a) 6

5b. No dorsal scutelike fleshy flaps on tips of digits 8

6a. Pattern of light stripes on dark background; terminal disks of outer two digits of fingers and toes slightly wider than digits; maxillary teeth present; flesh flesh colored or gray. 7

6b. No stripes in pattern; terminal disks of outer two digits of finger and toes two times width of digits; no maxillary teeth; flesh black . . . . *Dendrobates* (p. 382)

7a. Venter and underside of thighs white, although throat and chest may be black in adult males; flesh flesh colored . . . . *Colostethus* (p. 377)

7b. Venter and underside of thighs black, mottled with lighter color; flesh gray . . . . *Phyllobates* (p. 388)

8a. A distinct transverse fold across head just posterior to eyes (fig. 7.10c); one or two distinct fleshy ridges across posterior area of roof of mouth (fig. 7.10d) 9

8b. No transverse fold across head; no fleshy ridge across roof of mouth 11

9a. Only an inner metatarsal tubercle present 10

9b. Inner and outer metatarsal tubercles present (fig. 7.7) . . . . *Hypopachus* (p. 394)

10a. Toes with moderate webs; venter uniform . . . . *Nelsonophryne* (p. 396)

10b. Toes with basal webs; venter dark with distinct large white spots . . . . *Gastrophryne* (p. 392)

11a. A small digital groove around tips of at least the outer two fingers and toes, separating upper surface from subterminal pad; disks usually present on fingers, always on toes (fig. 7.10b) 14

11b. No digital groove completely around tips of fingers and toes; never any finger disks, usually no toe disks. . . 12

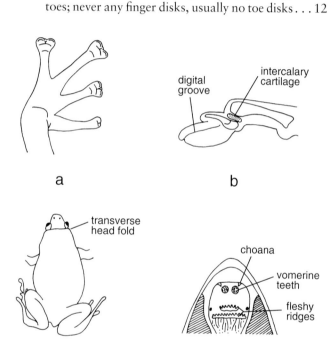

a

digital groove

intercalary cartilage

b

transverse head fold

choana

vomerine teeth

fleshy ridges

c

d

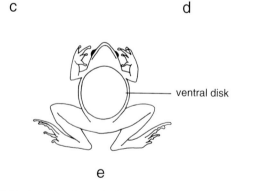

ventral disk

e

Figure 7.10. Specialized morphological features in anurans: (a) dorsal view of left hand in dendrobatid frogs showing scutelike fleshy flaps on fingertips; (b) lateral view of digit in hylid frogs showing skeletal structure and digital disk; (c) dorsal view of microhylid frog illustrating transverse head fold; (d) roof of an anuran mouth showing internal nares (choanae), vomerine tooth patches, and fleshy ridges; (e) ventral view of leptodactylid frog showing ventral disk.

12a. Webs present between toes; toes without lateral fringe; no ventral disk 13

12b. No webs between toes; often a lateral fringe on toes; a distinct ventral disk present (fig. 10e) . . . .
. . . . *Leptodactylus* (p. 213)

13a. A tympanum; teeth on upper jaw . . . . .*Rana* (p. 398)

13b. No tympanum; no teeth on upper jaw . . . .
. . . . *Atelopus* (p. 186)

14a. No ventral disk; finger webs usually present; an intercalary cartilage between distal and penultimate phalanges on each digit (fig. 7.10b) 15

14b. A ventral circular or V-shaped disk terminating on the posterior abdomen or femurs, respectively (fig. 7.10e); no finger webs; no intercalary cartilages . . . .
. . . . *Eleutherodactylus* (p. 226)

15a. Venter transparent in life, translucent in preservative so that white parietal pericardium and/or internal organs are visible; eyes directed forward at about a 45° angle (fig. 7.11a); tibiale and fibulare fused throughout length; terminal phalanges T-shaped, Y-shaped, or linear 16

15b. Venter usually not transparent or translucent, internal organs not visible; *if* transparent or translucent, fingers with vestigial webs or webs absent; eyes usually directed laterally (fig. 7.11b); tibiale and fibulare separate, except at tips; terminal phalanges claw shaped 18

16a. Viscera not covered by white pigment; liver trilobate; dorsum purple to lavender in preservative; bones dark green in life 17

16b. Viscera covered by white pigment; liver bulbous; dorsum white in preservative; bones pale green or usually white in life . . . . *Hyalinobatrachium* (p. 367)

17a. Adult males with humeral hook (fig. 7.11c); white parietal peritoneal sheath extending far posterior to hide most visceral organs . . . . *Centrolenella* (p. 360)

17b. No humeral hook; white parietal peritoneal sheath not extending far posterially beyond level of liver . . .
. . . . *Cochranella* (p. 362)

18a. Pupil of eye horizontal in bright light (in life) or in preservative (fig. 7.11d) 19

18b. Pupil of eye vertical in bright light (in life) or in preservative (fig. 7.11e). 26

19a. No series of bony occipital and frontal spines; no light-outlined dark markings on ventral surface of thighs 20

19b. A series of bony occipital and frontal spines in adults (fig. 7.11f); little or no finger webbing; light-outlined dark spots, bars, or rings on undersurface of thighs; lavender brown in life and in preservative . . . .
. . . . *Anotheca* (p. 295)

20a. No triangular dermal flap on upper eyelids. 21

20b. A triangular supraorbital flap on each eyelid (fig. 7.11g); no finger webbing. . . . . *Gastrotheca* (p. 293)

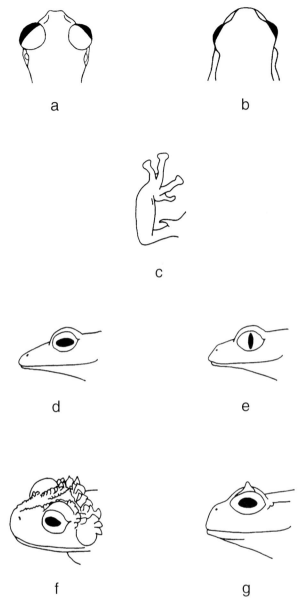

Figure 7.11. Specialized morphological features in frogs and toads; (a) eyes directed forward; (b) eyes directed laterally; (c) left arm of glassfrog showing humeral hook; (d) horizontal pupil; (e) vertical pupil; (f) lateral view of head of adult *Anotheca spinosa* showing bony occipital and frontal spines; (g) lateral view of head of *Gastrotheca cornuta* showing fleshy supraorbital flap.

21a. Skin of dorsal surfaces of head usually not co-ossified with skull roof; *if* co-ossified, fleshy fringes on outer limb margins (fig. 7.70a) and protuberant prepollex present; bones usually white 22

21b. Skin of dorsal surfaces of head co-ossified with exostosed (ridged and pitted) skull roof (fig. 7.99); no fleshy fringes on limbs, prepollex not protuberant; bones green in life . . . . *Osteopilus* (p. 338)

22a. Usually a definite web between toes I and II; *if* not, snout bluntly rounded in profile (fig. 7.6) 23

22b. Webbing between toes I and II reduced to a fringe along inner margin of toe II or absent (fig. 7.12a); snout protruding beyond lower jaw (fig. 7.6) . . . . *Scinax* (p. 342)

23a. No dense concentration of granular poison glands in occipital and frontal regions, as indicated by thickened skin; vocal sacs subgular, single, bilobate, or paired in adult males (fig. 7.12c,d,e); bones usually not green in life 24

23b. Dorsal integument thick and glandular; a dense zone of granular poison glands indicated by thickened, rugose skin in occipital and frontal regions (fig. 7.12b); paired lateral vocal sacs in adult males (fig. 7.12f); bones green in life . . . . *Phrynohyas* (p. 340)

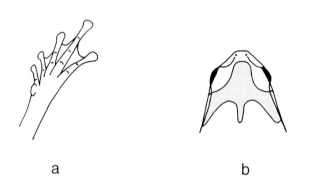

a   b

c   d

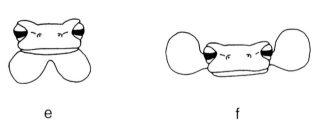

e   f

Figure 7.12. Specialized morphological features in anurans: (a) reduced webbing between toes I and II; (b) distribution of granular-poison glands on the upper head surface of *Phrynohyas venulosa*; external vocal sacs: (c) single (median) subgular; (d) bilobate subgular; (e) paired subgular; (f) paired lateral.

24a. Males with paired subgular vocal sacs (fig. 7.12e); depressor mandibulae arising from two slips, one from dorsal fascia, the other from posterior arm of squamosal; bones not green in life; adductor mandible externus superficialis present . . . . *Smilisca* (p. 348)

24b. Males with single, or sometimes bilobate, median vocal sacs (fig. 7.12c,d); depressor mandibulae arising from dorsal fascia only; bone color variable, usually not green in life; no adductor mandibulae externus superficialis 25

25a. Tadpoles with huge pendant, funnel-shaped oral disks; disk covered with large, widely spaced conical papillae; denticle rows 2-3/2–3; all rows short, about as long as width of mouth; red eyes in adults . . . . *Duellmanohyla* (p. 297)

25b. Tadpoles without pendant, funnel-shaped oral disks; disk not covered with large conical papillae; denticle rows various, when present, at least some rows longer than width of mouth; eyes usually not red in adults . . . .*Hyla* (p. 301)

26a. Considerable webbing on fingers and toes . . . . *Agalychnis* (p. 278)

26b. No webbing on fingers and toes . . . . *Phyllomedusa* (p. 286)

Note that the species of *Duellmanohyla*, *Hyla*, and *Smilisca* are in a single key (p. 288) so that it is not necessary to know the differentiating features of internal anatomy or their larvae to make identifications.

IDENTIFICATION OF THE LARVAE (TADPOLES) OF COSTA RICAN ANURANS

Of the 131 known native and two introduced species of frogs and toads known from Costa Rica, ninety-one of the former and one of the latter have or are thought to have free-living tadpoles. Family distributions are as follows: Rhinophrynidae (1), Bufonidae (14), Leptodactylidae (6), Hylidae (41 + 1), Centrolenidae (13), Dendrobatidae (8), Microhylidae (3), and Ranidae (5). Available data make it possible to include (76) of these species in the key presented below. The eggs of most of these species are deposited in aquatic situations, with the following exceptions: (1) two species of *Leptodactylus*, *L. labialis* and *L. poecilochilus*, lay their eggs in a foam-filled burrow (Heyer 1969), (2) the treefrog *Hyla ebraccata* lays its eggs on the leaves of broad-leafed herbaceous plants emerging from ponds (Duellman 1967a); (3) all members of the family Centrolenidae (Starrett 1960a), the hylid genera *Agalychnis* and *Phyllomedusa*, and *Hyla calypsa* (Lips 1996a,b) lay eggs on vegetation, usually leaf surfaces, a considerable distance above a stream or other body of water: and (4) all members of the family Dendrobatidae lay their eggs in moist terrestrial sites and transport the hatched tadpoles to water on the adult's back (Savage 1968b).

LARVAL STRUCTURE: KEY FEATURES

The characteristics of anuran larvae that may be used in taxonomic description have been detailed by Orton (1952), Altig (1970), Altig and Brandon (1971), Duellman (1970, 2001), Starrett (1973), and McDiarmid and Altig (1999a). Distinctions among preserved tadpoles generally can be made only in developmental stages 25 (the time of operculum closure) through 40 (immediately before front leg eruption), using the Gosner (1960) system. The synopsis that follows is based on tadpoles of these stages and the characterizations provided by the authors cited. The principal morphological features used for identification are illustrated (fig. 7.13). Sizes given in the text are mean total lengths of material in stages 35–40 unless otherwise indicated. They are meant to indicate general size categories as total length increases until about stage 40, when the tail begins to be reabsorbed. Generally stage 40 larvae are two to two and a half times as long as those at stage 25. Points of minor departure or preference include:

1. Denticle is used instead of labial tooth, since larval Anura have no true teeth (with enamel or dentine), but only a series of nonhomologous keratinized structures.
2. The jaws, beaks (called sheaths [= rhamphothecae] by Altig 1970; Altig and McDiarmid 1999), labia, and denticle rows are referred to as upper and lower because of their association with the upper and lower jaw structures common to most vertebrates; Altig (1970) uses anterior and posterior, respectively, since most tadpoles have ventrally located mouths (but see McDiarmid and Altig 1999).
3. The denticle rows are coded as A1, A2 (upper) and P1, P2 (lower) after Altig (1970), with A1 one being nearest the upper disk margin and P1 nearest the mouth. The denticle formula denotes the number of denticle rows above the mouth over the number below the mouth (e.g., three above, four below = 3/4).
4. A complete or entire oral disk refers to an unindented disk.
5. Description of mouth position and disk size follows the terminology of Duellman (1970; as modified by McDiarmid and Altig 1999b):

    Mouth position: dorsal, terminal, anteroventral, anteroventral, but directed ventrally or ventral.

    Oral disk size (= mouth size): small (less than 67% of body width); medium (greater than 67% but less than body width); large (greater than body width).
6. Other terminology:

    Body shape: ovoid (depth equal to width); robust (= mesomorphic) (depth slightly greater than width); compressed (depth 15% or more of width); depressed (depth less than width).

    Tail length: short (less than 60% of total length); moderate (60 to 67% of total length); long (more than 67% of total length).

The tadpoles of seventy-six native Costa Rican species and one introduced form have been described or illustrated in considerable detail. Reasonably accurate predictions as to the basic features of several other species can be made as follows:

1. Bufonidae: tadpoles of all known members of the genus *Atelopus* (Starrett 1967; Duellman and Lynch 1969) have the peculiar enlarged ventral disk found in *Atelopus varius*: all known Central American *Bufo*, except *Bufo holdridgei*, agree in having 2/3 denticle rows, an emarginate oral disk, and oral papillae across both upper and lower labia.
2. Centrolenidae: it is anticipated that the other Costa Rican tadpoles of the genera *Centrolenella*, *Cochranella*, and *Hyalinobatrachium* will agree with known species in having 2/3 denticle rows, the oral disk complete, a median vent tube, the spiracle posterior in position, and the denticle row above the mouth (A2) restricted to two short lateral segments.

Of the remaining taxa, thirty-nine native and one introduced species belong to the genus *Eleutherodactylus*. So far as is known, all members of this genus but one extralimital form lay terrestrial encapsulated eggs that undergo direct development into small frogs. The single exception is the ovoviviparous Puerto Rican *Eleutherodactylus jasperi* (Drewry and Jones 1976; Wake 1978). One introduced form and one

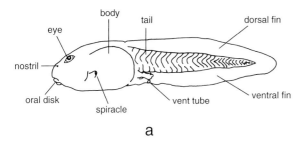

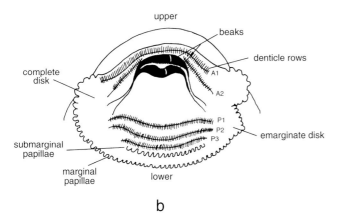

Figure 7.13. Morphological characteristics of anuran larvae (tadpoles): (a) lateral view; (b) oral disk and mouthparts.

Table 7.4. Costa Rican Anurans for Which
Larvae Are Unknown or Not Fully Described

*Atelopus chiriquiensis**	*Hyalinobatrachium chirripoi*
Atelopus senex	*Hyalinobatrachium pulveratum*
Bufo melanochlorus	*Hyalinobatrachium talamancae*
Centrolenella ilex	*Hyalinobatrachium vireovittatum*
Cochranella albomaculata	*Hyla miliaria*
Cochranella euknemos	*Hyla rufitela*
Crepidophryne epiotica	*Hyla xanthosticta*
Duellmanohyla lythrodes	

*Not fully described

other species *(Gastrotheca cornuta)* have no larval stage. Consequently the tadpoles of 84% of the native species having a free-living aquatic stage (seventy-six out of ninety-one) have been described.

The following key is slightly modified from the one published by Lips and Savage (1996a). It is based on available specimens and published data. It is designed to lead to identification when well-preserved material conspecific with described tadpoles is available. In most cases it is anticipated that examples of as yet undescribed tadpoles of known species will be correctly identified at the generic level. Larvae of the genera *Atelopus, Bufo, Centrolenella, Cochranella,* and *Hyalinobatrachium* fall into this category, so by elimination any other tadpole will probably be one of the species of *Hyla.*

The species of Costa Rican frogs listed in table 7.4 do not have described or fully described and illustrated tadpoles. Where material does not conform to the key or illustrations, you must seriously consider the possibility that you have discovered a previously missing immature stage.

KEY TO THE TADPOLES OF COSTA RICAN ANURANS

1a. No denticle rows 2
1b. Denticle rows present 10
2a. No jaw beaks (fig. 7.3S,X). 3
2b. Jaw beaks present (fig. 7.3A) 6
3a. One ventral spiracle; no barbels (fig. 7.3S) 4
3b. Two lateral spiracles; barbels present at corners of mouth (fig. 7.3X) . . . . *Rhinophrynus dorsalis* (p. 183)
4a. Spiracle opens near vent tube; oral flaps present (fig. 7.14a,b) 5
4b. Spiracle opens near midbelly; no oral flaps. . . . . *Nelsonophryne aterrima* (p. 396)
5a. Oral flaps without scallops or papillae (fig. 7.14a). . . . . *Gastrophryne pictiventris* (p. 393)
5b. Oral flaps scalloped or with papillae (fig. 7.14b) . . . . *Hypopachus variolosus* (p. 394)
6a. Tail terminating in a thin flagellum (fig. 7.15e) 7
6b. Tail not terminating in a thin flagellum 9
7a. Dextral vent; no rows of submarginal papillae 8
7b. Median vent; one row of submarginal papillae . . . . *Hyla ebraccata* (p. 313)
8a. Postorbital dark stripe absent or extending onto tail; no dark spots on tail . . . . *Hyla microcephala* (p. 316)

8b. Postorbital dark stripe restricted to body; dark spots on tail . . . . *Hyla phlebodes* (p. 318)
9a. Eyes lateral; large umbelliform oral disk . . . . *Colostethus flotator* (p. 379)
9b. Eyes dorsal; moderate oral disk, protuberant labia, velum covering anterior beak . *Hyla picadoi* (p. 336)
10a. No enlarged abdominal sucker 11
10b. An enlarged abdominal sucker (fig. 7.14c) . . . . *Atelopus varius* (p. 189)
11a. Denticle rows 1/0, 2/0; 01, 1/1; or 1/2 12
11b. Denticle rows 2/2 or more. 14
12a. Vent dextral; papillae present on upper disk margin 13
12b. Vent median; no papillae on upper disk margin. . . 76
13a. Denticle rows 1/1 or 2; small oral disk 77
13b. Denticle rows 1/0 or 2/0; umbelliform oral disk . . . . *Colostethus nubicola* (p. 380)
14a. Denticle rows 2/2 15
14b. Denticle rows 2/3 or more. 16
15a. A2 and P1 complete or with narrow median gap; oral disk bordered by minute papillae . . . . *Duellmanohyla rufioculis* (p. 298)
15b. Broad median gap in A2; oral disk bordered by one row of large papillae . . . . *Anotheca spinosa* (p. 295)
16a. Denticle rows 2/3 or 3/3; if 3/3 oral disk not covered by large papillae 17
16b. Denticle rows more than 2/3. 68
17a. Oral disk emarginate (fig. 7.13). 18
17b. Oral disk complete (nonemarginate) (fig. 7.13). . . 30
18a. Gaps in anterior and posterior labial papillae 19
18b. No gaps in posterior labial papillae. 25
19a. White saddles on tail musculature. 20

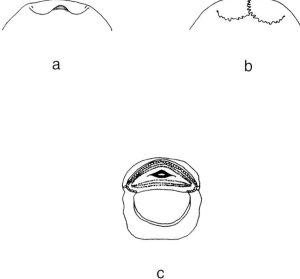

Figure 7.14. Specialized mouth region in some Costa Rican tadpoles: (a) oral flaps without scallops; (b) oral flaps with scallops; (c) oral disk and abdominal sucker in *Atelopus.*

19b. Tail musculature solid black or brown above 21

20a. Scattered dark pigment on caudal fin . . . .
. . . . *Bufo coccifer* (p. 203)

20b. Reticulate pattern on caudal fin . . . .
. . . . *Bufo valliceps* (p. 210)

21a. Body and tail dark above and below 22

21b. Body light below, dark above 23

22a. Tail fins suffused with dark pigment . . . .
. . . . *Bufo fastidiosus* (p. 195)

22b. Tail fins usually clear . . . . *Bufo marinus* (p. 199)

23a. All denticle rows of equal length 24

23b. Denticle row P3 shorter than other rows. 78

24a. No submarginal papillae on posterior disk area . . . . .
. . . . *Bufo coniferus* (p. 207)

24b. Few submarginal papillae on posterior disk area . . . .
. . . . *Bufo periglenes* (p. 202)

25a. Oral disk anteroventral; body ovoid; one row of papillae posterior to mouth 26

25b. Oral disk ventral; body depressed; two rows of papillae posterior to mouth 28

26a. Dorsal fin high, extending onto body 27

26b. Dorsal fin low, not extending onto body; body, tail,
and fins with large dark blotches . . . .
. . . . *Physalaemus pustulosus* (p. 224)

27a. Body brown, tail heavily pigmented with large spots
or splotches; well-developed lateral line system . . . . .
. . . . *Rana forreri* (p. 399)

27b. Body dark brown, tail with light brown punctations;
lateral line system indistinct . . . *Rana taylori* (p. 400)

28a. Tail and body with dark spots; P3 = P2 = P1 29

28b. Uniformly dark body and tail; P3 < P1, P2 . . . .
. . . . *Phyllobates vittatus* (p. 390)

29a. Dark tail spots small . . . .
. . . . *Colostethus talamancae* (p. 381)

29b. Dark tail spots large . . . *Phyllobates lugubris* (p. 389)

30a. Median vent 31

30b. Dextral vent 44

31a. A2 complete or barely interrupted medially 32

31b. A2 broadly interrupted medially 38

32a. Gap in papillae of anterior labium; oral disk unexpanded; no folds lateral to mouth 33

32b. Oral papillae complete across anterior labium; oral
disk greatly expanded; deep folds lateral to mouth . .
. . . . *Hyla pictipes* (p. 321)

33a. Tail with dark spots and blotches 34

33b. Tail without dark spots and blotches; uniformly dark
or with a few light spots 36

34a. Distinct ventral light spot posterior to oral disk; disk
anteroventral; P.3 more than half length of P1, P2
. . . . 35

34b. No distinct light spot posterior to oral disk; disk terminal; P3 half length of P1, P2 . . . .
. . . . *Leptodactylus pentadactylus* (p. 219)

35a. Body without dark flecks . . . .
. . . . *Leptodactylus labialis* (p. 221)

35b. Body with small dark flecks; gap in A2 and P1 . . . . .
. . . . *Leptodactylus poecilochilus* (p. 222)

36a. Median gap in A2 37

36b. All denticle rows complete . . . .
. . . . *Leptodactylus bolivianus* (p. 217)

37a. Oral papillae in two rows; body and tail musculature
uniform black . . . . *Dendrobates auratus* (p. 383)

37b. Oral papillae in one row below denticle row P3; body
and tail musculature black, with spiracle and areas lateral to eyes lighter . . . .
. . . . *Leptodactylus melanonotus* (p. 215)

38a. Spiracle much closer to posterior margin of body than
to eye; A2 restricted to two short segments near corners of mouth 39

38b. Spiracle about halfway between eye and posterior margin of body; A2 extending medially well beyond corners of mouth . . . . *Bufo holdridgei* (p. 196)

39a. P3 equal to or only slightly shorter than other posterior denticle rows 40

39b. P3 much shorter than other posterior denticle rows
. . . . *Centrolenella prosoblepon* (p. 361)

40a. Lower beak with large serrations; lateral serrations
distinctly larger than median ones 41

40b. Lower beak with small, fine serrations; all serrations
about equal in size 42

41a. Upper beak with small serrations; lower beak with large
blunt serrations . . . . . *Cochranella granulosa* (p. 363)

41b. Both beaks with large sharp serrations . . . .
. . . . *Cochranella spinosa* (p. 365)

42a. Tail musculature lacking dark blotches 43

42b. Anterodorsal area of tail musculature with dark
blotches . . . . . . *Hyalinobatrachium valerioi* (p. 374)

43a. Anterodorsal area of tail musculature with a narrow
dark stripe or row of spots along each side of tail fin
. . . . . . *Hyalinobatrachium colymbiphyllum* (p. 370)

43b. Anterodorsal area of tail fin uniformly covered with
small dark punctations . . . .
. . . . *Hyalinobatrachium fleischmanni* (p. 371)

44a. No gap in anterior marginal papillae 45

44b. Median gap in anterior marginal papillae 53

45a. Oral disk covered by a few large, widely spaced, conical papillae; denticle rows not extending laterally past
level of beak margins 46

45b. Oral disk not covered by enlarged papillae; denticle
rows extending laterally well beyond level of beak
margins 47

46a. A1 and A2 about same length . . . .
. . . . *Duellmanohyla rufioculis* (p. 298)

46b. A1 shorter than A2 . . . .
. . . . *Duellmanohyla uranochroa* (p. 299)

47a. No more than three rows of papillae posterior to
mouth; median gap in A2 48

47b. Four to seven rows of papillae below posterior tooth
rows; no median gap in A2; body and tail spotted with
dark 52

48a. Two to three rows of papillae anterior to mouth; body and tail with dark spots; body depressed, wider than deep 49

48b. One row of papillae anterior to mouth; body uniform, tail tan with dark spots 51

49a. Oral disk small, about two-thirds width of body. . . 50

49b. Oral disk large, as wide as body . . . . *Hyla debilis* (p. 320)

50a. Tail length 68–70% body length; snout truncate. . . . . *Hyla calypsa* (p. 310)

50b. Tail 57–66% body length; snout rounded . . . . *Hyla lancasteri* (p. 311)

51a. Spiracle lateral, tubular; body ovoid, depth equal to width; body uniform; tail tan with flecks and dashes of darker pigment that form crossbars on dorsal surface of tail musculature . . . . *Smilisca sordida* (p. 355)

51b. Spiracle ventrolateral, a slit without a tube; body depressed, slightly wider than deep; body uniform olive brown; caudal musculature tan, brown reticulations on both fins . . . . *Agalychnis calcarifer* (p. 279)

52a. Tail fin not extending onto body; lateral folds of disk lacking papillae . . . . *Hyla tica* (p. 324)

52b. Tail fin extending onto body; several large papillae in lateral folds of disk . . . . *Hyla rivularis* (p. 323)

53a. No gap in posterior marginal papillae; P3 not on extension 54

53b. Gap in posterior papillae; P3 mounted on short, vertically movable extension. . . *Scinax boulengeri* (p. 343)

54a. Spiracle ventrolateral 55

54b. Spiracle lateral 59

55a. Body robust, slightly deeper than wide; tail equal to or slightly deeper than body; tail fins low 56

55b. Body deep, at least 15% deeper than wide; tail much deeper than body; tail fins very high. . . . .*Agalychnis annae* (p. 278)

56a. P3 complete 57

56b. P3 interrupted medially . . . . *Agalychnis saltator* (p. 283)

57a. Beak serrations pointed 58

57b. Beak serrations blunt . . . .*Agalychnis callidryas* (p. 281)

58a. Tail fin not extending onto body . . . . *Agalychnis spurrelli* (p. 285)

58b. Tail fin extending onto body. . . . . *Phyllomedusa lemur* (p. 286)

59a. Tail not acuminate, tip not tapering to a long tip, round or pointed (fig. 7.15a,b,d) 60

59b. Tail acuminate or tapering to a long point (fig. 7.15c); tail fin extending onto body; oral disk anteroventral, small 66

60a. Tail fin not extending onto body; oral disk ventral, small 61

60b. Tail fin extending onto body; oral disk anteroventral 63

61a. Oral disk small 62

61b. Oral disk medium . . . . *Smilisca sila* (p. 354)

62a. Tail speckled, body lightly colored . . . . *Hyla pseudopuma* (p. 328)

62b. Tail spotted, body dark. . . . . *Hyla angustilineata* (p. 327)

63a. Body dark brown or black. 64

63b. Body pale brown to tan 65

64a. Body not tuberculate; no median gap in P1 . . . . *Smilisca baudinii* (p. 349)

64b. Body tuberculate; median gap in P1 . . . .*Hyla fimbrimembra* (p. 332)

65a. P3 much shorter than P1 and P2 . . . . *Smilisca puma* (p. 352)

65b. P3 equal in length to P1 and P2. . . . . *Smilisca phaeota* (p. 351)

66a. No gap in P2; P3 slightly shorter than P1 and P2. 67

66b. Median gap in P2; P3 extremely short, much shorter than P1 or P2 . . . . *Hyla loquax* (p. 308)

67a. Tail fins moderate, fin barely extends onto body . . . . *Scinax elaeochroa* (p. 345)

67b. Tail fins very high, with high arch; fin extends well onto body. . . . . *Scinax staufferi* (p. 347)

68a. Oral disk emarginate (fig. 7.13). 69

68b. Oral disk complete (nonemarginate) (fig. 7.13). . . 71

69a. Denticle rows 5–7/4–10 70

69b. Denticle rows 4/4. . . . . *Rana vaillanti* (p. 402)

70a. Denticle rows 5–7/7–10 79

70b. Denticle rows 5–6/4 . . . . *Rana vibicaria* (p. 403)

71a. Dextral vent. 72

71b. Median vent; gap in anterior marginal papillae; tooth rows 3–4/4–6 . . . . . . *Phrynohyas venulosa* (p. 340)

72a. No gap in upper marginal papillae 73

72b. Gap in upper marginal papillae. 74

73a. Disk surface with large papillae; denticle rows short, 3/3 . . . . *Duellmanohyla uranochroa* (p. 299)

73b. Disk surface covered by denticle rows; denticle rows long, 2–3/5 . . . . *Hyla legleri* (p. 331)

74a. Denticle rows 2/4. 75

74b. Denticle rows 6/4 . . . . . *Rana warszewitschii* (p. 404)

75a. Body tan to gray . . . . *Hyla rosenbergi* (p. 303)

75b. Body black . . . . . *Osteopilus septentrionalis* (p. 338)

76a. Tail tip rounded (fig. 7.15a); beaks huge, occupying most of oral disk; A1 short, restricted to area above middle of beak or lateral with large gap above mouth; papillae below mouth not widely separated from P1 . . . .*Dendrobates pumilio* (p. 386)

76b. Tail tip long and attenuate (fig. 7.15d); beaks large but not occupying most of oral disk; A1 extends to level of lateral margins of upper beak; papillae below mouth widely separated from P1 . . . . *Dendrobates granuliferus* (p. 384)

77a. Denticle rows 1/2 . . . . *Hyla zeteki* (p. 337)

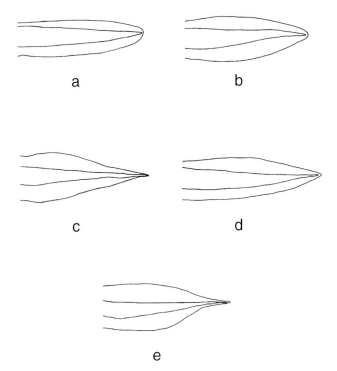

Figure 7.15. Characteristics of tails and fins in Costa Rican tadpoles: (a) low fins, rounded tip; (b) moderate fins, bluntly pointed tip; (c) deep fins, acuminate form; (d) moderate fins, pointed tip; (e) deep fins, tail tip a flagellum.

77b. Denticle rows 1/1 . . . . *Hyla picadoi* (p. 336)
78a. No submarginal papillae on lower labium; caudal fins clear . . . . *Bufo haematiticus* (p. 198)
78b. Submarginal papillae on lower labium; caudal fins mottled with dark pigment . . *Bufo leutkenii* (p. 208)
79a. Tail acuminate; tail with a longitudinal stripe; tail fins with dark blotches posteriorly; denticle rows at least 5/8 . . . . *Hyla palmeri* (p. 307)
79b. Tail bluntly pointed; dorsal fin and tail with numerous brown flecks; denticle rows 6–7/6–10 . . . . *Hyla colymba* (p. 306)

Illustrations of the larvae of Costa Rican anurans accompany the appropriate species accounts. These include a semidiagrammatic lateral view emphasizing general habitus, the most obvious features of coloration with a scale indicating the total length of a typical specimen at the Gosner (1960) stage illustrated, and a diagram of the mouthparts. These are composite renderings of available material and the illustrations in the literature cited below. Because colors change or fade in preservative, the figures primarily depict dark components of the pattern retained after preservation. The illustrations of mouthparts are drawn at a variety of scales to show major structural features and should not be interpreted as representing differences in size. Consult the following references for more detailed descriptions and illustrations of specific larvae.

INDEX TO PUBLISHED ILLUSTRATIONS OF COSTA RICAN TADPOLES

Agalychnis
   annae Duellman (1963a, 1970)
   calcarifer (Donnelly et al. 1987)
   callidryas (Starrett 1960a [as *helenae*]; Duellman 1970; Lee 1996)
   saltator (Duellman 1970)
   spurrelli (Duellman 1970)
Anotheca
   spinosa (Taylor 1954c; Robinson 1961 [as *coronata*]; (Duellman 1970)
Atelopus
   chiriquiensis (not fully described) (Lötters 1996)
   senex (unknown)
   varius (Starrett 1967)
Bufo
   coccifer (McDiarmid and Foster 1981)
   coniferus (Livezey 1986)
   fastidiosus (Lips and Krempels 1995)
   haematiticus (McDiarmid and Altig 1990)
   holdridgei (Novak and Robinson 1975)
   luetkenii (not completely described) (Haas and Köhler 1997; Köhler, Lehr, and McCranie 2000)
   marinus (Breder 1946; Savage 1960a)
   melanochlorus (unknown)
   periglenes (Savage 1966b)
   valliceps (Limbaugh and Volpe 1957; Lee 1996)
Centrolenella
   ilex (unknown)
   prosoblepon (Starrett 1960a)
Cochranella
   albomaculata (unknown)
   euknemos (unknown)
   granulosa (Starrett 1960a)
   spinosa (Starrett 1960a)
Colostethus
   flotator (Dunn 1924b; Ibáñez and Smith 1995)
   nubicola (Dunn 1924b; Savage 1968b; Ibáñez and Smith 1995)
   talamancae (Breder 1946; Savage 1968b)
Crepidophryne
   epiotica (unknown)
Dendrobates
   auratus (Breder 1946; Savage 1968b; Silverstone 1975)
   granuliferus (Van Wijngaarden and Bolaños 1992)
   pumilio (Starrett 1960a; Savage 1968b; Silverstone 1975)
Duellmanohyla
   lythrodes (unknown)
   rufioculis (Duellman 1970)
   uranochroa (Dunn 1924b; Duellman 1970)
Gastrophryne
   pictiventris (Donnelly, de Sá, and Guyer 1990)

Hyalinobatrachium
  chirripoi (unknown)
  colymbiphyllum (Jaramillo, Jaramillo, and Ibáñez 1997)
  fleischmanni (Starrett 1960a; Duellman and Tulecke 1960 [as *viridissima*]); Lee 1996)
  pulveratum (unknown)
  talamancae (unknown)
  valerioi (Starrett 1960a [as *reticulata*])
  vireovittatum (unknown)
Hyla
  angustilineata (Duellman 1970)
  calypsa (Lips 1996a,b)
  colymba (Dunn 1924b [as *albomarginata*]; Duellman 1970)
  debilis (Duellman 1970)
  ebraccata (Duellman 1970; Lee 1996)
  fimbrimembra (Savage 1981c)
  lancasteri (Starrett, 1960a [as *moraviensis*]; Duellman 1970)
  legleri (Duellman 1970)
  loquax (Lee 1996)
  microcephala (Duellman 1970; Lee 1996)
  miliaria (unknown)
  palmeri (Duellman 2001)
  phlebodes (Duellman 1970)
  picadoi (Robinson 1977)
  pictipes (Starrett 1966; Duellman 1970)
  pseudopuma (Starrett 1960a; Duellman 1970)
  rivularis (Starrett 1960a; Duellman 1970)
  rosenbergi (Breder 1946; Duellman 1970)
  rufitela (unknown)
  tica (Duellman 1970)
  xanthosticta (unknown)
  zeteki (Dunn 1937a; Starrett 1960a; Duellman 1970)
Hypopachus
  variolosus (Taylor 1942 [as *caprimimus?*]; Lee 1996)
Leptodactylus
  bolivianus (Heyer 1968)
  labialis (Heyer 1968; Lee 1996)
  melanonotus (Heyer 1968; Lee 1996)
  pentadactylus (Heyer 1968)
  poecilochilus (Heyer 1968)
Nelsonophryne
  aterrima (Donnelly, de Sá, and Guyer 1990)
Osteopilus
  septentrionalis (Duellman and Schwartz 1958 [as *Hyla*])
Phrynohyas
  venulosa (Zweifel 1964a; Duellman 1970; Lee 1996)
Phyllobates
  lugubris (Donnelly, Guyer, and de Sá 1990)
  vittatus (Savage 1968b [as *lugubris*]; Silverstone 1976)
Phyllomedusa
  lemur (Duellman 1970)

Physalaemus
  pustulosus (Noble 1927 [as *Eupemphix*]; Breder 1946 [as *Engystomops*]; Lee 1996)
Rana
  forreri (described here)
  vaillanti (Volpe and Harvey 1958 [as *palmipes*]; Hillis and de Sá 1988; Lee 1996)
  taylori (described here)
  vibicaria (Zweifel 1964b)
  warszewitschii (Starrett 1960a [as *warschewitschii*])
Rhinophrynus
  dorsalis (Orton 1943; Starrett 1960a; Lee 1996)
Scinax
  boulengeri (Duellman 1970)
  elaeochroa (Starrett 1960a; Duellman 1970)
  staufferi (Duellman 1970; Lee 1996)
Smilisca
  baudinii (Duellman and Trueb 1966; Duellman 1970; Lee 1996)
  phaeota (Duellman and Trueb 1966; Duellman 1970)
  puma (Duellman and Trueb 1966; Duellman 1970)
  sila (Duellman and Trueb 1966; Duellman 1970)
  sordida (Duellman and Trueb 1966; Duellman 1970)

Anurans at all life history stages are major contributors to maintaining ecosystem health. The larvae are important regulators of aquatic plant growth and a food source for other animals. The adults are prime predators on invertebrates, especially arthropods, and contribute significantly to the control of insect pests in cultivated areas. They are the primary food source for many invertebrates, birds, and other vertebrates. When humans remove too many anurans from the food web for food or through indiscriminate malicious killing, it can have serious consequences in collapsing a functional system. I hope this book will encourage readers to appreciate these sometimes maligned creatures both for their beauty and for their significant benefit to the environment.

Throughout the world and across the millennia humans have attributed great mystic power to frogs and toads, as they have to their archenemies the serpents. First there is their role as harbingers of spring in the temperate zone or of the rains in tropical environments. For months the earth has been frozen or seared by the tropical sun so as to be nearly dead; then frogs magically appear, calling to the heavens for the life-giving rain that will replenish the land. Although frogs are despised by many, this association marks the humble frog as a potent symbol of fertility, unrecognized but still appealing, as may be seen from the immense popularity of frog totems in the culture of our own prosaic time.

Further, the life cycle of the frog, where the fishlike pollywog miraculously metamorphoses into a froglet, has long evoked powerful feelings as symbolic of growth and transformation, the underlying theme of all great myths and their daughter religions. Finally, as a messenger of the self, the frog comes to call us—at least princesses and herpetologists—to the adventure of discovering both the world and ourselves.

Family Rhinophrynidae Günther "1858b," 1859

A single living species, moderate-sized and of toadlike habitus, is placed in this family. Two genera and a distinct species of *Rhinophrynus* are known as fossils from Paleocene to middle Eocene times of Wyoming and from the early Oligocene of Saskatchewan, Canada. *Rhinophrynus* is a specialized fossorial form that feeds on subterranean insects, especially ants and termites, and emerges on the surface only to reproduce for a brief period during the early rainy season. The genus shares several external similarities with other burrowing frogs from throughout the world that are placed in several families, including a stocky, rounded body; a small narrow head; small eyes; short arms and short, fat legs, enclosed for the most part by the skin of body; and an internal tympanum or none. Externally the species is distinct from all other American species in combining these features with the following: tip of snout covered with calluses; upper eyelids movable; two elongate spadelike inner metatarsal digging tubercles with a free margin and only four toes apparent, since the first toe is reduced to a rudiment that supports the distal metatarsal tubercle; prehallux supporting the proximal metatarsal tubercle. Other characteristics defining the family are eight ectochordal, modified opisthocoelous presacral vertebrae; no free ribs; sacrum with expanded diapophyses and bicondylar articulation with the coccyx, which lacks transverse processes; pectoral girdle arciferal, no omosternum or sternum; parahyoid bone present; tibiale and fibulare fused proximally and distally; no intercalary cartilages; sartorius muscle distinct from semitendinosus, tendon of latter penetrating gracilis muscle. Unlike that of all other anurans, the tongue is attached posteriorly but is free anteriorly and protrusible; there are no teeth on jaws or vomer; and the pupil is vertically elliptical. Amplexus is inguinal; larvae are type X, without beaks or denticles and with paired spiracles; the karyotype is 2N = 22, NF = 44.

The family is restricted to lowland regions from southern Texas to eastern Honduras and from Michoacán, Mexico, to Costa Rica.

Genus *Rhinophrynus* C. Duméril and Bibron, 1841

The unusual robust, globular body form, short, fat limbs that are enclosed in the body skin, small head, callused snout, small eyes with movable upper lids, and dorsally directed nostrils make the genus and its sole living species distinctive. The genus *Rhinophrynus*, which also includes one fossil species from the early Oligocene of Saskatchewan, Canada, differs from the other two recognized genera (fossil) placed in the family by having a reduced humeral crest

and the shaft of the femur strongly bowed and narrow directly proximal to the distal condyle, probably as a further specialization for burrowing. In all three genera the prehallux and distal phalanx of the first toe are modified as bony digging spades that in living *Rhinophrynus*, and presumably the fossil forms, support a pair of inner sharp-edged metatarsal tubercles. In association with this specialization the first toe is reduced so that the phalangeal formula of the foot is 1-2-3-4-3 (the formula in the fossils is unknown but probably is the same).

Rhinophrynus are unique among anurans in having the tongue attached posteriorly and free anteriorly and extensible. At rest the tongue is a flat triangular structure. During feeding the free portion is protruded as a rodlike tube stiffened by hydrostatic pressure. In this process it is projected forward along the vaulted buccal roof directly through the barely opened small mouth. Prey are then enfolded by the sticky, cup-shaped villous tongue tip and pulled back into the mouth.

Rhinophrynus dorsalis C. Duméril and Bibron, 1841

Cone-nosed Frog, Alma de Vaca

(plate 62; map 7.1)

DIAGNOSTICS: This moderate-sized to large frog (adults 50 to 89 mm in standard length) is so distinctive that there is little likelihood it will be confused with any other species. Superficially it resembles a rotund bag of jelly with five short protrusions (four limbs and the head). The presence of the callused snout, posteriorly placed and dorsally directed nostrils, and two sharp inner metatarsal digging tubercles (fig. 7.9) will immediately confirm identification.

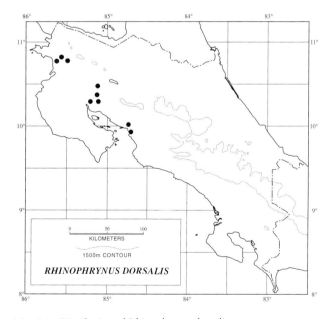

Map 7.1. Distribution of *Rhinophrynus dorsalis*.

DESCRIPTION: A moderately large species, adult females to 89 mm in standard length, adult males to 75 mm; body oval, limbs short and stocky, enclosed basally in body skin folds; skin relatively smooth, covered with widely separated pustules, loose when body not inflated; head small, broad, roughly trapezoidal in dorsal outline; snout truncate in dorsal outline; nostrils dorsal, nearer to eyes than snout tip; eyes small, both lids movable; pupil vertically elliptical; tympanum concealed; vocal slits present in males; vocal sacs fully distensible, paired, internal, lateral; arms short, fingers without subarticular tubercles or webs; hind limbs barely protruding from body skin; reduced first toe supporting distal of two digging metatarsal tubercles; other four toes connected by thick webbing; webbing formula: II 1½–2 III 2–3 IV 3½–2 V; no subarticular tubercles. Dorsum dark gray and maroon brown to dark brown, with pale yellow to orange spots or blotches and usually a vertebral stripe of the same color; ground color of head and limbs lighter than dorsum; venter uniform gray.

LARVAE (fig. 7.16): Large and superficially resembling a tiny catfish; distinctive in lacking beaks and denticles and having mouth bordered by eleven barbels; barbels begin to develop at a total length of 22 mm; large tadpoles equivalent to stage 40 are 45 mm in total length; two lateral spiracles and median vent tube; tail moderate; fins moderate; tail with thin terminal flagellum; blackish above, iridescent silvery gold below in life.

VOICE: The finely tuned call is an extremely loud, long-drawn-out whoop with a rising inflection at the end ("Wh-o-o-o-a"). Large choruses may be heard from a distance of 3 to 4 km, and up close the sound is deafening. The call lasts about 1.4 to 2 seconds, and the notes are repeated about 15 to 20 times a minute. There is no dominant frequency, but several major frequencies are 0.5 kHz, 1.3 kHz, and 1.8 kHz, and additional higher frequencies (to 2.5 kHz) are added toward the end of the call (Porter 1962; Fouquette 1969; Lee 1996).

SIMILAR SPECIES: The microhylid frogs *Gastrophryne pictiventris*, *Hypopachus variolosus*, and *Nelsonophryne aterrima* have five distinct toes and a smooth snout and do not have two enlarged and elevated inner metatarsal tubercles.

HABITAT: An inhabitant of the Lowland Dry Forest region of northwestern Costa Rica, where it is commonly found during the breeding season in roadside ditches, pastures, cultivated fields, and other open areas as well as in the forest.

BIOLOGY: Foster and McDiarmid (1983) have described the salient features of this species' biology in Costa Rica, and much of the following is based on that account. *Rhinophrynus dorsalis* is active nocturnally on the surface only during the reproductive period. It spends the rest of the year underground. Aboveground these frogs generally walk awkwardly, with the limbs splayed out at the sides. When disturbed they may hop 150 mm or more. *Rhinophrynus* burrow backward into the ground corkscrew fashion. Once they are below the surface, the sides of the burrow opening collapse to cover the entrance. Underground, this species makes a small, round chamber and inflates the body to wedge itself into the retreat. This behavior is apparently defensive, since to dislodge the frog one side of the chamber has to be completely removed. The chambers have been located at depths of 70 to 150 mm in the wet season, but doubtless the frogs dig deeper during the rest of the year. The chamber may be occupied for a single day during the rainy season as individual frogs return to the surface, apparently to feed, but others remain belowground as long as a month. In captivity (Fouquette and Rossman 1963) one individual stayed buried and inactive for two years without feeding. A complex of physiological factors, most importantly changes in renal function, use of fat stores, and metabolic reduction (Pinder, Storey, and Ultish 1992), are probable elements in such a long period of inactivity. Foster and McDiarmid (1983) noted that *R. dorsalis* does not secrete a protective cocoonlike covering while in the chamber as some other burrowers do (e.g., *Pternohyla fodiens* of Mexico), and individuals removed from underground are immediately active (not dormant as in estivating forms). This species eats primarily ants and termites. When underground, they are apparently capable of burrowing forward using their spatulate hands and forelimbs (Trueb and Gans 1983) to find ant or termite tunnels, although no one has observed this behavior. They probably penetrate the tough outer covering of the termite passageway (saliva-cemented soil) with their heavily callused snouts, and ant tunnels would be easily entered by the same means. The specialized tongue and

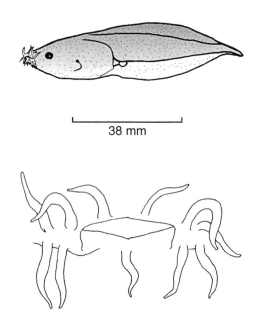

38 mm

Figure 7.16. Tadpole of *Rhinophrynus dorsalis*.

prey capture behavior of *R. dorsalis* have already been described in the generic account, but several other unique features of buccal morphology also seem to relate to food habits. These include the broad lips that effect a double closure along the wedge-shaped maxillary arch, a series of parallel grooves in the middorsal area of the roof of the mouth, and leaflike foldings of the esophageal lining. The first lets these frogs protrude the anterior portion of the tongue without a wide gape. The last two apparently expedite food processing and reduce the effects of the stings and bites of the principal prey (ants and termites). Bear in mind that this frog lacks teeth and swallows the small prey whole. Reproduction occurs during the first heavy rains in late May or early June. At the onset of the rainy season, some individuals move toward the breeding ponds or ditches, which are beginning to fill, then burrow into the ground there to await the reproductive period. During this interim, males may be heard calling from underground on afternoons with heavy rains. Breeding is explosive and usually occurs in a population once the ponds fill. In some years there are two breeding periods a few days to a few weeks apart during the heaviest rains. On a night when everything is right, the frogs move rapidly to the pond, with males often calling from the land en route. Early in the evening, males begin to congregate in the water, sitting partially submerged in the shallows and calling or inflating their bodies. Some call from deeper water while they float like balloons with limbs outstretched. As these males call the body rocks noticeably, and the head rises progressively out of the water until the call terminates. Later in the evening (more or less from 9:30 on) most males arriving at the breeding site are in pelvic amplexus with a female and carried on her back. During egg laying both frogs are submerged. Eggs are extruded one at a time, usually with six to twelve laid in a pulse. The single cream-colored eggs are demersal (sink to the bottom), but they become sticky and tend to adhere to one another in clumps. Clutch size ranges from 2,000 to 8,000. Each ovum is about 4.5 mm in diameter, and the egg plus the envelope is about 5 mm. The eggs hatch in a few days into filter-feeding larvae (ecomorph 4, table 7.2) that aggregate into schools ranging in diameter from about 100 mm to 1 m. Some schools are regularly arranged, but others form "boils" where individuals descend in a circular pattern in the center of the aggregation and ascend on its margins. Like most tadpoles, *R. dorsalis* may be cannibalistic at high densities. Metamorphosis is more or less simultaneous from any particular pond, and the froglets leave en masse. Metamorphs are mostly black with some small light spots. They may move by walking but more often jump (up to 100–200 mm). The tail is still present when they leave the water, but they are able to burrow by folding the tail over the upper body. The froglets seem to retain their social structure during this phase of the life history, maintaining aggregations for two weeks or more after transformation. Both the tadpoles and metamorphs are preyed on by a wide variety of vertebrates (fishes, turtles, birds, and aquatic invertebrates). The froglets are eaten especially by snakes that forage along the pond margins during the period following metamorphosis. A major defense is inflating the lungs and puffing up the body to stretch the loose skin, turning the seemingly pliable animal into a greatly enlarged and rather rigid object. When handled, the skin gives off large amounts of a sticky white, noxious secretion that may discourage predators. It seems more likely, however, that the secretion deters attacking soldier castes of termites and ants and protects the frog from bites and stings when penetrating the insects' passageways to obtain food. This frog is the legendary *uo* of Mayan mythology, whose calling causes Chac, the rain god, to empty his water-filled rain gourd onto the parched earth and restore its fecundity.

REMARKS: The karyotype is 2N = 22, NF = 44; all are macrochromosomes, but seven pairs are definitely larger than the four smaller pairs. Among the former, one is metacentric, three are submetacentric, and three are intermediate; among the smaller sets, three are submetacentric and one is nearly telocentric (Cole 1971; Bogart and Nelson 1976). Fouquette (1969) presented a summary of information on this species up to that time.

DISTRIBUTION: Lowlands from Michoacán, Mexico, to central Costa Rica, on the Pacific versant, and from southern Texas to western Honduras (with one record from northeastern Nicaragua) on the Atlantic slope; in Costa Rica only in the northwest, including the Nicoya Peninsula (0–300 m).

Family Bufonidae Gray, 1825

The toads, *Bufo* and related genera with a similar morphology, and a number of anurans having the frog habitus make up this large family (thirty-four genera, 410 species). Many of the species included have thick glandular skins, most often with pustular warts and well-developed concentrations of glandular tissue in the parotoid region to form discrete, small to very large glands. In others the skin is smooth, there are no wartlike pustules, and the parotoid glands are inconspicuous or absent. Many American and Asian species have well-developed cranial ridges on the skull that are evident externally as prominent crests covered by co-ossified integument.

Most bufonids are nocturnal, but some are diurnal or crepuscular. The majority of species are terrestrial or semifossorial, but some are fossorial and a few are scansorial or semiaquatic in habits. The following combination of characteristics technically defines the family: five to eight holochordal presacral vertebrae; no ribs; sacrum with dilated diapophyses and usually a bicondylar articulation with the coccyx; monocondylar or sacrum fused to vertebral column in taxa with reduced vertebral numbers; coccyx lacking transverse processes; pectoral girdle arciferal to firmisternal through fusion of epicoracoids; omosternum usually absent; bony sternum present; no parahyoid bone; tibiale and fibulare fused distally and proximally; no intercalary cartilages; unlike all other anurans, a rudimentary ovary (Bidder's organ) retained in adult males except in a few species; sartorius muscle distinct from semitendinosus, with tendon of latter penetrating gracilis muscle; tongue free posteriorly, highly protrusible; no teeth on jaws; pupil horizontally elliptical; amplexus usually axillary (rarely inguinal); larval type A with beaks and denticles and single sinistral spiracle; the karyotype is usually 2N = 22; 20 in one group of African *Bufo*.

The three Costa Rican genera of the family are unlikely to be confused with representatives of other anuran families. Costa Rican *Bufo* are quintessential toads, and both they and the allied *Crepidophryne* have parotoid glands and usually cranial crests. The species of *Atelopus*, which are more froglike in appearance than the other two genera and are brightly colored, differ from any other frogs that might be mistaken for them in lacking a tympanum and scutelike flaps on the tips of the digits.

The family has an extensive temperate, subtropical, and tropical distribution, being found on every continent and major island where anurans occur except Madagascar, New Guinea, Australia, New Zealand, and the islands in the Pacific oceanic region. One species *(Bufo marinus)* has been widely introduced by humans to some of the areas where the family does not occur naturally (New Guinea, Australia, Fiji, Hawaii, and other Pacific islands).

Most bufonids are oviparous, but they have a variety of reproductive patterns (1–2, 3–4, 17–19, 21, 34–35; see table 7.1) that occur in scattered genera (Duellman and Trueb 1986, 1994), including putative direct development (several South American genera), ovoviviparity (two species of African *Nectophrynoides*), and viviparity (two species of *Nimbaphrynoides*). So far as is known all Costa Rican species are oviparous, but *Crepidophryne epiotica* may undergo direct development.

Genus *Atelopus* C. Duméril and Bibron, 1841

Harlequin Frogs

This is a large genus (sixty-six species) of small to moderate-sized frogs, usually marked with bright and contrasting colors on the dorsum and often on the venter, or with the venter a uniform bright color. Many *Atelopus* are smooth-skinned, slim-bodied, long-limbed forms in which the arms are very slender. Others are more toadlike in appearance, with dorsal integument that is wrinkled, smooth but warty or pustulate, or else spinulate and warty, an ovoid body, and short limbs. The snout tends to be pointed and protuberant in all species, there is no tympanum, and the columella is usually lacking. The first digit of the hands and feet is very short, there is finger webbing at least between digits I and II; the toes are webbed and the palmar and plantar surfaces smooth. Some forms have discrete integumentary glands in the paro-

toid region and on the back and upper limb surfaces. All *Atelopus* tadpoles, so far as is known, have a distinctive suctorial disk posterior to the oral disk (ecomorph 12, table 7.2).

Atelopus is distinguished from other bufonid genera by the following combination of mostly internal features: seven presacral vertebrae (atlas fused to first trunk vertebra); pectoral girdle firmisternal (epicoracoid cartilages completely fused); phalangeal formula of hand usually 2-2-3-3, but thumb shortened or a phalanx lost (1-2-3-3); distal phalanges slightly knobbed; coccyx not expanded; anterior process of ceratohyal present; tensor fasciae latae muscle elongate; larvae with a large ventral sucking disk; eggs white.

The presence of toe webs, absence of a tympanum, teeth, and tarsal folds, in combination with the lateral nares and horizontal elliptical pupil, separate Costa Rican representatives of the genus from all other anurans from the republic.

Members of this genus are diurnal and prey on soft-bodied arthropods, primarily ants. In Costa Rica they are usually found near watercourses and tend to concentrate along stream banks and on rocks in the stream during the dry season (or at least they used to). These frogs are walker-hoppers that walk about the environment when foraging, seeking mates, or in amplexus. They hop mainly to escape potential predators.

Because of the conspicuous coloration of many forms, their diurnal habits, and their rather lackadaisical response to intruders, it is not surprising that studies (Kim et al. 1975; Daly and Spande 1986) have demonstrated high concentrations of the highly toxic alkaloid tetrodotoxin in the skin of several species, including *Atelopus chiriquiensis* and *Atelopus varius*. In addition their skins contain high concentrations of bufodienolides, a group of toxic steroid compounds, and *A. chiriquiensis* secretes large quantities of another toxic tetrodotoxin-like substance (chiriquitoxin). The alkaloids in the skin secretions are primarily neurotoxic, while the steroids are principally cardiotoxic. These compounds clearly serve an antipredator function, and the bright coloration of many *Atelopus* is putatively aposematic.

Lötters (1996) has recently published a compilation of data, primarily from the literature, on the taxonomy, biology, and distribution of members of this genus.

Atelopus ranges in the uplands and lowlands from Costa Rica to Pacific slope Ecuador throughout the northern Andes and disjunctly in northeastern Venezuela and the Guianas; it is also present in the Amazon basin and the Andes of Peru and Bolivia.

KEY TO THE COSTA RICAN HARLEQUIN FROGS OF THE GENUS *ATELOPUS*

1a. Distinct rostral-canthal, parotoid, dorsal, and limb glands (fig. 7.17b,c) 2

1b. No differentiated glandular areas (fig. 7.17a); dorsal

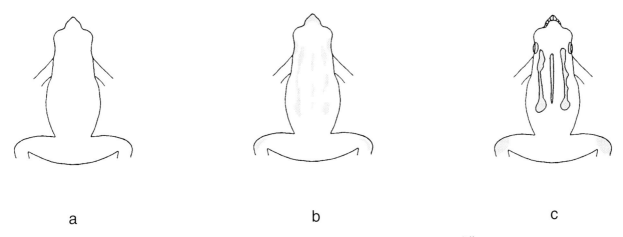

a　　　　　　　　　　b　　　　　　　　　　c

Figure 7.17. Glandular development in Costa Rican species of the genus *Atelopus*:. (a) *A. varius*, no differentiated glandular areas; (b) *A. chiriquiensis*, glandular areas weakly developed; (c) *A. senex*, glandular areas well developed.

ground color in life lime green, chartreuse, yellow, yellow orange, or red; slightly to heavily mottled with black or black blotches, bars, or spots (fig. 7.19). . . . .
. . . . *Atelopus varius* (p. 189)

2a. Glands well developed (fig. 7.17c); uniform black, blue gray, gray green, to black with green to pale orange spots and blotches or an orange suffusion in life . . . .
. . . .*Atelopus senex* (p. 188)

2b. Glands weakly developed (fig. 7.17b); uniform green, yellow, or rust to black with yellow or red lines, spots, or blotches in life. . . . . *Atelopus chiriquiensis* (p. 187)

Atelopus chiriquiensis Shreve, 1936

(plates 63–64; map 7.2)

DIAGNOSTICS: A moderate-sized frog, highly variable in coloration, with moderately developed parotoid and dorsal glandular concentrations and a well-developed glandular protuberance on the tip of the snout (fig. 7.17b).

DESCRIPTION: Adult males 28 to 34 mm in standard length, adult females 36 to 49 mm; habitus froglike, with long arms and legs; skin essentially smooth, but often with minute pustules dorsally; some large adults with well-developed small spicules along anterior flank; head angular, longer than broad; canthus rostralis well defined; weakly developed parotoid, dorsal, elbow, knee, and heel glandular concentrations; eyes moderate; no tympanum; ostia pharyngea small, slitlike, or absent; paired elongate vocal slits and a single internal subgular vocal sac in adult males; undersides of hands and feet wrinkled and fleshy; thenar, paired metatarsal and subarticular tubercles present; other tubercles on palms and soles otherwise obsolescent; thumb webbed basally, with brown nuptial pad in adult males; toes connected by extensive fleshy webbing; webbing formulas: males I 0–½ II ½–1 III 1–3 IV 3–½ V; females, I 0–¼ II ¼–½ III ½–2¼ IV 2¾–¼ V; no tarsal fold. Males usually uniform lime

green, chartreuse, yellow, or rust brown dorsally, occasionally with thin black reticulate pattern or underlying pattern lavender with fine dark vermiculations on a yellow ground color; females rarely similar to males in coloration, usually with black-outlined orange red dorsolateral stripe from snout to groin, but frequently broken into elongate red blotches or spots; in some examples considerable red spotting on dorsum; upper limb surfaces uniform; red mottling also often present on limb surfaces and posterior thigh surface; venter in both sexes usually uniform yellow, but sometimes completely covered by broad black-and-red reticulum in females; iris gold.

LARVAE: The tadpole has not been formally described, but Lötters (1996) included a photograph of the underside

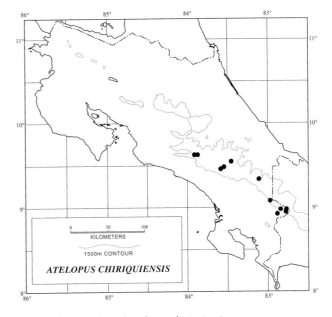

Map 7.2. Distribution of *Atelopus chiriquiensis*.

of a larva of this species. It is gastromyzophorous and has 2/3 denticle rows. The photograph does not allow the tadpole to be distinguished from that of *Atelopus varius*.

VOICE: The advertisement call was described (Jaslow 1979) as a buzz and as the pulsed call (Cocroft et al. 1990) and is repeated about 10 to 12 times a minute, lasting 315 to 478 milliseconds. The dominant frequency is 2.0 to 2.7 kHz and the pulse rate about 60 to 82 per second. Males of this species also give a short "chirp" or "twitter" when engaged in conspecific wrestling bouts, captured by a predator, or crowded in captivity, as well as a pure tone "whistle." Lindquist (1995) suggests that the pure tone call indicates a preconfrontational surrender to a dominant male.

SIMILAR SPECIES: (1) *Atelopus senex* has extremely developed glandular concentrations with well-developed parotoid glands (fig. 7.17c). (2) *Atelopus varius* lacks head, dorsal, and limb glands (fig. 7.17a).

HABITAT: At one time very common along stream margins in the upper portions of Lower Montane Wet Forest and Rainforest of the Cordillera de Talamanca.

BIOLOGY: *Atelopus chiriquiensis*, like other members of the genus, is diurnal and is generally found along watercourses. Males are territorial and use the advertisement call to maintain the calling site. Physical combat results when another male attempts to enter the resident male's territory and is not deterred by the call. A dominant male usually climbs or jumps on the intruder's back, and the lower male may become motionless until released and then retreat. Wrestling bouts also occur in which the males attempt to flip one another to the ground until one adopts the motionless, submissive position. The winner in these bouts usually gives a whistle or buzz call immediately after releasing the loser (Jaslow 1979). This same author noted a chirping call by males when crowded in captivity, probably corresponding to the release call of most anurans. Breeding apparently takes place in the early wet season (May to July). Females probably locate males by their advertisement calls. Once amplexus begins it may be days to months before egg laying occurs, and females are frequently seen carrying the grasping males on their backs. Lindquist and Swihart (1997) reported finding amplecting pairs along a stream course in February (dry season). In this instance solitary males attempted, but failed, to dislodge amplexed males. Mated pairs were observed to submerge on the stream bottom for fifteen to thirty minutes, but no eggs were laid. Two days later a female of a captive pair laid a complete clutch of 364 unpigmented eggs in two strings over a period of eight to ten hours during the night. The eggs formed a single massive clutch. Eggs averaging 2.05 mm in diameter were enclosed in an envelope averaging 3.63 mm in diameter. The skin of these brightly colored frogs contains a number of alkaloids, including several steroid compounds of the bufogenin series and two related highly toxic guanidines, tetrodotoxin and chiriquitoxin, in a ratio of about 30:70 (Kim et al. 1975). The amount of

these last two compounds in the skin of one frog is enough to kill 350 20 g mice and serves as a robust antipredator defense since, like some other amphibian toxins, they are not destroyed if ingested. It seems logical to conclude that the coloration serves a warning (aposematic) function. Pavelka, Kim, and Mosher (1977) reported that these toxins are also found in the eggs of *A. chiriquiensis*.

REMARKS: *Atelopus chiriquiensis* was not recognized until Shreve described it in 1926 from Panama. No Costa Rican localities for the species had been reported until Savage (1972a) pointed out that W. M. Gabb collected this frog from Cerro Utyum, Limón Province, in 1873 and Cope (1875) reported it under *A. varius*. Since 1972 the species has been taken at a number of other localities throughout the Cordillera de Talamanca. This form continued to be present in large numbers in southern Costa Rica when populations of *A. senex* and *A. varius* declined elsewhere between the reproductive seasons of 1987 and 1988. However, it suffered serious declines between 1992 and 1993 that have continued to the present. This is another species that showed spectacular declines in the 1990s (Lips 1998).

DISTRIBUTION: The lower montane zone of the Cordillera de Talamanca–Chiriqui axis of Costa Rica (1,800–2,500 m) and western Panama (1,400–2,100 m).

Atelopus senex Taylor, 1952a

(plates 65–66; map 7.3)

DIAGNOSTICS: The extreme development of glands on the head, dorsum, and limbs (fig. 7.17c) immediately distinguishes this moderate-sized frog from any other species lacking a tympanum, cranial crests, and tarsal folds.

DESCRIPTION: Adult males 28 to 32 mm in standard length, adult females 30 to 43 mm; habitus froglike with

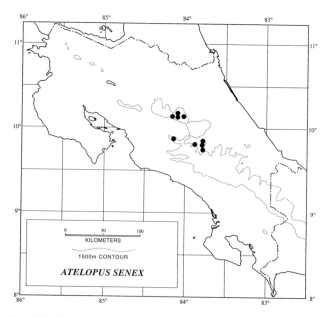

Map 7.3. Distribution of *Atelopus senex*.

long arms and legs; skin basically smooth with minute pustules dorsally in many examples and some well-developed spicules on anterior flanks in some large adults; head angular, longer than broad; canthus rostralis well defined; strongly developed glandular canthal, parotoid, dorsal, elbow, knee, and heel areas; eyes moderate; no tympanum; ostia pharyngea small, slitlike, or absent; paired elongate vocal slits and single internal subgular vocal sac in adult males; undersides of hands and feet wrinkled and fleshy; thenar, paired metatarsal and subarticular tubercles present; other tubercles on palms and soles obsolescent; thumb webbed basally, brown nuptial pad in adult males; toes connected by fleshy webbing; webbing formulas: males, I 0–½ II ½–1 III 1–3 IV 3–½ V; females, I 0–¼ II ¼–½ III ¾–2¾ IV 2¾–¼ V; no tarsal fold. Males tend to be uniform (bluish gray, blue green, to black) dorsally with parotoid and limb glands pink to cream; females may be similar to males but usually are patterned with contrasting dark and light (cream, lemon, or lime); some males with dorsal greenish suffusion, some gray females with dorsal rusty orange suffusion, often over most of upper surfaces; upper limb surfaces without transverse dark bars; venter uniform gray with or without black markings and sometimes suffused with yellow to orange.

LARVAE: Unknown.

VOICE: Not known.

SIMILAR SPECIES: (1) *Atelopus chiriquiensis* has weakly developed glandular concentrations (fig. 7.17b). (2) *Atelopus varius* lacks snout, parotoid, dorsal, and limb glandular areas (fig. 7.17a).

HABITAT: Stream margins in the upper portion of Premontane Rainforest and in Lower Montane Wet Forest and Rainforest.

BIOLOGY: These diurnal stream-breeding frogs formerly were found in great concentrations during the reproductive period (July to August). The extreme development of glandular areas suggests that they may be even more toxic than *A. chiriquiensis*. It is expected that they will resemble other Costa Rican *Atelopus* in most features of their natural history. This is one of the species that underwent serious declines between 1987 and 1988 and has never recovered. This toad was once abundant on the slopes of Volcán Barva within the Braulio Carrillo National Park but is now believed to be extinct in this region.

REMARKS: *Atelopus senex* is known from three geographic areas: the slopes of Volcán Barva (1,960–2,040 m); the Macizo de Cedral, south of the city of San José (2,150 m); and the upper Río Reventazón basin, south of the city of Cartago (1,280 to 1,320 m). The populations differ only in coloration (Savage 1972a), with the Barva examples tending to be uniform, the Cedral frogs markedly dimorphic (males uniform, females patterned), and those from the Río Reventazón drainage variable.

DISTRIBUTION: Slopes of Volcán Barva and the extreme

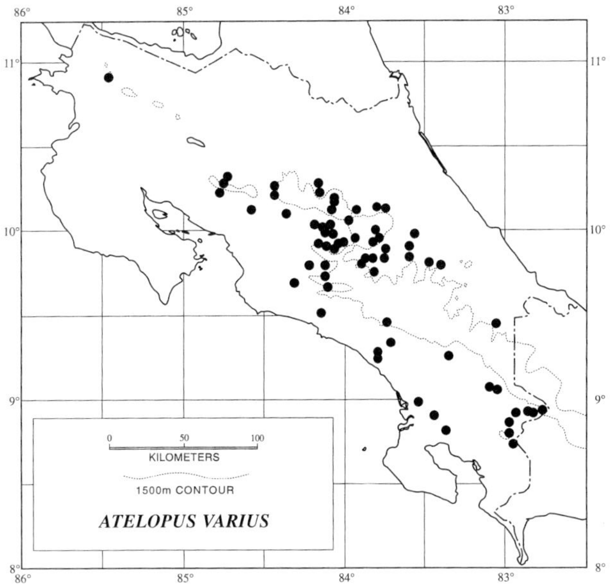

Map 7.4. Distribution of *Atelopus varius*.

northern slopes of the Cordillera de Talamanca (1,280–2,040 m).

Atelopus varius (Lichtenstein and von Martens, 1856)

(plates 67–69; map 7.4)

DIAGNOSTICS: A moderate-sized frog variously marked with a pattern of black and light (lime green, chartreuse, yellow, yellow orange, or red) and lacking glandular concentrations on the head, body, and limbs (fig. 7.17a), as well as lacking a tympanum, cranial crests, and tarsal folds.

DESCRIPTION: Adult males 27 to 39 mm in standard length, adult females 33 to 48 mm; habitus froglike with very long arms and legs; skin generally smooth but with minute pustules dorsally and anterior flank spicules in some examples; head longer than broad; canthus rostralis rounded; eyes moderate; no tympanum; ostia pharyngea slitlike or absent; paired elongate vocal slits and single internal subgular vocal sac in adult males; undersides of hands and feet wrinkled and fleshy; thenar, paired metatarsal and subarticular tubercles present; other tubercles on palms and soles obsolescent; thumb webbed basally, with brown nuptial pad in adult males; toes connected by moderately extensive fleshy webs; webbing formulas: males, I ¼–1 II ½–2 III 1¼–3¼ IV 3¼–1 V; females, I 0–¾ II ¼–1¾ III 1–3 IV 3–¼ V; no tarsal fold. Coloration extremely variable, ground color lime green, chartreuse, yellow, yellow orange, or red, marked with black blotches, bars, spots, or a reticulum or nearly uniform black; upper limb surfaces with dark transverse bars; ventral ground color light (usually yellow), often suffused with greenish or bright scarlet; males and females similarly colored to markedly sexually dichromatic; iris green.

LARVAE (fig. 7.18): Small, total length 12 mm at stage 36; body depressed with large ventral sucking disk; mouth ven-

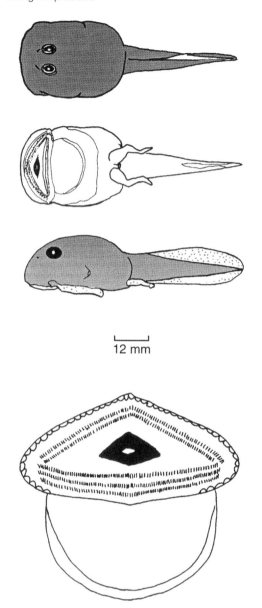

Figure 7.18. Tadpole of *Atelopus varius*.

tral; nares and eyes dorsal; spiracle lateral and sinistral; vent tube medial; tail short, tip rounded; oral disk very large, entire; beaks and 2/3 rows of denticles present; large papillae border oral disk except posteriorly; body uniform dark brown, except light ventral disk; tail musculature brown; tail fins covered by flecks, lines, or blotches of dark pigment.

VOICE: The advertisement call is a pulsed buzz lasting 370 to 460 milliseconds repeated about 12 times a minute. The dominant frequency is 2.2 to 2.5 kHz, frequency modulated and rising to 2.4 to 2.5 kHz at the end, and the pulse rate is 43 to 56 per second (Cocroft et al. 1990). This call apparently serves a territorial function. The "chirps" given by both males and females may serve as release calls or in other social interactions (see biology below).

SIMILAR SPECIES: (1) *Atelopus chiriquiensis* has a protuberant and glandular snout tip and weakly developed glandular concentrations on the head, body, and limbs (fig. 7.17b). (2) *Atelopus senex* has well-developed glands on the snout, parotoid, dorsal, and limb regions (fig. 7.17c).

HABITAT: Margins of high-gradient, rocky streams in Lowland Wet Forest and Premontane Moist and Wet Forests and Rainforest and the lower margin of Lower Montane Wet Forest and Rainforest.

BIOLOGY: The following account draws heavily on the papers by Crump (1986a, 1988), Crump and Pounds (1989), and Pounds and Crump (1987) on this species at Monteverde, Puntarenas Province, and my own field observations. The frogs are associated with small, fast-moving streams and are found along the banks and sitting out on rocks in the stream, although they rarely enter the water. At night they sleep in crevices or on low vegetation. They formerly occurred in great concentrations in the dry season (December to May). During the wet season (June to November), they were abundant but more widely dispersed into the adjacent forest. During the dry season, *A. varius* may average one per 1.8 m of stream, and 751 individuals were marked in one year along 200 m of stream. They feed on a wide variety of arthropods (small flies, wasps, ants, beetles, springtails, silverfish, bugs, caterpillars, and spiders). Males exhibit strong territoriality during the wet season; they call from preferred sites and may engage in calling duels with intruders. If the intruder continues to infringe on the resident male's territory and approaches within 150 mm, chasing, pouncing, and wrestling ensue until one male is pinned to the ground, becomes passive, and (after release) retreats. Males also have a "chirp" call that is given by one or the other after pinning or when a male first pounces on a female for amplexus. During the dry season, males show little aggressive behavior. Females also exhibit territorial behavior, primarily during the nonbreeding season (January to August), and chase other females from their preferred sites. They also chase males away immediately following the breeding season and during the several months when males attempt early amplexus. There appears to be no courtship in this species, since amplexus occurs without preliminaries, about two and a half months into the rainy season. As is apparently the case in all *Atelopus*, amplexus is prolonged, and females carry males about on their backs for many days to over a month. Sometimes a second male may join the amplectant pair; after an extended struggle, he is usually dislodged. Oviposition appears to take place from October to early December. The unpigmented eggs are laid in paired strings in the laboratory, each containing twenty or so eggs 1.6 mm in diameter, separated by partitions. The aquarium literature, however, reports a clutch of 950 eggs (Heselhaus and Schmidt 1994). The gelatinous envelope is scalloped and about 2.4 mm in diameter. The eggs are laid in water and probably attached to rocks (Starrett 1967). Hatching occurs in six days. Tadpoles of related species (Duellman and Lynch 1969) have been found adhering by their ventral suckers to rocks or pebbles in streams (ecomorph 12, table 7.2). The larvae probably feed on algae. Metamorphs are about 7 mm in standard length. The bright coloration of adults of this species appears to be aposematic, since the skin of one individual contains a substantial amount of tetrodotoxin (enough to kill 100 20 g mice; Kim et al. 1975). While apparently safe from many predators, they may be infested by the larvae of a parasitoid fly *(Notochaeta bufonivora)* that kills the frog within four days of infection (Crump and Pounds 1985). Unfortunately this species is no longer as abundant at any site as it was before 1988, and it has virtually disappeared at many localities, including Monteverde (Pounds and Crump 1994). The cause of these declines remains uncertain, as I discussed in the introduction to this chapter. It is important that populations near the Panama border in southwestern Costa Rica were seemingly unaffected in the 1980s but crashed between 1992 and 1993.

REMARKS: Taylor (1952a, 1955) reported three taxa, *Atelopus varius varius, A. varius ambulatorius,* and *A. loomisi* from Costa Rica that Savage (1972a) demonstrated were conspecific. The supposed differences among these forms are based on dorsal coloration and sexual dichromatism. Savage described coloration for fifteen population samples from Costa Rica and western Panama. Two others (his 8 and 9) are now regarded as representing a distinct species, *A. zeteki* (see below). In many populations males and females are similar in coloration, with a ground color that is either greenish, yellow, orange red, or red (plates 67–68). In others males have a greenish ground color (plate 69) while the females' light areas are yellow, yellow orange, or red. In some samples, males and some females have a greenish ground color and female coloration is variably yellow or orange red or, as in the population from the southwestern slopes of the Cordillera de Talamanca and adjacent areas, bright red. In preservative the bright colors fade over time— pale green and yellow fade to white, and orange red and red to yellow or white. The darker greens (many males) turn grayish to brown in preservative. The degree to which the dorsum is marked with black is also variable (fig. 7.19). It ranges from a broken black reticulum on a greenish background (a–c) to a black dorsum with a few light spots (k). Taylor's *A. v. ambulatorius* and *A. loomisi* are based on a male (FMNH 178270) and a female (KU 24746), respectively, from the population found on the north slope of Volcán Poás. The male has pattern b and the female has pattern e, as do other examples of the two sexes from that area. All *A. varius* from this sample have a green ground color. Ontogenetic change also occurs in some populations. In these frogs the juveniles are black and yellow above, and the yellow becomes suffused with bright red pigment during ontog-

eny to produce adults that are black, yellow, and red (typical of the Meseta Central Occidental samples) or black and red (southwestern Costa Rica specimens). The karyotype of *A. varius* is 2N = 22, with eleven pairs of metacentric chromosomes and the first six larger than the last five; NF = 44 (Duellman 1967b; Schmid 1980). Savage (1972a) regarded *Atelopus* from Coclé Province, Panama, as *A. varius,* but the presence of the unique zetekitoxin along with tetrodotoxin (in a ratio of about 96:4) seems to establish these frogs as belonging to a distinct species. So far as is known (Kim et al. 1975; Pavelka, Kim, and Mosher 1977), this is the most toxic species of *Atelopus;* the skin of one individual contains enough toxins to kill 1,200 20 g mice. In addition, frogs in this population are larger (adult males 35–41 mm in standard length, adult females 46–60 mm) than *A. varius* and have a distinctive coloration. Both males and females are similarly colored: golden yellow above, with three transverse black bands (juveniles and some adults), to uniform golden yellow above. The undersurface is bright yellow in all examples. It seems best to recognize this form as a distinct species, *Atelopus zeteki* Dunn, 1933b, which includes the famous golden frogs of El Valle de Antón, Panama.

DISTRIBUTION: Premontane and lower montane zones on both Atlantic and Pacific versants of the cordilleras of Costa Rica and western Panama; also on outlying ridges and hills down to 16 m at a few lowland sites; the lowland rainforest localities are all along high-gradient, rocky streams, in hilly areas; absent from the lowlands of the Pacific northwest (16–2,000 m).

Genus *Bufo* Laurenti, 1768

Toads, Sapos

This very large genus (229 species) shows considerable variation in coloration and details of morphology, but all toads are relatively compact in body form, short-limbed creatures having distinct parotoid glands and the upper body usually covered by warts and/or glands or spines. The nares are lateral, the pupil of the eye is horizontally elliptical, and an external tympanum is usually present; the fingers usually lack webs, and the toes are webbed at least basally.

In addition, the genus *Bufo* differs from other genera in the family, including several sharing the toad Gestalt, by the following combination of features: eight presacral vertebrae, atlas not fused to first trunk vertebra; pectoral girdle arciferal (epicoracoid cartilages not fused); phalangeal formula of hand 2-2-3-3, thumb not shortened; distal phalanges not expanded; columella usually present, sometimes absent; coccyx not expanded; no anterior process on ceratohyal; tensor fasciae latae short; eggs usually pigmented; larvae without sucking disk on abdomen. All known *Bufo* have a karyotype of 2N = 22 except for one African clade that has 20.

All Costa Rican *Bufo* are moderate- to giant-sized toads

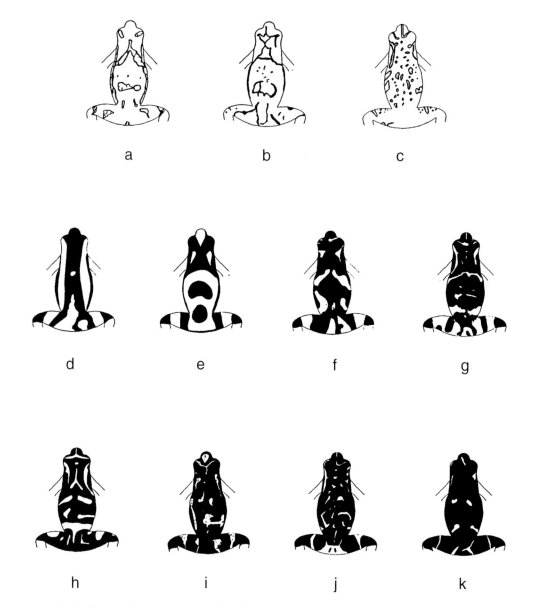

Figure 7.19. Variation in color pattern in Costa Rican *Atelopus varius:* (a–c) green ground color, typical of males on Atlantic slopes of Volcanes Poás and Barba; (d–k) ground color may be greenish, yellow, orange red, or red, typical of females in most dimorphic populations and males in nondimorphic populations; g, h, and j are the most common patterns.

and have very well defined parotoid glands and definite cranial crests (except in *Bufo haematiticus* and all very young toadlets) (fig. 7.20) that will consistently separate them from other sympatric anurans. They further differ from the allied *Atelopus* and *Crepidophryne* in having a well-developed thumb and usually lacking finger webbing. Toads are hopper-walkers that favor the hopping mode during most activities.

The skins of all *Bufo* contain numerous bufodienolides, also called bufogenins, which are steroid compounds having cardiotoxic effects and sometimes other effects as well. The most important of these are several bufotoxins and bufogenins that may be secreted in large quantities from the

very large skin glands, especially the parotoids. These compounds serve an antipredator function, since ingesting them may cause death. *Bufo* eggs and their jelly envelopes also are toxic to some predators (Licht 1968).

Bufo are found on every continent and continental island where Anura are known to occur except for Madagascar, the Seychelles, the Philippines, New Guinea, Australia, and New Zealand. *Bufo marinus* has been widely introduced to many areas, including those where *Bufo* do not occur naturally (Australia, New Guinea, New Zealand, and many Pacific islands, including Hawaii).

At least thirty-three mostly phenetically based species

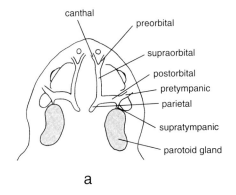

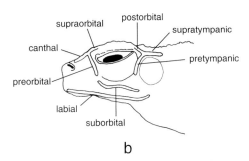

Figure 7.20. Parotoid glands and cranial crests in the Costa Rican toad genera *Bufo* and *Crepidophryne*: (a) dorsal view of head; (b) lateral view of head.

groups, twenty of them monotypic, have been proposed to accommodate the multitude of toads (Blair 1972b). The following putative groups may be recognized in Costa Rica (number of Costa Rican representatives in parentheses): *Bufo fastidiosus* group (two species), *B. guttatus* group (one species), *B. marinus* group (one species), *B. periglenes* group (one species), and *B. valliceps* group (five species).

Members of this genus generally are easy to hybridize in vivo in the laboratory, unlike some other anurans (e.g., hylid treefrogs). Artificial experimental hybridizations have been made between many species (Blair 1972a), and genetic relationships have been estimated from the degree of fertility, viability, and stages of development the offspring attain. In a number of cases hybrids were raised to sexual maturity and backcrossed to representatives of the parental species to further assess the compatibility of the genotypes. All Costa Rican species of *Bufo* except *B. fastidiosus* and *B. melanochlorus* have been included in such studies.

Artificial hybridization circumvents any premating isolating mechanisms, which must be very effective since I know of no specimens of Costa Rican toads that are putative interspecies hybrids. Two species, *Bufo haematiticus* and *B. holdridgei*, with hypertrophied testes, are unusual in that all interspecies crosses to them fail from lack of fertilization or normal early development. Blair (1972a) attributes this

situation to a yet unidentified compound produced by the testes.

Tadpole aggregations or schools of hundreds to thousands of individuals are typical of many *Bufo* species that breed in lentic waters. The schools form dense black mats in shallow water or on the bottom, with individuals usually touching one another. Movement of the schools has been described as "amoeboid," with several projections advancing in a particular direction. The aggregations are polarized; that is, individuals may shift their positions in the school, but its total orientation is not affected. When disturbed, the tadpoles scatter but regroup within a short time. The school structure and behavior described above are characteristic of *Bufo marinus* in particular but are also found in other Costa Rican *Bufo* except *B. coccifer*, *B. fastidiosus*, *B. holdridgei*, and *B. periglenes*. These last three forms lay their eggs in small puddles or depressions in the forest, so larval concentrations are small after hatching.

Recently metamorphosed and small juvenile toads under a year old are often encountered in large numbers in the field. These toadlets differ in appearance from older examples of the same species, especially in lacking cranial crests or having them poorly to partially developed. Metamorphs lack any suggestion of the crests and a tympanum, but they are partially developed in most species two to three months after transformation and approach the adult condition at four to six months of age. If these ontogenetic changes are kept in mind, most toadlets may be identified to species by using the key below but emphasizing the presence or absence of the tarsal fold, size and shape of the parotoid gland, and the characteristics (or absence) of digital, palmar, and plantar tubercles.

KEY TO THE COSTA RICAN TOADS OF THE GENERA *BUFO* AND *CREPIDOPHRYNE*

1a. A well-developed inner tarsal fold; parotoid gland very large, several times the area of upper eyelid. 2
1b. No tarsal fold; parotoid gland moderate to small, less than twice the area of upper eyelid 3
2a. Numerous pointed supernumerary tubercles under fingers and toes and similar accessory palmar and plantar tubercles; no black stripe from snout backward on side of head and body; prominent cranial crests (fig. 7.30d); adult males 85–145 mm in standard length, adult females 90–175 mm . . . . *Bufo marinus* (p. 199)
2b. No supernumerary tubercles on undersides of fingers and toes, no accessory palmar or plantar tubercles; a broad black stripe covers entire side of head and extends at least to posterior margin of parotoid gland; no cranial crests (fig. 7.30c); adult males 42–62 mm in standard length, adult females 50–80 mm . . . .
. . . . *Bufo haematiticus* (p. 198)

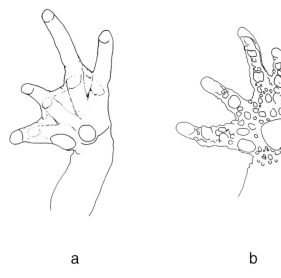

a   b

Figure 7.21. Palmar view of hands in toads of the genus *Bufo:*
(a) *B. fastidiosus,* showing webbed fingers and weakly developed subarticular tubercles; (b) *B. melanochlorus,* which lacks webs but has
definite subarticular tubercles.

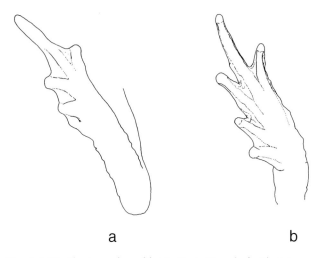

a   b

Figure 7.22. Plantar surface of feet in Costa Rican bufonids: (a) completely webbed foot of *Crepidophryne epiotica;* (b) partially webbed
foot of *Bufo periglenes.*

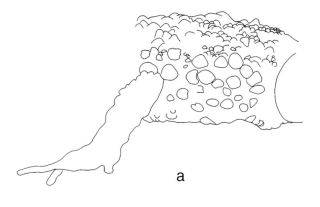

a

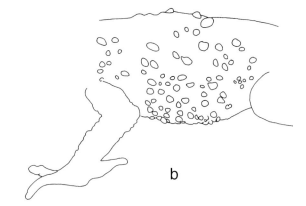

b

Figure 7.23. Lateral views of body in Costa Rican toads of the *Bufo
fastidiosus* group: (a) *B. fastidiosus;* (b) *B. holdridgei.*

3a. No tympanum; fingers and toes without well-developed
    subarticular tubercles (fig. 7.21a); hands and feet with
    fleshy pads; no parietal crests 4
3b. Tympanum present; definite subarticular tubercles pres-
    ent (fig. 7.21b); parietal crests present. 7
4a. Fingers not completely webbed, toes not completely
    enclosed by webs (fig. 7.22b); first toe well developed
    5
4b. Fingers (I–II–IV) and toes (I–II–III–V) nearly com-
    pletely enclosed by webbing; first toe greatly reduced
    and not distinguishable externally (fig. 7.22a); adult
    males 20–23 mm in standard length, adult females 26–
    29 mm. . . . . *Crepidophryne epiotica* (p. 211)

5a. No Eustachian tubes or ostia pharyngea; ground color
    black with parotoid glands and some warts reddish
    brown 6
5b. Eustachian tubes and ostia pharyngea present; males
    bright uniform orange (yellow in preservative), females
    black to olive green with large bright red spots; adult
    males 39–52 mm in standard length, adult females 42–
    56 mm . . . . *Bufo periglenes* (p. 202)
6a. Dorsum with several discrete, regular, longitudinal se-
    ries of enlarged warts (fig. 7.23a); cranial crests massive,
    elevated (fig. 7.30a); venter light, heavily mottled with
    black; adult males 43–52 mm in standard length, adult
    females 40–62 mm. . . . . *Bufo fastidiosus* (p. 195)
6b. Dorsum with scattered moderate-sized, beadlike warts
    (fig. 7.23b); cranial crests low, obsolescent (fig. 7.30b);
    venter light with some dark mottling on chest; adult
    males 32–52 mm in standard length, adult females 38–
    53 mm. . . . . *Bufo holdridgei* (p. 196)
7a. First finger equal to or longer than second; toes webbed
    only at base; supernumerary tubercles under fingers
    and toes raised, globular to pointed; enlarged series of
    lateral warts, if present, not tipped by elongate spines;
    dorsal ground color yellow to dark brown 8

7b. Second finger longer than first; toes with considerable webbing between toes II–III–IV; supernumerary tubercles under fingers and toes low, rounded; enlarged series of lateral warts tipped with elongate spines; dorsum usually yellow green to olive green; adult males 53–72 mm in standard length, adult females 76–94 mm. . . . . *Bufo coniferus* (p. 207)

8a. Parotoid gland small, area definitely less than area of upper eyelid 9

8b. Parotoid gland moderate, area equal to or exceeding area of upper eyelid 10

9a. A definite broad dorsolateral dark stripe, demarcated by a series of white warts above, along each side; parotoid gland triangular, very narrow and elongate, area about three-fourths area of upper eyelid; pretympanic, suborbital, and labial crests weak or absent; all crests usually brown (fig. 7.32c); adult males 43–65 mm in standard length, adult females 65–103 mm . . . . *Bufo melanochlorus* (p. 209)

9b. Body uniform, blotched, or striped; if dorsolateral dark stripe is indicated, a broad dorsal light stripe present along each side; parotoid gland round to oval, area one-half or less area of upper eyelid; all crests well developed, covered with black keratin (fig. 7.32b); adult males 75–96 mm in standard length, adult females 73–107 mm . . . . *Bufo luetkenii* (p. 208)

10a. A distinct light interorbital bar; all cranial crests well developed (fig. 7.30f); subarticular and supernumerary tubercles under fingers III and IV and toes III–IV–V bifid or double; adult males 45–65 mm in standard length, adult females 53–83 mm . . . . *Bufo coccifer* (p. 203)

10b. No light interorbital bar; no suborbital or labial crests (fig. 7.32d); a broad dark dorsolateral stripe bordered above by light, from parotoid to groin; subarticular and supernumerary tubercles under fingers and toes single; adult males 48–76 mm in standard length, adult females 57–82 mm. . . . *Bufo valliceps* (p. 210)

Bufo fastidiosus Group

Cranial crests present, small parotoid glands and hypertrophied testes in males, no tympanum, columella, Eustachian tubes, or ostia pharyngea. In addition to the two Costa Rican species described below, the only other member of the group is *Bufo peripetates* of western Panama (Savage 1972b; Savage and Donnelly 1992).

Bufo fastidiosus (Cope, 1875)

(plate 70; map 7.5)

DIAGNOSTICS: A moderate-sized toad with massive light-colored cranial crests (fig. 7.30a), several enlarged longitudinal series of dorsal warts, the dorsal ground color dark, and the warts usually rusty, pink, or red. These features, combined with the absence of the ear apparatus (no

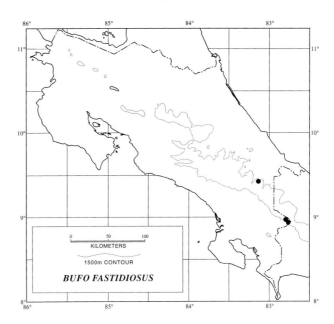

Map 7.5. Distribution of *Bufo fastidiosus*.

tympanum or middle ear), fleshy padlike hands and feet, and the skin of the head co-ossified to the frontoparietal and squamosal bones in adults, are distinctive.

DESCRIPTION: Adult males 43 to 52 mm in standard length, adult females 40 to 60 mm; upper surfaces with large and small discrete smooth warts; a single distinct series of smooth lateral warts; venter strongly tuberculate; head broader than long; snout subelliptical in dorsal outline; eyes moderate; skin on upper surface of head co-ossified to skull, massive canthal, supraorbital, postorbital, and bulbous supratympanic cranial crests; parietal crest weak; smooth, oblong parotoid gland moderate, about two-thirds the area of eyelid; a well-developed broad temporal plate with ornamentation (exostosis) continuous with supratympanic crest posterior to orbit; no vocal slits or sac in males; limbs short; hands and feet fleshy pads without subarticular, accessory palmar, or plantar tubercles; forelimbs hypertrophied in adult males; finger II longer than finger I; a discrete light nuptial pad on thumb in adult males; fingers with definite webbing that continues as a fringe along each finger; hind limbs relatively short; upper surface of tibial segment with huge pointed warts; toes connected by extensive fleshy webs; webbing formula: I 1–1 II 1–2 III 1–3 IV 3–2 V; inner metatarsal tubercle elliptical, outer elongate and larger than inner; no tarsal fold. Dorsum brown to black, nearly uniform or with warts and tubercles contrasting rusty, pink, or red; usually a narrow to broad middorsal light stripe; cranial crests light; no transverse dark limb bars; venter brown to black mottled with lighter pigment; iris brown.

LARVAE (fig. 7.24): Small, total length 16 mm at stage 36; body ovoid; mouth anteroventral, directed ventrally; nares and eyes dorsal; spiracle lateral and sinistral; vent tube medial; tail moderate; fins moderate; small caudal fin subovoid

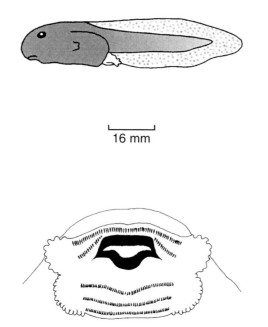

16 mm

Figure 7.24. Tadpole of *Bufo fastidiosus*.

to round at tip; oral disk medium, emarginate; beaks and 2/3 rows of denticles present; A2 with wide median gap above mouth; single row of large papillae border oral disk laterally; body uniform chocolate brown in life, venter gray; caudal musculature dark olive brown; fins translucent olive brown with darker peppering.

VOICE: Males lack an advertisement call but have a weak release trill.

SIMILAR SPECIES: (1) *Bufo holdridgei* has reduced cranial crests (fig. 7.30b) and beadlike warts on the body (fig. 7.23b) and limbs. (2) *Crepidophryne epiotica* has the first toe indistinguishable from the second.

HABITAT: The Premontane and Lower Montane Rainforest zones of southern Costa Rica and western Panama.

BIOLOGY: Most of the information summarized here is from Lips and Krempels (1995). This is a diurnal species that is found walking about the forest floor, under logs, and in open pastures during the reproductive season or after heavy rains during the dry season (February to March). *B. fastidiosus* is mainly fossorial and was excavated from leaf litter along stream banks in August. Occasional juveniles are found throughout the year on rocky stream margins. In a series of seven samplings along a 400 m section of the Río Cotón, above Las Tablas, Puntarenas Province, February to April, 116 gravid females, 84 adult males, plus 23 juveniles were captured. Of 187 adults toe clipped and released, only 4 were recaptured. The sex ratio was 0.58 (female biased). These data and those from litter plot samples indicate densities in excess of 1,100/ha near this stream course. *B. fastidiosus* is an explosive breeder, and reproduction takes place in ephemeral pools after heavy rains in late April to May. Males often amplex with other males but let go when the

grasped male gives a release call and vibration. Presumably females locate mates visually (Savage 1966b). Males arrive at the breeding site earlier than females. Amplexus is inguinal (Graybeal and Queiroz 1992). On arrival females are immediately amplexed by two to ten males, which form "balls" of toads in the pools. While males struggle to attain amplexus the females remain immobile, and they occasionally drown under the weight of the competing males. A single file of black-and-yellow eggs is laid in shallow pools in a weakly rosarylike string. There are about 80 to 90 eggs in a clutch, averaging 4.3 mm in diameter. A single outer envelope is present in addition to the vitelline membrane, and the eggs are not separated by partitions. Hatching occurs in four to five days and metamorphosis within sixty to seventy days of egg deposition. Postmetamorphic juveniles are 17 to 37 mm in standard length. This is another species of toad that has suffered incredible population declines since the 1980s (Lips 1998)

REMARKS: *Bufo holdridgei* of Volcán Barva, and *B. peripetates* of western Panama are closely related to *B. fastidiosus* (Savage 1972a; Savage and Donnelly 1992). *B. peripetates* is much larger than the Costa Rican species (adult males 56–78 mm in standard length, adult females 54–96 mm) but resembles *B. fastidiosus* in having massive, elevated cranial crests and the cephalic skin co-ossified to the roofing bones of the skull. The Panama toad is markedly sexually dichromatic, males nearly uniform yellow, females light tan to brown, often with a lateral dark stripe, and lacks ornamentation in the temporal area and large pointed warts on the lower hind limb. The nominal species *Ollotis coerulescens* Cope, 1875 (Costa Rica: Limón: "Pico Blanco" = Cerro Utyum) as recognized by Taylor (1952a) was synonymized by Savage (1972a). The name was based on juveniles now lost, whose description agrees closely in morphology and coloration with juvenile *B. fastidiosus*. It is apparent that Cope inadvertently switched the color description of *Crepidophryne epiotica* and *B. fastidiosus* in the 1875 paper, making it appear that *O. coerulescens* differed from the latter species in that regard.

DISTRIBUTION: Premontane and Lower Montane zones on both slopes of the southern Cordillera de Talamanca of Costa Rica and the Atlantic slope of immediately adjacent western Panama (760–2,100 m).

Bufo holdridgei Taylor, 1952a

(plate 71; map 7.6)

DIAGNOSTICS: A moderate-sized black to light brown toad with low cranial crests except for the supratympanic (fig. 7.30b) and having the parotoids and some of the dorsal and lateral warts reddish. This species lacks all elements of the outer and middle ear apparatus; it has fleshy, padlike hands and feet and small, smooth beadlike warts on the upper surfaces of the back and limbs.

DESCRIPTION: Adult males usually 32 to 46 mm (one

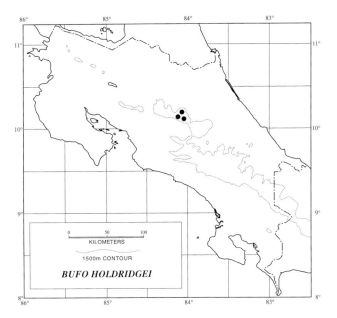

Map 7.6. Distribution of *Bufo holdridgei*.

51.7 mm) in standard length, adult females 38 to 53 mm; upper surfaces with large and small beadlike warts; a single distinct lateral series of smooth conical warts; venter strongly pustulate; head broader than long; snout subovoid in dorsal outline; eyes moderate; skin on upper surface of head co-ossified to roofing bones of skull; low canthal, supraorbital, postorbital, and parietal crests, supratympanic crest thickened; smooth globular parotoid gland less than two-thirds the area of upper eyelid; a well-developed narrow temporal plate with ornamentation (exostosis), continuous with supratympanic crest posterior to orbit; no vocal slits or sac in males; limbs short; hands and feet fleshy pads without subarticular, accessory palmar, or plantar tubercles; forelimbs hypertrophied in adult males; fingers I and II equal in length; a definite light nuptial pad on thumb in adult males; fingers with webs but no digital fringes; legs short, upper surfaces without large pointed warts; toes connected by fleshy web; webbing moderate; modal webbing formula: I 1–2 II 1–2½ III 1¾–3¼ IV 3¼–2 V; metatarsal tubercles obsolescent, round, outer largest, no tarsal fold. Dorsum black or dark to light brown; some reddish warts on body and limbs; supratympanic crests similarly colored; venter whitish with some dark mottling on chest; iris black.

LARVAE (fig. 7.25): Small, total length 23 mm at stage 36; body ovoid; mouth anteroventral, directed ventrally; nares and eyes dorsal; spiracle lateral and sinistral; vent tube medial; caudal fin rounded at tip; oral disk small, entire; beaks and 2/3 rows of denticles present; A2 with median gap above mouth; single row of large papillae borders oral disk laterally; body, upper tail musculature, and dorsal fin heavily pigmented with dark brown; ventral areas lighter.

VOICE: None.

SIMILAR SPECIES: (1) *Bufo fastidiosus* has massive light-colored cranial crests (fig. 7.30a) and large pointed warts on the upper surface of the tibial segment of the leg. (2) *Crepidophryne epiotica* has the toes so extensively webbed that only III is free of complete webbing (fig. 7.22a).

HABITAT: Lower Montane Rainforest of Volcán Barva.

BIOLOGY: This account is based principally on Novak and Robinson (1975) and my own observations. *Bufo holdridgei* is a fossorial species that is active during the daytime when aboveground. These toads may be found under surface debris within the forest during periods of heavy rains (March to November) but apparently concentrate in mossy stream banks during low-rainfall periods. Food includes a wide variety of arthropods, mostly spiders, larval lepidopterans, flies, and beetles, and adult beetles, Dermaptera, ants, and mites. The species is an explosive breeder that originally reproduced in forest floor pools but now also deposits eggs in man-made drainage ditches. The onset of reproductive activity follows heavy rains in early to mid-April and extends to late May. Males arrive at the breeding site somewhat earlier than females and may remain there for some time after the reproductive period. Males frequently amplex with individuals of other frogs, *Atelopus senex* and *Rana vibicaria*, using the same sites and with conspecific males. Conspecifics are released in response to the amplexed male's release vibrations, but other frogs are usually drowned, since they are grasped and submerged by several male *B. holdridgei* at once. Females are similarly besieged by two to eight males vying for success as soon as they reach the water. This struggling mass forms a compact ball of toads. Amplexus is inguinal, and females remain at the breeding site for less than an hour. There are always many more males than females

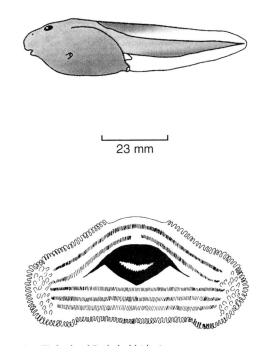

23 mm

Figure 7.25. Tadpole of *Bufo holdridgei*.

at the breeding site. At two pools monitored by Novak and Robinson (1975) using mark and recapture methods, 2,765 different males but only 30 females were recorded (April 17–26). Males, unlike females, leave the ponds and then return, and these figures suggest that females may not breed every year. The black-and-cream eggs are deposited in shallow pools or puddles as single files in two slightly crenate strings. The female leaves the site immediately after oviposition. Clutch sizes vary from 45 to 137. The eggs are 2.5 to 3.5 mm in diameter and eggs plus envelopes measure about 7 mm. There are two outer gelatinous layers; the inner one adhering to the vitelline layer is definitely crenate, but there are no partitions within the jelly layers. Hatching occurs in about twelve days, and metamorphosis may be estimated to take place in another twenty-seven days. Metamorphosing toadlets are about 7 mm in standard length and juveniles 15 to 38 mm in standard length. Sexual maturity is not attained until at least the third year of life. *B. holdridgei* samples contain a high proportion of individuals lacking part of a limb or an entire limb. Whether these deformities are the result of predation or parasitism or are congenital remains unknown. This species is one of those that has completely disappeared during the 1980s and may now be presumed to be extinct.

REMARKS: Blair (1972a) pointed out the unusual hypertrophied testes in males of this species, which he believed contain a substance that prevents fertilization or normal early development of eggs in attempted crosses with other *Bufo*. Adult males of *B. holdridgei* are usually under 47 mm (average 42) in standard length, but Novak and Robinson (1975) reported an exceptionally large example (UCR 5858) that measured 51.7 mm. Aside from size this toad is not remarkable, and its karyotype (from gonadal tissue) was identical to that recorded in Blair (1972b): 2N = 22, all chromosomes metacentrics or submetacentrics, six pairs larger than remaining five, with a secondary constriction on chromosome 1; NF = 44 (Bogart 1972). This toad is closely related to *Bufo fastidiosus* of southern Costa Rica and western Panama and to *B. peripetates* of western Panama. *B. holdridgei* differs from the latter in its smaller adult size (32 to 52 mm versus 54 to 96 mm), development of cranial crests (relatively low versus massive, elevated) and ornamentation (exostosis of the temporal plate versus plate not evident, no ornamentation).

DISTRIBUTION: Lower montane zones of Volcán Barva, Costa Rica (200–2,200 m).

Bufo guttatus Group

No cranial crests, large parotoid glands, a fully developed auditory apparatus (external tympanum, columella, Eustachian tubes, and ostia pharyngea). Other members of this group besides *Bufo haematiticus* are *Bufo blombergi* (eastern Ecuador and Colombia), *B. caeruleostictus* (western Ecuador), *B. glaberrimus* (Amazonian Colombia, Ecuador, and Peru and western Colombia), *B. guttatus* (Amazon basin), and *B. hypomelas* (western Colombia and Ecuador).

Bufo haematiticus Cope, 1862b

(plate 72; map 7.7)

DIAGNOSTICS: A moderate-sized to large toad having large parotoid glands (fig. 7.30c) and a distinct inner tarsal fold and lacking any cranial crests. This toad has a "dead leaf" color pattern, light above (leaf surface) with a broad dark lateral band (leaf shadow).

DESCRIPTION: Adult males 42 to 62 mm in standard length, adult females 50 to 80 mm; upper surfaces have small pustules and smooth warts; venter smooth anteriorly, granular posteriorly; head about as broad as long; snout subelliptical in dorsal outline; eyes large; parotoid glands very large, elongate, equal in area to side of head, covered with pustulate warts, continuous with lateral fold on body; height of external tympanum about one-half eye diameter; a single vocal slit (dextral or sinistral) and fully distensible single internal subgular vocal sac in adult males; limbs long; finger I longer than finger II; well-developed single subarticular tubercles present; no supernumerary or accessory palmar tubercles; a brown nuptial pad on thumb in adult males; no finger webs; single rounded subarticular tubercles on toes; no supernumerary or plantar tubercles; toes webbed only at base; inner metatarsal tubercle elongate, outer a small round bump; well-developed inner tarsal fold. Dorsum tan to purplish gray, uniform or with few to many moderate black spots and/or irregular burnt orange areas that may cover entire back; parotoids somewhat lighter; a lateral dark brown to black band outlined above by a narrow white line from tip of snout along side of head and parotoid along

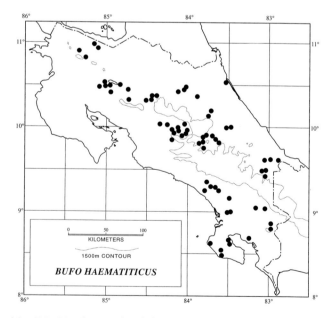

Map 7.7. Distribution of *Bufo haematiticus*.

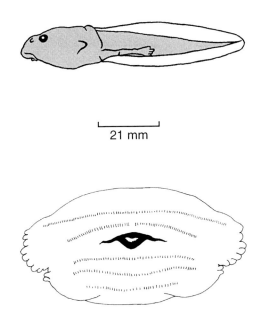

Figure 7.26. Tadpole of *Bufo haematiticus*.

21 mm

flank to groin; upper surfaces of limbs without transverse limb bars, although dark spots may be present; venter light beige to yellow, often grading to burnt orange posteriorly with dark brown markings, particularly on chest; soles dark brown; iris dark brown to black with gold flecks.

LARVAE (fig. 7.26): Small, total length 21 mm at stage 37; body ovoid, somewhat depressed; mouth anteroventral; nares anterolateral; eyes dorsal; spiracle lateral and sinistral; vent tube medial; tail long; fins low; caudal fin rounded at tip; oral disk small, emarginate; beaks and 2/3 denticles; broad gap in A2 above mouth; single row of labial papillae laterally; body uniform dark brown dorsally; venter clear; tail musculature dark; fins clear.

VOICE: No known advertisement call; the series of high-pitched peeps described by Ibáñez, Rand, and Jarmillo (1999) may be a release call.

SIMILAR SPECIES: There are none.

HABITAT: Forest floor of Lowland Moist and Wet Forests and Premontane Moist and Wet Forests; metamorphs and small juveniles typically found along stream courses under rocks.

BIOLOGY: A nocturnal "dead leaf" toad of the litter layer. Food consists mostly of small ants (averaging about twenty-one in each of eleven stomachs examined) and averaging 3.7 mm in length. Other arthropods averaged between 0.4 (mites) and 4.8 (beetles) in length (Lieberman 1986). Densities are low, being 21/ha at La Selva, Heredia Province, where the diet study was carried out. The species is an explosive breeder, with reproduction in the early to middle wet season (March to July). As breeding sites it prefers rocky pools (Scott 1983b) along the borders of forest streams and large rivers. The black-and-cream eggs are laid in single files in paired strings. Metamorphs are about 19 mm in standard

length; the smallest juveniles are 38 mm. Small, recently metamorphosed examples are easily identifiable by having a black venter sprinkled with glossy white spots in life. This pattern is retained in preservative, although the characteristic light spots lose their intensity.

REMARKS: The karyotype consists of 2N = 22 metacentric to submetacentric chromosomes, five pairs of which are markedly shorter than the others. Secondary constrictions are found on chromosomes 1, 4, and 7; NF = 44 (Bogart 1972).

DISTRIBUTION: Humid lowlands and premontane slopes on the Atlantic versant from eastern Honduras to northern Colombia and on the Pacific slope from Costa Rica to central Ecuador; in Costa Rica on the Atlantic and southwestern lowlands and adjacent premontane slopes, the Mesetas Central Oriental and Occidental, and the premontane Pacific slope in Guanacaste Province (20–1,300 m).

Bufo marinus Group

Cranial crests present, enormous parotoid glands, a fully developed auditory apparatus (external tympanum, columella, Eustachian tubes, and ostia pharyngea). In addition to the wide-ranging *Bufo marinus* that occurs in Costa Rica, other members of this group are *Bufo arenarum* (southern Brazil, Uruguay, and Argentina), *B. ictericus* (southern Brazil and Paraguay), *B. poeppigii* (uplands from Colombia to Bolivia), and *B. rufus* (central Brazil to Argentina).

Bufo marinus Linné, 1758

Cane Toad, Sapo Grande

(plate 73; map 7.8)

DIAGNOSTICS: The large size, enormous parotoid glands, cranial crests (fig. 7.30d), and tarsal fold make adults

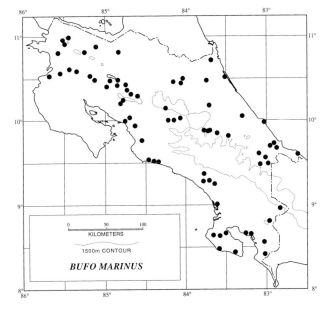

Map 7.8. Distribution of *Bufo marinus*.

of this very common species immediately recognizable. Juveniles are often mistaken for other toads, but the presence of the tarsal fold, in combination with pointed tubercles under the fingers and toes and on the palm and sole, is unique in Costa Rican toads.

DESCRIPTION: The largest Costa Rican amphibian and among the largest anurans in the world: adult males 85 to 145 mm in standard length, adult females 90 to 175 mm (to 238 mm in northeastern South America and 241 mm in Australia); females weigh up to 1.5 kg; upper surfaces relatively smooth, with scattered large and small warts; warts and tubercles keratinized with one or more spines and most numerous on limbs, often darkened in males; venter with small granular tubercles; head broader than long; snout rounded; eyes moderate; skin of head co-ossified to upper surface of skull; massive canthal, supraorbital, preorbital, postorbital, and pretympanic cranial crests; weak parietal and labial crests but no suborbital; parotoid glands very large, triangular, relatively smooth, equal to or exceeding area of side of head; height of external tympanum about one-half eye diameter; paired vocal slits and fully distensible single internal subgular vocal sac in adult males; arms and legs short; finger I longer than finger II; well-developed single subarticular tubercles and pointed supernumerary, accessory palmar, and plantar tubercles present; adult males with brown nuptial pads on upper surface of first and second and sometimes third fingers; inner and outer metatarsal tubercles subequal, oval; toes with some webbing; webbing formula: I 1–1½ II 1–2½ III 1½–3 IV 3½–1½ V; a distinct inner tarsal fold present. Dorsum of juveniles and females various shades of brown, mottled with irregular chocolate and beige blotches, large dark brown scapular blotches and narrow beige middorsal stripe usually present; adult males uniform cinnamon brown above; transverse dark bars on upper limb surfaces in juveniles; venter cream mottled with dusky pigment; iris brown.

LARVAE (fig. 7.27): Small, total length 24 mm at stage 37; body ovoid; mouth anteroventral, directed ventrally; nares dorsal; spiracle lateral, sinistral; vent tube medial; eyes dorsolateral; tail short; fins moderate; caudal fin pointed at tip; oral disk small, emarginate; beaks and 2/3 denticle rows present; broad gap in A2 above mouth; two rows of labial papillae laterally; body and upper tail musculature black, belly light; caudal fins with some dark flecks or clear.

VOICE: The advertisement call is a low-pitched trill reminiscent of the exhaust sound of a distant tractor (Conant and Collins 1991). The call is usually long, 10 to 20 seconds, with ten or so trills and has a dominant frequency between 0.63 and 0.8 kHz and a pulse rate of 15 to 20 per second (Easteal 1986; Lee 1996).

SIMILAR SPECIES: The five Costa Rican species of the *valliceps* group that might be confused with juvenile *B. marinus* lack tarsal folds.

HABITAT: All lowland and premontane zones, especially in disturbed areas and around human habitations, but ranging to 2,100 m in the lower montane zone in a few areas.

BIOLOGY: *Bufo marinus* is a "weed" species found in greatest densities in agricultural areas, but also in the urban sprawl of the Meseta Central and less commonly in undisturbed habitats. The account below is based primarily on the study of Zug and Zug (1979) in Panama and other observations on Central American populations (Zug 1983). The extensive literature on this species, which is regarded as a pest where it has been introduced by humans, is summarized by Easteal (1986). These toads are nocturnal and are most active during the wet season (April/May to July), but on nights with little rain. During the day they hide under objects in the environment, but they emerge near dark and are active for two or three hours. Apparently they regularly use one or a few feeding stations. Juveniles, unlike the adults, are active during the day. Dry season activity (November to March/April) is curtailed by danger of desiccation, and toads are most active whenever some light rain has fallen. Originally *B. marinus* was a denizen of secondary growth and open habitats, and it was and is rarely found in closed canopy forest tracts. However, the extensive clearing of forest for agricultural pursuits has opened up vast areas for toad colonization, and it may now be regarded as a human commensal. In some localities—for example, along ditches in the lowland banana plantations—almost incredible densities exceeding 300/ha can be found. In more natural open areas or along forest margins, Zug and Zug (1979) estimated densities of 25/ha and 10/ha, respectively. Typical

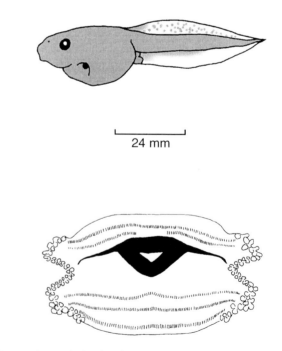

24 mm

Figure 7.27. Tadpole of *Bufo marinus*.

foraging areas average 160 m², with different sectors used on different nights. Critical thermal limits for the transformed individuals are minimum 10 to 12°C, maximum 41 to 42°C; tadpoles have a critical thermal maximum of 41.5 to 42.5°C. The species is an aggressive predator, known to eat almost anything that moves if it is not too large to catch and swallow. Among arthropods, it favors beetles and ants. Noxious and stinging animals, including wasps, spiders, and millipedes, are common food items, and many other arthropods have been recorded in the diet. Earthworms, slugs, snails, lizards, frogs, and juvenile *B. marinus* are less commonly consumed. They also eat canned and dry dog and cat food set out for family pets. Apparently they will also eat plant material when other food is unavailable (Zug, Lindgren, and Pippett 1975). Any differences in stomach contents between localities or seasons seem to reflect their catholic tastes. Most prey items are 2 to 10 mm in length, and fifty items are commonly ingested in an evening's foraging. In developed situations the toads are regularly seen sitting under street lights, neon signs, or other sources of artificial light, where they capture phototropic insects by the score. Opportunistic reproduction in *B. marinus* occurs sporadically beginning with the first rains of the season, and breeding congresses may be heard thereafter throughout the rainy season from April to November. Males call from the edge of almost any temporary and permanent lentic body of water. At sites with high densities of toads there is strong male-male competition for mates, and other males attempt to dislodge amplectant males, with minimal success. Amplexus is axillary, and both sexes float in the water. The black-and-cream eggs are laid in the shallows of any lentic body of water with little or no tree cover, including horse troughs, fountains, and swimming pools. The eggs are laid in paired strings and anchored to the bottom by the female, who winds them around submerged vegetation or other objects. Each string contains 2,500 to 12,500 eggs, 1.7 to 2 mm in diameter. Only a single gelatinous envelope lies outside the vitelline membrane, and it has fused with those of adjacent eggs to form the string. The egg coat is probably toxic (Licht 1968). There are no partitions between the eggs. The eggs hatch in thirty-six hours to four days (depending on water temperatures), and the larvae metamorphose in sixty to seventy-five days (thirty to eighty days in introduced populations). The small black tadpoles are very social, and congregations of hundreds or even thousands swim about together in the shallows at the breeding sites. They are also distasteful (Wassersug 1971). Experiments by Valerio (1971) showed that the tadpoles could survive for ten hours out of water. Metamorphs are 6 to 12 mm in standard length. Growth is extremely fast, and sexual maturity is reached in about one year, but the reproductive period appears to first occur in the second year of life. One captive specimen lived for sixteen years. The toxicity of the parotoid gland secre-tions, like the toxic and noxious properties of the eggs and larvae, serves as a potent antipredator defense. These toads are able to eject the secretion for a short distance as a fine spray, probably aimed at the eyes. Predators foolish enough to grab one of the toads suffer considerable discomfort or even death by ingesting the skin toxins. Domestic dogs are particularly likely to be victims, especially in areas where the toads have been introduced (Roberts et al. 2000). Mammals, birds, lizards, and snakes have all been reported as killed by mouthing or swallowing individuals of this species. However, both tadpoles and adults are known to be eaten by a variety of predators—the former by insects, crustaceans, and fishes, the latter by crocodilians, turtles, large lizards, snakes, birds, and mammals. The juveniles are much less toxic than adults and probably are more affected by predation. Dodds (1923) states that natives in eastern Sinaloa, Mexico, used the gland secretions to tip their arrows, but this has not been confirmed. Rabor (1952) reported that a man died after eating three *B. marinus* thinking they were edible frogs. He was chief of detectives in a town in the Philippines. So much for his powers of observation and deduction. Licht (1967) suggests that two human deaths likely followed ingestion of eggs of this species. *B. marinus* has been widely introduced into areas to which it is not native, generally in an attempt to reduce populations of insect pests in sugar plantations. This attempt had some success in Puerto Rico in the 1920s, although earlier introductions to the West Indies go back to the 1840s. They were subsequently introduced to Hawaii (1932), the Philippines (1934), Australia (1935), Florida (1940s), and many other sugarcane-producing areas. In most areas they are now considered pests, but the Australian introduction has had the most serious repercussions, and *B. marinus* has expanded its range to include most of coastal eastern tropical and subtropical Australia. The toads now breed and die in high numbers in the various water sources used by sheep and cattle herds, contaminating them by making the water toxic, with extreme economic results.

REMARKS: The name *marinus* is misleading and is based on an error in Seba (1734–65), who thought this toad could live in the sea. There is no evidence that this species has any ability to tolerate salt water, nor can its tadpoles survive in brackish water under natural conditions. Ely (1944) showed that eggs and embryos could develop in very low saline water at salinities of 9 or below (seawater average salinity is 35) in the laboratory. Linné (1758) based his original description on Seba, and the species is still incorrectly referred to as the marine toad in some popular accounts nearly three hundred years later. The karyotype is 2N = 22, all chromosomes metacentric or submetacentric; five large pairs, six small with a secondary constriction on chromosome 7; NF = 44 (Bogart 1972).

DISTRIBUTION: Lowlands and premontane areas and

occasionally into the lower montane zone from southeastern Texas and Sonora, Mexico, throughout Central America and to southwestern Ecuador and the Amazon basin of Peru and central Brazil in South America; successfully introduced throughout the West Indies (exclusive of Cuba), south Florida, Hawaii, eastern Australia, the Philippines, Taiwan, eastern New Guinea, Fiji, Samoa, and many other islands in the south and western Pacific; in Costa Rica (1–1,600 m).

Bufo periglenes Group

Cranial crests present, small parotoid glands, no tympanum and no columella, but Eustachian tubes and ostia pharyngea present. A single Costa Rican species forms this subdivision within *Bufo*.

Bufo periglenes Savage, "1966b," 1967

(plates 74–76; map 7.9)

DIAGNOSTICS: The bright yellow orange to red orange males and the red-spotted black to greenish females are unlikely to be mistaken for any other Costa Rican amphibian. The well-developed parotoid glands and cranial crests (fig. 7.30e) will confirm that specimens in hand are toads and not unusually colored harlequin frogs *(Atelopus)*.

DESCRIPTION: A moderate-sized toad, adult males 39 to 48 mm in standard length, adult females 42 to 56 mm; upper surfaces relatively smooth with dark-tipped granular warts; venter granular; head about as broad as long; snout pointed in dorsal outline in males, subelliptical in females; eyes large; skin of upper surface of head co-ossified to skull; low canthal, supraorbital, postorbital, and supratympanic cranial crests; parotoid gland elongate, inconspicuous

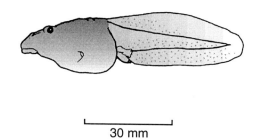

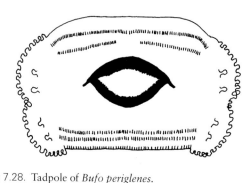

30 mm

Figure 7.28. Tadpole of *Bufo periglenes.*

warty; supratympanic crest, about same length as upper eyelid; temporal plate short and broad, without dermal ornamentation; no vocal slits or sac in males; no tympanum or columella, but reduced Eustachian tubes and ostia pharyngea present; limbs relatively long; fingers I and II equal, or II slightly longer than I; hands and feet fleshy pads without definite subarticular, accessory palmar, or plantar tubercles; brown male nuptial pad usually involves thenar tubercle as well as thumb and second finger; toes with lateral fringes and webbed at base; webbing formula: I 1–2 II 1¼–2¾ III 2–3½ IV 3¼–2 V; metatarsal tubercles globular; inner elongate, outer round, smaller; no tarsal fold. Males: uniform yellow orange, orange, or red orange above; venter lighter, mottled with dark; females: upper surfaces dirty greenish yellow, yellow olive, smoky black, or dark black with large bright red spots; venter yellowish green to flesh, usually heavily mottled with dark; juveniles: deep brown with bluish white spots; venter lighter; iris black with gold to copper outlining pupil.

LARVAE (fig. 7.28): Moderate-sized, total length 30 mm at stage 37; body ovoid, slightly depressed; mouth ventral; nares and eyes dorsal; spiracle lateral and sinistral; vent tube medial; tail short; fins moderate; caudal fin rounded at tip; oral disk small, emarginate; beaks and 2/3 rows of denticles; broad gap in A2 above mouth; two rows of papillae border oral disk laterally; uniform dark brown body, lighter below; tail musculature and anterior caudal fin similarly colored.

VOICE: It remains questionable whether *Bufo periglenes* emits an advertisement call. The male release call is a single short, low trilled chirp, accompanied by the typical release

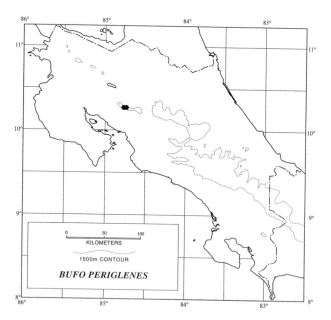

Map 7.9. Distribution of *Bufo periglenes.*

vibration of the body wall and hind limbs. Jacobson and Vandenberg (1991) reported a second call given by males, much less frequently heard, that may be an advertisement call. This vocalization sounded like the soft "tep-tep-tep" of wooden spoons clicking together and could be heard several meters away. The call duration was 500 milliseconds, and the dominant frequency ranged from 0.9 to 1.3 kHz.

SIMILAR SPECIES: There are none.

HABITAT: Lower Montane Rainforest.

BIOLOGY: The following summary is based on the accounts of Crump (1989a), Crump, Hensley, and Clark (1992), and Jacobson and Vandenberg (1991) and on observations over the years by M. Fogden, M. P. Hayes, J. A. Pounds, and me. B. periglenes is a secretive species that spends most of the year underground and comes to the surface to breed during one to three short periods near the end of the dry season (March to June). The toads are active diurnally, and in the period just before the breeding frenzy a few males may be seen moving about the forest floor, perhaps foraging. The species is an explosive breeder (plate 74), and depending on the timing of heavy thunderstorms, up to three bouts of about ten days each constitute the reproductive season. Most breeding sites are small, shallow pools (maximum sizes 0.5 m × 0.5 m, less than 20 mm deep) among the roots of wind-stunted trees along the continental divide, around the 1,500 to 1,620 m zone. The operational sex ratio is strongly male biased, with as many as nine hundred males to only ninety females present at the height of reproductive activity at the best-known Brilliante breeding site. In these circumstances it is not surprising that the males exhibit strong antagonistic behavior. Males amplex other males but usually respond to the release call and vibration. They also attack amplexing male-female pairs and form writhing masses of toad balls (plate 76), but the original male is rarely displaced. These activities continue at night, but at a reduced level. Amplexus is axillary (plate 75), and females generally remain at the breeding pools only during amplexus, which may last up to twenty-five hours. Females appear to leave the breeding pools shortly after oviposition and may not breed every year. The black-and-cream eggs are about 3 mm in diameter. They are deposited in single files in paired strings and hatch within several days. Metamorphosis requires four to five additional weeks. The tadpoles are facultatively nonfeeding and are able to attain transformation by subsisting on their yolk when no external food supply is available. Metamorphs are about 8 mm in standard length, and juveniles not showing any sign of adult coloration are 12 to 23 mm. Juveniles with evidence of the adult dichromatic coloration are between 24 mm and 33 mm in standard length. The bright coloration of this species is putatively aposematic, but no evidence from the field, terrarium, or biochemical analysis of skin toxins is available in support of this possibility. B. periglenes is one of the most

spectacular frogs to exhibit spectacular population declines during the 1980s. Since the early 1960s (Savage, unpubl.) "golden" toads have emerged from their retreats in large numbers through 1987 (Crump, Hensley, and Clark 1992). Regular monitoring of the population began that year, but in 1988 only one male was observed at the Brilliante site, and seven males and two females at a second locality 4 to 5 km away. In 1989, a single male was encountered after searches of areas where "golden" toads had been seen in previous years. At this time, although extensive monitoring and searching continues, no B. periglenes have been seen in subsequent years.

REMARKS: The common name "golden toad" applied to this anuran is a misnomer, since males are uniformly orange and the females blackish to greenish with red spots. Both the auditory apparatus and vocal ability of this species are greatly reduced, leading me to suggest (Savage 1966b) that the sexual dichromatism was related to mate recognition. This now seems an unlikely explanation, since female choice is not a factor in this situation, where any female arriving at the breeding site is immediately grasped by the nearest male and, if released, by another of the males scrambling to amplex.

DISTRIBUTION: Restricted to the lower montane zone on both slopes along the continental divide between Puntarenas and Alajuela Provinces, generally north and east of Monteverde, Costa Rica (1,500–1,620 m).

Bufo valliceps Group

Cranial crests present, small to moderate-sized parotoid glands and usually a fully developed auditory apparatus (tympanum, columella, Eustachian tubes, and ostia pharyngea). Besides the four representatives found in Costa Rica, other members of the group include *Bufo campbelli* (Belize, Honduras, Guatemala, and southern Mexico), *B. canaliferus* (southern Mexico and Guatemala), *B. cavifrons* (eastern Mexico), *B. cristatus* (upland central Mexico), *B. gemmifer* (Guerrero, Mexico), *B. ibarai* (Guatemala), *B. macrocristatus* (southern Mexico and Guatemala), *B. mazatlanensis* (western Mexico), *B. occidentalis* (Mexico), *B. turtelarius* (southern Mexico and Guatemala), and *B. tacanensis* (southern Mexico and Guatemala).

Bufo coccifer Cope, 1866a

(plate 77; map 7.10)

DIAGNOSTICS: A moderate-sized to large toad with a distinct light interorbital bar, a median dorsal light stripe, light-colored cranial crests (fig. 7.30f), and bifid or double subarticular (fig. 7.72b,c) and supernumerary tubercles under outer portions of the fingers and toes.

DESCRIPTION: Adult males 45 to 65 mm in standard length, adult females 53 to 83 mm; upper surfaces with mixture of large and small closely set warts; flanks covered with

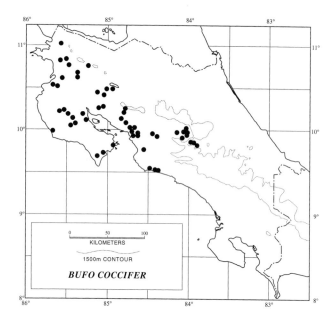

Map 7.10. Distribution of *Bufo coccifer*.

conical warts but no lateral series distinctly enlarged; venter finely granular; head broader than long; snout somewhat triangular in dorsal outline; eyes large; skin of head co-ossified to upper surface of skull; well-developed canthal, preorbital, supraorbital, postorbital, parietal, supratympanic, suborbital, and labial crests; pretympanic less well developed or absent; parotoid glands moderate, ovoid and diagonal, smooth or with few spicules; area of parotoid equal to or greater than area of upper eyelid; tympanum external, height about one-half eye diameter; adult males with large paired vocal slits and fully distensible single internal subgular vocal sac, round when calling; arms and legs short; finger I equal to or longer than finger II; well-developed sub-articular and supernumerary tubercles under fingers and toes; bifid or double on fingers III and IV, toes III–IV–V; brown nuptial pads on upper surfaces of first two fingers in adult males; toes with some webbing; webbing formula: I 1–2 II 1–3 III 2–3¼ IV 3½–1½ V; numerous accessory palmar and plantar tubercles present; inner metatarsal tubercle elongate, elevated outer rounded and flattened. Upper surfaces and flanks light tan to gray with series of dark brown dorsal blotches partially outlined by black; a definite light interorbital bar and light brown cranial crests; a narrow median stripe down back lying within a broader light stripe; a broad dorsolateral light stripe from parotoid to groin; broad lateral dark stripes; dark transverse bars on upper limb surfaces; undersides dull yellow, often with dark spotting in pectoral area; iris brown.

LARVAE (fig. 7.29): Small, total length 19 mm; body ovoid; mouth anteroventral, directed ventrally; nares dorsal; eyes dorsolateral; spiracle lateral, sinistral; vent tube medial; tail short; fins moderate; caudal fin bluntly rounded at tip; oral disk small, emarginate; beaks and 2/3 denticle

rows present; broad gap in A2 above mouth; one row of lateral labial papillae; body dark with small light spots, belly light; caudal musculature mottled with dark, upper surface with dark saddles alternating with light interspaces; scattered dark pigment in caudal fins.

VOICE: The call is a piercing buzz lasting 1.5 to 16.3 seconds with pulses (90 to 95 per second) not distinguishable to the human ear. The dominant frequency is 2.3 to 2.4 kHz for lower Central American examples (Porter 1965; Zweifel 1965).

SIMILAR SPECIES: (1) *Bufo luetkenii* has the cranial crests mounted by black keratin (fig. 7.32b) and lacks a light interorbital bar. (2) *Bufo melanochlorus* lacks a light interorbital bar and the middorsal pinstripe and has small triangular parotoid glands (fig. 7.32c). (3) *Bufo valliceps* has a series of strongly enlarged lateral warts and lacks a light interorbital bar and middorsal pinstripe.

HABITAT: Relatively open and disturbed areas within Lowland Dry Forest and Premontane Moist Forest, Wet Forest, and Rainforest zones, including pastures, roadside ditches, and gardens and vacant lots in urban areas.

BIOLOGY: This account is based mostly on McDiarmid and Foster (1981) and my own field observations. *Bufo coccifer* is relatively common and nocturnal and tends to favor open situations, although occasionally it is found in forest habitats. Little is known regarding activity aside from the reproductive period, but these toads probably remain below the ground surface during most of the dry season (Novem-

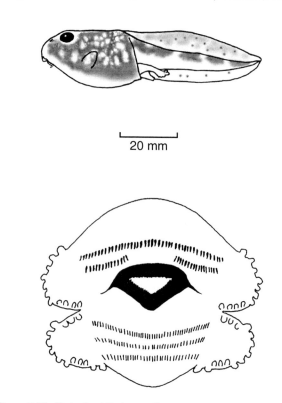

Figure 7.29. Tadpole of *Bufo coccifer*.

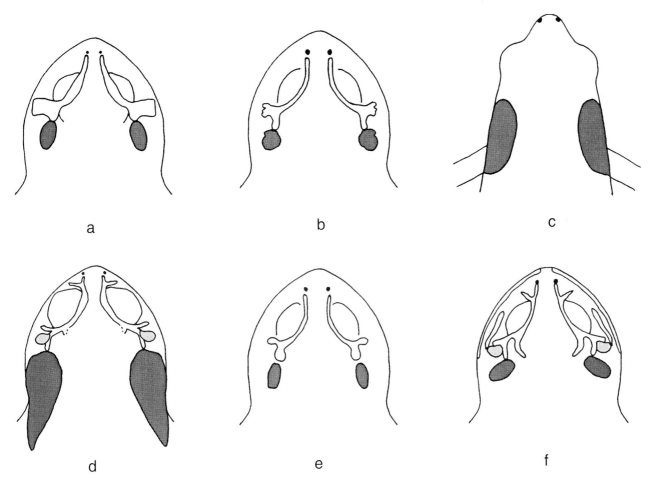

Figure 7.30. Characteristics of head structure in Costa Rican toads of the genus *Bufo*, dorsal view; all except (e) shown on a generalized outline of the head that does not accurately portray differences in head dimensions among taxa: (a) *B. fastidiosus*; (b) *B. holdridgei*; (c) *B. haematiticus*; (d) *B. marinus*; (e) *B. periglenes*; (f) *B. coccifer*. Dark shading = parotoid gland, light shading = tympanum.

ber to May). Mating takes place in shallow, usually temporary ponds or flooded fields. Males arrive at the breeding site during the first rains of the season but do not begin calling for three to four weeks. Males usually call submerged in the shallows on the edge of the pond but sometimes call from the bank. Some may also call from open water or dense low vegetation. The first females appear as the rainy season settles in, about the middle of May. The principal breeding season is from that time to mid-June, but some reproductive activity occurs into August. Amplexus is axillary, and egg laying takes place in the shallows with the female usually submerged on the bottom except for her head. The black-and-cream eggs are laid in two strings in single files of approximately 2,000 to 4,000 small eggs (1.2 to 1.37 mm in diameter). The strings are usually broken into 100 to 150 mm sections by the female's activity during egg laying. Metamorphosis apparently occurs in about thirty-five days. Unlike the tadpoles of some other Costa Rican species (e.g., *Bufo marinus*), those of *B. coccifer* tend to be solitary and do not form large aggregations. Metamorphs average about

8.5 mm in standard length. The smallest juveniles are about 20 mm in standard length. Adults are eaten by a variety of predators, most particularly the snake *Leptodeira annulata*, which feasts on a variety of anurans at breeding sites in the lowland forest. Larvae are heavily preyed on by the carnivorous tadpoles of *Leptodactylus labialis* in lowland northwestern Costa Rica.

REMARKS: It is often hard to distinguish among the species of the *Bufo valliceps* group except for the very distinctive *Bufo coniferus*. *Bufo coccifer* is the most easily recognized because it always has a light pinstripe running down the median raphe (furrow) of the back from the tip of the snout to the vent and a light interorbital bar. In this species the light pinstripe usually lies in the middle of a broader light stripe that extends paravertebrally for some distance. No light middorsal pinstripe is found in the other taxa, but all *Bufo melanochlorus* and *Bufo valliceps* and some *Bufo luetkenii* have a moderately broad middorsal light stripe at least anteriorly. *B. coccifer* further differs from its allies in having the flanks covered with conical warts of various sizes and

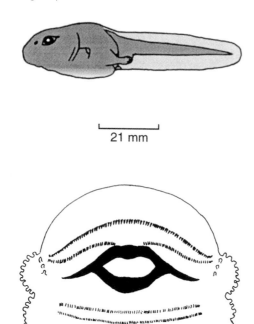

Figure 7.31. Tadpole of *Bufo coniferus*.

bifid or double subarticular (fig. 7.72b,c) and supernumerary tubercles under the outer digits. The other three species have the flanks covered with moderate low warts and a definite longitudinal series of much larger lateral warts (some *B. luetkenii* males lack the enlarged warts). Neither *B. luetkenii* nor *B. valliceps* has bifid or double digital tubercles, but a few examples of *B. melanochlorus* have bifid tubercles on some digits. The other three species differ most obviously in size of the parotoid gland, degree of cranial crest development, and details of coloration. Some female *B. luetkenii*, *B. melanochlorus*, and *B. valliceps* agree in having a broad lateral dark stripe running from below the parotoid to the groin. The dark stripe is bordered above by the enlarged lateral warts that form its light margin. *B. luetkenii* males lack the lateral dark stripe. The latter species has a very small oval parotoid gland but well-developed cranial crests, including a definite labial and weak suborbital, all topped by black keratin (fig. 7.32b). *B. melanochlorus* has a small triangular parotoid, and the pretympanic, suborbital, and labial crests are weak or absent (fig. 7.32c). The cranial crests are not as massive in *B. melanochlorus*, and only in the largest specimens are they black. *B. valliceps* has a large, well-

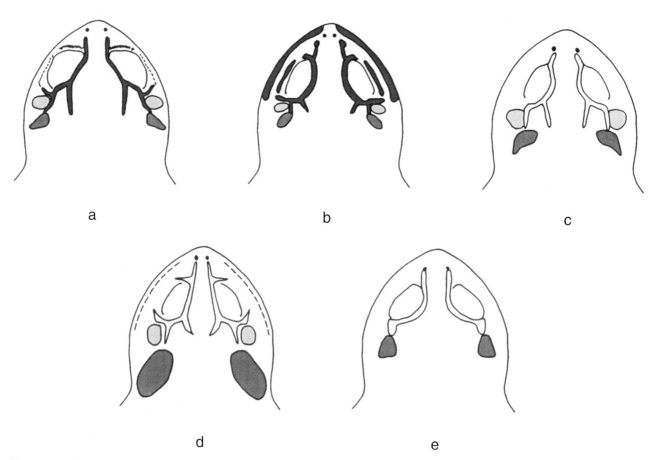

Figure 7.32. Characteristics of head structure in Costa Rican bufonids, dorsal view; shown on a generalized outline of the head that does not accurately illustrate differences in head shape and proportions among species: (a) *B. coniferus*; (b) *B. luetkenii*; (c) *B. melanochlorus*; (d) *B. valliceps*; (e) *Crepidophryne epiotica*. Dark shading = parotoid gland, light shading = tympanum.

developed pretympanic crest but lacks suborbital and labial crests (fig. 7.32d). As in *B. melanochlorus,* the cranial crests are less massive than in *B. luetkenii* and only rarely, in very large specimens, are they black. In patterned examples of these species, there is frequently a broad dorsolateral light stripe running from the parotoid to the groin. It contrasts most obviously with the dark lateral stripe in individuals having that pattern component. *B. coccifer* differs from the other forms in that the lateral dark stripe is tan and not demarcated by enlarged light-colored warts on its upper edge. Although there is considerable geographic variation in male call characteristics (Porter 1965; Zweifel 1965; McDiarmid and Foster 1981), there appears to be little reason to recognize *Bufo cycladen* Lynch and H. Smith, 1966, based on examples from Guerrero and Oaxaca, Mexico, that other authors regard as typical *B. coccifer.*

DISTRIBUTION: Subhumid Pacific lowlands and premontane areas from Guerrero, Mexico, to central Costa Rica and disjunctly in southwestern Panama, but also in several Atlantic drainage areas of Guatemala, El Salvador, Honduras, Nicaragua, and Costa Rica; in Costa Rica, restricted to the northwestern region and the Mesetas Central Oriental and Occidental (4–1,435 m).

Bufo coniferus Cope, 1862b

(plate 78; map 7.11)

DIAGNOSTICS: This moderate-sized to large toad is unlikely to be confused with any other anuran because of its many dark spinous warts and usually greenish coloration.

DESCRIPTION: Adult males 53 to 72 mm in standard length, adult females 76 to 94 mm; upper surfaces covered with widely spaced dark-tipped, pointed warts and spicules; flanks with a longitudinal series of large dark-tipped spinous warts; underside except abdomen covered by spinous tubercles; head broader than long, subovoid in dorsal outline; eyes large; skin of head co-ossified to upper skull bones; well-developed black canthal supraorbital, postorbital, parietal, and supratympanic and pretympanic crests; preorbital and suborbital crests variable, labial crest poorly indicated; parotoid gland highly variable in shape, elongate, ovoid, or triangular, often obscure or barely larger than adjacent warts, covered with spinous warts; height of external tympanum about one-half diameter of eye; adult males with one or two vocal slits and a fully distensible single internal subgular vocal sac; sac round when inflated; finger II longer than finger I; subarticular and supernumerary tubercles under fingers and toes low, rounded to flattened, usually single, bifid in some examples; dark brown nuptial pads on upper surfaces of fingers I and II in adult males; toes with moderate webbing; modal webbing formula: I ½–1½ II ½–2 III 1½–3 IV 3–1½ V; inner metatarsal tubercle small, outer tiny or absent. Dorsum usually yellow green to olive green, sometimes dull gray to brown, uniform or with contrasting dark and/or white blotches or with one or two gold spots; small juveniles (to 30 mm in standard length) usually green with light

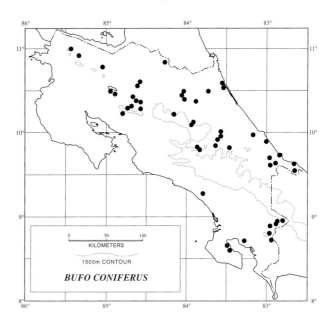

Map 7.11. Distribution of *Bufo coniferus.*

outlined red orange warts; venter dirty white; iris green and brown.

LARVAE (fig. 7.31): Small, total length 21 mm; body ovoid; mouth anteroventral, directed ventrally; nares dorsal; eyes dorsolateral; spiracle lateral, sinistral; vent tube medial; tip of caudal fin rounded; oral disk medium, emarginate; broad gap in A2 above mouth; one row of lateral labial papillae; body black above, gray below; caudal musculature uniform black; caudal fins mottled with dark pigment.

VOICE: A very long trill lasting 18 to 20 seconds; the dominant frequency is 1.0 kHz, and the trill rate is 39 to 40 per second (Porter 1966). Large choruses sound like gasoline-powered generators.

SIMILAR SPECIES: There are none.

HABITAT: A common toad of Lowland Wet and Moist Forest zones and less frequently found in Premontane Wet Forest and Rainforest and Lower Montane Wet Forest and Rainforest. Usually found in undisturbed forest and often climbing some distance above the ground on shrubs or trees.

BIOLOGY: A nocturnal scansorial toad often found several meters above the ground hiding or climbing in vegetation. This species feeds almost exclusively on ants (Toft 1981). *B. coniferus* breeds in shallow ponds or pools in the forest or nearby pastures during the dry season (December to April). Male choruses may be heard both at night and during the day in the reproductive season. The call site is at the edge of the pond or pool. Amplexus is axillary. The black-and-gray eggs are laid in paired strings in double files. The string consists of two gelatinous layers about 4.2 mm in diameter, and the eggs average 1.8 mm in diameter without partitions. The eggs hatch in about five days. Metamorphs are tiny, about 9 to 10 mm in standard length. Small juveniles are 14 mm.

REMARKS: Cope (1875) described *Bufo auritus* based on

two juvenile toadlets from the east coast region of Costa Rica. Taylor (1952a) recognized this taxon as valid but pointed out the junior homonymy of the name with *Bufo auritus* Raddi, "1822," 1823. Consequently Taylor proposed a replacement name, *Bufo gabbi*. I have examined one of the syntypes (USNM 30676), and there can be little question that it is a juvenile *B. coniferus*, since it has, among other shared features, the diagnostic conditions of the second finger being much longer than the first and the digital tubercles very weak. The karyotype is 2N = 22, with six large and five small pairs of chromosomes, all metacentrics or submetacentrics. A secondary constriction is present on chromosome 7; NF = 44 (Bogart 1972).

DISTRIBUTION: Humid lowlands and premontane slopes from eastern Nicaragua on the Atlantic versant and southwestern Costa Rica on the Pacific slope south to western Colombia and northwestern Ecuador (2–1,550 m).

Bufo luetkenii Boulenger, 1891

(plate 79; map 7.12)

DIAGNOSTICS: The uniform yellow brown to yellow green adult males of this large species are unlikely to be confused with any other toad. Females and juveniles are variable in coloration but have small parotoid glands, well-developed black cranial crests (fig. 7.32b), and usually middorsal and lateral light stripes.

DESCRIPTION: Adult males 77 to 96 mm in standard length, adult females 73 to 107 mm; upper surface of body and limbs covered by moderate-sized low warts; none or a few lateral warts enlarged; venter finely granular; head broader than long, broadly rounded in dorsal outline; eyes large; skin of head co-ossified to upper surface of skull; can-

thal, preorbital, supraorbital, postorbital, parietal, supratympanic, pretympanic, and labial crests well developed, suborbital less so, all capped by black keratin; parotoid gland very small, ovoid, area one-half or less that of upper eyelid; height of external tympanum about one-half diameter of eye; adult males with paired vocal slits and fully distensible single internal subgular vocal sac, round when calling; limbs short; finger I longer than finger II; well-developed single subarticular and supernumerary tubercles under fingers and toes; upper and medial surfaces of first and second finger in adult males covered by black nuptial pad that extends onto outer edge of thenar tubercle; toes with some webbing; webbing formula: I 1–2 II 1–3 III 2–3½ IV 3½–1½ V; inner metatarsal tubercle elevated, elongate, slightly larger than elongate outer tubercle. Dorsum of adult males and some females uniform pale yellow brown to yellow green; most females with dark brown, olive green, or rusty brown dorsal ground color, a broad yellow middorsal stripe at least anteriorly and a broad lateral dark stripe; light dorsolateral stripes sometimes present; juveniles and some females with black lineate spots and lines often forming chevrons in shoulder region; dark transverse bars on upper surface of limbs in patterned females and juveniles, absent in adult males; underside dirty yellow without conspicuous markings; throat of adult males discolored, yellowish green in life.

LARVAE: Moderate, total length 29 mm; body ovoid; mouth anteroventral, directed ventrally; nares dorsal; eyes dorsolateral; spiracle lateral, sinistral; vent tube medial; tip of caudal fin rounded; oral disk small, emarginate; broad gap in A2 above mouth; one row of lateral labial papillae; body black above, gray below; caudal musculature uniform black; upper caudal fin mottled with dark pigment, lower with scattered dark punctations. (Larvae of this species were discovered too late for preparation of a figure.)

VOICE: A short trill lasting about 4 seconds repeated after intervals of 1 to 4 seconds. The dominant frequency is 1.6 to 1.95 kHz with a trill rate of 21 per second (Porter 1966).

SIMILAR SPECIES: (1) *Bufo melanochlorus* has a triangular parotoid gland and weak pretympanic, suborbital, and labial crests (fig. 7.32c). (2) *Bufo valliceps* has larger parotoid glands that are about equal in area to the area of the upper eyelid. (3) *Bufo coccifer* has a light interorbital bar and a middorsal light pinstripe.

HABITAT: Primarily in Lowland Dry Forest but also marginally in Lowland Moist and Premontane Moist Forests. It commonly occurs in disturbed pastureland and other open areas.

BIOLOGY: These relatively common nocturnal anurans congregate during the first half of the rainy season (June to August) around shallow temporary ponds, ditches, and permanent or temporary stream courses. Males call from shore near the water's edge. Presumably they retreat to underground hideaways during the long dry season (November to

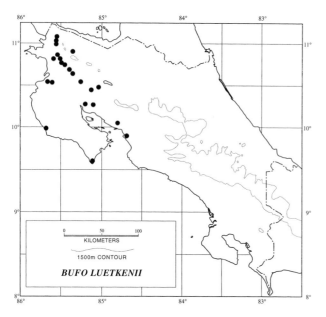

Map 7.12. Distribution of *Bufo luetkenii*.

May). Haas and Köhler (1997) reported on captive breeding and development. Two females laid about 1,000 and 5,600 eggs in two strings. Individual eggs were 1.5 mm in diameter and were surrounded by a 1.5 mm thickness of gelatinous material. The eggs hatched in three days, and the larvae were 5 mm in total length. They reached stage 38 thirty days later at a standard length of 34 mm. Metamorphosis began six days after that. Metamorphs are 9 to 12 mm in body length. Small juveniles are 20 mm in standard length.

REMARKS: See the account for *Bufo coccifer* for additional comparisons of features separating *B. luetkenii* from related species. This toad was originally described by Boulenger (1891) from the city of Cartago, Cartago Province, Costa Rica. All other known records of the species in the republic are from the Pacific Lowland Dry Forest region. Since it has never been taken again anywhere near Cartago, that locality must have been erroneously associated with the type specimen. *B. luetkenii* has often been confused with the allied species *B. valliceps*. Although recognizing the validity of *B. luetkenii*, Porter (1966, 1970) included many records of this form in his 1970 range map for *B. valliceps*. *B. luetkenii* is now known to have an extensive range along the Pacific lowlands from southern Chiapas, Mexico, to Costa Rica. All of Porter's *B. valliceps* locality records for this area, including El Salvador and western Nicaragua, are based on specimens of *B. luetkenii*. The karyotype is 2N = 22, consisting of six pairs of large and five pairs of smaller chromosomes, all metacentric or submetacentric, with a secondary constriction on chromosome 1; NF = 44 (Bogart 1972).

DISTRIBUTION: Pacific lowlands from Chiapas, Mexico, to central Costa Rica (6–300 m) and on the Atlantic versant in dry interior valleys of Guatemala and Honduras (to 1,300 m) and the upper Río San Juan drainage in Costa Rica (10–436 m).

Bufo melanochlorus Cope, 1877

(plates 80–81; map 7.13)

DIAGNOSTICS: Moderate-sized to large toads having well-developed cranial crests and a broad dark lateral stripe bordered above by light warts. This species differs from others exhibiting these features by having small triangular parotoid glands (smaller than the area of the upper eyelid) and the pretympanic crest weak and short or absent (fig. 7.32c).

DESCRIPTION: Adult males 43 to 65 mm in standard length, adult females 65 to 103 mm; upper surface of body smooth with scattered spicules; flank with series of moderately enlarged warts; upper surface of hind limbs with many large pointed warts; venter finely granular; head broader than long; dorsal outline of snout somewhat pointed in males, subelliptical in females; eyes large; skin of head co-ossified to upper surface of skull; canthal, supraorbital, postorbital, parietal, and supratympanic crests well developed, preorbital variable, pretympanic weak and short or absent, suborbital and labial weak to absent; parotoid gland trian-

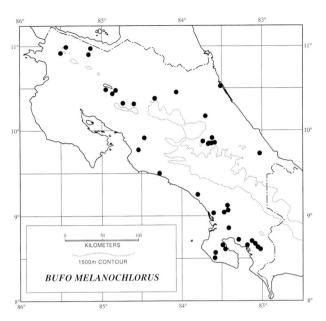

Map 7.13. Distribution of *Bufo melanochlorus*.

gular, small, about three-fourths area of upper eyelid; height of external tympanum about one-half diameter of eye; adult males with paired vocal slits and a single internal subgular vocal sac, fully distensible and round when calling; finger I longer than finger II; well-developed single subarticular and supernumerary tubercles under fingers and toes; occasionally first subarticular tubercle on finger III bifid; fingers I and II with dark brown nuptial pad on upper surface in adult males; toes with some webbing; webbing formula: I 1–2 II 1–3 III 2–3½ IV 3½–2 V; elevated inner metatarsal tubercle elongate, larger than rounded outer tubercle. Dorsum uniform rusty tan to brown with a light middorsal and often paired light dorsolateral stripes; dark dorsal spots and blotches often present in juveniles and females; dark lateral stripe from parotoid to groin, bordered above by white line that includes largest lateral warts; dark stripe broad in females and juveniles; upper limb surfaces with transverse dark bars; venter dull yellow; throat of adult male discolored, yellowish green in life; iris gold above, brownish gold below, separated by dark brown line.

VOICE: The call is similar to those of *Bufo luetkenii* and *Bufo valliceps*. It consists of a short trill lasting several seconds, repeated several times at intervals of a few seconds.

SIMILAR SPECIES: (1) *Bufo valliceps* has large parotoid glands (equal to or larger than upper eyelid) and a well-developed pretympanic crest (fig. 7.32d). (2) *Bufo luetkenii* has small round parotoid glands and well-developed pretympanic, suborbital, and labial crests (fig. 7.32b). (3) *Bufo coccifer* has a light interorbital bar and a light pinstripe down the center of the back.

HABITAT: Lowland Moist and Wet Forests and the lower portion of Premontane Wet Forest. Usually seen only during the breeding season in association with large streams or occasionally found in the forest litter.

BIOLOGY: These nocturnal anurans are dry season breeders (January to February) that reproduce in fairly large rocky-bottomed streams at low water. On the Pacific versant large choruses may be encountered at this time. Males call from within a 0.5 m or so of the water or sometimes from the shallows. Amplexus is axillary. The smallest juvenile found in the wet season within the forest is 22 mm in standard length.

REMARKS: See the account for *Bufo coccifer* for a more detailed comparison of the characteristics distinguishing *B. melanochlorus* from allied forms. This species apparently consists of two allopatric populations, one along the Atlantic versant and a second on the Pacific slope. Most of the available material from the Atlantic population consists of juvenile and subadults, while the Pacific series contains many large adults. The largest examples of *B. melanochlorus*, however, are two females, one from the Atlantic versant at Pavones, Cartago Province (Greding 1972a), and another from La Balsa, Alajuela Province, in central Costa Rica on the Pacific slope (KU 139995), 97 and 103 mm in standard length, respectively. The largest southwest Pacific specimen is also a female 93 mm in standard length. These additional data confirm Greding's conclusion that *B. melanochlorus* reaches a much larger size than Taylor's (1952a) account and previously reported collections suggested. Taylor (1952a) introduced the incorrect subsequent spelling *B. melanochloris* for this taxon, a course followed by most authors since that time. The correct original spelling *B. melanochlorus* is used here. The date of publication for this name is usually listed as 1878. However, a preprint dated August 15, 1877, was distributed before the publication in March 1878 of the volume containing this paper.

DISTRIBUTION: Lowlands and premontane slopes of the Atlantic versant and lowland Pacific southwest of Costa Rica (2–1,080 m).

Bufo valliceps Wiegmann, 1833

(plate 82; map 7.14)

DIAGNOSTICS: One of several moderate-sized to large toads having well-developed cranial crests and a distinct broad lateral dark stripe bordered above by light warts. *B. valliceps* differs most obviously from similar forms in having a much larger, oblong to triangular parotoid gland that is at least equal in area to the upper eyelid (fig. 7.32d).

DESCRIPTION: Adult males 48 to 76 mm in standard length, adult females 57 to 82 mm (to 130 mm in northern populations); upper surfaces of body wrinkled with scattered low warts often covered with spicules; flank with a strong continuous series of distinctly enlarged warts; upper surface of hind limbs with mixture of large and small conical warts; venter granular; head broader than long, broadly rounded in dorsal outline; eye large; skin of head co-ossified to upper skull; well-developed canthal, preorbital, supraorbital, postorbital, parietal, supratympanic, and pretympanic crests, labial crests weak, no suborbital crest; parotoid gland oblong to triangular, area equal to or greater than area of upper eyelid; height of external tympanum about three-fourths diameter of eye; adult males with paired vocal slits and a single internal subgular vocal sac, fully distensible and round when calling; limbs short; finger I longer than finger II; dark brown nuptial pad on first two fingers of adult males; well-developed single subarticular and supernumerary tubercles under fingers and toes; toes with some webbing; webbing formula: I 1–2 II 1–3 III 2–3½ IV 3½–2 V; inner metatarsal tubercle raised, ovoid, much larger than outer. Upper surfaces mostly uniform tan in adult males with scattered dark blotches on dorsum in females and juveniles; broad middorsal light stripe in all examples; dark lateral stripe from parotoid to groin, demarcated by white row of large warts above; dark stripe broader in juveniles and females; light dorsolateral stripes usually present; upper limb surface with dark transverse bars; venter dull yellowish; throat of adult male discolored, yellowish green in life; iris brown, pupil rimmed with yellow.

LARVAE (fig. 7.33): Small, total length 23 mm; body ovoid; mouth anteroventral, directed ventrally; nares dorsal; eyes dorsolateral; spiracle lateral, sinistral; vent tube medial; tail short; fins moderate; tip of caudal fin rounded; oral disk small, emarginate; beaks and 2/3 denticle rows present; broad gap in A2; one row of lateral labial papillae; body black, spiracle and venter clear; caudal musculature dark, dorsal margin with a series of dark saddles alternating with light, ventral half with some light spots; caudal fins heavily mottled with dark pigment.

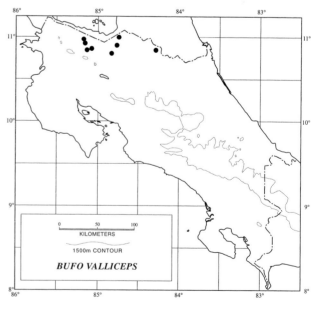

Map 7.14. Distribution of *Bufo valliceps*.

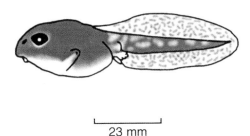

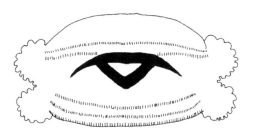

Figure 7.33. Tadpole of *Bufo valliceps*.

VOICE: A short trill lasting 2 to 6 seconds and repeated several times at intervals of 1 to 4 seconds. The dominant frequency is 1.4 to 2.1 kHz and the trill rate 30 to 40 per second (Porter 1966).

SIMILAR SPECIES: (1) *Bufo melanochlorus* has a small parotoid gland (less than area of upper eyelid), and the pretympanic crest is weak or absent (fig. 7.32c). (2) *Bufo luetkenii* has a small parotoid gland (less than area of upper eyelid) (fig. 7.32b) and usually no greatly enlarged lateral wart series. (3) *Bufo coccifer* has a light interorbital band and a narrow light pinstripe down the center of the back.

HABITAT: Lowland rainforests of extreme northern Atlantic slope Costa Rica.

BIOLOGY: In Costa Rica these uncommon nocturnal forest toads seem to be associated with streams. Reproduction usually takes place in shallow water in ponds or pools, although backwater areas of streams may be used. Amplexus is axillary. The black-and-white eggs are laid in paired strings in single or double files that sink toward the bottom. The gelatinous envelope consists of two layers, and the eggs are not separated by partitions. The eggs are small (1.2 mm in diameter) and closely packed. The diameter of the string is about 3 mm. Eggs of this species and their jelly coats are definitively toxic and unpalatable to some predators (Licht 1968). Hatching occurs in twenty-eight to forty-eight hours and metamorphosis in twenty-five to thirty days. Metamorphs are 7 to 12 mm in total length.

REMARKS: See remarks under *Bufo coccifer* for a detailed comparison of *B. valliceps* with related forms. *B. valliceps* only marginally reaches Costa Rica, where it is replaced by *B. melanochlorus* on the Atlantic versant. The two species may be sympatric in the extreme north-central portion of the country near the Río San Juan. It is also possible

that the allied *Bufo luetkenii* may overlap the range of one or both of these toads along the eastern foothills of the northernmost volcanoes in the Cordillera de Guanacaste. *B. valliceps* and *B. luetkenii* probably co-occur at some localities farther north in Central America, although the former is unknown over most of the primarily Pacific lowland range of the latter. The karyotype is 2N = 22 and consists of six pairs of large and four pairs of small metacentrics or submetacentrics and a pair of subtelocentric chromosomes with secondary constrictions on pairs 1, 3, and 8; NF = 42 (Bogart 1972).

DISTRIBUTION: Mostly in the lowlands, but also in premontane areas and marginally in the lower montane zone on the Atlantic versant from Louisiana to northern Costa Rica and on the Pacific slope around the Isthmus of Tehuantepec to central Guatemala (0–1,850); in Costa Rica known only from the lowlands in the upper portion of the Río San Juan drainage (40–495 m).

Genus *Crepidophryne* Cope, 1889

A single small, toadlike species is placed in this genus. It has cranial crests and parotoid glands, lacks an ear apparatus, has the fingers and toes enclosed in a web to form thick pads, and has the inner digit on both hands and feet greatly reduced.

Other characteristics definitive of the taxon, when taken in combination, are eight presacral vertebrae, atlas not fused to first trunk vertebra; pectoral girdle arciferal; epicoracoid cartilages not fused; phalangeal formula of hand 1-2-3-3 and foot 1-2-2-4-3; thumb greatly reduced; toe I virtually indistinguishable; distal phalanges not expanded; no tympanum, columella, Eustachian tubes, or ostia pharyngea; coccyx expanded; no anterior process on ceratohyal; tensor fasciae latae short; eggs and larvae not known.

Several Costa Rican *Bufo* resemble *Crepidophryne* in lacking a tympanum and have the hands and feet forming fleshy pads. However, they lack extensive finger webbing, the thumb is not reduced, and all five toes are distinct. The genus is known only from the Atlantic slope of the Cordillera de Talamanca in southeastern Costa Rica and adjacent western Panama and Volcán Miravalles in northwestern Costa Rica.

Crepidophryne epiotica (Cope, 1875)

(plate 83; map 7.15)

DIAGNOSTICS: A small to moderate-sized toad having well-developed cranial crests (fig. 7.32e), pointed warts on the upper thigh surface, and the hands and feet with the first digit reduced and enclosed in extensive webs to form fleshy pads. Toes I and II are not distinct from one another, and webbing nearly completely encloses them and toes III and V (fig. 7.22a).

DESCRIPTION: Adult male 23.4 mm in standard length, adult females 26 to 37.4 mm; dorsal surface relatively

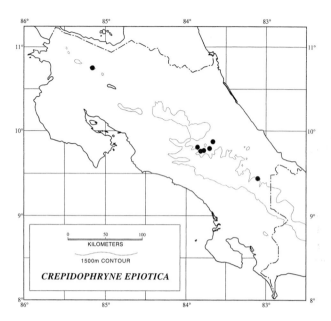

Map 7.15. Distribution of *Crepidophryne epiotica*.

smooth with widely spaced rounded warts; flanks smooth with widely spaced pointed warts, two distinct longitudinal lateral series of conical pointed warts; upper surface of thigh with numerous pointed warts; venter with spinous pustules; head broader than long, wedge-shaped in dorsal outline; nares lateral; eyes small, pupil horizontally elliptical; skin of upper head surfaces co-ossified to skull; well-developed canthal, supraorbital, postorbital, and supratympanic crests, parietal crest weak; small ovoid parotoid gland smooth, slightly smaller than area of upper eyelid; temporal plate well-developed, broad, and smooth; no vocal slits or sacs; no tympanum, columella, Eustachian tubes, or ostia pharyngea; limbs short; finger I shorter than finger II; hands and feet fleshy pads, undersides without definite tubercles; a light-colored nuptial pad in adult males on finger I; fingers webbed, webbing formula: I 0–1 II ½–3 III 3–1 IV; toes strongly webbed, webbing formula: I 0–0 II 0–½ III ¼–3 IV 3–1 V. Ground color of upper surfaces grayish to orangish, a dark lateral stripe from parotoid to groin; sometimes an orangish blotch just posterior to parotoids; no transverse dark limb bars; venter light brown with bluish cast.

LARVAE: None known.

VOICE: Unknown.

SIMILAR SPECIES: *B. holdridgei* and *B. fastidiosus* have five distinct toes and smooth warts on the upper thigh.

HABITAT: Undisturbed Premontane Rainforest.

BIOLOGY: This rare probably diurnal, fossorial toad has been collected only a few times. It is a denizen of deep forest and appears to be active aboveground for only a short time each year. Based on examination of a preserved female, Trueb (1971) reported that the eggs are unpigmented, small, and numerous.

REMARKS: The species was described as *Crepidius epioticus* by Cope (1875). Cope (1889) later realized that the name *Crepidius* was preoccupied by *Crepidius* Candeze, 1859 for a genus of beetles and proposed the replacement name *Crepidophryne*, a fact overlooked by Taylor (1952a) and Savage and Kluge (1961). A juvenile *Crepidophryne* (UCR 12877) was recently collected from the rim of Volcán Miravalles (2,000 m). This record extends the known range of the genus approximately 185 km to the northwest. The toadlet is 20 mm in standard length and differs most obviously from *C. epiotica* in the texture of the skin and degree of cranial crest development. It probably represents an undescribed species.

DISTRIBUTION: Humid premontane slopes of Atlantic versant southeastern Costa Rica and adjacent northeastern Panama (1,051–2,040 m).

Family Leptodactylidae Werner, 1896 (1838)

Another speciose family of frogs (fifty-four genera, 980 species) of tiny (12 mm) to large (220 mm) amphibians showing substantial diversity in morphology (e.g., froglike, toadlike, broad-headed, carnivorous, globular, fully webbed, aquatic, and big-disked arboreal forms) and occupying a wide variety of habitats. Most frogs in this family are terrestrial, semiaquatic, or arboreal, but some are completely aquatic, semifossorial, fossorial, or saxicolous.

Because of this diversity, it is not possible to characterize the family as a whole based on features of external morphology, although it is amply distinguished from other anurans by the following combination of attributes: eight usually holochordal (rarely stegochordal) presacral vertebrae; no ribs, sacrum with diapophyses usually cylindrical (rarely weakly dilated) and usually a bicondylar articulation to the coccyx; coccyx lacking transverse processes proximally (present in a few forms); pectoral girdle arciferal; a cartilaginous omosternum usually present, sometimes absent; a cartilaginous sternum present; no parahyoid bone; tibiale and fibulare fused distally and proximally; no intercalary cartilages; sartorius muscle distinct from semitendinosus and tendon of latter usually ventral to gracilis muscle (rarely penetrating it); tongue usually not highly protrusible; usually teeth on upper jaw; pupil usually horizontally elliptical (vertically elliptical in several forms); amplexus usually axillary (rarely inguinal); larval type A with beaks and denticles (absent in *Lepidobatrachus* of southern South America) and a single sinistral spiracle (paired and ventrolateral in *Lepidobatrachus*) (fig. 7.5b); 2N karyotype extremely variable, 18 to 36; polyploids 4N = 44 and 8N = 104 (King 1990).

The Leptodactylidae have an extensive distribution in the Americas ranging from Arizona and Texas, south to Chile and Argentina, including the West Indies. In practice, throughout this region frogs may be referred to this family by recognizing the characteristics of the genera occurring at a particular site or through a process of elimination. Generally speaking, if an anuran in hand cannot be placed in one or the other families known for a particular area, it is probably a leptodactylid.

In Costa Rica, frogs of this family are of four general types:

1. Froglike: smooth to rugose skin; long legs; no digital webbing; no digital disks: *Leptodactylus*
2. Froglike: smooth to rugose skin; long legs; toe webbing; complete digital disks: some *Eleutherodactylus*
3. Toadlike: warty skin or cranial crests; short limbs; no finger webs, no or slight toe webbing; no well-developed digital disks: *Physalaemus*, some *Eleutherodactylus*
4. Treefrog-like: smooth to rugose skin; long legs; no finger webs, no to some toe webbing; well-developed digital disks: many *Eleutherodactylus*

Leptodactylids of group 1 differ as follows from frogs of the families of Costa Rican frogs with which they might be misidentified (features for the indicated families in parentheses): no terminal scutes on upper surface of digits (present in Dendrobatidae); no webs (extensive toe webbing in Ranidae and *Atelopus* of the Bufonidae). Those of group 2 lack terminal scutes on the fingers and toes (present in Dendrobatidae) and have complete digital disk grooves (lacking in *Atelopus* and Ranidae) and a ventral disk (absent in Hylidae). Those of group 3 have a tarsal tubercle or complete digital grooves (absent in *Bufo*, *Crepidophryne*, and *Atelopus*, family Bufonidae). Frogs of group 4 differ as follows: no finger webbing, a definite ventral disk (usually finger webbing, no ventral disk, families Centrolenidae and Hylidae). In addition, representatives of the two treefrog families have intercalary cartilages between the last two phalanges of each digit; the terminal phalanges of hylids are claw shaped (no intercalary cartilages and most digits T-shaped in *Eleutherodactylus*).

The very narrow-headed burrowing frogs of the families Microhylidae and Rhinophrynidae each have a distinctive feature (a definite transverse fold across head and two elongate spadelike inner metatarsal tubercles, respectively) that separate them from any leptodactylid in the unlikely event that they are not immediately identified.

Members of the Leptodactylidae may be grouped into four subfamilies following Frost (1985), which is based on Lynch (1971, 1978). An alternative classification was proposed by Heyer (1975) but is not generally accepted (Duellman and Trueb 1986, 1994):

Ceratophryinae Tschudi, 1838 (two genera): large, broadheaded, carnivorous frogs with fanglike nonpedicellate teeth on upper jaw; sternum cartilaginous; a vertebral shield ossified to skin; transverse processes of anterior presacral vertebrae greatly expanded (northern South America to Argentina)

Telmatobiinae Fitzinger, 1843 (thirty-seven genera): a diverse group of frogs, usually having pedicellate teeth on upper jaw, sometimes jaws edentulous; sternum cartilaginous; no vertebral shield; transverse processes on anterior presacral vertebrae not greatly expanded (southern Texas, New Mexico, and Arizona and northwestern Mexico, throughout the Neotropics to southern Argentina and Chile, including the West Indies)

Hylodinae Günther, "1858b," 1859 (three genera): a small group of semiaquatic frogs having pointed pedicellate teeth on upper jaw; sternum cartilaginous (calcifying in most old adults); no vertebral shield; transverse processes on presacral vertebrae short (southeastern Brazil to Argentina)

Leptodactylinae Werner, 1896 (1838) (eleven genera): a diverse subfamily with blunt nonpedicellate teeth on upper jaw or sometimes jaw edentulous; sternum containing an osseous style or plate; no vertebral shield; transverse processes of anterior presacral vertebrae not expanded or shortened (Texas and Mexico to Argentina, including the West Indies)

As might be expected in such a large and ecologically diverse group, a considerable range of reproductive patterns occur in the family. These include 1–3, 5–8, 15–19, 21, 25, 31 (table 7.1).

Subfamily Leptodactylinae Werner, 1896 (1838)

Genus *Leptodactylus* Fitzinger, 1826

Members of this genus (sixty species) are quintessential frogs—narrow-waisted, long limbed, smooth-skinned, and narrow-headed—but they lack webs on the hands, and toe webbing is vestigial at most. The digits lack disks, although the toe tips may be somewhat expanded. All species have toe fringes at the time of metamorphosis, but these are lost in ontogeny by members of the *L. fuscus* and *L. pentadactylus* groups. The snout tends to be pointed, the nares are lateral, the eyes and tympanum are large, and the pupil of the eye is horizontal. Most species have smooth glandular concentrations that may form parotoid, ventrolateral, or lumbar glands or dorsolateral ridges, and these are variously present or absent in different species. These folds are often obscured in preserved material. Most *Leptodactylus* are moderate-sized frogs, but some are tiny, and several are large, with size range in adults 20 to 200 mm in standard length.

The genus *Leptodactylus* is distinguished from other genera in the subfamily by having a sternum with an elongate osseous style; xiphisternum expanded, symmetrical, shield shaped; maxillary and vomerine teeth present; vomers large, narrowly separated medially; maxillary arch complete; nasals large, narrowly separated medially; terminal phalanges knobbed; sacral diapophyses rounded; tongue emarginate, free posteriorly; fold of ventral disk not extending onto hind limbs; larvae with median vent, 2/3 denticle rows, and no papillae above the mouth. The presence of a large tympanum and the absence of toe webs, an enlarged tarsal tubercle, digital disks, and the absence of digital dermal scutes will distinguish *Leptodactylus* from all other sympatric Costa Rican anurans.

The genus has a wide geographic distribution over most of subtropical and tropical America exclusive of arid regions as follows: southern Texas and Sonora, Mexico, south through Central and South America to northwestern Peru west of the Andes and central Argentina and southern Bolivia east of the Andes; also northeastern Hispaniola, Puerto Rico, and other islands of the Puerto Rico Bank, several islands in the Lesser Antilles, and the islands of Trinidad, Tobago, Providencia, and San Andrés.

The genus *Leptodactylus* has been divided into four major groups by Heyer (1969, 1970, 1978, 1979, 1994), each represented by Costa Rican species (number of species in the republic in parentheses): *Leptodactylus melanonotus* group (one species), *L. ocellatus* group (one species), *L. pentadactylus* group (one species), and *L. fuscus* group (two species). The species *L. riveroi* (Colombia) is of uncertain affinities.

All members of this genus are prodigious leapers, exhibit axillary amplexus, build foam nests, and have nidicolous larvae. In addition, they produce substantial amounts of noxious skin secretions that contain several amines and peptides in studied species (Erspamer 1994). These form an effective antipredator defense, and you should be careful in handling these frogs because the secretions are particularly irritating to mucous membranes and may cause a skin rash.

KEY TO THE COSTA RICAN SPECIES OF THE GENUS *LEPTODACTYLUS*

1a. A distinct light longitudinal stripe along posterior surface of thigh (fig. 7.34); males with paired lateral external vocal sacs (fig. 7.12f); no horny spines on thumb 2
1b. Posterior surface of thigh marbled, without a distinct light longitudinal stripe; vocal sacs of males internal; adult males with horny spine or spines on thumb (fig. 7.37) 3
2a. Lower surface of tarsus and sole of foot smooth; vomerine teeth in almost transverse to arched series, usually entirely posterior to choanae; adult males 33–49 mm in standard length, adult females 32–50 mm . . . .*Leptodactylus poecilochilus* (p. 222)
2b. Lower surface of tarsus and sole of foot covered with prominent white tubercles (fig. 7.35a); vomerine teeth in transverse series entirely posterior to choanae (fig. 7.36c); adult males 24–36 mm in standard length, adult females 25–40 mm . . . . *Leptodactylus labialis* (p. 221)

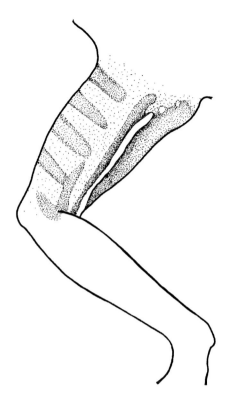

Figure 7.34. Leg of *Leptodactylus poecilochilus* showing white stripe on posterior thigh surface.

3a. Lower surface of tarsus tuberculate (fig. 7.35a); no lumbar glands; toes with prominent lateral fringes; males with two horny spines on thumb (fig. 7.37b); males never with horny spines on chest; standard length to 95 mm in adult males, to 120 mm in adult females 4
3b. Lower surface of tarsus smooth; distinct lumbar glands present; toes lacking prominent fringe, although sometimes with weak ridges; vomerine teeth in arched series lying between choanae, but extending posteriorly (fig. 7.36b); a single horny spine on thumb in adult males (fig. 7.37a); males with or without horny spines on chest; adult males 106–77 mm in standard length, adult females 118–85 mm . . . . *Leptodactylus pentadactylus* (p. 219)
4a. A pair of distinct dorsolateral glandular folds (fig. 7.36a); vomerine teeth at least partly extending between the choanae in arched series (fig. 7.36b); adult males with hypertrophied arms; tibia long, 48–54% of standard length; adult males 50–95 mm in standard length, adult females 70–120 mm . . . . *Leptodactylus bolivianus* (p. 217)
4b. No paired dorsolateral glandular folds; vomerine teeth entirely posterior to choanae, in transverse series (fig. 7.36c); male arm not noticeably hypertrophied; tibia short, 40–45% of standard length; adult males 30–

45 mm in standard length, adult females 35–55 mm
. . . .*Leptodactylus melanonotus* (p. 215)

Leptodactylus melanonotus Group

Thumb spines but no chest spines in adult males; no dorsolateral folds or lumbar glands; sphenethmoid complex not
calcified anteriorly; posterior squamosal spur overlapping
crista parotica; male humerus expanded; eggs black-and-
yellow, laid in foam nest floating on water (pattern 8, table
7.1); larvae aquatic. Non–Costa Rican species belonging to
this group are *Leptodactylus colombiensis* (Andes of Colombia), *L. dantasi* (Amazonian Brazil), *L. diedrus* (northwestern Amazonia), *L. griseigularis* (eastern Peru and Bolivia),
L. leptodactyloides (Amazon basin), *L. natalensis* (eastern
to southeastern Brazil), *L. nesiotus* (Trinidad), *L. pallidirostris* (northern South America), *L. pascoensis* (eastern Peru),
L. petersii (Amazon basin), *L. podicipinus* (southern Amazon basin to northern Argentina), *L. pustulatus* (central Brazil), *L. sabanensis* (southeastern Venezuela), *L. silvanimbus*
(Honduras), *L. validus* (Trinidad, Tobago, and southern
Lesser Antilles), and *L. wagneri* (Colombia to eastern Peru).

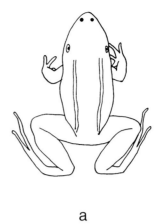

a

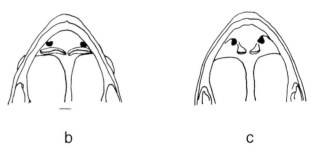

b   c

Figure 7.36. Diagnostic characteristics of Costa Rican anurans:
(a) paired dorsolateral glandular folds; roof of mouth: (b) vomerine
teeth in arched series; (c) vomerine teeth in linear series well posterior
to choanae.

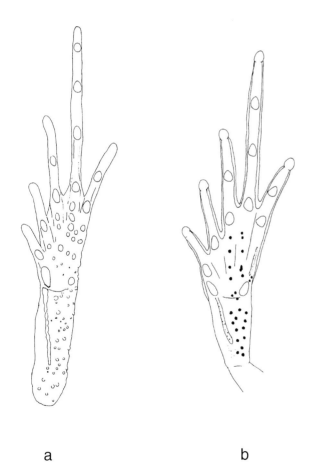

a   b

Figure 7.35. Tarsus and plantar surface of foot in Costa Rican frogs
of the genus *Leptodactylus*: (a) *L. labialis* with white tarsal and plantar tubercles; (b) *L. melanonotus* with black tarsal tubercles.

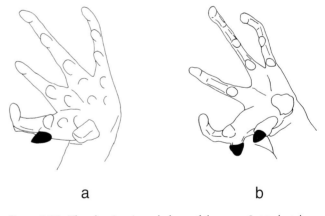

a   b

Figure 7.37. Thumb spines in male frogs of the genus *Leptodactylus*:
(a) single spine in *L. pentadactylus*; (b) two spines in *L. bolivianus*.

Leptodactylus melanonotus (Hallowell, "1860," 1861)

(plate 84; map 7.16)

DIAGNOSTICS: A rather robust, moderate-sized *Leptodactylus* having well-developed toe fringes and black-tipped
tubercles on the lower surface of the tarsus (fig. 7.35b); the

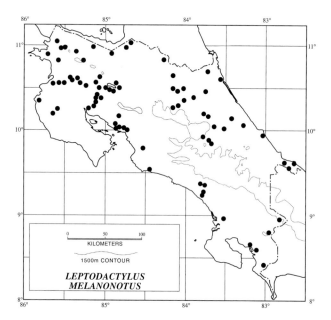

Map 7.16. Distribution of *Leptodactylus melanonotus.*

upper lip is barred or uniform, and brownish ventrolateral glands are usually present.

DESCRIPTION: Adult males 30 to 45 mm in standard length, adult females 35 to 55 mm; dorsum tuberculate; without dorsolateral folds and lumbar glands; venter smooth; head longer than broad; snout subelliptical to almost pointed in dorsal outline; eyes large; tympanum large, width one-half to two-thirds diameter of eye; paired elongate vocal slits and a single internal subgular vocal sac in adult males; vomerine teeth in short transverse series behind choanae; finger I longer than finger II; subarticular tubercles under fingers and toes round, globular; no supernumerary digital tubercles; definite fringe along toes; accessory palmar tubercles; two nuptial spines on thumb in adult males; forelimb not hypertrophied in males; thenar tubercle smaller than palmar; inner metatarsal tubercle elongate, outer much smaller; an inner tarsal fold present; lower surface of tarsus and usually sole of foot with black-tipped tubercles. Dorsum dark brown with obscure darker spots, blotches, bands, stripes, or uniform; interorbital light-outlined dark bar or triangle usually present; anterior margin of chin with light spots; upper surfaces of limbs with or without transverse dark bars; posterior thigh mottled; venter light to heavily suffused with dark pigment anteriorly; sometimes with small light spots; iris gold above, brown below.

LARVAE (fig. 7.38): Moderate-sized, total length 40 mm at stage 40; body ovoid; mouth anteroventral, directed ventrally; nares and eyes dorsal; spiracle lateral and sinistral; tail moderate; vent median; fins moderate; caudal fin rounded at tip; oral disk small, entire with beaks and 2/3 rows of denticles; A2 usually with narrow median gap above mouth; serrations on beaks small, blunt; double or triple row of

moderate papillae laterally, one or two rows ventrally, none above mouth; body and tail mostly black with light areas lateral to eye; scattered small light spots on tail; venter dark.

VOICE: The call is a single explosive short metallic "doink" repeated rapidly but irregularly at an average of 150 to 160 times a minute. Breaks between calls are 200 to 300 milliseconds. The note has two components, an initial untuned one, and the second that produces the *ck* sound, starting at 2.4 kHz and dropping smoothly to 2.0 kHz (Fouquette 1960; Heyer 1970; Straughan and Heyer 1976; Lee 1996).

SIMILAR SPECIES: (1) *Leptodactylus labialis* has the soles and underside of tarsal surfaces covered with white tubercles (fig. 7.35a) and a white longitudinal stripe on the posterior thigh surface (fig. 7.34). (2) *L. poecilochilus* has distinct dorsolateral folds, smooth soles and tarsal undersurfaces, and a light longitudinal stripe on the posterior thigh (fig. 7.34). (3) *L. bolivianus* is a much larger frog (adults more than 50 mm in standard length) and has paired dorsolateral folds (fig. 7.36a) and a white upper lip stripe; (4) All species of *Rana* have well-developed toe webbing.

HABITAT: Generally found near or in temporary ponds, roadside ditches, swamp margins, and marshes throughout the Lowland and Premontane Wet Forest zones.

BIOLOGY: This common nocturnal species is most abundant in the Lowland Dry Forest zone of northwestern Costa Rica. Although apparently active during most of the year in humid areas of the country, in the northwest foraging and mating are concentrated in the wet season (May to November). Both males and females congregate at the margins of shallow bodies of water. Breeding seems to occur primarily during the early heavy rains and sporadically thereafter to

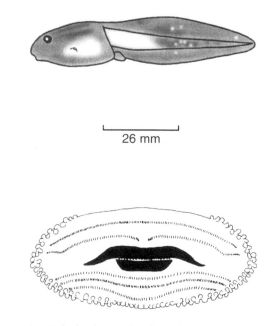

26 mm

Figure 7.38. Tadpole of *Leptodactylus melanonotus.*

August, although single males may call from time to time until the end of the wet season. Males call from near shore, often while hiding in vegetation mats, bases of grass clumps, or slight depressions and are extremely shy. Males are territorial. The "resident" male usually responds to an intruder by raising and lowering the posterior portion of the body by moving the hind legs to expose the ventrolateral glands. The nonresident male usually retreats or cowers. If he does not, the other male jumps on his back, and a tussle ensues with the defender biting his opponent or both biting each other. At this point the challenging male usually retreats (Brattstrom and Yarnell 1968). Amplexus takes place in the shallows, and the males' thumb spines aid in clasping. As the eggs are deposited, the male beats glandular secretions and the gelatinous envelope of the eggs into a foam by short, quick strokes of his hind legs. The foam is much the consistency and color of meringue and floats on the surface. The nest is usually deposited in shallow, often isolated, small pools or puddles or in the matted roots of aquatic vegetation. A given nest contains 1,000 to 2,000 small black-and-yellow eggs. Eggs are 1.0 to 1.5 mm in diameter. The nests are very sturdy, and having the eggs out of the water provides considerable defense against aquatic predators. Later in the season a heavy rain washes the advanced larvae out of the nest. In some cases the body of water may dry up without the foam's decomposing, and heavy rain may release the tadpoles as the breeding site refills. The dark-colored tadpoles of this species form dense schools containing hundreds of individuals. Small juveniles of this form having ventrolateral glands are about 16 mm in standard length.

REMARKS: The function of the ventrolateral glands, which are present in both males and females, remains unknown. In some individuals no glands are present, while in others they are restricted to the flanks. Usually glandular tissue is found on the prepectoral area, the flanks, and the chest, but in some examples the entire ventral surface is glandular. Similarly, glandular areas may be present on the lower portions of the thigh, the chin, and the undersides of the upper arms; the glands are usually brown to orange but in extralimital areas may be black or gray. Taylor's (1952a) description of *Leptodactylus maculilabris* from Costa Rica is based on examples of this species. Heyer (1968) provided a description of this form and its tadpole. The karyotype is 2N = 22, and there are two pairs each of large and four pairs of medium metacentric chromosomes, two pairs of medium and one pair of small submetacentrics, and one pair each of large and medium subtelocentrics: NF = 44 (Bogart 1974).

DISTRIBUTION: Lowlands and premontane zone from Sonora and Tamaulipas, Mexico, south through Central America to western Ecuador, but the range is disjunct in Costa Rica and Panama on the Pacific versant, with populations in northwestern Costa Rica, southwestern Costa Rica, and adjacent western Panama and in the Darién of Panama; in Costa Rica on both Atlantic and Pacific slopes

exclusive of the Pacific central region and the Meseta Central (6–1,440 m).

Leptodactylus ocellatus Group

Thumb but no chest spines in adult males; dorsolateral folds present but no lumbar glands; sphenethmoid complex not calcified anteriorly; posterior squamosal spur overlapping crista parotica; male humerus expanded; eggs black-and-yellow, laid in foam nest on top of water (pattern 8, table 7.1); larvae aquatic. Besides the Costa Rican representative of this group the following four species belong here: *Leptodactylus chaquensis* (southern Brazil to Argentina), *L. macrosternum* (Amazon basin to Paraguay), *L. ocellatus* (south America east of Andes), and *L. viridis* (eastern Brazil).

Leptodactylus bolivianus Boulenger, 1898a

(plate 85; map 7.17)

DIAGNOSTICS: A large dark-spotted frog having well-developed toe fringes, the lower surface of the tarsus with dark-tipped tubercles (fig. 7.35b), a pair of prominent dorsolateral folds (fig. 7.36a), white upper lip stripe, and the posterior thigh surface mottled to spotted with light and dark colors.

DESCRIPTION: Adult males 50 to 95 mm in standard length, adult females 70 to 120 mm; dorsum smooth; venter smooth; head longer than broad; snout subovoid to subelliptical in dorsal outline; eyes large; tympanum large, width one-half to two-thirds diameter of eye; vomerine teeth in arched series extending posteriorly from between choanae; paired elongate vocal slits and single internal subgular vocal sac in adult males; finger I longer than finger II; subarticular tubercles elongate, obtuse, and somewhat projecting; no supernumerary tubercles on digits; no accessory palmar

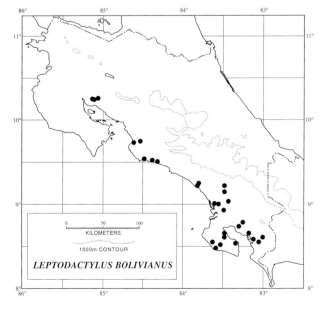

Map 7.17. Distribution of *Leptodactylus bolivianus*.

tubercles, but a few very small dark-tipped tubercles on sole of foot; two horny nuptial spines on thumb and forelimb hypertrophied in adult males; thenar tubercle smaller than usually bifid palmar tubercle; inner metatarsal tubercle elongate, somewhat raised, outer small, round; an inner tarsal fold; lower surface of tarsus with dark-tipped tubercles; dorsolateral folds dark, interdorsal fold field lighter, often reddish brown, usually with larger dark spots; flanks usually with light-outlined dark spots; dark canthal stripe extending posteriorly through eye and along supratympanic fold; anterior margin of chin suffused with melanophores; upper surface of arms spotted, legs with dark oblong spots on upper surface forming bars; posterior thigh surface mottled with dark and light colors; venter pale cream, immaculate, or suffused with brown pigment: iris gold.

LARVAE (fig. 7.39): Moderate-sized, total length 32 mm at stage 34; body ovoid; mouth anteroventral, directed ventrally; nares and eyes dorsal; spiracle lateral and sinistral; vent tube median; tail long; fins moderate; caudal fin rounded at tip; oral disk moderate, entire, with beaks and 2/3 rows of denticles present; A2 complete above mouth; beak serrations small, blunt; labial papillae absent above mouth, mostly in double rows; body with small light flecks; brown dorsally, venter lighter, slightly suffused with melanophores; tail mostly dark, with scattered light flecks; ventral fin lightest near vent tube.

VOICE: A somewhat melodic short whooping call, rising smoothly from a frequency of 0.4 to 1.3 kHz, lasting about 70 to 100 milliseconds and repeated 112 to 140 times a minute (Fouquette 1960; Straughan and Heyer 1976).

SIMILAR SPECIES: (1) The several species of *Rana* have well-developed toe webs. (2) *Leptodactylus pentadactylus* has distinct lumbar glands and the upper lip spotted or barred. (3) *L. poecilochilus* lacks toe fringes and has smooth

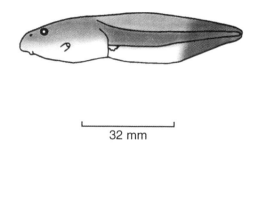

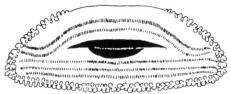

Figure 7.39. Tadpole of *Leptodactylus bolivianus*.

lower tarsal surfaces and a longitudinal white stripe on the posterior thigh surface (fig. 7.34). (4) The much smaller *L. melanonotus* (to 55 mm in standard length) lacks dorsolateral folds and has a barred or uniform upper lip.

HABITAT: Found near shallow bodies of water, including temporary ponds and roadside ditches in the Pacific Lowland Dry, Moist, and Wet Forest zones.

BIOLOGY: This relatively common large, principally nocturnal carnivorous frog feeds primarily on arthropods but eats almost any small animal prey. Adults are often found in or near burrow retreats, but there is no evidence that the frogs dig the burrows. Courtship and mating take place throughout the wet season (May to November). The species usually is associated with permanent bodies of water, but males call from temporary ponds and marshes as well, and reproduction may also occur there. Males call from the bank near the water or more usually while sitting in the shallows during the night and early morning hours. Amplexus takes place in shallow water, and the males use thumb spines to maintain their position. The female deposits 1,000 to 2,000 black-and-yellowish eggs in a foam nest one liter in volume. The nest floats on water but is usually hidden in vegetation, under debris, or in the mouth of a burrow. The white foam is produced as described in the accounts of *L. melanonotus* and *L. pentadactylus*. Subsequent heavy rains wash the tadpoles out of the nest into the water, where development is completed. The dark-colored tadpoles form dense schools containing hundreds of individuals. Small juveniles are about 20 mm in standard length. Males of this species have been observed to call from the foam nests and apparently defend the nest site against other males (Sexton 1962). Females seemingly remain close to their foam nests and tend the school of tadpoles once they hatch (Wells and Bard 1988; Vaira 1997), communicating with them through a stereotypic up-and-down pumping of the posterior part of the body.

REMARKS: This taxon is often referred to as *Leptodactylus insularum* (Barbour 1906) because there is doubt whether the frogs forming the basis for the original description of *L. bolivianus* are conspecific with the Central American and northern South American population. The name *L. insularum* is based on specimens (syntypes: MCZ 2424, 2444, 6901–2) from islands in the Bay of Panama, while the lectotype (MSNG 28875A) of *L. bolivianus* is from Bolivia. Heyer (1968, 1974) was unable to differentiate between the two nominal forms, and his conclusion is followed here. Heyer (1968) provided a description of this form and its tadpoles. The occurrence of this species on several offshore islands may be due to introductions, because they are a human food item in many areas of tropical America. The karyotype is 2N = 22. The size and type are as follows: three pairs of large, four pairs of medium, and one pair of small metacentric chromosomes; one pair of large, one pair of medium submetacentrics; one pair of medium subtelocentrics; NF = 44 (Bogart 1974).

DISTRIBUTION: Lowlands of Pacific slope Costa Rica, south through northern South America and east of the Andes to Bolivia, including Trinidad; also on Caribbean slope of Nicaragua and Providencia and San Andrés Islands (1–404 m).

Leptodactylus pentadactylus Group

Thumb spines and usually chest spines present in adult males; dorsolateral folds usually present; lumbar glands present; sphenethmoid complex calcified anteriorly; squamosal spur overlapping crista parotica; male humerus expanded; black-and-yellow eggs laid in foam on top of water (pattern 8, table 7.1) or in a burrow near water (pattern 10, table 7.1); larvae aquatic. Other species belonging to this group beside the Costa Rican representative are *Leptodactylus fallax* (Lesser Antilles), *L. flavopictus* (southeastern Brazil), *L. knudseni* (Amazon basin), *L. labyrinthicus* (Paraguay to Venezuela), *L. laticeps* (Paraguay, Bolivia, and Argentina), *L. lithonaetes* (upper Amazonian Colombia and Brazil), *L. myersi* (Guianas and northern Brazil), *L. rhodomystax* (Amazon basin), *L. rhodonotus* (Amazonian Peru and Bolivia), *L. rugosus* (eastern Colombia to the Guianas), *L. stenodema* (Amazon basin), and *L. syphax* (central to northeastern Brazil).

Leptodactylus pentadactylus (Laurenti, 1768)

Smoki Jungle Frog, Rana Ternero

(plate 86; map 7.18)

DIAGNOSTICS: Adults of this very large frog are unlikely to be mistaken for any other species because of size alone. The presence of paired dorsolateral folds (fig. 7.36a), distinct paired lumbar glands, and the barred or spotted upper lip will confirm most initial identifications. Adult males of this species in Costa Rica have spines on the chest. Juveniles are usually more brightly colored than adults and have toe ridges that are lost in older examples.

DESCRIPTION: Adult males 106 to 177 mm in standard length, adult females 118 to 185 mm; dorsum smooth; definite inguinal glands, entire flank often glandular; snout rounded to nearly semicircular from above; eyes large; tympanum large, width one-half to two-thirds diameter of eye; vomerine teeth in arched series extending from between choanae posteriorly; paired elongate vocal slits and single internal subgular vocal sac in adult males; finger I longer than finger II; subarticular tubercles round, globular, projecting; no supernumerary tubercles on digits; no accessory or plantar tubercles; a single nuptial thumb spine usually present and forelimb greatly hypertrophied in adult males; thenar tubercle ovoid, somewhat flattened, about half the size of more or less bifid palmar tubercle; inner metatarsal tubercle elongate, somewhat raised, outer small, round; an inner tarsal fold; lower surface of tarsus smooth; dorsum of adults uniform gray to reddish brown or spotted or barred with darker color; chin margin with light spots, often indistinct in large adults; upper surfaces of limbs uniform, mottled, or distinctly barred or striped with dark markings; posterior surface of thighs mottled with dark and light; venter dark, mottled with white to yellow or with distinct light spots; juveniles reddish, dark lip markings, thigh mottling, and overall coloration much more brilliant than in adults; iris gold silver above, dark brown below; eyeshine red.

LARVAE (fig. 7.40): Giant tadpoles, total length 83 mm at stage 40; body ovoid; mouth nearly terminal; nares and eyes dorsal; spiracle lateral and sinistral; vent medial; tail

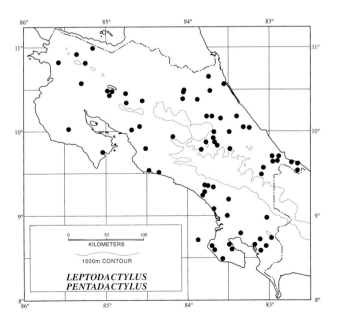

Map 7.18. Distribution of *Leptodactylus pentadactylus*.

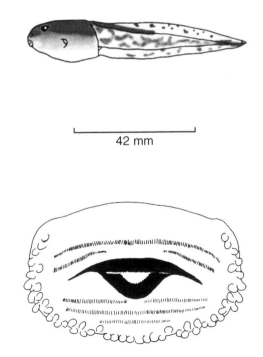

Figure 7.40. Tadpole of *Leptodactylus pentadactylus*.

long; fins low; caudal fin pointed; oral disk moderate, entire with beaks and 2/3 rows of denticles; a broad gap in A2 above mouth; beak serrations small and blunt to moderate and pointed; labial papillae variable, usually in one row under mouth, two rows laterally, always absent above mouth; body brown above, lighter below, not marked with dark pigment.

VOICE: The advertisement call is a very loud "whorup," repeated every 5 to 10 seconds, and can be heard from a distance of a kilometer or slightly more in open country. Call duration is 200 to 330 milliseconds with an irregular repetition rate up to 35 per minute. The dominant frequency is 0.2 kHz at the beginning, rising to 0.6 kHz at the end (Fouquette 1960; Heyer 1979; Straughan and Heyer 1976).

SIMILAR SPECIES: (1) *Leptodactylus bolivianus* lacks lumbar glands and has a light upper lip stripe and dark-tipped tubercles on the lower tarsal surface (fig. 7.35b); (2) The species of *Rana* lack lumbar glands and have toe webbing.

HABITAT: All lowland forests and marginally in Premontane Wet Forest and Rainforest. Found frequently at some distance from bodies of water or stream courses.

BIOLOGY: Adults of this rather common species are nocturnal and retreat into burrows, under logs, or among the interstices of tree roots during the day, but they may also hide under *campesino* houses. Juveniles may be found active during the day on the leaf litter in dense forests. These large frogs are voracious carnivores eating a wide variety of animal prey, including small or nesting birds, bats, snakes, lizards, and frogs and many larger arthropods (plate 87). Males are heard calling throughout the rainy season, and the call seems to function in territorial spacing as well as in courtship. Males often sit and call at the mouth of their burrows or other retreats and quickly retreat to safety if approached. Mating occurs over the course of the wet season (May to November). Breeding males call from the margins of ponds, marshes, swamps, or sometimes river backwaters, and mating usually takes place in these situations. However, it may be that some individuals use their burrows or other retreats, some distance from the water, as nest sites. Amplexus is axillary, and the thumb and chest spines and hypertrophied forearms of the males are used to clasp the females very tightly. A huge foaming mass (2 to 7 l in volume) is produced by backward-and-forward movements of the males' hind limbs to mix air with the egg gelatin and possible mucous secretions from the body (Heyer and Rand 1977). Nests are usually placed in dry depressions in seasonally flooded areas or in cavities in the ground near ephemeral bodies of water. About 1,000 light gray eggs are typically found in a single nest (Muedeking and Heyer 1976) and were about 2.5 mm in diameter at an early stage of development. Subsequent heavy rains wash the tadpoles out of the nests, usually into nearby bodies of water, but some may remain in flooded burrows or potholes. Experiments by Valerio (1971) showed that tadpoles of this species could live up to

156 hours out of water, so even if a site dries up they may survive until the next rain. The tadpoles are large, carnivorous (ecomorph 8, table 7.2), and cannibalistic (Heyer, McDiarmid, and Weigmann 1975). Initially they feed on the foam, but once washed into a stream or other body of water, they eat tadpoles and eggs of their own and other frog species, although they can grow and metamorphose while eating plant material (Vinton 1951). The time from hatching to metamorphosis is quite short, about twenty-eight days. Small juveniles are about 22 mm in standard length. In addition to the usual aquatic predators that eat frog eggs and their larvae, the eggs of this and other species having foam nests are probably preyed upon by spiders and the larvae of ephydrid flies (Villa, McDiarmid, and Gallardo 1982). Transformed *L. pentadactylus* exhibit several unusual antipredator defenses. When handled, they secrete immense amounts of mucus that not only makes them extremely difficult to hold but contains noxious compounds. The secretions induce heavy sneezing, swelling of the eyes, and irritation of mucous membranes by diffusion through the air, without direct contact. You need only be in the same room when someone else handles these animals to show the results. Direct contact with the secretion often produces a skin rash and stings any nicks or cuts on the hands, and the symptoms above are exacerbated if you rub your eyes, nose, or lips after touching the frog. A vertebrate predator that bites one of these frogs might well release it. In addition, other frogs placed in a collecting bag that previously contained *L. pentadactylus* and has not been thoroughly washed will be killed by the toxic residue. Villa (1969a) has described in detail the defensive behavior of this species, which may take full advantage of the threat of noxious secretions. When approached, individuals inflate their bodies and elevate themselves with all four limbs. This particularly exposes the glandular back and groin as the longer hind limbs lift the posterior part of the body above the head. The body is now raised and lowered with the longitudinal axis centered on the potential predator. This behavior may effectively repulse an enemy that has previously been exposed to the irritating amines and perhaps toxic peptides in the skin. When caught by a collector (or another predator), frogs of this species usually emit an ear-shattering high-pitched scream. Presumably this distress call warns other frogs of danger in the vicinity. Scott (1983c) had another interpretation, since he noted that caimans *(Caiman crocodilus)* were attracted by this call. He proposed that the call attracts large predators that in turn may attack the frog's assailant.

REMARKS: Metamorphs have fringelike ridges on the toes that become reduced through ontogeny. A low ridge may be present on the toes in examples up to 100 mm in standard length. Some authors, including Taylor (1952a), recognize subspecies within this form. In the most recent revision of the group, Heyer (1979) finds no basis for such distinctions. Heyer (1968) provided a description of Costa Rican material, including the larva. The karyotype is 2N = 22. The size and types are as follows: two large, four medium, and one small pair of metacentric chromosomes; two large and one medium pairs of submetacentrics; one medium pair of sub-telocentrics; NF = 44 (Bogart 1974).

DISTRIBUTION: Lowland and premontane slopes from northern Honduras on the Atlantic slope and northern Nicaragua on the Pacific south to Colombia and Ecuador; discontinuous with populations east of the Andes from Colombia and the Guianas through Amazonia (5–1,200 m).

Leptodactylus fuscus Group

No thumb or chest spines in males; dorsolateral folds usually present; lumbar glands usually absent; sphenethmoid complex calcified anteriorly; posterior squamosal spur not overlapping crista parotica; male humerus not expanded; eggs cream, laid in foam nest in incubating chamber evacuated by the male (pattern 14, table 7.1); larvae aquatic.

Extralimital species included in this group are *Leptodactylus albilabris* (Puerto Rico and Virgin Islands), *L. bufonius* (Bolivia to Argentina), *L. camaquara* (central Brazil), *L. cunicularius* (central Brazil), *L. didymus,* and *L. elenae* (central Brazil to Argentina), *L. dominicensis* (Dominican Republic), *L. furnarius* (central and southeastern Brazil), *L. fuscus* (Paraguay and Argentina), *L. geminus* (southern Brazil and Argentina), *L. gracilis* (southern Brazil to Argentina), *L. jolyi* (central and southeastern Brazil), *L. labrosus* (western Ecuador and Peru), *L. latinasus* (southern Brazil to Argentina), *L. longirostris* (Amazon basin), *L. marambaiae* (southeastern Brazil), *L. mystaceus* (Amazon basin), *L. mystacinus* (southern Brazil to Argentina), *L. notoaktites* (southeastern Brazil), *L. plaumanni* (southeastern Brazil), *L. spixii* (eastern Brazil), *L. tapiti* (central Brazil), *L. troglodytes* (northeastern Brazil), and *L. ventrimaculatus* (western Ecuador and Peru).

Leptodactylus labialis (Cope, 1877)

(plate 88; map 7.19)

DIAGNOSTICS: A moderate-sized species with two pairs of dorsolateral folds, a more or less evident white labial stripe, and tuberculate undersurfaces of the tarsi and feet (fig. 7.35a).

DESCRIPTION: Adult males 24 to 36 mm in standard length, adult females 25 to 40 mm; dorsum and venter smooth; head longer than broad; snout subelliptical to pointed from above; eyes large; tympanum large, width about three-fifths diameter of eye; vomerine teeth in short, transverse series behind choanae; paired elongate vocal slits and external lateral paired vocal sacs in adult males; finger I longer than finger II; subarticular tubercles round, pungent, projecting; no supernumerary digital tubercles; no nuptial pads or spines; forelimb not hypertrophied in adult males; thenar tubercle elongate, much smaller than round palmar tubercle; accessory palmar and plantar tubercles present; inner metatarsal tubercles small, elongate, outer smaller, conical; inner metatarsal fold present, usually ru-

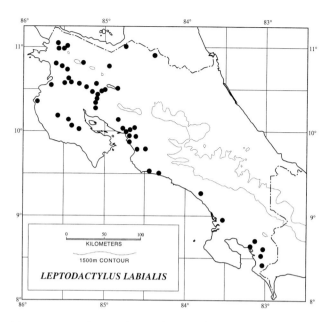

Map 7.19. Distribution of *Leptodactylus labialis.*

gose; lower surface of tarsus covered with many tubercles. Dorsum brown to grayish with large, often light-outlined dark blotches; chin banded with dark pigment; vocal sacs black; upper surfaces of limbs with dark bars or stripes; posterior thigh surface with a distinct white longitudinal stripe (fig. 7.34); venter white to yellowish, immaculate; undersides of limbs white to yellowish with or without scattered dark pigment; iris gold.

LARVAE (fig. 7.41): Moderately large, total length 40 mm at stage 40; body ovoid; mouth anteroventral, directed ventrally; nares and eyes dorsal; spiracle lateral and sinistral;

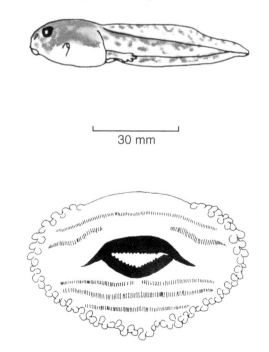

30 mm

Figure 7.41. Tadpole of *Leptodactylus labialis.*

vent tube medial; tail long; fins moderate; caudal fin tip rounded; oral disk small, entire with beaks and 2/3 denticle rows; very broad gap in A2 above mouth; beak serrations moderate, pointed; labial papillae usually in one row laterally and two rows below mouth, absent above mouth; body dark above, light below; body without pattern; tail blotched.

VOICE: The call is a series of "throw-up, throw-up" about 200 milliseconds in length with a rising inflection at the end, repeated continuously 72 to 114 times a minute, shifting from 1.0 to 2.2 kHz (0.6 to 1.2 in northern examples) in dominant frequencies (Fouquette 1960; Heyer 1971; Straughan and Heyer 1976; Lee 1996).

SIMILAR SPECIES: (1) Other *Leptodactylus* lack white tubercles on the sole and underside of the tarsus. (2) The species of *Rana* have toe webs.

HABITAT: Lowland forest zones of the Pacific versant and marginally on the northern Atlantic plains. Found near marshes, ponds, and any temporary lentic body of water during rainy periods. Most common in open and disturbed sites.

BIOLOGY: This very common species is primarily nocturnal, but males may call sporadically during the day. It feeds mostly on arthropods. Courtship and mating take place after heavy rains throughout the wet season (April/May to November). The males construct burrows and call from these sites just below the ground surface, making it extremely difficult for the field biologist to locate them. Females are attracted to a calling male, and breeding takes place in his burrow. A foam nest is produced as in other *Leptodactylus*. During egg fertilization the male whips the egg gelatin and his mucous body secretions into a meringuelike froth. From 72 to 250 eggs 1.6 to 2.5 mm in diameter are deposited per nest. After the eggs hatch and development proceeds, the larvae are flooded out of the burrow and nest directly by heavy rains or the advancing water from nearby ponds and finish development as free-swimming tadpoles. Metamorphosis occurs in about fifteen days. The smallest available juvenile is 13 mm in standard length.

REMARKS: Just metamorphosed froglets of this taxon have definite lateral toe ridges; these have disappeared in individuals over 30 mm in standard length. Representatives of this species have sometimes been referred to incorrectly as *Leptodactylus albilabris*. That form is restricted to Puerto Rico and the Virgin Islands and does not occur elsewhere. Heyer (1978) erroneously applied the name *Cystignathus (= Leptodactylus) fragilis* Brocchi, 1877 to this form, based on its supposed priority, and a number of subsequent workers followed his conclusion. Recently Dubois and Heyer (1992) demonstrated that *Cystignathus labialis* Cope, 1877 has priority over *C. fragilis* and restored the former usage. Although there is a general hiatus in the Atlantic versant range of *L. labialis* from Honduras to central Panama, as an exception it occurs in the upper Río San Juan drainage in northern Costa Rica. Heyer (1968) presented a review of this frog in Costa Rica. The karyotype is 2N = 22. It is com-

posed of one large pair, five medium pairs, and one small pair of metacentric chromosomes; one pair of large submetacentrics; and one pair of subtelocentrics: NF = 44 (Bogart 1974).

DISTRIBUTION: Lowlands on the Atlantic slope from extreme southern Texas to Honduras, disjunct in Costa Rica and from central Panama to northeastern Venezuela and from Colima, Mexico, to northern Colombia on the Pacific side (0–520 m).

Leptodactylus poecilochilus (Cope, 1862b)

(plates 89–90; map 7.20)

DIAGNOSTICS: A moderate-sized species having one to three pairs of dorsolateral folds, smooth soles and tarsi, and a longitudinal white stripe across the posterior thigh surface (fig. 7.34).

DESCRIPTION: Adult males 33 to 49 mm in standard length, adult females 32 to 50 mm; dorsum and venter smooth; head longer than broad; snout subovoid to elliptical in outline from above; eyes large; tympanum large, width one-half to two-thirds diameter of eye; vomerine teeth in strongly to weakly arched series, usually entirely behind choanae; paired elongate vocal slits and paired external lateral vocal sacs in adult males; finger I longer than finger II; subarticular tubercles round, obtuse, projecting; no supernumerary digital tubercles on hands and feet; no nuptial pads or spines; forelimb not hypertrophied in adult males; thenar tubercle elongate, subequal in size to palmar tubercle; accessory palmar tubercles present; inner metatarsal tubercle small, elongate, outer tiny; low, smooth inner metatarsal fold present. Dorsum brownish gray, pattern spotted or striped; dorsolateral folds same color, cream, or dark

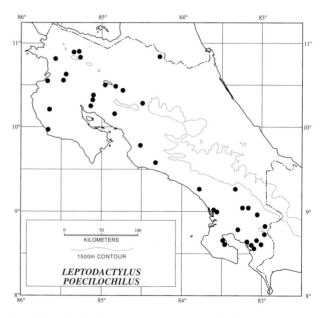

Map 7.20. Distribution of *Leptodactylus poecilochilus*.

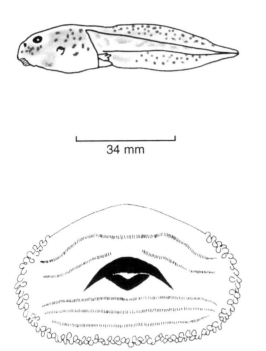

Figure 7.42. Tadpole of *Leptodactylus poecilochilus*.

brown; broad middorsal light stripe often present; dark canthal stripe and dark supratympanic fold present; upper lip banded or spotted with dark; chin sometimes bordered with dark pigment, usually immaculate; vocal sacs black; upper surface of legs with transverse dark bars; venter white; iris gold.

LARVAE (fig. 7.42): Moderate-sized, total length 36 mm at stage 40; body ovoid; mouth anteroventral, directed ventrally; nares and eyes dorsal; spiracle lateral and sinistral; vent tube medial; tail moderate; fins moderate; caudal fin pointed; oral disk small, entire with beaks and 2/3 denticle rows; broad gap in A2 above mouth; beak serrations small, usually blunt; labial papillae mostly in two rows, absent above mouth; body dark above, lighter below, blotched with dark pigment; tail blotched with dark spotting.

VOICE: A short whoop with the pitch rising rapidly, lasting 50 to 60 milliseconds and repeated about 100 times a minute. The dominant frequency shifts through the course of the call from 0.35 to 0.55 kHz (Fouquette 1960; Straughan and Heyer 1976).

SIMILAR SPECIES: (1) *Leptodactylus labialis* has white tubercles on the sole and tarsal undersurface (fig. 7.35a). (2) *Leptodactylus bolivianus* has toe fringes and dark-tipped tarsal tubercles (fig. 7.35b) and lacks the light thigh stripe; (3) *Leptodactylus melanonotus* has black-tipped tubercles on the lower tarsal surface (fig. 7.35b) and toe fringes. (4) The species of *Rana* have toe webs.

HABITAT: Marshes, swamps, temporary ponds, and ditches throughout the Pacific lowlands in Dry, Moist, and Wet Forest habitats and marginally in Lowland Moist Forest on the Atlantic versant near Laguna de Arenal and Pre-montane Rainforest near the southwestern Panama frontier. The preference of this species is for open and disturbed situations.

BIOLOGY: A relatively common nocturnal, basically arthropodivorous frog found in association with lentic waters. The species is an explosive early wet season breeder (May to June, depending on locale and year). Otherwise the life history and habits are similar to those of *Leptodactylus labialis*, but they have not been studied in detail. Small juveniles are 15 to 16 mm in standard length. Experiments by Valerio (1971) showed that tadpoles of this species could survive up to 140 hours out of water.

REMARKS: Heyer (1968) demonstrated that, contrary to Taylor (1952a), the nominal species *Leptodactylus quadrivittatus* (Cope 1893), based on a striped individual (plate 90) from Costa Rica (holotype now lost; Puntarenas: Buenos Aires), is a synonym of *L. poecilochilus* (Cope 1862b). The holotype (USNM 4347; Colombia: Antioquia: Turbo) of the latter is spotted, but the occurrence of intermediate patterned examples and the presence of the extremes and intermediates in metamorphs collected at the same time and place are conclusive evidence of conspecificity. Taylor (1952a) also used the name *Leptodactylus maculilabris* (Boulenger 1896b), another synonym of *L. poecilochilus*, for several series of *L. melanonotus* from Atlantic slope southeastern Costa Rica. Metamorphosing froglets of this species have low lateral toe ridges that are lost in individuals larger than 30 mm in standard length. Heyer (1968) provided a detailed description of this species and its tadpole.

DISTRIBUTION: Lowlands of the Pacific slope from northwestern Costa Rica to northern Colombia and then eastward along the Atlantic slope to northwestern Venezuela; Atlantic slope records also include the region near Laguna de Arenal and the San Carlos area in Costa Rica and the Canal Zone area of central Panama (3–1,150 m).

Genus *Physalaemus* Fitzinger, 1826

A group (forty species) of small to moderate-sized rather toadlike anurans, frequently with one or more pairs of enlarged glandular concentrations in the parotoid, flank, lumbar, or inguinal region. The genus is a diverse one in terms of morphology, but all species have lateral nares and a horizontally elliptical pupil and lack finger and toe disks and webbing.

The genus *Physalaemus* differs from other members of its subfamily in the following combination of characters: sternum with a broad osseous style; xiphisternum broad but irregular in shape; maxillary usually edentulous; vomers small, widely separated medially, usually edentulous; maxillary arch complete; nasals large, in broad contact medially; terminal phalanges knobbed; sacral diapophyses slightly dilated; tongue narrow, entire, and free posteriorly; no ventral disk; larvae with a dextral vent, 2/3 denticle rows, and no papillae above mouth.

The single Costa Rican species placed in this genus, *Physalaemus pustulosus*, is toadlike in appearance, having a tuberculate dorsum and parotoid and flank glands, but it differs from other superficially similar forms in the country in lacking toe webs and a tympanum and in having a conspicuous inner tarsal tubercle.

Physalaemus tend to be denizens of lowland open areas, grasslands, savannas, semiarid woodlands, and secondary growth, but in some places they occur in rainforest habitats. The range includes most of mainland subtropical and tropical America from eastern and southern Mexico to central Argentina, exclusive of the Pacific slope south of Ecuador.

Some frogs now placed in this genus were formerly referred to the nominal genera *Engystomops* or *Eupemphix*. Lynch (1970) demonstrated that these taxa were congeneric with *Physalaemus* and proposed that the four species groups could be recognized within the genus. Only the *Physalaemus pustulosus* group is represented in Costa Rica. Extralimital forms referred to this group include *Physalaemus coloradorum* (western Ecuador), *P. petersi* (Amazon basin), and *P. pustulatus* (western Ecuador and Peru).

Adult males of all species have nuptial thumb pads and external subgular vocal sacs. All members of the genus use axillary amplexus, build foam nests that float on bodies of water, and have nidicolous larvae (pattern 8, table 7.1).

Physalaemus pustulosus (Cope, 1864a)

Tungara Frog, Sapito Tungara

(plate 91; map 7.21)

DIAGNOSTICS: A small to moderate-sized anuran unmistakable because of its toadlike appearance, definite parotoid glands, and distinct outer tarsal tubercle (fig. 7.8).

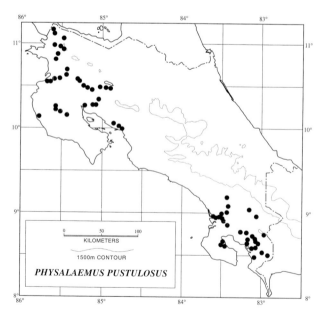

Map 7.21. Distribution of *Physalaemus pustulosus*.

DESCRIPTION: Adult males 25 to 34 mm in standard length, adult females 26 to 35 mm; body squat, limbs short; upper surfaces with wartlike glandular concentrations covered with granules and pustules, one or more of them spinous; parotoid glands well developed; head about as wide as long; snout short, somewhat pointed from above; eyes large; tympanum small, indistinct in females, hidden in males; no vomerine teeth; paired elongate vocal slits and single external subgular vocal sac in adult males; subarticular tubercles ovoid, conical, projecting, proximal ones especially strong; supernumerary digital tubercles present; brownish nuptial pads on thumb of adult males; thenar and palmar tubercles subequal, ovoid; accessory palmar and plantar tubercles present; no toe webs; metatarsal tubercle raised, pointed, inner much larger than outer; undersurface of tarsus covered with many tubercles; no inner tarsal fold or large tubercle; a conspicuous outer tarsal tubercle and projecting tarsal fold. Ground color brownish to gray above, nearly uniform or with dark brown blotches, a narrow middorsal cream to orange stripe or paired dorsolateral light stripes or a combination of these; warts on anterior part of body usually reddish, orange or yellow; often an orange spot between shoulders; male vocal sac black with a medium light line when inflated; chest dark; upper surface of forearm with one transverse dark bar, upper surface of leg with several transverse dark bars; venter dirty white, usually with large black spots posteriorly; iris tan or pale brown.

LARVAE (fig. 7.43): Small, total length 9.5 mm at stage 26, 20 mm at stage 37; body ovoid; mouth anteroventral; nares and eyes dorsal; spiracle lateral, sinistral; vent tube dextral; tail moderate; fins low; caudal fin tip rounded; oral disk small, emarginate, with beaks and 2/3 denticle rows; broad gap in A2 above mouth; beak serrations small, blunt; labial papillae in single row below and lateral to mouth, no papillae above mouth; body dark above, light below; tail musculature same color as body; tail fins mostly clear.

VOICE: The advertisement calls are complex. The simplest is a whine about 250 to 500 milliseconds long, falling from a frequency of 0.9 to 0.4 kHz and repeated about every 2 seconds. A short harsh "chuck" at about 2 kHz may be added to the whine if a calling male hears another male call, or a second similar "chuck" may be added. The release call is a series of clucks. A "mew" call is used when a male approaches another male too closely. A "glug" sound is emitted when a male with vocal sac inflated is startled and dives underwater (Rand 1983).

SIMILAR SPECIES: (1) Occasionally mistaken for a member of the genus *Bufo*, all of which have toe webs and lack the outer tarsal tubercle. (2) *Gastrophryne pictiventris* has a smooth dorsum and a transverse dermal fold just behind the eyes (fig. 7.10c).

HABITAT: Open and disturbed areas throughout the Lowland Dry, Moist, and Wet Forest and marginally in the Premontane Wet Forest of the Pacific versant. Commonly

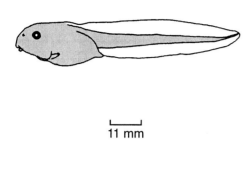

11 mm

Figure 7.43. Tadpole of *Physalaemus pustulosus*.

found around and in almost any natural or human-made temporary ponds, puddles, potholes, hoofprints, ditches, pastures, gardens, secondary growth and along forest edges or small permanent ponds or water catchments.

BIOLOGY: Rand (1983) and Ryan (1985) have discussed the biology and acoustic communication of this very common species in detail. Much of the summary below is based on those accounts. *Physalaemus pustulosus* is completely nocturnal both as larvae and as adults. It is a leaf litter denizen that hides during the day well below the litter surface and is never found sleeping at night. This frog is inactive most of the dry season (December to April) in Costa Rica and may burrow deep within the soil and estivate in the Lowland Dry Forest region of Costa Rica during this period. *P. pustulosus* is a hopping anuran, jumping only when disturbed. The species is an arthropod generalist with no strong preference for ants in spite of being edentulous. Like toads and most advanced anurans, it has a highly protrusible tongue that is free posteriorly. The tadpoles are mainly detritovores but sometimes eat floating eggs of other frogs. In Costa Rica breeding occurs early on the wet season (May to June) and sporadically thereafter until the dry season begins in November. Males call while inflated, floating on the surface of any available temporary or small permanent lentic body of water. Calling begins at dusk and generally slows down by 10:00 P.M. Calls alternate between a particular male and his nearest neighbors. Females usually come to the site only on the night they breed, and they actively locate the presumptive mate of choice, often swimming toward a par-

ticular male and passing up others. Isolated males usually give only the whine call described above, and this is sufficient for species recognition. In a chorus males add one to six "chucks" to the whine in response to movements of nearby frogs. Females are attracted preferentially to large males giving complex calls in which the "chucks" have a lower frequency (as negatively correlated with male size). Amplexus, which is axillary, usually takes place in the shallow water. Eggs are deposited over a period of half an hour to an hour, and the amplexing male repeatedly whips air into the egg jelly with his hind legs to produce the foam nest. The eggs apparently are unpigmented but later darken, and 80 to 450 are laid in a single floating nest. After breeding, the pair usually leaves the breeding site. Females may lay several clutches per wet season, potentially every six weeks (Davidson and Hough 1969). The eggs hatch in a day and a half to three days, and the tadpoles wiggle down through the foam into the water below. If the breeding site has dried up, the tadpoles collect under the nest and may survive up to five days, during which a rainstorm may refill the site and save them. The larvae metamorphose in about five to nine weeks at a length of 5 to 7 mm and reach breeding condition within two to three months in the laboratory. Small juveniles collected in the field are about 15 mm in standard length. Larvae are eaten by a variety of aquatic predators including *Leptodactylus pentadactylus* tadpoles and dragonfly nymphs. It appears that they are not distasteful to humans or natural predators (Wassersug 1971). Adults are consumed by snakes, *Bufo marinus*, *Leptodactylus pentadactylus*, small carnivorous mammals, and the frog-eating bat *Trachops cirrhosus*. The bats and possibly some other predators locate male frogs acoustically, and unfortunately the frogs are more likely to be caught when they are producing complex calls, the very ones most attractive to the females. The interactions between males and females and male frogs and bats are the basis for Ryan's classic studies of the balancing forces of natural (predation) and sexual selection (female choice) (summarized in Ryan 1985).

REMARKS: Before 1970 this frog was called *Eupemphix pustulosus* or *Engystomops pustulosus*. Lynch (1970, 1971) demonstrated that neither of the two nominal genera in which *P. pustulosus* had been placed by various workers could be distinguished from *Physalaemus*, resulting in the present nomenclature. Several recently collected specimens (UCR 5633–35, 9233) were taken at or in the vicinity of Siquirres, Limón Province, on the Atlantic lowlands. These probably represent an introduction, because no examples were ever seen in that well-collected region before the 1980s. The only other Costa Rican Atlantic versant record is from the Lake Nicaragua drainage at the north end of the Cordillera de Guanacaste at Los Inocentes, Guanacaste Province. Interestingly, *P. pustulosus* is not known to occur along the Pacific versant over a distance of 175 km between the Río Barranca, Puntarenas Province, and Puerto Cortés, Puntarenas Province, although suitable habitat seems to be present. The karyotype is 2N = 22 (Morescalchi 1973).

DISTRIBUTION: Atlantic slope Mexico in Veracruz and Campeche and the Pacific versant from Oaxaca, Mexico, to northwestern Costa Rica; disjunctly present in the Golfo Dulce region of southwestern Costa Rica and adjacent southwestern Panama and through western Panama; eastward from central Panama to Colombia; also on the Atlantic versant of northern Colombia and northwestern Venezuela; apparently a recent introduction to Atlantic lowland Costa Rica (3–930 m).

Subfamily Telmatobiinae Fitzinger, 1843

Genus *Eleutherodactylus* C. Duméril and Bibron, 1841

This, the largest genus of vertebrates (about six hundred species), includes a wide variety of morphologies and adaptations to specialized habitats. The species vary in size from tiny to large and may be quite toadlike in appearance, although most are froglike, and many of the latter have greatly expanded digital disks to superficially resemble the arboreal treefrogs.

Members of the genus have disks at least on the outer fingers and toes that are demarcated by a terminal transverse groove between the disk cover and pad and a well-developed circular to V-shaped ventral disk and lack finger webs. Features that in combination will separate them from other members of the subfamily are most terminal phalanges T-shaped; maxillary toothed; columella present; vomerine teeth usually present; nares laterally placed; a horizontally elliptical pupil; tongue free posteriorly, highly protrusible in some if not all species.

The differences in body form, the absence of parotoid glands, and the presence of digital disks will immediately separate species of *Eleutherodactylus* from most other Costa Rican anurans. One species of the genus *(E. gaigeae)* resembles members of *Phyllobates* (family Dendrobatidae) in coloration but lacks the paired dermal scutes on the digits that are characteristic of that family. *Eleutherodactylus,* especially those with expanded digital disks, are often confused with members of the treefrog family Hylidae. Many hylids have finger webs and some have vertically elliptical pupils, conditions never found in *Eleutherodactylus,* and all lack any indication of a ventral disk and have intercalary cartilages in the digits.

The genus has an extensive distribution over tropical America, from Mexico to western Ecuador and northern Argentina, the Greater and Lesser Antilles, and the Bahama Islands; one species has been introduced to Florida.

The relationships and classification of this immense array of species (more are described each year) remain enigmatic. The number of taxa, combined with the mosaic combination of the same features of external morphology in distantly related forms, continues to frustrate every attempt to bring

systematic order to the group. Those interested in the current stage of understanding should refer to Hedges (1989), Lynch (1986, 1993, 2000), Savage (1987), and Lynch and Duellman (1997), none of whose conclusions are very convincing, especially to the others. Lynch (1986) and Savage (1987) concur at least in recognizing two major clades within the genus: a South America–West Indian lineage (I) and a Central American lineage (II). These stocks differ most notably in the condition of the adductor mandibularis muscles, the former having only the externus superficialis present (e) and the latter only the posterior subexternus (s).

The South American lineage ranges from northeastern Honduras and northwestern Costa Rica to western Ecuador and northern Argentina and to the Greater and Lesser Antilles and Bahama Islands. The Central American lineage ranges from Mexico to southwestern Ecuador, Caribbean lowland Colombia, and disjunctly in the Cordillera Costeña of Venezuela. The genus *Hylactophryne* (three species) of the southwestern United States and adjacent Mexico is closely related to the Central American lineage, and the nominal genera *Syrrhophus* (fifteen species) of southern Texas to Guatemala and *Tomodactylus* (ten species) of upland Mexico seem allied to the South American stock. Both lineages I and II are represented in Costa Rica.

A number of attempts have been made to further divide the plethora of species-level taxa into meaningful or taxonomically useful groups, most notably by Lynch (1976), Savage (1987), Hedges (1989), and Lynch and Duellman (1997). The characterization of these infrageneric clusters, other than the two major clades, is equivocal, but their recognition allows for the association of similar and apparently related forms. Savage (1987) suggested that the many groups, subgroups, or assemblies of species recognized by these authors, often only on phenetic grounds, could be most efficiently organized into a system of *series* containing a number of species *groups*. These clusters of species represent a first approximation of possible relationships and a framework for further study.

Completely detailing of such a system and allocating species to it are beyond the scope of this book. The following arrangement identifies the principal series and groups found in Costa Rica within the overall scheme:

Clade I

fitzingeri series

    fitzingeri species group: *Eleutherodactylus andi, E. crassidigitus, E. cuaquero, E. emcelae, E. fitzingeri, E. longirostris, E. melanostictus, E. monnichorum, E. phasma, E. raniformis, E. rayo, E. talamancae* (Honduras to western Ecuador)

    gollmeri species group: *Eleutherodactylus chac, E. coffeus, E. gollmeri, E. laticeps, E. lineatus, E. mimus, E. noblei, E. rostralis* (Mexico to Panama)

    rugulosus species group: *Eleutherodactylus anciano, E. angelicus, E. aurilegulus, E. azueroensis, E. ber-*

kenbuschii, E. brocchi, E. emleni, E. escoces, E. fleischmanni, E. laevissimus, E. merendonensis, E. olanchano, E. pechorum, E. pozo, E. psephosypharus, E. punctariolus, E. ranoides, E. rugulosus, E. sandersoni, E. taurus, E. vocalis, E. vulcani (Mexico to western Panama)

biporcatus species group: *Eleutherodactylus aphanus, E. biporcatus, E. gulosus, E. megacephalus, E. rugosus* (Guatemala to western Ecuador and northern Venezuela)

bufoniformis species group: *Eleutherodactylus bufoniformis, E. necerus* (Costa Rica to western Ecuador)

rhodopis series

rhodopis species group: *Eleutherodactylus bransfordii, E. hobartsmithi, E. lauraster, E. mexicanus, E. podiciferus, E. persimilis, E. polyptychus, E. pygmaeus, E. rhodopis, E. saltator, E. sartori, E. stejnegerianus, E. underwoodi* (Mexico to Panama)

Clade II

conspicillatus series

gaigeae species group: *Eleutherodactylus gaigeae* (Costa Rica to northern Colombia and western Ecuador)

cerasinus series

cerasinus species group: *Eleutherodactylus cerasinus, E. crenunguis, E. labriosus, E. ocellatus, E. orpacobates, E. rubricundus, E. tenebrionis* (Nicaragua to western Ecuador)

martinicensis series

cruentus species group: *Eleutherodactylus altae, E. caryophyllaceus, E. cruentus, E. latidiscus, E. moro, E. operosus, E. pardalis, E. ridens, E. taeniatus* (Honduras to western Ecuador)

martinicensis species group: *Eleutherodactylus amplinympha, E. barlagnei, E. johnstonei, E. martinicensis, E. pinchoni,* (Lesser Antilles)

diastema species group: *Eleutherodactylus chalceus, E. diastema, E. gularis, E. hylaeformis, E. scolodiscus, E. tigrillo, E. vocator* (Nicaragua to western Ecuador)

Most *Eleutherodactylus* are capable of astonishingly long and nearly vertical leaps when disturbed. Locomotion otherwise consists of shorter jumps when on the ground or short leaps from one leaf, twig, or branch to another in arboreal forms.

All members of the genus, as far as is known, have axillary amplexus. With a single exception, all species in the genus for which reproduction is known lay encapsulated eggs in moist terrestrial or arboreal situations. They have no tadpole stage, and development is completed in the egg capsule (patterns 21 and 27, table 7.1) (plate 92). In *Eleutherodactylus jasperi* of Puerto Rico, females retain the eggs in the oviducts and give birth to little froglets (pattern 34, table 7.1). The large pigmentless eggs, 3 to 6 mm in diameter (including the outer capsule), are concealed under rocks, in depressions in the ground covered by soil or leaf litter, in moss banks, or adhering to the leaves of aerial plants. Clutch sizes vary, depending on species size, from 3 to 104, but most lay about 20 eggs at one time. The eggs are laid singly but form an adherent mass that in some species resembles a raspberry. It is suspected that females attend the eggs in many cases. Details of the life history remain unknown for most forms, and the eggs of only a few have been described.

Most species in the genus display marked sexual dimorphism in size. Adult males are generally much smaller (33 to 67% of the standard length of adult females in many forms), but some are only slightly smaller (70 to 85% of female standard lengths).

Adult males of many species have vocal slits and a single internal or external subgular vocal sac. Similarly, nuptial excrescences in the form of glandular brown or white nuptial pads may be present on the thumbs of adult males. These several features are species specific. The external auditory apparatus also shows considerable dimorphism. In most species the distinct external tympanum is round and one-third to one-half again as large as the more elliptical structure in females (fig. 7.47a,b). In those species with the tympanum covered by a layer of skin but still visible (indistinct), the structure is smaller in males and often less evident or completely concealed in females.

POLYMORPHISM AND POLYCHROMATISM IN *ELEUTHERODACTYLUS*

The recurrence of the same characters of external morphology in mosaic combinations in seemingly distantly related lineages has already been discussed above and by Savage (1987) and Lynch (1993) as a hindrance to recognizing natural groups in the genus. Another factor confounding species recognition and evaluation of evolutionary relationships is the extreme polymorphism and polychromatism that occurs within many *Eleutherodactylus*. That different states of a polymorphic or polychromatic series in one species may be stabilized (found in all individuals) in another further confuses the issue. The following summary describes only the most common and obvious polymorphic and polychromatic conditions.

Polymorphism is most common in characteristics of the dorsal skin texture and patterns of skin warts or knobs and ridges. Skin texture may be variously smooth, shagreened, granular, warty, tuberculate, or tuberculate with whitish tubercle tips. In some forms a range of textures occurs as individual variation but not this entire spectrum. Superimposed on the basic skin texture may be a series of distinct tubercles, particularly in the postorbital, suprascapular, and supraxillary areas (fig. 7.44). In some cases these fuse to form a series of short ridges between the arms (fig. 7.44b) or a W-shaped ridge pattern (fig. 7.44c). Short longitudinal ridges may also be present throughout the length of the body in other individuals (fig. 7.44a). These often fuse to form

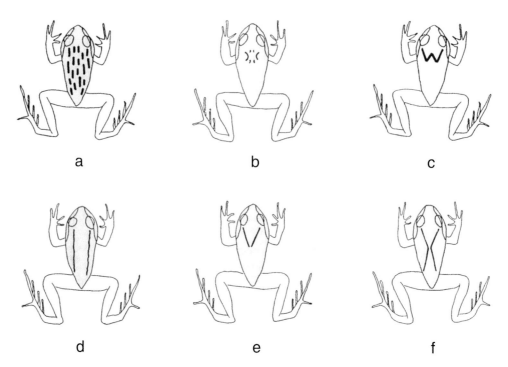

Figure 7.44. Glandular ridging patterns in frogs of the genus *Eleutherodactylus:* (a) short linear ridging; (b) scattered suprascapular tubercles and ridges; (c) W-shaped ridging; (d) paired dorsolateral ridges, sometimes discontinuous; (e) V-shaped ridging; (f) open hourglass ridging.

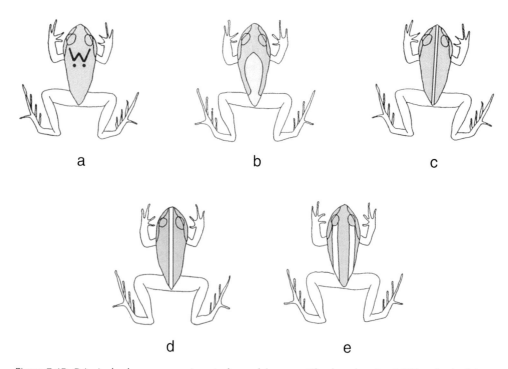

Figure 7.45. Principal color pattern variants in frogs of the genus *Eleutherodactylus:* (a) W and paired dot patterns; (b) picket pattern; (c) light middorsal pinstripe that may branch above cloaca and continue along posterior surface of each thigh; (d) broad light middorsal stripe; (e) paired dorsolateral light stripes. In some individuals the entire dorsum between the dorsolateral ridges is light in color.

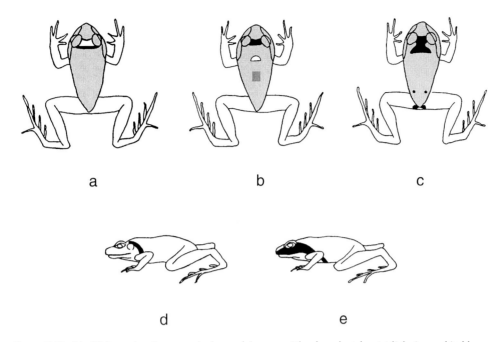

Figure 7.46. Modifying color characters in frogs of the genus *Eleutherodactylus*: (a) light interorbital bar bordered anteriorly and posteriorly by dark bar; one or both bordering bars usually absent; (b) dark interorbital bar, light dorsal shield mark, and dorsal dark blotch; (c) dark interorbital-suprascapular blotch, paired sacral dark spots, and dark seat patch; (d) supratympanic dark stripe; (e) dark eye mask.

a single pair of continuous dorsal ridges (fig. 7.44d) or several long dorsolateral and/or shorter dorsal ridges. In many cases the ridges form paired V-shaped ridges on the anterior part of the body (fig. 7.44e) or hourglass-shaped ridges extending from behind the eyes to the sacral area (fig. 7.44f).

The dorsal color pattern shows the greatest polychromatism. Examples within a single population may have a uniform or mottled pattern, usually with some dark warts or ridges, or a pattern of many discrete small spots. Individuals with a light middorsal pinstripe (fig. 7.45c), a broader middorsal stripe (fig. 7.45d), or paired dorsolateral stripes (usually associated with paired dorsolateral ridges) (fig. 7.45e) may also be present. Another variant has the light middorsal pinstripe forking at the vent and continuing along each posterior thigh surface. In some cases the entire dorsum or the upper head surface and dorsum are light and bordered by broad, dark longitudinal lateral stripes that may cover the entire flank. Frogs from populations showing several of these variations have often been described as different species.

Superimposed on these several basic patterns are variable modifying characters: interocular dark or light spots of several shapes (fig. 7.46a,b,c), dark or light body chevrons or crescents, suprascapular dark spots or a W- or H-shaped dark mark (fig. 7.45a), pelvic dark spots, and dark cloacal patches (fig. 7.46c). The head may have a dark eye mask or a supratympanic stripe, or these may be fused into one element (fig. 7.46d,e). The upper lip may be uniformly light or dark or variously barred (fig. 7.47b). In some species the eye

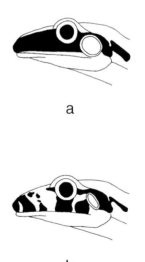

Figure 7.47. Head pattern and tympanum size and shape in frogs of the genus *Eleutherodactylus*: (a) dark eye mask; large round tympanum found in males; (b) dark lip bars; small elliptical tympanum found in females. Some taxa exhibit less common variants: the entire upper lip light, a distinct light upper lip stripe, or contrasting light areas between the first two dark lip bars. In a few species there is no sexual dimorphism in tympanum size and shape.

mask is always present (fig. 7.47a), in others it is usually present but may be broken up into a barred pattern in some individuals. In still others the dark eye mask is an occasional variant. The color of the iris, in life, may also show variation within a single species (e.g., gray or red).

KEY TO THE COSTA RICAN RAINFROGS OF THE GENUS *ELEUTHERODACTYLUS*

The great number of species, the variation discussed above, and the superficial resemblance among many forms create problems of identification even when specimens are in hand. As many as fourteen species of *Eleutherodactylus* may occur at a single site in lowland Costa Rica. However, once you become familiar with the local suite of species, their differences in size, coloration, male call, microhabitat preference, diel activity pattern, and general behavior will aid in recognition. In some cases you will usually have to examine individuals to distinguish among very similar forms. Juveniles may be especially difficult to place without use of a hand lens or microscope.

Although experienced fieldworkers soon learn to identify living animals at their study sites, preserved specimens have their own special difficulties. Freshly preserved examples that retain remnants of the life colors offer the fewest chances for misidentification. However, as the colors fade and the skin softens and tends to lose distinguishing warts, ridges, or tuberculation over time, individuals of different species begin to look more and more alike. Rather subtle difference in the structure of the hands and feet (disks, tubercles, folds, and webbing) are usually the only way to discriminate among forms that in life may be easily recognized as distinct.

In the following key, these last-named morphological characters are emphasized, but size, sexually dimorphic features of adult males, especially the presence or absence of vocal slits and nuptial thumb pads, and coloration in life are included, since they provide additional clues for positive diagnosis. In problematic cases a careful comparison of the appropriate species accounts usually will be definitive. In a few cases (especially juveniles), you may still need to compare the difficult specimen(s) with authoritatively identified preserved material.

1a. A thin membranous web extending at least to the level of the proximal subarticular tubercle on at least one side of toes I and III and usually on II and IV (fig. 7.48b–e) 2
1b. No toe webs (fig. 7.48a) 17
2a. Web between toes minimal, extending at most slightly distal to proximal subarticular tubercles on toes I and IV (fig. 7. 48b,c) 3
2b. Web between toes III and IV substantial, reaching distal subarticular tubercle on toe III (fig. 7.48d,e), or toes with fleshy fringes (fig. 7.49e) 14
3a. Dorsum smooth or with a few scattered warts and ridges, never tuberculate; tips of fingers, especially on III and IV, with distinctly expanded disks; no cranial crests; head narrow, 30–43% of standard length . . . 4
3b. Dorsum tuberculate; tips of fingers without well-developed expanded disks (fig. 7.51a), although digital

pads and disk grooves visible under magnification; distinct cranial crests on skull between eyes in large juveniles and adults not obvious externally; head broad, 44–58% of standard length; no nuptial thumb pads but vocal slits in adult males; adult males 28–50 mm in standard length, adult females 52–94 mm . . . . *Eleutherodactylus bufoniformis* (p. 256)
4a. Disks on outer two fingers (III and IV) much larger than disks on fingers I and II; outer finger and toe disks usually truncate, often emarginate (fig. 7.51f,g) 5
4b. Disks on outer two fingers (III and IV) only slightly, if at all, larger than disks on fingers I and II; outer finger and toe disks usually not truncate or emarginate 10
5a. Posterior thigh surface uniform gray white, yellowish brown, reddish brown, or purplish black 6
5b. Posterior thigh surface with distinct light spots and/or light vertical, lineolate markings visible to unaided eye 8
6a. Heel smooth (fig. 7.49a); posterior thigh surface gray white *or* yellowish brown to reddish, suffused with red in life 7
6b. Heel with a well-developed tubercle (fig. 7.49b); posterior thigh surface dark purple in life; no red in axilla and groin in life; no dark eye mask; nuptial thumb pads and vocal slits in adult males; adult males 37–45 mm in standard length, adult females 38–71 mm . . . . *Eleutherodactylus rayo* (p. 243)
7a. Posterior thigh surface yellowish brown to reddish, suffused with red in life; a definite dark eye mask (fig. 7.47a); axilla and groin usually red in life 39
7b. Posterior thigh surface gray white; no eye mask; no red in coloration in life; some disks on fingers and toes emarginate (fig. 7.51g); a weak inner tarsal fold; no plantar tubercles; adult males approximately 25–30 mm in standard length, adult female 48 mm . . . . *Eleutherodactylus phasma* (p. 242)
8a. Toe webbing usually extending to level of the proximal subarticular tubercles between toes I–II and somewhat beyond the proximal subarticular tubercles on toe IV (fig. 7.48c) 9
8b. Toe webbing barely extending to level of proximal margin of proximal subarticular tubercle on any toe (fig. 7.48b); finger disks greatly enlarged; pads on fingers III and IV emarginate (fig. 7.51g), broader than length of inner metatarsal tubercle; posterior thigh surface dark brown, with small bright yellow spots or vertical stripes; groin mottled; throat and undersides of hind limbs heavily marked with dark brown pigment; posterior undersurfaces bright yellow suffused with pink in life; nuptial thumb pads and vocal slits in adult males; adult males approximately 20–39 mm in standard length, adult females 33 to 44 mm . . . . *Eleutherodactylus cuaquero* (p. 239)
9a. Finger disks greatly enlarged (covers on III and IV emar-

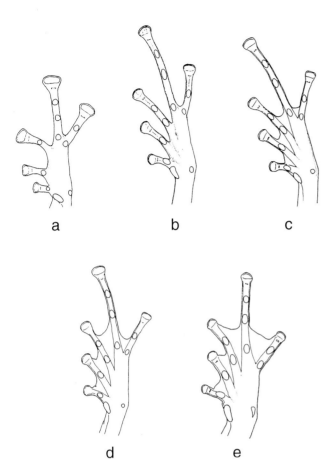

Figure 7.48. Characteristics of feet in frogs of the genus *Eleuthero-dactylus:* (a) no web, toe III much shorter than toe V; toe III much longer than toe V: (b) basal webbing; (c) moderate webbing; (d) extensive webbing; (e) very extensive webbing.

ginate) (fig. 7.51g) equal to or broader than length of inner metatarsal tubercle; posterior thigh surface dark chocolate brown, with discrete large bright yellow (in life) stripes and/or spots (fig. 7.50a); groin with bright yellow (in life) large spots and stripes; throat almost solid black to dark brown; undersurfaces of body and limbs heavily marked with dark brown spots or mottling; light areas on posterior undersurfaces bright yellow, almost always suffused with salmon red in life; nuptial thumb pads and vocal slits in adult males; adult males 40–55 mm in standard length, adult females 65–80 mm . . . . *Eleutherodactylus andi* (p. 237)

9b. Finger disks moderately enlarged (covers on III and IV barely indented), narrower than length of inner metatarsal tubercle; posterior thigh surface dark black to brown, with numerous small pale yellow spots in life (fig. 7.50b); groin mottled or uniform; throat almost immaculate to heavily mottled with dark brown; undersurface white, with a yellow cast posteriorly and under thighs, usually immaculate but sometimes weakly mottled with gray and rarely strongly mottled

with dark brown; nuptial thumb pads and vocal slits in adult males; adult males 23–35 mm in standard length, adult females 36–53 mm . . . .
. . . . *Eleutherodactylus fitzingeri* (p. 240)

10a. No heel tubercle (fig. 7.49a); disk pads rounded, never lanceolate; no eye mask 11

10b. One or two distinct heel tubercles (fig. 7.49b,c); disk pads on some digits lanceolate (fig. 7.51i); almost always with a distinct dark eye mask from snout through eye to tympanum and continuing downward and posteriorly well onto body at least as a narrow line beyond axilla (fig. 7.46e); posterior thigh surface uniform brown; no nuptial pads or vocal slits in adult males; adult males 30–37 mm in standard length, adult females 45–54 mm . . . .
. . . . *Eleutherodactylus gollmeri* (p. 245)

11a. Posterior thigh surface mottled, blotched, lineate, or uniform, never with numerous discrete light spots on a dark brown ground color 12

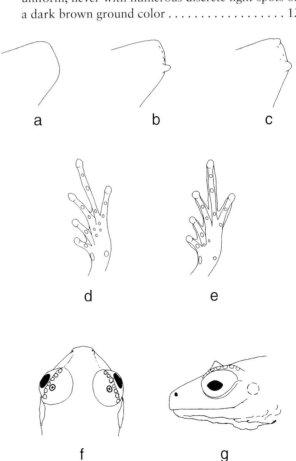

Figure 7.49. Special characteristics in frogs of the genus *Eleuthero-dactylus:* (a) smooth heel; (b) single heel tubercle; (c) paired heel tubercles; (d) toe webbing in *E. gollmeri;* (e) toe fringe and webbing in *E. mimus;* (f) dorsal view of head showing superciliary tubercles and enlarged supraocular tubercles; (g) lateral view of head showing superciliary tubercles and enlarged supraocular tubercle. In *E. caryophyllaceous* there is a single enlarged pointed superciliary tubercle on each upper eyelid.

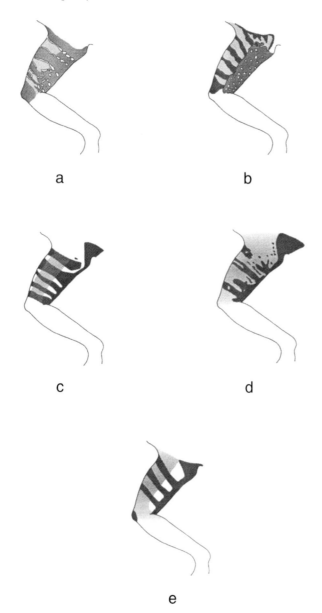

a

b

c

d

e

Figure 7.50. Coloration of the thigh in Costa Rican frogs of the genus *Eleutherodactylus:* (a) *E. andi;* (b) *E. fitzingeri;* (c) *E. rugosus;* (d) *E. megacephalus;* (e) *E. melanostictus.*

11b. Posterior thigh surface with numerous small discrete light spots on a dark brown ground color; venter of adults white to bright golden yellow in life; no nuptial pads or vocal slits in adult males; adult males 26–45 mm in standard length, adult females 40–74 mm . . . . *Eleutherodactylus ranoides* (p. 251)

12a. Toe disks definite, about one and a half to two times width of digit; posterior thigh surfaces uniform or suffused; basal toe webbing, considerably more than four phalanges free of webs on toe IV; venter of adults orange to bright red in life 13

12b. Toe disks weak, barely expanded; posterior thigh sur-

faces usually mottled or blotched; basal toe webbing, about four phalanges free of webs on toe IV (fig. 7.48b); venter of adults pale yellow; nuptial thumb pads and vocal slits in adult males; adult males 29–45 mm in standard length, adult females 38–75 mm . . . .*Eleutherodactylus fleischmanni* (p. 250)

13a. Finger disks definite, slightly expanded, about one and one-half width of digits on fingers III and IV; dorsum gray to brown; venter of adults orange to red; nuptial thumb pads present but no vocal slits in adult males; adult males 26–46 mm in standard length, adult females 38–75 mm . . . . *Eleutherodactylus angelicus* (p. 248)

13b. Finger disks definite, about twice width of digits on fingers III and IV; dorsum dark olive gray with purple cast in preservative; venter of adults tomato red; nuptial thumb pads and vocal slits present in adult males; adult males 26–46 mm in standard length, adult females 38–72 mm . . . . *Eleutherodactylus escoces* (p. 249)

14a. Posterior thigh surface uniform brown to reddish brown 15

14b. Posterior thigh surface with mottling of dark and light irregular to lineate markings 16

15a. No fleshy toe fringes (fig. 7.48d); disks on outer two fingers (III and IV) much larger than disks on fingers I and II; covers on outer disks truncate (fig. 7.51f); no distinctive eye mask; nuptial pads and vocal slits in adult males; adult males 20–31 mm in standard length, adult females 34–48 mm . . . . *Eleutherodactylus crassidigitus* (p. 238)

15b. Toes with fleshy fringes (fig. 7.49e); disks on outer two fingers (III and IV) only slightly if any larger than disks on fingers I and II; covers on outer disks rounded (fig. 7.51b); a distinct eye mask from snout through eye at least to axilla (fig. 7.46e); no nuptial pads or vocal slits in adult males; adult males 30–37 mm in standard length, adult females 45–58 mm . . . . *Eleutherodactylus mimus* (p. 246)

16a. Toe disks definitely to strongly expanded, about two times width of digit; venter of adults yellow in life; nuptial thumb pads and vocal slits in adult males; adult males 40–50 mm in standard length, adult females 55–81 mm . . . .*Eleutherodactylus punctariolus* (p. 250)

16b. Toe disks weak, barely wider than digit; venter of adults white in life; no nuptial pads, but vocal slits present in adult males; adult males 24–44 mm in standard length, adult females 40–80 mm . . . .*Eleutherodactylus taurus* (p. 252)

17a. Tips of fingers without expanded disks, although digital pads and grooves visible under magnification (fig. 7.51a). 18

17b. Tips of fingers with expanded disks. 26

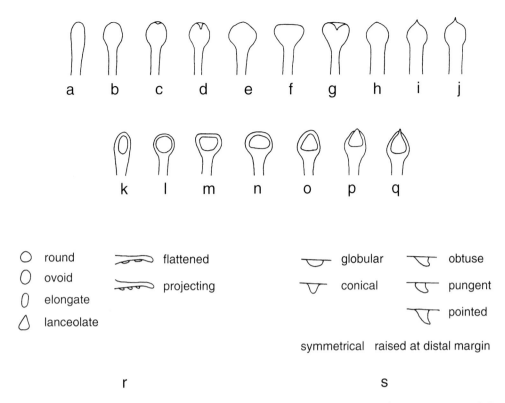

Figure 7.51. Variation in digital features in frogs of the genus *Eleutherodactylus*: Disk cover (a) unexpanded, even; (b) expanded, even, round; (c) expanded, indented, round; (d) expanded, notched, round; (e) expanded, even, palmate; (f) expanded, even, truncate; (g) expanded, emarginate, truncate; (h) expanded, spadate; (i) expanded, lanceolate; (j) expanded, papillate. Disk pad (k) even, elliptical; (l) even, ovoid; (m) even, truncate; (n) even, broadened; (o) even, triangular; (p) swollen; (q) cuspidate. Digital tubercles in outline (r) and profile (s).

18a. Fingers I and II subequal or finger II longer than finger I; no cranial crests; a weak inner tarsal fold or tubercle; head narrow, 30–43% of standard length 19

18b. Finger I distinctly longer than finger II; adults and all but smallest juveniles with paired cranial crests that are best developed in the occipital region; no inner tarsal fold or tubercle; venter smooth; head very broad, 47–54% of standard length.................. 24

19a. Venter coarsely areolate; subarticular, palmar and thenar tubercles of hand strongly elevated (fig. 7.52d,e); accessory palmar tubercles well developed; distal subarticular tubercles on toes III and IV projecting (fig. 7.52a,b); heel rugose or with a series of very small subequal warts; underside of tarsal segment rugose to warty; an inguinal gland usually evident; ventral disk V-shaped, extending posteriorly to level of femora; no vocal slits in adult males 20

19b. Venter smooth at least centrally; subarticular palmar and thenar tubercles of hand low and rounded (fig. 7.52c,f); usually no accessory palmar tubercles but, if present, weakly indicated; distal subarticular tubercles on toes III and IV rounded, low (fig. 7.52c,i);

a single, well-developed conical heel tubercle always present, occasionally one or two smaller tubercles as well (fig. 7.49b); usually no plantar tubercles but, if present, large, weak, rounded and few in number (fig. 7.52i); underside of tarsal segment smooth; venter smooth to granulate; no inguinal gland; ventral disk circular, posterior margin lying anterior to femurs; no nuptial pads, but vocal slits present in adult males; adult males 21–28 mm in standard length, adult females 23–40 mm .... *Eleutherodactylus podiciferus* (p. 259)

20a. Thenar and palmar tubercles about same size, much larger than subarticular tubercles under fingers II–III–IV (fig. 7.53a) 21

20b. Thenar tubercle much smaller than palmar, about same size as subarticular tubercles under fingers II–III–IV (fig. 7.53b); no vocal slits or nuptial thumb pads in adult males............................. 23

21a. All tubercles under fingers and toes raised, often pointed (fig. 7.52a,d); some accessory palmar tubercles pointed; supernumerary subarticular tubercles present; numerous, often pointed, large plantar tubercles (fig. 7.52g) 22

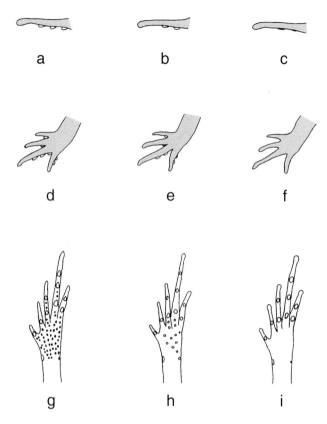

a   b   c

d   e   f

g   h   i

Figure 7.52. Characteristics of hands and feet in frogs of the *Eleutherodactylus rhodopis* group: upper row showing profile of toes, middle row profile of hands, lower row plantar surfaces; (a,d,g) typical features for most Costa Rican species in the group; (b, e, h) *E. underwoodi*; (c,f,i) *E. podiciferus*.

21b. Tubercles under fingers and toes relatively low and globular (fig. 7.52b,e); accessory palmar tubercles definitely so; usually no supernumerary subarticular tubercles (fig. 7.52h); a few large low, scattered plantar tubercles; nuptial pads in adult males; adult males 16–26 mm in standard length, adult females 18.4–30 mm . . . . *Eleutherodactylus underwoodi* (p. 262)
22a. Adult males with nuptial pads; adult males 13.3–23 mm in standard length, adult females 13.6–26 mm . . . . *Eleutherodactylus bransfordii* (p. 257)
22b. Adult males lack nuptial pads; adult males 15.3–26.6 mm in standard length, adult females 15.6–28.7 mm . . . *Eleutherodactylus polyptychus* (p. 260)
23a. Fingers I and II equal in length; adult males 12–18 mm in standard length, adult females 12.5–22.4 mm . . . . *Eleutherodactylus stejnegerianus* (p. 261)
23b. Finger I shorter than finger II; adult males 13.1–17.8 mm in standard length, adult females 13.3–23 mm . . . . *Eleutherodactylus persimilis* (p. 258)
24a. Distinct chalice- or hourglass-shaped suprascapular plicae (fig. 7.54); most of venter, undersurfaces of calf, thigh, and groin, and adjacent areas covered by large red, orange, or yellow (white in preservative and some

juveniles) spots on a black or dark brown ground color 25
24b. Dorsum essentially smooth, without plicae; venter, undersides of calf, thigh, groin, and adjacent areas pale yellow, mottled with dark brown; no nuptial pads or vocal slits in adult males; adult males probably 40–60 mm in standard length, adult females 72–103 mm . . . . *Eleutherodactylus gulosus* (p. 253)
25a. Posterior thigh surface marked with vertical black bars separated by bright red interspaces in live adults (white in juveniles) (fig. 7.50c); no laterosacral plicae or definite series of linearly arranged warts (fig. 7.54a); dorsum covered with warty tubercles and short tuberculate ridges; no vocal slits or nuptial pads; adult males 30–44 mm in standard length, adult females 35–69 mm . . . . *Eleutherodactylus rugosus* (p. 255)
25b. Posterior thigh surface variously mottled with dark or light or with some orange or yellow spotting, not marked with vertical black bars (fig. 7.50d); usually laterosacral plicae; if absent, a definite series of laterosacral warts (fig. 7.54b); dorsum relatively smooth, with a few scattered tubercles; no vocal slits or nuptial pads in adult males; adult males 30–43 mm in standard length, adult females 50–70 mm . . . . *Eleutherodactylus megacephalus* (p. 254)
26a. Toe III definitely shorter than toe V 27
26b. Toe III longer than toe V; tip of toe V not reaching distal border of subarticular tubercle on toe IV (fig. 7.48b); venter granulate; enlarged supraocular and heel tubercles; no accessory palmar tubercles; anterior, upper, and posterior thigh surfaces usually marked by continuous dark transverse bars (fig. 7.50e) that alternate with light areas (yellow, yellow green, orange, pink, magenta, or scarlet in life); tarsus smooth; snout never with a fleshy pustule at tip; no light spots outlined by dark in groin region; nuptial thumb pads and vocal slits present in adult males; adult males 35–43 mm in standard length, adult females 35–56 mm . . . . *Eleutherodactylus melanostictus* (p. 241)
27a. Tip of toe V reaching distal border of distal subarticular tubercle on toe IV (fig. 7.55a) 28
27b. Tip of toe V not reaching distal border of distal subarticular tubercle on toe IV (fig. 7.55b); no distinct enlarged supraocular or superciliary tubercle, although a series of small superciliary tubercles usually present along edge of eyelid and several small warts on upper surfaces; some subarticular tubercles, at least under toes III and IV, pungent or pointed (fig. 7.51s); tympanum thin, distinct; an enlarged heel tubercle present (fig. 7.49b); tarsus with 1–2 outer tubercles and a short inner tarsal fold or tubercle; a few low accessory palmar tubercles; dorsum granular; posterior thigh surface brown, with red in groin, on anterior and posterior thigh surfaces, and on underside of calf in life;

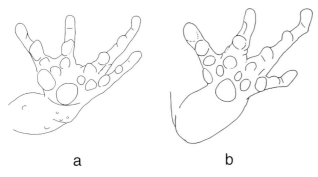

a   **b**

Figure 7.53. Palmar view of hands in the *Eleutherodactylus rhodopis* group: (a) thenar and palmar tubercles about the same size, much larger than subarticular tubercles; (b) thenar tubercle much smaller than palmar tubercle, about same size as subarticular tubercles.

no discrete dark and light markings in groin; dorsum light brownish marked with dark spots, blotches, or light stripe or nearly uniform; nuptial thumb pads and vocal slits present in adult males; adult males to 19–25 mm in standard length, adult females 25–35 mm . . . . *Eleutherodactylus cerasinus* (p. 264)

28a. A distinct well-developed pointed heel tubercle (fig. 7.49b); an enlarged supraocular or superciliary tubercle on upper eyelid (fig. 7.49f,g) 29

28b. Heel smooth (fig. 7.49a) or with a few small warts all of about the same size and similar to others scattered over upper surface of hind limb; no enlarged eyelid tubercle 30

29a. Disk covers on outer fingers (III and IV) greatly expanded, truncate, and emarginate (fig. 7.51g); an enlarged pointed or flaplike supraocular tubercle located mediad from free upper eyelid margin (fig. 7.49f,g); a pointed heel tubercle (fig. 7.49b); dorsum tuberculate; accessory palmar tubercles present; tarsus with a series of 2–4 tubercles along outer margin and a short inner tarsal fold or tubercle; snout rounded to truncate in dorsal outline (fig. 7.6b), sometimes with a fleshy pustule at end; dorsum gray, dark brown, or black, with spots, blotches, stripes, or uniform, not covered by discrete dark punctations; usually one to several bright golden yellow to orange spots in groin and on front of thigh in life, each outlined by dark pigment (fig. 7.58); nuptial thumb pads but no vocal slits in adult males; adult males 21–28 mm in standard length, adult females 25–42 mm . . . . *Eleutherodactylus cruentus* (p. 267)

29b. Disk covers on outer fingers (III and IV) moderately expanded, rounded (fig. 7.51b); an enlarged pointed superciliary tubercle located on free margin of upper eyelid; a large pointed tubercle on heel (fig. 7.49b); tarsus smooth; no accessory palmar tubercles; dorsum very smooth; snout pointed in dorsal outline (fig. 7.6b); no distinct pattern of light spots or dark and light bars

in groin or on thigh surfaces; dorsum pale yellow, pinkish, brownish or gray covered by discrete dark punctations, occasionally with some dark crossbars or longitudinal silver stripes; no nuptial pads but vocal slits present in adult males; adult males 21–24 mm in standard length, adult females 23–26 mm . . . . *Eleutherodactylus caryophyllaceus* (p. 266)

30a. Groin and anterior thigh surface marked with large light spots on a dark gray to black background . . 31

30b. Groin and anterior thigh surface without contrasting light spots or other pattern 32

31a. Large silvery white spots in groin and on anterior and posterior thigh surfaces; disks on outer fingers enlarged, equal to tympanum; no nuptial pads, but vocal slits present in adult males; adult males 16–19 mm in standard length, adult females 25–29 mm . . . . *Eleutherodactylus pardalis* (p. 269)

31b. Groin and anterior and posterior thigh surfaces with large light (coral in life) spots; disks on outer fingers moderate, not equal to tympanum; no nuptial pads but vocal slits in adult males; adult males 18–23.5 mm in standard length, adult females 20–27 mm . . . . *Eleutherodactylus altae* (p. 265)

32a. Disk pads of some fingers and toes, including finger III and toe IV, broadened or triangular (fig. 7.51o–q); some disk covers spadate, lanceolate, or papillate (fig. 7.51h–j) 33

32b. Disk pads and covers rounded (fig. 7.51b) 36

33a. Disk covers not lanceolate or papillate 34

33b. Disk covers lanceolate and papillate (fig. 7.51i,j); dorsum usually gray with definite dark markings; no nuptial pads but vocal slits present in adult males; adult males 14–16 mm in standard length, adult females 16.5–18 mm. . . . *Eleutherodactylus vocator* (p. 274)

34a. Most disk covers spadate (fig. 7.51h); most disk pads triangular (fig. 7.51o) 35

34b. Most disk covers palmate (fig. 7.51e); most disk pads expanded, not triangular; adults with vomerine teeth; no nuptial pads, but vocal slits present in adult males; adult males 16–21 mm in standard length, adult females 18–24 mm . . . .*Eleutherodactylus diastema* (p. 272)

35a. Dorsal pattern of discrete brown spots lined up in rows on yellow ground color; posterior thigh surface unpigmented yellow; venter yellow; vomerine teeth present in adults; no nuptial pads, but vocal slits present in adult males; adult males 16–18 mm in standard length, adult females about 19–21 mm . . . . *Eleutherodactylus tigrillo* (p. 274)

35b. Dorsum uniform gray tan or marked with dark blotches, often suffused with pink or red; posterior thigh surface pigmented, pink to reddish; venter pink; no vomerine teeth; no nuptial pads, but vocal slits present in adult males; adult males 19–22 mm in stan-

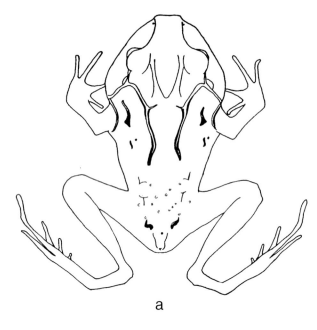

a

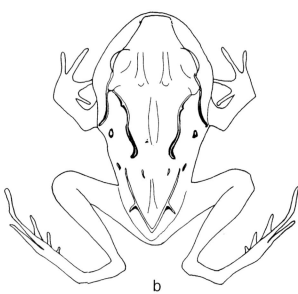

b

Figure 7.54. Dorsal view of Costa Rican frogs of the *Eleutherodactylus biporcatus* group: (a) *E. rugosus* with chalice-shaped suprascapular plicae; (b) *E. megacephalus* with hourglass-shaped suprascapular and laterosacral plicae. Note paired cranial crests in both species.

dard length, adult females 20–26 mm . . . .
. . . . *Eleutherodactylus hylaeformis* (p. 273)
36a. Venter coarsely areolate. 37
36b. Venter smooth; dorsum uniformly granulate; dorsum black with a pair of dorsal red stripes from eye to sacrum in life, stripes often disappearing in preservative; no nuptial pads or vocal slits in adult males; adult males 26–31 mm in standard length, adult females 32–44 mm . . . . . *Eleutherodactylus gaigeae* (p. 263)
37a. Posterior thigh surface uniform; dorsum pinkish or

uniform green in life; head or hind limbs with red orange or red coloration in life 38
37b. Posterior thigh surface marbled, stippled, or blotched with dark on tan ground color; dorsum tan to grayish tan; no red in coloration; no nuptial thumb pads, but paired vocal slits present in adult males; adult males 17–25 mm in standard length, adult females 17–35 mm . . . . . . *Eleutherodactylus johnstonei* (p. 271)
38a. Surface of upper eyelid with one to several enlarged, pointed supraocular tubercles; anterior and posterior surfaces of thigh, calf, and feet red in life; dorsum not green in life; no nuptial pads, but vocal slits present in adult males; adult males 16–19 mm in standard length, adult females 21–25 mm . . . .
. . . . *Eleutherodactylus ridens* (p. 269)
38b. Surface of upper eyelid smooth; no red on hind limbs in life; dorsal surfaces green, head red orange or with red orange markings in life; no nuptial pads, but vocal slits present in adult males; adult males 16.5–19.5 mm in standard length, adult females to 24–25 mm. . . . .
. . . . *Eleutherodactylus moro* (p. 268)
39a. Sole of foot with definite low plantar tubercles; inner tarsal fold present; dorsum smooth; upper lip light gray, yellow, or brown, same color as dorsum; a distinct longitudinal glandular ridge, continuous with supratympanic ridge, extending posteriorly at least one-third the length of body; no nuptial pads or vocal slits in adult males; adult males 43–48 mm in standard length, adult females 58–66 mm . . . .
. . . . *Eleutherodactylus noblei* (p. 247)

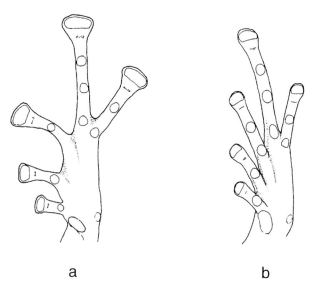

a   b

Figure 7.55. Foot characteristics in Costa Rican frogs of the *Eleutherodactylus cruentus* and *E. cerasinus* groups: (a) *E. cruentus* group, tip of toe V reaching distal subarticular tubercle on toe IV; (b) *E. cerasinus* group, tip of toe V not reaching distal subarticular tubercle on toe IV.

Plate 49. *Nototriton guanacaste;* eggs with developing embryos. Costa Rica: Guanacaste: Cerro Cacao, 1,500 m.

Plate 50. *Nototriton picadoi.* Costa Rica: Cartago: Tapantí National Park, about 1,500 m.

Plate 51. *Nototriton richardi.* Costa Rica: Heredia: Volcán Barva: Braulio Carrillo National Park, about 1,800 m.

Plate 52. *Nototriton tapanti;* holotype. Costa Rica: Cartago: Tapantí National Park, about 1,300 m.

Plate 53. *Oedipina alleni.* Costa Rica: Puntarenas: Corcovado National Park, about 40 m.

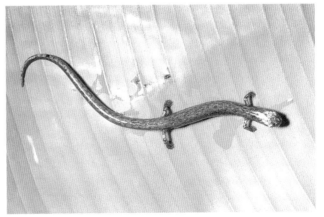

Plate 54. *Oedipina savagei.* Costa Rica: Puntarenas: Las Cruces Biological Station, about 1,300 m.

Plate 55. *Oedipina cyclocauda*. Costa Rica: Heredia: La Selva Biological Station, about 60 m.

Plate 56. *Oedipina cyclocauda*, view of head. Costa Rica: Heredia: La Selva Biological Station, about 60 m.

Plate 57. *Oedipina gracilis*. Costa Rica: Heredia: La Selva Biological Station, about 60 m.

Plate 58. *Oedipina gracilis;* view of head. Costa Rica: Heredia: La Selva Biological Station, about 60 m.

Plate 59. *Oedipina grandis*. Costa Rica: Puntarenas: Las Tablas, about 1,900 m.

Plate 60. *Oedipina* poelzi. Costa Rica: Alajuela: Monteverde Cloud Forest Preserve, about 1,600 m.

Plate 61. *Oedipina uniformis;* view of head. Costa Rica: Puntarenas: Monteverde, about 1,550 m.

Plate 62. *Rhinophrynus dorsalis.* Costa Rica: Guanacaste: Santa Rosa National Park, about 280 m.

Plate 63. *Atelopus chiriquiensis*, amplectant pair. Costa Rica: Puntarenas: Las Tablas, about 1,800 m.

Plate 64. *Atelopus chiriquiensis*, female. Panama: Chiriquí: Cerro Punta, about 1,600 m.

Plate 65. *Atelopus senex;* male. Costa Rica: Heredia: Volcán Barva: Braulio Carrillo National Park, about 2,300 m.

Plate 66. *Atelopus senex*, female. Costa Rica: Heredia: Volcán Barva: Braulio Carrillo National Park, about 2,300 m.

Plate 67. *Atelopus varius*. Costa Rica: Alajuela-Puntarenas: Monteverde Cloud Forest Preserve, about 1,600 m.

Plate 68. *Atelopus varius*. Costa Rica: Heredia: Volcán Barva: Braulio Carrillo National Park, about 1,000 m.

Plate 69. *Atelopus varius*, male. Costa Rica.

Plate 70. *Bufo fastidiosus*. Costa Rica: Puntarenas: Las Tablas, about 1,900 m.

Plate 71. *Bufo holdridgei*. Costa Rica: Heredia: Cerro Chompipe, about 2,100 m.

Plate 72. *Bufo haematiticus*. Costa Rica: Heredia: La Selva Biological Station, about 60 m.

Plate 73. *Bufo marinus.* Costa Rica: Guanacaste: La Pacifica, 40 m.

Plate 74. *Bufo periglenes,* males gathering by breeding pools. Costa Rica: Alajuela-Puntarenas: Monteverde Cloud Forest Preserve, about 1,600 m.

Plate 75. *Bufo periglenes;* amplectant pair. Costa Rica: Alajuela-Puntarenas: Monteverde Cloud Forest Preserve, about 1,600 m.

Plate 76. *Bufo periglenes;* cluster of males trying to mate with one female. Costa Rica: Alajuela-Puntarenas: Monteverde Cloud Forest Preserve, about 1,600 m.

Plate 77. *Bufo coccifer.* Costa Rica: Guanacaste: Santa Rosa National Park, about 280 m.

Plate 78. *Bufo coniferus.* Costa Rica: Alajuela: Peñas Blancas Valley, about 800 m.

Plate 79. *Bufo luetkenii.* Costa Rica: Guanacaste: Santa Rosa National Park, about 280 m.

Plate 80. *Bufo melanochlorus,* amplectant pair. Costa Rica: Puntarenas: Corcovado National Park, about 40 m.

Plate 81. *Bufo melanochlorus,* subadult. Costa Rica: Alajuela: Peñas Blancas Valley, about 800 m.

Plate 82. *Bufo valliceps.* Costa Rica: Alajuela: Upala, 48 m.

Plate 83. *Crepidophryne epiotica.* Costa Rica: Cartago: Tapantí National Park, about 1,800 m.

Plate 84. *Leptodactylus melanonotus.* Costa Rica: Puntarenas: Guacimal, about 370 m.

Plate 85. *Leptodactylus bolivianus*. Costa Rica: Puntarenas: Rincón de Osa.

Plate 86. *Leptodactylus pentadactylus*. Costa Rica: Heredia: La Selva Biological Station, about 60 m.

Plate 87. *Leptodactylus pentadactylus* in action, eating a *Smilisca phaeota*. Costa Rica: Puntarenas: Corcovado National Park, about 40 m.

Plate 88. *Leptodactylus labialis*. Costa Rica: Guanacaste: La Pacifica, 40 m.

Plate 89. *Leptodactylus poecilochilus*. Costa Rica: Puntarenas: Guacimal, about 370 m.

Plate 90. *Leptodactylus poecilochilus, "quadrivittatus"* color phase. Costa Rica: Puntarenas: Golfito, about 5 m.

Plate 91. *Physalaemus pustulosus;* calling male. Costa Rica: Guanacaste: Santa Rosa National Park, about 280 m.

Plate 92. *Eleutherodactylus* sp., froglets about to emerge from egg capsules. Costa Rica.

Plate 93. *Eleutherodactylus andi.* Costa Rica: Puntarenas: Monteverde, about 1,400 m.

Plate 94. *Eleutherodactylus crassidigitus.* Costa Rica: Puntarenas: Monteverde, about 1,400 m.

Plate 95. *Eleutherodactylus fitzingeri.* Costa Rica: Heredia: La Selva Biological Station, about 60 m.

Plate 96. *Eleutherodactylus melanostictus.* Costa Rica: Alajuela: Monteverde Cloud Forest Preserve, about 1,600 m.

Plate 97. *Eleutherodactylus phasma*; holotype. Costa Rica: Puntarenas: Las Tablas, 1,850 m.

Plate 98. *Eleutherodactylus talamancae.* Costa Rica: Heredia: La Selva Biological Station, about 60 m.

Plate 99. *Eleutherodactylus gollmeri.* Costa Rica: Limón: Comadre, 20 m.

Plate 100. *Eleutherodactylus mimus.* Costa Rica: Heredia: La Selva Biological Station, about 60 m.

Plate 101. *Eleutherodactylus noblei.* Costa Rica: Heredia: La Selva Biological Station, about 60 m.

Plate 102. *Eleutherodactylus angelicus*. Costa Rica: Puntarenas: Monteverde Cloud Forest Preserve, about 1,600 m.

Plate 103. *Eleutherodactylus escoces*. Costa Rica: Cordillera Central: Braulio Carrillo National Park, about 1,800 m.

Plate 104. *Eleutherodactylus catalinae*. Costa Rica: Puntarenas: Las Tablas, about 1,800 m; see p. 840

Plate 105. *Eleutherodactylus ranoides*. Costa Rica: Heredia: La Selva Biological Station, about 60 m.

Plate 106. *Eleutherodactylus taurus*. Costa Rica: Puntarenas: Rincón de Osa, about 40 m.

Plate 107. *Eleutherodactylus gulosus*. Panama: Chiriquí: Fortuna Forest Reserve, about 1,200 m.

Plate 108. *Eleutherodactylus megacephalus*; gray color phase. Costa Rica: Heredia: La Selva Biological Station, about 60 m.

Plate 109. *Eleutherodactylus megacephalus*; orange color phase. Costa Rica: Heredia: La Selva Biological Station, about 60 m.

Plate 110. *Eleutherodactylus bufoniformis*. Panama: Panama: near San Juan de Pequení: Chagres National Park, about 360 m.

Plate 111. *Eleutherodactylus bransfordii*. Costa Rica: Heredia: La Selva Biological Station, about 60 m.

Plate 112. *Eleutherodactylus persimilis*. Costa Rica: Limón: Guayacán, about 650 m.

Plate 113. *Eleutherodactylus podiciferus*. Costa Rica: Alajuela: Monteverde Cloud Forest Preserve, about 1,400 m.

Plate 114. *Eleutherodactylus stejnegerianus;* amplectant pair. Costa Rica: Puntarenas: Lagarto de Santa Elena, about 1,200 m.

Plate 115. *Eleutherodactylus underwoodi*. Costa Rica: Puntarenas: Monteverde, about 1,550 m.

Plate 116. *Eleutherodactylus gaigeae*. Panama: Coclé: El Copé, about 1,100 m.

Plate 117. *Eleutherodactylus cerasinus*. Costa Rica: Heredia: La Selva Biological Station, about 60 m.

Plate 118. *Eleutherodactylus altae*. Costa Rica: Heredia: La Selva Biological Station, about 60 m.

Plate 119. *Eleutherodactylus caryophyllaceus*. Costa Rica: Heredia: La Selva Biological Station, about 60 m.

Plate 120. *Eleutherodactylus cruentus*. Costa Rica: Alajuela: Monteverde Cloud Forest Preserve, about 1,600 m.

Plate 121. *Eleutherodactylus cruentus*. Costa Rica: Alajuela: Peñas Blancas Valley, about 800 m.

Plate 122. *Eleutherodactylus moro*. Panama: Colón: Cerro Brujo, Chagres National Park, about 800 m.

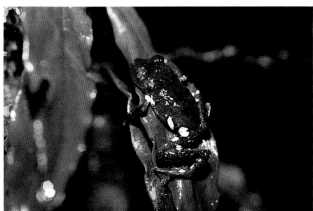

Plate 123. *Eleutherodactylus pardalis*. Costa Rica: Puntarenas: Las Cruces Biological Station, about 1,200 m.

Plate 124. *Eleutherodactylus ridens*. Costa Rica: Alajuela: Peñas Plancas Valley, about 800 m.

Plate 125. *Eleutherodactylus ridens;* amplectant pair. Costa Rica: Alajuela: Peñas Plancas Valley, about 800 m.

Plate 126. *Eleutherodactylus johnstonei.* Panama: Panama: Panama, about 35 m.

Plate 127. *Eleutherodactylus diastema.* Costa Rica: Puntarenas: Monteverde, about 1,400 m.

Plate 128. *Eleutherodactylus hylaeformis;* common color phase. Costa Rica: Alajuela: Monteverde Cloud Forest Preserve, about 1,600 m.

Plate 129. *Eleutherodactylus hylaeformis;* red color phase. Costa Rica: Alajuela: Monteverde Cloud Forest Preserve, about 1,600 m.

Plate 130. *Eleutherodactylus vocator.* Costa Rica: Puntarenas: Las Cruces Biological Station, about 1,300 m.

Plate 131. *Agalychnis annae*. Costa Rica: Puntarenas: Monteverde, about 1,400 m.

Plate 132. *Agalychnis calcarifer*. Costa Rica: Heredia: La Selva Biological Station, about 60 m.

Plate 133. *Agalychnis calcarifer*. Costa Rica: Heredia: La Selva Biological Station, about 60 m.

Plate 134. *Agalychnis callidryas*. Costa Rica: Heredia: La Selva Biological Station, about 60 m.

Plate 135. *Agalychnis callidryas*; night colors. Costa Rica: Heredia: La Selva Biological Station, about 60 m.

Plate 137. *Agalychnis saltator*. Costa Rica: Heredia: La Selva Biological Station, about 60 m.

Plate 136. *Agalychnis callidryas;* tadpoles; note vertical position in water column. Costa Rica.

Plate 138. *Agalychnis spurrelli*. Costa Rica: Limón: Guayacán, about 750 m.

Plate 139. *Agalychnis spurrelli* parachuting. Costa Rica: Limón: Guayacán, about 750 m.

Plate 140. *Phyllomedusa lemur;* night colors. Costa Rica: Alajuela: Monteverde Cloud Forest Preserve, about 1,500 m.

39b. Sole of foot smooth; no tarsal fold; dorsum granulate; upper lip marked with an enamel white longitudinal stripe in juveniles and some adults, broken up by darker pigment forming a more or less discontinuous white stripe or series of white spots in most adults, so that lip contrasts markedly with dark dorsal color; a glandular supratympanic ridge that curves posteriorly downward but does not extend posterior to axilla; nuptial thumb pads and vocal slits in adult males; adult males 21–30 mm in standard length, adult females 34–50 mm. . . . .
. . . .*Eleutherodactylus talamancae* (p. 244)

Eleutherodactylus fitzingeri Group

Costa Rican species of this group are froglike anurans with slender, relatively pointed heads. Other features include finger I usually longer than finger II; toe III longer than toe V; prominent or internal tympanum; vomerine odontophores usually triangular, posterior to choanae, and narrowly separated medially; vocal slits and nuptial thumb pads in adult males; vocal sac subgular and internal; disks on all digits expanded; some disks on fingers usually emarginate; subarticular tubercles symmetrical, projecting moderately; no supernumerary tubercles on hands and feet; accessory palmar tubercles obsolescent; no plantar tubercles; toes usually webbed at least basally; an inner tarsal fold; smooth or granulate venter.

Eleutherodactylus andi Savage, 1974b

(plate 93; map 7.22)

DIAGNOSTICS: A species of moderate-sized (males) to large (females), long-limbed arboreal frogs having bright

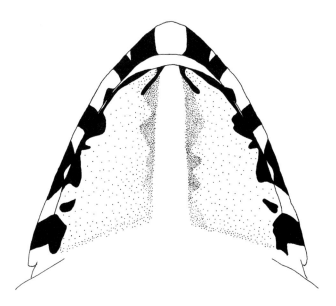

Figure 7.56. Throat of *Eleutherodactylus fitzingeri* showing characteristic midgular light stripe.

yellow spots and/or vertical lineolate markings on the posterior thigh (fig. 7.50a) and usually on the groin area, the undersurfaces of the body and limbs heavily marked with dark spots or mottling, and a distinct midgular cream stripe (fig. 7.56). In addition, the finger disks are greatly expanded (III and IV emarginate) (fig. 7.51g), and the posterior venter and the undersides of the hind limbs are bright yellow and usually suffused salmon red.

DESCRIPTION: Adult males 40 to 55 mm in standard length, adult females 86 to 80 mm; upper and ventral surfaces smooth; head about as wide as long; snout subelliptical in outline from above; nares lateral; eyes large; tympanum large, height about two-thirds diameter of eye in males, about one-half in females; paired vocal slits and a single internal subgular vocal pouch in adult males; finger I slightly longer than finger II; finger disks greatly expanded, covers truncate, emarginate; disks on fingers III and IV much larger than on fingers I and II; subarticular tubercles on fingers and toes ovoid, conical to obtuse in profile; no supernumerary tubercles; brownish nuptial thumb pads in adult males; thenar tubercle elongate, palmar bifid; accessory palmar tubercles obsolescent; heel smooth; toe disks large but smaller than largest finger disks, truncated, expanded, and emarginate; definite fleshy keels along margins of toes I and II; toes with moderate webs; webbing formula: I 2–2⅓ II 2–3¼ III 3–4 IV 4–2¾ V; no plantar tubercles; elongate inner metatarsal tubercle, much larger than small, rounded outer; definite elongate inner tarsal fold. Dorsum and upper surfaces of head a nearly uniform deep, rich chocolate brown; often a median light dorsal pinstripe; posterior thigh surface deep chocolate brown with yellow spots and/or vertical lines that may continue onto upper thigh surface; throat dark to dark spotted with a median cream stripe; iris coppery gold above

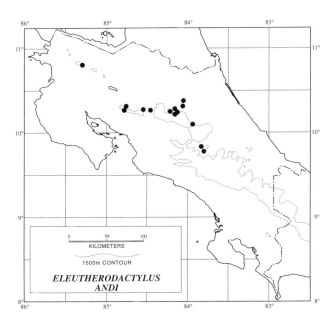

Map 7.22. Distribution of *Eleutherodactylus andi.*

to gold below with a horizontal dark brown line separating upper and lower halves.

VOICE: A deep guttural "glug" repeated several times a minute.

SIMILAR SPECIES: (1) *Eleutherodactylus fitzingeri* is a smaller species, adult males 23 to 35 mm in standard length, adult females 36 to 53 mm, with a rugose dorsum, nonemarginate finger disks (fig. 7.51f), and never a light middorsal pinstripe. (2) *Eleutherodactylus cuaquero* is also a smaller species (maximum standard length 48 mm) with only basal toe webs (fig. 7.48b) and the groin mottled.

HABITAT: An upland form found in Premontane Wet Forest and Rainforest of the Atlantic slope.

BIOLOGY: These uncommon nocturnal, arboreal frogs are usually found fairly close to watercourses. They have occasionally been collected from hiding places in tank bromeliads at least 5 m above the ground. During heavy rains toward the end of the dry season, males call from the stream banks, and the females descend from the trees for mating. The eggs are apparently deposited on the ground, after which the frogs return to the tree canopy. Small juveniles are 13 to 14 mm in standard length.

REMARKS: Savage (1974b) has reviewed the complicated nomenclature associated with this taxon. Günther (1900) described and figured a Costa Rican specimen of this form as *Lioyla guentheri* (= *Leiyla güntherii* Keferstein, 1868a). Taylor (1952a), with some uncertainty, used the name *Eleutherodactylus palmatus* (Boulenger, 1882) for another example of the same form. The latter name is based on the original descriptions of *Liyla rugulosa* Cope, 1869 (= *Eleutherodactylus rugulosus*) and *Leiyla güntherii* (= *Eleutherodactylus fitzingeri*) and cannot be applied to this species. Consequently the name *E. andi* was proposed based on fresh Costa Rican material. The karyotype is 2N = 22; five pairs of chromosomes are metacentric, one pair is submetacentric, three pairs are telocentric, and two pairs are subtelocentric; NF = 38 (DeWeese 1976).

DISTRIBUTION: Atlantic premontane slopes of the northern and central portions of the Costa Rican cordilleras (1,000–1,200 m).

Eleutherodactylus crassidigitus Taylor, 1952a

(plate 94; map 7.23)

DIAGNOSTICS: A small (males) to moderate-sized (females) species having the posterior thigh surface uniform, usually orange or red, with substantial webbing between toes II–III–IV and frequently lacking a median light gular stripe.

DESCRIPTION: Adult males 20 to 31 mm in standard length, adult females 34 to 48 mm; upper surfaces shagreened; venter smooth; head about as wide as long; snout subelliptical in dorsal outline; eyes large; tympanum large, height about two-thirds diameter of eye in males, about

three-fifths in females; paired vocal slits and single internal subgular vocal sac in adult males; finger I longer than finger II; finger disks moderately expanded, covers round on fingers I and II, truncate and larger on III and IV, nonemarginate; subarticular tubercles under digits of hands and feet round to ovoid, conical; no supernumerary tubercles; brownish nuptial thumb pads in adult males; thenar tubercle elongate, palmar much larger, round; accessory palmar tubercles obsolescent; heel smooth; toe disks smaller than largest finger disks; toe webbing substantial, reaching well beyond proximal subarticular tubercle on toe IV and at least to distal tubercle on outer margin of toe III; webbing formula: I 2–2 II 1½–3 III 2–3½ IV 4–2½ V; no plantar tubercles; elongate inner metatarsal and very small, round outer metatarsal tubercle; definite inner tarsal fold. Ground color of upper surfaces various shades of gray or brown, usually with pink, orange, or reddish cast, sometimes with partial or complete greenish overlay, usually with some dark spotting on glandular areas, often with a broad middorsal light stripe or rarely a pinstripe; some-times entire dorsum forming wide light field bordered by dark line or series of spots laterally; upper surface of limbs with transverse dark bars; posterior thigh surface usually uniform orange, red orange, or red, sometimes brown tinged with these colors; anterior surface of thigh and groin usually suffused with orange to red; upper lip uniform or barred; sometimes a dark eye mask present; throat uniform white to gray with or without a median light stripe; venter white to yellow, often with a greenish cast; iris gray to bronze above, separated from gray lower portion by brown to red horizontal stripe.

VOICE: The typical advertisement call of *E. crassidigitus* consists of a single two-part "chuck-chirp" given at spo-

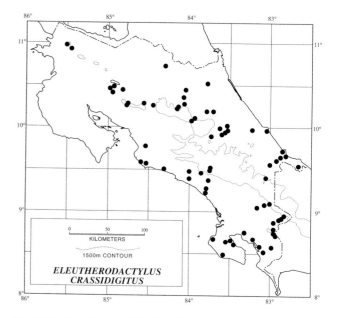

Map 7.23. Distribution of *Eleutherodactylus crassidigitus*.

radic intervals. Single chirps may also be given. The chirps are 100 to 200 milliseconds long, with the frequency spread from 1.4 to 4.5 kHz. The "chucks" are 20 to 130 milliseconds in duration with a similar frequency range (Lynch and Myers 1983).

SIMILAR SPECIES: (1) *Eleutherodactylus talamancae* has minimal toe webbing that extends at most slightly distal to the proximal subarticular tubercles on toes I and IV (fig. 7.48c), and juveniles have a definite white upper lip stripe. (2) *Eleutherodactylus rayo* has minimal toe webbing and a well-developed heel tubercle (fig. 7.49b). (3) *Eleutherodactylus mimus* has subequal finger disks that are palmate on fingers III and IV (fig. 7.51e) and fleshy toe fringes (fig. 7.49e).

HABITAT: In mature and second-growth forests and along forest margins, also in coffee fincas and brushy pastures throughout the Lowland Moist and Wet Forest and Premontane Wet Forest and Rainforest and Lower Montane Rainforest zones.

BIOLOGY: A very common species that is primarily nocturnal and forages on the forest floor but may be somewhat active during the day. It is characteristically seen perched on the leaf litter. At night it perches on low vegetation about 0.5 m above the ground, the usual call site for males. It may seek refuge in the leaf litter during the day and sometimes is flushed out by simply walking or kicking the litter. However, in a thirteen-month sampling of the leaf litter herpetofauna at La Selva, Heredia Province, only two *E. crassidigitus* were collected (Lieberman 1986). This suggests that most activities are carried out above the ground surface in the understory layer of the forest. Although the species is also found near small stream courses, it is not a strongly riparian form. Activity is year round but is definitely reduced during dry periods. Males call throughout the year but only occasionally and lazily during dry spells or in the dry season. Like its sympatric ally *E. fitzingeri*, males call occasionally at wide intervals on dark or rainy days. Similarly, males call with greatest frequency around dusk and infrequently later in the evening. Males are particularly vocal on evenings following heavy afternoon rains. The calls act to space males in the environment (they are territorial) as well as to attract mates. *E. crassidigitus* is most abundant at upland sites (900–1,500 m) but is less common than sympatric *E. fitzingeri* and *E. talamancae* at lowland localities. Food consists of arthropods; Lieberman (1986) reported on stomach contents of two examples containing two prey items each; prey ranged in length from 2.5 to 40 mm and included centipedes, orthopterans, dipterans, and isopods. In the original description of the species Taylor (1952a) records collecting a specimen (presumably a female) and 26 eggs from under a rock (pattern 21, table 7.1). The smallest juvenile is 12.1 mm in standard length. The principal predators on this species are a variety of frog-specialist snakes

(e.g., *Leptodeira* and *Liophis*) and the blunt-headed vine snakes of the genus *Imantodes*. This frog is probably also preyed on by predators that home in on the males' calls, especially the frog-eating bat *Trachops cirrhosis*. Lynch and Myers (1983) suggest that the concentration of calling at dusk that tapers off after dark may be a predator-avoidance defense against the last-named nocturnal predator. These authors also reported that a female of this species being eaten by a snake *(Liophis epinephalus)* emitted a high-pitched *Rana*-like distress call.

REMARKS: Dunn (1931a *et alibi*) applied the name *Eleutherodactylus longirostris* (Boulenger, 1898b) to Costa Rican and Panamanian examples of this taxon. Taylor (1952a) described *E. crassidigitus* from Costa Rica, but most authors continued to call Panamanian examples *E. longirostris*. Savage (1973b) recognized *E. crassidigitus* but, influenced by Dunn's identifications, thought *E. longirostris* might be a different, primarily Panamanian form that entered southwestern Costa Rica. Lynch and Myers (1983) demonstrated that *E. longirostris* is a valid species occurring in Colombia and extreme eastern Panama, distinct from *E. crassidigitus*. They further showed that all records of *E. longirostris* from Panama published before the Lynch and Myers opus are based on examples of *E. crassidigitus*. Similarly, all Costa Rican material confused with *E. longirostris* is composed of representatives of *E. crassidigitus*. The karyotype is apparently variable 2N = 22, composed of six pairs of metacentric chromosomes, two or three pairs of telocentrics, and two or three pairs of subtelocentrics; NF = 38, 40 (DeWeese 1976). The specimen identified as *E. longirostris* in the latter work is *E. crassidigitus*.

DISTRIBUTION: Humid lowland and premontane forest zones of Costa Rica to extreme northwestern Colombia (10–2,000 m).

Eleutherodactylus cuaquero Savage, 1980a
(map 7.24)

DIAGNOSTICS: A small (males) to moderate-sized (females) long-limbed frog having the posterior thigh surface marked with small bright yellow spots that may be fused to form vertical stripes and with basal toe webs extending no farther than the proximal margin of the proximal subarticular tubercle on any toe (fig. 7.48b). In addition, the finger disks are greatly expanded, truncate, and emarginate on fingers III and IV (fig. 7.51g), and a light midgular stripe is present (fig. 7.56).

DESCRIPTION: Adult male 39 mm in standard length, adult females 33 to 48 mm; upper surfaces and venter smooth; head as wide as long; dorsal outline of snout subovoid to subelliptical; nares lateral; eye large; tympanum large, height about two-thirds diameter of eye in males, one-third in females; paired vocal slits and single internal subgular vocal sac in adult male; finger I longer than finger II;

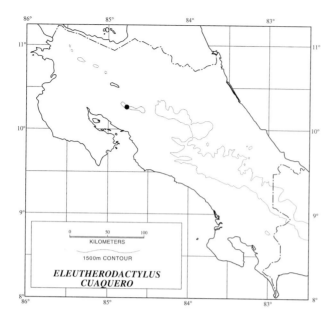

Map 7.24. Distribution of *Eleutherodactylus cuaquero*.

outer two finger disks greatly expanded, covers emarginate; subarticular tubercles on fingers and toes ovoid, globular to obtuse in profile; no supernumerary tubercles under fingers and toes; brownish nuptial thumb pads in adult males; thenar tubercle elongate, much smaller than bifid palmar tubercle; accessory palmar tubercles obsolescent; heel with several small tubercles; toe disks smaller than largest finger disks; only basal toe webbing; webbing formula: I 2^+–$2\frac{1}{2}$ II 2^+–$3\frac{1}{2}$ III 3^+–4^+ IV 4^+–3^- V; no plantar tubercles; inner metatarsal tubercle oval, outer round, much smaller; an inner tarsal fold. Upper surfaces of body and head dark brown with a few dark dorsal spots; limbs lighter with dark crossbars; posterior thigh surface dark brown with distinct small yellow spots or spots fused to form vertical stripes; throat and underside of hind limbs heavily marked with dark; a midgular light stripe; groin mottled; venter yellow; underside of hind limbs and groin bright yellow suffused with pink; iris upper half silver, lower brown.

VOICE: Unknown.

SIMILAR SPECIES: (1) *Eleutherodactylus andi* is much larger (adult males 40 to 55 mm in standard length, adult females 65 to 80 mm) with more toe webbing (web includes proximal subarticular tubercle on all toes) (fig. 7.48c) and usually large yellow groin spots. (2) *Eleutherodactylus fitzingeri* also has more toe webbing (similar to *E. andi*), smaller, nonemarginate finger disks (fig. 7.51f), a rugose dorsum, and a white to light yellow venter.

HABITAT: Lower limits of the Lower Montane Rainforest near Monteverde, Puntarenas Province.

BIOLOGY: A nocturnal form found on herbaceous vegetation 1.0 to 1.5 m above the ground.

REMARKS: Since its original collection (1964) and description (Savage 1980a), this species has been known from

only two adult females. A male (UCR 4418) from the continental divide (1,600 m) above Monteverde appears to belong here, although the disks of this individual are not obviously emarginate. It further lacks a midgular light stripe, and the thigh pattern, though barred and spotted with light, is not as vivid as the female's. This species' morphological similarity to the larger *E. andi* suggests that it is an arboreal canopy species, which may explain its seeming rarity.

DISTRIBUTION: Pacific slope lower montane zone of the Monteverde Forest Reserve, Puntarenas Province, Costa Rica (1,520–1,600 m).

Eleutherodactylus fitzingeri (O. Schmidt, 1857)

(plate 95; map 7.25)

DIAGNOSTICS: The light-spotted posterior thigh surface (fig. 7.50b), generally rugose dorsum, and moderate toe webbing (fig. 7.48c), with the midgular light stripe that is almost always present (fig. 7.56), will immediately separate this common small to moderate-sized species from all other similar forms.

DESCRIPTION: Adult males 23 to 35 mm in standard length, adult females 36 to 53 mm; dorsum rugose with many warts and some short ridging; venter smooth; head about as long as wide; snout subelliptical from above; eyes large; tympanum large, height about four-fifths diameter of eye in males, about three-fifths in females; paired vocal slits and single internal subgular vocal sac in adult males; finger I longer than finger II; finger disks moderately expanded; disk pads on I and II round, on III and IV truncate and larger, nonemarginate; subarticular tubercles on all digits ovoid, conical to obtuse in profile; no supernumerary tubercles under fingers and toes; brownish nuptial thumb

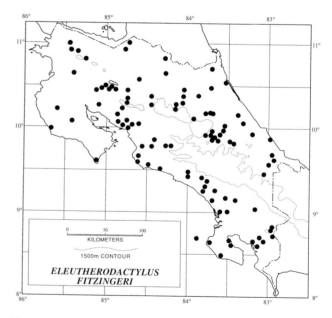

Map 7.25. Distribution of *Eleutherodactylus fitzingeri*.

pads in adult males; thenar tubercle elongate, palmar much larger, bifid; accessory palmar tubercles obsolescent; heel rugose; toe disks smaller than finger disks; toe webbing moderate; webbing formula: I 2⁻–2⁺ II 1¾–3⁻ III 2½–4⁻ IV 4⁻–2½ V; no plantar tubercles; inner metatarsal tubercle elongate, outer smaller, round; an inner tarsal fold. Dorsal ground color grayish or orangish brown, tan, or olive, often with a broad light gray, tan, yellow, or orange middorsal stripe; darker elements form a vague complex pattern and/ or a few blackish spots; posterior thigh surface brown to nearly black with well-demarcated light spots, best developed in juveniles at proximal end of femoral segment; upper lip usually barred; throat gray to dark gray brown, almost always with a median light stripe; upper surface of limbs with transverse dark bars; venter white, pale yellow, or greenish yellow; underside of hind limbs yellow to greenish yellow; iris divided by brown or red line, upper portion usually bronze, lower grayish; sometimes entire iris tan or gray.

VOICE: The advertisement call is a series (2 to 18) of short, harsh "clacks" sounding like two stones struck together, lasting about 2.5 seconds and repeated after some delay. The dominant frequency is about 3.0 kHz with a wide spread of energy between 1.5 to 4.0 kHz (Lynch and Myers 1983).

SIMILAR SPECIES: (1) *Eleutherodactylus andi* is a much larger species (adult males 40 to 55 mm in standard length, adult females 65 to 85 mm) having emarginate outer finger disks (fig. 7.51g), a smooth dorsum, and usually large yellow spots in the groin. (2) *Eleutherodactylus cuaquero* has the venter, underside of hind limbs, and groin bright yellow and the small yellow spots on the posterior thigh surface often fused into vertical stripes. (3) *Eleutherodactylus ranoides* lacks expanded, truncate outer finger disks and a light mid-gular stripe, and adult males lack vocal slits and nuptial pads.

HABITAT: This ubiquitous lowland species is characteristic of humid forests and disturbed or forest edge situations. *E. fitzingeri* is also found in similar habitats in Premontane Wet Forest and Rainforest and marginally in Lower Montane Wet Forest and Rainforest. In Lowland Dry Forest areas it is restricted to riparian gallery forests.

BIOLOGY: This very common frog is primarily nocturnal in activity. Both sexes are usually found at night in the understory vegetation or bushes along the forest edge 0.5 to 1.6 m above the ground. Males typically call from elevated positions on stumps, fallen logs, and low vegetation. Some activity also occurs during the day, and they may be seen sitting on or partially concealed in the leaf litter. Nevertheless, in a thirteen-month study of the leaf litter herpetofauna at La Selva, Heredia Province, only two *E. fitzingeri*, a very common frog at this site, were collected in the litter samples (Lieberman 1986). Food consists of arthropods other than ants. *E. fitzingeri* is much more common at lowland than upland sites, the converse of the sympatric *E. crassidigitus*,

which is most abundant at higher elevations. Although this species is often found along the margins of stream courses and may escape capture by jumping into the water and hiding on the bottom, it is not a particularly riparian form. Escape is by a series of long leaps (usually three), each in a different direction, ending in a dive under the leaf litter. Males tend to call with greatest frequency at dusk, particularly after heavy afternoon rains, and then only sporadically at night. Males also make widely spaced calls on dark or rainy days. The calls appear to serve a territorial function in spacing males as well as in attracting females. Unfortunately the calls also attract predators, including the frog-eating bat *Trachops cirrhosus*. Lynch and Myers (1983) suggested that the reduction of call frequency after dusk is a predator-avoidance mechanism. A single report (Dunn 1931c) of 44 eggs attended by an adult constitutes the only data on reproduction. These were found on June 2 under leaves (pattern 21, table 7). Although Dunn indicated that the attending parent was a male, size alone (51 mm) suggests it was a female (Lynch and Myers 1983). A hatchling is 11.1 mm in standard length, and small juveniles are 14 to 15 mm.

REMARKS: Most of the Nicaraguan and all of the Costa Rican frogs referred to as *Eleutherodactylus nubilis* by Noble (1918) and *Eleutherodactylus ranoides* by Taylor (1952a), respectively, are examples of *E. fitzingeri*. Noble (1924) misapplied the name *Eleutherodactylus longirostris* to Panamanian *E. fitzingeri*, and Cochran and Goin (1970) included Nicaraguan, Costa Rican, and Panamanian specimens of this species in their composite *Eleutherodactylus longirostris*. Savage (1974b) and Lynch and Myers (1983) provide more detailed accounts of the complicated history of the several names and their misapplications and detailed descriptions of *E. fitzingeri*. The usual rugosity of the dorsum with its projecting warts and ridging is obvious in living specimens but may disappear in preservative. The karyotype is 2N = 22, usually with four pairs of metacentric chromosomes, three pairs of submetacentrics, three pairs of telocentrics, and one pair of subtelocentrics. However, there is individual and geographic variation, especially in the number of metacentric and submetacentric pairs, so that the NF = 36, 38, or 40 (DeWeese 1976).

DISTRIBUTION: Lowland and premontane slopes from northeastern Honduras and northwestern Costa Rica, south to northwestern and central Colombia; marginally into the lower montane zone in Costa Rica as well (0–1,520 m).

Eleutherodactylus melanostictus (Cope, 1875)

(plate 96; map 7.26)

DIAGNOSTICS: A moderate-sized species that lacks toe webs and has large truncate, emarginate disks on fingers II and III and toes II–III–IV (fig. 7.51g). These frogs are unusual in that the distinct black thigh bars usually continue onto the posterior thigh surface (fig. 7.50e). The light spaces between the bars are usually bright yellow, greenish, orange,

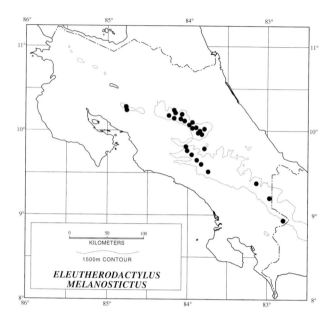

Map 7.26. Distribution of *Eleutherodactylus melanostictus.*

pink, magenta, or scarlet, although sometimes the light areas are obscured by a suffusion of dark pigment.

DESCRIPTION: Adult males 35 to 43 mm in standard length, adult females 35 to 56 mm; dorsum relatively smooth with scattered tubercles; venter granulate; head about as wide as long; snout subovoid to subelliptical from above; eyes large; a single large supraorbital tubercle and several smaller superciliary tubercles on upper eyelid; tympanum internal, indistinct, round in males, oval in females; vomerine teeth in two short transverse series, posterior to choanae; paired vocal slits and single internal subgular vocal sac in adult males; finger I shorter than finger II; finger disks greatly expanded; disk covers on fingers III and IV truncate, emarginate; subarticular tubercles on hands and feet ovoid, flattened to projecting; no supernumerary tubercles under digits; whitish nuptial pad on thumb in adult males; thenar tubercle large, elongate, cordate; palmar larger; accessory palmar tubercles obsolescent; a distinct enlarged tubercle on heel; toe disks smaller than largest finger disks, emarginate or not; no toe webs; no plantar tubercles; elongate inner metatarsal tubercle, outer small, round, indistinct; no tarsal fold. Upper surface pale tan to dark brown variously marked with green, yellow green, yellow, orange, magenta, or scarlet; a narrow tan middorsal pinstripe or a broader chestnut to reddish middorsal stripe often present or entire central area a very broad yellow to scarlet stripe; some examples with a few dark spots or short lines, a W-shaped suprascapular dark mark, or a series of dark marks and/or chevrons on back; middorsal pinstripe when present continuing as narrow light line on back of each thigh; flanks with oblique dark elongate blotches outlined by black to form tigerlike pattern; dorsal, anterior, and posterior thigh surfaces with continuous dark bars; light areas between bars on posterior thigh surface bright yellow, greenish yellow, orange, pink, magenta, or scarlet, sometimes suffused with dark pigment; throat and venter white to heavily marked or spotted with dark pigment; iris coppery gold above, dark brown below.

SIMILAR SPECIES: (1) *Eleutherodactylus cruentus* is a smaller species (adult males 23 to 28 mm in standard length, adult females 34 to 42 mm) that has toe V much longer than toe III (fig. 7.48a), very obvious accessory palmar tubercles, a strongly areolate venter and usually bright yellow to orange spots in the groin and on the anterior surface of the thigh (fig. 7.58). (2) *Eleutherodactylus rugosus* has a very broad head, has finger I much longer than finger II, and lacks expanded finger disks and a heel tubercle.

HABITAT: Relatively undisturbed situations in the upper portions of Premontane Wet Forest and Rainforest and throughout Lower Montane Wet Forest and Rainforest and Montane Rainforest areas.

BIOLOGY: An uncommon nocturnal frog that is usually found in dense forest stands on herbaceous vegetation 1.0 to 2.5 m above the ground that also are male calling sites. During the day individuals hide in bromeliads and under rocks and logs. Small juveniles are 16 to 17 mm in standard length.

REMARKS: Taylor (1952a) used the name *Eleutherodactylus platyrhynchus* (Günther, 1900) for his specimens of this species. Although he also included *Eleutherodactylus melanostictus* in his key and a redescription based on Cope's (1875) original description, Savage and DeWeese (1981) demonstrated that only one species was involved. The latter authors pointed out that samples from higher elevations (2,000 to 2,400 m) in the Cordillera Central and northern Talamancas differed from those from lower slopes (1,150 to 1,600 m) and in the Cordillera de Tilarán (1,500 to 1,600 m) as follows (features for lower-elevation populations in parentheses): snout rounded in profile (obtuse), disk on finger II small, round (expanded, palmate), toe disks small (large, emarginate), throat and venter dark (venter light), and posterior thigh interspaces yellow, yellow green, purplish, or orange (scarlet). Individuals from intermediate elevations and the southern Talamanca range bridge the apparent gap. The karyotype is 2N = 22, with five chromosome pairs metacentric, one pair submetacentric, one pair subtelocentric, and four pairs telocentric; NF = 36 (Savage and DeWeese 1981).

DISTRIBUTION: Humid forests in the upper premontane, lower montane, and montane zones of the cordilleras of Costa Rica and western Panama (1,150–2,700 m).

Eleutherodactylus phasma Lips and Savage, 1996b

(plate 97; map 7.27)

DIAGNOSTICS: A long-legged species known only from a single moderate-sized female that is immediately recognizable by its predominantly gray white coloration. Toe web-

Map 7.27. Distribution of *Eleutherodactylus phasma*.

bing in this form is basal, extending no farther than the proximal margin of the proximal subarticular tubercle on any toe (fig. 7.48b), some of the digital disks are indented (fig. 7.51c), and the posterior thigh surface is uniform gray white.

DESCRIPTION: Adult males estimated to be 25 to 30 mm in standard length, adult females 35 to 50 mm; upper and lower surfaces smooth; head slightly longer than broad; snout subovoid in dorsal outline; nares lateral; eyes large; tympanum distinct, elliptical, height slightly more than one-half diameter of eye in adult female; finger I longer than finger II; outer two finger disks greatly expanded, with palmate disk covers, indented on finger III; subarticular tubercles on all digits round and globular in profile; no supernumerary tubercles under digits; thenar tubercle large, elongate, smaller than very large bifid palmar tubercle; accessory palmar tubercles obsolescent; no heel tubercles; toe disks smaller than largest finger disks; only basal toe webbing; webbing formula: I 1–2 II 2–3 III 3–4 IV 4¼–2¾ V; no plantar tubercles; inner metatarsal tubercle large, elongate; outer metatarsal tubercle low, round, much smaller than inner; a weak inner tarsal fold. Dorsal ground color gray white; fifteen to twenty black spots coalesced into blotches in scapular region; a narrow black stripe from canthus along outer margin of eyelid to tympanum; tympanum mostly dark brown; isolated black blotches on one knee and tibiae; lateral, ventral, and thigh surfaces gray white; iris black.

VOICE: Not known.

SIMILAR SPECIES: There are none.

HABITAT: Lower Montane Rainforest.

BIOLOGY: The type and only known specimen was found on a rocky stream bank.

DISTRIBUTION: Known only from the type locality on

the bank of the Río Cotón at Finca Jaguar in the Las Tablas area, Puntarenas Province, near the border with Panama in southwestern Costa Rica (1,850 m).

Eleutherodactylus rayo Savage and DeWeese, 1979

(map 7.28)

DIAGNOSTICS: A rare moderate-sized to large (females larger) frog having a moderate amount of toe webbing (fig. 7.48c), a uniform purple posterior thigh surface, and an enlarged heel tubercle (fig. 7.49b).

DESCRIPTION: Adult males 37 to 45 mm in standard length, adult females 38 to 71 mm; dorsum and venter smooth; head about as long as wide; snout subelliptical from above; eye large; tympanum internal, indistinct, round in males, oval in females, its diameter about one-half diameter of eye; paired vocal slits and single internal subgular vocal sac in adult males; finger II longer than finger I; finger disks expanded; disks on fingers III and IV larger than on I and II with emarginate disk covers; subarticular tubercles on fingers ovoid and globular; no supernumerary tubercles on digits; whitish nuptial pads on thumbs of adult males; thenar tubercle large, elongate, palmar larger, ovate; no accessory palmar tubercles; a heel tubercle, toe disks smaller than finger disks, those on II–III–IV palmate, emarginate; toe webbing moderate; webbing formula: I 2⁻–2⁺ II 2–3 III 3–4⁺ IV 4⁺–3⁻ V; subarticular tubercles on toes ovoid and obtusely projecting; inner metatarsal tubercle elongate, outer round and much smaller; an inner tarsal fold. Ground color of upper surfaces dark brown with bluish purple cast; a white middorsal pinstripe or broader middorsal light stripe sometimes present; olive green blotches present on dorsum of some males; upper surfaces of limbs uniform or with transverse dark bars; posterior and ventral thigh surfaces uniform

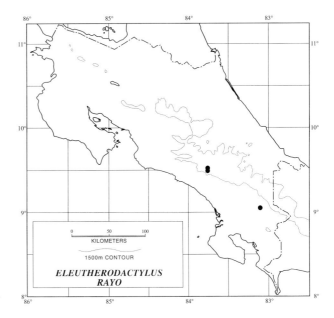

Map 7.28. Distribution of *Eleutherodactylus rayo*.

dark purplish; throat dark with median light stripe; venter light with heavy mottling of dark pigment; undersides of hind limbs and soles uniform purplish.

VOICE: Not known.

SIMILAR SPECIES: *Eleutherodactylus talamancae* has a dark eye mask, usually has a white upper lip stripe, and lacks a heel tubercle.

HABITAT: Stream margins in the upper portions of Premontane Wet Forest and Rainforest and the lower section of Lower Montane Rainforest.

BIOLOGY: The male holotype was captured during the day while calling from a bromeliad during a heavy rain. Other individuals were captured from under debris along a small montane stream. Others apparently were frightened from hiding places and were hopping along the stream banks or across the shallows.

REMARKS: The karyotype is 2N = 22, composed of five pairs of metacentric chromosomes, one pair of submetacentrics, one pair of subtelocentrics, and four pairs of telocentrics; NF = 36 (Savage and DeWeese 1979).

DISTRIBUTION: Upper premontane zone and lower portions of the lower montane belt in the Cordillera de Talamanca of Costa Rica (1,480–1,820 m).

Eleutherodactylus talamancae Dunn, 1931a

(plate 98; map 7.29)

DIAGNOSTICS: This elegant, long-limbed small (males) to moderate-sized (females) frog has the posterior thigh surface uniform yellowish to reddish brown suffused with red, has moderate toe webbing (fig. 7.48c), and never has a light midgular stripe. Juvenile individuals always have an enamel white stripe along the upper lip, and the stripe or white lip spots are present in many adults.

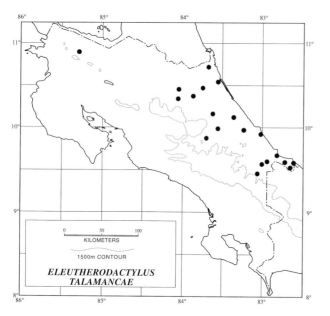

Map 7.29. Distribution of *Eleutherodactylus talamancae*.

DESCRIPTION: Adult males 21 to 30 mm in standard length, adult females 34 to 50 mm; dorsum finely tuberculate; head about as long as wide; snout subelliptical from above; eyes large; tympanum prominent, diameter about two-thirds diameter of eye in males, one-half in females; paired vocal slits and single internal subgular vocal sac in adult males; finger I longer than finger II; finger disks moderately expanded, disk cover on I and II rounded, on III and IV truncate, nonemarginate, and larger; subarticular tubercles on hands obtuse, on feet round and globular in profile; no supernumerary tubercles under digits; mottled brown and white nuptial pad on thumb in adult males; thenar tubercle elongate, palmar bifid, larger; heel smooth; toe disks smaller than largest finger disks; toe webbing moderate; webbing formula: I 2–2¼ II 2–3½ III 3–4⁺ IV 4⁺–3⁻ V; no plantar tubercles; inner metatarsal tubercle elongate, much larger than small round outer tubercle; no inner tarsal fold. Dorsal ground color brown, uniform, or with some small black spots; often with a broad light-outlined brown hourglass-shaped figure and/or a series of transverse dark brown stripes posteriorly; distinct dark eye mask present; a white upper lip stripe usually bordered above by a narrow golden line and one to three white oblique stripes on flank in juveniles; lip stripe retained in some adults or reduced to a series of light spots but absent in many large adults; dark transverse bars on upper surface of limbs; throat and venter white shading to yellow posteriorly; throat, groin, and thigh surfaces suffused with red to purple; venter immaculate, white to dusky in juveniles; upper portion of iris gold, lower dark brown.

VOICE: High-pitched single "mew" given sporadically.

SIMILAR SPECIES: (1) *Eleutherodactylus crassidigitus* has toe webbing that reaches well beyond the proximal subarticular tubercle on toe IV (fig. 7.48d) and a definite inner tarsal fold. (2) *Eleutherodactylus rayo* has a well-developed heel tubercle (fig. 7.49b) and inner metatarsal fold and lacks red pigmentation in the groin and thigh area. (3) *Eleutherodactylus gollmeri* has a tarsal fold, and its dark eye mask (when present) continues as a dark line well posterior to the axilla (fig. 7.46e). (4) *Eleutherodactylus noblei* has a tarsal fold and paired dorsolateral glandular ridges (fig. 7.36a).

HABITAT: Undisturbed Lowland Moist and Wet Forests of the Atlantic slope.

BIOLOGY: Frogs of this nocturnal species are usually found perched in low vegetation but hide in the leaf litter by day. When disturbed they escape in a series of bounding leaps, usually followed by freezing on top of or just under the uppermost leaf litter. Food consists primarily of a variety of small arthropods exclusive of hemipterans. In twenty-three stomachs analyzed, the average prey number per stomach was 1.9; prey lengths averaged 1.6 mm (ants) to 14 mm (beetle larvae) (Lieberman 1986). One specimen examined in that study had eaten a 14 mm anole (*Norops limifrons*). Density of *Eleutherodactylus talamancae* at La Selva, Heredia Province, where the diet study was carried out, averaged

43/ha. A hatchling of the species is 10.1 mm in standard length and a small juvenile is 13 mm.

REMARKS: The karyotype is 2N = 22, consisting of four pairs of metacentric chromosomes, two pairs of submetacentrics, one pair of subtelocentrics, and four pairs of telocentrics; NF = 36 (DeWeese 1976).

DISTRIBUTION: Humid lowland forests of the Atlantic versant from Nicaragua to eastern Panama (15–646 m).

Eleutherodactylus gollmeri Group

Members of this lineage in Costa Rica are elegant froglike forms with narrow and relatively pointed snouts. Other characteristics are finger I longer than finger II; toe III longer than toe V; prominent tympanum; vomerine odontophores triangular, posterior to choanae, and narrowly separated medially; no vocal slits or nuptial thumb pads; disks on all digits, but usually none are expanded and none are notched; subarticular tubercles projecting; no supernumerary tubercles on hands or feet; toes webbed; an inner tarsal fold; venter smooth.

These animals are often diurnal and characteristically bound through the forest in a zigzag pattern of long leaps when disturbed, using surprise, speed, and evasive action to escape danger. Because of their cryptic coloration, these frogs may pass unnoticed by an observer until they seemingly explode out of the leaf litter directly underfoot and disappear into the underbrush with graceful, arching leaps. Unlike most of their congeners, males of the *Eleutherodactylus gollmeri* group apparently lack the ability to vocalize. As a result their reproductive behavior remains unknown, although in most forms the eggs probably are laid in the leaf litter.

Eleutherodactylus gollmeri (Peters, 1863d)

(plate 99; map 7.30)

DIAGNOSTICS: A moderate-sized, long-legged diurnal frog with basal toe webbing (fig. 7.49d), the finger disks barely expanded and subequal, and one or two enlarged heel tubercles (fig. 7.49b,c). In addition the upper half of the iris is red in life, the posterior thigh surface is uniform rust brown, and most examples have a dark eye mask that extends beyond the axilla onto the flank (fig. 7.46e).

DESCRIPTION: Adult males 30 to 37 mm in standard length, adult females 45 to 54 mm; dorsum mostly smooth with some well-developed supratympanic and suprascapular tubercles; venter smooth; head about as long as broad; snout subelliptical from above; eyes large; tympanum large, height greater than diameter of eye in males, about equal in females; no vocal slits or sac; finger disks all about same size, outer two with lanceolate covers and cuspidate pads; subarticular tubercles on all digits ovoid, projecting, and obtuse (pungent in juveniles); no supernumerary tubercles; thenar tubercle elongate, palmar usually bifid, otherwise ovoid; no nuptial thumb pad; a few low accessory palmar tubercles; one or two distinct conical heel tubercles; toe disks barely

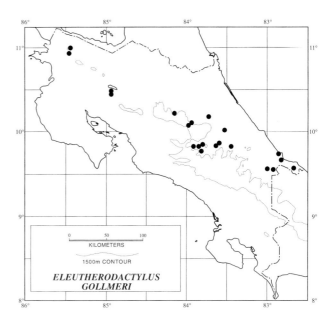

Map 7.30. Distribution of *Eleutherodactylus gollmeri*.

expanded, disk pads cuspidate; no fleshy fringe along toe margins; only basal toe webbing; toe webbing formula: I 2– 2½ II 2–2¾ III 3¼–4 IV 4–3 V; inner metatarsal tubercle raised, elongate, outer small, round; no plantar tubercles; an inner tarsal fold. Dorsum tan to dark brown with thin middorsal pinstripe, light-centered small suprascapular black spots, and paired short, dark paravertebral blotches usually present; dark eye mask usually present, sometimes broken up into dark lip bars; dark transverse bars on upper surface of limbs; posterior thigh surface uniform rust brown not suffused with red; seat patch mark usually a pair of black dots above and lateral to vent, sometimes spots fused or extending ventrally for short distances, discontinuous from dark area on posterioventral surface of thigh; definite dark stripe on anterior thigh surface continuing along anterior margin of lower limb; undersurface white, nearly immaculate; iris red above, brown below.

SIMILAR SPECIES: (1) *Eleutherodactylus mimus* has substantial toe webbing (fig. 7.49e) and lacks enlarged heel tubercles. (2) *Eleutherodactylus noblei* has the outer two finger disks greatly enlarged and the posterior thigh surface area suffused with red. (3) *Eleutherodactylus talamancae* lacks a tarsal fold, and its eye mask terminates near the axilla. (4) *Eleutherodactylus podiciferus* has a tarsal fold and lacks expanded digital disks (fig. 7.51b).

HABITAT: A leaf litter habitué in Lowland Moist and Wet Forests and Premontane Wet Forest and Rainforest.

BIOLOGY: Individuals of this relatively common species forage in the leaf litter during the day and hide under the surface leaves at night. When disturbed they escape with bounding leaps as they search for cover. Food consists of small arthropods exclusive of ants, principally orthopterans (Toft 1981).

REMARKS: *Eleutherodactylus gollmeri* was originally

thought to have been from Venezuela (Peters 1863d); Dunn and Emlen (1932) realized that it was a Central American form, and Rivero (1961) convincingly demonstrated that a mix-up of data was involved and the lectotype (ZMB 3168) of *Eleutherodactylus gollmeri* must have been collected in western Panama. Savage (1987) showed that the name *Eleutherodactylus lanciformis*, usually applied to examples of the species between 1878 and 1932, is a synonym. Unfortunately Dunn and Emlen (1932) and Meyer and Wilson (1971a) cite this species as ranging from Guatemala to Panama. As Savage (1987) pointed out all, Nicaraguan records are based on misidentified specimens of *Eleutherodactylus mimus* and *Eleutherodactylus noblei* and Guatemala and Honduras records of distinctive related forms, *Eleutherodactylus laticeps* (A. Duméril, 1853) and *Eleutherodactylus rostralis* (Werner, 1896). Although found at a number of lowland localities, *Eleutherodactylus gollmeri* is mainly a premontane form in Costa Rica (most records 640–1,500 m). Its ally *Eleutherodactylus mimus* is more lowland in distribution (15–640 m). In Panama, where only *Eleutherodactylus gollmeri* is found, it occurs regularly at lower elevations. The listing of this species from La Selva, Heredia Province (Lieberman 1986), is based on juvenile *Eleutherodactylus mimus*.

DISTRIBUTION: Lowland and premontane humid forests of eastern and central Panama (10–850 m) and Atlantic slopes of northwestern Panama and eastern Costa Rica; a few records for premontane Pacific slope Costa Rica (10–1,520 m).

Eleutherodactylus mimus Taylor, 1955

(plate 100; map 7.31)

DIAGNOSTICS: A moderate-sized long-limbed, narrow-headed frog having fringed toes, considerable toe webbing,

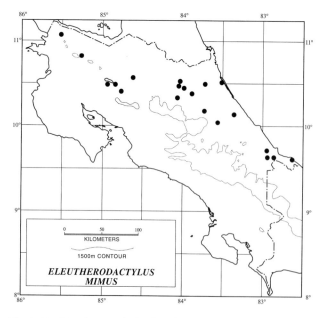

Map 7.31. Distribution of *Eleutherodactylus mimus*.

barely expanded finger disks (fig. 7.49e), no enlarged heel tubercles, a uniform brown posterior thigh surface, and a yellow iris in life. In addition, the dark eye mask continues well beyond the axilla onto the flank.

DESCRIPTION: Adult males 30 to 37 mm in standard length, adult females 45 to 58 mm; dorsum smooth, except for paired supratympanic and suprascapular tubercles; venter smooth; head about as broad as long; snout pointed in dorsal outline; eyes large; tympanum prominent, height about equal to diameter of eye in males, slightly less in females; no vocal slits or sac; finger disks subequal with rounded covers; outer two with swollen disk pads; subarticular tubercles on all digits round, globular in profile, projecting; no supernumerary tubercles; thenar tubercle ovoid to elongate, smaller than bifid palmar tubercle; no nuptial thumb pad; two or three low accessory palmar tubercles; no heel tubercles; toe disks barely expanded; toes with fleshy fringe; toes with substantial webbing; webbing formula: I 1–¼ II 1½–3½ III 2½–3¾ IV 3–2½ V; distinct elongate, raised inner and small round outer metatarsal tubercles; zero to three plantar tubercles; an inner tarsal fold. Dorsum tan to medium brown, essentially uniform or with a dark middorsal blotch; middorsal pinstripe invariably present; dark eye mask extending well beyond axilla onto flank; dark transverse bars on upper surface of limbs; posterior thigh surface uniform brown, not suffused with red; seat patch dark mark an inverted V or U, not continuous with dark posterioventral thigh surface; definite dark stripe along anterior surface of thigh to ankle; undersurface white, nearly immaculate; iris yellow above, brown below.

VOICE: No sound production.

SIMILAR SPECIES: (1) *Eleutherodactylus gollmeri* has red eyes, heel tubercles (fig. 7.49b,c), and basal toe webbing and lacks fleshy toe fringes (fig. 7.49d). (2) *Eleutherodactylus noblei* has the outer two finger disks much larger than those on fingers I and II, only basal toe webbing (fig. 7.49d), and the posterior thigh suffused with red. (3) *Eleutherodactylus crassidigitus* has disks on fingers III and IV truncate (fig. 7.51f) and much larger than on I and II and lacks fleshy toe fringes. (4) *Eleutherodactylus podiciferus* has a tarsal fold and lacks expanded digital disks (fig. 7.51b).

HABITAT: On the forest floor in undisturbed Lowland Moist and Wet Forest zones of the Atlantic slope.

BIOLOGY: A common, primarily diurnal species that forages on or just under the superficial layer of leaf litter. Because of their cryptic coloration these frogs pass unnoticed until they seemingly explode directly from underfoot and disappear to safety with graceful leaps. Food consists of a variety of small insects and centipedes. In fourteen stomachs analyzed from La Selva, Heredia Province, the number of food items averaged 1.3 per stomach. Average length for prey was 3.6 mm (termites) to 14 mm (orthopterans). Densities at this site are relatively low (Lieberman 1986), about 26/ha. The smallest juveniles are 14 mm in standard length.

REMARKS: Various workers, especially E. R. Dunn, confused juveniles of this form with *E. gollmeri*. Taylor (1955) was the first to recognize its distinctiveness, and Savage (1987) pointed out that several putative records of *E. gollmeri* from Nicaragua are based on examples of *E. mimus*. This species and *E. gollmeri* appear to have distinctive ecological requirements. Throughout most of its range *E. mimus* is principally a lowland form, found between 15 and 640 m in altitude. Although found at lower elevations in Panama where *E. mimus* does not occur, *E. gollmeri* in Costa Rica is primarily an upland frog found most abundantly between 640 and 1,500 m. Farther to the north where *E. gollmeri* does not range, *E. mimus* is found from 100 to 940 m in elevation. The two species are marginally sympatric at a few sites in Costa Rica but appear to replace each other ecologically over much of the country. The records of *E. gollmeri* (three specimens) at La Selva, Heredia Province, by Lieberman (1986) are based on juvenile *E. mimus*. These misidentifications are the source for Donnelly's (1994a, 206) reference to the species there. *E. gollmeri* is not known from the La Selva Field Station property, nor is it likely to be found there, and Donnelly (1994b, 381) does not include it in her definitive listing of the La Selva amphibian fauna. The karyotype is 2N = 20, composed of seven pairs of metacentric, one pair of submetacentric, and two pairs of telocentric chromosomes; NF = 36 (DeWeese 1976).

DISTRIBUTION: Slopes of the lowland and premontane Atlantic versant from northeastern Honduras and eastern Nicaragua (100–940 m) to southeastern Costa Rica (15–685 m).

Eleutherodactylus noblei Barbour and Dunn, 1921

(plate 101; map 7.32)

DIAGNOSTICS: A moderate-sized, long-legged frog having basal toe webbing (fig. 7.49d), the outer two fingers with disks much larger than those on fingers I and II, no heel tubercles, and the groin and uniform posterior thigh surface suffused with red. In addition, it lacks a dark seat patch mark, the dark eye mask continues onto the flank, and the iris is gold.

DESCRIPTION: Adult males 43 to 48 mm in standard length, adult females 58 to 66 mm; dorsum weakly granulate; dorsal tubercles poorly developed; venter smooth; a well-developed glandular dorsolateral ridge from eye nearly to groin; head about as broad as long; snout pointed in dorsal outline; eyes large; tympanum large, height about equal to diameter of eye in males, somewhat less in females; no vocal slits or sac; outer two finger disk covers palmate, much larger than round disks on fingers I and II; disk pads on III and IV broadened; subarticular tubercles on all digits rounded, obtuse, and projecting; no supernumerary digital tubercles; thenar tubercle elongate, about half as large as bifid palmar tubercle; no nuptial thumb pad; no heel tubercles; toe disks only slightly wider than toes; toe pads cuspidate; no fleshy fringe along toe margins; basal toe webbing; toe webbing formula: I 2⁺–2⁺ II 2⁺–3¾ III 3–4¼ IV

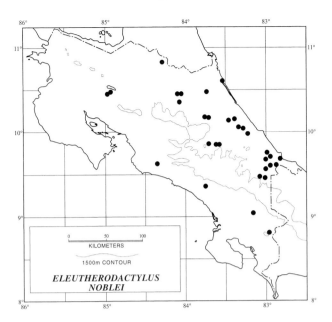

Map 7.32. Distribution of *Eleutherodactylus noblei*.

4¼–3 V; distinct elongate inner metatarsal and small round outer metatarsal tubercles; several (four to seven) distinct plantar tubercles; an inner tarsal fold. Dorsum light tan to dark brown, usually with a dark brown hourglass-shaped middorsal figure outlined by light pigment; usually a series of narrow dark dorsolateral and/or lateral longitudinal lines; body frequently covered by pink wash; usually no dark stripe along anterior face of hind limb; upper surface of limbs with dark transverse bars; no dark seat patch; uniform gray or brown posterior thigh suffused with red to bright scarlet; groin suffused with red; undersurface white, immaculate to heavily flecked with brown, especially on throat and limbs; upper iris gold, lower portion brown.

VOICE: Does not vocalize.

SIMILAR SPECIES: (1) *Eleutherodactylus talamancae* lacks a tarsal fold and any indication of a dorsolateral longitudinal ridge. (2) *Eleutherodactylus gollmeri* has heel tubercles (fig. 7.49b,c), and the outer finger disks are not enlarged. (3) *Eleutherodactylus mimus* has fleshy toe fringes and substantial webbing (fig. 7. 49e).

HABITAT: Lowland Moist and Wet Forests and Premontane Wet Forest and Rainforest in both primary forest and forest edge situations.

BIOLOGY: A relatively uncommon diurnal forest floor species.

REMARKS: Although Barbour and Dunn (1921) recognized and clearly defined this distinctive form, the latter author subsequently identified several juvenile *E. noblei* from Nicaragua and Costa Rica as *E. gollmeri*. Savage (1987) showed that the latter species is restricted to Panama and Costa Rica and that more northern records are based on misidentified examples of *E. mimus, E. noblei,* and *E. rostralis,* a species found only in northwestern Honduras and adjacent Guatemala.

DISTRIBUTION: Lowland and premontane humid Atlantic slope forests from northeastern Honduras to western Panama and Pacific versant Costa Rica and west-central Panama (4–1,200 m).

Eleutherodactylus rugulosus Group

Rather robust short, stout-limbed frogs with blunt snouts and the head about as long as wide; other distinctive features include finger I usually longer than finger II; toe III longer than toe V; vomerine odontophores triangular, posterior to choanae, and narrowly separated medially; distinct tympanum, vocal slits, and nuptial thumb pads present or absent in adult males; if vocal sac present, subgular and internal; disks on all fingers and toes barely to strongly expanded, none notched; subarticular tubercles flattened; no supernumerary tubercles on fingers or toes; accessory palmar and plantar tubercles obsolescent; toes webbed at least basally; an inner tarsal fold; venter smooth.

These frogs are essentially nocturnal as adults and feed while moving along the banks and over rocks and boulders in stream bottoms, except during periods of extreme high water in the wet season. All members of this group show marked sexual dimorphism, with the females often twice as large as the males and reaching lengths of nearly 100 mm in some populations. Breeding takes place on the ground along stream banks during the rainy season, but the males lack a breeding call, or the call is so weak as to be barely perceptible to the human ear above the roar of a stream or in driving rain. Although these frogs have a distinctive silver green eyeshine that lets collectors find them at night, the females' large size and collectors' inability to locate males by their calls makes the group one of the few frog taxa represented in scientific collections by disproportionately more females than males. Even with this imbalance between adult females and males, most of the material of these species in collections is composed of juveniles. Young tend to be active in the daytime and are the commonest frogs found under rocks, leaves, and other debris in stream bottoms and adjacent areas, particularly in the drier periods of the year. At most localities within the range of the group, any small, robust, rather short-legged gray or brown frog captured within 10 m of a rocky stream course will be a juvenile of this stock.

Eleutherodactylus angelicus Savage, 1975

(plate 102; map 7.33)

DIAGNOSTICS: Adult males of this common species are moderate-sized frogs lacking vocal slits but having nuptial thumb pads. Adult females are large. The species has finger and toe disks that are wider than the digits (fig. 7.51b) and only basal toe webbing (fig. 7.48b); the ventral color in adults is often bright yellow orange, orange, or red.

DESCRIPTION: Adult males 26 to 46 mm in standard length, adult females 38 to 75 mm; dorsum smooth to granulate; venter smooth; head about as wide as long; snout subovoid to rounded from above; eyes large; upper eyelid smooth to tuberculate; tympanum round and diameter two-thirds to three-fourths diameter of eye in males, oval and one-third to one-half in females; no vocal slits or sac; finger I longer than finger II; finger disk covers rounded, about one and a half times the width of digit on fingers III and IV; subarticular tubercles on hands and feet rounded, not projecting; no supernumerary tubercles under digits; whitish nuptial pad on thumbs in adult males; thenar and palmar tubercles elongate, former slightly larger; toe disks definite, about one and a half times width of toe on toes III–IV–V; basal toe webbing; toe webbing formula: I 2^{+}–$2\frac{3}{4}$ II 2^{+}–$3\frac{1}{2}$ III 3–$4\frac{1}{2}$ IV $4\frac{1}{4}$–3 V; large inner and very small outer metatarsal tubercles; a strong inner tarsal fold. Dorsum dark olive or brown to bluish gray, usually marked with obscure darker spots and/or blotches; upper surface of limbs uniform or with obscure transverse dark bars; posterior thigh surface uniform or with obscure light mottling; venter usually yellow orange to bright orange or red in adults (pale yellow in one population); iris gold above, brown below, separated by black line.

SIMILAR SPECIES: (1) *Eleutherodactylus escoces* has larger finger disks that are nearly twice the width of the digits on fingers III and IV, and vocal slits in adult males. (2) *Eleutherodactylus fleischmanni* never has a brightly colored venter and has barely expanded finger and toe disks.

HABITAT: These are riparian frogs found along stream banks in Premontane Wet Forest and Rainforest and Lower Montane Wet Forest and Rainforest.

BIOLOGY: *Eleutherodactylus angelicus* is active at night, foraging along the steep sloping banks of mountain streams. Amplexus has been observed in April and May. Hayes (1985) reported on the nest and female attendance of an egg clutch

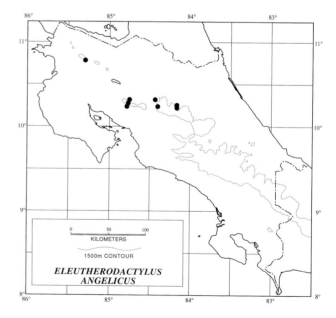

Map 7.33. Distribution of *Eleutherodactylus angelicus*.

for this species. The female laid the clutch of 77 loosely adherent eggs in a hole and then covered it with sand. She then sat on the covered clutch for several days. Individual eggs averaged 6.5 to 7 mm in diameter including the jelly layers, and the eggs were about 5 to 6 mm in diameter. Another female laid 51 eggs. Hayes indicated that females may lay up to three clutches a year and that the species probably uses internal fertilization. Small juveniles are 13 to 14 mm in standard length.

REMARKS: There is geographic, ontogenetic, and sexual variation in ventral coloration in *E. angelicus*. The venter is pale yellow in juveniles, and adults from the northern part of the range retain this color. On Volcán Poás adult females and large males have yellow orange to bright red venters. Although the secondary sex characters will always separate this species from *E. escoces* and *E. fleischmanni*, live and preserved juveniles and preserved adult females of the three forms are often difficult to classify. *E. angelicus*, however, has wider toe disks and less toe webbing than *E. fleischmanni* and narrower finger disks than *E. escoces*. The karyotype is 2 N = 20: four pairs of metacentric chromosomes, three pairs of submetacentrics, one pair of subtelocentrics, and two pairs of telocentrics; NF = 36 (DeWeese 1976). Savage (1975) provided a detailed description of this form and its relatives in lower Central America.

DISTRIBUTION: Humid forests of the premontane and lower montane slopes of the Cordillera de Tilarán (1,200– 1,600 m) and northern and eastern slopes of Volcán Poás (656–1,680 m), Costa Rica.

Eleutherodactylus escoces Savage, 1975

(plate 103; map 7.34)

DIAGNOSTICS: Adult males of this common species are moderate-sized and have nuptial thumb pads and vocal slits. Adult females are large frogs. All adults have the undersurfaces bright red, disks on fingers III and IV nearly twice as wide as digits, and only basal toe webbing (fig. 7.48b).

DESCRIPTION: Adult males 26 to 46 mm in standard length, adult females 38 to 72 mm; dorsum smooth to weakly granulate; venter smooth; head about as broad as long; snout rounded to subovoid from above; eyes large; upper eyelid smooth to granulate; tympanum, round and diameter two-thirds to three-fourths diameter of eye in males, oval and one-third to one-half in females; paired vocal slits and single internal subgular vocal sac in adult males; finger I longer than finger II; finger disk covers rounded, about twice as wide as digit on fingers III and IV; subarticular tubercles on hands and feet rounded, not projecting; no supernumerary tubercles under digits; whitish nuptial pad on thumbs in adult males; thenar and palmar tubercles elongate, former slightly larger; toe disks expanded, almost twice width of digit on toes III–IV–V; only basal toe webbing; webbing formula: I 2–2½ II 2–3½ III 3–4½ IV 4¼–3 V; large inner, very small outer metatarsal tubercles; a strong inner dorsal

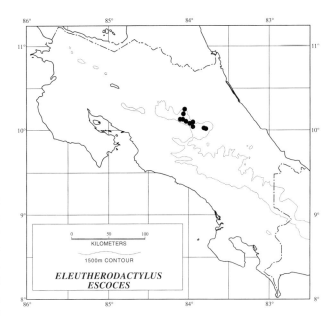

Map 7.34. Distribution of *Eleutherodactylus escoces*.

fold. Dorsum dark olive (dark bluish gray to purple in preservative), usually uniform; posterior thigh surface suffused to uniform, without definite pattern; venter bright tomato red in adults; iris gold above, black below.

VOICE: Unknown.

SIMILAR SPECIES: (1) *Eleutherodactylus angelicus* has finger and toe disks that are slightly wider than the digits, and adult males lack vocal slits. (2) *Eleutherodactylus fleischmanni* has a pale yellow venter in adults, and the finger and toe disks are barely wider than the digits.

HABITAT: Another riparian species usually found along stream banks in Premontane and Lower Montane Wet Forest and Rainforest.

BIOLOGY: A nocturnal stream bank forager. Small juveniles are 13 to 14 mm in standard length.

REMARKS: Ontogenetic variation in this species involves changes in ventral coloration. Small juveniles (under 30 mm in standard length) have the undersurface pale yellow. By the time of sexual maturity the venter has become bright red. Within another season all undersurfaces including the chest and limbs are tomato red as well. Juveniles and preserved adult females of *E. fleischmanni* and this form are difficult to separate. *E. escoces* has broader finger and toe disks and less webbing between toes III and IV than *E. fleischmanni*. In addition, living and preserved examples of the former are bluish gray to deep purple dorsally while the latter is gray to dark brown. The karyotype is 2N = 20, consisting of five pairs of metacentric chromosomes, two pairs of submetacentrics, one pair of subtelocentrics, and two pairs of telocentrics; NF = 36 (DeWeese 1976). Savage (1975) reviewed the *rugulosus* group in lower Central America and provided a detailed description of this form.

DISTRIBUTION: Humid forests of the premontane and lower montane northern slopes of Volcanes Barva, Irazú, and Turrialba of the Cordillera Central of Costa Rica (1,100–2,100 m) and on the southern slopes of these mountains in the lower montane zone (1,500–2,100 m).

Eleutherodactylus fleischmanni (Boettger, 1892a)

(map 7.35)

DIAGNOSTICS: Adult males are moderate-sized frogs having nuptial thumb pads and vocal slits. Adult females are large. This species has basal toe webbing (fig. 7.48b), barely expanded finger and toe disks, and a pale yellow venter.

DESCRIPTION: Adult males 29 to 47 mm in standard length, adult females 38 to 75 mm; dorsum smooth to granulate, venter smooth; head about as broad as long; snout subovoid to rounded from above; eyes large; upper eyelid granulate to tuberculate; tympanum round and diameter two-thirds to four-fifths diameter of eye in males, oval and one-third to one-half in females; single internal subgular vocal sac in adult males; finger I longer than finger II; finger disks rounded, barely expanded; subarticular tubercles on hands and feet rounded, not projecting; no supernumerary tubercles under digits; whitish nuptial pad on thumb in adult males; thenar and palmar tubercles elongate, former slightly larger than latter; toe disks weak, barely expanded; minimal toe webbing; webbing formula: I $2^--2\frac{1}{2}$ II $2^--3\frac{1}{2}$ III $3-4^+$ IV 4^+-3 V; large inner and very small outer metatarsal tubercles; a strong inner tarsal fold. Dorsum gray to dark brown, uniform or usually dark spotted; upper surface of limbs with obscure transverse dark bars; posterior thigh surface usually blotched or mottled but often marked with dark or light lines and/or obscure small light spots; venter pale yellow; iris dull gold.

SIMILAR SPECIES: (1) *Eleutherodactylus escoces* has larger finger and toe disks that are about twice as wide as the digits on fingers III and IV and toes III–IV–V and a bright red undersurface in adults. (2) *Eleutherodactylus angelicus* has larger toe disks that are about one and a half times as wide as the digits on toes III–IV–V and a bright yellow to red venter; adult males lack vocal slits.

VOICE: Unknown.

HABITAT: A riparian frog that is frequently found some distance from stream courses at disturbed sites. The habitat includes Premontane Moist and Wet Forests and Lower Montane Wet Forest.

BIOLOGY: A nocturnal forager. Small juveniles are 14 to 15 mm in standard length.

REMARKS: See remarks sections under *E. angelicus* and *E. escoces* for further information on distinguishing *E. fleischmanni* from these forms. Adults of those two species usually have brightly colored venters (yellow orange to red). Rust or red spots or blotches may be present dorsally in all three species. A reddish or rusty suffusion may cover the axilla and/or groin or extend along the flank in some examples of *E. angelicus* and *E. fleischmanni*. The throat may also be suffused with red in these frogs. In some large females of those two species, although there is considerable red on the throat, chest, flanks, and undersides of the limbs, the venter is pale yellow. Only in mature *E. escoces* are all the undersurfaces, including the venter, bright red. The nominal species *Eleutherodactylus engytympanum* and *Eleutherodactylus euryglossus* recognized as distinct taxa by Taylor (1952a) were shown by Savage (1975) to be synonyms of *E. fleischmanni*. The karyotype of this species is 2N = 20, consisting of four pairs of metacentric chromosomes, three pairs of submetacentrics, one pair of subtelocentrics, and two pairs of telocentrics; NF = 36 (DeWeese 1976).

DISTRIBUTION: Premontane to lower montane southern slopes of the Cordillera Central and the Mesetas Central Oriental and Occidental of Costa Rica and both slopes of the Cordillera de Talamanca of Costa Rica (1,050–2,286 m).

Eleutherodactylus punctariolus (Peters, 1863e)

(map 7.36)

DIAGNOSTICS: The moderate-sized adult males, large females, and small juveniles are immediately recognizable because of their large finger and toe disks, extensive toe webbing, toe fringes (fig. 7.48e, and yellow venters.

DESCRIPTION: Adult males 40 to 50 mm in standard length, adult females 55 to 81 mm; dorsum rugose; venter smooth; head about as wide as long; snout ovoid to rounded in dorsal view; eyes large; tympanum round and diameter two-thirds to eight-tenths eye diameter in males, one-third to two-fifths in females; vocal slits and single internal subgular vocal sac in adult males; finger I longer than finger II; finger and toe disk covers rounded, the largest at least twice width of digits; subarticular tubercles on hands and feet

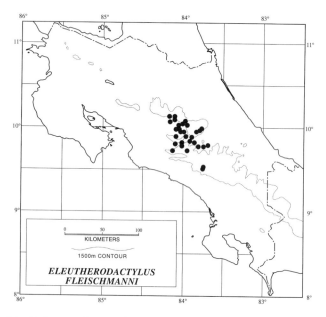

Map 7.35. Distribution of *Eleutherodactylus fleischmanni*.

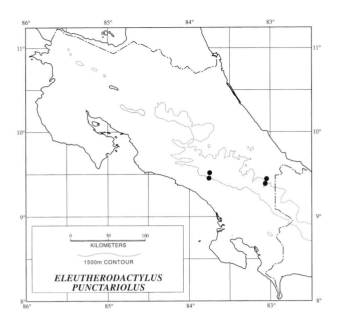

Map 7.36. Distribution of *Eleutherodactylus punctariolus*.

rounded, not projecting; no supernumerary tubercles under digits; whitish nuptial pad on thumb in adult males; thenar and palmar tubercles elongate, former larger than latter; extensive toe webbing; webbing formula: I 1½–2 II 1⅓–2¾ III 1¾–3⁺ IV 3¼–1½ V; large inner and small outer metatarsal tubercles; a very strong to flaplike tarsal fold. Dorsum brown, heavily spotted or blotched with darker markings; transverse dark bars on upper surface of limbs; posterior thigh surface mottled; venter pale yellow, heavily marked with brown.

VOICE: Not known.

SIMILAR SPECIES: *Eleutherodactylus taurus* has toe disks that are barely expanded and a white venter; adult males lack nuptial pads.

HABITAT: The spray zone on rocks, boulders, and cliff faces in the middle of moderate-sized cascading streams, rapids, or waterfalls from the upper portions of the Lowland Moist Forest into Premontane Wet Forest and Rainforest and Lower Montane Wet Forest.

BIOLOGY: An uncommon nocturnal frog that forages on rock faces in or immediately adjacent to steep torrential watercourses. Small juveniles are 16 to 17 mm in standard length.

REMARKS: In my review (Savage 1975) of lower Central American frogs of this group I pointed out slight differences in webbing and disk development among populations of this form. It now appears that *E. punctariolus* is a complex of several species (see addendum, p. 840). This frog was originally described as a treefrog (genus *Hyla*) by Peters (1863e) because of its large digital disks and extensive toe webs. The karyotype is 2N = 20, consisting of seven pairs of metacentric chromosomes, four pairs of subtelocentrics, and two pairs of telocentrics; NF = 36 (DeWeese 1976).

DISTRIBUTION: Stream courses within the upper portions of the lowland zone through the premontane belt on both slopes of the Cordillera de Talamanca of Costa Rica and western Panama to Pacific slope Panama, west of the Canal area (400–1,700 m).

Eleutherodactylus ranoides (Cope, "1885," 1886)

(plate 105; map 7.37)

DIAGNOSTICS: Adult males of this common form are small to moderate-sized frogs that are much smaller than the large females. Among members of the *rugulosus* group in Costa Rica, they are unique in having the dark posterior thigh surface marked with discrete small light spots and in lacking vocal slits, a vocal sac, and nuptial pads in males.

DESCRIPTION: Adult males 26 to 45 mm in standard length, adult females 40 to 74 mm; dorsum smooth to granulate; venter smooth; head about as wide as long; snout subovoid to rounded from above; eyes large; upper eyelid granulate to tuberculate; tympanum round and diameter two-thirds to seven-tenths of eye diameter in males, oval and one-third to one-half in females; finger I longer than finger II; disk covers rounded and definite, one and a half times width of digits on fingers III and IV and toes III–IV–V; subarticular tubercles on hands and feet rounded, not projecting; no supernumerary tubercles under digits; no nuptial pads; thenar and palmar tubercles elongate, former slightly larger than latter; modal webbing formula: I 2–2½ II 2⁻–3⁺ III 3–4 IV 4–2½ V; large inner and very small outer metatarsal tubercles; a strong inner tarsal fold. Dorsum dark olive green or olive brown, uniform or spotted or blotched with darker pigment; dark transverse bars on upper surface of limbs; posterior thigh surface with numerous

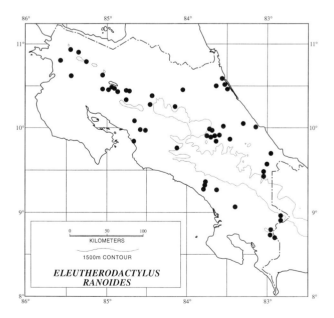

Map 7.37. Distribution of *Eleutherodactylus ranoides*.

sharply contrasting small light spots on dark brown; venter pale yellow or bright yellow gold; iris gold above, dark brown below.

VOICE: Apparently mute.

SIMILAR SPECIES: (1) *Eleutherodactylus fitzingeri* has large truncate disks on fingers III and IV (fig. 7.51f), a brown nuptial thumb pad in adult males, a bifid palmar tubercle, and usually a light midgular stripe (fig. 7.56). (2) *Eleutherodactylus fleischmanni* has barely expanded finger and toe disks, has vocal slits and nuptial pads in adult males, and lacks the characteristic thigh pattern of *E. ranoides*.

HABITAT: This species is usually associated with small streams in Lowland Moist and Wet Forests and Premontane Wet Forest, but it also ranges into drier areas along stream banks supporting gallery forest.

BIOLOGY: Like other members of the group, this riparian frog forages along streams at night and jumps into the water when disturbed. Small juveniles are 13 to 14 mm in standard length.

REMARKS: There is geographic variation in the ventral coloration in populations of this species in Costa Rica. Adult examples from both Atlantic and Pacific versants (Guanacaste and northern Limón Provinces) have pale yellow venters. Those from farther south into Panama have bright yellow gold venters. The names *Lithodytes ranoides* Cope (1886) and *Liohyla pittieri* Günther (1900) are based on examples having pale yellow and yellow gold venters, respectively. Taylor (1952a) included both taxa in his treatment of Costa Rican frogs as members of *Eleutherodactylus*, although he referred to no material other than the type of *Eleutherodactylus pittieri*. The specimens Taylor identified as *Eleutherodactylus ranoides* are actually examples of *Eleutherodactylus fitzingeri*, as I pointed out (Savage 1975). I used the name *Eleutherodactylus rugulosus* (Cope, "1869," 1870) for the Costa Rican frogs here called *E. ranoides* (Savage 1975, 1976, 1980b; Savage and Villa 1986). Current research indicates that *Eleutherodactylus rugulosus* as conceived at present (Savage 1975) consists of several distinctive species. Because the name is based on Mexican material (Cope 1869) and the lower Central American populations are not conspecific with the lectotype (USNM 29771; Mexico: Oaxaca: Tehuantepec) of *Eleutherodactylus rugulosus*, the name *E. ranoides* (lectotype: USNM 14179A; Nicaragua: Río San Juan: between El Castillo and San Juan del Norte) is applicable to the Costa Rican form. The karyotype of Costa Rican examples is 2N = 20, consisting of six pairs of metacentric chromosomes, one pair each of submetacentrics and subtelocentrics, and two pairs of telocentrics; NF = 36 (DeWeese 1976). The Mexican specimens referred to *E. rugulosus* by DeWeese, with 2N = 22, have been recognized as a distinct species, *Eleutherodactylus berkenbuschii* (Savage and DeWeese 1979).

DISTRIBUTION: Lowlands and premontane slopes from eastern Nicaragua on the Atlantic slope and northwestern Costa Rica on the Pacific versant to extreme western Panama, exclusive of the Golfo Dulce region of southwestern Costa Rica; in Costa Rica found at 10–1,300 m in the northwest, 1–116 m on the Atlantic slopes, and 500–1,220 m in the southwest.

Eleutherodactylus taurus Taylor, 1958a

(plate 106; map 7.38)

DIAGNOSTICS: Adult males of this species are small to moderate-sized anurans; adult females are large. The combination of expanded finger disks, toe fringes, moderate toe webbing, and barely expanded toe disks is distinctive within Costa Rican members of the *rugulosus* group. In addition, the venter is white and adult males have vocal slits but lack nuptial pads.

DESCRIPTION: Adult males 24 to 44 mm in standard length, adult females 40 to 80 mm; dorsum rugose; venter smooth; head about as wide as long; snout rounded to subovoid from above; upper eyelid granulate to tuberculate; eyes large; tympanum round and diameter two-thirds to three-fourths diameter of eye in males, oval and one-third to one-half in females; single internal subgular vocal sac in adult males; finger I longer than finger II; finger disk covers rounded, expanded, about twice width of fingers III and IV; subarticular tubercles on hands and feet rounded, not projecting; no supernumerary tubercles under digits; no nuptial thumb pads; thenar and palmar tubercles elongate, former slightly larger than latter; toe disks barely expanded; moderate toe webbing; webbing formula: I $^3/_4$–2^+ II $1^3/_4$–3 III $2^1/_4$–$2^2/_3$ IV $3^3/_4$–2 V; large inner and very small outer metatarsal tubercles; very strong inner tarsal fold. Dorsum gray, heavily blotched with dark pigment; transverse dark bars

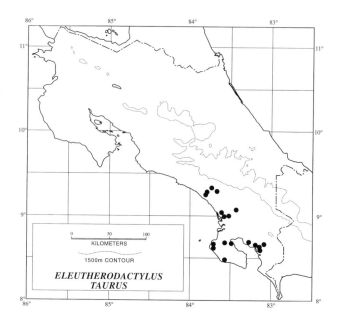

Map 7.38. Distribution of *Eleutherodactylus taurus*.

on upper surface of limbs; posterior thigh surface blotched with irregular light markings; venter dull white or with a light greenish yellow wash posteriorly in adults.

VOICE: Appears not to vocalize.

SIMILAR SPECIES: *Eleutherodactylus punctariolus* has disks that are at least twice as wide as the digits on fingers III–IV–V and a yellow venter; adult males have nuptial thumb pads.

HABITAT: In and around rocky streams in Lowland Moist and Wet Forests.

BIOLOGY: A relatively common nocturnal frog that forages out into the stream bed. The species is often found in cavities in the banks of deeply cut small streams and among fallen logs choking such watercourses. Small juveniles are 13 to 14 mm in standard length.

REMARKS: This distinctive form was first described by Taylor (1958a). A detailed description may be found in Savage (1975). The karyotype is 2N = 20 and consists of six pairs of metacentric chromosomes, one pair each of submetacentrics and subtelocentrics, and two pairs of telocentrics; NF = 36.

DISTRIBUTION: Lowland humid forests of southwestern Costa Rica and extreme southwestern Panama (25–525 m).

Eleutherodactylus biporcatus Group

These robust, thick-limbed, broad-headed toadlike anurans have a distinct pair of interorbital cranial crests visible in adults of most species. Additional distinguishing features are finger I longer than finger II; toe III longer than toe V; tympanum present; vomerine odontophores triangular, posterior to choanae, and narrowly separated medially; no vocal slits and no vocal sac or nuptial pads in adult males; disks present on all fingers and toes but not expanded, none notched; subarticular tubercles flattened or obtuse; no supernumerary tubercles on hands or feet; toe web absent; no inner tarsal fold; venter smooth.

All species are terrestrial, and adults are nocturnal. Juveniles, however, are active in the leaf litter during the day and are among the most common little frogs observed on the forest floor.

Eleutherodactylus gulosus (Cope, 1875)

(plate 107; map 7.39)

DIAGNOSTICS: A giant broad-headed frog having cranial crests in the frontoparietal region except in the smallest examples, having narrow finger disks (fig. 7.51a) and slightly expanded toe disks, and lacking toe webs and dorsal plicae. Juveniles are covered over the entire dorsal surface by numerous small white-tipped pustules.

DESCRIPTION: Adult males probably 40 to 60 mm in standard length, adult females 72 to 103 mm; dorsum smooth or with scattered low pustules in adults; no suprascapular plicae in adults, but these are weakly indicated anteriorly in juveniles; no V-shaped ridge or tubercle series in

Map 7.39. Distribution of *Eleutherodactylus gulosus*.

laterosacral region; head much broader than long (width 48–54% of standard length); large knobs at posterior tip of cranial crests and above supratympanic fold in large adults; head outline semicircular from above; snout short; eyes moderate; upper eyelid with widely spaced supraocular and a few or no superciliary warts; tympanum distinct, small, diameter one-third to one-half diameter of eye in juvenile males, two-fifths to one-half in females; no expanded finger disks, but digital pads separated from cover by a transverse groove; subarticular tubercles ovoid, flattened; thenar tubercle large, elongate; no nuptial thumb pads; palmar tubercle low, round, bifid to trifid; obscure accessory palmar tubercles; toe disks slightly expanded; no toe webs or fringes; inner metatarsal tubercle elongate, strongly compressed, outer small, obsolescent; plantar surface smooth; no tarsal fold. Dorsal surfaces uniform dark brown in adult females, adult males with obscure dark blotches posteriorly; groin uniform or marked like venter; upper limb surfaces uniform with transverse dark bars; posterior thigh surface uniform or with some light spots or dashes; seat patch discrete, not continuous with black on lower posterior thigh; undersurfaces of upper arms, hind limbs, and venter mottled or spotted with brown on yellow; *juveniles:* mottled with light and dark brown above; undersurfaces of limbs and posterior venter without markings; posterior thighs mostly mottled light brown and whitish yellow.

VOICE: Probably does not vocalize.

SIMILAR SPECIES: (1) *Eleutherodactylus bufoniformis* has definite toe webbing (fig. 7.48b). (2) *Eleutherodactylus megacephalus* has well-developed dorsal plicae (fig. 7.54b). (3) *Eleutherodactylus rugosus* has well-developed dorsal plicae (fig. 7.54a) and black and bright red posterior thigh markings in adults (fig. 7.50c).

HABITAT: Premontane Rainforest.

BIOLOGY: Probably similar to *Eleutherodactylus megacephalus*. The smallest available juvenile is 20 mm in standard length, but hatching size is certainly much smaller.

REMARKS: This form was long placed in the synonymy of *E. biporcatus*, but Savage and Myers (2002) recently documented its distinctiveness based on new material from western Panama. The holotype of this species (USNM 23590), a giant female 103 mm in standard length, was collected on Cerro Utyum, Limón Province, by the W. W. Gabb expedition to the Talamanca region in 1873 (Savage 1970). The specimen was said to have been captured at an elevation of 6,000 feet (1,829 m). Although this is an unlikely altitude for *E. megacephalus*, as Savage (1974a) pointed out, *E. gulosus* may well range into the lower montane zone based on its distribution in Panama. The karyotype is 2N = 20, consisting of seven pairs of metacentrics, two pairs of submetacentrics, and one pair of telocentric chromosomes; NF = 38 (S.-H. Chen, pers. comm.).

DISTRIBUTION: Humid premontane and possibly lower montane areas of the southern Cordillera de Talamanca of Costa Rica and western Panama (1,000–1,873 m).

Eleutherodactylus megacephalus (Cope, 1875)

(plates 108–9; map 7.40)

DIAGNOSTICS: This large broad-headed frog has narrow finger disks but slightly expanded toe disks and no toe webbing or toe fringes. Large juveniles and adults have paired cranial crests externally evident in the frontoparietal region, and examples of all ages have well-developed suprascapular plicae that form chalice- or hourglass-shaped ridging and also laterosacral plicae (fig. 7.54b). The posterior thigh surfaces are black below, spotted or mottled with light

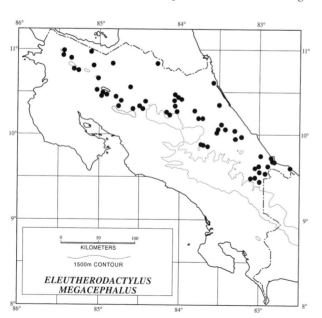

Map 7.40. Distribution of *Eleutherodactylus megacephalus*.

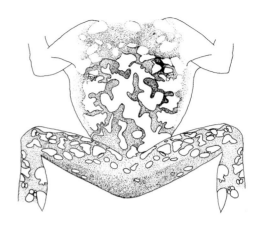

Figure 7.57. Ventral view of *Eleutherodactylus megacephalus* illustrating the typical color pattern of the venter and legs.

(fig. 7.50d). Juveniles and all but the largest adults have the venter, groin, and undersurfaces of hind limbs orange to red orange and heavily marked with dark brown spots or forming a reticulum around the light spots (fig. 7.57). Juveniles have the groin, posterior venter, and undersides of the legs red orange.

DESCRIPTION: Adult males 30 to 43 mm in standard length, adult females 50 to 70 mm; dorsum smooth to tuberculate between major ridges; flanks tuberculate to granulate; scattered paired dorsal and sacral enlarged tubercles, forming V-shaped ridges or rows of large tubercles; tubercles often tipped with white glandular tissue; large knobs at posterior tip of each cranial crest and above supratympanic fold in large adults; head much broader than long (width 44–55% of standard length); head outline semicircular from above; snout short; eyes moderate; upper eyelid with widely spaced supraocular and superciliary warts, one or two of former enlarged; tympanum distinct, large, diameter four-fifths to nine-tenths diameter of eye in males, two-fifths to one-half in females; no vocal slits or sac; no expanded finger disks, but digital pad separated from upper finger surface by a terminal transverse groove; subarticular tubercles ovoid, obtuse, slightly projecting; thenar tubercle low, oblong, palmar larger, bifid; no nuptial pads; low accessory palmar tubercles at base of fingers II and III; toe disks slightly expanded; no toe webs or fringes; inner metatarsal tubercle elongate, strongly compressed, terminal margin projecting; outer metatarsal tubercle small, obsolescent; plantar surface smooth; no tarsal fold. Dorsum gray tan to gray olive brown, individual tubercles and dorsal plicae bordered by black; groin marked like venter; lower half of posterior thigh surface and adjacent undersurface black, including V-shaped dark seat patch; faint dark bars on upper surface of thigh, when present, continuous with dark color on lower posterior thigh; groin, posterior venter, and undersurfaces of hands and hind limbs orange to red orange in juveniles, heavily reticulated with dark brown to black; in large adults venter may be yellow spotted with dark or reticulate enclosing white or yellow spots; soles black; iris black flecked with gold.

VOICE: Apparently mute.

SIMILAR SPECIES: (1) *Eleutherodactylus rugosus* has much more dorsal and cephalic ridging and has contrasting black bars and bright red (white in juveniles) interspaces on the posterior thigh (fig. 7.50c). (2) *Eleutherodactylus gulosus* lacks the suprascapular plicae in adults, but these are indicated anteriorly in juveniles. (3) *Eleutherodactylus bufoniformis* has definite toe webs.

HABITAT: Lowland Moist and Wet Forests and lower portions of the premontane belt in Premontane Wet Forest and Rainforest.

BIOLOGY: Adults are nocturnal, terrestrial frogs that hide in burrows during the day. At night they commonly sit at the burrow entrance and ambush passing prey. Juveniles are essentially diurnal and are among the most common leaf litter anurans. At La Selva, Heredia Province, densities of this form are 84/ha (Lieberman 1986). Food consists of a variety of arthropods, mostly ants (23%) and beetles (30%), but orthopterans (14%), isopods (9.3%), and various larvae (9.3%) are eaten in large numbers at La Selva. Large adults also eat small frogs and lizards. In twenty-one stomachs at La Selva the mean number of prey items was 2.2, and mean prey length varied from 2.2 mm (ants) to 16.9 mm (spiders). Large prey items to 34 mm were also eaten (Lieberman 1986). Eggs are apparently laid in the leaf litter. The smallest juveniles are 10 mm in standard length, which is near the size at hatching.

REMARKS: Taylor (1952a) used the name *Eleutherodactylus rugosus* for this species. *Eleutherodactylus biporcatus* has been applied to it by most workers, following Dunn (1931c). Savage and Myers (2002) demonstrated that the syntypes (ZMB 3222a,b,c, 3330) of the name *Strabomantis biporcatus* W. Peters, 1863e, are representatives of the Venezuelan species *Eleutherodactylus maussi* (Boettger 1893a). Therefore the earliest available name for this taxon is *Lithodytes megacephalus* Cope, 1875 (holotype: USNM 32579; an adult female; Limón: Cerro Utyum). The karyotype is 2N = 20, consisting of six pairs of metacentrics and two pairs each of submetacentric and telocentric chromosomes; NF = 36 (S.-H. Chen, pers. comm.).

DISTRIBUTION: Humid lowlands and lower premontane slopes on the Atlantic versant from extreme southeastern Honduras to western Panama. Restricted to the Atlantic slope in Costa Rica except for a few records just west of the continental divide in Guanacaste Province (1–1,200 m).

Eleutherodactylus rugosus (Peters, 1873a)

(map 7.41)

DIAGNOSTICS: This relatively common large, broad-headed, extremely rugose species is immediately distinguishable by having black bars alternating with bright scarlet (white in juveniles) interspaces on the posterior thighs

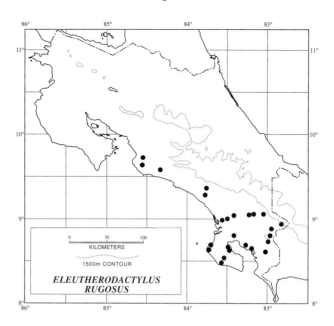

Map 7.41. Distribution of *Eleutherodactylus rugosus*.

(fig. 7.50c) and narrow finger disks, while lacking webs and fringes.

DESCRIPTION: Adult males probably 30 to 44 mm in standard length, adult females 35 to 69 mm; dorsum tuberculate with numerous large tubercles and tuberculate ridges, sometimes white tipped; a chalice- or hourglass-shaped ridging system present in suprascapular area, without longitudinal ridges laterally and in sacral region; smooth to pointed tubercles on flanks; head much broader than long (48–54% of standard length); paired cranial crests in frontoparietal region evident externally in all but smallest individuals, with large knobs at posterior tip and paired bony knobs above supratympanic fold in large adults; head outline semicircular from above; snout short; eyes moderate; upper eyelid with numerous warts, several supraocular warts enlarged; tympanum distinct, large, diameter four-fifths to nine-tenths diameter of eye in males, two-fifths to one-half in females; no vocal slits or sac; no expanded finger disks, but digital pad separated from disk cover by a terminal transverse groove; subarticular tubercles on all digits ovoid, not projecting; thenar tubercle low, oblong; palmar larger, bifid; no nuptial pads; low accessory palmar tubercles at base of fingers II and III; elongate accessory palmar tubercle at base of finger IV; toe disks slightly expanded; no toe webs or fringes; inner metatarsal tubercle elongate, strongly compressed terminal margin projecting; outer metatarsal tubercle well developed, raised, smaller than inner; plantar surface smooth; no tarsal fold. Dorsum gray or dark brown to nearly black; dorsal plicae and warts black or black tipped with white; groin with vivid complex black and white markings; upper surface of limbs with transverse dark bars; posterior thigh surface black below, including V-shaped seat patch, barred red (white in juveniles) and black above; bold black bars on upper anterior thigh surfaces continuous with segments on posterior thigh; anterior surface of thigh, undersurface of hind limbs, and venter usually with large white to red spots in a dark brown reticulum (fig. 7.57); juveniles with red orange in groin and on undersurfaces of hands, venter, and hind limbs, fading in large adults to yellow orange or yellow; soles black; iris black.

VOICE: Apparently mute.

SIMILAR SPECIES: (1) *Eleutherodactylus megacephalus* lacks contrasting black and red (or white) thigh bars and has much less dorsal ridging and tuberculation. (2) *Eleutherodactylus gulosus* lacks dorsal plicae in adults and red (or white) and black thigh markings.

HABITAT: Lowland Moist and Wet Forest and Premontane Wet Forest and Rainforest in southwestern Costa Rica.

BIOLOGY: Similar to *Eleutherodactylus megacephalus* as far as known. The smallest juvenile is 9.9 mm in standard length, which probably is near hatching size.

REMARKS: Taylor (1952a, 1955) used the name *Eleutherodactylus florulentus* for this species. After examining the holotypes of *Hylodes rugosus* W. Peters (1873a) (ZMB 7812; Panama: Chiriquí) and *Lithodytes pelviculus* Cope, 1877 (USNM 32326; west coast or Central America), Savage and Myers (2002) concluded that they were conspecific with Cope's (1893) *Lithodytes florulentus* (type lost; Costa Rica: Puntarenas: Boruca). The name *Eleutherodactylus rugosus* is thus correct for this species. The karyotype is 2N = 20, composed of seven pairs of metacentric chromosomes, two pairs of submetacentrics, and one pair of telocentrics; NF = 38 (DeWeese 1976).

DISTRIBUTION: Humid lowland and premontane zones of southwestern Costa Rica and adjacent southwestern Panama (10–1,220 m).

Eleutherodactylus bufoniformis Group

Another group of robust, thick-limbed toadlike animals having a pair of cranial crests in adults and the tips of the fingers not expanded into definite disks. Other features shared by members of the group are finger I longer than finger II; toe III much longer than toe V; tympanum present, indistinct; vomerine odontophores triangular, posterior to choanae, narrowly separated medially; vocal slits and a single internal subgular vocal sac but no nuptial pads in adult males; disks poorly developed but larger on toes; disk covers not notched; subarticular tubercles round to ovoid, not projecting; no supernumerary tubercles on hands and feet; no accessory palmar or plantar tubercles; toes webbed basally or no webbing; inner tarsal fold weak or absent; venter smooth.

Eleutherodactylus bufoniformis (Boulenger, 1896d)

(plate 110; map 7.42)

DIAGNOSTICS: A large broad-headed anuran immediately recognizable by having the dorsal surfaces covered by tubercles tipped with white glandular tissue, having toe webs

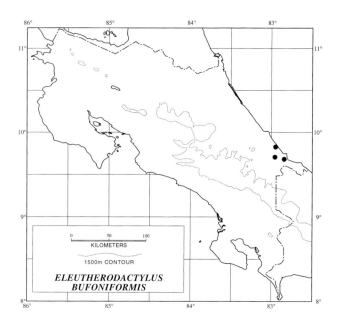

Map 7.42. Distribution of *Eleutherodactylus bufoniformis*.

and fringes, but not having the fingertips expanded into distinct disks (fig. 7.51a). In addition, large juveniles and adults have cranial crests on the frontoparietal bones that are not evident externally but can be felt through the skin.

DESCRIPTION: Adult males 51 to 59 mm in standard length, adult females 52 to 94 mm; dorsum tuberculate and ridged, each tubercle tipped with white glandular tissue, and also ridges in juveniles and some adults, with rest of upper surfaces having short rows of cream-colored glands; head much broader than long, width 44–58% of standard length; snout broadly rounded from above; eyes large; upper eyelid tuberculate; tympanum small, distinct, diameter one-half to three-fourths diameter of eye in males, two-fifths to one-half in females; paired vocal slits and a single median internal subgular vocal sac in males; no expanded finger disks, but digital pad separated from upper finger surface by a terminal transverse groove; subarticular tubercles round to ovoid on hands and feet; thenar tubercle elongate, smaller than low, bifid palmar tubercle; palm without accessory tubercles; toe disks expanded, rounded, about one and a half times width of digit; toes webbed basally; webbing formula: I 2–3 II 2–3¼ III 2½–4⁻ IV 4–2 V; inner metatarsal tubercle compressed, larger than tiny outer metatarsal tubercle; plantar surface smooth; a weak inner tarsal fold or none; dorsum brown with darker spots; upper surface of limbs with transverse dark bars; posterior thigh surfaces brown with small cream spots; venter white.

VOICE: A series of paused short low barklike notes followed by shorter knocklike ones; dominant frequency 0.5 to 1.2 kHz (Ibáñez, Rand, and Jarmillo 1999).

SIMILAR SPECIES: *Eleutherodactylus gulosus, E. megacephalus,* and *E. rugosus* lack toe fringes and webs and have a distinct tympanum.

HABITAT: Lowland Moist and Wet Forests of the southern Atlantic versant.

BIOLOGY: A rarely seen nocturnal riparian frog that feeds on insects other than ants.

DISTRIBUTION: Humid lowlands from southeastern Costa Rica to western Colombia; in Costa Rica known only from extreme southern Limón Province (15–20 m).

Eleutherodactylus rhodopis Group

These are small, short-limbed frogs with chunky, narrow heads. Other features include fingers I and II equal in length or finger I shorter than finger II; toe III much longer than toe V; tympanum prominent; vocal slits and nuptial pads present in some species; vocal sac, if present, single internal subgular; fingers and toes without expanded disks; fingers I and II with terminal transverse groove weak or absent; subarticular tubercles projecting or not; usually supernumerary tubercles on hands and feet; no toe webbing; no well-developed inner tarsal fold, although a short, weak ridge or tubercle may be present; venter coarsely areolate or smooth.

In lower Central America these small frogs are active diurnally and hop and skitter across the thin layer of fallen leaves on the forest floor. When disturbed they dive under the litter after two or three short hops. Males call from under the litter, and the terrestrial eggs are laid in moist, shady places near the calling site.

Ridging patterns (fig. 7.44) and coloration (fig. 7.45) in frogs of this group show extreme variation. Most individuals have granular to warty skin, a series of short ridges in the suprascapular area, a uniform or mottled to blotched ground color, a light dorsal shield mark of varying extent, a dark seat patch or pelvic dark spots, an interocular dark bar and a supratympanic dark stripe (fig. 7.46), and dark bars on the upper lip (fig. 7.47b). Other less common variants are illustrated (fig. 7.46). The several modifying color features may occur in different combinations with different ridging patterns and basic dorsal patterns.

Two species of this group, *Eleutherodactylus bransfordii* and *E. podiciferus,* have usually been recognized in Costa Rica following Savage and Emerson (1970). Miyamoto (1983) revived *Eleutherodactylus stejnegerianus* from synonymy in *E. bransfordii* based on biochemical evidence. More recent unpublished studies of karyological features indicate that *E. bransfordii* as currently understood is a complex of cryptic forms, several of which were recognized by Taylor (1952a). Unfortunately there are few obvious external features that consistently distinguish between individuals allocated to the various taxa. Adult size differences and the presence or absence of male nuptial pads provide the best indicators for species assignment among the most similar forms. Juveniles and females most often are identified only through association with males from the same locality.

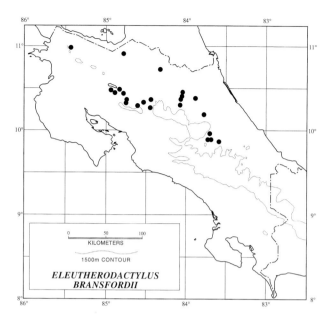

Map 7.43. Distribution of *Eleutherodactylus bransfordii.*

Eleutherodactylus bransfordii (Cope, "1885," 1886)

(plate 111; map 7.43)

DIAGNOSTICS: Very common tiny to very small brownish frogs having very obvious projecting tubercles on the hands and feet (fig. 7.52a,d,g), the thenar tubercle equal to or slightly smaller than the palmar tubercle (fig. 7.53a), the groin and hind limbs covered by a reddish suffusion, and nuptial pads present in adult males.

DESCRIPTION: Adult males 13.3 to 23 mm in standard length, adult females 13.6 to 26 mm; dorsum smooth, granular, or usually warty, or with a series of short suprascapular ridges or often parallel dorsolateral ridges; venter coarsely areolate; inguinal gland moderate, slightly larger than palmar tubercle; head longer than broad; snout subelliptical in outline from above; eyes large; tympanum distinct, large, diameter equal to diameter of eye in males, three-fourths in females, translucent; no vocal slits or sac; rudimentary unexpanded digital disks and grooves on fingers III and IV; toes with expanded disks; disk covers often pointed, disks often asymmetrical; subarticular tubercles pointed; supernumerary tubercles under most fingers and toes; thenar tubercle and palmar tubercle ovoid, thenar equal to or slightly smaller than palmar; whitish nuptial pads on base of thumb in adult males; fingers I and II equal in length; accessory palmar tubercles numerous, projecting; heel granular; no toe webbing; large elevated, elongate inner metatarsal tubercle; outer very small; supernumerary tubercles present on toes III and IV, often on others as well; numerous pointed plantar tubercles; weak inner tarsal fold; ventral disk V-shaped, extending posteriorly to level of femurs; upper surface brown to brownish cream, uniform, mottled with darker pigment or with paired dorsolateral light stripes; frequently with light shield marking in suprascapular area and/or dark seat patch; sometimes with narrow or broad middorsal light stripe, sometimes with an interocular dark bar or hourglass-shaped mark or pelvic dark spots; usually a supratympanic dark stripe and upper lip barred with light and dark; upper surface of thighs usually barred and covered by reddish suffusion; posterior surface of thigh uniform reddish brown; throat and chest yellowish, usually unpigmented; venter dull yellow; groin area reddish.

VOICE: A low chirp, repeated after a long pause.

SIMILAR SPECIES: (1) *Eleutherodactylus polyptychus* is a larger form (adult males 15.3 to 26.6 mm in standard length, adult females 15.6 to 28.7 mm), and adult males lack nuptial pads. (2) *Eleutherodactylus underwoodi* is larger (adult males 16 to 36 mm in standard length, adult females 18.4 to 30 mm), and the tubercles on the undersurface of the hands are low, smooth, and usually well separated from one another (fig. 7.52b,e).

HABITAT: Lowland Moist and Wet Forests, Premontane Wet Forest, and marginally in Premontane Rainforest.

BIOLOGY: These diurnally active forest frogs live among the fallen leaves on the forest floor. When disturbed they hop and skitter over the surface. At most lowland Atlantic sites this is the commonest leaf litter amphibian or reptile. At the La Selva Biological Station, Heredia Province, *E. bransfordii* has densities of 1,200/ha during the late dry season and first part of the rainy season (February to May), with much lower abundances in August to November (Lieberman 1986). Scott (1976) recorded densities of 345/ha in July and August. These frogs actively forage in the litter throughout the day. They eat a wide variety of prey including beetles, spiders, mites, and isopods in proportions significantly higher than found in the leaf litter. Orthopterans and ants are also common food items. Prey ranges from 0.6 mm (mites) to 13.3 mm (isopods and centipedes). The mean number of prey items per stomach examined was 2.2 (Lieberman 1986; Limerick 1976). Males call sporadically from under leaves, and courtship and mating appear to take place within the litter but breeding apparently takes place at night (Scott 1983d). Donnelly (1999) recorded oviductal eggs for this species at La Selva as 3–6.3–10. Because eggs at various stages of development were present, she concluded that the females oviposit several times annually. This species appears to be reproductively active throughout the year, although there may be variation in breeding intensity. The large flush of juveniles in the late dry season supports that notion.

REMARKS: Taylor (1952a) recognized *Microbatrachylus polyptychus* as a member of the Costa Rican fauna and tentatively included *Microbatrachylus bransfordii* as well, although questioning the identification of specimens referred to these two species (as *Eleutherodactylus*) from the republic by Barbour and Dunn (1921). The nominal species were thought to differ based on the presence of longitudinal dorsal ridges in *M. polyptychus* and their absence in *M. bransfordii*. Both taxa were described by Cope (1885) as *Lithodytes* from "Nicaragua" and had been collected by J. F. Bransford along the Río San Juan between Machuca and San Juan del Norte, Department of Río San Juan (Savage 1973a). Savage and Emerson (1970) examined the syntypic series, designated lectotypes (*Lithodytes bransfordii*: USNM 166895; *Lithodytes polyptychus*: USNM 166894), concluded they were based on the same species population, and selected the name *Eleutherodactylus bransfordii*. The discovery that two Atlantic lowland species in Costa Rica were included in these authors' *E. bransfordii* suggests that the name *E. polyptychus* may be applicable to one of them, a course tentatively followed here. Savage and Emerson also synonymized several other names under *E. bransfordii* that have subsequently been shown to be distinctive in molecular or karyological features. See remarks sections for the other species recognized below. The variation in ridging and color patterns in all taxa included in the *rhodopis* species group is extreme and led Savage and Emerson (1970) to several erroneous conclusions because the variants appeared in most nominal forms. *Eleutherodactylus bransfordii* occurs with *E. persimilis* at several localities and overlaps slightly with the range of *E. stejnegerianus* in the Laguna Arenal area of northwestern Costa Rica. *E. bransfordii* is replaced at higher elevations (above 1,000 m) by the very similar but larger *E. underwoodi*. The exact status of these two forms is ambiguous, and additional study is needed to clarify their systematic and distributional relationship to one another. The karyotype is 2N = 18, with six pairs of metacentric chromosomes, two pairs of submetacentrics, and one pair of subtelocentrics; NF = 36 (DeWeese 1976). Bogart (1970a,b) reported a 2N = 20, which DeWeese attributes to a possible misidentification.

DISTRIBUTION: Humid lowlands and adjacent premontane slopes on the Atlantic versant from eastern Nicaragua to central Costa Rica (60–880 m).

Eleutherodactylus persimilis Barbour, 1926
(plate 112; map 7.44)

DIAGNOSTICS: Common tiny to very small brownish frogs with finger II longer than finger I, prominent tubercles under the fingers and toes (fig. 7.52a,d,g), the thenar tubercle much smaller than the palmar (fig. 7.53b), a larger whitish inguinal gland, the groin and hind limbs covered by a reddish suffusion, and no nuptial pads in adult males.

DESCRIPTION: Adult males 13.1–17.8 mm in standard length, adult females 13.3–23 mm; dorsum irregularly granular, often with dorsolateral, suprascapular, or hourglass-shaped ridges; venter coarsely areolate; inguinal gland flattened, large, twice size of palmar tubercle; head longer than broad; snout subovoid to subelliptical from above; eyes large; tympanum distinct, translucent, diameter three-fourths diameter of eye in males, one-half in females; no

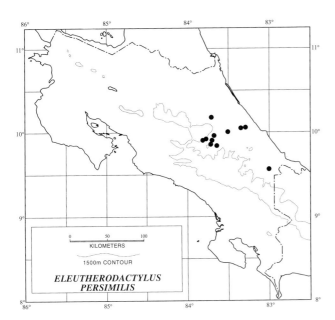

Map 7.44. Distribution of *Eleutherodactylus persimilis*.

vocal slits or sacs; finger II longer than finger I; no vocal slits or sac; expanded digital disks and grooves on fingers III and IV; toes with expanded disks; disk covers often pointed, disks often asymmetric; subarticular tubercles ovoid, obtuse, projecting; supernumerary tubercles on some toes; thenar and palmar tubercles ovoid, thenar much smaller than palmar; no nuptial pads; accessory palmar tubercles prominent; heel granular; no toe webs; numerous rounded plantar tubercles; inner metatarsal tubercle elongate, low, outer small, pointed; a barely evident inner tarsal fold; ventral disk V-shaped, extending posteriorly to level of femurs. Dorsum tan, usually uniform, sometimes mottled, very rarely with dorsolateral light stripes; frequently with narrow middorsal light stripe, rarely with broad one; usually without light shield mark; usually a dark seat patch, pelvic spots, and often dark interocular bar; usually a dark supratympanic stripe and dark lip bars; dorsal surfaces of thighs with dark bars and suffused with red; posterior surface of thighs uniform reddish brown; throat and chest pale yellow heavily pigmented with dark; groin area red.

VOICE: Unknown.

SIMILAR SPECIES: *Eleutherodactylus stejnegerianus* has fingers I and II about the same length.

HABITAT: Lowland Moist and Wet Forests and Premontane Wet Forest and Rainforest.

BIOLOGY: This frog, in common with other members of the group, is a diurnal leaf litter form. It probably resembles *E. bransfordii* in most features of its biology. Small juveniles about 6 mm in standard length.

REMARKS: Taylor (1952a) recognized this form as *Microbatrachylus persimilis*. Savage and Emerson (1970) placed it in the synonymy of *Eleutherodactylus bransfordii*. The recognition that *E. stejnegerianus* was distinct from *E. brans-*

fordii and that the type of *E. persimilis* (MCZ 11598; Costa Rica: Limón: Suretka) shared with the former the characteristic tiny thenar tubercle led to a reexamination of its status. *E. persimilis* differs from *E. bransfordii* in the size of the thenar tubercle and in lacking nuptial pads in adult males. Although very similar to *E. stejnegerianus* of the Pacific slope, it is readily distinguished by having finger II definitely longer than finger I (about equal in length in *E. stejnegerianus*). The nominal species *Microbatrachylus rearki* Taylor, 1952a (holotype: KU 31533; Costa Rica: Limón: Batán) appears to be based on large examples of this form having several interrupted dorsal and lateral folds. *E. persimilis* is syntopic with *E. bransfordii* at lower elevations in Atlantic slope central Costa Rica (600 to 700 m). It also occurs syntopically with *E. underwoodi* at higher elevations (1,100 to 1,200 m) in the same region. Both this species and *E. polyptychus* have been collected at Suretka, Limón Province (60 m).

DISTRIBUTION: Humid lowlands and premontane Atlantic slopes of central to southeastern Costa Rica (39–1,200 m).

Eleutherodactylus podiciferus (Cope, 1875)

(plate 113; map 7.45)

DIAGNOSTICS: A common small to moderate-sized tan, pink, or dark brown species distinguished from other members of the *rhodopis* group in Costa Rica by usually having a smooth venter, one to three distinct heel tubercles (fig. 7.49b,c), and the ventral disk circular with its posterior margin crossing the venter anterior to the femurs (fig. 7.10e). In addition this species lacks toe webbing and expanded digital disks, and fingers I and II are subequal in length.

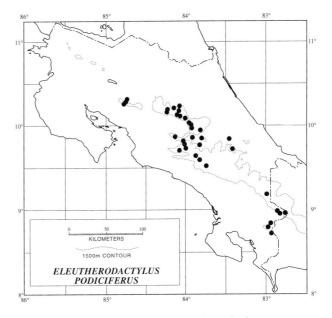

Map 7.45. Distribution of *Eleutherodactylus podiciferus*.

DESCRIPTION: Adult males 21 to 28 mm in standard length, adult females 23 to 40 mm; dorsum smooth or laterally warty with a pair of complete or interrupted chalice-shaped dorsal ridges or thin dorsolateral ridges; venter smooth, encroached on laterally by large granules; no inguinal gland; head about as broad as long, subovoid to subelliptical from above; eyes large; tympanum usually heavily pigmented, opaque; diameter of tympanum three-fourths that of eye in males, one-half in females; fingers I and II subequal; vocal slits and single internal subgular vocal sac in adult males; unexpanded disks and grooves on all fingers; disk covers rounded; expanded, spadate (fig. 7.51h) disks on toes; subarticular tubercles on digits flattened or rounded, low in profile; no supernumerary tubercles; usually no accessory palmar tubercles; no nuptial pads; thenar and palmar tubercles low, rounded, palmar much larger than thenar; heel with one to three conical tubercles; usually no plantar tubercles; inner metatarsal tubercle ovoid, outer round, inner three times size of outer; no inner tarsal fold. Dorsum tan, pink, or dark brown, nearly uniform or with some small dark spots or blotches or obscure middorsal or lateral dark stripes or dark blotches, frequently paired suprascapular and pelvic dark spots; dorsal or dorsolateral folds often light in color; a narrow light middorsal stripe sometimes present; a purple black eye mark usually present continuing above tympanum and downward behind axilla, often bordered above by narrow light line; area between dorsal folds usually forms lighter field contrasting with flank color; upper surface of thighs usually with dark bars; a purplish black, triangular, posterior seat patch usually present; sole of foot purple black; venter yellow, orangish, grayish, or olive, reticulated with tiny lighter spots; posterior venter and groin usually suffused with pink; posterior surface of thigh purple gray; iris coppery, brown, or bluish.

VOICE: Males appear to have distinct advertisement and courtship calls. The former consists of a "squeak" lasting about 44 milliseconds, with a dominant frequency at 2.7 kHz and repeated every 10 to 20 seconds. The courtship call is a "trill" and has a dominant frequency of 5.5 kHz and a duration of about 1 second with 8 to 9 pulses. The female's reciprocal call is also a "squeak" lasting about 58 milliseconds, with a dominant frequency at 3.1 kHz (Schlaepfer and Figeroa 1998).

SIMILAR SPECIES: This frog superficially resembles *Eleutherodactylus gollmeri* in size and coloration, but that species has expanded finger and toe disks and an inner tarsal fold.

HABITAT: Premontane Wet Forest and Rainforest, Lower Montane Wet Forest and Rainforest, and marginally in Montane Rainforest.

BIOLOGY: *Eleutherodactylus podiciferus* is a diurnal leaf litter inhabitant. It tends to be found in primary forest situations. Active juveniles are about 9 mm in standard length.

REMARKS: More than one valid species may be encompassed within the present concept of *E. podiciferus*. Frogs referred to this form from higher elevations reach the greatest adult size, and those from the Cordillera de Talamanca grow larger than those from the northern cordilleras. There is also a tendency for examples from lower elevations (1,100 to 1,200 m), here placed in *E. podiciferus*, to have more elevated subarticular tubercles and weak but larger plantar tubercles. The testes of male *E. podiciferus* are enclosed in a black peritoneum. Other *Eleutherodactylus* in Costa Rica have an unpigmented peritoneum around the testes.

DISTRIBUTION: Humid premontane and lower montane zones and lower portion of the montane belt on both slopes of the cordilleras of Costa Rica and adjacent western Panama (1,089–2,650 m).

Eleutherodactylus polyptychus (Cope, "1885," 1886)
(map 7.46)

DIAGNOSTICS: A common tiny to small brown frog having very obvious projecting tubercles on the hands and feet (fig. 7.52a,d,g), the thenar tubercle equal to or slightly smaller than the palmar tubercle (fig. 7.53a), the groin and legs suffused with red, and no nuptial pads in adult males.

DESCRIPTION: Adult males 15.3 to 26.6 mm in standard length, adult females 15.6 to 28.7 mm; dorsum usually warty, less often smooth or granular, usually with a series of short suprascapular ridges, rarely with other ridge patterns; venter coarsely areolate; inguinal gland moderate, about size of palmar tubercle; head longer than broad; snout ovoid to subelliptical in outline from above; eyes large; tympanum distinct, large, translucent, diameter two-thirds diameter of eye in males, one-half in females; no vocal slits or sacs; narrow, unexpanded digital disks and grooves on fingers III and IV; toes with expanded disks; disk covers often pointed,

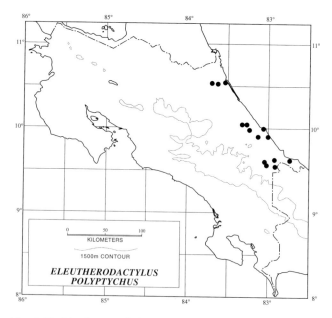

Map 7.46. Distribution of *Eleutherodactylus polyptychus*.

disks often asymmetric; subarticular tubercles projecting, often pointed; usually supernumerary tubercles under some fingers and toes; thenar and palmar tubercles ovoid, thenar equal to or slightly smaller than palmar; accessory palmar tubercles raised, some pointed; heel granular; no nuptial pads; fingers I and II of about equal length; heel granular; no toe webs; inner metatarsal tubercle raised, elongate, outer small, pointed; numerous pointed plantar tubercles; weak inner tarsal fold; ventral disk V-shaped, extending posteriorly to level of femurs. Dorsum grayish brown to tan, uniform, mottled with dark and rarely with dorsolateral or middorsal light stripes; light shield mark, interocular area dark, and dark seat patch marks often present; dark supratympanic stripe and dark lip bars; upper surface of thighs lightly barred, suffused with red; posterior surface uniform reddish brown; throat, chest, and venter pale yellow, with some scattered dark pigment on venter; groin area light red.

VOICE: Not known.

SIMILAR SPECIES: (1) *Eleutherodactylus bransfordii* is smaller (adult males 13.3 to 23 mm in standard length, adult females 13.6 to 26 mm), and adult males have nuptial pads. (2) *Eleutherodactylus underwoodi* has tubercles on the undersurface of the hands that are low, smooth, and usually well separated, and adult males have nuptial pads.

HABITAT: Lowland Moist and Wet Forests.

BIOLOGY: These diurnal leaf litter inhabitants probably resemble *E. bransfordii* in habits. They occur at lower elevations than that species and do not approach it in terms of local abundance. Small juveniles active in the litter are about 7 mm in standard length.

REMARKS: This form is closely related to *E. bransfordii* and *E. underwoodi* and has usually been included within the former species (Savage and Emerson 1970; Miyamoto 1983). It is larger than *E. bransfordii* and generally occurs at lower elevations than that species, and adult males lack the nuptial pads that are characteristic of *E. bransfordii* and *E. underwoodi*. I use the name *E. polyptychus* tentatively for this species rather than coining a new name. The lectotype (USNM 166894) and lectoparatype (USNM 14199) are both adult females, 25.5 and 24 mm in standard length (type locality, Nicaragua: Río San Juan: between Machuca and San Juan del Norte). Further study is needed to establish the status of this taxon.

DISTRIBUTION: Humid lowlands of Atlantic slope Nicaragua, Costa Rica, and adjacent northwestern Panama (2–60 m).

Eleutherodactylus stejnegerianus (Cope, 1893)

(plate 114; map 7.47)

DIAGNOSTICS: Very common tiny to very small brown frogs having very prominent projecting tubercles on the hands and feet (fig. 7.52a,d,g), the thenar tubercle much smaller than the palmar tubercle (fig. 7.53b), a red wash over the groin and hind limbs, and no nuptial pads in adult males.

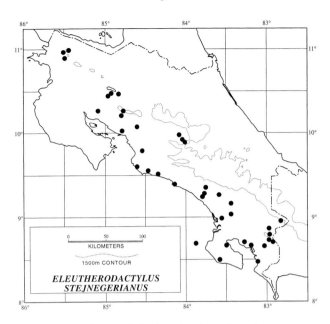

Map 7.47. Distribution of *Eleutherodactylus stejnegerianus*.

DESCRIPTION: Adult males 12 to 18 mm in standard length, adult females 12.5 to 22.4 mm; dorsum irregularly granular to warty, sometimes with suprascapular, dorsolateral, or hourglass-shaped ridges; venter coarsely areolate; inguinal gland large, two to three times size of palmar tubercle; head about as broad as long; snout ovoid to subelliptical in dorsal outline; eyes large; tympanum distinct, translucent, diameter three-fourths diameter of eye in males, one-half in females; no vocal slits or sac; narrow, unexpanded digital disks and groove on fingers III and IV; expanded disks on toes; disk covers frequently pointed, disks often asymmetric; subarticular tubercles globular, obtuse, projecting; supernumerary tubercles on some toes; thenar and palmar tubercles ovoid, thenar much smaller than palmar; no nuptial pads; fingers I and II equal in length; accessory palmar tubercles prominent; heel smooth or granular; no toe webs; numerous rounded plantar tubercles; inner metatarsal tubercle rounded, elongate, outer rounded, much smaller than inner; weak inner tarsal fold present; ventral disk V-shaped, extending posteriorly to level of femurs. Dorsum gray brown to tan, uniform, mottled, or rarely with dorsolateral light stripes or a broad or narrow middorsal stripe; usually a light shield mark, usually a dark seat patch; interocular dark bar usually present; usually a dark supratympanic stripe and upper lip barred with light and dark; upper surfaces of thighs usually with dark bars and covered by reddish suffusion; posterior surface of thigh uniform reddish brown; throat and chest yellowish, usually with dark mottling; venter dull yellow; groin area pale red.

SIMILAR SPECIES: *Eleutherodactylus persimilis* has finger II definitely longer than finger I.

VOICE: A single low squeak, repeated after a pause.

HABITAT: Lowland Moist and Wet Forests, Premontane

Wet Forest and Rainforest, gallery forests in the Lowland Dry Forest region, and on the Meseta Central Occidental within Premontane Moist Forest.

BIOLOGY: Another diurnal leaf litter species, probably having biological characteristics similar to those of *Eleutherodactylus bransfordii*. Densities for this species at some localities may be even higher than those recorded for *E. bransfordii* at La Selva. Scott (1976) reported densities for *E. stejnegerianus* at the Las Cruces Biological Station and Rincón de Osa, Puntarenas Province, as an incredible 4,586/ha and 431/ha, respectively. Small juveniles about 6 mm in standard length.

REMARKS: Taylor (1952a) recognized this form as *Microbatrachylus stejnegerianus*, but Savage and Emerson (1970) synonymized it with *E. bransfordii*. Miyamoto (1983) demonstrated on biochemical grounds that *Eleutherodactylus stejnegerianus* was a distinctive Pacific versant species genetically distinguished from Atlantic slope populations. Subsequently Chen (unpublished MS) showed that *E. bransfordii* (Atlantic) and *E. stejnegerianus* also differed significantly in karyology. Fortunately a morphological feature, the relative sizes of the thenar and palmar tubercles (see diagnostics for each species) is in agreement with the molecular and biochemical evidence. *E. stejnegerianus* further lacks nuptial pads in adult males, which are present in *E. bransfordii*. These two species are syntopic at a few sites near Laguna Arenal at elevations of 825 to 880 m in northwestern Costa Rica on the Atlantic slope of the Cordillera de Tilarán. Although the type specimen of this name is lost, there can be little question that Cope's (1893) description applies to a specimen of this species. *E. stejnegerianus* overlaps in range with *E. underwoodi* at Las Cruces, Puntarenas Province, in southwestern Costa Rica and about 900 m in elevation in the Cordillera de Tilarán.

DISTRIBUTION: Humid lowlands and premontane slopes on the Pacific versant of western Panama and Costa Rica, on the Meseta Central Occidental, and in gallery forests in the subhumid northwest and peripherally on the Atlantic lowlands near Laguna Arenal (3–1,330 m).

Eleutherodactylus underwoodi (Boulenger, 1896a)

(plate 115; map 7.48)

DIAGNOSTICS: A common tiny to small brown species having prominent but relatively low, smooth accessory palmar tubercles (fig. 7.52b,e,h), the thenar tubercle equal to or slightly smaller than the palmar tubercle (fig. 7.53a), the groin and the hind limbs suffused with red, and nuptial pads in adult males.

DESCRIPTION: Adult males 16 to 26 mm in standard length, adult females 18.4–30 mm; dorsum granular or warty; usually with only short suprascapular ridges, sometimes with paired ridging; venter coarsely areolate; inguinal gland moderate, somewhat larger than palmar tubercle;

head slightly broader than long; snout ovoid to subelliptical from above; eyes large; tympanum distinct, translucent; diameter equal to that of eye in males, three-fourths in females; no vocal slits or sac; narrow, unexpanded disks and grooves on fingers III and IV; expanded disks on toes; disk covers often pointed, disks often asymmetrical; subarticular tubercles projecting, low, ovoid, and globular; usually no supernumerary tubercles under digits; thenar and palmar tubercles low in profile, ovoid, thenar equal to or slightly smaller than palmar; white nuptial pads on base of thumb in adult males; fingers I and II about equal in length; accessory palmar tubercles low, globular, ovoid; heel with small warts; no toe webs; plantar tubercles few, round, widely spaced; inner tarsal tubercle low, elongate, outer very small, round; weak inner tarsal fold; ventral disk V-shaped, extending posteriorly to level of femurs. Dorsum brown to tan, rarely gray, uniform, mottled, or rarely with dorsolateral or middorsal light longitudinal stripes; usually with light shield mark, dark interocular mark, and dark seat patch; usually a distinct supratympanic dark stripe and dark bars on upper lip; upper surfaces of thighs with dark bars and suffused with red; posterior thigh surface uniform reddish brown; throat and chest yellowish, often heavily mottled with dark pigment; venter dull yellow; groin region suffused with red.

VOICE: A two-note call consisting of "squeak-squeak" followed by a long pause.

SIMILAR SPECIES: (1) *Eleutherodactylus bransfordii* has projecting, often pointed subarticular accessory palmar and plantar tubercles (fig. 7.52a,d,g) and is smaller (adult males 13.3 to 23 mm in standard length, adult females 13.6 to 26 mm). (2) *Eleutherodactylus polyptychus* has the tubercles on the hands and feet and under the fingers and toes,

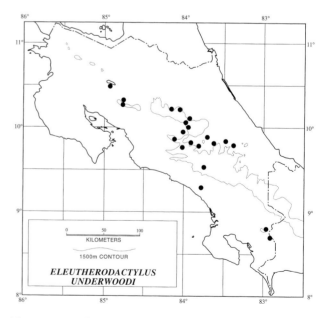

Map 7.48. Distribution of *Eleutherodactylus underwoodi*.

frequently pointed (fig. 7.52a,d,g), and adult males lack nuptial pads.

HABITAT: Premontane Wet Forest and Rainforest and marginally in Lower Montane Wet Forest.

BIOLOGY: This leaf litter denizen, like others in the *rhodopis* group, is diurnal. Little is known regarding its habits, but they are probably similar to those of other species in that group. Small juveniles moving around are about 7 mm in standard length.

REMARKS: Taylor (1952a) referred to this form as *Microbatrachylus underwoodi*, but Savage and Emerson (1970) synonymized it with *E. bransfordii*. It seems to be an upland cognate of *E. bransfordii*, but karyologically it differs markedly from that form and *E. stejnegerianus* (Chen, unpublished MS). It occurs sympatrically with the latter species at several sites at elevations (900–1,200 m) near the lower limit of the altitudinal range of *E. underwoodi*. This species is extremely similar to *E. bransfordii*, but as far as can be determined their ranges do not overlap. Additional study is clearly needed to clarify the status of *E. bransfordii* and *E. underwoodi* as well as their relations to the Atlantic lowland *E. polyptychus*. A specimen now assigned to this taxon had a karyotype of 2N = 18, consisting of six pairs of metacentric chromosomes, two pairs of submetacentrics, and one pair of subtelocentrics; NF = 36 (DeWeese 1976).

DISTRIBUTION: Humid premontane slopes of the Cordillera Central, Cordillera Costeña, Cordillera de Tilarán, and Cordillera de Talamanca; also marginally in lower montane belt (920–1,590 m). This species probably ranges into western Panama.

Eleutherodactylus gaigeae Group

Relatively short-limbed, narrow-waisted, and narrow-headed terrestrial frogs with a relatively pointed snout, prominent tympanum, and well-developed accessory palmar and plantar tubercles. Other features include finger I longer than finger II; toe III shorter than toe V; tip of toe V not reaching distal subarticular tubercle on toe IV; no vocal slits or nuptial pads in adult males; disks on all digits.

Eleutherodactylus gaigeae (Dunn, 1931a)

(plate 116; map 7.49)

DIAGNOSTICS: A small (males) to moderate-sized (females) frog very distinctive in having a black ground color dorsally, marked with vivid paired orange, orange red, to gold dorsolateral stripes. This coloration closely resembles that of the two very toxic Costa Rican species of the genus *Phyllobates*. This species differs from them most obviously in lacking a light upper lip stripe and paired scutes on the digital disk pads.

DESCRIPTION: Adult males 26 to 31 mm in standard length, adult females 32 to 44 mm; dorsum shagreened;

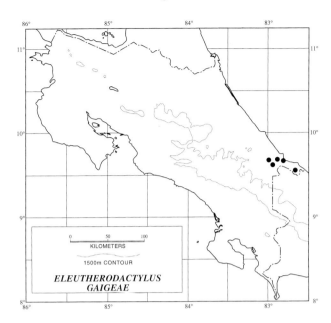

Map 7.49. Distribution of *Eleutherodactylus gaigeae*.

venter smooth; head about as wide as long; snout subelliptical from above; eyes moderate; tympanum distinct, height about two-thirds diameter of eye in males, slightly more than one-half in females; vomerine odontophores oblique, widely separated from one another medially; no vocal slits or sac in adult males; finger I longer than finger II; disk cover on outer two fingers slightly expanded, truncate; inner two barely expanded, round; subarticular tubercles on all digits ovoid, conical in profile; no supernumerary tubercles under any digit; no nuptial pads in adult males; thenar tubercle oval, palmar twice as large, bifid; accessory palmar tubercles distinct; heel smooth; toe disks expanded, barely smaller than largest finger disks; no toe webbing; inner metatarsal tubercle elongate, elevated, outer small, conical; plantar tubercles present; no tarsal fold. Dorsum black with pair of bright gold, orange, or orange red dorsolateral stripes from eye at least to sacrum, sometimes broken into short segments posteriorly; no transverse bars on limbs; venter uniform gray, or orange brown or gray flecked with silver white or pale blue or mottled with bluish green; iris bronze, orange brown, or orange, narrow horizontal black line through center.

VOICE: Does not vocalize.

SIMILAR SPECIES: (1) *Phyllobates vittatus* is smaller (adults 23 to 29 mm in standard length) and has a light upper lip stripe from under the eye to the shoulder and paired scutes on each disk cover (fig. 7.10a). (2) *Phyllobates lugubris* is even smaller (adults 19 to 24 mm in standard length) and has a light upper lip stripe from below the eye to the shoulder and paired scutes on each disk cover (fig. 7.10a).

HABITAT: Tropical Moist Forest in extreme southeastern Costa Rica. Found under surface debris and elsewhere in the leaf litter and in its range often associated with caves or rocky stream banks.

BIOLOGY: A relatively rare nocturnal, terrestrial species that feeds on small arthropods. The striking convergence in coloration between *E. gaigeae* and two members of the highly toxic genus *Phyllobates*, with which it shares parts of its range, suggests a mimetic complex. This species is sympatric with the similarly patterned *Phyllobates lugubris* in southeastern Costa Rica and adjacent northwestern Panama and with *Phyllobates aurotaenia* in northwestern Colombia. The resemblance is enhanced by behavior. Unlike most other *Eleutherodactylus*, *E. gaigeae* and *Phyllobates* use short hops as their principal means of locomotion on and through the upper layer of the leaf litter.

REMARKS: The principal objections to the mimic scenario are (Lynch 1985): (1) *Phyllobates* are diurnal, adult *E. gaigeae* are nocturnal. (2) Adults of the striped *Phyllobates* are smaller than adult *E. gaigeae*. (3) Several populations of *E. gaigeae* (Panama and Colombia) do not co-occur with striped *Phyllobates*. (4) Several populations (valley of the Ríos Sinú, Cauca, and Magdalena, Colombia) putatively referred to *E. gaigeae* based on morphological similarity consist of dull-colored frogs that do not resemble any *Phyllobates* or the "mimics" from the east and north. The first and second points are not conclusive. Juveniles of many predominantly nocturnal *Eleutherodactylus* are often diurnally active. In addition, juvenile *E. gaigeae* are about the same size as adults of the presumed *Phyllobates* models, and the coloration thus may be of greatest selective advantage to individuals at prereproductive stages. Moreover, the coloration probably is a releaser stimulus, so that the size of the mimic is irrelevant to the predator's response or the larger size of *E. gaigeae* may even enhance the effectiveness of the sign stimulus (larger equals greater intensity). The third objection may be the fairly recent result of extinction of the striped *Phyllobates* from areas between their current ranges, or the avoidance response of the predator(s) to the *Phyllobates–E. gaigeae* color pattern may be intrinsic. In the latter case *E. gaigeae* would have the potential for invading areas where striped *Phyllobates* do not occur once the aposematic coloration was established in the mimic. Finally, the absence of mimetic coloration in populations of "*E. gaigeae*" from east of the Cordillera Occidental in Colombia may represent a condition antecedent to the highly specialized pattern found elsewhere in the species' range. On the other hand, two species may be involved, one nonmimetic and one mimetic. Two issues thus remain to be resolved. Although circumstantial evidence points to an aposematic role for the color pattern of *E. gaigeae*, the data available are suggestive but not conclusive. Lynch's (1985) belief that both mimetic and nonmimetic populations of frogs that he assigned to *E. gaigeae* are conspecific also needs further review.

Dunn (1931a) originally described this species as *Lithodytes gaigei*. Taylor (1952a) used the original spelling but noted that perhaps it should be changed to *gaigeae* (the Latin feminine genitive), since the species was named for Helen Thompson Gaige. Lynch (1980, 1985), apparently influenced by Taylor's comment, used the feminine form. Subsequently Lynch (1996) completely muddied the waters by stating that although this frog was explicitly named to honor Mrs. Gaige, Taylor (1952a) emended the original spelling to *gaigeae* and that articles 32 and 33 of the *International Code of Zoological Nomenclature* required retention of the original spelling *(gaigei)*. Since Dunn indicated that he named the frog after Mrs. Gaige, under article 31.1.2 and article 32.5.1 the original spelling would have been incorrect, and the change to *gaigeae* is mandatory. Such a change is not an emendation but a correction of an incorrect original spelling (article 32.4). Lynch concluded that *E. gaigeae* was a homonym of *E. gaigei* that needed to be replaced, citing article 58 of the code. Unfortunately he misread the article, which lists variant spellings for species names that are deemed to be homonymous when used for different nominal taxa placed in the same genus. Item 12 in that list states that "use of *-i* or *-ii*, *-ae* or *-iae*, *-orum* or *-iorum*, *-arum* or *-iarum*" as the termination in a genitive-based name of a person constitutes such pairs of alternative suffix spellings for names. Lynch apparently thought that all of these terminations, if used with the same personal name, make the names homonyms. That is not what the article says. Thus *smithi* and *smithii* and *joanae* and *joaniae* are homonyms, but *smithi* and *smithae* are not homonyms and are available names. Nevertheless, since the correct original spelling for Dunn's frog is *E. gaigeae*, Lynch's conclusion regarding secondary homonymy in the case of *Syrrhophus gaigeae* Schmidt and Smith, 1944, now referred to *Eleutherodactylus*, is valid.

DISTRIBUTION: Atlantic lowlands from extreme southeastern Costa Rica to eastern Panama and on the Pacific versant in central Colombia; populations in the Ríos Sinú, Cauca, and Magdalena questionably placed here (following Lynch 1985); in Costa Rica known from the Valle de Talamanca and vicinity of Carbón, Limón Province (20–200 m).

Eleutherodactylus cerasinus Group

Members of this group have expanded digital disks, finger I usually shorter than finger II, toe V much longer than toe III, but the tip of the disk on toe V does not reach the distal subarticular tubercle on toe IV (fig. 7.55b). Other features shared by species in this group are: no supernumerary tubercles under fingers and toes, no toe webs, and venter coarsely areolate.

Eleutherodactylus cerasinus (Cope, 1875)

(plate 117; map 7.50)

DIAGNOSTICS: Males are small and females moderate-sized, long-legged frogs having a well-developed heel tubercle (fig. 7.49b), red in the groin and on the anterior and

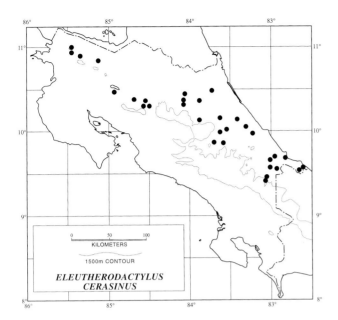

Map 7.50. Distribution of *Eleutherodactylus cerasinus*.

posterior thigh and calf, some subarticular tubercles pungent to pointed (fig. 7.51s), and the tip of toe V not reaching the ultimate subarticular tubercle on toe IV (fig. 7.55b).

DESCRIPTION: Adult males 19 to 25 mm in standard length, adult females 25 to 35 mm; dorsum weakly granular with evident scattered tubercles often forming W-shaped or H-shaped suprascapular ridging; head longer than broad; snout ovoid to subelliptical from above; snout long; eyes moderate; upper eyelid with several small tubercles; superciliary border with small tubercles; tympanum distinct, diameter about one-third diameter of eye, but annulus tympanicus not evident externally; vomerine odontophores oblique, widely separated from one another medially; vocal slits and single internal subgular vocal sac in adult males; finger and toe disks expanded, covers truncate on fingers III and IV, rounded on other digits; subarticular tubercles projecting, at least some on finger III and toes III and IV pungent to pointed; accessory palmar and plantar tubercles present; whitish nuptial thumb pads present in adult males; thenar tubercle on outer margin of base of thumb oval, smaller than more or less bifid palmar tubercle; inner metatarsal tubercle large, projecting, outer round, small; a weak, short inner tarsal fold or tubercle. Upper surfaces yellowish brown, usually marked with dark brown spots or blotches and with a W-shaped or H-shaped suprascapular dark marking, sometimes nearly uniform or rarely with a broad middorsal longitudinal yellow stripe; dark transverse bars on upper limb surfaces; posterior thigh surface brown, suffused with red; anterior thigh, underside of calf, and groin red; venter white; iris golden yellow above, copper ventrally.

VOICE: A single very short "tick," repeated after a long interval; dominant frequencies 2.2 to 4.9 kHz (Ibáñez, Rand, and Jarmillo 1999).

SIMILAR SPECIES: (1) *Eleutherodactylus ridens* lacks the large heel tubercle and pungent to pointed subarticular tubercles. (2) *Eleutherodactylus cruentus* has a large flaplike tubercle on the upper eyelid (fig. 7.49f,g) and lacks pungent to pointed subarticular tubercles.

HABITAT: Lowland Moist and Wet Forests and Premontane Wet Forest and Rainforest.

BIOLOGY: *Eleutherodactylus cerasinus* is often found moving about in the leaf litter during the day and perched on low vegetation at night. Escape behavior consists of three or four long leaps ending in a dive under the leaf litter. The species is common in premontane situations, with lower apparent densities at lowland sites. Food includes a variety of insects exclusive of ants (Lieberman 1986).

REMARKS: Taylor (1952a) lists *Eleutherodactylus peraltae* Barbour (1928) (holotype: MCZ 13601; Costa Rica: Cartago: Peralta) as a Costa Rican form based on the original description. Savage (1981b) demonstrated that this nominal taxon is conspecific with *E. cerasinus* (holotype: USNM 32572; Costa Rica: Limón: Cerro Utyum). He also pointed out that Dunn (1931c) correctly applied the name *E. cerasinus* to this species but later (Dunn in Zetek and Wetmore 1951) used the older name *Eleutherodactylus cruentus*. *E. cruentus* is a representative of another valid species (see account below).

DISTRIBUTION: Humid forests of the Atlantic lowlands and southern premontane slopes from Nicaragua to eastern Panama and marginally on the Pacific slope in northwestern Costa Rica (19–920 m).

Eleutherodactylus cruentus Group

Costa Rican members of this group are often mistaken for treefrogs because of their large digital disks. Other characteristics are finger I usually shorter than finger II; toe III much shorter than toe V; tip of toe V reaching distal subarticular tubercle on toe IV (fig. 7.55a); tympanum distinct, indistinct, or concealed; vomerine odontophores oblique, widely separated medially; vocal slits and nuptial pads present or absent in adult males; vocal sac, if present, single internal subgular; subarticular tubercles flattened; no supernumerary tubercles on hands and feet; no toe webbing; short inner tarsal fold or tubercle present or absent; venter coarsely areolate.

Most frogs of this group in lower Central America are found on the forest floor or on low vegetation. Males usually call from sites from ground level to a meter above the substrate, and most species are predominantly nocturnal.

Eleutherodactylus altae Dunn, 1942b

(plate 118; map 7.51)

DIAGNOSTICS: A beautiful tiny to small blackish frog with a bright deep coral red spot in the groin that usually extends onto the anterior thigh surface, a truncate snout

Map 7.51. Distribution of *Eleutherodactylus altae*.

(fig. 7.6c) with each nostril in a protuberance, and an indistinct tympanum.

DESCRIPTION: Adult males 18 to 23.5 mm in standard length, adult females 20 to 27 mm; dorsum granulate to tuberculate; head as wide as long; eyes moderate; upper eyelid rugose; tympanum indistinct, height less that one-half diameter of eye, oval in both sexes; annulus tympanicus visible through skin anteriorly and ventrally; no vocal slits or sac; finger and toe disk covers expanded, palmate to spadate, those on fingers III and IV and toes II–III–IV–V nearly twice width of digit; some toe pads triangular; subarticular tubercles on hands and feet, flattened, ovoid; no accessory palmar tubercles; no nuptial thumb pads in males; thenar tubercle elongate, on outer margin at base of thumb; much larger than rounded palmar tubercle; no enlarged heel tubercle; one to several small plantar tubercles; inner metatarsal tubercle elongate, flattened, outer round, much smaller; no tarsal fold. Dorsum dark gray to black; axilla coral; an elongate coral to red orange spot in groin area extending onto anterior surface of thigh, where it may be broken into several spots; often a coral spot or spots on posterior thigh surface; sometimes black vertical bars between light spots on thighs; venter heavily pigmented with dark brown; iris copper.

VOICE: Apparently mute.

SIMILAR SPECIES: In *Eleutherodactylus pardalis* the light groin and thigh markings are silvery white and the anterior thigh surface is always marked with black vertical bars.

HABITAT: In leaf litter and arboreal bromeliads in Lowland Wet Forest and Premontane Wet Forest and Rainforest areas.

BIOLOGY: Found active on low vegetation at night,

in bromeliads and leaf litter during the day. The smallest known juvenile is 12.9 mm in standard length.

REMARKS: Examples of this relatively uncommon species from Atlantic lowland sites have distinct black vertical bars separating the brightly colored areas on the anterior thigh surface and resemble *Eleutherodactylus pardalis* in pattern but not color. Those from upland areas lack the black bars. Preserved specimens from the lowlands are difficult to separate from *E. pardalis* because the coral to red color fades rapidly. *E. altae*, however, has somewhat smaller outer finger disks (smaller than the tympanum) than *E. pardalis* (equal to tympanum). The karyotype is 2N = 26, composed of six pairs of metacentrics, three pairs of submetacentrics, one pair of subtelocentrics, and three pairs of telocentrics; NF = 46 (DeWeese 1976).

DISTRIBUTION: Humid lowland and premontane zones of Atlantic slope northern and central Costa Rica, south to extreme northwestern Panama (60–1,245 m).

Eleutherodactylus caryophyllaceus (Barbour, 1928)

(plate 119; map 7.52)

DIAGNOSTICS: This elegant small, long-legged species exhibits considerable polychromatism but is distinctive in having a long, pointed snout (fig. 7.5b), a large, pointed heel tubercle (fig. 7.49b), a well-developed flaplike superciliary tubercle (fig. 7.49f,g), and no contrasting pattern in the groin and on the posterior thigh surface.

DESCRIPTION: Adult males 21 to 24 mm in standard length, adult females 23 to 26 mm; dorsum smooth; head longer than broad; snout pointed from above; eyes moderate; upper eyelid smooth with a distinct large superciliary tubercle; tympanum indistinct in females, hidden in males, annulus tympanicus not visible externally; vocal slits and

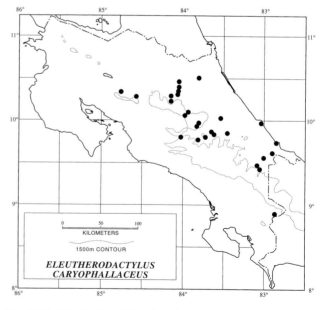

Map 7.52. Distribution of *Eleutherodactylus caryophyllaceus*.

single internal subgular vocal sac in adult males; finger and toe disks moderately expanded, covers rounded; subarticular tubercles on all digits oval, globular, not projecting; no accessory palmar or plantar tubercles; no nuptial thumb pads in males; thenar tubercle elongate, on outer margin at base of thumb; palmar tubercle much smaller, flattened, usually slightly bifid; inner metatarsal tubercle elongate, outer obscure, small, and rounded; no tarsal fold. Upper surfaces pale yellow, pink, brownish, gray, or dark green, covered by discrete dark punctations; occasionally with some dark crossbars, broad longitudinal silver stripes, or a broad dorsal dark field bordered by dark brown longitudinal stripes and lateral yellow pinstripes; no dark transverse bars on upper surface of limbs; posterior thigh surface uniform pale yellow; venter white with a few to many scattered dark punctations; iris gray, yellow, or bright red.

VOICE: Unknown.

SIMILAR SPECIES: *Eleutherodactylus ridens* lacks a heel tubercle, has one to several large supraocular tubercles, and has red on anterior and posterior surfaces of the thighs, calves, and feet.

HABITAT: Lowland Moist and Wet Forests, Premontane Wet Forest and Rainforest, and marginally into Lower Montane Rainforest.

BIOLOGY: Typically found on low vegetation within the forest. The species is relatively uncommon at lowland sites but common at many premontane locales. Food consists of a variety of arthropods, exclusive of ants. In five individuals from La Selva, Heredia Province, stomachs contained an average of 1.7 food items with average lengths of 1.0 mm (dipterans and isopods) to 17.0 mm (orthopterans) (Lieberman 1986). Females lay their eggs on leaf surfaces and brood them (Myers 1969a). Shortly after laying, the jelly envelope of the eggs turns white. Small juveniles are 12 to 13 mm in standard length. A recent hatchling measured 9 mm.

REMARKS: The karyotype is 2N = 32, consisting of all telocentric chromosomes; NF = 32 (DeWeese 1976).

DISTRIBUTION: Lowland, premontane, and lower portions of the lower montane belts from Atlantic slope northern Costa Rica to northern Colombia and along the Pacific versant from extreme southwestern Costa Rica to western Colombia (2–1,900 m).

Eleutherodactylus cruentus (Peters, 1873a)

(plates 120–21; map 7.53)

DIAGNOSTICS: An extremely variable small (males) to moderate-sized (females) species having a distinct large heel tubercle (fig. 7.49b), an enlarged supraocular tubercle (fig. 7.49f,g), the groin and anterior thigh surface usually marked with several bright golden yellow to orange spots separated by dark brown to black borders (fig. 7.58), and no pungent or pointed subarticular tubercles.

DESCRIPTION: Adult males 21 to 28 mm in standard length, adult females 25 to 42 mm; dorsum smooth to gran-

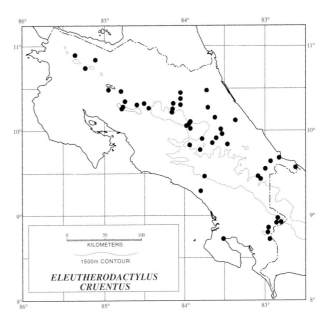

Map 7.53. Distribution of *Eleutherodactylus cruentus*.

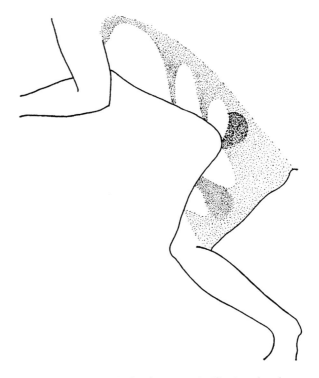

Figure 7.58. Flank and thigh color pattern in *Eleutherodactylus cruentus*; the light areas are usually gold to orange in life.

ulate with numerous smooth tubercles and/or with tubercles fused into low ridges; W-shaped or H-shaped suprascapular ridges often present; head about as broad as long; snout rounded to truncate from above, sometimes with a fleshy terminal pustule; eyes moderate; upper eyelid smooth or with several small tubercles in addition to large supraocular tubercle; tympanum oval, small, height about one-fifth to one-fourth eye diameter; indistinct in females, hidden under skin in males; annulus tympanicus partially evident in females, not visible in males; no vocal slits or sac; finger and toe disks expanded, covers truncate on fingers III and IV, and toes II–III–IV, nonemarginate and at least twice width of digits; subarticular tubercles small, ovoid, globular on all digits; low, small, conical accessory palmar and plantar tubercles present; whitish nuptial pads present on thumbs in adult males; thenar tubercle small, elongate, on outer margin at base of thumb; palmar tubercle larger, bifid to trifid; inner metatarsal tubercle large, ovoid, outer tiny; a short inner tarsal fold or tubercle. Upper surfaces gray, brown, brownish black, or nearly black; usually dark markings of various types including W-shaped or H-shaped figures on suprascapular region; sometimes nearly uniform above or with the upper head surface yellow or orange, or a broad yellow longitudinal middorsal stripe present; other examples with dorsum light brown, bordered by broad dark brown longitudinal lateral stripes; upper surface of limbs uniform to heavily barred with black; groin usually with large golden yellow to orange spots separated by dark brown to black areas, often in the form of oblique bars; groin area sometimes lacking light spots, groin and thighs sometimes suffused with pink, especially ventrally; irregular rose to red areas sometimes on upper surface of body; posterior thigh surface brown to black, uniform or with small irregular light spots; venter usually heavily mottled with dark pigment to almost uniform black; iris gold above, coppery brown below to uniform black.

VOICE: Apparently mute.

SIMILAR SPECIES: (1) *Eleutherodactylus cerasinus* consistently has red groin, thigh, and calf coloration and lacks a well-developed supraocular tubercle. (2) *Eleutherodactylus melanostictus* is much larger (adult males 35 to 43 mm in standard length, adult females 35 to 56 mm), usually has the dark bars on the legs extending completely across the posterior thigh surface (fig. 7.50e), where the light interspaces are bright yellow, orange, pink, magenta, or scarlet, although sometimes suffused with dark pigment, and has toe III much longer than toe V (fig. 7.48b).

HABITAT: Lowland Moist and Wet Forests, Premontane Wet Forest and Rainforest, and marginally into Lower Montane Wet Forest and Rainforest.

BIOLOGY: A nocturnal species usually found on low vegetation, up to a meter above the substrate in dense forest. It has also been collected during the day in thick moss banks in association with presumably conspecific egg masses. In a thirteen-month study of the leaf litter herpetofauna at La Selva, Heredia Province, only three individuals were captured, suggesting that the species hides above the ground during the day (Lieberman 1986). The only definite food items found through analysis of stomach contents in that study were a hymenopteran (not an ant) and a hemipteran, 1 and 6 mm in length, respectively. Eggs sometimes are laid in crevices of tree trunks but are not attended by either parent (Myers 1969a). The species is much more abundant at premontane sites than at lowland localities, and most samples contain large numbers of subadults. Small juveniles are 10 mm in standard length.

REMARKS: Savage (1981b) reviewed the confusions relating to the correct name for this species (holotype, ZMB 9811; Panama: Chiriquí) and its misapplication to *Eleutherodactylus cerasinus*. Because Dunn (in Zetek and Wetmore 1951) associated the name *E. cruentus* with the latter species, Taylor (1952a) described and named *Eleutherodactylus dubitus* from Costa Rica. The holotype of *E. dubitus* (KU 24942; Costa Rica: Alajuela: Isla Bonita) is conspecific with *E. cruentus* as recognized here. Taylor also included *Eleutherodactylus lutosus* (Barbour and Dunn 1921) in his treatment of Costa Rican frogs. This name is based on a juvenile male (holotype: MCZ 8023; Costa Rica: Cartago: Navarro) of *E. cruentus* (Savage 1981b). The karyotype of *E. cruentus* is 2N = 20, composed of five pairs of metacentrics, two pairs of submetacentrics, one pair of subtelocentrics, and two pairs of telocentrics; NF = 36 (DeWeese 1976).

DISTRIBUTION: Humid forests of lowland and premontane zones and lower portion of the lower montane zone from northern Costa Rica to western Colombia (40–1,800 m).

Eleutherodactylus moro Savage, 1965

(plate 122; map 7.54)

DIAGNOSTICS: A tiny (males) to small (females) green frog with a red head and red orange eyes, rounded digital disks (fig. 7.51b), and no enlarged supraocular and heel tubercles. In preservative, the nearly uniform pale white dor-

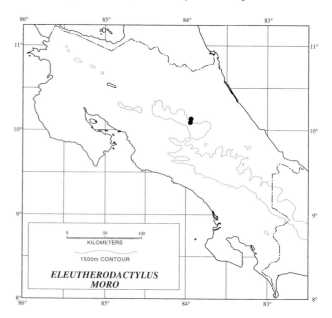

Map 7.54. Distribution of *Eleutherodactylus moro*.

sal coloration might cause this species to be confused with *Hyla colymba* or some of the glassfrogs (Centrolenidae), but these frogs have finger webs, which are absent in *E. moro*.

DESCRIPTION: Adult males 16.5 to 19.5 mm in standard length, adult females 24 to 25 mm; dorsum smooth; snout suboovoid; eyes moderate; upper eyelid smooth; tympanum internal, barely discernible; annulus tympanicus not evident externally; paired vocal slits and somewhat distensible single internal subgular vocal sac in adult males; disk covers on fingers III and IV and toes II–III–IV–V about one and a half times width of digits; subarticular tubercles on all digits small, round, not projecting; no definite accessory palmar tubercles, but several plantar tubercles present; no nuptial thumb pads in males; thenar tubercle elongate, on outer margin of base of thumb; palmar tubercle large, round; inner metatarsal tubercle flat, elongate, smaller outer tubercle rounded and conical; no tarsal fold. Dorsum green, head red above; no transverse dark bars on limbs; venter immaculate greenish yellow, transparent; peritoneum visible, white; pericardial sac visible, white; iris red orange.

VOICE: Unknown.

SIMILAR SPECIES: In life there are none. In preservative: (1) *Eleutherodactylus ridens* has one to several enlarged supraocular tubercles (fig. 7.49f,g), a distinct tympanum, and a definite dark supratympanic spot. (2) *Eleutherodactylus caryophyllaceus* has a well-developed heel tubercle (fig. 7.49b), and the dorsum and venter are punctuated with dark pigment.

HABITAT: Premontane Rainforest.

BIOLOGY: A rare nocturnal, probably arboreal species; most examples have been found hiding in epiphytic bromeliads during the day.

REMARKS: Savage (1965), in the original description, suggested that *Eleutherodactylus moro* might be allied to *E. diastema*. The differences in disk structure between the *diastema* species group (which see) and *E. moro* seem to preclude any close relationship. Lynch and Duellman (1997) placed *E. moro* in the *unistrigatus* species group along with all members of the *cruentus* group discussed in the present account.

DISTRIBUTION: Premontane zone of Atlantic slope central Costa Rica (1,245 m) and Pacific versant central Panama (550–975 m) to lowland western Colombia (above 50 m); know from localities near La Hondura, San José Province, in Costa Rica (1,245 m).

Eleutherodactylus pardalis (Barbour, 1928)

(plate 123; map 7.55)

DIAGNOSTICS: Tiny to small blackish frogs having a silvery white spot in the groin and the anterior thigh surface silvery white marked with black vertical bars, the snout truncate (fig. 7.6) with nostrils in protuberances, and an indistinct tympanum.

DESCRIPTION: Adult males 16 to 19 mm in standard

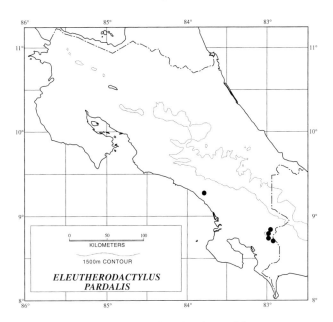

Map 7.55. Distribution of *Eleutherodactylus pardalis*.

length, adult females 25 to 29 mm; dorsum granulate to tuberculate; head as wide as long; eyes moderate; upper eyelid rugose; indistinct oval tympanum, height less than one-half diameter of eye in both sexes; annulus tympanicus visible through skin anteriorly and ventrally; no vocal slits or sac; finger and toe disk covers expanded, palmate to spadate, twice as wide as digits on fingers III and IV and toes II–III–IV–V; some disk pads triangular; subarticular tubercles ovoid, flattened on all digits; no distinct accessory palmar tubercles; no nuptial thumb pads in males; thenar tubercle elongate, on outer margin of base of thumb; palmar tubercle much smaller, rounded; no heel tubercle; one or several plantar tubercles; inner metatarsal tubercle elongate, flattened, much larger than tiny, round outer tubercle; no tarsal fold. Dorsum dark gray to black; axilla, groin, thigh, and lower limb spots silvery white; light thigh and calf spots separated by vertical black bars; venter heavily marked with dark brown pigment; iris coppery.

VOICE: Does not vocalize.

SIMILAR SPECIES: *Eleutherodactylus altae* has coral to red orange spots in the groin and on the anterior thigh. .

HABITAT: Premontane Rainforest zone.

BIOLOGY: A relatively uncommon, nocturnal form found on low vegetation within dense forest.

DISTRIBUTION: Premontane belt on Pacific slope southwestern Costa Rica (884–1,220 m) and Atlantic slope northwestern Panama (365–1,450 m) to eastern Panama (50–800 m).

Eleutherodactylus ridens (Cope, 1866a)

(plates 124–25; map 7.56)

DIAGNOSTICS: Tiny to small light-colored frogs having the anterior and posterior thigh surfaces, calves, and feet

Map 7.56. Distribution of *Eleutherodactylus ridens*.

red, round disk covers and pads (fig. 7.51b,l), the tip of the snout (fig. 7.6) usually extended into a fleshy point, one to several enlarged, pointed supraocular tubercles (fig. 7.49f,g), and no enlarged heel tubercle.

DESCRIPTION: Adult males 16 to 19 mm in standard length, adult females 21 to 25 mm; dorsum smooth with a few scattered tubercles, frequently several marking points of W-shaped suprascapular dark mark; head slightly broader than long; snout pointed; eyes moderate; upper eyelid with several tubercles, at least one enlarged, pointed; tympanum oval, larger and more distinct in females than males, height about one-third diameter of eye in former, less than one-third in latter; annulus tympanicus evident through skin in females, partially evident anteriorly and ventrally in males; paired vocal slits and a somewhat distensible single internal subgular vocal sac in adult males; disks expanded, rounded; disks on fingers II–III–IV and toes II–III–IV–V not twice width of digit; subarticular tubercles round, not projecting; accessory palmar and plantar tubercles; no nuptial thumb pads in males; thenar tubercle elongate, on outer margin of thumb base; palmar tubercle cordate to trifid; larger than thenar tubercle; heel smooth or with several small warts; inner metatarsal tubercle elongate, not raised, larger than round, outer metatarsal tubercle; no tarsal fold. Dorsum beige, yellowish tan, or pale gold, uniform or usually with a pinkish cast; a characteristic distinct dark brown supratympanic mark; a brown W-shaped suprascapular mark and/or chevron-shaped sacral mark present; upper limb surfaces marked with transverse dark bars; a light middorsal pinstripe sometimes present, or entire dorsum pale tan bordered laterally by broad dark brown longitudinal stripes; venter and vocal sac yellow, undersurfaces marked with scattered dark punctations; iris beige above, copper below.

VOICE: A short harsh trill; dominant frequency 5.2 to 6 kHz (Ibáñez, Rand, and Jarmillo 1999).

SIMILAR SPECIES: (1) *Eleutherodactylus caryophyllaceus* has a well-developed heel tubercle (fig. 7.49b), has no large supraocular tubercles, and lacks red groin and limb coloration. (2) *Eleutherodactylus moro* has most of the upper surfaces green and lacks supraocular tubercles, red groin and limb coloration, a dark supratympanic mark, and ventral punctations.

HABITAT: Lowland Moist and Wet Forests, Premontane Wet Forest and Rainforest, and Lower Montane Rainforest.

BIOLOGY: A very common nocturnal forager in low vegetation that often hides in the leaf litter during the day. Males call from leaf surfaces 0.5 to 1.5 m above the ground. Food consists of a variety of small arthropods, but especially spiders and ants. Stomach analysis of fifteen examples from La Selva, Heredia Province, averaged 2.1 prey items per stomach; prey length averaged 0.8 mm (ants) to 5.5 mm (one isopod) depending on taxonomic group (Lieberman 1986). Average density for *E. ridens* at this locale was 26/ha. Densities at premontane sites are noticeably higher than at lowland localities. The smallest juvenile measured 11 mm in standard length.

REMARKS: This species was originally described from Atlantic versant Nicaragua near the boundary with Costa Rica. The original account (Cope 1866a) emphasized the areolate venter, reddish dorsal color, and dark supratympanic mark characteristic of individuals associated with this name. Taylor (1952a, 690) was the first subsequently to apply the name *E. ridens* to this form, since Costa Rica examples agreed with Cope's original account. Examination of specimens from Honduras and Nicaragua confirms their conspecificity with Costa Rican examples. Although the type is lost, the brief type description cannot be associated with any other species. Barbour's (1928) *Syrrhopus molinoi* (holotype: MCZ 13051; Panama: Panama: Barro Colorado Island) is based on a 19 mm female that agrees in detail with Costa Rican specimens in lacking a well-developed heel tubercle and in having one to several enlarged, pointed supraocular tubercles and the finger disks expanded and rounded. Lynch (1980) concurs with this allocation based on an independent study of the situation and the holotype of *S. molinoi*. Dunn's (1931c) placement of the latter as a subspecies of *Eleutherodactylus lutosus* is clearly incorrect, as may be seen from the account of *E. cruentus* above. Although Taylor (1952a) recognized *E. ridens* in Costa Rica, he failed to associate it with the Panamanian frogs called *S. molinoi*. However, following Savage (1981b) the name *E. ridens* now is applied to Panama populations (Rand and Myers 1990). The karyotype for this species is 2N = 34, consisting of one pair of metacentric and sixteen pairs of telocentric chromosomes; NF = 36 (DeWeese 1976).

DISTRIBUTION: Humid lowlands and premontane slopes on the Atlantic versant from northern Honduras to

Panama, on the Pacific slope from southwestern Costa Rica to western Ecuador, and marginally in northwestern Costa Rica (15–1,600 m).

Eleutherodactylus martinicensis Group

Small to moderate-sized, rather treefroglike forms having well-developed finger and toe disks, paired vocal slits, and a single median external subgular vocal sac in adult males, the disk pads broadened or truncate, finger I shorter than finger II and toe III much shorter than toe V, the tip of toe V reaching the distal subarticular tubercle on toe IV. Other features of this group are tympanum distinct; no nuptial thumb pads in males; subarticular tubercles ovoid, not projecting; no supernumerary tubercles on hands and feet; toes usually without webs; inner tarsal fold indistinct or absent; venter coarsely areolate.

Eleutherodactylus johnstonei Barbour, 1914

(plate 126; map 7.57)

DIAGNOSTICS: A small to marginally moderate-sized dull-colored species that lacks digital webbing, inguinal glands, and distinct light spots in the groin and has distinct but small round finger and toe disks (fig. 7.51b), finger I longer than finger II, and toe III much shorter than toe V.

DESCRIPTION: Adult males 17 to 25 mm in standard length, adult females 17 to 35 mm; dorsum smooth to weakly tuberculate; venter strongly areolate; head slightly broader than long; snout truncate from above; eyes large; eyelid with many low, rounded tubercles; tympanum distinct, diameter about one-half diameter of eye, slightly larger in males than females; vomerine odontophores oblique, widely separated medially; paired vocal slits and a distensible single internal

Map 7.57. Distribution of *Eleutherodactylus johnstonei.*

subgular vocal sac in adult males; sac strongly granular when uninflated; finger I longer than finger II; finger disks expanded, rounded, less than twice width of finger; those on II–III–IV larger than on finger I; subarticular tubercles on hands and feet rounded and obtuse; no supernumerary tubercles under digits; no nuptial thumb pads in males; thenar and palmar tubercles large, about same size; a few low accessory palmar tubercles; heel tuberculate; largest toe disks about same size as finger disks, rounded; no toe webs; many small plantar tubercles; elongate inner metatarsal tubercle much larger than conical outer metatarsal tubercle; no tarsal fold. Dorsal ground color brown to gray tan, usually with one or two darker chevrons; frequently a narrow light middorsal pinstripe or broad paired dorsal stripes; posterior thigh surface marbled, stippled, or blotched on a dark brown to grayish tan ground; undersurface creamy; iris gold above, brownish below.

VOICE: The call is a two-note whistle repeated a maximum of 60 times per minute. The first note has a frequency near 2 kHz and a duration of 70 to 90 milliseconds. The second note is longer, with a duration of 180–270 milliseconds, rising sharply from about 3 kHz and continuing for some time at about 4 kHz. The interval between calls averages 1.2 seconds (Kaiser 1992).

SIMILAR SPECIES: (1) *Eleutherodactylus ridens* has red thighs, calves, and feet and several enlarged, pointed supraocular tubercles on the upper eyelid (fig. 7.49f,g). (2) *Eleutherodactylus cruentus* has large truncate and emarginate disks on the fingers (fig. 7.51g).

HABITAT: A "weed" species likely to be found in vacant lots, gardens, and greenhouses.

BIOLOGY: This summary is based on the species in its native Lesser Antilles habitat as reviewed by (Schwartz and Henderson 1991) and Kaiser and Hardy (1994a,b). A nocturnal, ecological generalist readily introduced as a stowaway on agricultural plant materials. Food consists principally of ants (44%), but spiders, leafhoppers, and springtails are also common prey. Males call from ground level up to 3 m from a variety of sites, including leaf surfaces, tree trunks, and stone walls, but also from piles of debris, especially coconut husks. Reproduction occurs primarily in the early rainy season (July–August), but the season probably extends from April through November. In the laboratory breeding takes place every two to three months. In courtship a female approaches a male that then follows her to a previously selected oviposition site. Females lay 10 to 30 large eggs averaging 3 mm in diameter under debris or in concealed places near ground level. One of the parents, usually the male, attends the clutch until it hatches (Bayley 1950). Hatching occurs in about two weeks. The fully developed young have an egg tooth on the upper jaw used to burst out of the egg capsule. Hatchlings are 2 to 3.5 mm in standard length; they lose the egg tooth and reach sexual maturity in one year.

REMARKS: In Costa Rica, known only from a single specimen (UCR 7184) taken in a park in San José City (1,200 m). It remains to be seen if a population has been established there as it has been in Panama City, Panama Province, Panama (Ibáñez and Rand 1990). The karyotype is 2N = 28, composed of four pairs of metacentric or submetacentric chromosomes and ten telocentric pairs; NF = 36 (Kaiser, Green, and Schmid 1994).

DISTRIBUTION: Native to the Lesser Antilles but introduced to Bermuda, Jamaica, Trinidad, Venezuela, Guyana, and recently to Panama and Costa Rica. It is also reputed to occur in greenhouses in Europe and North America, but these records may be based on other species (Kaiser and Hardy 1994b).

Eleutherodactylus diastema Group

Tiny to small frogs with short, stubby arms and hands and more or less triangular disk pads, although these may be only broadened in some forms, and a single external subgular vocal sac in adult males. Other features typifying the group are finger I shorter than finger II, toe III much shorter than toe V; tympanum distinct; some or all digital disks spadate, lanceolate, or papillate (fig. 7.51h–j; vocal slits and sacs but no nuptial thumb pads in adult males; subarticular tubercles flattened; no supernumerary tubercles on hands and feet; no toe webs; no tarsal fold or tubercle; venter coarsely areolate.

These small anurans are more often heard than seen. Males produce a characteristic "dink-dink-dink" call from hiding places at ground level to several meters up on any kind of plant cover, and the calls are among the most common background sounds at night in most humid forest situations in Costa Rica.

Eleutherodactylus diastema (Cope, 1875)

(plate 127; map 7.58)

DIAGNOSTICS: Rather nondescript tiny to small gray to brown frogs having broad, mostly palmate disk covers (fig. 7.51e) and broadened disk pads (fig. 7.51n) and the posterior thigh surface uniform and pigmented. In addition, this species lacks enlarged supraocular and heel tubercles, contrasting colors, and a vivid pattern in the groin or on the thighs.

DESCRIPTION: Adult males 16 to 21 mm in standard length, adult females 18 to 24 mm; dorsum smooth with a few scattered tubercles; head slightly broader than long; snout rounded from above, often with a slight fleshy bump at tip; eyes large, upper eyelid with a few small tubercles; tympanum distinct, diameter about one-third diameter of eye; annulus tympanicus visible except dorsally; vomerine odontophores oblique, widely separated medially; long paired vocal slits and a single external subgular vocal sac in adult males; finger and toe disks moderately expanded, most disk covers palmate, one or more spadate; disk pads mostly expanded; accessory palmar and plantar tubercles

present; no nuptial thumb pads in males; thenar tubercle elongate, on outer margin of thumb base; palmar tubercle smaller than thenar, round to cordate; inner metatarsal tubercle elongate, outer about same size, round. Dorsum various shades of gray to tan, often tinged with pink; nearly uniform or with dark brown interocular bar, suprascapular spots, or lumbar spots or large blotch; paired light longitudinal lateral stripes sometimes present; upper surface of legs with or without obscure transverse dark bars; posterior thigh surface brownish, pigmented; venter yellow to dark brown, always with many small dark punctations; vocal sac yellow; iris silver above, brown below.

VOICE: A series of well-spaced "dink" notes, each lasting about 205 milliseconds and repeated 30 to 40 times a minute, with a dominant frequency between 3.1 and 4.0 kHz (Fouquette 1960; Wilczynski and Brenowitz 1988).

SIMILAR SPECIES: (1) *Eleutherodactylus hylaeformis* has most digital disk covers spadate (fig. 7.51h) and most disk pads triangular (fig. 7.51o). (2) *Eleutherodactylus tigrillo* has a pattern of dark spots on a yellow background, unpigmented yellow posterior thigh surfaces, and mostly spadate disk covers (fig. 7.51h) and triangular disk pads (fig. 7.51o).

HABITAT: Lowland Moist and Wet Forests, Premontane Wet Forest and Rainforest, and marginally in Lower Montane Wet Forest and Rainforest.

BIOLOGY: These are extremely common nocturnal frogs. Males begin calling shortly after dusk from piles of dead leaves, hidden in heavily vegetated road cuts, on stream banks, or within bromeliads, but also from leaf surfaces, low shrubs, or small trees. Calling continues throughout the night, usually peaking between 10:00 P.M. and 2:00 A.M. Wilczynski and Brenowitz (1988) found that calling males are spatially distributed in a nonrandom pattern. Examples

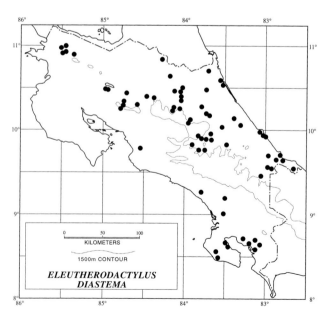

Map 7.58. Distribution of *Eleutherodactylus diastema*.

are often found resting in bromeliads or leaf litter in the daytime. Densities in leaf litter samples at La Selva, Heredia Province, were 37/ha (Lieberman 1986), but this vastly underestimates total densities, since most individuals call or hide above ground level. These agile little frogs either jump or use an accelerated walk (or run) to escape intruders (Scott 1983e). Food consists of a wide variety of arthropods, including a high proportion of ants (95%) at La Selva. The mean number of prey items in seventeen stomachs was 3.4, and average prey length ranged from 1.0 mm (spiders and dipterans) to 3 mm (orthopterans) at that site (Lieberman 1986). Reproduction occurs throughout the wet season (May to November). Taylor (1955) reported egg clutches laid in bromeliads and suggested that several females may deposit eggs in a single plant. A single female may lay up to 10 eggs that are about 4 mm in diameter. Dunn (1937a) suggested that males of this species attend the eggs. Ovaska and Rand (2001) described courtship and reproductive behavior in frogs referred to this species at Barro Colorado Island, Panama. Males show considerable calling site fidelity and have exclusive home ranges but may call from up to four different sites per night. The maximum distance between calling sites of individual males per night varies from about 1 to 8 m. Spacing between calling males is 5 to 50 m according to Wilczynski and Brenowitz (1988). Courtship was observed in July. When a female enters a male's territory he moves from the calling station toward her, calling continuously. He stops in front of her and lifts his legs up and down. The female then touches the male's distended vocal sac with her snout. The male then lunges and bumps into the female. This sequence is repeated nine to twelve times. The male then turns away and over a period of twenty-seven to forty-seven minutes leads the female to a nest site where she backs under him to establish amplexus. The nest site is usually concealed in bromeliads, in leaf petioles, under bark or fallen leaves, or in other cavities from 180 to 1,500 mm above the ground. One instance of female-female agonistic behavior was noted when a female evicted another female already in amplexus from a nest. Males are apparently polygynous and were noted as leading different females to the nest site on two or three occasions. Three to seven egg clutches in different stages of development were also found at several nest sites. Clutch sizes ranged from 11–15.7–19 in newly laid clutches to 7–11.9–16 in older clutches. Newly laid eggs (including the capsule) are 4.0 to 4.5 mm in diameter but expand to about 6.5 mm by absorbing water. Hatching occurred in 23 to 26 days. Males do not appear to attend the eggs and resume calling the next evening after oviposition. The smallest juvenile is 10.5 mm in standard length.

REMARKS: Some midelevation populations in Costa Rica and western Panama are larger in average size than lowland ones and differ subtly in coloration and in male advertisement call. At some intermediate-elevation sites E. hylaeformis and E. diastema (sensu lato) may be sympatric. Whether some of the higher-elevation E. diastema-like frogs represent additional undescribed species remains an open question requiring an in-depth analysis of acoustical differences. Bogart (1970b) and León (1969) reported karyotypes of 2N = 18 and 20, respectively. According to Bogart there are two pairs of metacentric, five pairs of submetacentric, and two pairs of subtelocentric chromosomes; NF = 36.

DISTRIBUTION: Humid lowlands and premontane slopes on the Atlantic versant from Nicaragua to eastern Panama; on the Pacific slope, marginally in the Cordillera de Guanacaste and Cordillera de Tilarán of Costa Rica, southwestern Costa Rica, and adjacent western Panama and from just west of the Canal area in Panama south to western Ecuador (1–1,620 m).

Eleutherodactylus hylaeformis (Cope, 1875)

(plates 128–29; map 7.59)

DIAGNOSTICS: Tiny to small gray to nearly black frogs, with the dorsal surfaces often tinged with pink or suffused with red, having spadate disk covers (fig. 7.51h) and triangular disk pads (fig. 7.51o) on most digits and the posterior thigh surface uniform and pigmented. Like other members of this group E. hylaeformis lacks enlarged supraocular or heel tubercles and a vivid contrasting pattern or color in the groin or on the thigh.

DESCRIPTION: Adult males 19 to 22 mm in standard length, adult females 20 to 26 mm; dorsum smooth or with a few scattered low tubercles; head slightly longer than broad; snout rounded from above, with a fleshy bump at tip; eyes large; tympanum distinct, diameter about one-third diameter of eye; annulus tympanicus evident except dorsally; no vomerine teeth; long paired vocal slits and single external subgular vocal sac in adult males; finger and toe disks moderately expanded, most disk covers spadate and most

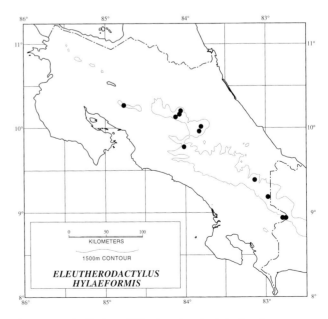

Map 7.59. Distribution of *Eleutherodactylus hylaeformis*.

pads triangular; subarticular tubercles round and flattened; accessory palmar and plantar tubercles obsolescent; thenar tubercle elongate, on outer thumb base; palmar larger rounded; inner metatarsal tubercle small, elongate, outer about same size, round, conical. Dorsum tan, gray, to almost black, often tinged with pink (males) or suffused with red (females), nearly uniform in most females, marked with dark interocular, suprascapular, or lumbar dark blotches in many males; occasionally a pink or yellow middorsal pinstripe; posterior thigh surface brownish, pigmented, often suffused with red; venter yellowish to tan punctuated with scattered dark pigment to uniformly dark brown with dark punctations; vocal sac yellow to pink; iris brown.

VOICE: A series of "dink-dink-dinks" similar to that of *Eleutherodactylus diastema* but of slightly lower pitch.

SIMILAR SPECIES: (1) *Eleutherodactylus diastema* has most disk covers palmate (fig. 7.51e) and most disk pads expanded (fig. 7.51n), not triangular. (2) *Eleutherodactylus tigrillo* is vividly spotted with small dark dorsal spots on a yellow ground and has unpigmented bright yellow posterior thigh surfaces.

HABITAT: Lower Montane Wet Forest and Rainforest.

BIOLOGY: A relatively uncommon nocturnal forager that is most often seen in low vegetation within dense forests.

DISTRIBUTION: Humid lower montane situations in the cordilleras of Costa Rica and western Panama (1,500–2,500).

Eleutherodactylus tigrillo Savage, 1997b

(map 7.60)

DIAGNOSTICS: A tiny species immediately recognizable by its pattern of brown spots on bright yellow orange, spadate disk covers (fig. 7.51h), and triangular disk pads (fig. 7.51o).

Map 7.60. Distribution of *Eleutherodactylus tigrillo*.

DESCRIPTION: Adult males 16 to 18 mm in standard length, adult females probably to 20 mm; dorsum smooth with widely spaced low pustules; head longer than broad; snout subovoid in dorsal outline, no bump at tip; eye moderate; upper eyelid with a few scattered tubercles; tympanum indistinct, diameter a little less than one-half diameter of eye; annulus tympanicus partially visible through skin anteriorly and ventrally; a few vomerine teeth on oblique odontophores, widely separated medially; large paired vocal slits and single external subgular vocal sac in adult males; finger and toe disks moderately expanded, disk covers spadate, pads triangular; subarticular tubercles on fingers and toe I weakly bifid, ovoid, flattened; no accessory palmar or plantar tubercles evident; no nuptial thumb pads in males; thenar tubercle somewhat elongate, larger than round palmar; inner metatarsal tubercle elongate, larger than round outer tubercle. Dorsum bright yellow orange with each pustule dark brown; posterior thigh surface unpigmented, bright yellow orange; no transverse dark bars on upper surface of limbs; venter yellow with many small dark punctations; vocal sac immaculate yellow; iris golden.

VOICE: Another "dink-dink-dink" call as in *Eleutherodactylus diastema*.

SIMILAR SPECIES: *Eleutherodactylus diastema* and *E. hylaeformis* lack the vivid spotted dorsal pattern and have pigmented posterior thigh surfaces.

HABITAT: Known from two localities near the Río Lari, Limón Province, in Premontane Rainforest.

BIOLOGY: The adult male holotype and paratype and only known examples were calling from low shrubs early in the evening.

DISTRIBUTION: Known only from the valley of the Río Lari, Limón Province, Costa Rica (400–440 m); it is to be expected to occur in western Panama.

Eleutherodactylus vocator Taylor, 1955

(plate 130; map 7.61)

DIAGNOSTICS: A tiny, rather dark-colored species, distinctive among all Costa Rican frogs in having mostly lanceolate finger and toe disks (fig. 7.51i).

DESCRIPTION: Adult males 14 to 16 mm in standard length, adult females 16.5–18 mm; dorsum shagreened; head longer than broad; snout subovoid from above without terminal fleshy bump; eyes moderate; upper eyelid essentially smooth; tympanum distinct, diameter one-fourth to one-third diameter of eye; annulus tympanicus visible except above; no vomerine teeth; single external subgular vocal sac in adult males; finger and toe disks barely expanded, disk covers on fingers and toes lanceolate, disk pads triangular; subarticular tubercles ovoid, flat; accessory palmar tubercles but obscure plantar tubercles; thenar tubercle rounded, larger than oval palmar tubercle; inner metatarsal tubercle low, elongate, larger than conical outer tubercle. Dorsum gray to nearly black with some scattered dark flecking, a dark suprascapular blotch and/or suprasacral blotch;

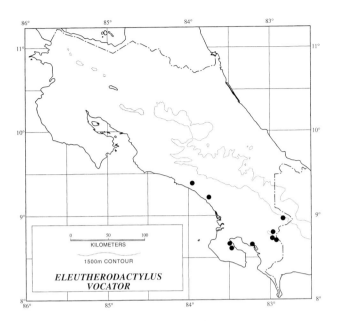

Map 7.61. Distribution of *Eleutherodactylus vocator*.

in most females and some males, dark blotches pronounced and a pair of oblique light lateral stripes extend from behind tympanum onto flank and from sacrum to groin on each side; a dark interorbital mark and area anterior to it usually light; upper surface of thigh with or without transverse dark bars; posterior thigh surface and venter heavily pigmented with dark punctations surrounding light spots; vocal sac pale yellow; iris brown.

VOICE: A low "tink-tink-tink."

SIMILAR SPECIES: *Eleutherodactylus diastema* and *E. hylaeformis* lack lanceolate disk pads.

HABITAT: Found in relatively pristine to disturbed situations in Lowland Wet Forest and Premontane Rainforest.

BIOLOGY: An essentially nocturnal, rather common species found on the ground and in low vegetation. Males call from hiding places on the ground or from shrubs up to 1 m above the substratum. On cloudy or rainy days males may be heard calling throughout the afternoon and early evening. Food consists of a variety of small arthropods (under 8 mm in length) including ants (Toft 1981). Reproduction occurs throughout the wet season (May to November). Females contain and presumably lay 4 to 6 large eggs at a time.

DISTRIBUTION: Very humid lowland and premontane areas from southwestern Costa Rica to western Colombia on the Pacific versant and from central Panama to northern Colombia on the Atlantic slope (2–1,220 m).

Family Hylidae Gray, 1825

Treefrogs

A large family (forty genera, 594 species) of primarily arboreal frogs that usually have smooth to rugose skins, long legs, and well-developed finger and toe disks. Many treefrogs also have finger webbing. All members of the Hylidae have a

short intercalary cartilage between the penultimate phalanx and claw-shaped terminal phalanx (fig. 7.10b) that helps them use the digital disks for climbing on vertical and other surfaces. Size variation is substantial: some species are tiny (to 17 mm in standard length), most are moderate, but some are relatively large (to 120 mm).

The definitive characteristics of the Hylidae, in addition to the digital ones, are eight holochordal, procoelous presacral vertebrae; no ribs; sacrum usually with dilated diapophyses, round, cylindrical in several genera, and with a bicondylar articulation to coccyx; coccyx lacking transverse processes; pectoral girdle arciferal; cartilaginous omosternum and sternum; no parahyoid bone; tibiale and fibulare fused proximally and distally; sartorius muscle discrete from semitendinosus, and tendon of latter inserts ventral to gracilis muscle; teeth present on upper jaw, usually pedicellate; pupil usually horizontally elliptical (vertically elliptical in subfamily Phyllomedusinae); nares lateral, amplexus axillary; larva type A, usually with beaks and denticles (latter absent in a few species) and single sinistral lateral to lateroventral spiracle; the karyotype is 2N = 22 to 30.

In Costa Rica any frog having finger webbing is a hylid. In addition, members of this family may be distinguished from other frogs in the republic that have finger disks as follows: no ventral disk (a ventral disk present in *Eleutherodactylus*, family Leptodactylidae); venter usually not transparent in life or translucent in preservative (transparent in life, translucent in preservative in Centrolenidae).

Hylid frogs are mainly nocturnal, scansorial, and arboreal in habits, but one North American genus *(Acris)* is essentially aquatic. The family ranges over the Americas from Canada to Ecuador (Pacific versant) and Argentina (Atlantic), including the Greater Antilles, and also occurs in Europe, northwestern Africa, southwestern Arabia, and the Middle East and in eastern Eurasia south to Vietnam. Three major subdivisions are recognized within the family:

Phyllomedusinae Günther, "1858b," 1859 (six genera): pupil vertically elliptical; tongue highly protrusible; pterorhodophores present; eggs laid above water, usually on vegetation; spiracle of larvae lateroventral

Hemiphractinae Peters, 1862a: pupil horizontally elliptical; tongue not highly protrusible; no pterorhodophores; most with direct development, eggs brooded by female and having two (usually) or one pair of specialized external gills; in forms with larvae, spiracle lateral (Panama to Ecuador on the Pacific slope and Costa Rica to Argentina on the Atlantic)

Hylinae Gray, 1825 (twenty-four genera): pupil horizontally elliptical; tongue not highly protrusible; pterorhodophores present or not; eggs usually deposited in water (above water on vegetation in a few species); spiracle of larvae lateral (North, Central, and South America exclusive of Pacific lowland Peru and Chile, also the Greater Antilles, Europe, northwestern Africa, southwestern Arabia, the Middle East, eastern Eurasia including Japan, and south to Vietnam)

The scansorial and arboreal hylids characteristically use a jumping-walking style of locomotion that is expedited by the structure and functioning of the digital disks. These anurans typically walk while clambering for short distances along branches or twigs or across leaf surfaces. For more rapid movement, particularly to avoid predators, they leap considerable distances from one branch, bush, or tree to another. Similar behaviors and digital specializations are found in arboreal and scansorial representatives of other families of "treefrogs" (Pelodryadidae, Centrolenidae, Hyperoliidae, and Rhacophoridae).

The digital disk is a complex structure supported at its base by the upwardly directed terminal phalanx. Externally it consists of an upper disk cover, separated from the lower disk pad by a circumferal groove, except proximally. The pad is composed of columnar (usually hexagonal) cells that are separated from one another and flattened distally. Interspersed among these cells are mucous pores providing sticky secretions to the pad surface. The offset provided by the intercalary cartilage allows the entire disk pad to be applied to the substrate. This improves traction between the specialized pad cells and any surface irregularity and helps the sticky pad adhere to any smooth surface by surface tension. Treefrogs are thus able to climb smooth surfaces vertically or move across them upside down.

The structure and function of digital disks in Costa Rican glassfrogs (Centrolenidae) is similar to that of hylids except that the terminal phalanges are T-shaped. *Eleutherodactylus* (Leptodactylidae) also have digital disks and T-shaped terminal phalanges on at least the outer fingers and toes (inner finger or fingers and inner toe may have knobbed terminal phalanges), but they lack intercalary cartilages.

One species of native ranid *(Rana warszewitschii)* also has digital disks (as do many Paleotropic species in the family), but the disk is bounded by a groove only laterally.

The dominant dorsal color of most hylid frogs without reference to darker or lighter markings, if present, is some shade of brown, green, or yellow. The browns and yellows and most contrasting markings are produced by different chromatophores or their interactions as previously described (chapter 4). As discussed below in the account of the family Centrolenidae, two systems are responsible for green colors. In some cases the green color in life is a structural one, and after preservation the frog turns bluish purple (e.g., *Agalychnis* and *Duellmanohyla uranochroa*). In others a green pigment accumulated in the skin dissolves on preservation, and the skin turns yellow or white (e.g., *Hyla colymba* and *H. rufitela*). This pigment is almost certainly biliverdin, which in both green (e.g., *H. colymba*) and nongreen (e.g., *Phyrnohyas venulosa*) frogs sometimes accumulates in the bones. See account of the Centrolenidae for details (p. 358).

Most species of hylid frogs have some ability to change the intensity of the dorsal ground color and pattern. Many exhibit conspicuously different colorations at night (active) and during the day (resting) (e.g., *Hyla ebraccata*, and *Phyllomedusa lemur*). Others show metachrosis not necessarily related to differences in light stimulation (e.g., *Smilisca baudinii*). In several forms males turn a brighter or different color during the breeding season (e.g., *Hyla elaeochroa* and *H. pseudopuma*).

Another interesting aspect of coloration, infrared reflectance in anurans, was originally documented by Cott (1940) for an Australian treefrog, *Pelodryas coeruleus* (family Pelodryadidae). Photographs taken with infrared-sensitive film showed that the frog matched the infrared reflectance of the green leaf background. Subsequently Schwalm, Starrett, and McDiarmid (1977) demonstrated that a variety of Neotropical frogs including two species of phyllomedusine hylids and two centrolenids were similar cryptic infrared reflectors. A variety of other anurans (*Bufo* spp., *Hyla* spp., and *Rana* spp.) absorbed infrared and stood out in contrast to the background leaves.

Schwalm (1981) later established that infrared reflectance occurred in many tropical green frogs of the families Hylidae, Centrolenidae, Pelodryadidae, Rhacophoridae, and Hyperoliidae. All twenty-nine species examined had the specialized, very large melanophores (pterorhodophores), unique to anurans, originally described by Taylor and Bagnara (1969). These cells contain complex melanosomes consisting of a eumelanin core surrounded by a fibrous mass of the red compound pterorhodin.

Krempels (1989) discovered that a wide variety (ninety-four species) of mostly tropical amphibians and reptiles (mainly lizards), exhibit infrared reflectance in the near infrared (NIR) range of wavelengths (760 to 900 nm). These are potentially visible to some animals but not to human beings, whose visible spectrum (VL) is from 400 to 800 nm. These wave lengths (NIR) are much more similar to visible radiation than to higher wavelengths of infrared (900 to 100,000 nm), which usually increase temperature when absorbed, and only those above 3,000 nm are emitted as heat via reradiation or metabolism by living organisms.

In this study Krempels found a significant relation between VL and NIR in that innocuous taxa were cryptic in both VL and NIR, whereas poisonous or venomous species were conspicuous in both. However, crypsis was found to involve high-intensity reflectance in arboreal amphibians and lizards and low-intensity reflectance in leaf litter forms matching NIR reflectance of the substrate. As far as is known, none of the amphibian low-intensity forms and none of the reptiles have the specialized pterorhodophores mentioned above. Most snakes were found not to reflect significant intensities of NIR; exceptions included coral snakes and some of their mimics.

These results suggest that NIR reflection may serve as protective or aposematic coloration or have a thermoregulatory function. In the latter case regulation is by reflection rather than absorption of the irradiance of NIR in high-incident light (e.g., forest canopy) or consistently warm environments. Evidence that known predators actually can "see" NIR, however, remains ambiguous.

The skins of phyllomedusine frogs contain an unusual and characteristic mixture of noxious and toxic peptides not found in any other group of amphibians. These include bradykinins, caerulins, tachykinins, sauvagine, and opioids. The last two peptides are unique to the subfamily (Erspamer 1994). Although it is not nearly as irritating as the skin secretions of some other Costa Rican amphibians, these frogs emit an acrid odor. After handling them you need to take care not to touch mucous membranes to prevent itching and swelling.

Reproductive patterns are diverse in this family and include patterns 1–4, 22–25, 31–33 in table 7.1. All species of *Gastrotheca* have brood pouches (patterns 30 and 32), and those exhibiting direct development are thought to have gill placentas that are involved in gaseous exchange between the embryo and the highly vascularized lining of the pouch.

KEY TO THE COSTA RICAN SUBFAMILIES OF THE TREEFROG FAMILY HYLIDAE

1a. Pupil of eye horizontally elliptical in strong light or preservative (fig. 7.11d) 2
1b. Pupil of eye vertically elliptical in strong light or preservative (fig. 7.11e) . . . . . . Phyllomedusinae (p. 277)
2a. Definite triangular dermal flaps on upper eyelids (fig. 7.11g) . . . . Hemiphractinae (p. 288)
2b. No triangular dermal flaps on upper eyelids. . . . . Hylinae (p. 288)

Subfamily Phyllomedusinae
Günther "1858b," 1859

KEY TO COSTA RICAN PHYLLOMEDUSINE TREEFROGS (FAMILY HYLIDAE)

1a. Considerable webbing on fingers and toes. 2
1b. No webbing on fingers or toes; adult males 30–41 mm in standard length, adult females 39–53 mm . . . . *Phyllomedusa lemur* (p. 286)
2a. Flanks uniform, no series of contrasting dark and light areas. 3
2b. Flanks with a contrasting pattern of light and dark areas (fig. 7.59). 5
3a. Web between fingers II and III originates proximal to distal subarticular tubercle of finger III; web between toes III and IV originates at or proximal to distal subarticular tubercle of toe IV; flank blue to purple in life, dark in preservative 4
3b. Web between fingers II and III originates at or distal to distal subarticular tubercle of finger III; web between toes III and IV originates distal to distal subarticular tubercle of toe IV; flanks orange in life, light in preserva-

tive; adult males 48–75 mm in standard length, adult females 81–87 mm. . . . . . *Agalychnis spurrelli* (p. 285)
4a. Web between toes IV and V originates distal to penultimate subarticular tubercle (second from disk) of toe IV; no dark transverse lines on dorsum; iris yellow to yellow orange in life; standard length of adult males 57–74 mm, adult females 67–84 mm. . . . . *Agalychnis annae* (p. 278)
4b. Web between toes IV and V originates at or proximal to penultimate subarticular tubercle (second from disk) of toe IV; dorsum often marked with wavy transverse lines; iris red in life; standard length of adult males 34–54 mm, adult females 52–66 mm . . . . *Agalychnis saltator* (p. 283)
5a. No calcar or heel; flanks with a series of light vertical bars on a dark field (fig. 7.59a); no transverse dark bars on upper surface of thigh; lower eyelid reticulate (fig. 7.60b); iris red in life; adult males 39–59 mm in standard length, adult females 51–71 mm. . . . . *Agalychnis callidryas* (p. 281)
5b. A triangular calcar on heel (fig. 7.60a); flanks with a series of dark vertical bars on light ground color (fig. 7.59b); a series of transverse dark bars on upper

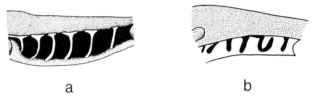

a   b

Figure 7.59. Color patterns in Costa Rican frogs of the genus *Agalychnis*: (a) *A. callidryas*, flank with light bars on a dark field; (b) *A. calcarifer*, flank with dark bars.

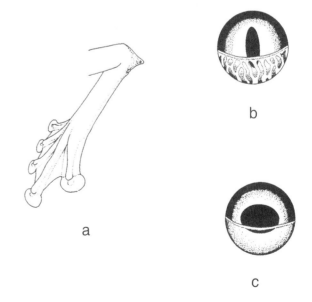

Figure 7.60. Diagnostic features of Costa Rican treefrogs (Hylidae): (a) triangular heel calcar; (b) reticulate lower eyelid; (c) nonreticulate lower eyelid.

surface of thighs; lower eyelid not reticulate (fig. 7.60c); iris gray in life; adult males 51–81 mm in standard length, adult females 61–87 mm . . . .
. . . . *Agalychnis calcarifer* (p. 279)

Genus *Agalychnis* Cope, 1864a

Treefrogs of this genus (eight species) are moderate to large forms having a green dorsum, large digital disks, and finger and toe webbing. They include some of the most colorful and most frequently photographed Costa Rican anurans. The first toe is shorter than the second and not opposable to the others, and the iris is bright red, yellow, or gray. The head is about as wide as long, with males having narrower heads than females, and the eyes and tympanum are large.

The genus *Agalychnis* is distinguished from other genera in the subfamily by definite finger and toe webs; a nonopposable big toe; a shallow skull (greatest depth less than 40% of length), weak quadratojugals, moderately ossified sphenethmoid; spatulate, strongly bifid teeth present on premaxillary, maxillary, and vomers; larvae deep bodied with terminal to nearly terminal mouth, a ventrolateral sinistral spiracle, 2/3 denticle rows, with or without papillae above the mouth.

The vertical pupil and large digital disks combined with the presence of finger and toe webs will distinguish these treefrogs from all other Costa Rican forms.

Savage and Heyer (1967, 1968) included all Costa Rican species of this genus in *Phyllomedusa* following Funkhouser (1957). Duellman (1968a, 1970) convincingly demonstrated the validity of *Agalychnis* as distinct from *Phyllomedusa*.

The genus ranges from southern Veracruz and southern Guerrero, Mexico, throughout Central America to northern Colombia and on the Pacific versant to western Ecuador; a single species, *A. craspedopus*, is known from lowland eastern Ecuador and Peru.

All members of the genus walk as their principal mode of locomotion but will leap from branch to branch as well.

So far as is known all species in the genus except *Agalychnis calcarifer* and *Agalychnis craspedopus* lay eggs on the upper leaf surfaces of moderately tall shrubs and trees over lentic waters (reproductive patterns 22 and 23, table 7.1). These last two species regularly deposit their eggs on leaves or other surfaces above water-filled depressions or cavities in fallen logs (reproductive mode 23, table 7.1). After hatching, in all cases the tadpoles (ecomorph 5, table 7.2) fall or are washed by heavy rains into water-filled depressions, puddles, ponds, or backwaters of slow-moving small streams.

In preserved specimens the green coloration becomes dull blue to purplish blue.

Agalychnis annae Duellman, 1963a

(plate 131; map 7.62)

DIAGNOSTICS: A moderate to large bright green treefrog having uniform blue flanks and anterior and posterior

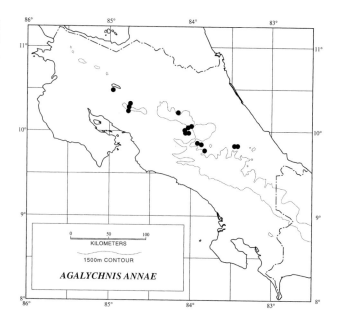

Map 7.62. Distribution of *Agalychnis annae*.

thigh surfaces and a yellow orange eye. It also has substantial webbing between fingers I–II–III.

DESCRIPTION: A slender species; adult males 57 to 74 mm in standard length, adult females 67 to 84 mm; dorsal surfaces smooth; venter weakly granulate; head longer than broad; snout subelliptical in dorsal outline; eyes large; lower eyelid reticulate; tympanum distinct, moderate, diameter about one-half diameter of eye; small paired vocal slits and slightly distensible single internal subgular vocal sac in adult males; finger disks large, disk on finger III equals tympanum; subarticular tubercles on fingers III and IV large, round, occasionally bifid; brown nuptial pad on base of thumb in adult males; fingers with substantial webbing; webbing formula: I 2–2½; II 1½–3⁻ III 2–2 IV; thenar pad moderately enlarged in males; no palmar tubercle; two small heel tubercles; weak dermal fold from heel to toe V; toes with substantial webbing; webbing formula: I 2–2⁺ II 1½–3 III 1½–2½ IV 2⁺–1½ V; inner metatarsal tubercle large, flat, elliptical, no outer tubercle; weak tarsal fold present. Most of upper surfaces uniform green; proximal dorsal portion of upper arm pink to lavender, distal portion blue; flanks and anterior and posterior thigh surfaces blue; upper surface of hands and feet green, orange, and blue; bright creamy yellow stripes along ventrolateral margin of forearm and tarsus and foot; venter creamy yellow to orange; colors darken at night to darker green and bluish purple; metamorphs lack blue coloring and turn reddish brown at night and in preservative; iris yellow orange.

LARVAE (fig. 7.61): Moderate-sized, total length 33 mm at stage 31; body deep; mouth anteroventral; nostrils dorsolateral, directed anterolaterally; eyes dorsolateral, directed laterally; spiracle lateroventral, sinistral; vent dextral; tail moderate; caudal fins high; tip of tail forms short, thin fla-

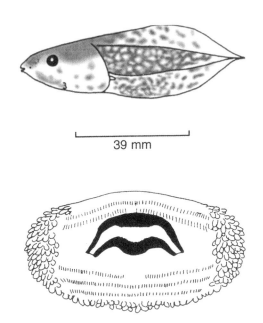

39 mm

Figure 7.61. Tadpole of *Agalychnis annae*.

gellum; oral disk small, entire, with beaks and 2/3 rows of denticles; A2 and P1 with slight median gaps; both beaks massive and serrate; disk bordered by papillae except above mouth; generally in two rows, but five to six rows lateral to mouth opening; body grayish brown above, bluish gray laterally and silvery blue below; caudal fins and musculature with brown flecks, increasing in amount during ontogeny; iris yellow.

VOICE: The call consists of a single-note "wor-or-op" repeated at intervals of 40 seconds to several minutes. The duration is 160 to 440 milliseconds, and the pulse rate is 38 to 50 per second. The dominant frequency is 1.0 to 1.3 kHz (Duellman 1963a).

SIMILAR SPECIES: (1) *Agalychnis saltator* has red eyes, and the dorsum has transverse dark green lines (diurnally) and is reddish brown at night. (2) *Agalychnis spurrelli* has fully webbed fingers and toes and orange flanks.

HABITAT: Premontane Moist and Wet Forests and Rainforest and marginally into Lower Montane Rainforest. This forest species is rather adaptable and may be found in disturbed sites including vacant lots and gardens within the metropolitan areas of the Meseta Central.

BIOLOGY: A nocturnal form that formerly was common in forest remnants throughout the heavily populated Central Valley. Males call throughout the year in rainforest areas, but in seasonal forests they call only during the rainy season (May through November). Calling males are usually stationed above ponds in overhanging vegetation. Initial amplexus takes place 3 to 10 m above the ground. The female, carrying the male, usually climbs down to the water source and then returns to the vegetation seeking a deposition surface from 350 mm to 3 m above the pond. The pale green eggs are laid in irregularly shaped jelly masses. Clutch

sizes are 45 to 126 eggs (Duellman 1970; Proy 1993a) each about 4 mm in diameter, including the envelope; ova are about 3.3 mm in diameter. As development proceeds the yolk becomes creamy tan. The eggs are usually laid on upper leaf surfaces but are also deposited on branches, vines, and other vegetation. Eggs hatch in five to seven days; the tadpoles wriggle to the surface of the jelly mass and flip themselves into the pond below or are washed there by heavy rains. Free-swimming larvae float in the midwaters with their heads up and bodies at about 45° to the water surface, maintaining this position by constant fluttering of the tail. Tadpoles raised in the laboratory metamorphosed in 247 days (Duellman 1970). Metamorphs are between 20 to 23 mm in standard length and have considerably less digital webbing than adults. Larvae are often found in swimming pools and garden fountains in the major upland cities of Costa Rica.

REMARKS: Taylor (1952a), following previous authors, referred to this species under the name *Agalychnis moreletti*. Duellman (1963a) demonstrated the distinctiveness of the Costa Rican form and described it as new. He further showed that *A. moreletti* was a valid form restricted to southern Mexico and adjacent upper Central America. The karyotype is 2N = 24 (León 1969).

DISTRIBUTION: Humid premontane and lower portions of lower montane zones on slopes of the cordilleras of northern and central Costa Rica (780–1,650 m).

Agalychnis calcarifer Boulenger, 1902

(plates 132–33; map 7.63)

DIAGNOSTICS: This strikingly marked moderate-sized to large dark green treefrog is instantly recognizable by having bold purple to black vertical bars on the orange yellow

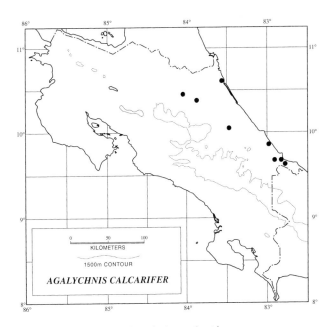

Map 7.63. Distribution of *Agalychnis calcarifer*.

flanks (fig. 7.59b), transverse purplish black bars on the limbs, and a prominent large calcar on the heel (fig. 7.60a). Juveniles lack the characteristic black bars, but the calcar is evident even in metamorphs.

DESCRIPTION: A relatively robust frog; adult males 51 to 81 mm in standard length, adult females 61 to 87 mm; dorsal surfaces and venter smooth; head broader than long; snout truncate from above, acute in profile; eyes large; lower eyelid not reticulate; tympanum distinct, large, diameter about three-fourths diameter of eye; small paired vocal slits and slightly distensible single subgular vocal sac in adult males; finger disks large, that of finger III as large as or larger than tympanum; dermal fringe from elbow to near tip of finger IV; subarticular tubercles sometimes bifid; brown spinous nuptial pad at base of thumb in adult males; fingers extensively webbed; webbing formula: I 2–2 II 1$^+$–2 III 1½–1¼ IV; thenar pad moderately enlarged; no palmar tubercle; large calcar on heel; narrow dermal fringe along outer edge of tarsus to disk of toe V; toes extensively webbed; webbing formula: I 1–1½ II 1–2 III 1–2$^-$ IV 2–1$^+$ V; inner metatarsal tubercle low, elliptical, no outer tubercle; short, weak tarsal fold present. Upper surfaces of head, body, forearm, and tibial segment of leg dark green, sometimes with bluish spots; upper surfaces of forearms, thighs, and flanks yellow to yellow orange; belly, ventral surface of limbs, hands, and feet and upper surfaces of hands and feet deep red orange, except for the outer digits on each; flanks, thighs, and tarsal segment of leg marked with bold purplish black bars; lower eyelid unpigmented; iris gray to purplish gray with yellow border; metamorphs are tan above but soon become grayish then bluish after a few months before turning green (Brian Kubicki, pers. comm.).

LARVAE (fig. 7.62): Large, 52 mm in total length at stage 37; body ovoid; mouth anteroventral; nostrils dorsolateral, directed anteriorly; eyes dorsolateral, directed laterally; spiracle lateroventral, sinistral; vent dextral; tail moderate; caudal fins reduced, low; tip of tail bluntly pointed; oral disk small, entire, with robust beaks and 2/3 rows of denticles; slight median gap in A2 and P1; both beaks serrate; disk completely bordered by papillae, mostly in three rows except above, where reduced to one row; body and caudal musculature olive green with brown reticulations on fins.

VOICE: Typically a single-note "whuunk." Sometimes a series of widely spaced notes. In a captive specimen (Myers and Duellman 1982) the single note lasted 100 milliseconds, with a pulse rate of 140 per second and maximum energy output about 0.8 kHz. The chuckle these authors noted is probably the release call (Marquis, Donnelly, and Guyer 1986).

SIMILAR SPECIES: There are none.

HABITAT: Undisturbed Lowland Moist and Wet Forests of the Atlantic slope.

BIOLOGY: This uncommonly seen nocturnal canopy species apparently moves downward toward favorable repro-

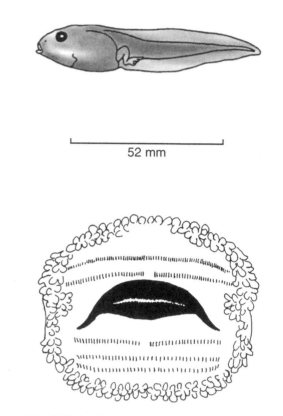

Figure 7.62. Tadpole of *Agalychnis calcarifer*.

ductive sites during the mating season. Calling males have been heard through most of the rainy season (March to October), sometimes from well up in trees to 8 m (Marquis, Donnelly, and Guyer 1986), but before mating they call from nearer the substrate, on vegetation 2 to 4 m above the ground. Unlike most other members of the genus, calling and breeding *A. calcarifer* are not found near either permanent or temporary ponds. In addition, the eggs are deposited on vegetation or on the bark of fallen trees over shallow water-filled cavities, depressions in the tree trunk or its buttresses (pattern 23, table 7.1). Other Costa Rican *Agalychnis* follow pattern 22. The same sites may be used for two or more years if not filled by falling debris, especially leaves. Eggs are laid throughout the year, including the dry season. Amplectant pairs have been observed in March, April, August, and September (Caldwell 1994; Roberts 1994a, 1995). During amplexus the female apparently hydrates her bladder in the water below the ultimate oviposition site as in other *Agalychnis*. After she climbs out of the water the eggs are laid and fertilized. Oviposition occurs in the morning (before 10:00) and is very rapid, lasting three to ten minutes. Clutch sizes are 10 to 54 eggs. The greenish blue ova are about 4 mm in diameter, but eggs plus envelopes are about 9 mm in diameter. The egg masses include 20 to 50% empty capsules. Within a short time the yolk becomes yellowish brown. When laid under field conditions, eggs hatch in five to ten days and fall or are washed by rain into the puddles below. Hatching is nonsynchronous and under laboratory conditions (Donnelly et al. 1987) may extend more than a month for a single clutch. Hatchlings are 17 to 33 mm in total length. In containers, tadpoles hang at the surface or float in midwater most of the time, but they also graze on the benthos (Caldwell 1994). Under natural conditions at La Selva, Heredia Province (Roberts 1994a), complete mortality of egg masses is high (50%); in masses that produced tadpoles, mean survival per clutch was 2 to 47%. A variety of effects contributed to mortality: embryos prematurely falling into the water, fungal infection or parasitism, infertile eggs, and in one case displacement by a land crab. Tadpoles are apparently preyed on by damselfly naiads and notonectids. Tadpoles appear to be cannibalistic on embryos that prematurely fall into the water. In laboratory-raised tadpoles metamorphosis occurred in a minimum of 211 days (Donnelly et al. 1987). Metamorphs are 23 to 27.5 mm in standard length.

REMARKS: Taylor (1952a, 1954a, 1955, 1958a) did not include this species in his review of Costa Rican frogs because it was not recorded from the country until the 1960s (Savage and Heyer 1967). The karyotype is 2N = 26 (Duellman and Cole 1965).

DISTRIBUTION: Humid lowlands of the Atlantic versant from eastern Honduras to central Panama and northern Colombia, and on the Pacific slope from eastern Panama to northwestern Ecuador (15–85 m).

Agalychnis callidryas (Cope, 1862d)

(plates 134–35; map 7.64)

DIAGNOSTICS: The combination of red eyes and dark blue to brownish flanks marked with white to creamy vertical bars (fig. 7.59a) will make recognition of this moderate-sized to large photogenic green frog instantaneous. In addition, the anterior and posterior thigh surfaces are uniform blue and the venter is white.

DESCRIPTION: A slender form; adult males 30 to 59 mm in standard length, adult females 51 to 71 mm (48 to 77 mm); dorsal surfaces and venter smooth; head longer than broad; head rounded in dorsal outline; snout obtuse in profile; eyes large, lower eyelid reticulate; tympanum distinct, moderate, diameter about one-half diameter of eye; small paired vocal slits and a slightly distensible single internal subgular vocal sac in adult males; finger disks large, disk on finger III equal to or slightly larger than tympanum; distal subarticular tubercles often bifid on fingers III and IV; gray brown spinous nuptial pad on base of thumb in adult males; fingers with substantial webbing; webbing formula: I 2–2½ II 2⁻–3⁻ III 2⁺–2⁺ IV; thenar pad large; no palmar tubercle; no heel tubercles or calcar; narrow dermal fold from heel to disk of toe V; toes substantially webbed; modal webbing formula: I 2–2⁺ II 2⁻–3 III 1½–2½ IV 2–1½ V; inner metatarsal tubercle large, ovoid, no outer tubercle; long, weak tarsal fold present; upper surfaces generally leaf green, sometimes with

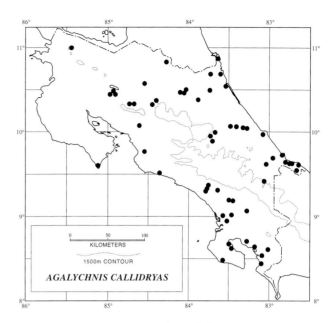

Map 7.64. Distribution of *Agalychnis callidryas*.

faint, narrow dark green transverse lines or small white spots on dorsum; flanks blue to purple or brown, marked with white or cream vertical bars; upper surfaces of upper arm and hands and feet orange or yellow orange except for outer digits on each; anterior, posterior, and ventral thigh surfaces blue or orange; venter creamy white; iris burgundy to ruby red; metamorphs are purplish brown at night, pale green in the day, have yellow eyes, and lack dark flank color; somewhat older juveniles resemble adults, are green, and do not change color at night.

LARVAE (fig. 7.63): Large, 48 mm in total length at stage 34; body robust; mouth anteroventral; nostrils dorsolateral, directed anteriorly; eyes dorsolateral, directed laterally; spiracle lateroventral, sinistral; vent dextral; tail moderate; caudal fins moderate; tip of tail a thin flagellum; oral disk small, entire, with robust beaks and 2/3 rows of denticles; A2 with slight median gap; both beaks with blunt serrations; papillae present lateral and ventral to month, mostly in two to three rows, none above mouth, usually only one row just lateral to the broad gap above mouth; body olive gray becoming bluish laterally and ventrally; caudal musculature grayish tan, fins heavily flecked with dark gray; iris bronze (plate 136).

VOICE: The call consists of single "chock" or double "chock-chock" repeated every 8 to 60 seconds. Call duration is 80 to 240 milliseconds with 180 to 200 pulses per second. The dominant frequency is 1.9 to 2.0 (1.7 to 2.4) kHz (Duellman 1970; Lee 1996).

SIMILAR SPECIES: Adults are unlikely to be mistaken for any other species. Juveniles might be confused with *Duellmanohyla uranochroa*, which has a horizontal pupil (fig. 7.11d) and an enamel white stripe along the upper lip, under the tympanum, to the groin (fig. 7.73c).

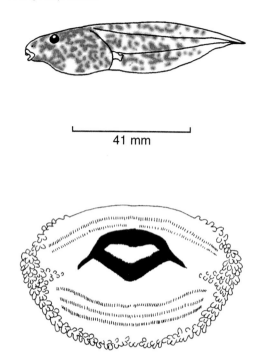

41 mm

Figure 7.63. Tadpole of *Agalychnis callidryas*.

HABITAT: Lowland Moist and Wet Forests and marginally into Premontane Wet Forest and Rainforest in a variety of situations, including disturbed habitats and mangrove forests.

BIOLOGY: The reproductive biology of this species has been studied in some detail by Pyburn (1963, 1970) in Mexico, by Duellman (1970) as summarized by Scott (1983f), and by Donnelly and Guyer (1994). Much of what follows is based on these accounts. *A. callidryas* is very common, nocturnal, and most active during the wet season, hiding in trees during the day and drier periods. Pyburn (1964) reported parachuting in this species, and Roberts (1994a) found in trials that it parachuted about as well as *Agalychnis spurrelli* (see that account). Roberts, however, never saw *A. callidryas* use parachuting in the field. Breeding takes place throughout the wet season (June to November) with peaks in June and sometimes October. Males are heard calling throughout this period, but courtship centers on nights after heavy afternoon rains. Males call from higher up in the trees on nights unsuitable for reproduction or call early in the evening before approaching the breeding sites. If conditions are right they descend to the margins of ponds or backwaters, where they begin serious calling later in the night. Calling sites are small trees and bushes near the water's edge from ground level to 3 m in height. Females then descend from the canopy. A gravid female apparently selects among calling males and approaches his calling site. After amplexus she usually carries the piggybacking male down to the water, where it is thought that she fills her bladder by absorption so its contents can be used in making the jelly mass sur-

rounding the eggs. She then climbs back up into the trees, with the male still in amplexus, searching for a suitable egg deposition site. Amplectant males are sometimes attacked by single males when the pair moves from water to the oviposition site. In some cases more than one male may amplect a female during egg laying, and multiple paternity has been confirmed by DNA fingerprinting (Orgeix and Turner 1995). Egg clutches have been found on both lower and upper leaf surfaces overhanging the water. Clutches of 11 to 104 eggs are then laid and fertilized. The egg mass is surrounded by clear jelly, and the ova are green. An egg and its envelope are about 3 mm in diameter. A female usually lays three to five clutches per night and may ovulate twice during the reproductive season. One female laid five clutches in one night for a total of 265 eggs (Pyburn 1970). Eggs may be laid before standing water is available in anticipation of an approaching storm. As the embryos develop the yolk becomes yellow. Hatching is nonsynchronous and occurs in five to eight days. Survival of eggs to the larval stage was 63% at La Selva, Heredia Province (Roberts 1994a). In heavy rains the tadpoles usually fall or wash into the water below, although some perish by landing on obstacles or the ground. Experiments by Valerio (1971) showed that the larvae may survive out of water for up to twenty hours, so even those falling on land may later be washed into a pond in a heavy rain. The larvae orient themselves with the head up near the surface of the water; they are suspension-rasper feeders that are suspended in the water column most of the time (plate 136). Pyburn (1963) reported that the sole survivor of a captive-raised clutch metamorphosed in seventy-nine days. Metamorphs are 18 to 19 mm in standard length. Juvenile perch heights are from near ground level to 0.65 m. Juveniles are most abundant in the early dry season (December to January) when no adults are present at the now dry breeding sites. Another flush of juveniles may appear in late August to early September. The eggs of this species are preyed on by the snakes *Imantodes inornatus* and *Leptodeira septentrionalis* (Donnelly and Guyer 1994; Michael Fogden, pers. comm.). Warkentin's (1995) claim that early hatching of the embryos is an antipredator device is without foundation. Her observation that embryos hatch at an early stage of development (at least five days old) and that some escape when the egg mass is attacked by *L. septentrionalis* while embryos in masses not so disturbed continue further development seems valid. However, embryos in the egg masses of leaf-breeding treefrogs and glassfrogs generally are tuned to be released by the physical stimulus of heavy showers. Once they reach a minimum developmental stage any movement of, or contact with, the egg mass (simulating the impact of raindrops) stimulates hatching. A strong wind or a human observer can initiate this response by jiggling the egg mass, which is precisely what a snake does when eating part of the mass. This process allows hatching to be delayed for several days if a suitable rain does not occur. In labora-

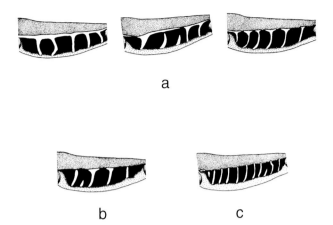

a

b   c

Figure 7.64. Variation in flank coloration in *Agalychnis callidryas:* (a) Atlantic slope Nicaragua and Costa Rica; (b) intermediate pattern; (c) southwestern Costa Rica and Panama.

tory tests Roberts (1994a) found that conspecific and heterospecific tadpoles and the turtle *Kinosternon leucostomum* found the eggs of this species palatable, but none of the syntopic freshwater fish ate them.

REMARKS: Taylor (1952a) recognized the material of this species available to him as *Agalychnis helenae* but also included *A. callidryas* in his list of Costa Rican anurans. The two forms were thought to be distinct in that *A. helenae* had the light lateral bars connected above by a light lateral stripe (fig. 7.64a) and deep blue flanks, upper forearms, anterior and posterior thighs, and undersurfaces of the thighs and calves. *A. callidryas* was distinguished from *A. helenae* by having the flanks and several limb surfaces orange rather than blue, and it lacked the light lateral stripe connecting the lateral light bars (fig. 7.64c). Savage and Heyer (1967) reviewed frogs of this complex from throughout the range of the two putative species. They found that the light lateral stripe was typically found in a high percentage of individuals from Atlantic slope Nicaragua and northeastern Costa Rica and was absent or rare in other populations. In addition, flank patterns intermediate between the two extremes are present in most populations. These include the lateral stripe being incomplete and one or more of the vertical bars having the dorsal tip expanded to form a T-shaped bar (fig. 7.64a). Duellman (1970) further documented variability in the flank pattern. These authors agreed that the complex appeared to consist of three geographic units:

A. Mexico to Honduras: no lateral stripe (fig. 7.64c), blue flanks, orange anterior and posterior thigh surfaces

B. Atlantic slope Nicaragua and Costa Rica: lateral stripe usually present (fig. 7.64a), flanks blue, anterior and posterior thigh surfaces blue

C. Panama and southwestern Costa Rica: lateral stripe rarely present (fig. 7.64c), flanks blue to purplish brown, anterior and posterior thigh surfaces blue, orange, or blue and orange

Examples from northwestern Costa Rica in the Tilarán region are most similar to unit B. Those from southwestern Nicaragua and northwestern Panama tend to be intermediate between B and C, with about 20% of the samples having the lateral stripe present (fig. 7.64a). These data support the conclusion that only one geographically variable species is involved. The karyotype is 2N = 26 (Duellman and Cole 1965), all chromosomes apparently metacentric or submetacentric; NF = 52 (Schmid et al. 1995).

DISTRIBUTION: Humid lowlands and premontane slopes on the Atlantic versant from southern Veracruz and northern Oaxaca, Mexico, to central Panama and northern Colombia and on the Pacific slope in southwestern Nicaragua and southwestern Costa Rica to eastern Panama; barely on the Pacific side of the continental divide near Tilarán, Guanacaste Province and tip of the Península de Nicoya in northwestern Costa Rica (1–970 m).

Agalychnis saltator Taylor, 1955

(plate 137; map 7.65)

DIAGNOSTICS: No other green Costa Rican treefrog has red eyes and uniform dark blue or purple flanks and anterior and posterior thighs.

DESCRIPTION: A moderate-sized species, adult males 35 to 54 mm in standard length, adult females 52 to 66 mm; dorsal surfaces smooth, venter granular; head longer than broad; snout subelliptical from above, bluntly rounded in profile; eyes large, lower eyelid reticulate; tympanum distinct, moderate, diameter about one-half diameter of eye; small paired vocal slits and slightly distensible single internal subgular vocal sac in adult males; finger disks large, disk on finger III slightly larger than tympanum; distal subarticular

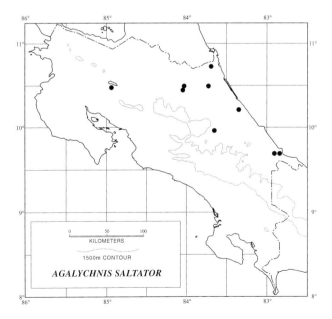

Map 7.65. Distribution of *Agalychnis saltator.*

tubercle on finger IV sometimes bifid; spiny brown nuptial pad on base of thumb in adult males; fingers with reduced webbing; webbing formula: I 2^+–3 II 2^-–3 III $2\frac{1}{2}$–2^+ IV; thenar pad hardly enlarged; no palmar tubercle; no heel tubercles or calcar; narrow dermal fold from heel to disk of toe V; toes with reduced webbing; webbing formula: I 2–2^+ II 2^-–3 III 2–3 IV 3^-–2^- V; many plantar tubercles; inner metatarsal tubercle large, ovoid, no outer tubercle; long, weak tarsal fold present. Dorsal surfaces generally green during day, reddish brown at night, sometimes with a few black-outlined pale yellow spots on dorsum; upper surface of upper arm pink to lavender proximally, blue distally; flanks and anterior and posterior thigh surfaces blue to bluish purple; in the diurnal mode thin transverse green lines may cross dorsum; nocturnally, dark transverse markings contrast with ground color; hands, feet, and venter orange; iris red; metamorphs are purplish brown at night, pale green during the day, with yellow eyes and white flanks and thigh surfaces.

LARVAE (fig. 7.65): Moderate-sized, 34 mm in total length at stage 32; body robust; mouth anteroventral; nares dorsolateral, directed anteriorly; eyes dorsolateral directed laterally; spiracle lateroventral, sinistral; vent dextral; tail moderate; caudal fins moderate; tip of tail tapering gradually to point; oral disk small, entire, with beaks and 2/3 rows of denticles; A2 and P3 with slight medial gap; both beaks serrate; one or two rows of papillae except above mouth; body grayish; body, fins, and tail musculature with fine grayish reticulations.

VOICE: Duellman (1970) described the call as a single or double "clack" repeated every 30 seconds to several minutes. Guyer and Donnelly (2002) reported that the only sounds

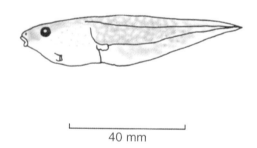

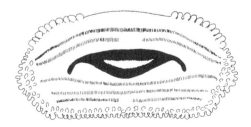

Figure 7.65. Tadpole of *Agalychnis saltator*.

this species emits are high-pitched peeps during breeding. I suspect this is a release call. Duellman's description and sonograph for this species are remarkably similar to those of *Agalychnis callidryas*. Roberts (1994b) stated that the only vocalizations by *A. saltator* males were soft squeaking and chuckling noises.

SIMILAR SPECIES: (1) The larger *Agalychnis annae* (adult males 57 to 74 mm in standard length, adult females 67 to 84 mm) has yellow orange eyes and substantial webbing between fingers I–II–III. (2) *Duellmanohyla uranochroa* has a horizontal pupil (fig. 7.11d) and lateral enamel white stripe (fig. 7.73c).

HABITAT: Undisturbed Lowland Moist and Wet Forests and marginally in adjacent Premontane Wet Forest and Rainforest.

BIOLOGY: The reproductive biology of *A. saltator* has been studied in detail by Proy (1992), Donnelly and Guyer (1994), and Roberts (1994a,b) at La Selva, Heredia Province, and their reports form much of the basis for the following account. This frog is a nocturnal canopy species. It is unlikely to be seen except during the reproductive period. Hand-over-hand locomotion is characteristic and is used to rapidly climb up vines into the canopy or to descend from it. This form also parachutes, spreading the limbs out so webs and digital disks are parallel to the substrate, and descends at angles up to 45°. Roberts recorded parachuting leaps from as high as 5 m above the landing site, traveling horizontal distances of as much as 4 m. *A. saltator* is an explosive seasonal breeder. Reproduction occurs several times between June and October following particularly heavy rains. Breeding takes place in the early morning, when large concentrations of 25 to 400 frogs, many in amplexus, congregate on vines overhanging seasonal ponds. The aggregation forms a writhing mass, with individual females grasped by one or more males while other males pull and kick trying to dislodge amplectant males or single males from the female or vines. During this period of frantic activity thousands of eggs are deposited in large masses plastered tightly to the moss and roots of epiphytes growing on the vines. The clutches of single mating pairs are mixed with those of others by the movements of paired and single frogs clambering over the vines and egg masses. Peak reproductive activity is at 7:00, but it continues until noon, when fewer than ten pairs are still out in the daylight. At night single and amplectant pairs are dispersed and found sitting on the leaves of aroids marginal to the pond from ground level to a height of 1.5 m. Other individuals probably descend from the canopy during the breeding frenzy, and females return there after oviposition. Unlike some other *Agalychnis*, the females do not hydrate the bladder before laying eggs. The same vines were favored oviposition sites during each reproductive period over three successive years. Clutch sizes from pairs brought to the laboratory that had not yet laid eggs were 25 to 311. Clutches from the field contained 21 to 72 eggs, and each egg was about 5 mm in diameter. The individual egg capsules are unusually sticky, and the ova are slate gray. The embryos are also slate gray and hatch at a total length of 8 to 16 mm after an average of 5.9 days. *A. saltator* shares its breeding site at La Selva with eleven other frog species. Three of these, *Agalychnis calcarifer, Agalychnis callidryas,* and *Hyla ebraccata,* lay eggs out of the water on vegetation, but the last two frogs do not include daytime egg laying in their reproductive cycle. Predators on the eggs include ants, warblers *(Phaeothylpis),* and the nocturnal snake *Leptodeira septentrionalis.* The diurnal snake *Leptophis ahaetulla* and the nocturnal snake *Imantodes inornatus* are attracted to the aggregations and eat adults. Roberts (1994a) reported predation by a hawk and capuchin monkeys as well. The smallest juvenile is 22 mm in standard length.

REMARKS: Taylor described this form in 1955, so it was not included in his 1952a monograph on Costa Rican frogs.

DISTRIBUTION: Humid lowlands and marginally on premontane slopes of the Atlantic versant from northeastern Honduras to southeastern Costa Rica (15–819 m).

Agalychnis spurrelli Boulenger, 1913

(plates 138–39; map 7.66)

DIAGNOSTICS: A moderate to large green treefrog having red eyes, bright orange flanks and anterior and posterior thigh surfaces, and extensively webbed hands and feet.

DESCRIPTION: Adult males 46 to 56 mm in standard length (66 to 75 mm), adult females 57 to 68 mm (62 to 93 mm); upper surfaces smooth; venter granular; head about as broad as long; snout subelliptical in dorsal outline, obtuse in profile; eyes large, lower eyelid reticulate; tympanum distinct, moderate, diameter equal to about one-half diameter of eye; short paired vocal slits and slightly distensible single internal subgular vocal sac in adult males; finger disks large, disk on finger III about equal to diameter of eye; distal subarticular tubercles under fingers III and IV sometimes bifid; spiny brown nuptial pads present on base of thumb in adult males; fingers extensively webbed; webbing formula: I 2⁻–2 II 1–1¾ III 1–1 IV; thenar tubercle greatly enlarged; no palmar tubercle; no heel tubercle or calcar; narrow dermal fold from heel to disk of toe V; toes extensively webbed, webbing formula: I 1–2⁻ II 1–2⁻ III 1–1½ IV 1½–I V; many plantar tubercles; inner metatarsal tubercle low, elliptical, no outer tubercle; moderately developed long tarsal fold present. Dorsal surfaces generally light green during the day and darker green at night, often with white dorsal spots that are usually outlined by black; upper surface of forearm, flanks, belly, anterior and posterior thigh surfaces, webs, and most digits orange; iris red; juveniles are green during the day but reddish brown at night.

LARVAE (fig. 7.66): Moderate-sized, 41 mm in total length at stage 35; body robust; mouth anteroventral; nares dorsolateral, directed anteriorly; eyes dorsolateral, directed laterally; spiracle lateroventral, sinistral; vent dextral; tail moderate; caudal fin reduced, low; tip of tail a thin flagellum; oral disk small, entire, with beaks and 1/3 rows of denticles; A2 with slight median indentation; both beaks serrate; two or three rows of papillae lateral and ventral to mouth; body dark bluish gray, venter paler; dark brown flecks on sides and anterior portion of fins; caudal musculature reticulated brown and gray anteriorly.

VOICE: A single low-pitched groan repeated every 10 to 17 seconds. Call duration is 340 to 400 milliseconds with a

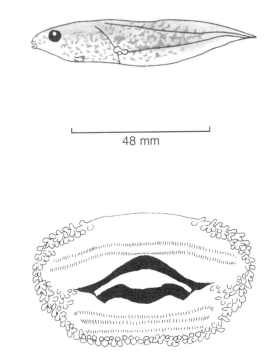

48 mm

Figure 7.66. Tadpole of *Agalychnis spurrelli.*

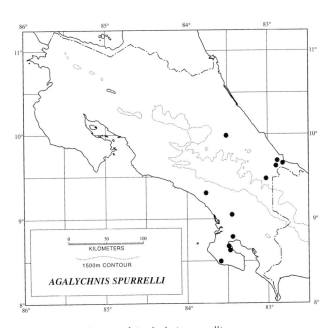

AGALYCHNIS SPURRELLI

Map 7.66. Distribution of *Agalychnis spurrelli.*

pulse rate of 60 to 90 per second. The dominant frequency is 0.44 to 0.7 kHz (Duellman 1970). The low "wuk! wuk! wuk!" reported by Scott and Starrett (1974) is probably a courtship call.

SIMILAR SPECIES: (1) Other red-eyed *Agalychnis (A. callidryas* and *A. saltator)* have the flanks blue to purplish brown rather than orange. (2) *Duellmanohyla uranochroa* has a horizontal pupil (fig. 7.11d) and an enamel white lateral stripe on each flank (fig. 7.73c).

HABITAT: Lowland Moist and Wet Forests and marginally in Premontane Rainforest, usually seen only at reproductive sites.

BIOLOGY: An essentially nocturnal canopy species. It resembles *A. saltator* in using hand-over-hand locomotion to rapidly descend and ascend vines and small branches when moving through the canopy or to and from the breeding site. *A. spurrelli* also uses parachuting leaps (sensu Oliver 1951) for rapid horizontal movement. In this behavior the hands and feet are spread out so the large webs and digital disks are parallel to the substrate. These frogs can maintain angles of descent up to 45° from the vertical for considerable distances, probably helped by their extensive webs. Trials where the frogs were released at a height of 4.5 m produced horizontal leaps of 1.5 to 4 m (median 2.2 m) (Scott and Starrett 1974). This species is an explosive rainy season breeder. The following is based principally on Scott and Starrett's account of a breeding aggregation near Rincón de Osa, Puntarenas Province. After or during heavy rains, hundreds to thousands of frogs congregate around the temporary ponds. Breeding takes place in the early morning (about 6:00) for the most part, but amplectant pairs were still laying eggs until about 9:00, when the direct sun struck the pond. A seething mass of frogs covered vegetation up to 10 m above the pond, but most were concentrated 1.5 to 3 m above the water. Males predominate, since once egg laying is complete females return to the canopy. Egg masses are deposited 1.5 to 8 mm above the water, but density is heaviest between 1.5 and 3 m. Thousands of eggs are deposited by the aggregation, mostly on the upper surfaces of leaves. Clutch size varies between 14 and 67 eggs. Male *A. spurrelli* engage in a scraping behavior that dislodges many eggs from the leaves. Hatching commences six days after oviposition and may continue for several days. The tadpoles "hang" in the midwaters with their heads pointed toward the surface most of the time, as in other *Agalychnis.* Metamorphs are 18 to 20 mm in standard length.

REMARKS: This species is not recorded by Taylor (1952a, 1954a, 1955, 1958a) from Costa Rica, since it was not collected there until the 1960s (Savage and Heyer 1967). Duellman (1970) regarded this species as closely allied to *Agalychnis litodryas* of eastern Panama to northwestern Ecuador. Roberts (1994a,b) pointed out the similarities in breeding behavior and locomotion between *A. spurrelli* and *A. saltator* and suggested a possible relationship.

DISTRIBUTION: Humid lowlands and lower portions of the premontane zone of southeastern and southwestern Costa Rica to northwestern Ecuador (15–750 m).

Genus *Phyllomedusa* Wagler, 1830

These small to large treefrogs (twenty-nine species) are usually green, but some change to purplish brown or orange tan at night. The digital disks are small, and both fingers and toes lack webs or have only rudimentary webbing. The first toe is variable in length relative to the second but is shorter than the second in the Costa Rican species. The toes are opposable in some South American forms having a long first toe. The eyes and tympanum are moderate. and the latter is distinct. The iris in the lone Costa Rican species is silvery white.

Phyllomedusa formerly also included nine South American species now variously allocated by Cruz (1990) to *Hylomantis* (two), *Phasmahyla* (four), and *Phrynomedusa* (three). The genus as currently understood is probably composite and putatively may be distinguished from other genera in the subfamily by finger and toe webs absent or rudimentary, quadratojugal robust to slender, sphenethmoid well ossified; spatulate, bifid teeth present on the premaxilla and maxilla and present on the vomer in some species but absent in others; larvae deep bodied with an anteroventral mouth, a lateroventral sinistral spiracle, 2/3 denticle rows, and no papillae above the mouth.

The vertical pupil and lack of finger and toe webbing will separate the Costa Rican member of the genus from any other frog known from the republic.

Phyllomedusa range from Costa Rica to western Colombia west of the Andes and to northern Argentina east of the Andes and on Trinidad.

Several species groups have been proposed within *Phyllomedusa,* but many taxa have not yet been allocated to group. Of the four groups recognized, the Costa Rican species has been referred to the *buckleyi* species group (Cannatella 1980). Other members are *Phyllomedusa buckleyi* (eastern Ecuador), *P. medinai* (northern Venezuela), *P. hulli* (eastern Peru), and *P. danieli* and *P. psilopygion* (western Colombia).

Members of this genus tend to be slower moving than the species of *Agalychnis* and often walk through or on vegetation. In all species studied to date eggs are laid on vegetation above bodies of water, and hatching tadpoles are washed by rain or fall into the ponds or puddles to complete development (pattern 22, table 7.1). In preservative these frogs are dull blue to dark purple.

Phyllomedusa lemur Boulenger, 1882

(plates 140–41; map 7.67)

DIAGNOSTICS: This moderate-sized anuran is the only green treefrog in Costa Rica having a vertical pupil (fig. 7.11e) and lacking finger and toe webs.

DESCRIPTION: Adult males 30 to 41 mm in standard length, adult females 39 to 53 mm; upper surfaces smooth; venter granular; head about as wide as long; snout short,

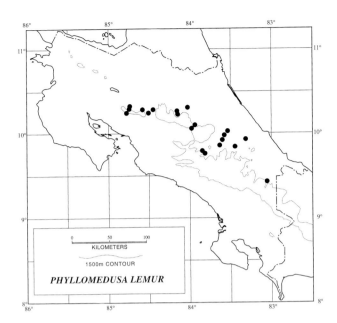

Map 7.67. Distribution of *Phyllomedusa lemur*.

truncate in dorsal outline, vertical in profile; eyes large; lower eyelid not reticulate; tympanum distinct, moderate, diameter equal to or less than one-half diameter of eye; elongate paired vocal slits and barely distensible internal median subgular vocal sac in adult males; finger disks small, disk on finger III slightly larger than tympanum; a weak ulnar fold from elbow to tip of finger IV; subarticular tubercles single; nonspinous brownish nuptial pad on base of thumb in adult males; thenar pad moderately enlarged; small round, flattened palmar tubercle; no well-developed heel tubercle or calcar, but a single tiny tubercle occasionally present; toe V with slight outer fringe; toes usually without webs, with slight webbing at most between toes III and IV; weak plantar tubercles present; inner metatarsal tubercle low, elliptical, no outer metatarsal tubercle or distinct tarsal fold. Dorsal surfaces mostly pale green by day, various shades of reddish brown, lavender brown, or orange tan at night; upper surface of forearm dark yellow; anterior and posterior thigh surfaces orange; flanks yellow to orange; venter creamy white; iris silvery white. In preservative, dorsal surfaces bluish purple to reddish brown, depending on when preserved. Small juveniles reddish brown in life and in preservative.

LARVAE (fig. 7.67): Moderate-sized, 42 mm in total length at stage 30; body ovoid; mouth anteroventral; nares dorsolateral, directed anterolaterally; eyes dorsolateral, directed laterally; spiracle lateroventral, sinistral; vent dextral; tail moderate; caudal fins reduced; tip of tail pointed; oral disk small, entire, with beaks and 2/3 rows of denticles; A2 with slight median gap; upper and lower beaks serrate; two or three rows of oral papillae lateral and ventral to mouth; body purplish gray; venter creamy yellow with a pinkish suffusion; venter, flanks, and tail musculature with dark flecks and mottling; fins mostly transparent with a few flecks anteriorly; iris gold.

VOICE: A very short "tick" repeated every 25 seconds. Call duration is 190 to 200 milliseconds, with a pulse rate of 39 to 41 per second. The dominant frequency is 1.4 to 3.1 kHz (Jungfer and Weygoldt 1994).

SIMILAR SPECIES: All species of *Agalychnis* have well-developed finger and toe webs, and the species of *Duellmanohyla* have well-developed toe webs.

HABITAT: Primarily undisturbed Premontane Wet Forest and Rainforest, but marginally in Lower Montane Wet Forest and Rainforest.

BIOLOGY: These uncommon nocturnal frogs are active throughout most of the year, with a reproductive season concentrated in the early rainy period, April to July. They sleep on the underside of leaves during the day. After dusk they begin to move about. They rarely jump but usually slowly climb or walk on leaves or twigs. They approach food items slowly and pounce on them from a distance of 50 to 60 mm, continuing to grip vegetation with their feet during predation. In addition to the advertisement call they emit a two- to four-pulse encounter call when another calling male is nearby. They also produce a release call. Aggressive behavior in males consists of clasping other males and unreceptive females and trying to push them off the perch. Amplectant pairs lay and fertilize eggs that are deposited on leaf surfaces, usually on top. Clutches of 15 to 30 green eggs were deposited in terrariums (Schulte 1977; Van Eijsden 1977; Jungfer and Weygoldt 1994) early in the morning. Females may produce two or three clutches in one night or may lay clutches on separate nights. An egg mass containing 70 eggs was reported by Cannatella (1980) for this species, but

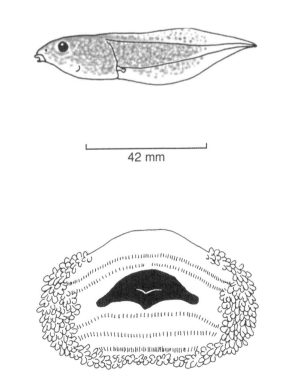

42 mm

Figure 7.67. Tadpole of *Phyllomedusa lemur*.

the clutch was probably that of an *Agalychnis annae* (Jungfer and Weygoldt 1994). The egg mass may consist of a single or double layer of eggs 3.0 to 3.5 mm in diameter within a gelatinous capsule 5 to 6 mm in diameter. Eggs hatch seven to fourteen days after fertilization. Embryos at this stage are light green and are washed or fall into the water below the oviposition site. Larger tadpoles become silvery gray, then tan, and finally olive gray during further development. The larvae are suspension-raspers feeding in the water column and on the bottom (ecomorph 4, table 7.2). Metamorphosis begins sixty-nine to ninety-eight days after fertilization. At this stage green pigment develops, and large tadpoles and metamorphs show color changes between day and night as adults do. Metamorphs captured in the field are 25 to 27 mm in standard length and lack the orange or yellow flank color of adults.

REMARKS: Specimens of this form from Costa Rica are somewhat smaller than those from Panama, and there appears to be an increase in size from west to east. Adult Costa Rican *P. lemur* are 30 to 35 mm in standard length for males, 30 to 45 mm for females; males from western Panama are 31 to 37 mm in standard length, and females are 38 to 47 mm. The largest known example is a female from extreme eastern Panama measuring nearly 53 mm in standard length. The smallest juvenile is 18.5 mm. The advertisement calls Duellman (1970) and Cannatella (1980) reported for this species appear to be based on encounter calls. The karyotype is 2N = 26 (León 1969).

DISTRIBUTION: Humid premontane and lower portions of lower montane zone predominantly on the Atlantic versant from the vicinity of Tilarán, Guanacaste Province, to western Panama; the disjunct Pacific slope records are from northwestern Costa Rica and southwestern, central, and extreme eastern Panama (440–1,600 m).

Subfamilies Hemiphractinae and Hylinae

Because of the external similarities among treefrogs of this family having a horizontal pupil, I have devised a single key that includes all species of the genera *Anotheca, Duellmanohyla, Gastrotheca, Hyla, Osteopilus, Scinax,* and *Smilisca*. In the cases of *Duellmanohyla, Scinax,* and *Smilisca* I have included separate keys to the species of those genera as well. If you are confident of the generic identification, proceed directly to the latter keys as appropriate. Extensive variation in some features has led me to double-key two species *(Smilisca puma* and *S. sila)* in the following key.

KEY TO COSTA RICAN HEMIPHRACTINE AND HYLINE TREEFROGS (FAMILY HYLIDAE)

1a. Lower surfaces of limbs without large light-outlined dark markings; no large knoblike or spiny cranial projections. 2
1b. Lower surfaces of limbs with sharply contrasting large,

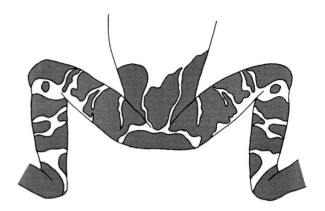

Figure 7.68. Ventral view of *Anotheca spinosa* showing characteristic color pattern of posterior venter and legs.

light-outlined dark rings, bars, or spots (fig. 7.68); little or no finger webbing; occipital, temporal, and supraorbital surface of head with numerous large spiny or knoblike cranial projections in adults (fig. 7.11f); adult males 59–69 mm in standard length, adult females 58–80 mm . . . . *Anotheca spinosa* (p. 295)
2a. No triangular dermal flaps on upper eyelids 3
2b. A triangular supraorbital flap on each eyelid (fig. 7.11g); no finger webbing, adult males 66–81 mm in standard length, adult females to 77 mm . . . . *Gastrotheca cornuta* (p. 294)
3a. Snout protruding beyond lower jaw (fig. 7.6c); web between toes I and II reduced to a fringe on toe II (fig. 7.12a); no finger webs 4
3b. Snout not protruding beyond lower jaw; web between toes I and II rarely reduced to a fringe on toe II. 6
4a. Posterior surface of thigh uniform, dusky or dull yellowish green in life; no dark spots in groin; adult males 24–35 mm in standard length, adult females 24–40 mm. 5
4b. Posterior surface of thigh with dark bars that alternate with light areas (greenish yellow to bright yellow in life, white in preservative) (fig. 7.83b); a dark spot or spots in groin; adult males 36–49 mm in standard length, adult females 44–52 mm . . . . *Scinax boulengeri* (p. 343)
5a. Dorsum tuberculate; finger disks equal to or slightly smaller than tympanum; snout pointed in dorsal outline (fig. 7.6b); vocal slits in males hidden by tongue; bones white; adult males 21–27 mm in standard length, adult females 21–32 mm . . . . *Scinax staufferi* (p. 347)
5b. Dorsum smooth to granular, never tuberculate; finger disks about one and a quarter times as large as tympanum; snout rounded in dorsal outline (fig. 7.6b); vocal slits in males not hidden by tongue; bones green in life; adult males 26–38 mm in standard length, adult females 30–40 mm . . . . *Scinax elaeochroa* (p. 345)
6a. No finger webs. 7

6b. Finger webs present 8

7a. No dorsolateral light stripe, dorsum not uniform; venter without dark peppering 35

7b. A dorsolateral light stripe from tip of snout to level of groin bordered below by a dark line, flank dark; dorsum uniform (fig. 7.69); venter peppered with brown flecks; adult males 30–34 mm in standard length, adult females 34–37 mm . . . .
. . . . *Hyla angustilineata* (p. 327)

8a. An extensive fleshy fringe along posteroventral margin of lower arm and leg (fig. 7.70a); prepollex protuberant in both sexes, with a terminal spine in some males 9

8b. No arm or leg fringes, prepollex protuberant or not 10

9a. Web of hand extending to disk on at least one digit; dorsum tubercular; prepollex with a terminal bony spine in adult males; adult males 57–110 mm in standard length, adult females 69–86 mm . . . .
. . . . *Hyla miliaria* (p. 334)

9b. Web of hand not extending to disk on any digit; dorsum smooth or nearly so; distal end of prepollex with horny black spines in adult males; adult male 79 mm in standard length, adult females 71–92 mm . . . .
. . . . *Hyla fimbrimembra* (p. 332)

10a. Skin on dorsal surface of head movable, not co-ossified with skull roof. 11

10b. Skin on dorsal surface of head immovable, co-ossified to exostosed (pitted and ridged) skull roof (fig. 7.99); bones green in life; adult males 27–89 mm in standard length, adult females 52–165 mm . . . .
. . . . *Osteopilus septentrionalis* (p. 338)

11a. A protuberant prepollex with a terminal spine in adult males, indicated by a nub in females; green with red digital webs *or* yellowish tan with a midcephalic dark stripe and/or vertical dark bars on flanks in adults; bones green in life 12

11b. No protuberant prepollex, no dark midcephalic stripe or dark vertical flank bars, never green with red digital webs 13

12a. Dorsal ground color yellowish tan to reddish tan or

Figure 7.69. Coloration of *Hyla angustilineata* on an outline of a generalized treefrog.

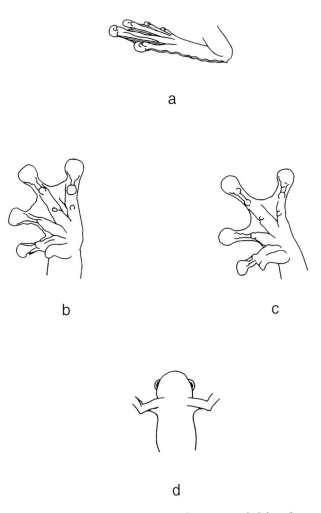

Figure 7.70. Distinctive characteristics of Costa Rican hylid treefrogs: (a) fleshy fringe along outer margin of tarsus and foot; (b) palmar view of hand of *Hyla rosenbergi* showing definite web between fingers I and II; (c) palmar view of hand of *Hyla rufitela* showing absence of web between fingers I and II; (d) ventral view showing axillary web or membrane.

gray in life and preservative; flanks usually marked with narrow vertical dark bars; midcephalic dark stripe usually present and continuing as middorsal stripe; digital webs brown; bones green in juveniles, not obviously so in adults; a basal web between fingers I and II (fig. 7.70b); tympanum two-thirds to three-fourths diameter of eye; adult males 60–79 mm in standard length, adult females 63–82 mm . . . .
. . . . *Hyla rosenbergi* (p. 303)

12b. Dorsal ground color green in life, yellowish brown in preservative; no vertical flank bars or midcephalic or middorsal dark stripes; digital webs red; bones green; no web between fingers I and II, although finger I with low fringe (fig. 7.70c); tympanum one-half diameter of eye; adult males 39–49 mm in standard length, adult females 46–55 mm . . . . . *Hyla rufitela* (p. 302)

13a. Posterior surface of thigh without dark bars; no dark

blotches in groin, although small punctations may be present 14

13b. Dorsal and posterior thigh surfaces contrastingly barred black and yellow (fig. 7.83a); upper surfaces smooth or with small, rounded tuberosities; one or more dark blotches in groin; adult males with smooth brown nuptial pad; adult males 27–34 mm in standard length, adult females 31–38 mm . . . .
. . . .*Hyla lancasteri* (p. 311)

14a. One odontoid at tip of lower jaw 15

14b. A pair of odontoids at tip of lower jaw (fig. 7.71b); virtually no finger webs; adult males 27–32 mm in standard length, adult females 34–35 mm . . . .
. . . . *Hyla picadoi* (p. 336)

15a. No dense concentration of mucous-poison glands in occipital and frontal region, as indicated by thickened skin; vocal sacs subgular, single or paired, in adult males 16

15b. Dorsal integument thick and glandular; a dense zone of mucous-poison glands indicated by thickened, rugose skin in occipital and frontal regions (fig. 7.12b); paired lateral vocal sacs in adult males (fig. 7.12f); bones green in life; adult males 70–101 mm in standard length, adult females 65–114 mm . . . .
. . . . *Phrynohyas venulosa* (p. 340)

16a. Dorsum strongly tuberculate or warty. 17

16b. Dorsum smooth or weakly granular 19

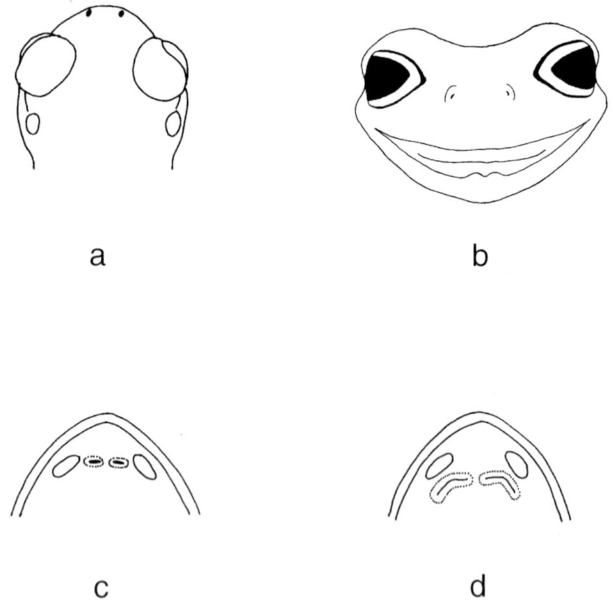

a   b

c   d

Figure 7.71. Diagnostic features of Costa Rican hylid treefrogs: (a) dorsal view of head of *Hyla zeteki* showing forward-directed eyes and dorsolaterally directed tympana; (b) frontal view of *Hyla picadoi* illustrating paired odontoid projections at tip of lower jaw; (c) transverse vomerine tooth patches lying between choanae; (d) acute, angulate vomerine tooth patches lying behind choanae. In one form (*Hyla debilis*) the transverse tooth patches lie near the posterior margins of the choanae.

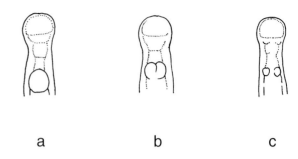

a   b   c

Figure 7.72. Undersurface of digit showing variation in subarticular tubercle character states in anurans: (a) single; (b) bifid; (c) double.

17a. Dorsum with low tubercles or warts, never with large pointed spines 18

17b. Head, body, and limbs with enlarged fleshy, spiny tubercles; nuptial pads of adult males composed of small black spines; adult males 26–35 mm in standard length, adult females 31–40 mm . . . .
. . . . *Hyla calypsa* (p. 310)

18a. Subarticular tubercles under fingers III and IV double or bifid (fig. 7.72b,c); posterior surface of thigh yellow, usually suffused with dark pigment; adult males 27–34 mm in standard length, adult females 33–92 mm . . . . *Hyla tica* (p. 324)

18b. Subarticular tubercles under fingers single (fig. 7.72a); posterior surface of thigh mottled with lighter color and brown to purple (fig. 7.74b); adult males 31–45 mm in standard length, adult females 32–46 mm . . . . *Smilisca sila* (p. 354)

19a. Eyes directed laterally (fig. 7.11b); tympanum vertical, directed laterally; dorsum frequently with pattern of lines or blotches. 20

19b. Eyes directed forward at about a 45° angle; tympanum oblique, directed dorsolaterally (fig. 7.71a); dorsum nearly uniform with a few dark punctations; a conspicuous narrow dark band on wrist; adult males 21–24 mm in standard length, adult females 24–27 mm . . . . *Hyla zeteki* (p. 337)

20a. A distinct dark lateral band from tympanum to level of axilla or extending farther posterior (fig. 7.73a,b) 21

20b. No distinct dark lateral band, although entire flank may be darker than dorsum or a narrow dark dorsolateral stripe may be present 23

21a. No axillary membrane; posterior surface of thigh suffused with dark pigment; groin usually mottled with dark and light pigment that may form a reticulate pattern; adult males 40–78 mm in standard length, adult females 50–90 mm 22

21b. Axillary membrane well developed (fig. 7.70d); posterior surface of thigh yellow orange in life, without dark pigment; groin uniform light; adult males 23–27 mm in standard length, adult females 30–35 mm . . . . *Hyla ebraccata* (p. 313)

the larval characters under the Campbell and Smith (1992) arrangement.

I have sympathy for any attempt to tease out the several major phyletic lines that are subsumed under the paraphyletic genus *Hyla*. I agree that certain lineages (e.g., the *uranochroa* species group) of *Hyla* having highly modified stream-adapted tadpoles form monophyletic clusters of closely related forms. There is some risk in recognizing genera based mostly on reproductive and larval features that are subject to great homoplasy; however, I follow Campbell and Smith (1992) in recognizing *Duellmanohyla*. The transfer of *Hyla legleri* to *Ptychohyla* seems to be on less secure grounds, and here it is retained in *Hyla*.

Genus *Anotheca* Smith, 1939a

A single large, strikingly patterned species having a co-ossified, casque head with a series of spiny projections dorsally as adults is placed in this genus. The fingers lack webs, the digital disks are large and rounded (as long as wide), the web between toes I and II is well developed, the eyes are moderate, and the tympanum is large. Both juveniles and adults have a series of large light-outlined dark bars, rings, or spots on the venter and underside of the legs.

Other features that in combination differentiate *Anotheca* from other genera in the subfamily are skull completely roofed; anterior arm of squamosal extending nearly or completely to maxilla; quadratojugal well developed, articulating with maxilla; sphenethmoid well ossified; narrowly spatulate, bifid teeth on premaxilla, maxilla, and vomers; no vocal slits or vocal sac; depressor mandibulae muscle with two parts, one from dorsal fascia and one from otic arm of squamosal (DFSQ); adductor mandibulae subexternus and externus superficialis present (s + e); widely dilated sacral diapophyses present; larvae with depressed body, small anteroventral mouth, lateroventral sinistral spiracle, 2/2 denticle rows, and labial papillae bordering entire oral disk; 2N = 24.

The distinctive coloration, absence of finger webs, and cephalic spines of adults make identification to genus and species easy.

The eggs of this genus appear to be deposited above the waterline in water-filled spaces in epiphytes or tree holes (pattern 24, table 7.1). The tadpoles are free swimming and obligately oophagous, feeding on unfertilized eggs laid by their mother (ecomorph 9, table 7.2).

The genus is found disjunctly on the Atlantic slope of Veracruz, Oaxaca, and Chiapas, Mexico, in eastern Honduras, Costa Rica, and western Panama, and in southwestern Costa Rica and western Panama on the Pacific versant.

Anotheca spinosa (Steindachner, 1864)

(plate 144; map 7.69)

DIAGNOSTICS: Adults with their sharp pointed and knoblike projections on the head and a huge tympanum

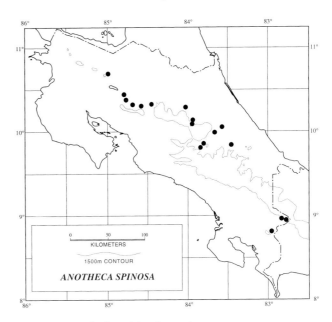

Map 7.69. Distribution of *Anotheca spinosa*.

(fig. 7.11f) are unmistakable. Juveniles lack the projections but share with the adults the color pattern of large black blotches, rings, and/or spots narrowly outlined by creamy white (fig. 7.68) and the large tympanum.

DESCRIPTION: A large, long-legged species; adult males 59 to 69 mm in standard length, adult females 58 to 80 mm; dorsum, upper limb surfaces, and venter smooth; head as broad as or broader than long; snout acutely rounded from above, truncate in profile; tympanum distinct, large, 70–80% of diameter of eye; annulus tympanicus prominent; no vocal slits or vocal sac; fingers long and slender with moderate-sized disks, disk on finger III two-thirds diameter of tympanum; distal subarticular tubercle on finger IV bifid or single; black spinous nuptial pad on thumb base in adult males; no finger webs; thenar pad moderately enlarged; palmar tubercle double; no heel tubercle or calcar; toes with moderate webbing; webbing formula: I 2^-–$2\frac{1}{4}$ II $1\frac{3}{4}$–3 III $1\frac{1}{3}$–3 IV 3^-–$1\frac{1}{2}$ V; inner metatarsal tubercle moderate, ovoid, outer small, conical; distinct inner tarsal fold present. Dorsal surfaces mostly tan or brown, often with a pinkish to reddish cast, marked with large light-outlined purplish dark brown to black blotches on flanks as extensions of ventral color; similar light-outlined black bars and some spots on surfaces of the limbs; side of head black with tan or gray vertical bar on lip just anterior to eye; juveniles somewhat more vividly marked than adults; iris bronze to coppery brown, with or without fine reticulations.

LARVAE (fig. 7.75): Moderate-sized, total length 28 mm at stage 33, 45 mm at stage 38; body stout, depressed; mouth anteroventral; nares dorsal; eyes small; eyes dorsolateral, directed dorsally; spiracle lateroventral, sinistral; vent dextral; tail short; caudal fins moderate; tip of tail rounded; oral disk small, entire, with immense beaks and 2/2 rows of denticles;

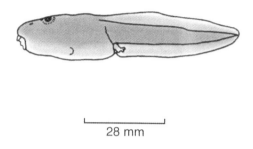

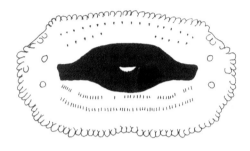

Figure 7.75. Tadpole of *Anotheca spinosa*.

denticles widely spaced on upper rows; A2 with large median gap; beaks with fine serrations; papillae around entire disk; body dark brown above, grayish below; tail musculature brown, fins tan.

VOICE: This species lacks vocal slits and sacs but produces a loud call carrying more than 100 m. The call is a "boop-boop-boop," each note lasting about 100 milliseconds with an interval of about 300 milliseconds between notes and about 180 notes per minute. The dominant frequency is 0.54 kHz, with several harmonics at higher frequencies (Duellman 1970; Jungfer 1996).

SIMILAR SPECIES: There are none.

HABITAT: Primarily in relatively undisturbed Premontane Rainforest, but marginally in Lowland Wet Forest habitats. Rarely seen except during the reproductive period.

BIOLOGY: Much of what is known about *Anotheca* biology comes from the study of captive animals (Jungfer 1996). That account forms the basis for much that follows, with field observations interspersed with behavior in the terrarium. Males of this rather uncommon nocturnal species are often heard calling while sitting in water in tree holes, bromeliads, or open bamboo internodes, although they are infrequently observed. They do not seem to stay at the calling site during the day. Amplexus takes place early in the morning. In captivity they choose water-filled containers and call from them at night, returning to the same container each night for a month or longer. In the terrarium amplexus took place when a female approached a calling male and, after sliding into the water in the container, was immediately clasped by him. She then dived under the water, bringing their heads close to the bottom with both cloacae above the water surface. The eggs were affixed mostly to the sides of the container at or above the water surface and fertilized by the male. Egg laying occurred during the day and lasted between two and four hours. Clutch sizes varied between 48 and 311 (mean about 158), with eggs ranging in diameter from 1.48 to 1.82 mm and surrounded by jelly envelopes of 0.2 to 0.3 mm. The eggs are dark gray and white. In captivity the larvae hatched after six to seven days; they are white and are about 10 mm in total length. At this stage the operculum has formed and the tadpoles are able to swim to the surface to breathe. In nature the eggs apparently are laid out of the water on the walls of tree holes or on leaves of bromeliads, and after hatching they slide into the water in the bottom of the hole or leaf axils. Survival to the larval stage in captivity was very low, which correlates well with the number of larvae found together in the field: two and three (Taylor 1954c), four (Robinson 1961), seven, eight, and eleven (Duellman 1970). Most of these tadpoles were unusual because their stomachs contained numerous undamaged frog eggs. In the terrarium the individual female returned to the same container of her first oviposition after five to seven days. She sat or floated on the water with the cloaca submerged and deposited large clutches of unfertilized eggs (48 to 194), which the tadpole(s) ate. The tadpoles grasped the first eggs with their beaks and sucked out the contents, leaving the egg capsules behind. During later feedings they ate the whole egg. Apparently the mother is stimulated to lay the nutritive eggs when the larvae touch, suck, or bite her. Metamorphosis took place in 60 to 132 days but was related to the number of tadpoles in the container. If for some reason a second clutch of fertilized eggs was laid in a container, the larvae that hatched were apparently eaten by the tadpoles from the first clutch. In most cases larvae of two size-classes developed from the same clutch, based on success in obtaining nutritive eggs. The smaller ones were not attacked by the larger tadpoles unless they were immobile from starvation or dead, in which case they were eaten. Some of the feeble larvae survived until the more successful large larvae metamorphosed. If they subsequently received nutritive eggs, they ultimately transformed. Metamorphs are 26 to 28 mm in total length.

REMARKS: *A. spinosa* tadpoles are obligate oophages that feed on conspecific eggs their mother provides. This same type of parental care is documented in the Jamaican hylid *Osteopilus brunneus* (Lannoo, Townsend, and Wasserzug 1986), several species of *Dendrobates* (see the accounts for that genus), and the Asian rhacophorid *Chirixalus eiffingeri* (Ueda 1986). Tadpoles of *Hyla zeteki* and *H. picadoi* of Costa Rica are also oophagous, and I suspect their mothers similarly nourish them with unfertilized eggs. *Osteocephalus oophagus* of the Amazon and Guyana region is obligately oophagous but feeds on fertilized eggs (Jungfer and Schiesari 1995). Taylor (1954c) was the first to report egg predation in *Anotheca*, based on examples from Moravia, Cartago Province. Starrett (1960a) confused the matter by associating an unspecialized tadpole from Costa Rica that inhabits slow streams with *A. spinosa*. Robinson (1961) contested this conclusion and confirmed Taylor's identification after allowing representative tadpoles from Veracruz, Mexico, to transform. He also confirmed that *A. spinosa* tadpoles are oophagous. Duellman (1970) reported on additional egg-filled tadpoles of this species but found parts of mosquito larvae in very small examples (stage 25). Jungfer (1996) noted that starving early-stage *A. spinosa* larvae being outcompeted for eggs by the larger size-cohort tried to gain nourishment by swallowing bits of bark picked from the walls of the container. It is likely that members of the smaller size-cohort, when starving, will try to eat anything available, including mosquito larvae. Although Duellman was unwilling to make an identification of the Starrett tadpoles (UNMN 118664), they conform very closely to those of *Smilisca phaeota*, a relatively common frog in the area of Turrialba, Cartago Province, where her tadpoles were collected. This species was known for many years before 1970 as *Anotheca coronata* (Taylor 1952a), but Duellman (1970) established that the two names applied to the same frog, and that *A. spinosa* has priority. Contrary to Duellman's account (1970) stating that no nuptial pads are found in this species, adult males have black nuptial thumb pads composed of small spinules.

DISTRIBUTION: Disjunctly in humid forests, primarily in the premontane zone, in eastern Mexico (800–2,068 m), northeastern Honduras (95 m), Atlantic slope Costa Rica and western Panama, and from southwestern Costa Rica to west-central Panama on the Pacific slope (350–1,330 m).

Genus *Duellmanohyla* Campbell and Smith, 1992

These small to moderately small stream-breeding frogs have bright red, bronze, or yellow irises and the dorsum uniform pale green, olive, reddish brown, or lichenose, with green or olive spots on a black background. Pale upper labial and lateral stripes are present in several species. In addition, the fingers and some of the toes have moderate webbing; dark nuptial thumb pads present or absent in adult males but no protuberant prepollices; no fleshy limb fringes, axillary membrane, or mental gland; vomerine teeth in linear series; quadratojugal may or may not touch maxilla. Tadpoles with low fins, long tails, large, pendant ventral oral disks that have short 2/2 to 3/3 denticle rows and large conical papillae on the disks external to the denticle rows; the karyotype is 2N = 24. Other species allocated to this group besides the three Costa Rican members are *Duellmanohyla salvavida* of Honduras and *D. soralia* of Honduras and Guatemala, *D. chamulae* and *D. ignicolor* of southern Mexico, and *D. schmidtorum* of southern Mexico and Guatemala. Oviposition is unknown for any species in this group, leading to speculation that the eggs are laid on vegetation over streams (Scott and Limerick 1983). This group represents an expansion of species from two to eight over the *uranochroa* species group of Duellman (1970) and is part of the *uranochroa* species group of Savage and Heyer (1968).

KEY TO THE COSTA RICAN TREEFROGS OF THE GENUS *DUELLMANOHYLA*

1a. White lip stripe greatly expanded under orbit (fig. 7.73c); snout relatively short, distance from eye to nostril less than diameter of eye; dorsum greenish brown to brown in life, brown in preservative 2

1b. White lip stripe not expanded under eye; snout moderately long, distance from eye to nostril equals diameter of eye; dorsum leaf green in life, bluish to purplish in preservative; adult males 31–37 mm in standard length, adult females 36–40 mm . . . . *Duellmanohyla uranochroa* (p. 299)

2a. Tympanum small, horizontal diameter 30–46% of eye length; throat and venter white in life; posterior surface of thigh usually suffused by dark pigment; adult males 25–30 mm in standard length, adult females 33–40 mm . . . . *Duellmanohyla rufioculis* (p. 298)

2b. Tympanum moderate, horizontal diameter 55–62% of eye length; throat and venter yellow in life; posterior surface of thigh not suffused with dark pigment; adult males 30–33 mm in standard length, adult females probably to 36 mm. . . . . *Duellmanohyla lythrodes* (p. 297)

Duellmanohyla lythrodes Savage, 1968a

(plate 145; map 7.70)

DIAGNOSTICS: A moderately small elegant, red-eyed, brown to light green treefrog having a continuous white upper labial–lateral strip from snout to groin expanded into a large suborbital spot (fig. 7.73c), a moderate-sized tympanum (diameter 55 to 63% diameter of eye), posterior thigh surface clear yellow, and a yellow venter.

DESCRIPTION: Adult males 30 to 33 mm in total length,

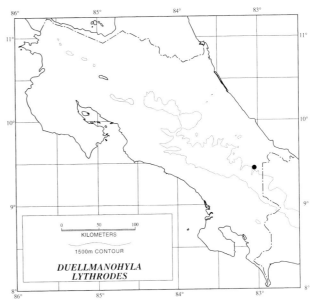

Map 7.70. Distribution of *Duellmanohyla lythrodes*.

adult females probably to 36 mm; upper surfaces smooth; venter granular; head wider than long; snout truncate from above; snout rounded in profile; eyes large; tympanum distinct, diameter 55 to 63% diameter of eye, smaller than disk on finger III; paired vocal slits and moderately distensible single internal subgular vocal sac in adult males; fingers relatively short and stout; finger disks moderate, width of disk on finger III about two-thirds diameter of eye; distal subarticular tubercles single on all fingers; nonspinous brown nuptial pad on base of thumb in adult males; fingers with minimal webbing, vestigial between fingers I and II; webbing formula: I 3–3 II 2¼–3½ III 3–2¼ IV; prepollical pad barely enlarged, elongate; no palmar tubercle; no ulnar ridge; toes moderately webbed; webbing formula: I 2–2½ II 1½–2½ III 1¾–2⁺ IV 2½–1½ V; inner metatarsal tubercle oblong, large, no outer tubercle; no tarsal fold. Dorsum light brown or light to bright green with definite metachrosis; eyelids and top of head with green and bronze flecks in brown phase; light stripe continuous from snout along head and side to groin, expanded into a large subocular spot; light stripes on margins or forearm, feet, and above cloaca; upper forearm yellow marked with oblique white stripe above; anterior and posterior surfaces of thighs yellow without dark pigment; no transverse dark bars on limbs; ventral surfaces yellow; vocal sac white; plantar surfaces with dark pigment; iris bright red.

LARVAE: Not known.

VOICE: A weak, bell-like "ping" in bursts of five to seven notes (Savage 1968a).

SIMILAR SPECIES: (1) *Duellmanohyla rufioculis* has white ventral surfaces, a tympanum smaller than the disk on finger III, and the posterior thigh usually suffused with dark pigment. (2) *Duellmanohyla uranochroa* has the tympanum smaller than the width of finger disk III, the sole of the foot immaculate, and no expanded suborbital spot. (3) *Hyla legleri* has a light upper lip stripe that terminates anterior to the tympanum when present, a light lateral stripe from axilla to groin, and no expanded suborbital light spot.

HABITAT: Found near streams in Lowland Wet Forest.

BIOLOGY: The holotype was calling from a tree about 3 m above a small stream. The other three known examples, also males, were perched in low vegetation along a shallow headwaters stream.

REMARKS: Savage (1968a) described this species based on a single adult male (LACM 28184; Limón: juncture of Río Lari and Rió Dipari). Although in the original description this locality was cited as being at 800 m, revised topographic maps show it to be at 440 m. Duellman (1970) placed *D. lythrodes* in the synonymy of *D. rufioculis*. Myers and Duellman (1982) resurrected the species after additional material from northwestern Panama confirmed the differences between it and *H. rufioculis* in tympanum size and ventral coloration.

DISTRIBUTION: Humid lowland forests of Atlantic ver-

sant southern Costa Rica and adjacent northwestern Panama (170–440 m).

Duellmanohyla rufioculis (Taylor, 1952a)

(plate 146; map 7.71)

DIAGNOSTICS: A small to moderate-sized brown or green, red-eyed treefrog having a continuous white upper lip–lateral stripe from snout to groin, expanded into a large suborbital light spot (fig. 7.73c), small tympanum (30–46% diameter of eye), posterior thigh surface dull tan or yellow, usually suffused with dark pigment, and a white venter.

DESCRIPTION: Adult males 25 to 30 mm in standard length, adult females 33 to 40 mm; upper surfaces smooth; venter granular; head about as wide as long; dorsal outline of snout truncate, snout profile rounded; eyes large; tympanum distinct, less than one-half diameter of eye, much smaller than finger disk III; paired vocal slits and slightly distensible single internal subgular vocal sac in adult males; fingers moderately long and slender; finger disks small, width of disk on finger III equal to diameter of eye; much larger than diameter of tympanum; distal subarticular tubercle on finger IV usually bifid; brown nonspinous nuptial pad on base of thumb in adult males; fingers with minimal webbing, vestigial between fingers I and II; webbing formula: I 3–3 II 2⁻–3 III 2½–2⁺ IV; prepollical pad barely enlarged; moderate, usually bifid palmar tubercle; ulnar ridge or warts present; toes moderately webbed; webbing formula: I 1⅔–2⁺ II 1–2½ III 1–2 IV 2–1 V; inner metatarsal tubercle large, ovoid, outer small or absent; weak tarsal fold. Upper surfaces dull brown, olive green, or rarely bright green, uniform or mottled with bronze, tan, or brown; light stripe continuous from snout along head and side to groin, usually

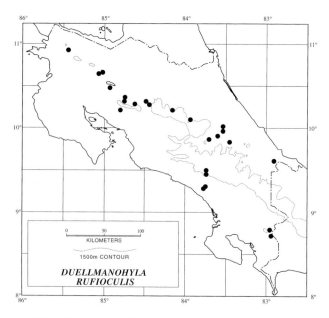

Map 7.71. Distribution of *Duellmanohyla rufioculis*.

expanded into large light suborbital spot and occasionally broken into spots on flank; light stripes on outer margin of forearm and tarsus; white stripe or paired white spots usually present above cloaca; upper arm dull tan with oblique white stripe above; anterior and posterior thigh surfaces dull tan or yellowish, usually suffused with dark pigment; no transverse dark bars on limbs; plantar surface suffused with dark pigment; ventral surfaces white, usually with suffusion of dark pigment in males; vocal sac white; iris bright red.

LARVAE (fig. 7.76): Moderately large, 40 mm in total length at stage 28; body somewhat depressed; mouth ventral; nostril directed anterodorsally; eyes directed dorsolaterally; spiracle lateral, sinistral; vent tube dextral; tail long; fins low; tail tip rounded; oral disk large, pendant, complete with moderately delicate beaks and 2/2–3 rows of denticles; denticle rows short, no longer than width of beaks; A2 and P1 with or without slight median gap; disk completely bordered by one row of tiny papillae; surface of disk covered by many large papillae; beaks with long, pointed serrations; dorsum dull olive brown; flanks paler with bluish white flecks; caudal musculature pale brown, heavily suffused with darker flecks and small spots that extend onto fins; iris red.

VOICE: A short rattle or "scraape" consisting of three to four short notes. The call group is repeated every 3 to 18 seconds, with note durations of 50 to 70 milliseconds and a dominant frequency of 2.0 to 2.6 kHz (Duellman 1970).

SIMILAR SPECIES: (1) *Duellmanohyla lythrodes* has yellow ventral surfaces, a moderate-sized tympanum that is larger than the disk on finger III, and the posterior thigh surface immaculate yellow. (2) *Duellmanohyla legleri* has a light upper lip stripe that terminates anterior to the tympanum when present, a light lateral stripe from axilla to groin, and no enlarged suborbital light spot. (3) *Hyla uranochroa* is bright green above and yellow below, has a large tympanum that is larger than the disk on finger III, has the posterior thigh surface immaculate deep yellow, and lacks an expanded suborbital spot.

HABITAT: Premontane Wet Forest and Rainforest and marginally in Premontane Moist Forest.

BIOLOGY: A nocturnal stream breeder. Males call throughout the year but are most active from August to December. The species tends to aggregate around small, shallow streams or seeps draining into streams. Calling males usually are 300 to 750 mm above the substrate. Characteristics of amplexus, oviposition, and eggs remain unknown for this species, but ten gravid females contained 82 to 142 eggs averaging 2.36 mm in diameter (Lang 1995). Tadpoles (ecomorph 11, table 7.2) are found in quiet pools, and when disturbed they hide under debris. The very specialized mouthparts help them adhere to rocks in the streams as the current increases in torrential rains. Metamorphs are 16 to 17 mm in standard length.

REMARKS: This common species is the most variable among Costa Rican red-eyed treefrogs. In addition to the

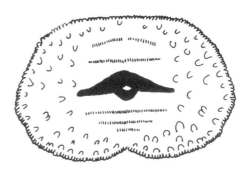

Figure 7.76. Tadpole of *Duellmanohyla rufioculis.*

variation in general coloration, which is affected by considerable metachrosis, the suborbital light spot may be obscured by dark pigment and the posterior thigh suffusion, and cloacal white marks are usually present but sometimes absent. The lateral white stripe is occasionally discontinuous on the flank as well. In most samples, maximum sizes do not exceed 28 mm in standard length for adult males and 35 mm for adult females. In a small series from El Tigre, Limón Province (CRE 290 1–3), two males are 29.4 to 30 mm in standard length and three females 38.5 to 39.9 mm. It may be that one or more cryptic species are subsumed under this name, although I can find no consistent difference among the several populations. Duellman (1970) reached a similar conclusion.

DISTRIBUTION: Humid premontane slopes of the cordilleras of Costa Rica (680–1,524 m).

Duellmanohyla uranochroa (Cope, 1875)

(plate 147; map 7.72)

DIAGNOSTICS: A beautiful moderate-sized bright green, red-eyed treefrog having a continuous white upper labial–lateral white stripe from snout to groin that is not greatly expanded below the eye, a large tympanum (diameter 66–71% diameter of eye), the posterior thigh surface clear yellow, and a deep yellow venter.

DESCRIPTION: Adult males 31 to 37 mm in standard length, adult females 36 to 40 mm; dorsal surfaces smooth; venter granular; head broader than long; dorsal outline of snout truncate, snout profile rounded; tympanum distinct,

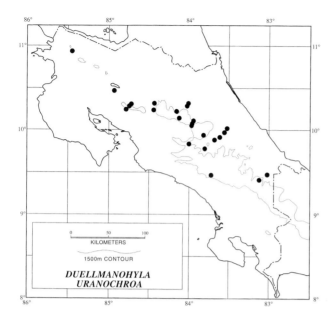

Map 7.72. Distribution of *Duellmanohyla uranochroa*.

large, diameter about 66–71% diameter of eye, larger than disk on finger III; paired vocal slits and a moderately distensible single internal subgular vocal sac in adult males; fingers moderately short and stout; finger disks moderate; width of disk on finger III about two-thirds of eye, much larger than tympanum; distal subarticular tubercle under finger IV often bifid, occasionally bifid under fingers II and III; brown nonspinous nuptial pad at base of thumb in adult males; finger webs minimal, vestigial between fingers I and II; webbing formula: I 3–3 II 2–3 III 2½–2⁺ IV; prepollical pad barely enlarged, bifid palmar tubercle usually present; ulnar ridge or warts present; toes moderately webbed; webbing formula: I 1⅔–2⁺ II 1–2½ III 1–2 IV 2–1 V; inner metatarsal tubercle large, outer small, round; weak tarsal fold present. Upper surfaces bright green; continuous light stripe from snout along side of head and body to groin, usually very slightly expanded toward suborbital rim; light stripes on margins of forearm, tarsus, and foot; white supracloacal stripe usually present; upper arm deep yellow with oblique white stripe; anterior and posterior thigh surfaces deep yellow, immaculate; no transverse dark bars on limbs; plantar surface immaculate; venter bright yellow; vocal sac yellow; iris bright red.

LARVAE (fig. 7.77): Moderately large, 42 mm in total length; body slightly depressed; mouth ventral; nostrils directed anterolaterally; eyes directed dorsolaterally; spiracle lateral, sinistral; vent tube dextral; tail long; fins low; tip of tail rounded; oral disk large, pendant, complete with delicate beaks and 2–3/2–3 rows of denticles; denticle rows short, no longer than width of mouth; no gap in P1; disk completely bordered by one row of minute papillae; surface of disk covered by large papillae; beaks with long, pointed serrations; dorsum and flanks olive brown, flecked with bluish

white; caudal musculature and fins cream tan with brown spots; minute white flecks; iris red.

VOICE: the call is unmistakable and consists of a long series of deep, melodic bell-like notes variously described as "boop-boop-boop-boop" or "ping'-ping'-ping'" with note repetition rates of 217 to 264 per minute. Note duration is 30 to 40 milliseconds with a pulse rate of 240 to 280 per second, and there are two emphasized dominant frequencies at about 1.7 and 1.8 kHz (Savage 1968a; Duellman 1970).

SIMILAR SPECIES: (1) Red-eyed *Agalychnis (A. callidryas, A. saltator,* and *A. spurrelli)* have a vertical pupil and do not have a continuous light stripe from snout to groin. (2) *Duellmanohyla lythrodes* has an enlarged suborbital light spot, smaller tympanum (55–63% diameter of eye) and considerable dark pigment on the plantar surface. (3) *Duellmanohyla rufioculis* has a white venter a much smaller tympanum (diameter much less than width of disk on finger III) and usually a large light suborbital stripe and dark pigment suffusion of the plantar and posterior thigh surfaces. (4) *Hyla legleri* has a light upper lip stripe that, if present, terminates anterior to the tympanum and the lateral light stripe running only from axilla to groin.

HABITAT: Characteristically found near streams in Premontane Wet Forest and Rainforest and marginally in Lower Montane Wet Forest and Rainforest.

BIOLOGY: The bell-like call of males of this beautiful species is one of the commonest nocturnal sounds in montane Costa Rica. During the day individuals hide in vegetation, especially leaf axils of epiphytes and terrestrial aroids. At night moderately dense congregations are found along

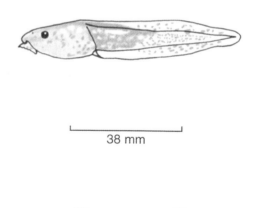

38 mm

Figure 7.77. Tadpole of *Duellmanohyla uranochroa*.

small, fast-moving mountain streams. The males call from dense vegetation several meters away from the stream and from 0.5 to 3 m above the substrate. Calls are heard throughout the year but only sporadically in the drier months (December to April). The largest choruses have been noted in May to June and September to November, suggesting two reproductive peaks in the wet season. No observations of amplexus or oviposition have been reported for this species, and the eggs have not been described. Tadpoles (ecomorph 11, table 7.2) are found in relatively quiet pools but adhere to large rocks in the stream bottom when the streams rise after heavy rains. Metamorphs are 16 to 17 mm in standard length.

REMARKS: Taylor (1952a) included this species in his volume on Costa Rican frogs and described a new red-eyed *Hyla* as *H. alleei* (holotype: FMNH 178234, an adult male: Costa Rica: Alajuela: Isla Bonita). Duellman (1966a) and Savage (1968a) showed that the supposed differentiating features of *H. alleei* were encompassed within the range of variation of *D. uranochroa* and synonymized Taylor's name with that species. In preservative, representatives of this taxon are bluish to purplish above, indicating that their green life color is structural (see discussion of coloration in hylids, p. 276). Dunn (1924b) reported eggs assumed to be of this species attached to a leaf overhanging the stream. The egg capsules contained well-developed tadpoles, but these were not raised to establish identity. The karyotype is 2N = 24 (León 1969).

DISTRIBUTION: Humid premontane and lower portions of the lower montane zones in the cordilleras of Costa Rica and western Panama (300–1,450 m); in Costa Rica on the Atlantic versant (656–1,740 m) and on the Pacific slope (880–1,600 m).

Genus *Hyla* Laurenti, 1768

This large genus (three hundred species) is an unnatural taxon defined by the absence of the derived characteristics diagnostic for other hylid genera. Members of the genus range from tiny forms (18 to 25 mm in standard length) to very large species (up to 132 mm).

Morphologies of adults are varied: the integument is smooth, tuberculate, or spinous dorsally, always granular ventrally, and rarely co-ossified with the skull roofing bones; the pupil is horizontally elliptical, the tympanum is usually distinct, and vocal slits and a single median internal or external subgular vocal sac generally are present in adult males; the fingers are usually webbed; the toes are always definitely webbed, and the finger disks are round (about as wide as long). In some species the enlarged prepollex of adult males protrudes and may be tipped with a spine. Skeletal features are also variable, but the sacral diapophyses are strongly dilated. The depressor mandibulae arises from the dorsal fascia (DF), and there is only a depressor mandibulae posterior subexternus present (s). Larvae are also variable; some lack

denticles, but in those having them they vary from 1/1 to 9/14, although most are 2/3. Karyotypes are 2N = 18, 22, 24, 30, or 34.

Coloration is also varied. Many species are gray or brown, some are green, others are nearly uniform yellow, and still others have striking and contrasting patterns. Posterior thigh surfaces are often boldly marked or brightly colored, and iris colors vary from yellow, bronze, or copper in most species to bright red in others.

As a reflection of this diversity, about fifty phenetically based species groups have been proposed within the genus (Frost 1985; Duellman 1993), with about twenty-six represented in North and Central America depending on a particular author's preference.

In the following accounts species are treated alphabetically by eleven species groups without any implication of intergroup phylogenetic relationships. Similar and presumably related species are included in each group as modified from the treatments of Savage and Heyer (1968) and Duellman (1970). Generally, difficulties in identification will involve discriminating among species belonging to the same species group. Once you become familiar with hylid frogs, positive identifications will be expedited by turning directly to the appropriate species group account.

So far as is known, all species have free-living aquatic larvae, but the eggs may be laid in ponds, streams, epiphytes, or tree holes, and in a few species on vegetation above a body of water or a stream course.

The genus as currently recognized has three distributional areas: Europe, northwestern Africa, the Middle East, and southern Arabia; East Asia and Japan; and the Americas from northern North America to western Ecuador and central Argentina, including the Greater Antilles.

Hyla albomarginata Group

Frogs of this group are moderate-sized and have a green dorsum in life (cream or pale yellow in preservative) with dark flecks and/or small white spots; considerable finger and toe webbing; no nuptial thumb pads in adult males, but a protruding prepollex tipped with a spine in some species including the Costa Rican taxon; no fleshy fringes on arms and legs; no axillary membrane; no mental gland; webbing and/or concealed surfaces or hind limbs pink to red in some forms; bones green in several species including the Costa Rican form; vomerine teeth in a curved to linear series just behind the choanae; quadratojugal articulates with maxilla; 2N = 24. Tadpoles have small oral disks with 2/4 denticle rows. Included species beside the single Costa Rica form are *Hyla albomarginata* (Atlantic slope from Colombia throughout Amazon basin to southeast Brazil); *H. pellucens* (western Colombia, Ecuador); *H. rubracyla* (western Colombia); and *H. arianae, H. arildae, H. albofrenata, H. albosignata,* and *H. musica* (southeastern Brazil).

Reproduction in this group is not specialized; the eggs are

deposited and the larvae develop in lentic situations (pattern 1, table 7.1). This group corresponds to the *H. albomarginata* species group of Duellman (1970) and Savage and Heyer (1968).

Hyla rufitela Fouquette, 1961

(plates 148–49; map 7.73)

DIAGNOSTICS: A spectacularly colored moderate-sized green treefrog having bright red digital webs, green bones, and a protuberant prepollex (fig. 7.70c).

DESCRIPTION: Adult males 39 to 49 mm in standard length, adult females 46 to 55 mm; dorsal surfaces smooth; venter granular; head about as broad as long; snout rounded to nearly semicircular from above, rounded in profile; eyes moderate; tympanum distinct, diameter two-fifths to one-half diameter of eye; paired vocal slits and a moderately distensible single external subgular vocal sac in adult males; fingers short and robust; finger disks moderate, width of disk on finger III about equal to diameter of tympanum; distal subarticular tubercle under finger IV usually single, sometimes bifid; prepollex protuberant with a definite spine in a retractable fleshy sheath in males, reduced to nub in females; no nuptial thumb pads in males; fingers moderately webbed, at best a trace between fingers I and II; webbing formula: I 3$^+$–3 II 1½–2$^+$ III 2$^-$–1¼ IV; thenar pad greatly enlarged; no palmar tubercle; distinct fleshy flap on heel; toes extensively webbed; webbing formula: I 1¼–2 II 1$^+$–2¼ III 1–2¼ IV 2$^+$–1 V; moderate low, elliptical inner metatarsal tubercle, small outer tubercle present or absent; weak inner tarsal fold. Dorsal surfaces lime green to bluish green with randomly scattered melanophores and usually scattered white or light bluish spots; occasionally a light stripe from tip of

snout onto flank; in preservative, cream with brown flecks and small dark brown spots; flanks yellowish olive; groin pale blue, bluish green, or red; no transverse dark bars on limbs; anterior and posterior thigh surfaces green to bright red; webbing of hands and feet red orange to tomato red; throat and undersurfaces of limbs green; venter creamy; juveniles are lime green anteriorly to yellowish green posteriorly, with a narrow red dorsal stripe from tip of snout over upper eyelid to level of the vent on each side; a broad lateral yellow stripe runs along the flank on each side; limbs green, without red markings, and digital webs red orange; iris silvery bronze.

LARVAE: Unknown. Duellman's (1970) account and figure are based on a tadpole of *Hyla rosenbergi*.

VOICE: A high-pitched series of clucks. Call groups may consist of nine to twenty-one notes repeated 22 to 63 times a minute. Note duration is about 50 milliseconds with a dominant frequency about 1.6 kHz, but there are usually two other harmonics out of a total of seven to nine that are emphasized (Duellman 1970).

SIMILAR SPECIES: (1) *Hyla palmeri* has a heel calcar. (2) *Hyla rosenbergi* is a tan to gray frog, usually with dark vertical lines on flanks and/or a midcephalic and/or middorsal dark stripe and definite webbing between fingers I and II (fig. 7.70b).

HABITAT: Lowland Moist and Wet Forests.

BIOLOGY: Another relatively uncommon nocturnal treefrog. Males may be heard calling throughout most of the year but are seldom seen outside the reproductive period. Reproduction takes place on suitable rainy nights during the wet season, with a peak during the heaviest rainy period (late August to October). Males call from dense vegetation near standing water within the forest and do not join the mixed-species choruses found around large temporary ponds at many lowland sites. Amplexus takes place in shallow vegetation-choked ponds and muddy pools, and the eggs are laid as a clear surface film. The eggs are black-and-cream and are about 1.8 mm in diameter; egg plus vitelline membrane is about 2.1 mm. Metamorphs are about 19 to 22 mm in standard length.

REMARKS: Taylor (1952a) used the name *Hyla albomarginata* for this species. Fouquette (1961) demonstrated that the Central American frogs are distinct from South American forms of this group. The prepollical spines found in all males of this species probably are used in male agonistic interactions, as in the gladiator frogs of the *Hyla boans* species group. The spine is usually concealed by the retractile sheath in living frogs. References in the literature to the variable occurrence of male prepollical spines are based on preserved material in which the sheath may or may not be retracted to expose the spine. The green coloration in this species is apparently produced by a pigment, probably biliverdin, that dissolves in formalin or alcohol when the animal is preserved. See the discussion of coloration in treefrogs

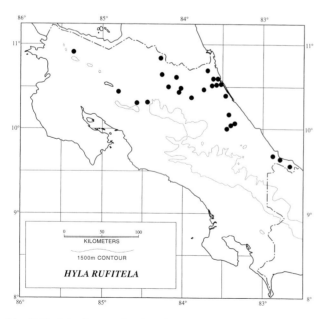

Map 7.73. Distribution of *Hyla rufitela*.

(p. 276) for additional information. All records of this species from the Pacific slope in Central America are based on larval and/or juvenile *Hyla rosenbergi*, but *H. rufitela* possibly occurs on the Pacific versant in northwestern Colombia. See remarks section under *H. rosenbergi*.

DISTRIBUTION: Humid lowlands from eastern Nicaragua to central Panama on the Atlantic slope (11–650 m).

Hyla boans Group

Gladiator Frogs

Large tan, brownish, or gray frogs, usually with darker irregular blotches on the dorsum and often with a narrow middorsal dark stripe. Transverse dark bars mark the limbs and narrow vertical lines or reticulations cover the thighs and flanks in most species. Other features include extensive finger and toe webs; no nuptial thumb pads, but a protruding prepollex whose spine may be exposed in adult males; no fleshy limb fringes, axillary membrane, or mental gland; vomerine teeth in curved series; quadratojugal articulates with maxilla; 2N = 24. Tadpoles robust, lentic forms with small oral disks and 2/4 denticle rows. Other than the single Costa Rican member, the following species belong to this group: *Hyla biobeba* (west-central Brazil); *H. boans* (eastern Panama and western Colombia throughout Amazon and Orinoco basins); *H. crepitans* (eastern Panama to central Brazil); *H. faber* (eastern Brazil to Argentina); *H. pardalis* (eastern and central Brazil to Argentina); *H. pugnax* (central Panama and northern Colombia); *H. wavrini* (Orinoco and Amazon basins).

Males of this group typically excavate basinlike nests in the mud adjacent to ponds or slow-moving streams, and the eggs are laid on the surface of the water inside the basin (pattern 3, table 7.1). The males aggressively defend their nests against other males, and they use the unsheathed prepollical spines to injure one another.

This group is equivalent to the *H. faber* species group of Savage and Heyer (1968) and the *H. boans* species group of Duellman (1970).

Hyla rosenbergi Boulenger, 1898b

(plates 150–51; map 7.74)

DIAGNOSTICS: A large long-snouted yellowish, tan to reddish tan, or gray treefrog having large light-colored eyes, a protuberant prepollex (fig. 7.70b), and usually narrow, wavy dark bars on the flanks and/or a dark midcephalic stripe that often continues onto the body as a middorsal stripe. Occasional examples lack the dark bars and dark stripe but are unlikely to be confused with any other form having a protuberant prepollex since they have the fingers less than fully webbed and lack fleshy fringes along the limb margin and green or red in their coloration. Juveniles lack lateral bars and a middorsal stripe and are covered with nu-

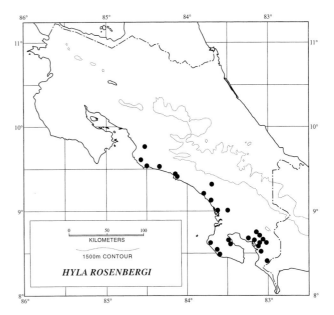

Map 7.74. Distribution of *Hyla rosenbergi*.

merous small dark spots. In addition, the bones of juveniles are obviously green.

DESCRIPTION: Adult males 60 to 79 mm (60 to 91 mm) in standard length, adult females 63 to 82 mm (60 to 95 mm); upper surfaces smooth; venter granular; head slightly longer than broad; snout subovoid to almost pointed from above, long, low, rounded in profile; eyes large; tympanum large, distinct, diameter two-thirds to three-fourths diameter of eye; paired vocal slits and moderately distensible single internal subgular vocal sac in adult males; fingers moderately short and robust; finger disks large, width of disk on finger III about three-fourths diameter of tympanum; subarticular tubercles single; prepollex protuberant, with terminal spine, usually concealed in retractable fleshy sheath in males, reduced to nub in females; no nuptial thumb pads in males; fingers extensively webbed; webbing formula: I 2–2$^+$ II 1–2$^-$ III 1½–1 IV; thenar pad greatly enlarged; palmar tubercle weak, flat, bifid; distinct heel flap sometimes present; toes extensively webbed; webbing formula: I 1–1⅓ II 1–1½ III 1–1½ IV 1⅔–1 V; inner metatarsal tubercle large, outer small or absent; distinct inner tarsal fold. Dorsum various shades of tan or gray, usually mottled with darker color; breeding females yellow; midcephalic dark stripe usually present, often continuing down middorsum; obscure dark, narrow vertical flank bars usually present, often interconnected to form irregular reticulum; legs with dorsal dark bars; narrow vertical pale dark bars often present on yellowish brown posterior thigh surface; ground color fades to pale gray or tan and dark markings less evident during day; webs yellowish tan to orange brown; flanks yellowish brown; venter pale bluish green, most intense in axilla and undersurface of thigh; iris pale gold above, silvery white below; juveniles, unlike adults, are covered with numerous

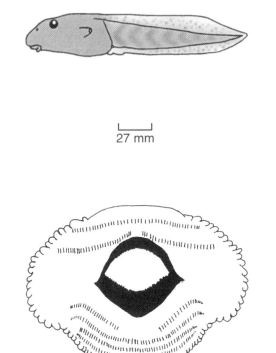

27 mm

Figure 7.78. Tadpole of *Hyla rosenbergi*.

small dark spots on tan ground color, without lateral dark bars or middorsal dark stripe; bones green.

LARVAE (fig. 7.78): Large, 27 mm in total length at stage 27; body robust; mouth anteroventral; nostrils and eyes directed dorsolaterally; spiracle sinistral, lateral; vent tube dextral; tail moderate; caudal fins moderate; tail tip pointed; oral disk entire, small, with moderate beaks and 2/4 denticle rows; A2 narrowly interrupted above mouth, P1 interrupted below mouth or not; beaks with pointed serrations; papillae in a single row around disk, absent on upper labium above mouth; body brown, venter creamy; caudal musculature brown; fins and musculature with dark brown flecks and blotches tending to form transverse bars dorsally.

VOICE: A loud series of "tonk-tonk-tonks" sounding like someone monotonously beating on a piece of wood. The call group consists of two to five notes, and the group is repeated 32 to 58 times a minute. Note duration is 50 to 70 milliseconds. The dominant frequency is 0.4 kHz, and the call may be heard from over 100 m away (Breder 1925; Duellman 1970; Kluge 1981). Kluge has described courtship, territorial, and encounter distress calls for this species in addition to the advertisement call. There is no release call.

SIMILAR SPECIES: (1) *Hyla fimbrimembra* and *Hyla miliaria* have fleshy flaps along the forearm and lower hind limb (fig. 7.70a) and *H. milaria* has fully webbed hands and feet. (2) *Hyla rufitela* is green with red webs, never has midcephalic, middorsal, or vertical flank stripes, and has essentially no web between fingers I and II (fig. 7.70c).

HABITAT: Lowland Moist and Wet Forests of the southwestern zone.

BIOLOGY: *Hyla rosenbergi* has been extensively studied in Panama by Kluge (1981). His account forms the basis for much of the material in this section (but also see Breder 1946). Like other members of the genus, this rather common species is nocturnal and eats a variety of arthropods. Males call throughout the year from vegetation or near water, including occasional periods during the dry season. However, during the reproductive periods males construct basins in the soft substrate near standing or slow-moving bodies of water. The nests are filled with water by seepage from these sources. In southwestern Costa Rica *H. rosenbergi* often uses grassy, temporary marshes for nest building, and the nest margins (ramparts) tend to be formed of matted grass rather than mud, as pointed out by Kluge (1981) and confirmed by my observations near Rincón de Osa, Puntarenas Province. At some more disturbed Costa Rican sites, the frogs use puddles and cattle footprints as substitutes for male-constructed nests (Höbel 1999b). Nests are 100 to 300 mm in diameter, with water depths of 5 to 90 mm. Males renovate their nests almost nightly, and they form their calling sites and the center of social interactions. Males sit on a small calling platform on the lower inside of one of the ramparts. The males occupy and defend the nests at night, but during the day they retreat to the nearby forest canopy to hide and rest. Early in the evening males begin to call and descend to their nests. At this time they are very aggressive, giving high-intensity territorial calls and encounter calls (chuckle, growl, mew, bark, or hiss). Territorial calls apparently space the males. Resident males give advertisement calls between 5:00 and 10:00, from the nest or near it. The males alternate calls with near neighbors, frequently forming duets that are expanded into small choruses as the evening progresses. The resident male gives encounter calls in response to visual contact with another male to indicate nest ownership. Males may also defend the nest site by rapid charges and chases without physical contact but with a range of territorial and encounter vocalizations. As a last resort a resident male will attack an intruding male by rushing or diving on him and knocking him away from the nest. A violent wrestling match ensues in which they jab and cut one another with the now unsheathed prepollical spines. Stabs are especially directed to the opponent's eyes and ears. During the breeding season most adult males probably receive wounds in these fights, and some suffer permanent damage (e.g., loss of eye or eardrum) or are killed. Females do not vocalize, are not involved in these aggressive displays, and are not attacked. They usually arrive at the chorus site about one hour after the males. Operationally males outnumber females at the breeding site, but the seasonal sex ratio is about 1:1. An interested female will approach and enter the nest of a calling male from a point opposite the male's back and move about the nest inspecting it. At this time the male shifts from his advertisement call to a lower-intensity, slightly slower courtship call. As the inspection continues the female usually bumps the male. Females are rather selective, and at least half the time they find the nest or the male unacceptable and move to another nest. If a female remains she continues to touch the male and eventually massages his head and body with her hands, chest, and chin. After this preamplexus ritual the male mounts the female and shifts to an axillary amplexus. The female then spends several hours renovating the nest before oviposition, which typically occurs early in the morning (2:00 to 5:00). During egg laying the female is completely submerged except for the cloaca. The male is essentially hunkered down on the female's back with his cloaca immediately above hers. Eggs are laid in small clusters every two to ten seconds and come to form a surface film of about 1,700–2,350–3,100 black-and-cream eggs. Oviposition takes about ten minutes. Eggs are initially surrounded by two envelopes and are 1.8 to 2.1 mm in diameter; the diameter of egg plus inner envelope is about 3.0 to 3.3 mm. Additional complete clutches (up to five more) may be deposited by the same female on average every 24 or 25 days but in different nests. The entire process of nest building, nest guarding, and reproduction is extremely dynamic. Nests were occupied 2.7 to 100 out of 100 nights. Many nests are used only one or two nights owing to shifting of choruses, heavy rain, and avoidance of nests already containing developmental stages. The same nest may be used by different males during the course of the breeding season. Male *H. rosenbergi* continue to defend the nest during the early stages of their offspring's development until hatching. The father alternately guards and patrols the nest during this period and chases away intruding males or engages them in combat. Hatching takes place 40 to 66 hours after fertilization, and stage 25 is reached at 150 to 165 hours. At later stages larvae mill about within the nest basin and often remain at the surface gulping air and releasing bubbles. Flooding of the nest releases the now active free-swimming larvae into the main body of water. Larvae from any one nest tend to form schools that persist for several hours. Most clutches suffer high mortality (from heavy rains and disturbance by intruding males). The dead eggs and embryos, however, provide additional nutrients for the survivors in the same nest. The larvae are benthic (ecomorph 1, table 7.2) and are often cannibalistic while confined to the nest. Metamorphosis occurs approximately forty days after fertilization, and metamorphs are about 21 mm in standard length. Sexual maturity is reached in about a year. The principal predators on adult *H. rosenbergi* appear to be the large carnivorous frog *Leptodactylus pentadactylus* and the snake *Leptodeira annulata*, but several frog-eating opossums are also putatively important. Some evidence suggests that *L. pentadactylus* and the opossums use vocal clues to locate male *H. rosenbergi*. Eggs and tadpoles apparently are eaten by pisaurid fishing spiders and

water bugs. The tadpoles of the frogs *L. pentadactylus* and *Physalaemus pustulosus* and *H. rosenbergi* are primary consumers of the early life stages. *L. pentadactylus* tadpoles eat eggs whole, and *H. rosenbergi* larvae frequently cannibalize embryos and the tails of siblings while in the nest. Cannibalistic tadpoles of this species grow faster than the others, so two size-classes are often present in the nest, although all are from the same egg clutch. Höbel (2000) found striking differences in the biology of this species at La Gamba, Puntarenas Province (north of Golfito), and that reported for Panama by Kluge (1981) as summarized above. Some of these relate to the ready availability of nests (= cattle footprints, small puddles) at La Gamba. The males at this site were generally "wimps." They did not show fighting behavior, stayed in the breeding chorus only a few nights (average 8.2 versus 18.3 in Panama), had a lower mating success (35% versus 57% in Panama), and did not guard the egg clutches after amplexus. Höbel also pointed out differences in aggressive call types (two types versus six types in Panama) and the courtship call. Clutch size was also smaller (513 to 1,231) at La Gamba, being about one-third the size reported from Panama (Kluge 1981). This may relate to the much smaller size of adults at La Gamba (males 24% larger, females 21% larger in Panama). These data suggest that the Costa Rica/western Panama population may represent an undescribed species distinct from *H. rosenbergi* from farther to the east (see remarks).

REMARKS: Taylor (1954a) was the first to record *H. rosenbergi* from Costa Rica. Material from the Golfo Dulce region and adjacent western Panama represents a population widely disjunct from those in central and eastern Panama. The latter Panama examples are larger than those farther to the west (adult males 72 to 91 mm in standard length, adult females 70 to 95 mm), with a smaller tympanum and toe disks in males and the heel calcar more often present (Kluge 1979). Juveniles have green bones, and this condition may persist in adults but is not readily observed because of their thick skin and heavy musculature. All records of *Hyla rufitela* from the Pacific slope of Costa Rica and western Panama are based on larvae and juveniles of *H. rosenbergi*. The juvenile coloration of the latter resembles the color in preservative of *H. rufitela*, causing all recent workers, Savage and Heyer (1968) and Duellman (1970) among them, to misidentify juvenile *H. rosenbergi* as *H. rufitela*. Duellman's (1970) description of the tadpole of *H. rufitela* is likewise based on the larva of *H. rosenbergi*, misidentified because a metamorph from the same clutch was erroneously thought to be *H. rufitela*. Duellman's (1970) comments on possible hybridization between *Hyla boans* and *H. rosenbergi* in central Panama are based on examples of the gladiator frog *Hyla pugnax* (Schmidt 1857), a species recognized and redescribed by Kluge (1979). The karyotype is 2N = 24 (León 1969).

DISTRIBUTION: Disjunctly in the humid lowlands of

southwestern Costa Rica and adjacent southwestern Panama, central and eastern Panama, northern Colombia, northwestern Ecuador, southwestern Colombia (10–100 m).

Hyla bogotensis Group

These are moderate-sized green or pale brown frogs that sometimes have dark flecking, often a narrow pale yellow dorsolateral stripe on each side, and green bones in some forms, including the Costa Rican species. Other characteristics are considerable finger and toe webbing; no nuptial thumb pads in males, protruding prepollex, fleshy limb fringes or axillary membrane; mental gland present in adult males; vomerine tooth patches acute and angulate, extending posterior to choanae; quadratojugal articulates with maxilla; 2N not known. Tadpoles elongate with low fins, large ventral oral disk, and 4–7/7–10 denticle rows. Species included besides the two Costa Rican forms are *Hyla bogotensis, H. callipeza, H. denticulenta, H. lynchi, H. piceigularis,* and *H. simmonsi* (Colombia); *H. alytolylax* (Ecuador); *H. torrenticola* (Colombia and Ecuador); *H. phyllognatha* (eastern Andean slopes from Colombia to Peru); and *H. jahni, H. lascinia,* and *H. platydactyla* (Venezuela). Duellman recognized this group (1970), then revised and expanded it (1972a, 1989).

Eggs are deposited and the tadpoles develop in torrential streams (pattern 2, table 7.1; ecomorph 11, table 7.2).

Hyla colymba Dunn, 1931a

(plate 152; map 7.75)

DIAGNOSTICS: A moderately small stream-breeding treefrog, usually green but sometimes pale brown, with a creamy yellow stripe running along the canthus rostralis over the upper eyelid to behind the tympanum; it has webbed fingers,

white ulnar, tarsal, and supracloacal stripes, a whitish mental gland on the chin in adult males and usually in adult females, and pale green bones.

DESCRIPTION: Adult males 31 to 37 mm in standard length, adult females 31 to 39 mm; upper surfaces smooth; venter smooth to weakly granular; head broader than long, head outline semicircular; snout truncate from above, bluntly rounded in profile; eyes moderate, tympanum indistinct, internal, covered by skin, diameter less than one-half diameter of eye; paired vocal slits and a moderately distensible single external subgular vocal sac in adult males; a round whitish mental gland usually present (all males and most females); fingers moderately short and robust; finger disks small, width of disk on finger III equals diameter of tympanum; distal subarticular tubercle under finger IV usually single, sometimes bifid; no nuptial thumb pads in males; fingers weakly webbed, vestigial between fingers I and II; webbing formula: I 2½–2½ II 2–3 III 2–2 IV; prepollical pad enlarged in males, with sharp free margin, not enlarged in females; small, conical palmar tubercle present; no heel calcar; toes extensively webbed; webbing formula: I 1⁺–1⅓ II ¾–1½ III 1–2 IV 2–¾ V; low, elliptical inner metatarsal tubercle, no outer tubercle; low inner tarsal fold present. Upper surfaces usually pale green with a faint yellow stripe along canthus rostralis over upper eyelid to above axilla; often with faint yellow flecks or scattered brown dots; faint white ulnar, tarsal, and supracloacal stripes present; no transverse dark bars on limbs; venter and mental gland white to cream; undersurfaces of limbs pale green; throat pale bluish green; bones pale green; toe webs greenish tan; iris pale brown.

LARVAE (fig. 7.79): Large, 37 mm in total length at stage 25; body ovoid; mouth ventral; nostrils directed dorsally; eyes directed dorsolaterally; spiracle lateral, sinistral; vent tube dextral; tail short; caudal fin moderate; tail tip pointed; oral disk large, emarginate, with moderate beaks and 6–7/7–10 denticle rows; row above mouth and P1 with slight median gap; beaks with blunt serrations; oral disk completely bordered by two rows of small papillae. Body bronze tan with lichenous gold flecks; caudal musculature tan with brown spots; small brown spots on dorsal fin but not on ventral fin; iris pale brown.

VOICE: A series of short, rapid high-pitched cricketlike "chirp-chirp-chirps" comprising 12 to 104 notes per call group at a call rate of 123 to 236 per minute. The interval between call groups is 1 to 3 seconds, and each note lasts about 500 milliseconds. The dominant frequency is about 3.6 kHz (Duellman 1970).

SIMILAR SPECIES: (1) *Hyla palmeri* lacks the dorsolateral postorbital light stripe and has a heel calcar (fig. 7.60a). (2) Juveniles of *Hyla rufitela* have an extensive web between fingers II and III and some red in their coloration. (3) *Hyla angustilineata* lacks finger webs and has narrow dorsolateral light stripes that continue to the groin area.

HABITAT: Small stream courses within the Premontane Wet Forest and Rainforest zones.

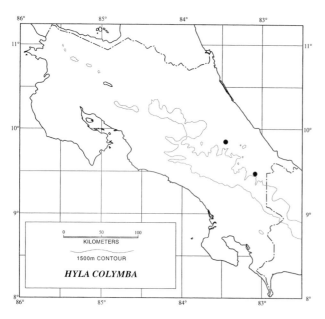

Map 7.75. Distribution of *Hyla colymba*.

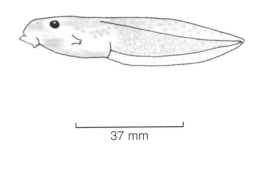

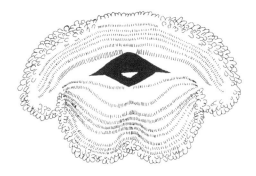

Figure 7.79. Tadpole of *Hyla colymba*.

BIOLOGY: This stream-breeding frog is a rarely seen nocturnal species. Males have been heard calling from beneath boulders in small, swift-moving streams in April, July, and December. The vocalizing frogs are very difficult to locate, and they stop calling for considerable periods at the slightest disturbance. Reproduction, however, probably is restricted to the drier times of the year. Tadpoles use the oral disk to cling more or less continuously to rocks in the swift current. Metamorphs are about 17 to 19 mm in standard length.

REMARKS: Taylor (1952a) described *Hyla alvaradoi* based on an example (KU 30886) of this form from Moravia de Chirripó, Cartago Province. Dunn (1931a) previously recognized the species as *Hyla colymba* (holotype: MCZ 10234) from material he collected in northwestern Panama (Bocas del Toro: La Loma), but he (1924b) had earlier called it *Hyla albomarginata*. Duellman (1966a) demonstrated that the two holotypes were conspecific, so Dunn's *H. colymba* is the correct name for this taxon. Duellman (1970) mentions an unusual individual from eastern Panama that he referred to this species. In life the color was yellowish tan with brown flecks above and the anterior portion of the head brown. No specimens from Costa Rica have this coloration. In preservative these frogs are pale tan to almost white; the dorsolateral light stripe and minute dark dorsal flecks usually are evident. This color change and the green bones strongly suggest that the pigment biliverdin is responsible for the green color in life (see discussion on coloration, p. 358). Dunn (1924b) noted a foamy mass of eggs found under a rock at the type locality and assumed it belonged to

this species. It seems likely that this is the case, since the males call from under rocks in the streams where the characteristic tadpoles co-occur.

DISTRIBUTION: Humid premontane Atlantic slopes of southeastern Costa Rica and western Panama (610–1,116 m) and Pacific slopes of central and eastern Panama (560–1,410 m).

Hyla palmeri Boulenger, 1908

(plate 153; map 7.76)

DIAGNOSTICS: A moderate-sized dark green frog with a heel calcar (fig. 7.60a), webbed fingers, white ulnar, tarsal, and supracloacal stripes, a whitish mental gland in both sexes, green bones, and pale orange to yellow toe webs.

DESCRIPTION: Adult males 36 to 45 mm in standard length, adult females 36–50 mm; upper surfaces smooth; venter smooth to granular posteriorly; head about as wide as long; head outline rounded, snout truncate from above and in profile; eyes moderate; tympanum indistinct, internal, covered by skin, diameter less than one-half diameter of eye; paired vocal slits and a moderately distensible single external subgular vocal sac in adult males; a round, whitish mental gland present in both sexes; fingers moderately short and robust; finger disks small, width of disk on finger III slightly larger than diameter of tympanum; distal subarticular tubercle under finger IV bifid; no nuptial thumb pads in adult males; fingers weakly webbed, vestigial between fingers I and II; webbing formula: I 2½–2½ II 2–3 III 2¼–2 IV; prepollical pad enlarged in males, with sharp free margin, not enlarged in females; small conical palmar tubercle present; dermal ridge on ventrolateral edge of forearm; a small triangular heel calcar; toes extensively webbed; webbing formula: I 1⁺–1¼ II 1–1½ III 1⁺–2 IV 2–1 V; inner tarsal

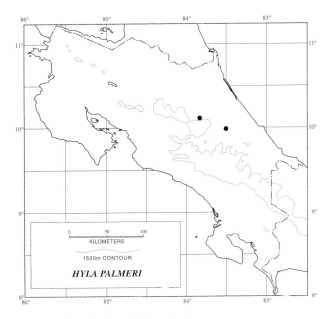

Map 7.76. Distribution of *Hyla palmeri*.

tubercle low, elliptical, no outer tubercle; low tarsal fold present; low dermal ridge on ventrolateral margin of tarsus. Dorsum dark green with irregular, sparse cream to white flecks; upper surface of limbs brown, but knees and elbows green; narrow yellowish white labial stripe; white to pale cream ulnar, tarsal, and supracloacal stripes present; no transverse dark bars on limbs; venter green, translucent, so white parietal peritoneum visible; mental gland green, sometimes with a yellowish suffusion; ventral surfaces or limbs with yellowish suffusion; toe webs pale orangish; bones green; iris silvery gray with reddish brown reticulations. In preservative, creamy white above and below, often spotted with brown on dorsum with or without a light labial stripe; plantar tubercles brown.

LARVAE: Large, total length 45.5 mm at stage 25; body ovoid; mouth ventral; nares and eyes directed dorsolaterally; spiracle lateral and sinistral; vent tube dextral; tail long; fins low; caudal fin acuminate; oral disk large, emarginate with beaks and at least 5/8 rows of denticles; one or two rows of labial papillae around entire labial disk; body brown with darker flecks; venter greenish silver; tail musculature tan with brown spots and a longitudinal stripe anteriorly; dorsum of tail orange with dark brown transverse bars; tail fins translucent with reddish brown flecks. (Larvae of this species were discovered too late for preparation of a figure.)

VOICE: A long series of short high-pitched whistles given singly and followed by paired notes for most of the rest of the call. The dominant frequency is about 2.5 kHz, each note lasts about 100 milliseconds, and the interval between calls is about 350 milliseconds (Ibáñez, Rand, and Jarmillo 1999).

SIMILAR SPECIES: (1) *Hyla colymba* has a light stripe from the eye to the axilla and lacks a heel calcar. (2) Juveniles of *Hyla rufitela* have an extensive web between fingers III and IV and red toe webs.

HABITAT: Premontane Rain Forest on the Atlantic slope.

BIOLOGY: *Hyla palmeri* is a rare nocturnal frog that breeds in rocky streams. Males call from rocks along the stream margins or from under rocks in the stream bottom. Otherwise habits are probably similar to those of *H. colymba*. The smallest metamorphs are about 20 mm in standard length.

REMARKS: Myers and Duellman (1982) resurrected this frog from the synonymy of the Amazonian *Hyla albopunctulata*. It was only recently found in Panama (Myers and Duellman 1982; Ibáñez, Rand, and Jarmillo 1999). D. C. Robinson thought the first (a female) specimen collected in Costa Rica (UCR 8152, San José: Parque Nacional Braulio Carrillo: near Rió Sucio) was a centrolenid. The green color, green bones, and white parietal peritoneal sheath contributed to that misidentification. Recent collections of this form from Guayacán, Limón Province (UCR 14155–61), confirm its occurrence in the republic. In preservative this frog resembles *H. rufitela* in having a creamy white to gray dorsal ground color often flecked with widely separated dark

brown spots, but *H. rufitela* lacks the heel calcar characteristic of *H. palmeri*.

DISTRIBUTION: Scattered lowland and premontane humid forest sites on the Atlantic slope from central Costa Rica (600–750 m) to central Panama and on the Pacific versant in the Chocó of western Colombia (100–420 m), northwestern Ecuador (550–920 m).

Hyla godmani Group

These moderate-sized frogs are pale gray, yellow, or tan to brown with the concealed leg surfaces bright red or yellow. Other definitive features include extensive finger and toe webbing; an extensive axillary web; no fleshy limb fringes, nuptial thumb pads in adult males, protruding prepollex or mental gland; vomerine teeth in transverse linear series between choanae; quadratojugal articulates with maxilla; 2N = 24. Tadpoles are lentic types with small oral disks and 2/3 denticle rows. Only one species (*Hyla godmani* of eastern Mexico) besides *Hyla loquax* belongs to this group. Reproduction in these frogs is not specialized; the eggs are deposited and the tadpoles develop in lentic situations (pattern 1, table 7.1).

This group has exactly the same composition as proposed by Duellman (1970).

Hyla loquax Gaige and Stuart, 1934

(plate 154; map 7.77)

DIAGNOSTICS: A moderate-sized robust light gray or yellow to tan treefrog unmistakable in having a yellow venter and bright red axilla, groin, anterior and posterior thighs, and webs.

DESCRIPTION: Adult males 33 to 45 mm in total length,

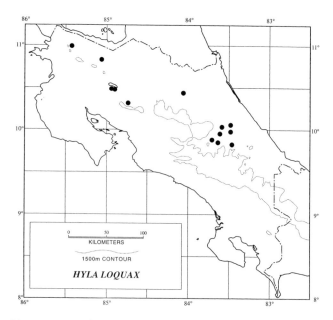

Map 7.77. Distribution of *Hyla loquax*.

adult females 38 to 47 mm; upper surfaces smooth; venter granular; head slightly wider than long; snout bluntly rounded from above and rounded in profile; eyes large; tympanum distinct, large, diameter about one-half to two-thirds of eye diameter in males, one-half or less in females; paired vocal sacs and moderately distensible single internal subgular vocal sac in adult males; extensive axillary membrane present; fingers short and robust; finger disks moderate, disk on finger III equal to or larger than diameter of tympanum; subarticular tubercles all single; no nuptial thumb pads in males; fingers moderately webbed; webbing formula I 2–3 II 1$^+$–2 III 2$^-$–1$^+$ IV; prepollical pad slightly enlarged; an elevated tripartite palmar tubercle; toes extensively webbed; webbing formula: I 1–1 II 1–1$^+$ III 1–1½ IV 1½–1 V; flat, elongate inner metatarsal tubercle, outer small, subconical; short, weak tarsal fold present. Upper surfaces usually yellow to reddish brown at night, yellow or pale gray, almost white, by day; some small dark brown flecks or mottling often present; no transverse dark bars on limbs; concealed surfaces and webs bright red; venter lemon yellow; iris reddish tan.

LARVAE (fig. 7.80): Large, 45 mm in total length at stage 25; body ovoid; mouth anteroventral; nostrils and eyes directed dorsolaterally; spiracle lateral, sinistral; vent tube dextral; tail moderate, acuminate; caudal fins high; tip of tail pointed; oral disk small, complete with moderate beaks; 2/3 rows of denticles; A2 broadly interrupted medially, P2 with a narrow median gap; P3 extremely short, much shorter

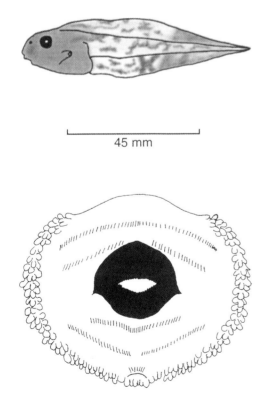

45 mm

Figure 7.80. Tadpole of *Hyla loquax*.

than width of beaks; beaks finely serrate; two rows of papillae around disk except above mouth, additional row or rows at level of angle of jaws; body light with dark spot between eyes and V-shaped dark mark from snout pointed toward spot between eyes; venter metallic silver or copper; caudal musculature mottled with silver; fins mottled with dark pigment becoming uniform gray on posterior two-thirds of tail; iris orange.

VOICE: A series of "kaaack" or "wonk" notes somewhat resembling the honking of geese. Notes are repeated at intervals of 900 milliseconds to 5 seconds, last 50 to 160 milliseconds, and have a repetition rate of 9 to 62 per minute; the dominant frequency is between 2.1 and 2.4 kHz, and the call is poorly modulated (Duellman 1970; Lee 1996).

SIMILAR SPECIES: The much smaller *Hyla phlebodes* (adult males 19 to 24 mm in standard length, adult females 23 to 28 mm) usually has a dark dorsolateral stripe, a dark canthal stripe (fig. 7.73e), and interorbital stripes and lacks red in its coloration.

HABITAT: Typically found in and around temporary ponds in Lowland Moist and Wet Forest zones and marginally in Premontane Wet Forest and Rainforest.

BIOLOGY: Males of this relatively uncommon nocturnal species call throughout the wet season, usually from emergent vegetation or leaves of floating plants in the deepest parts of the breeding ponds. Although not an explosive breeder, its reproduction peaks in the middle of the rainy season in July and early August. Amplexus also takes place in deep water, and eggs presumably of this species (but see remarks) form large gelatinous masses attached to vegetation near the water surface; one such clutch contained 250 eggs (Duellman 1970). The tadpoles are nektonic (ecomorph 2, table 7.2) and inhabit the deepest parts of the pond away from shore.

REMARKS: Duellman's (1970) illustrations and description of the larva of this species are based on misidentified material of a ranid frog easily recognizable by the extremely broad area between the mouth and the posterior margin of the lower jaw. Lee (1996) has recently described and illustrated the tadpole of *H. loquax*, which like other *Hyla* has a narrow lower jaw.

DISTRIBUTION: Humid lowlands and premontane slopes of the Atlantic versant from southern Veracruz, Mexico, to south-central eastern Costa Rica (60–1,116 m).

Hyla lancasteri Group

Frogs of this group are moderately small stream breeders having the upper surfaces mottled with shades of brown or copper and green or gray and the toes extensively webbed. Additional features characteristic of the group are fingers with moderate or reduced webbing; dark nuptial thumb pads present but no protuberant prepollical spines in adult males; no fleshy limb fringes, axillary membrane, or mental gland; vomerine tooth patches extending posterior to choanae, linear, transverse to obtusely angulate; quadrato-jugal articulates with maxilla; 2N not known. Tadpoles elongate with moderate-sized oral disks and 2/3 denticle rows. The only other species referred to this group besides the two Costa Rican forms is *Hyla insolita* of Honduras.

Most species in this group have an unspecialized reproductive pattern (pattern 2, table 7.1), but one species, *Hyla calypsa*, deposits its eggs on vegetation overhanging torrential streams (pattern 22, table 7.1).

Duellman (1970) proposed this group for *H. lancasteri*, the only species described up to that time.

Hyla calypsa Lips, 1996b

(plate 155; map 7.78)

DIAGNOSTICS: An eye-catching moderately small bright metallic green treefrog mottled or marked with dark olive green to brown blotches and having the upper surfaces covered by large tubercular spines.

DESCRIPTION: Adult males 26 to 36 mm in standard length, adult females 31 to 41 mm; upper surfaces of head, body, and limbs covered with large tubercular spines; females much spinier than males; venter strongly granular; head broader than long; snout short, truncate in dorsal outline and profile; eyes large; tympanum distinct, diameter about four-fifths diameter of eye; paired vocal slits and fully distensible single external subgular vocal sac in adult males; fingers short and thick; finger disks moderate, width of disk on finger III slightly smaller than diameter of tympanum; distal subarticular tubercle usually bifid under finger IV; male nuptial pad at base of thumb composed of large black spines; a series of warts along outer margin of forearm; webs of hands and feet strongly areolate; fingers weakly webbed, web vestigial between fingers I and II; webbing formula:

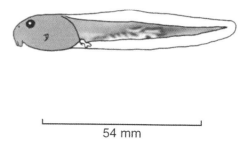

54 mm

Figure 7.81. Tadpole of *Hyla calypsa*.

I 2½–3 II 2–3 III 2¾–2¼ IV; prepollical pad moderately enlarged; no palmar tubercle; a series of spines along outer margin of tarsus and foot; toes extensively webbed; webbing formula: I 2–2¼ II 1⁺–1½ III 2¾–1½ IV 2¾–1½ V; inner metatarsal tubercle low, flat, elliptical, outer small, subconical; an inner tarsal fold present. Upper surfaces bright metallic green with few large dark olive green to brown blotches in males, dark mottling in females; lips barred green and dark brown; groin bright white with a few black spots; irregular transverse dark bars on upper surface of limbs; anterior and posterior thigh surfaces gray to white with small black spots; venter dirty white with numerous black flecks and one or two large black blotches; iris cream with irregular brown line around margin.

LARVAE (fig. 7.81): Large, 54 mm in total length at stage 37; body depressed; mouth ventral, nostrils dorsal; eyes dorsolateral; spiracle lateral, sinistral; vent tube dextral; tail long, fins low; tail tip rounded; oral disk small with small beaks and 2/3 rows of denticles; A2 with narrow gap above mouth; beaks with large, pointed serrations; disk bordered on all margins by two or three rows of small papillae, with many additional papillae lateral to mouth; a series of large submarginal papillae also present; body uniform olive brown suffused with orange; tail musculature same color as body with darker gray blotches on midline and anterior half of tail, posterior portion with large black spots; fins translucent.

VOICE: Usually a short one-note call rising in frequency at the end, lasting 110 to 330 milliseconds. The dominant frequency is about 2.9 kHz, with no additional harmonics.

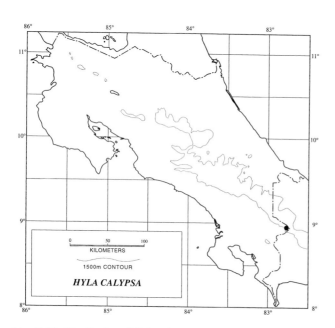

Map 7.78. Distribution of *Hyla calypsa*.

Sometimes two- or three-note calls are produced (Duellman 1970; Lips 1996a,b).

SIMILAR SPECIES: No other Costa Rican frog is covered by spines on the upper surfaces of the head, body, and limbs.

HABITAT: Stream banks in Lower Montane Rainforest.

BIOLOGY: Lips (1996a,b) studied the Costa Rican population of this species during twenty-four months over a four-year period, and the account below is based on her report. The species is a locally common nocturnal leaf-breeding treefrog that has a prolonged reproductive period (April to December). Individuals are present along torrential stream courses throughout the year, although some males call less frequently during the dry season (January to mid-April). Males are strongly territorial and have high calling-site fidelity, often calling from the same spot over a span of years. When calling they alternate calls with their nearest calling neighbor and usually call from vegetation 1 to 2 m directly above the stream. Their home ranges are small, usually about 6 m². Intermale spacing is relatively constant, averaging about 1.9 m. Females, when present, are found on lower vegetation, leave the area after oviposition, and also show considerable site fidelity. Most females returned to the same general area to breed each year. Females, juveniles, and a flush of males all arrived at the breeding stream two weeks before the start of the rainy season. Juveniles disappeared after two days, and females did not mate until the first rains. Amplexus and oviposition in this species take place near the male's calling site but usually somewhat lower. Females probably lay only one clutch per visit and up to three clutches during the breeding season. Clutches consist of 10 to 36 creamy yellow eggs deposited in a mass on leaf surfaces (pattern 22, table 7.1). Eggs average 3.6 mm in diameter and are contained in jelly capsules averaging 5.3 mm in diameter. The early embryos are light chocolate and hatch in 23 to 56 days after fertilization. Average total length of hatchlings was about 15 mm. On hatching, tadpoles fall or are washed by rain into the stream below. They are of the mountain brook ecomorph (guild 11, table 7.2), without an extremely specialized oral disk, and inhabit slow-moving portions of the stream. In the laboratory tadpoles metamorphosed in about 270 days after fertilization. Metamorphs are about 15 mm in standard length. The maximum life span based on recapture data was estimated as three years, with sexual maturity probably reached during the first year of life. This species, which was relatively abundant in the early 1990s, showed a substantial population decline from 1992 onward (Lips 1998).

REMARKS: Trueb (1968) originally described the Cerro Pando (Costa Rica–Panama border) population (1,920 m). However, she interpreted a population at 1,450 m as intermediate between this form and the lower-elevation species *Hyla lancasteri* and referred all examples to the latter taxon. Early in her study of this species Lips (1993e) also referred to this species as *H. lancasteri*. Lips (1996a,b) demonstrated

consistent differences in morphology, coloration, advertisement calls, and life history between *H. lancasteri*, including Trueb's 1,450 m sample, and the higher-elevation species and gave the latter the name *H. calypsa*. She further clarified the situation by showing that some of Trueb's published data were in conflict with Trueb's original field notes, which had led to the incorrect hypothesis of intergradation between the two taxa. Myers and Duellman (1982) also regarded these forms as conspecifics when reporting on material from western Panama. The population they noted from Fortuna, Chiriquí Province, was studied in some detail by Lips (1996b) and unequivocally referred to *H. lancasteri*.

DISTRIBUTION: Humid lower montane rainforest of the southern Cordillera de Talamanca, on Cerro Pando on the Pacific slope in Costa Rica and Atlantic versant in Panama, and on the Pacific slope in southwestern Panama (1,810–1,920 m).

Hyla lancasteri Barbour, 1928

(plate 156; map 7.79)

DIAGNOSTICS: This moderately small frog mottled with green and light brown has a contrasting pattern of black and bright yellow bars on the posterior thigh surface (fig. 7.83a), an extremely short and truncate snout (from above and in profile) (fig. 7.6b,c), and upper surfaces without spines.

DESCRIPTION: Adult males 27 to 34 mm in standard length, adult females 31 to 38 mm; upper surface of head, body, and limbs smooth or with scattered low, rounded warts; venter granulate; head slightly broader than long; snout very short and truncate from above and in profile; eyes moderate; tympanum distinct, diameter about one-half diameter of eye; paired vocal slits and fully distensible single external subgular vocal sac in adult males; fingers short and

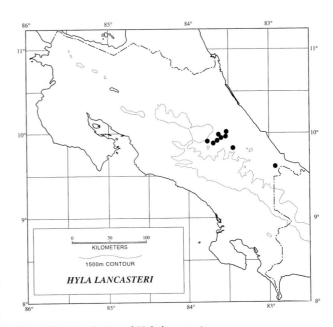

Map 7.79. Distribution of *Hyla lancasteri*.

thick; finger disks small, width of disk on finger III about equal to diameter of tympanum; distal subarticular tubercle on finger IV often bifid; black nuptial pad in adult males composed of tiny spinules on base of thumb; row of low warts along outer margin of forearm; webs of hands and feet smooth or barely areolate; fingers barely webbed, web vestigial between fingers I and II; webbing formula: I 3–3 II 1⅔–2⅔ III 2⅓–2 IV; prepollical pad moderately enlarged in males, not enlarged in females; low, bipartite or tripartite palmar tubercle; low warts along outer margin of tarsus and foot; toes extensively webbed; webbing formula: I 1½–2⁺ II 1–2⁺ III 1¼–2 IV 2–1 V; inner metatarsal tubercle low, flat, elliptical; no outer tubercle; weak, short inner tarsal fold present. Upper surfaces generally tan with pale to metallic green mottling; lips with dark bars; upper surface of upper arm uniform yellow; groin, anterior and posterior thighs, upper surfaces of thighs, and tarsal segment bright yellow with contrasting black bars; venter white with irregular dark spots medially and larger blotches laterally; iris reddish brown.

LARVAE (fig. 7.82): Moderate-sized, 43 mm in total length at stage 37; body depressed; mouth ventral; nostrils directed anterolaterally; eyes directed dorsolaterally; spiracle lateral, sinistral; vent tube dextral; tail short; fins moderate, tail tip rounded; oral disk small, with moderate beaks and 2/3 rows of denticles; A2 with narrow median gap; beaks with long, pointed serrations; two or three rows of small papillae completely around oral disk, many additional papillae lateral to mouth; series of large submarginal papillae also present; body uniform olive green brown suffused with orange yellow; tail suffused with orange, usually with

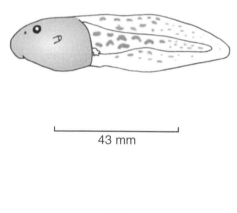

43 mm

Figure 7.82. Tadpole of *Hyla lancasteri.*

black markings along dorsal half to one-third of fins and musculature, appearing transversely banded from above.

VOICE: Usually a low-pitched, vibrating extended one-note call having an average duration of 200 to 350 milliseconds and repeated 7 to 19 times a minute. Two frequencies are emphasized, at about 1.6 and 2.2 kHz. Sometimes a second, somewhat squeaky note follows the first (Duellman 1970; Myers and Duellman 1982; Lips 1996a,b).

SIMILAR SPECIES: *Scinax boulengeri* is larger (adult males 36 to 49 mm in standard length, adult females 44 to 52 mm), has a long snout that is rounded in profile (fig. 7.6c), and lacks webs between the fingers.

HABITAT: Found in Premontane Wet Forest and Rainforest along small streams or sometimes near shallow pools in pastures or roadside ditches.

BIOLOGY: A relatively uncommon nocturnal stream-breeding treefrog that may be heard calling most of the year but reproduces primarily in the drier months (January to March) when stream flow is reduced. Males form breeding aggregations and call from low vegetation and rocks in or along streams, where they favor quiet backwater pools or pools in roadside ditches. Males alternate calls with their nearest calling neighbors. Eggs are laid in water in single long strands of 70 to 80 chocolate brown eggs, each separated from the others by partitions in the sticky string. The eggs average 2.6 mm in diameter. After two to three days the jelly breaks down, and the eggs sink to the bottom and develop among the debris on the bottom of the pool. Hatching occurs about nine to ten days after fertilization at a total length of 9 to 11 mm, and the chocolate brown larvae remain hidden on the bottom. The tadpoles characteristically use the oral disk to maintain position in the stream but are somewhat intermediate in features between ecomorphs 10 and 11 (table 7.2). They have the body and tail form typical of the former, but the oral disk structure of the latter (multiple rows of small papillae completely around the oral disk). Laboratory-reared larvae metamorphosed seventeen weeks after fertilization at a standard length of 13 to 15 mm.

REMARKS: This species was originally described by Barbour (1928) from a juvenile (MCZ 13062; Costa Rica: Cartago: Peralta). For a long time it was thought to be a synonym of *Hyla*, now *Scinax boulengeri*, following Dunn and Emlen (1932), because of the barred leg pattern. Subsequently Taylor (1952a) questioned this decision and described a putative new species with barred legs as *Hyla moraviensis* (KU 30284, an adult male; Costa Rica: Cartago: Moravia de Chirripó). After comparing the type specimens, Duellman (1966a) concluded that only one species was involved and rejected the notion that *H. lancasteri* had any relation to *S. boulengeri.* Trueb (1968) discussed the species at length, but as described in the account of *Hyla calypsa,* she included that taxon within the variational range of *H. lancasteri.* Lips (1996b) clarified the situation by describing the high-elevation leaf-breeding populations as *H. calypsa* to

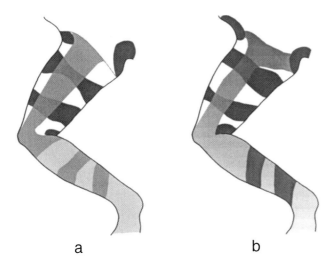

Figure 7.83. Thigh patterns in Costa Rican hylid treefrogs: (a) in *Hyla lancasteri* the light interspaces on the thighs are yellow; (b) in *Scinax boulengeri* the light interspaces on the thighs are yellowish green or yellowish orange.

distinguish them from the lower-elevation stream-breeding *H. lancasteri*. However, as documented by Trueb (1968), Duellman and Myers (1982), and Lips (1996a,b), there is an altitudinal cline in features of morphology and coloration within *H. lancasteri*. At low elevations (368 to 1,200 m in Costa Rica; 70 to 830 m in Panama) this species has smooth dorsal surfaces and distinct black and yellow bars on the thighs. At higher elevations in Panama (910 to 1,450 m) the upper surfaces are covered with low, rounded protuberances, and the thighs are grayish with some black markings. No specimens with this morphology and coloration have been taken in Costa Rica. All these frogs share the features of nuptial pads, advertisement calls, strings of eggs laid in water, and tadpole morphology, as compared with the lower montane species *Hyla calypsa*. That form also differs from *H. lancasteri* in having the upper surfaces covered by large tubercles and spines. Specimens originally referred to *H. lancasteri* by Lips (1993e) were later used as a partial basis in describing *H. calypsa* (Lips 1996b). Anderson (1991) reported a karyotype for *H. lancasteri* as 2N = 24, all metacentrics or submetacentrics; NF = 48. It is unclear whether her material was of this species or *H. calypsa*.

DISTRIBUTION: Humid premontane slopes of the Cordillera de Talamanca of Costa Rica and western Panama (368–1,200 m).

Hyla leucophyllata Group

This group includes a variety of frogs with brightly colored thighs or thigh patterns and often a bright white, yellow, or tan dorsum. All members of the group have tadpoles that lack denticles but have beaks, one row of labial papillae or none, and a tail fin that terminates in a long filament. Other features shared by species placed here are fingers

and toes with considerable webbing; usually no nuptial thumb pads and no protuberant prepollices in adult males; axillary membrane present; no fleshy limb fringes or mental gland; vomerine teeth in linear series; quadratojugal reduced or absent; 2N = 30. Three subgroups may be recognized within this group based on larval features and were considered separate species groups by Duellman (1970) and Duellman and Trueb (1989). These are:

Hyla leucophyllata subgroup: one row of large labial papillae; body violin shaped from above. Species included besides the single Costa Rican representatives are *Hyla elegans* (eastern and southeastern Brazil); *H. bifurca* and *H. triangulum* (upper Amazon basin); and *H. leucophyllata* (Amazon basin, the Guianas to southeastern Brazil).

Hyla microcephala subgroup: no oral papillae; body depressed. Other species included besides the two Costa Rican forms are *Hyla gryllata* (western Ecuador); *H. mathiassoni* (eastern Colombia); *H. rhodopepla* (upper Amazon basin); *H. robertmertensi* (southeastern Mexico to El Salvador); and *H. sartori* (southwestern Mexico).

Hyla parviceps subgroup: one row of large labial papillae; body ovoid. No members of this subgroup occur in Costa Rica; the range is from Panama south throughout the humid Neotropics.

Members of these subgroups share macrophagous tadpoles (ecomorph 7, table 7.2). They differ in reproductive patterns. That of *H. microcephala* and its allies and the *parviceps* subgroup is unspecialized (pattern 1, table 7.1), but members of the *H. leucophyllata* subgroup lay eggs on vegetation overhanging lentic habitats (pattern 22).

The *leucophyllata* species group is equivalent to the *leucophyllata* species group of Savage and Heyer (1968) sans *Hyla loquax*.

Hyla ebraccata Cope, 1874

(fig. 7.85a; plates 157–58; map 7.80)

DIAGNOSTICS: This moderately small frog is a jungle jewel much favored as a photographic subject because of its usually vivid dorsal coloration of one or more bold dark brown blotches, usually broadly bordered by white or yellow, a light yellow labial stripe that is usually expanded into a large suborbital light spot (fig. 7.73a), and bright yellow to orange thighs. Usually the dark dorsal figure is roughly hourglass-shaped or is broken into a large triangular occipital and a large irregular sacral blotch enclosing a yellow spot. In some individuals the dark markings are reduced to small spots or flecks on a yellow tan background or are uniform yellow tan above. This species has a well-developed axillary membrane (fig. 7.70d), extensive finger webbing, and usually a dark brown lateral band extending from the tympanum at least to the midbody.

DESCRIPTION: Adult males 23 to 27 mm in standard length, adult females 30 to 35 mm; upper surfaces smooth;

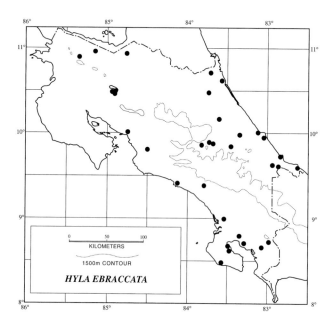

Map 7.80. Distribution of *Hyla ebraccata*.

venter granular; head slightly broader than long; snout short, truncate in dorsal outline, rounded to slightly acute in profile; eyes large; tympanum distinct, diameter one-fourth to two-fifths diameter of eye; paired vocal slits and fully distensible single external subgular vocal sac in adult males; extensive axillary web present; fingers short and broad; finger disks large, width of disk on finger III slightly greater than diameter of tympanum; distal subarticular tubercle under finger IV usually bifid; no nuptial pads; finger webs extensive; webbing formula: I 2–3 II 1$^+$–2$^+$ III 2$^+$–2$^-$ IV; prepollical pad poorly developed, elongate; palmar tubercle flat, bifurcate; toes extensively webbed; webbing formula: I 1–2$^-$ II ½–1 III 1–2 IV 2$^+$–1 V; inner metatarsal tubercle low, flat, elliptical, outer tiny; inner tarsal fold present. Dorsum pale yellow, yellowish tan, or white with dark brown markings (diurnal) or pale yellow tan with markings obscure (nocturnal); side of head dark brown with narrow white to yellow upper lip stripe expanded under eye to form suborbital spot; usually dark brown band along flank from eye to groin bordering lighter dorsal areas, often bordered above by narrow yellow stripe (fig. 7.85a); flanks, upper arm, hands, and feet pale yellow; thighs clear yellow to orange; vocal sac bright yellow; venter creamy white; iris reddish brown to bronze with red flecks.

LARVAE (fig. 7.84): Moderate-sized, 29 mm in total length at stage 36; body ovoid; mouth terminal; nostrils dorsolateral; eyes directly laterally; spiracle lateral, sinistral; vent tube median; tail long; fins high; tail with thin terminal flagellum; oral disk small, complete with large lower beak and thin upper beak; no denticles or upper labium; lower lip fleshy with one row of large papillae; beaks with fine serrations; dorsum dark brown mottled with tan; sides and an-

terior part of tail gold; venter white; fins marked with three large black spots dorsally and ventrally; area between spots bright red; tail flagellum barred with black; iris red centrally, bronze laterally (plate 159).

VOICE: A prolonged insectlike "wreek" sometimes followed by one to four secondary "click" notes. The primary notes have a duration of 90 to 290 milliseconds, a call rate of 5 to 25 times per minute, a pulse rate of 85 to 110 per second, and a dominant frequency of 2.5 to 3.4 kHz (Duellman 1970; Wells and Schwartz 1984a; Lee 1996).

SIMILAR SPECIES: (1) *Hyla microcephala* has reduced finger webbing, no broad lateral dark band from tympanum to at least midbody, and no suborbital light spot (fig. 7.73d). (2) *Hyla phlebodes* has reduced finger webbing, never has the bold brown and yellow pattern of most *H. ebraccata*, and lacks a suborbital light spot (fig. 7.73e).

HABITAT: Typically found around temporary ponds in Lowland Moist and Wet Forests and Premontane Wet Forest but also in disturbed situations in these zones and marginally in Premontane Rainforest.

BIOLOGY: These beautiful frogs are among the most commonly seen and heard species in the humid lowlands. At many sites males are heard calling nightly throughout the wet season. Calling males are usually heard between 6:00 and midnight, with a peak around 10:00. Breeding aggregations develop several times during the wet season between June and August. Individuals are not active during the day but hide in vegetation near the ponds. Males usually call from emergent vegetation in the pond or from bushes and trees overhanging it and are spaced 1 to 2 m apart. Calls

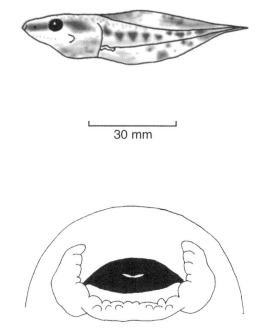

30 mm

Figure 7.84. Tadpole of *Hyla ebraccata*.

overlap (are synchronous) with the calls of both conspecific and heterospecific males. Males defend their calling sites from intruders by vocal challenges or physical contact. Generally encounters involve exchanges of aggression calls by males only a few centimeters apart. These calls are similar to the advertisement calls in number of notes, but because of their higher pulse rates (140 to 315 per second) they produce a different sound (a squeak). On occasion males engage in pushing contests or grapple with one another to maintain their calling sites. Aggression calls are also elicited by calling males of syntopic *Hyla microcephala* and *Hyla phlebodes*, especially the former, which breed at the same time and place as *H. ebraccata*. However, loud choruses of *H. microcephala* inhibit calling male *H. ebraccata*. Male *H. ebraccata* respond to the advertisement calls of both conspecific and heterospecific males by increasing their call rate and adding secondary notes to produce complex multinote calls. Females prefer three-note calls over less complex ones. Females usually move into the chorus between 6:00 and 10:00 P.M. A female may sit for several hours before selecting a mate. A female attracted to a particular calling male first tilts her head in his direction and jumps toward him, then tilts her head and jumps again, continuing this sequence until she is within 60 to 300 mm. At that time she rotates her flanks toward him. His call frequency and intensity increase as she approaches but cease when she rotates her body. Almost immediately he leaps toward her, positions himself parallel to her, and shortly mounts and amplexes her (Miyamoto and Cane 1980a). In dense choruses of this species, silent satellite males sit in spaces between calling males and intercept and amplex females en route to a calling male (Miyamoto and Cane 1980b). Axillary amplexus takes place out of the water, and eggs are laid above the water in a single layer on leaf surfaces, usually of herbaceous plants emergent in the pond. Clutch sizes are 15 to 296, and the eggs are yellow with a small black area at one pole. Embryos become light brown as they develop. Eggs containing embryos at the yolk-plug stage were 1.2 to 1.4 mm in diameter. On hatching the tadpoles are about 6 mm in total length; they wiggle out of the jelly mass and fall into the water. Egg mortality is principally from desiccation, and survival from egg to larvae at La Selva, Heredia Province, was 78% (Roberts 1994a). The pelagic, macrophagous tadpoles (ecomorph 7, table 7.2; plate 159) cruise through shallow weedy areas of the pond and transform in another four to six weeks. Experiments by Valerio (1971) showed that the tadpoles could survive out of water for only thirteen hours. Juveniles are found in greatest numbers in the early dry season (January), but a smaller flush appears in August to September (Donnelly and Guyer 1994). Eggs were shown in laboratory tests to be palatable to conspecific and heterospecific tadpoles and a turtle, *Kinosternon leucostomum* (Roberts 1994a). Co-occurring fishes did not eat eggs of this species.

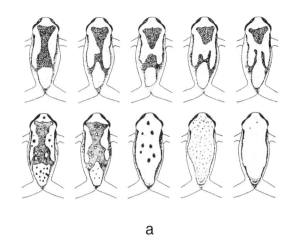

a

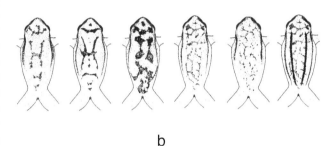

b

Figure 7.85. Dorsal color pattern variation in Costa Rican hylid treefrogs: (a) *Hyla ebraccata*; (b) *Hyla microcephala*.

REMARKS: Taylor (1954a) described *Hyla weyerae* based on a specimen (KU 34850) of this species from the Golfo Dulce region that had a uniform yellow dorsum. Duellman (1966b) pointed out that this occasional variant occurs throughout the range of *H. ebraccata* and synonymized *H. weyerae*. Wells and his associates (Wells and Greer 1981; Wells and Schwartz 1984a,b; Schwartz and Wells 1984a; Wells and Bard 1987; Wells 1989) have studied male vocalization in this species in great detail, and only the bare minimum of their results has been included in this account. Further data on vocalization are in the remarks section of the *H. phlebodes* account. The karyotype of *H. ebraccata* is 2N = 30 (Duellman and Cole 1965; Kaiser et al. 1996).

DISTRIBUTION: Disjunctly distributed in the humid lowlands and adjacent premontane slopes on the Atlantic slope: from southern Mexico to Belize; from northeastern Honduras to western Panama, on the Pacific versant: from central Costa Rica to western Panama; in eastern Panama and northern and western Colombia; and on both slopes in central Panama. In Costa Rica it occurs primarily in the humid lowlands on both versants but also ranges onto the humid premontane Atlantic slope (2–1,320 m).

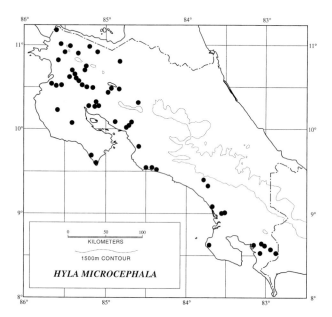

Map 7.81. Distribution of *Hyla microcephala*.

Hyla microcephala Cope, "1885," 1886

(fig. 7.85b; plate 160; map 7.81)

DIAGNOSTICS: A small slender yellowish tan treefrog having minimal finger webbing, a well-developed dorsolateral white stripe bordered below by a narrow dark brown stripe running from the eye above the tympanum past midbody (fig. 7.73d), and uniform yellow thighs, while lacking a suborbital light spot.

DESCRIPTION: Adult males 18 to 25 mm in standard length, adult females 24 to 31 mm; upper surfaces smooth; venter granular; head slightly longer than broad; snout rounded to subovoid from above, acutely rounded in profile; eyes moderate; tympanum indistinct, diameter one-third to three-fifths diameter of eye; paired vocal slits and fully distensible single external subgular vocal sac in adult males; an abbreviated axillary web present; fingers short and broad; finger disks small, width of disk on finger III about three-fourths diameter of tympanum; distal subarticular tubercles on fingers III and IV usually bifid; white nonspinous nuptial thumb pads present in breeding males but absent in juvenile and nonbreeding adult males; finger webs reduced, barely indicated between fingers I and II; webbing formula: I 3–3 II 1⅔–3⁻ III 2⁺–2⁺ IV; prepollical pad not enlarged; no palmar tubercle; toes extensively webbed; webbing formula: I 1–2⁻ II 1–2 III 1–2 IV 2–1 V; inner metatarsal tubercle large, oblong, outer obscure or absent; weak inner tarsal fold present. Dorsum tan to yellow with darker brown fragmented network of lines often fusing to form H-shaped mark or larger blotches; narrow lateral dark stripe from nostril to midbody, sacrum, or groin, bordered above by narrow white line (fig. 7.85b); coloration paler at night than during day; flanks tan to reddish brown; axilla, groin, and

thighs yellow; lower leg flecked with dark or with dark transverse bars above; vocal sac yellow; venter white; iris bronze.

LARVAE (fig. 7.86): Small, 24 mm in total length at stage 34; body ovoid; mouth terminal; nostrils dorsal; eyes directly laterally; spiracle lateral, sinistral; vent tube dextral; tail short to moderate; dorsal fin high, ventral fin low; tail with thin terminal flagellum; oral disk small, complete, with large lower beak, thin upper; no denticles or papillae; beaks finely serrate; upper surfaces tan; dark stripe from snout through eye, may extend onto tail; venter white; anterior half of tail unpigmented, posterior half deep orange in life; iris bronze.

VOICE: An insectlike "creek-eek-eek-eek" ("buzz-click-click-click"). The introductory note lasts 34 to 104 milliseconds, with pulse rates of 214 to 274 per second. The dominant frequencies are about 2.9 and 6.0 kHz. Call rates are 5 to 32 times per minute. Typically the primary note is given alone at the beginning of the call series. Later the primary note is followed by up to eighteen secondary notes (usually about four). These are usually biphasic and last 48 to 88 milliseconds (Schwartz and Wells 1984a, 1985).

SIMILAR SPECIES: (1) *Hyla phlebodes* lacks the narrow white dorsolateral stripe, and the dark dorsolateral stripe, if present, extends no farther posteriorly than the axilla (fig. 7.73e). (2) *Hyla ebraccata* has extensive finger webs, a broad dark band from tympanum at least to midbody, and a suborbital light spot (fig. 7.73a). Most *H. ebraccata* also have the back marked by large dark brown blotches.

HABITAT: Lowland Moist, Wet, and Dry Forests and Premontane Wet Forest and Rainforest zones. Not normally seen within the forest, but congregates around temporary

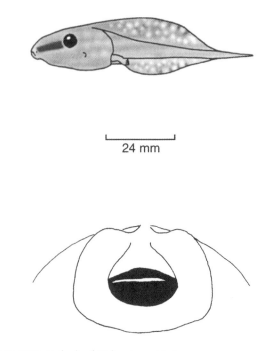

24 mm

Figure 7.86. Tadpole of *Hyla microcephala*.

ponds in open areas, especially secondary growth, pastures, and roadside ditches.

BIOLOGY: Nocturnal breeding aggregations of this very common species are formed throughout the wet season in response to heavy rains. Males call from perches on tall grass or reeds in or at the margins of the breeding site. Calling usually begins at dusk and continues until near dawn, with a peak between 9:00 and 11:00. Hundreds of males form choruses, particularly after torrential rains in the sub-humid seasonal forest zone in northwestern Costa Rica. In the humid lowlands of southwestern Costa Rica similar large choruses are commonly found in roadside ditches. In this region males may call singly throughout the rainy season but are inactive the rest of the year (January to February or March). Male advertisement calls of this species are generally synchronous. The addition of clicks and an increase in calling rates are responses to the calls of conspecific and heterospecific males *(H. ebraccata* and *H. phlebodes)*. The calls are unusual in that the different notes in the call group are given alternately with the notes of both conspecific and heterospecific nearby neighbors. Males of this species use an aggressive vocalization to space themselves from both conspecific and heterospecific intruders, especially *Hyla ebraccata* and *H. phlebodes* (Schwartz and Wells 1984a, 1985). This call has a higher pulse repetition rate than the advertisement call and lower dominant frequencies (2.6 and 5.6 kHz). Both conspecific advertisement and aggressive calls elicit increased aggressive calling. Aggressive calls with a long introductory buzz are used in close vocal or physical encounters with *H. ebraccata*. For additional detail on vocalizations in this species see Schwartz and Wells (1985) and the remarks section of the *H. phlebodes* species account below. As in all Central American hylids, amplexus is axillary and takes place in the shallows. Eggs are laid in small clumps, usually attached to vegetation near the water surface. Tadpoles are nektonic macrophages (ecomorph 7, table 7.2). Metamorphs are 16 to 17 mm in standard length.

REMARKS: Taylor (1952a) listed this frog as *Hyla underwoodi* and later referred other material to *H. microcephala* (1958a), since he regarded them as distinct species. Duellman and Fouquette (1968), Savage and Heyer (1968), and Duellman (1970) concluded that only one species was involved. In Mesoamerica geographic differences are evident in coloration in this species as follows:

Southwestern Costa Rica and Panama: the dorsolateral dark and light stripes do not extend posterior to the groin, usually a dark interorbital mark is present, and the tibial segment of the leg usually is marked with narrow transverse dark bars (the name *H. microcephala* Cope, "1885," 1886, syntypes: USNM 113473a,b; Panama: Chiriquí) is based on this population (fig. 7.85b).

Northwestern Costa Rica to Mexico: the dorsolateral dark and light stripes extend to the level of the groin, there is usually no dark interorbital mark, and the tibial segment

of legs has dark spots without transverse bars (the name *H. microcephala* Boulenger 1898c [*nec* Cope, 1866] and its replacement name, *H. underwoodi* Boulenger, 1899; syntypes: BMNH 1947.2.23.28–29; Costa Rica: Guanacaste: Bebedero) is based on frogs having these features (fig. 7.85b). Examples from the Pacific lowlands of central Costa Rica are intermediate in features between these extremes, indicating that only one species is involved. Fouquette (1968) pointed out that *Hyla misera* Werner, 1903a of South America was a synonym of *H. microcephala*. Lutz (1973) continued to recognize *H. misera* and several putative but poorly differentiated subspecies as distinct. Duellman (1974) regarded these as conspecific with *H. microcephala* although continuing to recognize five nominal subspecies (two from Mesoamerica). Although there is some question whether all the populations (Mexico to southern Brazil) represent the same species, that course is followed here. The geographic distribution pattern of *H. microcephala* and its ally *H. phlebodes* is unusual. From Mexico to eastern Nicaragua the former is an Atlantic slope species, but from Nicaragua south to eastern Panama it is a Pacific versant form. *H. microcephala* is found in both humid and subhumid habitats on the Atlantic slope, subhumid situations in southern Nicaragua and northwestern Costa Rica, and humid habitats in southwestern Costa Rica and Panama. *H. phlebodes* essentially replaces *H. microcephala* in the humid Atlantic lowlands of Costa Rica and Panama. The two species are sympatric at a few sites in southeastern Nicaragua, in northern Costa Rica on both slopes of the Cordillera de Guanacaste, and in central and eastern Panama. Savage and Heyer (1968) presented the following evidence for synonymizing *Hyla cherrei* Cope, 1894a (only known specimen lost; Costa Rica: Alajuela). The significant characteristics of the holotype of *H. cherrei* are vomerine teeth in two patches between very large choanae; snout not prominent, vertical in profile; canthus rostralis sharp, straight, angular; vertical lores; distance from eye to nostril slightly less than diameter of eye; interorbital space one and a half times width of upper eyelid; tympanum height equal to one-half diameter of eye; upper surfaces smooth; fingers almost without webbing; toes fully palmate; heel of extended leg reaches tip of snout; head and body pigmented above, probably with yellow, in abrupt contrast to the color of rest of body, from which it is separated by a narrow light stripe from eye to sacrum; undersurface straw color; upper surfaces of limbs pigmented, except humerus and femur. The type appears to be a female, since no mention was made of vocal sac, vocal slits, or nuptial thumb pads. Measurements of the type were given as 26 mm in standard length, length of leg 45 mm, and length of foot including tarsal segment 20 mm. The presence of a distinct continuous dorsolateral light stripe from the eye along the body to the sacrum is found among Costa Rican hylids only in *H. angustilineata*, some *H. ebraccata*, and *H. microcephala*. The only other Costa Rican treefrogs with distinct continuous light stripes are *H. legleri, Duellmano-hyla lythrodes, D. rufioculis,* and *D. uranochroa.* In these forms a lateral light stripe runs below the tympanum to the groin. These frogs further differ from the type of *H. cherrei* in having the dorsum dark in ground color (bluish or purplish to brown in preservative) and a rounded snout in profile. In other aspects of coloration the type of *H. cherrei* differs markedly from *H. angustilineata,* which has a dark green to blackish brown dorsum and numerous dark flecks on the undersurfaces. In these respects the questionable form resembles individuals of *H. microcephala* (for example, the type of *H. underwoodi*) and *H. ebraccata* (for example, the type of *H. weyerae*) that have a uniform light dorsum. In these two forms the undersurfaces are immaculate. *H. angustilineata* further differs from Cope's description in having the vomerine teeth extending posterior to the choanae, a rounded snout profile, and very little toe webbing. There seems to be ample evidence that the types of *H. cherrei* and *H. angustilineata* represent distinct species. *H. cherrei* does not appear to be conspecific with *H. ebraccata,* since the latter has considerable finger webbing, particularly between fingers I and II (webless in *H. cherrei*). It is apparent from the remarks above that *H. cherrei* is rather similar to *H. microcephala.* Cope's description of *H. cherrei* agrees with *H. microcephala* in all but two points: the description of toe webbing and the lack of pigmentation of the humerus and femur. Taylor (1952a, 853) interpreted the statement "pes fully palmate" to mean that *H. cherrei* had toes webbed to the disks. It seems that Cope's statement "manus almost without web; pes fully palmate" is meant to imply that the webs between the toes were well developed and not vestigial as on the fingers. It may be noted further that in his remarks on the distinctive characteristics of the frog, after the description, Cope does not mention toe webbing. If the toes were webbed to the disks, an unusual situation in *Hyla,* it is probable that Cope would have emphasized this feature as he did the webless fingers. The amount of pigment on the upper surface of both humerus and femur limb segments in *H. microcephala* is minimal in some individuals and may be further reduced by fading. It seems likely that Cope's example was similarly colored. Cope gives the type locality as "Alajuela," which could stand for either the town or the province. The collector was apparently Anastasio Alfaro, who later became director of the Museo Nacional de Costa Rica, but through some error he was listed as R. Alfaro (a possible misinterpretation of a Spanish script *A* as an *R*). *H. microcephala* is not known from either the city or the province of Alajuela, but it has been collected from Puntarenas Province, near the boundary with Alajuela Province (San José de San Ramón). Because as nearly as can be determined the description of *Hyla cherrei* agrees in all significant features with *H. microcephala,* and since the type is lost, it seemed appropriate to place the name as a synonym of the latter form. The name *H. cherrei* is available for use as a sen-

ior synonym over *H. underwoodi* Boulenger. If the population of *H. microcephala* from northwestern Costa Rica is recognized as racially distinct from populations in Panama and southwestern Costa Rica, as was done by Duellman and Fouquette (1968) and Duellman (1970), it would be called *H. microcephala cherrei.* The description of *H. cherrei* agrees in detail with the type description of *H. underwoodi,* which is also stated to have toes nearly entirely webbed and thighs lacking pigment. Southern populations thus are referable to *H. m. microcephala.* The karyotype is 2N = 30, all metacentric or submetacentric chromosomes; NF = 60 (Bogart 1973).

DISTRIBUTION: Subhumid and humid lowlands onto lower portions of premontane slopes from southern Veracruz, Mexico, on the Atlantic versant to the Great Lakes drainage of Nicaragua, Costa Rica, and disjunctly in central Panama; on the Pacific slope disjunctly from northwestern Costa Rica to eastern Panama; in South America in Atlantic drainage northern South America and at scattered localities in the Amazon basin to southern Brazil (7–780 m).

Hyla phlebodes Stejneger, 1906

(plate 161; map 7.82)

DIAGNOSTICS: Another moderately common small yellowish tan treefrog having minimal finger webbing, usually a narrow dorsolateral dark stripe that extends only to the axilla but without a white dorsolateral stripe bordering it above (fig. 7.73e), uniform yellow thighs, and no suborbital light spot.

DESCRIPTION: Adult males 19 to 24 mm in standard length, adult females 23 to 28 mm; upper surfaces smooth; venter granular; head about as wide as long; dorsal outline

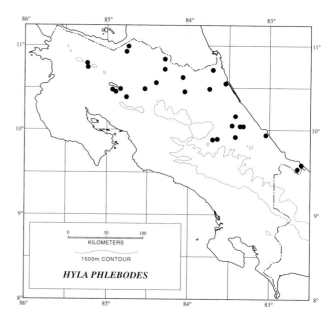

Map 7.82. Distribution of *Hyla phlebodes.*

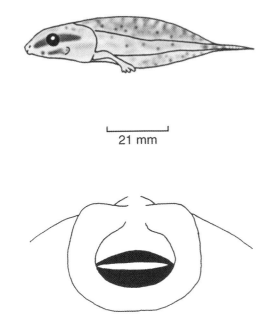

21 mm

Figure 7.87. Tadpole of *Hyla phlebodes*.

of snout subovoid; snout profile rounded; eyes moderate; tympanum distinct, diameter about one-third diameter of eye; paired vocal slits and fully distensible single external subgular vocal sac in adult males; distinct axillary membrane present; fingers short and stout; finger disks moderate, width of disk on finger IV equals diameter of tympanum; distal subarticular tubercle, occasionally bifid under finger IV; nonspinous white nuptial pad on base of thumb in adult males; finger webs reduced, vestigial between fingers I and II; webbing formula: I 3–3 II 1½–2 III 2½–2$^+$ IV; prepollical pad not enlarged; no palmar tubercle; toes extensively webbed; webbing formula: I 1$^+$–2$^-$ II 1–2 III 1–2 IV 2–1 V; inner metatarsal tubercle low, flat, elongate, no outer tubercle; no tarsal fold present. Dorsal surfaces pale yellowish tan with dull darker brown markings; narrow dark stripe from canthus rostralis through eye to axilla or sometimes to midbody present (75%) or absent (25%), not bordered by narrow light stripe above; interorbital dark bar present; flanks pale yellow; thigh yellow; tibial segment of leg with dark transverse bars on upper surface; vocal sac yellow; iris creamy bronze with brown flecks.

LARVAE (fig. 7.87): Small, 21 mm in total length; body ovoid; mouth terminal; nostrils directed anteriorly; eyes directed laterally; spiracle lateral, sinistral; vent tube dextral; tail long; dorsal fin high, ventral fin low; tail terminating in thin, long flagellum; oral disk small, complete, with large lower beak, narrow upper beak; no denticles or papillae present; beaks finely serrate; upper surfaces reddish tan mottled with brown; dark stripe from snout through eye; venter white, mottled with dark brown; caudal musculature heavily pigmented with tan; fins blotched with gray; iris orange tan, center red.

VOICE: Similar to the call of *H. microcephala*, being an insectlike "creek-eek-eek-eek" (buzz-click-click-click) that is higher in pitch and longer in duration than the call of syntopic *Hyla ebraccata*. Introductory notes have a duration of 42 to 94 milliseconds, with pulse repetition rates of 147 to 192 per second and a dominant frequency of about 4 kHz. Call rates are 5 to 22 times per minute. A secondary call of click notes lasts 24 to 56 milliseconds and varies from 0 to 28 times per minute following the primary note (Duellman 1970; Schwartz and Wells 1984b).

SIMILAR SPECIES: (1) *Hyla microcephala* has a narrow white stripe above the dorsolateral dark stripe, and the latter always extends at least to midbody (fig. 7.73d). (2) *Hyla loquax* is much larger (adult males 33 to 45 mm in standard length, adult females 38 to 47 mm), has bright red thighs and webs, and lacks a dorsolateral dark stripe. (3) *Hyla ebraccata* has extensive finger webs, a broad dark band from the tympanum onto the flank, a light suborbital spot, and usually large brown blotches on the back and lower limbs (fig. 7.73a).

HABITAT: Found in Lowland Moist and Wet Forests and marginally in Premontane Wet Forest and Rainforest, where it breeds in grassy temporary ponds, often in disturbed situations.

BIOLOGY: Males call from perches on grasses, reeds, or sedges at heights of 1 to 2 m, in or at the edge of breeding ponds. Individuals may be heard calling from time to time throughout the wet season (April to November). Reproduction usually takes place one or two nights after a heavy rain. The largest aggregations peak shortly after the first torrential rains of the season. The following summary of vocal communication is based primarily on the work of Schwartz and Wells (1984b). Males of this species usually synchronize, increase the call rate, and often add click notes to the advertisement call in response to calls of both conspecific and heterospecific males. Because of the structure of multinote calls (one to twenty-eight clicks) and the timing of male responses, there is little interference between notes in the long overlapping calls of any two *H. phlebodes*. Males are not as aggressive as those of *H. ebraccata* and *H. microcephala*, but they may emit a long aggressive call when a conspecific male approaches. The first note in this call has a higher pulse repetition rate (mean of 247 per second) than the advertisement call. Amplexus is axillary and takes place in the water. Eggs are deposited in small masses that float near the water surface and are usually attached to vegetation. Tadpoles are nektonic macrophages (ecomorph 7, table 7.2) and spend most of their time in submerged vegetation near the pond margin. Juveniles are found in greatest numbers in August to October (Donnelly and Guyer 1994). Small juveniles are 8 to 9 mm in standard length.

REMARKS: In Costa Rica *H. phlebodes* generally is restricted to the Atlantic slope and *H. microcephala* is found primarily along the Pacific versant. At some sites they are syntopic with *Hyla ebraccata,* and at one Atlantic locale (Guanacaste: Finca Jenkins, near San Bosco de Tilarán) the three co-occur at the same breeding site. *H. phlebodes* and *H. microcephala* are also syntopic in southeastern Nicaragua, and they use the same breeding sites as *H. ebraccata* at the same time in central and eastern Panama. Schwartz and Wells (1984a,b, 1985) and Wells and Schwartz (1984a,b) have studied in detail vocal communication in the three species in central Panama. All three species are responsive to one another's advertisement and aggressive calls. All show increased calling rates and complexity of advertisement calls when other males are calling. In this process they produce synchronous calls and often add secondary clicks. Aggressive calls in all three species have higher pulse repetition rates than the advertisement call and are highly variable in pulse repetition rates, rise time, and duration. These calls are emitted during close vocal or physical contact between conspecific males or between males of *H. ebraccata* and *H. microcephala.* *H. phlebodes* appears to be the least aggressive of the three species, perhaps owing to its relatively low densities at breeding ponds. The karyotype is 2N = 30 (Duellman and Cole 1965; Kaiser et al. 1996).

DISTRIBUTION: Humid lowlands of the Atlantic versant from southeastern Nicaragua, including Isla Grande de Maíz, to eastern Panama; on the Pacific lowlands from central Panama to western Colombia and barely across the continental divide into northwestern Costa Rica (20–620 m).

Hyla pictipes Group

Members of this group are small to moderate-sized stream-breeding frogs having the dorsum green, tan, or deep purple or mottled green and black, with some finger webbing and moderate toe webs. Dark brown nuptial thumb pads but no prepollical protrusion in adult males and the absence of fleshy limb fringes, axillary membrane, and mental gland are characteristic. Other features of this group include vomerine teeth in linear series; quadratojugal reduced or absent; 2N = 24. Tadpoles have low fins, long, rounded tail, large ventral oral disk, and 2/3 denticle rows. All species placed in this group are found in Costa Rica.

Oviposition site is unknown for any species in this group, leading to speculation that the eggs are deposited on vegetation over streams (Scott and Limerick 1983). The tadpoles live in torrential streams (ecomorph 11, table 7.2).

The group is equivalent to the *pictipes* and *rivularis* groups of Duellman (1970) and part of the *uranochroa* group of Savage and Heyer (1968).

Hyla debilis Taylor, 1952a

(plate 162; map 7.83)

DIAGNOSTICS: A relatively small treefrog that is grayish to olive green (night color) or green to bluish green (day color), uniform or with flecks of green, with a light sub-orbital spot, continuous to discontinuous golden yellow to white lateral stripe or elongate spot on the glandular ridge between the midlateral area and groin, and a small tympanum, one-third diameter of eye.

DESCRIPTION: Adult males 25 to 30 mm in standard length, adult females 27 to 32 mm; upper surfaces smooth; venter granular; head slightly longer than wide; dorsal outline of snout subelliptical to truncate, snout profile acute; eyes moderate; tympanum distinct, small, height one-third diameter of eye; paired vocal slits and moderately distensible single median external subgular vocal sac in adult males; an abbreviated axillary web present; finger I shorter than finger II; fingers short and broad; finger disks small, width of disks on fingers III and IV equal to diameter of tympanum; distal subarticular tubercle under finger IV usually bifid; brown spinous nuptial pad at base of thumb in adult males; finger webs reduced, barely indicated between fingers I and II; webbing formula: I 3–3 II 2⁻–3 II 2½–2 IV; accessory palmar tubercles small and low, no plantar tubercles; prepollical pad moderately enlarged; weak bipartite or tripartite palmar tubercle; weak row of warts on outer surface of forearm; toes moderately webbed; webbing formula: I 2⁻–2⁺ II 1⁺–2½ III 1½–2½ IV 2½–1⁺ V; inner metatarsal tubercle low, flat, ovoid, outer an indistinct bump; no inner tarsal fold; weak row of warts on outer edge of tarsus. Dorsal surfaces uniform or punctated, dull green, olive, tan, beige, or gray with enamel green flecks and/or brown mottling; well-developed light suborbital spot and creamy (juveniles) or dark stripe on canthus rostralis; light lateral stripe from groin to midlateral area, sometimes reduced to an elongate spot; no transverse dark bars on upper surface of limbs; flanks, groin, and posterior thigh surfaces bright yellow,

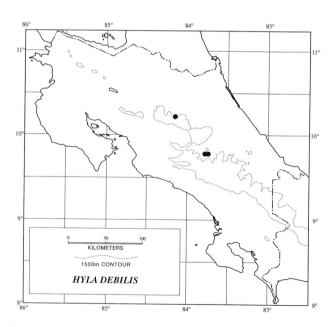

Map 7.83. Distribution of *Hyla debilis.*

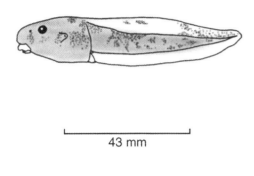

Figure 7.88. Tadpole of *Hyla debilis*.

43 mm

often suffused with dark pigment; venter white with many dark flecks; iris coppery brown or gray to dull yellow or orangish.

LARVAE (fig. 7.88): Large, 43 mm in total length at stage 28; body somewhat depressed; mouth ventral; nostrils dorsolateral; eyes directed dorsolaterally; spiracle lateral, sinistral; vent tube dextral; tail long; fins low; tip of tail rounded; oral disk large, complete, with robust beaks and 2/3 rows of denticles; A2 with slight median gap, P1 with or without slight median gap; rows of moderate-sized papillae border entire disk; additional submarginal papillae at level of mouth; beaks with moderately fine serrations; upper and lateral surfaces of body mottled dark brown and tan; venter mostly dark brown, gray midventrally; caudal musculature tan with darker flecks and spots; fins clear except for a few dark spots and suffusion of gray flecks posteriorly; iris gold centrally, gray peripherally.

VOICE: A very low series of weak cricketlike "chirp-chirp-chirps," repeated five to nine times with the call group lasting 100 to 200 milliseconds. There is a single pulse per call note, with a repetition rate for the call group of about 30 per minute and a dominant frequency of 4.8 to 5.4 kHz (Duellman 1970; Myers and Duellman 1982).

SIMILAR SPECIES: (1) *Hyla rivularis* has the outer two finger disks larger than the tympanum and lacks a suborbital light spot. (2) *Hyla tica* has tuberculate skin on the upper surfaces of the head, body, and limbs, rounded snout profile, disks on the outer two finger disks larger than the tympanum, and no suborbital light spot. (3) *Hyla xanthosticta* has large yellow spots in the groin and on the poste-

rior thigh surface. (4) *Duellmanohyla lythrodes*, *D. rufioculis*, and *D. uranochroa* have bright red eyes, a continuous enamel white lateral stripe along the upper lip back to the groin (fig. 7.73c), and an enamel white supracloacal stripe.

HABITAT: Found along small, clear streams in Premontane Rainforest.

BIOLOGY: This rare stream-breeding species has been collected on only six occasions. Males hide and call from dense vegetation overhanging small montane streams 100 mm to 1 m above the water. The weak call may barely be heard above the sound of flowing water. Duellman (1970) reported finding females near the stream and one asleep on a leaf in the forest 300 m away. *H. debilis* and *H. tica* have been taken together in Costa Rica during their breeding season from the same stream at the Río Quirí, Cartago Province (1,350 m). Tadpoles are of the mountain brook type and use their large oral apparatus to adhere nearly continuously to rocks against the stream flow. Metamorphs are 12 to 13 mm in standard length.

DISTRIBUTION: Humid premontane areas on the Atlantic slopes of the Cordillera Central and the Cordillera de Talamanca in Costa Rica and western Panama (910–1,450 m) and Pacific slope southwestern Panama (1,200–1,400 m).

Hyla pictipes Cope, 1875

(plate 163; map 7.84)

DIAGNOSTICS: A medium-sized frog, uniform purple black, brown, or green or green mottled with black, having discrete creamy white to bright yellow spots on the posterior thigh surfaces (fig. 7.74a), an acute snout profile (fig. 7.6c), a small tympanum, height one-third diameter of eye, and the venter heavily suffused with dark pigment.

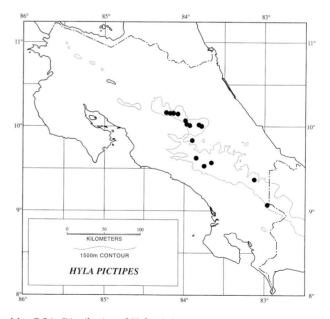

Map 7.84. Distribution of *Hyla pictipes*.

DESCRIPTION: Adult males 31 to 39 mm in standard length, adult females 40 to 45 mm; upper surfaces smooth; venter granular; head outline from above, semicircular with truncate snout; snout profile acute; eyes large; tympanum indistinct, small, about one-third diameter of eye; paired vocal slits and barely distensible single internal subgular vocal sac in adult males; no axillary web indicated; finger I shorter than finger II; fingers relatively long and robust; finger disks large, width of disks on fingers III and IV one and one-third to one and one-half times diameter of tympanum; distal subarticular tubercle under finger IV divided or bifid; brown nonspinous nuptial pad at base of thumb in adult males; finger webs basal, vestigial between fingers I–II–III; webbing formula: I 3–3 II 2⁻–3½ III 3⁻–2½ IV; prepollical pad greatly enlarged; palmar tubercle low, flat, often tripartite; low row of tubercles on outer surface of forearm; toes moderately webbed; webbing formula: I 1½–2½ II 1⁺– 2½ III 1⁺–2½ IV 2½–1⁺ V; inner metatarsal tubercle low, elongate, outer small, round; no tarsal fold. Upper surface leaf green, olive green, dark brown, or purple black, uniform or with some dark mottling; no transverse dark bars on upper limb surfaces; a narrow light upper lip stripe present or not; posterior thigh surfaces dark brown with small discrete yellow spots; usually with similar coloration on anterior surface of thigh, groin, and flank; venter uniform bright yellow or with dark marbling or nearly uniform dark; iris greenish to golden brown or dull copper.

LARVAE (fig. 7.89): Large, 43 mm in total length; body depressed; mouth ventral; nostrils dorsolateral; eyes dorsal, directed dorsolaterally; spiracle lateral, sinistral; vent tube median; tail short; fins low; tail tip rounded; oral disk large, complete with rather weak beaks and 2/3 complete rows of denticles; single row of marginal papillae borders disk; two rows of submarginal papillae above mouth, four to six rows of submarginal papillae below mouth; beaks with minute serrations; body black with golden lichenous markings; tail musculature tan, fins clear; iris black with gold flecks.

VOICE: A single low-pitched, pulsed "eeck" or sometimes two "eecks" repeated every twenty seconds to several minutes apart. Note duration is about 320 milliseconds; the pulse rate is 110 to 143 per second, and the dominant frequency is about 2.2 to 2.7 kHz (Savage 1968a; Duellman 1970).

SIMILAR SPECIES: (1) *Hyla xanthosticta* has a bronze tan stripe on the canthus rostralis and light stripes on the upper lip, forearm, and shank and above the cloaca. (2) *Hyla pseudopuma* has definite webbing between fingers II and III, the posterior thigh is uniform tan or brown, and the venter is never heavily suffused with dark pigment.

HABITAT: Found in or near clear torrential streams in Lower Montane and Montane Rainforests.

BIOLOGY: These rather common nocturnal stream-breeding, treefrogs are found in the wettest and the coolest of montane forests. Males appear to become reproductive during the dry season (December to April), but other reports suggest breeding begins during the first heavy rains of the year (March or April). Males call primarily from rocks in streams and along stream margins. At other times of the year they may also call from low vegetation near the stream. During the day *H. pictipes* hides under rocks on the stream margins. Amplexus and oviposition have not been observed in any species in the *pictipes* species group, and it is not known where early development takes place. Eight gravid females contained 100 to 149 eggs averaging 2.44 mm in diameter (Lang 1995). Tadpoles typically are found adhering to large rocks in cold mountain streams. They are powerful swimmers, but when disturbed they quickly swim against the current before adhering to the exposed surface of another rock or hiding in crevices between or under the rocks in the stream bottom. Metamorphs are 16 to 21 mm in standard length.

REMARKS: Taylor (1952a) listed *Hyla moesta*, *Hyla monticola*, and *H. pictipes*, all described by Cope in 1875 from "Pico Blanco," Costa Rica (= Cerro Utyum; see Savage 1970), as varieties of *Hyla punctariola*. Dunn (1940b) had demonstrated previously that *H. punctariola* was based on a juvenile *Eleutherodactylus* (see remarks section in the account of *Eleutherodactylus punctariolus*). Starrett (1966) reported rediscovery of frogs matching Cope's description, described the substantial color variation that led him to recognize three taxa, and concluded that only one species was involved. As first reviser, she selected *Hyla pictipes* as the valid name for this species. Duellman (1966a) concurred

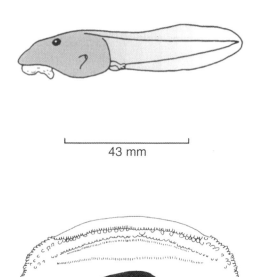

Figure 7.89. Tadpole of *Hyla pictipes*.

43 mm

with this decision and described putative sexual dichromatism in *H. pictipes*. He reported in that paper and later (Duellman 1970) that males had pale green, olive green, or olive tan upper surfaces mottled with darker colors whereas females are uniform pale green. Savage (1968a) pointed out that some males are uniform dark brown or purple black while others exhibit all other previously described color patterns including being uniform green. Females show similar variation, except none are dark brown to purple black. A specimen (UCR 7658) reputedly from Orosi, Cartago Province (1,051 m), is definitely a representative of this species. This locality is from a much lower elevation that any other record of the species, and I suspect there was a mix-up in data or that the frog was collected in the mountains to the south of this town. Another record from La Palma, San José Province (1,520 m) is also probably based on animals (USNM 75060–61, 75065) collected on the slopes above the La Palma Pass.

DISTRIBUTION: Humid lower montane and montane slopes of the Cordillera Central and Cordillera de Talamanca of Costa Rica (1,930–2,800 m).

Hyla rivularis Taylor, 1952a

(plate 164; map 7.85)

DIAGNOSTICS: A common moderate-sized treefrog, tan, yellowish tan, or pale gray, uniform or usually with irregular dark markings; thighs, upper arms, and venter creamy white, and the venter marked with dark brown or black flecks. In addition, the snout profile is acute (fig. 7.6c) and the tympanum small, height one-third diameter of eye.

DESCRIPTION: Adult males 29 to 34 mm in standard length, adult females 33 to 37 mm; upper surfaces smooth to weakly granular; venter granular; head broader than long; dorsal outline of snout subelliptical to truncate; snout profile acute; eyes moderate; tympanum hidden, small, height about one-third diameter of eye; paired vocal slits and fully distensible single external subgular vocal sac in adult males; slight axillary web; finger I shorter than finger II; fingers short and broad; finger disks large, width of disks III and IV larger, one-half to two times tympanum, distal subarticular tubercles divided or bifid under most fingers; tan to brown spinous nuptial pad on base of thumb in adult males; finger webs basal, vestigial between fingers II–III–IV; webbing formula: I 3–3 II 2$^+$–3 III 2⅔–2$^+$ IV; prepollical pad moderately enlarged; small bifid palmar tubercle; low ridge of warts on outer surface of forearm; toes moderately webbed; webbing formula: I 2$^-$–3 II 1$^+$–2⅔ III 1¼–2⅔ IV 2⅓–1¼ V; inner metatarsal tubercle ovoid, flat, outer small; no tarsal fold; thin outer dermal fold from heel to base of toe V. Dorsum and upper head surfaces uniform, punctate, lineate, or blotched gray, yellow brown, or tan; usually nearly uniform yellowish tan (night) or tan with darker markings (day); thin dark canthal-postocular dark stripe; green or turquoise on sides of some examples; thighs, upper arms, flanks, vocal sac, and venter creamy yellow; posterior thigh surface yellow or sometimes greenish, often suffused with melanophores; weak transverse dark bars sometimes present on upper surface of tibial segment; venter flecked with brown or black; iris bronze with fine black reticulations.

LARVAE (fig. 7.90): Moderate-sized, 40 mm in total length; body slightly depressed; mouth ventral; nostrils dorsolateral, directed anteriorly; eyes directed dorsolaterally;

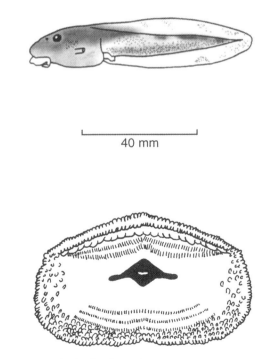

Figure 7.90. Tadpole of *Hyla rivularis*.

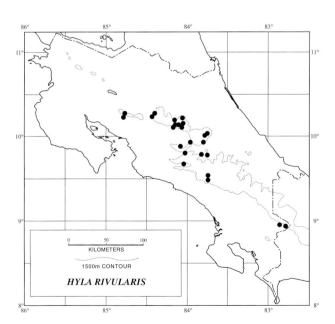

Map 7.85. Distribution of *Hyla rivularis*.

spiracle lateral, sinistral; vent tube median; tail long; fins low; tip of tail rounded; oral disk huge, complete, with moderate beaks and 2/3 denticle rows; upper labium bordered by one row of small marginal papillae and one row of moderate-sized and one row of large submarginal peglike papillae above mouth; lower labium with two or three rows of small marginal papillae and two or three rows of moderate-sized submarginal papillae; beaks with fine serrations; body brown with gold flecks; caudal musculature tan with dark brown blotches dorsally and flecks laterally; fins transparent with a few scattered brown flecks; iris greenish bronze.

VOICE: A series of low, high-pitched short "cheep-cheep-cheeps" similar to notes of a cricket. Each note contains three pulses. A call group consists of 12 to 137 notes, has a duration of 5 to 66 seconds, is repeated at intervals of about 10 to 70 seconds, and has a dominant frequency of 1.98 to 2.30 kHz. The call is of lower intensity and longer duration than the similar call of *Hyla debilis* (Savage 1968a; Duellman 1970).

SIMILAR SPECIES: (1) *Hyla tica* has tuberculate dorsal surfaces, a rounded to vertical snout profile (fig. 7.6c), usually a barred thigh, and a moderate-sized tympanum, height one-half diameter of eye. (2) *Hyla pseudopuma* has a moderate-sized tympanum, rounded snout profile (fig. 7.6c), and usually yellow groin and flank spots (fig. 7.74c,d). (3) *Smilisca sordida* has a rounded snout profile (fig. 7.6c), has substantial webbing between fingers II and III, and usually has the flank and posterior thigh surfaces mottled with lighter color and bluish purple (fig. 7.74b).

HABITAT: Seen most frequently along or in clear streams in Premontane Wet Forest and Rainforest and Lower Montane Wet Forest and Rainforest.

BIOLOGY: Males call at night during most of the year, concealed in herbaceous vegetation or bushes at the margin of or overhanging fast-moving mountain streams. They generally call deeper in the stream bank foliage than does *H. debilis*. Duellman (1970) suggested that reproduction is year-round because tadpoles in various stages of development are found throughout the year. It appears, however, that at most sites breeding peaks during the early part of the rainy season (April to June). Amplexus, oviposition, and eggs of this species have never been observed. Eight gravid females contained 76 to 104 eggs, averaging 2.36 mm in diameter (Lang 1995). The tadpoles use the very large oral disks to maintain their position in the stream by adhering to the tops and sides of large rocks. They may be seen at night moving across the rock's surface using the oral disk. When disturbed they often hide under rocks on the stream bottom, again maintaining position against the current with the oral funnel. Metamorphs are 13 to 15 mm in standard length.

REMARKS: This very common frog was described by Taylor (1952a). However, Dunn had recognized its validity many years earlier. Dunn, unlike his great rival Taylor, was very conservative in describing new forms and always made

an effort whenever possible to associate available names with species at hand. In this case he used the name *Hyla monticola* Cope, 1875, for specimens of *H. rivularis* that he collected on Volcán Barva, Heredia Province. Unfortunately the types of *H. monticola* turned out to be representatives of another montane species, *Hyla pictipes* (Starrett 1966; Duellman 1966a). See the remarks section under that species for further information on the history of these names. Myers and Duellman (1982) recorded a juvenile example (AMNH 10795) of *H. rivularis* that was bluish green above and yellow below in life (Panama: Bocas del Toro: eastern slopes of Cerro Colorado, 1,600 m). Costa Rican examples of *H. rivularis* sometimes have some green on the head or green flecks elsewhere in the coloration, and this is especially the case with juveniles. However, no Costa Rican examples have a primarily green dorsal coloration. The karyotype is 2N = 24 (León 1969).

DISTRIBUTION: Humid premontane and lower montane slopes of the cordilleras (Tilarán, Central, and Talamanca) of Costa Rica and adjacent western Panama (1,210–2,040 m).

Hyla tica Starrett, 1966

(plate 165; map 7.86)

DIAGNOSTICS: A small to moderate-sized tuberculate treefrog mottled with various shades of green and brown above and having a rounded snout profile (fig. 7.6c), a moderate tympanum, height one-half diameter of eye, transverse dark bars usually present on the upper surfaces of the limbs, and usually no dark flecking on the undersurface.

DESCRIPTION: Adult males 27 to 34 mm in standard length, adult females 33 to 42 mm; upper surface with many rounded warts and tubercles, especially in occipital and

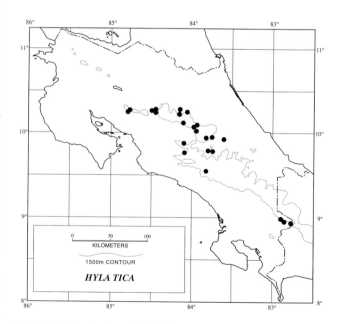

Map 7.86. Distribution of *Hyla tica*.

sacral regions; venter granular; head broader than long; dorsal outline of snout subelliptical to truncate; snout rounded to vertical in profile; eyes moderate; tympanum distinct, moderate, height about one-half diameter of eye; paired vocal slits and moderately distensible single external subgular vocal sac in adult males; moderate axillary web present; fingers short and broad; finger disks large, width of disks on fingers III and IV equal to or slightly larger than diameter of tympanum; distal subarticular tubercle under finger IV always divided, that under finger III divided or bifid; brown nonspinous nuptial pad at base of thumb in adult males; webs vestigial between fingers I and II; webbing formula: I 3–3 II 2–3 III 2⅓–2 IV; prepollical pad moderately enlarged; two small palmar tubercles; low ridge of warts along outer surface of forearm; toes extensively webbed; webbing formula: I 1⁺–2 II 1–2 III 1–2 IV 2–1 V; inner metatarsal tubercle low, ovoid, outer small; no tarsal fold; sometimes a few small tubercles along outer margin of tarsus. Some metachrosis occurs in this species; upper surfaces except upper arm mottled or blotched green and brown or tan to dark brown with scattered metallic green flecks; upper arm uniform tan to orange brown, usually with bold transverse dark bars on limbs; flanks mottled brown and yellow or with irregular light spots; posterior thigh surfaces tan to yellow (brown to orange brown extralimitally) usually suffused with brown; venter dull white; iris reddish bronze.

LARVAE (fig. 7.91): Moderate-sized, 38 mm in total length at stage 34; body depressed; mouth ventral; nostrils directed anterodorsally; eyes directed dorsolaterally; spiracle lateral, sinistral; vent tube dextral; tail very long; fins low; tip of tail rounded; oral disk huge, complete, with delicate beaks and 2/3 rows of denticles; one row of small marginal papillae completely around disk and one row of small and one of large submarginal papillae above mouth; three to six additional rows of small papillae below mouth; body brown above, silvery gray below; tail creamy tan with brown transverse marks dorsally and brown reticulations laterally; fins transparent with brown flecks on posterior half of tail; beaks with minute serrations.

VOICE: A series of three to five cricketlike "chirp-chirp-chirps" with the call repeated at intervals of about 1.5 seconds. Each note contains two to four pulses, and the call group lasts 380 to 600 milliseconds, with two emphasized frequencies at about 2.2 and 4.7 kHz (Savage 1968a; Duellman 1970).

SIMILAR SPECIES: (1) *Hyla rivularis* has smooth skin on the upper surfaces, an acute snout profile (fig. 7.6c), a small tympanum, height one-third diameter of eye, and rarely has transverse dark bars on the tibial segment of the leg. (2) *Hyla pseudopuma* has smooth dorsal surfaces; the groin is brown with discrete yellow spots (fig. 7.74c,d) or uniform blue and has numerous pigmented plantar tubercles. (3) *Smilisca sila* has small blue spots in the groin and on the posterior thigh surface (fig. 7.74b) and lacks divided or bifid subarticular

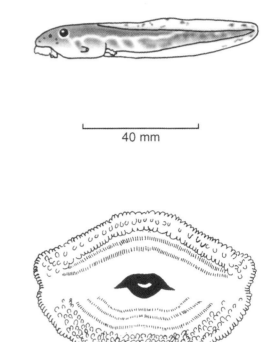

Figure 7.91. Tadpole of *Hyla tica*.

tubercles under the fingers. (4) *Hyla debilis* has a light suborbital spot, smooth dorsal integument, and an acute snout profile (fig. 7.6c).

HABITAT: Found near mountain streams in Premontane Wet Forest and Rainforest and very marginally in Lower Montane Wet Forest and Rainforest.

BIOLOGY: These common nocturnal stream-breeding frogs are active all year, but reproduction probably takes place in the dry season (February to April). Males usually call from vegetation overhanging the fastest stretches of the stream, 1 to 3 m above the water. Occasional males call from rocks in the stream. Individuals of *H. tica* generally are not concealed in vegetation like *H. debilis* and *H. rivularis* but sit openly on limbs, branches, or leaf surfaces. Amplexus and oviposition are unknown for this species. Starrett (1966) reported that females contained large (2 mm in diameter) unpigmented eggs, and (Lang 1995) found 190 to 194 eggs averaging 2.32 mm in diameter in three gravid females. These are probably deposited under rocks in streams. Tadpoles are adapted to fast-moving streams and adhere to the lee side of rocks with the large oral funnel. When disturbed they swim to other rocks or hide in debris on the bottom. Metamorphs are 17 to 19 mm in standard length.

REMARKS: This relatively common species was not discovered until the late 1950s (Starrett 1966). *Hyla tica* and *H. rivularis* have complementary altitudinal distributions, with the former found at lower elevations. The two occur in the same streams and overlap in breeding periods between 1,210 and 1,650 m. *H. pictipes* altitudinally overlaps (1,980 to 2,040 m) with *H. rivularis*, and the two have been taken from the same stream during their presumed breeding seasons. *H. pictipes* appears to be an ecological replacement for *H. tica* in the montane zone. The karyotype is 2N = 24 (León 1969).

DISTRIBUTION: Humid premontane areas and the lowest portion of the humid lower montane zone in the Cordillera de Tilarán, Cordillera Central, and Cordillera de Talamanca of Costa Rica and western Panama (1,100–1,650 m).

Hyla xanthosticta Duellman, 1968b

(plate 166; map 7.87)

DIAGNOSTICS: A small green treefrog with large bright yellow spots in the groin and on the posterior thigh, a broad bronze tan stripe on the canthus rostralis, and white stripes on the upper lip, forearm, and tarsal segment of the leg. In addition, the snout is acute in profile (fig. 7.6c) and the tympanum is small, height one-half diameter of eye.

DESCRIPTION: Known from a single adult female 29 mm in standard length; upper surfaces smooth; venter granular; head broader than long; dorsal outline of snout narrow, truncate, snout profile acute; eyes moderate; tympanum indistinct, height one-half diameter of eye; no axillary web; fingers long and rather slender; finger disks large, width of disks on fingers III and IV equal to tympanum; distal subarticular tubercle under finger IV bifid; webs much reduced, absent between fingers I and II; webbing formula: I 3–3 II 2½–3 III 3⁻–2½ IV; prepollical pad moderately enlarged; moderate, indistinct, round palmar tubercle; outer margin of forearm smooth; toes moderately webbed; webbing formula: I 2⁻–3 II 1½–2⅔ III 1½–3 IV 2½–1½ V; inner meta-

tarsal tubercle large, flat, outer tubercle small; no tarsal fold; dermal fold along outer edge of tarsus. Dorsal surfaces green except for upper arm and thigh; bronze tan stripe on canthus rostralis; flanks and thighs brown with discrete large yellow spots; upper arm and dorsal surface of thigh uniform tan; white stripe along upper lip, outer margin of forearm, and tarsus and above cloaca; venter pale yellow; iris gold with fine black reticulations, suffused with reddish centrally.

LARVAE: Unknown.

VOICE: Unknown.

SIMILAR SPECIES: (1) *Hyla debilis* has a suborbital light spot, and the posterior thigh surfaces are uniform yellow. (2) *Hyla pictipes* has small yellow spots on the posterior thigh surface (fig. 7.74a) and lacks a dark stripe on the canthus rostralis and light labial, forearm, and shank stripes.

HABITAT: Lower Montane Rainforest.

REMARKS: The only known specimen (KU 103772; Heredia: south fork of Río Las Vueltas, on base of Cerro Chompipe, southern slope of Volcán Barva) was collected from a leaf about 1 m above the ground. The forest in this area is composed primarily of large moss-covered oaks supporting many epiphytic plants, especially bromeliads. *H. xanthosticta* resembles juvenile *H. pictipes* in many ways, although Duellman (1968b) thought it was more closely related to *H. debilis*, *H. rivularis*, and *H. tica*. Both *H. pictipes* and *H. rivularis* are found in association along the Río Vueltas at the type locality of *H. xanthosticta*. Unfortunately this area, although now in the Braulio Carrillo National Park, is one where all anurans disappeared in the 1980s when amphibian populations declined. It thus seems unlikely that additional specimens of this rarity will be found.

DISTRIBUTION: Known only from Costa Rica at the type locality, on the south slope of Volcán Barva in the lower montane zone (2,100 m).

Hyla pseudopuma Group

These moderate-sized frogs may have a brown, green, or yellow dorsum that may be marked by darker blotches, and definite dorsolateral light or lateral dark stripes may be present. Additional features of these upland anurans are finger webbing present or absent; toes definitely webbed; dark brown nuptial thumb pads in adult males; prepollices not protuberant; no fleshy limb fringes, axillary membrane, or mental gland; vomerine teeth in linear series; quadratojugal articulates with maxilla; 2N = 24. Tadpoles robust, oral disk anteroventral with 2/3 rows of denticles. Reproduction in this group is not specialized; the eggs are deposited and the larvae develop in montane ponds or small water-filled depressions (pattern 1, table 7.1).

This group includes the species contained in Duellman's (1970) *H. pseudopuma* group and the subsequently described *Hyla graceae* Myers and Duellman (1982) from

Map 7.87. Distribution of *Hyla xanthosticta*.

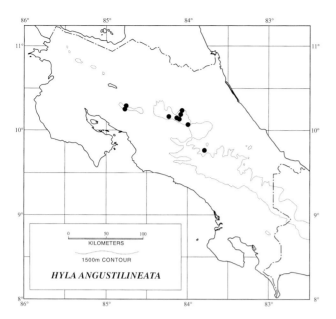

Map 7.88. Distribution of *Hyla angustilineata*.

Panama. I regard *Hyla infucata* Duellman, 1968a, of northwestern Panama, originally described as a subspecies of *H. pseudopuma*, as a valid species belonging to this group.

Hyla angustilineata Taylor, 1952a

(plates 167–68; map 7.88)

DIAGNOSTICS: This elegant moderately small treefrog has a dark dorsum (dark brown or green), a dorsolateral pinstripe on each side running from the tympanum to the groin that is bordered below by a broader dark brown band, and the venter heavily flecked with dark brown (fig. 7.69). These features, combined with the fact that it is one of only a few Costa Rican treefrogs lacking finger webs, makes it easily recognizable.

DESCRIPTION: Adult males 30 to 34 in standard length, adult females 34 to 37 mm; upper surfaces smooth; venter granular; head slightly longer than broad; dorsal outline of snout subelliptical, snout rounded in profile; eyes moderate; tympanum distinct, diameter two-fifths to one-half diameter of eye; paired vocal slits and moderately distensible single internal subgular vocal sac in adult males; fingers moderate, robust; finger disks large, width of disk on finger III equal to or slightly greater than diameter of tympanum; distal subarticular tubercle on finger IV usually bifid; brown nonspinous nuptial pad on base of thumb in adult males; no finger webs; prepollical pad moderately enlarged; large, elevated bifid palmar tubercle; toe webbing reduced; webbing formula: I 2$^+$–3$^-$ II 2$^-$–3⅓ III 2–3¼ IV 3–2$^-$ V; inner metatarsal tubercle large, flat, elliptical, outer tubercle small; no tarsal fold. Upper surfaces of head, body, and limbs pale brown or olive tan with bronze cast; numerous small

darker brown spots always on upper limb surfaces, usually on back as well; juveniles under 20 mm in standard length with upper surfaces of head and back bright green or brown and limb surfaces as in adults; white dorsolateral pinstripe edged with green from tip of snout along canthal ridge and outer margin of upper eyelid and above tympanum to level of groin, bordered below by dark brown longitudinal band from eye to groin; groin, anterior and posterior thigh surfaces, and undersurfaces of legs dull yellow; venter creamy white; all ventral surfaces heavily spotted with dark brown; iris reddish copper.

LARVAE (fig. 7.92): Moderate-sized, 35 mm in total length; body robust; mouth directed ventrally; nostrils directed anterolaterally; eyes directed dorsolaterally; spiracle lateral, sinistral; vent tube dextral; tail short; fins moderate; tip of tail broadly rounded; oral disk small, complete, with moderate beaks and 2/3 rows of denticles; A2 with narrow median gap; two rows of papillae around margin of disk, except for broad gap above mouth; beaks with short, pointed serrations; body mottled brown and bronze tan; tail creamy tan with brown markings; iris deep bronze.

VOICE: The call consists of two short pulsed, poorly, modulated notes repeated every 70 to 80 seconds. Each note lasts 70 to 140 milliseconds, with 90 pulses per second and a dominant frequency of 1.6 kHz (Duellman 1970).

SIMILAR SPECIES: Unlikely to be confused with any other treefrog having dorsolateral light stripes *(H. colymba, H. ebraccata, H. microcephala,* or *H. phlebodes)*, since these forms have webbed fingers and lack dark spotting on the ventral surfaces.

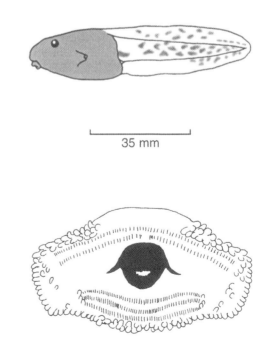

35 mm

Figure 7.92. Tadpole of *Hyla angustilineata*.

HABITAT: Found in Lower Montane Wet Forest and Rainforest and marginally in Premontane Rainforest.

BIOLOGY: An uncommon nocturnal treefrog that usually breeds in small puddles and other water-filled depressions within the forest. Males call from leaves and twigs a short distance above the forest floor. Reproduction apparently takes place early in the year (January to March), since many tadpoles are in pools in March and April and males call only sporadically at that time of year. Tadpoles are benthic (ecomorph 1, table 7.2). Metamorphs are 15 to 19 mm in standard length.

REMARKS: This rarely seen frog was described by Taylor (1952a) based on a series of specimens from La Palma, San José Province, collected by Manual Valerio in 1928 (holotype: USNM 75061, paratypes USNM 75061, 75065). Years earlier Dunn had recognized the distinctiveness of these animals and had given them to the National Museum of Natural History while keeping several more Valerio specimens (now ANSP 23976, 23985, 248223–28) in Philadelphia. Dunn was firmly committed to trying to associate previously proposed available names to apparent novelties from Central America rather than proposing new ones. In this case he applied the name *Hyla moesta* Cope, 1875, to these specimens. Unfortunately the holotype of *H. moesta* is an example of another species, *Hyla pictipes* (Starrett 1966; Duellman 1966a), so ironically the name proposed by Dunn's bête noire, Taylor, is valid. For additional details on the history of these names see the remarks section in the account on *H. pictipes*. The karyotype is 2N = 24 (León 1969).

DISTRIBUTION: Humid lower montane rain forests of the Cordillera de Tilarán, Cordillera Central, and Cordillera de Talamanca of Costa Rica and western Panama (1,500–2,040 m).

Hyla pseudopuma Günther, 1901

(fig. 7.94; plates 169–70; map 7.89)

DIAGNOSTICS: This rather nondescript, extremely common moderate-sized treefrog, usually brown but sometimes bright yellow (breeding males), shows considerable variation in morphology and coloration. The most distinctive features that in combination will distinguish it from other Costa Rica treefrogs are an acuminate snout profile (fig. 7.6c), numerous darkly pigmented plantar tubercles, groin usually marked with yellow spots that invade the dark brown to blue flanks from below (fig. 7.74c,d), anterior and posterior thigh surfaces usually tan or brown or with a few diffuse cream spots, fingers with considerable webbing, and a bilobate subgular vocal sac in adult males (fig. 7.12d). Some individuals of *H. pseudopuma* are most likely to be confused with members of the genus *Smilisca*, and immatures may be confused with adult *Hyla rivularis* and *Hyla tica*. See similar species section below to further differentiate these forms from *H. pseudopuma*.

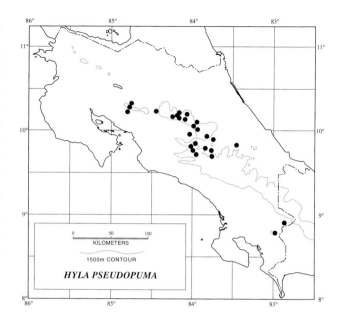

Map 7.89. Distribution of *Hyla pseudopuma*.

DESCRIPTION: Adult males 37 to 45 mm in standard length, adult females 41 to 52 mm; upper surfaces smooth; venter granular; head about as broad as long; snout rounded from above; snout acuminate in profile; eyes moderate; tympanum distinct, diameter two-fifths to two-thirds diameter of eye, largest in females; paired vocal slits and moderately distensible bilobate internal subgular vocal sac in adult males; fingers moderately long and robust; finger disks large, width of disk on finger III equals diameter of tympanum; distal subarticular tubercles under fingers III and IV often bifid; brown, nonspinous nuptial pads on base of thumb in adult males; fingers moderately webbed, only a trace between fingers I and II. Webbing formula: I 3–3 II 2⁻–3⁻ III 2⅓–2 IV; prepollical pad greatly enlarged, bulbous; small, elevated palmar tubercle; toes extensively webbed; toe webbing formula: I 1½–2½ II 1⁺–2½ III 1½–2½ IV 2½–1⁺ V; inner metatarsal tubercle small, flat, elliptical, no outer tubercle; weak inner tarsal fold. Dorsal pattern variable, nearly uniform or marked with dark brown spots or blotches on tan to brown ground color; breeding males bright yellow above with black or blue purple dorsolateral band along flanks and yellow anterior and posterior thigh surfaces; flanks usually dark brown to black with bright yellow spots or single yellow area in groin encroaching on lateral dark band from below; upper surface of thighs with or without transverse dark bars; anterior and posterior thigh surfaces usually tan or brown; venter creamy white; bold dark markings and yellow spots in diurnal color phase somewhat muted at night; iris deep bronze.

LARVAE (fig. 7.93): Moderate-sized, 31 mm in total length; body robust; mouth ventral; nostrils directed anterodorsally; eyes directed laterally; spiracle lateral, sinistral;

vent tube dextral; tail short; fins deep; tail tip round; oral disk small, complete with robust beaks and 2/3 rows of denticles; A2 with median gap; disk bordered by two rows of small papillae with broad median gap above mouth; beaks with small, pointed serrations; body dark brown above and below; caudal musculature brown; fins transparent and heavily flecked with brown; iris bronze.

VOICE: Call a single low, poorly modulated note repeated 45 times a minute, with a duration of 30 milliseconds and a pulse rate of 85 per second. The dominant frequency is 0.95 kHz. Sometimes the note is broken up into a series of short notes (Duellman 1970).

SIMILAR SPECIES: Variation in this species often leads to confusion with other usually brown treefrogs. Specimens without distinctive morphologies or coloration, especially juveniles, will often turn out to be *H. pseudopuma*. (1) *Smilisca puma* has the groin and flank with a fine venated pattern (fig. 7.74b), has white labial and tarsal-foot stripes, and lacks finger webs. (2) *Smilisca sordida* has the flank, groin, and posterior thighs usually marked with bluish or creamy tan flecks or larger bluish spots (fig. 7.73b), has more toe webbing (webbing formula I 1–1 II versus 1½–2 II in *H. pseudopuma*), and lacks bright yellow groin or flank spots or overall yellow coloration. (3) *Smilisca sila* has a truncate snout profile (fig. 7.6c), has the groin and posterior thighs marked with blue or creamy tan spots or flecks (fig. 7.74b), and lacks bright yellow in its coloration. (4) *Hyla rivularis* has a truncate snout profile (fig. 7.6c), usually heavy dark ventral peppering, and the tympanum one-third diameter of

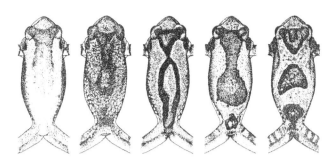

Figure 7.94. Variation in color pattern in *Hyla pseudopuma*, dorsal view. Frog on the left has typical pattern of breeding males.

eye. (5) *Hyla tica* has a strongly tuberculate dorsal surface and a truncate snout profile (fig. 7.6c).

HABITAT: Found in a variety of situations in Lower Montane Wet Forest and Rainforest and marginally into Premontane Moist and Wet Forests and Rainforest. Common within the woods, in secondary growth, meadows, and pastures, and around breeding sites within these habitats.

BIOLOGY: *Hyla pseudopuma* is among the commonest Costa Rican treefrogs. The following account summarizes the observations and experiments of Duellman (1970), Crump (1983a, 1984, 1986b, 1989b, 1990), and Crump and Townsend (1990) together with my own field observations. By day these frogs hide in vegetation, especially in ground and arboreal tank bromeliads or in the leaf axils of other plants. The species is an explosive pond breeder that uses water-filled depressions ranging in size from human or cow footprints to small ponds up to 15 m in diameter. Flooded pastures and roadside ditches also form breeding sites. Males call sporadically throughout the rainy times of the year from vegetation or near lentic situations. During reproductive periods large numbers of males congregate around breeding sites, call from near the water's edge, and are joined by a lower number of females. Reproduction follows heavy rains beginning early in the wet season (April or early May at most sites). At that time hundreds of individuals will be around large ponds, and the first wave of reproduction may be completed within twenty-four hours. These explosions occur at appropriate sites at several times scattered throughout the rainy season, although it is unclear if the same females return to lay additional egg clutches. Males exhibit a frenzy of scramble competition for mates. Mating by males is random with respect to size, with no apparent advantage to larger males (Crump and Townsend 1990). Amplexus is axillary, and amplectant pairs are found both in the water and on the shore. Males try to dislodge amplectant males, and a single female may become the center of a mating ball of two to sixteen competing suitors. Oviposition takes place in the water, and eggs are laid in multiple small masses (usually fewer than 500 eggs) and attached to emergent or submerged leaves, sticks, reeds, or

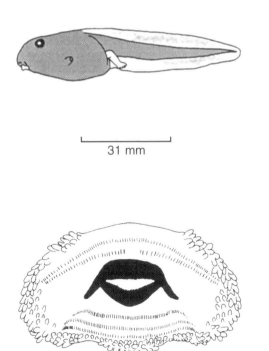

31 mm

Figure 7.93. Tadpole of *Hyla pseudopuma*.

blades of grass at several locations in the puddle or pond. Total clutch size is estimated as 1,800 to 2,500 eggs. Oviposition by a single female is completed within a few hours, and afterward she leaves the breeding site. The eggs in each mass remain discrete; they are about 1.7 mm in diameter and about 5 mm including the envelope. Eggs hatch in about twenty-four hours, and in captivity tadpoles metamorphosed in fifty-seven to eighty-one days (Duellman 1970). Metamorphs are 10.5 to 13 mm in standard length. Metamorphs are pale olive tan above with brown markings and pale tan spots, have silvery white upper lips and venter, and have barred hind limbs. Crump (1983a, 1989b, 1990) suggested that two factors, desiccation and food limitation, were especially important in the survival of the development stages. At many breeding sites the ephemeral pools often dry up within a week or two, and all the larvae in them may die. In a series of controlled experiments with eggs from a single mated pair, Crump (1989b) demonstrated that tadpoles of this species exhibit phenotypic plasticity in the rate of development. Under simulated drying conditions, development is accelerated and size at metamorphosis is significantly decreased. Tadpoles raised in shallow-water experiments, where crowding was a factor, took the longest to metamorphose, and size was also significantly decreased. Tadpoles raised in optimal hydric conditions took an intermediate amount of time to metamorphose and metamorphosed at a significantly larger size. Clearly, in a drying environment rapid metamorphosis is a plus, but the price is smaller size. Because both rapidly drying ponds and more persistent ones are used as reproductive sites, it is also advantageous for tadpoles in the latter case to remain in the water to eat and grow larger before transforming. Similar adaptability is to be predicted in other tropical frogs reproducing in heterogeneous aquatic environments. Tadpoles feed on plant material and detritus like typical lentic, benthic ecomorphs (ecomorph 1, table 7.2). However, these materials are limited at most H. pseudopuma sites. In ponds that persist longest, several age-classes may coexist as the result of successive bursts of breeding. In these cases Crump (1990) showed that under natural conditions older tadpoles with well-developed mouthparts (stage 25 on) eat recently laid conspecific viable eggs, smaller conspecific tadpoles (1983a), and heterospecific tadpoles (1990) (plate 171). In a controlled experiment she demonstrated paradoxically that H. pseudopuma tadpoles fed on conspecific tadpoles grow larger and metamorphose at a larger size than those fed heterospecifics. She argued that the costs of cannibalism in this case are reduced because so many pairs lay so many eggs at one time that it is unlikely most conspecifics encountered will be siblings or other close relatives. The mouthparts and intestinal tract of larval H. pseudopuma are not highly modified for carnivory or oophagy, so predation and cannibalism are probably opportunistic. Further exper-iments (Crump 1991) showed that females seem to assess the egg load of potential breeding sites, preferring ponds with deeper water and those lacking conspecific tadpoles.

REMARKS: Duellman (1968b) described a distinctive population allied to H. pseudopuma from the Atlantic slope of Cerro Pando, Bocas del Toco Province, Panama, having truncate instead of an acuminate snout profile and the axilla, groin, anterior and posterior thigh surfaces, and webbing of the feet bright red. Typical H. pseudopuma are found on the Pacific slope in western Panama and have been taken recently from the Pacific slope near Cerro Pando in Costa Rica. No examples appear to bridge the gap in characters between the two forms, and it seems appropriate to regard H. infucata as a separate species. In June 1961 I collected several H. pseudopuma males in breeding coloration from an aggregation on the campus of the University of Costa Rica in a suburb of San José City (1,160 m). This sample suggests that the species formerly may have been distributed on the upper portions of the Meseta Central Occidental.

DISTRIBUTION: Primarily in the humid lower montane zone of the Cordillera de Tilarán, Cordillera Central, and Cordillera de Talamanca of Costa Rica and Pacific slope western Panama, but marginally in humid areas of the premontane zone (1,120–2,340 m).

Hyla salvadorensis Group

These moderately small dark brown to olive green frogs are stream breeders with red or bronze irises and definite but moderate webbing of the hands and feet. Other features characteristic of the group are no fleshy limb fringes, axillary membrane, or mental gland; dark brown to black nuptial thumb pads in adult males but without protuberant prepollices; vomerine teeth in linear series; quadratojugal articulating to maxilla; 2N = 24. Tadpoles have long tails and large, ventral oral disks and 2–3/3–5 rows of denticles. Only one other species besides the endemic Costa Rican form belongs to this group, Hyla salvadorensis of El Salvador. These two species were included in Ptychohyla by Campbell and Smith (1992). According to these authors Hyla erythromma of southwestern Mexico is related to the species in this group, and they also placed it in Ptychohyla. The eggs of these frogs remain unknown, leading to speculation that eggs are deposited on vegetation over a stream (Scott and Limerick 1983). The tadpoles develop in small montane streams in the slower-moving shallows or backwaters (pattern 2, table 7.1; ecomorph 11, table 7.2).

The group as defined at present is the same as Duellman's (1970) H. salvadorensis species group. Savage and Heyer (1968) placed Hyla legleri in their Hyla uranochroa species group, an allocation not followed here.

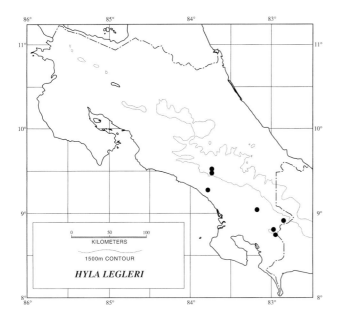

Map 7.90. Distribution of *Hyla legleri*.

Hyla legleri Taylor, 1958a

(plate 172; map 7.90)

DIAGNOSTICS: A moderate-sized red-eyed treefrog, dark brown to olive brown with a distinct lateral light stripe from axilla to groin and lacking a suborbital light spot.

DESCRIPTION: Adult males 31 to 37 mm in standard length, adult females 36 to 39 mm; upper surfaces smooth; venter granular; head about as long as wide; dorsal outline of snout subovoid to truncate, snout rounded in profile; eyes moderate; tympanum distinct, diameter 42 to 62% diameter of eye, smaller than disk on finger III; paired vocal slits and moderately distensible single internal subgular vocal sac in adult males; fingers rather short and stout; finger disks large, width of disk on finger III equal to diameter of eye, larger than diameter of tympanum; distal subarticular tubercle on finger IV sometimes weakly bifid; brown nonspinous nuptial pad on base of thumb in adult males; fingers moderately webbed, only trace between fingers I and II; webbing formula I 3–3$^+$ II 1½–2¼ III 2$^+$–2 IV; prepollical pad moderately enlarged; low bifid palmar tubercle present; row of ulnar tubercles or low fleshy ridge on outer margin of forearm; toes extensively webbed; webbing formula: I 1–2 II 1–2 III 1–2 IV 2–1 V; inner metatarsal tubercle large, oblong, outer small, round; weak tarsal fold present. Upper surfaces usually dark brown, uniform or mottled with small light, bronze, or green flecks; light stripe along upper lip when present, not enlarged into suborbital light spot under eye; cream white stripes from axilla to groin, on margins of forearm, feet, and above cloaca; upper forearm orange brown marked with oblique white stripe above; anterior and posterior surfaces of thighs orange brown to yellowish

tan, suffused with dark pigment; no transverse dark bars on upper limb surfaces; plantar surfaces heavily pigmented, dark; venter white; throat dark in adult males; iris deep red; small juveniles olive green above, white below with obvious lateral white stripes and red eyes.

LARVAE (fig. 7.95): Moderate-sized, 40 mm in total length; body ovoid; mouth ventral; nostrils directed anterolaterally; eyes directed dorsolaterally; spiracle lateral, sinistral; vent tube dextral; tail long; fins relatively low; tail tip rounded; oral disk moderate, complete with moderate beaks and 2–3/5 rows of denticles; when three rows present A3 fragmented; when two rows present, A1 usually with narrow median gap, row immediately above mouth sometimes with narrow median gap; two complete rows of small papillae around disk; many additional submarginal papillae at level of mouth; beaks serrate; body and tail pale brown with dark flecks and spots; transverse bars tending to form across upper surface of tail; venter creamy white; iris bronze turning reddish at stages 38 to 39.

VOICE: A single long, poorly modulated "wauk!" Notes are repeated 8 to 32 times a minute and have a duration of 230 to 400 milliseconds and a pulse rate of 89 to 125 per second. Two frequencies are usually emphasized at about 1.43 and 2.4 kHz (Savage 1968a; Duellman 1970).

SIMILAR SPECIES: In *Duellmanohyla lythrodes*, *D. rufioculis*, and *D. uranochroa* a continuous white stripe runs from the tip of the snout along the upper lip and under the tympanum to the groin (fig. 7.73c).

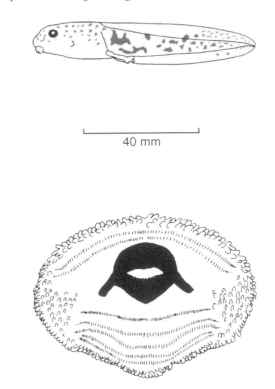

40 mm

Figure 7.95. Tadpole of *Hyla legleri*.

HABITAT: Usually found near streams in Premontane Wet Forest and Rainforest.

BIOLOGY: A moderately common nocturnal, rather shy stream breeder. Males call from February to July from thick vegetation no more than 1 m above the water. Occasional individuals have been seen calling from plant-covered rocks in the stream and ponds or water supplies at the Las Cruces Biological Station, Puntarenas Province. In captivity males are territorial and defended egg-laying cavities under an artificial waterfall (Proy 1993b). Eggs were attached to the roof of the cavity, usually below the waterline. Clutch size was 42 to 142 black-and-yellow eggs about 2 mm in diameter. Larvae metamorphosed one hundred days after egg deposition. Tadpoles are found in small, shallow streams where they rest and feed on the bottom. They are somewhat intermediate between ecomorphs 10 and 11 (table 7.2) and have a moderate-sized oral disk (10) but with papillae completely around the disk and denticle rows increased to 2–3/5 (11). Metamorphs have been found in June and July, but amplexus, oviposition, and eggs are unknown for this species. Metamorphs are 14 to 18 mm in standard length.

REMARKS: Recently Campbell and Smith (1992) claimed that the iris in *H. legleri* is bronze or copper. The original describer (Taylor 1958a) and subsequent fieldworkers in Costa Rica (Savage 1968a; Savage and Heyer 1968; Duellman 1970) all report red irises as characteristic of this species, including metamorphs. My observations indicate that the iris is more a ruby red than the bright scarlet of some *Agalychnis* and members of *Duellmanohyla*. Of course color is a perception, so two people may disagree on their use of the terms orange, red, bronze, or copper to describe a particular hue. Nevertheless, the consensus is that *H. legleri* has red eyes. You can test this statement by looking at Duellman's 1970 color plate (54:3) or by catching some of these frogs.

DISTRIBUTION: Humid premontane areas on the Pacific slopes of the Cordillera de Talamanca of Costa Rica and extreme western Panama (880–1,524 m).

Hyla tuberculosa Group

These moderate-sized to large denizens of the canopy are immediately recognizable by having immense hands and feet and scalloped fleshy fringes along the outer margins of the forearm and foot. The dorsal coloration is green or tan mottled (or not) with brown or dark green, and the toes are fully or nearly fully webbed. Other significant characteristics shared by the group members are adult males with a protuberant prepollex and a projecting spine or spiny nuptial thumb pads; cephalic skin co-ossified to skull (or not) in adults; no axillary membrane or mental gland; vomerine teeth in curved series; quadratojugal articulates with maxilla; 2N unknown. Tadpoles have a depressed body, small oral disk, and 2/3 denticle rows. Other species belonging to this group beside the two Costa Rican forms are *Hyla echi-*

nata (Oaxaca, Mexico); *H. minera* (Guatemala); *H. salvaje* (southwestern Honduras); *H. thysanota* (eastern Panama); *H. tuberculosa* (upper Amazon basin); and *H. valancifer* (eastern Mexico).

These frogs appear to lay their eggs in tree holes, where the tadpoles complete development (pattern 4, table 7.1), although the tadpole of one species, *Hyla fimbrimembra*, was collected in a roadside ditch (Savage 1981c).

Duellman (1970) and Savage and Heyer (1968) called this group the *Hyla miliaria* species group.

Hyla fimbrimembra Taylor, 1948b

(plate 173; map 7.91)

DIAGNOSTICS: A very large brown, casque-headed treefrog with large hands, feet, and finger disks, scalloped dermal flaps along the outer margins of the limbs, a protuberant prepollex, a supratympanic fold that curves downward behind the tympanum and then posteriorly to terminate above the axilla, and the web not extending to the base of the disk on any finger. Adults of this species have the skin co-ossified to the anterior skull bones and a minutely granular dorsum, but in juveniles there is no co-ossification and the dorsum is tuberculate.

DESCRIPTION: Adult male 79 mm in total length, adult females 71 to 92 mm; venter granular; no keratinized tips on ventral tubercles; head broader than long; head subovoid in dorsal outline, snout truncate from above and in profile; eyes large; tympanum distinct, diameter one-half to three-fourths diameter of eye; presumably paired vocal slits and apparently a single internal subgular vocal sac in adult males; fingers moderately long and robust; large finger disks, width of disk on finger III about equal to tympanum; distal

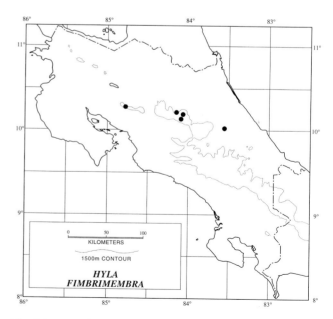

Map 7.91. Distribution of *Hyla fimbrimembra*.

subarticular tubercles single; fingers extensively webbed; webbing formula: I 2–2 II 2¼–2⁻ III 2¾–1 IV; prepollical pad elongate, rectangular, and terminus blunt; scattered small black spines on base of thumb and prepollex in adult male; palmar tubercle small, conical; scalloped dermal flap from elbow to base of disk on finger IV, scallops pointed; similar dermal flap from heel to base of disk on toe V; several pointed heel tubercles present; toes extensively webbed; webbing formula: I ½–2⁺ II 1–2⁺ III 1⁺–2 IV 2–1 V; inner metatarsal tubercle elongate, flat, outer small, conical; fringelike elevated tarsal fold. Adults: overall color brown with darker brown markings on flank and groin, which may be venated; edge of upper lip and tips of digits dark brown; venter and undersides of thighs lavender brown with cream marks; webs brown; flanks and groin tan with or without dark blotch in groin or uniform dark brown; posterior thigh surfaces tan; iris coppery brown; juveniles have lichenlike dorsal pattern of yellows, greens, and browns, showing marked metachrosis; white heel tubercles, limb fringes, and circumcloacal ornamentation (not present in adults); iris red orange.

LARVAE (fig. 7.96): Giant, total length 77 mm; body depressed, heavily tuberculate laterally; mouth directed ventrally; nostrils directed anterolaterally; eyes directed laterally; spiracle lateral, sinistral; vent tube dextral; tail long; fins low; tip of tail rounded; oral disk small with delicate beaks and 2/3 rows of denticles; A1, A2, and P1 with median gap; disk bordered by two to three rows of small papillae laterally, one row along lower margin of lower labium; papillae incomplete across upper labium; beaks serrate. In

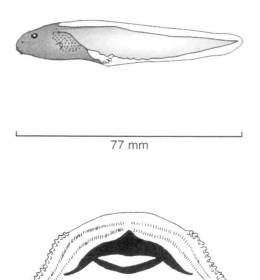

77 mm

Figure 7.96. Tadpole of *Hyla fimbrimembra*.

preservative: body uniform brown, except dirty white venter; black peritoneum visible through skin; tail musculature pinkish tan; caudal fins and overlying skin on tail transparent, with network of black pigment over myosepta.

VOICE: Not known.

SIMILAR SPECIES: (1) *Hyla miliaria* has fingers II–III–IV and all toes fully webbed, and the dorsal surfaces are tuberculate in adults. (2) *Hyla rosenbergi* almost always has a midcephalic dark stripe and thin dark vertical flank and groin markings and lacks fleshy flaps on the limbs.

HABITAT: Undisturbed Premontane and Lower Montane Rainforests.

BIOLOGY: All evidence points to this rarely seen or collected species' being a nocturnal canopy dweller. Although the only known tadpole was found in a roadside ditch, it is probable that *H. fimbrimembra* usually oviposits and the larvae develop in water-filled cavities in trees, as in other members of the *tuberculosa* species group. Other aspects of its biology probably are similar to the slightly better known features of *H. miliaria* (see that species account below). The smallest juvenile is 48 mm in standard length.

REMARKS: Taylor (1948b) described *Hyla richardi* and *Hyla fimbrimembra*, both from the slopes of Volcán Poás, Alajuela Province. Since the former name was preoccupied by *Hyla richardi* Baird, 1854, Taylor (1952a) renamed the species *Hyla richardtaylori*. The holotype of *H. richardtaylori* (FMNH 191783) is a juvenile female with a truncate snout but lacking cranial co-ossification. The holotype of *H. fimbrimembra* (FMNH 191784) is an adult female with rounded snout and cranial co-ossification. Savage and Heyer (1968) raised the possibility that the differences were due to ontogenetic variation, since the two specimens resembled one another in most other features, particularly finger and toe webbing. Duellman (1970), after a review of all fringe-limbed hylids known up to that time, confirmed that suggestion and synonymized the two nominal taxa. As first reviser he chose *H. fimbrimembra* as the correct name for the species. Subsequent collections now bring the total number of examples of *H. fimbrimembra* to one adult male, six females (three adults, three juveniles), and one tadpole. Duellman (1970) and Hayes, Pounds, and Robinson (1986) emphasized four features that in combination separate the species from other fringe-limbed forms: a supratympanic fold curving posteriorly behind the tympanum to the axilla; skin folds below the cloaca; sharply pointed scallops on the limb fringes; and narrow transverse dark bands on the limbs. Newly available material indicates some variability in several of these features. The skin folds below the cloaca are best developed in juveniles and are absent in the largest adult (CRE 7414; Heredia: Braulio Carrillo National Park: La Selva extension: northern slope of Volcán Barva, 1,800 m). The scallops in this specimen are bluntly rounded, not sharp, and the dark transverse leg bands are very broad. Savage (1981c) referred a single large tadpole (CRE 7015) to this species based on elimination, since the larvae of only five hylids from Costa Rica were undescribed at that time. This allocation was further supported by the tuberculate body of the tadpole, since tuberculation of the dorsal surfaces of the body in juveniles is characteristic of frogs of the *tuberculosa* group. In addition the tadpole was collected virtually at the type locality of *H. fimbrimembra* (Costa Rica: Alajuela: Isla Bonita, 1,200 m). The secondary sexual feature of the male prepollex is of great taxonomic significance in this group, but possible relationships of *H. fimbrimembra* have been obscure because no male had been collected until recently. It should be pointed out, however, that the tadpoles of the recently described *Hyla salvaje* of Honduras are strikingly similar to that of *H. fimbrimembra*, and adults also resemble the latter form (Wilson, McCranie, and Williams 1985). Males of *H. salvaje* have the prepollex and base of the thumb covered by a series of black-tipped horny spines. Several other species in the *tuberculosa* group have similar structures *(Hyla echinata, H. minera, and H. tuberculosa)*. Adult males of *Hyla miliaria* have a single sharp, bony spine on the free end of the prepollex and no black-tipped integumentary spines. *H. valancifer* has a blunt protuberant prepollex but lacks any spines. *H. thysanota* is known from a single female. The recently collected adult male of *H. fimbrimembra* (CHP 1036; Panama: Chiriquí: Cerro Horqueta) has scattered small black spines on the base of the thumb and prepollex, confirming the species relationship to the four other taxa in the group having similar secondary sexual features. It was not possible to definitely establish the presence of vocal slits or an internal vocal sac without damaging this specimen. Coloration in the juvenile *H. fimbrimembra* (CRE 4643), described in detail by Hayes, Pounds, and Robinson (1986) as a melange of yellow, greens, and browns, resembles that of adult *H. miliaria*. This example also showed remarkable color changes from an essentially reticulated green and dark brown appearance to yellow cream to tan, narrowly reticulated with dark and having green areas anteriorly and brown lower back and upper surface of legs. Duellman's (1970) figure has led to some confusion, since the illustration of the hand of *H. fimbremembra* (fig. 166A) is labeled *H. valancifer* and the hand of the latter (fig. 166C) is labeled *H. fimbrimembra*.

DISTRIBUTION: Humid premontane and lower montane slopes of the Cordillera de Tilarán, Cordillera Central, and Cordillera de Talamanca of Costa Rica and western Panama (750–1,900 m).

Hyla miliaria (Cope, "1885," 1886)

(plates 174–76; map 7.92)

DIAGNOSTICS: This spectacular species is a huge mottled brown and green frog with exceptionally large hands, feet, and digital disks, scalloped dermal flaps along the outer limb margins, a protuberant prepollex (with a spine in adult males), a supratympanic fold that runs directly downward

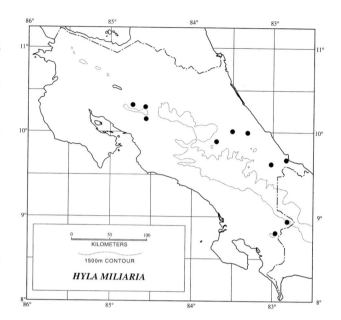

Map 7.92. Distribution of *Hyla miliaria*.

behind the tympanum, and the webs extending to the base of at least one finger disk. Osteoderms are present on the head in adults, so the skull roof and skin above it appear to be co-ossified. The dorsum is tuberculate in juveniles and adults, and adults have osteoderms middorsally on the body.

DESCRIPTION: Adult males 57 to 110 mm in standard length, adult females 69 to 86 mm; head wider than long; ventral tubercles of adults with keratinized tips; head rounded in dorsal outline; snout truncate from above; profile truncate; eyes large; tympanum distinct, diameter one-half to three-fourths diameter of eye; paired vocal slits and single moderately distensible internal subgular vocal sac; fingers moderately long and robust; huge finger disks, width of disk on finger III one and one-half times diameter of eye; distal subarticular tubercle on finger IV sometimes bifid; a terminal recurved prepollical spine and spine sheath in adult males; when arms are flexed, prepollical spine fits into depression on anteroventral margin of upper arm; fingers III and IV fully webbed; webbing formula: I 1½–2 II ½–¾ III 1–¾ IV; prepollical pad greatly enlarged, recurved terminally; palmar tubercle flat, semidivided; scalloped dermal flap from elbow to base of disk on finger IV; similar flap from heel to base of toe V; numerous small heel tubercles; toes fully webbed; webbing formula: I 1⁻–1 II ¾–1 III ¾–1¼ IV 1–¾ V; inner metatarsal tubercle elongate, elliptical; several small tubercles on distal outer edge of tarsus; distinct tarsal fold. Dorsal surfaces mottled brown and green with some white, orange tan, and metallic green areas or spots; showing marked metachrosis from predominantly green to predominantly brown; webs brown or brown and green; axilla, groin, and posterior thigh dull brown, with irregular paler areas on rear of thigh; broad dark transverse bars on upper surface of legs; venter tan or brown, sometimes with blotches of white; iris dark brown or bronze with reddish brown reticulations.

LARVAE: Not known.

VOICE: A single "gurrrrgh" repeated at widely separated intervals (Michael Fogden, pers. comm.).

SIMILAR SPECIES: (1) *Hyla fimbrimembra* has the webs on fingers not reaching the base of the disks, the toes three-fourths webbed, and the dorsum smooth in adults. (2) *Hyla rosenbergi* nearly always has a midcephalic dark stripe and vertical dark markings along the flanks and in the groin and lacks fleshy fringes on the limbs.

HABITAT: Undisturbed Lowland Moist Forest and Premontane Wet Forest and Rainforest.

BIOLOGY: Undoubtedly a nocturnal canopy species rarely seen or collected. Dunn (1943) and Duellman (1970) suggested that their large size, rugosity, thick skin, and the presence of osteoderms protect frogs of this group from desiccation in the treetops. This frog, and probably all members of the *tuberculosa* group, parachutes (Oliver 1951) using the large hands feet and disks and the limb fringes to increase the planing surface. When leaping from a perch, these frogs extend the forearms and legs, spread the fingers and toes wide, and arch the fingers upward, as beautifully illustrated in Duellman (1970). That author recounts this frog's parachuting under natural conditions and reports on experiments with a captive specimen. In the posture just described, the frog could remain airborne for about 3 m while losing 1 m of altitude. Circumstantial evidence indicates that these frogs call and breed in tree holes or other water-filled cavities. Taylor (1952a) collected a specimen of this species (the male holotype of *H. immensa*) from a water-filled hole in a tree 1.5 m above the ground. Michael Fogden reports hearing a call, later associated with this species, coming from high in a tree at night. The tree hole site was spotted during the day, but with no frog. Late in the afternoon Fogden waited on a ladder at the tree hole and captured the frog when it returned at dusk. Unlike most anurans, males of this species are larger than females. In at least some other such species (e.g., *Hyla boans* group) males physically defend their territory (oviposition sites) and use their prepollical spines to slash their rivals. Similar behaviors may well be typical of *H. miliaria*. In the related Honduras species *Hyla salvaje*, tadpoles were found in two water-filled cavities in tree trunks. In each of these instances an adult male was taken from the same cavity, the first along with the tadpoles, the second a year later (Wilson, McCranie, and Williams 1985). Tadpoles in one sample contained probably conspecific frog eggs, and it may be that all frogs in this species group are oophagous and perhaps are obligate egg eaters. The smallest juvenile is 40 mm in standard length, and the skin on its head is free from any sign of osteoderms.

REMARKS: This uncommon species was originally described by Cope (1885) from a locality between El Castillo and San Juan del Norte, Department of Río San Juan, Nicaragua (holotype: USNM 14193, an adult male). Subsequently Dunn (1943) described *Hyla phantasmagoria* (holotype: MLS 267) from north-central Colombia and Taylor (1952a) described *Hyla immensa* (holotype: KU 30404; Cartago: Centro Agronomico Tropical de Investigación y Enseñanza) from Costa Rica, also based on male examples. Duellman (1970) compared these three specimens with one male and one female he collected in western Panama and concluded that only one taxon was involved. It now seems questionable that the type of *H. phantasmagoria* is conspecific with *H. miliaria* (W. Duellman, pers. comm.). Sixteen specimens of this bizarre frog are now in collections. In preserved adult males the prepollical bony spine is variously exposed, strongly indicating that, as in the *Hyla boans* group, the sheath around the spine is retractable. Adults of this species superficially appear to have the skull roofing bones co-ossified with the overlying integument. As Duellman (1970) pointed out, osteoderms are present and closely compacted on the head of large *H. miliaria*, making the skin on top of the head immovable but not co-ossified to the skull (pseudoco-ossification).

DISTRIBUTION: Humid lowlands and premontane slopes from southeastern Nicaragua to southeastern Costa Rica on the Atlantic versant (20–900 m) and on the Pacific versant in humid premontane areas of southwestern Costa Rica and western Panama (1,000–1,330 m).

Hyla zeteki Group

These small tan, yellowish brown to yellowish green frogs breed high in the canopy, usually in tank bromeliads, and have reduced finger and toe webbing. Other characteristics include adult males with white nuptial thumb pads, without protuberant prepollices; axillary membrane present; no fleshy limb fringes or mental gland; vomerine teeth in linear patches entirely posterior to choanae; quadratojugal present and articulating with maxilla or absent; 2N unknown. Tadpoles robust, with reduced fins and small anterodorsal oral disks without denticles or in 1/1–2 rows.

The eggs in this group are deposited in bromeliads, where the larvae undergo further development. Dunn's (1937a) account indicates that the eggs of *H. zeteki* are laid out of the water on the leaves of bromeliads (pattern 24, table 7.1), as in *Anotheca spinosa*. The tadpoles are highly specialized in having the denticles widely spaced in a reduced number of rows. One species, *Hyla zeteki*, is obligately oophagous (ecomorph 9, table 7.2), and the other, *Hyla picadoi*, is putatively an egg eater. It is probable that the mothers of both species feed their tadpoles unfertilized eggs, as in *Anotheca* and some dendrobatids.

Both species assigned to this group occur in Costa Rica, and it corresponds to the *H. zeteki* groups of both Duellman (1970) and Savage and Heyer (1968).

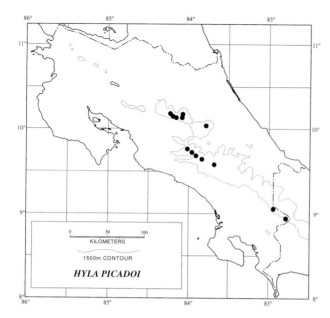

Map 7.93. Distribution of *Hyla picadoi*.

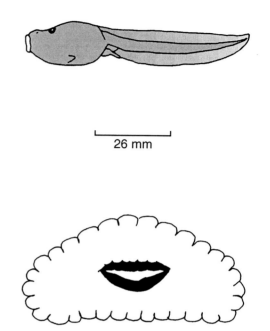

26 mm

Figure 7.97. Tadpole of *Hyla picadoi*.

Hyla picadoi Dunn, 1937a

(plate 177; map 7.93)

DIAGNOSTICS: A moderately small yellowish to orangish tan treefrog having a pair of odontoids at the symphysis of the lower jaw (fig. 7.71b), the tympanum vertical and indistinct to hidden, and vestigial finger webs.

DESCRIPTION: Adult males 27 to 32 mm in standard length, adult females 34 to 35 mm; skin smooth; venter strongly granular; head broader than long, with bulging muscles in temporal region; from above, snout truncate with a median point, snout acute in profile; tympanum indistinct to hidden, diameter less than one-half diameter of eye; paired vocal sacs and barely distensible single internal subgular vocal sac in adult males; fingers short and robust; finger disks moderate, width of disk on fingers III and IV one and one-half times diameter of tympanum; distal subarticular tubercle on finger IV usually bifid; white nonspinous nuptial pad on base of thumb in adult males; finger webs vestigial; webbing formula: I 3–3 II 2–3 III 3⁻–2⁺ IV; prepollical pad moderately enlarged; palmar tubercle moderate, bifid; toe webbing reduced, vestigial between toes I–II–III; webbing formula: I 2–3 II 2⁻–3 III 1⅔–3 IV 3–2⁻ V; inner metatarsal tubercle large, ovoid, outer small or absent; strong tarsal fold present. Dorsal surfaces metallic gold orange, yellowish tan, to olive brown with darker markings; sides of head and supratympanic fold usually marked with dark pigment that continues as narrow, wavy dark lateral line along flank, sometimes reaching groin; top of head usually with dark blotch, sometimes extending posteriorly as irregular narrow middorsal stripe; temporal areas sometimes orange; no transverse dark bars on upper surfaces of limbs, although dark reticulation sometimes present; a narrow dark wrist band usually present; venter white; iris reddish copper.

LARVAE (fig. 7.97): Moderate-sized, 26 mm in total length stage 31; body depressed, stout, guitar-shaped from above; mouth anterior; nostrils dorsal; eyes directed dorsolaterally; spiracle lateroventral, sinistral; vent tube dextral; tail long; fins low; tail tip pointed; oral disk moderate, complete with beaks and 0/0 or possibly 1/1 rows of denticles; disk bordered by single row of large papillae; upper beak smooth, lower beak robust, with moderate serrations; lower beak not visible without spreading disk; uniformly gray except light venter and transparent fins.

VOICE: Not known. A loud call probably of this species is often heard from high in large trees, especially in the giant oak forests.

SIMILAR SPECIES: *Hyla zeteki* is smaller (adult males 21 to 24 mm in standard length, adult females 24 to 27 mm), has considerable finger webbing, and has a single odontoid at the tip of the lower jaw and the tympanum distinct and directed dorsolaterally (fig. 7.71a).

HABITAT: Lower Montane and Montane Rainforests.

BIOLOGY: Males of this uncommonly seen species appear to call from high in the canopy throughout the rainy season. Most examples have been collected from bromeliads. Tadpoles, adults, and metamorphs were recently found together in bromeliads. The first time the leaves were gently jiggled, tadpoles swam to the surface (K. R. Lips, pers. comm.). The morphology and habits of the tadpoles of this species strongly suggest that they, like the allied *H. zeteki*, are obligately oophagous (ecomorph 9, table 7.2). The smallest juvenile is 18.4 mm in standard length.

REMARKS: Robinson (1977) originally described the tadpole of this species as lacking denticles. Later Wassersug (in Lannoo, Townsend, and Wasserzug 1987) indicated that

1/1 rows of denticles are present. Recently collected material confirms the absence of denticles in larvae of this species. Since the presence or absence of denticles may be a variant in this taxon, *H. picadoi* is double keyed in the tadpole key (p. 178).

DISTRIBUTION: Humid lower montane and montane slopes of the Cordillera Central and Cordillera de Talamanca of Costa Rica and western Panama (1,920–2,770 m).

Hyla zeteki Gaige, 1929

(plate 178; map 7.94)

DIAGNOSTICS: An uncommonly found small tan, yellowish brown, or pale yellowish green treefrog having the eyes directed forward at about a 45° angle, the tympanum indistinct and directed obliquely dorsally (fig. 7.71a), a distinct dark black band across the wrist, a single odontoid at tip of lower jaw, and moderate webbing between fingers II–III–IV.

DESCRIPTION: Adult males 21 to 24 mm in standard length, adult females 24 to 27 mm; dorsum smooth; venter strongly granulate; head broader than long, with bulging muscle in temporal region;, dorsal outline of snout truncate, snout profile acute; tympanum small, distinct, diameter one-third that of eye; paired vocal slits and a slightly distensible single internal subgular vocal sac in adult males; fingers relatively short and stout; finger disks fairly large, width of disks on fingers III and IV one and one-half times diameter of tympanum; distal subarticular tubercle on finger IV often bifid; white nonspinous nuptial pads on base of thumb in adult males; finger webs minimal, vestigial between fingers I and II; webbing formula: I 3–3 II 2–3 III 3⁺–2⁺ IV; prepollical pad greatly enlarged, bulbous; palmar tubercle flat, bifid; toes moderately webbed; webbing formula: I 2–3 II

21 mm

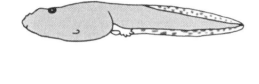

Figure 7.98. Tadpole of *Hyla zeteki*.

2⁻–3⁺ III 2–3 IV 3⁻–2 V; large oblong, inner metatarsal tubercle; outer low, small or absent; weak tarsal fold present. Upper surfaces yellowish tan to yellowish green, lacking distinct pattern except conspicuous narrow transverse dark band across wrist; venter unpigmented so that silvery parietal peritoneum visible; other undersurfaces pale yellow to yellowish green; iris light dull red, red brown, or pinkish bronze.

LARVAE (fig. 7.98): Small, 21 mm in total length; body stout, depressed, guitar shaped from above; mouth anterodorsal; nostrils dorsal; eyes directed dorsolaterally; spiracle lateroventral, sinistral; vent tube dextral; tail moderate; fins very low; tip of tail rounded; oral disk moderate, complete with very broad beaks and 1/2 rows of denticles; outer row completely encircling mouth; denticles large, widely spaced; single row of large, widely spaced marginal papillae; an irregular row of submarginal papillae; upper beak smooth, massive lower beak finely serrate; beaks partially hidden in natural state; tadpole uniform gray.

VOICE: The call remains unknown.

SIMILAR SPECIES: *Hyla picadoi* is larger (adult males 27 to 32 mm in standard length, adult females 34 to 35 mm), has vestigial finger webs, has paired odontoids at tip of lower jaw (fig. 7.71b), and usually has lips, dorsal area, and supratympanic fold covered by dark pigment.

HABITAT: Associated with epiphytic bromeliads in Premontane Wet Forest and Rainforest and the lower portion of Lower Montane Rainforest.

BIOLOGY: Dunn (1937a) reported on the collection of adults, eggs, and tadpoles of *H. zeteki* in bromeliads between La Palma and La Hondura, San José Province. Five adults,

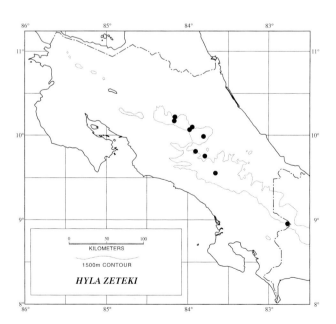

Map 7.94. Distribution of *Hyla zeteki*.

nine tadpoles, and five eggs were recovered from the eleven bromeliads on a single downed tree and from one brome-liad each from two standing trees nearby (within 7.5 m). Single plants contained one or two adults, three tadpoles (in one case of different size-classes), two or three eggs, and one frog and a tadpole. The eggs were laid on the outside leaves above the waterline in the bromeliads' cups (pattern 24, table 7.1). Two were on the same leaf in one plant, and three were on a different leaf in another. The tadpoles' stomachs contained conspecific eggs. Comparing these data with the now better known biology of another Costa Rican hylid, *Anothca spinosa*, which also lays eggs in bromeliads and is obligatorily oophagous, suggests a similar life history. In both species the eggs are deposited above the waterline, only one or a few tadpoles are present, and two size-classes from the same clutch are often found together. In *Anotheca* the female returns and lays unfertilized nutritive eggs for the tadpoles to eat. The co-occurrence of an adult and a tadpole together in *H. zeteki* suggests similar behavior (see species account of *A. spinosa* for more details). One gravid female reported by Taylor (1958a) and Duellman (1970) contained 24 large, heavily pigmented eggs.

REMARKS: A specimen of this form is the basis of Pi-cado's (1913) identification of "*Hylella fleischmanni*" as a bromeliad associate. In the original description Gaige (1929) reported that in her thirteen specimens some had a narrow red middorsal stripe and/or some red elsewhere in the coloration. None of the subsequently collected examples from Costa Rica or Panama (Myers and Duellman 1982) have red markings. Duellman's (1970) figure (fig. 153) of the mouthparts of the larva was printed upside down.

DISTRIBUTION: Humid premontane areas or marginally in the lower montane zone of the Cordillera Central and Cordillera de Talamanca of Costa Rica and western Panama (1,200–1,804 m).

Genus *Osteopilus* Fitzinger, 1843

Moderate-sized to giant treefrogs with the cephalic integu-ment co-ossified to the skull to form a casque in adults. They may be uniform brown; brown, gray, or olive with dis-tinct darker markings (blotched, reticulate, or mottled); or else mottled, marbled, or variegated with greens, grays, and brown with a distinct pattern. The finger disks are round, fingers have reduced webbing, and toes are moderately webbed; the eyes are large, and the tympanum is large and distinct with a prominent annulus tympanicus.

Other characteristics shared by members of the genus include no fleshy fringes on limbs; single external or inter-nal subgular vocal sac in adult males; nuptial thumb pads in adult males and prepollices not protuberant; no fronto-parietal fontanelle; quadratojugal articulating with maxilla; sphenethmoid present; teeth on premaxilla, maxilla, and vo-mer; depressor mandibulae with two parts, one from dorsal fascia, one from the annulus tympanicus (DFAT); adductor mandibulae externus superficialis present (e); sacral diapo-

physes broadly dilated. Tadpoles with elongate tails, small anteroventral mouths, lateral sinistral spiracle, 0/1 or 2/4– 5 denticle rows, and papillae absent above mouth; 2N = 24 or 34.

The combination of co-ossified cephalic integument, green bones, and absence of fleshy limb fringes and a pro-tuberant prepollex will separate this genus from all other Costa Rican treefrogs.

This is an Antillean genus; one species, *Osteopilus septen-trionalis*, has been introduced to Costa Rica, Puerto Rico, St. Croix, St. Thomas, and southern Florida. *O. brunneus* is endemic to Jamaica and *O. dominicensis* to Hispaniola, and *O. septentrionalis* occurs naturally in Cuba, the Cayman Is-lands, and the Bahamas. The last two species lay their eggs opportunistically in lentic situations (pattern 1, table 7.1). *O. brunneus*, on the other hand, deposits its eggs in bro-meliads, and the larvae are obligately oophagous (pattern 4, table 7.1; ecomorph 9, table 7.2), feeding on unfertilized eggs of their own species probably deposited by the mother, as in some *Dendrobates* (Lannoo, Townsend, and Wasser-zug 1987) and *Anotheca* (Jungfer 1996).

Osteopilus septentrionalis (C. Duméril and Bibron, 1841)

(plate 179; map 7.95)

DIAGNOSTICS: A large pale green, ashy gray, pale brown, or reddish brown treefrog marked with bold dark reticula-tions or elongate blotches and having very large finger disks. In adults of this species the skin on top of the head is co-ossified with the skull to form a definite casque (fig. 7.99). The bones are green, the fingers are webbed basally, and the flanks and posterior thigh surface are reticulated cream or yellow and dark brown to black.

DESCRIPTION: Adult males 27 to 89 mm in standard

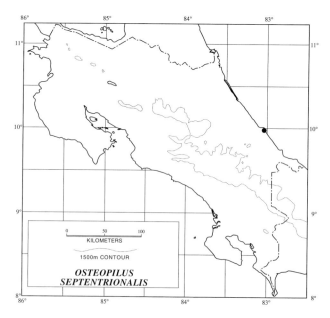

Map 7.95. Distribution of *Osteopilus septentrionalis*.

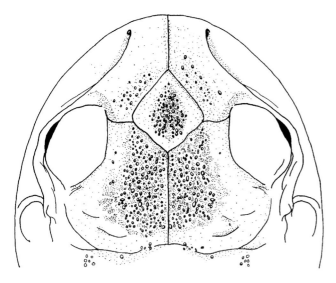

Figure 7.99. Dorsal view of casque head of adult *Osteopilus septentrionalis* showing skin co-ossified to exostosed (pitted) skull roof.

length, rarely over 75 mm, adult females 52 to 165 mm; sexual dimorphism pronounced, females (mean = 75 mm) much larger than males (mean = 55 mm); dorsal surfaces usually tuberculate; venter strongly areolate; head broader than long; snout rounded in dorsal outline, truncate in profile; eyes moderate; tympanum large, distinct, diameter one-third to three-fourths that of eye; annulus tympanicus prominent; paired vocal slits and slightly distensible single internal subgular vocal sac with lateral extensions in adult males; finger and toe disks very large, round, disk on finger III equal to or greater than diameter of tympanum; subarticular tubercle single; palm and sole granular; dark brown nonspinous nuptial pad at base of thumb in adult males; fingers with basal webbing; webbing formula: I 2½–2¾ II 2–3 III 2¼–2¾ IV; thenar pad moderately enlarged; no palmar tubercle: toes moderately webbed; webbing formula: I ½–2 II 1–2¼ III 1¼–2¾ IV 2¾–1 V; inner metatarsal tubercle moderate, elongate, no outer tubercle; tarsal fold present. Dorsal ground color variable from light green or gray to tan or reddish brown marked with bold darker reticulations or elongate blotches; flanks and upper limb surfaces with dark transverse bars; flanks and posterior thigh surfaces reticulated with light and dark; creamy white spots usually enclosed in flank and thigh reticulum; venter creamy white; throat often suffused with gray in males; bones green; iris dull yellow.

LARVAE (fig. 7.100): Large, 32 mm in total length at stage 25; body ovoid; mouth anteroventral; nostrils dorsolateral; eyes dorsolateral, directed dorsally; spiracle lateral, sinistral; vent tube dextral; tail moderate; caudal fins high; tail tip pointed; oral disk small, entire with slender beaks and 2/4–5 rows of denticles; A2 and P1 with median gaps; papillae present laterally and ventrally; body black, caudal musculature grayish brown; fins with scattered melanophores.

VOICE: A rasping snarl with great variability in range. The commonest call consists of a pulsed call lasting 300 to 400 milliseconds, often preceded by a shorter variable series of notes lasting a total of about 150 milliseconds. This element has a variable series of harmonics with the dominant frequency rising from about 2.1 kHz to 2.5 kHz at the end. The pulse rate for the second portion of the call is 25 to 40 per second, again with a broad series of harmonics. The combined elements form a call given at intervals of 0.9 second (Blair 1958; Duellman and Crombie 1970). The frog also produces a warning shriek or scream when attacked by a predator.

SIMILAR SPECIES: (1) *Phrynohyas venulosa* lacks a casque head and reticulate flanks and thighs. (2) *Smilisca baudinii* has a broad dark postorbital to axilla mark and lacks a casque head and green bones. (3) *Smilisca phaeota* has distinct white labial, forearm, and tarsal stripes and white bones and lacks a casque head.

HABITAT: In Costa Rica known only from the urban center of Puerto Limón, Limón Province, where it has become established and is common in city parks.

BIOLOGY: An essentially nocturnal treefrog that usually hides during the day. In its native Cuba and the Bahamas it is found in a variety of generally humid habitats but also in disturbed areas and in and around human habitations. Because of this propensity it is a typical tramp species that humans have incidentally introduced to a number of Caribbean islands and the mainland. In these circumstances it is a "weed" that has become established in heavily populated and disturbed areas. The following account provides general information on the species based on Schwartz and Henderson (1991) and on the behavior of the species in southern

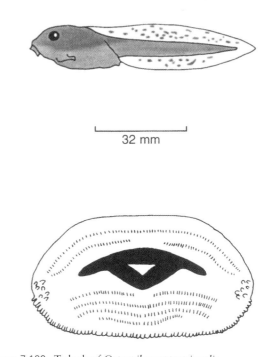

32 mm

Figure 7.100. Tadpole of *Osteopilus septentrionalis*.

Florida. This frog is most commonly seen or heard in breeding aggregations. Individuals are also found throughout most of the year in a variety of places, both under surface objects and hiding in plants, often in gardens. In urban areas they are found around swimming pools and occasionally in toilet bowls, to which they gain access through the soil stack (plumbing air vent) and drain. *Osteopilus septentrionalis* is an explosive breeder that opportunistically carries out reproductive activity throughout the wet season after heavy rains. Males call in large choruses from a variety of elevated sites, primarily in bushes and trees. They also call singly and sporadically throughout the wet seasons, which are at somewhat different times in various areas of introduction. Breeding takes place in temporary ponds, flooded areas, roadside ditches and drainage ditches, or even swimming pools and decorative fountains. Females follow the males to the breeding site after one to two hours and remain in trees or on other elevated objects 3 or 4 m above the ground for some time before joining them in the water. Choruses usually cease about midnight, and egg laying is completed in the early morning hours. Clutches contain up to 130 eggs about 3 mm in diameter including the envelopes (measured twenty-four hours after deposition) laid in a thin sheet on the water surface. Eggs hatch in twenty-four to thirty hours, and the embryos are about 2.6 to 2.8 mm in total length. Metamorphs are 11 to 14 mm in standard length and have pale green to light tan dorsums and a moderately broad cream to yellow dorsolateral stripe on each side. The tadpoles (ecomorph 2, table 7.2) feed by rasping surfaces and are cannibalistic in crowded conditions. The skin secretions of these frogs are very irritating to mucous membranes, and they should be handled with care.

REMARKS: Before 1974 this species was usually called *Hyla septentrionalis*, but Trueb and Tyler (1974) established the validity of *Osteopilus* at that time. Duellman and Crombie (1970) provide a review of this form. *O. septentrionalis* was introduced into southern Florida probably in the nineteenth century and to Puerto Rico more recently. It was first seen in Costa Rica in the mid-1980s and probably arrived in the Caribbean port city of Puerto Limón, Limón Province, as a tramp passenger on a ship docking there. The karyotype is 2N = 24; ten chromosome pairs are metacentrics or submetacentrics, and two pairs are subtelocentrics; NF = 48 (Cole 1974).

DISTRIBUTION: Low and moderate elevations throughout Cuba and the Isla de Juventud, on the Cayman Islands, and on several of the Bahama Islands north of the Mayaguana Passage and Great Inagua Island. Introduced to the Islands of Puerto Rico, St. Croix, and St. Thomas and to the Florida Keys and southern Florida and Costa Rica at Puerto Limón (2 m).

Genus *Phrynohyas* Fitzinger, 1843

Members of this genus (five species) are large, short-legged, robust tan, brown, or gray frogs with exceptionally glandular and generally pustulate skin on the dorsum, having finger webs, large, rounded (as long as wide) finger disks, well-developed webbing between toes I and II, and large eyes and tympanum. Adult males are distinctive among Central American treefrogs in having paired lateral external vocal sacs lying behind the angle of the jaws that are balloon shaped when fully inflated and dark nuptial thumb pads.

Other features that in combination distinguish frogs of this genus from other hylids are bones green; skull completely roofed but not co-ossified with cephalic integument; anterior arm of squamosal not reaching maxilla; quadratojugal articulating with maxilla; sphenethmoid well ossified; bifid, spatulate teeth present on premaxilla, maxilla, and vomer; depressor mandibulae arising as single mass from dorsal fascia (DF); both adductor mandibulae subexternus and externus superficial present (s + e); sacral diapophyses broadly dilated; larvae robust, with small anteroventral mouth, lateral sinistral spiracle, 3–4/4–6 denticle rows and lacking labial papillae above mouth; 2N = 24.

In Costa Rica adult males are readily referred to this genus by reason of the unique paired, lateral vocal sacs. Females and juveniles may be confused with some species of *Hyla* or *Smilisca*, but unlike those forms they have green limb bones visible through the skin.

Species of *Phrynohyas* emit large quantities of a white volatile, noxious alkaline skin secretion when handled. This substance is sticky, insoluble in water, and extremely irritating to mucous membranes. When rubbed in the eyes it causes sharp pain and temporary blindness lasting several hours. It is highly probable that the secretion serves as an antipredator defense. Take great care when collecting frogs of this group. Other animals should never be placed in a collecting bag with them or in an unlaundered bag that previously was used to transport *Phrynohyas* or valuable specimens, may die. Since they are not immune to their own toxins, many *Phyrnohyas* placed together in a bag also may die.

The genus is found in Mesoamerica, primarily in the subhumid lowlands from Tamaulipas and Sinaloa, Mexico, to Nicaragua on the Atlantic versant and eastern Panama on the Pacific slope; in South America the range includes both humid and subhumid areas from northern Colombia and Venezuela, and southward east of the Andes to Bolivia, northeastern Argentina, and eastern Brazil.

The eggs of most species in this genus are laid in lentic, usually temporary, bodies of water (pattern 1, table 7.1), and the tadpoles are free-swimming. At least one South American form, *Phrynohyas resinifictrix*, lays its eggs in tree holes (pattern 4, table 7.1), and the tadpoles are known to eat conspecific eggs on occasion (Jungfer 1996).

Phrynohyas venulosa (Laurenti, 1768)

(plates 180–81; map 7.96)

DIAGNOSTICS: A large, rather robust tan, reddish brown, or pale gray treefrog usually marked with a large dark dorsal blotch covering most of the back, though occa-

Figure 7.101. Tadpole of *Phrynohyas venulosa*.

sionally the dorsum is uniform in color. Adult males are unmistakable because of their paired lateral external vocal sacs behind the angles of the jaws (fig. 7.12f). The species is further distinguished from other treefrogs it may be confused with by lacking dark vertical lip bars and having uniform, not mottled or reticulate, flanks and green bones.

DESCRIPTION: Adult males 70 to 101 mm in standard length, adult females 93 to 114 mm; skin thick and glandular dorsally, smooth to tuberculate in adults, strongly tuberculate in juveniles; throat, venter, and posteroventral surfaces of thighs strongly areolate; head broader than long; snout bluntly rounded from above, bluntly rounded to nearly truncate in profile; eyes moderate; tympanum large, distinct, diameter one-half to four-fifths that of eye, slightly larger in females than in males; paired vocal slits and fully distensible paired lateral external vocal sacs in adult males; sacs evident behind angles of jaws when deflated; fingers short and robust; finger disks large, round, width of disk on finger III greater than diameter of tympanum; distal subarticular tubercle under finger IV usually bifid; tan nonspinous nuptial pad at base of thumb in adult males; fingers moderately webbed; webbing formula: I 2^+–3^- II $1\frac{1}{2}$–$2\frac{1}{2}$ III $1\frac{1}{3}$–2^+ IV; thenar tubercle pad moderately enlarged; no palmar tubercle; toes extensively webbed; webbing formula: I 1–2^- II 1–2 III 1–2 IV 2^-–1 V; inner metatarsal tubercle moderate, ovoid, outer small, conical; distinct tarsal fold present. Dorsal surfaces yellowish tan, tan, reddish brown, or pale gray; usually with a dark olive tan to brown blotch that covers most of dorsum, sometimes with light spots or blotches interrupted by transverse band of ground color; usually with distinct transverse dark bars on upper surfaces of limbs; some individuals uniform brown above; no distinctive flank markings; venter dirty light brown to cream; deflated vocal

sacs dark brown to black, pale brown or olive brown when inflated; iris golden bronze flecked with black.

LARVAE (fig. 7.101): Moderately large, 41 mm in total length; body robust; mouth anteroventral; nostrils anterolateral; eyes directed laterally; spiracle lateral, sinistral; vent tube median; tail moderate; caudal fins high; tip of tail pointed; oral disk small, entire with slender beaks and 3–4/4–6 denticle rows; A1 and A2 present only laterally, usually fragmented; slight median gap in row above mouth (A2 or A3), P1 sometimes with narrow median gap; papillae present laterally and ventrally; lowermost row or two rows on lower jaw usually fragmented; beaks finely serrate; body brown above; venter creamy tan with brown flecks; dark brown lateral longitudinal stripe; fins with minute dark punctations; longitudinal dark stripe on legs; iris pale brown.

VOICE: A loud growl regularly repeated at short intervals. Note duration is 230 to 400 milliseconds repeated 42 to 67 times a minute, and the call is well modulated, with 12 to 14 harmonics and a dominant frequency of 1.39 to 2.5 kHz (Duellman 1970; Lee 1996).

SIMILAR SPECIES: Adult males are unique among Costa Rican hylids in having paired lateral vocal sacs. (1) *Smilisca baudinii* has white bones, a mottled flank pattern, and a broad postorbital to axilla dark mark. (2) *Smilisca phaeota* has a venate flank pattern (fig. 7.73b). (3) *Osteopilus septentrionalis* has a casque head (fig. 7.99) and a reticulate pattern on the flanks and thighs. (4) *Smilisca sila* has prominent warts along the posteroventral margin of the forearm and a dark brown to black reticulum enclosing bluish flecks

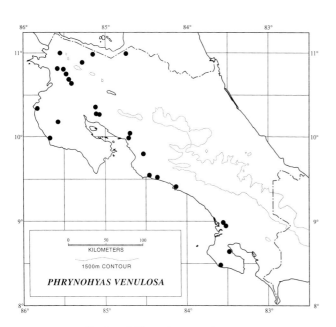

Map 7.96. Distribution of *Phrynohyas venulosa*.

in the groin and on the posterior thigh surface (fig. 7.74b) and lacks green bones.

HABITAT: Lowland Dry, Moist, and Wet Forests. Usually adults are seen only at reproductive sites, but they are sometimes found in hiding places. This species is active at night in the dry season.

BIOLOGY: These nocturnal arboreal frogs are most commonly encountered around temporary ponds during the breeding season (in the rainy season). In the dry season they hide in bromeliads, in tree holes, under bark on living trees or snags, or in the sheaths of heliconias or bananas and are uncommonly seen. At night they emerge and perch on branches or other plant surfaces to forage. Like some *Agalychnis* and members of the *Hyla tuberculosa* group, the species has well-developed parachuting ability (Cott 1926). In one of Cott's field experiments an individual of *Phrynohyas venulosa* dropped from a height of 43 m landed 27 m away on a horizontal line. *P. venulosa* breeds explosively after heavy rains. Males usually call while sitting or floating in shallow water. The inflated vocal sacs resemble balloons floating above the sides of the head. Amplexus takes place in the water, and the eggs are laid as a surface film covering about 1.5 m². The eggs plus envelope are each about 3.5 mm in diameter (Zweifel 1964a). Larvae from eggs collected in the field hatched in one day at total lengths of 3.8 to 4.4 mm (stage 18). Metamorphosis took place at thirty-seven to forty-seven days (Zweifel 1964a; Pyburn 1967). Metamorphs retain the larval dark leg stripe, have green bones, and are about 13 to 17 mm in total length. The leg stripe is lost and adult coloration appears a few days after metamorphosis. The larvae are nektonic (ecomorph 2, table 7.2) and feed by rasping surfaces. The thick glandular skin of this frog is richly supplied with granular glands that secrete copious amounts of a noxious or toxic mucus. The glands are concentrated in the interorbital, occipital, supratympanic, and anterior dorsal regions and in the dorsal pustules. McDiarmid (1968) showed that there are seasonal differences in gland development. In the wet season in Costa Rica (late June on) the glands along the side of the neck are not noticeably enlarged, but they are hypertrophied at the height of the dry season (February and March) and expand to cover most of the tympanum. The glandular secretions, though volatile and apparently poisonous, may also reduce the skin's permeability to water and limit dehydration in the drier times of the year. In some South American *Phrynohyas* these secretions are used to line tree holes that serve as refuges during the dry season (Goeldi 1907; Vellard 1948). This may be a further step in reducing desiccation used by Costa Rican members of the genus. As with all *Phrynohyas*, take care in handling these frogs because the skin secretions produce extreme irritation, swelling, and pain if they are accidentally rubbed in the eyes or touch other mucous membranes (Smith 1941a; Duellman 1956; Shannon and Humphrey 1957; Neill and Allen 1959; Janzen 1962). In Belize

this species is called the "pepper treefrog" because it may stimulate sneezing (Meyer and Foster 1996).

REMARKS: The correct name for this species has had a tortuous history fully detailed by Duellman (1970). Controversy existed over whether the name *Acrodytes* should be used as the generic name and whether the species should be called *Phrynohyas zonata* or *Acrodytes venulosa* as Taylor (1952a) did. The matter was resolved by action of the International Commission on Zoological Nomenclature (Opinion 520, 1958), which confirmed the nomenclature followed here. A plethora of synonyms are based on representatives of this taxon (Duellman 1970; McDiarmid 1968). Only a single species seems to be represented in Mexico and Central America, since the purported differences in coloration among named forms are subject to considerable individual variation. Differences in size emphasized by Duellman (1956) among the nominal forms are bridged by samples from intermediate geographic areas as well (McDiarmid 1968). Nevertheless, *P. venulosa* from Costa Rica and Panama are larger (maximum standard length for males 101 mm, females 114 mm) than elsewhere in Central America (male maximum 87 mm, female 93 mm). The largest examples are from the rainforests of the Golfo Dulce region, contradicting McDiarmid's (1968) hypothesis that large size (lower surface-to-volume ratio) in this species is an adaptation to semiarid environments. Duellman (1971) reviewed the genus in South America, considerably expanding the known range of *P. venulosa* but without change in the status of Central American populations. The karyotype is 2N = 24, consisting of all metacentric or subtelocentric chromosomes; NF = 48 (Bogart 1973).

DISTRIBUTION: Lowlands of Mesoamerica from Sinaloa, Mexico, to eastern Panama on the Pacific slope and on the Atlantic versant from Tamaulipas, Mexico, to northern Nicaragua, with populations contiguous with those of the Pacific slope in northern Costa Rica and central Panama; in South America south to western Ecuador, east to Venezuela, Trinidad, and Tobago and south through the Guianas to northeastern Brazil in the Amazon basin to Bolivia and thence to northern Argentina; in Costa Rica throughout the Pacific lowlands and in the Río San Juan (Atlantic) drainage in the north (2–285 m).

Genus *Scinax* Wagler, 1830

These small to moderate-sized rather drably colored treefrogs (seventy-six species) have the web between toes I and II absent or reduced to a fringe along the inner margin of toe II, the finger disks palmate (wider than long), and fingers with webs reduced or absent. The eyes are moderate, and the tympanum is large and distinct.

Other features that in combination distinguish the genus from others in the subfamily are frontoparietal fontanelle variably exposed; integument of head not co-ossified to skull; anterior arm of squamosal not reaching maxilla;

quadratojugal slender, articulating with maxilla; sphenethmoid well ossified; small, blunt, pedicellate teeth on premaxilla, maxilla, and vomer; depressor mandibulae arising from dorsal fascia (DF); only the adductor mandibulae posterior subexternus present (s); sacral diapophyses weakly dilated; vocal slits and vocal sac present in adult males, usually single external subgular but paired lateral external subgular sac present in a few species; white nuptial thumb pads in adult males; larvae deep bodied with small anteroventral mouth, lateral, sinistral spiracle, tail often terminating in a thin flagellum, 2/3 rows of denticles, and labial papillae usually absent above mouth; 2N = 24.

The characters of the fingers and toes serve to separate species of *Scinax* from other genera of the family found in Costa Rica. These frogs were usually placed in the *Hyla rubra* group before 1977 (León 1969; Duellman 1970). At that time Fouquette and Delahoussye (1977) proposed that the *rubra* group represented a genus distinct from *Hyla* and used the name *Ololygon* Fitzinger, 1843. This decision was based primarily on sperm tail morphology. According to these authors *Ololygon* had double-filamented tails whereas most other hylids had single ones. However, a variety of other frogs (bufonids, some leptodactylids, pelodryadids, and centrolenids) also have double-filamented tails, as does the presumably close relative of this genus, *Sphaenorhynchus*.

The genus ranges from Guerrero and Tamaulipas, Mexico, to Uruguay and Argentina and extreme northwestern Peru.

Pombal and Gordo (1991) established that *Scinax* Wagler, 1830, is the proper name for the genus, since it antedates *Ololygon*. Duellman and Wiens (1992) accepted that conclusion, further reviewed the characteristics of the genus, and recognized seven species groups within it. Most of the groups were pointed out by Fouquette and Delahoussye (1977) based on differences in sperm head morphology. Three of the species groups are represented in Costa Rica, each by a single form. Because the characterization of the species groups remains incomplete and because presumed defining features are interspecifically variable and many named taxa have not been assigned to a group, I have eschewed detailed discussion or description of intrageneric divisions.

When Duellman and Wiens (1992) resurrected *Scinax* for this genus, they stated that the name was from the Greek *skinos*, meaning quick or nimble, and was feminine. Köhler and Böhme (1996) disputed that interpretation, since there is no such word in Greek. They properly point out that *scinax* is the Latinized form of the Greek *skinax*, as is clear from Wagler's (1830) original description. They further conclude that the generic name *Scinax* is masculine and, following article 31.2 (*International Code of Zoological Nomenclature*, 1985), changed the termination of species epithets to agree in gender with the generic name.

The Greek word *skinax* (plural *skinakos*) is actually an adjective that is used to modify both masculine and feminine nouns with no change in termination. In the case of the generic name, *scinax* is an adjective used as a noun ("the nimble one"), and contrary to Köhler and Böhme (1996) is not "clearly" of masculine gender. Article 30.1.4.2 of the current code (1999) states that when a generic name is based on a word of variable gender and is not clearly indicated as feminine by the author establishing the name, it is to be treated as masculine. Peters's (1872) reference to *Scinax bipunctata* notwithstanding, *Scinax* must be regarded as of masculine gender.

Frogs of this genus usually breed in ponds or swamps, often of a temporary nature (pattern 1, table 7.1). One species group in South America is a stream breeder (pattern 2) and another lays eggs in terrestrial bromeliads (pattern 4) (Duellman and Wiens 1992). Tadpoles are free-living.

Members of this genus characteristically sit head down on vertical surfaces. Jungfer (1987) illustrated the peculiar ability of these frogs to bend the first finger and first toe posteriorly at a 90° angle to their axes. He suggested that this feature provides traction when they perch head down.

KEY TO THE COSTA RICAN TREEFROGS OF THE GENUS *SCINAX*

1a. Posterior surface of thigh uniform, dusky or dull yellowish green in life; no dark spots in groin; adult males 24–35 mm in standard length, adult females 24–40 mm 2
1b. Posterior surface of thigh with dark bars that alternate with light areas (bright yellow to light green in life, white in preservative) (fig. 7.83b); a dark spot or spots in groin; adult males 36–49 mm in standard length, females 44–52 mm . . . . *Scinax boulengeri* (p. 343)
2a. Dorsum tuberculate; finger disks equal to or slightly smaller than tympanum; snout pointed in dorsal outline (fig. 7. 6b); vocal slits in males hidden by tongue; bones colorless; adult males 21–29 mm in standard length, adult females 21–32 mm . . . . *Scinax staufferi* (p. 347)
2b. Dorsum smooth to granular, never tuberculate; finger disks about 1.25 times as large as tympanum; snout rounded in dorsal outline (fig.7.6b); vocal slits in males not hidden by tongue; bones green in life; adult males 26–38 mm in standard length, adult females 30–40 mm . . . . *Scinax elaeochroa* (p. 345)

Scinax boulengeri (Cope, 1887)

(plate 182; map 7.97)

DIAGNOSTICS: A very attractive moderate-sized strongly tuberculate long-snouted treefrog lacking finger webs and having the posterior thigh marked with black bars alternating with yellow or green interspaces (fig. 7.83b). This relatively common species is unlikely to be confused with any

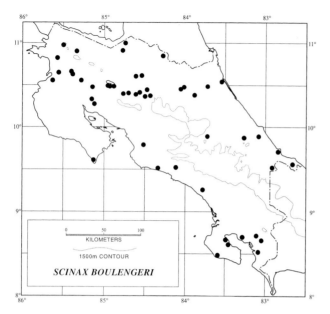

Map 7.97. Distribution of *Scinax boulengeri*.

other treefrog. Because of its "schnozz," OTS students have come to know it as the Jimmy Durante frog.

DESCRIPTION: Adult males 36 to 49 mm in standard length, adult females 42 to 53 mm; upper surfaces covered with small tubercles; venter granulate; head much longer than broad; snout subovoid to semielliptical from above, profile acuminate, the long snout projecting beyond leading edge of lower jaw; eye moderate; tympanum large, diameter two-thirds to four-fifths diameter of eye in males, seven-tenths to four-fifths in females; paired vocal slits and a moderately distensible single external subgular vocal sac in adult males; fingers long and slender; finger disks large, truncate, width of disk on finger III equal to diameter of tympanum; subarticular tubercles under fingers and toes round and single; a white nonspinous nuptial pad on base of thumb in adult males; no finger webbing, but definite fleshy fringe on fingers; thenar tubercle elongate, flat; palmar tubercle large, flat, tripartite; one or two heel tubercles; toes III–IV–V extensively webbed; webbing formula: I $2-2^+$ II $1-2$ III $1-2$ IV 2^+-0 V; webbing reduced to flaps on inner margin of toe I and outer margin of toe II; inner metatarsal tubercle low, flat, round, outer indistinct, small, conical, or absent; weak tarsal fold present. Dorsal surfaces grayish tan or dull green with slightly darker markings; upper surfaces of limbs with transverse dark bars; groin pale green with dark spots or mottling; anterior and posterior thigh surfaces and inner edge of tarsal segment greenish yellow to orange yellow marked with black vertical bars; venter creamy white; iris dull bronze to goldish.

LARVAE (fig. 7.102): Moderate-sized, 34 mm in total length; body compressed; mouth anterior, ventral; nostrils directed anterolaterally; eyes directed laterally; spiracle lateral, sinistral; vent tube dextral; caudal fins very high; dor-

sal fin extending onto body to posterior level of eye; tail pointed; oral disk small, entire, with massive, strongly serrate beaks, 2/3 rows of denticles; A2 and P2 with slight median gaps; P3 a short row of elongate denticles mounted on movable extension of lower lip; large blunt papillae on lateral margins of upper and lower lips; in addition, several small conical papillae, internal to lateral papillae of lower lip; no papillae on dorsal or ventral margins of lips; body silver yellow; large dark midcaudal spots present; tail musculature cream; fins transparent.

VOICE: A single low-pitched guttural "wraak" repeated every ten seconds to several minutes. The call is lower in pitch than that of *S. elaeochroa* and has a pronounced throaty quality. Note duration is 240 to 470 milliseconds with a pulse rate of 80 to 120 per second with two harmonics about equally emphasized at 1.6 and 2.8 kHz (Duellman 1970).

SIMILAR SPECIES: *Hyla lancasteri* has a short, truncate snout (fig. 7.6c) and finger webbing.

HABITAT: Lowland Dry, Moist, and Wet Forests and marginally in Premontane Wet Forest. Most commonly found near breeding sites.

BIOLOGY: A nocturnal, arboreal form that is usually found on bushes, low trees, stumps, and fallen logs near the forest floor but also in secondary growth or in isolated trees and shrubs in pastures or other open areas (Scott 1983g). Males call from sites close to the ground with the head facing down and cocked away from the substrate or from the edges of temporary ponds. The species is a prolonged

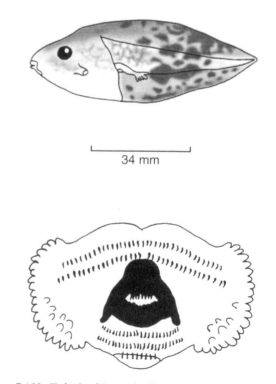

34 mm

Figure 7.102. Tadpole of *Scinax boulengeri*.

breeder that arrives at the breeding sites shortly after the first heavy rains of the season. However, most males remain in the chorus one to six nights, although some stay much longer (Bevier 1997), up to fifty-one nights for one male. Males have strong call-site fidelity. Males call throughout the year in humid areas but only during the wet times of the year (May to November) in dry forest habitats. Breeding peaks appear to occur in May to June and again in August, but reproduction may take place at any time during the rainy season. Eggs are laid in ponds (pattern 1, table 7.1). Clutches of 600 to 700 eggs 1.5 to 1.6 mm in diameter were laid in captivity. These eggs hatched in about thirty-six hours. Metamorphosis was completed after forty to eighty-eight days (Jungfer 1987). The tadpoles are nektonic and apparently feed on small crustaceans and other aquatic insect larvae and on developing conspecific and heterospecific embryonic frogs that they flip into the mouth area with a catapultlike extension of the lower lip, which is mounted with long, pointed denticles (P3). Metamorphs are about 15 mm in standard length. The heaviest predation on tadpoles of this species is by other tadpoles, probably including conspecifics (Roberts 1994a). Co-occurring fish species do not eat eggs of this form under laboratory conditions. Juveniles are rarely seen except in the latter part of the wet season and then in small numbers when few adults are active.

REMARKS: *Scinax boulengeri* is a member of the *Scinax rostratus* group (Duellman 1972a; Duellman and Wiens 1992). Most species in this group share the specialized oral features of the tadpole with *S. boulengeri*. Taylor (1952a) called this form *Hyla boulengeri*, and many post-1977 publications have cited it as *Ololygon boulengeri*. Contrary to Duellman's monograph (1970), adult males of this species have whitish nuptial pads on the base of the thumb.

DISTRIBUTION: Humid lowlands and lower edges of the premontane zone from Nicaragua to central Panama on the Atlantic versant and Pacific slope subhumid lowlands of northwestern Costa Rica south through the humid lowlands of southwestern Costa Rica to northwestern Ecuador (1–700 m).

Scinax elaeochroa (Cope, 1875)

(plate 183; map 7.98)

DIAGNOSTICS: A moderately small yellow, yellowish green, or yellowish tan to olive green treefrog that may be nearly uniform in color or more usually has dark dorsal and limb markings. This species has visible dark green bones, a protruding snout (fig. 7.6c), and a uniform thigh that is similar in color to the upper surfaces of the body, and it lacks finger webs and dorsolateral light stripes.

DESCRIPTION: Adult males 26 to 38 mm in standard length, adult females 30 to 40 mm; dorsal surfaces usually smooth, sometimes with small tubercles; venter granular; head longer than broad; head outline from above rounded with a terminal point; snout profile protruding, tip rounded;

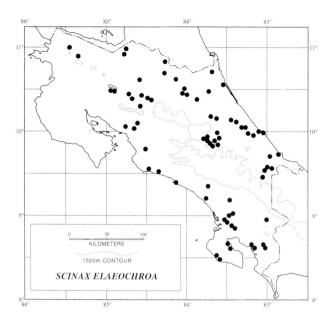

Map 7.98. Distribution of *Scinax elaeochroa*.

eyes moderate; tympanum distinct, but upper edge obscured by supratympanic fold, diameter sexually dimorphic, 44–68% of eye diameter in males, 49–69% in females; paired vocal slits and moderately distensible single external subgular vocal sac in adult males; fingers moderately long and slender; finger disks large and truncate, width of disk on finger III greater than diameter of tympanum; subarticular tubercles under all digits single, round; a whitish nonspinous nuptial pad present on base of thumb in adult males; no finger webs; thenar tubercle elongate; palmar tubercle cordate to nearly tripartite; no heel tubercles or calcar; toes III–IV–V moderately webbed; webbing formula: I 2–2$^+$ II 1–2 III 1–2 IV 2–1 V; web absent between toes I and II, except fringe on outer edge of toe II; inner metatarsal tubercle low, ovoid, outer elongate; no tarsal fold. Ground color of dorsal surfaces yellow green, pale yellow, or tan with faint darker markings, flanks yellow (night) or olive green to tan with dark dorsal lineate blotches or spots and pale yellow flanks (day); upper surface of limbs obscurely to strongly barred with darker color; groin usually bright yellow; posterior thigh surface yellowish green with melanophores; venter orange, pale yellow, or white, some individuals with bluish gray throats and axillary regions; vocal sac usually yellow; iris bronze or copper with gray to goldish suffusion.

LARVAE (fig. 7.103): Moderate-sized, 30 mm in total length; body compressed; mouth anteroventral; nostrils directed anterodorsally; eyes directed laterally; spiracle lateral, sinistral; vent tube dextral; tail moderate; tail fins moderate; tail fins tapering to a long, pointed tip; oral disk small, entire, with robust, finely serrate beaks and 2/3 denticle rows; A2 with a gap; single row of papillae lateral to upper lip and along ventral disk margin; lower lip with additional small lateral papillae; body yellowish tan with gray mottling;

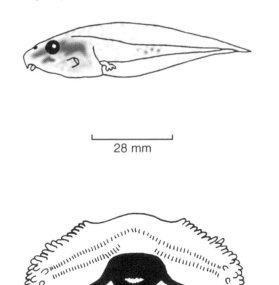

28 mm

Figure 7.103. Tadpole of *Scinax elaeochroa.*

venter mostly white; a brown stripe from snout through eye; tail musculature and fins marked with small grayish brown spots; iris gold.

VOICE: The call is a series of short "waack" notes, usually two to fifteen, but sometimes up to ninety-five notes are produced. The notes are a higher pitch than the call of *Scinax boulengeri* and lack its throaty quality. When not in a chorus or on overcast days, lazy single calls or several "waack" calls may be emitted from hiding places. Call groups cover 1.5 to 48 seconds, with two harmonics emphasized at about 1.5 and 2.9 kHz, and have an average pulse rate of 40 to 50 per second (Duellman 1970).

SIMILAR SPECIES: The smaller *Scinax staufferi* (adult males 21 to 29 mm, adult females 21 to 32 mm) has a definitely pointed snout, is more tuberculate, and lacks green bones.

HABITAT: Generally Lowland Moist and Wet Forests but also in the lower reaches of Premontane Wet Forest and Rainforest.

BIOLOGY: This is one of the commonest frogs throughout lowland rainforests. Like other hylids it is primarily nocturnal. Males may be heard calling throughout most of the year from hiding places in tree holes and other cavities. Early in the wet season they begin to congregate around temporary ponds, where they usually call from vegetation or bushes emergent from the pond or on shore near the water. After the first torrential rains that flood the reproductive sites there is an explosive increase in calling males (Duellman 1970; Donnelly and Guyer 1994). Soon hundreds to thousands appear and call continually and synchronously

day and night for twenty-four to forty-eight hours, producing an awful din. At the same time males turn a brilliant yellow overall, and the vocal sac becomes yellow orange. During this time the males scramble, leap about on the vegetation, and grapple with one another in apparent competition for calling sites, but some are forced to call from the shore or while floating in the water. During this chaotic period females join the action, and amplexus takes place in the water. After this frenetic bout of reproductive activity, most of these frogs disperse just as quickly as they appeared. Eggs are laid in a mass adjacent to vegetation in the pond or adhering to it and hatch in about 1.6 days. Early embryos orient themselves vertically with the mouth pointed toward the water surface, gradually sinking to the bottom and then swimming back to the surface (Duellman 1970). Tadpoles with mouthparts stay in the shallow water near the pond margins and hide in vegetation (ecomorph 2, table 7.2). Metamorphs are about 13 mm in standard length. Large numbers of juveniles are encountered in the humid lowlands during the driest time of the year, January and February. During the rest of the year juveniles are rarely seen, suggesting that sexual maturity is reached in about one year. *Scinax elaeochroa* is primarily a lowland species, but it has been collected at up to 1,000 to 1,200 m on the Atlantic slope at Juan Viñas, El Silencio de Sitia Mata, and Pacuare, all in Cartago Province. On the Pacific versant it has a continuous range from southwestern Panama to Parrita, Puntarenas Province. It has also been taken at 400 to 500 m in altitude in the foothill region in one or two small streams in the Río Barranca drainage system about 60 km north-northwest of Parrita. That area was at one time part of a moist forest zone along the western foothills that was continuous with the forests of the Pacific lowlands to the south. The slope environment was heavily deforested by the beginning of the twentieth century, and *S. elaeochroa* also occurs in other moist forest fragments between Parrita and the Río Barranca. The karyotype is 2N = 24 all chromosomes are metacentric or submetacentric; NF = 48 (Anderson 1991).

REMARKS: *S. elaeochroa* is a member of the phenetically based *Scinax ruber* group (Duellman 1972b; Duellman and Wiens 1992). Taylor (1952a) referred to this species as *Hyla elaeochroa*. He (1958a) later described *Hyla dulcensis* (holotype: KU 32168) from southwestern Costa Rica, but Duellman (1966a) concluded that this is a synonym of *S. elaeochroa*. Most post-1977 references to this species used *Ololygon elaeochroa* for this taxon until Duellman and Wiens (1992) agreed that the generic name *Scinax* had priority. Although the generic name *Scinax* must be considered a masculine noun, this does not affect the termination of the name *elaeochroa*, which is a noun in apposition. It is from the Greek *elaeo-*, olive and *-chroa*, skin. Adult males of this species have whitish nuptial thumb pads, although Duellman (1970) stated they were absent. The isolated population in the Chocó of Colombia referred to this species by

Ruiz, Ardilla, and Lynch (1996) probably is another distinct species.

DISTRIBUTION: Humid lowland and lower premontane forests from eastern Nicaragua to northwestern Panama on the Atlantic versant and on the Pacific slope from southwestern Panama to the Río Barranca drainage of central Costa Rica (1–1,200 m, mostly below 700 m).

Scinax staufferi (Cope, 1865a)

(plate 184; map 7.99)

DIAGNOSTICS: A small, rather tuberculate dull-colored treefrog that has irregular longitudinal dark blotches on the back and a pointed, protruding snout (fig. 7.6b,c) but lacks finger webs, distinctive thigh markings, and dorsolateral light stripes.

DESCRIPTION: Adult males 21 to 29 mm in standard length, adult females 21 to 32 mm; dorsal surfaces tuberculate; venter granulate; head longer than broad; snout pointed from above, protruding in profile, tip rounded; eyes moderate; tympanum distinct, sometimes with upper margin obscured by supratympanic fold, diameter of tympanum 48–67% of diameter of eye in males, 48–71% in females; paired vocal slits and fully distensible single external subgular vocal sac in adult males; fingers relative short and stout; finger disks moderate, slightly truncate, width of disk on finger III about equal to diameter of tympanum; subarticular tubercles under all digits single, often indistinct; a whitish nonspinous nuptial pad on thumb base in adult males; no finger webs; thenar tubercle elongate; palmar tubercle large, bifid; no heel tubercles or calcar; toes III–IV–V moderately webbed; webbing formula: I 2^+–$2\frac{1}{2}$ II $1\frac{2}{3}$–$2\frac{1}{2}$ III 1^+–$2\frac{1}{3}$ IV 2^+–1^+ V; webbing vestigial between toes I

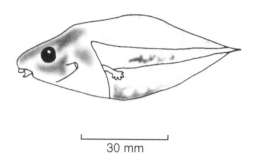

30 mm

Figure 7.104. Tadpole of *Scinax staufferi*.

and II; inner metatarsal tubercle large, oblong, outer small, rounded, without melanophores, in contrast to adjacent area of foot; no tarsal fold. Dorsal ground color usually gray but sometimes various shades of brown; dorsum markings darker than ground, usually elongate blotches and/or spots, sometimes forming continuous dark longitudinal stripes; upper limb surfaces usually with dark bars, sometimes blotched; flanks creamy white to creamy tan; posterior thigh surface dull brown to dusky with melanophores; venter dirty white; iris brown with gold flecks or bronze with brownish suffusion.

LARVAE (fig. 7.104): Moderately small, total length 28 mm; body compressed; mouth anteroventral; nostrils directed anterodorsally; eyes directed laterally; spiracle lateral, sinistral; vent tube dextral; tail short; tail fins very high; tip pointed; oral disk small, entire, with moderately robust beaks and 2/3 rows of denticles; A2 interrupted above mouth; P1 sometimes interrupted below mouth; small marginal papillae border lips except above mouth; numerous small submarginal papillae on lateral portion of lower labium; body pale; venter white; tail similarly colored with dark reticulations; fins transparent; iris pale gold.

VOICE: The call consists of a series of short nasal "ah-ah-ah-ah" notes usually repeated two to thirty times or more (rarely). The notes have a duration of 130 to 230 milliseconds and a pulse rate of 100 to 130 per second. Two nearly equally emphasized harmonics have frequencies of about 1.74 and 3.06 kHz (Duellman 1970; Lee 1996).

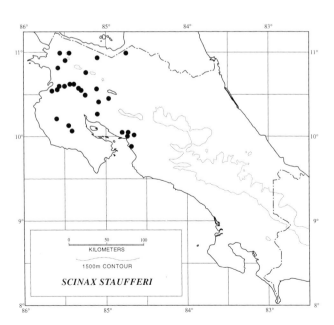

Map 7.99. Distribution of *Scinax staufferi*.

SIMILAR SPECIES: *Scinax elaeochroa* is larger (adult males 26 to 38 mm in standard length, adult females 30 to 40 mm); the snout outline is rounded above but with a point on the tip, and the bones are green.

HABITAT: Primarily in the Lowland Dry Forest region, where it is found in the breeding season around temporary shallow ponds, flooded pastures, and roadside ditches. Also in the Lowland Moist Forest in the upper Río San Juan drainage.

BIOLOGY: This very common nocturnal species is semi-terrestrial and is usually found in the dry season (November to May) near or only a few centimeters above the substrate, hiding in leaf axils of herbaceous plants. Reproduction takes place in shallow temporary bodies of water throughout most of the wet season after heavy showers. Males usually call from low bushes, herbs, and grasses 50 to 800 mm above the ground at the margin of the pond, marsh, or ditch, but they sometimes call while sitting on the ground or from the shallows. The breeding congregations include numerous males that produce a raucous chorus. Amplexus and egg deposition take place in shallow water. Egg masses are in small clumps. Tadpoles (ecomorph 2, table 7.2) are nektonic forms that feed by rasping vegetation surfaces.

REMARKS: *Scinax staufferi* belongs to the *Scinax staufferi* phenetically based group (Duellman and Wiens 1992). Taylor (1952a) referred to this species as *Hyla staufferi*, although its occurrence in the republic was confirmed only later (Taylor 1954a). Much of the post-1977 literature used *Ololygon staufferi* for this form until *Scinax* was shown to have priority (Duellman and Wiens 1992). León (1969) and Duellman (1970) recognized (as *Hyla*) two completely allopatric subspecies: *S. s. staufferi* (Cope, 1865a) of Central America and Mexico and *S. s. altae* (Dunn, 1933c) of Pacific versant Panama. The two forms are separated geographically by the humid lowland forests of the Golfo Dulce region of Costa Rica and extreme southwestern Panama. The rather consistent differences between the two in color pattern, toe webs, and calls suggest that they may represent distinct species. *S. s. altae* is distinguished from other *S. staufferi* (after Duellman 1970) (characters for the nominate form in parentheses) by having a longitudinal dark stripe on the tibial segment (barred or blotched); usually distinct, complete dark dorsolateral stripes (blotches or irregular dorsolateral stripes); toes with reduced webbing, "40%" webbed (moderate webbing, "75%"); and a shorter call with the lower emphasized harmonic at about 2 kHz (longer, lower harmonic at about 1.74 kHz). The Panama population is distributed disjunctly in western to central Panama. The species is discontinuously distributed on the Atlantic slope of northern Central America in savanna islands within rainforests in northern Guatemala and in the isolated dry valleys of Guatemala and Honduras. The karyotype of this species is 2N = 24; all chromosomes are metacentric or sub-metacentric; NF = 48 (Anderson 1991). Adult males have whitish nuptial thumb pads, contra Duellman's (1970) statement that none are present.

DISTRIBUTION: Subhumid to semiarid lowlands from southern Tamaulipas and Guerrero, Mexico, to extreme northeastern Costa Rica on the Atlantic versant and central Costa Rica on the Pacific slope; also disjunctly on the Pacific lowlands of western to central Panama (4–550 m).

Genus *Smilisca* Cope, 1865a

Treefrogs of this genus (six species) are moderate-sized to large anurans having a blotched or barred dorsal pattern in various shades of brown or green and the flanks mottled, spotted, reticulated, or venated. Adult males have paired external subgular vocal sacs that form two bulbous protrusions when fully inflated and dark nuptial thumb pads. In most species the fingers and toes are definitely webbed, the finger disks are round (about as wide as long), the eyes are moderate, and the tympanum is large, distinct, and with a prominent annulus tympanicus.

Additional features that in combination separate this genus from related ones are frontoparietal fontanelle usually present; skull not co-ossified to integument or exostosed; quadratojugal articulating with maxilla; sphenethmoid large, well ossified; spatulate, strongly bifid teeth on premaxilla, maxilla, and vomer; depressor mandibulae with two parts, one from dorsal fascia, one from otic arm of squamosal (DFSQ); adductor mandibulae posterior subexternus and externus superficialis present (s + e); sacral diapophyses broadly dilated; larvae ovoid to robust with small anteroventral to ventral mouths, lateral, sinistral spiracle, 2/3 denticle rows, and papillae usually absent above mouth; 2N = 24.

Although adult males are generally easy to recognize because of the vocal sac, juveniles and female *Smilisca* may be most often confused with some Costa Rican species of the genus *Hyla*. Generally the posterior surface of the thigh is solid brown or bluish to purple or reticulated with light spots, and the groin is reticulated or venated with bluish purple to brown or marked with discrete light spots in *Smilisca*. Only two species of *Hyla* have coloration resembling that described above. These are *Hyla pictipes*, a high-elevation form never found in sympatry with any *Smilisca*, and *Hyla pseudopuma*. The descriptions of these forms should be consulted when there is doubt regarding a particular juvenile or female specimen.

Duellman and Trueb (1966) and Duellman (1970) recognized two poorly differentiated groups within the genus, the *Smilisca baudinii* group and the *S. sordida* group. Both groups are represented in Costa Rica.

Eggs of this genus are deposited in lotic or lentic situations depending on the species. All larvae are relatively generalized free-living forms.

Smilisca is an essentially Middle American genus and

ranges through lowland and premontane areas from southern Texas and Sonora, Mexico, to north-central Colombia and northwestern Ecuador.

KEY TO THE COSTA RICAN TREEFROGS OF THE GENUS *SMILISCA*

1a. A dark brown to black postorbital band encompassing tympanum (fig. 7.73b)......................2

1b. No dark postorbital band involving tympanum 3

2a. Flanks with dark venation (fig. 7.73b); white labial stripe and white stripe from heel along tarsal segment to foot present; no lip bars; usually no light suborbital spot; no series of prominent warts along posteroventral margin of lower arm; adult males 40−66 mm in standard length, adult females 50−78 mm ....
.... *Smilisca phaeota* (p. 351)

2b. Flanks cream; groin with dark reticulations; no white labial or tibial stripes; lips usually barred, and light suborbital spot usually present (fig. 7.73a); a series of prominent warts along posteroventral margin or arm; adult males 47−75 mm in standard length, adult females 56−90 mm .... *Smilisca baudinii* (p. 349)

3a. Considerable webbing between fingers II−III−IV, extending to or proximal to distal subarticular tubercle on finger IV; a distinct series of warts along lower arm; no light stripe from heel to foot; dorsum uniform or with irregular dark blotches, bands, or spots, never with a pair of elongate dark blotches from head to sacrum; usually no interorbital dark bar or spot, but if present, faint 4

3b. Practically no webbing between fingers; no distinct series of warts along posteroventral margin of forearm; a distinct light stripe, bordered below by a dark line, from heel to foot; dorsum with two elongate dark blotches running from head to sacrum that anastomose at one to several points across midline (fig. 7.73f); a distinct interorbital dark bar or spot; adult males 32−38 mm in standard length, adult females 40−46 mm.........
.... *Smilisca puma* (p. 352)

4a. Snout short, truncate in profile (fig. 7.6c); web reduced between fingers I and II, modal formula I 1−2⁻ II; dorsum strongly tuberculate in females; blue spots on flanks and posterior thigh surface; vocal sacs in breeding males dark gray or brown; adult males 31−45 mm in standard length, adult females 44−62 mm........
.... *Smilisca sila* (p. 354)

4b. Snout long, rounded to obtuse in profile (fig. 7.6c); web extensive between fingers I and II, modal formula I 1−1⁺ II; dorsum smooth; cream or pale blue flecks on flanks and posterior thigh surface; vocal sacs of breeding males white; adult males 32−54 mm in standard length, adult females 56−64 mm ....
.... *Smilisca sordida* (p. 355)

Smilisca baudinii Group

Moderately large frogs capable of considerable color change from tan through olive green to green and marked with contrasting dark blotches or bars. Skull well ossified, the supraorbital margin of the frontoparietals irregular or having a distinct projecting process; male advertisement calls consist of a series of similar-sounding notes. The squamosal articulates with the maxilla in two members of this group *(S. baudinii* and *S. cyanosticta)*. *S. cyanosticta* of southern Mexico and Guatemala is the only extralimital member of this group.

Smilisca baudinii (C. Duméril and Bibron, 1841)

(plate 185; map 7.100)

DIAGNOSTICS: A very common moderate-sized to large, rather chubby, short-legged treefrog usually having a dorsal pattern of large dark spots and/or blotches, a cream flank, groin boldly reticulated with brown or black, and usually a broad dark postorbital band running back to the axilla (fig. 7.73a) and a contrastingly barred upper lip with a distinct light suborbital area (fig. 7.73c). Metachrosis is extreme, and individuals may be brown, very dark gray, gray green, pale green, yellow, tan, or gray depending on circumstances, sometimes uniform but usually with persistent dark markings outlined by black.

DESCRIPTION: Adult males 47 to 75 mm in standard length, adult females 56 to 90 mm; dorsal surfaces smooth; venter granulate; head broader than long; snout rounded from above, snout profile bluntly rounded; eyes moderate; tympanum large, distinct, diameter about one-half to two-thirds diameter of eye; paired vocal slits and fully distensible paired gray external subgular vocal sacs in adult males;

Map 7.100. Distribution of *Smilisca baudinii*.

fingers moderately long and stout; finger disks moderate, width of disk on finger III nearly equals diameter of tympanum; distal subarticular tubercle on finger IV single or bifid; brown nonspinous nuptial pad on base of thumb in adult males; fingers slightly webbed; webbing formula: I 2^+–$2\frac{1}{2}$ II 2^-–3 III 2^+–2^+ IV; thenar tubercle moderately enlarged; palmar tubercle tripartite; a prominent row of warts along posteroventral margin of forearm; toes extensively webbed; webbing formula: I 1^+–2 II 1–2 III 1^+–2^+ IV 2–1 V; inner metatarsal tubercle raised, very large, elliptical; outer metatarsal tubercle small, round or absent; inner tarsal fold present. Dorsal surfaces generally pale green, olive green, or pale brown, uniform or usually with dark olive green, brown, or dark brown markings, respectively; dark blotches or spots usually outlined by black; side of head pale green or tan, usually marked with vertical dark bars and a cream suborbital bar or spot; broad dark brown to black postorbital stripe usually present, sometimes reduced to narrow line below supratympanic fold; flanks cream with dark mottling, especially evident in groin; upper limb surfaces with dark transverse bars; posterior thigh surfaces brown with small pale yellow spots; venter white (yellow in some extralimital cases); iris bronze with black reticulations.

LARVAE (fig. 7.105): Moderate-sized, about 22 mm in total length at stage 30; body robust; mouth anteroventral; nostrils dorsolateral; eyes dorsolateral; spiracle lateral, sinistral; vent tube dextral; tail short; caudal fins moderate; tail tip pointed; oral disk small, entire, with moderate-sized beaks and 2/3 denticle rows; A2 with large gap above mouth; beaks serrate; papillae in two rows lateral and ven-

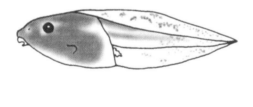

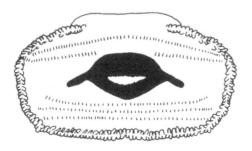

22 mm

Figure 7.105. Tadpole of *Smilisca baudinii*.

tral to mouth with numerous submarginal papillae immediately lateral to mouth; papillae with a broad gap above mouth; dorsum dark brown with crescent-shaped light mark anterior to base of tail, usually forming irregular light stripe continuous with tail stripe on upper posterior body; venter transparent, with scattered brown flecks anterolaterally; caudal musculature pale brown with longitudinal dark lines down middorsal and midlateral areas anteriorly to produce dorsolateral light stripe more or less continuous with light stripe on body; rest of tail flecked with brown; fins transparent, with brown flecks and spots; iris bronze.

VOICE: A series of blurred "heck" or "keck" notes reminiscent of starting up an old car. Two to fifteen notes make up a call group, with note durations of 90 to 130 milliseconds. Call groups are spaced from fifteen seconds to several minutes apart, and note repetition rate is about 300 to 350 per minute. Notes have pulse rates of 140 to 195 per second, and two frequencies at about 0.35 and 2.5 kHz are emphasized (Duellman 1970; Conant and Collins 1991; Lee 1996).

SIMILAR SPECIES: (1) *Smilisca phaeota* has white lip and tibial stripes and fine dark venation on the flanks (fig. 7.73b). (2) *Phrynohyas venulosa* has green bones and uniform flank and groin areas; males of this species have paired lateral vocal sacs as well (fig. 7.12f). (3) *Osteopilus septentrionalis* has a casque head (fig. 7.99).

HABITAT: Common in Lowland Dry Forest areas but also occurring in Lowland Moist and Wet Forests, often in association with disturbed and second-growth sites and rarely in Premontane Rainforest. Typically found around temporary ponds and flooded fields.

BIOLOGY: Like other members of the genus, this is a nocturnal arboreal frog that favors lower levels of the forest, where it takes refuge during the dry season in tree holes, under bark, in the axils of plants, and in the outer sheaths of banana plants. Members of the species also seem to use their own or mammals' burrows, fallen logs, and boards as diurnal or dry-season retreats. McDiarmid and Foster (1987) report that a Costa Rican representative of this species formed an antidesiccation cocoon similar to those found in several other anurans. The parchmentlike covering consisted of thirty-eight to forty layers of stratum corneum separated by a mucoid substance. Unfortunately the exact habitat of this frog is unknown, although it was found in a roadbed late in the dry season (April). It could not be determined if it was estivating in a tree or underground. *Smilisca baudinii* is an explosive breeder that uses shallow temporary bodies of water including roadside ditches as breeding sites, but it will also attempt to breed in cisterns and swimming pools. Both sexes emit high-pitched distress calls with the mouth wide open when attacked by a predator or handled roughly. Males usually call from the edge of the water but sometimes call from the shallows or low vegetation. Large congregations are typically heard after the first heavy rains (May), although scattered males may call throughout the wet season. These aggregations usually contain hundreds to thousands of frogs. At many places *S. baudinii* breed in ponds separate from communal ponds containing multispecies assemblages. In these choruses males alternate calls with near neighbors to form duets. Amplexus takes places in the shallows, and females lay 2,500 to 3,500 eggs in a surface film. The black-and-cream eggs are about 1.3 mm in diameter, 1.5 mm including the vitelline membrane. The tadpoles are pelagic (ecomorph 2, table 7.2) and feed in shallow water. Metamorphosis takes place about fourteen to twenty days after fertilization. There are no outer egg envelopes. Metamorphs are 12 to 15.5 mm in standard length.

REMARKS: There is considerable intraspecific variation in size, proportions, and coloration in *S. baudinii*. Costa Rican examples are near the average size for the species, with adult males to 68 mm in standard length and adult females to 72 mm compared with those in Sinaloa, Mexico, where the largest individuals occur. The frogs' hind limbs are also slightly longer in Atlantic versant Costa Rica than on the Pacific slope. The coloration of some Sinaloa specimens resembles that of *Smilisca phaeota* in having a white stripe on the foot and a white, but discontinuous, labial stripe (Duellman 1970). The posterior thigh in these individuals is creamy white, strongly marked with black. These variants have not been noted in Costa Rican examples. The species *Hyla manisorum* (holotype: KU 34927) described by Taylor (1954a) from Batán, Limón Province, is based on a female *S. baudinii*. The inner metatarsal tubercle in this species is raised and very large, at least twice as long as the width of the disk on toe IV when compared with other *Smilisca* (equal to or shorter than disk width). The tubercle is spadelike in samples from northern Mexican and the Central American Pacific slope. In other populations it is less protuberant and has a more rounded margin. When comparative material is at hand, this feature will immediately distinguish *S. baudinii* from its congeners. The karyotype is 2N = 24, with all metacentric or submetacentric chromosomes; NF = 48 (Anderson 1991).

DISTRIBUTION: Humid, subhumid, and semiarid lowlands and marginally on premontane slopes from southern Sonora, Mexico, and extreme southeastern Texas to southeastern and southwestern Costa Rica (1–750 m).

Smilisca phaeota (Cope, 1962d)

(plates 186–88; map 7.101)

DIAGNOSTICS: A common moderate-sized to large, relatively long-legged pale green to tan treefrog, uniform or with dark olive green to dark brown dorsal blotches. It is most distinctive in having a white stripe on the outer margin of the tibial segment of the leg, a white labial stripe, and the flanks marked with a fine dark brown to black venation (fig. 7.73b). As in *Smilisca baudinii*, the color change repertoire of individuals of this species is extensive, but the

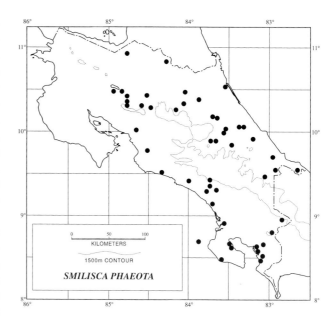

Map 7.101. Distribution of *Smilisca phaeota*.

venation of the flanks provides an invariable identification characteristic.

DESCRIPTION: Adult males 40 to 66 mm in standard length, adult females 50 to 78 mm; dorsal surfaces smooth; venter granular; head about as wide as long; snout bluntly rounded from above, rounded in profile; eyes moderate; tympanum distinct, diameter 63–85% of eye diameter in males, 75–90% in females; paired vocal slits and fully distensible paired gray external subgular vocal sacs in adult males; fingers moderately long and broad; finger disks moderate, width of disk on finger III about two-thirds diameter of tympanum; distal subarticular tubercle on finger IV sometimes bifid; nonspinous tan nuptial pads on base of thumb in adult males; fingers slightly webbed; webbing formula: I 2^+–2^+ II 2^-–3^- III $2\frac{1}{2}$–2 IV; thenar pad moderately enlarged; palmar tubercle large, tripartite; toes extensively webbed; webbing formula: I 1–2 II 1–2 III 1^+–2^+ IV 2–$\frac{1}{2}$ V; inner metatarsal tubercle low, flat, elliptical, and relatively small, outer tubercle absent; inner tarsal fold relatively short. Dorsal surfaces pale green or tan, uniform or with dark olive green to dark brown markings, usually with dark brown markings; loreal region pale green, demarcated by dark stripe on canthus rostralis above and white labial stripe below; dark brown to black postorbital mark from eye backward beyond axilla; flanks with fine dark venation; a light stripe on outer edge of forearm and finger IV and along outer margin of tibial segment, usually onto toe V; upper limb surfaces with dark transverse bars; usually a white supracloacal transverse stripe present; posterior thigh surface pale brown to gray with dark flecks or sometimes small light spots; venter creamy white; iris bronze with fine black reticulum.

LARVAE (fig. 7.106): Moderate-sized, 30 mm in total

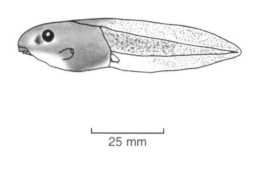

25 mm

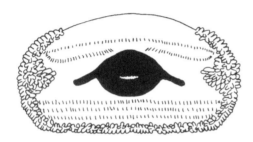

Figure 7.106. Tadpole of *Smilisca phaeota*.

length at stage 36; body robust; mouth anteroventral; nostrils dorsolateral; eyes dorsolateral; spiracle lateral, sinistral; vent tube dextral; tail short; caudal fins moderate; tail tip rounded; oral disk small, entire, with well-developed beaks and 2/3 denticle rows; A2 with moderate gap above mouth; beaks bluntly serrate; papillae in two rows lateral and ventral to mouth, many additional rows lateral to mouth; broad gap in papillae above mouth; dorsum pale brown with a light posterior crescent-shaped marking; venter transparent, with scattered brown flecks; caudal musculature creamy tan with brown spots; fins transparent with brown flecks and blotches; iris bronze.

VOICE: The call is a low vibrant growl ("wawk") consisting of one or two notes lasting 100 to 450 milliseconds and repeated at intervals of twenty seconds to several minutes. The pulse rate is 100 to 130 per second, with the dominant frequency at 0.33 to 0.5 kHz (Duellman 1970).

SIMILAR SPECIES: (1) *Smilisca baudinii* lacks the reticulate flanks and white stripes on the lip, forearm, and leg. (2) *Smilisca puma* is smaller (adult males 32 to 38 mm in standard length, adult females 40 to 40 mm) and has little or no finger webbing. (3) *Osteopilus septentrionalis* has a casque head and green bones and lacks white labial and limb stripes. (4) *Phrynohyas venulosa* also lacks these features of coloration and has green bones.

HABITAT: Lowland Wet and Moist Forests into the Premontane Wet Forest zone, including disturbed areas, secondary growth, and pastures, and marginally in Lowland Dry Forest.

BIOLOGY: A nocturnal treefrog that is active throughout the wet season in humid areas. Its breeding is prolonged, and males call throughout most of the year whenever it rains. Individuals usually hide during the day in low vegetation near water. Males call from secluded spots at the edge of the breeding site, usually hidden by an overhanging leaf. The species favors small ponds, shallow streamlets, or ditches as mating and oviposition sites. Sometimes they are found near large ponds or stream backwaters, but apparently they do not breed regularly in such situations. Clutches of 1,600 to over 2,000 black-and-cream eggs are laid in a surface film at the base of emergent vegetation in the shallows. The tadpoles are nektonic (ecomorph 2, table 7.2). Experiments by Valerio (1971) showed that the tadpoles could survive up to twenty-four hours out of water. Metamorphs are 13 to 17 mm in standard length.

REMARKS: Taylor (1952a) used the name *Hyla phaeota* for this species, following the consensus at that time. Starrett (1960b) placed it in *Smilisca*, as confirmed by Duellman and Trueb (1966).

DISTRIBUTION: Humid lowland and premontane forests from Atlantic slope northeastern Honduras to northern Colombia and on the Pacific versant from central Costa Rica to southwestern Panama and from El Valle de Antón, Coclé Province, Panama, to western Ecuador (2–1,116 m).

Smilisca sordida Group

Moderate-sized brown to tan frogs marked with darker blotches and/or bars, which exhibit changes in intensity of color but do not turn green. These anurans are characterized by a tendency for reduction in the ossification of several skull bones, the supraorbital margin of the frontoparietals is smooth, and the squamosal does not contact the maxilla. In two species of this group the male advertisement call consists of a primary note followed by secondary notes (*Smilisca puma* and *S. sila*). All three species referred to this group occur in Costa Rica.

Smilisca puma (Cope, "1884," 1885)

(plate 189; map 7.102)

DIAGNOSTICS: A moderate-sized tan treefrog that has a white lip stripe and a white stripe along the outer margin of the tibial segment of the leg but lacks finger webs. The dorsal pattern is distinct in being composed of paired elongate dark brown to green blotches that usually anastomose at one to several points across the midline (fig. 7.73f).

DESCRIPTION: Adult males 32 to 38 mm in standard length, adult females 40 to 46 mm; dorsal surfaces smooth; venter granular; head slightly longer than broad; snout rounded to subelliptical in dorsal outline, rounded in profile; eyes small; tympanum distinct, diameter one-half to seven-tenths of eye diameter; paired vocal slits and fully distensible paired grayish brown external subgular vocal sacs in adult males; fingers short and stout; finger disks moder-

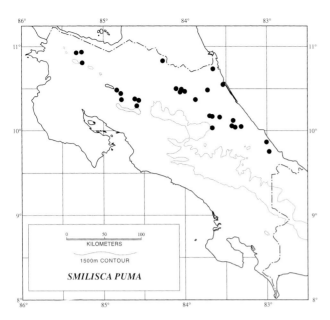

Map 7.102. Distribution of *Smilisca puma*.

ate, width of disk on finger III about equal to diameter of tympanum; distal subarticular tubercle on finger IV rarely slightly bifid; light brown nonspinous nuptial pad on base of thumb in adult males; fingers without webbing or with only a trace; thenar pad barely enlarged; palmar tubercle low, flat, indistinct; toes weakly webbed; webbing formula: I 2^-–$2\frac{1}{2}$ II 1^+–$2\frac{1}{2}$ III $1\frac{1}{3}$–$2\frac{1}{3}$ IV $2\frac{1}{3}$–1^+ V; inner metatarsal tubercle small, low, elliptical, outer absent; inner metatarsal fold present. Dorsal surfaces golden tan with a broad, irregular dark brown to dark green stripe from each eye, usually transversely anastomosed to form H-shaped figure and/or broken into spots posteriorly; narrow white stripe on upper lip; anterior flank with fine dark venation, inguinal region white mottled with dark brown; light stripe on outer margin of forearm and tarsal segment; transverse supracloacal white stripe always present; upper surfaces with narrow transverse dark lines; posterior thigh surface dark brown; venter creamy white; iris deep bronze.

LARVAE (fig. 7.107): Moderate-sized, 24 mm in total length at stage 34; body ovoid; mouth anteroventral; nostrils and eyes dorsolateral; spiracle lateral, sinistral; vent tube dextral; tail short; caudal fins moderate; tail tip rounded; oral disk small, entire, with strong beaks and 2/3 denticle rows; A2 with moderate gap above mouth; beaks finely serrate; papillae with wide gap above mouth, one row on upper labium, two on lower, with many additional rows at angle of jaws; dorsum olive brown with greenish tan flecks; tail fins pale brown with greenish gold flecks; dark reticulations on musculature and at fin bases anteriorly, dark flecks posteriorly; iris bronze.

VOICE: A low "squawk" usually followed by one or more secondary rattling notes. The primary notes last 60 to 350 milliseconds, and the secondaries last 100 to 470 milliseconds. Primary notes have pulse rates of 187 to 240 per second and dominant frequencies at about 0.74 and 1.897 kHz (Duellman 1970).

SIMILAR SPECIES: (1) *Smilisca phaeota* is much larger (adult males 40 to 66 mm in standard length, adult females 50 to 78 mm) and has fine dark venation along the flank and in the groin (fig. 7.73b). (2) *Smilisca sordida* has substantial finger webbing, the flanks mottled dark brown and cream, greenish, or blue, the posterior thigh surface dark brown with blue, green, or tan flecks (fig. 7.74b), and no white labial and limb stripes. (3) *Hyla pseudopuma* has substantial webs between fingers II–III–IV and lacks white labial and limb stripes.

HABITAT: A denizen of relatively undisturbed Atlantic Lowland Moist and Wet Forests.

BIOLOGY: This is an uncommonly seen nocturnal treefrog that is usually found only near breeding sites. Males call throughout the rainy season on the Atlantic lowlands (February through September), from shallow water and low bushes, usually hidden in the vegetation. They alternate calls with nearby neighbors, and the choruses are formed of duets, trios, and quartets. Breeding, however is restricted to periods of heavy rain from June through August. Reproductive sites are usually small, shallow temporary pools or ponds within the forest, but more open sites are also used (Duellman 1967a). Tadpoles are apparently benthic forms (ecomorph 1, table 7.2). Metamorphs are about 12.5 mm in standard length.

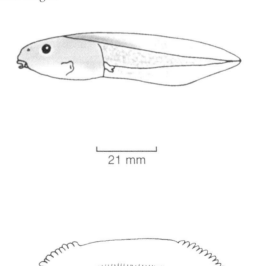

21 mm

Figure 7.107. Tadpole of *Smilisca puma*.

REMARKS: This species was described as *Hyla puma* by Cope ("1884," 1885) and was known for many years only from the holotype (USNM 13735; "Nicaragua"). E. R. Dunn in his unpublished review of the lower Central American herpetofauna recognized the validity of *H. puma* and included it in the Costa Rica biota based on specimens from La Castilla (ANSP 23686) and Monteverde, Limón Province (MCZ 7836). However, he thought that *H. puma* was a synonym of three older names, *Hyla molitor* (lectotype: NHMW 16494), *H. molitor marmorata* (holotype: KM 1010/1342), and *H. splendens* (holotype: KM 1008/1340), all supposedly collected in western Panama by Josef Warszewicz and described by Oskar Schmidt (1857). Dunn is doubtless the source for Cochran's (1961) listing of *H. puma* as a synonym of *H. molitor*. Savage and Heyer (1968) and Savage (1970) concluded that a number of Warszewicz's specimens most likely originated from Colombia, Peru, or Bolivia, where he collected immediately after his stay in Panama. Savage and Heyer, Duellman, and the late Charles F. Walker all agreed (Duellman 1970) that the three Schmidt names were based on non–Central American material. The type of *H. splendens* is a large (51 mm) female *Gastrotheca* but does not resemble any known member of that genus from Central America or Colombia. Dunn's archrival E. H. Taylor (1952a) also recognized the validity of *H. puma* but thought it was a newly discovered species, which he described and named *Hyla wellmanorum*. Duellman and Trueb (1966) in their revision of the genus *Smilisca* conclusively demonstrated that only a single species was involved, for which the name *S. puma* has priority.

DISTRIBUTION: Humid Atlantic lowlands of Costa Rica and adjacent Nicaragua (15–520 m).

Smilisca sila Duellman and Trueb, 1966

(plate 190; map 7.103)

DIAGNOSTICS: A moderate-sized gray, tan, to reddish brown treefrog having darker dorsal markings, a truncate snout profile, and the flanks and posterior thigh surface brown with bluish flecks (fig. 7.74b). In addition, individuals of this species have tuberculate upper surfaces and a tuberculate tympanum; adult males have gray vocal sacs, a series of prominent warts along the posteroventral margin of the lower arm, and single subarticular tubercles under fingers III and IV.

DESCRIPTION: Adult males 31 to 45 mm in standard length, adult females 44–62 mm; upper surfaces tuberculate; venter granular; head wider than long; dorsal outline of head round to semicircular; snout truncate from above and in profile; eyes relatively large; tympanum distinct, tuberculate, diameter about one-half diameter of eye; paired vocal slits and fully distensible paired gray external vocal sacs in adult males; fingers moderately long and stout; finger disks moderate; width of disk on finger IV about two-thirds diameter of tympanum; subarticular tubercles single; black

nonspinous nuptial pad on base of thumb in adult males; fingers webbed, but only a trace of webbing between fingers I and II; webbing formula: I 3–3 II 1⅔–2½ III 2–2⁻ IV; thenar pad moderately enlarged; palmar tubercle indistinct, triangular; toes extensively webbed; webbing formula: I 1–2⁻ II 1–2⁻ III 1–2⁻ IV 2⁻–1 V; inner metatarsal tubercle small, low, flat, outer absent; inner tarsal fold present. Dorsal surfaces various shades or gray and brown with darker brown markings; green metallic flecks also present in many examples; upper limb surfaces with transverse dark bars; groin and posterior thigh surface a brown reticulum enclosing blue flecks; venter creamy white; iris variable, pale brown to grayish brown with or without bronze suffusion and dark brown or black reticulations.

LARVAE (fig. 7.108): Large, 25 to 31 mm in total length at stage 25; body ovoid; mouth ventral; nostrils and eyes dorsolateral; spiracle lateral, sinistral; vent tube dextral; tail moderate; caudal fins moderate; tail tip pointed; oral disk moderate, entire, with well-developed beaks and 2/3 denticle rows; A2 with narrow gap above mouth; beaks bluntly serrate; papillae in one or two rows with many additional rows on lateral portion of lower lip; no papillae in area of upper lip above mouth; body colors in preservative: gray brown above, paler below with dark brown dorsal spots and light lateral flecks; caudal musculature pale tan with brown flecks; dark brown flecks on dorsal fin and posterior half of ventral.

VOICE: A low squawk usually followed by one or more rattling secondary notes. The primary notes last 60 to 280 milliseconds and the secondary 140 to 480 milliseconds. Primary notes have pulse rates of 97 to 120 per second and dominant frequencies at about 0.9 and 2.2 kHz (Duellman 1970).

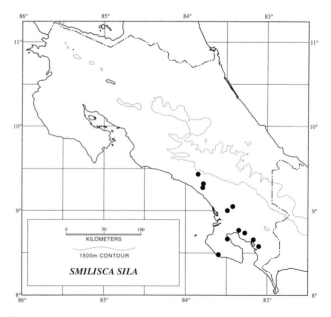

Map 7.103. Distribution of *Smilisca sila*.

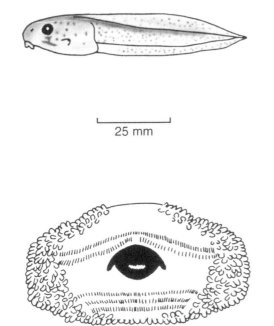

Figure 7.108. Tadpole of *Smilisca sila*.

25 mm

SIMILAR SPECIES: (1) *Smilisca sordida* has a smooth dorsum and tympanum, and the web between toes I and II reaches the base of the distal phalanx on toe II; adult males have white vocal sacs, and where the two species are sympatric the snout is obtuse (fig. 7.6c) in profile. (2) *Phrynohyas venulosa* has green bones and lacks prominent warts along the posteroventral margin of the forearm and dark brown to black reticulations in the groin and on the posterior thigh surfaces that enclose bluish to creamy tan flecks. (3) *Hyla tica* has the distal subarticular tubercle under finger IV divided and that under finger III divided or bifid (fig. 7.72b,c). (4) *Hyla pseudopuma* has smooth dorsal surfaces and tympanum and a rounded snout profile (fig. 7.6c) and lacks the reticulate groin and posterior thigh surface, although large yellow spots may be present along the flank and in the groin (fig. 7.74c,d).

HABITAT: Found primarily in or near shallow streams having a considerable gradient and rocky bottoms and/or margins in Lowland Moist and Wet Forests and rarely in Premontane Rainforest.

BIOLOGY: A rather common nocturnal treefrog that is active throughout the year. Males call from rocks in the streams or on the banks and sometimes from low vegetation overhanging the water. Breeding is restricted to the dry season (December to February or March) when water levels are down. Occasional rains appear to stimulate reproductive activity several times per season. Males of this species produce synchronous one- to four-note advertisement calls. Complexity of call is determined by vocal interactions with other males. Complex calls probably increase the frogs' predation risk, especially from the frog-eating bat *Trachops cir-*

rhossus, which locates its food acoustically and preferentially responds to complex frog calls over simple ones. Like other frogs, *S. sila* probably uses visual cues to detect hunting bats and employs a variety of behaviors to reduce predation (Ryan and Tuttle 1981; Ryan, Tuttle, and Rand 1982; Nunes 1988). These include calling from near waterfalls that increase ambient noise; decreasing the frequency of calling and the proportion of complex calls; calling more frequently at dusk; and reducing calling periods. On nights with low ambient illumination, call synchrony itself may confuse the bats. Experiments also showed that the calls of *S. sila* decreased in number and complexity as a response to overhead flight of a model of the frog-eating bat. On nights with greater ambient illumination males of *S. sila* call more often and with greater complexity, have a longer calling period, and hide under leaves less frequently. The benthic tadpoles are found in clear pools and use the clasping oral apparatus to maintain position (ecomorph 10, table 7.2). Metamorphs are 13 to 16 mm in standard length.

REMARKS: This frog was confused with its ally *Smilisca sordida* for many years until Duellman and Trueb (1966) clarified its status as a distinct species. Some geographic variation occurs in the coloration of the flanks and posterior thigh surface. Generally, in frogs from Costa Rica and western Panama the blue coloration is subdued compared with that in specimens from eastern Panama. The flanks are brown with pale blue flecks, and the posterior thigh surfaces are reticulated brown, enclosing small blue flecks. As one moves farther east through Panama the light areas become spots, the reticulations get darker, and the spots become a darker blue.

DISTRIBUTION: Humid lowlands from southwestern Costa Rica to eastern Panama on the Pacific slope and on the Atlantic versant in central Panama and northern Colombia (10–970 m).

Smilisca sordida (Peters, 1863e)

(plates 191–92; map 7.104)

DIAGNOSTICS: A rather common, nondescript moderate (males) to moderately large (females) treefrog that is gray, tan, greenish brown, reddish brown, or gray marked with darker blotches and is often confused with a variety of other forms. It differs from other *Smilisca* in the following combination of features: webbed fingers and extensively webbed toes (most toes have only the distal phalanx free of web); snout obtuse to rounded in profile (fig. 7.6c); dorsal surfaces and tympanum smooth; flanks and posterior thigh surfaces marked with bluish white or cream flecks or bluish white spots in a darker ground color or boldly mottled black and bluish white (fig. 7.74b); a prominent series of warts along posteroventral margin of lower arm; no white labial or tibial stripe; no suborbital light spot.

DESCRIPTION: Adult males 32 to 54 mm in standard length, adult females 56 to 64 mm; dorsal surfaces smooth;

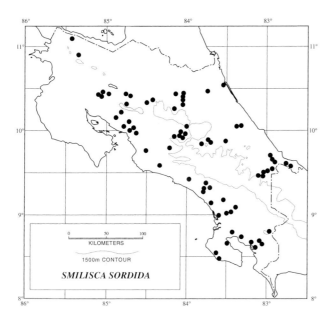

Map 7.104. Distribution of *Smilisca sordida*.

venter granular; head longer than broad; dorsal outline of snout subovoid to subelliptical, snout bluntly rounded to obtuse in profile; eyes moderate; tympanum distinct, diameter about one-half diameter of eye; paired vocal slits and slightly distensible paired white internal subgular vocal sacs in adult males; fingers short and stout; finger disks large, diameter of disk on finger III equals diameter of tympanum; subarticular tubercles under finger IV single or bifid; light brown nonspinous nuptial pad on base of thumb in adult males; fingers webbed, but only a trace between fingers I and II; webbing formula: I $3-3^+$ II 1^+-2^+ III 2^--1^- IV; thenar pad enlarged, no palmar tubercle; toes extensively webbed; webbing formula: I $1-1^+$ II $1-1\frac{1}{2}$ III $1\frac{1}{4}-2\frac{1}{4}$ IV $1\frac{1}{2}-1$ V; inner metatarsal tubercle large, oblong, outer small, indistinct or absent; distinct inner tarsal fold. Dorsum yellowish brown, reddish brown, gray, green brown, to dark brown, uniform or with darker blotches, some of them dark green; in some examples dark markings form large transverse bands; flanks mottled dark brown and cream, greenish gray, or bluish gray; upper limb surfaces with obscure to well-developed transverse dark bars; posterior thigh surfaces dark brown with pale blue, bluish green, or tan flecks; venter white; iris silvery, yellowish, or bronze with some black reticulation.

LARVAE (fig. 7.109): Moderate-sized, 32 mm in total length; body ovoid; mouth ventral; nostrils and eyes dorsolateral; spiracle lateral, sinistral; vent tube dextral; tail moderate; caudal fins moderate; tail tip pointed; oral disk large, entire, with well-developed beaks and 2/3 denticle rows; A2 with narrow median gap; beaks bluntly serrate; papillae in two rows lateral and ventral to mouth, many additional rows at corner of mouth, none above mouth; body tan; belly pale tan with whitish to silvery tint; caudal musculature tan

with dull reddish flecks and dashes forming alternating dark and light areas along midline; fins flecked with dark; iris bronze.

VOICE: A "wrink" repeated up to six times with a rising inflection at the end. The call sounds like someone rapidly running a fingernail along a pocket comb, most closely spaced teeth last. The call group is repeated at intervals of twelve seconds to several minutes, and each "wrink" lasts 180 to 450 milliseconds, with a pulse rate of 78 to 135 per second and two emphasized frequencies at about 1.20 and 2.7 kHz (Duellman 1970).

SIMILAR SPECIES: *S. sordida* is a rather variable species and difficult to identify. Treefrogs, especially juveniles, without highly distinctive morphologies or coloration will often turn out to be this form. (1) *Smilisca sila* has a truncate snout profile (fig. 7.6c), tuberculate dorsum and tympanum, and the web between fingers I and II not reaching to the base of the first digit on finger II. (2) *Hyla pseudopuma* lacks the reticulate groin and posterior thigh pattern, although yellow spots may be present along the flank and in the groin (fig. 7.74c,d), and has one or two phalanges on every toe free of webbing. (3) *Smilisca puma* has no finger webs or only vestigial ones.

HABITAT: Usually found in or near shallow, low-gradient rocky streams in Lowland Moist and Wet Forests in some gallery forests in the Lowland Dry Forest zone and in Premontane Moist and Wet Forests.

BIOLOGY: This common nocturnal treefrog is found throughout the republic in humid situations, and males may be heard calling sporadically outside the breeding season.

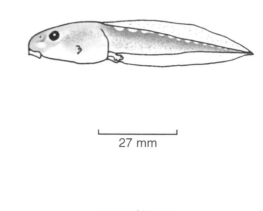

27 mm

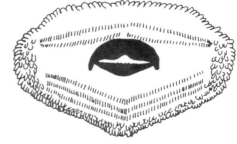

Figure 7.109. Tadpole of *Smilisca sordida*.

Although it usually is found on low vegetation when not on rocks or in the stream, *S. sordida* often hides during the day in bromeliads. Breeding takes place in the dry season, which may be relatively long, as in northwestern Costa Rica and on the Meseta Central (November to May), or relatively short on the Atlantic slope or in southwestern Costa Rica (January to February or March). At these times the streams are clear and at low water, and males primarily call from rocks or gravel bars in the watercourse. Males of this species call alternately with their nearby neighbors, and it is usual for the calls to have a domino effect as males call in sequence along a section of the streambed. Apparently stimulated by occasional heavy showers, females move to the stream, forming large breeding congregations. One to several of these aggregations may take place at a particular site during the dry period, depending on the frequency of showers, but it is not known if females lay more than one clutch annually. Amplexus takes place near shore or in the water. The eggs are deposited singly on the bottom of shallow pools. They adhere to one another to form masses of 20 to 50 eggs (James 1944). The benthic larvae live in shallow, clear backwaters and use the large oral disk to clasp submerged rock surfaces to maintain stability (ecomorph 10, table 7.2). Metamorphs are 13 to 16 mm in standard length.

REMARKS: The species called *Hyla gabbi* (syntypes: USNM 30658–59; Costa Rica: Limón: near Sipurio) and *Hyla nigripes* (syntypes: USNM, 30685–86; Costa Rica: Limón: Cerro Utyum) by Taylor (1952a), both described by Cope (1875), are based on Costa Rican examples of this species. Duellman and Trueb (1966) demonstrated that these nominal forms were conspecific with *S. sordida*, which was originally described from western Panama. Specimens of *S. sordida* are also the basis for Taylor's (1958a) reports of *Hyla monticola* and *H. gabbi* from Costa Rica. *H. monticola* is a synonym of *H. pictipes*. Considerable geographic variation occurs in this species as follows: Atlantic lowlands: snout short, blunt, rounded in profile; two to six tibial dark bars usually present, sometimes indistinct; little bluish white to tan flecking in groin, and flecking reduced or occasionally absent on posterior thigh surface. Meseta Central: snout moderately short, blunt, rounded in profile; two to eight tibial dark bars usually present, sometimes indistinct; numerous large pale blue flecks or round spots on posterior half of flanks, groin, and posterior thigh surface. Southwestern Pacific lowlands: snout long, obtuse in profile; four to six distinct tibial dark bars; bold mottling of black and bluish white on flanks, numerous bluish white flecks on posterior thigh surface. Representatives from northwestern Costa Rica, including those from gallery forests, resemble those of the Atlantic lowlands. Duellman and Trueb's (1966) record of this species for the Lake Nicaragua drainage (MCZ 7924–25) is based on a confusion of localities. These examples are from Santa Cecilia, Limón Province, a locality confused with San Cecilio, Guanacaste Province,

by these authors and others, including me (Savage 1974b; Savage and Heyer 1968). Recent collections, however, confirm that *S. sordida* occurs along several streams draining into Lake Nicaragua in Alajuela and Guanacaste Provinces. The disjunct record of this taxon from the Río Magdalena valley of Colombia (Ruiz, Ardilla, and Lynch 1996) awaits confirmation. The karyotype is 2N = 24, with all chromosomes metacentric or submetacentric; NF = 48 (Bogart, 1973).

DISTRIBUTION: Humid forests in the lowlands and on premontane slopes from northeastern Honduras to northwestern Panama on the Atlantic slope and in southwestern Costa Rica and adjacent western Panama and El Valle de Antón, Coclé Province, in west-central Panama; also in Pacific slope gallery forests in the subhumid region of northwestern Costa Rica (2–1,525 m).

Family Centrolenidae Taylor, 1951a

Glassfrogs

The representatives of this family (four genera, 137 species) include some of the most beautiful and delicate of anurans. Most are tiny to small (up to 30 mm in standard length) green frogs with large eyes that are directed forward at about a 45° angle and have the parietal peritoneum white and/or internal organs visible through the transparent ventral integument. Two species of the Colombian genus *Centrolene*, however, are much larger (*C. geckoideum* to 84 mm in standard length and *C. paezorum* to 44.5 mm) and have small eyes, as does the other smaller species (*C. acanthidiocephalum*).

Centrolenids, like the hylids and frogs of several other families (some microhylids and all hyperoliids), have a short intercalary cartilage between the penultimate and terminal phalanges. Well-developed disks and webbing are present on both fingers and toes, the disks are supported by T-shaped terminal phalanges, and the venter is coarsely areolate. The name glassfrogs is derived from the transparent venter through which, in all Costa Rican species, a white parietal peritoneal sheath, some or all of the viscera, or the heart may be seen in living specimens.

Additional features that define the family are eight holochordal procoelous presacral vertebrae; no ribs; sacrum with dilated diapophyses and bicondylar articulation to coccyx; coccyx without transverse processes; pectoral girdle arciferal; no omosternum, but a cartilaginous sternum present; no parahyoid bone; tibiale and fibulare usually fused throughout length (but see Sanchiz and Riva 1993); a dilated medial process on third metacarpal (unique in anurans); sartorius muscle distinct from semitendinosus and tendon of latter inserting ventral to gracilis muscle; teeth present on upper jaw; tongue not highly protrusible; pupil horizontally elliptical; amplexus axillary; larva type A with beaks and denticles and single sinistral spiracle. The karyotype is 2N = 20.

Among Costa Rican frogs, centrolenids are most likely to be confused with green members of the Hylidae or with *Eleutherodactylus moro.* Hylids usually have the eyes oriented laterally. Only one member of the family in the republic *(Hyla zeteki)* has a transparent ventral integument, and two green species *(Hyla colymba* and *H. palmeri)* have a white parietal peritoneal sheath also characteristic of some glassfrogs. *H. zeteki* has minimal finger webs and a dark black band across the wrist (absent in centrolenids). *H. colymba* has a yellowish stripe along the canthus rostralis over the eyelid to behind the tympanum, and *H. palmeri* has a heel calcar. Neither of these distinctive features occurs in centrolenids. *E. moro* has laterally directed eyes and lacks any digital webbing.

The structure and function of the digital disks in this family are similar to those of the Hylidae (see that family account for details), except that in centrolenids the terminal phalanges are T-shaped, not claw shaped.

The family is found in humid habitats from Mexico to Peru on the Pacific slope and to Bolivia and northeastern Argentina on the Atlantic versant but is absent from the lower Amazon basin.

Ruiz and Lynch (1991) have recently reviewed the generic classification of the family. Their proposal recognizes two derived lineages as *Centrolene* and *Hyalinobatrachium.* The former is characterized solely by the presence of humeral hooks on the arms of adult males, the latter by the presence of a bulbous, nontrilobate liver. *Cochranella* in this arrangement is probably not monophyletic, since its species share no derived features. In previous works (Savage 1967, 1973b, 1976, 1980b; Savage and Villa 1986) I included all Costa Rican species in a single genus, *Centrolenella.* Since the Ruiz and Lynch arrangement has considerable merit, I follow it here except that I regard the three small-eyed species of *Centrolene* from Colombia and Ecuador mentioned above as distinct from the large-eyed taxa placed in that genus by those authors. In this book these latter species are retained as the genus *Centrolenella.*

At least two chromatophore arrangements apparently are responsible for the green coloration in glassfrogs (Goin and Goin 1968; Starrett and Savage 1973; Schwalm 1981), as indicated by color in preservative. In *Centrolene, Centrolenella,* and *Cochranella* the color in life seems to be structural. It is produced by reflection of shorter wavelengths (greens, blues, and purples) from the iridophore layer. Longer wavelengths (oranges, yellows) are absorbed or reflected (reds) by special large and complex melanophores (pterorhodophores) also found in some other treefrogs. See the account for the family Hylidae for further details on these specialized melanophores.

In the three genera mentioned above, when reflected light passes through the superficial xanthophore layer the blues, reds, and purples are mostly absorbed while the greens are transmitted, so the frog is seen as green. On preservation, the fatty yellow materials in the iridophores are dissolved or destroyed and the reflected blues, reds, and purples are transmitted, so the skin appears purple to lavender to the unaided eye. Dried skins that have not been fixed in preservative also change from green to purplish in a short time. Preserved specimens exposed to strong light for a long time will fade to white. Examples that have been in preservative for many years may also pale, but this may be the effect of the original fixative.

In *Hyalinobatrachium* the green pigment biliverdin (Barrio 1965; Goin and Goin 1968; Starrett and Savage 1973; Jones 1967) is accumulated in the skin rather than in the chromatophores. On preservation the pigment is dissolved, so the skin appears pale yellow or white, although scattered purplish and black (melanophores) punctations may be present. Significantly, members of this genus lack substantial numbers of pterorhodophores. Dried skins of fresh examples of most species in this genus resemble preserved specimens, but *Hyalinobatrachium pulveratum* skins retain their bright green color.

Barrio (1965) also found high quantities of the pigment biliverdin in the bones, muscles, eggs, and lymph of several hylid and pseudid frogs. Jones (1967) confirmed that the green pigment in the skin of most *Hyalinobatrachium* species is biliverdin, and this is probably the substance responsible for the green bones in the other genera. In Costa Rica members of the genus *Hyalinobatrachium* have green eggs, and these almost certainly contain biliverdin.

White pigment is variously deposited in the parietal peritoneum, the parietal pericardium, and the visceral peritoneum of the liver, intestinal tract, and urinary bladder in centrolenids. Except for those species in which the parietal peritoneum (peritoneal sheath) extends far posterior, some or all of the white-covered organs may be seen through the ventral wall in life. The white pigment does not dissolve in fixative, and its presence may be determined by dissecting frogs that have been preserved for fifty years or more.

Glassfrogs are nocturnal, epiphyllous, and arboreal. In all species for which the reproductive biology is known, eggs are laid out of the water in gelatinous circular masses on vegetation overhanging fast-moving streams (plate 209), or in one case *(C. geckoideum)* on rocks above the stream. Males of some species attend the eggs, but in all cases when tadpoles hatch they are washed or fall into the stream (pattern 6, table 7.2). Because of these habits, the tadpoles have extremely high concentrations of hemoglobin and large blood sinuses and usually appear pink or red (plate 193). It is interesting that biliverdin is an oxidation product of hemoglobin and may be deposited in the skin, bones, or oocytes of centrolenids as a waste product during metamorphosis.

The two genera *Centrolenella* and *Cochranella* differ primarily in that the adult males of the former have bony hooks on the humerus. For that reason the following key covers the species of both genera to expedite identification of females and juveniles. Remember that in life the white parietal peritoneal sheath extends far posterior in *Centrolenella*, concealing most of the viscera, whereas in *Cochranella* it extends only to the level of the liver, so most of the digestive tract is visible. These same relationships can be determined in preserved specimens with careful dissection.

KEY TO THE COSTA RICAN GLASSFROGS OF THE GENERA *CENTROLENELLA* AND *COCHRANELLA*

1a. Snout vertical to rounded in profile (fig. 7.110a–d) 2
1b. Snout strongly obtuse in profile (fig. 110e,f); no humeral hook in adult males 5

2a. Dorsum uniform or with a few large light or dark spots or heavily spotted with dark; no fleshy ridge along posterior margin of lower arm 3
2b. Dorsum marked with numerous light spots; distinct fleshy fold along posterior margin of lower arm; no humeral hook in adult males . . . .
. . . . *Cochranella albomaculata* (p. 365)
3a. No free prepollex or prepollical spine; a humeral hook in adult males (fig. 7.11c); dorsum smooth; adult males to 27 mm in standard length, adult females to 32 mm
. . . . 4
3b. A free prepollex or prepollical spine; no humeral hook in adult males; dorsum weakly granular; adult males to 20 mm in standard length, adult females to 23 mm
. . . . *Cochranella spinosa* (p. 365)
4a. Webs between toes III–IV–V not reaching base of disk on toes III and V, not reaching distal subarticular tu-

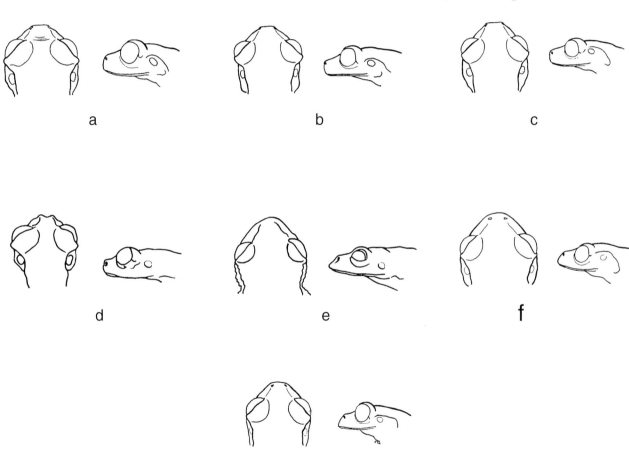

a   b   c

d   e   f

g

Figure 7.110. Diagnostic features of head morphology in Costa Rican glassfrogs (Centrolenidae), dorsal and lateral views: (a) *Cochranella albomaculata* T, N; (b) *C. spinosa* R to T, B; (c) *Centrolenella prosoblepon* T, N; (d) *C. ilex* T, S; (e) *Cochranella euknemos* O, B; (f) *C. granulosa* O, VB; (g) *Hyalinobatrachium pulveratum* O, N. (Snout profile: O = obtuse, R = round, T = truncate; interorbital space: B = broader than eye diameter, N = narrower than eye diameter, SN = slightly narrower than eye diameter, VB = much broader than eye diameter.)

bercle on toe IV; a protruding humeral hook present in adult males and some females; snout subovoid in dorsal outline (fig. 7.110c); nostrils not elevated on protuberant ridges; dorsal color uniform or with numerous small dark spots . . . .
. . . . *Centrolenella prosoblepon* (p. 361)

4b. Webs between toes III–IV–V reaching nearly to base of disk on toes III and V, reaching distal subarticular tubercle on toe IV; humeral hook present in adult males, hidden in musculature; snout semicircular in dorsal outline (fig. 7.110d); nostrils elevated on protuberant ridges; dorsum uniform or with a few large light spots, never marked with numerous dark spots . . . .
. . . .*Centrolenella ilex* (p. 360)

5a. A distinct fleshy fringe along posterior margin of lower arm and lower leg and foot (fig. 7.70a); upper surface with scattered tubercles; dorsal coloration with distinct light spots. . . . . *Cochranella euknemos* (p. 363)

5b. No fleshy fringe along arms or legs; dorsum strongly granulate, without raised tubercles; dorsum uniform or with a few black spots . . . .
. . . .*Cochranella granulosa* (p. 363)

Genus *Centrolenella* Noble, 1920

These diminutive frogs (thirty-five species) are most distinctive in having eyes larger than the disk on finger III, humeral hooks in adult males, and the parietal peritoneum and the pericardium white. The liver is trilobate and unpigmented. In all Costa Rican species the bones are green in life, vomerine teeth are present, and the white peritoneal sheath extends far posteriorly to hide most of the internal organs. In preservative the skin is purple to lavender. Adult males of all known *Centrolenella* have paired vocal slits and a single external vocal sac that is fully distensible.

Members of this genus are found on vegetation surfaces along watercourses. The eggs are laid on unshaded surfaces, usually on the upper surface of leaves. The eggs are uniform black in Costa Rican species. No species in this genus is known to brood its eggs.

The genus is composed of two species clusters; only one, the *Centrolenella prosoblepon* group, characterized by the absence of white pigment on the organs of the digestive tract, occurs in Costa Rica. Besides the native *C. ilex* and *C. prosoblepon*, extralimital taxa placed in this group are *Centrolenella andina* (Venezuela), *C. antioquiensis* (Colombia), *C. audax* (Ecuador), *C. azulae* (Peru), *C. ballux* (Colombia and Ecuador), *C. buckleyi* (Venezuela to Ecuador), *C. fernandoi* (Peru), *C. grandisonae* (Colombia and Ecuador), *C. guanacara* (Colombia), *C. heloderma* (Ecuador), *C. hesperia* (Peru), *C. huilensis* (Colombia), *C. lemniscata* (Peru), *C. mariae* (Peru), *C. medemi* (Colombia), *C. muelleri* (Peru), *C. notosticta* (Colombia), *C. petrophila* (Colombia), *C. pipilata* (Ecuador), *C. puyoensis* (Ecuador), *C. quindiana* (Colom-

bia), *C. robledoi* (Colombia), *C. scirtetes* (Ecuador and Colombia), and *C. tayrona* (Colombia).

This genus ranges from eastern Honduras to Venezuela and on both slopes of the Andes to northern Peru.

Centrolenella ilex Savage, 1967

(plate 194; map 7.105)

DIAGNOSTICS: Prominent protuberant nostrils (fig. 7.110d) are characteristic of this medium-sized dark green centrolenid that lacks any dorsal dark spots. The bones are dark green in life, and a white parietal peritoneal sheath and distinct white lip stripe are present. The toes are extensively webbed so that the webbing reaches the distal subarticular tubercle on toe IV and to, or nearly to, the base of the disks on toes III and V. Adult males of this form have a sharp, pointed humeral hook embedded in the arm musculature that is not visible externally.

DESCRIPTION: Adult males 27 to 29 mm in standard length, adult females 28 to 34 mm; upper surfaces smooth; head slightly broader than long; head basically semicircular in dorsal outline, interrupted by prominent protuberant nostrils; snout profile truncate (fig. 7.110d); interorbital space slightly less than diameter of eye; eyes large, protuberant; tympanum round, indistinct, directed obliquely upward; vomerine teeth in transverse rows between choanae, separated medially; finger I longer than finger II; finger and toe disks truncate; subarticular tubercles small, smooth, usually double on fingers I and IV; no supernumerary tubercles, but distinct accessory palmar and plantar tubercles present; no web between fingers I and II, webbing vestigial between fingers II and III; webbing formula for outer fingers: II $1\frac{3}{4}$–$3\frac{1}{4}$ III $1\frac{1}{2}$–$1\frac{1}{4}$ IV; thenar tubercle weak, elon-

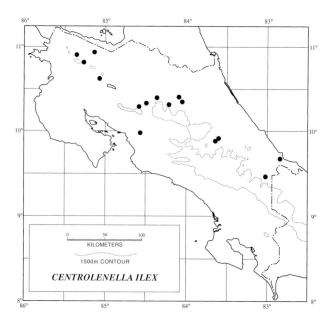

Map 7.105. Distribution of *Centrolenella ilex*.

gate; palmar tubercle broader than thenar but elongate; well-developed white nuptial pad on dorsal and outer lateral surfaces of thumb base in adult males; toes extensively webbed; webbing formula: I 1–1$^+$ II 1–1¾ III 1–2 IV 1½–1 V; inner metatarsal tubercle elongate, no outer metatarsal tubercle, tarsal fold, or tubercle. Dorsum uniform deep green or with a few large light spots (in preservative only); white stripe along upper lip; undersurface dull white; pericardium white (not visible); digestive tract unpigmented; iris ivory gray with black reticulum.

LARVAE: Unknown.

VOICE: A single high-pitched "click" repeated at intervals of several minutes.

SIMILAR SPECIES: *Centrolenella prosoblepon* does not have prominent protuberant nostrils (fig. 7.110c), and the toe webs are less extensive, those on toe IV not reaching the distal subarticular tubercle; it frequently has the dorsal surfaces flecked with black, and adult males have a protuberant humeral hook (fig. 7.11c).

HABITAT: Lowland Moist and Wet Forests and Premontane Rainforest.

BIOLOGY: The eggs are black and are deposited on upper leaf surfaces overhanging a stream.

REMARKS: This species was originally described from Costa Rica (Savage 1967), and records extending its known range were published by Starrett and Savage (1973) for Panama and by Hayes and Starrett (1980) for Nicaragua and Colombia.

DISTRIBUTION: Scattered localities in humid lowland and perhumid premontane areas from eastern Nicaragua to western Panama and western Colombia; also from the caldera of the Río Grande de Tárcoles of central Costa Rica on the Pacific versant (250–900 m).

Centrolenella prosoblepon (Boettger, 1892a)

(plate 195; map 7.106)

DIAGNOSTICS: This small very common species is light green, usually with many small black spots dorsally, dark green bones in life, and a white parietal peritoneal sheath. It lacks protuberant nostrils (fig. 7.110c) and has moderate toe webbing, the web not reaching distal subarticular tubercle on toe IV or base of disks on toes III and IV. Adult males of this species have an obvious projecting blunt humeral hook (fig. 7.11c).

DESCRIPTION: Adult males 21 to 28 mm in standard length, adult females 25 to 31 mm; upper surfaces smooth to finely shagreened; head broader than long; head round from above; snout truncate in profile; nostrils slightly protuberant or not; interorbital space narrower than diameter of eye; eyes large, protuberant; tympanum indistinct, directed obliquely upward; vomerine teeth in slightly diagonal patch between choanae, separated medially; finger I longer than finger II; finger and toe disks truncate; subarticular tu-

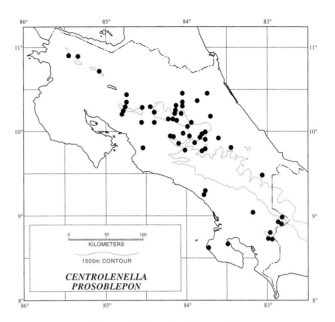

Map 7.106. Distribution of *Centrolenella prosoblepon*.

bercles small, round; no supernumerary, accessory palmar, or plantar tubercles; no webbing between fingers I and II, webbing vestigial between II and III; webbing formula for outer fingers: II 2–3¾ III 2½–2 IV; thenar tubercle elongate; palmar tubercle round, large; well-developed white nuptial pad on outer lateral and dorsal surfaces of thumb base in adult males; toes moderately webbed; webbing formula: I 1$^+$–2½ II 1½–1¼ III 1¼–2¼ IV 2½–1½ V; inner metatarsal tubercle small, flat, elongate, outer small, ovoid; no tarsal fold or tubercle. Dorsal surfaces green, uniform or usually with black flecks; yellow to transparent flesh below; pericardium white (not visible); digestive tract unpigmented; iris ivory gray with fine black reticulum.

LARVAE (fig. 7.111): Small, total length 12.3 mm at stage 25; body elongate, somewhat depressed; mouth ventral; nares and eyes dorsal; spiracle midlateral, posterior, sinistral; vent tube long, median; tail long; fins very low; caudal fin reduced, rounded at tip; oral disk complete, with medium-sized beaks and 2/3 rows of denticles present; A2 with great gap above mouth; lower beak serrate laterally; single row of labial papillae, absent above mouth; black on hatching, becoming pale brown above and lighter below but appearing bright red; brown dorsal and lateral streaks on musculature and tail fins anteriorly, uniform posteriorly except fin margins.

VOICE: A rasping series, usually of three notes, "dik, dik, dik," followed by a pause and repeated. Call rate is low, 1 to 43 per hour, note duration is 1.5 to 3 seconds, and the dominant frequency is 5.3 to 6.0 kHz (Jacobson 1985).

SIMILAR SPECIES: *Centrolenella ilex* has extremely protuberant nostrils lying on a distinct ridge (fig. 7.110d), and the web reaches the distal subarticular tubercle on toe IV.

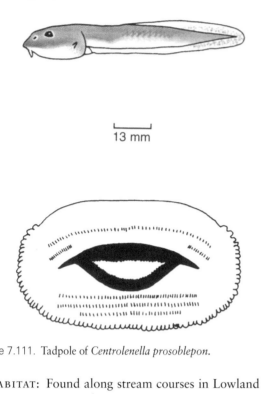

Figure 7.111. Tadpole of *Centrolenella prosoblepon*.

HABITAT: Found along stream courses in Lowland Wet and Moist Forests and Premontane Moist and Wet Forests and Rainforest and marginally in Lower Montane Wet Forest and Rainforest.

BIOLOGY: This account draws heavily on information in Jacobson (1985) and Hayes (1991) as well as on my own observations. *Centrolenella prosoblepon* is a nocturnal species that usually is found in low vegetation along the margins of fast-moving streams. Males usually call from the upper surfaces of leaves throughout the year but are most vocal during the wet season (May to November). The males are strongly territorial, and the advertisement call is used to space them along the stream at intervals averaging 3.2 m. If an intruder moves into an occupied territory, the resident male gives a rapid condensed series of "dik-dik-diks." If the intruder approaches too closely, a fight ensues with both males dangling upside down while holding onto the vegetation with their legs and grappling with their arms. This combat may last three to thirty minutes. It concludes when the loser drops from the fight site or signals submission by flattening his body against the leaf substrate, then moves out of the territory. In this species males initiate amplexus when a gravid female approaches the call site. After the male jumps on the female's back and grasps her, he continues calling. Amplexus continues for thirty-eight to eighty-five minutes, and the pair stays in place or moves up to 2 m away. Oviposition occurs in the early morning before sunup (from 2:30 on). All eggs are laid on upper leaf surfaces, moss-covered rocks, or branches from 0 to 3 m above the ground. Egg clutches may be laid throughout the year, but reproduction peaks in August to September. The black eggs are laid in a single layer

in a thin film of jelly about 50 mm in diameter; clutch size averages about 20 eggs, with diameters including the envelope averaging 10 mm. The females remain with the eggs for a time, but once they leave after the first few minutes or hours, neither parent returns to the clutch. Observed females returned to the canopy after oviposition. They probably remain there or disperse some distance into the adjacent forest during nonreproductive periods, since they are not found along the stream margin at other times. Embryos are black but quickly turn yellow and then appear red by the time they fall into the stream. Small juveniles are 14 mm in standard length.

REMARKS: Taylor (1952a, 1958b) used the name *Centrolene prosoblepon* for this species. He reports on one male lacking vomerine teeth. The humeral hook may be developed in occasional females. The karotype is 2N = 20 (Duellman 1967b).

DISTRIBUTION: Humid lowland, premontane, and lower levels of the lower montane zones from Atlantic slope eastern Honduras and Pacific versant Costa Rica to northern Colombia and western Ecuador (20–1,900 m).

Genus *Cochranella* Taylor, 1951a

Members of this genus (sixty-seven species) are small anurans having the eyes larger than the disk on finger III and the parietal peritoneum and the pericardium white but lacking humeral hooks in males. The white peritoneal sheath in this genus extends no farther posteriorly than the level of the liver, so most of the viscera are visible in life. The liver is trilobate and unpigmented, and in preservative the skin appears lavender. In all Costa Rican species, the bones are green and vomerine teeth are present. Adult males of all known species of the genus have paired vocal slits and a single external subgular vocal sac.

These frogs are found on vegetation along stream margins and, as far as is known, usually lay their eggs on the unshaded or weakly shaded upper surface of leaves. Studied species do not brood their eggs.

The genus is composed of three divisions, the *Cochranella granulosa*, *Cochranella ocellata*, and *Cochranella spinosa* species groups. The first and last are represented in Costa Rica. *Cochranella* ranges from Honduras to the Guianas, Amazonian Brazil, Peru, and Bolivia east of the Andes and to northern Peru west of the Andes.

Cochranella granulosa Group

Frogs of this group have white pigment on the organs of the digestive tract. Extralimital species are *Cochranella castroviejoi* (Venezuela), *C. daidalea* (Colombia), *C. ramirezi* (Colombia), *C. resplendens* (Ecuador and Colombia), *C. savagei* (Colombia), and *C. solitaria* (Colombia). As far as is known the eggs of Costa Rican members of this group are black-and-white.

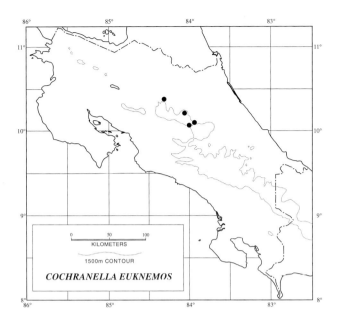

Map 7.107. Distribution of *Cochranella euknemos*.

Cochranella euknemos (Savage and Starrett, 1967)

(map 7.107)

DIAGNOSTICS: A rare moderate-sized deep blue green, light-spotted glassfrog having distinct crenulate fringes along the posteroventral margin of the lower arm and lower leg and foot (fig. 7.70a), a snout obtuse in profile (fig. 7.110e), dark green bones, a white parietal peritoneum and pericardium, and visible white digestive tract.

DESCRIPTION: Adult males 21 to 25 mm in standard length, adult females 26 to 32 mm; upper surfaces granular with isolated raised tubercles; head longer than broad; head rounded in dorsal outline; snout much longer than orbital diameter, obtuse in profile; interorbital space much broader than diameter of orbit; eyes large, not protuberant; tympanum indistinct, round, directed laterally; vomerine teeth in transverse series between choanae, separated medially; finger I longer than finger II; finger and toe disks truncate; subarticular tubercles tiny; no supernumerary tubercles, but two accessory palmar and three plantar tubercles; no web between fingers I and II, webbing vestigial between fingers II and III; webbing formula for outer fingers: II 2–2¾ III 2¼–1¾ IV; thenar tubercle elongate, large, palmar tubercle rounded, large; well-developed white nuptial pad on dorsal and outer lateral surfaces of thumb in adult males; toes moderately webbed; webbing formula: I 1–1½ II 1–2 III 1–2 IV 2–1 V; inner metatarsal tubercle elongate, large; no outer metatarsal or tarsal tubercles, no tarsal fold; a distinct scalloped fringelike ridge below vent. Dorsal surfaces deep bluish green, with numerous bright yellow to yellowish white spots; ivory lip stripe present; arm and leg fringes ivory white; undersurfaces of limbs yellowish white; white

parietal peritoneal sheath anteriorly, reddish liver, green kidneys, and white digestive tract visible posteriorly; pericardium white (not visible); iris ivory gray.

LARVAE: Unknown.

VOICE: One to three rapidly repeated harsh "creep-creep-creep" notes that blur together; dominant frequency 3.8 to 4.6 kHz (Savage and Starrett 1967; Ibáñez, Rand, and Jarmillo 1999).

SIMILAR SPECIES: No other Costa Rican frog closely resembles this beautiful species. (1) *Cochranella albomaculata* has a similar color pattern, a short truncate to rounded snout profile, a narrow interorbital space (fig. 7.110a), and no fleshy fringe along the lower leg and foot. (2) *Hyalinobatrachium pulveratum* has substantial webbing between fingers II and III and lacks crenulate limb fringes.

HABITAT: Premontane Moist and Wet Forests and Rainforest.

BIOLOGY: According to Hayes (1991), males call from small trees or bushes overhanging torrential streams in the rainy season (May to November). The black-and-white oviductal eggs are laid in a large jelly mass suspended from the upper surface of a leaf, hanging over the tip to form a gelatinous drip tip that ensures the mass a steady water supply. There is no long-term parental care in this species.

REMARKS: The specimens observed by Hayes and reported on by Hayes, Pounds, and Timmermann (1989) and by Pounds and Fogden (2000) appear to represent an undescribed species distinct from *C. euknemos* (plate 196).

DISTRIBUTION: Scattered lowland and premontane localities from central Costa Rica (840–1,500 m) and Panama (90–1,270 m) to western Colombia (100–200 m).

Cochranella granulosa (Taylor, 1949a)

(plate 197; map 7.108)

DIAGNOSTICS: Small uniform dark green frogs, usually with scattered large black spots, having an obtuse snout in profile (fig. 7.110f), dark green bones, a white stripe on the upper lip, a white parietal peritoneal sheath, a white pericardium, and white digestive tract. This moderately uncommon species has no evidence of arm or leg fringes.

DESCRIPTION: Adult males 22.5 to 29 mm in standard length, adult females 29 to 32 mm; upper surfaces strongly granulate; head broader than long; head subovoid in dorsal outline; profile of snout obtuse; snout moderate, about equal to diameter of orbit; interorbital space much broader than diameter of eye; eyes large, not protuberant; tympanum indistinct, round, directed obliquely upward; vomerine odontophores round, between choanae, only a few teeth present, separated medially; finger I slightly longer than finger II; finger and toe disks truncate; no supernumerary, accessory palmar, or plantar tubercles; subarticular tubercles tiny, round; no web between fingers I and II, webbing vestigial between II and III; webbing formula for outer fingers:

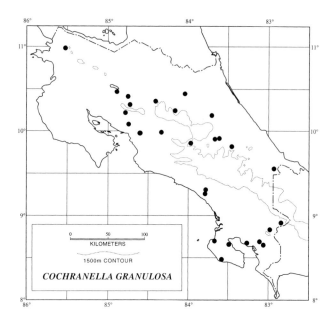

Map 7.108. Distribution of *Cochranella granulosa*.

II 1¾–3½ III 2–1½ IV; thenar tubercle elongate, large; palmar tubercle oval, large; well-developed white nuptial pad on dorsal and outer lateral margin of thumb base in adult males; toes moderately webbed; webbing formula: I 1½–2½ II 1–2 III 1–2 IV 2–1 V; inner metatarsal tubercle elongate, outer small, weak, or absent; no tarsal fold or tubercle. Dorsal surface uniform deep blue green or usually with scattered black spots; venter white to flesh below; white parietal peritoneal sheath covers anterior internal organs; pericardium (not visible) and digestive tract white; liver reddish; iris pale gray gold.

LARVAE (fig. 7.112): Small, total length 11 mm at stage 25; elongate, body somewhat depressed; mouth ventral; nares and eyes dorsal; spiracle midlateral, posterior, sinistral; vent tube median; tail long; caudal fins reduced; tail round at tip; oral disk complete, moderate; beaks and 2/3 rows of denticles present; A2 with great gap above mouth; lower beak with blunt serrations; a single row of labial papillae, absent above mouth; black on hatching, becoming pale brown above and lighter below but appearing bright pink or red; tail musculature colored like body; fins translucent with a few dark spots posteriorly.

VOICE: A call of two or three rapidly repeated harsh notes, "creep-creep-creep," repeated after several minutes; dominant frequency 4 to 4.5 kHz (Savage and Starrett 1967; Ibáñez, Rand, and Jarmillo 1999).

SIMILAR SPECIES: *Hyalinobatrachium pulveratum* has the webbing between fingers II and III almost as extensive as between fingers III and IV (fig. 7.114b), a narrow interorbital space (less than diameter of orbit), pale green bones, and a white, bulbous liver.

HABITAT: Lowland Moist and Wet Forests and Premontane Wet Forest.

BIOLOGY: This is a strictly nocturnal species that usually calls from trees 5 to 10 m above fast-moving streams. Eggs are laid in a single layer at these sites in jelly masses about 20 by 35 mm. The black-and-white eggs are about 1.5 mm in diameter, 3 mm including the envelope. Clutches consist of 49 to 60 eggs. The larvae at hatching are black. The egg mass hangs over the leaf margin to form a drip tip that ensures a constant flow of water over the larvae (Starrett 1960a; McDiarmid 1975). There is no long-term parental care in this species. The smallest subadult juvenile available is 23 mm in standard length.

REMARKS: The figure legends in McDiarmid (1975, figs. 2 and 3) are transposed, and the male labeled *Centrolenella fleischmanni* is *C. granulosa*. Taylor (1949a) originally described this species as a member of *Centrolenella*. He later (1951a) made it the type of his new genus *Cochranella*. Savage (1967) returned it to *Centrolenella*, a procedure followed by subsequent workers until the Ruiz and Lynch (1991) revision. The karyotype is 2N = 20 (Duellman 1967b).

DISTRIBUTION: Humid lowland and premontane slopes from Atlantic slope eastern Honduras to central Panama and on the Pacific versant in humid upland or gallery forest situations from northern Costa Rica to southwestern Panama (40–1,500 m).

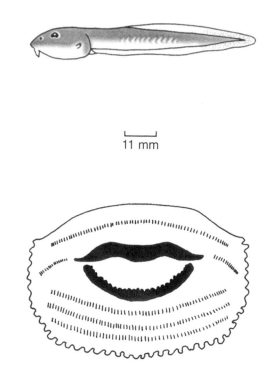

Figure 7.112. Tadpole of *Cochranella granulosa*.

Plate 141. *Phyllomedusa lemur;* day colors. Costa Rica: Alajuela: Monteverde Cloud Forest Preserve, about 1,500 m.

Plate 142. *Gastrotheca cornuta;* male. Costa Rica: Limón: Guayacán, about 600 m.

Plate 143. *Gastrotheca cornuta;* female with eggs in pouch. Costa Rica: Limón: near Río Blanco de Liverpool, about 300 m.

Plate 144. *Anotheca spinosa.* Costa Rica: Alajuela: Peñas Blancas Valley, about 800 m.

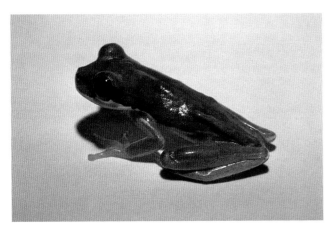

Plate 145. *Duellmanohyla lythrodes.* Panama: Bocas del Toro: Río Changuinola, near Quebrada El Guabo, 90 m.

Plate 146. *Duellmanohyla rufioculis.* Costa Rica: Alajuela: Peñas Blancas Valley, about 800 m.

Plate 147. *Duellmanohyla uranochroa*. Costa Rica: Alajuela: Monteverde Cloud Forest Preserve, about 1,600 m.

Plate 148. *Hyla rufitela*. Costa Rica: Heredia: La Selva Biological Station, about 60 m.

Plate 149. *Hyla rufitela*. Costa Rica: Heredia: Puerto Viejo, about 35 m.

Plate 150. *Hyla rosenbergi*. Costa Rica: Puntarenas: Rincón de Osa, about 30 m.

Plate 151. *Hyla rosenbergi;* day color. Costa Rica: Puntarenas: Rincón de Osa, about 30 m.

Plate 152. *Hyla colymba*. Panama: Coclé: El Copé National Park, about 1,100 m.

Plate 153. *Hyla palmeri*. Costa Rica: Limón: Guayacán, about 750 m.

Plate 154. *Hyla loquax;* amplectant pair. Costa Rica: Heredia: La Selva Biological Station, about 60 m.

Plate 155. *Hyla calypsa*. Costa Rica: Puntarenas: Las Tablas, about 1,900 m.

Plate 156. *Hyla lancasteri*. Costa Rica.

Plate 157. *Hyla ebraccata;* amplectant pair, male with *"weyerae"* color phase. Costa Rica: Heredia: La Selva Biological Station, about 60 m.

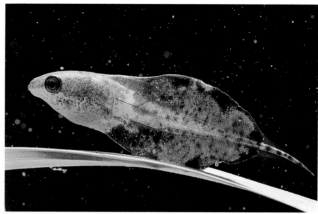

Plate 159. *Hyla ebraccata;* tadpole showing possibly aposematic coloration. Costa Rica: Heredia: La Selva Biological Station, about 60 m.

Plate 158. *Hyla ebraccata;* amplectant pair and eggs being deposited on vegetation. Costa Rica: Heredia: La Selva Biological Station, about 60 m.

Plate 160. *Hyla microcephala.* Costa Rica: Guanacaste: La Pacifica, about 40 m.

Plate 161. *Hyla phlebodes;* calling male. Costa Rica: Heredia: La Selva Biological Station, about 60 m.

Plate 162. *Hyla debilis.* Panama: Chiriquí: Fortuna Forest Reserve, about 1,100 m.

Plate 163. *Hyla pictipes*. Costa Rica: Heredia: Volcán Barva.

Plate 164. *Hyla rivularis*. Costa Rica: Alajuela: Monteverde Cloud Forest Preserve, about 1,600 m.

Plate 165. *Hyla tica*. Costa Rica: Alajuela: Monteverde Cloud Forest Preserve, about 1,600 m.

Plate 166. *Hyla xanthosticta;* holotype. Costa Rica: Heredia: Rama Sur Río Las Vueltas, 2,100 m.

Plate 167. *Hyla angustilineata;* adult. Costa Rica: Alajuela: Monteverde Cloud Forest Preserve, about 1,600 m.

Plate 168. *Hyla angustilineata;* metamorphosing froglet. Costa Rica: Alajuela-Puntarenas: Monteverde Cloud Forest Preserve, about 1,600 m.

Plate 169. *Hyla pseudopuma;* breeding male coloration. Costa Rica: Puntarenas: Monteverde, about 1,400 m.

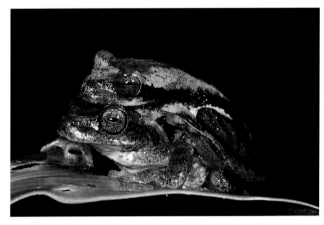

Plate 170. *Hyla pseudopuma;* amplectant pair. Costa Rica: Puntarenas: Monteverde, about 1,400 m.

Plate 171. *Hyla pseudopuma;* tadpoles cannibalizing a conspecific tadpole. Costa Rica: Puntarenas: Monteverde, about 1,400 m.

Plate 172. *Hyla legleri.* Costa Rica: San José: near Alfombra, about 780 m.

Plate 173. *Hyla fimbrimembra.* Costa Rica: Heredia: Braulio Carrillo National Park, about 1,800 m.

Plate 174. *Hyla miliaria.* Costa Rica: Alajuela: Peñas Blancas Valley, about 800 m.

Plate 175. *Hyla miliaria*. Costa Rica: Limón: Comadre, 20 m.

Plate 176. *Hyla miliaria;* view of head. Costa Rica: Limón: Comadre, 20 m.

Plate 177. *Hyla picadoi*. Costa Rica: Heredia: Volcán Barva.

Plate 178. *Hyla zeteki*. Costa Rica: Cartago: Tapantí National Park, about 1,500 m.

Plate 179. *Osteopilus septentrionalis.*

Plate 180. *Phrynohyas venulosa*. Costa Rica: Guanacaste: Santa Rosa National Park, about 280 m.

Plate 181. *Phrynohyas venulosa*; amplectant pair. Costa Rica: Guanacaste: Santa Rosa National Park, about 280 m.

Plate 182. *Scinax boulengeri*. Costa Rica: Heredia: La Selva Biological Station, about 60 m.

Plate 183. *Scinax elaeochroa*. Costa Rica: Heredia: La Selva Biological Station, about 60 m.

Plate 184. *Scinax staufferi*. Costa Rica: Guanacaste: Santa Rosa National Park, about 280 m.

Plate 185. *Smilisca baudinii*. Costa Rica: Guanacaste: Santa Rosa National Park, about 280 m.

Plate 186. *Smilisca phaeota*. Costa Rica: Puntarenas: Rincón de Osa, about 30 m.

Plate 187. *Smilisca phaeota*; calling male. Costa Rica: Puntarenas: Corcovado National Park.

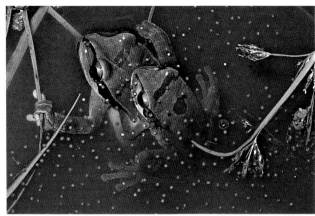

Plate 188. *Smilisca phaeota*; amplectant pair and recently oviposited eggs. Costa Rica: Puntarenas: Corcovado National Park, about 40 m.

Plate 189. *Smilisca puma*. Costa Rica: Heredia: La Selva, about 60 m.

Plate 190. *Smilisca sila*. Costa Rica: Puntarenas: Rincón de Osa, about 10 m.

Plate 191. *Smilisca sordida*. Costa Rica: Alajuela: Peñas Blancas Valley, about 800 m.

Plate 192. *Smilisca sordida*. Costa Rica: Limón: Cahuita, 5 m.

Plate 193. Centrolenid tadpoles; note red appearance. Colombia.

Plate 194. *Centrolenella ilex*. Costa Rica: Heredia: Rara Avis, about 700 m.

Plate 195. *Centrolenella prosoblepon;* male with humeral hook. Costa Rica: Alajuela: Monteverde Cloud Forest Preserve, about 1,600 m.

Plate 196. *Cochranella* sp. Costa Rica: Puntarenas: San Luis Valley, 840 m.

Plate 197. *Cochranella granulosa*. Costa Rica: Alajuela: Peñas Blancas Valley, about 800 m.

Plate 198. *Cochranella albomaculata*. Costa Rica.

Plate 199. *Cochranella spinosa*. Costa Rica: Heredia: La Selva Biological Station, about 60 m.

Plate 200. *Hyalinobatrachium pulveratum*. Costa Rica: Heredia: La Selva Biological Station, about 60 m.

Plate 201. *Hyalinobatrachium chirripoi*. Panama: Darién: Estación Pirre, Darién National Park.

Plate 202. *Hyalinobatrachium chirripoi;* ventral view showing bare heart and white viscera. Panama: Darién: Estación Pirre, Darién National Park.

Plate 203. *Hyalinobatrachium colymbiphyllum*. Costa Rica: Puntarenas: Monteverde Cloud Forest Preserve, about 1,600 m.

Plate 204. *Hyalinobatrachium colymbiphyllum*. Costa Rica: Puntarenas: Monteverde Cloud Forest Preserve, about 1,600 m.

Plate 205. *Hyalinobatrachium colymbiphyllum;* ventral view showing bare heart and white viscera. Costa Rica: Puntarenas: Rincón de Osa, about 60 m.

Plate 206. *Hyalinobatrachium fleischmanni;* amplectant pair. Costa Rica: Alajuela: Monteverde Cloud Forest Preserve, 1,400 m.

Plate 207. *Hyalinobatrachium fleischmanni;* amplectant pair. Costa Rica: Puntarenas: Monteverde, about 1,400 m.

Plate 208. *Hyalinobatrachium valerioi.* Costa Rica: Alajuela: Peñas Blancas Valley, about 800 m.

Plate 209. *Hyalinobatrachium valerioi;* mated pair with two egg clutches. Costa Rica: Heredia: La Selva Biological Station, about 60 m.

Plate 210. *Hyalinobatrachium vireovittatum.* Costa Rica: Alajuela: Peñas Blancas Valley, about 800 m.

Plate 211. *Colostethus flotator*. Costa Rica: Puntarenas: Rincón de Osa, about 30 m.

Plate 212. *Colostethus flotator*. Panama: Bocas del Toro: Río Changuinola.

Plate 213. *Colostethus nubicola*. Panama: Panama: Cerro Campana, about 860 m.

Plate 214. *Colostethus talamancae*. Costa Rica: Puntarenas: Rincón de Osa, about 30 m.

Plate 215. *Dendrobates auratus*. Costa Rica: Puntarenas: Rincón de Osa, about 40 m.

Plate 216. *Dendrobates granuliferus*. Costa Rica: Puntarenas: Corcovado National Park, about 40 m.

Plate 217. *Dendrobates granuliferus;* female transporting tadpole. Costa Rica: Puntarenas: Corcovado National Park, about 40 m.

Plate 218. *Dendrobates pumilio.* Costa Rica: Heredia: La Selva Biological Station, about 60 m.

Plate 219. *Dendrobates pumilio;* calling male on his territory. Costa Rica: Heredia: La Selva Biological Station, about 60 m.

Plate 220. *Dendrobates pumilio;* males in combat. Costa Rica: Heredia: La Selva Biological Station, about 60 m.

Plate 221. *Dendrobates pumilio;* female transporting tadpole. Costa Rica: Heredia: La Selva Biological Station, about 60 m.

Plate 222. *Phyllobates lugubris;* male transporting tadpoles. Costa Rica: Heredia: La Selva Biological Station, 60 m.

Plate 223. *Phyllobates vittatus*. Costa Rica: Puntarenas: Corcovado National Park, about 40 m.

Plate 224. *Gastrophryne pictiventris;* amplectant pair. Costa Rica: Heredia: La Selva Biological Station, about 60 m.

Plate 225. *Hypopachus variolosus*. Costa Rica: Puntarenas: Monteverde, about 1,400 m.

Plate 226. *Nelsonophryne aterrima*. Panama: Chiriquí: Fortuna Forest Reserve, about 1,000 m.

Plate 227. *Rana forreri*. Costa Rica: Puntarenas: Monteverde, about 1,400 m.

Plate 228. *Rana taylori*. Costa Rica: Alajuela: near Laguna Arenal, about 500 m.

Plate 229. *Rana taylori*. Costa Rica: Cartago: Tapantí National Park, about 1,500 m.

Plate 230. *Rana vaillanti*. Costa Rica: Alajuela: Upala, 48 m.

Plate 231. *Rana vaillanti*. Costa Rica: Guanacaste: Taboga, 4 m.

Plate 232. *Rana vibicaria*. Costa Rica: Alajuela: Monteverde Cloud Forest Preserve, about 1,600 m.

Plate 233. *Rana warszewitschii*. Costa Rica: Puntarenas: Monteverde, about 1,400 m.

Plate 234. *Basiliscus basiliscus;* adult male. Costa Rica: Guanacaste: La Pacifica, 40 m.

Cochranella spinosa Group

Frogs of this group lack white pigment on any organs of the digestive tract and have extensive webbing between fingers III and IV (Ruiz and Lynch 1995). The following extralimital species are placed here: *Cochranella adiazeta* (Colombia), *C. amertarsia* (Colombia), *C. croceopodis* (Peru), *C. duidaeana* (Venezuela), *C. euhystrix* (Peru), *C. flavopunctata* (Ecuador), *C. geijskei* (Suriname), *C. helenae* (Venezuela), *C. megistra* (Colombia), *C. midas* (Ecuador and Peru), *C. ocellifera* (Ecuador), *C. orejuela* (Colombia), *C. oyampiensis* (French Guyana and Suriname), *C. punctulata* (Colombia), *C. ritae* (Brazil), *C. riveroi* (Venezuela), *C. saxiscandens* (Peru), *C. spiculata* (Peru), *C. susatami* (Colombia), *C. tangarana* (Peru), and *C. xanthocheridia* (Colombia).

Cochranella albomaculata (Taylor, 1949a)

(plate 198; map 7.109)

DIAGNOSTICS: A moderately common small green frog with numerous small silver cream or pale yellow spots on the upper surfaces, having a white stripe on the upper lip and along the lower limb margins, a round snout profile (fig. 7.110a), dark green bones, a white parietal peritoneal sheath and pericardium, and a distinct fleshy ridge along the posterior lower margin of the forearm. This species has no white pigment on the digestive tract.

DESCRIPTION: Adult males 20.5 to 29 mm in standard length, adult females 22 to 32 mm; upper surfaces granular, with widely scattered low tubercles; head about as broad as long; head rounded from above, truncate to round in profile; snout shorter than diameter of orbit; interorbital space narrower than diameter of eye; eyes large, protuberant; tympa-

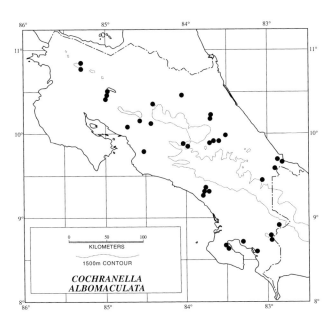

Map 7.109. Distribution of *Cochranella albomaculata*.

num indistinct, directed obliquely upward; vomerine teeth in two transverse rows between choanae, separated medially; finger I longer than finger II; finger and toe disks truncate; subarticular tubercles round, moderate-sized; no supernumerary, accessory palmar, or plantar tubercles; web between fingers I and II vestigial; webbing formula for outer fingers: II 2^+–3½ III 2–1½ IV; thenar tubercle large, elongate; palmar tubercle large, round; well-developed white nuptial pad on dorsal and outer lateral margin of thumb base in adult males; a distinct fleshy fold along outer margin of forearm; toes moderately webbed; webbing formula: I 1–2 II 1–2 III 1¼–2^+ IV 2–1 V; inner metatarsal tubercle elongate, outer small or absent; no tarsal fold or tubercle. Dorsal surfaces bluish green with a silver, cream, or yellow spot on each low tubercle; many light spots on sides of neck; undersurfaces of limbs creamy white; white parietal peritoneal sheath not extending far posteriorly, pericardium white; iris gray gold with black reticulum.

LARVAE: Unknown.

VOICE: A single short "dik," widely spaced, that may be repeated several times.

SIMILAR SPECIES: (1) *Centrolenella euknemos* has a long, obtuse snout (fig. 7.110e), a broad interorbital space, and a crenulate fringe along the lower leg and foot (fig. 7.70a). (2) *Hyalinobatrachium pulveratum* has substantial webbing between fingers II and III (fig. 7.114b), an obtuse snout profile (fig. 7.110g), and no white line along the outer arm margin.

HABITAT: Lowland Moist and Wet Forests and Premontane Wet Forest and Rainforest.

BIOLOGY: Males call at night from low vegetation near fast-moving streams. In the oviduct and shortly after laying, the eggs are black-and-white. The smallest juvenile is 17 mm in standard length.

REMARKS: The karyotype is 2N = 20 (Duellman 1967b)

DISTRIBUTION: Humid lowlands and premontane slopes from north-central Honduras to western Colombia; in Costa Rica in the humid lowlands and on the slopes of the cordilleras (20–1,500 m).

Cochranella spinosa (Taylor, 1949a)

(plate 199; map 7.110)

DIAGNOSTICS: This uncommonly seen small uniformly green frog is distinctive within the family in having a free prepollex at the base of the thumb that supports an exposed spine in adult males. Other diagnostic features are snout profile round to truncate (fig. 7.110b), dark green bones, and only the parietal peritoneal sheath and heart white.

DESCRIPTION: Adult males 17.8 to 20 mm in standard length, adult females 20 to 23 mm; upper surfaces minutely shagreened; head about as wide as long, truncate in profile; snout about equal to diameter of orbit; interorbital space broader than diameter of eye; eyes large, protuberant;

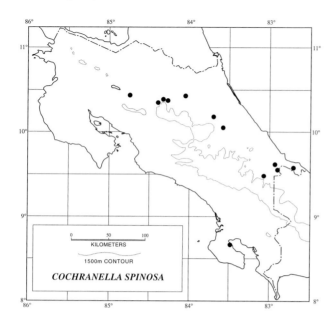

Map 7.110. Distribution of *Cochranella spinosa*.

tympanum indistinct, directed obliquely upward; vomerine teeth in two small patches immediately medial to choanae, widely separated from one another; finger I longer than finger II; finger and toe disks truncate; subarticular tubercles small; no supernumerary tubercles, but accessory palmar and plantar tubercles present; free prepollex and prepollical spine project externally in large males; web between fingers I and II vestigial; webbing formula for outer fingers: II 2–3 III 2–1¾ IV; thenar tubercle large, elongate; palmar tubercle large, round; white nuptial pad on dorsal and outer lateral margin of thumb base in adult males; toes moderately webbed; webbing formula: I 1–2 II 1–2 III 1–2 IV 2–1 V; inner metatarsal tubercle elongate, outer absent; no inner tarsal fold. Dorsal surfaces uniform green; undersurfaces of limbs whitish; iris gray ivory with black reticulum.

LARVAE (fig. 7.113): Tiny on hatching, total length about 11 mm at stage 25; body elongate, depressed; mouth ventral; nares and eyes dorsal; spiracle midlateral, posterior, sinistral; vent tube median; tail long; caudal fins reduced; tail rounded at tip; oral disk complete, moderate; beaks and probably 2/3 rows of denticles; A2 probably with great gap above mouth; both beaks with sharp serrations; a single row of papillae laterally and ventrally, none above mouth; tail musculature mostly brown except light anterior ventral area; fins with scattered dark spots (see remarks).

VOICE: Similar to *Cochranella euknemos* and *C. granulosa*. A series of harsh "creep-creeps" followed by a considerable pause, then repeated; dominant frequency 6.8 to 7.2 kHz (Ibáñez, Rand, and Jarmillo 1999).

SIMILAR SPECIES: (1) *Centrolenella prosoblepon* has a white parietal peritoneal sheath concealing most of the internal organs, the toe web reaching nearly to the base of the disks on toes III and V, and no enlarged prepollex. (2) *Centrolenella illex* has a white parietal peritoneal sheath, a smooth dorsum, protuberant nostrils, and no enlarged prepollex.

HABITAT: Lowland Moist and Wet Forests.

BIOLOGY: Males call throughout the wet season (May to October) from low vegetation near small streams. Females lay green eggs on the underside of vegetation, unlike other members of the genus reported to date (Starrett 1960a). Eggs are laid in a single layer in loose jelly, 18 to 25 at one time. Eggs plus capsules are 5 to 7 mm in diameter. Embryos are gray at hatching; later stages are light brown above, paler below. The smallest juvenile is 16 mm in standard length.

REMARKS: Taylor (1949a) originally described this form as a *Centrolenella* and later made it the type of a new genus, *Teratohyla* (Taylor 1951a). Savage (1967) returned it to *Centrolenella*, and the recent revision by Ruiz and Lynch (1991) placed it in the resurrected genus *Cochranella*. Starrett's (1960a) description of egg deposition site and egg color places it at variance with other members of *Cochranella*, which lay black-and-white eggs on the tips of leaves. Her description is of captive-hatched and -raised tadpoles. These lacked denticles, although they had 1/3 ridges where denticles usually develop. This is probably anomalous, produced by the artificial diet, and tadpoles of this species likely develop 2/3 denticle rows like other centrolenids, with a very short A2 lying at the corners of the mouth opening. Lynch and Ruiz (1996) did not find nuptial pads in males

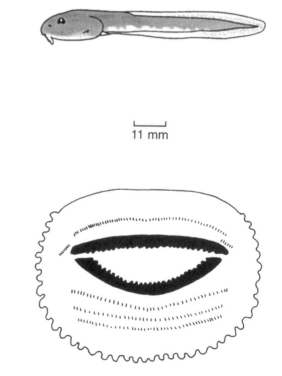

11 mm

Figure 7.113. Tadpole of *Cochranella spinosa*.

of this species. Adult Costa Rica males, however, have well-developed nuptial pads.

DISTRIBUTION: Humid Atlantic lowlands from Costa Rica to eastern Panama, thence south on the Pacific versant to western Ecuador (20–560 m).

Genus *Hyalinobatrachium* Ruiz and Lynch, 1991

These little frogs include some of the most commonly encountered streamside species. The genus (thirty-three species) is characterized by having eyes larger than the disk on finger III, the parietal peritoneum transparent, and the liver bulbous and covered by white pigment, as are one or more of the organs of the digestive tract and usually the pericardium. Component species further lack vomerine teeth and humeral hooks in males, and the bones are usually white. In preservative the skin loses its green color so these frogs appear yellowish white. Adult males have paired vocal slits and fully distensible single median external subgular vocal sacs.

In all species studied to date the eggs are green in the oviduct and when laid. As cleavage takes place the embryos become whitish yellow, but the yolk retains a green cast for some time. Males of all species studied brood the eggs.

Ruiz and Lynch (1991) recognized three subdivisions within the genus based on bone color and white pigmentation of the internal organs. The *Hyalinobatrachium pulveratum* species group and the *Hyalinobatrachium fleischmanni* species group are represented in Costa Rica. The *Hyalinobatrachium parvulum* species group in southeastern Brazil differs from the others by having white pigment on the urinary bladder.

This is a widely distributed genus ranging from Veracruz and Guerrero, Mexico, to Ecuador and to southeastern Brazil, Bolivia, and northeastern Argentina east of the Andes.

Frogs of this genus are small, elusive creatures that usually call from low vegetation overhanging streams at night, with both sexes hiding among streamside vegetation by day, well camouflaged by their coloration. In some disturbed areas, including vacant lots within large cities, certain species may be common among stands of ginger lilies (family Zingiberaceae, especially *Costus* and *Hedychium*) in and around small streams or even rivulets only a few millimeters deep. During the wet season males call nearly every evening, and the voice typically is a series of high-pitched single notes, "seet" or "wheet," repeated several times.

KEY TO THE COSTA RICAN GLASSFROGS OF THE GENUS *HYALINOBATRACHIUM*

1a. Snout rounded or truncate in profile (fig. 7.115); no vomerine teeth; no fleshy fold on outer margin of arm; bones white in life 2
1b. Snout obtuse in profile (fig. 7.110g); vomerine teeth present; a fleshy fold along outer margin of lower arm

and tarsal segment of leg; bones pale green in life; webs between fingers II and III almost as extensive as between fingers III and IV. . . . .
. . . . *Hyalinobatrachium pulveratum* (p. 368)
2a. Web between fingers II and III not nearly as extensive as between fingers III and IV, restricted to base of fingers (fig. 7.114a) 3
2b. Web between fingers II and III almost as extensive as between fingers III and IV (fig. 7.114b) . . . .
. . . . *Hyalinobatrachium chirripoi* (p. 369)
3a. Dorsum without a distinct straight middorsal green stripe bordered on either side by yellow stripes; color green with small yellow spots or reticulated with green and yellow in life; uniform white to yellow or with a dark reticulum in preservative 4
3b. Dorsum with a distinct straight middorsal green stripe bordered on either side by yellow stripes in life; dark stripe clearly indicated by purplish to brownish pigment in preservative; pericardium colorless . . . .
. . . . *Hyalinobatrachium vireovittatum* (p. 375)
4a. Dorsal pattern of small to moderate yellow spots on a green background in life; almost uniform yellowish to whitish, although often with minute punctations of brown or purplish in preservative 5
4b. Dorsal pattern in life a broad reticulum of green with black punctations bordering yellow spots; black punctations clearly demarcate reticulum in preservative; pericardium white. . . . .
. . . . *Hyalinobatrachium valerioi* (p. 374)
5a. Nostrils in strongly protuberant, fleshy swellings; tympanum distinct; pericardium colorless. 6
5b. Nostrils open through very slightly indicated fleshy swellings (fig. 7.115d); tympanum concealed; pericardium white. . . . .
. . . . *Hyalinobatrachium fleishmanni* (p. 371)

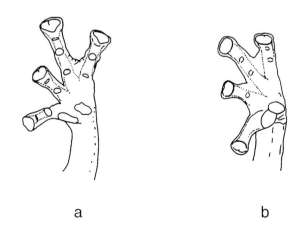

a   b

Figure 7.114. Left hands of glassfrogs (Centrolenidae): (a) vestigial webbing between fingers II and III; (b) substantial webbing between fingers II and III.

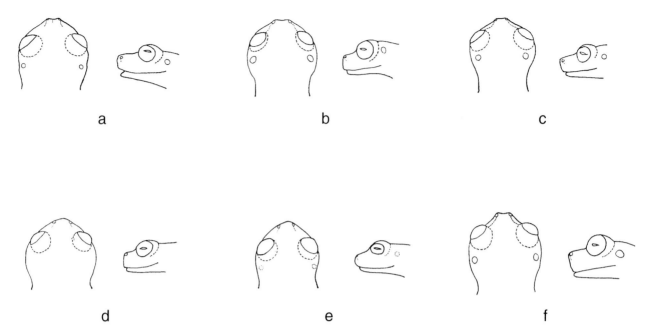

Figure 7.115. Diagnostic characters of head structure in Costa Rican glassfrogs, genus *Hyalinobatrachium*, dorsal and lateral views: (a) *H. chirripoi;* (b) *H. vireovittatum;* (c) *H. valerioi;* (d) *H. fleischmanni,* note snout rounded when viewed from above, nostrils not protuberant; (e) *H. talamancae,* note snout subovoid when viewed from above, rounded in profile; (f) *H. colymbiphyllum.* In all species except *H. fleischmanni* and *H. talamancae* there is a distinct elevated intercanthal platform terminating with the nostrils. The tympanum may be hidden (*H. fleischmanni*), indistinct (*H. talamancae*), or distinct (all other species).

6a. Head viewed from above rounded, with a weakly pointed snout (fig. 7.115e); canthus rostralis weak; no elevated intercanthal platform; eyes not protuberant . . . . *Hyalinobatrachium talamancae* (p. 373)

6b. Head viewed from above truncate (fig. 7.6b) with an indentation between nostrils; canthus rostralis strong; an elevated intercanthal platform (fig. 7.115f); eyes protruding laterally well beyond level of lip margin . . . . *Hyalinobatrachium colymbiphyllum* (p. 370)

Hyalinobatrachium pulveratum Group

This group is characterized by having pale green bones in life and vomerine teeth. The only other species besides *H. pulveratum* belonging to the group is *Hyalinobatrachium antisthenesi* of Venezuela. The latter species has sex determination of the type where the male is heteromorphic (Schmid, Steinlein, and Feichtinger 1989).

Hyalinobatrachium pulveratum (Peters, 1873a)

(plate 200; map 7.111)

DIAGNOSTICS: An uncommon small yellow-spotted, lime green frog having an obtuse snout profile (fig. 7.110g), light green bones, translucent parietal peritoneum, and white pericardium, liver, and digestive tract.

DESCRIPTION: Adult males 22 to 29 mm in standard length, adult females 23 to 33 mm; upper surfaces sha-

greened; head longer than broad; snout truncate from above; obtuse in profile; snout about equal to diameter of orbit; interorbital space narrower than diameter of orbit; eyes large, protuberant; tympanum indistinct, directed laterally; vomerine teeth in two small patches between choanae, widely

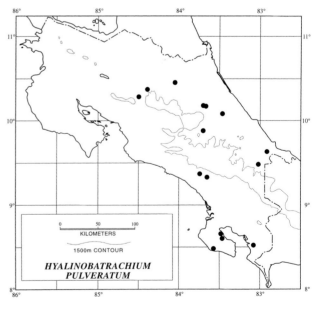

Map 7.111. Distribution of *Hyalinobatrachium pulveratum.*

separated medially; fingers I and II subequal; finger and toe disks truncate; a fleshy fold along outer margin of lower arm; subarticular tubercles small, round; no supernumerary, accessory palmar, or plantar tubercles; webs between fingers I and II vestigial; substantial web between fingers II and III; webbing formula for outer fingers: II 1½–3 III 1½–1¼ IV; thenar tubercle elongate; palmar tubercle round; white nuptial pad on dorsal and outer surfaces of thumb base in adult males; a fleshy fold along outer margin of tarsal segment of leg; toes moderately webbed; webbing formula: I 1–2⁺ II 1¼–2¼ III 1¼–2½ IV 2½–1¼ V; inner metatarsal tubercle elongate, outer absent; no tarsal fold. Dorsal surfaces green with small white dots; venter transparent; heart, liver, and digestive tract white; iris silver gray.

LARVAE: Unknown.

VOICE: Similar to *Centrolenella prosoblepon*, a rasping series of "dik-dik-diks" but in a slower cadence; dominant frequency 5.7 to 6.2 kHz (Savage and Starrett 1967; Ibáñez, Rand, and Jarmillo 1999).

SIMILAR SPECIES: (1) *Cochranella albomaculata* has a distinct white line on the outer margin of the forearm and little webbing between fingers I and II. (2) *Cochranella euknemos* has white fringes along the outer edge of the forearm, lower leg, and foot (fig. 7.70a) and little webbing between fingers I and II.

HABITAT: Lowland Moist and Wet Forests.

BIOLOGY: Like other Costa Rican centrolenids, males call from vegetation along fast-moving streams throughout the rainy season (May to October). It is not known whether males brood the eggs as in the *Hyalinobatrachium fleischmanni* group. Eggs are laid on the upper surfaces of leaves according to Villa (1984a). He did not specify if the egg mass was attached to the leaf tip as in *Cochranella granulosa*, but in Costa Rica this is the case. The eggs are green and laid in clutches of 44 to 80.

REMARKS: Taylor (1952a, 1958b) referred to this species as *Cochranella pulverata*. Savage (1967) placed it in *Centrolenella*, but the Ruiz and Lynch (1991) revision allocated it to *Hyalinobatrachium*. *H. pulveratum* differs from other members of the latter genus in Costa Rica in remaining lavender in preservative for a longer time before turning white and in the fact that dried skins of fresh specimens remain bright green. In life the dorsal coloration is formed by widely scattered chromatophores and a diffuse green pigment. The chromatophores change to purple in preservative, and the diffuse pigment is dissolved or destroyed. This suggests that the diffuse pigment is not biliverdin, since that compound is lost from dried skins of other *Hyalinobatrachium*. Lynch and Ruiz (1996) did not find nuptial pads in males of this species. Adult Costa Rican males, however, have well-developed nuptial pads.

DISTRIBUTION: Humid lowlands on the Atlantic versant from north-central Honduras and on the Pacific slope from southwestern Costa Rica to northern Colombia (2–560 m).

Hyalinobatrachium fleischmanni Group

These frogs have white bones and usually lack vomerine teeth (absent in all Costa Rican species). Extralimital members of this group include *Hyalinobatrachium aureoguttatum* (Colombia), *H. bergeri* (Bolivia), *H. cardiacalyptum* (Honduras), *H. crurifasciatum* (Venezuela), *H. crybetes* (Honduras), *H. duranti* (Venezuela), *H. eccentrum* (Venezuela), *H. esmeralda* (Colombia), *H. fragilis* (Venezuela), *H. iaspidiensis* (Venezuela), *H. ibama* (Colombia), *H. lemur* (Peru), *H. loreocarinatum* (Venezuela), *H. munozorum* (Ecuador and Peru), *H. orientale* (Venezuela), *H. ostracodermoides* (Venezuela), *H. pallidum* (Venezuela), *H. pellucidum* (Ecuador), *H. petersi* (Colombia), *H. pleurolineatum* (Venezuela), *H. ruedai* (Colombia), and *H. taylori* (Guianas and Venezuela)

Hyalinobatrachium chirripoi (Taylor, 1958b)

(plates 201–2; map 7.112)

DIAGNOSTICS: A small yellow-spotted green frog having an extensive web between fingers II and III (fig. 7.114b), the snout indented above and rounded to truncate in profile (fig. 7.115a), and a white liver and digestive tract but a clear pericardium.

DESCRIPTION: Adult males 24 to 26 mm in standard length, adult females probably somewhat larger; upper surfaces shagreened; head broader than long; snout truncate and slightly indented from above, rounded in profile; loreal distance longer than diameter of orbit; nasal region markedly swollen and nares opening in fleshy protuberances that lie on distinct raised ridges; eyes not protuberant; interorbital space about equal to snout length; tympanum indistinct, directed dorsally; finger I longer than finger II;

Map 7.112. Distribution of *Hyalinobatrachium chirripoi*.

finger and toe disks truncate; subarticular tubercles small, rounded; no supernumerary, accessory palmar, or plantar tubercles; webbing formula for outer fingers: II 1½–3 III 2¼–1 IV; thenar tubercle elongate; palmar tubercle smaller, rounded; white nuptial pad of scattered, separate glands on dorsal and outer surfaces of thumb base in adult males; toe webs extensive; webbing formula: I 1½–2 II 1–2 III 1–2¼ IV 2½–1 V; elongate inner metatarsal tubercle; no outer or tarsal fold. Dorsal surfaces green; venter transparent, liver and digestive tract white; iris gold.

LARVAE: Unknown.

VOICE: Unknown

HABITAT: Lowland Moist Forest.

SIMILAR SPECIES: *Hyalinobatrachium colymbiphyllum* lacks distinct light dorsal spots and has little webbing between fingers II and III (fig. 7.114a).

REMARKS: This rare species was originally described by Taylor (1958b) as *Cochranella*, referred to *Centrolenella* by Savage (1967), and assigned to its present genus by Ruiz and Lynch (1991).

DISTRIBUTION: Lowlands of southeastern Costa Rica, eastern Panama, and western Colombia (60–100 m).

Hyalinobatrachium colymbiphyllum (Taylor, 1949a)

(plates 203–5; map 7.113)

DIAGNOSTICS: A small yellow-spotted green frog having protuberant eyes, a markedly truncate snout profile, and the nares opening through fleshy protuberances lying on distinct raised ridges (fig. 7.115f). From above the snout is truncated with a distinct indentation between the nostrils, and the pericardium is colorless.

DESCRIPTION: Adult males 23 to 27 mm in standard length, adult females 24.5 to 29 mm; upper surfaces shagreened; head broader than long; snout truncate, indented in dorsal view; profile truncate; loreal distance greater than orbital diameter; nasal region markedly swollen and nares opening in fleshy protuberances lying on distinct raised ridges; eyes protuberant; interorbital space wider than snout length; tympanum indistinct, directed dorsally; finger I longer than II; finger and toe disks truncate; subarticular tubercles small, rounded; no supernumerary, accessory palmar, or plantar tubercles; webbing vestigial between fingers I and II, reduced between outer fingers; webbing formula: II 2–3¼ III 1¾–2 IV; thenar tubercle elongate; palmar tubercle smaller, round; white nuptial pad of scattered separate glands present on outer surface of thumb base in adult males; toe webs extensive; formula I 1½–2 II 1–2 III 1–2¼ IV 2½–1 V; inner metatarsal tubercle elongate, no outer tarsal fold. Upper surfaces lime green; venter transparent; undersides of limbs white; liver and digestive tract white; iris gold.

LARVAE (fig. 7.116): Small, 9.5 mm at stage 25; body elongate, depressed; mouth ventral; spiracle low, posterior, sinistral; vent tube median; tail long; tail fins reduced; tail tip pointed; oral disk complete, moderate; beaks and 2/3 rows of denticles present; A2 with very large gap above mouth; lower beak with fine serrations; single row of papillae around disk, except above; dorsum clear chocolate, light below, with blood making undersurface appear pink; tail musculature with dark punctations on anterior dorsal surface, marking myosepta and in groove between epimeric and hypomeric musculature; fins clear.

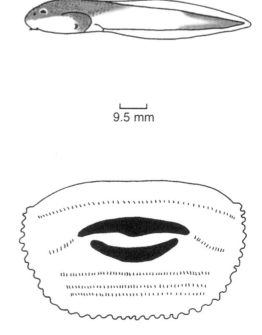

9.5 mm

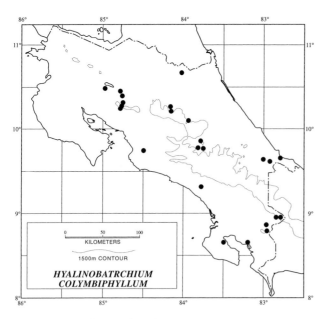

Map 7.113. Distribution of *Hyalinobatrachium colymbiphyllum*.

Figure 7.116. Tadpole of *Hyalinobatrachium colymbiphyllum*.

VOICE: Call a single sustained, rather musical trill lasting 600 to 800 milliseconds, having a dominant frequency of 4.4 to 4.6 kHz and repeated after a pause (Starrett and Savage 1973).

SIMILAR SPECIES: *Hyalinobatrachium chirripoi* has definite light dorsal spots and an extensive web between fingers II and III (fig. 7.114b).

HABITAT: Lowland Moist and Wet Forests, Premontane Wet Forest and Rainforest, and marginally in Lower Montane Wet Forest and Rainforest.

BIOLOGY: McDiarmid (1978) and Hayes (1991) studied this rather common species in some detail, and the following account includes material from their reports. Like other members of the family, *Hyalinobatrachium colymbiphyllum* is a nocturnal insectivore. Males call at night from the undersides of leaves along small streams during the rainy season (May to November). They are strongly territorial and call from the same site or nearby throughout the reproductive period. Calling both spaces the males and attracts females. Call frequency increases three- to fourfold if a nonresident male enters the territory. If the intruder does not retreat, physical contact ensues until one male (usually the resident) presses the other's venter to the leaf. When released the defeated male retreats from the territory. Eggs of this species are laid on the underside of the calling-site leaf. The male continues to call after the eggs are deposited, and clutches at different stages of development laid by from one to eight females may be attended by the father. Clutch size averages about 50 eggs (maximum 75), each egg having a diameter, exclusive of the gelatinous envelope, of about 1.5 mm. Males spend the night calling and visiting the egg masses. From time to time the male climbs on an egg mass and touches the eggs with ventral, pelvic, and thigh areas. Hayes (1991) calls this behavior hydric brooding, since its function is to hydrate the eggs and the clutch jellies are noticeably swollen afterward. Males of *H. colymbiphyllum* move away from the stream near daylight and rest on the underside of leaves during the day, but they return to the calling site and egg masses at night. McDiarmid (1978) reported a high incidence of predation at Rincón de Osa, Puntarenas Province, by a diurnal wasp that removes a single egg from the mass and flies off. Entire egg masses are depleted in this way, apparently by repeated visits from the same wasp or another individual. He attributed the higher egg mortality in this species than in the syntopic *Hyalinobatrachium valerioi* to the absence of diurnal guarding of the egg masses in *H. colymbiphyllum*.

HABITAT: Lowland Moist and Wet Forests, Premontane Wet Forest and Rainforest, and marginally in Lower Montane Rainforest.

REMARKS: Taylor (1949a) described this species as a member of *Centrolenella* but later (1951a, 1952a) referred it to *Cochranella*. Savage (1967) returned it to *Centrolenella*, but the Ruiz and Lynch (1991) revision elevated the

C. fleischmanni species group to generic rank as *Hyalinobatrachium*.

DISTRIBUTION: Humid lowlands, premontane slopes, and lower areas of the lower montane belt on the Atlantic slope of Costa Rica and on the Pacific slope marginally in northwestern Costa Rica and from southwestern Costa Rica to western Colombia; also on the Atlantic versant in central Panama and northern Colombia (6–1,580 m).

Hyalinobatrachium fleischmanni (Boettger 1893b)

(plates 206–7; map 7.114)

DIAGNOSTICS: An extremely common small yellow-spotted frog characterized by having the nares in very slightly indicated swellings (fig. 7.115d), the tympanum concealed, and the pericardium white.

DESCRIPTION: Adult males 19 to 28 mm in standard length, adult females 23 to 32 mm; upper surfaces shagreened; head broader than long; snout slightly pointed in dorsal view, truncate in lateral profile; loreal distance less than orbital diameter; nares open through slight swellings; eyes not protuberant; interorbital space wider than snout length; tympanum concealed; finger I longer than II; digital disks truncate; subarticular tubercles small, round; no supernumerary, accessory palmar, or plantar tubercles; webbing vestigial between fingers I and II, reduced between fingers II and III but extensive between III and IV; webbing formula: II 2–3½ III 2½–2 IV; thenar tubercle elongate, palmar small, round; no nuptial pads in adult males; toe webs extensive; webbing formula: I 1½–2 II 1–2 III 1–2¼ IV 2½–1 V; inner metatarsal tubercle small, round, flat, no outer tubercle; no tarsal fold. Upper surfaces lime green with pale yellowish or yellow green spots; undersides of

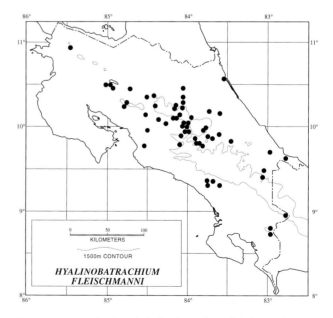

Map 7.114. Distribution of *Hyalinobatrachium fleischmanni*.

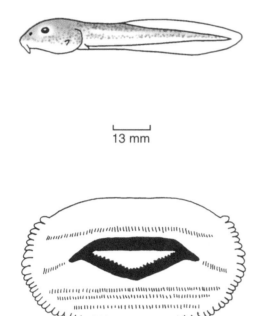

Figure 7.117. Tadpole of *Hyalinobatrachium fleischmanni.*

limbs flesh-colored; venter transparent; pericardium, liver, and digestive tract white; iris gold.

LARVAE (fig. 7.117): Small, 9 to 15 mm in total length on hatching (stage 25); body elongate, depressed; mouth ventral; spiracle midlateral, posterior, sinistral; vent tube median; tail long; tail fins reduced; tail tip rounded; oral disk complete, moderate; beaks and 2/3 rows of denticles present; A2 with very large gap above mouth; lower beak with fine serrations; a single row of papillae around disk, except none above denticle rows; whitish on hatching, becoming brownish above, cream below but appearing bright pink or red; tail musculature with small dark spots, fins clear.

VOICE: One of the calls most often heard in the rainy season. The call consists of a single "wheet," rising at the end and lasting 150 to 300 milliseconds. The dominant frequency is 3.8 kHz to 4.5 kHz at the beginning and 4.8 kHz to 5.3 kHz at the end. Call rates are 4 to 19 per minute (Starrett and Savage 1973; Greer and Wells 1980; Jacobson 1985).

SIMILAR SPECIES: *Hyalinobatrachium talamancae* has a rounded snout profile, protuberant narial swellings (fig. 7.114e), and the tympanum faintly indicated.

HABITAT: Lowland Moist and Wet Forests, Premontane Moist and Wet Forests and Rainforest, and marginally into Lower Montane Wet Forest. Ubiquitous and often common even in pastures and other cleared sites, where it occurs along small streams or rivulets when some leafy vegetation remains.

BIOLOGY: This is among the most studied Central American amphibians. Materials in this section are based primar-ily on the accounts of Greer and Wells (1980), Villa (1984a), Jacobson (1985), and Hayes (1991). Males call beginning from up to two hours before dusk on cloudy afternoons, usually from under leaves along stream courses. They may be heard at night at all times of the year but only sporadically or not at all during the dry season (December to May). Females are found along the streams at night only during the reproductive season (May to October) and return to the canopy or nearby forest after mating. The males are strongly territorial, and the advertisement call spaces males at high-density sites 0.5 to 1.5 m apart. It also attracts females. Call sites are near the ground to a height of 2.5 m. If the male territory is substantially encroached on, the resident male utters a short "mew" interspersed with the advertisement call. If these encounter calls do not work physical combat ensues, and the intruder is usually pinned venter down for some time and withdraws when released. Females remain two days at most in the riparian habitat, apparently searching for mates. When a female approaches a male closely he mixes "mews" and "wheets." After amplexus the female moves about in the male's territory carrying him on her back, presumably looking for an oviposition site. Oviposition is rapid, taking no more than five minutes, during which the male rubs his legs over the female's sides. Eggs are almost always laid on the underside of leaves overhanging the stream, at least 0.5 m above the water or bank and within 1 m of the stream margin. Clutches vary from 10 to 50 eggs, and only one clutch is usually found under one leaf. Females leave the oviposition site within two hours of egg deposition. Males remain close to the egg masses and exhibit hydric brooding behavior. When they visit the clutches at night they bring the ventral posterior and thighs into contact with the eggs and apparently empty the bladder over them. This behavior also occurs immediately after the female is released from amplexus. Brooding takes place at night, but irregularly throughout the period of embryonic development, each bout lasting seventeen to forty-six minutes. Brooding is infrequent, averaging about 15% of the nights during the embryonic stage. The number of nights when brooding takes place increases with drier conditions, but initial hydration of the jelly of the egg mass seems the principal function. During the day the males rest, usually on the undersides of leaves a short distance away from the calling and oviposition sites, but they do not attend the eggs. The hypothesis that guarding or brooding the eggs is an antipredator behavior has been effectively refuted by Hayes (1991). At fertilization, eggs are about 2 mm in diameter. The embryos remain green until about stage 19, when they pale but the yolk retains a green cast. The embryonic interval is fourteen to thirty-four days, and hatching may be delayed at stage 25 for up to ten days until there is a sufficiently heavy rain to wash the larvae into the stream. Hatching larvae are pinkish brown to brilliant red. The tadpoles are fossorial and burrow into benthic debris down to 200 mm. Metamorphs are 9.5 to 12 mm in standard length and are found throughout the wet season. Many organisms doubtless feed on these frogs, especially ctenid spiders. Hayes (1983) noted egg or embryo predation by phalangers *(Prionostemma)* and a gryllid *(Paroecanthus)*. However, the most important causes of egg or embryo mortality are the combined effects of fungal infection and predation by frogfly larvae (21 to 88%) or desiccation (9 to 72%), depending on year and site. The frogflies are members of the family Drosophilidae and were identified in one study as members of the genus *Zygothrica* (Villa 1977, 1978), although other genera may be implicated. According to Villa (1984a), in the maggot stage the flies are obligate predators on frog eggs. Hayes (1991), on the other hand, suggests that the fungal infection precedes the laying of fly eggs on the jelly mass. Since the adult drosophilids feed on fungus, the frog eggs may be one of several potential oviposition sites for the flies. Hayes's (1991) study convincingly demonstrated that the brooding behavior reduced mortality in the early stages of development and during periods when environmental humidity was reduced. However, frequent brooding of a clutch led to higher morality by the fungusfrogfly complex, suggesting that overwatering may encourage infection and predation.

REMARKS: Taylor (1951a) referred this species to his new genus *Cochranella*, and it is so listed in his monograph on Costa Rican frogs (1952a). Savage (1967) placed it in *Centrolenella*, an allocation followed by subsequent authors until Ruiz and Lynch (1991) made it the type of their new genus *Hyalinobatrachium*. Taylor (1952a, 1958b) used the color of the eye tunic in preservative, shape of the digital disks, and presence of folds lateral to the vent as a basis for recognizing several putative species within this genus. In centrolenids the outer surface of the eyeball, posterior to the cornea, is covered by the cartilaginous sclera that underlies the upper eyelid. In life this membrane appears white. On preservation, as the green pigments of the integument fade, the eye tunic may be seen through the upper eyelids as a definite white area. Taylor (1952a) noted that in preservative some individuals lacked the white tunic area, and he used this difference as a basis for separating several nominal species. The presumed differences (white versus black eye tunics) are artifacts, for depending on the nature and time of death, strength of preservative, exposure to light, and other factors, the white may disappear on preservation so the area seen through the upper eyelid appears black from the color of the lining of the eyeball. In all members of the *fleischmanni* group the scleral portion of the eye tunic is white in life. After death, some individuals in any large sample lose the white color, whereas others retain it for years. The other two features mentioned above are variable or are artifacts of preservation (Starrett and Savage 1973). Consequently these authors concluded that *Hylella (Cochranella) chry-*

sops, Cochranella decorata, and *C. millepunctata* as recognized by Taylor (1958a) were based on variant individuals of *H. fleischmanni.* The karyotype is 2N = 20 (Duellman and Cole 1965).

DISTRIBUTION: Humid lowlands and premontane and lower parts of the lower montane slopes from Guerrero and Veracruz, Mexico, to Suriname and western Ecuador; in Costa Rica widely distributed except in the subhumid northwestern lowlands, higher montane areas, or near the coast (103–1,680 m).

Hyalinobatrachium talamancae (Taylor, 1952a)

(map 7.115)

DIAGNOSTICS: A small yellow-spotted green frog having a somewhat pointed snout from above, a rounded snout profile, nostrils in protuberant swellings, a very indistinct tympanum, and the pericardium colorless (fig. 7.114e).

DESCRIPTION: Adult males 22 mm in standard length; upper surfaces shagreened; head broader than long; snout weakly pointed in dorsal outline, rounded in profile; loreal distance less than orbital diameter; nares in protuberant, fleshy swellings; eyes not protuberant; interorbital space about equal to snout length; tympanum barely indicated, directed dorsally; finger I longer than finger II; digital disks truncate; subarticular tubercles small, round; no supernumerary, accessory palmar, or plantar tubercles; webbing vestigial between fingers I and II but extensive between other fingers; webbing formula for outer fingers: II 2–3¼ III 2¾–2 IV; thenar tubercle elongate, palmar small, round; no nuptial pads in adult males; toe webs extensive; webbing formula: I 1½–2 II 1–2 III 1–2¼ IV 2½–1 V; inner metatarsal tubercle small, round, flat; no outer metatarsal tubercle or

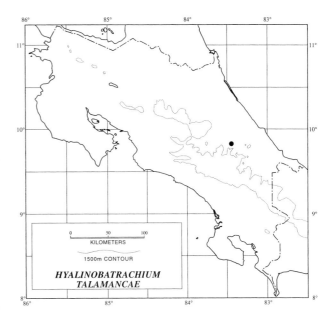

Map 7.115. Distribution of *Hyalinobatrachium talamancae.*

tarsal fold. Upper surfaces green with moderate-sized yellowish spots; undersides of limbs flesh-colored; venter transparent; iris gold and black.

LARVAE: Unknown.

VOICE: Unknown.

SIMILAR SPECIES: *Hyalinobatrachium fleischmanni* usually has a truncate snout profile, lacks protuberant narial swellings, and has a concealed tympanum (fig. 7.114d).

HABITAT: Known from a single Premontane Wet Forest site.

REMARKS: Taylor (1952a) originally described this species from Moravia de Chirripó, Cartago Province, as *Cochranella talamancae* and compared it with what is now known as *Hyalinobatrachium valerioi*. He (1958b) referred four additional specimens from the same locality to the species and again compared it with *H. valerioi*. Starrett and Savage (1973) reexamined the type and three of the four other specimens assigned by Taylor to this species. They found no similarity to the reticulate pattern of *H. valerioi* in these frogs, which differ only subtly from examples of *Hyalinobatrachium fleischmanni*. These latter authors referred the species to *Centrolenella*, but Ruiz and Lynch's (1991) revision led to the generic allocation adopted here.

DISTRIBUTION: Known only from Moravia de Chirripó in the premontane zone of southeastern Costa Rica (1,116 m).

Hyalinobatrachium valerioi (Dunn, 1931a)

(plates 208–9; map 7.116)

DIAGNOSTICS: The pattern of this species, large yellow spots surrounded by a strong reticulum of green pigment and dark punctations, is distinctive among Costa Rican centrolenids. Like other members of the genus, this is a small species that differs from several somewhat similar forms in having the liver, pericardium, and digestive tract white.

DESCRIPTION: Adult males 19.5 to 24 mm in standard length, adult females 22.5 to 26 mm; upper surfaces shagreened; head broader than long; snout truncate from above and in profile; loreal distance longer than orbital diameter; nares opening through elevated protuberances on distinct ridges; eyes not protuberant; interorbital space broader than snout length; tympanum indistinct, directed dorsally; finger I longer than finger II; digital disks truncate; small, subarticular tubercles round; no supernumerary or plantar tubercles, a few weak accessory palmar tubercles; webbing vestigial between fingers I–II–III; webbing formula for outer fingers: III 2^{+}–2 IV; thenar tubercle elongate; palmar tubercle round; white nuptial pad of a few separate glands on outer margin of thumb base in adult males; toes extensively webbed; webbing formula: I 1½–2 II 1–2 III 1–2¼ IV 2½–1 V; inner metatarsal tubercle elongate, no outer tubercle; no tarsal fold. Upper surfaces appearing yellow, overlain by green reticulum; scattered dark punctations in reticulum; venter transparent; pericardium, liver, and digestive tract white; iris gold.

LARVAE (fig. 7.118): Small, about 12 mm in total length at hatching (stage 25); body elongate, depressed; mouth ventral; spiracle posterior, midlateral; vent tube median; tail

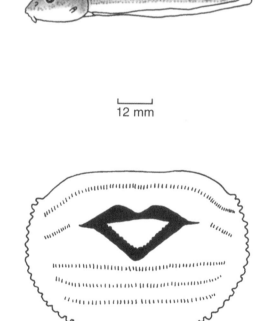

12 mm

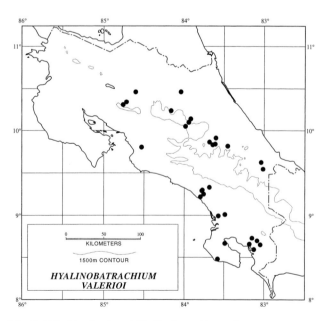

Map 7.116. Distribution of *Hyalinobatrachium valerioi*.

Figure 7.118. Tadpole of *Hyalinobatrachium valerioi*.

long; tail fins reduced, low; tail tip rounded; oral disk complete, moderate; beaks and 2/3 rows of denticles present; A2 with very large gap above mouth; lower beak with fine serrations; single row of papillae on lateral and ventral margins of disk; color light on hatching; body with dorsal brown blotches, lower surfaces lighter but appearing bright red; tail musculature with brown blotches, fins clear.

VOICE: A high-pitched "seet" lasting about 200 to 250 milliseconds with a dominant frequency initially of 7.0 kHz rising to 7.5 kHz and then dropping again to 7.2 kHz, repeated every 7 to 10 seconds (Starrett and Savage 1973).

SIMILAR SPECIES: *Hyalinobatrachium vireovittatum* has a middorsal green stripe bordered by a yellow lateral stripe on each side and a colorless pericardium.

HABITAT: Lowland Moist and Wet Forests and Premontane Wet Forest and Rainforest.

BIOLOGY: A rather uncommon nocturnal insectivore with strong territorial behavior in the males. Although male *H. valerioi* call from under leaves, they have been collected up to 6 m above the stream brooding egg masses. The following account incorporates data from McDiarmid and Adler (1974), McDiarmid (1975), and Hayes (1991). Males call throughout the wet season (May to late October). Like *Hyalinobatrachium fleischmanni*, they maintain spacing with this call but emit a squeak if another male intrudes on their territory. Physical combat may ensue if the intruder does not retreat. Both males may squeak during the interaction. Once one of the males (usually the nonresident) is pinned venter down, he is held for a while, then leaves. Females are attracted by the advertisement call, and mating follows, with the green eggs deposited in a single-layer jelly mass about 20 mm in diameter on a leaf undersurface (plate 209). Clutch sizes average about 35, with 40 being the upper limit. The eggs, including the envelopes, are 3 to 4 mm in diameter; the egg itself is about 2 mm in diameter. Males continue to advertise and may end up with as many as seven clutches, from seven different females, at various stages of development under one leaf. The males of this species attend the eggs both during the day and at night. At night the male continues to call and engage in the ventral hydric brooding behavior described in the account of *H. fleischmanni*. During the day they sit with one hand on the edge of the egg mass, facing the eggs, and they may adopt this guarding position from time to time at night. Only males are involved in parental care of the eggs; the females leave the streamside shortly after mating. McDiarmid (1978) attributed the higher survivorship to hatching in this species than in the syntopic *Hyalinobatrachium colymbiphyllum* to the diurnal "guarding" behavior. The egg masses at his study site, Rincón de Osa, Puntarenas Province, are preyed on by diurnal wasps, which remove eggs one at a time and carry them away, presumably to a nest, until the egg mass is de-

pleted. *H. colymbiphyllum* males do not stay with the egg masses during the day. He further noted that the coloration of *H. valerioi* closely resembles an egg mass and may serve a cryptic purpose as a decoy that confuses a visually hunting enemy, which may approach or attempt to attack the "guarding" males instead of the eggs. Although no observations have been made to confirm this hypothesis, it may be that the male merely chases the predators away or eats them. The smallest juvenile is 16.5 mm in standard length.

REMARKS: This species was described as *Centrolene valerioi* by Dunn (1931a) and placed in *Cochranella* by Taylor (1952a, 1958b). Taylor (1958a) described a putative new form as *Cochranella reticulata*, while comparing the reticulated *C. valerioi* with *Cochranella talamancae*. Starrett and Savage (1973) concluded that *C. reticulata* could not be distinguished from *C. valerioi* and referred to the species as *Centrolenella valerioi*. This conclusion was followed by other workers until the generic name change by Ruiz and Lynch (1991). It is possible that more than one species with a reticulate pattern is included under this name. Reticulate frogs lacking white pigment in the pericardium have been reported from several localities. The name *H. reticulata* (Taylor 1958a) may be applicable to these centrolenids, or they may represent one or more undescribed taxa.

DISTRIBUTION: Humid lowlands and premontane slopes of northeastern and southwestern Costa Rica and disjunctly in west-central Panama and western and north-central Colombia (6–1,500 m).

Hyalinobatrachium vireovittatum (Starrett and Savage, 1973)

(plate 210; map 7.117)

DIAGNOSTICS: This rather rare very small green-and-yellow-striped centrolenid lacking white pericardial pigment is unlikely to be confused with any other frog.

DESCRIPTION: Adult males 21.5 to 23 mm in standard length, adult females about 24 to 25 mm; upper surface shagreened; head broader than long; snout truncate from above and in profile; loreal distance greater than orbital diameter; nares lying in swollen protuberances on distinct raised ridges; eyes not protuberant; interorbital space broader than snout length; tympanum indistinct, directed dorsally; finger I longer than finger II; digital disks truncate; subarticular tubercles small, round; no supernumerary, accessory palmar, or plantar tubercles; webbing vestigial between fingers I–II–III; outermost finger web well developed; webbing formula for outer fingers: II 2–3¼ III 1¾–2 IV; thenar tubercle elongate; palmar tubercle round; no nuptial pad in adult males; toes extensively webbed; webbing formula: I 1½–2 II 1–2 III 1–2 IV 2½–2 V; inner metatarsal tubercle elongate, no outer tubercle; no tarsal fold. A middorsal longitudinal green stripe bordered on each side by a paravertebral yellow

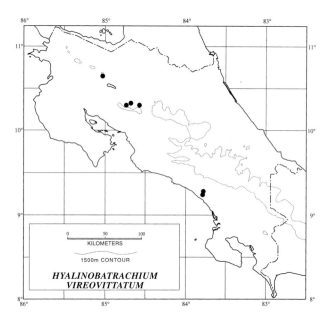

Map 7.117. Distribution of *Hyalinobatrachium vireovittatum.*

stripe; sides and upper surfaces of head and limbs green with moderate yellow spots; venter transparent; liver and digestive tract white; heart unpigmented; iris gold.

LARVAE: Unknown.

VOICE: A rising whistled "wheet" lasting 400 to 500 milliseconds with a dominant frequency of 5 kHz initially and rising to 5.4 kHz. The call is repeated several times in a row, much as in *Hyalinobatrachium fleischmanni* (Starrett and Savage 1973).

SIMILAR SPECIES: No other Costa Rican centrolenid has a longitudinal middorsal green stripe.

HABITAT: Premontane Wet Forest and Rainforest.

BIOLOGY: Similar to *H. fleischmanni* as far as is known. Males call from the undersides of leaves, where the green eggs are laid. Males perform ventral hydric brooding at night but do not guard the eggs during the day.

REMARKS: Starrett and Savage (1973) described this species as *Centrolenella vireovittata.* Ruiz and Lynch's (1991) revision placed this form and all others belonging to the *fleischmanni* species group (Savage 1967; Starrett and Savage 1973) in the new genus *Hyalinobatrachium.*

DISTRIBUTION: Known from the premontane zone at scattered localities from the slopes of Volcán Tenorio, Guanacaste Province, to near Barú, Puntarenas Province, in southwestern Costa Rica, exclusive of the Cordillera Central and Cordillera de Talamanca, and from west-central Panama (800–1,100).

Family *Dendrobatidae* Cope, 1865b

A small family (six genera, 187 species) of anurans found only in humid tropical regions of the Americas from Nicaragua south to Ecuador and the Amazon basin and southeast-

ern Brazil. Dendrobatids are mainly diurnal, semiaquatic or terrestrial, often highly toxic, and brightly colored. Most species in the family are relatively small, but the recently described nocturnal, aquatic *Aromobates nocturnus* from Venezuela reaches a maximum standard length of 62 mm. All members of the family have smooth venters. Like treefrogs (Hylidae), glassfrogs (Centrolenidae), and *Eleutherodactylus,* members of this family have digital disks divided by a transverse terminal groove into a dorsal disk cover and ventral disk pad. They are distinctive among anurans in having the disk cover divided longitudinally into a pair of scutelike flaps. Other characteristics are eight holochordal, procoelous presacral vertebrae; no ribs; sacrum with cylindrical diapophyses (dilated in *Aromobates*) and bicondylar articulation to coccyx: coccyx usually with transverse processes proximally; pectoral girdle firmisternal; a cartilaginous omosternum and bony sternum present; no parahyoid bone; tibiale and fibulare fused proximally and distally; no intercalary cartilages; sartorius muscle distinct from semitendinosus, tendon of later pierces gracilis muscle; upper jaw with or without teeth, tongue highly protrusible; pupil horizontally elliptical; amplexus cephalic or independent; larva type A, with keratinized beaks and denticles and single sinistral spiracle. The karyotype is 2N = 18 to 24. For intrafamilial taxonomy I have followed Myers (1987), Schulte (1989), and Myers et al. (1991) regarding generic limits. Schulte (1989) and Myers, Paolillo, and Daly (1991) questioned the validity of the nominal genera *Allobates* and *Phobobates* Zimmermann and Zimmermann, 1988, and synonymized them with *Epipedobates* Myers (1987) of South America. Myers, Paolillo, and Daly (1991) also refrained from subdividing *Colostethus,* although shortly thereafter La Marca (1992) erected *Mannophryne* and then *Nephelobates* (La Marca 1994) for species of this genus. Coloma's (1995) placement of the Venezuelan *Aromobates* Myers, Paolillo, and Daly (1991) within *Colostethus* is not followed in the present treatment of dendrobatids.

Four genera of this family, *Dendrobates, Epipedobates, Minyobates,* and *Phyllobates,* secrete toxic alkaloids and are rarely preyed on by the usual frog predators, and they share aposematic coloration (Daly, Myers, and Whittaker 1987). Until recently it was thought that the toxins were manufactured by glands in the frogs' skins. It has now been documented (see summary in Myers et al. 1995) that some and perhaps all of the toxins are the result of diet. Wild-caught frogs of these genera show marked declines in toxicity when kept in captivity, and captive-reared individuals are nontoxic. In addition, some alkaloids were found to accumulate in the skin when added to the diet or when leaf litter insects were fed to captive-raised frogs. This suggests that these dendrobatids have sophisticated metabolic pathways for accumulating the toxins, concentrating them in the granular skin glands, and protecting themselves from toxic effects. These biochemical processes doubtless have a genetic basis, so accumulation of different toxins by different taxa remains phylogenetically significant.

Most species in this family belong to the genus *Colostethus* (106 species), which like *Aromobates* are relatively dull dark brown to tan, usually with some lighter markings. The representatives of the other four genera are brightly colored (red, green, blue, yellow, or orange), highly toxic forms.

Although capable of jumping when disturbed, dendrobatids usually move about in a series of short hops. As far as is known, they lay small clutches of unpigmented to grayish eggs in moist terrestrial and arboreal sites (patterns 15, 18, and 26, table 7.1). Either males or females attend the developing eggs or visit them from time to time, and with a few exceptions they carry the tadpoles on their backs to water, where the free-swimming larvae complete their development.

This is a strictly Neotropical family that ranges from eastern Nicaragua south to southeastern Brazil and throughout the Amazon basin east of the Andes, including Trinidad, Tobago, and Martinique, throughout the Andes to northern Peru, and west of the Andes from Colombia to central Peru.

Genus *Colostethus* Cope, 1866a

A large group (106 species) of small to moderate-sized diurnal frogs having various shades of brown as the dorsal ground color, relatively small finger and toe disks, and usually longitudinal dorsolateral and/or lateral light stripes. No member of the genus has finger webs or vomerine teeth, the nares are lateral, the pupil of the eye is horizontally elliptical, and the tympanum is small and indistinct. Most species placed in the genus are tiny to small (14 to 27 mm in standard length), but some are moderate-sized, reaching maximum standard lengths of 35 to 38 mm. The genus *Colostethus* (sensu lato) differs from other genera of dendrobatids in the following combination of features: teeth present on upper jaw; disk of finger III less than twice width of digit; toes with or without basal webbing; no lipophilic alkaloid skin toxins; larvae with median or dextral vent tube, oral disk indented laterally with 2/3 denticle rows or without denticles and marginal papillae and umbelliform (ecomorph 3, table 7.2). Members of the genus are terrestrial or riparian forms that occur in a variety of habitats from lowland sites to 4,000 m in altitude. Most are found in humid forests, but others occur in upland meadows and subparamo to paramo conditions. The range includes much of the Neotropical region from Costa Rica south in the lowlands and in the Andes into Peru, to northern and Andean Venezuela, the Guianas, Trinidad, Tobago, and Martinique, throughout the Amazon basin and eastern and southern Brazil.

Adult males of all species have paired vocal slits and a single internal subgular vocal sac. Amplexus is cephalic in all species of the genus whose behavior has been described. In an unpublished dissertation based on a phenetic analysis of external features, Edwards (1974) divided the genus into twenty-one informal clusters of species. Rivero (1988) proposed that there are eight recognizable species groups within the genus. Myers, Paolillo, and Daly (1991) identified a cluster of northern South American species having a throat collar or bar as a possible monophyletic group within the genus. They indicated that *Colostethus* (sensu stricto) consisted of species in which the third finger is widened or swollen in adult males. They further suggested that the generic name *Hyloxalus* Jiménez de la Espada (1871) is probably applicable to the remaining species of *Colostethus* (sensu lato).

La Marca (1992) subsequently proposed the new generic name *Mannophryne* for the collared species recognized as a group by Myers, Paolillo, and Daly (1991). Coloma (1995) concluded that none of these groupings were strongly supported and retained *Colostethus* (sensu lato) plus the putative monotypic genus *Aromobates* (Myers, Paolillo, and Daly 1991) within an all-inclusive *Colostethus*.

While Coloma's paper was in press La Marca (1994) described another new genus, *Nephelobates*, that Kaplan (1997) associated with *Mannophryne* and some *Hyloxalus* species. Kaplan further concluded that other *Hyloxalus* and *Colostethus* (sensu stricto) differed from the remaining nontoxic dendrobatids in lacking neopalatine bones. Whatever final taxonomic arrangement is accepted, two Costa Rican species *(C. flotator* and *C. nubicola)* belong to *Colostethus* (sensu stricto), which I regard as distinct from *Aromobates*, contrary to Coloma (1995). The third, *C. talamancae*, would be referred to *Hyloxalus* under the Myers, Paolillo, and Daly (1991) scheme. In these circumstances I recognize *Colostethus* (sensu lato) for the purposes of this account.

KEY TO THE COSTA RICAN SPECIES OF THE GENUS *COLOSTETHUS*

1a. A strong lateral light line usually running obliquely from groin to eye, sometimes incomplete and interrupted near level of midbody; finger III swollen in adult males (fig. 7.120) 2

1b. No oblique lateral light line anterior to groin (fig. 7.119c), although a dorsolateral light line extends from sacrum to eye; throat black in adult males; finger III not swollen in males; adult males 17–24 mm in standard length, adult females 16–25 mm . . . . *Colostethus talamancae* (p. 381)

2a. Venter white; throat gray in adult males; usually no bars on upper surface of calf; oblique lateral light line complete or incomplete (fig. 7.119a,b); adult males 14.4–16.5 mm in standard length, adult females 14.8–18 mm . . . . *Colostethus flotator* (p. 378)

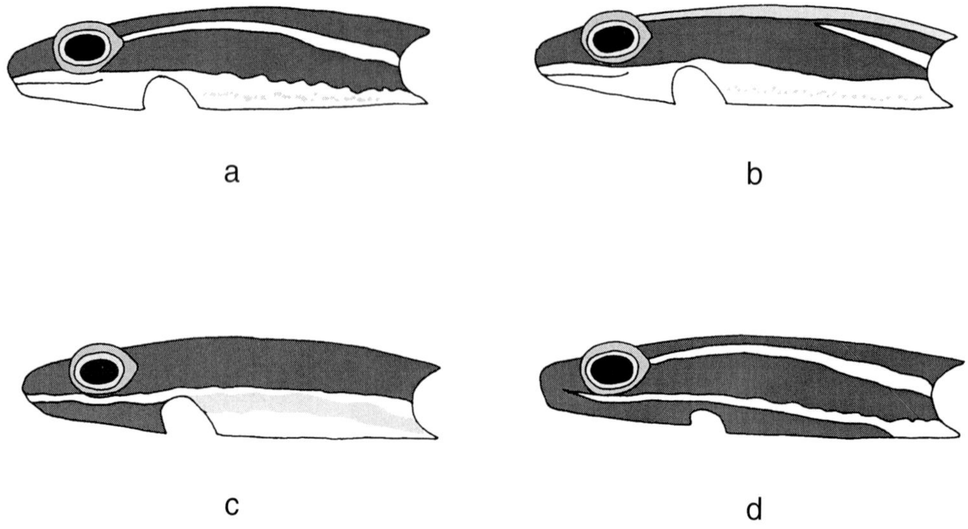

a

b

c

d

Figure 7.119. Color patterns in Costa Rican frogs of the genus *Colostethus:* (a) *C. flotator* from southwestern Costa Rica, oblique light lateral stripe from eye to groin; (b) *C. flotator* from southeastern Costa Rica, short oblique light lateral stripe; (c) male *C. talamancae*, no light lateral stripe, throat black; (d) male *C. nubicola*, oblique light lateral stripe from eye to groin, throat and chest black.

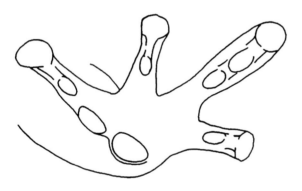

Figure 7.120. Palmar view of left hand showing swollen finger III found in male *C. flotator* and *C. nubicola*.

2b. Venter yellow; throat black in adult males; calf usually with dark bars or spots on upper surface; oblique lateral light line complete (fig. 7.119d); adult males 16–20 mm in standard length, adult females 17.5–23 mm
.... *Colostethus nubicola* (p. 380)

Colostethus flotator (Dunn 1931a)

(plates 211–12; map 7.118)

DIAGNOSTICS: A tiny species having an oblique lateral light stripe originating in the groin that runs partway or all the way to the upper eyelid (fig. 7.119a,b) and a white venter; adult males have a gray throat.

DESCRIPTION: Adult males 14.4 to 16.5 mm in standard length, adult females 14.8 to 18 mm; dorsum shagreened; head slightly longer than broad; snout truncate from above; eyes moderate; paired vocal slits, a single external subgular vocal sac and swollen finger III in adult males; finger I longer

than II; subarticular tubercles ovoid to elongate on all digits, obtuse in profile; no supernumerary tubercles under digits; no distinct accessory palmar or plantar tubercles; thenar tubercle elongate, low, slightly smaller than round palmar tubercle; no toe webs; inner and outer metatarsal tubercles small, ovoid and round, respectively; weak tarsal fold and inner tarsal tubercle present. Dorsum dark brown; flanks black with cream oblique lateral stripe; no dorsolateral light stripe; a ventrolateral light stripe indicated in life but not bordered below by dark pigment and not evident in preservative; no dark transverse bars on clear upper thigh surface; light thigh surfaces orange; calf usually without dark bars on upper surface; testes dark gray or black.

LARVAE: (fig. 7.121): Moderate-sized, total length 12.5 mm at stage 25; oral disk umbelliform; body ovoid; mouth anteroventral; nares lateral; eyes dorsolateral; spiracle low, lateral, sinistral; vent tube dextral; tail long; fins low; tail tip bluntly pointed; oral disk large, complete; serrate beaks but no denticles present; large papillae on surface of oral disk; brown on dorsum, cream below, with scattered dark pigment on tail musculature and fins.

VOICE: The call is heard during the day and is a more or less continuous series of slow, weak "peet-peet-peets" lasting 7 to 50 seconds. Mean note repetition rate is about 180 per minute; the initial dominant frequency of the call is 4.8 to 5.9 kHz rising to 5.1 to 6.9 kHz in samples from Panama (Ibáñez and Smith 1995).

SIMILAR SPECIES: (1) *Colostethus nubicola* has a yellow venter, a black throat in males, and usually dark calf markings. (2) *Colostethus talamancae* has a broad dorsolateral light stripe on each side but lacks an oblique lateral stripe (fig. 7.119c). (3) *Phyllobates lugubris* is mostly black

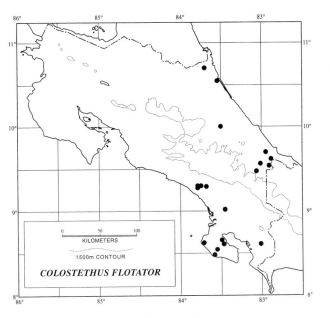

Map 7.118. Distribution of *Colostethus flotator*.

on the undersurfaces, has paired dorsolateral light stripes, and lacks an oblique lateral light stripe.

HABITAT: Relatively undisturbed Lowland Moist and Wet Forests and marginally into Premontane Rainforest.

BIOLOGY: These common terrestrial, diurnal frogs are leaf litter habitués. They are active all year but tend to concentrate near stream courses in the drier months. Males call throughout the day on the forest floor, usually from under leaf litter but most commonly during early morning and late afternoon after showers. Calls are used to maintain territories. As an intruder approaches a resident male, the resident accelerates the advertisement call. If the intruder persists, the resident initiates a wrestling match by jumping on him. After a quick pinning or a few falls, the dominant male sits on top of the loser, pressing him to the ground. The loser then abandons the disputed territory. The diet consists of a variety of small arthropods, with larvae forming a high proportion of the prey (Toft 1981). Mating occurs in the leaf litter, where the eggs are laid and hatched. Males transport two to ten tadpoles on their backs to streams, where they complete development. The mouthparts of the larvae are specially modified for feeding near the surface of the water. The oral disk is expanded, lacks denticles, and is usually directed dorsally. In an aquarium, the tadpoles swim along beneath the surface film and, by slight movements of the edges of the oral disk, create a current that flows toward the mouth. The papillae over the surface of the disk apparently aid in sorting particles out of the water. The tadpoles can orient the umbrellalike oral disk ventrally to attach to rocks or other substrate features when resting (ecomorph 3, table 7.2). The smallest juveniles are 9 mm in standard length.

REMARKS: Taylor (1952a) used the name *Phyllobates nu-*

bicola flotator for this taxon following Dunn (1933b). Savage (1968b) established the distinctiveness of *Colostethus* and *Phyllobates* but used the name *Colostethus nubicola* for this species. Subsequent research by Ibáñez and Smith (1995) demonstrated that the upland *C. nubicola* and this lowland form are not conspecific and occur sympatrically at several localities in Panama. Nevertheless, the geographic variation described for lowland populations by Savage (1968b) and for intermediate elevations in Panama by Ibáñez and Smith (1995) suggests that one or more unnamed species may be involved. Atlantic and Pacific versant Costa Rican examples differ most obviously in that the former have an incomplete lateral light stripe (plate 212) rather than a complete one (plate 211) as found elsewhere in the range of *C. flotator*. Lowland and midelevation samples from west-central Panama differ in ventral coloration, and there is subtle variation in the males' calls and size, with the latter most closely resembling southwestern Costa Rican samples. Taylor (1954a) listed and discussed the possibility that *Phyllobates pratti* (*Colostethus pratti*) might be represented in Costa Rica. Atlantic slope Costa Rican *C. flotator* may be confused with

12.5 mm

Figure 7.121. Tadpole of *Colostethus flotator*. Center figure shows rotation of mouth for surface feeding in this species and *C. nubicola*.

C. pratti, which also has an incomplete lateral light stripe. Savage (1968b), through a *lapsus*, incorrectly mentioned *C. pratti* as occurring in Costa Rica but listed no definite localities. The species does range into northwestern Panama and ultimately may be found in southeastern Costa Rica. *C. pratti* has a broad dorsolateral light stripe, and the lateral light stripe nearly parallels it (no dorsolateral light stripe and a strongly oblique lateral stripe in *C. flotator*).

DISTRIBUTION: Humid lowlands on the Atlantic slope from Costa Rica to east-central Panama and on the Pacific versant in southwestern Costa Rica, west-central Panama, and possibly extreme eastern Panama and adjacent Colombia (10–865 m).

Colostethus nubicola (Dunn, 1924b)

(plate 213; map 7.119)

DIAGNOSTICS: A tiny to very small frog having an oblique lateral line extending from the groin to the upper eyelid, a yellow venter, and males with a black throat (fig. 7.119d).

DESCRIPTION: Adult males 16 to 20 mm in standard length, adult females 17.5 to 23 mm; dorsum shagreened; head slightly longer than wide; snout truncate from above; eyes moderate; paired vocal slits, a single external subgular vocal sac and swollen finger III in adult males; finger I longer than finger II; subarticular tubercles ovoid to elongate, obtuse in profile; no supernumerary tubercles under digits; no distinct accessory palmar or plantar tubercles; thenar tubercle elongate, smaller than round palmar tubercle; no toe webs; inner and outer metatarsal tubercles small, ovoid and round, respectively; weak tarsal fold but inner tarsal tubercle present. Dorsum dark brown; flanks black with a complete

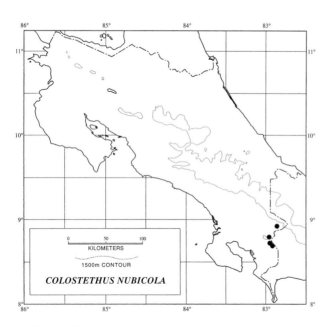

Map 7.119. Distribution of *Colostethus nubicola*.

11 mm

Figure 7.122. Tadpole of *Colostethus nubicola*; also see figure 7.121.

oblique golden cream lateral line from groin to upper eyelid; no dorsolateral light stripe; ventrolateral golden cream stripe, bordered below by dusky to black in males, continuing onto upper lip, reduced to a pinstripe below eye; in females and juveniles ventrolateral stripe not bordered below by dark pigment, not evident in preservative; venter and throat pale yellow to yellow in females and juveniles; throat black in males, black often extending onto chest and abdomen; upper surface of thigh uniform, suffused with dark pigment; light thigh surfaces red orange; calf usually with dark spots or bars on upper surface; testes white or pale gray.

LARVAE (fig. 7.122): Small, total length 11 mm at stage 25; oral disk umbelliform, large; body ovoid; mouth anteroventral; nares lateral; eyes dorsolateral; spiracle low, sinistral; vent tube dextral; tail long; fins low; tail tip rounded; oral disk large, emarginate; serrate beaks and 1/0 or 2/0 short denticle rows present; large papillae on surface of oral disk; brown on dorsum, cream below; scattered dark pigment on tail musculature and dorsal fin.

VOICE: The advertisement call is a rapid long "peet-peet-peet . . . peet" lasting about 2 seconds with a 2 second break between calls. Mean note repetition is about 510 per minute. The initial dominant frequency is 4.0 to 4.6 kHz rising to 4.3 to 5.5 kHz in samples from Panama (Ibáñez and Smith 1995).

SIMILAR SPECIES: (1) *Colostethus flotator* has a white venter, a gray throat in males, and usually no dark calf markings. (2) *Colostethus talamancae* has a dorsolateral light stripe on each side but lacks an oblique lateral stripe (fig. 7.119c). (3) *Phyllobates lugubris* is black on the limb undersurfaces and lacks an oblique lateral light stripe.

HABITAT: Premontane Wet Forest and Rainforest.

BIOLOGY: A moderately common diurnal terrestrial species favoring leaf litter situations. Males call during the day, early in the morning, late in the afternoon, or during very cloudy or rainy periods. Food consists of a variety of small arthropods. Reproduction probably takes place all year in Costa Rica. Eggs are laid in the leaf litter, and males carry the one to eleven tadpoles on their backs to small streams to complete metamorphosis. The large umbelliform oral disk probably functions in the same fashion described above for *C. flotator*.

DISTRIBUTION: Humid premontane forests in southwestern Costa Rica and adjacent western Panama (1,050–1,600 m) south to eastern Panama in lowland and premontane zones (200–860 m); possibly occurring in northern Colombia.

Colostethus talamancae (Cope, 1875)

(plate 214; map 7.120)

DIAGNOSTICS: A tiny to very small anuran having a definite pair of dorsolateral light stripes, no oblique lateral stripes, and males with a black throat (fig. 7.119c).

DESCRIPTION: Adult males 17 to 24 mm in standard length, adult females 16 to 25 mm; dorsum shagreened; head slightly longer than broad; snout truncate from above; eyes moderate; paired vocal slits and a single external subgular vocal sac in adult males; finger III not swollen in males; finger I longer than finger II; subarticular tubercles mostly round and low on all digits; no accessory palmar or plantar tubercles; thenar tubercle low, elongate, slightly smaller than

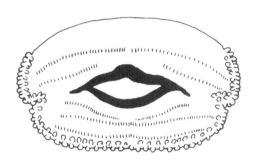

12 mm

Figure 7.123. Tadpole of *Colostethus talamancae*.

round palmar tubercle; no toe webs; inner and outer metatarsal tubercles small, ovoid; weak tarsal fold and inner tarsal tubercle present. Dorsum dark brown; dorsolateral light stripe from sacrum onto upper eyelid; flanks black, bordered below by broad ventrolateral light stripe that runs from leg to upper lip, edged below by dusky to black pigment; venter and throat white in females and juveniles; throat and chest of adult males black; light thigh surfaces beige suffused with orange; thigh and calf barred with dark above; testes white.

LARVAE (fig. 7.123): Small, total length 12 mm at stage 26; oral disk not umbelliform; body ovoid; mouth anteroventral; nares lateral; eyes dorsolateral; spiracle low, lateral, sinistral; vent tube dextral; tail long; fins very low; tail tip bluntly pointed; oral disk small, emarginate; beaks and 2/3 denticle rows present; broad gap in A2 above mouth; marginal papillae row interrupted above mouth; dark brown on dorsum, slightly lighter below; tail light brown with heavy blotching of darker pigment on musculature and fins.

VOICE: A variable series of slightly harsh "peet-peet-peets" lasting 150 to 180 milliseconds, rising in pitch and followed by a pause and repeated eight to twenty times. The dominant frequency starts at 3.4 to 3.5 kHz and rises to 4.4 to 5.0 kHz (Savage 1968b; Edwards 1974).

SIMILAR SPECIES: (1) *Phyllobates lugubris* and *P. vittatus* lack ventrolateral stripes, and the undersurfaces of the legs are covered with black. (2) *Colostethus flotator* and *C. nubicola* have oblique lateral light stripes (fig. 7.119a,b,d).

HABITAT: Lowland Moist and Wet Forests and marginally into the Premontane Wet Forest and Rainforest belts.

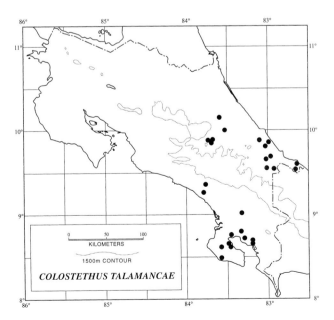

Map 7.120. Distribution of *Colostethus talamancae*.

BIOLOGY: A common diurnal denizen of the forest floor that is active throughout the year, although less so in the driest months. Males call throughout the day from the forest floor, usually sitting on top of the leaf litter. Calling tends to be concentrated at times of low light intensity—early morning, late afternoon, and cloudy or rainy periods—often peaking during the afternoon or after showers. Prey includes a variety of small arthropods (under 9 mm in length), including ants (Toft 1981). Mating occurs in the litter, apparently throughout the year but concentrated in wetter months. Males transport eight to twenty-nine tadpoles on their backs from the litter where early development takes place to water-filled depressions in small streams.

REMARKS: Taylor (1952a) used the name *Phyllobates talamancae* for this species. The karyotype is 2N = 24, composed of six large and six smaller pairs of chromosomes: nine pairs of metacentrics, two pairs of submetacentrics, and one pair of subtelocentrics; NF = 48 (Bogart 1991).

DISTRIBUTION: Humid lowlands and lower premontane slopes from northeastern and southwestern Costa Rica (2–703 m) south to western Colombia (2–820 m).

Genus *Dendrobates* Wagler, 1830

This genus (twenty-five species) includes some of the most brilliantly colored tropical poison frogs. The genus is composed of tiny to moderate-sized forms usually having a black ground color with bright markings or black blotches or large or small black spots on a bright red, orange, yellow, green, or blue ground. Most species have broad finger and toe disks and finger I shorter than finger II. All lack finger and toe webs and teeth on the upper jaw. The nares are lateral, the pupil of the eye is horizontally elliptical, the tympanum is small and indistinct, and the venter is smooth. Species placed in the genus range in adult size from 13.5 to 50 mm in standard length. Other characteristics that in combination will separate *Dendrobates* from other genera in the family are disk on finger III usually two or more times width of digit; no palatine bones; lipophilic alkaloid skin toxins include 3,5 disubstituted indolizidines but no batrachotoxins; larvae with median vent tube, oral disk not indented laterally, 0/1, 1/1, 2/1, or 2/3 rows of denticles and no papillae above mouth.

Four species groups may be recognized within *Dendrobates* (Silverstone 1975, as modified by Myers and Daly 1976, 1980; Myers 1982a; and Myers, Daly, and Martínez 1984). Two of these, the *Dendrobates pumilio* and *Dendrobates tinctorius* groups, occur in Costa Rica.

All frogs of this genus and the allied genera *Epipedobates*, *Minyobates*, and *Phyllobates* secrete a variety of lipophilic alkaloid skin toxins. These genera also exhibit warning coloration and are collectively known to be aposematic (Myers et al. 1995).

Take care in handling them, since the skin toxins are very irritating if you accidentally touch your skin, eyes, or mucous membranes after contact. Washing your hands will usually prevent such accidents, but do not be surprised if any scratch or cut tingles or burns for some time after you handle these frogs. Do not place other organisms in the same container with *Dendrobates* or they will probably be poisoned and die. Overcrowding specimens of *Dendrobates* in a collecting bag usually kills them as well.

These frogs are found primarily in humid lowland forests, but several occur above 1,000 m to a maximum of 1,400 m. The genus is found from southeastern Nicaragua south to western Ecuador, west of the Andes, and east of the Andes in the Guianas and the Orinoco and Amazon basins.

Adult males have paired vocal slits and a moderately distensible single internal subgular vocal sac. There is no amplexus in the species of this genus studied to date. After courtship the members of a pair sit facing away from one another with their cloacal regions juxtaposed during fertilization of the eggs. Silverstone's (1975) outstanding treatment of this taxon, although somewhat outdated, is a valuable guide. Note that the species referred to the *Dendrobates minutus* species group, except *D. quinquevittatus*, are now placed in the genus *Minyobates* (Myers 1987) and that a number of new species have been described in both genera since 1975.

KEY TO THE COSTA RICAN POISON FROGS OF THE GENUS *DENDROBATES*

1a. Dorsum usually bright red, rarely blue, green, or yellow in life, often with black flecks; adults 19–24 mm in standard length. 2
1b. Dorsum and venter bright bluish green to green in life, with large black blotches, bands, or stripes; adults 25–42 mm in standard length. . . . .
. . . . *Dendrobates auratus* (p. 383)
2a. Dorsum extremely granular, uniform bright red or rarely green or yellow in life; hind limbs and venter usually bright jade green; adults 19–22 mm in standard length. . . . . *Dendrobates granuliferus* (p. 384)
2b. Dorsum smooth, a uniform bright red (rarely blue) or flecked with black in life; hind limbs and venter usually red, dark blue black, or purple; adults 17.5–24 mm in standard length. . . . . . . *Dendrobates pumilio* (p. 386)

Although readily distinguished by coloration in life, frogs of this genus quickly become nearly uniform black in preservative. However, the contrasting dark and light areas of living individuals are usually indicated in preserved *D. auratus*, particularly if the specimen is immersed in water. Size and the texture of the dorsal skin should let you differentiate the other species from *D. auratus* and from one another except for poorly preserved examples.

Dendrobates tinctorius Group

A cluster of five species with the color in life never red but black and yellow, green, or blue, stripes usually absent but sometimes present, lacking an omosternum and having a buzz call. This group ranges from Nicaragua to western Colombia and the lower Amazon basin. Extralimital species include *Dendrobates azureus* of Suriname, *D. galactonotus* from the lower Amazon basin of Brazil, *D. tinctorius* from the Guianas and adjacent Brazil, and *D. truncatus* of Colombia.

Dendrobates auratus (Girard, "1854," 1855)

(plate 215; map 7.121)

DIAGNOSTICS: Small to moderate-sized blue, blue green, green, or yellow green frogs with black, brown, or bronze spots, bands, or stripes, large finger disks, and no webs between fingers or toes.

DESCRIPTION: Adult males 25 to 40 mm in standard length, adult females 27 to 42 mm; upper surfaces smooth; head longer than wide; snout truncate to slightly rounded from above; eyes large; finger I shorter than finger II; subarticular tubercles low, ovoid to elongate; no supernumerary tubercles under digits; accessory palmar tubercles present; no plantar tubercles; thenar tubercle small, elongate; palmar tubercle much larger, round; inner metatarsal tubercle small, elongate, outer small, rounded, larger than inner; a definite inner tarsal fold; inner tarsal tubercle weak or absent. Dorsum and venter a mixture of spots, bands, blotches, or stripes of black, brown, or bronze and of blue on various shades of green.

LARVAE (fig. 7.124): Moderate-sized, total length 30 mm

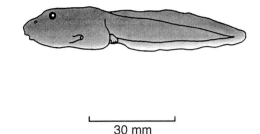

30 mm

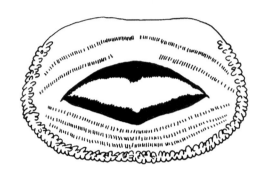

Figure 7.124. Tadpole of *Dendrobates auratus*.

at stage 34; body robust; mouth anteroventral, directed ventrally; nares and eyes dorsal; spiracle low, lateral, sinistral; median vent tube; tail long; fins moderate; tail tip bluntly pointed; oral disk small, complete; beaks and 2/3 rows of denticles present; broad gap in A2 above mouth; two rows of marginal papillae, no papillae above mouth; uniformly black.

VOICE: The call is a low, slurred buzzing sound lasting about 2 to 4 seconds, usually consisting of three to five notes followed by a 5 second pause and repeat. The dominant frequency is 3.5 kHz through most of the call (Myers and Daly 1976).

SIMILAR SPECIES: There are none.

HABITAT: Undisturbed Lowland Moist and Wet Forests.

BIOLOGY: This section combines the reports of Dunn (1941a), Eaton (1941), the summary in Silverstone (1975), Wells (1978), Zimmerman (1986), and Summers (1989, 1990) with my own observations. This common beautiful and active frog is an obvious forest floor denizen. It is a diurnal forager that moves by a characteristic series of hops, stopping briefly and then hopping away again when disturbed. It escapes by a series of hops, following an irregular course until it reaches a hiding place. The species also climbs well, and individuals have been observed in the canopy as high as 45 m above the ground (D. Bickford, pers. comm.). It eats a wide variety of small arthropods (under 8 mm in length), especially ants (Toft 1981). Males call infrequently from holes or hollows near tree bases throughout the day,

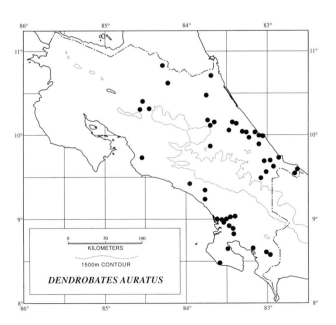

Map 7.121. Distribution of *Dendrobates auratus*.

but most commonly in the early afternoon. Males vocalize to establish territories, and wrestling matches occur where their territories overlap. Both males and females are territorial. Females court the males and wrestle other females and chase them from their territories. Individual males mate with up to six females. Female aggression is directed toward preventing other females from mating with a particular male. Courtship in this species lasts up to several hours. Both sexes are active in touching and hopping around one another. Mating is typical for the genus (Wells 1978). Eggs are laid in the leaf litter. Males keep the egg clutches moist and may continue to court females while caring for clutches from four different females (Summers 1989). The clutch is small (5 to 13 eggs). On hatching, the tadpoles crawl onto the male's back and are carried to small still-water sources such as crevices or holes in fallen logs, curled-up dead leaves, the pods or shells of fallen fruit, fallen palm fronds, or tree holes or bromeliads up to 20 m or so above the ground. Individual males often repeatedly place tadpoles in the same site. In captivity one female laid multiple clutches every eight to ten days during the breeding period of several months. Eggs hatch in ten to sixteen days and transform in thirty-nine to eighty nine days (mean sixty-six) depending on the degree of crowding, with more rapid development at lower tadpole densities. Juveniles are about 13 mm in standard length, and they reach sexual maturity in about fifteen months. In captivity the larvae are omnivorous. Summers (1989) and Caldwell and Arajúo (1998) reported attacks on heterospecific tadpoles and aquatic arthropods in field experiments, and the latter authors hypothesize that *D. auratus* may be cannibalistic when crowded. A recent report (Master 1998) indicates that motmots *(Baryphthengus* and *Momotus)* eat this species with apparent immunity to its toxins. Myers, Daly, and Malkin (1978) discount Breder's (1946) hearsay report that Amerindians used the skin of this species for poisoning darts.

REMARKS: Coloration in *D. auratus* is rather variable. Generally the dark blotches are ebony, but at some localities they are suffused with a metallic light pigment so that they appear bronze to dull gold. In some individuals a light suffusion gives the entire body a gold or silvery cast that is heightened after preservation and is maintained until the specimens turn solid black after several months. Specimens of this kind apparently formed the basis for the species name proposed by Girard (1854). Silverstone (1975) reported on an extralimital variant from one locality in central Panama, specimens of which lack green and are dark brown with tan markings. One of the plates in Silverstone's opus (1975) is marred by wrongly reproduced color for *D. auratus;* the light color should be green, not bluish white (Silverstone in Myers et al. 1995). The karyotype is 2N = 18 and consists of six large pairs and three small pairs of chromosomes, all metacentrics; NF = 36 (Rasotto, Cardellini, and Sala 1987).

DISTRIBUTION: Humid lowlands from southern Nicaragua on the Atlantic slope and southwestern Costa Rica on the Pacific versant to northwestern Colombia; in Costa Rica 2–601 m; in eastern Panama to 800 m.

Dendrobates pumilio Group

This group of three species usually has red as the dominant color in life, no stripes in the pattern, an omosternum present, and a harsh, often constantly repeated chirp call in adult males. A single species of the group, *Dendrobates speciosus* of western Panama, is extralimital to Costa Rica.

Dendrobates granuliferus Taylor, 1958a

(plates 216–17; map 7.122)

DIAGNOSTICS: A small, short-limbed species with an extremely granular dorsum and usually uniform bright red upper surfaces, except for the generally bright jade hind limbs; however, some examples are uniform green or yellow on the body and arms and others have the upper surface of the arms uniform red.

DESCRIPTION: Adults 18 to 25 mm in standard length; dorsum strongly granular, upper hind limb surfaces granular; head about as broad as long; snout short, broadly rounded from above; eyes moderate; finger I shorter than finger II; subarticular tubercles ovoid, low; no supernumerary tubercles under digits; no accessory palmar or plantar tubercles; elongate thenar tubercle much smaller than round palmar tubercle; inner metatarsal tubercle elongate, much smaller than round outer tubercle; weak tarsal fold; small tarsal tubercle absent or weak. Upper surfaces of head and arms usually uniform red; hind limbs, venter, and nonred areas turquoise to jade green, often with some black spotting; dorsal surfaces rarely all green or yellow, sometimes all red; throat red in females, green or black in males.

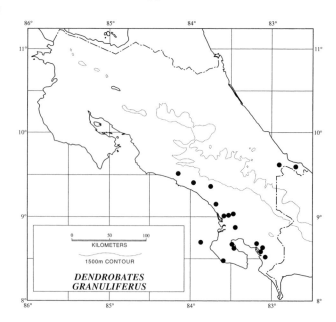

Map 7.122. Distribution of *Dendrobates granuliferus.*

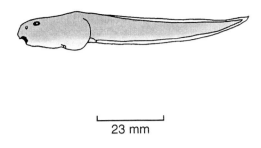

23 mm

Figure 7.125. Tadpole of *Dendrobates granuliferus*.

LARVAE (fig. 7.125): Small, total length 25 mm at stage 39; body depressed; mouth anteroventral, directed ventrally; nares and eyes dorsal; spiracle low, sinistral; vent tube median; tail long; fins very reduced; tail tip pointed; oral disk small, complete; very large beaks with pointed serrations and 1/1 or 0/1 rows of denticles present; A1 often interrupted by gap; one row of very large marginal papillae, two at level of mouth opening; papillae not continuous above.

VOICE: A harsh series of unmusical "chirp-chirp-chirp" calls. Calls last 2 to 39 seconds with two to three notes per second, and intervals between calls are 5 to 15 seconds. The dominant frequency is 3.4 to 4.6 kHz (Myers and Daly 1976; Myers et al. 1995).

SIMILAR SPECIES: *Dendrobates pumilio* has a smooth dorsum and lacks bright green in its coloration (but see remarks).

HABITAT: Relatively undisturbed Lowland Moist and Wet Forests.

BIOLOGY: This account summarizes published material by Goodman (1971), Crump (1972, 1983b), Silverstone (1975), Meyer (1992, 1993, 1996), Van Wijngaarden and Bolaños (1992), and my own field observations. *Dendrobates granuliferus* is a relatively common diurnal forest frog with strong arboreal tendencies. It may be seen hopping across or sitting on the leaf litter but frequently is found on vegetation, tree trunks, rock faces, or moss banks along fast-moving streams. Activity is seasonal with few individuals seen during the dry season (December to January). Calling males are heard from May to November during the rainy times of the year. Reproduction probably occurs throughout this period. Activity of males and females is highest from

July to November but is influenced by the beginning of the first rains. Males call primarily from vegetation from 0.1 to 2.0 m above the ground during the early morning hours (5:30 to 9:00) and again in the late afternoon (peaking at 5:30). Males are strongly territorial. Resident males wrestle with intruding male conspecifics, and the loser retreats. Food consists of many small arthropods, but especially ants. Reproduction takes place on the ground. During courtship, the male leads the female to an oviposition site in the litter. The female characteristically rubs the male's head and chin with her head and raises her vent in front of him. Fertilization follows, with the two frogs facing in opposite directions and bringing only their vents into contact. The female lays two to four eggs at one time and produces multiple clutches. Males attend the eggs and may move from one egg clutch to another. The recently hatched tadpoles (plate 217) crawl onto the females' backs and are transported to water-containing axils of plants, particularly *Diffenbachia*. Females then return from time to time and deposit unfertilized eggs in the water as food for the obligatorily oophagous tadpoles (ecomorph 9, table 7.2) The smallest juveniles are 13 mm in standard length.

REMARKS: Although the species was recently cited as *D. granulifera* (Duellman 1993), there seems no basis for this change (Myers et al. 1995). Article 32.3 of the *International Code of Zoological Nomenclature* (International Commission on Zoological Nomenclature 1999) seems applicable in requiring retention of Taylor's (1958a) original spelling. Different populations within this species are recognizable by coloration. Frogs from south of the Río Grande de Térraba usually have the red extending posteriorly only to the sacral region. Farther north the red usually extends above the cloaca. Northern frogs tend to have more red on the arms than southern ones. In the extreme northern populations from Río Damitas, Puntarenas Province, near Quepos, most of the upper surfaces are green with the hind limbs turquoise and the hands blue. Frogs in another northern population are uniform yellow (Mason Ryan, pers. comm.). Myers et al. (1995) have recently reported that Silverstone's (1975) plate has the color of *D. granuliferus* incorrectly reproduced. The limbs of this frog should be blue green, not light blue (Silverstone in Myers et al. 1995). Myers et al. (1995) reported on a population of this species from the Atlantic versant completely disjunct from the previously known range of *D. granuliferus* in southwestern Costa Rica. At this site, Río Sand Box, a tributary of the Río Sixaola, Limón Province, it is sympatric with *Dendrobates pumilio*, an Atlantic slope endemic. Although this is a well-collected area visited by E. R. Dunn in the 1930s, E. H. Taylor in the 1950s, J. M. Savage in the 1960s, and D. C. Robinson in the 1970s and 1980s, none of these or other collectors during that period obtained this species from the region. However, two specimens undeniably of *D. granuliferus* were collected at the site in 1989 and another four in 1990. Eberhard Meyer (pers.

comm.), who has carried out the most extensive studies on this species over several years, doubts that these animals represent a natural population. He notes that beginning in the 1980s the Servicio de Fauna Silvestre (Costa Rica) confiscated many collections of live dendrobatids illegally collected for sale to terrarium enthusiasts in Europe. Their policy was to release the contraband animals at or near the presumed capture locality. The Río Sand Box area was well known as a site for *D. pumilio*. Wildlife personnel had little reason to discriminate among red frogs, and some of Pacific slope origin may well have been mixed in at the time of release. Certainly the collector or exporter was unlikely to inform the agents involved about correct identifications or the origins of confiscated specimens. E. Meyer further pointed out that many local populations of *D. granuliferus* differ in details of coloration, and the Sand Box sample exactly matches a population from the Golfo Dulce region. Contravening this argument is a specimen from Sixaola, Limón Province (BMNH 1956.1.6.54), collected in 1955 or 1956. Silverstone (1975) noted that it was nearly as granular as *D. granuliferus*. Myers et al. (1995) did not examine this example but regard it as evidence for the natural occurrence of *D. granuliferus* in sympatry with *D. pumilio*. "Sixaola" could refer either to the town of that name (8 m) or more probably to the general region, since data with the specimen in question indicated a collection elevation of 240 m. The Río Sand Box examples reportedly are from "approximately 200 meters above sea level." The karyotype is 2N = 20, composed of six large pairs and four small pairs of chromosomes: six pairs of metacentrics, two pairs of submetacentrics, and one pair each of telocentrics and subtelocentrics; NF = 38 (Rasotto, Cardellini, and Sala 1987).

DISTRIBUTION: Lowlands of southwestern Costa Rica, adjacent southwestern Panama, and possibly disjunctly in southeastern Costa Rica (20–100 m).

Dendrobates pumilio O. Schmidt, 1857

(plates 218–21; map 7.123)

DIAGNOSTICS: Tiny to small rather robust red (rarely blue) frogs with a smooth dorsum and the upper surfaces of the hind limbs red, blue, purple, or black.

DESCRIPTION: Adults 17.5 to 24 mm in standard length; upper surfaces smooth to shagreened; head about as broad as long; snout truncate or broadly rounded; eyes moderate; finger I shorter than finger II; subarticular tubercles low, ovoid; no supernumerary, accessory palmar, or plantar tubercles; small elongate thenar tubercles and larger round palmar tubercles; inner and outer metatarsal tubercles distinct, small, subequal; inner tarsal tubercle usually indicated. Upper surfaces strawberry red (rarely blue), uniform or with black flecks; limb surfaces similar to dorsum, red or with black flecks or invaded posteriorly by darker hind limb color; throat red in females, with a buff center in adult males.

LARVAE (fig. 7.126): Small, total length 16 mm at stage 30; body depressed; mouth anteroventral, directed ventrally; nares and eyes dorsal; spiracle low, sinistral; vent tube median; tail long; fins very low; caudal fin rounded; oral disk small, complete; large beaks with large pointed serrations and 1/1 rows of denticles; one row of very large marginal papillae; no papillae above mouth; body dark brown above, lighter ventrally; tail light brown with scattered darker punctations.

VOICE: A series of harsh "chirp-chirp-chirp" calls. Call lengths are 5 to 32 seconds with 5 to 9 notes per second, and the dominant frequency is in the range of 3 to 4 kHz. The calls are repeated after a pause of up to 20 seconds (Myers et al. 1995).

SIMILAR SPECIES: *Dendrobates granuliferus* has extremely granular dorsal body and hind limb surfaces and usually bright green hind limbs.

HABITAT: Undisturbed forest, cacao plantations, and abandoned clearings, but not including banana plantation sites, within the Lowland Moist and Wet Forest zones.

BIOLOGY: This charming little frog is among the most studied and photographed of tropical amphibians because of its diurnal habits, bright color, unusual reproductive biology, and high population density. The following account is based mostly on the research of Brust (1990, 1993), Donnelly (1987, 1989a,b,c, 1991), Lieberman (1986), Limerick (1980), Pröhl (1997b), and Weygoldt (1980, 1987). The field studies of this species cited above were carried out primarily at the Organization for Tropical Studies (OTS), La Selva Biological Station, and I have included additional information from student field problem reports and the unpublished

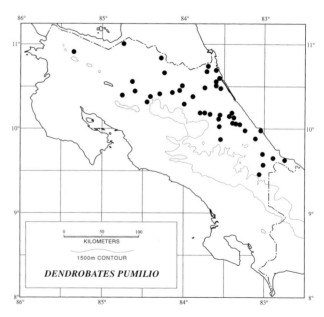

Map 7.123. Distribution of *Dendrobates pumilio*.

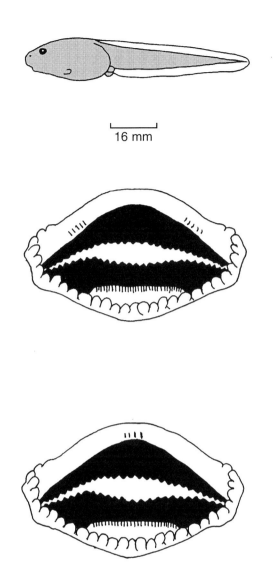

16 mm

Figure 7.126. Tadpole of *Dendrobates pumilio;* note variation in denticle rows.

master's theses of John T. Kitasako and Sandra Limerick as well as from my own notes. Like other *Dendrobates,* this is a diurnal species. It is typically seen moving through or over the leaf litter, interspersing a series of short, jerky hops with walking to cross obstacles or when approaching food items. *Dendrobates pumilio* is active throughout the year. Males are territorial and call from fallen logs, stumps, and the bases of trees usually less than 1 m above the ground. At La Selva, home ranges are 2 to 16 m² for males and 6 to 16 m² for females. Male territories range from 0.24 to 4.78 m². Pröhl (1997b) reported territory sizes of 1 to 24.5 m² at Hitoy-Cerre National Park, Limón Province, but these figures are probably for home ranges. Males use the call to establish and maintain the territory. If an intruder enters, the resident's call rate increases significantly. The intruder usually calls back. If approached more closely, the resident leaps onto

the other male and they may wrestle for up to 20 minutes, standing on their hind legs and grasping one another (plate 220). After one or more takedowns, one male (usually the resident) pins the other venter down, and in a few minutes the loser is released and leaves the territory. Females, non-calling males, and juveniles are not attacked by the resident male. Females are also territorial. Population densities are high for this species, being 827/ha during the wettest months and somewhat lower during November and December. Food consists of small arthropods, especially ants (50% of food items) and mites (40%); juveniles eat proportionally more mites than adults. Mean number of prey items per individual is high (34.8) because of the large amount of small prey (55% of prey items). Average prey lengths for ants and mites in stomachs were 1.4 and 0.7 mm, respectively. Most prey items are under 3 mm long. Mating appear to occur throughout the year. Courtship is initiated when a receptive female approaches a calling male, and it lasts 10 to 180 minutes. The male leads the female to an oviposition site in the leaf litter. Some mutual tactile stimulation takes place before egg laying and fertilization, but there is no amplexus and the two face away from one another and assume a vent-to-vent posture. Clutch sizes are 3 to 5 eggs in the field but higher in terrariums (5 to 20), and the female may lay a clutch every week or so. The following details are based on observations of captive frogs (Weygoldt 1980) and confirmed by field study of a Panamanian population by Brust (1990, 1993). The male returns, usually daily, to moisten the eggs. Females also attend the nest. The eggs hatch in approximately seven days (Limerick 1980) (fourteen to seventeen days in vivariums). At that time the mother returns and carries the tadpoles, one to four at a time (plate 221), placing each in a separate axil of water-filled bromeliads or other plants that are devoid of other tadpoles. The same female returns to each tadpole from time to time during morning hours and lays one to five unfertilized eggs (twenty and forty over time) that serve as food for the larvae (ecomorph 9, table 7.2). Young tadpoles feed by biting a hole in the jelly envelope and sucking out the contents. Older ones often eat the jelly too. Apparently the occasional tadpole that a male transported to a bromeliad is not fed and dies, since the larvae are obligatory oophages. Adult males in captivity often try to eat the fertilized eggs of other males, but in nature this is probably prevented by aggressive resident fathers. Weygoldt (1980) and Brust (1990, 1993) further point out a complex "begging behavior" by the larva that stimulates the female to oviposit in the occupied axil. In this behavior the female backs into the axil and touches the water surface with her vent. The tadpole signals its presence by stiffening its body, placing its head near the female's vent, and vibrating. Similar mother-tadpole interactions are also now known in *Dendrobates granuliferus* (E. Meyer, pers. comm.). In both species the mothers feed only their own offspring even if solicited by another frog's larva. The tadpoles transform in forty-three to fifty-two days, and the froglets are about 10 mm in standard length and uniformly maroon. They attain sexual maturity in about ten months.

REMARKS: Taylor (1952a, 1958a) referred to this species as *Dendrobates typographicus*. Savage (1968b) established the synonymy of that name with *D. pumilio* and was followed by subsequent workers. Daly and Myers (1967) briefly outlined the interesting color variation in this species. The following data are based on living material from Costa Rica combined with the data on Panamanian populations published by Daly and Myers (1967). Along the Nicaraguan–Costa Rican lowlands there is a gradual clinal change in the hind limb color and the amount of dorsal and ventral black flecking. At the level of Matagalpa, Department of Matagalpa, Nicaragua, *D. pumilio* are essentially solid scarlet with bright purple hind limbs and no more than a few scattered dark flecks on the dorsum or venter. In the region of northeastern Costa Rica, 15 km north of Ciudad Quesada, Alajuela Province, and at La Selva, Heredia Province, the coloration is similar to Nicaraguan examples, but with slightly more dark flecks. At La Lola, Limón Province, the frogs have black hind legs and considerable dark flecking. At Pandora, Limón Province, the frogs are red orange with less black on the hind limbs and considerable dark flecking, some with small black spots. At Suretka, Limón Province, near the border with Panama, the frogs are brick red with the hind limbs black and some black spotting dorsally and ventrally. Daly and Myers (1967) and Myers and Daly (1983) reported the situation shown in table 7.4 in adjacent Bocas del Toro Province, Panama.

Surprisingly, a single blue frog of this species was found for the first time at La Selva, Heredia Province, in 1999, supporting the concept that the color variants in Panama represent a single species. Perhaps this variation is related to the extensive intra- and interspecific variability in alkaloid toxins in this species (Myers et al. 1995). Could dietary uptake be involved in the color variation too? The karyotype of *D. pumilio* is 2N = 20, composed of six pairs of metacentric, two pairs of submetacentric, and one pair each of subtelocentric and telocentric chromosomes; NF = 38 (Bogart 1991).

Table 7.5.   Color Variation in Panamanian *Dendrobates pumilio*

Place	Dorsum	Venter	Dorsal Black Flecking
Almirante Mainland opposite Isla Split Hill	Dull red	Red	None
Isla Colón	Dark blue	Powder blue	None to some
Isla Bastimentos	Green	Yellow	Small spots
Isla Bastimentos	Pale green	White	Small spots
Cayo Nancy	Red orange	White	Small spots
Isla Shepard	Red	Red	None
Isla San Cristóbal	Olive	Yellow	Present
	Red	Red	Present

DISTRIBUTION: Atlantic versant, humid lowlands, and premontane slopes in eastern Nicaragua (0–940 m) south through the lowlands of Costa Rica and northwestern Panama (1–495 m).

Genus *Phyllobates* Bibron, 1840

These beautiful, brightly colored frogs include the most toxic members of the family. Only five species are placed in this genus; three have a pattern of light stripes on a black ground color *(Phyllobates aurotaenia, P. lugubris,* and *P. vittatus),* and in two the dorsum is essentially uniform yellow *(P. terribilis)* or yellow to orange *(P. bicolor).*

These species share the presence of narrow finger and toe disks and finger I longer or equal to finger II. All lack finger and toe webs. The nares are lateral, the eye is horizontally elliptical, the tympanum is concealed, and the venter is smooth. Adult males have paired vocal slits and a single external subgular vocal sac. Adults range in standard length from 19 to 47 mm.

Additional features that in combination define the genus are disk on finger III no more than one and one-half times width of digit; teeth present on upper jaw; no palatine bones; lipophilic alkaloid skin toxins include 3,5 disubstituted indolizidines and the steroid batrachotoxins; larvae with dextral vent tube, oral disk indented laterally, 2/3 rows of denticles, and no papillae above mouth.

Silverstone's (1976) fine monograph on the genus has been revised by the removal of all but the five species mentioned above to the genus *Epipedobates* (Myers 1987) and the description of several new forms in the latter genus since 1976. I follow Schulte (1989) and Myers, Paolillo, and Daly (1991) in regarding *Allobates* and *Phobobates* (Zimmerman and Zimmerman 1988) as synonyms of *Epipedobates.*

The batrachotoxins characteristic of this genus are among the most powerful naturally occurring nonprotein poisons. They are very potent cardiotoxins, so be careful in handling Costa Rican members of the genus. Although not in the same toxicity league as *Phyllobates terribilis* of Colombia, their secretions can be very irritating to skin, eyes, and mucous membranes and possibly fatal if ingested. The secretions will almost immediately cause burning in any small scratch or cut, but it is doubtful that sufficient poison could be absorbed this way to cause serious harm. The cautions about placing other animals in a container with poisonous frogs or overcrowding them in a collecting bag, detailed in the generic account of *Dendrobates,* also apply here.

These aposematic dendrobatids are found in humid lowland or premontane forests, with one species, *Phyllobates bicolor* of western Colombia, recorded from as high as 1,525 m. The genus ranges from Nicaragua to southwestern Colombia. There is no amplexus in species of this genus as far as is known. Fertilization is accomplished by vent-to-vent contact, with the male and female facing in opposite directions.

KEY TO THE COSTA RICAN POISON FROGS OF THE GENUS *PHYLLOBATES*

1a. Light areas on upper surfaces of limbs marbled gold, yellow, or yellow green in life; dorsolateral stripe narrow, bordering upper eyelid laterally; adult males 18.5–21 mm in standard length, adult females 20–24 mm . . . . *Phyllobates lugubris* (p. 389)

1b. Light areas on upper surfaces of limbs mostly blue green in life; dorsolateral light stripe broad, nearly as wide as upper eyelid; adult males 22.5–26 mm in standard length, adult females 26–31 mm . . . . *Phyllobates vittatus* (p. 390)

Like *Dendrobates*, members of this genus lose most of their distinctive colors in preservative and may become a nearly uniform black. Usually the striped pattern can still be seen in well-preserved Costa Rican examples, and the differences in adult size between the two species are definitive.

Phyllobates lugubris (O. Schmidt, 1857)

(plate 222; map 7.124)

DIAGNOSTICS: A tiny to small striped species with narrow gold, yellow, yellow orange, or orange dorsolateral stripes and the upper limb surfaces marbled with similar colors.

DESCRIPTION: Adult males 18.5 to 21 mm in standard length, adult females 20 to 24 mm; dorsal surfaces shagreened; head longer than broad; snout rounded, truncate from above; eyes large; finger I longer than finger II; subarticular tubercles ovoid, low; no supernumerary, accessory palmar, or plantar tubercles; thenar and palmar tubercles round, low, thenar much smaller; inner metatarsal tubercle

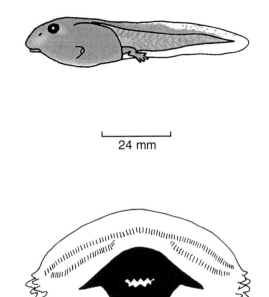

24 mm

Figure 7.127. Tadpole of *Phyllobates lugubris*.

elongate, outer round, subequal; a weak tarsal fold or tubercle present. Dorsum and head black; a complete narrow dorsolateral stripe from dorsal base of thigh onto upper eyelid and along canthus rostralis to tip of snout, continuous around snout with stripe from other side; stripes narrow, usually only along outer margin of upper eyelid; a gold, cream, whitish, or bluish stripe from above arm insertion to under eye; dorsal surfaces of limbs marbled black and gold, yellow, or yellow green; venter and undersurfaces of limbs marbled black and blue, greenish, white, or silver and a ventrolateral light stripe present on flank; iris jet black.

LARVAE (fig. 7.127): Small, 24 mm in total length; body depressed; mouth directed ventrally; nares and eyes dorsal; spiracle low, sinistral; vent tube dextral; tail moderate; fins low; tip of tail rounded; oral disk small, emarginate; serrate beaks and 2/3 rows of denticles present; A2 row with median gap; one row of large papillae, two rows laterally below mouth; papillae not continuous above mouth; body dark brown dorsally, grading to light brown ventrally; caudal musculature brown; some brown blotches on dorsal fin.

VOICE: Very low rasping trill followed by an interval, then repeated.

SIMILAR SPECIES: (1) *Phyllobates vittatus* is larger (adult males more than 21 mm in standard length, adult females more than 25 mm), always has blue green on the upper limb surfaces, and has broad light dorsolateral stripes. (2) *Eleutherodactylus gaigeae* is much larger (adults 26 to 44 mm in standard length) and lacks digital scutes and a light lip stripe.

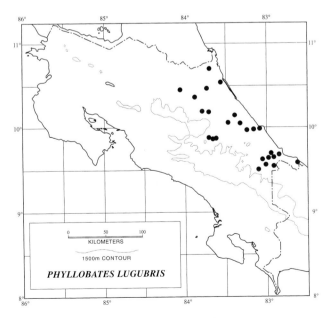

Map 7.124. Distribution of *Phyllobates lugubris*.

HABITAT: Lowland Wet and Moist Forests and marginally in Premontane Wet Forest.

BIOLOGY: These elegant little diurnal frogs are relatively infrequently encountered in the leaf litter. They move about primarily by a series of short hops interspersed with walking. Unlike the red *Dendrobates,* they show no strong territorial behaviors. In captivity (Zimmerman 1982; Weygoldt 1987) courtship lasts several days, and both partners are active, frequently touching one another. Egg clutches are not guarded, but the male moistens them while they are among dry leaves. The eggs hatch in nine to fourteen days. Several days later the male transports one or a few tadpoles (plate 222) at a time on his back to a water source. Transformation occurs about sixty days later, and the froglets are about 12 mm in standard length. They attain sexual maturity in about ten months.

REMARKS: Savage (1968b) included *P. vittatus* as a synonym of *P. lugubris,* but Silverstone (1976) demonstrated that both are valid species. Much of Savage's account refers to *P. vittatus,* especially the description of the voice and larvae. Ibáñez, Jaramillo, and Solís (1994) recorded this species in Panama well east of previously known localities. If the single specimen represents a naturally occurring population, it may prove not to be conspecific with *P. lugubris.* The karyotype is 2N = 24, composed of six large and six small pairs of chromosomes: seven pairs of metacentrics, three pairs of submetacentrics, and two pairs of telocentrics; NF = 44 (Rasotto, Cardellini, and Sala 1987).

DISTRIBUTION: Humid lowlands and marginally in the premontane zone of the Atlantic versant from extreme southeastern Nicaragua to northwestern Panama; a single record from just west of the Panama Canal (10–601 m).

Phyllobates vittatus (Cope, 1893)

(plate 223; map 7.125)

DIAGNOSTICS: A moderately common small striped frog with the upper limb surfaces mostly blue green and having broad gold, orange, or red orange dorsolateral stripes.

DESCRIPTION: Adult males 22.5 to 26 mm in standard length, adult females 26 to 31 mm; dorsal surfaces shagreened; head longer than broad; snout rounded, truncate from above; eyes large; finger I longer than finger II; subarticular tubercles ovoid, low; no supernumerary, accessory palmar, or plantar tubercles; thenar and palmar tubercles round, low, thenar much smaller; inner metatarsal tubercle oblong, outer round, subequal; a weak tarsal fold or tubercle present. Dorsum and head black; a complete broad gold, orange, or red orange dorsolateral stripe from dorsal base of thigh across upper eyelid along canthus nostralis to tip of snout, continuous across snout with stripe from other side of body and usually covering most of upper eyelid; a white stripe from above arm continuing on lip to under eye; venter and dorsal surface of limbs mostly blue green; undersides of limbs black with pale blue green marbling, forming a ventrolateral light stripe on each flank.

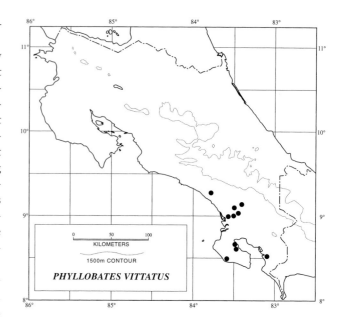

Map 7.125. Distribution of *Phyllobates vittatus.*

LARVAE (fig. 7.128): Moderate-sized, 30 mm in total length; body depressed; mouth ventral; nares and eyes dorsal; spiracle low, sinistral; tail moderate; fins moderate; tail tip rounded; oral disk small, emarginate; serrate beaks and 2/3 rows of denticles present; A2 row with median gap; two complete rows of oral papillae below mouth; papillae not continuous above mouth; body, tail, and fins uniform dark brown, lighter ventrally.

VOICE: A low rasping trill lasting 2 to 6 seconds, repeated after a pause (Savage 1968b; Silverstone 1976).

SIMILAR SPECIES: (1) *Phyllobates lugubris* is smaller (adult males under 22 mm in standard length, adult females under 25 mm), with narrow dorsolateral light stripes, and the upper limb surfaces are often yellow. (2) *Eleutherodactylus gaigeae* is usually larger (adults 26 to 44 in standard length) and lacks a light lip stripe.

HABITAT: Lowland Moist and Wet Forests, especially near streams.

BIOLOGY: This brief account summarizes information in Savage (1968b), Silverstone (1976), Zimmermann (1982), and Weygoldt (1987). This bright striped species is an obvious diurnal component of the forest floor fauna. Like other members of the genus, it preys on a variety of small arthropods, including ants. Locomotion involves walking mixed with a series of rapid hops. There appears to be no territorial behavior. Males produce the advertisement call as described above and emit a second call when courtship is in progress. The latter consists of a series of two to five chirps dropping in pitch, followed by another series after a pause. Courtship in captivity lasts from one to seven days and involves much moving about and tactile contact initiated by both partners. Fertilization is by vent-to-vent contact, with the two frogs facing away from one another. Eggs are usually laid on leaves above the ground. Clutch sizes are 7 to

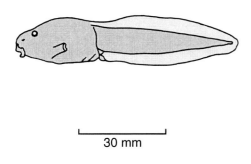

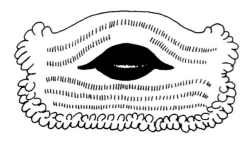

Figure 7.128. Tadpole of *Phyllobates vittatus.*

21 eggs, and captive females lay clutches every one or two weeks during the three-month breeding cycle. The eggs are not guarded, but the male moistens them one to three times a day. Eggs hatch in thirteen to seventeen days. Some few days later the male approaches the tadpoles, and one or more crawl onto his back. In nature he then carries them to a water source such as a small pool or a fallen palm petiole to continue development. Males carry one to thirteen tadpoles at one time, but not all the tadpoles from a clutch are transported together. Metamorphosis takes place about forty-five days later, and juveniles are about 13 mm in standard length. They reach sexual maturity in about ten months.

REMARKS: Savage (1968b) included this species in his concept of *P. lugubris.* Silverstone (1976) established the distinctiveness of the two forms. Savage's description of tadpoles and calls and notes on life history all refer to *P. vittatus,* not *P. lugubris.* In his account on *P. lugubris,* Silverstone (1976) inadvertently repeated Savage's description of the *P. vittatus* vocalization. Unfortunately, no adequate description of the call of the former species has been published.

DISTRIBUTION: Lowlands of the Golfo Dulce region of southwestern Costa Rica; expected to occur in immediately adjacent southwestern Panama (20–550 m).

Family Microhylidae Günther "1858b," 1859

The seventy genera (345 species) making up this family include a variety of terrestrial, scansorial, and arboreal frogs (with expanded digital tips), but some are toadlike and many forms are semifossorial, smooth skinned, short legged, and narrow headed. Most are small to moderate in size, but a few species reach 105 mm in standard length. All microhylids

have one to three transverse dermal folds running across the palate anterior to the pharynx, except for two genera (*Melanobatrachus,* of India and *Hoplophryne* of eastern Africa). This feature is found in no other group of anurans.

The family is further characterized by the following combination of attributes: eight holochordal, procoelous, presacral vertebrae, or eighth presacral vertebra biconcave and sacrum biconvex; no ribs; sacrum with broadly dilated diapophyses and bicondylar articulation to coccyx (fused in a few genera); coccyx lacking transverse processes; pectoral girdle firmisternal; usually no omosternum; a cartilaginous sternum; no parahyoid bone; tibiale and fibulare fused proximally and distally; intercalary cartilages in one genus; sartorius muscle distinct from semitendinosus, the tendon of the latter passing dorsal to the gracilis muscle; usually no teeth on upper jaw; tongue usually highly protrusible; pupil horizontal or round; amplexus usually axillary, but in robust species male adheres to posterior part of female; larva type S, no beaks (cornified beaks present in *Scaphiophryne* of Madagascar) or denticles (upper and lower jaws with cornified denticles in *Otophryne* of northern South America); a single posterior midventral spiracle (opening sinistral in *Scaphiophryne* and *Otophryne*) and nares not perforating until just before metamorphosis. The karyotype is 2N = 22 to 28.

Costa Rican frogs of this family are unlikely to be mistaken for those belonging to other families because of their rotund body, short limbs, and very short, narrow heads demarcated by a transverse fold just behind the small eyes. *Rhinophrynus dorsalis* has a similar habitus, but unlike that species (characteristics in parentheses) the native microhylids have a single large inner metatarsal tubercle (two large keratinized inner metatarsal tubercles) and the snout is covered by smooth skin (tip of snout covered with calluses).

The general morphology of Costa Rican microhylids and *Rhinophrynus* are typical of secretive, essentially fossorial forms that feed more or less exclusively on termites and ants. These anurans emerge aboveground for very brief periods during heavy rains to reproduce. The rest of the year they live in termitaria, or the tunnels of termites and ants. The head shape is an adaptation for breaking into the passageways of the food species and plugging them, and the reduced eyes are a correlate of their lightless, subterranean existence. These frogs give off mucous secretions that apparently protect them from attack by their potential prey.

The highly specialized feeding mode of microhylids is unique among anurans. In common with other group D frogs, most have highly protrusible tongues (a few genera in this family have fully attached tongues). In all other frogs and toads having this feature, the head and body are aimed at the prey. Microhylids, however, can control the lateral direction of tongue flipping through an angle of at least 90°, independent of the orientation of the head and body, aiming the tongue, not the animal as a whole, at the food item.

The family has an extensive circumtropical distribution

(Africa, Madagascar, Asia, northern Australia, and the Americas), with some temperate zone representation in East Asia and the United States.

The classification of the Microhylidae remains controversial, since there is reason to suspect that the system proposed by Parker (1934) is based in a number of cases on convergent and shared primitive features (Savage 1973d). Duellman and Trueb (1986, 1994) criticized the modifications suggested by Savage and essentially retained Parker's scheme as followed by Zweifel (1971). The subfamilal classification offered below attempts to balance these differing views and follows Blommers-Schlösser (1975), Wassersug (1984), and Wassersug and Pyburn (1987) in grouping the Madagascar *Scaphiophryne* with the microhylids:

Scaphiophyrinae Laurent, 1946 (two genera): a single sphenethmoid bone; vomer not divided; edentulous; no intercalary cartilages; eighth presacral vertebra biconcave; sacrum biconvex; palatal folds present; middle ear and Eustachian tubes present; larvae free-swimming (Madagascar)

Dyscophinae Boulenger, 1882 (two genera): sphenethmoids paired; vomer not divided; upper jaw and vomer dentate; eighth presacral vertebra biconcave; sacrum biconvex; palatal folds present; middle ear and Eustachian tubes present; larvae free-swimming (Madagascar, Southeast Asia)

Cophylinae Cope, 1889 (seven genera): sphenethmoid bones usually paired (single in one genus); vomer divided into two parts, with anterior bordering choana, posterior underlying palatine; teeth present on upper jaw, upper jaw and vomer, or vomer only, or jaw and vomer edentulous; no intercalary cartilages; all vertebrae procoelous; palatal folds present; middle ear and Eustachian tubes present; eggs laid in nest in litter, hatching into nonfeeding tadpoles that complete development in nest (pattern 19, table 7.1) (Madagascar)

Asterophyrinae Günther, "1858b," 1859 (eighteen genera): sphenethmoid bones paired; vomer not divided; usually edentulous; vertebrae procoelous or eighth presacral biconcave and sacrum biconcave; palatal folds present; middle ear and Eustachian tubes present; no larval stage (direct development) (Indonesia to the Philippines, New Guinea, and northern Australia)

Brevicipitinae Bonaparte, 1850 (five genera): no sphenethmoid bones; vomer not divided; edentulous; no intercalary cartilages; eighth presacral vertebra biconcave, sacrum biconvex; palatal folds present; middle ear and Eustachian tubes present; no larval stage (direct development) (eastern and southern Africa)

Melanobatrachinae Noble, 1931 (three genera): sphenethmoid fused with parasphenoid bone; no vomer; edentulous; no palatal folds; all vertebrae procoelous; no middle ear, columella, or Eustachian tubes; eggs laid in water in axils or holes of plants, hatching into nonfeeding tadpoles that complete development in situ (pattern 6, table 7.1) (Tanzania and southern India)

Phrynomerinae Noble, 1931 (one genus): paired sphenethmoid bones; anterior (choanal) portion of vomer present, posterior portion absent; edentulous; intercalary cartilage between penultimate and terminal phalanges on all digits; eighth presacral vertebra biconcave, sacrum biconvex; palatal folds present; middle ear and Eustachian tubes present; free-swimming larvae (sub-Saharan Africa).

Otophryninae Wassersug and Pyburn, 1987 (one genus): paired sphenethmoid bones; vomer small, posterior portion lost; edentulous; no intercalary cartilage; eighth presacral vertebra biconcave, sacrum biconvex; palatal folds present; middle ear and Eustachian tubes present; free-swimming larvae with an elongate sinistral spiracle (siphon) (northern South America)

Microhylinae Günther, "1858b," 1859 (nine genera): paired sphenethmoid bones (absent in one genus); vomer small, usually divided and posterior portion lost; edentulous; no intercalary cartilages; eighth presacral vertebra usually biconcave and sacrum biconvex, all vertebrae rarely procoelous; palatal folds present; middle ear and Eustachian tubes present; usually with free-swimming larvae, two genera lay eggs in nests, which hatch into nonfeeding larvae and complete development there (pattern 19, table 7.1) (United States to Argentina and Bolivia, temperate East Asia to India and Sri Lanka, Indonesia, and the Philippines)

All Costa Rican species belong to the Microhylinae and have free-swimming, suspension feeding tadpoles (ecomorph 4, table 7.2).

Genus *Gastrophryne* Fitzinger, 1843

This genus contains five species of small to moderate-sized frogs. It shares the characteristic specialized habitus and physiognomy described in the family account above with all other American genera included in the subfamily. Externally, the following combination of features occur in the genus: tympanum hidden, tips of digits flattened dorsoventally or not, one or two metatarsal tubercles present, and toes with at most basal webbing. None of these features distinguish all species of *Gastrophryne* from those of the allied genera *Hypopachus* and *Nelsonophryne*. The following combination of internal and larval features, however, is distinctive. Skeletal: pectoral girdle lacking clavicles and procoracoids; no palatine or postchoanal portion of vomer; maxillary in contact with quadratojugal; terminal phalanges simple; larval: vent and spiracle juxtaposed; upper lips with pendant flaps covering lower lips; flaps without scallops or papillae. The single Costa Rican species of *Gastrophryne* differs from *Hypopachus* by having only one metatarsal tubercle (versus two in all species of the latter genus) and from *Nelsonophryne*, which has moderate toe webbing and unflattened toe tips.

The karyotypes of two North American species, *Gas-*

trophryne carolinensis (Morescalchi 1968) and *G. olivacea* (Bogart and Nelson 1976), are known. Both have 2N = 22 and NF = 42; there are six pairs of macrochromosomes, of which four are metacentrics and two are submetacentrics plus four pairs of metacentrics and one telocentric pair of smaller chromosomes.

The genus ranges from the southern United States south on the Atlantic slope to Costa Rica and from southern Arizona to El Salvador on the Pacific versant.

Gastrophryne pictiventris (Cope, "1885," 1886)

(plate 224; map 7.126)

DIAGNOSTICS: This species is instantly recognizable by having a vivid ventral pattern of large white spots or blotches on a dark brown ground combined with a smooth skin.

DESCRIPTION: A small to moderate-sized species (adult males 25 to 31 mm in standard length, adult females 27 to 38 mm), body rotund, limbs short, skin smooth above and below with many scattered pustules dorsally; head small, narrow, triangular in dorsal outline; snout subelliptical in dorsal outline, protruding in profile; nostrils lateral, almost at tip of snout, far distant from orbit; eyes small; pupil round; no vomerine teeth; single external subgular vocal sac and paired vocal slits in males; arms short; fingers with single subarticular tubercles; no finger webbing; thenar and palmar tubercles present, otherwise palm smooth; tips of outer toes dorsoventrally flattened; a rounded, low inner metatarsal tubercle and single subarticular tubercles present; no plantar tubercles or tarsal fold; basal toe webbing; webbing formula: I 1–2½ II 2–3½ III 3–4 IV 4–2½ V; gravid females have pericloacal spicules; breeding males have spicules on chin. Dorsum light brown to grayish brown, uniform or with faint inverted V-shaped dark mark; dark brown to

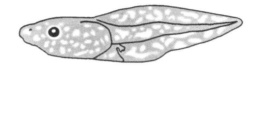

Figure 7.129. Tadpole of *Gastrophryne pictiventris*.

black lateral band from tip of snout above nostril through middle of eye above shoulder to groin, bordered above by light pinstripe; obscure, narrow transverse dark bars on upper surface of femur; venter dark brown with vivid pattern of large white spots or blotches; iris light brown.

LARVAE (fig. 7.129): Small, total length 20.5 mm at stage 33; no beaks or denticles; upper lips with smooth pendant flaps; mouth terminal, small; nares dorsal; eyes lateral; spiracle opens ventral to median vent tube; tail short; fins moderate; tail tip bluntly pointed; dorsum brown, uniform or mottled with light; lateral and anterior ventral areas brown, mottled with white; caudal musculature brown with an irregular white stripe suggested; fins transparent, mottled with brown and white.

VOICE: A single prolonged nasal "baa" or "whaaa" (Nelson 1973).

SIMILAR SPECIES: (1) *Hypopachus variolosus* has two raised shovel-like metatarsal tubercles. (2) *Nelsonophryne* has a uniform yellowish brown venter. (3) *Rhinophrynus dorsalis* has two shovel-like metatarsal tubercles and a uniform gray venter. (4) *Physalaemus pustulosus* has the dorsum covered with large wartlike pustules.

HABITAT: An inhabitant of Lowland Moist and Wet Forests, rarely found except during the breeding season.

BIOLOGY: *Gastrophryne pictiventris* is a secretive forest floor species that is present in considerable numbers in the leaf litter at all times of the year. These frogs are primarily jumpers but may hop when foraging for food. Although it was long thought to be a truly rare frog, studies at the La Selva Biological Station showed it to be among the five most common species in the leaf litter herpetofauna (Lieberman 1986), with a density of 145/ha. This species forages at night and hides in the leaf litter during the day. It favors undisturbed forest sites and is not likely to be found in open or secondary growth situations. This species is a classic ex-

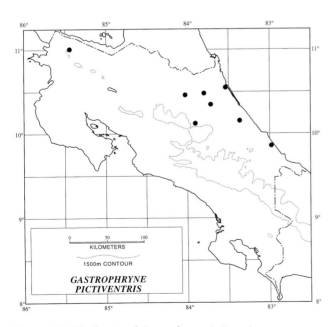

Map 7.126. Distribution of *Gastrophryne pictiventris*.

ample of leaf resemblance (Cott 1940). The color, depressed body, dorsal pattern, and broad lateral dark band conspire to create the illusion of a dead leaf and its shadow when the frog is motionless in the surface litter. Food consists largely of ants (85%), with beetles (4%) and termites (6%) also well represented. An average of 9.8 prey were present in the twenty-four stomachs examined. These frogs eat relatively few but large prey. Ants up to 9 mm in length (mean = 4.3 mm) and an earthworm 18 mm in standard length were significant large prey items. *G. pictiventris* is an explosive breeder that congregates at the onset of the rainy season around shallow temporary ponds, flooded areas within the forest, and swales near the forest edge. Males float at the edge of the pond when calling, and the calls are synchronous. Chin pustules are present in breeding males, and gravid females have pericloacal pustules. Males also have a definite abdominal nuptial gland. Calling males have been noted in May. Eggs are laid as a surface film. Mature tadpoles were collected from the same pond in August.

REMARKS: For many years the members of the genus *Gastrophryne*, including *G. pictiventris*, were referred to the genus *Microhyla* following Parker (1934), Dunn (1949), and Taylor (1952a). Carvalho (1954) demonstrated that the two genera were distinct and restricted the use of *Microhyla* to Asian species, a course followed by subsequent workers. Nelson (1972) published a revision of the genus and presented a description based on a relatively small sample of *G. pictiventris* in that and a subsequent account (Nelson 1973).

DISTRIBUTION: Lowlands of southeastern Nicaragua and adjacent Atlantic slope Costa Rica (1–500 m).

Genus *Hypopachus* Keferstein, 1867

Sheep Frogs

Only two species are currently recognized in this genus (Nelson 1973, 1974). Externally they closely resemble some members of the genus *Gastrophryne* but differ in always having two raised shovel-like metatarsal tubercles, unflattened toe tips, and some toe webbing. Technically *Hypopachus* may be separated from *Gastrophryne* by having clavicles and procoracoid bones (absent in the latter). Other characteristics that in combination distinguish *Hypopachus* from other American microhylids include skeletal: no palatine or postchoanal portion of prevomer; maxillary in contact with quadratojugal; terminal phalanges simple; and larval vent and spiracle juxtaposed; upper lips with pendant flaps covering lower lips; flaps with scallops or papillae.

The sole Costa Rican *Hypopachus* (*H. variolosus*) differs most obviously from the co-occurring species of *Gastrophryne* and *Nelsonophryne* by having two metatarsal tubercles (one in the other two taxa).

Hypopachus is found from southern Texas to Honduras on the Atlantic versant with populations in the Great Lakes region of Nicaragua and the Meseta Central Oriental of Costa Rica and from southern Sonora, Mexico, to Costa Rica on the Pacific slope.

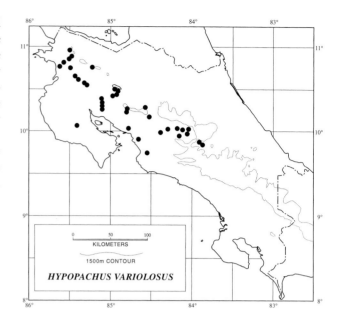

Map 7.127. Distribution of *Hypopachus variolosus*.

Hypopachus variolosus (Cope, 1866a)

(plate 225; map 7.127)

DIAGNOSTICS: This species is readily identified by its two enlarged metatarsal tubercles and anteriorly and laterally placed nostrils.

DESCRIPTION: A moderate-sized species (adult males 33 to 39 mm in standard length, adult females 30 to 53 mm); body rotund, limbs short; skin smooth, somewhat thickened on upper surfaces, with many scattered tiny pustules; head small, narrow, triangular in dorsal outline; snout subelliptical in dorsal outline; snout rounded in profile; nostrils lateral, almost at tip of snout, far forward of orbit; eyes small; pupil round; no vomerine teeth; single external subgular vocal sac and paired vocal slits in adult males; fingers with single subarticular tubercles; thenar and palmar tubercles present, otherwise palm smooth except for tiny pustules; no webbing between fingers; tips of toes not flattened; a pair of raised sharp metatarsal tubercles and single subarticular tubercles present; no plantar tubercles or tarsal fold; a thick basal web between toes; webbing formula: I 1–2½ II 1¼–3 III 2¾–4 IV 4½–2 V; gravid females have pericloacal spicules; breeding males have spicules or pustules on chin, edges of fingers and toes, and usually more pustules on dorsum; a large subcircular nuptial gland on the chest and anterior abdomen also present in males. Dorsum purplish to lavender brown to light chocolate; an irregular pair of dark blotches originate in the groin and variously extend onto dorsum; several light areas indicated by complete stripes, bands, or rows of dots from occiput to groin, eye to groin, and groin along anterior thigh surface; other hairline light markings down midorsum, on legs, throat, arms, and venter poorly developed or absent in Costa Rican examples (present elsewhere in the species range); postorbital dark stripe from

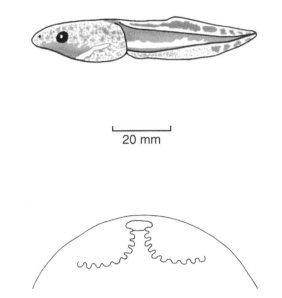

20 mm

Figure 7.130. Tadpole of *Hypopachus variolosus*.

behind eye continuing posteriorly above arm to groin, bordered anteriorly by light stripe; legs with narrow wavy transverse dark bars on upper surface; venter usually dusky brown with many cream spots (uniformly light in some extralimital populations); vocal sac blackish; iris black.

LARVAE (fig. 7.130): Moderate-sized, total length 28 mm at stage 36; no beaks or denticles; scalloped or papillate pendant flaps on upper lip; mouth terminal, small; nares dorsal; eyes lateral; spiracle opens ventral to median vent tube; tail moderate; fins moderate; tail tip bluntly pointed; dorsum nearly uniform brown, slightly lighter laterally; venter brownish with irregular light spots; tail musculature and fins spotted and marbled with brown; lateral light stripe along base of tail.

VOICE: The advertisement call of this species is a loud single prolonged bleat (Nelson 1966a). The duration is about 2 to 3.6 seconds and the dominant frequency is 2.05 to 2.9 kHz in the Costa Rican population (0.8 to 7.2 seconds and 1.05 to 3.6 kHz overall, respectively (Nelson 1973; Lee 1996).

SIMILAR SPECIES: (1) *Gastrophryne pictiventris* has a vivid ventral pattern of white spots on a dark background and only an inner metatarsal tubercle. (2) *Nelsonophryne aterrima* is uniform black above and has only an inner metatarsal tubercle. (3) *Rhinophrynus dorsalis* has two large and sharp inner metatarsal tubercles, a callused snout, and dorsally placed nostrils.

HABITAT: This secretive but relatively common species is found under rocks, logs, and other debris in Lowland Dry and Moist Forests and Premontane Moist and Wet Forests when it is aboveground. During most of the year it leads a subterranean existence.

BIOLOGY: A nocturnal, fossorial species that lives in burrows, hollows under trees, and other cavities. Locomotion

in primarily by hopping. This is a "sliding" backward burrower that uses the modified metatarsal tubercles for digging. Once a burrow reaches a sufficient depth, the opening closes over at the surface. During the wet season the burrows are usually only a short distance below the surface, but at other times of the year they may be up to 1 m underground. The species tends to favor open areas dominated by grasses, so it is frequently found in disturbed situations, especially pastures. Occasional examples are dug up or plowed up during agricultural work (Nelson 1966a). Food consists primarily of ants, along with a few other kinds of arthropods. *Hypopachus variolosus* is an explosive breeder that primarily uses pools, marshy sites, or standing water in pastures and agricultural fields. Breeding usually commences shortly after the first heavy rains of the wet season. Breeding males have spicules or pustules on the chin, and gravid females have pericloacal spicules. Early on the males, like several *Leptodactylus*, call from excavations in the water near shore. As densities increase males call from more exposed positions, usually resting their hands on vegetation or other objects. At the peak of the breeding activity, with high male densities, males swim rapidly about the breeding site, call incessantly, and indiscriminately clasp, at least momentarily, each other or any other frog, toad, or other object that might be a female. The synchronous choruses produce rhythmic bursts of sound. When amplexus occurs, the nuptial gland secretions cause the male's venter to adhere tightly to the dorsum of the female, and a vigorous struggle is required to separate them once breeding is complete. The 600 to 800 pigmented eggs (black upper pole, yellow lower pole) are deposited as a loose mass on the surface of the water. The egg and single gelatinous envelope are about 2.5 to 4 mm in diameter, and the egg is 1 mm. The eggs hatch in about twelve hours, and metamorphs appear about thirty days later. The principal predators on these anurans are the several frog-eating snakes, especially *Leptodeira* and *Coniophanes*, which seem undeterred by the heavy mucous secretions emitted by *Hypopachus*. These secretions are irritating to human skin, mucous membranes, and eyes and often stimulate sneezing spells when the frogs are handled or are in a collecting bag nearby. Rather than an antipredator defense, it seems more likely that the secretions discourage stings or bites by the principal prey, ants and termites, which the frogs attack in their underground tunnels, nests, and runaways.

REMARKS: Although a plethora of species had been described in this genus, Nelson (1973, 1974) recognized only two valid taxa for the twelve names then available. His analysis left unresolved whether the nominal upland form, *H. barberi*, was really distinct from *H. variolosus*, since no diagnostic feature unequivocally separates these population systems. The karyotype is 2N = 22 (León 1969).

DISTRIBUTION: Lowland and lower premontane zones from southern Texas to Honduras on the Atlantic versant and from Sinaloa, Mexico, to central Costa Rica on the

Pacific slope, including the Great Lakes region of Nicaragua and the Meseta Central of Costa Rica; in Costa Rica restricted to the northwest and the Meseta Central Oriental and Meseta Occidental (4–1,435 m).

Genus *Nelsonophryne* Frost, 1987

This genus contains only two species of moderate-sized frogs restricted in distribution to lower Central America and northwestern South America. External features characteristic for the genus are tympanum hidden, tips of toes not dilated, one inner metatarsal tubercle, and toes with some webbing. The following combination of internal and larval features will separate *Nelsonophyrne* from other American microhylids. Skeletal: pectoral girdle lacking clavicles and procoracoids; palatine present but postchoanal portion of prevomer absent; maxillary in contact with quadratojugal; terminal phalanges simple; larval: spiracle opens ventral to gut; no labial flaps.

The Costa Rican species of *Nelsonophryne*, *N. aterrima*, differs from *Hypopachus* in having a single metatarsal tubercle instead of two and from *Gastrophryne pictiventris* by having considerable webbing between the toes and undilated toe tips. *Nelsonophryne aequatorialis* of western Ecuador is the only other member of this genus, and unlike *N. aterrima* it has a dorsal pattern of contrasting dark markings on a lighter ground and only basal toe webbing.

Nelsonophryne aterrima (Günther 1900)

(plate 226; map 7.128)

DIAGNOSTICS: The essentially uniform, glossy black coloration of the upper surfaces combined with the moderately large size, short, narrow head, and definite toe webbing are distinctive.

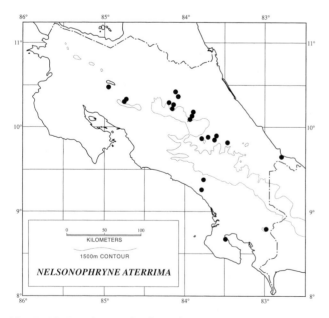

Map 7.128. Distribution of *Nelsonophryne aterrima*.

DESCRIPTION: A moderately large frog (adult males 43 to 61 mm in standard length, adult females 46 to 67); body oblong, no distinct waist; arms short; legs relatively long; skin thick, leathery, mostly smooth, but with some scattered pustules dorsally; head small, narrow, essentially triangular with subovoid snout from dorsal view; snout rounded in profile; nostrils lateral and near tip of snout, far anterior to orbit; eyes small, pupil round; no vomerine teeth; no vocal sac or slits; fingers with single flattened subarticular tubercles; thenar and palmar tubercles present, otherwise palm smooth; no finger webs; tips of toes not flattened; flat inner metatarsal tubercle, no outer tubercle; single flattened subarticular tubercles; no plantar tubercles or tarsal fold; toe webs and fringes present; toe webbing formula: I 2–2½ II 3–3 III 2½–3½ IV 3½–1 V; adult males have white spicules or pustules on chin, edges of fingers, toes, and webs; gravid females have pericloacal white pustules; whitish nuptial pad involving thumb, second finger, and third finger in males; a large nuptial gland on abdomen and chest in males. Upper surfaces black (purple brown in preservative); venter lighter, uniform brownish; iris black.

LARVAE (fig. 7.131): Large, total length 44 mm at stage 37; mouth terminal, small; no beaks, denticles, or labial flaps; nares dorsal; eyes dorsolateral; spiracle opens ventral to gut; vent tube medial; tail fin bluntly pointed; dorsum brown; venter transparent; caudal musculature white; fins transparent.

SIMILAR SPECIES: (1) *Gastrophryne* has a vivid ventral pattern of white spots on a dark background. (2) *Hypopachus variolosus* has two large metatarsal tubercles and a variegated or reticulate ventral pattern. (3) *Rhinophrynus dorsalis* has two large, sharp inner metatarsal tubercles and a callused snout.

VOICE: Not known; possibly lacks an advertisement call.

HABITAT: Usually found under fallen logs and other surface debris in Premontane Wet Forest and Rainforest but also known from Lowland Tropical Wet Forest.

BIOLOGY: A secretive, rarely encountered, nocturnal species that spends much time under surface litter on the forest floor. However, J. Campbell (pers. comm.) has collected this form at a depth of 800 mm in the ground. Gravid females have been collected from June to August, corresponding to the early part of the rainy season. The black-and-yellow eggs are apparently laid in shallow depressions in the forest floor. Fully developed oviductal eggs (without envelopes) are about 1.5 mm in diameter. Between 70 and 80 eggs are present in the oviducts. Tadpoles were collected in western Panama in October (Donnelly, de Sá, and Guyer 1990) along with a metamorph (11.3 mm in standard length). The smallest juveniles (21 mm in standard length) from Costa Rica were collected with gravid females in June and August.

REMARKS: Parker (1934), followed by Dunn (1949) and Taylor (1952a), referred this species to the genus *Microhyla*.

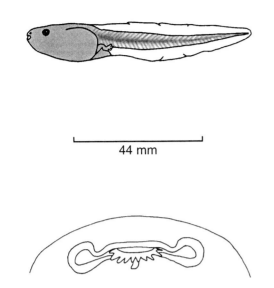

44 mm

Figure 7.131. Tadpole of *Nelsonophryne aterrima*.

Carvalho (1954) revived the generic name *Glossostoma* for the two species now placed in *Nelsonophyrne*. Frost (1987) proposed the generic name *Nelsonophyrne*, since *Glossostoma* is preoccupied by a name for a genus of flatworms. The karyotype is 2N = 26 (León 1969).

DISTRIBUTION: Lowland and premontane zones from Costa Rica to montane Colombia and northwestern Ecuador; in Costa Rica found primarily on the lower Atlantic slopes of the cordilleras and in the lowlands and on slopes of the southwestern region (20–1,350 m).

Family Ranidae Gray, 1825

A widely distributed and diverse family (fifty-one genera, 746 species) of mainly semiaquatic or terrestrial, usually long-legged, smooth to rugose-skinned frogs with substantial toe webbing. They vary in standard length from tiny (20 mm) to giant (300 mm). American and European members of the family all are typical in these features. Elsewhere in the family's range, although froggy species predominate, many are robust to toadlike (including the broad-headed carnivorous *Pyxicephalus* of Africa), and others are arboreal forms that superficially resemble treefrogs or scansorial members of the leptodactylid genus *Eleutherodactylus*.

The family is characterized as follows: eight holochordal presacral vertebrae, one to seven procoelous, eighth biconcave and sacrum biconvex; no ribs; sacrum with cylindrical diapophyses and bicondylar articulation to coccyx, which lacks transverse processes; pectoral girdle firmisternal; omosternum present, usually bony; sternum present, usually bony; no parahyoid bone; tibiale and fibulare fused proximally and distally; no intercalary cartilages; sartorius muscle distinct from semitendinosus, tendon of latter passing dorsal to gracilis muscle; teeth present on upper jaw; tongue highly protrusible; pupil horizontally elliptical; amplexus axillary; larva type A, keratinized beaks and denticles and single sinistral spiracle present. The karyotype is 2N = 16 to 20.

Costa Rican ranid frogs are sterotypical *Rana* in habitus, having narrow heads, long arms and legs, large eyes and tympanum, and considerable toe webbing. They are semiaquatic or terrestrial in habits and unlikely to be mistaken for any other species except members of the genus *Leptodactylus*, which completely lack toe webs. Other froglike forms have a ventral disk *(Eleutherodactylus)*, have paired scutelike flaps on the upper surface of the tip of the digits (Dendrobatidae), or lack a tympanum (no ventral disk, no flaps on digits, but a tympanum present in native ranids).

The family is well represented on the northern continents, Africa, Madagascar, and tropical Asia to the Philippines and Fiji, but only one species ranges into northeastern Australia, and three others are found in northern South America. This family has been classified into seven subdivisions following Dubois (1992), as modified by Ford and Cannatella (1993). Dubois's arrangement is based primarily on Clarke's (1981) research on African species and must be considered tentative until a broader study is undertaken. Since the status of the several groups remains questionable, as emphasized by Inger (1996), and since all American species belong to a single subfamily (the Raninae), I have not presented characterizations but merely list the subfamilies here.

Dicroglossinae Anderson, 1871 (thirteen genera): sub-Saharan Africa and Asia, south and east to the Philippines and Fiji Islands

Petropedatinae Noble, 1931 (thirteen genera): sub-Saharan Africa

Ptychadeninae Dubois, 1986, 1987 (three genera): Africa, Madagascar, and adjacent islands

Pyxicephalinae Bonaparte, 1850 (two genera): sub-Saharan Africa

Raninae Gray, 1825 (ten genera): Africa, Eurasia to northern Australia, the Americas to northwestern Peru, eastern Brazil, and northern Bolivia

Ranixalinae Dubois, 1986, 1987 (nine genera): India, Sri Lanka, and the Malay Peninsula

Tomopterninae Dubois, 1986, 1987 (one genus): India, Madagascar, and sub-Saharan Africa

As this list shows, ranids occur everywhere frogs are found except for New Zealand, most of Australia, and southern South America.

Members of this family usually deposit small pigmented eggs in lentic waters and have benthic (generalized) tadpoles, but there are many species with lotic-clasping larvae (ecomorph 11, table 7.2). One Asian genus *(Amolops)* has gastromyzophorous larvae (ecomorph 12, table 7.2), and some African forms have nonfeeding tadpoles that complete development in the terrestrial nest (pattern 19, table 7.1). Several Paleotropic genera lack a larval stage and have direct development.

Subfamily Raninae Gray, 1825

Technically this subfamily may be defined by the following suite of features: skull not exostosed; nasals reduced, not in contact with one another, moderately to widely separated; otic plate of squamosal narrow; palatines present; parasphenoid narrow, anterior tip rounded; omosternum forked or not; clavicles straight, not in contact with anterior margin of coracoid; sternal style short, broad; eighth vertebra not fused to sacrum; distal end of terminal phalanges knoblike or notched.

Genus *Rana* Linné, 1758

Members of this large genus (222 species) are definitive frogs, narrow waisted, long limbed, smooth skinned and narrow headed with substantial toe webbing. The snout is usually pointed, the nares are lateral, the eyes are large and have horizontal pupils, and the tympanum is large and distinct. Most species have smooth glandular dorsolateral folds. *Rana* are generally large, but adults range from 35 mm to 203 mm in standard length. They are distinguished from other genera in the subfamily by the following combination of features: base of omosternum not forked; anterior process of vomer long, passing dorsal to the maxilla-premaxilla articulation, and anterior end of maxilla convex; larvae with dextral vent and emarginate oral disk. The presence of the large tympanum, toe webs, and dorsolateral folds and the absence of finger disks and digital dermal scutes will distinguish species of this genus from all other Costa Rican frogs.

The genus as understood at present is found in Africa except for the Saharan region, throughout Eurasia, south through India, Southeast Asia, and the East Indies to the Philippines, New Guinea, and northern Australia, and in North, Central, and South America to extreme northwestern Peru, eastern Brazil, and northern Bolivia.

The considerable number of species contained in *Rana* has frustrated efforts to establish intrageneric relationships and a well-supported infrageneric classification. Dubois (1992) recognized thirty-three subgenera within *Rana* based at best on insufficient evidence (Inger 1996). Hillis (1985) and Hillis and Davis (1986) recognized nine species groups in the Americas using morphological, allozyme, and DNA data. Two of these, the *Rana berlandieri* and *Rana palmipes* species groups, are represented in Costa Rica.

Frogs of this genus are usually diurnal and use leaping for rapid movement.

KEY TO THE COSTA RICAN SPECIES OF THE GENUS *RANA*

1a. Posterior surface of thigh uniform or mottled; no grooves around margins of toe tips (fig. 7.132b–d) 2

1b. Posterior surface of thigh with large yellow spots or bars; well-defined toe disks demarcated by marginal grooves (fig. 7.132a); an inner and a small outer metatarsal tubercle . . . . *Rana warszewitschii* (p. 404)

2a. Tympanum equal to or larger than eye; posterior surface of thigh heavily mottled; no red in coloration in life; toes pointed or enlarged; only an inner metatarsal tubercle 3

2b. Tympanum smaller than eye; posterior surface of thigh uniform red in life, with a few faint small black flecks or light areas, uniform in preservative; ventral areas red in life; toe tips enlarged; two metatarsal tubercles, outer tiny, obscure (fig. 7.132b) . . . . *Rana vibicaria* (p. 403)

3a. Webs not reaching base of toe pads on any toes, at least one full phalanx free of web on all toes (fig. 7.132d);

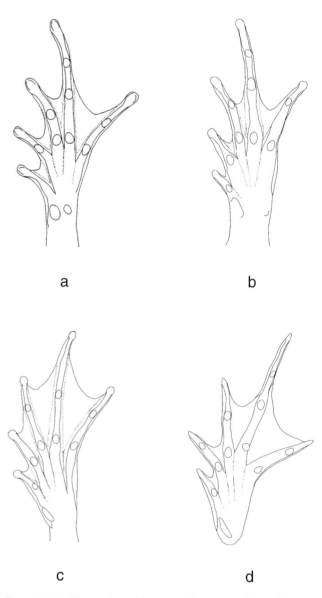

Figure 7.132. Plantar view of the feet of Costa Rican frogs of the genus *Rana*: (a) *R. warszewitschii*, note lateral grooves on toe disks; (b) *R. vibicaria*; (c) *R. vaillanti*, note complete webbing and single inner metatarsal tubercle; (d) *R. taylori*, note pointed toe tips.

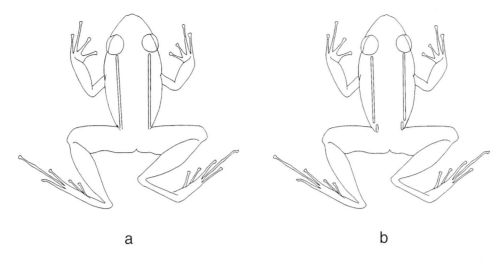

a   b

Figure 7.133. Dorsal view of Costa Rican ranid frogs: (a) paired dorsolateral folds continuous; (b) paired dorsolateral folds discontinuous and offset posteriorly.

dorsum smooth or rugose, usually covered by large blotches outlined by lighter color 4

3b. Webs on toes I–II–III–V reaching to, or near, base of toe pad on at least one margin of toe (fig. 7.132c); dorsum denticulate, uniform, or with moderate to small dark spots, particularly posteriorly. . . . . *Rana vaillanti* (p. 402)

4a. Dorsolateral folds usually continuous or nearly so (fig. 7.133a); vocal sacs darkly pigmented in adult males; large Müllerian ducts present in adult males . . . . *Rana forreri* (p. 399)

4b. Dorsolateral folds usually discontinuous posteriorly and posterior portion usually inset medially (fig. 7.133b); vocal sacs not pigmented in adult males; no Müllerian ducts in adult males. . . *Rana taylori* (p. 400)

Rana berlandieri Group

Characterized by tips of toes obtusely pointed, not expanded; skin smooth or rugose, never denticulate; sacral and presacral vertebrae not fused. In addition to the Costa Rican forms the other species in this group are (Hillis 1988): *Rana berlandieri* (Texas to Nicaragua), *R. magnaocularis* (western Mexico), *R. miadis* (Corn Islands, Nicaragua), *R. neovolcanica* (western trans-Mexican volcanic belt), *R. spectabilis* (eastern trans-Mexican volcanic belt and Sierras Oriental and Madre del Sur), *R. taloci* (Valle de Mexico), and *R. yavapaiensis* (southwestern United States and northwestern Mexico). The group as a whole ranges from Texas and the southwestern United States to western Panama (Hillis 1988).

Rana forreri Boulenger, 1883a

(plate 227; map 7.129)

DIAGNOSTICS: A large, long-legged brownish, greenish gray, to metallic green frog with a distinct, essentially con-

tinuous pair of dorsolateral folds (fig. 7.133a), large light-outlined, sepia-colored dorsal spots, and webbed feet.

DESCRIPTION: Adult males 65 to 90 mm in standard length, adult females 85 to 114 mm; skin of upper and ventral surfaces smooth, sometimes with prominent smooth tubercles or short ridges between smooth dorsolateral folds; head longer than broad, snout subelliptical from above; tympanum large, about equal in diameter to eye, not sexually dimorphic; paired round vocal slits and dark-colored paired external lateral vocal sacs in adult males; sacs lying in a slit under the tympanum at margin of lower jaw when deflated; finger I longer than finger II; fingers obtusely pointed; subarticular tubercles round; no supernumerary, accessory palmar, or plantar tubercles; no lateral ridges on fingers; brown-

Map 7.129. Distribution of *Rana forreri.*

ish nuptial pad on outer surface of thumb base and forelimb hypertrophied in large adult males; thenar tubercle elongate, palmar cordate, bifid to double; tips of toes obtusely pointed, not expanded; toes extensively webbed, webs between toes II–IV–V deeply incised; webbing formula: I 1–1½ II 2–2¾ III 1½–3 IV 3–1¼ V; inner metatarsal tubercle elongate, no outer tubercle; a low inner tarsal ridge; tarsal surfaces smooth. Dorsum green gray with large dark spots; no dark eye mask; supralabial light stripe incomplete; posterior thigh mottled dark and light; upper surfaces of limbs marked like dorsum, sometimes with only small dark spots; undersurface white (venter yellow extralimitally), with some melanistic patches; iris gold above, black below.

LARVAE (fig. 7.134): Giant, 82 mm in total length; body ovoid; mouth ventral; eyes and nares dorsal; spiracle midlateral, near level of eye, sinistral; vent tube dextral; tail moderate, tail fins deep; tail tip bluntly pointed; oral disk small, emarginate, with beaks and 2/3 rows of denticles present; large gap in A2 above mouth; beaks finely serrate; one row of papillae around lower half of disk, one row on upper lateral portion, none above mouth; body brown, tail including fins heavily pigmented with large spots.

VOICE: A short guttural trill usually composed of two notes each lasting about 2.5 seconds; dominant frequencies are 0.7 to 1 kHz, and the pulse rate is 21 to 24 per second (Frost 1982).

SIMILAR SPECIES: (1) *Rana taylori* has the dorso-

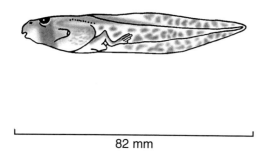

82 mm

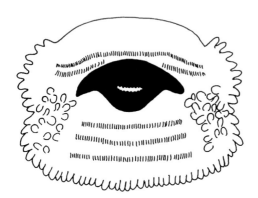

Figure 7.134. Tadpole of *Rana forreri*.

lateral folds discontinuous posteriorly and inset medially (fig. 7.133b). (2) Species of *Leptodactylus* lack toe webs.

HABITAT: Ponds and marshy situations in Lowland Dry Forest and marginally into the Lowland Wet and Premontane Forest zones.

BIOLOGY: This common species is active both during the day and at night. Males call while floating on the water surface, and the oval paired lateral vocal sacs are very evident. Mating takes place at night in the water during the wet season (May to November), frequently in temporary ponds. The black-and-white eggs are laid in a plinth containing about a thousand that is usually attached to submerged plant parts or debris. The frogs escape from intruders by leaping into the water and swimming rapidly to hide under objects on the bottom. If they are already in water, they use a similar rapid swim and dive. Small juveniles are 45 mm in standard length.

REMARKS: This species is a member of the *Rana pipiens* complex (Hillis 1988) and is one of at least two Costa Rican forms subsumed under the name *R. pipiens* by previous workers (Taylor 1952a; Savage 1973b, 1976, 1980b; Savage and Villa 1986). Although it was once regarded as a race of *Rana berlandieri*, Frost (1982) and Hillis (1985) recognized *R. forreri* as distinct because of reproductive incompatibility with *R. pipiens*.

DISTRIBUTION: Mostly subhumid and semiarid lowlands and marginally into more humid premontane areas from southern Sonora, Mexico, along the Pacific versant to central Costa Rica, where it is found at a few humid lowland localities (12–840 m).

Rana taylori Smith, 1959

(plates 228–29; map 7.130)

DIAGNOSTICS: A large, relatively short legged brownish to gray green frog having distinct but discontinuous dorsolateral folds with the posterior portion offset medially (fig. 7.133b), large, often elongate, light-outlined dorsal spots, and webbed feet (fig. 7.132d).

DESCRIPTION: Adult males 61 to 78 mm in standard length, adult females 71 to 88 mm; skin of upper surfaces and venter smooth, sometimes with prominent smooth tubercles and short ridges between smooth dorsolateral folds; head longer than broad; snout pointed from above; paired round vocal slits and paired light-colored lateral external vocal pouches in adult males; sacs lying in a slit under the tympanum at edge of lower jaw when deflated; tympanum large, about equal in diameter to eye, not sexually dimorphic; finger I longer than finger II; fingers obtusely pointed; subarticular tubercles round; no supernumerary, accessory palmar, or plantar tubercles; no lateral ridge on fingers; brownish nuptial pad on outer surface of thumb base and forearm hypertrophied in large adult males; thenar tubercle elongate; palmar irregular, tending to bifid or trifid; tips of toes obtusely pointed, not expanded; toes extensively webbed;

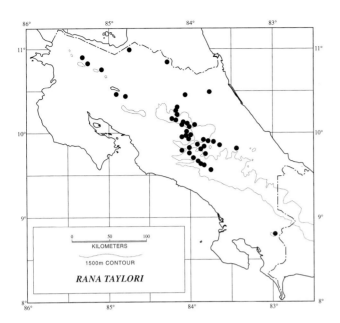

Map 7.130. Distribution of *Rana taylori*.

webs between toes II–III–IV deeply incised; webbing formula: I 1–1½ II 2–2¾ III 1½–3 IV 3–1¼ V; inner metatarsal tubercle elongate, no outer tubercle; a low inner tarsal ridge; surface of tarsus smooth. Dorsum and upper surface of limbs gray tan with large, often elongate dark spots; no dark eye mask; supralabial light stripe incomplete; posterior thigh surface mottled dark and light; undersurface white; iris gold above, brown below.

LARVAE (fig. 7.135): Giant, 82 mm in total length; body ovoid; mouth ventral; eyes and nares dorsal; spiracle midlateral, near level of eye, sinistral; vent tube dextral; tail moderate; tail fins deep; tail tip bluntly pointed; oral disk small, emarginate, with beaks and 2/3 rows of denticles, sometimes a third short anterior row lateral to beak; large gap in A2 above mouth; beaks finely serrate; one row of papillae around lower half of disk, one row on upper lateral portion, none above mouth; body dark brown; tail including fins light brown with numerous light punctations.

VOICE: A short chuckle-like, guttural trill composed of two notes lasting about 3 seconds with a pulse rate of 16 to 18 per second; dominant frequencies 0.5 to 2.2 kHz (Greding 1972b).

SIMILAR SPECIES: (1) *Rana forreri* has essentially continuous dorsolateral folds. (2) Species of *Leptodactylus* lack toe webs.

HABITAT: Ponds, swamps, marshes, and backwaters in Lowland Wet Forest and Premontane Moist and Wet Forests and Rainforest.

BIOLOGY: This is a semiaquatic species that is active both during the day and at night. *Rana taylori* is primarily an upland species in Costa Rica. It is recorded only from scattered localities below 1,100 m on the Atlantic lowlands, and only a few individuals have been seen or collected at

these sites. The following observations consequently are based on upland *R. taylori*. Males call while floating on the water, and the oval paired lateral vocal sacs are very evident. Mating takes place at night during the wet season (May to November). The black-and-white eggs are laid in a plinth, containing about a thousand, that is usually attached to grass or reed stems or other submerged vegetation or bottom debris. These frogs usually jump into the water or swim under debris on the bottom to escape. Metamorphs are about 25 mm in standard length.

REMARKS: *R. taylori* is another member of the *Rana pipiens* (leopard or grass frog) complex. It was originally confused by Taylor (1952a) with *Rana maculata*, a member of the *Rana palmipes* species group, which does not occur south of Nicaragua. Smith (1959) named Taylor's three specimens *Rana taylori* (holotype: FMNH 103210; Cartago: Peralta). Savage (1973b, 1976, 1980b) and Savage and Villa (1986) used the name *R. "pipiens"* for this form and *R. forreri*. Although several authors used the name *R. taylori* for Atlantic slope *R. "pipiens,"* Hillis (1985) was the first to document differences establishing the specific status of the two Costa Rican forms as distinct from other species in the *R. pipiens* complex. The situation is complicated by the fact that the montane population of the *R. pipiens* complex from the Cordillera de Talamanca (1,700–3,200 m) probably represents an undescribed relatively small species distinct from other Costa Rican forms and another not yet described species from western Panama, which appears to range into extreme southwestern Costa Rica (Hillis 1988).

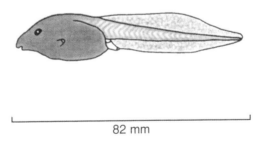

82 mm

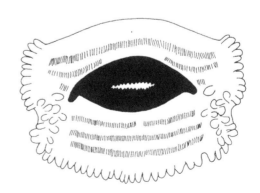

Figure 7.135. Tadpole of *Rana taylori*.

The Talamanca frogs will key out to *R. taylori*, but they are smaller (adult males 53 to 65 mm in standard length, adult females 60 to 75 mm) and occur at elevations between 1,700 and 2,950 m in Costa Rica. They are not included in the description of *R. taylori* above. The members of this complex from the Meseta Central (below 1,700 m) and Cordillera Central from 1,500 to 1,862 m are also problematic, but they are indistinguishable from typical *R. taylori* and so are included in that species here.

DISTRIBUTION: At scattered localities on the humid Atlantic lowlands from eastern Nicaragua to southeastern Costa Rica and in the humid premontane and lower montane areas of upland Costa Rica, including the Meseta Oriental and Meseta Occidental and probably the Cordillera Central (60–1,862 m).

Rana palmipes Group

Members of this group have the tips of the toes with the pad flattened or expanded and bordered laterally by a transverse groove, skin denticulate in some species, and sacral and presacral vertebrae fused. Besides the three Costa Rican representatives, the group includes *Rana bwana* (northwestern Peru and southwestern Ecuador), *R. juliani* (Belize), *R. maculata* (southern Mexico to Nicaragua), *R. sierramadrensis* (southern Mexico), and *R. palmipes* (lowlands of South America, east of the Andes). The group is strictly tropical in distribution and ranges from Veracruz and Oaxaca, Mexico, to northwestern Peru, eastern Brazil, and Bolivia.

The webbing formula given for species of the *palmipes* group in the Hillis and Sá (1988) revision differs markedly from those presented here. They apparently based their formula on a literal interpretation of Savage and Heyer's original formulation (1967) and counted only segments demarcated by a subarticular tubercle. These *Rana* and most other anurans lack definite subarticular tubercles under the joint between the penultimate and distal phalanges, and consequently Hillis and Sá counted these two segments as one. In addition, they did not take into account the revision in the formula proposed by Myers and Duellman (1982) whereby a zero is recorded when the web reaches the tip of the digital disk, not its base. Thus, for example, their formula I 0–0 II 0–1 III 0–1 IV 0–0 V would convert to I 1–1 II 1–2 III 1–2 IV 1–1 V under the revised system used in this book. See the section of key characters in this chapter for further details on webbing formulas.

Rana vaillanti Brocchi, 1877

(plates 230–31; map 7.131)

DIAGNOSTICS: A large frog with the dorsum between the prominent paired dorsolateral folds denticulate, tan to brown, usually with a greenish cast anteriorly, the denticles with white tips, a black border along the outer margin of the two folds, few if any dorsal dark spots, and the feet fully webbed (fig. 7.132c).

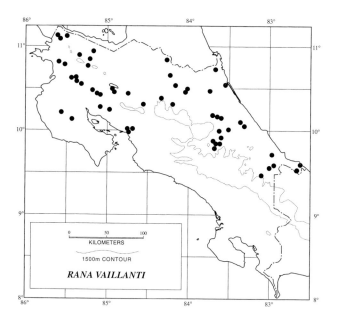

Map 7.131. Distribution of *Rana vaillanti*.

DESCRIPTION: Adult males 67 to 94 mm in standard length, adult females 76 to 125 mm; skin of upper surfaces denticulate; denticles with white tips; venter smooth; head longer than broad; snout pointed from above; tympanum large, equal to or larger than eye diameter, not sexually dimorphic; paired round vocal slits and paired internal subgular vocal sacs in adult males; finger I slightly longer than finger II; tips of fingers slightly swollen; subarticular tubercles oblong under fingers, definitely oblong under toes; no finger webs, but definite lateral ridge on each finger; no supernumerary, accessory palmar, or plantar tubercles; yellowish nuptial pad on dorsolateral surface of thumb and forearm hypertrophied in adult males; thenar tubercle elongate, palmar cordate to bifid; dorsal surface of shank with longitudinal rows of ridges of white-tipped denticles; tips of toes expanded; toes fully webbed; webbing formula: I 1–2 II 1–2 III 1–1½ IV 2–1¼ V; inner metatarsal tubercle elongate, no outer tubercle; a weak tarsal ridge but undersurface of tarsus denticulate. Dorsum brown to tan in adults, usually becoming green anteriorly; juveniles usually with green head and flanks; often with scattered dark brown markings posteriorly; dorsolateral fold marked by a black stripe laterally; no dark face mask; usually no supralabial light stripe, or present only posterior to eye; legs with transverse dark bars on upper surface; venter cream to yellow; iris gold with black flecks.

LARVAE (fig. 7.136): Giant, 80 mm in total length; body ovoid; mouth ventral; eyes and nares dorsal; spiracle low, near level of eye, sinistral; vent tube dextral; tail moderate; tail fins deep; tail tip rounded; oral disk moderate, emarginate, with beaks and 4/4 rows of denticles; A2 to A4 interrupted medially above or by mouth; P1 interrupted medially below mouth; beaks finely serrate; three lateral rows of

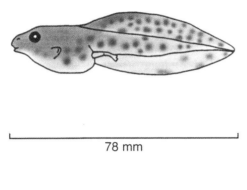

females, and 6.4% juveniles with little change throughout a year of sampling, suggesting year-round reproduction. If disturbed when ashore, *R. vaillanti* leaps into the water and hides on the bottom or under bordering or floating vegetation.

REMARKS: Many authors including Boulenger (1920), Taylor (1952a), Savage (1973b, 1976, 1980b), and Savage and Villa (1986) regarded this species as part of *Rana palmipes*. Hillis and Sá (1988) separated the two on a morphometric basis. In their paper the species are primarily distinguished by tibia length as a percentage of interorbital width: *R. palmipes* less than 600%, *R. vaillanti* more than 600%.

DISTRIBUTION: Lowlands of the Atlantic versant from southern Veracruz, Mexico, to northern Colombia and on the Pacific slope from Ecuador to western Panama and disjunctly in northwestern Costa Rica and western Nicaragua and southern Oaxaca and adjacent Chiapas, Mexico; in Costa Rica on the Atlantic versant and on the Pacific slope north of the Río Barranca, Puntarenas Province; unknown from the humid lowlands of southwestern Costa Rica and adjacent southwestern Panama (1–700 m).

Rana vibicaria (Cope, 1894a)

(plate 232; map 7.132)

DIAGNOSTICS: A large frog having the axilla, groin, and posterior thigh surface and often the undersurface of the thigh bright red, a black face mask, a definite light lip stripe, and webbed feet (fig. 7.132b).

DESCRIPTION: Adult males 60 to 73 mm in standard length, adult females 66 to 92 mm; skin of upper surfaces shagreened; venter smooth; head longer than broad; snout pointed from above; tympanum distinct, smaller than eye

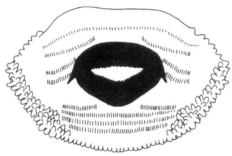

Figure 7.136. Tadpole of *Rana vaillanti*.

papillae, none above mouth; body dark above and laterally, with irregular dark spots; tail musculature and fins with large dark spots.

VOICE: A series of grunts each about 0.2 seconds long at intervals of 2 to 11 seconds. Each grunt has 5 or 6 pulses with a dominant frequency of 1 kHz (Greding 1972b, 1976; Lee 1996).

SIMILAR SPECIES: (1) *Rana vibicaria* has a substantial amount of red coloration on the hind limbs, and the dorsal surface of the shank is smooth. (2) Species of *Leptodactylus* lack toe webs.

HABITAT: Associated with lentic waters within the Lowland Dry, Moist, and Wet Forests and marginally in Premontane Rainforest.

BIOLOGY: A common semiaquatic frog usually seen near the water's edge either on land or resting in the shallows (Robinson 1983a). It is active both during the day and at night and often floats in the water among vegetation with only the head emergent. If you see a green-headed frog in any of these situations, it will be this species. *Rana vaillanti* is a sit-and-wait predator that eats a wide variety of prey including arthropods, small fishes, frogs (including small conspecifics), occasional birds, and possibly small mammals. Males call sporadically from floating vegetation or from shore during the day and more often at night during the reproductive season. Mating occurs at least from June to August. The calls appear to serve a spacing (territorial) function as well as attracting gravid females. In a study in a lake in Veracruz, Mexico, Rameríz, Vogt, and Villarreal (1998) found that the population contained 41.5% males, 52.4%

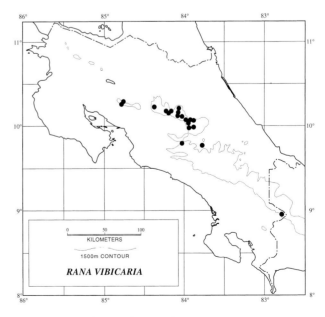

Map 7.132. Distribution of *Rana vibicaria*.

diameter, not sexually dimorphic; no vocal slits or vocal sac; finger I longer than finger II; tips of at least fingers II and III flattened; subarticular tubercles round to oblong on fingers, oblong on toes; no finger webbing or ridges; no supernumerary, accessory palmar, or plantar tubercles; light brownish nuptial pad on dorsolateral surface of thumb and forearm hypertrophied in adult males; thenar tubercle elongate, palmar low, cordate to bifid; dorsal surface of tibia smooth to shagreened; tips of toes expanded; toe webs extensive; webbing formula: I 1½–2⁺ II 1⁺–3 III 2–3¼ IV 3½–2½; inner metatarsal tubercle elongate; outer metatarsal tubercle round, tiny, and obscure; a weak tarsal ridge and underside of tarsus smooth. Upper surfaces gold, yellow brown, brassy to sooty brown to green, often with many small dark spots and/or blotches; juveniles usually green above; a black lateral border stripe along light dorsolateral folds; a black face mask and supralabial light stripe; no transverse dark bars on upper surface of limbs; axilla, groin, posterior surface of limbs, and often undersurface of thigh bright red; iris green.

LARVAE (fig. 7.137): Giant, 70 mm in total length; body ovoid; mouth ventral; eyes and nares dorsal; spiracle low, near level of eye, sinistral; vent tube dextral; tail moderate, tail fins deep; tail tip bluntly rounded; oral disk small, emarginate with beaks and 5–6/4 rows of denticles; A2 to A5 or A6 interrupted medially above or by mouth; P1 interrupted by slight gap below mouth; beaks finely serrate; three rows of papillae lateral to mouth, one row below, one row lateral to upper beak, none above mouth; body and tail brown, lighter below; tail fins heavily spotted with dark.

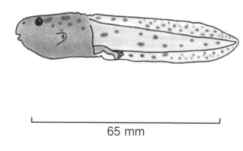

65 mm

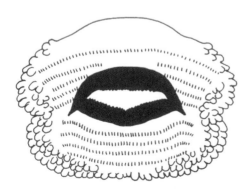

Figure 7.137. Tadpole of *Rana vibicaria*.

VOICE: A low harsh trill lasting 300 to 900 milliseconds composed of 9 to 30 pulses and usually given as two to five irregularly spaced calls 10 to 12 times a minute. Pulse rates are 26.6 to 30 per minute, and frequencies are spread over a 0.4 to 3 kHz range, but 0.8 kHz is dominant. One to three untrilled calls may follow a trilled one but are less frequent. The function of the second call is unclear, but it is similar in duration and frequencies to the trilled one (Zweifel 1964b; Greding 1972b).

SIMILAR SPECIES: (1) *Rana vaillanti* has the dorsum tuberculate, the upper surface of the shank with white-tipped longitudinal denticulate ridges, and no red in its coloration. (2) Species of *Leptodactylus* have toe webs.

HABITAT: Lower Montane Wet Forest and Rainforest and lower portions of Montane Wet Forest and Rainforest.

BIOLOGY: These semiaquatic forest frogs are locally rather common and may be active both during the day and at night. They prefer dense woods but may also be found near bodies of water in clearings or pastures. Calling and mating take place at night, but amplexus may continue into the daylight hours. Males generally call from vegetation in the water. The call is of very low amplitude because of the lack of a vocal sac. Choruses may include up to a hundred or so calling individuals, but even these are inaudible from more than 100 m. Breeding sites include shallow ponds or puddles or slow-moving seeps or backwaters of very small streams. The breeding season coincides with the early rainy season (May to July), although Zweifel (1964b) suggests there may be year-round reproduction. Eggs are laid in globular masses about 100 mm in diameter and attached to vegetation or the substrate. The black-and-cream eggs average 2.5 mm in diameter and have two jelly envelopes, so eggs with envelopes average 5.86 mm in diameter.

REMARKS: Villa (1988c) summarized information on this species based primarily on the literature.

DISTRIBUTION: Humid forests of the lower montane and lower portions of the montane belts in the Cordillera de Tilarán, Cordillera Central, and Cordillera de Talamanca of Costa Rica and western Panama (1,500–2,700 m).

Rana warszewitschii (O. Schmidt, 1857)

(plate 233; map 7.133)

DIAGNOSTICS: These handsome moderate-sized frogs have bright yellow spots or bars on the posterior thigh surface, expanded toe tips bordered laterally by a transverse groove, toe webs (fig. 7.132a), a black face mask and light supralabial stripe, and usually some green on the dorsum, with the undersurfaces of the limbs red.

DESCRIPTION: Adult males 37 to 52 mm in standard length, adult females 45 to 63 mm; skin of upper surfaces shagreened to weakly denticulate, denticles white-tipped when present; venter smooth; head longer than broad, very narrow, and pointed in dorsal outline; tympanum distinct, smaller than eye, not sexually dimorphic; no vocal slits or

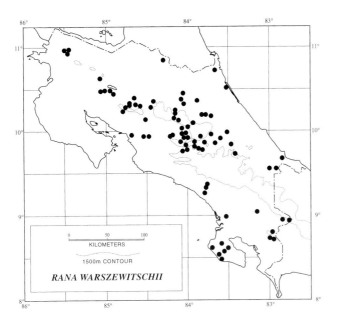

Map 7.133. Distribution of *Rana warszewitschii*.

sacs; finger I longer than finger II; tips of at least fingers II–III–IV slightly flattened; subarticular tubercles round under fingers except elevated, elongate one on thumb; oblong on toes; no finger ridges or webbing; no supernumerary, accessory palmar, or plantar tubercles; a yellowish nuptial pad on dorsolateral surface of thumb and forearm hypertrophied in adult males; thenar tubercle elongate, palmar cordate; dorsal surface of shank shagreened to weakly denticulate; tips of toes with expanded disks bordered laterally by a transverse groove; toes moderately webbed; modal formula: I 1⁺–2⁺ II 1½–2½ III 1½–3 IV 3–2 V; elongate inner and smaller round outer metatarsal tubercles; a weak short, tarsal ridge; underside of tarsus smooth. Upper surfaces golden tan to dark brown with a few to many green flecks that may be fused into larger green patches; juveniles mostly metallic green on back; flanks dark brown to nearly black; dorsolateral folds gold to pale yellow; black face mask continuous with dark flank color, bordered below by supralabial light stripe; transverse dark bars on upper surface of legs; posterior thigh with two to four large bright yellow spots or bars; similar yellow spots may be present in axilla; venter bright yellow to yellow orange becoming red on undersurfaces of legs in all but small juveniles; throat and venter heavily mottled with black in males; iris gold above and brown below.

LARVAE (fig. 7.138): Giant, 115 mm in total length; body robust; mouth ventral; eyes dorsolateral; nares dorsal; spiracle midlateral, sinistral; vent tube dextral; tail short; tail fins low; tip pointed; oral disk moderate, complete, with beaks and 6/4 denticle rows; A2–A6 interrupted medially or by mouth; P1 interrupted below mouth; upper beaks serrate; two rows of papillae laterally, one row below mouth,

none above; body uniformly dark; tail fins and musculature with large dark blotches.

VOICE: The call is a slow, low soft trill lasting 120 to 350 milliseconds, composed of 4 to 9 pulses; pulse rate 16.6 to 32 per second with frequencies spread over 0.016 to 2.08 kHz (Greding 1972b).

SIMILAR SPECIES: There are none.

HABITAT: Gallery forests in the Lowland Dry Forest zone, Lowland Moist and Wet Forests, Premontane Moist and Wet Forests and Rainforest into Lower Montane Wet Forest and Rainforest.

BIOLOGY: These rather common frogs are forest inhabitants, generally encountered near small streams. They are primarily diurnal and are found wherever patches of forest remain, even within large cities. Tadpoles are found in small streams and are among the most commonly collected Costa Rican anuran larvae.

REMARKS: The specific epithet was misspelled as *R. warschewitschii* by most workers before Hillis and Sá (1988), who reinstated the original correct spelling adopted here. Villa (1990a) presented a summary of available data on this species primarily based on Zweifel (1964b) and Hillis and Sá (1988).

DISTRIBUTION: Humid lowlands on the Atlantic versant from northeastern Honduras to central Panama, both slopes of the cordilleras of Costa Rica and western Panama, lowlands of southwestern Costa Rica, and eastern Panama and gallery forests in nonpeninsular northwestern Costa Rica (1–1,740 m).

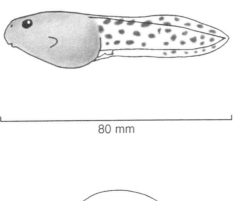

80 mm

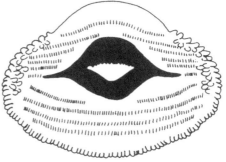

Figure 7.138. Tadpole of *Rana warszewitschii*.

3

Living Reptiles

8 | Reptiles
Class Reptilia Laurenti, 1768

As usually defined, this group contains those tetrapod vertebrates that cannot be considered amphibians since they have an amniotic egg, or birds since they lack feathers, or mammals since they lack hair and mammary glands. Recent study (Gauthier, Kluge, and Rowe 1988) indicates that the "reptiles" actually constitute two major evolutionary lineages. One, the Synapsida, includes a whole series of now extinct forms that is represented today by living mammals (class Mammalia). The second (class Reptilia) is composed of all living nonsynapsid amniotes, including the tuataras, lizards, snakes, turtles, and crocodilians and those volant reptiles usually called birds. The earliest remains of reptiles in the fossil record are from 300 million years ago in the Pennsylvanian period, and many major groups of now extinct animals (including ichthyosaurs, phytosaurs, pterosaurs, and dinosaurs) are also members of this class. The accounts that follow are restricted to descriptions and discussion of extant nonvolant reptiles, the several groups traditionally considered the subject of herpetology. The excellent field guide by Stiles, Skutch, and Gardner (1989) will provide interested readers with information on living Costa Rican volant reptiles.

The living nonvolant reptiles are frequently portrayed as a small company of survivors from the reptilian legions that were a dominant component of the earth's fauna during Mesozoic times (230 to 65 million years ago). Nevertheless, although many major stocks within the class are now extinct, several recent groups, especially lizards and snakes, are conspicuous inhabitants of a wide variety of ecological situations in the warmer regions of the world. The continuing success of the group is attested in that there are approximately 7,900 living species of these kinds of reptiles, placed in 999 genera, compared with about 4,700 living mammals. Living species of volant reptiles (Aves) number about 9,700.

Because of the great diversity of extinct reptiles and their radiation into a wide array of environments, today's remnant stocks are only distantly related to one another. In addition, the unique derived characters of soft anatomy and osteology that Gauthier, Kluge, and Rowe (1988) used to diagnose the Reptilia are shared with that subgroup known as birds. Consequently it is possible to distinguish nonvolant reptiles from other tetrapods only by a combination of shared primitive features and the absence of several derived characters found in birds and mammals. Nonvolant reptiles may be characterized as follows without recourse to the significant but relatively obscure internal features that technically diagnose the class as a whole:

1. Ectothermic: dependent on environmental heat sources (feature shared with amphibians; birds and mammals are endothermic); body temperature fluctuating, temperatures regulated by gross behavioral adjustments
2. Lower jaw composed of several bones (as in amphibians and birds; a single bone, the dentary, is found in mammals)
3. Middle ear region with a single ossicle, the columella (as in some amphibians and all birds; two middle ear ossicles, the operculum and columella, are found in most amphibians; three middle ear bones occur in mammals: the stapes [homologous to the columella], incus, and malleus)
4. Two systemic arteries (as in amphibians; a single systemic artery is found in birds and mammals)
5. Cleidoic eggs surrounded by a complex series of protective layers, including a leathery or calcareous shell in oviparous forms, and containing a rich food supply in the form of yolk (shared with birds and egg-laying mammals)
6. A series of protective extraembryonic membranes (serosa or chorion, amnion, and allantois) that appear during development (fig. 8.1) and are lost at hatching or birth (shared with birds and mammals)
7. No hair or feathers
8. No mammary glands

The essential differences between reptiles and amphibians are associated with features of their life history. Whereas amphibian eggs are virtually without protection against desiccation and must be deposited in water or very moist terrestrial situations, the eggs of reptiles are deposited on land, usually buried in the soil or rotting vegetation, out of direct sunlight. In most amphibians the eggs hatch into a gilled, free-swimming larval stage that later undergoes a rapid metamorphosis into a lunged terrestrial form (see the section on Amphibia for modifications of this general pattern). Reptiles have no stage with functional gills; either they hatch fully formed from the egg or, in a number of lizards and snakes, the egg is retained inside the mother's body during development and she gives birth to living young (is viviparous).

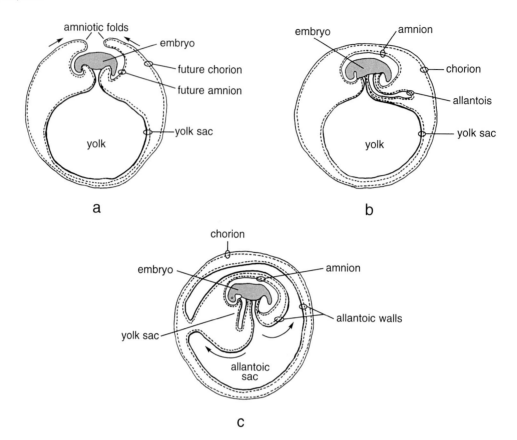

Figure 8.1. Development in an amniotic egg. Each extraembryonic membrane is bilaminar and composed of a layer of mesoderm (dotted line) and either ectoderm (solid line) or endoderm (heavy line). The amnion, chorion, allantois, and yolk sac are characteristic of reptiles and mammals. A yolk sac is also present during development in many fishes and some amphibians but is a different, quadrilaminar structure.

Although most amphibians and reptiles may be distinguished from one another by morphological features (especially skeletal), a number of fossil forms are transitional between the two classes. This fact, coupled with the great diversity of extinct groups within both stocks, makes comprehensive definition of the classes equivocal without reference to life history data. Fortunately, living reptiles may be distinguished from all living amphibians by sharing the following features:

1. Body covered with horny or bony scales or plates and/or modified into an encasing shell (the sea turtle *Dermochelys* lacks epidermal scales as an adult)
2. Skin dry to touch in life
3. Claws on digits (absent in *Dermochelys*)
4. Usually one occipital condyle (three in the snake *Bolyeria*)
5. Two or more sacral vertebrae in limbed forms
6. Teeth nonpedicellate
7. Ribs curved, long, extending into flank musculature

Living reptiles are placed in five orders (fig. 8.2). One of these, the Squamata, contains most extant species and is usually divided into two suborders. The classification followed here is based on DeBraga and Rieppel (1997):

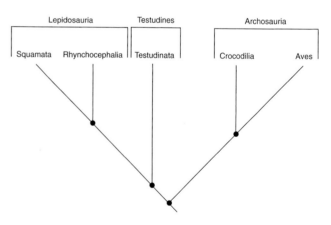

Figure 8.2. Phylogenetic relationships of the living orders of reptiles.

Class Reptilia Laurenti, 1768
 Subclass Eureptilia Olson, 1947
  Superorder Lepidosauria Haeckel, 1866
   Order Rhynchocephalia Günther, 1867: tuataras
   Order Squamata Oppel, "1810," 1811
    Suborder Sauria: lizards
    Suborder Serpentes: snakes

Superorder Testudines Linné, 1758
  Order Testudinata: turtles
Superorder Archosauria Romer, 1945
  Order Crocodilia: crocodilians
  Order Aves: birds

For convenience the Costa Rican reptiles (lizards, snakes, turtles, and crocodilians) are reviewed in separate sections in the order indicated. The order Rhynchocephalia, represented by two living lizardlike species, the tuataras (genus *Sphenodon*) of New Zealand, does not fall within the scope of this book.

Amphisbaenians (about 132 species) are known from Africa, Arabia, the Middle East, Spain and Portugal, South America, Florida (USA), western Mexico, the Greater Antilles (Cuba, Hispaniola, and Puerto Rico), and the Virgin Islands. Although sometimes placed in a separate suborder, these animals are best considered highly specialized lizards. One species, *Amphisbaena fuliginosa*, ranges from tropical South America to central Panama, and there is a possibility the genus might be found in Costa Rica. In that event, examples would key out to couplet 6b (Serpentes) in the key to the major groups of the Costa Rican herpetofauna (p. 102). Amphisbaenians are readily separated from all snakes by having large, quadrangular scales that form regular rings (annuli) around the body and tail (ventral scales much larger than dorsal scales, or dorsal scales rounded posteriorly, in Costa Rican nonmarine snakes).

Although all oviparous reptiles are characterized by having cleidoic eggs (Witschi 1956), these exhibit considerable variation in the nature of the outer protective covering or shell (Packard, Packard, and Boardman 1982). The shell may consist of as many as three layers: one or two flexible, fibrous ones adjacent to the albumin (egg white) and an outer rigid, calcareous one. The tuataras, most lizards, and as far as is known all snakes have flexible-shelled eggs either without a calcareous layer or with a very thin one. These eggs are strongly affected by the hydric environment and have higher hatching success and produce larger hatchlings when adequate moisture is available. Because of the uptake of moisture and growth of the embryo, a slight but definite increase in size takes place in this kind of egg during incubation. The flexible-shelled eggs laid by some turtles (Cheloniidae, Chelydridae, and most of the Emydidae) have a definite calcareous layer that is usually about the same thickness as the shell membrane it is attached to. These eggs are relatively independent of the hydric environment, although larger hatchlings are produced when adequate moisture is available. The rigid-shelled eggs of other turtles, a few lizards (all members of the Dibamidae and Gekkonidae), and crocodilians have a well-developed calcareous layer that is much thicker than the shell membrane, as in birds. These eggs are essentially independent of the hydric environment. Among living reptiles only some members of the order Squamata are viviparous.

Living nonvolant reptiles control their temperature by selective behavioral exploitation of thermal differences in the environment in order to maintain body temperature during activity within a narrow range. Postural adjustments, basking (heliothermy), differential exposure of parts of the body to solar radiation, or contact with the substrate (thigmothermy) and active selection of microclimatic differentials are typical behaviors that allow free-ranging animals to maintain a characteristic and restricted range of body temperatures (the activity temperature range). While much higher and lower temperatures may be tolerated or avoided by reduced activity before organisms incur lethal effects, the activity temperature range characterizes the body temperatures at which most routine activities are carried out. Recorded activity temperatures for turtles range between 8 and 37.8°C, with means ranging between 25 and 35°C; for the tuatara between 6.2 and 24.5°C, mean 12.8°C; for lizards between 11 and 46.4°C, means 15.8 to 40.5°C; for snakes 9.0 to 38°C, means 14.8 to 30.3°C; and for crocodilians about 23 to 35°C (Brattstrom 1965; Avery 1982).

Reptiles are covered by an integument that is essentially impermeable to water, and respiration is carried out primarily in the lungs, where carbon dioxide is replaced in the capillary blood by inhaled oxygen and then eliminated when air is exhaled. In reptiles (exclusive of turtles) the relatively long lungs lie in the body cavity. As the intercostal muscles elevate the elongate ribs the trunk cavity expands and air is sucked into lungs, which expand into the available space. Conversely, compression of the rib cage forces air out of the lungs. Since turtles cannot move their ribs, they use another system involving specialized abdominal or abdominal and pectoral muscles to expand and contract the lungs. In a few aquatic turtles auxiliary respiratory organs of highly vascularized tissue in the buccal, pharyngeal, or cloacal area allow for gaseous exchange with the surrounding water.

The sense organs that provide reptiles with information regarding the external environment are similar to those found in other amniotes. These include chemoreceptors (taste and olfaction); mechanoreceptors (tactile, equilibrium, gravity source, acceleration, and hearing); thermoreceptors; and photoreceptors. The lateral line system of mechanoreceptors found in fishes and in larval and most strictly aquatic adult amphibians does not occur in amniotes, whether aquatic or not. Because of the significant diversity among the major groups of reptiles in chemoreception, hearing, thermoreception, and photoreception, these features will be described under the individual major groups. Similarly distinctive aspects of locomotion, food and feeding, and reproductive biology will be treated in the same fashion.

9 | Squamates
Order Squamata, Oppel, 1811

LIZARDS AND snakes are the living representatives of this order, which first appeared in the fossil record about 245 million years ago (Upper Permian). The earliest forms were lizardlike creatures in which the diapsid skull condition (two temporal openings on each side, a quadratojugal-jugal arch and a rigid quadrate), typical of more primitive forms and today found in the tuatara and crocodilians, has been drastically modified. In primitive lizards the bony arch bordering the lower temporal opening was lost through loss of the quadratojugal bone so that the quadrate became movable to swing the lower jaw forward and raise the anterior portion of the skull during feeding. Further modification of this kinetic skull has allowed the development of specialized feeding habits in advanced lizards, amphisbaenians, and snakes and will be discussed later in appropriate sections of this book.

The distinctive features of the order are:

1. Skull with one or no temporal fossa; kinetic; no quadratojugal; jugal reduced or absent
2. Quadrate movable
3. No secondary palate
4. Paired penes (two hemipenes) in males
5. Egg tooth or teeth covered with enamel (a true tooth)
6. Urinary bladder present or absent
7. Teeth not socketed (pleurodont or acrodont)
8. Subclavian artery arising from the systemic artery

Lizards, amphisbaenians, and snakes share many biological features. All these groups except strictly aquatic forms use behavior to regulate body temperature. Heliothermic lizards and most snakes have preferred body temperatures between 28 and 35°C, but a number of lizards maintain activity temperatures of 38 to 42°C. Nocturnal, fossorial, and shade-loving squamates tend to have lower activity temperatures, with most in the 20 to 26°C range.

Squamates obtain water by drinking or licking moisture off plants or other objects when it is available. A number of species that occur in xeric environments are almost completely dependent on water of metabolism. Salt secretory glands are present in some lizards, a number of marine and estuarine snakes, and sea turtles (see details in sections on these groups). Nitrogenous wastes are excreted almost exclusively as insoluble uric acid.

All squamates are characterized by a distinctive pair of specialized organs of chemoreception that lie below the olfactory bulb and open through separate ducts into the roof of the mouth. These structures, the Jacobson's (vomeronasal) organs, are largest in snakes and monitor lizards and smallest among arboreal lizards. Snakes and many kinds of lizards practice tongue flicking, where the tongue is frequently protruded (often for long periods) and waved in the air or lightly pressed against the substrate. Chemical particles obtained in this fashion from the air, water, soil, or objects in the environment are transferred via the ducts of the Jacobson's organs to the sensory cells in the organs proper. It has long been thought and repeatedly stated that when the tongue is retracted the tips are rubbed against or inserted into the duct openings to effect the transfer.

This may be the case in many lizards. However, a study by Gillingham and Clarke (1981) demonstrated that in one snake species the ventral surface of the tongue touches the anterior processes of the sublingual plicae during tongue retraction, and the processes are then elevated into the openings of the Jacobson's organs, presumably to effect the final transfer. They concluded that this method of transfer is likely typical of all snakes and monitor lizards. Young (1990) experimentally confirmed that elevating the anterior lingual processes is the mechanism for stimulus transfer and that tongue tips are not inserted into the Jacobson's organs in snakes and monitor lizards. The elongate and forked tongues of these forms are obviously designed to obtain frequent samples of chemicals from the environment, but many lizards have shorter and less protrusible bifurcate tongues yet also exhibit tongue flicking. Halpern and Borghjid (1997) questioned the role of the anterior lingual processes as the sole mechanism of stimulus transfer in snakes, since their experiments showed that the anterior processes are not essential for stimulus transfer to the vomeronasal organs. It thus remains unclear whether the mechanism described by Gillingham and Clarke (1981) and Young (1990) is universal among squamates or whether other methods of transfer are involved.

Although most squamates are oviparous, many species bear living young, a situation not duplicated in other modern reptile groups. In these forms the oviduct does not secrete the outer organic covering (shell) around the egg as in oviparous squamates. In addition, a variety of placental connections are established between the uterine wall of the parent (the maternal component) and the extraembryonic nutritive and protective structures of the developing fetus (the embryonic component). These connections mediate the flow of water, oxygen, and in some cases nutrients from the parent to the embryo and the removal of water, nitrogenous wastes, and carbon dioxide from the embryo to the parent. They vary in complexity from simple apposition of fetal and maternal vascular systems to extensive interdigitation of fetal and maternal tissues that are much thickened and vascularized.

Four principal types of reptile placentas may be recognized (Stewart and Blackburn 1988) and are named for the embryonic contribution to the connection. The first, the choriovitelline placenta, forms as a vascularized area around the embryo derived from the yolk sac and functions only early in development. A second kind of placenta, the omphaloplacenta, is unique to squamates and forms at the abembryonic pole through development of a cleft in the yolk sac to form an isolated yolk mass. The outer bilaminar covering of the latter contacts the uterine wall and mediates physiological exchange between the parent and the yolk sac proper. This kind of placenta occurs in most viviparous lizards. A further specialization of this basic arrangement produces the third placental type. In this structure, typical of all snakes studied to date and some lizards, the allantois invades the yolk cleft and fuses with the peripheral isolated yolk mass to form a complex omphalallantoic placenta (fig. 9.1). Finally, all viviparous reptiles have a chorioallantoic placenta, of exactly the same kind found in eutherian mammals, formed around the fetus by the fusion of the chorionic and allantoic membranes. In viviparous species the chorioallantoic placenta develops over the dorsal portion of the developing embryo. Yolk sac placentas may form around the developing embryo and be transitory (replaced by the chorioallantoic placenta) or form abembryonically (the omphaloplacenta) and persist until birth. Omphalallantoic placentas are also persistent structures, so that viviparous reptiles have two coexisting, functional placentas (an abembryonic one and the chorioallantoic one).

Among oviparous squamates hatching is expedited by a single median egg tooth (paired in geckos) that is used to pierce the surrounding egg shell. Typically the tooth has a hollow base that fits onto the premaxillary bone, and its tip projects forward from the anterioventral region of the snout. Generally, in snakes and many lizards it cuts a slit through the shell, but in at least some lizards after a preliminary puncture the opening is enlarged by the head. The egg

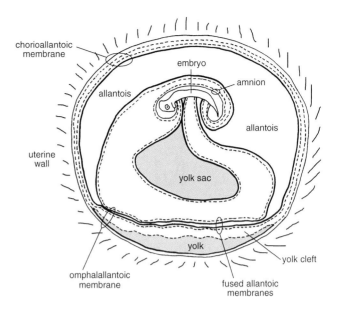

Figure 9.1. Chorioallantoic and omphalallantoic placentas of squamate reptiles. The embryonic portion of the former consists of fused chorionic and allantoic membranes and is quadrilaminar (has four layers). The embryonic portion of the latter is a complex structure consisting of seven layers.

teeth are shed within a few hours or a day or so after hatching. In viviparous squamates the egg teeth are variously nonfunctional, rudimentary, or absent, and the snout is used to rupture the protective extraembryonic membranes immediately after birth.

Many squamates use tail breakage (urotomy) as an antipredator defense. When a predator attempts to seize the intended prey, the prey sheds its tail, which may thrash about from reflex action. While the predator concentrates on the severed tail, the prey quietly escapes. Two major kinds of urotomy besides traumatic loss are recognized: pseudautotomy, applicable to forms having a high incidence of tail loss, intervertebral breakage, no capacity for spontaneous separation, and no tail regeneration (some lizards and snakes); and autotomy, descriptive of taxa having intravertebral breakage, autonomous separation, and tail regeneration (most lizards). The terms nonspecialized pseudautotomy and specialized pseudautotomy are used to differentiate the condition in those forms with moderately long to long tails but no other modification for tail breakage from that in those with a specialized caudal morphology (e.g., the snake genera *Enulius, Scaphiodontophis,* and *Urotheca*).

10 | Lizards
Suborder Sauria Macartney, 1802 (Lacertilia)

Mᴇᴍʙᴇʀs ᴏꜰ the suborder occur in a wide variety of habitats (terrestrial, arboreal, fossorial, semiaquatic, freshwater, and marine), include nocturnal, crepuscular, and diurnal species, and exhibit a wide range of food habits (insectivorous, omnivorous, herbivorous, and carnivorous). This diversity is reflected in the considerable modification of structure and body form away from the basic tetrapod morphology of four well-developed limbs, five fingers and toes, a slightly depressed body, and a long tail. A number of lizard families containing snakelike forms have one to several structural features that resemble the conditions of the Serpentes so closely as to make separation of the two groups on any single criterion impossible. Most lizards differ from snakes as follows:

1. Usually with limbs
2. Usually a pelvis
3. Usually with movable eyelids
4. Usually with an external ear opening
5. Usually with a pectoral girdle
6. Cranium open anteriorly, bounded by a tectum
7. Mandibles usually suturally united

By contrast, all snakes lack functional limbs, movable eyelids, external ear openings, and a pectoral girdle and have the braincase enclosed anteriorly by a bony septum; the halves of the mandible are not articulated, and though some primitive snakes have a pelvic girdle and hind limb rudiments, these are reduced to mere vestiges. All amphisbaenian lizards lack hind limbs and have at best a vestige of the pelvis; further, they lack movable eyelids and external ear openings and have the eyes covered by skin, but they have the cranium slightly open anteriorly and the mandibles suturally united. The pectoral girdle is vestigial in most amphisbaenians, but members of the Mexican genus *Bipes* (two species) have rudimentary forelimbs and a pectoral girdle. Most workers now regard these animals as highly specialized lizards.

In Costa Rica there is no difficulty in separating lizards from snakes, since there are no limbless, snakelike lizards in the republic. There the inexperienced person might confuse lizards only with salamanders or young crocodilians. Salamanders lack epidermal scales and scutes and never have more than four fingers; they also lack an external ear opening and claws on the digits (Costa Rican lizards have scales and an external ear opening; most have five fingers,

and all have claws on fingers and toes). Crocodilians have a longitudinal cloacal opening, some teeth exposed when the mouth is closed, and webbed toes (lizards have a transverse cloacal opening, and the teeth are concealed when the mouth is closed; three genera of Costa Rican lizards—*Basiliscus, Lepidodactylus,* and *Thecadactylus*—have toe webs.

There are approximately 459 genera and 4,700 species of living lizards placed into twenty-eight families as follows (fig. 10.1):

Infraorder Iguania Cope, 1864b
  Family Agamidae (agamids)
  Family Chamaeleonidae (chameleons)
  Family Corytophanidae (helmeted lizards)
  Family Crotaphytidae (collared and leopard lizards)
  Family Hoplocercidae (hoplocercids)
  Family Iguanidae (iguanas)
  Family Opluridae (Madagascar iguanians)
  Family Phrynosomatidae (sceloporines)
  Family Polychrotidae (anoloids)
  Family Tropiduridae (tropidurines)
Infraorder Gekkota Camp, 1923
  Family Eublepharidae (eyelash geckos)
  Family Gekkonidae (geckos)
  Family Pygopodidae (Australasian geckos and flap-foots)
Infraorder Scincomorpha Camp, 1923
  Family Xantusiidae (night lizards)
  Family Lacertidae (lacertids)
  Family Scincidae (skinks)
  Family Dibamidae (dibamids)
  Family Cordylidae (girdle-tailed lizards)
  Family Teiidae (macroteiids)
  Family Gymnophthalmidae (microteiids)
  Family Bipedidae (limbed amphisbaenians)
  Family Amphisbaenidae (worm lizards)
  Family Trogonophidae (desert ringed lizards)
  Family Rhineuridae (rhineurids)

(These last four families are commonly called amphisbaenians.)

Infraorder Anguimorpha Fürbringer, 1900
  Family Xenosauridae (xenosaurs)
  Family Anguidae (anguids)
  Family Helodermatidae (beaded lizards)
  Family Varanidae (monitor lizards)

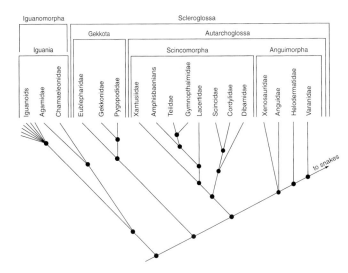

Figure 10.1. Phylogenetic relationships among living lizards. Several often recognized monophyletic suprafamilial taxa are also indicated.

Lizards essentially similar to those living today appear in the Upper Jurassic, about 160 million years ago, but the earliest squamates from the Upper Permian (245 million years ago) are generally regarded as primitive lizards.

The great diversity within the Sauria makes diagnosis of the major divisions difficult. Many characters that seem definitive for a particular evolutionary line may occur as convergences in a few representatives of other stocks. Consequently, in the systematic sections that follow I emphasize diagnoses at the family level and provide only a summary of the most important features characteristic of the four infraorders. For readers interested in greater detail, the papers by Estes, Queiroz, and Gauthier (1988) and Frost and Etheridge (1989) are invaluable references. Note in this regard that all lizards, amphisbaenians, and snakes have procoelous vertebrae, except for most of the geckos (families Gekkonidae and Pygopodidae), which have amphicoelous ones. Note also that the jaw dentition is acrodont (the spaces around and between the bases of the teeth are filled with bone, so that the teeth appear to sit on the jaw margin) in the Agamidae, Chamaeleonidae, and some amphisbaenians. Pleurodont dentition, in which the teeth sit on a ridge on the inner base of the jawbones and have their lingual surface exposed, is found in all other squamates.

Lizards are among the most widespread of vertebrates and occur on all continents except Antarctica, on all major (and many minor) continental islands in warmer latitudes, and on many oceanic islands, including the myriad islands and atolls of the tropical and subtropical Pacific. Their northern limits of distribution extend beyond the Arctic Circle to near 70° north latitude in Scandinavia (Europe), to near 62° N in eastern Siberia (Asia), and to about 51° N in British Columbia but only about 45° N in Ontario on the North American continent. Southern limits reach to land's end: 36° S

(Africa), 40° S (Australia), and 57° S (South America). In addition, they are found in Tasmania (near 42° S) and in extreme southern New Zealand (to about 46°30′ S).

A considerable number of lizard families are extralimital to Costa Rica: the Agamidae (forty-five genera) of Africa, temperate and tropical Asia, to New Guinea and Australia and the Varanidae (two genera), with a similar range but reaching Fiji and some other Pacific islands; Chamaeleonidae (six genera), Africa, Madagascar, India and Sri Lanka, and extreme southwestern Europe; Lacertidae (twenty-nine genera), Africa and temperate and tropical Eurasia; Trogonophidae (four genera), deserts of northern Africa to Iran; Cordylidae (eleven genera), Africa and Madagascar; Opluridae (two genera), Madagascar; Pygopodidae (twenty-one genera), Australia and New Zealand; Dibamidae (two genera), Mexico and tropical Southeast Asia to New Guinea and the Philippines; Crotaphytidae (two genera), southwestern North America; Bipedidae (one genus), western Mexico; Helodermatidae (one genus), southwestern United States to Guatemala; Xenosauridae (two genera), southern Mexico and northern Central America and southeastern China; Hoploceridae (three genera), eastern Panama and South America; Tropiduridae (thirteen genera), South America, the West Indies, and the Galápagos Islands; Amphisbaenidae (nineteen genera), northern West Africa and sub-Saharan Africa, southwestern Europe and the Middle East, the West Indies, and the Neotropics from Panama south; Rhineuridae (one genus) (Florida).

General aspects of osmoregulation (maintaining water balance) and processes of excretion in lizards are covered in the sections on squamates. A considerable number of lizard families include species with glands posterior to the nasal sac that secrete salts into that structure. The several families of geckos and the Cordylidae, Anguidae, and Helodermatidae lack nasal salt glands.

Chemoreception involving the olfactory sac and Jacobson's (vomeronasal) organs (described previously for squamates in general) is used for sampling airborne particles and is characterized by tongue flicking. In addition, many species use the tongue tips to carry chemical particles from the substrate into the mouth and thence to the openings of the Jacobson's organs. In most lizard families olfaction plays an important part in locating food and mates, though the extent varies among groups.

Hearing is an important sense in most lizards, but the auditory apparatus exhibits degenerate evolution in many stocks, especially among semifossorial and fossorial forms. In the usual condition there is a tympanum that may be superficial or sunken into an external ear cavity, an air-filled middle ear, a cartilaginous extracolumella, an ossified columella, and an inner ear. In some groups the tympanum is lost completely, the extracolumella is reduced or lost, the columella abuts the quadrate as in snakes, and the middle ear cavity is reduced or absent, but many intermediates exist between the extremes. In forms with a greatly reduced auditory apparatus, substratum-borne sound vibrations are carried to the inner ear via the lower jaw–quadrate–columella route. Because of the great variability in ear structure and its impact on hearing ability, it is difficult to generalize regarding vibratory sensitivity. In lizards with fully developed ears, excellent sensitivity ranges from 300 to 3,000 Hz. In those with modified auditory apparatus, sensitivity to higher frequencies is reduced, and in those with the greatest reduction in the ear, hearing is poor even for low-frequency vibrations. Auditory communication among lizards is of considerable significance only in several geckos, as will be discussed in the appropriate family accounts below. Vocalizations in other lizards consist primarily of grunts and hisses.

Reception of visual stimuli also varies considerably in lizards. Most lizards have well-developed eyes and movable eyelids, with the lower eyelid mainly involved in closing the eye. A transparent nictitating membrane that moistens, cleans, and protects the eye is usually present. Some forms have a transparent window in the lower eyelid. A number of families are characterized by having the eyelids fused to form a fixed transparent brille or spectacle (as in most snakes) and lack a nictitating membrane, and a few have the eyes covered by a thickened, nontransparent scale. In the Old World chameleons (Chamaeleonidae) the eye is covered by a thick, granular lid, the opening is reduced to a small transverse slit, the nictitating membrane is absent, and the bulging eyes may be rotated independently of one another 180° horizontally and 90° vertically.

Most lizards are diurnal, have color vision, and depend on vision to a considerable extent in all activities. As in birds, accommodation in all lizards is effected by the ciliary muscle, which squeezes the very soft lens, deforming its shape to change focal length. The pupil is round in all lizards, but in nocturnal forms it becomes reduced to a vertical slit in bright light. In diurnal species the retina is composed of two types of cones (a major single cone and a double cone) and often a second type of minor single cone that mediates color vision and vision under high light intensities; that is, they have pure cone vision. Yellow droplets are present in the major single and double cones in diurnal species, but the droplets may be variously colorless or absent in burrowing or nocturnal forms. Nocturnal geckos are unusual in having all rod vision, with one kind of single rod, double rods, and a second unique kind of double rods. It is thought that the latter are not homologous to the rods of other tetrapods but are independently derived from the cone cells of diurnal lizards (Walls 1942; Underwood 1970).

Many lizards have some ability to achieve binocular vision through lateral movements. The chameleons, of course, have complete binocularity. Color vision is well established for diurnal lizards, which have a high preference for green and the ability to discriminate red, orange, yellow, green, blue, and violet.

The paired camera eyes described above for lizards are homologous to those found in other vertebrates. In addition, many lizards have a single dorsal photoreceptor, the parietal eye, on the posterior region of the head. This eye is derived from the paraphysis (parapineal organ), a dorsally projecting portion of the brain diencephalon. The organ typically has a dorsally placed lens and a retina composed principally of secretory cells that is exposed to light through an opening in the skull (the parietal foramen). Frequently the skin over the organ is modified to form a clear cornea. This structure is often incorrectly termed a "pineal" eye. Turtles, snakes, and crocodilians lack parietal eyes, but one is present in the tuatara (order Rhynchocephalia). The parietal eye is definitely light sensitive, and photic information is relayed to the brain. However, it is unclear what function the eye performs. It has been suggested that it regulates the time lizards spend in surface activity, thyroid function, metabolic rates, and/or thermoregulatory behaviors, since in parietalectomized examples these functions increase significantly (Quay 1979).

Most lizards exhibit some ability to change color (metachrosis) by the expansion and contraction of pigment-containing cells (chromatophores) in the dermis. The arrangement of the chromatophores and their functioning is essentially similar to that described for amphibians (chapter 4). The basic color pigments in the cells are brown black (melanin), yellow, red, and silver white (guanine). Blue and green coloration is based on layering of the pigmented cells and structural colors and is produced by light refraction and scattering off the guanine. Changes in color occur in response to both external stimuli (especially light, temperature, and humidity) and internal stimuli (e.g., excitement, fatigue, nutrition, and reproductive state). This ability is under neuronal or hormonal control. In thermoregulation it regulates the absorption of solar energy by decreasing the amount when lightened and increasing it when darkened. In addition, several groups undergo remarkable short-term color changes related to concealment, territoriality, or reproduction, involving brown, green, blue, red, or yellow hues. Visual stimuli seem to have a role in changes related to thermoregulation, concealment, and territorial behavior.

When lizards with well-developed limbs are moving slowly, their locomotion is typical of most tetrapods. This basic walk involves a lateral sequence in which a forefoot strikes the ground after the hind foot on the same side. However, most lizards move in short bursts of speed (skittering) or may run for considerable distances. In these forms of locomotion the gait is a trot, with the forelimb and hind limb on opposite sides of the body thrust backward and striking the substratum in unison. A number of lizards also exhibit bipedal locomotion, either running with the forelimbs off the ground at a slight diagonal angle or walking or running in a vertical position. Fast-moving lizards have been timed at 29 km/h. Forms with the limbs reduced or absent generally use serpentine locomotion, but some favor rectilinear locomotion (both described in more detail under snakes; see chapter 11). All lizards use serpentine motion when swimming.

Many arboreal lizards appear to slow their descent through the air after leaping from a high perch. Oliver (1951) recognized two types of locomotion for such species: gliding and parachuting. Only members of the Asian genus *Draco* are true gliders with a specialized volplaning surface between the arms and legs. These lizards can maintain themselves at an angle of more than 45° to the vertical during the glide. Although not studied in detail, most Costa Rican members of the polychrotid genera *Dactyloa* and *Norops* and the gecko *Thecadactylus* apparently retard the angle of descent by spreading out the limbs, hands and feet, and digital pads and orienting the venter downward and parallel to the ground to form a living parachute. The angle of descent in these cases is not more than 45° to the vertical.

The structure of the limbs, hands, and feet varies widely, from a basic pattern of well-developed limbs through limb and digit reduction to complete loss. The early tetrapod phalangeal formula for the hand (2-3-4-5-3) and foot (2-3-4-5-4) is retained in many forms, but reductions are common. The order of reduction is variable, since the inner or outer digits or both may be lost or all digits may be reduced or lost, leaving a budlike limb. Digit and limb reduction or loss is generally positively correlated with elongation of the body and is associated with semifossorial or fossorial habits.

Lizards are basically carnivores, with most species eating insects and other arthropods. However, some lizards are omnivores, a few are herbivores, and some specialize on large vertebrate prey (birds and mammals), fishes, mollusks, or crustaceans.

There are three principal feeding modes: sight feeders are attracted by movement or color and simply grab their usually small prey; others use olfaction to sample prey items before eating them, seizing the item with the mouth, sampling it with the Jacobson's organs, and either spitting it out (if distasteful or inanimate) or eating it; others sample potential food items by tongue flicking before taking them into the mouth. Old World chameleons have a protrusible tongue that they shoot out for considerable distances to catch insect prey on the sticky, clublike tip. Finally, two species of lizards (family Helodermatidae) can introduce venom into their prey through a series of grooved fangs in the lower jaw. The delivery system in these forms is clumsy, since venom flows from the glands in the lower jaw by capillary action (there is no muscular means of injection) and the lizard must hold on tightly for some time to introduce venom into a victim. Saurians exhibit several styles of foraging behavior. Many species are active, opportunistic feeders, moving about the environment in search of prey in the style characteristic of "taste" feeders (scincomorphs and anguimorphs), which use sight and chemotracking to locate prey. Most sight feeders

(iguanians and gekkotans), however, are sit-and-wait predators that remain motionless while visually searching for prey. In many cases (e.g., anolines) the lizard moves to the prey item and captures it, then returns to the same perch site or observation point and continues searching. In a relatively few forms, including the Costa Rican *Corytophanes* and the chameleons, the predator remains motionless at one site and waits for the prey to approach. The lizard then captures the prey without moving from its perch.

Various lizards are preyed on by almost every known kind of carnivore, particularly snakes, many kinds of birds, and small mammals (cats, both domestic and feral, weasels, civets, wild dogs, foxes, etc.). Large mammalian carnivores and eagles are important predators on large iguanas and monitor lizards, as are boas on the former and pythons on the latter. Crocodilians feed on both large and small aquatic species. A number of carnivorous lizards prey primarily on other saurians. Large spiders, especially in the Tropics, often catch small lizards in their webs and eat them. Except where there is specific information regarding predation on a Costa Rican species, I have not treated this subject in the species accounts for lizards, but the remarks above may be taken to apply generally depending on the size of the lizard.

The precloacal and femoral pores (collectively called precloacal organs) are part of a holocrine secretory gland unique to lizards and amphisbaenians. A glandular secretion is produced by germinal cells lining the base of the gland, which lies within the dermis. The secretion is pushed outward as additional cells arise from the germinal region to form a secretion plug that fills the duct and its external pore. This secretion, which probably produces olfactory stimulation, appears to play some role in courtship. Precloacal organs occur in all genera in the families Crotaphytidae, Hoplocercidae, Iguanidae, Phrynosomatidae, Xantusiidae, and Cordylidae. They are absent in all members of the Chamaeleonidae, Scincidae, Dibamidae, Xenosauridae, Anguidae, Helodermatidae, and Varanidae and in all snakes. They are variably present in all other families of lizards and amphisbaenians. A series of a different kind of holocrine secretory glands forms the unique and obvious posterior ventral escutcheon typical of males of the sphaerodactyline gecko genera *Gonatodes*, *Lepidoblepharis*, and *Sphaerodactylus*.

Many species of lizards show sexual dimorphism in size, coloration, or details of scalation. Usually males are larger or differently or more brightly marked; they often have precloacal and femoral pores when the females lack them or have enlarged postcloacal scales. The base of the tail is often distinctly swollen in males compared with females because of the hemipenes. You can evert these organs in live individuals of most species by gently squeezing with the thumb and forefinger pushing toward the cloacal opening at a point just behind the swollen tail base. In animals without caudal swelling, use the same technique, pressing down and forward 20 to 25 mm posterior to the vent. Another way to check if your specimen is a male is to hold the lizard by its head and carefully probe the rear edge and sides of the cloaca with a small blunt object (paper clip, wooden matchstick, dissecting probe, etc.), thrusting the probe posteriorly. In males the probe should pass into the opening of the inverted hemipenis. Do not be too vigorous, since the probe may damage the specimen or lead to incorrect sexing of a female. In preserved examples you may have to dissect the tail base to confirm the presence of hemipenes or examine the gonads.

Fertilization in all lizards is internal. The paired hemipenes function exactly as in snakes, and details on the processes of eversion are described in the sections on those organisms. Note that these organs are very short in lizards and that their structure has not been as extensively described or used for establishing systematic relationships as it has in snakes. Consequently the considerable variation in hemipenial characteristics is reviewed in the section on snakes and the same terminology is applied in describing the structures in lizards, with one exception: in many lizards the hemipenes are covered by thin transverse laminate folds. This condition has not been reported for any snake.

Unfortunately the snake terminology is at variance with that adopted for some structures in the most notable review of lizard hemipenial morphology to date (Böhme 1988). The confusion arises because Cope (1894bc, 1896) used the term plicae for two different structures (Savage 1997a). He applied the term to a series of longitudinal membranous folds in some snakes that disappear when the organ is everted (Dowling and Savage 1960) and also to the thick, fleshy transverse folds that Dowling and Savage called flounces and Böhme (1988) termed paryphasmata. The latter author used plicae as the descriptor for the laminate folds found in some lizards. To reduce the confusion produced by using plicae in three different contexts, Savage (1997a) proposed that the laminate folds be called petala (sing. petalum). That and the term flounces are used in the following accounts to differentiate between the two distinctive types of transverse hemipenial folds (see chapter 11).

Since the number of lizard species for which the hemipenes have been described is relatively small and the differences between related forms may be trivial at best, I usually provide a general description of these organs in the family accounts. When there are marked differences in this feature among Costa Rican genera, they are noted in the generic diagnoses.

A considerable period of courtship precedes mating, with visual stimuli most important in diurnal groups, chemical ones in secretive stocks, and vocalizations in nocturnal geckos. In mating the male grasps the female from above; they intertwine their tails and bring their cloacas together to allow intromission of one of the hemipenes. Most lizards lay eggs. In some the shell is calcareous and brittle, but in many it is surrounded by a thickened, parchmentlike envelope that contains a small amount of lime. Eggs are deposited in a variety of sites where there is adequate moisture and minimal temperature fluctuation: under debris or rocks, in the ground, in sand, humus, or fallen logs, and in tree holes and aerial plants. The eggs are usually abandoned, but a few species remain with their eggs. A considerable number of lizards are viviparous and have placental connections mediating physiological exchange between the embryo and the mother's oviductal lining. Ovoviviparity does not occur in reptiles (Stewart and Blackburn 1988).

Territorial behavior is typical of most lizards, and males in various groups may defend their territories against conspecific males by visual or vocal stimuli or by actual fighting.

Many lizards exhibit tail autotomy, the ability to break off a portion of the tail by muscular contraction. Autotomy is a predator defense, since the reflex movements of the separated portion of the tail may continue for some time afterward and the predator frequently continues to attack the severed tail while the otherwise intact lizard escapes. Breakage is accomplished by intravertebral septa present in all but the anteriormost caudal vertebrae. After a time the tail regenerates, but the new tissue is supported by a rod of cartilage, not bone. The tail may be shed an additional time or two if the early break(s) was distal to retained autotomous vertebrae. The cartilaginous rod cannot be shed independently if no autotomous vertebrae are present in the base of the tail.

As in snakes, to permit growth, from time to time a new outer layer of skin is proliferated by the synchronous differentiation of cells over the entire surface of the body. The new skin underlies the outer dead layer of the old skin, which separates from the new one and is then shed. Because lizards generally are not as streamlined as snakes, the old skin tends to become worn and torn and may be shed in flakes or patches. Like many amphibians, geckos usually eat the sloughed skin.

LIZARD IDENTIFICATION: KEY FEATURES

The complex scalation that forms the outer protective covering of the body of lizards provides a wide array of identifying characteristics (fig. 10.2). These features and those of coloration and size are emphasized in the keys and diagnostics sections. The following paragraphs summarize the most substantive differences among lizard groups in squamation and other characteristics that may not be familiar to the general reader or where the differences mentioned are not self-explanatory.

Head scalation: The head may be covered by small granular scales, by a mixture of small and large (tubercular) granular scales, by scales of variable size (none granular), most of them platelike, or by large symmetrical plates. Terminology for head structures is given on the accompanying illustrations (figs. 10.2, 10.5).

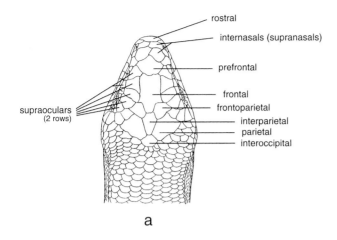

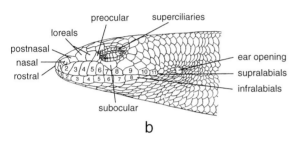

Figure 10.2. Terminology for head scalation in lizards: (a) dorsal view; (b) lateral view.

Ventral scalation: There is considerable diversity in the scales covering the venter. These may consist of numerous small rounded or rectangular, semi-imbricate, or juxtaposed scales or small granules. In others the scales, though numerous, are moderately large round, notched, or pointed (mucronate), imbricate (overlapping) or semi-imbricate, and smooth or keeled (fig. 10.3). Finally, the scalation may be composed of large smooth imbricate cycloid scales or large rectangular juxtaposed plates.

Dorsal scalation: The dorsum often is covered by small granular scales, a mixture of small granules and larger tubercles, or small to large rounded or pointed juxtaposed scales that are imbricate or semi-imbricate, keeled or smooth. Several forms have large smooth imbricate cycloid dorsals, and others have large rectangular smooth or keeled dorsal plates (fig. 10.3).

Eyes and eyelids: In most lizards the eye is closed by the movable lower eyelid that is usually covered by pigmented scales. In some forms a thinned unpigmented area as large as several scales forms a translucent disk in the lower eyelid. In some lizards the disk is without scale divisions and completely transparent. In other species the eyelids are fused and consequently immovable and form a completely transparent spectacle. The pupil is either round or vertically elliptical in bright light or preservative. A parietal eye that is located in the interparietal plate and de-

limited by a clear lenslike structure is a characteristic of many Costa Rican lizards.

Tympanum: An external exposed or recessed tympanum is present in all Costa Rican lizards except *Bachia blairi*.

Throat structure: A number of genera are characterized by a complete transverse fold and groove marking the posterior boundary of the gular region (fig. 10.4). The gular fold is demarcated by a change of squamation, usually with enlarged scales on the fold and smaller ones in the groove, followed by larger ventral scales or plates. Several genera have the hyoid apparatus modified so that the gular area can be expanded to form a vertical fanlike gular appendage or dewlap. The dewlap is usually best developed or found only in males, and dewlap size and coloration are species specific (fig. 10.5).

Helmet: The head is produced dorsally or posteriorly to form one or more flaps, one of which overhangs the neck and is continuous with a dorsal crest in subadult and adult members of the family Corytophanidae. In silhouette this structure resembles the helmet of a medieval knight, hence the name.

Crests: Scales on the midline of the neck, body, and tail may be enlarged to form low to very well developed crests. In some cases the crests are restricted to males.

Tail structure: The tail in most Costa Rican lizards is cylindrical to subcylindrical in cross section, but in semiaquatic forms it is somewhat to greatly compressed laterally. Squamation of the tail may be like that of the body or may differ from it. In some forms the caudal and subcaudal scales are the same size and form complete rings (whorls) around the tail. In others the whorls or scales may be of different sizes, or the number of whorls of small and large scales and their sequences down the tail may differ among related species (fig. 10.15). In still others the caudal and subcaudal scales are different in size or shape and amount of keeling. A pair of enlarged postcloacal scales may be also present in males of some species (fig. 10.3).

Limb structure: The shape and scalation of the digits are significantly different in various groups of lizards. Typically the digits are subcylindrical and the scales on the undersides (lamellae) are narrow and keeled throughout, but in several groups the keels are lost and in others the lamellae on the basal or distal portion of the digits are expanded and smooth. The tips of the digits may be further modified in geckos and their allies, and the claws may be concealed or retractile within a sheath (fig. 10.6d). A basal web between the fingers and toes may be present, or in some genera the three outer toes may have a scaly fringe (fig. 10.3).

Glandular structures: A series of precloacal or femoral pores are found on the ventral surface and underside of the hind limbs; in several species the two sets of pores, if both are present, may form a continuous series running from one leg across the precloacal area to the other leg. These may

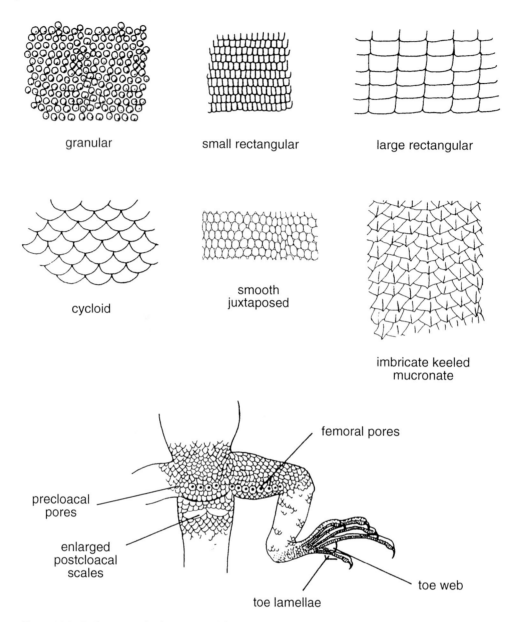

granular   small rectangular   large rectangular

cycloid   smooth juxtaposed   imbricate keeled mucronate

Figure 10.3. Scale types and other structural features in lizards.

be restricted to males, or in females they may be fewer, smaller, and less well developed (fig. 10.3). In sphaerodactylines adult males have a series of enlarged glandlike scales on the posterior ventral surface and often on the adjacent undersurface of the femur. The distribution and number of escutcheon scales is characteristic at the species level (fig. 10.29).

Integumentary pockets: One species each of the genera *Norops* and *Sceloporus* has a specialized integumentary pocket associated with the limb insertions. In the former the pocket forms a deep circular pit in the axillary region, and in the latter it has a slitlike opening lying behind the femur and extending obliquely downward toward the vent (fig. 10.20).

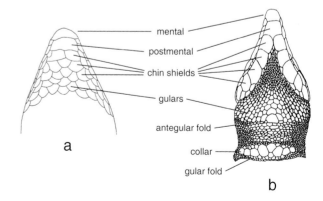

Figure 10.4. Terminology for scalation of chin, throat, and neck region in lizards: (a) a skink, *Mabuya*; (b) a macroteiid, *Ameiva*.

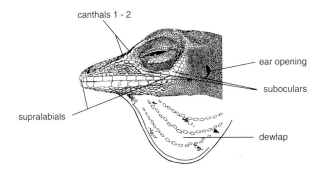

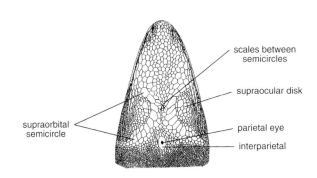

Figure 10.5. Special terminology for head scalation and other structural features in anole lizards: (a) lateral view; (b) dorsal view.

Coloration: Pattern and coloration are important features that are helpful in species identification. In some forms sexual dimorphism and ontogenetic changes complicate field identification, although females tend to retain a juvenile pattern where there is a marked sexual difference in adults. Overall coloration of the upper surfaces of the body, limbs, and tail may be uniform or variously marked with dark or light spots, blotches, bands, or stripes or combinations of these components (fig. 3.1). Gular col-

oration and especially the color(s) of the dewlap, when present, are often diagnostic. In a few species adult males (and in some cases females) have distinctive brightly colored lateral belly patches. In other forms the tail may be brightly and contrastingly colored compared with the body (at least in juveniles). Differences in iris color may also aid in identification.

In the following sections and elsewhere in this book, reference to lizard size refers to total length as follows:

Tiny: under 75 mm
Small: 75 to 200 mm
Moderate: 200 to 550 mm
Large: 550 to 1,200 mm
Very large: 1,200 to 2,500 mm
Giant: greater than 2,500 mm

The approximate maximum total length is given in each species account. However, since many lizard species frequently lose their tails and then regenerate a new but shorter one, the maximum standard length (head plus body length [snout to vent length]) is also indicated. Because of the great range of size in lizards, this feature alone will eliminate many species from consideration when you make a preliminary identification.

The largest Costa Rican lizards are *Ctenosaura similis* (maximum total length about 1,400 mm) and *Iguana iguana* (maximum total length 2,010 mm). These species are much smaller than the largest living lizard, the Komodo dragon or ora, *Varanus komodoensis* of Komodo Island, the western part of Flores Island, and several other East Indian islands between Komodo and Flores. This lizard attains a total length of approximately 3 m (Auffenberg 1981) and a weight of up to 250 kg. The smallest lizards are members of the gecko genus *Sphaerodactylus*; several West Indian species have maximum total lengths of about 40 mm.

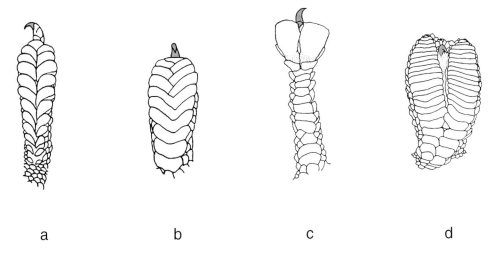

a   b   c   d

Figure 10.6. Structure of digits in genera of geckos having expanded digital lamellae, ventral view: (a) *Hemidactylus;* (b) *Lepidodactylus;* (c) *Phyllodactylus;* (d) *Thecadactylus* note retractile claw.

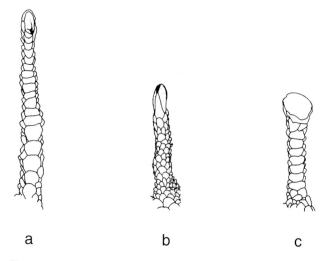

Figure 10.7. Structure of digits in genera of geckos with unexpanded distal lamellae, ventral view: (a) *Gonatodes;* (b) *Lepidoblepharis,* note retracted claw; (c) *Sphaerodactylus,* claw retracted.

In general, large species of lizards live longer than smaller forms, many of which live only one or two years. Only a few species under captive conditions reach an age of twenty to thirty years. However, Auffenberg (1981) estimated that the largest living lizard, the Komodo dragon (*Varanus komodoensis*), attains an age of fifty years in the wild.

KEY TO THE GENERA OF COSTA RICAN LIZARDS

1a. No movable eyelids 2
1b. Movable eyelids present. 10
2a. Head covered primarily by granular scales (fig. 10.3) 3
2b. Head covered by enlarged plates (fig. 10.2). 9
3a. A series of greatly expanded lamellae under fingers and toes; digits markedly compressed dorsoventrally, never cylindrical in cross section (fig. 10.6); pupil vertically elliptical in bright light or preservative (fig. 10.9a) . . . 6
3b. Lamellae not greatly expanded under fingers and toes; digits not compressed dorsoventrally but cylindrical or subcylindrical in cross section (fig. 10.7); pupil round (fig. 10.9b) 4
4a. Claws retractile into a sheath; terminal scales of digits much larger than other scales (fig. 10.7b,c) 5
4b. Claws not retractile into a sheath; terminal scales of digits about same size as other scales (fig. 10.7a) . . . . *Gonatodes* (p. 489)
5a. Tips of digits distinctly asymmetrical and dorsolaterally compressed; claw retractile into a sheath between one extremely enlarged and flattened terminal scale and similar but smaller scales (fig. 10.7c); a supraorbital spinelike scale on the superciliary margin (fig. 10.9b) . . . . *Sphaerodactylus* (p. 492)

5b. Tips of digits symmetrical, subcylindrical in cross section; claw retractile into a sheath surrounded by five essentially equal-sized scales and one smaller terminal scale (fig. 10.7b); no supraorbital spinelike scale, although one to several slightly projecting scales present along superciliary margin . . *Lepidoblepharis* (p. 491)
6a. Terminal lamellae of toes about the same size as other lamellae; all lamellae divided (fig. 10.6a,b,d) 7
6b. Terminal lamellae of toes much larger than other lamellae; basal lamellae single (fig. 10.6c); no precloacal or femoral pores . . . . *Phyllodactylus* (p. 487)
7a. Claws at tip of dilated area, retractile into distinct longitudinal groove (fig. 10.6d); no precloacal or femoral pores . . . . *Thecadactylus* (p. 488)
7b. Claws not retractile, on slender compressed terminal phalanx (fig. 10.6a,b); males with a continuous series of precloacal and femoral pores (fig. 10.3) 8
8a. Inner digits with a claw; terminal phalanges free, arising angularly from expanded portion of digit (fig. 10.6a); digits not webbed. . . . *Hemidactylus* (p. 482)
8b. Inner digit on fingers and toes clawless; terminal phalanges of outer four digits united with expanded portion (fig. 10.6c); digits webbed (fig. 10.3). . . . . *Lepidodactylus* (p. 485)
9a. Four fingers; no gular fold; infralabials and postmentals distinct (fig. 10.8a); pupil round (fig. 10.9b) . . . . *Gymnophthalmus* (p. 521)
9b. Five fingers; a gular fold present; infralabials and postmentals fused into a single series (fig. 10.8b); pupil vertically elliptical in bright light or preservative (fig. 10.9a). . . . . *Lepidophyma* (p. 498)
10a. Venter covered by large rectangular plates (fig. 10.3) 11
10b. Venter covered by large cycloid scales or numerous small to granular rectangular scales (fig. 10.3) . . . 18
11a. No strong lateral fold or series of small granules separating lateral from ventral scales 12

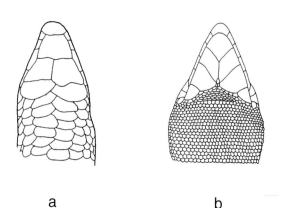

Figure 10.8. Scalation of chin and throat region: (a) *Gymnophthalmus,* infralabials and postmentals distinct; (b) *Lepidophyma,* infralabials and postmentals fused (supralabials visible).

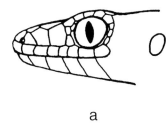

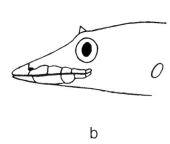

b

Figure 10.9. Lateral view of lizard heads: (a) *Lepidophyma*, showing vertical pupil; (b) *Sphaerodactylus*, showing round pupil and superciliary spinelike scale above eye.

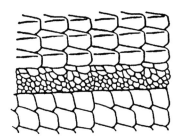

Figure 10.10. Body scalation in the anguid lizard genus *Mesaspis* in lateral view; keeled lateral scales separated from smooth ventrals by a lateral fold containing smaller scales.

11b. A strong lateral fold enclosing a series of small scales that separate ventral from lateral scales (fig. 10.10) 24
12a. Scales on head and venter smooth; a distinct gular fold present 13
12b. Scales on head and venter keeled; no gular fold . . . . *Leposoma* (p. 522)
13a. Some or all dorsals enlarged scales or plates (fig. 10.3); nasals not in contact on upper surface of head posterior to rostral, separated by a large shield or several smaller plates (fig. 10.34) 14
13b. Dorsal scales small granules (fig. 10.3); nasals in contact with one another on midline posterior to rostral scale (fig. 10.31). 17
14a. Some dorsal scales keeled (fig. 10.3) 16
14b. Dorsal scales smooth (fig. 10.3). 15
15a. An ear opening (fig. 10.2); five fingers and toes . . . . *Anadia* (p. 518)

15b. No ear opening; no more than four fingers or toes . . . . *Bachia* (p. 519)
16a. Dorsal scutellation a mixture of small smooth flattened and rounded scales, enlarged tubercles, and enlarged keeled scales (fig. 10.3); a pair of weak longitudinal crests on tail . . . . *Neusticurus* (p. 523)
16b. Dorsal scutellation homogeneous, all keeled (fig. 10.3); no longitudinal crests on tail . . . . *Ptychoglossus* (p. 524)
17a. A fleshy lingual sheath enclosing base of tongue from posterior margin of scaly portion of tongue to larynx when tongue retracted into mouth; posterior margin of tongue rounded and only slightly notched (fig. 10.11a); collar scales small and irregular or no more than one row of enlarged scales on gular collar (fig. 10.32a) . . . . *Ameiva* (p. 508)
17b. No fleshy lingual sheath enclosing base of tongue; posterior margin of tongue deeply emarginate (fig. 10.11b); two to four regular rows of enlarged scales on gular collar (fig. 10.4b) . . . .*Cnemidophorus* (p. 514)
18a. Head covered by large plates or scales, never by granular scales (fig. 10.2); pupil round (fig. 10.9b) 19
18b. Head covered by a mixture of large and small granular scales; pupil vertically elliptical in bright light and preservative (fig. 10.9a). . . . . *Coleonyx* (p. 481)
19a. Body covered by uniform cycloid scales (fig. 10.3); head covered by enlarged plates (fig. 10.2) 20
19b. Body not covered by cycloid scales; venter covered by small noncycloid scales (fig. 10.3); head covered by a mixture of small and enlarged platelike scales. 25
20a. Paired supranasals (internasals) present; two frontoparietals (fig. 10.12b,c) 21
20b. No supranasals, a single large frontonasal (internasal); one frontoparietal (fig. 10.12a) . . . .*Sphenomorphus* (p. 504)

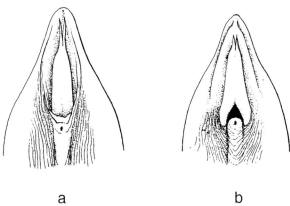

a   b

Figure 10.11. Floor of mouth in macroteiid lizards: (a) *Ameiva*, showing fleshy lingual sheath; (b) *Cnemidophorus*, showing absence of sheath and presence of deeply notched tongue.

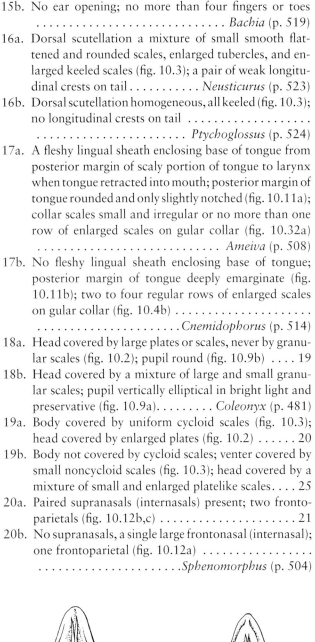

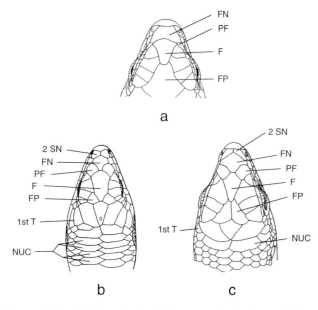

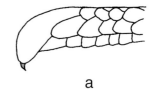

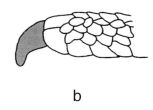

Figure 10.12. Head scalation in skink genera, dorsal view: (a) *Sphenomorphus*, note single frontonasal (internasal) and single frontoparietal; (b) *Eumeces*, note paired supranasals, enlarged first temporal and four pairs of nuchals; (c) *Mabuya*, note paired supranasals, small first temporal, and one pair of nuchals. Abbreviations: F = frontal, FN = frontonasal (internasal), FP = frontoparietal, NUC = nuchal, PF = prefrontal, SN = supranasal, T = temporal.

Figure 10.13. Claws in anguid lizards: (a) *Diploglossus*, covered permanently by sheath of scales; (b) *Celestus*, exposed for entire length.

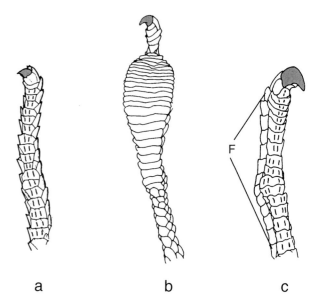

Figure 10.14. Structure of fingers and toes in iguanian lizards showing digital lamellae and claws, ventral view: (a) lamellae narrow, keeled; (b) lamellae under two medial phalanges expanded, smooth (anole lizards); (c) lamellae keeled, scaly fringe (F) present on toes only *(Basiliscus)*.

21a. A median interoccipital scale bordering interparietal posteriorly; two rows of supraoculars (fig. 10.2); more than 32 scale rows around midbody 22

21b. No median interoccipital scale bordering interparietal posteriorly; one row of supraoculars (fig. 10.12b,c); no more than 32 scales around midbody. 23

22a. Claws permanently covered by a sheath of scales, with only the claw tip exposed at most (fig. 10.13a) . . . . .
. . . . *Diploglossus* (p. 530)

22b. Claws exposed for entire length (fig. 10.13b). . . . . . .
. . . . *Celestus* (p. 527)

23a. No translucent disk in lower eyelid; four pairs of nuchal scutes; primary temporal markedly larger than lateral scales at midbody (fig. 10.12b) . . . .
. . . . *Eumeces* (p. 501)

23b. A translucent disk in lower eyelid; one pair of nuchal scutes; primary temporal about same size as lateral scales at midbody (fig. 10.12c). . . . *Mabuya* (p. 502)

24a. Twelve longitudinal rows of ventral scales . . . .
. . . . *Mesaspis* (p. 533)

24b. Ten longitudinal rows of ventral scales. . . . .
. . . . *Coloptychon* (p. 532)

25a. Lamellae narrow, keeled throughout (fig. 10.14a) . . . .
. . . . 26

25b. Lamellae of medial portion of digits expanded, smooth (fig. 10.14b). 31

26a. A gular fold present (fig. 10.4b); a dorsal crest . . . 27

26b. No gular fold present; no dorsal crest 30

27a. Femoral pores present (fig. 10.3). 28

27b. No femoral pores. 29

28a. Caudal scales arranged in whorls of large scales separated by small scales (fig. 10.15); no greatly enlarged scale below angle of jaw. . . . . . . *Ctenosaura* (p. 434)

28b. Caudal scales not arranged in whorls of large and small scales; a greatly enlarged scale below angle of jaw . . . . *Iguana* (p. 437)

29a. Three outer toes with a narrow, scaled fringe (fig. 10.14c); anterior superciliary scales imbricate and elongate, strongly overlapping (fig. 10.16a) . . . .
. . . . *Basiliscus* (p. 426)

29b. Outer toes without a scaled fringe; superciliary scales quadrangular, not overlapping (fig. 10.16b). . . . . *Corytophanes* (p. 432)

30a. Third and fourth toes about equal in length; more than 40 lamellae under fourth toe. . . . . *Polychrus* (p. 445)

30b. Third toe much shorter than fourth; fewer than 30 lamellae under fourth toe . . . . *Sceloporus* (p. 440)

31a. All or most autotomic caudal vertebrae without transverse processes; if present, short and laterally directed 32

31b. Forward-directed transverse processes present on all autotomic caudal vertebrae . . . . . . *Norops* (p. 457)

32a. Parietal foramen in frontoparietal suture or absent; parietal ridges U-shaped . . . . *Dactyloa* (p. 453)

32b. Parietal foramen in anterior part of parietal; parietal ridges Y-shaped . . . . *Ctenonotus* (p. 452)

The principal differences between the genera *Ameiva* and *Cnemidophorus* and among the genera *Ctenonotus, Dactyloa,* and *Norops* are internal, so I provide a single key to species of the first two genera (p. 507) and a single one for the last three (p. 449).

Most lizards are small, innocuous forms, but large and moderate-sized Costa Rican species can inflict painful bites

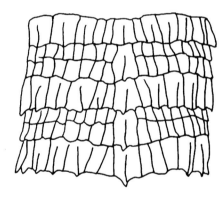

Figure 10.15. Dorsal view of section of tail in the spiny-tailed iguana, genus *Ctenosaura*, showing alternating whorls of large and small scales (up toward head).

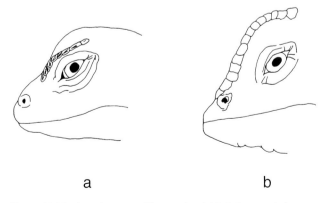

a   b

Figure 10.16. Anterior superciliary scales: (a) imbricate and elongate *(Basiliscus);* (b) juxtaposed and quadrangular *(Corytophanes).*

if not handled carefully. Although local folklore may ascribe poisonous attributes to some species, *there are no poisonous or venomous lizards native to Costa Rica!* The only venomous lizards are members of the genus *Heloderma* (two species) of the beaded lizard family Helodermatidae, found in the southwestern United States, over most of western Mexico, and in two Atlantic versant valleys in extreme southern Mexico and Guatemala.

Iguanian Lizards (Infraorder Iguania)

This group includes a diverse series of predominantly diurnal sight feeders (with color vision) that are mostly terrestrial, semiarboreal, or arboreal in habits and prey on arthropods. Some species eat larger animal prey, and several are omnivores or herbivores. The marine iguana *(Amblyrhynchus cristatus)* of the Galápagos Islands is a semiaquatic algae eater.

Members of this group usually have rather robust bodies, short necks, distinct heads, well-developed limbs, and moderate to long tails. The body is covered with noncycloid granular juxtaposed or imbricate scales, which may be smooth, tuberculate, keeled, or with spiny projections. The ventral scales may be as small as the dorsals but are usually larger, juxtaposed, and rectangular. The head is usually covered with small scales, although the supraoculars and interparietal (containing the parietal eye cornea) may be somewhat enlarged, but most members of the Phrynosomatidae have rather large and symmetrical head shields. Technically iguanians may be separated from other major lizard stocks by the combination of the following features:

1. Tongue slightly notched (less than 10%); foretongue not differentiated; tongue usually covered with villi anteriorly and conical and/or reticulate papillae posteriorly; prey caught on surface of tongue; tongue protrusion restricted to extension in most iguanians (e.g., Iguanidae and Agamidae), but in Chamaeleonidae whole tongue protrusible, tip club shaped, unnotched; tongue mucocytes mostly serous and seromucous
2. Postfrontal reduced to subtriangular element on orbital rim or absent
3. Postorbital and temporal arches and supratemporal foramen present; jugal present
4. Septomaxillae flat and small, not forming a midline crest; Jacobson's organs small
5. Jaw tooth replacement basal, each tooth with a well-developed resorption pit
6. A single egg tooth in oviparous species
7. Notochord not persistent in adults
8. Rectus abdominis lateralis muscle absent
9. Tibial epiphysis unnotched

Among other features, the families Agamidae and Chamaeleonidae differ from other iguanians in having acrodont teeth that are fused to the underlying bone in adults. The former has a wide range in Africa, temperate and tropical Eurasia (absent from most of Europe), and Australia and adjacent islands. The Chamaeleonidae occur in India and Sri Lanka, Madagascar and associated islands (Seychelles and Mascarenes), Africa, southwestern Arabia, the Middle East, and southern Spain.

Other families belonging to this stock but not treated in this book are the Crotaphytidae (two genera) of the western United States and adjacent Mexico, the Hoplocercidae (three genera) of eastern Panama and tropical South America, the Tropiduridae (thirteen genera) of South America and the Antilles, and the Opluridae (two genera) of Madagascar and the Comoro Islands.

Family Corytophanidae Fitzinger, 1843

Helmet Lizards

Members of this small family are moderate-sized to large long-tailed semiaquatic, semiarboreal, and arboreal lizards that are unique in having a posteriorly extended skull crest formed by the parietal bone that supports one or more prominent head flaps or a parietal casque. Three genera— *Basiliscus* (four species), *Corytophanes* (three species), and *Laemanctus* (two species)—are recognized within the group. The family is restricted to the Neotropics, and the distribution is centered on Central America. *Basiliscus* ranges from Mexico to northwestern Venezuela and western Ecuador, *Corytophanes* is found from Mexico to Colombia, and *Laemanctus* is found on the Atlantic lowlands from Mexico to Nicaragua and on the Pacific slope at the Isthmus of Tehuantepec.

Besides the specialized head structures, helmeted lizards have the following obvious external features: body laterally compressed; eye large; pupil round; tympanum large, not recessed; interparietal scale not markedly enlarged; dorsal and ventral scales smooth or keeled, juxtaposed to imbricate; no femoral pores; no spinulate scale organs; infradigital lamellae with a single keel or tuberculate; gular fold complete medially.

In addition to these features the family is distinctive among iguanians in the following combination of mostly osteological characters: parietal foramen in frontal or frontoparietal suture *(Laemanctus)*; supratemporal on lateral surface of supratemporal process of parietal; nuchal endolymphatic sacs not penetrating neck musculature; no labial blade on coronoid; Meckel's groove closed (open in some *Basiliscus*); lateral teeth on jaws triconodont; caudal autotomy septa present or absent *(Corytophanes* and *Laemanctus)*; transverse processes weak or absent on posterior trunk and autotomous vertebrae; no posterior coracoid foramen; sternal fontanelles very small or absent; postxiphisternal inscriptional ribs short; vertebrae with zygosphenes and zygantra; nasal chamber with short vestibule and well-developed concha; hemipenes slightly bilobed, with a simple sulcus spermaticus, apical region calyculate; no colic septa.

Most members of this family are oviparous and lay cream to white eggs with flexible shells, but *Corytophanes pericarinatus* is viviparous.

Genus *Basiliscus* Laurenti, 1768

Garrobos, Basilisks, Jesus Christ Lizards

The members of this genus (four species) are large lizards characteristically found in riparian situations, where they forage in bushes or trees or along the margins of the watercourse or in it. When disturbed they usually escape by diving into the stream or by running across the water surface on their hind legs. Costa Rican males are immediately recognizable because of the well-developed fleshy flap extending from the head and the dorsal and caudal crests. The fleshy head appendage and crests are much less obvious in juvenile males, and females lack body and tail crests, but all ages and both sexes share the following features: compressed body and long, laterally compressed tail; well-developed limbs; anterior superciliary scales elongate, strongly overlapping; outer three toes with a narrow, scaled fringe; an elongate Y-shaped parietal bone with a posteriorly directed blade that supports the head appendage (best developed in adults); parietal foramen in frontal bone; clavicles fenestrate; dorsal scales small, rhombic, smooth or keeled; ventral scales larger, square and smooth or rounded, imbricate, and keeled; a slightly indicated gular pouch in males; infradigital lamellae with a single keel; caudal autotomy septa present but partially ossified during ontogeny. Among New World lizards *Basiliscus* is likely to be confused only with the other members of this family, *Corytophanes* and *Laemanctus*. Both of these genera have a distinctive modified head structure. In *Corytophanes* the parietal bone is greatly expanded laterally to form a casque, and distinct raised ridges border the casque on each side from the loreal to the occipital region, where they fuse into a single median crest. In *Laemanctus* the scales on the upper surface of the head are partially fused to the underlying bone to form a flattened casque that projects slightly posteriorly to overhang the neck region. Neither of these genera have scaly flaps or fringes along any toe margins. Other features of *Laemanctus* include superciliary scales not quadrangular, some slightly overlapping; an elongate Y-shaped parietal bone; parietal foramen in frontoparietal suture; dorsal and ventral body scales large, keeled, imbricate; infradigital lamellae with single knoblike keels; no caudal autotomy. Other features for *Corytophanes* are described in the account of that genus below.

All members of the genus *Basiliscus* exhibit remarkable ontogenetic transformations in morphology, culminating in the extraordinary development of the head crest and, in three of the species, of the dorsal and caudal crests characteristic of adult males. Head crest development involves the posterior extension and expansion of the median crest of the parietal bone and the elaboration of a dorsally or posteriorly expanded vertical fleshy flap or flaps supported by it. Development of these structures is entirely postembryonic, and hatchlings and small juveniles lack any indication of them. As species of *Basiliscus* grow, the parietal crest begins to project slightly, and a fleshy ridge develops around it and begins to extend posteriorly over the nuchal region. This degree of development persists in adult females. In males the crest continues to expand and gradually forms a large thin vertical parietal blade that provides support for the greatly enlarged fleshy cephalic ornamentation typical of adult males. Dorsal and caudal crest development follows a similar ontogenetic trajectory in males of *Basiliscus basiliscus* and *B. plumifrons*. Small juveniles and small adult females show no evidence of these crests, which involve the gradual elongation of the neural spines of the dorsal and anterior caudal vertebrae and the elaboration of a fleshy vertical fin covered by thin smooth scales. The crests, when fully developed, have neural spines that may be 50 to 60 mm long and curve gently posteriorly. The dorsal crest begins in the nuchal region and terminates in the lumbar area; the caudal crest is restricted to the anterior two-thirds of the tail in these forms. In *B. vittatus* the dorsal fin is lower than in the other two taxa, and only the largest males have some elongate neural spines. The caudal crest consists of a series of enlarged serrate scales in males of this species. *Basiliscus* is found throughout the tropical lowlands and on premontane slopes from Jalisco and Tamaulipas, Mexico, to central Ecuador and northwestern Venezuela. One species, *B. vittatus*, has been introduced to southern Florida.

KEY TO THE COSTA RICAN SPECIES OF THE GENUS *BASILISCUS*

1a. Ventral scales smooth; ventral surface of third phalanx of fourth toe with a series of very small scales that separate one row of large lateral scales from two rows of large medial scales; three or four chin shields almost always in contact with infralabials; usually an odd number (one or three) of scale rows between supraorbital semicircles, occasionally two rows; head crest of adult males rounded in outline or with a posteriorly directed lobe; ground color often bright to dull green in life, often bluish gray in preservative 2

1b. Ventral scales keeled; ventral surface of third phalanx of fourth toe covered by two rows of large scales, usually in contact; one or two chin shields in contact with an infralabial; usually two rows of scales between supraocular semicircles, occasionally one or three rows; head crest of adult and large juvenile males (92–140 mm in standard length) single, triangular in outline (fig. 10.17a); ground color brown to olive, often with a distinct light (reddish to yellow in life) dorsolateral stripe from eye to shoulder or beyond . . . . *Basiliscus vittatus* (p. 431)

2a. Occipital scales large, as large as or larger than enlarged

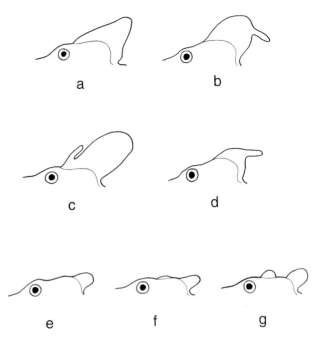

Figure 10.17. Head crests in Costa Rican lizards of the genus *Basiliscus*: (a) male *B. vittatus*; (b) adult male *B. basiliscus*; (c) adult male *B. plumifrons*; (d) large juveniles and adult female *B. basiliscus*; posterior projection may be absent; (e–g) Ontogenetic series of head crest development in juvenile male *B. plumifrons*; adult females have head crest like (e).

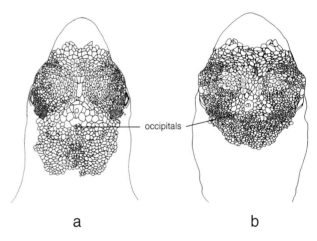

Figure 10.18. Dorsal view of head in Costa Rican lizards of the genus *Basiliscus*: (a) *B. basiliscus*; (b) *B. plumifrons*; note differences in size of occipital scales.

supraocular scales, abruptly differentiated from small scales on posterior dorsal surface of head; supraorbital semicircles usually separated by one scale row, occasionally by two rows (fig. 10.18a); head crest single in adult and large juvenile males (75–250 mm in standard length) and in large females (75–194 mm in standard length) (fig. 10.17b,d); throat of juveniles (37–80 mm in standard length) with three distinct longitudinal light stripes (fig. 10.19b); ground color of adults brown,

a   b

Figure 10.19. Throat coloration in juvenile lizards of the genus *Basiliscus*: (a) *B. plumifrons*; (b) *B. basiliscus*.

olive, or bronze in life and in preservative; juveniles pale green in life, gray in preservative; never with a series of distinct large light spots along sides, although two pairs of light lateral stripes are usually present . . . . *Basiliscus basiliscus* (p. 428)
2b. Occipital scales much smaller than enlarged supraocular scales, gradually grading into small scales on posterior dorsal surface of head; supraorbital semicircles almost always separated by three to five scale rows, rarely by one or two rows (fig. 10.18b); head crest of adult males (122–210 mm in standard length) comprising two lobes (fig. 10.17c); crest of large juvenile males (109–21 mm in standard length) consisting of a posterior flap and an anterior ridge or of two distinct flaps (fig. 10.17f,g); adult females (146–74 mm in standard length) with a single crest; throat of juveniles and adults uniform, usually dark, sometimes with a diffuse light median spot or stripe; ground color of adults dark bluish green in life, dark bluish gray to purple in preservative; dorsum of very small juveniles (44 mm in standard length) reddish to golden brown (in life and preservative), usually with a green head in life; dorsum of larger juveniles and small adults bright green in life (bluish gray in preservative); adults usually with one or two series of large light spots along sides . . . . *Basiliscus plumifrons* (p. 430)

Basiliscus basiliscus (Linné, 1758)

(Plates 234–35; map 10.1)

DIAGNOSTICS: This is the most commonly encountered large lizard in Costa Rica. Individuals with well-developed dorsal and caudal crests differ from other crested species in having a rounded or pointed head crest (fig. 10.17b,d). Juveniles and females have smooth ventrals and large occipital scales (fig. 10.18a) and a pattern of dorsal dark crossbands and usually paired longitudinal lateral light stripes, but it may be necessary to capture them in order to distin-

guish them from their allies *Basiliscus vittatus* (keeled ventrals) and *B. plumifrons* (small occipitals). Identification is rarely a problem, since *B. basiliscus* does not occur with *B. vittatus* and is sympatric with *B. plumifrons* only in southwestern Costa Rica. Individuals of the latter species are very distinct from *B. basiliscus* in being bright green or brown with a bright green head (very small juveniles).

DESCRIPTION: A large lizard, adult males to 900 mm in total length (standard length 130 to 250 mm) and 600 g, adult females to 610 mm in total length (standard length 135 to 194 mm) and 300 g, most examples 430 to 520 mm in total length; tail long, 70 to 75% of total length; head crest single, rounded in outline or with a posteriorly directed fleshy lobe (adult males), barely suggested in small juveniles (less than 75 mm in standard length); a middorsal series of large and small scales present in females and juveniles; adult males with a high, finlike dorsal crest covered with thin scales and supported by 16 to 18 elongate neural spines (to 60 mm in height); a separate similar, somewhat lower tail fin supported by 22 to 23 elongate neural spines present in adult males; supraorbital semicircles usually separated by one scale row, occasionally by two rows; occipital scales large, as large as or larger than largest supraocular scales, abruptly differentiated from small scales on posterior upper surface of head; three or four chin shields almost always touch infralabials; ventral surface of third phalanx of fourth toe with a series of very small scales that separate one row of large lateral scales from two rows of large medial scales. Adults: body brown, olive, or bronze, with darker crossbands and usually cream to yellow light stripes along lips and a pair of similarly colored lateral stripes. Juveniles: yellowish green with brown crossbands, lip and lateral stripes usually

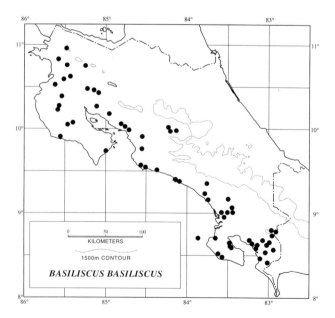

Map 10.1. Distribution of *Basiliscus basiliscus*.

present, all markings more vivid than in adults; throat with three distinct longitudinal light lines (fig. 10.19b); iris brown to bronze in all age-classes.

SIMILAR SPECIES: (1) *Basiliscus plumifrons* has the head (very small juveniles) or head and body bright green, and the throat lacks distinct light stripes (fig. 10.19a); adult males have a bipartite crest (fig. 10.17c). (2) *Basiliscus vitattus* has keeled ventral scales and a triangular head crest in adults (fig. 10.17a). (3) Species of *Ctenosaura similis* have whorls of large spiny scales around the tail (fig. 10.15). (4) *Iguana iguana* has a very large and conspicuous scale below the tympanum. (5) *Polychrus gutturosus* is bright green in ground color, lacks any indication of cephalic, dorsal, and caudal crests, and has a very long and attenuate tail that is round in cross section.

HABITAT: Along watercourses within Lowland Dry, Moist, and Wet Forests and Premontane Moist and Wet Forests, including gallery forests and secondary growth along streams in highly disturbed agricultural areas.

BIOLOGY: Our knowledge of the biology of this very common species has been greatly enhanced by the excellent work of Van Devender (1975, 1978, 1982a,b, 1983), and much of what follows is derived from his reports, which are based on populations in northwestern Costa Rica. These semiaquatic, diurnal lizards spend most of their time basking, foraging, and resting along streams. At night they sleep on perches from near ground level to 20 m high; adults tend to favor perches near or above a watercourse. Diurnal active body temperatures are 25 to 34°C (Fitch 1973a). They are especially conspicuous and numerous (200 to 400 per hectare) in the Dry Forest zone of northwestern Costa Rica. The name Jesus Christ lizard derives from their ability to run across the surface of water using erect bipedal locomotion with the arms held at the sides of the body. Smaller individuals are most adept at this behavior and may actually stroll on the water for some distance before sprinting for 10 to 20 m to seize prey or escape intruders. Because of their greater mass, large adults are less adroit and usually sink after enthusiastic splashing bursts of a few meters. In the drier areas of Costa Rica an observer may see as many as ten to fifteen of these lizards running down a small stream at one time as they are flushed out of the riparian vegetation. Snyder (1949) showed that water walking or strolling in these animals is facilitated by the large feet and specialized flattened toe fringes characteristic of all members of the genus. Basilisks are good swimmers and can swim underwater or remain submerged for considerable periods. The species is omnivorous, feeding mostly on active prey, including many kinds of insects and other arthropods, lizards, snakes, birds, mammals, and freshwater shrimp and fishes (Barden 1943; Van Devender 1983), but also eating flowers and fruits of many common streamside plants (e.g., *Ardisia*, *Muntingia*, *Cordia*, *Spondias*, *Manilkara*, *Psidium*, *Ficus*, and *Sloanea*). Juveniles are primarily insectivorous but may eat some small

fish. Pough (1973) and Fleet and Fitch (1974) indicated that herbivory increased with age but that adults fed on whatever was available. Hatchling iguanid lizards *(Iguana* and *Ctenosaura)* are a favorite food in northwestern Costa Rica. Breeding begins in March and females lay five to eight clutches of 2 to 18 eggs during the succeeding ten months. The eggs average 24 × 13 mm. Reduced reproduction is evident in January and February. Smaller and younger females lay fewer eggs; incubation averages about ninety days, and hatchlings are about 2 g in weight and 37 to 43 mm in standard length (average about 40 mm). In wetter areas reproduction appears to be prolonged, with reduced activity in the drier times of the year. Females reach sexual maturity at about twenty months of age at a standard length of 135 mm. Males exhibit a size-determined dominance hierarchy. Large males often attack younger ones, and males are often three to four years old before becoming fully integrated into the breeding system, although they reach sexual maturity sometime during the second year of life. Many males live to be four to six years of age (one was known to be seven years old), although females probably do not live as long. Raptors are the principal predators on basilisks, but opposums *(Philander* and *Chironectes)* and snakes are known to capture sleeping individuals at night.

REMARKS: Some authors divide this species into two geographic races, one ranging from Costa Rica to northwestern Colombia and the other restricted to northeastern Colombia and adjacent Venezuela. Adult males of the latter population have a head crest with a very low profile and a longer posterior projection than in northern examples (Maturana 1962; Lang 1989), but intermediate conditions occur in Panama and northwestern Colombia. Juveniles and young females of this species from Costa Rica have been mistakenly identified as the South American form, *Basiliscus galeritus*, by Taylor (1956) and Lang (1989). The latter differs from *B. basiliscus* most significantly in the sexually dimorphic features of the males. Adult male *B. galeritus* have a single high, rounded head crest and lack the well-developed dorsal and caudal fins and their elongate neural spine supports that are typical of *B. basiliscus* and *B. plumifrons*. Van Devender's (1983) comment that *B. galeritus* differs from *B. basiliscus* by lacking skin webbing between the dorsal rays is completely erroneous, since the neural spines are never elongate in the former. Instead, the middorsal scales are arranged in a regular series of enlarged triangular scales separated at regular intervals by two to four small intervening scales to form a low, saw-toothed ridge that continues onto the basal one-third of the tail. A similar ridged arrangement occurs in females and juveniles. *B. galeritus* resembles *B. basiliscus* in having large occipital scales and the supraocular semicircles separated by one scale (rarely two). Although the middorsal scale rows of other female and juvenile *Basiliscus* are also composed of large and small scales alternating, at least anteriorly, in a regular fashion, none approaches the distinct pattern found in *B. galeritus*. Lang (1989) stated that he examined numerous examples of the latter species from Costa Rica during the course of his study, and Köhler (1993a)—apparently based on Lang—perpetuated this error. I have been unable to locate a single specimen that confirms the occurrence of *B. galeritus* in Costa Rica or western Panama. As far as I can determine, the species is found only in northern and western Colombia and perhaps in eastern Panama. Although Lang's (1989) and Köhler's (1993a) distribution maps imply that *B. basiliscus* occurs on the Atlantic slope in southeastern Costa Rica and western Panama, I know of no confirmed records of the species from that region. The karyotype is 2N = 36, with six pairs of metacentrics and twelve pairs of microchromosomes; NF = 48 (Paull, Williams, and Hall 1976).

DISTRIBUTION: Lowlands and adjacent slopes on the Pacific versant from southwestern Nicaragua to northwestern Colombia, and on the Atlantic versant from central Panama to northwestern Venezuela, including the Meseta Central Occidental in Costa Rica (1–1,200 m).

Basiliscus plumifrons Cope, 1875

(plates 236–38; map 10.2)

DIAGNOSTICS: Adult males of this species, with their bipartite head crest (fig. 10.17c), well-developed dorsal and caudal crests, bright green coloration, and red eyes, are so striking in visual impact as to never be confused with any other animal. Females and juveniles are somewhat less impressive than the dragonlike males but are distinctive in their bright green coloration (only the head is bright green in very small juveniles). Coloration alone should distinguish female and juvenile examples of this species from other *Ba-*

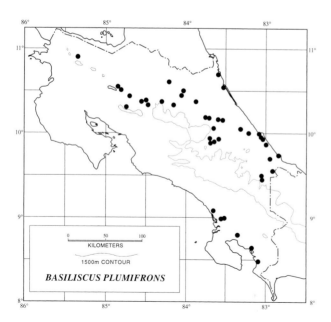

Map 10.2. Distribution of *Basiliscus plumifrons*.

siliscus, but they are further distinctive in having smooth ventral scales (keeled in *B. vitattus*) and occipital scales much smaller than the enlarged supraoculars (as large as or larger than in *B. basiliscus*) (fig. 10.18).

DESCRIPTION: A large lizard; males to 920 mm in total length (standard length in adults 122 to 250 mm); females 510 mm (standard length in adults 146 to 174 mm); most adults 750 to 800 mm in total length; tail long, 72 to 75% of total length; head crest in males bipartite, consisting of a narrow anterior flap and a broad, rounded posterior midcephalic flap (adults and some large juveniles) or an anterior ridge and a rounded posterior flap (many large juveniles); adult females and all but the smallest juveniles have a single flaplike crest; adult males with a high, finlike dorsal crest supported by 15 to 17 elongate neural spines (to 50 mm in height) covered by small scales; a separate similar, somewhat lower tail fin supported by 15 elongated neural spines also present in adult males; supraorbital semicircles usually separated by three to five scales (rarely one or two); occipital scales small, much smaller than enlarged supraocular scales, gradually grading into small scales on posterior head surface; three or four chin shields almost always contact infralabials; ventral surface of third phalanx of fourth toe with a series of very small scales that separate one row of large lateral scales from two rows of large medial scales. Head and body bright green, tending toward blue green in the largest individuals, usually with one or two rows of light spots (pale green, blue, yellow, or white) along flanks in adults; no transverse or longitudinal light or dark bands on head, body, or throat; females and juveniles usually more drab in coloration, tending to olive brown with lateral black bars; smallest juveniles (44 mm in standard length) reddish to golden brown with bright green heads; all individuals with yellowish green venter, brightest in juveniles; iris red in adult males, yellow in females and juveniles.

SIMILAR SPECIES: (1) *Basiliscus basiliscus* is brown or yellowish green, usually marked with dark crossbands and/or longitudinal light stripes; adult males have a single head crest (fig. 10.17b,d). (2) *Basiliscus vitattus* has keeled ventrals and is brown to olive in overall color with dark dorsal crossbands and/or longitudinal light stripes; adult males lack dorsal and caudal fins. (3) *Ctenosaura similis* has whorls of large spines around the tail (fig. 10.15). (4) *Iguana iguana* has a very large scale below the tympanic region. (5) *Polychrus gutturosus* lacks cephalic, dorsal, and caudal crests and has a very long and attenuate tail that is round in cross section.

HABITAT: Lowland Moist and Wet Forests and into Premontane Wet Forest, primarily along stream courses with considerable stands of trees still intact.

BIOLOGY: The habits of this common diurnal, semiaquatic, semiarboreal lizard are generally similar to those described for *B. basiliscus*. *B. plumifrons* is shier than its congener and less inclined to leave the vicinity of water.

Hirth (1963a) studied food habits of this lizard at Tortuguero, Limón Province, and reported that small juveniles appear to be primarily insectivorous but that larger examples (135 mm or more in standard length) contained a considerable amount of plant material (mostly seeds, fruits, and leaves). Examples of this species also ate shrimps, crabs, and a small free-tailed bat (Molossidae). Reproduction appears to take place throughout the year but is centered on the wet season (May to September). Clutches consist of 4 to 17 eggs about 22 × 14 mm. Hatching times in captivity were fifty-five to seventy-five days (Köhler 1993a). Hatchlings were 35 to 42 mm in standard length.

REMARKS: Van Devender (1983) and Lang (1989) incorrectly state that this species is restricted to the Caribbean slope. It also occurs in the rainforests of Pacific versant southwestern Costa Rica (as shown on Lang's figure 48) and probably in adjacent western Panama. It does not range into northwestern Costa Rica as erroneously shown on Köhler's map (1993a).

DISTRIBUTION: Humid lowlands on the Atlantic versant from eastern Honduras to western Panama and on the Pacific slope in southwestern Costa Rica and probably adjacent southwestern Panama; also in the premontane zone in central Costa Rica in the Río Reventazón valley (1–775 m).

Basiliscus vittatus Weigmann, 1828

(plate 239; map 10.3)

DIAGNOSTICS: Adult males are unlikely to be misidentified, since they have a single cephalic crest with a triangular outline (fig. 10.17a) and lack the high fins found in other Costa Rican Basiliscus. Females and juvenile males are distinct from the sympatric B. plumifrons in their dull brown to

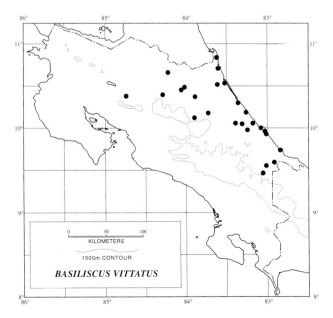

Map 10.3. Distribution of Basiliscus vittatus.

olive coloration marked with dark crossbands and longitudinal light stripes. This species also has keeled ventral scales and lacks a head casque, characters not found in combination in any other large Costa Rican lizard.

DESCRIPTION: A moderate-sized species; adult males to 590 mm in total length (80 to 170 mm in standard length); adult females to 470 mm in total length (83 to 132 mm in standard length); most adults 430 to 520 mm in total length; tail about 75% of total length; head crest single, triangular in outline in adult males, a fleshy ridge in females and juveniles; middorsal ridge of large and small scales in females and juveniles, in adult males raised into a low dorsal fin covered with small scales and supported by slightly elongate neural spines (20 to 25 mm high); serrate caudal crest of adults consisting of elongate pointed scales, best developed in males; supraorbital semicircles usually separated by two scales, sometimes one or three; occipital scales moderate, smaller than enlarged supraoculars but abruptly differentiated from posterior head scales; one or two chin shields in contact with infralabial; ventral surface of third phalanx of fourth toe covered by two rows of large scales, usually in contact. Ground color brown to olive, frequently with dark crossbands and two pairs of longitudinal light stripes along posterior side of head and along body; stripes yellow or upper stripe reddish posteriorly; pattern obscured in large adults; iris yellow or greenish yellow, suffused with black, with a yellow ring around pupil.

SIMILAR SPECIES: (1) Basiliscus basiliscus has smooth ventral scales and a rounded and/or pointed head crest (fig. 10.17b,d). (2) Basiliscus plumifrons has the head (small juveniles) or the head and body bright green, smooth ventral scales, and a rounded or bipartite head crest (fig. 10.17c,e,f,g). (3) Ctenosaura similis has whorls of spiny scales around the tail (fig. 10.15). (4) Iguana iguana has a very large conspicuous scale below the tympanum. (5) Polychrus gutturosus has a bright green ground color, lacks any indication of cephalic, dorsal, or caudal crests, and has a very long and attenuate tail that is round in cross section.

HABITAT: Lowland Moist and Wet Forest zones of the Atlantic versant including pastures and other disturbed areas; prefers more open habitats than other Basiliscus and is very common in coconut groves and other areas bordering sandy beaches.

BIOLOGY: This common lizard is much more terrestrial than its congeners and is frequently found at considerable distances from water. Hirth (1963b) and Fitch (1973a,b) have studied B. vittatus in some detail in Costa Rica, and much of the following information is derived from their accounts. It is frequently seen on the ground or on piles of debris and is less inclined to seek shelter in trees when disturbed than are other basilisks. The species is strictly diurnal and heliothermic and resembles B. basiliscus in most features of its biology, including bipedal locomotion across water. At night these lizards sleep under debris near the ground and in low vegetation (up to 3 m above the ground). Body temperatures during normal diurnal activity range from 33.1 to 37.5°C and are usually several degrees higher than ambient air temperature; nocturnal sleeping lizards had body temperatures of 22.5 to 26°C (Brattstrom 1965). Young of this species are exclusively carnivorous and feed principally on insects and spiders, with terrestrial crustaceans an important prey component in animals from coastal margins. Adults eat a wide variety of animal prey and considerable amounts of plant material (grasses, seeds, fruits, and stems). *B. vittatus* is eaten by many carnivores. It is particularly vulnerable to birds of prey (e.g., the great black hawk *Buteogallus urubitinga* and the white hawk *Leucopternis albicollis*; Gerhardt, Harris, and Vásquez 1993) because of its preference for open areas or forest fringes. The main reproductive period begins from mid-February to March and continues to October, but some eggs are laid throughout the year. Females probably lay four or five clutches per year of 2 to 18 eggs about 17 × 11 mm. Incubation time is fifty to seventy days. Hatchlings are 32 to 43 mm in standard length. These lizards become sexually mature at about six months, in females at a standard length of about 83 to 87 mm and in males at about 85 mm. Maximum life expectancy in the field is two to three years, but one captive animal lived for nearly six years (Conant and Downs 1940).

REMARKS: Lang (1989) erroneously included most of Panama and northwestern Colombia in the range of this species. Köhler (1993a) erroneously included southwestern Nicaragua, Pacific slope Costa Rica, and southwestern Panama within the range of this species on his map. The karyotype is 2N = 36, with six pairs of metacentric macrochromosomes and twelve pairs of telcocentric microchromosomes; NF = 48 (Gorman, Atkins, and Holzinger 1967).

DISTRIBUTION: Lowlands and premontane slopes from Jalisco and southern Tamaulipas, Mexico, south through Central America to northwestern Colombia, exclusive of Pacific versant southern Nicaragua, Costa Rica, and southwestern Panama; also introduced into southern Florida; confined to the Atlantic versant in Costa Rica (0–120 m).

Genus *Corytophanes* Boie in Schlegel, 1826

The helmetlike head casque with paired raised lateral crests that fuse posteriorly to form a single serrate midcephalic crest is unique to the American forms of this genus (three species) of arboreal lizards. Other features that characterize the genus include body and tail strongly compressed; limbs well developed; superciliary scales quadrangular, not overlapping; toes without scaly fringe; expanded parietal bone underlying casque and extending posteriorly as a thin blade that supports the midcephalic crest; casque and crests present in all age-classes and both sexes; parietal foramen in frontal bone; clavicles without fenestrations; dorsal scales irregular and unequal, larger ones weakly to strongly keeled, others smooth; ventral scales strongly keeled, slightly im-

bricate; an extensible gular appendage; infradigital lamellae with a single keel; no caudal autotomy septa or autotomy.

The genus ranges from Mexico (San Luis Potosí and Veracruz on the Atlantic slope and Oaxaca on the Pacific) south to northwestern Colombia.

Corytophanes cristatus (Merrem, "1822")
Camaleón, Perro Zompopo

(plates 240–41; map 10.4)

DIAGNOSTICS: The distinctive helmet-shaped casque with raised lateral ridges that join to form the medial dorsal cephalic crest instantly identify this species.

DESCRIPTION: A moderate-sized species (to 360 mm in total length, 120 mm in standard length), adult females (80 to 120 in standard length) average larger than males (adults 99 to 120 mm in standard length); tail long, 65 to 72% of total length; a well-developed head casque mounted by a Y-shaped crest that continues as a median serrate nuchal crest along posterior margin of head; upper head scales smooth, small; irregular occipital scales small; scales of supraorbital semicircles smooth; no squamosal spine above tympanum; gular scales small, keeled, arranged in longitudinal rows separated by granules; a midgular crest of large triangular scales; a serrate dorsal crest of large scales continuous with nuchal crest; no caudal crest. Ground color olive to rich chocolate or reddish brown; head with distinct black lines radiating from eye across light-colored upper lip onto infralabials; dorsum with irregular darker brown to black spots or crossbars; a white spot on elbow and at underside of arm near axilla; often with some green on head, body, limbs, and tail; venter light brown with a few darker and lighter markings; iris coppery gold.

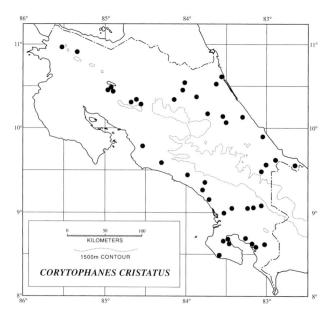

Map 10.4. Distribution of *Corytophanes cristatus*.

SIMILAR SPECIES: There are none in Costa Rica.

HABITAT: Lowland Rainforests and Premontane Wet Forest and Rainforest, primarily in undisturbed areas.

BIOLOGY: *Corytophanes cristatus* is a strictly arboreal, nonheliothermic lizard. Its physical resemblance to African chameleons is striking, and like them it is a sit-and-wait predator. These lizards carry the sit-and-wait approach to the extreme, remaining completely motionless, usually on a vertical perch with the head pointed upward, for hours at a time until some animal approaches close enough to be snapped up. They apparently descend from their perches to feed on the forest floor as well. On the ground, they frequently use erect, bipedal locomotion, running upright for 2 to 10 m. They prey on large, generally slow-moving arthropods, primarily lepidopterous and coleopterous larvae and orthopterans, but they also feed on small lizards of the genus *Norops* (Andrews 1983). At least some of these are eaten during the brief forays to the forest floor. Andrews (1979a) concluded that they capture prey infrequently, perhaps only every day or so. The coloration of *C. cristatus* and the behaviors described above combine to produce the epitome of passive concealment as an antipredator defense. Davis (1953) described in some detail the rigid or cataleptic freezing of body position this species uses to eliminate all minor movements. Although they are relatively common animals, it is unusual to see one in the field during the day, and most examples are located while sleeping at night on their vertical perches (usually tree trunks). The first line of defense against intruders is camouflage, followed by catalepsy in response to a specific localized stimulus. If these fail and a lizard becomes excited or threatened, it compresses its body, erects its nuchal crest, expands its gular pouch, and bobs its head. These actions, combined with presenting the broadened lateral area to the threatening agent, increase the lizard's apparent size and may discourage the perceived aggressor. If all else fails, *C. cristatus* attacks and bites. Fighting among captive males follows this same sequence. Bock (1987) discovered a female at Rara Avis, Heredia Province, on June 25 that had dug a shallow nest (50 mm in diameter, 95 mm deep) in the ground that contained 5 eggs. When disturbed she displayed the defensive behaviors described above, although it is unclear whether she was guarding the eggs. Bock suggested that this species may use the head casque as a scoop for nest evacuation, but this needs confirmation. Some support for the idea is provided by Lazcano and Góngora's (1993) observation of a female that had soil on her head after depositing 5 eggs in a shallow nest. There is evidence (Fitch 1970) that breeding and egg laying occur throughout the year. Taylor (1956) found a female ovipositing a clutch of 6 eggs in a nest dug in the ground on September 9 at Golfito, Puntarenas Province. Clutch size is 4 to 11; average egg dimensions are 12 × 25 mm; hatchlings are about 30 to 37 mm in standard length. Eggs laid in captivity had an incubation time of 115 to 155 days (Ream 1965; Köhler, Ritt-

mann, and Ihringer 1994). This species is erroneously reputed to be venomous by many Costa Ricans.

REMARKS: The other members of this genus differ from *C. cristatus* as follows: *Corytophanes hernandezi* (eastern and southwestern Mexico, to the Yucatán Peninsula in Mexico, Guatemala and Belize) has the nuchal crest not continuous with the dorsal crest, the upper head scales keeled or striated, and a prominent lateral squamosal spine above the tympanum. *Corytophanes percarinatus* (Isthmus of Tehuantepec to northern Guatemala and El Salvador) has the scales on the upper head surface distinctly keeled or rugose and a prominent lateral squamosal spine above the tympanum. The karyotype is 2N = 36, with six pairs of metacentrics and twelve pairs of microchromosomes; NF = 48 (Schwenk, Sessions, and Peccinini-Seale 1982).

DISTRIBUTION: Lowland and premontane areas on the Atlantic slope from central Veracruz and the western margin of the Yucatán Peninsula in Mexico south to northwestern Colombia; also on the Pacific versant in southwestern Costa Rica and Panama and marginally in northwestern Costa Rica (2–1,640 m).

Family Iguanidae Gray, 1827

Iguanas

A family of mostly large to very large omnivorous or strictly herbivorous lizards usually having a well-developed dorsal crest and/or a tail covered with whorls of large spiny scales. Eight genera may be placed in the family: *Dipsosaurus* (one species) and *Sauromalus* (five species) of the southwestern United States and adjacent Mexico; *Iguana* (two species), with a broad range in the Neotropics; *Ctenosaura* (twelve species) of Mexico and Central America; *Cyclura* (eight species) in the West Indies; *Amblyrhynchus* (one species) and *Conolophus* (two species) of the Galápagos Islands; and *Brachylophus* (two species) of the Fiji and Tonga Islands.

Iguanas share the following general features: body laterally compressed, cylindrical, or depressed (*Sauromalus*); eye large, pupil round; tympanum large to very large; interparietal scale not enlarged; dorsal scales small, usually keeled; ventrals larger than dorsals, smooth or slightly keeled; femoral pores present; no spinulate scale organs; infradigital lamellae with two or three keels; gular fold complete medially.

Additional (mostly osteological) features characteristic of iguanids include parietal foramen in frontoparietal suture or frontal bone (*Dipsosaurus*); supratemporal on medial side of supratemporal process of parietal; nuchal endolymphatic glands not penetrating neck musculature; labial blade of coronoid large; Meckel's groove closed; lateral teeth on jaws triconodont or with up to 15 to 20 small cusps; caudal autotomy septa usually present, with a pair of transverse processes one set anterior and the other posterior to septum; two paired transverse processes on posterior vertebrae retained in forms lacking autotomy (genera from Galápagos and Fiji and *Iguana delicatissima*); posterior coracoid foramen present; sternal fontanelles very small or absent; post-xiphisternal inscriptional ribs variable, short or long and confluent medially; vertebrae with zygantra and zygosphenes; nasal chamber with long S-shaped vestibule and concha; hemipenes slightly bilobate, with a simple sulcus spermaticus, apical region calyculate; colic septa present. All members of this family are oviparous and lay cream to white eggs with flexible shells.

Genus *Ctenosaura* Wiegmann, 1828

Spiny-tailed Iguanas

Members of this genus are moderate to large reptiles (to about 1.3 m in total length) characterized by having the caudal scales arranged in whorls of large spiny scales that are separated by whorls of small smooth or keeled scales, a condition unique among lizards occurring on the mainland of the Americas. Other features that in combination define the genus include a distinct dorsal crest of enlarged scales; dorsal scales small, smooth or keeled; upper head scales small, smooth; one subocular scale elongate, much larger than others; a gular pouch or a pendulous gular appendage present (well developed in *Ctenosaura palearis* and *C. melanosterna*, weakly developed in *C. bakeri*); infradigital lamellae usually not fused at their bases to form comblike structures (conspicuous combs present under proximal phalanx of digit III in *C. defensor*); lateral teeth on jaws triconodont or usually four cusped; caudal autotomy septa present but fused over by bone during ontogeny in one species *(C. acanthura)*. Queiroz (1987a) referred all species formerly placed in the nominal genus *Enyaliosaurus* Gray to *Ctenosaura*, and I follow his conclusion here. Lizards formerly included in *Enyaliosaurus* have a patch of enlarged, strongly keeled scales on the anterodorsal surface of the lower hind limb, as indicated by an asterisk (*) in the following discussion. The largest species included in this genus—*C. acanthura* (eastern Mexico), *C. bakeri** (Isla Utila, Honduras), *C. hemilopha* (northwestern Mexico and southern Baja California), *C. melanosterna* (north-central Honduras), *C. oedirhina** (Islas Roatán and Santa Elena, Honduras), *C. palearis* (Atlantic versant northeastern Guatemala), *C. pectinata* (western Mexico), and *C. similis* (southeastern Mexico to Panama)—have long tails (55 to 73% of total length), more than 30 caudal vertebrae, and spinous whorls proximally but not distally. The smaller forms (to about 300 mm in total length)—*C. clarki** (Michoacán, Mexico) and *C. defensor** (northern Yucatán Peninsula, Mexico)—have short tails (44 to 52% of total length), fewer than 30 caudal vertebrae, and a tail that is spinous for its entire length. *C. quinquecarinata**, which occurs in Costa Rica, and the recently described *C. flavidorsalis** of Honduras and El Salvador and *C. alfredschmidti** of Campeche, Mexico, are intermediate in size and have moderately long tails but resemble the large species in cau-

dal features. Beside the review of the systematics of this genus by Queiroz (1987b), Köhler (1993b, 1995a,b,c, 1996) and Köhler and Streit (1996) have published several papers constituting a general revision of the group. Köhler and Klemmer (1994) and Köhler (1995d) also described two new species, *C. flavidorsalis* and *C. alfredschmidti*, respectively, and Buckley and Axtell (1997) described *C. melanosterna*.

The genus ranges over the lowlands of tropical and subtropical Mexico (including Baja California) to eastern Nicaragua and adjacent islands and the valley of the Great Lakes in Nicaragua on the Atlantic versant and to Panama on the Pacific slope; also on the Atlantic slope in southeastern Costa Rica to central Panama; *C. pectinata* has been introduced to southern Florida and extreme southern Texas.

KEY TO THE COSTA RICAN SPECIES OF THE GENUS *CTENOSAURA*

1a. Tail dorsolaterally compressed proximally; a patch of enlarged, strongly keeled scales on anterodorsal surface of lower leg; dorsal crest interrupted at sacrum . . . . *Ctenosaura quinquecarinata* (p. 434)

1b. Tail subcylindrical; no patch of enlarged, strongly keeled scales on anterodorsal surface of lower leg; dorsal crest continuous from nape onto tail. . . . . *Ctenosaura similis* (p. 435)

Ctenosaura quinquecarinata (Gray, 1842)

(plate 242; map 10.5)

DIAGNOSTICS: This species can be confused only with small examples of its larger congener *Ctenosaura similis*, since they are the only Costa Rican lizards with spiny tails. *C. quinquecarinata* is immediately distinguishable by its

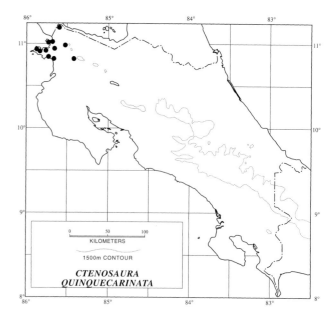

Map 10.5. Distribution of *Ctenosaura quinquecarinata*.

proximally flattened spiny tail and patch of enlarged, spiny scales on the anterodorsal surface of the lower hind limb (tail cylindrical and no scale patch in *C. similis*).

DESCRIPTION: A moderate-sized species (to 475 mm in total length, adult males 115 to 185 mm in standard length), males larger than females (female maximum total length about 320 mm, adults 100 to 154 mm in standard length); tail moderate, 52 to 57% of total length in adults; a distinct middorsal crest of enlarged scales; postmental scales usually two; no gular appendage; total femoral pores 8 to 18, minute in females; tail greatly flattened, covered with whorls of greatly enlarged spines proximally; first ten whorls separated from one another by one row of small scales dorsally and laterally. Ground color olive green, usually with darker crossbands indicated and a series of light olive green spots and blotches interrupting the bands on the dorsum; dorsal crest with alternating dark and light bands anteriorly; upper surfaces of limbs spotted with black and/or olive green; upper surface of head and neck uniform; lateral head surface and lower jaw black; most of undersurfaces grayish white, throat black; gular pouch of males black; juveniles with a bright green ground color on upper surfaces of head, body, limbs, and tail.

SIMILAR SPECIES: (1) *Ctenosaura similis* has a cylindrical tail and lacks enlarged spinous scales on the anterodorsal margin of the lower hind limb. (2) *Iguana iguana* lacks tail spines and has a greatly enlarged smooth scale below the tympanum.

HABITAT: The species is restricted to the Lowland Dry Forest region in extreme northwestern Costa Rica in relatively open woodlands.

BIOLOGY: *Ctenosaura quinquecarinata* is a relatively, uncommon diurnal, semiarboreal omnivore that is often found on stumps and trees and uses tree hollows for refuge (Villa and Scott 1967). A captive female laid five eggs that were 26.6–29.2 × 16.7–18.5 (Werler 1970).

REMARKS: This species is currently known from extreme northwestern Costa Rica, primarily near the coast in Santa Rosa National Park, Guanacaste Province. A specimen (UCR 7407) kept for some time in the little zoo at La Pacifica, Guanacaste Province, was erroneously labeled as being from there but was originally collected in the Santa Elena area of the park. There remains some question whether the several allopatric populations grouped under this name (Gicca 1983) are conspecific. Gicca's (1983) and Köhler's (1993b) distribution maps erroneously indicate a record of this species from north-central Honduras, apparently based upon a confusion with *C. palearis*, which occurs in the Aguan valley. The population from the Atlantic versant upper Comayagua valley in Honduras formerly included in this species has recently been described as a distinct form, *C. flavidorsalis* Köhler and Klemmer, 1994; examples from the interior valleys of El Salvador are also referable to *C. flavidorsalis*.

DISTRIBUTION: Disjunct in semiarid and subhumid Pacific slope lowland forests as follows: southern Oaxaca, Mexico, near the Isthmus of Tehuantepec (2–100 m); west-central Nicaragua (100–600 m); northwestern Costa Rica (1–145 m).

Ctenosaura similis (Gray, 1831a)

Iguana Negra, Black Ctenosaur, Garrobo, Gallina de Palo

(plates 243–44; map 10.6)

DIAGNOSTICS: The long cylindrical and spinose tail of this species makes it unmistakable. Juveniles might be initially mistaken for *C. quinquecarinata* or young *Iguana iguana*, but the tail differences (proximally flattened in the former, without spines in the latter) make recognition easy. In addition, *C. similis* lacks the patch of enlarged spines on the anterodorsal margin of the lower hind limb (characteristic of *C. quinquecarinata*) and the greatly enlarged smooth scale found below the tympanum in *I. iguana*.

DESCRIPTION: A large lizard reaching a total length of 1.3 m and a weight of 2 kg in adult males (adult males 239 to 489 in standard length), adult females smaller than males (to 870 mm in total length, 204 to 400 mm in standard length, 1 kg in weight); tail moderate, 60 to 67% of total length in adults but about 70% in small green juveniles; a distinct middorsal crest of enlarged scales best developed in adult males; postmental scales usually four; no gular appendage; total femoral pores 6 to 18; tail cylindrical or slightly compressed, covered with whorls of large spines; first ten whorls separated from one another by two or more rows of small scales dorsally and laterally. Adults: yellowish, gray, or tan with a series of four to six broad, ill-defined dark dorsal crossbands, usually separated along midline by

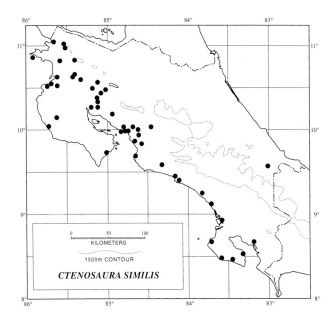

Map 10.6. Distribution of *Ctenosaura similis*.

definite yellow or cream interspaces, dark bands often split by light; limbs similarly banded; often with some orange or red spots, especially in breeding season; flanks and limbs spotted with dark and light; tail marked with alternating dark and light bands; pattern, especially on anterior part of body, becoming obscure in large examples; head and anterior portion of body and forelimbs black in large males; tail pattern becoming obscured in large individuals. Juveniles: bright green or pale tan with dark reticulations on hatching, but turning bright green shortly thereafter; green color persisting for about a year or until the lizard is about 120 mm in standard length, when the banded pattern appears.

SIMILAR SPECIES: (1) *Ctenosaura quinquecarinata* has a greatly flattened tail proximally. (2) *Iguana iguana* has a very large smooth scale below the tympanum.

HABITAT: Lowland Dry and Moist Forests in relatively open and/or severely disturbed situations, especially pastures. Ctenosaurs may be found in vacant city lots, gardens, cemeteries, and junk piles near cultivated fields. They prefer locations with rocky outcrops, fallen logs, or standing hollow snags and are often found in abundance along rock walls and fence lines, using them for basking or shelter. They also occur along the entire Pacific coastline in association with sandy beaches but are not found in the adjacent rainforest habitats. The only Atlantic lowland record is from some distance inland.

BIOLOGY: This diurnal, semiarboreal lizard frequently occurs in considerable numbers at favored sites and is easy to observe. Fitch and Henderson (1978), Fitch and Hackforth-Jones (1983), and Van Devender (1982b) have carried out substantial field studies on the species in Costa Rica, and their reports and Köhler (1993b) provide much of the material summarized here. Adult ctenosaurs are heliotherms and spend most of their waking time basking. Preferred body temperatures are 36 to 37°C, and as might be expected, adults do not ordinarily emerge from their shelters on rainy or overcast days or at night. Individuals are active at all times of the year and spend most of the day on the ground. The green juveniles perch in low vegetation at night, but larger lizards rest or hide in various shelters (burrows, rock piles, or hollow trees and fence posts) during the day and sleep in them at night. These lizards excavate their own burrows using their well-developed sharp, curved claws and may construct rather complicated tunnel systems (Burger and Gochfeld 1991). Typical burrows have a single entrance; they are 200 to 400 mm below ground level and 1 to 2 m long, with room for the lizard to turn around in the twisting main tunnel or its branches. Nest burrows are much longer and include many side tunnels. An elevated perch near the shelter is required for basking and display. Ctenosaurs of all ages and sexes are territorial, with defense centered on each individual's shelter and perch. Males exhibit the strongest territorial behavior, which varies seasonally. However, there is frequently a dominance hierarchy among males so that many may occur close together with minimal interaction. Actual fighting is unusual, and they use a head-bobbing display to establish territorial rights. The body may be laterally flattened, the head uplifted, the gular pouch expanded, and the mouth opened to impress any intruder. These actions are also used in courting females. A male and female may share a shelter and perch for considerable periods, but the males are polygynous and may wander off to associate with a new female at any time. Adult individuals rarely forage or move very far from their home site, and most activities take place within 1 to 10 m of the retreat. Young ctenosaurs feed primarily on insects, while larger individuals take increasing amounts of plant material. Adults are essentially opportunistic omnivores eating large amounts of vegetation during the rainy season and primarily flowers (e.g., *Tabebuia*) and fruits (especially *Spondias*) in the dry season. These are supplemented by animal food, including insects, spiders, marine crabs, rodents, bats, lizards, lizard eggs (including those of their own species), frogs, small birds, and perhaps smaller ctenosaurs. Breeding begins in December or January and continues into February. Eggs are laid mostly in March in burrows in open, sunny places, frequently along with the eggs of other females and sometimes with those of *Iguana iguana* (Mora 1989). Each clutch is separated into its own incubation chamber. Nest burrows may be extensive (11 to 200 m in length) and complexly twisted. Females produce one clutch a year; clutch size varies from 12 to 88 eggs and is correlated with female size. Hatching occurs from April to July but is concentrated near the beginning of the wet season in May. Hatchlings are 48 to 58 mm in standard length. Some males attain sexual maturity as early as the first year, but most reach that state in the second year. Hatchlings are very different from adults, not only in coloration but in being much more slender, active, and nervous; their heads are shorter and their tails are much longer (about two and a quarter to two and a half times the standard length) than in adults (one and a half to two times). At first they are terrestrial, but they soon begin to spend most of their time climbing in low trees or bushes. They are heavily preyed on by hawks, jays, basilisks, and probably small carnivorous mammals. The snake *Trimorphodon biscutatus* is a principal predator, and another snake, *Loxocemus bicolor,* eats ctenosaur eggs (Mora 1987). Adult *Ctenosaura* are more difficult to attack than the young, since they rush for shelter when approached; if cornered they lash out with the spiny tail in a most effective defense, and they bite if caught. Ctenosaurs are prized as food by humans in Mexico and most Central American countries, where they are sold alive by the thousands (along with *Iguana iguana*) in the central markets of all the larger towns (Fitch, Henderson, and Hillis 1982). The meat is preferred to that of the iguana and is said to have medicinal value, especially for sexual dysfunction. While it may have been an important food item for *campesinos* in the past, it does not appear to be heavily hunted today in the republic, and much hunting seems to be recreational. Unlike the situation elsewhere in Central America, there appears to be no commercial exploitation of this species in Costa Rica. The hunted animals may be shot, noosed, or dug out of their burrows, often with the aid of trained dogs.

REMARKS: No thorough study of geographic variation has been carried out on this form, although some authors follow Barbour and Shreve (1934) in recognizing the population on Isla San Andrés (Colombia) as distinctive in coloration, a view rejected by Schwartz and Henderson (1988). The range of this species is usually stated as including the entire Atlantic versant of Central America and Panama as reported and mapped by Köhler and Streit (1996). Since the species is restricted to marine strand areas with rocky outcrops on the Pacific slope from central Costa Rica southward to central Panama, it seems logical that it may have a similar distribution on the Caribbean coast. However I have been able to locate only one authentic record for the species on the Atlantic versant of Costa Rica and none for northwestern Panama or most of eastern Nicaragua. The Costa Rican specimens are from well inland in the valley of the Río Telire (KU 56060–56061; Limón: Suretka, 60 m). I have not seen *C. similis* at other Caribbean localities. This suggests that these ctenosaurs may be absent from most of the Atlantic lowlands from northeastern Honduras to Panama, probably because this stretch of coast consists primarily of sandy beaches with few if any rocky areas. Fitch and Henderson (1978) mentioned seeing a roadkill of this species on the old Carretera Interamericana at Río Itiquís, Alajuela Province (862 m). This may have been an animal killed elsewhere and thrown onto the road by someone returning from the Pacific coastal area. It is possible that *C. similis* formerly ranged or still ranges up the Río Grande de Tárcoles valley to the western margin of the Meseta Central Occidental.

DISTRIBUTION: Lowlands and lower premontane areas from the Isthmus of Tehuantepec eastward and southward on the Pacific versant to central Costa Rica and then southward in coastal areas to central Panama and on the Atlantic versant to eastern Honduras, with scattered records in Costa Rica and central Panama, including the coastal islands of Honduras, Isla de Maíz Grande (Nicaragua), and Islas San Andrés and Providencia (Colombia); in Costa Rica primarily on the Pacific lowlands but possibly or formerly ranging up the valley of the Río Grande de Tárcoles to the western portion of the Meseta Central Occidental and with one Atlantic lowland record (1–765 m).

Genus *Iguana* Laurenti, 1768

These very large and conspicuous arboreal lizards (to somewhat over 2 m in total length) are among the most characteristic and familiar inhabitants of the Neotropical forests. Their most obvious distinguishing characteristics are the well-developed serrate, middorsal crest composed of tall, curved spines that are especially prominent in adults, a large pendulous fan bordered along its free margin by large pointed scales, and enlarged smooth scales covering the upper surface of the snout. In addition, the tail is covered by whorls of even-sized small scales that are keeled but never spinous. In *Iguana* (two species) the dorsal crest continues onto the proximal one-third of the tail; dorsal scales small, keeled; subocular scales all about same size; lateral teeth on jaws serrate with numerous small cusps (as many as 15 to 20 per tooth); caudal autotomy septa present *(I. iguana)* or absent *(I. delicatissima)*. Etheridge (1967) has pointed out that in *I. iguana* the septa are fused over during ontogeny so that large examples secondarily lack them.

The genus ranges over the Neotropical lowlands from Mexico to Ecuador and Bolivia, Paraguay, and south-central Brazil; also in the Lesser Antilles as far north as Anguilla in the Virgin Islands and on several other Caribbean islands. One species, *I. iguana*, has been introduced into southern Florida.

Iguana iguana (Linné, 1758)

Iguana, Green Iguana, Iguana Verde

(plates 245–46; map 10.7)

DIAGNOSTICS: Always readily recognizable, even at some distance, by the presence of a greatly enlarged smooth scale near the posterior margin of the head below the tympanum on the side of the throat.

DESCRIPTION: This is the largest species of Central American lizard, with males reaching a total length of 2,010 mm (adults 250 to 580 mm in standard length), and weighing up to 4 kg; adult females smaller (to 1,440 mm in total length, 236 to 411 in standard length) and weighing up to 2.9 kg; tail moderate to long, somewhat compressed, attenuate and

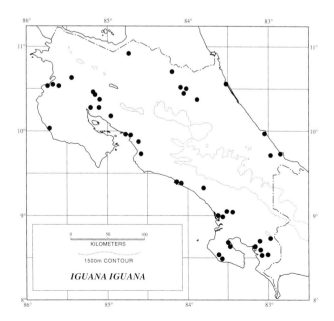

Map 10.7. Distribution of *Iguana iguana*.

whiplike, about 65 to 72% of total length; a distinct mid-dorsal crest of enlarged scales (10 to 40 mm high in adult males) continues onto tail; a pendulous extensible gular appendage, larger in males, has a midventral series of enlarged triangular scales anteriorly; total femoral pores 18 to 46; tail covered by uniform small keeled scales, slightly larger inferiorly. Most adults green, often with obscure wavy transverse dorsal crossbands that may have light borders; bluish green shoulder stripes and black abdominal stripes (most obvious in displays); undersurfaces pale green; head green, gray, or white with a few dark markings; tail banded with dark; very large males uniform gray, gold, or tan; excited animals lighter in color; iris orange tan. During the breeding season males develop broad dorsal areas of gold to red orange, and the largest males become nearly uniform gold to red orange; juveniles bright green above; undersurfaces yellow in all age-classes.

SIMILAR SPECIES: (1) Species of *Basiliscus* lack the greatly enlarged smooth scale below the tympanum. (2) Species of *Ctenosaura* have spinous whorls of scales around the tail and lack the enlarged smooth lateral head scale. (3) *Polychrus gutturosus* lacks nuchal and dorsal crests and has a very long, attenuate tail that is round in cross section.

HABITAT: Lowland rainforests, both in the forest and along forest margins, and gallery forests and riparian vegetation in the Lowland Dry Forest zone. Large solitary adults are usually spotted in trees, often near the canopy 20 to 30 m above the ground at forest edges.

BIOLOGY: This rather common diurnal species is much more arboreal than the black ctenosaur *(Ctenosaura similis)*, is active throughout the year, and is a heliotherm with preferred body temperature of 33 to 38°C. Adults spend most of their time in trees, foraging and basking during the day and sleeping near the basking perch at night. Young spend much time on the ground or in low vegetation during the day but climb to a height of 2 to 3 m to sleep at night. Young iguanas, unlike young ctenosaurs, are primarily herbivorous, eating leaves, flowers, and fruits, although they occasionally take insects. Adults are almost exclusively herbivorous, feeding principally on leaves but occasionally eating small animals. Iguanas are usually quadrupedal but may accelerate to an upright bipedal position with the arms held out at right angles to the sides (typical of juveniles) or against the sides (usual in adults). Iguanas are excellent swimmers and may be found far from the shore, even in the ocean. They often jump into the water to avoid intruders or predators, usually submerging and swimming with graceful undulations of the body and long tail, the limbs held against the sides. Individuals of this species are given to violent and explosive movements when disturbed, and the heavy adults frequently lose their footing in such maneuvers and come crashing down to the forest floor, a surprise to both the field biologist who is startled by the thrashing in the trees and to the slightly stunned iguana that lands on the ground! Important studies on the biology of iguanids have been carried out by Burghardt and Rand (1982) and their associates (especially Dugan 1982) in Panama and by Van Devender (1982b) in Costa Rica. Much of the information summarized here is from the published work of these scientists. Adult iguanas occur singly or in groups of twelve to twenty. Groups are generally composed of one large and up to three medium-sized males (320 to 360 mm in standard length), several smaller males (260 to 310 mm in standard length), and four to six females. Except during the breeding season, aggressive interactions between individuals are minimal, but there is a clear dominance hierarchy among males based on size. Large males establish territories beginning in mid-October, just before the breeding season, and defend both the territory and the females within it against other males. They patrol the territory, but it is maintained primarily by an increase in display bouts. Display includes extending the gular appendage, bobbing the head, and compressing the body when approached by or approaching an intruder. This may be followed by strutting, circling, hissing, and tail lashing. Fighting may occur on occasion, consisting first of a face-off or head to neck contact, then pushing or leaning on the opponent, followed by mounting and neck biting. Sometimes the sequence may be repeated several times if the males are evenly matched. Ultimately the neck biting causes the mounted male to suddenly turn darker and flee. Interactions between females are infrequent but include extending the dewlap, compressing the body, circling, hissing, and snapping. During the breeding season each mating territory contains one to four females, and the polygynous male courts all of them. A courtship period of at least four weeks precedes copulation, and usually a male courts and is temporarily bonded to a particular female for about two weeks before mating occurs. Typically, courtship involves the male's approaching the female from behind and performing a rapid, low shudder-bob; the female then moves her tail to one side, and the male bobs again. This behavior may be repeated many times. If the female is receptive the male continues shudder-bobbing, mounts the female from behind, bites her neck, and twists his tail under hers to insert a hemipenis. Large males tolerate small males (since they resemble females in size and coloration), and the small males may breed opportunistically with females when the dominant male is occupied. Medium-sized males are driven off the large male's territory. The actual breeding season is concentrated in the first six weeks of the dry season (mid-November to January). Females descend from the trees to nest in the open, usually in banks or beaches along rivers or other forest-edge situations, sometimes migrating as far as 3 km to suitable nesting sites. In Costa Rica eggs are laid from March to April in a nest burrow the female constructs using her sharp, curved claws. Burrows typically are about 1 or 2 m long and 300 to 600 mm below the surface and are large enough to allow the female to turn around. However, communal nests of considerable complexity (up to 24 m of tunnels, multiple entrances, and individual incubation chambers) may be constructed and shared by as many as eight females (Rand and Dugan 1983). Communal nesting also occurs with *Ctenosaura similis* at Palo Verde, Guanacaste Province, in similarly large and complex burrows (up to 200 m in extent) (Mora 1989). Females apparently produce a single clutch of 9–35.5–71 eggs, and clutch size is correlated with female size. Incubation time is 65 to 115 days, and in Costa Rica hatching occurs from April to June. Hatchlings are 70 to 80 mm in standard length. Young iguanas spend considerable time basking on the ground or in clumps of low vegetation. At night they move into low branches and tend to aggregate in clusters of ten to twenty, some of them in physical contact. Sexual maturity usually is attained in the third year (at a standard length of about 260 mm), but some individuals may breed at the end of two years. The eggs of iguanas are eaten by the snake *Loxocemus bicolor* (Mora 1987) and probably by many small mammals (Rand and Robinson 1969). Young iguanas are preyed on by a wide variety of animals including the boas of the genera *Boa* and *Corallus*, the colubrid snake *Oxybelis aeneus*, basilisks (*Basiliscus* spp.), a number of raptors and other birds (e.g., toucans, *Rhamphastos*; anis, *Crotaphaga*) and small mammalian carnivores. Adults are preyed on by crocodiles *(Crocodylus acutus)*, king vultures *(Sacroramphus papa)*, and probably large hawks, cats (*Felis* spp.), coatis *(Nasua narica)*, and tayras *(Eira barbara)* (Greene et al. 1978). The principal line of defense against predation is camouflage and staying motionless. Other defenses include lunging, snapping, head shaking, and tail lashing, and ultimately violent flight, which often involves jumping or falling to the ground or diving into a body of water. Since iguanas and their eggs are prized food items in Latin America, their numbers have been greatly reduced by human predation. Unlike other Central American countries, Costa Rica has no recent history of commercial exploitation of this food source, but the eggs (considered a great delicacy) and meat are harvested locally. Dogs are usually used to find nests. The Costa Rican government has established a project near Orotina, Puntarenas Province, designed and operated by Dagmar Werner, an authority on large lizard biology, as an experiment in raising *I. iguana* commercially (Cohn 1989). Eggs are collected, incubated, and hatched, and the young iguanas are maintained in wire coops. With a constant rich supply of food, especially fruits, the lizards grow and mature twice as rapidly as in the field. When they reach adulthood they may be released or marketed. The government hopes this operation will provide sufficient eggs and meat to satisfy the market and reduce hunting of wild populations.

REMARKS: No extensive analysis of geographic variation in this species has been undertaken, and the purported differences (Dunn 1934) between northern (Mexico to Costa Rica) and southern populations have been shown to be the result of individual variation (Lazell 1973). The only other member of the genus, *Iguana delicatissima* from the Lesser Antilles, has a small subtympanic scale that is separated from the tympanum by more than twelve small scales (scale greatly enlarged and separated from the tympanum by no more than twelve small scales in *I. iguana*). A record of this species (KU 67224; Puntarenas: Agua Buena, 1,000 m) from Premontane Wet Forest in southwestern Costa Rica is suspect, since elsewhere in Costa Rica it is not known from above an elevation of 500 m. The karyotype of *I. iguana* is 2N = 34, with six pairs of metacentrics and eleven pairs of microchromosomes; NF = 46 (Cohen, Huang, and Clark 1967).

DISTRIBUTION: Lowlands from Sinaloa and Veracruz, Mexico, to Ecuador on the Pacific versant and northern Bolivia, Paraguay, and south-central Brazil on the Atlantic slope; also in the Lesser Antilles out to the Virgin Islands and on Isla de Maíz Grande (Nicaragua) and Islas San Andrés and Providencia (Colombia); in Costa Rica throughout the lowlands (2–500 m); introduced to southern Florida.

Family Phrynosomatidae Fitzinger, 1843

Sceloporine Lizards

A moderate-sized family (ten genera and 125 species) of mostly small insectivorous diurnal lizards that are terrestrial, semiarboreal, arboreal, or saxicolous in habits. Members of the group lack dorsal crests and expanded digits and skull crests, although the species of one genus, *Phrynosoma* (the horned lizards), usually have the head protected by large spiny scales underlain by bony projections. The family comprises ten genera, mostly occurring in the western and southwestern United States and adjacent Mexico, but one *(Sceloporus)* ranges south into Central America to western Panama and another *(Phrynosoma)* into southern Canada. Phrynosomatid lizards share the following general features: body depressed; eye moderate, pupil round; ear opening moderate to large, recessed, and usually with auricular scales protecting it, although ear occasionally absent *(Cophosaurus* and *Holbrookia)*; interparietal usually a large shield (not enlarged in the sand lizard genera *Uma* and *Phrynosoma*); dorsal scales smooth or keeled, small to large, often imbricate and mucronate; ventral scales usually smooth and larger than dorsals; femoral pores present; no spinulate scale organs; infradigital lamellae with three to five keels; gular fold usually complete medially (gular fold lacking in *Sceloporus*).

The features listed below, while mainly osteological, provide additional characterization of this family: parietal foramen in frontoparietal suture; supratemporal sits on lateral side of supratemporal process of frontal; nuchal endolymphatic glands not penetrating nuchal musculature; labial blade of coronoid weakly developed or absent; Meckel's groove open; lateral teeth on jaws triconodont; caudal autotomy septa present, with transverse process anterior to them (no septa, but a single transverse process in *Phrynosoma*); no posterior coracoid fenestra; sternal fontanelle enlarged, median; postxiphisternal inscriptional ribs short; nasal chamber vestibule long, straight, supported by elongated septomaxillae, no concha; hemipenes single, with a simple sulcus spermaticus, apical region trilobate, median lobe smooth; lateral lobes calyculate proximally; truncus calyculate; colic septa absent.

Most members of the family lay cream to white eggs with flexible shells, but several species of *Phrynosoma* and *Sceloporus* are viviparous and have a transitory yolk sac placenta in addition to a well-developed chorioallantoic one.

Genus *Sceloporus* Wiegmann, 1828

Members of this large genus (seventy-nine species) are small to moderate-sized lizards (under 350 mm in total length) having a depressed body, moderately long limbs, keeled and imbricate dorsal scales, and a large interparietal shield and lacking a gular fold, a gular appendage, enlarged middorsal scales, and whorls of enlarged spiny caudal scales. Other characteristics that further define this group are elongate and imbricate superciliary scales; no head crests or enlarged cephalic spines; posterior maxillary teeth triconodont; three sternal ribs; caudal autotomy septa present. All *Sceloporus* are diurnal, and many species are scansorial or arboreal forms that perch on fence posts, hence the common name "fence lizard" often applied to members of the genus in the United States. However, many species are primarily terrestrial, and a few are saxicolous in habits.

The classic study of *Sceloporus* by Hobart M. Smith (1939b) concluded that the genus comprised two major lineages: small-scaled, small-sized *Sceloporus* and large-scaled, large-sized *Sceloporus*. Each of these lineages contained a number of distinctive species groups. Subsequent work that focused on karyology, by Hall (1973) and others as reviewed by Sites et al. (1992), retained these divisions and most of Smith's species groups, although the phenetic study of Larsen and Tanner (1974) based on morphological features supported a contradictory arrangement. The recent phylogenetic analysis of molecular and morphological evidence (Wiens and Reeder 1997) showed that Smith's two divisions were artificial and proposed a reclassification or reassignment of a few species groups and species. I follow their classification here. Of the twenty-two species groups Wiens and Reeder recognized, only three occur in Costa Rica. *Sceloporus* has its greatest diversity in Mexico and the southwestern United States but ranges over a wide geographic and altitudinal range in North and Central America from the northeastern United States and Washington State to western Panama.

KEY TO THE COSTA RICAN SPECIES OF THE GENUS *SCELOPORUS*

1a. No postfemoral dermal pocket; 28–38 rows of dorsal scales down middle of back from interparietal shield to level of posterior margin of hind limb 2
1b. Postfemoral dermal pocket present (fig. 10.20b); 48–

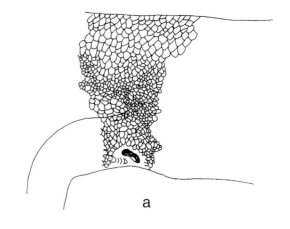

a

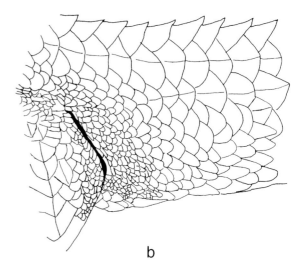

b

Figure 10.20. Specialized integumentary structures in iguanian lizards: (a) axillary (posthumeral) pocket in *Norops humilis*; (b) postfemoral pocket in *Sceloporus variabilis;* direction of head to the left.

59 rows of dorsal scales down middle of back from interparietal shield to level of posterior margin of hind limb; dorsal ground color auburn in males, gray brown in females; usually a distinct dorsolateral light stripe running down each side from eye to base of tail; a series of 10–12 paired dark dorsal spots that lie between dorsolateral light stripes when latter present; throat pink (in life), lateral belly patches pink in life (pinkish to white in preservative) bordered by dark blue in adult males; total femoral pores 16–22; adults to 70 mm in standard length; enlarged postcloacal scales in males . . . . *Sceloporus variabilis* (p. 443)

2a. Total femoral pores 22–34; ventral scales notched; dorsal ground color malachite blue, olive green, or olive brown; no dorsolateral light stripes; black collar on sides of neck extending from above shoulders across posterior throat region; throat blue, lateral belly patches blue with black median border in adult males; throat and lateral belly areas of mature females blue; adults to 91 mm in standard length; enlarged postcloacal scales in males . . . . *Sceloporus malachiticus* (p. 441)

2b. Total femoral pores 6–12; ventral scales pointed; dorsal ground color dark reddish to seal brown; a distinct dorsolateral white stripe running down each side from above ear opening onto tail, suffused with reddish brown posteriorly; no black collar; throat and venter dirty white without blue patches; adults to 57 mm in standard length; no enlarged postcloacal scales . . . . *Sceloporus squamosus* (p. 442)

Sceloporus formosus Group

These are large-scaled, large-sized *Sceloporus* with a karyotype of 2N = 22 and brightly colored ventral patches in males. Members of this group (thirteen species) range from Mexico to Panama. Included besides the lone Costa Rican representative are *S. adleri, S. cryptus, S. formosus, S. internasalis, S. salvini, S. stejnegeri, S. subpictus,* and *S. tanneri* (Mexico); *S. taeniocnemis* (Mexico and Guatemala); *S. lunnaei* and *S. smaragdinus* (Guatemala); and *S. acanthinus* (Guatemala and El Salvador).

Sceloporus malachiticus Cope, 1864a

(plate 247; map 10.8)

DIAGNOSTICS: The conspicuous bright turquoise individuals of this small species are frequently seen basking or head bobbing on fence posts, walls, or rocks, and they are unmistakable. Unfortunately these lizards display considerable color change at lower body temperatures, and under other conditions they may be dark gray to black. Nevertheless, they are unlikely to be confused with other forms because of their overall spiny appearance and absence of longitudinal light stripes in the dorsal pattern.

DESCRIPTION: A small species (to 190 mm in total length) standard length 67 to 98 mm in males, 64 to 94 mm in females; tail moderately short, about 50% of total length; head scales smooth; frontal divided transversely; supraoculars in two irregular series, one or more supraoculars often in contact with median head shields; one canthal; keeled, mucronate scales on upper surfaces of body, limbs, and tail; ventral and proximal subcaudal scales smooth, ventrals notched; lateral scales about three-fourths as large as dorsals and more strongly mucronate and denticulate; scales on posterior surface of thigh not granular; enlarged postcloacals present in males; no postfemoral dermal pocket; 30 to 38 rows of dorsal scales down middle of back between interparietal shield and posterior margin of hind limb; scales around midbody 28 to 45; total femoral pores 22 to 34. Dorsum varying from bright bluish green through greenish tan to gray or black; juveniles and females with nine to ten pairs of dorsal dark spots; a definite black collar on sides of

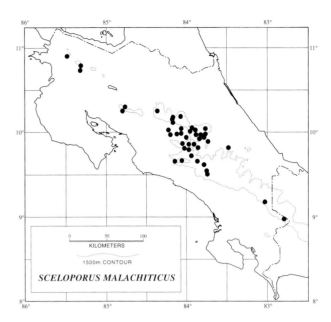

Map 10.8. Distribution of *Sceloporus malachiticus.*

neck; throat blue in adults; adult males with black-bordered longitudinal dark blue ventral areas, blue without black borders in females; blue areas least developed in juveniles.

SIMILAR SPECIES: (1) *Sceloporus squamosus* has a pair of distinct longitudinal dorsolateral white stripes. (2) *Sceloporus variabilis* usually has a pair of distinct longitudinal dorsolateral light stripes on a tan background and a postfemoral dermal pocket.

HABITAT: This species is very common in relatively open and sunny situations, including boulders or outcrops in clearings or pastures, fencerows, logs, woodpiles, rock walls, tile roofs, gardens, and rough-barked solitary trees within the Premontane, Lower Montane, Montane, and Subalpine zones. It is frequently seen within the urban areas of both the Meseta Central Oriental and Meseta Central Occidental.

BIOLOGY: Fitch (1973a,b) and Robinson (1983b) have provided recent summaries for the biology of this lizard, and their reports combined with my field observations form the basis for the following review. *Sceloporus malachiticus* is heliothermic, and its diel activity pattern closely follows the sun, especially at higher elevations (Vial 1984). Early in the day individuals occupy a basking site in the open, but as they warm up they shift to the shady side of the perch. At this time they usually adopt a vertical position with the head pointed downward and descend rapidly to the ground to pounce on any passing prey. During activity, body temperatures range from 20 to 35°C and are 7 to 15°C above ambient air temperature. Juveniles are more terrestrial than adults, but adults are more terrestrial at higher elevations than in the premontane zone. Food consists of arthropods, primarily insects. The species is territorial and defends the central core of the rather small home range (estimated to be only a few square meters). The home ranges of a male and several females may overlap, but individual males are widely spaced, and most territorial encounters are between head-bobbing juveniles. *S. malachiticus* is viviparous. At elevations below 1,500 m reproduction appears to be seasonal, with mating occurring in June, ovulation beginning in early September, and parturition occurring in late November and continuing until early March (Marion and Sexton 1971). At higher elevations births seem to occur throughout the year. Litter sizes are six to twelve. Neonates are about 28 to 30 mm in standard length at birth and reach sexual maturity at eight to nine months of age at lower elevations and at eighteen months at higher ones. These lizards are preyed on by a variety of birds, snakes, and probably small mammals. They are extremely wary, cannot be easily approached, and usually move away from an intruder within 10 to 15 m. At closer distances these lizards scramble to hide in crevices or holes in or near the perch site.

REMARKS: Stuart (1971) demonstrated the distinctiveness of this taxon and separated it from several other valid but similar species *(S. acanthinus, S. internasalis, S. smaragdinus,* and *S. taeniocnemis)* from upper Central America and southeastern Mexico that others had included in *S. malachiticus.* Nevertheless, a problem remains regarding examples from the area between the ranges of unquestioned *S. malachiticus* in lower Central America and the related *S. smaragdinus* in Guatemala, since individual populations variously resemble one or the other species in one or another feature with no consistent geographic pattern. However, most authors place the representatives of this group from El Salvador, Honduras, and Nicaragua in *S. malachiticus,* a course I reluctantly follow here pending a reanalysis of the problem. The karyotype (2N = 22) consists of five pairs of metacentric and one pair of submetacentric macrochromosomes and five pairs of microchromosomes, with XY sex chromosomes indistinct; NF = 34 (Sites et al. 1992).

DISTRIBUTION: Premontane, lower montane, montane, and subalpine zones of Costa Rica and western Panama (600–3,800 m); also reported from the lowland and premontane zones in El Salvador, western Honduras, and Nicaragua.

Sceloporus siniferus Group

Small-scaled, small-sized *Sceloporus* with a karyotype of 2N = 34 and lacking brightly colored ventral patches. Other forms included in this group besides *Sceloporus squamosus,* which occurs in Costa Rica, are *Sceloporus carinatus* and *S. siniferus* (Mexico and Guatemala) and *S. cupreus* (Mexico).

Sceloporus squamosus Bocourt, 1874

(plate 248; map 10.9)

DIAGNOSTICS: This small mostly terrestrial lizard is likely to be confused only with its ally *Sceloporus variabilis* where they occur together in Costa Rica. *S. squamosus* is

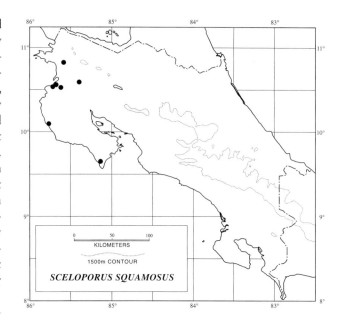

Map 10.9. Distribution of *Sceloporus squamosus.*

distinctive in having a reddish brown head and in lacking dorsal dark spots and bright-colored ventral patches.

DESCRIPTION: A small lizard (total length to 174 mm; standard length of adults 45 to 57 mm); tail long, 65 to 68% of total length; head scales strongly rugose; anterior and posterior sections of frontal longitudinally divided; usually five (four to six) large supraoculars separated from median head shields by a (usually) complete row of small scales; one canthal; scales on upper surfaces of body, limbs, and tail finely keeled, mucronate; ventrals smooth to weakly keeled, pointed; subcaudals strongly keeled; lateral scales about two-thirds size of dorsals, mucronate; neither dorsals nor laterals denticulate; scales on posterior surface of thigh small but granular; postcloacal scales not enlarged; no postfemoral dermal pockets; 28 to 37 rows of transverse dorsal scales (between interparietal and base of tail); 29 to 46 scales around midbody; total femoral pores 6 to 12. Dorsal ground color dark reddish to seal brown, head a darker red; a pair of longitudinal dorsolateral white stripes from ear to tail, suffused with reddish brown posteriorly; undersurfaces uniform dirty white without blue or pink display patches.

SIMILAR SPECIES: (1) *Sceloporus variabilis* does not have a reddish brown head and has a postfemoral dermal pocket (fig. 10.20b) and pink or pink-and-blue ventral color patches in adult males. (2) *Sceloporus malachiticus* lacks dorsolateral stripes.

HABITAT: The species is locally common and restricted to the Lowland Dry Forest area of northwestern Costa Rica, where it is found in open, disturbed areas (barren fields, roads, and trails).

BIOLOGY: The species is a diurnal, mainly terrestrial form that sometimes climbs onto boulders, into the bases of trees, or into low bushes.

DISTRIBUTION: Lowlands and premontane slopes of southeastern Mexico along the Pacific versant to northwestern Costa Rica; also semiarid valleys of Atlantic versant Guatemala and Honduras; in Costa Rica found only below 50 m in altitude.

Sceloporus variabilis Group

Small-scaled, small-sized *Sceloporus* having a karyotype of 2N = 34 and brightly colored ventral patches in males. In addition to the wide-ranging *Sceloporus variabilis*, which is found in Costa Rica, other members of the group are *Sceloporus couchii*, *S. parvus*, and *S. smithi* (Mexico), *S. cozumelae* (outer Yucatán Peninsula), and *S. chryostictus*. *S. chryostictus* is the only member of this group lacking a distinct postfemoral pocket.

Sceloporus variabilis Wiegmann, 1834a

(plates 249–50; map 10.10)

DIAGNOSTICS: Any small terrestrial lizard lacking a distinctly reddish brown head and having distinct paired dorsolateral light stripes and/or many pairs of dorsal dark spots will be recognizable as this species. In addition, adult males of this form have a pink throat and blue-and-pink lateral belly patches; all individuals have a definite postfemoral dermal pocket (fig. 10.20b) and dorsal scales five to six times as large as the lateral ones.

DESCRIPTION: A small species (adult males to 175 mm in total length, 57 to 75 mm in standard length; adult females 44 to 68 mm in standard length); tail moderate, 55 to 60% of total length; head scales rugose; anterior section of frontal longitudinally divided, posterior section often divided into two or more scales; usually five (four to six) large supraoculars, usually separated from medium-sized head

scales by a complete row of small scales; single canthal; scales on upper surfaces of body, limbs, and tail keeled, mucronate; ventrals smooth, notched; subcaudal scales keeled, mucronate; dorsal scales five to six times as large as laterals, not denticulate; scales on posterior surface of thigh granular; postcloacal scales not enlarged; a distinct postfemoral dermal pocket present; 48 to 61 (38 to 72) transverse rows of dorsal scales; scale rows around midbody (between interparietal shield and level of vent) 53 to 68 (39 to 91); total femoral pores 15 to 25 (14 to 34). Head reddish brown, other dorsal areas auburn in males, gray brown in females; usually a pair of light, longitudinal, dorsolateral stripes running from eyes to base of tail; females and juveniles with ten to twelve pairs of dark dorsal spots between light stripes; dark spots obscure or imperceptible in large males; usually a dark spot on each shoulder; undersurfaces dirty white with a pink throat and paired, blue-outlined pink ventrolateral display patches in adult males (adult females with patches in some extralimital populations).

SIMILAR SPECIES: (1) *Sceloporus squamosus* has a reddish head, has lateral scales about two-thirds the size of the dorsals, and lacks a postfemoral dermal pocket. (2) *Sceloporus malachiticus* lacks dorsolateral light stripes and a postfemoral dermal pocket and is usually some shade of green, gray, or black.

HABITAT: This lizard is found on dry, open ground including gullies, washes, and pastures and especially along sandy beaches in the Lowland Dry Forest zone; also marginally in Premontane Moist Forest.

BIOLOGY: This very common species is a diurnal heliotherm that spends most of its time on the ground, although it often climbs on or hides in small trees, logs, and fence posts. It emerges from nocturnal resting places rather late in the morning. Activity-level body temperatures are clustered between 31 and 35°C (range 27.5 to 39.5°C) and are 3 to 5°C above ambient air temperature (Fitch 1973a). It spends most of the day basking, with intermittent retreats to shady places to avoid higher temperatures. Individuals center their activities on preferred basking sites (dry pieces of wood, logs, tree trunks, driftwood, or fence posts) that also provide hiding or resting places. Forays are usually only of a few meters, but they may shift to other home sites 15 to 25 m away. Fitch (1973a) noted that spring tides seriously disturb these lizards, which are common denizens of beach areas and hide in and feed on beach wrack. When approached they run a short distance (3 to 6 m) if on the ground or move to the opposite side of the perch site. If further disturbed they may hide in a hole or crevice or dash to a nearby clump of brush or grass. Often the lizard then quickly burrows into the sandy or soft substrate with a series of lateral wiggling movements. Food consists of a variety of small arthropods, especially insects. *Sceloporus variabilis* is oviparous and lays multiple clutches of 1-3-5 eggs during the year. Gravid females are present from May to early January, and hatching

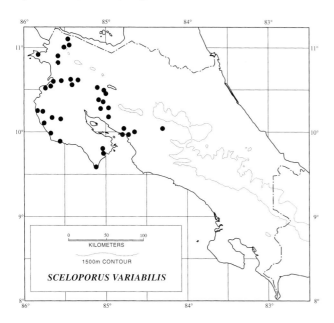

Map 10.10. Distribution of *Sceloporus variabilis*.

occurs throughout the year. However, the highest reproductive rates are found in July to November, corresponding to the heaviest rainfall period. Hatchlings are 25 to 28 mm in standard length.

REMARKS: Variation in this form has led to the recognition of several geographic races based on nonconcordant clinal trends in meristic and color features, as follows (Sites and Dixon 1982):

A. Dorsolateral light stripes broad, usually 2½ scale rows wide, prominent in both sexes and extending well onto tail; 71 to 91 scales around midbody; 54 to 69 transverse rows of dorsal scales: lowland Oaxaca, Mexico, west of Isthmus of Tehuantepec
B. Dorsolateral light stripes narrow, usually ½–1–½ scale rows wide, best developed in males and rarely extending onto tail.
   1. 54 to 72 transverse rows of dorsal scales; 64 to 73 scales around body: southern Texas to southern Nuevo León and southern Tamaulipas, Mexico
   2. 50 to 60 transverse rows of dorsal scales, 58 to 70 scales around body: southern Tamaulipas, Mexico, southward in dry habitats to Honduras and Nicaragua, exclusive of most of the Yucatán Peninsula on the Atlantic versant, and to Costa Rica on the Pacific slope
   3. 38 to 50 transverse rows of dorsal scales, 39 to 55 scales around body: Atlantic slope southern Veracruz and northeastern Oaxaca, Mexico, to Belize

Intermediate populations connect these rather poorly differentiated geographic subdivisions. All Costa Rican examples belong to population B2. Mather and Sites (1985) provide a redescription and summarize pertinent literature on this species. For a contrary view see Smith, Pérez-Higareda, and Chiszar (1993), who recognize *Sceloporus smithi* and *Sceloporus teapensis* as species distinct from *S. variabilis*. A record (Stafford 1998) from above the Pitilla Station, Guanacaste National Park, is from the Atlantic slope at 650 m, and additional specimens from this essentially Lowland Wet Forest area are needed to confirm the occurrence of this species. The karyotype of this species is $2N = 34$, with six pairs of metacentrics and submetacentrics and eleven pairs of microchromosomes (as many as ten pairs biarmed and one large subtelocentric pair with a secondary constriction); it has a male XY sex chromosome system; $NF = 46$ (Cole 1978).

DISTRIBUTION: Semiarid and subhumid lowlands and premontane areas from southern Texas to Nicaragua, exclusive of most of the Yucatán Peninsula on the Atlantic versant, and from Oaxaca, Mexico, to Costa Rica on the Pacific slope; in Costa Rica found in the Pacific northwest region and in the caldera of the Río Grande de Tárcoles to the border of the Meseta Central Occidental (1–800 m).

Family Polychrotidae Fitzinger, 1843

Anoloid Lizards

This is a diverse family (about 391 described species) of mostly semiarboreal and arboreal lizards of small to large size. Most of the species belong to one of several genera in which the digital lamellae are expanded and smooth (an arboreal adaptation), the hyoid apparatus is modified to support the extensible gular fan (dewlap) typically used in male displays but sometimes appearing in females as well, and the tongue shows specializations (reticular papillae on foretongue, no conical papillae on hindtongue) not found in any other iguanians except agamids and chameleons. Included in the anoline radiation are the following: *Chamaeleolis* (four species), *Chamaelinorops* (one species), *Ctenonotus* (thirty-seven species), and *Xiphosurus* (five species), endemic to the West Indies; *Anolis* (sixty species), essentially a West Indian endemic with one species in the southeastern United States; *Phenacosaurus* (ten species), in mountains from western Venezuela to Peru; *Norops* (143 species) of the mainland Neotropics and the western West Indies; and *Dactyloa* (sixty-one species) of lower Central America and South America.

Many authors place all eight putative genera into a single genus *Anolis* (Poe 1998; Jackman, Larson, Queiroz, et al. 1999). Others (e.g., Williams 1989) would retain *Chamaeleolis*, *Chamaelinorops*, and *Phenacosaurus* as distinct but place the other five in *Anolis* (sensu lato).

Another cluster of four genera (*Diplolaemus*, *Enyalius*, *Leiosaurus*, and *Pristidactylus*) comprises mostly moderate-sized terrestrial short-tailed, chubby forms and occurs in central and southern South America, primarily in Amazonia and southeastern Brazil (*Enyalius*) and the temperate region of Argentina and Chile. The genus *Enyalius* is arboreal, as are some species of *Pristidactylus*. In addition, three genera (*Anisolepis*, *Polychrus*, and *Urostrophus*) composed of moderate-sized to large arboreal lizards are South American, with *Polychrus* ranging north to Honduras. These two groups of genera appear to represent an early radiation that included the precursor of the anolines.

The anoloids are difficult to diagnose based on external features because of their historical and ecological diversity (terrestrial, arboreal, saxicolous). The following features are found in most of the genera: ear opening moderate to small, tympanum recessed; eye moderate, pupil round; interparietal small (sometimes absent in *Polychrus*); usually no femoral pores (present in *Polychrus*); usually spinulate scale organs present (absent in *Polychrus*); infradigital lamellae usually smooth (variable in some genera, keeled in *Polychrus*); gular fold complete (lacking in *Polychrus* and anoles).

The following internal characteristics, in combination, are definitive for the family: parietal foramen usually in frontoparietal suture (in parietal in some anoles, absent in some

Polychrus); supratemporal sits on lateral side of supratemporal process of parietal; nuchal endolymphatic glands penetrating nuchal musculature and filled with endolymph rich in calcium carbonate; labial blade of coronoid variable, from well developed to absent; Meckel's groove closed; lateral teeth on jaws triconodont; caudal autotomy septa present or absent with transverse processes posterior to septa or absent; posterior coracoid foramen small or absent; sternal fontanelles small or absent; postxiphisternal inscriptional ribs long, confluent medially; nasal chamber vestibule relatively short and straight, concha present or absent; hemipenes usually strongly bilobed, sometimes single, sulcus spermaticus simple or bifurcate, apical region calyculate or spinous (*Polychrus*); colic septa absent.

As far as is known all members of this family are oviparous and lay yellowish white soft-shelled eggs.

Genus *Polychrus* Cuvier, "1817," 1816

These moderately large (to 700 mm in total length), arboreal lizards (seven species) are unlikely to be confused with any others because of their bright green coloration; compressed body without nuchal or dorsal crests; extremely attenuate moderate to very long tails (two to three and a half times standard length, three or more times in Costa Rican species) that are round in cross section and lack whorls of spiny scales; and third and fourth fingers and third and fourth toes of equal length. Other distinguishing features of *Polychrus* are digits compressed, with keeled, unexpanded lamellae; anterior upper head surface covered with large polygonal scales; upper and lower eyelids partially fused into a cone with narrow opening at tip; eyes independently rotatable; no transverse gular fold; a definite extensible saclike gular appendage, sometimes with a median serrate crest; femoral pores present; no caudal autotomy or autotomy septa; posterior caudal vertebrae without transverse processes; no palatal teeth; angular bone and splenial bones present; no jaw sculpturing; parietal roof trapezoidal; parietal ridges U-shaped; occipital not overlapped by parietal half funnel; interclavicle arrow shaped; formula for inscriptional ribs (see glossary) 10:0, 9:2, 9:0, or 8:0; karyotype 2N = 26 to 30 with XY sexual heteromorphism (Gorman, Atkins, and Holzinger 1967; Gorman, Huey, and Williams 1969); substantial ability for color change (metachrosis).

The genus ranges in the lowlands from northern Argentina, Paraguay, southern Bolivia, and eastern and southeastern Brazil east of the Andes and from northwestern Peru to northwestern Honduras.

Polychrus gutturosus Berthold, 1845

(plates 251–52; map 10.11)

DIAGNOSTICS: This large blunt-headed lizard is immediately recognizable because of the extremely long round tail that is more than three times the length of the head and

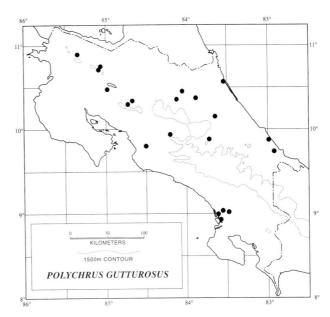

Map 10.11. Distribution of *Polychrus gutturosus*.

body. In addition, it lacks toe pads, nuchal, dorsal, and caudal crests, a large subtympanic scale, and tail spines. Individuals observed in the field are usually bright green in general coloration, but depending on mood or other factors they may change to a dull brown. Most examples have a distinct large light spot below the posterior margin of the orbit.

DESCRIPTION: Total length to 700 mm, females much larger (125 to 170 mm in standard length) than males (83 to 135 mm in standard length); limbs moderately long; upper head scales smooth, weakly granulate or striated, greatly enlarged on snout; symmetrical scales in supraorbital semicircles, separated by one scale; gular scales oval, striated, much larger than ventrals, those on gular pouch widely separated by granular skin; gular pouch moderate, extending posterior to axilla in males, to axilla in females, without midventral serrated ridge; dorsal, lateral, and ventral scales about same size; dorsals and laterals unicarinate, ventrals multicarinate; toes III and IV about same length; 12 to 36 total femoral pores; tail round, thin, covered with homogeneous unicarinate scales. Head, dorsum, venter, and tail green, usually with some distinctive bright yellow to reddish orange markings, including a middorsal light stripe, oblique narrow transverse stripes on flanks, and/or longitudinal lateroventral stripes; frequently large, distinct postocular spot; sometimes dark green transverse bands on body and tail; green areas change to dull dark brown under changing conditions of temperature, light, and mood; venter somewhat lighter green than upper surfaces; dewlap magenta with large green scales forming oblique horizontal stripes; iris coppery brown; lining of throat bluish black.

SIMILAR SPECIES: (1) Species of *Basiliscus* have the fourth toe much longer than the third, often have head, dorsal, and caudal crests (fig. 10.17), and lack femoral pores. (2) *Iguana iguana* has a large smooth subtympanic scale. (3) *Dactyloa frenata* has a very long head, distinct expanded toe pads, and the fourth toe much longer than the third.

HABITAT: Undisturbed Lowland Moist and Wet Forests and marginally along stream courses into adjacent Premontane Moist Forest.

BIOLOGY: A strictly diurnal and arboreal lizard that is restricted to humid forests and rarely seen. One example was taken at a height of 40 m. The hands and feet are used to grasp small branches and twigs, and locomotion is slow and deliberate with the limbs moving alternately. Individuals move higher in the tree canopy by pulling themselves upward by the forelimbs. They move downward by hanging from a branch by the hind limbs and tail and then dropping to a lower one. Like *Corytophanes*, these lizards may remain totally still for long periods, often hanging by one or more limbs and/or the tail in awkward or bizarre positions. However, they are capable or rapid movements both in vegetation or on the ground if threatened by a predator. The species is a sit-and-wait predator that, unlike *Corytophanes*, slowly advances on the prey once it is spotted and feeds primarily on relatively large arthropods but also eats leaves, flowers, fruits, and seeds. When disturbed *Polychrus* threatens with a wide-open mouth and extends the dewlap. Roberts (1997) reported copulation in this species in May at the La Selva Biological Station, Heredia Province. The lizards were in a tree 2 m above the ground, and the male was on top of the female with their tails entwined. The male was observed earlier in the morning but was found copulating at noon and mating continued until 2:48 P.M. A female with palpable eggs had been noted by another investigator in July, suggesting a rainy season reproductive period (May to December) at this locale. Taylor (1956) noted a female containing nine ovarian eggs, four in the right ovary and five in the left. The eggs are apparently laid in the leaf litter on the ground. A juvenile (CRE 8411) hatched from an egg measured 57 mm in standard length. Another small juvenile was 53.5 mm in standard length.

DISTRIBUTION: Lowland and lower premontane evergreen forests from northwestern Honduras and western Costa Rica to northwestern Ecuador; on both Atlantic and Pacific versants in Costa Rica but not present in the northwest Pacific slope subhumid area (6–700 m).

ANOLES

Morphological Specializations and Behavioral Ecology

The next three genera described in this book, *Ctenonotus*, *Dactyloa*, and *Norops*, and their allies *Anolis* (sensu stricto) and *Xiphosurus* form a speciose and ecologically diverse radiation of Neotropical lizards that until recently were placed in the single all-inclusive genus *Anolis* (sensu lato). Guyer and Savage (1986, 1992) have presented arguments for the classification adopted here, but see Williams (1989) and Jackman, Larson, Queiroz, et al. (1999) for alternative views. The members of this radiation are collectively called anoles.

In spite of the diversity of anole species, members of the five genera share many morphological, ecological, and behavioral features, while others exhibit remarkable convergences among species in these characteristics across generic limits. The rest of this section will discuss the most significant of these features to provide an introduction to anole biology and to avoid redundancy in the individual species accounts.

The single most obvious feature of anole morphology is the expanded digital pads composed of smooth transverse lamellae on the undersurface of each finger and toe (except in *Norops annectans* and *N. onca*). The lamellae are covered in turn by many microscopic spinelike to setiform projections that apparently have an adhesive or prehensile function in locomotion, especially on vertical surfaces (Peterson and Williams 1981). The digital structure of most anoles (including all Costa Rican species) is further distinctive in that the compressed terminal phalanx rises above and slightly proximal to the end of the expanded portion of the digit. There is some variation in the degree of expansion and number of expanded lamellae in Costa Rican species. Highly arboreal forms have greatly expanded digital pads, whereas terrestrial and semiterrestrial species tend to have only slightly expanded ones. Large species also have more lamellae and more expanded ones than smaller forms.

Another distinct feature of these lizards is the extensive, laterally compressed gular appendage (dewlap) of males (and some females) that is often bright and contrasting in color and used in territorial and courtship displays.

All anoles are diurnal. Some are heliotherms that regulate their temperature by alternating rounds of basking with retreats to cover or shade. Others tend to be shade species, do not bask, and are thermoconformers. Many anoles have considerable ability to change color, usually in a range from green to gray or brown. Most species are primarily arboreal. They are also pure pursuer predators that do not actively forage but visually scan the environment for prey of suitable size. When it finds a food item a lizard moves quickly to snap it up and then typically returns to its perch. The term "foraging" as used throughout this section thus refers to the combination of scanning and pursuit behaviors. All anoles feed principally on arthropods that they locate visually from the perch and eat without olfactory testing. Most species are active predators, but some forms remain motionless until the prey approaches. Average prey size is positively correlated with lizard size in all species studied to date; small anoles eat small prey, large anoles eat small and large prey. In contrast to many other lizards, anoles readily desiccate and need to drink frequently. Because of this requirement, confining anoles in a collecting bag even for a few hours usually kills them, as many field biologists have discovered to their chagrin.

The myriad anole species may be divided into ecologic groups based on preferred perch height and size and usual foraging and defensive behaviors, as modified from Williams (1972, 1983), Fitch (1973a), Losos (1992), and Beuttell and Losos (1999): (1) large canopy species that live, forage, and court in the treetops and are aggressive defenders against intruders; (2) species that forage on twigs in the canopy and use crypsis primarily as a defense; (3) trunk-crown species that occupy the upper trunk and tree crown area, forage in the crown, and run up the tree when approached; (4) trunk or low tree species that forage on the trunk or crown and scurry around the trunk when disturbed; (5) trunk-litter species that perch on lower trunks, sit and wait for prey moving on the leaf litter, and escape by moving down and around the perch or onto the ground.

Other ecologic types include (1) low bush-grass species that forage on the perch vegetation and flee downward when approached; (2) bush-litter species that perch on low shrubs, bushes, and small trees, forage on the ground, and escape downward when disturbed from the perch but often run to a tree and climb the trunk when frightened while on the ground; (3) terrestrial species that spend most of their time on or near the ground, feed there, and run and hide at ground level if disturbed; (4) saxicolous species that tend to perch on boulders, forage on the ground, and escape downward if approached; (5) riparian species that live and forage along stream margins and move toward the water or jump into it to escape.

Williams (1972, 1983) has documented a strong correlation in morphology among those West Indian species (regardless of generic allocation) having the same ecologic role and has characterized them as ecomorphs. Morphological congruence does not seem to be as strongly coupled with ecology in mainland forms, and the term ecomorph is not applied to the species covered in this book. Generally, in all predominantly arboreal anoles, perch height is positively correlated with size both inter- and intraspecifically, so that smaller species or smaller individuals of a particular species tend to be found closer to ground level.

There is strong sexual dimorphism in most anole species. Males often differ from females in having nuchal or caudal crests, or both, and enlarged postcloacal scales (fig. 10.3), and the two sexes often differ in overall color or pattern (fig. 10.21). The crests in most Costa Rican anoles, when present, are not prominent. They are most fully developed in the introduced species *Ctenonotus cristatellus* (caudal), and *Dactyloa insignis* and *D. microtus* (nuchal). In other forms

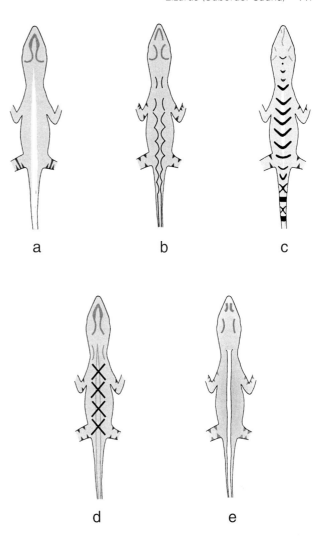

Figure 10.21. Variation in dorsal patterns in female anoles; not all patterns present in every species: (a) middorsal light stripe; (b) diamond; (c) dark chevrons; (d) dark X marks; (e) light stripe bordered by narrow dark stripes.

they are indicated by one or two rows of vertebral scales that are larger or project conspicuously above the paravertebral scales. The nuchal crest is usually most obvious when it is elevated by contracting the neck muscles during territorial and courtship displays. In many species males are markedly larger than females, but in some the reverse is true, while in others the males are only slightly larger than females or are of nearly the same size (Fitch 1976). The most striking sexually dimorphic feature of anoles remains the throat fan, dewlap, or gular appendage (fig. 10.5). The dewlap is on the ventral midline of the throat and can be extended by the hyoid musculature. It is made rigid by the hyoid bones and when not extended is folded up against the throat. This structure is variously developed in males and characteristically colored. In most females the dewlap is small or absent, but in some it may be almost as large as in the males.

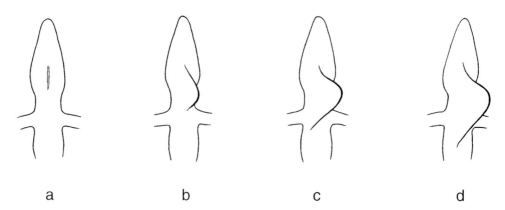

Figure 10.22. Variation in dewlaps in anole lizards, ventral view: (a) very small; (b) small; (c) moderate; (d) large.

For this account dewlap size (fig. 10.22) is defined as follows:

Very small, posterior extent well anterior to the axilla
Small, posterior extent to front limb insertion
Moderate, posterior extent to slightly beyond front limb insertion
Large, posterior extent to one-fourth to one-half distance between axilla and groin

Females may have large or small dewlaps. In some species in which females have no dewlaps the throat may be uniform (plain) in color, but in others the medial throat region is colored differently. In males the dewlap is used for aggressive displays to defend the male's territory and in courtship behavior. Females with large dewlaps may also use them in territorial defense. Each species has a characteristic display action pattern of aggression that may involve nuchal and dorsal crest erection, tail lashing, mouth gaping, tongue protrusion, inflation of the throat, push-ups, and head bobbing as well as display of the dewlap. Typically, the displaying male presents the side of his body to the intruder and compresses it to increase its height and his apparent size. If the display fails to intimidate, the resident male may chase and bite the trespassing male. Females have feeding territories, and a male's territory encompasses those of one to several females.

When a male sees a female he gives the dewlap display some distance away, then struts or hops toward her. Often she flees a short distance one or more times at his approach, but if she is gravid she may initially or ultimately flatten herself on the substrate. The male then grasps her by the neck, and their tails intertwine so they can appose their cloacas and complete copulation. Courtship and copulation usually occur off the ground near a favorite perch site, but in more terrestrial forms they may take place on fallen logs, rocks, or the ground.

As far as is known all species of anoles normally lay a single egg at a time from alternate ovaries. The egg may be deposited rather casually without any pretense of nest construction, often in the leaf litter, but some forms use other more protected sites. Many anoles are reproductive throughout the year, producing an egg as often as every one to four weeks. In areas of seasonal rainfall, egg production may be reduced at the height of the dry season, and a few forms have been found to be seasonal breeders where there is a long dry season.

A wide variety of organisms including spiders, mantises, predaceous katydids, lizards, many birds, and small mammals include anoles in their diet. However, the most important anole predators are the many species of semiarboreal and arboreal snakes found in the Neotropics.

Anole Identification: Clues and Cautions

Because the species of these genera resemble one another in external features but differ primarily in skeletal characteristics, they are included in a single key. The most important of these features in Costa Rican species (should a herpetologist wish to confirm them) are *Ctenonotus:* caudal autotomy septa present, arrow-shaped interclavicle, no transverse processes on autotomic caudal vertebrae; *Dactyloa:* no caudal autonomy septa or transverse processes, T-shaped interclavicle; *Norops:* caudal autotomy septa and transverse processes, T-shaped interclavicle.

The various features of head and body scalation used in the key and descriptions are illustrated in figures 10.3 and 10.5. These characters taken in combination will usually provide correct identifications. However, the large number of species and the superficial resemblances among many of the smaller forms often create problems even when specimens are in the hand. Little difficulty should be encountered identifying live adult males in the field, since these lizards usually will display when approached by a human. Consequently the size and coloration of the dewlap are the most important field marks for species recognition. For this reason dewlap color is included for each anole in the following key, and dewlap color and size are emphasized in the diagnostics section of each species account. Sometimes individual males are reluctant to spread their dewlap, but it is often possible to stimulate a display by remaining quiet at a distance and slowly moving the forefinger or a vertically flattened hand up and down within the lizard's field of vision.

Unfortunately, field identification of females and juveniles remains a problem. The best course of action, when studying or visiting a new site, is to capture some individuals to confirm identifications. However, once you become familiar with the local suite of anoles (a maximum of eight at any one site) the differences in perch preference, coloration, and general behavior will make visual recognition of most species routine.

Difficulties (especially for preserved specimens) are most common in attempting to identify moderately short-legged anoles, in which the tip of the fourth toe of the adpressed hind limb usually reaches to the orbit but sometimes somewhat farther forward. In my experience the most frequent misidentifications are of examples of *Norops altae, N. intermedius, N. lemurinus,* and *N. tropidolepis,* which exhibit considerable variability in scutellation. If you have a questionable anole from within the geographic range of any of these forms, carefully comparing it with the appropriate species accounts usually will resolve matters. In a few cases you may still need to compare the difficult example with authoritatively identified preserved material.

Another important character is the number of lamellae under the fourth toe (fig. 10.14b). In the key the total number of lamellae under that toe is given for each species. In the species accounts both the total number of lamellae and those under phalanges 2 and 3 are given. The latter count includes those lamellae that are definitely expanded as compared with those at the very base of the toe and on the terminal segment. For simplicity these are referred to as expanded or enlarged lamellae.

One final caution is in order before using the key and in identifying anoles. To see details of scutellation you often need a hand lens or microscope, particularly to ascertain rugosity and degree of keeling and to count scales. In the case of preserved material it is best to dry off the scales to be examined with a towel before trying to determine whether a particular character state is present. This is mandatory when differentiating among the several conditions of the ventral scales, since keeling is usually obscured on wet scales. To avoid some of these difficulties Williams et al. (1995) devised a computerized "key" to Costa Rica and Panama anoles, which may be of interest to serious students.

KEY TO COSTA RICAN ANOLES, GENERA *CTENONOTUS, DACTYLOA,* AND *NOROPS*

1a. Ventral scales strongly keeled, often imbricate and/or mucronate (fig. 10.3) 2

1b. Ventral scales smooth or some (usually near midbody) tuberculate or weakly keeled with posterior margin rounded 13

2a. No axillary pocket 3

2b. A deep tubelike axillary pocket opening through a pore (fig. 10.20a); 7–12 enlarged keeled dorsal scale rows; 23–30 lamellae under fourth toe; no nuchal or caudal crest; standard length to 44 mm; no enlarged postcloacal scales; dorsal ground color brown; dewlap of adult males in life deep red with the free margin yellow . . . .
. . . . *Norops humilis* (p. 460)

3a. Legs long; when hind limb adpressed along side of body, fourth toe reaches to anterior margin of orbit or beyond 4

3b. Legs short; when hind limb adpressed along side of body, fourth toe reaches no farther than between shoulder and orbit, usually no farther than tympanum . . . 6

4a. Dorsal and lateral scales keeled, imbricate and usually mucronate (fig. 10.3); several rows of somewhat enlarged keeled dorsal scales 5

4b. Dorsal and lateral scales grading into each other, smooth, flattened, and juxtaposed to one another (fig. 10.3); most dorsal scales pentagonal or hexagonal; 33–41 lamellae under fourth toe; no nuchal or dorsal crest; dorsal ground color olive to olive brown; standard length to 96 mm; no enlarged postcloacal scales; dewlap of adult males in life greenish yellow . . . .
. . . . *Norops capito* (p. 476)

5a. Pads under fourth toe with only slightly expanded lamellae, at most about twice as wide as those on base of digit; usually four to six scales separating supraorbital semicircles, rarely only three; dorsal ground color dark brown; 31–38 lamellae under fourth toe; no nuchal or caudal crest; standard length to 55 mm; no enlarged postcloacals; dewlap small, base extending only to level of arm in adult males (fig. 10.22b); dewlap of adult males plum red to purple; iris yellow in life . . . .
. . . . *Norops tropidolepis* (p. 467)

5b. Pads under fourth toe with greatly expanded lamellae, several times wider than those on base of digit; two or three scales separating supraorbital semicircles; dorsal ground color dull brick red or dark olive in life, pale yellow to tan in preservative, with numerous small punctations; standard length to 95 mm; enlarged postcloacals present in males; 34–43 lamellae under fourth toe; low nuchal and dorsal crests in males; dewlap large, base extending beyond axilla onto venter in adult males; dewlap of adult males in life pink orange to olive green to black, often with a reddish hue; iris blue in life
. . . . *Norops woodi* (p. 479)

6a. Interparietal plate surrounded by moderate-sized scales (fig. 10.5), separated from supraorbital semicircles by four or fewer scales 7

6b. Interparietal plate surrounded by minute scales, sepa-

rated from supraorbital semicircles by six to nine scales; suboculars separated from supralabials; supraorbital semicircles separated by two to four scales; 29–33 lamellae under fourth toe; nuchal and caudal crests present in males; standard length to 71 mm; dorsal ground color brown with darker spots and light blotches, suffused with green to olive in life; body marked with a distinct broad dorsolateral light stripe from shoulder to groin; no enlarged postcloacals; dewlap of adult males in life red orange with oblique yellow stripes . . . .
. . . . *Norops aquaticus* (p. 458)

7a. Enlarged dorsal scales strongly keeled, not flattened or juxtaposed 8

7b. Enlarged dorsal scales (smooth and weakly keeled) in 18–24 rows, flattened and rounded, pentagonal or hexagonal and usually juxtaposed; suboculars separated from supralabials; supraorbital semicircles separated by one or two scales; 30–38 lamellae under fourth toe; nuchal and caudal crests in males; standard length to 85 mm; usually a distinct light dorsolateral stripe from neck at least to midbody; dorsal ground color dark brown; enlarged postcloacals in males; dewlap of adult males in life burnt orange . . . .
. . . . *Norops oxylophus* (p. 472)

8a. Legs short; when hind limb adpressed along side of body, fourth toe reaches farther than shoulder, usually to tympanum; scales of supraocular and frontal series usually keeled; no scattered elevated wartlike scales on flanks and lateral area of venter 9

8b. Legs very short; when hind limb adpressed along side of body, fourth toe reaches no farther than shoulder; scales of supraocular and fontal series smooth to rugose; scattered elevated wartlike scales on lower flanks and adjacent area of venter; 6–14 definitely enlarged median rows of keeled dorsal scales; supraorbital semicircles in contact; all anterior head scales mostly smooth and somewhat flattened; dorsal and lateral scales not tuberculate; suboculars in contact with supralabials; 24–33 lamellae under fourth toe; nuchal and caudal crests in males; standard length to 54 mm; gray, brown, or cream with darker markings in life and preservative; enlarged postcloacal scales in males; dewlap of adult males in life bone white . . . .
. . . . *Norops intermedius* (p. 462)

9a. Dorsal scales not tuberculate, four or more median rows enlarged, keeled; lateral scales small, usually granular; dorsal ground color brown to gray; lining of throat flesh colored 10

9b. Dorsal and lateral scales raised, tuberculate, and keeled; two median rows of dorsal scales slightly enlarged; supraorbital semicircles separated by one to three scales; some anterior head scales (nasal-rostral area) with tuberculate keels; suboculars separated from supralabials or in contact with them; 35–47 lamellae under fourth

toe; nuchal crest in males; standard length to 98 mm; usually bright green in life; purplish brown in preservative; lining of throat black; no enlarged postcloacal scales; dewlap of adult males in life white basally, mostly powder blue with a red orange free margin . .
. . . . *Norops biporcatus* (p. 475)

10a. Pattern without lateral light stripes 11

10b. A lateral light stripe bordered above and below by a black stripe along each side of the body; 38–46 lamellae under fourth toe; no nuchal or caudal crest; standard length to 49 mm; no enlarged postcloacal scales; supraorbital semicircles separated by two to four rows of scales; dorsal ground color olive to brown; dewlap of adult males in life amber . . . .
. . . . *Norops townsendi* (p. 466)

11a. Pads under fourth toe with greatly expanded lamellae, several times wider than lamellae on base of digit 12

11b. Pads under fourth toe with only slightly expanded lamellae, only about twice as wide as those on base of digit; suboculars separated from supralabials by a row of scales; 26–33 lamellae under fourth toe; nuchal and caudal crests in males; standard length to 57 mm; no enlarged postcloacal scales; dorsal ground color gray to olive brown; dewlap large, base extending well back of axilla onto venter in adult males (fig. 10.22d); dewlap of adult males in life burnt orange to red orange with an anterior white area and free margin pink to yellow . . . . *Norops cupreus* (p. 458)

12a. Suboculars usually separated from supralabials by a row of scales, occasionally these scales partially fused to suboculars, but still evident; 31–38 lamellae under fourth toe; no nuchal or caudal crest; standard length to 79 mm; no enlarged postcloacal scales; dorsal ground color brown; dewlap small, base extending to level of arm insertion in adult males (fig. 10.22b); dewlap of adult males in life dark red . . . .
. . . . *Norops lemurinus* (p. 463)

12b. Suboculars in contact with supralabials; 27–33 lamellae under fourth toe; nuchal crest in both sexes; standard length to 52 mm; no enlarged postcloacal scales; dorsal ground color pale brown to gray; dewlap moderate, extending to level of axilla in adult males (fig. 10.22c); dewlap of adult males in life yellow to orange with a central purple round spot . . . .
. . . . *Norops sericeus* (p. 465)

13a. Legs short; when hind limb adpressed along side of body, fourth toe reaches to between shoulder and posterior margin of orbit; suboculars usually in contact with supralabials 14

13b. Legs long; when hind limb adpressed along side of body, fourth toe reaches to anterior margin of orbit or beyond; suboculars usually separated from supralabials by one or two rows of scales 21

14a. A parietal eye present 15
14b. No parietal eye; 41–49 lamellae under fourth toe; nuchal and caudal crests present; standard length to 111 mm; brown and cream in coloration, often with a greenish cast in life; enlarged postcloacals in males; dewlap of adult males in life pink with white scales . . . . *Dactyloa microtus* (p. 455)
15a. Lateral scales granular to conical, not flattened and juxtaposed; adults to 79 mm in standard length . . .16
15b. Lateral and dorsal scales flattened, juxtaposed, and somewhat elongate; adults to 160 mm in standard length; 51–58 lamellae under fourth toe; nuchal crest present; brown and white, often with a greenish cast in life; enlarged postcloacals in males; dewlap of adult males in life dark red . . . . *Dactyloa insignis* (p. 455)
16a. Lining of throat black; sliver of bluish tissue at corner of mouth; legs very short; when hind limb adpressed along side of body, fourth toe rarely reaches beyond shoulder 17
16b. Lining of throat flesh colored; no sliver of bluish tissue at corner of mouth; legs short; when hind limb adpressed along side of body, fourth toe reaches well beyond shoulder 19
17a. No paired bony parietal protuberances; scales along dorsal edge of tail enlarged and carinate 18
17b. A pair of small bony parietal protuberances posterior and lateral to interparietal plate; scales along dorsal surface of tail small but carinate; 30–35 lamellae under fourth toe; no nuchal or caudal crest, although a paired series of enlarged carinate subcaudal scales present (fig. 10.23a); standard length to 47 mm; dorsum brown with a series of large white funguslike blotches; bluish gray tissue at corner of mouth; no enlarged postcloacal scales; dewlap of adult males in life red with a few white scales . . . . *Norops fungosus* (p. 477)
18a. Some middorsal granules keeled; ventrals keeled in adults; a dark interorbital bar; 33 lamellae under fourth toe; nuchal crest but no caudal crest in adult males (fig. 10.23b); standard length to 57 mm; dorsum brown with darker brown bands and blotches, flanks with definite irregular reticulation; bright blue

tissue at corner of mouth; enlarged postcloacal scales in males; dewlap of adult males in life red, with white scales . . . . *Norops vociferans* (p. 478)
18b. Middorsal granules smooth; ventrals smooth; no dark interorbital bar; 34–41 lamellae under fourth toe; nuchal crest in adult males; distinct dorsal caudal crest in adults (fig. 10.23c); standard length to 79 mm; dorsum and flanks brown with darker brown markings; back and especially flanks with a reticulum of dark lines and lichenose light areas; bluish gray tissue at corner of mouth; no enlarged postcloacal scales; dewlap of adult males Bing cherry red-purple . . . . *Norops pentaprion* (p. 477)
19a. Dorsum gray to brown, in life and in preservative; uniform or with some dark spots or blotches; venter light 20
19b. Dorsum pale green in life; distinctly yellow with numerous small dark punctations in preservative; venter usually covered by a dark and light lichenous reticulum; ventrals smooth; 29–31 lamellae under fourth toe; no nuchal or caudal crest; standard length to 41 mm; no enlarged postcloacal scales; dewlap of adult males in life bright orange . . . . *Norops carpenteri* (p. 469)
20a. Dewlap of adult males orange in life, moderate, extending slightly posterior to axilla (fig. 10.22c); no dewlap in females; no enlarged postcloacals in males; 28–37 lamallae under fourth toe; no nuchal or caudal crest; standard length to 52 mm . . . . *Norops altae* (p. 468)
20b. Dewlap of adult males pink with an orange anterior margin; large, extending onto chest (fig. 10.22d); females with small dewlap, same color as throat; males with enlarged postcloacal scales; 31–32 lamellae under fourth toe; standard length to 60 mm. . . . . *Norops pandoensis* (p. 473)
21a. Supraorbital semicircles separated by one to six scales 22
21b. Supraorbital semicircles in contact; caudal crest with bony rays in males; suboculars in contact with supralabials; standard length to 75 mm; ground color light gray to brownish green in life, gray or brown in pre-

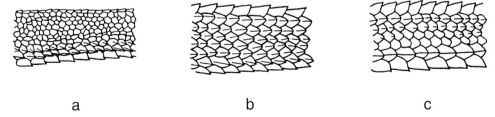

a   b   c

Figure 10.23. Caudal scalation in adult Costa Rican anoles, lateral view: (a) *Norops fungosus*; (b) *N. vociferans*; (c) *N. pentaprion*.

servative; no enlarged postcloacals; dewlap of adult males in life greenish yellow with free margin of burnt orange to red . . . . . . *Ctenonotus cristatellus* (p. 452)

22a. Fewer than 45 lamellae under fourth toe; supraocular disk with definitely enlarged scales medially that contrast with small lateral scales; standard length to 73 mm 23

22b. Lamellae under fourth toe 50–58; supraocular disk comprises a series of small scales that gradually increase in size medially; suboculars separated from supralabials; nuchal but no caudal crest; standard length to 143 mm; green in life, lavender in preservative; enlarged postcloacals in males; dewlap of adult males cream white in life . . . . *Dactyloa frenata* (p. 454)

23a. Lamellae under fourth toe 28–35; no dewlap in females; lining of throat flesh colored 24

23b. Lamellae under fourth toe 37–43; females with a large dewlap; supraorbital semicircles separated by one or two scales; no nuchal or dorsal crest; standard length to 75 mm; body green in life, purple in preservative, with a series of chevron-shaped marks on back and flanks; lining of throat black; enlarged postcloacal scales in males; dewlap of adult males in life pale orange with a series of faint green longitudinal lines, white along anterior margin . . . . *Dactyloa chocorum* (p. 456)

24a. Canthals unicarinate; interparietal at least two-thirds size of ear opening 25

24b. Canthals one and two multicarinate; interparietal about one-third size of ear opening; suboculars usually separated from supralabials; 28–35 lamellae under fourth toe; no nuchal or caudal crest; standard length to 50 mm; dorsum and flanks dark brown in life and in preservative; no enlarged postcloacal scales; dewlap small, extending posteriorly to level of arm insertion in adult males (fig. 10.22b); dewlap of adult males red orange in life, often with a central yellow spot; no dewlap in females but an orange spot on throat in life . . . . *Norops pachypus* (p. 464)

25a. Male dewlap small, base extending to level of arm insertion in adult males (fig. 10.22b); no dewlap in females, throat plain; interparietal and ear opening about equal; suboculars usually in contact with supralabials; dorsal ground color gray brown to olive; 28–32 lamellae under fourth toe; no nuchal or caudal crest; standard length to 43 mm; enlarged postcloacals in males; dewlap of adult males white in life with a basal orange spot . . . . *Norops limifrons* (p. 470)

25b. Male dewlap large, base extending well beyond axilla onto venter (fig. 10.22d); no dewlap in females, but an orange spot on throat in life; interparietal smaller than ear opening; suboculars usually separated from supralabials below eye; dorsal ground color light yellow to reddish brown; 29–35 lamellae under fourth toe; cau-

dal crest in males; standard length to 57 mm; enlarged postcloacals in males; dewlap of adult males orange in life . . . . *Norops polylepis* (p. 473)

Genus *Ctenonotus* Fitzinger, 1843

Members of this genus are small to moderate-sized (maximum total length approximately 210 mm; maximum standard length 70 mm) diurnal lizards with expanded smooth digital lamellae, no transverse gular fold, the fourth toe much longer than the third, no femoral pores, and the tail one and a half to two and a quarter times the standard length. Other (mostly osteological) characteristics of *Ctenonotus* include no palatal teeth; no angular bone; a splenial bone present; lower jaw often heavily sculptured in males; parietal roof triangular; parietal ridges Y-shaped; parietal forming a half funnel overlapping occipital bone; interclavicle arrow shaped; inscriptional rib formula 3:1 or 2:2; caudal autotomy septa present, posterior caudal vertebrae without transverse processes; XY or usually XYY sexual heteromorphism in the karyotype.

Ctenonotus is indigenous to the West Indies, but two species have been introduced to several sites on the mainland of North and Central America, including *Ctenonotus cristatellus* to Costa Rica. The genus definitely contains two allopatric series (Guyer and Savage 1992), the *cristatellus* and *bimaculatus* series. The *cristatellus* series is characterized by (features for the *bimaculatus* series in parentheses): extensive sculpturing of lower margin of dentary bone in adult males (no jaw sculpturing); inscriptional rib formula 2:2 (3:1); presacral vertebrae 23 (24).

The *cristatellus* series (nineteen species) occurs in Puerto Rico, Hispaniola, and the Virgin and Bahama Islands, and the *bimaculatus* series (ten species) is restricted to the northern Lesser Antilles. A third group, the *cybotes* series (eight species) of Hispaniola, is also currently included in this genus, although its relation to the other two stocks is ambiguous. Members of the *cybotes* series have extensive jaw sculpturing in adult males, an inscriptional rib formula of 2:2, 24 presacral vertebrae, and no sexual heteromorphism in the karyotype. For a complete listing of species referred to the three series see Savage and Guyer (1989).

Ctenonotus cristatellus (C. Duméril and Bibron, 1837)

(plate 253; map 10.12)

DIAGNOSTICS: Adult males of this small lizard are easily recognizable because they have a distinct caudal crest supported by bony extensions of the neural arches and the dewlap is uniform mustard or greenish yellow, with the free margin burnt orange to reddish. Females have a well-developed but smaller dewlap (fig. 10.22a) and a low caudal crest. The species has smooth ventral scales and long hind legs, and the scales of the supraorbital semicircles are keeled and in contact.

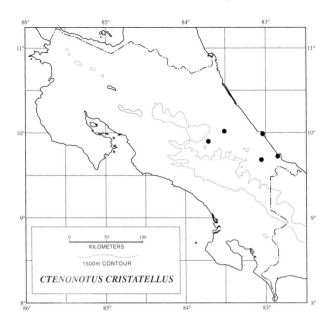

Map 10.12. Distribution of *Ctenonotus cristatellus*.

DESCRIPTION: A moderate-sized anole, total length to
205 mm in large males (standard length in adult males 50
to 75 mm), females averaging smaller (adults 39 to 73 mm
in standard length); tail moderate, about 60 to 65% of total length; hind limbs long; when adpressed along side of
body, tip of fourth toe reaches to anterior margin of eye
or beyond; anterior upper head scales flattened, smooth to
weakly rugose; circumnasal or anterior nasal separated from
rostral by no more than one scale; supraorbital semicircles
in contact; suboculars separated from supralabials by one
row of scales; ventrals smooth, much larger than dorsals;
postcloacal scales not enlarged; lateral and dorsal scales
about same size, small, conical to granular, gradually grading into one another; 38 to 40 lamellae under fourth toe (20
to 22 greatly enlarged); tail compressed, basal half of tail in
adult males with a definite caudal crest supported by bony
rays forming a fin in large males; distinct, widely spaced
whorls of larger keeled caudal scales separated by smaller
keeled ones; adult male dewlap extending no farther posteriorly than front limb insertion in males; dewlap present
in females. Dorsal ground color usually dull brown with a
series of transverse dark bars in males and overlapping
diamond-shaped blotches in many females; some females
with broad middorsal dark stripe indicated; venter greenish
yellow; iris dark brown.

SIMILAR SPECIES: (1) *Norops capito* has a very short
head, a uniform greenish yellow dewlap, and keeled ventrals. (2) *Norops woodi* has a dark olive green to black dewlap, a dull brick red or dark olive ground color, and keeled
ventrals.

HABITAT: In Parque Vargas, Puerto Limón, this species
is found on giant fig trees.

BIOLOGY: *Ctenonotus cristatellus* is a heliothermic species that is very commonly seen in the Parque Vargas in
Puerto Limón, Limón Province, and less frequently elsewhere in the city. Recently they have been collected in Costa
Rica from Cahuita, Guayacán, and Valle de Rosas, Limón
Province, and Turrialba, Cartago Province. This lizard conforms to the trunk-ground ecomorph of Williams (1972,
1983), perching on the lower trunks of trees (1 to 3 m above
the ground), where it sits and waits for prey, principally arthropods, moving along the ground. It makes a quick scurry
off the perch to capture prey. When disturbed, these lizards
move upward to escape by hiding in the tangle of roots from
the fig trees.

REMARKS: A member of the C. *cristatellus* series (nineteen species) of Hispaniola, Isla Mona, Puerto Rico, and the
Virgin and Bahama Islands (Guyer and Savage 1986). The
karyotype is 2N = 26 (in females), consisting of six pairs of
large and two pairs of smaller metacentrics, eight pairs of
microchromosomes, and a pair of X chromosomes; NF = 42
(males are 2N = 27 with XXY heteromophism) (Gorman,
Thomas, and Atkins 1968).

DISTRIBUTION: Native to Puerto Rico, the Virgin Islands, and associated smaller islands; introduced to the Dominican Republic, southern Florida, and Costa Rica. Known
in Costa Rica from Puerto Limón and Cahuita and several
other localities in Limón Province and Turrialba, Cartago
Province, on the Atlantic slope (2–750 m).

Genus *Dactyloa* Wagler, 1830

Anoles referred to this genus (sixty-one species) include the
largest mainland species (total length to 480 mm, standard
length to 160 mm), which occur primarily in the rainforest
canopy, but also many small to moderate-sized forms. Species of *Dactyloa* share the following features: expanded
smooth digital lamellae, fourth toe much longer than third;
no transverse gular fold; no femoral pores; no caudal autotomy or septa (present in *roquet* series); posterior caudal
vertebrae without transverse processes; and tail one and a
half to two and a quarter times standard length. Other characteristics typical of the genus are no palatal teeth; no angular bone; splenial bone present; no jaw sculpturing; parietal roof trapezoidal; parietal ridges U-shaped; occipital not
overlapped by parietal half funnel; interclavicle T-shaped or
arrow shaped; inscriptional rib formula 4:0, 5:0; karyotype
2N = 32, 34, or 36, six pairs of biarmed macrochromosomes
and ten, eleven, or twelve pairs of microchromosomes, no
sexual heteromorphism (Gorman and Atkins 1967; Williams et al. 1970). All species in this genus have the ability
for some color change. *Dactyloa* contains several distinctive
lineages (Guyer and Savage 1992), the extralimital *aequatorialis* (nine species), *laevis* (three species), *roquet* (nine
species), and *tigrina* (seven species) series, and the *latifrons*
(twelve species) and *punctata* (twenty species) series that are
represented in Costa Rica. One species, *Dactyloa agassizi*

from Malpelo, Colombia, has not been allocated to a series. For a complete listing of species by species group within all series, see Savage and Guyer (1989).

This is a South American genus with a wide distribution on that continent and adjacent islands, from western Ecuador and southeastern Brazil northward, with several species occurring in Panama and Costa Rica.

Latifrons Series

These are "giant" anoles (Dunn 1937c), all exceeding 300 mm in total length as adults (more than 100 mm in standard length) and having a T-shaped interclavicle. They are highly arboreal, and some are rarely seen, since they spend most of their time in the canopy of intact forests. In addition to the three Costa Rican species described below, the other giant anoles of this genus may be grouped as follows (Williams 1988; Arosemena, Ibáñez, and Sousa 1991): (1) green long-legged forms: *D. casildae* (Panama), *D. latifrons* (Colombia and Panama), *D. princeps* (Ecuador), *D. purpurescens* (Colombia), and *D. squamatula* (Venezuela); (2) green short-legged species: *D. apollinaris*, *D. danieli*, and *D. propinqua* (Colombia); (3) mostly brown short-legged species: *D. fraseri* (Colombia and Ecuador). Of the Costa Rican species, *D. frenata* belongs to the first group and *D. insignis* and *D. microtus* belong to the third.

Dactyloa frenata (Cope, 1899)

(plates 254–55; map 10.13)

DIAGNOSTICS: A giant green anole (to 430 mm in total length) marked by light-centered oblique green bands on the flanks and having an enormous, uniformly cream white dewlap in adult males. This long-legged species is unlikely to be confused with any other anole except the smaller *Dactyloa chocorum* (maximum total length 280 mm), which has some definite enlarged supraoculars and the suboculars in contact with the supralabials.

DESCRIPTION: Total length to 430 mm in adult males (standard length 121 to 143 mm), adult females much smaller (110 to 118 mm in standard length), tail long, 65 to 70% of total length; hind limbs long; when adpressed along side of body, tip of fourth toe reaches posterior margin of orbit or beyond; anterior upper head scales flattened and mostly smooth; interparietal shield large, and parietal eye present; circumnasal or anterior nasal separated from rostral by one scale; an elongate anterior superciliary scale; supraorbital semicircles separated by one to five scales; suboculars separated from supralabials by one or two rows of scales; two or three rows of slightly enlarged keeled middorsal scales; other dorsals and laterals small, about same size, juxtaposed or subimbricate, smooth or weakly keeled; ventrals much larger than dorsals, smooth with rounded margins, juxtaposed to imbricate; scales along anterior margin of upper thigh surface unicarinate; low nuchal crest; postcloacal scales enlarged in males; 50 to 58 lamellae under fourth toe (22 to 28 greatly enlarged); tail subcylindrical, without a crest; dewlap of males extremely large, extending posteriorly about one-third length of body, smaller in females but well developed. Upper surfaces green with four or five light-centered, oblique dark bands on each flank; tail similarly banded; venter essentially yellowish.

SIMILAR SPECIES: (1) *Dactyloa chocorum* is smaller and does not have light-centered body bands. (2) *Polychrus gutturosus* has a short, blunt head, an extremely long tail three or more times standard length, and no expanded subdigital lamellae.

HABITAT: Relatively undisturbed forests in Lowland Moist and Wet Forests and lower portions of the Premontane Wet Forest zone.

BIOLOGY: *Dactyloa frenata* is uncommonly seen in Costa Rica. It is a thermoconformer that uses tree trunks for perches but actively feeds on the leaf litter once it finds a prey item. Although the data of Scott, Wilson, Jones, et al. (1976) and Losos et al. (1991) at Barro Colorado Island, Panama Province, Panama, indicate that *D. frenata* perches 0.3–2.1–6 m above the ground, observations in Costa Rica suggest that large males frequently perch at the 4 to 5 m level. These lizards remain relatively motionless for long periods with the body parallel to the trunk and the head raised and pointed downward. This position allows a lizard to survey the environment for prey, conspecifics, and predators. Males consistently perch higher on the trunks than females, and Scott et al. (1976) believed this reflects a different focus in females (oriented toward the food supply) than in males (oriented toward territorial and courtship activities). Territory sizes on Barro Colorado Island average about 800 m² for males, 438 m² for females, and 177 m² for juveniles.

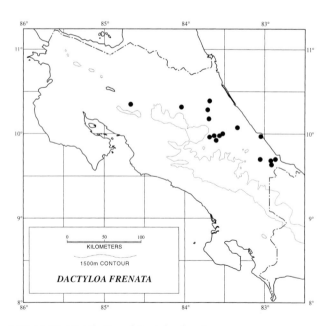

Map 10.13. Distribution of *Dactyloa frenata*.

Food at that site consists of a variety of arthropods and their larvae, especially beetles and orthopterans. Egg laying appears to be concentrated in the dry season (December to April) in central Panama, but reproduction occurs throughout the wet season at many sites. Females place the eggs in a shallow subsurface nest.

REMARKS: The specimens (KU 34245–46) referred to *Anolis purpurescens* by Taylor (1956) are *Dactyloa frenata*. Most earlier accounts on this species use the name *Anolis frenatus*.

DISTRIBUTION: Lowland and lower premontane areas of Atlantic slope Costa Rica, south to northern Colombia and the upper reaches of the Río Magdalena of that country; also in central and eastern Panama on the Pacific versant (2–819 m).

Dactyloa insignis (Cope, 1871)

(plates 256–57; map 10.14)

DIAGNOSTICS: A short-limbed giant anole (to 464 mm in total length) with a series of dark oblique bands (green to brown) on a yellowish gray to tan ground color and a distinct large, light-centered, dark-outlined preaxillary blotch or postorbital dark spot. The enormous dewlap of adult males is primarily orange red, with or without several horizontal green or white bars and dark spotting; the free margin is white to greenish.

DESCRIPTION: Total length to 464 mm in males (adult standard length 129 to 160 mm), adult females smaller (126 to 135 mm in standard length); tail long, 64 to 70% of total length; hind limbs short; when adpressed to side of body, tip of fourth toe reaches at most to between shoulder and tympanum; anterior upper head scales flattened, smooth; interparietal scale large, parietal eye present; circumnasal or

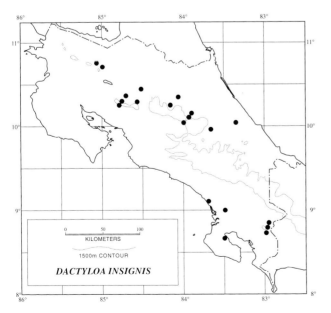

Map 10.14. Distribution of *Dactyloa insignis*.

anterior nasal separated from rostral by one or two scales; no elongate anterior superciliary scale; supraorbital semicircles separated; some enlarged suboculars; supraoculars in contact with supralabials; one to three slightly enlarged middorsal scale rows; dorsals and laterals smooth, conical, and juxtaposed; ventrals smooth, juxtaposed, rounded, much larger than dorsals; scales along anterior margin of upper thigh surface smooth; a distinct nuchal crest; postcloacal scales enlarged in males; 51 to 58 lamellae under fourth toe (23 to 27 greatly expanded); tail strongly compressed, without a crest; dewlap of males very large, extending posteriorly about one-third of body length, smaller but well developed in females. Upper surfaces gray yellow to light tan with three oblique dark bands, containing light spots (or composing a reticulum) on flanks; light and dark lip bars and a light-centered, dark-outlined preaxillary spot (absent in some extralimital samples); some examples have a distinct dark postorbital spot instead; dark areas varying from green to dark brown depending on mood of individual lizard; tail marked like body; venter cream with numerous small dark spots; female dewlap pale green; iris brown with blue margin or uniform blue.

SIMILAR SPECIES: (1) *Dactyloa microtus* has a strongly developed dark-outlined postorbital light stripe that continues posteriorly through the axilla and lacks an enlarged interparietal scale and parietal eye. (2) *Norops biporcatus* is usually bright green and has strongly keeled ventrals.

HABITAT: An uncommonly seen canopy species restricted to undisturbed midelevation Premontane Wet Forest and Rainforest, Lowland Wet Forest, and Lower Montane Wet Forest and Rainforest.

BIOLOGY: This species appears to be a denizen of the forest canopy.

REMARKS: As pointed out by Etheridge (1959) and confirmed by Savage and Talbot (1978), but contrary to Peters and Donoso-Barros (1970), *Diaphoranolis brooksi* Barbour, 1923, is based on a juvenile of this species (MCZ 16297) from eastern Panama. Note that in Panamanian examples of *D. insignis* the lip bars are weakly developed or absent and that the dark bands are reticulate. In addition, the preaxillary spot is absent in all examples from eastern and central Panama. The name *Anolis insignis* was used for this species before 1987.

DISTRIBUTION: Premontane and lower montane humid forests and lowland forests on both slopes of the cordilleras of Costa Rica; lowland rainforests of northwestern Panama (Atlantic slope) and from southwestern Costa Rica to eastern Panama (Pacific versant) (4–1,600 m).

Dactyloa microtus (Cope, 1871)

(plate 258; map 10.15)

DIAGNOSTICS: A short-legged giant anole (to 331 mm in total length) with a light yellow or tan ground color marked by a series of light brown-centered dark bands that may

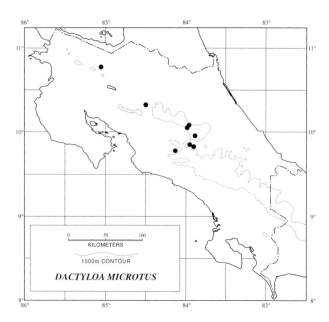

Map 10.15. Distribution of *Dactyloa microtus*.

change to green on flanks and a definite postorbital dark-bordered light stripe. The large dewlap of adult males is pale salmon with several vertical rows of white scales.

DESCRIPTION: Total length to 331 mm (adult male standard lengths 95 to 111 mm); females smaller than males (95 to 104 mm in standard length); tail long, 65 to 68% of total length; hind limbs short; when adpressed to side of body, tip of fourth toe not reaching farther than between shoulder and tympanum; anterior head scales smooth, flattened; no enlarged interparietal scale, no parietal eye; circumnasal or anterior nasal separated from rostral by one scale; one slightly elongate superciliary scale; supraorbital semicircles separated by one or two scales; some enlarged supraoculars; suboculars in contact with supralabials; no enlarged middorsal scales, lateral scales mostly flattened and smooth; ventrals smooth, rounded, much larger than dorsals; scales along anterior margin of upper thigh surface multicarinate; a distinct nuchal crest; postcloacal scales enlarged in males; 41 to 49 lamellae under fourth toe (20 to 22 expanded); tail laterally compressed without a crest; dewlap of male large, extending posteriorly about one-fourth distance from axilla to groin; smaller but well developed in females. Upper surfaces a light yellowish to tan marked with three oblique transverse bands on flanks; bands with lighter centers; light interspaces divided by irregular linear, transverse series of dark spots; some metachrosis, with green becoming more evident in dark bands depending on mood; a definite dark-outlined light postorbital stripe from lower posterior corner of eye to beyond axilla; tail marked with alternating dark and light bars; venter yellow, essentially unmarked by dark pigment; female dewlap pinkish.

SIMILAR SPECIES: (1) *Dactyloa insignis* lacks a postocular light stripe and has a well-developed parietal eye.

(2) *Norops biporcatus* is usually bright green and has keeled ventrals.

HABITAT: Undisturbed rainforests of the premontane zone and marginally into similar forests of the upper lowland belt.

BIOLOGY: Apparently a canopy species.

REMARKS: This rare species is now known from ten examples from Costa Rica and four from Panama. Before 1987 it was referred to as *Anolis microtus*.

DISTRIBUTION: Premontane forests and upper portions of adjacent lowland zone on both slopes from northwestern Costa Rica to southwestern Panama (425–1,500 m).

Punctata Series

Members of this series are small to moderate-sized lizards with an arrow-shaped interclavicle. Only a single member of this stock ranges into Costa Rica, but species in the series occur in the humid tropical forests of South America both east and west of the Andes as follows:

Northwestern South America (west of Andes)
1. Ecuador: *D. fasciata*, *D. festae*, *D. gemmosa*, *D. nigrolineata*
2. Colombia: *D. calimae*, *D. chocorum* (to Costa Rica), *D. huilae*, *D. santamartae*
3. Ecuador and Colombia: *D. chloris* (to Panama), *D. peraccae*

Northern South America
1. Eastern Colombia and western Venezuela: *D. jacare*, *D. nigropunctata*
2. Eastern Venezuela: *D. deltae*

Amazon drainage
1. Peru: *D. albimaculatus*, *D. boettgeri*, *D. dissimilis*
2. Colombia: *D. caquetae*, *D. vaupesiana*
3. Venezuela to Peru and Brazil: *D. transvalis*

Widespread
1. Amazon basin, northeastern and eastern Venezuela, the Guianas, and eastern Brazil: *D. punctata*

Dactyloa chocorum (Williams and Duellman, 1967)
(plate 259; map 10.16)

DIAGNOSTICS: A very distinctive moderate-sized, long-legged green arboreal anole that may be uniform in dorsal color or have darker green oblique flank bars. Both sexes lack nuchal and caudal crests, and the male dewlap is mostly orange and marked with oblique, forward-directed greenish lines. The dewlap of females is also large (fig. 10.22d) but is mostly green, bordered by white and often marked with oblique white lines. Members of this species further differ from other green forms in having smooth ventrals, definitely enlarged medial supraoculars, and the supraorbital semicircles separated by one or two scales.

DESCRIPTION: Total length to 275 mm (standard length 60 to 79 mm in adult males, 54 to 73 mm in adult females); tail very long, 70 to 73% of total length; hind limbs long;

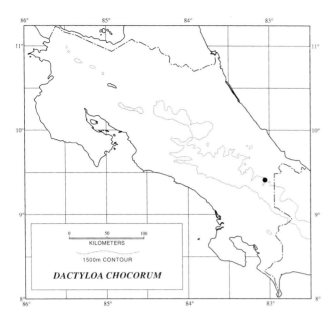

Map 10.16. Distribution of *Dactyloa chocorum*.

when adpressed along side of body, tip of fourth toe reaches to anterior margin of orbit or beyond; anterior head scales moderate to small, smooth; an enlarged interparietal and parietal eye present; nasal separated from rostral by one scale; an elongate anterior superciliary; several enlarged supraoculars separated by one to three scales; suboculars in contact with supralabials; no enlarged middorsal scales; dorsal and lateral scales small, granular to conical; ventrals much larger than dorsals, smooth, rounded; no nuchal crest; enlarged postcloacal scales in males; 37 to 43 lamellae under fourth toe (17 to 20 greatly expanded); tail slightly compressed, without a crest; dewlap of both sexes very large, extending posteriorly about one-third of distance between axilla and groin. Upper surfaces uniform green or with darker green oblique transverse bands; tail marked with dark and light bands; venter blue gray with light spots; in addition to the dewlap features mentioned in the diagnostics, the anterior edge is white and the base greenish in males and yellowish to gray in females; lining of throat black; iris brown.

SIMILAR SPECIES: (1) *Dactyloa frenata* is a larger anole with the transverse body bands having definite light centers and all supraocular scales small. (2) *Norops biporcatus* is usually a uniform green and has short legs and keeled ventrals.

HABITAT: In Costa Rica known only from the Lowland Wet Forest zone on the southeast Atlantic slope.

BIOLOGY: A diurnal trunk-crown species that is rarely seen in Costa Rica. An egg belonging to this form was found under loose bark (Williams and Duellman 1967) and measured 16 × 12 mm. The hatchling lizard that emerged on July 26 was 30 mm in standard length, 92 mm in total length. When held, this anole emits a few squeaks or a series of them (Myers 1971a).

REMARKS: This form was originally described as *Anolis chocorum,* and that name was used for it before 1987.

DISTRIBUTION: Lowlands of southeastern Costa Rica and adjacent western Panama (100–300 m) on the Atlantic versant, the Darién region of eastern Panama (100–1,100 m), and the Chocó of western Colombia (220 m) on the Pacific slope.

Genus *Norops* Wagler, 1830

Norops (143 species) is the largest genus of anoles and contains small, moderate-sized, and large species that include terrestrial, scansorial, semiaquatic, semiarboreal, and arboreal forms. The genus is characterized as follows: expanded smooth subdigital lamellae present, except in two forms (*N. annectans* and *N. onca* of South America), fourth toe much longer than third, no transverse gular fold, no femoral pores, caudal autotomy with autotomy septa lying anterior to anteriorly directed transverse processes, and tail one and a half to two and a quarter times standard length. Other features that in combination diagnose the genus include no palatal teeth; no angular bone; no splenial bone; no lower jaw sculpturing; parietal roof triangular; parietal ridges V- or Y-shaped; parietal forming a half funnel overlapping occipital bone; interclavicle T-shaped; inscriptional rib formula usually 2:2, 3:1, or 3:2, sometimes 1:3 or 1:2, rarely 2:1.

The studies of Etheridge (1959), Williams (1976a,b), Lieb (1981), and Guyer and Savage (1986, 1992) indicate that most species of this large genus may be placed in seven major subdivisions (series) based on skeletal, external morphological, and karyological features as follows: *auratus* series containing seventy-two Mexican, Central American, and South American forms; *fuscoauratus* series composed of twenty-seven Mexican, Central American, and South American species; *grahami* series consisting of seven species on Jamaica and Grand Cayman Island; *meridionalis* series containing a single species from southern Brazil and Paraguay; *onca* series composed of two species from northern South America; *petersii* series consisting of twelve forms in Mexico, Central America, and South America; and *sagrei* series made up of fifteen species from Cuba, Jamaica, the Bahamas, and Little Cayman Island, but one species introduced to the mainland of North and Central America.

The *auratus, fuscoauratus,* and *petersii* series are represented in Costa Rica. Technically these three series differ only in their basic inscriptional rib formulas (which, however, exhibit rare variations within a few species). For a complete listing of species by species group within each series see Savage and Guyer (1989).

All species now placed in this genus were referred to *Anolis* before 1987, and some authors continue to do so.

As seen above, *Norops* is primarily a Mexican and Central American genus with a few representatives ranging as far south as central Bolivia, Brazil, and Paraguay and with a moderate radiation in Jamaica, Cuba, the Bahamas, and the Cayman Islands.

Auratus Series

This is the largest, most heterogeneous, and systematically most difficult group within *Norops,* having a characteristic inscriptional rib formula of 2:2 or 2:1. Most members of the series are small lizards, and in Costa Rica they include terrestrial and riparian forms and others that are denizens of the lower strata within the forest.

Norops aquaticus (Taylor, 1956)

(plate 260; map 10.17)

DIAGNOSTICS: An elegant little lizard found along stream courses and readily distinguished by color alone. The dorsum, limbs, and tail are marked with broad transverse brown bands on a pale yellowish green ground color, a light greenish yellow stripe runs from under the eye some distance or down the whole length of the body, and the iris is blue. The large dewlap of adult males (fig. 10.22d) is red orange, usually with oblique forward-directed yellow stripes.

DESCRIPTION: A moderate-sized anole (to 188 mm in total length), adult males (57 to 71 mm in standard length) larger than females (adults 59 to 62 mm in standard length); tail moderately long, 60 to 65% of total length; hind limbs short; when adpressed against body, tip of fourth toe reaches no farther than posterior margin of orbit; scales on upper surface of snout large, keeled; scales of prefrontal region small, keeled; circumnasal or anterior nasal separated from rostral by two rows of scales; supraorbital semicircles separated by two to four scales; semicircle and enlarged supraocular scales with strong median keel; suboculars separated

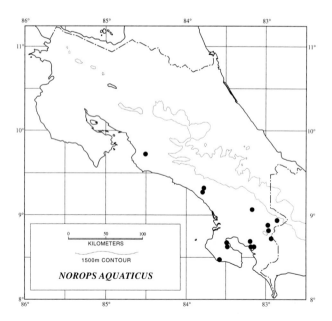

Map 10.17. Distribution of *Norops aquaticus.*

from supralabials by one or two rows of scales; two to four enlarged keeled middorsal scale rows; laterals small, subimbricate or conical; ventrals keeled, imbricate, much larger than all dorsals except middorsals; a distinct nuchal crest in males; no enlarged postcloacal scales; 29 to 33 lamellae under fourth toe (14 to 19 definitely expanded); tail strongly compressed, a distinct low serrate crest in males, midcaudal scales enlarged in females; dewlap of male large, extending posteriorly about one-fourth the way between axilla and groin; no dewlap, throat unpigmented in females. Dorsal ground color pale yellowish green marked by dark dorsal brown bands; a pale yellowish green longitudinal stripe runs under eye and down flank, usually to groin; numerous small pale greenish yellow spots scattered over dorsal and lateral surfaces; tail similarly marked; venter yellowish green; iris blue.

SIMILAR SPECIES: *Norops oxylophus* lacks light spotting and has light-edged black bands on the tail and a coppery iris.

HABITAT: Restricted to stream courses within Lowland Wet Forest and Premontane Wet Forest and Rainforest.

BIOLOGY: *Norops aquaticus* is an abundant diurnal thermoconforming species. The lizards spend most of their time along stream margins perching and feeding on rocks, logs, or low vegetation and often swimming or hopping out to boulders in the stream to feed. When approached they dive into the water and swim for several meters, emerging to hide under log jams or accumulations of debris or in cavities under the bank or rocks. They occur along relatively small slow-moving streams in the lowlands, but at higher elevations are found near rocky mountain brooks of high gradient and water flow. Individuals are frequently encountered at night sleeping on small branches along stream margins. Fitch, Echelle, and Echelle (1976) suggested that this lizard reproduces year round, as do many other tropical anoles.

REMARKS: The relation of this distinctive species to other *auratus* series anoles remains unclear (Lieb 1981; Savage and Guyer 1989). The karyotype of this species is 2N = 30, composed of six pairs of metacentrics and one pair of submetacentrics and eight pairs of microchromosomes; NF = 44 (Lieb 1981).

DISTRIBUTION: Lowlands and premontane slopes of southwestern Costa Rica and adjacent southwestern Panama (30–1,170 m).

Norops cupreus (Hallowell, "1860," 1861)

(plate 261; map 10.18)

DIAGNOSTICS: A very common small, short-legged gray to brown lizard found in the drier portions of Costa Rica including the Meseta Central, where it is often seen even in the most urbanized areas on fence posts and low vegetation in coffee plantations, vacant lots, and gardens. The adult males are unmistakable in having a very large (fig. 10.22d) brightly colored dewlap that is burnt orange to red orange

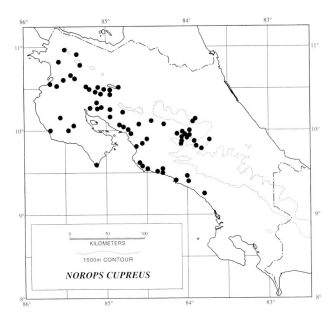

Map 10.18. Distribution of *Norops cupreus*.

basally and has a pinkish free margin fading into a yellow to white area at its anteriobasal margin with the throat. Females lack a dewlap, but the throat may have a spot of color. Other features that taken in combination characterize this species are toe lamellae only moderately expanded, ventrals strongly keeled, and four median rows of dorsal scales enlarged and keeled.

DESCRIPTION: Total length to 170 mm in adult males; adult males (38 to 57 mm in standard length) larger than females (adults to 51 mm in standard length); tail moderately long, 60 to 67% of total length; hind limbs short; when adpressed against side of body, tip of fourth toe reaches no farther than center of eye; scales on upper surface of snout small, keeled; prefrontal scales larger, keeled; circumnasal or anterior nasal separated from rostral by two scale rows; supraorbital semicircles separated by one to three scales; semicircle and enlarged supraocular scales with strong median keel; suboculars usually separated from supralabials by one row of scales, sometimes partially fused to suboculars; four median rows of dorsal scales enlarged, keeled; other dorsal and lateral scales small, granular; ventrals much larger than dorsals, keeled, imbricate, rounded posteriorly; a low nuchal crest in males; no enlarged postcloacal scales; digital pads only slightly expanded; 26 to 33 lamellae under fourth toe (11 to 15 expanded); tail subcylindrical; a low caudal crest in males; dewlap of males very large, extending posteriorly halfway between axilla and groin; no dewlap, throat usually unpigmented or with a pinkish or orange spot or mottled with gray in females. Dorsum olive gray to dark brown, uniform or marked with narrow to broad light (whitish, tan, yellow to orange) middorsal stripe or with a series of dark middorsal black marks, often in the form of chevrons; venter dull white; iris coppery.

SIMILAR SPECIES: (1) *Norops sericeus* has a dark blue to purplish round spot on the reddish orange male dewlap, a definite small dewlap is present in females (fig. 10.22a), and the suboculars are in contact with the supralabials. (2) *Norops limifrons* has a white and orange dewlap in adult males, has smooth or weakly keeled ventrals, and lacks enlarged middorsal scales. (3) *Norops polylepis* has a uniform orange dewlap in adult males, has smooth ventrals, and lacks enlarged middorsal scales.

HABITAT: *Norops cupreus* is found in a wide variety of situations including disturbed and converted Premontane Moist Forest areas on the Meseta Central and marginally in Premontane Wet Forest. Especially abundant in gallery forest sites along stream courses within the Lowland Dry Forest region. It also occurs in dense vegetation bordering beach frontage and at scattered sites within humid forests.

BIOLOGY: This very common lizard has been extensively studied in Costa Rica by Henry Fitch and associates (Fitch 1973a,b,c, 1975; Fitch, Echelle, and Echelle 1972), and these accounts and those of Fleming and Hooker (1975) provide much of the detail that follows. This lizard is characteristic of areas that have a marked, hot dry season. During cooler periods of the year individuals bask, but in the dry season they tend to hide during the day and are active near sunset. Body temperatures of active lowland individuals range from 18.8 to 34.2°C and are only slightly above ambient air temperature. At higher elevations basking is more frequent, and body temperatures of active lizards were somewhat higher than ambient, although lower than those from lowland sites (Fitch 1973a). Lizards of this species spend a good deal of their time on the ground foraging and are generally found where there is abundant leaf litter. They also forage in small trees, vines, bushes, and trash piles, and their perch sites include these elevated areas as well as fence posts, walls, and grass clumps. In the hot, dry times of the year perches are on or near the ground (0 to 400 mm), with little difference between the sexes in preference. In the wet season male perch sites are much higher, as much as 10 m above the ground. Territory sizes average 35 m² for males and 8 m² for females. The species occurs in very high densities (over 1,900/ha) in favored habitats during the dry season. As males move into more arboreal sites in the wet season, densities at ground level decrease to about 500/ha, of which 20 to 30% are adults. Food consists of a variety of arthropods, predominantly ants, lepidopteran larvae, and spiders, with ants making up the bulk of the diet in the dry season and caterpillars in the wet season. Other prey items include (in decreasing order of frequency) orthopterans, isopods, termites, cockroaches, ticks, and cicadas, and wasps and snails are eaten occasionally. Prey items range in size from 1 to 17 mm, and there is considerable overlap in the size of prey taken by males and by females. In the dry season males feed on larger prey than females, and the stomachs of both sexes contain more food items in the dry season (males one to twenty-two, females none to thirteen) than in the wet (males none to eight, females none to eleven) (Fleming and Hooker 1975). Social interactions are limited in the dry season, but large males become increasingly aggressive in the wet season and spend much time patrolling, displaying, and courting. Male reproductive behavior is intense from April through June and decreases thereafter. This behavior corresponds to the distinctly seasonal reproductive cycle of this species. Territorial males are 42 mm or longer in standard length, and smaller males are excluded from interactions with females by the aggressive large males. Egg production is concentrated between late April or May and October, with essentially no reproduction during the dry season and both tails of the rainy season. On average females lay a single egg every ten days, and some may lay an egg every seven or eight days. Hatchlings are about 18 mm in standard length and appear in large numbers beginning in mid-July. In June only adults and large juveniles make up the population, but a flush of hatchlings appears in July to August, and by November to December both small juveniles and larger lizards are well represented. The proportion of juveniles then gradually decreases to zero until the next summer's hatchling flush. Both sexes attain sexual maturity during the first year of life at a standard length of about 37 to 38 mm.

REMARKS: *Norops cupreus* is thought to be most closely related to *N. dolfusianus* (Mexico and Guatemala) and the putative species, *N. villai* (Isla de Maíz Grande, Nicaragua), which formed the *cupreus* species group (Lieb 1981; Savage and Guyer 1989). Fitch, Echelle, and Echelle (1972) and Fitch and Seigel (1984) analyzed geographic variation in this form and recognized a number of races based on various combinations of mean differences in size and proportions, scale counts, and coloration in females. Most of these features vary in a discordant clinal fashion, but two peripheral populations differ from the others in male dewlap color. Males from the far northern segment of the range (Guatemala and El Salvador) have a mostly pink dewlap with a basal orange spot and are now recognized as a distinct species, *Norops macrophallus* (Werner 1917), following Köhler and Kreutz (1999). Males from the central montane region and lowland Atlantic drainage Nicaragua have a uniform dark brown dewlap. This feature, the distinctly smaller body, scalation (138-160-179 scales around midbody versus 98-120-152 in other *N. cupreus*), and the more arboreal habits suggest that this population represents a different species than *N. cupreus*, for which the name *N. dariense* Fitch and Seigel, 1984 is appropriate. Records of *N. cupreus* from the lower montane zone (ANSP 21920–21; Costa Rica: San José: Tablazo, 1,600–1,900 m; UCR 3008; San José: Finca Echandi, 1,740 m) may be based on animals from lower elevations. A questionable specimen collected in the nineteenth century (AMNH 16353) is listed as being from Puerto Viejo, Heredia Province, and another (AMNH 67228) was more recently obtained from Río Sarapiquí in the same area. The species is known from the Atlantic slope in the Río Reven-

tazón valley and on the north side of the pass between Volcanes Barva and Irazú. No specimens from the next pass north (Desengaño) or the lowlands along the Río Sarapiquí have been taken in the past half century, so these records are questionable. The karyotype of *N. cupreus* is 2N = 40, consisting of twelve pairs of macrochromosomes and eight pairs of microchromosomes (Lieb 1981).

DISTRIBUTION: Lowland and premontane zones of Pacific versant southern Nicaragua to central Costa Rica as far south as the Río Barú, Puntarenas Province, including the Meseta Central Occidental; scattered Atlantic versant records for Honduras and Nicaragua (may be based on *N. dariense*) and Costa Rica, including the Meseta Central Oriental (1–1,435 m).

Norops humilis (Peters, 1863b)

(plate 262; map 10.19)

DIAGNOSTICS: This extremely abundant very small brown mainly terrestrial forest dweller is easily identified by the deep axillary pore and the eight to ten rows of enlarged keeled middorsal scale rows. The dewlap of adult males is mostly deep red with a bright yellow free margin.

DESCRIPTION: Total length to 114 mm, females (adults 32 to 44 mm in standard length) somewhat larger than males (adults 29 to 45 mm); tail moderately long, 60 to 63% of total length; hind limbs long; when adpressed along side of body, tip of fourth toe reaches to anterior margin of orbit or beyond; scales on upper surface of snout small, elongate, keeled; scales in prefrontal area larger, keeled; circumnasal or anterior nasal separated from rostral by one or two scales; supraorbital semicircles separated by two or three scales; semicircle and enlarged supraocular scales with strong median keel; suboculars separated from supralabials

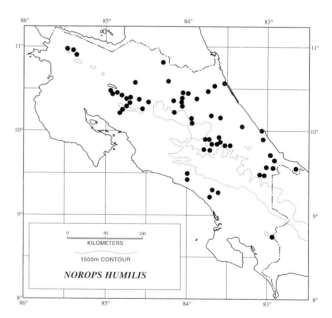

Map 10.19. Distribution of *Norops humilis*.

by one row of scales; eight to ten median rows of dorsal scales enlarged, keeled; other dorsal and lateral scales small, granular; ventrals smaller than enlarged dorsals, keeled, imbricate, rounded; no nuchal crest; no enlarged postcloacal scales in males; digital pads barely expanded; 23 to 30 lamellae (10 to 14 slightly expanded) under fourth toe; tail compressed; no caudal crest; dewlap of males moderate, extending slightly posterior to axilla; no dewlap in females, but throat has a patch of red color. Dorsum brown to olive brown, often with dark markings forming broad middorsal bronze or dark stripe, paired dark dorsolateral blotches, or a middorsal series of hourglass-shaped figures; usually a broad dark band across prefrontal region; venter light brown; iris brown with gold flecks.

SIMILAR SPECIES: (1) *Norops tropidolepis* usually has a curved black mark between the eyes and another on the nape and often a distinct light diagonal subocular stripe that slants posteriorly. All dorsal scales are smaller than the ventrals in this species, and males have a uniform deep plum to burgundy dewlap and the tail is distinctly swollen at the base. (2) *Norops pachypus* has mostly smooth ventrals, head markings, and a swollen tail base as in *N. tropidolepis* and a uniform red orange dewlap, often with a central yellow spot.

HABITAT: Most commonly observed on or in the leaf litter on the forest floor, particularly in the vicinity of tree buttresses in Lowland Moist and Wet Forests and Premontane Wet Forest and Rainforest; marginally present in Lower Montane Wet Forest and Rainforest at a few sites.

BIOLOGY: *Norops humilis* has been the subject of several detailed ecological studies in Costa Rica (Fitch 1973a,b, 1975; Talbot 1976, 1977, 1979; Corn 1981; Guyer 1986, 1987, 1988a,b), and these accounts form the basis for the following summary. The species is a rainforest habitué that is most common in shaded areas near the base of large trees, although locally abundant in abandoned cacao plantations. These lizards are thermoconformers and do not bask. Body temperatures of active lizards from the Atlantic lowlands are 20–21.5–27°C and are about 1°C above ambient air temperature. These temperatures are distinctly low for a lowland species. Lizards of this species spend a considerable portion of their time on the ground and are most abundant where there is deep leaf litter. They are trunk-litter foragers that perch head downward on trees, stumps, or logs from near ground level up to 1.5 m (average about 600 mm). Males usually have higher perches than females. When approached *N. humilis* usually darts around and down the perch to hide in the litter or under low vegetation. The species is strictly diurnal, and males spend about 20% of the day displaying or engaging in other social interactions and 70% on the foraging perch site or briefly and actively pursuing visually located prey. Females are more active foragers (88% of their time) but are less involved in social activities (1% of their time). Most male social interaction (52%) is centered on territory maintenance. *N. humilis* is a sedentary species, and individuals rarely venture more than a few meters from the preferred perch and often return to the same site day after day. In especially favorable habitats (e.g., abandoned cacao plantations) densities are high (500 to 1,100/ha), but in undisturbed forest they average about 350/ha. Home range sizes average about 30 m² for males and 23 m² for females, unlike the situation in most anoles, where females' home ranges usually are one-fourth to one-fifth the size of males'. *N. humilis* preys on a wide variety of arthropods, principally araneid spiders and isopods. Other common dietary components (in decreasing order of frequency) are hemipterans, caterpillars, beetles, dipterans, termites, and centipedes. In twenty-six stomachs analyzed at La Selva, Heredia Province, there was an average of 1.9 prey items per stomach, and mean prey length was 2.9 mm (Lieberman 1986). Although ants are included in the diet, they are selectively avoided, whereas other food items are eaten in proportion to what is available within the size range (0 to 20 mm) of food items known to be taken. In this species prey size increases as lizards become larger, with no significant seasonal differences in size or number (zero to eight) of prey items in stomachs. Aggressive displays, courtship, and reproduction occur throughout the year, but the greatest activity is in the heavy rainy period (May through November). Females frequently have a large egg in each oviduct and may lay a single egg as often as every seven days. Hatchlings are 16 to 17 mm in standard length and reach sexual maturity within one year. There is a slight drop in the relative abundance of juveniles in the March to June period of lower rainfall. This is a short-lived species, with few individuals surviving through two breeding seasons.

REMARKS: *Norops humilis* and its close relatives, *N. compressicauda* (southeastern Mexico), *N. notopholis* (western Colombia), *N. tropidonotus* (eastern Mexico to eastern Nicaragua), and *N. uniformis* (eastern Mexico to northern Guatemala), form the *humilis* species group. For many years *N. uniformis* was regarded as a geographic race of *N. humilis*, and Taylor (1956) referred four lizards (KU 40922–25) from the slopes of Volcán Tenorio, Guanacaste Province, to *Anolis humilis uniformis*. In addition he described a new form, *A. h. marsupialis*, from southwestern Costa Rica. Meyer and Wilson (1971b) concluded after a review of available material that *N. uniformis* was a valid species restricted in distribution to Atlantic slope southern Mexico and northern Guatemala, characterized by a bright red dewlap with a more or less purple anterior spot. Contrary to some subsequent authors (Villa, Wilson, and Johnson 1988), *N. uniformis* does not occur in Honduras, and its range is separated from that of *N. humilis* in extreme northeastern Honduras by about 350 km. In his description of the presumed Costa Rican *uniformis* Taylor (1956) implied that his specimens had the dewlap color typical of the Mexican and Guatemalan form. As pointed out by Meyer and Wilson (1971b) and confirmed by my own examination, only one of Taylor's examples (KU 40925) is a male, and its dewlap is red with the anterior margin yellow, as in all other Costa Rican examples of *N. humilis*. Taylor's *A. h. marsupialis* is based on specimens from the Pacific versant of southwestern Costa Rica that were said to be larger than Atlantic slope material. Additional collections confirm that males (to 44 mm in standard length) and females (to 45 mm in standard length) from the southwest area are larger than in other populations (to 43 mm in both sexes). However, no other difference separates these populations, and it seems best to regard them as forming a single species. Whether their ranges are discontinuous or connected by intermediate populations in Panama remains to be determined. Instead of the typical inscriptional rib formula of 2:2, Etheridge (1959) noted an unusual amount of variation in the formulas in individuals of this form, 8% having 1:2, 4% having 3:1, and 40% having 2:1 (not found in any other anole). The karyotype in this species is 2N = 40, consisting of twelve pairs of macrochromosomes and sixteen pairs of microchromosomes (Gorman 1973).

DISTRIBUTION: Humid forests of the Atlantic lowlands and premontane slopes and marginally in the lower montane zone from extreme eastern Honduras to western Panama (2–1,500 m); upper portion of lowland zone and premontane areas of southwestern Pacific versant Costa Rica (880–1,200 m), south on both slopes to central Panama (350–1,200 m); marginally on the Pacific slope in northwestern Costa Rica (690–1,600 m).

Norops intermedius (Peters, 1863b)

(plate 263; map 10.20)

DIAGNOSTICS: A small, short-legged usually pale-colored arboreal lizard of the mountain slopes having strongly keeled ventrals and elevated wartlike white scales on the lower

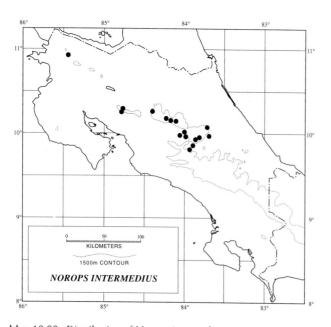

Map 10.20. Distribution of *Norops intermedius*.

flanks and lateral areas of the venter. Males have a small ivory to bone white dewlap and a low but obvious nuchal crest.

DESCRIPTION: Total length to 145 mm, adult males and females about same size (39 to 54 mm in standard length); tail moderate, 60 to 62% of total length; hind limbs very short; when adpressed along side of body tip of fourth toe reaches to shoulder; scales on upper surface of snout small and rugose or flattened and smooth; scales in prefrontal area large, flattened, usually smooth, sometimes rugose; circumnasal or anterior nasal separated from rostral by one scale row; supraorbital semicircles in contact; semicircle and supraocular scales smooth; suboculars and supralabials in contact; six to fourteen median rows of enlarged, keeled, dorsal scales, grading into small tubercular laterals; scattered elevated, wartlike white scales on lower flank and lateral ventral area; imbricate ventrals larger than dorsals, strongly keeled and bluntly pointed posteriorly; nuchal crest present in males; enlarged postcloacals in males; 24 to 33 lamellae (14 to 17 strongly expanded) under fourth toe; tail compressed distally; caudal crest in males; male dewlap small, extending posteriorly to level of forelimb insertion; no dewlap in females and throat plain. Dorsal ground color usually gray, tan, or cream but capable of change to nearly black; often with four pairs of dark brown or black dorsolateral blotches that may be broken up and fused to form a lichenous pattern; females frequently with broad middorsal cream to orange stripe; tail usually banded with brown; venter dirty white; iris beige or coppery.

SIMILAR SPECIES: *Norops lemurinus* has keeled supraocular scales and lacks scattered elevated lateral and lateroventral scales; males have a uniform bright red dewlap, and the dewlap is large in females.

HABITAT: An inhabitant of Premontane Moist and Wet Forests and Rainforest and Lower Montane Wet Forest and Rainforest; very adaptable to disturbed conditions, found along pasture margins, near dwellings, and in gardens and vacant lots in the metropolitan area of central Costa Rica.

BIOLOGY: Fitch (1973a, 1975) has published excellent summaries of the biology of this common species, and these accounts, supplemented by my own field observations, form the basis for the following. *Norops intermedius* is a strictly diurnal and essentially arboreal species that perches and forages between 0.3 and 2 m above the ground on low trees, bushes, or the trunks of taller trees. These lizards are often seen in open areas when favorable perches are available. As expected from the description above, they are heliothermic baskers that are more likely to have an extended diel activity period on overcast days and retreat to cover earlier on clear, sunny days. Body temperatures of active lizards range from 17.4 to 33.7°C (preferred temperatures 25 to 28°C) and are usually 1 to 3°C above ambient air temperature. When approached *N. intermedius* usually escapes by running to the side of the perch opposite an intruder and then upward.

It will however, dive into dense vegetation at ground level near the perch if it is available. This species does not exhibit strong territoriality, and several individuals may co-occur within a few meters. The species tends to be sedentary, and individual lizards show high fidelity to their perch sites, using them for days at a time or shifting to another nearby. Movements by adults within home ranges are minimal (less than 10 m from original capture sites). However, juveniles and adults seeking new ranges may move a considerable distance (up to 50 m in forty-five days). Reproduction in this form is markedly seasonal, although males may be seen displaying throughout the year. Egg laying begins near the end of the dry season (late April or early May), and a female will lay one egg every two to three weeks until the end of the wet season (November). Hatchlings are about 18 mm in standard length, and females become reproductive in about four months. These figures probably apply to males as well, since there is no sexual dimorphism in size. Juveniles are very abundant in November to December, but by June the population consists solely of adults. During August to September both adults and small juveniles and hatchlings are present in large numbers. This is a short-lived form, with few individuals surviving through two breeding seasons.

REMARKS: Meyer and Wilson (1971b) argue that the nominal species *N. laeviventris* (eastern Mexico to western Guatemala), *N. nannodes* (western Guatemala to Honduras), and *N. intermedius* are representatives of a single form, contra Stuart (1955). The three species are completely allopatric, and the populations in Costa Rica and Panama are separated by some 580 km from the nearest Honduras record. However, examples of this stock have recently been reported from Nicaragua by Fitch and Seigel (1984). Their sample resembles *N. intermedius* more closely than lizards from farther north and suggests that at least two species are involved. The enigmatic *N. salvini*, ostensibly from Guatemala, appears to be another possibly valid species belonging to this complex (Stuart 1955; Lieb 1981). Consequently I continue to recognize *N. intermedius* as distinct pending further study, as did Lieb (1981) in his summary of Central American anoles. For those wishing to follow Meyer and Wilson (1971b), the oldest available name for this form is *Norops laeviventris* (Wiegmann, 1834a). The karyotype is 2N = 40, with twelve pairs of macrochromosomes and eight pairs of microchromosomes (Lieb 1981).

DISTRIBUTION: Humid premontane and lower montane forests on the slopes of the cordilleras of Costa Rica and western Panama; possibly in north-central Nicaragua (1,160–1,950 m).

Norops lemurinus (Cope, 1861b)

(plate 264; map 10.21)

DIAGNOSTICS: A moderate-sized, short-legged gray arboreal anole having strongly keeled ventrals and lacking lateral light stripes and elevated wartlike white lateral scales.

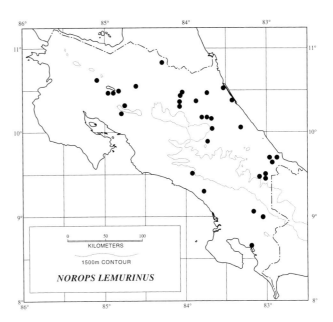

Map 10.21. Distribution of *Norops lemurinus*.

Males have a small (fig. 10.22b) dull dark red dewlap with scattered black scales and low nuchal and caudal crests. Females have slightly smaller white dewlaps that at some localities are colored like the males'.

DESCRIPTION: Total length to 228 mm, adult males and females about same size (59 to 79 mm in standard length); tail moderate, 60 to 65% of total length; hind limbs short; when adpressed against body, tip of fourth toe reaches beyond shoulder, usually to tympanum; scales on upper surface of snout usually keeled and/or rugose, sometimes smooth; scales in prefrontal area large, usually keeled or rugose, sometimes smooth; circumnasal or anterior nasal separated from rostral by one scale row; supraorbital semicircles often in contact but usually separated by one scale row; semicircle scales usually with low keel; enlarged supraocular scales strongly keeled; suboculars and supralabials separated by one row of scales; about twenty median rows of enlarged keeled scales (median four to six rows largest); lateral scales granular; ventrals rounded posteriorly, subimbricate, strongly keeled, and often mucronate, much larger than dorsals; a low nuchal crest in males; no enlarged postcloacals in males; 31 to 38 lamellae (15 to 20 strongly expanded) under fourth toe; tail compressed distally; a weak caudal crest in males; male dewlap small, extending no farther posteriorly than to level of forelimb insertion; female dewlap relatively large. Dorsal ground color brown, often with a yellow wash on sides and venter, usually with four or five large rectangular dark blotches; some female examples with four or five diamond-shaped dark blotches, other females have a broad gold to yellow middorsal band or a narrow midline stripe; tail usually banded with dark brown; venter white; iris coppery gold or brown.

SIMILAR SPECIES: *Norops intermedius* has smooth supraocular scales and elevated wartlike light scales on the flanks; the dewlap is white in males and absent in females.

HABITAT: In Costa Rica this species inhabits lowland rainforests but marginally occurs in the lower portion of Premontane Wet Forest and Rainforest zones. It seems to be a forest shade species.

BIOLOGY: *Norops lemurinus* is a strictly diurnal and arboreal species that perches on tree trunks 0.5 to 3 m above the ground and when disturbed escapes by fleeing upward along the trunk, often to considerable heights (about 5 m). It is a shade species and thermoconformer with a mean activity temperature of 21°C, about 1.9°C above ambient air temperature (Fitch 1975). These lizards occur in low densities and appear to have strong perch site fidelities and small home ranges. They are essentially solitary, and there are no published observations of intraspecific interactions in the field. However, C. Guyer informs me that a single male and single female usually occur together. Food includes a considerable variety of arthropods. Adults prey extensively on beetles, orthopterans, and caterpillars, but juveniles favor smaller and softer food items (e.g., flies and bark lice). Prey size ranges up to 8 mm in length, and three or four prey items usually are found in stomachs containing food. Larger lizards eat larger prey. The principal reproductive activity occurs from May through December, with a definite decline during the drier times of the year from January to March.

REMARKS: According to Lieb (1981), *Norops lemurinus* apparently is most closely related to *N. pachypus* and *N. tropidolepis* of Costa Rica and Panama and to *N. vittigerus* of western Colombia and eastern Panama. It may be that material called *N. lemurinus* from eastern Panama actually is *N. vittigerus*, or the latter may prove to be the southern representative of *N. lemurinus*. Lizards that have been referred to this rather variable species form three major geographic divisions: (1) Atlantic slope Mexico to northern Honduras; (2) Pacific slope Chiapas, Mexico, Guatemala, and El Salvador; and (3) southern Honduras to Panama. Examples from the first population usually have the supraorbital semicircles separated by two or three scales (80%), while in the other two the supraorbital semicircles are usually in contact or, farther to the south, separated by one scale (Stuart 1948). In Costa Rica the semicircles are rarely in contact. These trivial and variable differences (zero to two scale separation in populations 1 and 2; zero to three in 3) do not provide a sound basis for recognizing geographic races. Note, however, that representatives of the two northern populations occur in dry forest and savanna habitats, whereas those from the southern one inhabit closed-crown rainforests. Köhler (1999b) has recently recognized the lizards in population 2 as a distinct species, *Norops serranoi*, principally based on hemipenial and dewlap differences from *N. lemurinus*. It may be that population 1 will prove to be specifically distinct as well. The karyotype of female *N. lemurinus* is 2N = 38, with eleven pairs of macrochromosomes and eight pairs

of microchromosomes; the macrochromosomes include one pair of metacentrics, eight pairs of submetacentrics, and two pairs of subtelocentrics; NF = 40 (Lieb 1981).

DISTRIBUTION: Lowlands and premontane slopes of the Atlantic versant from central Veracruz to western Panama and disjunctly on the Pacific versant in southwestern Costa Rica (13–970 m) and adjacent Panama, marginally on the Pacific slope in northwestern Costa Rica (480–570 m).

Norops pachypus (Cope, 1875)

(map 10.22)

DIAGNOSTICS: A small, long-legged brown semiarboreal lizard resembling *N. tropidolepis* in having a subocular and one or more postocular light stripes radiating from the eye and usually having curved black bands between the eyes and on the nape, but differing from it in having mostly smooth ventrals and in male dewlap color. The dewlap in *N. pachypus* is small (fig. 10.22b) and orange red, usually with a central yellow spot, while females lack a dewlap but have a central orangish spot on the throat.

DESCRIPTION: Total length 160 mm, no sexual dimorphism in size (standard length in adults 40 to 50 mm); tail moderately long, 60 to 65% of total length, markedly swollen at base in males; hind limbs long; when adpressed against body, tip of fourth toe reaches at least to between eye and tip of snout; scales on upper surface of snout and on prefrontal region small, mostly multicarinate; circumnasal or anterior nasal separated from rostral by two or three scales; supraorbital semicircles separated by four to six keeled scales; semicircle and enlarged supraocular scales strongly multicarinate; subocular and supralabial scales usually separated by one or two scales (rarely three); four to eight middorsal scale rows slightly enlarged and keeled; lateral scales

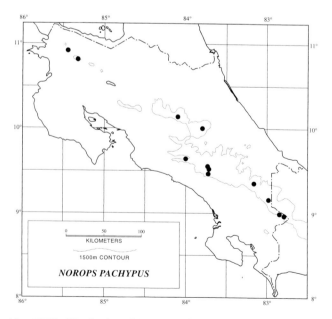

Map 10.22. Distribution of *Norops pachypus*.

granular; most ventral scales smooth, some anteriorly or along midline weakly keeled, rounded, semi-imbricate, much larger than dorsals; no nuchal or caudal crest; no enlarged postcloacals; digital pads only slightly expanded, 28 to 35 lamellae (12 to 14 expanded) under fourth toe; tail subcylindrical; male dewlap small, extending posteriorly to level of front limb insertion. Dorsal ground color light tan to dark black depending on mood; dorsum uniform, with middorsal dark spots or pairs of spots extending posteriorly to level of front limb insertion. Some females with broad middorsal light stripe or a series of light or dark diamond-shaped blotches; venter dull yellow speckled with dark; iris coppery.

SIMILAR SPECIES: (1) *Norops tropidolepis* has distinctly keeled ventrals and a purple red dewlap in males. (2) *Norops humilis* has enlarged middorsals that are larger than the ventrals and a distinct axillary pore; it lacks a nape band, and the male dewlap is deep red bordered on the free margin by bright yellow.

HABITAT: *Norops pachypus* is found in areas of high rainfall, fog, and cloud cover in the upper portion of Premontane Wet Rainforest communities and throughout the lower montane zone in Wet Forest and Rainforest formations. It is common in moss banks, clumps of ground epiphytes, and other vegetation and is expert at hiding in them. Moss-covered trees, stumps, and logs are favored sites both within and on the margins of primary and secondary forests.

BIOLOGY: Very little is known about the habits of this diurnal forest species, but in essential activities it probably resembles its close ally *N. tropidolepis* (which see).

REMARKS: This species is closely allied to *N. tropidolepis,* and their status as distinct species has been controversial. The holotype (USNM 30683) of *Anolis pachypus* Cope, 1875 is from the slopes of Cerro Utyum in the Cordillera de Talamanca of southern Costa Rica and has smooth ventral scales. The holotype (BMNH 85.3.24.11) of *A. tropidolepis* Boulenger, 1885 is from Volcán Irazú (probably Finca Irazú) in the Cordillera Central and has strongly keeled ventrals. Similarly, the name *A. curtus* Boulenger, 1898d, from La Estrella, Cartago Province, in the northern Cordillera de Talamanca is based on an example (BMNH 95.7.13.2) having keeled ventrals. Dunn (1930b) regarded the three as representing a single taxon that he called *A. pachypus.* Walters (1953) reported *A. pachypus* from Volcán Barú in western Panama. Taylor (1956) preferred to recognize the three species as valid. The situation has been confused by variation in the keeling of the ventral scales in this complex. Individuals from lower elevations in the Cordillera de Guanacaste, Cordillera de Tilarán, and Cordillera Central have most ventral scales strongly keeled, as do examples from the extreme northern portions of the Cordillera de Talamanca. Specimens from higher elevations in the Cordillera Central and over most the Talamanca range have mostly smooth ventrals, but some have many weakly keeled ones. Field observations of male dewlap color have influ-

enced my present treatment of these anoles. Males from most areas of the Talamanca range from populations with weakly keeled or mostly smooth ventrals have a red orange dewlap that usually has a central yellow spot. The name *N. pachypus* is appropriately applied to these populations. Males from the Cordillera de Guanacaste and Cordillera de Tilarán have deep purple red dewlaps and strongly keeled ventrals, as do those from lower elevations in the Cordillera Central and some localities in the extreme northern Cordillera de Talamanca. These animals are considered to represent a distinct species, *N. tropidolepis,* and *A. curtus* appears to be based on the same form. At elevations above 2,000 m in the Cordillera Central anoles of this group have mostly smooth ventrals, and males' dewlaps are orange red. For these reasons I include them in *N. pachypus,* although further work is needed to establish their relation to the Talamanca species and *N. tropidolepis.* On a broader scale, these two species are placed with *N. lemurinus* of Mexico and Central America and *N. vittigerus* of eastern Panama and Colombia in the *lemurinus* species group (Lieb 1981).

DISTRIBUTION: Lower montane zone of the Cordillera Central of Costa Rica (2,000–2,500 m) and upper premontane areas and the lower montane belt of the Cordillera de Guanacaste and Cordillera de Talamanca of Costa Rica (1,500–2,400 m) and western Panama (1,372–1,560 m).

Norops sericeus (Hallowell, 1856)

(plate 265; map 10.23)

DIAGNOSTICS: A small, short-legged, slender arboreal anole with strongly keeled middorsal and ventral scales and a dorsal coloration that has a silky sheen. Males have a moderate-sized (fig. 10.22c) reddish orange dewlap with a large blue to indigo central spot, and females have a small dewlap

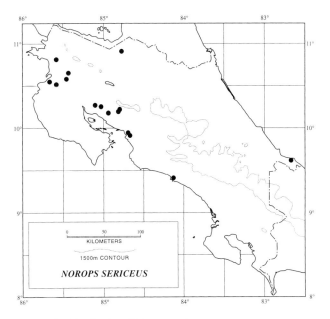

Map 10.23. Distribution of *Norops sericeus.*

with a trace of color. Both sexes also have a definite nuchal crest.

DESCRIPTION: Total length to 162 mm, males somewhat larger (adults 40 to 52 mm in standard length) than females (adults 36 to 47 mm); tail moderately long, 67 to 70% of total length; hind limbs short; when adpressed against body, tip of fourth toe reaches to well beyond shoulder, usually to tympanum; scales on upper surface of snout small, keeled; scales in prefrontal area large, smooth; circumnasal or anterior nasal separated from rostral by one scale row; supraorbital semicircles usually separated by one scale row, sometimes by none or two; semicircle scales smooth or weakly keeled; enlarged supraocular scales keeled; suboculars and supralabials in contact; six to eight rows of middorsal scales enlarged and keeled; lateral scales tubercular; ventrals imbricate, strongly keeled, usually mucronate, much larger than dorsals; definite nuchal crests in both sexes; no enlarged postcloacals; 27 to 33 lamellae (13 to 16 greatly expanded) under fourth toe; tail subcylindrical, attenuate; male dewlap moderate, extending slightly posterior to axilla. Dorsum usually pale gray to bronze, uniform or with a series of six or seven middorsal dark spots; some females with broad light middorsal stripe; tail often weakly banded with brown; venter pale yellow to white.

SIMILAR SPECIES: *Norops cupreus* has only moderately expanded lamellae under the fingers and toes, the male dewlap is bicolored but without a central dark spot, no dewlap occurs in females, and the suboculars are usually separated from the supralabials by one scale.

HABITAT: An inhabitant of the Lowland Dry Forest zone in open areas or along the edges of the woodlands, including pastures and beach margin vegetation; marginally in Lowland Moist Forest zone near beaches; fence posts are common perch sites.

BIOLOGY: Fitch (1973a, 1975) reported in depth on his studies of this common species, and this account is based substantially on them. *Norops sericeus* is strictly diurnal and perches 0.3 to 3 m above the ground, usually on tree trunks or the stems of shrubs. It is often sympatric with *N. cupreus* in Costa Rica, but it has much lower densities and tends to spend more time on the perch and farther above the ground. The species is a heliothermic basker with the preferred activity temperature in the 32 to 33°C range, one or two degrees above the ambient air temperature. When approached these lizards usually flatten themselves on the trunk or stem, where their coloration blends well with the bark. If further disturbed they generally move higher up the surface and around the perch, away from the intruder. Available information indicates that Pacific versant (Mexico to Costa Rica) populations of this species have an April to September (wet season) breeding period, with reproduction stopping in the dry season (December to April) to parallel the pattern in *N. cupreus*. The situation may be different in populations on the Atlantic slope in Mexico and upper Central America, where a longer breeding season might be anticipated.

REMARKS: Lee (1980) analyzed variation in *N. sericeus* and concluded that there are no distinct geographic subdivisions within the species. He noted that the magnitude of sexual dimorphism varies geographically and that in slope samples there appears to be no dimorphism. In addition, Lee concluded that the nominal species *Anolis kidderi* and *Anolis ustus* of the Yucatán Peninsula are synonyms of *N. sericeus*. Lee (1983) also redescribed and summarized pertinent literature on this species. This form occurs on the Atlantic versant of Costa Rica near the upper Río San Juan. A record from extreme southeastern Costa Rica (UCR 11114; Limón: Cocles) may be an accidental introduction. Lieb (1981) regarded *N. sericeus* as the sole member of the *sericeus* species group. The karyotype of this species is 2N = 40, with twelve pairs of macrochromosomes and eight pairs of microchromosomes (Lieb 1981).

DISTRIBUTION: Lowland and premontane zones from Tamaulipas, Mexico, to northeastern Nicaragua and Isla de Maíz Grande and extreme northern Costa Rica near the Río San Juan on the Atlantic versant, and on the Pacific slope from the Isthmus of Tehuantepec to the vicinity of Quepos, Costa Rica; in Costa Rica essentially restricted to the northwest Pacific lowlands but also found only in back beach vegetation along the coast south of the Río Grande de Tárcoles (2–90 m).

Norops townsendi (Stejneger, 1900)

(map 10.24)

DIAGNOSTICS: This small, short-legged olive to brown anole is endemic to Isla del Coco and is one of only two lizards found on the island. The other is a ground gecko, *Sphaerodactylus pacificus*. *Norops townsendi* is distinctive

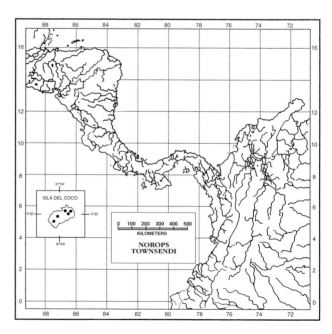

Map 10.24. Distribution of *Norops townsendi*.

in having a light lateral stripe along each flank, bordered above and below by a black stripe, in combination with strongly keeled ventrals, the supraorbital semicircles separated by two to four scales and the suboculars and supralabials in contact. Males have a moderate-sized (fig. 10.22b) dull amber dewlap; females lack any sign of a dewlap or throat color.

DESCRIPTION: Total length to about 130 mm, males (40 to 49 mm in standard length) larger than females (41 to 46 mm in standard length); tail moderately long, about 60% of total length; hind limbs short; when adpressed against body, tip of fourth toe reaches to between shoulder and tympanum; scales on upper surface of snout small, keeled; scales in prefrontal area small, rugose or keeled; circumnasal or anterior nasal in contact with rostral; supraorbital semicircles separated by two to four small rugose scales; semicircle scales keeled; enlarged supraoculars keeled; suboculars and supralabials in contact; four to six middorsal scale rows enlarged, keeled; lateral scales granular; ventrals imbricate, keeled, bluntly pointed, larger than middorsals; a weak nuchal crest in males; no enlarged postcloacal scales; 38 to 46 lamellae (17 to 20 strongly expanded) under fourth toe; tail cylindrical; male dewlap extending slightly posterior to axilla. Dorsum uniform olive or with a few small dark blotches; golden yellow stripe, bordered by dark stripes above and below, from shoulder to or near groin; venter yellowish.

SIMILAR SPECIES: There are none.

HABITAT: Found at all levels and on all types of vegetation in the unique rainforests on Isla del Coco.

BIOLOGY: The species is extremely abundant on Isla del Coco, and several males are often seen displaying within a meter or two of one another. They use any perch including rocks, abandoned human shelters, or low herbaceous plants from near ground level to over 10 m in height. When approached, or after losing a territorial bout, a lizard usually escapes by running along or rapidly jumping onto tree trunks, logs, or branches. The species appears to be a diurnal forest (shade) thermoconformer with activity body temperatures in the range 28.7–30.9–32.6°C (Carpenter 1965). Almost all examples of this form have broken or regenerated tails, suggesting a very high predation level, acute aggression, or inept predators.

REMARKS: The relation of this anole to mainland forms remains obscure.

DISTRIBUTION: Isla del Coco (1–634 m).

Norops tropidolepis (Boulenger, 1885)

(plates 266–67; map 10.25)

DIAGNOSTICS: A small, long-legged brown semiarboreal lizard distinctive in usually having a subocular light stripe and one or more postocular ones radiating from the eye, usually with black bands between the eyes and on the nape, and strongly keeled ventrals. Males have a small dewlap (fig. 10.22b) that is deep purplish red to burgundy, and

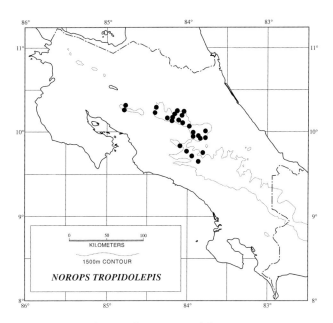

Map 10.25. Distribution of *Norops tropidolepis.*

though females lack a dewlap, the throat has a reddish central area.

DESCRIPTION: Total length to 165 mm, no sexual size dimorphism (standard length 43 to 59 mm); tail moderate, 60 to 65% of total length; hind limbs long; when adpressed against side of body, tip of fourth toe reaches to anterior margin of orbit or beyond; scales on upper surface of snout and prefrontal area small, mostly multicarinate; circumnasal or anterior nasal separated from rostral by two scales; supraorbital semicircles usually separated by four to six (occasionally three or seven) keeled scales; semicircle and enlarged supraocular scales strongly multicarinate; subocular and supralabial scales usually separated by one scale, rarely by two or none; four to eight rows of middorsal scales somewhat enlarged and keeled; lateral scales granular; ventral scales rounded, semi-imbricate, and keeled at least on the chest and anterior abdomen, much larger than dorsals; no nuchal or caudal crest; no enlarged postcloacals; lamellae of digital pad only slightly expanded, 31 to 38 (13 to 16 expanded) under fourth toe; tail subcylindrical; male dewlap small, extending slightly posteriorly to level of front limb insertion. Dorsal ground color usually medium to dark brown, capable of considerable color change to light tan or nearly black; dorsum uniform, with middorsal dark spots or pairs of spots; some females with broad middorsal light stripe or series of light or dark diamond-shaped blotches; venter dull white speckled with dark; iris coppery.

SIMILAR SPECIES: (1) *Norops pachypus* has mostly smooth ventrals, and the red orange dewlaps of males often have a central yellow spot. (2) *Norops humilis* has the greatly enlarged middorsals larger than the ventrals, has a distinct axillary pore, and lacks a dark nape band. The male dewlap in this form is deep red, bordered on its free margin by bright yellow.

HABITAT: *Norops tropidolepis* is a very common species in the Lower Montane Wet and Rainforest zones, both in dense forest and in disturbed areas where there is substantial epiphyte growth on downed trees and their remnants. It may also be encountered along pasture margins in the upper portion of Premontane Wet Forest and Rainforest and in clump grasses and along road cuts where there is matted vegetation.

BIOLOGY: Fitch (1972, 1973a,b, 1975) reported extensively on this species, and much of the following is derived from his accounts. *N. tropidolepis* is characteristic of the cloud-draped slopes of the Costa Rican cordilleras where fog, light to heavy rain, and overcast skies occur almost daily. The species is diurnal, a thermoconformer, and is active throughout the year at rather low body temperatures (12.4 to 26.5°C) that are similar to ambient air temperature. These lizards are frequently found foraging on the ground, on matted low-lying vegetation, or among ground-rooted epiphytes. They also climb moss-covered trees or shrubs up to 2 m above the ground. Their movements are rather slow compared with those of other anoles, but alarmed lizards immediately seek cover within the interstices of the complex vegetational cover. Individuals of this species are very sedentary and usually remain close to a home site. Occupation and defense of perch sites are uncommon in this species. Although the lizards occur in high densities and although both sexes exhibit aggressive behaviors in captivity, intermale interactions are uncommon in the field. They eat a variety of insects in the 6 to 20 mm size range. Breeding and reproduction take place throughout the year. Females produce a single egg from alternate ovaries every twenty to thirty days. Hatchlings are about 21 mm standard length. They attain sexual maturity at eight to nine months, and many individuals survive into the second year of life.

REMARKS: Boulenger (1898d) described *Anolis curtus* from the northeast slope of the Cordillera de Talamanca, and Taylor (1956) used this name for a single example (KU 4837) from Volcán Barva. These lizards fall within the range of variation seen in *N. tropidolepis*, although they have a high number of scales separating the supraorbital semicircles (six or seven) and two rows of scales separating the suboculars from the supralabials, a relatively uncommon condition. This species is closely allied to and very similar to *N. pachypus*. There remains some doubt as to the distinctiveness of the two forms (see remarks under *N. pachypus*), but as understood here *N. tropidolepis* generally occurs at lower elevations than *N. pachypus* and barely ranges into the Cordillera de Talamanca. Examples from the Cordillera de Guanacaste appear to be *N. pachypus*. These species are apparently members of the *lemurinus* species group that also includes the wide-ranging *N. lemurinus* of Mexico and Central America and *N. vittigerus* of Colombia and eastern Panama. The karyotype of *N. tropidolepis* is 2N = 40, with twelve pairs of macrochromosomes and eight pairs of microchromosomes (Lieb 1981).

DISTRIBUTION: Lower montane and extreme upper portions of the premontane zone in the Cordillera de Tilarán, and Cordillera Central and the northern Cordillera de Talamanca of Costa Rica (1,220–2,506 m).

Fuscoauratus Series

Members of this series usually have an inscriptional rib formula of 3:1 and are placed in two subdivisions. The *fuscoauratus* species group of lower trunk-bush-litter anoles includes the following nominal forms: *Norops altae*, *N. carpenteri*, *N. exsul*, *N. fortunensis*, *N. kemptoni*, *N. limifrons*, *N. pandoensis*, and *N. polylepis* of lower Central America; *N. albii*, *N. antonii*, *N. lemniscatus*, *N. maculiventris*, *N. mariarum*, *N. medemi*, and *N. tolimensis* of northwestern South America; *N. bocourtii* and *N. trachyderma* of the upper Amazon basin of Peru and *N. scapularis* of that region and adjacent Bolivia; *N. ortonii* of the Guianas, Amazon basin, and eastern Brazil; *N. gibbiceps* of the Guianas and northern Venezuela; and *N. fuscoauratus*, which has a wide range east of the Andes, in eastern Brazil, and in northern South America north to Panama.

A second cluster comprising riparian species, the *lionotus* species group, ranges from Nicaragua to northwestern Ecuador and includes the following forms: *Norops oxylophus* (Nicaragua and Costa Rica), *N. lionotus* (western Panama), *N. poecilopus* (eastern Panama and northwestern Colombia), *N. rivalis* (northwestern Colombia), *N. macrolepis* (southwestern Colombia), and *N. lynchi* (northwestern Ecuador).

Norops altae (Dunn, 1930b)

(plate 268; map 10.26)

DIAGNOSTICS: A small, short-legged pale yellow, beige to light brown, or rusty-colored anole with a distinct light upper lip and some weakly keeled to tuberculate ventral scales but lacking a definite zone of enlarged dorsal scales. The moderate male dewlap (fig. 10.22c) is uniform dull orange.

DESCRIPTION: Total length to 156 mm, adult males (32 to 52 mm in standard length) larger than females (adults 30 to 47 mm in standard length); tail long and attenuate, 65 to 70% of total length; hind limbs short; when adpressed against body, tip of fourth toe reaches to between shoulder and orbit, usually only to tympanum; scales on upper surface of snout small, keeled; scales on prefrontal area large, smooth; circumnasal or anterior nasal separated from rostral by one scale; supraorbital semicircles separated by one or two smooth scales; semicircle scales smooth; some enlarged supraocular scales weakly keeled; subocular and supralabial scales usually in contact, sometimes separated by one scale; dorsal and lateral scales small, grading into one another; ventral scales rounded, larger than dorsals, some weakly keeled to tuberculate; no nuchal or caudal crest; no

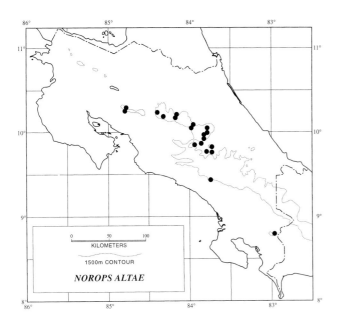

Map 10.26. Distribution of *Norops altae*.

enlarged postcloacals; 28 to 37 lamellae (13 to 17 greatly expanded) under fourth toe; tail subcylindrical; male dewlap moderate, extending posteriorly beyond axilla; no dewlap and throat plain in females. Dorsal ground color pale yellow, beige, light brown, or rusty, uniform or marked with small rectangular spots or faint bands; some females with X-shaped saddles or middorsal stripe; venter pale yellow and white, often overlain by dark flecks; iris beige yellow.

SIMILAR SPECIES: (1) *Norops limifrons* is a long-legged species with smooth ventrals, and males have a white dewlap with a large orange basal spot. (2) *Norops polylepis* has long legs, smooth ventrals, and the supraorbital semicircles separated by two to four scales. (3) *Norops vociferans* has a dark interorbital bar, darkly pigmented throat lining, and very short legs. (4) *Norops pandoensis* males have the dewlap rose with a marginal red orange spot, and females have a small plain dewlap (fig. 10.22a); males have enlarged postanal scales.

HABITAT: *Norops altae* is found in Premontane Wet Forest and Rainforest and Lower Montane Wet Forest and Rainforest.

BIOLOGY: This uncommon species is diurnal in habit and perches on the lower trunks of trees and shrubs but forages mostly on the ground. It reaches sexual maturity at about 32 mm in standard length.

REMARKS: Taylor's (1956) *Anolis achilles* (holotype: KU 40634; Costa Rica: San José: La Palma) is based on an example of this form from near the type locality. Examination of skeletal (inscriptional rib formula 3:1) and external features establishes this species as a member of the *fuscoauratus* species group.

DISTRIBUTION: Humid forest areas of the premontane

and lower montane zones of the Cordillera de Tilarán, Cordillera Central, and Cordillera de Talamanca of Costa Rica (1,220–2,000 m).

Norops carpenteri (Echelle, Echelle, and Fitch, 1971)

(plate 269; map 10.27)

DIAGNOSTICS: This elegant very small green semiarboreal anole is unlikely to be mistaken for any other species. The moderate-sized male dewlap (fig. 10.22b) is a uniform bright orange, and though females lack a dewlap they have an orange spot in the middle of the throat. In preservative the green turns to a very pale yellow and the upper body surfaces are marked with numerous dark punctations.

DESCRIPTION: Total length to 118 mm, no marked sexual dimorphism (standard length in adults 35 to 45 mm); tail moderately long, 60 to 65% of total length; hind limbs short; when adpressed alongside of body, tip of fourth toe reaches tympanum or slightly beyond; scales on upper surface of snout small, keeled, those of prefrontal area larger, smooth; circumnasal or anterior nasal separated from rostral by one scale; supraorbital semicircles separated by two or three smooth scales; semicircle scales smooth to rugose; enlarged supraoculars weakly keeled; subocular and supralabial scales in contact; dorsal scales smooth, granular; ventrals small, smooth, larger than dorsals; no nuchal or caudal crest; no enlarged postcloacals; digital pads definitely expanded, 29 to 34 lamellae under fourth toe (13 to 16 expanded); tail rounded; male dewlap moderate, extending slightly posterior to axilla. Upper surfaces yellow green to slate green (turning yellow in preservative) with many fine darker reticulations, uniform in males, with some middorsal X-shaped or narrow chainlike marks in females; venter

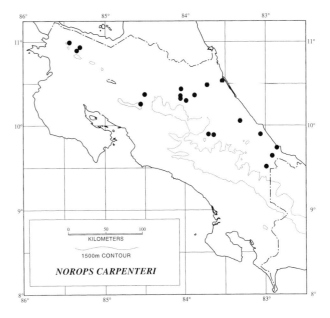

Map 10.27. Distribution of *Norops carpenteri*.

white; a white ring completely surrounds the eye, or a half ring borders lower eye; iris light brown.

SIMILAR SPECIES: *Norops carpenteri* cannot be confused with any other small anole in life. *Norops limifrons* and *Norops polylepis* have longer legs, are brown to gray in preservative, and lack the punctate pattern characteristic of preserved *Norops carpenteri*.

HABITAT: This species occurs only in relatively undisturbed lowland rainforests of the Caribbean slope.

BIOLOGY: *Norops carpenteri* is a diurnal thermoconformer that prefers the shaded forest interior (Fitch 1975; Fitch, Echelle, and Echelle 1976). Although the lizards are sometimes found on boulder-strewn stream banks and in vegetation tangles in swamp forests, they are usually seen on low perches (0.5 to 1 m above the ground) or occasionally on the ground. This species is much less abundant that the similarly sized syntopic *Norops limifrons*, is less likely to be found along forest margins and disturbed or second-growth situations than *N. limifrons*, and may spend more time higher up on tree trunks. Corn (1981) regarded *N. carpenteri* as a tree-crown species, since he usually found individuals among recently felled trees. He also believed that examples found at ground level were there because of their drop-and-freeze escape behavior. Individuals appear to favor lichen-covered exposed surfaces, where their cryptic coloration is most effective. This species feeds on a variety of small (to 10 mm in length) arthropods, especially araneid spiders and beetles and their larvae. Juveniles are unusual in eating many small snails. This anole probably breeds throughout the year, and gravid females have been taken in February, April, August, September, and October.

REMARKS: *Anolis procellaris* (Myers 1971a), originally described from western Panama (holotype: KU 113452; Veraguas: near mouth of Río Concepción), appears to be a synonym of *N. carpenteri*.

DISTRIBUTION: Humid forests of the Atlantic versant on the premontane slopes of Nicaragua (to 1,100 m) and the lowlands of eastern Costa Rica and northwestern Panama (4–682 m).

Norops limifrons (Cope, 1862c)

(plate 270; map 10.28)

DIAGNOSTICS: An extremely abundant very small, long-legged gray brown anole with a white upper lip, smooth or weakly keeled ventral scales, small granular dorsal and lateral scales, and rugose or keeled scales in the prefrontal region. The male dewlap is dull white with a distinct basal orange spot.

DESCRIPTION: Total length to 156 mm; no sexual dimorphism in size (adults 33 to 45 mm in standard length); tail moderate, 62 to 66% of total length; hind limbs long; when adpressed against body, fourth toe reaches between eye and snout; scales on upper surface of snout and on prefrontal region small, rugose or keeled; circumnasal or anterior nasal in contact with rostral; supraorbital semicircles separated by one to four (usually two or three) rugose or keeled scales; semicircle and enlarged supraocular scales strongly keeled; subocular and supralabial scales usually in contact, sometimes separated by one scale; dorsal and lateral scales granular; ventrals rounded, usually smooth (sometimes weakly keeled), much larger than dorsals; no nuchal or caudal crest; enlarged postcloacals in males; digital pads definitely expanded, 28 to 32 lamellae (13 to 16 expanded) under fourth toe; tail round; male dewlap small, extending posteriorly only to front of limb insertion; no dewlap in females. Dorsum gray brown to olive; males uniform or with middorsal black dots; some females colored like males, others with a broad to narrow pale cream, yellow, or pink dorsal stripe, often outlined by black or with a series of diamond-shaped middorsal dark blotches; venter white to washed with pale yellow; iris coppery gold to brown.

SIMILAR SPECIES: (1) *Norops polylepis* is a brown lizard usually with a yellowish to reddish cast, with smooth prefrontal scales and the suboculars usually separated from the supralabials; the males have a large (fig. 10.22d) solid orange dewlap. (2) *Norops altae* has short legs, some weakly tuberculate to weakly keeled ventrals, and a burnt orange dewlap. (3) *Norops carpenteri* is green in life and yellow with dark punctations in preservative. (4) *Norops cupreus* has strongly keeled ventral scales.

HABITAT: Ubiquitous in humid lowland areas and lower portions of the premontane belt, especially along forest margins, in secondary growth, back beach areas, plantations of cacao, banana, coconut, coffee, or pejibaye, and in palm gardens, other disturbed areas, and even relatively open pastures.

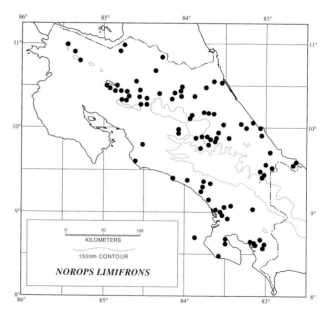

Map 10.28. Distribution of *Norops limifrons*.

BIOLOGY: The life history and ecology of this species have been extensively studied in Costa Rica by Corn (1981), Fitch (1973a,b, 1975), Fitch, Echelle, and Echelle (1976), and Talbot (1976, 1977, 1979) and in Panama by Andrews (1979b, 1982, 1989), Andrews and Rand (1974, 1982), and Andrews, Rand, and Guerrero (1983). Much of what follows is based on their reports. *N. limifrons* is a "weed" (or gap) species that has its highest population densities in relatively open or disturbed areas formerly covered by forests. In Costa Rica it is restricted to humid bioclimatic sites and is not found in the dry forest region of the northwestern lowlands. These lizards bask in direct sunlight to some extent, and body temperatures for active lizards are concentrated in the 23 to 30°C range (21–28.9–32°C) and are slightly higher than the ambient air temperature. These anoles are both scansorial and terrestrial and spend considerable time foraging on the ground, often at some distance from perch sites. They use a slightly elevated object for a perch site, including tree trunks, fallen logs, stems of saplings and herbaceous plants, leaf surfaces, vines, and grass stems, and usually sit head downward 0.25–0.75–5 m above the ground. Males usually have higher perches than females, and juveniles usually are found closest to the ground surface. Lizards of this species are agile and swift. When approached they tend to drop down onto the forest floor and move in quick jumps to other vegetation or hide in the leaf litter. At other times they leap from stem to stem, run over leaf surfaces, or clamber up a trunk or stem, where they freeze or run downward again. These lizards are strictly diurnal, and males spend about 10% of the day displaying or engaging in other social interactions and 80% of the time on the perch site or actively pursuing prey. Females forage about as long and are involved in social behavior about 5% of the time. A single male and one female tend to be bonded for four to six months and are usually found within 1 to 2 m of one another, even moving together for 20 m or so to set up new territories. The bond is reinforced by frequent display between the two lizards, and social interaction centers on this relationship (55% of total male social interaction time versus 29% for male-male interactions). Densities are relatively high in preferred habitats (about 400/ha) but are lower in shadier sites (abandoned cacao plantations, about 200/ha) and within forest stands (about 100/ha). Andrews (1979b), however, reported a density of 1,980/ha in one cacao plantation. Andrews and Rand (1982) documented marked fluctuations in adult densities for this species over a ten-year period at a single site on Barro Colorado Island, Panama. In some years densities were near 170/ha and in others about 1,143/ha. Home ranges average about 23 m², with no apparent differences between the sexes. *N. limifrons* eats a variety of arthropods, especially araneid spiders and beetles, but avoids ants. Other common food items are orthopterans, dipterans, homopterans, and lepidopteran larvae. Average number of prey in twenty-three stomachs analyzed at

La Selva, Heredia Province, was 2.1, averaging between 0.5 mm (mites) and 8.6 mm (dipteran larvae) (Lieberman 1986). Aside from the avoiding ants, they eat food items in proportion to what is available within the size range (0–30 mm) of prey known to be taken. Prey size increases as the lizard grows. It appears that adult females tend to consume larger prey at the end of the dry season and to contain more food items per stomach in the middle of the rainy season. Otherwise there are no seasonal differences in size or number of items per stomach (none to ten). Aggressive and courtship displays and reproduction occur throughout the year, but egg production is reduced in the drier times (December to March). Females lay a single egg every seven to ten days during the wet season or about every three weeks during the drier times of the year. Hatchlings are 18 to 19 mm in standard length and reach sexual maturity in less than four months. Essentially this is an annual species in which very few individuals live longer than one year. There is a peak in the relative abundance of juveniles in the rainiest part of the year (August to October).

REMARKS: The members of the *fuscoauratus* species group form a notoriously difficult complex. It remains unclear whether *Norops limifrons,* as usually understood (Peters and Donoso-Barros 1970), constitutes a single species or a series of slightly differentiated but distinctive ones. Workers (Fitch, Echelle, and Echelle 1976; Villa et al., 1988) have recognized a northern species, *Norops rodriguezii* (Caribbean slope Mexico east of the Isthmus of Tehuantepec to Honduras) and a southern one, *N. limifrons* (Honduras to eastern Panama). It is also clear that populations of the latter from Honduras to western Panama differ from those from central and eastern Panama (Fitch 1973a). In addition, Taylor (1956) described, and Peters and Donoso-Barros (1970), Fitch (1973a), and Fitch, Echelle, and Echelle (1976) recognized, a member of this complex from southwestern Pacific versant Costa Rica and adjacent western Panama as a separate taxon, *Anolis (= Norops) biscutiger* (holotype: KU 40771; Costa Rica: Puntarenas: Golfito). The relation of these various forms to *N. fuscoauratus,* which ranges into eastern Panama, also remains unclear. This problem is not without some significance, since a large body of ecological data has been published on the central and eastern Panamanian form usually placed in *N. limifrons* (Sexton, Heatwole, and Meseth 1963; Sexton 1967; Sexton et al. 1971; Andrews 1979b, 1982; Andrews and Rand 1974, 1982; Andrews, Rand, and Guerrero 1983). This is not the appropriate place to attempt to resolve the situation. However, C. Guyer and I have been unable to find any consistent differences between Atlantic and Pacific slope populations of this complex in Costa Rica and adjacent western Panama, so I do not recognize *N. biscutiger.* Should the central and eastern Panamanian form prove to be distinct and not conspecific with the holotype of *N. limifrons* (ANSP 79001; Panama: Veraguas: Cocuyos), the first available name for the Costa Rican and western Panamanian populations is *Anolis pulchripes* Peters, 1873b (lectotype: ZMB 7827; Panama: Chiriquí). The nominal species *Norops godmani* (Boulenger, 1885; Costa Rica: Cartago: Irazú) is based on representatives of this species with weakly keeled ventral scales but otherwise identical in scutellation and coloration. Examples with keeled ventrals are also known from the vicinity of Tapantí and Turrialba, Cartago Province, and near Bajo La Hondura, San José Province. Most of these captures are from higher altitudes, primarily above 1,100 m, but several from near Turrialba are from about 600 m. Although I previously recognized this form as distinct (Savage 1973b, 1976, 1980b; Savage and Villa 1986), I now agree with C. Guyer's conclusion that it is an individual variant. The karyotype is 2N = 40, comprising twelve pairs of macrochromosomes and eight pairs of microchromosomes (Gorman 1967, 1973; Lieb 1981).

DISTRIBUTION: Lowlands and lower portions of the premontane zone on the Atlantic versant from eastern Honduras to central Panama and on the Pacific slope from central Costa Rica to eastern Panama (1–1,340 m).

Norops oxylophus (Cope, 1875)

(plate 271; map 10.29)

DIAGNOSTICS: A moderately large, short-legged chocolate brown semiaquatic anole having a pair of cream-colored lateral stripes. It is unlikely to be confused with any other Costa Rican anole except *Norops aquaticus*, which differs in having blue eyes and a definite green cast to its coloration in life. The iris is coppery in *N. oxylophus*, and the male dewlap is uniform burnt orange.

DESCRIPTION: Total length to 243 mm, males (adults 59 to 85 mm in standard length) larger than females (adults 56

to 68 mm in standard length); tail moderate, 60 to 65% of total length; hind limbs short; when adpressed along side of body, tip of fourth toe reaches between shoulder and posterior margin of orbit; scales on upper surface of snout moderate, in prefrontal area larger, rugose; circumnasal or anterior nasal separated from rostral by one or two scales; supraorbital semicircles separated by one or two scales; some semicircle and supraocular scales enlarged; subocular and supralabial scales separated by one or two rows of scales; enlarged dorsal scales in 18 to 24 rows, flattened and rounded, pentagonal or hexagonal, mostly juxtaposed, smooth or weakly keeled; laterals much smaller than dorsals; ventrals strongly keeled, about same size as dorsals; dorsal crest and enlarged postcloacals in males; digital pads expanded, 30 to 38 lamellae (13 to 17 expanded) under fourth toe; tail compressed; caudal crest in males; male dewlap large, extending posteriorly onto venter about one-third distance from axilla to groin; no dewlap, throat plain in females. Upper surface dark brown with an olive cast; dorsum usually marked with dark bands, as are limbs and tail; a distinct cream stripe runs from above shoulder posteriorly about two-thirds length of body (rarely obscured or absent in females); venter cream, often with yellow (or orangish in adult males) wash; iris brown.

SIMILAR SPECIES: *Norops aquaticus* has blue eyes, a greenish cast to its coloration, and only two to four rows of enlarged middorsal scales.

HABITAT: *Norops oxylophus* is a riparian species found along moderate-sized, fairly rapid streams in lowland rainforests, gallery forests, and the lower portions of the Premontane Wet Forest and Rainforest zones.

BIOLOGY: Fitch (1973a,b, 1975) and Corn (1981) have studied this common species to some extent, and their material is incorporated as appropriate in the following account. Vitt, Zani, and Durtsche (1995) found similar features for a Nicaraguan population. This handsome anole is rarely found far from stream banks and prefers boulders, fallen logs, or piles of stream drift as perch sites. In addition, *N. oxylophus* is a thermoconformer that seeks shady sites and is most likely to be found along stream courses overhung by the tree canopy. Active body temperatures for these lizard center on 26°C, range from 19 to 28°C, and are near ambient air temperatures. Males display from any slightly elevated object near the stream course. *Norops oxylophus* of both sexes and all age groups forage primarily along the stream margin but do not hesitate to climb on partially submerged rocks, logs, or other materials in the stream in search of food. To reach these objects (or if frightened) these anoles dive into the stream and swim underwater for several meters. When avoiding an intruder they often swim to the far side of a screening object and remain submerged with only the head exposed until it is safe to emerge. Food consists primarily of stream margin arthropods, mostly araneid spiders, beetles (adults and larvae), homopterans, and dipterans. The lizards generally eat in proportion to what is available in the

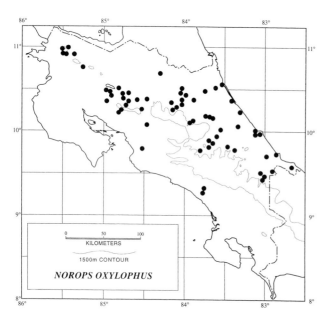

Map 10.29. Distribution of *Norops oxylophus*.

size range of 1.1 to 25 mm, but they avoid ants. As with other anoles, larger individuals usually take larger prey. *N. oxylophus* is unusual in that large adults continue to eat a high proportion of small prey items. This species appears to breed throughout the year, since young of various sizes are always present. Eggs are laid in nests under moss, and several females may use the same nest concurrently (C. Guyer, pers. comm.). The smallest hatchlings are 26 mm in standard length.

REMARKS: This species has been called *Anolis* (or *Norops*) *lionotus* in most recent publications. That name was based on an example, now apparently lost, from Cocuyos (an abandoned mine) on the Río Santiago, Veraguas Province, Panama (the locality has also been spelled Cucuyas or Cucuyos). Williams (1984) concluded that examples from central Panama were conspecific with the type of *N. lionotus* but specifically distinct from the related populations from Nicaragua and Costa Rica included under *N. lionotus* by other workers. The first available name for the northern form is *Anolis oxylophus* Cope, 1875, originally collected by the W. M. Gabb party (Savage 1970) in southeastern Costa Rica (syntypes: USNM 30556–30557). The presumed differences in scutellation between what may now be called *Norops lionotus* and *N. oxylophus* are slight, and C. Guyer and I suspect that the two nominal forms represent a single species and will be shown to intergrade in western Panama. In that eventuality the name *N. lionotus* has priority for the populations discussed here. Köhler (2000) goes further with this notion and without comment lists *N. oxylophus* as a synonym of *Norops lionotus* (Cope 1861b). The karyotype for *N. oxylophus* is 2N = 40, with twelve pairs of macrochromosomes and eight pairs of microchromosomes (Gorman 1973; Lieb 1981; both based on Costa Rican material).

DISTRIBUTION: Humid lowlands and premontane slopes of Atlantic versant eastern Honduras, Nicaragua, and Costa Rica, and extreme northwestern Panama; on the Pacific slope in gallery forests in the lowlands of northwestern Costa Rica and in the Cordillera Costeña, near San Isidro de El General, Puntarenas Province, in the southwest (20–1,200 m).

Norops pandoensis Savage and Guyer, "1998," 1999

(map 10.30)

DIAGNOSTICS: A small, short-legged yellowish brown anole with a light upper lip, ventral scales usually smooth but occasionally weakly keeled, and no definite zone of enlarged dorsal scales. The large male dewlap (fig. 10.22d) is pink with an orange margin.

DESCRIPTION: Total length to 161 mm, males (adults 49 to 53 in standard length) smaller than females (adults 49 to 60 mm in standard length); tail long, 61 to 66% of total length; hind limbs short; when adpressed against body, tip of fourth toe reaches between shoulder and orbit, usually only to tympanum; scales on upper surface of snout moderate, smooth to rugose; scales in prefrontal area flattened;

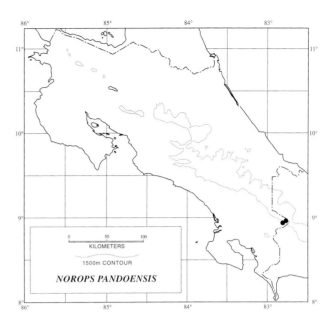

Map 10.30. Distribution of *Norops pandoensis*.

circumnasal or anterior nasal separated from rostral by one or two scales; supraorbital semicircles separated by one or two scales; semicircle scales keeled; enlarged supraoculars; subocular and supralabial scales in contact; dorsal and lateral scales small, grading into one another; ventral scales smooth, occasionally weakly keeled; no nuchal or caudal crest; postcloacals enlarged in males; 31 to 32 lamellae (15 to 16 greatly expanded) under fourth toe; tail slightly compressed; male dewlap large, extending posteriorly onto venter; small unpigmented dewlap in females. Dorsum yellow tan, sometimes with obscure darker markings; some females with yellow middorsal longitudinal stripe bordered by dark brown; venter pale yellow to white; iris orange red.

SIMILAR SPECIES: (1) *Norops altae* has the adult male dewlap orange, no enlarged postcloacal scales in males, and no dewlap in females. (2) *Norops limifrons* is a long-legged species, and males have a white dewlap with a large basal orange spot. (3) *Norops polylepis* has long legs and solid orange dewlap in males and no dewlap in females.

HABITAT: Lower Montane Rainforest.

BIOLOGY: A somewhat common diurnal species found perched on the lower trunks of trees and shrubs but usually foraging on the ground.

DISTRIBUTION: Lower montane humid forests of the Cerro Pando area of southwestern Costa Rica and adjacent Panama (1,350–1,950 m).

Norops polylepis (Peters, "1873b," 1874)

(plates 272–73; map 10.31)

DIAGNOSTICS: A small, long-legged yellowish brown to reddish brown anole with smooth ventrals, small granular dorsal and lateral scales, and smooth prefrontal scales. This lizard is most likely to be mistaken in the field for the somewhat smaller *N. limifrons* in the area of sympatry because

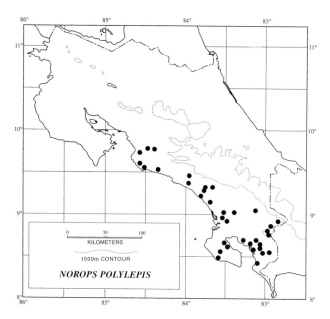

Map 10.31. Distribution of *Norops polylepis*.

they overlap in adult size and perch height preference. *N. polylepis* is a forest shade species, whereas *N. limifrons* prefers more open sunlight microhabitats. The male dewlap in *N. polylepis* is large (fig. 10.22d) and solid orange.

DESCRIPTION: Total length to 177 mm, males (adults 45 to 57 mm in standard length) larger than females (adults 41 to 53 mm); tail long, 65 to 70% of total length; hind limbs long; when adpressed against body, tip of fourth toe reaches to orbit; scales on upper surface of snout small, keeled; scales on prefrontal area small, smooth; circumnasal or anterior nasal separated from rostral by one scale; supraorbital semicircles separated by two to four smooth scales (usually by three); semicircle and enlarged supraocular scales keeled; subocular and supralabial scales usually separated by one scale, sometimes in contact; dorsal and lateral scales granular; ventrals rounded, smooth, much larger than dorsals; low caudal crest and enlarged postcloacals in males; digital pads definitely expanded, 29 to 35 lamellae (13–16 expanded) under fourth toe; male dewlap large, extending posteriorly about halfway between axilla and groin; no dewlap in females, but a spot of orange on throat. Dorsum brown with a yellowish or reddish cast, uniform or with dark median blotches in both sexes, but some females with diamond-shaped middorsal blotches or a broad middorsal light stripe often bordered with dark; frequently a distinct dark spot on side of neck; limbs and tail uniform or with dark bands; venter cream white, heavily pigmented with black in some females; iris brown.

SIMILAR SPECIES: (1) *Norops limifrons* is gray brown to olive; it has rugose to keeled prefrontal and intersemicircle scales, and the small male dewlap is cream with an orange basal spot. (2) *Norops altae* has short legs and some ventrals keeled or tuberculate. (3) *Norops cupreus* has strongly keeled

ventrals, short legs, and a bicolor or tricolor male dewlap. (4) *Norops carpenteri* has short legs and is green in life.

HABITAT: This species occurs only in the lowland rainforests and Premontane Wet Forest and Rainforest on the Pacific versant.

BIOLOGY: Andrews (1971, 1983) studied this very common species in detail at Rincón de Osa, Puntarenas Province, and her report provides much of the information summarized below. *Norops polylepis* is strictly diurnal and very abundant. It is a thermoconformer with recorded activity temperatures of 20.9–27.8–31.2°C. Most individuals are seen perched on saplings or herbaceous vegetation, most frequently on leaves 0.75 to 3 m above the ground in the shady understory of intact forests. Their foraging behavior is similar to that of other shrub-ground anoles and consists of sitting on a vertical surface, usually head downward, and scanning the surrounding area (in this case the forest floor) for prey of suitable size. When they see a prey item they dart down to grab it. These lizards are commonly encountered at night asleep on leaf surfaces. Males have higher perches (mean 920 mm wet season, 720 mm dry season) than females (mean 490 mm wet season, 420 mm dry season), and both sexes perch closer to the ground in the dry season (February to April). Males are very territorial and spend most of their time (about 50%) displaying or fighting with other males, about 10% in courtship or mating, and 20% in foraging. Males use higher perches when interacting with other males and lower ones for foraging and courtship. Females forage about 86% of the day and are involved in social interactions with males about 6% of the time. As in most anoles, adult males have fairly large territories (average 30 m²), but female territories are much smaller (average 7 m²). The territory of an adult male overlaps those of one to three adult females, but the territories of juveniles of both sexes do not overlap with those of other individuals. Densities are relatively high in closed-canopy forest (about 290-300/ha). Food includes the usual variety of arthropods, especially larvae of orthopterans and lepidopterans. The food size ranges up to 20 mm or more, with adult females eating larger prey than adult males. The feeding rate is about one food item per hour. Reproduction is continuous throughout the year but slows during the dry season. Females lay one egg every seven to fourteen days. Incubation time is about fifty days. Courtship and mating intensify at the onset of the wet season (May to June), and there is a major flush of hatchlings (about 19 mm in standard length) in August and September. They reach sexual maturity in three to four months. This is an annual species, and only a few individuals survive to a second year of life. Andrews (1983) notes that motmots, trogons, several snakes and the large lizards *Corytophanes cristatus* and *Norops capito* prey on this anole.

REMARKS: The karyotype of this species is 2N = 40, comprising twelve pairs of macrochromosomes and eight pairs of microchromosomes (Gorman 1973; Lieb 1981).

DISTRIBUTION: Lowlands and premontane slopes of Pacific versant southwestern Costa Rica and adjacent southwestern Panama (1–1,330 m).

Petersii Series

These are arboreal lizards including the largest species of *Norops* as well as some small and moderate-sized forms. The series is characterized by an inscriptional rib formula of 3:2. Two clusters of allied taxa are usually recognized. The first (the *petersii* subseries) is composed of rather large species having strongly keeled ventrals and moderate to long hind limbs. The second (the *pentaprion* subseries) contains a group of very short-legged small to moderate-sized forms that usually have smooth ventrals and a black lining to the throat.

Members of the *petersii* subseries include *Norops biporcatus* (Mexico to western Ecuador), *N. capito* (Mexico to Panama), *N. loveridgei* (Honduras), *N. petersii* (Mexico to Honduras), and *N. woodi* (Costa Rica and Panama). The following species constitute the *pentaprion* subseries: *Norops fungosus* and *N. vociferans* (Costa Rica and Panama); *N. ibague* and *N. sulcifrons* (Colombia), and *N. pentaprion* (Mexico to Colombia).

Norops biporcatus (Wiegmann, 1834a)

(plates 274–76; map 10.32)

DIAGNOSTICS: A large, short-legged anole having strongly keeled ventrals and tuberculate dorsal and lateral scales. Although the species is capable of considerable color change from green to dark brown, most active individuals are bright green. The male dewlap is tricolor: basally white, mostly powder blue, and the free margin red orange. The lining of the throat is black.

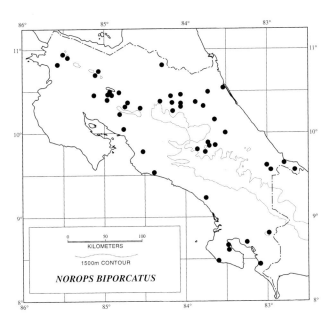

Map 10.32. Distribution of *Norops biporcatus*.

DESCRIPTION: Total length to 330 mm (adults 73 to 99 mm standard length), without sexual dimorphism in size; tail long, attenuate, 70 to 75% of total length; hind limbs very short; when adpressed along side of body, fourth toe usually reaches to tympanum, sometimes to orbit; scales on upper surface of snout area large, tuberculate, often keeled; scales in prefrontal area large, rugose to tuberculate; circumnasal or anterior nasal separated from rostral by two or three scale rows; supraorbital semicircles usually separated by two or three (rarely one) tuberculate scales; semicircle scales tuberculate; enlarged supraocular scales tuberculate to keeled; subocular and supralabial scales in contact or separated by one scale row; dorsal and lateral scales large, with tuberculate keels; ventrals strongly keeled, mucronate, larger than dorsals; a distinct nuchal crest in males; no enlarged postcloacals; 35 to 47 lamellae (18 to 25 greatly expanded) under fourth toe; tail subcylindrical; male dewlap small, extending posteriorly only to the forelimb insertion; dewlap almost as large in females, with some bluish color. Dorsum usually uniform bright green in life (brownish purple in preservative) but capable of color change to dark brown; occasionally with light spots or dark blotches; venter yellowish white (light gray in preservative) with many brown flecks and spots; iris gold.

SIMILAR SPECIES: (1) *Dactyloa insignis* has a yellow to tan dorsal ground color, strongly marked with dark bands and smooth ventrals. (2) *Dactyloa microtus* has a yellow to tan ground color, a well-developed postorbital light stripe, and smooth ventrals.

HABITAT: Usually found in relatively undisturbed situations in Lowland Moist and Wet Forests and Premontane Wet Forest and Rainforest. It also occurs in gallery forests along rivers in the Lowland Dry Forest zone.

BIOLOGY: *Norops biporcatus* is a strictly diurnal arboreal lizard that basks in direct sunlight and forages in the trunk-canopy zone. Individuals usually occur singly and in low densities. The species feeds on a wide variety of arthropod prey, but small individuals tend to eat mostly beetles, and large examples have a high proportion of large ants in their diets (Corn 1981). They also eat smaller anoles (Taylor 1956). As in other species, larger individuals eat larger prey, and prey length ranges up to 50 mm. Stomachs typically contain zero to eleven prey items. No detailed information on reproduction in this species is available, but Fitch (1975) indicated that eggs may be laid year-round. Corn (1981), on the other hand, found evidence of a decline during the short dry season on the Atlantic lowlands of Costa Rica, suggesting that there may be local differences in the reproductive cycle in this species, especially between sites having a long dry season and those where rainfall is more evenly distributed throughout the year. Hatchlings are 25 to 26 mm in standard length. When handled, these lizards may emit one squeak or a series.

REMARKS: Taylor's (1956) description of the male dewlap color (white or slightly greenish) involves a transposition of data for the condition in *Norops capito*. According to Williams (1966), specimens from the southern portion of the species' range (western Colombia and Ecuador) differ from other populations in having the scales surrounding the interparietal smaller than the middorsal scales or equal in size, a small round auricular opening, and a black-edged male dewlap (northern examples have the scales surrounding the interparietal larger than the middorsal scales, a large oval auricular opening, and a male dewlap edged with red orange). Both sexes have six pairs of metacentrics and seven pairs of microchromosomes, but there is XXY heteromorphism in males (2N = 30 in females, 29 in males); NF = 42 in females (Gorman 1973).

DISTRIBUTION: Lowlands and premontane areas of the Atlantic versant from Chiapas, Mexico, to northern Venezuela; at scattered sites in western Nicaragua and northwestern Costa Rica and more or less continuously from southwestern Costa Rica to western Ecuador (2–1,220 m).

Norops capito (W. Peters, 1863b)

(plate 277; map 10.33)

DIAGNOSTICS: This large, very long-legged anole is unlikely to be confused with any other allied form because of its extremely short head. It is always immediately distinguishable in the hand, since it is the only anole from the region having smooth, flattened, and juxtaposed dorsal and lateral scales (fig. 10.3). The small male dewlap is greenish yellow.

DESCRIPTION: Total length to 266 mm, females (adults 83 to 96 mm in standard length) larger than males (adults 78 to 90 mm in standard length); tail moderately long, ex-

tremely attenuate, 60 to 65% of total length; hind limbs very long; when adpressed along side of body, tip of fourth toe usually reaches tip of snout or beyond; scales on upper surface of snout small, elongate, keeled; scales in prefrontal region large, smooth; circumnasal or anterior nasal separated from rostral by two or three rows of scales; supraorbital semicircles separated by one or two usually smooth (sometimes tuberculate) scales; semicircle scales tuberculate; enlarged supraocular scales keeled; suboculars and supralabials separated by one or two rows of scales; dorsal scales hexagonal or pentagonal, smooth, juxtaposed, gradually grading into similar, slightly smaller laterals; ventrals larger than dorsals, imbricate, keeled, and mucronate; no nuchal or caudal crest; no enlarged postcloacals; 33 to 41 (14 to 18 greatly expanded) lamellae under fourth toe; tail round; male dewlap very small, restricted to throat, even smaller and plain in females. Dorsum greenish yellow, olive, to olive brown, uniform or mottled, frequently with three dark jagged narrow or broad transverse bands on body, sometimes a narrow or broad middorsal light stripe in females; a distinct V-shaped dark interorbital band almost always present; venter light, usually heavily mottled with dark pigment; tail usually banded with dark brown; iris coppery brown.

SIMILAR SPECIES: *Norops woodi* has a blue iris, an obviously elongate snout, and keeled dorsals; the male dewlap is red to orange to dark olive green to black.

HABITAT: Relatively undisturbed Lowland Moist and Wet Forests and Premontane Wet Forest and Rainforest.

BIOLOGY: This common forest anole is diurnal and essentially arboreal, perching on tree trunks, usually 0.25 to 2 m above the ground. It typically forages by making rapid forays to the ground within 2 or 3 m of the perch and quickly returning. *Norops capito* is a shade species and appears to be a thermoconformer. When first approached these lizards freeze, but if the observer comes closer they jump down, run across the ground, and scurry up and around another tree trunk to a height of 1 or 2 m. They are frequently associated with very large and heavily buttressed trees and effectively use the buttresses in escape behavior. Individuals appear to occur singly in low densities and are widely spaced, with territories about 15 m² for both sexes. Males are strongly territorial and use displays and attacks to expel other males intruding on their space. This species preys on a variety of arthropods (especially araneid spiders, orthopterans, and caterpillars) but also eats other lizards and often feeds on slugs. Prey size ranges to about 60 mm, with larger individuals eating larger prey. Stomachs analyzed contained 0–1–3 prey items. Reproduction appears to take place throughout the year (Fitch 1975; Corn 1981). Hatchlings are about 25 mm in standard length.

REMARKS: Skeletal material of this species confirms its placement in the *petersii* series, which is characterized by an inscriptional rib formula of 3:2. The karyotype (2N = 40)

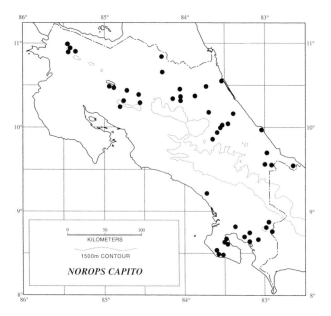

Map 10.33. Distribution of *Norops capito*.

Plate 235. *Basiliscus basiliscus;* juvenile. Costa Rica: Guanacaste: La Pacifica, 40 m.

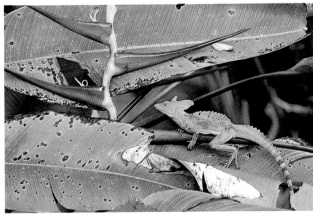

Plate 236. *Basiliscus plumifrons;* male. Costa Rica: Heredia: La Selva Biological Station, about 60 m.

Plate 237. *Basiliscus plumifrons;* male. Costa Rica: Heredia: La Selva Biological Station, about 60 m.

Plate 238. *Basiliscus plumifrons;* female. Costa Rica: Heredia: La Selva Biological Station, about 60 m.

Plate 239. *Basiliscus vittatus;* adult male. Costa Rica: Limón: Tortuguero, about 3 m.

Plate 240. *Corytophanes cristatus.* Costa Rica: Heredia: La Selva Biological Station, about 60 m.

Plate 241. *Corytophanes cristatus.* Costa Rica: Heredia: La Selva Biological Station, about 60 m.

Plate 242. *Ctenosaura quinquecarinata.* Costa Rica: Guanacaste: Santa Rosa National Park, about 100 m.

Plate 243. *Ctenosaura similis;* adult male. Costa Rica: Guanacaste: Santa Rosa National Park, 280 m.

Plate 244. *Ctenosaura similis;* adult female. Costa Rica: Guanacaste: Santa Rosa National Park, 280 m.

Plate 245. *Iguana iguana;* adult male. Costa Rica: Heredia: La Selva Biological Station, about 60 m.

Plate 246. *Iguana iguana;* juvenile. Costa Rica: Heredia: La Selva Biological Station, about 60 m.

Plate 247. *Sceloporus malachiticus;* male. Costa Rica: Puntarenas: Monteverde, about 1,400 m.

Plate 248. *Sceloporus squamosus.* Guatemala: Jalapa.

Plate 249. *Sceloporus variabilis;* male. Costa Rica: Guanacaste: Santa Rosa National Park, about 280 m.

Plate 250. *Sceloporus variabilis;* female. Costa Rica: Guanacaste: Santa Rosa National Park, about 280 m.

Plate 251. *Polychrus gutturosus.* Costa Rica: Alajuela: Peñas Blancas Valley, about 800 m.

Plate 252. *Polychrus gutturosus;* view of head. Costa Rica: Alajuela: Peñas Blancas Valley, about 800 m.

Plate 253. *Ctenonotus cristatellus*. Costa Rica: Limón: Cahuita, about 4 m.

Plate 254. *Dactyloa frenata*; male. Costa Rica: Limón: Guayacán, about 650 m.

Plate 255. *Dactyloa frenata*; view of head. Costa Rica: Limón: Comadre, 20 m.

Plate 256. *Dactyloa insignis*; male. Costa Rica: Puntarenas: Monteverde, 1,400 m.

Plate 257. *Dactyloa insignis*; male, view of head. Costa Rica: Puntarenas: Monteverde, 1,400 m.

Plate 258. *Dactyloa microtus*; female. Costa Rica.

Plate 259. *Dactyloa chocorum.* Panama: Panama: near San Juan de Pequení: Chagres National Park, about 360 m.

Plate 260. *Norops aquaticus.* Costa Rica: Puntarenas: Las Cruces, about 1,200 m.

Plate 261. *Norops cupreus;* male: Costa Rica: Guanacaste: Santa Rosa National Park, about 280 m.

Plate 262. *Norops humilis;* male. Costa Rica: Heredia: La Selva Biological Station, about 60 m.

Plate 263. *Norops intermedius;* male. Costa Rica: Puntarenas: Monteverde, about 1,400 m.

Plate 264. *Norops lemurinus.* Costa Rica: Heredia: La Selva Biological Station, about 60 m.

Plate 265. *Norops sericeus;* male. Costa Rica: Guanacaste: Santa Rosa National Park, about 280 m.

Plate 266. *Norops tropidolepis;* male. Costa Rica: Alajuela: Monteverde Cloud Forest Preserve, about 1,600 m.

Plate 267. *Norops tropidolepis;* female. Costa Rica: Alajuela: Monteverde Cloud Forest Preserve, about 1,600 m.

Plate 268. *Norops altae;* male. Costa Rica: Alajuela: Monteverde Cloud Forest Preserve, about 1,600 m.

Plate 269. *Norops carpenteri.* Costa Rica: Heredia: La Selva Biological Station, about 60 m.

Plate 270. *Norops limifrons.* Costa Rica: Heredia: La Selva Biological Station, about 60 m.

Plate 271. *Norops oxylophus;* male. Costa Rica: Heredia: La Selva Biological Station, about 60 m.

Plate 272. *Norops polylepis;* male. Costa Rica: Puntarenas: Corcovado National Park, about 40 m.

Plate 273. *Norops polylepis;* female. Costa Rica: Puntarenas: Rincón de Osa, about 40 m.

Plate 274. *Norops biporcatus;* male. Costa Rica: Heredia: La Selva Biological Station, about 60 m.

Plate 275. *Norops biporcatus;* male. Costa Rica: Heredia: La Selva Biological Station, about 60 m.

Plate 276. *Norops biporcatus;* male displaying to female. Costa Rica: Heredia: La Selva Biological Station, about 60 m.

Plate 277. *Norops capito*; male. Costa Rica: Heredia: La Selva Biological Station, about 60 m.

Plate 278. *Norops pentaprion*; male. Costa Rica: Heredia: La Selva Biological Station, about 60 m.

Plate 279. *Norops vociferans*. Panama: Chiriquí: Río Candela, about 1,200 m.

Plate 280. *Norops woodi*; male. Costa Rica: Puntarenas: Monteverde, about 1,400 m.

Plate 281. *Norops woodi;* same male as in Plate 280, showing dewlap color change. Costa Rica: Puntarenas: Monteverde, about 1,400 m.

Plate 282. *Coleonyx mitratus*. Costa Rica: Guanacaste: Santa Rosa National Park, about 280 m.

Plate 283. *Hemidactylus garnotii*. Costa Rica: San José: San José, about 1,200 m.

Plate 284. *Lepidodactylus lugubris*. Costa Rica: Puntarenas: Golfito, about 3 m.

Plate 285. *Phyllodactylus tuberculosus*. Costa Rica: Guanacaste: Santa Rosa National Park, about 280 m.

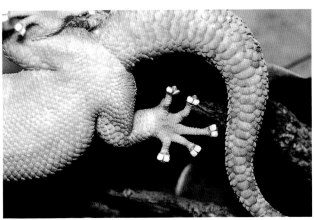

Plate 286. *Phyllodactylus tuberculosus,* showing toe pads from underside. Costa Rica: Guanacaste: Santa Rosa National Park, about 280 m.

Plate 287. *Thecadactylus rapicauda*. Costa Rica: Heredia La Selvia Biological Station, about 60 m.

Plate 288. *Gonatodes albogularis;* male. Costa Rica: Guanacaste: La Pacifica, 40 m.

Plate 289. *Gonatodes albogularis;* female. Costa Rica: Guanacaste: Santa Rosa National Park, about 280 m.

Plate 290. *Lepidoblepharis xanthostigma.* Costa Rica: Heredia: La Selva Biological Station, about 60 m.

Plate 291. *Sphaerodactylus graptolaemus.* Costa Rica: Puntarenas: Corcovado National Park, about 30 m.

Plate 292. *Sphaerodactylus homolepis;* male. Costa Rica: Heredia: La Selva Biological Station, about 60 m.

Plate 293. *Sphaerodactylus homolepis;* juvenile. Costa Rica: Heredia: La Selva Biological Station, about 60 m.

Plate 294. *Sphaerodactylus millepunctatus.* Guatemala: Izabal.

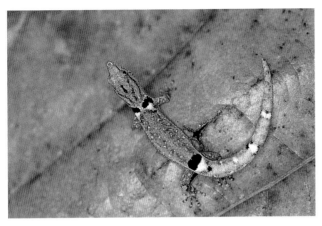

Plate 295. *Sphaerodactylus millepunctatus*. Honduras: Cortés: Matamoros, about 10 m.

Plate 296. *Lepidophyma flavimaculatum*. Costa Rica: Heredia: La Selva Biological Station, about 60 m.

Plate 297. *Lepidophyma reticulatum*. Costa Rica: Puntarenas: Las Cruces, about 1,200 m.

Plate 298. *Eumeces managuae*. Costa Rica: Guanacaste: Río Sandillal at Route 1, about 50 m.

Plate 299. *Mabuya unimarginata*. Costa Rica: Puntarenas: Monteverde, about 1,400 m.

Plate 300. *Sphenomorphus cherriei*. Costa Rica: Heredia: La Selva Biological Station, about 60 m.

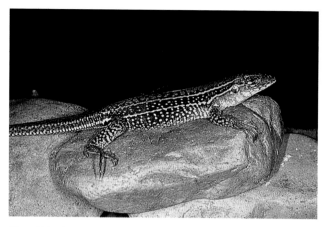

Plate 301. *Ameiva ameiva*. Panama: Panama: San Juan de Pequení, Chagres National Park.

Plate 302. *Ameiva festiva;* male. Costa Rica: Heredia: La Selva Biological Station, about 60 m.

Plate 303. *Ameiva festiva;* view of head Costa Rica: Heredia: La Selva Biological Station, about 60 m.

Plate 304. *Ameiva festiva;* juvenile. Costa Rica: Heredia: La Selva Biological Station, about 60 m.

Plate 305. *Ameiva leptophrys;* adult male. Costa Rica: Puntarenas: Rincón de Osa, about 40 m.

Plate 306. *Ameiva leptophrys;* juvenile. Costa Rica: Puntarenas: Rincón de Osa, about 40 m.

Plate 307. *Ameiva quadrilineata;* adult male. Costa Rica: Puntarenas: Rincón de Osa, about 40 m.

Plate 308. *Ameiva undulata.* Costa Rica: Puntarenas: Santa Rosa National Park, about 280 m.

Plate 309. *Cnemidophorus deppii.* Costa Rica: Guanacaste: Santa Rosa National Park, 280 m.

Plate 310. *Cnemidophorus lemniscatus.* Florida.

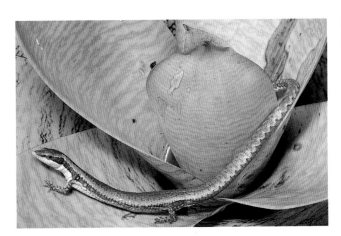

Plate 311. *Anadia ocellata.* Costa Rica: Alajuela: Peñas Blancas Valley, about 800 m.

Plate 312. *Anadia ocellata* at lunch. Costa Rica: Alajuela: Peñas Blancas Valley, about 800 m.

Plate 313. *Bachia blairi*. Costa Rica: Puntarenas: Corcovado National Park, about 20 m.

Plate 314. *Gymnophthalmus speciosus*. Costa Rica: Guanacaste: Santa Rosa National Park, about 280 m.

Plate 315. *Leposoma southi*. Costa Rica: Puntarenas: Sirena, Corcovado National Park, about 5 m.

Plate 316. *Neusticrurus apodemus*. Costa Rica: Puntarenas: Palmar Norte, about 30 m.

Plate 317. *Ptychoglossus plicatus*. Costa Rica: Cartago: Tapantí National Park, about 1,200 m.

Plate 318. *Celestus cyanochloris*. Costa Rica: Puntarenas: Monteverde, about 1,400 m.

Plate 319. *Celestus hylaius*. Costa Rica: Heredia: La Selva Biological Station, about 60 m.

Plate 320. *Celestus hylaius;* view of head. Costa Rica: Heredia: La Selva Biological Station, about 60 m.

Plate 321. *Diploglossus bilobatus;* female. Costa Rica: Heredia: La Selva Biological Station, about 60 m.

Plate 322. *Diploglossus bilobatus;* adult male. Costa Rica: Puntarenas: Monteverde, about 1,400 m.

Plate 323. *Diploglossus bilobatus;* female with eggs. Costa Rica: Monteverde area.

Plate 324. *Diploglossus monotropis*. Costa Rica: Heredia: La Selva Biological Station, about 60 m.

Plate 325. *Diploglossus monotropis;* view of head. Costa Rica: Heredia: La Selva Biological Station, about 60 m.

Plate 326. *Coloptychon rhombifer;* preserved specimen. Costa Rica: Puntarenas: near Golfito, about 10 m.

Plate 327. *Mesaspis monticola;* male above, female below. Costa Rica: Cordillera de Talamanca: Cerro de la Muerte, about 3,000 m.

Plate 328. *Mesaspis monticola;* female giving birth. Costa Rica: San José: Cerro de la Muerte, about 3,000 m.

Plate 329. *Anomalepis mexicanus.* Costa Rica: Guanacaste: Los Angeles de Tilarán, 436 m.

Plate 330. *Helminthophis frontalis.* Costa Rica: San José: San José, about 1,200 m.

is composed of twelve pairs of macrochromosomes and eight pairs of microchromosomes (Gorman 1973).

DISTRIBUTION: Humid lowlands and premontane slopes on the Atlantic versant from Tabasco, Mexico, to eastern Panama; on the Pacific slope from southwestern Costa Rica (10–1,200 m) through Panama; also barely reaching the Pacific slope in northwestern Costa Rica (700–1,230 m).

Norops fungosus (Myers, 1971b)

(map 10.34)

DIAGNOSTICS: A small, very short-legged anole with white funguslike markings and a pair of scale-covered parietal protuberances on the posterior upper head surface. The male dewlap is red with a few rows of white granules.

DESCRIPTION: Total length 103 mm, standard length 47 mm (in only known examples); tail moderate, 52 to 54% of total length; hind limbs very short; when adpressed along body, fourth toe does not reach shoulder; scales on upper surface of snout and in prefrontal region large, smooth; circumnasal or anterior nasal separated from rostral by one scale; supraorbital semicircles separated by one or two smooth scales; supraocular scales smooth; dorsal and lateral areas covered by smooth, flat granules; ventrals smooth, larger than dorsal; no nuchal crest; no enlarged postanal scales; 30 lamellae (17 expanded) under fourth toe; tail moderately compressed with keeled scales; no caudal crest; male dewlap moderate, extending posteriorly to slightly beyond axilla; dewlap condition unknown for females. Dorsal ground color medium brown with irregular blotching of white funguslike spots; tongue light yellow; lining of mouth and throat black, turning bluish gray at corners of mouth; tail with indefinite white rings; iris brown.

SIMILAR SPECIES: (1) *Norops vociferans* lacks parietal protuberances and funguslike white spotting in the pattern and has a dark interorbital bar. (2) *Norops pentaprion* is much larger, lacks parietal protuberances, and has a distinct caudal crest in adults.

HABITAT: Undisturbed Premontane Rainforest.

REMARKS: This distinctive and rare form is known from four males and one juvenile. The holotype (KU 113451) is from a clearing (Campo Mojica) along a trail on the north slope of Cerro Pando, Bocas del Toro Province, Panama (1,450 m). The second example (CRE 57) is from El Silencio de Sitia Mata, Cartago Province (1,200 m). The other examples are from Chirripó Grande, on the trail to Llano Bonito, San José Province (UCR 11912, male), and Tapantí, Cartago Province (UCR 9477, male; 11367, juvenile). The adult specimens have almost precisely the same characteristics and a standard length of 47 mm.

DISTRIBUTION: Humid premontane zone of southeastern and southwestern Costa Rica and adjacent western Panama (1,200–1,600 m).

Norops pentaprion (Cope, 1862c)

(plate 278; map 10.35)

DIAGNOSTICS: A small to moderate-sized, very short-legged anole having a flattened body, a variable pattern including a reticulum of dark lines and lichenose light areas, and smooth, granular ventrals. Male dewlap Bing cherry red-purple.

DESCRIPTION: Total length to 178 mm, males (adults 70 to 79 mm in standard length) larger than females (adults 57 to 63 mm in standard length); tail moderately short, about 52 to 57% of total length; hind limbs very short; when

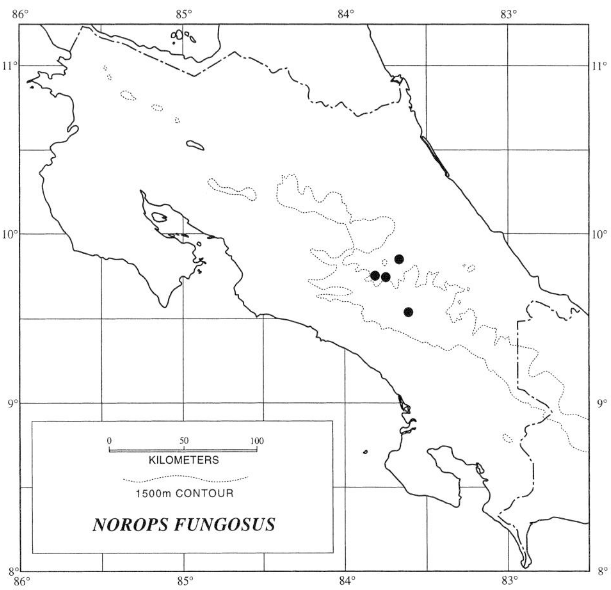

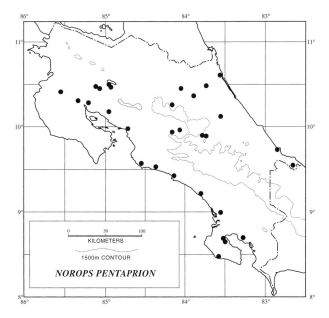

Map 10.34. Distribution of *Norops fungosus*.

Map 10.35. Distribution of *Norops pentaprion*.

adpressed to side of body, fourth toe reaches to shoulder; scales on upper surface of snout smooth, small laterally, large medially; scales in prefrontal area large, flattened, and smooth; circumnasal or anterior nasal separated from rostral by zero to three scales; supraorbital scales rugose; supraocular scales smooth; suboculars and supralabials in contact; dorsal and lateral scales small, smooth granules (a few middorsal scales rarely keeled); ventrals smooth, much larger than dorsals; a distinct nuchal crest in adult males; no enlarged supraorbital; semicircles separated by one smooth scale; postcloacal scales; 34 to 41 (18 to 22 greatly expanded) lamellae under fourth toe; tail moderately compressed, a distinct serrate caudal crest in all but smallest examples, most evident basally, larger in males. Male dewlap moderate, extending posteriorly slightly beyond axilla; female dewlap large, only slightly smaller than in males. Dorsal ground color brown to pale ashy white with a variable reticulum of dark lines and lichenose light areas, often with a greenish cast; tail marked like body; capable of extensive color change and in light phase appears white at a distance; tongue yellowish, lining of throat black turning bluish gray at corners of mouth; iris bronze.

SIMILAR SPECIES: (1) *Norops vociferans* has a dark interorbital bar, bright blue tissue at the corners of the mouth, and keeled ventrals in adults. (2) *Norops fungosus* is a smaller species (to 103 mm in total length) with paired parietal protuberances and the mouth and throat lined with black.

HABITAT: In all lowland forest zones (Dry, Moist, and Wet), even when much modified, as long as occasional trees are present. Marginally, in adjacent Premontane Moist and Wet Forest areas.

BIOLOGY: *Norops pentaprion* is a highly arboreal diurnal canopy species that occurs in low densities. Although they sometimes perch on tree trunks, these lizards spend most of their time in tree crowns. They are heliothermic and bask in the morning, late afternoon, and on overcast days. Often present in undisturbed forest situations, where they bask on the canopy surface 10 to 30 m above the ground, they are also regularly found in more open situations in disturbed or cleared areas. In the latter habitat they are found most commonly in widely scattered tall bushes or any tree that has been spared. In dry forest areas they tend to be found in trees along watercourses but often are present in isolated trees in pastures. If approached while on an exposed surface (e.g., a tree trunk), they flatten the body and sidle away from the observer much like the forest-dwelling gecko *Thecadactylus rapicauda*. They take a wide variety of arthropod prey (Corn 1981), but small flying insects (beetles and flies) make up a disproportionate portion of the diet. Stomachs examined contained zero to thirteen prey items. Prey size range (up to 9 mm) reflects this preference as well and is much lower than would be expected based on the relatively large size of these lizards. It seems likely that this food habit

relates to the canopy ecology of this form, which seems to have specialized on prey taxa that are abundant in tree crowns. Pérez-Higareda, Smith, and Chiszar (1997) reported that in Mexico they eat *Eugenia* fruits and cannibalize hatchlings. Reproduction appears to be continuous throughout the year in rainforest areas but may show a different pattern in western Costa Rica, where it may stop during the long dry season (November to May). Hatchlings are about 23 mm in standard length.

REMARKS: Myers (1971b) pointed out that examples referred to this species from the Yucatán Peninsula and Honduras lack caudal crests. This species may have once had a wider range on the Meseta Central Occidental, as suggested by an old record (FMNH 5627; Costa Rica: San José: San José). The karyotype of *N. pentaprion* is 2N = 28, composed of four pairs of metacentrics, two pairs of submetacentrics, and eight pairs of microchromosomes; NF = 40 (Lieb 1981).

DISTRIBUTION: Lowlands and marginally onto the premontane slopes from the Isthmus of Tehuantepec on the Atlantic versant to Colombia; also on the Pacific versant in Chiapas, Mexico, and Costa Rica to northwestern Colombia (2–900 m).

Norops vociferans (Myers, 1971b)

(plate 279; map 10.36)

DIAGNOSTICS: A small, very short-legged anole having a distinct V-shaped interorbital dark bar, a bright blue sliver of color at the corners of the mouth, keeled ventrals (in adults), and no bony parietal projections. The male dewlap is red.

DESCRIPTION: Total length to 132 mm (adults 35 to

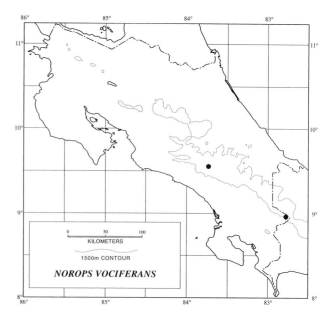

Map 10.36. Distribution of *Norops vociferans*.

57 mm in standard length); tail moderate, 55 to 57% of total length; hind limbs very short; when adpressed along side of body, tip of fourth toe reaches no farther than shoulder; scales on upper surface of snout small, some keeled; scales of prefrontal region large, smooth; circumnasal or anterior nasal separated from rostral by two rows of scales; supraorbital semicircles separated by zero to two smooth scales; semicircular and enlarged supraocular scales smooth; suboculars and supralabials in contact; small, mostly smooth, granular dorsal and lateral scales. Some middorsal scales weakly keeled; ventrals smooth, smaller than dorsals, larger than laterals, smooth in juveniles, keeled in adults; a nuchal crest in adult males; enlarged postcloacals in males; 33 (15 to 20 expanded) lamellae under fourth toe; tail moderately compressed without a crest, although middorsal row of caudal scale enlarged (fig. 10.23); dewlap of males moderate, extending posteriorly to near axilla; female dewlap slightly smaller than in males. Dorsal ground color brown to gray brown with definite but irregular darker reticulation and usually one to three dark chevrons on trunk; usually a butterfly-shaped dark marking on base of tail; venter white with a few dark spots; lining of throat black; iris brown.

SIMILAR SPECIES: (1) *Norops pentaprion* is larger, lacks the interorbital dark bar, and has grayish blue tissue at the corners of the mouth and smooth ventrals. (2) *Norops fungosus* is a smaller lizard (to 103 mm in total length) having bony parietal protuberances and the mouth and throat lining black.

HABITAT: In Costa Rica known only from two localities at relatively undisturbed sites in the Lower Montane Rainforest zone.

BIOLOGY: This species squeaks when handled (Myers 1971b).

DISTRIBUTION: Known from two Costa Rican localities on the south slope of the Cordillera de Talamanca, in the lower montane zone; also from extreme southwestern Panama in premontane and lower montane areas (1,402–1,830 m).

Norops woodi (Dunn, 1940b)

(plates 280–81; map 10.37)

DIAGNOSTICS: A large, extremely slender, long-legged anole having bright blue eyes, an attenuate tail, and strongly keeled ventral scales. Male dewlap dark olive green to black but capable of undergoing considerable lightening with mood changes to expose orange to red pigment.

DESCRIPTION: Total length to 345 mm; males (adults 78 to 95 mm in standard length) larger than females (adults 61 to 86 mm in standard length); tail long, 68 to 72% of total length; hind limbs very long; when adpressed against body, tip of fourth toe reaches beyond orbit; scales on upper surface of snout large, elongate, keeled; scales in prefrontal region large, keeled; circumnasal or anterior nasal separated from rostral by one or two scales; supraorbital semicircles

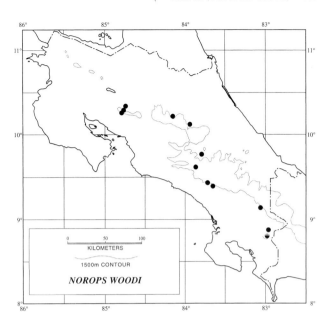

Map 10.37. Distribution of *Norops woodi*.

separated by two or three small rugose to keeled scales; semicircle and supraocular scales keeled; subocular and supralabial scales usually separated by one scale, rarely in contact; four to twelve rows of keeled middorsal scales often somewhat enlarged; lateral scales smaller than dorsals, granular and smooth; ventrals strongly keeled, mucronate, larger than dorsals and laterals; distinct nuchal crest in adult males; enlarged postcloacals present in males; 34 to 43 (14 to 19 greatly expanded) lamellae under fourth toe; tail compressed, attenuate; no caudal crest; male dewlap moderate, extending posteriorly to level of axilla; very small dewlap in females. Dorsum light or dark olive, dark brown, or blackish brown, often with a brick red suffusion; frequently with dark green or dark brown crossbands and/or many small black punctations; in preservative pale tan with numerous black punctations; venter light olive to brown or reddish brown; iris blue.

SIMILAR SPECIES: *Norops capito* has a very short head, a coppery iris, large smooth dorsal scales, and a pale green male dewlap.

HABITAT: In wooded areas and disturbed sites on trees within Premontane Wet Forest and Rainforest and Lower Montane Wet Forest and Rainforest areas; marginally in the Premontane Moist Forest zone at a few sites.

BIOLOGY: This common diurnal scansorial lizard forages on tree trunks; perch sites occur from near ground level to a height of 4.5 m. Individuals are often seen along the forest edge and on nonnative trees *(Cupressus)* used as windbreaks along mountain roads. When approached they flatten against the trunk and may slowly sidle out of sight around the tree. They do not try to escape by rapid dashes up, down, or around the trunk and are easily captured by hand. Year-round breeding is indicated by available data

(Fitch 1975). Hatchlings are about 31 mm in standard length.

REMARKS: *Norops woodi* was originally described from El Volcán, Chiriquí Province, Panama, by Dunn (1940b). Taylor (1956) recorded examples from southwestern Costa Rica from near the type locality. He also collected a series of similar lizards from farther north in the Cordillera Central (Heredia: Isla Bonita) that he described as a distinct subspecies, *N. woodi attenuatus*. The two forms were said to differ primarily in male dewlap color. In the northern form the dewlap was described as dark olive to blackish, whereas the southern lizards were said to have a tricolor dewlap (bluish white base, central amber band, and pink orange margin). Fitch, Echelle, and Echelle (1976) elevated *attenuatus* to species status based on the presumed differences in dewlap color, slightly larger size, reddish dorsal coloration, and number of scale rows around the midbody. Material now at hand indicates that all of these features are variable in both northern and southern samples so that individuals typical of "*attenuatus*" in coloration and scalation occur with typical *woodi* in the south and vice versa. Most of the confusion in this case arises from individual variation in coloration and the fact that these lizards also have a considerable repertoire of color changes. The general dorsal color varies from olive to reddish brown to nearly black depending on mood or temperature. Light and dark spots and dark banding also appear and disappear on the same individual. The dewlap color also shows metachrosis, so that the same specimen may have a primarily pink orange, yellow green, or black dewlap at any one time or exhibit one of several intermediate phases (plates 280–81). Consequently it is apparent that only a single species of anole is represented, for which the earliest name is *N. woodi*, and *N. attenuatus* must be considered a junior synonym. The karyotype is N = 30 or 32 with seven pairs of macrochromosomes and seven or eight pairs of microchromosomes (Lieb 1981).

DISTRIBUTION: Premontane and lower montane slopes of the cordilleras of Costa Rica and adjacent western Panama (1,150–2,500 m).

Geckos and Flap-foots (Infraorder Gekkota)

Most geckos are nocturnal sight feeders that are terrestrial, semiarboreal, or arboreal. Most feed on arthropods, but some large species eat larger prey. Typical geckos are rather delicate, thin-skinned lizards with stout, dorsoventrally depressed bodies, distinct short necks and heads, well-developed limbs, short tails, and the tips of the digits greatly expanded. Some are diurnal, however, and one family of geckos contains a series of snakelike forms, the flap-foots of Australia. Most four-limbed geckos are covered with small granular scales (sometimes intermixed with larger tubercles) on both the head and body, but some geckos have enlarged head plates and/or cycloid body scales. Flap-foots have variable cephalic scalation but consistently large and regularly arranged body scales. The following features are characteristic of this evolutionary line:

1. Tongue slightly notched (less than 10%); foretongue not differentiated; tongue covered anteriorly by peg-shaped papillae, posteriorly by plicae; tongue protrusion usually restricted to extension; prey apprehended by jaws; tongue monocytes mostly mucoserous
2. Postfrontal semilunate, forked medially and clasping frontoparietal suture
3. No postorbital or temporal arches; no supratemporal foramen; no external process on the extracolumella; jugal reduced
4. Septomaxillae dorsally expanded, convex, and forming midline crest; Jacobson's organs enlarged
5. Jaw tooth replacement basal, each tooth with a well-developed resorption pit
6. Two egg teeth in oviparous species
7. Notochord usually persistent in adults
8. Rectus abdominis lateralis muscle absent
9. Tibial epiphysis notched

In addition, all gekkotans exhibit eye-licking behavior to clean the cornea or brille and typically have clutch sizes of one or two eggs. All geckos and flap-foots, as far as is known, are oviparous except for two New Zealand genera *(Hoplodactylus* and *Naultinus)* and one genus from New Caledonia *(Rhacodactylus)*. The omphaloplacenta of these forms is well developed and persists until hatching. Of the three families now placed in this group, the Pygopodidae (Australasian geckos and flap-foots) of Australia (excluding Tasmania), New Caledonia, Loyalty Islands, New Zealand, and New Guinea (two species) and New Britain are extralimital.

Family Eublepharidae Boulenger, 1883b
Eyelash Geckos

The members of this family (six genera) are terrestrial nocturnal forms with widely disjunct distributions. The presence of movable eyelids with enlarged scales on the margin (hence eyelash geckos) and nictitating membrane separates them from all other gecko groups, and the absence of expanded digital lamellae distinguishes them from most others. As in most geckos, the head is covered with small, granular scales or a mixture of granules and tubercles. The dorsal scalation is homogeneous and covered with granular scales or with a mixture of granules and larger tubercles (which may be keeled), and the ventral scales are somewhat larger than the dorsals and usually imbricate. The eyes are moderate and the pupil is vertically elliptical. The ear opening is moderate and deeply recessed.

The following combination of features characterizes this family: movable eyelids; procoelous vertebrae; no digital expansions; nuchal endolymphatic sacs not penetrating nuchal musculature; a single premaxillary derived from two ossification centers, usually notched posteriorly; frontal process of the prefrontals in contact medially; a single parietal; supratemporal usually present; splenial present; a stapedial foramen; auditory meatal muscle bordering posterior and ventral meatal margins; cloacal sacs and bones present; femoral and/or precloacal pores usually present in males, sometimes in females; pupil vertically elliptical; egg shell leathery; complex vocalizations of chirps and clicks; hemipenes bilobed, calyculate, without flounces or petala, sulcus spermaticus simple.

The family is distributed as follows: *Coleonyx* (seven species) of the southwestern United States to Costa Rica; *Hemitheconyx* (nine species) and *Hoplodactylus* (two species) of Africa; *Eublepharis* (four species) of arid Asia from India to central Iraq; *Goniurosaurus* (nine species) from islands in the Gulf of Tonkin and the Ryuku Islands, Japan; *Aeluroscalabotes* (one species) of Thailand through Malaysia, Borneo, and Sanana Island.

Genus *Coleonyx* Gray, 1845

The members of this genus (seven species) are small to moderate-sized, delicate nocturnal lizards most readily recognized by the combination of movable eyelids, vertically elliptical pupils, small granular scales or a mixture of granular and small tuberculate scales on the upper head surface, and strongly dark-banded body and tail. Other distinguishing features include digits narrow, without expanded lamellae; dorsal scales uniformly granular or a mixture of granules and enlarged tubercles; ventral scales flat, imbricate, and smooth; digital lamellae uniform, imbricate, and in a single row; claws partially or entirely hidden by two shell-like lateral scales and a pointed terminal scale; precloacal pores present in males.

Characters that further separate this genus from other eublepharids are posterior margin of rostral V-shaped, extending posterior to nostrils; posterior fringe scales on eyelid elongate and conical; upper surface of eyelids covered by small granules; terminal digital scales laterally compressed, not entirely enclosing claw; a complete second ceratobranchial arch.

Members of this genus are strange-appearing nocturnal lizards believed to be extremely toxic by most uneducated and many educated people in Latin America. Even touching a specimen of *Coleonyx* is said to be fatal. Although this folklore is quaint, it is totally without foundation, for these delicate and shy geckos are completely harmless.

The genus may be divided into two clusters based on dorsal scalation: *Coleonyx brevis* (Chihuahua Desert), *C. fasciatus* (northwestern Mexico), and *C. variegatus* (Greater Sonora Desert) are covered by uniform small granules; *C. elegans* (tropical Mexico to El Salvador and the Yucatán Peninsula), *C. mitratus* (Guatemala to Costa Rica), *C. reticulatus* (Big Bend region of Texas), and *C. switaki* (southern California and Baja California) have a heterogeneous mixture of dorsal granules and larger tubercles.

The generic range is discontinuous in three areas, the Greater Sonoran Desert and adjacent western Mexico, the Chihuahua Desert, and the tropical lowlands from Veracruz and Guerrero, Mexico, to Honduras (Atlantic) and Costa Rica (Pacific).

Coleonyx mitratus (Peters, 1863a)

Escorpión Tobobo, Camaleón Breñero

(plate 282; map 10.38)

DIAGNOSTICS: The combination of vertical pupil (fig. 10.9a) and movable eyelids separates this small species from all other Costa Rican lizards. The mixture of small granular scales and enlarged tubercules on the body and the pattern of alternating broad dark brown and narrow light bands on the body and tail are also distinctive.

DESCRIPTION: A small species (to 190 mm in total length); males (adults 55 to 91 mm in standard length) smaller than females (adults 60 to 97 mm standard length); tail moderate (50 to 54% of total length); circular in cross section; upper surface of head covered by a mixture of small granular scales and larger tubercles; prominent enlarged eyelid fringe scales; neck and dorsum covered by small granules interspersed with larger tubercles in 21 to 23 irregular rows; venter covered with flat, imbricate scales; 5 to 9 precloacal pores in males (and analogous pits in females), the two series forming a continuous V; cloacal spurs longitudinally flattened and ridged, weakly developed, hardly larger than adjacent tubercles; scales of claw sheath short, claws conspicuous; 16 to 18 lamellae under fourth toe; tail covered by rings of imbricate scales, enlarged ventrally. Dorsum with a pattern of round, solid dark bands separated by much narrower

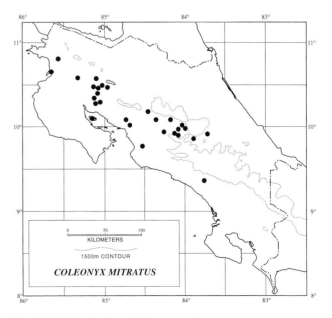

Map 10.38. Distribution of *Coleonyx mitratus*.

clear light ones in juveniles; bands in adults invaded laterally by light pigment to produce indented bands and mottled flanks; sometimes all of anterior bands becoming mottled as well; a light nuchal loop terminating at orbits on each side, best developed in juveniles; a horseshoe-shaped light rostral marking; tail banded with dark and undergoing ontogenetic change similar to that for body; undersurface immaculate white.

SIMILAR SPECIES: There are none.

HABITAT: This lizard is most frequently recorded from the Lowland Dry Forest region of northwestern Costa Rica, but it also is found in Premontane Moist Forest and marginally in Premontane Wet Forest.

BIOLOGY: This secretive form is usually discovered under debris on the ground during the day or occasionally walking about at night. Cope (1879) reported that J. Zeledón found it in anthills near the city of San José, San José Province. Dial and Grismer (1992) reported mean standard metabolic rate, evaporative water loss ratio, and temperature preference as 0.057 ml/g/hr, 2.557 mg/g/hr, and 25.7°C, respectively.

REMARKS: *Eublepharis dovii* Boulenger (1885) is a synonym of *Coleonyx mitratus* and was based on an example from "Panama." No other examples have been collected in that country, although it is possible that the species occurs there. Taylor (1956) had some reservations about whether this form occurred in Atlantic versant Costa Rica. Recently collected examples from the Meseta Central Oriental suggest that the record for Turrialba, Cartago Province (Klauber 1945), that Taylor questioned probably is valid.

DISTRIBUTION: Atlantic lowlands of northeastern Guatemala and northwestern Honduras; Pacific lowlands from Guatemala to southwestern Costa Rica, including the Premontane Mesetas Central Occidental and Oriental and the Valle de General (7–1,435 m). Questionably ranging into Panama.

Family Gekkonidae Gray, 1825

Geckos

A large and diverse family (sixty-nine genera, about 795 species) comprising mostly small to moderate-sized species of nocturnal insectivorous lizards. However, a number of genera are diurnal, and several of the larger species may eat small vertebrates. The family has a circumtropical, circumsubtropical distribution and is found on every continent and across the Pacific Islands from Australia to Hawaii. Most members of the family have some or most of the digital lamallae smooth and greatly expanded. In actuality the ventral surfaces of the lamellae are covered by a great many tiny bristles or setae. Each seta has numerous (100 to 1,000) tiny suction cups at its tip (visible under a scanning electron microscope) that make it possible for geckos to climb smooth vertical surfaces or run along the ceiling. All geckos of this family have the eyelids fused into a transparent, immovable

brille (spectacle) that lies over the cornea and protects the eye. Most are covered with small granular scales or a mixture of small granules and larger tubercles on the upper surfaces of the head and body, and with rounded or hexagonal imbricate scales ventrally. The eye is moderate, and the pupil is vertically elliptical or round. The ear opening is moderate to large, and the tympanum is deeply recessed.

The following features in combination characterize the family: no movable eyelids, a brille present; vertebrae usually amphicoelous, sometimes procoelous; nuchal endolymphatic sacs usually penetrating nuchal musculature and filled with an endolymph rich in calcium carbonate; a single premaxillary derived from one ossification center; frontal processes of prefrontals not in contact medially; auditory meatus muscle bordering posterior and ventral meatal margin or absent; usually paired parietals; supratemporal absent; splenial present or absent; stapedial foramen present or absent; cloacal sacs and bones usually present in males; egg shell calcareous, brittle; complex vocalization present or absent; hemipenes bilobate, calyculate, without flounces or petala, sulcus spermaticus simple.

In Costa Rica two major evolutionary lines variously treated as subfamlilies (Kluge 1967) or tribes (Kluge 1987) are represented and are briefly characterized as follows:

Gekkonines (sixty-four genera): digits usually expanded; splenial usually present; stapedial foramen absent; auditory meatal muscle usually present; femoral and/or precloacal pores usually present in males; no escutcheon scales; pupil vertically elliptical or round; chirping and/or clicking sound produced; clutch almost always of two eggs, but sometimes one, three, or four (distribution as for family)

Sphaerodactylines (five genera): digits not expanded; splenial absent; stapedial foramen present; no auditory meatal muscle; no femoral or precloacal pores; escutcheon scales present in males of most genera; no cloacal sacs or bones; pupil usually round; no chirping or clicking sounds produced; clutch a single egg (Neotropics)

The eggs of all members of this family are soft and flexible when laid but harden on exposure to air.

Subfamily Gekkoninae Gray, 1825

In rural areas throughout Latin America, these geckos and their allies, including *Coleonyx* (family Eublepharidae), are thought to be highly poisonous. They are reputed to sting or inject venom with their tails, and people believe that even touching one of them will lead to an unpleasant and rapid death. Actually these delicate creatures are neither venomous nor poisonous and are totally harmless.

Genus *Hemidactylus* Oken, 1817

Gekkos of this species (seventy-four species) are mostly small, delicate arboreal nocturnal forms between 100 and

200 mm in total length, but one species (*Hemidactylus giganteus* of India) is moderate-sized with a total length of 235 mm. These lizards have greatly expanded digits with transverse, divided lamellae, and all digits have a slender, compressed clawed terminal phalanx that is free from the expansion for part of its length. They lack movable eyelids, have a vertical elliptical pupil, and have most of the upper head surface covered by granular scales. Other features that in combination characterize this taxon are dorsal scales granular, uniform or mixed with larger tubercles; ventral scales smooth, rounded, and imbricate; no supraorbital spinelike scale; claws not retractile; no webs between digits; precloacal and femoral pores in males; postcloacal slits and bones in males.

Hemidactylus has a wide range from southern Europe, Africa, and South Asia to Polynesia and tropical and subtropical America.

Several species in the genus including the all-female triploid parthenogen *Hemidactylus garnotii* and the bisexual diploid *H. frenatus* are human commensals. They have become widely distributed to many subtropical and tropical port areas, probably as stowaways on boats and ships.

Hemidactylus garnotii seems to have been an East Indian species that now ranges into subtropical and tropical portions of South Asia, throughout Oceania to the Hawaiian Islands and to southern Central America, the Bahamas, and Florida. *H. frenatus* is probably a Southeast Asian and Indian species that has expanded its range to East Africa, Oceania, St. Helena, and the mainland of Mesoamerica,

eastern Texas, and southern Florida. Other species were likely carried to the Americas in a similar fashion from West Africa or the Canary Islands *(Hemidactylus brooki* and *H. mabouia)* and the Mediterranean region *(Hemidactylus turcicus)*.

KEY TO THE COSTA RICAN GECKOS OF THE GENUS *HEMIDACTYLUS*

1a. Tail with denticulate fringe (fig. 10.24a); two or three pairs of chin shields, posterior pair(s) not in contact with infralabials (fig. 10.25a) . . . . *Hemidactylus garnotii* (p. 485)

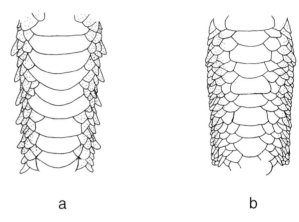

a   b

Figure 10.24. Ventral view of basal portion of tails in geckos of the genus *Hemidactylus*: (a) *H. garnotii*; (b) *H. frenatus*.

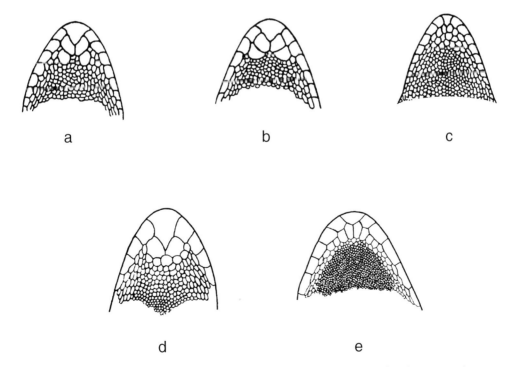

Figure 10.25. Chin scalation in geckos having expanded digital lamellae: (a) *Hemidactylus garnotii*; (b) *Hemidactylus frenatus*; (c) *Lepidodactylus lugubris*; (d) *Phyllodactylus tuberculosus*; (e) *Thecadactylus rapicauda*.

1b. Tail without denticulate fringe, although widely spaced ventrolateral spines present (fig. 10.24b); two pairs of chin shields, both pairs in contact with infralabials (fig. 10.25b) . . . . *Hemidactylus frenatus* (p. 484)

Hemidactylus frenatus Schlegel, 1836

(map 10.39)

DIAGNOSTICS: This small gecko is likely to be confused only with *Hemidactylus garnotii* and *Lepidodactylus lugubris*. It differs most obviously from *H. garnotii* in lacking a denticulated fringe on the tail, having instead a series of widely spaced pointed tubercles (fig. 10.24b). It is distinguished from *L. lugubris* in having all digits clawed and the claw-bearing terminal phalanx free from the digital expansion (fig. 10.6a). Distinctive features that in combination will separate it from other geckos are large divided digital pads, the claws nonretractile, and the digits without basal webbing.

DESCRIPTION: Total length to 135 mm, males (adults 52 to 65 mm in standard length) larger than females (adults 50 to 60 mm in standard length); dorsal scales heterogeneous, with a number of small tubercles (in two to eight uneven rows at midbody) interspersed among small granular scales; two pairs of large chin shields present and in contact with infralabials; ventral scales flattened, rounded, imbricate; 26 to 36 enlarged precloacal and femoral pores in continuous series in males; 9 to 10 expanded lamallae under fourth toe; tail flattened, most caudals small but interspersed by six longitudinal rows of keeled tubercles; median subcaudals transversely enlarged plates. Grayish brown, pinkish brown to red brown or dark brown above, uniform or with darker flecks or blotches or occasionally broken longitudinal stripes or short lines; dark streak from tip of snout through eye to ear; venter yellow often speckled with brown, underside of tail pale red in adults.

SIMILAR SPECIES: (1) *Hemidactylus garnotii* has a denticulated lateral tail fringe (fig. 10.24a) and the posterior pair of chin shields separated from the infralabials (fig. 10.25a). (2) *Lepidodactylus lugubris* lacks claws on the inner finger and toe. (3) *Thecadactylus rapicauda* is much larger, with retractile claws (fig. 10.6d) on the digits and a well-developed web between the digits. (4) *Phyllodactylus tuberculosus* has the terminal lamellae on each digit greatly expanded to form a leaflike tip (fig. 10.6c).

HABITAT: In Costa Rica *H. frenatus* has been found mostly in several coastal areas, where it has been incidentally introduced by shipping. Collections from two inland sites in Guanacaste Province suggest a wider and possibly expanding distribution in the republic. It is commonly found in human habitations throughout most of its range.

BIOLOGY: This very common gecko is often active during the day but more active at night. It feeds on insects, especially those attracted to electric lights. Individuals move rapidly up or down vertical surfaces and upside down across ceilings in houses. The species is very vocal, and adults of both sexes emit a multiple chirp call consisting of five to fifteen "chucks" over a period of 1 to 3.7 seconds (Marcellini 1971, 1974, 1977). They may give the call before or after any actual or anticipated change of state, but it is most frequently given by males. All but the smallest juveniles give a single chirp call when grasped by a predator or another gecko. Finally, a chirp call consisting of a very rapid series of short chirps ("similar to the rattle of a high-speed Teletype machine") is given by territorial males, usually just before they attack another male. *H. frenatus* is active at body temperatures of 26 to 30°C. Pairs of eggs (rarely, only one egg is laid) are usually deposited in crevices in walls or beams, in the thatch of roofs, or under ground debris. Many females use the same communal nesting sites, so eggs at various stages of development and old egg shells may be found together. Eggs are about 8 × 12 mm and hatch in forty-five to ninety days depending on temperature. Hatchlings are 19 to 20 mm in standard length and reach sexual maturity at about 45 mm in standard length.

REMARKS: *Hemidactylus frenatus* has only very recently invaded Costa Rica, and all records of its occurrence in the republic are post-1990. King (1978) reported that Australian geckos referred to this species had a karyotype of 2N = 40, consisting of five pairs of metacentrics, two pairs of submetacentrics, and thirteen pairs of telocentrics; NF = 54. Makino and Momma (1949) described the karyotype for *H. frenatus* in Taiwan as 2N = 46 (all pairs apparently telocentric), NF = 46. This leaves open the possibility that more than one species is represented under a single name, since Darvesky, Kupriyanova, and Roschin (1984) found

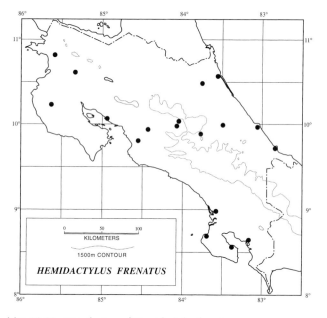

Map 10.39. Distribution of *Hemidactylus frenatus*.

Vietnamese examples to be similar to Australian lizards in karyotype. In addition, Moritz and King (1985) reported a 3N = 60 individual from Tarawa (Micronesia) that they thought was an autotriploid spontaneously derived from the sympatric diploid (2N = 40) population.

DISTRIBUTION: Southern India and Sri Lanka; Burma to southern China and through the Malay Peninsula to the Philippines. Other portions of the range are probably due to accidental human introductions and include New Guinea and the Solomon Islands, northern Australia, Samoa, Guam, and the Hawaiian Islands, islands adjacent to southern China and in the Indian Ocean, Madagascar, East Africa, the Cape of Good Hope, St. Helena, the Mediterranean region, Mexico, Central America, Panama, Texas, and Florida. In Costa Rica the species is known from several localities on both the Pacific and Caribbean lowlands including: Golfito, Puerto Jiménez, and Palmar Norte, Puntarenas Province; Liberia, Santa Cruz, and Santa Rosa National Park, Guanacaste Province; Cahuita, Puerto Limón, Siquirres, and Tortuguero, Limón Province; and on the Meseta Central Occidental (4–1,231 m).

Hemidactylus garnotii C. Duméril and Bibron, 1836

(plate 283; map 10.40)

DIAGNOSTICS: A small gecko with a sharply denticulated lateral fringe on the strongly depressed tail (fig. 10.24a) and claws on the inner fingers and toes. Other features that in combination will separate it from other geckos are large divided digital pads and nonretractile claws (fig. 10.6a). All known examples of this taxon are females.

DESCRIPTION: Total length to 135 mm (adults 50 to 65 mm in standard length); dorsum with small flat granular

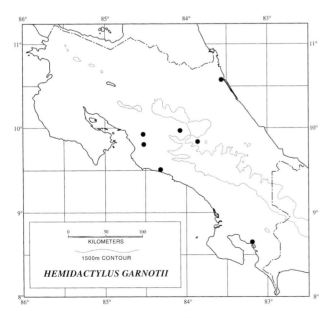

Map 10.40. Distribution of *Hemidactylus garnotii*.

scales except along sides, where small tubercles present; two or three pairs of large chin shields, posterior pair(s) not in contact with infralabials; ventral scales smooth, rounded, imbricate; 30 to 40 enlarged precloacal and femoral scales; 11 to 14 expanded lamallae under fourth toe; tail flattened with denticulate fringe; subcaudals transversely enlarged scutes. Dorsum uniform grayish tan or marbled with darker brown, usually with small whitish spots of various sizes and shapes; venter yellow, underside of tail pale red in adults.

SIMILAR SPECIES: (1) *Hemidactylus frenatus* lacks a denticulated tail fringe (fig. 10.24b), and the chin shields are all in contact with the infralabials (fig. 10.25b). (2) *Lepidodactylus lugubris* lacks claws on the inner finger and toe. (3) *Thecadactylus rapicauda* is much larger and has retractile claws (fig. 10.6d). (4) *Phyllodactylus tuberculosus* has the expanded terminal lamallae of the digits forming a leaflike tip.

HABITAT: Known only from the Meseta Central and the Lowland Wet Forest zone.

BIOLOGY: *Hemidactylus garnotii* is an invasive lizard expected to be found in Costa Rican port cities and urban centers, since it has successfully colonized these kinds of areas throughout its expanding range. It is essentially nocturnal and has habits similar to those described for *H. frenatus*. Individuals are territorial and emit a loud squeak when attacking conspecifics.

REMARKS: This species was first collected in a suburb of San José City, San José Province, in 1992. Additional records are to be expected as parthenogenesis expedites reproduction and range expansion. Kluge and Eckardt (1969) demonstrated that this species is an all-female triploid with a karyotype of 3N = 70, with two metacentrics, three submetacentrics, and one subtelocentric and sixty-four telocentric rods; NF = 75.

DISTRIBUTION: Southeast Asia, the East Indies to the Philippines, most islands in the tropical Pacific Ocean to Hawaii; introduced to St. Helena, Mesoamerica, the Bahamas, and southern Florida. In Costa Rica originally known only from a suburb of San José City, San José Province, and now collected from many localities (2–1,160 m).

Genus *Lepidodactylus* Fitzinger, 1843

Members of this genus (twenty-five species) are small, delicate arboreal nocturnal geckos reaching total lengths between 72 and 140 mm as adults. The most obvious distinguishing characteristics are the greatly expanded digits with transverse or oblique entire or divided lamellae; the outer four digits with a slender compressed and clawed terminal phalanx that unites with the dilated area above; the absence of eyelids; the presence of granular scales on the upper head surface; and the vertically elliptical pupil. Other features that in combination distinguish the genus include dorsal scales uniformly granular; ventral scales flat, rounded, juxtaposed, subimbricate or imbricate; no supraorbital spinelike scale; first digit without a claw in most species; claws not retractile; a basal web between digits in most species; precloacal pores and often femoral pores present in males; postcloacal slits and cloacal bones present in males.

Lepidodactylus is an Indo-Australian genus that ranges from Sri Lanka, northern India, Burma, Southeast Asia, and Malaysia through the East Indies to the Philippine and Solomon Islands and over much of Oceania to the Hawaiian Islands. One species, *Lepidodactylus lugubris,* has been accidentally introduced by humans into New Zealand, southern Florida, lower Central America, Colombia, and Ecuador, and its extensive distribution, especially on the islands of the South Pacific, also may have been expedited by human migration and exploration.

Lepidodactylus lugubris (C. Duméril and Bibron, 1836)

(plate 284; map 10.41)

DIAGNOSTICS: This distinctive little gecko is readily identified, since it has large digital pads, nonretractile claws on the outer digits (fig. 10.6b), the first digit clawless, and basally webbed digits.

DESCRIPTION: Total length to 105 mm; males (adults 35 to 39 mm in standard length) smaller than females (adults 31 to 45 mm in standard length); tail moderate, 52 to 55% of total length; digits webbed basally; dorsal scales homogeneous, granular; chin shields small, several in contact with infralabials (fig. 10.25c); ventral scales flat, rounded, imbricate; 28 to 35 enlarged precloacal and femoral scales in a continuous series in females; 10 to 26 precloacal and femoral pores in a continuous series in males; 11 to 15 expanded lamellae under fourth toe; tail flattened beneath with somewhat serrate margin, covered with small flat scales; subcau-

dal scales somewhat enlarged. Dorsum pale light pinkish gray to light brown, sometimes uniform but usually with prominent pair of brown spots just anterior to forelimbs; often with scattered small dark spots and/or irregular lines or chevrons; a wide dark band from snout through eye to forelimb; tail usually with dark markings; venter creamy white.

SIMILAR SPECIES: (1) The species of *Hemidactylus* have claws on all digits and definitely expanded midsubcaudal scales (fig. 10.24). (2) *Thecadactylus rapicauda* is much larger and has retractile claws on the digits (fig. 10.6d) and a very well developed web between the digits. (3) *Phyllodactylus tuberculosus* has distinct large tuberculate dorsal scales, claws on all digits, and the terminal lamellae of each digit greatly expanded to form a leaflike tip (fig. 10.6c).

HABITAT: In Costa Rica this species has been seen around and in human habitations at Golfito, a maritime town on the Golfo Dulce, on the Península de Osa, and near Quepos, Puntarenas Province. It is similarly associated with human developments throughout its natural range. In Asia, Indo-Malaysia, Australia, and Oceania, *L. lugubris* is commonest in the coastal zone and is most frequently encountered foraging or hiding on mangrove, coconut, and screw pine *(Pandanus)* trees.

BIOLOGY: A common nocturnal, scansorial species that often feeds on insects attracted to lights inside buildings. These lizards are excellent climbers and frequently perch on vertical surfaces or may be seen running upside down across ceilings. Most individuals of this species are diploid females (2N = 44) that reproduce parthenogenetically (Moritz and King 1985), but all-female triploid (3N = 66) clones are also known. Males are rare, but the triploids are probably produced by the insemination of diploid parthenogens. Pairs of eggs (about 8.5 × 7 mm) are laid off the ground under the bark of mangroves, in the leaf axils of palms, bananas, and screw pines, and frequently in the thatch on human-constructed structures. Hatchlings are about 16 mm in standard length.

REMARKS: The status of the many populations subsumed under the name *L. lugubris* remains equivocal. Some insular isolates of this complex are regarded as distinct species, but others are not. Any decision regarding the status of these populations is complicated by the prevalence of parthenogenetic reproduction and the absence of detailed study. The origin and relationships of populations that are probably the result of recent human transport, as in the Costa Rican case, are enigmatic in present circumstances but are referred to *L. lugubris* at this time. This species also occurs on several of the islands in the Bocas del Toro region of northwestern Panama, and it may be expected to appear in coastal localities in nearby southeastern Costa Rica. The karyotype in diploid populations comprises twenty-two pairs of telocentrics, and an additional set is present in the triploids to constitute NF = 44 and 66, respectively (Mortiz and King 1985).

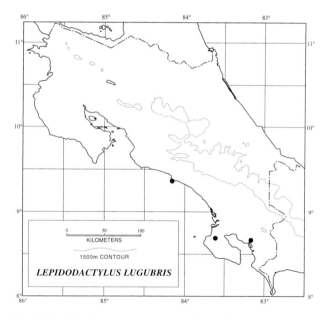

Map 10.41. Distribution of *Lepidodactylus lugubris.*

DISTRIBUTION: From Sri Lanka, northern India, Burma, Southeast Asia, and Malaysia through the Indo-Australian archipelago to the Philippine Islands, northern Australia, and Oceania to the Hawaiian Islands; introduced into New Zealand, the Galápagos Islands, southwestern Costa Rica (13–200 m), central Panama, western Colombia, and western Ecuador on the Pacific slope of the Americas and on the Atlantic versant in southern Florida, southeastern Nicaragua, and northwestern Panama.

Genus *Phyllodactylus* Gray, 1828

This large genus (forty-five species) of small to moderate-sized (to 212 mm in total length) nocturnal geckos is most trenchantly distinguished from other genera by having the terminal subdigital lamellae of the digits enlarged to form a pair of leaflike pads and the other subdigital lamellae narrow and in a single row on most of the digit but in two or three rows just proximal to the greatly expanded terminal pair. The characteristic toe pads combined with the absence of movable eyelids and the presence of granular head scales and the vertically elliptical pupil will separate these lizards from all other Costa Rican forms. Other features that aid in recognizing *Phyllodactylus* include claws present, not retractile; granular dorsal scales, sometimes homogeneous but usually mixed with rows of enlarged tubercles; ventral scales smooth and imbricate; no supraorbital spinelike scale; no precloacal or femoral pores; postcloacal slits and cloacal bones present in males.

For most of the past hundred years, the generic name *Phyllodactylus* was applied to species from southern Europe and southwestern Asia, sub-Saharan Africa and Madagascar, Southeast Asia, Australia, and adjacent islands, and the Americas. Bauer, Good, and Branch (1997) recently separated the Old World species previously included in *Phyllodactylus* into eleven discrete genera and restricted the latter name to forms occurring in the Americas. Within this clade the *Phyllodactylus tuberculosus* group (Dixon 1964) contains the following species in addition to *P. tuberculosus*, which is found in Costa Rica: *Phyllodactylus xanti* (southern California and Baja California); *P. angelensis*, *P. apricus*, *P. bugastrolepis*, *P. partidus*, *P. santacruzensis*, and *P. tinklei* (islands in the Gulf of California, Mexico); *P. homolepidurus* and *P. lanei* (western Mexico); *P. muralis* (southern Mexico); *P. insularis* (Half Moon Cay, Belize); and *P. palmeus* (Roatan Island, Honduras). The American species occur primarily in semiarid to arid regions along the Pacific coast in extreme southern California, Baja California, and western Mexico to Costa Rica and from Ecuador to northern Chile and also the Galápagos Islands; on the Caribbean slope in Belize and Guatemala; on islands off Honduras and northern Colombia and Venezuela; and on the Dutch Leeward Islands, Barbados, and Puerto Rico.

The American species exhibit a considerable range of ecological preferences. Most are scansorial and usually saxicolous, although they often climb trees and shrubs where boulders or rocky outcrops are unavailable. Several of the western South American species, however, are completely terrestrial.

Phyllodactylus tuberculosus Wiegmann, "1834b," 1835

(plates 285–86; map 10.42)

DIAGNOSTICS: The most obvious features of this moderate-sized gecko are the expanded digital tips (fig. 10.6c) and prominent keeled tubercles on the body. No other Costa Rican species lacking eyelids has these characteristics.

DESCRIPTION: Total length to 170 mm (to 212 mm in one extralimital population); males (adults 54 to 80 mm in standard length) smaller than females (adults 72 to 83 mm in standard length; to 100 mm extralimitally); tail moderate, 50 to 53% of total length; dorsum with 12 to 17 longitudinal rows of enlarged, keeled tubercles; 6 to 8 longitudinal rows of enlarged keeled tubercles on base of tail; a single large chin shield on each side in contact with first infralabial (fig. 10.2d); ventral scales rounded, imbricate; 11 to 16 lamellae under fourth toe. Dorsum fawn to dark brown with lateral paired series of small gray to black spots; similar spotting on head, limbs, and tail; proximal one-fourth of tail colored like body, distal portion usually marked with alternating dark and light bands; ventral margins or entire venter lemon yellow (bright ocher in one extralimital population), as are undersides of limbs and tail.

SIMILAR SPECIES: (1) *Thecadactylus rapicauda* lacks enlarged tuberculate scales and has the entire digit forming a large pad (fig. 10.6d), retractile claws, and the digits webbed at the base. (2) *Lepidodactylus lugubris* has the claws on a slender compressed terminal phalanx (fig. 10.6b), uniformly

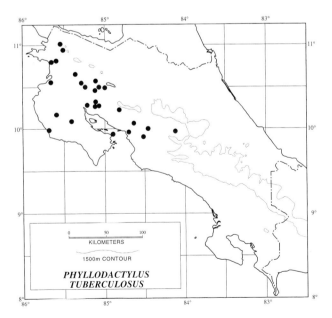

Map 10.42. Distribution of *Phyllodactylus tuberculosus*.

smooth granular dorsal scales, and femoral and precloacal pores in males. (3) The species of *Hemidactylus* have the claws on a slender compressed terminal phalanx (fig. 10.6a), precloacal and femoral pores in males, and enlarged precloacal and femoral scales in females.

HABITAT: Primarily found on rocky outcrops and boulders but also around cave and tunnel entrances, under bridges or rock fences, and in human habitations in the Lowland Dry Forest region. The species also occurs in similar situations along the caldera of the Río Grande de Tárcoles in the Premontane Moist Forest belt of the Meseta Central Occidental.

BIOLOGY: These relatively common strictly nocturnal scansorial insectivores favor rocky surfaces as a substrate but also may be found around human habitations. The two eggs that form a clutch are placed in crevices or on the undersurfaces of rocks. This species probably has continuous reproduction (Dixon 1964), since mature oviductal eggs are found in females throughout the year. Hatchlings are about 23 mm in standard length.

REMARKS: Taylor (1956) used the name *Phyllodactylus eduardofischeri* for examples of this species, but Dixon (1960) demonstrated that all mainland Central American populations belong in *P. tuberculosus*. As understood at present, the species consists of four geographic components differing slightly in adult size, ventral coloration, and scale count averages (Dixon 1964): (a) western Mexico (Sonora to Jalisco); (b) southwestern Mexico (Guerrero to Chiapas) and western Guatemala; (c) Pacific slopes of Central America and semiarid basins of central Guatemala; (d) Belize. Populations a and d are completely allopatric to one another, but b and c are connected by intermediates. Two related putative species have been recognized from the Caribbean islands of Half Moon Cay off the coast of Belize *(Phyllodactylus insularis)* and the Bay Islands off Honduras *(Phyllodactylus palmeus)* (Dixon 1960, 1968). How much this species ranges onto the Valle Central of Costa Rica is questionable. Several old records indicate that it occurred or may still occur "near San José," San José Province (ca. 1,100 m). Two examples (MCZ 15470–71) are also supposed to have been collected at Navarro, Cartago Province (1,100 m), on the Meseta Central Oriental. Since a few other forms (e.g., *Bufo coccifer, Norops cupreus*) that are essentially denizens of the Lowland Dry Forest region have similar distributions on the Meseta, it may well be that *P. tuberculosus* occurs on the Atlantic slope as well.

DISTRIBUTION: Pacific slope lowlands and adjacent slopes from central Sonora, Mexico, to central Costa Rica; definitely known from the Atlantic versant of central Guatemala and Belize. In Costa Rica found throughout the northwestern lowlands on the premontane Meseta Central Occidental and possibly on the Meseta Central Oriental (2–1,160 m).

Genus *Thecadactylus* Oken, 1817

Thecadactylus is a monotypic genus of moderately large (to 225 mm in total length) strictly arboreal and generally nocturnal geckos. Lizards of this genus are distinctive in having large toe pads divided below by a distinct longitudinal groove into which the claw retracts and having the digits webbed. These features and the absence of movable eyelids, the presence of a vertically elliptical pupil, and the dorsal scalation of small granules will immediately distinguish this genus from all other Costa Rican lizards. Other characteristics of the genus are subdigital lamellae divided; ventral scales imbricate, smooth; no supraorbital spine; no precloacal or femoral pores; and postcloacal slits and cloacal bones usually present in males.

The genus and single species have a wide range, primarily in lowland forests from southeastern Mexico and western Costa Rica to Ecuador and northern Brazil, including the Amazon basin, south to Bolivia and from Trinidad and Tobago through the Lesser Antilles to the Virgin Islands.

Thecadactylus rapicauda (Houttuyn, 1782)

Escorpión Tobobo

(plate 287; map 10.43)

DIAGNOSTICS: The largest Costa Rican gecko, this arboreal lizard may be immediately recognized by the combination of very large digital pads, digits enclosed in a well-developed fleshy web, and retractile claws at the tips of the fingers and toes (fig. 10.6d). No other lizard species in the Neotropics lacking movable eyelids shares these features.

DESCRIPTION: A moderate-sized lizard to 225 mm in total length; females (adults 95 to 126 mm in standard length)

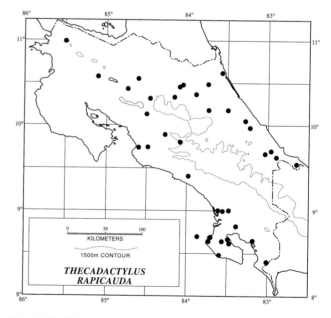

Map 10.43. Distribution of *Thecadactylus rapicauda*.

slightly larger than males (adults 93 to 125 in standard length); tail short, 42 to 46% of total length, rarely unregenerated; dorsal scales small, granular; two large and several smaller chin shields in contact with infralabials (fig. 10.25e); ventral scales round, imbricate; claw on inner digits very small or absent; 18 to 25 lamellae under fourth toe. Dorsum light to dark gray or nearly black with black and white markings; capable of changing color to become pale pinkish gray to almost white at night; conspicuous white postorbital stripes running posteriorly to ear opening; labials white, often with black margins; in juveniles the tail weakly banded; venter pale tan; tongue blue; lining of mouth orange; iris olive green with black streaks.

SIMILAR SPECIES: (1) *Lepidodactylus lugubris* is a much smaller species with nonretractile claws (fig. 10.6b) and claws on only the outer four digits. (2) *Phyllodactylus tuberculosus* has large and obvious tubercles on the body and tail and the terminal lamellae on each digit expanded into a leaflike tip (fig. 10.6c). (3) The species of *Hemidactylus* have some definite tubercles and spines laterally and on the tail and have nonretractile claws (fig. 10.6a).

HABITAT: Principally an inhabitant of undisturbed lowland rainforests but also found at intermediate elevations on the Pacific versant in remnants of Premontane Moist Forest.

BIOLOGY: Although usually considered a nocturnal species, *Thecadactylus rapicauda* is often seen on tree trunks or buttresses during daylight. When approached at these times it rapidly retreats into a convenient hiding place. It favors similar vertical sites for foraging and social interactions at night and spends most of the day resting in hollow trees, tree holes, or other cavities or crevices or in fallen logs or palm leaf axils. *Thecadactylus rapicauda* is a thermoconformer with activity temperatures of 24.2 to 28.6°C. These common geckos emerge and begin feeding at dusk, and they forage well up into the forest canopy (to 40 m). They eat a wide variety of arthropod prey, especially grasshoppers, katydids, and roaches. In a food study in Amazonia, Vitt and Zani (1997) found that orthopterans constituted the principal prey items (mean 82% by volume, 63% by number). They eat relatively large prey, length 4.7–5.9–42 mm. Females lay eggs at night on tree trunks or under bark and are unusual within the family in laying only one 16.7–18.3–21.3 × 11.2–12.9–14.7 mm egg at a time from alternate ovaries. Like many geckos *Thecadactylus rapicauda* eats its own shed skin. This species emits a high-pitched, rapidly repeated series of fifteen to twenty-five "chick-chick-chick" notes from time to time. These vocalizations appear to have a social function and probably are involved in territorial behavior. A tail-waving display appears to have a similar function, although it may be used to call predators' attention to the tail. Parachuting (Vitt and Zani 1997) and tail autotomy are characteristic antipredator defenses in this species.

REMARKS: The karyotype has been reported as 2N = 44 (Soma, Beçak, and Beçak 1975) or 2N = 42 (McBee et al. 1984). According to the latter authors there is one pair of large submetacentric chromosomes and twenty pairs of telocentric ones that grade in size from small to medium; NF = 44.

DISTRIBUTION: Lowlands on the Atlantic versant from Chiapas, Mexico, southward throughout Central America, including the outer Yucatán Peninsula and Isla Maíz Grande (Nicaragua), and northern South America to central Maranhão, Brazil, on the east and throughout the Amazon basin to northern Bolivia to the west; also from Trinidad and Tobago through the Lesser Antilles to St. Croix, Virgin Islands; on the Pacific slope primarily in the lowlands from central Costa Rica to western Ecuador. In Costa Rica found in the humid lowlands of the Atlantic versant and the southwest Pacific area, but also in the premontane zone on the Pacific versant in northwestern and central Costa Rica along the western border of the Meseta Central Occidental (1–1,052 m).

Subfamily Sphaerodactylinae Underwood, 1954

The Costa Rican sphaerodactyline geckos are all very small lizards and, except for brightly colored adult male *Gonatodes*, are not readily identified by someone unfamiliar with them without having an individual in hand. The most important distinguishing features of the three genera involve the structure of the toe tips and whether a small spinelike superciliary scale is present *(Sphaerodactylus)* or not *(Gonatodes* and *Lepidoblepharis)*. Kluge (1995) has established the relationships among the genera of this group and presented a detailed analysis of the evolution of ungual (claw) sheath characters. When coloration is ambiguous, differentiating the species of *Sphaerodactylus* usually requires checking the size of the subcaudal scales and whether there are one or two supranasal scales. Any small gecko should be examined for these several characteristics and compared with the following descriptions to ensure accurate identifications.

Genus *Gonatodes* Fitzinger, 1843

These small (to 115 mm in total length) diurnal scansorial lizards (sixteen to nineteen species in the genus) are most readily distinguished from other geckos by having undilated digits, the distal two or three phalanges of the digits angulate, and the claws not retractile and between a pair (superior and inferior) of terminal scales. Characteristics that further differentiate *Gonatodes* from other Costa Rican lizards are the absence of movable eyelids; the absence of a spinelike superciliary scale; the uniform granular scalation on the dorsum and upper head surface; the flat, rounded, imbricate, and smooth ventral scales; and the round pupil (said to be vertical in one extralimital form; Peters and Donoso-Barros 1970).

Lizards of this genus are widely distributed in the Neotropics from Chiapas, Mexico, to western Ecuador and Venezuela and on adjacent islands and thence south to Pará, Brazil, and the upper Amazon region from Colombia to Mato Grosso, Brazil, and northern Bolivia; also the West Indian islands of Grand Cayman, Cuba, Jamaica, and Hispaniola; one species *(Gonatodes albogularis)* has been introduced into southern Florida.

Gonatodes albogularis (C. Duméril and Bibron, 1836)

(plates 288–89; map 10.44)

DIAGNOSTICS: Adult males of this small diurnal arboreal gecko cannot be mistaken for any other lizard. They have bright orange heads, blackish to gray brown bodies with bluish lateral spots, a conspicuous blue line on the white supralabials, and a white tail tip (on complete tails). Females and juveniles are likely to be confused only with other small geckos *(Lepidoblepharis* and *Sphaerodactylus),* and they differ from them in lacking retractile claws and enlarged terminal scales on the digits (fig. 10.7a).

DESCRIPTION: Total length to 113 mm with no sexual dimorphism in size (adults 36 to 48 mm in standard length); tail moderate, 50 to 58% of total length; pupil vertical; two supranasals; scales on upper surfaces of body and head granular; caudals granular; ventrals flat, imbricate, larger than dorsals; median subcaudals large, slightly smaller than ventrals, about twice as wide as adjacent scales; numerous enlarged escutcheon scales in males on venter and underside of thighs; 18 to 22 lamellae under fourth toe. Dorsum of males blackish to gray brown with many bluish lateral spots; dorsum of juveniles and females pale grayish brown mottled with irregular black markings; venter tan to whitish; iris brown.

HABITAT: This gecko is found throughout the lowlands in Dry, Moist, and Wet Forests in every conceivable situation including pastures, roadside fences, fallen logs, trash piles, and human-made structures.

SIMILAR SPECIES: (1) *Lepidoblepharis xanthostigma* is found in leaf litter. It has enlarged terminal scales on the digits and retractile claws (fig. 10.7b). (2) The species of *Sphaerodactyclus* have a spinelike scale on the superciliary margin (fig. 10.9a), enlarged terminal lamellae on the digits, and retractile claws (fig. 10.7c).

BIOLOGY: Fitch (1973a,b) has described the ecology and behavior of this gecko in some detail, and the following account is based to a considerable extent on his reports. Although widely distributed throughout the lowlands, *G. albogularis* occurs in greatest densities in the subhumid Lowland Dry Forest region of northwestern Costa Rica. These lizards are most active late in the day, when they forage in deep shade. They are very shy and move about on any inclined or vertical surface from ground level to about 4 m above the ground. They rarely are motionless and dart from one place to another, frequently disappearing into holes, under bark, or into any available crevice to emerge again and sprint to a new hiding place. If flushed from cover they usually spiral around the tree trunk or other substrate and dash to another retreat. Adult males are territorial and chase smaller males away from a relatively small defended area (about 10 to 20 m²). Adult males maintain spacing by an aggressive display in which they lower the tail directly forward over the back and wave it up and down in a jerky motion that emphasizes the white tail tip. If approached closely by a rival male, they may also raise the body, lower the head, and twitch the head laterally. If these behaviors do not discourage an intruding male, the resident male may rush toward the opponent, striking him and trying to bite him. In most situations individuals are dispersed, with one or two per tree, fence post, or other vertical site. However, as Fitch (1973a) noted, large trees (especially *Ficus*) often have sizable colonies of ten to thirty or forty individuals. Reproductive activity appears to be influenced by food availability. On the Atlantic slope, where there is no marked dry season, females produce eggs throughout the year with some decrease in December and January, laying them one at a time from alternate ovaries. In the Dry Forest region of the Pacific versant, reproduction markedly declines during the dry season (November through March). Atlantic slope populations consequently have individuals of all sizes and ages present in unvarying proportions throughout the year, whereas those from the northwest have marked fluctuations in proportions. At the beginning of the dry season in the latter area, all age-groups are present in the population. By the end of the dry season when the breeding season begins, mostly

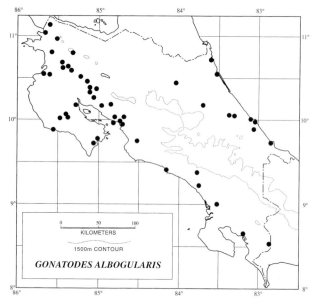

Map 10.44. Distribution of *Gonatodes albogularis.*

adults and large juveniles are present. A flush of hatchlings occurs from July to August, so by the end of the wet season all age groups are again represented. Hatchlings are 17 to 19 mm in standard length; they attain sexual maturity at about six months of age.

REMARKS: Taylor (1956) and other authors applied the name *Gonatodes fuscus* to this form. Vanzolini and Williams (1962) demonstrated that South American lizards referred to *G. fuscus* are conspecific with *G. albogularis*. There seems little basis for recognizing geographic races in this species, although differences in adult male coloration in life may have some geographic fidelity. Stuart (1963) was of the opinion that all West Indian records are the result of human introductions.

DISTRIBUTION: Lowlands of Central America from Pacific slope Chiapas, Mexico, and eastern Guatemala (Atlantic) south on both coasts to northwestern Colombia and western Venezuela and adjacent islands; Cuba, Jamaica, Grand Cayman, and Hispaniola in the Antilles; introduced to southern Florida (2–530 m).

Genus *Lepidoblepharis* Peracca, 1897

Members of this genus (fifteen species) are small diurnal denizens of the leaf litter. They are separated from other geckos by having undilated digits, the distal two or three phalanges on the toes angulate, and the claws retractile into a sheath of five equal-sized elongate scales and one smaller scale. Other distinguishing features include no movable eyelids, round pupil, no spinelike superciliary scale, uniform granular scales on dorsum and upper surface of the head, and ventral scales smooth, flattened, rounded, and imbricate.

The genus ranges from eastern Nicaragua south to western Ecuador, and east of the Andes it occurs in Colombia, Ecuador, Peru, and the Amazon basin to near the mouth of the Rio Amazon in Brazil.

Lepidoblepharis xanthostigma (Noble, 1916)

(plate 290; map 10.45)

DIAGNOSTICS: A very small diurnal lizard that is seldom seen but is often heard rustling between leaves in the litter at ground level. Most examples have a pair of more or less distinct longitudinal lateral light stripes (often discontinuous), a dark-outlined light nuchal band, and/or one or two definite large light-outlined dark spots on the upper surface of the sacral and postcloacal regions. The claws are retractile into a sheath of equal-sized elongate scales (fig. 10.7b), and there is no superciliary spinelike scale in this gecko.

DESCRIPTION: Total length to 90 mm, both sexes about same size (adults 25 to 45 mm in standard length); tail moderately short, 52 to 55% of total length; pupil round; two supranasals; scales on upper surfaces of body and tail uniformly granular; caudals flat, subimbricate, larger than

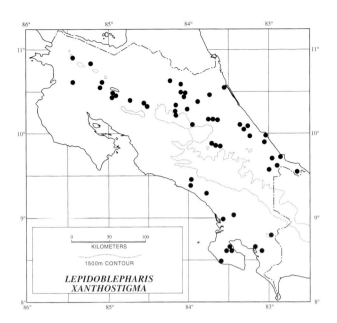

Map 10.45. Distribution of *Lepidoblepharis xanthostigma*.

dorsals; ventrals much larger, imbricate; median subcaudal scales greatly enlarged, three to four times as wide as adjacent scales; numerous escutcheon scales on venter and a few on thighs on males. Coloration extremely variable, dorsum brownish lavender, uniform or with light stripes and/or spots and/or dark spots or blotches; nuchal light color almost always indicated; sacral and precloacal light-outlined dark spots almost always present; venter pale lavender marked with many dark punctations posteriorly; throat marked with small brown spots and reticulations; iris bronze.

SIMILAR SPECIES: (1) *Gonatodes albogularis* is scansorial, lacks enlarged terminal scales on the digits, and has nonretractile claws (fig. 10.7a). (2) The species of *Sphaerodactylus* have a spinelike superciliary scale (fig. 10.9a), asymmetric claw-sheath scales, and a large terminal subdigital pad (fig. 10.7c).

HABITAT: *Lepidoblepharis xanthostigma* is very common and ubiquitous in the leaf litter of Lowland Moist and Wet Forests and marginally into Premontane Wet Forest and Rain Forest wherever some tree cover remains and in modified habitats including cacao plantations. It frequently may be found in second growth and also in the gallery forests of the Dry Forest region of northwestern Costa Rica.

BIOLOGY: These lizards carry out their activities under the fallen debris, especially dead leaves, on the forest floor during the day. They may be flushed out by an observer moving along a trail or through the forest but quickly dive back under cover. *L. xanthostigma* is extremely common, with densities as high as 88/ha in undisturbed forest and 548/ha in abandoned cacao plantations (Lieberman 1986) on the Atlantic lowlands. Food consists primarily of small spiders, with high proportions of isopods and mites also included in the diet. Zero to four (mean 2.3) food items are usually present per stomach, and the prey lengths range up to about 12 mm, averaging between 1.3 (dipterans) and 11.6 mm (centipedes) (Lieberman 1986). Females lay a single egg from alternate ovaries; reproduction on the Atlantic slope appears to occur throughout the year. Small hatchlings are 15 mm in standard length.

DISTRIBUTION: Humid lowland and adjacent premontane localities from southeastern Nicaragua and northwestern Costa Rica to northern Colombia; in Costa Rica, also in gallery forests of the northwest (2–1,360 m).

Genus *Sphaerodactylus* Wagler, 1830

This is a large genus (ninety species) comprising tiny and small secretive essentially diurnal lizards ranging in total adult length from 26 to 94 mm. Most of the species are terrestrial forms and are found under any available ground cover, but others are saxicolous or semiarboreal. Species of *Sphaerodactylus* differ from other geckos in sharing the following combination of features: digits not dilated, but tips distinctly dorsolaterally compressed and asymmetric, the distal phalanges not angulate, the claws retractile into a sheath between one greatly enlarged subterminal pilose pad and several much smaller scales, and a distinct spinelike superciliary scale present. Additional features that aid in distinguishing these geckos from other Costa Rican lizards are the absence of movable eyelids; the upper head surface covered by small granular scales; dorsals granular, smooth, or keeled; some extralimital forms with large flat dorsal scales that may be smooth or keeled and juxtaposed or imbricate; small flattened rounded imbricate ventrals; and a round pupil. Hass (1991) regarded most mainland species of this genus as constituting a distinct clade, the *Sphaerodactylus lineolatus* section. In addition to the three Costa Rican members this section includes *Sphaerodactylus dunni* (northeastern Honduras), *S. heliconiae* (Colombia), *S. molei* (Venezuela, Trinidad and Tobago, and Guyana), and *S. scapularis* (western Ecuador and Colombia). *Sphaerodactylus glaucus* (southern Mexico to western Honduras) and *S. millepunctatus,* which occur in Costa Rica, were not placed in either the mainland or West Indian (the *sputator* section) groups recognized by Hass (1991).

Sphaerodactylus ranges over the Neotropical mainland from southern Veracruz, Mexico, through northern Venezuela to the Guianas on the Atlantic versant and on the Pacific slope at the Isthmus of Tehuantepec and from Nicaragua to northwestern Ecuador. Numerous species also occur in the West Indies and the Bahama Islands, and several others on the nearshore Caribbean islands; two species have been introduced to the Florida Keys, and another *(Sphaerodactylus notatus)* occurs in southern Florida, on Cuba, and in the Bahamas.

KEY TO THE COSTA RICAN GECKOS OF THE GENUS *SPHAERODACTYLUS*

1a. Median subcaudal scales not forming a continuous series of enlarged scales, some equal to, others broader than adjacent scales (fig. 10.26a)..............2

1b. Median subcaudal scales forming a definitely enlarged series running most of length of tail (fig. 10.26b) ... 3

2a. Dorsum of adults and some juveniles brown, with very small dark spots and often a diffuse broad dark nuchal band; other juveniles with distinct light-outlined dark bands on body and tail (fig. 10.28f); head marked with dark stripes or a reticulum of light spots outlined by dark in adult males (fig. 10.28a); head and neck of adult females with a series of narrow dark stripes (fig. 10.28b); male escutcheon comprising 35–79 unpigmented glandular scales restricted to the precloacal ventral region (fig. 10.29b); adults to 33 mm in standard length ....*Sphaerodactylus homolepis* (p. 494)

2b. Dorsum of adults marked with definite large dark blotches, spots, or mottling (fig. 10.28c), usually several moderate-sized light spots present as well; dorsum of juveniles uniform brown; a narrow nuchal dark band bordered posteriorly by a light area usually present; head marked with a broad dark stripe that passes along canthus through eye, bordered above in the postorbital area by a distinct broad light stripe; male escutcheon comprising 10–16 unpigmented glandular scales restricted to a small area in the precloacal ventral region (fig. 10.29b); adults to 47 mm in standard length ....*Sphaerodactylus pacificus* (p. 496)

3a. Dorsum with dark brown reticulum or with distinct dark-outlined longitudinal light stripes (fig. 10.28d); no sacral dark band or spot; a single supranasal (fig. 10.27a); male escutcheon comprising 22–61 glandular

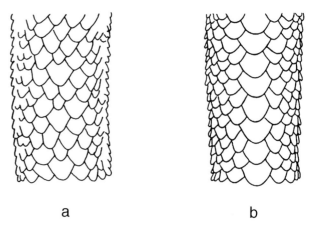

Figure 10.26. Subcaudal scalation near base of tail in the gecko genus *Sphaerodactylus:* (a) median subcaudals a mix of large and smaller scales; (b) median subcaudals forming a longitudinal series of enlarged scales.

Figure 10.27. Arrangement of supranasal scales in geckos of the genus *Sphaerodactylus:* (a) one large supranasal; (b) two supranasals.

scales restricted to the precloacal ventral region (fig. 10.29a); adults to 32 mm in standard length . . . . *Sphaerodactylus graptolaemus* (p. 494)

3b. Dorsum usually uniform or with many small dark spots; usually a light-outlined sacral dark band or large spot (fig. 10.28e,g); two supranasals (fig. 10.27b); male escutcheon comprising 22–138 glandular scales in a continuous series from the precloacal ventral area out on the underside of both legs to the knee (fig. 10.29c); adults to 31 mm in standard length . . . . *Sphaerodactylus millepunctatus* (p. 495)

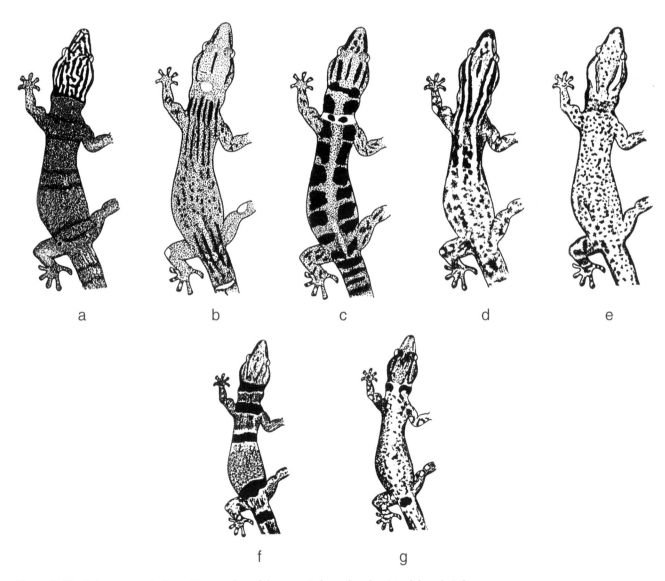

Figure 10.28. Color patterns in Costa Rican geckos of the genus *Sphaerodactylus:* (a) adult male *S. homolepis;* (b) adult female *S. homolepis;* (c) adult *S. pacificus;* (d) adult female *S. graptolaemus;* (e) adult *S. millepunctatus;* (f) juvenile *S. homolepis;*, some adult females retain this pattern; (g) juvenile *S. millepunctatus.*

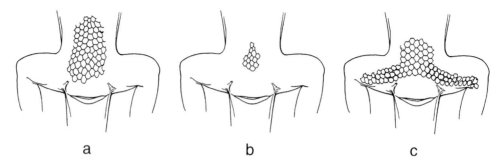

Figure 10.29. Escutcheon scale patterns in male geckos of the genus *Sphaerodactylus:*. (a) *S. graptolaemus;* (b) *S. homolepis* and *S. pacificus;* (c) *S. millepunctatus.*

Sphaerodactylus graptolaemus Harris and Kluge, 1984

(fig. 10.28d; plate 291; map 10.46)

DIAGNOSTICS: A very small secretive (partially nocturnal), scansorial species having paired lateral light stripes or a dark brown reticulate dorsal pattern and lacking sacral dark spots. In addition, the subcaudals form a continuous series of greatly enlarged scales (fig. 10.26b) and there is a single large supranasal scale (fig. 10.27a).

DESCRIPTION: Total length to 67 mm (adults 24 to 32 mm in standard length); tail moderate, 52% of total length; pupil round; scales on upper surface of body small, oval, flat, keeled; caudals flat, imbricate, smooth except on base of tail; ventrals round, smooth, imbricate, much larger than dorsals; median subcaudals three to four times as wide as adjacent scales and forming a continuous series; 11 to 16 lamellae under fourth toe; escutcheon scales of males restricted to precloacal venter. Dorsum tan with a dark brown reticulum or bold cream lateral stripe bordered by dark from eye to axilla or indicated as far posteriorly as base of tail; no large sacral dark marking.

SIMILAR SPECIES: (1) The much larger *Sphaerodactylus pacificus* does not have stripes or a dark dorsal reticulum. (2) *Sphaerodactylus homolepis* does not have a continuous series of enlarged subcaudal scales (fig. 10.26a). (3) *Sphaerodactylus millepunctatus* has two supranasals (fig. 10.27b) and a uniform or spotted dorsum. (4) *Lepidoblepharis xanthostigma* has symmetrical claw sheaths and no subdigital pads (fig. 10.7b). (5) *Gonatodes albogularis* has nonretractile claws, and the terminal digital lamellae are about the same size as the other lamellae (fig. 10.7a).

HABITAT: Humid lowland forests of the Pacific southwest region.

REMARKS: Taylor (1956) called examples of this species *Sphaerodactylus lineolatus* and *Sphaerodactylus* sp. I applied the name *Sphaerodactylus continentalis* to it (Savage 1973b) but later used *S. lineolatus* (Savage 1976, 1980b). Harris and Kluge (1984) demonstrated the distinctiveness of this uncommon form from its allies *Sphaerodactylus homolepis* (Atlantic slope southeastern Nicaragua, Costa Rica, and Panama) and *S. lineolatus* (west-central Panama to Colombia).

DISTRIBUTION: Lowlands of southwestern Costa Rica and adjacent southwestern Panama (5–700 m).

Sphaerodactylus homolepis Cope, "1885," 1886

(fig. 10.28a,b,f; plates 292–93; map 10.47)

DIAGNOSTICS: Adult males of this tiny semiarboreal diurnal gecko are unmistakable in having the head striped, spotted, or reticulated with bright lemon yellow. Most juveniles are distinctive in having the body crossed by four light-edged dark bands that fade during ontogeny but are usually indicated in adults as well. Some juveniles and adult females are creamy brown with small dorsally dark spots and may be allocated to this species if they lack a continuous series of enlarged median subcaudals (fig. 10.26a) but have a single supranasal (fig. 10.27a).

DESCRIPTION: Total length to 69 mm, adults 25 to 33 mm in standard length; tail rather short, 48 to 52% of total

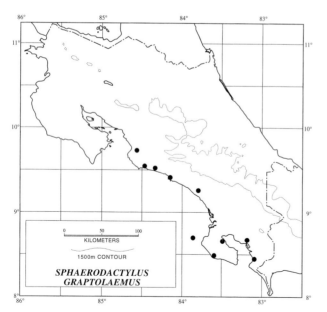

Map 10.46. Distribution of *Sphaerodactylus graptolaemus.*

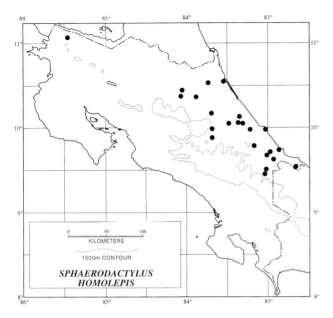

Map 10.47. Distribution of *Sphaerodactylus homolepis*.

length; pupil round; scales on upper surface of body rhomboid, juxtaposed, and strongly keeled; caudals flat, imbricate, smooth except at base of tail; ventrals round, smooth, imbricate, larger than dorsals; median subcaudals not forming a continuous series of enlarged scales but repeating a sequence of large and small scales (one small, one large, two small); escutcheon scales of males restricted to precloacal ventral area; 10 to 12 lamellae under fourth toe. Dorsum cream to light brown; body and tail strongly banded in most juveniles (fig. 10.28f), bands fading during ontogeny but often still indicated in adults; other juveniles and adult females with numerous small dark spots; adult males with striped, spotted, or reticulate pattern of dark brown and bright lemon yellow on head and tail; some very narrow dark lines on head and neck in most adult females; venter light yellow; iris golden.

SIMILAR SPECIES: (1) The much larger *Sphaerodactylus pacificus* shows no sexual dichromatism; it has a dorsal pattern of large dark blotches, large dark spots, or mottling in adults, and the juveniles are uniform brown. (2) *Sphaerodactylus graptolaemus* has a striped or reticulate dorsal pattern and a continuous series of expanded subcaudals (fig. 10.26b). (3) *Sphaerodactylus millepunctatus* usually has a dorsal pattern of small spots, and those with an overall uniform dorsal color have definite light-outlined dark sacral spots or blotches. All examples of this species have a continuous series of enlarged subcaudals (fig. 10.26b). (4) *Lepidoblepharis xanthostigma* has symmetrical claw sheaths and no subdigital pads (fig. 10.7b). (5) *Gonatodes albogularis* has nonretractile claws, and the terminal digital lamellae are about the same size as the other lamellae (fig. 10.7a).

HABITAT: Lowland Moist and Wet Forests of the Atlantic slope.

REMARKS: The remarkable individual, ontogenetic, and sexually dichromatic variation in this uncommon species and its similarity to related forms have caused considerable confusion. Dunn (1940c) regarded this species as conspecific with *Sphaerodactylus lineolatus* (now regarded as a strictly Panamanian form) and stated that he had seen six examples from Nicaragua and four from Costa Rica with definite localities. The former are representatives of *Sphaerodactylus millepunctatus*, a species that ranges from southeastern Mexico to Costa Rica, but the Costa Rica specimens are *S. homolepis*. Taylor (1956) recognized *S. homolepis* Cope, 1886 (based on a banded juvenile, USNM 14207, from along the Río San Juan, Nicaragua) and *Sphaerodactylus imbricatus* Andersson, 1916 (an adult female and two adult males, GNM 1326a–c, from Costa Rica). Although Taylor had material of both adult males and females, he considered the possibility that two species might be represented in Andersson's type series. I referred to this species as *Sphaerodactylus lineolatus* (Savage 1973b, 1976) and as *S. homolepis* (Savage 1980b). Harris and Kluge (1984) resolved the matter by recognizing *S. lineolatus* (west-central Panama to Colombia) and *S. homolepis* as distinct, although they suggested that the two might intergrade in western Panama. Taylor's (1956) records of *S. lineolatus* from Costa Rica are based on *S. millepunctatus*.

DISTRIBUTION: Humid Atlantic lowlands from southeastern Nicaragua to west-central Panama (2–600 m).

Sphaerodactylus millepunctatus Hallowell, "1860," 1861
(fig. 10.28e,g; plates 294–95; map 10.48)

DIAGNOSTICS: A tiny secretive diurnal terrestrial gecko in which the dorsum is marked with small dark spots or is

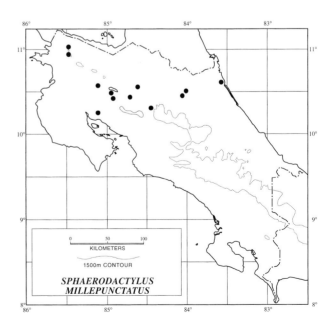

Map 10.48. Distribution of *Sphaerodactylus millepunctatus*.

nearly uniform tan and usually has a light-outlined sacral dark band or spot. Besides the differences in coloration, it may be distinguished from similar species by a continuous series of greatly enlarged median subcaudal scales (fig. 10.26b) and two supranasals (fig. 10.27b).

DESCRIPTION: Total length to 62 mm (adults 24 to 31 mm in standard length); tail short, 48 to 50% of total length; pupil round; scales on upper surface of body oval, flattened, keeled, imbricate; caudals rhomboid, flat, imbricate, smooth except at base of tail; ventrals round, smooth, imbricate, larger than dorsals; median subcaudals enlarged in a continuous series; escutcheon scales of adult males forming a bell-shaped area on venter and usually extending onto undersides of thighs; 10 to 11 lamellae under fourth toe. Dorsum light brown with distinct nuchal and sacral light-outlined dark spots or bands in juveniles; adult pattern of small dark dots on uniform light brown ground color; sacral dark spots or bands retained in most adults; dark-outlined light postorbital stripes in some examples; head of adult males reticulate; iris tan.

SIMILAR SPECIES: (1) *Sphaerodactylus graptolaemus* has light dorsal stripes or a reticulate dorsal pattern. (2) *Sphaerodactylus homolepis* does not have a continuous series of expanded subcaudals (fig. 10.26a). (3) The much larger *Sphaerodactylus pacificus* has a dorsal pattern in adults of large dark blotches or spots or mottling; juveniles have a uniform brown dorsum as in some *S. millepunctatus,* but neither adults nor juveniles have definite light-outlined sacral bands or spots. (4) *Lepidoblepharis xanthostigma* has symmetrical claw sheaths and no subdigital pads (fig. 10.7b). (5) *Gonatodes albogularis* has nonretractile claws, and the terminal digital lamellae are about the same size as the other lamellae (fig. 10.7a).

HABITAT: Found under tree bark or on tree trunks, on the ground, under debris and near human habitations in the Lowland Moist and Wet Forests and Premontane Wet Forest of northern Costa Rica.

REMARKS: Sexual dichromatism and ontogenetic change in coloration led in the past to some confusion as to the status of this uncommon form. Dunn (1940c) referred to four and six examples of this species from definite localities in Costa Rica and Nicaragua, respectively, under the name *Sphaerodactylus lineolatus* (a Panamanian form). Taylor's (1956) account of *S. lineolatus* is based on three examples of *S. millepunctatus* from Tenorio and one from Tilarán, Guanacaste Province. He correctly referred another specimen from Tenorio (KU 40504) to *S. millepunctatus.* Earlier I used the name *Sphaerodatylus continentalis* for this form (Savage 1973b), but later I recognized it as *S. millepunctatus* (Savage 1976, 1980b). This allocation was confirmed in the excellent revision of mainland members of the genus by Harris and Kluge (1984), and they amply distinguished it from other related forms.

DISTRIBUTION: Lowlands of the Atlantic versant from the Isthmus of Tehuantepec to northern Costa Rica, including Isla Cozumel, Quintana Roo, Mexico, and the Great Lakes region of Nicaragua; also in the humid lowland and gallery forests of northwestern Costa Rica and marginally in the premontane zone at a few sites in Costa Rica (37–657 m).

Sphaerodactylus pacificus Stejneger, 1903

(fig. 10.28c; map 10.49)

DIAGNOSTICS: Among the very largest members of its genus, this species is characterized by a pattern of irregular large to smaller lineate dark blotches on a lavender brown ground color in adults. Juveniles are uniform brown but may be identified by having a single supranasal (fig. 10.27a) and lacking a continuous series of enlarged subcaudal scales (fig. 10.26a) and definite light-outlined dark sacral (lineate) bands or spots. This species is restricted to Isla del Coco, and the only other lizard occurring there is the anole *Norops townsendi.*

DESCRIPTION: Total length to 98 mm (standard length of adults 48 to 49 mm); tail rather short, 48 to 52% of total length; pupil round; scales on upper surface of body rhomboid, keeled, juxtaposed, smooth except on base of tail; ventrals rounded, imbricate, smooth, much larger than dorsals; median subcaudals forming a repetitive sequence of large and small scales (one small, one large, two small); escutcheon scales of males forming a small oval region on the precloacal ventral area; 13 to 15 lamellae under fourth toe. Dorsum uniform tan in juveniles with large irregular dark blotches or lineate spots or mottling in adults; sometimes

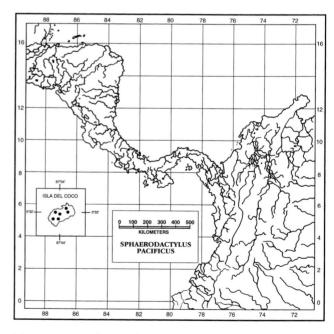

Map 10.49. Distribution of *Sphaerodactylus pacificus.*

with white spotting; narrow nuchal dark band, marked by light posteriorly, usually indicated; a broad dark stripe along canthus through eye bordered above in postorbital area by a light stripe.

SIMILAR SPECIES: (1) *Sphaerodactylus millepunctatus* usually has the dorsum marked with small dark spots and has a distinct light-outlined dark sacral band or spot(s) in juveniles. (2) The smaller *Sphaerodactylus homolepis* has light-outlined dark bands on the body in many cases (obscured in most adults) or the dorsum marked with small dark spots and the head yellow with a dark reticulum (adult males) or brown marked with narrow darker lines (adult females).

HABITAT: This species is restricted to Isla del Coco, where it is mainly found in the Lowland Wet Forest Zone and is especially abundant near the shoreline under debris and around the bases of coconut palms. It may, however, be widely distributed over the rest of the island, where the densities may be lower but have not been well sampled.

BIOLOGY: A nocturnal species (Carpenter 1965). Taylor (1956) reported that eggs and empty shells of this species were found hidden in the roots of coconut palms, usually 50 or so mm above the ground. Eight to ten empty shells were found in one nook and empty shells and eggs together in others. Six eggs alone were found in one site. Gravid females have a single oviductal egg, as is typical for the genus. The smallest known juvenile (probably a hatchling) measures 17.5 mm in standard length.

REMARKS: Taylor (1956, fig. 5) indicates total lengths for three illustrated specimens of this species as 121 to 129.5 mm. Elsewhere he stated that the maximum standard length is 47 mm. In material available to me the maximum total length recorded was 94 mm, which is consistent with the latter figure. The longest examples of this species are 47 mm in standard length but lack complete tails. Extrapolating from these values and assuming a maximum tail length (52% of total length), the largest examples would be about 98 mm in total length—well below Taylor's high figures, which appear to be in error.

DISTRIBUTION: Isla del Coco.

Scincomorph Lizards (Infraorder Scincomorpha)

This group is the most diverse of lizard stocks, and consequently it is difficult to generalize regarding habits or habitus. Many species have the limbs reduced, and several families contain limbless, snakelike forms. These lizards characteristically have enlarged, symmetrical shields on the head. Body scalation shows much variability, but cycloid scales are typical of several included families and are also found in a few additional genera of other families. In those families lacking cycloid scales the ventrals consist of enlarged juxtaposed rectangular plates or smaller elongate, more or less quadrangular scales. The following features technically define this group:

1. Tongue slightly (less than 10%) or deeply (20 to 50%) notched; foretongue not retractile; tongue usually covered with lingual scales and/or plicae; tongue protrusion usually oscillatory; prey apprehended with jaws; tongue mucocytes mostly mucoserous
2. Postfrontal with few exceptions semilunate, forked medially and clasping frontoparietal suture
3. Usually postorbital and temporal arches; supratemporal foramen open, closed, or absent
4. Septomaxillae dorsally expanded, convex, and forming midline crest; Jacobson's organs enlarged
5. Jaw tooth replacement usually basal with large resorption pits, sometimes alternating with reduced resorption pits
6. A single egg tooth in oviparous species
7. Notochord not persisting in adults
8. Rectus abdominis lateralis muscle present
9. Tibial epiphysis notched
10. Usually no internal process on the extracolumella; no extracolumella muscle

As understood at present, ten lizard families are referred to this infraorder, and an eleventh, the Dibamidae, probably belongs here. The following families are extralimital to the area covered by this book and will not be further discussed: the Lacertidae of Africa and Eurasia; the Cordylidae of Africa and Madagascar; the snakelike Dibamidae of Southeast Asia to the Philippines and New Guinea and also eastern Mexico; and the amphisbaenian families (Amphisbaenidae, Bipedidae, Rhineuridae, and Trogonophidae).

Family Xantusiidae Baird, 1859a

Night Lizards

A small family (four genera) of tiny to moderate-sized insectivorous or omnivorous lizards that are mostly nocturnal and/or secretive. Night lizards superficially resemble geckos in lacking movable eyelids and having the dorsum covered with small granular scales or a mixture of granules and larger tubercles. Xantusiids are immediately distinguished from geckos by having large symmetrical head shields and large rectangular juxtaposed ventral plates. Other external features shared by the members of this family include body depressed, limbs moderate, tail short; eye small, covered by a transparent spectacle and lacking a nictitating membrane; pupil vertically elliptical; ear opening moderate, tympanum exposed and slightly recessed; femoral pores usually present; a complete gular fold; digits not expanded and lamellae keeled. The group is a New World endemic with one genus (*Cricosaura*, two species) found in Cuba, another (*Lepidophyma*, seventeen to nineteen species) from Mexico to Panama, one (*Klauberina*, one species) on the Channel Islands off southern California, and a fourth (*Xantusia*, six species) in western North America.

The following combination of features characterizes the family: premaxilla single; vomer single; postfrontal and postorbital fused into a single element; no supratemporal fenestra; parietal foramen usually present; jaw tooth replacement basal, with large resorption pits; jaw teeth conical or triconodont; Meckel's groove closed and fused; a small intercentrum associated with each dorsal vertebra; caudal autotomy, with autotomy septa passing through or behind a single pair of nondiverging, nonconverging transverse processes; extracolumella with or without internal process; no extracolumella muscle; no dorsal, ventral, or separable cephalic osteoderms; nasal plates not separated by frontonasal(s); tongue weakly notched; foretongue covered with peglike papillae, hindtongue with plicae; tongue protrusion limited to extension; hemipenes slightly bilobed, apical regions pustulate, each with a terminal disk, truncus flounced, no calyces or petala, and a simple sulcus spermaticus.

The family has been divided into two subfamilies as follows (Crother, Miyamoto, and Presch 1986):

Xantusiinae: parietal foramen present; nostril at juncture of nasal, postnasal, and first supralabial; gulars enlarged; caudal scales homogeneous; 12 to 16 longitudinal rows of ventral scales (*Xantusia* and *Klauberina*)

Lepidophyminae: no parietal foramen; nostril at nasal-postnasal juncture; gulars same size as pregulars; caudal scales heterogeneous in size (*Cricosaura* and *Lepidophyma*)

Members of this family exhibit eye-licking behavior. They are "taste" feeders that take any small object that might be food into the mouth, sample it with the Jacobson's organs, and then spit out distasteful or inedible items. As far as is known, all Xantusiidae are viviparous and have both a well-developed omphaloplacenta and a chorioallantoic placenta.

Genus *Lepidophyma* A. Duméril, 1851

The genus is composed of seventeen to nineteen species of small to moderate size that are unique among all American lizards in lacking movable eyelids and having vertically elliptical pupils, enlarged head shields, five digits on all limbs, and the dorsal and lateral surface of the body covered by a heterogeneous mixture of granules and enlarged tuberculate scales. Characteristics that in combination distinguish *Lepidophyma* from other xantusiids are one frontonasal scale, two frontal and two parietal scales and bones; supraocular scales reduced to a fleshy ridge projecting from side of frontal above eye; caudals arranged in whorls of large and small scales; orbit completely roofed over by frontal bone; squamosal contacting parietal bone; and clavicle perforate.

Smith (1973) divided the genus into five species groups, but there is some overlap in distinguishing characteristics among several of them (Bezy 1984; Bezy and Camarillo 1992). The two Costa Rican species are unambiguous members of the *Lepidophyma flavimaculatum* group, to which the following extralimital taxa unquestionably belong: *Lepidophyma micropholis* and *L. occulor* (northeastern Mexico); *L. tarascarae* (southern Mexico); *L. mayae* (Guatemala); and *L. smithii* (Guatemala and El Salvador).

Night lizards belonging to this genus occur from Michoacán and Tamaulipas, Mexico, south over most of tropical Mesoamerica, on the Atlantic versant to central Panama exclusive of the outer portions of the Yucatán Peninsula, and on the Pacific slope to Guatemala and El Salvador and disjunctly in central and southwestern Costa Rica; also uplands of northwestern Costa Rica.

KEY TO THE COSTA RICAN SPECIES OF THE GENUS *LEPIDOPHYMA*

1a. Gular area covered by reticulum of dark pigment enclosing series of rounded light spots (fig. 10.30a).....
.... *Lepidophyma reticulatum* (p. 500)
1b. Gular area uniformly light or with small dark spots (fig. 10.30b) .. *Lepidophyma flavimaculatum* (p. 498)

Lepidophyma flavimaculatum A. Duméril, 1851

(plate 296; map 10.50)

DIAGNOSTICS: A moderate-sized species immediately recognizable in the field as *Lepidophyma* in having a large smooth head that sharply contrasts in appearance with the tuberculate body and a dorsal pattern of large round light spots on a dark brown to black ground color. It may be distinguished from its congener *Lepidophyma reticulatum* only by differences in throat coloration. This species lacks the reticulate gular pattern enclosing large round light spots typical of *L. reticulatum* (fig. 10.30a).

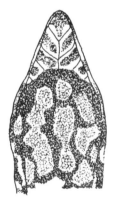 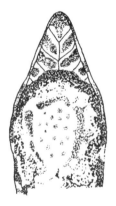

a   b

Figure 10.30. Coloration of throat in Costa Rican species of the night lizard, genus *Lepidophyma:*. (a) *L. reticulatum;* (b) *L. flavimaculatum.*

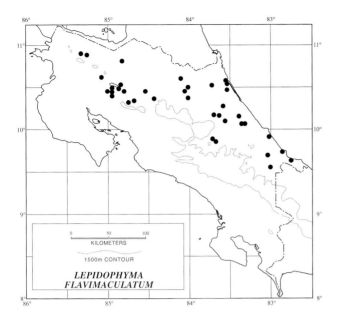

Map 10.50. Distribution of *Lepidophyma flavimaculatum*.

DESCRIPTION: Total length to 245 mm (adults 60 to 102 mm in standard length); tail moderate, 57 to 60% of total length; a definite parietal foramen visible externally; dorsum covered with a heterogeneous mixture of small granules and enlarged tubercles; lateral tubercles arranged in 23 to 33 vertical rows; enlarged paravertebral scales in two discontinuous rows; transverse rows of enlarged keeled caudal scales in continuous series, separated by three or four rows of small scales; enlarged rectangular ventrals in eight to ten longitudinal rows; 26 to 33 femoral pores; 26 to 31 (25 to 31) lamellae under fourth toe, 14 to 22 (12 to 22) divided. Dorsal ground color dark brown to black with large light round spots; upper and lateral surfaces of head light brown; throat light or with small dark spots; venter cream with some brown markings; iris dark brown.

SIMILAR SPECIES: (1) *Lepidophyma reticulatum* has large cream spots on the throat (fig. 10.30a). (2) *Neusticurus apodemus* has movable eyelids, small flattened and rounded gular scales, and the caudals a mixture of keeled tubercles arranged in discontinuous transverse rows and smaller, smooth scales.

HABITAT: A secretive denizen of undisturbed Lowland Moist and Wet Forests that is usually found under or in fallen logs or other debris on the forest floor. Although found primarily on the Atlantic versant including the lower reaches of the Premontane Wet Forest zone, it also occurs marginally on the northwest Pacific slope in the Premontane Rainforest and Lowland Moist Forest.

BIOLOGY: *Lepidophyma flavimaculatum* has been variously considered a nocturnal, crepuscular, or secretive diurnal species. It is apparently active early in the evening but has also been seen during the day under conditions of low illumination. Since most examples are discovered when their hiding places in or under surface debris are disturbed, it is difficult to evaluate densities and how active they may be during daylight. The species is insectivorous and eats fairly large arthropods, including centipedes (up to 12.5 mm in length) and spiders as well as many ants (mean 2 mm). Telford and Campbell (1970) pointed out that the population of this form in central Panama seemed to be composed exclusively of females. Bezy (1989) concluded after examination of extensive material that all Panamanian and most Costa Rican populations referred to this species are all female (diploid) parthenogens that reproduce asexually. However, some populations in extreme northeastern Costa Rica have males present in small numbers. This lizard is strictly viviparous and produces four to six young at a time during the wet season (May to November), with birth sometimes spread over two or three days. The smallest neonates are 35 mm in standard length, and Telford and Campbell (1970) estimated that they do not reach sexual maturity until twenty months of age.

REMARKS: The Costa Rican and Panamanian lizards referred to this species belong to a complex of morphologically, karyologically, and electrophoretically identical populations that range north along the Atlantic versant to southern Veracruz and Oaxaca, Mexico. Populations in this complex from Honduras northward regularly include males and are probably bisexual reproducers (Bezy 1989). Whether the few males in northern Costa Rican populations are involved in reproduction remains to be determined. The northern (Honduras to Mexico) bisexual populations differ in coloration and average scale counts from the southern ones here placed in *L. flavimaculatum*. Unfortunately, there are so few individuals from the intervening area (Nicaragua) that any conclusion regarding genetic continuity or noncontinuity is impossible. The name *L. flavimaculatum* A. Duméril, 1851 (holotype: MNHNP 782; Guatemala: Petén Province) is based on an example of the northern group. Should further study establish that the southern populations ought to be recognized as a distinct species, the name *Lepidophyma obscurum* Barbour, 1924 (holotype: MCZ 17747; Panama: Panama Province: Río Chilibrillo) is available. Taylor's (1955, 1956) specimens variously referred to *L. flavimaculatum* or *L. obscurum* or described as the new species *L. anomalum* and *L. ophiophthalmum* are of this form. The karyotype is 2N = 38, with nine pairs of macrochromosomes (two pairs of metacentrics, one pair of submetacentrics, five pairs of subtelocentrics, and one pair of telocentrics) and ten pairs of microchromosomes; NF = 44 (Bezy 1972).

DISTRIBUTION: Humid lowlands of the Atlantic versant from Oaxaca and southern Veracruz, Mexico, to central Panama; in Costa Rica also marginally on the Pacific slope in the premontane and upper lowland zones in the area of Tenorio and Tilarán in Guanacaste Province (2–690 m).

Lepidophyma reticulatum Taylor, 1955

Lagartija Pintada

(plate 297; map 10.51)

DIAGNOSTICS: A moderate-sized lizard distinct from all other Costa Rican species except *Lepidophyma flavimaculatum* in having a large smooth head that contrasts markedly in texture with the tuberculate body and a dorsal pattern of large round light spots on a dark brown ground color. It differs from *L. flavimaculatum* in having a gular reticulum surrounding large oval light spots instead having the throat uniformly light or with small dark spots (fig. 10.30).

DESCRIPTION: Total length to 220 mm (adults 60 to 100 mm in standard length); tail moderate, 57 to 59% of total length; a definite parietal foramen visible externally; dorsum covered with a heterogeneous mixture of small granules and large tubercles; lateral tubercles arranged in 28 to 33 vertical rows; enlarged paravertebral scales in two discontinuous rows; transverse rows of enlarged keeled caudal scales in continuous series, separated by three or four rows of small scales; enlarged, rectangular ventrals in eight to ten longitudinal rows; 25 to 43 femoral pores; 26 to 31 lamellae under fourth toe, 15 to 20 divided. Dorsum dark brown to black marked with large round light spots; upper and lateral surfaces of head light brown; throat with large cream spots surrounded by a dark brown to black reticulum; venter cream with some brown smudges; iris dark brown.

SIMILAR SPECIES: (1) *Lepidophyma flavimaculatum* lacks large light spots set into a dark reticulum on the throat (fig. 10.30b). (2) *Neusticurus apodemus* has movable eyelids, small flattened and rounded gular scales, and the caudals a mixture of keeled tubercles arranged in discontinuous transverse rows and smaller smooth scales.

HABITAT: This secretive epigeal species occurs in Lowland Moist and Wet Forests and Premontane Moist and Wet Forests. Most examples have been taken on the Pacific versant from under or in fallen logs or under debris in relatively undisturbed forest or along stream courses where some tree cover remains after the adjacent forest has been cleared.

BIOLOGY: All known examples of this species are diploid females. It probably has habits similar to those described for *L. flavimaculatum*.

REMARKS: This species has been taken in sympatry with *L. flavimaculatum* 4 to 5 km north-northeast of Tilarán (probably near Finca San Bosco, 640 m), Guanacaste Province. In earlier accounts (Savage 1973b, 1976, 1980b; Savage and Villa 1986) I included this species, originally described by Taylor (1955) (holotype: KU 36245; Costa Rica: Puntarenas: Agua Buena, near the Panama border, 1,060 m), in *L. flavimaculatum*. The consistent and distinctive throat coloration is the only univariate feature that separates *L. reticulatum* from *L. flavimaculatum*, and Bezy (1989) convincingly argues for the validity of the former based on coloration, multivariate analysis, and absence of intergradation. The karyotype is identical to that of *L. flavimaculatum*: 2N = 38, composed of nine pairs of macrochromosomes (two pairs of metacentrics, one pair of submetacentrics, two pairs of subtelocentrics, and one pair of telocentrics) and ten pairs of microchromosomes; NF = 44 (Bezy 1972).

DISTRIBUTION: Humid lowlands and premontane slopes of southwestern Costa Rica and probably adjacent southwestern Panama; Pacific slope premontane zone of the Meseta Central Occidental and the Cordillera de Tilarán to the Atlantic versant near Laguna de Arenal (10–1,250 m).

Family Scincidae Gray, 1825

Skinks

A large family (about 125 genera and 1,250 species) of mostly diurnal fast-moving small to large terrestrial and semifossorial lizards. Some, however, are nocturnal or crepuscular, a few are semiaquatic, and others are scansorial, but only one species (*Corucia zebrata* of the Solomon Islands) is arboreal. Most skinks are insectivorous, but a few are omnivores, and some larger species eat small vertebrates. Skinks usually have cylindrical bodies, the limbs short, reduced, or absent, and the head not distinct from the neck and covered with enlarged symmetrical shields. The body and tail are generally covered by smooth, glossy imbricate cycloid scales (some extralimital genera have enlarged keeled and/or spiny dorsal and caudal scales). In addition, skinks have small eyes; most have movable eyelids, a few have only a lower eyelid, and in some the eyelids may be fused to form a transparent spectacle, and the pupils are round; the ear opening is usually small and slitlike with the tympanum deeply recessed, but some extralimital forms have the tympanum exposed and superficial and others lack

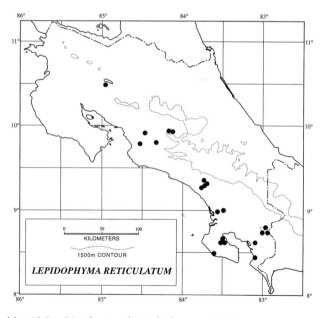

Map 10.51. Distribution of *Lepidophyma reticulatum*.

ear openings and tympana; all skinks lack femoral pores and gular folds. A considerable number of skinks have the limbs variously reduced, but remnants of the pectoral and pelvic girdles are present even in limbless species. The family is found on all continents, major islands, and island groups throughout warm temperate, subtropical, and tropical regions except for northeastern Europe, southern South America, and Cuba.

The following combination of features technically characterizes the family: usually paired premaxillae; vomers usually paired; postfrontal and postorbital usually distinct; supratemporal fenestra usually more or less closed by expansion of postfrontal supratemporal arch, sometimes absent; parietal foramen usually present; jaw tooth replacement basal, with large resorption pits; jaw teeth conical or bicuspid or with spheroid or compressed crowns; Meckel's groove usually open, but sometimes closed and fused; no intercentra associated with dorsal vertebrae; caudal autotomy usually present with autotomy septa usually passing through a single pair of nondiverging, nonconverging transverse processes or rarely anterior to them; extracolumella without internal process; no extracolumella muscle; dorsal, ventral, and separable cephalic osteoderms present; foretongue arrowhead shaped, slightly notched; most of tongue covered by scale-like papillae that have the posterior (free) edge serrate and posterior limbs of tongue covered by plicae; tongue protrusion oscillatory; hemipenes with flounces or petala or nude, usually without calyces, single or bilobed with a simple or bifurcate sulcus spermaticus.

The Costa Rican members of this family are elongate, short-limbed lizards with the head not distinct from the neck and have the body and tail covered with glossy to iridescent cycloid scales underlain by osteoderms. They are unlikely to be confused with any other lizard in the area covered here except for species of the anguid genera *Celestus* and *Diploglossus*. Skinks lack the enlarged median interoccipital scale that borders the interparietal posteriorly in these anguids but instead may have a pair of enlarged nuchals (occipitals) lying posterior to the interparietal and parietal shields.

Skinks are placed in four subfamilies characterized below:

Scincinae Gray, 1825 (thirty-five genera): includes both limbed and elongate limbless forms with moderately long tails (more than 30% of total length); ear opening present or absent; frontal bones paired; palatine bones usually separated on midline; palatal ramus of pterygoid almost always extends to posterior border of infraorbital vacuity, which is bordered by palatine anteriorly; Meckel's groove usually open; movable eyelids usually present, sometimes fused to form a transparent spectacle; one or usually two pairs of enlarged precloacal scales (widely distributed in North and Central America, around the Mediterranean Sea, in Africa, Madagascar, East Asia and the Philippines, and disjunctly in India and Sri Lanka).

Lygosominae Mittleman, 1952 (about eighty-five genera): includes mostly limbed species showing various degrees of limb reduction (only a few lack limbs) with short to moderately long tails (more than 30% of total length, usually more than 50%); ear opening present or absent; a single frontal bone; palatines usually in contact along midline; palatal ramus of pterygoid extends to posterior margin of infraorbital vacuity, which is bordered anteriorly by palatine; Meckel's groove open, closed, or fused; movable eyelids usually present but sometimes fused to form a transparent spectacle; at least one pair of enlarged precloacal scales (the Americas, Africa, and Madagascar, temperate, subtropical, and tropical Eurasia from southeastern Europe to Korea, the Indo-Australian archipelagoes, Australia, New Zealand, and Oceania).

Acontinae Gray, 1845 (three genera): elongate, limbless, very short tailed (tail less than 25% of length) burrowers lacking an ear opening; frontal bones paired; palatine bones separated medially; palatal ramus of pterygoid not reaching posterior margin of infraorbital vacuity, which is bordered by palatine; Meckel's groove closed and fused; eyes small; movable transparent or translucent lower eyelid present, eyelids fused to form immovable transparent spectacle, or eyes vestigial and covered by head shields; a single transversely enlarged precloacal scale (disjunct in Kenya and southern Africa)

Feyliniinae Camp, 1923 (two genera): elongate, limbless burrowers with short tails (about 33% of total length), lacking an ear opening; frontal bones paired; palatine bones separated medially; palatal ramus of pterygoid extends far forward to exclude palatine from infraorbital vacuity; Meckel's groove open; eyes vestigial and covered by scales; precloacal scales subequal (central and West Africa)

Most species of skinks are oviparous, but many (about 40%) are viviparous. There is considerable variation in the types of placentas present and their development. Transient and persistent yolk sac placentas or omphalallantoic ones may be present in different species along with simple to complex chorioallantoic placentas. The eggs are cream to white with flexible shells, but many have definite calcareous deposits. Skinks are "taste" feeders that test possible food items with the tongue tips and sample the particles so obtained with the Jacobson's organs before taking anything into the mouth.

Subfamily Scincinae Gray, 1825
Genus *Eumeces* Wiegmann, 1834a

This large genus (forty-eight species) is composed of small to large species that differ externally from other skinks primarily in details of head scalation. Species of *Eumeces* share the following features that in combination will distinguish them from superficially similar forms: nostril pierced in nasal; large paired supranasals (internasals); a single frontonasal posterior to supranasals; prefrontals and frontoparietals present; a single interparietal; no median interoccipital scale bordering the interparietal posteriorly; the eyelids well developed, with the lower one scaly; enlarged precloacal scales. The members of this genus are typical skinks in having the body covered by uniform imbricate cycloid scales, one row of supraocular scales, a distinct and deeply recessed tympanum, pentadactyl limbs, exposed claws, and transverse lamellae under the digits. Most members of this genus are oviparous, but some are viviparous. In all egg-laying forms, so far as is known, the female remains with the eggs throughout incubation.

Taylor (1935) established the intrageneric classification for the genus by recognizing fifteen species groups. The single Costa Rican form belongs to the *Eumeces schwartzei* group, which contains two extralimital species, *Eumeces altamirani* (Michoacán, Mexico) and *E. schwartzei* (Yucatán Peninsula). Griffith, Ngo, and Murphy (2000) proposed that these species should be placed in a new genus, *Mesoeumeces*. All members of this stock are confined to arid to subhumid lowland habitats.

Eumeces has a fragmented Northern Hemisphere distribution in North and Central America; the semiarid and desert regions extending from North Africa to Pakistan, exclusive of the southern Arabian Peninsula; and Japan and East Asia south to Vietnam.

Eumeces managuae Dunn, 1933

(plate 298; map 10.52)

DIAGNOSTICS: A large diurnal skink; the only Costa Rican lizard with enlarged head shields (fig. 10.12b) and the body covered with smooth cycloid scales, having a dorsal pattern of eleven to fifteen dark brown longitudinal stripes on a lighter brown background.

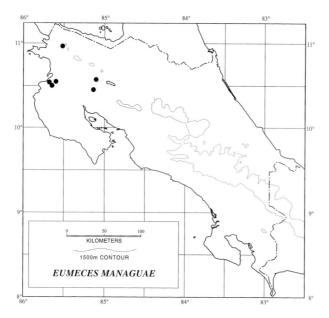

Map 10.52. Distribution of *Eumeces managuae*.

DESCRIPTION: A moderate-size lizard to 282 mm in total length (to 125 mm in standard length); tail moderate, 56 to 59% of total length; limbs very short, widely separated when adpressed along side of body; one frontoparietal, a postnasal, and one postmental present; three presuboculars between second loreal and first enlarged subocular supralabial; large auricular lobule scales; four supraoculars; four pairs of expanded occipitals or nuchals; median middorsal scales greatly expanded, about four times wider than adjacent scales; 17 rows of scales around midbody; 67 to 70 transverse rows of dorsal scales (between interparietal to above vent); subcaudals greatly expanded; two discrete enlarged heel plates and one enlarged plantar scale; 12 to 14 keeled lamellae under fourth toe. Dorsum greenish brown to yellowish brown; dark brown spots on many scales lined up to form eleven to fifteen longitudinal stripes that continue onto tail as five stripes; limbs similarly striped above; head heavily spotted with dark brown above and on sides; palms and soles light in color; venter immaculate white to cream.

SIMILAR SPECIES: There are none.

HABITAT: This diurnal species is uncommon and usually found under rocks in the Lowland Dry Forest and marginally in Moist Forest areas of extreme northwestern Costa Rica. All sites where the species has been collected have been highly modified by humans.

REMARKS: Reeder (1990) redescribed this species and summarized the pertinent literature.

DISTRIBUTION: In subhumid to semiarid lowland areas from extreme southwestern Honduras to northwestern Costa Rica (40–530 m).

Lygosominae Mittleman, 1952

Genus *Mabuya* Fitzinger, 1826

This large (eighty-five species) and geographically widespread genus is represented by fifteen nominal species in the Neotropics, but a number of others will probably be recognized on further study. The American lizards of this stock are moderate sized and share the following features: nostril pierced in nasal; large paired supranasals (internasals); a single frontonasal posterior to supranasals; prefrontals and one or two frontoparietals present; paired interparietals or a single one; no median interoccipital scale bordering interparietal(s) posteriorly; the eyelids well developed, and the lower one with a translucent disk; enlarged precloacal scales. The members of this genus are quintessential skinks in having the body covered with uniform imbricate cycloid scales, one row of supraocular scales, a distinct and deeply recessed tympanum, pentadactyl limbs, exposed claws, and transverse lamellae under the digits. All the American species of this genus in which the reproductive mode has been studied are viviparous and have a persistent omphaloplacenta as well as a chorioallantoic placenta. Both oviparous and viviparous species occur in Africa and Asia.

Mabuya is primarily tropical and subtropical in distribution and occurs over much of Africa and on Madagascar and the Seychelles; in Pakistan and India and over much of South Asia south through the Indo-Malaysian archipelago to the Philippines and New Guinea; in Mexico to Peru and central Argentina; and throughout the Antilles exclusive of Cuba to the southern Bahamas.

Mabuya unimarginata Cope, 1862c

(plate 299; map 10.53)

DIAGNOSTICS: A moderate-sized, relatively long-legged skink having the dorsum some shade of brown, usually with a metallic cast, contrasting sharply with the dark chocolate flanks, which may be bordered above and/or below by light stripes, the upper lip light, a pair of enlarged nuchals (occipitals) (fig. 10.12c) and a transparent disk in the lower eyelid.

DESCRIPTION: Total length to 246 mm, females (adults 62 to 91 mm in standard length) average larger than males (adults 56 to 91 mm in standard length); tail moderate, 62 to 65% of total length; limbs overlapping or barely separate when adpressed along side of body; two frontoparietals, one postnasal, and one postmental; three presuboculars between second loreal and enlarged subocular supralabial; no large auricular lobe scales; four supraoculars; one pair of expanded occipitals or nuchals; dorsal and lateral scales all about same size, mostly smooth or with one or two striations (two or three low ridges); 26 to 32 scales around midbody; 53 to 61 transverse rows of dorsal scales (between interparietal to above vent); ventrals same size as dorsals, smooth; palms and soles covered with small tubercular and juxtaposed scales, proximal row somewhat larger than oth-ers; subcaudals not expanded; 12 to 16 smooth lamellae under fourth toe. Dorsum light to coppery brown, uniform or with small dark spots that sometimes form longitudinal rows or fuse into stripes; flanks dark brown, forming a longitudinal stripe very distinct from dorsal field; usually a light, dark-bordered stripe separating dorsal and flank colors; dark flank stripe continuing forward across eye to tip of snout; a white stripe runs from tip of snout along labials, below eye and ear opening, almost always continuing as a dark-bordered light stripe separating dark flank from venter; upper dorsolateral light stripes often obscure or absent, lower rarely obscure; palms and soles black; ventral surface greenish white, bluish white, or white; iris black.

SIMILAR SPECIES: (1) *Sphenomorphus cherriei* is a much smaller skink that has one frontoparietal plate and the upper lip speckled with dark pigment and lacks both enlarged nuchals (occipitals) and a transparent disk in the lower eyelid. (2) The species of *Celestus* lack lateral white stripes and have a single large prefrontal shield (fig. 10.37a), and the frontoparietals are not in contact (fig. 10.12b). (3) *Diploglossus bilobatus* lacks longitudinal light stripes and has all but the tips of the claws concealed in a scaly sheath (fig. 10.13a). (4) *Mesaspis monticola* has a lateral fold (fig. 10.10) and several rows of the rhomboidal to rectangular middorsal scales strongly keeled.

HABITAT: This very adaptable lizard is found in both humid and subhumid climatic regions in undisturbed and modified Lowland Dry, Moist, and Wet Forests and Premontane Moist and Wet Forest and Rainforest habitats. It ranges marginally into Lower Montane Wet Forest at a few sites. Individuals of this species are found on fallen logs, in pastures, in banana and cacao plantations, among coconut stands, and on fences, rock walls, and bridge abutments as well as in forest stands and second growth.

BIOLOGY: These skinks are very common and strictly diurnal but are secretive and extremely wary. They are never far from a safe retreat in a crevice or hollow log, under bark, or in some other animal's burrow. When approached they immediately and swiftly escape to a hiding place. They are mostly terrestrial but frequently and rapidly climb onto fallen logs or tree trunks for some distance above the ground. Individuals, especially gravid females, are often seen basking on any inclined surface, particularly early in the day. Food consists primarily of arthropods, especially spiders and orthopterans. The species is viviparous, and four to six young are usually born at one time. Evidence (Fitch 1973a) is equivocal regarding the reproductive season, which appears to differ under different climatic regimens. On the Atlantic versant limited data indicate that births are concentrated in March, just before the heavy rains. In the Pacific northwest region reproduction appears to be concentrated in the middle of the rainy season (July to August). The height of breeding apparently occurs three months before these peaks, in December or early January on the Atlantic and in April in

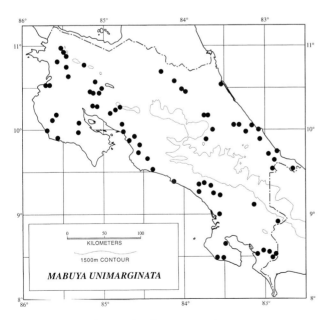

Map 10.53. Distribution of *Mabuya unimarginata*.

the northwest. Available information suggests that in rainforest areas some reproduction may occur throughout the year, but in the strongly seasonal northwest Dry Forest there may be a nonreproductive period of two or three months. Neonates are about 35 mm in standard length and reach sexual maturity at about twelve months of age.

REMARKS: Dunn (1936a) included all Mexican and Central American *Mabuya* within his concept of *Mabuya mabouya*. As visualized by Dunn, this species also ranged south to Ecuador on the Pacific slope of South America, over most of the lowland northern and eastern areas of that continent to southeastern Brazil and Bolivia, and throughout the Antilles, exclusive of Cuba, to the Bahamas. Taylor (1956) regarded the Mesoamerican populations as distinct from *M. mabouya* and recognized three species in the region: an Atlantic slope form, *Mabuya alliacea;* a Pacific slope form presumably ranging from Mexico to Costa Rica, *Mabuya brachypoda;* and a Pacific slope Costa Rica and Panama species, *Mabuya unimarginata.* Taylor's action was partially based on the perception that Dunn's concept of *M. mabouya* was too inclusive, as evinced by the range of variation in scalation and coloration, and on his skepticism that one species would have such a wide geographic range. The latter point has been borne out by recent studies (Rebouças-Spieker 1981a,b; Avila-Pires 1995) demonstrating that Amazonian and northern South American skinks called *M. mabouya* by Dunn and subsequent authors represent several species. Contrary to Taylor (1956), there does not appear to be more than one species of *Mabuya* in Costa Rica. The presumed differences between the putative species are all bridged by individual variation within them or by intermediate conditions in individuals from populations geographically intermediate between their presumptive ranges. My decision to regard the Mesoamerican populations of this complex as distinct is somewhat arbitrary. It is based primarily on the fact that the northern South American and Antillean lizards are larger (to 103 mm in standard length), have fewer transverse rows of dorsal scales, and differ in coloration in life compared with examples from Costa Rica and Panama. Only further study will satisfactorily resolve whether the northeastern South American and Antillean populations (*M. mabouya* sensu stricto) and the Mesoamerican ones are connected by intermediate populations in northwestern South America and the status of Colombia and Ecuador members of this complex. Until that time I propose to use *M. unimarginata* Cope, 1862 (holotype lost, but from Panama, no other data) for the Mexican and Central American populations, since that name has priority over *M. alliacea* Cope, 1875, and *M. brachypoda* Taylor, 1956.

DISTRIBUTION: Lowlands and adjacent slopes from Colima and Veracruz, Mexico, to Panama (1–1,500 m).

Genus *Sphenomorphus* Fitzinger, 1843

This taxon as now recognized includes a great number of species (more than a hundred), many of which are unrelated to one another and are placed in it as a matter of convenience because of similarity in external morphology. Only three small species of American skinks were referred to this genus by Greer (1970)—*Sphenomorphus assatus* (Mexico and northern Central America), *S. cherriei* (Mexico to Panama), and *S. incertum* (Guatemala)—and Myers and Donnelly (1991) have added a fourth, *S. rarus* (western Panama). The Neotropical forms share the following features that in combination distinguish them from other American skinks or other lizards they might be confused with: nostril pierced in the nasal; a single large frontonasal (internasal) between nasals; no supranasals; paired prefrontals and a single frontoparietal; a single interparietal; no median interoccipital scale bordering the interparietal posteriorly; eyelids well developed, lower one scaly or with a translucent disk; enlarged precloacal scales. Like other Neotropical skinks, these lizards have the body covered with uniform imbricate cycloid scales, a single row of supraocular scales, a distinct recessed tympanum, pentadactyl limbs, digits with exposed claws, and transverse lamellae under the digits. The Neotropical species of *Sphenomorphus* in which reproduction is known *(S. assatus* and *S. cherriei)* are oviparous. Most Old World forms are viviparous.

Except for the three Mesoamerican forms, other species placed in this composite genus range from Pakistan, India, and eastern China throughout Southeast Asia and the Indo-Australian region to New Guinea, Australia, and the Philippines.

Sphenomorphus cherriei (Cope, 1893)

Chirbala Lisa

(plate 300; map 10.54)

DIAGNOSTICS: A small diurnal skink inhabiting the leaf litter, having the upper lip spotted with dark pigment, the dorsum bronze, lacking enlarged nuchals (occipitals) (fig. 10.2a) and usually lacking a transparent disk in the lower eyelid.

DESCRIPTION: Total length to 178 mm, females (adults 49 to 68 mm in standard length) larger than males (adults 50 to 58 mm in standard length); tail moderate, 59 to 62% of total length; limbs overlapping or barely separated when adpressed along side of body; one frontoparietal; one postnasal and a single postmental; two presuboculars between second loreal and enlarged subocular supralabial; no enlarged auricular lobule scales; four supraoculars; no expanded nuchal or occipital scales; dorsal and lateral scales all about same size, smooth; 29 to 34 (26 to 34) scales around midbody; 60 to 72 (54 to 72) transverse rows of dorsal scales (between interparietal to above vent); ventrals same size as dorsals, smooth; palms and soles covered with small smooth flattened scales, proximal row largest; subcaudals not expanded; 13 to 19 smooth lamellae under fourth toe. Dorsum bronze, appearing uniform but with many tiny black dots; a black stripe begins behind nostril and passes through

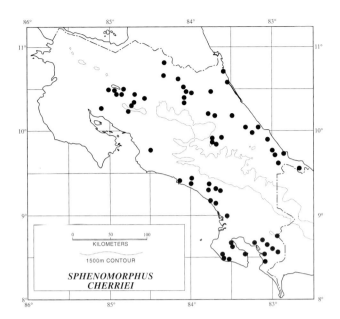

Map 10.54. Distribution of *Sphenomorphus cherriei*.

the orbit, widens postorbitally, and continues onto neck or farther onto the body, where it becomes diffuse posteriorly; bordered below by a narrow cream light stripe that also tends to disappear posteriorly; flanks often with many small yellow spots; upper lips heavily speckled with dark pigment; palms and soles dark brown; ventral surface immaculate yellow; sometimes with salmon suffusions under the hind limbs and tail base; upper surface of tail blue to black in small juveniles, bronze in other individuals; undersurface of tail yellowish but tip gray; iris black.

SIMILAR SPECIES: (1) *Mabuya unimarginata* is a much larger skink having an immaculate white stripe along the upper lip, two frontparietals, a pair of enlarged nuchals (occipitals) (fig. 10.12c), and a transparent disk in the lower eyelid. (2) *Gymnophthalmus speciosus* has a red tail, has only four fingers, and lacks movable eyelids. (3) The species of *Celestus* are much larger lizards that have a single large prefrontal shield and paired frontonasals (fig. 10.37a). (4) *Diploglossus bilobatus* is a much larger lizard that has the claws mostly hidden within a scaly (fig. 10.13a) sheath and a single large prefrontal shield (fig. 10.37a). (5) *Mesaspis monticola* is also a larger lizard with lateral (fig. 10.10) folds and several keeled rows of middorsal scales.

HABITAT: This forest floor species prefers forest edge and disturbed situations to intact forests, primarily within Lowland Moist and Wet Forest regions. It also occurs marginally in Premontane Wet Forest and Rainforest habitats and gallery forest corridors in the eastern portions of the Lowland Dry Forest area.

BIOLOGY: Fitch (1973a,b, 1983) has studied this common skink in considerable detail, and his accounts and that of Lieberman (1986) provide much of the data summarized below. *Sphenomorphus cherriei* is diurnal and typically found in leaf litter, carrying out most of its activities

under cover of surface debris. Like the gecko *Lepidoblepharis xanthostigma* and the several microteiids inhabiting the leaf litter, these lizards are more likely to be heard rustling under the leaves than to be seen. At some localities they may also occur in low grassy or herbaceous vegetation, where they have places of concealment under debris or in holes or crevices. Diel activity is centered on midmorning and midafternoon, and *S. cherriei* avoids high midday temperatures. These lizards rarely appear until the morning sun warms their hiding places, and after emergence they may bask for brief periods in patches of sunlight. Body temperatures during active periods are usually 25 to 27°C, about the same as the environment. These lizards are extremely furtive and almost always have a portion of the body or tail partially concealed under a leaf or similar object, whether resting, basking, or moving about. If approached while on the surface, they immediately dart under cover, fold the limbs against the body, and "swim" through the litter with a serpentine movement to reach some hiding place. Since the temporary hideout may consist merely of a single leaf or other superficial surface debris, stomping or kicking the litter or striking the surface with a stick or machete will often flush them out. The lizard will then dash away 0.5 to 2 m and dive under cover again. When you walk on the forest floor these skinks frequently pop out from underfoot and seek new hiding places as described above. Densities average 38/ha in suitable forest habitats, with higher values (104/ha) in abandoned cacao plantations (Lieberman 1986). Fitch (1983) estimated that home ranges average about 300 m² for males and 476 m² for females. This situation is unusual in that in most lizards males have larger home ranges than females. There appears to be no home site or definite territorial behavior. Food consists of a variety of arthropods (to 20 mm in length), especially spiders (mean 2.4 mm), isopods (mean 6.1 mm), and homopterans (mean 16.5 mm), but the species avoids mites, ants, and beetles almost completely. Stomachs usually contain no more than two relatively large food items. Reproduction may be continuous throughout the year, but there is a definite decline during January to February in the rainforest climates of the Atlantic versant. Fitch (1973b) noted a longer period of low reproduction (January to April) near Turrialba, Cartago Province (602 m), and it seems likely that the relative length of the dry season at different sites affects the reproductive pattern. One to three (usually two) eggs are laid at a time, and a female probably lays several clutches each year. Hatchlings (22 mm in standard length) are very large (about 40% of adult length). They grow rapidly and reach sexual maturity in seven to eight months. Fitch (1983) reported that a *Norops lemurinus* had swallowed one of these lizards whole.

REMARKS: This species had been referred under the generic names *Leiolopisma*, *Lygosoma*, and *Scincella* by various authors during the past fifty years until Greer (1970) placed it in *Sphenomorphus*. Taylor (1956) called it *Leiolopisma cherriei*. Several geographic races have been recognized by previous authors (Smith 1939a; Smith and Taylor 1950; Taylor 1956; Stuart 1963). Purported differences among the several forms included the number of dorsal scales and the arrangement of temporal scales. When large series are examined, individual and sexual dimorphism in the dorsal counts and individual variation in the temporal scale patterns used to differentiate two Costa Rican races, *S. cherriei cherriei* (Pacific) and *S. cherriei lampropholis* (Atlantic), provide no means for distinguishing geographic subdivisions over most of the species' range. The one exception is those specimens from the outer two-thirds of the Yucatán Peninsula that have 26 to 28 scales around the midbody and 54 to 60 transverse rows of dorsal scales versus 30 to 34 and 59 to 72, respectively, in other samples. Stuart (1940) noted intermediate conditions in these characters (in reality, only one feature is involved—the size of the dorsal scales) from an intermediate location, suggesting a clinal decrease in dorsal scale number from the base of the peninsula northward to its tip. Myers and Donnelly (1991) reported that rare examples (2 out of 193) from Costa Rica have a translucent disk in the lower eyelid.

DISTRIBUTION: Lowlands and adjacent premontane areas from central Veracruz, Mexico, on the Atlantic versant and Costa Rica on the Pacific slope to extreme western Panama; in Costa Rica throughout the Atlantic versant and in the Pacific southwest but also occurring in gallery forests along streams in the northwest Pacific region (2–1,170 m).

Family Teiidae Gray, 1827

Macroteiid Lizards, Chirbalas

A moderate-sized family (nine genera and 116 species) of medium to very large insectivorous, carnivorous, or herbivorous diurnal lizards with relatively long, narrow heads that are barely distinct from the neck, well-developed limbs, and long tails. Most species are terrestrial, but two South American genera *(Crocodilurus* and *Dracaena)* are semiaquatic. In addition, the head is covered with large symmetrical plates and the dorsum by granules or larger imbricate scales or a mixture of granules and enlarged juxtaposed rectangular plates or smaller elongate, more or less quandrangular scales; the scales are never cycloid. The ear opening is moderate to large, and the tympanum is exposed and slightly recessed. The eyes are moderate, with movable eyelids and round pupils. Femoral pores and two complete gular folds are present.

The family is restricted to the Americas, ranging from the northern United States to central Chile and central Argentina, including the West Indies. Six of the genera occur only in South America.

Technically the Teiidae are characterized as follows: premaxilla single; vomers paired; postfrontal small, absent, or fused to postorbital; supratemporal fenestra present; parietal foramen absent in adults; jaw tooth replacement usually basal, with large resorption pits, sometimes alternating with reduced resorption pits; jaw teeth heterodont, conical anteriorly and conical, biconodont, triconodont, or molariform posteriorly; Meckel's groove open or closed without fusion ventrally; no intracentra associated with dorsal vertebrae; caudal autotomy present in most forms; autotomy septa of anterior autotomic vertebrae typically passing between double pair of diverging transverse processes, a series of posterior vertebrae with single pair of transverse processes anterior to septum; followed by the most posterior vertebra lacking processes; both anterior and posterior autotomic vertebrae with single posterior processes anterior to the septum in one genus *(Teius)* or no processes on autotomic vertebra *(Dicrodon)*; extracolumella with an internal process; no extracolumella muscle; no dorsal, ventral, or separable cephalic osteoderms; nasal plates not separated by frontonasals; tongue deeply bifurcate for 40 to 50% of length; most of tongue covered by flattened scalelike papillae arranged in regular diagonal rows, posterior limbs of tongue covered by plicae; a single-layered tongue sheath sometimes present; tongue protrusion oscillatory; truncus and apical region of hemipenes petalate, without calyces or flounces, single or usually slightly bilobed, with a simple sulcus spermaticus.

The combination of movable eyelids, enlarged head plates, granular dorsals, and large rectangular ventrals will distinguish Costa Rican representatives of the family from all other lizards found there.

Two subfamilies (Presch 1974, 1983) are recognized within the Teiidae and are characterized as follows:

Teiinae Gray, 1827 (five genera): postfrontal and postorbital fused; pterygoid flange high; quadrate process of pterygoid expanded; dorsal squamosal process present; clavicles expanded and perforate; scapular foramen present

Tupinambinae Presch, 1974 (four genera): prefrontal and postorbital usually separate; pterygoid flange low; quadrate process of pterygoid not expanded; dorsal squamosal process absent; clavicle simple, rod shaped, with or without hooks

The Costa Rican genera of this family belong to the Teiinae, which also include the herbivorous *Dicrodon* (three species) of western Ecuador and Peru. Members of the other subfamily (all South American) include the large (to 1.4 m) tegus *(Tupinambis)*, and smaller (to 1 m) *Callopistes*, which are carnivores that prey mainly on small vertebrates, and the semiaquatic *Crocodilurus* and *Dracaena*. The latter is a snail specialist and has the lateroposterior teeth molariform to aid in smashing the prey's shell.

Macroteiids are "taste" feeders that sample possible food items with the tongue tips, then transfer the particles to the openings of the Jacobson's organs for evaluation before taking anything into the mouth. All macroteiids are oviparous and lay eggs with cream to white flexible shells.

The two genera represented in Costa Rica, *Ameiva* and *Cnemidophorus*, differ only in the morphology of the tongue and related structures. Superficially the species placed in these genera resemble one another in overall proportions, scale characteristics, and habits. In addition, most exhibit considerable ontogenetic and sexual variation in coloration, making immediate identification difficult. Consequently the following key is based on external features, designed to distinguish among the species of both genera known or likely to occur in the country.

KEY TO THE COSTA RICAN SPECIES OF *AMEIVA* AND *CNEMIDOPHORUS*

1a. Ventrals in 8 longitudinal rows 2
1b. Ventrals in 10–12 longitudinal rows; 4–5 light stripes on body in juveniles and adult females, stripes completely replaced by light spots in adult males; to 197 mm in standard length . . . . *Ameiva ameiva* (p. 508)

2a. A single parietal on each side; usually one interparietal, sometimes two (fig. 10.31b,c) 3
2b. Two parietals on each side (rarely incompletely divided); one interparietal (fig. 10.31a); 6–10 narrow light stripes on body in juveniles and adult females, stripes replaced by light spots on flanks in adult males; to 113 mm in standard length . . . . *Cnemidophorus lemniscatus* (p. 516)
3a. Gular scales posterior to level of enlarged central scales, abruptly and markedly smaller than anterior gular scales (fig. 10.32b,c) 4
3b. Anterior and posterior gular scales subequal in size; a gradual reduction in scale size radiating out from enlarged central gular scales (fig. 10.32d,e) 6
4a. Interparietal shield in broad contact with frontoparietals and parietals (fig. 10.31c); prefrontal in contact with

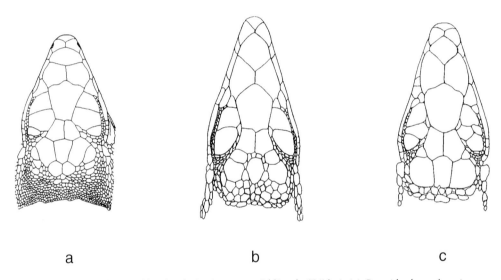

a b c

Figure 10.31. Dorsal view of head scalation in macroteiid lizards (Teiidae): (a) *Cnemidophorus lemniscatus*, five shields in parietal region; (b) *Ameiva leptophrys*, three large shields in parietal region and parietal separated from frontoparietals and interparietal shields by small scales; (c) *A. quadrilineata*, three large shields in parietal region in contact with frontoparietals and parietals in contact with interparietal shield.

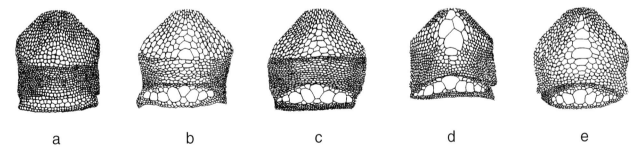

a b c d e

Figure 10.32. Throat and neck region in macroteiid lizards (Teiidae): (a) *Ameiva ameiva*, no enlarged scales on gular collar, no enlarged median gular scales, scales between antegular and gular folds very small; (b) *Ameiva leptophrys*, enlarged scales on gular collar, small scales between folds abruptly contrasting in size with enlarged median gular scales; (c) *Ameiva quadrilineata*, enlarged scales on gular collar, small scales between folds abruptly contrasting with enlarged median gular scales; (d) *Ameiva festiva*, enlarged scales on gular collar, scales between folds gradually decreasing in size away from greatly enlarged median gular scales; (e) *Ameiva undulata*, enlarged scales on gular collar, scales between folds gradually decreasing in size away from enlarged median gular scales.

a   b

Figure 10.33. Scales (brachials) on posterior surface of upper arm in macroteiid lizards (Teiidae): (a) *Cnemidophorus*, several enlarged rows; (b) *Ameiva*, a single row of greatly enlarged scales.

postnasal; middorsal dark field usually with even margins and without faint dark irregular crossbars..... 5

4b. Interparietal shield separated from broad contact with frontoparietals and/or parietals by one to several rows of small scales (fig. 10.31b); prefrontal nearly always separated from postnasal; middorsal dark field usually with scalloped margins and marked by faint irregular dark crossbars; to 133 mm in standard length ....*Ameiva leptophrys* (p. 511)

5a. Several rows of enlarged brachial scales (fig. 10.33a); dorsal pattern of 5–10 longitudinal light stripes, 5 or more retained in adults; to 92 mm in standard length .... *Cnemidophorus deppii* (p. 515)

5b. One row of enlarged brachial scales (fig. 10.33b); usually a pair of continuous well-defined lateral and more or less continuous ventrolateral light stripes running length of body maintained throughout life; never a middorsal light stripe; to 88 mm in standard length .... *Ameiva quadrilineata* (p. 512)

6a. A distinct vertebral light stripe in all but very large adults; flanks marked with numerous light short longitudinal lines or spots; to 129 mm in standard length ....*Ameiva festiva* (p. 510)

6b. No vertebral light stripe; flanks marked by distinct vertical dark and light bars in adults; to 129 mm in standard length ....*Ameiva undulata* (p. 513)

Genus *Ameiva* Meyer, 1795

A large genus (thirty-three species) of moderate-sized to moderately large (to 640 mm in total length) active terrestrial and diurnal lizards with well-developed limbs and long tails. In addition, the species of *Ameiva* have the snout pointed, a homogeneous dorsal scalation of small smooth granules, large smooth ventrals, and the tongue not deeply notched posteriorly and retractile into a basal sheath. Other distinguishing features for the genus include pupil of eye reniform (kidney-shaped), indented ventrally; femoral pores present, no precloacal pores; a definite collar bounded by anterior and posterior gular folds; a single row of enlarged brachials; five fingers and toes; outer margin of fourth toe not serrate; pterygoid teeth usually absent; premaxillary teeth conical; posterior teeth on jaws compressed, biconodont or triconodont; single pair of processes posterior to

autotomy septa on anterior series of posterior autotomic caudal vertebrae.

Among Costa Rican members of the genus differences in dorsal color patterns may be used to separate the several species from one another and the species of *Cnemidophorus*. Using these characters is complicated by ontogenetic change and by differences between the sexes. The basic pattern in juveniles consists of a series of narrow longitudinal light stripes separating darker fields. Typically there is a broad middorsal field bordered on each side by a lateral light stripe. The lateral light stripe originates on the superciliary scales above the eye and continues posteriorly onto the side of the tail. The area lateral to this light stripe is dark brown to black and is bordered below by a ventrolateral light stripe. The ventrolateral light stripe runs from behind the eye through the ear and above the forelimb insertion and continues to the groin area or slightly beyond onto the tail. A ventrolateral dark field lies between this stripe and the margin of the venter and may be uniform in color or marked with light spots. The area between the lateral and ventrolateral light stripes is much darker than the middorsal area. A vertebral light stripe may also be present in addition to the paired lateral and ventrolateral stripes. In some species (*Ameiva festiva*) the lateral and ventrolateral stripes are discontinuous and form a series of light dashes and spots running parallel to the body axis in both juveniles and adults.

This pattern changes with age. Adult females of all species tend to retain light stripes to a greater extent than males. The ventrolateral stripe frequently becomes discontinuous or is lost entirely; light spots or bars often develop in the lateral dark field, and the ventrolateral field usually becomes marked with light or dark spots or blotches and/or vertical bars in females. Small males often resemble adult females in coloration. Adult males of some species lose all indications of light stripes and have numerous light spots over the entire body or a pattern of distinct vertical dark and light bars on the flanks.

Because of their similarities and the range of variation in external morphology and coloration in *Ameiva* and *Cnemidophorus*, it may be hard to decide which genus a particular specimen belongs to. Costa Rican *Ameiva*, however, never have more than one row of greatly enlarged collar scales and never more than five longitudinal light stripes on the body.

The genus *Ameiva* has a wide distribution at lower elevations in subtropical and tropical America, from Nayarit and Tamaulipas, Mexico, south to Ecuador west of the Andes and to northern Argentina east of the Andes; also found throughout the Antilles and the Bahama Islands; one species (*Ameiva ameiva*) has been introduced into southern Florida.

Ameiva ameiva (Linné, 1758)

(plate 301; map 10.55)

DIAGNOSTICS: Adult males of this large lizard are unmistakable because of their dorsal pattern of numerous greenish yellow to orange spots on a blue green to brown ground color. Juveniles and females have two narrow creamy white lateral and two ventrolateral stripes and a broader dull gray green to brown vertebral stripe. In adult females the areas between the stripes are marked with numerous light spots. The combination of five large shields in the parietal row (fig. 10.31a) combined with a longitudinal ventral count of 10 to 12 will conclusively separate *A. ameiva* from all other macroteiids in Central America.

DESCRIPTION: Total length to 640 mm; adult males 90 to 197 mm and adult females 80 to 157 mm in standard length; tail long, 65 to 70% of total length; head not distinct from neck except in large males with hypertrophied jaw muscles; head plates smooth; nostril in nasal suture; postnasals touch a prefrontal; a single frontal; four supraoculars; surpraorbital semicircles extending no farther than posterior margin of frontal; inner parietals in broad contact with interparietal and frontoparietals; five large plates in parietal row; anterior gular scales much larger than posterior ones; scales in midgular patch only slightly enlarged, irregular; collar scales on posterior gular fold in two rows, subequal in size to posterior gular scales; central scales on posterior surface of forearm slightly enlarged; ventrals in 10 to 12 longitudinal rows; 29 to 40 (25 to 41) femoral pores; 26 to 36 lamellae under fourth toe. Juveniles: dorsum marked by pair of lateral and ventrolateral narrow cream to yellow longitudinal stripes; often a broader dull brown vertebral stripe as well; area between lateral and vertebral stripes with many small whitish spots; area below lateroventral light stripes with similar small light spots; tail brown; females: middorsal area marked with irregular, incomplete dark bars, and white spotting increased on sides; adult males: stripes replaced by pattern of numerous greenish to orange spots;

back often some shade of green in all ontogenetic stages; throat orange in reproductive adults; venter white to pale blue; tail brown to bluish in adults; upper surface of thighs mottled with light and dark in smaller examples, spotted in large males; iris light brown in all stages.

SIMILAR SPECIES: (1) Juvenile *Cnemidophorus* have nine to ten light stripes on the body and greatly enlarged gular collar scales (fig. 10.4b). (2) Juvenile *Ameiva quadrilineata* lack any suggestion of light dorsal or lateral spotting and have only three enlarged shields in the parietal region (fig. 10.31c).

HABITAT: Elsewhere in its range this species is found in open areas including grasslands and savanna habitats. The single Costa Rican record is from a banana plantation in the Lowland Moist Forest area of the extreme southwestern part of the country.

BIOLOGY: Like other members of the genus, *Ameiva ameiva* is a sun-loving terrestrial species that feeds principally on a variety of arthropods. Heatwole (1966) suggested that the distribution of this lizard is expanding in association with land clearing, and the species may have only recently dispersed into southwestern Panama and Costa Rica in association with the establishment of extensive banana cultivation. Clutch size is three to five, with four the usual complement. Hatchlings appear in large numbers in the wet season (May to November).

REMARKS: The color pattern of adult males of this species emphasizes light spots, whereas that of females is a combination of light spots and longitudinal light stripes. Peters and Donoso-Barros (1970) recognized nine mainland geographic races, and Schwartz and Henderson (1988, 1991) recognized an additional western Caribbean insular form based on putative differences in adult coloration. Vanzolini (1986), however, concluded that there was little basis for recognizing any of these forms, presumably because of the substantial individual, ontogenetic, and sexual variation. Echternacht (1971) noted that individuals of this species from central Panama retain a distinct vertebral light stripe as adults that is lacking in those from western Panama (and in the single Costa Rican example). The coloration described above applies to Costa Rican and western Panama examples. Variations that occur elsewhere in the range of the species include dorsum and head grass green, changing to brown anteriorly in large adult males; flanks brown, tan, or blue; light spots cream yellow, greenish to bluish, often outlined by dark; spots or ocelli regularly or irregularly arranged continuously across back or interrupted by a broad brown or green dorsal field; ocelli often fusing to form a dark reticulum on head, neck, and anterior part of body in adults; throat and chest pale blue, spotted or suffused with black pigment; venter white, cream, gray, blue, or orange. *A. ameiva* is apparently absent from the heavily forested areas of eastern Panama and adjacent northwestern Colombia (Heatwole 1966; Echternacht 1971). The karyotype of this species is 2N = 50, comprising thirteen pairs of telocen-

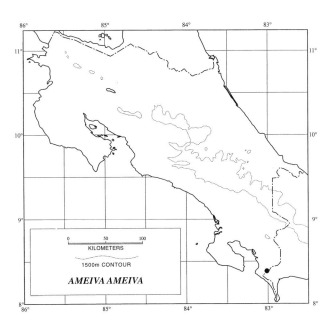

Map 10.55. Distribution of *Ameiva ameiva*.

tric macrochromosomes and twelve pairs of microchromosomes; NF = 76 (Gorman 1970).

DISTRIBUTION: In the lowlands from extreme southwestern Costa Rica along the Pacific versant to central Panama, where it occurs on both coasts in populations disjunct from those in northern and central Colombia; found more or less continuously from Colombia and northern Venezuela southward over most of South America to southern and southeastern Brazil, Paraguay, and northern Argentina; also on Trinidad and Tobago, the southern lesser Antilles, and Swan Island, Islas San Andrés, and Providencia; introduced to southern Florida; known in Costa Rica from Laurel, Puntarenas Province (23 m).

Ameiva festiva (Lichtenstein and von Martens, 1856)

(plates 302–4; map 10.56)

DIAGNOSTICS: All examples of this moderate-sized species except the largest males have a well-developed vertebral stripe. In juveniles the stripe is bright blue and extends down the center of the tail nearly to the tip. During ontogeny it becomes greenish blue and finally yellow in adults and gradually disappears from posterior to anterior on the tail. All examples from Costa Rica have the dorsolateral area marked by numerous short longitudinal yellow dashes or spots arranged in two or three parallel rows along the length of the body.

DESCRIPTION: Total length to 345 mm; males (adults 85 to 129 mm in standard length) larger than females (adults 78 to 129 mm in standard length); tail long, 65 to 70% of total length; head not distinct from neck; head shields smooth; nostril in nasal suture; postnasals in contact with a prefrontal; a single frontal; usually three supraoculars; three large plates in parietal region; parietals in broad contact with interparietal, and usually in contact with frontoparietals but often separated from them by small scales; supraorbital semicircle scales extending no farther forward than posterior margin of frontal; anterior gular scales grading in size into posterior ones; scales in midgular series very large, irregularly arranged, much larger than enlarged collar scales; enlarged collar scales in a single row on gular fold; one row of greatly enlarged central scales on posterior surface of forearm; eight longitudinal rows of ventrals; 32 to 55 femoral pores; 23 to 34 lamellae under fourth toe. Juveniles (plate 304): middorsal field dark brown with a blue vertebral stripe running from tip of snout almost to tip of tail, fading to white or yellow with age; tail entirely bright blue in smallest examples; paired rows of lateral and ventrolateral yellow to orange discontinuous longitudinal stripes in smallest examples, fragmented into dashes or spots in larger juveniles; undersurface grayish white; adult females: similar to juveniles but vertebral stripe becoming sinuous and turning yellow to orange anteriorly and becoming faint on posterior body and tail in large examples and only lateral series of spots present in some cases; throat turquoise; venter pale bronze; adult males: vertebral bluish white to blue stripe becoming yellow and sinuous, gradually fading with age to disappear in largest examples; lateral and ventrolateral rows of lines present posteriorly only in largest males, or only lateral series indicated; chin and venter usually pale blue; throat and chest orange in reproductive individuals; tail brown in adults; upper surface of thigh not distinctly light spotted; iris brown with gold flecks in all stages.

SIMILAR SPECIES: (1) *Ameiva undulata* never has a vertebral light stripe and lacks the series of light lateral and ventrolateral dashes. (2) *Ameiva quadrilineata* is smaller, always lacks a vertebral light stripe, and usually has a continuous pair of lateral and/or ventrolateral light stripes retained throughout life.

HABITAT: An inhabitant of forested regions in the humid lowlands, primarily in relatively undisturbed areas of Lowland Moist and Wet Forests but also in banana, cacao, and coconut plantations. Marginally in lower portions of Premontane Wet Forest and Rainforest at a few sites.

BIOLOGY: *Ameiva festiva* is a very common, extremely active and nervous lizard that constantly moves across the ground with quick, jerky movements. The head and flicking tongue are always probing and sampling the substrate environment. Both the snout and forelimbs are used in intermittent bouts of digging in search of food. Although it is strictly diurnal and basically a forest species, its activity is concentrated in the midmorning on sunny days. The various aspects of the biology of the species in Costa Rica have been reported on by Smith (1968a,b), Hillman (1969), Fitch (1973a,b), and Echternacht (1983) and in Nicaragua by Vitt and Zani (1996) and their observations form the basis for parts of the following account. These lizards attain high activity temperatures of 34–37.7–40.5°C by basking along the edge of clearings or in sunny spots within the forest. On

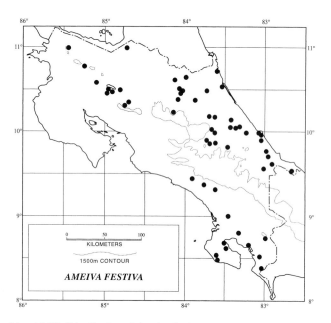

Map 10.56. Distribution of *Ameiva festiva*.

cloudy or rainy days they seldom emerge from their retreats down burrows or under logs or debris, and they normally retire early in the afternoon even on sunny days. During the wet season they may remain inactive for long periods (a week or so) during the heaviest rains. When approached they run quickly away and disappear under cover. Food consists of a variety of small arthropods, primarily orthopterans and araneid spiders, but they also eat small amphibians and probably other lizards. Food sizes range up to 40 mm in length, and stomachs (with food items) contain an average of four items. Reproduction appears to occur throughout the year, with greatest egg production and hatching success during the rainy season (July through September). Eggs are laid about eighty days after fertilization. Females lay three or four clutches of one to four eggs per year. Hatchlings are about 37 mm in standard length and reach sexual maturity between three six months after hatching.

REMARKS: As Echternacht's (1971) analysis shows, there is little basis for recognizing geographic subdivisions within this species, since most variation involves nonconcordant clinal trends, often interrupted by erratic variation in values for one or more populations. Examples from the Atlantic slopes of Mexico and most of Central America to western Panama have the lateral dark field between the lateral and ventrolateral rows of light dashes black and the vertebral stripe relatively narrow. Taylor (1956) regarded these populations as typical *Ameiva festiva*. Lizards of this species from southwestern Costa Rica, Pacific slope western and eastern Panama and Colombia, east of the Cordillera Central, have a broad vertebral stripe (sometimes as broad as the entire middorsal area); the dark lateral area is brown, and the ventrolateral row of dashes is often absent or indicated only posteriorly. Taylor (1956) described a new form, *A. f. occidentalis*, based on Costa Rican examples with this coloration. Examples of *A. festiva* from western Colombia retain a narrow vertebral stripe with even margins as adults, and the lateral and ventrolateral stripes may be continuous or consist of spots. Echternacht (1971) noted a population from Isla Escudo de Veraguas, Bocas del Toco Province, off northwestern Panama, in which the adult pattern is obscured by melanin.

DISTRIBUTION: Humid lowlands on the Atlantic versant from Tabasco, Mexico, to northern Colombia and on the Pacific slope from western Costa Rica and southwestern, central, and eastern Panama to western Colombia; in Costa Rica found throughout the Atlantic and southwest Pacific lowlands and marginally on the Pacific slope in the northwest (2–1,200 m). Also in Premontane Wet Forest or Rainforest at some sites.

Ameiva leptophrys (Cope, 1893)

(plate 305–6; map 10.57)

DIAGNOSTICS: A moderately large lizard characterized by having a broad middorsal area with distinct scalloped margins that is lighter than the flanks and usually marked

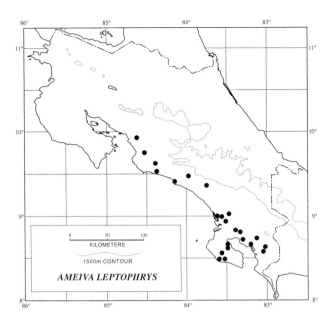

Map 10.57. Distribution of *Ameiva leptophrys*.

with faint irregular dark crossbars and lacking a light vertebral stripe. Unlike other macroteiids in Central America, *A. leptophrys* has many irregular, small scales in the parietal region that prevent contact between the parietals and interparietal and/or the interparietal(s) and frontoparietal scales (fig. 10.31b).

DESCRIPTION: Total length to 439 mm, males larger (adults 83 to 133 mm in standard length) than females (adults 81 to 129 mm in standard length); tail long, 65 to 70% of total length; head not distinct from neck except in very largest males; nostril in nasal suture; head shields smooth; postnasals usually not in contact with a prefrontal; a single frontal; usually three supraoculars; three or four enlarged scales in parietal area (two parietals and one or two interparietals); one large to many small scales in the parietal area, separating parietals from frontoparietals; scales in supraorbital semicircles usually extending no farther than posterior margin of frontal, sometimes reaching anterior frontal margin; anterior gular scales much larger than posterior gulars; largest scales in midgular series very large, about same size as enlarged collar scales; enlarged collar scales in single row on posterior gular fold; central scales on posterior surface of forearm small or slightly enlarged and in one or two rows; ventrals in eight longitudinal rows; 34 to 61 femoral pores; 24 to 35 lamellae under fourth toe. Juveniles (plate 306): lateral light stripe bordering broad middorsal field on each side undulating at least posteriorly; lateral dark field black, bordered below by discontinuous narrow light ventrolateral stripe; undersurface white cream or coppery; females: similar to juveniles but with discontinuous, indications of lateral dark bars on flanks; adult males: strongly marked with dark bars that extend from the metallic brown middorsal field to venter; separated by irregular yellowish spots and bars; light stripes reduced or lost;

ventral surface of body and tail white, cream, or coppery or sometimes blue; chin and throat yellow to brick red in reproductive males; tail brown in all stages; upper surface of thighs same color as dorsum, uniform or mottled with darker pigment in all age-groups.

SIMILAR SPECIES: (1) *Ameiva undulata* does not have the scales of the parietal region greatly fragmented (fig. 10.31a), and the broad middorsal field has even margins. (2) *Ameiva quadrilineata* usually retains lateral and ventrolateral light stripes throughout life and has only three large scales in the parietal area (fig. 10.31c).

HABITAT: *Ameiva leptophrys* is a rainforest species most commonly found where some relatively undisturbed Lowland Moist or Wet Forest stands or patches remain.

BIOLOGY: This strictly diurnal terrestrial lizard forages in the shadiest areas of the forest. It is heliothermic and basks in patches of sunlight within the forest or under dense low vegetation on forest margins or along trails. *A. leptophrys* is most active in midmorning after warming up. These common lizards are less nervous than sympatric *A. festiva* and may be approached within 1 m or so before bolting. They are frequently seen foraging around fallen logs or tree buttresses and may seek cover in these locations when disturbed. Body temperatures for active lizards are 35–37.2–41°C (Hillman 1969). Small individuals are most active, forage later in the morning than adults, and have their activity peak near midday. Food consists primarily of small arthropods (6 to 30 mm in length). Hatchlings are about 42 mm in standard length and appear throughout the wet season (May to November).

REMARKS: The Costa Rican–southwestern Panamanian population is completely allopatric to those from farther east in Panama.

DISTRIBUTION: Lowland humid forests of southwestern Costa Rica and adjacent southwestern Panama, the Península de Azuero, and on both Atlantic and Pacific slopes in central and eastern Panama to western Colombia; the northernmost record (USNM 37754) is from Orotina, Alajuela Province, Costa Rica (2–703 m).

Ameiva quadrilineata (Hallowell, "1860," 1861)

(plate 307; map 10.58)

DIAGNOSTICS: This is the smallest species of macroteiid in Costa Rica. Typically it has a pair of well-developed continuous narrow longitudinal lateral and ventrolateral light stripes and a dark brown to black lateral field between them. The middorsal area is light and its margins are even, but there is never a distinct vertebral light stripe. In addition, the posterior gular scales are much smaller than the anterior gulars and abruptly differentiated from them (fig. 10.32c).

DESCRIPTION: A moderate-sized lizard, to 283 mm in total length; males (adults 66 to 88 mm in standard length) larger than females (adults 62 to 82 mm in standard length); tail long, 67 to 69% of total length; head not distinct from

neck, except in large males; nostril in nasal suture; head shields smooth; postnasals in contact with a prefrontal; a single frontal; usually three supraoculars; an enlarged interparietal and two large parietals; parietals in broad contact with interparietal and frontoparietals; scales in supraorbital semicircles extending no farther than posterior margin of frontal; anterior gular scales much larger than posterior gulars; largest scales in midgular patch smaller than enlarged collar scales; enlarged collar scales in a single row on posterior gular fold; central scales on posterior surface of forearm moderately enlarged, irregular or in one row; ventrals in eight longitudinal rows; 18 to 42 femoral pores; 25 to 34 lamellae under fourth toe. Juveniles: vividly marked with two pairs (lateral and ventrolateral) of continuous narrow longitudinal yellow stripes; middorsal field light brown, with even margins; lateral fields black; venter coppery to white; tail blue; females: similar to juveniles but often with darker flecking in middorsal field; venter white to coppery; throat usually with some yellow; males: lateral field rust, flanks brown; ventrolateral stripe may be discontinuous; ventrolateral dark field with grayish reticulations in lower half; venter white, coppery, or blue; throat region with or without heavy dark spotting in all stages; throat yellow to bright orange in reproductive males; tail brown in adults; upper surface of thighs without light spots in all cases.

SIMILAR SPECIES: (1) The much larger *Ameiva undulatus* usually has the flanks strongly barred, lacks longitudinal light stripes except in very young examples, and has anterior and posterior gular scales not abruptly differentiated in size (fig. 10.32d). (2) *Ameiva leptophrys* has scalloped margins to the middorsal field and many small scales in the parietal region (fig. 10.31b). (3) *Ameiva festiva* usually has

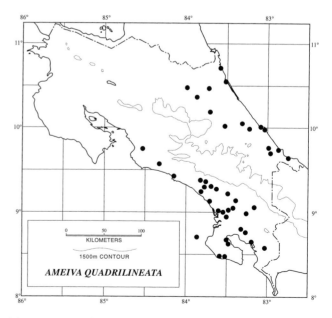

Map 10.58. Distribution of *Ameiva quadrilineata*.

a light vertebral stripe, and the lateral and ventrolateral stripes are broken up to form a series of short linear spots. (4) *Ameiva ameiva* has numerous light spots on the sides of the body, and the collar scales are not greatly enlarged (fig. 10.31a).

HABITAT: Open areas including cacao and coconut groves in Lowland Moist and Wet Forest areas. Rarely seen in heavy forest but common along forest margins and frequently seen along roadsides and trails and in clearings.

BIOLOGY: *Ameiva quadrilineata* has been the object of field studies by Hirth (1963b), Smith (1968a,b), Hillman (1969), and Fitch (1973a,b). Their data have been integrated with my observations to form the basis of the following account. This very common heliophilic species is an obvious component of the biota throughout the humid lowlands. It prefers much more open areas than *Ameiva festiva* or *Ameiva leptophrys*, is less shy and secretive than *Ameiva undulata*, and has longer diel activity periods than *A. undulata*. It is active primarily in the morning. These lizards usually bask in open areas between 7:00 to 9:00 and begin foraging toward the end of this time. Activity increases throughout the morning and peaks between 11:00 A.M and 1:00 P.M. Usually there is little afternoon activity on cloudy or rainy days, but on sunny days there may be a period of activity in midafternoon. *A. quadrilineata* is a dynamic forager, moving rapidly about with jerky movements and constantly probing and digging in the leaf litter or other substrate. Body temperatures of active lizards are in the 34.6 to 40°C range and are several degrees above the ambient air temperature. Home ranges average about 450 m² for adult males and 190 m² for adult females and, females tend to favor a particular home site (Hirth 1963b; Fitch 1973a). Densities range from 32 to 64/ha. Like other *Ameiva*, this species eats mostly small arthropods, especially spiders, orthopterans, beetles, and ants. At a beach site where Hirth (1963b) carried out his study, amphipods were a major component of the diet. Food size ranges up to 30 mm in length. Reproduction in *A. quadrilineata* is continuous throughout the year, with the egg laying at a peak during the rainy season and depressed in the dry season. Differences in the timing of peak egg production (May to June on the Atlantic, October on the Pacific) were noted by Fitch (1973b). Clutch size is usually two eggs and varies from one to three. Several clutches are produced annually. Hatchlings are about 30 mm in standard length and usually reach sexual maturity at about six months. Hirth (1963b), however, mentions a sexually mature female thought to be two months of age. Individuals of this species rarely survive more than two breeding seasons.

DISTRIBUTION: Humid lowlands from extreme southeastern Nicaragua to northwestern Panama on the Atlantic slope and from the Carara area, Puntarenas Province, Costa Rica, to the Península de Azuero in western Panama on the Pacific versant (0–1,050 m).

Ameiva undulata (Wiegmann, 1834a)

(plate 308; map 10.59)

DIAGNOSTICS: A moderately large lizard marked with contrasting light and dark bars along the flanks in adults and a broad brownish to greenish middorsal field. There is no vertebral light stripe, but juveniles have a pair of continuous light lateral stripes and a pair of more or less continuous ventrolateral ones that may be retained in adult females. The anterior and posterior gular scales gradually grade into one another in size in this species (fig. 10.32e).

DESCRIPTION: Total length to 416 mm, males (adults 93 to 129 mm in standard length) larger than females (adults 78 to 111 mm in standard length); tail long, 67 to 70% of total length; head not distinct from neck except in large males; nostril in nasal suture; head shields smooth; postnasals in contact with a prefrontal; a single frontal; usually three supraoculars; one or two enlarged interparietals and two enlarged parietal shields; parietals in broad contact with interparietal and frontoparietals; scales in supraorbital semicircles extending no farther forward than posterior margin of frontal; anterior and posterior gular scales grading into one another in size; median pregulars moderately to greatly enlarged, regularly or irregularly arranged; a single row of enlarged collar scales on posterior gular fold; one row of greatly enlarged central scales on posterior surface of forearm; ventrals in eight longitudinal rows; 26 to 34 femoral pores; 24 to 38 lamellae under fourth toe. Juveniles: middorsal field yellowish or greenish brown; a pair of continuous light lateral longitudinal stripes and a pair of usually discontinuous ventrolateral ones; venter white to bluish; tail blue; females: dorsolateral stripes present or restricted to anterior part of body; ventrolateral stripe usually present;

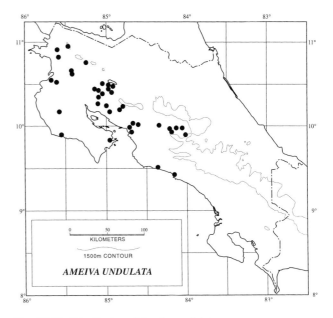

Map 10.59. Distribution of *Ameiva undulata*.

lateral area between stripes vertically barred with bluish white and dark; adult males: longitudinal light stripes reduced or absent; flanks marked from middorsal field to venter by alternating bluish to greenish white and dark bars; venter white to blue; throat and neck red in breeding males; tail brown in adults; upper surface of thighs without light spots in all age-groups.

SIMILAR SPECIES: (1) *Ameiva quadrilineata* retains light stripes throughout life, lacks strongly barred flanks, and has the small posterior gulars abruptly differentiated from the larger anterior ones (fig. 10.32c). (2) *Ameiva leptophrys* has the margins of the middorsal field scalloped and many small scales in the parietal region (fig. 10.31b). (3) *Ameiva festiva* usually has a vertebral light stripe, has two or more rows of lineate spots on the sides, and lacks lateral dark and light bars.

HABITAT: Found in relatively open situations in the drier regions of the country, primarily in Lowland Dry Forest and Premontane Moist Forest areas but marginally in the Lowland Moist Forest and Premontane Wet Forest zones. *Ameiva undulata* occurs in open woodlands, forest edges, secondary growth, fields, and pastures, on coffee, banana, and pineapple plantations, along roadsides, and in vacant lots and gardens.

BIOLOGY: A very common strictly terrestrial diurnal species that like its congeners is a heliotherm, basking to bring the body to an appropriate activity temperature. It forages primarily in the late morning (9:00 to noon) on sunny days and moves about in a brisk, jerky fashion, constantly probing and digging in the leaf litter or other substrate. These lizards rest during the warmest part of the day and usually remain hidden in the afternoon. During the long, hot dry season adults often are quiescent for several days or weeks and are seen infrequently, although juveniles may still be active. Body temperatures of active lizards are usually in the 33 to 41°C range and several degrees above the ambient air temperature (Fitch 1973a). Food consists of small arthropods. Reproduction apparently occurs during all months of the year in populations from the Meseta Central, but there is a definite decline in egg production in the dry season (December to April). In the drier lowlands of northwestern Costa Rica egg production virtually ceases from January to May. During the reproductive period clutches of two to seven eggs are laid every few weeks. Hatchlings are about 35 mm in standard length.

REMARKS: This species occurs within the limits of the capital city, San José, San José Province, and may once have had a wider distribution on the Meseta Central Occidental. There is no reason to doubt the assignment (Taylor 1956; Echternacht 1971) of *Cnemidophorus amivoides* Cope, 1894 (holotype: AMNH 163116) to the synonymy of *A. undulata* or its type locality (Cartago: La Carpintera), 10 km east of San José. Although as many as twelve geographic races have been recognized within this species (Smith and Laufe 1946), Echternacht (1971) has shown that most of the differences in meristic features they are based on reflect clinal changes and are frequently nonconcordant or often show chaotic variation. Color pattern variation is extensive in this species, but all populations have bars or blotches on the flanks in adult males except for those from the Gulf coast of Tabasco and Campeche and adjacent Chiapas, Mexico. These lizards retain light stripes as adults, but intermediates are found to the south. Many other samples exhibit considerable intrapopulational variation in lateral patterns, ranging from having light blotches in the lateral dark field and spots, blotches, or bars in the ventrolateral fields to having continuous bars from the middorsal field to the venter. This last pattern is characteristic of most Central American examples. Females are similarly variable and usually retain longitudinal light stripes but have a pattern of light and dark spots or bars in the lateral field. Because of noncongruence between scalation and coloration and the substantial within-populational variability, there seems to be no basis for formal recognition of geographic divisions within this species.

DISTRIBUTION: Lowlands and adjacent premontane areas from Tamaulipas and Nayarit, Mexico, south to Nicaragua (including Isla de Maíz Grande) on the Atlantic versant and central Costa Rica on the Pacific slope; in Costa Rica confined to the Pacific slope and Meseta Central and occurring near the coast as far south as Quepos, Puntarenas Province (2–1,300 m).

Genus *Cnemidophorus* Wagler, 1830

Another speciose genus (fifty species) of macroteiid lizards whose members have pointed snouts, well-developed limbs, and long tails. These are mostly moderate-sized lizards reaching somewhat over 500 mm in maximum total length, but most are in the 200 to 400 mm range. All are active terrestrial and strictly diurnal forms that have a homogeneous complement of small granular dorsal scales, large smooth ventrals, and the tongue deeply notched posteriorly and not retractile into a basal sheath. Other distinguishing features are pupil of eye reniform, indented ventrally; femoral pores present; a definite gular collar bordered by anterior and posterior gular folds; two to five rows of large brachials; five fingers and toes; outer margin of fourth toe not serrate; pterygoid teeth present; premaxillary teeth conical; posterior teeth on jaws compressed longitudinally, biconodont or triconodont; single pair of processes posterior to autotomy septa on anterior series of posterior autotomic vertebrae.

As with their close allies in the genus *Ameiva*, most differences between Costa Rica individuals of this genus involve dorsal coloration. The color pattern of juveniles consists of a series of narrow longitudinal light stripes separated by dark brown fields. These may include paired paravertebral, dorsal, dorsolateral, lateral, and ventrolateral light stripes (ten stripes), or the paravertebrals may be fused into a single vertebral stripe (nine stripes). The vertebral or paravertabrals do not reach the enlarged head shields and terminate on the rump. The dorsals originate at the median edges of the parietal plates, run down the back, and usually fuse into a single stripe that continues onto the tail. The dorsolateral stripes arise on the level of the lateral margins of the parietal plates and continue onto the dorsal surface of the tail. The lateral stripes run from the superciliary scales above the eye posteriorly onto the sides of the tail. The ventrolaterals originate below the eye, pass through the ear, and continue to the groin. During ontogeny the ventrolateral stripes become obscure or may be replaced by a series of light spots, or the ventrolateral and lateral stripes may both break up into spots.

Members of this genus and the species of *Ameiva* are very similar in external features. Most species in both genera also undergo considerable ontogenetic change in coloration, and the sexes usually show substantial dichromatism. These factors often complicate ready determination of genus, although careful comparison with the illustrations and descriptions provided in this book aids in secure identification. Any macroteiid from Costa Rica having two or more rows of enlarged collar scales or a body pattern of six to ten light longitudinal stripes is a *Cnemidophorus*.

Of the recognized species in this genus, nineteen are all-female parthenogenetic forms (Wright 1993; Cole and Dessauer 1993). The parthenogens may be diploids that have originated by hybridization between two bisexual species (nine forms) or triploids that were formed by a hybrid diploid backcrossing to one of the parental bisexual species (ten forms). Several additional putative but as yet unnamed and undescribed parthenogenetic species are also known (Wright 1993).

Four principal species groups are recognized within *Cnemodophorus*, based primarily on a few details of scutellation and chromosome morphology (Lowe et al. 1970; Wright 1993). The genus has an extensive range from Maryland, Wisconsin, Minnesota, the Dakotas, Oregon, and Idaho southward over North and Central America to western Colombia and northern Venezuela and thence south to Bolivia and northern Argentina; also found on Trinidad and Tobago and several Caribbean islands.

Cnemidophorus deppii Group

These macroteiids have a karyotype of 2N = 52 and lack large metacentric or submetacentric chromosomes. All species are bisexual diploids and include two extralimital members of the group in addition to the Costa Rican representative: *Cnemidophorus ceralbensis* (Gulf of California and Baja California, Mexico) and *C. lineatissimus* (western Mexico).

Cnemidophorus deppii Wiegmann, 1834a

(plate 309; map 10.60)

DIAGNOSTICS: This handsome species is a moderately small form marked with nine to ten narrow yellowish longitudinal light stripes in juveniles and females. In adult males the lower stripes are replaced by spots, and the vertebral and dorsal ones may be obscured. These lizards also have three enlarged shields in the parietal region (two parietals and one interparietal) (fig. 10.31c) and usually only three supraocular plates.

DESCRIPTION: Total length to 295 mm, males (adults 58 to 93 mm in standard length) larger than females (adults 69 to 81 mm in standard length); tail long, 67 to 70% of total length; head not distinct from neck; head shields smooth; nostril anterior to nasal suture; a single frontal; postnasals in contact with a prefrontal; usually three supraoculars; an enlarged interparietal and two large parietals; parietals in broad contact with interparietal and paired frontoparietals; scales in supraorbital semicircles extending no farther forward than posterior margin of frontal (extending to anterior margin of frontal in some extralimital examples); anterior gular scales very large, much larger than posterior ones; midgular scales somewhat enlarged (about same size as anterior gulars), larger than enlarged collar scales; two to four regular rows of collar scales on posterior gular fold; central scales on posterior surface of forearm small to slightly enlarged; ventrals in eight longitudinal rows; 29 to 44 femoral pores. Juveniles and females: dorsum dark brown with nine or ten yellowish longitudinal stripes; tail blue, venter white; adult males: middorsal field suffused with green; paravertebral light stripes fused to form vertebral stripe; dorsal and vertebral stripes may be obscured; ventrolateral stripes broken up into bluish white spots; lateral stripe often yellow, partially discontinuous or forming a chain of bluish white spots in largest adults; upper stripes greenish; flanks rust red; tail bluish gray; throat blue to black; venter black; underside of hind limbs blue to black; upper surface of thighs with distinct light spots in juveniles and adults.

SIMILAR SPECIES: (1) Adult male *Cnemidophorus lemniscatus* have no more than four light stripes on the body,

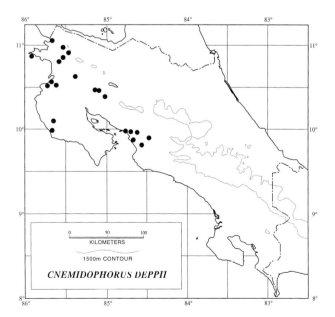

Map 10.60. Distribution of *Cnemidophorus deppii*.

and all examples of that species have five shields in the parietal region and four or five supraoculars (fig. 10.31a). (2) *Ameiva ameiva* does not have greatly enlarged collar scales on the posterior gular fold and has five shields in the parietal region (fig. 10.31a).

HABITAT: This lizard is an inhabitant of subhumid areas and prefers open situations, primarily in the Lowland Dry Forest, but also occurs marginally in Lowland Moist Forest in the caldera of the Río Grande de Tárcoles. It is very common along back beach areas, in open sandy flats, along streambeds, and in pastures and grassy areas throughout northwestern Costa Rica.

BIOLOGY: This very active, nervous, and quick-footed species is strictly diurnal and terrestrial. Like other members of the genus, when active it is constantly poking, probing, and scratching the substrate and moves from place to place in rapid bursts. Fitch (1973a,b) studied *Cnemidophorus deppii* in Costa Rica in some detail. A one-month (March) study of this species on a sandy beach in Nicaragua (Vitt et al. 1993) is mostly concordant with Fitch's reports. These accounts form the basis for what follows. As might be expected from its behavior, it is a heliotherm that maintains an activity temperature 9 to 12°C above ambient air temperature in a range from 29 to 43°C, but mostly about 40°C. Individuals reach these temperatures by basking, but they move in and out of the shade as the morning progresses. Activity is intense in the late morning, but most adults retire to underground retreats by midday. These lizards may be approached closely, but once alarmed they dash to a shelter or burrow with remarkable speed, often taking a zigzag route or even doubling back and then diving to safety. The home range appears to be relatively small (approximately 400 m²). Food consists primarily of small arthropods, especially termites, spiders, and orthopterans. The orthopterans are initially located visually, but the others are dug up from the substrate. Reproduction in this species is seasonal in most areas, with high egg production from late May through October and virtually none in the dry season (December through April). In these circumstances adults make up most of the population from March to October. They are then replaced by the much more numerous juveniles during November through January, and by February both juveniles and adults are well represented. Hatching occurs after about sixty days of incubation. In moister areas egg production continues through the dry season, but at a reduced rate, and the population contains balanced numbers of juveniles and adults throughout the year. Clutches range from one to four eggs (usually three), and four to six clutches are probably laid each year. Eggs are about 14 × 8 mm. Hatchlings are about 31 mm in standard length and reach sexual maturity in five to six months. Individuals of this species rarely survive more than two breeding seasons.

REMARKS: Ontogenetic change in coloration in this species is relatively straightforward. The smallest juveniles usually have a single vertebral light stripe that later divides into two paravertebral stripes. This pattern (ten light stripes) is characteristic of larger juveniles and adult females. In males the paravertebral and dorsal stripes become obscured as the area between the dorsal stripes becomes green. The ventrolateral stripes more or less simultaneously break up into a series of light spots, and the lateral stripe may become indistinct. There seems little reason to recognize races in this species, since all color patterns described by Duellman and Wellman (1960) occur in the Costa Rican samples. The all-female diploid forms *Cnemidophorus cozumela*, *Cnemidophorus maslini*, and *Cnemidophorus rodecki* of the Yucatán Peninsula and adjacent islands appear to have been produced by hybridization between *C. deppii* and *Cnemidophorus angusticeps*, a member of the *sexlineatus* group (McCoy and Maslin 1962; Fritts 1969). Although Taylor (1956) recognized *Cnemidophorus alfaronis* Cope, 1894a (holotype: AMNH 16315; Costa Rica: Alajuela: San Mateo), it is clear that this name is based on a juvenile *C. deppii* with four supraoculars. The karyotype of this species is 2N = 52, with fourteen pairs of telocentrics and twelve pairs of microchromosomes; NF = 52 (Gorman 1970, 1973).

DISTRIBUTION: Lowlands from Veracruz and Michoacán, Mexico, south on the Atlantic versant to the northeastern Great Lakes region of Nicaragua, exclusive of most of the Yucatán Peninsula, the rainforests of Guatemala, Belize, Honduras, and Nicaragua, and on the Pacific slope to central Costa Rica (2–530 m).

Cnemidophorus lemniscatus Group

These lizards are characterized by a basic karyotype of 2N = 50, having a pair of large submetacentric chromosomes. Most species are bisexual diploids, one is a parthenogenetic diploid (P), and one is a parthenogenetic triploid (3N = 75). Included species besides *Cnemidophorus lemniscatus* are *Cnemidophorus arubensis* (Aruba Island, Dutch West Indies), *C. cryptus* (P; Venezuela and Brazil); *C. gramivagus* (Colombia and Venezuela); *C. lacertoides* (Brazil and Argentina); *C. leachei* and *C. longicaudus* (Argentina); *C. murinus* (Curaçao and Bonaire Islands, Dutch West Indies), *C. nigricolor* (islands off Venezuela), *C. ocellifer* (Brazil, Bolivia, and Argentina); *C. pseudolemniscatus* (3N; Suriname); *C. serranus* (Argentina); *C. vanzoi* (St. Lucia); and *C. vittatus* (Bolivia).

Cnemidophorus lemniscatus (Linné, 1758)

(plate 310)

DIAGNOSTICS: The brilliant and gaudy color of adult males of this moderate-sized species immediately distinguishes them from any other Central American lizard. In these males the sides of the head are grass green to sky blue, the flanks are rust red to yellow brown with rows of white spots, the limbs and tail are grass green to blue above and bright blue below, and the brown middorsal field is bordered by green stripes. Juveniles and females have a pattern of nine or ten light-colored longitudinal stripes on a dark ground color. All examples have five large shields in the parietal area and four or five supraocular plates (fig. 10.31a).

DESCRIPTION: Total length to 330 mm, males (adults 60 to 113 in standard length) larger than females (adults 50 to 93 in standard length); tail long, 66 to 70% of total length; head not distinct from body; head shields smooth; nostril in nasal suture; postnasal contacts a prefrontal; a single frontal; four or five supraoculars; an enlarged interparietal and two pairs of enlarged parietal shields; inner parietals in broad contact with interparietal and paired frontoparietals; scales in supraorbital semicircles extending no farther forward than posterior margin of frontal; anterior gular scales larger than posterior ones, grading gradually into them in size; midgular scales not enlarged; three or four rows of enlarged collar scales on posterior gular fold; central scales on posterior surface of forearm small, uniform; ventrals in eight longitudinal rows; 30 to 58 (30 to 66) femoral pores; 30 to 37 lamellae under fourth toe. Juveniles and females: middorsal field and lower flanks brown; nine to ten light longitudinal light stripes; lateral dark fields between light stripes black; upper surface of hind limbs with white spots; venter white; side of head orange in females; adult males: middorsal field brown, bordered by yellow or green dorsal stripes that are separated from green lateral stripes by a black field; flanks rust red to yellowish brown with three longitudinal rows of round spots; side of head grass green to sky blue; upper surfaces of limbs grass green to blue; undersurfaces of head, chest, limbs, and tail sky blue; venter white.

SIMILAR SPECIES: (1) Adult male *Cnemidophorus deppii* retain five or six light stripes on the body, and all examples of that species have three enlarged shields in the parietal area (fig. 10.31c) and usually have only three supraoculars. (2) *Ameiva ameiva* lacks greatly enlarged collar scales on the posterior gular fold (fig. 10.32a) and has ten to twelve longitudinal rows of ventrals.

HABITAT: This species has not been collected in Costa Rica but occurs in Honduras to the north and Panama to the east and may range into Costa Rica. Elsewhere in Central America *C. lemniscatus* is restricted to humid lowland areas, primarily near the Caribbean coast.

BIOLOGY: A terrestrial heliophilic species often found on beaches and similar in habits to *Ameiva quadrilineata* and *Cnemidophorus deppii*.

REMARKS: There is a major hiatus in the known range of this conspicuous species between northeastern Honduras (Echternacht 1968) and central Panama. It has never been collected in Costa Rica. The record for Granada, Granada Province, Nicaragua (Burt 1931) has recently been shown by Köhler (1999a) to be based on examples of *Cnemidophorus deppii* (FMNH 1948, eight specimens). It is possible that this lizard ranges along the Atlantic coast of Nicaragua into extreme northeastern Costa Rica, and I include it to aid in identification should a reader of this book capture an example in the republic. The lizards long referred to *C. lemniscatus* were remarkable in having both bisexual and unisexual (all-female parthenogenetic) populations rather randomly distributed in northern Brazil (Vanzolini 1970) and Suriname (Hoogmoed 1973). The Amazonian unisexuals are diploids (Peccinini-Seale and Frota-Pessoa 1974; Peccinini-Seale 1989), and those from Suriname are triploids (Serena 1985; Dessauer and Cole 1989). The karyotype of bisexual individuals is 2N = 50, comprising thirteen pairs of macrochromosomes (twelve pairs of telocentrics, one pair of subtelocentrics) and twelve pairs of microchromosomes (NF = 50; Lowe et al. 1970), but there are two cytotypes (D and E), from northern Brazil and Suriname and from the central Amazon basin, respectively. The karyotypes of the diploid unisexuals differ slightly among populations— some (A, B) 2N = 48, one (C) 2N = 50—and in numbers of heteromorphic macrochromsomes, and that of the diploid unisexuals is virtually the same. In the triploid unisexuals the karyotype is 3N = 75 (thirty-nine macrochromosomes and thirty-six microchromosomes). Dessauer and Cole (1989) suggested that the triploid populations are a single lineage derived from hybridization between bisexual *C. lemniscatus* and a cryptic bisexual species followed by a backcross by the hybrid to *C. lemniscatus*. Sites et al. (1990) and Vyas et al. (1990) supported this hypothesis based on extensive cytological and biochemical evidence suggesting that the diploid unisexuals arose from hybridization between cytotypes D and E (now regarded as distinct species) and the triploids from a backcross between the unisexual C population and population D. Cole and Dessauer (1993) have now confirmed this explanation and formally recognized four species in the complex. Populations D and E (bisexual diploids) are recognized as *C. lemniscatus* (Brazil, Suriname, Venezuela, and presumably other northern populations) and *C. gramivagus* (Amazonian Brazil), respectively. The diploid unisexual populations (A, B, C) that arose by hybridization between the two bisexual species were named *C. cryptus*. Finally, the triploid unisexual (E), produced by a backcross between *C. cryptus* and *C. lemniscatus*, is recognized as *C. pseudolemniscatus*.

DISTRIBUTION: Lowlands of Central America from extreme southeastern Guatemala to northeastern Honduras on the Atlantic versant and on the Pacific slope in central and eastern Panama; in South American from northern Colombia and Venezuela (including its offshore islands) south, east of the Andes, to the Amazon basin of northern Brazil; also Swan Island and Islas Roatón, Utila, Providencia, and San Andrés and Trinidad and Tobago; introduced into southern Florida; not yet collected in Costa Rica.

Family Gymnophthalmidae MacClean, 1974

Microteiid Lizards

A large family (thirty-four genera and 184 species) of small to medium-sized mostly diurnal secretive subfossorial or fossorial insectivores. Some forms, however, are arboreal or semiaquatic. The diversity of this group makes a statement regarding general features difficult; however, the body is usually slender and often elongate, the limbs are short to greatly reduced, the tail is long, and the head is covered with large symmetrical plates. The dorsal scales show a great diversity of homogeneous small to large scales that may be juxtaposed, imbricate, or cycloid, smooth or keeled, or a mixture of several kinds; the ventrals are usually large and quadrangular and may be smooth or keeled. The ear opening is moderate to small, and the superficial to recessed tympanum is exposed or both are absent; the eyes are small, the pupil is usually round, and the eyelids are usually movable but may be fused to form a transparent spectacle. Femoral pores are usually present at least in males, and a gular fold may be present or not.

The family is endemic to the Neotropics from central Argentina and northwestern Peru northward, including the Lesser Antilles as far as Guadeloupe, but only one genus (Gymnophthalmus) ranges north of Costa Rica into southern Mexico.

The gymnophthalmids are further characterized by the following features: postfrontal and postorbital present; supratemporal fenestra present or more frequently roofed over by postorbital and squamosal; parietal foramen absent; jaw tooth replacement usually basal, with large resorption pits sometimes alternating with small resorption pits; lateral jaw teeth biconodont to triconodont; Meckel's groove open or closed and fused; no intracentra associated with dorsal vertebrae; caudal autotomy, with autotomy septa on anterior vertebrae passing between a double pair of diverging transverse processes, more posterior vertebrae variable in form, and the short terminal series without processes; extracolumella with an internal process; no extracolumella muscle; no dorsal, ventral, or separable cephalic osteoderms; nasal plates separated by one or two frontonasals; tongue deeply bifurcate for 40 to 50% of length; most of tongue covered by flattened scalelike papillae arranged in regular diagonal rows; posterior limbs of tongue plicate; tongue protrusion oscillatory; hemipenes usually petalate, rarely flounced or nude, never calyculate, somewhat bilobed and with a simple sulcus spermaticus. The combination of enlarged head shields, enlarged noncycloid dorsal and ventral scales, and the absence of a lateral fold and series of small scales separating the dorsal from ventral scales will separate five of the Costa Rican microteiid genera from all other lizards in the republic. The sixth genus, Gymnophthalmus, has cycloid body scales but has only four fingers and lacks movable eyelids.

Microteiids use oscillatory tongue movements to obtain chemical particles from potential prey items and convey them to the Jacobson's organs. If the stimulus is positive, they take the prey into the mouth and swallow it. So far as is known all members of this family are oviparous and lay eggs with flexible yellowish to white shells. The microteiids were formerly included in the Teiidae, but Presch (1980, 1983) has validated the Gymnophthalmidae and established osteological differences among the genera included.

Genus *Anadia* Gray, 1845

Members of this genus (twelve species) are small, elongate lizards with pointed heads, large recessed ear openings, well-developed pentadactyl limbs, claws on all digits, and a long cylindrical tail. They differ from other Central American microteiids in sharing the following combination of features: head plates smooth; ventrals smooth and quadrangular or rounded and imbricate; dorsal scales smooth and homogeneous in size, larger ones slightly smaller than ventrals; scales forming complete annuli around body; nasals separated by a single frontonasal (internasal). Other distinguishing external characteristics include two prefrontals; two frontoparietals; no enlarged interoccipital; pupil round; usually a translucent disk in lower eyelid composed of several scales (pigmented in one extralimital form); tongue covered by rhomboidal imbricate scalelike papillae; a gular fold and collar; collar scales larger than other gulars; femoral pores present in males, present or absent in females.

The following osteological characters also define the genus: frontal enclosing olfactory tract ventrally; supratemporal fenestra open; postorbital and postfrontal present and distinct; extrastapedial present; no pterygoid teeth; anterior teeth on jaws conical, lateral ones laterally compressed, biconodont and triconodont; glossohyal single, second ceratobranchial absent; lacrimal present; a secondary palate; twenty-five presacral vertebrae; no parasternum; clavicle flattened and with a hook; two xiphisternal ribs; a single pair of processes anterior to autotomy septa on anterior series of posterior autotomic caudal vertebrae.

The genus occurs in the Andes of Colombia, Ecuador, and Venezuela and at lower elevations from Costa Rica to southwestern Ecuador and northeastern Venezuela. In his review of the genus Oftedal (1974) recognized five species groups based on slight differences in scalation and, except for the *ocellata* group (four species), comprising one or two forms. The validity of this extremely fine-graded arrangement is questionable, and consequently it is not detailed here.

Anadia ocellata Gray, 1845

(plate 311; map 10.61)

DIAGNOSTICS: This beautiful little long-tailed lizard having a pointed head and the body covered above and below by smooth flattened quandrangular and juxtaposed scales is unlikely to be mistaken for any other species (fig. 10.34a). Although the coloration is variable, there is always a series of distinct white spots along the flank. The anteriormost of these spots, above the forelimb insertion, is outlined by black to form an ocellus, and some or all of the more posterior white spots may also form ocelli.

DESCRIPTION: Total length to 216 mm (adults 50 to 75 mm in standard length); tail long, 65 to 70% of total length, attenuate; limbs short, well developed; nasal partially or completely divided; three supraoculars; suboculars unequal in size, one extending downward between fourth and

Map 10.61. Distribution of *Anadia ocellata*.

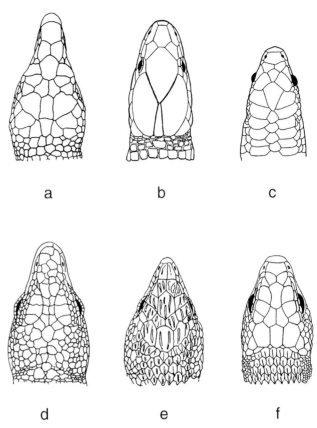

Figure 10.34. Dorsal views of heads of the Costa Rican genera of microteiid lizards (Gymnophthalmidae): (a) *Anadia;* (b) *Bachia;* (c) *Gymnophthalmus;* (d) *Neusticurus;* (e) *Leposoma;* (f) *Ptychoglossus.*

fifth supralabials; 14 to 20 transverse rows of smooth, flat rhomboidal to quadrangular gular scales; large interparietal and large occipitals; 27 to 32 scale rows around midbody; 51 to 59 transverse rows of dorsals; femoral and precloacal pores in a continuous series, 8 to 11 in males, 2 to 6 in females; caudals smooth, juxtaposed, and quadrangular, in uniform whorls around tail. Middorsal area metallic olive, brown, or bronze; a narrow, somewhat irregular dull light to bright white stripe runs from above each eye down body onto side of tail; stripe bordered below by dark lateral field that originates on head, passes through eye, and continues onto tail; dark field with three to eight light spots; the anteriormost, above arm insertion, is outlined by black, as may be one or more posterior ones; middorsal field may be uniform or heavily marked with small dark spots or larger rectangular blotches; venter clear cream to yellow; hemipenes bilobate with comblike rows of calcareous spinules (Charles Myers, pers. comm.).

SIMILAR SPECIES: (1) *Neusticurus apodemus* has a mixture of small and large keeled tubercular dorsal scales. (2) *Mesaspis monticola* is much larger, with a distinct lateral fold and a number of dorsal scale rows with well-developed keels.

HABITAT: In the tree canopy among mosses and epiphytes in Premontane Wet Forest and Rainforest.

BIOLOGY: This relatively rare upland lizard is completely arboreal. It apparently hides in the heavy moss growths on the limbs of trees in high rainfall premontane forests and has been collected several times from bromeliads, which also serve as shelter. *Anadia ocellata* is insectivorous (plate 312).

REMARKS: Taylor (1955, 1956) used the name *Anadia metallica* for this species and recognized three geographic races from various upland areas of Costa Rica: *A. m. metal-*

lica (Pacific slope central), *A. m. arborea* (northwestern), and *A. m. attenuata* (Atlantic slope central). Oftedal (1974) discovered that the type specimen (BMNH 1946.8.2.2) of *A. ocellata* Gray, 1845 (type locality: "Tropical America"), was conspecific with the holotype of *Chalcidolepis metallica* Cope, 1875 (holotype: USNM 30568; Alajuela: Cerros de Aguacate), and used the former name for this lizard. Oftedal also demonstrated that all three of Taylor's (1955, 1956) putative races were based on individual variation within a single species. The description of the hemipenis of this form by Presch (1978) is based on a partially everted organ.

DISTRIBUTION: Humid forested areas on both slopes in the premontane zone of the cordilleras of Costa Rica and western Panama (520–1,200 m).

Genus *Bachia* Gray, 1845

The small lizards of this semifossorial genus (fifteen species) have elongate, snakelike bodies and tails and very small eyes; although they lack external ear openings, they have movable eyelids, have rudimentary forelimbs and hind limbs with no more than four fingers (zero to four) and toes (zero, three, or four), with or without claws, or the hind limbs may be absent. The scalation on both head and body, though species specific, has a wide range of variation in the genus. The head shields are smooth, and paired prefrontals, a frontonasal (internasal), and an elongate interparietal may be present but are often absent. The ventral scales are usually smooth, quadrangular, and juxtaposed, but one Amazonian species *(Bachia panoplia)* has keeled lanceolate and imbricate ventrals. The dorsals may be hexagonal or quadrangular, smooth, and juxtaposed or elongate, keeled, and imbricate. Other external features include no frontoparietal; no enlarged interoccipital or occipitals; scales in complete annuli around the body; pupil round; lower eyelid with an unsegmented translucent disk; tongue covered with rounded imbricate scalelike papillae in oblique rows; a distinct postcephalic groove demarcated by small scales present; a weakly developed gular fold and collar may be present; one or two precloacal pores in males. The following osteological features further characterize this genus: frontal not enclosing olfactory tract ventrally; supratemporal fenestra open; postorbital and postfrontal present and distinct; no extrastapedial; no pterygoid teeth; all teeth conical; no lacrimal; no secondary palate; glossohyal single, continuous with basihyal; Meckel's groove closed; 38 to 55 presacral vertebrae; a parasternum with 8 to 10 inscriptional ribs; clavicle rod shaped, without a hook; single pair of processes anterior to autotomy septa on anterior series of posterior autotomic caudal vertebrae; hemipenes slightly bilobed, nude (Presch 1978), but this may be based on partially everted organs.

Four species groups are recognized within *Bachia* following Dixon (1973). The single Costa Rican species belongs to the *heteropa* group, having the dorsal scales hexagonal, smooth, imbricate and lateral scales smooth, quadrangular, juxtaposed. Other extralimital species are *Bachia pallidiceps* (eastern Panama) and *B. guianensis* and *B. heteropa* (northern Venezuela).

Bachia ranges from southwestern Costa Rica through northern Colombia and Venezuela and the Guianas south through the upper Amazon basin to Bolivia and northern Paraguay; also on Trinidad and Tobago, the Grenadines, and Grenada.

Bachia blairi (Dunn, 1940b)

(plate 313; map 10.62)

DIAGNOSTICS: This little lizard superficially looks like a small black snake, but the ridiculously short arms and legs and the regular annuli of large quadrangular to hexagonal scales that circle the body and tail prevent one from confusing it with any other Costa Rican animal (fig. 10.34b).

DESCRIPTION: A small lizard reaching a total length of 160 mm (adults 50 to 63 mm in standard length); tail moderate, 60 to 63% of total length; a frontonasal (internasal), two prefrontals, and an elongate interparietal present; two supraoculars; frontal large, in contact with frontonasal (internasal); gulars smooth, imbricate, rectangular; scales on weakly developed gular collar larger than gulars, smooth and rectangular; dorsals hexagonal, smooth, juxta-

posed; ventrals and laterals quandrangular, smooth, juxtaposed; 24 to 25 scales around midbody; 43 to 45 transverse rows of dorsals; two supraoculars, two superciliaries; two precloacal pores in males, none in females; four clawed fingers and three clawed toes; scales on tail smooth, juxtaposed, elongate, and hexagonal in uniform whorls around tail; subcaudals slightly smaller and less elongate than caudals. Body and tail uniform dark brown to black, with the chin and throat lighter in adults; hatchlings light brown with a yellowish venter.

SIMILAR SPECIES: Unique in Costa Rica.

HABITAT: A semifossorial leaf litter inhabitant of undisturbed rainforests of the Golfo Dulce region.

BIOLOGY: All available examples were collected from the leaf litter, usually within tree buttresses or from trench traps in the forest floor. McDiarmid and Deweese (1977) reported that one example contained remains of a small elongated beetle and a lepidopteran larva. They also hatched an egg collected in early July after twenty-nine days. Similar eggs were also found in the leaf litter. The hatchling was 28 mm in standard length, and a hatchling 27 mm in standard length was also captured in the leaf litter. A female (61 mm in standard length) captured in July contained an egg in each oviduct.

REMARKS: This species was first described by (Dunn 1940b) based on a single specimen (ANSP 21773) from Puerto Armuelles, Chiriquí Province, Panama. Dixon (1973) synonymized it with *Bachia pallidiceps* Cope, 1862c (holotype: ANSP 4324, now lost, Colombia: Chocó: Río Truando region) and called the species *Bachia pallidiceps* following Vanzolini (1961). McDiarmid and DeWeese (1977) showed that recently collected material from Costa Rica was con-

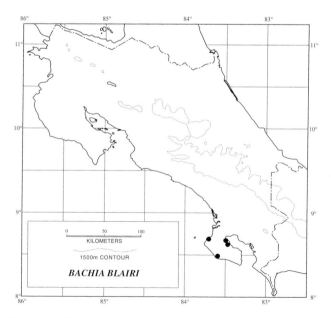

Map 10.62. Distribution of *Bachia blairi*.

specific with Dunn's (1940b) holotype but differed substantially from material of *B. pallidiceps* from eastern Panama and adjacent Colombia, and they concluded that two species were involved. These authors showed that *B. pallidiceps* is a more slender form than *B. blairi* and has a color pattern that includes a middorsal dark stripe and a pair of lateral light stripes rather than the uniform pattern of the latter species.

DISTRIBUTION: Lowlands of the Golfo Dulce region of southwestern Costa Rica and adjacent western Panama; in Costa Rica known only from the Península de Osa (2–40 m).

Genus *Gymnophthalmus* Merrem, 1820

These small, slender lizards (eight species) have cylindrical bodies, short, small limbs, recessed ear openings, only four fingers (inner absent), and no movable eyelids, and the body is covered above and below by broad cycloid imbricate scales. These features separate them from all other Central American lizards and will, in combination with the following characteristics, distinguish them from other microteiids: head shields smooth; dorsal scales usually smooth, posterior ones keeled in one extralimital form (*Gymnophthalmus pleii* of the Lesser Antilles); scales forming an uninterrupted series around body; nasals separated by a single frontonasal (internasal). Other features include two prefrontals; no frontoparietals; a single supraocular on each side; no enlarged interoccipital or occipitals; two pairs of chin shields in contact medially; pupil round; tongue covered by rhomboidal imbricate scalelike papillae in oblique rows; no gular fold or collar; digits clawed; femoral pores present in males.

Osteological features further define the genus as follows: frontal not enclosing olfactory tract ventrally; supratemporal fenestra open; postorbital and postfrontal present, separate; extrastapedial present; a lacrimal present; a secondary palate present; Meckel's groove open; pterygoid teeth present; all teeth conical; glossohyal single, continuous with basihyal, second ceratobranchial present; 26 to 32 presacral vertebrae; two xiphisternal ribs; no parasternum; clavicle rod shaped, without a hook; single pair of processes anterior to autotomy septa on anterior series of posterior autotomic caudal vertebrae; hemipenes slightly bilobed, papillate or flounced.

Rodrigues (1991) has recently reviewed the status of microteiids lacking movable eyelids. He placed *Gymnophthalmus rubricauda* (northern Bolivia to Argentina) in a monotypic new genus, *Vanzosaura*, characterized by having three pairs of chin shields in contact medially and two supraoculars on each side.

One of the species currently recognized in this genus, *Gymnophthalmus underwoodi* of northeastern South America and several of the Lesser Antilles, is a diploid all female parthenogenetic form (Cole et al. 1990; Cole, Dessauer, and Markezich 1993). It apparently arose through hybridiza-

tion between the recently described *Gymnopthalmus cryptus* Hoogmoed, Cole, and Ayarzaguena 1992 and *G. speciosus.* This genus has an extensive distribution from the Isthmus of Tehuantepec through Central America and northern South America to northeastern Brazil; also on Trinidad and Tobago and in the Lesser Antilles to Guadeloupe.

Gymnophthalmus speciosus (Hallowell, "1860," 1861)

(plate 314; map 10.63)

DIAGNOSTICS: This very small shiny skinklike lizard is immediately recognizable by having a bright red tail and only four fingers (all clawed) and lacking movable eyelids (the eyelids are fused to form a transparent brille or spectacle as in snakes) (figs. 10.8a, 10.34c).

DESCRIPTION: Total length to 122 mm (adults 33 to 44 mm in standard length); tail moderately long, 63 to 68% of total length; limbs very short; nasal entire; one large supraocular; frontal small, not in contact with frontonasal (internasal); gular scales smooth, imbricate, cycloid; body scales smooth; vertebral and paravertebral scales enlarged, larger than laterals or ventrals; 13 to 15 (usually 15) scale rows around midbody; 31 to 38 transverse rows of dorsals; 4 to 13 femoral and precloacal pores in males, none in females; tail covered by smooth imbricate cycloid scales in uniform whorls around tail; subcaudals slightly smaller than caudals. Dorsum metallic bronze brown; light tan lateral stripe from above eye to near midbody or to base of tail; broad lateral black field running from snout to tail base; venter cream in smaller individuals, heavily suffused with black in adult males; tail bright red in juveniles, usually red orange in adults if tail not regenerated. Everted hemipenes slightly bilobed, truncus and apex flounced (Presch 1978).

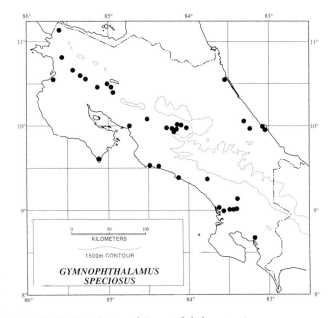

Map 10.63. Distribution of *Gymnophthalmus speciosus.*

SIMILAR SPECIES: *Sphenomorphus cherriei* has light supralabials sprinkled with small dark spots and lacks enlarged vertebral and paravertebral scales on the body; the tail is not red.

HABITAT: This species prefers open situations, often near human habitations. The species is most common in Lowland Dry Forest and Premontane Moist Forest areas. It also is locally abundant at a number of coastal sites on both the Pacific and Caribbean shorelines, where it favors disturbed situations, especially stands of coconut. It may also be found in pastures or clearings within the rainforest zone of the southwest Pacific region.

BIOLOGY: A diurnal thermophilic species, locally common, that is most active near midday. At other times it seeks shelter in burrows, under surface debris, or in rock walls. Lizards of this species are often encountered in gardens in the metropolitan area of the Meseta Central Occidental, and they were probably common in the coffee fincas that formerly covered much of that area. *Gymnophthalmus speciosus* feeds primarily on small insects but also eats other arthropods. Courtship and mating in Panama (Telford 1971) occur in the dry season (December to March). Egg laying and hatching take place from February to May, and no sexually mature individuals are present from March to October, indicating a complete annual population turnover. Clutch size is one to four eggs, and most females lay two or three clutches. Hatchlings are 16 to 19 mm in standard length and reach sexual maturity at about seven months of age.

REMARKS: The karyotype of this species is 2N = 44, comprising three pairs of submetacentrics, six pairs of telocentrics, one pair of submetacentric to subtelocentric chromosomes, and twelve pairs of microchromosomes; NF = 50 (Cole et al. 1990). These authors point out that the karyotype of the all-female parthenogen *G. underwoodi*, from northeastern South America, contains a perfect, complete haploid complement of chromosomes from putative South American *G. speciosus*. The other haploid complement is very different, leading them to propose that *G. underwoodi* arose from hybridization between *G. speciosus* and a species that at that time was unrecognized and undescribed but subsequently was discovered (Hoogmoed, Cole, and Ayarzaguena 1992) and named *Gymnophthalmus cryptus*. As pointed out by Cole et al. (1990) and by Vanzolini and Morato de Carvalho (1991), there remains a question whether *G. speciosus* of Central America is conspecific with the taxon called by that name in northern South America, but this does not affect the conclusions regarding the hybrid origin or the parental sources of the karyotype of *G. underwoodi*.

DISTRIBUTION: Lowlands and adjacent premontane slopes from the Isthmus of Tehuantepec (Pacific) and northeastern Guatemala (Atlantic) southward to Colombia, Venezuela, and the country of Guyana; found on both Atlantic and Pacific lowlands and on the Meseta Central Occidental in Costa Rica (2–1,222 m).

Genus *Leposoma* Spix, 1825

The members of this genus (thirteen species) are small forms with well-developed pentadactyl limbs, clawed digits, a large superficial tympanum, and cylindrical bodies and tails. They are immediately distinguished from all other Costa Rican lizards by having large keeled head shields and dorsal scales and elongate imbricate mucronate keeled and quadrangular ventral scales. Additional features that in combination differentiate *Leposoma* from other microteiids include body scales in transverse and oblique rows, forming an uninterrupted series around body, or not; ventrals keeled or smooth, truncate or pointed, imbricate; nasals separated by one or two frontonasals (internasals); paired prefrontals and frontoparietals present; no enlarged interoccipital or occipitals; pupil round; distinct gular fold; collar present or not; if present, collar scales keeled, mucronate; femoral pores present or absent and precloacal pores present in males; precloacal pores present or absent in females. Osteological characteristics that further distinguish this genus are frontal not enclosing olfactory tract ventrally; supratemporal fenestra roofed over; postorbital and postfrontal present and separate; extrastapedial present; a secondary palate in some species; Meckel's groove closed; pterygoid teeth absent; anterior teeth on jaws conical, lateral ones laterally compressed and triconodont; glossohyal in two parts, proximal portion continuous with basihyal, no second ceratobranchial; 25 presacral vertebrae; two xiphisternal ribs; no parasternum; clavicle rod shaped without hook; a single pair of processes anterior to autotomy septa on anterior series of posterior autotomic caudal vertebrae.

One recognized all-female species, *Leposoma precarinatum* of the Guianas and adjacent Brazil and Venezuela, appears to be a parthenogen (Uzzell and Barry 1971; Hoogmoed 1973).

Leposoma occurs from Costa Rica through Panama to western Colombia and from southwestern and northeastern Venezuela and the Guianas to eastern Brazil; also in the upper Amazon basin of Colombia, Ecuador, Peru, and Bolivia.

Leposoma southi Ruthven and Gaige, 1924
(plate 315; map 10.64)

DIAGNOSTICS: This small leaf litter lizard is immediately recognizable because of its strongly keeled head plates and large heavily keeled spiny (mucronate) more or less quadrangular to rectangular dorsal and ventral scales (fig. 10.34e).

DESCRIPTION: Total length to 103 m (standard length of adults 24 to 41 mm); tail moderately long, 63 to 67% of total length; limbs short, well developed; two frontonasals (internasals); three supraoculars; lower eyelid with a translucent disk composed of two to four scales; pregulars flat and quadrangular; gular scales keeled, mucronate, imbricate; gular fold indistinct; lateral nuchal scales large conical tubercles; ventrals in longitudinal and transverse rows; 20

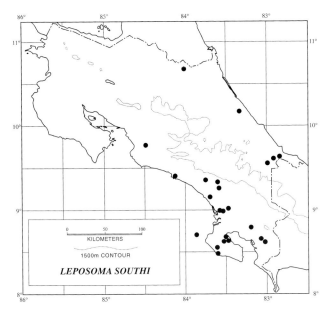

Map 10.64. Distribution of *Leposoma southi*.

to 26 scales around midbody; 27 to 33 transverse rows of dorsals; males with 8 to 14 femoral pores and 4 precloacal pores; females with 2 precloacal pores; tail covered with keeled mucronate imbricate essentially quadrangular scales in uniform whorls around tail. Dorsum dark brown; a pair of broad tan or cream lateral stripes often present running from head onto tail, most obvious in juveniles; venter cream, immaculate.

SIMILAR SPECIES: *Ptychoglossus plicatus* has smooth head shields (fig. 10.34f), gulars, and ventral scales.

HABITAT: On and under leaf litter in Lowland Moist and Wet Forests at relatively undisturbed sites.

BIOLOGY: This common denizen of the lowland rainforests is more often heard skittering through the forest litter than seen. It is strictly diurnal and feeds on arthropods, mostly small insects. Courtship, mating, egg laying, and hatching occur throughout most of the rainy season (May to November). The usual clutch size appears to be two, and the eggs are about 8 × 4.7 mm. A female captured July 25 laid eggs that day. Hatchlings are about 16 mm in standard length.

REMARKS: Taylor (1949b) described *Leposoma bisecta* (holotype: KU 23801; San José: 1.5 km east of El General) based on an example of this form (Ruibal 1952). In 1954 he also described *L. southi orientalis* (holotype: KU 3612; Limón: Volio, now Uatsi). Taylor (1956) reluctantly accepted the synonymy of *L. bisecta* with *L. southi* Ruthven and Gaige (1924) (holotype: UMMZ 48065; Panama: Chiriquí: Progreso) but continued to recognize Atlantic *(southi)* and Pacific slope *(orientalis)* races. The purported differences between these forms are individually variable in the larger series now available, and these nominal taxa are not recognizable.

DISTRIBUTION: Humid lowlands of eastern and southwestern Costa Rica to western Colombia (2–703 m).

Genus *Neusticurus* C. Duméril and Bibron, 1839

A genus (eleven species) of small semiaquatic lizards having pointed heads, well-developed pentadactyl limbs, clawed digits, cylindrical bodies, and compressed tails that have a series of enlarged scales forming a pair of caudal crests. Other characteristics that in combination differentiate them from other microteiids include head shields smooth; ventrals usually flattened, quadrangular or rounded posteriorly; imbricate dorsals uniformly small or a heterogeneous mixture of keeled tubercles and smaller scales; nasals separated by one or two frontonasals (internasals); two prefrontals present or prefrontal-frontonasal area covered by irregular scales; two to four frontoparietals; no enlarged interoccipital or occipitals; pupil round or reniform; tongue covered with rhomboidal imbricate scalelike papillae; lower eyelid with a translucent disk (undivided or composed of several scales) or pigmented; tympanum large, superficial to recessed; gular fold and collar well developed; collar scales not larger than adjacent scales; femoral and precloacal pores in males, precloacal pores and often femoral pores in females.

Osteological features that further characterize this genus are frontal enclosing olfactory tract ventrally; supratemporal fenestra open; postorbital and postfrontal present, separate; extrastapedial present; no lacrimal; secondary palate present in some species; Meckel's groove closed; pterygoid teeth absent; anterior teeth on jaws conical, lateral ones biconodont and laterally compressed; glossohyal single, continuous with basihyal, second ceratobranchial absent; 25 presacral vertebrae; two xiphisternal ribs; no parasternum; clavicle flattened, with a hook; single pair of processes anterior to autotomy septa on anterior series of posterior autotomic caudal vertebrae.

Neusticurus ranges from southwestern Costa Rica to north-central Colombia and western Ecuador and east of the Andes in the upper Amazon basin, from Colombia to Bolivia; also from southeastern Venezuela through the Guianas to northern Brazil.

Neusticurus apodemus Uzzell, 1966

(plate 316; map 10.65)

DIAGNOSTICS: A small, semiaquatic species having a pointed head, the dorsal scalation consisting of a mixture of small flat scales and enlarged keeled tuberculate ones that line up to form four weak keels, a pair of weak discontinuous keels on the slightly compressed tail, the ventrals flat and quadrangular, outer two rows weakly keeled and flat, rounded gular scales (fig. 10.34d).

DESCRIPTION: Total length to 157 mm, (adult males 40 to 57 mm in standard length, adult females 45 to 65 mm in standard length); tail slightly compressed, moderate, 60 to

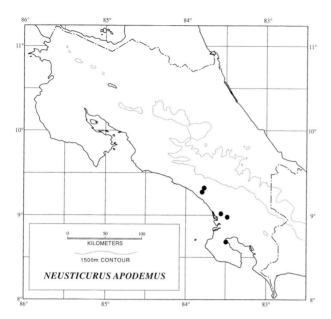

Map 10.65. Distribution of *Neusticurus apodemus*.

65% of total length; limbs short, well developed; frontonasal-prefrontonasal area covered by irregular scales; frontals in two rows; three or four large supraoculars usually bordered laterally by a series of much smaller ones; lower eyelid with a translucent disk composed of several scales; tympanum large, superficial; gular scales smooth, flattened, and juxtaposed, with rounded margins; gular fold indistinct; lateral scales a mixture of small flat scales and large tubercles; femoral and precloacal pore series continuous, 26 to 30 in males; 4 precloacal pores in females; scales on dorsal and lateral surface of tail arranged in whorls of large keeled tubercles separated by two rows of irregular flat scales; whorls of large scales interrupted laterally by interdigitated smaller smooth ones; four longitudinal rows of rectangular juxtaposed subcaudals, outer row on each side weakly keeled on tail base. Dorsum and flanks brown, sides darker with bright to dull dark-outlined light ocelli; a distinct posttympanic ocellus and one above forelimb insertion; most of upper arm covered by yellowish white area; venter mostly dark in adults, but midventer often clear pale orange in large females; chin red orange in large males.

SIMILAR SPECIES: (1) *Anadia ocellata* has smooth homogeneous platelike dorsal scales and lacks any suggestion of caudal crests. (2) The species of *Lepidophyma* have small granular gular scales and the enlarged keeled caudal scales in continuous transverse series separated by three or four transverse rows of regularly arranged smaller keeled scales. Members of that genus also lack movable eyelids.

HABITAT: A riparian form found along stream courses in undisturbed Lowland Wet Forest and Premontane Rainforest.

BIOLOGY: This rare lizard is found in leaf litter during the day but appears to be primarily active in the early evening along the margins of small streams. When approached it jumps into the stream and swims until it finds a hiding place, usually under a rock. The smallest known juvenile is 32 mm in standard length.

DISTRIBUTION: Apparently endemic to the lowlands and the lower premontane zone of southwestern Costa Rica, but may occur in adjacent southwestern Panama (30–884 m).

Genus *Ptychoglossus* Boulenger, 1890

Another group (fifteen species) of small microteiids with cylindrical bodies and tails, short but well-developed pentadactyl limbs, clawed digits, and large recessed tympanum. They differ from other members of the family by having smooth head shields; smooth juxtaposed quadrangular ventrals; imbricate keeled dorsals; scales forming complete annuli around body; no gular fold. Other characteristic features include nasals separated by frontonasal (internasal); two prefrontals or no prefrontals; two frontoparietals; no enlarged interoccipital or occipitals; pupil round; a translucent disk composed of several scales in lower eyelid; tongue covered by two longitudinal rows of oblique, transverse plicae; precloacal and femoral pores present; hemipenes tubular with semicircular flounces, each with an external band of noncalcareous hooklike papillae; distal flounces forming broad collar around bulbous or auricular tip that may be nude or usually papillate.

Defining osteological features are frontal enclosing olfactory tract ventrally; supratemporal fenestra roofed over; postorbital and postfrontal fused into single element; extrastapedial present; lacrimal present; no secondary palate; Meckel's groove open; no pterygoid teeth; anterior teeth on jaws conical; lateral ones laterally compressed, biconodont and triconodont; glossohyal in two parts, proximal portion continuous with basihyal, second ceratobranchial present; 25 presacral vertebrae; two xiphisternal ribs; no parasternum; clavicle flattened with a hook; single pair of processes anterior to autotomy septa on anterior series of posterior autotomic caudal vertebrae.

Ptychoglossus ranges from Costa Rica to central and western Colombia and occurs disjunctly in upper Amazonian Brazil, Colombia, Ecuador, and Peru and from northern Venezuela through the Guianas.

Ptychoglossus plicatus (Taylor, 1949b)

(plate 317; map 10.66)

DIAGNOSTICS: This relatively large microteiid is immediately recognized by the combination of elongate keeled and mucronate dorsal scales, smooth quadrangular ventral scales, and smooth head shields (fig. 10.34f).

DESCRIPTION: Total length to 195 mm (adults 38 to 66 mm in standard length); tail moderate, 60 to 65% of

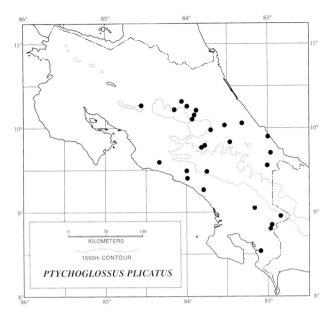

Map 10.66. Distribution of *Ptychoglossus plicatus*.

total length; limbs short, well developed; prefrontals present; four supraoculars; gulars large, rectangular (usually wider than long), imbricate, smooth; gular fold distinct, collar scales slightly larger than gulars, rectangular (longer than wide); dorsals elongate, hexagonal, keeled, and mucronate; 32 to 39 scales around midbody; 28 to 31 transverse rows of dorsal scales; 18 to 29 femoral and precloacal pores in males forming a continuous series; 0 to 3 precloacal pores in females; no femoral pores in females (0 to 3 in Panama females); caudal scales smaller than dorsals, keeled, elongate, imbricate, and mucronate; subcaudals rectangular, convex, smooth. Hemipenes with 23 to 28 flounces; lateral portion of tip with large conical protuberance, medial portion with fingerlike and bilobed projections, all papillate. Juveniles: dorsum and flanks light brown; paired light stripes on neck; underside immaculate cream; females: middorsal field brown, uniform or with a few dark spots, bordered by a light lateral stripe running to midbody or onto tail; stripes sometimes discontinuous, usually bordered below by dark chocolate; underside mostly clear with some black spots on chin; males: light stripes restricted to neck; middorsal field heavily marked with rectangular black spots; flanks dark chocolate; underside heavily marked with rectangular or triangular black spots.

SIMILAR SPECIES: *Leposoma southi* has keeled mucronate ventrals and gular scales and keeled head shields (fig. 10.34e). Other Costa Rican microteiids lack keeled dorsal scales.

HABITAT: A common leaf litter species found in Lowland Moist and Wet Forests and Premontane Wet Forest and Rainforest.

BIOLOGY: These restless little diurnal lizards forage on the leaf litter surface and when approached rapidly "snake" under the litter.

REMARKS: Taylor (1949b) originally placed this form in the genus *Alopoglossus*, which differs from *Ptychoglossus* in having the body scales in oblique (laterals) and transverse (dorsals) rows, the loreal scale separated from the supralabials by the frenocular, and the upper arm surface covered by keeled scales (the body scales are in transverse rows, the loreal scale touches a supralabial, and the upper arm scales are smooth in *Ptychoglossus*). Ruibal (1952) reassigned this species to *Ptychoglossus*, but some doubt remains as to the distinctiveness of these nominal genera (Ayala and Harris 1984). The genus and this species have recently been reviewed by Harris (1994), who only tentatively referred Colombian material to this form, which occurs in adjacent eastern Panama. *Ptychoglossus myersi* of extreme eastern Pacific slope Panama is a cryptic species superficially indistinguishable from *P. plicatus*. They differ, however, in hemipenial features (only 9 to 12 flounces in *P. myersi* versus 23 to 28 in *P. plicatus*).

BIOLOGY: A diurnal, secretive leaf litter lizard. The smallest known juvenile is 25 mm in standard length.

DISTRIBUTION: Lowlands and premontane slopes of eastern and southwestern Costa Rica (35–1,540 m) to northwestern Colombia (400–1,890 m).

Anguimorph Lizards (Infraorder Anguimorpha)

Anguimorph lizards are all carnivorous species, ranging in size from small insectivorous lizards to the large beaded and monitor lizards that prey on small to large vertebrates. Several genera of the family Anguidae are limbless and snakelike. Members of this stock may have small to medium or beadlike head scales, while others have enlarged plates. There is also considerable variation in body scalation, with granular, beadlike, and cycloid types present in different forms. The infraorder may best be recognized by the following combination of features:

1. Tongue moderately to deeply notched (from 10% to more than 50%); tongue covered with villi, foretongue retractile into hindtongue, or tongue smooth, retractile into a basal sheath; tongue protrusion oscillatory; prey apprehended by jaws; tongue mucocytes mostly mucoserous

2. Postfrontal semilunate, forked medially and clasping frontoparietal suture

3. Usually postorbital and temporal arches; no supratemporal foramen

4. Septomaxillae dorsally expanded, convex, and forming a midline crest; Jacobson's organs enlarged

5. Jaw tooth replacement alternate (with a few exceptions), reduced resorption pits present or completely absent

6. A single egg tooth in oviparous species
7. Notochord not persisting in adults
8. Rectus abdominis lateralis muscle present
9. Tibial epiphysis notched
10. No internal process on extracolumella; no extracolumella muscle

Only one family (Anguidae) of this group is represented in Costa Rica. The others, Xenosauridae of southern China, Mexico, and Guatemala; Helodermatidae of southwestern North America, Mexico, and Guatemala; and Varanidae of Africa, Asia, New Guinea, and Australia will not be considered further.

Family Anguidae Gray, 1825

Anguid Lizards

The lizards of this family (fifteen genera and 112 species) are small to moderate-sized mainly insectivorous terrestrial, fossorial, and arboreal forms that are mostly diurnal and rather secretive. Some however, feed on larger animal prey, including birds' eggs and small birds and mammals. The body is cylindrical and elongate, the limbs are short, reduced, or absent (but remnants of the limb girdles are present), and the head is not distinct or only slightly distinct from the neck and covered with large symmetrical plates. The tail is long and fragile. The body and tail are usually covered by rectangular-rhomboidal scales arranged in definite longitudinal series, but some species have cycloid scales; the dorsal scales may be smooth, striated, or keeled. The eyes are moderate to small with movable eyelids and round pupils; moderate to small ear openings are usually present, and the tympanum is recessed, but both the opening and the tympanum are absent in some forms; the digits, when present, are subcylindrical, and the lamallae lack keels; femoral pores and gular folds are absent.

Anguids occur in North, Central, and South America and the Greater and Lesser Antilles, northwestern Africa, Europe, and subtropical and tropical Asia to Sumatra and Borneo.

The following additional features formally characterize the family: premaxilla single; vomer single; postfrontal and postorbital present; supratemporal fenestra open or roofed over; postfrontal expanded; parietal foramen small or absent; jaw tooth replacement alternate, with small resorption pits; jaw teeth tuberculate, conical, or recurved; Meckel's groove open; no intercentra associated with dorsal vertebrae; caudal autotomy present, with fracture septa usually passing between two pairs of converging or parallel transverse processes or sometimes passing anterior to the processes; extracolumella without internal process, no extracolumella muscle; dorsal, ventral, and separable cephalic osteoderms present; tongue moderately notched (10 to 20% of length); foretongue retractile into hindtongue, entire tongue covered with villi; tongue protrusion oscilla-

tory; hemipenes covered with smooth or spinous papillate flounces, without calyces or petala, single or bilobed and with a simple sulcus spermaticus.

The members of this family found in Costa Rica are unlikely to be confused with any other lizards except microteiids or skinks. No microteiid or skink has a lateral fold as found in *Coloptyochon* and *Mesaspis*. No Costa Rican microteiid has small uniform cycloid body scales as in *Celestus* and *Diploglossus*, and no skink has an enlarged median interoccipital scale bordering the interparietal posteriorly as found in the last-named genera. Living anguids are placed in four subfamilies that may be briefly characterized as follows:

Diploglossinae Cope, 1864b (five genera): elongate, usually limbed lizards (*Ophiodes* has only hind limb rudiments) lacking a lateral fold, body scales cycloid and about same size; supratemporal bone short; frontal bone with parallel sides; large cephalic osteoderms present; ear opening present or absent (Central and South America and the West Indies)

Gerrhonotinae Tihen, 1949 (six genera): limbed forms having a lateral fold containing small scales separating enlarged dorsals from ventral scales; supratemporal bone elongate; frontal hourglass shaped; large cephalic osteoderms present; ear opening present (North and Central America)

Anguinae Gray, 1825 (three genera): elongate, limbless forms (some with hind limb rudiments); a lateral fold containing small scales separating enlarged dorsal and ventral scales present or not; body scales squarish to rounded; supratemporal bone short; frontal bone with parallel sides; large cephalic osteoderms present; ear opening distinct, small, or absent (North America, Europe, subtropical and tropical Asia to Borneo)

Anniellinae Cope, 1864b (one genus): elongate, limbless burrowing lizards lacking a lateral fold and having the body covered with uniform cycloid scales; frontal bone with parallel sides; cephalic osteoderms small; no ear opening (western North America)

Anguids are mostly viviparous and have both yolk sac and chorioallantoic placentas in species studied in detail. In oviparous forms the eggs have flexible cream to white shells. These lizards constantly sample the environment with the tongue tips, which carry particles to the Jacobson's organs, "tasting" prey items before they grasp and swallow them.

Subfamily Diploglossinae Cope, 1864b

Considerable confusion has existed regarding the use of the generic names *Celestus* and *Diploglossus*. Boulenger (1885) recognized a single genus *(Diploglossus)* and was followed in this by Dunn (1939b) and Underwood (1959) and most recently Campbell and Camarillo (1994). Others, beginning with Stejneger (1904) and including Taylor (1956), Wilson,

Porras, and McCranie (1986), Savage and Lips (1993), and McCranie and Wilson (1996), regarded the two as distinct based on exposure of the claws, which are enclosed in a claw sheath in *Diploglossus* and exposed in *Celestus*.

Straham and Schwartz (1977) used putative differences in osteoderm canal structure to recognize two genera, each containing species with or without exposed claws. Variation and ontogenetic changes in the osteoderm characters led subsequent authors to reject them as diagnostic (Wilson, Porras, and McCranie 1986; Campbell and Camarillo 1994; Savage and Lips 1993).

In the present account *Celestus* and *Diploglossus* are considered to represent two lineages based on the differences in claw sheaths.

Three other genera in addition to the two described here are usually recognized within the subfamily. *Ophiodes* (four species from Bolivia and southern Brazil to south-central Argentina) are elongate lizards that lack forelimbs, have the hind limbs reduced to clawless flaps, and have the auricular opening small and hidden by scales. The nominal Hispaniolan genera, *Sauresia* (two species) and *Wetmorena* (one species), have only four digits on each limb and the claws hidden within a scaly sheath. In addition, members of the latter genus lack an ear opening.

Genus *Celestus* Gray, 1839

These small to moderate-sized, skinklike lizards (twenty-seven species) have short pentadactyl limbs with exposed claws on all digits, large head shields including an enlarged interoccipital, and the body covered by small uniform striated cycloid scales. Other characteristics include two pairs of large internasals (supranasals) present and preventing contact between frontonasal and rostral; three (one fronto-parietal and two prefrontals) or one (fused frontonasal and prefrontals) plates in frontonasal-prefrontal region; supra-oculars in two rows, inner row of much larger plates; a pair of frontoparietals; a large interoccipital plate.

Costa Rican species of the genus are distinguished most readily from the several skinks *(Eumeces, Mabuya,* and *Sphenomorphus)* by having an enlarged interoccipital and two rows of supraoculars.

Celestus ranges on the mainland from Veracruz, Mexico, to Costa Rica (ten species) and is found on Jamaica (eight species), Hispaniola (eight species), and Navassa Island (one species); one Jamaican species also occurs on Grand Cayman Island, and one recognized form *(Celestus striatus)* is of unknown provenance.

As far as is known, members of this genus are viviparous.

KEY TO THE COSTA RICAN LIZARDS OF THE GENUS *CELESTUS*

1a. 73–77 transverse rows of ventral scales; 65–73 transverse rows of dorsal scales 2

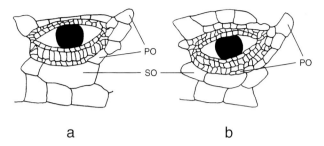

Figure 10.35. Relation between postocular (PO) and subocular (SO) scales in anguid lizards: (a) postoculars and suboculars continuous; (b) postoculars and suboculars juxtaposed.

1b. 84–92 or more transverse rows of ventral scales; 76–81 transverse rows of dorsal scales; dorsal and lateral pattern includes scattered light-tipped black scales that sometimes form short longitudinal or vertical lines; side of neck without vertical dark and light bars; a pair of light longitudinal dorsolateral and dark ventrolateral stripes . . . . *Celestus hylaius* (p. 529)

2a. Suboculars and postoculars continuous (fig. 10.35a); 10–12 precloacal scales; body pattern of scattered black scales . . . . *Celestus cyanochloris* (p. 527)

2b. Suboculars and postoculars juxtaposed (fig. 10.35b); 8 precloacal scales; body pattern of scattered black scales dorsally and definite black-margined light bars on flanks . . . . *Celestus orobius* (p. 529)

Note that the specimen, now lost, reported as *Celestus steindachnerii* by Cope (1893) from Boruca, Puntarenas Province (550 m), remains an enigma, as pointed out by Taylor (1956). Cope described it as having two prefrontals, 36 rows of scales around the midbody, and the body scales smooth anteriorly but striated and lacking a median keel posteriorly and on the tail. The first listed feature eliminates from consideration all mainland Central American species except *Celestus montanus* (Honduras). The second eliminates *C. montanus* and all known Costa Rican forms as possible conspecifics as well. In addition, *C. cyanochloris* and *C. orobius* have well-developed median keels on the caudal scales. If the presence of two prefrontals is regarded as a possible anomaly, the scale count and the structure of the posterior dorsal and caudal scales are similar to those of *Celestus enneagrammus* (Mexico). Even if the locality data for Cope's example are erroneous, identity with *C. enneagrammus* seems a remote possibility. Consequently Savage and Lips (1993) concluded, in agreement with Taylor, that an undescribed species of *Celestus* may await rediscovery from southwestern Costa Rica.

Celestus cyanochloris Cope, 1894

(fig. 10.36c; plate 318; map 10.67)

DIAGNOSTICS: A moderate-sized, short-limbed skinklike lizard having the body covered with small uniform shiny cycloid scales, the upper caudal scales with a median

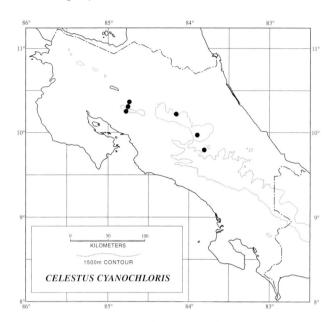

Map 10.67. Distribution of *Celestus cyanochloris*.

keel, the digits with exposed claws (fig. 10.13b), an enlarged interoccipital scale, and a brown ground color without light lateral bars.

DESCRIPTION: Total length to 238 mm (adults 67 to 99 mm in standard length); tail moderate, 53 to 59% of total length; head not distinct from neck; two pairs of internasals (supranasals) meeting on midline; frontonasal and prefrontals fused into one large shield; five median supraoculars; suboculars and postoculars continuous; ten or eleven supralabials; an azygous postmental; five pairs of enlarged chin shields; 32 to 34 rows of scales around midbody; 65 to 73 transverse rows of dorsal scales; 73 to 77 transverse rows

of ventrals; 10 to 12 enlarged precloacal scales; 20 to 25 lamellae under fourth toe. Dorsal ground color tan, becoming darker brown posteriorly, with black speckling; tail reddish brown; venter lime green, chin and throat bluish; underside of tail orange; iris dark brown.

SIMILAR SPECIES: (1) *Celestus orobius* has light bars on the flanks (fig. 10.36a), the postoculars and subocular juxtaposed (fig. 10.35b), and eight enlarged precloacal scales. (2) *Celestus hylaius* has the postoculars and subocular juxtaposed (fig. 10.35b) and light and dark longitudinal stripes in the dorsal pattern (fig. 10.36b). (3) *Diploglossus bilobatus* has all but the tips of the claws hidden in a scaly sheath (fig. 10.13a). (4) *Mabuya unimarginata* has enlarged paired nuchal (occipital) scales but no enlarged interoccipital shield (fig. 10.12c). (5) *Sphenomorphus cherriei* has two prefrontals and a frontonasal (fig. 10.12a) and lacks an enlarged interoccipital shield.

HABITAT: Upper portion of the premontane belt in Premontane Moist and Wet Forests and Rainforest and marginally into Lower Montane Rainforest.

BIOLOGY: A rare secretive terrestrial form that climbs into low bushes on occasion. The smallest known juvenile is 54 mm in standard length.

REMARKS: Taylor (1956) was the first to collect additional examples of this form after its description by Cope (1894a). Although Taylor thought that the holotype (originally MNCR 217) was lost, it is extant (now AMNH 16290; Cartago: Volcán Irazú). Taylor inadvertently called this species *Celestus chrysochorus* in his discussion and in his key to the genus. Straham and Schwartz (1977) included this species in *Diploglossus*.

DISTRIBUTION: Humid forests of the premontane and lower montane zones of the Cordillera de Tilarán, Cordi-

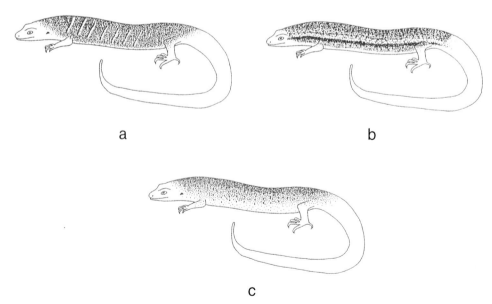

a

b

c

Figure 10.36. Color patterns in Costa Rican anguid lizards of the genus *Celestus:* (a) *C. orobius;* (b) *C. hylaius;* (c) *C. cyanochloris.*

Map 10.68. Distribution of *Celestus hylaius*.

llera Central, and Cordillera de Talamanca of Costa Rica (1,200–1,710 m).

Celestus hylaius Savage and Lips, 1993

(fig. 10.36b; plates 319–20; map 10.68)

DIAGNOSTICS: A small, short-legged skinklike lizard having the body covered with small uniform cycloid scales, the upper caudal scales without a median keel, the digits with exposed claws, an enlarged interoccipital shield, and ground color olive with black flecking and dorsolateral yellow green stripes and black ventrolateral stripes.

DESCRIPTION: Total length to 198 mm (adults 70 to 107 mm in standard length); tail moderate, 57 to 60% of total length, frequently regenerated; head not distinct from neck; two pairs of internasals (supranasals) meeting on midline; prefrontals and frontonasal fused into a single large plate; usually five median supraoculars, sometimes six; suboculars and postoculars juxtaposed; nine to eleven supralabials, usually nine or ten; an azygous postmental; five pairs of enlarged chin shields; 31 to 33 scales around midbody; 76 to 81 transverse rows of dorsals; 84 to 92 transverse rows of ventrals; 10 to 12 enlarged precloacals; 22 to 27 lamellae under fourth toe. Dorsal ground color coppery brown with greenish cast and numerous black flecks that coalesce to form an indistinct ventrolateral stripe along each side; a greenish dorsolateral stripe on each side; labials and venter bright yellow green; iris rusty brown.

SIMILAR SPECIES: (1) *Celestus cyanochloris* has the postoculars and subocular continuous (fig. 10.35a) and lacks light and dark longitudinal stripes in the pattern. (2) *Celestus orobius* has dark-edged vertical bars on the flanks (fig. 10.36a). (3) *Diploglossus bilobatus* has all but the tips of the claws hidden in a scaly sheath (fig. 10.13a). (4) *Ma-*

buya unimarginata has enlarged nuchal (occipital) scales but no enlarged interoccipital shield (fig. 10.12c). (5) *Sphenomorphus cherriei* has two prefrontals and a frontonasal (fig. 10.12a) and lacks an enlarged interoccipital plate.

HABITAT: An inhabitant of relatively undisturbed Lowland Wet and Moist Forests, where it has been found in leaf litter and among recently felled trees.

BIOLOGY: This relatively rare secretive species apparently is diurnal and terrestrial, though Corn (1981) thought it might be arboreal and possibly active at night. The smallest juvenile examined is 38.5 mm in standard length.

DISTRIBUTION: Lowlands of the Atlantic versant of eastern Costa Rica (40–430 m).

Celestus orobius Savage and Lips, 1993

(fig. 10.36a; map 10.69)

DIAGNOSTICS: A small, short-limbed shiny skinklike lizard having the body covered with small uniform cycloid scales, the upper caudal scales with a median keel, the digits with exposed claws, an enlarged interoccipital shield, and a pattern of dark-edged vertical light bars on the flanks.

DESCRIPTION: Total length to approximately 197 mm (standard length to 83 mm); head not distinct from neck; two pairs of internasals (supranasals) meeting on midline; frontonasal and prefrontals fused into a single large shield; five median supraoculars; suboculars and postoculars juxtaposed; nine supralabials; an azygous postmental; five pairs of enlarged chin shields; 30 scales around midbody; 66 transverse rows of dorsal scales; 75 transverse rows of ventrals; 8 enlarged precloacal scales; 21 to 22 lamellae under fourth toe. Upper surface olive brown with some darker brown scales evenly scattered on neck, body and tail; defi-

Map 10.69. Distribution of *Celestus orobius*.

nite black-edged light bars on flanks; venter uniform gray (in preservative).

SIMILAR SPECIES: (1) *Celestus cyanochloris* has the postoculars and suboculars continuous (fig. 10.35a), lacks light bars on the flanks, and has ten to twelve enlarged precloacal scales. (2) *Celestus hylaius* lacks light bars on the flanks and has light and dark longitudinal stripes in the dorsal pattern (fig. 10.36b). (3) *Diploglossus bilobatus* has all but the tips of the claws hidden in a scaly sheath (fig. 10.13a). (4) *Mabuya unimarginata* has enlarged paired nuchal (occipital) scales and lacks an enlarged interoccipital plate (fig. 10.12c). (5) *Sphenomorphus cherriei* has three shields in the prefrontal region (two prefrontals and a frontonasal) (fig. 10.12a) and lacks an enlarged interoccipital plate.

HABITAT: Lower Montane Rainforest of the Cordillera de Talamanca.

REMARKS: *Celestus orobius* Savage and Lips, 1993, is known from a single specimen (LACM 138540).

DISTRIBUTION: The single known example is from the Pacific slope of the Cordillera de Talamanca near the Carretera Interamericana between Hortenisa and Florencia, San José Province, (1,500–2,000 m).

Genus *Diploglossus* Wiegmann, 1834a

Another genus (eleven species) of small to large skinklike forms having short pentadactyl limbs, large head shields, and the body covered by small uniform striated cycloid scales. They differ from all other similar Costa Rican lizards, including members of the allied genus *Celestus*, in having the claws on all digits hidden within a scaly sheath so that only the claw tips are exposed. Additional features that in combination will separate this genus from others include two pairs of large internasals (supranasals) present, preventing contact between frontonasal and rostral; three (one frontonasal and two prefrontals) or one (fused frontonasal and prefrontals) plates in frontonasal-prefrontal region; supraoculars in two rows, inner row of much larger shields; a pair of frontoparietals; a large interoccipital plate.

The genus occurs disjunctly on the mainland in Central America and western South America, Nicaragua to Ecuador (three species) and southeastern Brazil, and Amazonian Peru and Bolivia (two species). Single endemic species are also found on the islands of Malpelo, Montserrat, and Puerto Rico, and three species occur on Cuba.

The mainland and Cuban species of this genus are apparently oviparous, but *D. pleii* of Puerto Rico is viviparous (Greer 1967).

KEY TO THE COSTA RICAN SPECIES OF THE GENUS *DIPLOGLOSSUS*

1a. Two prefrontals and a single median frontonasal plate, the latter separated at posterolateral margins from contact with the supraoculars (fig. 10.37b); dorsal body

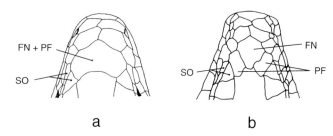

Figure 10.37. Scalation of the prefrontal region in Costa Rican anguid lizards: (a) single large plate (FN + PF = fused frontonasal and prefrontals) in contact with supraoculars (SO) on each side; (b) two prefrontals (PF) and a single median frontonasal (FN) that is not in contact with supraoculars (SO).

scales with a pronounced median keel; a series of dark-edged, narrow, light transverse bands across dorsum somewhat obscured in large adults; adults 150–200 mm in standard length . . . . *Diploglossus monotropis* (p. 531)

1b. Prefrontals and frontonasal fused to form a single large plate that touches supraoculars on each side (fig. 10.37a); dorsal body scales strongly striated, but median keel weak or absent; dorsum uniform black to brown, often weakly checkered or reticulated with black; standard length 62–99 mm in adults . . . . *Diploglossus bilobatus* p. 530)

Diploglossus bilobatus (O'Shaughnessy, 1874)

(plates 321–22; map 10.70)

DIAGNOSTICS: A small to moderate-sized, short-limbed shiny skinklike lizard covered with small cycloid scales that may have a weak median keel, having the claws hidden within a scaly sheath so that only the tips are exposed (fig. 10.13a) and an enlarged interoccipital shield and lacking well-developed median keels on the caudal scales. In addition, there are no dark dorsal crossbands, but well-developed light-centered black bars may be present on the flanks.

DESCRIPTION: Total length to 240 mm; males (adults 74 to 99 mm in standard length) larger than females (adults 62 to 79 mm in standard length); tail moderate, 57 to 60% of total length, usually regenerated or incomplete in most examples; head not distinct from neck; two pairs of internasals (supranasals) meeting on midline; frontonasal and prefrontals fused into one large shield; five enlarged supraoculars; suboculars and postoculars juxtaposed; ten supralabials; one postmental; three pairs of enlarged chin shields; 36 to 39 rows of scales around midbody; 11 to 13 lamellae under fourth toe. Dorsum uniform dark brown to black in juveniles, brown and often checkered or reticulated with black in adults; flanks with a definite greenish cast in adults; middorsal field often bordered by thin continuous to discontinuous light line bordered in turn laterally by narrow dark line; golden to greenish spots or ocelli indicated on

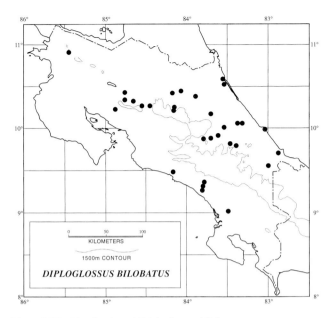

Map 10.70. Distribution of *Diploglossus bilobatus*.

flank, set in vertical dark bars or completely obscured by dark bars in adult males; venter gray in juveniles, greenish in adults, with a pink to salmon suffusion over undersides of body, limbs, and tail base in large adults.

SIMILAR SPECIES: (1) The species of *Celestus* have the claws exposed (fig. 10.13b). (2) *Mabuya unimarginata* has the claws exposed, a pair of enlarged nuchal scales, and two prefrontal shields and lacks an enlarged interoccipital plate (fig. 10.12c). (3) *Sphenomorphus cherriei* has exposed claws and lacks an enlarged interoccipital plate. (4) *Mesaspis monticola* has several middorsal rows of rhomboidal to rectangular strongly keeled dorsal scales (fig. 10.10) and exposed claws.

HABITAT: Lowland Moist and Wet Forests and Premontane Wet Forest and Rainforest on the forest floor.

BIOLOGY: The species is relatively uncommon, secretive, diurnal, and semifossorial. Two males were captured while fighting in the center of a trail. Taylor (1956) found a female under a log apparently brooding six eggs (plate 323). A female, which escaped, was found with fifteen eggs on June 3; thirteen eggs hatched on July 6–8. Two hatchlings, the smallest of which was 28 mm in standard length, were preserved August 19.

REMARKS: As reported by Taylor (1956) and Myers (1973), the body scales of this species (and probably of other *Celestus* and *Diploglossus*) are smooth in juveniles and strongly striated in adults. In addition, adults may have a weak median keel on most body scales or on the lateral scales only or no keeling.

DISTRIBUTION: Lowlands and adjacent premontane slopes on the Atlantic versant of Costa Rica and northwestern Panama and on the Pacific versant of southwestern Costa Rica (2–1,360 m).

Diploglossus monotropis (Kuhl, 1820)

El Escorpión Coral

(plates 324–25; map 10.71)

DIAGNOSTICS: The moderate-sized to large adult males of this species are unmistakable with their bold dorsal pattern of alternating broad, black-edged dark banks and narrow light interspaces, orange to bright red venter and flanks, and orange iris. Juveniles and females are vividly banded as well, but the venter is yellow in small specimens and gradually turns orange in adult females. The species is further characterized by having all but the tips of the claws concealed in a scaly sheath (fig. 10.13a) and the dorsal, lateral, and caudal scales heavily striated and with a pronounced median keel.

DESCRIPTION: Adult males of this species to 537 mm in total length, males (adults to 215 mm in standard length, but usually in 175 to 190 mm range), larger than females (adults 150 to 188 mm in standard length); tail moderate, 57 to 60% of total length, rarely complete or unregenerated in adults; head slightly distinct from neck in large males; two pairs of internasals (supranasals) meet on midline; two prefrontals and a frontonasal present; five enlarged supraoculars; suboculars and postoculars continuous; ten supralabials; a large azygous postmental; three pairs of enlarged chin shields; 37 to 40 scales around midbody; 8 to 10 lamellae under fourth toe. Dorsum with a pattern of alternating broad olive brown to blackish bands and narrow black-edged white, gray, or greenish ones; light bands usually continuing onto flanks, often only black edged on anterior margin, sometimes broken up into ocelli; occasionally only the black vertical line indicated; light areas on flanks orange to red; dorsal dark bands usually much darker than contin-

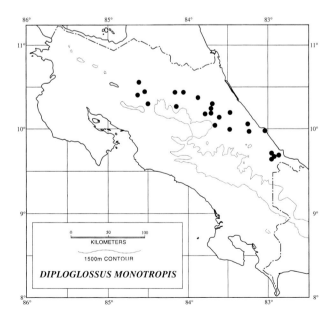

Map 10.71. Distribution of *Diploglossus monotropis*.

uations on flanks, sometimes olive brown throughout; tail with alternating broad dark bands and black-edged light interspaces; undersurface yellowish in juveniles; venter and underside of tail changing to orange or red in adults; iris rusty red.

SIMILAR SPECIES: *Coloptychon rhombifer* lacks orange or red in its coloration and has lateral folds (fig. 10.10), smooth rhomboidal to rectangular dorsal and ventral scales, and exposed claws (fig. 10.13b).

HABITAT: A forest floor species restricted to the Lowland Moist and Wet Forests of the Atlantic versant.

BIOLOGY: A secretive diurnal terrestrial species that is rarely encountered. As implied by the common name (which translates as the coral snake scorpion) this harmless lizard is believed to be deadly poisonous and is much feared by most Costa Ricans because it is large and colored like a coral snake. The smallest juvenile is 117 mm in standard length.

DISTRIBUTION: Humid Atlantic lowlands of southern Nicaragua, Costa Rica, and western Panama and both slopes from Panama to Colombia and western Ecuador (2–320 m).

Subfamily Gerrhonotinae Tihen, 1949

In the most recent review of the gerrhonotines, Good (1987, 1988) recognized four other genera besides the two described below as belonging to the subfamily. These include *Gerrhonotus* (two species), ranging from central Texas to Guatemala; *Elgaria* (six species), from British Columbia to the tip of Baja California and from Arizona and New Mexico to Jalisco, Mexico, with an isolate in Neuvo León, Mexico; *Barisia* (three species), upland areas of Mexico, west of the Isthmus of Tehuantepec; and *Abronia* (twenty-eight species), mountains from eastern and central Mexico to El Salvador and Honduras. Except for some of the species of *Elgaria* that occur along the Pacific coast of temperate North America, these are primarily upland lizards that do not range into the tropical lowlands in Mexico or Central America.

Genus *Coloptychon* Tihen, 1949

A moderate-sized elongate lizard having short pentadactyl limbs, clawed digits, a long tail, large head shields, large smooth rhomboidal dorsal and ventral scales, and a distinct lateral fold. The genus is further characterized by having a pair of anterior internasals, a pair of supranasals followed by paired posterior internasals, the dorsal scales in 16 to 18 longitudinal rows, and the ventrals in 10 longitudinal rows. The presence of the lateral fold separates this lizard from all other Costa Rican species except *Mesaspis monticola*. Nothing is known of the osteology of this genus.

The single representative of *Coloptychon* occurs at low elevations in southwestern Costa Rica and adjacent southwestern Panama.

Coloptychon rhombifer (Peters, 1876)

(plate 326; map 10.72)

DIAGNOSTICS: This moderate sized broad-headed and elongate species is unlikely to be confused with any other lizard except its ally *Diploglossus monotropis*. *C. rhombifer* differs most obviously from the last-named species in having dark-edged dark crossbands on the body and tail, exposed claws on the digits (fig. 10.13b), and large rhomboidal to quadrangular dorsal and ventral scales.

DESCRIPTION: Total length to 353 mm; tail 66% of total length; head distinct from neck, bluntly pointed; two or three postrostrals border rostral; two anterior internasals meeting on midline or separated by an azygous median scale; two medially expanded supranasals and two posterior internasals; frontonasal present, separated from supranasals by internasals; two prefrontals in contact on midline; one or two frontals; three to five suboculars, continuous with postoculars; five to six scales in each vertical row of temporals; two postmentals; five pairs of enlarged chin shields; dorsal scales in 16 to 18 longitudinal rows; ventrals in 10 longitudinal rows. Ground color of upper surface tan; body marked with eight triangular to rhomboidal black-edged brown crossbands; top of head uniform; tailed banded like body; venter with narrow brown posteriorly pointed chevrons.

SIMILAR SPECIES: *Diploglossus monotropis* has a pattern of dark crossbands alternating with narrow light lines, orange to red flanks, small cycloid scales on the body, and all but the tips of the claws hidden in a scaly sheath (fig. 10.13a).

HABITAT: In Costa Rica known only from the lowland rainforests of the southwest Pacific versant.

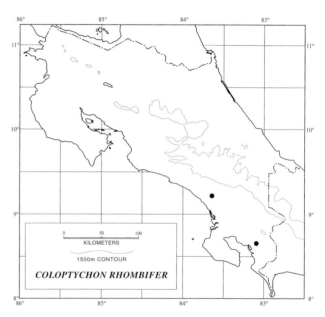

Map 10.72. Distribution of *Coloptychon rhombifer*.

REMARKS: This rare lizard was originally described by W. Peters as *Gerrhonotus rhombifer* from "Chiriqui," Panama (holotype ZMB 8655). Tihen (1949) erected the genus *Coloptychon* for the species, and Good (1988) confirmed its distinctiveness.

DISTRIBUTION: Humid lowlands of southwestern Costa Rica and adjacent western Panama; in Costa Rica known only from the Río General drainage system and from near Golfito, southern Puntarenas Province (50–500 m).

Genus *Mesaspis* Cope, 1877

The six species of this montane genus are moderate-sized, rather stocky lizards having short pentadactyl limbs, clawed digits, a long tail, large rhomboidal dorsal and ventral scales, the several middorsal rows strongly to weakly keeled, and a distinct lateral fold. Other features that in combination will distinguish the genus from other gerrhonotines are a pair of supranasals and one or two pairs of internasals present, the dorsal scales in 14 to 20 longitudinal rows, the ventrals in 12 longitudinal rows, and marked sexual dichromatism. The presence of a lateral fold separates this species from all other Costa Rican forms except *Coloptychon rhombifer*.

Good (1987) provided osteological diagnoses of the gerrhonotine genera based primarily on detailed and subtle differences in bone shapes. He recognized three species groups of *Mesaspis* (Good 1988) based on external characteristics; one, the *Mesaspis moreletii* group, is represented in Costa Rica.

All species of *Mesaspis* are high montane forms, and the generic distribution is disjunct in Oaxaca and upper and lower Central America, with no two species occurring in sympatry. As far as is known the species of this genus (confirmed for *M. gadovii*, *M. monticola*, and *M. moreletii*) are viviparous.

Extralimital taxa belonging to this group include *Mesaspis moreletii* (Chiapas, Mexico, to Nicaragua) and *M. viridiflava* (Oaxaca, Mexico).

Mesaspis monticola (Cope, 1877)

Dragón, Lagartija de Altura

(plate 327; map 10.73)

DIAGNOSTICS: Adult males of this moderate-sized species are instantly recognizable by having all the dorsal surfaces dark brown to black and marked with numerous greenish yellow to turquoise flecks and dashes. Juveniles and females are dull in color but may be distinguished from similar forms in having five to seven rows of strongly keeled middorsal scales.

DESCRIPTION: Total length to 236 mm, males (adults 63 to 88 mm in standard length) slightly larger than females (adults 63 to 84 mm in standard length); tail 61 to 63% of total length; head somewhat distinct from neck in males,

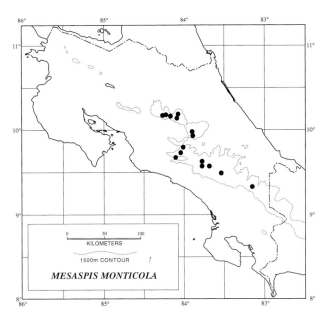

Map 10.73. Distribution of *Mesaspis monticola*.

bluntly pointed; two postrostrals bordering rostral; two anterior internasals, two supranasals, and sometimes two small posterior internasal remnants restricted to lateral margin of snout; a frontonasal present in contact with supranasals; two prefrontals not meeting on midline; a single frontal; usually two suboculars, rarely three, continuous with postoculars; no more than four in any vertical row of temporal scales; one postmental; four pairs of enlarged chin shields; 16 longitudinal rows of dorsals at midbody and 12 longitudinal rows of ventrals. Juveniles and females: broad light to dark brown dorsal field; usually with a dark discontinuous medial stripe that terminates at various points on body or continues nearly to tip of tail; dorsal field usually bordered by a more or less distinct narrow light stripe and a dark lateral one on upper margin of dark brown flank; underside dull gray with numerous dark marks; adult males: upper surface black or dark brown, heavily mottled with green to yellow flecks and dashes; middorsal and lateral dark stripes sometimes present; undersurface dull gray, heavily marked with dark spots.

SIMILAR SPECIES: (1) *Diploglossus bilobatus* has the body covered with small cycloid scales and has all but the tips of the claws hidden in a scaly sheath (fig. 10.13a). (2) *Anadia ocellata* is a much smaller lizard that is covered with smooth flat scales and lacks paired supranasal and internasal scales (fig. 10.34a). (3) *Mabuya unimarginata* has the body covered with small cycloid scales. (4) *Sphenomorphus cherriei* is a much smaller species covered by small cycloid scales.

HABITAT: The species occurs in abundance in Lower Montane Rainforest, Montane Wet Forest and Rainforest, and Subalpine Pluvial Paramo zones. It is usually found on the ground in association with fallen logs, stumps, woodpiles, boards, loose bark, moss mats, grass clumps, or rock piles. It most commonly occurs in forest edge or gaps and is most abundant in open areas such as grasslands and pastures, disturbed situations including those near human habitation, and above timberline in the paramo.

BIOLOGY: Various aspects of the biology of this species have been reported by Fitch (1973a), Vial (1975), and Vial and Stewart (1985, 1989), and their observations form the basis for the following account. *Mesaspis monticola* is an essentially terrestrial diurnal species that tends to spend most of its time foraging or resting under surface litter. These lizards raise their body temperatures by basking on rocks or logs on sunny mornings. Activity temperatures are in the 20.2 to 27.2°C range, several degrees above that of the substrate. Fluctuations in temperatures at the high elevations where this species occurs are extreme both between day and night and also from day to day depending on cloud cover. Nighttime temperatures may drop to nearly 0°C and rise to 24°C during a single twenty-four-hour period. Consequently activity is concentrated in daytime periods when the sun is shining, and the lizards may be inactive and hidden in their retreats for several days at a time. Food is primarily arthropods, but they occasionally eat juvenile salamanders *(Bolitoglossa pesrubra* or *B. subpalmata)* and the young of their own species. The species is strictly viviparous and has a transitory yolk sac placenta that is replaced by an omphalallantoic placenta. Ovulation occurs in late June to September, and there is an extended gestation period of over five months, with parturition in May to June of the following year. In the laboratory females not obviously pregnant on capture in October produced young about 14.5 to 17.5 weeks later, in January and early February (Wicknick 1993). Vial and Stewart (1985) provide evidence that females reproduce biennially. There appears to be some parental care by females, since groups of young are often found in close association with the presumed mother (plate 328). The number of young varies from two to ten, averaging about five. Neonates are 24 to 27 mm in standard length. In some areas of Costa Rica the *campesinos* swear that this harmless lizard is very venomous, inflicting stings with its tail.

REMARKS: This species has been referred to by many recent authors under the name *Gerrhonotus monticola* or *Barisia monticola*. It is a member of the *Mesaspis moreletii* group (Good 1988) that also includes *M. moreletii* (Chiapas, Mexico, to Nicaragua), characterized by having the dorsal scales in 18 to 20 longitudinal rows, and *M. viridiflava* (Oaxaca, Mexico), with 14 rows. The karyotype of *M. monticola* is 2N = 30, comprising two pairs of metacentrics, seven pairs of telocentrics, and six pairs of microchromosomes; NF = 34 (Bury, Gorman, and Lynch 1969).

DISTRIBUTION: Humid areas of the upper portions of the lower montane zone and montane and subalpine belts of the cordilleras of Costa Rica and extreme western Panama (1,800–3,800 m).

11 | Snakes
Suborder Serpentes Linné, 1758 (Ophidia)

Snakes are essentially specialized limbless lizards that probably evolved from burrowing forms and now have expanded from the subterranean life to occupy a wide variety of terrestrial, arboreal, and aquatic (both marine and freshwater) habitats. Snakes are distinguished from lizards by consistently having the following character states:

1. No external functional limbs
2. Pelvis reduced to a vestige or absent
3. No movable eyelids
4. No external ear openings
5. No pectoral girdle
6. Cranium encased in a bony capsule, not open anteriorly
7. Mandibles connected at symphysis by an elastic ligament

Other features shared by all snakes and only occasionally found in lizards include no parietal foramen, zygantra and zygosphenes on vertebrae, and no urinary bladder. In addition, and most important, the upper temporal skull arch and the jugal bone are absent in snakes so that the quadrate and lower jaw are very loosely attached to the cranium. This gives the jaw even greater mobility than in lizards (although the arch and jugal may be reduced or absent in a few saurians). The flexible jaw arrangement combined with the elastic connection between the two halves of the lower jaw is what allows most snakes to move each mandible independently when swallowing prey. By moving first one mandible and then the other forward under the prey, the snake literally pulls itself around the food and engulfs it. The jaw movements are enhanced by single rows of recurved teeth on the jaws and usually on the palate as well, which help move the prey item into the esophagus. Although not all snakes eat large prey that require this behavior, even those feeding on smaller animals have the characteristic jaw arrangement, although the primitive blind snakes cannot thrust one mandible ahead of the other. No lizard has both the ligamentous connection between the mandibles and the modified temporal quadrate region typical of snakes, and no lizard can engulf prey in the characteristic manner of the serpent.

Snakes differ from amphisbaenian lizards (features for the latter in parentheses) in lacking a pectoral girdle (present but reduced), in having the mandibular rami connected by an elastic ligament (united by a suture), in having zygantra and zygosphenes on the vertebrae (absent), in having the left lung reduced or absent (right lung reduced), and in having body scutellation comprising uniform small cycloid or granular scales or with enlarged ventral scutes (scales quadrangular, forming rings that encircle body).

The 442 genera and approximately 2,900 species of living snakes are placed in nineteen families as follows (fig. 11.1):

Infraorder Scolecophidia Cope, 1864b
  Family Anomalepididae (blind worm snakes)
  Family Typhlopidae (blind snakes)
  Family Leptotyphlopidae (thread snakes)
Infraorder Alethinophidia Nopsca, 1923
  Family Anomochilidae (stumpheads)
  Family Aniliidae (coral pipe snake)
  Family Cylindrophiidae (Asian pipe snakes)
  Family Uropeltidae (shieldtails)
  Family Xenopeltidae (sunbeam snake)
  Family Loxocemidae (dwarf boa)
  Family Boidae (boas and pythons)
  Family Ungaliophiidae (ungaliophiids)
  Family Xenophidiidae (xenophiids)
  Family Bolyeriidae (Round Island snakes)
  Family Tropidophiidae (wood snakes)
  Family Acrochordidae (file snakes)
  Family Colubridae (harmless and rear-fanged snakes)
  Family Atractaspididae (mole vipers)
  Family Elapidae (cobras, coral snakes, sea snakes, and
    allies)
  Family Viperidae (adders and vipers)

Note that the name Alethinophidia is often attributed to Hoffstetter (1955), but I follow McDiarmid, Campbell, and Toure (1999) in crediting Nopsca (1923).

The origin of snakes has long been controversial (see Rieppel 1988 for a review). Did they arise from burrowing lizards, or are they derived from marine varanoid lizards? The burrowing hypothesis has had many supporters, especially since Walls's (1942) demonstration of the great differences in eye morphology in lizards and snakes. However, Caldwell and Lee (1997) and Lee and Caldwell (1998) proposed that the earliest undoubted snake (*Pachyrhachis problematicus*) known up to that time seems to have been a marine species. *Pachyrhachis* has a well-developed pelvis and hind limbs and shows affinities to Cretaceous marine varanoid lizards (the mosasaurs). This elongated form (164 presacral

535

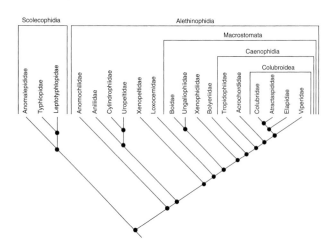

Figure 11.1. Phylogenetic relationships among living families of snakes. Several often recognized monophyletic suprafamilial taxa are also indicated.

vertebrae) was over 1 m in total length. The age of the fossil is about 100 Ma (lowermost late Cretaceous).

The oldest known snake remains (vertebrae only) are from the earlier Cretaceous, about 120 Ma (Rage and Richter 1994). Werner and Rage (1994) also report that a variety of snakes, including colubroids, were present in the middle Cretaceous, about 95 Ma. Zaher (1998) questioned the basal position of *Pachyrhachis* among snakes and suggested that it is an advanced snake allied to the Macrostomata (fig. 11.1), but see Lee (1998) for an opposing view. Recently a second limbed species, *Haasiophis terrasanctus*, was discovered in the same fossil bed in Israel that yielded *Pachyrhachis* (Tchernov et al. 2000). Tchernov and his coauthors conclude that both genera are advanced snakes most closely related to boas and their allies among living forms. Greene and Cundall (2000) nicely contrast the issues raised by these discoveries in seeking an understanding of snake evolution.

Snakes are found on all continents except Antarctica but are most diverse in numbers and adaptive radiations in tropical and subtropical regions. Their northern limits of distribution are above the Arctic Circle near 67° N latitude in Scandinavia (Europe), nearly to the Arctic Circle (67° N) in eastern Siberia (Asia), and to near 60° N in western Canada and 49° N in eastern Canada. Southern limits are the tips of South Africa (35° S) and Australia (40° S), Tasmania (42° S), and southern Argentina (45° S). There are no native snakes on New Zealand and most oceanic islands. Highly venomous snakes of the families Elapidae and Viperidae are absent from Madagascar, the Greater Antilles, and Ireland.

Ten snake families are extralimital to the area covered in this book: the Anomochilidae (one genus) in Malaysia and the Indo-Malayan archipelago; the Uropeltidae (eight genera) in India and Sri Lanka; the Cylindrophiidae (one genus) of tropical Asia; the Tropidophiidae of Panama, South America, and the West Indies (two genera); the Xenopelti-

dae (one genus) in Southeast Asia and the Indo-Malayan archipelago; the Xenophidiidae (one genus) of the Malay Peninsula and northern Borneo; the Bolyeriidae (two genera) on Round Island near Mauritius in the Indian Ocean; the Aniliidae (one genus) in Amazonian and Guyanian South America; the Acrochordidae (one genus) from India, Sri Lanka, and Southeast Asia to the Philippines, northern Australia, New Guinea, and the Bismark and Solomon Islands; and the Atractaspididae (twelve genera) of Africa and the Middle East.

Snakes resemble other squamates in most aspects of maintaining water balance and excretion. A sublingual salt-secreting gland is found in the coastal-dwelling file snakes and in sea snakes. The gland empties via multiple ducts into the tongue sheath, and salt is expelled when the tongue is extended.

Olfaction is a very significant sense in snakes, both in terms of particles inhaled through the nostrils to the olfactory sac and those brought to the Jacobson's (vomeronasal) organs by the tongue tips, as described in chapter 9 on squamate reptiles. Most snakes, especially nocturnal, semifossorial, and fossorial forms, locate food and mates principally through chemoreception. In many species well-developed cloacal glands at the base of the tail secrete odoriferous substances. They appear to be especially important in helping males track females during the reproductive season, but they may have other functions in social communication as well.

All snakes have the auditory apparatus reduced. There is no external ear or tympanum, the middle ear is reduced, and the osseous columella articulates to the quadrate bone laterally. A well-developed inner ear is present. Much of a snake's hearing involves low-frequency vibrations that are transmitted from the substrate to the inner ear via the lower jaw–quadrate–columella route. Aerial sounds are received by the skin and muscles on the side of the head and transmitted to the quadrate and thence via the columella to the inner ear. The snake ear is a low-frequency receptor most sensitive to airborne and substrate vibrations in the 100 to 500 Hz range (Wever 1978). Auditory communication is obviously minimal in snakes, and their vocal repertoire consists of hisses.

Vision is important in almost all snakes, and their eyes appear especially designed to perceive movement. Snakes' eyes however, are the most novel among the reptiles and resemble those of lizards in no substantial way, as Walls (1942) pointed out many years ago. The differences between lizards' and snakes' eyes appear to be due to an initial modification for nocturnal vision (in snake ancestors) followed by simplification associated with burrowing habits (in primitive snakes or their ancestors). Subsequently snakes appear to have returned to the surface, and the eye became reelaborated from the simplified remnants. Consequently a whole series of eye structures in snakes have a different embryological origin or are composed of entirely different tissue than those of lizards. In addition, snakes have developed a unique type of double cones in the retina.

Apparently, as an adaptation to burrowing the eyelids were lost (Scolecophidia) or became fused into a transparent, immovable brille or spectacle that covers and protects the cornea. The brille is derived primarily from the lower eyelid and is shed along with the rest of the skin. There is no nictitating membrane. Most snakes lack muscles to change the shape of the lens, and they focus by moving the lens toward or away from the retina. The few forms known to change the lens shape do so by means of the pupillary sphincter, as in most turtles and unlike any lizard. Many snakes have a duplex retina of rods and cones and others have pure rod retinas, but many diurnal species have only cones. None of the visual cells contain oil droplets. Recall that rods are the effective photoreceptor cells under conditions of low light intensity and that cones have that function in bright light and mediate color vision in vertebrates. Snakes' eyes may have as many as three kinds of cones, including the unique kind of double cone already mentioned (Underwood 1970). These differences correlate generally with ecology. Many burrowing and nocturnal forms have pure rod vision, and most diurnal ones have pure cone vision, but there are also many forms having both types of retinal cells. Binocular vision occurs in a number of arboreal snakes. Color vision is probably absent even in forms with a substantial number of cones (Walls 1942; Smythe 1975).

Four basic patterns of locomotion are found in snakes, and a particular individual may use several at different times. The most familiar type is serpentine curvilinear. In this pattern the snake moves forward by throwing out lateral undulations of the body and pushing the sides of the ventrals against any irregularity in the surface. Fast-moving snakes using this type of locomotion have been timed at 11 km/h. This style of movement is also used in swimming. Snakes using rectilinear locomotion move forward in a straight line, without any lateral undulations, by producing wavelike movements in the belly plates. This behavior is especially characteristic of large, heavy-bodied snakes (boas and vipers). Laterolinear locomotion or sidewinding is used primarily on smooth or yielding surfaces and is very complex. In essence, the snake anchors a portion or portions of the body in the substratum and laterally lifts the rest to a new position. When the lifted part is anchored again, the portions of the body left behind are lifted to the new position. By constantly lifting and anchoring alternate parts, the snake moves laterally. Concertina locomotion resembles the expansion and contraction of that musical instrument, with the head extended first and the looped curves of the body then brought forward. It is used primarily in moving through a burrow or in climbing vertical surfaces such as tree trunks.

Snakes of the arboreal Asian colubrid genus *Chrysopelea*, erroneously called "flying snakes," are able to behaviorally reduce the speed of descent after an initial leap. Once in the air they stretch out lengthwise, flatten and broaden the body, and draw in the venter to produce a concave surface. Since the angle of descent does not deviate from the vertical by more than 45°, Oliver (1951) regarded this form of locomotion as parachuting. In contrast, gliding locomotion involves morphological specialization, typically a fleshy fold between or connecting the arms and legs (a patagium), as in the flying squirrel and lizards of the genus *Draco*. Gliders are able to maintain an angle of descent greater than 45° to the vertical. Contrary to popular belief, no Costa Rican snake can fly or glide, although some may prove to be capable of parachuting locomotion.

Snakes are carnivores that feed on a wide variety of animal prey. Many species eat a broad range of vertebrates, but a substantial number specialize exclusively on a limited set of prey items, including one or more of the following: earthworms, slugs and snails, crustaceans, centipedes, scorpions, spiders, termites or ants, fish, other snakes, lizard eggs, and birds' eggs. Many forms also eat carrion. Insects are often eaten by young snakes that as adults eat larger prey. Because of the flexible jaw arrangement previously described, snakes can engulf prey that is very large in proportion to themselves. Consequently, larger species tend to prey on large vertebrates, especially lizards, birds, and mammals. Snakes that eat insects or other small animals feed more often than those eating larger prey. Most snakes may go several weeks or months between meals. However, during warm periods these species may feed every few days if prey is available.

These reptiles use three methods of catching and killing prey. Many forms simply grab the prey in the mouth, hold it with the sharp recurved teeth, and swallow it. Constrictors grab the animal and hold it while coiling their bodies around the victim and then, in the case of endothermic prey, quickly squeeze the thoracic cage to produce circulatory arrest, causing rapid tissue hypoxia of vital organs. According to Hardy (1994b), suffocation occurs simultaneously but is not the proximate cause of death. Similarly, ectothermic prey may be effectively immobilized by several minutes of nonlethal constriction. Other snakes use venom to kill the prey. Grab-and-swallow and constriction methods are used by a number of both relatively primitive and specialized species.

Among the most advanced snakes, specialized grooved fangs and venom glands increase the efficiency of predation. In many genera of the family Colubridae, toxic secretions from a modified supralabial salivary gland (Duvernoy's gland) may flow into a wound when prey (or predators) are bitten (McKinstry 1983). A considerable number of genera showing no dental specializations have a Duvernoy's gland that opens through a duct into the mouth near the posteriormost maxillary teeth. Others have one to several (usually two) small to greatly enlarged grooved fangs at the posterior end of each maxillary (opisthoglyphous condition) that are connected by a duct to the Duvernoy's gland. The venom delivery apparatus is relatively inefficient in these forms, since there are no special muscles to force the venom into the prey and the snake usually must chew the intended victim to ensure envenomation. Small rear-fanged forms are consequently harmless to humans because they cannot get their mouths around any part of a person's body to bring the fangs into contact. Some large Old World species, however, can cause serious injury or death. No such cases have been reported for Costa Rica, but unpleasant reactions to the venoms of snakes of the genera *Coniophanes, Conophis, Erythrolamprus, Leptodeira,* and *Urotheca* have been reported when people injudiciously handled these animals and were bitten.

Much more dangerous to humans than the rear-fanged snakes are those forms nominally having two hollow (canaliculate) fangs at the front of each maxillary (proteroglyphous condition), although usually only one fang is fixed, immobile, and functional on each side owing to periodic shedding. The venom gland is a modified salivary gland in these forms, and special muscles force the venom down through the hypodermic-like teeth into the victim. Members of the family Elapidae (the cobras, kraits, coral snakes, Australian venomous snakes, and sea snakes) have this venom delivery system. Many humans die or are seriously injured when bitten by these snakes, especially in Asia and to a lesser extent in Africa and Australia. The venomous coral snakes of Costa Rica, all members of the genus *Micrurus,* and the sea snake *(Pelamis)* are the only snakes in the republic having this type of venom apparatus. Because their fangs are relatively short and reportedly require some chewing to inject the venom, most recorded bites come from careless handling of a recently captured or captive coral snake that may envenomate the victim with a single quick bite.

By far the most effective venom delivery system is found in the vipers, and since many members of this stock are large, they are extremely dangerous to humans. In these snakes the maxillary bone is very short and has a single very long, curved hollow fang (solenoglyphous condition). Since the maxillary can be rotated when the fangs are not advanced to the biting position, they may be folded up against the roof of the mouth. The usual biting sequence involves a strike of blinding speed in which the mouth opens and the fangs are rotated and pointed slightly forward or directly downward to stab the victim on impact. Simultaneously a large amount of venom is injected into the wounds. In Costa Rica almost all serious injuries and deaths from snakebite are caused by the pitvipers of the genera *Atropoides, Bothrops, Bothriechis,* and *Lachesis.* Most bites from these snakes occur during land clearing, and they usually are on the hands or lower extremities.

Snake venoms are complex mixtures primarily of proteins, some of them enzymes. The individual proteins are each capable of producing at least one pharmacological effect that is deleterious to living animals. The more lethal venom components appear to be peptides or certain nonenzymatic pro-

teins, but the enzymes, which may digest tissue or otherwise disrupt the normal functioning of cell structures and metabolic pathways, certainly contribute to the overall destructive effect of the venom. The polypeptide fractions of snake venoms include compounds that tend to have systemic effects that may include blockage of nerve transmission, local tissue damage, breakdown of red blood cells, convulsions, and cardiac arrest. Other fractions affect the clotting of blood cells or produce hemorrhaging. Several to many of the components mentioned above may be present in the venom of a particular species and will usually produce a complex of pharmacological effects or symptoms. Both the venoms and their actions are variable within taxonomic groups, and any attempt to generalize regarding the principal effects for a family, subfamily, or even some genera is misleading (Russell 1983). Generally speaking, the rear-fanged colubrid snakes have the least complex venoms and the vipers have the most complex, with as many as twenty components. The venoms of elapids are somewhat less complex than those of vipers. Envenomation serves three functions: to immobilize and kill prey; to partially digest prey; and to defend the snake against predators, especially those larger than itself. The chief uses of venom are in obtaining and digesting relatively large animals that the hungry snake otherwise could not capture or kill without sustaining potentially serious damage.

Although a serious snakebite accident is a very rare event (about twenty-two per hundred thousand people per year in Costa Rica, according to Bolaños 1984), it is not a very pleasant experience and proves fatal 3.3% of the time. For this reason I urge you to exercise reasonable caution when looking for snakes in the field (Watch where you put your hands and feet!) and to avoid touching or handling any snake if you are not sure whether it is harmless or venomous. Finally, in case a snakebite does occur, chapter 1 outlines first aid and emergency handling of the victim.

A wide variety of animals are natural enemies of snakes. Aside from humans, who may eat snakes or merely kill them from fear or for pleasure, principal predators include many other snakes (see below), raptors (Gerhardt, Harris, and Vásquez 1993), many large and small carnivorous mammals (domestic and feral cats, weasels, mongooses, shrews, etc.), and wild pigs and peccaries. The large carnivorous monitor lizards of Africa, Asia, and Australia also take their toll. Boas and pythons are often victims of large predators, especially members of the cat family and various crocodilians. Aquatic and semiaquatic species are preyed on by crocodilians, fishes, turtles, and many wading birds. Large carnivorous frogs of several families also include snakes in their diets. A number of large tropical spiders are known to eat snakes among other prey caught in their webs.

In Costa Rica the colubrids *Clelia* and *Erythrolamprus* and several of the venomous coral snakes of the genus *Micrurus* feed exclusively or primarily on other snakes. Another colubrid, *Lampropeltis triangulum*, includes snakes in its diet elsewhere in its range and may do so in Costa Rica. With these generalities in mind, in the species accounts I make reference only to particularly interesting or unusual instances of predation on Costa Rican snakes.

One of the more interesting features of squamate reptiles important in snake systematics is the presence of paired, saclike penes (each one is a hemipenis) at the base of the tail in males. These structures open from the cloaca and are everted during courtship and mating by being filled with lymph and blood, aided by propulsor muscles at the base and by relaxation of the retractor muscles. In this process the hemipenis is turned inside out so that the structures lining the organ in situ are now on the external surface (fig. 11.2). The everted organ has a groove on the surface (the sulcus

spermaticus) that runs all or most of the penis length to provide a channel for carrying the sperm to the opening of one of the oviducts when the organ is inserted into the female's cloaca. Only one of the two hemipenes is inserted at one time, although both may be everted. Invagination is accomplished by reduction in fluid pressure, relaxation of the propulsor muscles, and contraction of the retractor muscles that lie inside the everted organ and attach to the outer connective sheath of the retracted organ.

The structure and ornamentation of the hemipenes have been used as systematic characters, especially in colubrid snakes. For this reason their condition, where known, will be succinctly described in the species accounts, with a reference to a published description or illustration. The terms defined here are also used to describe lizard hemipenes (chapter 10).

The hemipenes may be single (fig. 11.2b), bilobed, or divided. In the bilobed state the organ is divided at the apex for less than the length of the undivided basal portion (fig. 11.2a,c). A divided hemipenis has the undivided basal segment equal to or shorter than the apical portion (fig. 11.2d). In terms of general shape the organ may be subcylindrical, attenuate, bulbous, cirrate, or clavate. The sulcus spermaticus may be simple (fig. 11.2b,c) or bifurcate (fig. 11.2a,d). If the sulcus is simple the hemipenis is termed symmetrical when the sulcus lies in the center of the organ and the right and left hemipenes are mirror images of one another. In this case the sulcus is semicentripetal. The contrasting asymmetric condition has the simple sulcus almost always terminating on the right lobe of both right and left hemipenes, and the hemipenes are mirror images only at their bases (fig. 11.2c). The position of the sulcus on these organs may be semicentrolineal (confined to the sulcate surface) or semicentrifugal (extending beyond the sulcate surface onto the side of the organ). Where the sulcus is bifurcate three conditions are recognized:

1. Centripetal, where the branches of the sulcus diverge minimally and extend up the center of the hemipenis to lie on the facing sides of the lobes of the bilobed organ (fig. 11.3b)
2. Centrolineal, where the branches of the sulcus diverge moderately and lie on the same surface of the hemipenis as the point where the sulcus bifurcates (fig. 11.2a)
3. Centrifugal, where the branches of the sulcus diverge outward from the center to terminate on opposite sides of the organ (fig. 11.3a,c)

The organ may be naked or covered with ornamentation, including flounces, calyces, papillae, or spines, or a combination of several kinds. In addition the tip of the hemipenis may be differentiated from the base or modified in several ways.

Most hemipenes include a naked basal region, the pedicel, and an ornamented central zone, the truncus. The distal

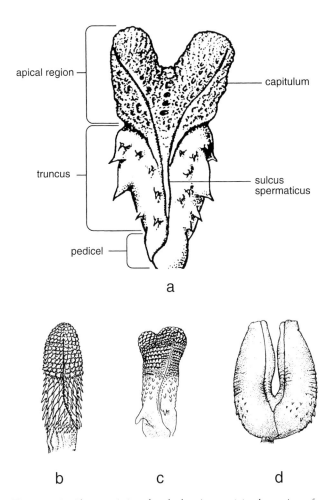

apical region —

truncus —

pedicel —

— capitulum

— sulcus spermaticus

a

b   c   d

Figure 11.2. Characteristics of snake hemipenes: (a) sulcate view of a bilobed capitate hemipenis having a bifurcate, centrolineal sulcus spermaticus with principal regions of the organ indicated; (b) a single noncapitate hemipenis having a simple sulcus spermaticus and apical calyces and spines on the truncus *(Chironius)*; (c) an asymmetric bilobed noncapitate hemipenis with a simple sulcus spermaticus and apical calyces and small spines on the truncus *(Coluber)*; (d) a divided hemipenis with a bifurcate sulcus spermaticus with flounces on the truncus and spines on the pedicel.

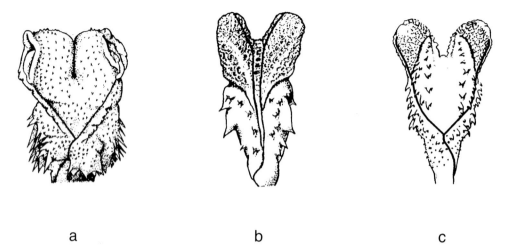

a   b   c

Figure 11.3. Specialized features of snake hemipenes: (a) centrifugal sulcus spermaticus and apical disks *(Erythrolamprus)* on bilobed organ; (b) centripetal sulcus spermaticus on bilobed unicapitate organ; (c) centrifugal sulcus spermaticus on bilobed bicapitate organ *(Oxyrhopus)*.

portion of the organ or apical region may or may not have ornamentation distinct from that of the truncus (fig. 11.2a).

Many species have a distinct naked groove, or a pair of grooves in some bilobed taxa, separating the differentiated distal portion of the hemipenis (hemipenes) from the proximal section(s). In capitate hemipenes the groove(s) completely encircle the organ or each lobe, except where the sulcus spermaticus passes through the groove(s) (fig. 11.2a). The term for the differentiated apical region(s) of such organs is capitulum (pl. capitula). In semicapitate hemipenes the naked groove(s) only partially separate(s) the distal portion or the apical areas of the lobes from the rest of the organ. In bilobed, semicapitate hemipenes the sulcus spermaticus forks proximal to the base of the bilobate region.

Capitate hemipenes are of two distinct types. Unicapitate organs, whether single or bilobed, have a single capitulum. The sulcus spermaticus may be simple or bifurcate, and it forks proximal to or within the capitulum. Bicapitate organs are bilobed and have a separate capitulum on the distal portion of each lobe. In this kind of hemipenis the sulcus spermaticus forks proximal to the base of the bilobate region (fig. 11.3c).

Other snakes may have a well-developed papilla or a fleshy elongation extending from the apex of the organ or from one or both of its lobes. These structures are called terminal awns. Finally, a number of forms have well-defined, more or less round, unornamented areas at the terminus of each branch of the sulcus that are referred to as disks (fig. 11.3a). Descriptive terminology generally follows Dowling and Savage (1960), Myers and Campbell (1981), and Zaher (1999).

Most structural features are best seen in fully everted hemipenes. However, many character states may be determined by dissection (in situ) of the retracted organ by cutting down the midline and spreading the tissue flat. The descriptions and illustrations cited are based on everted organs when these are available, but descriptions of preparations in situ are also referenced. Members of the same genus usually have similar organs, but where there are substantial species differences within a group I will include references to specific characters and illustrations. In a few cases I include brief descriptions of the hemipenes of certain species that have not previously been described, and these are indicated by catalog numbers. If no reference is given it means the hemipenes have not been described and no material is available for description. Although I include the descriptions of penial structure for completeness, all but the most serious students of snake systematics may skip over these esoteric details when attempting to identify Costa Rican snakes.

Most species of snakes exhibit sexual dimorphism in size, but which sex is larger is variable. Females usually are larger than males, have relatively longer bodies and shorter tails than males, and consequently have more ventral and fewer subcaudal scutes. Where size differences have been documented, the larger sex is mentioned in the generic or species accounts. Probably in most cases females are larger because they carry eggs or fetuses and males have longer tails to accommodate their penes. In some forms the tails are much thicker basally in males than in females, also reflecting penial location. Males of a number of species have enlarged tubercles on the chin or keeled supracloacal scales, or the keels in the supracloacal region are tuberculate. In boas the pelvic spurs are usually more prominent in males than in females.

If these obvious external features are not present in live males of a particular species, another way to see if the specimen is a male is to use the cloacal probing technique described for lizards in chapter 10. This involves carefully probing the posterior and lateral margins of the cloaca with a small blunt instrument, preferably the standard probe in a dissecting kit, and gently thrusting it posteriorly. If the specimen is a male the probe will pass into the opening of the hemipenis for some distance. You may need to dissect preserved specimens to determine sex by making a slit just off the midline of the underside of the tail to see if the hemipenes are present. Otherwise you can make an incision through the ventral body wall, from about the posterior one-half to two-thirds of the body length, to examine the gonads.

Fertilization in all snakes is internal. A long and involved courtship precedes mating, including the male's pursuing the female and much tactile stimulation. During mating the male usually grasps the female's neck in his mouth; their tails and often their bodies are intertwined, and the cloacae are brought together to allow intromission of one hemipenis.

Most snakes lay eggs (are oviparous), but a considerable number give birth to living young (are viviparous). The eggs are surrounded by a whitish or yellowish parchmentlike covering that contains a small amount of lime (calcium oxide) and corresponds to the hard calcareous shell found in most other reptiles, including birds. Fertilization is internal and occurs before the secretion of the tough, protective outer covering. Eggs may be retained in the oviduct for some time after fertilization before being laid. The eggs are oval and usually about twice as long as broad, but some are much more elongate. When laid they frequently stick together because of an adhesive fluid secreted by the oviduct. Eggs are deposited in a variety of sites where there is some moisture and temperature fluctuations are minimal or in decomposing plant material where the temperature remains above that of the ambient air and is relatively constant. Typical locations include sand or soil, decaying leaf litter, hollow or rotting logs, abandoned burrows of other organisms, and ant and termite nests. A few species construct nests of decaying vegetation or make nests in the ground, while others coil themselves around the eggs to protect them and keep the temperature stable. Contrary to many accounts there are no known ovoviviparous snakes, in which the eggs are retained in the oviduct and hatch there, and the snakelets are born alive. All live-bearing snakes are truly viviparous, since they have relatively simple placental connections (zones of physiological exchange) between the developing embryo and the lining of the maternal oviduct (Stewart and Blackburn 1988).

Snakes show little territorial behavior, but males of some (perhaps all) boas and their allies, colubrids, elapids, and vipers exhibit a ritualized combat that was for many years erroneously thought to be courtship behavior. In this activity two males lift the front of the body off the ground. The bodies usually remain parallel to one another while the heads variously make pushing, hovering, or pinning movements so that they push one another to the ground, become momentarily intertwined, then begin again. Vipers raise the body vertically and maintain that position until one individual topples the other, only to ascend and start anew. Males may continue these bouts for several hours. The familiar caduceus—two intertwined serpents symbolizing the medical profession—is a representation of this behavior that goes back to classical Greece.

Many snakes, especially venomous ones, are reputed to be extremely aggressive and to pursue human intruders. This canard is vastly exaggerated by imagination. The first line of defense for most species is concealment, the next flight. If these defenses fail they may use various antipredator defenses, most often including tail vibrating, neck spreading, gape opening, and striking with a closed mouth. Many snakes will strike or bite when closely approached or handled, but none will frequently or consistently "chase" a human who is retreating. A number of snakes, including the species of the American snake genera *Coluber, Coniophanes, Dendrophidion, Drymobius, Enulius, Nerodia, Scaphiodontophis, Thamnophis,* and *Urotheca,* have a high incidence of tail breakage (urotomy). Breakage in all these cases is intervertebral, tail regeneration does not occur, and a calcified cap and skin develop over the exposed portion of the vertebra as the wound heals. This situation stands in marked contrast to that found in most lizards (nineteen of twenty-eight families), where intravertebral breakage occurs across a fracture plane and the tail regenerates around a cartilaginous rod. Intravertebral fracturing of the vertebrae is autonomous, but intervertebral breakage requires physical resistance. In at least some snakes *(Coluber, Dendrophidion, Drymobius, Nerodia,* and *Thamnophis)* intervertebral breakage is "self-induced," for when its tail is grasped the snake rotates its body longitudinally in one direction until its tail snaps off. These two modes of urotomy have been defined earlier (p. 413) as autotomy (most lizards) and pseudautotomy (snakes and some lizards).

Among American snakes having a high percentage of tail breakage, three genera—*Enulius, Scaphiodontophis,* and *Urotheca*—have an extremely fragile, thickened, and very long tail (more than 35% of total length in adults). This peculiar caudal morphology contrasts markedly with that in other pseudautotomous genera, where the tail is slender and tapering. Recent studies (Slowinski and Savage 1995; Savage and Slowinski 1996) indicate that the specialized tail morphology of *Enulius, Scaphiodontophis,* and *Urotheca* allows for multiple breaks and escapes during the life of individual snakes (specialized pseudautotomy). In the other genera multiple breaks are unusual, although they have a high frequency of tail loss.

Another unusual attribute of snakes is skin shedding. All vertebrates shed their skin as growth takes place, usually more or less continuously or in small patches. Snakes usually shed the skin in one piece. As the snake prepares to shed, the eyes and general color become pale or cloudy owing to a milky fluid secreted between the old skin and the new one. After several days in this premolt state, during which time vision is impaired, the snake rubs its head and neck against rough objects until the old skin catches and is peeled off inside out. Shedding may occur once to several times a year.

In general large species of snakes are longer lived than small forms. Many species apparently attain ages of twelve to fifteen years under natural conditions. Captive examples of some large species have lived twenty-five to twenty-nine years.

SNAKE IDENTIFICATION: KEY FEATURES

The accurate identification of snakes depends to a great extent on differences in squamation, at least until you become familiar with individual species, which most often will be recognizable from form and coloration. The principal features of scalation used in the keys and the diagnostics sections of snake species accounts are illustrated (figs. 11.4 and 11.5). The following paragraphs are designed to help you apply these in the identification process.

Head scalation: The size, shape, and arrangement of the head scales or shields are of great importance in identifying snake taxa. The several families of blind burrowing snakes generally have the upper and lateral surfaces of the head covered by enlarged smooth, shiny shields that look much like the smaller scales on the body (same size and shape in one extralimital genus). In some snakes the upper head surface is covered with many small scales that are about the same size as the dorsal scales and may be smooth or have a median keel, and the supralabial and infralabial series

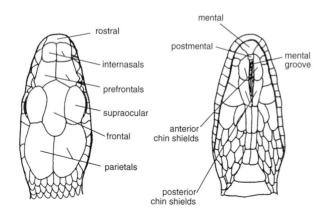

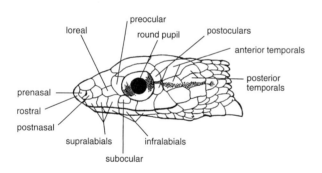

Figure 11.4. Basic terminology for snake head shields; this diagram shows the basic colubrid (Colubridae) complement of enlarged plates on the dorsal head surface.

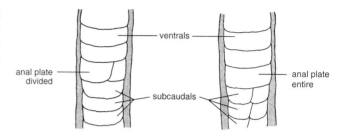

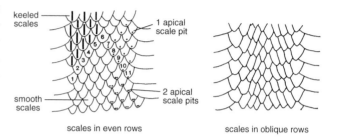

Figure 11.5. Terminology for snake body scalation.

consist of enlarged symmetrical shields. Most Costa Rican snakes have large smooth symmetrical head shields (fig. 11.4), but some forms have a mixture of enlarged shields and small smooth or keeled scales on the upper head surface. In most taxa with symmetrical shields the arrangement is very constant. However, rare anomalies involving partial or complete fusion or fragmentation of the head shields may occur and cause difficulty in identification.

The throat region also allows differentiation in scale features. In some snakes the gular area is covered with small scales like those on the body. In others a definite mental groove runs down the midline of the throat and usually is bordered laterally by one or two pairs of enlarged chin (genial) shields (fig. 11.4).

Other head specializations: A series of thermoreceptors lodged in distinct depressions in the sutures between the labials occur in some forms, and others have indistinct pits between the supralabials (fig. 11.8a). A single large loreal pit lies between the eye and nostril on the lateral surface of the head in all pitvipers (family Viperidae) (fig. 11.8b).

General body scalation: In the blind burrowing snakes the body is covered by smooth cycloid scales arranged in an alternating pattern of both longitudinal and transverse rows, and there are no enlarged ventral scutes. In most other Costa Rican snakes the abdominal region is covered by a series of greatly expanded ventral scutes so that the dorsal squamation and ventral squamation are distinct (fig. 11.5). In the sea snake *(Pelamis)* some slightly enlarged ventrals are present, but others are similar to the body scales in size and shape.

Dorsal scalation: Where there is differentiation of dorsal and ventral scalation, the dorsals are much smaller than the ventrals (three or more times as wide as dorsals), but there is only one transverse row of dorsals for each ventral (fig. 11.5). Dorsal scales are typically hexagonal and lanceolate, with some species having quadrangular (diamond shaped), rectangular, or ovate ones. The scales may be smooth or may have a central ridge (keel) (fig. 11.5) or several shallow grooves (striae). The vertebral row may be definitely enlarged. In snakes lacking ventrals the number of transverse scale rows (counted down the middle of the back) is usually used as a systematic feature. The dorsal scales usually are arranged evenly in straight rows (longitudinal rows parallel, transverse rows diagonal to body axis) in most snakes (fig. 11.5). Some snakes, however, have the scales arranged in oblique rows (longitudinal rows straight, transverse rows curved (fig. 11.5).

Scale row formulas: The number of dorsal scale rows is among the most diagnostic features for many snake genera and species. In species without enlarged ventral scutes, scale rows are counted around the body, whereas in other snakes the scale row count is made transversely from the ventral tip on one side to a ventral tip on the other (fig. 11.5). In the latter case the scales are usually in an odd number of rows across the body, but some genera have them in even numbers. Many forms have the same number of scale rows across the neck (counted one head length posterior to the head), midbody, and posteriorly (counted one head length anterior to the vent). Others have different numbers in two or more of these regions. If there is a change in the number of scale rows from one region to the next, that change is called a scale reduction if the number is smaller or a scale increase if it is larger. These features of body scalation are denoted in this book by a dorsal scale formula. A typical formula would be 15-19-17, indicating the number of scales across the neck, midbody, and posterior body, respectively. A formula such as 19–15 indicates the count on the neck and midbody (19) and posterior body (15), since there is no scale reduction between the first two regions. When only a single count is given the number of scale rows is identical in all three regions. When there is variation for a particular area, that variation is indicated as scale rows 17 to 19. A similar system applies to record the number of scale rows around the body in forms lacking ventral shields. As with other scutellation features, rare examples of some species will exhibit aberrant scale row numbers or scale row formulas.

Scale pits: Many snakes have one or two small apical pore-like pits on some or all dorsal scales (fig. 11.5). These structures are often difficult to see, even under magnification. Most herpetologists, when in doubt, remove the outer layer of the scale, let it dry, and then examine it for the pit(s).

Ventral, anal, and subcaudal scalation· The number of enlarged ventral scutes corresponds to the number of trunk vertebrae in snakes. The first ventral is the most anterior plate intercepted by the first row of dorsal scales on each side. The enlarged anal scute covering the cloacal opening is not counted as a ventral. This structure is single (entire) or divided (fig. 11.5). The condition is usually constant in a genus or species but may vary. In addition, occasional specimens of forms that almost always have entire anal plates will exhibit the divided condition and vice versa. In all snakes having expanded ventral scutes the subcaudal scales are enlarged. In most snakes they form paired shields, but in a few the subcaudals are single (entire). In a few others a mixture of single and paired scutes may be present, and in one case there may be four or five subcaudals per tail segment posteriorly (figs. 11.5, 11.19).

Coloration: Although there is some ontogenetic change in pattern and color in some species, most forms can be rather consistently identified by coloration, or at least placed in the correct genus. Sexual dichromatism is also minimal. Dorsal patterns, whether uniform, spotted, banded, ringed, or striped (fig. 3.1), are useful features if you bear in mind that the juveniles of some genera have a dark blotched pattern that disappears in adults (table 11.1). Color and pattern of the ventral and subcaudal regions are also often diagnostic.

Table 11.1. Identification of Costa Rican Colubrine Snakes Having Blotched or Banded Patterns as Juveniles

	Blotched/Banded		Scales	Scale Rows	Blotches or Bands	Subocular	Venter
	Juveniles	Adults					
Coluber mentovarius	X	X	S	17	D	X	Immaculate
Dendrophidion nuchale	X	—	K	17	L	—	Dark marks on anterior margin of ventrals
Dendrophidion paucicarinatum	X	—	K	17	L	—	Dark marks on posterior margin of ventrals
Dendrophidion percarinatum	X	X	K	17	L	—	Immaculate
Dendrophidion vinitor	X	X	K	17	L	—	Immaculate
Drymarchon corais	X	—	S	17	D	—	Some dark lines
Drymobius margaritiferus	X	—	K	17	D	—	Dark marks on posterior margin of ventrals
Drymobius melanotropis	X	—	K	17	D	—	Dark marks on posterior margin of ventrals
Drymobius rhombifer	X	X	K	17	D	—	Immaculate
Elaphe triaspis	X	—	K	29–39	L/B	—	Immaculate anteriorly
Leptophis riveti	X	X	K	15	G	—	Immaculate
Mastigodryas melanolomus	X	—	S	17	D	—	Stippled to immaculate
Pseustes poecilonotus	X	—	K	19 25	L/B	—	Suffused with brown
Scaphiodontophis annulatus	X	—	S	17	D	—	Flecked or suffused with brown
*Stenorrhina degenhardtii**	X	X	S	17	D	—	Dark markings
Tantilla supracincta	X	—	S	15	D	—	Immaculate, red in adults

S = smooth; D = dark; L = light; K = keeled; G = green in life; L/B = light outlined by black; * = some individuals.

The pattern of the head and neck—especially distinctive markings, extent of a dark head cap, and the presence of light or dark nuchal bars or collars—is informative, but light nuchal markings may disappear with age in some species. The color of the iris, lining of the mouth and throat, and tongue often aid in identification as well.

Costa Rican snakes range in size from less than 200 mm to 4.5 m in total length. This feature is among those most useful in eliminating other much larger or smaller snakes from consideration.

In the following sections and elsewhere in this book, descriptions of snake size refers to total length according to the following ranges:

Tiny: under 200 mm
Small: 200 to 650 mm
Moderate: 650 to 1,500 mm
Large: 1,500 to 3,500 mm
Very large: 3,500 to 4,500 mm
Giant: greater than 4,500 mm

The largest of living serpents are members of the family Boidae. The reticulated python *(Python reticulatus)* of tropical Asia attains a length of 10.1 m and a weight of about 91 kg. The South American anaconda *(Eunectes murinus)* is somewhat shorter at 9.4 m but is heavier bodied, weighing about 135 kg at the maximum. The king cobra *(Ophiophagus hannah)* of India and Southeast Asia is the largest venomous snake, reaching a length of 5.5 m. The largest Costa Rican snakes include the following forms:

Boa constrictor (Boidae): up to 4.5 m, but rarely seen over 3.5 m (boa or béquer)
Drymarchon corais (Colubridae): up to 3 m (zopilota)
Lachesis melanocephala and *L. stenophrys* (Viperidae): up to 3.8 m (bushmasters or cascabelas mudas), venomous and very dangerous snakes

The smallest snakes are several members of the genus *Leptotyphlops*, with maximum total lengths about 112 mm. *Helminthophis frontalis* is the smallest Costa Rican snake, with a maximum total length of 160 mm.

KEY TO THE GENERA OF COSTA RICAN SNAKES

Note that because of intrageneric variation in some features, several genera are double or triple keyed below and more than one couplet may lead ultimately to the same genus (e.g., couplet 27 may be reached from 26a, 44b, and 46a).

1a. Ventral scutes markedly enlarged, at least three times size of dorsal scales (fig. 11.5) 7
1b. Ventral scales not noticeably enlarged, equal in size to dorsal scales or slightly larger 2
2a. Tail cylindrical or subcylindrical in cross section, not distinctly compressed laterally 3

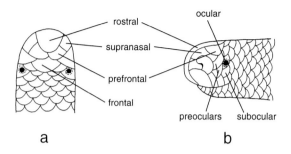

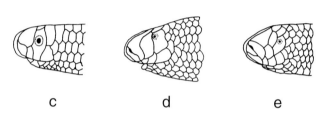

Figure 11.6. Specialized terminology for head scalation in primitive blind burrowing snakes (Scolecophidia): (a) dorsal view; (b) lateral view; (c) lateral view of *Leptotyphlops* showing ocular scale extending to upper lip; (d) lateral view of *Typhlops* showing the one preocular condition; (e) lateral view of *Liotyphlops* with two preoculars.

2b. Tail not cylindrical or subcylindrical, distinctly compressed laterally . . . . *Pelamis* (p. 714)
3a. Ocular scale separated from lip by one to several scales (fig. 11.6b); 18–28 scale rows around body 4
3b. Ocular scale bordering lip (fig. 11.6c); 14 scale rows around body . . . . *Leptotyphlops* (p. 558)
4a. Rostral meeting frontal, separating prefrontals or supranasals from one another (fig. 11.7a,b) 5
4b. Rostral separated from frontal; prefrontals or supranasals in contact (fig. 11.7c,d) 6
5a. Prefrontals fused with supranasals (fig. 11.7a); a preocular; no subocular (fig. 11.6d) . . *Typhlops* (p. 556)
5b. Prefrontals and supranasals distinct (fig. 11.7b); two preoculars; no subocular . . . . . *Liotyphlops* (p. 555)
6a. Upper head plates rounded (fig. 11.7c); one subocular (fig. 11.6b) . . . . *Helminthophis* (p. 554)
6b. Upper head plates angular (fig. 11.7d); two suboculars . . . . *Anomalepis* (p. 553)
7a. A well-developed loreal pit present (fig. 11.8b) . . . 74
7b. No loreal pit (fig. 11.8a) 8
8a. Snout, prefrontal, and frontal areas covered by large symmetrical shields (fig. 11.4) 12
8b. Frontal and parietal areas covered by small irregular scales 9
9a. Supralabials separated from orbit by suboculars (fig. 11.8a); dark blotches in pattern 10
9b. Supralabials border orbit; adults uniform, young with dark blotches . . . . *Epicrates* (p. 567)
10a. Dorsal scales smooth; subcaudals entire (fig. 11.5); snout without enlarged internasals, covered with small

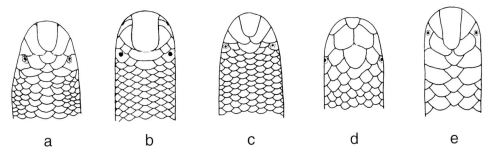

Figure 11.7. Dorsal view of head scalation in Costa Rican primitive blind burrowing snakes (Scolecophidia): (a) *Typhlops*; (b) *Liotyphlops*; (c) *Helminthophis*; (d) *Anomolepis*; (e) *Leptotyphlops*.

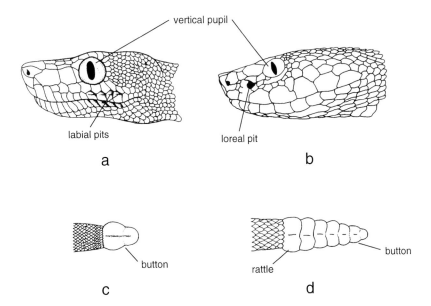

Figure 11.8. Special features in snakes: (a) lateral view of boid head showing labial pits and vertical pupil; (b) lateral view of pitviper head showing loreal pit and vertical pupil: (c) button on tail of newborn rattlesnake *(Crotalus)*; (d) rattle of rattlesnake after six skin sheddings.

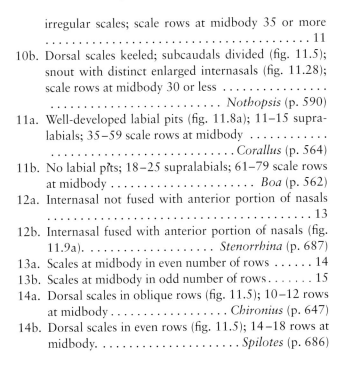

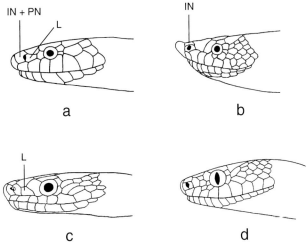

Figure 11.9 Distinctive features of lateral head scalation and pupil shape in snakes: (a) *Stenorrhina* with round pupil, a loreal, and fusion of internasal and prenasal scales; (b) *Loxocemus* with round pupil, upturned rostral, no loreal, and internasal separate from nasal scale; (c) *Elaphe* with round pupil and a single loreal; (d) *Trimorphodon* with vertical pupil and several loreal scales.

irregular scales; scale rows at midbody 35 or more
.... 11
10b. Dorsal scales keeled; subcaudals divided (fig. 11.5); snout with distinct enlarged internasals (fig. 11.28); scale rows at midbody 30 or less ....
.... *Nothopsis* (p. 590)
11a. Well-developed labial pits (fig. 11.8a); 11–15 supralabials; 35–59 scale rows at midbody ....
.... *Corallus* (p. 564)
11b. No labial pits; 18–25 supralabials; 61–79 scale rows at midbody .... *Boa* (p. 562)
12a. Internasal not fused with anterior portion of nasals
.... 13
12b. Internasal fused with anterior portion of nasals (fig. 11.9a). .... *Stenorrhina* (p. 687)
13a. Scales at midbody in even number of rows 14
13b. Scales at midbody in odd number of rows........ 15
14a. Dorsal scales in oblique rows (fig. 11.5); 10–12 rows at midbody.................. *Chironius* (p. 647)
14b. Dorsal scales in even rows (fig. 11.5); 14–18 rows at midbody. .... *Spilotes* (p. 686)

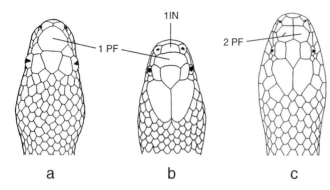

Figure 11.10. Distinctive features of upper head plates in Costa Rican snakes: (a) *Ungaliophis* showing single prefrontal shield; (b) *Hydromorphus* showing single internasal and prefrontal shields; (c) *Micrurus* with two internasals and two prefrontals.

15a. A single prefrontal (fig. 11.10a,b) 16
15b. A pair of prefrontals (fig. 11.4) 18
16a. 15–17 scale rows at midbody 17
16b. 21–23 scale rows at midbody . . *Ungaliophis* (p. 569)
17a. Nostrils directed laterally; loreal separated from eye by preocular (fig. 11.4); ventrals 118–57. 65
17b. Nostrils directed dorsally; loreal usually bordering eye (fig. 11.10b); ventrals 157–86 . . . .
. . . . *Hydromorphus* (p. 605)
18a. Scales at midbody in more than 29 rows 19
18b. Scales at midbody in fewer than 29 rows 20
19a. A prominent upturned rostral; no loreal, prefrontal meets supralabials (fig. 11.9b); scales smooth . . . . . .
. . . . *Loxocemus* (p. 560)
19b. Rostral not upturned; loreal present, prefrontal separated from supralabials (fig. 11.9c); posterior body scales keeled. . . . . *Elaphe* (p. 663)
20a. A loreal, sometimes fused with posterior part of nasal, separating prefrontal from supralabials (fig. 11.4) 28
20b. No loreal, prefrontal touching one or more supralabials (fig. 11.11d) 21
21a. Scale rows at midbody 15; anal divided (fig. 11.4); nasal usually divided 22
21b. Scale rows at midbody 17, or if 15, anal entire; a single nasal (fig. 11.11d) . . . . *Oxybelis* (p. 675)
22a. Anal divided (fig. 11.5) 23
22b. Anal entire (fig. 11.5). 24
23a. Dorsal scales keeled (fig. 11.5); scale rows around body one head length anterior to vent two to four fewer than around midbody 37
23b. Dorsal scales smooth (fig. 11.5), except a few above vent in some males; scale rows one head length anterior to vent not more than one fewer than around midbody. 24
24a. Scales in 17 or more rows at midbody; apical scale pits present (fig. 11.5). 26

24b. Scales in 15 rows at midbody; no scale pits (fig. 11.5) 25
25a. Pattern of alternating black and/or red, yellow, or white rings, sometimes evident only on venter (fig. 11.47), never striped . . . . *Micrurus* (p. 707)
25b. Pattern without alternating black and light rings, often striped . . . . *Tantilla* (p. 690)
26a. Venter essentially unmarked, except where dorsal coloration impinges on tips of ventrals; dorsum uniform, with very small dark spots or banded; pupil vertically elliptical in strong light or preservative (fig. 11.9d); subcaudals 60 or more. 27
26b. Venter heavily marked by black rings to nearly uniform black, dorsal pattern of alternating rings that encircle body (fig. 11.47) *or* uniform black; pupil round (fig. 11.4); subcaudals usually fewer than 60 . . . .
. . . . *Lampropeltis* (p. 665)
27a. Dorsal pattern consisting of dark bands on a lighter ground color (fig. 3.1) . . . . *Oxyrhopus* (p. 572)
27b. Dorsum uniform black or uniform reddish with small black scale tips and a single nuchal black band . . . . .
. . . . *Clelia* (p. 572)
28a. Strong keels on several uppermost dorsal scale rows to most scale rows over posterior third or more of body length (fig. 11.5) 29
28b. Dorsal scales smooth (fig. 11.5), except a few above vent in some examples or very weakly keeled in 17–17 rows. 40
29a. Anal entire (fig. 11.5). 30
29b. Anal divided (fig. 11.5) 34
30a. Dorsal scales in evenly arranged rows; 15 to 21 scale rows at midbody; 5 uppermost to all dorsal rows keeled (fig. 11.5) 31

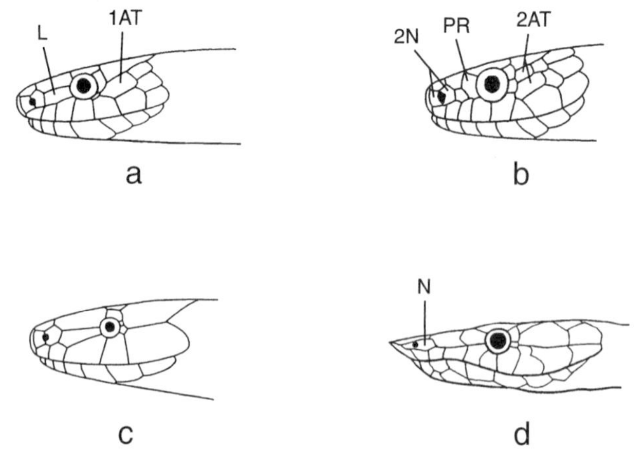

Figure 11.11. Distinctive features of lateral head scalation in Costa Rican snakes: (a) *Ninia* with a loreal and single anterior temporal but no preocular; (b) *Drymobius* with two nasals, a loreal, a preocular, and two anterior temporals; (c) *Geophis hoffmanni* with a single loreal, no preocular, and no temporals; (d) *Oxybelis fulgidus* having a single nasal.

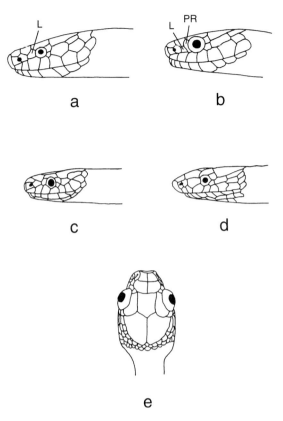

Figure 11.12. Distinctive features of head scalation in Costa Rican snakes. Lateral views: (a) *Thamnophis proximus*, which lacks a preocular; (b) *Erythrolamprus* with a loreal and preocular; (c) *Enulius sclateri* having a loreal and postnasal but no preocular; (d) *Tantilla* with a loreal but no preocular. Dorsal view: (e) *Imantodes* showing blunt head, narrow neck, and bulging eyes.

30b. Dorsal scales in obliquely arranged rows; usually 23–25 scale rows at midbody, sometimes 21; only 3–13 uppermost dorsal scale rows keeled (fig. 11.5) . . . . *Pseustes* (p. 679)
31a. No preocular (fig. 11.12a) 32
31b. A preocular present (fig. 11.12b) 33
32a. Scales in 15 rows at midbody; no anterior temporal (fig. 11.11c) 41
32b. Scales in 17–21 rows at midbody; an anterior temporal (fig. 11.11a) . . . . *Ninia* (p. 615)
33a. 17 or fewer scale rows at midbody and on neck . . . 35
33b. 19 scale rows at midbody or on neck . . . . *Thamnophis* (p. 699)
34a. 23 or fewer scale rows at midbody 35
34b. More than 23 scale rows at midbody . . . . *Trimorphodon* (p. 697)
35a. 17 or fewer scale rows at midbody 37
35b. More than 17 scale rows at midbody 36
36a. Nostrils directed upward; no rings in pattern; usually 2 preoculars and 2 loreals (fig. 11.13a,c); ventrals 127–51; no apical scale pits (fig. 11.5) . . . . *Tretanorhinus* (p. 635)

36b. Nostrils directed laterally; ringed pattern (fig. 11.47); a preocular and a loreal (fig. 11.13b,d); ventrals 239–46; apical scale pits present (fig. 11.5) . . . . *Rhinobothryum* (p. 680)
37a. Scale rows at midbody 17 38
37b. Scale rows at midbody 15 or fewer . . . . *Leptophis* (p. 668)
38a. Nasal divided, usually completely (fig. 11.11b) . . . 39
38b. A single undivided nasal scale (fig. 11.11d) 21
39a. Apical area of hemipenes covered by large smooth or scalloped calyces or by flounces; maxillary teeth 30–50 . . . . *Dendrophidion* (p. 652)
39b. Distal one-half to one-third (including apical area) of hemipenes covered with small, papillate calyces; maxillary teeth 22–34 . . . . *Drymobius* (p. 660)
40a. Anal entire (fig. 11.5) 41
40b. Anal divided (fig. 11.5) 50
41a. Parietal separated from supralabials by temporals 42
41b. Parietal touches at least one supralabial; no anterior temporal (fig. 11.11c) . . . . *Geophis* (p. 599)
42a. Loreal borders orbit; number of scale rows at midbody not more than one greater than the number one head length anterior to vent; no apical scale pits (fig. 11.5) 43
42b. 1 or 2 preoculars separate loreal from bordering orbit; number of scale rows at midbody usually two or more

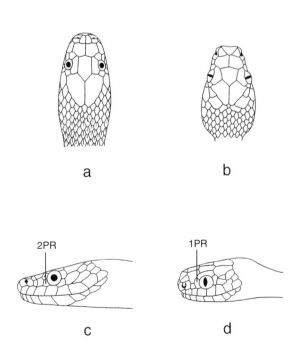

Figure 11.13. Distinctive features of head scalation in Costa Rican snakes. Dorsal views: (a) *Tretanorhinus* showing dorsally directed nostrils and internasals; (b) *Rhinobothryum* with laterally directed nostrils and internasals separated by rostral. Lateral views: (c) *Tretanorhinus* with two preoculars and two loreals; (d) *Rhinobothryum* with one preocular and one loreal.

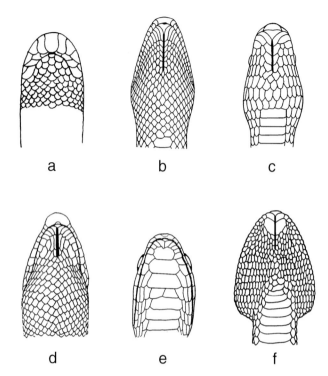

Figure 11.14. Features of gular scalation in Costa Rican snakes: (a) generalized scolecophidian; note anteriorly placed mouth and lack of mental groove; (b) a generalized boid, note absence of enlarged gular scales; (c) *Ungaliophis* with three pairs of enlarged scales bordering the mental groove; (d) *Loxocemus* with a pair of small but elongate gular scales bordering the mental groove; (e) *Dipsas* bicolor with three broad unpaired chin shields and no mental groove; (f) a generalized pitviper with a pair of large chin shields and a mental groove. See figure 11.3 for condition in most colubrid snakes.

greater than one head length anterior to vent; apical scale pits usually present 44

43a. Supralabial just posterior to orbit greatly enlarged (fig. 11.36); chin shields paired, first pair elongate; a more or less well developed mental groove . . . .
. . . . *Sibon* (p. 627)

43b. No greatly enlarged supralabial; chin shields single or paired; if paired, rectangular; no mental groove (fig. 11.14e) . . . . *Dipsas* (p. 596)

44a. 17 or more scale rows at midbody; nasal divided, usually completely 45

44b. 15 scale rows at midbody; a single nasal 27

45a. Pupil vertically elliptical in strong light or preservative (fig. 11.9d) 46

45b. Pupil round (fig. 11.4) 47

46a. Vertebral scales of about the same size as paravertebral rows; uniform black or red dorsum in life (pale brown in preservative), with a small black spot on tip of each dorsal scale or with alternating red (in life) and broad black bands; head not completely white, although a light nuchal collar bordered by a narrow black nuchal collar often present 27

46b. Vertebral scales distinctly enlarged (fig. 11.15); red in

life (pale brown in preservative) with small black blotches or narrow bands on dorsum; head red or white; a broad black nuchal collar . . . .
. . . . *Tripanurgos* (p. 576)

47a. 17 scale rows at midbody 49

47b. 19 or more scale rows at midbody. 48

48a. Pattern uniform or with crossbands or rings (fig. 3.1); dorsal scales in even rows (fig. 11.5) 26

48b. Pattern blotched (fig. 3.1); dorsal scales in oblique rows (fig. 11.5) . . . . *Xenodon* (p. 582)

49a. Subcaudals more than 90 37

49b. Subcaudals fewer than 90 . . . . *Drymarchon* (p. 658)

50a. No apical scale pits (fig. 11.5) 51

50b. Apical scale pits present (fig. 11.5) at least on anterior 20 dorsal scale rows 66

51a. Scale rows at midbody at least two more than number one head length anterior to vent 52

51b. Scale rows at midbody not more than one greater than number one head length anterior to vent. 55

52a. Pattern of light and/or dark stripes *or* uniform (fig. 3.1) 53

52b. Pattern of flecks, spots, or blotches (fig. 3.1); pupil vertically elliptical in strong light or preservative (fig. 11.9d) 59

53a. Usually 2–3 anterior temporals (fig. 11.16a), if one anterior temporal present, dorsum dark brown with 2–4 narrow longitudinal white stripes or with dark smudges on tips of ventrals at midbody. 54

53b. One anterior temporal (fig. 11.16b); dorsal pattern

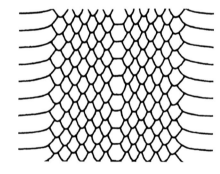

Figure 11.15. Dorsal view near midbody in snakes having a distinctly enlarged middorsal scale row.

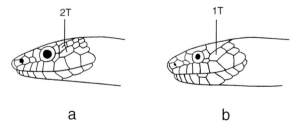

a   b

Figure 11.16. Lateral view of head scalation: (a) *Conophis* showing two primary temporals; (b) *Coniophanes* with one anterior temporal.

a

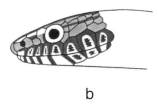

b

Figure 11.17. Lateral view of head color patterns: (a) *Conophis lineatus*; (b) *Crisantophis nevermanni*.

without narrow longitudinal white stripes or dark smudges on tips of ventrals, although other dark ventral markings may be present . . . . *Coniophanes* (p. 591)

54a. Supralabials immaculate or with dark lower borders (fig. 11.17a); dorsum gray to brownish with 6–13 narrow longitudinal dark stripes at midbody; lateral profile of head acute, with decurved rostral projecting well beyond tip of lower jaw (fig. 11.25); rostral concave below . . . . *Conophis* (p. 584)

54b. Supralabials bordered with black above (fig. 11.17b); dorsum dark with 2–4 narrow longitudinal white stripes; lateral profile of head rounded, with normal rostral not projecting much beyond tip of lower jaw; rostral not concave below . . . . *Crisantophis* (p. 586)

55a. Tail disproportionately thickened for most of its length (fig. 11.18a), long (30–48% of total length) and fragile (about 50% of specimens with part of tail missing) 56

55b. Tail short, less than 30% of total *or* if 30% or more of total length, slender and tapering to tip (fig. 11.18b) 58

56a. Dorsal pattern uniform, striped, or with black rings (fig. 3.1) 57

56b. Dorsum with tricolor pattern of black, red, and yellow or white bands at least anteriorly or bicolor with black and light bands (fig. 11.47) . . . . *Scaphiodontophis* (p. 681)

57a. Two scales (loreal and postnasal) between eye and nostril separating prefrontal from supralabials (fig. 11.12c); dorsum uniform dark brown or black, with a light nuchal collar or neck and head mostly white; 1

or 2 apical scale pits; 15 or 17 scale rows at midbody . . . . *Enulius* (p. 587)

57b. Three scales (preocular, loreal, and postnasal) between eye and nostril separating prefrontal from supralabials (fig. 11.11b); pattern of black and light rings (fig. 11.47), light stripes, or uniform dark brown or black, without a nuchal light collar; no apical scale pits; 17 scale rows at midbody . . . *Urotheca* (p. 640)

58a. Dorsal pattern includes flecks, spots, dark bands, or rings (fig. 3.1) 59

58b. Dorsal pattern striped or uniform (fig. 3.1) 62

59a. Head distinct from neck or not, but not greatly enlarged; eyes not markedly enlarged or bulging out from head 60

59b. Head greatly enlarged, neck extremely narrow and elongate; eyes very large and bulging out from head (fig. 11.12e) . . . . *Imantodes* (p. 606)

60a. Pattern with dark bands (fig. 11.47); ventrals 160 or fewer 24

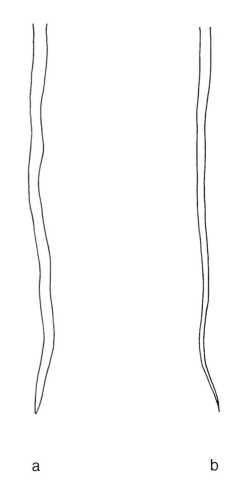

a   b

Figure 11.18. Distinctive caudal morphologies in long-tailed Costa Rican snakes: (a) disproportionately thickened and fragile tail found in the genera *Enulius, Scaphiodontophis,* and *Urotheca;* (b) slender and gradually tapering tail seen in *Coluber, Coniophanes, Dendrophidion, Leptophis, Oxybelis, Rhadinaea,* and *Thamnophis.*

60b. Pattern includes black rings (fig. 11.47); ventrals 170 or more 61

61a. Only a loreal present (no preocular) (fig. 11.12d) 25

61b. Both a preocular and a loreal present (fig. 11.12b) . . . . *Erythrolamprus* (p. 578)

62a. Two scales (loreal and postnasal) between eye and nostril separating prefrontal from supralabials (fig. 11.12c). 63

62b. Three scales (preocular, loreal, and postnasal) between eye and nostril separating prefrontal from supralabials (fig. 11.11b) 65

63a. No canthal ridge; venter light; head shields without irregular dark markings, although light spots may be present; dorsum uniform or with dark and/or light stripes, no lateral light dots; usually a nuchal light collar or sometimes head nearly completely white . . . 64

63b. A definite canthal ridge; venter uniformly dark; upper head shields marked with irregular dark lines on a gray to rust ground color (fig. 11.24); dorsum essentially uniform dark brown or black, often with a series of small lateral white dots; no nuchal light collar . . . . *Amastridium* (p. 583)

64a. Dorsum reddish brown, usually with dark and/or light longitudinal stripes, sometimes uniform; head never mostly white above, although a nuchal light collar and small light spots often present on dorsal surface of head; nuchal collar often outlined by black, usually somewhat suffused with darker pigment if it extends forward beyond posterior tips of parietals; dorsal scales in 15 rows at midbody; rostral plate acuminate, not enlarged nor markedly protruding beyond mouth opening; no apical scale pits 25

64b. Dorsum uniform black or brown; a light nuchal collar usually present, but without small light spots on dorsal surface of head; dorsal scales in 15–17 rows at midbody, if 15 the parietals to almost entire upper head surface clear white; rostral plate may protrude markedly beyond mouth opening; apical scale pits present 57

65a. A single prefrontal (fig. 11.38) *or* if two prefrontals, scale rows 17 at midbody, venter immaculate red in life, and ventrals more than 150. . . . . *Trimetopon* (p. 636)

65b. Two prefrontals (fig. 11.4), but never in combination with 17 scale rows at midbody, immaculate red venter, and more than 150 ventrals. . . . . *Rhadinaea* (p. 622)

66a. Scale rows at midbody not more than one greater than number one head length anterior to vent. 67

66b. Scale rows at midbody at least two more than number one head length anterior to vent 69

67a. Dorsal pattern consisting of flecks, spots, blotches, or rings (fig. 3.1) 68

67b. Dorsal pattern consisting of stripes or uniform (fig. 3.1) 63

68a. Pattern consisting of flecks, spots or blotches (fig. 3.1) 55

68b. Pattern consisting of rings (fig. 11.47) . . . . *Scolecophis* (p. 685)

69a. No more than 17 scale rows at midbody. 70

69b. 19 or more scale rows. . . . . *Leptodeira* (p. 610)

70a. Dorsal scale rows reduced to 15 at about one head length anterior to vent; no subocular below preoculars 71

70b. Dorsal scale rows reduced to 13 at about one head length anterior to vent; a small subocular below preocular (fig. 11.4) . . . . *Coluber* (p. 651)

71a. Pattern consisting of dark blotches or bands or with 2–4 more or less well developed lateral light stripes (fig. 3.1). 72

71b. Pattern with a distinct pair of dark, dorsal stripes bounding a broad light middorsal area (fig. 3.1); more than 140 subcaudals. . . . . *Leptodrymus* (p. 667)

72a. Ventrals 200 or more; pattern consisting of spots, speckles, or blotches (fig. 3.1) 59

72b. Ventrals fewer than 196; adults with pale stripes or with dark transverse bands that extend completely across back; juveniles may have dark dorsal blotches (fig. 3.1). 73

73a. Dorsal pattern of adults composed of two or four pale lateral stripes; juveniles with dark alternating dorsal and lateral blotches (fig. 3.1); ventrals 160–95; subcaudals 85–110 . . . . *Mastigodryas* (p. 673)

73b. Dorsal pattern banded (fig. 3.1); ventrals 139–52; subcaudals 51–63 . . . . *Liophis* (p. 580)

74a. No rattle or "button" on tail, although an enlarged scale or spine may be present. 75

74b. A rattle (fig. 11.8c) or "button" (fig. 11.8d) on tail . . . . *Crotalus* (p. 735)

75a. Posterior subcaudals in 1–2 rows (fig. 11.19a) . . . 76

75b. Posterior subcaudals in 4–5 rows (fig. 11.19b) . . . . *Lachesis* (p. 729)

76a. Dorsal surface of head covered by mostly small irregular scales, including frontal and parietal region . . . 77

76b. Dorsal surface of head covered by large symmetrical plates including paired parietals and a single frontal (fig. 11.4) . . . . *Agkistrodon* (p. 718)

77a. Subcaudals single (in one row) (fig. 11.4) or some divided (in two rows); 8–11 supralabials; ventrals plus caudals fewer than 240 78

77b. Most subcaudals divided (in two rows) over entire length of tail (fig. 11.19a); 7 supralabials; ventrals plus caudals 250–88; dorsal pattern a series of dark triangular lateral blotches that often fuse on midline to form hourglass-shaped marks . . . . *Bothrops* (p. 726)

78a. Tail prehensile; dorsal ground color usually bright green or yellow. . . . . *Bothriechis* (p. 723)

78b. Tail not prehensile; dorsal ground color never bright green or yellow 79

79a. Frontal and parietal region covered by small imbricate

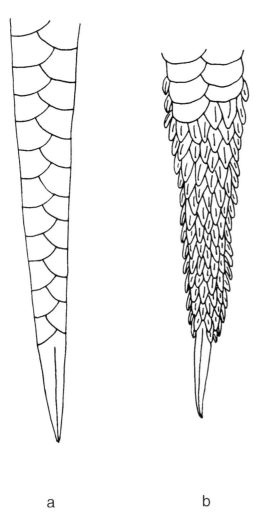

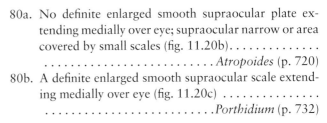

80a. No definite enlarged smooth supraocular plate extending medially over eye; supraocular narrow or area covered by small scales (fig. 11.20b)..............
.........................Atropoides (p. 720)
80b. A definite enlarged smooth supraocular scale extending medially over eye (fig. 11.20c) ....
.........................Porthidium (p. 732)

Remember that in Costa Rica any reptile lacking a shell, limbs, movable eyelids, and external ear openings is a snake. A few elongate lizards with reduced limbs, especially *Bachia*, might be mistaken briefly for snakes, but all Costa Rican lizards have four limbs. The only other members of the herpetofauna that are snakelike are the limbless burrowing amphibians called caecilians (Gymnophiona). These forms lack epidermal scales, have virtually no tail, and are encircled by a series of skin folds that give them the appearance of large earthworms. All snakes are scaly, have a well-developed tail (although it is very short in the three blind snake families), and lack skin folds.

Although tropical lands are reputed to teem with snakes, most species are rarely seen, and a field biologist working in relatively undisturbed areas may go for days without seeing a single one. Exaggeration concerning the size of snakes sighted in casual encounters is notorious. Since size is an important feature in identification, remember that to even the calm uninitiated observer a snake looks 50 to 60% larger than it actually is. Estimates of snake sizes by people who are superstitious, frightened, or upset are usually at least twice the actual length.

Throughout the world the serpent is the most revered and feared of all animals and evokes extremely strong emotions of attraction or disgust. The symbolic effect of these creatures has so influenced humans that in many cultures snakes are objects of either worship or persecution, for the serpent is a powerful androgynous symbol emphasizing the masculine (phallic) in its form and, as the swallower, representing the female organ as well. In addition, its wondrous ability to slough its skin to renew its beauty symbolizes growth, change, the mastery of time, and the mystery of rebirth.

Figure 11.19. Diagnostic caudal morphologies in pitvipers: (a) paired subcaudals; (b) *Lachesis* with posterior subcaudals in four or five rows.

mostly keeled scales; scales on upper surface of snout keeled (fig. 11.20b,c) 80
79b. Definite enlarged smooth frontal and parietal plates; scales on upper surface of snout mostly smooth (fig. 11.20a)................. *Cerrophidion* (p. 728)

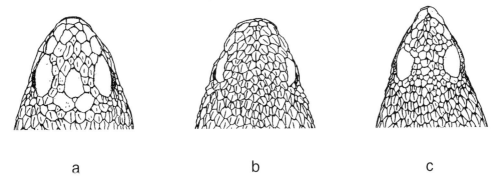

Figure 11.20. Scalation of upper anterior head surface in Costa Rican pitvipers: (a) *Cerrophidion godmani* with smooth enlarged frontal and parietal plates; (b) *Atropoides nummifer* with narrow supraocular plates; (c) *Porthidium nasutum* having definite enlarged supraocular plates.

Through its darting tongue and lethal, burning poison the snake symbolizes the equality of death, and in its unflickering gaze the wisdom of the ages (Campbell 1964). It is no wonder, then, that at the vulgar level of existence unfounded superstitions and lies abound about these creatures, whose limbless movements made them messengers of the gods in ancient times. The common libels that must be repeatedly dispelled are briefly mentioned below.

Snakes do not hypnotize or charm their prey, even though they have an unflickering stare (since they lack movable eyelids).

Snakes cannot sting or inject venom with their tongues or tails, nor is their breath poisonous, nor do they leave balls of white spit on leaves or the substrate.

Most snakes arc not venomous.

Snakes, like all other organisms, may die at any time of the day or night, not only after sundown.

Snakes that are cut in two or decapitated die and do not regenerate the missing parts to produce two individuals.

Because of their mouth and jaw structure, snakes are incapable of sucking; thus they cannot "milk" a cow or any other mammal.

Snakes do not swallow their young or take them into their mouths in time of danger.

Snakes do not form pairs that live or travel together, nor does the "mate" of a snake that is killed come in search of the slayer.

No snakes have functional limbs, but under duress males may evert the hemipenes, which are sometimes mistaken for legs.

Garlic or horsehair ropes do not repel any snake.

Neither the skins nor the internal organs of snakes have any curative powers for human diseases.

No snake places its tail in its mouth to form a hoop and then bowls itself over the ground at high speeds; significantly, a serpent biting its own tail—the uroboros of ancient cosmologies—is a powerful symbol of wholeness and completion in many cultures.

Solórzano (1993) reviewed many commonly held but erroneous beliefs well established in Costa Rican folklore. His list, of course, includes most of those above, and others that have been attributed to particular species will be mentioned in the species accounts to follow.

Partially because of the venomous capabilities of some species in the group, most people view snakes as frightful, dangerous, and disgusting creatures. In Costa Rica most *campesinos* kill any snake on sight, and people's inordinate and often irrational fear of even the smallest harmless snake leads to exaggerated accounts of danger and narrow escapes. The vast majority of Costa Rican snakes are inoffensive, shy, and totally harmless animals that prefer escape to discovery and are incapable of hurting a human. Most of these much maligned reptiles are important predators that help maintain natural ecosystems and keep down populations of rats, mice, and insect pests. Normally they avoid contact with humans by concealment or flight. Although snakes evoke strong feelings of irrational fear and revulsion in many people, these should not be the basis for wanton and irresponsible killing. I hope the information in this book will give readers a new appreciation for these fascinating animals, which deserve respect as a part of humankind's natural heritage and need to be protected from our irrational impulses.

Primitive Blind Burrowing Snakes (Infraorder Scolecophidia)

Representatives of the three families of wormlike, burrowing snakes resemble one another very much in overall general appearance. All have smooth, shiny cycloid scales on the body and tail, without any hint of enlargement of the scales on the underside. The eyes are reduced to small dark spots under the head scales, do not function to form an image, and are questionably light sensitive, hence the name blind snakes. However, the center of the scale over the eye is transparent, suggesting some ability to sense light intensity. The retina contains only rods. The small, curved mouth is well behind and below the snout tip (fig. 11.14a). The body is uniform in diameter throughout its length, while the tail is very short and ends in a conical spine.

Many species in the group appear to be commensals with ants or termites, living in their hosts' subterranean nests and feeding on workers, larvae, and pupae and on soft-bodied arthropod commensals in the same nest. Some members of the Typhlopidae appear to construct their own burrows and feed on a variety of soft-bodied arthropods and their larvae. Although they are sometimes found under debris on the ground surface, the close association of these snakes with their hosts' underground nests or their own burrows means they are rarely seen. Occasional individuals are found moving about on the surface at night, usually after heavy rains.

When disturbed, all blind burrowing snakes emit an odoriferous and, to most humans, objectionable secretion from the cloacal glands. Studies by Gehlbach, Watkins, and Reno (1968) show that in one North American species *(Leptotyphlops dulcis)* the secretion is used to repel attacking ants when the snake first enters a colony. It seems likely that this is the function of the secretion in other species as well. This form and one species of *Typhlops* have been shown to follow the pheromone trails of ant prey (Gehlbach, Watkins, and Kroll 1971).

In addition to features of external appearance, blind burrowing snakes are defined, and distinguished from other snakes, by a series of characteristics (mostly osteological) associated with their fossorial lives and specialized feeding habits:

1. Quadrate bone long and inclined downward and forward to attach with lower jaw; supratemporal bone vestigial or absent

2. Lower jaw shortened, much shorter than skull

3. Dentary bones cannot be thrust forward one ahead of the other

4. Eye covered by a large scute that does not conform to shape of eye (brille absent); one type of visual cell (single rods) in retina

5. No teeth in roof of mouth, on palatine, or on pterygoid bones

6. No nasofrontal joint

7. No mental groove or differentiated chin shields

8. Dorsal vertebrae lacking neural spines and hypapophyses

Because of the superficial similarity among the blind burrowing snakes that often makes it difficult to identify individual snakes to family or genus, I include a supplementary key to this group below. This key will help you identify any known blind snake from the Americas to genus.

KEY TO NEW WORLD GENERA OF BLIND BURROWING SNAKES

1a. Nasal and ocular separated from lip by one to several scales (fig. 11.6d,e); 18–20 scale rows around body 2

1b. Lower nasal, and usually ocular, bordering lip (fig. 11.6c); 14 scale rows around body. . . . . *Leptotyphlops* (LEPTOTYPHLOPIDAE)

2a. Prefrontals fused with large supranasal plates that meet frontal behind rostral (fig. 11.7a). . . . . *Typhlops* (TYPHLOPIDAE)

2b. Prefrontals and supranasals distinct plates, or head covered by small scales (ANOMALEPIDIDAE) 3

3a. Head covered by large plates, distinctly larger than those on body; nostril between two large plates 4

3b. Head covered by small uniform scales, indistinguishable from those on body; nostril between two small nasals . . . . *Typhlophis*

4a. Upper head plates rounded (fig. 11.7b,c). 5

4b. Upper head plates angular; polygonal prefrontals in contact with each other, separating rostral from pentagonal frontal (fig. 11.7d) . . . . *Anomalepis*

5a. Prefrontals in contact, separating rostral from frontal (fig. 11.7c) . . . . *Helminthophis*

5b. Prefrontals separated by rostral, which touches the frontal (fig. 11.7b) . . . . *Liotyphlops*

Family Anomalepididae Taylor, 1939a

Blind Worm Snakes

The family comprises four genera: *Anomalepis* (four species), *Helminthophis* (three species), *Liotyphlops* (seven species), and *Typhlophis* (one species). The distribution of the group is restricted to the Neotropics from lower Central America and northwestern South America south to southern and southeastern Brazil, Paraguay, and northeastern Argentina. *Anomalepis* is found in lower Central America,

northwestern South America, and the Amazon drainage of Peru. *Helminthophis* is known from Costa Rica and northern Colombia and Venezuela. *Typhlophis* occurs along the Atlantic coast of South America from Trinidad and the Guianas to southern Brazil. *Liotyphlops* occurs throughout the range of the family. All species included are very small snakes, none exceeding 380 mm in total length.

In addition to the features of external morphology that are given in the key (p. 553) to separate this family from the Typhlopidae and Leptotyphlopidae, it is distinctive among blind snakes in having the following combination of characters:

1. Maxillary toothed, mobile on the skull, and oriented transverse to it

2. Dentary small, with one to three teeth; lower jaw long, slender, hinged

3. Hyoid apparatus an M-shaped element (fused basihyal and first ceratobranchials)

4. A tracheal lung present; no left lung

5. Pelvic vestiges, if present, composed of cartilaginous elements

The family shares the maxillary condition and the presence of a tracheal lung with the Typhlopidae and agrees with the Leptotyphlopidae in having teeth on the lower jaw. However, the small tooth-bearing dentary of the anomalepidids resembles the tiny element in the typhlopids, whereas the dentary in the leptotyphlopids is a large bone with an enlarged anterior dentigerous process. As far as is known all members of the family are oviparous (Blackburn 1985).

Genus *Anomalepis* Jan and Sordelli, 1860

(fig. 11.7d)

The blind worm snakes of this genus (four species) share the following diagnostic features that will separate them from the other three genera in the family: enlarged head scales, with those on the upper head surface forming polygonal plates; prefrontal plates meet on the midline to prevent any contact between the rostral and prefrontal; a roughly pentagonal frontal, with an ovate posterior margin. In addition there are 22 to 28 scale rows around the middle of the body, and the tail has a terminal spine. *Typhlophis*, which is not represented in Costa Rica, differs from *Anomalepis* by having small cycloid scales on the upper head surface so that prefrontal and frontal plates are not distinctive, but the tail terminates in a spine. Differences between *Anomalepis* and the other genera of the family represented in Costa Rica are provided in the appropriate generic accounts.

The members of the genus are secretive leaf litter dwellers or fossorial and insectivorous forms that have been collected only a few times at widely scattered localities in the Neotropics: Costa Rica, central Panama, central Colombia, northwestern Ecuador, and the extreme upper Amazonian drainage of northern Peru.

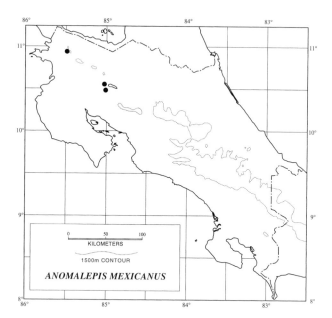

Map 11.1. Distribution of *Anomalepis mexicanus*.

Anomalepis mexicanus Jan and Sordelli, 1860

(plate 329; map 11.1)

DIAGNOSTICS: A tiny uniformly glossy brown snake, without enlarged ventral scales and with the minute eyes hidden under the head scales.

DESCRIPTION: A rare secretive species (to 180 mm in total length); scale rows around midbody 22; dorsal scales from head to tip of tail 267 to 272; a pair of enlarged precloacal scales; tail short about same length as head. Uniform brown.

SIMILAR SPECIES: (1) *Helminthophis frontalis* has the head and neck area pinkish. (2) *Liotyphlops albirostris* has the snout or occasionally the whole head light. (3) *Typhlops costaricensis* has the venter yellow and 20 scales around midbody. (4) *Leptotyphlops ater* has a light tail tip and 14 scale rows around midbody. (5) The caecilian genera *(Dermophis, Gymnopis,* and *Oscaecilia)* lack epidermal scales.

HABITAT: Taken under leaf litter in remnants of Pacific-Atlantic Lowland Moist Forest.

REMARKS: The original description (Jan and Sordelli 1860) attributes the species to Mexico, hence the specific name. Taylor (1939a, 1951b) thought the form in Costa Rica and Panama was distinctive from *A. mexicanus* and called it *Anomalepis dentatus.* Most subsequent workers follow Dunn (1932a, 1941b) in regarding the lower Central American specimens as conspecific with the supposed Mexican holotype. Since the species has never again been collected in Mexico, it seems likely that some mixup in locality data was involved. However, given their secretive habits and the rarity of anomalepidids in collection, there remains an outside possibility that the species may be collected elsewhere in Central America or even in Mexico. Villa (1983)

lists this species as occurring in Nicaragua without documenting specimens or exact localities.

DISTRIBUTION: Foothills of northwestern Costa Rica and lowlands of central Panama. In Costa Rica known only from the Pacific drainage in Guanacaste Province (350–500 m).

Genus *Helminthophis* Peters, 1860

(fig. 11.7c)

Members of the genus (three species) share the following distinctive characteristics that will distinguish them from other genera of blind worm snakes: enlarged, nonpolygonal head plates that have rounded, posterior margins; prefrontal plates meeting on the midline to separate rostral and prefrontal; frontal enlarged, rounded, and similar in appearance to dorsal scales, but about twice as large. Scale rows around midbody 20 to 24; tail with a terminal spine. *Helminthophis* differs from the extralimital *Typhlophis* in head scutellation, since the upper head scales in the latter genus are similar in size and shape to the dorsal scutes. Detailed differences distinguishing the other two Costa Rican genera from *Helminthophis* are given in their generic accounts.

Species in this genus are secretive, burrowing forms that superficially resemble small, slender earthworms. They are known from disjunct areas in Costa Rica and southwestern Panama, northern Colombia, and northern Venezuela.

Helminthophis frontalis (Peters, 1860)

(plate 330; map 11.2)

DIAGNOSTICS: A tiny species with the head and neck pinkish in life, lacking enlarged ventral scales, and with the minute eyes hidden under scales.

DESCRIPTION: Maximum total length 160 mm; uniform glossy black above and below with the head and neck and usually cloacal region pinkish yellow in life (yellow to white in preservative). Scale rows around midbody 22; dorsal scales from head to tip of tail 470 to 480; no enlarged precloacal scales; tail short, about as long as head.

SIMILAR SPECIES: (1) *Anomalepis mexicanus* has the head colored like the body. (2) *Liotyphlops albirostris* has a light spot on the snout and the rostral and frontal in contact. (3) *Typhlops costaricensis* has the undersides of the posterior venter and tail yellow. (4) *Leptotyphlops ater* has a yellow tail tip and only 14 scale rows around the midbody. (5) The caecilian genera *(Dermophis, Gymnopis,* and *Oscaecilia)* lack epidermal scales.

HABITAT: A probably common burrowing form sometimes found under debris on the surface in Lowland Moist and Premontane Moist Forest areas.

REMARKS: One specimen (CRE 6283) labeled "Cartago" is probably from the city of Cartago (1,435 m) and establishes the species' occurrence on the Meseta Central Oriental.

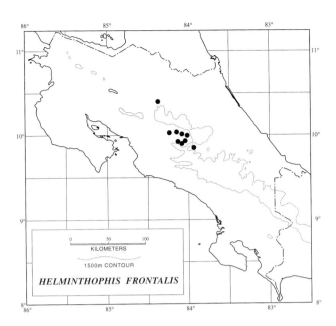

Map 11.2. Distribution of *Helminthophis frontalis*.

DISTRIBUTION: The Meseta Central and the area near La Florencia, Alajuela Province, on the Atlantic slope of Costa Rica and southwestern Panama (96–1,435 m).

Genus *Liotyphlops* Peters, 1881

(figs. 11.6e, 11.7b)

The seven species of *Liotyphlops* share the following suite of features that separate them from other anomalepidids: enlarged, nonpolygonal head plates that have rounded posterior margins; the enlarged rostral meeting a broad and short frontal plate to completely separate the prefrontals; scale rows at midbody 22 to 28; tail with a terminal spine. *Liotyphlops* differs from *Typhlophis* of South America by having enlarged head scales (scales on upper head surface same size and shape as dorsals in *Typhlophis*) and from the other Costa Rican genera in the enlarged rostral and peculiar frontal scales.

The hemipenes have been described for only one species *(Liotyphlops albirostris)*. The everted organ is simple, symmetrical, somewhat bulbous, capitate, and ornamented with spines. The pedicel is nude, but moderate to large spines are present on the asulcate surfaces below the capitulum margin and on the capitulum; several large spines project from the apex. The sulcate surface has spinules just below the capitulum but is otherwise nude. The sulcus spermaticus appears to be centrolineal and simple, although the describers (Myers and Trueb 1967), perhaps influenced by the effects of a collapsed tissue artifact they illustrated, suggest that it may bifurcate distally.

These secretive forms apparently hide in the leaf litter or burrow into the soil. The genus is known from Costa Rica, Panama, Colombia, western Ecuador, northern Venezuela, Isla Bonaire, Suriname, east-central and southeastern Brazil, and Paraguay.

Liotyphlops albirostris (Peters, 1857)

(plate 331; map 11.3)

DIAGNOSTICS: A small rare species with a light spot on the snout, lacking enlarged ventral scales and with the minute eyes hidden under scales.

DESCRIPTION: Total length to 223 mm; glossy black above and below. Scale rows around midbody 22; dorsal scale rows from head to tip of tail 370 to 455; no enlarged precloacal scales; tail short, about as long as head.

SIMILAR SPECIES: (1) *Anomalepis mexicanus* has the head colored like the body. (2) *Helminthophis frontalis* has the head and neck pinkish in color, and the rostral and frontal scales are separated by the prefrontals. (3) *Typhlops costaricensis* has the undersides of the posterior body and tail yellow. (4) *Leptotyphlops ater* has 14 scale rows around the midbody. (5) The limbless amphibians *(Dermophis, Gymnopis,* and *Oscaecilia)* lack epidermal scales.

REMARKS: Some Panamanian and Colombian specimens of this species have the light area on the snout expanded so that the entire head is light. Dixon and Kofron (1984) in their generic revision review scutellational and geographic variation for this species and mention color changes from dark brown (slightly lighter below) to whitish that occur during the various stages of skin shedding. Dixon and Kofron (1984) agree with Dunn (1932a) in placing *Typhlops emunctus* Garman (1883) as a synonym of *L. albirostris.* Taylor (1951b) regarded the former as a distinct species and mentioned as a dubious record for Costa Rica an example of this nominal taxon at the United States National Museum

Map 11.3. Distribution of *Liotyphlops albirostris*.

collected by J. A. McNeil. This specimen (USNM 23748) was originally labeled as from "Chirique, Costa Rica." Chirique is an obvious transliteration of Chiriquí, the name of a western province of Panama, bordering Costa Rica. The confusion probably arose because this area, originally part of the Province of Cartago in colonial times, was claimed by Costa Rica during all of the nineteenth century. In any event, Dunn (1932a) and Dixon and Kofron (1984) correctly list this snake as being from Panama. Contrary to Dixon and Kofron, Taylor did not question the occurrence of *Liotyphlops* in Costa Rica, but only the provenance of the McNeil specimen.

DISTRIBUTION: The species is definitely known from the lowlands of western and central Panama and northern Colombia, southward into the Río Magdalena valley, western Ecuador, northern Venezuela, and Isla Bonaire. A population from Mérida, Venezuela (1,640 m), was questionably referred to this species by Dixon and Kofron (1984). The Panama records (ZMB 8565, USNM 23748, Chiriquí; and ZMB 9529, Veragua) probably are from the southwestern portion of that country, where the species is definitely known to occur near the Costa Rican frontier in the (Finca) Blanco district (MCZ 31540). The only specimen from Costa Rica (SMF 7005a) lacks additional data, but because of the lowland distribution of the species and the Panama records, it is anticipated that it will be rediscovered, probably in lowland southwestern Costa Rica.

Family Typhlopidae Fitzinger, 1826

Blind Snakes

Members of this family occur in sub-Saharan Africa and Madagascar; from the Balkans and Greece southward across Asia to southern China, through the Indo-Malayan archipelago to New Guinea and Australia and throughout Oceania to the Hawaiian Islands; in the Neotropics from tropical Mexico south through Central and South America (east of the Andes) to southeastern Brazil and central Argentina, the Greater and Lesser Antilles, and the Bahama Islands. Six genera are currently recognized: *Acutotyphlops* (four species) of New Guinea and the Solomon Islands, *Cyclotyphlops* of Sulawesi (one species); *Ramphotyphlops* (forty-eight species), with a range centered on the Indo-Australian area; *Rhinotyphlops* (twenty-eight species) of sub-Saharan Africa; *Typhlops* (124 species), with representatives in most areas where the family occurs except for Australia and Oceania and the arid regions of sub-Saharan Africa; and *Xenotyphlops* (one species) of Madagascar.

The distributional picture is further complicated by the extensive range of the all-female triploid (3N) species *Ramphotyphlops braminus*, which occurs throughout much of tropical Africa and Asia and the Pacific islands and has been introduced into Egypt, Mexico, Guatemala, Florida, Louisiana, Massachusetts, and the Lesser Antilles. It seems likely that much of the distribution of this Indian or Asian species has been a relatively recent dispersion, probably aided inadvertently by human transport of soil either by itself or as a medium for living plants. Most species are small, with a total length less than 400 mm, but one African species reaches a total length of 900 mm. Except for a single record of viviparity (Smith 1943), the family is apparently oviparous (Blackburn 1985).

Typhlopids are distinguished from other blind snakes by the external features included in the key (p. 553) and the following combination of features:

1. Maxillary toothed, mobile on the skull, and transverse to it
2. Dentary very small, toothless; lower jaw not hinged
3. Hyoid apparatus (fused basihyal and first ceratobranchials) wishbone shaped
4. Usually a tracheal lung, left lung vestigial or absent
5. Pelvic vestiges usually consisting of usually ossified ischia, with cartilaginous ilia or pubes sometimes present

The first of these characters also occurs in the anomalepidid blind snakes, and typhlopids share the condition of the hyoid with leptotyphlopids.

Genus *Typhlops* Oppel, "1810," 1811

(figs. 11.6d, 11.7a)

Members of this wide-ranging genus differ from the other five genera placed in the family by having simple hemipenes without a terminal awn, a relatively unenlarged and undifferentiated rostral plate that lacks papillae, and three to five precloacal scales. *Acutotyphlops* has a V-shaped, pointed lower jaw. *Cyclotyphlops* has four rostrals and a large circular frontal scale. The rostral plate of *Xenotyphlops* is papillate. In *Ramphotyphlops* the greater portion of the hemipenis is a solid protrusible structure that lies coiled within a connective tissue sheath until everted (Robb 1960) and corresponds to the terminal awn found in some other snakes. *Rhinotyphlops* has a greatly enlarged rostral with a sharp horizontal edge running across its lateral and anterior surfaces at the level of the nostrils, simple hemipenes, and one precloacal scale. *Typhlops* have 18 to 44 scale rows around the midbody, and the tail ends in a terminal spine.

Branch (1986) illustrated the very short, simple hemipenis, ornamented with a few oblique ridges laterally, characteristic of *Typhlops*, but there is great variability in the genus in ornamentation (nude, papillate, or spinous), and some species have a capitulum.

Of the 120 species placed in *Typhlops*, 34 occur in the Neotropics, with 24 endemics in the Antilles. The introduced all-female blind snake *Ramphotyphlops braminus* is known from several localities in tropical and subtropical America (Mexico, upland Guatemala, the Lesser Antilles, and southern Florida) (Dixon and Hendricks 1979) and has the potential to be introduced elsewhere in subtropical and tropical America. *R. braminus* is much smaller (total length to 173 mm) than the single species of *Typhlops* found in Costa Rica and has many fewer (300 to 338) transverse rows of scales from head scales to tail tip. *Typhlops* occurs from Greece across the arid regions of Eurasia, throughout India and eastward to southern China, and then southward through Southeast Asia and the Indo-Australian region to the Philippine Islands and New Guinea. It also is found in humid regions of sub-Saharan Africa and on Madagascar. In the Americas members of the genus are found from eastern tropical Mexico south to Honduras; in central Nicaragua and northern Costa Rica; northern Venezuela; tropical South America (east of the Andes) to southeastern Brazil and central Argentina; Trinidad, the Antilles, and the Bahama Islands.

Typhlops costaricensis Jiménez and Savage, 1962

(plate 332; map 11.4)

DIAGNOSTICS: Uniformly glossy gray to dark brown dorsally and laterally, lighter ventrally, lacking enlarged ventral scales, and with the minute eyes hidden under the scales.

DESCRIPTION: A moderately large (to 360 mm in total length) species of blind snake, tip of snout, throat, center of venter, underside of tail, and tail spine cream, yellow, or beige, with the anterior two-thirds of venter sometimes suffused with dark pigment. Nasal suture complete, nasal in contact with rostral; scale rows around midbody 20; dorsal scales from head scales to tail tip 394 to 415, no enlarged precloacal scales; tail short, about as long as head.

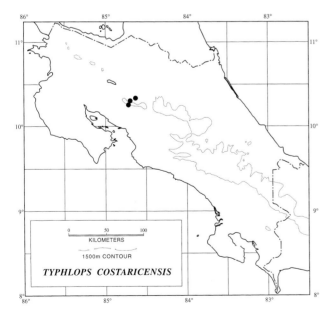

Map 11.4. Distribution of *Typhlops costaricensis*.

SIMILAR SPECIES: (1) *Anomalepis mexicanus* is uniform in color and has enlarged polygonal scales on top of the head. (2) *Helminthophis frontalis* has a light head and neck. (3) *Liotyphlops albirostris* has only a light snout spot, and the large rostral and frontal are in contact. (4) *Leptotyphlops ater* has 14 scale rows around the middle of the body. (5) The caecilians of the genera *Oscaecilia*, *Dermophis*, and *Gymnopis* lack epidermal scales.

HABITAT: This relatively rare burrower is known to occur in Lowland (Atlantic) Wet Premontane Wet and Lower Montane Wet Forests in Costa Rica.

BIOLOGY: Two examples (CRE 4623–24) from Monteverde, Puntarenas Province, Costa Rica, were regurgitated by a coral snake *(Micrurus nigrocinctus)*.

REMARKS: Dixon and Hendricks (1979) provided a detailed account of this species, and Villa (1988a) used it for his redescription.

DISTRIBUTION: Premontane zone of Pacific slope Honduras; Atlantic versant of Nicaragua and northeastern Costa Rica; premontane and lower montane zones (Pacific versant) in the vicinity of Monteverde, Puntarenas Province, Costa Rica (540–1,500 m).

Family Leptotyphlopidae Stejneger, "1891," 1892

Thread Snakes

The thread snakes are short, extremely slender blind burrowing forms that usually are less than 160 mm in total length, although a few reach 300 to 350 mm. These slender snakes have 13 to 15 scale rows around the body compared with 18 to 44 in other blind snakes. The family occurs in tropical and subtropical areas of western North America, Central and South America, the Lesser Antilles, Hispaniola, Africa, southern Arabia, and through the Middle East to Pakistan.

Two genera are included in the family: *Leptotyphlops* (eighty-six species), which has the same range as the family, and the West African *Rhinoleptus* (one species).

The species of this family are readily separated from other blind burrowing snakes because of their slender habitus and other features outlined in the key (p. 553). In addition the thread snakes share the following combination of characters:

1. Maxillary toothless, sutured to the braincase and parallel to it, immobile
2. Dentary large, with four or five teeth; jaw not hinged
3. Hyoid apparatus (fused basihyal and first ceratobranchials) wishbone shaped
4. No tracheal lung, left lung absent
5. Pelvic vestiges usually present, including ossified ischia, pubes, and femurs

The thread snakes share the hyoid character state with typhlopids and the toothed dentary with anomalepidids. The size and shape of the dentary, large and with an enlarged anterior dentigerous process in leptotyphlopids versus a tiny sliver of bone in the blind worm snakes, is distinctive for each group. So far as is known the family is oviparous (Blackburn 1985).

Genus *Leptotyphlops* Fitzinger, 1843

(figs. 11.6c, 11.7e)

The only other genus placed in the family Leptotyphlopidae is the monotypic *Rhinoleptus* from West Africa, which is differentiated from *Leptotyphlops* by having the snout terminating in a conical point with a sharp cutting edge that is visible from above. No *Leptotyphlops* exhibits this condition.

The genus has the same distribution as given for the family. In the Americas there are thirty-six recognized mainland species, one in the Lesser Antilles, four in Hispaniola, and one in the Bahama Islands.

Leptotyphlops ater Taylor, "1939b," 1940

(plate 333; map 11.5)

DIAGNOSTICS: A tiny snake, black above, chocolate brown below, with tail tip yellow in life, lacking enlarged ventral scales and with the minute eyes hidden under scales.

DESCRIPTION: A delicate thread snake (to 185 mm in total length) with a silvery cast in life; each scale with dark edges that tend to form indistinct dark longitudinal lines; tip of snout, lips, throat, subcaudal area, and the spiny tail tip yellow in life. The color pattern and the following scale characteristics will distinguish this species from other American *Leptotyphlops*: supraoculars large, larger than frontal; supraoculars separated from supralabials by oculars; prefrontal fused with rostral, which extends posteriorly on top

of head to behind level of eyes; ocular bordering lips; 14 scale rows around midbody, 12 around tail; dorsal scales from head to tip of tail 250 to 260; an enlarged precloacal scale; tail long, several times longer than the head. The hemipenes are similar in shape to those found in other members of the *L. goudotti* complex (Peters and Orejas-Miranda 1970, fig. 3) except that the terminal two-thirds is more robust; pedicel bulbous, naked, and sharply demarcated from terminal region by several large flounces on asulcate side; area distal to the large flounces covered by small flounces, then calyces, and finally spinulate calyces; the sulcus spermaticus is single and runs to tip of the symmetrical organ.

SIMILAR SPECIES: (1) Other blind burrowing snakes (*Anomalepis, Helminthophis, Liotyphlops,* and *Typhlops*) have 20 or more scales around midbody. (2) The limbless amphibians of the genera *Dermophis, Gymnopis,* and *Oscaecilia* lack epidermal scales.

HABITAT: Usually discovered under small stones or surface debris in Lowland Dry and Premontane Moist Forest.

BIOLOGY: Probably a common species, but uncommonly seen because of its secretive habits. Although little is known regarding the habits and ecology of the Costa Rican species, it seems likely that it resembles other ant and termite commensals in the genus (Scott 1983h). *Leptotyphlops dulcis* of the southwestern United States is such a commensal and feeds on the soft-bodied stages of its insect hosts and other soft-bodied arthropods. Sometimes it eats the prey whole by rotating the tooth-bearing half of each lower jaw synchronously in and out of the mouth (Kley and Brainerd 1999). At other times it breaks off the prey's head and apparently sucks out the body contents to leave a hollow shell of chitin (Reid and Lott 1963; Punzo 1974). *Leptotyphlops goudotti* is known to have a similar feeding style (Smith 1957). *L. dulcis* also exhibits a highly specialized behavior that permits it to coexist in the nests of its hosts (Gehlbach, Watkins, and Reno 1968). When a snake is newly introduced into an ant colony it is usually attacked. It then elevates the free margins of the scales to form a barrier resembling a pinecone. At the same time the animal excretes fecal material and secretes an odoriferous substance from the cloacal glands. The two materials are smeared over the body with the tail and are repellent to the attackers. The ants eventually cease to bite or sting the snake as it acquires the odor of the colony, although there may be several attacks before the intruder is fully accepted. Between attacks and once the snake is safe from them, it lowers the scales to form the glossy, low-friction surface typical of all worm snakes. This same species has been shown to follow pheromone trails of army ants (Watkins, Gehlbach, and Baldridge 1967; Gehlbach, Watkins, and Kroll 1971). It seems likely that the Costa Rican species has similar habits (a fruitful area for further study). Scott (1983h) notes that *L. ater* occasionally congregates in small colonies, sometimes in association with the small colubrid snake *Enulius flavitorques*. This behavior is reminis-

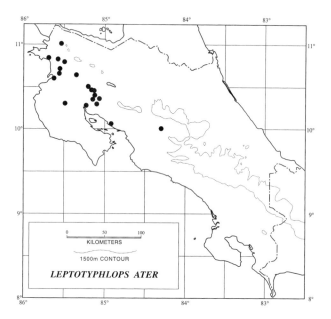

Map 11.5. Distribution of *Leptotyphlops ater*.

cent of that reported by Slevin (1950) for an allied species, *Leptotyphlops humilis*, which was found in large numbers with considerable numbers of the secretive burrowing colubrid *Sonora semiannulata* (numbers on four sampling days between 1940 and 1947 were 13:56, 99:26, 36:27, and 4:6). These concentrations may be associated with communal nesting in female thread snakes, which has been reported for *L. dulcis* (Tennant 1984).

REMARKS: Most recent references to this species follow Peters and Orejas-Miranda (1970) in regarding it as a race of *Leptotyphlops goudotti* (C. Duméril and Bibron). The Costa Rican and Nicaraguan populations of this complex are completely allopatric to the distribution of *goudotti*, are unique in having the rostral-prefrontal fusion (prefrontal present in *goudotti*), and are not connected to *goudotti* by a zone of intergradation. For these reasons *L. ater* is regarded as a distinct species. In spite of Peters and Orejas-Miranda's (1970) comments and Villa's (1990c) redescription to the contrary, I regard *Leptotyphlops nasalis* Taylor, "1939b," 1940, known from a single specimen (USNM 79947) from Managua, Department of Managua, Nicaragua, as a synonym of *L. ater*. The holotype of *L. ater* (USNM 79947) is from the same place. The type of *L. nasalis* differs from *L. ater* solely in having the supraocular and upper nasal on each side fused into a single plate while having the elongate, fused rostral-prefrontal typical of *L. ater*. Dunn and Saxe (1950) reached a similar conclusion, and as first revisors selected *ater* over *nasalis* for the species, since both names were proposed by Taylor (1939b) in the same article.

DISTRIBUTION: Lowlands of western Nicaragua and northwestern Costa Rica and inland in the valley of the Río Grande de Tárcoles to near La Garita, Alajuela Province (12–560 m).

Boas, Pythons, and Their Allies and Advanced Snakes (Infraorder Alethinophidia)

To most people the members of this group are the familiar serpents, since few individuals have seen examples of the primitive blind burrowing snakes or, if they have, recognized them as snakes. The present stock includes species ranging in size from the huge pythons (*Python reticulatus* of India and Southeast Asia, which reach a length of 10.1 m) and boas (*Eunectes murinus* of South America, to 9.4 m) to tiny burrowing forms (several burrowing or litter inhabitants in several families to 150 to 200 mm in total length). In addition the group has undergone a radiation into a great variety of habitats to produce a number of strictly fossorial, arboreal, freshwater, and marine groups.

The diversity of adaptations associated with the wide range of sizes and habitat specializations makes ready definition of the group as a whole based on external features difficult. However, the vast majority of the species in this taxon have the scales on the upper surfaces polygonal, with the posterior tip bluntly pointed; the ventral scutes enlarged and much broader than the dorsals; the subcaudals forming one or more series of enlarged scutes that are much larger than the caudal scales; the eye not covered by the head scales; and the mouth large and not set well back of the snout tip. Although there is some variation in these features elsewhere in the world, all Costa Rican snakes of this stock may be distinguished invariably from the primitive blind burrowing snakes by this set of features. Other features defining this stock and technically distinguishing them from the blind burrowing snakes are:

1. Quadrate bone short and inclined vertically or long and inclined backward; supratemporal bone present, usually long and mobile
2. Lower jaw only slightly shorter than skull
3. Dentary bones capable of being thrust forward independently of one another
4. Eye almost always covered by a large scute (= brille) that conforms to eye shape; two to four kinds of visual cells in retina (single rods, two kinds of single cones, and double cones)
5. Palatal teeth almost always present on palatine and pterygoid bones
6. A nasofrontal joint that allows engulfment of relatively large prey
7. A mental groove and one or two pairs of elongate chin shields usually present
8. Dorsal vertebrae usually with neural spines; hypapophyses usually present on anterior dorsal vertebrae and often on posterior ones as well

In addition the hyoid apparatus is composed of paired cornuae forming an inverted V-shaped or elongate, paired structure. Contrasting features for the blind burrowing snakes may be found on pages 552–53.

Most of the differences among the more primitive families of the group (boas, pythons, and their allies) and the advanced snakes (acrochordids, atractaspidids, colubrids, elapids, and vipers) involve increasing specialization in feeding mechanisms, reflected in the structure of skull, jaw, and dentition. The changes show a transition from the Anomochilidae, Cylindrophiidae, Aniliidae, and Uropeltidae, the least specialized, through boas and pythons, which are intermediate, to advanced snakes.

In the four primitive families the quadrate is short and vertical and the immobile supratemporal is vestigial or small and embedded in the skull. In addition, the upper jaw and snout are not movable on each other or the cranium, and each mandible is rigid, with a small coronoid bone at the juncture of the dentary and angular. These conditions allow little flexibility or distensibility of the mouth and restrict these forms' ability to engulf large prey. The genus *Xenopeltis*, the sunbeam snake of tropical Asia (Xenopeltidae), has the upper jaw condition just described, but the lower jaw lacks a coronoid and has great flexibility at the joint between the dentary and angular. The boas and pythons have a considerable amount of independent movement between the bones of the snout, jaws, and palate. In addition the mouth is proportionally larger and the entire head is elongated. In these forms the quadrate is somewhat longer and slightly inclined backward, and the supratemporal is elongate and mobile on the skull or quadrate or both. This arrangement allows both bones to swing sideways to greatly extend the distance between the hinges of the lower jaw and permit the snake to engulf very large prey relative to its size. In acrochordid and colubrid snakes the skull is very kinetic, and the lower jaws are very flexible and without a coronoid bone. In the most advanced snakes, the elapids and vipers, the jaw apparatus is further modified to ensure envenomation of prey. The highly evolved venom delivery systems of these groups are not connected by intermediate stages to the gland, skull, and dentition conditions found in the other advanced snakes.

KEY TO COSTA RICAN SNAKES OF THE FAMILIES LOXOCEMIDAE, BOIDAE, AND UNGALIOPHIIDAE

The following key will serve to distinguish the species covered in the present volume:

1a. Snout, prefrontal, and frontal areas covered by large symmetrical shields (figs. 11.4, 11.28); scale rows at midbody fewer than 40 2

1b. Frontal and parietal areas covered by small irregular scales; scale rows at midbody more than 40 3

2a. No loreal (fig. 11.9b), two frontals; subcaudals paired (fig. 11.5); scales in 31–35 rows at midbody . . . .
. . . . *Loxocemus bicolor* (p. 560)

2b. A loreal, one large prefrontal (fig. 11.10a); subcaudals single (fig. 11.5); scales in 21–23 rows at midbody
. . . . *Ungaliophis panamensis* (p. 569)

3a. Supralabials separated from orbit by suboculars (fig. 11.8a); snout covered with small irregular scales to somewhat enlarged plates; obvious dark blotches in pattern 4

3b. Supralabials bordering orbit; snout covered with symmetrical shields; pattern of adults uniform, young with dark blotches. . . . . *Epicrates cenchria* (p. 567)

4a. A well-developed series of labial sensory pits (fig. 11.8a); supralabials 11–15; 35–59 scale rows at midbody. . . 5

4b. No labial sensory pits; supralabials 18–25; 61–79 scale rows at midbody . . . . *Boa constrictor* (p. 563)

5a. Nasals separated from each other (fig. 11.21a); subcaudals 80–86 . . . . *Corallus annulatus* (p. 564)

5b. Nasals in contact with each other (fig. 11.21b); subcaudals 101–18 . . . . *Corallus hortulanus* (p. 566)

a   b

Figure 11.21. Scalation of upper anterior head surface in Costa Rican snakes of the genus *Corallus:* (a) *C. annulatus* with the nasals separated by the internasals; (b) *C. hortulanus* with supranasals in contact on the midline.

Family Loxocemidae Cope, 1861c

Dwarf Boa

A single species of moderate-sized snakes is placed in this family, which occurs only in western Mexico and Central America. This group shares many features with primitive boids, including the relatively rigid lower jaw with a coronoid bone present and the slightly elongated quadrates and elongated supratemporals that permit maximum unhinging of the lower jaws for engulfing large prey. Additional shared characteristics are pelvic vestiges, premaxillary teeth, and supraorbital bones present; two functional lungs (the left more than 50% length of right) but no tracheal lung; a brille; inverted V-shaped hyoid composed of the first ceratobranchials, and expanded ventral plates present. Distinctive features of the family include a short stapes, head covered by large plates, and no postorbital bone.

In Costa Rica, *Loxocemus* is readily distinguishable from the blind burrowing snakes in having enlarged ventrals and subcaudal scutes, noncycloid scales on the body and tail, well-developed functional eyes, and a mental groove (fig. 11.14d) (dorsal and ventral areas covered by homogeneous cycloid scales, eyes covered by head scales, and no mental groove in the blind burrowing snakes). Similarly, it may be separated from the several vipers, which always have a well-developed loreal sensory pit (no loreal pit in *Loxocemus*). *Loxocemus* superficially resembles several colubrid genera because it has enlarged head plates, but it has 31 to 35 scale rows at midbody (30 or fewer in colubrids and terrestrial elapids), and males have distinct spurs (not present in these other groups).

Genus *Loxocemus* Cope, 1861a

A single species constitutes this monotypic genus and family. It may be separated from other snake genera by the following combination of features: postfrontal bones and premaxillary teeth present; head covered with large symmetrical plates including paired internasals, paired prefrontals, a frontal, and a prominently upturned rostral; no labial pits; scales on head, body, and tail smooth; anterior maxillary and mandibular teeth enlarged, gradually decreasing in size posteriorly; no hypapophyses on posterior dorsal vertebrae. Other distinctive characteristics of the genus are nostrils in a divided nasal; dorsal scales in 31 to 35 rows at midbody; anal and subcaudals divided. *Loxocemus* is unlikely to be confused with any other Costa Rican snake because of the prominent upturned rostral and divided anal and subcaudals. The everted hemipenis (Dowling 1975b) is weakly bilobed, clavate; sulcus spermaticus bifurcates near tip of organ, centrolineal; basal half of organ naked, distal portion flounced, with a distinct distal disk at tip of each branch, sulcus terminating in disk.

The genus ranges along the western coast of tropical Mexico and Central America to northern Costa Rica, with isolated populations in arid valleys of Atlantic slope Guatemala and Honduras.

Loxocemus bicolor Cope, 1861a

(figs. 11.9b, 11.14d; plate 334; map 11.6)

DIAGNOSTICS: The combination of the prominently upturned rostral, essentially uniform dark dorsum, and absence of a loreal pit will distinguish this snake from all other Costa Rican species.

DESCRIPTION: A moderate-sized snake reaching 1.53 m in total length (most adults under 1 m); head not distinct from neck; body cylindrical; tail short (10 to 14% of total length), not prehensile; eye small, with vertically elliptical pupil; no loreal, prefrontal in contact with supralabials; 9 to 11 supralabials, two entering orbit; 11 to 14 infralabials;

a pair of small but elongate gular scales bordering a distinct mental groove; ventrals 252 to 270 (234 to 270); subcaudals 44 to 52 (39 to 52); pelvic spurs present in both sexes but usually not visible in females. Dorsal surfaces of head, body, and tail uniform lavender brown to gray brown, sometimes with a little irregular white spotting; labials white to cream with some dark spotting on anterior ones; tip of tail light; venter immaculate white or cream to uniform brown.

SIMILAR SPECIES: (1) *Hydromorphus* has dorsally directed nostrils, a loreal, and 15 to 17 scale rows at midbody. (2) Species of *Stenorrhina* have 17 scale rows and the internasal fused to the anterior portion of the nasal.

HABITAT: Apparently restricted to the drier portions of the Lowland Dry Forest; also recorded from coastal beaches.

BIOLOGY: A semifossorial nocturnal constrictor that feeds on small mammals and frogs in captivity (Scott 1969). Greene (1983a) found two *Cnemidophorus* and two small rodents as prey items in three examples from Mexico and one from Costa Rica. Mora and Robinson (1984) recorded predation by this species on eggs of the sea turtle *Lepidochelys olivacea* in Costa Rica. Mora (1987) subsequently reported predation by *Loxocemus* on the eggs of the large iguanas *Ctenosaura* and *Iguana* in the Palo Verde area of Guanacaste Province. One individual contained shells of thirty-two *Ctenosaura* eggs and another twenty-three shells of *Iguana* eggs. *Loxocemus* apparently enters the nesting tunnels shared by the two iguanas in search of eggs. When feeding on sea turtle eggs this snake loops its body around the eggs in typical constrictor fashion before engulfing them whole. *Ctenosaura* eggs are first bitten, pushed against the body with the mouth, and then swallowed. Some eggs may be broken in the process, but most are swallowed whole. The feeding process has not been described for *Iguana* eggs. Later Mora (1991) recorded predation by this species on juvenile *Ctenosaura*. It seems likely, however, that reptile eggs constitute a major food source for *Loxocemus*. Male combat with biting is characteristic of this species (Shine 1994). This snake usually burrows in loose soil or leaf mold during the day but moves about on the surface at night and is apparently oviparous (Blackburn 1985).

REMARKS: Roldán (1985) misidentified a Panamanian *Nothopsis rugosus* as this species. *Loxocemus* does not occur in Panama (Peréz-Santos, Moreno, and Gearhart 1993; Solís, Ibáñez, and Jaramillo 1996). Nelson and Meyer (1967) reviewed variation in this species. The karyotype is 2N = 36, consisting of eight pairs of macrochromosomes (four pairs metacentric, four pairs telocentric) and ten pairs of microchromosomes; NF = 44 (Fishman, Mitra, and Dowling 1972).

DISTRIBUTION: Subhumid lowlands and adjacent premontane slopes on the Pacific versant from Nayarit, Mexico, to northwestern Costa Rica; scattered Atlantic versant localities in Chiapas, Mexico, Guatemala, and Honduras.

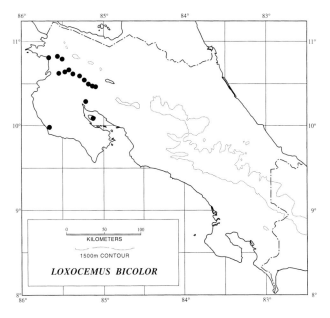

Map 11.6. Distribution of *Loxocemus bicolor*.

Restricted to the northwestern lowlands in Costa Rica (20–100 m).

Family Boidae Gray, 1825

Boas and Pythons

This moderate-sized family is composed of sixteen very diverse genera (sixty-six species) that are mostly medium-sized (650 m to 1,500 mm) to very large, with four species reaching a total length of over 5.5 m. However, several forms are relatively small (480 to 500 mm in total length) secretive creatures. The family is widely distributed in the Tropics and subtropics in Africa, Arabia, Madagascar, India, South and Southeast Asia, the Indo-Australian archipelago eastward in the South Pacific to Samoa, and southward over most of Australia; North, Central, and South America and the southern Lesser and the Greater Antilles and Bahama Islands; and also in temperate areas of Pacific coastal North America, Eurasia from Greece and North Africa to central Asia, India, and Sri Lanka, and southwestern China.

Because of the long evolution of this stock dating from some 100 Ma in the Cretaceous, it is difficult to present a definition applicable to all the diverse living subgroups placed in the family. Primitive features characteristic of the Boidae and found in most species include a relatively rigid lower jaw with a coronoid bone present; a postorbital bone; a rudimentary pelvis and vestigial hind limbs that are visible externally as small spurs just lateral to the cloacal opening (absent in some females); a long series of palatal teeth; premaxillary teeth present; a long stapes; and a functional left lung that may be 75% as long as the right lung but no tracheal lung. All have a brille and expanded ventral plates and an inverted V-shaped hyoid apparatus. All boids have the somewhat elongated quadrates and elongated supratemporal bones that allow for maximum unhinging of the lower jaws while swallowing large prey, as described above (p. 560). All members of the family use constriction to subdue large prey. In this style of feeding, the snake grasps the prey in its mouth while wrapping it in coils of the muscular body. It then gradually tightens the coils until the victim's circulation stops and it suffocates.

Three subdivisions (subfamilies) are recognizable within the family, based primarily on departures from the general pattern of characteristics just described:

Pythoninae, Fitzinger, 1826 (eight genera; twenty-six species): a supraorbital bone; prefrontals in contact or narrowly separated; labial pits usually present within boundaries of labial scales; oviparous (Africa, South Asia, the Indo-Australian archipelago, and Australia)

Boinae, Gray, 1825 (five genera; twenty-seven species): no supraorbital bone; prefrontals in contact or narrowly separated; labial pits, when present, usually lying on borders between labial scales; viviparous (tropical America, Madagascar, Sulawesi, and New Guinea to Samoa)

Erycinae Gray, 1825 (three genera, fourteen species): usually no supraorbital bones (present in *Charina reinhardtii*); prefrontals widely separated; no labial pits; viviparous, except *Charina reinhardtii*, which is oviparous (Africa, southwestern Asia to eastern Europe, western China, India, Sri Lanka, and western North America)

In Costa Rica members of the family are readily distinguishable from the blind burrowing snakes in having enlarged ventrals and subcaudal scutes, noncycloid scales on the body and tail, well-developed functional eyes, and a mental groove (fig. 11.14b) (the blind burrowing snakes have dorsal and ventral areas covered by homogeneous cycloid scales, eyes covered by head scales, and no mental groove). Similarly, boids are separated from the several vipers, which always have a well-developed loreal sensory pit (no loreal pit in boids). There is little difficulty in distinguishing even juveniles of the boas, *Boa, Corallus,* and *Epicrates,* from colubrid snakes and the elapid coral snakes (features for the latter two groups in parentheses) since these boas have the upper head surface covered by small scales (enlarged plates), the dorsal scale rows at midbody 39 or more (30 or less), and spurs lateral to the vent in both sexes (no spurs). Of course none of the boas may be confused with the elapid marine snake *Pelamis,* which lacks enlarged ventrals and has a laterally compressed tail.

Subfamily Boinae Gray, 1825

Genus *Boa* Linné, 1758

This genus (four species) is distinguished from other members of the subfamily by the following combination of features: head covered by small scales; no labial pits; scales on head, body, and tail smooth; maxillary and mandibular teeth distinctly enlarged anteriorly and gradually decreasing in size posteriorly; no hypapophyses on posterior dorsal vertebrae. Other distinctive characteristics of the genus include nostril surrounded by two or three small scales; upper nasals separated by a series of small scales; dorsal scales small, in 59 to 87 rows at midbody; anal entire; all or most subcaudals entire. The uniformly small head scales will separate this genus from other related Costa Rican genera of the family. The upper head surface is covered by large plates in *Loxocemus* and *Ungaliophis;* there are distinct enlarged internasals, prefrontals, and supraoculars in *Epicrates;* and the supranasals are in contact or separated by only two relatively large scales in *Corallus.* The absence of labial pits will further distinguish *Boa* from *Corallus* and *Epicrates,* which have them. The everted hemipenis is bilobed and subcylindrical; sulcus spermaticus bifurcate, centrolineal; pedicel and truncus naked for about first half of length, followed distally by five flounces on undivided portion of organ and five to seven flounces on each lobe; distal half of each lobe covered with smooth calyces (Cope 1900; Vellard 1946; Branch 1981).

Kluge (1991) includes in this genus the Madagascar species previously placed in *Acrantophis* (two species) and *Sanzania* (one species).

The genus has a wide range throughout the lowlands of subtropical and tropical America from northern Mexico to northwestern Peru, Uruguay, and central Argentina, including the Lesser Antilles; also in Madagascar and on Réunion Island.

Boa constrictor Linné, 1758

Boa, Boa Constrictor, Béquer

(plates 335–36; map 11.7)

DIAGNOSTICS: This heavy-bodied snake is distinct in having the following combination of characteristics: no labial pits, a dark brown middorsal stripe running from the snout onto the neck, head surface covered by small smooth scales, the dorsal scales small and smooth, and the subcaudals single (fig. 11.5).

DESCRIPTION: A giant among snakes, adults reaching a total length of 4.5 m (most adults 2 to 3 m); head distinct from neck; body somewhat compressed, stout; tail short (10 to 15% of total length) and prehensile; eye small, with vertically elliptical pupil; 2 or 3 nasal scales; no distinct loreal; 18 to 25 supralabials; 17 to 25 infralabials; eye separated from supralabials by 1 or 2 small scales; no enlarged chin shields, distinct mental groove bordered by small scales; 61 to 79 scale rows at midbody (55 to 97); ventrals 225 to 258 (225 to 283); subcaudals 48 to 70 (35 to 70); pelvic spurs visible. Ground color light brown to gray; head usually with a median elongate dark mark, sometimes bifurcate posteriorly or forming cross between eyes; side of head usually with

a distinct downward-slanting dark postorbital stripe that is continuous through eye with a distinct broad dark brown blotch that extends from lip nearly to nostril; usually a narrow vertical dark brown bar below eye; dorsum with 22 to 35 dark brown or black dorsal blotches and similarly colored smaller lateral blotches; dorsal blotches usually with an elongate cream or yellow spot laterally and frequently vertebrally; lateral blotches usually with light centers; dorsal blotches frequently fusing to create a chain of dark figures surrounding a series of oval light areas on middle of back at least anteriorly; lateral blotches may be variously fused to dorsal blotch chain and to one another; tail usually banded with black, yellowish, and red; venter gray with paired lateral dark spots; subcaudal area with a series of median black spots; iris beige, transected by a narrow dark brown horizontal bar.

SIMILAR SPECIES: (1) Other boids *(Corallus* and *Epicrates)* have some enlarged head scales. (2) The venomous pitvipers all have a distinct loreal pit (fig. 11.8b) and keeled dorsal scales (fig. 11.5). (3) *Nothopsis* has keeled dorsal scales in 26 to 30 rows at midbody and divided subcaudals (fig. 11.5).

HABITAT: Forests, shrublands, and secondary growth in Lowland Dry, Moist, and Wet Forests and in Premontane Moist and Wet Forests.

BIOLOGY: Greene (1983b) has summarized much of the information available on this relatively common snake, and most of what follows is from that review. These snakes are constrictors and eat a variety of prey including birds, large lizards, bats, rats and mice, opossums, agoutis, rabbits, young deer, coatis, ocelots, and juvenile tree porcupines. A record exists of one trying to eat a tree anteater. Contrary to Costa Rican folklore, they cannot kill or eat animals as large as cattle. Mating has been recorded from August to March and birth from March to August. Boas are viviparous and give birth to ten to sixty-four living young, each about 350 to 450 mm long, primarily in the wet season (May to November). There is no parental care. Sexual maturity occurs apparently at a length of 1.5 to 2 m. Captive boas often live for twenty-five to thirty-five years, with the record thirty-eight years and ten months. Boas are semiarboreal, though adults are often found on the ground; the species is most active at night, but large examples are frequently seen moving about during the day. Large birds (e.g., the giant black hawk *Buteogallus urubitinga;* Gerhardt, Harris, and Vásquez, 1993), mammals, and crocodilians are their chief predators. Many Costa Ricans believe this harmless species is venomous, but only between 6 P.M. and 6 A.M. It is also erroneously reputed to give birth to other distantly related venomous snakes (e.g., *Bothrops asper, Micrurus* spp.) and to mate with these and some nonvenomous forms (e.g., *Leptophis ahaetulla*) and produce hybrids (Solórzano 1993). Many Costa Ricans are extremely afraid of the béquer, since they believe there is a high risk of strangulation in handling

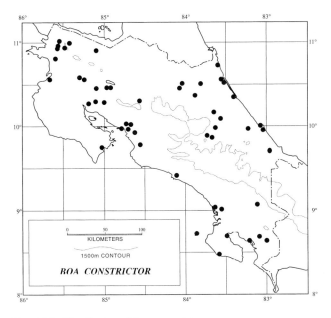

Map 11.7. Distribution of *Boa constrictor.*

one of these snakes. Because of their great strength, it is not impossible that a very large boa could constrict the neck and suffocate an unwary or foolish human who draped it around his shoulders. Humans are much too large to be food items for this species, so such a behavior by a boa is highly unlikely.

REMARKS: Between 1901 and 1951 this species was often referred to as *Constrictor constrictor*. Forcart (1951) reestablished the current usage. No adequate study of individual and geographic variation in this wide-ranging and common species has ever been published. Most accounts derive from Stull's (1932) description of a new race from southern Brazil and adjacent southeastern Bolivia. Langhammer's (1983) popular paper drawn primarily from that source contributes little to current knowledge, as pointed out by Price and Russo (1991). These authors, however, compound the problem by describing a new form from northwestern Peru and proposing that several insular taxa deserve species status. Their conclusions remain dubious, since aside from examining specimens from western Peru they derived all their comparisons from the same literature sources used by Langhammer. The following poorly diagnosable geographic variants are often regarded as races of this boa that intergrade in geographically intermediate areas:

A. Mexico, Central America, and western South America to northern Pacific slope of Peru: 55 to 78 scale rows at midbody, 22 to 35 dark dorsal blotches, and 225 to 253 ventrals
B. Northern Colombia and Venezuela south over the Guianas and the Amazon basin: 81 to 95 scale rows at midbody, 15 to 22 dark dorsal blotches, and 234 to 252 ventrals
C. Central, eastern, and southeastern Brazil and adjacent eastern Paraguay: 71 to 79 scale rows at midbody, 22 to 30 dark dorsal blotches, and 226 to 237 ventrals
D. Southern Bolivia, Paraguay, and Argentina: 80 to 97 scale rows at midbody, 22 to 30 dorsal dark blotches, usually interconnected to form a reticulate pattern with many small yellow spots, and 227 to 258 ventrals

Several insular forms have also been recognized, primarily based on high ventral counts or slight differences in color pattern that occur elsewhere as occasional individual variants in mainland populations. The most distinctive of these are from Dominica and St. Lucia in the Lesser Antilles (Lazell 1964). The karyotype is 2N = 36, consisting of eight pairs of macrochromosomes (four metacentric, four subtelocentric) and ten pairs of microchromosomes; NF = 44 (Beçak, Beçak, and Nazareth 1962).

DISTRIBUTION: Lowlands and adjacent lower slopes from Sonora and Tamaulipas, Mexico, south to northwestern Peru on the Pacific versant and throughout eastern South America to central Argentina; also the Lesser Antilles. In Costa Rica, in the lowlands and on premontane slopes throughout the republic (2–1,360 m).

Genus *Corallus* Daudin, 1803

Tree Boas

Members of this genus (six species) are distinctive in sharing the following features, when compared with other boas: anterior surface of head covered by small scales or somewhat enlarged plates; posterior head surface covered by small scales; all posterior supralabials and infralabials with deep sensory pits between them; scales on head, body, and tail smooth; anterior maxillary and mandibular teeth greatly enlarged and abruptly distinguished from the small posterior teeth; no hypapophyses on posterior dorsal vertebrae. Other characters of the genus are nostrils surrounded by two to three nasals, the upper usually enlarged; upper nasals separated by two internasal shields or in contact behind rostral (rarely slightly separated by rostral and tips of internasals); small dorsal scales in 30 to 70 rows at midbody; anal entire; subcaudals entire. The presence of labial pits will separate the two Costa Rican *Corallus* from all other snakes in the republic except their relative *Epicrates*. The latter genus differs from *Corallus* in having moderately enlarged shields over the anterior surface of the head, including a pair of internasals, a pair of anterior prefrontals, often an azygous prefrontal, and frequently a single supraocular. Most important, *Epicrates* has shallow labial pits (well-developed labial pits in *Corallus*).

Corallus is an arboreal genus that ranges through the lowlands of the Neotropics from eastern Guatemala to western Ecuador, the Amazon basin, and the southern Lesser Antilles. Two extralimital mainland species are recognized. *Corallus caninus* (the emerald tree boa) of northeastern South America and the Amazon basin is bright green as an adult (juveniles are red or yellow) with narrow white crossbands and has 186 to 219 ventrals and 62 to 84 subcaudals. *Corallus cropanii* of southeastern Brazil has only 20 to 32 dorsal scale rows at midbody, 51 to 53 subcaudals, and 36 to 38 dark body blotches. This species was originally described in a monotypic genus, *Xenoboa* Hoge, "1953," 1954, but Kluge (1991) synonymized that genus with *Corallus*. Henderson (1993a,b,c,d,e) has recently provided a review of *Corallus* and its included species. Members of this genus are viviparous.

KEY TO THE COSTA RICAN SPECIES OF THE GENUS *CORALLUS*

1a. Nasals separated by internasals (fig. 11.21a); subcaudals 80–86 . . . . *Corallus annulatus* (p. 564)
1b. Supranasals in contact (fig. 11.21b); subcaudals 101–18 . . . . *Corallus hortulanus* (p. 566)

Corallus annulatus (Cope, 1875)

(plate 337; map 11.8)

DIAGNOSTICS: No other Costa Rican snake has deep sensory pits between the labials except *Corallus hortulanus*.

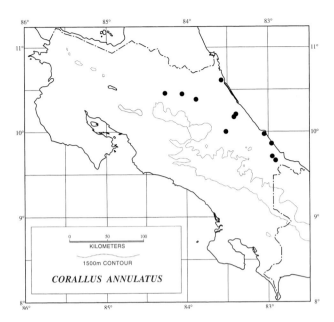

Map 11.8. Distribution of *Corallus annulatus*.

C. annulatus has nasals separated (in contact in *C. hortulanus*) and fewer subcaudals (80 to 86).

DESCRIPTION: This moderate-sized species attains a length of about 1.5 m; head distinct from neck; body definitely compressed, relatively slender; tail moderate (17 to 20% of total length), prehensile; eye moderate, with vertically elliptical pupil; supranasals enlarged, separated from one another by rostral and two internasals; no enlarged prefrontals; 3 distinct loreals, 4 scales between postnasals and eye; 14 to 15 supralabials; 18 to 19 infralabials; sensory pits between posterior supralabials and posterior infralabials; eye separated from supralabials by row of scales; no enlarged chin shields, a distinct mental groove bordered by one or two pairs of moderate-sized scales anteriorly; 50 to 55 (50 to 57) scale rows at midbody; ventrals 252 to 269 (251 to 269); subcaudals 80 to 86 (76 to 88); visible pelvic spurs. Ground color cinnamon, light brown, or dark grayish brown in adults; neonates brick red, charcoal gray, or tan to khaki, gradually turning to adult condition at about one year of age; upper head surface without definite dark markings, usually uniform; lips much lighter than rest of head, without dark markings; a distinct dark postorbital stripe runs obliquely downward from lower posterior rim of orbit; often a narrower dark stripe running posteriorly from upper eye rim for a short distance; dorsum usually marked with dark brown saddle-shaped blotches anteriorly that extend down sides to ventrals; centers of lateral extensions of blotches often contain large white to reddish brown spots, especially on posterior part of body and tail; in some examples middorsal portions of dark blotches have light central areas; occasionally blotches fuse to form a dull light stripe down middle of back, bordered by light-centered dark blotches on either side; ground color darkens with age, and

pattern becomes indefinite in large examples; underside pale buff with pinkish cast, usually suffused with some dark pigment and with dark pigment along free margin of ventrals and some dark spots on tips of ventrals; tongue black, iris light brown.

SIMILAR SPECIES: (1) *Corallus hortulanus* has the scales on the anterior head surface (internasals and prefrontals) somewhat enlarged, the supranasals in contact behind the rostral (fig. 11.21b), and many subcaudals (101 to 118). (2) *Boa constrictor* has the head covered by very small scales and lacks labial pits. (3) *Epicrates cenchria* lacks the strongly compressed body, lacks suboculars so that the supralabials border the orbit, and has weakly developed labial pits. (4) All pitvipers have a distinct loreal pit and keeled dorsal scales (fig. 11.8b). (5) *Nothopsis* has 26 to 30 rows of keeled scales at midbody and divided subcaudals.

HABITAT: Lowland (Atlantic) Moist and Wet Forests.

BIOLOGY: The species is an uncommonly seen arboreal nocturnal constrictor. It feeds primarily on vertebrates, especially hatchling, nesting, or roosting birds and bats. Male combat has been reported for this boa (Shine 1994). Captive breeding (Spataro 1996) indicates a gestation period of six to seven months. Litter sizes are six to fifteen, and neonates are 422 to 470 mm in total length. A small juvenile is 512 mm in total length.

REMARKS: There appears to be no consistency in the arrangement of the anterior head shields that Rendahl and Vestergen (1940, 1941) used to characterize geographic races of this species, as Peters (1957) suggested. Some question remains as to the status of the single example from eastern Ecuador (Orcés and Matheus 1988) purported to represent this species (Rendahl and Vestergren 1941), since all other known specimens of *C. annulatus* from South America are from west of the Andes. Pérez-Santos and Moreno's (1988) records of this species from Leticia, Amazonas Division, Colombia, are based on *C. hortulanus*. Henderson's (1993b) inclusion of Honduras in the range of this species was based on examples thought to be of dubious provenance, obtained by a commercial collector, and was not accepted by Wilson and McCranie (1994). However, Smith and Acevedo (1997) reported an authentic record of the species from eastern Guatemala near the Honduras border. J. Campbell (pers. comm.) informs me that the snakes Henderson (1993b) reported were actually collected in northwestern Honduras. This suggests that *C. annulatus* probably occurs in suitable habitats throughout northern Honduras. In the older literature (1901–51) this form is often called *Boa annulata*. Forcart (1951) reestablished nineteenth-century usage by again calling the genus *Corallus*.

DISTRIBUTION: Lowlands from extreme eastern Guatemala south through Atlantic drainage Nicaragua, Costa Rica, and western Panama to central Panama, eastward through Pacific versant Panama, and south through western Colombia to western Ecuador. In Costa Rica restricted to the Atlantic lowlands (2–56 m).

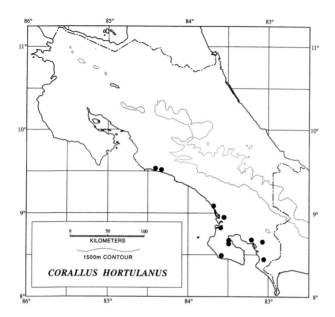

Map 11.9. Distribution of *Corallus hortulanus*.

Corallus hortulanus (Linné, 1758)

(plate 338; map 11.9)

DIAGNOSTICS: Only *Corallus annulatus*, among Costa Rican snakes, shares the deep labial pits with this species. *C. hortulanus* has supranasals in contact (separated in *C. annulatus*) and has a higher subcaudal count (101 to 118).

DESCRIPTION: A large species reaching a total length of nearly 2 m (most adult examples 1 to 1.25 m); sexes equal in length; head distinct from neck; body definitely compressed, relatively slender; tail moderate (17 to 20% of total length), prehensile; eye moderate, with vertically elliptical pupil; supranasals in contact with one another behind rostral; a pair of large prefrontals; 1 to 4 distinct loreals; 2 to 5 scales between eye and postnasals; 11 to 15 supralabials; 16 to 18 infralabials; sensory pits between posterior supralabials and posterior infralabials; eye separated from supralabials by row of scales; no enlarged chin shields, a distinct mental grove bordered by one or two pairs of moderate-sized scales anteriorly; scales in 35 to 49 (30 to 63) rows at midbody; ventrals 253 to 283 (250 to 299), subcaudals 101 to 118 (94 to 137); pelvic spurs visible. Ground color varies from yellow or light brown through brownish gray to gray; upper head surface with a distinct median dark stripe in the parietal region; distinct postorbital dark stripe slants obliquely downward from lower posterior rim of orbit toward or to angle of jaw; narrower dark line running from upper rim of eye onto temporal area; dorsum usually marked with two series of obscure dark brown diamond-shaped to round spots, usually with light centers and narrow yellow margin; dark portions of spots often narrowly connected across middle of back to form butterfly-shaped figures; marking most distinctive posteriorly and in juve-

niles; tail spots dark brown to black with yellow centers; some examples with dull markings or nearly uniform in dorsal color; venter clear white to yellow anteriorly, with some dark spotting at tips of ventrals; spotting more extensive posteriorly, and dark pigment often forms lines along free margin of each ventral; underside of tail invaded by extensions of caudal blotches to form an irregular pattern of dark and light areas anteriorly, but posterior third mostly dark brown or black. Everted hemipenis slightly bilobed, clavate; sulcus spermaticus bifurcate, centrifugal; pedicel naked, truncus with five wavy flounces, and apical area covered with papillae (Branch 1981).

SIMILAR SPECIES: (1) *Corallus annulatus* has most of the upper head surface covered by small scales (no enlarged prefrontals), the supranasals separated by two internasals (fig. 11.21a), and 80 to 86 subcaudals. (2) *Boa constrictor* has the head covered by small scales and lacks labial pits. (3) *Epicrates cenchria* lacks the compressed body, lacks suboculars so that the supralabials border the orbit, and has weakly developed labial pits and fewer than 100 subcaudals. (4) The venomous pitvipers have a distinct loreal pit and keeled dorsal scales (fig. 11.8b). (5) *Nothopsis* has keeled scales in 26 to 30 rows at midbody and divided subcaudals.

HABITAT: Relatively undisturbed Lowland (Pacific) Moist and Wet Forests.

BIOLOGY: The species is a relatively common arboreal nocturnal constrictor that feeds primarily on birds and mammals. Juveniles and small adults mostly eat young, nesting, and roosting birds and bats and also some lizards at mainland sites, but lizards are predominant food items for West Indian juveniles (Henderson 1993f). Larger adults (above 750 mm in standard length) apparently prey principally on mammals (e.g., rats, bats, mice, opossums) (Henderson 1993f). Mole (1924) reported mating in Trinidad during February, with birth of twenty to thirty living young the following August. Fitch (1970) recorded ten young from another female.

REMARKS: References to this species in the literature appear under several names: *Boa* and *Corallus* (*cookii, enydris,* and *hortulanus*). *Boa hortulana* and *Boa enydris* were described originally on the same page in Linné (1758). Duméril and Bibron (1883–44) and Boulenger (1893) recognized that a single species was involved and used the name *hortulanus* for it. As pointed out by McDiarmid, Toure, and Savage (1996), this usage predates any other in which the conspecificity of the two nominal taxa was noted or where *enydris* was used in preference to *hortulanus* (article 24.2, *International Code of Zoological Nomenclature*). Forcart (1951) established a taxonomy that regarded *cookii* as a Central American and northern South American race of this species. The difference between that nominal form and examples from farther south involves scalation: dorsal scale rows 35 to 49 in the northern race, 47 to 63 in southern examples (Henderson 1993e). Henderson (1997) proposed an entirely different arrangement and recognized four species in the "*hortulanus* complex": *Corallus cookii* (St. Vincent Island), *C. grenadensis* (Grenada Bank), *C. hortulanus* (South America, south of the Orinoco River), and *C. ruschenbergerii* (Central America and northern South America). The mainland forms are differentiated on much the same scutellation features used by Forcart, but with a change of name for the Central American and northern South American populations. The extreme color pattern variation in this species shows no consistent geographic correlation with the scale count differences, and all mainland samples have paired dark-outlined light spots as the predominant pattern. The only difference between Henderson's two putative mainland species that does not show substantial overlap remains the number of dorsal scale rows at midbody. For these reasons I continue to regard the mainland boas of this complex as representing a single variable species. Two minor errors in Henderson's treatment should be noted. The correct spelling is *cookii*, not *cooki* (article 31, *International Code of Zoological Nomenclature*). The name *Xiphosoma ruschenbergerii* Cope dates from 1875, not 1876, based on when the preprint of the paper proposing that name was distributed. Although it has been recorded from Nicaragua by several authors (Stimson 1969; Peters and Orejas-Miranda 1970; Lancini 1979; Villa 1983), I know of no authenticated records of the species from that country, and the range of *C. hortulanus* in Costa Rica (restricted to the southwestern Pacific lowlands) makes it unlikely that it occurs in Nicaragua. A single Costa Rican specimen (UCR 6221) listed as from the Río Macho dam, Cartago Province (an Atlantic slope locality at 1,114 m), has all the diagnostic features of *C. hortulanus*. This record is suspect, since no other examples are known from the Atlantic drainage of the republic or from such a high altitude, and the specimen may be an escaped or released pet. The karyotype is 2N = 40, consisting of ten pairs of macrochromosomes (two pairs metacentric, eight pairs telocentric) and ten pairs of microchromosomes; NF = 44 (Gorman and Gress 1970).

DISTRIBUTION: The humid lowlands of Pacific versant southwestern Costa Rica and western Panama, both slopes of central and eastern Panama and northern Colombia, throughout northern South America, and southward to southern Brazil and through the Amazon basin to Bolivia; Trinidad and Tobago and the southern Lesser Antilles (3–200 m).

Genus *Epicrates* Wagler, 1830

The genus contains ten moderate to very large species, only one of which occurs on the mainland in Central and South America. Members of the genus differ from other boas in sharing the following features: anterior upper surface of head covered by relatively large scales, including a pair of internasals and a pair of anterior prefrontals; posterior head region covered by irregular scales but one or a pair of small frontals and one to three supraoculars often present; scales on head, body, and tail smooth; anterior maxillary and mandibular teeth greatly enlarged and abruptly differentiated from the small posterior teeth; no hypapophyses on posterior dorsal vertebrae. Other characters of the genus are nostril surrounded by two or three nasals, the upper ones in contact behind rostral; postnasals separated by two large internasals that do not touch rostral; small dorsal scales in 20 to 69 rows at midbody; anal entire; subcaudals entire; usually no labial pits; *Epicrates angulifer* and *E. cenchria* with shallow pits between supralabials (also present between infralabials in the latter). The presence of labial pits separates *Epicrates* from other Costa Rican snakes except the two species of *Corallus*. The latter genus has well-developed sensory pits in the supralabials and infralabials (shallow pits between anterior supralabials and infralabials in *Epicrates*). Species of this genus are generally terrestrial in habits but often climb into bushes or trees to hunt or rest. Nine species of *Epicrates* occur in the Greater Antilles and the Bahama Islands, and the single mainland form has a wide range in the lowlands from Costa Rica to north-central Argentina. So far as is known, all included species are viviparous. The Cuban boa *(Epicrates angulifer)* is the largest member of the genus and reaches a length of 4 m.

Epicrates cenchria (Linné, 1758)

Rainbow Boa

(plates 339–40; map 11.10)

DIAGNOSTICS: The presence of shallow labial pits between the anterior labials combined with the cylindrical body will distinguish this species from all other Costa Rican

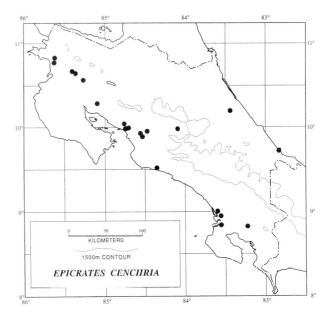

Map 11.10. Distribution of *Epicrates cenchria*.

forms. Rainbow boas are among the most beautiful snakes in the republic, both as strikingly patterned juveniles and as iridescent, uniformly colored adults. The very appropriate common name derives from the shimmering, multicolored effect produced by light reflected from the body surface of the dark brown adults.

DESCRIPTION: A large form reaching a total length of 2.2 m (most examples less than 1.5 m); adults 1,080 mm or more; females longer than males; head distinct from neck; tail short (10 to 14% of total length), prehensile; eye moderate, with vertically elliptical pupil; a single large loreal, two scales between postnasal and eye; 11 to 15 supralabials; 13 to 17 infralabials; orbit not separated from supralabials by suboculars, two supralabials border orbit; sensory pits between anterior supralabials and all infralabials; no enlarged chin shields, a distinct mental groove bordered by moderate-sized scales; 45 to 53 (39 to 56) scale rows at midbody; ventrals 225 to 245 (214 to 271); subcaudals 47 to 58 (34 to 58); pelvic spurs visible. Juveniles: Ground color cream to tan, often with an orange tinge; upper head surface with three dark brown longitudinal stripes, one from snout tip to parietal region and a similar stripe running from the upper posterior rim of each eye posteriorly almost onto neck; lips light, supralabials with a few dark markings; a distinct postorbital dark stripe from lower rim of eye to near angle of jaw; dorsal dark brown to reddish brown markings form a chainlike pattern down center of back, with the lighter ground color appearing as large cream to light tan oval spots or ocelli; a series of short irregular narrow longitudinal light lines or discrete light crescent-shaped spots on flanks, bordered above and below by dark lines, spots, or crescents; ventrolateral area marked with a complex pattern of dark or light spots, blotches, or ocelli; underside cream or white, sometimes with small brown spots laterally. Adults: Colors gradually darken with age, and light areas become suffused with dark pigment until all upper surfaces of head, body, and tail become a uniform rich, dark, iridescent brown (plate 339); no dark head markings, labials and underside white; change (plate 340) from juvenile to uniform coloration completed by the time individual is about 0.6 m in total length; iris tan. Everted hemipenis bilobed; sulcus spermaticus centrifugal; pedicel nude, truncus with flounces, apical one-third covered with papillae (Vellard 1946; Branch 1981).

SIMILAR SPECIES: (1) *Boa constrictor* lacks labial pits and a loreal. (2) The species of *Corallus* have strongly compressed bodies and well-developed pits in the posterior labials (fig. 11.8a). (3) *Elaphe triaspis* has large symmetrical plates over the entire dorsal head surface (fig. 11.4). (4) *Nothopsis* has keeled scales in 26 to 30 rows at midbody and divided subcaudals (fig. 11.5).

HABITAT: Relatively undisturbed Lowland Dry, Moist, and Wet Forests and Premontane Moist Forest.

BIOLOGY: *Epicrates cenchria* is an uncommonly seen terrestrial nocturnal constrictor that may use bushes and low branches of trees when hunting or as resting sites. Food consists of birds, their eggs, small mammals, and on occasion frogs. In Trinidad Mole (1924) observed mating in October and January, with live young produced in July from the October mating. Brood size is about eighteen, and in captivity neonates are 435 to 490 mm in total length (Murphy, Barker, and Tyson 1978).

REMARKS: A number of putative geographic races have been recognized within this species, primarily based on differences in coloration (Amaral 1954; Abalos, Baez, and Nader 1964). Two principal adult color patterns have been defined, but there is considerable individual variation:

A. Middorsal area marked by a series of narrow dark brown or black rings, S-shaped marks (fused rings), or saddle-shaped marks that contrast with the middorsal ground color; a lateral series of white crescents bordered by dark brown or black above and below to form ocelli that alternate with the middorsal marks and a series of smaller ventrolateral dark spots that alternate with the ocelli

B. Middorsal area marked by dark-outlined, light-centered spots or ocelli that may be paired or fused across dorsum; a lateral series of dark spots bordered, at least above, by light color and a series of ventrolateral dark spots that may also be bordered with light

Snakes having color pattern A occur from Central America and northern South America south through the Guianas and the Amazon basin to Bolivia and in southeastern Brazil. Central and northern South American populations have this pattern type as juveniles but lose it as adults, whereas it is retained in adults from other areas in South America. Aside from this feature, other purported geographic subdivisions are based on minor differences in scalation and coloration. Snakes having pattern B are found on Marajó Island at the mouth of the Amazon in Brazil and in dry habitats in the northeast region and Planalto Central of Brazil, south to northern Argentina and western Ecuador. In the southernmost populations the black borders of the spots usually fuse with one another laterally to form a broad longitudinal dark band. In other populations each dorsal spot is discrete, but the lateral, light-edged dark spots usually fuse to form a distinct light-bordered dark stripe anteriorly. Relatively minor differences in head markings and scale counts have been used to further subdivide populations having this pattern. All Costa Rican *Epicrates* show pattern A as juveniles, and all adults are uniform in color. The karyotype is 2N = 36, consisting of eight pairs of macrochromosomes (four pairs metacentric, four telocentric) and ten pairs of microchromosomes; NF = 44 (Gorman and Gress 1970).

DISTRIBUTION: Humid and subhumid lowlands from northwestern and eastern Costa Rica, south through Panama, northern Colombia, and Venezuela to north-central Argentina and western Ecuador. In Costa Rica on both Atlantic and Pacific slopes and the Meseta Central Occidental (0–1,200 m).

Family Ungaliophiidae McDowell, 1987

This family consists of two genera (three species) of small to moderate-sized strictly Neotropical snakes. They share many features with *Loxocemus* and the Boidae, including those skull and jaw specializations associated with swallowing large prey; they have a postorbital bone and a rudimentary pelvis (spurs present in most males, absent in females), a brille, expanded ventral plates and an inverted V-shaped hyoid. The family is distinctive among boas and their allies in lacking supraorbital bones and premaxillary teeth, having the left lung vestigial or absent, having the coronoid bone reduced or absent, and possessing a tracheal lung.

The only Costa Rican representative of the family *(Ungaliophis)* is readily distinguishable from the blind burrowing snakes in having enlarged ventrals and subcaudal scutes, noncycloid scales on the body and tail, and well-developed functional eyes (dorsal and ventral areas covered by homogeneous cycloid scales and eyes covered by head scales in the blind burrowing snakes). *Ungaliophis* are separated from the several vipers, which always have a well-developed loreal sensory pit (no loreal pit in boids and their allies). They superficially resemble several colubrid genera because they have enlarged head plates, but males have distinct spurs (not present in other groups), and the species of *Ungaliophis* have a single prefrontal shield, a condition they share with only two Costa Rican colubrid genera *(Hydromorphus* and some *Trimetopon)*. These forms have 15 to 17 scale rows at midbody (whereas *Ungaliophis* have 21 to 23), and both have a uniform gray to black dorsum (*Ungaliophis* are blotched).

The family occurs on the Pacific versant of Mexico in upland Oaxaca and lowland Chiapas; also on the Pacific slope of Guatemala, Honduras, and southwestern Costa Rica; disjunct populations occur on the Atlantic versant in Chiapas, Mexico, Nicaragua, Costa Rica, and Panama to northwestern Colombia.

Genus *Ungaliophis* Müller, 1880

The genus comprises two relatively rare arboreal species that differ from other boas by sharing the following combination of character states: upper surfaces of snout, prefrontal, and frontal areas covered with large symmetrical plates, including a single large prefrontal and a single loreal; no labial pits; a mental groove (fig. 11.14c); scales on head, body, and tail smooth; anterior maxillary and mandibular teeth longest, former gradually decreasing in size posteriorly, latter abruptly differentiated; no hypapophyses on posterior dorsal vertebrae. Other distinctive features include a single nasal followed by a postnasal, loreal, and preocular; a pair of internasals (fig. 11.10a); dorsal scales in 19 to 25 rows at midbody; anal entire; subcaudals entire. The hemipenes

(Bogert 1968) are relatively long, bilobed, symmetrical; sulcus spermaticus bifurcate, dividing near base of organ, and centrolineal; basal area naked, replaced by smooth calyces distally. The genus differs from the only other one in the family as follows (features for *Ungaliophis* in parentheses): the genus *Exiliboa* (one species) of Oaxaca, Mexico, has a single enlarged internasal and a pair of prefrontals (paired internasals and a single prefrontal).

Although they were long placed in the Tropidophiidae, Zaher (1994a) has conclusively demonstrated the affinities of the two genera with the Boidae. The tropidophiids *(Trachyboa* and *Tropidophis)* differ from the ungaliophiids in several features of jaw musculature, in cephalic circulatory pathways, and especially in having the hyoid apparatus comprising paired cornuae that form elongate branches directed posteriorly. This latter condition is found in "higher" snakes (Acrochordidae, Colubridae, Atractaspididae, Elapidae, and Viperidae) in contrast to all other alethinophidians, which have the inverted V-shaped condition.

The single large prefrontal shield will immediately distinguish *Ungaliophis* from other Costa Rican boids.

The genus ranges from eastern Chiapas, Mexico, south through Central America to northwestern Colombia.

Ungaliophis panamensis K. Schmidt, 1933a

(plate 341; map 11.11)

DIAGNOSTICS: The pattern of alternating light-outlined paravertebral dark blotches is distinctive. In addition, no other Costa Rican snake has a single large prefrontal shield (fig. 11.10a) while lacking elongate chin shields (fig. 11.14c).

DESCRIPTION: A small snake reaching a total length of 670 mm; head distinct from neck; body compressed; tail

Map 11.11. Distribution of *Ungaliophis panamensis*.

short (8.5 to 9.5% of total length), prehensile; eye moderate, with vertically elliptical pupil; internasals in contact; loreal present; 7 to 9 supralabials, with 2 or 3 bordering orbit; 9 to 11 infralabials; first infralabials and three pairs of enlarged scales border mental groove (fig. 11.14c); 23 scale rows at midbody (19 to 25); ventrals 237 to 254 (226 to 254); subcaudals 41 to 48; pelvic spurs usually present in both sexes but not visible in females; maxillary teeth 13 to 15. Ground color light brown with dark brown punctations, spots, and lines; usually an internasal dark band; a large dark brown spot in the frontal-parietal area that may continue posteriorly as a pair of broad stripes; a distinct narrow dark line along canthus; a similar line from upper posterior rim of eye onto temporal area; distinct postorbital dark stripe from middle of eye down to angle of jaw; 50 to 55 pairs of large angular paravertebral dark (light-outlined) blotches, with apexes pointing toward vertebral line; some blotches confluent across back to form hourglass- or butterfly-shaped figures, but usually alternating or offset on the two sides; an incomplete lateral row of dark spots sometimes fusing to form irregular longitudinal stripes; 40 to 50 lateroventral brown spots that extend onto or across venter posteriorly; underside yellow, heavily speckled with large dark chocolate brown areas on underside of tail, sometimes fusing into a median longitudinal stripe.

SIMILAR SPECIES: (1) All pitvipers have a distinct loreal pit and keeled dorsal scales (fig. 11.8b). (2) *Nothopsis* has keeled dorsal scales (fig. 11.5). (3) The species of the colubrid genera *Stenorrhina*, *Trimorphodon*, and *Xenodon* have enlarged parietal shields and paired prefrontals (fig. 11.4).

HABITAT: A rare species known only from the canopy of undisturbed Lowland Moist and Wet Forests and Lower Montane Rainforest.

BIOLOGY: The following brief notes are based on data summarized in Bogert (1968) and Corn (1974). The species appears to be a nocturnal arboreal snake that hides under epiphytic growth or in bromeliads. In captivity individuals ate mice and anole lizards. Apparently this snake is restricted to the canopy layer in undisturbed forest, and like other denizens of that zone it remains rarely collected (seventeen known specimens, five without locality data) and barely known biologically. The other member of the genus is viviparous.

REMARKS: The only other species placed in the genus, *Ungaliophis continentalis*, is known from eastern and central Chiapas, Mexico, to Pacific versant Guatemala and Honduras and Atlantic drainage northern Nicaragua (Köhler 1997). It differs from *U. panamensis* (features for the latter in parentheses) by having the internasals separated by the broad contact between the rostral and prefrontal (internasals in contact behind rostral), 12 maxillary teeth (13 to 15), and the dorsal dark blotches ovoid (triangular).

Bogert (1968) reviewed the genus. Villa and Wilson (1990) provided an updated account of the genus and species based

primarily on the detailed descriptions and data in Bogert (1968) and Corn (1974).

DISTRIBUTION: Evergreen forests of the Atlantic lowlands of southeastern Nicaragua, Costa Rica, and Panama and northwestern Colombia. One record (UCR 7580) from the rim of Volcán Orosí in Costa Rica in the lower montane zone and several from the southwestern Pacific lowlands of the republic, including the records of Merahtzakis (1988) and Krywicki (2001), (41–1, 487 m).

Family Colubridae Oppel, "1810," 1811

Harmless and Rear-Fanged Snakes

The family contains the greatest number of snake genera (about 287) and species (1,800) and comprises nearly two-thirds of the world's ophidians. Everywhere that snakes occur, excluding the Australian continent and the Pacific Ocean islands, colubrids are the dominant ones. The family exhibits a wide variety of ecologic adaptations, with fossorial, semifossorial, terrestrial, arboreal, freshwater aquatic, and marine forms. Like all snakes the group is strictly carnivorous, and though most eat frogs, lizards, and small mammals, some specialists feed primarily or exclusively on fish, crustaceans, earthworms, snails and slugs, birds and their eggs, reptile or frog eggs, or insects and other arthropods.

Within this diverse stock, modifications in structure to meet life requirements make it difficult to provide a general definition of the family. However, all members of the Colubridae share the following advanced features: a very flexible lower jaw that lacks a coronoid bone; no remnant of the pelvic structure; no premaxillary teeth; left lung vestigial or absent; enlarged ventrals present; brille present; quadrate elongate, posteriorly inclined; supratemporal usually elongate; no prefrontal bone; the hyoid apparatus composed of paired cornuae that form a structure with paired, elongate, parallel branches directed posteriorly; in addition colubrids lack a compressed, paddlelike tail and lack hollow venom-injecting fangs. The diversity of genera, parallelism in adaptive features associated with different specific ecological roles, and the streamlined and simplified morphology of colubrids have made it difficult to cluster genera into meaningful subdivisions. Dowling and Duellman (1974–78) suggest that four subfamilies be recognized, but other authors (see citations below) recognize five to seven major subgroups. Zaher (1999) presents evidence that four subfamilies may be recognized in the Americas based on hemipenial features. See Whistler and Wright (1989) for an insightful critique of the entire situation. Since the stability of any of these arrangements is questionable, I have decided to place the Costa Rican genera in informal groups of apparently allied stocks, based in part on Dowling (1975a), Dowling et al. (1983), Jenner (1981), Jenner and Dowling (1985), Dowling and Jenner (1988), and Dowling (2002) and influenced by Cadle (1984a,b,c, 1988). In this scheme groups are recognized as follows, primarily for convenience:

Xenodontines (six genera): moderate-sized to large (to 2.5 m) terrestrial or semiarboreal snakes with a pair of enlarged (usually grooved) teeth on the posterior end of the maxillary that are usually separated from the anterior teeth by a gap (diastema); Duvernoy's gland present; lacking hypapophyses on the posterior dorsal vertebrae; with symmetrical, noncapitate and disked (fig. 11.3a) or bicapitate hemipenes and a bifurcate sulcus spermaticus (fig. 11.3c). These snakes are included in the South American xenodontines of Cadle (1984a) and the Xenodontidae of Dowling and Jenner (1988) and Dowling (2002). Zaher (1994b, 1999) argues that a subset of Cadle's South American xenodontines and the North American genera *Farancia* and *Heterodon* form a monophyletic (natural) group distinguished by shared hemipenial features. The included genera have enlarged lateral spines on the truncus and enlarged calyces or a nude area on the asulcate surface of the truncus and apical lobes. The implication that Cadle's South American xenodontines are polyphyletic further urges caution in establishing formal recognition of the various lineages.

Amastridines (five genera): small to moderate-sized semifossorial or terrestrial snakes with two enlarged (usually grooved) teeth on the posterior end of the maxillary that may or may not be separated from the anterior teeth by a diastema; Duvernoy's gland present in genera with grooved fangs, probably present in others; three of the five genera (absent in *Conophis* and *Enulius*) placed in this group have broad but not recurved hypapophyses on all dorsal vertebrae; with symmetrical, noncapitate hemipenes and a bifurcate sulcus spermaticus (fig. 11.2).

Dipsadines (twelve genera): a wide variety of adaptive types including small semifossorial, terrestrial, semiaquatic, and aquatic forms without grooved teeth and a series of small to moderate-sized semifossorial, terrestrial, semiarboreal, and arboreal species with a pair of enlarged grooved fangs on the posterior end of the maxillary that are separated by a diastema from the anterior teeth; in nine of the genera Duvernoy's gland is known to be present, and it probably occurs in the others; most genera in this group lack hypapophyses on the posterior dorsal vertebrae, but one genus has them; hemipenes symmetrical, usually unicapitate, with the bifurcate sulcus spermaticus forking within the capitulum (noncapitate or semicapitate with a bifurcate sulcus in a few species) or capitate with a simple sulcus (figs. 11.2, 11.3). Essentially the Central American xenodontines of Cadle (1984b), the Dipsadinae of Dowling and Jenner (1987) or the family Dipsadidae of Dowling and Jenner (1988) and Dowling (2002). Cadle (1984b) argued for the monophyly of this group based on hemipenial attributes, especially if the sulcus spermaticus forks within the capitulum, although some included taxa (e.g., *Imantodes*) have a simple sulcus spermaticus.

Colubrines (nineteen genera): a diverse group of small to large (up to 3 m in total length) semifossorial, terrestrial, semiarboreal, and arboreal snakes. One series of genera (thirteen) lacks grooved fangs, but another cluster has one to five enlarged grooved fangs that are usually separated from the other maxillary teeth by a diastema; Duvernoy's gland absent in four genera, definitely present in eleven, and unknown but probably present in four; all but one genus (*Scaphiodontophis*) lack hypapophyses on the posterior dorsal vertebrae; hemipenes single or bilobed, symmetrical or asymmetrical, usually noncapitate, and always with a single sulcus spermaticus (fig. 11.2c). The subfamily Colubrinae of most workers includes these genera (Zaher 1999). Dowling and Jenner (1988) and Dowling (2002) recognize this stock as a separate restricted family Colubridae.

Natricines (one genus): two moderate-sized semiaquatic species (genus *Thamnophis*) with the posterior two or three maxillary teeth somewhat larger than the anterior ones, but not grooved or separated from them by a diastema; Duvernoy's gland present; distinctive short, recurved hypapophyses on all dorsal vertebrae; single noncapitate symmetrical hemipenes having a simple sulcus spermaticus (fig. 11.2). The subfamily Natricinae of most authors (Zaher 1999) includes these genera. Dowling and Jenner (1988) and Dowling (2002) recognized the group as having family status (Natricidae), as have some previous researchers.

As may be seen from the listing above, Costa Rican colubrids form a diverse subset of the ophidian fauna. Differentiation of the genera in this family is based to a considerable extent on characteristics of head scutellation, maxillary dentition, and hemipenial structure, although other scutellation features (dorsal scale row form and number and anal and subcaudal scale form) are important. In this regard the basic colubrid complement of head shields referred to in many of the generic descriptions that follow consists of a large rostral; paired internasals; paired prefrontals; a single large frontal; a single large supraocular over each orbit; and large paired parietals (fig. 11.4). Any deviation from this basic pattern is described in detail in the generic accounts.

Costa Rican colubrid snakes are immediately separated from members of the three blind snake families by having enlarged ventral and subcaudal scutes, functional eyes that are not hidden under the head plates, large dorsal scales, and usually a mental groove on the underside of the head (fig. 11.4) (no enlarged ventral, subcaudal, or chin scales; eyes vestigial, hidden under the head scales; and no mental groove in the blind snakes). There is no difficulty in distinguishing colubrids from the pitvipers (family Viperidae), since the latter have a distinct loreal pit on the side of the head, between and slightly below the level of the eye and nostril (this pit is lacking in all colubrids).

It is more difficult to invariably separate Costa Rican

colubrids from members of the families Boidae and Elapidae. Most colubrids may be distinguished from most boids in the republic in having the snout, frontal, and parietal areas covered by large symmetrical plates and/or having fewer than 30 scale rows at midbody (frontal and parietal areas covered by small irregular scales and/or scale rows at midbody 31 or more). Two boids (*Loxocemus* and *Ungaliophis*) and one colubrid (*Nothopsis*) cause some difficulty in immediate placement. The two boid allies have large and symmetrical head plates, and the colubrid has the head covered with small scales except for the enlarged internasal plates. Since *Loxocemus* has a single pair of elongate chin shields (fig. 11.14d), 31 to 35 dorsal scale rows at midbody, and the prefrontal meeting the supralabials, it cannot be confused with any colubrid except *Elaphe* (with two pairs of elongate chin shields and the prefrontal separated from the supralabials). Species of *Ungaliophis* have a single prefrontal plate (fig. 11.10a), have 21 to 23 scale rows at midbody, and lack elongate chin shields (fig. 11.14c). The only Costa Rican colubrids with a single prefrontal (*Hydromorphus* and some *Trimetopon*) have elongate chin shields and 15 to 17 scale rows at midbody. The colubrid *Nothopsis* is distinct from all Costa Rican boids in having keeled dorsal scales. It resembles *Loxocemus* in having a pair of elongate chin shields and divided subcaudals and *Ungaliophis* in having 30 or fewer scale rows at midbody (26 to 30 in *Nothopsis*, 21 to 23 in *Ungaliophis*). However, both *Loxocemus* and *Ungaliophis* have smooth dorsal scales.

Costa Rican colubrids are easily separable from the elapid sea snakes *Pelamis* because of their paddle-shaped tails and barely enlarged ventral scales, neither of which occur in any other Costa Rican snake. In the republic it may be very difficult to distinguish colubrids from the terrestrial elapids (*Micrurus*). In fact the only definitive way to make this distinction is to examine the maxillary dentition. No Costa Rican colubrid has hollow fangs on the upper jaw (*Micrurus* and *Pelamis* have hollow fangs on the front of the jaw). All Costa Rican *Micrurus* but one population of *Micrurus alleni* have a definite ringed pattern throughout life, and all have 15 scale rows at midbody and a single square scale (a loreal or preocular) separating the prefrontal from the supralabials. A key distinguishing colubrids that resemble coral snakes from the highly venomous species of *Micrurus* will be found below (p. 706). Until you can make a positive identification, it is best to assume that any snake with a ringed pattern and/or bright red, orange, and yellow (or even white) markings alternating with black may be a venomous coral snake and act accordingly!

Xenodontines

The Central American representatives of this stock comprise two clusters of genera that differ primarily in hemipenial features. *Clelia*, *Oxyrhopus*, and *Tripanurgos* have bicapitate hemipenes (fig. 11.3c), whereas *Erythrolamprus*, *Lio-phis*, and *Xenodon* lack capitation (and distal calyces) but have a disklike structure surrounding the terminus of each sulcus spermaticus ramus (fig. 11.3a). The distributions of these genera are centered on South America. The first three are closely allied to a series of other South American genera sharing the unique hemipenial structure: *Drepanoides*, *Phimophis*, *Pseudoboa*, *Rhachidelus*, and *Siphlophis*. *Saphenophis* and *Tropidodryas* do not belong here (Myers and Cadle 1994), contrary to Jenner and Dowling (1985). The second cluster shares the peculiar disked hemipenes with another group of South American genera: *Lystrophis*, *Umbrivaga*, and *Waglerophis* (Dixon 1980 regards the nominal genera *Dromicus*, *Leimadophis*, and *Lygophis* as synonyms of *Liophis*). Zaher (1999) regards all the genera mentioned as belonging to the subfamily Xenodontinae. So far as is known, all members of this stock are oviparous.

Genus *Clelia* Fitzinger, 1826

A generalized colubrid genus (six species) in terms of external morphology, with the basic arrangement of enlarged head shields; nasal completely divided; a single preocular; usually a single loreal (sometimes absent); pupil vertically elliptical; two pairs of chin shields; dorsal scales smooth with two apical pits (or pits absent), in 17 to 19 rows at midbody; vertebral scale row usually not enlarged; anal usually single; subcaudals divided or with some undivided scutes; maxillary teeth subequal, 10 to 15 plus 2 moderately enlarged grooved fangs separated from anterior teeth by a diastema; Duvernoy's gland present; anterior mandibular teeth not markedly larger than posterior ones; no hypapophyses on posterior dorsal vertebrae. The combination of smooth scales in 17 to 19 rows at midbody, single preocular, divided nasal, single anal, and coloration will separate members of the genus from other snakes in Costa Rica. The everted hemipenis is moderate (extending to subcaudal 10 to 12), bicapitate; sulcus spermaticus bifurcate, centrifugal, and dividing proximal to the capitulate lobes; usually spinous proximally; capitula calyculate (Jenner and Dowling 1985; Zaher 1996, 1999).

The juvenile coloration is unique in having the dorsal surfaces of the body and tail red with a single black nuchal band separated from the black upper head surfaces by a light collar; venter immaculate. Adults are uniform glossy bluish black with an essentially unmarked venter.

The bright red juveniles of the two Costa Rican species are considered to be viboras de sangre (blood snakes) whose bite results in bleeding over the entire body surface and death. This in an unfounded belief lacking any element of fact. Zaher (1996) recently provided a revision of this genus.

Clelia range from the tropical lowlands of Atlantic and Pacific Mexico and at scattered upland sites south through Panama to central Ecuador on the Pacific versant and through northern South America and the southern Lesser Antilles, southward east of the Andes to central Argentina (41° S).

KEY TO THE COSTA RICAN SPECIES OF THE GENUS *CLELIA*

1a. Dorsal scales in 19 rows at midbody . . . .
. . . . *Clelia clelia* (p. 573)
1b. Dorsal scales in 17 rows at midbody . . . .
. . . .*Clelia scytalina* (p. 574)

Clelia clelia (Daudin, 1803)

Zopilota, Tiznada, Vibora de Sangre (Juvenile)

(plates 342–43, map 11.12)

DIAGNOSTICS: Juveniles of this species and *Clelia scytalina* are the only Costa Rican species appearing to have a uniform red body and tail (actually there is a small black spot on each scale) in combination with a light nuchal collar that crosses the parietal shields and is bordered posteriorly by a black band (fig. 11.22a). The adults are uniform bluish black with a cream venter and 19 rows of scales at midbody.

DESCRIPTION: A large snake reaching a total length of 2.6 m; females longer than males (to 1.8 m); head somewhat distinct from neck; body cylindrical; a moderately long tail (16 to 20% of total length); eye moderate; supralabials 7 to 8, with 2 bordering orbit; 7 to 9 infralabials; 2 postoculars; temporals 2 + 2 or 2 + 3; smooth dorsal scales in 17 to 19-19-17 rows, with two apical pits; ventrals 198 to 247; subcaudals 65 to 93 (70 to 98). Dorsum of adults uniform bluish black above; venter ivory with some invasion of dark pigment along the upper portions of the ventrals; scattered dark markings often present on posterior ventrals and under the tail; iris slate black. Juveniles uniform reddish above with a black spot at the tip of each scale; upper head shields

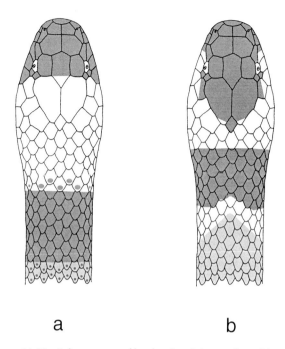

a   b

Figure 11.22. Color patterns of head and neck in two Costa Rican snakes with red dorsal coloration; patterns are shown on a generalized template and do not accurately depict differences in the arrangement of head plates for the two taxa: (a) *Clelia*, juveniles, black head cap and nuchal band separated by white to yellow nuchal collar; (b) *Ninia sebae*, black head cap involves most of parietal shields, black nuchal saddle demarcated posteriorly by yellow band.

black followed by a white, cream, or yellow collar three to five scale rows in longitudinal extent and bordered posteriorly by a black nuchal band (plate 343). During ontogeny the black areas expand to produce the adult coloration, but the venter of juveniles is immaculate cream except near the tail tip, which has some black markings.

SIMILAR SPECIES: Juveniles: (1) *Ninia sebae* has the black nuchal band bordered by yellow posteriorly (fig. 11.22b). (2) Red or orange *Stenorrhina freminvillii* lack a nuchal light collar. Adults: (1) *Chironius grandisquamis* has 10 scale rows at midbody. (2) *Clelia scytalina* has 17 scale rows at midbody. (3) *Lampropeltis triangulum* has the venter heavily marked by black that extends completely across, or the belly is uniform black.

HABITAT: Ubiquitous in the lowlands; often found in disturbed areas, secondary growth, and along roadsides as well as in forested tracts. Occurs in Lowland Dry, Moist, and Wet Forests and marginally in Premontane Moist Forest.

BIOLOGY: This relatively common large active terrestrial snake, also called musarana or ratonera elsewhere in Latin America, usually forages at night but is also active diurnally. It feeds primarily on other snakes, including various large pitvipers *(Atropoides, Bothrops, Crotalus,* and *Porthidium),* but eats many lizards and mammals as well. The species is much appreciated by *campesinos* because of its ophiophagous (snake-eating) habits. Terciopelos *(Bothrops asper)* are

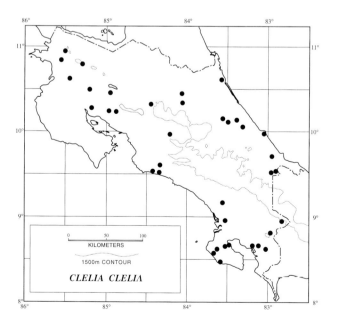

Map 11.12. Distribution of *Clelia clelia*.

regularly fed to *Clelia* at the antivenin-producing Instituto Clodomiro Picado in San Isidro de Coronado each Friday afternoon in a public demonstration. Although equipped with grooved rear fangs and the associated venom glands used in feeding, this species also constricts its prey. When attacking a large viper the zopilota usually strikes and grasps the neck of the victim and then constricts the helpless prey in a series of body coils. This snake has natural immunity to the venoms of pitvipers (Cerdas and Lomonte 1982). As far as is known (Scott 1983i) this species rarely bites in defense, and I know of no Costa Rican instances of human envenomation by it. Courtship and mating in a captive pair involved initial female aggression. Shortly afterward the male crawled over the female's back, lying parallel to her body, raised his tail, and pressed his cloaca against her body. Copulation followed, with one hemipenis, and lasted eighteen minutes. There was also increased tongue flicking by both snakes. Oviposition from the captive mating occurred 47 days later. Incubation time in captivity was 117 to 120 days (Martínez and Cerdas 1986). Ten to twenty-two eggs 46 to 66 mm long by 32 to 39 mm in diameter are laid during the late rainy season (August to November). Hatchlings from the captive mating were about 310 to 490 mm in total length.

REMARKS: The several Costa Rican examples referred to *Clelia clelia clelia* by Taylor (1951b, 1954b) are representatives of *C. scytalina*.

DISTRIBUTION: Atlantic lowlands of Guatemala south on the Atlantic versant throughout Central America through Colombia, northern Venezuela, Trinidad, and Grenada, and south from the Guianas to central Bolivia, and on the Pacific slope from Costa Rica to southern Peru. Found throughout the lowlands and marginally onto the premontane Meseta Central Occidental of Costa Rica (2–913 m).

Clelia scytalina (Cope "1866b," 1867)

(plates 344–45; map 11.13)

DIAGNOSTICS: The vividly red body and tail combined with the light nuchal collar that crosses the parietal shields, demarcated posteriorly by a black band, constitute a juvenile coloration shared only with *Clelia clelia* (fig. 11.22a). Adults of both species are uniform bluish black with a cream venter, but *C. scytalina* has only 17 scale rows at midbody.

DESCRIPTION: This poorly known species is very similar to *Clelia clelia* in most respects. A moderate-sized to large snake, probably smaller than *C. clelia*, but adult females over 1,800 mm in total length, longer than males (to 1,540 mm); head somewhat distinct from neck; body cylindrical; a moderately long tail (15 to 22% of total length); eyes moderate; loreal present or absent; supralabials 7 to 8, with 2 bordering orbit; 7 to 9 infralabials; 2 postoculars; temporals 2 + 2 or 2 + 3; smooth dorsal scales in 17-17-17 rows; no apical pits (two extralimitally); ventrals 202 to 228; subcaudals 75 to 92. Upper surfaces of adults bluish black; venter white; iris black (plate 344). Juveniles uniform pale red above with black spot on each scale tip; upper head shields black followed by a white to yellow collar three to five scale rows in longitudinal extent, bordered behind by black nuchal band; venter cream (plate 345).

SIMILAR SPECIES: Juveniles: (1) *Ninia sebae* has the black nuchal band bordered by yellow posteriorly (fig. 11.22b). (2) Red or orange *Stenorrhina freminvillii* lacks a nuchal light collar. Adults: (1) *Chironius grandisquamis* has 10 scale rows at midbody. (2) *Clelia clelia* has 19 scale rows at midbody. (3) *Lampropeltis triangulum* has the venter heavily marked with black (bands or uniform).

HABITAT: Near streams in Lowland (Atlantic) Moist Forest and in Premontane and Lower Montane Wet Forest and Rainforest.

REMARKS: This relatively rare species was confused by Taylor (1951b, 1954b) with *C. clelia*, and all specimens he cited under the latter name are typical *C. scytalina*. All specimens from Costa Rica and western Panama lack apical pits and may be conspecific with *Clelia equatoriana* of eastern Panama to western Ecuador rather than *C. scytalina* from Mexico and upper Central America as proposed by Zaher (1996). The issue is confused further by Zaher's (1996) reference to a single Costa Rican specimen (AMNH 111267), which he regarded as typical *C. scytalina*. The single records of this taxon from southwestern Colombia (Pérez-Santos and Moreno 1988) and northwestern Ecuador (Pérez-Santos and Moreno 1991) need verification and probably are based on specimens of *Clelia equatoriana*, which is well known from these areas.

Map 11.13. Distribution of *Clelia scytalina*.

DISTRIBUTION: At scattered localities in Mexico from Veracruz on the Atlantic versant and Jalisco on the Pacific slope, then southward from the Isthmus of Tehuantepec at scattered sites in the Pacific slope uplands to western Panama, and on the Atlantic slope in humid lowland and premontane zones in Costa Rica (60–1,900 m).

Genus *Oxyrhopus* Wagler, 1830

An unspecialized colubrid genus (twelve species) with the basic arrangement of enlarged head shields; nasal completely divided; a loreal; one or two preoculars; pupil vertically elliptical; two pairs of chin shields; dorsal scales smooth with zero to two apical pits; dorsal scales in 15 to 19 rows at midbody; vertebral row not enlarged; anal usually entire; subcaudals usually divided; subequal maxillary teeth 10 to 15 plus two slightly enlarged grooved fangs separated from anterior teeth by a diastema; Duvernoy's gland present; anterior mandibular teeth markedly enlarged; no hypapophyses on posterior dorsal vertebrae. The combination of smooth scales in 19 rows at midbody, vertebral scale row not enlarged, preoculars present, divided nasal, and the banded coloration separate *Oxyrophus* from other Costa Rican genera.

The everted hemipenes are moderate (about 10 to 12 subcaudals in length), bicapitate, with the sulcus spermaticus bifurcate, centrifugal, and dividing proximal to the capitulate lobes; spinous proximally; capitula calyculate, with smooth apical cups (Jenner and Dowling 1985; Zaher 1999).

The genus is found on the Atlantic lowlands of Mexico from Veracruz, then south on both slopes through Central America to central Argentina east of the Andes and on the Pacific versant to central Peru.

Oxyrhopus petolarius (Linné, 1758)

(plates 346–47; map 11.14)

DIAGNOSTICS: This unmistakable species has a pattern of dark bands on a red orange to red dorsum, an unmarked cream-colored venter, no enlargement of the vertebral scale row, and 19 scales at midbody.

DESCRIPTION: A moderate-sized snake reaching a total length of 1,200 mm; adult females (980 to 1,200 mm) longer than adult males (920 to 1,130 mm); head distinct from neck; body compressed; tail moderate (19 to 24% of total length); eye moderate; preoculars usually not in contact with prefrontal; 2 postoculars; supralabials 7 to 9, with 2 bordering orbit; 9 to 10 infralabials; temporals usually 2 + 3; dorsal scales in 19 rows at midbody, reduced to 17 anterior to vent, with two apical pits; anal single; ventrals 192 to 244 (189 to 244); subcaudals 79 to 117 (79 to 123). Dorsal surfaces red with a series of 19 to 36 (17 to 59) black body bands that are narrowly outlined by yellow in juveniles and some adults; several bands sometimes fused for short

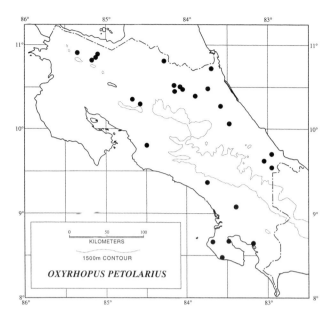

Map 11.14. Distribution of *Oxyrhopus petolarius*.

distances along middle of back to form zigzag figure; often considerable black spotting on scales in red bands; tail similarly marked above; nuchal area yellow, suffused with red in adults; head dark; venter cream, usually immaculate cream but sometimes with black markings posteriorly and/or under the tail; iris dark brown to black; tongue gray.

SIMILAR SPECIES: Adults: (1) *Tripanurgos compressus* has the vertebral scale row enlarged (fig. 11.15) and a very long black nuchal band. (2) All other red-and-black bicolored snakes in Costa Rica have substantial black rings or other extensive dark markings on the venter (fig. 11.47; table 11.2). Juveniles: (1) *Scaphiodontophis venustissimus* has broad yellow or white bands separating black and red areas. See key to coral snakes (p. 706), figure 11.47, and table 11.2.

HABITAT: Lowland Moist and Wet Forests.

BIOLOGY: An uncommon terrestrial nocturnal oviparous snake that feeds mostly on lizards and small mammals but also eats frogs. Like *Clelia* this species is a venomous form that may also constrict its prey. According to Fitch (1970) egg clutch size is 5–7.3–10 for samples from Amazonian Peru. Test, Sexton, and Heatwole (1966) reported that a female from Venezuela laid seven eggs on June 1 and 2 and that two hatched September 1 and 2. Campbell (1998) reported a clutch of ten eggs from Guatemala, 31 to 34 mm long and 18 to 19 mm in diameter, and estimated hatching time as about ninety days after oviposition.

REMARKS: Bailey (in Peters and Orejas-Miranda 1970) recognized three major geographic subdivisions in this species based primarily on coloration. Subsequently Bailey (in Wilson and Meyer 1985) suggested that four races might be involved but gave no further clarification. Examples from

Table 11.2. Color Patterns in Costa Rican Venomous Coral Snakes and Their Mimics

	BIB	BIR	BIR*	BIM	TSR	TDb	TZs	TZ	TM	TD	Venter
Dipsas articulata								X			Ringed
Dipsas bicolor							X				Ringed
Dipsas tenuissima*		X									Ringed
Erythrolamprus bizona										W/Y	Ringed
Erythrolamprus mimus								X			Ringed
Lampropeltis triangulum										W	Ringed
Leptodeira rubricata	X										Red
Liophis epinephalus	X										Dark markings
Oxyrhopus petolarius	X						rarely				Cream
Rhinobothryum bovallii									X		Ringed
Scaphiodontophis annulatus	j								X		Cream
Scolecophis atrocinctus								X			Cream
Sibon annulatus*		X			X					R	Ringed or with lateral bands
Sibon anthracops							X	X	X		No red
Sibon dimidiatus				X						R	Ringed or with lateral bands
Tantilla Supracincta	j				X						Red in adults
Tripanurgos compressus	X										Cream/white pink suffusion
Urotheca euryzona		X	X					X			No red
Micrurus alleni				X					X		Ringed
Micrurus clarki									X		Ringed
Micrurus nigrocinctus									X		Ringed
Micrurus mipartitus		X	X								Ringed

Note: See figure 11.47 for explanation.

*No red, orange, or pink dorsally; j = juveniles only; R = black-red-black dyad; W = black-white-black dyad; Y = black-yellow-black dyad.

Amazonia to Argentina and in the Chocó of Colombia have a few (10 to 24) dark body bands that are distinctly wider than the light interspaces, and large examples are often completely black. Samples from the Orinoco drainage of northern South America have 24 to 59 (usually more than 35) dorsal dark bands that are slightly wider than the light interspaces, which are one to two scale rows in length. Melanistic adults are rare in this sample. Snakes of this species from Mexico, Central America, northern Colombia, and western Ecuador have 17 to 36 dark body bands that are separated by light interspaces more than two scale rows in length. Uniformly black individuals have not been reported from Mexico or Central America, and Costa Rican *Oxyrhopus* are typical of this population. For most of the present century this species has been called *Oxyrhopus petola* after Amaral (1926), but Smith, Williams, and Pérez-Higareda (1986) chose *O. petolarius* as the correct name. See Bailey (1986) for a contrary view. This matter hinges on whether the subsequent spellings of *pethola* and *petalarius* by Linné (1766) and Gmelin (1788) are unjustified emendations or incorrect subsequent spellings (article 35, *International Code of Zoological Nomenclature*). If the former, the conclusion of Smith, Williams, and Pérez-Higareda (1986) prevails; if the latter is correct then Bailey's interpretation is valid. Taylor called this species *C. petolarius* (1951b) and *C. petolaria* (1954b). The karyotype of this species is 2N = 46, with eight pairs of macrochromosomes and fifteen pairs of microchromosomes; pair 1 is metacentric, pairs 4 and 6 are submetacentrics, and all others are telocentric; NF = 52; pair 4 shows a heteromorphism in females with the Z submetacentric and the W telocentric (Beçak and Beçak 1969).

DISTRIBUTION: Tropical lowlands from Veracruz, Mexico, on the Atlantic versant and Costa Rica on the Pacific slope through Central America to western Ecuador and throughout northern South America and the Amazon basin to Brazil, Peru, and Bolivia. In Costa Rica, restricted to the humid lowlands of the Atlantic versant and the central and southwest Pacific areas (5–700 m).

Genus *Tripanurgos* Fitzinger, 1843

A monotypic genus having a compressed body and the basic complement of enlarged head shields; nasal completely divided; a single loreal, one or two preoculars; pupil vertically elliptical; two pairs of chin shields; dorsal scales smooth with two apical pits, in 19-19-15 rows; vertebral row enlarged; anal entire; subcaudals divided; subequal maxillary teeth 13 to 15 plus two grooved fangs separated by a diastema; Duvernoy's gland present; anterior maxillary teeth longer than fangs; anterior mandibular teeth enlarged; no hypapophyses on the posterior dorsal vertebrae. The distinctive coloration of the single member of the genus, red in life with a very long black nuchal band and small black body blotches or narrow bands, and the enlarged vertebral scale row will distinguish the genus from others found in Costa Rica.

The everted hemipenes are relatively short (reaching to subcaudal 10), bicapitate, with the sulcus spermaticus bifurcate, centrifugal, and dividing proximal to the capitulate lobes; spinous proximally, with many spines on truncus; capitula with papillate calyces (Jenner 1981; Zaher 1999).

The genus occurs in southeastern, western, and central Brazil, from Bolivia and Paraguay in the Amazon basin to eastern Colombia, and from near the mouth of the Amazon and the Guianas throughout the Orinoco basin; also in northwestern South America to southwestern Costa Rica.

Tripanurgos compressus (Daudin, 1803)

(plate 348; map 11.15)

DIAGNOSTICS: The combination of an enlarged vertebral scale row (fig. 11.15), the very long postnuchal black band, immaculate venter, compressed body, and 19 scale rows at midbody is distinctive.

DESCRIPTION: A moderate-sized snake reaching a total length of somewhat over 1 m; head distinct from neck; tail moderate (20 to 24% of total length); eye large; supralabials 8 or 9, with 1 or 2 bordering orbit; 9 infralabials; 2 postoculars; temporals 2 + 3 or 2 + 2 + 3; smooth dorsal scales with two apical pits; ventrals 228 to 258; subcaudals 110 to 125. Dorsal surfaces red to reddish brown with 25 to 35 small black blotches or narrow bands; tail similarly marked above; upper head shields reddish to dark brown (white in some juveniles), followed by a yellow occipital collar that may be suffused with red pigment and a very long black nuchal collar (two to three times longer than head length); tip of tail often black; venter cream or white with a rose suffusion, without dark markings; iris burnt orange.

SIMILAR SPECIES: (1) *Oxyrhopus petolarius* lacks enlarged vertebral scales and has the upper head shields black. (2) All other Costa Rican snakes with the body bicolored red and black have substantial black rings or other dark markings on the venter. See key to coral snakes (p. 706) and table 11.2.

HABITAT: Undisturbed Lowland Pacific Wet Forest.

BIOLOGY: A rare nocturnal terrestrial species that feeds on lizards that it envenomates and/or constricts. Oviparous, presumably laying eggs with some frequency in the nests of parasol ants *(Atta)*. A set of eggs from such a site was reported by Riley and Winch (1985) from Trinidad. Apparently two clutches were represented, supposedly from two females, since six eggs measured 27 × 15 mm and six were 29.5 × 15 mm. The eggs were collected May 1, and five of the larger set hatched May 6–8; mean length of the hatchlings was 289 mm when they were preserved two weeks later.

REMARKS: Zaher (1999) recognized *Tripanurgos* as a valid genus. Zaher and Prudente (1999) synonynmized *Tripanurgos* with *Siphlophis* based on Zaher's (1994) unpublished Ph.D. dissertation. Because the evidence for their action has not been published, I tentatively regard *Tripanurgos* as a valid genus.

DISTRIBUTION: A rather disjunct range in southeastern Brazil, western and central Brazil, and adjacent Bolivia and Paraguay north to Colombia, southern Venezuela, and the Guianas, and the mouth of the Amazon, also, western Colombia and eastern Panama, with a disjunct population in southwestern Panama and adjacent Costa Rica. In the latter area known from the humid lowland evergreen forests of the Golfo Dulce region (0–80 m).

Genus *Erythrolamprus* F. Boie, 1826

The five species belonging to this genus are coral snake mimics with alternating rings of black, red, and usually yellow or white around the body. The body form is that of a generalized, cylindrical colubrid, but the tail is short (10 to 13% of total length). Members of the genus have the basic colubrid complement of head shields; an undivided nasal; a loreal; one or two preoculars; pupil round (fig. 11.12b); two pairs of chin shields; dorsal scales smooth in 15-15-15 rows, no apical pits; anal divided; subcaudals divided; subequal maxillary teeth 10 to 15 plus two feebly enlarged fangs that may be strongly or weakly grooved or without grooves; no diastema; Duvernoy's gland present, mandibular teeth subequal; no hypapophyses on posterior dorsal vertebrae. The tricolor ringed pattern of Costa Rican *Erythrolamprus* separates them at once from most other colubrid genera. Other genera in this family in the republic having similar coloration may be distinguished from *Erythrolamprus* as follows: *Dipsas* lack chin shields, *Lampropeltis* have 19 to 21 middorsal scale rows, and *Rhinobothryum* have keeled scales.

The everted hemipenes in *Erythrolamprus* are moderate (12 to 13 subcaudals in length), bilobed, with a bifurcate, centrifugal sulcus spermaticus; noncapitate; covered with spines that are enlarged on the basal third of the organ; with a terminal disk on each lobe (Dowling 1975b; Zaher 1999) (fig. 11.3a).

The genus is found in the lowland and slope zones from Atlantic versant Honduras and northwestern Costa Rica south to northwestern Ecuador (west of the Andes) and

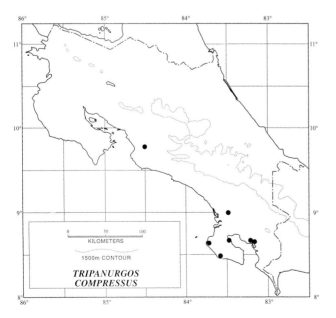

Map 11.15. Distribution of *Tripanurgos compressus*.

from northern South America throughout the Amazon basin south to Bolivia and extreme northeastern Argentina.

KEY TO THE COSTA RICAN SPECIES OF THE GENUS *ERYTHROLAMPRUS*

1a. Some infralabials and throat scales marked with black; nuchal region marked by two distinct black collars separated by a narrow yellow or white collar (fig. 11.23a); paired black body rings usually clearly separated by an intervening yellow or white ring; each pair of black rings broadly separated from next pair by red ring; supralabials bordered by black along their posterior margin; venter without a more or less continuous black stripe . . . . *Erythrolamprus bizona* (p. 578)

1b. Infralabials and throat scales immaculate; a single black nuchal collar crosses portions of parietals and may be split laterally by white or yellow (fig. 11.23b); black body rings not paired, most black rings split dorsally or laterally by white or yellow areas; each black ring broadly separated from next black ring by white-red-white rings; supralabials light, dark, or spotted with dark, but not margined by black along posterior borders; usually with an irregular more or less continuous midventral black stripe connecting black rings . . . . *Erythrolamprus mimus* (p. 579)

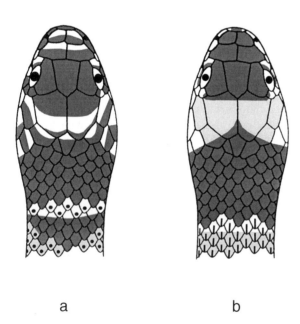

a   b

Figure 11.23. Color pattern of head and neck in Costa Rican species of the genus *Erythrolamprus;* patterns are shown on generalized template for the genus and do not accurately depict differences in the arrangement of the head plates for the two taxa: (a) *E. bizona,* two distinct black collars present; (b) *E. mimus,* a single black collar involving posterior tips of parietal plates. Light shading indicates red on head and first red ring on body.

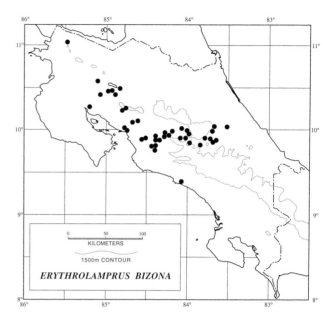

Map 11.16. Distribution of *Erythrolamprus bizona.*

Erythrolamprus bizona Jan, 1863

(fig. 11.23a; plate 349; map 11.16)

DIAGNOSTICS: A tricolored coral snake mimic in which the red rings on the body and tail border the black ones and each pair of black rings is separated by a white or yellow ring (i.e., has a pattern of dyads). *E. bizona* has a black ring bordering the light nuchal collar anteriorly and posteriorly. Only *Lampropeltis triangulum* and *Tantilla supracincta* among Costa Rican snakes have similar dorsal patterns, but tricolor individuals of the former have only one nuchal black ring, and the latter species has a uniform venter (table 11.2).

DESCRIPTION: A moderate-sized species (up to 1 m) in total length; head scarcely distinct from neck; tail short (10 to 13% of total length); eye moderate; 2 postoculars; 7 supralabials, with 2 bordering orbit; infralabials 7 to 9; temporals 1 + 2 or 1 + 1 + 2; ventrals 180 to 204; subcaudals 45 to 62; distinct grooves on fangs. The dorsal and caudal coloration consists of long red rings separated by dyads (two black rings separated by a white or yellow ring), of which there are 12 to 16½ on the body; often one or more white scales border black rings and partially separate them from contact with red rings; tips of scales in red and white/yellow rings are black; upper surface of the head mostly black, supralabials light with black along their margins; tips of the parietals usually red, sometimes suffused with black followed by a narrow yellow ring and then a nuchal dyad of two black bands separated by a yellow one; some infralabials and the throat marked with black; venter ringed like dorsum, sometimes with black pigment in the red rings along the midline; iris black; tongue black.

SIMILAR SPECIES: (1) *Lampropeltis triangulum* has 19 to 21 middorsal scale rows. (2) *Tantilla supracincta* lacks rings. (3) All other tricolor (black, red, yellow/white) snakes in Costa Rica have the black rings separated from the red rings by a light ring (fig. 11.47; table 11.2). See key to coral snakes (p. 706).

HABITAT: Pacific Lowland Dry and Moist Forests, Lowland Atlantic Moist Forest, Premontane Moist and Wet Forests, and Lower Montane Wet Forest.

BIOLOGY: An uncommon diurnal secretive terrestrial snake. It feeds principally on other snakes (plate 349) but eats lizards as well. One example (UCR 9229) from Escazú, San José Province, regurgitated a *Hydromorphus concolor* after capture; another (CRE 6243) from near Tilarán, Guanacaste Province, was eating a *Tantilla supracincta* when captured. The species is oviparous. This snake dorsoventrally flattens the neck and anterior trunk region when disturbed (Myers 1986), a maneuver that increases its apparent size from in front or behind. Presumably this behavior discourages some predators. Bites of this species may cause pain and some swelling in humans.

REMARKS: Taylor (1951b, 1954b) used the incorrect subsequent spelling, *E. bizonus*, for this species, an action followed by most subsequent authors. However, the original orthography is *bizona* (Jan 1863), a noun in apposition to the generic name. The specimen Villa, Wilson, and Johnson (1988) figured as a representative of this species is an *Erythrolamprus mimus*. The karyotype (based on one female) is 2N = 28, with no abrupt differentiation between macrochromosomes and microchromosomes; pairs 1, 3, 5, and 9 are metacentrics; 2, 7, 8, 10 are submetacentrics; 6 is subtelocentric; 4 shows a heteromorphism, with Z a little larger, but both Z and W are submetacentrics; the location of the centromere could not be determined for chromosomes 11, 12, 13, 14; NF = 44 (Gutiérrez, Solórzano, and Cerdas 1984).

DISTRIBUTION: Costa Rica south to Colombia and northern Venezuela; northern to central Pacific lowlands and adjacent slopes, the Meseta Central, and the valley of the Río Reventazón on the Atlantic versant (8–1,450 m) in Costa Rica; widely disjunct from the nearest records for southwestern Panama.

Erythrolamprus mimus (Cope, "1868b," 1869)

(fig. 11.23b; plates 350–51; map 11.17)

DIAGNOSTICS: This tricolored coral snake mimic is distinguishable from all others in the republic in having most of the black rings on the body split dorsally by white areas. It is one of several species of coral snakes and their mimics having the single black rings separated from the red ones by broad light rings (it has a pattern of monads).

DESCRIPTION: A moderate to large species with a total length up to 650 mm; head scarcely distinct from neck; tail

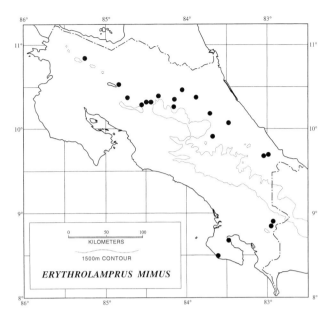

Map 11.17. Distribution of *Erythrolamprus mimus*.

short (10 to 13% of total length); eye moderate; supralabials 5 to 7, with 2 bordering orbit; 7 to 10 infralabials; temporals 1 + 2 or 1 + 1 + 2; ventrals 171 to 183 (179 to 199); subcaudals 42 to 56; grooves on fangs indistinct or absent. Dorsal and caudal pattern composed of moderately long red rings separated by a monad (one black ring, bordered on each side by a light, usually white, ring), of which there are 12 to 18; black rings usually split dorsally or laterally by light (usually white) areas; some black rings may be offset on two sides of body; scales in red and light areas tipped with black; supralabials black, black spotted, or immaculate, but never bordered posteriorly by black; light parietal area usually suffused with red; a single nuchal black collar (three to six scales in longitudinal extent along midline) crosses posterior tips of parietals and sometimes is split laterally by light color, followed by a narrow light ring and a long red ring; infralabials and throat scales immaculate; venter ringed like dorsum (without white splitting black rings), with a more or less continuous black stripe running along the midventer in most examples; iris black.

SIMILAR SPECIES: (1) All other tricolor ringed or banded snakes in Costa Rica that have the black markings separated from the red ones by light markings lack white areas that partially split the black rings or bands. (2) The remaining tricolor ringed or banded snakes in the republic have the black markings in contact with the red ones (fig. 11.47; table 11.2). See key to coral snakes (p. 706).

HABITAT: Lowland Atlantic Moist and Wet Forests and Pacific Wet Forest; Premontane Rainforest.

BIOLOGY: An uncommon terrestrial diurnal secretive species that feeds principally on other snakes (plate 351) but eats lizards on occasion and is oviparous. Neck flattening as described in the account of *Erythrolamprus bizona* also occurs in this species (Myers 1986). In addition, *E. mimus* is known to coil tightly and display the tail (Myers 1986), perhaps an aposematic (warning) behavior. Its bite causes pain and swelling in humans.

REMARKS: Elsewhere in its range this species shows consistent geographic variation in coloration. Examples from Honduras and Nicaragua have a tricolor dyad pattern with the black rings in contact with the red rings, and the black rings have light centers. The head cap and nuchal black band are as in Costa Rican examples. Specimens from eastern Panama, western Colombia, and Ecuador have a tricolor monad pattern with the solid black rings separated from the red ones by white rings; in some individuals there are white ventrolateral spots in the posterior black rings. In the upper Amazon basin of Ecuador and Peru the pattern is bicolor black and red with a very narrow black collar (one scale row long along midbody), or the collar is absent or reduced to a series of spots (Dunn and Bailey 1939). The example figured by Villa, Wilson, and Johnson (1988) is a representative of this species, not *E. bizona* as labeled.

DISTRIBUTION: Lowlands and premontane slopes on the Atlantic versant from Honduras through Panama; southwestern Costa Rica and western Panama, western Colombia and Ecuador, and northwestern Venezuela; a disjunct population occurs in the upper Amazon region of eastern Peru and Ecuador. In Costa Rica, known from evergreen forests of the Atlantic plains and slopes and the Golfo Dulce region (1–1,200 m).

Genus *Liophis* Wagler, 1830

As currently defined this large (forty-one species) and rather diverse genus has generalized body proportions and the basic colubrid complement of head shields, but there is a broad range of variation in many other features. Species belonging to the genus range from 0.5 to 1.5 m in total length and have short to long tails (13 to 31% of total length); nasal entire; usually a loreal (sometimes reduced or absent); one or two preoculars; pupil round; two pairs of chin shields, dorsal scales smooth in 15, 17, or 19 rows at midbody, with zero to two apical pits; anal and subcaudals divided; maxillary teeth 8 to 28 plus two enlarged ungrooved fangs, sometimes separated from anterior maxillary teeth by a diastema; Duvernoy's gland present; mandibular teeth subequal; no hypapophyses on posterior dorsal vertebrae. The single Costa Rican species of *Liophis* resembles a number of diurnal terrestrial genera *(Coluber, Leptodrymus,* and *Mastigodryas),* but its coloration (usually dark dorsal bands with red in the interspaces and red venter) will separate it from them.

The everted hemipenes are short to moderately long (8 to 12 subcaudals in length), bilobed, with a bifurcate, centrifugal sulcus spermaticus; noncapitate; covered with spines over entire surface, largest on pedicel; with a terminal disk on each lobe (Zaher 1999).

The genus *Liophis* as redefined by Dixon (1980, 1989) includes a variety of forms often placed in separate genera *(Dromicus, Leimadophis,* and *Lygophis).* As reconstituted this assemblage now contains a heterodox series of species that share a complex of primitive features not found in other allied xenodontines having disked hemipenes that are partially covered with spines and lack calyces. Whether or not Dixon's concept of *Liophis* constitutes a monophyletic group, present knowledge does not permit recognizing clearly defined subdivisions within the assemblage based on the characteristics formerly used to separate *Dromicus, Leimadophis, Liophis,* and *Lygophis.* These include presence of a long or short diastema in the maxillary dentition (or its absence), presence of one or two apical scale pits (or their absence), number of scale rows and their pattern of reduction posteriorly, coloration, and the presence or absence of basal hooks on the hemipenes. In addition, although there might be some merit in continuing to recognize *Leimadophis* and *Lygophis* for certain clusters within *Liophis,* the demonstration by Maglio (1970) that *Dromicus* Bibron, 1843, is congeneric with, and has priority over, *Leimadophis* Fitzinger, 1843 (*fide* Smith and Grant 1958), further complicates the matter. Since *Dromicus* was used primarily for Antillean and Galápagos Island species before Maglio's study and not for forms from the mainland of Central and South America, replacing *Leimadophis* with *Dromicus* at this point would be an added confusion. It therefore seems best to use *Liophis* in the broad sense defined by Dixon until more detailed studies confirm his conclusions or require the resurrection of *Dromicus (= Leimadophis)* and *Lygophis,* since *Liophis* was used for many of the mainland forms before Dunn (1932b,c, 1944d) placed them in *Leimadophis* or *Lygophis.*

Liophis ranges over much of the Neotropics from Costa Rica to northern Peru (on the Pacific slope) and northern South America, including the Andes, and thence to the Lesser Antilles and south throughout the eastern portions of the continent to extreme south-central Argentina.

Liophis epinephalus Cope, 1862a

(plates 352–53; map 11.18)

DIAGNOSTICS: This common species has smooth dorsal scales and a red venter marked with large black spots (fig. 11.34g–i) or occasionally with black bands. There is some red in the dorsal coloration as well, except in large adults. Most other species having red venters lack the dark markings or have alternating red and black rings that encircle the body (fig. 11.47; table 11.2).

DESCRIPTION: A moderate-sized species reaching 800 mm in total length; head distinct from neck; body cylindrical; tail moderate (17 to 27% of total length); eye moderate; a loreal; usually one preocular, sometimes two; 2 postoculars; usually 8 supralabials (occasionally 9 or 10), with 2 bordering orbit; infralabials usually 10, ranging from

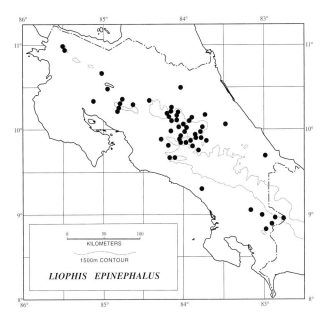

Map 11.18. Distribution of *Liophis epinephalus*.

8 to 11; temporals usually 1 + 2, sometimes no anterior temporal, sometimes 1 or 3 posterior temporals; 17-17-15 (17-17-17) dorsal scale rows; ventrals 139 to 152 (128 to 191); subcaudals 51–63 (44–80); maxillary teeth 15 to 17 plus two fangs separated from the anterior teeth by a diastema. Upper surfaces of body and tail reddish, marked with black bands, 32 to 49 on body; black bands tapering ventrally; red scales usually tipped or suffused with black; pattern most evident in juveniles, since red becomes completely suffused with darker pigment in some adults, especially on the anterior or posterior halves of the body; upper and lateral head shields black, but supralabials, infralabials, and throat light; chin and throat yellow green grading into bright red on venter; venter with various amounts of black, square or rectangular markings, usually offset from one another on each side, rarely forming a series of more or less continuous black bands across the belly; ventral black markings rarely continuous with dorsal ones; tongue black. The hemipenes are short, about 8 subcaudals in length.

SIMILAR SPECIES: (1) *Ninia psephota* has keeled scales (fig. 11.5).

HABITAT: A relatively common form usually found in marshy or riparian situations in relatively open areas, pastures, and secondary growth as well as in a variety of forest types (Lowland Moist and Wet Forests; Premontane Moist and Wet Forest and Rainforest; Lower Montane Wet Forest and Rainforest); an upland and lowland record from the Río Tenorio, Guanacaste Province, demonstrates its occurrence in gallery forest in northwestern Costa Rica.

BIOLOGY: A diurnal and terrestrial snake that feeds mainly on frogs and toads (Sexton and Heatwole 1965; Michaud and Dixon 1989). In captivity this species has been

fed toxic frogs of the genera *Atelopus, Dendrobates,* and *Phyllobates* with no apparent ill effects (Myers, Daly, and Malkin 1978), but the frogs may have lost their toxicity in captivity. Sexton and Heatwole (1965) reported gravid females from Panama containing five or six eggs in June. This form exhibits horizontal neck spreading (Myers 1986) to expose the red or orange and light blue skin between the scales, which presumably deters some predators. Although Pough's (1988) interpretation of this behavior as a warning seems to be correct, his statement that individuals of this species are normally solid colored and show red coloration only when the neck and anterior body are inflated is not. Most examples have considerable red banding on the body that is visible whether or not the animal is excited. Some large adults are almost uniformly dark, and in these individuals the brightly colored interscale skin of the neck and anterior body is exposed only during neck flattening.

REMARKS: Dunn (1937b) and Taylor (1951b, 1954b) used the name *Leimadophis taeniurus juvenalis* for Costa Rican representatives of *Liophis epinephalus.* The snakes from Costa Rica (Bebedero, Guanacaste Province, and Irazú, Cartago Province) that Wettstein (1934) identified as *Liophis cobella* are examples of *L. epinephalus.* The Bebedero record is the only one from lowland northwestern Costa Rica and is probably from the gallery forest along the Río Tenorio. Dixon (1983) has described the substantial geographic differences in this variable species. However, he used the incorrect subsequent spelling *L. epinephelus* following Boulenger (1894). Elsewhere in its range *L. epinephalus* may lack red coloration; may have an olive, leaf green, or brown dorsal ground color with two to four black stripes on the posterior portion of the body and tail; may be spotted, streaked, flecked, or banded with dark pigment; and may have the venter immaculate white, yellow, pinkish or reddish, black-and-yellow checkered, or dusky or uniform black. Dixon used the relative constancy of the presence or absence of tail stripes, dorsal pattern, and ventral coloration in combination with average differences in ventral and subcaudal counts to recognize eight races within the species. This treatment is difficult to interpret: there are many intermediates (intergrades); as many as three nominal forms occur in virtual sympatry; and one putative subspecies, the high Andean taxon *Liophis epinephalus bimaculatus,* contains individuals whose color variation encompasses that typical for five other forms. Dixon's treatment confirms the position of Wilson and Brown (1953) and Savage and Heyer (1967) that using subspecies is not an effective way to describe geographic variation. Snakes agreeing in most features with the Costa Rican population occur in extreme southwestern Panama, but those from farther east in Panama have a black lateral tail stripe and very little black marking on the venter. The karyotype of this snake is as follows (based on males only): 2N = 28, without an abrupt differentiation between macrochromosomes and microchromosomes; pairs 1, 3, 5,

7, and 10 are metacentrics; 2, 4, and 8 are submetacentrics; and 6 and 9 are subtelocentrics; the position of the centromere in the other chromosome pairs could not be determined; NF = 44 (Gutiérrez, Solórzano, and Cerdas 1984).

DISTRIBUTION: Slopes of the cordilleras of Costa Rica and western Panama and the eastern and southwestern lowlands of Costa Rica southward through the lowlands of central Panama to northern Peru on the Pacific slope and at higher elevations (to 3,500 m) in the Andes of Colombia, Ecuador, and Venezuela. In Costa Rica found in humid portions of the lowland, premontane, and lower montane zones throughout the country, including the Meseta Central, and in gallery forest in the northwestern region (2–2,100 m).

Genus *Xenodon* Boie, 1827

The four members of the genus *Xenodon* are relatively short, stocky, heavy-bodied, short-tailed (13 to 15% of total length) snakes with a triangular head that is very distinct from the body. They have the basic colubrid complement of head shields, nasal completely divided, a loreal, one or two preoculars, pupil round; two pairs of chin shields; dorsal scales smooth, with one apical pit; scales arranged in oblique rows; 19 to 21 scale rows at midbody; anal divided or entire; subcaudals divided; maxillary much shortened, rotatable; maxillary teeth increasing in size posteriorly, 6 to 15 plus two enlarged ungrooved fangs separated from anterior series by a diastema; Duvernoy's gland present; mandibular teeth subequal; no hypapophyses on posterior dorsal vertebrae. The distinctive viperlike head shape combined with the blotched pattern and round pupil will distinguish *Xenodon* from other Costa Rican snakes.

The everted hemipenis is bilobed, with a bifurcate, centrifugal sulcus spermaticus; spinous over entire surface with largest spines basally; no calyces; terminal disks present at tip of each organ around terminus of each sulcus branch (Jenner 1981) or not (Myers 1986; Zaher 1999).

The genus ranges through the tropical lowlands from Veracruz and Guerrero, Mexico, through Central America to central Pacific slope Ecuador, and through northern South America southward east of the Andes to southern Brazil and northeastern Argentina.

Xenodon rabdocephalus (Wied, 1824)

Terciopelo Falsa

(plate 354; map 11.19)

DIAGNOSTICS: The viperlike appearance and the smooth scales arranged in oblique rows (fig. 11.5) make this species easy to recognize.

DESCRIPTION: A moderate-sized snake that reaches 800 mm in total length; head distinct from neck; body depressed; tail short (14 to 15% of total length); eye large; supralabials 7 to 9, with 2 bordering orbit; infralabials 9 to 12; temporals most frequently 1 + 2 or 1 + 3; dorsal scales

in 19 rows at midbody reduced to 17 anterior to vent; anal single; ventrals 138 to 151 (124 to 153); subcaudals 44 to 48 (35 to 52). Dorsum brown, usually with a series of 11 to 16 hourglass-shaped blotches; each blotch dark brown laterally and grayish brown medially, outlined by a pale brown border with some white areas; in some examples the dark blotches are variously broken up into a series of smaller blotches, especially on the posterior part of the body, or the small blotches may be fused across the midline; head brown above, usually with a dark-outlined light lyre-shaped figure originating at the internasals with a branch running posteriorly above the eye and downward to the angle of the jaw; in some large specimens the light areas are reduced, suffused, or absent, but the dark outline persists as a canthal and/or postocular dark stripe; supralabials light brown, edged with dark; throat cream with very little dark pigment, which is usually restricted to scale margins; venter cream to beige with much light brown stippling and often with light brown irregular blotches as well; underside of tail cream, with little or no dark scale margins, contrasting sharply with venter; iris gray brown; tongue dark brown. The hemipenes of this species are long and slender, with both lobes covered by medium-sized spines (Zaher 1999).

SIMILAR SPECIES: (1) All pitvipers have a vertically elliptical pupil and a distinct labial pit (fig. 11.8b). (2) *Leptodeira nigrofasciata* has solid transverse dark dorsal bands. (3) *Trimorphodon biscutatus* has a vertically elliptical pupil (fig. 11.9d). (4) *Ungaliophis panamensis* has the dark dorsal markings outlined by light.

HABITAT: Widespread in Lowland Moist and Wet Forests and in Premontane Moist and Wet Forests and marginally in Premontane Rainforest.

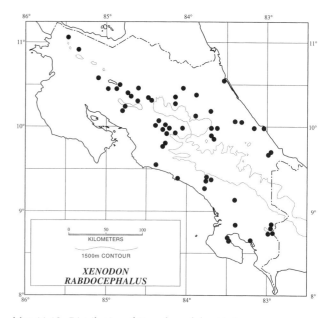

Map 11.19. Distribution of *Xenodon rabdocephalus*.

BIOLOGY: Snakes of the genus *Xenodon* prey almost exclusively on toads (genus *Bufo*), and this species is no exception. The specialized maxillary and feeding behavior of *Xenodon* is especially effective in eating these prey items. A standard defensive maneuver in toads is to increase their body size by inflating the lungs. *Xenodon* first grasps the toad using the anterior teeth but quickly moves the swollen victim farther back into the mouth. At this point it rotates the maxillaries medially so the large fangs stab the toad, puncturing the body cavity and lungs to release the air and deflate the animal. It then engulfs the prey in the usual fashion. These snakes also eat frogs, and tadpoles have been taken from their stomachs. Because of the flexibility of the upper jaw apparatus they can also rotate the fangs laterally on occasion. It is rather disconcerting to have this happen when you are holding the snake directly behind the head; the fangs protrude from the corners of the mouth, and the animal attempts to move its head to scratch or penetrate your hand. The species is relatively common, diurnal, riparian, and oviparous. Clutches contain nine or ten eggs and are laid in the rainy season. Gerhardt, Harris, and Vásquez (1993) recorded predation on this species by the great black hawk *(Buteogallus urubitinga)* in Guatemala. The color pattern of *X. rabdocephalus* resembles that of the venomous terciopelo *(Bothrops asper)*. The coloration combined with the aggressive defensive behavior of this species, which hisses and strikes with an open mouth at the slightest provocation (plate 355), has led to the suggestion that it is a Batesian mimic of the venomous species (Pough 1988). In addition, this snake exhibits horizontal neck spreading or flattening when disturbed.

REMARKS: Taylor (1951b, 1954b) used the names *Xenodon colubrinus* and *X. bertholdi* for examples of this form in Costa Rica. Only a single species, *X. rabdocephalus*, seems to occur in Mexico and Central America, and the presumed distinguishing features Taylor attributed to Costa Rican populations are encompassed within the known variation for that form. As Wilson and Meyer (1985) pointed out, there is a clinal increase in ventrals from north to south in this species, which precludes recognizing geographic races, since no other feature separates individuals from northern and southern populations. The karyotype of this snake is as follows: 2N = 34, with eleven pairs of macrochromosomes and six pairs of microchromosomes; pairs 6, 9, and 11 are metacentrics; 1, 5, are 7 are submetacentrics; 10 is subtelocentric; 2, 4, and 8 are telocentrics; 3 shows a female heteromorphism, the Z is submetacentric, and the small W is submetacentric; NF = 48 (Gutiérrez, Solórzano, and Cerdas 1984).

DISTRIBUTION: Lowlands of tropical Mexico from Veracruz and Guerrero, south through Central America to northwestern Ecuador, west of the Andes, Amazonian Colombia, and Ecuador, the Guianas and the upper Amazon portion of Brazil, Peru, and Bolivia; most records are from humid forest areas, so that the distribution is discontinuous on the Pacific versant of Central America. In Costa Rica on the humid lowland and premontane slopes of the Atlantic versant, on the Meseta Central (Occidental and Oriental), in the southwestern Pacific region and marginally on the Pacific slope in the northwest (1–1,200 m).

Amastridines
Genus *Amastridium* Cope, "1860c," 1861

A genus (two species) differing in morphology from generalized colubrids in having a distinct canthal ridge and moderately long tail (25 to 32% of total length). Other defining features include the basic colubrid complement of head shields; nasal completely divided; a loreal that may be fused with postnasal; a single preocular; pupil round; two pairs of chin shields; dorsal scales 17-17-17, smooth, with two apical pits on scales in nuchal region; keels present on supracloacal scales in adult males; anal and subcaudals divided; maxillary teeth 11 to 7 plus two enlarged fangs that may be grooved or not and separated from the subequal anterior teeth by a diastema; Duvernoy's gland probably present; mandibular teeth subequal; hypapophyses present on posterior dorsal vertebrae. The absence of a loreal, the presence of 17 scale rows at midbody, and the canthal ridge distinguish the genus from others in Costa Rica (although members of the genus from northern areas of Central America and Mexico have a loreal). Wilson (1988) provided a review of the genus, updating the Wilson and Meyer (1969) account. Zaher (1999) places this genus in the subfamily Dipsadinae.

The everted hemipenis is single, slightly clavate, noncapitate, with a bifurcate, centrolineal sulcus spermaticus; with basal spines (two enlarged hooks at very base) and calyces distal from the level where the sulcus forks (Wilson and Meyer 1969).

The genus has a discontinuous distribution from upland Nuevo León (Atlantic versant) and Oaxaca (Pacific versant), Mexico, to Belize and Honduras and from extreme southeastern Nicaragua south on the Atlantic slope to central Panama; also southwestern Pacific slope Costa Rica and adjacent western Panama.

Amastridium veliferum Cope, "1860c," 1861
(fig. 11.24; plate 356; map 11.20)

DIAGNOSTICS: Any moderate to small unstriped, dark gray to black snake with a distinct rusty colored head will be readily identified as this species.

DESCRIPTION: A small to moderate-sized snake attaining a total length of 724 mm; head distinct from neck; body cylindrical; eye moderate; supralabials 7, with 2 bordering orbit; 9 infralabials; 2 postoculars; temporals 1 + 2; ventrals 111 to 129 in males, 126 to 134 in females; subcaudals 70 to 86 in males, 69 to 85 in females; ventrals plus subcaudals 196 to 202. Upper surfaces of body and tail dark

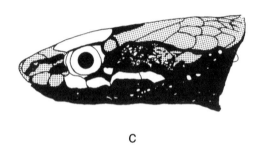

a   b

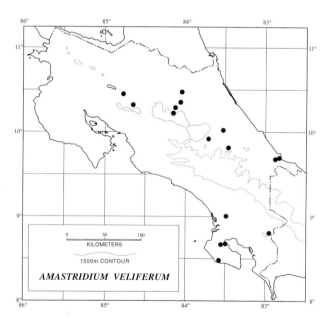

c

Figure 11.24. Cephalic coloration of *Amastridium veliferum*: (a) dorsal view; (b) ventral view; (c) lateral view. Light shading indicates rusty reddish color of head plates.

Map 11.20. Distribution of *Amastridium veliferum*.

gray to black, with a series of very small white spots every four or five scale rows (on fourth, fifth, or sixth scale row on each side); superficially head and nape appear rust colored, but head shields are marked with dark brown to black; a pointed extension of the dorsal ground color indents the posterior margin of the nape coloration; sides of head mostly dark but often with more or less continuous dark-bordered white stripes along the loreal region and supralabials; infralabials and throat dark brown to black, each scale marked with a white spot or only a few white spots present; venter and underside of tail dark gray to black with some light spots anteriorly; iris rust or dark brown; tongue black.

SIMILAR SPECIES: (1) *Geophis hoffmanni* and *G. zeledoni* lack the rust-colored head and nape and white spots on the throat and dorsum.

HABITAT: Undisturbed forests including Lowland Moist (exclusive of the Pacific-Atlantic type) and Wet Forests and Premontane Wet Forest and Rainforest.

BIOLOGY: A relatively uncommon diurnal semifossorial snake that is found in the leaf litter. It feeds on small frogs mostly of the genus *Eleutherodactylus* and is oviparous. Wilson and Robinson (1971) described an apparent pseudohermaphrodite of this snake from Costa Rica (UCR 2819) having oviducts and ovaries and one hemipenis.

REMARKS: Wilson and Meyer (1969, 1985) and Wilson (1988) regarded the populations of this genus from upper Central America and Mexico as conspecific with *A. veliferum*. Smith (1971) maintained that the northern populations constitute a separate species, *Amastridium sapperi* Werner, 1903b, since they are characterized by having a separate loreal and much higher segmental counts (ventrals 146 to 158 in males, 144 to 170 in females; subcaudals 84 to 86 in males, 79 to 85 in females; ventrals plus subcaudals 229 to 250). These differences and the substantial range gap (650 km) between the nearest records of *A. veliferum* in extreme southeastern Nicaragua and putative *A. sapperi* from western Honduras lead me to accept Smith's interpretation. The holotype (SMF 9227a) of *Fleischmannia obscura* Boettger, 1898, is a representative of this species and labeled as being from "San Jose, Costa Rica." It almost certainly is not from the capital city or environs, since no other Meseta Central specimen has been collected.

DISTRIBUTION: Lowlands of Atlantic slope southeastern Nicaragua, Costa Rica, Panama, and lowlands and premontane slopes of southwestern Costa Rica and western and central Panama (2–1,200 m).

Genus *Conophis* Peters, 1860

A Mesoamerican genus of striped snakes containing two or three species (Wellman 1963) having a generalized colubrid morphology and head shields. *Conophis* are distinguished from other members of the family by the following combination of features: rostral decurved; nasal completely divided; a loreal; usually one preocular (rarely two); pupil round (fig. 11.16a); two pairs of chin shields; dorsal scales smooth in 19-19-17 rows, without apical pits; anal and sucaudals divided; maxillary teeth 8 to 12 plus two grooved fangs; a distinct diastema; prediastemal teeth subequal; Duvernoy's gland present; anterior mandibular teeth largest; no hypapophyses on posterior dorsal vertebrae. Technically this genus can be separated from *Coniophanes* and *Crisantophis* only by details of dentition, vertebrae, and hemipenes, but Costa Rican *Conophis* usually have two or three anterior temporals, six to eleven dark and several light longitudinal stripes, and the supralabials not marked by dark pigment along the upper margins (one anterior temporal and two to four dark longitudinal dark stripes or only two longitudinal white stripes in *Coniophanes;* two to four longitudinal white stripes and the supralabials marked by black above in *Crisantophis*). Zaher (1999) places *Conophis* in the subfamily Xenodontinae.

The everted hemipenis is short (8 subcaudals in length), slightly bilobed, and noncapitate, with a bifurcate, centrolineal sulcus spermaticus; large basal hooks present; truncus covered by smaller spines; apical region covered by spinulate flounces (Myers 1986). *Conophis vittatus* has small apical lobes (Jenner 1981; Zaher 1999).

The genus occurs in tropical lowlands from Veracruz, Mexico, to eastern Honduras and the Yucatán Peninsula on the Atlantic versant and from the lowlands of Nayarit, Mexico, through lowland and premontane Pacific slope Mexico, and thence south through Central America to Costa Rica.

Conophis lineatus (C. Duméril, Bibron, and A. Duméril, 1854)

Guarda Camino

(fig. 11.25; plate 357; map 11.21)

DIAGNOSTICS: This species has six to eleven dark longitudinal stripes usually alternating with six or more white or tan areas (or stripes) and has at most the lower one-third of the supralabials brown. The pattern combined with a dorsal scale row formula of 19-19-17 will distinguish this snake from all others in Costa Rica.

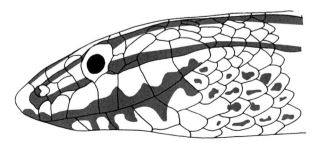

Figure 11.25. Head scalation and color pattern of *Conophis lineatus*. Light areas white, dark areas dark brown.

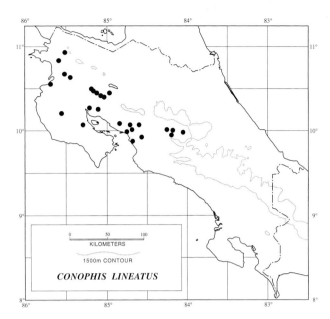

Map 11.21. Distribution of *Conophis lineatus*.

DESCRIPTION: A moderate-sized snake with a maximum total length of 1,167 mm; head moderately distinct from neck; body cylindrical; tail moderate (20 to 24% of total length); supralabials 7 or 8, with 2 bordering orbit; infralabials 8 to 11, usually 9 or 10; 2 or 3 postoculars, usually 2; anterior temporals usually 2 or 3 (rarely one), posterior temporals 1 to 3; ventrals 159 to 178 (149 to 180); subcaudals in males 67 to 80, in females 60 to 72 (56 to 72); ventrals plus subcaudals 224 to 247 (222 to 251). Upper surface of body and tail with a variable number of light and dark longitudinal stripes (six to thirteen white or tan stripes and six to eleven dark stripes, evident at midbody); middorsal area usually tan; head dark brown above, a distinct dark stripe from snout along side of head through eye and continuing as a lateral dark stripe along body; supralabials immaculate white or with dark lower margins; throat immaculate; venter white, usually with a series of more or less prominent dark spots on tips of ventrals.

SIMILAR SPECIES: (1) Some examples of several species of striped *Rhadinaea (R. calligaster, R. godmani,* and *R. serperaster)* superficially resemble this form but have the supralabial borders and upper margins marked with dark pigment. (2) *Coniophanes bipunctatus* has two rows of large conspicuous dark spots on the ventrals (fig. 11.29a). (3) *Coniophanes fissidens* has two or three dark longitudinal stripes. (4) Juveniles of *Crisantophis nevermanni* have dark markings on the upper portion of the supralabials (fig. 11.17b).

HABITAT: Found in secondary growth and open areas, including pastures and roadsides, in Lowland Dry and Premontane Moist Forests.

BIOLOGY: This rather common diurnal terrestrial snake

feeds primarily on lizards (principally *Ameiva* and *Cnemidophorus*) and the eggs of ground-nesting birds such as doves and the ground cuckoo (Scott 1983j). It is active even at midday while foraging. When approached it usually takes cover under bushes or down small mammal burrows. In captivity it will also eat frogs, snakes, and small mammals (Wellman 1963). When attacking prey this quick-moving snake slashes the victim with the enlarged rear fangs to introduce the venom and holds on until the venom takes effect. Similarly, representatives of this species deliver a slashing bite to predators or to humans foolish enough to carelessly handle one. The bite causes pain, substantial edema, and hematoma. Symptoms last for two to several days (Johanbocke 1974; Wellman 1963; Lee 1996), and the initial bite may bleed for an hour or more, suggesting that the venom may have anticoagulant properties (Scott 1983j; D. Janzen, pers. comm.). The bite of this species is extremely painful and dangerous, although not as life threatening as those from the coral snakes, sea snakes, and vipers. Cook's (1984) record of a *Stenorrhina* bite is based on a misidentification of the present species. Little is known about reproduction, but one female laid two eggs and seemed to have two more in her body when captured and then released in February 1986. Campbell (1998) reported clutches of four to six eggs. The decurved and concave rostral suggests that *C. lineatus* burrows into the substrate, and it frequently escapes capture by entering the burrows of other animals.

REMARKS: The considerable variation in details of the color pattern in this species shows little consistency, contrary to some authors (e.g., Wellman 1963), and does not form a basis for recognizing geographic subdivisions. Wilson and Meyer (1985) regarded the nominal species *Conophis pulcher* of Chiapas, Mexico, Guatemala, and Honduras as conspecific with *C. lineatus*. Scott (1983j) recorded *C. lineatus* from the Península de Osa of Costa Rica, based on the oral report of Daniel Janzen, who knew the species well from his extensive fieldwork in northwestern Costa Rica. Elsewhere in its range in western Central America the species is restricted to seasonally deciduous forest or semiarid habitats. The Osa record is from a lowland evergreen forest site and awaits confirmation. However, it seems likely that is based on some other striped snake. The records of this form from Panama (Auth 1994a) and northern Colombia (Pérez-Santos and Moreno 1988) seemed dubious on the face of it. Fortunately, Auth (1994b) quickly realized his mistake in confusing *Liophis lineatus* with this form. Later Auth et al. (1998) showed that the Colombian record was also based on *L. lineatus*.

DISTRIBUTION: Lowlands and premontane slopes from Veracruz, Mexico, to eastern Honduras on the Atlantic versant and Oaxaca, Mexico, on the Pacific slope possibly to southwestern Costa Rica, including upland areas of western Nicaragua belonging to the Atlantic drainage system. In Costa Rica in the northwestern Pacific lowlands and premontane Meseta Central Occidental, with a single questionable sight record from the Pacific southwest on the Península de Osa (2–1,160 m).

Genus *Crisantophis* Villa, 1971

This monotypic genus was long confused with *Conophis*, from which it differs primarily in dentition, vertebrae, hemipenes, rostral morphology, and coloration. In form members of *Crisantophis* are generalized colubrids and have the basic colubrid complement of head shields. They may be characterized by the following combination of features: nasal divided; a loreal; one preocular; pupil round; two pairs of chin shields; dorsal scales smooth, without apical pits, in 19-19-17 rows; anal and subcaudals divided; maxillary teeth 13 to 15 plus two laterally compressed grooved fangs; a diastema; prediastemal and mandibular teeth increasing in size posteriorly; Duvernoy's gland present; hypapophyses present on posterior dorsal vertebrae. The combination of narrow light longitudinal stripes and black marks on the intersupralabial sutures immediately distinguishes them from other Costa Rican snakes. Zaher (1999) tentatively refers this genus to the Dipsadinae.

The everted hemipenis is moderately long (13 to 15 subcaudals in length), bilobed, and subcylindrical with a bifurcate, centripetal sulcus spermaticus; noncapitate, with pedicel and truncus covered with spinules; tipped with small terminal awns (Villa 1971).

The genus is found along the Pacific lowlands of Central America from Guatemala to central Costa Rica.

Crisantophis nevermanni (Dunn, 1937b)
(plate 358; map 11.22)

DIAGNOSTICS: The adult coloration of two to four conspicuous narrow light longitudinal stripes on the dark brown ground color, the black marking along the upper margins of the supralabials (fig. 11.17b), and the presence of distinct dark spots on the tips of the ventrals are the principal identifying marks that will separate *Crisantophis* from other Costa Rican colubrids. Juveniles are pale with three to five dark stripes, have the upper portion of the supralabials dark, and have dark smudges on each ventral tip.

DESCRIPTION: A moderate-sized snake (maximum total length 828 mm); head moderately distinct from neck; body cylindrical; tail moderate (20 to 24% of total length); eye small; supralabials 8, with 2 bordering orbit; infralabials 10; 1 postocular; temporals usually 2 + 2, sometimes 1 or 3 in either position; ventrals 173 to 183; subcaudals in males 82 to 89, in females 71 to 76; ventrals plus subcaudals 250 to 263. Upper surfaces of juveniles relatively pale with three to five dark stripes (a middorsal stripe and one to two lateral ones on each side); larger examples have the paler areas becoming darker and fusing with the stripes to produce a uniform dark brown or black ground color with a lateral

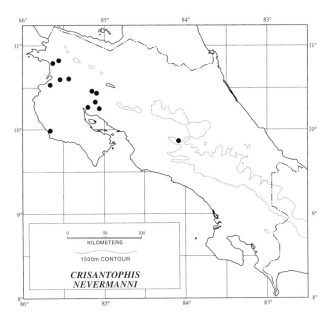

Map 11.22. Distribution of *Crisantophis nevermanni*.

and/or dorsolateral white stripe on each side; upper surface of head dark brown or black with a light stripe from tip of snout over eye to become continuous with the dorsolateral stripe; side of the head dark, with supralabials white and marked above and usually along their sutures by dark brown or black; infralabials white, marked along sutures by dark brown or black; throat immaculate except for a few dark flecks or heavily pigmented on chin; venter immaculate except for definite black spots, which are restricted to ventral tips.

SIMILAR SPECIES: (1) The several striped species of *Rhadinaea* and *Urotheca* lack the discrete dark spots on the tips of the ventrals, although the dorsal coloration may continue onto the ventral tips, and have no scale row reduction on the posterior body. (2) *Conophis lineatus* has six to eleven dark stripes and lacks dark markings on the upper margins of the supralabials (fig. 11.17a). (3) *Coniophanes bipunctatus* has a pair of large bold spots on each ventral, well mediad from the ventral tips (fig. 11.29a). (4) *Coniophanes piceivittis* has two broad light longitudinal stripes on the black body and an immaculate venter.

HABITAT: Found primarily in Lowland Dry Forest, but one record is from a Premontane Moist Forest; tends to be associated with stream courses, even during the dry season when streams dry up.

BIOLOGY: An uncommon diurnal and terrestrial form. It usually feeds on frogs and toads, but one adult contained a partially digested juvenile of its own species. A captive female laid three eggs in May and contained seven more (Villa 1969a,b).

REMARKS: For many years this species was placed in the genus *Conophis*. Villa (1971) demonstrated that *Conophis* and *Crisantophis* are remarkably distinct in many features

of internal and external morphology and concluded that the superficial resemblance in striped coloration was insignificant in evaluating their relationship. Villa (1988b) presented a full description and summarized the literature on the genus and its single species. As Wellman (1963) pointed out, Taylor's (1955) record of this species (KU 35630) is based on a specimen of *Conophis lineatus* taken well within the known range of the latter form. The holotype of *C. nevermanni* (ANSP 22423) is supposed to be from Río Poás de Asserí (1,385 m), 1.5 km northwest of Asserí on the north slope of the mountains just south of San José. All other known examples from Costa Rica are from the northwest in Guanacaste Province. It is possible that *C. nevermanni* ranges southward into Puntarenas Province and onto the Meseta Central Occidental, but no voucher specimens except the holotype support this notion. The five examples of this species in the Vienna Museum from Bebedero, Guanacaste Province, were originally misidentified by Wettstein (1934) as *Coniophanes imperialis*. They were subsequently referred to *Coniophanes piceivittis* by Dunn (in Bailey 1938) and by Taylor (1951b). Dunn (1937), however, had previously revised his identification.

DISTRIBUTION: Pacific lowlands from Guatemala to northwestern Costa Rica (4–50 m) and possibly onto the Meseta Central Occidental in the latter country (1,385 m).

Genus *Enulius* Cope, "1870," 1871

A genus of small, slender snakes (five species) with the basic colubrid complement of head shields. *Enulius* are separated from other members of the family by the following combination of characteristics: nasal divided; a loreal; usually no preoculars; pupil round; two pairs of chin shields; dorsal scales smooth, in 15 to 17 rows at midbody, with one or two apical pits; anal and subcaudals divided; maxillary teeth 3 to 5 plus two compressed and enlarged ungrooved postdiastemal teeth; prediastemal teeth increasing in size posteriorly; Duvernoy's gland probably present, but not confirmed; mandibular teeth small, equal; no hypapophyses on posterior dorsal vertebrae. Members of this genus in Costa Rica are essentially black or dark brown above with a light nuchal collar that may be enlarged to include almost the entire head. This pattern serves to distinguish them from most other small colubrid snakes in the republic. The pattern combined with the presence of a loreal, apical pits, and a divided anal separates them from all other Costa Rican snakes. Zaher (1999) tentatively refers *Enulius* to the subfamily Dipsadinae.

The everted hemipenes are short (7 to 10 subcaudals in length) and slightly to distinctly bilobed, with a bifurcate, centrifugal sulcus spermaticus; noncapitate, ornamented with spines, and lacking calyces (Jenner 1981).

The genus *Enulius* is one of several snake genera (including *Scaphiodontophis* and *Urotheca*) that have a disproportionately long thickened but fragile tail (fig. 11.18a). Nearly

50% of these snakes have a portion of the tail missing, and tail loss appears to be a predator escape device (specialized pseudautotomy). The predator apparently becomes confused by the reflex thrashing of the fractured tail segment while the snake seeks to escape. The tail is not regenerated, but because of its great length it may be broken several times during the life of an individual.

The genus ranges along the Pacific versant from Sinaloa, Mexico, through Central America to Colombia and onto the Atlantic slope in Honduras (including the Islas de la Bahía), Nicaragua, Costa Rica, Panama, and Colombia.

KEY TO THE COSTA RICAN SPECIES OF THE GENUS *ENULIUS*

1a. Dorsal scales with one apical pit; 165–216 ventrals; usually 17 dorsal scale rows, occasionally 15; snout distinctly shovel shaped in profile (fig. 11.26)..........
...................*Enulius flavitorques* (p. 588)
1b. Dorsal scales with two apical pits; 129–51 ventrals; 15 dorsal scale rows; snout blunt in profile (fig. 11.12c)
.... *Enulius sclateri* (p. 589)

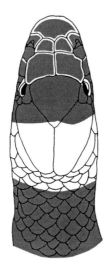

a

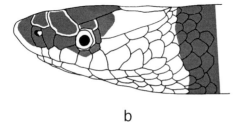

b

Figure 11.26. Head scalation and color pattern of *Enulius flavitorques:* (a) dorsal view; (b) lateral view.

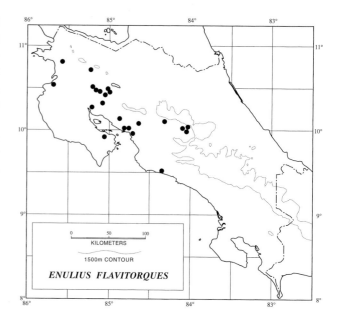

Map 11.23. Distribution of *Enulius flavitorques.*

Enulius flavitorques (Cope, "1868b," 1869)

(fig. 11.26; plate 359; map 11.23)

DIAGNOSTICS: This small smooth-scaled brown snake with a yellowish nuchal collar (fig. 11.33a) is readily distinguished from similar species by having a very long tail (that is frequently incomplete) and a projecting (shovel-shaped) snout.

DESCRIPTION: A small, elegant snake (maximum total length 500 mm); head barely distinct from neck; body cylindrical; tail long (30 to 41% of total length); eye small; rostral somewhat enlarged, projecting; no preocular (rarely present in examples from Panama and Colombia); 2 postoculars; supralabials 7 to 8, usually 7, with 2 bordering orbit; infralabials 6 to 8, usually 7; temporals 1 + 2; dorsal scales 15 or 17 at midbody, with a single apical pit; ventrals 165 to 216; subcaudals 85 to 121. Upper surfaces dark brown to black; venter cream with light color extending onto second or third dorsal scale row; a pale cream or yellow nuchal band usually present (present in Costa Rican examples). The everted hemipenis is bilobed, covered with minute uniform spines over most of length, distal area naked with a pair of large papillae in which the branches of the sulcus terminate.

SIMILAR SPECIES: (1) *Enulius sclateri* has the snout blunt (fig. 11.27). (2) The several species of *Geophis* with light nuchal collars as juveniles do not have the light ventral coloration extending onto the lower body scales. (3) Similarly colored species of *Ninia* (fig. 11.33c,d) have keeled dorsal scales (fig. 11.5). (4) *Trimetopon* with a nuchal light collar have a single prefrontal shield (fig. 11.38).

HABITAT: A denizen of Lowland Dry and Premontane Moist Forests; may be found under stones or debris in

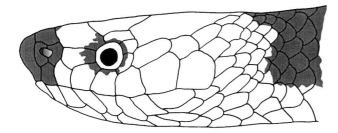

Figure 11.27. Head scalation and color pattern of *Enulius sclateri*, lateral view.

pastures or natural openings in the thickets or between tree clumps.

BIOLOGY: A rather uncommon fossorial diurnal form, often found in ant and termite nests in association with *Leptotyphlops ater*. It feeds primarily on small snake and lizard eggs (Scott 1983h), puncturing the egg shells with its greatly enlarged fangs and swallowing the contents. The species is oviparous.

REMARKS: There is some variation in the nuchal light collar in this species. Most Central American examples have a light collar extending from the middle of the parietals to the second or third scale row on the body. Sometimes the nuchal region and most of the head are light colored, or the nuchal collar may be reduced to several light spots, but these variants are not known from Costa Rica. In Mexico the nuchal light collar is absent, and this condition occurs as an occasional variant in Central America (reported in single specimens from Guatemala, Nicaragua, and Panama). Although *E. flavitorques* usually has the dorsal scales in 17 rows throughout the length of the body, a few examples from Nicaragua and Costa Rica have 15 rows at midbody. This variation makes suspect the status of *Enulius oligostrichus* Smith, Arndt, and Sherbrooke, 1967, from Sinaloa (McDiarmid and Bezy 1971) and Nayarit, Mexico, which differs from *E. flavitorques* by always having (in the four known specimens) 15 middorsal scale rows and somewhat fewer ventrals (152 to 163).

DISTRIBUTION: Pacific lowlands and premontane slopes from Jalisco, Mexico, to Colombia; also from the Atlantic drainage in the uplands and lowlands of Honduras, the Great Lakes region of Nicaragua, and northern Colombia. In Costa Rica commonly found in the Pacific northwestern lowlands and Meseta Central Occidental (5–1,250 m), but ranging along the coast to the vicinity of Parrita in the southwestern Pacific zone.

Enulius sclateri (Boulenger, 1894)

(fig. 11.27; plate 360; map 11.24)

DIAGNOSTICS: Most examples of this small black snake have the upper surfaces of the head and neck pure white except for the tip of the snout. In many other examples the tip of the frontal, all of the parietals, and several scale rows on

the neck are white (fig. 11.33b). The first of these head patterns occurs in no other species. The second is also found in *E. flavitorques*, but the rounded snout of *E. sclateri* will separate it from its congener.

DESCRIPTION: A small robust snake (maximum total length 550 mm); head barely distinct from neck; body cylindrical; tail long (36 to 42% of total length); eye small; snout blunt in profile, rostral not protruding; supralabials 7, with 2 bordering orbit; 6 infralabials; no preocular; 2 postoculars; temporals 1 + 2; dorsal scales in 15-15-15 rows, with two apical pits; ventrals 129 to 151; subcaudals 96 to 100. Upper surfaces dark brown or black; throat and venter white; undersurface of tail white to gray, sometimes with small dark blotches or a median dark line; nuchal area and head white except for dark around orbit and on snout (internasals, nasals, rostral, and first supralabials) that sometimes extends posteriorly to cover all or part of the frontal and anterior edges of parietals; often with some white areas on prefrontals even in dark-headed examples; iris and tongue black. The everted hemipenis is slightly bilobed; truncus with large spines, apical area with widely scattered small spinules and pedicel covered by numerous small spines on sulcate surface (McCranie and Villa 1993).

SIMILAR SPECIES: (1) *Enulius flavitorques* has a shovel-shaped snout (fig. 11.26). (2) *Geophis* with a light nuchal collar have keeled scales (fig. 11.5) over at least the last third of the body or the nuchal collar not involving all of the parietal shields. (3) Similarly colored species of *Ninia* (fig. 11.33c,d) have keeled scales (fig. 11.5). (4) *Trimetopon* with a light nuchal collar have a single prefrontal shield (figs. 11.33g, 11.38).

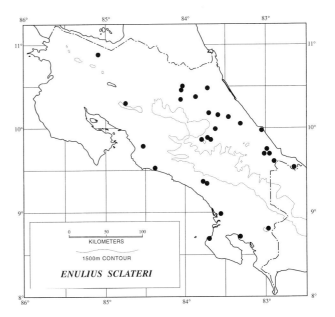

Map 11.24. Distribution of *Enulius sclateri*.

HABITAT: Most common in undisturbed Lowland Moist and Wet Forests, occasional in Premontane Wet Forest and Rainforest.

BIOLOGY: A relatively uncommon diurnal fossorial and oviparous species that feeds on small reptile eggs. It punctures the egg shells with its enlarged rear fangs and swallows the contents.

REMARKS: McCranie and Villa (1993) erected a new monotypic genus, *Enuliophis*, for this species because of its unusual hemipenes. In all other features *E. sclateri* clusters with *Enulius* when compared with other colubrid snakes. Since all evidence indicates that *E. sclateri* is the sister species to all other *Enulius*, taxonomic efficiency is best served by avoiding monotpyic genera and including the known species in an inclusive taxon. Zaher (1999) amply demonstrates the variation in hemipenes that may occur within a genus. Lips (1993d) confirmed the occurrence of this form in southwestern Costa Rica.

DISTRIBUTION: Evergreen forests of the Atlantic lowlands and premontane slopes from Nicaragua to central Colombia and in southwestern Costa Rica to eastern Panama on the Pacific versant (4–700 m).

Genus *Nothopsis* Cope, 1871

A distinctive monotypic genus of small rainforest snakes having the general habitus of a boa and highly modified head scalation. Unlike the condition found in most other colubrid snakes, *Nothopsis* have the upper head surface covered by numerous small, mostly tuberculate scales. This feature in combination with the following separates the genus from all others in the family: nasal partially divided below nostril; loreal area occupied by 5 to 10 smooth scales; 1 or 2 preoculars; pupil round; a single pair of chin shields; dorsal scales keeled, in 26 to 30 rows at midbody, without apical pits; anal entire; subcaudals divided; maxillary teeth 19 to 21 plus two slightly enlarged, ungrooved postdiastemal teeth, longest anteriorly; Duvernoy's gland probably present; several anterior mandibular teeth shortened, rest decreasing in length posteriorly; hypapophyses present on posterior dorsal vertebrae. The presence of keeled head and body scales and the absence of a loreal pit immediately separate *Nothopsis* from all other Costa Rican snakes. Zaher (1999) tentatively refers this genus to the subfamily Dipsadinae.

The everted hemipenes are short (6 subcaudals in length) and simple with a centrolineal, bifurcate sulcus spermaticus; semicapitate; pedicel covered with tiny spicules that near the sulcus grade into small papillate calyces, which in turn extend along and between the sulcus branches to tip; large spines lie proximal to the sulcus fork and extend distally as a paired series of smaller spines along outer edge of each hemipenial lobe. Dunn and Dowling (1957) illustrated a retracted organ and Zaher (1999) an everted one.

Nothopsis ranges from eastern Honduras through lower Central America and western Colombia to western Ecuador.

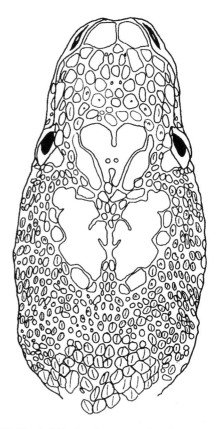

Figure 11.28. Head of *Nothopsis rugosus*, dorsal view.

Nothopsis rugosus Cope, 1871
(fig. 11.28; plate 361; map 11.25)

DIAGNOSTICS: No other species lacking a loreal pit has the head covered with small keeled scales.

DESCRIPTION: A small, slender snake with a maximum total length of 433 mm; head distinct from slender neck; body somewhat compressed; tail long, averaging 45% of total length in adults (range for all examples 32 to 52%); eye small, dorsally placed, and visible from above; upper surface of head covered for the most part by small keeled or tuberculate scales; internasals complete and in contact with one another; frontal usually divided and outer margins broken up into small scales; parietals with outer margins broken off as small scales and separated from one another and the frontals by one to three rows of small scales; eye surrounded by a ring of about 16 scales that may be slightly enlarged, of which 2 or 3 are preoculars, 2 or 3 supraoculars, and 2 or 3 postoculars; supralabials 9 to 13, separated from orbit by one or two rows of small scales (suboculars and lorealabials); infralabials 11 to 16; temporals smooth to tuberculate in 7 to 10 rows, smaller than dorsal scales; dorsal scales heavily keeled, 26 to 30 rows in neck region and at midbody, reduced to 22 to 26 rows anterior to vent; commonest dorsal scale formulas are 26-26-22, 26-28-24, 28-28-24; ventrals 152 to 162; subcaudals 95 to 112 in males,

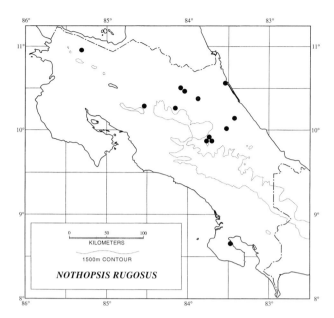

Map 11.25. Distribution of *Nothopsis rugosus*.

81 to 101 in females; ventrals plus subcaudals 235 to 268. Dorsal surfaces yellowish or cream colored with a series of 31 to 38 pairs of large subtriangular dorsolateral dark brown or black blotches; blotches often fused across midline to form hourglass-shaped figures, but usually blotches alternate over a part of body; lower margins of blotches are open and contain a series of small lateral rounded or rhomboidal dark spots; pattern continuing onto anterior part of tail; head uniform black or dark brown; venter cream or yellow; tips of each ventral with small dark spot, free margin irregularly stippled and streaked with brown.

SIMILAR SPECIES: (1) All species of *Boa, Corallus, Epicrates, Loxocemus,* and *Ungaliophis* have smooth dorsal scales (fig. 11.5). (2) All pitvipers have a loreal pit (fig. 11.8b).

HABITAT: Restricted to undisturbed rainforests (Lowland Moist and Wet Forests and Premontane Wet Forest and Rainforest).

BIOLOGY: A rather uncommon diurnal snake that preys on frogs and salamanders (Scott 1969; Cadle and Greene 1993); a semifossorial leaf litter inhabitant. It is apparently oviparous.

REMARKS: As Dunn and Dowling (1957) pointed out, *Nothopsis affinis* Boulenger, 1905 (Ecuador), and *N. torresi* Taylor, 1951b (Costa Rica), are conspecific with *N. rugosus.* The characteristics of the holotype of *N. torresi* (KU 28719) emphasized by Taylor (18 prediastemal maxillary teeth and the scale reduction formula of 26-26-22) fall well within the variability for *N. rugosus.* Similarly, Taylor's (1954b) reference to a second specimen of *N. torresi* (KU 31946) is based on a typical *N. rugosus.*

DISTRIBUTION: Evergreen lowland and premontane forests from eastern Honduras to eastern Panama on the Atlantic versant and on the Pacific slope from southwestern Costa Rica through Panama and western Colombia to northwestern Ecuador. In Costa Rica on the Atlantic slope and in the Golfo Dulce region (2–830 m).

Dipsadines

The Costa Rica genera of this stock may be placed in five clusters of more or less closely related taxa based principally on hemipenial features (Jenner 1981; Fernandes 1995). Because the members of each of these subgroups resemble one another in habitus, external features, or coloration (Cadle and Greene 1993; Campbell and Smith 1998), it seems useful to list them here with their extralimital allies. Generally, when you have difficulty classifying a snake belonging to this group after making an initial generic placement, you should compare the specimen in question with the descriptions of other genera included in a particular subgroup. In this way you can most readily confirm or revise the generic determination.

The major subdivisions of diapsadine snakes are as follows:

I. *Coniophanes-Rhadinaea-Trimetopon-Urotheca: Rhadinophanes* and *Tantalophis* of the Sierra Madre del Sur of Mexico and *Chapinophis* of Guatemala may belong here or with group II (terrestrial and semifossorial forms)

II. *Geophis-Ninia: Adelphicos* of Mexico and upper Central America and probably *Chersodromus* of Mexico belong in this group (semifossorial and fossorial snakes)

III. *Hydromorphus* and *Tretanorhinus* (a semiaquatic and an aquatic genus, respectively)

IV. *Dipsas-Sibon: Sibynomorphus* of central and southeastern South America and *Tropidodipsas* of Mexico and northern Central America (snail- and slug-eating snakes).

V. *Imantodes-Leptodeira: Cryophis* of the Sierra Juárez of Oaxaca, Mexico, *Eridiphas* of Baja California, Mexico, and *Hypsiglena* of central and western Mexico and the southwestern United States (terrestrial and arboreal forms)

Zaher (1999) places all these genera except *Chapinophis* in the subfamily Dipsadinae. He did not examine *Chapinophis*, which probably will prove to be a dipsadine.

So far as is known all dipsadine snakes are oviparous.

Genus *Coniophanes* Hallowell, in Cope, 1860b

A generalized colubrid genus (fourteen species) in terms of external morphology, with the basic arrangement of enlarged head shields; nasal partially or completely divided; a loreal; one or two preoculars; pupil round; two pairs of chin shields; dorsal scales smooth, without apical pits; in 17 to 25 scale rows at midbody (with a reduction anterior to the vent); anal and subcaudals divided; 8 to 15 subequal maxillary teeth separated by a diastema from two grooved fangs; Duvernoy's gland present; mandibular teeth subequal; no hypapophyses on posterior dorsal vertebrae. The combination of a loreal, divided anal, smooth scales without apical pits, scale row reduction anterior to the vent, and striped color pattern will separate *Coniophanes* from other Costa Rican genera. Members of this genus have a high incidence of tail loss in adults, about 40% in most samples having broken or healed tails. As is characteristic of unspecialized pseudautotomy, the tail shows no specialized morphology other than elongation, and multiple tail breaks are unusual (Mendelson 1992).

The hemipenis varies among the several species groups. All have a bifurcate sulcus spermaticus; several groups have a single spinous organ with a calyculate capitulum; others have the organ slightly bilobed, spinous, and with a calyculate capitulum; one group has the hemipenis divided, spineless, calyculate, flounced, and noncapitate (Bailey 1938; Jenner 1981). The genus *Coniophanes* is closely related to *Rhadinaea, Trimetopon,* and *Urotheca.* If you have trouble with species identification, check the accounts for these three genera.

Bailey (1938) divided the genus into five poorly differentiated species groups, each containing no more than two species. The three Costa Rican species belong to different groups. Recently described extralimital species (Myers 1966; Campbell 1989; Cadle 1989) do not significantly affect Bailey's arrangement.

Coniophanes range in the lowlands from extreme southern Texas and Sinaloa, Mexico, south through Central America and western South America into northwestern Peru.

KEY TO THE COSTA RICAN SPECIES OF THE GENUS *CONIOPHANES*

1a. Dorsal scales in 17 to 21 rows at midbody; belly with a black spot or spots mediad to tip of most ventrals (fig. 11.29); ventrals 110–45; dorsum brownish to gray, with a narrow (less than one scale row in width) pair of dorsolateral light stripes or a distinct pair of broad (two or more scale rows in width) dorsolateral dark stripes 2

1b. Dorsal scales in 23–25 rows at midbody; belly nearly immaculate; ventrals 153–74; dorsum black, with two broad (two scale rows in width) light dorsolateral stripes . . . . *Coniophanes piceivittis* (p. 595)

2a. A large dark spot on the lateral third of most ventrals (in addition to the small dark spots on ventral tips), forming a paired series of conspicuous dark spots along the otherwise clear venter (fig. 11.29a); dorsum with a broad (about six scale rows in width) middorsal dark area separated from the dorsolateral dark stripe on each side by a lighter area; supralabials mottled black and white . . . . *Coniophanes bipunctatus* (p. 592)

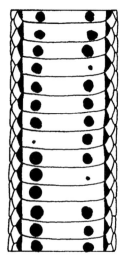
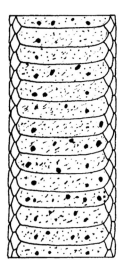

a   b

Figure 11.29. Ventral color pattern at midbody in two Costa Rican species of the snake genus *Coniophanes:* (a) *C. bipunctatus;* (b) *C. fissidens.*

2b. Ventrals irregularly marked with numerous black speckles and small spots, some of which may tend to form rows along lateral third of ventrals in some individuals, but never with a paired series of conspicuous large dark spots along the venter (fig. 11.29b); dorsum usually with a narrow middorsal dark stripe or a series of dots and a pair of narrow dorsolateral dark stripes that are at best faintly separated from dorsal and lateral ground colors; supralabials spotted and speckled, with a white longitudinal stripe bordered above by black posterior to orbit. . . . . *Coniophanes fissidens* (p. 593)

Coniophanes bipunctatus Group

Snakes of this group lack a temporal light stripe, have large regular dark spots on the ventrals, and have the hemipenes slightly bilobed, spinous, and capitate. Two species are included in the group, *Coniophanes bipunctatus* and *Coniophanes quinquevittatus* of Veracruz, Mexico, to northern Guatemala.

Coniophanes bipunctatus (Günther, 1858a)

(plate 362; map 11.26)

DIAGNOSTICS: Among the many striped species of snakes in Central America, this is the only one with a pair of large dark spots on each ventral and the tip of each ventral marked with a small spot (fig. 11.29a).

DESCRIPTION: A moderate-sized snake (maximum total length 750 mm); head distinct from neck; tail moderate (29 to 36% of total length); eye moderate; one preocular; two

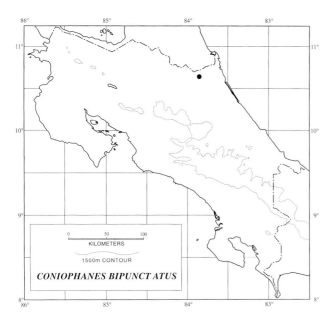

Map 11.26. Distribution of *Coniophanes bipunctatus*.

postoculars; supralabials 7 to 9 (usually 8), with 2 bordering orbit; infralabials 7 to 11 (usually 9 or 10); temporals 1 + 1, 1 + 2, or 1 + 3; dorsal scales in 21 rows at midbody reducing to 17 or 19 anterior to vent; supracloacal keels present in males; ventrals 124 to 145; subcaudals 72 to 101; maxillary teeth 13 to 15 + 2. Ground color pale, dark, or rust brown; a lateral dark stripe poorly developed on scale rows 3 and 4, 5 and 4, and 5 or 3 only; a diffuse dark brown middorsal stripe; venter with a pair of distinct large dark brown or black spots on each ventral scute; venter otherwise cream anteriorly and/or medially but suffused with pinkish orange laterally and/or posteriorly; iris rust brown.

The everted hemipenis is short (about 6 subcaudals in length); single and unicapitate, with a centrolineal, bifurcate sulcus spermaticus that forks slightly below the capitulum; large basal spines and distal calyces and flounces are present (Myers 1969b).

SIMILAR SPECIES: (1) *Coniophanes fissidens* has numerous small dark spots and speckles on each ventral (fig. 11.29b) and a pair of narrow dorsal dark stripes. (2) Adult *Crisantophis nevermanni* have two or four conspicuous dorsal white stripes. (3) *Conophis lineatus* has a series of six to eleven dorsal dark stripes.

HABITAT: Swampy areas in Lowland Wet Forest.

BIOLOGY: A semiaquatic snake that eats fish (Myers 1969b) and frogs and is mostly nocturnal. Smith (1940) reported a clutch of five eggs for this species in Mexico.

REMARKS: This species has long been expected to be found in Costa Rica, since it was known from Nicaragua and northwestern Panama. A recently collected male example (UCR 13039) from 38 km northeast of Puerto Viejo (Tres Marías), Heredia Province, confirms its presence in

Costa Rica. Additional specimens are likely to be found in marshes and swamps near the Río San Juan and near the coast on the Atlantic versant.

DISTRIBUTION: Lowlands and premontane slopes on the Atlantic versant from Veracruz, Mexico, south through Central America through the Great Lakes region of Nicaragua to northwestern Panama (10 m).

Coniophanes fissidens Group

Members of this group lack a temporal light stripe, have the ventrals immaculate or with small dark spots, and have the hemipenes single, spinous, and capitate. Three species are allocated to this group: *Coniophanes fissidens*, *Coniophanes andresensis* (San Andrés Island, Colombia), and *Coniophanes alvarezi* (Chiapas, Mexico).

Coniophanes fissidens (Günther, 1858a)

(plate 363; map 11.27)

DIAGNOSTICS: This common, rather nondescript snake is best recognized by the absence of broad and/or conspicuous light or dark stripes that markedly contrast with the ground color. A weakly differentiated pair of thin lateral dark stripes usually emphasized above by light dashes is present, and the venter is marked with dark speckles and spots (fig. 11.29b).

DESCRIPTION: A small to moderate-sized snake reaching a total length of 800 mm, adult females (300 to 800 mm) longer than adults males (300 to 630 mm); head distinct from neck; tail long (33 to 42% of total length in males, 26 to 38% in females); eye moderate; one or two preoculars, usually one; usually two postoculars, sometimes one; supralabials 8 (7), with 2 bordering orbit; infralabials 9 or 10;

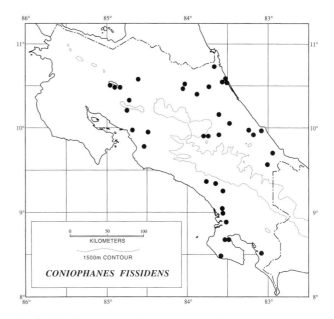

Map 11.27. Distribution of *Coniophanes fissidens*.

temporals 1 + 2, 1 + 2 + 3, or 1 + 3; dorsal scales variable, anterior count 19 to 23, midbody 17 to 21, but usually 21-21-17 or 21-19-17 (sometimes reduced to 15 posteriorly); supracloacal keels present on adult males; ventrals 110 to 127 (109 to 146); subcaudals 63 to 80 (53 to 103); maxillary teeth 12 to 15 + 2. Ground color brown, often with an orange cast; a narrow dark lateral stripe on scale rows 4 and 5 and lower edge of row 6 on each side, sometimes interrupted, and in most examples fading imperceptibly into rather dark flanks below; scale areas immediately above stripe often light; often a series of dark dots on the vertebral scale row; labials stippled with dark pigment, usually with a dark-centered ocellus on most labials; chin with moderate-sized dark spots; venter white to yellow, sometimes with an orange cast; usually several rows of small dark spots along ventrals and much dark stippling on ventrals and subcaudals; iris reddish brown; tongue black.

The everted hemipenes are short (about 6 subcaudals in length), single, and unicapitate, with a centrolineal, bifurcate sulcus spermaticus; large basal spines present; capitulum has papillate calyces on distal and sulcate surface and spinulate calyces on asulcate surface (Bailey 1938; Myers 1969b).

SIMILAR SPECIES: (1) *Coniophanes bipunctatus* has a pair of large dark spots on each ventral (fig. 11.29a) and a pair of broad lateral dark stripes. (2) *Rhadinaea godmani* and *R. serperaster* have immaculate venters. (3) *Rhadinaea calligaster* has the supralabials heavily margined with black. (4) *Rhadinaea pulveriventris* has a distinct median black streak on the neck that bifurcates on the nape. (5) *Tantilla vermiformis* has an immaculate venter. (6) Striped *Trimetopon* have essentially immaculate venters except for the ventral tips that are encroached on by the dorsal color.

HABITAT: An inhabitant of lowland rainforests, upland semideciduous forests, and rainforests of the following types: Lowland Moist and Wet Forests, Premontane Moist and Wet Forests.

BIOLOGY: This denizen of the forest floor is active during the day and early evening. It feeds on small prey that it envenomates with the grooved rear fangs. In a detailed analysis of stomach contents in 256 examples of this species from Pacific slope Chiapas, Mexico, and Guatemala, Seib (1985a) found it to be a generalist in food consumption. Frogs of the genus *Eleutherodactylus* and their egg clutches composed about 50% of the prey items, but it also ate other small vertebrates (frogs and toads, lizards of the genera *Norops* and *Sphenomorphus*, leaf litter snakes, and a salamander of the genus *Bolitoglossa*) and reptile eggs as well as conspecifics. Juveniles are at least partially insectivorous. At La Selva, Heredia Province, a similar pattern is evident, with *Eleutherodactylus* and their egg clutches making up at least 50% of the diet items. *Coniophanes fissidens* is among those rear-fanged snakes that can inflict a painful bite on humans. The bite often numbs the bitten part and causes substantial

swelling that may persist for several days. The species, like others in the genus, is oviparous and lays one to seven eggs from April through July (Zug, Hedges, and Sunkel 1979). Hatchlings are 111 to 123 mm in total length and attain sexual maturity during the third year after hatching. *C. fissidens* shows a high frequency of broken tails (about 40% in most samples).

REMARKS: Individual, ontogenetic, and geographic variation in this species is extensive, especially in details of color pattern. Fisher (1969) recognized five mainland races and another on San Andrés Island (Colombia) based on scutellation and coloration. The features of these nominal forms show intergradation at intermediate localities on the mainland and considerable individual variation in head and ventral coloration, at least in some populations. The several characteristics and their states according to Fisher's analysis are as follows:

1. Supralabial light stripe that is bordered above by black: R = continues to rostral; E = ends at posterior margin of eye
2. Supralabial/infralabial dark markings (all stippled to some degree): O = each labial often or usually with a dark-centered ocellus; SM = each labial usually with a dark smudge or spot; N = labials stippled but without dark ocellus, smudge, or spot
3. Lateral dark stripes: L = a single lateral dark stripe or row of spots on each side; D = a lateral dark stripe or row of spots and a dorsolateral row of dark spots on each side
4. Venter (always some dark stippling on tips of ventrals): C = center clear; S = heavily stippled, often with row of small dots; W = usually some central stippling and some small spots; SP = distinct small dark spots; NSP = numerous distinct small dark spots
5. Scale rows: usually 19; usually 21
6. Supralabials: usually 7; usually 8

The several populations have the following features using the abbreviations indicated above; features in italics were regarded as unique to a population (by Fisher 1969):

I. Atlantic slopes of Mexico, west of the Isthmus of Tehuantepec (R, O, L, *W*, 19, 7)
II. Pacific slopes of Mexico, west of the Isthmus of Tehuantepec (R, O, L, C, 19, 8)
III. Atlantic slopes of Mexico, east of the Isthmus of Tehuantepec, Guatemala, Belize, Honduras, Nicaragua, and disjunct on both versants in central Panama (R, N, L, SP, 21, 8)
IV. San Andrés Island (R, N, L, *S*, 19, 8)
V. Pacific slopes of Mexico, east of the Isthmus of Tehuantepec south to Honduras; also at some Atlantic drainage localities in Honduras (Wilson and Meyer 1985) (*SM*, D, C, 21, 8)
VI. Costa Rica, western and extreme eastern Panama, west-

ern Colombia, and western Ecuador (*E*, O, L, *NSP*, 21, 8)

Material from Costa Rica now available agrees well with subdivision VI in always having the supralabial stripe terminating at the posterior margin of the eye and a single lateral dark stripe and in usually having the indicated scale counts. However, while most examples have ocellate labials and numerous definite dark ventral spots, there is much more individual variation in these features than suggested by Fisher's analysis. Some examples have the labials marked with numerous small dark spots, and others are uniformly lightly stippled with dark pigment. Ventral spotting is also variable, with some individuals approaching the W, SP, and S conditions. There are ontogenetic changes in coloration in Costa Rican examples. Juveniles (the smallest example is 155 mm in total length) lack labial and ventral dark stippling and have fewer, smaller, and less intense dark labial, chin, and ventral spots than larger examples. These markings all become more fully developed as size increases.

DISTRIBUTION: Lowlands and premontane zones from San Luis Potosí (Atlantic versant) and Michoacán (Pacific slope), Mexico, southward throughout Central America exclusive of southwestern and western Nicaragua to central Ecuador. In Costa Rica found on the Atlantic slope, on the Meseta Central (Oriental and Occidental), and in southwestern Costa Rica (2–800 m).

Coniophanes piceivittus Group

These snakes have a distinct light temporal stripe; venter immaculate; broad light dorsal stripes; hemipenes slightly bilobed, spinous, and capitate. Usually two species are recognized as belonging to this group—*Coniophanes piceivittis* and *Coniophanes schmidti* of the Yucatán Peninsula in Mexico, Guatemala, and Belize—but see remarks below.

Coniophanes piceivittus Cope, "1869," 1870

(plate 364; map 11.28)

DIAGNOSTICS: The only species in Costa Rica with a pair of broad longitudinal dorsolateral light stripes on a black background and the white ventral color continuing onto the lowermost 2½ scale rows.

DESCRIPTION: A small snake (maximum total length 571 mm); head distinct from neck; tail long (31 to 39% of total length); eye moderate; 1 or 2 preoculars, usually 2; 2 postoculars; supralabials 7 or 8, with 2 bordering orbit; infralabials 9 or 10; temporals 1 + 2; dorsal scales 23 to 25 anteriorly, usually in 25 (rarely 23) rows at midbody and reduced to 17 to 21 anterior to vent; adult males with supracloacal keels; ventrals 153 to 174; subcaudals 78 to 115; maxillary teeth 8 to 11 + 2. Body with a bold pattern of dark and light areas forming three black stripes (broad middorsal and paired lateral) and two dorsolateral light stripes; undersurfaces cream, essentially immaculate; tongue black.

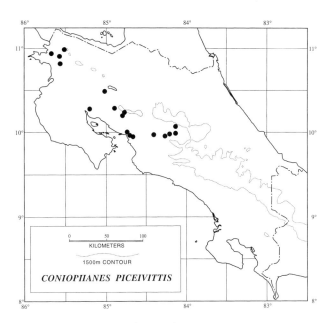

Map 11.28. Distribution of *Coniophanes piceivittis*.

The everted hemipenes are short (about 8 subcaudals in length), unicapitate, and slightly bilobed with a centrolineal bifurcate sulcus spermaticus; covered with moderate spines basally; capitulum calyculate.

SIMILAR SPECIES: (1) *Crisantophis nevermanni* has the supralabials marked by dark pigment (fig. 11.17b). (2) *Conophis lineatus* has six to eleven narrow dark dorsal and lateral stripes. (3) Striped species of *Urotheca* have narrow light stripes and the ventral tips dark.

HABITAT: Uncommon and found in deciduous, semideciduous, and seasonal evergreen forest areas, including Lowland Dry and Moist Forest and Premontane Moist Forest and marginally in Premontane Wet Forest.

REMARKS: Specimens called *Coniophanes imperialis* from Bebedero, Guanacaste Province, and subsequently referred to *C. piceivittus* by Dunn (in Bailey 1938) and Taylor (1951b) are examples of *Crisantophis nevermanni* (Dunn 1937b). Wilson and Meyer (1985), following the equivocal comments of Neill and Allen (1960), regarded *Coniophanes schmidti* of the Yucatán Peninsula as a geographic variant of *C. piceivittus*. I do not follow that course here, since intergradation between the two forms has not been conclusively demonstrated. *C. schmidti* has a narrow lateral dark stripe (one-half to one and one-half scale rows wide) that fades into the ground color lateroventrally. It occurs from the tip of the Yucatán Peninsula south through Belize into central Petén, Guatemala, and northern Chiapas, Mexico.

DISTRIBUTION: Lowlands and premontane areas from southern Tamaulipas and Guerrero, Mexico, to the Isthmus of Tehuantepec, thence eastward and southward through Central America to northwestern Costa Rica; also Atlantic drainage uplands of central Honduras and western Nicaragua. In Costa Rica restricted to the seasonal forests of northwestern Costa Rica and the Meseta Central Occidental (10–1,305 m).

Genus *Dipsas* Laurenti, 1768

A highly evolved genus of arboreal colubrid snakes (thirty species) that have a specialized jaw apparatus designed for feeding on slugs and snails. The head is short, blunt, extremely prominent, and very distinct from the narrow neck and strongly compressed body. The eyes are large and protuberant over the lip line and clearly visible from below. The basic complement of enlarged head shields is present, but these snakes lack a mental groove and the chin shields are rectangular rather than elongate as in most other snakes. Additional features characterizing the genus include nasal variable (undivided, partially divided, or completely divided); a loreal, bordering orbit; one or rarely two preoculars above loreal or no preocular present; pupil vertically elliptical; usually two to four pairs of rectangular chin shields, sometimes three unpaired ones; dorsal scales smooth in 11 to 19 rows (13 to 19 at midbody), without apical pits; anal entire; subcaudals divided; 10 to 26 maxillary teeth that are subequal or increase slightly in length posteriorly; Duvernoy's gland present; mandibular teeth decreasing in length posteriorly; no hypapophyses on posterior dorsal vertebrae. The absence of a mental groove and the large rectangular chin shields will distinguish members of this genus from all other Costa Rican snakes.

The everted hemipenes are short (7 to 8 subcaudals in length); slightly bilobed, with a bifurcate, centrolineal sulcus spermaticus; unicapitate, covered with small spines over truncus and large spines basally and terminating in a calyculate capitulum.

Members of this genus and its allies (*Sibon* of Mesoamerica and northern and northwestern South America, *Sibynomorphis* of central and southern South America, and *Tropidodipsas* of tropical Mexico and northern Central America) are all specialized cochleophagous predators. *Dipsas* and *Sibynomorphis* share a unique set of modifications in the jaw apparatus that allows them to grasp and swallow their slimy prey. These features include the inward orientation of the teeth on the upper jaw; the shortening of the pterygoid bone and the reduction or loss of a connection between that element and the jaw hinge so that the pterygoid no longer moves in unison with the lower jaw as in most snakes; the reduction or absence of pterygoid teeth; the presence of a functional hinge in the lower jaw so that the anterior portion can move independently of the rigid posterior part and the tip of the other mandible; a modification in the muscle responsible for closing the mouth (adductor mandibulae), which has a special elongate portion that inserts on the anterior part of the mandible. Apparently (Dunn 1951; Mertens 1952; Peters 1960) these last two features allow the snake to insert the tip of each mandible independently into the shell of a snail and then hook the recurved teeth into its soft body.

Typically these snakes grab the snail's soft body near the shell opening. As the snail retracts, the snake's lower jaws are pulled into the shell. Contraction of the specialized adductor muscles alternately retracts the mandibles to extract the snail gradually and neatly from its protective covering (Gans 1975; Sazima 1989). The tooth-bearing elements of the upper jaw fold inward during this process and slide over the shell, probably to prevent damage to the teeth. The other modifications appear to be designed to hold small slippery, wiggling prey in the mouth during the feeding process and effectively move them to the back of the mouth by independent operation of each mandible and pterygoid. These features seem to be equally effective in deshelling snails and in swallowing slugs and soft-bodied insects (occasionally found in the stomachs of these snakes; Beebe 1946).

Sibon and *Tropidodipsas* lack the extreme modifications characteristic of the other two genera, except for the alignment of the maxillary teeth, and appear to have a more primitive method of feeding on soft-bodied prey than their allies. These differences are reflected externally, since *Sibon* and *Tropidodipsas* usually have a more generalized colubrid head shape and always have a more or less well-developed mental groove. The head is short and blunt in *Dipsas*, and the mental groove is entirely absent in *Dipsas* and *Sibonymorphus*. All *Dipsas* are nocturnal arboreal snakes.

Peters (1960) recognized seven species groups within this taxon. All Costa Rican members of the genus belong to the *Dipsas articulata* group that also includes *Dipsas brevifacies* (Yucatán Peninsula of Mexico, Belize, and Guatemala), *D. gaigeae* (Colima, Mexico), *D. gracilis* (western Ecuador), *D. maxillaris* (Tabasco, Mexico), *D. temporalis* (northwestern South America and eastern Panama), and *D. viguieri* (Panama and northern South America).

The genus *Dipsas* has an apparently fragmented range in western Mexico (Colima and Jalisco) and the Yucatán Peninsula; it then occurs southward in the humid lowlands from eastern Honduras on the Atlantic slope and southwestern Costa Rica on the Pacific through Panama, and thence over most of tropical and subtropical South America (except for the Pacific slopes of Peru and Chile) to Amazonian basin Bolivia and Peru and southern Brazil, Paraguay, and extreme northern Argentina. Costa Rican representatives of the genus are rare in collections, probably because they are arboreal inhabitants of the rainforest canopy. Available samples include (number from Costa Rica in parentheses): *Dipsas articulata* 16 (10); *D. bicolor* 14 (12); and *D. tenuissima* 9 (7).

KEY TO THE COSTA RICAN SPECIES OF THE GENUS *DIPSAS*

1a. Two to four pairs of chin shields; dorsal pattern consisting of alternating dark and light rings completely encircling body 2
1b. Three large unpaired chin shields (fig. 11.14e); tricolor dorsal pattern consisting of alternating dark and white

rings, the latter split by red bands in life. . . . .
. . . . *Dipsas bicolor* (p. 597)

2a. Bicolor; posterior light rings heavily streaked with dark pigment on dorsum; one preocular, usually three postoculars. . . . . *Dipsas tenuissima* (p. 598)

2b. Tricolor red, cream, and dark brown to black rings; posterior light rings not, or only lightly, spotted with dark pigment on dorsum; usually no preocular, usually two postoculars. . . . . *Dipsas articulata* (p. 597)

Dipsas articulata (Cope, 1868a)

(map 11.29)

DIAGNOSTICS: The absence of a mental groove (fig. 11.14e), the rectangular chin shields, and the relatively unspotted light rings separate this form from all other tricolored Costa Rican snakes.

DESCRIPTION: A moderate-sized snake (maximum total length 712 mm); tail long (about 30% of total length); head distinct from neck; eyes large; nasal divided or not; no or rarely one preocular above loreal; 2 to 4 postoculars; sometimes 1 or 2 subpreoculars; supralabials 9 or 10, with usually 3, sometimes 4 bordering orbit; 11 or 14 infralabials; temporals usually 2 + 2 or 2 + 3; two to four pairs of chin shields; dorsal scale rows 15-15-15, vertebral row sometimes moderately enlarged; ventrals 196 to 214 (167 to 214); subcaudals 120 to 135 (116 to 135) in males, 108 to 113 (108 to 115) in females; ventrals plus subcaudals 309 to 334 (281 to 334). Tricolor monad pattern of alternating red–cream–dark–cream–red rings; 14 to 18 black or brown body rings; occasionally some irregular brown spotting in light rings; head mostly reddish brown with a light nuchal collar posterior to parietals; tail marked like body.

SIMILAR SPECIES: (1) *Dipsas tenuissima* is heavily

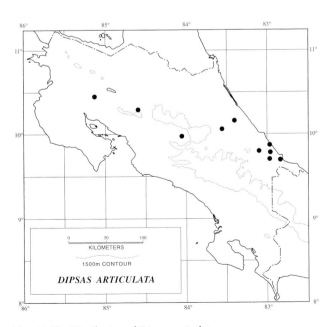

Map 11.29. Distribution of *Dipsas articulata*.

streaked and spotted with dark pigment in the light rings. (2) *Dipsas bicolor* has red bands in the light dorsal interspaces. (3) *Sibon anthracops* has red saddles in the light dorsal interspaces. (4) Other ringed species of *Sibon* have 26 to 45 body rings and bands. (5) *Micrurus mipartitus* has a cylindrical body with the head not distinct from the neck and has small, nonprotuberant eyes. See key to coral snakes (p. 706), figure 11.47, and table 11.2.

HABITAT: A rarely seen, apparently canopy-inhabiting species found in undisturbed Lowland Atlantic Moist and Wet Forests, but one example (UCR 7695) is from a gallery forest (Guanacaste: La Pacifica) within the Dry Forest zone of northwestern Costa Rica. Another record (AMNH 17370) said to be from the city of San José is questionable, but in light of the La Pacifica specimen the species may have occurred there in earlier times.

BIOLOGY: A strictly nocturnal, snail-eating snake. Schumacher (1996) reported on captive breeding. Two juveniles hatched from an undiscovered clutch. A second clutch consisted of three eggs, and a third laid fifty-three days later also was of three eggs. Eggs were 13–15 × 24–39 mm. The eggs in the second clutch hatched after eighty-five days. Hatchlings were about 180 mm in total length.

REMARKS: Schumacher (1996) misidentified his captive snakes as *Dipsas tenuissima*.

DISTRIBUTION: Lowland evergreen forests of Atlantic slope southeastern Nicaragua, Costa Rica, and northwestern Panama (4–100 m); one recorded from Pacific northwestern Costa Rica (40 m).

Dipsas bicolor (Günther, 1895)

(plates 365–66; map 11.30)

DIAGNOSTICS: The tricolor coloration of dark and light rings with red areas in the light rings, in combination with the absence of a mental groove and the unpaired rectangular chin shields (fig. 11.14e), will distinguish this form from all other snakes treated in this book.

DESCRIPTION: A small snake reaching a total length of 482 mm; tail very long (about 35% of total length); head distinct from neck; eye large; nasal entire or partially divided; one or two preoculars, if one preocular present, it lies above loreal; when two preoculars present, lower one appears to result from an anomalous vertical division of loreal; 2 to 4 postoculars; supralabials 9 or 10, usually 10, with 4 bordering orbit; infralabials 10 to 12; temporals 1 to 2 + 2 to 3 + 3 to 4; postmental sometimes present; three unpaired chin shields; dorsal scale rows 15-15-15, vertebral row sometimes slightly enlarged; ventrals 186 to 199; subcaudals 129 in male, 111 in female, with complete tails; ventrals plus subcaudals 299 to 328. Dorsal and ventral coloration consisting of alternating white and 14 to 17 reddish brown to black rings; white rings with a red saddle dorsally and sometimes stippled with brown.

SIMILAR SPECIES: (1) *Sibon annulatus* has a mental groove and 26 to 41 dark body rings and bands. (2) *Sibon*

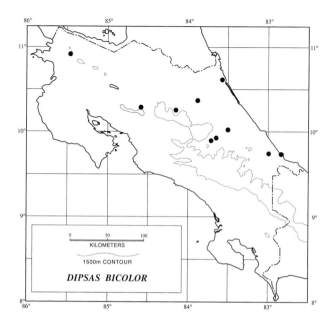

Map 11.30. Distribution of *Dipsas bicolor*.

anthracops has a mental groove and 19 to 25 black body rings. (3) *Sibon dimidiatus* has a mental groove and 30 to 45 body rings and bands. (4) *Micrurus mipartitus* has a cylindrical body, small, nonprotuberant eyes, and the head not distinct from the neck. See key to coral snakes (p. 706), figure 11.47, and table 11.2.

HABITAT: Apparently a rare species restricted to the canopy of undisturbed Lowland Atlantic Moist and Wet Forests and Premontane Rainforest.

REMARKS: This species was described as a *Neopareas* (Günther, 1895; type locality: Nicaragua: Chontales: Chontales Mine) from a preserved specimen (BMNH 94.10.1.39, male) that had lost the red dorsal saddles in preservative, hence the name *bicolor*. Brattstrom and Howell (1954) described a freshly preserved example as a new species, *Neopareas tricolor* (type locality: Honduras: El Paraíso: Arenal), because of the presence of the red saddles in life. There now seems little question that both snakes represent the same rare species (Peters 1960; Wilson and Meyer 1985). The last authors included Pacific versant Costa Rica in the range of this form based on an example (ZMH 5503) labeled San José, Costa Rica. This locality is suspected to be a point of shipment, not the collection site, since all other known records are from Atlantic locales.

DISTRIBUTION: Lowland and premontane evergreen forests on the Atlantic versant of southeastern Honduras, Nicaragua, and Costa Rica and marginally on the Pacific slope in northwestern Costa Rica (4–1,100 m).

Dipsas tenuissima Taylor, 1954b

(plate 367; map 11.31)

DIAGNOSTICS: The heavily pigmented light rings, combined with the absence of a mental groove (fig. 11.14e) and

the paired rectangular chin shields, make this form easy to recognize.

DESCRIPTION: A small snake reaching a total length of 555 mm; tail long (about 30% of total length); head distinct from neck; eyes large; nasal entire or partially divided; one preocular above loreal; usually three postoculars; a subocular; supralabials 8, with 2 bordering orbit; infralabials 9 or 10; temporals 2 + 3 + 3; two pairs of chin shields; dorsal scale rows 15-15-15, vertebral row enlarged; ventrals 225 to 228; subcaudals 118 to 128; ventrals plus subcaudals 346 to 353; pattern of alternating rings of cream or white and dark chocolate brown; light rings heavily streaked and spotted with dark pigment; upper head surfaces dark brown, no nuchal light collar.

SIMILAR SPECIES: (1) *Dipsas articulata* has few or no dark speckles or spots in the light rings. (2) *Dipsas bicolor* has red in the light dorsal interspaces. (3) *Sibon anthracops* has red saddles in the light dorsal interspaces. (4) Other ringed species of *Sibon* have the first pair of chin shields elongate. (5) Species of *Imantodes* lack rings around the body. (6) *Micrurus mipartitus* has a cylindrical body, small, nonprotuberant eyes, and the head not distinct from the neck. See key to coral snakes (p. 706), figure 11.47, and table 11.2.

HABITAT: Probably a canopy species; found in the rainforests of the Golfo Dulce (Lowland Pacific Wet Forest) and marginally in Premontane Rainforest.

REMARKS: One might suspect that the two allopatric taxa *D. tenuissima* and *D. viguieri* of eastern Panama are conspecific geographic variants of a single form. The differences between them are trivial (modal number of pre- and postoculars and supralabials and details of coloration),

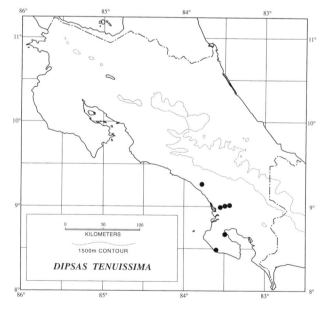

Map 11.31. Distribution of *Dipsas tenuissima*.

especially considering the amount of intraspecific variation in other species in this genus (Peters 1960).

DISTRIBUTION: The Gulfo Dulce rainforest lowland zone of southwestern Costa Rica and adjacent southwestern Panama (40–970 m).

Genus *Geophis* Wagler, 1830

A speciose genus (thirty-nine species) of small (to 400 mm in total length) secretive fossorial or semifossorial snakes showing considerable variation in scutellation, maxillary dentition, and hemipenial characters. Most members of this stock have the basic colubrid arrangement of enlarged head shields, but some forms have the internasals fused with the prefrontals or the supraoculars fused with the frontal or parietals. Other features characterizing the genus are head not distinct from neck; body cylindrical; tail short to moderate (7.4 to 22% of total length); nasal completely divided; a loreal; no, one, or two postoculars; pupil round; one or two pairs of chin shields; dorsal scales smooth or keeled, with or without paired apical pits, in 15 or 17 rows; anal entire; subcaudals divided; 6 to 17 maxillary teeth, subequal or increasing or decreasing in length posteriorly; no diastema; Duvernoy's gland present; mandibular teeth also variable in length; no hypapophyses on posterior dorsal vertebrae. The combination of entire anal, presence of a loreal, paired prefrontals, distinct pre- and postnasals, contact between the parietal and supralabial(s), and absence of a dorsal scale reduction will separate Costa Rican *Geophis* from other colubrid snakes known from the republic.

The hemipenes are 6 to 14 subcaudals in length and when everted may be simple or distinctly bilobed, with the sulcus spermaticus bifurcate and centrolineal; the pedicel bears numerous minute spinules, a few large spines, and a naked pocket (Downs 1967; Jenner 1981).

Members of this genus feed primarily on earthworms but also eat soft-bodied arthropod larvae (Cadle and Greene 1993).

Downs (1967) divided *Geophis* into seven species groups. Two of these groups are represented by Costa Rican forms: the *Geophis championi* and *Geophis sieboldi* groups.

Geophis range from Tamaulipas, Sinaloa, and Chihuahua, Mexico, southward throughout Central America to western and central Colombia, from the lowlands to montane situations.

KEY TO THE COSTA RICAN SPECIES OF THE GENUS *GEOPHIS*

1a. Supraocular shields distinct; parietal separated from orbit and prefrontal shield (fig. 11.30a); color of rostral and prenasals similar to that of adjacent plates, not contrasting with them 2

1b. No supraoculars; parietal shield enters orbit and meets prefrontal (fig. 11.30b); rostral and prenasal sometimes

whitish to yellow and strongly contrasting with dark adjacent plates 6

2a. Uppermost dorsal scales keeled (fig. 11.5), at least on posterior half of body and on tail 3

2b. Dorsal and caudal scales smooth (fig. 11.5), except for some faintly keeled scales above vent 5

3a. Dorsal scales on anterior half of body smooth; dorsum uniform dark; venter mostly dark, at least posteriorly 4

3b. Dorsal and caudal scales (exclusive of neck) distinctly keeled (fig. 11.5); dorsum uniform dark or often with light (usually red in life) lateral blotches, crossbands, or a lateral light stripe; venter yellowish white, sometimes immaculate, usually at least posterior ventrals with about equal parts dark (anteriorly) and light (posteriorly) producing a banded appearance . . . . *Geophis brachycephalus* (p. 602)

4a. Snout rounded; rostral barely produced posteriorly between internasals; frontal shorter than interparietal suture (fig. 11.30a,c); mental rounded . . . . *Geophis talamancae* (p. 604)

4b. Snout pointed; rostral markedly produced posteriorly; frontal one-third longer than interparietal suture (fig. 11.30d); mental pointed . . . *Geophis ruthveni* (p. 601)

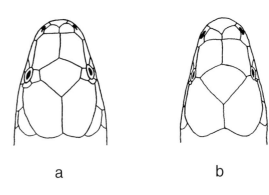

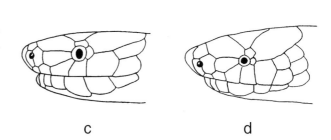

Figure 11.30. Diagnostic features of head scalation in Costa Rican snakes of the genus *Geophis*: (a) dorsal view of *G. brachycephalus* showing presence of supraocular scales; (b) dorsal view of *G. godmani*, which lacks supraocular scales; (c) lateral view of *G. brachycephalus* showing rounded snout; (d) lateral view of *G. ruthveni* illustrating pointed snout.

5a. Two supralabials posterior to orbit; supralabials usually 6, sometimes 5; ventrals plus subcaudals 178–91; venter mostly black with a few scattered irregular light markings in adults, faintly indicated in juveniles . . . . *Geophis zeledoni* (p. 604)

5b. One supralabial posterior to orbit; supralabials usually 5, sometimes 4 (fig. 11.11c); ventrals plus subcaudals 147–68; ventrals immaculate yellow white or with dark pigment only along anterior edges . . . . *Geophis hoffmanni* (p. 603)

6a. Uppermost dorsal scales keeled at least on posterior third of body and on tail (fig. 11.5); ventrals 122–33; subcaudals 41–46; color of rostral and prenasals similar to that of adjacent plates and not contrasting with them . . . . *Geophis downsi* (p. 600)

6b. Dorsal scales smooth (fig. 11.5); ventrals 132–45; subcaudals 26–38, color of rostral and prenasals usually whitish to yellow, strongly contrasting with dark color of adjacent plates . . . . *Geophis godmani* (p. 600)

Geophis championi Group

This stock is characterized by having the rostral extending posteriorly between the internasals, 7 to 9 maxillary teeth (first tooth at tip of bone), and short (8 to 10 subcaudals in length) simple noncapitate hemipenes. Four species are included in the group, with only *Geophis championi* of southwestern Panama extralimital to this account.

Geophis downsi Savage, 1981d

(map 11.32)

DIAGNOSTICS: One of several very similar small uniformly dark brown to black burrowing snakes, distinctive

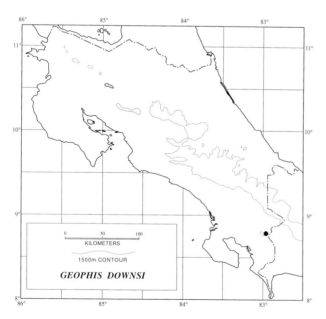

Map 11.32. Distribution of *Geophis downsi*.

in having keeled dorsal scales on the posterior third of the body and in lacking supraocular shields (fig. 11.30b) and a nuchal light collar.

DESCRIPTION: A rare small snake reaching 248 mm in total length; eye very small; tail moderate (17 to 22% of total length); no supraoculars; no postoculars; supralabials 5 or 6, with 2 or 3 bordering orbit; infralabials 5 to 6; anterior chin shields slightly longer than broad, posterior chin shields short; temporals 0 + 1; dorsal scales in 15 scale rows, strongly keeled on posterior portion of body and tail; without apical pits; ventrals 122 to 125 in males, 131 to 133 in females; subcaudals 45 to 46 in males, 41 in one female; ventrals plus subcaudals 168 to 174. Dorsum uniform dark brown or black, upper surface of head slightly lighter; venter white in juveniles, each ventral banded with black along anterior margin in adults.

SIMILAR SPECIES: (1) *Geophis godmani* usually has the snout white to yellow, the venter bright yellow in adults, and smooth dorsal scales (fig. 11.5). (2) *Geophis ruthveni* has a light nuchal collar and supraocular and postocular shields. (3) *Ninia psephota* has keeled scales over the entire body and the venter checkered with dark pigment (fig. 11.34g–i).

HABITAT: Found under surface debris in Premontane Rainforest.

DISTRIBUTION: Known only from Las Cruces, Puntarenas Province, in the premontane zone of Pacific southwestern Costa Rica (1,200 m).

Geophis godmani Boulenger, 1894

(plate 368; map 11.33)

DIAGNOSTICS: This relatively common species might be confused with several other small species having a uniform dark brown to black dorsal coloration, smooth scales, and no nuchal collar. However, it is readily separated from similar forms by lacking supraocular shields (fig. 11.30b). In addition, most examples have the tip of the snout light in color.

DESCRIPTION: A small snake (401 mm maximum known total length); eye small; tail short to moderately long (about 17% of total length in males, about 13% in females); no supraoculars; postocular present or absent; supralabials 6, with 2 or 3 bordering orbit; infralabials 6; anterior chin shields squarish, posterior pair barely enlarged; temporals 0 + 1, dorsal scales in 15 smooth rows, without apical pits; ventrals 132 to 143 in males, 133 to 145 in females; subcaudals 34 to 38 in males, 26 to 28 in females; ventrals plus subcaudals 162 to 180. Dorsum uniform dark brown or black; snout tip white; venter white in juveniles, bright yellow in adults, uniform or with scattered dark spots or irregular transverse bands.

SIMILAR SPECIES: (1) *Geophis downsi* has the snout colored similarly to the rest of the head and dorsum and has keeled scales on the posterior body and tail (fig. 11.5). (2) *Geophis ruthveni* has the snout colored like the upper

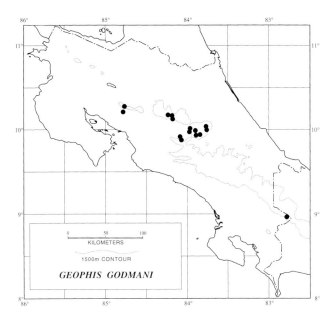

Map 11.33. Distribution of *Geophis godmani.*

head surface and body and the venter marked with transverse bands. (3) *Ninia psephota* has keeled scales (fig. 11.5) and the venter checkered with dark pigment (fig. 11.34g–i).

HABITAT: Most commonly collected under logs or rocks in pastures cleared from Lower Montane Wet Forest and Rainforest, but also found in these forests as well; marginally in Premontane Wet Forest.

REMARKS: For some time I thought that a specimen (BMNH 1913.7.19.144) of this species labeled "Escazú," a place a few kilometers west of the city of San José at 1,100 m, was almost certainly collected from the mountain ridge just to the south, sometimes called the Fila de Escazú (Cerros de Candelaria). However, another example (UCR 7269) was collected recently at San Rafael de Escazú (1,033 m), suggesting that a population of *G. godmani* may be established in the Escazú area. The other lowest locality for the species in Costa Rica is at 1,734 m, and no other examples are known from the Meseta Central.

DISTRIBUTION: Lower montane evergreen forests of the cordilleras of Costa Rica and western Panama (1,600–2,100 m) and apparently in the Escazú area (premontane zone) on the southern Meseta Central Occidental of Costa Rica (1,000–1,100 m).

Geophis ruthveni Werner, 1925

(map 11.34)

DIAGNOSTICS: One of several small uniform dark brown to black snakes having a light nuchal collar (in juveniles) but differing from the others by having keeled scales (fig. 11.5) on only the posterior half of the body and tail.

DESCRIPTION: A rare small snake attaining a total length of 260 mm; eye very small; tail short to moderate (about

18% of total length in males, about 14% in females); supraoculars present; a postocular; supralabials 6, with 2 bordering orbit; infralabials 5 to 7; anterior chin shields almost as broad as long, slightly longer than posterior ones; temporals 0 + 1; dorsal scales in 15 rows, strongly keeled on posterior half of body and tail, with paired apical pits; ventrals 122 to 126 in males, 130 to 142 in females; subcaudals 37 to 41 in males, 32 to 35 in females; ventrals plus subcaudals 163 to 176. Upper surfaces of head, body, and tail dark brown to bluish black; a light collar involves supralabial 6, posterior parts of parietals, and first 3 to 4 rows of nuchal scales in juveniles; venter mostly dark, with each ventral having a light posterior margin; iris and tongue black.

SIMILAR SPECIES: (1) *Geophis brachycephalus* often is marked with red blotches, crossbands, or lateral stripes and has a rounded snout and a higher number for the ventral plus subcaudal count (169 to 186 in Costa Rica). (2) *Geophis downsi* lacks a light nuchal collar and supraocular (fig. 11.30b) and postocular shields. (3) *Geophis godmani* usually has a light snout and lacks supraocular plates (fig. 11.30b). (4) *Geophis talamancae* has a rounded snout and small frontal. (5) *Ninia celata* has keels on all body and tail scales (fig. 11.5). (6) *Ninia psephota* has keels on all body and tail scales (fig. 11.5) and the venter checkered with dark pigment (fig. 11.34g–i). (7) Species of *Enulius* have smooth dorsal scales (fig. 11.5).

HABITAT: Under surface debris or rocks in Lowland Atlantic Moist Forest and Premontane Rainforest.

REMARKS: Although the holotype (NHMW 16508) was reported as being from "Sarapigui, Brasil" (Werner 1925), Dunn (in Savage 1960b) and Downs (1967) concur that this is a transcription, probably of Sarapiquí, Costa Rica. The

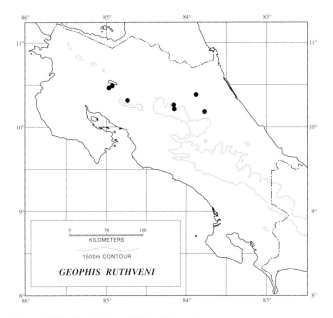

Map 11.34. Distribution of *Geophis ruthveni.*

Sarapiquí region was visited by many collectors over the years (Savage 1974a), with most examples labeled with that name coming from Cariblanco de Sarapiquí (830 m). However, the old oxcart road from Puerto Viejo to Heredia, across the Cordillera Central, passes through Cariblanco, so collectors often worked all along this road up to the pass (1,936 m) near Vara Blanca. *G. ruthveni* has been taken at Cinchona (1,360 m) on this road and most recently near Cariblanco (853 m). Downs (1967) showed that the type specimen for this name is of the same species as available Costa Rican material. The record of this species from the Río Peñas Blancas valley, Alajuela Province (Hayes, Pounds, and Timmermann 1989), is based on an example (CRE 7481) of *Geophis brachycephalus*, but Michael Fogden informs me that it occurs there.

DISTRIBUTION: Evergreen forests of the northeastern Atlantic lowlands and adjacent premontane slopes of Costa Rica (85–1,360 m).

Geophis sieboldi Group

A cluster of allied species distinct in having the rostral not projecting posteriorly between the internasals; 8 to 15 maxillary teeth (anterior tip of bone pointed, toothless), and short slightly bilobed, unicapitate hemipenes (6 to 10 subcaudals in length). This group ranges from upland and Pacific drainage tropical Mexico to northwestern Colombia. Extralimital species include *Geophis laticollaris, G. petersi, G. pyburni, G. russatus, G. sallaei,* and *G. sieboldi* from Mexico; *G. nasalis* from Pacific slope Chiapas, Mexico, and Guatemala, *G. dunni* from Nicaragua, and *Geophis nigroalbus* from eastern Panama and Colombia.

Geophis brachycephalus (Cope, 1871)

(plates 369–70; map 11.35)

DIAGNOSTICS: This extremely variable species often has substantial red areas in the dorsal coloration and most of the scales on the body keeled (fig. 11.5). Because of the variation in coloration, positive identification of uniformly dark brown to black individuals or juveniles with light (not red) nuchal collars may require counting the dorsal scale rows (15 in this species) and/or ventrals.

DESCRIPTION: A small snake (known maximum total length 460 mm); females average longer than males; eye moderate; tail short to moderate (about 19% of total length in males, about 14% in females); supraoculars present; 1 postocular; supralabials 6, with 2 bordering orbit; infralabials 6 or 7; anterior chin shields short and broad, posterior ones smaller than anterior shields; temporals 0 + 1; dorsal scales in 15 rows, smooth on neck, keeled over most of the body and tail, with paired apical pits; ventrals 131 to 148 (119 to 148) in males, 135 to 145 (123 to 153) in females; subcaudals 36 to 48 (35 to 51) in males, 30 to 39 in females; ventrals plus subcaudals 169 to 186 (154 to 185). Dorsal ground color brown, bluish gray, or black; some ju-

veniles with a light nuchal collar; dorsum uniform, with an irregular red orange to brick red lateral stripe or red lateral blotches; blotches sometimes confluent across back to form red crossbars; light components of pattern usually obliterated by dark pigment in large adults; venter flesh white to gray, occasionally immaculate but usually banded with dark pigment along the anterior edge of each ventral at least on the posterior part of the body; iris and tongue black.

SIMILAR SPECIES: (1) *Geophis ruthveni* lacks any red coloration, has a pointed snout, has keeled scales only on the posterior half or third of body (fig. 11.5), and has total segmental counts of 163 to 167 (ventrals plus subcaudals). (2) *Geophis talamancae* has smooth dorsal scales (fig. 11.5) on the anterior half of the body. (3) *Ninia psephota* has the venter checkered with black (fig. 11.34g–i) and 17 dorsal scale rows. (4) *Ninia celata* sometimes has a light nuchal collar (fig. 11.33c) and 19 dorsal scale rows. (5) *Enulius* species have smooth dorsal scales (fig. 11.5).

HABITAT: Commonly found under surface litter, logs, and rocks in pastures and disturbed sites, as well as in the forest proper in areas supporting or formerly supporting Lowland Wet Forest and Premontane and Lower Montane Moist and Wet Forest and Rainforest. Rarely found in the lowlands, and on the Atlantic versant it always occurs far inland.

BIOLOGY: A very common species. Fitch (1970) reported on gravid females from Volcán Barva, Heredia Province, taken in February and nongravid females and a hatchling collected in March from the pass between Volcanes Barva and Poás. Sasa (1993) collected gravid females from under tree trunks or in loose soil at Las Nubes on the northwest slope of Volcán Irazú, San José Province. Six were kept in

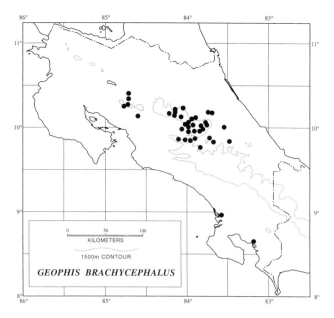

Map 11.35. Distribution of *Geophis brachycephalus*.

the laboratory and laid clutches of three to six (mean 4.2) eggs averaging about 23 × 10 mm. Another female laid four eggs on January 22, three of which hatched 109 days later. The hatchlings were 118 to 129 mm in total length.

REMARKS: Variation in this extremely variable species has been reviewed by Downs (1967). Taylor's records (1951b, 1954b) under the names *Geophis bakeri*, *G. brachycephala*, *G. dolicocephala*, and *G. moesta* are based on specimens of *G. brachycephalus*. Downs (1967) placed *Geophis nigroalbus* Boulenger, 1908 (Colombia: Department of Valle: Pavas) in the synonymy of *G. brachycephalus* and placed Colombian specimens in the latter species. I remain dubious about the status of specimens from eastern Panama and Colombia, some or all of which will be referred to *G. nigroalbus*, a probably valid form (see Downs 1967, note 146). For this reason I regard *G. brachycephalus* as restricted to Costa Rica and western Panama.

DISTRIBUTION: Lowland, premontane, and lower montane zones of Costa Rica and western Panama. In Costa Rica on the northeastern Atlantic and southwestern Pacific lowlands and in the Cordillera Central and Cordillera de Talamanca (13–2,115 m).

Geophis hoffmanni (Peters, 1859)

(plates 371–72; map 11.36)

DIAGNOSTICS: This is another of several essentially uniform dark brown to black snakes (juveniles with a light nuchal collar) with smooth dorsal scales in 15 rows. It is a common and widespread form that differs from other similar species by having the following additional features: a short tail, lower two dorsal scale rows same color as dorsum, and relatively few ventral scales (118 to 130).

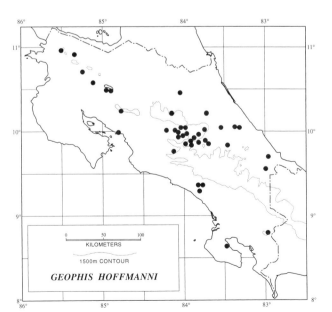

Map 11.36. Distribution of *Geophis hoffmanni*.

DESCRIPTION: A small snake with a maximum known total length of 300 mm, females average longer than males; eye small; tail short (13 to 15% of total length); supraoculars present; usually 1 postocular; supralabials usually 5, rarely 4, with 2 bordering orbit, only 1 posterior to orbit; infralabials 6; two pairs of distinct chin shields; no elongate temporals; dorsal scales in 15 rows, all smooth except a few above vent, with paired apical pits; ventrals 118 to 130 (117 to 130) in males, 122 to 125 (122 to 135) in females; subcaudals 28 to 37 in males, 24 to 32 in females; ventrals plus subcaudals 147 to 168. Dorsum uniform dark brown to black, with a yellowish nuchal collar in small juveniles; venter cream, white, yellowish white, or gray, immaculate or with dark pigment on anterior edge of each ventral and dorsal coloration encroaching onto tips of ventrals; subcaudals similarly colored; iris black.

SIMILAR SPECIES: (1) *Geophis zeledoni* has the venter uniformly dark or with scattered irregular light markings on a dark ground color. (2) Other similar *Geophis* have at least one-half to one-third of the body and the tail with keeled scales (fig. 11.5). (3) *Trimetopon simile* has a single prefrontal (fig. 11.38) and consistently a broad nuchal light collar. (4) Other *Trimetopon* have at least vaguely indicated dorsal dark stripes. (5) *Amastridium veliferum* has a chestnut head and nape. (6) *Enulius* have at least the tip of the frontal and all of the parietals involved in the nuchal collar (fig. 11.33a,b).

HABITAT: A very common species likely to be found under any surface litter, trash, logs, or rocks in pastures, secondary growth, or coffee fincas or within disturbed or undisturbed forests; frequently present in gardens on the Meseta Central. Occurs in areas that currently support or were formerly covered by a variety of forest vegetation (Lowland Moist and Wet Forests, Premontane Moist and Wet Forests and Rainforest, and Lower Montane Wet Forest).

REMARKS: *Geophis acutirostris* Taylor, 1954b (type locality: Costa Rica: Cartago: Cot), was regarded by Downs (1967) as an aberrant *G. hoffmanni*. In the type and only known specimen (KU 34670) of the former the internasals are fused with the prefrontals and there is no postocular. Both conditions are found in individuals typical of *G. hoffmanni* in all other respects. Since these kinds of fusions are not unusual in burrowing snakes, there seems little reason to question Downs's conclusion. Since complete or partial fusions of the head shields occur with some frequency in the large sample Downs examined (seventy-three snakes), this potential should be taken into account in identifying *G. hoffmanni*, its congeners, and other fossorial or semifossorial species.

DISTRIBUTION: Lowland and upland areas from eastern Honduras southward on the Atlantic versant to central Panama; also Pacific central and southwestern Costa Rica and adjacent southwestern Panama. In Costa Rica the species occurs on the Atlantic and Pacific southwestern lowlands and the Meseta Central (Oriental and Occidental) and the slopes of the Cordillera de Tilarán, Cordillera Central, and Cordillera de Talamanca in the premontane and lower montane zones (18–670 m).

Geophis talamancae Lips and Savage, 1994

(map 11.37)

DIAGNOSTICS: This form is uniform black in dorsal coloration and has a rounded snout and keeling restricted to the dorsal scales on the posterior half of the body and on the tail.

DESCRIPTION: A small form (218 mm in total length); eye small, pupil round; tail short (16% of total length); supraoculars present; 2 postoculars; 6 supralabials, with 3 or 4 bordering orbit; 6 infralabials; anterior chin shields slightly longer than broad, posterior pair short; no elongate temporals; 15 dorsal scale rows without apical pits; dorsal scales on anterior half of body smooth, those on posterior body and tail keeled; ventrals in female 138; subcaudals in female 33; ventrals plus subcaudals 171. Dorsum uniform charcoal gray to black; venter light anteriorly, mostly dark posteriorly, subcaudals black.

SIMILAR SPECIES: (1) *Geophis brachycephalus* has most of the dorsal scales keeled (fig. 11.5) and usually is marked with light (often red) bars, spots, and/or stripes. (2) *Geophis hoffmanni* and *G. zeldoni* have keeled dorsal scales only in the area near the vent. (3) *Geophis ruthveni* has a markedly pointed snout and a large frontal (fig. 11.30b,d). (4) *Ninia celata* often has a light nuchal collar (fig. 11.33c) and 19 dorsal scale rows. (5) *Ninia psephota* has a checkered venter (fig. 11.34g-i) and 17 dorsal scale rows. (6) *Enulius* species have a light collar or prominent light areas on the head (fig. 11.33a,b) and smooth dorsal scales (fig. 11.5).

HABITAT: Known from a single specimen in the Lower Montane Rainforest.

REMARKS: As pointed out in the original description (Lips and Savage 1994), there is a slight possibility that the type (CRE 5343, an adult female) had been accidentally transported from elsewhere in the Cordillera de Talamanca to the type locality (Puntarenas: Las Tablas). Another specimen similarly colored had been captured a year earlier but escaped. The only other *Geophis* from the area is *G. godmani*, which has a bright yellow venter.

DISTRIBUTION: Known only from the type locality at 1,880 m in the Cordillera de Talamanca of southwestern Costa Rica.

Geophis zeledoni Taylor, 1954b

(map 11.38)

DIAGNOSTICS: In this relatively rare species the venter is uniformly dark gray to black or heavily marked with those colors and is readily distinguished from the several other small snakes that have uniformly dark brown or black dorsums and light venters. *G. zeledoni* is also distinct from other similar species that have dark bands across each ventral or tessellated ventral markings.

DESCRIPTION: A small species (417 mm in maximum total length); eye moderate, pupil round; tail moderate (17 to 20% of total length); supraoculars present; 1 or rarely 2 postoculars; supralabials usually 6, sometimes 5, with 2 bordering eye, 2 posterior to orbit; 6 infralabials, sometimes 5 or 7; anterior chin shields short, scarcely longer than broad, posterior pair small, scarcely distinguishable from gulars; no elongate temporals; 15 rows of smooth scales, with paired apical pits; scales on posterior 20% of body and on base of tail lightly keeled; ventrals 141 to 146 in males,

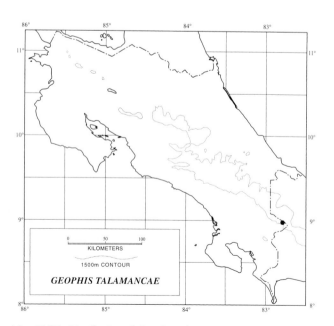

Map 11.37. Distribution of *Geophis talamancae*.

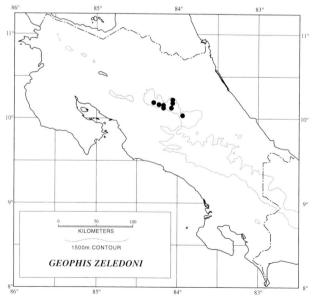

Map 11.38. Distribution of *Geophis zeledoni*.

Plate 331. *Liotyphlops albirostris*. Panama: Colón: Gamboa, about 26 m.

Plate 332. *Typhlops costaricensis*. Costa Rica: Puntarenas: Monteverde, about 1,400 m.

Plate 333. *Leptotyphlops ater*. Costa Rica: Guanacaste: La Pacifica, about 40 m.

Plate 334. *Loxocemus bicolor*. Costa Rica: Guanacaste: Santa Rosa National Park, about 280 m.

Plate 335. *Boa constrictor*. Costa Rica: Heredia: La Selva Biological Station, about 60 m.

Plate 336. *Boa constrictor*. Costa Rica: Heredia: La Selva Biological Station, about 60 m.

Plate 337. *Corallus annulatus*. Costa Rica: Heredia: La Selva Biological Station, about 60 m.

Plate 338. *Corallus hortulanus*. Costa Rica: Puntarenas: Corcovado National Park, about 40 m.

Plate 339. *Epicrates cenchria*; adult. Costa Rica: Guanacaste: Santa Rosa National Park, about 280 m.

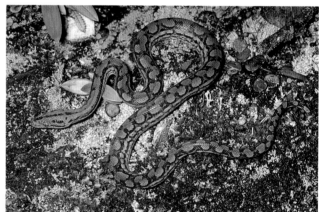

Plate 340. *Epicrates cenchria*; young adult. Costa Rica: Guanacaste: Santa Rosa National Park, about 280 m.

Plate 341. *Ungaliophis panamensis*. Costa Rica: Heredia: La Selva Biological Station, about 60 m.

Plate 342. *Clelia clelia*; adult. Costa Rica: Heredia: La Selva Biological Station, about 60 m.

Plate 343. *Clelia clelia;* juvenile. Costa Rica: Heredia:
La Selva Biological Station, about 60 m.

Plate 344. *Clelia scytalina;* adult. Costa Rica: Alajuela:
Monteverde Cloud Forest Preserve, about 1,600 m.

Plate 345. *Clelia scytalina;* juvenile. Costa Rica: Alajuela:
Monteverde Cloud Forest Preserve, about 1,600 m.

Plate 346. *Oxyrhopus petolarius.* Costa Rica: Heredia:
La Selva Biological Station, about 60 m.

Plate 347. *Oxyrhopus petolarius;* view of head. Costa Rica:
Heredia: La Selva Biological Station, about 60 m.

Plate 348. *Tripanurgos compressus.* Costa Rica: Puntarenas:
Corcovado National Park, about 40 m.

Plate 349. *Erythrolamprus bizona* eating a *Leptodeira septentrionalis*. Costa Rica: Alajuela: Monteverde Cloud Forest Preserve, about 1,600 m.

Plate 350. *Erythrolamprus mimus*. Costa Rica: Alajuela: Peñas Blancas Valley, 800 m.

Plate 351. *Erythrolamprus mimus* killing a *Dendrophidion vinitor*. Costa Rica: Heredia: La Selva Biological Station, about 60 m.

Plate 352. *Liophis epinephalus*. Costa Rica: Alajuela: Monteverde Cloud Forest Preserve, about 1,600 m.

Plate 353. *Liophis epinephalus;* view of head. Panama.

Plate 354. *Xenodon rabdocephalus*. Costa Rica: Heredia: La Selva Biological Station, about 60 m.

Plate 367. *Dipsas tenuissima*. Costa Rica.

Plate 368. *Geophis godmani*. Costa Rica: Alajuela: Monteverde Cloud Forest Preserve, about 1,600 m.

Plate 369. *Geophis brachycephalus*. Costa Rica: Puntarenas: Monteverde, about 1,400 m.

Plate 370. *Geophis brachycephalus*. Costa Rica: San José: Las Nubes, about 1,800 m.

Plate 371. *Geophis hoffmanni*. Costa Rica: Heredia: La Selva Biological Station, about 60 m.

Plate 372. *Geophis hoffmanni*; lateral view of head. Costa Rica: Heredia: La Selva Biological Station, about 60 m.

Plate 373. *Hydromorphus concolor.* Costa Rica: Alajuela: Peñas Blancas Valley, about 800 m.

Plate 374. *Imantodes cenchoa.* Costa Rica: Puntarenas: Rincón de Osa, about 40 m.

Plate 376. *Imantodes gemmistratus.* Costa Rica: Guanacaste: Santa Rosa National Park, about 280 m.

Plate 375. *Imantodes cenchoa* killing a *Norops;* spider just passing by. Costa Rica: Heredia: La Selva Biological Station, about 60 m.

Plate 378. *Imantodes inornatus:* Costa Rica: Puntarenas: Corcovado National Park, about 30 m.

Plate 377. *Imantodes inornatus:* Costa Rica: Puntarenas: Corcovado National Park, about 30 m.

Plate 379. *Leptodeira annulata.* Costa Rica: Heredia: La Selva Biological Station, about 60 m.

Plate 380. *Leptodeira rubricata.* Costa Rica: San José: Tárcoles, 2 m.

Plate 381. *Leptodeira nigrofasciata.* Costa Rica: Guanacaste: La Pacifica, 40 m.

Plate 382. *Leptodeira nigrofasciata;* view of head. Costa Rica: Guanacaste: near Liberia, 40 m.

Plate 383. *Leptodeira septentrionalis* eating eggs of a leaf frog, *Agalychnis*. Costa Rica: Puntarenas: Cartago, about 40 m.

Plate 384. *Leptodeira septentrionalis;* juvenile, note white nuchal collar. Costa Rica: Heredia: La Selva Biological Station, about 60 m.

Plate 385. *Ninia celata*. Costa Rica: Limón: Guayacán, about 750 m.

Plate 386. *Ninia maculata*. Costa Rica: Heredia: La Selva Biological Station, about 60 m.

Plate 387. *Ninia psephota*. Costa Rica: Heredia: Braulio Carrillo National Park, about 1,800 m.

Plate 388. *Ninia sebae*. Costa Rica: Heredia:
La Selva Biological Station, about 60 m.

Plate 389. *Rhadinaea calligaster*. Costa Rica: Alajuela:
Monteverde Cloud Forest Preserve, about 1,600 m.

Plate 390. *Rhadinaea decorata*. Costa Rica: Heredia:
La Selva Biological Station, about 60 m.

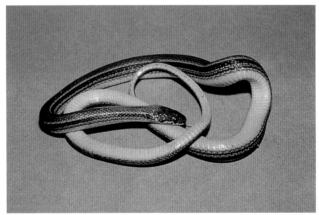

Plate 391. *Rhadinaea godmani*. Costa Rica: Cartago:
Tres Ríos, 1,345 m.

Plate 392. *Rhadinaea serperaster*. Costa Rica: Puntarenas:
Monteverde, about 1,400 m.

Plate 393. *Rhadinaea pulveriventris*. Costa Rica.

Plate 394. *Sibon annulatus*. Costa Rica: Heredia: La Selva Biological Station, about 60 m.

Plate 395. *Sibon* sp. Costa Rica: Limón: Guayacán, about 650 m.

Plate 396. *Sibon anthracops*. Costa Rica: Guanacaste: Santa Rosa National Park, about 280 m.

Plate 397. *Sibon dimidiatus*. Costa Rica: Puntarenas: Buenos Aires area, 350 m.

Plate 398. *Sibon argus*. Costa Rica: Limón: Guayacán, about 550 m.

Plate 399. *Sibon longifrenis*. Costa Rica: Heredia: La Selva Biological Station, about 60 m.

Plate 400. *Sibon nebulatus*. Costa Rica: Heredia: La Selva Biological Station, about 60 m.

Plate 401. *Tretanorhinus nigroluteus*. Honduras: Yoro: north of Lago de Yojoa, about 750 m.

Plate 402. *Trimetopon gracile*. Costa Rica: Alajuela, Monteverde Cloud Forest Preserve, about 1,600 m.

Plate 403. *Urotheca decipiens*. Costa Rica: Puntarenas: Monteverde, about 1,400 m.

Plate 404. *Urotheca decipiens*. Costa Rica: Heredia: La Selva Biological Station, about 60 m.

Plate 405. *Urotheca euryzona*; adult. Costa Rica: Heredia: La Selva Biological Station, about 60 m.

Plate 406. *Urotheca euryzona*; juvenile. Costa Rica: Heredia: La Selva Biological Station, about 60 m.

Plate 407. *Urotheca guentheri*. Costa Rica: Alajuela: Monteverde Cloud Forest Preserve, about 1,600 m.

Plate 408. *Urotheca pachyura*. Costa Rica: Cartago: Tapantí National Park, about 1,200 m.

Plate 409. *Chironius carinatus*. Costa Rica: Limón: Valle de Estrella, about 35 m.

Plate 410. *Chironius exoletus*. Costa Rica: Puntarenas: Monteverde, about 1,400 m.

Plate 411. *Chironius grandisquamis*. Costa Rica: Heredia: La Selva Biological Station, about 60 m.

Plate 412. *Coluber mentovarius*. Costa Rica: Guanacaste: Palo Verde National Park, about 10 m.

Plate 413. *Dendrophidion nuchale*. Costa Rica: Puntarenas: San Luis Valley, about 800 m.

Plate 414. *Dendrophidion nuchale*. Costa Rica: Limón: Guayacán, about 650 m.

Plate 415. *Dendrophidion nuchale*; view of head. Costa Rica: Puntarenas: Cajón, 687 m.

Plate 416. *Dendrophidion vinitor* sleeping at night. Costa Rica: Heredia: La Selva Biological Station, about 60 m.

Plate 417. *Dendrophidion paucicarinatum*. Costa Rica: Alajuela: Monteverde Cloud Forest Preserve, about 1,600 m.

Plate 418. *Dendrophidion percarinatum*. Costa Rica: Puntarenas: Rincón de Osa, about 40 m.

Plate 419. *Drymarchon corais*. Costa Rica: Puntarenas: Monteverde, about 1,400 m.

Plate 420. *Drymobius margaritiferus*. Costa Rica: Puntarenas: Monteverde, about 1,400 m.

Plate 421. *Drymobius melanotropis;* adult. Costa Rica: Alajuela: Peñas Blancas Valley, about 800 m.

Plate 422. *Drymobius melanotropis;* juvenile. Costa Rica: Alajuela: Peñas Blancas Valley, about 800 m.

Plate 423. *Drymobius rhombifer*. Costa Rica: Heredia: Horquetas, 50 m.

143 to 150 in females; 39 to 46 subcaudals in males, 33 to 43 in females; ventrals plus subcaudals 178 to 191. Dorsum uniform grayish to brownish black or marked with some light spots and narrow light bands and a light lateral stripe; venter predominately black, uniform or with irregular light blotches or diffuse mottled areas, never banded. Probably a narrow nuchal light ring in juveniles.

SIMILAR SPECIES: (1) *Geophis hoffmanni* has the venter light or with dark pigment only on the anterior edges of the ventrals. (2) Other similar *Geophis* have at least one-half to one-third of the body and the tail strongly keeled (fig. 11.5). (3) *Amastridium veliferum* has a chestnut head and neck. (4) The species of *Enulius* have part or all of the parietals involved in the nuchal ring (fig. 11.33a,b).

HABITAT: Under rocks and litter in pastures or forest remnants in the Lower Montane Wet Forest and Rainforest communities.

REMARKS: With the exception of the unusual specimen described below, all known examples of this species are uniform in dorsal coloration. A single snake (UF 10438) from above the Tres Ríos hydroelectric station on Volcán Irazú, Cartago Province (1,830 m), agrees with other material of *G. zeledoni* in having the dorsal and caudal scales smooth except in the vent region, 6 supralabials, 146 ventrals, and 33 subcaudals. In coloration this example resembles some specimens of *G. brachycephalus* in having light markings on the dorsum. There is a faint light nuchal collar, light spots, and narrow light bands on the anterior third of the body, light spots above the precloacal and vent area, and a light longitudinal stripe running down each flank. One suspects that the light areas were red or orange in life. The venter appears to be uniformly dark.

DISTRIBUTION: Known only from the lower montane zone on Volcanes Barba, Poás, and Irazú, Cordillera Central, Costa Rica (1,830–2,100 m).

Genus *Hydromorphus* Peters, 1859

A semiaquatic genus (two species) of moderate-sized snakes having the nostrils directly dorsally and an odd number of prefrontal head shields (one or three). Other features that in combination will distinguish *Hydromorphus* from other colubrid genera are: nasal entire or partially divided below the nostril; a loreal; usually one preocular, sometimes two or preoculars absent; pupil round (fig. 11.10b); two pairs of chin shields; one to several median gulars; mental groove variously developed, usually running between first infralabials and anterior chin shields, often extending part or entire length of chin shields, rarely evident only between first infralabials; tubercles on snout and chin; dorsal scales smooth in 15 to 17 rows at midbody, with two apical pits; supracloacal keels in only known specimen (a male) of *Hydromorphus dunni* (CAS 78939); anal and subcaudals divided; maxillary teeth 13 to 14, increasing in length posteriorly, no diastema or enlarged fangs; Duvernoy's gland probably present; mandibular teeth subequal.

The everted hemipenes are short (reaching to subcaudal 11), and distinctly bilobed, with a bifurcate, centrolineal sulcus spermaticus; unicapitate, with two large basal hooks and large spines over the pedicel; capitulum spinous, calyculate distally (Crother 1989).

The genus is known from the Atlantic versant of Guatemala, Honduras, Nicaragua, Costa Rica, and western Panama, Pacific versant Costa Rica and western Panama, and on both slopes in central Panama.

Hydromorphus concolor Peters, 1859

(plate 373; map 11.39)

DIAGNOSTICS: The dorsally oriented nostrils (fig. 11.10b), smooth scales, and cylindrical body will separate this species from all other snakes from Costa Rica.

DESCRIPTION: A moderate-sized snake (total length to 797 mm), with a broadly rounded head that is barely distinct from body; body cylindrical; tail short to moderate (15 to 19% of total length in males, 13 to 15% in females); eye very small; rostral large, almost twice as broad as high; usually a single internasal, sometimes only partially fused or rarely two internasals; 1 prefrontal; a loreal, usually bordering orbit but excluded from orbit when 2 preoculars present; 0 to 2 preoculars; postoculars almost always 2, very rarely 1 or 3; temporals usually 1 + 2, rarely 1 + 1; supralabials 5 to 7, usually 6, with 1 bordering orbit; infralabials 6 to 9, usually 8 or 9; dorsal scales 15 to 21–15 to 17–15 to 17, usually 19-17-15; no supracloacal keels in males; ventrals 157 to 186; subcaudals 31 to 54; upper surfaces usually uniform grayish brown, somewhat paler on scale rows 1 and 2; some individuals have scattered small dark spots, others have paired dorsolateral dark blotches on body and tail, and a rare variant has distinct paired dark stripes on

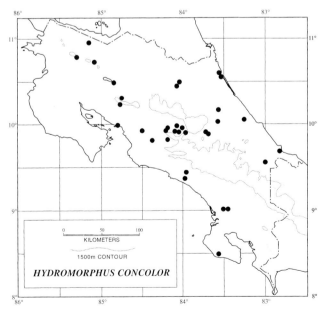

Map 11.39. Distribution of *Hydromorphus concolor.*

each lateral surface on the anterior portion of body; venter cream, usually with a few indistinct brownish smudges posteriorly, but sometimes smudges more extensive or lined up to form an irregular midventral dark line anteriorly; iris coppery brown.

SIMILAR SPECIES: (1) *Loxocemus bicolor* has paired prefrontals (fig. 11.4), a single pair of chin shields, and laterally directed nostrils. (2) *Tretanorhinus nigroluteus* has keeled scales (fig. 11.5).

HABITAT: Found in or near streams or freshwater marshes in Lowland and Premontane Moist and Wet Forests. Regularly found on the campus of the University of Costa Rica.

BIOLOGY: So far as is known, these uncommonly seen snakes are diurnal and semiaquatic and have been collected from small streams, under litter on stream bottoms, or moving about on land during heavy rains. Two captive females (UCR 4966, 10167) deposited four and seven oval eggs (32.8–44.8 × 9.3–10.6 mm) June 22–23 and in August. Two eggs hatched December 11 from the first clutch and one from the second hatched in early November (Savage and Donnelly 1988; Solórzano et al. 1989). It apparently feeds on small aquatic prey. One example (UCR 9229: San José Province: Escazú) was regurgitated by an *Erythrolamprus bizona*.

REMARKS: Nelson (1966b) recognized three species within the genus, but *Hydromorphus clarki* Dunn (1942c) from central Panama was synonymized with *H. concolor* by Savage and Donnelly (1988). *Hydromorphus dunni* Slevin (1942) from Boquete, Provincia de Chiriquí, Panama, is known from a single male example (CAS 78939) characterized by three prefrontals, 15-15-13 scale rows, and keeled supracloacal scales and appears to be a valid form. Villa (1990b) provided a redescription and summarized published data on the genus and its two species. The karyotype of *H. concolor* (based on a single female) is 2N = 46, composed of eight pairs of macrochromosomes and fifteen pairs of microchromosomes; pairs 1 and 7 are subtelocentric, 2, 3, 5, 6, and 8 are telocentric; 4 shows a marked ZW (probable sexual) heteromorphism, the former telocentric, the latter subtelocentric; NF = 46 (Solórzano, Gutiérrez, and Cerdas 1989).

DISTRIBUTION: Atlantic lowlands and premontane slopes from Guatemala to western Panama and Pacific versant central and southwestern Costa Rica to central Panama. In Costa Rica widely distributed on the Atlantic lowlands and adjacent premontane slopes; on the Pacific versant from Puntarenas, Puntarenas Province, south to Panama, including the Meseta Central Occidental (1–1,500 m).

Genus *Imantodes* C. Duméril, 1853

Blunt-headed Vine Snakes, Bejuquillas

A distinctive genus (six species) of very slender, elongate snakes having a short, blunt head that is trenchantly separated from the neck and possessing large, protuberant eyes (fig. 11.12e). In addition to its distinctive habitus, the genus *Imantodes* is further distinguished from others in the family Colubridae by the following combination of features: body strongly compressed; nasal divided; a loreal; 1 to 3 preoculars; pupil vertically elliptical (fig. 11.9d); two pairs of elongate chin shields; dorsal scales smooth, in 15 to 19 rows at midbody, lacking apical pits; anal entire or divided; subcaudals divided; maxillary teeth 10 to 22 plus two shallowly to strongly grooved fangs posterior to diastema; Duvernoy's gland present; prediastemal teeth subequal; anterior mandibular teeth longest; no hypapophyses on posterior dorsal vertebrae. The striking physiognomy of these snakes, elongate and compressed body, long tail (27 to 34% of total length), and presence of a mental groove will separate them from all other Costa Rican forms.

The everted hemipenes are very short (4 to 8 subcaudals in length), simple, with an undivided sulcus spermaticus, large basal hooks, spines, and a calyculate capitulum (Myers 1982b).

Imantodes range over much of the lowlands and lower adjacent slopes of the Neotropics from Sonora and southern Tamaulipas, Mexico, south to northwestern Ecuador and Bolivia, Paraguay, and extreme northeastern Argentina.

KEY TO THE COSTA RICAN SPECIES OF THE GENUS *IMANTODES*

1a. Dorsal pattern usually consisting of large dark blotches, rarely unblotched; ventrals more than 223 2
1b. Dorsal pattern consisting of small dark spots and speckles, sometimes lined up to form very narrow (no more than one scale row long) crossbars; fewer than 220 ventrals. . . . . *Imantodes inornatus* (p. 609)
2a. Scales in vertebral row greatly enlarged, 3 to 5 times as wide as laterals (fig. 11.31a); 31–52 (usually 48 or fewer) dark dorsal body blotches; subcaudals 146–83 . . . . *Imantodes cenchoa* (p. 606)
2b. Scales in vertebral row 1.2 to 2.5 times as wide as laterals (fig. 11.31b,c); 41–73 (usually 49 or more) dark dorsal body blotches or rarely unblotched; subcaudals 129–54 . . . . *Imantodes gemmistratus* (p. 608)

Imantodes cenchoa (Linné, 1758)

(plates 374–75; map 11.40)

DIAGNOSTICS: The long, thin, compressed body combined with the blunt head and bulging eyes (fig. 11.12e) of this species are common to all members of the genus. The conspicuously enlarged vertebral scale row and the reddish to chestnut dorsal blotches (fig. 11.31a) readily distinguish *I. cenchoa* from other *Imantodes*.

DESCRIPTION: A moderate-sized, very elongate and slender snake, attaining a maximum total length of 1,250 mm; females longer than males; body strongly compressed, neck

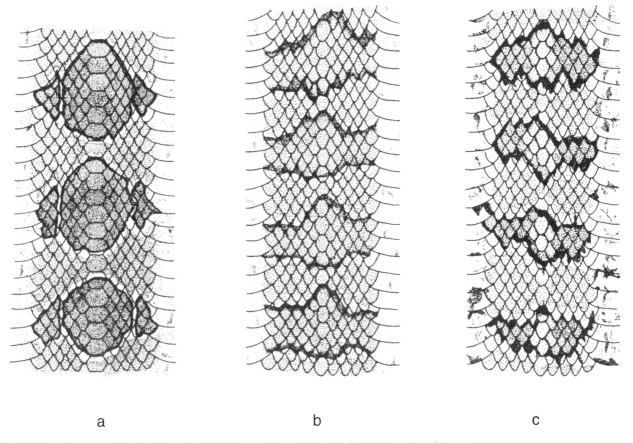

a   b   c

Figure 11.31. Midbody scalation and color pattern in Costa Rican snakes of the genus *Imantodes*: (a) *I. cenchoa*, note greatly enlarged vertebral scale row; (b) *I. gemmistratus*, usual pattern; (c) *I. gemmistratus*, variant pattern, note characteristic light spots in center of the dorsal dark bands.

very slender and distinct from head; head and protuberant eyes conspicuous; tail long (27 to 34% of total length); usually 1, sometimes 2, rarely 3 preoculars; 1 to 4 postoculars, usually 2; supralabials 7 to 9, usually 8, with 2 or 3 bordering orbit; infralabials 9 to 12, usually 10; temporals variable (4 to 12 scales per side), usually 2 + 3; dorsal scales 17-17-17 or 19-17-17, vertebral row conspicuously enlarged (three to five times width of lateral scales); anal divided; ventrals 233 to 278 (232 to 288) in males, 228 to 274 in females; subcaudals 147 to 183 (147 to 195) in males, 146 to 179 (134 to 179) in females; ventrals plus subcaudals 383 to 431. Dorsum pale to medium brown with 31 to 52 large chestnut brown (definitely reddish in juveniles) dorsal body blotches or saddles; venter whitish but heavily spotted or dotted with dark brown that sometimes forms median dark stripe; iris variable, usually pale greenish gray or light brown, sometimes yellowish gray or grayish tan, usually uniform but sometimes with vertical dark line through pupil or with fine black reticulum; tongue brown or gray with light tips on fork. Hemipenes very short (4 to 5 subcaudals in length) with evidence of capitation on asulcate side, which is spinous to the tip (Myers 1982b).

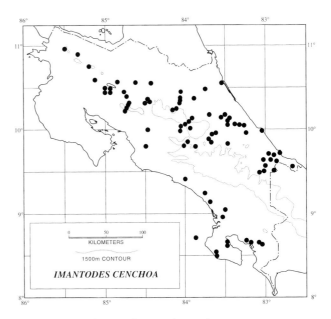

Map 11.40. Distribution of *Imantodes cenchoa*.

SIMILAR SPECIES: (1) *Imantodes gemmistratus* has the vertebral scale row only slightly enlarged and lacks the large chestnut blotches (fig. 11.31b,c). (2) *Sibon argus* and *Sibon longifrenis* have distinct ocelli in the dorsal pattern. (3) Other *Sibon* have the dorsal dark markings continuing onto the venter. (4) The species of *Dipsas* are ringed with dark and lack a mental groove. (5) The species of *Leptodeira* have 15 or more scale rows at midbody.

HABITAT: Occurs in forested areas, secondary growth, coffee fincas, fruit orchards, and banana plantations in localities originally covered by Lowland Moist and Wet Forests and Premontane Moist and Wet Forest and Rainforest; also found in the lower margins of Lower Montane Wet Forest and Rainforest.

BIOLOGY: This very common species is nocturnal and arboreal but may be seen prowling on the forest floor. The snakes sleep during the day hidden in hollow sections of trees and other plants, coiled in bromeliads, under ground debris (Myers 1982b), or even in artificial bromeliads (Young 1982). Frogs, especially *Eleutherodactylus*, and anole lizards *(Dactyloa* and *Norops)* (plate 375) are major food items, and they constrict large specimens. The data of Zug, Hedges, and Sunkel (1979) suggest that gravid females may be found in all months of the year in Costa Rica, indicating a continuous reproductive cycle. Females attain sexual maturity at a body length of about 620 mm, at about two years of age. Clutch sizes are small, with one to eight eggs laid, but usually two or three. Eggs are 21 to 38 mm in length, and the diameter is about a third of the length. Hatchlings are about 340 to 370 mm in total length.

REMARKS: For some years Scott (1969, 1983k) and Savage (1973b, 1976, 1980b) believed that *Imantodes cenchoa* and *I. gemmistratus* were probably conspecific and intergraded in northwestern Costa Rica. Following the stimulus of Myers (1982b), who demonstrated the distinctiveness of the two forms in Panama, we (Savage and Scott 1987) reexamined all Costa Rican material and confirmed that two species are involved. The two forms were clearly diagnosed and recognized by Taylor (1951b, 1954b), but additional specimens with intermediate scale and body blotch counts confused the issue for many years. The two species can be distinguished unequivocally in Costa Rica by the degree of vertebral scale enlargement and dorsal pattern. *I. cenchoa* occurs in evergreen lowland and semideciduous and evergreen premontane forests, principally on the Atlantic versant and in southwestern Pacific Costa Rica, but ranges into Lower Montane Rainforest at some locales. *I. gemmistratus* is relatively rare in those regions but occurs primarily in the subhumid forests of northwestern Pacific Costa Rica and the Meseta Central Occidental. The karyotype of this species is 2N = 36, with eight pairs of macrochromosomes and ten pairs of microchromosomes; pairs 1, 2, 3, 5, and 7 are submetacentrics; 6 and 8 are telocentrics; 4 shows a ZW heteromorphism in females, with the Z metacentric and the

somewhat shorter W submetacentric; NF = 48 (Gutiérrez, Solórzano, and Cerdas 1984).

DISTRIBUTION: In the lowlands and on premontane slopes from southern Tamaulipas (Atlantic) and Oaxaca (Pacific), Mexico, south through Central America (exclusive of Pacific slope El Salvador, Honduras, and northwestern Costa Rica) to central Ecuador (Pacific) and Bolivia, Paraguay, and northeastern Argentina (Atlantic). In Costa Rica the species occurs in forests of the Atlantic lowlands and premontane slopes and closely adjacent Pacific slopes (Premontane and lower portion of Lower Montane life zones), including the Meseta Central Occidental and the Montes del Aguacate, and in southwestern Pacific Costa Rica (2–1,500 m).

Imantodes gemmistratus Cope, 1861c

(plate 376; map 11.41)

DIAGNOSTICS: This species differs most obviously from most snakes in having an elongate, compressed body and a distinct blunt head with bulging eyes (fig. 11.12e). It differs from other members of the genus, which share these features, in having tan to brown dorsal bands (often with a light spot in the center) (fig. 11.31b,c) or rarely being uniform in color.

DESCRIPTION: A moderate-sized elongate and slender snake (maximum total length 880 mm); body compressed; short, blunt head distinct from neck; eyes large, protuberant; tail moderate to moderately long (24 to 28% of total length); usually 1 preocular, rarely 2; usually 2 (1 to 3) postoculars; supralabials usually 8, sometimes 7 or 9, with 3 bordering orbit; infralabials 9 to 11, usually 10; temporals variable (3 to 9 scales per side), usually 2 + 3, often 1 + 2 or

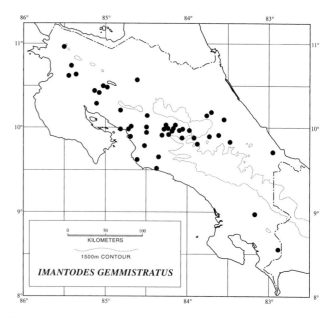

Map 11.41. Distribution of *Imantodes gemmistratus*.

1 + 3; dorsal scales 17-17-17 or 19-17-17, vertebral row moderately enlarged (1.2 to 2.5 times width of lateral scales); anal divided; ventrals 231 to 249 (201 to 278) in males, 225 to 252 (198 to 274) in females; subcaudals 138 to 147 (109 to 147) in males, 129 to 154 (104 to 154) in females; ventrals plus subcaudals 353 to 404 (302 to 404). Dorsal ground color pale brown to tan, often with a yellowish cast; almost always with dark dorsal blotches or bands, rarely nearly uniform in dorsal color; 41 to 73 (34 to 73) relatively narrow light brown dorsal blotches or bands; usually 49 or more dark bands with lower numbers usually the result of fusions between adjacent bands; often bands with a light spot in center that may form ocelli in posterior blotches; venter white, heavily speckled with brown; iris gray, suffused with brown; tongue brown with white tips on fork. Hemipenes very short (5 subcaudals in length), with no evidence of capitulum on asulcate surface, which is spinous to the tip (Myers 1982b).

SIMILAR SPECIES: (1) *Imantodes cenchoa* has the vertebral scales greatly enlarged and usually has 31 to 48 large chestnut blotches (fig. 11.31a). (2) *Sibon argus* and *Sibon longifrenis* have a series of dark-bordered lateral ocelli on the first three scale rows. (3) Other *Sibon* have the dorsal dark markings continuing onto venter. (4) The species of *Dipsas* are ringed with dark and lack a mental groove. (5) The species of *Leptodeira* have nineteen or more scale rows at midbody.

HABITAT: Most abundant in rather open situations including shrublands and secondary growth in Lowland Dry Forest and Premontane Moist Forest environments. Also found occasionally in Lowland Moist and Wet Forests, Premontane Wet Forest, and possibly Lower Montane Rainforest.

BIOLOGY: This very common essentially arboreal species occurs in more open situations than other members of the genus and is most abundant in seasonal deciduous to semideciduous forests and secondary growth in Costa Rica. It is typically found a meter or so above the ground in low trees or shrubs and fairly often moving across the ground. It may constrict its prey, which consists primarily of sleeping anole lizards it captures at night. Gravid females with two eggs (15 × 30–34 mm) were taken in Panama during May and September, and one Costa Rican example contained six eggs (Myers 1982b).

REMARKS: Myers (1982b) and Savage and Scott (1987) have reviewed the status of this form in Panama and Costa Rica, respectively, and support Taylor's (1951b, 1954b) conclusion that *Imantodes cenchoa* and *I. gemmistratus* are distinct species. Savage (1973b, 1976, 1980b) and Scott (1969, 1983k) earlier had thought that intergradation occurred between the two taxa because several Costa Rican populations of *I. cenchoa* have low ventral counts that overlap the variation in *I. gemmistratus*. The record of this species from Varablanca, Heredia Province (UCR 3323), is possibly from

the road to that locality but at a lower elevation. Varablanca (1,935 m) is much higher than the next highest locality record for this species from Cartago, Cartago Province (1,435 m). Some question remains whether all populations from Mexico and upper Central America referred to this species are conspecific (Savage and Scott 1987). The karyotype is 2N = 36, with eight pairs of macrochromosomes and ten pairs of microchromosomes; pairs 1, 2, 5, and 7 are submetacentrics; 6 and 8 are telocentrics; 4 shows a ZW heteromorphism in females, with the Z metacentric and the W submetacentric; NF = 44 (Gutiérrez, Solórzano, and Cerdas 1984).

DISTRIBUTION: From southern Sonora and northern Veracruz, Mexico, southward through lowland and premontane zones to the Yucatán Peninsula on the Atlantic versant and to eastern Panama on the Pacific slope; also scattered Atlantic slope records from Honduras, Costa Rica, Panama, and the Magdalena valley of Colombia. In Costa Rica, primarily in the northwestern Pacific lowlands (2–500 m) and adjacent slopes (to 1,400 m), including the Meseta Central Occidental (to 1,435 m or possibly higher) and at scattered localities in the Atlantic lowlands and southwestern Costa Rica (2–1,435 m; possibly to 1,935 m).

Imantodes inornatus Boulenger, 1896d

(plates 377–78; map 11.42)

DIAGNOSTICS: The coloration of numerous small black spots and speckles is not found in any other snake with a blunt head, bug eyes (fig. 11.12e), and a strongly compressed body.

DESCRIPTION: An elongate moderate-sized snake (up to 1,035 mm in total length); body strongly compressed, neck

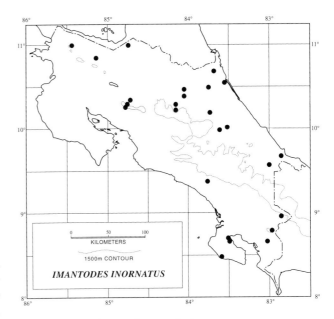

Map 11.42. Distribution of *Imantodes inornatus*.

very slender and distinct from neck; head and protuberant eyes conspicuous; tail moderate to long (21 to 31% of total length); usually 1 preocular, rarely 2; usually 2 postoculars, rarely 3; supralabials 7 to 9, usually 8, with 2 or 3 bordering orbit; infralabials 8 to 11, usually 10; temporals variable (3 to 6 scales per side), usually 2 + 3; dorsal scales usually 17-17-17, rarely 17-19-17 in females, 17-17-15 or 17-17-13 in males; vertebral scale row moderately enlarged (1.2 to 1.5 times midlateral scales); anal entire (33%) or divided (67%); ventrals 199 to 218 in males; 196 to 212 in females; subcaudals 166 to 132 in males, 110 to 122 in females; ventrals plus subcaudals 316 to 334. Dorsum yellowish to light brown with pattern of small black spots and speckles that often form very narrow crosslines dorsally and alternate with similar lateral markings; upper head surfaces marked with black speckles; venter yellow to pale orange, heavily speckled with black, usually a thin midventral dark stripe; iris pale yellow brown, tan, light brown, orangish, or greenish gray; tongue black with white tips on fork. Hemipenes short (6 to 8 subcaudals in length), proximal half with large spines; capitulum with a deep uninterrupted overhang on asulcate side (Myers 1982b).

SIMILAR SPECIES: (1) *Bothriechis schlegelii* has distinct supraciliary hornlike spines above the eyes, a labial pit (fig. 11.50), and a prehensile tail. (2) Other species of *Imantodes* almost always have a strongly blotched or banded pattern and always lack the black speckling on the head and body.

HABITAT: Usually restricted to relatively undisturbed rainforest; most commonly found in Lowland Moist and Wet Forests but occasionally in Premontane Wet Forest and Rainforest and Lower Montane Wet Forest.

BIOLOGY: This moderately common snake is a very arboreal nocturnal form that feeds primarily on frogs, sleeping anole lizards, and recently deposited eggs of the frogs *Agalychnis* and *Hyla ebraccata*, which lay their eggs on vegetation out of the water. Like other members of the genus, it constricts large prey. Its color pattern is similar to that of the yellow varieties of *Bothriechis schlegelii* and may represent mimicry. Myers (1982b) reports on a gravid female from Panama that contained four eggs about 20 mm in length.

DISTRIBUTION: Lowlands and adjacent premontane slopes on the Caribbean versant from northwestern Honduras to eastern Panama; on the Pacific slope in southwestern Costa Rica and adjacent southwestern Panama; and from extreme eastern Panama to central Ecuador. In Costa Rica nearly restricted to the evergreen forests of the Atlantic and Pacific southwest lowlands, with a few records from premontane areas near Monteverde, Puntarenas Province; Cariblanco, Alajuela Province; and Las Cruces, Puntarenas Province (5–1,450 m).

Genus *Leptodeira* Fitzinger, 1843

Members of this genus (nine species) are generalized colubrids without striking modifications in body morphology or head scalation; nasal divided; a loreal; one or two preocu-

lars; eye large, with a vertically elliptical pupil; two pairs of elongate chin shields; dorsal scales smooth in 17 to 25 at midbody with a reduction anterior to the vent; two apical scale pits; supracloacal keels rarely present; anal and subcaudals divided; 8 to 18 subequal maxillary teeth followed after a diastema by two enlarged grooved fangs; Duvernoy's gland present; anterior mandibular teeth slightly enlarged; no hypapophyses on posterior dorsal vertebrae. Females are longer than males. The combination of the features listed and the pattern of dark blotches, spots, or bands will separate representatives of this genus from all other Costa Rican snakes.

The everted hemipenes in *Leptodeira* are short (6 to 10 subcaudals in length), single, symmetrical, with a forked sulcus spermaticus that divides within the capitulum (Dowling 1975b), and have large basal and medial spines and a calyculate capitulum. Duellman (1958) divided the genus into four weakly defined species groups; three of them occur in Costa Rica.

The genus occurs in the lowlands and on adjacent mountain slopes throughout the Americas from extreme southern Texas in the United States and Sonora, Mexico, south through tropical Mexico and Central America, to northern Chile on the Pacific coast, and to southeastern Brazil on the Atlantic versant, including the entire Amazon basin and then south to Paraguay and northern Argentina; also the islands of Aruba, Margarita, Trinidad, and Tobago in the southern Caribbean Sea.

KEY TO THE COSTA RICAN SPECIES OF THE GENUS *LEPTODEIRA*

1a. Dorsal pattern with a series of 23 or more dark spots, blotches, or bands on a red, gray, or brown ground color, *or* blotches fused, forming an irregular zigzag figure down center of back; dorsal markings not forming light and dark bands that extend completely across dorsum. 2

1b. Dorsal pattern a series of 10–21 dark brown bands on a light ground color forming a series of regularly alternating dark and light transverse bands extending completely across body . . . .
. . . . *Leptodeira nigrofasciata* (p. 613)

2a. Ventrals fewer than 185; a median dark nape stripe usually running from parietal shields to connect with first dark dorsal blotch or band (fig. 11.32b,c); distinct paired nuchal dark spots on side of neck anterior and lateral to first dorsal dark marking, sometimes fused to first dorsal dark markings *or* fused with one another across parietal area and with the nape stripe and first dark dorsal band to produce a figure eight pattern on dorsal neck region 3

2b. Ventrals 186–211; median dark nape stripe, if present, usually extending only a few scales posterior to parietal shields, not connecting with first body blotch; nape

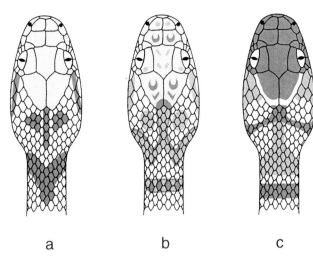

a   b   c

Figure 11.32. Color pattern of head and neck in Costa Rican snakes of the genus *Leptodeira;* patterns are shown on a generalized template for this genus and do not accurately depict differences in scalation among the species: (a) *L. septentrionalis*, note incomplete dark nape stripe; (b) *L. annulata* showing complete dark nape stripe connecting to first dark body blotch and distinct paired nuchal dark spots; (c) *L. rubricata* showing fusion of dark nape stripe, nuchal dark spots, and first dark body blotch.

stripe rarely connected with fused dark nuchal blotches to form a transverse band across head just posterior to parietals (fig. 11.32a); dorsal dark markings not outlined by a lighter color, continuous with dorsal ground color . . . . *Leptodeira septentrionalis* (p. 614)

3a. Dark body blotches 23–51; pattern consisting of dark spots or blotches on gray or brown ground color *or* blotches fused to form an irregular zigzag figure down center of back; light interspaces cream to light tan in life; a median dark nape stripe usually runs from parietal shields to connect with first dark dorsal blotch or spot; paired dark nuchal spots distinct, rarely fused with first dark dorsal marking (fig. 11.32b); venter cream in life; dorsal dark markings brown, usually outlined by a lighter color that separates them from ground color; ventrals 151–81 . . . . . . *Leptodeira annulata* (p. 611)

3b. Dark body bands 57–67; pattern consisting of regular alternating dark and light transverse bands; light ground color red in life; dark nuchal blotches, median nape stripe, and first dark body band fused to form a figure eight pattern on dorsal neck region (fig. 11.32c); venter red in life; dorsal dark bands black, not outlined by a lighter color, contiguous with dorsal ground color; ventrals 177–82 . . . . *Leptodeira rubricata* (p. 612)

Leptodeira annulata Group

Snakes of this group have a dorsal pattern of dark blotches on lighter ground color, maxillary teeth 7 to 18 + 2, the hemipenes usually with a cup-shaped depression in the capitulum, and crenulate calyces. In addition to *Leptodeira*

annulata and *Leptodeira rubricata*, the following taxa are placed in this group: *Leptodeira bakeri* (Aruba Island, Dutch West Indies), *L. frenata* (Veracruz, Mexico to Belize), and *L. maculata* (Sinaloa to Guerrero, Mexico).

Leptodeira annulata (Linné, 1758)

(plate 379; map 11.43)

DIAGNOSTICS: Any small to moderate-sized brown-blotched snake lacking extensive ventral markings and having smooth dorsal scales (fig. 11.5) and a vertically elliptical pupil is likely to be this common species. The combination of a considerable number of scale rows (21 to 25) at midbody and a long nape stripe that begins on the parietals and extends posteriorly to almost always fuse with the first body blotch is also distinctive (fig. 11.32b).

DESCRIPTION: A moderate-sized snake reaching a total length of 1,038 mm; females (264 to 1,038 mm) longer than males (200 to 907 mm); body cylindrical (cylindrical to compressed), head distinct from neck; tail length variable (20 to 42% of total length), but moderately long in the Costa Rican population (25 to 30% of total length); 1 to 4 preoculars, usually 2; 1 to 4 postoculars, usually 2; supralabials 7 to 9, usually 8, almost always with 2 bordering orbit, rarely 3; infralabials 8 to 12, usually 10; temporals usually 1 + 2, rarely 1 + 1 or 1 + 3; dorsal scale rows 21 to 25 (17 to 25) at midbody, with a reduction to 15 to 19 (11 to 19) anterior to vent; males occasionally with weakly keeled supracloacal scales; vertebral scale row not enlarged (not to somewhat enlarged); ventrals 151 to 181 (151 to 204); subcaudals 59 to 89 (54 to 102). Dorsum cream to grayish brown with a series of 23 to 51 (21 to 56) dark brown to black dorsal blotches; lateral intercalary spots present (present or absent); dorsal blotches sometimes fused to form an irregular

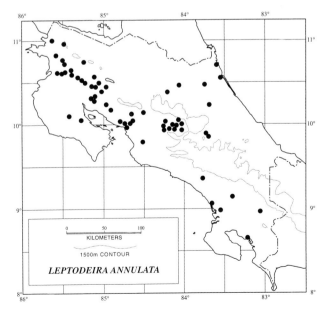

Map 11.43. Distribution of *Leptodeira annulata*.

zigzag figure on center of back; head and neck pattern variable, a single dark nape stripe present in Costa Rican examples that usually connects parietals and first dark dorsal blotch or band; paired dark nuchal spots present or not (in Costa Rican specimens the spots are distinct or rarely fused with the first body blotch); venter cream, immaculate; iris gray brown to olive tan. The everted hemipenes are short (7 to 10 subcaudals in length), with a deep cup-shaped depression in the capitulum and the latter covered with crenulate calyces (Duellman 1958; Underwood 1967).

SIMILAR SPECIES: (1) *Leptodeira septentrionalis* usually lacks the dark nape stripe, or if present the stripe extends only a few scale rows posterior to the parietal shields and is not fused with the first body blotch (fig. 11.32a). (2) *Leptodeira rubricata* has a red ground color and 57 to 67 transverse dark bands. (3) *Drymobius rhombifer* has keeled scales (fig. 11.5). (4) *Trimorphodon biscutatus* has three loreals (fig. 11.45) and the posterior dorsal scales weakly keeled. (5) *Xenodon rabdocephalus* has considerable dark markings ventrally.

HABITAT: Most commonly found in semiarid and seasonal semideciduous and deciduous forest and savanna areas in Central America. Usually occurs near slow-moving streams or standing water in Lowland Dry Forest and Premontane Moist Forest environments in Costa Rica, but occasionally found in Lowland Moist and Wet Forests and Premontane Wet Forest.

BIOLOGY: In Central America this nocturnal very common species is primarily terrestrial but readily enters water. Food consists of frogs and toads that, if small enough, it grabs and swallows. With larger prey items the snake strikes, grabs the victim, and rapidly chews with the mouth open until the grooved fangs penetrate. It then holds the frog until its struggles cease. During the wet season or the frog breeding season, *L. annulata* is found in some numbers with congregations of pond-breeding anurans *(Bufo, Hyla, Leptodactylus, Smilisca,* and *Rhinophrynus)*. These snakes literally gorge themselves during this period, eating as many as thirty to forty small frogs in an evening. At other times they climb low vegetation to search for food or to rest. The smallest juvenile is 179 mm in total length. The venom of *L. annulata* is strong enough to cause a severe reaction in humans.

REMARKS: This wide-ranging species shows substantial variation in body form (compressed in the Amazonian populations), scalation, and coloration. Duellman (1958) has written a monograph on the genus and provides extensive information on variation in *L. annulata*. The following listing succinctly diagnoses the major geographic subdivisions of this species as recognized by Duellman (1958):

1. Mexico: no nape stripe; lateral intercalary dark spots present; dark body blotches 21-39.2-48; preoculars two; vertebral scale row not enlarged, body cylindrical
2. Central America: dark nape stripe (fig. 11.32b); lateral intercalary dark spots present; dark body blotches 23-

35.7-56; preoculars two; vertebral scale row not enlarged; body cylindrical
3. Northern South America: two parallel longitudinal dark bars on nape; lateral intercalary spots present; dark body blotches 31-39-51; preoculars two; vertebral scale row not enlarged; body cylindrical
4. Amazon basin and coastal Brazil: no nape bars or stripes; no lateral spots; dark body blotches 25-33.4-50; usually one preocular; vertebral scale row enlarged; body compressed
5. Paraguay and northern Argentina: two parallel longitudinal dark nape bars; lateral intercalary spots present; dark body blotches 33-39.2-46; usually two preoculars; vertebral scale row not enlarged; body cylindrical

Populations intermediate between the several geographic forms occur in intermediate geographic areas. Mexican and Central American *L. annulata* are relatively short-bodied, robust snakes compared with their more arboreal congener *Leptodeira septentrionalis*, reflecting the terrestrial habits of the former.

All the Costa Rica specimens referred to this species by Taylor (1951b, 1954b) are actually representatives of *L. septentrionalis*. Examples Taylor listed under the names *Leptodeira ocellata* and *L. rhombifera* are *L. annulata*. Although relatively rare in such habitats, this species is now known from several Atlantic lowland sites (e.g., Barra del Colorado, Roxana, and Tortuguero, Limón Province; Río Cuarto, Alajuela Province; La Selva, Heredia Province; and Turrialba, Cartago Province) and Pacific southwestern sites (e.g., Palmares, San José Province; Palmar and Gromaco, Puntarenas Province) in Costa Rica.

DISTRIBUTION: From southern Tamaulipas and Guerrero, Mexico, south through much of lowland and premontane Central America (exclusive of the forests of Atlantic slope southern Mexico and Guatemala and the Yucatán Peninsula) to Ecuador on the Pacific versant and throughout most of South America east of the Andes to Bolivia, Paraguay, northern Argentina, and southeastern Brazil, including the islands of Margarita, Trinidad, and Tobago in the Caribbean; the species is uncommon in the evergreen forests of Atlantic drainage Honduras, Nicaragua, Costa Rica, and Panama. In Costa Rica found principally in the northwestern lowlands and adjacent slopes and on the Meseta Central Occidental, with scattered records from the Atlantic slope and the southwestern Pacific lowlands (4–1,400 m).

Leptodeira rubricata (Cope, 1893)

(plate 380; map 11.44)

DIAGNOSTICS: The red dorsal and ventral ground color and the banded pattern combined with the cylindrical body and 21 to 23 smooth scale rows at midbody will readily distinguish this snake from all others in the republic.

DESCRIPTION: A small to moderate-sized species attain-

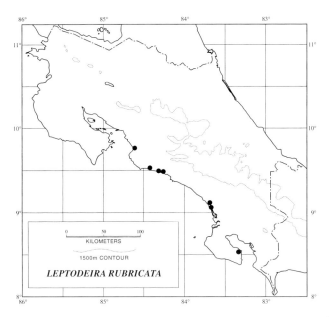

Map 11.44. Distribution of *Leptodeira rubricata*.

ing a maximum total length of 700 mm; body cylindrical; head distinct from neck; tail length moderate (about 20% of total length); preoculars 2; postoculars 2; supralabials 8, with 2 bordering orbit; infralabials 9 to 10; temporals 1 + 2; dorsal scale rows at midbody 21 to 23, with a reduction to 17 or 19 anterior to vent; ventrals 177 to 182; subcaudals 82 to 97. Dorsum red, with a series of 57 to 67 dark brown to black transverse bands across back that extend about halfway down sides toward venter; dark nuchal blotches, median nape stripe, and first dark dorsal band fused to form a complex figure eight pattern on dorsal neck region; venter red, heavily dusted with brown.

SIMILAR SPECIES: (1) *Liophis epinephalus* has the venter with prominent black marks. (2) *Oxyrhopus petolarius* has a cream venter and 19 scale rows at midbody. (3) *Tripanurgos compressus* has a compressed body and the vertebral scale row enlarged (fig. 11.15). (4) *Scolecophis atrocinctus* has black rings around the body. See key to coral snakes (p. 706), figure 11.47, and table 11.2.

HABITAT: Found only in Pacific mangrove forests.

BIOLOGY: This unusual nocturnal species is a semiarboreal snake that occurs only in mangroves and feeds principally on small crabs. Its relative rarity in collections derives from its specialized habitat.

REMARKS: Although Dunn (1936b) and Duellman (1958) regarded this form as a variant of what is now called *Leptodeira annulata*, Taylor (1951b) recognized that it was distinct. Floyd Downs and the late Douglas C. Robinson have prepared a manuscript (as yet unpublished) that convincingly establishes *L. rubricata* as a valid species. Most of the data presented here are from that account.

DISTRIBUTION: Mangrove swamps of the Pacific coast from Tárcoles, Puntarenas Province, south through south-

western Costa Rica into adjacent southwestern Panama (0–5 m).

Leptodeira nigrofasciata Group

Members of the group have a dorsal pattern of broad dark bands on a lighter ground color, maxillary teeth 10 to 13 + 2, and the hemipenes lacking a naked cup in the capitulum that is covered by spinulate calyces. Only *Leptodeira nigrofasciata* belongs here. Dowling and Jenner (1987) concluded that *Pseudoleptodeira latifasciata* of Colima to Puebla and Guerrero, Mexico, which Duellman (1958) placed in this group of *Leptodeira*, properly forms a distinct monotypic genus.

Leptodeira nigrofasciata Günther, 1868a

(plates 381–82; map 11.45)

DIAGNOSTICS: The pattern of broad straight-edged transverse dorsal bands that extend completely across the body on a light ground color immediately separates this species from all others.

DESCRIPTION: A small snake reaching a total length of 581 mm; females (240 to 581 mm) longer than males (202 to 486 mm); body cylindrical; head distinct from neck; tail moderate to long (23 to 38% of total length); preoculars 1 to 3, usually 2; postoculars 1 or 2, usually 2; supralabials 7 to 9, usually 8, with 2 (rarely 3) bordering orbit; infralabials 9 to 11, usually 10; temporals usually 1 + 2, rarely 1 + 1; dorsal scale rows 17 to 19 at midbody, with a reduction to 15 or 17 anterior to vent; vertebral scale row not enlarged; ventrals 168 to 186 (161 to 196); subcaudals 50 to 75 (54 to 76). Dorsum cream to grayish tan with 13 to 18 broad straight-edged dark brown to black crossbands on

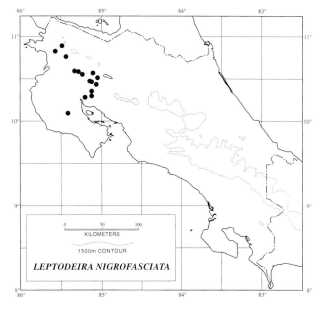

Map 11.45. Distribution of *Leptodeira nigrofasciata*.

body that extend to ventrals; several bands often fused posteriorly across midline to produce an irregular dark figure with the remnants of the light interspaces forming alternate even-margined narrow light lateral bars within the dark figures; top of head including parietals dark, a light nape band separating head cap from first body band; venter cream to tan with dark markings restricted to lateral margins. The everted hemipenes are short (8 or 9 subcaudals in length), with spinulate calyces on the capitulum; the pedicel is covered with small spines, followed on the truncus by large hooked spines that extend to the capitulum (Duellman 1958).

SIMILAR SPECIES: (1) *Urotheca euryzona* has a ringed pattern. (2) *Micrurus mipartitus* has a ringed pattern.

HABITAT: Restricted to Lowland Dry Forest.

BIOLOGY: This uncommon species is nocturnal and terrestrial and is often encountered moving across paved roads at night. Duellman (1958) reported the presence of a microteiid lizard *(Gymnophthalmus speciosus)* in the stomach of a snake of this species. The smallest juvenile is 214 mm in total length.

REMARKS: Wilson and Meyer (1985) noted two unusual variants in material from Honduras. In one snake the loreal is fused with the prefrontal on both sides of the head, and in another the dorsal scale row formula is 17-17-17. Dunn's (1936b) record of *L. nigrofasciata* from Turrialba, Cartago Province, Costa Rica, seems to be in error, since the species is known elsewhere in the republic only from the Pacific northwest (also see Scott 1969).

DISTRIBUTION: Primarily in subhumid to semiarid areas of the lowlands and adjacent lower slopes on the Pacific versant from Guerrero, Mexico, to northwestern Costa Rica; also on the Caribbean versant in northwestern Honduras. In Costa Rica restricted to the Pacific northwest (7–400 m).

Leptodeira septentrionalis Group

Species referred to this group have a dorsal pattern of dark spots on a lighter ground color, maxillary teeth 13 to 18 + 2, and the hemipenes without a naked cup in the capitulum that is covered by crenulate calyces. In addition to *Leptodeira septentrionalis*, only *Leptodeira splendida* of Sonora to Michoacán and Puebla, Mexico, belongs to this group.

Leptodeira septentrionalis (Kennicott in Baird, 1859b)

(plates 383–84; map 11.46)

DIAGNOSTICS: Another common species with a pattern of brown dorsal blotches, smooth dorsal scales (fig. 11.5), a virtually unmarked venter, and a vertically elliptical pupil (fig. 11.32a). In this form the nape stripe, if present (fig. 11.32a), does not connect to the first dark body blotch.

DESCRIPTION: A moderate-sized snake reaching a maximum total length of 1,055 mm; females (339 to 1,055 mm) larger than males (340 to 965 mm); body cylindrical (cylin-

drical to compressed); head distinct from neck; tail length moderate to long (26 to 38% [23 to 41%] of total length); preoculars 1 to 3, usually 3; postoculars 1 to 3, usually 2; supralabials 8 or 9, usually 8, with 2 or 3 bordering orbit; infralabials 9 to 12, usually 10; temporals usually 1 + 2, rarely 1 + 3; dorsal scale rows at midbody almost always 21 or 23 (sometimes 19), with a reduction to 17 or 15 anterior to the vent (rarely 11, 13, or 19); ventrals 186 to 211 (170 to 211); subcaudals 73 to 107 (60 to 107). Dorsum cream to reddish tan (no red tint in Costa Rica), with 38 to 70 (20 to 70) dark brown to black blotches; lateral intercalary dark spots present (present or absent); head and neck pattern variable, median nape stripe, if present, short in Costa Rican examples, not connected to first body blotch; sometimes lateral nuchal dark spots fused across parietal region and with nape stripe; many juveniles have a pure white horseshoe-shaped collar that begins behind the eyes and passes across the head and neck behind the parietals (plate 384); venter cream, sometimes suffused with orange and with some brown flecking; iris cinnamon, reddish tan, to yellowish gray. The everted hemipenes are short (7 to 9 subcaudals in length) and covered with numerous spines, and the capitulum lacks a cup-shaped depression and is covered by crenulate calyces (Duellman 1958).

SIMILAR SPECIES: (1) *Leptodeira annulata* has a long dark nape stripe that usually connects to the first body blotch, and the lateral nuchal dark spots are almost always distinct or, rarely, fused with the first dark body blotch (fig. 11.32b). (2) *Leptodeira rubricata* has a red ground color and a complex dark figure eight marking formed from the fusion of the dark nape stripe, the nuchal spots, and the first body blotch (fig. 11.32c). (3) *Drymobius rhombifer* has

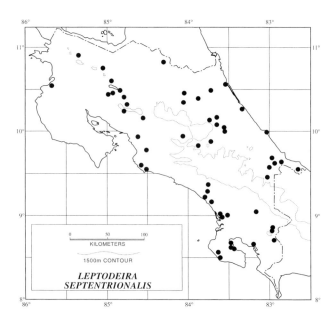

Map 11.46. Distribution of *Leptodeira septentrionalis*.

keeled scales (fig. 11.5). (4) *Trimorphodon biscutatus* has three loreals (fig. 11.45), and the posterior dorsal scales are weakly keeled.

HABITAT: Commonly encountered in Lowland Moist and Wet Forests but occasionally found in Dry Forest, Premontane Moist and Wet Forests and Rainforests.

BIOLOGY: In Central America this very common nocturnal and semiarboreal species is found primarily in humid evergreen forests. It forages at night on the ground and in the lower bush-sapling-tree layer of the forest. Food consists primarily of frogs and toads, although small lizards have also been recorded from stomach contents (Duellman 1958). This species also feeds commonly on the eggs of the hylid treefrog genus *Agalychnis* (Smith 1943; Duellman 1958; Fogden and Fogden in Greene 1997) (plate 383). These frogs lay their eggs in gelatinous masses on the branches or leaves of bushes or trees overhanging ponds or slow-moving streams. Numerous observations of *L. septentrionalis* eating the eggs of *Agalychnis callidryas* at La Selva, Heredia Province, have been made over many years and suggest that this is a common and widespread habit. In addition, *L. septentrionalis* feeds on the eggs of the treefrog *Hyla ebraccata*, a species that deposits its eggs on leaves, stems, and branches of low vegetation emergent from ponds or swamps. Clutches of four to thirteen eggs have been reported for captive animals from Central America or presumed to be from there (Duellman 1958; Haines 1940), with mean egg dimensions of 26 × 11.5 mm. Incubation time in captivity varied between seventy-nine (Duellman 1958) and ninety days (Behler and King 1979). One captive female (Haines 1940) isolated from males produced clutches of seven and six eggs twenty-six months apart and another thirteen eggs fourteen months later, indicating long-term sperm storage and delayed fertilization. The smallest juvenile is 164 mm in total length. *L. septentrionalis* is relatively docile and rarely bites when handled. When it does, it may inflict a painful but not serious bite on humans with its grooved rear fangs.

REMARKS: This species exhibits considerable geographic variation in body shape, scalation, and coloration. The following information (primarily from Duellman 1958) summarizes available data for the principal populational units:

1. Eastern Mexico and Texas: lateral intercalary spots absent; dark body blotches 20–26.2–35; preoculars three, midbody scale rows 23; body slightly compressed
2. Western Mexico and Central America: lateral intercalary spots present; dark body blotches 38–53.6–70; preoculars usually three; midbody scale rows usually 23; body slightly compressed
3. Panama and northwestern South America: lateral intercalary spots present; dark body blotches 30–39.9–52; preoculars usually two; midbody scale rows usually 21; body compressed.
4. Northwestern Peru: lateral intercalary spots present;

dark body blotches 34–42.1–49; preoculars usually two; midbody scale rows usually 21; body cylindrical

Intermediate conditions occur in zones lying between the various major geographic subunits characterized above. However, Campbell (1998) argues that the snakes of this series from southern Mexico to Costa Rica (on both coasts) should be recognized as a separate species, *Leptodeira polysticta* Günther, 1895. The situation remains confused, since the name *Leptodeira ornata* (Bocourt, 1884), based on an example from eastern Panama, has priority over *L. polysticta*. If the populations from Panama and adjacent South America and the rest of Central America are conspecific. as Duellman (1958) concluded, and distinct from the form in Texas and northeastern Texas (*L. septentrionalis* of Campbell 1998), Bocourt's name would apply. Taylor's (1951b, 1954b) Costa Rican specimens identified as *L. annulata* are all representatives of *L. septentrionalis*. In addition, his 1954 record of *Hypsiglena* from the republic is based on a specimen (KU 31930) of *L. septentrionalis* (Duellman 1958). The genus *Hypsiglena* is not known to occur south of southwestern Mexico and is not expected to be found in Central America.

DISTRIBUTION: Lowlands and premontane slopes on the Atlantic versant from extreme southern Texas through Mexico and Central America to northern Colombia and in disjunct areas on the Pacific slope from Sinaloa, Mexico, to El Salvador; also Pacific northwestern Costa Rica to adjacent southwestern Panama and thence more or less continuously southward through western South America to northwestern Peru. In Costa Rica found on the Atlantic versant and in western Pacific areas and on adjacent forested slopes (2–1,200 m).

Genus *Ninia* Baird and Girard, 1853

A group of small semifossorial snakes (eight species) with the basic colubrid complement of head shields and the following distinctive combination of features: nasal entire or divided; a loreal; usually no preocular, very rarely one; pupil vertically subelliptical (fig. 11.11a); two pairs of elongate chin shields; adult males with tubercles on anterior supralabials, mental, anterior infralabials, and chin shields; dorsal scales keeled and strongly striated, in 17 to 19 rows at midbody, without a scale reduction posteriorly; no apical scale pits; anal usually entire, sometimes divided; subcaudals divided; maxillary teeth 15 to 18, subequal in size; condition of Duvernoy's gland unknown, probably present; mandibular teeth subequal; hypapophyses present on posterior dorsal vertebrae. The combination of keeled scales in 17 to 21 rows, a loreal, paired prefrontals (and the usual conditions of no preocular and an entire anal), and the absence of stripes in the coloration will separate *Ninia* from other Costa Rican colubrid snakes.

The everted hemipenes are short and simple, have a

centrolineal, bifurcate sulcus spermaticus, and are unicapitate with basal spines and a calyculate capitulum (Dowling 1975b; Jenner 1981).

The genus ranges from San Luis Potosí (Atlantic slope) and Oaxaca (Pacific slope), Mexico, southward through Central America to Guyana and the upper Amazon basin and Pacific lowland Ecuador.

KEY TO THE COSTA RICAN SPECIES OF THE GENUS *NINIA*

1a. 19 scale rows at midbody; ventrals plus subcaudals fewer than 210; usually 7–8 supralabials 2
1b. 17 scale rows at midbody; ventrals plus subcaudals usually more than 210 (190–238); 5–6 supralabials; dorsum uniform dark brown to black or banded with narrow light crossbars; often with an incomplete light nuchal collar indicated (fig. 11.33d); venter tessellated with black and light (red to yellow or whitish in life) (fig. 11.34g–i) . . . . *Ninia psephota* (p. 619)
2a. Venter immaculate or with a few scattered dark markings 3
2b. Venter variegated with dark and light areas (fig. 11.34a–f); dorsum brown with a series of black crossbars and without a distinct light nuchal collar; ventrals plus subcaudals 168–204 . . . *Ninia maculata* (p. 618)
3a. Dorsum uniform black; a nuchal light collar involving the parietals present or absent; ventrals plus subcaudals 158–71 . . . . *Ninia celata* (p. 617)
3b. Dorsum red in life (reddish brown in preservative), uniform; a distinct yellow nuchal collar lies posterior to the parietals, bordered by black posteriorly (fig. 11.22b); ventrals plus subcaudals 196–206 . . . . *Ninia sebae* (p. 620)

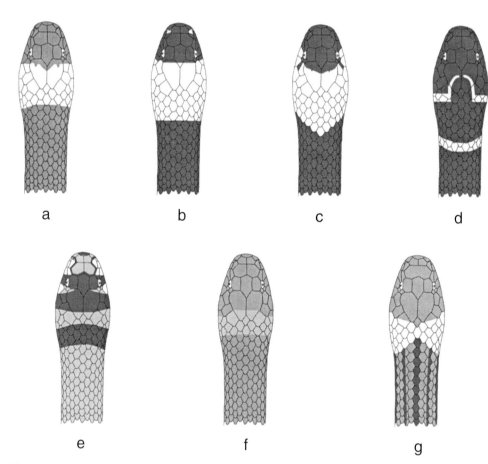

Figure 11.33. Diagnostic features of head and neck color patterns in some Costa Rican snakes having light nuchal collars or bars; patterns are figured on a generalized template of head plates and do not accurately depict differences in scalation among the taxa illustrated: (a) *Enulius flavitorques* illustrating broad yellow nuchal collar; (b) *E. sclateri*, dark-headed example; (c) *Ninia celata*, some examples lack nuchal collar; (d) *Ninia psephota* with narrow anterior nuchal bars; (e) *Tantilla alticola* with narrow light brown nuchal collar between dark brown bands; (f) *T. schistosa* with narrow light brown nuchal collar not demarcated by darker bands; (g) *Trimetopon pliolepis* with light brown head shields and yellow nuchal collar. Light shading indicates light brown; darker shading indicates dark brown or black.

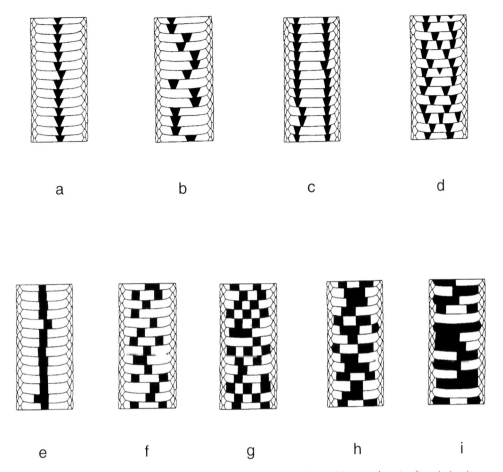

a   b   c   d

e   f   g   h   i

Figure 11.34. Variation in ventral coloration at midbody in some Costa Rican snakes: (a–f) variation in *Ninia maculata*; (g–i) variation in *Liophis epinephalus* and *Ninia psephota*.

Ninia celata McCranie and Wilson, 1995

(plate 385; map 11.47)

DIAGNOSTICS: One of several small species with a uniform dark dorsum (fig. 11.33c) and immaculate venter. It is distinctive among these forms in having keeled scales in 19 rows at midbody.

DESCRIPTION: A small snake reaching a total length of 450 mm; head slightly distinct from neck; body cylindrical; tail very long (44 to 47% of total length); eye small; nasal entire; a loreal, bordering orbit; usually no preocular, rarely 1; postoculars 1 or 2 (usually 2); supralabials 7–7, with 1 or 2 (usually 2) bordering orbit; infralabials 8–8; temporals 1 + 2; 19 dorsal scale rows; ventrals 123 to 130; subcaudals 43 to 45 in males, 33 to 40 in females, ventrals plus subcaudals 158 to 171. Dorsal coloration black; a well-developed yellow to red orange nuchal band across posterior two-thirds of parietals, temporals and 4 to 5 dorsal scale rows in some examples, otherwise head, neck, body, and tail uniform black; chin and underside of tail mostly dark brown to black; tips of ventrals black; venter immaculate white.

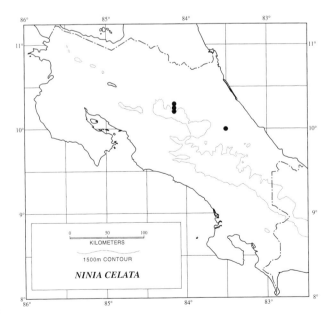

Map 11.47. Distribution of *Ninia celata*.

SIMILAR SPECIES: (1) *Ninia psephota* has the nuchal light bars, if present, involving only the tips of the parietals (fig. 11.33d) and 17 dorsal scale rows. (2) Species of *Enulius* have smooth dorsal scales (fig. 11.5) in 15 to 17 rows. (3) Species of *Geophis* with some keeled scales (fig. 11.5) and a nuchal light collar have the scales in 15 rows. (4) *Trimetopon* have smooth scales and the ventrals unmarked excepts for the tips.

HABITAT: Taken from under debris in undisturbed Premontane Rainforest.

REMARKS: Dunn (1935), Taylor (1951b, 1954b), and Savage and Lahanas (1991) called Costa Rican examples of this species *Ninia atrata*. McCranie and Wilson (1995) demonstrated that three distinct forms could be recognized from material formerly referred to that species: *N. atrata* (eastern Panama to Trinidad, Tobago, and western Ecuador); *N. celata* (western Panama and Costa Rica); and *N. espinali* (El Salvador and western Honduras). *N. atrata* differs from *N. celata* most obviously in having reduced parietals and more ventrals and subcaudals (ventrals plus subcaudals 177 to 223 versus 158 to 171 in *N. celata*). *N. espinali* also has higher segemental counts (ventrals plus subcaudals 192 to 215) than *N. celata*. Dunn's (1935) reference to the species' occurring in the city of Cartago, Cartago Province (as followed by Savage and Lahanas 1991), is based on four examples of *Ninia maculata* (BMNH 71.11.22.23–26) (McCranie and Wilson 1995). Although the presence of a light nuchal collar was thought to be diagnostic of this species, recently collected examples (UCR 12634, 13875, both adult females) from Alto Guayacán, Limón Province, are uniform black above. Taylor's (1951b) listing of this species from the city of San José, San José Province, is based on a misidentification of two examples (USNM 38147–48) of *Ninia maculata*. This species was called *Ninia lansbergii* by Burger and Werler (1954) and a few other authors. The International Commission of Zoological Nomenclature (1963) has established the priority of the name *Ninia atrata* over *N. lansbergii*.

DISTRIBUTION: Northern slopes of the Cordillera Central and the Alto Guayacán area of central Limón Province in Costa Rica and upland northwestern Panama (750–1,360 m).

Ninia maculata (Peters, 1861)

(plate 386; map 11.48)

DIAGNOSTICS: This small banded snake has a reddish to dark brown ground color, a distinct black nuchal collar, strongly keeled scales (fig. 11.5), and the venter marked with distinct black spots, dashes, or lines of black and light (fig. 11.34a–f).

DESCRIPTION: A small snake (maximum total length 352 mm); females (adults 233 to 352 mm) larger than males (adults 249 to 308 mm); head distinct from neck; body subcylindrical, with a middorsal ridge; tail moderate (20 to

25% of total length); eye small; nasal divided; a loreal, bordering orbit (some extralimital examples have small preoculars, and one has 2 preoculars excluding loreal from the orbit); 2 postoculars; supralabials 7, with 2 bordering orbit; infralabials 7; temporals 1 + 2; dorsal scales in 19 rows; ventrals 125 to 155; subcaudals 45 to 63 (44 to 63) in males, 44 to 63 (34 to 63) in females; ventrals plus subcaudals 168 to 204 (168 to 206). Dorsum reddish tan to dark brown with a series of up to 54 transverse black markings on back that are frequently reduced to a few black scales or spots posteriorly; sometimes entire body suffused with dark pigment to obscure pattern; frequently a few irregular red spots and/or short dashes present; venter white to yellow with black quadrangular or triangular spots; if present, spots scattered or tending to form paired lines or, rarely, a single midventral stripe; underside of tail uniform black or heavily marked with dark pigment.

SIMILAR SPECIES: (1) *Ninia psephota* is uniform or with very narrow light crossbands and broad dark bands and has 17 scale rows at midbody. (2) Other *Ninia* lack dark crossbands or spotting on the body.

HABITAT: Under leaf litter and other debris in Lowland Moist and Wet Forests and Premontane Moist and Wet Forest and Rainforest and marginally in Lower Montane Wet Forest.

BIOLOGY: This very common diurnal inhabitant of the leaf litter is one of the more abundant snakes in the upland areas of Costa Rica and is commonly found in gardens in the city of San José. *Ninia maculata* feeds on earthworms and slugs for the most part (Cadle 1992) but also eats soft-bodied insects and insect larvae (Burger and Werler 1954). Fitch (1970) found six gravid females (187 to 233 mm in body

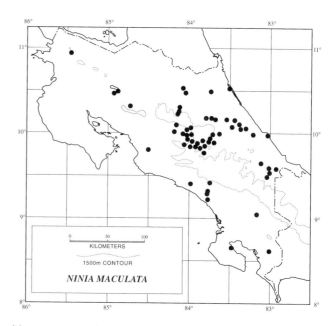

Map 11.48. Distribution of *Ninia maculata*.

length) taken on June 2 and August 30, probably all from Volcán Turrialba, Cartago Province, Costa Rica. One was dissected and contained five eggs. Young (111 to 168 mm in body length) were noted as having been taken June through August, leading Fitch to suggest that breeding occurs during most of the year. Snakes of this species flatten the entire body when alarmed, especially the head. They may then feign death by extruding the tongue and failing to right themselves when when flipped onto the back.

REMARKS: Several Costa Rica populations of this taxon have been variously recognized as distinct species (Taylor 1951b, 1954b) or subspecies (Burger and Werler 1954) based on coloration and segmental counts. Taylor (1951b) listed a Pacific slope and upland (both Atlantic and Pacific drainages) species *(Ninia maculata)* and an Atlantic lowland one *(Ninia tessellata)*. Burger and Werler (1954) pointed out that examples from the area near Turrialba, Cartago Province, geographically intermediate between the ranges of Taylor's two species, were intermediate in the features that Taylor thought distinguished between them. Burger and Werler (1954) concluded that the two nominal forms intergraded with one another and were best considered as geographic races of a single species. Taylor (1954b), on the other hand, regarded the Turrialba area specimens as representing two valid species that occurred sympatrically. Additional material now at hand (Savage and Lahanas 1991) supports the concept that a single species is involved. The presumed differences in body band counts reflect the point at which the body bands are reduced to spots. In the upland series the bands are replaced by small spots about halfway down the back, although occasionally the posterior third of the body and tail lack black markings. The number of transverse bands plus spots in snakes from the Meseta Central Occidental varies (0–11.1–24). In examples from the southwestern Pacific lowlands, the bands occur over the entire length of the body but number 21 to 31. In the Atlantic lowland samples definite bands are present along most of the back (26-40-54). Material (38 specimens) now available from the Turrialba area is intermediate between Atlantic lowland and Pacific drainage examples, having 1–21.6–38 crossbands. This series is also intermediate in total segmental scale counts (ventrals plus subcaudals 179–185.5–197) between the Atlantic lowland (168–181.5–190) and upland Costa Rica series (189–194.7–202). Pacific lowland examples agree closely with upland Costa Rican snakes in this feature (187–193.5–198). The variation in coloration, the overlap in segmental counts, and the evidence of geographic intergradation are consistent with the conclusion that only a single species is represented. Savage and Lahanas (1991) regard *Ninia pavimenta* (of Atlantic slope Guatemala), usually considered a race of *N. maculata*, as a valid species, characterized by high segmental counts (ventrals plus subcaudals 212-216-220) and a distinctive coloration of narrow light-outlined dark crossbands.

DISTRIBUTION: Evergreen forests of the Atlantic lowlands and premontane slopes of northeastern Honduras, Nicaragua, Costa Rica, and western Panama; lowland and premontane zones of Pacific Costa Rica and western Panama; also central and eastern Panama. In Costa Rica on the Atlantic lowlands and slopes, the Meseta Central, and the southwestern Pacific lowlands (36–1,800 m).

Ninia psephota (Cope, 1875)

(plate 387; map 11.49)

DIAGNOSTICS: The combination of a strongly tessellated black and light venter (fig. 11.34g–i) combined with the uniform dark or light-banded dorsum and keeled scales (fig. 11.5) is found in no other Costa Rican snake.

DESCRIPTION: A small snake reaching a total length of 494 mm; head distinct from neck; body cylindrical, with a sharp middorsal ridge; tail moderate (25 to 27% of total length); eye small; nasal divided; a loreal, bordering orbit; 2 postoculars; supralabials 6 or 7, usually 6, with 2 bordering orbit; infralabials 6; temporals 1 + 2; dorsal scale rows 17; ventrals 149 to 163 (139 to 163); subcaudals 60 to 77 in males (51 to 77), 60 to 77 (51 to 77) in females; ventrals plus subcaudals 217 to 238 (190 to 238). Dorsum gray, dark brown, to black, uniform or marked with dark bands separated by light interspaces (or crossbands); usually with an incomplete narrow light nuchal collar indicated; venter checkered with black and red, yellow, or white; iris and tongue black.

SIMILAR SPECIES: (1) *Ninia maculata* has narrow dark crossbands or spots on a lighter ground color, is never uniform in color, lacks the narrow light crossbands, and has 19 dorsal scale rows. (2) *Geophis hoffmanni* has smooth

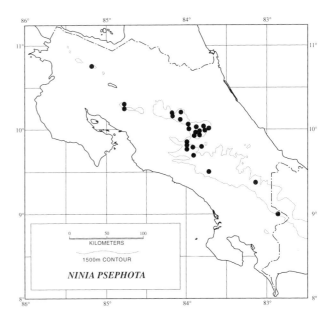

Map 11.49. Distribution of *Ninia psephota*.

scales (fig. 11.5). (3) *Geophis downsi* and *Geophis godmani* lack supraocular shields (fig. 11.30b). (4) *Geophis brachycephalus* lacks a nuchal light collar and has 15 scale rows at midbody. (5) *Liophis epinephalus* has smooth dorsal scales.

HABITAT: Found under rocks, logs, and other surface debris in Lower Montane Wet Forest and Rainforest; occasional in Premontane Wet Forest and Montane Wet Forest; often taken in pastures in areas formerly covered by these forest types.

BIOLOGY: Slevin (1942) reported this snake as common under debris on coffee plantations near Boquete, Provincia de Chiriquí, Panama, but it is relatively uncommon in Costa Rica. Stomachs from the Panama snakes contained a considerable quantity of beetle remains, but like other members of the genus the species probably eats mostly earthworms and slugs (Cadle and Greene 1993).

REMARKS: Taylor (1954b) recognized three species of *Ninia* from Costa Rica having 17 scale rows at midbody, based on *Catostoma psephota* Cope, 1875 (type locality: Costa Rica: Limón: Cerro Utyum, 1,525–2,135 m); *Streptophorus oxynotus* Werner, 1910 (type locality: Costa Rica: Alajuela: Cariblanco, 830 m); *Ninia cerroensis* Taylor, 1954b (type locality: south slope of Cordillera de Talamanca on Carretera Interamericana, near Finca Jardín, 2,286 m). These putative taxa appear to be based on variants of a single species, since there are no consistent scale features that distinguish among them (Savage and Lahanas 1991). The holotype (USNM 61971) of *N. psephota* is uniformly black dorsally and was tessellated with black and red ventrally. Material from western Panama (Slevin 1942) agrees in coloration with the type but has lower ventral counts than other samples in this series. The holotype (ZMH 818) of *N. oxynota* had definite dark body bands separated by narrow light bands, and the belly was tessellated with black and white. Most available specimens from the northern Talamaca, Central, and Tilarán ranges of Costa Rica resemble the type of *N. oxynota* in coloration. The holotype of *N. cerroensis* (KU 31935) appears to be an essentially uniformly colored snake that had died, whose epidermis began to slough off in preservation (it may have been on the verge of shedding when captured). The description of each scale as having a transparent lobe or extension, lacking striations, and the color description that says the underlying scale color is black support this view. The venter is variegated with the dark dorsal color but is not strongly tessellated. There is a narrow yellow collar (in the same position as in other *Ninia* having 17 scale rows and a collar). The types of *N. psephota* (a male with 162 ventrals and 73 subcaudals), *N. oxynota* (a female with 158 ventrals and 68 subcaudals), and *N. cerroensis* (a male with 160 ventrals and 60 subcaudals) are similar in scale counts and well within the range of variation for other Costa Rican samples. Consequently they are regarded as representatives of a single valid species. In this regard it should be noted that samples from western Panama placed in *N. psephota* have markedly lower ventral counts (139 to 153 in males, 146 to 153 in females) than the type of the name or other Costa Rican specimens. In all other respects the Panamanian snakes agree with the type of *N. psephota,* and it seems probable that the differences in ventral counts represent a south to north cline from lower to higher. Coloration also varies along this cline, with samples from southwestern Panama having red or pink and black ventral tessellations and uniform dorsums, and those from northern Talamanca and other northern regions having yellowish to white and black checkered or variegated ventrals and definite dorsal dark and light bands. The holotype of *N. psephota* has the coloration of the southern type, has a ventral count within the variation for northern samples, and comes from a site geographically intermediate between Panamanian and central and northern Costa Rican localities for the species. The type locality for *S. oxynotus* Werner, 1910 is probably in error, since Cariblanco is at an elevation of 830 m and the species is known elsewhere in Costa Rica from sites no lower than 1,400 m. Cariblanco lies at the foot of the old oxcart road that passes across the Cordillera Central between Volcanes Barba and Poás through the Pass of Desengaño, and collectors frequently worked along this road. Since *N. psephota* is known from near the pass at about 1,900 m, it seems likely that Werner's specimen came from a higher elevation along the road and was incorrectly labeled Cariblanco.

DISTRIBUTION: Evergreen forests of the upper portion of the premontane and the lower montane zones of the cordilleras of Costa Rica and western Panama; from western Panama (1,200–1,700 m) in the south through the Cordillera de Talamanca and Cordillera Central (1,300–2,771 m) to the Cordillera de Tilarán at Monteverde de Puntarenas (1,400–1,520 m) and Volcán Miravalles, Guanacaste Province, Costa Rica (about 2,000 m).

Ninia sebae (C. Duméril, Bibron, and A. Duméril, 1854)
Culebrilla del Café

(plate 388; map 11.50)

DIAGNOSTICS: The uniform red body and the black upper head surfaces, including the parietals (fig. 11.22b), will immediately identify this species.

DESCRIPTION: A small snake with a maximum total length of 386 mm; head distinct from neck; body subcylindrical with a median dorsal keel; tail moderate (20 to 24% of total length); eye small; nasal divided; a loreal, bordering orbit; almost always no preocular, rarely a small one above loreal; postoculars 1 to 3, usually 2; supralabials 6 to 8, usually 7, with 2 bordering orbit; infralabials 5 to 8, usually 7; temporals usually 1 + 2; dorsal scales 19; ventrals 132 to 148 (131 to 156); subcaudals in males 55 to 66 (41 to 74), in females 51 to 59 (37 to 65); ventrals plus subcaudals 196–206. Dorsum some shade of red, top of head including parietals

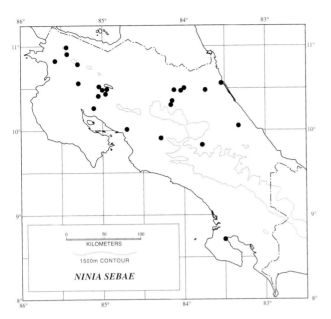

Map 11.50. Distribution of *Ninia sebae*.

black with a yellow collar followed by a broader black nuchal saddle that is separated from the red dorsal color by a yellow band; body without spots or crossbars (black spots or crossbars present in northern Central America); venter usually immaculate white, yellow, or pinkish; iris black.

SIMILAR SPECIES: (1) Juveniles of the two species of *Clelia* have a white nuchal collar that involves most of the parietal shields (fig. 11.22a). (2) Red and orange *Stenorrhina freminvillii* lack a light nuchal collar.

HABITAT: Found in forest litter in Lowland Moist and Wet Forests, Premontane Wet Forest, and gallery forests of the Lowland Dry Forest region.

BIOLOGY: A moderately common secretive nocturnal leaf litter inhabitant. Landy et al. (1966) found a caecilian in the stomach of a specimen from Chiapas, Mexico. Greene (1975) reported that this species feeds on earthworms, slugs, and small land snails. In eastern Guatemala (Stuart 1948) it reaches sexual maturity at a total length of about 250 mm. Females taken between March and April were gravid. Gaige (1936) recorded gravid females from Yucatán, taken on July 13 and 14. Meyer (1966) found an example in northern Honduras that contained three eggs. Bohuslavek (1996) found two eggs of this species in a cavity in the ground under a log in Belize in January. They were 16 × 9 mm and 18 × 8 mm. One egg hatched in March, and the hatchling snake was 123 mm in total length. Greene (1975) reported on examples from Veracruz, Mexico, laying clutches of one, two, and three eggs between August 28 and September 6. The elongate eggs were smooth and creamy white and averaged 22 × 7.8 mm. The eggs hatched seventy-five to seventy-nine days later. In Chiapas, southern Mexico, Burger and Werler (1954) reported eggs of this species found under a log on August 20 that hatched during the next four days to

produce seven hatchlings at total lengths of 123 to 128 mm. Alvarez del Toro (1960) mentioned that in Chiapas, Mexico, a clutch consists of three or four eggs. Campbell (1998) reported that eggs were 20–31 × 7–9 mm in several clutches from Chiapas, Mexico, and hatchlings were 85 to 135 mm in total length. He noted that females may lay eggs in communal nests. Schmidt (1932) found examples of this species in the stomachs of coral snakes (*Micrurus nigrocinctus*) from Guatemala and Honduras. One was palped from the stomach of an *Erythrolamprus mimus* at La Selva, Heredia Province. Greene (1975) reported that individuals of *N. sebae* flattened the entire body dorsoventrally, raised the anterior third or half of the body vertically, and cocked the head slightly downward when startled. If touched on the body the snakes moved with jerky movements forward or to the side so that the dorsal red coloration always faced the threatening stimulus. If the head or neck was touched they hid the head under the flattened and loosely coiled body. If handled, they thrashed the body and tail and extruded the cloacal contents but did not attempt to bite. Greene (1975) suggested that this behavior may mimic the tail coiling response of coral snakes when they are disturbed, with the black and yellow nuchal bands and the red dorsal coloration in *N. sebae* resembling the underside of the sympatric venomous coral snake *Micrurus diastema*. This species is one of those called vibaro de sangre by many Costa Ricans. The bite of "blood snakes" is erroneously claimed to cause death by profuse bleeding through the entire skin of the victim.

REMARKS: Schmidt and Rand (1957) and Wilson and Meyer (1985) discussed in detail the geographic variation in coloration and segmental counts for this species. Whereas the first two authors thought they saw consistent patterns in these features, Wilson and Meyer (1985) found much greater variability than previously documented in their large population samples from Honduras. The dorsal pattern in *N. sebae* may be uniform red except for the nuchal collar; may be composed of distinct narrow black bands (usually outlined by yellow, but often not) that extend over the length of the body or any portion of it; may be made up of narrow black crossbars alternating with or flanking paravertebral spots or bars; or may be formed of numerous dark spots lined up in four to six rows along the back; and several pattern components may occur on the same individual. In Costa Rica and Nicaragua all known examples lack dorsal black bands or spots. Specimens from Honduras and El Salvador exhibit a wide range of color patterns, from uniform to spotted, with the pattern extending to varying degrees down the body. Material from the northern Yucatán Peninsula comprises about equal numbers of patternless and patterned snakes, and those from Atlantic slope Mexico and Guatemala are nearly as variable as the Honduras–El Salvador sample. Pacific drainage *N. sebae* from southern Mexico and Guatemala are generally spotted. Although Schmidt and Rand (1957) emphasized mean segmental count differences as another factor distinguishing among geographic subunits within the species, Wilson and Meyer (1985) pointed out that the slight differences reflect clinal variation, one cline showing a general decrease in ventral and subcaudal counts from south to north and another a decrease from northern Honduras northwestward through El Salvador, western Guatemala, and southwestern Mexico. The specimen from Panama (UMMZ 57971) listed by Dunn (1935) as this species is *N. maculata*.

DISTRIBUTION: Lowlands and premontane slopes from Veracruz (Atlantic) and Oaxaca (Pacific), Mexico, southward over Central America, including the Yucatán Peninsula, to southern Costa Rica (4–1,100 m).

Genus *Rhadinaea* Cope, 1863

A genus of rather slender, elegant diurnal and terrestrial snakes (thirty-four species) that are generalized in morphology and retain the basic colubrid complement of head shields. The genus exhibits considerable variation but is characterized by the following combination of features: head barely distinct from neck; body cylindrical; nasal variable, usually undivided; a loreal; one or two preoculars, rarely fused with loreal; pupil round; two pairs of elongate chin shields; dorsal scales smooth, in 15 to 21 rows at midbody (keeled in some *Rhadinaea decorata*), but strong keels may be present above and/or immediately anterior to the vent; scales without apical pits except on the neck in a few individuals; anal divided; subcaudals divided; maxillary teeth 11 to 23 (usually with several posterior teeth enlarged) or 10 to 24 plus two enlarged fangs with or without a distinct diastema; anterior teeth subequal; Duvernoy's gland present; mandibular teeth subequal; no hypapophyses on the posterior vertebrae. In Costa Rica these snakes are distinctive in having a divided anal, smooth scales in fewer than 23 rows at midbody, a loreal, and preoculars combined with either a striped or uniform dorsal pattern.

The everted hemipenis is short to moderately long (5 to 14 subcaudals in length), simple or slightly bilobed, with a centrolineal, bifurcate sulcus spermaticus; most species have unicapitate organs, a few are semicapitate, and one (*Rhadinaea calligaster*) lacks a capitulum; the central third of the organ is covered by large spines, and the pedicel is covered with very small spines; calyces of the capitulum are papillate or spinulate (Myers 1974).

Myers (1974) divided this genus into eight species groups, one of which (the *lateristriga* group) is now placed in the genus *Urotheca* (Savage and Crother 1989). Cadle (1984b) noted, based on biochemical data, that the *Rhadinaea brevirostris* group (seven species) of South America (east of the Andes) was not closely related to other *Rhadinaea*. DiBernardo (1992) placed these species and several others (*E. amoena, E. cynopleura, E. melanostigma,* and *E. undulata*) in *Echinathera* Cope, 1894b. Myers and Cadle (1994) resurrected the name *Taeniophallus* Cope, "1894c," 1895 for

brevirostris and its allies and proposed a new genus *Psomophis* for *P. joberti, P. genimaculatus,* and *P. obtusus*. Zaher (1999) tentatively regarded *Echinathera* and *Taeniophallus* as dispadines. Four of the remaining six lineages in *Rhadinaea* are represented in Costa Rica, the *Rhadinaea calligaster, R. decorata, R. godmani,* and *R. vermiculaticeps* species groups.

The genus *Rhadinaea* as constituted at present occurs in humid forested areas from Nuevo León and Sinaloa, Mexico to southwestern Ecuador; a single disjunct species (*Rhadinaea flavilata*) is found in the southeastern United States.

KEY TO THE COSTA RICAN SPECIES OF THE GENUS *RHADINAEA*

1a. 17 dorsal scale rows at midbody 3
1b. 19 to 21 dorsal scale rows at midbody 2
2a. 19 dorsal scale rows at midbody; ventrals plus subcaudals 228–38 . . . . *Rhadinaea serperaster* (p. 626)
2b. 21 dorsal scale rows at midbody; ventrals plus subcaudals 253–64 . . . . *Rhadinaea godmani* (p. 625)
3a. No distinct postocular or temporal light marking, although center of postocular plate may be light; subcaudals 46–80 4
3b. A distinct light line, stripe, ocellus, or wedge-shaped marking just behind upper posterior margin of eye, extending horizontally toward neck or obliquely across temporal area toward angle of mouth or side of neck; subcaudals 85–124 . . . . *Rhadinaea decorata* (p. 624)
4a. A median black stripe on neck that expands and bifurcates at the nape; supralabials white except where crossed by the lateral black stripe; belly immaculate or heavily dotted with black, but no evidence of distinct line of spots or stripe down middle of belly; ventrals 119–34 . . . . *Rhadinaea pulveriventris* (p. 626)
4b. No distinctive bifurcate median dark stripe on neck, nape, or head; supralabials light, bordered by black; a line of small black spots or a solid stripe down middle of belly; ventrals 141–56 . . . .
. . . . *Rhadinaea calligaster* (p. 622)

Rhadinaea calligaster Group

This stock is characterized by having the body variably dark striped, sometimes almost uniform; the supralabials boldly margined with black; the last fang offset; slightly bilobed, noncapitate hemipenes with papillate calyces. Only a single species is placed in this group.

Rhadinaea calligaster (Cope, 1875)

(plate 389; map 11.51)

DIAGNOSTICS: This small relatively common high-elevation species is readily recognized by the bold black markings on the supralabial margins and the line of black

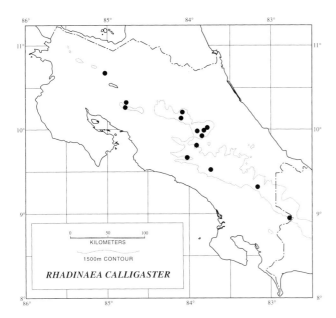

Map 11.51. Distribution of *Rhadinaea calligaster*.

figures (from small spots to large triangles or half-moons) or stripe down the middle of the ventrals. The dorsal pattern is variable (uniform to heavily striped), but no other species with smooth dorsal scales (fig. 11.5) in 17 rows at midbody has the supralabial and ventral coloration of this form.

DESCRIPTION: A small snake (most adults 300 to 500 mm in total length), maximum total length of 513 mm; females much longer than males (largest male 440 mm); tail moderate to long (20 to 29% of total length in males, 18 to 25% in females); eye moderate; 1 preocular; 2 postoculars; supralabials 7 to 9, usually 8, with 2 bordering orbit; infralabials 7 to 9, usually 8; temporals usually 1 + 1, occasionally 1 + 2, rarely 1 + 1 + 2; dorsal scales in 17 rows; supracloacal keels present in some males; ventrals 141 to 156; subcaudals 51 to 68 in males, 46 to 58 in females; ventrals plus subcaudals 205 to 209; maxillary teeth 15 to 19 + 2. Dorsal ground color light brown, olive, or dull green; head uniform or mottled, with supralabials yellow and heavily margined with black; dorsum may be essentially uniform in color, each scale having a somewhat lighter center, but always with a middorsal blackish stripe or row of small spots at least anteriorly; many individuals have additional paired dorsal, dorsolateral, and lateroventral dark stripes of varying widths; in some individuals light centers of scale rows 1, 2, and 3 strongly contrasting with dark ground color to form a series of oval spots; venter bright yellow to greenish yellow in adults, white or yellowish green in juveniles; ventrals tipped with black; midventral black markings in form of triangles or half-moons at base of each ventral, often forming median longitudinal stripe or transversely expanded across ventrals; iris brown to black; tongue black with white tips on fork. Everted hemipenis short (7 to 10 subcaudals in length), with distal third covered by papillate calyces; trun-

cus covered by large spines and pedicel by small spinules (Myers 1974).

SIMILAR SPECIES: (1) Other *Rhadinaea* and species of *Urotheca* lack the dark markings down the middle of the venter. (2) Species of *Coniophanes* lack black-margined supralabials.

HABITAT: This snake is most commonly collected under logs and surface debris in pastures and other disturbed sites. It may well be equally common in the surrounding primary forests, but the density of the vegetation and tangle of the understory make discovering individuals difficult. All Costa Rican examples are from areas originally or currently covered by Lower Montane Wet Forest or Rainforest.

BIOLOGY: This common species spends most of its time hidden under rocks, moss, ground litter, and fallen logs, although it may be found moving about during the day. Myers (1974) reported finding a communal nest of decaying palm fronds at 1,810 m on Cerro Pando in western Panama that contained thirty-eight eggs. The eggs ranged from 20 × 10 mm to 43 × 8 mm and were variously adhered to one another. The nest was discovered on May 9 and contained four adult female snakes; one female had recently oviposited, but the other three contained oviductal eggs; two other gravid females were collected in the nest on May 12 and May 30. The five gravid females contained three to five oviductal eggs well within the size range of those in the nest. These females would have contributed an additional nineteen eggs, and Myers suggested that as many as ten females had already laid eggs in the nest when it was first discovered. Seventeen eggs from the nest hatched between August 5 and 15; the rest spoiled or were opened for study. Another female from the same mountain (taken at 2,260 m), captured May 31, had recently laid eggs. Of two female examples from the south fork of the Río Las Vueltas, Heredia Province, Costa Rica (2,100 m), captured March 29 to 30, one seemed to have laid eggs recently and the second contained two eggs (29 × 8 and 32.5 × 7 mm) that appeared ready to be laid (Myers 1974).

REMARKS: According to Dunn's notes, Wettstein's (1934) record of *Rhadinaea pulveriventris* from Volcán Irazú (1,500 m), mentioned by Taylor (1951b, 116), is based on a specimen of *Rhadinaea calligaster*. Myers (1974) reached a similar conclusion.

DISTRIBUTION: Humid lower montane forests in the Cordillera de Tilarán, Cordillera Central, and Cordillera de Talamanca of Costa Rica and on Volcán Tenorio in the Cordillera de Guanacaste (1,500–2,100 m) to extreme western Panama (1,220–2,439 m).

Rhadinaea decorata Group

Members of this stock (eleven species) share the following distinctive features: variably striped with dark on a brown ground color; a pale postocular marking on temporal or outer parietal region; last fang offset; single, capitate hemipenes. Ten members of this group are montane forms confined to the sierras of eastern and southern Mexico: *Rhadinaea bogertorum* (Oaxaca), *R. cuneata* (central Veracruz), *R. forbesi* (central Veracruz), *R. gaigeae* (northeastern sierras), *R. hesperia* (Sinaloa to Guerrero and Morelos), *R. macdougalli* (Oaxaca), *R. marcellae* (San Luis Potosí), *R. montana* (Nuevo León), *R. myersi* (Oaxaca), and *R. quinquelineata* (Puebla). The single wide-ranging species *Rhadinaea decorata*, which occurs from eastern Mexico to Ecuador, is the only representative of the group in Costa Rica.

Rhadinaea decorata (Günther, 1858a)

(plate 390; map 11.52)

DIAGNOSTICS: Any small brown snake with a pair of longitudinal light stripes, distinct light postocular marks, and an immaculate orange to red posterior venter found on a lowland forest floor will be a member of this common species. Some examples (southwestern Costa Rica) have white venters.

DESCRIPTION: A small snake reaching a total length of 470 mm (most adults under 400 mm); males longer than females (more than 444 mm); tail long (35 to 47% of total length); eye moderate; 1 or 2 preoculars and frequently a subpreocular; usually 2 postoculars, sometimes 1 or 3; supralabials almost always 8, rarely 9, with 2 bordering orbit; infralabials 8 to 11, usually 10; temporals usually 1 + 2, but may be 1 + 1, 1 + 1 + 2 or have the upper secondary temporal divided vertically or the primary temporal fused to upper secondary temporal; dorsal scales in 17 rows throughout (very rarely fewer on neck or anterior to vent); scales usually smooth, although obviously striated posteriorly in many examples; rarely some or most dorsal scales with keels;

moderately strong keels frequently present in pre- and supracloacal regions; one or two apical scale pits on nuchal scales in some examples; ventrals 111 to 125 (110 to 134); subcaudals 85 to 109 (85 to 124); maxillary teeth 17 to 21 + 2 (17 to 25 + 2). Dorsal ground color a shade of brown; head marked with a black-edged postocular light spot, spots, or stripe; a narrow blackish line along upper border of supralabials, which are otherwise immaculate white or weakly dotted with dark along sutures; sides darker brown than dorsum and two areas separated by a narrow light dorsal stripe bordered by a thin black stripe below; anteriorly light stripe bordered above by a thin black line as well; median dark stripe one to five scale rows wide often present, but variable in intensity and definition; linear markings absent on most of tail length; venter and underside of tail white (southwestern Costa Rica) or salmon to red turning yellow or white anteriorly; usually a series of small black spots or dashes across tips of ventrals (sometimes confluent) that forms a solid black line on ends of subcaudals; venter otherwise immaculate, subcaudal area usually so, but some irregular dark speckling sometimes present; iris variable, upper sector tan to orange red, lower part brown to reddish brown; tongue dark brown to black. Everted hemipenes short to moderately long, 6 to 13 (5 to 14) subcaudals in length; single, unicapitate, with a bifurcate, centrolineal sulcus spermaticus; pedicel with spinules, truncus covered with 15 to 33 (15 to 50) spines, 4 or 5 of which are large hooks; capitulum with papillate calyces distally, spinulate ones proximally.

SIMILAR SPECIES: (1) *Urotheca guentheri* has a pattern of paired lateral and ventrolateral longitudinal white stripes. (2) Other striped *Urotheca* lack the distinctive black-outlined light postocular spots or stripe. (3) *Coniophanes fissidens* lacks the distinctive black-outlined postocular light spots or stripe.

HABITAT: Usually found in relatively undisturbed Lowland Moist and Wet Forests and Premontane Wet Forest and Rainforest. Occasional in secondary growth.

BIOLOGY: This common species is active in the daytime and is commonly found foraging on or resting in or under the litter on the forest floor. Food consists mostly of small *Eleutherodactylus* frogs and their terrestrial eggs, although it also eats salamanders and small lizards. Scott (1983l) emphasized the lack of knowledge on the biology of this common species.

REMARKS: Myers (1974) discussed geographic variation in coloration, segmental counts, maxillary dentition, and hemipenes. The variation in shape and size of the latter is especially interesting. Two morphologically distinct types occur in Costa Rica. Atlantic lowland males have an elongate organ (11 to 13 subcaudals in length) with three-fourths of the length covered by spinules and 28 to 33 spines on the truncus. In other examples the organ is 6 to 9 subcaudals in length and bulbous; the spinulate region is only about

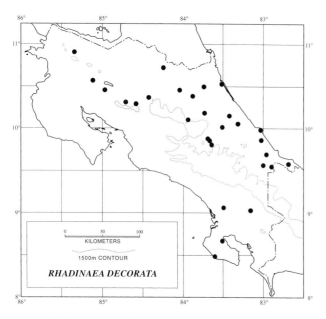

Map 11.52. Distribution of *Rhadinaea decorata*.

one-third the length of the organ, which has 15 to 18 central spines.

DISTRIBUTION: Humid broadleaf evergreen forests in the lowland and premontane zones from southeastern San Luis Potosí south to southern Oaxaca and northern Chiapas in Mexico; Atlantic slope Guatemala, Nicaragua, Costa Rica, and Panama; southwestern Costa Rica to Ecuador. Unknown from El Salvador, Honduras, and Colombia. The species is widespread on the Atlantic versant and is known from several southwestern Pacific localities in Costa Rica (2–1,200 m). It also ranges from the Atlantic slope across the low continental divide in northern Costa Rica onto the Pacific versant (395–561 m), but not onto the lowlands or lower slopes in the Pacific northwest.

Rhadinaea godmani Group

The fifteen species composing this group usually have the body striped or with light dashes on dark scales; usually an oblique pale bar from the lower part of the eye to the mouth, a U-shaped dark mark on the rostral, and dark-edged supralabials; last fang not offset; diastema absent or short and variable in position; hemipenes single or slightly bilobed, with a naked basal pocket, capitate. The members of this stock range in montane situations from Veracruz and Oaxaca, Mexico, south to western Panama. The highest concentration of species (twelve) occurs in northern nuclear Central America: *Rhadinaea anachoreta* (Guatemala), *R. hannsteini* (Chiapas, Mexico, and Guatemala), *R. hempsteadae* (Guatemala), *R. kanalchutchan* (Mexico and Guatemala), *R. kinkelini* (Guatemala to Nicaragua), *R. lachrymans* (Chiapas, Mexico, and Guatemala), *R. montecristi* (El Salvador and Honduras), *R. pilonaorum* (Guatemala and El Salvador), *R. posadai* (Guatemala), *R. rogerromani* (Nicaragua), *R. schistosa* (Oaxaca and Veracuz, Mexico), *R. stadelmani* (Guatemala), and *R. tolpanorum* (northern Honduras). Only two members of the group, *Rhadinaea godmani* and *Rhadinaea serperaster,* are found in lower Central America.

Rhadinaea godmani (Günther, 1865)

(plate 391; map 11.53)

DIAGNOSTICS: One of several small snakes having one or more series of dark stripes on a pale to medium brown background. *R. godmani* is most easily distinguished from similarly patterned snakes by having 21 rows of scales at midbody and an immaculate venter (white in juveniles, yellow in adults).

DESCRIPTION: A small snake attaining a maximum length of 568 mm; males longer than females (to 544 mm); tail moderate to long (23 to 32% of total length); eye moderate; usually 1 preocular, rarely 2; no subpreocular; usually 2 postoculars, very rarely 1; supralabials usually 8, rarely 7 or 9, with 2 bordering orbit; infralabials 8 to 10, usually 9; temporals usually 1 + 2, but some individuals have 1 + 1

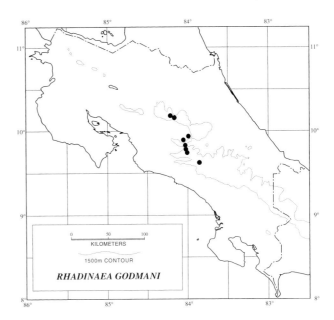

Map 11.53. Distribution of *Rhadinaea godmani.*

or 1 + 2 + 3 or have the secondary temporals vertically divided; dorsal scale rows almost always 21 throughout, sometimes reduced anterior to vent; supracloacal keels in some males and females; ventrals 169 to 180 (156 to 186); subcaudals 73 to 94 (71 to 95); ventrals plus subcaudals 253 to 264; maxillary teeth 18 to 20 (16 to 20), the last three to five heavier and somewhat larger than the others. Upper surfaces pale to medium brown with darker longitudinal stripes on body and tail; often a pale tan nuchal collar (one scale row wide), usually interrupted by the dark vertebral stripe; usually a black-bordered postocular light bar that slants obliquely from eye to mouth; a dark vertebral stripe and usually one to three dark stripes on each side; underside of head, venter, and subcaudal surfaces cream to bright yellow, immaculate or with scattered dark punctations or a median dark streak or row of dots under tail; ventral tips often involved in lateroventral dark stripe; iris light brown, tongue black. The everted hemipenis short, 7 to 9 subcaudals in length in Costa Rican examples (short to moderate elsewhere, 7 to 13 subcaudals long); very slightly bilobed and semicapitate with a centrolineal, bifurcate sulcus spermaticus; pedicel covered with spinules, truncus with about 50 small to medium-sized spines that surround a nude pocket on the asulcate side; distal calyces of capitulum papillate, proximal ones spinulate (Myers 1974).

SIMILAR SPECIES: (1) *Rhadinaea serperaster* has 19 scale rows at midbody.

HABITAT: This rather common snake is found under logs and surface debris in Lower Montane Wet Forest and Rainforest or in pastures or other areas where the forest has been removed.

BIOLOGY: A female of this species (CRE 2983) from Tres Ríos, Cartago Province (1,345 m), laid five eggs.

REMARKS: The nominal species *Rhadinaea altamontana* Taylor (1954b), from the Cordillera de Talamanca, was synonymized with *R. godmani* by Myers (1974). The holotype (KU 30962) of *R. altamontana* was collected in the northern Cordillera de Talamanca (Costa Rica: Cartago: National Forest Reserve off the Carretera Interamericana) at about 2,200 m. It is unusual in having no distinctive lateral stripe, and the dorsal ground color is pale tan. Myers regarded this specimen as conspecific with *R. godmani* because typical examples were known from both the north and south in the Cordillera de Talamanca–Chiriquí massif and the scutellation of this example fell within the variation for *R. godmani*. Recently a second specimen (CRE 7510) having the *R. altamontana* color pattern was obtained from a somewhat higher elevation (Cartago: Los Lagos, about 2,600 m) near the type locality. This snake has 178 ventrals and 86 subcaudals, and except for the color pattern it agrees in every respect with examples of *R. godmani*. McCranie and Wilson (1992) provided an updated description and illustration of the species.

DISTRIBUTION: Highland areas from southern Oaxaca and Chiapas, Mexico, south to southeastern Honduras (1,450–2,650 m); the lower montane zone of the Cordillera Central and Cordillera de Talamanca of Costa Rica (1,300–2,600 m) and adjacent western Panama (1,200–2,200 m).

Rhadinaea serperaster Cope, 1871

(plate 392; Map 11.54)

DIAGNOSTICS: Another small snake with a complex pattern of body stripes on a pale tan to brown ground color. This form differs from others with a similar pattern in having 19 scale rows at midbody.

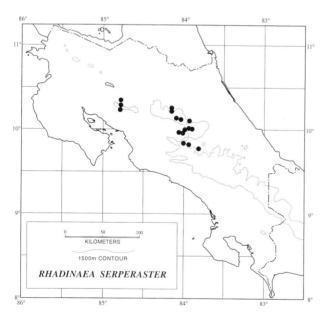

Map 11.54. Distribution of *Rhadinaea serperaster*.

DESCRIPTION: A small species, maximum total length to 445 mm (longest male 368 mm); tail moderate to long (23 to 28% of total length); eye moderate; usually 1 preocular, rarely 2; no subpreocular; 2 postoculars; supralabials 8, with 2 bordering orbit; infralabials usually 8, occasionally 9; 1 + 2 temporals; dorsal scales in 19-19-19 rows, supracloacal keels present on some males and females; ventrals 156 to 172; subcaudals 66 to 79; ventrals plus subcaudals 228 to 238; maxillary teeth 16 to 17, the last five somewhat enlarged and heavier than others. Middorsal area light to medium brown; lateral areas tan, with distinct longitudinal dark stripes on body and tail; no nuchal collar; a black-bordered postocular light bar slanting obliquely from eye to mouth; a dark vertebral stripe and two or three lateral dark stripes on each side; underside cream (in preservative), immaculate or with a few dark markings posteriorly or under tail; ventral tips dark, often involved in lateroventral stripe; iris brown; tongue black. The retracted hemipenis is short, 7 subcaudals in length, apparently very slightly bilobed, unicapitate, with a bifurcate and probably centrolineal sulcus spermaticus; pedicel covered with minute spinules, truncus covered by 50 or so small to medium-sized spines that surround a nude pocket on the asulcate side; calyces of capitulum spinulate (Myers 1974).

SIMILAR SPECIES: (1) *Rhadinaea godmani* has 21 scale rows at midbody.

HABITAT: Found in Premontane Moist and Wet Forests and Lower Montane Wet Forest and Rainforest environments, under moss on logs or under debris.

REMARKS: Dunn's (1947) listing of this species from western Panama is based upon examples of *R. godmani* (Myers 1974).

DISTRIBUTION: Upper part of the premontane and the lower montane life zones of the Cordillera de Tilarán, Cordillera Central, and northern Cordillera de Talamanca of Costa Rica (1,200–2,200 m).

Rhadinaea vermiculaticeps Group

The three species *Rhadinaea pulveriventris* (Costa Rica and Panama), *R. sargenti* (Panama), and *R. vermiculaticeps* (Panama), placed in this group share the following diagnostic features: some dark stripes or bands (usually a middorsal dark band); last fang offset; hemipenes (unknown in *Rhadinaea pulveriventris*) single, unicapitate (with papillate calyces), and with a naked basal pocket. The group is restricted to upland areas of lower Central America.

Rhadinaea pulveriventris Boulenger, 1896d

(plate 393; map 11.55)

DIAGNOSTICS: This rare, handsome species is unique in having the combination of a pattern of paired narrow dark stripes (best developed anteriorly) and a median black streak on the neck that expands and bifurcates on the nape to form a wineglass-shaped marking.

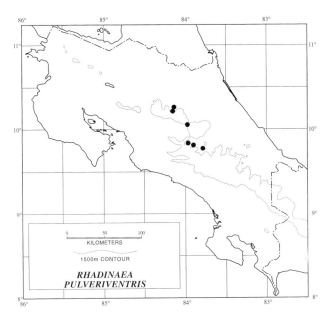

Map 11.55. Distribution of *Rhadinaea pulveriventris*.

DESCRIPTION: A small species reaching 502 mm in total length in females, no adult males known; tail long (26 to 32% of total length); eye moderate; 1 preocular; no subpreocular; 2 postoculars; supralabials 8; infralabials 8 to 11, usually 10, with 2 bordering orbit; temporals 1 + 2; dorsal scales in 17-17-17 rows; weak supracloacal keels in one female; ventrals 119 to 134; subcaudals 63 to 80; ventrals plus subcaudals 194 to 199; maxillary teeth 18 to 19 followed after diastema by two enlarged fangs, last fang offset laterad. Dorsum brown, with longitudinal dark stripes on body, less distinct posteriorly and on tail; black stripe from rostral through eye continuous with lateral stripe, bordered on head and neck by a pale light field or stripe; a wineglass-shaped dark marking on neck and nape in form of a median dark neck stripe continuous with a dark bifurcate nape area; a single narrow and somewhat diffuse longitudinal dark stripe along each side of body; ventral surface immaculate to heavily dotted with black, or sometimes some black pigment on tips of ventrals; iris reddish tan above, darker below; tongue reddish brown with black tips.

SIMILAR SPECIES: (1) *Coniophanes fissidens* has no distinctive dark marking on the neck and nape and 19 to 21 scale rows at midbody. (2) *Leptodrymus pulcherrimus* has a pair of broad (on 2½ scale rows) longitudinal dark stripes and no median nuchal dark marking.

HABITAT: Found in Premontane and Lower Montane Rainforests.

BIOLOGY: Myers (1974) reported on two clutches, one of four eggs (UMMZ 117651) and the other of five (UMMZ 117652), that were hatched at the University of Michigan.

REMARKS: The specimen (MCZ 15261) of this species from Navarro, Cartago Province (1,100 m), was probably collected up the Río Navarro at a higher elevation above

the village of that name. The nine hatchlings (UMMZ 117651–52) of this species mentioned above are listed as from 14.1 miles north of Varablanca, Heredia Province (4,500 feet [1,372 m]). The recorded elevation for these snakes is probably correct. The distance (22.5 km) is patently incorrect. The road running north from Varablanca drops to an altitude of 1,372 m between the waterfall on the Río La Paz Grande and Cinchona, about 6.5 km straight line north-northwest of Varablanca or 8.5 km by road. A distance of 22.5 km by road north of Varablanca would be out of the mountains and down on the Atlantic lowlands near 500 m in altitude.

DISTRIBUTION: The upper portion of the premontane and adjacent lower montane zones in the Cordillera Central and Cordillera de Talamanca in Costa Rica (1,372–1,600 m) and immediately adjacent western Panama 1,000–1,600 m).

Genus *Sibon* Fitzinger, 1826

One of the four genera of specialized snail- and slug-eating snakes confined to the Neotropics (*Dipsas*, *Sibynomorphus*, and *Tropidodipsas* are the others). Members of *Sibon* (ten species) have a head that is very distinct from the narrow neck and compressed body and large eyes that are not markedly protuberant in most forms but definitely so in one species, *Sibon argus*. The basic colubrid complement of enlarged head shields is present, but the penultimate supralabial is in contact with the postoculars and primary and secondary temporals and is greatly enlarged and higher than the other labials. These features, in combination with the following characteristics, will distinguish the genus from other colubrids: nasal entire, partially divided, or divided; a loreal; 1 preocular or none; 2 or 3 (rarely) postoculars; no suboculars; pupil vertically elliptical; usually three but sometimes two or four pairs of chin shields, first pair elongate, crescent-shaped; mental groove more or less developed; dorsal scales smooth, in 13 or 15 rows, without apical pits; anal entire; subcaudals divided; 12 to 17 maxillary teeth, anterior ones largest, the rest subequal; mandibular teeth decreasing in length posteriorly; no hypapophyses on posterior dorsal vertebrae. The presence of a mental groove, the lack of a dorsal scale reduction, loreal bordering the orbit but no preocular in all but one species (the extralimital *Sibon sanniolus*), a divided anal, and the smooth scales will separate Costa Rican *Sibon* from other slender snakes with large heads and eyes.

The everted hemipenes are short (7 to 8 subcaudals long); single with a bifurcate, centrolineal sulcus spermaticus; unicapitate; four large and several smaller hooked basal spines present; a spinous truncus and a capitulum covered by spinulate calyces.

Taylor (1951b, 1954b) placed most of the Costa Rican species now referred to *Sibon* in the genus *Dipsas*. Kofron (1985) has argued that members of the genus *Tropidodipsas* ought to be placed in *Sibon* because of resemblance in cranial osteology, hemipenes, and diet. He further suggested that *Sibon carri* (El Salvador and Pacific versant Honduras) and *Tropidodipsas fischeri* (Mexico, Guatemala, El Salvador, and Guatemala) are not congeneric with his inclusive *Sibon* assemblage. Wallach (1995) presented evidence that supports recognizing *Sibon* (including *S. carri*) and *Tropidodipsas* (six species) as separate genera, with the status of "*T.*" *fischeri* unresolved. The matter is of little consequence here, since all Costa Rican forms are unequivocal *Sibon* (sensu stricto). Peters (1960) recognized three species groups within the genus, all of them represented in Costa Rica. All members of the genus are nocturnal and arboreal. *Sibon*, as defined by Wallach (1995), range through the lowlands and into the premontane zone from Nayarit (Pacific slope) and Veracruz and Oaxaca (Atlantic versant), Mexico, south to northern South America, including Pacific versant Colombia and Ecuador, Atlantic drainage Colombia, Venezuela, the Guianas, adjacent Brazil, and upper Amazonian drainage Ecuador. Smith (1982) presented reasons why the name *Sibon* should be regarded as masculine, and I follow his reasoning here.

KEY TO THE COSTA RICAN SPECIES OF THE GENUS *SIBON*

1a. 15 scale rows at midbody 2
1b. 13 scale rows at midbody; a pair of infralabials in contact posterior to mental, so that no accessory postmental scale lies between paired chin shields and mental (fig. 11.35a); ventrals 166–88; 16–31 black body rings alternating with light (red dorsally in life) rings . . . . . .
. . . . *Sibon anthracops* (p. 630)
2a. Dark components of dorsal pattern continue well onto or across venter (fig. 11.37a); enlarged penultimate supralabial not bordering orbit (fig. 11.36a). 3
2b. Dark components of dorsal pattern extending only to tips of ventrals, venter mostly light with dark dashes (fig. 11.37b,c); enlarged penultimate supralabial bordering orbit (fig. 11.36b,c). 5
3a. No infralabials in contact behind mental (fig. 11.35b,c); often with a single large, a single small, *or* paired small accessory postmental scales that separate mental from first pair of chin shields 4
3b. First pair of infralabials in contact behind mental so that no accessory postmental scales lie between paired chin shields and mental (fig. 11.35a); ventrals 174–90; pattern of irregular dark dorsal bands or rings on a light (white, yellow, or pink) ground color that is usually heavily invaded by dark spots and streaks. . . . .
. . . . *Sibon nebulatus* (p. 634)
4a. A single large postmental (fig. 11.35b); dorsal pattern of red, orange, or light bands or rings that are much lighter than interspaces; center of bands or rings much lighter than margins; ventrals 178–200 . . . .
. . . . *Sibon dimidiatus* (p. 631)

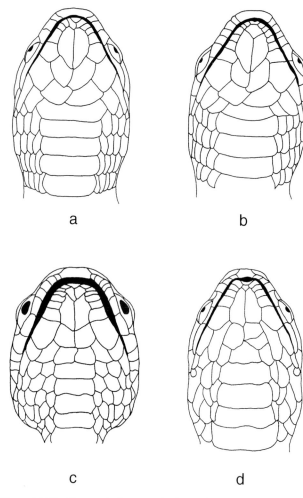

a   b

c   d

Figure 11.35. Diagnostic features of gular scalation in Costa Rican snakes of the genus *Sibon*: (a) *Sibon nebulatus* showing condition where the first pair of infralabials meets posterior to mental scale; (b) *S. dimidiatus* with a single large postmental preventing infralabial contact; (c) *S. annulatus* showing paired postmentals preventing infralabial contact; (d) *S. longifrenus* with a single small postmental preventing infralabial contact.

4b. Usually with a pair of postmentals, sometimes partially fused or reduced to a tiny single scale (fig. 11.35c); dorsal pattern of reddish, brownish, or gray bands or rings that are darker than light pale yellow, gray, or greenish interspaces; bands or rings essentially unicolor; ventrals 161–84. . . . . *Sibon annulatus* (p. 629)
5a. Eyes protuberant, snout blunt (fig. 11.36b); ventrals 192–201; subcaudals 115–21 . . *Sibon argus* (p. 632)
5b. Eyes nonprotuberant, snout acuminate (fig. 11.36c); ventrals 151–73; subcaudals 82–103. . . . .
. . . . *Sibon longifrenis* (p. 633)

Sibon annulatus Group

These snakes have an enlarged penultimate supralabial separated from the orbit by the lower postocular and a dorsal pattern of dark spots or of dark bands or rings that extend onto or across venter. *Sibon sanniolus* (Yucatán Peninsula)

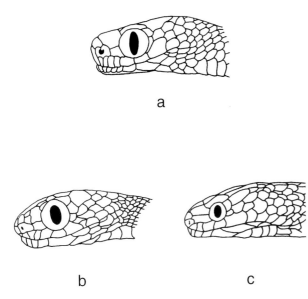

a

b   c

Figure 11.36. Lateral view of heads in Costa Rican snakes of the genus *Sibon*: (a) *S. dimidiatus* showing large eye and enlarged supralabial separated from orbit by lower postocular; (b) *S. argus* with large eye and enlarged supralabial bordering orbit; (c) *S. longifrenus* with small eye and enlarged supralabial bordering orbit.

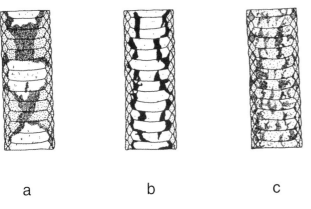

a   b   c

Figure 11.37. Ventral color patterns at midbody in Costa Rican snakes of the genus *Sibon*: (a) ringed pattern in *S. annulatus*; (b) zigzag pattern on lateral edges of ventrals in *S. argus*; (c) spotted and stippled pattern in *S. longifrenus*.

is the only other species in this group besides the three forms found in Costa Rica, *Sibon annulatus, Sibon anthracops,* and *Sibon dimidiatus.*

Sibon annulatus (Günther, 1872)

(plate 394; map 11.56)

DIAGNOSTICS: Any snake with a compressed body, distinct head, large eyes, a mental groove, and a pattern of reddish, brownish, or gray bands and rings on a lighter ground color will be immediately recognizable as this species, except for occasional juvenile examples of *Sibon dimidiatus. S. annulatus* may be distinguished from the questionable examples of *S. dimidiatus* by usually having two postmental

plates, or these are sometimes partially fused to one another or reduced to a tiny scale (fig. 11.36c). In *S. dimidiatus* there is a single large postmental plate (fig. 11.36b).

DESCRIPTION: A small snake (total length to 557 mm); tail very long (32 to 37% of total length); nasal entire or partially divided; no preoculars; a loreal, bordering orbit; usually 2 postoculars, occasionally 3; supralabials 7 or 8 (rarely 9), with 2 bordering orbit; 7 to 9 infralabials, usually 9, none in contact behind mental; temporal formula usually 1 + 2, rarely 2 + 2 or 2 + 3; a single small or two small or large postmentals (sometimes partially fused), followed by two to four pairs of enlarged chin shields, two of which border mental groove; dorsal scale rows 15-15-15, vertebral row scarcely or moderately enlarged; ventrals 161 to 184 (161 to 192); subcaudals 108 to 135. Dorsal pattern of 26 to 41 dark-outlined reddish, brownish, or gray rings and bands that are darker and much broader than the light pale yellow, gray, or greenish interspaces; rings and bands essentially unicolor; interspaces heavily suffused with dark pigment laterally, clear dorsally and toward venter; venter yellowish; dorsal dark markings narrowing abruptly on first scale row but continuing on venter, usually forming rings or often alternating on the two sides and becoming dark brown or black; tail similarly colored; iris dark reddish brown.

SIMILAR SPECIES: (1) *Sibon dimidiatus* has a single large postmental (fig. 11.35b) and dorsal body bands that are much lighter than the interspaces. (2) *Dipsas articulata* has 14 to 18 reddish brown body rings. (3) *Dipsas bicolor* has 14 to 17 dark body rings. (4) *Dipsas tenuissima* has a preocular (fig. 11.4). (5) *Micrurus mipartitus* has a cylindrical body, the head not distinct from the neck, and small eyes (fig. 11.48a). See key to coral snakes (p. 706), figure 11.47, and table 11.2.

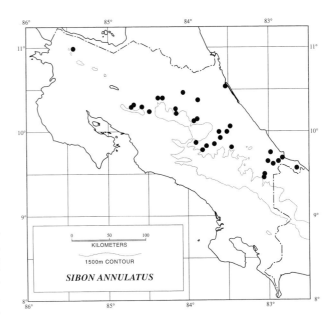

Map 11.56. Distribution of *Sibon annulatus.*

HABITAT: Relatively undisturbed Atlantic Lowland Moist and Wet Forests, Premontane Wet Forest and Rainforest.

REMARKS: This uncommon species is very similar to *Sibon dimidiatus* (sensu Peters 1960 and Wilson and Meyer 1985) and to the snakes referred here to that form from the uplands and southwestern Pacific versant Costa Rica. Peters (1960) suggested that the two nominal forms may represent a single species, a view Kofron (1990) has echoed. Typical *S. annulatus* occurs along the Atlantic slope of eastern Panama and Costa Rica, with a single record from Pacific versant Panama (FMNH 238196, Coclé: El Valle de Antón). These snakes usually have the postmental divided into two large scales or, less commonly, reduced to a single small scale that may be divided; they have 161 to 184 ventrals and the color pattern described above. *S. dimidiatus* is completely allopatric to *S. annulatus*, with a range from Veracruz, Mexico, to Atlantic slope Nicaragua on the Atlantic slope with an outlying population in the Pacific lowlands of Guatemala and Honduras. Examples of this stock have a single large postmental scale, and ventrals number 178 to 200. A Costa Rican population agreeing with *S. dimidiatus* in scale features and usually differing from *S. annulatus* in coloration is regarded here as *S. dimidiatus*, as discussed under the latter name, where I summarize reasons for recognizing the two taxa as distinct species. Peters (1960) demonstrated that *Leptognathus pictiventris* Cope, 1875, is a synonym of *S. annulatus*. An apparently undescribed species of *Sibon* (plate 395) occurs with *S. annulatus* at some localities on the Atlantic versant in Costa Rica.

DISTRIBUTION: Lowland and premontane evergreen forests of the Atlantic versant of eastern Honduras, Nicaragua, Costa Rica, and western Panama and the Darién region of Pacific slope eastern Panama; a single record from a similar habitat on Pacific slope western Panama (2–1,300 m).

Sibon anthracops (Cope, 1868a)

(plate 396; map 11.57)

DIAGNOSTICS: No other Costa Rican snake has a compressed body, distinct head, and large eyes combined with a dorsal pattern of alternating black and light rings that are red dorsally and 13 scale rows at midbody.

DESCRIPTION: A small snake reaching a total length of 544 mm, with a moderate to long tail (24 to 31% of total length); nasal divided; no preocular; a loreal, bordering orbit; usually 2 postoculars, rarely 3; supralabials 6 to 8 usually 7, almost always with 2 bordering orbit; infralabials 7 to 9, usually 8, first pair in contact behind mental; temporals 1 + 2, rarely 2 + 2 or 1 + 3; no postmental; three pairs of chin shields, the first two pairs bordering mental groove; dorsal scales 13-13-13, rarely 13-15-13, vertebral row not enlarged; ventrals 166 to 188; subcaudals 73 to 91. Dorsum tricolor (cream, red, and black) with 16 to 31 black rings, some with two halves alternating or only one half of dorsal ring present; interspaces between black rings cream marginally and suffused with red over most of dorsal surface and often lateral surfaces as well; interspaces horizontally streaked and dashed with black; black rings narrowing laterally and continuing around body, although some are offset so that the two halves alternate, or rarely incomplete across venter on posterior one-half to two-thirds of body; ventral interspaces cream; tail marked like body; anterior portion of head black with a distinct light nuchal collar that may or may not include tips of parietal shields; iris black.

SIMILAR SPECIES: (1) The species of *Dipsas* lack elongate chin shields and a mental groove (fig. 11.14e). (2) *Scolecophis atrocinctus* has a divided anal (fig. 11.5), and the supralabial under the first temporal is not greatly enlarged. (3) *Urotheca euryzona* has a divided anal (fig. 11.5) and 17 rows of dorsal scales. (4) *Micrurus mipartitus* has a very short, stubby tail and small eyes and lacks the enlarged supralabial under the first temporal (fig. 11.48a). (5) Other ringed species of *Sibon* lack red saddles in the light dorsal interspaces and have one or two postmentals (fig. 11.35b–d). (6) Other ringed snakes in Costa Rica have red on the venter. See key to coral snakes (p. 706), figure 11.47, and table 11.2.

HABITAT: Widespread in Lowland Dry Forest environments and occurring marginally in Pacific Lowland Moist Forest and Premontane Moist Forest.

BIOLOGY: Although this relatively common species, like others in the genus, is essentially arboreal, it is often found moving about on the ground or dead on the road at night. It is a somewhat more robust form than other Costa Rican species, a possible correlate of less arboreal habits. Kofron (1987) confirmed that this species feeds primarily on slugs.

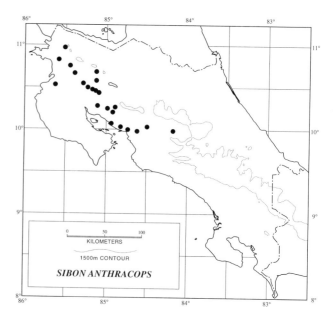

Map 11.57. Distribution of *Sibon anthracops*.

He reported three females with potential clutches of three eggs, one of which (from El Salvador) would have laid the eggs in July.

REMARKS: The holotype of *Sibymorphus ruthveni* Barbour and Dunn, 1921 (MCZ 15549), is unusual in having 15 dorsal scale rows over the posterior quarter of the body. In all other respects this specimen from Montes del Aguacate, Alajuela Province, agrees with typical *S. anthracops*.

DISTRIBUTION: Semiarid and subhumid lowlands from Guatemala through western Nicaragua (Atlantic and Pacific drainages) and northwestern Pacific Costa Rica to the Meseta Central Occidental; a single record from semiarid north-central (Atlantic versant) Honduras (4–915 m).

Sibon dimidiatus (Günther, 1872)

(plate 397; map 11.58)

DIAGNOSTICS: The compressed body, distinct head, and large eyes combined with the presence of a mental groove, a single large postmental (fig. 11.35b), and an adult dorsal pattern of red, orange, or light, dark-outlined rings and bands that are lighter than the interspaces and usually continue onto or across the venter (fig. 11.37a) readily distinguish this species. Some juveniles of this form resemble *Sibon annulatus* in coloration, but that species has paired postmentals (fig. 11.35c) or a single tiny one.

DESCRIPTION: A small to moderate-sized snake (total length to 678 mm); males much longer than females; tail moderately to very long (30 to 33% of total length); nasal entire or partially divided; usually no preocular (a small one present on one side in one non–Costa Rican specimen); a loreal, bordering orbit; 2 postoculars; supralabials 7 to 9 (usually 8), with 2 bordering orbit; infralabials 9 or 10, none in contact behind mental; temporals usually 1 + 2, rarely 1 + 3 or 2 + 3; a single large postmental followed by three or four pairs of chin shields, the first two pair bordering the mental groove; dorsal scales 15-15-15, vertebral row may be slightly enlarged on posterior portion of body; ventrals 178 to 200; subcaudals 107 to 125 (106 to 126). Dorsal pattern of 30 to 45 (23 to 45) red, orange, or light rings and bands, usually outlined by dark pigment; rings and bands usually lighter than interspaces and sometimes uniform in color; dorsal light-centered figures usually continuing well onto venter or crossing it; venter cream, usually with a series of pale-centered dark-outlined alternating lateral blotches that lie opposite dorsal interspaces and are confluent with them (corresponding to ventral ground color between extensions of dorsal bands or rings); tail marked like body; iris brown, tongue black.

SIMILAR SPECIES: (1) *Sibon annulatus* has paired postmentals (fig. 11.35c) or a single small one and dorsal bands that are uniform in color. (2) *Sibon nebulatus* has well-developed continuations of the dark dorsal bands on the venter. (3) Species of *Imantodes* lack pale-centered light lateroventral spots and rings or bands on venter. (4). *Micrurus mipartitus* has a cylindrical body, the head not distinct from the neck, and small eyes. See key to coral snakes (p. 706), figure 11.47, and table 11.2.

HABITAT: Undisturbed Pacific Lowland Wet Forest, Premontane Moist and Wet Forest and Rainforest, and marginally into Lower Montane Wet Forest.

BIOLOGY: Several examples of this uncommon snake have been collected from tank bromeliads, where they appeared to be hiding during the day. Three recent hatchlings were found together in leaf litter on November 14 at Monteverde, Puntarenas Province, suggesting that the eggs are laid on the forest floor. McCoy (1990) reported that a female from Belize laid two eggs.

REMARKS: Costa Rican specimens referred to this species agree with *Sibon dimidiatus* from Mexico to Nicaragua in having a single large postmental and a high number of ventrals and in adult coloration. The northern population is separated from the nearest known Costa Rican record by approximately 200 km (between Nicaragua: Matagalpa: Hacienda Rosa de Jericó (= Jericó) and Puntarenas: Cerro Cacao). Other Costa Rican examples are from the southwestern Pacific lowlands and premontane slopes. This form is closely allied to *Sibon annulatus* of Atlantic slope Costa Rica and eastern Panama (both versants). Peters (1960) suggested that *S. annulatus* and *S. dimidiatus* may be conspecific, since they seemed to be completely allopatric and differ primarily in the postmental character states. At that time no putative *S. dimidiatus* were known from Costa Rica, but recently collected examples from the Pacific versant leave little doubt that both forms occur in the republic. However, the two are allopatric, since *S. annulatus* is restricted to the Atlantic versant and *S. dimidiatus* to Pacific drainage sites.

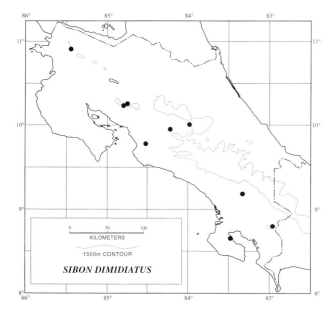

Map 11.58. Distribution of *Sibon dimidiatus*.

S. annulatus consists of two allopatric populations, one from the Atlantic versant of Costa Rica and western Panama and a second from El Valle de Antón, Coclé Province, on the Pacific slope of Panama. *S. dimidiatus* comprises several allopatric populations as well: Atlantic lowlands and premontane slopes from Veracruz, Mexico, to Nicaragua; Pacific slopes of Guatemala; Pacific slopes of Honduras; and Pacific slopes of Costa Rica. Wilson and Meyer (1985) have pointed out that the color pattern features emphasized by Peters (1960) and used here to characterize the two species in Costa Rica have considerably more variability in *S. dimidiatus* than previously thought. Peters (1960) recognized his Guatemalan specimens of *S. dimidiatus* as a distinct race *(S. dimidiatus grandoculis)* because they had a pattern similar to that of *S. annulatus* (unicolor dark bands). Wilson and Meyer (1985) showed that this feature is variable in adult *S. dimidiatus* from Honduras and that juveniles have unicolor bands. The color pattern differences appear to consistently separate the adults of the two species in Costa Rica and Panama. One juvenile specimen (CRE 4678; Costa Rica: Puntarenas Province: Monteverde) was called *S. annulatus* by Hayes, Pounds, and Timmerman (1989) at my suggestion. This example, if properly identified, would establish sympatry of *S. annulatus* with *S. diminiatus*, which is also known from Monteverde. The questionable specimen has a single moderate-sized postmental, 195 ventrals, and 34 black-bordered uniform reddish brown body blotches that are darker than the interspaces. The color pattern is similar to that found in adult *S. annulatus* and differs from that typical of slightly larger juvenile *S. dimidiatus* in Costa Rica. This snake agrees in characteristics with *S. dimidiatus* (high ventral count) and *S. annulatus* (coloration and a single postmental that is smaller than in *S. dimidiatus*) and might be regarded as a representative of either form. The intermediate nature of the postmental and the combination of high ventral count (at 195, eleven more than for *S. annulatus*) and the color pattern, which resemble the condition in *S. annulatus*, suggest possible intergradation between the two putative species. I would favor the latter interpretation except that four additional examples from Monteverde (CRE 4604A, 4604B, 7210A, 7210B) are unquestionably *S. dimidiatus*. In these circumstances it seems best to regard this example as an unusually colored *S. dimidiatus* based on the resemblances to that form in scutellation, while admitting other possible interpretations.

DISTRIBUTION: Evergreen forests in the lowlands and on the premontane slopes of the Atlantic versant from Oaxaca and Veracruz, Mexico, through Central America to eastern Nicaragua; in similar habitats on the Pacific versant in Guatemala, Honduras, and Costa Rica; in Costa Rica known from Cerro Cacao, the Cordillera de Tilarán, the Meseta Central Occidental, and the Pacific southwestern region (40–1,600 m).

Sibon argus Group

Members of the group have an enlarged penultimate supralabial bordering the orbit, the dark components of dorsal markings extending only to tips of ventrals, and a dorsal pattern of ocellate markings. Both species assigned to this group, *Sibon argus* and *Sibon longifrenis*, occur in Costa Rica.

Sibon argus (Cope, 1875)

(plate 398; map 11.59)

DIAGNOSTICS: The dorsal pattern of dark-bordered ocelli combined with the compressed body, distinct, blunt head, and large and protuberant eyes (fig. 11.36b) distinguish this rare species from all other forms.

DESCRIPTION: A small snake (maximum total length 690 or more); tail long (28 to 33% of total length); nasal divided; no preoculars; a loreal, bordering orbit; 2 or 3 postoculars; supralabials 6 to 9, usually 7, with 3 bordering orbit; infralabials 6 to 9, first pair usually not in contact behind mental; temporals 1 + 2 or, rarely, 2 + 3; no postmental; three pairs of chin shields, first two bordering mental groove; dorsal scales 15-15-15, vertebral row not enlarged; ventrals 192 to 201 (182 to 201), subcaudals 115 to 121 (112 to 121). Dorsal ground color greenish gray with a paravertebral series of 30 to 55 light, dark-bordered yellow to orangish ocelli, usually alternating at middorsal axis and usually fused across midline (absent anteriorly in some specimens); a lateral series of yellow to orangish black-bordered ocelli on first three scale rows; a lateral ocellus below each paravertebral one; an additional white lateral ocellus corresponding to each interspace between the paravertebral figures suggested; venter cream with much dark mottling to

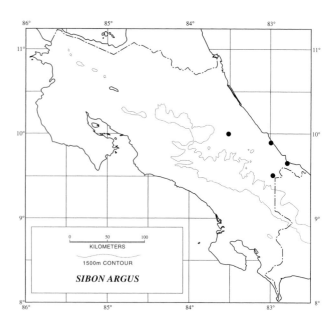

Map 11.59. Distribution of *Sibon argus*.

produce a zigzag along margins of ventrals that meets edges of lateral ocelli; mottling heaviest posteriorly and under tail; upper head surfaces vermiculated with black and light areas; supralabials yellow with dark spots; iris greenish gray.

SIMILAR SPECIES: (1) *Sibon longifrenis* has 160 to 173 ventrals and 90 to 103 subcaudals. (2) *Sibon dimidiatus* has the dorsal bands extending well onto venter or crossing it (fig. 11.37a).

HABITAT: Known only from Atlantic Lowland Rainforest sites.

REMARKS: For many years *S. argus* was known from a single juvenile male (USNM 30656) from near the Panama border in southeastern Costa Rica at Sipurio, Limón Province. Because of the similarity in pattern between this species and *Sibon longifrenis* (known until recently from only three female examples), it seemed possible that a single species was involved or that the holotype of *S. argus* was an aberrant. Recently collected material (five additional specimens), as reported elsewhere (Savage and McDiarmid 1992), fully confirms the validity of this taxon as distinct from all other members of the genus.

DISTRIBUTION: Known from lowland evergreen forests of southeastern Costa Rica (15–550 m), the Darién region of eastern Panama (550 m), and one locality on the Pacific slope of western Panama (600 m).

Sibon longifrenis (Stejneger, 1909)

(plate 399; map 11.60)

DIAGNOSTICS: The distinctive dorsal pattern of black-bordered ocelli in combination with the compressed body, distinct head, and moderate eyes separates this form from all other Costa Rican species except *Sibon argus*. *Sibon longifrenis* differs from *S. argus* in having the head slightly distinct from the neck, nonprotuberant eyes (fig. 11.36c), and many fewer ventrals (151 to 173 [192 to 201 for *S. argus*]) and subcaudals (82 to 103 [115 to 121]).

DESCRIPTION: A small species (total length to 624 mm); tail moderately long (about 28 to 33% of total length); nasal partly or completely divided; usually no preocular, but a small one below loreal sometimes present; a loreal, bordering orbit; 2 or 3 postoculars; supralabials 7 to 9, usually 7, with 3 bordering orbit; infralabials 6 to 9, none in contact behind mental; temporals 1 + 1 or 2, 2 + 2 or 3; when two primary temporals present, lower one appears to have split off from enlarged supralabial; usually a single small postmental (absent in one example) followed by two or three pairs of chin shields (the first two pairs border mental groove); dorsal scales in 15-15-15 rows, vertebral row not enlarged; ventrals 151 to 173, subcaudals 82 to 103. Dorsal ground color grayish green, with 33 to 44 black-bordered pinkish tan ocelli that form a single figure across the middorsal axis or are paired on the paravertebral areas and alternate in position (single ocelli 28 to 35; paired ocelli 5 to 8); a series of black-bordered lateral pinkish ocelli present on first three scale rows on each side, each directly below a dorsal ocellus; interspaces between lateral ocelli clear white; upper surfaces irregularly variegated with brown or black spots or lines; tail banded; venter mostly bright yellow with dark brown spotting and stippling that tends to form lineate markings laterally and/or on midventer; iris olive green speckled with dark pigment.

SIMILAR SPECIES: (1) *Sibon argus* has 192 to 201 ventrals and 115 to 121 subcaudals. (2) *Sibon dimidiatus* has extensions of the dorsal rings and bands well onto or across the venter (fig. 11.37a). (3) *Bothriechis schlegelii* has a series of distinctive hornlike scales above the eyes and a loreal pit (fig. 11.50).

HABITAT: Relatively undisturbed Atlantic Lowland Moist and Wet Forests and Premontane Wet Forest and Rainforest.

BIOLOGY: The coloration of this rare species closely resembles that of the greenish variety of *B. schlegelii* and may represent mimicry.

REMARKS: Although Peters (1960) described *Sibon argus* as having two lateral ocelli for each dorsal one, there is no essential difference between *Sibon longifrenis* and *S. argus* in this regard. Both have dark-outlined light lateral ocelli directly below the dorsal ones. In both species the interspaces between the lateral ocelli are clear white, and Taylor (1951b) described them as like ocelli (they are bordered by the same dark margin that forms the outline of the tan lateral ocelli). The second series of ocelli that Peters (1960) described as characteristic for *S. argus* corresponds to the white interspaces typical of *S. longifrenis* that have yellowed

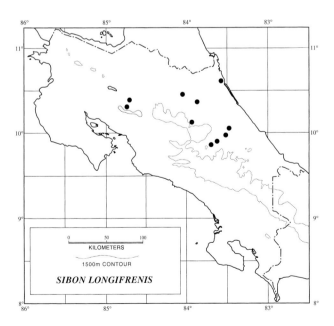

Map 11.60. Distribution of *Sibon longifrenis*.

in preservative. *Dipsas costaricensis* Taylor, 1951b, is based on an example of this species (KU 25703) lacking a post-mental (Peters 1960).

DISTRIBUTION: Lowland and lower premontane zone evergreen forests of Atlantic versant northeastern Honduras, Costa Rica, and adjacent western Panama (60–750 m).

Sibon nebulatus Group

Snakes referred to this group have an enlarged penultimate supralabial that is separated from the orbit by a postocular and a variable dorsal pattern with much white spotting, variegation, and irregular dark markings. In addition to *Sibon nebulatus*, which occurs in Costa Rica, *Sibon carri* (El Salvador and Honduras) and *Sibon dunni* (Ecuador) have been referred to this group.

Sibon nebulatus (Linné, 1758)

(plate 400; map 11.61)

DIAGNOSTICS: This common species shares the compressed body, distinct head, and large eyes with other members of the genus and with species of *Dipsas* and *Imantodes*. None of these other forms has a pattern similar to that of *Sibon nebulatus*, which consists of irregular dark bands and spots bordered by white on a lighter ground color and with the dark bands extending onto or across the venter.

DESCRIPTION: A moderate-sized species (maximum recorded total length to 830 mm); sexes about same length; moderate to long tail (22 to 27% of total length); nasal divided; usually no preoculars so that loreal borders orbit; 1 to 4 postoculars, usually 2; supralabials 5 to 9, usually 7 or 8, with 2 bordering orbit; infralabials 6 to 10, usually 8 or 9, first pair in contact behind mental; temporals usually 1 +

2; no postmental; three pairs of chin shields, first two pair bordering well-developed mental groove; dorsal scales in 15-15-15 rows, vertebral row moderately enlarged; ventrals 174 to 190 (159 to 200); subcaudals 79 to 106 (75 to 114) in males; 72 to 93 (164 to 100) in females. Dorsum pale brown, brownish gray, to grayish black with 38 to 42 irregular narrow dark brown to black bands that extend onto venter; dorsal portion of bands usually bordered by white; bands tend to cross body axis diagonally and alternate with one another ventrally, where they often reach the midline and occasionally form complete rings; bands sometimes broken into lateral spots; venter cream; iris gray, speckled with dark pigment.

SIMILAR SPECIES: (1) *Sibon annulatus* has broad dark bands that are wider in longitudinal extent than the interspaces. (2) *Sibon dimidiatus* has light-centered dark dorsal bands that extend well onto or across the venter (fig. 11.37a). (3) Species of *Imantodes* lack dark bands on the venter. (4) All *Leptodeira* have an essentially immaculate venter.

HABITAT: Primarily an inhabitant of relatively undisturbed Lowland Moist and Wet Forests, but also found in secondary growth, in gallery forests within the Lowland Dry Forest area, and marginally ranging into Premontane Moist and Wet Forest and Rainforest.

BIOLOGY: An arboreal nocturnal form that feeds on snails and slugs. Campbell (1998) reported that egg clutches of three to nine eggs 20 to 26 mm long are laid in the rainy season in Guatemala. Hatchlings from these eggs were 222 to 234 mm in total length. He also noted an unusual aggregation of this species near Turrialba, Cartago Province, where seven individuals were found in one tree, apparently feeding on the many small snails crawling over its limbs and branches. Other trees nearby lacked snails and snakes.

REMARKS: Old records of this species from the city of San José (MNCR 414, now lost; ZMH 5502) and Mina, Río Torres (AMNH 17311), both in San José Province, in the Meseta Central Occidental, suggest that this species formerly occurred there. No recently collected examples are known from that area. There is considerable variation in coloration in this wide-ranging species (Peters 1960). Some populations have broad regular dark blotches anteriorly, and others have the markings obscured by extensive suffusions of dark pigment to leave the white borders of the dorsal dark bands as the conspicuous components of the pattern. Costa Rican examples of the species are relatively constant in basic pattern (no differentiation of anterior bands, interspaces not obscured by melanin deposits, and with irregular dorsal bands). However, there is great variation in detail because of the irregularity of the dark bands and the distribution of white scales that usually border them. Nevertheless, this is one of the most easily recognizable snakes in Costa Rica, and any dark-colored moderate-sized snake with irregular banding and with scattered speckling of white dorsally will turn out to be *Sibon nebulatus*.

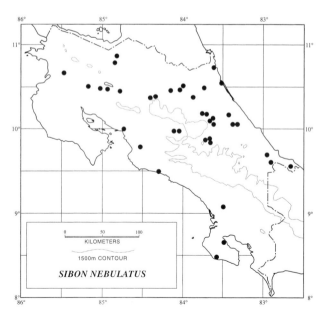

Map 11.61. Distribution of *Sibon nebulatus*.

There is also considerable variation in segmental counts in this form (Peters 1960), but Central American and northern South American samples are similar in the number of ventrals and subcaudals. Extremes in the range of variation occur in the Cauca valley, Colombia, and the Andes of Venezuela near Mérida for low ventral numbers (159 to 173) and in northwestern South America for high numbers (178 to 200). Subcaudal ranges are low for the Cauca valley (77 to 84 in males, 64 to 71 in females), for Mérida (75 to 101 in males, 70 to 85 in females), and for material from the Río Magdalena and Río Porce valleys of Colombia (85 to 103 in males, 75 to 82 in females).

DISTRIBUTION: Lowlands and adjacent premontane areas on the Atlantic versant from central Veracruz, Mexico, to Amazonian Ecuador, the Guianas, and northern Brazil; on the Pacific versant from Nayarit, Mexico, to southern Guatemala, with an isolated population in southwestern Costa Rica and adjacent Panama; also in eastern Panama south through Pacific drainage Colombia and Ecuador. In Costa Rica the species occurs in evergreen forests of the Atlantic versant and the Golfo Dulce region, but it is also known from a few Pacific slope sites on the Meseta Central Occidental and in northwestern Costa Rica (3–1,160 m).

Genus *Tretanorhinus* C. Duméril, Bibron, and A. Duméril, 1854

The members of this genus are aquatic snakes (four species) that are found in lowland freshwater streams or mangrove forest areas. The principal morphological modification associated with these habits is the dorsally directed nostrils, since the body proportions and head shield complement are characteristic for many nonaquatic colubrids, except for occasionally having one or three prefrontals and more than one loreal. The following additional combination of features typifies the genus: head slightly distinct from neck; body cylindrical; nasals completely or partially divided; one or two loreals; one to three preoculars; pupil vertically elliptical; two pairs of elongate chin shields; dorsal scales striated and keeled, in 19 to 21 rows at midbody, without apical pits; anal divided; subcaudals divided; 27 to 30 subequal maxillary teeth; no diastema or enlarged fangs; mandibular teeth subequal; no hypapophyses present on posterior dorsal vertebrae. The keeled scales and nostril placement alone will separate this genus from all other Costa Rican snakes.

The everted hemipenis in *Tretanorhinos variabilis* (Pinou and Dowling 1994) is single and about 10 subcaudals long; unicapitate with a centrolineal bifurcate sulcus spermaticus; capitulum calyculate; truncus with small spines followed proximally by two to four large hooks.

Members of the genus are found on Cuba, Isla de la Juventud, and Grand Cayman Island in the Antilles and on the mainland in the lowlands from southern Veracruz, Mexico, to central Panama and northern Colombia on the Atlantic versant and then southward on the Pacific slope to western Ecuador.

Tretanorhinus nigroluteus Cope, 1861c

Cativo

(plate 401; map 11.62)

DIAGNOSTICS: This rare (in Costa Rica) aquatic snake with dorsally directed nostrils (fig. 11.13a,c) and keeled dorsal scales is unlikely to be confused with any other species.

DESCRIPTION: A medium-sized snake, up to 885 mm in total length; females much longer than males; tail moderately long (19 to 22% of total length); eye small; usually 2 loreals, sometimes 1; 1 to 3 preoculars, usually 2; 2 postoculars; supralabials 7 to 9, usually 8, with 1 bordering orbit; infralabials 9 to 11, usually 10; temporals 1 + 2; tubercles on chin of adult males; dorsal scales variable, most frequently 21-21-19 or 21-21-17 (sometimes 21-19-17 to 19), without apical pits; ventrals 127 to 151 (entire range); subcaudals 56 to 82 (entire range). Dorsal ground color pale to dark brown (black in some extralimital examples), extending to third scale row; pattern usually a double series of small dark blotches or paravertebral blotches that may be fused across back to form narrow transverse bars over a portion or entire length of body; venter usually orange to red, sometimes cream or tan, with dark ventral tips and some heavy dark markings, especially near midline, that may form a midventral stripe.

SIMILAR SPECIES: (1) *Hydromorphus concolor* has smooth scales (fig. 11.5). (2) *Pelamis platurus* has a paddle-shaped tail.

HABITAT: Expected to be found in slow-moving fresh

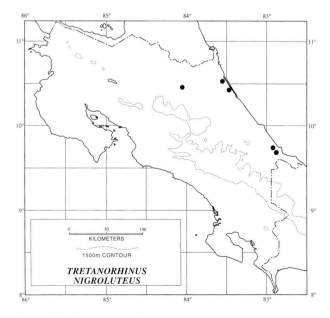

Map 11.62. Distribution of *Tretanorhinus nigroluteus*.

water, tree swamps (freshwater and brackish), and mangrove forests in Atlantic Lowland Moist and Wet Forest environments, based on occurrences elsewhere in Central America and the few records from Costa Rica.

BIOLOGY: This primarily aquatic species is essentially nocturnal and is rarely seen during the day, when it remains submerged or hides among aquatic vegetation. Little is known regarding other aspects of its biology, and the following comes mostly from Villa's (1970) and Wilson and Hahn's (1973) observations and summaries of previous reports, none of which are based on Costa Rican specimens. The species often lies at the surface with only the snout above water and rapidly dives to the bottom to hide under rocks or other cover. Several have been collected crawling on roads at night, but there are no observations of them lying on stream banks out of the water. Food consists of small fishes, tadpoles, and frogs. Feeding may involve active swimming, but sometimes these snakes lie on the bottom and try to catch approaching fishes with a quick sideways snap of the head. Three gravid females (date of collection unknown for one, the others obtained in January and July) contained six to nine well-developed eggs and three to four small ones (18–20 × 10.2–11 mm) and confirm that the species is oviparous. Predators include wading birds (Conant 1965), turtles (Villa 1973), and possibly crabs (Villa 1970), but this is not documented.

REMARKS: Dunn (1939b) commented on the absence of records for this species from Costa Rica. *Tretanorhinus nigroluteus* is now known to be rather common elsewhere along the Atlantic slope of southern Mexico and Central America, but it remains a rarity in Costa Rica.

DISTRIBUTION: Lowlands from southern Veracruz, Mexico, southward on the Atlantic versant, exclusive of the northern two-thirds of the Yucatán Peninsula, to central Panama and northern Colombia, but ranging onto the Pacific slope and into the premontane zone in Honduras (to 750 m) and Nicaragua (1,200 m). In Costa Rica known from a few sites near the coast or inland; expected in tributaries of the Río San Juan (0–100 m).

Genus *Trimetopon* Cope, "1884," 1885

Members of this genus are tiny to very small semifossorial snakes, all under 300 mm in total length. Of the six species placed in the genus, three *(Trimetopon barbouri, T. slevini, and T. viquezi)* have the basic complement of colubrid head shields (fig. 11.4), and three *(T. gracile, T. pliolepis, and T. simile)* have the prefrontals fused into a single plate (fig. 11.38). The following combination of features is characteristic: head scarcely distinct from neck; body cylindrical; nasal usually single, rarely partially divided; one loreal or none; one preocular; pupil round; two pairs of elongate chin shields (posterior ones usually shortest); dorsal scales smooth in 15-15-15 or 17-17-17 rows, without apical

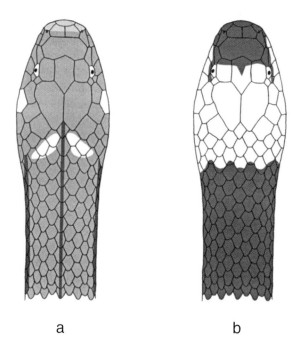

a   b

Figure 11.38. Head and neck patterns in snakes of the genus *Trimetopon*: (a) *T. gracile* showing paired nuchal light spots; (b) *T. simile* showing broad nuchal light collar; patterns are illustrated on a generalized template of head shields and do not accurately portray details of head scalation for these species.

pits; supracloacal keels present in most males; anal divided; subcaudals divided; maxillary teeth 11 to 12, gradually increasing in size posteriorly, with or without a short diastema separating the last two moderately enlarged teeth from the others; Duvernoy's gland probably present; mandibular teeth subequal; no hypapophyses on posterior vertebrae.

Externally these snakes closely resemble some species of the allied genus *Rhadinaea*, with which they share the divided anal, smooth scales, the presence of a loreal (usually) and preocular, and a dark striped or uniform pattern. No *Rhadinaea* have a single prefrontal or 15 scale rows at midbody. Those *Rhadinaea* that have 17 scale rows at midbody and two prefrontals either have the venter spotted or heavily marked with dark blotches *(Rhadinaea calligaster)* or have fewer than 150 ventral scales and a pair of dark dorsal stripes *(Rhadinaea decorata)*. *Trimetopon slevini* and *T. viquezi*, with 17 scale rows, have immaculate red venters, more than 150 ventrals, and no light dorsal stripes.

The hemipenes in this genus are short to moderately long (5 to 12 subcaudals in length), single or bilobed, with a bifurcate, centrolineal sulcus spermaticus, and are semicapitate (a portion of the apical area has a capitate appearance) or unicapitate. The apical area is calyculate or not, and spines cover the central sector (Myers 2002).

The genus is known from Costa Rica and Panama, with only *T. barbouri* extralimital to the former country.

KEY TO THE COSTA RICAN SPECIES OF THE GENUS *TRIMETOPON*

1a. Prefrontals fused forming a single plate (fig. 11.38); venter white or yellow in life 2

1b. Prefrontals separate, paired (fig. 11.4); venter red in life 4

2a. Dorsal scales in 15 rows at midbody 3

2b. Dorsal scales in 17 rows at midbody; dorsum with longitudinal dark stripes; a narrow yellow collar (fig. 11.33g), usually broken above by a vertebral dark stripe, confluent with throat color (usually on both sides of neck) . . . . *Trimetopon pliolepis* (p. 637)

3a. Body with indications of dark stripes; a narrow nuchal collar or paired nuchal spots present, involving head shields and not confluent with throat color; posterior portion of head dark (fig. 11.38a); ventrals 140–54 . . . . *Trimetopon gracile* (p. 637)

3b. Body uniform; a broad nuchal collar involving posterior portion of head and first few nuchal scales and confluent with throat color (fig. 11.38b); ventrals 118–19 . . . . *Trimetopon simile* (p. 638)

4a. Paired nuchal light spots present (fig. 11.38); dorsum with a vertebral dark stripe and two or three lateral dark stripes . . . . *Trimetopon slevini* (p. 639)

4b. No nuchal light spots; dorsum with a middorsal and a single dark stripe on each side . . . . *Trimetopon viquezi* (p. 639)

Trimetopon gracile (Günther, 1872)

(plate 402; map 11.63)

DIAGNOSTICS: The combination of a pair of light nape spots (fig. 11.38a) (or a nuchal light bar), an obscurely striped dorsal pattern, a single prefrontal (fig. 11.38a), and 15 scale rows at midbody will distinguish this relatively uncommon snake from all others known from Costa Rica.

DESCRIPTION: A small species reaching 293 mm in total length; tail moderate (22 to 27% of total length); 1 large prefrontal; 1 loreal; 1 postocular; supralabials usually 7, rarely 8, with 2 bordering orbit; infralabials usually 7, rarely 8; temporals 1 + 1; dorsal scales in 15 rows; ventrals 140 to 154; subcaudals 51 to 64; maxillary teeth 12 to 13, gradually increasing in size posteriorly, no diastema. Brown above with dark lateral stripes that may be obscured and only vaguely indicated, with or without narrow dark stripe down vertebral scale row; usually two yellow nape spots that sometimes fuse across midline to form a nuchal bar that is not confluent with throat color; top and sides of head brown, punctated with yellow; short oblique postocular light stripe often present; venter pale yellow or white, immaculate or with scattered brown markings; iris brown, tongue brown. The retracted hemipenis is bilobed, semicapitate, with apical calyces (Myers 2002).

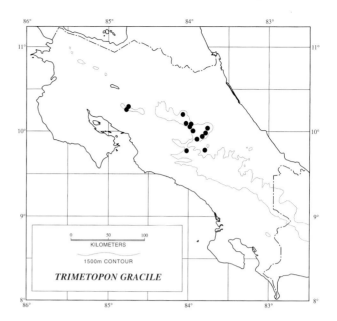

Map 11.63. Distribution of *Trimetopon gracile*.

SIMILAR SPECIES: (1) *Trimetopon pliolepis* usually has the light nuchal collar broken at the midline but confluent with the throat color laterally (fig. 11.33g). (2) *Trimetopon simile* has a light nuchal collar that involves most of the parietals (fig. 11.38b). (3) *Trimetopon slevini* and *T. viquezi* have red venters.

HABITAT: Undisturbed situations at the upper margins of Premontane Wet Forest and Rainforest and in Lower Montane Wet Forest and Rainforest.

BIOLOGY: A specimen (CRE 46) of this species from Volcán Turrialba, Cartago Province (1,950 m), contained a salamander *(Oedipina uniformis)*.

REMARKS: An example of this form listed as from Turrialba (646 m), Cartago Province, at the British Museum (BMNH 95.7.13.17.) probably came from the slope of Volcán Turrialba at a higher elevation. The lowest altitudinal record for the species otherwise is 8.7 km northeast of the Tapantí bridge, Cartago Province (about 1,300 m).

DISTRIBUTION: Principal cordilleras (Tilarán, Central, and Talamanca) of Costa Rica (1,300–2,286 m).

Trimetopon pliolepis Cope, 1894a

(map 11.64)

DIAGNOSTICS: The combination of a dark-striped dorsal pattern, a light nuchal collar (fig. 11.33g) usually interrupted on the midline but confluent with the throat color, white venter, and a single prefrontal shield distinguishes this snake from other Costa Rican forms.

DESCRIPTION: A small species attaining a maximum total length of 277 mm; tail moderate to moderately long (22 to 27% of total length); 1 large prefrontal; a loreal; 1 postocular; supralabials usually 8, rarely 7, with 2 bordering

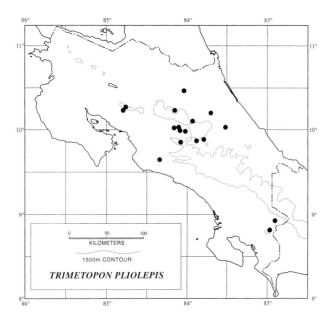

Map 11.64. Distribution of *Trimetopon pliolepis*.

orbit; infralabials usually 8, rarely 7, temporals usually 1 + 1, very rarely 1 + 2, or 2 + 1; dorsal scales in 17 rows; ventrals 137 to 157; subcaudals 58 to 73; maxillary teeth 10 to 11 + 2, gradually increasing in size posteriorly but last two after diastema slightly enlarged. Brown above with middorsal dark stripe; usually three or four dark stripes on each side of body, which may be ill-defined, interrupted, or obscured by dark pigment; yellow tan nuchal collar present, usually broken medially by dark vertebral stripe and usually confluent with light throat color (rarely not confluent on one side); head brown, conspicuously or inconspicuously spotted with yellow, sometimes with distinct diagonal subocular or postocular light stripe; venter white or yellow, immaculate or with some small dark spots; subcaudals immaculate to heavily spotted on midline; iris cinnamon brown; tongue dark brown to black. The everted hemipenes are single, semicapitate, with a calyculate capitulum (Myers 2002).

SIMILAR SPECIES: (1) *Trimetopon gracile* has paired nuchal light spots or a narrow light nuchal bar that is not confluent laterally with the light throat color (fig. 11.38a). (2) *Trimetopon simile* has a broad light collar that involves most of the parietals (fig. 11.38b). (3) *Trimetopon slevini* and *Trimetopon viquezi* have red venters. (4) *Tantilla schistosa* has a uniform dorsum (fig. 11.33f) and a red venter. (5) *Tantilla reticulata* has distinct dark and light dorsal stripes.

HABITAT: Under surface debris and logs in relatively undisturbed Lowland Moist and Wet Forests, Premontane Moist and Wet Forests and Rainforest, and marginally into Lower Montane Wet Forest.

BIOLOGY: A relatively common leaf litter species that eats small lizards.

REMARKS: Whereas Dunn (1937b) regarded this form as a synonym of *Trimetopon gracile,* Taylor (1951b, 1954b) pointed out the principal difference between them (number of scale rows) and established the distinctiveness of this form. All subsequent workers have validated Taylor's conclusions.

DISTRIBUTION: In Costa Rica on the northeastern Atlantic lowlands and adjacent mountain slopes and Pacific premontane areas of the Cordillera de Tilarán, Cordillera Central, Meseta Central, northern Cordillera de Talamanca, and the southwest; also in adjacent southwestern Panama (60–1,600 m).

Trimetopon simile Dunn, 1930c

(map 11.65)

DIAGNOSTICS: This relatively rare snake is one of a number of Costa Rican species with a uniform dark brown or black dorsum, an immaculate yellowish venter, and a light nuchal collar involving the parietal shields (fig. 11.38b). *T. simile* differs from other species having these features in the following combination of characteristics: 1 prefrontal (fig. 11.38b), smooth dorsal scales (fig. 11.5) in 15 rows, and dark dorsal coloration extending onto tips of ventrals.

DESCRIPTION: A tiny snake (maximum total length 176 mm); tail long (30 to 33% of total length); 1 large prefrontal; 1 loreal; 1 postocular; supralabials 7, with 2 bordering orbit; infralabials 7; temporals 1 + 1; dorsal scales in 15 rows; ventrals 118 to 119; subcaudals 72 to 68; maxillary teeth 10, slightly increasing in size posteriorly. Dorsum uniform blackish brown; broad light nuchal collar involving most of the parietal shields and several scale rows

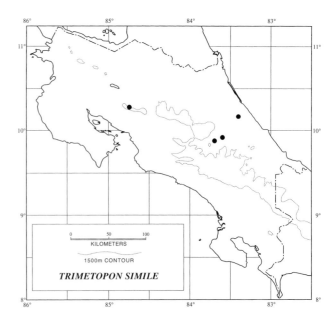

Map 11.65. Distribution of *Trimetopon simile*.

on neck, complete or barely interrupted by dark pigment on midline; venter yellowish white. The retracted hemipenis is bilobed, semicapitate, and apparently without calyces (Myers 2002)

SIMILAR SPECIES: (1) *Enulius sclateri* has two prefrontals, no preocular, and 2 postoculars (fig. 11.27). (2) *Trimetopon pliolepis* has the light nuchal collar not or barely involving the parietals (fig. 11.33g). (3) *Trimetopon gracile* has light nuchal spots (fig. 11.38a) or a narrow light nuchal bar that barely involves the parietals. (4) *Enulius flavitorques* has the ventral tips and first one to three scale rows on the body light in color. (5) Juvenile *Geophis ruthveni* have keeled dorsal scales (fig. 11.5). (6) *Ninia celata* has keeled dorsal scales (fig. 11.5) in 19 rows.

HABITAT: Known from the type locality in Atlantic Lowland Moist Forest, two sites in Premontane Wet Forest, and one place in Lower Montane Rainforest.

BIOLOGY: Myers (2002) pointed out that there is just a single ovary (the right) in the only known female (CRE 2922) and suggested that this may be an evolutionary loss related to the tiny size. Alternatively, he thought it might be a congenital anomaly.

REMARKS: The holotype (MCZ 15263) was reported to have two postocular shields in the original description (Dunn 1930c), an error copied by Taylor (1951b), Peters and Orejas-Miranda (1970), Savage (1980b), and Savage and Villa (1986). Myers (2002) reexamined this specimen and confirmed that it has one postocular.

DISTRIBUTION: Known only from the lowland and premontane zones of the lower Río Reventazón valley in central Atlantic slope Costa Rica (60–660 m) and in the lower montane zone in the region near Monteverde, in the Cordillera de Tilarán (1,500 m).

Trimetopon slevini Dunn, 1940b

(map 11.66)

DIAGNOSTICS: This small relatively rare dark brown snake is one of several Costa Rican forms with dark dorsal stripes, a bright red belly, and light nuchal markings. *T. slevini* differs from similar forms in having pale nuchal light spots instead of a definite light nuchal collar.

DESCRIPTION: A small species (maximum total length 294 mm); tail moderate (19 to 24% of total length); 2 prefrontals; loreal present or absent (present in all Costa Rican examples); 2 postoculars; supralabials 6 to 7, usually 7, with 2 bordering orbit; infralabials 7 to 8; temporals 1 + 1, rarely 1 + 1 + 1; dorsal scales in 17 rows; ventrals 152 to 165; subcaudals 43 to 64; maxillary teeth 11 to 12, increasing in size posteriorly, with or without a small diastema anterior to last two teeth. Dark brown above with a narrow vertebral dark stripe and three or four pairs of lateral dark stripes that are barely indicated or discontinuous; a pair of small nuchal light spots; ventral surfaces of body and tail

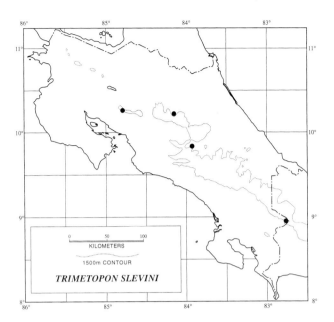

Map 11.66. Distribution of *Trimetopon slevini*.

red, subcaudals sometimes punctuated with brown; iris dark brown; tongue black. The retracted hemipenis is single, calyculate, and semicapitate (Myers 2002).

SIMILAR SPECIES: (1) *Trimetopon viquezi* lacks light nuchal spots. (2) *Trimetopon gracile* has a pale yellow or white venter and 15 rows of dorsal scales. (3) *Tantilla schistosa* has a uniform dorsum (fig. 11.33f) and 15 scale rows. (4) *Trimetopon pliolepis* has a single prefrontal (fig. 11.33g) and a yellowish venter.

HABITAT: Premontane Wet Forest and Rainforest and Lower Montane Rainforest.

DISTRIBUTION: Premontane and marginally into lower montane areas of the Costa Rican cordilleras (1,200–1,800 m) and their extension into southwestern Panama 120–1,825 m).

Trimetopon viquezi Dunn, 1937b

(map 11.67)

DIAGNOSTICS: Dark dorsal stripes, a red venter, and the absence of light nuchal spots or a collar will separate this form from all other small Costa Rican snakes.

DESCRIPTION: A small species (total length 240 mm or more, tail incomplete) known solely from the type specimen, now lost; 2 prefrontals; a loreal; 2 postoculars; 7 supralabials, two bordering eye; 8 infralabials; temporals 1 + 1; dorsal scales in 17 rows; ventrals 161; subcaudals 33 (incomplete tail). Black above with a middorsal and a pair of lateral dark stripes; venter and underside of tail red. Hemipenis bilobed, "non-capitate" (probably equals semicapitate), calyculate (Dunn 1937b).

SIMILAR SPECIES: (1) *Trimetopon slevini* has light nuchal spots. (2) All other *Trimetopon* have light nuchal mark-

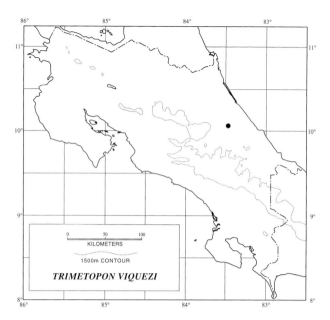

Map 11.67. Distribution of *Trimetopon viquezi*.

ings and a yellowish to white venter. (3) *Tantilla vermiformis* has a single middorsal dark stripe and light-colored parietal tips (fig. 11.44g).

HABITAT: Atlantic Lowland Moist Forest.

REMARKS: This apparently valid species has been collected only once, from Siquirres, Limón Province. Myers (2002) regarded it as a relative of the more upland *Trimetopon slevini*. A recently discovered photograph of an unusual *Trimetopon* from La Selva, Heredia Province, may represent this species. This snake had paired prefrontals, 17 scale rows, and a striped pattern and lacked small nuchal light spots, which agrees with the original description of *T. viquezi*. Unlike the holotype and only known specimen, formerly in the collection of the Museo Nacional de Costa Rica and now lost, the La Selva example had a white venter.

DISTRIBUTION: Known from a single locality in the Atlantic lowlands of central Costa Rica (62 m).

Genus *Urotheca* Bibron, 1843

A genus of rather slender, usually diurnal terrestrial snakes having very long tails (35 to 47% of total length in adults) that are disproportionately thick but very fragile and consequently are frequently incomplete (about 50% of most samples). Most *Urotheca* (seven species) have a brown ground color that may be marked with distinct longitudinal white stripes, but two species are coral snake mimics typically with a pattern of alternating black and light rings encircling the body. In the coral snake mimics the rings may be bicolor (black with white, pink, or red) or tricolor. Tricolored populations may have black and white rings around the body and the white rings suffused with yellow, pink, or red dorsally; have black, red, and yellow rings around the

body (one red ring for each black one); or have a tricolor triad pattern (three black rings for each red ring). An additional variant in one of the species has the black markings reduced to dorsal bands. Members of the genus have the basic colubrid complement of head shields, the head slightly distinct from neck; body cylindrical; nasal variable, sometimes completely divided; a loreal separated from the orbit by one or two preoculars and frequently by a subpreocular; pupil round; two pairs of elongate chin shields; dorsal scales smooth in 17-17-17 rows, except that supracloacal keels may be present in some male *U. guentheri*; scales without apical pits; anal divided; subcaudals divided; maxillary teeth 11 to 19 + 2, with a distinct diastema, fangs usually not grooved (usually with weak to moderate grooves in examples of *U. guentheri*); ultimate fang offset laterad; teeth anterior to diastema subequal; Duvernoy's gland present; mandibular teeth subequal; no hypapophyses on the posterior vertebrae. In Costa Rica these snakes are distinctive in having the combination of smooth scales in 17 rows throughout, a divided anal, a loreal and preoculars, and the long, thick, and often broken tail (fig. 11.18). Two other Costa Rican genera, *Enulius* and *Scaphiodontophis*, have similarly long and disproportionately thickened and fragile tails. *Enulius* have a uniform dark dorsum and no preocular scales; *Scaphiodontophis* have a tricolor coral snake mimic pattern of bands, not rings.

The everted hemipenes of *Urotheca* are short to moderately long (7 to 12 subcaudals in length), simple to slightly bilobed, with a centrolineal, bifurcate sulcus spermaticus; organ unicapitate with the capitulum calyculate and the calyces with papillate micro-ornamentation; a large naked pocket within the capitulum on the asulcate side; pedicel of hemipenes covered by small spinules, truncus by large spines and hooks (Myers 1974; Savage and Crother 1989).

The species previously referred to the nominal genus *Pliocercus* and the *lateristriga* group of *Rhadinaea* were recently shown (Savage and Crother 1989) to constitute a distinct genus for which *Urotheca* is the earliest available name. The snakes appear to thrash the long and fragile tail as a predator-avoidance behavior (Greene 1973). When a portion of the tail breaks off, reflex action continues the thrashing, and the predator is decoyed into chasing the separated caudal segment while the snake sneaks away. The tail breaks intercentrally and is not regenerated, so the elongate tail allows for several breaks (and escapes) over the life of an individual snake, with each succeeding loss further shortening the tail (specialized pseudautotomy). In these snakes (unlike lizards with tail autonomy) the tail does not break in response to slight tactile stimuli but requires some physical resistance to separate the segments. Although Wilson (1968) and Muñoz (1987) suggested there is intravertebral fracture in *Urotheca* (as in many lizards), nothing in the available data supports that idea (see Savage and Crother 1989 for a full discussion).

Members of this genus, although usually lacking grooved fangs, have the Duvernoy's venom gland well developed. In at least one case (Seib 1980) a person bitten on the hand by *Urotheca elapoides* suffered severe pain and swelling that persisted in modified form for several days. You should obviously take care in handling individuals of this genus.

Urotheca occurs from Tamaulipas and Oaxaca, Mexico, south through northern and western Colombia to western Ecuador, northern Venezuela, and eastern Colombia.

KEY TO THE COSTA RICAN SPECIES OF THE GENUS *UROTHECA*

1a. Dorsum uniform or with definite longitudinal light stripes 2
1b. Pattern of black and light rings that encircle body. . . . . *Urotheca euryzona* (p. 642)
2a. No ocelli on head or neck; body uniform or with one or two pairs of light longitudinal stripes 3
2b. A pair of pale ocelli on head, one on each parietal plate and sometimes additional ocelli on neck; two pairs of longitudinal light stripes on body . . . . *Urotheca guentheri* (p. 645)
3a. Body uniform brown without stripes. 4
3b. One or two distinct narrow white stripes on each side of body, ... 5
4a. Supralabials heavily speckled with brown; head cap much lighter than dorsum, extending four scale rows onto neck 6
4b. Supralabials suffused with brown, or white above and blackish brown below; head cap somewhat lighter than dorsum, extending two scale rows onto neck . . . .*Urotheca myersi* (p. 646)
5a. Supralabials weakly to heavily spotted with brown; upper surfaces of head yellow brown (often reddish brown in life), much lighter than brown dorsum; usually with a distinct narrow light stripe on each side of body; usually no nuchal light collar 6
5b. Supralabials sometimes dark above and immaculate white below, but usually with dark pigment along lower margins forming a conspicuous light longitudinal stripe across several supralabials; upper surfaces of head brown, usually of the same color as dorsum; usually with a pair of distinct light stripes on each side of body; often with a nuchal light collar. . . . . *Urotheca decipiens* (p. 641)
6a. Head cap extending only one or two scale rows onto neck with a distinct longitudinal white stripe on first scale row; supralabials with brown sutures or weakly spotted with brown; temporals usually 1 + 1 . . . . *Urotheca pachyura* (p. 646)
6b. Head cap extending three or four scale rows onto neck; dorsum often uniform, sometimes with a distinct longitudinal white stripe on first scale row; supralabials

heavily speckled with dark; temporals 1 + 2 . . . .*Urotheca fulviceps* (p. 644)

Because the tail is so frequently missing in these species, tail lengths have been estimated as a percentage of total length for a particular specimen based on average tail length from individuals with complete tails.

Urotheca decipiens (Günther, 1893)

(plates 403–4; map 11.68)

DIAGNOSTICS: A small dark brown or black snake with a distinct narrow light stripe running the length of the body on the first scale row on each side of the body. Some examples have a second pair of dorsolateral light stripes as well. Most examples have a broad nuchal light collar that crosses the parietal, but in examples lacking the collar the upper and lower margins of the supralabials are heavily pigmented to produce a black-outlined supralabial light stripe. These features, combined with the absence of postocular or parietal ocelli, will separate this form from other striped species that have 17 rows of dorsal scales at midbody and a white venter.

DESCRIPTION: A small species reaching a total length of 569 mm (longest male 490 mm with very tip of tail missing); tail very long (35 to 46% of total length); eye moderate; usually 2 preoculars (sometimes 1) and a subpreocular; 2 postoculars; supralabials 8 to 9, usually 8, with 2 bordering orbit; usually 10 infralabials, sometimes 9; temporals usually 1 + 1, rarely 1 + 2, or 1 + 1 + 1 (one side); dorsal scales in 17 rows; no supracloacal keels; ventrals 122 to 140; subcaudals 90 to 121; maxillary teeth 15 to 17 + 2. Dorsum dark brown to black with two or four longitudinal light stripes; head brown above, somewhat lighter to same color

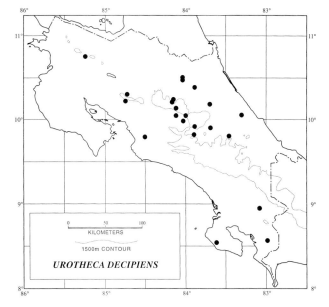

Map 11.68. Distribution of *Urotheca decipiens*.

as dorsum, often with an orangish, pinkish, dull brown, tan, or yellow nuchal collar that crosses the parietals and continues for one to two scale rows onto neck, sometimes interrupted on midline; in examples lacking the collar, black pigment is concentrated on the upper and lower margins of the supralabials to produce a black-bordered light supralabial stripe; no ocelli or short bars or stripes on sides of neck; a longitudinal light stripe bordered below by dark brown to black running along scale row 1 on each side of body; some individuals with a second dorsolateral light tan stripe on each side; number of light stripes, presence or absence of nuchal collar, and supralabial stripes occur in all combinations; venter white; dorsal ground color encroaches on ventral tips throughout length of body; sometimes with scattered brown or black spots or the spots arranged in a medial line on the belly and/or underside of tail; iris dark brown to reddish brown; tongue black, with tips of fork white. The everted hemipenis is short to moderate (8 to 12 subcaudals in length) with a small capitulum, truncus with large spines (45 to 84); basal half covered by minute spines; sulcus spermaticus bifurcates immediately after entering capitulum; small nude area on asulcate side of central portion bordered by a basal spine below; spines around nude area are not largest on organ (Myers 1974)

SIMILAR SPECIES: (1) *Urotheca guentheri* has parietal ocelli. (2) *Urotheca pachyura* usually lacks a light nuchal collar and never has four lateral light stripes or a supralabial light stripe. (3) Striped individuals of *Urotheca fulviceps* lack the nuchal light collar and supralabial light stripe and have the supralabial plates heavily spotted and speckled with dark.

HABITAT: Undisturbed Lowland Moist and Wet Forests and Premontane Moist and Wet Forests and Rainforest.

BIOLOGY: This common species is active during the day on the forest floor.

REMARKS: Dunn (1938, 1942c) regarded this form as a subspecies of *Rhadinaea pachyura* (now *Urotheca pachyura*) and provided the only Pacific slope records for upland Costa Rica, from San José and La Palma, both in San José Province. Although Myers (1974) was equivocal regarding the San José specimen ("town or province?") the two British Museum specimens from the La Palma pass area confirm that this form occurs in suitable habitat on the Pacific slopes of the Cordillera Central. It is very likely that the "San José" specimen (MCZ 28071) was taken near La Palma, which was a favorite collecting site readily accessible from San José in the nineteenth and early twentieth centuries (see Savage 1974a for more detail). Taylor (1951b) followed Dunn's (1942c) treatment, but then (Taylor 1954b) recognized *Rhadinaea decipiens* as distinct and described a new subspecies, *R. decipiens rubricollis*, from Volcán Poás. Myers (1974) identified the type (KU 31956) of the last name as a juvenile *R. pachyura* and demonstrated the distinctiveness of *R. decipiens*, *R. fulviceps*, and *R. pachyura*, all of

which Dunn (1942c) regarded as subspecies of *R. pachyura*. The three species were referred to *Urotheca* by Savage and Crother (1989).

DISTRIBUTION: Humid lowland and premontane forests of Atlantic slope Costa Rica and western Panama, central and southwestern Pacific Costa Rica, southeastern Panama, and northern Colombia. In Costa Rica also ranging onto the Pacific slope on the Meseta Central and between Volcanes Barba and Irazú (15–1,500 m).

Urotheca euryzona (Cope, 1862a)
(plates 405–6; map 11.69)

DIAGNOSTICS: This coral snake mimic differs from most Costa Rican snakes in having the body encircled by alternating black and white rings, with the latter often suffused above with yellow, pink, or red pigment. The combination of a black head cap extending back onto the parietal shields, smooth dorsal scales in 17 rows at midbody, three plates between the eye and nostril, divided anal, and elongate chin shields will separate it from other similarly colored Costa Rican species.

DESCRIPTION: A small to moderate-sized snake reaching 795 mm in total length; tail very long (38 to 43% of total length); eye moderate; usually 2 preoculars and a subpreocular, but sometimes 3 or 1 preocular, and the subpreocular may be absent; postoculars usually 2, sometimes 3; supralabials usually 8, sometimes 9, with 2 bordering orbit; infralabials 9 to 11, usually 9 or 10; temporals usually 1 + 1, rarely 1 + 2; 17 dorsal scale rows; no supracloacal keels; ventrals 131 to 142 (118 to 142); subcaudals 90 to 122; maxillary teeth 18 to 19 plus two ungrooved fangs. Dorsal pattern of alternating black and light rings; 13½ to

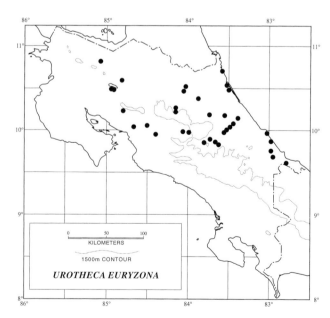

Map 11.69. Distribution of *Urotheca euryzona*.

22 (13½ to 24½) black body rings; 4½ to 13½ (4½ to 18½) black tail rings; anterior upper and lateral surfaces of head mostly black with a broad light band across parietals (very narrow or interrupted on the midline in samples from eastern Panama and southward); light rings white, yellow, or bright red, sometimes with only a yellow or red suffusion dorsally and becoming white or yellow ventrally; ventral light areas usually yellow anteriorly grading into red posteriorly (but white in specimens from eastern Panama and South America); iris black. The everted hemipenis is short (7 subcaudals long), slightly bilobed, with a small capitulum, about 30 moderate to large spines present on truncus; pedicel covered by small spinules; sulcus spermaticus bifurcates immediately on entering capitulum (Savage and Crother 1989).

SIMILAR SPECIES: (1) *Micrurus mipartitus* has the black head cap restricted to the area well forward to the middle of the eyes, only two plates between the eye and nostril, and 15 dorsal scale rows. (2) *Scolecophis atrocinctus* has a light line running from the supralabials across the prefrontals and has 15 dorsal scale rows. (3) *Sibon anthracops* has a light nuchal ring that extends no farther forward than the tips of parietals and 13 dorsal scale rows. (4) *Dipsas bicolor* lacks a mental groove and elongate chin shields. (5) *Rhinobothryum bovallii* has complete rings of three colors around the body and weakly keeled scales. (6) Tricolor *Lampropeltis triangulum* has complete rings of three colors around the body and an undivided anal plate. See key to coral snakes (p. 706), figure 11.47, and table 11.2.

HABITAT: A forest floor denizen found in relatively undisturbed areas of Lowland Pacific-Atlantic Moist Forest, Atlantic Moist and Wet Forests, and Premontane Moist and Wet Forests and Rainforest.

BIOLOGY: This uncommon species is one of several snakes forming the coral snake mimic guild. Individuals of this mostly diurnal form are usually seen moving about on the forest floor but probably spend most of the time resting or foraging under surface debris. *Urotheca euryzona* seems to feed primarily on small amphibians, and several species of the salamander genus *Bolitoglossa* have been reported from stomachs of the allied Mexican and upper Central American coral snake mimic *Urotheca elapoides*. It is presumed to be oviparous, since its close relative *U. elapoides* is known to lay eggs (Greene 1969). Known predators include the laughing falcon *(Herpetotheres cachinnas)* as reported by Pough (1964), and one specimen in the CRE collection came from the stomach of a venomous coral snake, *Micrurus nigrocinctus*. Greene and McDiarmid (1981) have described the extensive geographic variation in coloration in this form and its usually tricolor ringed relative *U. elapoides*, of Mexico and upper Central America. The variation correlates closely with the coloration of venomous coral snakes that occur sympatrically with *Urotheca*. At some sites the coral snake model and the mimic are bicolor, at others

they are tricolor ringed, and at still others the patterns are of the tricolor triad type, with a very close match in details of the pattern as well. In most instances red is a primary component in the coloration of both model and mimic. Greene and McDiarmid (1981) concluded, based on this evidence and similar concordance in color patterns between venomous coral snakes and other presumed mimics, that coral snake mimicry by relatively innocuous or harmless species is the best explanation of available data. Smith (1975), Smith (1977), Brodie (1993), and Brodie and Janzen (1995) have demonstrated a generalized innate avoidance response to coral snake patterns in naive and free-ranging avian predators. The level of resemblance between *Urotheca* and several species of *Micrurus* implies a much finer distinction by some predators than was made by the birds in those studies. This situation parallels the geographic variation and congruence shown by coral snake mimics and their models in *Scaphiodontophis* (Savage and Slowinski 1996) and appears to represent the second step in the two-step evolution of Batesian mimicry (Pough 1988). Savage and Crother (1989) have shown that individual and geographic variation in coloration in *U. euryzona* corresponds to similar variation in the bicolor ringed venomous coral snake *Micrurus mipartitus*, which occurs in sympatry with the former over much of their ranges. As I pointed out above, Costa Rican *U. euryzona* have the light rings completely red or pink, or the rings are white or yellow with the upper three-fourths to one-fourth red, or the dorsal area is yellow, or the entire ring is white. In most *M. mipartitus* the light rings are pink or red, or the rings are white to cream with the pink restricted to the upper scale rows (Savage and Vial 1973), exactly as in many *U. euryzona*. In some *M. mipartitus* the light rings are completely white or with cream dorsally. In addition, *U. euryzona* shows a north-to-south change in coloration, with a loss of red in the color pattern, so that in eastern Panama and South America the species is bicolor black-and-white ringed. *M. mipartitus* shows a similar geographic shift in coloration, and in eastern Panama and western South America it is also black-and-white ringed, but unlike *Urotheca*, retains a red head band and red tail rings. *U. euryzona* thus combines two powerful predator avoidance systems: the fragile tail (typical of all *Urotheca*) and coral snake mimicry. A study of this unusually modified species offers a unique research opportunity in behavioral ecology.

REMARKS: This species has been called *Pliocercus euryzonus* by most previous authors. Savage and Crother (1989) revived the name *Urotheca* for the coral snake mimics *Urotheca elapoides* and *U. euryzona* and those species placed in the *Rhadinaea lateristriga* group by Myers (1974). They demonstrated that all lower Central American specimens previously assigned to *Pliocercus*, including *Pliocercus dimidiatus* Cope, 1865a, *P. annellatus* Taylor, 1951b, and *P. arubricus* Taylor, 1954b, are variants of *U. euryzona*. Although Savage and Crother (1989) were dubious that this species occurred east of the Andes, Pérez-Santos and Moreno (1988) report it from the Orinoco drainage of eastern Colombia and Brazil. Myers and Cadle (1994) and Smith, Wallach, and Chiszar (1995) argue for retaining *Pliocercus* as a separate genus based on coloration alone, a course obviously not followed here and one not to be applauded for the mischief it would cause if widely applied as a basis for generic differentiation (e.g., in *Lampropeltis, Lystrophis, Tantilla*). Smith and Chiszar (1996) recognize six (in the key) or seven (in the text) species of *Pliocercus* based principally on coloration differences shown by Savage and Crother (1989) to be individually variable. In addition, they recognize pattern classes as named geographic races in *Pliocercus bicolor* (two), *P. elapoides* (four), and *P. euryzonus* (two). Their statement (p. 32) that the newly described subspecies of *P. euryzonus* from Panama is intermediate in all characters between Colombian *P. euryzonus* and "*P. dimidiatus*" from Costa Rica supports the taxonomy adopted in this account. Wilson and McCranie (1997) have recently provided an enlightened critique of the Smith and Chiszar (1996) paper, further indicating that at the most only two valid species can be recognized in the "*Pliocercus*" complex.

DISTRIBUTION: Humid forests of the lowland and premontane zones from northern Nicaragua south on the Atlantic versant to central Panama and on both slopes from central Panama to western Colombia and Ecuador and the Orinoco drainage of eastern Colombia; also a few records from lowland and humid premontane forest sites on the Pacific versant of Costa Rica; absent from southwestern Costa Rica and adjacent southwestern Panama (4–1,460 m).

Urotheca fulviceps (Cope "1885," 1886)

(map 11.70)

DIAGNOSTICS: Most representatives of this species have a uniform brown dorsum, but a few have a pair of narrow light stripes or a row of white dashes on the first scale row. The presence of smooth scales (fig. 11.5) in 17 rows at midbody, the absence of nuchal light spots or a light collar, and the immaculate venter will separate *U. fulviceps* from all similar snakes except its close relatives in the genus *Urotheca*. It differs from these species by the following combination of features: head cap reddish brown, contrasting with the darker brown dorsum and extending three to four scale rows onto neck, and supralabials heavily speckled with dark.

DESCRIPTION: A small to moderate-sized species reaching 649 mm in total length (largest female 461 mm); tail very long (39 to 45% of total length); eye moderate, usually 2 preoculars (sometimes 1) and a subpreocular; 2 postoculars; supralabials 8; infralabials 9, occasionally 8 or 10, with 2 bordering orbit; temporals 1 + 2 (rarely 1); dorsal scales in 17 rows; no supracloacal keels; ventrals 136 to 143; subcaudals 98 to 122; maxillary teeth 13 to 16 + 2. Dorsum light to dark brown, usually uniform, but sometimes with an ill-defined light stripe on first scale row on each side; head

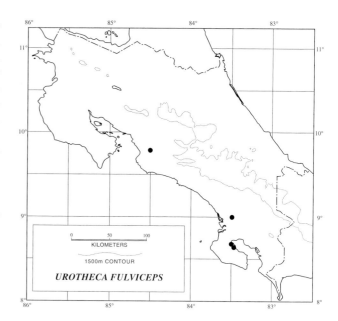

Map 11.70. Distribution of *Urotheca fulviceps*.

and neck reddish in life, yellow brown in preservative, with light color of upper head surface continuing three to four scale rows onto neck; some dark speckling on top of head; no ocelli or short dark-outlined light stripe on side of neck; a poorly defined dark stripe along upper edge of supralabials; supralabials and infralabials white, but with dense pattern of blackish specks or spots and dark pigment along sutures; venter white, immaculate; iris reddish brown, with golden tinting above; tongue black with yellowish gray tips. The everted hemipenis is short (to subcaudal 7), with a well-developed calyculate capitulum; 28 to 44 small to medium slightly recurved spines below capitulum; pedicel with only a few minute spinules; sulcus spermaticus bifurcating within lower third of capitulum; small nude area on asulcate side of truncus flanked by two rows of two or three spines (Myers 1974).

SIMILAR SPECIES: (1) *Urotheca myersi* lacks the heavy speckling of dark pigment on the labials. (2) *Urotheca pachyura* has the light head cap extending only one to two scale rows onto the neck. (3) Species of *Enulius* have a light nuchal collar (fig. 11.33a,b) or most of the head white (fig. 11.27) and the first two rows of scales the same color as the ventrals. (4) *Urotheca decipiens* has two or four longitudinal light stripes, and often a nuchal light collar or a dark-bordered light supralabial stripe.

HABITAT: Undisturbed Lowland Wet Forest of the Pacific southwest.

BIOLOGY: A rare denizen of the forest floor that may be active during the day or early evening (Myers 1974).

REMARKS: This species is included in the Costa Rican herpetofauna based on several examples collected in the area near Rincón on the Península de Osa, from near Palmar Norte and Carara, Puntarenas Province. They agree in all details with Myers's (1974) description of the species, especially in the reddish head coloration in life. The nearest records for *U. fulviceps* in Panama are in the Canal Zone area some 375 km to the east.

DISTRIBUTION: Disjunct in the lowlands of southwestern Costa Rica, central Panama, eastern Panama, northern Colombia, western Ecuador, and northwestern Venezuela (0–600 m); records for Colombia suggest that the species ranges into the premontane zone, but Myers (1974) questions some high altitude records (800–2,023 m) for that country. In Costa Rica known only from the Península de Osa (20 m), Cajón, and Carara National Park, Puntarenas Province (80 m).

Urotheca guentheri (Dunn, 1938)

(plate 407; map 11.71)

DIAGNOSTICS: This beautiful little snake is readily distinguished by having two pairs of distinct longitudinal light stripes on the body and tail; an immaculate bright orange to red venter; a distinct black-bordered light ocellus on each parietal plate; and usually a similar ocellus on each side of the neck, although this ocellus may fuse with the dorsolateral light stripe, which then appears to have an enlarged terminus.

DESCRIPTION: A small to moderate-sized form (maximum total length about 670 mm, 540 mm in largest female); tail very long (36 to 43% of total length in adults; one juvenile female, CRE 277, has the tail 27% of total length); eye moderate; usually 1 preocular (rarely 2), usually a subpreocular; 2 postoculars; supralabials usually 7, often 8, rarely 6, usually with 2 bordering orbit; infralabials 8; temporals usually 1 + 1, occasionally 0 + 1; dorsal scales in 17 rows;

supracloacal keels in some males; ventrals 135 to 176; subcaudals 82 to 110; maxillary teeth 11 to 13 + 2, fangs usually weakly to strongly grooved. Dorsum light to dark brown with a faint gray brown middorsal stripe and two pairs of light stripes bordered below by black on each side of body; anterior terminus of dorsolateral stripe usually expanded to form a circular or elongate white spot, or spot broken off to form an ocellus; a light ocellus on outer edge of parietal shield behind upper preocular plate; supralabials and infralabials mostly white, with a blackish line along upper margins of former; venter orange to red, immaculate; iris coppery above, dark brown to black below; tongue black with white tips on fork. The everted hemipenis is short (6 to 7 subcaudals in length); capitulum large; 27 to 29 small to large spines on truncus; pedicel with tiny spines; sulcus spermaticus bifurcates near distal margin of capitulum; nude area on asulcate side of truncus flanked by two rows of two or three spines (Myers 1974).

SIMILAR SPECIES: (1) *Urotheca decipiens* lacks parietal ocelli and has a white venter. (2) *Crisantophis nevermanni* has the supralabial margins heavily marked with dark pigment (fig. 11.17b), a white venter, and 19 scale rows at midbody.

HABITAT: Relatively undisturbed Lowland and Premontane Moist and Wet Forests; also marginally in Premontane Rainforest and Lower Montane Wet Forest and Rainforest.

BIOLOGY: *Urotheca guentheri* is relatively uncommon and usually found moving about during the day on the forest floor. It feeds primarily on frogs. Taylor (1954b) captured a juvenile by digging around the base of a rotting tree.

REMARKS: Dunn (1938) proposed the name *Rhadinaea guentheri* as a replacement for *Trachymenis decipiens* Günther, 1895, a junior secondary homonym of *Ablabes decipiens* Günther, 1893, created by Dunn's inclusion of both taxa in *Rhadinaea*. At the same time, Dunn described *Rhadinaea persimilis* as new. Myers (1974) concluded that *R. guentheri* and *R. persimilis* represented a single species, and as first reviser selected the former name to have precedence over the latter. Taylor (1951b, 1954b) used the name *R. persimilis* for this form. Dunn's (1938) record of this species from Cartago, Cartago Province (1,435 m), was based on a specimen in the collection of the Colegio de San Luis Gonzaga in Cartago and is no longer extant. I regard this locality as questionable because no other specimens have been collected in or near this city and because specimens in high school collections often are labeled with a general locality. Myers (1974) listed a specimen (LSUMZ 8761) from the southern slope of the Cordillera de Talamanca as having been taken at 6,900 feet (2,103 m) some 12 miles south of "Georgiana Motel." This altitude seems to be erroneous, since any locality that distance (19.3 km) south of La Georgina (the correct name for the "motel") would lie well below 2,000 m. The specimen in question was doubtless taken near the Carretera Interamericana. A location 19.3 km

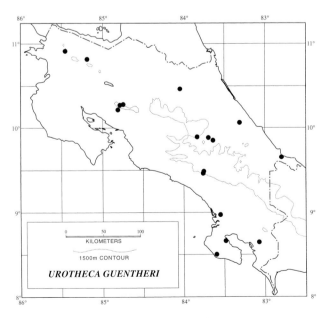

Map 11.71. Distribution of *Urotheca guentheri*.

south of La Georgina on this road lies near the Río Paynor at 1,540 m. The LSUMZ example most likely was collected near this stream, not at a higher elevation.

DISTRIBUTION: Lowland and premontane zones and lower portions of the lower montane zone from northeastern Honduras south to northwestern Panama on the Atlantic versant; southwestern Costa Rica and adjacent southwestern Panama and disjunct west-central Panama on the Pacific slope; also marginally on the premontane Pacific slopes of the cordilleras of northwestern Costa Rica (25–1,600 m).

Urotheca myersi Savage and Lahanas, 1989

(map 11.72)

DIAGNOSTICS: One of several small uniform brown snakes lacking conspicuous distinguishing features, but distinctive from most other similar forms in having smooth scales (fig. 11.5) in 17 rows at midbody, no nuchal light spots or collar, no distinct light head markings, and an immaculate venter. It differs from its closest allies in the genus *Urotheca* in having the head cap much lighter than the dorsum and extending only one to two scale rows onto the neck, a vague but definite narrow black nuchal collar bordering the posterior margin of the head cap, and a bright yellow venter in adults.

DESCRIPTION: A small species reaching a total length of 349 mm; tail very long (41% of total length); eye moderate; 1 preocular and a subpreocular; usually 2 preoculars (occasionally an additional small one); 8 supralabials, with 2 bordering orbit; usually 10 infralabials, occasionally 11; temporals usually 1 + 1, occasionally 1 + 2; dorsal scales in 17 rows, no supracloacal keels; ventrals 132 to 138, sub-

caudals 119 in example with nearly complete tail (possibly 120 to 123 since tail tip missing); maxillary teeth 16 to 17 plus two ungrooved fangs. Dorsum uniform brown to gray, with white dashes on scale row 1 anteriorly in one specimen; top of head lighter than dorsum, with head cap coloration extending one to two scale rows onto neck; light area demarcated posteriorly by an indistinct narrow dark collar; supralabials mostly light with some dark pigment along dorsal or ventral margins; venter canary yellow, immaculate. The retracted hemipenis is relatively short (to subcaudal 10), with a small capitulum; truncus with 40 or more small to medium spines; more than one-half of organ covered only with minute spinules; sulcus spermaticus bifurcates near margin of capitulum; nude area on asulcate side of truncus bordered by a spine below (Myers 1974).

SIMILAR SPECIES: (1) *Urotheca fulviceps* has the light head cap extending three or four scales onto the neck and is heavily marked with dark speckles and dots on the supralabials. (2) The species of *Enulius* have a light nuchal collar (fig. 11.33a,b) or most of the head white (fig. 11.27), and the light ventral color extends two scale rows onto the body on each side.

HABITAT: Lower Montane Wet Forest and Rainforest in the Cordillera de Talamanca.

REMARKS: The male holotype (LACM 137604) of *U. myersi* was placed as *Rhadinaea species inquirenda* by Myers (1974). With the discovery of a second example (a female, CRE 10095), Savage and Lahanas (1989) described it as a new taxon in the genus *Urotheca*. Lips (1993g) reported on two additional examples, one the first record for Panama.

DISTRIBUTION: Known only from the lower montane zone on the southern slope of the Cordillera de Talamanca of Costa Rica and adjacent western Panama (1,500–2,255 m).

Urotheca pachyura (Cope, 1875)

(plate 408; map 11.73)

DIAGNOSTICS: This is another of several dark brown to black snakes with a distinct narrow light stripe running the length of the body on scale row 1. The only species it may be confused with besides other members of the genus *Urotheca* have more than 17 scale rows at midbody. This relatively uncommon form may be separated from its striped congeners by having the head cap lighter than the body and extending one to two scale rows onto the neck and in lacking a nuchal collar (in adults) and distinct dark-bordered light markings on the sides of the head and neck.

DESCRIPTION: A moderate-sized species reaching a maximum total length of about 676 mm (largest female about 590 mm); tail very long (about 41% of total length); eye moderate; usually 1 preocular (sometimes 2) and a subpreocular; 2 postoculars; supralabials 8, with 2 bordering orbit; infralabials 9 to 10; temporals 1 + 1 (rarely 1 + 2); dorsal scales in 17 rows; no supracloacal keels; ventrals 130

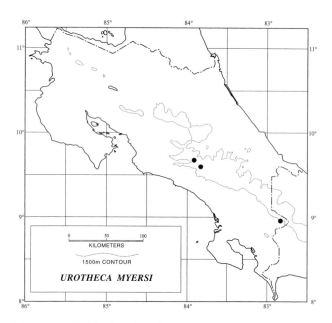

Map 11.72. Distribution of *Urotheca myersi*.

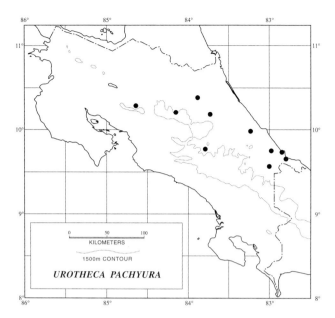

Map 11.73. Distribution of *Urotheca pachyura*.

to 138; subcaudals 104 to 124; maxillary teeth 15 to 17 plus two ungrooved fangs. Dorsum medium to dark brown with a pair of light dorsolateral longitudinal stripes; upper head surface reddish brown, color extending one to two scale rows onto neck, sharply contrasting with body coloration (on some examples head color extends farther back on vertebral scale row); nuchal light collar across parietals and on one and one-half scale rows on neck in juveniles; supralabials marked with dark pigment above, some dark along sutures or occasionally with medial spots; venter white, immaculate or with a few scattered dark specks. The retracted hemipenis is short (about 10 subcaudals in length), capitulum small; about 46 small to medium spines proximal to capitulum; pedicel and truncus (60% of organ) covered by spinules; sulcus spermaticus bifurcates about halfway into capitulum; nude pocket on asulcate surface of truncus flanked by two rows of spines, each row containing three spines (Myers 1974).

SIMILAR SPECIES: (1) Specimens of *Urotheca decipiens* that lack a nuchal light collar and have only one pair of light dorsal stripes have a conspicuous dark-bordered light supralabial stripe. (2) *Urotheca fulviceps* has the light head cap extending three to four scale rows onto the neck and the labials heavily marked with black speckles and spots.

HABITAT: Undisturbed Atlantic Lowland Moist and Wet Forests and Premontane Rainforest.

REMARKS: Taylor (1951b) followed Dunn (1938) in treating *Rhadinaea pachyura* as the nominate subspecies of a polytypic taxon that included *Rhadinaea pachyura decipiens*. Subsequently Taylor (1954b) raised the latter form to specific status and described a new subspecies *R. decipiens rubricollis* from the Cordillera Central. Myers (1974) synonymized *rubricollis* with *R. pachyura* and confirmed the

distinctiveness of *R. decipiens*. Both taxa are now placed in the genus *Urotheca* according to Savage and Crother (1989).

DISTRIBUTION: Atlantic lowland and premontane zones of Costa Rica and northwestern Panama; one record in the premontane zone of Pacific slope western Panama (4–1,360 m).

Colubrines

The Costa Rican genera of this lineage may be placed in four general divisions. The first of these includes the active, mostly diurnal racers and their semiarboreal and arboreal allies: *Chironius, Coluber, Dendrophidion, Drymarchon, Drymobius, Leptodrymus, Leptophis, Mastigodryas, Pseustes*, and *Spilotes*. All of these lack grooved fangs, although most have Duvernoy's gland (definitely absent in *Spilotes*, condition not known for *Dendrophidion* and *Pseustes*). A second group includes two genera *(Elaphe* and *Lampropeltis)* of moderate-sized constrictors (other Costa Rican colubrines are grab-and-swallow predators) that lack both grooved fangs and Duvernoy's gland. Also lacking grooved fangs and Duvernoy's gland is the coral snake mimic *Scaphiodontophis*, which is unique among American colubrines in having double-socketed teeth and hypapophyses on the posterior vertebrae. The remaining genera *(Oxybelis, Rhinobothryum, Scolecophis, Stenorrhina, Tantilla*, and *Trimorphodon)* have one or more grooved fangs on the posterior part of the maxillary and a Duvernoy's gland (condition not known in *Scolecophis*, but probably present). All members of the colubrine stock in Costa Rica are probably oviparous.

Genus *Chironius* Fitzinger, 1826

Members of this genus (thirteen species) are large (to nearly 3 m in total length), powerful, and active diurnal snakes that are unique among New World taxa in having the dorsal scales arranged in even numbers of oblique rows. Other characteristics include presence of the basic colubrid complement of enlarged head plates; body somewhat compressed; nasal entire; a loreal (very rarely none or two); usually 1 preocular, sometimes 2; eye large, with a round pupil; two pairs of elongate chin shields; dorsal scales smooth or most rows keeled or keels restricted to paravertebral rows; scales in 10 to 12 rows at midbody; usually one apical pit (but sometimes two or none), pits usually restricted to dorsal scale rows on neck and/or paravertebral rows on tail but may be present on paravertebral rows throughout length of body or on neck and paravertebrals of the posterior body and tail; anal entire or divided; subcaudals divided; 24 to 51 maxillary teeth that gradually increase in size posteriorly; Duvernoy's gland present; mandibular teeth subequal or decreasing in length posteriorly; no hypapophyses on posterior vertebrae.

The short to long everted hemipenes (5 to 13 subcaudals in length) are single and symmetrical, with a simple semicentripetal sulcus spermaticus; a basal naked pocket present;

pedicel with spinules or naked; truncus covered with many spines and four to six large hooks; distal half calyculate (Vellard 1946; Wiest 1978; Dixon, Wiest, and Cei 1993) (fig. 11.2b).

Wiest's 1978 unpublished dissertation was incorporated into a monograph on the entire genus (Dixon, Wiest, and Cei 1993). Much of the revision is verbatim from Wiest's manuscript, so either work may be consulted for more information on *Chironius*.

The genus has a wide range in the lowlands and onto the adjacent slopes on the Atlantic versant from Honduras southward through most of South America to southern Brazil, central Bolivia, and adjacent areas of Uruguay, Paraguay, and northern Argentina; on the Pacific versant it ranges from Costa Rica to southern Ecuador; also occurs on the islands of Trinidad and St. Vincent (Lesser Antilles).

KEY TO THE COSTA RICAN SNAKES OF THE GENUS *CHIRONIUS*

1a. Dorsal scale rows 12 at midbody; dorsum weakly variegated, uniform or with one or more longitudinal dark or light stripes 2
1b. Dorsal scale rows 10; dorsum uniformly dark or with alternating dark and light crossbands. . . . .*Chironius grandisquamis* (p. 650)
2a. Dorsum uniform green without longitudinal stripes; subcaudals 133–44 . . . . . *Chironius exoletus* (p. 649)
2b. Dorsum brown, olive green, olive gray, or gray, usually with a light middorsal stripe in adults; juveniles with a conspicuous dark lateroventral longitudinal stripe involving tips of ventrals and first scale row along posterior part of body and tail; subcaudals 118–29 . . . . *Chironius carinatus* (p. 648)

Chironius carinatus (Linné, 1758)

(plate 409; map 11.74)

DIAGNOSTICS: This species is unlikely to be confused with any other, except for some examples of *Chironius exoletus* and *C. grandisquamis*, which also have very large, obliquely arranged dorsal scales (fig. 11.5) in an even number of rows. The dorsal color of *C. carinatus* is dull (brown, olive, or gray) compared with that of the green *C. exoletus*. In addition the former has a bright yellow venter in adults and most juveniles, whereas in *C. exoletus* the venter is greenish. *C. grandisquamis* has the dorsum banded or uniform black, and the venter is white in juveniles and white anteriorly and black posteriorly in adults.

DESCRIPTION: A large species reaching a total length of 2,054 mm (most examples 1.5 m or less); males longer than females; head very distinct from neck; tail very long (30 to 40% of total length); 2 postoculars (rarely 3); supralabials 8 to 10, usually 9, with 2 or 3 bordering orbit; infralabials 8 to 12, usually 11; temporals 1 + 2, rarely 1 + 1 or 1 + 3;

dorsal scale rows 16 to 12-12-8 to 10, paravertebral rows keeled, very strongly so in males; sometimes 4 to 6 dorsalmost rows keeled in females; apical scale pits restricted to scales on neck, usually a single pit, sometimes two; ventrals 143 to 165 (143 to 169); subcaudals 118 to 129 (108 to 139); anal divided; maxillary teeth 29 to 36 (29 to 39). Adults some shade of olive green to gray dorsally, usually with a light middorsal stripe and/or distinct yellow flecking or distinct spots on first scale row on anterior body and tail; juveniles brown to brownish gray with obscure brownish crossbands, usually edged by a lighter color, with definite lateroventral dark stripes, changing ontogenetically to olive green or olive gray; supralabials, throat, and venter bright yellow in adults and some juveniles (others have dusky to white venters); iris grayish brown; tongue black above, grayish below. The everted hemipenis is 11 to 13 subcaudals long, and the pedicel lacks spinules (Wiest 1978; Dixon, Wiest, and Cei 1993).

SIMILAR SPECIES: (1) *Chironius exoletus* is green with a greenish venter and lacks light stripes and spots.

HABITAT: Lowland Moist and Wet Forests, secondary growth, and more or less disturbed areas within these formations, especially near stream courses, and the landward side of mangrove forests.

BIOLOGY: These relatively common active predators are often seen moving about on the ground during the day, but they sleep during the night well above the ground (to 15 m) in bushes and trees. Because of this habit they are restricted to areas of primary forest or perturbed sites where shrubs and trees persist, including orchards and plantations. Available data indicate that Costa Rican representatives of this species feed exclusively on anurans (particularly frogs of the genera *Eleutherodactylus, Hyla,* and *Leptodactylus*), as is

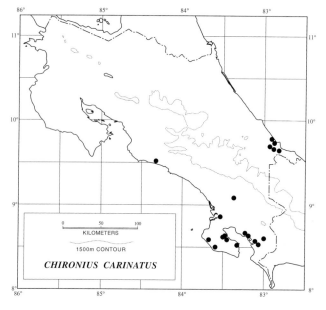

Map 11.74. Distribution of *Chironius carinatus.*

the case elsewhere in its range (Duellman 1978; Cunha and Nascimeto 1982, 1983; Wiest 1978; Dixon, Wiest, and Cei 1993). Oviductal eggs from non–Costa Rican examples numbered three to ten and averaged 28 × 10 mm (Wiest 1978). The smallest known juvenile is 433 mm in total length. When threatened, snakes of this species raise the anterior part of the body off the ground and attempt to bluff the perceived enemy by holding the mouth open and hissing. If this is not effective they then inflate the neck region, exposing the whitish to light blue skin, and flatten it dorsoventrally to form a hood, while whipping the tail back and forth. Similar behaviors doubtless are exhibited by other members of the genus.

REMARKS: Wiest (1978) and Dixon, Wiest, and Cei (1993) have described the geographic variation in this species:

1. Populations from Brazil, the Guianas, eastern Venezuela, and Trinidad: uniform dark olive brown dorsum and distinct yellow spots on first scale row on tail
2. Populations from northern and eastern Colombia and central and western Venezuela: dorsum with a pair of broad lateral reddish brown or brownish stripes on anterior part of body as well as a greenish to bluish middorsal stripe and lacking light spots on the tail
3. Populations from western Ecuador and Colombia to Costa Rica: dorsum usually with a yellowish middorsal stripe and distinct yellow spots on the first scale row of the anterior part of the body and on the tail

The specimens Taylor (1951b) referred to this species are representatives of *Chironius exoletus*, which has long been confused with *C. carinatus*.

DISTRIBUTION: Widespread in the lowlands from southern Costa Rica south to Ecuador on the Pacific slope and throughout northern South America, the Guianas, the lower Amazon basin, and northeastern Brazil and Trinidad. In Costa Rica restricted to the southwestern Pacific area and the southeastern Atlantic zone (2–200 m).

Chironius exoletus (Linné, 1758)

(plate 410; map 11.75)

DIAGNOSTICS: This green snake is distinct from other Costa Rican species that are primarily green in having large, obliquely arranged dorsal scales (fig. 11.5) in even numbers of rows. It may be separated from its congener *Chironius carinatus* by coloration, since that form has brown, olive, or gray dorsal coloration and usually a light middorsal stripe and a bright yellow venter (sometimes white in juveniles). The venter in *C. exoletus* is greenish white (juveniles) to green.

DESCRIPTION: A moderate-sized species reaching a total length of 1,545 mm (most examples 1 m or less); head very distinct from neck; tail moderate to very long (28 to 44% of total length); usually 2 postoculars, rarely 3; supra-

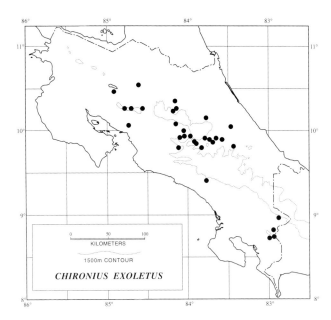

Map 11.75. Distribution of *Chironius exoletus*.

labials 7 to 10 (usually 9), with usually 3 (sometimes 2 or 4) bordering orbit; infralabials 9 to 13, usually 10 or 11; temporals usually 1 + 2, sometimes 1 + 1, 1 + 1 + 1 or 1 + 1 + 2; dorsal scale rows 12 to 16-12-8 to 10, usually 12-12-8, paravertebral rows keeled in adults, weakly keeled in females; apical pits on scales in neck region and sometimes on paravertebral scales on tail or posterior body and tail; usually one pit per scale, sometimes two; ventrals 132 to 154 (123 to 162); subcaudals 133 to 144 (111 to 160); anal usually divided (entire in 22% of available samples); maxillary teeth 24 to 34. Dorsal surfaces dark green; juveniles sometimes with narrow light crossbands; venter greenish white in juveniles, pale green in adults; lips and chin yellow; iris pale yellowish brown to black with yellowish rim around pupil; tongue red or orange with gray tips. The everted hemipenes are 5 to 11 subcaudals long, and the pedicel may be naked or covered with spinules (Wiest 1978; Dixon, Wiest, and Cei 1993).

SIMILAR SPECIES: (1) All other uniform green snake species in Costa Rica have an odd number of dorsal scale rows. (2) *Chironius carinatus* has a brown, olive, or gray dorsum and usually a bright yellow venter (whitish in some juveniles).

HABITAT: A rather common inhabitant of Lowland Moist and Wet Forests and Premontane Moist and Wet Forests and Rainforest; marginal in Lower Montane Wet Forest and Rainforest.

BIOLOGY: A conspicuous, relatively uncommon diurnal species, usually seen foraging on the ground or in low bushes. It spends the night sleeping in bushes or trees up to 4 m above the ground. It feeds primarily on frogs, especially hylids, but also eats lizards (Wiest 1978; Dixon, Wiest, and Cei 1993) like other members of the genus (Dixon and Soini

1977) such as *C. carinatus*. Oviductal eggs from non–Costa Rican examples ranged between four and twelve per female and averaged 30.4 × 12.8 mm. The smallest known individual is 310 mm.

REMARKS: This snake has long been regarded as a variant of *C. carinatus*, but Hoge, Romano, and Cordeiro (1976–77) demonstrated that *C. exoletus* was a valid species occurring sympatrically with *C. carinatus* in eastern Amazonian Brazil. Wiest (1978) and Cunha and Nascimiento (1982, 1983) provided additional evidence confirming that two species are involved. Most examples of *C. exoletus* from Costa Rica have previously been identified as *C. carinatus*, including those reported under the latter name by Taylor (1951b). Scott (1969) realized that two species were included in Costa Rican material previously referred to *C. carinatus* but was uncertain about the correct name to apply to the green species here called *C. exoletus*. There is no question that *C. exoletus* and *C. carinatus* are distinct taxa. There remains some question regarding the application of the name *C. exoletus* to the green species described in this account, since the color variation in *C. exoletus* (sensu lato) is extensive. Wiest (1978) and Dixon, Wiest, and Cei (1993) noted that examples from eastern South America were brown to olive (some with a light middorsal stripe), upper Amazon examples were olive to dark green, and those from Bolivia had dark crossbands and a distinct postocular dark stripe. The populations in western South America, Panama, and Costa Rica referred to this species appear to be completely allopatric with those from east and south of the Andes. In these circumstances, since no other features consistently separate geographic populations of *C. exoletus*, I use this name for Costa Rican examples.

DISTRIBUTION: Lowlands and both Atlantic and Pacific premontane and lower montane areas of Costa Rica, south through western South America to Ecuador, northern South America, the Guianas, and the Amazon basin to Peru, southern Brazil, and extreme northeastern Argentina (40–1,600 m).

Chironius grandisquamis (Peters, 1868)

(plate 411; map 11.76)

DIAGNOSTICS: This species is unique among Costa Rican snakes in having large, obliquely arranged dorsal scales (fig. 11.5) in 10 rows at midbody. No other species in the area that is uniformly black above, as are adults of *C. grandisquamis*, or banded with brown and white (juveniles) has even numbers of oblique scale rows.

DESCRIPTION: A very large species, total length to 2,718 mm, largest female 1,995 mm; head very distinct from neck; tail long to very long (33 to 43% of total length); 2 postoculars (rarely 3); supralabials usually 9, sometimes 8 or 10, usually with 3 bordering orbit, occasionally 2; infralabials 9 to 12, usually 10; temporals variable, usually 1 + 1 or 1 + 2, sometimes 1 + 1 + 1 or 1 + 1 + 2; dorsal

scales usually 10-10-8, sometimes 12-10-8; paravertebral rows keeled (strongly in males, weakly in females) and usually six to eight uppermost scale rows keeled in males; apical pits single, restricted to scales on neck or neck and paravertebral rows above vent; ventrals 151 to 167; subcaudals 129 to 155; anal divided; maxillary teeth 35 to 45. Dorsal surfaces in adults glossy black, rarely peppered with white spots; juveniles banded with brown and yellow or white dorsally; head dark except for lower half of supralabials and chin, which are white; venter white in juveniles, white anteriorly in adults, becoming black on posterior half of body and under tail; iris black; tongue black. The everted hemipenis is 7 to 9 subcaudals in length and has many spinules on the pedicel (Wiest 1978; Dixon, Wiest, and Cei 1993).

SIMILAR SPECIES: (1) Uniform black examples of *Lampropeltis triangulum* have smooth scales in 19 to 21 evenly arranged rows. (2) Adult examples of *Clelia* have smooth scales in 19 evenly arranged rows, and the undersurfaces of the body and tail are light throughout. (3) *Dendrophidion paucicarinatum* juveniles have dark ventral markings.

HABITAT: Lowland Moist and Wet Forests and Premontane Wet Forest and Rainforest.

BIOLOGY: A rather common active diurnal forager on the forest floor that rests at night well off the ground (to 5 m) in bushes and trees. These snakes often evade predators by escaping into streams or rivers, since they are powerful swimmers. They feed mostly on frogs, primarily *Eleutherodactylus*, and one specimen contained a salamander (*Bolitoglossa* sp.) (Wiest 1978; Dixon, Wiest, and Cei 1993). They are known to eat the rather toxic *Leptodactylus pentadactylus*, and Scott (1969) noted that they also eat small lizards, birds, and mice. A female captured in Panama laid an egg on May 21, and fourteen more were removed from the oviducts

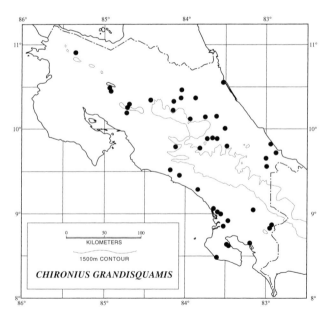

Map 11.76. Distribution of *Chironius grandisquamis*.

the next day. The three preserved eggs averaged 39 × 26 mm (Wiest 1978; Dixon, Wiest, and Cei 1993). The smallest known example is 392 mm in total length.

REMARKS: Wettstein (1934) pointed out that the nominal species *Chironius melas* was conspecific with *C. grandisquamis*. Although Taylor (1951b, 1954b) recognized both forms as valid based on the presumed differences in number of keeled rows of scales, Scott (1969) placed both in the synonymy of the wide-ranging Amazonian species *Chironius fuscus*. Wiest (1978) confirmed Wettstein's conclusion that the supposed diagnostic characters for *C. grandisquamis* versus *C. melas* are sexually dimorphic and reduced the latter to synonymy. He further demonstrated the distinctiveness of *C. fuscus* and *C. grandisquamis* in coloration and scalation. *C. fuscus* is bright green, olive green, or brown with a yellow venter; it has a single anal plate and 116 to 126 subcaudals. In contrast, *C. grandisquamis* is banded with black and light or uniform black dorsally, the venter is white or white anteriorly and black posteriorly, the anal plate is divided, and subcaudals number 129 to 155.

DISTRIBUTION: Lowlands and premontane slopes from Atlantic versant Honduras to northern Colombia, and from Pacific versant central Costa Rica to Ecuador (2–1,600 m).

Genus *Coluber* Linné, 1758

A genus of slender, active diurnal racers (nine species) with the basic colubrid complement of enlarged head shields and the following combination of characteristics: nasal divided or entire; one loreal; two preoculars; eye large, pupil round; two pairs of elongate chin shields; dorsal scales smooth in 15 to 17 rows at midbody and a reduction anterior to vent; two apical pits; anal divided; subcaudals divided; 12 to 23 maxillary teeth that increase in length posteriorly; Duvernoy's gland present; mandibular teeth enlarged anteriorly; no hypapophyses on posterior dorsal vertebrae. The combination of the listed features, the characteristic dark mottling on the side of the head (and usually the chin, throat, and anterior ventrals), and having only two supralabials bordering the orbit (in Costa Rica) will immediately separate this genus from other colubrids described in this book. The everted hemipenes are short to moderate (7 to 14 subcaudals in length), bilobed, asymmetrical, with a simple, semicentrolineal sulcus spermaticus; pedicel naked distally, followed by three large spines; truncus covered by numerous spines (75 to 135); apical one-quarter to one-third calyculate (Ortenburger 1928; Schätti 1987) (fig. 11.2c). Schätti (1987) showed that these New World racers are distinct from Old World forms (thirty-three species) often referred to *Coluber*. The genus as defined here corresponds to Schätti *Coluber* (sensu stricto), since the type species is the North American racer *Coluber constrictor*. Schätti also conclusively demonstrated that the putative genus *Masticophis* cannot be separated from *Coluber* and that all species often placed in the

former genus are properly allocated to *Coluber*. Determination of the correct generic name(s) for Palearctic racers awaits the completion of ongoing research on their interrelationships. However, it is clear that none of them are congeneric with the New World forms.

Three species groups are recognized within *Colober* (sensu stricto) following Ortenburger (1928). Of these only the *Coluber flagellum* species group is represented in Costa Rica, by *Coluber mentovarius*. Other species placed in this group include *Coluber anthonyi* (Clarion Island, Mexico) and *Coluber flagellum* (southern United States and northern Mexico).

Coluber has a wide distribution from British Columbia, Canada, and the northern United States southward throughout Mexico and Central America to Nicaragua on the Atlantic versant and central Costa Rica on the Pacific; isolated populations occur on Pacific slope western Panama and in northern Colombia and Venezuela (Atlantic drainage), including Isla Margarita.

Coluber mentovarius (C. Duméril, Bibron, and A. Duméril, 1854)

Savenera

(plate 412; map 11.77)

DIAGNOSTICS: The distinctive mottled pattern of the side of the head (and usually the chin, throat, and anterior ventrals), combined with the smooth dorsal scales (fig. 11.5), separates this species from most other forms. The dorsal pattern is rather variable (uniform, obscurely banded, or lineate), but only some lineate examples of *Mastigodryas melanolomus* are likely to be confused with *C. mentovarius*. The condition of the supralabials, with one bordering the

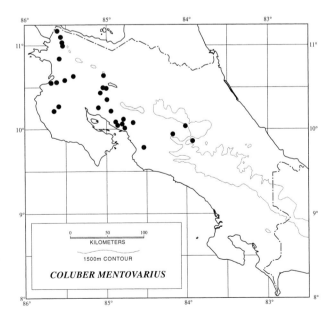

Map 11.77. Distribution of *Coluber mentovarius*.

orbit in Costa Rican *C. mentovarius,* three in *C. melanolomus,* will consistently distinguish between them.

DESCRIPTION: A large species reaching a total length of 2,527 mm (most examples less than 1.5 m); body cylindrical; head distinct from neck; tail long (32 to 36% of total length but often incomplete); 2 preoculars; 2 postoculars; supralabials 7 (6 to 8), with 1 or 2 bordering orbit; infralabials 8 to 11; temporals 1 + 2 to 3 + 4; smooth dorsal scales in 19-17-13 rows, with two apical pits; ventrals 181 to 205 (166 to 205); subcaudals 102 to 123 (95 to 126). Dorsal ground color tan or brown to grayish; adults uniform or with a dark brown spot on each scale; dark spots tend to form narrow longitudinal dark stripes or indistinct, irregular dark, crossbands or stripes; sides of head and usually chin, throat, and anterior ventrals mottled with dark pigment; venter immaculate pale yellow anteriorly grading to pink on posterior third of body and underside of tail; juveniles similarly colored but with one or two light longitudinal stripes on each side of neck that may continue well back onto body; iris bronze, with gold ring around pupil; tongue brown.

SIMILAR SPECIES: (1) Uniform individuals of *Pseustes poecilonotus* have obliquely arranged keeled scales (fig. 11.5). (2) *Mastigodryas melanolomus* juveniles have alternating middorsal and lateral dark blotches; adults have two pairs of narrow light longitudinal stripes on each side, and three supralabials border the orbit. (3) *Drymarchon corais* has subocular dark stripes. Also see table 11.1.

HABITAT: Open areas (including pastures and roadsides) and thickets within Lowland Dry Forest and Premontane Moist Forest and marginally in Premontane Wet Forest.

BIOLOGY: This swift, elegant diurnal snake is frequently seen foraging during the early part of the day. It often climbs into bushes or low trees to rest or search for prey. It feeds primarily on the diurnal ground lizards of the genera *Ameiva* and *Cnemidophorus* but also eats the common bush anole *Norops cupreus.* Guyer and Laska (1996) report an attack by a member of this species on a juvenile *Boa constrictor* in Guanacaste Province. It probably also eats birds' eggs and hatchlings, rodents, and an occasional adult bird, like most other members of the genus. Hatchlings and small juveniles feed on arthropods, mainly insects. Mexican examples of the species laid seventeen eggs from March 23 to 25 (Werler 1951) and twenty in early spring (Alvarez del Toro 1960). Clutch size ranges from sixteen to thirty eggs slightly over 50 mm long, based on these reports and Minton and Minton (1991). Gerhardt, Harris, and Vásquez (1993) recorded this species as a prey item of the great black hawk *(Buteogallus urubitinga)* in Guatemala. This species exhibits unspecialized pseudautotomy as a antipredator defense.

REMARKS: This form has been called *Masticophis mentovarius* in most recent literature and in Taylor (1951b). Johnson (1977, 1982) provided a detailed analysis of geographic variation for the species based on slight differences

in color pattern and scalation. The principal variants and their geographic distributions are summarized as follows:

1. Dorsum in adults brown, with or without darker spots or stippling; juveniles similar but with four or five light longitudinal stripes on neck; with or without a few dark spots on labials and chin; seven supralabials: north-central Venezuela and Isla Margarita

2. Dorsum in adults pale brown, with many obscure, irregular dark stripes and bands (Venezuela and Colombia) or spots (Panama); juveniles similar but with anterior pattern more pronounced; chin and venter without dark spotting; seven supralabials: northwestern Venezuela and adjacent northeastern Colombia and western Panama

3. Dorsum in adults brown to gray brown; juveniles brown with two pale stripes (rarely one) on each side of neck; spotting on chin and anterior ventrals present or absent; seven supralabials: drier areas of Central America from Costa Rica to the Isthmus of Tehuantepec (on both Atlantic and Pacific slopes) and Atlantic versant Mexico to San Luis Potosí

4. Dorsum in adults blue gray to brown, with or without dark stripes or spots; juveniles with dark bands anteriorly; chin and venter immaculate; eight supralabials: Pacific versant Mexico west of the Isthmus of Tehuantepec to southern Sonora, including the Islas de Tres Marías and disjunct in eastern Chihuahua

In Costa Rica, as expected, available material generally agrees with features for population 3. There is considerable variation, however, and some examples approach the adult coloration found in populations 1 and 2.

DISTRIBUTION: In Mesoamerica in the drier lowlands and adjacent slopes from southern Sonora, southern San Luis Potosí, and northern Veracruz, Mexico, to northwestern Costa Rica on the Pacific slope; with Atlantic slope populations in the Yucatán Peninsula, Guatemala, Belize, Honduras, and northeastern Nicaragua; disjunct populations to the south occur in central western Panama (Pacific drainage); Atlantic slope northeastern Colombia and adjacent western Venezuela; north-central Venezuela; and Isla Margarita, Venezuela. In Costa Rica, found in the subhumid lowlands of northwestern Costa Rica and the Meseta Central (Oriental and Occidental) (1–1,435 m).

Genus *Dendrophidion* Fitzinger, 1843

A genus (nine species) of relatively conspicuous diurnal semiarboreal racers with the following suite of characteristics: a generalized colubrid complement of enlarged head plates; body cylindrical; nasal divided; one loreal; one preocular; eye large, pupil round; two pairs of elongate chin shields; keels present at least on the uppermost seven dorsal scale rows; dorsal scales in 17-17-15 or 15–15 rows; two apical scale pits; anal entire or divided; subcaudals divided; 30 to 50 maxillary teeth subequal, or last three or four somewhat larger than others; condition of Duvernoy's gland unknown; mandibular teeth subequal; no hypapophyses on posterior dorsal vertebrae. This combination of features separates the genus from other Costa Rican colubrids except species of the genus *Drymobius*, as will be discussed below.

The everted hemipenes are short (7 to 9 subcaudals in length), single, symmetrical and have a simple semicentripetal sulcus spermaticus. In some species the pedicel is nude or covered with spinules, the truncus is covered by large spines, and the apical area bears scalloped flounces. In others the pedicel is nude, but there are three or four large basal hooks proximal to the truncus, which is covered by spines. The apical area is composed of scalloped or smooth calyces or smooth or spinulate flounces. The very tip of the organ is naked in all species (Dunn 1933b; Stuart 1932; Roze 1966).

Some species of the genus *Drymobius* agree in general external features with *Dendrophidion*. Technically the two genera differ only in hemipenial structure, since the presumed differences in maxillary dentition (Stuart 1932; Roze 1966) are variable (Dunn 1933b; Chippaux 1986). The hemipenes of *Drymobius* differ from those of *Dendrophidion* in having the apical area covered with small, papillate calyces. Since only a few users of this book will wish to make a generic determination on this basis, both taxa have been integrated into a single key based on external characteristics (p. 653). In a recent review Lieb (1988) recognized two subdivisions within the genus, the *Dendrophidion dendrophis* (five species) and *D. percarinatum* (four species) species groups, based principally on differences in scalation and hemipenial structure. Members of the genus appear to feed exclusively on frogs. These snakes use unspecialized pseudautotomy as an antipredator defense. *Dendrophidion* ranges throughout the lowlands and adjacent slopes on the Atlantic versant from southern Veracruz, Mexico, to northeastern Brazil and in the Amazon basin to northern Bolivia and on the Pacific versant from Costa Rica to Ecuador.

KEY TO THE COSTA RICAN SPECIES OF THE GENERA *DENDROPHIDION* AND *DRYMOBIUS*

1a. Caudal scale row reduction, from 8 to 6 rows, around tail occurs anterior to subcaudal 24 2
1b. Caudal scale row reduction, from 8 to 6 rows, around tail occurs posterior to subcaudal 24. 6
2a. Ventrals 140–72 3
2b. Ventrals 179–95; dorsum uniform in adults, juveniles with some markings; subcaudals 119–39; ventrals dark edged on free margin (fig. 11.39a) . . . . *Dendrophidion paucicarinatum* (p. 656)
3a. Dorsum uniform, striped, reticulate, or with narrow crossbands, never with distinct large blotches. 4
3b. Dorsum with large dark rhomboidal blotches, with light centers, on a lighter ground color; a series of small

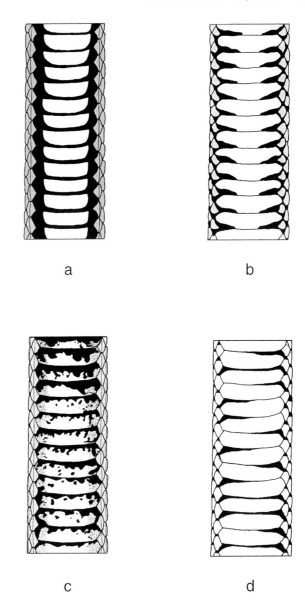

a   b

c   d

Figure 11.39. Ventral coloration at midbody in several Costa Rican snakes: (a) *Dendrophidion paucicarinatum* with distinct dark lateral and posterior margins on ventrals that are continuous with one another; (b) *Drymobius margaritiferus* with ventrals edged with black on lateral posterior margins; (c) adult *Dendrophidion nuchale* with ventrals heavily marked with dark pigment; (d) *Drymarchon corais* with mixture of complete and incomplete narrow dark lines on posterior margin of ventrals.

black ventrolateral blotches; anterior venter immaculate, becoming mottled with gray posteriorly; ventrals 145–63; subcaudals 84–108 . . . . *Drymobius rhombifer* (p. 662)
4a. Subcaudals 100 or more; dorsum reticulate, striped, crossbanded, or uniform brown; keels on vertebral and paravertebral scale rows light or of the same color as the rest of scales. 5
4b. Subcaudals 90–96; living adults uniform green (bluish

in preservative); juveniles green with black and white interscale flecking that forms narrow crossbands dorsally; keels on vertebral and paravertebral scale rows usually black; ventrals dark edged on posteriolateral margins (fig. 11.39b) ....
.... *Drymobius melanotropis* (p. 661)

5a. Dorsal pattern of adults reticulate, each scale with a yellow center bordered by dark green to black; small juveniles with dark crossbands anteriorly but reticulated posteriorly and with distinct light semicircular nuchal spots; ventrals 139–50; subcaudals 100–121; ventrals dark edged at least on posteriolateral margins (fig. 11.39b) ..... *Drymobius margaritiferus* (p. 660)

5b. Dorsum soft brown, usually with narrow dark-edged light crossbands anteriorly and dark longitudinal stripes posteriorly; dorsum occasionally uniform tawny brown; ventrals 153–66; subcaudals 146–55; venter immaculate. .... *Dendrophidion percarinatum* (p. 657)

6a. Subcaudals 111–28; ventrals 148–65; vivid dark-edged light crossbands on neck more than one scale row wide; venter immaculate white or pale yellow; anal entire .... *Dendrophidion vinitor* (p. 655)

6b. Subcaudals 140–60; ventrals 160–75; neck usually uniform green, sometimes reddish brown or with faint dark-edged light crossbands less than one scale row wide; venter of adults with suffusions of dark pigment on the medial portions of the ventral scales (fig. 11.39c); anal divided or entire (fig. 11.39c) ....
................*Dendrophidion nuchale* (p. 654)

Dendrophidion dendrophis Group

Snakes referred to this group have the dorsal scales strongly keeled in adults and the hemipenes with large basal spines or hooks. In addition to *Dendrophidion nuchale* and *Dendrophidion vinitor*, which occur in Costa Rica, *Dendrophidion dendrophis* (northern Amazonia) and an undescribed species (southern Amazonia) compose this group (Lieb 1988).

Dendrophidion nuchale (Peters, 1863c)

(plates 413–15; map 11.78)

DIAGNOSTICS: The two-toned adults are unlikely to be misidentified, having an overall coloration usually with the anterior dorsum bright green and the posterior part of the body and the tail dark brown or black, or the body green or reddish brown and the tail reddish brown to coral red. Juveniles resemble the young of some other species of *Dendrophidion*, but the obscure dark nuchal collar (if present) of juvenile D. *nuchale* and the presence of dark crossbands throughout the length of the body are usually diagnostic. D. *nuchale* juveniles, with dark-edged light crossbands on the neck, closely resemble some individuals of *Dendrophidion percarinatum*. When such juveniles have a single anal plate

(fig. 11.5) they may with some certainty be placed in D. *nuchale* (all D. *percarinatum* have divided anals). If the examples have a divided anal, identification must be based on the level at which the caudal scales reduce from eight to six (posterior to subcaudal 24 in D. *nuchale*, anterior to this point in D. *percarinatum*).

DESCRIPTION: A moderate to large snake reaching a total length of 1,530 mm (most examples less than 1 m); head distinct from neck; tail very long (35 to 40% of total length); nasal divided; postoculars 2 or 3, usually 2; supralabials 9 to 10, usually 9, with 3 bordering orbit; infralabials 9 to 12, usually 10; temporals usually 2 + 2; scales keeled, strongly in adults; scales in 17-17-15 rows; two apical scale pits; anal entire or divided; ventrals 160 to 175 (153 to 175); subcaudals 140 to 160 (132 to 163). Adults: usually bright green on anterior dorsum, with posterior body and tail dark brown or black with rows of pale yellow ocelli or tail pink to red; sometimes with dark crossbands that enclose ocelli on posterior third of body; a dark nuchal collar (usually obscured in preservative); venter white or pale yellow anteriorly, becoming pink under tail, with definite dark markings (fig. 11.39c). Juveniles: nuchal collar faint, dorsal ground color brown; dark crossbands present throughout length of body, usually containing light ocelli; dark-edged light crossbands on neck, if present, less than one scale row wide; ventrals immaculate white in juveniles, heavily suffused and marked with dark pigment medially in adults. The everted hemipenis has two to four large basal hooks, the truncus covered by large spines and on the apical section by large scalloped calyces.

SIMILAR SPECIES: (1) *Dendophidion vinitor* juveniles have dark-edged light crossbands on the neck that are more

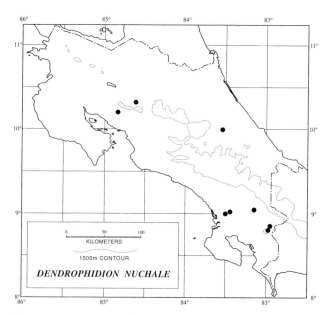

Map 11.78. Distribution of *Dendrophidion nuchale*.

than one scale row wide. (2) *Dendrophidion paucicarinatum* juveniles have narrow dark and pale gray crossbands, no light ocelli, and the ventrals marked with dark pigment. (3) *Dendrophidion percarinatum* juveniles have dark-edged light bands on the anterior part of the body and the caudal reduction from eight to six scale rows anterior to the level of subcaudal 24. (4) *Chironius carinatus* lacks dark-edged ventrals. (5) Adult *Drymobius melanotropis* are uniform green dorsally, and the banded juveniles have distinct dark marks on the posteriolateral edges of the ventrals (fig. 11.39b). (6) *Drymarchon corais* has dark subocular stripes (fig. 11.40a). Also see table 11.1.

HABITAT: Relatively undisturbed Lowland Wet and Moist Forests and Premontane Wet Forest and Rainforest; marginally in Premontane Moist Forest. Often found in riparian remnants or gallery forest situations in otherwise cleared land.

BIOLOGY: A relative uncommon diurnal terrestrial species that feeds on frogs, mostly *Eleutherodactylus*. It lays about six eggs per clutch (Campbell 1998). A small hatchling is 425 mm in total length.

REMARKS: In most recent literature this species has been called *Dendrophidion clarkii*, but Lieb (1988, 1991a) documented the synonymy of that name with *D. nuchale* (Peters, 1863c), a form long confused with its ally *Dendrophidion dendrophis*, of Amazonian and Guianan South America. Geographic, ontogenetic, and individual variation is substantial in this form, which has a disjunct distribution in Central America and northern South America. The more northern populations (Guatemala, Belize, and Honduras) and those from northern Costa Rica have a pink to coral red tail (plate 413). Southern Costa Rican examples and those from western Panama and northwestern South America have dark brown or black tails. Those from northern Venezuela tend to have posterior crossbanding with vivid ocellation in adults. Lieb (1988) points out some other differences between each pair of these populations. The more northern populations (those with pink or red tails) have lower subcaudal counts (mean 143.4) than those with dark brown or black tails from lower Central America and northwestern South America (mean 149.6). Venezuelan examples resemble the more northern sample (mean 142.3). The Venezuelan population also lacks keeling on the first dorsal scale row on each side in all cases and has a lower ventral count (152 to 160) than the northern sample with red or pink tails (166 to 171) and the dark-tailed series from the intermediate area (162 to 175). It seems likely that two or three distinct species are subsumed under this name.

DISTRIBUTION: Lowland and premontane zones of Atlantic versant Guatemala, Belize, and northwestern Honduras, northeastern Costa Rica and northwestern Panama; and western Costa Rica, southwestern Panama, and northwestern Colombia to Ecuador on the Pacific slope; also a pair of disjunct populations in northern Venezuela. In Costa

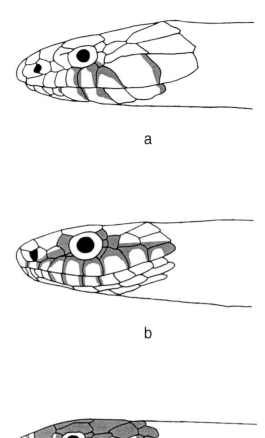

a

b

c

Figure 11.40. Dark head markings in some large Costa Rican colubrid snakes, lateral view: (a) *Drymarchon corais*; (b) *Pseustes poecilonotus*; (c) *Spilotes pullatus*. Shading indicates dark brown or black markings.

Rica, known from both slopes of the Cordillera de Tilarán in the premontane zone, at similar elevations in central Costa Rica on the Atlantic versant, and from the lowlands and premontane slopes of the Pacific southwest (75–1,220 m).

Dendrophidion vinitor Smith, 1941

(plate 416; map 11.79)

DIAGNOSTICS: The presence of vivid dark-edged light bands (two to three scale rows in width) on the neck in concert with the immaculate venter will separate this species from all other Costa Rican snakes except *Dendrophidion percarinatum*. *D. vinitor* differs from *D. percarinatum* (features for the latter in parentheses) in having broader light crossbands on the neck (less than one scale row wide) and a single anal plate (anal divided) (fig. 11.5). Occasional juve-

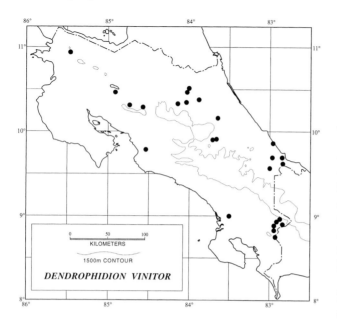

Map 11.79. Distribution of *Dendrophidion vinitor*.

nile *Dendrophidion nuchale* have light crossbands on the neck, but they are narrow as in *D. percarinatum*.

DESCRIPTION: A moderate-sized form reaching a total length of 980 mm; head distinct from neck; tail very long (35 to 37% of total length); nasal divided; postoculars 2; supralabials 8 to 10; usually 9, with 3 bordering orbit; infralabials 9 to 11, usually 10; temporals usually 2 + 2; all but first row (on each side) of dorsal scales strongly keeled; 17-17-15 scale rows; two apical scale pits; anal entire; ventrals 148 to 165; subcaudals 111 to 128. Adults and juveniles similar in coloration, upper surfaces brown to brownish gray, with vivid dark-edged narrow light crossbands anteriorly, light bands on neck two to three scale rows wide; interscale skin and/or light bands on neck red or orange to bluish white; usually some red to orange areas in dorsal coloration; similar bands may be present posteriorly or grade into pattern of dark crossbands, ocelli, or dark longitudinal stripes posteriorly; upper surface of head reddish brown, brown, or greenish; underside immaculate white (juveniles) or yellow (adults); iris black with gold flecks; tongue brown. The everted hemipenis has large basal hooks, large spines on truncus, and spinulate flounces on the apical region.

SIMILAR SPECIES: (1) *Dendrophidion percarinatum* has the light crossbands on the neck less than one scale row wide and a divided anal plate. (2) Juveniles of *Dendrophidion nuchale* usually lack light bands on the neck, or if present, they are less than one scale row in width. Also see table 11.1.

HABITAT: Undisturbed Lowland Moist and Wet Forests and marginally into Premontane Wet Forest and Rainforest. Also found in riparian situations and isolated remnants of these forest types in areas cleared for other purposes.

BIOLOGY: This somewhat common diurnal frog eater readily climbs into bushes or trees when foraging or escap-

ing. The smallest hatchling is 230 mm in total length. It apparently inflates the anterior portion of the body to expose the interscale colors as a antipredator defense.

REMARKS: Although this species most closely resembles *D. percarinatum* in coloration, the consistent difference in anal plate condition (single in *D. vinitor* versus divided in *D. percarinatum*) has prevented misidentification since Smith (1941b) established that the two species were distinct from *Dendrophidion dendrophis* of South America. On the other hand, some confusion of the upland species *Dendrophidion paucicarinatus* (anal plate variably single or divided) and *Dendrophidion vinitor* was created by Scott's (1969) inclusion of Atlantic lowland *D. vinitor* in his concept of *D. paucicarinatum*. Lieb (1988, 1991b) has satisfactorily clarified the status of all Costa Rican forms of the genus, and I follow his treatment here.

DISTRIBUTION: Disjunct in rainforests of the Atlantic lowlands and foothills of southern Veracruz, Mexico; Chiapas, Mexico, and Guatemala; Nicaragua, Costa Rica, and western Panama. In similar environments on the Pacific versant in southwestern Costa Rica and adjacent western Panama (15–1,360 m).

Dendrophidion percarinatum Group

Members of this species group lack keels on the first dorsal scale row; the keeling is reduced or weak in adults, and the hemipenes are without basal spines or hooks. *Dendrophidion paucicarinatum* and *Dendrophidion percarinatum* are representatives that occur in Costa Rica. *Dendrophidion bivittatum* and *Dendrophidion brunneum*, both from Colombia, are the other members of the group.

Dendrophidion paucicarinatum (Cope, 1894a)

(plate 417; map 11.80)

DIAGNOSTICS: Adults of this relatively common species have a uniform dull greenish or brown dorsum and may be confused with several other similarly colored Costa Rican species, except that this form has the posterior margin of each ventral marked with dark pigment. Juvenile *D. paucicarinatum* are banded with dark and gray and have dark pigment along the posterior margin of each ventral (fig. 11.39a) to separate them from other juvenile *Dendrophidion*.

DESCRIPTION: A moderate-sized snake, up to 1,400 m in total length (most examples less than 1 m); head distinct from neck; tail very long (32 to 35% of total length); nasal divided; postoculars 2 or 3, usually 2; supralabials 8 to 10, usually 9, with 3 bordering orbit; infralabials 9 to 12, usually 10; temporals usually 2 + 2; dorsal scale keeling reduced in adults, often restricted to upper 7 to 13 rows; 17-17-15 scale rows; two apical scale pits; anal usually divided; ventrals 179 to 195; subcaudals 119 to 139. Adults: uniform olive green, gray green, or occasionally brown; underside yellow, with distinct dark posterior borders to each ventral and subcaudal. Juveniles: ground colors similar to adults,

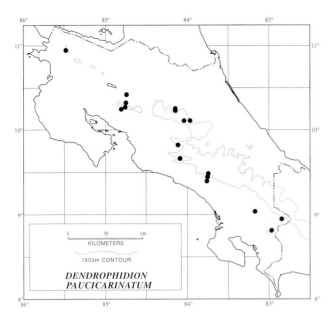

Map 11.80. Distribution of *Dendrophidion paucicarinatum.*

with narrow dark (black or brown) and pale gray crossbands (less than two scale rows wide); underside white to yellowish with dark markings as in adults; iris dark brown to black, upper one-fourth lighter, tongue dark brown to black. The everted hemipenis lacks large basal hooks, and the apical area is covered with smooth flounces.

SIMILAR SPECIES: (1) Juvenile *Dendrophidion nuchale* lack dark markings on the venter. (2) *Dendrophidion percarinatum* and *D. vinitor* have immaculate venters. (3) *Liophis epinephalus* has the venter checkered with black and orange, yellow, or red. (4) *Drymobius melanotropis* has the dark ventral markings restricted to the posteriolateral margins of each ventral (fig. 11.39b) or at most weakly indicated medially. Also see table 11.1.

HABITAT: Undisturbed Premontane Wet Forest and Rainforest and Lower Montane Wet Forest and Rainforest and riparian remnants of these habitats, where the forests have been converted to other uses.

REMARKS: The record of this upland species from Sara (Madre de Dios), Limón Province (AMNH 36450), in the Atlantic lowlands at about 40 m in elevation is suspect. All other examples have been taken at localities above 1,000 m. Other records from the Atlantic lowlands for this species reported by Scott (1969) are based on examples of *Dendrophidion vinitor*, which resemble some examples of *D. paucicarinatum* in having a single anal plate but differ from them in always having an immaculate venter. *D. paucicarinatum* consistently has each ventral marked with dark pigment posteriorly. Taylor's (1951b, 1954b) records for *Drymobius chloroticus* are based on examples of this species (Wilson 1970). Lieb (1991c) redescribed *D. paucicarinatum* and provided a figure of a specimen from "Cichona," Alajuela Province, an error for Cinchona.

DISTRIBUTION: Humid premontane and lower montane zones in the mountains of Costa Rica and western Panama (1,040–1,500 m).

Dendrophidion percarinatum (Cope, 1893)

(plate 418; map 11.81)

DIAGNOSTICS: Most individuals of this form are readily identified by the dark-edged light bands on the anterior part of the body combined with the immaculate venter. The immaculate white upper lip, keeled dorsal scales, and immaculate venter will separate the occasional uniform brown or obscurely striped adults from other snakes with these patterns. This species differs from examples of the similarly marked *Dendophidion vinitor* in having the light bands less than one scale row in width and a divided anal plate (fig. 11.5) (versus light bands two to three scale rows in width and a single anal in *D. vinitor*). It may be necessary to determine the level of the caudal scale row reduction from eight to six in order to discriminate between this form and the occasional juvenile *Dendrophidion nuchale* having light crossbands on the neck, unmarked ventrals, and a divided anal plate (anterior to the level of subcaudal 24 in *D. percarinatum*, posterior to that level in *D. nuchale*).

DESCRIPTION: A moderate-sized species attaining a total length of 1,175 mm; head distinct from neck; tail very long (35 to 45% of total length), very often incomplete; nasal divided; postoculars 2; supralabials 8 to 10, usually 9, with 3 bordering orbit; infralabials 9 to 11, usually 10; temporals usually 2 + 2; keels on dorsal scales strongest on upper rows, weaker laterally; first scale row on each side smooth; scales in 17-17-15 rows; two apical pits; anal divided; ventrals 153 to 166 (153 to 169); subcaudals 146 to

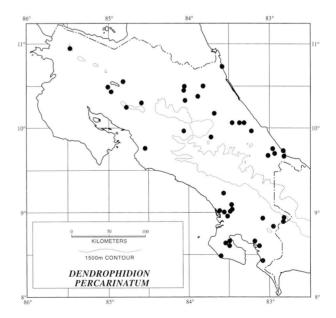

Map 11.81. Distribution of *Dendrophidion percarinatum.*

155 (133 to 164). Upper surfaces brown, sometimes uniform or with indistinct dark longitudinal stripes; usually crossbanded with dark-edged narrow (less than one scale row wide) light bands anteriorly, grading into dark crossbands containing light ocelli posteriorly; often with light longitudinal stripes posteriorly and on tail, or sometimes tail uniform; venter immaculate white anteriorly becoming pale yellow posteriorly; tongue brown. The hemipenis lacks basal hooks, and the apical area is covered by large scalloped calyces and the truncus by spines.

SIMILAR SPECIES: (1) *Dendrophidion vinitor* has light neck bands two to three scale rows in width and a single anal plate. (2) Juveniles of *Dendrophidion nuchale* with light crossbands on the neck and a divided anal plate (fig. 11.5) have the caudal scale reduction from eight to six posterior to the level of subcaudal 24. (3) *Dendrophidion paucicarinatum* has each ventral marked posteriorly with dark pigment (fig. 11.39a). Also see table 11.1.

HABITAT: Relatively undisturbed lowland Moist and Wet Forests and lower margins of adjacent Premontane Wet Forest and Rainforest and marginally in Premontane Moist Forest; also occurring in isolated patches of these forests, especially near stream courses on otherwise cleared land.

BIOLOGY: A relatively common terrestrial diurnal frog-eating form.

REMARKS: A single example of this species (AMNH 17374) has been recorded from the city of San José on the Meseta Central. Since no other specimens are known from the Pacific uplands in central Costa Rica, I regard this locality as possibly in error, perhaps based on the point the specimen was shipped from. Lieb (1988, 1996) has redescribed this species and summarized the literature on it.

DISTRIBUTION: Rainforests of the lowlands and foothills of Atlantic slope northern Honduras south to eastern Panama and on Pacific versant southwestern Costa Rica, Panama, Colombia, and Ecuador. In Costa Rica the species also barely ranges onto the northwestern Pacific slope through the low pass near Laguna Arenal (4–1,200 m).

Genus *Drymarchon* Fitzinger, 1843

This monotypic genus is represented in Costa Rica by the largest known species of colubrid snake (approaching 3 m in total length), the diurnal terrestrial racer *Drymarchon corais*. The genus has the basic colubrid complement of enlarged head shields and the following distinguishing features: nasal entire; one loreal; one preocular; eye large, pupil round; two pairs of elongate chin shields, dorsal scales smooth or a few weakly keeled; scales rows 17 to 19-17-13 to 15, with two apical pits; anal entire; subcaudals divided; 17 to 21 subequal maxillary teeth; Duvernoy's gland present; anterior mandibular teeth slightly longer than posterior ones; no hypapophyses on posterior dorsal vertebrae. These features and the unique pattern of dark lines radiating from

below the eye along the margins of the supralabials makes identification easy.

The everted hemipenes are slightly bilobed and asymmetric, with a simple semicentrolineal sulcus spermaticus; basal area naked, truncus covered by small spines, distal surface by fringed calyces (Dowling and Duellman 1974–78; McCranie 1980).

The genus occurs disjunctly in the southeastern United States and from southern Sonora and Chihuahua, Mexico, to northern Peru (on the Pacific versant), and from southern Texas south through Mesoamerica, northern South America, and most areas east of the Andes to Paraguay, southern Brazil, and extreme northern Argentina.

Drymarchon corais (Boie, 1827)

Cribo, Zopilota

(plate 419; map 11.82)

DIAGNOSTICS: All Costa Rican examples of this species have a series of three dark lines radiating from the orbit along the supralabial sutures, a similar dark line between the last two supralabials, the supralabials anterior to the eyes unmarked (fig. 11.40a), and usually a distinct diagonal black mark on each side of the neck. These marks are evident at all sizes, including the obscurely banded hatchlings.

DESCRIPTION: A large species with a known maximum total length of 2,950 mm (most adults 1.5 to 2 m); body cylindrical, robust; head distinct from neck; tail moderate (17 to 24% of total length); 2 postoculars; supralabials 7 to 9, usually 8, penultimate scale wedge shaped and in contact with an anterior temporal or lower postocular or both (variable in some non–Costa Rican populations); usually

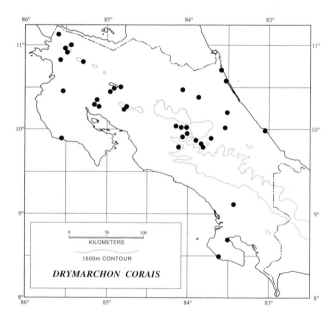

Map 11.82. Distribution of *Drymarchon corais*.

2 supralabials bordering orbit, sometimes 3; infralabials 7 to 10, usually 8; temporals usually 2 + 2, sometimes 1 or 3 posterior ones; supracloacal keels in adult males; ventrals 188 to 215 (182 to 222); subcaudals 56 to 88 (55 to 88). Adults uniform pale brown to pale gray dorsally or becoming black on the extreme posterior part of body and entire tail; venter pale brown throughout or becoming darker brown or black posteriorly and under tail; sides of body and lateral posterior edges of ventrals often marked with narrow lines, especially in juveniles; these lines frequently continuous along posterior margin of ventrals across venter (fig. 11.39d); hatchlings with obscure dark crossbands anteriorly; iris black, paler above than below.

SIMILAR SPECIES: (1) Juveniles of *Mastigodryas melanolomus* have all the supralabials heavily marked with dark pigment. (2) *Pseustes poecilonotus* has oblique keeled dorsal scale rows (fig. 11.5) and lacks the subocular dark stripes. (3) *Dendrophidion nuchale* has immaculate supralabials. (4) *Coluber mentovarius* lacks subocular stripes. (5) *Spilotes pullatus* has most or all the supralabial sutures marked with black (fig. 11.40c) and has keeled scales (fig. 11.5). Also see table 11.1.

HABITAT: Found in a wide variety of situations, but most commonly near riverbeds, swamps, and seasonal marshes, in Lowland Moist and Wet Forests and Premontane Moist and Wet Forests and Rainforest; marginal in Lower Montane Wet Forest.

BIOLOGY: This uncommon powerful racer usually forages on the ground but sometimes climbs low shrubs and trees. In Costa Rica this snake is usually diurnal. It feeds on a wide variety of prey including fishes, frogs and toads, turtles, lizards, snakes (including its own species), mammals, birds, and birds' eggs (Scott 1983m). Hatchlings are 430 to 660 mm in total length. Duellman 1963b reported an individual 2,950 mm in total length eating a 1,683 mm *Boa constrictor* in Guatemala. The same *Drymarchon* contained an adult pitviper *(Atropoides nummifer)* 953 mm in total length. When threatened, *Drymarchon* lift the anterior part of the body off the ground and attempt to bluff the intruder away by holding the mouth open and hissing. If this is not successful these snakes vertically compress and inflate the neck, hiss, and vibrate the tail, which may produce a rattling sound in dry leaves. Male combat including biting occurs in this species (Shine 1994). These snakes are oviparous (Fitch 1970) and lay four to eleven eggs in temperate and subtropical areas. Carson (1945) reported sperm storage and delayed fertilization for over four years. This species was among the prey of the great black hawk *(Buteogallus urubitinga)* in Guatemala (Gerhardt, Harris, and Vásquez 1993).

REMARKS: Substantial geographic variation has been described for this species (McCranie 1980). The principal variants, as defined principally by adult coloration, are:

1. South America from Venezuela southward over most of the continent east of the Andes: anterior dorsum dark brown or black; yellowish, brownish, or orange posteriorly; venter yellow white; no dark subocular stripes radiating from eye; 199 to 222 ventrals; 65 to 87 subcaudals

2. Northwestern Venezuela and northwestern Peru north to central Costa Rica on the Pacific slope and to central Veracruz, Mexico, on the Atlantic versant: dorsum and venter pale brown to pale gray anteriorly, becoming black on the posterior part of body and tail; definite dark subocular stripes and diagonal dark stripes on neck; 191 to 225 ventrals; 59 to 88 subcaudals. The form on Isla Margarita, off Venezuela, described (Roze 1959) as lacking subocular stripes probably belongs here (Lancini 1979)

3. Central Veracruz, Mexico, to southern Texas: dorsum brownish anteriorly, black posteriorly, with some suggestion of spots or bands; venter gray anteriorly, becoming black posteriorly; definite dark subocular stripes; 182 to 196 ventrals; 55 to 65 subcaudals; consistently 14 scale rows at vent (almost always 15 rows at vent in other populations)

4. Southern Sonora to eastern Chiapas, Mexico, on the Pacific slope, southern portion of the Mexican Plateau and an adjacent area on Atlantic slope of the Sierra Madre Oriental: dorsum and tail usually uniform black; anterior venter light, usually reddish, posterior venter and underside of tail black; subocular dark stripes present; 186 to 209 ventrals; 59 to 82 subcaudals

5. Guatemala to northwestern Costa Rica: dorsum pale brown; venter pale brown anteriorly becoming darker under tail; subocular stripes present; 192 to 206 ventrals; 56 to 77 subcaudals

6. Southeastern United States: dorsum and venter blue black, except for reddish to cream suffusion on chin, throat, and side of head; subocular dark stripes present; 184 to 195 ventrals; 63 to 70 subcaudals

In Costa Rica most examples fit the features listed for population 2. Specimens from the northwestern Pacific area agree with those listed for population 5. The karyotype is 2N = 36, with eight pairs of macrochromosomes (seven pairs metacentric, one pair telocentric) and ten pairs of microchromosomes with ZW heteromorphism in the fourth pair in females; NF = 50 (Beçak 1965).

DISTRIBUTION: Lowlands and premontane slopes from southern Sonora, Mexico, and southern Texas to extreme northwestern Peru on the Pacific versant, to southern Brazil, Paraguay, and extreme northern Argentina, including most of South America east of the Andes and Isla Margarita, Trinidad, and Tobago on the Atlantic slope; disjunct populations in southeastern Georgia, Florida, and southern Alabama (2–1,600 m).

Genus *Drymobius* Fitzinger, 1843

A group of active diurnal terrestrial racers (four species) with the basic colubrid complement of enlarged head shields and the following additional characteristics: head distinct from neck; body cylindrical; nasal divided; one loreal; one preocular; eye large, pupil round (fig. 11.11b); two pairs of elongate chin shields; dorsal scales keeled, in 17-17-15 rows; two apical scale pits; anal divided; subcaudals divided; maxillary teeth 22 to 36, gradually becoming larger and stouter posteriorly; Duvernoy's gland present; mandibular teeth subequal; no hypapophyses on posterior dorsal vertebrae. The combination of keeled scales in 17 rows at midbody, divided anal, and the presence of a loreal will distinguish the genus from all others in Costa Rica genera except *Dendrophidion*. Technically *Drymobius* differs from *Dendrophidion* in having the apical region of the hemipenes covered by small papillate calyces and 22 to 34 maxillary teeth (apical area of hemipenes covered by flounces or smooth or scalloped calyces and maxillary teeth 30 to 50 in *Dendrophidion*). Since most readers of this book probably will not care to determine the condition of the hemipenes and dentition, the species of these two genera have been combined into a single key (p. 653) for convenient identification.

The moderately short everted hemipenes (8 to 11 subcaudals in length) of *Drymobius* have the following features: slightly bilobed, symmetrical, with a simple semicentripetal sulcus spermaticus; pedicel nude or covered with small scattered spinules; truncus covered with spines, and apical area covered with papillate calyces (Stuart 1932; Villa 1967; Wilson 1970, 1975a).

The species of this genus use unspecialized pseudautotomy as an antipredator defense. Wilson (1975a) summarized characters and pertinent literature for *Drymobius*.

The genus ranges from southern Texas and Sonora, Mexico, over most of lowland and premontane Mesoamerica through western South America to Ecuador and from northern South America to the Guianas and upper Amazonian Brazil, Peru, and Bolivia. Note that the distribution map for this genus provided by Wilson (1975a) is patently in error. No *Drymobius* occur in the arid Pacific lowlands of southern Ecuador and Peru (contra Wilson), and *Drymobius rhombifer* ranges much farther to the east in the northern Amazon basin than shown.

See key to species of *Dendrophidion* and *Drymobius* (p. 653).

Drymobius margaritiferus (Schlegel, 1837)

(plate 420; map 11.83)

DIAGNOSTICS: This beautiful species is probably the most commonly seen snake in Costa Rica. Adults are immediately distinguished from all other species by the reticulate coloration alone. The coloration appears to be green, with each dorsal scale marked with a yellow to orangish cen-

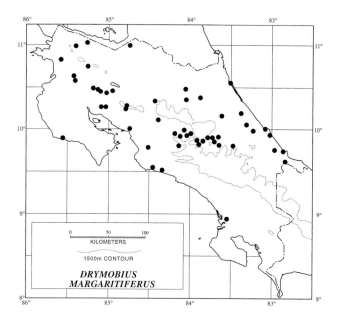

Map 11.83. Distribution of *Drymobius margaritiferus*.

tral area, and each scale is edged with dark green to bluish green grading to black at the posterior tip. The overall pattern is a reticulum of small discrete light spots on a greenish to black background. Juveniles have dark crossbands on the anterior part of the body, but the posterior portion has the distinctive reticulate coloration of the adults.

DESCRIPTION: A moderate-sized species (total length to 1,339 mm, most adults 760 to 1,000 mm); tail long (26 to 30% of total length); 2 postoculars; supralabials 8 to 10, usually 9, with 3 bordering orbit; infralabials 9 to 12, usually 10; temporals 2 + 1 to 2 + 3; ventrals 139 to 150 (137 to 158); subcaudals 100 to 121 (100 to 138). Adults reticulate with a yellow to orange streak or spot on center of each dorsal scale, which otherwise are dark green to black; juveniles with dark and yellow crossbands anteriorly, but reticulate pattern posteriorly and definite semicircular nuchal light spots; top of head often with a Y-shaped yellow mark, with arms of Y pointed posteriorly; supralabials edged with dark; a definite dark temporal stripe; venter and subcaudal area cream to yellow with each ventral edged with black posteriorly, at least on lateral margins (fig. 11.39b); iris rust brown.

SIMILAR SPECIES: (1) Juvenile banded examples of *Mastigodryas melanolomus* entirely lack the reticulate coloration and have no distinctive nuchal light spots. Also see table 11.1.

HABITAT: Ubiquitous in all but the most humid lowland and premontane zones, but favoring open areas, especially on forest edges and in clearings, riparian sites, secondary growth, pastures, and roadsides. Present in Lowland Dry, Moist, and Wet Forests, Premontane Moist and Wet Forests, and rarely in Lower Montane Wet Forest.

BIOLOGY: This species is active primarily during the daytime as it forages on the ground. Principal food items are frogs and toads of several genera (Duellman 1963b; Meyer 1966; Seib 1984, 1985b; Smith 1943; Stuart 1935), but it also eats lizards, small snakes, reptile eggs, and small mammals. Juveniles also include insects in their diet. Gaige, Hartweg, and Stuart (1937) reported a gravid female from Nicaragua taken in August, and Stuart (1948) found females from Guatemala with eggs in May and June and a female about to lay eggs in July (Stuart 1943). A Mexican female laid seven eggs on April 22, two of which hatched on June 9 and 11, and two eggs were laid on July 29 (Werler 1949). Two clutches of four and five eggs from Costa Rica, laid in February and March, respectively, hatched after sixty-four to sixty-eight days, in April and June (Solórzano and Cerdas 1987). The eggs averaged 35.6 × 53.4 mm, and the hatchlings were 225 to 275 mm in total length. This is another species preyed on by the great black hawk *(Buteogallus urubitinga)* in Guatemala (Gerhardt, Harris, and Vásquez 1993).

REMARKS: The considerable geographic variation in coloration in this widespread species has been summarized by Wilson (1974) and Wilson and Meyer (1985). The major geographic pattern is as follows:

1. Populations of the Atlantic versant from southern Texas to northeastern Nicaragua: dorsal scales blue on anterior margin, black posteriorly, with a central yellow to orange area; a distinct dark temporal stripe; subcaudals usually edged with black posteriorly
2. Populations of the Pacific versant of Mexico from Sonora to the Isthmus of Tehuantepec: entire border of each dorsal scale black, central area blue or yellow; a distinct dark temporal stripe; subcaudals usually edged with black posteriorly
3. Populations of the Pacific versant from the Isthmus of Tehuantepec to El Salvador: diffuse, stippled gray brown spots on the dorsal scales; no dark temporal stripe; subcaudal surface white

No one has critically analyzed color variation in this species from Honduras to northern Colombia. Villa (1967) regarded specimens from both versants in Nicaragua as identical to those in population 1 and described a distinctive variant from Great Corn Island (Nicaragua) in which the juvenile pattern is retained in all known adults. All material from Costa Rica (both Atlantic and Pacific localities) and western Panama agrees with the description of examples from population 1. The karyotype for this form (based on males only) is 2N = 36, with eight pairs of macrochromosomes and ten pairs of microchromosomes; pairs 1, 3, 4, and 5 are metacentrics; 2 is submetacentric; and 6, 7, and 8 are subtelocentrics; NF = 52 (Gutiérrez, Solórzano, and Cerdas 1984).

DISTRIBUTION: Widespread in the lowlands and onto adjacent slopes on the Atlantic versant from Texas to Costa Rica and on the north coast of Colombia; on the Pacific versant from southern Sonora, Mexico, south through Central America and Panama. In Costa Rica in all lowland and premontane areas (1–1,500 m).

Drymobius melanotropis (Cope, 1875)

(plates 421–22; map 11.84)

DIAGNOSTICS: Adults are uniform bright green snakes, with the venter mostly lemon yellow but with dark margins on the posteriolateral edges of each ventral (fig. 11.39b) and sometimes dark speckles on the rest of the posterior margins of each ventral. Juveniles have narrow black crossbands or black spots on a bright green ground color (plate 422). These features combined with 17 scale rows at midbody will distinguish *Drymobius melanotropis* from all other Costa Rican snakes.

DESCRIPTION: A moderate-sized species reaching a total length of 1,300 mm; tail long (29 to 30% of total length); 1 postocular; supralabials 9, with 3 bordering orbit; infralabials 10 to 11, usually 10; temporals 2 + 2 or 2 + 3; ventrals 150 to 161, subcaudals 90 to 96. Adults essentially uniform green above, frequently with black keels on three middorsal scale rows; dorsal green color extending onto outer one-fourth of each ventral scale; venter white anteriorly becoming lemon yellow posteriorly; posterior edge of each ventral edged with dark at least on posteriolateral margins; juveniles with narrow black and light crossbands on a green background; iris golden brown to coppery flecked with gray; tongue black.

SIMILAR SPECIES: (1) *Dendrophidion paucicarinatum* has the entire posterior margin of each ventral and subcaudal heavily marked with dark pigment (fig. 11.39a). (2) *Dendrophidion nuchale* has a tail uniformly pink, black,

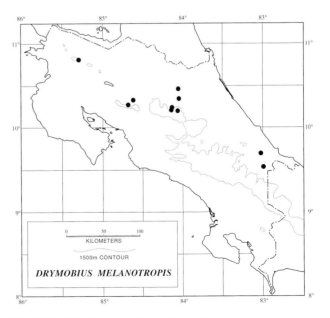

Map 11.84. Distribution of *Drymobius melanotropis*.

or brown or body marked with bands or ocelli, and banded juveniles lack dark-edged ventrals. (3) *Chironius exoletus* has the lateral scales smooth and the venter pale green. (4) Uniform green species of *Leptophis* have a definite dark postocular stripe and often a preocular dark stripe as well (fig. 11.41). (5) Green species of *Oxybelis* have obliquely arranged dorsal scales (fig. 11.5). Also see table 11.1.

HABITAT: Relatively undisturbed Atlantic slope Lowland Moist and Wet Forests and Premontane Rainforest.

REMARKS: Although relatively rare in collections, the species is often encountered along the Atlantic slopes of the Cordillera Central and Cordillera de Tilarán. Taylor's records (1951b, 1954b) for the related species *Drymobius chloroticus* are based on examples of *Dendrophidion paucicarinatum* (Wilson 1970). *D. chloroticus* is now known to be restricted to Mexico and upper Central America, ranging southward to Nicaragua, and does not occur in Costa Rica. It seems likely that the two nominal taxa, *D. chloroticus* and *D. melanotropis*, will ultimately prove to be geographic variants of a single species, although Wilson (1970, 1975b) reported virtual sympatry for the two: *D. chloroticus* at a locality 19 km north of Matagalpa, Department of Matagalpa, Nicaragua, and *D. melanotropis* from the city of Matagalpa. The presumed differences between the two are trivial (*chloroticus*: no black keels on the dorsal scales, subcaudals 107 to 125; *melanotropis*: black keels on the three uppermost dorsal scales, subcaudals 91 to 96). The degree to which the keels are darkened in Costa Rican examples is now known to be variable, so this feature does not reliably separate the nominal species. The name *melanotropis* (Cope, 1875) predates *chloroticus* (Cope, "1885," 1886), so the name of the species in Costa Rica would remain unchanged in any event. Wilson (1975b) provided a summary of then current knowledge of this species. My note (Savage 1974a) regarding the provenance of the specimen (KU 31973) of this species from Estrella reported by Taylor (1954b) and Wilson (1970) is in error. The example is from Limón Province, as confirmed by other Atlantic lowland collections, not, as I suggested, from La Estrella, Cartago Province.

DISTRIBUTION: Atlantic lowlands and premontane slopes from eastern Honduras through southeastern Costa Rica (35–1,000 m).

Drymobius rhombifer (Günther, 1860)

(plate 423; map 11.85)

DIAGNOSTICS: Unlikely to be mistaken for any other racer because of the distinctive light-centered dark-outlined rhomboidal dorsal blotches. These markings resemble those of the harmless *Xenodon rabdocephalus* and the dangerous poisonous viper *Bothrops asper*. The presence of keeled dorsal scales arranged in nonoblique rows (fig. 11.5) will separate *D. rhombifer* from *Xenodon*, and the absence of a labial pit will separate it from *B. asper*.

DESCRIPTION: A moderate-sized species attaining a to-

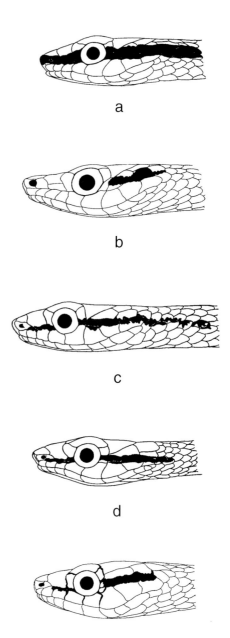

Figure 11.41. Diagnostic features of head scalation and coloration in Costa Rican snakes of the genus *Leptophis*: (a) *L. mexicanus*, a loreal present and a well-developed pre- and postocular dark stripe that is continuous with dark body stripe; (b) *L. depressirostris*, a loreal present, no preocular dark stripe, and a short postocular dark stripe; (c) *L. riveti*, no loreal, a narrow pre- and postocular stripe that does not reach tip of snout or continue any great distance onto body; (d) *L. nebulosus*, no loreal, a narrow pre- and postocular dark stripe that does not reach tip of snout or extend or continue posteriorly past neck; (e) *L. ahaetulla*, no loreal, pre- and postocular dark stripe does not reach tip of snout or extend onto neck or body.

tal length of 1,200 mm; tail long (30 to 34% of total length); 2 postoculars; supralabials 9, with 3 bordering orbit; infralabials 8 to 10, usually 9; temporals 2 + 2; ventrals 145 to 163; subcaudals 84 to 108. Dorsum gray, marked with about 20 light-centered rhomboidal dark brown blotches

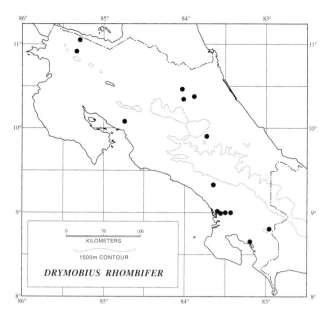

Map 11.85. Distribution of *Drymobius rhombifer*.

that are outlined by black; a series of definite small ventro-lateral black blotches present on tips of ventrals and adjacent dorsal scales; dorsal markings becoming somewhat indistinct in large adults and partially obscured by darker pigment; tail similarly marked; labials almost uniform gray; most of anterior ventral surface immaculate, without dark posterior edges, becoming mottled with gray posteriorly, most pronounced under tail; juveniles similar to adults in ventral coloration or with considerable mottling on anterior ventrals as well; iris bronze.

SIMILAR SPECIES: (1) *Xenodon rabdocephalus* has smooth, obliquely arranged dorsal scales (fig. 11.5). (2) *Bothrops asper* has a labial pit (fig. 11.8b) and 25 to 29 dorsal scale rows at midbody. Also see table 11.1.

HABITAT: Undisturbed Lowland Moist and Wet Forests and Premontane Wet Forest and Rainforest.

DISTRIBUTION: Widespread in tropical America from Nicaragua south through Central America, Colombia, Venezuela, and the Guianas and over much of the upper Amazon basin to southern Peru and northern Bolivia on the Atlantic versant and from central Costa Rica to western Ecuador on the Pacific slope. Relatively rare in Costa Rica but known from the Atlantic and southern Pacific lowlands and adjacent premontane slopes, with two records from the Pacific-Atlantic Moist Forest zone at the northern end of the Cordillera de Guanacaste (13–1,200 m).

Genus *Elaphe* Fitzinger, 1833

As currently understood this is a diverse group (thirty to forty species) of small to moderate-sized snakes that use constriction to subdue their prey. The genus is represented in the New World by seven species, one of which ranges into Costa Rica. Other species occur over most of the Eurasian landmass. The following features characterize this genus: basic colubrid complement of enlarged head plates: nasal divided; loreal present, large or very small, or absent; one preocular; eye medium to large, pupil round; two pairs of elongate chin shields; dorsal scales smooth or keeled in 19 to 39 rows at midbody, with a posterior scale row reduction; two apical scale pits; anal divided or entire; subcaudals divided; maxillary teeth 14 to 24, subequal; no Duvernoy's gland; mandibular teeth gradually increasing in size anteriorly; no hypapophyses on posterior dorsal vertebrae. These features combined with hemipenial characteristics reviewed below separate the genus from other colubrines. The single Costa Rican species has a very high number of scale rows at midbody (29 to 39), and this, in combination with the presence of enlarged head plates and two pairs of elongate chin shields, will separate it from all other species known from the republic.

There seems little question that *Elaphe* (sensu lato) is probably polyphyletic, and Dowling and Fries (1987) and Dowling and Price (1988) attempted to partition the genus into at least four genera based principally on hemipenial characters. In this scheme American members of the genus would be placed as follows (characteristics of everted hemipenes):

Elaphe: asymmetric bilobed hemipenes with a simple semicentrolineal sulcus spermaticus, nude pedicel, spinous truncus, and apical area covered by scalloped calyces (*Elaphe flavirufa, E. guttata, E. obsoleta,* and *E. vulpina* and twenty-three Old World forms)

Bogertophis: asymmetric single clavate hemipenes with a simple semicentrolineal sulcus spermaticus, pedicel spinulate, truncus spinous, and the distal portion covered by scalloped calyces (*Elaphe rosaliae* and *E. subocularis*)

Senticolis: asymmetric single subcylindrical hemipenes with a simple semicentripetal sulcus spermaticus, large basal hooks grading into an area of large spines, but about the distal half of the organ covered by papillate calyces (*Elaphe triaspis*)

These authors concede that their newly restricted *Elaphe* as defined above remains polyphyletic, and though the hemipenial differences between it and the other two nominal American genera are impressive, such differences are not unusual within a single large genus of snakes (e.g., *Dendrophidion* and *Coniophanes*). For this reason I have not used the name *Senticolis* for the Costa Rican species pending a reanalysis of species groupings within *Elaphe* (sensu lato).

In the Old World the genus has an extensive range in suitable habitats throughout the temperate and semiarid areas of Eurasia, India, the Indo-Malayan archipelago, China, and Japan. In the Western Hemisphere the genus ranges from Arizona and the northeastern United States south to Costa Rica, with one species (*Elaphe rosaliae*) restricted to California and Baja California, Mexico.

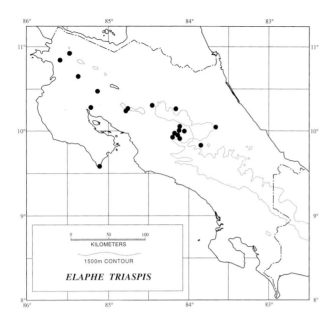

Map 11.86. Distribution of *Elaphe triaspis*.

Elaphe triaspis (Cope, 1866a)

(plates 424–26; map 11.86)

DIAGNOSTICS: Adults are uniform brown to golden brown (gray, olive green, to lime green extralimitally), often with indications of dark blotching on the head, and have an immaculate venter. These features in combination with the faintly keeled middorsal scale rows and 29 to 39 scale rows at midbody (27 to 39 extralimitally) will separate them from all other Costa Rican snakes. The juveniles have forty-five or more black-outlined reddish brown dorsal blotches, which in combination with the aforementioned features of scalation will distinguish them from other small blotched snakes with heads distinct from the neck.

DESCRIPTION: A moderate-sized snake reaching a total length of 1,222 mm; body cylindrical; head distinct from neck; tail very long (31 to 37% of total length; moderate to long for entire species, 23 to 37% of total length); eye medium; one loreal; postoculars usually 2, rarely 3; supralabials 8 to 10, usually 8, with 2 bordering orbit; infralabials 10 to 12, usually 11; temporals usually 3 + 3 or 3 + 4, but sometimes 1 or 2 anterior temporals or 5 posterior ones (fig. 11.9c); dorsal scales smooth except for 3 to 11 faintly keeled middorsal rows; dorsal scales in 27 to 33–27 to 39–19 to 25 rows; ventrals 247 to 264 (241 to 282); anal divided; subcaudals 107 to 115 (87 to 126). Juveniles tan with 45 to 51 (43 to 73) black-outlined reddish brown blotches (plate 425); adults (over about 800 mm in total length) uniform brown to golden brown (blotched as in juveniles in some non–Costa Rican areas); venter immaculate cream to pinkish anteriorly with some dark mottling posteriorly and under the tail in some examples; dark head markings, usually evident throughout life although obscured in adults,

include a transverse internasal-frontal bar, a longitudinal supraorbital-parietal bar that may continue onto neck, a frontoparietal spot, a posterior parietal spot on suture between parietals that continues onto nape, and a longitudinal stripe, often continuous with supraorbital-parietal bar on each side of neck; iris beige.

SIMILAR SPECIES: Adults: (1) Examples of *Liophis epinephalus* with a uniform dorsal coloration have tessellated venters. (2) Adult *Epicrates cenchria* have smooth scales in 45 to 53 rows at midbody. (3) All pitvipers have strongly keeled dorsal scales, a vertically elliptical pupil, and a loreal pit (fig. 11.8b). Juveniles: (1) *Leptodeira annulata* and *Leptodeira septentrionalis* have vertically elliptical pupils and smooth scales (fig. 11.5) in 19 to 25 rows at midbody. (2) *Mastigodryas melanolomus* juveniles have smooth scales (fig. 11.5) in 17 rows at midbody. (3) *Trimorphodon biscutatus* has a vertically elliptical pupil and fewer than twenty-five dark body blotches. Also see table 11.1.

HABITAT: Lowland Dry and Wet Forests, Premontane Moist and Wet Forest and Rainforest, and Lower Montane Wet Forest; in drier habitats usually found in riparian or gallery forest situations.

BIOLOGY: This relatively rare species is terrestrial but frequently climbs shrubs and trees when foraging, usually at night. The species kills its prey by constriction. Food consists primarily of small mammals, but it also eats birds and their eggs. Censky and McCoy (1988) reported on reproduction for the northern Yucatán Peninsula population of *E. triaspis*. Although the climate is strongly seasonal in the area (with a long dry season between November and April) this species appeared to be a continuous breeder. Females with body lengths of over 624 mm were sexually mature and laid clutches of three to seven eggs that were 35 to 42 mm long. Clutch size did not increase with body size. Captures of subadult snakes (300 to 500 mm in body length) were evenly distributed throughout the year. Sexual maturity was attained at about twenty-four months.

REMARKS: Although Taylor (1951b) listed *Elaphe flavirufa* for Costa Rica, that species is not known to occur south of Nicaragua. Juvenile *E. triaspis* sometimes are misidentified as *E. flavirufa* because they resemble the latter form in having dark-outlined reddish dorsal blotches. *E. triaspis* usually has more dorsal blotches (43 to 73) than *E. flavirufa* (29 to 46) and has three supralabials bordering the orbit (two in *E. flavirufa*). Three geographic subdivisions of *E. triaspis* may be recognized as follows (Dowling 1960; Price 1991), but intermediates connect these geographic units:

1. Arizona and Tamaulipas, Mexico, south to Chiapas, Mexico: adults uniform gray green to lime green; juveniles with 53 to 69 body blotches and the median frontoparietal band with a central light area reaching forward to the tip of the frontal shield

2. Yucatán Peninsula and adjacent Guatemala: adults and

juveniles with 42 to 55 body blotches and the median frontoparietal band with a light area at about the midpoint of the parietals

3. Guatemala to Costa Rica: adults uniform brown to golden brown; juveniles with 45 to 51 body blotches and with a central elongate light area in the frontoparietal band and the band usually broken at the frontoparietal suture

All Costa Rican examples conform to the features listed for population 3. Price (1991) provided a review of characteristics and pertinent literature updating Dowling's (1960) treatment.

DISTRIBUTION: Lowlands and adjacent slopes from Tamaulipas, Mexico, and Arizona to central Costa Rica. In that country known from the lowlands, premontane and lower montane zones in northern Costa Rica, the Meseta Central Occidental and the Atlantic versant foothills of the Cordillera Central (10–1,500 m).

Genus *Lampropeltis* Fitzingerm 1843

A genus of moderate-sized terrestrial constrictors (eight species) characterized by the basic colubrid complement of head shields; head not or only slightly distinct from neck; body cylindrical; nasal divided; zero to two loreals, usually one; zero to two preoculars, usually one; eye moderate, pupil round; two pairs of elongate chin shields; dorsal scales smooth, in 17 to 27 rows at midbody; two apical scale pits; anal entire; subcaudals mostly divided; 12 to 20 maxillary teeth, subequal or last two slightly enlarged; no Duvernoy's gland; mandibular teeth increasing in size posteriorly; no hypapophyses on posterior dorsal vertebrae. The only Costa Rican member of the genus may be separated from other snakes in the republic by having smooth scales in 19 to 23 rows at midbody and a single anal plate and in being uniform black dorsally and ventrally or ringed with black, red, and white.

The hemipenes are clavate, asymmetric with a simple sulcus spermaticus and a semicentrifugal sulcus or bilobed with a semicentrolineal sulcus; pedicel naked or with spinules; truncus spinous; apical area calyculate (Blanchard 1921; Dowling and Savage 1960; Dowling and Duellman 1974–78; Williams 1988).

Blaney (1973) provided a review of generic features and pertinent literature. He recognized two poorly defined species groups in this genus: the *getulus* group (two species) and the *triangulum* group (six species). In addition to *Lampropeltis triangulum*, which occurs in Costa Rica (but see remarks under that species), other species included in the latter division are *Lampropeltis alterna* (Mexico and Texas), *L. mexicana* and *L. ruthveni* (Mexico), *L. pyromelana* (Mexico and southwestern United States), and *L. zonata* (Pacific slope United States and Baja California, Mexico). *Lampropeltis* ranges from southeastern Canada and the

northern United States south throughout Mexico and Central America to Ecuador and northern Venezuela.

Lampropeltis triangulum (Lacépède, 1788)

(plates 427–28; map 11.87)

DIAGNOSTICS: Individuals of this species have the body encircled by red, black, and white rings forming dyads (two black rings, separated by a white ring, alternating with each red ring) or this pattern changing ontogenetically to a uniform black dorsal and ventral coloration. A dyad pattern is found in no other Costa Rican snake except *Erythrolamprus bizona* (fig. 11.47; table 11.2). Tricolor *Lampropeltis triangulum* have the white nuchal collar bordered posteriorly by a black ring, but in *E. bizona* the light (white or yellow) nuchal collar lies between an anterior black nuchal band and the first black ring. Uniform black *L. triangulum* might be confused with adults of *Chironius grandisquamis*, which have the dorsum uniform black and only the posterior ventral and subcaudal surfaces black. *L. triangulum* has 19 to 23 scale rows at midbody; *C. grandisquamis* has 10.

DESCRIPTION: As understood at present (Williams 1988), an extremely variable species with adults reaching a maximum total length of 500 mm in some populations and over 1,900 mm in others; maximum length for Costa Rican examples 1,600 mm; males longer than females; tail short (12 to 16% of total length [11 to 17%]); head slightly distinct from neck; 1 loreal or rarely 0 (0 to 2); 1 preocular or rarely 2 (0 to 2); 2 postoculars, rarely 1 (1 to 4); supralabials usually 7, rarely 8 (6 to 9), usually with 2 bordering orbit; infralabials 8 to 11, usually 9 (7 to 11); temporals variable, usually 1 + 2 or 2 + 3; usually 21-21-19, 21-23-19 or 23-23-19 dorsal scale rows, 19 to 23 (17 to 23) at midbody; ventrals 216 to 242 (152 to 244); subcaudals 42 to 63 (31

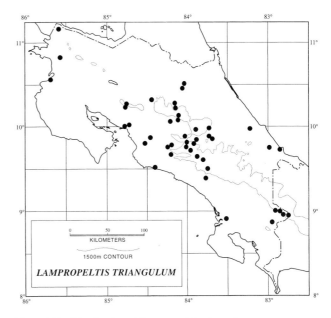

Map 11.87. Distribution of *Lampropeltis triangulum*.

to 63). Maxillary teeth 11 to 15, last two largest. Basic coloration of juvenile Costa Rican examples tricolor, with more or less complete rings of red, black, and white or yellow; each red ring separated from next by two black rings that are separated from one another by a white or yellow ring, forming a dyad (black-light-black); this pattern retained in some adults, but most examples with red and white or yellow areas invaded by black ontogenetically so that tricolor pattern variously obscured or so that adults bicolor (red and black) or uniform black above and below; red scales tipped with black in most examples, some black spotting on other light scales; black markings on venter of tricolor individuals highly irregular, often only a few complete black rings; elsewhere in range dorsal patterns may be of gray or brown blotches, red blotches, black-bordered red bands or tricolor, with red rings and black bands and/or rings; ventral patterns may be immaculate white, checkered black and white, or tricolor ringed; iris reddish; tongue black. The everted hemipenis is short to moderate (5 to 11 subcaudals in length), bilobed with a semicentrolineal sulcus spermaticus, pedicel naked, truncus with spines, apical region calyculate (Williams 1988).

SIMILAR SPECIES: (1) *Erythrolamprus bizona* has a pair of black nuchal collars (fig. 11.23a) and 15 scale rows at midbody. (2) *Chironius grandisquamis* has the throat and anterior part of the body white in adults that are uniform black above. (3) Species of *Clelia* have a cream venter. Also see key to coral snakes (p. 706), figure 11.47, and table 11.2.

HABITAT: Known from Lowland Dry, Moist, and Wet Forests, Premontane Moist and Wet Forests and Rainforest, and Lower Montane Wet Forest and Rainforest.

BIOLOGY: A usually nocturnal or crepuscular terrestrial constrictor that in Costa Rica is often active during the day within rainforests. It feeds mainly on small mammals (plate 428), birds and their eggs, other snakes, and lizards, based on data from tropical specimens of the species (Williams 1988). Male combat with biting has been recorded for temperate populations. Clutches consist of five to sixteen eggs that hatch in thirty-five to fifty days. Presumably Costa Rican examples have similar reproductive habits. Gerhardt, Harris, and Vásquez (1993) reported predation on this snake by the great black hawk *(Buteogallus urubitinga)* in Guatemala. A relatively rare snake in the republic.

REMARKS: Williams (1988) has reviewed the extensive variation in the snakes currently placed in *Lampropeltis triangulum* and essentially repeated and summarized it again (Williams 1994). As a consequence he recognized twenty-five geographic races based primarily on head and body pattern differences. Wilson and Meyer (1985) discussed these aspects of coloration and speculated on possible evolutionary trends. Three general patterns may be recognized within this complex of populations:

1. Blotched pattern: dorsum with brown or reddish brown
   (bright chestnut red to red in juveniles), dark-outlined
   blotches on a lighter ground color (gray, tan, yellowish or white); venter checkered: northeastern United States and south-central Canada
2. Simple banded pattern: dorsum with black-bordered red to orange bands extending onto ventral tips on a yellow to white ground color; midventral surface light or variously encroached on by the dorsal bands: Louisiana and the Great Plains west to the fringes of the Great Basin
3. Dyad pattern: dorsum with red bands and/or rings separated from one another by a black-white (or red or yellow)-black dyad; venter variously marked but usually with some complete rings of red, white or yellow, and black, although black often encroaching onto other colors to produce an irregular ventral pattern: southeastern United States and Mexico to northern South America

Most of the fourteen head patterns Williams (1988) delimited appear to be parts of a continuum in variation ranging from uniform black through a number of intermediate conditions to having most of the prefrontals and the rest of the head anterior to them white. In populations from several areas of the United States the head is mostly red or orange with black variously covering most of the parietal shields or restricted to a band crossing the parietal tips. In Costa Rica most examples have the head uniformly black. Juveniles usually have the snout black followed by a white V-shaped figure and the rest of the upper head surface black. Some adults have a broader light area between the black snout and black head cap. There seems to be little geographic constancy in the elements of these patterns in the republic, with snakes from all areas tending toward a uniformly black head and snout as adults. Williams (1988) indicated that the Atlantic and Pacific lowland populations in Costa Rica belonged to two geographic races based on head pattern and details of body pattern. These forms were described as ranging from Costa Rica north to northwestern Honduras on the Caribbean slope and to El Salvador on the Pacific versant. Atlantic examples were supposed to have a white V-shaped figure on an otherwise black head and a higher number of red rings (19–22.8–28). The Pacific lowland race was defined as having the head mostly black with a broad white band crossing the prefrontals and fewer red body rings (13–17.1–26). With the acquisition of additional material these features now appear to show substantial individual variation. Snakes with both kinds of head patterns are now known from both Atlantic and Pacific slopes and on individuals having both high and low numbers of red rings. In addition, several examples show a remnant of the light V-shaped marking, while others exhibit a head pattern not described by Williams (1988) in which the head is mostly black except for a narrow transverse white mark across each internasal. Williams also regarded the upland Costa Rican population in which adults become nearly uniform black as a third geographic race. It is now clear that the tendency toward obfuscation of the tricolor pattern with age by expansion of black pigment occurs at lowland sites as well, although it reaches its fullest expression at higher elevations. These data suggest that there are more appropriate ways to describe variation in this species than by recognizing rather arbitrarily defined geographic races. On the other hand, it appears likely that one or more of the very distinctive forms now referred to *Lampropeltis triangulum* may well represent distinct species. Until these extralimital systematic problems are resolved, that name must be applied to Costa Rican examples of the complex. It is interesting that the highly venomous coral snake *Micrurus alleni* exhibits color pattern variation that parallels the situation in *L. triangulum*. *M. alleni* from the Atlantic lowlands of Nicaragua and Costa Rica is a tricolor snake with a pattern of monads (one black ring, bordered on each side by a white or yellow ring, separating each red ring from the next one). However, the southwestern Pacific area of Costa Rica supports a population of *M. alleni* in which the juveniles have a pattern of tricolor monads that changes ontogenetically to bicolor black and yellow in adults (Savage and Vial 1973). Small adults of *L. triangulum* with a bicolor pattern (where the white rings have been invaded by black) and large uniform black adults occur sympatrically with examples of these bicolor *M. alleni* in southwestern Costa Rica and adjacent western Panama.

DISTRIBUTION: Southeastern Canada and the United States east of the Great Basin south through Mexico and Central America to Ecuador and northern Venezuela. Atlantic and Pacific lowland, premontane and lower montane zones of Costa Rica (4–2,460 m).

Genus *Leptodrymus* Amaral, 1927

A monotypic genus of elongate, medium-sized, diurnal racers having the basic colubrid complement of enlarged head shields and the following additional features: head distinct from neck; body cylindrical; nasal divided; a loreal; one preocular; eye large, pupil round; two pairs of elongate chin shields; dorsal scales smooth in 17-17-15 rows, with two apical pits; anal plate divided; subcaudals divided; 17 to 19 subequal maxillary teeth followed after a diastema by three slightly enlarged ones; condition of Duvernoy's gland not known, presumably present; mandibular teeth subequal; no hypapophyses on posterior dorsal vertebrae. The genus is most easily recognizable among Costa Rican snakes by the unique color pattern of broad dorsal black stripes on a beige to cream ground color.

The retracted hemipenes are asymmetric, single, with a simple semicentrolineal sulcus spermaticus; the pedicel and truncus are covered by spines; there is a single large spine on each side of the sulcus distally, and the apex is covered by calyces (Bogert 1947).

Leptodrymus ranges through the Pacific lowlands and premontane slopes from Guatemala to central Costa Rica and onto the Atlantic versant in Guatemala and northwestern Honduras.

Leptodrymus pulcherrimus (Cope, 1874)

(plate 429; map 11.88)

DIAGNOSTICS: No other Costa Rican snake has a green head and a broad black dorsal stripe on each side running from eye to tail tip.

DESCRIPTION: A moderate-sized to large species (maximum total length 1,600 m); tail very long (33 to 36% of total length); 3 postoculars; supralabials 9 to 10, usually 9, with 3 bordering orbit; infralabials 10 to 12, usually 10; temporals usually 2 + 2 + 2, but often one temporal in any position or rarely some fusions; 17 scale rows at midbody; ventrals 195 to 210; subcaudals 145 to 152. Middorsal field reddish beige to cream with a pair of black stripes (two plus two half-scale rows wide) running from eyes to tip of tail; lateral areas grayish; top of head and neck and tail green; underside immaculate white to cream with a faint pink suffusion on subcaudal area; iris brown, darkest across equator; tongue red, tips of fork black.

SIMILAR SPECIES: (1) *Coniophanes piceivittis* has the middorsal area brownish black to black and a pair of dorsolateral light stripes.

HABITAT: Pacific Lowland Dry and Moist Forests and Premontane Moist Forest, usually in open situations.

BIOLOGY: A common terrestrial snake, but often seen foraging in bushes and small trees.

REMARKS: All examples except one (USNM 38150) from the Río Pacacua (Pacaca) near Colón, San José Province (about 800 m) are from northwestern Costa Rica.

DISTRIBUTION: Pacific lowlands onto premontane slopes from Guatemala to the Meseta Central Occidental of Costa Rica and on the Atlantic slope in the Motagua valley, Guatemala, and northwestern Honduras (2–850 m).

Map 11.88. Distribution of *Leptodrymus pulcherrimus*.

Genus *Leptophis* Bell, 1825

A group (nine species) of semiarboreal elongate snakes having the dorsal color predominantly bright green and/or bronze. Members of the genus have the basic series of enlarged head shields found in most colubrids and the following combination of definitive features: relatively narrow and elongate head distinct from neck; body slender, cylindrical; anal divided; one loreal or none; usually one preocular, sometimes two; eye large, pupil round; two pairs of elongate chin shields; keeling of dorsal scales variable from all rows keeled to only paravertebral rows on posterior one-fourth of body keeled (keels rarely absent in juvenile females); dorsal scales in 15-15-11 obliquely arranged rows, with one apical pit on most scales, some with two pits; anal plate usually divided; subcaudals divided; 18 to 36 maxillary teeth increasing in size posteriorly and last two to four somewhat enlarged; Duvernoy's gland present; mandibular teeth longest anteriorly, decreasing in size posteriorly; no hypapophyses on posterior dorsal vertebrae. The green and/ or bronze coloration in combination with 15 scale rows at midbody, divided anal, and immaculate venter will separate this genus from all other Costa Rican colubrids.

The hemipenis is very short to moderate in length (5 to 9 subcaudals long), asymmetric and bilobed, with a simple semicentrolineal sulcus spermaticus. The proximal portion is covered by spines and the distal area by papillate calyces (Vellard 1946; Oliver 1948; Dowling and Savage 1960).

All *Leptophis* are oviparous and lay three to ten eggs (Oliver 1948). Taylor (1951b, 1954b) used the name *Thalerophis* for this genus following Oliver (1948). Savage (1952) established that *Leptophis* is the correct name for these snakes.

The genus ranges over most of the lowland and premontane areas of the Neotropics from Sonora and Tamaulipas, Mexico, south to Ecuador on the Pacific slope and Bolivia and northern Argentina on the Atlantic versant; also on the islands of Trinidad and Tobago.

KEY TO THE COSTA RICAN SPECIES OF THE GENUS *LEPTOPHIS*

1a. Loreal plate separating prefrontal from supralabials (fig. 11.41b); usually a broad dark stripe that extends from tip of snout through the eye to become continuous with lateral body stripe or with only a discrete broad postocular dark stripe 2
1b. No loreal (fig. 11.41c–e); narrow preocular dark stripe, if present, not reaching tip of snout; a narrow postocular dark stripe extending onto body a distance not greater than length of head 3
2a. Keels present on dorsal scales except outer (lower) row; keels present on most of scales on upper surface on tail; adults usually with a pair of broad greenish blue to dark blue (green in life) longitudinal lateral stripes separated by a light (bronze in life) middorsal area; usually a dis-

tinct broad dark stripe from eye through nostril to tip of snout; usually a broad postocular stripe continuous with dark lateral body stripe when present (fig. 11.41a) . . . . *Leptophis mexicanus* (p. 671)
2b. Keels present only on paravertebral dorsal scale rows and occasionally on an adjacent row of dorsal scales; no keels on scales on upper surface of tail; adults with an essentially uniform green dorsum in life (bluish in preservative), without definite dark lateral stripes, although very narrow dark paravertebral stripes are often present; no dark stripe running from eye to tip of snout; a short, broad postocular stripe (fig. 11.41b) . . . .*Leptophis depressirostris* (p. 670)
3a. No keels on first row of dorsal scales; adults uniform or with dark stripes dorsally; juveniles striped or banded dorsally, predominately green 4
3b. Keels present on all dorsal scales; adults and juveniles with a series of narrow transverse dark dorsal bands; all ages predominately bronze. . . *Leptophis riveti* (p. 672)
4a. Dorsum in adults and juveniles with a pair of dark green or bluish lateral stripes, about two scale rows wide anteriorly, separated by a light (bronze in life) middorsal area. . . . . *Leptophis nebulosus* (p. 672)
4b. Dorsal color in adults essentially uniform green in life, bluish in preservative, green with dark bands in juveniles . . . . *Leptophis ahaetulla* (p. 668)

Leptophis ahaetulla (Linné, 1758)

Lora Falsa

(plate 430; map 11.89)

DIAGNOSTICS: A snake characterized by having a uniformly green dorsum, pale green venter, and a narrow postocular black stripe that extends backward a short distance

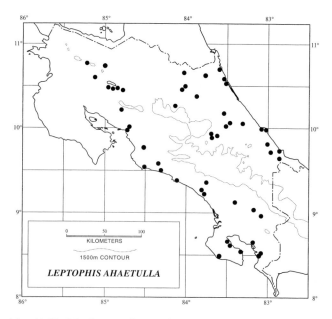

Map 11.89. Distribution of *Leptophis ahaetulla*.

onto the neck and lacking a loreal. It usually has a narrow preocular black stripe as well (fig. 11.41e). Although a slender species, it differs from *Oxybelis brevirostris* in not having an extremely narrow and elongate head, body, and tail and in having a divided anal plate (fig. 11.5).

DESCRIPTION: A large snake attaining a maximum total length over 2,250 mm (most examples less than 1.5 m); males longer than females; tail very long (35 to 41% of total length) but incomplete in 60 to 65% of adult examples; no loreal; postoculars 1 to 4, usually 2; supralabials 7 to 10, usually 8, with 2 bordering orbit; infralabials 7 to 12, usually 10 or 11; temporals variable but usually 1 + 2; keels usually present on all but first scale row on each side (Pacific slope) or usually only on upper two rows and/or vertebral row with prominent keels (Atlantic slope) (keels sometimes absent on anterior three-quarters of body in non–Costa Rican samples); weak keels or no keeling on posterior caudal scales; anal divided (often entire in some South American populations); ventrals 152 to 182 (147 to 183); subcaudals 153 to 189 (137 to 189); maxillary teeth 18 to 25 (18 to 29). Dorsum uniform green or marked with thin dark stripes along dorsal scale keels in adults, marked with narrow dark green to bluish crossbands in young; narrow postocular black stripe extending no farther than one head length onto nuchal area; narrow preocular black stripe, if present, not reaching level of nostril (may reach nostril in other populations); venter and subcaudal area pale green; iris yellow; tongue black. The everted hemipenes are very short (5 to 6 subcaudals long) with 5 to 6 large basal spines (0, 2, 4 to 6 extralimitally) (Vellard 1946).

SIMILAR SPECIES: (1) *Leptophis depressirostris* has a short broad postocular black stripe and a loreal plate and lacks a preocular black stripe (fig. 11.41b). (2) *Leptophis riveti* has narrow green bands on the dorsum throughout life and a bronze dorsal and ventral ground color. (3) *Oxybelis brevirostris* has a very thin black stripe that runs from the nostril to the neck and has an extremely elongate head. (4) Species of *Chironius* lack the postocular black stripe. (5) Species of *Dendrophidion* lack the postocular black stripe.

HABITAT: Most abundant in relatively undisturbed Lowland Moist and Wet Forests but also found marginally in Premontane Moist and Wet Forest and Rainforest and marginally in gallery forests of the Lowland Dry Forest region.

BIOLOGY: These common diurnal snakes usually are seen foraging in trees and shrubs. Sleeping or resting frogs appear to be their favored food items, but arboreal lizards, grasshoppers, nestling birds, and birds' eggs have also been found in their stomachs (Oliver 1948). Clutches consist of one to five eggs. Rand (1969) reported finding, in central Panama, three groups of eggs in a mass of bromeliads that had fallen from a tree 10 to 20 m above the ground. One set of two eggs hatched snakes of this species. The eggs were 29 × 12 mm, and the hatchlings were 240 and 242 mm in total length. The other groups of four and three empty egg

shells were presumed to be of this species. Sexton and Heatwole (1965) previously reported a female containing three large eggs. Hatchlings appear throughout the wet season (May to November). When frightened these snakes raise the anterior part of the body off the substrate, hold the mouth open, exposing the flesh-colored lining and blue tongue sheath, and hiss. If a threat persists they inflate the neck region, exposing the underlying iridescent greenish yellow skin. If handled they bite without hesitation.

REMARKS: This wide-ranging species exhibits extensive geographic and individual variation in keeling of the dorsal scales and rather consistent geographic stability in dorsal pattern (Oliver 1948). In most populations keels are usually present (at least on the posterior one-third to one-quarter of the body) on all but the lowest scale row on each side. In most examples from Atlantic slope Central America and many individuals from Colombia, northern South America, and the Amazon basin, prominent keels are found only on the uppermost two scale rows on each side. In addition, there tends to be an ontogenetic increase in degree of keeling and sexual dimorphism in keeling as well (stronger and more extensive keeling in adult males). The following major geographic components may be recognized based on color pattern, with intermediate conditions found in contact areas between the various subdivisions:

1. Populations with a uniform green dorsum, with or without black keels on scales: Mexico, south to northwestern Venezuela and western Ecuador
2. Populations with the dorsum green and the head shields and each dorsal scale heavily edged with black: disjunct in northwestern Ecuador and Gorgona Island, Colombia; upper Amazon basin of Ecuador, Peru, Brazil, and Bolivia
3. Populations with the middorsal area contrasting in color with the flanks
   a. Middorsal area bronze, flanks green: the Guianas to eastern Brazil
   b. Middorsal area green, flanks light: Venezuela and eastern Colombia south through the central Amazon basin
4. Populations with the anterior dorsum green, changing to bronze posteriorly: disjunct in headwaters between Río Negro and Río Orinoco drainages; southeastern and south central Brazil to northern Argentina

Oliver (1948) recognized a number of geographic races within these major subdivisions, primarily based on the development of the postocular black stripe (whether it was broad, narrow, or absent). In addition, the preocular dark stripe shows some geographic fidelity in how far it extends along the upper edge of the supralabial sutures (whether it terminates posterior to the level of the nostril, under the nostril, or anterior to it). There is no consistent correlation between scale keeling, dorsal pattern, and pre- and postocular stripe development (e.g., uniform green populations may have much or little keeling and preocular and/or postocular dark stripes). According to Oliver (1948) there is some geographic consistency in most populations in the number of enlarged basal spines present on the hemipenes. Those from north-central and eastern Venezuela, the main parts of the Amazon basin, and its Colombian and Bolivian drainage areas lack enlarged basal spines. Those from the upper Amazon region of Peru have two somewhat enlarged spines, and other populations have four to six somewhat to definitely enlarged basal spines. There are five or six definitely enlarged basal spines on the hemipenes of males from Mexico, Central America, and northwestern South America. In all the indicated features Costa Rican examples fall within the variation exhibited in populations included in group 1 above.

DISTRIBUTION: Lowlands and premontane areas from central Veracruz and Oaxaca, Mexico, on the Atlantic slope through Central and South America to northern Argentina; on the Pacific versant from Costa Rica to southwestern Ecuador. Throughout the lowlands and adjacent premontane slopes in Costa Rica, but restricted to gallery forests east of the Río Tempisque in the lowlands of northwestern Costa Rica (1–1,400 m).

Leptophis depressirostris (Cope, "1860d," 1861)

(plate 431; map 11.90)

DIAGNOSTICS: A species with a uniform green dorsum or often green with a pair of paravertebral black stripes and a pale green venter, but differing from similar forms in having a broad black postocular stripe and a loreal and in lacking a preocular black stripe (fig. 11.41b).

DESCRIPTION: A moderate-sized species reaching a total length of about 1,500 mm; tail very long (36 to 40% of total length), but about 80% of specimens with incomplete

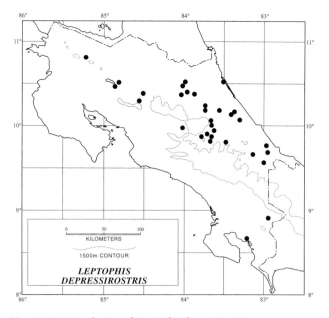

Map 11.90. Distribution of *Leptophis depressirostris*.

tails; a loreal; usually 2 postoculars, occasionally 1; supralabials 7 to 10, usually 9, with 2 bordering orbit; infralabials 9 to 11, usually 10; temporals usually 1 + 2, rarely 1 + 1 or 1 + 3; keels restricted to paravertebral rows on body, absent on posterior caudal scales; anal usually divided, very rarely entire; ventrals 144 to 158; subcaudals 158 to 170; maxillary teeth 33 to 36. Dorsum uniform green or with pair of paravertebral black stripes; a broad postocular dark stripe occupying most of area on one or two temporal scales (may be reduced to spots in extralimital examples), barely extending onto neck; no preocular dark stripe (present in some non–Costa Rican specimens); underside of body and tail pale green; tongue blue; iris yellow. The retracted hemipenis is very short (6 subcaudals long), without enlarged basal spines (Oliver 1948).

SIMILAR SPECIES: (1) *Leptophis ahaetulla* has a narrow postocular dark stripe and sometimes a narrow preocular stripe and lacks a loreal plate (fig. 11.41e). (2) *Oxybelis brevirostris* has a very thin black stripe that runs from the nostril to the neck and an extremely elongate head. (3) *Chironius exoletus* lacks a postocular dark stripe. (4) Species of *Dendrophidion* lack a postocular dark stripe.

HABITAT: Lowland rainforests of both slopes in relatively undisturbed sites and adjacent Premontane Wet Forest and Rainforest. One record for Premontane Rainforest in the extreme southwestern Pacific region.

BIOLOGY: This uncommon diurnal species preys primarily on sleeping frogs and anole lizards. When disturbed it attempts to bluff a potential predator by raising the anterior part of the body off the substrate and holding the mouth open as wide as possible (plate 432). If this fails to impress the intruder, the snake inflates the neck region and exposes the interscale skin, which is dark green. The overall impression created is of a larger-than-life, vicious snake, and the behavior probably discourages some of the snake's enemies from attacking. If handled this snake delivers a slashing bite by rotating the upper jaw so that the rear fangs project. There is apparently an anticoagulant in the buccal gland secretions that causes profuse bleeding in humans, but there are no other adverse effects. Dundee and Liner (1974) collected three eggs of this species together with four old egg shells from a bromeliad approximately 3.5 m high on the Atlantic slope. The eggs were 35 to 48 mm in length, and the hatchlings were 186 to 198 mm in total length.

REMARKS: A single record (USNM 11266) for the city of San José is questionable, since no other examples have been taken on the Meseta Central Occidental and the species does not occur elsewhere in similar habitats on the Pacific slope of Costa Rica. The example (KU 31962) Taylor (1954b) referred to *Leptophis aeruginosus* (a name placed in the synonymy of *L. depressirostris* by Oliver 1948) is unquestionably *L. depressirostris*.

DISTRIBUTION: Lowlands and adjacent premontane slopes from northeastern Nicaragua on the Atlantic versant to central Panama and from southwestern Costa Rica to western Ecuador on the Pacific slope (4–1,120 m).

Leptophis mexicanus C. Duméril, Bibron, and A. Duméril, 1854

(plates 433–34; map 11.91)

DIAGNOSTICS: The midbody pattern of alternating longitudinal bands of bronze or gold and bright green is found only in this species and its ally, *Leptophis nebulosus*, among Costa Rican forms. The middorsal area is usually bronze, bordered on each side by broad green stripes, which in turn are bordered by a narrow yellow stripe below. In addition this species usually has a broad preocular black stripe that runs posteriorly from the snout tip and continues as a broad postocular stripe from the eye along the anterior one-half to two-thirds of the body (fig. 11.41a). In *Leptophis nebulosus* the narrow preocular black stripe begins behind the nostril, and the postocular stripe, also narrow, continues posteriorly only onto the neck (fig. 11.41d).

DESCRIPTION: A moderate-sized snake (to 1,269 mm in total length); tail very long (36 to 40% of total length), about 40% of tails incomplete; a loreal; usually 2 postoculars, rarely 1 or 3; 8 or 9 supralabials, usually 8, with 2 bordering orbit; 9 to 11 infralabials, usually 10; temporals variable, usually 1 + 2 but rarely 1 + 1, 1 + 1 + 2, or other combinations; keels on all but outermost dorsal scale row on each side; keels on median two rows of posterior caudal scales; anal divided; ventrals 145 to 161 (145 to 174); subcaudals 141 to 166 (140 to 181); maxillary teeth 20 to 25. Middorsal area bronze to gold, bordered on each side by a broad green longitudinal band, bordered in turn below by a narrow yellow stripe; head green above, usually with a distinct broad lateral black stripe running from tip of snout

posteriorly through eye and continuous with dark stripe on body; in some examples preocular black stripe narrow, no lateral dark stripes present, and most of dorsum and lateral areas green (plate 433); undersides gray or beige, often heavily suffused with pink to salmon; iris yellow above, black below; tongue black, with tips of fork green to blue. The retracted hemipenis is short (8 to 9 subcaudals in length), with four enlarged basal spines.

SIMILAR SPECIES: (1) *Leptophis nebulosus* has the postocular black stripe narrow and terminating on the neck and lacks a loreal plate (fig. 11.41d).

HABITAT: Lowland Dry Forest habitats, especially in gallery forests and adjacent Premontane Moist and Wet Forests, and marginally in Lowland Wet Forest of the northeastern Atlantic slope, and the Meseta Central Occidental.

BIOLOGY: This common form is similar to other members of the genus in being a diurnal forager that feeds primarily on sleeping treefrogs. Individuals may also eat the arboreal nocturnal gecko *Thecadactylus rapicauda* (which, like the treefrogs, rests or sleeps during the day), anoles, small snakes, birds' eggs, and tadpoles (Henderson 1982). Censky and McCoy (1988) reported on reproduction for the northern Yucatán Peninsula population of this species. The ovaries were quiescent during much of the dry season (November to March). The reproductive season is March or April to late October or early November. The region has a strongly seasonal climate, and eggs are laid during the height of the wet season (July to August or September). Females with body lengths over 541 mm were sexually mature and had clutches of two to six eggs 37 to 43 mm in length. Clutch size did not increase with body size. The smallest specimen (262 mm in body length) was collected in October, and females attained sexual maturity in about eighteen months. When threatened this species raises the anterior part of the body off the substrate, hisses, and opens the mouth wide to expose the buccal lining (plate 434). The mouth is usually kept wide open until the potential predator retreats or the snake is attacked.

REMARKS: Several geographic variants have been recognized within this species (Henderson 1976). Examples from the extreme northern part of its range in Tamaulipas, Mexico, have the postocular stripe terminating on the neck and black spotting on the lateral scales on the anterior part of the body (Mertens 1942, 1943). Those from Big Cay Bokel, Belize (Henderson 1976), are uniform green, with the postocular stripe weakly developed and terminating on the head. More widely distributed are the following two dorsal patterns, which both have the middorsal area grayish gold to bronze and the area below the lateral black stripe green, bordered by white, gray, or yellow below:

A. Lateral dark stripe when present usually restricted to upper margin of scale row 2 and lower margin of scale row 3 on posterior third of body: central Veracruz and

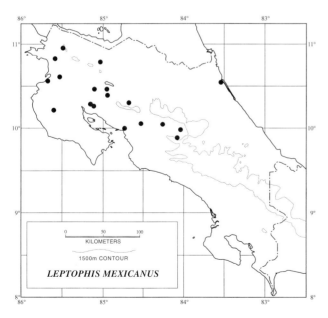

Map 11.91. Distribution of *Leptophis mexicanus*.

Oaxaca, Mexico, to Costa Rica, exclusive of the outer part of the Yucatán Peninsula

B. Lateral dark stripe on upper margin of scale row 2, all of scale row 3, and lower margin of row 4 on posterior third of body: outer part of Yucatán Peninsula

There are also slight average differences in ventral (V) and subcaudal (SC) counts among these populations as follows: Tamaulipas, Mexico (V171/SC158); Belize (V182/SC177); population A (V159/SC157); population B (V168/SC169). Costa Rican examples agree in all respects with others belonging to pattern A above.

DISTRIBUTION: Atlantic lowlands and premontane slopes from Tamaulipas, Mexico, to extreme northeastern Costa Rica, with disjunct populations on the Pacific versant from Oaxaca and Chiapas, Mexico, to El Salvador; also northwestern Costa Rica including the Meseta Central Occidental (4–1,160 m).

Leptophis nebulosus Oliver, 1942

(plate 435; map 11.92)

DIAGNOSTICS: The pattern of alternating longitudinal bands of bronze or gold and bright green is found only in this species and *Leptophis mexicanus*. *Leptophis nebulosus* has a narrow preocular black stripe that begins near the nostril and a narrow postocular black stripe that terminates on the neck (fig. 11.41d), whereas *L. mexicanus* usually has a broad preocular stripe that begins on the tip of the snout and a broad postocular stripe that is continuous with a lateral stripe coursing down the length of the body (fig. 11.41a).

DESCRIPTION: A moderate-sized species (maximum total length 600 mm); tail very long (37 to 41% of total length),

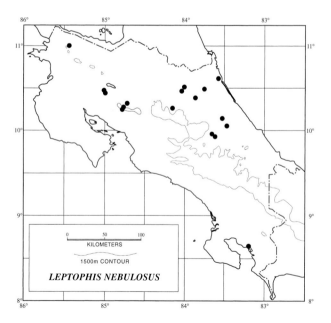

Map 11.92. Distribution of *Leptophis nebulosus*.

about 40% of tails incomplete; no loreal; usually 2 postoculars, rarely 1; 8 or 9 supralabials, with 2 bordering orbit; usually 10 infralabials, occasionally 11; temporals 1 + 2; keels on all but outermost scale rows on body and on posterior caudal scales; anal divided; ventrals 150 to 160; subcaudals 146 to 151; maxillary teeth 27 to 28. Middorsal area bronze to gold, bordered on each side by a broad green longitudinal band, bordered in turn below by a narrow yellow stripe; head green above; narrow black stripe running from near level of nostril through eye onto neck; venter white to beige, often heavily suffused with pink to salmon, iris yellow; tongue black with extreme tips green. Hemipenes short (7 caudals long), without enlarged basal spines.

SIMILAR SPECIES: (1) *Leptophis mexicanus* usually has a black stripe running from the eye down half or more of the body and a loreal plate (fig. 11.41a).

HABITAT: Lowland Moist and Wet Forests, Premontane Moist and Wet Forest and Rainforest, and marginally in Lower Montane Wet Forest and Rainforest.

BIOLOGY: A common diurnal species that specializes in finding and eating sleeping frogs. When approached this snake holds the mouth wide open to expose the lavender buccal lining and blue tongue sheath and intimidate potential predators.

DISTRIBUTION: Atlantic lowlands from extreme northeastern Honduras through Costa Rica, where it ranges onto the premontane slopes; also on the Pacific versant in the lowlands of southwestern Costa Rica and on the slopes of the Cordillera de Tilarán (13–1,600 m).

Leptophis riveti Despax, 1910

(map 11.93)

DIAGNOSTICS: The distinctive pattern of dark green bands on a coppery ground color and uniform bronze venter are unique among Costa Rican snakes.

DESCRIPTION: A moderate-sized snake reaching a maximum length about 1 m; tail very long (30 to 41% of total length), about 50% of tails incomplete; no loreal; 1 or 2 postoculars; 8 supralabials, with 2 bordering orbit; infralabials 9 to 11; usually 10; temporals variable, usually 1 + 2; keels usually present on all dorsal scale rows; prominent keels on two median caudal rows for most of tail length; anal divided; ventrals 133 to 155, angulate with a moderate keel; subcaudals 140 to 145; maxillary teeth 26 to 29. Dorsum bronze with narrow bright green bands; head green above; narrow preocular black stripe not reaching level of nostril; narrow postocular black stripe extending onto neck (fig. 11.41c); undersurfaces of body and tail bronze; iris yellow interrupted by median black bar. The retracted hemipenes are short (6 subcaudals long), without enlarged basal spines.

HABITAT: Undisturbed Pacific Lowland Moist Forest and Premontane Rainforest.

SIMILAR SPECIES: (1) Juvenile *Leptophis ahaetulla* are

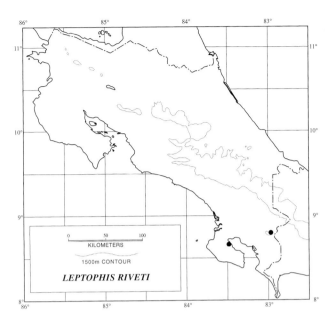

Map 11.93. Distribution of *Leptophis riveti.*

predominately green, not bronze, and the outermost scale row on each side is not keeled. Also see table 11.1.

BIOLOGY: This rare snake apparently feeds primarily on sleeping treefrogs during the day.

REMARKS: Scott's (1969) record (CRE 851) of this form from the Cordillera de Tilarán (Monteverde, Puntarenas Province) is based on an example of *Leptophis ahaetulla* having 180 ventrals and 165 subcaudals (tail incomplete).

DISTRIBUTION: The lowland and premontane zones of southwestern Costa Rica and central Panama, south to western and upland Ecuador and Amazonian Peru; also Trinidad. Known from two Pacific versant localities in Costa Rica (4–1,200 m).

Genus *Mastigodryas* Amaral, "1934," 1935

A genus (eleven species) of terrestrial diurnal racers that share the following combination of features: the basic colubrid complement of enlarged head shields; head distinct from neck; body cylindrical or slightly compressed; nasal divided; a loreal; usually one preocular, rarely two; eye large, pupil round; two pairs of elongate chin shields; dorsal scales smooth in 15–15 or 17–15 rows; two apical pits on each scale; anal divided; subcaudals divided; 10 to 26 subequal maxillary teeth or last two or three larger and preceded by a short diastema; Duvernoy's gland present; mandibular teeth subequal; no hypapophyses on posterior dorsal vertebrae. *Mastigodryas* is distinguished from other genera of smooth-scaled racers in Costa Rica *(Coluber, Drymarchon,* and *Leptodrymus)* by having a divided anal (entire in *Drymarchon),* 15 scale rows anterior to the anus (13 in *Coluber),* and two postoculars (three in *Leptodrymus)* and in coloration as detailed in the species account below.

The hemipenes are short (10 to 11 subcaudals long), asym-

metric, bilobed, with a simple semicentrolineal sulcus spermaticus. The pedicel is nude, the truncus is spinous, and the apical half is covered with papillate calyces (Stuart 1932; Vellard 1946). Members of this stock were referred to the genus *Dryadophis* Stuart, 1939, in the period from 1939 to 1970, when Peters and Orejas-Miranda (1970) accepted the priority of *Mastigodryas* Amaral, "1934," 1935. Since that time some authors have persisted in using *Dryadophis* for this genus following Smith and Larsen (1974). Their arguments are not persuasive, since Peters examined the type species *(Mastigodryas danieli)* and confirmed the synonymy. Consequently, *Mastigodryas* is used here. Stuart (1941) recognized four species groups within this genus. The *melanolomus* group contains the only Costa Rican representative. The only other species definitely assignable to this group is *Mastigodryas dorsalis* of Guatemala, Honduras, and Nicaragua. However, *Mastigodryas cliftoni* of northwestern Mexico may belong here.

The genus has a wide range over most of the continental Neotropics from Sinaloa, Mexico, on the Pacific versant to central Peru and on the Atlantic slope from Tamaulipas, Mexico, to northern Argentina, including the islands of Trinidad, Tobago, Grenada, St. Vincent, and the Grenadines.

Mastigodryas melanolomus (Cope, 1868a)

(plates 436–37; map 11.94)

DIAGNOSTICS: The pattern of alternating dark dorsal and lateral blotches and the boldly marked labials, chin, and throat immediately identify juveniles. Superficially the juveniles resemble some individuals (juveniles and adults) of several species of *Dendrophidion,* but the latter have a pattern of light and/or dark crossbands (not of alternating blotches) and lack bold labial, chin, and throat markings. Adults have

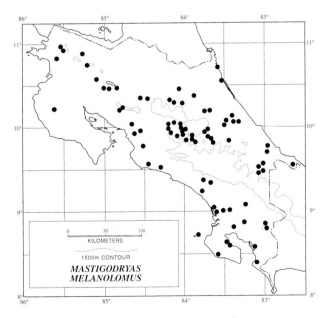

Map 11.94. Distribution of *Mastigodryas melanolomus.*

one or two pairs of dull light stripes on a brown body, lack distinctive light head or nuchal markings, and have an essentially immaculate venter. Other Costa Rican snakes that share these features and have smooth scales in 17 rows at midbody are smaller forms under 700 mm in total length. *Mastigodryas* with a total length of 700 mm or less still exhibit the juvenile color pattern at least on the anterior part of the body.

DESCRIPTION: A moderate-sized snake reaching a maximum total length of about 1,500 mm; males larger than females; tail moderately long in females to very long in males (24 to 35% of total length); postoculars 2; supralabials 8 to 10, usually 9, with 3 bordering orbit; infralabials 9 to 11; temporals usually 2 + 2, variable; dorsal scales in 17-17-15 rows; ventrals 160 to 195; subcaudals 85 to 110 (85 to 136). Adults: dorsum light to medium brown (bluish gray in preservative), with a dull cream to orangish dorsolateral stripe on scale rows 4 and 5 on each side and often with a similarly colored lateral stripe on scale rows 1 and 2; sometimes with suggestion of blotched juvenile pattern, especially on neck; labials without bold dark markings; infralabials, chin, and throat yellowish to red, usually mottled with brownish gray; undersides of body and tail immaculate cream, beige, bright salmon, or red. Juveniles: dorsum with series of 42 to 54 vivid dark brown to black rectangular body blotches, separated by lighter interspaces and alternating with brown lateral blotches (one on each side) that border light interspaces below (plate 437) (dark crossbands extralimitally); labials, chin, and throat heavily marked with dark brown to black to form a bold pattern of white to gray-centered squarish blotches; venter and underside of tail stippled with gray in smallest examples, often suffused with pink in larger juveniles; iris coppery above, brown below; tongue black.

SIMILAR SPECIES: (1) Juveniles of *Coluber mentovarius* lack alternating dorsal and lateral blotches. (2) *Dendrophidion* species that include banded individuals lack alternating dorsal and lateral blotches and bold dark labial, chin, and throat markings. Also see table 11.1.

HABITAT: Ubiquitous in all lowland and premontane forests and marginally into lower montane forests including extensively modified lands (pastures and farmlands); uncommon in the Dry Forest region.

BIOLOGY: The principal prey of this rather common species is lizards, especially anoles, that it finds during diurnal foraging on the ground; it also eats small snakes, reptile eggs, nesting birds, and small mammals (Cadle and Greene 1993; Seib 1984). Juveniles may feed on insects. At night *M. melanolomus* sleeps several meters above the ground on tree limbs Censky and McCoy (1988) reported on the reproductive cycle in the northern Yucatán Peninsula population of this species. In this strongly seasonal area eggs are laid during the wet season (July through October). The vitellogenic cycle was estimated as being twelve months or more, with production in alternate years. Females of a body length of

580 mm or more were considered to be mature and had clutches of two to five eggs that were 42 to 55 mm long. Clutch size increased with body size. The smallest individual (260 mm in body length) was taken in January, and females were estimated to require at least eighteen months to attain sexual maturity.

REMARKS: *Mastigodyras melanolomus* is one of the most commonly encountered Costa Rican snakes. The marked ontogenetic change in coloration involves the gradual expansion of the dorsal blotches to eliminate the light interspaces and the elongation and fusion of the lateral blotches to form dark fields bordering the lateral light stripes. In addition the bold head markings become muted and obscured with age. Another remarkable change occurs in preserved adults. The dorsal ground color turns to bluish gray, and the light stripe frequently becomes the same color so that the snake appears a uniform color above. Elsewhere in the species' range some adults are unicolor. So far as can be determined, based on available field notes, slides, and recollections, all Costa Rican adults of the species are striped in life, but it may be that an occasional uniform brown example will be found. The species exhibits considerable geographic variation in coloration and shows clinal trends in subcaudal counts as follows:

1. Juveniles with narrow light dorsal crossbands; a pair of dorsolateral light stripes in small adults, large adults may be unicolor; subcaudals 98 to 136, increasing from north to south: Tamaulipas to western Honduras
2. Juveniles with narrow light dorsal crossbands; each dorsal scale in adults bordered by black to produce a reticulate pattern; subcaudals 116 to 130: Yucatán Peninsula and northern Guatemala
3. Juveniles with alternating dorsal and lateral blotches; adults with one or two pairs of dorsolateral stripes; subcaudals 85 to 110: Atlantic slopes from Honduras to Panama
4. Juveniles with alternating dorsal and lateral blotches (in known examples); adults unicolor; subcaudals 103 to 118, increasing from south to north: Nayarit, Mexico, to western Guatemala, including the Tres Marías Islands

The differences in coloration intergrade in areas intermediate in location to the principal geographic sectors listed above, and there is extensive overlap in subcaudal count ranges. Stuart (1941) and Smith (1942) proposed recognizing a total of seven geographic races based on these slight differences that are further obscured by changes in coloration on preservation. These proposals appear to have little taxonomic value, since individual and ontogenetic variation within the several putative subdivisions blurs the distinctions among them. Taylor (1954b) described a new species in this genus as *Dryadophis sanguiventris* from the Golfo Dulce area of southwestern Costa Rica and separated it from what is now called *Mastigodryas melanolomus* because of its red belly and labials. Many Costa Rican examples of the genus have red venters (the labials may be red or not) but differ in no other way from other specimens having yellowish or beige undersurfaces. Red-bellied and light-bellied individuals have been taken at the same locality on both Atlantic and Pacific slopes of the republic and are otherwise indistinguishable. Consequently Taylor's name is regarded as a synonym of *M. melanolomus*. Pérez-Santos and Moreno (1988) record this species from two localities on the slopes of the central Andes of Colombia (1,125 to 1,260 m) but state that it occurs more widely in that country. These records need confirmation.

DISTRIBUTION: Lowlands and premontane slopes on the Atlantic versant from Tamaulipas, Mexico, to western Panama; on the Pacific slope from Nayarit, Mexico, to western Guatemala and in Costa Rica and western and central Panama; marginally in the lower montane zone in Costa Rica (4–1,760 m).

Genus *Oxybelis* Wagler, 1830

These extraordinarily thin elongate and narrow-headed arboreal snakes are unlikely to be confused with any other genus (four species) in Mesoamerica. Five species were formerly recognized as forming the genus (Keiser 1974, 1989). Machado (1993) erected a new genus (*Xenoxybelis*) for two Amazonian forms (*X. argenteus* and *X. boulengeri*) because of striking differences in hemipenial morphology that place them with the South American xenodontine assemblage (Zaher 1999). Unlike *Oxybelis* (sensu stricto), these snakes have a bifurcate sulcus spermaticus, distal spinulate calyces, and semicapitate hemipenes. Note that the sulcus is centrolineal, not semicentripetal as stated by Machado (1993; see her fig. 1) for *Xenoxybelis*. Villa and McCranie (1995) recently added a fourth species to *Oxybelis* as recognized by Machado (1993). See the account of *Oxybelis fulgidus* for comments on this new form. Members of the genus have the basic colubrid complement of enlarged head shields, but in correlation with the attenuation of the head, the internasals and the frontal or the supraoculars and parietal are extremely narrow and elongate compared with those of other snakes (the prefrontals also may be narrowed). Additional features that characterize the genus include head distinct from neck; body slender and compressed; nasal entire or partially divided, elongate; usually no loreal (sometimes present); almost always one preocular, rarely two; eye small, pupil round; two pairs of elongate chin shields; dorsal scales keeled or smooth, in 15 to 17 rows at midbody with a reduction anterior to vent; no apical scale pits; anal divided or entire; 14 to 27 maxillary teeth, subequal except for posterior three to five grooved fangs that are somewhat enlarged and sometimes preceded by a short diastema; Duvernoy's gland present; anteriormost mandibular teeth small, followed by enlarged teeth that decrease in size posteriorly; no hypapophyses on posterior dorsal vertebrae.

The everted hemipenes are short (6 to 10 subcaudals in length), asymmetric, bilobed or single, with a simple semicentrolineal sulcus spermaticus, usually with proximal spines and the distal half covered with papillate calyces (Underwood 1967; Keiser 1974).

Vision in this genus is binocular, and the eyes are directed forward rather than laterally as in most snakes. As a correlate to the eye position, a narrow groove runs forward from each orbit to the snout.

Oxybelis occurs over a wide region of subtropical and tropical, lowland and premontane zones in the Americas. It ranges from southern Arizona and Coahuila, Mexico, south to northern Peru on the Pacific versant and to Bolivia, northeastern Argentina, and southern Brazil east of the Andes, including Trinidad and Tobago.

KEY TO THE COSTA RICAN SPECIES OF THE GENUS *OXYBELIS*

1a. 17 scale rows at midbody; anal plate almost always divided; supralabials usually 8–10 2
1b. 15 scale rows at midbody; anal plate single; supralabials usually 6; dorsal ground color green in life, purplish to dark brown in preservative; ventrals fewer than 187 . . . . *Oxybelis brevirostris* (p. 677)
2a. Prominent paired white or yellow stripes along sides of ventrals for full length of body (fig. 11.42a); dorsal ground color green in life, bluish to purplish in preservative; ventrals 198 or more . . . . *Oxybelis fulgidus* (p. 678)

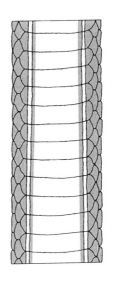 

a   b

Figure 11.42. Ventral color patterns at midbody in two green Costa Rican snakes: (a) *Oxybelis fulgidus* showing paired light stripes on ventrals; (b) *Bothriechis lateralis* illustrating paired light lines along tips of ventrals and first scale row.

2b. No paired light stripes along ventrals or, if present, indicated only on anterior half of body; dorsal ground color gray brown in life and preservative; ventrals fewer than 195. . . . . *Oxybelis aeneus* (p. 676)

Oxybelis aeneus (Wagler, 1824)

Narrow-headed Vine Snake, Bejuquilla

(plates 438–39; map 11.95)

DIAGNOSTICS: The species is unmistakable. No other snake has an elongate, narrow head, an attenuate brown or gray body, and a black-lined mouth that is exposed in a threatening display during nearly every encounter with humans. In addition, the eyes are directed forward, and there is a slight depression from each orbit to the tip of the snout.

DESCRIPTION: A moderate-sized to large species (adult total length 900 to 1,700 mm) with an elongate pointed head and extremely long and slender body and tail; tail very long (35 to 45% of total length), incomplete in about 50% of individuals; a loreal; postoculars 2; supralabials 6 to 10, usually 8 or 9, usually with 3 bordering orbit, rarely 2 on both sides; infralabials 6 to 11, usually 8 or 9; temporals variable, usually 1 + 2; dorsal scales usually in 17-17-13 (sometimes reduced to 15 or 12 rows), smooth or weakly keeled, with keels most evident on scale rows 7 to 9 on each side at midbody; anal plate almost always divided, rarely entire; ventrals 177 to 194 (173 to 205); subcaudals 165 to 197 (158 to 203); maxillary teeth 14 to 23 (14 to 27). Dorsum gray to brown, often with tints of green, yellow, orange, red, or black or obvious iridescence, usually peppered with light and black spots; sometimes with yellow to reddish ventrolateral stripes that are restricted to anterior half of body; usually a dark preocular stripe that runs posteriorly through

eye and continues onto body a short distance; venter similar in color to dorsum posteriorly, but anterior venter, throat, and chin white to yellow; venter sometimes with dark median stripe and/or dark lateral stripes; iris yellow to beige; tongue brown; lining of mouth blue black. Hemipenes single, with basal spines and distal calyces (some Mexican samples lack spines on organs).

SIMILAR SPECIES: (1) Other *Oxybelis* are predominately green.

HABITAT: Any situation, including secondary growth and other disturbed areas, where there are low shrubs, tall grass, or trees that can be climbed, within Lowland Dry, Moist, and Wet Forests and immediately adjacent lower portions of Premontane Moist and Wet Forests; in Dry Forest habitats it is most common in semideciduous forest remnants and in gallery forest situations.

BIOLOGY: *Oxybelis aeneus* is a common diurnal snake that forages in low vegetation, from half a meter to a meter or so above the ground to several meters in height. Prey consists of a variety of small vertebrates, but especially anole lizards (Keiser 1967), and includes birds, some frogs, mammals, and some insects (Henderson 1982). The grooved fangs introduce a mild venom into the prey. One human who was bitten (Crimmins 1937; erroneously identified as *Oxybelis fulgidus*) suffered local swelling and blistering. The species, however, rarely bites when handled. Except for rapid escape, the snake moves by steretoped oscillation of the usually elevated head and anterior body imposed on the curvilinear pattern of forward locomotion (Fleischman 1985). This behavior produces a swaying movement closely resembling the motion of windblown vegetation (branches, twigs, and vines) and appears to have a cryptic function in masking the snake's movement from potential prey. Snakes of this species often protrude the tongue rigidly and hold it motionless for several seconds while foraging. The possible functions of this behavior have been reviewed by Henderson and Binder (1980), but its significance remains enigmatic. These snakes are effectively camouflaged by their coloration and usually remain stretched out motionless and undetected, resembling a small vine. If this behavior fails and an intruder approaches too closely, they will open the mouth and expose its blue black lining while extending the gape to the maximum (plate 439) and then hiss and make short rapid strikes without biting hard (Scott 1983n). At night they sleep in loose coils with the head down, 2 to 5 m above the ground. Censky and McCoy (1988) described the reproductive cycle for samples of this species from the seasonally dry forest area of northern Yucatán. Reproductive activity in females was compressed into a twelve- to thirteen-week period from March to June, with eggs laid at the wettest time of year (July to September). Females with a body length of 650 mm or more were sexually mature and laid clutches of three to five eggs 30 to 45 mm in length. Clutch size increased with body size. The smallest female (290 mm in

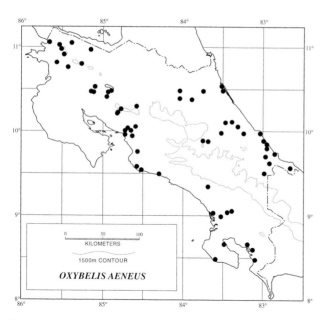

Map 11.95. Distribution of *Oxybelis aeneus*.

body length) was collected in October. Juveniles reached about 450 mm in body length in one year, so sexual maturity appeared to be reached at twenty-four months. Goldberg (1998) recorded clutches of three to eight eggs for females from western and southern Mexico. Sexton and Heatwole (1965) reported a clutch of four eggs of this species found in a depression in leaf litter in Panama on July 1 that hatched twelve days later. Scott (1983n) noted that these snakes may reduce activity and retire to hollow trees and other moist places in the dry season in the Dry Forest region of northwestern Costa Rica.

REMARKS: Keiser (1974, 1982) has reviewed the considerable individual and geographic variation in this wide-ranging species. Among the most noteworthy geographically localized differences are those in scale keeling, hemipenial structure, and to some extent coloration. In North and Central American examples, about 50% have weakly keeled scales, and keeling is present in about 25% of those from South America. In most males of *O. aeneus* the hemipenes are covered basally over a short sector by spines and have the distal half calyculate. However, in males from Sinaloa, Zacetacas, and the Tres Marías Islands, Mexico, basal spines are reduced or absent and the organs are attenuate. Coloration shows extensive individual variation superimposed on the basic gray to brown dorsum and the similarly colored but lighter posterior venter and underside of tail. The skin between the scales is white to yellow, orange, red, tan, brown, or black. Many examples have a checkerboard pattern of black and light on the neck and anterior body, some others have a distinctive mottled appearance. In some specimens the dark ventral colors extend forward onto the anterior ventrals, in others the chin, throat, venter, and subcaudal areas are white to yellow. Specimens from North and Central America usually have the labials and chin white to yellow (or rarely red) and unmarked by dark pigment. South American examples usually have the labials and chin peppered, spotted, or mottled with dark pigment. Bogert and Oliver (1945) thought that two major geographic subdivisions of the species could be recognized based on relative eye size in adults. Populations from Arizona and Mexico appeared to have smaller eyes relative to snout length than those from Central and South America. Keiser (1974) showed, however, that there is a clinal increase in relative eye size from north to south in Mexico, relatively high values for Central America, a clinal decrease from Panama to western Ecuador and northern Venezuela, and a general increase south through the Amazon basin and Brazil. Moreover, scattered thoughout the range of the species are populations that do not fit these patterns. For example, individuals from Amazonian Venezuela, Peru, Bolivia, and the Mato Grosso of Brazil have relatively small eyes. Consequently he concluded that this feature did not provide a basis for recognizing geographic races. In Costa Rican samples the mean of the ratio of eye diameter to snout length expressed as a

percentage is 37% compared with 32.5% for examples from Arizona-Mexico and 39.8% for the population with the relatively largest eyes, from Panama. The range for the species as a whole in this feature is 25.4% to 50.5%, with population means varying between 29.8% and 39.8%.

DISTRIBUTION: Lowland and premontane zones from Arizona and Coahuila, Mexico, south to northern Peru on the Pacific versant and Bolivia, east of the Andes; also on the islands of Trinidad and Tobago. In Costa Rica in the lowlands, except the driest areas of the Pacific northwest, onto adjacent premontane slopes, exclusive of the Meseta Central Occidental (2–1,340 m).

Oxybelis brevirostris (Cope, "1860d," 1861)
(plate 440; map 11.96)

DIAGNOSTICS: This relatively uncommon uniform green narrow-headed, elongate vine snake is unlikely to be confused with any other form except its larger ally *Oxybelis fulgidus. Oxybelis brevirostris* differs most obviously from the latter species in having an acuminate snout and lacking a pair of ventral light stripes (in *O. fulgidus* the snout is pointed and projecting and ventral light stripes are present). *O. brevirostris* differs from all other Costa Rican species that have the dorsum and venter green in its attenuate head and elongate body and tail. The prefrontals and the frontal plate are extremely long and narrow compared with those of any other snake in the republic, except the gray to brown *Oxybelis aeneus* (which has extremely elongate internasal, prefrontal, frontal, and supraocular shields) and *O. fulgidus* (which has extremely elongate internasal and frontal shields).

DESCRIPTION: A moderate-sized species attaining a total length of 1,200 mm, with an elongate, bluntly acuminate

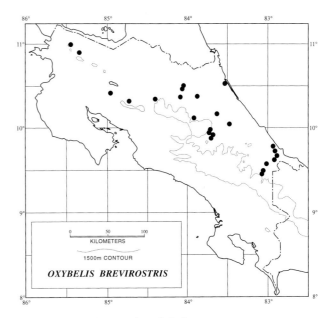

Map 11.96. Distribution of *Oxybelis brevirostris.*

head and an extremely attenuate body and tail; tail very long (36 to 42% of total length); tail frequently incomplete; no loreal; postoculars 1 to 2; supralabials 6 to 7, usually 6, with 2 bordering orbit; infralabials 7 to 8, usually 7; temporals usually 1 + 2; dorsal scales 15-15-13, smooth or faintly keeled; anal plate entire; ventrals 166 to 186; subcaudals 159 to 179. Dorsal surfaces uniform leaf green; a narrow dark stripe from nostril through eye to neck; labials yellow green; venter bright yellow green anteriorly, lime posteriorly; lining of mouth flesh colored; iris pale yellow to gold bisected by dark head stripe. Hemipenes single, with basal spines.

SIMILAR SPECIES: (1) *Oxybelis fulgidus* has a projecting pointed snout (fig. 11.11d) and a pair of ventral light stripes (fig. 11.42a). (2) Uniformly green *Leptophis* have strong keels on some dorsal scales and a divided anal (fig. 11.5). (3) *Drymobius melanotropis* has a yellow belly and 17 scale rows at midbody.

HABITAT: Lowlands and lower margins of premontane forests of the Caribbean slope; also found in a Premontane Wet Forest on the Pacific slope of the Cordillera de Tilarán.

BIOLOGY: Frogs and lizards are major food items of this active diurnal forager. Guyer and Donnelly (2002) reported on a female containing three oviductal eggs.

REMARKS: A single example (ANSP 5245), presumably from the city of San José (1,160 m), constitutes a questionable record. The species has not been otherwise recorded from the Meseta Central or from an altitude higher than 800 m.

DISTRIBUTION: Lowlands and the adjacent lower premontane belt on the Atlantic versant from eastern Honduras to central Panama and on the Pacific slope of eastern Panama, western Colombia, and western Ecuador; a single Pacific versant record in Costa Rica from near Tilarán, Guanacaste Province, otherwise restricted to the Atlantic versant in the republic (4–800 m).

Oxybelis fulgidus (Daudin, 1803)

Falsa Lora

(plate 441; map 11.97)

DIAGNOSTICS: One of the most distinctive members of the country's snake fauna, unmistakable because of the uniform leaf green dorsal coloration combined with the elongate head and narrow, pointed snout that projects well beyond the mouth opening (fig. 11.11d). In addition the green venter is marked by a narrow yellowish white stripe on each side (fig. 11.42a), a feature unique among Costa Rican snakes.

DESCRIPTION: A large species reaching a total length of 2,000 mm, with an elongate, pointed head, projecting snout, and attenuate body and tail; tail very long (32% to 42% of total length), frequently incomplete; no loreal; usually 2 postoculars, sometimes 1; supralabials 9 to 12, usually 10,

with 3 bordering orbit; infralabials 9 to 12, usually 10; temporals usually 1 + 2, occasionally 1 + 1 or 2 + 3; dorsal scales 17-17-13, keeled; anal divided (rarely entire in extralimital examples); ventrals 198 to 217; subcaudals 139 to 186. Dorsal surfaces bright green; venter vivid, pale green to chartreuse, with paired yellowish white stripes that continue onto underside of tail; lining of mouth flesh colored; iris yellow, tongue green. Hemipenes bilobed, with basal spines (Underwood 1967).

SIMILAR SPECIES: (1) *Oxybelis brevirostris* has a bluntly acuminate snout and lacks ventral yellow stripes. (2) Uniformly green *Leptophis* do not have pointed and projecting snouts (fig. 11.41) and lack the ventral yellow stripes. (3) *Drymobius melanotropis* lacks the pointed and projecting snout and ventral yellow stripes. (4) *Bothriechis lateralis* has a labial pit (fig. 11.8b) and has light stripes on the first scale row, not on the venter (fig. 11.42b).

HABITAT: Gallery forests within the Lowland Dry Forest and Premontane Moist Forest areas. Elsewhere in relatively undisturbed Lowland Moist and Wet Forests, Premontane Wet and Moist Forests and Rainforest, and the lower margins of Lower Montane Wet Forest and Rainforest of northern Costa Rica.

BIOLOGY: An uncommon diurnal form that forages in low vegetation and small trees. It eats numerous frogs, lizards, birds, and an occasional mammal. Connors (1989) reported a female (provenance unknown) that laid ten eggs in May, which hatched in 105 to 106 days. Hatchlings were 335 to 360 mm in total length. Gerhardt, Harris, and Vásquez (1993) list this snake as a prey item of the great black hawk *(Buteogallus urubitinga)* in Guatemala. This species exhibits rigid tongue protrusion, as described in the account

Map 11.97. Distribution of *Oxybelis fulgidus*.

of *O. aeneus*, when stalking prey. It bites if handled, and the mild venom injected by the rear fangs may cause pain and swelling (Campbell and Lamar 1989).

REMARKS: Keiser (1969) reported on an unusual coloration for examples from the Bay Islands off Honduras, in which the dorsum is mustard yellow and the venter pale yellow. Villa and McCranie (1995) named this population, restricted to Isla Roatán, *Oxybelis wilsoni* because of its coloration and higher subcaudal counts. Whether this is a valid species, a localized morph, or an unusual genetic variant that may appear elsewhere in the species' range is unclear, but no yellow *Oxybelis* have been reported from elsewhere. The karyotype of *O. fulgidus* is 2N = 34 (Beçak, Cameiro, and Beçak 1971).

DISTRIBUTION: Lowlands and adjacent slopes on the Pacific versant from the Isthmus of Tehuantepec to eastern Panama and on the Atlantic versant in southern Veracruz, Mexico, and then south from the Yucatán Peninsula through Central America and throughout South America, east of the Andes to northern Bolivia and extreme northeastern Argentina. In Costa Rica in the northwestern lowlands, Cordillera de Tilarán, Meseta Central, and northeastern lowlands (1–1,600 m).

Genus *Pseustes* Fitzinger, 1843

A group (three species) of semiarboreal diurnal snakes that have the dorsal scale rows obliquely arranged and the uppermost 3 to 19 rows keeled. In addition, the genus has the basic colubrid complement of head shields; head distinct from neck; body compressed; a single nasal; a loreal; one preocular; one or no subocular; eye large, pupil round; two pairs of elongate chin shields; dorsal scales in 19 to 27 rows at midbody, with a reduction in number posteriorly; keeling on all but first scale row on each side to only thre to four dorsalmost scales keeled; two apical scale pits; anal entire; subcaudals divided; 15 to 21 maxillary teeth increasing in size posteriorly; Duvernoy's gland present; anterior mandibular teeth somewhat longer than posterior ones; no hypapophyses on posterior dorsal vertebrae. The high number and oblique arrangement of the dorsal scale rows, the entire anal, and the restriction of keeling to the middorsal area (in the Costa Rican species) will distinguish the genus from similar snakes treated in this book.

The retracted hemipenes are short (8 to 9 subcaudals in length), symmetrical, single, clavate, with a simple semicentripetal sulcus spermaticus. The pedicel is nude, the truncus spinous, and the distal third to half calyculate, with a naked pocket within the calyculate area on the asulcate surface (Amaral 1929c).

The genus occurs over most of the Neotropical region at low to moderate elevations from San Luis Potosí, Mexico, on the Atlantic versant south to extreme northern Argentina and on the Pacific slope in Oaxaca, Mexico, and from Costa Rica to western Ecuador.

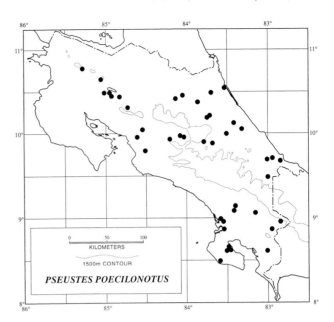

Map 11.98. Distribution of *Pseustes poecilonotus*.

Pseustes poecilonotus (Günther, 1858a)

(plate 442; map 11.98)

DIAGNOSTICS: The extensive individual and ontogenetic variation in this species makes immediate recognition of unusually marked examples challenging. Most adults are obscurely banded, have most of the dorsal scales edged by dark pigment (brown, black, or dark green) and the centers yellow and/or orange, pink, or reddish to form a variegated and multicolored pattern. Others have a bold pattern of black bands on a bright yellow ground, and still others are nearly uniform black. The juveniles are strikingly different from adults as well, having well-developed distinctive dark-edged orangish bands on the body and dark lateral head stripes. The combination of oblique dorsal scales, with the keeling restricted to the uppermost rows, single anal (fig. 11.5), a preocular (fig. 11.40c), and the high number of scale rows (21 to 25 at midbody) and a posterior reduction will distinguish even the most nondescript or confusing examples of this species from all other Costa Rican species.

DESCRIPTION: A large snake attaining a maximum total length of 2,400 mm; females longer than males; tail moderately long (26 to 32% of total length); postoculars usually 2 (occasionally 1 or 3); supralabials 6 to 10, usually 8 or 9, usually with 3 bordering orbit (rarely 2); infralabials 11 to 14; temporals 1 + 2, 2 + 2, or 2 + 3; dorsal scales in 21 to 25 rows at midbody (19 to 27), modal formula 23-25-15; ventrals 206 to 220 (181 to 220); subcaudals 117 to 147 (95 to 147). Adults: Dorsal coloration extremely variable; ground color brown, olive, greenish, yellow, or various shades of orange or gray, often uniform, but frequently with pale or dark spots on each scale; usually with some indication of dark brown to black transverse bands that in some cases are vivid and in others obscure; many individuals with both light spotting and obscure bands present; sometimes black bands expanded to almost obliterate lighter ground color, in which case upper head surface and anterior dorsum usually uniform black; labials usually light, but supralabial sutures sometimes marked with black (fig. 11.40b); ventral surfaces yellow, suffused with brown, with black spots laterally or free margins of ventrals marked with black laterally or underside of tail to entire undersurface black. Juveniles: dorsum with a series of irregular black-outlined orange brown bands on yellow to orange ground color; head with a broad orange brown dark-outlined stripe from nostril through eye and onto neck; yellow venter suffused with brown; ontogenetic changes involve expansion, to various degrees, of black pigment into bands or the obliteration of the bands to produce a more or less uniform or light dotted pattern, loss of eye stripe, and expansion of dark pigment on undersurface; iris light brown above, dark brown below; tongue black.

SIMILAR SPECIES: (1) *Spilotes pullatus* has 14 to 18 scale rows at midbody, all of them keeled (fig. 11.5) and arranged in even rows. Also see table 11.1.

HABITAT: Common in a variety of situations within Lowland Moist and Wet Forests and Premontane Moist and Wet Forests and Rainforest.

BIOLOGY: This common diurnal species is usually seen foraging in low trees or bushes or moving rapidly across open areas on the ground. It feeds primarily on birds and their eggs but also eats small mammals. When approached these snakes carry out an elaborate threatening behavior, hissing, opening the mouth wide, compressing the body laterally, and inflating the neck. Frequently they do this while elevated in shrubs or bushes, and *Pseustes* will often strike from this position if the bluff fails to deter a potential predator.

REMARKS: The extreme color variation in this species does not appear to show consistent geographic patterns, contrary to Amaral (1929c), Smith (1943), and Peters and Orejas-Miranda (1970), whose attempts to define subspecies based on differences in adult coloration are contradictory. Almost all of the several color patterns that these authors claim distinguish geographic races are represented in Costa Rican samples and show no geographic fidelity in the republic. The most common coloration in Costa Rican *Pseustes* is a bluish gray ground color, with red spots on the scales or entire scales red and usually also having a suggestion of dark bands. Other individuals have a similar pattern with light yellow, orange, pinkish, or greenish light scale centers, or entire scales of several of these colors (including red) may be present. At the other extreme of color variation are individuals boldly banded with yellow and black or with the yellow reduced to a few light spots. These black or yellow-and-black individuals have the upper head and nuchal area uniform black. Other individuals exhibit one or the other of

the color patterns described from elsewhere in the species' range, except that none have distinctive narrow paravertebral dark stripes or a completely reticulate dorsum. These last two features are supposed to be characteristic of adults from Mexico, exclusive of the Yucatán Peninsula, and adjacent Guatemala, according to Amaral (1929c) and Smith (1943). Completely uniform *P. poecilonotus* are also rare in Costa Rica, which contrasts markedly with the situation in the Amazon basin, where large adults (1,000 mm or more in total length) usually are uniform brown dorsally. The nominal species *Pseustes shropshirei* (Barbour and Amaral 1924) appears to be based on a typical obscurely banded and light-spotted example of Panamanian *P. poecilonotus*, as pointed out by Amaral (1929c). Taylor (1951b) thought that two species of *Pseustes* occurred in Costa Rica. One was characterized by having most of the dorsal scales marked with light centers combined with obscure banding. The second had distinct dark bands posteriorly and the head and neck black. Although Taylor (1951b) regarded the two as distinct, since both pattern types were found near Turrialba, Cartago Province, he inadvertently referred the light-spotted example (KU 25173) to *P. poecilonotus* and the black-headed one (KU 8738) to *P. shropshirei*. He reversed this arrangement in 1954. Peters and Orejas-Miranda (1970) seemed to have recognized *P. shropshirei* as valid based on Taylor's papers. They apparently used Taylor's (1951b) original identifications and characteristics to distinguish the two nominal forms in their keys. Under this arrangement black-headed individuals with strong black and yellow dorsal banding were erroneously called *P. shropshirei* and recognized as a distinct species. As correctly pointed out by Taylor (1954b), these snakes are properly regarded as *P. poecilonotus*. Taylor's (1954b) concept of *P. schropshirei* was based on the obscurely banded and light-spotted individuals described above as the commonest Costa Rican variant of *Pseustes*. There seems little question that only a single species of *Pseustes* is represented by Costa Rican material. In conclusion, both light-spotted and black-headed color phases of this species also occur in Panama and adjacent areas of Colombia.

DISTRIBUTION: Lowlands and premontane slopes of the Atlantic versant from San Luis Potosí, Mexico, to Amazonian Peru, Bolivia, and Brazil; also on the Pacific versant in Oaxaca, Mexico, central and southern Costa Rica, Panama, and western Colombia and Ecuador. In Costa Rica in lowland and premontane areas of the Atlantic slope and in the Tilarán area, on the Meseta Central Occidental, and in the southwest on the Pacific slope (2–1,330 m).

Genus *Rhinobothryum* Wagler, 1830

Snakes of this genus (two species) are unusual among arboreal forms in having a pattern of bicolor (black and red) or tricolor (red-white/yellow-black) rings to make them members of the coral snake mimic guild. Other significant features that in combination distinguish the genus from other colubrids include basic colubid complement of enlarged head shields; body compressed; nasal divided; a loreal; one preocular; eye large, pupil vertically elliptical (fig. 11.13b,d); two pairs of elongate chin shields; dorsal scales smooth on anterior part of body, with uppermost 5 to 9 rows keeled posteriorly, in 21-21-17 or 21-19-15 to 17 rows; two apical pits present; anal divided; subcaudals divided; 11 to 14 plus 2 enlarged grooved maxillary teeth, increasing in size posteriorly; Duvernoy's gland present; anterior mandibular teeth largest; no hypapophyses on posterior dorsal vertebrae. The coral snake pattern, compressed body, and large distinct head combined with the upper head shields black but bordered by white or cream are absolutely diagnostic.

The retracted hemipenes are apparently symmetrical, single, and cylindrical and have a simple semicentripetal sulcus spermaticus. The pedicel is nude, the truncus is spinous, and the distal third is covered with papillate calyces based on *Rhinobothryum lentiginosum* (Cope, 1894c).

The genus ranges from southern Honduras on the Atlantic versant southward in humid lowland forest areas over most of tropical South America east of the Andes (exclusive of southern Brazil) and from central Panama to western Ecuador on the Pacific slope.

Rhinobothryum bovallii Andersson, 1916

(plate 443; map 11.99)

DIAGNOSTICS: This species is unlike any other tricolor (red-white/yellow-black) ringed snake in having the enlarged black shields on top of the head edged with white or cream.

DESCRIPTION: A large-sized snake reaching a total length of 1,760 mm; head very distinct from neck; tail moderate (22 to 27% of total length); 1 preocular (fig. 11.13d);

2 or 3 postoculars; supralabials 8, with 2 bordering orbit; infralabials 10; temporals 2 + 2 or 3 + 3; dorsal scales in 21-19-17 or 21-21-17 rows, 5 uppermost rows keeled posteriorly; ventrals 239 to 246, angulate and keeled laterally; subcaudals 115 to 125. Dorsal pattern of tricolor rings arranged in monads (one black ring for each red ring); each black ring separated from red rings by a narrow white or yellowish ring; most scales in red rings tipped with black; 11 to 16 black body rings; tail patterned like dorsum; enlarged head shields black outlined by white or cream, temporal area red with black spots.

SIMILAR SPECIES: (1) Tricolor venomous coral snakes *(Micrurus)* do not have the enlarged black head shields outlined by light (fig. 11.48). (2) *Erythrolamprus mimus* has most of its black rings split dorsally or laterally by white areas. See key to coral snakes (p. 706), figure 11.47, and table 11.2.

HABITAT: Undisturbed Lowland Moist and Wet Forests of the Atlantic versant.

BIOLOGY: This rare serpent appears to be a canopy inhabitant. Scott (1969) reported that in captivity the species ate anole lizards.

REMARKS: The only other member of the genus, *Rhinobothryum lentiginosum*, occurs over much of tropical South America east of the Andes (exclusive of southeastern Brazil) and is a bicolor (red-black) ringed species that has the uppermost 7 to 9 rows of dorsal scales keeled on the posterior part of the body.

DISTRIBUTION: Atlantic lowlands from southern Honduras to western Venezuela; Pacific lowlands from central western Panama to western Ecuador (4–550 m).

Genus *Scaphiodontophis* Taylor and Smith, 1943

A single variable species is placed in this extraordinary genus, which has a tricolor (red-yellow/gray-black) banded coral snake pattern at least on the anterior part of the body. In addition to their distinctive coloration these snakes have very long tails (34 to 49% of total length in adults) that are disproportionately thick but very fragile and consequently are frequently incomplete (about 70% of most samples). Representatives of the genus are further distinguished from other snakes by the following combination of characteristics: the basic colubrid complement of enlarged head shields; body cylindrical; nasal divided; a loreal; one preocular; eye moderate, pupil round; two pairs of elongate chin shields; dorsal scales smooth, in 17 rows; no apical pits; anal divided; subcaudals divided; 39 to 59 hinged teeth on maxillary, many of them spatulate and appearing to have enamel only on their leading surface; anterior 8 to 10 and posterior 5 or 6 maxillary teeth shorter than most teeth, which are arranged in groups of two long teeth and one small or one long and one small over most of maxillary; no diastema; mandibular teeth similarly arranged; tooth replacement on jaws simultaneous; no Duvernoy's gland; hypapophyses present

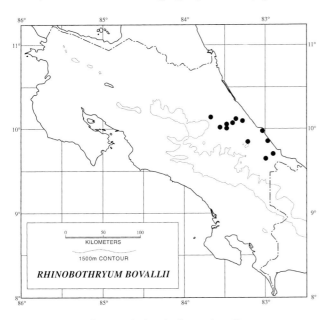

Map 11.99. Distribution of *Rhinobothryum bovallii*.

on posterior dorsal vertebrae. The restriction of the dorsal coral snake pattern to the upper surfaces (no rings crossing venter) and the long and thick but often broken tail will distinguish *Scaphiodontophis* from all other Costa Rican snakes. *Enulius* and *Urotheca* have similarly long and disproportionately thickened and fragile tails, but the former have uniform dorsums and the only similar species of *Urotheca* has a pattern of black rings.

The everted hemipenes are moderate (10 to 12 subcaudals in length), symmetrical, single, and cylindrical and have a simple semicentripetal sulcus spermaticus. The pedicel is covered by scattered very small spines, followed distally by 11 to 15 large spines and 40 to 50 smaller spines. The distal third of the organ is covered with papillate calyces (Taylor and Smith 1943; Morgan 1973; Zaher 1999).

The genus is found in the lowlands and on premontane slopes from Tamaulipas and Oaxaca, Mexico, south to northern and central Colombia, exclusive of the subhumid to semiarid zones of Pacific versant Central America.

Scaphiodontophis annulatus (C. Duméril, Bibron, and A. Duméril, 1854)

(plates 444–46; map 11.100)

DIAGNOSTICS: The tricolor banded pattern of monads (each red band separated by a yellow-black-yellow series of bands) will immediately distinguish this species from all other snakes in Costa Rica except some *Oxyrhopus petolarius*. *S. annulatus* differs from that form in having a white snout in juveniles, a light (white to brown) interocular band in adults, and the black head cap extending several scale rows onto the neck (the black head cap covers all of the enlarged head shields but does not extend onto the neck in

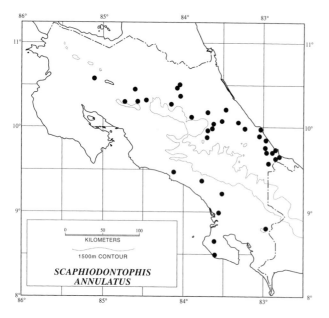

Map 11.100. Distribution of *Scaphiodontophis annulatus*.

juvenile and adult *O. petolarius*). Juveniles of *S. annulatus* superficially appear to be bicolor black and yellow.

DESCRIPTION: A moderate-sized species (to 920 mm in total length); females longer than males, largest examples with complete tails 795 mm and 740 mm in total length, respectively; standard length of adult males 215 to 435 mm, adult females 205 to 480 mm; body cylindrical; head slightly distinct from neck; tail long (41 to 49% of total length in males, 33 to 45% in females); tail extremely fragile, 60 to 70% of adults with missing or broken tails; a loreal; 2 postoculars; supralabials 8 to 10, with 3 bordering orbit, usually fourth to sixth, sometimes third to fifth, rarely fifth to seventh; infralabials 9 to 12, usually 10; temporals almost always 1 + 2; ventrals 127 to 155 (127 to 166); subcaudals 105 to 121 (105 to 145) in males, 93 to 105 (93 to 126) in females; ventrals plus subcaudals 234 to 294. Coloration is extremely variable in this species (see remarks), and the following applies only to Costa Rican examples. Tricolor pattern on body and tail mostly of monads, each red band separated from the next by yellow-black-yellow bands; number of black bands on body five to sixteen; black bands often offset; sometimes a few dyads or triads present; in a few cases tricolor pattern interrupted on posterior body by a brownish gray segment; snout white in juveniles, black in adults; black head cap involving parietals and several rows of nuchal scales followed by a yellow nuchal collar that usually borders first red band; sometimes separated from it by a narrow black nuchal collar; venter immaculate dull yellow, flecked with darker pigment or heavily suffused with brown, especially on ventral tips; subcaudals immaculate in very small individuals and with large median black spots in subadults and adults; tongue banded black and yellow, tips of fork yellow; iris brown; small juveniles heavily suffused with black pigment (banded black and yellow), red bands obscured by black; otherwise like adults.

SIMILAR SPECIES: (1) Tricolor specimens of *Oxyrhopus petolarius* have all of the enlarged upper head shields black. (2) Tricolor specimens of *Lampropeltis triangulum* have a pattern of dyads with the red and black bands in contact with one another and the red bands not in contact with the yellow bands. See key to coral snakes (p. 706), figure 11.47, and tables 11.1, 11.2.

HABITAT: This relatively uncommon snake is a terrestrial form that is found in Lowland Moist and Wet Forests and Premontane Wet Forest and Rainforest.

BIOLOGY: *Scaphiodontophis annulatus* is one of several Costa Rican snakes forming a coral snake mimic guild. The species is a diurnal leaf litter habitué. Adults typically sit motionless with the head and anterior one-fourth to one-third of the body raised above the substratum and most of the body and tail hidden under the surrounding litter waiting for approaching prey. In Costa Rica, food consists almost exclusively of the common leaf litter skink *Sphenomorphus cherriei*, although elsewhere in its range it also eats other skinks, microteiid lizards, and frogs (Alvarez del Toro 1960, 1973, 1982; Savitsky 1981, 1983; Henderson 1984). These snakes are known to constrict larger prey items under captive conditions, but their usual modus operandi is a quick strike as the skink passes within range. The lizard is then swallowed headfirst while alive with incredible speed. Henderson (1984) recorded times from when captive examples first grasped the prey to when the lizard was no longer visible as ranging from 2.8 to 16.9 seconds. The highly specialized jaw apparatus, the spatulate and hinged teeth, hypertrophy of the levator anguli oris muscle at the angle of the jaws, and the binding of the dentaries by a nonelastic ligament anteriorly are modifications that apparently expedite rapid prey ingestion. As Savitsky (1981, 1983) pointed out, the morphological specializations are designed for processing hard-bodied prey (durophagy). In *Sphenomorphus* and other skinks each of the scales on the body, limbs, and tail is underlain by a small osseous plate (an osteoderm). The scales and osteoderms overlap to allow movement but form an essentially continuous bony protective armor. When *Scaphiodontophis* is feeding, the adaptations for durophagy allow the teeth, especially the spatulate ones, to hook under the free, trailing edges of the scale osteoderms. The hinged teeth then fold toward the back of the mouth to protect them against breakage as the prey is moved back. The other jaw and muscle modifications aid in this unusual method of ingestion and in restricting the lateral movement of the living prey. Juveniles (up to about 200 mm in standard length) have the snout a glossy white and the red bands suffused by black pigment. Campbell (1998) reported courtship behavior in this species from two observations in Guatemala. Three or four males were found entwined around a single female in both cases. The species is oviparous and lays clutches of one to twelve eggs, usually within the decomposing leaf litter or under rotten logs or in soil depressions. Eggs are about 33 to 35 mm long. Hatchlings are about 122 to 152 mm in standard length. Like two species of the genus *Urotheca (U. elapoides* and *U. euryzona), Scaphiodontophis* exhibits a predator avoidance complex involving both ready loss of the long and fragile tail and mimicry of the coloration found in the venomous coral snakes. These snakes appear to thrash their tails when threatened or touched (Henderson 1984; pers. obs.) When part of the tail breaks off, reflex action continues the thrashing, and the predator tends to focus on this decoy while the snake escapes. The processes involved are the same as described for the genus *Urotheca* (p. 640); see that account for further detail. The Costa Rican snakes of the genus *Enulius* also use this antipredator defense. Let me emphasize that in all these cases the breaks occur intervertebrally (pseudautotomy) as detailed by Savage and Crother (1989) and not intravertebrally as in most lizards (autotomy), contrary to Wilson (1968) and Muñoz (1987). Slowinski and Savage (1995) demonstrated that multiple tail breaks regularly occur during the life of individuals of this species. Presumably the thick but long and fragile tail permits multiple breaks, allowing for repeated escapes from predators. *S. annulatus* and other snakes *(Enulius, Sibynophis,* and *Urotheca)* having a similarly modified caudal morphology allowing multiple tail breaks are the only known examples of specialized pseudautotomy (Slowinski and Savage 1995). *S. annulatus* is sympatric with three species of monadal-patterned venomous coral snakes in Costa Rica. In general the pattern on the body and tail most closely resembles that of Atlantic lowland *Micrurus nigrocinctus* (ten to sixteen monads on the body), but some individuals could be mistaken for *Micrurus alleni, M. clarki,* or Pacific lowland *M. nigrocinctus.* The experiments of S. Smith (1975, 1977), Brodie (1993), and Brodie and Janzen (1995) demonstrated that naive and free-ranging avian predators have an innate generalized avoidance response to coral snake patterns. The level of resemblance between monadal *Scaphiodontophis* and these *Micrurus* implies a much finer distinction by some predators than is made by birds in these studies. This situation parallels the geographic variation and congruence shown by coral snake mimics and their models in *Urotheca* (Greene and McDiarmid 1981; Savage and Crother 1989) and appears to represent the second step in the two-step evolution of Batesian mimicry (Pough 1988). Like many coral snakes, *S. annulatus* exhibits head thrashing as well as tail thrashing when touched or held. It seems probable that these behaviors emphasize the aposematic coloration so that a potential predator will be warned of a venomous bite by a coral snake or misled about the danger from the mimic *S. annulatus.* In some areas extralimital to Costa Rica many examples of *S. annulatus* have the tricolor pattern only on the anterior part of the body, as described below. Some workers have doubted the efficacy of the mimicry in these cases. However, recall that the characteristic diurnal foraging posture in this species differentially exposes the head and anterior body region, and it seems likely that aposematic coloration, even if restricted to that region, will confer an adaptive advantage (Savage and Slowinski 1996).

REMARKS: Variation in coloration in *S. annulatus* over its entire geographic range is extreme, and at one time as many as eight species were recognized in the genus based on characteristics of body and tail and on head and nuchal patterns. Recently Savage and Slowinski (1996) completed an extensive study of variation in *Scaphiodontophis* that supported the conclusion of Smith et al. (1986) that only a single species is involved. As these authors pointed out, snakes of this genus are among the most variable in coloration of any vertebrate. Almost every individual from any area differs in details of coloration from every other one, and at one time individuals known to be from the same egg clutch would have been placed in a different species than some of their siblings or their mothers. Although there is some geographic consistency in general color pattern, almost every individual has one or more anomalous elements or combination of elements in the pattern. The basic coloration of the body and tail always contains one or more tricolor (red-yellow/gray-black) components. Generally one of two principal component types predominates. In many populations the red bands are separated by black-yellow/gray-black elements to form a dyadal pattern (two black bands for each red one). In others the red bands are separated by yellow-black-yellow elements to form a monadal pattern (one black band for each red one). The tricolor pattern may cover the entire body and tail or be variously replaced by a brown to gray ground color that is usually marked with black-dotted longitudinal stripes. Four major head and nuchal patterns are also recognizable:

A: black head cap involving posterior tips of frontal and supraoculars; nuchal collar red bordered posteriorly by first body dyad

Du: black head cap involving posterior half of frontal and supraoculars; nuchal region and anterior portion of body red, no black nuchal collar

V: black head cap not involving frontal or supraoculars; nuchal collar yellow, bordered posteriorly by red

Z: black head cap involving posterior tips of frontals and supraoculars; light collar yellow, bordered posteriorly by a single black nuchal collar

Geographic variation in *S. annulatus* may be summarized as follows:

1. Northern populations: subcaudals 123 to 145 in males, 110 to 126 in females; subcaudals usually lacking large medial black spots
   a. Atlantic and Pacific slope Mexico to El Salvador: dyadal pattern over entire length of body and tail or variously replaced by brown to gray ground color on tail, posterior portion of body and tail, or over most of body and tail; 3-6-17 dyads on body; Z head and nuchal pattern
   b. Atlantic slope Guatemala, Belize, and Honduras: dyadal pattern restricted to anterior portion of body; 1–3.4–6 dyads on body; A head and nuchal pattern
2. Southern populations: subcaudals 105 to 121 in males, 93 to 105 in females; subcaudals usually with large medial black spots
   c. Nicaragua, Costa Rica, and western Panama: monadal pattern, usually over entire length of body and tail; 5–10.5–16 monads on body; V head and nuchal pattern
   d. Central to eastern Panama: dyadal pattern over entire body and tail or restricted to anterior part of body or that region and tail; 4-8-12 dyads on body; Z head and nuchal pattern
   e. Colombia: pattern a mixture of dyadal and monadal elements over entire length of body and tail, on anterior portion of body and tail, or with a short precloa-

cal tricolor area as well; total number of monads, dyads, and tetrads on body 2-6-11; Du head and nuchal pattern

Variation in body and tail coloration is much greater than this summary suggests, and this fact, together with the presence of intergrading populations between populations a and b, c and d, and d and e and trends indicating probable intergradation between b and c, supports the concept that only one extremely variable species is involved (Savage and Slowinski 1996). Note that Costa Rican examples of this taxon have usually been referred to as *Scaphiodontophis venustissimus* by most workers. As I have pointed out in previous sections, Costa Rican snakes of this species undergo remarkable ontogenetic changes in coloration. All juvenile *Scaphiodontophis* (under 215 mm from snout to vent) have the rostral, internasals, prefrontals, most or all of the frontal and supraoculars, and in some populations most of the parietal shields bright white (plate 445). Some individuals retain the white area into adulthood, but in most cases dark pigment is laid down over the snout as growth proceeds, beginning with the appearance of yellowish to light brown areas near the internasal-prefrontal sutures. In time this color expands to completely cover first the internasals and prefrontals, then the rostral, and finally (in some populations) the anterior tips of the frontal and supraocular plates. As growth continues, black pigment becomes concentrated in the brown areas so that most snakes with a snout-vent length of more than 300 mm have the black snout characteristic of adults of this genus. In addition, Costa Rican juveniles under 200 mm in standard length superficially appear to have a bicolor dorsal pattern of narrow white and broad black bands. As I noted in a previous section, closer inspection reveals that the black bands consist of two parts, an anterior lighter area in which basically red scales are heavily suffused with black pigment and a solid black posterior one. The solid black bands are bordered posteriorly by a single row of white scales that form the narrow light bands. The pattern is an alternating series (anterior to posterior) of suffused red-black–narrow white bands. As growth proceeds, the influence of the black suffusing pigment is reduced and gradually concentrates on the tips of the red scales to fully reveal the underlying red pigment. In addition, both the red and suffusing black pigments gradually recede from both margins of the red band to expose additional light scales posterior to the solid black region and a second light zone along its anterior edge. As the light bands develop to adult dimensions, the light zones become yellow and the characteristic tricolor monadal pattern is established. The yellow nuchal collar typical of subadult and adult examples of this population develops in a similar fashion. It first appears as several white scales bordering the black head cap posteriorly. These scales are followed posteriorly by an area that becomes red later on but is so heavily suffused with black pigment at this stage as to obscure all but a hint of the underlying red pigment. With further development the obscuring black pigment becomes restricted to the scale tips and a distinct red band is unveiled. A little later and more or less simultaneously, both red and black pigments recede from the nuchal region and the nuchal yellow band is exposed.

DISTRIBUTION: Lowlands and adjacent premontane Atlantic slopes from Tamaulipas, Mexico, to northern and central Colombia; on the Pacific versant from west of the Isthmus of Tehuantepec in Oaxaca, Mexico, to Honduras and from southwestern Costa Rica to eastern Panama; in Costa Rica found throughout the humid lowlands of the Atlantic slope and the southwest; also a single record for the Pacific slope on Volcán Tenorio in northwestern Costa Rica (2–830 m).

Genus *Scolecophis* Fitzinger, 1843

A single species of small brightly colored semifossorial snakes belonging to the coral snake mimic guild is placed in this genus. The pattern consists of alternating black and light rings (red to orange dorsally, yellow laterally and on the venter). The following additional combination of features characterizes the genus: the basic colubrid complement of enlarged head plates; nasal divided; a loreal (rarely absent); a preocular; eye small, pupil round; two elongate pairs of chin shields; dorsal scales smooth in 15-15-15 rows; one apical scale pit; anal divided; subcaudals divided; 13 to 15 small, equal maxillary teeth followed by two somewhat enlarged grooved fangs; Duvernoy's gland probably present; mandibular teeth equal; no hypapophyses on posterior dorsal vertebrae. No other Costa Rican snake with a coral snake pattern has both loreal and preocular plates, elongate chin shields, and the dorsal scales in 15 rows at midbody.

The everted hemipenes are short (about 10 subcaudals in length), simple, symmetrical, and with a single sulcus spermaticus. The pedicel is naked, and two basal hooks on the asulcate surface demarcate the base of the truncus, which is covered with spines that are largest on the asculate surface. The apical region is covered by spinulate calyces.

The genus ranges along the Pacific slope of Central America from Guatemala to central Costa Rica and onto the western fringe of the Atlantic drainage uplands of Honduras.

Scolecophis atrocinctus (Schlegel, 1837)

(plates 447–48; map 11.101)

DIAGNOSTICS: The pattern of alternating black and light rings completely encircling the body, in which the upper portions of the light rings are red to orange and the lateral and ventral portions are yellow, is unusual. The dorsal coloration, in combination with the dorsal head pattern consisting of a black snout and a light (yellow to orangish) prefrontal bar followed by a black head band that extends

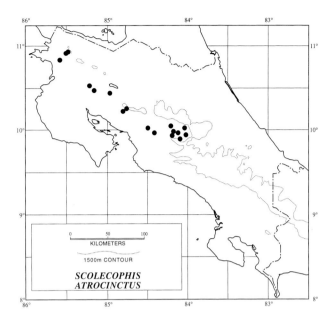

Map 11.101. Distribution of *Scolecophis atrocinctus*.

almost to the posterior tips of the parietal shields, is absolutely diagnostic. Only four other Costa Rican snakes might be confused with this form. Most *Urotheca euryzona* have a similar dorsal color pattern, but the black head cap reaches only onto the anterior portion of the parietals. Some *Micrurus mipartitus* have a dorsal pattern similar to that of *Scolecophis atrocinctus*, but the black head cap is restricted to the area anterior to the middle of the orbits. *Dipsas bicolor* and *Sibon anthracops* have red on the dorsal portion of the light (yellow) rings, and the head is mostly black and without a prefrontal light bar.

DESCRIPTION: A small snake reaching a total length of 450 mm; body cylindrical; head slightly distinct from neck; tail short to moderate (14 to 19% of total length); a small loreal, rarely absent on one side; 2 postoculars; supralabials 7, with 2 bordering orbit; infralabials 6 to 7, almost always 7; temporals 1 + 1; ventrals 184 to 196 (181 to 197); subcaudals 46 to 53 (45 to 54). Pattern on body of 30 to 37 (30 to 45) black rings that alternate with light rings that are red to orange red dorsally and yellow laterally and ventrally; lateral scales in light rings usually heavily tipped with black that sometimes gives lateral and ventral areas a bluish cast; tail similarly marked with five to eight black rings; iris black; tongue black.

SIMILAR SPECIES: (1) *Urotheca euryzona* lacks a prefrontal light bar, and the black head cap extends only onto the anterior portion of the parietal shields. (2) *Micrurus mipartitus* lacks a prefrontal light bar, and the black head cap is restricted to the dorsal surface of the head anterior to the middle of the orbits. (3) *Dipsas bicolor* lacks a prefrontal light bar and has the dorsal red saddles separated from the black body rings by a narrow light area. (4) *Sibon anthracops* lacks a prefrontal light bar. See key to coral snakes (p. 706), figure 11.47, and table 11.2.

HABITAT: Moister portions of the Lowland Dry Forest and immediately adjacent areas of Pacific slope Lowland Moist Forest and Premontane Moist and Wet Forests, including lower margin of Lower Montane Wet Forest in northern Costa Rica.

BIOLOGY: This secretive snake is relatively rare, but one example was collected as it was climbing a small tree. These snakes appear to prey exclusively on centipedes. In a series of eight examples, three contained centipede remains. Harry Greene (pers. comm.) saw a specimen of *S. atrocinctus* attacking and eating a large centipede during the day near Monteverde, Puntarenas Province. When disturbed *Scolecophis* thrash spasmodically from side to side in a defensive display similar to that exhibited by the venomous coral snakes *(Micrurus)*.

REMARKS: Douglas C. Robinson captured a single specimen (UCR 3364) at La Selva Biological Station, Heredia Province, in the Atlantic lowlands. This was probably an escape, since no other examples are known from Atlantic versant Costa Rica and the La Selva site has been intensively inventoried by many biologists over the past thirty-five years.

DISTRIBUTION: Foothills and slopes of the Pacific versant from Guatemala to central Costa Rica; one record for the western uplands of Honduras lies on the Atlantic slope. In Costa Rica restricted to the Pacific lowlands, premontane, and lowermost lower montane zones of the northwestern area and the Mesta Central Occidental (100–1,530 mm).

Genus *Spilotes* Wagler, 1830

A single species of large, powerful diurnal terrestrial to semiarboreal racer is included in this genus. *Spilotes* is immediately distinguishable from most other snakes by having the dorsal scales in an even number of rows (14 to 20) at midbody. In addition the genus has the basic colubrid complement of enlarged head shields and the following combination of distinctive characters: head distinct from neck; nasal entire; one loreal or none; usually one preocular, rarely two; eye moderate to large, pupil round; two pairs of elongate chin shields; dorsal scales keeled, with a reduction in number anterior to vent; two apical scale pits; anal entire; subcaudals divided; 19 to 22 equal maxillary teeth; no Duvernoy's gland; anterior mandibular teeth longest; no hypapophyses on posterior dorsal vertebrae. The combination of keeled scales in even numbers of nonoblique rows will separate *Spilotes* from all other Costa Rican snake genera.

The everted hemipenes are symmetrical, single, and have a simple semicentripetal sulcus spermaticus. Only the very short pedicel is nude, and most of the proximal half of the organ is covered with moderate spines. The rest of the hemipenes is covered by spinulate calyces (Chippaux 1986).

The genus ranges over most of the tropical and subtropical lowlands and adjacent slopes on the Atlantic versant

from Tamaulipas, Mexico, to Paraguay and extreme northeastern Argentina and on the Pacific slope from the Isthmus of Tehuantepec to western Ecuador.

Spilotes pullatus (Linné, 1758)

Mica, Zumbadora

(fig. 11.40c; plate 449; map 11.102)

DIAGNOSTICS: Any moderate-sized to large snake marked with alternating oblique or transverse bands of yellow or white and black dorsally and ventrally is most likely to be this species. Some adults are nearly uniform black and may be confused with large examples of *Chironius grandisquamis*. *S. pullatus* has evenly arranged keeled dorsal scales in 14 to 18 rows at midbody, while the dorsal scales are obliquely arranged in 10 rows at midbody, and at most only the upper 6 to 8 scale rows are keeled in *C. grandisquamis*. Some *Pseustes poecilonotus* have a yellow and black coloration similar to that of many *S. pullatus*. This species differs from *P. poecilonotus* (characteristics for the latter in parentheses), in having most of the evenly arranged dorsal scales keeled (fig. 11.5) and in 14 to 18 (14 to 20) rows at midbody (obliquely arranged, keeling only on uppermost scale rows, and scales in 19 to 25 rows at midbody). Juveniles of *S. pullatus* superficially resemble a number of banded Costa Rican snake species but can be distinguished from all other similar forms in having keeled dorsal scales in 14 to 18 rows at midbody.

DESCRIPTION: A very large species reaching a total length of 2,650 mm, most examples 1 to 2 m; body laterally compressed, robust; head distinct from neck; tail moderate (24 to 27% of total length); 1 or 2 postoculars; supralabials 6 to 9, usually with 2 bordering orbit, sometimes 3; infra-

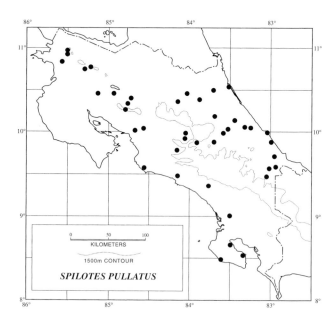

Map 11.102. Distribution of *Spilotes pullatus*.

labials 6 to 10; temporals variable, usually 1 + 1, often 0 + 1 or 1 + 2; 16 to 20 scale rows on neck; ventrals 200 to 226 (198 to 241); subcaudals 115 to 134 (100 to 142). Adults: extremely variable in coloration, dorsum and tail usually marked with transverse or oblique alternating bands of yellow or white and black; when light color predominates, black markings form bands, streaks, bars, and/or spots of various sizes and in variable combinations (pattern reticulate with a yellow spot on each black-edged scale in some non–Costa Rican examples); when black color dominates, light bands, streaks, bars, or spots usually present but posterior one-half to one-third of body and all of tail often uniform black and some adults almost entirely black; venter and underside of tail sometimes mostly light in color, usually with irregular alteration of black and light areas or sometimes nearly uniform black or uniform black on posterior venter and tail; juveniles: dorsal pattern of quadrangular dark blotches separated by definite narrow light crossbands (oblique anteriorly, transverse posteriorly and on tail); venter light, anterior ventrals edged with black; ontogenetically light areas expand in width and number in predominately light examples; in many examples light areas then become invaded by black, producing irregular light bands, streaks, and spots and in some large examples obscuring light color almost entirely; iris brown to black; tongue black.

SIMILAR SPECIES: (1) *Chironius grandisquamis* has oblique dorsal scales (fig. 11.5) in 10 rows at midbody. (2) *Pseustes poecilonotus* has oblique dorsal scale rows in 19 to 25 rows at midbody. (3) *Drymarchon corais* has smooth dorsal scales (fig. 11.5) in 17 rows at midbody.

HABITAT: In riparian situations in Lowland Dry Forests; forests, secondary growth, pastures, and most other situations within Lowland Moist and Wet Forests and Premontane Moist and Wet Forests.

BIOLOGY: A moderately common species most frequently seen moving about on the ground or foraging in shrubs or low trees, where it preys on lizards, birds' eggs, and nestling birds, but these appear to supplement a diet composed primarily of mammals (Cadle and Greene 1993). Although racerlike in form, *Spilotes*, unlike other colubrines with this gestalt, constricts its prey. Egg clutches are seven to ten, and hatchlings are about 500 mm in total length. In Guatemala, Gerhardt, Harris, and Vásquez (1993) found this snake to be preyed on by the great black hawk *(Buteogallus urubitinga)*. Its defensive display is most impressive and includes opening the gape wide, hissing, and greatly inflating the neck region to expose the dark gray to black skin. This behavior typically occurs when the snake is hanging down from an arboreal perch but may also be used to intimidate potential predators that approach it on the ground.

REMARKS: The extreme and complicated individual color variation in representatives of this species makes suspect attempts to recognize major geographic races based on differences in pattern (Amaral 1929d; Peters and Orejas-Miranda 1970). Most of the supposed geographically consistent differences in color pattern are represented as individual variants within Central American samples and show no geographic fidelity in Costa Rica, contrary to Taylor (1951b, 1954b). The situation is further complicated by ontogenetic changes that modify the juvenile pattern of dark quadrangular blotches separated by narrow transverse light lines into the several rather distinctive adult patterns. The end points of these changes produce some individuals that are mostly light with black markings of various sizes and shapes, snakes that may have a pale vertebral stripe and a reticulate pattern (a variant not known from Costa Rica), and examples that are mostly black with light markings of various sizes and shapes or nearly uniform black. Almost every imaginable intermediate between the extremes in this spectrum of patterns occurs among adults or is represented in various subadult and/or small adult *Spilotes*. For these reasons it is obvious that our current understanding of variation is insufficient to allow a rational review of geographic trends, if indeed any exist. The karyotype of this species is 2N = 36, with eight pairs of macrochromosomes and ten pairs of microchromosomes and a ZW heteromorphism in females (Beçak 1965).

DISTRIBUTION: Atlantic lowlands and premontane slopes from Tamaulipas, Mexico, south through Central America and northern South America to Bolivia, Paraguay, and extreme northeastern Argentina; on the Pacific versant from the Isthmus of Tehuantepec to western Ecuador. Both slopes of Costa Rica exclusive of the driest portions of the northwestern region (2–1,435 m).

Genus *Stenorrhina* C. Duméril, 1853

Members of this unusual genus (two species) are specialized predators on large arthropods, especially scorpions and tarantulas. The group is unique among Central American taxa in having each internasal shield fused with an anterior nasal, the rostral shield prominent, and the snout flattened and shovel-like (fig. 11.9a). Otherwise the basic colubrid complement of enlarged head plates is present. Additional characteristics of the genus include nasal divided below nostril; loreal present or absent; one preocular; eye small, pupil round; two elongate pairs of chin shields; dorsal scales smooth in 17-17-17 rows; no apical scale pits; anal divided; subcaudals divided; 10 to 21 equal maxillary teeth followed after a diastema by two feebly enlarged grooved fangs; Duvernoy's gland present; mandibular teeth equal; no hypapophyses on posterior dorsal vertebrae. The fusion of the internasal and anterior nasal will distinguish the genus from other Costa Rican snakes.

The everted hemipenes are moderately long (11 to 20 subcaudals in length), single, symmetrical, with a simple semicentripetal sulcus spermaticus. The organ may have basal hooks or not but is covered by spines proximally, and the apical region is calyculate.

Stenorrhina ranges from Guerrero and southern Veracruz, Mexico, to north-central Venezuela and western Ecuador.

KEY TO THE COSTA RICAN SPECIES OF THE GENUS *STENORRHINA*

1a. Dorsum with distinctive large dark blotches, or uniform; if dorsum uniform, the free margin of each ventral darker than rest of scute, sometimes with a midventral dark stripe; ventrals 153–68; ventrals plus subcaudals 189–205; hemipenes of males long, 17–20 subcaudals in length, with several large basal hooks .... *Stenorrhina degenhardtii* (p. 688)

1b. Dorsum with a pattern of narrow longitudinal dark stripes on a lighter ground color, at least anteriorly, or uniform; if dorsum uniform, ventrals immaculate, without dark pigment along free margin or with a few scattered dark spots along midline; ventrals 164–83; ventrals plus subcaudals 203–17; hemipenes of males short, 11–13 subcaudals in length, without large basal hooks .... *Stenorrhina freminvillii* (p. 689)

Stenorrhina degenhardtii (Berthold, 1845)

(plate 450; map 11.103)

DIAGNOSTICS: The enlarged rostral, shovel-shaped snout, and fusion of the internasal to the anterior nasal will immediately separate this form from any other Costa Rican snake having a blotched or uniformly colored dorsum except uniformly colored examples of its congener *Stenorrhina freminvillii*. In uniformly colored *S. degenhardtii* the ventrals have extensive dark markings, whereas in *S. freminvillii* having a uniform dorsum the venters are immaculate or with a few scattered dark marks.

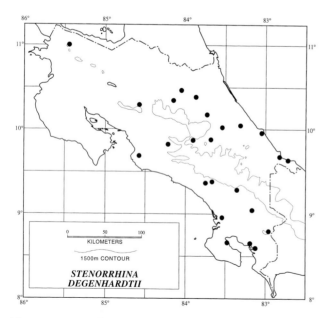

Map 11.103. Distribution of *Stenorrhina degenhardtii*.

DESCRIPTION: A moderate-sized species reaching a maximum total length of 800 mm (small adults about 500 mm); body stout, cylindrical; head not distinct from neck; tail moderately long in males (19 to 25% of total length), short in females (13 to 18% of total length); loreal usually lacking (sometimes present), and prefrontals usually in contact with supralabials; 2 postoculars; supralabials 7, with 2 bordering orbit; 7 to 9 infralabials, usually 7; temporals usually 1 + 2, occasionally 1 + 1, 1 + 3, or 2 + 3; ventrals 153 to 162 (136 to 168) in males, 156 to 168 (155 to 174) in females; subcaudals 42 to 50 (39 to 50) in males, 30 to 42 in females; ventrals plus subcaudals 189 to 205. Dorsal ground color brown to gray brown; uniform or with a series of twenty to thirty light-outlined quadrangular blotches alternating with smaller lateral blotches or variously spotted with dark; no dark temporal stripe; venter marked with dark at least along free margin of each ventral, sometimes heavily spotted with dark or with a median dark stripe (venter with little or no dark spotting in some non–Costa Rican individuals); iris light brown. Hemipenes long (17 to 20 subcaudals in length), with large basal hooks.

SIMILAR SPECIES: (1) *Stenorrhina freminvillii* has an immaculate venter or only a few scattered ventral dark marks, is usually marked with dark dorsal stripes, and never has a blotched or spotted dorsal pattern. Also see table 11.1.

HABITAT: In Atlantic and southwestern Pacific Lowland Moist and Wet Forests and Premontane Wet Forest; a few marginal records from Premontane Moist Forest and Rainforest.

BIOLOGY: This uncommon secretive semifossorial snake appears to be diurnal. It apparently feeds primarily on scorpions and tarantulas but also eats other spiders, crickets, grasshoppers, and insect larvae. Small juveniles are somewhat less than 250 mm in total length.

REMARKS: Several authors (Smith and Taylor 1945; Roze 1966; Peters and Orejas-Miranda 1970) have recognized geographic races within this species based on the presence or absence of a temporal dark stripe and differences in ventral coloration. Others regard the differences as reflecting individual variation (Duellman 1963b; Stuart 1963) or are unconvinced that the geographic pattern has been properly analyzed (Wilson and Meyer 1985). The three putative forms are said to differ as follows:

1. Mexico and Guatemala: no temporal dark stripe; venter with black spots

2. Panama, western and northern Colombia, western Ecuador, and northwestern Peru: a temporal dark stripe; venter with dark markings

3. Northern Venezuela: no temporal dark stripe; venter grayish with light spots

As pointed out by Wilson and Meyer (1985) for Honduran examples and confirmed by examination of Costa Rican specimens, the ventral coloration is extremely variable. Some

Central American individuals have an almost unmarked venter, others have a midventral row of brown to black spots, others have a dark mark across the free margin of each ventral, and in some the midventral dark markings are lined up to form a definite stripe. All examples lack a temporal dark stripe. Most juveniles are marked with dark crossbars, but some are uniform, and barred examples apparently may lose their markings with age. This variation suggests that the presumed differences among geographic divisions are based primarily on individual color variation. The specimen (USNM 9774) that Taylor (1951b) called *Stenorrhina degenhardtii apiata* is an example of *S. freminvillii*.

DISTRIBUTION: Lowlands and premontane slopes from southern Veracruz, Mexico, to north-central Venezuela, on the Atlantic versant; on the Pacific versant in upland Honduras and from central Costa Rica to northwestern Peru. In Costa Rica restricted to the humid lowlands and adjacent lower premontane slopes of the Atlantic slope and southwestern Pacific region (2–1,435 m).

Stenorrhina freminvillii C. Duméril, Bibron, and A. Duméril, 1854

(plate 451; map 11.104)

DIAGNOSTICS: *Stenorrhina freminvillii* cannot be confused with any other Costa Rican snake except *Stenorrhina degenhardtii*, with which it shares the enlarged rostral, shovel-shaped snout, and fusion of the internasal with the anterior nasal. Only uniformly colored specimens of the two species are likely to be hard to identify. Examples of *S. freminvillii* are often striped, never spotted or blotched, may be bright red or reddish orange in dorsal coloration, and have unmarked venters or a few dark spots along the midline (*S.*

degenhardtii are never striped, often have dark dorsal spots or blotches, never have a reddish orange dorsal ground color, and have the venters in uniformly colored individuals extensively marked with dark).

DESCRIPTION: A moderate-sized snake, total length of adults 490 to 800 mm; adult males 328 to 477 mm in standard length, adult females 317 to 473 mm); body stout, cylindrical; head not distinct from neck; tail moderate (15 to 18% of total length); loreal usually present, sometimes fused with postnasal; prefrontals usually separated from supralabials; 1 to 3 postoculars, usually 2; supralabials 6 to 8, usually 7, usually 2 bordering orbit; usually 7 infralabials, sometimes 6 or 8; temporals usually 1 + 2, sometimes 1 + 3; ventrals 164 to 176 (157 to 176) in males, 172 to 183 (163 to 183) in females; subcaudals 34 to 39 (33 to 47) in males, 30 to 38 (24 to 42) in females; 203 to 217 ventrals plus subcaudals; Dorsum gray, brown, reddish brown, reddish orange, or red (orange to red in some non–Costa Rican specimens), uniform or with one to five narrow longitudinal dark stripes that may be discontinuous along length of body; large adults usually uniform dorsally; a dark temporal stripe usually present; venter usually immaculate, sometimes with scattered dark spots along midline; tail immaculate or often with a dark stripe down midline; iris brown to black; tongue red basally, tips of fork black. Hemipenes moderate (11 to 13 subcaudals long), without large basal spines.

SIMILAR SPECIES: (1) *Stenorrhina degenhardtii* may have a dark spotted or blotched dorsum and are never striped; uniformly colored examples have extensive dark ventral markings. (2) Juvenile *Clelia* have a light nuchal collar. (3) *Ninia sebae* has a light nuchal collar.

HABITAT: Lowland Dry Forest and Premontane Moist and Wet Forests of the northern Pacific slopes; also at scattered localities in lowland Moist and Wet Forests and Premontane Moist Forest on the northern Atlantic slopes.

BIOLOGY: An uncommon secretive fossorial species that nevertheless is diurnal. Food consists primarily of scorpions and tarantulas (Censky and McCoy 1988). These authors described the reproductive cycle for the species from the seasonally dry forest area of the northern Yucatán Peninsula. Conditions in that region closely resemble those for the lowland sites where *S. freminvillii* occurs in Costa Rica. Females were reproductively active between August and April and laid eggs during the driest part of the year, between October and April. In November females with a body length of 448 mm or more were sexually mature and had clutches of five to nineteen eggs 15 to 28 × 10 mm. Clutch size increased with body size, and it appears that two clutches are produced annually. The smallest female hatchling (129 mm in body length) was captured in March. Juveniles reach a length of 490 mm in one year, so sexual maturity is attained in eleven to twelve months. A mostly digested example and four eggs of this species were found in the stomach of a coral snake (*Micrurus nigrocinctus*, CRE 3022). The red individ-

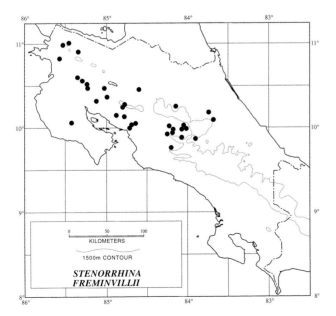

Map 11.104. Distribution of *Stenorrhina freminvillii*.

uals of this species are one of the kinds of snakes called viboras de sangre in Costa Rican folklore. According to this erroneous belief, anyone bitten by a "blood snake" rapidly dies from loss of blood through the skin surface.

REMARKS: No consistent pattern of geographic variation occurs in this species, contra Smith and Taylor (1945). The supposed differences in color pattern these authors used are based on individual variation, as pointed out by Stuart (1963) and Wilson and Meyer (1985) among others. The Costa Rican specimen (USNM 9774) that Taylor (1951b) called *Stenorrhina degenhardtii apiata* is *S. freminvillii.* Contrary to Peters and Orejas-Miranda (1970) and Villa, Wilson, and Johnson (1988), this species is not known from Panama.

DISTRIBUTION: An inhabitant of the semiarid and sub-humid lowlands and premontane slopes on the Atlantic versant from the Isthmus of Tehuantepec to central Honduras and from Guerrero, Mexico, south to the Meseta Central Occidental of Costa Rica on the Pacific slope; also scattered records in Costa Rica on the Meseta Central Oriental and northeastern Atlantic lowlands near the base of the Cordillera Central and Cordillera de Tilarán (3–1,435 m).

Genus *Tantilla* Baird and Girard, 1853

A large genus (fifty-six species) of mostly tiny or small (adults about 500 to 727 mm in total length) secretive fossorial snakes most commonly found under surface debris and in the leaf litter. The following combination of features characterize the genus: the basic colubrid complement of enlarged head shields; body cylindrical; head not distinct from neck; nasal divided; usually no loreal, but prefrontal may be separated from supralabials by the postnasal or an elongate postnasal (loreonasal) scale and the preocular (fig. 11.43); usually one preocular (absent in one non–Costa Rican species); eye small, pupil round; two pairs of chin shields, posterior ones short; dorsal scales smooth, in 15-15-15 rows; no apical scale pits; anal divided; subcaudals divided; 10 to 21 equal anterior maxillary teeth separated by a diastema from two enlarged, grooved fangs; Duvernoy's gland present; mandibular teeth equal; no hypapophyses on posterior dorsal vertebrae. No other small Costa Rican snakes have smooth scales in 15 rows, a divided anal, a preocular, and no loreal.

The hemipenes are symmetrical, single, with a simple semicentripetal sulcus spermaticus. The organ may be non-capitate or semicapitate with spines and/or hooks and calyces or capitate with spines and calyces (Cole and Hardy 1981).

Members of the genus feed primarily on small arthropods, primarily centipedes. Wilson (1982) summarized generic features and literature and provided a key to recognized species. That review was supplemented with additional records by Wilson and McCranie (1984) and a revised checklist and key by Wilson (1999).

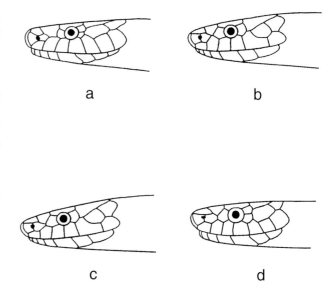

Figure 11.43. Variation in lateral head scalation in Costa Rican snakes of the genus *Tantilla:* (a) postnasal and preocular widely separated by prefrontal, no loreal; (b) postnasal and preocular narrowly separated by prefrontal, no loreal; (c) postnasal and preocular barely in contact, no loreal; (d) nasoloreal and preocular in broad contact.

Tantilla occurs over a wide range of altitudes and habitats, including lowland to montane sites throughout most of warm temperate North America, Mexico, Central America, and South America to central Peru on the Pacific slope and to southern Peru, Bolivia, Uruguay, and east-central Argentina on the Atlantic versant; also on Trindad and Tobago.

However, no one has made a comprehensive attempt to establish intrageneric relationships or a definitive clustering of allied forms into species groups. Several phenetically based groupings of similar species have been proposed by Wilson and Meyer (1971, 1981), Wilson and Mena (1980), and Wilson (1999). Where applicable, these subdivisions are indicated in the species accounts below.

KEY TO THE COSTA RICAN SPECIES OF THE GENUS *TANTILLA*

1a. Dorsum uniform brown or gray or with light and/or dark longitudinal stripes 2

1b. Dorsum red (in life) in adults with distinct black cross-bands having yellow centers, sometimes broken up on anterior part of body forming black-outlined yellow spots; juveniles bicolor with narrow yellowish bands between broad black bands; ventrals plus subcaudals 190–215. . . . . *Tantilla supracincta* (p. 696)

2a. Dorsum essentially uniform brown or gray, sometimes with a vague middorsal and lateral light lines or an obscure middorsal dark stripe 3

2b. Dorsum boldly marked with distinct single middorsal dark stripe at least anteriorly or definite light and/or dark longitudinal stripes 4

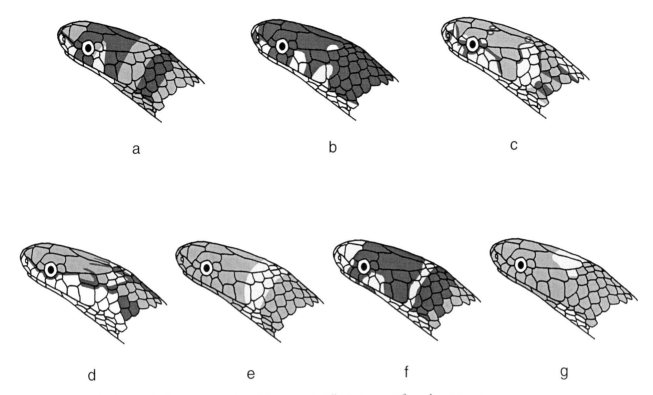

Figure 11.44. Head coloration in Costa Rican snakes of the genus *Tantilla*. Patterns are figured on a generalized template of head plates for the genus and do not accurately depict scalation differences among the species: (a) *T. alticola*; (b) *T. armillata*; (c) *T. reticulata*; (d) *T. ruficeps*; (e) *T. schistosa*; (f) *T. supracincta*; (g) *T. vermiformis*. Dark shading indicates dark brown or black; medium shading, light brown; light shading, yellow or red *(T. supracincta)*; no shading, white.

3a. Subcaudals 42–57; a definite postocular light spot on supralabials, usually a preocular light spot as well (fig. 11.44a); nuchal light collar complete or interrupted on midline; ventrals plus subcaudals 174–206 (fig. 11.44a). . . . . *Tantilla alticola* (p. 691)

3b. Subcaudals 28–36; no postocular or preocular light spots on supralabials (fig. 11.44e); nuchal light collar complete; ventrals plus subcaudals 164–75 . . . . *Tantilla schistosa* (p. 695)

4a. Ventrals plus subcaudals 200–241; ventrals 144–74 5

4b. Ventrals plus subcaudals 140–50; ventrals 115–29; a middorsal dark stripe; no nuchal light collar (fig. 11.44g) . . . . *Tantilla vermiformis* (p. 697)

5a. A middorsal dark stripe present; nuchal light collar reduced to a pair of spots centered on last supralabials or restricted to parietals, although a light postnuchal collar may be present 7–8 scale rows posterior to parietal shields; ventrals plus subcaudals 200–226 6

5b. Dorsum with a light middorsal stripe and several dark stripes; nuchal collar well developed but interrupted on midline by a dark area (fig. 11.44c); ventrals plus subcaudals 219–41. . . . . *Tantilla reticulata* (p. 693)

6a. Top of head dark brown or black with a distinct small light spot on each parietal or parietal and postparietal (fig. 11.44b); ventrals 163–74; subcaudals 48–59; a well-developed light postnuchal collar (interrupted on midline by black) located 4–8 rows posterior to parietal shields (fig. 11.44b). . . . *Tantilla armillata* (p. 692)

6b. Top of head light brown, no distinct small light parietal spot, although remnant of light nuchal collar may be indicated variously on parietals, postparietals, and temporals (fig. 11.44d); ventrals 142–56; subcaudals 64–83; rarely with a light postnuchal collar suggested (fig. 11.44d) . . . . *Tantilla ruficeps* (p. 694)

Tantilla alticola (Boulenger, 1903)

(plate 452; map 11.105)

DIAGNOSTICS: This relatively rare species is readily distinguishable by its immaculate red venter, narrow light nuchal collar (sometimes broken on the midline), and postocular light spot (fig. 11.44a) combined with an essentially uniform brown dorsum (fig. 11.33e). The presence of 15 rows of smooth dorsal scales will separate this form from somewhat similar small fossorial snakes of other genera that share some of these features.

DESCRIPTION: A small species (maximum total length 327 mm); tail moderately long (20 to 27% of total length); postnasal and preocular widely separated, barely separated, or in narrow contact with one another so that prefrontal

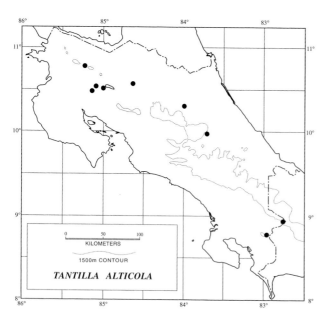

Map 11.105. Distribution of *Tantilla alticola*.

broadly meets, barely meets, or is separated from second supralabial, respectively (fig. 11.43a–c); 2 postoculars; 6 to 7 supralabials, usually 7, with 2 bordering orbit; infralabials 6; temporals 1 + 1 or 1 + 1 + 1; ventrals 132 to 149; subcaudals 42 to 57 in males (42 to 60), 44 (42 to 57) in females; ventrals plus subcaudals 172 to 206 (160 to 206). Dorsal ground color brown, sometimes with faint paling of color on middorsal scale row and rows 3 and 4 or 4 and 5; snout area may be lighter than rest of upper head or suffused with dark pigment; nuchal collar narrow, involving tips of parietals, one to one and one-half scale rows wide or interrupted by dark pigment on midline; usually a distinct preocular light spot on supralabials; a distinct light postocular spot present; venter immaculate bright red.

SIMILAR SPECIES: (1) *Tantilla schistosa* lacks pre- and postocular light spots (fig. 11.44e) and has fewer subcaudals (28 to 36). (2) *Trimetopon gracile* has obscure dorsal dark stripes and a beige to white venter.

HABITAT: Under surface debris in Lowland Moist Forest and Premontane Wet Forest and Rainforest and marginally in Lower Montane Rainforest; one record from a gallery forest site within the Lowland Dry Forest.

REMARKS: This rare species was long confused with *Tantilla schistosa*, but Wilson (1982) amply distinguished the two. The holotype (KU 30995) of Taylor's (1951b) *Tantilla costaricensis* is a male *T. alticola* with pre- and postocular light spots, the internasal area light, 142 ventrals, and an incomplete tail. Wilson (1982) emphasized the light internasal and preocular markings found in many specimens of this species to distinguish it from *T. schistosa*. Additional material indicates a greater variability in head pattern than Wilson described. Two Costa Rican examples (MCZ 15303 from Miravalles, Guanacaste Province, and CRE 7317 from

Cascante Camp, Braulio Carrillo National Park, Heredia Province) are typical *T. alticola* except that internasal and preocular light spots are absent. Wilson (1986) provided a redescription and literature summary for this species.

DISTRIBUTION: Lowland and premontane humid forest areas on Atlantic slope Nicaragua and Costa Rica, Pacific slope southwestern Costa Rica and southwestern Panama (55–1,600 m), and lowland to lower montane sites in western Colombia (70–2,743 m); one record in gallery forest in northwestern Costa Rica (40 m).

Tantilla armillata Cope, 1875

(plate 453; map 11.106)

DIAGNOSTICS: This beautiful little snake is readily separated from all other Costa Rican species by having a striped pattern and the dark brown to black head cap coloration continuing for three and one-half to seven scale rows onto the body (fig. 11.44b), where it is followed by a narrow light postnuchal collar that is interrupted by dark on the middorsal scale row.

DESCRIPTION: A small species reaching a maximum total length of 490 mm; tail moderate (18 to 23% of total length in males, 15 to 21% in females); postnasal and preocular broadly separated (13%) (fig. 11.43a), narrowly separated (40%) (fig. 11.43b), barely in contact (34%) (fig. 11.43c), or broadly in contact (13%) (fig. 11.43d) so that prefrontal does not meet second supralabial or barely or broadly meets it; 2 postoculars; supralabials 5 to 8, usually 7, with 2 bordering orbit; infralabials 4 to 7, usually 6; temporals 1 + 1 or 1 + 1 + 1; ventrals 163 to 167 (155 to 174) in males, 169 to 174 (158 to 177) in females, subcaudals 51 to 59 (46 to 59) in males, 48 to 52 (42 to 60) in females; ventrals plus subcaudals 209 to 221. Dorsum dark brown, with a narrow

Map 11.106. Distribution of *Tantilla armillata*.

dark brown middorsal stripe and usually a yellow stripe on scale rows 3 and 4; head cap dark brown to black, extending well onto neck, followed by a narrow yellow postnuchal collar (interrupted dorsally by middorsal dark stripe), one small pale spot on posterior part of each parietal, sometimes involving a postparietal scale; preocular, postocular, and nuchal light spots, light spots on supralabials and adjacent scales; venter immaculate whitish; iris brown; tongue black. A partially everted short (about 7 subcaudals long) hemipenis of one specimen (CRE 4631) lacks basal hooks and appears to be semicapitate, with the truncus covered by spines and the apical area by papillate calyces.

SIMILAR SPECIES: (1) *Tantilla ruficeps* lacks the distinct light spots on the head (fig. 11.44d) and has a prominent dark stripe just below the lateral light stripe.

HABITAT: Found under rocks, logs, debris, and leaf litter in Lowland Dry and Moist Forests and Premontane Moist and Wet Forests, often in pastures or disturbed sites.

REMARKS: A member of the *melanocephala* group (Wilson and Mena 1980; Wilson 1999). Wilson and Mena (1980) and Wilson (1992) regarded this rather common species and its presumed close relative, *Tantilla ruficeps*, as geographic variants of the (according to them) wide-ranging and variable *Tantilla melanocephala* (Linné), originally based on a South American snake. *T. armillata* and *T. ruficeps*, as recognized here, were shown by Wilson and Mena to be discrete geographic forms characterized by consistent differences in coloration and segmental counts. Individuals of *T. armillata* have color pattern A (in Wilson and Mena's system), a high number of ventrals (155 to 174 in males, 158 to 177 in females), and a low number of subcaudals (46 to 59 in males, 48 to 52 in females) and range from Guatemala to central Costa Rica, primarily in subhumid to semiarid areas on the Pacific versant. *T. ruficeps* has color pattern B (in Wilson and Mena's system), 142 to 148 ventrals in males, 143 to 156 in females and 64 to 83 subcaudals in males, 59 to 75 in females and occurs in humid habitats from eastern Nicaragua and southwestern Costa Rica through western Panama. The two forms, whose ranges meet and overlap slightly in central Costa Rica, show no sign of intergradation in coloration or scale counts at that juncture, and every specimen from the republic may be assigned without question to one or the other species. Wilson and Mena (1980) believed the two were conspecific because *Tantilla* presumed to be members of this complex from central and eastern Panama and some from northern and western Colombia were said to resemble *T. armillata* in coloration and *T. ruficeps* in segmental counts. They thus concluded that *T. armillata* and *T. ruficeps* intergrade not where the ranges of the two forms meet in Costa Rica, but over a substantial area east of the range of *T. ruficeps* and 500 km from the nearest *T. armillata* record to the west! Besides this incongruity, it appears that although the "intergrades" resemble *T. ruficeps* in segmental counts, their col-

oration, while perhaps most similar to that of *T. armillata*, differs markedly from both in usually having a definite light nuchal collar involving the posterior area of the parietals and adjacent portions of the posterior temporal and one postparietal scale. In the other nine features of coloration listed by Wilson and Mena (1980) that show differences, the "intergrades" resemble *T. armillata* in four conditions and *T. ruficeps* in three and have two unique conditions. One wonders why the cephalic and nuchal pattern of the "intergrades" was not considered as distinctive as the other six patterns delineated by these authors for their concept of *T. melanocephala*. Whatever the status of the central and eastern Panamanian and Colombian snakes that Wilson and Mena called intergrades, the fact remains that in Costa Rica, where the ranges of *T. armillata* and *T. ruficeps* meet, the two behave as distinct species. *T. armillata* is further distinguished from *T. ruficeps* (features for the latter in parentheses) by having a whitish venter and in lacking hooks on the hemipenis (venter red, hemipenial hooks present). In view of these data, the populations referred to *T. melanocephala* from Guatemala to western Panama are considered to constitute two distinct species, *T armillata* and *T. ruficeps*, that are marginally sympatric in central Costa Rica. Taylor's (1954b) record for *T. armillata* from southwestern Costa Rica is based on an example (KU 31960) of *T. ruficeps*.

DISTRIBUTION: Drier areas of the lowlands and premontane slopes on the Pacific versant from Guatemala to the Meseta Central Occidental of Costa Rica and in the Atlantic drainage of Honduras (mostly in the uplands), also upland Nicaragua and the Meseta Central Oriental of Costa Rica (4–1,435 m).

Tantilla reticulata Cope, 1860a
(plate 454; map 11.107)

DIAGNOSTICS: This strikingly marked little snake has a pattern of alternating light and dark longitudinal stripes and a broad light nuchal collar that is interrupted dorsally by a dark stripe (fig. 11.44c). These features will immediately separate it from all other Costa Rican snakes.

DESCRIPTION: A small species attaining a maximum total length of 312 mm; tail moderate (21 to 24% of total length); postnasal and preocular variously (barely or broadly) in contact or separated, so that prefrontal variously (barely or broadly) separated from or in contact with second supralabial (fig. 11.43); generally postnasal meeting preocular on one (14%) or both (55%) sides; 2 postoculars; supralabials 7, with 2 bordering orbit; 6 infralabials; temporals 1 + 1; ventrals 158 to 159 in males, 162 to 173 in females; subcaudals 60 to 67 in males, 59 to 70 in females; ventrals plus subcaudals 219 to 241. Multilineate dorsal color pattern with a broad middorsal light stripe and two or three pairs of narrow dark stripes; dark pigment on anteriolateral margins of scales in light areas often produces reticulate appearance; head cap light brown, bordered by dark

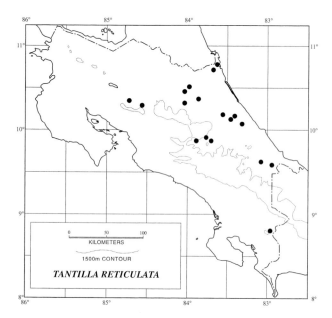

Map 11.107. Distribution of *Tantilla reticulata*.

brown to black posteriorly, followed by a broad white or yellow nuchal collar broadly interrupted dorsally by extension of head cap and bordered posteriorly by dark pigment; pre- and postocular spots present; venter immaculate yellow except for dark stripe running length of body on ventral tips. The everted hemipenis (USC 1265) very short (about 6 subcaudals in length), noncapitate, and with numerous hooks.

SIMILAR SPECIES: (1) *Tantilla ruficeps* has a middorsal dark stripe.

HABITAT: Under leaf litter and moss in undisturbed Atlantic versant Lowland Moist and Wet Forests and Premontane Wet Forest and Rainforest; one record from Premontane Rainforest in extreme southwestern Costa Rica. Possibly also occurring in Premontane Moist Forest (see remarks).

BIOLOGY: A rare snake usually found under leaf litter on the forest floor, but one example was taken 6.1 m above the ground from under moss on a tree limb (Wilson and Meyer 1971).

REMARKS: This species has been cited (Wilson and Meyer 1971; Wilson 1982, 1985a) as occurring in Nicaragua based on an example (USNM 19565) from "Colorado Junction." This locality is where the Río Colorado diverges from the Río San Juan and lies on the Costa Rican side of the boundary with Nicaragua. However, the species does occur in Nicaragua, as authenticated by an example (AMNH 12698) from Backas Creek, Department of Atlántico Sur, eastern Nicaragua. *Tantilla virgata* listed by Taylor (1951b) is a synonym of *T. reticulata*. The types (BMNH 1946.1.8.73–76) of *Microdromus virgatum* (Günther 1872) are purportedly from Cartago, Cartago Province. They were probably collected at a lower elevation on the Atlantic slope. A record for the city of San José, San Jose

Province, is the only specimen (ANSP 3364) recorded from the Meseta Central Occidental. Wilson and Meyer (1971) placed this form in the *taeniata* species group. Wilson (1985a) provided a redescription and literature review for this taxon.

DISTRIBUTION: Lowlands and lower reaches of premontane slopes from eastern Nicaragua and southwestern Costa Rica to western Colombia (10–1,200 m); possibly occurring at higher elevations on the Meseta Central of Costa Rica (to 1,345 m).

Tantilla ruficeps (Cope, 1894a)

(plate 455; map 11.108)

DIAGNOSTICS: A relatively common small variably striped species that always has a narrow middorsal dark stripe and a pair of lateral light stripes that are bordered below by a dark stripe, the nuchal collar reduced to a pale area on each side of the nape or absent (fig. 11.44d), and a red venter. An additional pair of narrow light stripes and one or two additional pairs of dark stripes may be present.

DESCRIPTION: A small snake with a maximum total length of 500 mm; tail moderate to long (23 to 31% of total length); postnasal and preocular usually widely separated from one another, and prefrontal broadly contacting second supralabial (fig. 11.43a) (sometimes narrowly separated with a narrow prefrontal-supralabial contact) (fig. 11.43b); 2 postoculars; supralabials 7, with 2 bordering orbit; usually 6 infralabials, temporals 1 + 1 or 1 + 1 + 1; ventrals 142 to 148 in males, 143 to 156 in females; subcaudals 65 to 83 in males, 66 to 75 (59 to 75) in females; ventrals plus subcaudals 206 to 226. Dorsum pale brown, with a narrow middorsal dark stripe, one or two pairs of lateral light stripes variously bordered above or below by dark stripes;

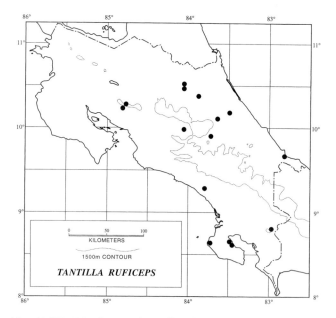

Map 11.108. Distribution of *Tantilla ruficeps*.

Snakes (Suborder Serpentes) 695

head cap tan; remnants of nuchal collar vaguely indicated as light areas on parietal, postparietal, and temporal shields or absent, never continuous across neck; dark nuchal band often indicated, rarely bordered posteriorly by indication of light postnuchal collar; supralabials light except along upper margins, or pre- and postocular and nuchal light spots present; anterior venter white, center of venter immaculate red posteriorly; tips of ventrals with black mark, often fusing to form ventral stripes; iris dark brown; tongue black. The everted hemipenis is moderately short (about 7 subcaudals in length), noncapitate, with basal hooks (CRE 4645).

SIMILAR SPECIES: (1) *Tantilla reticulata* has a middorsal light stripe and distinct nuchal collar divided by a middorsal bar (fig. 11.44c). (2) *Tantilla armillata* has the dark head cap continuing back onto the neck, where it is bordered by a light postnuchal collar and has distinct small parietal light spots (fig. 11.44b). (3) *Trimetopon* lack light lateral stripes.

HABITAT: Under leaf litter and other debris in undisturbed Lowland Moist and Wet Forests, Premontane Moist and Wet Forests, and marginally in Lower Montane Wet Forest.

BIOLOGY: An essentially diurnal arthropodivorous secretive snake (Cadle and Greene 1993).

REMARKS: As I pointed out in discussing the status of *Tantilla armillata*, Wilson and Mena (1980) and Wilson (1992) in their reviews of the *Tantilla melanocephala* species group regarded this form and *T. ruficeps* as geographic variants of the supposedly wide-ranging and variable species *Tantilla melanocephala*. As demonstrated above, the two forms are amply distinguished, are sympatric in distribution where their ranges meet, and behave as distinct species in Costa Rica. For this reason I treat the two as separate species in this account. Far (500 km) to the south and east of the range of *T. armillata*, in central Panama, typical *T. ruficeps* are replaced geographically by snakes that have similar segmental counts but a color pattern that somewhat resembles that of *T. armillata*. Wilson and Mena (1980) regarded these individuals and similar ones from Colombia as indicating intergradation between *T. armillata* and *T. ruficeps*. They suggest that the entire range of the latter form (eastern Nicaragua and southwestern Costa Rica to western Panama) lies between the range of the former (Guatemala to the central Pacific versant of Costa) and that of the "intergrades" in eastern Panama, a most unusual interpretation to say the least! It does seem likely that the "intergrades" from central and eastern Panama and western Colombia are conspecific with *T. ruficeps*. It is also possible that these geographic variants are part of a population system of distinctive allopatric forms, connected by intermediates, that constitute a single widely distributed species for which the earliest name is *T. melanocephala*, as proposed by Wilson and Mena (1980). However, the extreme differences in segmental counts and coloration within that population sys-

tem, the apparent sympatry of several morphs in Colombia, and those authors' failure to recognize the uniqueness of *T. armillata* lead me to question their conclusions. Since the issue cannot be resolved here, I have treated *T. ruficeps* as a valid species restricted in distribution to Nicaragua, Costa Rica, and western Panama. Others may wish to regard it as an allopatric variant of *T. melanocephala* as I did in earlier publications (Savage 1980b; Savage and Villa 1986). The snake (KU 31960) identified as *T. armillata* by Taylor (1954b) from the Dominical Road southwest of San Isidro de El General, San José Province, near Alfombra, is *T. ruficeps*.

DISTRIBUTION: Lowlands and premontane zone of the Atlantic versant of Nicaragua and Costa Rica; Pacific slope of the Cordillera de Tilarán, the Meseta Central, southwestern Pacific areas of Costa Rica, and western Panama (1–1,600 m).

Tantilla schistosa (Bocourt, 1883)

(plate 456; map 11.109)

DIAGNOSTICS: This snake is one of a number of little Costa Rican snakes with a uniform brown dorsum and a complete light nuchal collar (fig. 11.33f). It is separated from similar forms by having an immaculate red venter and lacking pre- and postocular light spots (fig. 11.44e).

DESCRIPTION: A very small species (maximum total length 350 mm); tail length moderate in males (17 to 20% of total length), short in females (12 to 16%); loreonasal almost always meeting preocular and separating prefrontal from second supralabial (fig. 11.43d); postoculars 1 or 2, usually 2; supralabials usually 7, sometimes 6, with 2 bordering orbit; infralabials 5 or 6, usually 5 (usually 6 north

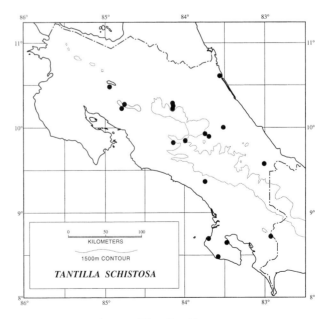

Map 11.109. Distribution of *Tantilla schistosa*.

of Nicaragua); temporals 1 + 1 or 1 + 1 + 1; ventrals 130 to 141 (121 to 141) in males, 117 to 141 (117 to 146) in females; subcaudals 33 to 36 (33 to 42) in males, 28 to 36 (26 to 40) in females; ventrals plus subcaudals 164 to 175 (151 to 186). Dorsum uniform brown, with or without slight paling of middorsal scale row; a complete light nuchal collar crossing parietals and extending one-half to two scales onto neck (divided on midline by dark pigment in some Mexican examples), often suffused with brown pigment; no well-defined pre- or postocular light spots; venter immaculate red, with some encroachment of dorsal coloration onto ventral tips. Hemipenes moderate (about 8 subcaudals in length), semicapitate, with basal hooks (CRE 4507).

SIMILAR SPECIES: (1) *Tantilla alticola* has postocular and usually preocular light spots on the supralabials (fig. 11.44a) and has more subcaudals (42 to 57). (2) *Tantilla vermiformis* has a dark middorsal stripe and lacks a nuchal collar (fig. 11.44g). (3) *Trimetopon slevini* has obscure dark dorsal stripes.

HABITAT: Under debris in undisturbed lowland Moist and Wet Forests and Premontane Wet Forest and Rainforest.

BIOLOGY: This rare species feeds primarily on centipedes and also insect larvae (Seib 1985b).

REMARKS: Wilson (1982) clarified the status of this species and pointed out the confusion surrounding Costa Rican specimens referred by previous authors to *T. schistosa* that are representatives of *Tantilla alticola*. He further demonstrated that individual and clinal variation in this form precludes recognition of geographic races as attempted by Smith (1962). Wilson (1987a) provided a summary of previous publications on this species, but it is largely redundant with his account of 1982. The holotype (KU 30995) of the nominal species *Tantilla costaricensis*, described by Taylor (1954b) and regarded by Smith (1962) as a race of *T. schistosa*,, is actually an example of *T. alticola*. An example (UCR 7601) from Cangrejal de Acosta, San José Province, is unusual in having a vague middorsal dark stripe on the posterior half of the body. It is clearly aberrant, since the anal and first six subcaudals are undivided.

DISTRIBUTION: Lowlands and premontane slopes of the Atlantic versant from Veracruz and Oaxaca, Mexico, to Costa Rica and from Costa Rica to central Panama on the Pacific versant (40–1,400 m).

Tantilla supracincta (Peters, 1863c)

(plates 457–58; map 11.110)

DIAGNOSTICS: A member of the coral snake mimic guild with a banded dorsal pattern of black-light-black figures (dyads) on a red ground color in adults and a uniform red venter. Other similarly colored Costa Rican snakes have the black markings forming rings that completely encircle the body. Juveniles have narrow cream to yellow, often offset bands separated by much broader black bands.

DESCRIPTION: A small species attaining a maximum

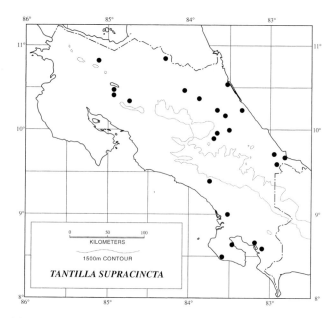

Map 11.110. Distribution of *Tantilla supracincta.*

total length of 590 mm; tail moderately long (19 to 24% of total length); loreonasal and preocular usually in contact, prefrontal usually separated from second supralabial (fig. 11.43d); 2 postoculars; supralabials 7, with 2 bordering orbit; infralabials 6; temporals 1 + 1 or 1 + 1 + 1; ventrals 138 to 151 (138 to 155) in males, 142 to 148 (141 to 153) in females; subcaudals 52 to 62 (52 to 65) in males, 61 to 65 (52 to 65) in females; ventrals plus subcaudals 190 to 215. Dorsum red in adults with 11 to 16 dyad (black-light-black) bands that are frequently offset and/or divided by black on midline and rarely restricted to anterior part of body; tail marked like posterior body; black head cap extends from posterior tips of prefrontals to posterior tips of parietals, bounded posteriorly by complete or medially divided light nuchal collar; most of snout and area anterior to eye light; definite large postocular light spot present (fig. 11.44f); venter and underside of tail red in adults; juveniles are black with narrow cream to yellow dorsal markings; iris brown; tongue black. The everted hemipenis is short (5 to 6 subcaudals long), noncapitate, with hooks (CRE 3697).

SIMILAR SPECIES: (1) *Erythrolamprus bizona* has the black and light dyads encircling the body. (2) Tricolor examples of *Lampropeltis triangulum* have the black and light dyads encircling the body. See key to coral snakes (p. 706), figure 11.47, and tables 11.1 and 11.2.

HABITAT: Under logs and other debris in undisturbed Lowland Moist and Wet Forests and the lower margins of Premontane Wet Forest and Rainforest.

BIOLOGY: These rather uncommon small snakes appear to eat only centipedes. One example (CRE 6243B) caught in the process of being devoured by an *Erythrolamprus bizona* contained a large centipede. Another was palped from the stomach of a specimen of *Clelia clelia* at La Selva, Heredia Province (E. D. Brodie III, pers. comm.).

REMARKS: This species parallels *Scaphiodontophis annulatus*, a fellow coral snake mimic, in having an ontogenetic change in coloration. Juveniles of *T. supracincta* are bicolor black and cream to yellow with the black bands as long as the combined extent of the black and red bands in adults. Apparently the underlying red bands are obscured at this stage by an overlying layer of black pigment (plate 458). With growth the suffusing black is reduced and concentrated toward the middle of each scale or nearly lost as the broad red bands become more evident. In this process the black is reduced to narrow bands on either side of the frequently offset yellow bands. The name *Tantilla annulata* Boettger, 1892b, was applied to this species before 1987. At that time L. D. Wilson (1987c) showed that the holotype (ZMB 4791) of *Tantilla supracincta* from Ecuador was conspecific with the holotype of *T. annulata* (Lubeck Museum) from Nicaragua. A redescription based on Wilson's 1982 account of this species is provided by Wilson (1985b) under the name *T. annulata*.

DISTRIBUTION: Lowlands and lower margins of the premontane slopes of the Atlantic versant from Nicaragua to central Panama; on the Pacific slope in the Cordillera de Tilarán and southwestern lowlands of Costa Rica south to western Ecuador; not yet recorded from Colombia (2–850 m).

Tantilla vermiformis (Hallowell, "1860," 1861)

(map 11.111)

DIAGNOSTICS: This little snake is rather nondescript in appearance but may be distinguished from similar forms by lacking a nuchal collar (fig. 11.44g) and striping except for a single narrow disjunct middorsal dark stripe.

DESCRIPTION: A tiny species (maximum total length 157 mm); tail short (9.6 to 15% of total length); loreonasal and preocular usually in contact, prefrontal usually not meeting second supralabial (fig. 11.43d) (preocular sometimes fused with prefrontal and supraocular in some Nicaragua examples); 2 postoculars; 7 supralabials, with 2 bordering orbit; 6 infralabials; temporals 1 + 1; ventrals 115 to 123 in males, 120 to 129 in females; subcaudals 23 to 28 in males, 19 to 24 in females; ventrals plus subcaudals 140 to 150. Dorsum uniform pinkish to brown with narrow disjunct median dark stripe; head brown with tan to yellowish mark on parietal or parietal and one postparietal; no distinct light spots on supralabials; venter immaculate white to pale yellow (pinkish in Nicaraguan specimens). The everted hemipenis is short (6 to 7 subcaudals long), capitate, with truncus covered by small spines (lacks hooks) and more distally by spinulate calyces; apical region nude (Van Devender and Cole 1977).

SIMILAR SPECIES: (1) *Tantilla schistosa* has a nuchal collar (fig. 11.44e) and a uniform brown dorsum.

HABITAT: Under rotting logs or stones or in the soil in Lowland Dry Forest and adjacent Lowland Moist Forest areas, including pastures, of northwestern Pacific slope Costa Rica.

BIOLOGY: The following information is primarily from Van Devender and Cole 1977). The species is relatively uncommon and usually found in the upper soil surface (3 to 5 cm) in the wet season, but it apparently retreats farther into the soil during the dry season. One female contained a single large (5 mm) egg in June. Small (smallest 71 mm) juveniles were taken in August. These data suggest that these snakes reproduce in the wet season and reach sexual maturity in one year. Food consists of small soft-bodied insects.

REMARKS: Van Devender and Cole (1977) and Wilson (1982, 1987b) provided detailed descriptions of this species. Wilson and Meyer (1981) and Wilson (1999) refer this form to the *calamarina* species group. The karyotype is 2N = 36, comprising three pairs of metacentric chromosomes, two pairs of submetacentrics, one pair of subtelocentrics, two pairs of submetaratrics or subtelocentrics, and ten pairs of michrochromosomes; NF = 46 or 50 (Van Devender and Cole 1977).

DISTRIBUTION: Pacific lowlands from El Salvador to northwestern Costa Rica (40–520 m).

Genus *Trimorphodon* Cope, 1861c

Members of this group (two species) are usually among colubrines in having the lateral head scales fragmented and variably numerous. The following combination of features characterize the genus: basic colubrid complement of enlarged dorsal head shields; body compressed; nasal divided; two to five (usually two to three) loreals; two to five preoculars (usu-

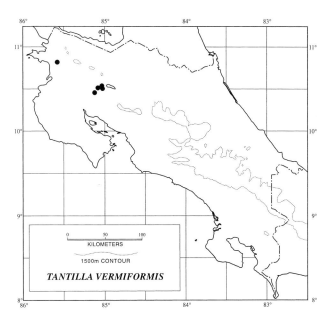

Map 11.111. Distribution of *Tantilla vermiformis*.

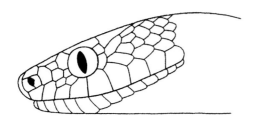

Figure 11.45. Lateral view of head scalation in *Trimorphodon* showing presence of three preoculars and two loreals.

ally three or four); eye moderate to large, pupil vertically elliptical; two pairs of elongate chin shields; dorsal scales smooth anteriorly, slightly keeled posteriorly, in 21 to 27 rows at midbody, with a scale row reduction posteriorly; two apical scale pits; anal divided or entire (almost always divided in Costa Rican examples); subcaudals divided; 10 to 12 maxillary teeth, anterior two or three much longer than others, which decrease in size to a diastema followed by one or two enlarged grooved fangs; Duvernoy's gland present; anterior mandibular teeth enlarged; no hypapophyses on posterior vertebrae. The presence of two or more loreals (fig. 11.45), combined with a blotched pattern, vertical evenly arranged dorsal scale rows in 23 or more rows at midbody, and the divided anal will separate the Costa Rican representatives of this genus from all other snakes found in the republic.

The everted hemipenes are moderate to long (14 to 25 subcaudals in length), single, symmetrical, attenuate, and noncapitate, with a simple semicentripetal sulcus spermaticus; pedicel naked or with small spines, proximal part of truncus covered with spines, followed by a spiny area (naked in some examples), several large spinulate calyces, and an apical enlarged calyx or naked pocket bordered by papillate spinules (Scott and McDiarmid 1984a).

Scott and McDiarmid (1984a) provided a description and literature review for the genus. *Trimorphodon* ranges from western Texas through much of the southwestern United States, in Baja California, and along the west coast and southern uplands of Mexico, south to northwestern Costa Rica.

It is absent from the Atlantic lowlands of Mexico and Central America except for a few scattered records in Guatemala, Honduras, and Nicaragua.

Trimorphodon biscutatus (C. Duméril, Bibron, and A. Duméril, 1854)

(fig. 11.45; plate 459; map 11.112)

DIAGNOSTICS: Any blotched snake with a definite dark chevron or lyre-shaped marking (with the apex directed anteriorly) on the posterior dorsal surface of the head is almost certainly of this species. Occasional specimens of other snakes that show an approach to this head pattern have round pupils, entire anal plates, and a single loreal scale and

may have all or most scales in strongly keeled oblique rows (figs. 11.4 and 11.5) or a loreal pit (fig. 11.8b).

DESCRIPTION: A large snake reaching a maximum total length of 1,660 mm; females longer than males; head distinct from neck; tail short to moderate (13 to 19% of total length); usually 2 or 3 loreals; usually 3 preoculars, often 4; usually 3 postoculars, often 4; supralabials 7 to 10, usually 9 often 8, with 2 bordering orbit; infralabials 11 to 14; dorsal scale rows at midbody 23 to 27 (21 to 27), posteriorly 17 to 21 (15 to 21), anteriorly 23 to 27 (21 to 27); anal divided (92%) or entire (8%) in Costa Rican examples (in most other populations anal divided or usually divided; anal entire in 92% of examples from southern California); ventrals 245 to 279 (215 to 279); subcaudals 75 to 102 (72 to 102) in males, 72 to 92 (61 to 92) in females. Dorsal ground color gray brown, tan, or brown; pattern of 18 to 24 complex dark and light figures consisting of two dark blotches (each usually partially split by a light area) partially or completely separated by a light area to form duplex (dark-light-dark) and quadruplex figures (dark-light-dark-light-dark-light-dark); always several quadruplex blotches in lower Central American specimens (in other populations single or duplex blotches predominant); dark chevron or lyre-shaped dark mark on upper head surface beginning on frontal shield with arms diverging posteriorly to behind angle of mouth (absent in some extralimital examples); venter yellow to white with moderate-sized dark spots laterally and considerable dark stippling; iris brown.

SIMILAR SPECIES: (1) *Elaphe triaspis* juveniles lack quadruplex markings and have 43 to 73 single body blotches; (2) *Xenodon rabdocephalus* has obliquely arranged dorsal scale rows (fig. 11.5) and a round pupil (fig. 11.4).

HABITAT: Usually found in or near rocky outcrops in

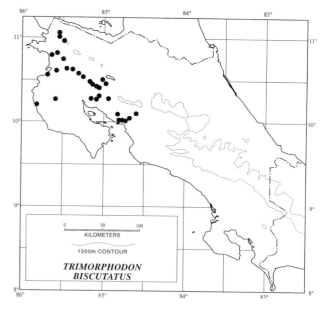

Map 11.112. Distribution of *Trimorphodon biscutatus*.

the Lowland Dry Forest and marginally into Lowland Moist Forest in the Cordillera de Tilarán in northwestern Costa Rica.

BIOLOGY: The species is relatively common, nocturnal, and terrestrial, although it sometimes climbs low shrubs and trees. In Costa Rica (based on stomach contents) juveniles feed on lizards and adults eat a variety of birds and small mammals, including bats. Small prey are envenomated, and larger prey are also constricted. A captive female from Colima, Mexico, was courted in November and laid twenty eggs the following March (Werler 1951). Egg clutches of seven to twenty eggs have been reported for this species in Arizona (Goldberg 1995).

REMARKS: Gehlbach (1971), Scott and McDiarmid (1984b), and Wilson and Meyer (1985) have described and discussed geographic variation in this snake. The principal differences among geographic subdivisions these authors emphasize involve the condition of the anal plate, ventral counts, nature and number of dorsal blotches, and details of head coloration. As Wilson and Meyer (1985) pointed out, the variation in ventrals shows a clinal pattern of gradual increase in numbers from north to south. Similarly, primary blotch numbers show a clinal increase from east to west and a decrease southward. Consequently, use of these features to distinguish among adjacent populations becomes arbitrary. The following subdivisions have been recognized based on other characteristics:

1. Western Texas, southern New Mexico, and northern Chihuahua, Mexico: mostly duplex blotches, some single; no light head chevron; light nape spots present or absent; no light snout bar; anal divided
2. Western New Mexico, Arizona, and southeastern California south to southern Sonora, Mexico, with disjunct populations in Utah, Nevada, and northern Arizona: mostly duplex body blotches, some single; light head chevron not connected to interocular bar; light nape spots present or absent; snout mottled; anal divided
3. Southern California and northern Baja California, Mexico: duplex body blotches; light head chevron usually connected to interocular bar; no light nape spots; snout pale or with light bar; anal usually entire (75%)
4. Southern and central Baja California, Mexico: duplex body blotches; light head chevron connected to interocular light bar; no light nape spots; snout pale or with light bar; anal divided
5. Sinaloa to western Chiapas, Mexico: most body blotches single, some duplex blotches in some examples; light head chevron not connected to interocular bar; light nape spots; snout pale, often with light bar; anal usually divided
6. Chiapas, Mexico, to Costa Rica: mostly quadruplex body blotches, some duplex ones; light head chevron not connected to interocular bar; no light bar on snout; anal usually divided

The slight differences in coloration between contiguous geographic segments and the variation found in these features within some of them indicate a mosaic rather than a concordant pattern of variation and presumably of character evolution. In these circumstances there seems little to be gained by formally recognizing these arbitrarily defined units, contrary to the conclusion of Scott and McDiarmid (1984b). The karyotype is 2N = 38, consisting of eight pairs of macrochromosomes (four pairs metacentric and four telocentric) and eleven pairs of microchromosomes, but lack a clear division between size categories, based on a male; NF = 46 (Bury, Gress, and Gorman 1970).

DISTRIBUTION: From Utah and Nevada southeastward to western Texas and north-central Mexico and southwestward to California and Baja California; in the lowlands of western Mexico and Central America to northwestern Costa Rica; also in a few dry areas in the Atlantic lowlands of Guatemala, Honduras, and Nicaragua (2–560 m).

Natricines
Genus *Thamnophis* Fitzinger, 1843

A diverse group (twenty-four species) of small (250 mm) to moderately large (1,450 mm) diurnal semiaquatic snakes that feed primarily on fishes, tadpoles, and frogs, although some forms specialize on slugs and earthworms. The genus is characterized by the following suite of features: the basic colubrid complement of enlarged head shields; nasal divided; a loreal (rarely absent); one to three preoculars; eye small to moderate, pupil round; two pairs of enlarged chin shields; dorsal scales keeled in 15 to 21–17 to 23–15 to 17 rows; no apical pits; anal entire; subcaudals divided; 14 to 34 maxillary teeth, last few usually abruptly enlarged, without a diastema or grooved fangs; Duvernoy's gland present; mandibular teeth subequal; short, recurved hypapophyses on all dorsal vertebrae. Among Costa Rican snakes the combination of keeled dorsal scales in 19 rows at midbody, entire anal, and a loreal will separate *Thamnophis* from all other snakes with the basic colubrid head shield complement.

The everted hemipenes are moderate to very long, single, and symmetrical, with a simple semicentripetal sulcus spermaticus not expanded apically or markedly expanded; the organ is noncapitate and is covered with spines and hooks only; there are no calyces.

So far as is known, all members of this stock are viviparous and have omphalallantoic and chorioallantoic placentas (Stewart and Blackburn 1988). The karyotype is 2N = 36, with no abrupt change from macrochromosmes to microchromosomes; females are heteromorphic (ZW) in the fifth chromosome pair (Baker, Mengden, and Bull 1972). All members of the genus use unspecialized pseudautotomy as a antipredator defense.

Recognition of intrageneric relationships in *Thamnophis* is in flux. There is no question that the ribbon snakes *(Tham-*

nophis proximus and *Thamnophis sauritus)* are not closely allied to *Thamnophis marcianus* and its close relatives. However, the vague limits of the *Thamnophis sirtalis* and *Thamnophis radix* groups and the ambiguous position of many recognized species as to group precludes further statements of group composition. Rossman, Ford, and Seigel (1996) disappointedly do not contribute to clarifying this situation.

The genus ranges over most of North and Central America except the most arid sections from British Columbia (57° N), the Northwest Territory (60° N), and Quebec (49° N), Canada, south to Costa Rica.

KEY TO THE COSTA RICAN GARTER SNAKES OF THE GENUS *THAMNOPHIS*

1a. Vertical black marks present along some supralabial sutures; definite dark spots or crossbars on venter (fig. 11.46); lateral light stripe rarely indicated, but if present confined to third scale row on anterior part of body; usually without a vertebral light stripe . . . .
. . . . *Thamnophis marcianus* (p. 700)
1b. Supralabials without vertical black marks; venter immaculate; lateral light stripe on dorsal scale rows 3 and 4, at least anteriorly; a vertebral light stripe present
. . . . *Thamnophis proximus* (p. 701)

Thamnophis marcianus (Baird and Girard, 1853)

(map 11.113)

DIAGNOSTICS: The dorsal checkerboard pattern of three rows of alternating black blotches on each side of the midline combined with the prominent black vertical stripes on

Figure 11.46. Ventral color pattern at midbody in *Thamnophis marcianus.*

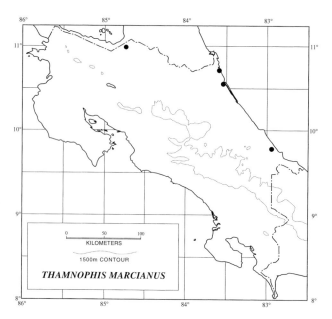

Map 11.113. Distribution of *Thamnophis marcianus.*

several supralabials unequivocally distinguishes this species from all other Costa Rican snakes.

DESCRIPTION: A moderate-sized species (total length of adults 450 to 1,050 mm); females longer than males; head distinct from neck; body cylindrical, chunky; tail long (24 to 27% of total length [20 to 27%]); 1 preocular (often 2 in some extralimital populations); postoculars usually 3 or 4, may be 2 or 5; supralabials 6 to 9, usually 8, with 2 bordering orbit; infralabials 8 to 12, usually 10; temporals variable, usually 1 + 2; dorsal scales 19-19-17 (19 to 21 on neck and/or midbody in extralimital populations); ventrals 138 to 148 (134 to 173); subcaudals 68 to 79 (56 to 83). Dorsum usually without longitudinal light stripes in Costa Rica (vertebral and lateral light stripe present in many extralimital populations), marked with checkerboard pattern of black blotches on flanks; large postoccipital black blotch preceded by upward-directed postrictal crescent of light pigment and deeply indented posteriorly along midline; distinct pre-, sub-, and postorbital vertical black marks on supralabial sutures; contrasting pre- and postocular light spots on side of head; skin between dorsal scales light orange to reddish; venter cream to orangish marked with definite dark spots or crossbars; iris reddish brown; tongue red with black tips on fork. Everted hemipenes without expanded apex.

SIMILAR SPECIES: (1) *Thamnophis proximus* does not have the supralabials boldly marked with dark lines and lacks the pre- and postocular and postrictal light markings.

HABITAT: Riverine situations in Lowland Moist and Wet Forests along the lower reaches of streams in the Río San Juan drainage and just inland from the Atlantic coast.

BIOLOGY: A rare snake in Costa Rica, semiaquatic and usually diurnal. One example was collected in the morning under a windrow of vegetable debris, wood, coconut husks, and such accumulated just above the waterline of Lake Nicaragua (Shreve and Gans 1958). Little else is known about tropical populations of this species. In extratropical areas it eats frogs, toads, tadpoles, and earthworms, supplemented by fish and salamanders. Females give birth to five to thirty-one snakelets that are 200 to 235 mm in total length. This species flattens the neck when disturbed, exposing the orange or red interscale skin in a bluffing display.

REMARKS: Two examples of this species were first collected in lower Central America in 1883 by Carl Bovallius (Andersson 1916) at Granada, Department of Granada, Nicaragua. Dunn (1940c) described these specimens (NHMG 1194A and B) as a new species, *Thamnophis bovallii*. Shreve and Gans (1958) reported additional examples from Nicaragua and Costa Rica. Rossman (1971) reviewed the status of Mexican and Central American lowland garter snakes not referable to *Thamnophis proximus* and concluded that the nominal species *Thamnophis praeocularis* of the Atlantic lowlands of northern Central America and *T. bovallii* were conspecific with *Thamnophis marcianus*. According to Rossman's concept, the species occupies an essentially continuous range in the southwestern United States and northern Mexico and has a disjunct distribution of scattered localities in southern Mexico and Atlantic slope Central America. Rossman recognized three major subdivisions within *T. marcianus* as follows:

A. North America (United States and Mexico west of the Isthmus of Tehuantepec): maximum number of dorsal scale rows 21 (very rarely 19); maxillary teeth 22 to 25; vertebral stripe present, lateral stripes evident, bright and distinct (except in southern Mexico)
B. Central America: maximum number of dorsal scale rows 19; maxillary teeth 26 to 31; lateral stripes indistinct or absent
   1. Upper Central America: broad vertebral stripe present, dull; two rows of small rounded spots on ventrals
   2. Lower Central America: usually no vertebral stripe; ventral spots large, often expanded into crossbars

Based on these data one is tempted to regard the North and Central American components in this arrangement as distinct, allopatric species. Rossman (1971) and Rossman, Ford, and Seigel (1996) argued that only a single species is involved because the isolated Isthmus of Tehuantepec population agrees in almost all features with examples of *T. marcianus* from the United States and northern Mexico, and material from Chiapas, Tabasco, and Veracruz, Mexico, is intermediate between northern *T. marcianus* and Central American specimens. These presumed intergrades agree with northern snakes in maxillary tooth numbers but resemble the Central American sample in dorsal scale row counts and coloration. In these circumstances it seems best to accept Rossman's conclusion pending detailed study of the various isolated allopatric populations referred to *T.*

marcianus. The karyotype is 2N = 36, consisting of seventeen pairs of macrochromosomes and one pair of microchromosomes; a ZW heteromorphism occurs in females (Baker, Mengden, and Bull 1972).

DISTRIBUTION: The southwestern United States and northern Mexico south from Kansas and New Mexico to northern Veracruz and Zacatecas on the east and from southern Arizona and California to Sonora on the west; a disjunct population at the Isthmus of Tehuantepec on the Pacific versant; at scattered localities in Central America on the Atlantic versant in southern Mexico, Belize, and Honduras and in the Great Lakes of Nicaragua and the Río San Juan drainage of Nicaragua and Costa Rica to coastal southeastern Costa Rica (2–30 m).

Thamnophis proximus (Say, 1823)

(plate 460; map 11.114)

DIAGNOSTICS: The presence of vertebral and lateral (at least anteriorly) light stripes, yellow orange labials that lack conspicuous dark markings, and the immaculate venter will readily separate this form from other Costa Rican snakes.

DESCRIPTION: A moderate-sized species reaching a maximum total length of about 1,250 mm, adults from 350 mm; females longer than males; body cylindrical, slender; tail long (28 to 31% of total length [26 to 33%]); 1 preocular; usually 3 postoculars, sometimes 2, 4, or 5; supralabials 7 to 9 (6 to 9), usually 8, with 2 bordering orbit; infralabials 8 to 11 (8 to 12), usually 10; temporals usually 1 + 2, sometimes other combinations from 1 + 1 to 2 + 3 (fig. 11.12a); dorsal scale rows almost always 19-19-17 (rarely 21 somewhere on body); ventrals 141 to 164 (141 to 181); subcaudals 91 to 104 (91 to 131) in males, 82 to 98 (82 to 124)

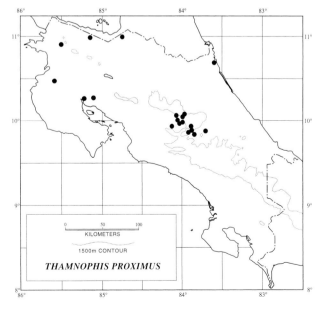

Map 11.114. Distribution of *Thamnophis proximus*.

in females. Dorsum olive brown (dark brown to olive gray in species as a whole); with broad gray tan vertebral stripe (grayish tan, gold, or bright red in extralimital populations), a pair of narrow lateral pale yellow stripes and a pair of narrow ventrolateral dark stripes (absent, narrow, or broad extralimitally); labials yellow orange (greenish white, greenish yellow, or yellow orange elsewhere in range); venter immaculate pale yellow to pale green. Everted hemipenes are short (about 7 to 8 subcaudals long), without an expanded apex.

SIMILAR SPECIES: (1) *Thamnophis marcianus* has a checkerboard dorsal pattern and a bold pattern on the supralabials.

HABITAT: Near streams, freshwater marshes, and swamps in open areas, including pastures, in Lowland Dry Forest and Premontane Moist and Wet Forests of the Meseta Central and into the lower margins of Lower Montane Wet Forest on the southern and western slopes of Volcán Irazú; also in the floodplains of the Río San Juan and Río Colorado in extreme northern Costa Rica.

BIOLOGY: This relatively rare species is active mostly during the day and occasionally climbs low bushes. It feeds primarily on fishes, tadpoles, and frogs. Females are mature at about 350 to 360 mm. An example captured in April in Belize gave birth to six young in July (Neill 1962). The hatchlings were 160 to 170 mm in body length. When approached closely it frequently opens its mouth wide in a threatening display.

REMARKS: Rossman (1963) analyzed variation in detail for this wide-ranging species and recognized six geographic races based solely on manipulation of different combinations of dorsal ground color, vertebral stripe width and color, labial color, and development of the ventrolateral dark stripe—features that vary independently. Considerable intraracial variation also occurs in some of these features. In addition, an extensive geographic zone (greater in area than the range of several of the putative races that border it) in the central United States is occupied by snakes showing nearly every possible combination of color characteristics described for the species. This case again points out the inadequacy of the subspecies concept for describing populational variation. Rossman (1970) essentially repeats his conclusions (1963) in his redescription and summary of published information on the species. Snakes from Central America belonging to this species were usually called *Thamnophis sauritus chalceus* or *Thamnophis sirtalis chalceus* before Rossman's revision. The karyotype is N = 36, consisting of seventeen pairs of macrochromosomes and one pair of microchromosomes, with a ZW heteromorphism in females (Baker, Mengden, and Bull 1972).

DISTRIBUTION: The central United States from Ohio, Lake Michigan, and southern Wisconsin and the Mississippi Valley westward through the Great Plains and southward through the Atlantic lowlands and adjacent uplands to Belize, and disjunctly distributed to the Meseta Central Ori-

ental of Costa Rica; disjunct populations on the Pacific slope from Guerrero to Chiapas, Mexico, El Salvador, Honduras, and Nicaragua south to the Meseta Central Occidental of Costa Rica; apparently disjunctly distributed in the republic with populations in northwestern and northeastern lowland areas and the premontane Meseta Central and adjacent lower margin of the lower montane zone on Volcán Irazú (2–1,751 m); in Mexico the species is found from near sea level to an altitude of 2,400 m.

Family Elapidae F. Boie, 1827

Coral Snakes, Cobras, Kraits, Mambas,
Australian Elapids, and Sea Snakes

All members of this family (approximately 65 genera and 327 species) are venomous, with a characteristic pair of rigid short hollow fangs on the anterior end of the maxillary. The fangs exhibit a groove along their anterior face where the infolded lips of what was an open groove earlier in development have met to form the enclosed venom-conducting channel. Externally the terrestrial and arboreal genera of the family resemble harmless and rear-fanged colubrid snakes, but the completely marine forms are greatly modified in morphology for life in the sea. In internal features, other than those associated with venom delivery, the family is very similar to the Colubridae (flexible lower jaw, without coronoid bone; no pelvic structures; no premaxillary teeth; no left lung; elongate posteriorly inclined quadrate and elongate supratemporal; hyoid composed of paired cornuae that form a structure with elongate posteriorly directed branches; no prefrontal bone; short recurved hypapophyses on all dorsal vertebrae). Externally all elapids lack scale pits, and terrestrial genera have a squarish scale between the eye and the nasal scale that separates the prefrontal from the supralabials. This scale is usually called a preocular (Peters 1964), but in the terminology I prefer it is properly called a loreal (Savage 1973c). If my definition is followed, all nonmarine elapids have a loreal and lack a preocular. However, in order to reduce any confusion in the keys and in the description of Costa Rican snakes, I specify the number of scales between the eye and nostril and their relation to the prefrontal and supralabials. The family ranges over most of Africa into the Middle East and Arabia, through tropical, subtropical, and temperate Asia to southern Korea and Japan, and southward through the Indo-Australian archipelago and Australia to Tasmania; in the Americas from the southern United States to southern Argentina; marine forms occur in the Indo–West Pacific region and on all coasts of Australia, with one species ranging across the Pacific to the western coasts of tropical America. The family is not represented on Madagascar or the Antilles, and marine forms are absent from the Atlantic Ocean. Members of this family are all dangerous to humans and include some of the most venomous and frequent agents of crippling or lethal bites. In the Old World the African mambas, the African and Asian cobras, and the Asian kraits are responsible for perhaps 400,000 serious bites a year, with 30,000 to 40,000 being fatal. Australia is the only region where venomous snakes (all elapids) outnumber harmless forms, but most of the toxic forms are small and relatively innocuous, and the number of bites recorded for the more dangerous species is small by comparison with those in Asia and Africa. The American coral snakes are dangerous because of the nature of their venoms, but bites and deaths are relatively rare (in Costa Rica, about eight reported bites per year, with deaths uncommon). Sea snakes also contribute few cases of bites or deaths worldwide.

Although there have been several attempts to divide the family into subfamilies, no scheme has been universally agreed on (see McDowell 1987; Mengden 1985). The following classification incorporates McDowell's (1967, 1969a,b, 1970) ideas into that advanced by Dowling and Duellman (1974–78) as a convenient arrangement:

Subfamily Elapinae Boie, 1827 (nine genera): palatine erectile; rear of venom gland bends downward behind corner of mouth; no tracheal lung (India and Sri Lanka to northern Australia and the Solomon Islands; southern North America to northern Argentina)

Subfamily Bungarinae Eichwald, 1831 (ten genera): palatine erectile; venom gland oval to suboval, without downward bend posteriorly; no tracheal lung (Africa, Arabia, Asia, and Indo-Malayan archipelago)

Subfamily Acanthophiinae Dowling, 1975a (thirty-five genera): palatine nonerectile; no tracheal lung; nostrils placed forward and/or lateral, no nasal valves (New Guinea and Australia to Fiji Islands)

Subfamily Hydrophiinae Fitzinger, 1843 (eleven genera): palatine nonerectile; a tracheal lung present; nostrils placed dorsally, valvular (Indian Ocean to the coasts of Australia and Fiji; one species reaches the coasts of tropical America)

Terrestrial and arboreal species of this family are mostly oviparous, but several are viviparous, as are all the marine forms except the members of the genus *Laticauda* of the western Pacific and Indo-Malayan waters.

In Costa Rica we find one genus of the first group *(Micrurus)* and one of the fourth group *(Pelamis)*. The latter genus, with its compressed, paddle-shaped tail, valvular, dorsally placed nostrils, and only slightly enlarged ventrals (about twice as large as the dorsals), cannot be mistaken for any other snake. This is especially so since the single species is totally marine and is restricted to the Pacific Ocean. As a practical matter no Costa Rican member of the families Anomalepididae, Typhlopidae, Leptotyphlopidae, Boidae, or Viperidae has a pattern that includes black rings encircling the body as in all of the indigenous venomous coral snakes referred to *Micrurus*. Most colubrid species that re-

semble coral snakes in scutellation (enlarged ventrals and large head shields) lack this color pattern as well. Harmless and rear-fanged species that resemble the venomous coral snakes in having a pattern that includes black rings (with or without red rings) or black and red bands may be separated from the species of *Micrurus* by using the key provided in a following section (p. 706). It must be emphasized that on a worldwide basis terrestrial and arboreal elapids and coral snakes may be separated definitively from the myriad genera of colubrid snakes only by examining the upper jaw to determine whether hollow anterior fangs are present.

The venomous coral snakes include sixty species placed in two genera: *Micrurus* (fifty-nine species; southern United States to Peru, Bolivia, and Argentina), and *Micruroides* (one species; southwestern United States and northwestern Mexico). Campbell and Lamar (1989) presented descriptions of all then recognized species of venomous coral snakes and color photographs for most of them. Roze's (1996) book provides a comprehensive coverage of biology, venoms, systematics, and distribution for this group. It also includes color photographs of many forms.

Remember, no brightly colored snake (or indeed no dull colored one) whose identity is in doubt should be carelessly approached or handled.

KEY TO THE AMERICAN GENERA OF ELAPIDAE

1a. Tail cylindrical in cross section; nostrils lateral; ventral scales broad, several times broader than dorsals; scales on body and tail rounded posteriorly, imbricate; 0–1 tooth present on maxillary posterior to fangs 2

1b. Tail strongly compressed, paddle shaped; nostrils superior; ventral scales small, about twice as broad as dorsals; scales on body and tail hexagonal or squarish, juxtaposed; 7–11 posterior maxillary teeth separated from the fangs by a diastema . . . . *Pelamis* (p. 714)

2a. Two pairs of chin shields; fangs present only on maxillary; body pattern various, but *if* monadal, a black nuchal collar present . . . . *Micrurus* (p. 707)

2b. One pair of chin shields; fangs on maxillary followed by a single small tooth; body pattern of monads but no black nuchal collar . . . . *Micruroides*

Almost nothing is known about the reproductive biology of coral snakes. So far as is known, all species of venomous coral snakes are oviparous, with clutches of two to nine eggs reported (Fitch 1970).

BEHAVIORAL ECOLOGY OF THE VENOMOUS CORAL SNAKES

The American coral snakes resemble one another closely in most aspects of their biology, which includes several special and interesting motifs. This section provides a summary overview of this subject to introduce readers to these snakes and avoid repetition in the species accounts. For additional details see the excellent papers by Greene (1973, 1984, 1988), Greene and McDiarmid (1981), Greene and Seib (1983), Pough (1988), and Roze (1982) and Roze's (1996) book on coral snakes. As might be expected from their general pattern of distribution, coral snakes occur in a wide variety of habitats in the lowlands and at moderate elevations from deserts to tropical rainforests. These snakes are both diurnal and nocturnal, and some South American species are semiaquatic.

Most species are secretive inhabitants of the forest litter and forage on the ground or in the leaf litter, moving slowly and poking their heads under debris and apparently using chemoreception to recognize their prey. Once they find the prey, they grasp it and hold in the mouth, using a chewing motion or a quick bite to puncture the victim with the short fangs and introduce venom into its system. They hold the envenomated animal until it stops moving, then swallow it as other snakes do, almost always headfirst. Most coral snakes are strictly ophiophagous, but some feed on other elongate vertebrates including fish, caecilians, lizards, and amphisbaenians as well as snakes. At least one population of *Micruroides euryxanthus* eats mainly thread snakes *(Leptotyphlops)*. Several species feed principally on freshwater fishes. Another South American form appears to specialize on onychophorans *(Peripatus)*, attenuate invertebrates allied to arthropods, and amphisbaenids. Snakes commonly reported as prey are members of the genera *Adelphicos*, *Coniophanes*, *Geophis*, *Ninia*, and *Urotheca*—typical denizens of the leaf litter—as well as other venomous coral snakes. Some coral snakes cannibalize their own species. Known predators on coral snakes are primarily birds (see Brugger 1989) and snake-eating snakes. There is evidence that envenomation by coral snakes may kill large raptors, but there are records of successful predation on these snakes (Smith 1969).

Contrasting patterns of alternating black and light bands or rings are found in many species of snakes. In these forms, including the venomous coral snakes, the pattern serves a cryptic function and helps to hide them from predators by (1) disrupting the impression of the snake's elongate shape and making it difficult to perceive when motionless; (2) creating an optical illusion that a snake is stationary when it is moving rapidly, since the viewer's eyes may focus on the bands and the snake seems to suddenly disappear out from under them; or (3) creating an optical illusion that the moving snake is unicolor (because of the rate of light-dark transitions across the visual field); when the snake becomes motionless the predator's search image is keyed to an elongate, unicolor object, but the sharply contrasting bands now conceal the snake's elongate form (Brattstrom 1955; Jackson, Ingram, and Campbell 1976; Pough 1976). These effects are

probably most pronounced under conditions of low illumination (i.e., at night or inside a forest) and where the background has considerable heterogeneity. However, as we shall see below, the addition of bright colors to the basic pattern has other effects as well.

Many species of coral snakes exhibit defensive behaviors that are keyed to reducing predation. Some or all of these may be used by a single individual and include hiding the head under the body coils, flattening the body, coiling the tail to expose the bright colors on the underside, and waving the tail about (plate 465). In addition, the snake may withdraw its head from hiding and swing it from side to side with the mouth open and ready to bite any object it touches; snap its body back and forth almost spasmodically; strike aggressively with its tail or wave it to divert attention from the head; and discharge the contents of the cloaca, with or without a series of loud popping sounds.

Although long and widely debated, there now seems little question that the bright coral snake color patterns and most of the behaviors described above are important antipredator defenses; they are aposematic, acting through the distance receptors of potential predators to threaten and warn them that they may be subject to a serious, venomous bite (Greene 1988; Pough 1988). It is also clear that a series of harmless or mildly toxic colubrid snakes with coloration identical or similar to that of various venomous coral snakes attain considerable protection from predators as mimics of the highly toxic species. This probably represents a case of Batesian mimicry, since the coloration of the coral snakes (the model) warns the predator of a potential danger while that of the mimic deceives the predator into avoiding a nonexistent danger. However, because some of the mimics are mildly toxic, it has been suggested that they are Müllerian mimics, since they also constitute a real danger to a predator and the color patterns they share with the venomous coral snakes mutually reinforce avoidance in predators.

Predators' avoidance of coral snake color patterns probably in some cases involves learning based on an unpleasant experience with a coral snake. Because of their short fangs and the need to chew to ensure envenomation, coral snakes frequently inject only small and not necessarily deadly amounts of venom, so predators probably often survive bites (e.g., Smith 1969). The carefully designed experiments of Smith (1975, 1977), however, demonstrated innate avoidance of the general coral snake pattern by two potential avian predators. Naive motmots *(Eumomota)* and kiskadees *(Pitangus)* presented with painted wooden dowels attacked those painted in a variety of solid colors, including red and yellow, and neutrally banded ones (i.e., without red and yellow). They also pecked at dowels with red and yellow longitudinal stripes. But when the birds were presented with dowels painted with tricolor monads (the most common coral snake pattern) they fled, giving high-intensity alarm calls. The birds also fled but gave lower-intensity alarm calls when dowels painted with red and yellow bands were introduced into their cage. Some individuals pecked a dowel painted with red, yellow, and black stripes, but at a significantly lower frequency than for a neutral or red and yellow striped model.

These experiments demonstrate that the birds are capable of discriminating among a number of different color patterns, but in practice they generalize their response broadly. Although they discriminated between monadal, bicolor (red and yellow) banded, and red-yellow-and-black striped patterns, they completely avoided the first two and showed less inclination to attack the third. Thus predators such as motmots and kiskadees will avoid potential prey that bears even a slight resemblance to a venomous coral snake.

Subsequently Brodie (1993) and Brodie and Janzen (1995), using plasticine snake replicas, showed that free-ranging avian predators display generalized avoidance of patterns resembling those of coral snakes. These studies convincingly established the reality of aposematic coloration in coral snakes and their mimics. The generalized response is what Pough (1988) regarded as the first step in the two-step evolution of Batesian mimicry. That mimicry is involved is further attested to by the precise resemblances in coloration between individual mimic species and specific venomous coral snakes and the concordant parallel geographic variation between models and mimics (Greene and McDiarmid

1981; Roze 1982, 1996; Savage and Crother 1989; Savage and Slowinski 1996). These levels of resemblance imply a much finer distinction by some predators than in the general avoidance response and seem to represent the second step in the two-step process leading to Batesian mimicry.

Vitt's (1992) novel hypothesis that bicolored (red and black) millipedes form the model for mimicry by both venomous and nontoxic coral snakes has been successfully refuted by the field experiments of Brodie and Moore (1995).

A substantial number of American colubrid genera have representatives that belong to this mimetic complex (Dunn 1954), here referred to as the coral snake mimic guild, characterized by having a pattern of alternating black and red (and/or yellow or white) dorsal bands or rings. In Costa Rica members of the guild include snakes of the following genera: *Dipsas, Erythrolamprus, Lampropeltis, Leptodeira, Liophis, Oxyrhopus, Rhinobothryum, Scaphiodontophis, Scolecophis, Sibon, Tantilla, Tripanurgos,* and *Urotheca.* These species may be distinguished from the venomous coral snakes by the use of the accompanying key and table 11.2. Terminology for color patterns in the key, table 11.2, and figure 11.47 follows Savage and Slowinski (1990, 1996).

The following key is designed to identify as readily as possible any Costa Rican snake having the coral snake pattern of alternating dorsal bands or rings encircling the body, without dissection of the jaw to determine the nature of the venom apparatus. Some snakes called *corales* by Costa

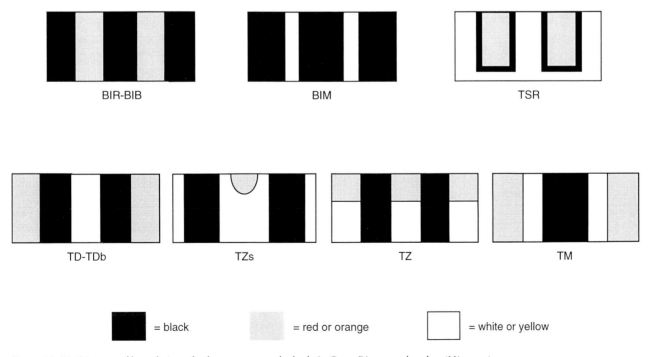

Figure 11.47. Diagram of lateral view of color patterns on the body in Costa Rican coral snakes *(Micrurus)* and their mimics: BI = bicolored, T = tricolored; M = monads, D = dyads; all patterns include black rings except for BIB and TDb, which have black bands, and TSR. See text for details.

Ricans are omitted, since they have the entire body red (especially, *Ninia sebae*, some *Stenorrhina freminvillii*, and juveniles of the two species of *Clelia*, a rear-fanged group); a considerable portion of the venter orange, pink, or red; a number of red spots on the body; or a nuchal red area and cannot be confused with the venomous coral snakes or the possibly toxic rear-fanged forms with coral snake patterns. Even a novice should be able to make a quick and accurate identification with this key in order to ensure immediate and appropriate treatment of a toxic bite. Life colors are emphasized for this reason, since living or recently dead snakes will be available for study.

KEY TO THE CORAL SNAKES OF COSTA RICA

1a. Pattern bicolor (BI), of alternating black and light (white, yellow, orange, rust, pink, or red) bands or rings; sometimes light areas orange to red above becoming yellow or white laterally and ventrally 2

1b. Pattern tricolor (T), of alternating black and light bands or rings of two colors (always clearly evident ventrally), white or yellow and reddish brown or red 13

2a. Pattern of alternating black and light rings that completely encircle body (BIR) 3

2b. Dorsal pattern of alternating black and light bands (BIB) 10

3a. Anal plate single 4

3b. Anal plate divided 7

4a. One or two pairs of elongate chin shields and a distinct mental groove 5

4b. No elongate chin shields or mental groove . . . .
. . . . *Dipsas tenuissima* (p. 398)

5a. Dorsal scale rows at midbody 13–15 6

5b. Dorsal scale rows at midbody 19 or more . . . .
. . . . *Lampropeltis triangulum* (p. 665)

6a. Dorsal scale rows at midbody 13 . . . .
. . . . *Sibon anthracops* (p. 630)

6b. Dorsal scale rows at midbody 15 . . . .
. . . . *Sibon annulatus* (p. 629)

7a. 15 dorsal scale rows at midbody 8

7b. Dorsal scale rows at midbody 17 . . . .
. . . . *Urotheca euryzona* (p. 642)

8a. Two scales (a loreal and postnasal) on side of head between eye and nostril 19

8b. Three scales (a preocular, loreal, and postnasal) on side of head between eye and nostril . . . .
. . . . *Scolecophis atrocinctus** (p. 685)

9a. Black head cap restricted to an area well forward of the middle of the orbit . . . *Micrurus mipartitus*** (p. 712)

9b. Black head cape extending posteriorly to cover interorbital area and reaching posteriorly onto parietal shields
. . . . *Micrurus alleni*** (p. 709)

10a. Venter light, almost immaculate 11

10b. Venter with a considerable amount of dark pigmentation, at least along the posterior margin of each ventral 12

11a. Some shields on upper head surface black 25

11b. Shields on upper head surface light (red or white); 17 dorsal scale rows at midbody . . . .
. . . . *Tripanurgos compressus** (p. 577)

12a. Venter with numerous bold black marks and blotches; 17 dorsal scale rows at midbody . . . .
. . . . *Liophis epinephalus* (p. 580)

12b. Venter with dark stippling at least along free margin of ventrals; 19 or more scale rows at midbody . . . . . .
. . . . *Leptodeira rubricata* (p. 612)

13a. No elongate chin shields or mental groove; anal plate entire 14

13b. Elongate chin shields and a mental groove present; anal plate divided 15

14a. Two to four pairs of chin shields; tricolor pattern of complete rings (TM) . . . . . *Dipsas articulata* (p. 597)

14b. Three large unpaired chin shields; tricolor dorsal pattern consisting of dark and light rings, the latter split by red bands (TZs); venter bicolor dark and white . .
. . . . *Dipsas bicolor* (p. 597)

15a. Dorsal scales smooth, except for a few above the cloacal region in some forms 16

15b. Dorsal scales keeled . . . .
. . . . *Rhinobothryum bovallii** (p. 681)

16a. Dorsal scale rows 13–15 at midbody 17

16b. Dorsal scale rows 17–21 at midbody 23

17a. Black rings bordered on either margin by yellow or white rings to completely separate black rings from red rings (TM) 18

17b. Black rings (TD) or bands (TDb, TSR) in broad contact with red areas 22

18a. Black rings not split by white or yellow areas dorsally
. . . . 19

18b. Most black rings split dorsally by white areas (TM)
. . . . *Erythrolamprus mimus* (p. 579)

19a. Red rings encircle body; scale rows 15 at midbody
. . . . 20

19b. Red areas variable (TZs, TZ, TM); scale rows 13 at midbody . . . . *Sibon anthracops* (p. 630)

20a. Parietal shields usually with considerable areas not covered by black head cap; if black head cap covering all or most of parietals, mental and several anterior infralabials mostly black and light lateral head plates not outlined by black 21

20b. Parietal shields completely covered by black head cap; mental and infralabials never covered with black, although some dark peppering often present and light lateral head plates outlined by black . . . .
. . . . *Micrurus clarki*** (p. 710)

21a. Black head cap not or barely involving parietal shields
.... *Micrurus nigrocinctus*** (p. 710)
21b. Black head cap extending as an oblong or lanceolate figure on both parietals along interparietal suture, rarely expanded to cover most or all of parietals ....
.... *Micrurus alleni*** (p. 709)
22a. Some black rings encircling body 27
22b. Pattern of black bands that do not encircle body (TDb)
.... *Tantilla supracincta** (p. 696)
23a. Dorsal scales at midbody, 19 or more; anal plate entire
.... 24
23b. Dorsal scales at midbody 17; black bands separated from red areas by yellow or white bands (TM); anal plate divided .. *Scaphiodontophis annulatus* (p. 682)
24a. Black rings in broad contact with red rings; light rings separated from red rings by black rings (TD).......
.... *Lampropeltis triangulum* (p. 665)
24b. Black bands narrowly separated from red bands by light areas (TZs) ... *Oxyrhopus petolarius** (p. 575)
25a. Venter cream; more than 15 scale rows at midbody
.... 26
25b. Venter red; 15 scale rows at midbody............
.... *Tantilla supracincta** (juveniles) (p. 696)
26a. All enlarged shields on upper surface of head black; anal plate entire ... *Oxyrhopus petolarius** (p. 575)
26b. Black head cap not involving frontal or supraocular shields; anal plate divided....................
... *Scaphiodontophis annulatus* (juveniles) (p. 682)
27a. Black and light areas usually forming rings, red to orange areas usually enclosed by black border, red not crossing ventrals; tail long; 107–35 subcaudals (TD, BIR)...................................... 28
27b. Pattern of rings completely encircling body (TD); tail relatively short; 45–62 subcaudals..............
.... *Erythrolamprus bizona** (p. 578)
28a. Reddish areas between black elements of pattern lighter than interspaces.... *Sibon annulatus* (p. 629)
28b. Reddish areas between black elements of pattern darker than interspaces ... *Sibon dimidiatus* (p. 631)

* Rear fanged, possibly toxic to humans.
** Highly venomous coral snakes.

Human envenomation by venomous coral snakes in Costa Rica is uncommon, with about eight bites reported in a typical year (Bolaños 1984). Deaths from bites by coral snakes are consequently rare. However, the possibility of death from an untreated bite is high (50%) because of the complex and potent venom. In addition there is considerable heterogeneity among the venoms of different species, so a specific monovalent antivenin should be used. Treatment, even if begun some time after a bite, is usually successful. Bites by *Micrurus mipartitus* appear to be particularly severe, and a number of deaths from untreated bites by other species occur when someone, often a child, handles a coral snake either carelessly or in ignorance of its toxicity.

Subfamily Elapinae

Genus *Micrurus* Wagler in Spix, 1824

Highly Venomous Coral Snakes

Snakes belonging to this genus (fifty-nine species) range from small to large forms (350 to 1,220 mm maximum total length); there is usually a bicolor or tricolor ringed pattern, but some individuals or species have a uniform red dorsum or may be marked with black spots or bands. All Costa Rican *Micrurus* have the body and tail encircled by black rings that alternate with rings of one or two different colors, yellow or white and/or red (at least indicated ventrally). In addition, they always have the prefrontal plate separated from the supralabials by a large squarish shield lying anterior to the orbit and touching the postnasal (two plates on the side of the head between the orbit and nostril). Other features include upper surface of head covered by enlarged shields, a rostral, paired internasals, prefrontals, supraoculars, and parietals and a single frontal (fig. 11.10c); body cylindrical; head not distinct from neck; nasal divided; eye small, pupil subcircular; two pairs of elongate chin shields, separated from the mental; dorsal scales smooth in 15 rows throughout; no apical scale pits; supracloacal keels on some individuals; anal divided; subcaudals divided; no maxillary teeth posterior to canaliculate fangs; mandibular teeth enlarged anteriorly; hypapophyses on posterior vertebrae. In Costa Rica *Micrurus* is likely to be confused only with colubrid snakes having enlarged head shields and the body banded or ringed with black alternating with light-colored bands or rings.

Harmless or mildly toxic snakes with coral snake coloration having the prefrontal in contact with one or more supralabials or the prefrontal separated from the supralabials and three scales (preocular, loreal, and nasal) on the side of the head between the orbit and nostril differ from *Micrurus*, which has only a single large squarish scale and the postnasal separating the prefrontal from the supralabials. No other Costa Rican species with an arrangement of lateral head scales similar to *Micrurus* has a pattern resembling that of the genus except some species of *Sibon*. Examples of the latter genus have an elongate, laterally compressed body, prominent eyes, the head distinct from the neck, and a very long tail.

Slowinski (1991, 1995) has shown that four species groups may be recognized within this genus (penial features for everted organs):

Micrurus fulvius group (thirty-eight species): anterior chin shields in contact with mental; body pattern variously bicolor or tricolor but usually of monads (red rings separated from one another by light-black-light rings) hemipenes elongate, slender, and strongly bilobed, and each lobe with a spiny terminal awn and a hypertrophied lateral fold on the naked basal pocket (southeastern United States to Peru, Bolivia, and Argentina)

Micrurus mipartitus group (two or three species): anterior chin shields in contact with mental; body pattern bicolor; hemipenes moderately bilobed, without terminal awns or hypertrophied lateral fold on the basal naked pocket (Nicaragua to southern Ecuador and northern Venezuela)

Micrurus collaris group (three species): anterior chin shields separated from mental by first infralabials; body pattern of modified monads, mostly black above with light rings only on head, neck, and tail; hemipenes moderately bilobed without terminal awns or hypertrophied lateral fold on the naked basal pocket (Amazonia and the Guianas)

Micrurus spixii group (fifteen species): anterior chin shields in contact with mental; body pattern of triads (each red ring separated by black-light-black-light-black rings, rarely indicated only ventrally; hemipenes short, slightly bilobed, without terminal awns or hypertrophied lateral fold on the naked basal pocket (Panama to western Peru and central Argentina)

The genus ranges from the southeastern United States to Texas on the east and from Sonora, Mexico, on the west, south over most of the Neotropical region to northwestern Peru on the Pacific versant, and to Bolivia and southern Argentina east of the Andes in South America.

KEY TO THE COSTA RICAN SPECIES OF VENOMOUS CORAL SNAKES, GENUS *MICRURUS*

1a. Black head cap including rostral, nasals, prefrontals, a substantial part of supraoculars, and frontal and often extending onto parietals (fig. 11.48b–f); usually a tricolor pattern of alternating black and light rings of two colors (white or yellow and red), sometimes evident only on venter 2

1b. Black head cap including rostral, nasals, prefrontals, and rarely anterior edges of supraoculars and frontal (fig. 11.48a); bicolor dorsal pattern of alternating black and light rings (white, yellow, pink, or red in life); light areas often yellow to red above, becoming white laterally . . . . *Micrurus mipartitus* (p. 712)

2a. Parietal shields usually with considerable areas not covered by black head cap; if black head cap covers all or most of parietals, mental and several anterior infralabials mostly black; light lateral head plates not outlined by black 3

2b. Parietal shields completely covered by black head cap; mental and infralabials never covered with black, although some dark peppering is often present; light lat-

eral head plates outlined by black (fig. 11.48f); ventrals plus subcaudals 241–57 . . . . *Micrurus clarki* (p. 710)

3a. Black head cap not, or barely, involving parietal shields (fig. 11.48b–d); ventrals plus subcaudals 222–68 . . . . *Micrurus nigrocinctus* (p. 710)

3b. Black head cap extending as an oblong or lanceolate figure on both parietals along interparietal suture, rarely expanded to cover most or all of parietals (fig. 11.48e); ventrals plus subcaudals 256–81 . . . . *Micrurus alleni* (p. 709)

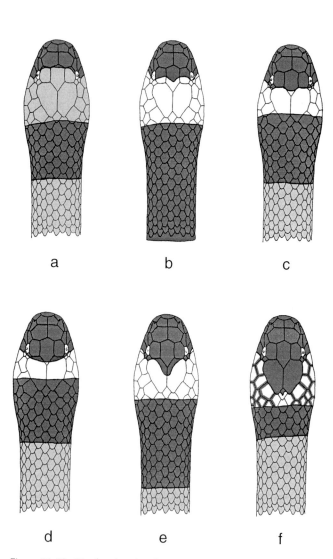

a   b   c

d   e   f

Figure 11.48. Head and neck color patterns in Costa Rican venomous coral snakes *(Micrurus)*; patterns are figured on a generalized template of head plates for the genus and do not accurately depict differences in scalation among the species illustrated: (a) *M. mipartitus*; (b) *M. nigrocinctus* from Atlantic slope; (c) *M. nigrocinctus* from southwestern Costa Rica and areas intermediate between northern Atlantic and Pacific slope populations in Nicaragua; (d) *M. nigrocinctus* from northwestern Costa Rica; (e) *M. alleni*; (f) *M. clarki*.

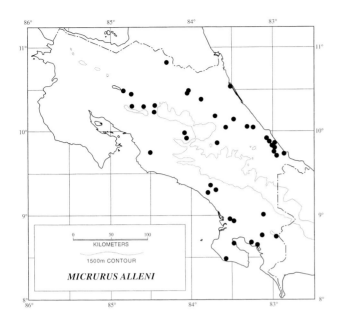

Map 11.115. Distribution of *Micrurus alleni*.

Micrurus fulvius Group

Micrurus alleni K. Schmidt, 1936

(plate 461; map 11.115)

DIAGNOSTICS: A relatively uncommon tricolor (black, red, and yellow or white) or bicolor (black and yellow) ringed species that usually has the solid black head cap forming an oblong or lanceolate figure along the interparietal suture (but rarely has it expanded to cover most of the parietals), the lateral head plates not outlined by black, and the nuchal black ring never crossing the parietal tips (fig. 11.48e). Bicolor examples have a tricolor ventral pattern.

DESCRIPTION: A moderate-sized species; most adults 500 to 700 mm in total length (maximum for females 1,165 mm, males to 800 m); tail short (12 to 19% of total length in males, 7.5 to 11% in females); postoculars 2; supralabials 7, with 2 bordering orbit, infralabials 7; temporals 1 + 1; anal divided; ventrals 209 to 224 in males, 221 to 244 in females; subcaudals 47 to 59 in males, 32 to 43 in females; ventrals plus subcaudals 256 to 281; supracloacal keels present in adult males and very large adult females. Body pattern bicolor dorsally (black and yellow and tricolor ventrally (black, red, and yellow) or tricolor, ringed by monads (light-black-light) and red; scales in red rings marked with black, sometimes completely obscuring red on dorsum in adults; black head cap extending down interparietal suture or rarely expanded over most of parietals; black nuchal ring not crossing parietals; light lateral head scales between head cap and nuchal ring not edged with dark; eye black; tongue black. The everted hemipenes moderately long (about 14 subcaudals in length), strongly bilobed, each lobe terminating in a long awn that has a single spine at the tip,

and the lateral fold on the basal naked pocket hypertrophied (CRE 2970).

SIMILAR SPECIES: (1) *Micrurus nigrocinctus* has the black head cap not or barely reaching the anterior edge of the parietals, and it never extends posteriorly along the interparietal suture (fig. 11.48b–d). (2) *Micrurus clarki* has the light lateral head scales outlined by black, and the black head cap always covers the entire surface of the parietals (fig. 11.48f). (3) *Micrurus mipartitus* has the black head cap restricted to an area well forward of the middle of the orbits (fig. 11.48a) and has a bicolor dorsal and ventral pattern. (4) *Rhinobothryum bovallii* has the upper head shields edged with white. (5) *Scaphiodontophis annulatus* is banded, not ringed by black. (6) *Urotheca euryzona* has a bicolor venter. (7) *Erythrolamprus mimus* has the black rings partially split by white. (8) *Oxyrhopus petolarius* is banded, not ringed. See key to coral snakes (p. 706), figure 11.47, and table 11.2.

HABITAT: Lowland Moist and Wet Forests and Premontane Moist and Wet Forests.

BIOLOGY: Roze (1996) states that this species feeds mostly on the eeliform freshwater fish *Synbranchus marmoratus* and some lizards.

REMARKS: Although long confused with its ally *M. nigrocinctus*, *M. alleni* is a distinctive and valid species that occurs sympatrically with the former at many localities. Taylor's description of *Micrurus alleni richardi* (1951b) and his references to *Micrurus nigrocinctus alleni*, *Micrurus nigrocinctus yatesi*, and *Micrurus richardi* are based on examples of this species. The completely allopatric Atlantic and Pacific populations of *M. alleni* exhibit marked differences in coloration (Savage and Vial 1973). Atlantic examples usually have the light rings white and retain the typical tricolor monad pattern throughout life. Pacific versant individuals, however, undergo a marked ontogenetic change in coloration. The juveniles have a tricolor monad pattern, and the light rings are bright yellow. With increasing size the black spots on each of the dorsal scales in the red rings expand to nearly or completely obscure the red coloration. Most adults are essentially bicolor, with alternating black and yellow bands dorsally. The venter retains a tricolor pattern, although black pigment invades the red ventral areas to a considerable extent. *M. alleni* is sympatric with the harmless *Lampropeltis triangulum* (a member of the coral snake mimic guild) at moderate elevations (1,200 to 1,400 m) in southwestern Panama and adjacent Costa Rica. When small it too has a tricolor pattern (although composed of dyads, not monads) that gradually changes to a bicolor one through invasion of the light rings by black to produce a black and red ringed snake. In this case the black pigment continues to expand, and adult *L. triangulum* from this area are nearly uniform black. Savage and Vial (1973) interpreted the concordance between juvenile (tricolor) and bicolor *M. alleni* and *L. triangulum* as evidence of the aposematic significance of coral snake mimicry. A specimen (KU 25190) of this species reported by Taylor (1954b) from the Lower Montane Rainforest zone at Boquete Camp, San José Province (2,000 m), is from 400 to 500 m higher than any other record and may have been collected at a lower elevation (Savage and Vial 1973). Gutiérrez and Bolaños (1979) described the karyotype for this species based on two males from Atlantic lowland Costa Rica. The diploid number is 34, with ten pairs of macrochromosomes and seven pairs of microchromosomes and a secondary constriction in the first pair of macrochromosomes. Pairs 2, 3, 6, 7, 9, and 10 are metacentrics, 8 is a submetacentric, and 1, 4, and 5 are telocentrics; NF = 48.

DISTRIBUTION: Atlantic lowlands and adjacent premontane slopes from eastern Honduras to western Panama (1–656 m) and the Pacific lowlands and premontane and lower montane zones of southwestern Costa Rica and adjacent Panama (2–1,500 m); also from the Meseta Central Occidental of Costa Rica (to 1,620 m) and the lowlands of eastern Panama on the Pacific slope.

Micrurus clarki K. Schmidt, 1936

(plate 462; map 11.116)

DIAGNOSTICS: A coral snake with a tricolor monad pattern (each black ring separated from the red rings by light rings) having the solid black head cap including all the parietals, the light lateral head scales between cap and nuchal collar edged with dark, and the nuchal black ring never crossing the parietal tips (fig. 11.48f).

DESCRIPTION: A rare relatively small coral snake (maximum known female total length 832 mm); most adults 380

to 600 mm; tail short (15 to 16% of total length in males, 10 to 13% in females); postoculars 2, supralabials 7, with 2 bordering orbit; 7 infralabials; temporals 1 + 1; no supracloacal tubercles; anal divided; ventrals 190 to 199 in males, 208 to 221 in females; subcaudals 53 to 58 in males, 33 to 44 in females; ventrals plus subcaudals 241 to 257. Tricolor pattern of black, yellow, and red rings arranged in monads; each scale in light rings with black tips; 13 to 20 black body rings; solid black head cap including all the parietals.

SIMILAR SPECIES: (1) *Micrurus alleni* has the light lateral head scales between the head cap and nuchal ring unmarked with dark pigment (fig. 11.48a). (2) *Micrurus nigrocinctus* has all or most of the parietals light in color (fig. 11.48b–d). (3) *Rhinobothryum bovallii* has the upper head shields outlined by white. (4) *Scaphiodontophis annulatus* is banded, not ringed. (5) *Erythrolamprus mimus* has the black rings partially split by white. (6) *Oxyrhopus petolarius* is banded, not ringed. See key to coral snakes (p. 706), figure 11.47, and table 11.2.

HABITAT: Lowland Moist and Wet Forests.

DISTRIBUTION: Lowlands of extreme southwestern Costa Rica and adjacent western Panama (5–500 m) and extreme eastern Panama and western Colombia (5–700 m) on the Pacific versant; central (both slopes) and Atlantic slope eastern Panama (2–100 m).

Micrurus nigrocinctus (Girard, "1854," 1855)

Coral Macho

(plates 463–66; map 11.117)

DIAGNOSTICS: A coral snake with a ringed tricolor monad pattern having the solid black head cap extending posteriorly to the level of the orbits and sometimes involving the anteriormost part of the parietals and the nuchal black ring often involving the posterior tips of the parietals (fig. 11.48b–d).

DESCRIPTION: A medium-sized coral snake (maximum female total length 1,150 mm, most adults 500 to 750 mm); tail short to moderate in males (13 to 17% of total length), short in females (8 to 13% of total length); postoculars 2; supralabials 7, with 2 bordering orbit; infralabials 7; temporals 1 + 1 or 1 + 2; supracloacal scales keeled in adult males; anal divided; ventrals 176 to 196 (176 to 217) in males, 190 to 227 (190 to 230) in females; subcaudals 43 to 58 (42 to 58) in males, 29 to 44 in females; ventrals plus subcaudals 222 to 268 (211 to 288). Dorsal pattern of tricolor monads, each black ring separated from red ring by yellow ring on each side (elsewhere in the range some examples are bicolor black and red); 10 to 24 (10 to 29) black rings on body; black head cap reaching no farther posteriorly than anterior tips of parietals; nuchal black ring crossing parietals or not; light lateral head scales between head cap and nuchal ring without dark borders, eye black; tongue black.

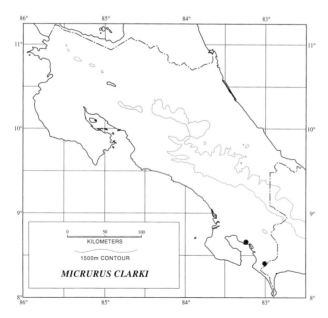

Map 11.116. Distribution of *Micrurus clarki*.

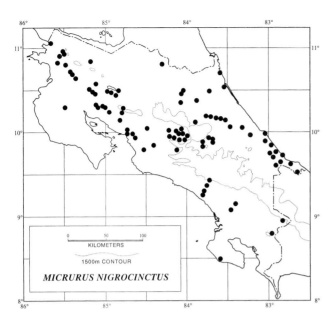

Map 11.117. Distribution of *Micrurus nigrocinctus*.

The everted hemipenes long (10 to 12 subcaudals long), strongly bilobed, each lobe terminating in a long awn that has a spine at tip, bifurcate, centripetal sulcus spermaticus; most of organ covered with thin, enlarged spines, lateral fold on basal naked pocket hypertrophied (Savitsky 1979).

SIMILAR SPECIES: (1) *Micrurus alleni* has the black head cap extending well onto the parietals along the interparietal suture. (2) *Micrurus clarki* has the parietals solid black. (3) *Rhinobothryum bovallii* has the enlarged upper head shields outlined by white. (4) *Scaphiodontophis annulatus* is banded, not ringed by black. (5) *Erythrolamprus mimus* has the black rings partially split by white. (6) *Oxyrhopus petolarius* is banded, not ringed. See key to coral snakes (p. 706), figure 11.47, and table 11.2.

HABITAT: A common snake in all lowland and premontane forest areas, often found under debris, in pastures, on coffee fincas, and in vacant lots and gardens in urban areas; also in the lower portion of the Lower Montane Wet Forest zone in a few areas.

BIOLOGY: Food items taken from the stomachs of this common species include caecilians, many kinds of lizards (small *Iguana* and *Ctenosaura, Ameiva, Cnemidophorus, Gymnophthalmus, Sphenomorphus,* and *Mabuya*), and snakes *(Anomalepis, Helminthophis, Coniophanes, Geophis,* and *Ninia)* (Roze 1996). Reproductive patterns for *M. nigrocinctus* appear to differ for Atlantic and Pacific slope populations (Solórzano and Cerdas 1988a), based on analysis of 269 examples collected for the Instituto Clodomiro Picado, including ten live gravid females (eight Pacific, two Atlantic). Mating occurs in the early dry season (November to January) in Pacific versant snakes and presumably from January to April in the Atlantic population. Oviposition takes place in February to March (Pacific) and March to June (Atlantic) with clutch sizes of 5–7.9–14 and 5–6.7–8, respectively. Hatching (plate 466) took place in February and March (Pacific) and August (one Atlantic clutch), with incubation times of forty-seven to eighty-one days and sixty days, respectively. Eggs were 22.1–32.3 × 36–45 mm, and hatchlings were 168 to 212 mm (Pacific) and 173 to 189 mm (Atlantic) in total length. According to many Costa Ricans, only the male is venomous. Others believe that all members of this species are males, the reputed females being differently colored. Both beliefs are untrue, and the snakes presumed to be nonvenomous females are doubtless harmless coral snake mimics (see table 11.2).

REMARKS: The substantial individual variation shown by this species in coloration and scutellation has been analyzed in detail for Honduras by Wilson and Meyer (1972) and for Nicaragua and Costa Rica by Savage and Vial (1973). The former authors eschewed recognizing geographic races within the species because of the variation was discordant, and the latter authors did so because the simplicity implied by recognizing races obfuscated the complex pattern of variation. Others (Roze 1967, 1982, 1996; Villa 1984b) have continued to use trinomials for as many as six or seven subdivisions within the species. In Costa Rica differences in color pattern and scutellation exist between three geographic areas as follows:

Atlantic slope: black head cap extending posteriorly to level of orbits so that part of supraoculars and frontal light in color; nuchal black ring not crossing parietals, covering 8 to 13 scale rows (fig. 11.48b); black body rings 10 to 14; ventrals plus subcaudals 233 to 245.

Northwestern Costa Rica: black head cap extending posteriorly onto parietals so that all of supraoculars and frontal black; nuchal black ring crossing parietal, covering 4 to 7 scale rows (fig. 11.48d); black body rings 12 to 22; ventrals plus subcaudals 248 to 269.

Southwestern Costa Rica: black head cap covering supraoculars and most of frontal, but not parietals; nuchal black ring crossing parietals or not, covering 6 to 8 scale rows (fig. 11.48c); black body rings 15 to 19; ventrals plus subcaudals 237 to 244

Taylor (1951b, 1954b) called the Atlantic population *Micrurus nigrocinctus mosquitensis* and that from the southwestern region *M. nigrocinctus nigrocinctus,* following Schmidt (1933b). He also described the northwestern population as a new species, *Micrurus pachecoi,* but his specimens are conspecific with the putative form *M. nigrocinctus melanocephalus* of other authors including Roze (1967).

As pointed out by Savage and Vial (1973) and Roze (1982), the Pacific coastal populations gradually intergrade in central Costa Rica, and snakes essentially like those from southwestern Costa Rica are found along that versant to Colombia. Most Pacific and Atlantic lowland examples of this species can be instantly separated from one another based on coloration. Pacific slope specimens, in addition to the differences in head and nuchal pattern, usually have much narrower red rings and fewer black rings than Atlantic examples. Snakes of this species from the Laguna de Arenal region and the Meseta Central (both drainages) suggest limited intergradation between Atlantic and Pacific populations. While some examples intermediate in coloration (e.g., Atlantic head cap, Pacific body pattern) are found at these sites, most individuals are exclusively Atlantic or exclusively Pacific in coloration, although intermediate in scutellation. Farther to the north in Nicaragua, typical Pacific slope examples occur on that versant and in the Great Lakes region. Most of the Atlantic lowlands in Nicaragua are inhabited by snakes similar to those found on that versant in Costa Rica. However, over a broad zone in central Nicaragua individual snakes of this species may be one of three morphs: like Pacific versant examples, like snakes found in the lowlands of eastern Nicaragua and Costa Rica, and a distinctive intermediate morph. All three morphs may be found at a single locale. Savage and Vial (1973) suggested that this represents a case of secondary intergradation between formerly isolated populations that have developed some degree of genetic isolation and subsequently have come back into contact followed by limited interbreeding. Gutiérrez and Bolaños (1979, 1981) have shown that in Costa Rica, Atlantic and Pacific slope snakes of this species are significantly different in karyology. The former has a diploid number of 30, the latter 26, 27, or 28, with the principal difference between them being in the number of microchromosomes (fourteen versus ten, eleven, or twelve), and the condition of the eighth pair of macrochromosomes (telocentric versus submetacentric), respectively. Both populations are similar in the following: chromosome pairs 2 and 3 are metacentrics; 1, 6, and 7 are submetacentrics; 4 and 5 are telocentrics; females have a heteromorphism in the sixth pair, with the Z being submetacentric and the W telocentric. Their descriptions are based on forty-one specimens having the Pacific karyotype (NF = 36 or 38 in males; 34 or 35 in females) and seventeen showing the Atlantic karyotype (NF = 42 in males, 40 in females). Unfortunately although these authors state that morphologically intermediate examples always have one or the other of the two karyotypes, they do not describe the characteristics of the intermediates or their geographic provenance. These data support the idea that Costa Rican Atlantic and Pacific populations of *M. nigrocinctus* are behaving like different species. It remains to be seen what karyologic studies of snakes of this species from the zone of intermediary in Nicaragua may tell us about their relationships. It is worth noting that Atlantic slope *M. nigrocinctus* and *M. alleni* resemble one another very closely in dorsal pattern, although the former has yellow rings and the latter white. Both have a broad light head band, a broad nuchal black ring, and broad black and white or yellow rings. This similarity in coloration may reinforce the aposematic effect of the pattern, since either a learned or an innate response

would lead a predator to avoid both species (i.e., they are Müllerian mimics). Janzen (1980) has speculated that the caterpillar of the moth *Pseudosphinx tetrio* (family Sphingidae) and the turtle *Rhinoclemmys pulcherrima* are coral snake mimics in northwestern Costa Rica, where only *M. nigrocinctus* occurs. The caterpillar is about 100 mm long, ringed with black and yellow, and has the head and anteriormost segment on the body red. When touched it thrashes back and forth much like the defensive display of coral snakes and bites (unusual behavior for a caterpillar). Consequently it may be avoided by the same predators that innately or through empathetic learning avoid coral snake patterns, making it a Batesian mimic.

In addition, the caterpillar may make itself distasteful or toxic by ingesting secondary compounds from its host plant, which would make it a Müllerian mimic. The turtle, on the other hand, resembles coral snakes only in having red, black, and yellow markings. The pattern is in no way similar to the ringed pattern of coral snakes, and individuals of this species are not particularly obvious, nor do they employ any display to call attention to the coloration. Merely having red in the coloration does not a mimic make!

DISTRIBUTION: Lowlands and premontane areas from southeastern Oaxaca, Mexico (Pacific), and northwestern Honduras (Atlantic) southward to northern Colombia; found in Costa Rica over the lowlands, premontane slopes, including the Meseta Central, and marginally into the lower montane zone (0–1,500 m).

Micrurus mipartitus Group
Micrurus mipartitus (C. Duméril, Bibron, and A. Duméril, 1854)

Gargantilla

(plates 467–69; map 11.118)

DIAGNOSTICS: A ringed snake with an essentially bicolor dorsal pattern in which the light rings are various shades of red, yellowish, or white dorsally but may become lighter laterally and ventrally. The solid black head cap is restricted to the enlarged head shields lying in front of the eye, and the nuchal black ring does not cross the parietals (fig. 11.48a).

DESCRIPTION: A moderate-sized, rather slender species (maximum total length 1,130 mm, average adults 500 to 800 mm); tail very short, without strong sexual dimorphism (5 to 8% of total length); postoculars 2; supralabials 7, with 2 bordering orbit; infralabials 6 to 8, usually 7; temporals 1 + 1, 1 + 2, or rarely 0 + 1; weak supracloacal keels in males; anal divided; ventrals 237 to 244 (197 to 275) in males, 256 to 274 (222 to 335) in females; subcaudals 31 to 38 (27 to 34) in males, 23 to 29 in females; ventrals plus subcaudals 264 to 274 in males, 288 to 305 in females (range for both sexes in species as a whole 264 to 353). Bicolor black and white, cream, pink, or red (usually black and red); some scales in light rings tipped with black; 41 to 63 (34 to 80) black body rings; black head cap terminates anterior to

niles have red rings that may be retained in adults but often change with age to pink and less often to cream or white. Leenders, Beckers, and Strijbosch (1996) reported capturing snakes of all three color variants at one locality in Costa Rica. The revival (Roze 1982) of the nominal Colombian species *Micrurus multiscutatus* Rendahl and Vestergren (1940) solely because it has red rings, whereas sympatric *M. mipartitus* has white ones, should be evaluated in light of these data. Roze (1982, 1996) argued that two species, *M. multifasciatus* of central Panama to Nicaragua and *M. mipartitus* of eastern Panama and northern and western South America, may be recognized within the taxon here called *M. mipartitus*. The two species were said to differ in coloration, *M. multifasciatus* supposedly having red rings and *M. mipartitus* white or yellow ones. Since this feature is variable ontogenetically and individually in Costa Rica, there seems little validity to Roze's proposal, although it appears that there are more red- or pink-ringed individuals of this group in Central America and more white- or yellow-ringed snakes in South American samples. Roze (1970) used the color of the light rings and geographic variation in ventrals to recognize six geographic races within *M. mipartitus*. The ventral counts show clinal variation, with a gradual increase in number from north to south (Nicaragua to Ecuador) and east to west (north-central Venezuela to western Colombia). Individual and geographic variation in coloration and lack of concordance between color and ventral counts do not support Roze's arrangement. The local individual variation and general geographic trend in coloration for *M. mipartitus* are paralleled in the harmless coral snake mimic *Urotheca euryzona*, which is sympatric with *M. mipartitus* over most of its range. In Costa Rica *U. euryzona* has the light rings entirely red, pink, or white, the rings are white or yellow with the upper three-quarters to one-quarter red, or the upper portion is yellow. This range of variation closely matches individual variation in Costa Rican *M. mipartitus*, and as in the coral snake model, individuals having red or pink rings predominate. On the geographic scale individuals of both species usually have red rings in western and central Panama, but examples of *U. euryzona* from eastern Panama and South America have white rings. *M. mipartitus* shows a similar change, since almost all examples from eastern Panama and South America have white or yellow rings on the body, although the head band and light tail rings are red. The precise resemblance between these two species in individual and geographic variation is among the types of evidence supporting coral snake mimicry by harmless forms. Taylor et al. (1973) state that this species occurs in the Pacific lowlands of Guanacaste Province, based at least in part on an erroneous interpretation of locality data. One snake in question is listed in the Instituto Clodomiro Picado records as from Colonia la Libertad de Liberia, Guanacaste Province. This village was recently established by settlers from the city of Liberia. However, the site is actually on the Atlantic slope, approximately 16 km north-northeast of

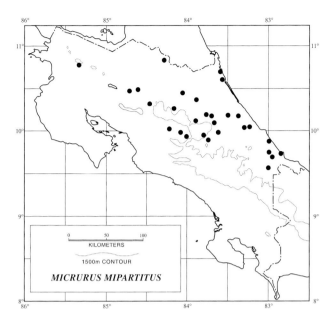

Map 11.118. Distribution of *Micrurus mipartitus*.

level of orbits (between the orbits in some extralimital populations); black nuchal ring not crossing parietals (crossing parietal tips elsewhere); area between black nuchal ring and head cap immaculate white, cream, pink, or red; iris black, tongue black. The everted hemipenes very short (about 6 subcaudals long), strongly bilobed, covered by moderate, robust spines, with small apical spines, sulcus spermaticus centripetal; no lateral fold in basal naked pocket.

SIMILAR SPECIES: (1) Dorsally bicolor examples of *Micrurus alleni* have the black head cap extending onto the parietals at least along the interparietal suture (fig. 11.48e). (2) *Urotheca euryzona* has most of the upper head shields black, including most of the parietals. (3) *Sibon anthracops* has 16 to 31 black body rings. (4) Bicolor individuals of *Lampropeltis triangulum* have three scales (preocular, loreal, and postnasal) between the eye and nostril (fig. 11.9c). See key to coral snakes (p. 706), figure 11.47, and table 11.2.

HABITAT: Undisturbed Atlantic Lowland Moist and Wet Forests and Premontane Moist and Wet Forests and Rainforest.

BIOLOGY: A relatively uncommon snake. A captive female from the Atlantic versant laid eighteen eggs on January 28. The eggs were 18–26 × 12–14 mm, and five hatched on March 31. The hatchlings averaged 169 m in total length (Solórzano and Cerdas 1988b).

REMARKS: The color of the light rings in this species is matched closely by that of the harmless *Urotheca euryzona*, although in that form cream, yellow, or red frequently is replaced by yellow or white laterally and ventrally. In Costa Rican *M. mipartitus* the light rings most commonly are entirely red or pink, the rings are white to cream with the pink restricted to the upper scale rows, or the rings may be cream dorsally or entirely white. There appears to be ontogenetic color change in Costa Rican members of this species. Juve-

Volcán Rincón de la Vieja in Alajuela Province. I know of no valid record of this species from Pacific lowland Costa Rica, although it occurs just west of the continental divide at some places in the Cordillera de Guanacaste (about 800 m). Savage and Vial (1973) list *M. mipartitus* from Monteverde, Puntarenas Province (1,400 m). No additional examples of this form have subsequently been collected there or at similar elevations. The snake in question (CRE 7246) was probably taken at a lower elevation to the east of this locale and was not properly labeled. The karyotype (from a single Atlantic slope male) has a diploid number of 34, with seven pairs of macrochromosomes, ten pairs of microchromosomes, and a secondary constriction in the second pair of macrochromosomes. Pairs 1, 3, 4, and 5 are metacentrics, 2 and 6 are submetacentrics, and 7 is a telocentric; NF = 34 (Gutiérrez and Bolaños 1979).

DISTRIBUTION: Central American lowlands and premontane slopes on the Atlantic versant from Nicaragua to central Panama; a few records from just west of the continental divide in northwestern Costa Rica and from the Meseta Central Occidental; Pacific versant eastern Panama south in the lowlands and on the mountain slopes of Colombia and Ecuador (possibly reaching northern Peru); Caribbean drainage Colombia to north-central Venezuela (2–1,160 m).

Subfamily Hydrophiinae

Sea Snakes

Genus *Pelamis* Daudin, 1803

This monotypic genus is the only marine snake that ranges into Costa Rican waters. In addition to the strongly compressed, paddle-shaped tail, small ventrals, superior nostrils, and juxtaposed scales (a combination found in no other American snake), *Pelamis* has the following characteristics: upper head surface covered by large regular shields (a rostral, nasals, prefrontals, supraoculars, frontal, and parietals); body short, stout, much compressed; head narrow, not distinct from neck; snout elongate; nasals meeting dorsally, undivided; no loreal; one or two preoculars; eye moderate, pupil round; tongue short, only forked part protrusible; no definite chin shields, throat covered by somewhat elongated small scales; no mental groove; dorsal scales hexagonal or squarish, mostly smooth, but lowermost rows with two or three small tubercles; maximum number of scale rows 49 to 67; no apical pits; ventrals small, divided by a median furrow, or same size as adjacent scales; precloacals somewhat enlarged; 7 to 11 maxillary teeth separated from caniculate fangs by a diastema; mandibular teeth equal; posterior vertebrae with hypapophyses. No other Costa Rican snake resembles the sole member of this genus, whose paddle-shaped tail is immediately diagnostic.

The mostly spinous hemipenes are symmetrical and weakly bilobed; the centripetal sulcus spermaticus divides near the tip of organ, which is covered by small papillae (McDowell 1972).

The karyotype is 2N = 38, consisting of ten pairs of macrochromosomes and nine pairs of microchromosomes; pairs 1 and 2 are metacentrics, 3 is a subtelocentric, and 4 to 9 have the centromere in a terminal position. Females have a slightly heteromorphic pair of sex chromosomes (metacentrics), one slightly smaller and with a more submedian centromere; NF = 44 (Gutiérrez and Bolaños 1980).

The genus ranges throughout the waters of the Indo–Pacific region from the Red Sea and Indian Ocean throughout the Indo-Australian area south to Tasmania and New Zealand and north to southern Siberia and across the tropical Pacific Ocean to extreme southern California, Ecuador, and possibly off extreme northern Peru; Taylor's (1951b) mention of Chile as the southern limit of the genus in the Americas is clearly in error. The extreme northern and southern records for the genus represent expatriates, because permanent breeding populations are generally restricted to subtropical and tropical waters where the surface temperature rarely is below 18°C.

Pelamis platurus (Linné, 1766)

Culebra del Mar

(plates 470–72; map 11.119)

DIAGNOSTICS: The compressed, paddle-shaped tail, which is yellowish to bright canary and almost always marked with black spots or bars, is absolutely diagnostic.

DESCRIPTION: A moderate-sized snake (adult total length 600 to 1,140 mm; usually no more than 900 mm, but most adults average 750 mm in the eastern Pacific); females longer than males; tail short (9 to 13% of total length); 2 or

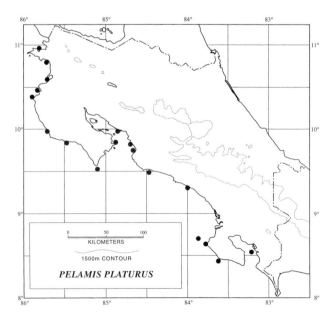

Map 11.119. Distribution of *Pelamis platurus*.

3 postoculars; 1 or 2 suboculars usually present; 7 to 11 supralabials, with usually 1 but sometimes 0 or 2 bordering orbit; infralabials 9 to 13; temporals small, variable, usually 2 or 3 anteriorly; dorsal scales in 44 to 67 rows around midbody; scale tubercles much larger in males than females; ventrals 254 to 465; subcaudals 39 to 66. Body pattern extremely variable, basically with a very broad to very narrow dark brown or black dorsal stripe, a bright yellow lateral stripe, and a brownish ventral stripe, but dark areas often broken into bars or spots, or sides and venter uniform cream, yellow, or brown, rarely almost yellow or black or dark areas with light spots; tail usually yellow marked with black spots, bars, or bands, rarely uniform yellow.

SIMILAR SPECIES: There are none.

HABITAT: A strictly marine snake ranging far out to sea in the Pacific Ocean; dead or dying individuals are often found on beaches along the Pacific coast. This snake is seen in substantial numbers off northwestern Costa Rica in the dry season (December to April).

BIOLOGY: This species is strictly aquatic and the only truly pelagic sea snake. It is most commonly found 1 to 20 km offshore, often in congregations of many hundreds of snakes in drift lines or slicks (Kropach 1975). The drift lines—bands of debris 3 to 6 m wide and frequently many kilometers long—are maintained for long periods by wind, waves, and temperature fronts. *Pelamis* usually are passive drifters, but when they come in contact with the drift line they remain there (much like the debris) to feed on the concentration of fishes attracted by the abundance of invertebrates and smaller fishes. The diet of this species is exclusively fishes of a great many families that gather under the snakes in the drift line. When feeding *Pelamis* swim slowly backward on the surface with the mouth slightly ajar. Small fish line up under the snake, and as they gradually fall back under the head, it strikes out laterally and somewhat ventrad to bite the prey (Pickwell 1972). The neurotoxic venom quickly (in less than a minute) immobilizes the prey, which it then usually swallows headfirst. Mating occurs at sea and may be seasonal off Costa Rica, with the apparent peak in July. *Pelamis* are viviparous and have placental connections between embryos and their mother. The gestation period is somewhere between six and eight months. Broods consist of one to six snakelets that are 220 to 260 mm in total length at birth. Males attain sexual maturity at about 500 mm in total length and females at 645 to 765 mm (Kropach 1975). The coloration in this form may reduce its visibility from above and below (countershading), or it may be disruptive. It has also been postulated that the bright, contrasting coloration of *Pelamis* is aposematic, since there is no record of natural predation on it. In laboratory experiments carried out in Panama, predatory fishes avoided the snakes almost absolutely. Atlantic fishes, on the other hand, attacked *Pelamis* under similar conditions, although only 11% were successful (Kropach 1975). Since *Pelamis* do not occur in the Atlantic, these data seem to support the hypothesis of aposematic coloration. The ostentatious coloration of the tail may lure prey, deflect a predator's attention from the head, or both. The snake sheds its skin by tying itself into a knot and then pulling its body through the knot to leave the skin behind (Voris 1983). Sea snakes can inflict serious injury or death on humans, because their venom contains an extremely virulent neurotoxin. It is unclear how dangerous to humans a bite from this snake may be. Bolaños (1984) recorded no bites from Costa Rica. Kropach (1972) reported six symptomless bites from Panama; they may have been "dry" bites. In a case reported by Solórzano (1995), a herpetologist was bitten on the hand while collecting *P. platurus* in Costa Rica. The victim's reaction included some pain, some edema, and slight hematoma, but there were no serious effects that required medical treatment. Anecdotal reports of death from bites by this species from the eastern Pacific remain unconfirmed (Culotta and Pickwell 1993). These snakes should not be casually approached in the water or carelessly handled if caught in a fisherman's net or found washed up on the beach. It is a common belief among many Costa Rican fisherman and others that this species is actually an eel. Conversely, some people insist that any eeliform marine or freshwater fish is a snake. Obviously these confusions are not based on a sound knowledge of biology.

REMARKS: Color variation in this species is extensively described by Tu (1976), and Pickwell and Culotta (1980) provided a useful redescription and summary of the literature on the species. Villa (1962, 1984b) and Alvarez del Toro (1973) reported a second species of sea snake, *Laticauda colubrina*, in waters offshore from Mexico, El Salvador, and Nicaragua. These records are suspect because no specimens are available for study, and the counts Villa (1962) gives for the two examples he examined (19 scales rows at midbody) are different from those known for the species (21 to 25 scale rows).

DISTRIBUTION: Breeding populations occur in tropical and subtropical seas in the Indian Ocean from the Horn of Africa to South Africa, eastward through the Bay of Bengal, from the Indo-Australian archipelago northward to southern Japan, southward along both coasts of Australia, and thence westward to Mexico, Central America, and northern South America. Expatriates are known from the Red Sea, Siberia, and southern California on the north and from Tasmania, New Zealand, and presumably extreme northern Peru. In Costa Rica restricted to the Pacific in open ocean and coastal waters and bays.

Family Viperidae Oppel, "1810," 1811

Adders and Vipers

The members of this family (32 genera and about 225 species) are the most highly efficient predators among snakes. Their effective venom delivery apparatus, the blinding speed of the lethal strike, and the skull modifications that allow them to maximize their gape for swallowing combine to ensure capture, killing, and engulfment of large food items. The family is characterized by the very short, rotatable maxillary bone that bears two fang sockets and at times two elongate hollow curved fangs (usually only one present). The fangs show no sign of a groove along the anterior face (unlike those of the elapids). During development the venom canal forms from an originally open groove that closes through the infolding of the lateral lips along the groove's margin, and a suture line is visible on the face of the functional fang. When the snake's mouth is closed, the fangs lie folded against the roof of the mouth. During the strike the movable maxillary bone and the fangs rotate 90 to 100° forward so that on impact the fangs stab the victim and the venom is simultaneously injected. The venom delivery system, although evolved for predation, may also be used for defense and makes the adders and vipers extremely dangerous, capable of causing death or serious injury in humans.

Most vipers have a short, stout body that enables them to swallow proportionately very large prey, and a broad, arrow-shaped head that is very distinct from the neck. Most species are small to moderate sized, but several exceed 2 m in length, and the bushmasters *(Lachesis)* are very large, reaching somewhat over 3.5 m. Many vipers are terrestrial, a few are semiaquatic, and a considerable number are arboreal forms, some with prehensile tails. Most vipers are viviparous, with well-developed omphalallantoic and chorioallantoic placentas. Among American species only the bushmaster is oviparous.

Aside from the characteristics associated with venom delivery, this family closely resembles the Colubridae and Elapidae in internal features: flexible lower jaw; no coronoid bone; no pelvic structures; no premaxillary teeth; no left lung or a rudimentary one; tracheal lung present; elongate, posteriorly inclined quadrate and elongate supratemporal; hyoid composed of paired cornuae that form a structure with elongate, posteriorly directed branches; no prefrontal bone; elongate hypapophyses on all dorsal vertebrae.

The family is widely distributed on all continents except Australia and is absent from Madagascar, New Guinea and adjacent islands, Oceania, the Greater Antilles, western Peru, and Chile. Although the number of humans bitten worldwide by members of this family each year (about 250,000, with perhaps 15,000 deaths) is lower than for elapids, the vast majority of crippling and lethal snakebites in the Americas are from vipers. In Costa Rica about five hundred serious viper bites are reported each year. There are about seven or eight deaths annually, almost always caused by bites of the terciopelo or fer-de-lance, *Bothrops asper.* Any viper, however, can cause a painful or permanently damaging injury, which at the least is a very unpleasant experience and is definitely to be avoided!

The members of the family are placed in three subdivisions as follows:

Subfamily Azemiopinae Liem, Marx, and Rabb, 1971 (monotypic): no loreal sensory pit or cavity in the maxillary bone; anteroventral medial wing and posteroventral medial process of prefrontal well developed; levator anguli oris muscle absent; cerebral artery and posterior Vidian canal pass through a common foramen (upper Myanmar, southern China, and Tibet)

Subfamily Viperinae Oppel, "1810," 1811 (thirteen genera): no loreal sensory pit or cavity in the maxillary bone; anteroventral medial wing of prefrontal poorly developed, posteroventral process absent; levator anguli oris muscle present; separate foramina for cerebral artery and posterior Vidian canal (Africa, Eurasia, and Arabia through Iraq, Iran, and Pakistan to India and Sri Lanka; Myanmar, Thailand, and at scattered localities in the East Indies to near Flores Island)

Subfamily Crotalinae Oppel, "1810," 1811 (eighteen genera): a loreal pit and a deep lateral cavity in the maxillary bone to hold pit organ; anteroventral medial wing of prefrontal poorly developed, posteroventral process absent; levator anguli oris muscle present; separate foramina for cerebral artery and posterior Vidian canal (the Americas, exclusive of the Greater Antilles, northern Lesser Antilles, western Peru, and Chile; from extreme eastern Europe across Asia to China and Japan, south through China, Southeast Asia, Malaysia, and Indonesia to Timor and the Philippines; India and Sri Lanka

In Costa Rica eight genera of the subfamily Crotalinae or pitvipers are represented. These forms are most easily distinguished from all other snakes in the republic by having a well-developed loreal pit between the eye and nostril, a feature they share with all members of the subfamily. *Any snake with a loreal pit is immediately identifiable as venomous!* The pit serves as the opening to a well-developed sensory organ that is sensitive to radiant heat (the infrared portion of the nonvisible spectrum) impinging on its diaphragm. Experimental evidence indicates that this device is used to locate prey that are warmer or cooler than the background environment. The pit organs are sensitive to temperature differences of about 10°C, depending on distance from the source, and this combined with their stereoscopic coverage of the field of reception makes them a significant device for detecting and locating prey or predators.

Unfortunately, the presence of the loreal pit is not obvious in the smaller species of pitvipers or in juvenile examples of large ones. Often it can be seen only through close examination of dead or caged examples. If you see snakes in the field that you suspect are pitvipers, do not approach closely enough to see the pit. Handle recently killed pitvipers with great care, since reflex biting can lead to envenomation. The head of a decapitated pitviper can inject venom into a careless human several hours after being severed from the body!

Plate 424. *Elaphe triaspis*; adult. Costa Rica: Puntarenas: Monteverde, about 1,400 m.

Plate 425. *Elaphe triaspis*; juvenile. Costa Rica: Puntarenas: Monteverde, about 1,400 m.

Plate 426. *Elaphe triaspis*; adult. Costa Rica: San José: Dulce Nombre, 1,340 m.

Plate 427. *Lampropeltis triangulum*. Costa Rica: Puntarenas: Monteverde, about 1,400 m.

Plate 428. *Lampropeltis triangulum* swallowing a spiny pocket mouse (*Heteromys*). Costa Rica: Heredia: La Selva Biological Station, about 60 m.

Plate 429. *Leptodrymus pulcherrimus*. Costa Rica: Puntarenas: San Luis, about 700 m.

Plate 430. *Leptophis ahaetulla.* Costa Rica: Heredia: La Selva Biological Station, about 60 m.

Plate 431. *Leptophis depressirostris.* Costa Rica: Heredia: La Selva Biological Station, about 60 m.

Plate 432. *Leptophis depressirostris,* showing threat display. Costa Rica: Heredia: La Selva Biological Station, about 60 m.

Plate 433. *Leptophis mexicanus;* uncommon color phase with reduced black dorsal striping. Costa Rica: Guanacaste: Santa Rosa National Park, about 280 m.

Plate 434. *Leptophis mexicanus;* view of head, showing threatening behavior. Costa Rica: Guanacaste, about 20 m.

Plate 435. *Leptophis nebulosus.* Costa Rica: Puntarenas: Monteverde, about 1,400 m.

Plate 436. *Mastigodryas melanolomus;* adult sleeping at night. Costa Rica: Heredia: La Selva Biological Station, about 60 m.

Plate 437. *Mastigodryas melanolomus;* juvenile. Costa Rica: Heredia: La Selva Biological Station, about 60 m.

Plate 438. *Oxybelis aeneus*. Costa Rica: Guanacaste: La Pacifica, about 40 m.

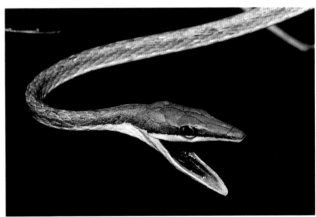

Plate 439. *Oxybelis aeneus*, showing threat display. Costa Rica: Heredia: La Selva Biological Station, about 60 m.

Plate 440. *Oxybelis brevirostris*. Costa Rica: Alajuela: Peñas Blancas Valley, about 800 m.

Plate 441. *Oxybelis fulgidus*. Costa Rica: Puntarenas: Monteverde, about 1,400 m.

Plate 442. *Pseustes poecilonotus.* Costa Rica: Heredia: La Selva Biological Station, about 60 m.

Plate 443. *Rhinobothryum bovallii.* Costa Rica: Limón: Guayacán, about 550 m.

Plate 444. *Scaphiodontophis annulatus;* adult, Atlantic slope population. Costa Rica: Heredia: La Selva Biological Station about 60 m.

Plate 445. *Scaphiodontophis annulatus;* juvenile. Costa Rica: Heredia: La Selva Biological Station, about 60 m.

Plate 446. *Scaphiodontophis annulatus;* southwestern Pacific population. Costa Rica: Puntarenas: Palmar Sur, 25 m.

Plate 447. *Scolecophis atrocinctus.* Costa Rica: Puntarenas: Monteverde, about 1,400 m.

Plate 448. *Scolecophis atrocinctus*; view of head. Nicaragua.

Plate 449. *Spilotes pullatus*. Costa Rica: Heredia: La Selva Biological Station, about 60 m.

Plate 450. *Stenorrhina degenhardtii*. Costa Rica.

Plate 451. *Stenorrhina freminvillii*. Costa Rica: San José: Brasil, 880 m.

Plate 452. *Tantilla alticola*. Costa Rica: Puntarenas: Las Tablas, about 1,600 m.

Plate 453. *Tantilla armillata*. Costa Rica: Cartago: Tapantí National Park, about 1,200 m.

Plate 454. *Tantilla reticulata*. Costa Rica: Alajuela: Peñas Blancas Valley, about 800 m.

Plate 455. *Tantilla ruficeps*. Costa Rica: Puntarenas: Monteverde, about 1,400 m.

Plate 456. *Tantilla schistosa*. Costa Rica: Alajuela: Peñas Blancas Valley, about 800 m.

Plate 457. Tantilla supracincta. Costa Rica: Alajuela: Peñas Blancas Valley, about 800 m.

Plate 458. *Tantilla supracincta;* subadult showing transition from bicolor to tricolor pattern. Costa Rica: Limón: Comadre, 20 m.

Plate 459. *Trimorphodon biscutatus*. Costa Rica: Guanacaste: Santa Rosa National Park, about 280 m.

Plate 460. *Thamnophis proximus.* Costa Rica: Guanacaste: Santa Rosa National Park, about 285 m.

Plate 461. *Micrurus alleni.* Costa Rica: Heredia: La Selva Biological Station, about 60 m.

Plate 462. *Micrurus clarki.* Costa Rica: Puntarenas: Punta Gallardo, about 20 m.

Plate 463. *Micrurus nigrocinctus.* Costa Rica: Puntarenas: Monteverde, about 1,400 m.

Plate 464. *Micrurus nigrocinctus;* view of head. Costa Rica: Heredia: La Selva Biological Station, about 60 m.

Plate 465. *Micrurus nigrocinctus,* showing antipredator behavior of tail waving and hiding head under coils. Costa Rica: Heredia: La Selva Biological Station, about 60 m.

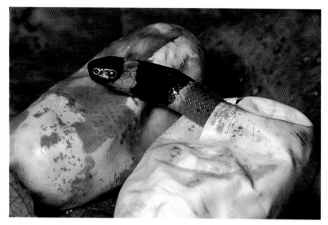

Plate 466. *Micrurus nigrocinctus* hatching from egg. Mother from Costa Rica: Cartago: San Diego de Tres Ríos, about 1,200 m.

Plate 467. *Micrurus mipartitus*. Costa Rica: Heredia: La Selva Biological Station, about 60 m.

Plate 468. *Micrurus mipartitus*. Costa Rica: Limón: San Andrés, 10 m.

Plate 469. *Micrurus mipartitus;* view of head. Costa Rica: Limón: San Andrés, 10 m.

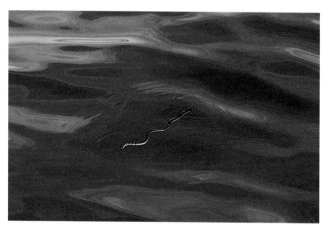

Plate 470. *Pelamis platurus*. Costa Rica: Guanacaste: Bahía de las Culebras.

Plate 471. *Pelamis platurus* on the beach. Costa Rica: Guanacaste: Bahía de las Culebras.

Plate 472. *Pelamis platurus;* view of head. Costa Rica: Guanacaste: Bahía de las Culebras.

Plate 473. *Agkistrodon bilineatus.* Costa Rica: Guanacaste: Santa Rosa National Park, about 285 m.

Plate 474. *Atropoides nummifer;* brown phase. Costa Rica: Cartago: Turrialba, about 620 m.

Plate 475. *Atropoides nummifer;* gray phase. Costa Rica: San José: San José, about 1,200 m.

Plate 476. *Atropoides picadoi.* Costa Rica: Limón: Siquírres, about 62 m.

Plate 477. *Bothriechis lateralis;* adult. Costa Rica: Alajuela: Monteverde Cloud Forest Preserve, about 1,600 m.

Plate 478. *Bothriechis lateralis;* juvenile. Costa Rica: Alajuela: Monteverde Cloud Forest Preserve, about 1,600 m.

Plate 479. *Bothriechis nigroviridis.* Costa Rica: Alajuela: Monteverde Cloud Forest Preserve, about 1,600 m.

Plate 480. *Bothriechis schlegelii.* Costa Rica: Heredia: La Selva Biological Station, about 60 m.

Plate 481. *Bothriechis schlegelii.* Costa Rica: Heredia: La Selva Biological Station, about 60 m.

Plate 482. *Bothriechis schlegelii;* brown color phase. Costa Rica: Heredia: La Selva Biological Station, about 60 m.

Plate 483. *Bothriechis schlegelii*; oropel color phase coiled among palm fruits. Costa Rica: Heredia: La Selva Biological Station, about 60 m.

Plate 484. *Bothriechis schlegelii*; female oropel giving birth to a litter; note color variation in neonates. Costa Rica: Heredia: La Selva Biological Station, about 60 m.

Plate 485. *Bothriechis "supraciliaris"* Costa Rica: Puntarenas: Las Cruces Biological Station, about 1,200 m.

Plate 486. *Bothrops asper*. Costa Rica: Heredia: La Selva Biological Station, about 60 m.

Plate 487. *Cerrophidion godmani*. Costa Rica: San José Province.

Plate 488. *Lachesis melanocephala.* Costa Rica: Puntarenas: Península de Osa.

Plate 489. *Lachesis stenophrys.* Costa Rica: Heredia: La Selva Biological Station, about 60 m.

Plate 490. *Lachesis stenophrys* hatching from egg. Mother from Costa Rica: Limón: Comadre, 20 m.

Plate 491. *Porthidium nasutum.* Costa Rica: Heredia: La Selva Biological Station, about 60 m.

Plate 492. *Porthidium nasutum* swallowing an *Eleutherodactylus megacephalus;* note use of fangs to manipulate frog into mouth. Costa Rica: Heredia: La Selva Biological Station, about 60 m.

Plate 493. *Porthidium ophryomegas.* Costa Rica: Guanacaste: Santa Rosa National Park, about 280 m.

Plate 494. *Porthidium volcanicum;* holotype. Costa Rica: Puntarenas: Ujarrás de Buenos Aires, 570 m.

Plate 495. *Crotalus durissus.* Costa Rica: Guanacaste: Santa Rose National Park, about 280 m.

Plate 496. *Kinosternon leucostomum.* Costa Rica: Heredia: La Selva Biological Station, about 60 m.

Plate 497. *Kinosternon scorpioides.* Costa Rica: Guanacaste: Palo Verde Biological Station, about 10 m.

Plate 498. *Dermochelys coriacea.* Costa Rica: Guanacaste: Las Baulas National Park.

Plate 499. *Caretta caretta.*

Plate 500. *Chelonia agassizii*; mating. Costa Rica: Guanacaste: Bahía de las Culebras.

Plate 501. *Chelonia agassizii*; nesting. Mexico: Michoacán.

Plate 503. *Eretmochelys imbricata*; nesting.

Plate 502. *Chelonia mydas*. Costa Rica: Limón: Tortuguero National Park.

Plate 504. *Lepidochelys olivacea*. Costa Rica: Guanacaste: Playa Nancite, Santa Rosa National Park.

Plate 505. *Lepidochelys olivacea*; nesting. Costa Rica: Guanacaste: Playa Nancite: Santa Rosa National Park.

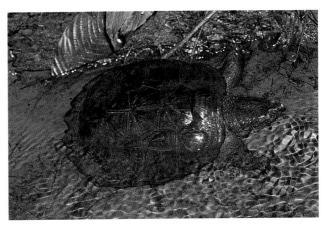

Plate 506. *Chelydra serpentina*. Costa Rica: Heredia: La Selva Biological Station, about 60 m.

Plate 507. *Chelydra serpentina*; view of head. Costa Rica: Heredia: La Selva Biological Station, about 60 m.

Plate 508. *Chelydra serpentina*. Costa Rica: San José: Río General, about 600 m.

Plate 509. *Rhinoclemmys annulata*. Costa Rica: Limón: Guayacán, about 550 m.

Plate 510. *Rhinoclemmys funerea*. Costa Rica: Heredia: La Selva Biological Station, about 60 m.

Plate 511. *Rhinoclemmys pulcherrima*. Costa Rica: Guanacaste: Santa Rosa National Park, about 280.

Plate 512. *Chrysemys ornata*. Costa Rica: Limón: Puerto Viejo, about 5 m.

Plate 513. *Caiman crocodilus*. Costa Rica: Heredia: La Selva Biological Station, about 60 m.

Plate 514. *Caiman crocodilus;* view of head. Costa Rica: Heredia: La Selva Biological Station, about 60 m.

Plate 515. *Crocodylus acutus*. Costa Rica: Puntarenas: Río Grande de Tárcoles, about 15 m.

Plate 516. *Crocodylus acutus;* view of head. Costa Rica: Limón: Tortuguero, about 3 m.

Other external features characteristic of all Costa Rican pitvipers include a lanceolate or arrow-shaped head that is much wider than the neck and very distinct from it, a vertically elliptical pupil, a single pair of elongate chin shields, and strongly keeled dorsal scales. Several harmless or mildly venomous Costa Rican snakes resemble pitvipers in head shape and pupil shape, including the species of the following genera: (Boidae) *Boa, Corallus, Epicrates;* (Ungaliophiidae) *Ungaliophis;* (Colubridae) *Dipsas, Imantodes, Leptodeira,* and *Sibon.* Members of these taxa all have smooth dorsal scales. The little leaf litter snake *Nothopsis ruqosus* and juveniles of *Elaphe triaspis* superficially resemble some Costa Rican pitvipers, but unlike them have round pupils. *Trimorphodon biscutatus* has a dark V-shaped or lyre-shaped mark (apex directly anteriorly) on the head and neck, and the dorsal scales are smooth anteriorly and weakly keeled posteriorly. *Xenodon rabdocephalus* has coloration and markings similar to those of the common and very dangerous terciopelo, *Bothrops asper.* The harmless species, however has smooth oblique dorsal scales and a round pupil. Table 11.3 will help you differentiate these snakes from pitvipers.

A number of harmless Costa Rican colubrids of the genera *Chironius, Dendrophidion, Drymobius, Leptophis,* and *Oxybelis* that are mostly green might be mistaken for similarly colored members of the aboreal pitvipers of the genus *Bothriechis.* All of the harmless green species are slender to attenuate snakes with round pupils and long thin tails, whereas the vipers are robust forms with vertically elliptical pupils and short prehensile tails.

Individuals of all of the taxa mentioned above may be initially misidentified as pitvipers, particularly by a novice or nonherpetologist. It is unlikely, however, that a pitviper will be erroneously identified as one of these species except for the boas and the species of *Trimorphodon* and *Xenodon.* Remember that only pitvipers have a loreal pit and that boas, *Trimorphodon,* and *Xenodon,* unlike pitvipers, have smooth dorsal scales (weakly keeled posteriorly in *Trimorphodon*). Table 11.3 summarizes features besides the presence or absence of the loreal pit that will help identify snakes often confused with pitvipers. In the field most observers will have neither the time nor the inclination to check out these details, so treat any snake even superficially resembling a pitviper circumspectly until its identity is firmly established.

The generic arrangement in the key follows the system proposed by Burger (1971), as accepted by Smith and Smith (1976) and Campbell and Lamar (1989) and modified by Werman (1992) and Campbell and Lamar (1992).

KEY TO THE GENERA OF AMERICAN PITVIPERS (FAMILY VIPERIDAE: SUBFAMILY CROTALINAE)

1a. Rattle or button present (fig. 11.8c,d); posterior 9 or 10 caudal vertebrae enlarged and fused into shaker.... 2
1b. Rattle and button lacking (fig. 11.19); posterior caudal vertebrae separate and not enlarged 3
2a. Top of head with large plates anteriorly, usually 9, including a single frontal and a pair of large symmetrical parietals in contact with one another ...... *Sistrurus*
2b. Top of head with scales and plates of varying size anteriorly; frontal and parietals, if present, separated by small scales.................... *Crotalus* (p. 735)
3a. Distal subcaudals in 1 or 2 rows (fig. 11.9a)....... 4
3b. Distal subcaudals in 4 or 5 rows (fig. 11.9b); tail not prehensile .... *Lachesis* (p. 729)
4a. Top of head with scales and plates of varying size anteriorly; if enlarged frontal and parietals present, 2–3 canthals on each side and/or parietals separated by small scales or parietals and frontal bordered by small scales 5
4b. Top of head covered with large plates anteriorly, usually 9, including a pair of prefrontals, a single frontal, and a pair of large symmetrical parietals in contact with one another; tail not prehensile..... *Agkistrodon* (p. 718)
5a. Tail prehensile 6
5b. Tail not prehensile 8

Table 11.3. Identification of Costa Rican Snakes Likely to Be Confused with Pitvipers

Upper Head Surface Covered by Large Symmetrical Shields		Frontal and Parietal Areas or Entire Upper Head Surface Covered by Small Scales	
Pupil Round	Pupil Vertically Elliptical	Pupil Round	Pupil Vertically Elliptical
Chironius K	*Ungaliophis* S	*Nothopsis* K	*Boa* S
Dendrophidion K	*Dipsas* S		*Corallus* S
Drymobius K	*Imantodes* S		*Epicrates* S
Elaphe K	*Leptodeira* S		*Atropoides* K
Leptophis K	*Sibon* S		*Bothriechis* K
Oxybelis S or K	*Trimorphodon* wk		*Bothrops* K
Stenorrhina S	*Agkistrondon* K		*Lachesis* K
Xenodon S	*Cerrophidion* K		*Porthidium* K
			Crotalus K

*= pitvipers; S = smooth dorsal scales; K = keeled dorsal scales; wk = posterior dorsal scales weakly keeled.

6a. Subcaudals divided throughout or entire anteriorly and divided posteriorly (fig. 11.5) 7
6b. Subcaudals entire (fig. 11.5). . . . *Bothriechis* (p. 723)
7a. An elongate erect hornlike supraciliary scale above eye; subcaudals divided throughout . . . . . *Ophryacus*
7b. No elongate, erect hornlike supraciliary scale above eye; subcaudals divided throughout or only posteriorly . . . . *Bothriopsis*
8a. Subcaudals divided throughout (fig. 11.5) . . . . *Bothrops* (p. 726)
8b. Subcaudals mostly entire (fig. 11.5). 9
9a. Frontal and parietal region covered by small imbricate and mostly keeled scales; scales on upper surface of snout keeled 10
9b. Definite enlarged smooth frontal and parietal plates; scales on upper surface of snout mostly smooth (fig. 11.20a) . . . . *Cerrophidion* (p. 728)
10a. No definite enlarged smooth supraocular plate extending medially over eye; supraocular narrow or area covered by small scales (fig. 11.20b). . . . . *Atropoides* (p. 720)
10b. A definite enlarged smooth supraocular scale extending medially over eye (fig. 11.20c) . . . . *Porthidium* (p. 732)

The genera of this subfamily may be placed in three informal subdivisions and will be treated by these groupings in the following accounts:

1. Agkistrodontines: *Agkistrodon* of North and Central America and *Calloselasma, Deinagkistrodon, Gloydius,* and *Hypnale* of temperate and tropical Asia
2. Lachesines: *Atropoides* of Mexico and Central America, *Trimeresurus* of Asia, *Bothriechis* of Central America and northern South America, *Bothriopsis* of South America, *Bothrops* of tropical America, *Cerrophidion* of Mexico and Central America, *Lachesis* of Central and South America; *Ophryacus* of Mexico, and *Porthidium* of Mexico to northern South America
3. Crotalines: *Crotalus* of North, Central, and South America and *Sisturus* of North America

All the American genera of pitvipers share the following features in addition to those listed above: head very distinct from neck; eye small, pupil vertically elliptical; a single pair of enlarged chin shields bordering the mental groove; dorsal scales keeled, with a posterior reduction in the number of rows; anal entire; hypapophyses on all body vertebrae.

The chromosome complements of all Costa Rican pitvipers, except that of *Agkistrodon bilineatus,* have been described in detail (Gutiérrez, Taylor, and Bolaños 1979; Solórzano and Cerdas 1986; Solórzano, Gutiérrez, and Cerdas 1988). All have a diploid (2N) number of 36, composed of eight pairs of macrochromosomes and ten pairs of microchromosomes with the fourth pair (ZW) heteromorphic in females. This karyotype is characteristic of all pitvipers stud-

ied to date. Most Costa Rican species have pairs 1, 3, 4, and 7 metacentric, 2 and 5 submetacentric, 6 and 8 subtelocentric, and the W chromosome submetacentric and shorter than the Z; NF = 48. Slight differences in this pattern occur in *Cerrophidion godmani* (chromosome 8 metacentric and W subtelocentric; NF = 48), *Bothriechis lateralis* (chromosome 8 submetacentric; NF = 50), and *Lachesis* (W chromosome subtelocentric; NF = 46). *Agkistroden bilineatus* (based on a single non–Costa Rican male) has 2N = 36, with eight pairs of macrochromosomes and ten pairs of microchromosomes; NF = 52 (Baker, Mengden, and Bull 1972).

Agkistrodontines

Genus *Agkistrodon* Beauvois, 1799

This genus comprises four species: the copperhead *(Agkistrodon contortrix)* and the cottonmouth or water moccasin *(Agkistrodon piscivorous)* of the eastern and central United States and the cantils *(Agkistrodon bilineatus* and *A. taylori)* of Mexico and Central America. Members of the genus have the upper surface of the head covered by nine unkeeled enlarged shields, as in most colubrid snakes: paired internasals, prefrontals, supraoculars, and parietals, and a single frontal. The following additional features characterize the taxon: body cylindrical; nasal pore present; dorsal scales in 23 to 25 rows at midbody; two apical scale pits; tail not prehensile; subcaudals entire anteriorly, divided posteriorly; margin of maxillary cavity with a definite projection; palatine rounded or triangular dorsally, with three to five teeth. The enlarged head shields are unique in Costa Rican pitvipers.

The hemipenes are deeply divided (forked) and have a centrolineal bifurcate sulcus spermaticus. Large basal hooks are present, and the distal two-thirds of the organ is covered with small spinulate calyces (Burger 1971).

Members of this genus are viviparous and range over much of the eastern and central United States and from northern Mexico south to Belize on the Atlantic versant and to northwestern Costa Rica on the Pacific slope. Many authors (see Gloyd and Conant 1990) include a number of Asian species within *Agkistrodon.* Others (Hoge and Romano 1978–79) place them in a separate genus, *Gloydius,* an arrangement supported by the biochemical findings of Knight, Densmore, and Rael (1992), Kraus, Mink, and Brown (1996), and Cullings et al. (1997) and followed here.

Agkistrodon bilineatus (Günther, 1863)

Cantil, Castellana

(plate 473; map 11.120)

DIAGNOSTICS: The facial markings of two light longitudinal lines, one running along the canthus to over the eye and usually continuing to the angle of the jaw and the second forking downward from the first anterior to the nostril and running along the supralabials, are distinctive. In some large adults the upper stripe may be intermittent or absent

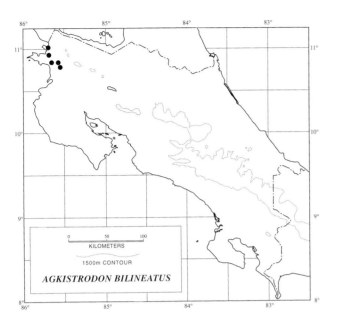

Map 11.120. Distribution of *Agkistrodon bilineatus.*

posterior to the eye. No other pitviper in Costa Rica has these head patterns, which combined with the symmetrical enlarged upper head shields are absolutely diagnostic.

DESCRIPTION: A moderate-sized snake, adults more than 500 mm in total length (maximum total length about 1,380 mm); tail moderate (16 to 21% of total length); upper head shields symmetrical, enlarged; smooth; 1 loreal; 2 or 3 preoculars, postoculars and suboculars 3 to 5; supralabials 8; infralabials 10 to 12; temporals irregular, first and second rows smooth; dorsal scales in 25 to 29–23 (rarely 21)-19 rows; ventrals 128 to 135 (128 to 144); subcaudals 50 to 68 (46 to 68). Dorsum gray, tan, or brown to very dark brown; 11 to 17 darker, often reddish crossbands on body, more or less outlined with light or light spots (obscured in large adults in some extralimital populations); head dark with upper canthal (usually) and lower supralabial longitudinal light stripes; midventer essentially unpatterned posteriorly (dark with light spotting in some extralimital examples); juveniles reddish to bright red with tip of tail bright sulfur yellow; iris orange to brown, upper half paler; tongue orange, red, or orange pink, tips yellow. The everted hemipenes are short (7 to 9 subcaudals in length) with 25 to 32 large basal spines; spinulate calyces cover distal two-thirds of lobes (Campbell and Lamar 1989).

SIMILAR SPECIES: (1) *Cerrophidion godmani* lacks the distinctive facial light lines. (2) *Porthidium nasutum* lacks the distinctive facial light lines and has most of the head covered by small keeled scales.

HABITAT: Restricted to Lowland Dry Forest environments in extreme northwestern Costa Rica.

BIOLOGY: This rare species is terrestrial and is relatively inactive during the day. It estivates during the severe dry season. Juveniles of *Agkistrodon bilineatus* apparently use the brightly colored tail as a lure to attract food organisms

(Neill 1960; Pough 1988) by lifting it well off the ground, holding it vertically, and wiggling the tip. Tail wiggling seems to be evoked only in the presence of prey, and under laboratory conditions frogs and anole lizards were attracted by this behavior. The brightly colored tail becomes darkened with age (probably by the end of the first eighteen months of life). Adults from Costa Rica are known to feed on small mammals and ctenosaurs. Juveniles' prey includes small lizards *(Mabuya unimarginata* and *Ameiva undulata)* and frogs (*Hypopachus variolosus* and *Leptodactylus* spp.) (Solórzano et al. 1999). In other tropical areas grasshoppers, frogs, snakes, and rodents have been reported as diet items (Gloyd and Conant 1990), with insects and frogs probably eaten only by juveniles. Male combat occurs in this species (Shine 1994). Births apparently occur between late May and August, the first half of the rainy season in Costa Rica. Brood sizes of eight to twenty have been reported for non–Costa Rican populations. The smallest neonate is 230 mm in total length. Although the snake is potentially dangerous because of its size, venom yield, and toxicity, there are no known bites of humans by this species in Costa Rica (Bolaños 1984). Villa (1962) mentions two cases of bites in Nicaragua, however.

REMARKS: Conant (1984) was the first to describe Costa Rican examples of the cantil. Gloyd and Conant (1990) provide an excellent review of geographic variation in this species and recognize a series of four apparently completely allopatric races based on slight differences in segmental counts and adult coloration. The differences used for this purpose are so subtle that they need no repetition here. Were it not for the disjunct distributions of the populations, most workers might regard these differences as portions of north to south clines or based on a higher percentage of individual variants in one area than in another. However, Parkinson, Zamudio, and Greene (2000) recognized the disjunct northeastern Mexican population as a separate species, *A. taylori,* based on biochemical and morphological differences from other cantils.

DISTRIBUTION: Disjunctly distributed in semiarid and subhumid situations as follows: (1) lowlands of the northern and eastern Yucatán Peninsula; (2) lowlands of southern Sonora to Guerrero, Mexico; (3) Río Grijalva valley, central Chiapas, Mexico; (4) Río Chixoy (Negro) valley, central Guatemala; (5) savannas of central El Petén, Guatemala; (6) Pacific lowlands from Chiapas, Mexico, to El Salvador; (7) Pacific lowlands of Honduras, Nicaragua (also including the Great Lakes region), and extreme northwestern Costa Rica (20–285 m).

Lachesines

All members of this clade except *Lachesis* were usually referred to the genus *Bothrops* until recently. In a paroxysm of lumping, Savage (1980b) and Savage and Villa (1986) placed all Central American "*Bothrops*" except *Bothrops asper* in the genus *Bothriechis* as a response to Burger's

(1971) unpublished fragmentation of "*Bothrops*" into five groups. The Savage and Villa (1986) scheme was followed by nobody else. The current generic arrangement accepts Burger's (1971) recognition of five genera (Campbell and Lamar 1989) and adds two more (Campbell and Lamar 1992) and Werman (1992). Two of these, *Bothriopsis* (South America) and *Ophryacus* (Mexico), are extralimital to Costa Rica.

The phylogenetic relationships of the genera remain controversial, with substantial conflict among morphological and several molecular data sets (Vidal et al. 1999; Werman 1992, 1999; Salomão et al. 1999). Until this issue is resolved it seems best to maintain the current arrangement for Neotropical pitvipers.

Genus *Atropoides* Werman, 1992

A small genus (three species) of very heavy-bodied pitvipers often referred to as the "jumping" pitvipers and formerly placed in the genus *Bothrops* and more recently in *Porthidium*. Werman (1992) proposed the name *Atropoides* and demonstrated the distinctiveness of this lineage, which is characterized by the following combination of features: upper head surface covered by small imbricate and keeled scales; supraoculars reduced in size and narrowed longitudinally or broken up into small scales; body cylindrical, extremely robust; nasal pore present; dorsal scales in 23 to 31 rows at midbody; two apical pits or none; subcaudals entire except in *A. picadoi*, which may have some divided; tail not prehensile; margin of maxillary cavity with a definite projection; palatine triangular dorsally with three or four teeth.

Hemipenes slender, strongly bilobed, with a bifurcate centrolineal sulcus spermaticus and spinous basal area gradually replaced by apical calyces (Burger 1971). The two Costa Rican species of this genus are unlikely to be confused with any other terrestrial pitviper because of their short, stocky habitus; they also lack large supraoculars, but that feature is not very useful for recognition if you stumble across an *Atropoides* on the trail.

All *Atropoides* are viviparous and range in humid forest situations from San Luis Potosí, Mexico, to central Panama on the Atlantic versant and from Oaxaca, Mexico, disjunctly on the Pacific versant to El Salvador and from northwestern Costa Rica to southwestern Panama. The single extralimital species, *Atropoides olmec*, is known only from the Los Tuxtlas region in southern Veracruz, Mexico.

KEY TO THE COSTA RICAN PITVIPERS OF THE GENUS *ATROPOIDES*

1a. Ventrals 138–55; prenasal in broad contact with rostral; prelacunal separated from second supralabial by one subfoveal row (fig. 11.49b); head scales moderately keeled; dorsal pattern consisting of diamond-shaped dark blotches, often fused on midline to form a zigzag pattern . . . . *Atropoides picadoi* (p. 722)

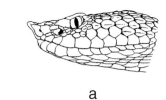

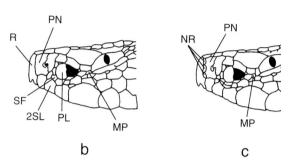

Figure 11.49. Diagnostic features of head scalation in some Costa Rican pitvipers, lateral view: (a) *Porthidium nasutum* showing upturned rostral and fleshy flap; (b) Loreal region of *Atropoides picadoi* showing separation of prelacunal (PL) from second supralabial (2SL) by one subfoveal (SF) and median preocular (MP) bordering orbit; (b) *Atropoides nummifer* showing nasorostrals separating the prenasal (PN) from contact with the rostral (R), two rows of subfoveals, and the median preocular separated from border of the orbit.

1b. Ventrals 114–35; prenasals separated from rostral by 2 or 4 small intervening scales (nasorostrals); prelacunal separated from second supralabial by at least two subfoveal rows (fig. 11.49c); head scales strongly keeled; dorsal pattern consisting of diamond-shaped blotches or saddles, sometimes fused on midline to form a zigzag pattern or uniform black . . . . *Atropoides nummifer* (p. 720)

Atropoides nummifer (Rüppell, 1845)

Mano de Piedra

(plates 474–75; map 11.121)

DIAGNOSTICS: This extremely stout medium-sized pitviper sometimes has dark spots on the supralabials below the pit and eye, a dorsal pattern of dark diamond-shaped blotches with the lateral points directed downward, and smaller lateral blotches aligned with the dorsal diamonds. Sometimes the dorsal blotches are fused along the midline to form a zigzag line, or the dorsal and lateral blotches may be fused; occasional individuals are uniform black. Definite rectangular ventrolateral dark spots are also present. Differentiating this species from its close relative *Atropoides picadoi* usually requires examining the scales around the loreal pit (fig. 11.49) or counting the ventrals.

DESCRIPTION: A moderate-sized species (total length to 870 mm, most adults 400 to 600 mm), females much larger than males; tail short (9 to 15% of total length); all scales on top of head strongly keeled, tubercular; supraoculars

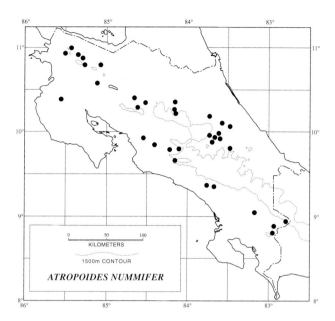

Map 11.121. Distribution of *Atropoides nummifer.*

small, long and narrow or broken up into small scales; 1 loreal; 3 preoculars; 1 subocular; 1 to 4 postoculars; supralabials usually 10 or 11, sometimes 8, 9, or 12; infralabials 10 to 14; temporals keeled; dorsal scales with tuberculate keels in 25 to 33–23 to 31–17 to 23 rows (mode at midbody 25); keels of vertebral and paravertebral scale rows strongly tuberculate, forming a dorsal crest on anterior part of body; ventrals 114 to 135 (114 to 138); subcaudals 22 to 39. Entire dorsum brown with 15 to 19 large dark brown diamond-shaped dorsal blotches, 14 to 28 smaller lateral blotches that may coalesce with the dorsal blotches or one another, and rectangular ventrolateral blotches; occasional examples mostly black; dark spots may be present below pit and eye; dark postocular stripe present; venter almost immaculate to heavily mottled with brown; tip of tail yellow in juveniles; underside of tail mostly dark. Hemipenes slender, with each lobe bearing 35 to 50 medium-sized spines proximally; apical area covered with scalloped to spinulate calyces (Campbell and Lamar 1989).

SIMILAR SPECIES: (1) *Atropoides picadoi,* a somewhat less robust species with weaker keeling on the scales, has the postocular dark stripe outlined by dark pigment and 138 to 155 ventrals. (2) Species of *Lachesis* are much more slender and lighter-colored snakes that lack dark spots on the supralabials and have four or five rows of subcaudals distally (fig. 11.19b).

HABITAT: Primarily found in Premontane Wet Forest and Rainforest, but also in lowland rainforests on the Atlantic versant and gallery forests within the Lowland Moist Forest zone of the Pacific slope.

BIOLOGY: Generally a relatively terrestrial, usually nocturnal pitviper that feeds mostly on small mammals. Juveniles have been reported from tree holes (Alvarez del Toro

1973) up to 3 m above the ground. Neonates eat small lizards and insects and may use the tail to lure prey (March 1929). Campbell (1998) reported finding a crayfish in the stomach of one juvenile. Adults feed principally on small mammals. Picado (1931) and Rokosky (1941) record broods of five to twenty-seven young. Murphy and Mitchell (1984) report captive birth for Guatemalan examples in January, with fourteen young measuring 166 to 189 mm in total length. In Costa Rica, Atlantic versant females give birth between August and November and Pacific slope females give birth between March and June (Solórzano 1989). The former period corresponds to the time of heaviest rains, whereas the latter is from the end of the dry season to the time of early rains. The number of young ranged between fourteen and thirty-four, and the number is positively correlated with maternal size based on a sample of seventeen females. Picado (1931) reported a lower limit of five neonates for one brood, and Taylor (1951b) recorded twenty-seven embryos from one female. Neonates were 181 to 220 mm in total length in a sample of seventy individuals (Solórzano 1989). Campbell (1998) reported litters of thirteen to thirty-six snakelets born from August to November in Atlantic slope Guatemala that were 145 to 200 mm in total length. March (1929) noted litters of eight to fifteen in Honduras. He reported the smallest gravid female as having a total length of 280 mm. Duellman (1963b) reported that a large *Drymarchon* had eaten a 953 mm individual of this species. *A. nummifer* was also among the prey of the great black hawk *(Buteogallus urubitinga)* in Guatemala (Gerhardt, Harris, and Vásquez 1993). Both juveniles and adults of this species are not inclined to flee when approached; they usually form a defensive coil with the body, raise the head, and threaten intruders with an open mouth display. They normally strike only as a last resort when closely approached or touched. Chiszar and Radcliffe (1989) reported that, unlike most other terrestrial vipers, *A. nummifer* holds the prey in the mouth after the strike rather than releasing it. They also pointed out that a fold of skin covers the pit and part of the eyes, apparently to protect them from the struggles of the dying prey. Although potentially dangerous, *P. nummifer* is rarely involved in serious human envenomations (about three bites per year are reported in Costa Rica) because of the small quantity of venom injected and its relatively mild effect. These snakes are aggressive. Their vicious strikes have given them the name "jumping" viper, although they are incapable of jumping and do not push more than one-third of the body forward and off the ground when they strike.

REMARKS: Most reports of this species from Costa Rica are under the name *Bothrops nummifer* or *Porthidium nummifer.* The Costa Rican vernacular name for this species, mano de piedra, refers to the resemblance between this short, stout snake and the stone traditionally used to grind corn on a metate. The record of this species from the Península de Nicoya (Taylor et al. 1973) is suspect. This specimen (ICP, no longer extant) was purported to be from Carrillo de

Guanacaste, a canton with Filadelphia (17 m) as its capital. This area borders the Río Tempisque. It may be that the example was carried downstream from its normal habitat or that there is a confusion in locality data. Attempts to recognize geographic races within this species have been based exclusively on individual variation in coloration (Wilson and Meyer 1985).

DISTRIBUTION: Disjunctly found on mountain slopes from San Luis Potosí to Oaxaca, Mexico; lowlands and premontane slopes on the Atlantic versant from southern Mexico to Belize and Honduras and in northern Nicaragua; on the Pacific versant in Chiapas, Mexico, Guatemala, and El Salvador; and on both slopes from northern Costa Rica to central Panama (50–1,400 m).

Atropoides picadoi (Dunn, 1939a)

Mano de Piedra

(plate 476; map 11.122)

DIAGNOSTICS: This species is a somewhat longer, more slender, darker colored and less strongly keeled version of *Atropoides nummifer*. Most examples have a dark subocular dark spot on the supralabials. The pattern is of dark dorsal diamonds that have the lateral tips pointing downward and smaller lateral dark spots aligned with the dorsal ones. Definite squarish ventrolateral dark spots are also present. Sometimes the dorsal blotches are fused to form a zigzag dorsal dark stripe, and occasionally the lateral spots are fused with the dorsal diamonds. Although these features separate *A. picadoi* from most other pitvipers, you may need to examine the scales around the loreal pit (fig. 11.49) and count the ventrals to distinguish it from *A. nummifer*.

DESCRIPTION: A moderate-sized snake reaching a maxi-

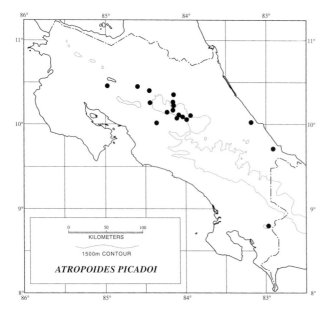

Map 11.122. Distribution of *Atropoides picadoi*.

mum total length of 1,202 mm; tail short (9 to 12% of total length); most scales on top of head weakly keeled; internasals and canthal smooth; supraoculars smooth, partially fragmented or broken into small scales; 1 loreal; 3 preoculars; 1 elongate subocular; 2 to 4 postoculars; supralabials 9 to 10; infralabials 9 to 13; dorsal scales keeled, not tuberculate, in 23 to 29 rows at midbody (mode 25), usually 27 anteriorly, usually 19 posteriorly; no obvious dorsal crest produced by keeling on vertebral and paravertebral scale rows; ventrals 138 to 155; subcaudals 30 to 40, entire or with 2 to 13 in various positions along tail divided (in about 50% of sample). Dorsum dark brown with 25 to 34 dark diamond-shaped blotches and 26 to 34 smaller lateral blotches; small ventrolateral dark spots present; sometimes dorsal and lateral blotches fused and/or dorsal blotches fuse along midline to form a zigzag middorsal dark stripe; top of head dark brown to auburn with irregular darker markings consisting of a pair of parenthesis-shaped marks and a few spots in juvenile females or a few dark spots in juvenile males; dark spots usually present below pit and eye; postocular dark stripe outlined by darker color; tail colored like body in subadults and adults but pale yellow in young; venter white to yellow with an irregular checkerboard pattern that becomes darker posteriorly; underside of tail uniform dark brown; iris dark brown to black.

SIMILAR SPECIES: (1) *Atropoides nummifer* has tuberculate keeling on head and body scales, a distinct vertebral crest, a solid-color postocular stripe, and 114 to 135 ventrals. (2) Species of *Lachesis* are lighter colored, lack ventral dark blotches, and have the postocular dark stripe uniformly colored when present.

HABITAT: Atlantic Lowland Moist and Wet Forests and Premontane Wet Forest and Rainforest; also marginally in Lowland Moist Forest on the Pacific slope.

BIOLOGY: Solórzano (1990) reports that births occur primarily during the height of the rainy season (August to November), but one litter was produced in April. Females are sexually mature at about 650 mm in total length. Ten litters consisted of 15-28-45 young, and litter size is positively correlated with female size. Newborns are 185 to 246 mm in total length. The species is relatively uncommon throughout its range. Very few bites have been attributed to this species and its venom yield is small, but a bite from a large individual may be life threatening.

REMARKS: This form has usually been called *Bothrops picadoi* or *Porthidium picadoi* by previous workers. Martínez (1983) and Solís (1991) have recently established its occurrence in Panama.

DISTRIBUTION: Atlantic slope Costa Rica in the lowlands and on the premontane slopes of the Cordillera Central and Cordillera de Talamanca; the type specimen is from La Palma, San José Province, just over the continental divide on the Pacific slope; also reported from the Meseta Central Occidental, Pacific slopes of the Cordillera de Tilarán, and the Cordillera de Talamanca in Costa Rica, and from montane southwestern Panama (15–1,750 m).

Genus *Bothriechis* Peters, 1859b

This genus (seven species) of moderate-sized vipers includes some of the most beautiful snakes in Mesoamerica. Any pitviper from central Panama north through the rest of Central America that has a prehensile tail belongs to this genus, and most examples of *Bothriechis* have a bright green or yellow dorsal ground color. Other features that distinguish the genus include upper surface of head covered by small keeled scales or medium to large mostly smooth plates; supraoculars definitely enlarged, smooth; internasals enlarged or not, usually smooth; if large smooth shields present, small scales present between frontal and supraoculars; body laterally compressed; dorsal scales in 17 to 25 rows at midbody; no apical scale pits; subcaudals entire; margin of maxillary cavity with a definite projection; palatine triangular dorsally, with four teeth.

The everted hemipenes (6 to 9 subcaudals in length) are deeply divided (forked), capitate, and covered by a mixture of large and small spines basally. The sulcus spermaticus is bifurcate and centrolineal, and the capitulum is covered by papillate calyces, in contrast to the spinous proximal region (Burger 1971; Campbell and Lamar 1989).

Crother, Campbell, and Hillis (1992) provide a detailed phylogenetic analysis and biogeographic hypothesis for the genus. All *Bothriechis* are found in the lowland, premontane, and lower montane zones from southern Mexico to western Venezuela and western Ecuador, exclusive of the subarid and subhumid Pacific coastal regions of Central America and Panama.

KEY TO THE COSTA RICAN PITVIPERS OF THE GENUS *BOTHRIECHIS*

1a. No elongate pointed supraciliary scales between supraocular scale and orbit 2
1b. 2–3 elongate pointed more or less erect hornlike supraciliary scales between supraocular scale and orbit (fig. 11.50); dorsum bright yellow, green, silver, to dark gray green or charcoal, usually contrasting dark blotches or spots, rarely almost uniform or with longitudinal light or dark stripes . . . .
. . . . *Bothriechis schlegelii* (p. 725)

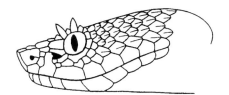

Figure 11.50. Lateral view of the head of *Bothriechis schlegelii* illustrating the hornlike superciliary scales above the eye.

2a. A distinct light line running length of body along first scale row on each side (fig. 11.42b); scales on upper surface of snout keeled; tip of tail green or yellow; dorsum and upper head surface green; juveniles with dark blotches on a brown background . . . .
. . . . *Bothriechis lateralis* (p. 723)
2b. No distinct light stripe along first scale row; scales on top of snout smooth; tip of tail black; dorsum and head green with black markings; no light paravertebral vertical bars . . . . *Bothriechis nigroviridis* (p. 724)

Bothriechis lateralis Peters, 1862b

Lora

(plates 477–78; map 11.123)

DIAGNOSTICS: No other snake in Costa Rica except the harmless *Oxybelis fulgidus* resembles adults of this species in having a bright green dorsum and a light stripe on each side of the body (fig. 11.42b). That species is an extremely slender form that has a very narrow head, lacks a loreal pit, and is unlikely to be confused with this viper. The blotched juveniles of *Bothriechis lateralis* (plate 478) are also distinctive in having ventrolateral light stripes.

DESCRIPTION: A moderate-sized species (maximum total length 950 mm, adults usually less than 700 mm); tail moderately short (14 to 19% of total length); upper head covered mostly with small keeled scales; smooth internasal, 2 canthals, and relatively large supraocular on each side; loreal single; 2 preoculars; usually 2 suboculars; 1 to 3 postoculars, usually 2; 9 to 12, supralabials, usually 10; infralabials 11 to 13; temporals mostly keeled; dorsal scales 21 to 23 (mode 23) at midbody, outer row usually smooth; ventrals 157 to 171; subcaudals 54 to 70. Adults: dorsum bright

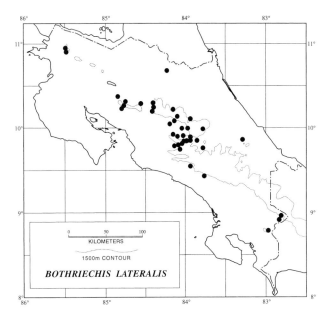

Map 11.123. Distribution of *Bothriechis lateralis*.

green with vertical yellow bars, often bordered with black or bluish; light ventrolateral stripes present; tail tip green or yellow; venter pale green; iris yellow green; tongue green; juveniles: dorsum brown with darker paravertebral blotches, often yellow edged (plate 478); ventrolateral light stripes present; tail tip yellow, or chartreuse; iris bronze; tongue green. The everted hemipenis has 10 to 12 enlarged basal spines and lacks enlarged mesial spines facing crotch (Campbell and Lamar 1989).

SIMILAR SPECIES: (1) *Bothriechis nigroviridis* has a considerable amount of black pigment in the pattern and lacks ventrolateral light stripes. (2) *Oxybelis fulgidus* lacks a loreal pit, is attenuate, and has the entire upper head surface covered by smooth shields.

HABITAT: Relatively undisturbed Premontane Moist and Wet Forest and Rainforest and Lower Montane Wet Forest and Rainforest.

BIOLOGY: The adults of this relatively uncommon species are arboreal. Campbell (1983) and Murphy and Mitchell (1984) observed captive females descending to the ground to give birth (six to thirteen young; total length 196 to 255 mm). Campbell suggested that in nature the young remain on the forest floor after birth (thus their cryptic coloration). Juveniles of this species feed primarily on leaf litter anoles and *Eleutherodactylus* frogs and use the brightly colored tail to lure them. Humans are often bitten on the hand by these snakes when clearing land, cutting trails, or logging. Bolaños (1984) reported about forty-one bites in a typical year in Costa Rica, with no known deaths, but often serious permanent damage to the hands. The amount of venom carried averages about 15 mg dry weight. An average bite thus injects about 0.5 cc of venom.

REMARKS: The ontogenetic color change becomes noticeable at about six months of age. By ten months the snakes are dull lime green, and the yellow areas in the dorsal pattern become more prominent and dark margined. At eighteen months or so the overall color is bluish green and the iris is yellow. The snakes attain typical adult coloration by about the end of two years (Campbell 1983). This species is usually called *Bothrops lateralis* in the literature.

DISTRIBUTION: Premontane and lower montane zones of the cordilleras of Costa Rica and western Panama (700–1,950 m).

Bothriechis nigroviridis Peters, 1859

(plate 479; map 11.124)

DIAGNOSTICS: A bright green arboreal pitviper unlike any other in Costa Rica in having considerable amounts of dark flecking and spotting in the pattern and a black tail tip.

DESCRIPTION: A medium-sized species to 937 mm in total length (adults usually about 500 mm), adult females (317 to 937 mm) longer than adult males (200 to 596 mm); tail moderately long (15 to 20% of total length); anterior upper surface of head covered by small smooth scales; in-

ternasals and canthals (2 on each side) about same size as other head scales, supraoculars somewhat larger; 1 loreal; 2 preoculars; usually 2 suboculars; 1 to 3 postoculars, usually 2; 9 to 10 supralabials usually 10; infralabials 9 to 12; temporals mostly keeled; dorsal scales 17 to 21 (usually 19) at midbody; first scale row on each side smooth; ventrals 139 to 158; subcaudals 47 to 58. Dorsum usually dark green, rarely yellowish green; each dorsal scale marked with some black pigment, usually on margins, to give flecked or mottled appearance, pattern sometimes appearing as black blotches with green centers; no light longitudinal stripes; tail tip black (pale yellow in juveniles); venter pale green; iris bronze or yellow but color obscured by heavy black speckling; tongue black. The everted hemipenis has 16 to 24 enlarged basal spines and enlarged mesial spines facing crotch (Campbell and Lamar 1989).

SIMILAR SPECIES: (1) *Bothriechis lateralis* has dorsoventral light stripes (fig. 11.42b), and adults have yellow bars in the dorsal pattern; juveniles are brown with darker blotches.

HABITAT: Undisturbed Premontane Wet Forest and Rainforest and Lower Montane Wet Forest and Rainforest.

BIOLOGY: This is a nocturnal and arboreal species, usually seen in shrubs or trees 200 mm to 3 m above the ground. Solórzano (1997) described courtship in a captive pair, which was similar to that of other pitvipers. He also provided data indicating that reproduction is timed so that births occur at the height of the rainy season (August to December). His report indicates that females in Costa Rica give birth to six to eight snakelets 165 to 177 mm in total length. Neonates resemble the adults in coloration except that the tail tip is pale yellow. Picado (1931) earlier recorded a female's giving birth to five young measuring 220 mm in total

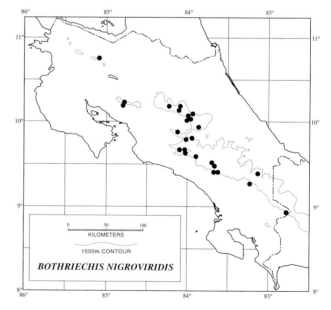

Map 11.124. Distribution of *Bothriechis nigroviridis*.

length. Murphy and Mitchell (1984) also report captive birth of two stillborn snakes from a Costa Rican mother in September.

REMARKS: This relatively uncommon form has been called *Bothrops nigroviridis* by most workers. A record (ICP, no longer extant) from Ujarras, San José Province (570 m), is probably based on an individual collected at a higher elevation. It does range down to 700 m in altitude in the vicinity of San Isidro de El General, San José Province. Bolaños (1984) reported that this species rarely envenomates humans, probably because it is most common in areas of low human density.

DISTRIBUTION: Premontane and lower montane zones of the cordilleras of Costa Rica and western Panama (700–2,400 m).

Bothriechis schlegelii (Berthold, 1845)

Eyelash Viper, Bocaracá, Oropel, Toboba de Pestañas

(plates 480–84; map 11.125)

DIAGNOSTICS: The distinctive hornlike scales above the eye that resemble eyelashes, combined with the presence of a loreal pit (fig. 11.50), immediately identify this variable arboreal species.

DESCRIPTION: A moderate-sized pitviper (maximum total length 820 mm); adult females (347 to 820 mm) longer than adult males (375 to 687 mm); tail short to moderate (13 to 19% of total length); upper surface of anterior head covered by small keeled scales; internasals 4 or 5 (on each side), canthals, and supraoculars larger, smooth; 1 to 3, usually 2, flexible flattened elongate superciliary scales on margin of eyelid; loreal divided; 1 preocular; 1 or 2 suboculars, usually 1; 2 or 3 postoculars, each with a raised process; 8 to 10 supralabials, usually 8; 10 to 13 infralabials;

temporals heavily keeled; dorsal scales 25 to 27–21 to 25–17 to 21 (middorsal mode 23), outermost row on each side smooth; ventrals 137 to 169; subcaudals 42 to 64. Dorsal ground color bright yellow, pink, green, olive green, silver, to dark gray green or charcoal; yellow morph almost uniform or with sprinkling of black, green, and/or red dots; others usually with black outlined light (red, orange, yellow, silver, pale green) spots, blotches, or crossbands; rarely with a middorsal dark stripe; tail tip yellow to green; venter pale yellow or more often stippled or mottled with dark or distinct dark blotches or stripes present; iris yellow to beige; tongue brown. The everted hemipenes has 16 to 24 enlarged basal spines and enlarged mesial spines facing crotch (Campbell and Lamar 1989).

SIMILAR SPECIES: (1) *Imantodes inornatus* has large protuberant eyes and a blunt head and lacks the hornlike supraciliary scales and loreal pit. (2) *Sibon longifrenis* lacks the hornlike supraciliary scales and loreal pit (fig. 11.36c).

HABITAT: Any wooded area or secondary growth, including low shrubby vegetation, in Lowland Moist and Wet Forests and Premontane Wet Forest and Rainforest; marginal at a few Premontane Moist Forest sites.

BIOLOGY: A nocturnal arboreal species that feeds on a variety of small vertebrates, frogs, lizards (especially anoles), birds, and mammals, including bats). The function of the enlarged supraciliary scales is unknown, but they may deflect vegetation from the eyes as the snake moves through dense cover (Cohen and Myers 1970). Costa Ricans apply the vernacular name oropel (literally gold skin), the Spanish word for tinsel or glitter, to the yellow phase of *B. schlegelii* (plates 483–84). The other color phases (plates 480–82) usually are called bocaraca. There are no known examples intermediate between the oropel phase and the darker variants, leading Picado (1931) to suggest that the yellow snakes are reproductively isolated from the others. Antonio (1980) reported on another unusual variant that was predominantly pink or rose. This specimen was crossed with a female that was green with reddish blotches, which produced twenty offspring, ten similar in coloration to each parent. Both snakes were from the same general area in Honduras, and they had been in captivity twenty months (male) and twelve months (female). This led Villa (1984b) to propose that the male was a homozygous recessive and the female was heterozygous. By implication the oropel phase would also be a homozygous recessive. Both the oropel and pinkish color phases occur discontinuously in this species from Honduras to Panama. Antonio (1980) described captive reproductive behavior for the parents of the brood discussed above. Courtship and mating took place off the ground on a branch in a cage. The male's swayed his head from side to side until he touched the female, then used forward thrusts to arrive in a dorsal position. The tails of both snakes hung down from the branch and were intertwined until the cloacas came into contact. The right hemipenis was used during copulation,

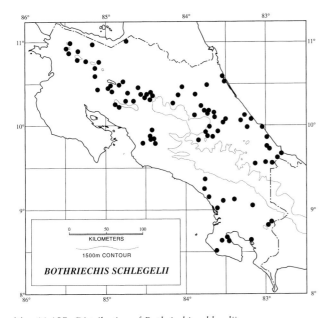

Map 11.125. Distribution of *Bothriechis schlegelii*.

which lasted somewhat under three hours. Gestation was from February 24 to August 9. The hatchlings were 171 to 215 mm in total length after seventeen days and used their brightly colored tails (yellow, pink, or white) to attract frogs that they then ate. Picado (1931) and Blody (1983) reported on other captive broods of eighteen and six to seventeen young, respectively. The latter suggested that more than one brood may be produced in one year (plate 484). Murphy and Mitchell (1984) recorded captive births of twelve and nineteen young (174 to 225 mm in total length). Gerhardt, Harris, and Vásquez (1993) report predation on this snake by the great black hawk *(Buteogallus urubitinga)* in Guatemala. One captive example lived more than sixteen years (Bowler 1977). This species is responsible for many snakebites, particularly on the face, upper body, and hands, for it commonly rests during the day in low vegetation 1 to 1.5 m above the ground. In a typical year about ninety to one hundred bites are reported in Costa Rica, and though deaths are rare (three to six in some years according to Seifert 1983), long-term impairment of the hands frequently occurs (Bolaños 1984). The amount of venom carried averages about 12 mg dry weight, so an average bite injects about 0.5 cc of venom.

REMARKS: The great individual variation in coloration has precluded any attempt to fragment the species into subspecies. This form has usually been called *Bothrops schlegelii* by most workers. Werman (1984b) concluded that *Bothrops schlegelii supraciliaris* Taylor (1954b), regarded by some as a separate species (Stuart 1963; Peters and Orejas-Miranda 1970), is based on a slightly aberrant Costa Rican individual of this form (KU 31997; San José: mountains near San Isidro de El General). However, Solórzano et al. (1998) consider *B. supraciliaris* to be a valid form that occurs in southwestern Costa Rica, generally at higher elevations (800 to 1,700 m) than *B. schlegelii* (plate 485). That the specimens assigned to *B. supraciliaris* resemble snakes referred to *B. schlegelii* from Colombia and Ecuador (Campbell and Lamar 1989) in squamation and coloration means that the status of *B. supraciliaris* needs further study.

DISTRIBUTION: Lowland and premontane areas on the Atlantic slope from northern Chiapas, Mexico, Belize, and Guatemala to extreme northwestern Venezuela; and on the Pacific versant in southern Costa Rica and southwestern Panama, central Panama, and eastern Panama to western Ecuador. In Costa Rica primarily on the Atlantic slope and in the southwestern region, but a few records lie on the upper Pacific slope of the Cordillera de Guanacaste and the Cordillera de Tilarán (5–1,530 m).

Genus *Bothrops* Wagler, 1824

This large, primarily South American genus (thirty-two species) is represented by one species in Mexico and Central America. Members of this stock include some of the most dangerous venomous snakes in the American tropics be-

cause of their aggressiveness, quickness, long fangs, and large volume of venom they can inject. The following features in combination will separate *Bothrops* from other pitvipers: upper surface of head covered by small keeled scales; supraoculars large, smooth; body triangular; head triangular in dorsal outline; nasal pore present; dorsal scales in 19 to 33 rows at midbody; 0 or 2 apical scale pits; tail not prehensile; subcaudals divided; margin of maxillary cavity an uninterrupted curve; palatine bifurcate with two to five teeth.

The everted hemipenes are short (8 to 11 subcaudals long) and deeply divided (forked), with a bifurcate centrolineal sulcus spermaticus. The penial lobes are subcylindrical or moderately to strongly attenuate; 35 to 50 spines cover the proximal one-half to two-thirds of each lobe but may extend farther proximally; apical portion of lobes usually covered by papillate calyces, but sometimes smooth, scalloped, or spinulate calyces present. The pedicel may be smooth or covered with spinules (Burger 1971; Campbell and Lamar 1989).

Members of this genus are viviparous and have a wide distribution in the Neotropics from Tamaulipas, Mexico, on the Atlantic versant to Patagonia, and from Costa Rica to western Ecuador on the Pacific slope; they also range from the Lesser Antilles to Martinique, and an isolated population of one species *(B. asper)* occurs on the Pacific versant of Chiapas, Mexico, and Guatemala.

Bothrops asper (Garman, "1883," 1884)

Terciopelo, Rabo Amarillo (Juveniles)

(plate 486; map 11.126)

DIAGNOSTICS: This large and dangerous snake is distinctive among pitvipers in Costa Rica in having light-outlined dark triangular lateral blotches (base lateral, apex dorsal) that often fuse across the back to form butterfly-shaped figures.

DESCRIPTION: A large species reaching a maximum total length of about 2,500 mm; adult females (800 to 2,500 mm) longer than adult males (800 to 2,200 mm), most adult examples 1,200 to 1,800 mm; tail short (12 to 15% of total length); upper head scales on anterior part of head small, keeled; canthals usually 1 (1 or 2 per side) and supraoculars enlarged, smooth; 1 to 3 internasals, somewhat enlarged, usually smooth; 1 to 3 loreals; usually 2 preoculars; 1 to 3 suboculars; 0 to 4 postoculars; usually 7 supralabials, sometimes 8 or 9; 8 to 12 infralabials, usually 10; temporal scales keeled; dorsal scales 25 to 29–23 to 33–19 to 21 (middorsal rows usually 25 to 29); all dorsal scale rows keeled; ventrals 184 to 215 (161 to 249); subcaudals 54 to 77 (46 to 86). Dorsal ground color olive green, pale, or dark gray or brown; 15 to 35 triangular dorsolateral dark markings on each side (base lateral, apex dorsal), outlined by light, often fused across back to form butterfly-shaped

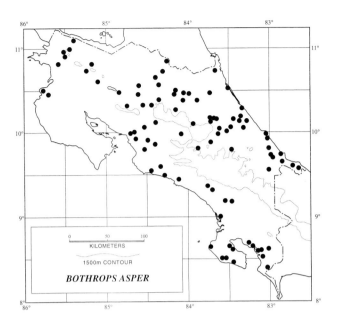

Map 11.126. Distribution of *Bothrops asper.*

dark figure; sometimes triangles offset on two sides of body; a pair of moderate-sized lateral dark spots below and just off the base of each triangle, sometimes fused with triangle; a series of small dark ventrolateral dark spots (six or so per triangle) evenly spaced below other spots; a distinct postocular dark stripe from eye backward and downward to posterior margin of mouth, demarcated by a light line above; labials, chin, and throat immaculate; central area of venter with or without scattered dark pigment; all pattern components usually more vivid in juveniles; tip of tail in juvenile and most subadult males bright yellow to orange, but usually yellow obscured by darker pigment in females and most adult males; iris beige to brown; tongue brown. Hemipenial lobes moderately attenuate, with 50 to 70 moderately enlarged spines proximally; calyces papillate (Campbell and Lamar 1989).

SIMILAR SPECIES: (1) Species of *Lachesis* have the dark dorsal diamonds with the tips pointing ventrally or continuing as a dark line down the flanks, and the postocular dark stripe, if present, is not separated from the dorsal head color by a light stripe. (2) *Xenodon rabdocephalus* lacks a loreal pit, has the upper head surface covered by large shields, and has smooth dorsal scales (fig. 11.5) and a round pupil (fig. 11.4). (3) *Boa constrictor* lacks a loreal pit and has smooth scales on the head and body. (4) *Trimorphodon biscutatus* lacks a loreal pit and has large plates covering the head surface (fig. 11.45). (5) Juveniles of *Elaphe triaspis* lack the loreal pit and have large head shields and a round pupil (fig. 11.9b).

HABITAT: Widespread and common in Lowland Moist and Wet Forests; especially abundant on banana plantations and in other human modified environments where rat populations have exploded. Less abundant in Lowland Dry For-

est, where the species is most likely to be found in riparian situations. Also ranging into the lower portions of Premontane Moist and Wet Forest and Rainforest zones.

BIOLOGY: These active and aggressive vipers are mostly terrestrial and nocturnal, but juveniles are often found in low trees and shrubs, and diurnal foraging also occurs (Scott 1983o). As young they eat a wide variety of terrestrial vertebrates, mainly frogs and lizards, and arthropods, but adults in Costa Rica feed primarily on small mammals, especially rodents and opossums. The brightly colored tail tip probably is used to lure prey, as in several other pitvipers, and gives the species one of its vernacular names, rabo amarillo (yellow tail). However, it is present only in males (Solórzano and Cerdas 1989), although young of both sexes undulate the tail tip (Tryon 1985a). Male combat with biting has been reported for the related *B. atrox* (Shine 1994). Hirth (1964) reported gravid females of this species with forty-eight to eighty-six embryos, and Scott (1983o) noted a litter of seventy-five. According to the study of Solórzano and Cerdas (1989) based on 647 Costa Rican adults, mating is seasonal, September to November on the Pacific versant versus March on the Atlantic, with births from April to June and September to November, respectively. Litter size was 5–18.6–40 for twenty-seven Pacific snakes and 14–41.1–86 for sixteen Atlantic ones. Number of neonates correlates positively with the size of the mother. Neonates were 280 to 346 mm in total length. Females in these samples reached sexual maturity at a total length of about 1,100 mm, and males did so at about 1,000 mm. Tryon (1985b) and Campbell (1998) report slightly different values for neonates, 300 to 327 mm and 275 to 350 mm in total length, respectively. This is the most dangerous venomous snake in Central America because of its size, length of fangs, venom yield, and generally nervous behavior. About half of all venomous snakebites suffered by humans in Costa Rica are from this species, and most deaths are caused by its bites. In a typical year there are about 220 *Bothrops* bites in Costa Rica, resulting in fifteen to twenty deaths as well as many serious injuries. The amount of venom carried averages about 458 mg dry weight, but one large individual produced 1,530 mg (Bolaños 1984). An average bite injects 1 to 2 cc of venom, and large snakes are capable of injecting 5 or 6 cc. The venom contains a definite anticoagulant component (Alvarado and Gutiérrez 1988). Lomonte et al. (1982) report that injecting blood serum of *Clelia clelia* reduces local effects from *B. asper* bites. Further study (Lomonte 1990) indicates that neonate serum from *Clelia* neutralizes the hemorrhagic action of *B. asper* venom.

REMARKS: This species is often regarded as conspecific with its close ally *Bothrops atrox* of South America. Hoge (1965), Hoge and Romano (1978–79) listed *B. asper* and *B. atrox* as separate species because they believed they were completely allopatric, nonintergrading forms. Villa (1984b) placed *B. asper* as a geographic race of *B. atrox*, and Wilson and Meyer (1985) called all snakes in this complex *B. atrox*. Aragón, Bolaños, and Vargas (1977) documented significant differences in several carbohydrate fractions of the venoms of these putative species. Most important, the carbohydrate fraction (percentage in dry base) was 1.3 to 1.65% in Central American samples and 0.6 to 1% in Amazonian snakes. An example from west of the Andes had 2% neutral carbohydrates, and although Aragón, Bolaños, and Vargas (1977) referred to it as *B. atrox,* both in biochemistry and in geographic source it is *B. asper*. The venom differences led Aragón, Bolaños, and Vargas (1977) to regard *B. asper* and *B. atrox* as distinct even though they confused the issue by calling the Atlantic drainage Colombian example *Bothrops atrox* and an Amazonian snake *Bothrops colombiensis*. Much impressed with the importance of the venom characters and the need to keep samples separated for the preparation of specific antivenins, I have continued to regard *B. asper* as a valid form. Under this concept *B. atrox* ranges over much of South America east of the Andes and *B. asper* occurs throughout Colombia and Ecuador west of the Andes. The nominal species *B. colombiensis* is apparently restricted to north-central Venezuela and northeastern Colombia and is allopatric to *B. atrox* in the semiarid region of northwestern Venezuela. *B. colombiensis* was synonymized with *B. atrox* by Johnson and Dixon (1984), but it may be a population of *B. asper*. Recent studies of venom (Aragón and Gubensek 1981) and reproductive biology (Sólorzano and Cerdas 1989) indicate marked differences between Atlantic and Pacific populations of *B. asper* in Costa Rica. *B. asper* apparently once ranged well onto the Meseta Central Occidental, based on several old records (ANSP, MNCR, SMF) from the vicinity of the city of San José. Human malice doubtless led to its extirpation from most of the metropolitan area by the 1940s. I know of no recent collections from east of Atenas, Alajuela Province (616 m), on the western Meseta. The vernacular name terciopelo (velvet skin) refers to the soft, velvetlike appearance of the color pattern in the living snake.

DISTRIBUTION: Lowland and premontane zones on the Atlantic versant from Tamaulipas, Mexico, to Colombia and on the Pacific versant from Costa Rica to western Ecuador, with a disjunct population in southern Chiapas, Mexico, and adjacent Pacific slope Guatemala (1–1,200 m).

Genus *Cerrophidion* Campbell and Lamar, 1992

This recently described genus of small terrestrial montane pitvipers contains four species (Campbell and Lamar 1992; López, Vogt, and Torre 1999). The most distinctive feature of the genus is the large smooth head scales of the frontal and parietal region. *Cerrophidion barbouri* of Guerrero, Mexico, has all the head scales formed of enlarged smooth flattened shields to approach the condition in *Agkistrodon,* but *Cerrophidion petlalcalensis* (Veracruz, Mexico), *Cerrophidion tzotzilorum* (Chiapas, Mexico), and *Cerrophidion godmani* (Mexico and Central America) have enlarged smooth scales irregularly mixed with small keeled ones in the frontal and parietal regions. Unlike *Agkistrodon* and most colubrid snakes (nine large supracephalic shields), *C. barbouri* has four internasals or four canthal shields, for a total of eleven or more plates or scales in the parietal to internasal region on top of the head. Other features include supraoculars large smooth plates; body cylindrical; nasal pore present; dorsal scales in 17 to 25 rows at midbody; 0 or 2 apical pits; subcaudals entire; tail not prehensile; margin of maxillary cavity with a definite projection; palatine triangular dorsally, with three or four teeth. The single high-elevation Costa Rican species of *Cerrophidion* does not co-occur with any other terrestrial pitviper and may always be distinguished from them in any case by the unique head scalation.

The everted hemipenes are strongly bilobed, with sulcus spermaticus bifurcate and centrolineal and spinous truncus replaced by calyces distally.

Members of this genus are viviparous, and their ranges are disjunct in montane situations in the isolated ranges of Veracruz, Guerrero, Oaxaca, and Chiapas, Mexico, and southward through Central America to western Panama, with a major hiatus in central to southern Nicaragua.

Cerrophidion godmani (Günther, 1863)

Toboba de Altura

(plate 487; map 11.127)

DIAGNOSTICS: This small dark-blotched montane species differs from other similarly colored Costa Rican pitvipers in having most of the upper head shields enlarged and smooth (fig. 11.20a).

DESCRIPTION: A moderate-sized species (maximum total length 822 mm); males average larger than females (to 500 mm in total length); tail short (10 to 13% of total length); upper head shields mostly enlarged and smooth; frontal and parietals bordered by smaller scales; 2 to 4 internasals; 2 or 3 canthals; a loreal; 1 upper and usually 2 lower preoculars; 1 to 4 suboculars, usually 1; 2 to 5 postoculars; 8 to 10 supralabials, usually 9; 9 to 12 infralabials, usually 10; temporals lightly keeled or smooth; dorsal scales except outer row keeled, in 23 to 25–21 to 23–19 to 21 (21 to 25–19 to 23–17 to 21) usually 23–21–19; ventrals 130 to 148; subcaudals 22 to 36, entire. Dorsal ground color gray with 20 to 46 greenish brown to brown dorsal and lateral blotches; blotches outlined by dark brown, which may in turn be bordered by white; dorsal blotches sometimes fuse on midline to form an irregular broad wavy dark stripe; postocular dark stripe bordered below by light stripe; venter cream anteriorly punctuated with gray posteriorly; underside of tail orangish; in juveniles entire tail pale orange; iris brown or gray. Hemipenes short (6 to 10 subcaudals

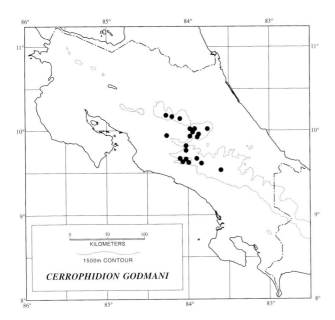

Map 11.127. Distribution of *Cerrophidion godmani.*

long), with 12 to 40 large spines on proximal one-third of each penial lobe; terminal portion of lobes calyculate (Campbell and Lamar 1989).

SIMILAR SPECIES: (1) *Agkistrodon bilineatus* has a distinct canthal light stripe. (2) *Trimorphodon biscutatus* lacks the loreal pit and has the anterior dorsal scales smooth (fig. 11.45). (3) Juveniles of *Elaphe triaspis* lack the loreal pit and have round pupils (fig. 11.9b). (4) *Ungaliophis panamensis* lacks the loreal pit and has smooth dorsal scales and the dorsal blotches outlined by white.

HABITAT: Primarily Lower Montane Wet Forest and Rainforest and Montane Rainforest zones, but also found along streams in the upper margin of the Premontane Wet Forest on the Meseta Central Occidental. Campbell and Sólorzano (1992) suggest that the species tends to occur in areas of localized moderate rainfall.

BIOLOGY: Campbell and Sólorzano (1992) provided much previously unknown data for this relatively uncommon species, derived in part by examining 240 snakes from Costa Rica, and the following paragraph summarizes important aspects of their study as it applies in that country. This is a diurnal terrestrial form that feeds primarily (49% of prey items) on small mammals, especially rodents *(Oryzomys, Reithrodontomys,* and *Mus)* and the shrew *Cryptotis nigrescens.* Reptiles make up 24% of the diet, with snakes *(Geophis godmani, G. brachycephalus,* and *Ninia maculata)* and the lizards *Mesaspis monticola* and *Sceloporus malachiticus* and several species of *Norops* as common prey items. Arthropods constitute a considerable portion of the diet (19%) in juvenile *P. godmani,* and birds and amphibians are occasional prey. Mating apparently takes place principally during the rainy season (April to Novem-

ber) but continues into the early dry season (December). Males exhibit combat behavior (Shine 1994) and are very aggressive in male-male and male-female interactions. The species appears to have a biennial reproductive cycle in nuclear Central America and probably in Costa Rica. Females are sexually mature at about 340 mm. Litters consisted of 2–5.5–12 young, and numbers are not related to female size in lower Central America. Neonate males are 179 to 204 mm in total length, and females are 143 to 191 mm.

REMARKS: Most previous authors have referred to this species as *Bothrops godmani* and more recently as *Porthidium godmani.* Taylor et al. (1973) indicated that it occurs throughout the Cordillera Central. Although I have been unable to locate voucher specimens from Volcán Barba and Volcán Poás and none are listed by Campbell and Sólorzano (1992), it seems very probable that *P. godmani* occurs at higher elevations on these mountains. Two putative records of this species from the Meseta Central (San José: Jaris de Mora, 820 m, and San Antonio de Coronado, 1,330 m), based on examples formerly at the Instituto Clodomiro Picado but now lost, are dubious but have been indicated on the distribution map (11.127).

DISTRIBUTION: Disjunct in montane situations from Oaxaca, Mexico, to Nicaragua and in the Cordillera Central and Cordillera de Talamanca in Costa Rica to extreme western Panama (1,460–2,875 m).

Genus *Lachesis* Daudin, 1803

Bushmasters

The three species referred to this genus are the largest pitvipers and largest venomous snakes in the Americas. The genus is characterized as follows: upper surface of head mostly covered by small scales or granules; small scales in the frontal and prefrontal region smooth, keeled, or tuberculate; supraoculars and internasals larger than other head scales, smooth; body triangular, head relatively narrow, blunt; large nasal pore; dorsal scales tuberculate, in 31 to 40 rows at midbody; two apical scale pits; tail not prehensile, with an elongate pointed terminal scale; subcaudals in two series anteriorly, in four or five rows posteriorly; anterior margin of maxillary concave; palatine triangular dorsally, with two to four teeth.

The short hemipenes (about 9 subcaudals long) are strongly divided, with a bifurcate centrolineal sulcus spermaticus. The pedicel is partially covered with small spines and the lobes with 65 to 80 moderate recurved spines proximally. The apical, conical tips (one-fifth of each lobe) are covered with calyces having smooth ridges (Campbell and Lamar 1989).

This genus is oviparous, unlike all other American pitvipers. It is found in the lowlands from eastern Nicaragua south on the Atlantic versant to Bolivia and southeastern Brazil and on the Pacific slope in southwestern Costa Rica to western Ecuador, but there are substantial disjunctions in the range in Panama, in much of Venezuela, and in northeastern and central Brazil (Vial and Jiménez-Porras 1967; Campbell and Lamar 1989).

KEY TO THE COSTA RICAN PITVIPERS OF THE GENUS *LACHESIS*

1a. Upper surface of head black; ventrals 209–16 . . . .
. . . . *Lachesis melanocephala* (p. 730)
1b. Upper surface of head light; ventrals 198–209 . . . . . .
. . . . *Lachesis stenophrys* (p. 731)

Lachesis melanocephala Solórzano and Cerdas, 1986

Matabuey, Cascabela Muda

(plate 488; map 11.128)

DIAGNOSTICS: A very large, relatively slender pitviper, distinctive in having a rather broad and blunt head that is uniform black above and the tail tip covered by small scales above and below and terminating in a long, pointed spine. The dorsal color pattern is characteristic in having large dark (but often light-centered) diamonds along the midline with the lateral tips pointed ventrally and usually continuing downward as dark lines and the anterior lateral blotches forming vertical bars.

DESCRIPTION: This species is one of the largest venomous snakes in the Americas, with a maximum total length of 3,900 mm; adults commonly 2 m or more. Size alone makes it dangerous! Tail very short (8 to 10% of total length); upper head shields mostly small, smooth or tuberculate; 1 pair of small smooth internasals; supraoculars large, smooth; 1 to 3 oval canthals; a loreal; 2 or 3 preoculars; 2 or 3 small

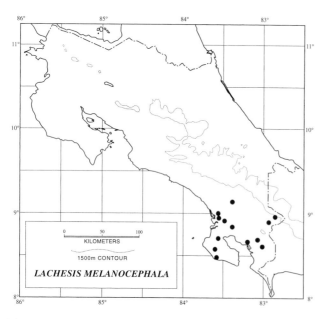

Map 11.128. Distribution of *Lachesis melanocephala.*

suboculars; 3 or 4 postoculars; 7 to 9 supralabials; 12 to 14 infralabials; temporals obtusely keeled; dorsal scales in 36 to 40 rows at midbody; all dorsal scales keeled; vertebral and uppermost 2 or 3 rows of paravertebral scales with very strong tuberculate keels, in adults appearing as a middorsal ridge or low crest; ventrals 209 to 216; subcaudals 43 to 56. Anterior surface of palatine bone concave. Dorsum light, deep yellow; neonates yellowish; supralabials usually immaculate, rarely with scattered dark pigment; 20 to 32 more or less diamond-shaped dark brown body blotches with points of diamond pointing anteriorly and posteriorly on the midline and toward venter on flanks; blotches frequently with light centers; lateral tips of diamonds often extending down sides as narrow dark lines; upper surface of head black and fused with postocular stripe; neonates with a thin light postocular stripe within black head cap; venter light yellowish to creamy with a pink cast; iris red brown; tongue black.

SIMILAR SPECIES: (1) *Lachesis stenophrys* has an unspotted upper head surface. (2) *Bothrops atrox* is much darker overall, with triangular light-outlined lateral dark markings that have their apexes pointed dorsally or are fused across the midline to form butterfly-shaped figures, and has a definite light postocular stripe. (3) *Atropoides picadoi* is a robust viper usually having a subocular dark spot, and blotched examples have dorsal, lateral, and lateroventral dark blotches and mostly entire subcaudals (fig. 11.5). (4) *Boa constrictor* lacks a loreal pit and has the body and head covered with small smooth scales. (5) *Atropoides nummifer* is an extremely stout-bodied snake, usually with dark spots below the pit and eye; blotched examples usually have distinct dorsal blotches (sometimes fused with one another), a series of rounded or oval lateral dark spots and lateroventral dark blotches, and the subcaudals mostly entire (fig. 11.5).

HABITAT: Undisturbed Lowland Moist and Wet Forests; also near the Panama border in Premontane Wet Forest.

BIOLOGY: These relatively uncommon terrestrial nocturnal snakes feed principally on small rodents, especially spiny rats (Echimyidae). They are sit-and-wait predators that feed every few weeks. Usually they move about the forest floor very little, remaining near a single resting site between meals. Females lay ten to twelve eggs that require about a sixty-day incubation. This bushmaster is encountered relatively infrequently but is rather aggressive. Bolaños (1984) indicates that in a typical year only two to four bites by the species of *Lachesis* are reported in Costa Rica. The amount of venom injected by the largest adult is about 5 cc. The venom is extremely toxic, and even with early treatment the mortality rate is estimated to be 75% (Bolaños 1984). Aragón (1988) characterizes the venom of this species as more coagulant and less proteolytic than that of *L. stenophrys.*

REMARKS: Although Sólorzano and Cerdas described

this taxon as a subspecies of *Lachesis muta* of South America, Zamudio and Greene (1997) demonstrated that three valid species were subsumed under that name. They concluded, based on biochemical, morphological, and behavioral features, that *L. melanocephala* is a distinct species. Sólorzano and Cerdas (1986) emphasized the substantial differences in coloration between this light yellow, black-headed species from southwestern Costa Rica and the pale grayish yellow to beige Atlantic bushmasters that have no black spots on the upper head surface and a black postocular stripe. Neonate *L. melanocephala* have a narrow, light postocular stripe within the black head cap, but this darkens and fuses with the cap during ontogeny. The Pacific population, which doubtless ranges into adjacent southwestern Panama, appears to be completely allopatric to other *Lachesis* populations. Snakes similar to those found on the Atlantic slope of Costa Rica occur in Atlantic versant Nicaragua and Panama, in Pacific versant central and extreme eastern Panama, and in north-central Colombia and Pacific versant Colombia and Ecuador. South American *Lachesis* exhibit considerable variation in coloration, but none have a black head cap.

DISTRIBUTION: Humid Pacific slope of southwestern Costa Rica and adjacent western Panama (13–1,330 m).

Lachesis stenophrys Cope, 1875

Matabuey, Cascabela Muda

(plates 489–90; map 11.129)

DIAGNOSTICS: A very large, relatively slender pitviper distinctive in having a rather broad and blunt head lacking dark spots, and having a distinct postocular dark stripe and the tail tip covered by small scales above and below and terminating in a long, pointed spine. The dorsal color pattern is characteristic in having large dark (but often light-centered) diamonds along the midline with the lateral tips pointed ventrally and usually continuing downward as dark lines and the anterior lateral blotches forming vertical bars.

DESCRIPTION: This species is one of the largest venomous snakes in the Americas, with a maximum total length of 3,900 mm; adults commonly 2 m or more. Size alone makes it dangerous! Tail very short (8 to 10% of total length); upper head shields mostly small, smooth or tuberculate; 1 pair of small, smooth internasals; supraoculars large, smooth; 1 to 3 oval canthals; a loreal; 2 or 3 preoculars; 2 or 3 small suboculars; 3 or 4 postoculars; 7 to 9 supralabials; 12 to 14 infralabials; temporals obtusely keeled; dorsal scales in 33 to 38 rows at midbody; all dorsal scales keeled; vertebral and uppermost 2 or 3 rows of paravertebral scales with very strong tuberculate keels, in adults appearing as a middorsal ridge or low crest; ventrals 198 to 209; subcaudals 36 to 49. Anterior surface of palatine bone even. Dorsum pale yellowish gray to beige; neonates orange with dark reddish markings; supralabials usually immaculate, rarely with scattered dark pigment; 20 to 37 more or less diamond-shaped dark brown body blotches with points of diamond pointing anteriorly and posteriorly on the midline and toward venter on flanks; blotches frequently with light centers; lateral tips of diamonds often extending down sides as narrow dark lines; upper surface of head light and postocular dark stripe present, not bordered above by contrasting light stripe; venter light yellowish to creamy with a pink cast; iris red brown; tongue black.

SIMILAR SPECIES: (1) *Lachesis melanocephala* has the upper head surface uniform black. (2) *Bothrops atrox* is much darker overall, has triangular light-outlined lateral dark markings that have their apexes pointed dorsally or are fused across the midline to form butterfly-shaped figures, and has a definite light postocular stripe. (4) *Boa constrictor* lacks a loreal pit and has the body and head covered with small smooth scales. (2) *Atropoides nummifer* is an extremely stout-bodied snake, usually with dark spots below the pit and eye, and blotched examples usually have distinct dorsal blotches (sometimes fused with one another), a series of rounded or oval lateral dark spots and lateroventral dark blotches, and the subcaudals mostly entire (fig. 11.5). (3) *Atropoides picadoi* is a robust viper usually having a subocular dark spot, and blotched examples have dorsal, lateral, and lateroventral dark blotches and the subcaudals mostly entire (fig. 11.5).

HABITAT: Undisturbed Lowland Moist and Wet Forests on the Atlantic slope.

BIOLOGY: These relatively uncommon terrestrial nocturnal snakes feed principally on small rodents, especially spiny rats (Echimyidae). They are sit-and-wait predators that feed every few weeks. Usually they move about the forest floor very little, remaining near a single resting site

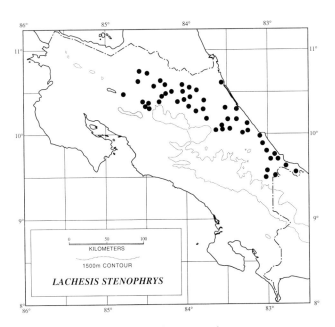

Map 11.129. Distribution of *Lachesis stenophrys*.

between meals. Females lay ten to twelve eggs that require about sixty days' incubation (plate 490). This bushmaster is encountered relatively infrequently in Costa Rica. Although it has a reputation for aggressiveness, this may be exaggerated because of the awesome size of adults, which are obviously dangerous. This species is much calmer than its relative *L. melanocephala*. Aragón (1988) reported that its venom is less coagulant and more proteolytic than that of *L. melanocephala*. Bolaños (1984) indicates that in a typical year only two to four bites by *Lachesis* species are reported in Costa Rica. The amount of venom injected by the largest adult is about 5 cc. The venom is extremely toxic (Bolaños, Muñoz, and Cerdas 1978), and even with early treatment the mortality rate is estimated to be 75% (Bolaños 1984). The common name matabuey (bullock killer) is self-explanatory in this context.

REMARKS: Zamudio and Greene (1997) have recently established that this taxon, long synonymized with the South American *Lachesis muta*, is a valid form. Morphology and biochemical evidence amply support its distinctiveness from *L. melanocephala* and *L. muta*, both of which are allopatric in distribution to each other and to *L. stenophrys*. *L. muta* consists of two allopatric populations: Amazon basin south to Peru and Bolivia and eastern Venezuela and the Guianas; and southeastern Brazil. This form has relatively high ventral counts (213 to 247) and a head pattern of small to large dark spots and either a narrow dark postocular stripe with a light upper border or a broad dark postocular stripe as in *L. stenophrys*.

DISTRIBUTION: Lowland humid forests from eastern Nicaragua to north-central Colombia on the Atlantic versant, continuing onto the Pacific slope in central Panama and eastern Panama to northwestern Ecuador (2–670 m).

Genus *Porthidium* Cope, 1871

Hog-nosed Pitvipers

This genus (seven species) includes the hog-nosed pitvipers. Significant features are body cylindrical; nasal pore present; dorsal scales in 21 to 27 rows at midbody; two apical scale pits or none; subcaudals entire; margin of maxillary cavity with a definite projection; palatine triangular dorsally, with three or four teeth.

The hemipenes are strongly bilobed, and the centrolineal sulcus spermaticus is bifurcate. There is a gradual transition from the spinous truncus to the calyculate apical region (Burger 1971), and the lobes terminate in a long awn containing a naked apical disk (Campbell and Lamar 1992).

The enigmatic *Porthidium melanurum* of Mexico has usually been assigned to this genus, but Gutberlet (1998) has shown that it belongs in *Ophryacus*.

The hog-nosed vipers are viviparous and occur disjunctly in subhumid to semiarid lowland sites from Colima, Mexico, to Costa Rica, and in more humid areas from Chiapas, Mexico, to western Ecuador.

KEY TO THE COSTA RICAN PITVIPERS OF THE GENUS *PORTHIDIUM*

1a. Tip of snout strongly upturned; a single canthal; internasals greatly elongate (figs. 11.49a, 11.20c) 2
1b. Tip of snout rounded or slightly upturned; two canthals; internasals about as wide as long; ventrals 156–76. . . . . *Porthidium ophryomegas* (p. 733)
2a. Ventrals 128–44; tip of snout usually with a fleshy flap forming a proboscis-like extension (fig. 11.49a); 15–28 dark dorsal blotches or bands. . . . . *Porthidium nasutum* (p. 732)
2b. Ventrals 156–65; tip of snout without a fleshy flap; 43–50 dark dorsal blotches . . . . *Porthidium volcanicum* (p. 734)

Porthidium nasutum (Bocourt, 1868a)

Hog-nosed Viper, Tamagá

(plates 491–92; map 11.130)

DIAGNOSTICS: The sharply upturned snout that is usually elongated into a proboscis (fig. 11.49a) is distinctive of this moderately small, stout-bodied snake among Costa Rican forms, and in combination with the number of dorsal dark markings (15 to 23) distinguishes it from all other species of pitvipers.

DESCRIPTION: A rather small to moderate-sized species reaching a maximum total length of 635 mm; adult females (412 to 635 mm), longer than adult males (300 to 500 mm); tail short (10 to 14% of total length in males, 9 to 12% in females); most scales on top of head small, keeled; 2 elongate smooth internasals; rostral and internasals bent upward to form proboscis; 1 smooth canthal on each side; supra-

Map 11.130. Distribution of *Porthidium nasutum*.

oculars large, smooth; a loreal; 2 to 4 preoculars, usually 3; 1 to 4 suboculars; 1 to 4 postoculars, usually 3; 8 to 11 supralabials, usually 9 or 10; usually 12 infralabials, sometimes 9, 10, or 13; temporals keeled; dorsal scale rows 23 to 29–21 to 27–17 to 21, usually 25-23-19, all keeled; ventrals 128 to 144 (123 to 145); subcaudals 27 to 41 in males, 24 to 35 in females; anal entire. Dorsal ground color various shades from yellow brown (females) to dark brown or gray (males), with 15 to 22 (13 to 23) dark body blotches narrowly outlined with white; blotches occasionally expanded into crossbands; a narrow vertebral light stripe (white, orange, or orangish tan) separating blotches, which sometimes alternate on the two sides; top of head often with arrow-shaped dark mark pointing anteriorly; distinct dark stripe from lower corner of eye to jaw angle, demarcated above by postocular light stripe and supralabial light stripe below; undersurfaces heavily stippled with brown; tail of juveniles yellowish white, iris gray. Hemipenial lobes with mix of 35 to 65 small and large spines (including 2 to 6 basal hooks) proximally; apical area covered by spinulate flounces and terminating in an awn (Campbell and Lamar 1989).

SIMILAR SPECIES: (1) *Porthidium ophryomegas* has a rounded or slightly upturned snout and lacks a proboscis, two canthals, and nonelongate internasals. (2) *Porthidium volcanicum* has many more dorsal dark markings (43 to 50) and ventrals (156 to 165). (3) *Agkistrodon bilineatus* has a distinct light canthal stripe and enlarged smooth head shields.

HABITAT: A common species often found in places modified by human activity. Widespread in Lowland Moist and Wet Forest zones and marginally in Premontane Wet Forest.

BIOLOGY: Like other Costa Rican pitvipers, this terrestrial form is primarily nocturnal and hides in a variety of sites (under rocks, logs, rubbish, in tree holes) during the day. In captivity, neonates ate earthworms (Picado 1931). Juveniles eat a variety of small arthropods and kill and eat frogs and toads; they may use caudal luring to attract prey. Adults eat mostly mammals but are known to eat birds and members of their own species (Porras, McCranie, and Wilson 1981; Greene 1997) (plate 492). Male combat with some biting is known for this viper (Shine 1994). These snakes will sometimes climb low bushes or trees. I found a mating pair at night during March about 1.5 m off the ground at the La Selva Biological Station. In Costa Rica litters of six to eighteen young have been reported. Neonates are about 120 to 145 mm in total length. Gerhardt, Harris, and Vásquez (1993) list this as a prey item of the great black hawk *(Buteogallus urubitinga)* in Guatemala. Bites of humans by this species are common (about fifty per year reported), but no deaths have been recorded in Costa Rica. The amount of venom injected is very small. However, fatal envenomations, probably untreated, have been reported in Colombia (Daniel 1949).

REMARKS: This species has been called *Bothrops nasu-* *tus* by most authors. Taylor's (1951b) record of *Bothrops lansbergii* is based on an example of *P. nasutum*. An old record (AMNH 17306) for the city of San José is questionable, since no other example is known from the Meseta Central Occidental. Wilson and McCranie (1984) provided a detailed account for this species based mostly on Porras, McCranie, and Wilson (1981). Unlike other Middle American species of hog-nosed vipers, *P. nasutum* is restricted to rainforest environments. Allied species of this stock are found primarily in semiarid to arid habitats and are mostly allopatric in distribution as follows: *Porthidium hespere* (Colima, Mexico); *P. dunni* (Isthmus of Tehuantepec, Oaxaca, Mexico); *P. yucatanicum* (northern Yucatán Peninsula); *P. volcanicum* (southwestern Costa Rica); *P. ophryomegas* (Pacific lowlands from Costa Rica to Guatemala and dry valleys of Atlantic lowland Guatemala and Honduras); and *P. lansbergii* (eastern Panama to northern Venezuela).

DISTRIBUTION: Lowlands and foothills from northern Chiapas, Mexico, to eastern Panama, on the Atlantic slope, and Costa Rica, central Panama, and eastern Panama to northwestern Ecuador on the Pacific versant. In Costa Rica on the Atlantic and southwestern Pacific lowlands, with a few records for higher elevations and on the Pacific slope near Tilarán (2–900 m).

Porthidium ophryomegas (Bocourt, 1868a)

Toboba Chinga, Toboba Gato

(plate 493; map 11.131)

DIAGNOSTICS: This relatively common species and *Porthidium nasutum* and *P. volcanicum* are the only Costa Rican pitvipers with a thin light vertebral stripe combined

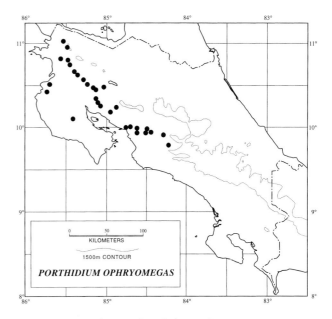

Map 11.131. Distribution of *Porthidium ophryomegas*.

with regular dorsal blotches and bands. *P. ophryomegas* does not have an upturned snout and definite proboscis or a light postocular stripe.

DESCRIPTION: A moderately small species (total length to 800 mm, most adults 400 to 500 mm); tail short (11 to 15% of total length); most scales on upper surface of head small, keeled; 2 flat smooth internasals; 2 smooth canthals on each side; supraoculars large, smooth; a loreal; 3 preoculars; usually 1 long slender subocular, sometimes 2; 2 or 3 (1 to 4) postoculars; usually 9 supralabials, sometimes 8, 10, or 11; infralabials 9 to 13, usually 11 or 12; temporals keeled; all dorsal scales keeled, in 21 to 28–23 to 28–19 rows (middorsal mode 25); ventrals 156 to 176; subcaudals 39 to 42 in males, 23 to 37 in females. Entire dorsum gray to brown with a thin light (white, cream, orange, or tan) vertebral stripe; 21 to 41 pairs of dorsal dark blotches (usually outlined by white) that are usually offset on the two sides, sometimes fused to form a wavy dorsal stripe; two or three additional rows of small dark blotches on flanks; rarely, entire dorsal pattern obscured by dark pigment and vertebral light stripe broken; postocular dark stripe present or absent, if present, not bordered by light stripe above; venter white, usually heavily mottled with dark pigment that may form transverse dark lines; underside of tail densely mottled with dark proximally, much lighter distally.

SIMILAR SPECIES: (1) *Porthidium nasutum* has a strongly upturned snout, usually a definite proboscis (fig. 11.49a), a light postocular stripe, and 128 to 144 ventrals. (2) *Porthidium volcanicum* has more dorsal blotches (43 to 50 pairs) and elongate internasals.

HABITAT: Primarily found in Lowland Dry Forest, but it ranges into Lowland Moist Forest and marginally into Premontane Moist Forest on the Pacific slope of northern and central Costa Rica.

REMARKS: This species has usually been referred to as *Bothrops ophryomegas* in other publications on Costa Rica. Although it is often listed as ranging into western Panama based on Amaral (1929b), this is doubtful. The southernmost record for *P. ophryomegas* in Costa Rica is in San José Province, 250 km northwest of the Panama frontier. This snake is found in subhumid to semiarid habitats, and the rainforests of southwestern Costa Rica and western Panama seem a highly unlikely environment for it. Human envenomations by *P. ophryomegas* are few (three or four reported annually). The small amount of venom it injects makes the bite innocuous, although unpleasant.

DISTRIBUTION: Pacific lowlands and premontane areas from Guatemala to central Costa Rica; disjunctly on the Atlantic versant in eastern Guatemala and northern Honduras; also in the Great Lakes drainage in Nicaragua. In Costa Rica found in the Pacific northwest area, the lower margin of the Meseta Central Occidental, and the upper Candelaria valley (7–1,095 m).

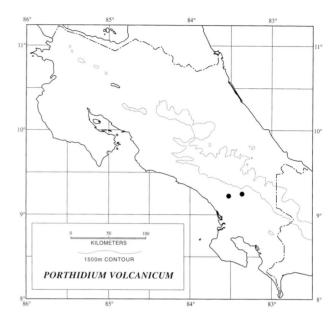

Map 11.132. Distribution of *Porthidium volcanicum*.

Porthidium volcanicum Solórzano, "1994," 1995

(plate 494; map 11.132)

DIAGNOSTICS: A hog-nosed pitviper with numerous (43 to 50) dark dorsal blotches, elongate internasals, and two canthals on each side.

DESCRIPTION: A small to moderate-sized snake attaining a maximum total length of 536 mm (males apparently smaller); tail short (10 to 15% of total length); most scales on top of head small, keeled; 2 elongate smooth internasals; rostral and internasals upturned; 1 smooth canthal on each side; supraoculars large, smooth; a loreal; usually 3 preoculars; 1 or 2 suboculars; 1 to 3 postoculars; 9 or 10 supralabials; 11 to 13 infralabials; temporals keeled; dorsal scale rows 19–25 to 26–19, all keeled; ventrals 156 in male, 159 to 165 in females; subcaudals 35 in male, 25 to 29 in females; anal entire. Dorsal ground color gray to brown, with 43 to 50 pairs of dark blotches; smaller irregular lateral blotches lie below most dorsal blotches; a narrow vertebral white stripe separates the halves of blotches; distinct postocular light stripe from upper surface of rostral over supraocular and curving downward posterior to eye onto neck; venter dull brown with scattered darker punctations.

SIMILAR SPECIES: (1) *Porthidium nasutum* usually has a fleshy flap on the tip of the snout (fig. 11.49a) and fewer dark dorsal markings (15 to 22). (2) *Porthidium ophryomegas* lacks elongate internasals, has one canthal on each side, and has fewer dorsal dark blotches (21 to 41 pairs). (3) *Agkistrodon bilineatus* has enlarged head shields.

HABITAT: Known only from a few localities in a disturbed area of the Pacific Lowland Moist Forest.

BIOLOGY: A rare form. Adult females have total lengths between 310 and 536 mm.

REMARKS: This recently described species (Solórzano 1994) is closely related to *Porthidium lansbergii* that ranges from eastern Panama to northern Colombia and northeastern Venezuela with an apparently relict population in western Ecuador. *P. langsbergii* differs from *P. volcanicum* primarily in having fewer dorsal dark blotches (16 to 25 versus 43 to 50 in the latter) and lower ventral counts (139 to 161).

DISTRIBUTION: Known only from the Valle de General, Puntarenas Province, in southwestern Costa Rica (400–570 m).

Crotalines

Genus *Crotalus* Linné, 1758

Rattlesnakes

The uniquely American rattlesnakes are represented in Costa Rica by a single member of the genus *Crotalus* (thirty-one species). Only one other snake genus, the pygmy rattlesnakes *(Sistrurus),* has the tail modifications characteristic of this stock. All species of rattlesnakes are viviparous and are born with a laterally compressed keratinized prebutton on the tip of the tail. A few days after birth the snake sheds it skin and loses the prebutton, which is replaced by the first rattle segment, or button. Subsequent sheddings (usually several per year) each add a segment to the rattle, and a complete rattle string has the button retained at the tip. A juvenile rattlesnake with only a prebutton or button cannot produce the characteristic rattle sound by vibrating the tail, since this requires two or more overlapping segments. However, their venom delivery system is operational at birth. One species in the genus, *Crotalus catalinensis* from an island in the Gulf of California, has a poorly developed rattle composed of a single segment that does not produce a sound. Occasionally individuals of other species are found that lack rattle strings because of wear or injury, but these snakes may be recognized because they will have a squared-off tail tip (not tapered as in other pitvipers). The species of *Sistrurus* differ from other rattlesnakes in having the upper head shields enlarged anteriorly, including a single frontal and paired parietals. In *Crotalus* the head scutellation may include scales and shields of varying sizes, but there are always several scales in the frontal region, and the parietals, if enlarged, are asymmetric and not in contact with one another. Besides the rattle, the following are significant characteristics of this genus: body cylindrical; nasal pore absent; dorsal scales in 21 to 31 rows at midbody; two apical scale pits; subcaudals usually entire throughout; tail not prehensile; palatine triangular or knobbed anteriorly, with zero to four teeth.

The hemipenes are short (6 to 10 subcaudals in length) and deeply forked, with a centrolineal, bifurcate sulcus spermaticus; pedicel and truncus are spinous, apical areas covered by small papillate calyces (Klauber 1956; Dowling 1975b).

The genus *Crotalus* ranges over a vast area from southwestern Canada and the eastern United States to central Argentina, but it is absent from the Atlantic lowlands of eastern Honduras, Nicaragua, and Costa Rica, southwestern Costa Rica, Panama, Pacific drainage South America, and much of the Amazon basin.

Crotalus durissus Linné, 1758

Tropical Rattlesnake, Cascabel

(plate 495; map 11.133)

DIAGNOSTICS: No other Costa Rican snake has the tail modified into a rattle (only a prebutton or button is present in very young rattlesnakes) (fig. 11.8c,d). The distinctive color pattern of paired elongate paravertebral stripes on the anterior part of the body is unique for indigenous pitvipers, but occasionally large adults lack these stripes.

DESCRIPTION: A large snake (maximum total length 1,800 mm); adult males (825 to 1,800 mm) much longer than adult females (802 to 1,580 mm); tail short (6.5 to 10% of total length); 2 large smooth internasals; 2 large smooth prefrontals; supraoculars large, smooth; a pair of large smooth plates in anterior frontal region; loreal divided horizontally into 2 scales; preoculars 2, upper sometimes divided vertically; 2 to 4 suboculars; 1 to 3 postoculars; supralabials 13 to 17 (11 to 18); infralabials 13 to 19 (12 to 21); upper temporals keeled, lower ones more or less smooth; all dorsal scales keeled, vertebral rows keeled and tuberculate, forming an obvious dorsal ridge; scale rows at midbody 27 to 31, usually 29 (25 to 33); ventrals 161 to 181 (159 to 192) in males, 172 to 186 (171 to 195) in females; subcaudals 28 to 34 (25 to 34) in males, 20 to 26 (19 to 26)

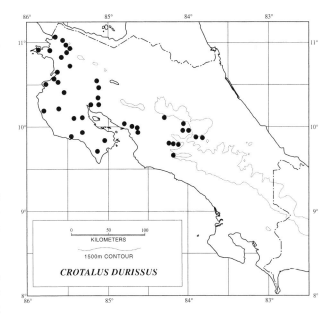

Map 11.133. Distribution of *Crotalus durissus.*

in females; ventrals plus subcaudals 192 to 209 (192 to 217). Dorsal ground color yellowish or light tan, becoming darker in large adults; usually a pair of dark brown paravertebral stripes anteriorly followed by 20 to 30 (18 to 35) light-outlined dark diamond-shaped blotches that usually have somewhat lighter centers; small light-outlined lateroventral dark spots present below diamonds, sometimes fusing with them (pattern variously obscured or broken up in adults in some South American populations); posterior part of body and tail uniform gray to black in adults (tail ringed in juveniles); pattern less intense in large examples as lateral areas darken and blotches become lighter; paravertebral dark stripes continue well onto neck; postocular dark stripe usually present to angle of mouth, not as prominent as paravertebral stripes; venter cream to buff with punctations at tips of ventrals and increasingly darker posteriorly; proximal rattle segment tan in young, gray in adults.

SIMILAR SPECIES: There are none.

HABITAT: Lowland Dry Forest, margins of Lowland Moist Forest in the northwest, Premontane Moist and marginally in Premontane Wet Forest and possibly in Lower Montane Moist Forest.

BIOLOGY: This relatively common species is most abundant in the more or less open country of the Pacific northwest and Península de Nicoya. Although often discovered resting during the day, the species is mostly nocturnal. It feeds primarily on mammals (Klauber 1972), but Picado (1931) reports that in Costa Rica the large spiny-tailed iguana *(Ctenosaura similis)* is a regular prey item. Male combat occurs in *C. durissus* (Shine 1994). Solórzano and Cerdas (1988c) report that Costa Rican rattlesnakes mate during the early dry season (December and January), and births occur in the early rainy season (May, June, and July) to give a gestation period of about six months. Females are sexually mature at about 1,200 mm in total length. Fifteen litters consisted of 14–22.9–35 young, but other workers have reported up to forty-seven neonates in a litter. Litter size is positively correlated with female length. Newborn rattlesnakes were 275 to 430 mm in total length. The tropical rattlesnake is very aggressive when disturbed, coiling and raising its head vertically well off the ground. If rattling and hissing do not deter the potential predator, the snake may uncoil and move toward the intruder. Striking from the coiled position is usual, but as with all vipers it does not have to coil before biting. Although *C. durissus* venom is often reported to be especially toxic because of a neurotoxic component (Klauber 1972; Scott 1983p), Bolaños et al. (1981) found no evidence of neurotoxic effects in twenty-one bite cases in Costa Rica. However, the venom of South American snakes referred to this species has a serious neurotoxic effect. Only about five cases of bites by this species are reported in a typical year in Costa Rica (Bolaños 1984), and serious permanent damage or death is rare. This reflects the relatively low venom yield (somewhat less than 1 cc from

a very large example) and the fact that the venom is much less hemorrhagic and myonecrotic than that of other Central American pitvipers. Costa Rican folklore erroneously claims that if a rattlesnake is placed in a guitar it will produce beautiful sounds instead of rattling.

REMARKS: In his monumental treatment of the rattlesnakes Klauber (1956) recognized six geographic races of *C. durissus*, principally based on color, and Hoge and Romano (1978–79) listed seven more as recognized by Hoge (1965) and by Harris and Simmons (1977). As Chippaux (1986) pointed out, in South America, at least, studies to date do not justify the Hoge and Romano (1978–79) arrangement, which is also based mostly on slight differences in coloration with little regard for individual variation. Whereas Klauber (1956) regarded most mainland South American rattlesnakes as constituting a single race differing from the Central American populations only in details of neck striping, Hoge and Romano (1978–79) split them into eight forms. The problem is further complicated because in South America the range of *C. durissus* is disjunct in subhumid, semiarid, and arid regions dominated by shrublands or savannas, and each isolate differs slightly from the others. McCranie (1993), although following the Hoge and Romano (1978–79) arrangement in his review of the species, expressed reservations regarding characterization of these races. In these circumstances it is not possible to describe details or to evaluate the significance of color variation in South American samples. In general, almost all South American rattlesnakes resemble those from Central America in having dark paravertebral stripes on the neck and anterior body. In one population on Aruba Island off the north coast of Venezuela, this pattern fades ontogenetically so that it is only obscurely evident in adults. In another (sometimes recognized as a separate species, *Crotalus vergrandis*) from northeastern Venezuelan savannas, the typical pattern is obliterated by scattered small white spots in adults. In all other South American rattlesnakes the light stripe bordering the paravertebral stripes below is sharply contrasted with the dark lateral color, and the dorsal diamonds are not in strong contrast with the lateral interdiamond area. In Mexican and Central American examples of *C. durissus* having dark neck stripes, the lateral light edge does not contrast or barely contrasts with the lateral ground color, and the dorsal diamonds contrast markedly with the lateral ground color. Other features shared by typically striped South American rattlesnakes include paravertebral neck stripes 1 to 3 scale rows wide; scale rows at midbody 25 to 33; ventrals 163 to 182; 2 internasals, 2 prefrontals. In Mexico and Central America three geographically subdivisions may be recognized as follows (Klauber 1952):

Tamualipas, southeastern San Luis Potosí, and northern Veracruz, Mexico: neck pattern a black band (about 7 scale rows wide) irregularly spotted and longitudinally streaked with light; scale rows at midbody 25; ventrals 184 to 195; 2 internasals, 2 prefrontals

Southern Mexico, west of the Isthmus of Tehuantepec: neck pattern of paired narrow (1 scale row wide), even-edged dark paravertebral stripes; scale rows at midbody 27 to 33, ventrals 170 to 188; usually more than 4 scales in the internasal-prefrontal region

Central and southern Veracruz, Mexico, and Central America: neck pattern of paired narrow (2 to 3 scale rows wide), even-edged dark paravertebral stripes; scale rows at midbody 27 to 29; ventrals 170 to 191; 2 internasals, 2 prefrontals

The Central American and South American populations are completely separated from one another across the Isthmus of Panama and northwestern Colombia. As Bolaños (1984) pointed out, the venoms of these two groups are markedly distinct. In the former the most important components are typically myonecrotic, hemorrhagic and coagulating, whereas neurotoxins and a hemolitic fraction predominate in the latter. Should these differences prove consistent, they could provide a basis for recognizing the two as distinct species. If that were to occur the South American taxon would be known either as *Crotalus terrificus* Laurenti (Klauber, 1956) or *Crotalus dryinus* Linné (Hoge, 1965). The rattlesnake was once much more common on the Meseta Central Occidental of Costa Rica than it is today. Human population pressure and the indiscriminate slaying of rattlers has led to their disappearance near and in the metropolitan areas where they once occurred. The species formerly was found in the Cartago-Paraíso area of the Meseta Central Oriental (Picado 1931) as well, but I know of no recent records from east of the continental divide. Rattlesnakes are rumored to persist in the small Lower Montane Moist Forest zone (now pretty much cleared for dairy ranching) on the southern slope of Volcán Irazú in the region of Tierra Blanca and Llano Grande, Cartago Province. The karyotype is 2N = 36, consisting of eight pairs of macrochromosomes and ten pairs of microchromosomes and having a ZW heteromorphism in females (Beçak 1965).

DISTRIBUTION: Disjunctly in the lowlands to premontane slopes as follows: (1) Nuevo León, southern Tamualipas, southeastern San Luis Potosí, eastern Queretero, and northern Veracruz, Mexico; (2) disjunctly on the Atlantic versant from central Veracruz, Mexico, to eastern Honduras; more or less continuously on the Pacific versant from Michoacán, Mexico, to central Costa Rica, including the Great Lakes region of Nicaragua; (3) northeastern Colombia and northwestern Venezuela; (4) northern Venezuela and the Orinoco drainage of Colombia and Venezuela to the Guianas; (5) scattered grasslands in the lower Amazon basin; (6) northeastern Brazil and extreme southeastern Peru to central Argentina. In Costa Rica in the Pacific slope lowland and premontane zones of the northwest, including the Meseta Central Occidental, dry valleys immediately to the south, and adjacent slopes (into the lower montane zone marginally); formerly on the Meseta Central Oriental (Atlantic slope) as well (1–1,796 m).

12 | Turtles
Order Testudinata Oppel, 1811 (Testudines, Chelonia)

So DISTINCTIVE is the shell encasing their bodies that turtles are unlikely to be confused with any other organism. The shell comprises a dorsal carapace (usually made up of dermal plates fused to the vertebrae and expanded ribs) and a ventral plastron (usually composed of dermal plates fused to remnants of the interclavicle, clavicles, and gastralia). In one species, the leatherback turtle (*Dermochelys coriacea*), the bony support of the shell is much reduced, the dermal plates are replaced in the carapace by a mosaic of polygonal osteoderms (epithecals) that are not fused to the vertebrae or ribs, and the plastron lacks dermal plates and epithecals. The soft shell of this turtle is composed primarily of connective tissue and covered by skin in adults. Other distinctive features of the order include:

1. Pectoral girdle internal to ribs
2. Skull without temporal fossae (anapsid), frequently emarginate posteriorly
3. Immovable quadrate
4. Sometimes a secondary bony palate
5. A single median penis in males
6. A horny (keratinized) egg tooth
7. Urinary bladder
8. Subclavian artery arising from the carotid or stapedial artery
9. Greatly enlarged perilymph-filled sac around the columella
10. Horny beak covering jaws
11. No teeth in living forms

The earliest fossil records of this order are of Upper Triassic age and date to about 210 million years ago.

In most turtles the shell bones are overlaid by a series of cornified epidermal shields (fig. 12.1). The sutures between the shields and underlying bones do not coincide, and features of both epidermal and bony structures (fig. 12.2) are used in classifying turtles. Other features emphasized for this purpose include details of the skull, beak, structure of the limbs, and coloration.

There are about 290 species of living turtles placed in eighty-nine genera and twelve families (fig. 12.3) as follows:

Suborder Pleurodira Cope, "1868c," 1869 (side-neck turtles)
  Family Pelomedusidae
  Family Chelidae

Suborder Cryptodira Cope, "1868c," 1869 (hidden-neck turtles)
  Superfamily Chelydroidea
    Family Chelydridae (snapping turtles)
  Superfamily Trionychoidea Fitzinger, 1826 (1819)
    Family Kinosternidae (mud and musk turtles)
    Family Dermatemydidae (Mesoamerican river turtle)
    Family Carretochelyidae (pitted-shell turtle)
    Family Trionychidae (soft-shelled turtles)
  Superfamily Chelonioidea Gray, 1825
    Family Dermochelyidae (leatherback turtles)
    Family Cheloniidae (sea turtles)
  Superfamily Testudinoidea Rafinesque, 1815
    Family Platysternidae (big-headed turtle)
    Family Emydidae (freshwater, marsh, and forest turtles and their allies)
    Family Testudinidae (land tortoises)

All Costa Rican turtles are cryptodires and, unlike the side-neck turtles, retract the head vertically into the shell by throwing the neck into an S-shaped curve to pull the head directly backward. In side-necked turtles the neck is curved horizontally to place the head and neck under the margin of the carapace. Sea turtles, although they are cryptodires, have the shell somewhat reduced so that the head cannot be fully retracted under the anterior margin of the carapace.

Turtles occur on every continent except Antarctica and are adapted to a wide variety of habitats, including deserts and oceans, in addition to freshwater and many terrestrial ones. Their northern limits of distribution are near 58° N in Europe, 48° N in Asia, and 51° N in America. Southern limits are at land's end in Africa and Australia and to 40° S in Argentina.

Families extralimital to the area covered in this book are (1) the pleurodiran Chelidae (ten genera) found in South America, Australia, and New Guinea and the Pelomedusidae (five genera) found in South America, sub-Saharan Africa, Madagascar, Mauritius, and the Seychelles Islands and (2) the following cryptodires: the Dermatemydidae (one species); two families of soft-shelled turtles, the Carretochelyidae (one species) of New Guinea and northern Australia, and the Trionychidae (fourteen genera) of North America, Africa, subtropical and tropical Asia to New Guinea, and northward in East Asia to Siberia and Japan; Platysternidae (one species) of southeastern China, Thailand, and Myanmar; and the land tortoises, the Testudinidae (sixteen genera) found in warmer areas on all continents except Australia (also absent from Central America and the East and West Indies) and on Madagascar and adjacent islands. However, one fossil form of uncertain generic status of the last-named family is known from the Oligocene of Costa Rica, and several fossils are known from the West Indies (Williams 1950; Auffenberg 1974).

Aquatic and semiaquatic freshwater turtles, when resting or active in the water, have body temperatures closely following the temperature of the surrounding water. In diurnally active terrestrial species and semiaquatic forms that leave the water to bask, feed, migrate, or lay eggs, the body temperature may be elevated above the ambient temperature. Basking in the sun while out of the water or by expos-

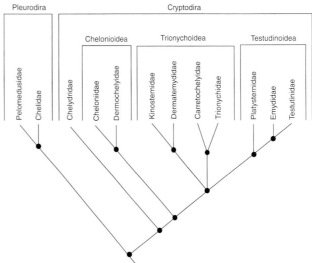

Figure 12.3. Phylogenetic relationships among living families of turtles. Several often recognized monophyletic suprafamilial taxa are also indicated.

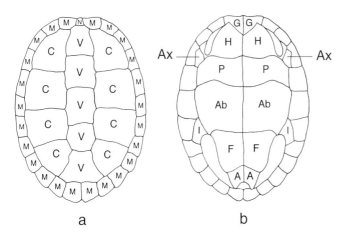

Figure 12.1. Epidermal scale terminology for turtle shells. (a) Carapace: C = costals, M = marginals, N = nuchal, V = vertebrals. (b) Plastron: A = anals, Ab = abdominals, Ax = axillaries, F = femorals, G = gulars, H = humerals, I = inguinals, P = pectorals.

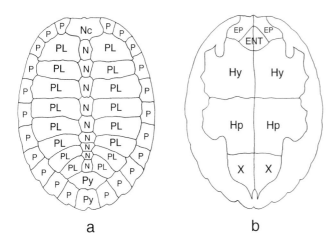

Figure 12.2. Bone terminology for turtle shells. (a) Carapace: N = neurals, Nc = nuchal, P = peripherals, PL = pleurals, Py = pygal. (b) Plastron: ENT = entoplastron, EP = epiplastra, Hy = hyoplastra, Hp = hypoplastra, X = xiphiplastra.

ing the carapace while floating in the water to raise the body temperature primarily expedites digestion and is characteristic of many turtle species. All marine turtles appear to have the capacity to maintain body temperatures that are 1° to 3°C above that of the water. The marine leatherback turtle *(Dermochelys),* however, is known to maintain core body temperatures as much as 18°C above the temperature of the surrounding water by a countercurrent heat exchange in each flipper.

In most amniotes, movements of the rib cage draw air into the lungs and force it out. Since turtles' ribs are expanded, involved in the rigid carapace, and immovable, that mode of breathing is impossible for them. In addition, the large lungs are attached to the carapace dorsally and laterally. Typically turtles exhale by contracting modified abdominal muscles that force the viscera upward to compress the lungs. When these muscles relax, the viscera fall down to their original position, and since they are attached to the lungs by a connective tissue sheath, the lungs expand and draw in air. In the fully terrestrial tortoises, both inhalation and exhalation are mediated by muscular activity. In these forms an anterior (pectoral) and a posterior (abdominal) muscle force the viscera upward, and conversely, another set (a posterior abdominal and the anterior serratus) expands the visceral cavity, allowing the viscera to settle back downward during each breathing cycle.

Turtles drink water, but some terrestrial forms depend primarily on water from metabolism. Marine turtles and the estuarine diamondback terrapin *(Malaclemys)* of eastern North America secrete excess salt from an enlarged and specialized lachrymal gland posterior to the eye to maintain the body's water balance. Excretion of nitrogenous wastes in turtles is accomplished in a variety of ways depending on their environment. In strictly aquatic species the principal nitrogenous waste products are ammonia (20 to 25%) and urea (20 to 25%), which are highly soluble in water and excreted as a very dilute urine. Semiaquatic turtles have lower fractions of ammonia (6 to 15%) and higher fractions of urea (40 to 60%) in their considerably more concentrated urine. Terrestrial turtles from humid environments excrete a highly concentrated urine with urea as its main constituent. Terrestrial turtles that occur under xeric conditions excrete most of the nitrogenous wastes in the form of urates (20 to 50% of total urinary nitrogen) that are precipitated out in the bladder to form a solid (insoluble uric acid) excretory product.

The olfactory sense in turtles is acute, but Jacobson's (vomeronasal) organs, so important in chemoreception in squamate reptiles, are not present. The turtle's characteristic pulsating throat movements apparently cause airflow into and out of the mouth and nostrils, where the olfactory tract and taste buds detect different airborne chemical particles. Locating food and other turtles and avoiding predators have an olfactory component in these animals. In addition, turtles are able to track animal food items or other turtles in aquatic situations using olfaction. Most turtles have specialized scent-producing glands (often called musk glands) near the cloaca or elsewhere on the body. Chemical trails left by these glands on land or in the water are readily followed and appear to be especially important in maintaining social aggregations and during courtship. Turtles' tongues are broad and soft, cannot be protruded, and do not play an important role in chemoreception of distant objects.

Hearing is also very important in the life of turtles. All living forms have a well-developed auditory apparatus including an external tympanum that lies on the side of the head, well behind the eye at about the level of the corner of the mouth, an air-filled tympanic cavity surrounding the ear cartilage and ossicle (extracolumella and columella), a pericapsular recess, and a uniquely complicated inner ear. Sensitivity is good to excellent at low-frequency ranges, with maximum sensitivity at 100 to 700 Hz in most cryptodires and 100 to 1500 Hz in pleurodires. Vocalization in turtles consists mostly of hissing and grunting, so hearing does not seem to play a major role in social communication.

Turtles are very visual animals that in most cases require eyes that function well underwater as well as in the air. The eyelids of all turtles are tilted so that the opening between them rises posteriorly. Floating turtles bring the head to the surface of the water with the inclined eye opening parallel with the water surface. The eye, as in amphibians, is covered by the lower eyelid when closed. A transparent nictitating membrane may be drawn back across the eyeball to clean and moisten it or to cover it during submergence without obstructing vision. The eyes of most turtles attain perfect focus in or out of the water by special modifications including an almost gelatinous lens that is shaped by contraction of the iris sphincter. The lens projects through the pupil to allow the iris to grip it during accommodation so that the pupil essentially remains open at all times. Only in sea turtles is this not the case, and consequently they lack the ability to focus when ashore at night. Because the pupil does not close in most forms, it is not surprising that turtle retinas are made up primarily of cones, those types of photosensitive cells important for daylight and color vision. There are both major single cones and double cones present and also, at least in some turtles, a second kind of single minor cone, all virtually identical to those found in lizards. These visual cells contain oil droplets, either colored (red, orange, or yellow) or colorless. The retina has no or very few rods, the cells responsible for vision at low light intensity and at night in diurnal species. In nocturnal forms the number of rods is increased, but the retina is still formed primarily of cones (Walls 1942; Underwood 1970). Lateral eye motion is coordinated for binocular vision, particularly in carnivorous species, but vertical movements of each eye are independent. Turtles have color vision with spectral limits about the same as in humans. The most important hues seem to be orange, green, and violet.

On land or in the water, most turtles move by alternating leg movements. Turtles walk slowly on land. Most terrestrial species use the typical tetrapod walk in which only one foot is off the ground at any one time and the forefoot strikes the ground after the hind foot on the same side, followed by the hind foot on the opposite side striking the ground (lateral sequence gait), or they may also use a walking trot (forelimbs and hind limbs on opposite sides of the body striking the ground in unison). In swimming, freshwater turtles use alternating limb motions analogous to those of a trot and/or a diagonal sequence where the forefoot pushes against the water after the hind foot on the opposite side (Zug 1971). Sea turtles swim by thrusting both forelimbs simultaneously backward and downward and then rotating them forward in a circular motion to prepare for the next stroke. Aquatic and semiaquatic forms have webs between the toes so the thrusts propel the turtle swiftly through the water. To walk on land these turtles use much the same motion as terrestrial species. Sea turtles have the front limbs modified into strong paddle-shaped flippers that are used in tandem in swimming, but they too use the alternating pattern when crawling on land.

The structure of the limbs reflects these differences in locomotion. A reduced phalangeal formula, hand and foot each 2-3-3-3-3 (from inner to outer digit), occurs in most semiaquatic and aquatic forms (i.e., the hands and feet are short), and all Costa Rican freshwater and terrestrial species have five claws on the forelimbs and four on the hind limbs (claw absent on outer toe). Terrestrial forms (tortoises) have greater reductions (extreme of 2-2-2-2-1) associated with their short, broad feet. The digits of sea turtles are long and slender with the basic 2-3-3-3-3 components or a loss of the phalanx from the inner toe. The aquatic soft-shelled turtles have increased toe lengths by adding phalanges to toe IV (four or five) and V (four to six).

Turtles locate food principally by olfaction and vision, and most species are omnivores. Several forms are strict herbivores, while a number are predators on fishes and large crustaceans or mollusks. The beaks (tomia) of turtles have sharp cutting edges that are usually serrate and often have sharp longitudinal ridges on the crushing plates (alveolar surfaces). The turtle uses these structures together with the front legs to tear food into pieces small enough to swallow. In forms specializing on large crustaceans or mollusks, the head is very large (increased jaw musculature) and the crushing plates are very broad to help crack the strong outer covering.

Adult turtles of the largest species are well protected by the shell from predators other than humans, and the shell doubtless discourages predation on most species except when the turtles are quite small. Crocodilians and large fishes, however, are important predators on freshwater turtles, even large ones, which they may swallow whole. Sharks appear to be major predators on small sea turtles and even bite the limbs off adults. Turtle eggs are particularly large packets of nutrients and are eaten by any predator that can dig them up, including a large variety of carnivorous mammals (especially bears, raccoons, and their allies skunks and dogs) and monitor lizards. Once a nest is opened, hawks, vultures, and other scavengers eat the exposed eggs. Crabs that burrow into the nest also appear to be important predators on sea turtle eggs. Hatchling turtles are, of course, much more vulnerable to predation than adults and are eaten opportunistically by almost every known kind of vertebrate predator on land or in the water. As described in detail in the account on *Lepidochelys olivacea,* on the way from the nest to the ocean they are attacked by crabs, birds, large lizards, and carnivorous mammals, and once in the water, by crocodilians, voracious sharks, and large fishes.

Sexual dimorphism is marked in most turtle species, with males smaller than females. Males of many species have concave plastrons whereas females have convex ones, and the tail is usually longer in males. Fertilization is internal. Mating may occur in the water or on land, and the male mounts the female from the rear, usually holding on to her carapace with his forelimbs, wraps his tail around hers, and inserts the penis into her cloaca from below. The penis of male turtles varies in details of structure among the main groups. In mating the penis is extruded, not everted as in squamates. The basic organ consists of a proximal shaft, derived embryologically from the ventromedial floor of the cloaca and continuous with it, and a terminal glans penis lying near the vent. The organ is usually described in a relaxed condition from the dorsal aspect (Zug 1966). A seminal groove runs posteriorly from the urethral opening to the highly folded glans. The groove may be single throughout its length, or

it may be bifurcate, doubly bifurcate, or trifurcate, with the median branch further bifurcated. As many as three (external, medial, internal) plicae or only one or two are present on the glans. During erection, when the spongy tissue in the organ is first filled with blood the penis is extruded posteriorly. As tension increases the penis bends ventrally and slightly forward. The seminal groove and glans are thus in a ventral or posteroventral position during copulation. Reduction of blood pressure combined with contraction of a retractor muscle withdraws the organ into the male's cloaca after mating. All turtles are oviparous and lay their eggs on land even if the rest of the life history is strictly aquatic.

Among the most interesting aspects of the biology of turtles is the occurrence of temperature-dependent sex determination (see review in Janzen and Paukstis 1991). This phenomenon has been reported to occur in forty-seven species representing most turtle families and all the genera of turtles found in Costa Rica. In some species eggs incubated at higher temperatures (typically above 30°C) produce females, while low temperatures (24° to 27°C) produce males. Intermediate temperatures yield hatchlings of both sexes. In others, however, 100% males are produced at some intermediate temperatures, 100% females at high and low temperatures, and hatchlings of both sexes at transitional temperatures along this gradient (Ewert and Nelson 1991). It appears that the critical temperature-sensitive stage in gonadogenesis occurs during the middle third of incubation, and the location of the nest site determines sex ratios in natural populations. Because of these effects, considerable attention has focused on conservation practices involving transportation and storage of turtle eggs (especially those of sea turtles), which may lead to very skewed sex ratios (Morreale et al. 1982) in released hatchlings.

Turtles are among the longest-lived vertebrates, and some giant land tortoises may live at least five hundred years. Smaller species may live fifty to seventy years in captivity. Many small and medium-sized forms probably live about twenty to twenty-five years under natural conditions. Contrary to popular belief, you cannot determine a turtle's age by counting the concentric rings on the shields of the shell.

TURTLE IDENTIFICATION: KEY FEATURES

The following paragraphs describe the characteristics used in the keys and diagnostics sections that require a technical terminology or might be confusing to the general reader.

Shell: The number and arrangement of the epidermal scutes are frequently diagnostic at the family and generic levels (fig. 12.1). The shape of the shell generally and the shape of the plastron, whether it is cruciform or ovate and the degree to which it covers the soft parts, are also useful features. In some forms there are ligamentous hinges between the pectoral and abdominal scutes and the abdominal and femoral scutes. The structure and scutellation of the bridge between the carapace and the plastron may also be distinctive, and in most families the bridge is covered by the margins of the abdominals or the abdominals and pectorals. In sea turtles an additional series of scutes (the inframarginals) cover the bridge, and in some species there is a pore at the posterior margin of each inframarginal.

Head shields: The dorsal surface of the head is covered with smooth skin in most forms, but distinct head shields occur in the snapping turtle *(Chelydra)* and sea turtles (family Cheloniidae) (fig. 12.6). In the latter group differences are evident in the number of prefrontals lying between the nasal and the single frontal and paired supraocular arcs. Usually there are two pairs of prefrontals (often with a single middorsal one bordering all four), but in *Chelonia* there is a single pair. The scutes on the lower jaw (inframandibulars) vary in numbers and size in those turtles where they are present. The shields lie posterior to the jaw sheath on the labial and ventral surfaces. The number bordering the tomium (one, two, or several) and their relative sizes are diagnostic (figs. 12.7, 12.18).

Jaws: The horny keratinized beaks (rhamphothecas) often exhibit differences in the structure of the shearing edge (tomium) and the crushing (alveolar) surface. In some turtles the upper tomium is level or concave at the jaw symphysis, but often there is a definite notch or a strongly developed hook. In *Dermochelys* the central notch at the symphysis lies between two pointed bony projections. The lower jaw frequently has a single large projection at the symphysis, and sometimes this is flanked by another one on each side. The tomium may be smooth or serrate over most of its length (fig. 12.18). The alveolar surfaces may be narrow or broad and often have a median ridge (or occasionally two ridges) that may or may not be strongly denticulate (fig. 12.9).

Limbs and claws: Sea turtles have paddle-shaped limbs (fig. 12.4) with rigid wrist and ankle joints. Other Costa Rican turtles have pentadactyl generalized testudinate limbs with movable wrist and ankle joints. Claws are present on all fingers and four toes in the latter case but are reduced to one or two in most sea turtles and are absent in *Dermochelys*.

Coloration: Color and color pattern are frequently distinctive among related species, but ontogenetic changes, sexual dimorphism, and melanism complicate use of these features. Generally, colors and pattern are most vivid in hatchlings and juveniles. Increased production of dark pigment with age often overlays and dulls the brighter colors, and large males of some species are consequently without much pattern, are melanistic, and bear little resemblance to females of the same age.

Size: There are marked size differences among chelonians, but in most genera the sexes tend to be about the same

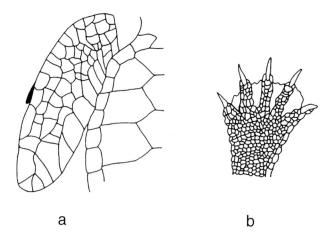

Figure 12.4. Forelimb structure of turtles: (a) paddle-shaped flipper of sea turtles; (b) hand of freshwater or terrestrial turtles.

size. The standard length measurement for turtles is the straight-line carapace length down the midline.

In the following sections and elsewhere in this book, reference to the size of turtles uses standard length, as follows:

Small: less than 200 mm
Moderate: 200 to 350 mm
Large: 350 to 700 mm
Very large: 700 to 1,400 mm
Giant: more than 1,400 mm

The largest Costa Rican turtle (also the largest living turtle) is the marine leatherback *Dermochelys coriacea* at 2,190 mm. The largest species found in the fresh waters of the republic or on land is the semiaquatic snapping turtle *Chelydra serpentina* at 494 mm, and the smallest is *Kinosternon angustipons* at 120 mm, making it among the smallest in the world.

KEY TO THE GENERA OF COSTA RICAN TURTLES

1a. Limbs paddle shaped; no more than two claws on hands and feet (fig. 12.4a) 2

1b. Limbs not paddle shaped; five claws on hands, four on feet (fig. 12.4b) 6

2a. Shell covered by epidermal scutes; claws present. . . . 3

2b. Shell covered by leathery skin or a mosaic of small scales (fig. 12.5); no claws, except in some hatchlings . . . . *Dermochelys* (p. 750)

3a. Two pairs of prefrontals (fig. 12.6b); usually two claws on hands and feet. 4

3b. One pair of prefrontals (fig. 12.6a); usually one claw on hands and feet . . . . *Chelonia* (p. 756)

4a. Five or more pairs of costals (fig. 12.15a,c) 5

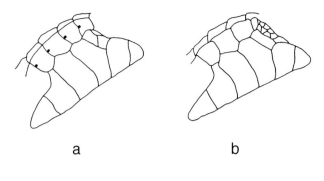

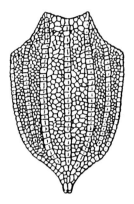

Figure 12.5. Dorsal view of carapace of hatchling *Dermochelys*.

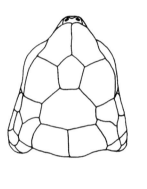
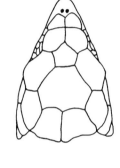

a   b

Figure 12.6. Dorsal surface of sea turtle heads. (a) *Chelonia* showing two prefrontals; (b) *Eretmochelys* showing four prefrontals.

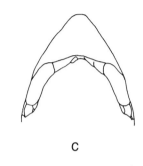
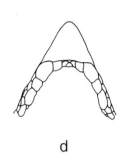

c   d

Figure 12.7. Ventral view of sea turtle bridges: (a) *Lepidochelys* with four inframarginals, all with pores; (b) *Caretta* with three inframarginals, all lacking pores. Ventral view of sea turtle jaws: (c) *Lepidochelys* with large inframandibular scale behind beak; (d) *Caretta* with several inframandibular scales behind beak.

4b. Four pairs of costals (fig. 12.16c) . . . .
. . . . *Eretmochelys* (p. 759)
5a. Bridge usually with four pored inframarginal shields (fig. 12.7a); a single large inframandibular scale behind beak on lower jaw, followed by a small one (fig. 12.7c)
. . . . *Lepidochelys* (p. 760)
5b. Bridge usually with three inframarginal shields all lacking pores (fig. 12.7b); a series of several (3–7) inframandibular scales behind beak on lower jaw (fig. 12.7d)
. . . . *Caretta* (p. 755)
6a. Plastron with nine to eleven plates (fig. 12.8) 7
6b. Plastron with twelve plates (fig. 12.1b) 8
7a. Plastron cruciform, with four pairs of plates and an unpaired one, or five pairs; length of plastron not more than 75% of carapace length (fig. 12.8a) . . . .
. . . . *Chelydra* (p. 762)
7b. Plastron ovate, with five pairs of plates and an unpaired one; length of plastron at least 80% of carapace length (fig. 12.8b) . . . . *Kinosternon* (p. 744)
8a. Alveolar surface of upper jaw without a ridge; head and neck spotted or obscurely lined with dark on light ground color . . . . *Rhinoclemmys* (p. 764)

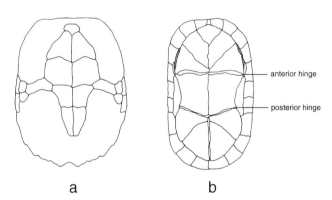

a   b

Figure 12.8. Costa Rican aquatic turtle plastrons: (a) cruciform plastron of *Chelydra*; (b) hinged plastron of *Kinosternon*.

8b. Alveolar surface of upper jaw with a ridge (fig. 12.9); head and neck with prominent bright yellow and black stripes on dark green or greenish black ground color
. . . . *Chrysemys* (p. 769)

Some confusion exists regarding vernacular names used for reptiles of this order. In British usage no single name covers all members of this group, and the word turtle is applied only to marine forms. Likewise, strictly land-dwelling spe-

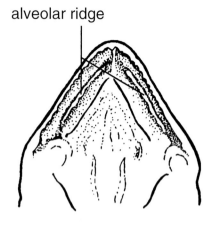

Figure 12.9. Ventral view of upper jaw showing tomium and alveolar ridge in turtles (see text for description).

cies are called tortoises and semiaquatic, freshwater and estuarine forms are terrapins. American usage, which is followed here, applies the word turtle to all representatives of the order. Tortoise is reserved for turtles of the strictly terrestrial family Testudinidae, and other forms are simply called turtles, including the terrestrial box turtles and strictly aquatic soft-shelled and marine forms. Some semiaquatic species may also be called terrapins (from the British usage), sliders, or cooters (from *kuta*, a West African word for turtle, imported to the southern United States during the early days of slavery).

The books by Cornelius (1986) and Acuña (1998) are excellent accounts of the marine and continental turtles, respectively, of Costa Rica. They have provided much of the information on the biology of Costa Rican species presented in the following accounts, and you should consult them for further details on life history and habits.

Suborder Cryptodira

The classification outlined earlier (p. 739, fig. 12.3) follows Gaffney and Meylan (1988) except for the placement of the Platysternidae. Ernst, Lovich, and Barbour (1994) proposed a different arrangement in which the positions of the Trionychoidea and Chelonioidea are reversed and the Chelydridae (plus Platysternidae) are allied with the Testudinoidea. Because there is no consensus the sequence of family accounts below is a compromise between the two.

Family Kinosternidae Gray, 1869 (1857)

Mud and Musk Turtles

The family comprises four genera: *Claudius* (one species), *Kinosternon* (eighteen to twenty species), *Staurotypus* (two species), and *Sternotherus* (three species). The family ranges from the eastern United States and Canada west to the southwestern United States and southward through Central America to northern Bolivia and Argentina. *Sternotherus* are re-stricted to eastern and central North America and *Claudius* and *Staurotypus* to southern Mexico and northern Central America. *Kinosternon* occur throughout the range of the family. The common names for turtles of this group derive from their generally inconspicuous color and aquatic habits and their habit of expelling a strong musky secretion from special paired glands near the shell's bridges when disturbed.

As with other turtle families, the most important definitive features are internal, and the Kinosternidae are distinguished as follows:

1. Entoplastron present *(Claudius* and *Staurotypus)* or absent
2. Seven to nine bones in the plastron
3. Nuchal bone with costiform processes
4. Twenty-three marginal plates (twenty-two marginals plus nuchal)
5. Inframarginal shields present, sometimes separating the marginals from the pectorals
6. Skull strongly emarginate
7. Vertebrae procoelous in part, one biconvex cervical vertebra, eighth cervical vertebra concave in front, the concavity usually double
8. Caudal vertebrae procoelous
9. Penis with a trifurcate seminal groove and the medial branch bifurcate; only external and medial plicae present on the glans

Genus *Kinosternon* Spix, 1824

The mud turtles of this genus are easily differentiated from the other three living genera placed in the family in having a noncruciform plastron, with one or a pair of well-developed hinges and covered by ten or eleven epidermal shields. *Claudius* lacks any hinge in the cruciform plastron and has the plastron covered by eight shields. *Staurotypus* have a cruciform plastron with a single anterior hinge between the abdominal and pectoral shields and has seven or eight plastral plates. *Sternotherus* are distinct from *Kinosternon* in having rectangular pectoral shields.

The same suite of plastral features distinguishes *Kinosternon* from all other turtles in Costa Rica. No other turtle in the republic has eleven (five paired and one unpaired) plastral shields, trapezoidal pectorals, and a pair of plastral hinges (one between the pectoral and abdominal plates, the second between the abdominals and femorals) (fig. 12.8b). The hinge structure makes it possible for the two movable plastral lobes to be retracted dorsally to effectively cover the head and limbs when the turtle is disturbed.

The genus is composed of eighteen to twenty species that are found in lowland sites from Canada to northern Argentina. The species are usually aquatic but often wander away from water, especially at night. They tend to frequent permanent slow-moving or lotic waters where there is a soft bottom. All the species are omnivores, taking about equal amounts of plant and animal food. Although they are generally not brightly colored or obvious and rarely climb out of the water to bask (they prefer to sit in the shallows), these are the most commonly seen turtles in Central America.

KEY TO THE COSTA RICAN SPECIES OF MUD TURTLES, GENUS *KINOSTERNON*

1a. Beak on upper jaw strongly hooked in adults (fig. 12.10a), usually hooked or broadly concave (fig. 12.10b) at symphysis in juveniles; tip of tail clawlike in males and at least horn covered in females; length of bridge more than 20% of carapace length; lobes of plastron expanded to conceal or nearly conceal soft parts (fig. 12.8b); adults to 175 mm in carapace length 2

1b. Beak on upper jaw even at symphysis (fig. 12.10c), neither hooked nor concave; tip of tail soft and unmodified in both sexes; length of bridge 20% or less of carapace length; lobes of plastron narrow so that soft parts greatly exposed (fig. 12.11); carapace flattened or unicarinate; dorsum of head dark, lateral and ventral areas light without distinct stripes or dark mottling; plastron golden yellow in life; definite patches of opposable

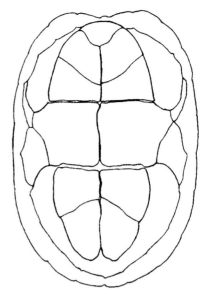

Figure 12.11. Reduced plastron of the turtle *Kinosternon angustipons* that leaves much of the soft parts exposed.

spines on inner surfaces of thigh and calf in adult males (fig. 12.13); adults to 120 mm in carapace length . . . .*Kinosternon angustipons* (p. 746)

2a. Two or three keels on carapace, in adults usually indicated only posteriorly (fig. 12.12b); head mottled with dark and light markings; light head markings usually reddish or orangish in life; plastron nearly always orangish in life; no patches of opposable spines on inner surface of thigh and calf in adult males . . . . *Kinosternon scorpioides* (p. 748)

2b. Carapace never with more than one keel (unicarinate) (fig. 12.12a); head of most adults with small light dots on a dark ground color, sometimes uniformly light; juveniles and some adults with a distinct broad tempo-

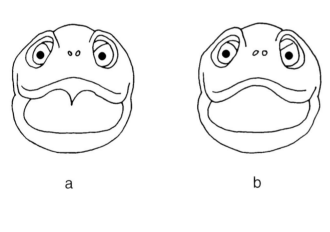

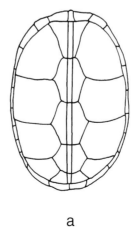

Figure 12.10. Frontal views of turtle heads showing types of upper jaw beaks: (a) hooked; (b) concave; (c) even; (d) notched.

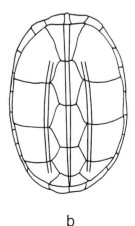

Figure 12.12. Carapace ridging in turtles: (a) unicarinate; (b) tricarinate.

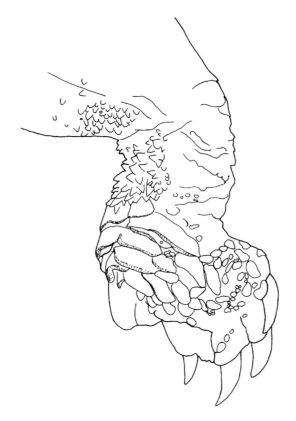

Figure 12.13. Patches of opposable thigh and calf spines in male *Kinosternon leucostomum*.

ral light stripe running obliquely from eye downward over tympanum onto neck; light areas on head yellowish in life; plastron usually yellowish in life; eight or nine patches of small opposable spines on inner surface of thigh and calf in adult males (fig. 12.13) . . . .
. . . . *Kinosternon leucostomum* (p. 747)

Kinosternon angustipons Legler, 1965

(map 12.1)

DIAGNOSTICS: A very small species (adult males to 115 mm in carapace length, females to 120 mm), readily recognized by having a relatively small plastron so that much of the soft parts are exposed (fig. 12.11), a head without strongly contrasting markings, and an unhooked upper beak that is even at the symphysis (fig. 12.10c).

DESCRIPTION: Carapace essentially smooth, without keels in adults, with weak middorsal keel in young; posterior margin of carapace smooth; plastron double-hinged, notched posteriorly; axillary and inguinal shields usually separate pectorals from any contact with marginals; beak margins smooth; one to three pairs of chin barbels; toes webbed. Carapace dark brown, plastron dark yellow with dark brown areas along interlaminal seams; ground color of soft parts gray to brown, with snout and upper eyelid paler than rest of head; ventral areas lighter, tending toward yel-

low; iris brown with yellow flecks. Tails of males extend well beyond margin of carapace, tails of females barely reach posterior margin of carapace; tails of both sexes with tip soft and unmodified; definite opposable patches of spines on inner surfaces of thigh and calf in adult males.

SIMILAR SPECIES: (1) Other *Kinosternon* have the upper beak hooked or concave at the symphysis (fig. 12.10a,b), the plastron covering most of the soft parts when closed (fig. 12.8b), and distinct, contrasting dark and light markings on the sides of the head and neck. (2) *Chelydra serpentina* has a cruciform plastron (fig. 12.8a) and a long tail.

HABITAT: Shallow permanent freshwater swamps and slow algae-filled backwaters of streams in lowland Atlantic rainforests (Moist and Wet Forests).

BIOLOGY: Practically nothing is known about the life of this rare swamp dweller, aside from its similarity to its congeners in food habits; it feeds primarily on vegetable matter and opportunistically eats insects and other small animals. Reproduction apparently occurs from May to August. Legler (1966) hypothesized that *K. angustipons* lays one to four eggs (often singly) and more than one clutch a year, based on examination of the ovaries of examples in the type series. One female from Panama contained a shelled oviductal egg in July that was laid in captivity the following June. The egg was elliptical, white, and 40 × 22 mm. Acuña (1998) indicates that adults are known to be eaten by the several Costa Rican representatives of the cat family (*Felis* spp., *Panthera* spp.) and the iguana *(Iguana iguana)*.

REMARKS: Iverson (1980) summarized pertinent information on the morphology and distribution of this species.

DISTRIBUTION: Lowlands of Atlantic versant Costa Rica and southeastern Nicaragua to northwestern Panama (2–260 m)

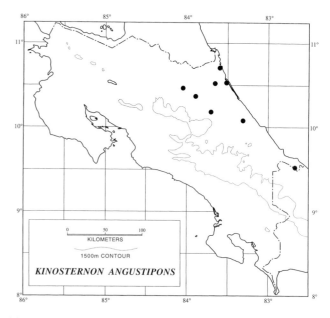

Map 12.1. Distribution of *Kinosternon angustipons*.

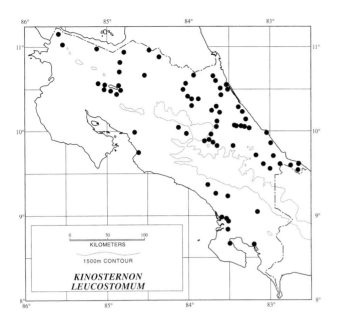

Map 12.2. Distribution of *Kinosternon leucostomum*.

Kinosternon leucostomum (A. Duméril and Bibron, 1851)

Tortuga Amarilla

(plate 496; map 12.2)

DIAGNOSTICS: Any Costa Rican turtle with a pair of broad yellow stripes on the side of the head and neck belongs to this species. The stripes are separated by a dark brown postorbital stripe, and the lower light stripe is continuous with the essentially unmarked upper beak. Many adults of this species, however, have the light areas on the head invaded by dark pigment so that the head is speckled. These latter individuals have a unicarinate (fig. 12.12a) or smooth carapace and a weakly hooked beak (fig. 12.10a) to distinguish them from other Costa Rican species of *Kinosternon*.

DESCRIPTION: Small turtles, males somewhat larger than females, with maximum carapace lengths of 174 mm and 160 mm, respectively. Carapace relatively smooth, except for growth ridges in subadult and small adult examples, smoothly rounded posteriorly, with a distinct low middorsal keel that may disappear in large adults; plastron double-hinged, not incised posteriorly, essentially covering soft parts when closed; axillary and inguinal scutes often in contact so that pectorals and marginals may or may not touch; beak margins smooth; one or more pairs of barbels on chin and throat; toes webbed. Carapace dark brown to black, plastron yellow, often with brown and/or seams marked with brown, sometimes almost entirely brown or black; soft parts generally grayish brown above and yellowish white below. In juveniles and smaller adults, head with a broad deep brown middorsal dark stripe that runs from above eye and continues obliquely downward onto neck, bordered laterally by a broad yellow temporal stripe that is continuous from nostril; a broad deep brown postorbital stripe bor-

ders temporal stripe below; beak yellow, and a yellow stripe runs from angle of jaw back onto neck, below postorbital dark stripe; head stripes tend to disappear with age as dark speckling invades light stripes and brown stripes become diffuse to produce a pattern of small light dots on a dark ground color; iris yellow, divided horizontally and vertically by brown stripes. Males have slight concavity at posterior bridge level and tails that extend well beyond carapace margin; females have tails that reach only to carapace margin; tails of both sexes have a terminal spine; adult males with patches of opposable small spines on inner surface of thigh and calf.

SIMILAR SPECIES: (1) *Kinosternon angustipons* has a reduced plastron that leaves much of the soft parts exposed even when closed (fig. 12.11), and the upper beak is unhooked, even at the symphysis (fig. 12.10c). (2) *Kinosternon scorpioides* has a tricarinate carapace (fig. 12.12b). (3) *Chelydra serpentina* has a cruciform plastron (fig. 12.8a) and a very long tail.

HABITAT: Found in a wide variety of aquatic situations including freshwater marshes, permanent ponds, small streams, the margins of slow-moving rivers, and freshwater swamp forests. Individuals are often encountered moving about on land far from watercourses and in temporary ponds. Typical of Lowland Moist and Wet Forests but also found in Premontane Moist and Wet Forest environments. Also known from a few streams in the Lowland Dry Forest area of central Costa Rica.

BIOLOGY: Acuña (1998) provides considerable data on this common species in Costa Rica, and his account, together with those of Moll and Legler (1971) and Medem (1962) for populations in Panama and Colombia, respectively, are the basis for most of the following. The species is mostly nocturnal and forages at night in shallow waters. It is an omnivore and eats a variety of foods, including algae, grasses, leaves, seeds, snails, freshwater crustaceans, insects, tadpoles, and possibly small reptiles and mammals, although the latter two may be carrion (Vogt and Guzman 1988). Villa (1973) recorded the species feeding on dead fish and carrion and predation on the aquatic snake *Tretanorhinus nigroluteus* in Nicaragua. Courtship and mating have recently been described by Acuña (1998). As in *Kinosternon scorpioides*, there are two mating seasons, one in April and another in July, although Moll and Legler (1971) thought that reproduction was continuous at the population level. Morales and Vogt (1997) reviewed Moll and Legler's material and concluded that there were three consecutive months when females were without oviductal eggs. This is concordant with observations of two breeding seasons. Breeding activities take place in very shallow water during the afternoon, and the female aggressively initiates courtship. She mounts the male for short periods, dismounts, then returns a number of times. After each dismount, the female positions herself in front of the male or to the side and vibrates her head in the water. After several tries, the male begins to respond with similar head vibrations. There follows a period of alternating approaches and retreats by both sexes, during which they face one another and rub snouts or nuzzle the base of the neck or arm or both. Finally the male begins to stroke the female's carapace with the right front foot, and as she becomes quiet he mounts her from above and behind and grasps the posterior part of her shell with the claws of the right hind foot. The two left limbs remain on the substrate. The male then doubles his tail under the female's to allow for insertion of the penis. Copulation lasts about twenty minutes, after which the male dismounts and the female moves away. Eggs are laid about four months after mating in July and October in Costa Rica, in clutches of one to five; they have hard, brittle shells and are oblong, averaging about 37 × 19 mm. No nest is dug, and the eggs are half buried in a shallow depression or covered by leaf litter. Three eggs hatched at normal environmental temperatures (20° to 33°C) under artificial conditions required 126 to 148 days to hatch. Hatchlings are about 25 to 30 mm in standard length. Sexual maturity is reached at about 102 mm (females) and 118 mm (males) in carapace length. Predators on this species include the coyote *(Canis latrans)*, the cougar *(Felis concolor)*, and the jaguar *(Panthera onca)*. A number of raptors, vultures, and the coyote eat many eggs and hatchlings. Juveniles are occasionally sold as pets in Costa Rica. Ewert and Nelson (1991) reported temperature-dependent sex determination in this species, with all females produced at incubation temperatures above 27°C. Vogt and Flores (1992) found that no temperature produced 100% of either sex in southeastern Mexico.

REMARKS: Berry (1978) has demonstrated intergradation between the more or less distinctive forms recognized by previous authors based on slight differences in details of the shell scutes, shell height, plastron size, male secondary sex characteristics, and coloration. His data clearly document the existence of a number of north-south character clines that make any attempt at delimiting geographic races within this species ambiguous. Berry's analysis fully supports his conclusion that the nominal species *Kinosternon postinguinale* and *K. spurrelli* are synonyms of *K. leucostomum*. The name *K. postinguinale* was frequently used for Costa Rican and Panamanian members of this species before 1980. Similarly, *K. spurrelli* was applied to Colombian and Ecuadorian examples. The karyotype is 2N = 56, comprising seven pairs of metacentric and submetacentric and six pairs of telocentric and subtelocentric macrochromosomes and fifteen pairs of microchromosomes; NF = 70 (Moon 1974).

DISTRIBUTION: Lowland and premontane slopes from southern Veracruz, Mexico, south on the Atlantic versant to eastern Panama and adjacent northern Colombia; on the Pacific slope from central Costa Rica and the Meseta Central to central Ecuador (1–1,200 m).

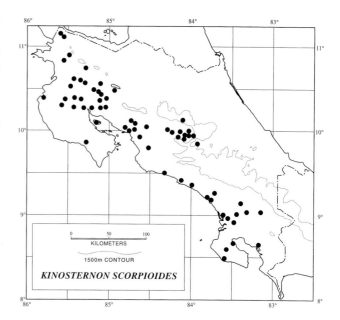

Map 12.3. Distribution of *Kinosternon scorpioides*.

Kinosternon scorpioides (Linné, 1766)

Tortuga Candado

(plate 497; map 12.3)

DIAGNOSTICS: The bi- or tricarinate carapace (fig. 12.12b) will distinguish this species from its congeners. In addition, most adults have a strongly hooked upper beak (best developed in males) (fig. 12.10a), and most individuals have a mottled or spotted head pattern; the light areas of the head are usually red or orange, and the plastron is usually orange.

DESCRIPTION: A small species (to 270 mm in carapace length); carapace relatively rugose in young, but becoming smooth with age, smoothly rounded posteriorly, with a series of three flattened keels (one middorsal, two lateral) in small examples; in larger specimens keels usually indicated posteriorly, or middorsal keel disappears to give shell a flattened appearance in cross section; plastron double-hinged, often weakly notched posteriorly and essentially covering the soft parts when closed, concave in males; axillary and inguinal scutes present but rarely in contact, so that pectorals and marginals usually meet; beak margins smooth; two or three pairs of chin and throat barbels; toes webbed. Carapace brown to black; plastron yellow or, more usually, orange and often partially or completely suffused with brown or black; soft parts grayish brown above, lighter below, with much dark flecking and mottling; light areas on head usually bright red to orange, but sometimes yellow; rarely head pattern consists of large light spots, or they coalesce to form a pair of broad light stripes separated by a dark brown stripe on side of head posterior to eye and continuing onto neck; beaks usually red or orange, spotted and vertically streaked with dark pigment; iris brown. Males with plastron slightly concave at posterior bridge and tail that extends well beyond carapace margin; females with tail barely extending to that margin; tails of both sexes with a terminal spine; no patches of opposable spines on inner surface of thigh and calf in males. Males have larger heads than females.

SIMILAR SPECIES: (1) Other Costa Rican *Kinosternon* have an unkeeled or unicarinate carapace (fig. 12.12a) and lack orange or red in their head coloration. (2) *Chelydra serpentina* has a cruciform plastron (fig. 12.8a) and a very long tail.

HABITAT: Occurring in many freshwater situations: marshes, permanent ponds, streams, rivers, and swamp forests, in Lowland Dry, Moist, and Wet Forests and Premontane Moist Forest.

BIOLOGY: Aside from its general resemblance to other mud turtles in habits, little is known of the life of this common species. It appears to prefer situations where there is considerable aquatic vegetation. Teska (1976) reported substantial overland movements by individuals of this mainly aquatic turtle during the dry season in Guanacaste Province, when most marshes and ponds dry up. It is a carnivorous form feeding mostly on insects, worms, snails, and fishes but is also capable of attacking larger prey. It seizes prey in its powerful jaws and tears it apart with the aid of the forefeet. The species is also a scavenger (Pritchard and Trebbau 1984). Courtship and mating take place twice a year in Costa Rica, during January and July (Acuña 1998). Sexton (1960) reported on mating for Venezuelan specimens of this form. The male initially pursues the female, and if she is receptive he extends his neck and makes an apparently olfactory investigation by touching her tail with his snout. After some preliminary nipping of the female's shell, the male then grasps the edge of her carapace with all four feet and the hooked tail. His head is now fully extended and the female's is retracted within her shell. During this phase of courtship and copulation, the male's neck and head remain fully extended, and he bites the female on the snout if she tries to move her head out of the retracted position. Afterward, in one case, the female dislodged the male, rapidly moved around, and bit his head. In Costa Rica eggs are laid shortly after mating in February and March and from August to November. One to six eggs are laid. They have hard, brittle shells and are about 40 × 19 mm. The eggs are laid in nests excavated at the bases of grass clumps 5 to 200 m away from water. Eggs are 4.4–5.8 mm × 2.7–3.4 mm (Castillo 1986). Incubation times vary from 115 to 128 days. Hatchlings are 30 to 40 mm in carapace length. The smallest mature female known is 122 mm in standard length. Adults are known to be eaten by the coyote *(Canis latrans)* and the several large and small Costa Rican cats. Eggs are heavily preyed on by coyotes, and hatchlings and juveniles are eaten by coyotes, the cats, raptors, and vultures.

REMARKS: This wide-ranging species exhibits considerable ontogenetic and individual variation that makes it hard to interpret geographic variation. The following major geographic subdivisions are recognized by Berry (1978) and by Pritchard and Trebbau (1984):

A. Mesoamerican populations: plastron completely covering soft parts when closed; anal notch slight or absent
1. Northern populations (Mexico, Guatemala, Belize, and El Salvador): posterior plastral hinge straight; axillary and inguinal scutes usually not in contact; carapace wide, high domed; head marbled or speckled with red, orange, yellow, or cream on a slate gray to black ground color
2. Río Grande de Chiapas drainage basin population (Mexico and Guatemala): posterior plastral hinge curved posteriorly; axillary scute absent or axillary abdominal seam incomplete; carapace wide, moderately high; head marbled or dotted with gray or yellow on an olive to slate gray ground color
3. Honduran, Nicaraguan, Costa Rican, and Panamanian populations: posterior plastral hinge curved posteriorly; axillary and inguinal scutes usually in contact; carapace wide, high domed; head spotted and/or reticulated with red, orange, yellow, or cream on a dark gray to brown ground color
B. South American populations: plastron not completely covering soft parts when closed; axillary and inguinal scutes usually in contact
1. Northern populations: posterior plastral hinge curved posteriorly; carapace narrow, depressed; anal notch usually distinct; head spotted or mottled with buff or cream on a gray to brown ground color
2. Southern population (Bolivia and Argentina): posterior plastral hinge curved posteriorly; carapace narrow, depressed; anal notch slight or absent; head spotted or mottled with buff to cream on a gray to brown ground color

These populations, except the last, are connected by zones of intergradation, where turtles intermediate in the cited differences are found. The population (B2) found in southeastern Bolivia and adjacent northern Argentina is completely allopatric to other populations, the nearest of which occurs 1,700 km to the northwest in Amazonian Peru. Cunha (1970) described a peculiar small population of this species restricted to a single flattened mountaintop, the Sierra dos Carajas (700 m) in the state of Pará, Brazil. Examples of this form are much smaller than *K. scorpioides* from the adjacent lowlands (maximum carapace length 130 mm and 178 mm, respectively) and have weakly developed lateral keels on the carapace, the posterior plastral hinge straight, and the carapace narrow but high domed. All Costa Rican material clearly belongs to population A3. The name *Kinosternon albogulare* is used in some of the older literature for this species in Costa Rica. The three examples reported by Acuña, Castaing, and Flores (1984) from Laguna Fraijanes

(1,650 m) on the southern slope of Volcán Poás may have been introduced. Elsewhere in Costa Rica the species is not found above 1,425 m. The karyotype is 2N = 56, comprising seven pairs of metacentric and submetacentric and six pairs of telocentric and subtelocentric macrochromosomes and fifteen pairs of microchromosomes; NF = 70 (Bickham and Baker 1976).

DISTRIBUTION: Lowlands and lower premontane areas from eastern Oaxaca, Mexico, south on the Pacific versant throughout Central America; from Tamaulipas, Mexico, south on the Atlantic slope to Guatemala and Belize, in the lowlands of eastern Honduras and Atlantic drainages of Nicaragua, south through northern South America to northern Peru and northern Brazil on the east; an isolated population in southern Bolivia and northern Argentina; absent from most of Atlantic versant Costa Rica and Panama and Pacific versant Colombia and Ecuador. In Costa Rica known only from the Pacific versant lowlands and the Meseta Central Occidental and Oriental (2–1,425 m).

Family Dermochelyidae Baur, 1888 (1825)

Leatherback Turtle

The members of the single living species in this family are so distinctive that they cannot be mistaken for any other organism. They are essentially marine, but the females come ashore on sandy beaches to lay their eggs. The leatherback is the largest of living turtles, and large adults attain a total length of 2,190 mm. All but the youngest examples of the species lack epidermal shields, so the carapace and plastron are covered by soft skin over a layer of connective tissue. This feature alone will immediately separate the family from all other sea turtles and from all freshwater and terrestrial turtles found in Costa Rica. The hatchlings are also unique in having the carapace and plastron covered by a series of

small polygonal scales that are lost during the first nine months of life (figs. 12.5, 12.14). All other sea turtles have large, well-developed scutes on the shell that persist through adulthood. No freshwater or terrestrial turtle has scalation similar to that of the young leatherback. The family is widely distributed in all the oceans of the world except in the Arctic and Antarctic.

The Dermochelyidae are defined by the following additional features:

1. Limbs modified into paddles
2. No entoplastron
3. Other plastral bones greatly reduced
4. Skull fully roofed by bone with a barely protruding supraoccipital process (fig. 12.17a)
5. Frontal bone excluded from the orbit
6. Head only partially retractable into shell
7. Only one biconvex cervical vertebra (usually the fourth), one or no cervical joints double, eighth centrum simply or doubly concave in front
8. No secondary bony palate
9. Internal nares lateral to alveolar surface of vomer
10. Penis with an undivided seminal groove; only a medial plica on the glans

Genus *Dermochelys* Blainville, 1816

The monstrous adults of the single species in the genus reach up to 2,190 mm in carapace length and 916 kg in weight and are easily recognized at a distance (either on the beach or in the water), since the naked carapace has a series of distinctive longitudinal keels (fig. 12.15b). These keels are also present in subadults and are represented by a series of light scales in longitudinal rows in the youngest juveniles (fig. 12.14). Unlike all other turtles, *Dermochelys* have the carapace lacking neural, pleural, and peripheral bones, the

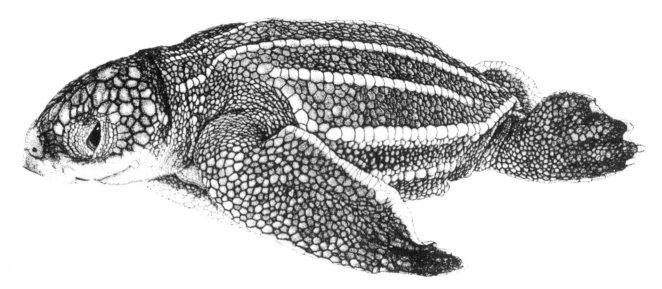

Figure 12.14. Hatchling of the sea turtle *Dermochelys coriacea.*

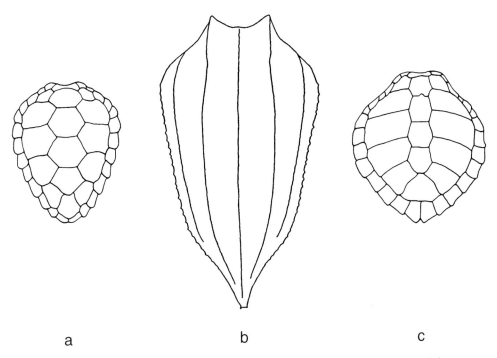

a   b   c

Figure 12.15. Carapaces of Costa Rican sea turtles: (a) *Caretta*; (b) *Dermochelys*; (c) *Lepidochelys*.

ribs not fused to the carapace but embedded in a layer of cartilage, and the surface of the carapace covered by skin over a layer of connective tissue that overlies a mosaic of small bones (epithecals) found nowhere else in the order. The plastron even lacks epithecals.

The genus is found in all tropical, subtropical, and temperate oceanic areas.

Dermochelys coriacea (Vandelli, 1761)

Leatherback, Baula

(fig. 12.14; plate 498; map 12.4)

DIAGNOSTICS: Adults of this giant reptile (to 1,830 mm in carapace length and to a weight of about 680 kg in Costa Rican waters) are unlikely to be confused with any other species because of size alone. The seven distinct knobby dorsal keels (a marginal pair, two lateral pairs, and a single serrate medial one) will immediately distinguish most leatherbacks from any other turtles (fig. 12.15b). Many subadults also have five weak ventral longitudinal ridges on the plastron, but these are almost or completely obliterated in large adults. In small juveniles the keels are represented by seven dorsal and five ventral longitudinal rows of polygonal raised yellowish or white scales that are unique features of this species (fig. 12.14). All but the youngest examples lack most keratinized epidermal structures, so large juveniles, subadults, and adults have no scutes on the shell, no scales on the head or body, and no claws on the digits. Although horny beaks are present on the jaws in young and adult leatherbacks, they are relatively feeble and thin.

DESCRIPTION: Most adult Costa Rican examples are 1,500 to 1,750 mm (Atlantic) or 1,280 to 1,510 mm (Pacific) in carapace length (most reports of larger individuals are probably exaggerations); upper jaw in larger specimens with a large triangular cusp on each side below nostril (fig. 12.10d); lower jaw with a medium bony hook that fits into a medium notch at upper jaw symphysis; deep notch present in the upper jaw just behind each cusp; in very young

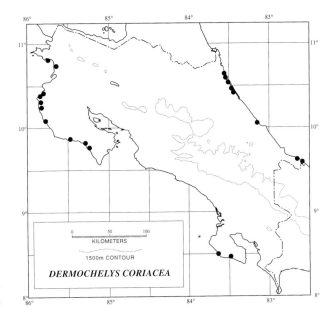

Map 12.4. Distribution of *Dermochelys coriacea*.

turtles, cusps sharply pointed; rarely a single claw present on hand in hatchlings. Upper surfaces dark brown or black, usually uniform but sometimes with paler marbling on back or with longitudinal rows of small dots; usually with pale white, cream, or pink spots and blotches laterally and on throat and neck; ventral surfaces mostly pale pink and white; females usually with a pink area on top of head; hatchlings blue black, trimmed with white, finely beaded in appearance, with seven dorsal and five ventral longitudinal white stripes along raised scale ridges. Tail longer in males than in females; when hind limbs stretched back they barely reach cloaca in males, whereas in females tail tip lies about halfway down legs.

SIMILAR SPECIES: All other marine turtles have claws and have large epidermal scutes covering the shell. Juveniles of these other species resemble the adults in the complement and arrangement of shields on the shell and lack the peculiar polygonal scales found in very young leatherbacks.

HABITAT: Marine, coming ashore on sandy beaches to nest.

BIOLOGY: Leatherback turtles range far out to sea and are found as solitary expatriates far from their nesting grounds. So far as is known, *Dermochely* feeds primarily on jellyfish and a few other soft-bodied pelagic invertebrates, and the weak beaks seem adapted to capturing little else (Bleakney 1965; Brongersma 1969). Long backward-projecting spines (50 to 70 mm) in the mouth and throat help hold the soft-bodied prey during swallowing. Individuals of this species are most likely to be seen in the water or on the beach near nesting sites. In Costa Rica isolated or sporadic nesting occurs along both coasts, but major colonial nesting areas are found near Boca de Matina and Gandoca beach, Limón Province, on the Caribbean shore and at Playa Naranjo, Guanacaste Province, on the Pacific side. Mating occurs in the sea before the females come ashore. The nesting season is generally from February through July on the Caribbean shore (Carr and Ogren 1959; Pritchard and Trebbau 1984; Chacón et al. 1996). On Pacific beaches nesting extends from October to March, with greatest activity in November and December (Cornelius 1986). As with other sea turtles, the female digs a concealing body pit with the forelimbs before evacuating the egg cavity with the hind limbs. The nest cavity itself is much deeper than for other marine turtles because the hind flippers are much longer in *Dermochelys*. Chacón et al. (1996) recorded 530 nests at Gandoca from February 25 to July 15. Nest densities were 12.8 per 100 m of beach, and the mean for nesting females was 5.12 per night. The flexible-shelled eggs are laid one to three at a time until about forty-five to one hundred yolked eggs are deposited, along with many (about 30% of total clutch) yolkless ones (Carr and Ogren 1959; Cornelius 1986; Fitch 1985). Yolked eggs are spherical, white, and soft-shelled, averaging about 51 mm in diameter; yolkless eggs are 0.3 to 45 mm in diameter. The female then fills the nest with sand

using the hind limbs, packs it down, and conceals it by throwing sand that buries the eggs to a depth of about a meter. This makes it extremely difficult for most natural predators to locate the deeply buried nest. In addition the female usually excavates several false body pits several meters away from the nest. Frequently she makes several 360° turns as she attempts to orient herself toward the sea and finally returns to the water. Nesting usually is at night. Renesting in this species occurs four to six times in a season at intervals of seven to eleven days. Temperature-dependent sex determination is characteristic of the leatherback (Mrosovsky, Dutton, and Whitmore 1984; Mrosovsky et al. 1985). Incubation times vary from fifty to seventy-five days, and the hatchlings (64 to 76 mm in carapace length) make a vertical climb through the sand to the surface to emerge as a group in an explosion of activity. The hatchlings then orient toward the sea and race for the water. Leatherback turtles intermediate in size between hatchlings and adults are virtually unknown, suggesting a pelagic existence in the open ocean during the missing growth period. A wide variety of mammal and bird predators eat the eggs, and beach crabs, sharks and large fishes feed on the hatchlings. Juvenile leatherbacks are eaten by sharks, but the adults have no predators other than humans. This species makes transoceanic migrations between feeding grounds and nesting beaches. Swimming locomotion is similar to that described for other sea turtles (p. 754). Swimming speeds are also comparable (1.9 to 9.3 km/h). Leatherback turtles have a number of unique anatomical and physiological features that allow them to penetrate cold north temperate waters. These characteristics also have significance in dives into deep, cool waters. *Dermochelys* are able to maintain core body temperatures 3° to 18°C above ambient water temperatures (Frair, Ackman, and Mrosovsky 1972) because of their great mass, thick, oily, nonvascular carapace, fatty flesh, and most important, a very effective countercurrent heat exchange mechanism in the base of each flipper (Greer, Lazell, and Wright 1973). Heat is conserved by the relatively low ratio of surface area to volume; the insulating properties of the skin, carapace, and flesh; and the redirection of blood away from the flippers. At sea these turtles and their allies in the family Cheloniidae are usually submarine and surface only briefly to breathe. Their lungs function as the major oxygen store, and unlike most diving mammals they have no special physiological adaptations for increased oxygen capacity. Submergence for up to twenty minutes is aerobic, but longer dives of at least three hours depend on anaerobic respiration and the special ability of the turtle brain to function in the absence of oxygen (Lutz and Bentley 1985). Mrosovsky (1987) indicates that leatherbacks may reach depths of at least 1,200 m in deep dives. It has been suggested (Davenport 1988; Davenport and Balazs 1991) that they pursue and feed on bioluminescent medusae, siphonophores, and tunicates at aphotic depths when there are few surface-dwelling jellyfish. Although seemingly docile on land, when molested or captured at sea a large leatherback is very dangerous, using flippers, jaws, and weight to advantage. Pritchard and Trebbau (1984) provide an excellent summary of the biology of this species, and Cornelius (1986) describes the Playa Naranjo population in some detail.

REMARKS: Pritchard (1980) provides an overview of the morphology, distribution, and pertinent literature for this species.

DISTRIBUTION: Worldwide in tropical and subtropical seas, but often straggling very far to the north (Alaska, Iceland, the North Cape, Bering Sea) and south (Chile, Argentina, South Africa, South Australia, New Zealand) in cooler waters. Nesting is almost entirely restricted to tropical beaches. Known to breed and nest on both Atlantic and Pacific coasts of Costa Rica, especially at the legally protected beaches near Boca de Matina and Tortuguero, Limón Province, and Playa Ostional and Playa Naranjo, Guanacaste Province, and on the outer coast of the Península de Osa.

Family Cheloniidae Gray, 1825

Sea Turtles

This family of marine turtles comprises five genera and seven species. Four of the genera and five of the species are found in Costa Rican waters. The family occurs in all tropical and subtropical seas, and several species range into temperate waters as well.

The family is technically defined based on the following characteristics:

1. Limbs modified into paddles
2. Entoplastron present
3. Skull fully roofed over by bone, only slightly emarginate (fig. 12.17b–e)
4. Head not retractable into shell

5. One biconvex cervical vertebra (usually the fourth), one or no cervical joints double, eighth centrum simply or doubly concave anteriorly
6. A secondary bony palate
7. Internal nares posterior to alveolar surface of vomer
8. Penis with an undivided seminal groove; only a medial plica on glans
9. Twenty-three to twenty-nine marginal plates (twenty-two to twenty-eight marginals plus nuchal)

Unlike the other sea turtle family (Dermochelyidae), this group has epidermal scales that persist through life on the shell (figs. 12.15a,c, 12.16), head, and limbs, one or two claws on each limb, and horny beaks on the jaws.

Differences in details of the skull and skeletal support of the shell form the principal basis for classification of the genera within the family. Since skulls are frequently found on the beach, more details concerning the characteristics of this unit are provided in the descriptions that follow than for other groups covered in this book (fig. 12.17). Although the character states of the shields on the head and shell are very useful in distinguishing sea turtles, care must be taken when identifying hatchlings and posthatchlings, since there are frequently numerous shell anomalies.

All members of the family are most likely to be observed in the water just offshore or on the beach during the breeding and nesting seasons. The little hatchlings may be extraordinarily abundant some weeks after the beginning of the nesting season. Because there is an enormous literature on the subject of sea turtle biology, especially on reproduction (see Bjorndal 1995 and Lutz and Musick 1997 for recent summaries and entry into the literature), I will make no attempt to fully review these matters in the following species accounts. Outstanding coverage of the biology of all Costa Rican forms based on their total geographic range is contained in Ernst, Lovich, and Barbour's (1994) account of

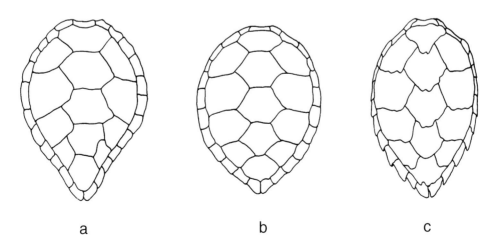

a   b   c

Figure 12.16. Carapaces of Costa Rican sea turtles: (a) *Chelonia agassizii;* (b) *Chelonia mydas;* (c) *Eretmochelys.*

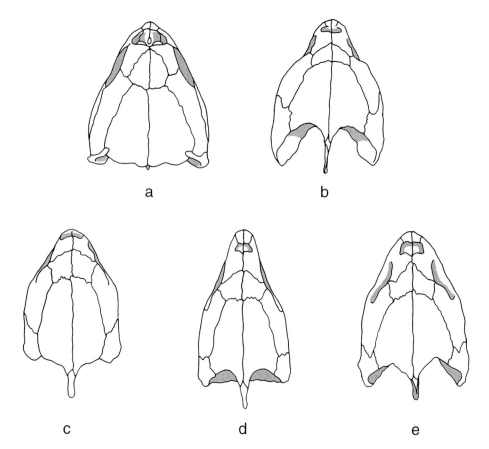

Figure 12.17. Dorsal views of skull of Costa Rican sea turtles: (a) *Dermochelys;* (b) *Caretta;* (c) *Chelonia;* (d) *Eretmochelys;* (e) *Lepidochelys.*

United States and Canadian turtles. Here is a brief summary of major elements of the biology common to all Costa Rican sea turtles.

Courtship and mating appear to take place offshore from the nesting grounds shortly before nesting, although there is a possibility of sperm storage and delayed egg fertilization (two to three years after copulation). Nesting behavior involves the female's coming ashore and laboriously moving to the nest site, excavating a body pit with her forelimbs and the nest with her hind limbs, then ovipositing, covering the eggs and packing the sand over them, concealing the nest by flipping sand, and returning to the sea. These activities usually occur at night. Many females lay multiple clutches (one to eleven) at about two-week intervals during a season and nest only every two to four years. The hatchlings emerge from the nest in a communal exodus about sixty days after nesting and scurry into the sea. They then scatter and appear to become pelagic drifters, feeding on the ocean surface, but little is known about them, since the turtles that reappear inshore are already about 350 mm in carapace length. After this long maturation period of five to twenty years (the "lost years"), the juveniles migrate to a coastal shallow-water area. There they join adults who have mi-

grated from the nesting beaches, and both feed on benthic organisms. They may attain sexual maturity by this time, but the age at first breeding is about twenty to fifty years. Both sexes return from the benthic feeding ground every two to eight years to the original beach where they were hatched, to begin the cycle again. During this process individual turtles migrate great distances (3,000 km or more), sometimes at sustained speeds of 50 to 90 km/day, but usually at about 20 to 40 km/day. The locomotion of all species is essentially similar. The paddle-like anterior flippers provide the propulsive force for swimming, reminiscent of the wing movement of birds in flight. The flippers are moved simultaneously downward and backward and then forward and upward along an axis inclined at 40° to 70°. The flipper tip usually inscribes a figure eight (with its axis inclined) to accentuate the thrust provided by the downstroke. The hind flippers primarily provide stability and regulate turning.

The brains of turtles contain particles of the magnetic mineral magnetite. Hatchlings are imprinted by the earth's magnetic field as they leave the nest. Since the angle of inclination of the magnetic field varies with latitude, adults return to the natal beach to breed using the magnetic field for navigation (Lohmann et al. 1997). Precise location of the beach may also involve olfactory imprinting on the water or beach substrate (Grassman et al. 1984).

Temperature-dependent sex determination has been verified for *Dermochelys, Caretta, Chelonia mydas, Eretmochelys,* and both species of *Lepidochelys,* and it probably occurs in all extant sea turtle species. Members of the family Cheloniidae have a limited ability to raise the temperature of active tissue up to 8°C above ambient water temperatures when swimming (Standora and Spotila 1985). Respiration in these sea turtles during submergence and diving and associated physiological adjustments are similar to those described earlier for *Dermochelys* (p. 752). All sea turtles spend most of their time submerged and may make dives that last up to three hours. A wide variety of mammals, birds, fishes, and invertebrates feed on eggs and hatchlings.

The family is found in tropical, subtropical, and warm temperate seas worldwide. Stragglers of all species except *Natator depressa* are often found in cooler waters well north and south of their usual ranges during the summer months. *N. depressa* is restricted in distribution to the northern coastal region of Australia and the Gulf of Papua.

Genus *Caretta* Rafinesque, 1814

This monotypic genus is separated from other sea turtles by having the frontal bone usually excluded from the orbit; the suture between the prefrontal bones about equal to or shorter than the one between the frontals (fig. 12.17b); and the alveolar surface of the lower jaw greatly expanded at the symphysis and with a median toothlike projection (fig. 12.18d), blunted by wear, in large adults. Externally, the presence of five pairs of costal shields (fig. 12.15a) and the very small intergular shield (which may be absent) will separate this genus from *Chelonia* and *Eretmochelys,* which have four pairs of costals, the nuchal scute in contact with the first costal shields, and a large intergular. *Caretta* usually have three pairs of inframarginals (although sometimes four to five may be present on one side or the extra scutes are quite small) and usually lacks pores on any of these shields, whereas *Lepidochelys* usually have four pairs of inframarginals (rarely three), each with a posteriorly placed pore (fig. 12.7).

The genus is found in tropical and subtropical seas with stragglers ranging into cooler waters.

Caretta caretta (Linné 1758)

Loggerhead Turtle, Cahuama, Caguama

(plate 499; map 12.5)

DIAGNOSTICS: The distinctive reddish brown coloration of the carapace and dorsal and lateral surfaces of the head and limbs in adults is the best field character for recognizing this very large species (carapace length to 2,130 mm, weight to 454 kg). Juveniles are also distinct from other sea turtles in being brown above and whitish, yellowish, or tan below.

DESCRIPTION: Most adult examples measure 700 to 1,000 mm in carapace length and weigh 70 to 160 kg. Carapace relatively elongate, with widest point well anterior to midpoint, without imbricate shields, serrate, but unnotched posteriorly; hatchlings with three dorsal keels, middorsal one with a series of knobs or spinelike protuberances that correspond to posterior section of each vertebral scute; keels and knobs lost in adults; marginals in 11 to 15 pairs, usually 12 to 13; inframarginals usually 3, often 4, sometimes

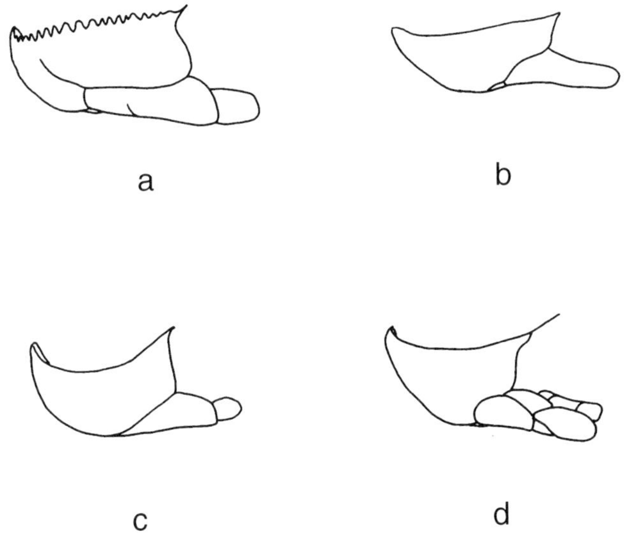

Figure 12.18. Lateral view of lower jaw of Costa Rican sea turtles showing form of tomium and number of inframandibular scales: (a) *Chelonia;* (b) *Eretmochelys;* (c) *Lepidochelys;* (d) *Caretta.*

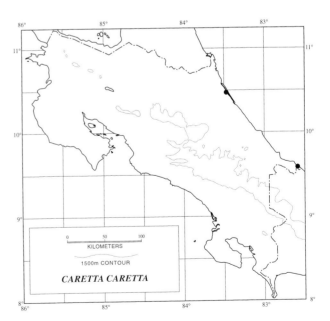

Map 12.5. Distribution of *Caretta caretta.*

5 or 6; rarely 1 or 2 with a pore; plastron covered by 6 pairs of shields and a small unpaired intergular, which may be present or not; two pairs of prefrontal scales sometimes form a median suture, but more often additional azygous scale or scales irregularly wedged between them; other times the additional scale is between the posterior pair so that 5 prefrontals are present; beak margins smooth, lower beak followed by a series of 3 to 7 inframandibular scales of various sizes (fig. 12.7d); two claws on both forelimbs and hind limbs. Lateral head scales reddish centrally, yellowish on margins, and plastron yellowish white to yellowish brown or orange. Males with tail extending far beyond the margin of the carapace, barely reaching it in females.

SIMILAR SPECIES: (1) *Chelonia* and *Eretmochelys* have the first costals separated from the nuchal (fig. 12.16). (2) *Lepidochelys* are olive green to gray dorsally and have the lower beak followed by a single large scale and then a smaller one (fig. 12.18c).

HABITAT: Marine but coming ashore to nest on sandy beaches. There are only a few records of nests of this species in Costa Rica, all from near Tortuguero and Gandoca beaches, Limón Province.

BIOLOGY: Pritchard and Trebbau (1984) summarize the data available on the biology of this turtle. Food includes a wide variety of benthic invertebrates and fishes, for the loggerhead is the most carnivorous of sea turtles and is able to smash the protective skeletons of crustaceans and the shells of mollusks with its powerful jaws and broad crushing surfaces (Mortimer 1981). It also eats jellyfish, salps, and algae. On land these turtles are relatively docile, but at sea they are aggressive and dangerous to handle when disturbed because of their large and powerful heads and jaws. The species does not usually nest in Costa Rica, although it does so in Nicaragua and Panama. Courtship and mating take place offshore and at a considerable distance from the nesting beach. Nesting is usually at night, and about 115 (60 to 200) spherical white soft-shelled eggs form a clutch; each egg is about 40 mm in diameter. This form renests three or four times a season, and the females migrate to the nesting beach every two or three years. Incubation is forty-six to sixty-two days; hatchlings are 41 to 48 mm in carapace length. Sexual maturity is attained in four years. This species exhibits temperature-dependent sex determination (Yntema and Mrosovsky 1980), with males increasingly produced at temperatures above 29°C (Mrosovsky 1983).

REMARKS: Dodd (1990a,b) has summarized current knowledge on variation, distribution, and pertinent literature on this species. The karyotype is 2N = 58 (Nakamura 1949).

DISTRIBUTION: Worldwide in tropical and subtropical seas, but rare along the Pacific coast of Central America; wandering far to the north (Washington State, Nova Scotia, Scandinavia, Japan) and south (Chile, Argentina, South Africa, New Zealand); no definite known records from the Pacific side of Costa Rica. Nesting grounds are generally peripheral to the tropical zone.

Genus *Chelonia* Brongniart, 1800

The genus may be distinguished from other sea turtles by having the frontal bone usually forming part of the orbital margin; the suture between the prefrontal bones shorter than the one between the frontals (fig. 12.17c); and the alveolar surface of the lower jaw at the symphysis narrow and without a distinct toothlike projection at the tip (fig. 12.18a). The genus *Chelonia* may be separated from its allies by having four pairs of costal shields (fig. 12.16a,b) on the carapace, the nuchal scute not in contact with the first costal shields, a large intergular shield, and four pairs of inframarginals (none with pores).

Hirth (1980a) provided an overview of the genus but did not recognize *C. agassizii* as distinct from *C. mydas* (Hirth 1980b). *Chelonia* occur in tropical and subtropical waters throughout the world, and both recognized species are found in Costa Rica.

KEY TO COSTA RICAN SPECIES OF GREEN TURTLES, GENUS *CHELONIA*

1a. Upper surfaces of shell tan to olive; upper soft parts gray; plastron light lemon yellow; no indentation in margin of shell above hind limbs (fig. 12.16b); intermediate scales on anterior and posterior margins of forelimb relatively large, three or four between third posterior scale (counting from joint) and anterior series (fig. 12.19a) . . . . *Chelonia mydas* (p. 758)
1b. Upper surfaces of shell and upper soft parts usually heavily suffused with black to produce a blackish, dark green, or dark olive brown color; plastron with extensive areas of bluish gray; usually an indentation in margin of shell above hind limbs (fig. 12.16a); intermediate scales lying between enlarged dorsal scales on anterior and posterior margins of forelimb small, five or six between third posterior scale and anterior series (fig. 12.19b) . . . . *Chelonia agassizii* (p. 756)

These characteristics will separate mature individuals; hatchling green turtles of the two species are identical in appearance. The carapace is dark blue black, and the underside is pure white except for a band of dark pigment along the underside of the marginals. Consequently only the difference in the paddle scalation may be used to distinguish between the hatchlings.

Chelonia agassizii Bocourt, 1868b

Pacific Green Turtle, Tora, Tortuga Verde

(plates 500–501; map 12.6)

DIAGNOSTICS: Adult turtles of this relatively small species (900 to 1,000 mm in carapace length and 125 kg in

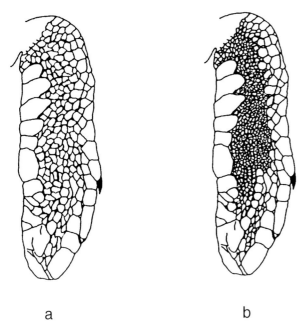

a   b

Figure 12.19. Flipper differences in Costa Rican species of the genus *Chelonia*: (a) *Chelonia mydas*; (b) *Chelonia agassizii*.

weight) are predominantly greenish to olive on the upper surfaces of the soft parts and carapace, often suffused with black pigment; plastron with extensive gray areas; and head scales dark, without light margins. The carapace is usually highly arched with steep sides, prominently tapered, indented above the hind limbs, smooth, heart shaped, without imbricate shields and not serrate or notched behind. In hatchlings the dorsal coloration is dark blue black; the plastron is white except for a band of dark pigment along each side; and the carapace has one prominent middorsal keel and the plastron has two.

DESCRIPTION: Twelve pairs of marginals; inframarginals four, without pores; plastron with six pairs of shields and a large unpaired intergular; upper beak with vertical ridges on inner margin; lower beak serrate, followed by a single large inframandibular scale and then a small one (fig. 12.18a); usually one claw on each flipper. Males with long tails that extend beyond carapace margin, with flattened corneous tip; females with tails that reach no farther than carapace margin.

SIMILAR SPECIES: (1) *Chelonia mydas* is essentially brown above and pale yellow below as an adult. (2) *Eretmochelys imbricata* usually has a serrate, notched carapace (fig. 12.16c) and two claws on each flipper.

HABITAT: Nests on sandy beaches, otherwise strictly marine.

BIOLOGY: Relatively little is known regarding the life of this species, although we can infer that it is similar in most regards to its much better known congener *Chelonia mydas*. Since sea grasses *(Thalassia)*, the principal food of the Atlantic green turtle *(Chelonia mydas)*, are not abundant along the coasts of Pacific Costa Rica, the Pacific species probably feeds mainly on algae supplemented by a diet of soft-bodied invertebrates. Nesting has been reported at Playa Naranjo in Guanacaste Province (Cornelius 1976, 1986), where the nests are well above the high tide line and usually in the shade of the trees in the upper beach zone. The peak nesting months are October to March, with occasional nesting in all months of the year. The eggs are white, round, and soft-shelled; clutch size varies from 67 to 107 eggs.

REMARKS: The eastern Pacific populations have been variously regarded as a distinct species or subspecies by different workers (see Pritchard and Trebbau 1984, for a review). Pritchard (1983, 1999) made the case for specific status based on shell and color characteristics and the smaller adult size of *C. agassizii*. Karl and Bowen (1999) and Bowen and Karl (1999) took issue with Pritchard's conclusion based on a mitochondrial DNA study and regard the Pacific black sea turtles as conspecific with *C. mydas*. Kamezaki and Matsui (1995) analyzed morphometrics of the skulls of green turtles and concluded that the eastern Pacific population is distinct from other populations. However, they declined to recognize *C. agassizii* as a species, preferring to regard it as a subspecies of *Chelonia mydas*. Since that conclusion is based solely on variation in skull measurements, I do not follow it here.

DISTRIBUTION: Tropical and subtropical waters of the eastern Pacific Ocean, including the Galápagos Islands and probably the Hawaiian archipelago, straying north to Alaska and Canada and south to southern Chile; possibly widespread in the central and southern Pacific Ocean, where it seems to occur sympatrically with *Chelonia mydas* at several sites. Only a few definite records from the Pacific coast of Costa Rica.

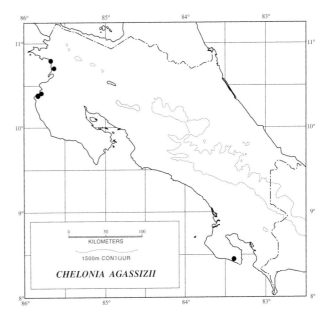

Map 12.6. Distribution of *Chelonia agassizii*.

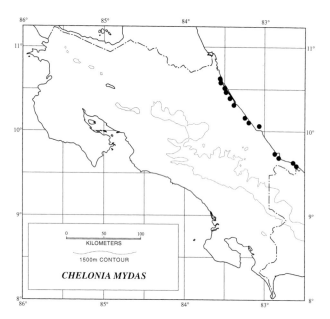

Map 12.7. Distribution of *Chelonia mydas*.

Chelonia mydas (Linné, 1758)

Atlantic Green Turtle, Tortuga Blanca

(plate 502; map 12.7)

DIAGNOSTICS: A very large turtle (to 1,530 mm in carapace length and 295 kg). Features of the coloration and shell provide the best means for distinguishing the green turtle from other sea turtles. Adult green turtles (the name comes from the greenish body fat) are brown in general coloration, sometimes shaded with olive and often with radiating dark mottling, wavy lines, or large brown to black blotches on each carapace shield; plastron pale yellow. The head scales have light edges. In juveniles the carapace is dark blue black and the plastron white except for a band of dark pigment along each side. Hatchlings have one middorsal keel (two lateral ridges obscurely suggested in some) and two ventral keels.

DESCRIPTION: Adult Atlantic green turtles usually are 900 to 1,220 mm in carapace length and weigh from 113 to 200 kg; maximum records (1,530 mm; 295 kg or more) for size are from non–Costa Rican waters. Carapace relatively low, not prominently tapered, not indented, smooth, heart shaped, without imbricate shields, not serrate or notched behind; marginals in 12 pairs; inframarginals 4, without pores; plastron covered with 6 pairs of shields and a single large intergular; upper beak with vertical ridges on inner margin and lower beak serrate; lower beak followed by a single large inframandibular scale and then a small one (fig. 12.18a); usually one claw on each flipper. Males with long tails and a flattened corneous tip that extends beyond carapace margin; females with short untipped tail that reaches no farther than margin of carapace.

SIMILAR SPECIES: (1) *Chelonia agassizii* is nearly black above and usually has dark markings on the plastron.

(2) *Eretmochelys imbricata* usually has a serrate, notched carapace with imbricate shields (fig. 12.16c) and two claws on each flipper.

HABITAT: Strictly marine, except that females nest on sandy beaches.

BIOLOGY: The green turtle is much esteemed for its eggs and delicious flesh. It is also the source of the "calipee" and "calipash," derived from the cartilaginous portions of the shell, that are the ingredients of green turtle soup. This esteem has led to the near extinction of green turtles in several parts of the world. More is known about the biology of this species than for any other sea turtle (indeed for any other turtle). Much of our knowledge is based on the work of Archie Carr and his associates, who have been studying this form at Tortuguero, Limón Province, since 1955 (Carr and Giovannoli 1957). Their many papers, particularly the summaries by Hirth (1971) and Carr (1983), form the basis for most of what follows. The species is primarily herbivorous and forages on sea grass pastures in relatively shallow inshore waters. The principal food is turtle grass *(Thalassia testudinum)*, or other sea grasses and algae when that is unavailable. The Tortuguero beach is the major nesting ground for the species in the Caribbean. Fluctuations in the number of turtles that nest each year are considerable; for example, 428 females nested along the 8 km research beach in 1979 and 3,192 the next year. Of more than 20,000 turtles tagged on that beach, none has ever been recorded as nesting elsewhere. The turtles seem to be imprinted on the beach where they are born and return repeatedly (but usually not every year) to nest there. The interval between egg-laying seasons is two to four years, but most individuals are on a three-year breeding cycle. Mating occurs off the nesting beach, but it remains unclear whether this occurs before or after egg laying and whether there is delayed fertilization with the current year's mating providing fertilized eggs several years later. Females lay clutches of 18 to 193 (average 112) round white soft-shelled eggs from June through October or November, almost always at night. The eggs are 40 to 50 mm in diameter. Multiple clutches (two to five) are often laid approximately twelve days apart. Incubation time is about forty-five to sixty-five days. Spotila et al. (1987) have discussed in detail the effects of the nesting site on temperature-dependent sex determination in this species. They report sex ratios for hatchling green turtles at Tortuguero beach for three distinct thermal zones based on exposure. The low beach belt (high water up to sparse vegetation) produced 72% females; midbeach zone (sparse vegetation to boundary of dense forest), 87% females; high beach (area of dense vegetation), 7.4% females. These values correlate closely with differences in sand temperatures in the nests at 30 to 50 cm below the surface, where temperatures during the critical temperature-sensitive stage in gonadogenesis produced males if less than 28.5°C and those greater than 30.3°C produced mostly females. The sex ratio for adults on this beach in the year of the study (1977) was sixty-seven females to thirty-three males. Immediately after emerging the hatch-lings (average carapace length 51 mm) swim out to sea, where they may take up a pelagic existence, possibly in as-sociation with sargassum rafts. At a weight of about 0.5 kg they reappear inshore and feed. At about 40 kg they seem to take up residence in or near sea grass pastures. At about 60 kg (an age of ten to fifteen years) they return to Tor-tuguero to breed and renew the cycle. Data from green turtles raised in captivity in the Cayman Islands (Wood and Wood 1993) indicate that growth is rapid during the first twelve and a half years, when juveniles attain an aver-age weight of 156 kg and an average carapace length of 1,100 mm. Turtles gain on average 3 kg per year thereafter. Mean age at sexual maturity was estimated to be sixteen years, at a weight of about 187 kg and a carapace length of 1,115 mm. The differences between these values and com-parable ones for Tortuguero turtles may be due to interpop-ulational variation. It is more likely that the effects of cul-tural conditions, especially a regulated and assured diet and restricted movement, contribute to larger sizes in captive turtles than in those of the same age from natural popula-tions. Tagged adults from Tortuguero are generally found feeding off Nicaragua, but individuals have been found in significant numbers off Yucatán, Mexico, Panama, Colom-bia, and Venezuela. Because of the decimation of its popu-lations by humans, the species borders on extermination and is regarded as endangered. Costa Rican law prohibits tak-ing the eggs of any green turtle except for local consump-tion, and most of the Tortuguero beach is fully protected within a national park. For a popular but informative ac-count of the work at Tortuguero beach, read Archie Carr's classic book *So Excellent a Fishe* (1967). Pritchard and Treb-bau (1984) provided an excellent review of the biology of the species throughout its range.

REMARKS: Hirth's (1980b) account on *Chelonia mydas* includes information relating to characteristics, distribution, and pertinent literature for both *C. agassizii* and this spe-cies. The karyotype is 2N = 56, composed of seven pairs of metacentric and submetacentric and five pairs of telocentric and subtelocentric macrochromosomes and sixteen pairs of microchromosomes; NF = 70 (Bickham et al. 1980).

DISTRIBUTION: Tropical and subtropical waters of the Atlantic Ocean and associated seas; Indian Ocean and western to central Pacific Ocean; straying into cooler wa-ters to the north (Massachusetts-Europe-Japan) and south (Argentina–South Africa–New Zealand). Nesting beaches are concentrated in tropical areas. Apparently sympatric with *Chelonia agassizii* in western Mexico, the Galápagos Islands, and Papua New Guinea (Pritchard and Trebbau 1984) and probably in the Hawaiian Islands.

Genus *Eretmochelys* Fitzinger, 1843

This monotypic genus may be separated from other sea turtles by having the frontal bone entering the orbit; the su-ture between the prefrontal bones equal to or longer than the suture between the frontals (fig. 12.17d); and the alveolar surface of the lower jaw very narrow, concave, not pointed (fig. 12.18b), and pitted near the symphysis. *Eretmochelys* are distinct in having four pairs of costal shields (fig. 12.16c), the nuchal shield not in contact with the first costal, a large intergular shield, four pairs of inframarginals (none with pores), and two pairs of frontal scales. In most examples of the species the carapace shields are imbricate, a condition found in no other adult sea turtle.

The genus is found in all tropical and subtropical seas and often ranges into cooler waters as well.

Eretmochelys imbricata (Linné 1766)

Hawksbill, Carey

(plate 503; map 12.8)

DIAGNOSTICS: All but the young and very oldest of adults of this species are immediately separated from other sea turtles by having the shields of the carapace imbricate and the carapace margin strongly serrate and notched pos-teriorly (fig. 12.16c). Very old adults and the young have nonimbricate shields on the carapace, and in the former the carapace margins are often smooth with age. The presence of four costal shields, two pairs of prefrontals (fig. 12.6b), and usually two claws on each flipper and, in the young, the presence of a single middorsal and two plastral ridges will identify difficult specimens.

DESCRIPTION: Most adults less than 1,000 mm in cara-pace length (to 1,140 mm) but may weigh up to 127 kg; usu-ally 750 to 900 mm in carapace length and 40 to 75 kg in weight. Carapace narrow and elongate; 12 pairs of margin-als, inframarginals 4, without pores; snout long and narrow, upper beak somewhat hooked but not cusped; both upper and lower beak smooth; a single large inframandibular scale

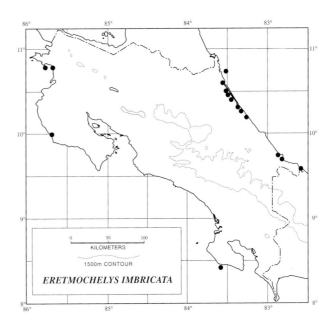

Map 12.8. Distribution of *Eretmochelys imbricata*.

behind lower beak (fig. 12.18b). Overall coloration brown dorsally, with some radiating darker streaks of black and brown on lighter background of carapace shields in small adults; hatchlings dark brown to black above and below except for light brown margins on keels, shell, paddles, and sides of neck; plastron clear yellow or yellow orange, sometimes with dark markings on each scute. Plastron slightly concave in males, which have long tails extending well beyond the plastral margin; tail in females reaches to plastral margin or slightly beyond.

SIMILAR SPECIES: (1) *Chelonia* have two claws on each flipper and one pair of prefrontal scales (fig. 12.6a). (2) *Caretta* and *Lepidochelys* have five or more pairs of costal shields and the nuchal in contact with the first costal (fig. 12.15a,c).

HABITAT: Marine but coming ashore on sandy beaches to nest.

BIOLOGY: The carey was heavily hunted for many years for its dense carapace shields that, when highly polished, were the tortoiseshell of commerce. Although the tortoiseshell trade has been essentially eliminated by competition from plastics, the hawksbill is still heavily exploited for eggs, and small stuffed hawksbills are sold as curios in many places. It is now strictly forbidden to import any hawskbill or part thereof into many countries. This turtle has been reported to feed primarily on encrusting organisms, including sponges, tunicates, bryozoans, mollusks, and calcareous algae, which it scrapes off the substrate. However, recent studies (Meylan 1984, 1988) based on digestive tract samples from widely scattered localities in the Caribbean indicate that *Eretmochelys* specialize on sponges. Of the food items present, 93% of the total dry weight consisted of sponges. The diet here and elsewhere in the species' range is mostly demosponges that contain numerous siliceous (glasslike) spicules that compose up to 50% of their dry weight (Meylan 1990). It remains unclear whether organic skeletal materials (spongin and collagen fibrils) in the sponges are digestible, but in any case, large numbers of sponges must be eaten to provide nutrition for these large, active swimmers from such a low-energy source. Nesting in this species is usually opportunistic and solitary and apparently occurs over most of the year. Occasional hawksbills nest on the green turtle beach at Tortuguero, Limón Province and south to Panama, and on scattered beaches on the Pacific side. Unlike the heavy females of other species of sea turtles, *Eretmochelys* walk with a typical tetrapod gait when on land. The females at the Tortuguero beach lay 53 to 206 (average 161) white spherical soft-shelled eggs about 38 mm in diameter. They appear to have multiple nestings with twelve to fifteen days between them. Nesting is thought to occur every two to three years. The small hatchlings (38 to 46 mm in carapace length) disappear into the sea to return inshore about a year later. While at sea they feed epipelagically. On their return they are 230 to 250 mm in carapace length

and begin feeding on the benthos. Pritchard and Trebbau (1984) provide a fine review of the biology of the hawksbill worldwide.

REMARKS: Although some authors have recognized Atlantic and Pacific geographic races within the species, this appears to be based on an inadequate interpretation of available data. As Pritchard and Trebbau (1984) pointed out, the presumed differences between these forms are not consistently present throughout the major ocean basins, but some apparently distinguishable, within-basin variants appear to have restricted geographic ranges. Consequently I make no attempt here to summarize the sketchy data and currently unresolved problem of geographic variation in this species. The karyotype is 2N = 56, with seven pairs of metacentrics and submetacentrics, five pairs of telocentrics and subtelocentrics, and sixteen pairs of microchromosomes; NF = 70 (Bickham and Carr 1983).

DISTRIBUTION: Worldwide in warm seas and reaching temperate oceans in the Northern Hemisphere and South Africa and New Zealand in the Southern Hemisphere. Nesting is mostly restricted to tropical beaches, with a few subtropical exceptions (Florida and China). The species is known from only a few definite records on the Pacific coast of Costa Rica and is uncommon in Atlantic waters of the republic. Small numbers nest at Tortuguero and between Puerto Limón and the Panama boundary and at scattered beaches on the Pacific coast.

Genus *Lepidochelys* Fitzinger, 1843

The genus is distinctive within the family in the following combination of features: frontal bone entering the orbit; suture between the prefrontal bones much shorter than the one between the frontals (fig. 12.17e); and alveolar surface of the lower jaw very broad at the symphysis and pointed, but often blunted by wear (fig. 12.18c). Other features that distinguish the genus from other sea turtles are the presence of five or more pairs of costal scutes on the carapace, the contact between the nuchal shield and the first costal (fig. 12.15c), a very small intergular scute (which may be absent), usually four (rarely three) pairs of pored inframarginals (fig. 12.7a), and two pairs of frontal scales.

The genus is composed of a species from the Gulf of Mexico (*Lepidochelys kempi*) that occurs sporadically northward to New England and Europe and a second wide-ranging form (*Lepidochelys olivacea*) found in the warmer waters of the Indian, Pacific, and southern Atlantic Oceans.

Lepidochelys olivacea (Eschscholtz, 1829)

Olive Ridley, Lora, Carpintera

(plates 504–5; map 12.9)

DIAGNOSTICS: Details of the shell shields (fig. 12.15c) remain the most reliable way to identify adults of this dorsally olive green to gray species. It has a very broad carapace (often as broad as long) that is widest near the midpoint of the shell length. Number of costal shield pairs is usually six to nine (sometimes five) and of the inframarginals four (rarely three), all usually with a posterior pore (fig. 12.7a). Hatchlings are black above and dark brown below and have three dorsal and two plastral keels, usually enlarged into knoblike structures, on each vertebral, costal, and plastral shield.

DESCRIPTION: Adults of this small sea turtle are up to 790 mm in carapace length and weigh up to 60 kg. Shields on carapace not imbricate, carapace margin smooth but notched posteriorly; marginals in 12 to 14 pairs; plastron with 6 pairs of shields, and a small unpaired intergular may be present or absent; 2 pairs of prefrontal scales; beak margins smooth; lower beak followed by a single larger inframandibular scale and then a small one (fig. 12.18c); two claws on both flippers in young, but often one in adults. Ventrally white to pale yellow. Males with tail extending well beyond carapace margin, females to margin.

SIMILAR SPECIES: (1) Adult *Caretta* are reddish brown above, and both adults and juveniles have three to seven inframandibular scales. (2) *Chelonia* and *Eretmochelys* have four pairs of costals, and the first costal is separated from the nuchal shield.

HABITAT: Marine except when females come ashore to nest on sandy beaches.

BIOLOGY: An excellent overview of the biology of this turtle is provided by Pritchard and Trebbau (1984). Its food is mainly decapod crustaceans but includes echinoderms, mollusks, fishes, and algae. The most striking feature of its life history is the formation of massive nesting aggregations or *arribadas*. Literally thousands of females congregate offshore during the nesting season near Playa Nancite and Playa Ostional, Guanacaste Province. At both of these short (1.3 km) beaches as many as 45,000 females will come ashore to nest on a single night during the July to December breeding season. As many as 150,000 individuals will nest during a four- to eight-day period. Then there will be a quiet period, with other *arribadas* following at unpredictable intervals, usually about two to four weeks apart. Mass nesting occurs during January to March, and smaller *arribadas* occur from April to June at both beaches. The females lay 30 to 168 (average 120) round white soft-shelled eggs about 40 mm in diameter. Interesting periods are erratic, as reflected in the unpredictability of the *arribadas*, but multiple nestings occur. However, most *Lepidochelys*, unlike other sea turtles, apparently nest annually. This species has temperature-dependent sex determination (Morreale et al. 1982). Hatchlings emerge in 49 to 72 days and are 40 to 46 mm in carapace length. Predation on the eggs and hatchlings on the beaches is documented by Cornelius (1986) in his extensive study of this species. Of the approximately 100,000 nests produced during an *arribada*, only about 5% hatch. Surprisingly, of the 15,000 nests that survive predation, erosion, and other hazards, 85% do not contain successful eggs. Chief egg predators at Nancite include the coati (or *pizote*), black vulture, raccoon, and coyote. According to Cornelius, these predators remove about 16% of the nests. The hatchlings suffer an incredible loss through predation on their way to the sea. Principal predators on the beach include various raptors and shorebirds, the mammals previously cited as egg eaters, ghost crabs, spiny-tailed iguanas, crocodiles, and especially frigate birds. Once the hatchlings are in the sea, carnivorous fishes and sharks vie with the crocodiles and frigate birds for their share of the feast. Sharks are probably the most effective predators on larger olive ridleys at sea. Tagged *Lepidochelys* from the Costa Rican beaches have been recovered from nearly every country in the eastern Pacific region from Mexico to Peru. One was found over 2,400 km west of Costa Rica in the open ocean. Nevertheless, little is known regarding the life of this species away from nesting beaches, and what happens between the hatchling and juvenile stages remains a mystery.

REMARKS: Zug and Ernst (1994) reviewed the genus based on a literature compilation. Their account suffers from not clearly discriminating between the nesting beaches and other records for this species and those for the allied *Lepidochelys kempii* in the western Atlantic. This objection is partially met by their treatment of *L. olivacea* (Zug, Ernst, and Wilson 1998).

DISTRIBUTION: Tropical and subtropical portions of the Indo–West Pacific region and the eastern Pacific but wandering to the coasts of Chile, Alaska, South Africa, and New Zealand; in the Atlantic Ocean between Mauritania and Guinea in West Africa and from northeast Cuba to the Guianas in the western Atlantic. Nesting seems to be restricted

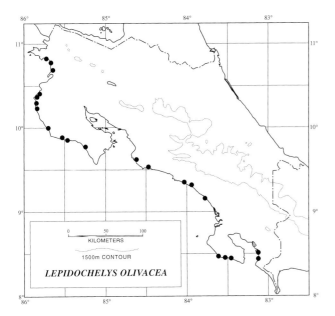

Map 12.9. Distribution of *Lepidochelys olivacea*.

to tropical beach sites. Found along the Pacific coast of Costa Rica; no member of the genus occurs along the Atlantic coast of the republic.

Family Chelydridae Agassiz, 1857 (1839)

Snapping Turtles

The large turtles that constitute this family are well known for their aggressive and predaceous behavior. There are two living species, each placed in its own genus. Both species are primarily aquatic but may venture onto land to bask or in search of another pond or stream. The alligator snapper, *Macroclemys temmincki*, is found in the central United States from southern Missouri, Kansas, Illinois, and Indiana to the Gulf of Mexico. *Chelydra serpentina* ranges from eastern and central Canada south to northern Mexico and from southern Veracruz, Mexico, to northwestern South America.

The family is defined in terms of the following (mostly internal) characteristics:

1. Entoplastron present
2. Nine bones in plastron
3. Nuchal bone with lateral (costiform) processes
4. Twenty-five marginal plates (twenty-four marginals plus nuchal), all separated from the pectoral shields
5. Skull strongly emarginate
6. Vertebrae procoelous in part, joints usually with a double articulation (concavity usually double on front surface of eighth vertebra), but only one biconvex cervical vertebra
7. Tail vertebrae mostly opisthocoelous
8. Penis with an undivided seminal groove; external, medial, and internal plicae present on glans

In Costa Rica the single member of the family may be distinguished from other freshwater and terrestrial turtles by having the plastron reduced, cruciform, and covered by four pairs of shields (pectorals, abdominals, femorals, and anals) and an unpaired one (a gular) or, rarely, by five pairs of shields (with paired gulars) and a lack of contact between the marginals and pectoral shields (fig. 12.8a). In addition, the plastron is without hinges and is rounded posteriorly.

Genus *Chelydra* Schweigger, 1812

A monotypic genus separated from its only living ally *Macroclemys* (features for the latter genus in parentheses) by having the head covered by soft skin (smooth symmetrical plates), underside of tail with large plates more or less in a double row (numerous rounded scales), no series of supramarginals separating the costals from the marginals (three to five supramarginals present), and the eyes dorsolateral in position and visible from above (lateral, not visible from above).

The generic range is coincident with that of the family and the single species discussed below.

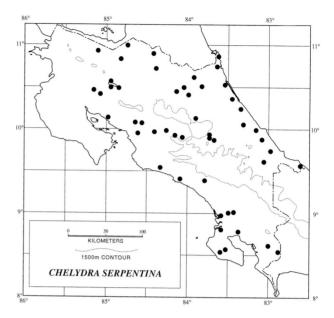

Map 12.10. Distribution of *Chelydra serpentina*.

Chelydra serpentina (Linné, 1758)

Snapping Turtle, Tortuga Lagarto

(plates 506–8; map 12.10)

DIAGNOSTICS: A large freshwater species reaching 494 mm in length and weighing up to 34 kg. This large-headed turtle with a strongly hooked (fig. 12.10a) smooth upper beak, two large chin barbels, usually followed by one to two smaller pairs, a long tail (as long as the carapace or longer), and reduced plastron (fig. 12.8a) cannot be mistaken for any other Costa Rican species.

DESCRIPTION: Carapace strongly tricarinate except in an occasional old, smooth-worn example; each keel usually ends with a knob or tubercle; posterior margin of brown to olive brown carapace strongly serrate; plastron reduced, not hinged, cruciform, leaving much of the soft parts exposed, rounded behind; toes webbed; tail with three rows of tubercles above. Carapace dark brown to black; plastron yellow and without markings; head dark brown, as are exposed upper surfaces of limbs and tail; numerous light warts on neck and limbs; soft parts gray to brownish gray above, light below; iris brown. Males larger than females, to 484 mm versus 390 mm in carapace length. Sexes not markedly different in other features, although vent situated more posteriorly in males than in females and males with a somewhat smaller concave plastron and narrower bridges than females.

SIMILAR SPECIES: *Chelydra* cannot be confused with other Costa Rican turtles because of the very long, stout tail and the reduced cruciform plastron (fig. 12.8a).

HABITAT: The snapper probably occurred originally in any permanent body of water throughout the lowlands of Costa Rica, including freshwater swamps, large permanent ponds, large streams, rivers, and both brackish and freshwater swamp forests. Currently uncommon, it is found in Lowland Dry, Moist, and Wet Forests and Premontane Moist and Wet Forest environments.

BIOLOGY: Because of their large size, aggressive demeanor, and prized flesh, snappers have suffered unusually from human pressures and are rarely encountered. In the water these turtles are relatively inoffensive, and when stepped on they will usually lie still and retract the neck. They often bury themselves to ambush their prey, including a wide variety of frogs (and their eggs and larvae), fishes, crustaceans, mollusks, and other invertebrates. They also eat snakes, birds, and small mammals. According to Acuña (1998), in Costa Rica about 35% of their diet consists of fishes. Aquatic plants compose a considerable portion of the diet as well. The "snap" is a very rapid, straightforward head lunge, culminating in rapid closing of the jaws, and is used to capture active prey as well as for defense. Snappers are poor swimmers and usually move about on the bottom with a characteristic slow-motion walk. When moving overland they hold the body up off the ground, and progress is slow. On land, snappers will strike and snap repeatedly when disturbed or cornered. Frequently they will stand with the hindquarters elevated and jaws agape before making a lunging snap. The species is primarily nocturnal, is rarely seen basking, and seems to move from one body of water to another overland, mostly at night. The species prefers deep ponds, lakes, and the mouths of rivers and hides on the bottom much of the time. The breeding season is from April to November. It is presumed that mating is similar to that of the species elsewhere in its range as outlined below (Medem 1962; Pritchard 1979). Mating occurs in the water; the male grasps the margins of the female's rather flattened carapace with the claws of his four feet and then twists his tail around the female's until their vents are in contact. Eggs are ovoid, hard (calcareous), but flexible, weigh between 5 and 15 g, and are deposited in a nest that the female excavates in the ground using her hind legs, sometimes at a considerable distance from water. Reproduction occurs during the wet season from April to November. Eleven to eighty-three eggs (mean twenty-five) are laid at one time. The eggs are about 28 to 33 mm in diameter. *Chelydra* are among those freshwater turtles known to have temperature-dependent sex determination (Yntema 1976; Vogt and Flores 1992). In the temperate zone eggs incubated at temperatures above 30°C or below 20°C become females. Males were produced from eggs developing at intermediate temperatures. In southeastern Mexico all males were produced in eggs incubated at 25°C, and increasingly higher percentages of females resulted from 27°C upward. Hatching times of 80 to 90 days are typical of northern populations of snappers (Carr 1952) but range between 55 and 125 days in Costa Rica (Acuña 1998). Hatchlings are 24 to 31 mm in length and are aggressive biters and snappers at birth. Few observations of predation on hatchlings in Costa Rica have been reported, but a wide variety of birds and mammals as well as crocodilians doubtless prey on them. The large snake *Drymarchon corais* is also known to feed on hatchlings (Acuña 1998). Larger snappers have little to fear from most predators except this snake and humans, but caimans *(Caiman crocodilus)* will occasionally attack and eat small ones. Coyotes *(Canis latrans)* and armadillos *(Dasypus)* are known to prey on the eggs and young. Raptors and vultures take eggs from nests opened by coyotes and armadillos and also eat many hatchlings. Human predation is intense, since the flesh is much favored for soups and stews. The eggs are also widely eaten and are often sold by enterprising vendors as sea turtle eggs. The hatchlings are often captured at birth and sold as pets.

REMARKS: Ernst, Gibbons, and Novak (1988) and Gibbons, Novak, and Ernst (1988) have summarized the systematics, distribution, and pertinent literature for this turtle. The species as currently understood occurs in two geographically disjunct segments, one ranging over the eastern and central United States and adjacent parts of southern Canada and the second from southern Mexico to Ecuador, although there are no records for most of Nicaragua. Feuer (1966) and Medem (1977) assigned all available material to this one species, but further study may recognize one or more of the following as distinct from *Chelydra serpentina* of temperate North America. Currently four major geographic variants may be characterized:

A. Eastern and central North American populations, exclusive of peninsular Florida and extreme south-central Georgia: neck tubercular or rugose; anterior plastral lobe usually less than 40% of carapace length
B. South-central Georgia and peninsular Florida: neck with long spinelike appendages; anterior plastral lobe usually less than 40% of carapace length
C. Atlantic versant southeastern Mexico, Guatemala, and Honduras: neck with long spinelike appendages; anterior plastral lobe usually more than 40% of carapace length
D. Western Ecuador and Colombia, Panama, Costa Rica, and (presumably) Nicaragua: neck covered with low rounded warts; anterior plastral lobe usually more than 40% of carapace length

Costa Rican material is typical of form D. The karyotype is 2N = 52, comprising seven pairs of metacentric and submetacentric and fifteen pairs of telocentric and subtelocentric macrochromosomes and fourteen pairs of microchromosomes; NF = 86 (Haiduk and Bickham 1982).

DISTRIBUTION: The United States and Canada, east of the Rocky Mountains, south through Texas and New Mexico and presumably into extreme northern Mexico (Morafka 1977; for a contrary view see Smith and Smith 1979); then in the lowlands from central Veracruz, Mexico, south on the Atlantic versant to northern South America and in the lowlands on the Pacific slope from northwestern Costa Rica and the Meseta Central Occidental to central Ecuador (4– 1,164 m).

Family Emydidae Schmid, 1819

Freshwater, Marsh, and Forest Turtles and Their Allies

The largest family of living turtles, the Emydidae contain thirty genera and 110 species of freshwater, semiterrestrial, and terrestrial species. Most forms are found in eastern North America and Southeast Asia. A few species are found south of the equator in Indonesia, and the genus *Chrysemys* ranges to Argentina. The family is absent from Africa south of the Mediterranean rim, Madagascar, Australia, and New Guinea.

Technically, the group is defined by the following features:

1. Entoplastron present
2. Nine bones in plastron, nuchal bone without lateral processes
3. Twenty-five marginal plates (twenty-four marginals plus nuchal), with some in contact with pectoral shields
4. Skull strongly emarginate
5. Vertebrae procoelous in part, usually two biconvex vertebrae, eighth centrum concave anteriorly, concavity usually double
6. Penis with an undivided seminal groove; external, medial, and internal plicae present on the glans

The turtles of this family are placed in two subfamilies following McDowell (1964) that are defined briefly below (Gaffney and Meylan 1988, advocate recognizing the two groups as distinct families):

Batagurinae (twenty-three genera): angular bone excluded from contact with Meckel's cartilage; basioccipital with a strong lateral tuberosity; joint between fifth and sixth cervical vertebrae simple, with a single condyle; supracaudal scutes extending forward onto suprapygal bone (from the Mediterranean region over most of tropical and subtropical Asia to eastern China and Japan; tropical America)

Emydinae (seven genera): angular bone forms floor of the canal for Meckel's cartilage; basioccipital without a strong lateral tuberosity; joint between fifth and seventh centra with paired condyles; supracaudal scutes fall short of suture between pygal and suprapygal (Europe, northwestern Africa, western Asia; North, Central, and South America)

In Costa Rica members of the family are immediately separated from other freshwater and terrestrial turtles by having six paired plates covering the plastron (gulars, humerals, pectorals, abdominals, femorals, and anals) and some contact between the marginals and pectoral shields. In ad-

dition, the plastron is well developed (not cruciform as in the Chelydridae), without hinges, and notched posteriorly.

Subfamily Batagurinae Gray, 1869
Genus *Rhinoclemmys* Fitzinger, 1835

Neotropical Forest Turtles

The genus differs from others in the family by a combination of skull characters and in having the following:

1. Large plastron without a hinge
2. Hexagonal neural bones short-sided behind
3. Simple articulation between the fifth and sixth cervical vertebrae
4. Alveolar surface of the upper jaw narrow and ridgeless
5. Head, neck, and limbs usually with vivid and contrasting colors

The only other Central American turtle genus *Rhinoclemmys* might be confused with is *Chrysemys*. The two differ most obviously in the presence of the ridge on the broad alveolar surface of the upper jaw in *Chrysemys* (alveolar surface narrow and lacking a ridge in *Rhinoclemmys*). *Rhinoclemmys* usually have a complex pattern of light and dark spots and stripes on the sides of the head and neck and red and orange coloration prevalent on the soft parts and shell. The light coloration in *Chrysemys* is mostly yellow without any red markings, and the lateral head and neck pattern is of dark and light stripes on a greenish ground color.

The nine Neotropical forest turtles are mostly terrestrial to aquatic forms that range from southern Sonora, Mexico, to northern Ecuador on the Pacific versant and east and south in South America through Colombia, Venezuela, and the Guianas to northeastern Brazil; on the Atlantic slope of Central America one species *(R. areolata)* occurs in extreme southern Veracruz and throughout the Yucatán Peninsula, and others occur in the area from extreme northeastern Honduras to Colombia.

These turtles were placed in the genus *Geoemyda* for many years, and much of the literature on them used that name. The generic names *Nicoria* and *Callopsis* have also been applied to these species. McDowell (1964) has been followed by all subsequent authors in recognizing this genus as distinct. Ernst (1978) has reviewed the genus (as *Callopsis*) and summarized (1981a) its characters, distribution, and pertinent literature.

KEY TO THE COSTA RICAN SPECIES OF THE GENUS *RHINOCLEMMYS*

1a. Almost entire plastron covered with a dark brown or black figure; underside of marginals at bridge mostly dark, no marginals with complex dark C-, U-, or V-shaped figure (fig. 12.20b,c) 2

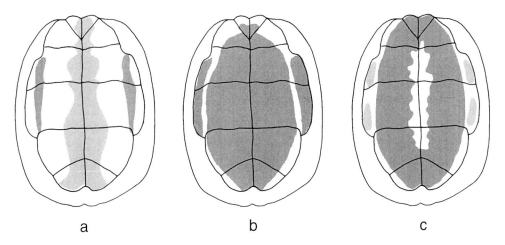

Figure 12.20. Plastron coloration in Costa Rican turtles of the genus *Rhinoclemmys*: (a) *R. pulcherrima*; (b) *R. annulata*; (c) *R. funerea*.

1b. Only center of plastron with a dark brown or black figure on a light background (figure faded in large adults) (fig. 12.20a); underside of each marginal marked with complex dark C-, U-, or V-shaped figures that have a red center in life . . . .
. . . . *Rhinoclemmys pulcherrima* (p. 767)

2a. A distinct free fleshy flap along outer margin of foot; toes nearly fully webbed; upper jaw notched (fig. 12.10d) or even (fig. 12.10c) at symphysis; plastron with a definite light central area in adults (fig. 12.20c), usually a uniform dark plastral figure in juveniles . . . .
. . . . *Rhinoclemmys funerea* (p. 766)

2b. No free fleshy flap on outer margin of foot; toes weakly webbed at base; upper jaw with a median hook (fig. 12.10a); plastron nearly completely dark in adults and juveniles, rarely with a light central area (fig. 12.20b)
. . . . *Rhinoclemmys annulata* (p. 765)

Rhinoclemmys annulata (Gray, 1860)

Brown Forest Turtle, Tortuga Café de Terrestre

(plate 509; map 12.11)

DIAGNOSTICS: This moderate-sized forest turtle (to a carapace length of 204 mm) is easily recognized when compared with other members of the genus by lacking a fleshy flap along the outer margin of the foot and having little or no toe webbing, a hooked, unnotched beak on the upper jaw (fig. 12.10a), and the plastron and bridge dark throughout life (fig. 12.20b).

DESCRIPTION: Carapace smooth in young and roughened in adults, with a definite but often flattened blunt middorsal keel and serrate posterior margin that often becomes rounded from wear; plastron not hinged, notched posteriorly. Carapace tan to uniform black; in most examples yellow or orange markings present on costals and vertebrals,

often forming a series of lines radiating from the dorsoposterior corner of each shield; middorsal keel usually yellow; each marginal with a rectangular to square yellow or orange blotch on upper surface and a similar blotch below, although often triangular; midseam of plastron sometimes yellow; upper surfaces of head and soft parts gray brown to dark brown and yellowish below; larger examples tend to melanism, with uniform dark carapaces and head markings obscured; head and neck striped with cream to red; a distinct supratemporal stripe runs from orbit slightly downward to continue onto neck; another stripe runs from upper anterior margin of eye to tympanum, where it meets a similar stripe from upper jaw (these often fuse and continue as

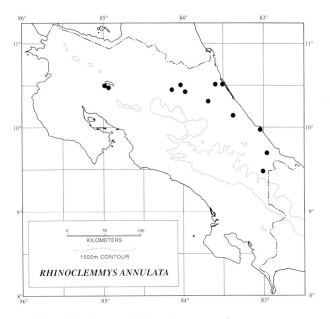

Map 12.11. Distribution of *Rhinoclemmys annulata*.

one stripe onto neck); often a light line from upper anterior margin of eye to tip of snout; iris brown; chin yellowish and often mottled with dark spots; forelimbs with pattern of dark stripes, hind limbs not; tail with yellow stripes above. Males with concave plastrons and longer tails than females (vent lying beyond carapace); no significant difference in size between sexes.

SIMILAR SPECIES: (1) *Rhinoclemmys funerea* has a fleshy flap along the outer margin of the foot and has large black spots on the lower jaw and chin. (2) *Rhinoclemmys pulcherrima* has a plastral dark figure usually restricted to the center of the plastron and the carapace, head, neck, and forelimbs usually marked with bright stripes, spots, and ocelli. (3) *Chrysemys ornata* has distinct ocellate markings on the carapace and a plastral figure consisting of concentric dark and light lines.

HABITAT: The forest floor in relatively undisturbed Lowland (Atlantic) Moist and Wet Forests and occasionally Premontane Wet Forest.

BIOLOGY: *Rhinoclemmys annulata* is diurnal and terrestrial in habit, being most active in the morning (7:00 to noon). Food consists of a wide variety of plant material, including leaves, seedlings, and fallen fruits. Moll and Jansen (1995) found a high proportion of ferns in the stomachs of *R. annulata* at Tortuguero, Limón Province. They also presented evidence for seed dispersal of a variety of plants by this form. During courtship, as described by Acuña (1998), the male initially begins to follow the female at some distance as she moves about. After pursuing the female for some time, he mounts her carapace from behind and nips or bites her head while attempting copulation. If the female tries to escape, the male salivates on her head, again places his tail under her carapace, and inserts his penis in her cloaca. Egg laying apparently occurs throughout the year. The female deposits one or two elliptical eggs (about 70 × 37 mm) in a shallow nest or simply places them on the ground and covers them with leaves. Hatchlings average 63 mm in carapace length. Raptors and vultures are known predators on the eggs and young of *R. annulata*, but other mammals doubtless eat them as well.

REMARKS: Ernst (1978, 1980a) provided detailed redescriptions and a summary of pertinent literature for this species. Ernst (1983a) cites the range of this form as extending to over 1,500 m in Costa Rica, based on USNM 8217 from "Talamanca." As I have discussed elsewhere (Savage 1970), the term Talamanca refers to a general region in southeastern Costa Rica, and this specimen is almost certainly from the lowlands, or Baja Talamanca. All other known localities for the species are below 950 m in elevation.

DISTRIBUTION: Lowland forests on the Atlantic versant from eastern Honduras through Nicaragua, Costa Rica, and Panama to the Pacific versant of eastern Panama, western Colombia, and north and central Ecuador (2–920 m).

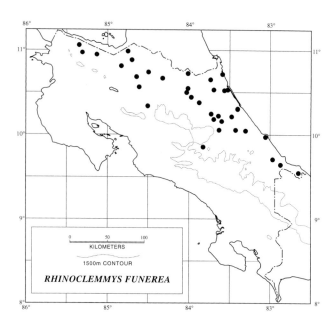

Map 12.12. Distribution of *Rhinoclemmys funerea*.

Rhinoclemmys funerea (Cope, 1875)

Black River Turtle, Tortuga Negra del Río

(plate 510; map 12.12)

DIAGNOSTICS: A moderate-sized turtle (to 375 mm in carapace length), immediately distinguishable from its congeners by having a free fleshy flap along the outer margin of the foot, toes almost fully webbed, beak on upper jaw notched (often with a cusp on either side) (fig. 12.10d) or even at the symphysis, the plastral figure and bridge dark in juveniles, with a light central plastral area in adults (fig. 12.20b,c).

DESCRIPTION: Carapace smooth in juveniles and smooth to rugose in adults, with a middorsal keel that may be worn flat with age; posterior margin serrate in young specimens, usually rounded by wear in adults; plastron not hinged, notched posteriorly. In adults carapace black to dark brown, in juveniles costals tinged with yellow; shield seams of plastron yellow in large specimens, dark in juveniles and most young adults; soft parts yellowish, variously marked with black; strong tendency toward melanism so that carapace and soft parts become darker with age, obscuring much of pattern; head and neck striped with dark olive green to black, demarcating a broad yellow stripe from eye to above tympanum; two other yellow stripes run from eye and corner of mouth to tympanum; iris brown; large black spots on lower jaw and chin; forelimbs yellow with black punctations and stripes and black outer border; hind limbs black to dark gray or brown on outer surfaces; tail with dark stripes below. Males with concave plastrons and longer tails than females (vent beyond the carapace margin); females with flat plastrons and vent beneath the carapace; no size difference between sexes.

SIMILAR SPECIES: (1) Other species of *Rhinoclemmys* lack a fleshy flap along the outer margin of the foot, have little or no toe webbing, and lack large black spots on the lower jaw and chin. (2) *Chrysemys ornata* has distinct ocellate markings on the carapace, and the plastral figure comprises a complex series of concentric dark and light lines.

HABITAT: Found in freshwater marshes, tree swamps, ponds, streams, and rivers in Lowland Atlantic Moist and Wet Forests.

BIOLOGY: Although often seen basking on logs or other debris or foraging near the water's edge during the day, individuals of *R. funerea* are active at night, when they may move some distance onto land. Food in the wild consists of a wide variety of plant material, but they will accept meat in captivity (Ernst 1983b), suggesting they may be partial scavengers. Moll and Jansen (1995), however, regard this as a strictly herbivorous species that plays a role in seed dispersal of terrestrial plants found near watercourses. Sexual maturity is reached at about 200 mm in plastron length. Acuña (1998) reports that reproduction occurs from March through August. During courtship the male pursues the female through the water, and when she stops or slows he swims to her side, presents his extended head and neck, and rapidly vibrates his head up and down (Iverson 1975). Egg laying occurs one to four times a season, with average clutch size of three eggs. The oblong eggs have a hard, brittle shell and average 68 × 35 mm. Eggs are deposited in a shallow nest in the ground and covered by leaves (Monge, Morera, and Chávez 1988). Eggs that Moll and Legler (1971) incubated at 20° to 35°C hatched in 98 to 104 days. Hatchlings are 55 to 60 mm in carapace length. Crocodilians *(Crocodylus* and *Caiman)* are primary predators on juveniles and also occasionally eat adults. Coyotes *(Canis latrans)* dig up the nests and eat the eggs as well as juveniles. Raptors and vultures commonly eat exposed eggs and attack and eat many hatchlings. Acuña (1998) reports an incident where a snake *(Lampropeltis triangulum)* appears to have eaten an egg of this species. The same author notes that the Amerind communities of the Atlantic lowlands in Costa Rica eat adults of this turtle. They also use a few adults of both sexes in a religious ritual during October of each year. They regard the eggs, hatchlings, and juveniles as sacred and do not eat them, but they include eggs of other turtles in their diet. The species attains sexual maturity at about 200 mm in carapace length.

REMARKS: Ernst (1978, 1980b) provided detailed descriptions and a review of pertinent studies on this species. A single record from the upper Río Reventazón in Costa Rica (Loaiza, Cartago Province, 990 m) (Acuña, Castaing, and Flores 1984) may be based on a released specimen. This may also be the case for the Pacific localities mapped by Acuña

(1998) from the upper Río Grande de Tárcoles drainage. These records are the basis of Acuña's (1994) inclusion of *R. funerea* in his discussion of the turtles of the Meseta Central. The karyotype is 2N = 52, comprising eight pairs of metacentric and submetacentric and five pairs of telocentric and subtelocentric macrochromosomes and thirteen pairs of microchromosomes; NF = 68 (Bickham and Carr 1986).

DISTRIBUTION: Atlantic lowlands from extreme southeastern Honduras to central Panama (2–600 m).

Rhinoclemmys pulcherrima (Gray, "1855," 1856)

Red Turtle, Tortuga Roja

(plate 511; map 12.13)

DIAGNOSTICS: The Costa Rican representatives of this small species (males to 159 mm in carapace length, females to 200 mm) are among the world's most beautiful turtles. The carapace, head, neck, and forelimbs are usually marked with bright red, orange, and/or yellow stripes, spots, and ocelli. Other distinguishing features are the absence of a fleshy flap along the outer margin of the foot, little or no toe webbing, a notched beak on the upper jaw that may have a cusp on either side (fig. 12.10d), the plastral figure restricted to the center of the plastron, and the bridge with contrasting dark and light areas (fig. 12.20a).

DESCRIPTION: Carapace smooth in young but rugose in adults, with definite middorsal keel and serrate posterior margin that may be rounded from wear; plastron not hinged, notched posteriorly. Costal shields marked with red, orange, or yellow ocelli bordered by black; vertebrals marked with one or two U-shaped black marks that point forward and

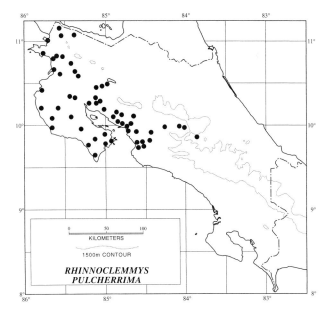

Map 12.13. Distribution of *Rhinoclemmys pulcherrima*.

alternate with red, orange, or yellow areas demarcated by broad black lines and/or ocelli; underside of most marginals marked with complex dark C-, U-, or V-shaped figures with red centers in life; bridge yellow toward carapace, with a sinuous or straight transverse black bar separating light area from light portion of plastron; plastral figure restricted to center of plastron and may be uniform brown but often with contrasting dark and light areas; plastral seams sometimes dark; upper surfaces of soft parts and head usually olive, as are lower surfaces of soft parts in most cases, but latter sometimes rufous to yellow; head with red stripes, several on top of head and one along upper jaw to tympanum; another from nostril to eye and several from eye to tympanum; iris gray to bluish green, bronze, or reddish; jaws and chin yellow and often with red stripes, large black spots, and/or red-centered ocelli; forelimbs covered with large red to yellow scales and rows of black spots; hind limbs uniform or with obscure light and dark stripes; tail yellow to red with dark dorsal stripes. Males with concave plastrons and longer tails (vent beyond the carapace margin) than females, which have the carapace flat and vent beneath carapace.

SIMILAR SPECIES: (1) Other *Rhinoclemmys* have most of the plastron covered by a dark figure (fig. 12.20b,c), the marginals at the bridge mostly dark, and no bright ocellate markings laterally on the carapace. (2) *Chrysemys ornata* has fully webbed feet.

HABITAT: Occurs in aquatic and terrestrial environments within deciduous and semideciduous forest, where it favors moist situations and in particular gallery forests. Found in Lowland Dry and Pacific Moist Forests and Premontane Forest.

BIOLOGY: *Rhinoclemmys pulcherrima* is a semiterrestrial turtle and takes to water readily, especially in the dry season. The species is mostly diurnal, but activity may be restricted to early morning or evening in the hot dry season. These turtles are omnivores but seem to prefer plant food (Ernst 1983c). Available reproductive behavior data on the species are for two captive specimens from Guatemala (Christensen 1975) and from nesting observations on two captive females in San José, Costa Rica (Monge, Morera, and Chávez 1988). No involved courtship preceded mating in these turtles, and the Guatemalan female laid four clutches of three to five eggs each from September through December. Eggs incubated at 18.3° to 29.4°C hatched in 115 to 186 days. The Costa Rican females dug nests about 100 mm deep with their hind limbs, laid one to three eggs, then covered them with dirt. Females tamp the dirt with the hind limbs or entire posterior body and then add twigs and other debris with movements of the hind limbs. The reproductive season lasts from May to December (Acuña 1998). Under natural conditions nests are constructed near grass clumps or roots of shrubs and trees where the soil is soft. According to Castillo (1986), eggs are about 52 × 32 mm. Eggs are

sometimes preyed on by the larvae of the phorid fly *Megaselia scalaris* (Acuña and Hanson 1990). Janzen (1980) speculated that this form is a harmless mimic of the coral snake *Micrurus nigrocinctus* because of the concentric red, yellow, and black markings on the shell and chin. No data support this claim. Predation on this species is most severe by the crocodile *(Crocodylus acutus)* and caiman *(Caiman crocodilus)*, which eat hatchlings, juveniles, and occasional adults. The usual suite of mammal predators (especially *Canis latrans*) and bird predators (hawks and vultures) eat eggs and hatchlings, and the coyote will also prey on small adults. Costa Ricans eat the eggs but not the flesh of this turtle. The young are much favored as pets because of their brilliant colors, although adults are often captured for the same purpose.

REMARKS: This species is found in three disjunct geographic areas on the Pacific coast of Mexico and Central America. Four rather distinctive forms show geographic fidelity in variation within the species as follows:

A. Northern populations
   1. Western Mexico (Sonora to Colima): carapace usually uniform brown, occasionally with a faint reddish stripe on each costal; bridge completely dark; an unbroken central dark figure on plastron
   2. Guerrero, Mexico: carapace with a single dark-bordered red or yellow spot on each costal; bridge black, divided by a yellow bar; central dark figure on plastron usually forked on gulars and anals
B. Southern populations
   1. Oaxaca, Mexico, to Lake Managua, Nicaragua: carapace with bright black-bordered yellow to red stripes or ocelli on each costal; bridge completely dark; an unbroken central dark figure on plastron
   3. Lake Nicaragua to central Costa Rica: carapace with bright black-bordered yellow to red ocelli on each costal; bridge black, divided by a yellow bar; central dark plastral figure usually divided on gulars and anals

Intermediates between A1 and A2 are found in central Nicaragua between the two great lakes (Managua and Nicaragua). Costa Rican examples are typical of the southernmost population (B2). Dunn's (1930a) record for Las Concavas (1,360 m), near the city of Cartago, may be based on a released specimen, since it is the only known Atlantic slope record for Costa Rica. Ernst (1978, 1981b) provided detailed descriptions and a review of previous studies on this beautiful turtle. The karyotype is 2N = 52, comprising nine pairs of metacentric and submetacentric and five pairs of telocentric and subtelocentric macrochromosomes and twelve pairs of microchromosomes; NF = 70 (Bickham and Carr 1986).

DISTRIBUTION: Subhumid lowlands and premontane slopes of the Pacific versant of Mexico from Sonora to

Colima, in Guerrero, and from eastern Oaxaca, Mexico, to central Costa Rica, including the Meseta Central Occidental (2–1,160 m); there are inland Atlantic drainage records for southern Guatemala, Honduras (many), and Nicaragua.

Subfamily Emydinae Schmid, 1819

Genus *Chrysemys* Gray, 1844

Sliders and Painted Turtles

Differences in details of the skull form the major basis for classifying turtles within the family Emydidae but need not be enumerated in detail here (see McDowell 1964). In addition to a unique combination of skull characteristics, *Chrysemys* differ from other genera in the family in having the following:

1. Large plastron without a hinge
2. The hexagonal neurals not short-sided behind
3. Double articulation between the fifth and sixth cervical centra
4. Alveolar surface of upper jaw broad and with a distinct ridge
5. Sides of the head and neck and limbs marked with light and dark stripes on a green ground color

The genus *Chrysemys* differs from *Rhinoclemmys*, its nearest ally in Central America, most obviously in the last two features. The light color in *Chrysemys* is mostly yellow, but often there is orange in the postorbital stripe on the head, whereas reds and oranges predominate on the light soft parts and shell of most *Rhinoclemmys* or the lateral surfaces of the head and neck are yellowish with distinct dark spots and stripes.

The genus comprises thirteen to twenty-two species (depending on how the *Chrysemys scripta* complex is treated) of aquatic turtles that range over much of eastern North America, through the southern half of Baja California, Mexico, and along both coasts of Middle America from Sonora, Mexico, on the west and the Rio Grande on the east to northern South America. Several isolated populations occur in northeastern and southeastern Brazil, and another isolate ranges over southern Brazil, Uruguay, and northeastern Argentina. The genus also occurs on the Greater Antilles, Grand Cayman Island, and the Bahamas. One species, *Chrysemys scripta*, especially the red-eared form, is very popular in the pet trade. Releases of captive individuals have led to the establishment of breeding colonies outside its natural range, in the northern United States (Michigan, New Jersey, Pennsylvania, Maryland, near Washington, D.C.), peninsular Florida, Arizona, California, Hawaii, Guadeloupe, Great Britain, Scotland, Germany, the Netherlands, France, Spain, Israel, South Africa, Marianas Islands, South Korea and Japan. Many of the species, including the Costa Rican form, are often placed in the genus *Pseudemys*, and

much published information uses that name. McDowell (1964), Rose and Weaver (1966), and Weaver and Rose (1967) demonstrated that *Chrysemys* and *Pseudemys* were closely allied and placed all species previously assigned to the two genera within *Chrysemys*. Vogt and McCoy (1980) presented evidence purporting to establish the distinctiveness of *Pseudemys* and suggested that it could be divided into two subgroups, *Pseudemys* proper and *Trachemys*. Ward (1984) and Seidel and Smith (1986) took this suggestion one step further and recognized three genera: *Chrysemys* (one species), *Pseudemys* (five or six species), and *Trachemys* (five to sixteen species). This arrangement was followed by Gaffney and Meylan (1988), Seidel and Jackson (1990), Ernst (1990), and Seidel and Ernst (1996), among others. Unfortunately this issue cannot be resolved based on current information, since it is unclear how these turtles are related to other emydine genera, especially *Deirochelys*, *Malaclemys* (including *Graptemys*), and *Emydoidea*. Consequently I continue to use *Chrysemys* for this stock. If two genera were recognized within the assemblage, the Costa Rican species would be placed in *Pseudemys*. If the tripartite arrangement of Seidel and Smith (1986) was followed, the Costa Rican turtles would be referred to *Trachemys*.

Chrysemys ornata (Gray, 1831b)

Tropical Slider, Tortuga Resbaladora

(plate 512; map 12.14)

DIAGNOSTICS: This moderate-sized largely aquatic turtle (adult males 120 to 340 mm, females 158 to 600 mm in carapace length) has a notched beak on the upper jaw (fig. 12.10d) and fully webbed feet, and the plastral figure in the

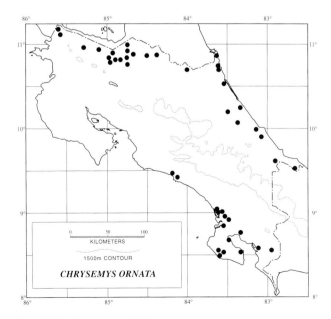

Map 12.14. Distribution of *Chrysemys ornata*.

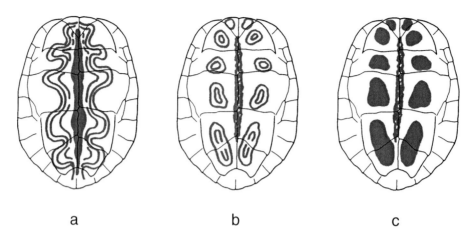

a   b   c

Figure 12.21. Color variation in plastron of the emydid turtle *Chrysemys ornata*: (a) basic pattern found in hatchlings and many adults; (b) modified pattern in some adults; (c) rare pattern in some large adults, concentric circles of pattern (b) filled with dark color. In the very largest adults much of the pattern may be lost.

center of the shell usually comprises a series of concentric dark and light lines (fig. 12.21).

DESCRIPTION: Carapace slightly rugose, with a definite middorsal keel in young that is reduced or absent in adults and with serrate posterior margin; plastron not hinged, notched posteriorly. Carapace greenish gray with definite ocelli on lateral scutes and often on vertebrals; middorsal keel marked with black posteriorly on each scute to form part of ocellus, when latter is present; underside of most marginals marked with distinct yellow and black ocelli; bridge yellow toward plastron but bordered by a complex figure composed of wavy parallel black and yellow lines along carapace margin; light shell markings particularly vivid in juveniles; soft parts greenish gray and marked with distinct black and yellow stripes; melanistic tendency in some large adults seldom obscures pattern of carapace and soft parts, and plastral figure never completely obliterated; a distinct yellow to orange supratemporal stripe expands slightly above ear and runs from orbit onto neck, another stripe runs from lower posterior margin of eye obliquely downward onto neck, and one extends from nostril downward and backward to beak margin under eye; many black-edged narrow light stripes run along top and sides of head and neck; iris yellowish green divided by a black horizontal line; chin marked with light stripes and elongate blotches; limbs with pattern of dark and light stripes; tail with paired yellow stripes above that fuse into a single middorsal stripe posteriorly and paired yellow stripes below. Males without concave plastrons or greatly elongated foreclaws and with elongate and upturned snout.

SIMILAR SPECIES: (1) *Rhinoclemmys annulata* and *R. funerea* lack any hint of an ocellate pattern on the lateral carapace scutes. (2) *Rhinoclemmys pulcherrima* has little or no toe webbing and has the upper jaw, head, neck, and forelimbs usually marked with bright spots and ocelli.

HABITAT: *Chrysemys ornata* is most commonly seen sleeping during the day in large numbers on the margins of mats of vegetation along the banks of slow-moving lowland rivers and streams but is found wherever there is permanent water and substantial aquatic, especially submergent, vegetation. Usually found in open or cleared areas along forest fringes in Lowland Moist and Wet Forests.

BIOLOGY: The species is essentially aquatic and nocturnal. Although they are omnivores as adults, these turtles feed primarily on aquatic plants (Moll and Legler 1971; Moll and Moll 1990; Acuña 1998), but juveniles eat invertebrates to a large extent. The species reaches sexual maturity at 125 to 135 mm in plastron length in males and 240 to 260 mm in females. They exhibit breeding behavior from September to December, but females apparently lay eggs throughout the year. Mating takes place in the water. During the initial phase of courtship the male faces the female and tickles her cheeks with the claws on his forefeet. She then submerges, and he swims after her until he is positioned on top of her carapace. He then grasps the shell with his claws and brings his tail downward to lie next to her cloaca. Copulation follows, lasting an hour to an hour and a half (Acuña 1998). Eggs are leathery, flexible and ellipsoidal, averaging 42 × 28 mm; eight to thirty-five form a single clutch, which is laid in a hole the female digs in the forest 3 m to 1.6 km from the home stream or pond. A female may lay one to six clutches per season. The flask-shaped nest is sealed after oviposition. Incubation time ranges from 69 to 123 days, when raised at temperatures averaging 28°C (Acuña 1998; Cabrera et al. 1996). Hatchlings are 30 to 40 mm in carapace length. Locally *Chrysemys* may be extremely abundant, and hundreds may be seen basking along the margins of major rivers in the course of a one- or two-hour boat ride. They are subject to considerable human predation in Costa Rica, which may account for their apparent absence from some areas of seemingly suitable habitat. The eggs are especially prized as food, and the juveniles are heavily harvested for the pet trade. A wide variety of other predators (mammals, birds, lizards, and even insects) eat the eggs of this species, and birds, crocodilians, and fish eat many hatchlings. Adult *Chrysemys* are relatively safe from all predators except crocodilians and humans (Moll and Legler 1971).

REMARKS: This form is a member of the wide-ranging *Chrysemys scripta* complex and is often regarded as conspecific with *C. scripta* (Moll and Legler 1971; Legler 1990; Smith and Smith 1979; Pritchard 1979; Iverson 1992). Weaver and Rose (1967) and Ward (1984) regarded *C. ornata* and some other named taxa as constituting species distinct from *C. scripta,* which, according to this view, is restricted to extreme northeastern Mexico and the central and eastern United States. I follow Weaver and Rose in their use of *C. ornata* for most Mexican and Central American sliders, although Ward (1980, 1984) believes that some of these populations are worthy of specific recognition. Under this interpretation, *Chrysemys gaigeae* of the Rio Grande and central and northwestern Mexico, *Chrysemys callirostris* of northern Venezuela, *Chrysemys adiutrix* of eastern Brazil, and *Chrysemys dorbigni* of southern Brazil, Uruguay, and Argentina, are regarded as species distinct from both *C. ornata* and *C. scripta.* Within this concept of *C. ornata*, substantial geographic variation is known. Legler (1990) recognized ten races within what I regard as *C. ornata*, but a number of these variants remain to be described. Several of these distinctive forms are completely allopatric and not connected by intergradation to other geographically adjacent populations (Iverson 1992; Legler 1990). Consequently some variants included in *C. ornata* under the arrangement followed here may constitute valid species as suggested by Ward's (1984) remarks. For these reasons I have eschewed any attempt to review geographic variation in Mesoamerican populations, but if you are interested see the summaries by Ernst

(1990) and Legler (1990), who place all populations in this complex in a single species. The names *Pseudemys ornata* and *Pseudemys scripta* are frequently applied to this species in the literature and have been used in many papers on Costa Rican turtles. Recently many authors have used the name *Trachemys ornata* or *Trachemys scripta* for this turtle (see Gibbons 1990 and the section on the genus above for a discussion of generic names). The karyotype is 2N = 50, with eight pairs of metacentrics and submetacentrics, five pairs of telocentrics and subtelocentrics, and twelve pairs of microchromosomes; NF = 66 (Stock 1972).

DISTRIBUTION: Lowlands of Mexico from Tamaulipas on the Atlantic versant and Sinaloa on the Pacific slope south through Central America on both coasts (except in extreme western Nicaragua and northwestern Costa Rica) to northern and western Colombia and northwestern Venezuela. In Costa Rica the species is found in lowland evergreen forest areas on the southwestern Pacific slope and Atlantic versant, especially in the drainage of Lake Nicaragua and the Río San Juan (2–60 m).

Extinct Turtles

Two fossil species of turtles have been described from Costa Rica. *Geochelone costaricensis* (Segura 1944) (family Testudinidae), from Milla 52, near Tunel Camp, Cartago Province, is of late Oligocene to early Miocene age, with a mosaic of primitive and advanced features making its placement in one or the other of the recognized subgenera ambiguous (Coto and Acuña 1986).

Acuña and Laurito (1996) recently described *Rhinoclemmys nicoyana* from the bed of the Río Nacaome, near Barra Honda (Nacaome), Puntarenas Province, dated as late Pleistocene. The authors suggest that this turtle is most closely related to similar semiaquatic forms from northern South America rather than its congeners in Costa Rica.

13 | Crocodilians
Order Crocodilia Gmelin, "1788," 1789 (Loricata)

THE EIGHT genera and twenty-three species of living crocodilians (alligators, caimans, crocodiles, the gharial, and the false gharial) are the last remnants of a major evolutionary radiation dating back to the Upper Triassic, some 210 million years ago. These large, voracious, and essentially aquatic creatures are the closest living relatives of the dinosaurs and birds. It is unlikely that anyone would mistake a heavily armored adult crocodilian for any other kind of animal. Juveniles, however, are sometimes confused with their distant allies the lizards (Sauria), which they superficially resemble. Young crocodilians may be distinguished from all lizards by having a number of the teeth exposed when the jaws are closed and a cloacal opening consisting of a longitudinal slit (teeth not exposed and cloacal slit transverse in lizards).

Other distinctive features of the order include:

1. Skull with two temporal fossae (diapsid)
2. Immovable quadrate
3. Bony secondary palate
4. Single median penis in males
5. A horny (keratinized) egg tooth
6. No urinary bladder
7. Socketed teeth
8. Subclavian artery arising from carotid artery
9. Only the ilium and ischium involved in the pelvic acetabulum for the femur
10. A four-chambered heart and an oblique septum separating the lungs and heart from the peritoneal cavity

A number of these unique features (the secondary palate, four-chambered heart, oblique septum) are intimately associated with aquatic life. The first, in conjunction with special valvular flaps at the posterior end of the mouth cavity, allows the crocodilian to breathe even with the head mostly underwater or when the mouth is open holding prey. The structure of the heart and oblique septum contribute to increased efficiency of respiration. Other aquatic adaptations include valvular external nostrils that are closed during submersion, a recessed eardrum that can be covered by a skin flap underwater, webbed toes, and the long, compressed muscular tail that propels the crocodilian through the water with strong lateral thrusts. The eyes (with a vertically elliptical pupil) and nostrils are mounted on raised areas of the head so the animal can see and breathe without exposing much of its body above water.

Modern species of the order are usually placed in two families (although some authors recognize a third family for the alligators and caimans). Others place the false gharial *(Tomistoma)* in the family Gavialidae (Densmore 1983; Densmore and White 1991; Poe 1996), but see Brochu (1997). The classification used here is as follows (fig. 13.1):

Family Crocodylidae
  Subfamily Alligatorinae: alligators and caimans (south eastern China and the Americas)
  Subfamily Crocodylinae: crocodiles and false gharial (Africa, Madagascar, Asia, Australia, and the Americas)
Family Gavialidae: gharial (northern portions of the Indian subcontinent)

Most crocodilians occur in lowland tropical and subtropical areas of Africa, Madagascar, Asia, Australia, and the Americas. The two species of alligators are the only crocodilians found in warm temperate areas in the southern United States and southeastern China. Several crocodilians range into brackish water, and some are passively carried or even venture out to sea for considerable distances. Most, however, are found in fresh water.

Differentiation among species of crocodilians is usually based on details of the skull, scutellation, and coloration. Skulls are frequently found in the field at sites of crocodile hunters' camps, and often skins or stuffed animals are offered for sale, although hunting and selling them are now heavily regulated. For these reasons I include more details than usual for skull features, and I also emphasize characteristics of prepared or commercial skins to aid in their identification. Remember that the epidermal scutes on the dorsum are underlain by thick bony plates (osteoderms) and that some alligators and crocodiles and all caimans have ventral osteoderms (called buttons in the commercial trade) as well. There may be one or two interlocking buttons under each ventral scute. In crocodiles and gharials the scutes on the underside of the body, at least in the gular region, contain a distinct porelike structure near the posterior margin of each scute; these were formerly thought to be glands but are now known to be integumentary sense organs (ISOs). Both osteoderms and ISOs are clearly indicated on tanned "belly" skins.

Crocodilians bask to elevate or maintain body temperatures above that of the surrounding medium. As water or air temperatures drop, adults' core temperatures remain higher than the ambient temperature by several degrees as the result of their large size, which retards heat loss.

As might be expected, crocodilians drink water and ingest it incidental to feeding. Some species lack any special salt-secreting glands such as are found in several other reptile groups (including birds). The saltwater crocodile *(Crocodylus porosus)* and freshwater crocodiles, however, have specialized salt secretory glands on the tongue, but these are lacking in alligators and caimans. Nitrogenous wastes are excreted primarily in the form of soluble ammonia (about 25% of urinary nitrogen) and urates (70%) that are precipitated out into insoluble uric acid in the cloaca, producing a mixed excretory product (liquid and solid).

The untrue notion that crocodiles "cry and sobbe" to entice human prey to come near them is an ancient one. Sir John Mandeville in an account of his mostly fictitious *Travels* (published about 1370) writes of the common proverb *(crocodili lachrimae)* that a crocodile will "sob, cry and weepe" so that a person taking pity on him approaches and "suddenly he destroyeth him." Edmund Spenser in *The Faerie Queene* poetically reinforced this false image:

. . . a cruell craftie crocodile,
Which in false grief hyding his harmful guile,
Doth weepe full sore, and sheddeth tender tears
The foolish man, that pities all the while
His mourneful plight, is swallowed up unawares
Forgetful of his owne, that mindes another's tears.

Merriam-Webster's Collegiate Dictionary defines crocodile tears as "false or affected tears: hypocritical sorrow." As at least one author has noted, the only tears are the victim's.

Olfaction in crocodilians is not used in a major way in locating prey in the aquatic environment, but it appears to play a significant role in social communication. All crocodilians have two pairs of specialized glands that secrete materials directly into the environment. One pair consists of single submandibular glands near the posterior inner margin of each half of the lower jaw. The second pair is in the cloaca. When the animal is excited the submandibular glands secrete a mildly odoriferous mucus. The cloacal glands are best developed in large males and secrete a mucuslike substance with a strong odor. It is thought that these secretions are used in courtship or that the cloacal glands are used in marking territories (Medem 1981). Adult crocodilians lack Jacobson's (vomeronasal) organs.

Hearing appears to be acute in members of this group. The auditory apparatus consists of a recessed tympanum covered by a pair of flaps that may be completely closed when the animal is submerged, an air-filled middle ear containing a cartilaginous extracolumella and osseous columella, and the inner ear. Hearing is excellent in low and middle frequencies, with maximum sensitivity from 100 to 6,000 Hz. Several vocalizations are important aspects of social communication in crocodilians, and the range and sensitivity of audition are partially related to these behavioral traits.

Vision appears to play an important role in the life of crocodilians, but the eye is basically adapted for nocturnal vision. The upper eyelid (unlike that of other reptiles) is mobile and covers most of the eye when the lids are closed. The transparent nictitating membrane covers the edge during submergence without impeding sight. Correlated with the specializations for nocturnal vision, the accommodation mechanisms are reduced, and crocodilians doubtless see only crude images. The pupil is reduced to a vertical slit in bright light to increase acuity during the day. Similarly, the retina has a higher percentage of rods than cones (Underwood 1970). The visual cells include two kinds of cones (single and double) and single rods, all lacking oil droplets. Spectral limits are about the same as in human beings, but crocodilians lack color vision.

The hands and feet of crocodilians are modified for their semiaquatic habits. In this group fingers IV and V (the outermost) are short and unclawed, so the hand has a phalangeal formula of 2-3-4-4-3. The foot has a phalangeal formula of 2-3-4-4 with toe V represented by a metapodial nub. During swimming the hands are folded back along the sides of the body, and the webbed feet serve as stabilizers.

While crocodilians are well adapted for an aquatic existence, they are also very mobile on land. They travel considerable distances from one aquatic site to another or to and

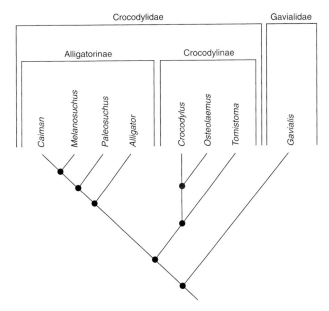

Figure 13.1. Phylogenetic relationships among living crocodilians.

from their nests, walking on all fours with the head and body lifted well off the ground while the tail drags behind. The basic movement is a trot, with the forelimb and hind limb on opposite sides of the body striking the ground at the same time. The trot may be converted to a high-speed gallop (3 to 17 km/h) when the animal is particularly excited, contrary to the popular idea that crocodilians must slither laboriously along when moving about on land. Slithering, in which the animal slides on its belly while thrusting simultaneously with all four feet, is usually used to reenter the water after basking, but it may be also used for slow or very rapid movement over relatively short distances.

Crocodilians are all predatory carnivores that usually feed on a variety of prey, but some are primarily fish eaters. The crocodilian head is designed for capturing and killing prey. The adductor muscles that close the mouth by pulling the lower jaw upward with a powerful snap are extremely well developed. The depressor muscles that open the mouth by lowering the mandible are relatively weak. Consequently relatively little force is required to keep a crocodilian's mouth closed, a fact exploited by hide hunters and others, who handle these animals regularly. In these cases the mouth is usually bound tightly closed; in "alligator wrestling" the wrestler keeps the mouth closed with a forearm grip under the animal's throat. For the reasons outlined in the final paragraph of this section, there is no need for you to test the relative strength of depressor and adductor jaw muscles. Take my word for it!

Sexual dimorphism in size is marked. In all species males are much larger than females, up to twice as large in most forms. Crocodilians mate in the water, but lay the eggs in a nest on land. Fertilization is internal. Courtship is usually initiated by females after male-male aggressive interactions and involves vocalizations, tactile stimulation, and swimming. The male mounts the female's back and positions his tail and cloaca underneath her tail, then inserts the anteriorly curved penis into her cloaca. The maternal instinct is strong in all species, and females guard the nest and help the hatchlings reach the water. The young usually remain near their mother for several weeks, since she provides protection from predators, including other crocodilians. Apparently all crocodilians have temperature-dependent sex determination (Ferguson 1985; Janzen and Paukstis 1991). In those cases studied to date, incubation temperatures of 30° to 34°C are critical. Individual species' responses to these temperatures are different, although 100% males usually are produced at the higher temperature and 100% females at the lower one. Crocodilians are relatively long-lived (twenty to forty or more years) and have an extended reproductive period (ten to thirty or more years). The largest living crocodilian is *Crocodylus porosus* of southern Asia, Australia, and adjacent areas, known to reach a maximum total length of 8,000 mm and a weight of 1,500 kg.

Several other features shared by all crocodilians are often the subject of popular curiosity. One is basking individuals' habit of remaining essentially immobile with the mouth partially or completely open for long periods. Folklore has it that these giant beasts are capturing flies and other small insects. It is now thought that this behavior has a thermoregulatory function and helps cool the crocodilian's body by evaporation. This behavior may also be a social signal related to dominance, since it occurs during rains, at night, and during basking. Another feature is the presence of rocks (gastroliths) in the stomachs of adult crocodilians. Apparently all members of the group regularly ingest sand, mud, mollusk shells, gravel, or rocks, depending on the size of the animal. As in some birds and other reptiles, these hard objects help fragment food to expedite digestion.

Caution: Only two species of crocodilians occur in Costa Rica: the spectacled caiman *(Caiman crocodilus)* and the American crocodile *(Crocodylus acutus)*. Although caimans generally are more docile than the vicious and aggressive American crocodile, both species should be treated with respect. Always remember that crocodilians are extremely dangerous reptiles, and even small individuals can cause serious injury from their rapid, snapping bites. Large examples can maim or kill relatively large animals, and some species attack terrestrial prey, which they grasp in their powerful jaws, drag underwater, and drown. The jaws, the numerous sharp teeth, and the habit of twisting and rolling underwater to further stun and confuse the prey all contribute to crocodilians' well-deserved reputation as truly dangerous creatures. On land or in shallow water, the muscular tail is an effective defensive weapon, which can inflict serious damage when thrashed from side to side. Most injuries from crocodilians in Costa Rica are caused by carelessly handling or by approaching small caimans. It is best to maintain a respectful distance when observing these animals, and you should be especially careful around even a small American crocodile.

CROCODILIAN IDENTIFICATION: KEY FEATURES

Conformation and scutellation are the principal characteristics used in identifying crocodilians, but features of the skull and dentition are also of value because skulls are often found where hunters have killed and skinned these animals. The external features included in the keys and descriptions that might cause some difficulty in interpretation are:

Head region: Snout shape and the presence of a transverse preorbital ridge or preorbital warts are characteristic (fig. 13.2).
Scutellation: The number and arrangement of postoccipital and nuchal scutes are particularly diagnostic and may be used to identify tanned skins (fig. 13.2).
Tail: Features that are definitive include the pattern of the

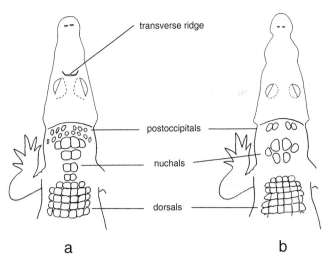

Figure 13.2. Dorsal view of head and anterior part of body in crocodilians: (a) *Caiman crocodilus*; (b) *Crocodylus acutus*. Note differences in scalation.

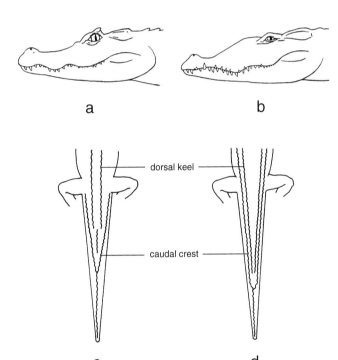

Figure 13.3. Head and keeling in crocodilians: (a) lateral view of head of *Caiman crocodilus*; (b) lateral view of head of *Crocodylus acutus*; (c) dorsal view of posterior body, tail, and keels in *C. crocodilus*; (d) dorsal view of posterior body, tail, and keels in *C. acutus*.

caudal crests (fig. 13.3c,d) and the number of whorls (around the tail) of caudal scales interrupted laterally or ventrally by small scales.

Coloration: Juveniles usually are more vividly colored, and the pattern is more distinct than in adults. Large adults tend to be nearly uniform in dorsal and caudal coloration.

In Costa Rica the ground color is light brown, olive brown, or light yellow *(Caiman)* or gray brown, olive green, or gray green *(Crocodylus)*.

Size and sexual dimorphism: There is definite sexual dimorphism in size, and females are smaller than males. Most adult Costa Rican *Caiman* are about 1,200 to 1,750 mm, and *Crocodylus* are 3,000 to 4,000 mm. All measurements for these animals in this book are for total length.

The sex of adult live crocodilians may be determined by digitally palpating the cloaca. Since the cloaca otherwise is smooth-walled, it is obvious when a penis is present. Pressing on the venter just anterior to the vent will cause the penis to be everted. This procedure requires that the animal be restrained and positioned ventral side up.

KEY TO COSTA RICAN CROCODILIANS

1a. An elevated crescent-shaped, transverse ridge just anterior to eyes on top of head, completely ossified in adults, partially in juveniles (fig. 13.2a); no finger webs; a pair of dorsal keels continuing onto tail between lateral caudal crests, but the dorsal keels fused posteriorly to form a single short crest between lateral pair (fig. 13.3c); fourth tooth on lower jaw fits into a fossa in upper jaw at premaxillary-maxillary suture and is not visible when mouth is closed (fig. 13.3a) . . . .*Caiman crocodilus* (p. 776)

1b. No crescent-shaped transverse preocular ridge (fig. 13.2b); fingers webbed; a pair of dorsal keels continuing onto tail between lateral caudal crests not fusing to form a single median crest posteriorly (fig. 13.3d); fourth tooth on lower jaw lies laterally in an open groove at premaxillary suture and is visible when mouth is closed (fig. 13.3b) . . . . *Crocodylus acutus* (p. 778)

Family Crocodylidae Cuvier, 1807

Alligators, Caimans, Crocodiles, and False Gharial

The largest living family of crocodilians, with twenty-two species. Two subfamilies are usually recognized, the Alligatorinae of east-central China and southeastern North America to Paraguay and Argentina and the Crocodylinae of subtropical and tropical Africa, Madagascar, Asia, Australia, extreme southeastern North America, and Mexico to northern South America and the Greater Antilles. The family is characterized as follows:

1. Total number of teeth on each upper jaw twenty-one or fewer, on each lower jaw twenty-two or fewer
2. Length of snout not more than three and a half times basal width
3. Nasal bones in contact with premaxillaries
4. Postorbital column attached to medial head of jugal
5. Foramen magnum bounded by supraoccipital

The extremely attenuate snouted monotypic Gavialidae of the northern rivers of the Indian peninsula, by contrast, have:

1. Twenty-six or more upper and twenty-four or more lower jaw teeth
2. Length of snout more than three and a half times basal width
3. Nasals separated from the premaxillaries
4. Postorbital bone attached to the head of jugal
5. Foramen magnum not bounded by the supraoccipital

Subfamily Alligatorinae Cuvier, 1807

Genus *Caiman* Spix, 1825

Spectacled and Broad-nosed Caimans

This genus is one of three Caiman genera found in the New World tropics. (The others are *Melanosuchus* and *Paleosuchus* of South America.) Caimans are allied to the Chinese and American alligators (genus *Alligator*), and the four genera differ in skull, scutellation, and color. As a group (the Alligatorinae) they differ from crocodiles in having

1. The fourth maxillary tooth the longest
2. The fourth mandibular tooth fitting into a fossa in the upper jaw so as not to be visible laterally when the mouth is closed
3. Teeth of the lower jaw internal to those of the upper jaw and hidden when the mouth is closed

In some large adult caimans, including the Costa Rican species, the fourth mandibular tooth wears a hole in the upper surface of the premaxillary bone and becomes exposed dorsally when the mouth is closed. A similar perforation may also develop to expose the first mandibular tooth in large adults.

Caiman is readily distinguished from alligators and other caiman genera (*Melanosuchus* and *Paleosuchus* of South America) by having at least eight enlarged nuchal tubercles grouped together, double osteoderms (buttons) underlying the scutes on the midbelly, five premaxillary teeth on each side, an interorbital crest, and the upper eyelid only partially ossified but with its surface wrinkled and tuberculate. Some recent evidence (Poe 1996) suggests that *Melanosuchus* should be synonymized into *Caiman*.

Two to four species are recognized in this genus, which ranges in the lowlands from southern Mexico to western Ecuador and northern Argentina; one species, *Caiman crocodilus*, has been introduced to southern Florida, eastern Puerto Rico, and Isla de la Juventud (Isle of Pines) off Cuba.

Caiman crocodilus (Linné, 1758)

Spectacled Caiman, Lagarto, Baba, Babilla, Guajipal, Caiman

(plates 513–14; map 13.1)

DIAGNOSTICS: In Central America this species is immediately recognized by an obvious, elevated transverse ridge just anterior to the eyes on the top of the head (fig. 13.2a).

DESCRIPTION: Adult males 1,100 to 2,750 mm, females 1,100 to 2,200 mm; large males weigh about 65 kg; hatchlings 210 to 260 mm in total length; length of snout one and a half to two times its basal width; 14 to 16 maxillary teeth; 18 to 20 mandibular teeth; premaxillaries often perforate in adults to receive anterior pair of teeth of lower jaw; five transverse series of nuchal scales (two or three with four scales); upper eyelids elevated into tubercle; postoccipital scutes in two or three indistinct rows; a single ventral collar of enlarged scales; 18 to 19 transverse rows of dorsal scutes, 8 to 13 longitudinal rows; 20 to 27 transverse rows of ventral scutes; fingers without webs; no integumentary sensory organs in ventral scales, but ventrals heavily pitted in tanned skins; rows of large keeled oval scales alternating with a series of beadlike scales laterally; whorls of large scales on the base of tail not interrupted laterally by small scales. Dorsum uniform dusky to olive brown in large adults, olive brown to light yellow with dark brown crossbands on sides of tail in small adults and juveniles; belly uniform cream to white; faint dark bars on lips in examples under 350 mm; iris gold in life; hatchlings light brown or yellow with dark crossbands on back and tail.

SIMILAR SPECIES: (1) The only other crocodilian in Costa Rica is *Crocodylus acutus*, which lacks a transverse preorbital ridge (fig. 13.2b), lacks ventral osteoderms, and has integumentary sensory organs on most ventral scales.

HABITAT: Occurs throughout the humid lowlands in rivers, streams, ponds, freshwater swamps, swamp forests, and occasionally near the ocean (in mangrove swamps or estuaries in brackish water). It is most common in high rainfall regions (Lowland Moist and Wet Forests) but also is found in Lowland Dry Forest.

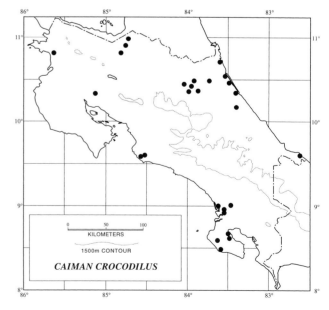

Map 13.1. Distribution of *Caiman crocodilus*.

BIOLOGY: The species usually is most active at night and hunts then, but it is often seen swimming or basking during the day. Body temperatures average 30°C in active specimens (Stanton and Dixon 1975, 1977). Caimans eat a wide variety of food, with the young eating mostly aquatic arthropods (insects and crustaceans) and adults mainly fishes and frogs. They also feed on carrion, including the corpses of large birds and moderate-sized mammals. Reproduction in the caiman has been extensively studied in Venezuela (Stanton and Dixon 1977) and Colombia (Medem 1981). These works, the study of nesting ecology for Caño Negro, Alajuela Province (Allsteadt and Vaughan 1988; Allsteadt 1994), and the summary for Costa Rica by Dixon and Stanton (1983) form the basis for the following paragraphs. Ovarian development and courtship begin about ninety days before nesting. Breeding seems to precede the start of the rainy season (February to March) in most of Costa Rica but is later in the northwest (June to August) and continues for several months. Courting males establish and patrol aquatic territories and display by arching the tail above the water, with the head out of the water. Intruder males are often chased and bitten, although when males of similar size give the tail display near each other the smaller one usually leaves. When receptive females approach displaying males, a brief courtship and copulation underwater follow. Vocalization plays little part in courtship in this species. The female constructs a nest of plant materials and soil immediately adjacent to the nearest water or as much as 200 m away about six weeks before peak water levels are reached. Nests are about 1,170 mm long, 1,000 mm wide, and 440 mm high. They maintain a relatively constant temperature (30.5 to 36°C) and high (70%) relative humidity inside. The female lays eggs in the nest two to six days after construction begins, from mid-June to August. The calcareous, hard-shelled elliptical to spherical eggs are about 65 × 44 mm and average 64 g. Clutch size is fifteen to forty, with an average of twenty-seven to twenty-nine. Incubation lasts about 73–82–90 days. Females tend and repair the nest during the incubation period, and vocalizing by the hatchlings within the nest stimulates her to open the nest and release the young. The adult also helps young to the nearest water. Postnatal care persisted for up to four months in the seasonally dry llanos of Venezuela (Stanton and Dixon 1977). In wetter situations parental care may continue longer, since the young will remain near the nest site for up to eighteen months after hatching. At Caño Negro juveniles formed pods of twenty to sixty individuals during the dry season. The larger pods included juveniles derived from several clutches. Pod members were usually young of the year (six months) or from the previous year (eighteen months), but some pods contained juveniles from several years. Alvarez del Toro (1974) noted that males of this species in Mexico are involved in nest tending and opening as in some other crocodilians. The young reach sexual maturity at a total length of somewhat over 1,000 mm, which they may attain in five to six years. A wide variety of mammals, including humans, prey on the eggs of this species, and large waterbirds are probably the principal predators of the young. Adults are safe from almost any predator except anacondas (Eunectes), large felids, and humans. Hatchlings are 191 to 260 mm in total length and weigh 35 to 49 g.

REMARKS: This species is often called Caiman sclerops in the literature (Medem 1981, 1983), but C. crocodilus must be the accepted name (Smith and Smith 1977). There is some controversy regarding the status of several distinctive geographic variants of caimans usually referred to this species (Groombridge 1987). Differences among them involve skull dimensions (especially snout shape, length, and width) and coloration. One of these populations, the yacare caiman, sometimes recognized as a distinct species (Caiman yacare), is characterized by a moderately broad snout (snout width at first maxillary tooth equal to or greater than basal width of snout), flanks covered by four or five rows of round nearly unkeeled scales separated by chainlike interscales, and black adult coloration. It is restricted to the Río Paraguay system of Brazil and Paraguay. A relatively narrow-snouted form, the Río Apaporis caiman (Caiman apaporensis), possibly best regarded as a separate species, occurs in the river of that name in the upper Amazon basin of southeastern Colombia. Although it agrees in general with the wide-ranging Amazon form of Caiman crocodilus, the Río Apaporis population has the snout strikingly elongate, narrow, and parallel-sided anteriorly; the pterygoid in at most narrow contact with the palatal foramen; the head with dark spots; the limbs dark gray or black; and the dorsum bright yellow (in life). The more broadly distributed form has the pterygoid in broad contact with the palatal foramen, no head spotting, and the limbs and body olive green (in life). The Costa Rican populations agree with others from Mexico and Central America and with those north and west of the Andean backbone of South America in having the snout relatively flat and broad (width of snout at level of fourth maxillary tooth greater than distance from that tooth to tip of snout) and in having a generally brown, olive brown, or yellow ground color (in life). Other populations in this complex have the snout width at the fourth maxillary tooth equal to or less than the distance from that tooth to the tip of the snout and differ in coloration from the caimans from Middle America and northern South America. Whether the several components of this complex are connected by intermediate populations is open to question, as is the constancy of the distinguishing features. All these forms differ from the other species in the genus, Caiman latirostris of northern Argentina, Uruguay, and Paraguay and adjacent parts of southern Brazil and eastern Bolivia. C. latirostris has a very short, compressed snout (length less than basal width); caudal whorls of scales interrupted laterally by small scales; three transverse series of nuchal scales (one series with four tubercles); ventral collar consisting of two rows of not noticeably enlarged scales; and 12 to

14 maxillary teeth. The several geographic variants of *C. crocodilus*, including the possibly distinct species *C. apaporensis* and *C. yacare*, conform to the character states for these features given in the description of *C. crocodilus* above. The karyotype (2N = 42) of *Caiman crocodilus* comprises a graded series of eleven pairs of telocentrics (four large pairs), two pairs of submetacentrics, and eight pairs of small metacentrics; NF = 62. There is no sexual heteromorphism (Cohen and Gans 1970).

DISTRIBUTION: Pacific lowlands from the Isthmus of Tehuantepec, Mexico, to southern Ecuador; Atlantic lowlands from extreme eastern Honduras to northern Colombia, Venezuela, and the Guianas and throughout the Amazon basin to eastern Peru and central Brazil; also introduced to southern Florida, eastern Puerto Rico, and Isla de la Juventud off Cuba. In Costa Rica only in the lowlands (1–200 m).

Subfamily Crocodylinae Cuvier, 1807

Genus *Crocodylus* Laurenti, 1768

Crocodiles

Crocodylus (twelve species), the pygmy crocodile *(Osteolaemus)* of Africa, and the Asian false gharial *(Tomistoma schlegelii)* constitute a distinctive subgroup (the Crocodylinae) within the family when compared with the alligators and caimans. The three genera share the following features that will distinguish them from the latter stock:

1. Fifth maxillary tooth the longest
2. Fourth mandibular tooth passes into a laterally open groove at the premaxillary-maxillary suture and is visible laterally when the mouth is closed
3. The teeth of the posterior upper and lower jaws interdigitate when the mouth is closed, and some teeth in both jaws are visible laterally

Crocodylus and *Osteolaemus* are distinguished from *Tomistoma* (characters for the latter in parentheses) in having not more than nineteen teeth on each side of the upper jaw, the splenial bone not forming part of the mandibular symphysis and the nuchal, and dorsal scales definitely separated (twenty or twenty-one teeth, splenial reaching symphysis, and nuchal and dorsal scutes forming a continuous unit). *Osteolaemus* is very similar to *Crocodylus*, but the former has a short, broad snout, with length only slightly exceeding basal width (snout with length markedly exceeding basal width) and differs from *Crocodylus* in minor details of the skull, scutellation, and color.

Twelve species, four in the New World, constitute *Crocodylus*, which ranges over tropical Africa, Madagascar, Asia, Australia, and associated islands to Fiji and also southern Florida, Mexico, to northern South America and the Greater Antilles in the Americas.

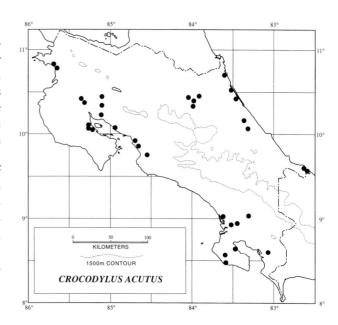

Map 13.2. Distribution of *Crocodylus acutus*.

Crocodylus acutus (Cuvier, 1807)

American Crocodile, Lagarto Negro, Cocodrilo

(plates 515–16; map 13.2)

DIAGNOSTICS: This species differs from all other New World crocodilians in having the following combination of features: a blunt, smooth wart in front of each eye in adults, the dorsal scales irregularly arranged, the snout long and pointed (length two or more times basal width), and the subcaudal scale rings of uniform-sized scales but with some inclusions of individual scales in the ventral-lateral area between the rings.

DESCRIPTION: Total length in adults 2,300 to 7,300 mm in males, 2,300 to 4,000 mm in females, most adults 3,000 to 4,000 mm; those 3,000 mm in total length weigh about 100 kg, those 4,000 mm weigh 270 kg; a 6,250 mm male should weigh about 1,100 kg; hatchlings are 250 to 300 mm; 13 to 14 maxillary and 15 mandibular teeth; both young and older adults with a median hump-like swelling anterior to orbits (fig. 13.3b); at least 4 nuchal scutes in two juxtaposed pairs; postoccipital scutes 2 to 4, in one transverse row; a single ventral collar; 16 to 17 transverse rows of dorsal scutes in 4 to 6 longitudinal rows; transverse ventral scale rows 25 to 35; fingers webbed; ventral scales with integumentary sense organs. Dorsum olive green, tan gray, or gray green with dark crossbands, flecks, and blotches on back and tail that become obscure in old adults; juveniles with gray ground color; venter unmarked, whitish or yellow white; iris greenish in life.

SIMILAR SPECIES: (1) The only other crocodilian in Costa Rica is *Caiman crocodilus*, which has a prominent preorbital transverse ridge (fig. 13.2a), has double ventral osteoderms, and lacks integumentary sense organs on the ventral scutes.

HABITAT: Typically occurs in larger rivers and streams and often in brackish water near the mouths of rivers. This crocodile may be found in salt marshes, freshwater marshes, mangrove swamps, and swamp forests (brackish and freshwater) as well as streams and rivers in Lowland Dry, Moist, and Wet Forests. Elsewhere in its range individuals sometimes venture into the open ocean.

BIOLOGY: Crocodiles formerly ranged over much of lowland Costa Rica, but because of hunting for hides and danger to humans, they had been eliminated from many areas by the 1960s. Conservation efforts in the 1980s and 1990s have seen the return of these animals to many areas. The species spends most of its time in the water and is most active at night but often basks along riverbanks during the day. Crocodiles frequently excavate subterranean caves in the banks of waterways; though these can be entered only from underwater, they are mostly above water level and are used as refuges. The food of the young includes aquatic insects, small fishes, and amphibians, but larger animals prey on crustaceans, snails, fishes, frogs, turtles, iguanas, caimans, birds, and small mammals. The now very uncommon large adults will attack large prey including domestic animals and humans. The frequency and success of such incidents in times past were probably exaggerated, and I know of no cases of unprovoked attacks in Costa Rica from the 1960s to the mid-1990s. As conservation efforts have begun paying off, however, the increase in the numbers of crocodiles has led to more interactions between crocodiles and humans. One of these, in the fall of 1997, led to the tragic death of a man fishing near the mouth of the Río Tivives, Puntarenas Province. Two more fatal attacks took place later in the 1990s near the mouth of the Rió Grande de Tárcoles. Little is known of crocodile biology in the republic, and most of what follows is extrapolated from Alvarez del Toro (1974) and Medem (1981, 1983) for Mexico, Colombia, and the rest of South America, respectively, and from the recent studies of Sasa and Chaves (1992), Sánchez, Bolaños, and Piedra (1996), and Bolaños, Sánchez, and Piedra (1997). Courtship and breeding seem to occur before the wetter periods of the year, probably from March to May in most of Costa Rica, but perhaps later in the Pacific northwest. Adults have definite territories and use sounds and physical attacks to maintain them. The species is rather vocal, and crocodiles produce a variety of sounds for various purposes. During the breeding season they often emit low grunts as well as loud roars. Copulation occurs in the water, and the female usually digs a hole in sand or soil to form a nest for the approximately twenty to sixty eggs that she then lays and covers with soil and vegetation. In some cases a mound nest may be constructed, similar to that of the spectacled caiman but lower. The eggs are calcareous, elliptical to spherical, usually about 65 × 40 mm, and weigh about 56 g. The nest is 40 to 70 cm deep and about 50 to 60 cm in diameter. A temperature of about 29°C is maintained during incubation. The female tends the nest for about seventy-five to ninety days, after which the voices of the hatchlings stimulate her to open the nest with her hands and snout. She then picks them up one at a time and carries them to the water in her mouth. She remains with the hatchlings for a considerable period, presumably to protect them from aquatic predators. Hatchlings average a little over 240 mm in total length. Growth is relatively rapid during the first few years of life. Both sexes are sexually mature at total lengths of about 2,300 mm. Animals of this size are believed to be seven to nine years old. The eggs and hatchlings are eaten by a variety of small carnivorous mammals and birds. Adults are generally safe from most predators except the jaguar, very large sharks, and humans. The karyotype (2N = 32) of *Crocodylus acutus* comprises three pairs of telocentrics, four pairs of submetacentrics, and four pairs of large and five pairs of small metacentrics; NF = 58. There is no sexual heteromorphism (Cohen and Gans 1970).

DISTRIBUTION: Lowlands from Sinaloa, Mexico, to extreme northern Peru on the Pacific slope and from Tamaulipas, Mexico, south to Colombia and northern Venezuela on the Atlantic versant; also in southern Florida and the Greater Antilles (Cuba, Jamaica, and Hispaniola). In Costa Rica only in the lowlands (0–200 m).

4 | Biogeography and Evolution

14 | Ecological Distribution of the Herpetofauna

THE SCIENCE of biogeography is concerned with describing the current patterns of distribution of organisms and developing and testing hypotheses as to how the patterns arose through time. In this chapter I will describe the major distributional patterns for the herpetofauna of Costa Rica and relate these patterns to ecological factors, especially climate, expressed in temperature and availability of moisture, vegetation, and geography.

ECOLOGICAL PATTERNS

Bioclimates and Plant Formations

The bioclimates of Costa Rica as characterized by distinctive plant formations were described and discussed in chapter 2. This system is followed in tables 14.1 to 14.8 to indicate the occurrence of each species in a particular plant formation. Tables 14.1 to 14.5 also list the known altitudinal limits for each species. Bear in mind that a species' ecological range tends to have fuzzy boundaries, so occasional individuals may be found in transitional zones between adjacent plant formations. Other factors that depend on local conditions, especially wind direction, cloud cover, and soil permeability, may influence the altitudinal range of a particular plant formation locally so that some amphibians and reptiles may be found at lower or higher elevations than expected from a strict correlation of bioclimate with elevation.

The following abbreviations are used in the tables to designate specific plant formations:

Lowland: Dry Forest (D), Moist Forest (LM), Wet Forest (LW)

Premontane: Moist Forest (PM), Wet forest (PW), Rainforest (PR)

Lower Montane: Moist Forest (LMM), Wet Forest (LMW), Rainforest (LMR)

Montane: Wet Forest (MW), Rainforest (MR)

Subalpine: Pluvial Paramo (SP)

In the case of lowland and some premontane zones, Pacific (P) and Atlantic (A) slope forest are treated separately.

In all tables in chapters 14 and 15 that display percentage values, these have been rounded to the nearest 1% except for very low values. Greater precision would be misleading given the state of knowledge of distributions.

Distinctive differences in the number of species found in each plant formation are a combination of ecologic valence, especially temperature and humidity regimens, and historical constraints. It is expected that any organism that has access to a particular situation and can survive in it will occur in that plant formation. Ecological assemblages thus consist of organisms that happen to be there.

Nevertheless, the differences in ecological valence lead to some predictable patterns. Crocodilians are restricted to the lower portion of the lowlands (below 200 m). Caecilians and salamanders are absent from Lowland Dry Forest. Aside from marine species, turtles occur only in lowland and premontane zones, below 1,300 m in altitude. Salamanders and anurans are most diverse in Premontane Rainforest, lizards in Lowland Moist and Wet Forests and snakes in the Atlantic Lowland Forests (table 14.6). Similar trends are indicated when one compares the proportion of each major group with the total herpetofauna by plant formation (table 14.7).

The highest total number of species occurs in Atlantic Lowland Moist Forest (185) and Premontane Rainforest (195), warm and wet and cool and wet environments, respectively (table 14.7). In Lowland Dry Forest and Pacific Premontane Moist Forest environments there are three times as many reptile species as amphibian species (table 14.8). The areas occupied by Atlantic Premontane Moist and Lower Montane Moist formations are so small that it is not surprising they have very low species numbers—forty-seven and two, respectively. These two formations aside, in most plant formations reptile species constitute 61 to 67% of the fauna. In Premontane Rainforest and Lower Montane Rainforest amphibians make up 54 to 61% of the total fauna, and in Montane Rainforest they make up 77%. The relative proportions of the two groups shift from high proportions of reptiles to higher proportions of amphibians along a gradient from warmer to cooler conditions. Drier and cooler environments have reduced numbers of species, with extremes in the Lowland Dry Forest (73 species) and Subalpine Pluvial Paramo (3 species).

Analysis of the distribution of amphibians and reptiles by major systematic group (table 14.6) shows that the greatest diversity of salamanders (64%) and anurans (60%) live in Premontane Rainforest. Lizard diversity is greatest in the lowland rainforests (45 to 49%). The greatest diversity of snake species is in Lowland Wet Forest (45 to 57%) and

Table 14.1. Distribution of Costa Rican Caecilians and Salamanders by Plant Formation and Altitude

	D	LM		LW		PM		PW		PR	LMM	LMW	LMR	MW	MR	SPR	Elevation (m)
		P	A	P	A	P	A	P	A								
Caecilians																	
Dermophis gracilior		X						X	X	X			X				404–2,000
Dermophis parviceps		X		X	X			X		X							40–1,220
Gymnopis multiplicata	G	X	X	X	X	X	X	X	X			X					1–1,400
Oscaecilia osae				X													3
Totals 4*	0	3	1	2	2	2	1	3	2	2	0	1	1	0	0	0	
Salamanders																	
Bolitoglossa alvaradoi			X	X						X							15–1,116
Bolitoglossa cerroensis															X		2,530–2,990
Bolitoglossa colonnea		X	X	X	X					X							2–1,245
Bolitoglossa compacta													X				1,650–1,980
Bolitoglossa diminuta										X							1,555
Bolitoglossa epimela								X		X							775–1,550
Bolitoglossa gracilis										X							1,225–1,280
Bolitoglossa lignicolor		X		X						X							2–884
Bolitoglossa minutula													X				1,810–2,660
Bolitoglossa nigrescens													X		X		1,900–3,000
Bolitoglossa pesrubra											X		X	X	X	X	1,875–3,620
Bolitoglossa robusta							X			X	X		X				500–2,048
Bolitoglossa schizodactyla										X							780
Bolitoglossa sooyorum													X		X		2,355–3,000
Bolitoglossa striatula		X	X	X		X				X							2–1,052
Bolitoglossa subpalmata										X	X	X	X	X	X		1,245–2,900
Nototriton abscondens					X		X			X			X				1,010–2,500
Nototriton gamezi										X			X				1,550–1,650
Nototriton guanacaste													X				1,400–1,580
Nototriton major										X							1,100–1,200
Nototriton picadoi										X		X					1,200–2,200
Nototriton richardi										X			X				1,370–1,800
Nototriton tapanti										X							1,300
Oedipina alfaroi			X							X							19–850
Oedipina alleni		X		X		X											2–880
Oedipina altura													X				2,286–2,320
Oedipina carablanca					X												60–260
Oedipina collaris					X												600
Oedipina cyclocauda			X		X												60–500
Oedipina gracilis			X		X					X							3–710
Oedipina grandis													X				1,810–1,950
Oedipina pacificensis		X		X						X							5–730
Oedipina paucidentata													X				2,286
Oedipina poelzi								X		X			X				775–2,050
Oedipina pseudouniformis					X			X	X	X							19–1,253
Oedipina savagei										X							1,200–1,400
Oedipina uniformis						X		X	X	X							750–2,150
Totals 37*	0	5	6	4	8	2	1	6	4	23	0	4	15	2	5	1	

*Not included in totals; G = gallery forest.

Lowland Moist and Pacific Premontane Moist and Wet Forests and Premontane Rainforest (46 to 55%). Turtle diversity is greatest in Atlantic Lowland Moist and Wet Forests (55%). Similar trends are reflected by the proportion of species in each major taxonomic group in the total herpetofauna by plant formation (table 14.7).

As expected based on total numbers of species, the highest proportion of species relative to the herpetofauna in each plant formation (table 14.8) live in the Atlantic lowland humid forests and Premontane Rainforest: Atlantic Moist Forest (47%), Atlantic Wet Forest (47%), and Premontane Rainforest (50%).

Temperature (Altitude)

It is usual in reports on ecology to chart changes in diversity along an elevational gradient. No one seriously believes that altitude has a direct effect on distributions, since the slight changes in air pressure are not likely to substantially affect an organism's physiology. Essentially, altitudinal gradients reflect changes in temperature, moving from warmer ones at low elevations to cooler ones at higher altitudes.

The known altitudinal distributions for Costa Rican amphibians and reptiles are given in tables 14.1 to 14.5. The number of species that are found in the biotemperate limited

Table 14.2. Distribution of Costa Rican Anurans by Plant Formation and Altitude

	D	LM		LW		PM		PW		PR	LMM	LMW	LMR	MW	MR	Elevation (m)
		P	A	P	A	P	A	P	A							
Rhinophyrnus dorsalis	X											X	X			0–300
Atelopus chiriquiensis												X	X			1,810–2,500
Atelopus senex										X		X	X			1,280–2,040
Atelopus varius				X		X	X	X	X	X		X	X			16–2,000
Bufo coccifer	X					X	X		X	X						4–1,550
Bufo coniferus		X	X	X	X			X	X	X		X	X			2–1,550
Bufo fastidiosus									X	X			X			760–2,100
Bufo haematiticus		X	X	X	X	X	X	X	X							20–1,300
Bufo holdridgei													X			2,000–2,200
Bufo luetkenii	X	X	X			X										6–436
Bufo marinus	X	X	X	X	X	X	X	X	X	X		X				1–1,600
Bufo melanochlorus		X	X	X	X		X		X							2–1,080
Bufo periglenes													X			1,500–1,620
Bufo valliceps			X		X											40–495
Crepidophryne epiotica										X						1,051–2,040
Eleutherodactylus altae										X	X					60–1,245
Eleutherodactylus andi					X				X	X						830–1,500
Eleutherodactylus angelicus						X		X	X	X		X	X			656–1,680
Eleutherodactylus bransfordii			X		X				X	X						6–880
Eleutherodactylus bufoniformis			X													15–20
Eleutherodactylus caryophallaceus			X		X				X	X			X			2–1,880
Eleutherodactylus cerasinus			X		X				X	X						19–590
Eleutherodactylus crassidigitus		X	X	X	X			X	X	X						2–2,000
Eleutherodactylus cruentus			X		X				X	X		X	X			40–1,800
Eleutherodactylus cuaquero													X			1,520–1,600
Eleutherodactylus diastema		X	X	X	X				X	X		X	X			1–1,620
Eleutherodactylus escoces										X			X			1,150–2,100
Eleutherodactylus fitzingeri	G	X	X	X	X			X	X	X		X	X			1–1,520
Eleutherodactylus fleischmanni						X	X	X	X			X				1,050–2,286
Eleutherodactylus gaigeae			X													20–200
Eleutherodactylus gollmeri			X		X				X	X						10–1,520
Eleutherodactylus gulosus													X			910–1,600
Eleutherodactylus hylaeformis												X	X			1,500–2,500
Eleutherodactylus johnstonei						I										1,160
Eleutherodactylus megacephalus			X		X				X	X						1–1,200
Eleutherodactylus melanostictus									X	X		X	X		X	1,150–2,700
Eleutherodactylus mimus			X		X					X						15–685
Eleutherodactylus moro										X						900–1,245
Eleutherodactylus noblei			X		X			X	X	X						4–1,200
Eleutherodactylus pardalis										X						884–1,200
Eleutherodactylus persimilis			X		X				X	X						39–1,200
Eleutherodactylus phasma													X			1,850
Eleutherodactylus podiciferus								X	X	X		X	X		X	1,089–2,650
Eleutherodactylus polyptychus			X		X											2–60
Eleutherodactylus punctariolus			X					X		X						400–1,700
Eleutherodactylus ranoides	G	X	X		X			X	X							1–1,300
Eleutherodactylus rayo								X		X			X			1,480–1,820
Eleutherodactylus ridens		X	X	X	X			X	X	X			X			15–1,600
Eleutherodactylus rugosus		X		X				X	X	X						10–1,220
Eleutherodactylus stejnegerianus	G	X		X		X		X		X						3–1,330
Eleutherodactylus talamancae			X		X											15–646
Eleutherodactylus taurus		X	X													5–525
Eleutherodactylus tigrillo					X											400–440
Eleutherodactylus underwoodi										X	X					920–1,590
Eleutherodactylus vocator			X							X						2–1,220
Leptodactylus bolivianus	X	X		X												1–404
Leptodactylus labialis	X	X	X	X	X											1–650
Leptodactylus melanonotus	X	X	X	X	X			X	X							6–1,440
Leptodactylus pentadactylus	X	X	X	X	X			X	X	X						5–1,200
Leptodactylus poecilochilus	X	X	X	X				X		X						3–1,150
Physalaemus pustulosus	X	X	X													3–1,050
Agalychnis annae						X	X	X	X	X			X			780–1,650
Agalychnis calcarifer			X		X											50–85
Agalychnis callidryas		X	X	X	X			X	X	X						1–970
Agalychnis saltator			X		X		X			X						15–819
Agalychnis spurrelli		X	X	X						X						15–750
Anotheca spinosa					X					X						350–1,330
Duellmanohyla lythrodes					X											440

continued

Table 14.2. continued

	D	LM		LW		PM		PW		PR	LMM	LMW	LMR	MW	MR	Elevation (m)
		P	A	P	A	P	A	P	A							
Duellmanohyla rufioculis				X				X	X	X						680–1,524
Duellmanohyla uranochroa								X		X		X	X			656–1,600
Gastrotheca cornuta					X											300–700
Hyla angustilineata										X		X	X			1,100–2,040
Hyla calypsa													X			1,810–1,920
Hyla colymba								X		X						1,116–1,200
Hyla debilis										X						1,200–1,400
Hyla ebraccata		X	X	X	X			X		X						2–1,320
Hyla fimbrimembra								X		X			X			750–1,900
Hyla lancasteri								X		X						368–1,200
Hyla legleri										X						880–1,524
Hyla loquax			X		X			X		X						60–1,116
Hyla microcephala	X	X	X	X	X	X				X						7–780
Hyla miliaria			X					X		X						20–1,330
Hyla palmeri										X						600–750
Hyla phlebodes			X					X		X						20–620
Hyla picadoi													X		X	1,920–2,770
Hyla pictipes													X		X	1,930–2,800
Hyla pseudopuma								X	X	X		X	X			1,120–2,340
Hyla rivularis								X	X	X		X	X			1,210–2,040
Hyla rosenbergi		X		X												10–100
Hyla rufitela			X		X											1,650
Hyla tica								X		X		X	X			1,100–1,650
Hyla xanthosticta													X			2,100
Hyla zeteki								X	X	X						680–1,804
Osteopilus septentrionalis			I													
Phrynohyas venulosa	X	X	X	X												2–285
Phyllomedusa lemur								X	X	X		X	X			440–1,600
Scinax boulengeri	X	X	X	X	X				X							1–700
Scinax elaeochroa		X	X	X	X			X	X	X						1–1,200
Scinax staufferi	X	X														4–550
Smilisca baudinii	X	X			X					X						1–750
Smilisca phaeota	G	X	X	X	X			X	X	X						2–1,116
Smilisca puma			X		X					X						15–520
Simisca sila		X		X												10–970
Smilisca sordida	G	X	X	X	X	X		X	X							1–1,525
Centrolenella ilex			X		X					X						250–900
Centrolenella prosoblepon		X	X	X	X	X		X	X	X		X	X			20–1,900
Cochranella albomaculata		X	X	X	X			X	X	X						20–1,500
Cochranella euknemos						X		X		X						840–1,500
Cochranella granulosa						X	X	X								40–1,500
Cochranella spinosa			X	X	X											20–560
Hyalinobatrachium chirripoi			X													60–100
Hyalinobatrachium colymbiphyllum		X	X	X	X	X				X		X	X			6–1,560
Hyalinobatrachium fleischmanni		X	X		X	X	X	X	X	X		X				103–1,730
Hyalinobatrachium pulveratum		X	X	X	X											2–560
Hyalinobatrachium talamancae										X						1116
Hyalinobatrachium valerioi		X	X	X	X			X		X						6–1,500
Hyalinobatrachium vireovittatum								X		X						880–1,100
Dendrobates auratus		X	X	X	X											2–601
Dendrobates granuliferus		X		X												20–100
Dendrobates pumilio			X		X											1–495
Phyllobates lugubris			X		X			X								20–601
Phyllobates vittatus		X		X												20–550
Gastrophryne pictiventris			X		X											1–500
Hypopachus variolosus	X					X	X	X				X				4–1,435
Nelsonophryne aterrima			X							X						20–1,350
Colostethus flotator		X	X	X						X						10–865
Colostethus nubicola								X		X						1,100–1,200
Colostethus talamancae		X	X	X	X				X	X						2–703
Rana forreri	X	X				X	X									12–1,311
Rana taylori					X	X	X		X	X		X				60–1,862
Rana vaillanti	X	X	X		X					X						1–700
Rana vibicaria												X	X	X	X	1,500–2,700
Rana warszewitschii	G	X	X	X	X	X	X	X	X	X		X	X			1–1,750
TOTALS 131 + 2*	18	44	65	42	60	19	12	41	56	79	0	27	38	1	5	

*Not included in totals: G = gallery forest; I = introduction.

Table 14.3. Distribution of Lizards by Plant Formation and Altitude

		LM		LW		PM		PW									Elevation
	D	P	A	P	A	P	A	P	A	PR	LMM	LMW	LMR	MW	MR	SPR	(m)
Basiliscus basiliscus	X	X		X		X		X									1–1,200
Basiliscus plumifrons			X		X				X								1–775
Basiliscus vittatus			X		X												1–120
Corytophanes cristatus		X	X	X	X			X	X	X							2–1,640
Ctenosaura quinquecarinata	X																1–145
Ctenosaura similis	X	X	X														1–176
Iguana iguana	G	X	X	X	X												1–500
Sceloporus malachiticus						X	X	X	X		X	X	X	X	X	X	146–3,800
Sceloporus squamosus	X																2–46
Sceloporus variabilis	X					X											1–800
Polychrus gutturosus		X	X	X	X	X											6–700
Ctenonotus cristatellus			I														2–750
Dactyloa chorcorum					X												300
Dactyloa frenata			X		X				X								2–819
Dactyloa insignis		X	X					X		X		X	X				4–1,600
Dactyloa microtus								X	X								425–1,500
Norops altae								X	X	X		X	X				1,100–2,000
Norops aquaticus			X					X		X							30–1,170
Norops biporcatus	G	X	X	X	X			X	X	X							5–1,220
Norops capito		X	X	X	X			X		X							2–1,230
Norops carpenteri			X		X												4–684
Norops cupreus	X			X		X	X	X	X								1–1,425
Norops fungosus										X							1,200–1,600
Norops humilis			X		X			X	X	X		X	X				2–1,600
Norops intermedius						X	X	X	X	X		X	X				1,160–1,950
Norops lemurinus		X	X	X	X				X	X							13–970
Norops limifrons		X	X	X	X	X	X	X	X	X							1–1,340
Norops oxylophus			X		X	G		X	X	X							20–1,200
Norops pachypus										X		X	X				1,500–2,400
Norops pandoensis													X				1,350–1,950
Norops pentaprion	X	X	X	X	X												2–900
Norops polylepis		X		X				X		X							1–1,300
Norops sericeus	X	X															2–90
Norops townsendi					X			IC		IC							1–634
Norops tropidolepis								X		X		X	X				1,220–2,506
Norops vociferans										X							1,700
Norops woodi						X		X		X		X	X				1,150–2,500
Coleonyx mitratus	X					X	X		X								10–1,435
Hemidactylus frenatus	I	I	I	I	I	I	I										4–1,231
Hemidactylus garnotii				I	I	I			I								2–1,160
Lepidodactylus lugubris				I													13–200
Lepidoblepharis xanthostigma	G	X	X	X	X			X		X							2–1,360
Phyllodactylus tuberculosus	X					X											2–1,160
Thecadactylus rapicauda		X	X	X	X												1–1,052
Gonatodes albogularis	X	X	X	X	X												2–530
Sphaerodactylus graptolaemus		X	X														5–700
Sphaerodactylus homolepis				X	X												2–600
Sphaerodactylus millepunctatus	G	X	X		X			X									37–657
Sphaerodactylus pacificus			IC														1–375
Lepidophyma flavimaculatum		X	X		X				X	X							2–690
Lepidophyma reticulatum		X		X		X		X									10–1,250
Eumeces managuae	X	X															40–530
Mabuya unimarginata	X	X	X	X	X	X		X		X		X					2–1,500
Sphenomorphus cherriei	G	X	X	X	X	X		X	X	X							2–1,170
Ameiva ameiva		X															33
Ameiva festiva		X	X	X	X			X	X	X							2–1,200
Ameiva leptophrys		X		X													2–703
Ameiva quadrilineata		X	X	X	X												0–1,050
Ameiva undulata	X	X				X		X									2–1,300
Cnemidophorus deppii	X	X															2–530
Anadia ocellata								X	X	X							520–1,200
Bachia blairi		X		X													2–40
Gymnophthalmus speciosus	X	X	X	X	X	X											2–1,222
Leposoma southi		X	X	X	X												2–703
Neusticurus apodemus			X							X							30–884
Ptychoglossus plicatus		X	X	X	X			X		X							35–1,540
Celestus cyanochloris						X		X		X			X				1,200–1,710
Celestus hylaius			X		X												4–430
Celestus orobius													X				1,500–2,000
Coloptychon rhombifer			X		X												50–500
Diploglossus bilobatus		X	X	X	X				X	X							83–1,360
Diploglossus monotropis			X	X													2–320
Mesaspis monticola													X	X	X	X	
TOTALS 69 + 4*	16	34	33	30	32	15	5	26	20	27	1	10	11	2	2	2	

IC = Isla del Coco.

*Not included in totals: G = gallery forest; I = introduction.

Table 14.4. Distribution of Costa Rican Snakes by Plant Formation and Altitude

	LD	LM P	LM A	LW P	LW A	PM P	PM A	PW P	PW A	PR	LMM	LMW	LMR	MW	MR	Elevation (m)
Anomalepis mexicanus		X														350–500
Helminthopis frontalis			X			X	X									96–1,435
Liotyphlops albirostris		X														—
Typhlops costaricense						X		X	X			X				540–1,500
Leptotyphlops ater	X					X										2–560
Loxocemus bicolor																20–100
Boa constrictor	X	X	X	X	X			X	X							2–1,360
Corallus annulatus		X			X											2–156
Corallus hortulanus				X	X											3–200
Epicrates cenchria	X	X	X	X	X											0–1,200
Ungaliophis panamensis		X	X	X	X								X			4–1,487
Amastridium veliferum		X	X	X	X				X	X						2–1,200
Chironius carinatus		X	X	X	X											2–200
Chironius exoletus			X		X	X	X	X	X	X		X	X			40–1,600
Chironius grandisquamus		X	X	X	X	X		X	X	X						2–1,600
Clelia clelia	X	X	X	X	X		X	X		X						2–913
Clelia scytalina			X		X							X	X			60–1,900
Coluber mentovarius	X					X	X	X								1–1,435
Coniophanes bipunctatus					X											10
Coniophanes fissidens		X	X	X	X	X			X							2–800
Coniophanes piceivittus	X					X	X									10–305
Conophis lineatus	X					X										2–1,160
Cristantophis nevermanni	X					X										4–1,385
Dendrophidion nuchale		X	X	X		X		X		X						75–1,220
Dendrophiodion paucicarinatum			X			X		X		X		X	X			1,040–1,500
Dendrophiodion percarinatum		X	X	X	X	X		X	X	X						4–1,200
Dendrophidion vinitor		X	X	X	X			X	X	X						15–1,360
Dipsas articulata	G	X		X												4–100
Dipsas bicolor		X		X						X						4–1,100
Dipsas tenuissima		X		X						X						4–970
Drymarchon corais	X	X	X	X		X	X	X	X	X		X				2–1,600
Drymobius margaritiferus	X	X	X		X	X	X	X	X			X				1–1,500
Drymobius melanotropis			X		X					X						35–1,000
Drymobius rhombifer		X	X	X	X			X	X	X						13–1,200
Elaphe triaspis	X			X	X	X		X		X		X				10–1,500
Enulius flavitorques	X				X											5–1,250
Enulius sclateri		X	X	X	X			X		X						4–700
Erythrolamprus bizona	X	X	X	X		X	X	X	X			X				8–1,450
Erythrolamprus mimus		X	X	X						X						1–1,200
Geophis brachycephalus				X	X	X	X	X	X	X						13–2,115
Geophis downsi										X						1,200
Geophis godmani								X				X	X			1,000–1,600
Geophis hoffmanni		X	X	X	X	X	X	X	X	X		X				18–1,670
Geophis ruthveni							X			X						851–1,360
Geophis talamancae													X			1,880
Geophis zeledoni												X	X			1,830–2,100
Hydromorphus concolor		X	X	X	X	X	X	X	X							1–1,500
Imantodes cenchoa		X	X	X	X	X	X	X	X	X		X	X			2–1,500
Imantodes gemnistratus	X	X	X	X	X	X	X	X		X						2–1,435
Imantodes inornatus		X	X	X	X			X		X		X				5–1,450
Lampropeltis triangulum	X	X	X		X	X	X	X	X	X		X	X			4–2,460
Leptodeira annulata	X	X	X	X	X	X			X							4–1,400
Leptodeira nigrofasciata	X															7–400
Leptodeira rubricata		MN		MN												0–5
Leptodeira septentrionalis	X	X	X	X	X	X		X	X	X						2–1,200
Leptodrymus pulcherrimus	X	X				X										2–850
Leptophis ahaetulla	G	X	X	X	X	X		X	X	X						1–1,400
Leptophis depressirostris			X	X	X			X	X	X						4–1,120
Leptophis mexicanus	X			X	X	X										4–1,160
Leptophis nebulosus			X	X		X		X		X		X	X			13–1,600
Leptophis riveti		X								X						4–1,200
Liophis epinephalus	G	X	X	X	X	X	X	X	X	X		X	X			2–2,100
Mastigodryas melanolomus	X	X	X	X	X	X		X		X	X	X				4–1,760
Ninia celata										X						750–1,360
Ninia maculata		X	X	X	X	X	X	X	X	X		X				36–1,800
Ninia psephota								X				X	X	X		1,300–2,771
Ninia sebae	G	X	X	X	X											4–1,100
Nothopsis rugosus		X	X	X	X				X	X						4–830

Table 14.4. continued

	LD	LM P	LM A	LW P	LW A	PM P	PM A	PW P	PW A	PR	LMM	LMW	LMR	MW	MR	Elevation (m)
Oxybelis aeneus	X	X		X		X		X	X							2–1,340
Oxybelis brevirostris			X		X	G		X	X	X						4–800
Oxybelis fulgidus	G					X		X				X	X			1–1,600
Oxyrhopus petolarius		X	X	X	X											5–700
Pseustes poecilonotus			X	X	X	X		X	X	X						2–1,320
Rhadinaea calligaster												X	X			1,220–2,439
Rhadinaea decorata			X	X	X			X		X						2–1,200
Rhadinaea godmani												X	X			1,200–2,600
Rhadinaea pulveriventris										X			X			1,372–1,600
Rhadinaea serperaster						X	X	X	X			X	X			1,220–2,200
Rhinobothryum bovallii			X		X											4–550
Scaphiodontophis annulatus	X	X	X	X	X			X		X						2–830
Scolecophis atrocinctus	X					X		X				X				100–1,530
Sibon annulatus			X		X				X	X		X				2–1,300
Sibon anthracops	X	X				X										30–915
Sibon argus			X		X											15–550
Sibon dimidiatus					X	X		X		X		X				40–1,600
Sibon longifrens					X		X	X		X						60–660
Sibon nebulatus	G	X	X	X	X	X			X	X						3–1,160
Spilotes pullatus	G	X	X	X	X	X	X	X	X							2–1,435
Stenorrhina degenhardtii		X	X	X	X		X	X	X	X						2–1,435
Stenorrhina freminvillii	X		X	X	X	X		X								3–1,435
Tantilla alticola	G	X	X					X	X	X						40–1,600
Tantilla armillata	X	X				X	X	X								4–1,435
Tantilla reticulata			X		X		X	X		X						10–1,345
Tantilla ruficeps		X	X	X	X	X		X	X	X		X				1–1,600
Tantilla schistosa			X	X	X			X	X	X						40–1,400
Tantilla supracincta		X	X	X	X			X	X			X				2–850
Tantilla vermiformis	X	X														40–520
Thamnophis marcianus			X		X							X				2–30
Thamnophis proximus	X					X	X	X	X							2–1,751
Tretanorhinus nigroluteus			X		X											0–100
Trimetopon gracile						X				X		X	X			1,300–2,286
Trimetopon pliolepis		X	X	X	X	X		X	X	X		X				60–1,600
Trimetopon simile			X						X				X			60–1,500
Trimetopon slevini										X						1,200–1,800
Trimetopon viquezi			X													62
Trimorphodon biscutatus	X	X														2–560
Tripanurgos compressus				X												0–80
Urotheca decipiens		X	X	X	X	X		X	X	X						15–1,500
Urotheca euryzona		X	X		X	X	X	X	X	X						2–1,460
Urotheca fulviceps				X												20–80
Urotheca guentheri					X	X	X	X		X	X		X			25–1,600
Urotheca myersi												X	X			1,500–2,255
Urotheca pachyrua			X		X							X				1–1,200
Xenodon rabdocephalus		X	X	X	X	X		X	X	X						4–1,360
Micrurus alleni		X	X	X	X	X			X							1–1,556
Micrurus clarki				X												5
Micrurus mipartitus			X			X	X	X	X	X						2–1,160
Micrurus nigrocinctus	X	X	X	X	X	X	X	X	X			X				1–1,500
Pelamis platurus	M	M		M												0
Agkistrodon bilineatus	X															20–285
Atropoides nummifer	G	X	X			X	X	X	X	X						50–1,400
Atropoides picadoi		X	X		X	X	X			X						15–1,750
Bothriechis lateralis						X		X		X		X	X			700–1,950
Bothriechis nigroviridis								X		X		X	X			700–2,400
Bothriechis schlegelii		X	X	X	X	X		X	X	X						5–1,530
Bothrops asper	G	X	X	X	X	X				X						1–1,200
Cerrophidion godmani								X				X	X		X	1,460–2,875
Crotalus durissus	X	X				X		X				X				1–1,796
Lachesis melanocephala		X		X		X		X				X				2–670
Lachesis stenophrys			X		X											2–670
Porthidium nasutum		X	X	X	X			X	X							2–900
Porthidium ophryomegas	X	X				X										7–1,095
Porthidium volcanicum		X														418–570
Totals 133*	33	65	73	60	75	60	26	66	52	63	1	38	24	1	1	

*Not included in individual totals: G = gallery forest; M = marine; MN = mangrove.

Table 14.5.   Distribution of Costa Rican Turtles and Crocodilians by Ocean or Plant formation and Altitude

| | MAR | | | LM | | LW | | PM | | PW | | Elevation |
	P	A	D	P	A	P	A	P	A	P	A	(m)
Turtles												
Kinosternon angustipons					X		X					2–260
Kinosternon leucostomum			X	X	X	X	X		X		X	1–1,200
Kinosternon scorpiodes			X	X		X		X	X			2–145
Dermochelys coriacea	X	X										—
Caretta caretta	X	X										—
Chelonia agassizii	X											—
Chelonia mydas		X										—
Eretmochelys imbricata	X	X										—
Lepidochelys olivacea	X											—
Chelydra serpentina			X	X	X	X	X	X			X	4–1,164
Chrysemys ornata												2–60
Rhinoclemmys annulata					X		X				X	2–920
Rhinoclemmys funerea					X		X					2–260
Rhinoclemmys pulcherrima			X	X		X		X				2–1,160
Totals 14	5	4	4	4	5	4	5	3	2	0	3	
Crocodilians												
Caiman crocodilus			X	X	X	X	X					1–200
Crocodylus acutus			X	X	X	X	X					0–200
Totals 2			2	2	2	2	2					

Table 14.6.   Percentage of Native Species of Each Major Taxonomic Group Living in Each Plant Formation

| | N | D | LM | | LW | | PM | | PW | | PR | LMM | LMW | LMR | MW | MR | SP |
			P	A	P	A	P	A	P	A							
Caecilians	4	0	75	25	50	50	50	25	75	25	50	0	25	25	0	0	0
Salamanders	37	0	14	17	11	22	5	3	17	11	64	0	11	41	12	14	0
Anurans	131	17	34	51	32	46	15	9	32	43	60	0	21	29	0.8	39	0
Lizards	69	23	49	48	45	45	21	7	36	29	39	1.4	7	16	3	3	3
Snakes	132	25	49	55	45	57	46	20	50	39	47	0.8	29	13	0.8	0.8	0
Turtles	8	44	44	55	4	55	22	22	0	33	17	0	0	0	0	0	0
Crocodilians	2	100	100	100	100	100	0	0	0	0	0	0	0	0	0	0	0
Total*	383	73	157	185	144	184	101	47	142	137	195	2	80	89	6	13	3

Exclusive of marine species.

*Not included in totals: Gallery forest; introductions; mangrove.

Table 14.7.   Percentage of Native Species by Major Taxonomic Group in Total Costa Rican Herpetofauna Living in Each Plant Formation

| | N | D | LM | | LW | | PM | | PW | | PR | LM | LM | LM | M | MR | SAR |
			P	A	P	A	P	A	P	A							
Caecilians	4	0	0.8	0.3	0.5	0.5	0.5	0.3	0.8	0.5	0.5	0	0.3	0.3	0	0	0
Salamanders	37	0	1.3	1.6	1.0	2	0.5	0.2	1.6	1	6	0	1	4	0.5	1.3	0.2
Anurans	131	5	12	17	11	16	5	3	11	15	21	0	7	10	0.2	1.3	0
Lizards	69	4	9	9	8	8	4	13	7	5	7	0.2	1.3	3	0.5	0.5	0.5
Snakes	132	9	17	19	16	19	16	7	16	14	16	0.2	10	6	0.2	0.2	0.2
Turtles	8	1	1	1.3	1	1.3	0.5	0.52	0	0.7	0	0	0	0	0	0	0
Crocodilians	2	0.5	0.5	0.5	0.5	0.5	0	0	0	0	0	0	0	0	0	0	0
Total*	383	73	157	185	144	184	101	47	142	137	195	2	80	89	6	13	3

Exclusive of marine species.

*Not included in percentages: Gallery forest; introductions; mangrove.

altitudinal zones show some interesting differences between the two groups and between Pacific and Atlantic versants (figs. 14.1 and 14.2). On these figures the zones are identified by the following abbreviations: Lowland (L), Premontane (P), Lower Montane (LM), Montane (M), and Subalpine (S), and the circumscribing biotemperatures (BT°) are indicated.

Analysis of the distributions of amphibians (fig. 14.1) shows that the number of species on the Pacific lowlands (60) is significantly lower than on the Atlantic (81). This may reflect the relative small number of amphibian species (18) in the subhumid northwestern lowland region. The number of species increases markedly in the premontane belt on both Pacific (121) and Atlantic (114) slopes and declines sharply

Table 14.8. Percentage of Native Species in Total Costa Rican Herpetofauna by Class Living in Major Marine and Terrestrial Habitats

	MAR			LM		LW		PM		PW		PR	LMM	LMW	LMR	MW	MR	SAR
	P	A	D	P	A	P	A	P	A	P	A							
Amphibians																		
N	0	0	18	52	72	48	70	23	14	50	62	105	0	32	54	3	10	1
% fauna in plant formation	0	0	25	33	39	33	38	23	30	35	45	54	0	40	61	50	77	33
% total fauna	0	0	5	13	18	12	18	6	4	13	16	27	0	8	14	0.7	3	0.2
Reptiles																		
N	6	4	55	105	113	96	114	78	33	92	75	90	2	48	35	3	3	2
% fauna in plant formation	100	10	75	67	61	67	62	77	70	65	55	46	100	60	40	50	23	67
% total fauna	1.6	1	14	26	29	25	29	20	9	24	19	23	0.5	12	9	0.7	0.7	0.5
N = 390																		
Totals	6	4	73	157	185	144	184	101	47	142	137	195	2	80	89	6	13	3
%	1.6	1	19	40	47	37	47	26	12	37	35	50	0.5	21	23	1.5	3	0.7

Including marine species.

Not included in totals: Gallery forest; introductions; mangrove.

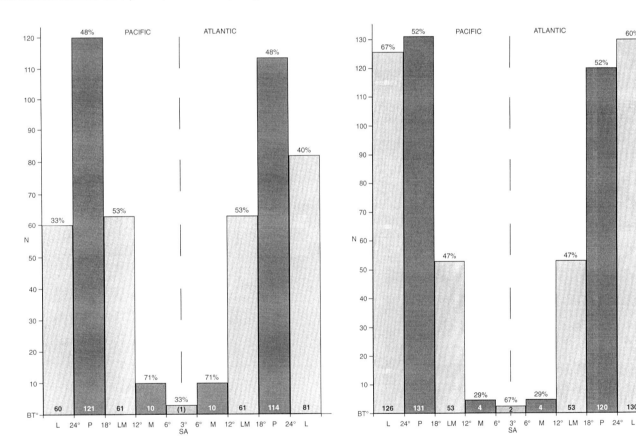

Figure 14.1. Distribution of amphibians by altitudinal zone. See text for abbreviations. Numbers of species and percentage of amphibians in total herpetofauna in each zone as delimited by BT°.

Figure 14.2. Distribution of reptiles by altitudinal zone. See text for abbreviations. Numbers of species and percentage of reptiles in total herpetofauna in each zone as delimited by BT°.

by 50% into the lower montane zones (61 species). Further reductions occur in the montane (10 species) and subalpine (1 species) zones. Species numbers are essentially symmetrical on Pacific and Atlantic slopes above the 500 to 600 m boundary between the lowland and premontane zones.

A similar analysis of the vertical distributions of reptiles (fig. 14.2) shows a different pattern. The number of species on the Pacific lowlands (126) is about the same as on the Atlantic lowlands (130). The premontane areas, unlike the situation with the amphibians, contain about the same number of species on the two versants (Pacific 131, Atlantic 120) but are about equal in number to the adjacent lowlands. Marked reductions in numbers occur between premontane and lower montane (53) zones and there are further reductions at lower biotemperatures and elevations (montane 4, subalpine 2). Species numbers are nearly symmetrical on the two versants at all elevations. The proportion of amphibians to reptiles is also symmetrical in all zones except for the lower percentage of amphibians to reptiles (33 to 67%) on the Pacific lowlands. Other proportions are premontane 48 to 52%, lower montane 53 to 47%, montane 71 to 29%, and subalpine 33 to 67%.

The peak diversity of the herpetofauna is in the premontane belt, which contains 65% of the total species complement for the country. This may be due in part to the overlap in vertical range between the upper limit of lowland species distributions and the lower limit of highland species ranges, but this zone is also home to a high number of endemics that do not range outside its limits.

Available Moisture

The second ecological variable that appears to be a major determinant of biodiversity is available moisture. Most studies express this factor in terms of mean annual precipitation as affected by differences in seasonality. As discussed

in chapter 2, the amount of rainfall has varying effects on the moisture available in an environment, depending on temperature parameters. One measure of environmental moisture availability is the relation between potential and actual evapotranspiration for a particular climatic regimen (fig. 2.5). For example, an annual precipitation of 1,500 mm in an area having a mean annual BT° = 22° will produce a PET ratio of 1.5. That is, the annual potential evapotranspiration exceeds the actual annual precipitation by 50%. However, in an area having a mean annual BT° = 9° and the same annual rainfall (1,500 mm), the PET ratio will be 0.35, meaning that the actual precipitation exceeds the potential evapotranspiration by 65%. Thus the first case produces a Lowland Dry Forest and the second a Montane Wet Forest.

Using this system, Costa Rica is represented by four humidity provinces (fig. 2.5) defined as follows with abbreviations to be used in figures 14.3 and 14.4:

Subhumid (SUH): PET ratio between 1 and 2
Humid (HUM): PET ratio between 0.5 and 1
Perhumid (PER): PET ratio between 0.25 and 0.5
Superhumid (SUP): PET ratio < 0.25.

The number of species found in each humidity province in Costa Rica is displayed graphically (figs. 14.3 and 14.4). Among amphibians, as might be predicted, the greatest number of species (122) are found in areas having superhumid

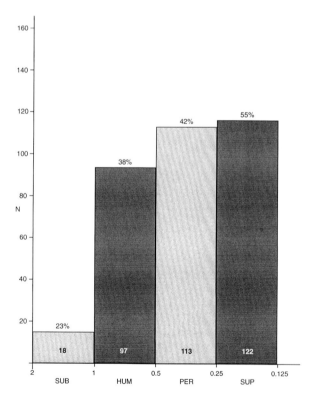

Figure 14.3. Distribution of amphibians by humidity province. See text for abbreviations. Numbers of species and percentage of amphibians in total herpetofauna in each zone as delimited by PET ratios.

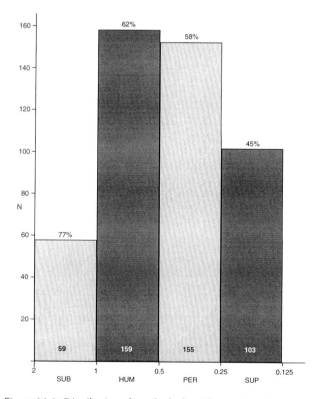

Figure 14.4. Distribution of reptiles by humidity province. See text for abbreviations. Numbers of species and percentage of reptiles in total herpetofauna in each zone as delimited by PET ratios.

conditions, where actual precipitation exceeds potential evapotranspiration by more than 75% annually. Nevertheless, there does not appear to be a significant difference in diversity between superhumid environments and perhumid ones (113 species). Lower amounts of available moisture produce marked reductions in numbers of species: 97 in humid situations and only 18 species in subhumid ones.

A similar analysis of reptile distributions by humidity province (fig. 14.2) reveals a different pattern. The highest reptile diversity (159 species) is in humid environments, but species diversity does not differ significantly from that in perhumid ones (155 species). There is a significant drop in species number in superhumid areas (103 species) and subhumid ones (59 species). Thus both high amounts of available moisture and relatively lower amounts are associated with reduced numbers of reptile species.

The relative proportions of amphibians to reptiles by humidity province reflects their different responses to drier or wetter conditions. Reptiles make up to 77% of the species assemblage in subhumid environments and 62% in humid ones. In superhumid environments amphibians make up 55% of the assemblage, and in perhumid ones reptiles predominate (58%). Given that there are substantially more reptiles than amphibians in the total Costa Rican herpetofauna, the latter figure probably represents a near balance of the two groups in the perhumid province. Basically, situations where precipitation greatly exceeds potential evapotranspiration support the greatest number of amphibians. Those with the PET ratio higher than 0.5 contain a disproportionate number of reptile species. Perhumid conditions are transitional, having about equal proportions of amphibians and reptiles, with a slight edge to the latter.

15 | Geographic Distribution
Historical Units, Faunal Areas, Endemism, and General Patterns

THE HERPETOFAUNA of Costa Rica lives between two continents and between two great seas that were connected by a broad seaway through most of Cenozoic times. The Isthmian Link between North and South America, formed mostly by present-day Costa Rica and Panama, has only recently been established to allow a mixing of major faunal elements from the two continents. Present patterns of distribution, however, are not simply products of the development of the recent land bridge, since a much earlier one connected the continents during the beginning of the Cenozoic, with effects still reflected in the composition of the isthmian herpetofauna.

For these reasons, any attempt to explain the biogeography of amphibians and reptiles in the region must recognize that the Costa Rican herpetofauna is embedded within a broader Tropical Mesoamerican faunal unit. This fauna formerly ranged much farther to the north but has become restricted to more southern latitudes by cooling and drying trends from the Oligocene onward and today is found in the tropical lowlands and premontane slopes from Tamaulipas and Sinaloa, Mexico, southward over southern Mexico and all of Central America. In addition, it mixes in a complex fashion with representatives of the northern or Nearctic herpetofauna on the slopes of the mountains bordering the central plateau of Mexico and with that of South America, principally in northern and northwestern areas of that continent.

COMPOSITION OF THE HERPETOFAUNA

The herpetofauna of the Tropical Mesoamerican unit comprises the following numbers of families and genera (tables 15.2, 15.3): Caecilians 1/4, salamanders 1/13, anurans 9/43, lizards 16/47, snakes 9/97, turtles 5/9, crocodilians 1/2; total 42/215. The numbers of families and genera of amphibians and reptiles living in Central America are caecilians 1/4, salamanders 1/8, anurans 9/38, lizards 14/41, snakes 11/79, turtles 5/9, crocodilians 1/2; total 40/181.

The genera of the Tropical Mesoamerican unit may be placed into one of four major groupings based on their distribution:

1. Widespread tropical: tropical genera found throughout the Middle and South American tropics with equally strong species differentiation in both regions

2. South American: genera with centers of distribution and differentiation in South America

3. Tropical Middle American: genera with centers of distribution and differentiation in tropical Mexico and Central America

4. Extratropical North American: genera with centers of distribution and differentiation in extratropical Mexico or the United States

A number of distinctive patterns of distribution within the four major groupings are evident and provide a basis for evaluating the composition of the Central American herpetofaunas as shown in table 15.1.

These data demonstrate that the Central American herpetofauna is composed primarily of genera with one of two major distribution patterns. The first (I) includes those genera with a tropical Middle American distribution that predominate on the lowlands and slopes in Mexico from Sinaloa and Tamaulipas southward and at all elevations east of the Isthmus of Tehuantepec to central Panama. The other (II) comprises genera with a South American distribution pattern and is most fully represented in the Central American region in Panama. Of the 215 genera in tropical Mesoamerica, 55% have their distribution centered there, 29% are South American, and 10% are extratropical (Nearctic) in pattern. North of Costa Rica only 21 genera (10%) are South American groups, while in eastern Panama 62 genera are South American (55% of the herpetofauna in the area). These data and the distribution of the 65 families of American amphibians and reptiles (table 15.2) are the basis (Savage 1966a, 1982) for recognizing the Middle American tropical assemblage as a biogeographic unit equivalent to the Nearctic and Neotropical units of traditional biogeography (Darlington 1957).

Within the tropical Middle America region seven major herpetofaunal assemblages may be recognized:

1. Eastern and Western Lowland Herpetofauna: a wide-ranging fauna, the most diverse and the richest in species composition in the region, found along the Atlantic lowlands from Tamaulipas, Mexico, to central Panama, with more or less isolated segments at moderate elevations along the Pacific slopes of Guatemala and in the Golfo Dulce region on the Pacific lowlands of southwestern Costa Rica and extreme western Panama.

Table 15.1. Distribution of Tropical Mesoamerican Genera of Living Amphibians and Reptiles

1. Widespread Tropical (8)

Eleutherodactylus	Mabuya
Bufo	Leptotyphlops
Phrynohyas	Micrurus
Hyla	Bothrops

2. South American (62)
   A. Northern Limit of Range in Panama (22)

Caecilia	Echinosaura
Protopipa	Prionodactylus
Rhamphophryne	Amphisbaena
Epipedobates	Trachyboa
Minyobates	Atractus
Chiasmocleis	Diaphorolepis
Elachistocleis	Phimophis
Hemiphractus	Pseudoboa
Pleurodema	Siphliophis
Morunasaurus	Bothriopsis
Enyalioides	Chelonoides

   B. Northern Limit of Range in Costa Rica (19)

Oscaecilia	Neusticurus
Gastrotheca	Anadia
Nelsonophryne	Anomalepis
Phyllobates	Helminthiophis
Colostethus	Liotyphlops
Phyllomedusa	Epicrates
Atelopus	Liophis
Dactyloa	Tripanurgos
Bachia	Ptychoglossus
Leposoma	

   C. Northern Limit of Range Between Costa Rica and Guatemala (9)

Centrolenella	Corallus
Cochranella	Erythrolamprus
Dendrobates	Nothopsis
Polychrus	Lachesis
Diploglossus*	

   D. Northern Limit of Range in Mexico (12)

Leptodactylus**	Gymnophthalmus
Physalaemus	Typhlops
Scinax	Clelia
Hyalinobatrachium	Oxyrhopus
Ameiva	Xenodon
Gonatodes	Caiman

3. Tropical Middle American (119)
   A. Endemics (44)

Gymnopis	Anotheca
Bradytriton	Abronia
Cryptotriton	Coloptychon
Dendrotriton	Loxocemus
Ixalotriton	Adelphicos
Nototriton	Amastridium
Nyctanolis	Chapinophis
Pseudoeurycea	Crisantophis
Atelophryniscus	Conophis
Crepidophryne	Hydromorphus
Duellmanohyla	Leptodrymus
Plectrohyla	Scolecophis
Ptychohyla	Symphimus
Triprion	Tantillita
Corytophanes	Trimetopon
Aristelliger*	Tropidodipsas
Laemanctus	Atropoides
Ctenosaura	Cerrophidium
Lepidophyma	Porthidium
Celestus*	Claudius
Mesaspis	Staurotypus
Xenosaurus	Dermatemys

B. Northern Limit of Range in Extratropical North America (19)

Rhinophrynus	Drymarchon
Hypopachus	Drymobius
Gastrophryne	Ficimia
Syrrhophus	Oxybelis
Coleonyx	Leptodeira
Phyllodactylus	Rhadinaea
Heloderma	Tantilla
Cnemidophorus	Trimorphodon
Gerrhonotus	Kinosternon
Coniophanes	

C. Southern Limit of Range in Northern and/or Northwestern South America (21)

Dermophis	Ninia
Oedipina	Rhinobothryum
Agalychnis	Scaphiodontophis
Smilisca	Sibon
Basiliscus	Stenorrhina
Lepidoblepharis	Tretanorhinus
Thecadactylus	Urotheca
Ungaliophis	Bothriechis
Dendrophidion	Rhinoclemmys
Enulius	Crocodylus
Geophis	

D. Southern Limit of Range in Amazon Basin or Farther South (12)

Bolitoglossa	Chironius
Norops	Imantodes
Iguana	Leptophis
Sphaerodactylus	Mastigodryas
Boa	Pseustes
Dipsas	Spilotes

E. Endemic Genera in Tropical to Subtropical Mexico (23)

Chiropterotriton	Adelophis
Lineatriton	Chersodromus
Parvimolge	Cryophis
Thorius	Conopsis
Hylactophryne***	Geagras
Pachymedusa	Manolepis
Pternohyla***	Pseudoficimia
Anelytropsis	Rhadinophanes
Barisia	Sympholis
Bipes	Tantalophis
Exiliboa	Toluca
Ophryacus	

4. Extratropical North American (34)
   A. Southern Limit of Range in Tropical Mexico (18)
      1) Southern Limit of Range in Central or Southern Mexico (10)

Scaphiopus	Gyalopion
Sisturus	Hypsiglena
Phrynosoma	Rhinocheilus
Urosaurus	Sonora
Ophisaurus	Salvadora

      2) Southern Limit of Range Marginally Tropical (8)****

Siren	Holbrookia
Notophthalmus	Anolis
Callisaurus	Arizona
Dipsosaurus	Micruroides

   B. Southern Limit of Range in Central America (10)

Terrapene	Storeria
Sceloporus	Nerodia
Eumeces	Elaphe
Sphenomorphus	Thamnophis
Pituophis	Agkistrodon

   C. Southern Limit of Range in South America (6)

Rana	Crotalus
Coluber	Chelydra
Lampropeltis	Chrysemys*****

Note: The lizard genus *Phyllodactylus* and the snake genera *Coniophanes* and *Drymobius* range south to western Peru; the lizard genus *Cnemidophorus* and the snake genera *Drymarchon*, *Leptodeira*, *Oxybelis*, and *Tantilla* range to the Amazon basin or farther south.

* Also in Antilles

** Also in Antilles and reaches southern United States.

*** Occurs in southern United States

**** The eight genera below are not treated further in this book and have been included here only for the sake of completeness.

***** Includes *Pseudemys* and *Trachemys*

Table 15.2. Distribution of New World Families
of Amphibians and Nonmarine Reptiles

Restricted to One Geographic Region (28)		
Nearctic (12)	Tropical Mesoamerica (6)	South America (10)
Cryptobranchidae	Rhinophrynidae	Rhinatrematidae
Sirenidae	Dibamidae	Typhlonectidae
Rhyacotritonidae	Xenosauridae	Pipidae *
Proteidae	Loxocemidae	Rhinodermatidae
Amphiumidae	Ungaliophiidae	Brachycephalidae
Dicamptodontidae	Dermatemydidae	Allophrynidae
Salamandridae		Pseudidae
Ambystomatidae		Aniliidae
Ascaphidae		Pelomedusidae
Pelobatidae		Chelidae
Crotaphytidae		
Trionychidae		

Occurring in Two Regions (16)		
Nearctic-Tropical Mesoamerica (5)	South America-Nearctic (1)	Tropical Mesoamerica-South America (10)
Phrynosomatidae	Testudinidae *	Caeciliidae
Eublepharidae		Centrolenidae
Bipedidae		Dendrobatidae
Xantusiidae		Corytophanidae
Helodermatidae		Hoplocercidae
		Gymnophthalmidae
		Amphisbaenidae
		Anomalepididae
		Typhlopidae
		Trophidophiidae *

Occurring in All Three Regions (21)		
Plethodontidae	Iguanidae	Elapidae
Leptodactylidae	Polychrotidae **	Viperidae
Bufonidae	Gekkonidae	Kinosternidae
Hylidae	Teiidae	Chelydridae
Michrohylidae	Scincidae	Emydidae
Ranidae	Anguidae	Crocodylidae
	Leptotyphlopidae	
	Boidae	
	Colubridae	

* Reaching Eastern Panama from South.

** One Nearctic species.

2. Pacific Lowland Herpetofauna: a fauna associated with semiarid to subhumid climatic conditions, ranging along the Pacific lowlands from northern Sinaloa in Mexico to the Golfo de Nicoya region and Meseta Central of Costa Rica; includes the subhumid and semiarid assemblages of Atlantic drainage valleys in Chiapas, Mexico, Guatemala, and Honduras and the uplands of Honduras and Nicaragua; characterized by a predominance of lizard and snake species and virtual absence of salamanders.

3. Mexican Highland Herpetofauna: an assemblage restricted to the Sierras of tropical Mexico.

4. Nuclear Highland Herpetofauna: an assemblage restricted to the cool, moist habitats of the Chiapas, Guatemala, and Honduras highlands.

5. Talamancan Herpetofauna: a fauna with a well-developed amphibian complement, occurring in the humid environments of highland Costa Rica and western Panama.

6. Panamanian Herpetofauna: a fauna associated with disjunct subhumid lowland habitats from eastern Panama, along the Pacific versant, to the Chiriquí region of western Panama; showing affinities to the herpetofaunas of northern lowland Colombia and Venezuela that are associated with subhumid to arid conditions along the Caribbean lowlands.

7. Chocoan Herpetofauna: a South American fauna, extremely rich in species composition, found along the Pacific lowlands from northern Ecuador through Colombia and entering eastern Panama, where it is found in the Darién region and along the Caribbean versant.

The tropical subhumid to semiarid areas of the northern Yucatán Peninsula (Lee 1996) are something of an enigma under this scheme, since they are characterized by a mixture of taxa having affinities to both faunas 1 and 2.

The approximate geographic limits of the assemblages are indicated in figure 15.1. The distribution by genus of all members of the Tropical Mesoamerican herpetofauna is presented in tables 15.3 to 15.7.

Table 15.8 shows the proportion of genera of each major group to the total Tropical Mesoamerican complement of genera in each assemblage and the proportion each makes up within a particular assemblage. The numbers of genera of caecilians, turtles, and crocodilians are so low as to have no significant effect on comparisons. However, note that turtles and crocodilians are absent from the highland faunas and caecilians are virtually absent.

Other comparisons are somewhat distorted by the unequal number of amphibian genera (60) in relation to reptile genera (155), with a heavy predominance of snake genera (97). Salamanders make their major contribution to the faunas of upland regions. Anurans and lizards contribute nearly equally to all assemblages except that the number of lizard genera is disproportionally greater in the Mexican Highland fauna, with the reverse most striking in the Talamancan Highland fauna. In most faunas the combined proportions of amphibian plus lizard genera are about equal to that of snake genera alone. The most striking exceptions are in the Pacific Lowland assemblage (amphibians plus lizards 39%, snakes 54%), the Mexican Highland assemblage (amphibians plus lizards 44%, snakes 56%), and the Talamancan Highland fauna (amphibians plus lizards 60%, snakes only 40%).

In terms of diversity by taxonomic class, amphibians reach their zenith in the Talamancan Highland fauna (19 genera, 47% of the assemblage) and reptiles in the humid East and West Lowland assemblage. However the Pacific Lowland, Mexican Highland, and Chocoan faunas also have very high values for the reptile contributions to their faunas, but many

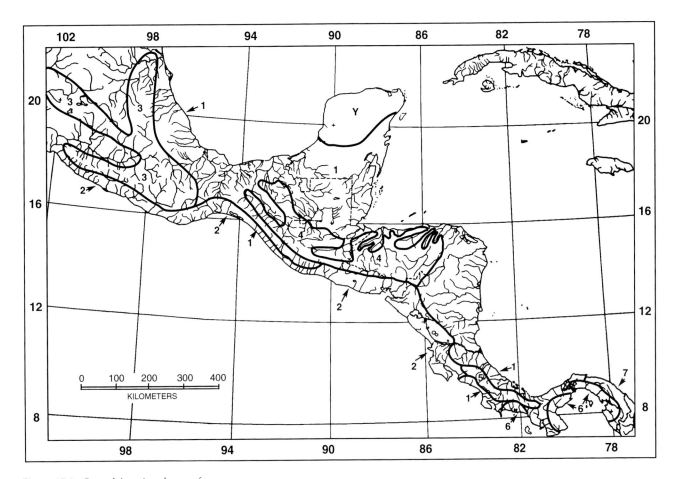

Figure 15.1. Central American herpetofaunas.

Table 15.3. Distribution of Caecilian and Salamander Genera in Tropical Mesoamerica

	Humid East and West Lowland	Pacific Lowland	Mexican Highland	Nuclear Highland	Talamancan Highland	Panamanian	Chocoan
Gymnophiona (4)							
Caecilia	X				X		X
Dermophis	X	X					X
Gymnopis	X						X
Oscaecilia	X					X	X
Total	4	1			1	1	4
Caudata (13)							
Bolitoglossa	X		X	X	X		X
Bradytriton				X			
Chiropterotriton			X				
Cryptotriton			X	X			
Dendrotriton				X			
Ixalotriton				X			
Lineatriton			X				
Nototriton					X		
Nyctanolis				X			
Oedipina	X	X		X	X		X
Parvimolge			X				
Pseudoeurycea			X				
Thorius			X				
Total	2	1	7	7	3	0	2

Table 15.4. Distribution of Anuran Genera in Tropical Mesoamerica

	Humid East and West Lowland	Pacific Lowland	Mexican Highland	Nuclear Highland	Talamancan Highland	Panamanian	Chocoan
Anura (43)							
Protopipa						X	
Rhinophrynus	X	X				X	X
Scaphiopus		X					
Eleutherodactylus	X	X	X	X	X		X
Hylactophryne			X				
Leptodactylus	X	X					
Physalaemus	X	X				X	X
Pleurodema						X	
Syrrhopus	X	X					
Atelophryniscus				X			
Atelopus	X				X		X
Bufo	X	X	X	X	X	X	X
Crepidophryne				X			
Rhamphophryne							X
Agalychnis	X			X	X	X	X
Anotheca	X		X				
Duellmanohyla			X	X	X		
Gastrotheca	X						X
Hemiphractus	X						X
Hyla	X	X	X	X	X	X	X
Pachymedusa		X				X	
Phrynohyas	X	X					
Phyllomedusa					X		X
Plectrohyla			X	X			
Pternohyla		X	X				
Ptychohyla			X	X			
Scinax	X	X				X	X
Smilisca	X	X				X	X
Triprion	X	X					
Phyllobates	X						
Centrolenella	X				X		X
Cochranella	X				X		X
Hyalinobatrachium	X				X		X
Colostethus	X				X		X
Dendrobates	X				X		X
Epipedobates					X		
Minyobates	X						X
Chiasmocleis	X						
Gastrophryne	X	X					
Elaschistocleis						X	
Hypopachus		X		X			
Nelsonophryne	X				X		
Rana	X	X	X	X	X	X	X
TOTAL	27	17	10	11	15	12	20

Table 15.5. Distribution of Lizard Genera in Tropical Mesoamerica

	Humid East and West Lowland	Pacific Lowland	Mexican Highland	Nuclear Highland	Talamancan Highland	Panamanian	Chocoan
Sauria (47)							
Basiliscus	X	X				X	X
Corytophanes	X						X
Laemanctus	X						
Ctenosaura	X	X				X	
Iguana	X	X				X	X
Morunasaurus	X						
Enyalioides							X
Dactyloa	X						X
Norops	X	X	X	X	X	X	X
Polychrus	X						X
Phyrnosoma		X	X				
Sceloporus		X	X	X	X		
Urosaurus		X					
Coleonyx	X	X					
Aristelliger	X						
Phyllodactylus	X	X					
Thecadactylus	X						X

Table 15.5. continued

	Humid East and West Lowland	Pacific Lowland	Mexican Highland	Nuclear Highland	Talamancan Highland	Panamanian	Chocoan
Gonatodes	X	X				X	X
Lepidoblepharis	X						X
Sphaerodactylus	X	X				X	X
Lepidophyma	X	X	X	X			
Ameiva	X	X		X		X	X
Cnemidophorus	X	X	X				
Anadia					X		X
Bachia	X					X	
Echinosaura	X						X
Gymnophthalmus	X	X				X	X
Leposoma	X						
Neusticurus	X						
Prionodactylus							X
Ptychoglossus	X						X
Amphibaenea	X					X	X
Bipes		X					
Eumeces	X	X	X	X			
Mabuya	X	X	X			X	X
Sphenomorphus	X		X	X			
Anelytropsis	X		X				
Celestus	X		X	X	X		
Diploglossus	X						X
Abronia			X	X			
Barisia			X				
Coloptychon	X						
Gerrhonotus			X				
Mesaspis			X	X	X		
Ophisaurus	X						
Xenosaurus			X	X			
Heloderma		X					
TOTAL	34	19	15	10	5	11	20

Table 15.6. Distribution of Snake Genera in Tropical Mesoamerica

	Humid East and West Lowland	Pacific Lowland	Mexican Highland	Nuclear Highland	Talamancan Highland	Panamanian	Chocoan
Serpentes (97)							
Anomalepis	X						
Helminthophis							X
Liotyphlops	X					X	
Typhlops	X		X	X	X		
Leptotyphlops	X	X	X	X		X	
Loxocemus		X					
Boa	X	X				X	X
Corallus	X					X	X
Epicrates	X	X				X	X
Trachyboa							X
Exiliboa			X				
Ungaliophis	X	X					
Adelophis		X					
Adelphicas				X			
Amastridium	X				X		
Atractus							X
Chapinophis				X			
Chersodromus			X				
Chironius	X					X	
Clelia	X	X		X	X		X
Coluber	X	X	X	X		X	
Coniophanes	X	X	X				X
Conophis		X					
Conopsis			X				
Crisantophis		X					
Cryophis			X				
Dendrophidion	X				X		X
Diaphorolepis							X
Dipsas	X	X					X
Drymarchon	X	X	X	X		X	X

continued

Table 15.6. continued

	Humid East and West Lowland	Pacific Lowland	Mexican Highland	Nuclear Highland	Talamancan Highland	Panamanian	Chocoan
Drymobius	X	X	X	X		X	X
Elaphe	X		X	X			
Enulius	X	X				X	X
Erythrolamprus	X	X					X
Ficimia	X	X					
Geargas	X						
Geophis	X	X	X	X	X		X
Gyalopion			X				
Hydromorphus	X				X		
Hypsiglena		X	X				
Imantodes	X	X				X	X
Lampropeltis	X	X	X	X	X	X	X
Leptodeira	X	X	X			X	X
Leptodrymus		X					
Leptophis	X	X	X	X	X		X
Liophis	X				X		X
Manolepis		X	X				
Mastigodryas	X	X		X		X	X
Nerodia	X						
Ninia	X	X	X	X	X		X
Nothopsis	X						X
Oxybelis	X	X	X	X	X	X	X
Oxyrhopus	X						X
Phimophis						X	
Pituophis			X	X			
Pseudoboa						X	X
Pseudoficimia		X	X				
Pseustes	X						X
Rhadinaea	X		X	X	X		X
Rhandinophanes			X				
Rhinobothryum	X						X
Rhinocheilus		X					
Salvadora		X	X				
Scaphiodontophis	X	X					X
Scolecophis		X					
Sibon	X	X				X	X
Siphlophis						X	X
Sonora		X	X				
Spilotes	X	X				X	X
Sternorrhina	X	X					X
Storeria	X		X				
Symphimus	X	X					
Sympholis	X	X					
Tantalophis		X	X				
Tantilla	X	X	X			X	X
Tantillita	X	X					
Thamnophis	X	X	X	X	X		
Toluca			X				
Tretanorhinus	X						
Trimetopon	X				X		
Trimorphodon	X	X	X	X			
Tripanurgos	X						X
Tropidodipsas	X	X	X	X			
Urotheca	X	X					X
Xenodon	X	X				X	X
Micrurus	X	X	X	X		X	X
Agkistrodon	X	X					
Atropoides	X	X			X		X
Bothriechis	X			X	X		
Bothriopsis		X					X
Bothrops	X					X	X
Cerrophidium			X	X	X		
Crotalus	X	X	X	X			
Lachesis	X						X
Ophryacus			X				
Porthidium	X		X			X	X
Sisturus			X				
Total	62	51	41	24	16	25	45

Table 15.7. Distribution of Turtle and Crocodilian Genera in Tropical Mesoamerica

	Humid East and West Lowland	Pacific Lowland	Mexican Highland	Nuclear Highland	Talamancan Highland	Panamanian	Chocoan
Testudinata (9)							
Claudius	X						
Kinosternon	X	X				X	X
Staurotypus	X	X					
Dermatemys	X						
Chelydra	X						X
Chrysemys	X	X				X	X
Rhinoclemmys	X	X					X
Terrapene	X						
Chelonoides						X	
TOTAL	8	4				3	4
Crocodilia (2)							
Caiman	X	X				X	X
Crocodylus	X	X				X	X
TOTAL	2	2				2	2

Table 15.8. Generic Composition of the Tropical Mesoamerican Herpetofaunas

	Humid East and West Lowland	Pacific Lowland	Mexican Highland	Nuclear Highland	Talamancan Highland	Panamanian	Chocoan
Caecilians (4)							
% total herpetofauna	2	<1	0	0	<1	<1	2
% assemblage	3	1	0	0	3	2	4
Salamanders (13)							
% total herpetofauna	<1	<1	3	3	1	0	<1
% assemblage	1	1	10	14	8	0	2
Anurans (43)							
% total herpetofauna	13	8	5	5	7	6	10
% assemblage	19	17	14	21	38	22	21
Lizards (47)							
% total herpetofauna	16	9	7	5	2	5	9
% assemblage	23	20	21	20	13	20	21
Snakes (97)							
% total herpetofauna	29	24	19	11	8	12	21
% assemblage	45	54	56	49	40	46	47
Turtles (9)							
% total herpetofauna	4	2	0	0	0	1.5	2
% assemblage	6	4	0	0	0	6	2
Crocodilians (2)							
% total herpetofauna	<1	<1	0	0	0	<1	<1
% assemblage	1	2	0	0	0	4	2
Total amphibians							
N (60)	33	19	17	18	19	13	26
% total herpetofauna	16	9	8	8	9	6	12
% assemblage	24	19	23	35	47	24	27
Total reptiles							
N (155)	106	76	56	34	21	41	71
% total herpetofauna	50	36	26	16	10	19	34
% assemblage	76	81	77	65	53	76	73
Grand total							
N (215)	139	95	73	52	40	54	97
% total herpetofauna	66	44	34	24	19	26	46

fewer genera. Amphibians contribute the least to the Pacific Lowland assemblage (19%) and reptiles the least to the Talamancan Highland fauna (40%).

The herpetofauna of Costa Rica is the meeting ground of three of these assemblages: The Eastern and Western Lowland fauna, the Pacific Lowland fauna, and the Talamancan fauna. The first ranges over the Atlantic lowlands and the Pacific lowlands of southwestern Costa Rica. The second is found on the lowlands of the northwestern region. The third lives in the Cordillera Central and the Cordillera de Talamanca. These major units have distributions broadly correlated with physiographic and climatic parameters and contain a mixture of genera with different histories. Therefore comparisons of the composition of these contemporary units cannot by themselves elucidate the process by which they were assembled over time.

GENERALIZED TRACKS AND HISTORICAL SOURCE UNITS

The raw data of historical biogeography are the distributions (or tracks) of individual species in space (geographical ecology) and time. Because each species has its own set of peculiar ecological requirements and its own unique evolutionary history, each species has a discrete nonrandom ecogeographic distribution. As a consequence, no species is universally present, and many species have very small or unique tracks.

The first level of generalization in biogeography is based on the recognition that in spite of the unique nature of individual species distributions, many individual tracks are concordant and show a common pattern. Determining the patterns (generalized tracks) involving the coincident distribution of many species or several monophyletic groups (genera, families, etc.) of species is the fundamental first step in biogeographic analysis.

The second level of generalization is to recognize the several disjunct adjoining or distant clusters of distributions that form nodes or track components along the generalized track. These components may be regarded as defining the geographic limits of major modern biotas, characterized by a high degree of endemism.

A third level of generalization attempts to identify the historical source units (ancestral biotas) that contributed to the modern biotas. In any given region, the biota may have been derived from several historical source units at different times, but usually the dominant source unit has developed in situ and is a component of a major generalized track.

In my 1982 paper on the biogeography of Central American herpetofauna I was able to discern three major general tracks for tropical Mesoamerica based on track analysis:

1. The North American–Central American track is a generalized track that includes North America, the Mexican lowlands and montane uplands, Central America, and the Greater Antilles (fig. 15.2a). South American portions of this track extend to Ecuador and Argentina but represent dispersal after the reconnection of Central and South America in the Tertiary.
2. The South American–Caribbean track is a generalized track including South America, the Greater and Lesser Antilles, and the Bahamas (fig. 15.2b). Mexican and Central American portions of this track represent dispersal from South America after establishment of the Isthmian Link in the Pliocene.
3. The Middle American–Caribbean track is a generalized track including the lowlands of Mexico, Central America and the Greater Antilles, and the Bahamas (fig. 15.2c). The portions of this track that extend to Ecuador and southern Brazil represent post-Miocene dispersal across the Isthmian Link.

A fourth track, the Western North American–Central American track, is a generalized track including western North America, Mexico, and Central America north of Panama (fig. 15.2d). A portion of this track, extending into South America, represents the dispersal of two genera (Cnemidophorus and Crotalus) across the Isthmian Link in late Cenozoic, followed by differentiation into a few species each. This track is represented by only a few taxa of reptiles in the tropical Mesoamerican region.

Each of these tracks is characterized by two features. First, they are composed of a string of components (areas of endemism) that represent a once more wide-ranging generalized fauna that has been fragmented by vicariance events and differentiated in situ. Second, they also reflect the recent emergence of the isthmian land connection between Central and South America that provided a corridor for concordant dispersal of many taxa from south to north and north to south in what has been called the great American biotic interchange (Marshall et al. 1982; Stehli and Webb 1985).

An analysis of distributional data, geologic, climatological, and vegetational correlates and changes, together with an assessment of phylogenetic relationships, led me (Savage 1982) to conclude that the four general tracks described above correspond to four historical herpetofaunal source units whose taxa have had an ancient and continuing association with one another. Genera and a few subgeneric groups whose distributions coincide with a particular track were grouped together as a primary historical unit or element. In the present context the following three elements are recognized as having made significant contributions to the Tropical Mesoamerican herpetofauna:

Old Northern Element: derivative stocks of originally extratropical (subtropical–warm temperate) groups distributed more or less continuously and circumpolarly in early Tertiary but forced southward and fragmented into several more or less disjunct components as a result of increased cooling and aridity trends and mountain building in the late Cenozoic. This unit comprises taxa having long-term Laurasian affinities. Typical members of this element, including the "hanging" Middle American relicts, the frog family Rhinophrynidae, the turtle family Dermatemydidae, and the lizard families Xantusiidae, Xenosauridae, and Helodermatidae, were widespread over much of North America to 40° N in the early Tertiary. As I pointed out in 1966 and as was confirmed by Rosen (1978), the Central American component of this stock has been disjunct from other components for most of later Tertiary and Quaternary times and evolved in situ in Middle America.

South American Element: derivatives of a generalized tropical American biota that evolved in situ in isolation in South America during most of the Cenozoic and must be considered a recent contributor to Middle American faunal diversity. The affinities of this unit are Gondwanan.

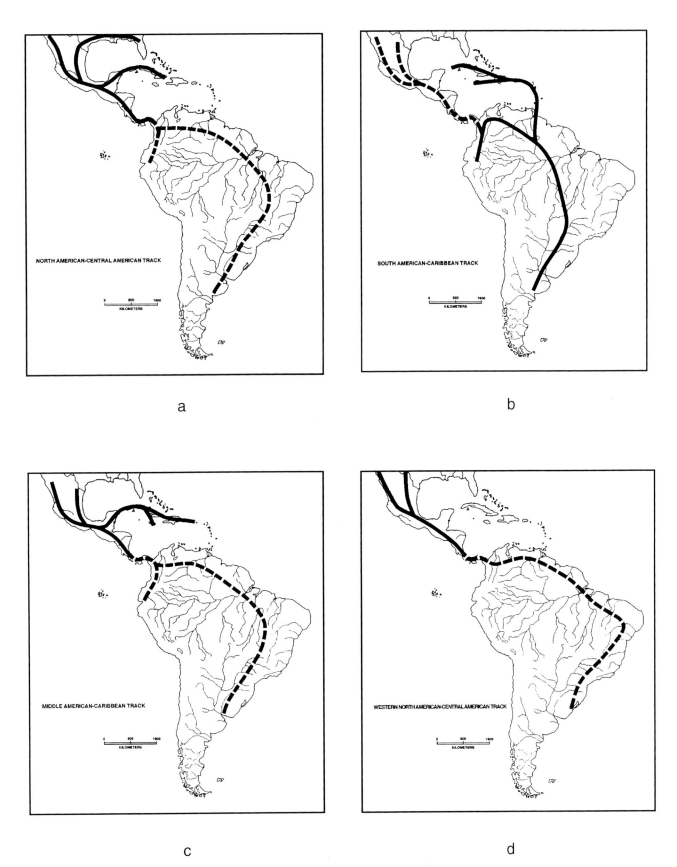

Figure 15.2. General tracks (historical source units) in the Tropical Mesoamerican region: (a) North American–Central American; (b) South American–Caribbean; (c) Middle American–Caribbean; (d) Western North American–Central American.

Table 15.9. Component Genera of Principal Historical Units of the Tropical Mesoamerican Herpetofauna

Old Northern (90)	Middle American (66)	South American (64)
Salamanders (13)	Caecilians (2)	Caecilians (2)
Bolitoglossa	*Dermophis*	*Caecilia*
Bradytriton	*Gymnopis*	*Oscaecilia*
Chiropterotriton		
Cryptotriton		
Dendotriton		
Ixalotriton		
Lineatriton		
Notrotriton		
Nyctanolis		
Oedipina		
Paravimolge		
Pseudoeurycea		
Thorius		
Frogs and Toads (3)	Frogs and Toads (18)	Frogs and Toads (24)
Rhinophrynus	*Atelophrynisus*	*Protopipa*
Scaphiopus	*Bufo* (pt.)	*Atelopus*
Rana	*Crepidophryne*	*Bufo* (pt.)
	Eleutherodactylus	*Rhamphophryne*
	(*Craugaster*)	*Eleutherodactylus*
	Hylactophyrne	*Leptodactylus*
	Syrrhophus	*Physalaemus*
	Agalychnis	*Pleurodema*
	Pachymedusa	*Gastrotheca*
	Anotheca	*Hemiphractus*
	Duellmanohyla	*Hyla* (pt.)
	Hyla (pt.)	*Phrynohyas*
	Plectrohyla	*Phyllomedusa*
	Pternohyla	*Scinax*
	Ptychohyla	*Centrolenella*
	Triprion	*Cochranella*
	Gastrophryne	*Hyalinbatrachium*
	Hypopachus	*Dendrobates*
	Nelsonophryne	*Epipedabates*
		Colostethus
		Minyobates
		Phyllobates
		Chiasmocleis
		Elachistocleis
Lizards (19)	Lizards (11)	Lizards (18)
Phrynosoma	*Basiliscus*	*Dactyloa*
Sceloporus	*Corytophanes*	*Enyalioides*
Urosaurus	*Ctenosaura*	*Morunasaurus*
Coleonyx	*Iguana*	*Polychrus*
Lepidophyma	*Laemanctus*	*Lepidoblepharis*
Cnemidophorus	*Norops*	*Phyllodactylus* (pt.)
Bipes	*Phyllodactylus* (pt.)	*Thecadactylus*
Eumeces	*Gonatodes*	*Amevia*
Mabuya	*Sphaerodactylus*	*Anadia*
Sphenomorphus	*Aristelliger*	*Bachia*
Anelytropsis	*Celestus*	*Echinosaura*
Abronia		*Gymnophthalmus*
Barisia		*Leposoma*
Coloptychon		*Neusticurus*
Gerrhonotus		*Ptychoglossus*
Mesaspis		*Prionodactylus*
Ophisaurus		*Amphisbaena*
Heloderma		*Diploglossus*
Xenosaurus		
Snakes (47)	Snakes (34)	Snakes (18)
Leptotyphlops (pt.)	*Boa*	*Anomalepis*
Loxocemus	*Exiliboa*	*Helminthophis*
Adelophis	*Ungaliophis*	*Liotyphlops*
Chironius	*Adelphicos*	*Typhlops*
Coluber	*Amastridium*	*Leptotyphlops* (pt.)
Conopsis	*Atractus*	*Corallus*
Dendrophidion	*Chapinophis*	*Epicrates*

Table 15.9.   continued

Old Northern (90)	Middle American (66)	South American (64)
Drymarchon	*Chersodromus*	*Trachyboa*
Drymobius	*Coniophanes*	*Clelia*
Elaphe	*Conophis*	*Erythrolamprus*
Ficimia	*Crisantophis*	*Liophis*
Gyalopion	*Cryophis*	*Oxyrhopus*
Lampropeltis	*Diaphorolepis*	*Phimophis*
Leptodrymus	*Dipsas*	*Pseudoboa*
Leptophis	*Enulius*	*Siphlophis*
Mastigodryas	*Geagras*	*Tripanurgos*
Nerodia	*Geophis*	*Xenodon*
Oxybelis	*Hydromorphus*	*Micrurus* (pt.)
Pituophis	*Hypsiglena*	
Pseudoficimia	*Imantodes*	
Pseustes	*Leptodeira*	
Rhinobothryum	*Manolepis*	
Rhinocheilus	*Ninia*	
Salvadora	*Nothopsis*	
Scaphiodontophis	*Rhadinaea*	
Sonora	*Rhadinophanes*	
Spilotes	*Scolecophis*	
Stenorrhina	*Sibon*	
Storeria	*Tantalophis*	
Symphimus	*Tretanorhinus*	
Sympholis	*Trimetopon*	
Tantilla	*Tropidodipsas*	
Tantillita	*Urotheca*	
Thamnophis	*Micrurus* (pt.)	
Toluca		
Trimorphodon		
Agkistrodon		
Atropoides		
Bothriechis		
Bothriopsis		
Bothrops		
Cerrophidium		
Crotalus		
Lachesis		
Ophryacus		
Porthidium		
Sisturus		
Turtles (8)		Turtles (1)
Claudius		*Chelonoides*
Kinosternon		
Staurotypus		
Dermatemys		
Chelydra		
Chrysemys		
Rhinoclemmys		
Terrapene		
	Crocodilians (1)	Crocodilians (1)
	Crocodylus	*Caiman*

Middle American Element: derivative groups of a generalized tropical American biota isolated in tropical North and Central America during most of the Cenozoic; developed in situ north of the Panamanian Portal and restricted by mountain building and climatic change in the late Cenozoic to Middle America. This element comprises genera that are primarily tropical Mesoamerican in distribution and have their closest allies either in the region or in South America but are mostly endemic to Central America and Mexico. Available evidence indicates that members of this unit or their ancestors had a more extensive range in North America in the early Tertiary, when humid warm climates occurred as far north as the region of what is now Montana, Wyoming, Utah, Colorado, and the Dakotas but became restricted southward to tropical Mesoamerica by climatic change in the late Cenozoic.

These units correspond to those I discussed earlier (Savage 1982), but with substantial revision in generic content (table 15.9) based on the most recent findings on phylogenetic relationships, especially for "iguanid" lizards (Frost and Etheridge 1989) and colubrid snakes (Jenner 1981;

Cadle 1984c, 1987; Zaher 1999). As well as more clearly defining the several elements, these studies have led me to abandon the idea of a separate Young Northern Element. The taxa formerly placed in that unit have proved to be a subset of the Old Northern Element that developed in response to the challenge of physiographic and climatic revolution in the middle latitudes of western North America and Mexico from the Oligocene to the present.

As now understood, the Old Northern Element is represented by four principal components in the Americas: Eastern North American, Western North American, Southwestern North American, and Central American. The Central American Component consists of derivatives of the Old Northern Element taxa that became associated with tropical conditions and isolated from the ancestral unit by cooling and drying trends from the Oligocene onward. Members of this component evolved in situ in Mesoamerica with the autochthonous Middle American Element during the rest of the Cenozoic.

The Southwestern North American Component contains most of the genera previously placed in the Young Northern Element (Savage 1966a, 1982). These genera are derivatives of Old Northern lineages that evolved in the southwest as a result of the orogenic and climatic drying trends in post-Oligocene times that produced the semiarid to desert environments of the region. Only a few tropical Mesoamerican taxa are distributionally associated with this component.

GEOGRAPHIC PATTERNS IN COSTA RICA

As I pointed out earlier (Savage 1975) and as was confirmed by Lynch and Duellman (1997), the Holdridge (1967) system is too sophisticated for sketching broad geographic correlates of animal distribution. Although this system, as discussed in chapter 2, is without peer in defining major tropical bioclimates, most species of amphibians and reptiles tend to occur over wider geographic areas that usually encompass two or more bioclimates/plant formations (tables 14.1 to 14.5).

A preliminary analysis of geographic distribution for Costa Rica, influenced in part by the unpublished manuscripts of Norman J. Scott Jr. (1969) for snakes and Marvalee H. Wake (1964) for lizards, identified nine putative geographic areas as high centers of diversity differing from the others in herpetofaunal composition. These faunal areas and their humidity province(s), representative plant formations, and approximate altitudinal limits are listed below:

Lowlands
> Northwest Pacific (NW): subhumid: Dry Forest; 0 to 600 m.
> Southwest Pacific (SW): humid and perhumid: Lowland Moist and Wet Forests; 0 to 600 m

Atlantic (A): humid and perhumid: Lowland Moist and Wet Forests; 0 to 500 m
Foothills and uplands
> Pacific Slope (PS): perhumid and superhumid: Premontane Wet and Rainforests; 600 to 1,600 m
> Meseta Central Occidental (MOC): humid: Premontane Moist Forest; 600 to 1,500 m
> Meseta Central Oriental (MOR): humid: Premontane Moist Forest; 500 to 1,500 m
> Atlantic Slope (AS): perhumid and superhumid: Premontane Wet Forest and Rainforest; 500 to 1,500 m
Highlands
> Cordillera Central (CC): humid, perhumid, and superhumid: Lower Montane and Montane Moist and Wet Forest and Rainforest; about 1,500 to 3,343 m
> Cordillera de Talamanca (CT): humid, perhumid, and superhumid: Montane Moist and Wet Forests and Rainforest and Subalpine Pluvial Paramo; about 1,500 to 3,840 m

These geographic units are mapped (fig. 15.3) and are used as the basis for evaluating the history of the Costa Rican herpetofauna. The abbreviations listed above are used throughout on appropriate tables and figures. Table 15.10 summarize species distributional data by major systematic group from tables 14.1 to 14.6 for each faunal area.

Not surprisingly, the distribution patterns by faunal area (figs. 15.4 and 15.5) reflect the same ecological trends discussed in chapter 14. The greatest overall diversity is found in humid lowland and slope areas. The number of reptile species in the subhumid Northwest Pacific area is three times that for amphibians. The number of amphibians spe-

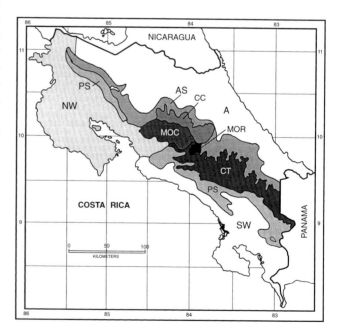

Figure 15.3. Putative faunal areas for the Costa Rican herpetofauna.

Table 15.10.   Number of Species by Major Taxonomic Group and Percentage Contribution to Each Faunal Area

		NW	SW	A	PS	MOC	MOR	AS	CC	CT
Caecilians	N	0	4	2	3	2	1	3	0	1
	%	0	2	1	2	2	2	2	0	1
Salamanders	N	0	5	9	19	2	1	24	5	10
	%	0	2	4	10	2	2	13	7	15
Anurans	N	18	47	73	66	19	12	63	36	30
	%	25	27	36	35	19	25	33	47	45
Subtotal	N	18	57	84	88	23	14	90	41	40
	%	25	33	41	46	24	30	48	53	60
Lizards	N	16	41	34	29	14	5	26	9	7
	%	22	24	17	15	14	11	14	12	10
Snakes	N	33	68	91	75	59	26	70	26	20
	%	45	40	39	39	60	55	37	34	30
Turtles	N	4	4	5	0	2	2	3	0	0
	%	5	2	2	0	2	4	2	0	0
Crocodilians	N	2	2	2	0	0	0	0	0	0
	%	3	1	1	0	0	0	0	0	0
Subtotal	N	55	115	122	104	75	33	99	35	27
	%	75	67	59	54	76	70	52	46	40
Grand total	N	73	172	206	192	98	47	189	76	67

cies exceeds that for reptiles at higher elevations (Cordillera Central and Cordillera de Talamanca areas).

In summary (table 15.10), the total number of species decreases along gradients from humid to drier conditions (e.g., SW to NW areas) and generally from warmer to cooler situations (e.g., AS to CC). The relative proportions change from a predominance of reptiles to amphibians from wet-ter to drier conditions (e.g., SW, 67:33 to NW, 75:25) and to cooler and wetter situations (e.g., AS, 52:48 to CT, 40:60). Because species distributions so strongly correlate with current ecological parameters, another approach seems required to analyze historical patterns.

As a consequence I have chosen to emphasize the contri-bution of genera and a few subgeneric groups to each of the

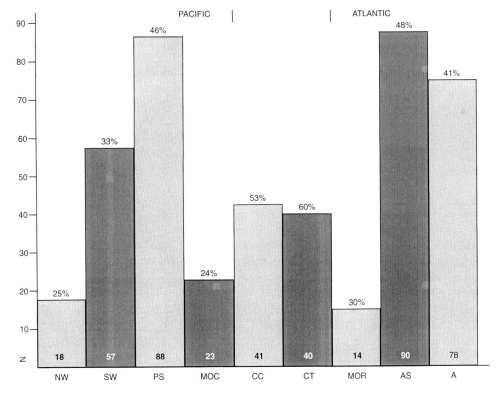

Figure 15.4. Composition by number of species of amphibians in each faunal area; % indicates contribu-tion of amphibians to total herpetofauna of each area.

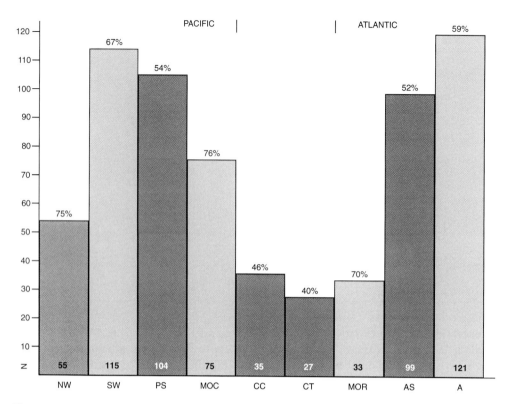

Figure 15.5. Composition by number of species of reptiles in each faunal area; % indicates contribution of reptiles to total herpetofauna of each area.

Table 15.11. Distinctive Mesoamerican Clades of the Anuran Genera *Bufo* and *Hyla*

Middle American Element	South American Element
Bufo fastidiosus group	*Bufo guttatus* group
Bufo periglenes group	*Bufo marinus* group
Bufo valliceps group	*Hyla albomarginata* group
Hyla godmani group	*Hyla boans* group
Hyla lancasteri group	*Hyla leucophyllata* group
Hyla pictipes group	*Hyla tuberculosa* group
Hyla pseudopuma group	
Hyla salvadorensis group	
Hyla zeteki group	

faunal areas as a method of establishing the broad historical picture. To carry out this analysis I have plotted the occurrence of each Costa Rica generic or subgeneric group in the faunal areas by their historical source units (elements and components) described earlier in this chapter.

Several currently recognized amphibian genera contain distinctive clades that are assigned to different historical units for this analysis. These include the Middle American *(Craugaster)* and South American (sensu stricto) divisions of the genus *Eleutherodactylus* and Central (*palmipes* group) and Eastern North American (*pipiens* group) components of the genus *Rana*. In addition, the species groups shown in table 15.11 within the composite genera *Bufo* and *Hyla* are treated as separate units.

The distribution of each taxonomic unit (genus or species group) by faunal area is presented in tables 15.12 to 15.15. Table 15.16 summarizes the data for number of genera and their proportional contribution to each faunal area by historical source unit. The contributions of Eastern (E) and Southwestern (SW) Components of the Old Northern Element to the Costa Rican herpetofauna are minimal, so further analysis will emphasize the Central American Component (CA) of the Old Northern Element and Middle American (MA) and South American (SA) Elements. Abbreviations listed above are used for the historical elements and components in table 15.16 and subsequent tables and figures.

The relative proportions (as percentages) of these three units are displayed diagrammatically in figure 15.6. The relative proportions of the predominant historical unit(s) in each area will be used as the basis for determining their historical characteristics and relationships in the following paragraphs.

These data indicate a distinctly different history for the Northwestern faunal area compared with other lowland areas. The Northwestern area is characterized by Central and Middle American genera in about equal proportions, 36:34%, respectively. The Southwestern and Atlantic areas are characterized by similar proportions of South American and Middle American taxa, 37:33% and 31:37%, respectively, with South American ones in slightly greater numbers

Table 15.12. Distribution of Genera and Species Groups of the Central American Component (Old Northern Element) by Faunal Area

	NW	SW	A	PS	MOC	MOR	AS	CC	CT
Salamanders (3)									
Bolitoglossa		X	X	X			X	X	X
Nototriton					X	X	X	X	X
Oedipina		X	X	X	X	X	X	X	X
Subtotal	0	2	2	2	2	2	3	3	3
Anurans (2)									
Rhinophrynus	X		X						
Rana palmipes group	X	X	X	X	X	X	X	X	X
Subtotal	2	1	2	1	1	1	1	1	1
Total amphibians (5)	2	3	4	3	3	3	4	4	4
Lizards (7)									
Coleonyx	X	X		X	X	X			
Lepidophyma		X	X	X	X	X			
Eumeces	X								
Mabuya	X	X	X	X			X		
Sphenomorphus		X	X	X			X		
Coloptychon		X							
Mesaspis								X	X
Subtotal	3	5	3	4	2	2	2	1	1
Snakes (25)									
Loxocemus	X								
Chironius		X	X	X	X		X		
Dendrophidion		X	X	X			X		
Drymarchon	X	X	X	X	X	X	X		
Drymobius	X	X	X	X	X	X	X		
Elaphe	X			X	X		X		
Lampropeltis	X	X	X	X		X	X	X	X
Leptodrymus	X				X				
Leptophis	X	X	X	X	X		X	X	X
Mastigodryas	X	X	X	X	X		X	X	
Oxybelis	X	X	X	X	X		X		
Pseustes		X	X	X	X		X		
Rhinobothryium			X				X		
Scaphiodontophis		X	X	X					
Scolecophis	X				X				
Spilotes	X	X	X	X	X	X	X		
Stenorrhina		X	X	X	X	X	X		
Tantilla	X	X	X	X	X	X	X		
Trimorphodon	X								
Atropoides		X	X	X	X		X	X	X
Bothriechis		X	X	X	X	X	X	X	X
Bothrops	X	X	X	X	X	X	X		
Cerrophidium								X	X
Lachesis		X	X	X					
Porthidium	X	X	X	X			X		
Subtotal	15	18	19	19	17	8	18	6	5
Turtles (2)									
Kinosternon	X	X	X	X	X	X	X		
Rhinoclemmys	X		X		X	X			
Subtotal	2	1	2	1	2	2	1	0	0
Total reptiles (34)	20	24	24	24	21	12	21	6	5
Grand total (39)	22	27	26	27	24	15	25	10	9

in the southwest and Middle American ones on the Atlantic. These latter differences may be trivial.

The highland areas are dominated by Middle American taxa (CC = 38%; CT = 43%), with a stronger representation of South American genera (28%) in the Cordillera de Talamanca and of Central American genera (29%) in the Cordillera Central. The two slope faunal areas are similar in having Middle American dominance (PS = 36%; AS = 41%), with considerable Central American influence (PS = 32%; AS = 30%).

The Meseta Central Occidental area is unique in having a high proportion of Central American (40%) to Middle

Table 15.13.   Distribution of Genera and Species Group of Eastern and Southwestern North American Components (Old Northern Element) by Faunal Areas

	NW	SW	A	PS	MOC	MOR	AS	CC	CT
Eastern North American									
Anurans (1)									
Rana pipiens group	X		X	X	X	X	X	X	X
Subtotal	1	0	1	1	1	1	1	1	1
Total amphibians (1)	1	0	1	1	1	1	1	1	1
Snakes (3)									
Coluber	X								
Thamnophis	X		X		X	X		X	
Agkistrodon	X								
Subtotal	3	0	1	0	1	1	0	1	0
Turtles (2)									
Chelydra	X	X	X	X	X		X		
Chrysemys		X	X						
Subtotal	1	2	2	1	2	0	1	0	0
Total reptiles (5)	4	2	3	1	2	1	1	1	0
Grand total (6)	5	2	4	2	3	2	2	2	1
Southwestern America									
Lizards (2)									
Sceloporus	X			X	X	X	X	X	X
Cnemidophorus	X				X				
Subtotal	2	0	0	1	2	1	1	1	1
Snakes (1)									
Crotalus	X			X	X	X			
Subtotal	1	0	0	1	1	1	0	0	0
Total reptiles (3)	3	0	0	2	3	1	1	1	1
Grand total (3)	3	0	0	2	3	2	1	1	1

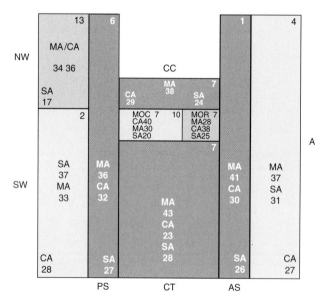

Figure 15.6. Diagram of composition of Costa Rican herpetofauna by percentage of genera of each major historical unit in faunal area; values in upper right hand corner are for E and SW Components (Old Northern Element).

American (30%) and South American genera (20%). The Meseta Central Oriental area is nearly identical to the Meseta Central Occidental area in faunal percentages: MA 28: CA 38: SA 25 and MA 30: CA 40: SA 20, respectively.

ENDEMISM

Another parameter of history is the degree of endemism each faunal area exhibits. A high level of endemism suggests a longer period of isolation for an area than for others with lower levels. Unfortunately the areas of endemism for lower Central America know no political boundaries, so that many species restricted in distribution to one faunal area in Costa Rica may extend into adjacent Panama or Nicaragua or farther to the north.

As I pointed out in my earlier paper (Savage 1982), the Northwest Pacific faunal area is the southern portion of a lowland faunal unit that includes lowland areas in western Nicaragua, Honduras, and El Salvador. The Southwest Pacific faunal area is the greater part of an area that includes the lowlands of extreme western Chiriquí Province, Panama. The Atlantic Lowland faunal area of Costa Rica is the

Table 15.14.   Distribution of Genera and Species Groups of the Middle American Element by Faunal Area

	NW	SW	A	PS	MOC	MOR	AS	CC	CT
Caecilians (2)									
Dermophis		X		X			X		X
Gymnopis		X	X	X			X		
Subtotal	0	2	1	2	0	0	2	0	1
Anurans (19)									
Bufo fastidiosus group				X			X	X	X
Bufo periglenes group								X	
Bufo valliceps group	X	X	X	X	X	X	X		
Crepidophryne				X					X
Eleutherodactylus (Craugaster)		X	X	X	X	X	X	X	X
Agalychnis		X	X	X			X		X
Anotheca			X	X			X		
Duellmanohyla				X			X		
Hyla godmani group			X	X			X		
Hyla lancesteri group			X	X					X
Hyla pictipes group				X			X	X	X
Hyla pseudopuma group				X			X	X	X
Hyla salvadorensis group				X					
Hyla zeteki group							X	X	X
Phrynohyas	X	X							
Smilisca	X	X	X	X	X	X	X		
Gastrophryne			X						
Hypopachus	X				X	X			
Nelsonophryne		X	X	X			X	X	X
Subtotal	3	6	9	14	3	3	12	7	9
Total amphibians (21)	3	8	10	16	3	3	17	7	10
Lizards (9)									
Basciliscus	X	X	X						
Corytophanes		X	X	X			X		
Ctenosaura	X	X	X						
Iguana	X	X	X						
Norops	X	X	X	X	X	X	X	X	X
Phyllodactylus	X				X				
Gonatodes	X	X	X						
Sphaerodactylus		X	X	X					
Celestus			X	X			X	X	X
Subtotal	6	7	8	4	2	1	3	2	2
Snakes (21)									
Loxocemus	X								
Boa	X	X	X	X			X		
Ungaliophis		X	X						
Amastridium		X	X				X		
Coniophanes	X	X		X	X		X		
Conophis	X				X				
Crisantophis	X				X				
Dipsas		X	X						
Enulius	X	X	X		X		X		
Geophis		X	X	X	X	X	X	X	X
Hydromorphus		X	X	X	X	X			X
Imantodes	X	X	X	X	X	X	X		
Leptodeira	X	X	X	X	X		X		
Ninia	X	X	X	X	X	X	X	X	X
Nothopsis		X	X				X		
Rhadinaea		X	X		X	X	X	X	X
Sibon	X	X	X	X	X	X	X		
Tretanorhinus			X						
Trimetopon			X	X		X	X	X	X
Urotheca		X	X	X	X		X		
Micrurus	X	X	X	X	X	X	X		
Subtotal	11	16	17	11	13	7	15	4	5
Crocodilians (1)									
Crocodylus	X	X	X						
Subtotal	1	1	1	0	0	0	0	0	0
Total reptiles (31)	18	24	26	15	15	8	18	6	7
Grand total (52)	21	32	36	31	18	11	35	13	17

Table 15.15. Distribution of Genera and Species Groups of the South American Element by Faunal Area

	NW	SW	A	PS	MOC	MOR	AS	CC	CT
Caecilians (1)									
Oscaecilia		X							
Subtotal	0	1	0	0	0	0	0	0	0
Anurans (20)									
Atelopus		X		X	X	X	X	X	X
Bufo guttatus group		X	X	X	X	X	X		
Bufo marinus group	X	X	X	X	X	X	X		
Eleutherodactylus (s.s.)		X	X	X			X	X	X
Leptodactylus	X	X	X	X			X		
Physalaemus	X	X							
Gastrotheca							X		
Hyla albomarginata group		X	X						
Hyla boans group		X							
Hyla bogotensis group			X						
Hyla leucophyllata group	X	X	X	X			X		
Hyla tuberculosa group			X	X			X	X	
Phyllomedusa				X			X	X	X
Scinax	X	X	X						
Centrolenella		X	X	X	X	X	X		X
Cochranella		X	X	X			X		X
Hyalinobatrachium		X	X	X	X	X	X		X
Colostethus		X	X				X		X
Dendrobates		X	X						X
Phyllobates		X	X						
Subtotal	5	16	15	11	5	5	13	4	8
Total amphibians (21)	5	17	15	11	5	5	13	4	8
Lizards (12)									
Dactyloa									
Polychrus		X	X	X	X				
Lepidoblepharis		X	X	X			X		
Thecadactylus		X	X	X	X				
Ameiva	X	X	X	X			X		
Anadia				X			X	X	X
Bachia		X							
Gymnophthalamus	X	X	X		X				
Leposoma		X	X						
Neusticurus		X							
Ptychoglossus		X	X	X			X		X
Diploglossus		X	X	X		X	X		
Subtotal	2	10	8	8	3	3	5	1	2
Snakes (13)									
Anomalepis				X					
Helminithophis			X		X	X			
Liotyphlops									
Leptotyphlops	X								
Typhlops				X				X	
Corallus		X	X						
Epicrates	X	X	X		X				
Clelia	X	X	X				X	X	
Erythrolamprus		X	X	X	X	X	X		
Liophis		X	X	X	X	X	X	X	X
Oxyrhopus		X	X						
Tripanurgos		X							
Xenodon		X	X	X	X		X		
Subtotal	3	8	8	4	4	2	4	3	1
Crocodilians (1)									
Caiman	X	X	X						
Subtotal	1	1	1	0	0	0	0	0	0
Total reptiles (26)	6	19	17	12	7	5	9	4	3
Grant total (47)	11	36	32	23	12	10	22	8	11

Table 15.16. Number and Percentage of Assemblage and Percentage of Total Genera and Species Groups by Faunal Area

Faunal Area	CA	E	SW	MA	SA	Total
NW						
N	22	5	3	21	11	62
% A	36	8	5	34	17	100
% T	15	3	2	14	8	42
SW						
N	27	2	0	32	36	98
% A	28	2	0	33	37	100
% T	18	1.4	0	22	25	66
A						
N	26	4	0	36	32	97
% A	27	4	0	37	31	100
% T	18	3	0	25	21	67
PS						
N	27	2	2	31	23	85
% A	32	2	2	36	27	100
% T	18	1.4	1.4	21	16	58
MOC						
N	24	3	3	18	12	60
% A	40	5	5	30	20	100
% T	17	2	2	12	8	41
MOR						
N	15	2	2	11	10	40
% A	38	5	5	28	25	100
% T	10	1.4	1.4	8	7	27
AS						
N	25	2	1	35	22	85
% A	30	2	1.2	41	26	100
% T	17	1.4	0.6	24	15	58
CC						
N	10	2	1	13	8	34
% A	29	6	3	38	24	100
% T	7	1.4	0.6	9	5	23
CT						
N	9	1	1	17	11	39
% A	23	3	3	43	28	100
% T	6	0.6	0.6	12	8	27

A = % of faunal area.

T = % of total herpetofauna.

central part of an area that includes the lowlands of Bocas de Toro Province, Panama, the lowlands of eastern Nicaragua, and extreme northeastern Honduras.

Similarly, the Pacific Slope and Atlantic Slope faunal areas are part of areas that extend along the slopes of Chiriquí and Bocas de Toro Provinces, respectively, in western Panama. The Talamancan Highland faunal area would also include the continuation of the Cordillera de Talamanca into western Panama.

For this reason table 15.17, which summarizes the degree of species endemism in the Costa Rican herpetofauna, gives two values for each faunal area. One indicates the number of species known only for Costa Rica. The second records the number of otherwise Costa Rican species endemic to a particular area, including those portions of a continuous faunal area that extend outside the country.

Species endemism is highest in the Cordillera de Talamanca faunal area (27%), with lower endemism in the

Table 15.17. Degree of Species Endemism in Costa Rican Herpetofauna by Faunal Area

Faunal Area	N in Area	CR N	Endemics %	Area N	Endemics %
NW	73	0	0	16	22
SW	172	4	2.3	17	10
A	206	8	4	27	13
PS	192	4	2.1	7	3.6
MOC	98	0	0	0	0
MOR	47	0	0	0	0
AS	189	9	4.8	13	7
CC	76	10	13	10	13
CT	67	11	16	18	27
TOTAL N + % total fauna	383	46	12	106	28

Note: See text for explanation of Costa Rican versus area endemics.

Northwestern, Cordillera Central, and Atlantic Lowland areas: 22, 13, and 13%, respectively. The Southwestern area has a lower but essentially equal value, 10%. The absence of endemics from the two Meseta Central faunal areas suggests that their recognition as discrete units is suspect.

GENERAL PATTERNS

The analysis above indicates that three general patterns of distribution reflect different histories for groups in the faunal areas. It should not be surprising that the autochthonous Middle American Element and Central American Component are major contributors to most of these faunas. The robust representation of South American Element taxa emphasizes the role of the isthmian region as an area of fairly recent, geologically speaking, mixing of tropical Middle American and South American biotas.

The three patterns are as follows:

I. Faunal areas in which a combination of Middle American and South American elements make up 65% or more of the herpetofauna (SW, A)
II. Faunal areas in which a combination of Middle American Element and Central American Component taxa make up 65% or more of their composition, with Middle American genera predominating (PS, AS, CC)
III. A faunal area where Middle American Element and Central American component genera are nearly equal in contribution (NW)

The areas constituting the Meseta Central Occidental and Meseta Central Oriental are anomalous as the only putative areas in which Central American genera (40:30%) substantially outnumber Middle American (30:28%) ones. They also resemble the NW area in having relatively low numbers of South American Element representatives.

The Cordillera de Talamanca area is unique and ambiguous as to placement, since it could be referred to pattern I (MA = 43% + SA = 28% = 71%) or pattern II (MA = 43% + CA = 23% = 66%).

The degree of species endemism within these areas sheds further light on the situation. The high degree of endemism combined with its unique combination of taxa strongly suggests a long-term isolation of the NW fauna from the others. The relatively high level of endemism in SW, A, and CT faunal areas seems to indicate separate histories within a single pattern (I).

Within pattern II, there appears to be no basis for separating out PS and AS as distinctive faunal areas, since they are similar in generic composition and proportional representation of historical units. The Cordillera Central is not supported as a separate faunal area but appears to be at one end of a gradient in historical unit contributions from

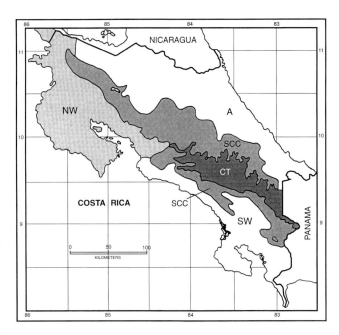

Figure 15.7. Faunal areas for the Costa Rican herpetofauna.

PS + AS → MOC → MORC → CC. If PS + AS + CC are combined the level of species endemism is 15%, similar to that for other recognizable faunal areas.

The Meseta Central Occidental (MOC) and Meseta Central Oriental (MOR) have no endemic species and most closely resemble NW in generic composition. As Scott (1969) pointed out long ago, they appear to form a geographic transitional area between patterns III (MA = 34% + CA = 36% = 70%) and II, that is, the NW and upland areas. The Meseta Central Occidental more closely resembles the Northwest region in proportions of historical contributions (MA = 30% + CA = 40% = 70%) and the Meseta Central Oriental (MA = 28% + CA = 38% = 66%) the uplands (PS = AS + CC): on balance both Meseta faunas are best regarded as part of the Upland Fauna because of the greater similarity in generic composition compared with that of the Northwestern region.

In summary, the following discrete recognizable faunal areas (fig. 15.7) appear to have had separate histories and are the biogeographic areas whose history will be traced in chapter 16:

Lowland
    Pacific Northwest (NW)
    Southwest (SW)
    Atlantic (A)
Upland/Highland
    Montane Slopes and Cordillera Central (SCC)
Highland
    Cordillera de Talamanca (CT)

16 | Development of the Herpetofauna

Central America as a land-positive region has a long and complex geologic and climatological history spanning some 90 million years from the late Cretaceous to the present (fig. 16.1). To explain the evolution of the Costa Rican herpetofauna it is first necessary to describe something of this broader history in order to establish the context. The following paragraphs highlight the most important events in earth's history that shaped the region and in turn are responsible for the patterns of distribution described in chapters 14 and 15.

PALEOGEOGRAPHIC BACKGROUND

Mobile Plates, Blocks, and Island Arcs

Among the most exciting scientific discoveries of the second half of the nineteenth century was the realization that the earth's outer layer consists of a number of rigid, mobile plates riding on a deeper elastic, nearly liquid plastic layer. The plates underlying the oceans are about 65 km deep, and those making up the continents are as deep as 140 km (box 16.1). The uppermost portion of the plates forms the earth's rocklike crust. Oceanic crust is very dense, highly magnetized, and relatively thin, about 5 km thick. Continental crust is much lighter, less magnetized, and much thicker than oceanic crust—about 35 to 40 km thick. The two kinds of crust are also composed of different kinds of rocks. In the course of geologic history continental and oceanic crust have maintained their integrity, and interactions between the various plates at their boundaries produced many of the most prominent features of the earth's geography. In addition, the plates have not remained static in position through time; in the Permian what we now recognize as continents formed a single continent, Pangaea, whose constituent plates have separated and drifted apart over the intervening 225 million years.

The geography of Central America is the result of a complex geologic development over the past 75 million years involving the interactions of five of these mobile tectonic plates (fig. 16.2). The current structure of the land portions of the area consists of four primary crustal blocks:

1. Mayan block: mostly continental crust with its southern border at the Motagua fault system in Guatemala

2. Chortis block: mostly continental crust with its southern boundary at the Santa Elena fault in northern Costa Rica
3. Chorotega block: accretionary crust, with its southeastern border at the Gatún fault
4. Chocó block: accretionary crust with its boundary with the South American plate at the Romerol fault

The Mayan, Chortis, and Chorotega blocks are bordered on the Pacific margins by the Middle American trench, where oceanic crust is being subducted under the lighter continental and accretionary crust. The Chocó block is similarly bordered on the Pacific by the Colombia trench subduction zone (box 16.2). It should be noted that before about 25 million years ago (Ma) the Cocos and Nazca plates were part of the Farallon plate (Atwater 1989).

The following paragraphs and accompanying figures (figs. 16.3 to 16.6) present a summary of the major geologic events affecting the origins of the extant herpetofauna, based primarily on a synthesis of data from Coates and Obando (1996), Donnelly et al. (1990), Escalante (1990), Frisch, Meschede, and Sick (1992), Kerr and Iturralde (1999), Lucas (1986), Mann (1995), Marshall and Sempre (1993), Pindell and Barrett (1990), Rage (1978, 1981, 1986, 1995), and Savage (1982).

The initial fragmentation of the supercontinent Pangaea into a northern land mass, Laurasia, and a southern one, Gondwanaland, was essentially completed by the middle of the Jurassic epoch, about 160 Ma (Barron et al. 1981).

Box 16.1

Principal Types of Crusts Forming Upper Layers of Earth's Surface

Oceanic: formed mainly at midoceanic ridges and hot spots

Continental: formed by mantle differentiation, magmatism, sedimentation, and metamorphism; mostly ancient Precambrian crystalline rock

Accretionary: formed from a mixture of rock types principally at a subduction zones

---

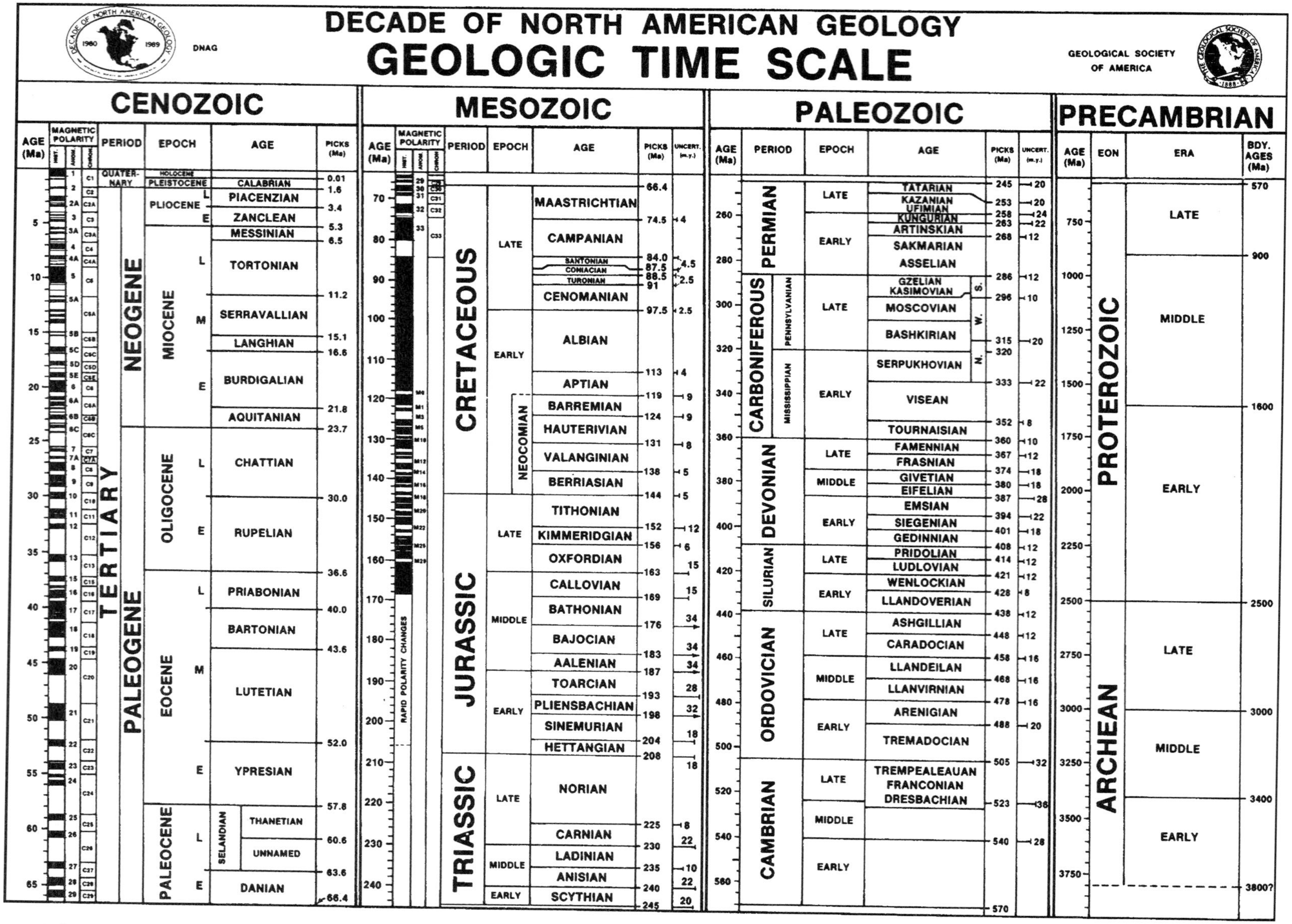

DECADE OF NORTH AMERICAN GEOLOGY
GEOLOGIC TIME SCALE

Figure 16.1. Geologic time scale.

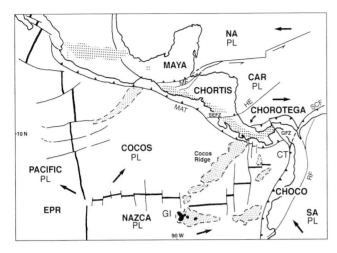

Figure 16.2. Tectonic features of Mesoamerica referred to in the text. Crustal blocks: Chocó, Chorotega, Chortis, Maya. Faults (F): R = Romeral, SC = South Caribbean. Fracture zones (FZ): G = Gatún, M = Motagua, SE = Santa Elena. Tectonic plates (PL): CAR = Caribbean, NA = North American, SA = South American. Trenches (T): C = Colombian, MA = Middle American; GI = Galápagos Islands, HE = Hess Escarpment, EPR = East Pacific Rise.

By 140 Ma the southern continent began to fragment, and by 80 Ma South America had become fully separated from Africa by seafloor spreading along the Mid-Atlantic ridge. Thus South America was completely isolated from other landmasses (fig. 16.3). This series of events doubtless set the stage for the origin of the extant families of amphibians and reptiles and was responsible for their association, blurred somewhat by later dispersals, with either Laurasia or South America.

During this time North and South America were separated by a wide proto-Caribbean seaway created by seafloor spreading of the Atlantic system (Pindell and Barrett 1990). This gap, however, was ultimately replaced by an accretionary land bridge that developed from activity along the southwestern margin of the Caribbean plate. As the plate moved generally northeastward by the late Cretaceous, an ancient Isthmian Link came to connect the two continents (fig. 16.4). This isthmus (the Proto-Antilles) was composed of rock that ultimately would fragment into the Greater Antilles.

At this time the Maya block was in its present position as the principal component of what would become eastern Mexico and northern nuclear Central America, and the Chortis block lay well to the west. Far to the southwest were a series of volcanic islands that would later coalesce into the Chorotega and Chocó blocks, forming by subduction of the Farallon plate under the Caribbean plate.

By the end of the Paleocene the land bridge between North and South America had undergone fragmentation, and some parts were submerged as the Caribbean plate continued its

Box 16.2

Important Tectonic Features of the Central American Region

Plate margins are the areas of greatest tectonic activity, marked by extensive volcanism, numerous earthquakes, and faults and mountain building.

Seafloor spreading: occurs where plates are diverging; magma rises at a midoceanic ridge through a rift in the crust and spreads equally and slowly at right angles on both sides of the rift to harden and form new oceanic crust; for example, the East Pacific Rise and the boundary between the Cocos and Nazca plates

Subduction: occurs where plates are converging on one another and the denser oceanic crest subducts under the continental crust on the opposing plate to form a deep ocean trench; for example, the Middle American and Colombian trenches

Transform faults: occur at right angles to midoceanic ridges as plates grind past one another in fits and starts along plate margins to reduce strain; for example, faults across the East Pacific Rise

Hot spots: occur where magma erupts through the oceanic crust to produce strings of volcanoes as a plate moves over the hot spot; for example, the Galápagos Islands.

northeasterly journey (Marshall and Sempere 1993). These events eliminated any terrestrial connection between the landmasses and completely isolated the faunas of North and South America during most of the Cenozoic. By the middle Eocene the Chortis block and the lower Central American island arc were also moving eastward, and the former became sutured to the Maya block by the end of that epoch, about 38 Ma (fig. 16.5).

The Oligocene and early Miocene saw the Chorotega and Chocó block volcanic islands moving into the now narrowed oceanic gap between the Chortis block and northern South America (fig. 16.6). These islands became more extensive as subduction under the now bordering Caribbean plate added additional volcanoes to the arc. This narrowing gap would be successively closed from the late Miocene to the late Pliocene (fig. 16.7) by the increasing uplift of the area through subduction of the Cocos ridge under the Chorotega block and the collision of the Chocó block with South America (fig. 16.8).

Complete closure of the Isthmian Portal between the Pacific Ocean and the Caribbean Sea was effected in the middle

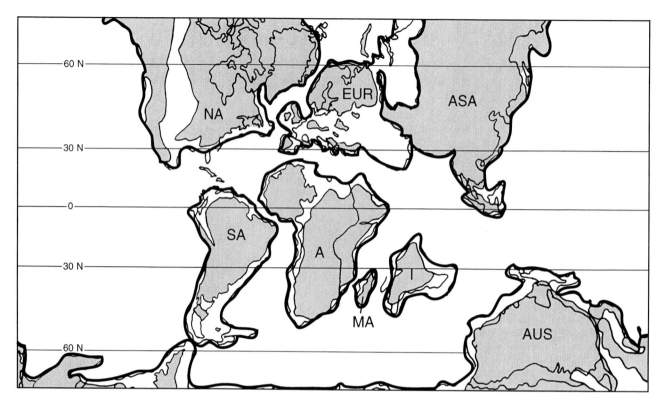

Figure 16.3. Relationships of continents approximately 80 Ma; shading indicates land-positive areas. Note complete isolation of South America.

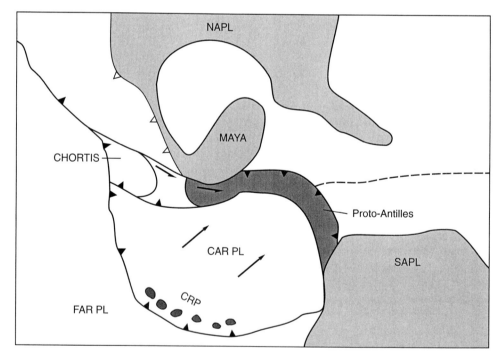

Figure 16.4. Mesoamerican region in the late Cretaceous–early Paleocene. Plates (PL): CAR = Caribbean, FAR = Farallon, NA = North American, SA = South American. Location of Chortis and Maya blocks. CRP = magmatic arc forming part of present-day Costa Rica and Panama. Dotted line equals approximate boundary between NAPL and SAPL.

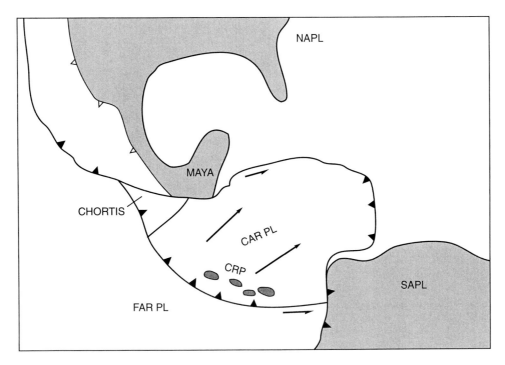

Figure 16.5. Mesoamerican region in the Eocene. Abbreviations as in figure 16.4.

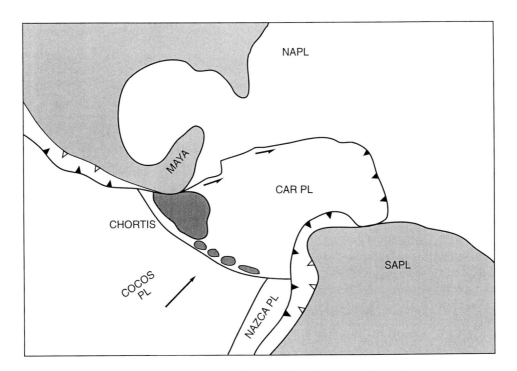

Figure 16.6. Mesoamerican region in the early Miocene. Abbreviations as in figure 16.4.

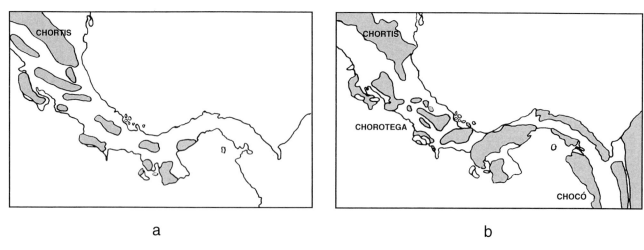

Figure 16.7. Isthmian Link area in the middle (a) and late (b) Miocene.

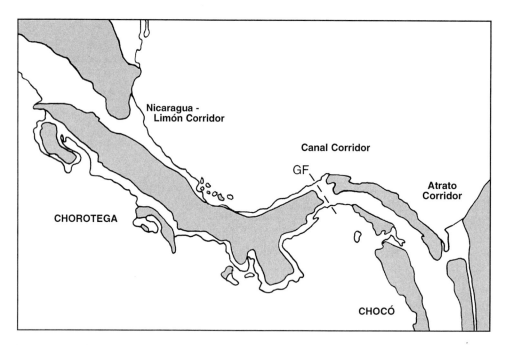

Figure 16.8. Isthmian Link area in the Pliocene. Chorotega and Chocó blocks sutured at Gatún fault (GF); principal marine barriers indicated.

Pliocene, 3.4 to 3.1 Ma (Coates and Obando 1996). At this time a continuous land corridor was established between the Chortis block region and South America. However, other evidence (Cronin and Dowsett 1996) indicates that the closure was temporary and that some exchange between Atlantic and Pacific marine faunas occurred between 2.8 and 2.5 Ma. This would explain the apparent two pulses of dispersal of northern terrestrial forms into South America and South American forms into lower Central America in late Pliocene and Pleistocene times as discussed below.

Uplands, Mountains, and Volcanoes

The Mesoamerican region is characterized by a complex series of elevated areas. Many of these support endemic taxa in their herpetofaunas. They are also often invoked as dispersal routes for various faunal components. Thus the distribution and timing of uplift and possible previous connections of these highlands with other areas are basic to understanding current faunal distribution patterns (Cserna 1989; Ferrusquía 1993; Ortega, Sedlock, and Speed 1994

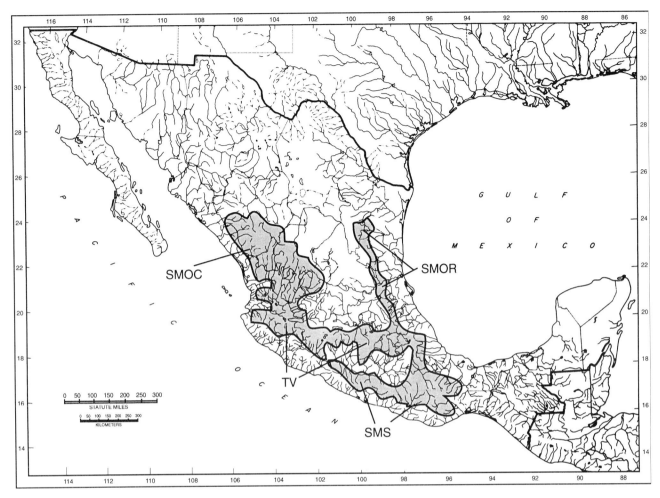

Figure 16.9. Upland areas of tropical Mexico, east of the Isthmus of Tehuantepec: SMOC = Sierra Madre Occidental, SMOR = Sierra Madre Oriental, SMS = Sierra Madre del Sur, TV = Transverse Volcanic Range.

for Mexico; Weyl, 1980 for Central America) (figs. 16.9 and 16.10).

Principal Upland Areas

Sierra Madre Oriental (Mexico): paralleling the Gulf coast, originally uplifted during the Laramide orogeny under influence of the now extinct Farallon–North American subduction zone; minimal Cenozoic volcanism; most recent uplift Miocene-Pliocene

Sierra Madre del Sur (Mexico): originally formed in the early Paleogene and probably related to subduction in the extinct Farallon–North American subduction zone; minimal Cenozoic volcanism; most recent uplift in the Miocene–Pliocene

North Central American Sierras: a series of subparallel ranges forming an arc open to the north extending from Chiapas, Mexico, across Guatemala and Honduras to northern Nicaragua; probably originally formed during suturing of the Chortis block to the Mayan block and

transfer of the former to the Caribbean plate; minimal Cenozoic volcanism; Pliocene uplift.

Tertiary Volcanics

Sierra Madre Occidental (Mexico): originally formed above former Farallon–North American plate subduction zone; extensive Tertiary volcanism; Miocene uplift

Central American: lying eastward of the Quaternary volcanic chain from Guatemala to southern Nicaragua; related to movements of the Chortis block; extensive volcanism from Miocene through Pliocene; Pliocene uplift

Costa Rica–Panama: mountains from central Costa Rica to eastern Panama originally formed as an island arc at Cocos–Caribbean subduction zone; extensive volcanism through Miocene; Miocene uplift

Quaternary Volcanics

Pacific Volcanic Chain: from Mexico-Guatemala border along Pacific versant to central Costa Rica; formed by

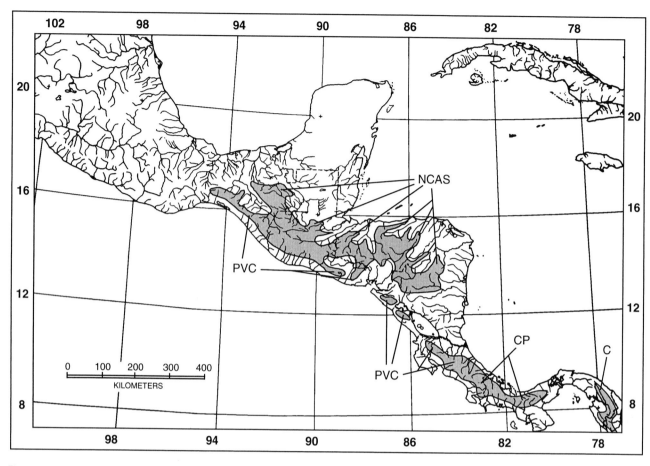

Figure 16.10. Upland areas of Central America: CPC = Costa Rica–Panama cordilleras, NCAS = Northern Central American sierras, PVC = Pacific volcanic chain. The Central American Quaternary volcanic chain lies generally to the east of the PVC and partially underlies it.

subduction at Middle American trench; many still active volcanoes often sitting on top of earlier volcanic ranges; uplift continuing today

Transverse Volcanic Belt (Mexico): formed by subduction of Cocos plate under North American plate; many still active volcanoes; uplift continuing today

PALEOCLIMATES AND VEGETATION

In addition to drifting landmasses and the uplift of mountain masses, the current distribution of the herpetofaunas has been shaped by changing climatic factors through the Cenozoic. Estimates of past climates are based principally on analyses of oxygen and carbon isotope composition, fossil faunal (especially marine) assemblages, fossil pollen profiles, and fossil plant assemblages (Frakes, Francis, and Syktus 1992; Graham 1994, 1996, 1997, 1999a,b,c; Burnham and Graham 1999). Terrestrial angiosperm foliar physiognomy reflects to a substantial extent climatic conditions, so fossil floras are invaluable indicators of climate, regardless of taxonomic composition, and palynofloras often provide the only estimates of past climate as reflected by vegetation.

In this section I will briefly review the major climatic shifts that have affected the herpetofauna and describe in broad strokes the changes in vegetation that occurred in our area of study. Frakes, Francis, and Syktus (1992) recognized two climate modes, periods during which similar climates prevail, as characteristic of Phanerozoic times. Cool modes are periods of global refrigeration, ranging from those in which glaciation occurs to those where ice is formed at high latitudes only during the winter. Warm modes are periods when climates are globally warm with little or no polar ice. Late Cretaceous and Paleocene times were among the very warmest and most humid of times in earth's history (fig. 16.11). The polar regions were free of ice, with temperatures warm enough to allow forests and associated reptile and amphibian faunas to live there. Mean global temperatures were 6°C higher than today. Sea levels were high in the late Cretaceous, about 200 m above current levels, and large areas of the continents were covered by extensive seaways (Vail, Mitchum,

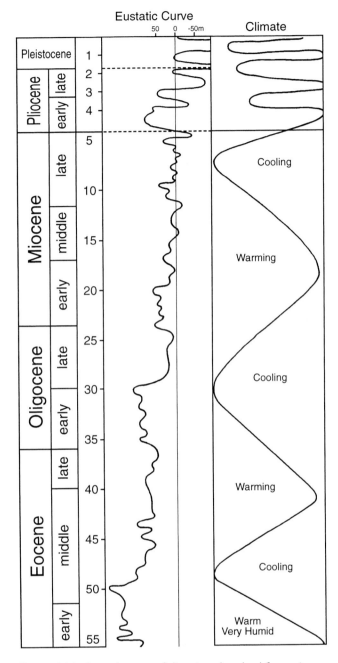

Figure 16.11. General pattern of climatic and sea level fluctuations, Eocene to present.

By the time the Proto-Antillean Isthmus began to break up in the late Paleocene, tropical and subtropical evergreen forests ranged north to 60 to 65°, temperate broadleaf forests to 70°, and a previously minor component of the Arcto-Tertiary geoflora, conifer forest, began developing around the pole (fig. 16.12a).

The onset of the overall cooling trend for the rest of the Cenozoic began in the Eocene at about 50 to 55 Ma. This early cool period saw displacement of the vegetational belts southward, and a number of temperate forests components ranged even farther southward along the developing uplands of western North America and eastern Mexico. At lower elevations a transitional zone appears to have been present between subtropical evergreen and temperate evergreen forests at about 50° N (fig. 16.12b). Although there were later warm modes in the Cenozoic with some shifting north and south of climate-vegetation zones, tropical and subtropical vegetation became displaced farther and farther south throughout the rest of the era (fig. 16.12c) (Wolfe 1985).

Superimposed on the cooling trend was a drying trend that was associated with the mountain building of the western North American cordilleras and the increasing continentalization of climate. The drying trend is already noticeable in the Oligocene, since broadleaf deciduous forests dominated eastern North America and subtropical deciduous vegetation covered most of what is now the southwestern United States and adjacent Mexico (fig. 16.13a). The latter vegetation corresponds to the Madro-Tertiary geoflora of Axelrod (1958).

The early Miocene (fig. 16.13b) was a period of warming and increasing aridity. This coupled with uplift of the Sierra Madres in Mexico completed the isolation of tropical Mesoamerica and its highlands from biotas to the north by the development of semiarid to desert vegetation across America west of the Mississippi and south through central Mexico. The late Miocene and earliest Pliocene (6 to 4.8 Ma) formed a period of severe cyclic glaciations in which average annual temperatures in Middle America were depressed as much as 6°C below present ones. Continuing drying trends, however, prevented any substantial movement southward of temperate forests, so that they came to be compressed between the taiga and scrubland-grassland belt (fig. 16.14).

After a brief interval of warming and a rise in sea level, the rest of Pliocene, from the initial closure of the Panamanian seaway (3.1 Ma) throughout the Quaternary, saw repeated cooling and warming episodes as polar glaciers waxed and waned. Interglacial periods in tropical Mesoamerica would have climate and vegetation patterns similar to those at present. Glacial periods saw temperature depressions of 5 to 6°C so that upland vegetation was displaced downward and mixed with lowland communities. Contrary to earlier authors (Clapperton 1993; Haffer 1969; Webb 1985), in lower Central America and the Amazon at least,

and Thompson 1977). By the Paleocene the seaways had gradually receded. Vegetation over most of what is now the Americas consisted of tropical and subtropical broadleaf evergreen forests (rainforest). Temperate forests including broadleaf evergreen and deciduous formations grew surrounding the poles. The tropical assemblage and the north temperate one have been recognized as the Neotropical Tertiary and Arcto-Tertiary geofloras, respectively (Axelrod 1958; Axelrod and Ting 1960).

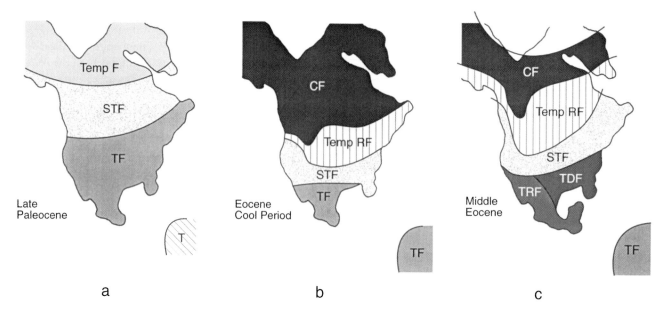

Figure 16.12. General distribution for forest vegetation in the early Paleocene to middle Eocene: CF = conifer forest, STF = subtropical rainforest, TDF = tropical deciduous forest, TF = tropical rainforest, Temp F = temperate forest, Temp RF = temperate rainforest.

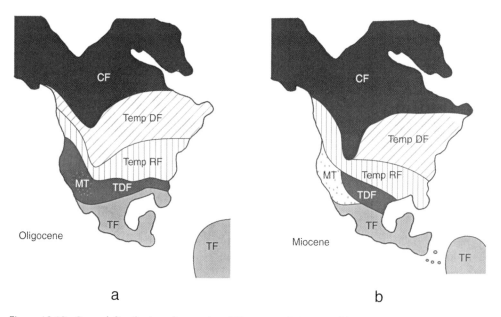

Figure 16.13. General distribution of vegetation, Oligocene to Miocene. Abbreviations as in figure 16.12. Temp DF = temperate deciduous forest. Subtropical vegetation, not shown because of scale, forms transition between tropical and temperate formations. MT = Madro-Tertiary woodland, chaparral, and scrub vegetation.

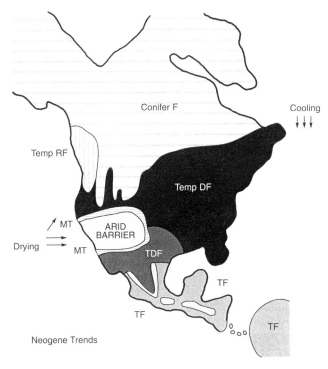

Figure 16.14. Neogene trends in vegetation produced by increased cooling and drying and continentalization of climates. Abbreviations as in figures 16.12 and 16.13.

interglacial intervals were not periods of increased dryness (Colinvaux 1993, 1996). I will return to this matter later in the discussion of the biogeographic patterns for montane amphibians and reptiles.

DISPERSALS, VICARIANCE, AND FAUNAL ASSEMBLAGE: THE BIG PICTURE

The history of the Tropical Mesoamerican herpetofauna includes a series of concordant dispersal events interspersed with periods of fragmentation of formerly continuous ranges followed by diversification within the now separate areas (vicariance). The earliest major vicariance to affect the present-day fauna was the gradual fragmentation of Pangaea into northern (Laurasia) and southern (Gondwanaland) landmasses in the middle Jurassic (about 160 Ma). During this period of separation all extant families of amphibians and reptiles probably evolved as further fragmentation led to the present pattern of continental units (table 16.1). For this discussion, that initial separation is called vicariance V_0.

Sometime in the late Cretaceous the accretionary island arc described above came to lie between North and South America and through uplift formed an isthmian connection between the two continents (the Proto-Antilles). This connection lasted 5 to 10 million years and mediated a major

Table 16.1. Historical Sources for Suprageneric Groups of the Mesoamerican Herpetofauna

Group	Laurasian	Pangaean	Gondwanan
Caecillians			Caeciliidae
Salamanders	Plethodontidae		
Frogs and Toads	Rhinophrynidae		Pipidae
	Ranidae		Bufonidae
			Leptodactylidae
			Hylidae
			Centrolenidae
			Dendrobatidae
			Microhylidae
Lizards	Phrynosomatidae		Corytophanidae
	Eublepharidae		Iguanidae
	Gekkoninae (pt.)		Hoplocercidae
	Xantusiidae		Polychrotidae
	Scincidae (pt.)		Sphaerodactylinae
	Dibamidae		Gymnophthalmidae
	Teiidae		Gekkoninae (pt.)
	Bipedidae		Scincidae (pt.)
	Anguidae		Amphisbaenidae
	Xenosauridae		
	Helodermatidae		
Snakes	Loxocemidae	Typhlopidae	Anomalepididae
	Ungaliophiidae	Leptotyphlopidae	Boidae
	Colubridae (pt.)		Tropidophiidae
	Elapidae		Colubridae (pt.)
	Viperidae		
Turtles	Kinosternidae		
	Dermatemydidae		
	Chelydridae		
	Emydidae		
Crocodilians	Crocodylidae (pt.)		Crocodylidae (pt.)

dispersal event (D_1) with apparent extensive faunal exchange between the two continents. All evidence points to an ancient continuity and essential similarity of a generalized tropical herpetofauna that ranged over tropical North, Middle, and South America in the late Cretaceous–Paleocene times. Descendants of this assemblage are represented today by the South and Middle American Elements defined in chapter 15. These elements are associated with the Neotropical-Tertiary geoflora defined above, which at this time ranged far to the north (70°), and its derivatives throughout the Cenozoic. To the north of this fauna ranged a temperate, Laurasian-derived unit associated with the Arcto-Tertiary geoflora, now represented by Old Northern Element herpetofaunal taxa.

By the end of the Paleocene the Proto-Antillean Isthmus became fragmented as the Caribbean plate moved northeastward to separate the continents and isolated northern and southern fragments of the generalized tropical herpetofauna in North and South America to constitute a second major vicariance event (V_1). Differentiation in situ within the two fragments during the next 54 million years created the distinctive tropical elements of the two herpetofaunas that became intermixed with the establishment of the new Isthmian Link in the Pliocene. Phylogenetic analysis of several groups of xenodontine snakes (Cadle 1985), the coral snakes of the genus *Micrurus* (Slowinski 1991, 1995), the frogs of the genus *Eleutherodactylus* (Lynch 1986; Savage 1987), and the anole lizards (Guyer and Savage 1992) strongly support the significance of this vicariance (fig. 16.15).

In the Eocene, probably related to the Eocene cool pe-

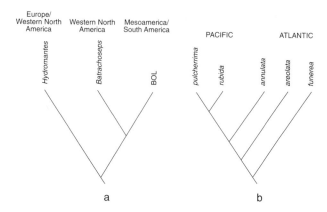

Figure 16.16. Phylogenetic relationships: (a) bolitoglossine salamanders illustrating incorporation of one lineage (BOL = *Bolitoglossa* plus twelve genera) into the Central American biogeographic component (after Wake and Larson 1987); (b) Mesoamerican batugarine turtles of the genus *Rhinoclemmys* showing vicariance event V_3.

riod and emerging uplands in western North America and Mexico, a substantial number of Old Northern groups became integrated with tropical taxa in Mexico (dispersal event D_2). Most prominent among these organisms were plethodontid salamanders and colubrine snakes, but several freshwater turtles and a variety of lizards are also represented (table 15.9). As the former continuity between the region and what is now the eastern and far western United States was affected by mountain building and the subsequent cooling and drying trends for the rest of the Cenozoic, these components became disjunct in Mexico (Axelrod 1975; Graham 1997). This disjunction (Rosen 1978) allowed differentiation of the Central American Component of the Old Northern Element (fig. 16.16a), which from the Oligocene onward evolved in association with the Middle American Element.

Thus the initial organization of what was to become the Mesoamerica herpetofauna involved a pair of vicariance events: complete geographic isolation from South America and fragmentation and isolation of the Central American Component from its northern congeners, by a combination of physiographic and climatic factors. By the Oligocene, most of the genera or their ancestors, which now form the Old Northern and Middle American Elements (table 15.9), were present in the region.

A momentous physiographic development, the uplift of the main mountain axis of Mexico and Central America, created one major dispersal event (D_3) and two important additional vicariance events. This process seems to have had a north-to-south sequence, with the Sierra Madres of Mexico present as upland areas beginning in the Oligocene, and the highlands of Nuclear Central America developing in the Miocene. The final sequence of uplift was in lower Central America leading to the closure of the Panamanian Portal

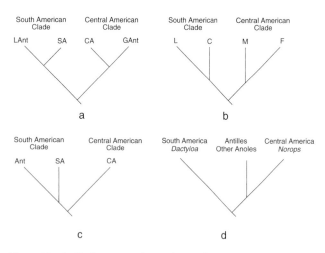

Figure 16.15. Phylogenetic relationships reflecting separation of North and South America in the early Paleogene: (a) xenodontine snakes; (b) venomous coral snakes (genus *Micrurus*): C = *collaris* group, F = *fulvus* group, L = *lemniscatus* group, M = *mipartitus* group; (c) frogs of the genus *Eleutherodactylus*; (d) anole lizards. Ant = Antilles, G = Greater, L = Lesser.

in the Pliocene. As the uplift proceeded, Middle American Element and Central American Component taxa dispersed southward (D₃) over the emerging landmass. A primary variance effect of the uplift was to gradually fragment what was a rather homogeneous Mesoamerican herpetofauna into several geographic assemblages, most notably in the lowlands, where the orogenic effects on the northeast trade winds produced a partial rain shadow along the Pacific versant from Sinaloa, Mexico, to Costa Rica. This process led to the replacement of humid conditions and evergreen vegetation by subhumid to semiarid climates and deciduous and thorn forest formations. As pointed out in chapter 14, many species and most genera of lowland groups in Central America are found on both the Pacific and Caribbean coastal strips. Duellman (1966b, 1988) and I (Savage 1966a, 1982) also emphasized the relative homogeneity of the herpetofauna on each lowland versant. Nevertheless, a considerable number of sister species pairs reflect the impact of the mountain barrier. An example supporting the significance of this vicariance event (V₃) is provided by the turtle genus *Rhinoclemmys* (fig. 16.16b) based on the study of Lahanas (1992).

As the mountains were uplifted, the distributions of certain other groups, perhaps originally associated with the low uplands of earlier times, became fragmented onto the three major highland areas today constituting the backbone of Middle America. This fragmentation has led in some cases to the development of endemic montane isolates from ancestors with a formerly continuous north-to-south range. However, this explanation does not seem satisfactory in all cases, and I will return to the problem of montane speciation in a later section.

The final major factor in shaping the herpetofauna of Central America was the complete emergence of the Panamanian Isthmus in the Pliocene to directly connect North and South America. Reconnection led to the dispersal (D₄) of many South American Element genera northward and permitted immigration into South America by Old Northern and many Middle American stocks. Concordant dispersal events (D₄), have brought sixty-five living generic-level taxa of clearly South American origin across the Isthmus to contribute to the Central American herpetofauna (table 15.9). Most of these groups are restricted to the region from eastern Panama to Costa Rica, so that the South American influence is minimal over most of Mesoamerica. Similarly, the greatest number of Old Northern and Middle American generic level taxa and species are found in northwestern South America, but many range southward to the Amazon basin or beyond.

The recent herpetofaunas of Central America, except those in eastern Panama, are based on a fundamental core of autochthonous Middle American groups whose history in the region goes back at least to the early Tertiary. Coexisting with this unit throughout the region are a series of en-

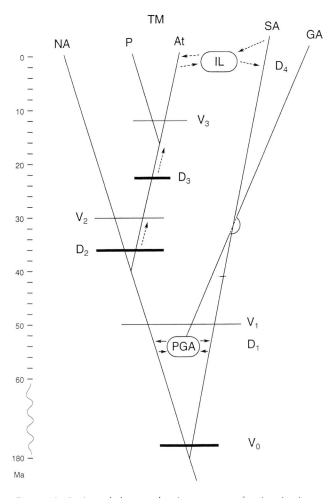

Figure 16.17. Area cladogram showing sequence of major vicariance (V) and dispersal (D) events responsible for present biogeographic patterns for the terrestrial herpetofauna: At = Atlantic, GA = Greater Antilles, IL = Isthmian Link, NA = North America, P = Pacific, PGA = Proto–Greater Antilles, TM = Tropical Mesoamerica.

demic derivative stocks of Old Northern relationships that have been in the region from Eocene-Oligocene times onward. Only in Panama and Costa Rica do South American Element taxa contribute significantly to the fauna and predominate in eastern Panama.

To summarize, the events leading to the present pattern for Mesoamerica are as follows (figs. 16.17 and 16.18):

Vicariance (V₀): breakup of Pangaea

Dispersal (D₁): from the south over the Proto-Antillean Isthmus

Vicariance (V₁): breakup of the Proto-Antillean Isthmus

Dispersal (D₂): invasion and integration of Old Northern fauna with the tropical herpetofauna

Vicariance (V₂): climate changes introducing a semiarid to arid barrier between temperate North America and tropical Mesoamerica

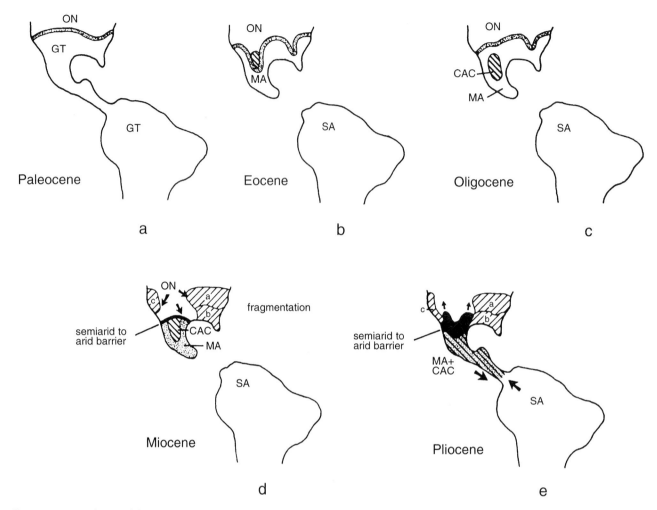

Figure 16.18. Evolution of the Tropical Mesoamerican herpetofauna showing distribution of historical source units over time: CAC = Central American Component, GT = Generalized Tropical Fauna, MA = Middle American Element, ON = Old Northern Element, SA = South American Element.

Dispersal (D₃): concordant population of the emerging Isthmian Link by Mesoamerican and Central American faunal units from the north

Vicariance (V₃): uplift of Mesoamerican highlands

Dispersal (D₄): from the south after emergence of the Panamanian Isthmus; this event subsumes two pulses that occurred about 1 million years apart

The Costa Rican Herpetofauna and the Closure of the Panamanian Portal

By the end of the Paleogene the land-positive portions of the Maya and Chortis blocks were populated by a variety of Mesoamerican Element and Central American Component genera, many shared with tropical Mexico. The continuing east-northeastward movement of the Caribbean plate by this time had brought the Costa Rica–Panama island arcs into

approximately their present locations. Uplift of a series of gradually emerging islands characterized this period for the Chorotega block, but the islands were widely separated from South America by a deep and wide seaway (fig. 16.7). The earliest of these islands are represented today by a series of Pacific versant peninsulas: Santa Elena, Nicoya, Herradura, Osa, Burica, Soná, and Azuero.

At this same time, although the Chortis and Chorotega blocks were sutured together at their margins, the contiguous Nicaragua depression and the Tempisque, San Carlos, and northern Limón basins formed a significant marine barrier to dispersal onto the islands from the north. Coates and Obando (1996) suggested that this barrier continued as an elongated marine connection between the Pacific Ocean and Caribbean Sea into the Pliocene (3 Ma). Other reconstructions (Van Andel et al. 1971; Donnelly 1989; Iturralde and MacPhee 1999; Perfit and Williams 1989; Pindell and Barrett 1990) indicate a land-positive connection between the Chortis and Chorotega blocks in the late Miocene (10 Ma). It seems likely that the connection was completed at the earlier time, as indicated by the late Miocene (6 Ma) mammal fossils of strictly northern affinities from central Panama (Whitmore and Stewart 1965). However, eustatic fluctuations may have opened and closed this and other interoceanic connections at various times before final closure of the Isthmian Portal.

In the Middle Miocene (ca. 15 Ma) the Panama portion of the arc began to be compressed by the South American plate, at which time the Chocó block was formed and a series of land-positive areas emerged in the diminishing gap between South America and the Chorotega block. These areas formed islands, continuing the isthmian archipelago southward, and today are represented by the Serranías San Blas–Darién and Majé of western Panama, the Serranía Sapo-Baudo of western Panama and northwestern Colombia, and the Dabeiba region of the Sierra Occidental of Colombia.

In addition, beginning about 5 Ma, the Cocos ridge began to subduct under the central portion of the Chorotega block to uplift the Cordillera de Talamanca. Thus by the early Pliocene a continuous land connection formed a dispersal corridor between the southern portion of nuclear Central America and what is now Costa Rica and western Panama. These events provided a landscape of coastal lowlands with a rapidly rising upland region that ultimately formed a continuous montane barrier 2,000 m in altitude across the Isthmus. Paleobotanical and paleoclimatic data (Graham 1987a,b, 1999a,b,c; Savin 1977; Savin and Douglas 1985) indicate that mid-Miocene climates were much the same as today at these latitudes (ca. 10° N), and the principal vegetation present was humid evergreen forests. However, Graham (1987a) noted that a substantial number of genera represented groups now found in midaltitude situations. These facts suggest that the basic habitats used by the modern herpetofauna were in place, except for the dry forest component, by the mid-Miocene (5 to 10 Ma). It is little wonder then that Mesoamerican Element and Central American Component taxa dominate all areas on the Chorotega and Chocó blocks, since they had the earliest and exclusive opportunity to invade the emerging Isthmus from the north.

The final chapters in this process involved complete closure of the seaway by further uplift of the Chocó block. The first such episode occurred in the Pliocene, when sea level was 100 m lower than today (Vail and Hardenbol 1979), about 3.4 to 3 Ma. This event led to the concordant dispersal of South American Element taxa into lower Central America. These earliest invaders are probably represented by the few genera of this unit now ranging north to Guatemala or beyond.

Later, warming temperatures and rising sea level appear to have again connected the Caribbean Sea and the Pacific

via the Atrato region for a time, with final closure dated at about 2 Ma (Cronin and Dowsett 1996). This resulted in a second pulse of dispersal that continues to this day. Although I have included them within dispersal event D_4 (fig. 16.17), they may be thought of as D_{4a} and D_{4b}. It is these dispersals that are responsible for the substantial representation of South American taxa in Costa Rican faunal areas other than the northwest region.

The temporary opening and final closure of the seaway in the Pliocene also may explain a puzzling aspect of South American biogeography. Many Mesoamerican and Central American unit genera have relatively few South American representatives, and their distributions are restricted to northern South America. Others, however, have wider distributions and considerable species richness on the continent (e.g., *Bolitoglossa*, *Norops*, and *Bothrops*). It appears that there were also two pulses of dispersal southward, mirroring those just described for South American genera dispersing northward across the Isthmus. Middle American genera showing wide distributions and considerable speciation in South America probably represent the first wave of migrants across the Isthmus—for example, the salamanders of the genus *Bolitoglossa* (Hanken and Wake 1985) and some colubrine snakes (*Atractus* and *Sibynomorphus*, Cadle 1985), and perhaps the pitvipers of the genus *Bothrops*. Those with more restricted ranges and less diversity would have arrived in the second, more recent pulse after the second and final closure of the seaway.

The two-step closure model would also explain the frequent occurrence of endemic species of Mesoamerican Element and Central American Component genera in the Chocó region of western Colombia or northern South America (e.g., the lizards of the genus *Basiliscus* and turtles of the genus *Rhinoclemmys*). The ancestors of such forms would have migrated across the first Pliocene land connection and become isolated from Central American populations. During separation, allopatric speciation occurred to produce the South American form(s). In some cases, after the final Pliocene closure, dispersal has brought the descendant species into sympatry. Examples include *Basiliscus basiliscus* of Central America and northern South America and *B. galeritus* of the Chocó and *Rhinoclemmys annulata* and *R. nausuta*, with the latter species in the Chocó. In other cases they may continue to be allopatric: *Kinosternon angustipons* (Atlantic slope Costa Rica and western Panama) and *K. dunni*. The South American genus *Phyllobates* mirrors this situation in reverse with allopatric species in Costa Rica and Panama (*Phyllobates lugubris* and *P. vittatus*) and their sister species (*P. aurotaenia*) in the Chocó. All are probably isolated fragments of an ancestor that crossed over the first Pliocene land bridge.

Several authors have argued that the presence and diversity of Middle American taxa in South America, here associated with the first Pliocene dispersal pulse, shows they must have arrived there by island hopping along the magmatic arc before the emergence of the Isthmian Link in the Pliocene. Hanken and Wake (1985) proposed that salamanders of the genus *Bolitoglossa* made such a journey in the middle Miocene about 18 Ma based on a molecular clock derived from genetic distances. Zamudio and Greene (1997) evoke a similar scenario for colubrine snakes and *Lachesis*. Using mtDNA data, they established a molecular clock that estimated the time of divergence between Central American and South American *Lachesis* as 6 to 18 Ma, again before completion of the Isthmian Link. Given the unreliability of molecular clock estimates (Scherer 1990; Avise et al. 1992; Avise 1994; Gaut 1998), the evidence for island hopping dispersal raises the issue whether dispersal could have occurred before the Pliocene by successive, sporadic connections of islands in the arc. The two-closure model of the portal also needs further verification to satisfactorily explain these and other distribution patterns.

By the late Pliocene most components of the present herpetofauna of Costa Rica and western Panama were in place. Mesoamerican Element and Central American stocks had arrived from the north during the Miocene to inhabit the emerging Isthmian Link as it gradually extended toward South America. Uplift of the Talamanca massif had provided an upland center for differentiation of these same lineages. Simultaneously, it split the lowland area into Pacific and Atlantic regions, where unique faunal assemblages would develop. Final closure of the Panamanian seaway had led to the migration of South American Element taxa northward to contribute eventually to herpetofaunal differentiation on both Pacific and Atlantic versants and in the uplands as well.

The Problem of Tropical Montane Herpetofaunas

Three principal upland areas occur in tropical Middle America: the Mexican Sierra Madres, the highlands of nuclear Central America, and the highlands of Costa Rica and western Panama. The first two are separated from one another by the low-lying Isthmus of Tehuantepec and the last two by the Nicaragua depression. No evidence suggests that these regions were ever connected by montane corridors at any time in their histories, yet closely related taxa and sometimes disjunct populations of arguably the same species occur on two or more of these highlands.

Three principal hypotheses have been proposed to account for these patterns:

At times of maximum Ice Age cooling, a temperature depression of about 6°C would allow dispersal through the lowlands from one upland area to another

The upland taxa are derivatives of widespread lowland taxa that independently invaded each upland area from below

The ancestral taxa were originally widespread lowland ones that were carried upward and fragmented as each highland was uplifted

Recent studies on fossil pollen profiles by Colinvaux (1993, 1996) have clarified the issue at least as it relates to climate and vegetation in the Quaternary of Lower Central America. His principal conclusions are:

During glacial periods temperatures were depressed by 6 to 8°C from Guatemala through Panama, and glaciers formed at the highest elevations in the Cordillera de Talamanca

During these periods there was not a simple downward depression of the vegetational zones by 800 m, a condition that would have eliminated all components of lowland vegetation

Rather, the distributions of upland genera were compressed downward, where they became mixed with lowland taxa

The lowlands of the region were covered during glacial periods by these humid forests, with no indication of Ice Age increase in aridity

During interglacial periods the upland taxa were sorted out by moving back up the cordilleras as temperatures returned to levels equivalent to those at present

There have been repeated cycles of these events in Quaternary times

Although this pattern is documented for the Quaternary, an earlier period of severe glaciation cycles that occurred in latest Miocene to earliest Pliocene times (6 to 4.8 Ma) probably showed a similar sequence of events. One may assume that, like the flora, herpetofaunal assemblages responded to long-term temperature changes.

It therefore seems likely that the populating of the mountains of nuclear Central America by upland genera from the Sierras of Mexico occurred during a late Miocene to early Pliocene glacial period. Upland taxa became mixed with lowland ones at this time and were able to disperse across the Isthmus of Tehuantepec. During a subsequent warming period the upland genera were sorted out of the mixed fauna by moving into the newly available montane habitats. An example of the result of such a process is the distribution of the lizard genus *Abronia* (fig. 16.19).

Similar subsequent cycles would lead to a sequential invasion from north to south of more southern uplands throughout the Pliocene. An example of this pattern is the distribution of the snakes of the *Rhadinaea godmani* group (fig. 16.20).

The formation of the Quaternary volcanic chain in Central America would expedite this process during the several cooling and warming cycles. A general model for montane speciation in the region (fig. 16.21) requires an initial dispersion event followed by cycles of temperature depression (glacial periods) and release (interglacial periods). The initial period of such a sequence is probably responsible for the

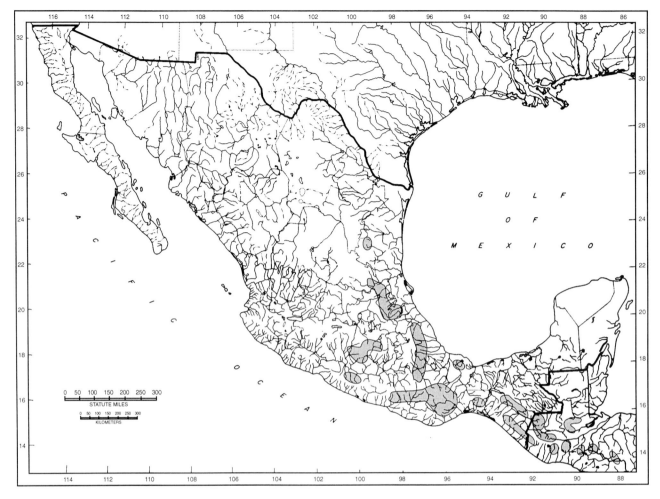

Figure 16.19. Distribution of the lizard genus *Abronia*.

differentiation of sister species in isolated ranges such as the Cordillera Central and Cordillera de Talamanca. An example of this process is the genus *Nototriton* in Costa Rica (fig. 16.22).

Evolution of the Lowland Dry Forest Herpetofauna

The northwestern region of Costa Rica forms the southern terminus of a continuous corridor of thorn woodland and semiarid, deciduous, and semideciduous forests and patches of savanna that extends from Sinaloa, Mexico, southward along the Pacific coast. The Meseta Central Occidental represents a transitional zone between the Lowland Dry Forest of Guanacaste and northern Puntarenas Provinces and more humid formations. Climates and vegetation similar to those found along the Pacific coastal plain also occur on the outer portion of the Yucatán Peninsula and in the rain shadow valleys on the Atlantic versant of nuclear Central America (fig. 16.23).

The disjunct distribution of several amphibians and rep-

tiles in the separate portions of these habitats suggests former continuity. An example is the distribution of the spiny-tailed iguanas of the *Ctenosaura quinquecarinatus* group (fig. 16.24).

Elements of these vegetational formations are of Neotropical-Tertiary geofloral types, but this complex probably represents a transition between the increasingly drier and cooler core of the Madro-Teritiary flora and tropical evergreen formations (Axelrod 1958). By late Oligocene times the relation between these might be visualized as a broad belt of tropical evergreen forest bordered on the north by tropical deciduous forest. The latter in turn was bordered by the horseshoe-shaped enclave of developing Madro-Tertiary vegetation in what is now northern Mexico and the southwestern United States (fig. 16.14). This element would continue to expand over western North America for the rest of Cenozoic to produce the familiar semiarid live oak–conifer woodland, chaparral, arid subtropical scrub, desert grasslands, and subdesert and desert formations.

In the Miocene the northern limits of tropical conditions

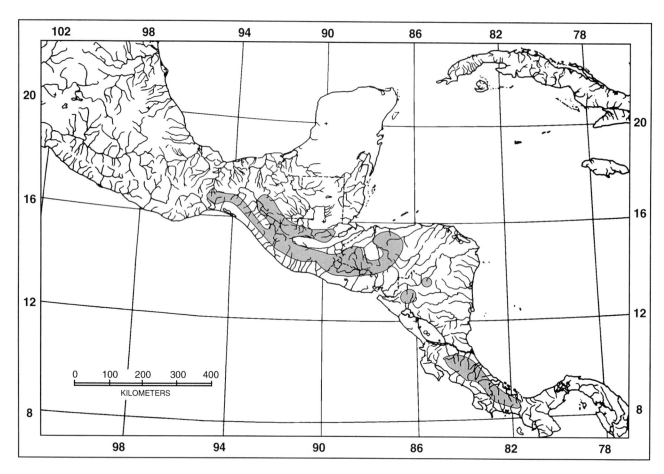

Figure 16.20. Distribution of the snakes of the *Rhadinaea godmani* group.

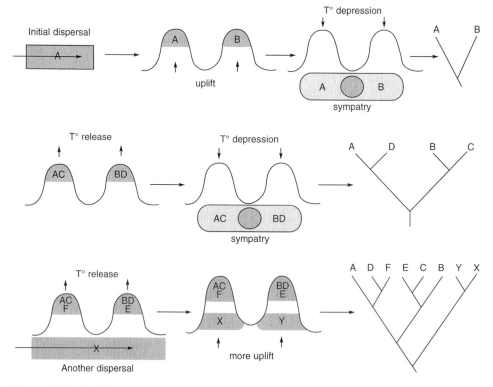

Figure 16.21. Model for montane speciation involving cycles of temperature depression and release beginning in the Pliocene; letters indicate species, cladograms indicate phylogenetic results.

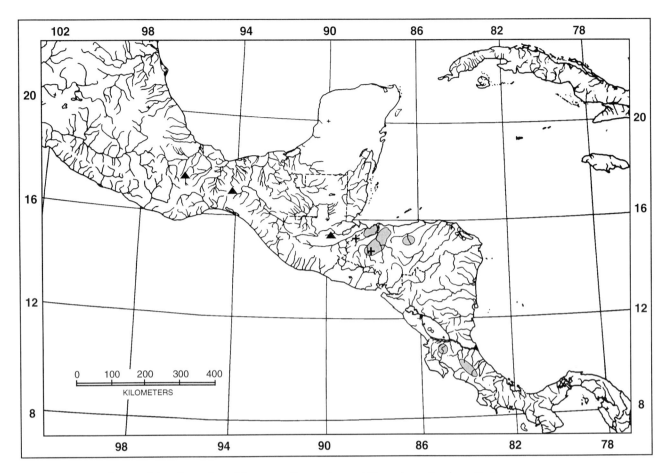

Figure 16.22. Distribution of the salamanders of the genera *Cryptotriton* and *Nototriton*: triangles = *Cryptotriton*; shading = *Nototriton*; cross = sympatry between the two genera.

were being restricted farther and farther south. At the same time the uplift of the Mexican Sierra Madres Oriental and Occidental and the Mexican Plateau drove an upland wedge southward that by the mid-Pliocene fragmented the lowland region into Pacific and Atlantic sections connected only across the Isthmus of Tehuantepec.

Herpetofaunal associates of the tropical deciduous forest belt were isolated at this time in three areas: along the western margin of the Gulf of Mexico; the outer margin of the Yucatán Peninsula, but doubtless extending more to the south than now; and western coastal Mexico. An example of a clade now found disjunctly in all three areas are the cantils *Agkistrodon bilineatus* and *A. taylori* (fig. 16.25).

During the rest of Neogene times the rain shadow effect of the rising nuclear Central American mountains combined with the Pacific climatic regimen saw the expansion southward of the subhumid to semiarid formations to replace humid ones for the most part in the western lowlands, producing considerable endemism over time. The herpetofauna of this corridor is more strongly influenced by members of the Central American Component than any other represented in Costa Rica. This fauna probably reached Costa Rica rel-

atively recently, during the time of Quaternary volcanism and after the uplift of the Nicaragua depression just to the north.

The distribution of this fauna in the Atlantic slope rain shadow valleys also appears to represent late Neogene fragmentation of once continuous ranges. Stuart (1954) long ago proposed a subhumid corridor involving dispersal across intermediate pine-oak forest barriers from the Pacific lowlands at the Isthmus of Tehuantepec through the Grijalva valley of Chiapas, Mexico, in the Atlantic drainage and into the semiarid Río Negro and Río Motagua valleys of Atlantic slope Guatemala (see also Wilson and McCranie 1998).

Another explanation seems more parsimonious. It is likely that these four areas and the dry Sula and Aguan valleys of Honduras were part of a more or less continuous corridor during glacial maxima in the Pliocene. Subsequent uplift of the Sierra Madre of Chiapas, the transverse ranges of Guatemala, and the Sierra de Oma and Sierra de Sulaco in Honduras appear to have fragmented this corridor into its present components and allowed differentiation of some endemic taxa.

The other isolated dry valleys of more southern Honduras

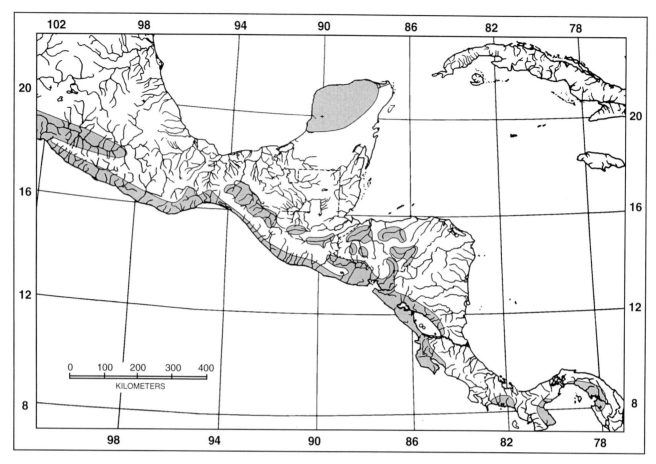

Figure 16.23. Distribution of subhumid to arid climatic conditions in tropical Mesoamerica supporting deciduous forests, thorn forests, thorn woodlands, and savanna. Areas of these types in Panama show little faunal similarity to those from Costa Rica northward.

were probably connected across the low early Pliocene continental divide with the Pacific slope dry corridor. Again Pliocene orogeny, reinforced by Quaternary volcanism, would have isolated the Atlantic sector from the Pacific sector. Although there clearly were considerable expansions of dry forest conditions into areas where evergreen forest now predominates in upper Central America during the Pleistocene, it is unlikely that the various isolated valleys were relinked as in their Pliocene continuities.

Webb (1977, 1978) and Webb and Rancy (1996) proposed that a continuous "savanna" corridor extended through Central America and across the emerging Isthmian Link and served as a major pathway for movements of fauna, as indicated by mammalian fossils, between the Americas in the late Pliocene to mid-Pleistocene. Their concept of "savanna" was broad and was applied to all subhumid to semiarid habitats now represented in Central America, including those discussed in the previous paragraphs. Colinvaux (1996) recently has demonstrated that such a corridor could develop only during glacial periods but found no support in the fossil plant record that one could have existed. All evidence

from herpetofaunal distributions indicates that the Pacific lowland subhumid-semiarid corridor never extended farther south than its present location near 10° N in Costa Rica.

Colinvaux (1993, 1996) has further discredited the notion, first proposed by Haffer (1969), that at times of glacial maxima in the Quaternary lowland evergreen forests and their faunas were replaced in tropical America, except in a few refugia, by dry-adapted formations. According to Haffer and others (Whitmore and Prance 1987), present-day dry forests and related associations have contracted ranges that will expand and reconnect during the next glacial expansion. The Pleistocene alternation of rainforest (interglacial) and dry forest (glacial) expansion and contractions was postulated as being responsible for numerous speciation events. As described above, the distribution of dry forest habitats and their faunal associates are the result of longterm changes in the physiography and climate of Central America in the Neogene, not slight Pleistocene fluctuations. Most important in this regard was the uplift of the central mountain backbone of the region that emphasized the rain shadow effect on the Pacific slope and fragmented Atlantic

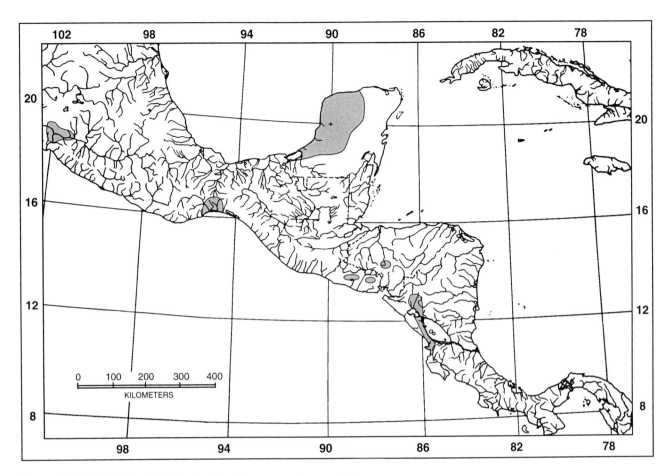

Figure 16.24. Distribution of the lizards of the *Ctenosaura quinquecarinatus* group.

dry forest habitats from those on the Pacific and from one another. The strong evidence for these vicariant processes as primary for dry forest development adds support to Colinvaux's refutation of the refugium hypothesis.

WHY THERE ARE SO MANY SPECIES BETWEEN TWO CONTINENTS, BETWEEN TWO SEAS

The present Isthmian Link is relatively recent geologically, having its initial connection to the southern end of the Chortis block between 6 and 10 Ma. Its final emergence as a continuous landmass between North and South America was not permanently effected until the late Pliocene, as recently as 2 Ma. Most of the species in the herpetofauna of Costa Rica and Panama or their ancestors could have migrated onto the link no sooner than 6 or 7 Ma from the north and no more than 3 Ma from the south. Subsequent rapid speciation must have occurred to produce the astonishing diversity characteristic of the region today (table 16.2).

It is not surprising that at the species level the greatest diversity is represented by the representatives of the autochthonous Mesoamerican Element and the coevolving Cen-

tral American Component that developed in situ in tropical Middle America throughout the Tertiary (table 16.3). Fully 70% of the species found in Costa Rica and elsewhere on the Link are derivatives of these historical units. There may have been some minimal overwater dispersal to the island precursors of the Isthmus in the early Miocene, but it is clear that the central core of the herpetofauna dispersed from nuclear Central America after the island chain became connected via the Chortis block. Without question, the availability of the many unoccupied ecological niches for amphibians and reptiles led to substantial differentiation. Within the Central American herpetofauna 156 species are endemic to Costa Rica and western Panama. Of these endemics, 54% are Mesoamerican Element species and 29% are Central American Component representatives. Of the total Costa Rican–western Panamanian herpetofauna, Mesoamerican Element forms endemics make up 20% and Central American Component 11% of the species.

The second principal factor contributing to differentiation was the continuing uplift and volcanic activity on the emerging Isthmus (Gardner et al. 1987). These processes opened new habitats and fragmented old ones. Volcanic eruptions and lava flows constantly modified topography

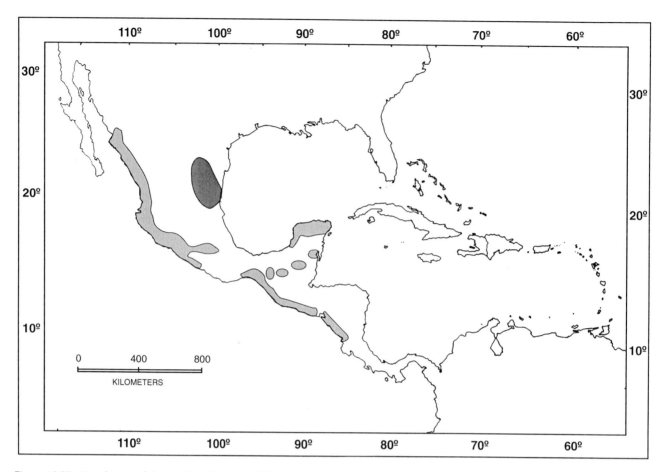

Figure 16.25. Distribution of the cantils *Agkistrodon bilineatus* and *A. taylori*, light and darker shading, respectively.

Table 16.2. Comparison of Species Numbers and Relative Species Richness (ISR) for Selected Geographic Areas

Geographic Unit	N	Area (km²)	ISR
United States	547	7,828,080	0.0007
Mexico	997	1,972,545	0.0005
California	126	441,015	0.03
Florida	122	151,670	0.08
Cuba	158	114,524	0.13
Guatemala	325	108,889	0.29
Oaxaca	355	95,364	0.37
Panama	399	75,474	0.53
Costa Rica	383	50,900	0.75

Introduced and marine species not included.

Index of Species Richness (ISR) = $N/km^2 \times 100$.

and produced a constantly changing mosaic of isolated areas where speciation could occur. The orogenic influence cannot be overemphasized as a factor in the diversification of the many upland endemics in the region.

The fluctuations of cooling and warming cycles in the latter part of the Cenozoic, in association with the continued uplift of the cordilleras, provided additional impetus for speciation. At times of glacial maximum the compression of altitudinal zones produced environments in the lowlands most closely resembling those of midelevation forests (500 to 1,600 m) at present. This situation allowed cool-adapted members of the herpetofauna to migrate between separate mountain masses and volcanoes, where during times of warmer conditions the corridor was broken and each allopatric montane population had the opportunity for differentiation. The several glacial cycles created multiple opportunities for this process to be repeated, with potential for further speciation events (fig. 16.21).

A fourth factor in the assemblage of the diverse Costa Rican herpetofauna was the reconnection of Central America to South America by the completed Isthmian Link. This resulted in the invasion of Central America by South American Element taxa long isolated from any relatives to the north. In Costa Rica and western Panama, endemics derived from this source make up 12% of the endemics and 4% of the total species in the herpetofauna. Some of these species doubtless are descendants of ancestors that crossed the first bridge that closed the seaway, and others probably arrived after the second closure proposed by Cronin and Dowsett (1996). In either case they arrived after the Mesoamerican

Table 16.3. Numbers (Upper Figure) of Species and Percentage (Lower Figure) of Total Costa Rican Herpetofauna by Historical Source Unit

Group	Old Northern			Mesoamerican	South American	Totals
	CAC	EAC	WAC			
Caecilians				3	1	4
				0.7	0.25	1
Salamanders	37	—	—	—	—	37
	10			0		9
Anurans	4	2	—	66	59	131
	1	0.5		17	15	34
Total amphibians	41	2	—	69	60	172
	10	0.5		18	15	45
Lizards	8	—	4	36	21	69
	2		1	9	6	18
Snakes	51	4	1	60	6	133
	13	1	0.25	16	4	35
Turtles		2	—	—	—	8
	1.6	0.5				2
Crocodilians	—	—	—	1	1	2
				0.25	0.25	0.5
Total reptiles	65	6	5	97	38	211
	17	1.6	1.3	26	10	55
Grant total	106	8	5	166	98	383
%	28	2.1	1.3	43	26	100

Seven marine species and six introduced species not included.

Element and Central American Component units were well in place and make up a significant but smaller percentage (26%) of the total species in the fauna.

The final act, bringing our story to its close, was the rather recent movement of dry forest taxa into northwestern Costa Rica and onto the Meseta Central. Late Quaternary volcanism was the principal factor in reducing the effects of the northeast trade winds on the Pacific slope. Increasingly xeric conditions in the northwest, unmitigated by monsoonal flow from the Southern Hemisphere, created an opportunity for northern forms to penetrate the formerly humid lowlands. The recency of these events is demonstrated by the absence of any species strictly endemic to this region of Costa Rica and the close resemblance of its herpetofauna to that of the subhumid to semiarid lowlands from Guatemala south.

The population of the Meseta Central Occidental by dry forest species may imply a recent ongoing dispersal event. On the other hand, both mesetas may have been covered by dry forest earlier in history, and perhaps their continuing uplift has merely carried dry forest taxa upward in the process.

The former hypothesis suggests that the subset of dry forest amphibians and reptiles on the Meseta are good dispersers with more to follow. The second hypothesis is that they are holdovers and are gradually slipping off the Pacific side from the Meseta Central Occidental as uplift continues, a process that will ultimately lead to their extinction on the Meseta Central Oriental.

Documentation of herpetofaunal diversity for Costa Rica remains incomplete. Recognition and description of species new to science by several authors is ongoing, and some of these works will doubtless appear while this book is in press. Still other species will be discovered in the future in under-explored areas of the republic as access increases and habitat destruction accelerates. *Vaya bien* to all of you searchers on this rich coast. May you too emerge from the forest of interlocking opposites on the Pass of Desengaño while following the trail of the golden frog to new and exciting discoveries. *¡Pura vida!*

Addendum

The following species were added to the herpetofauna of Costa Rica while this book was in the editorial process. Several are newly recognized taxa, and others have been resurrected from synonymy. It was not possible at that late date to include them in the keys or to prepare maps of their distributions. These additions bring the number of valid native amphibians to 178 and of native reptiles to 218. The total herpetofauna is now 396 native species plus 2 introduced frogs and 4 introduced lizards for a grand total of 402 species. These additions do not modify the patterns and conclusions presented in chapters 14, 15, and 16.

Amphibia
Gymnophiona
Family Caeciliidae

Savage and Wake (2001) clarified the status of Costa Rican caecilians referred to the genus *Dermophis*. They recognized one additional form previously confused with *D. mexicanus* (Savage and Wake 1972) or *D. gracilior* (this book) and two synonymized with *D. parviceps* (Savage and Wake 1972) but tentatively included with *D. gracilior* on map 5.1 in this book.

Dermophis costaricense Taylor, 1955

(plates 19–20)

DIAGNOSTICS: A relatively robust, moderate-sized species (to 387 mm in total length), plumbeous in dorsal color, having the eye visible (fig. 5.3c), 107 to 117 primary annuli, 74 to 96 secondaries, and 186 to 208 total annuli. In addition, the head is the same color as the body.

DISTRIBUTION: Premontane slopes of the Atlantic versant of Costa Rica (1,000–1,300 m).

COMMENT: All Atlantic slope localities indicated on map 5.1 are for this species.

Dermophis glandulosus Taylor, 1955

DIAGNOSTICS: A relatively robust, moderate-sized species (to 405 mm in total length), plumbeous in dorsal color, having the eye visible (fig 5.3c), 91 to 106 primary annuli, 37 to 60 secondaries, and 132 to 159 total annuli. In addition, the head is the same color as the body.

DISTRIBUTION: Lowland, premontane, and lower montane slopes of southwestern Costa Rica and southwestern Panama and lowlands of eastern Panama and northern Colombia (404–2,000 m).

COMMENT: This form is sympatric with *D. gracilior* and possibly *D. parviceps* in the uplands of southwestern Costa Rica and adjacent Panama.

Dermophis gracilior (Günther, 1902)

COMMENT: Characteristics for *D. gracilior* as distinct from *D. costaricense* are detailed in the remarks section for this species (p. 117). As now understood *D. gracilior* is known from only three premontane localities in southwestern Costa Rica and adjacent western Panama (980–1,200 m). It is putatively sympatric with *D. glandulosus* and *D. parviceps* in this region.

Dermophis occidentalis Taylor, 1955

DIAGNOSTICS: A relatively robust, small species (to 235 mm in total length) having the eye visible (fig. 5.3c), 95 to 112 primary annuli, 29 to 37 secondaries, and 126 to 129 total annuli. In addition, the head is lighter in color than the body.

DISTRIBUTION: Lowlands and premontane slopes of southwestern Costa Rica onto the Meseta Central Occidental (365–970 m).

COMMENT: This species generally occurs at lower elevations than *D. glandulosus* and *D. gracilior*. *D. occidentalis* and *D. glandulosus* are probably sympatric southwest of San Isidro de El General, San José Province.

Caudata
Family Plethodontidae

Bolitoglossa marmorea (Tanner and Brame, 1961)

DIAGNOSTICS: A moderate-sized (adults 128 to 134 mm in total length), long-legged, big-footed, purplish brown salamander marked with yellow flecks and small yellow spots. The species has moderate digital webs with subterminal pads (fig. 6.34m). Adult males 59 to 66 mm in standard length, adult females 60 to 72 mm in standard length; tail moderately long, 48 to 56% of total length; maxillary teeth 48 to 80; vomerine teeth 22 to 38; 0 to ½ costal folds separating adpressed limbs in males, 1½ to 2½ in females; proportions as percentages of standard length: head width 13 to 15% in adult males, 14 to 16% in adult females;

leg length 28 to 29% in adult males, 25 to 30% in adult females.

DISTRIBUTION: Lower montane and montane slopes of the southern Cordillera de Talamanca-Barú of Costa Rica and western Panama (1,920–3,444 m).

COMMENT: Biochemical evidence convinced David Wake (pers. comm.) that this species is distinct from the rather morphologically similar *Bolitoglossa sooyorum* of the northern Cordillera de Talamanca. *B. marmorea* was originally collected under rocks in the crater of Volcán Barú in western Panama. Wake, Brame, and Duellman (1973) reported that it was collected at night moving about on mossy trunks and limbs of trees in the Cerro Pando area in Panama on the Costa Rica–Panama border. The late Douglas Robinson (pers. comm.) reported similar observations for the species in extreme southwestern Costa Rica. This salamander may rest during the day under rocks but apparently emerges at night to forage on arborescent vegetation. In Costa Rica, known only from Lower Montane Rainforest on the western slope of Cerro Pando (UCR 8366–69), as indicated by the locality record near the Panama border on the distribution map for *B. sooyorum* (map 6.14). It will key out (p. 137) to *Bolitoglossa sooyorum*, an allied species in the *Bolitoglossa subpalmata* group.

Anura

Family Leptodactylidae

Campbell and Savage (2000) reviewed the entire *Eleutherodactylus rugulosus* group and resurrected six names from synonymy while describing ten new species. Their study requires some changes in the treatment presented earlier.

Eleutherodactylus catalinae Campbell and Savage, 2000

(plate 104)

DIAGNOSTICS: Adult males are moderate-sized frogs (to 49 mm in standard length) having nuptial thumb pads and vocal slits. Females are large (to 75 mm). This species has moderately strong toe webbing (I 2–2½ II 2–¾ III 2¾–4 IV 4–2½ V), with a fleshy fringe or broad flange on all toes, the disks on fingers III and IV and toes one and a half times the width of the digits, and the venter white to cream with some to considerable brown mottling. The posterior thigh surface is brown with distinct small yellow spots.

DISTRIBUTION: Streams in the premontane and lower montane humid forests of extreme southwestern Pacific slope Costa Rica and adjacent western Panama (1,219–1,800 m).

COMMENT: Panamanian examples of this species were referred to *E. fleischmanni* by Savage (1975). The later species is readily distinguished from *E. catalinae* because it lacks toe fringes or flanges and has less toe webbing.

Eleutherodactylus obesus (Barbour, 1928)

COMMENT: Members of this species conform to the description of *Eleutherodactylus punctariolus* presented earlier in this book (p. 251). All Atlantic slope localities indicated on map 7.36 are for *E. obesus*. As now understood *E. punctariolus* is restricted to mountain areas in Coclé, Veraguas, and eastern Chiriquí and Bocas del Toro Provinces, Panama. *E. obesus* differs from *E. punctariolus* as detailed by Campbell and Savage (2000) and has a karyotype of $2N = 22$ compared with the latter with $2N = 24$ and *E. rhyacobatrachus* with $2N = 20$.

Eleutherodactylus rhyacobatrachus Campbell and Savage, 2000

DIAGNOSTICS: A member of the *Eleutherodactylus punctariolus* complex (p. 251). Males moderate-sized (to 50 mm in standard length), having nuptial thumb pads and vocal slits. Females large (to 81 mm). This species has large finger and toe disks (twice the width of digits on fingers III and IV and toes III–IV–V), toe fringes, and extensive toe webbing (I 1¾–2 II 1½–3 III 2–3½ IV 3⅓–1¾ V) and a strong inner tarsal fold. The venter is pale yellow, usually heavily marked with brown pigment, and the posterior thigh surface is mottled with pale yellow and brown.

DISTRIBUTION: Premontane and lower montane southern slopes of the Cordillera Talamanca-Barú of Costa Rica and western Panama (950–1,800 m).

COMMENT: All Pacific slope localities on map 7.36 are for this species. The karyotype ($2N = 20$) reported by De Weese (1976) for *E. punctariolus* is from a specimen of this form. *E. rhyacobatrachus* is sympatric with *E. fleischmanni* at intermediate elevations along the southern slopes of the Cordillera de Talamanca in Costa Rica. It is also sympatric with *E. catalinae* at many upland localities in western Panama.

Glossary

The definitions provided below give the meanings of terms used in this book as they apply to living amphibians and reptiles. Features used in keys and species accounts defined in chapter 3, included in the sections on key features in chapters 5–7 and 10–13, or illustrated in the text usually have not been included. Users of this book primarily interested in species identification should rely on those sources to familiarize themselves with the minimum set of technical or specialized descriptors necessary to successfully use the keys or to fully understand the species accounts.

The glossary emphasizes features characterizing the major groups covered in this book in order to clarify for the serious student the material in the general accounts. For this reason morphological, including osteological, characters predominate, but systematic, developmental, ecological, genetic, and geological terminology that might otherwise be confusing or unintelligible is also included.

English nouns and adjectives derived from the scientific names of taxa covered in this book are not included, since their meanings are obvious (e.g., anurans, anoles, boids). To reduce redundancy, adjectival forms of many nouns defined in this glossary (or vice versa) are not separately defined (e.g., cranium is listed but not cranial; osteological but not osteology). In the case of skeletal features, the term element is used where a structure may be cartilaginous in some taxa but ossified (bone) in others.

The treatment is neither exhaustive nor complete. For fuller coverage of general biological terms, refer to Lincoln, Boxshall, and Clark (1987, 1998) or Martin, Ruse, and Holmes (1996). A very thorough coverage of the definitions of most terms used in herpetology is provided by Peters (1964).

Abdomen (n.): Generally, the ventral area of the body of vertebrates lying between the pectoral and pelvic appendages. Specifically, the cavity lying posterior to the lungs' peritoneal cavity.

Abductor (n.): A muscle drawing a segment of the body away from the midline or a part of the body.

Abembryonic (adj.): Away from the embryo.

Abrogate (v.): To abolish or nullify.

Acetabulum (n.): Depressions in the pelvic girdle, where the proximal limb bone (femur) articulates.

Acrodont (adj.): Used to describe a tooth attached to the apical crest of a jawbone.

Acuminate (adj.): Tapering to a slender point (fig. 7.15c).

Adductor (n.): A muscle drawing a segment of the body toward the midline or a part of the body.

Adductor mandibulae (n.): The muscle or muscles closing the mouth in amphibians and reptiles.

Advertisement call (n.): The species-specific call of a male anuran that serves to establish his territory or attract females to the breeding site.

Albinistic (adj.): Pertaining to or having little or no pigment; appearing white.

Alkaloids (n.): A series of complex bitter and often toxic organic bases.

Allantois (n.): One of four extraembryonic membranes found in amniotes. During development the allantois stores nitrogenous wastes.

Allochthonous (adj.): Originating outside a particular area or habitat but now living there.

Allopatric (adj.): Of or pertaining to taxa occupying different disjunct geographic areas.

Allopolyploid (n.): A hybrid organism having the chromosome complement of two different species, as in some hybrid frog and lizard species.

Alternate tooth replacement (n.): Tooth replacement where the old tooth is replaced by a new one developing behind it.

Alternating calls (n.): Calls by male frogs and toads spaced not to overlap with the calls of nearby males.

Amino acids (n.): A series of nitrogen-containing molecules, twenty of which commonly combine to form proteins.

Amnion (n.): One of four extraembryonic membranes found in amniotes forming a closed fluid-filled sac around the embryo in reptiles, birds, and mammals.

Amniote (n.): Any vertebrate in which an amnion is present in the egg or placenta. Reptiles, birds, and mammals are amniotes.

Amoeba (n.): A genus of naked unicellular organisms that move and change shape by extending parts of their protoplasm.

Amoeboid (adj.): Resembling an amoeba in form or movement.

Amphicoelous vertebra (n.): A vertebra with concave anterior and posterior faces.

Amplectant (adj.): Of or pertaining to amplexus.

Amplex (v.): To preform amplexus.

Amplexus (n.): The sexual embrace or clasping of a female anuran by a male.

Anal plate(s) (n.): The large transverse scute or pair of scutes covering the cloacal opening in snakes; this is a misnomer, since snakes lack an opening (anus) for the digestive tract separate from the cloaca.

Anamniote (n.): Any vertebrate lacking an amnion, chorion, and allantois in the egg or placenta. Amphibians and all fish-like vertebrates.

Anapsid (n.): A type of reptile skull lacking temporal foramina; i.e., the postorbital region is covered by bone, as in sea turtles (fig 12.17a).

Angiosperm (n.): A flowering plant.

Angular bone (n.): A dermal bone forming the posteroventral portion of the lower jaw in most reptiles.

Animal pole (n.): The center of the region of the egg that normally develops into the integument and nervous system. In amphibians it usually has a high concentration of dark pigment in contrast to the vegetal region, which is yellow to white; in reptiles it is the clear region of the egg that will develop into the blastodisc.

Annular groove (n.): A groove in the integument of caecilians that borders and defines an annulus. Dermal scales are present in the annular grooves of some species of caecilians.

Annulus (pl. annuli) (n.): A ring or ringlike structure or marking.

Anterior (adj.): The anatomical direction pertaining to or toward the front (head) end. Synonyms: cranial, rostral.

Anteroventral (adj.): A position intermediate between anterior and ventral; oriented forward and downward.

Antiphonal alternation (n.): Alternation of calling by two neighboring anuran males.

Anuran (adj.): Of or pertaining to the Anura or to members of that order.

Aphotic (adj.): Of or pertaining to an environment having no sunlight.

Apical (adj.): Of, at, or constituting the apex.

Aplacental viviparity (n.): Viviparity in which the embryo receives nutrition from the mother without a placental connection between them.

Aposematic (adj.): Of or relating to an antipredator defense that calls attention to the prey (e.g., bluffing, warning, or mimicry).

Arciferal pectoral girdle (n.): An anuran pectoral girdle in which the two epicoracoid cartilages are free and overlapping.

Areolate (adj.): Covered with large, globular, closely set prominences, as on the skin of the venter in some anurans.

Articulation (n.): The soft-tissue junction lying between and connecting two bones; when movable, the articulation forms a joint.

Astragalus. See Fibulare

Asynchronous breeding (n.): Females arriving at the breeding site irregularly through the breeding period and sometimes returning to mate several times.

Atlantal processes (n.): Paired lateral processes on the atlas that articulate with the occipital condyles in living amphibians.

Atlas (n.): The vertebra that articulates to the cranium.

Atmosphere (n.): The gaseous envelope surrounding the earth.

Attenuate (adj.): Elongate, slender, or thin; sharply tapering toward the distal end.

Auctorum (n.): Of authors.

Audiospectrogram (n.): The graphic representation of an anuran vocalization produced by a sonograph consisting of a sonogram (frequency against time) and a spectrogram (amplitude against frequency).

Autochthonous (adj.): Originating within a particular area or habitat.

Autotomy (n.): Autotomous separation of the tail, usually through intravertebral breakage followed by tail regeneration. Typical of most lizards exhibiting tail breakage.

Autotomy septum (n.): The region of a vertebra where fracture occurs during autotomy. Synonym: fracture septum.

Available name (n.): Any name proposed for a taxon that meets the requirements of the *International Code of Zoological Nomenclature.*

Axilla (n.): The posterior angle formed with the body by insertion of the forelimbs.

Azygous (adj.): Unpaired. The term is usually applied to the unpaired median head scales of snakes and lizards.

Backcross (n.): An individual or individuals resulting from backcrossing.

Backcrossing (n.): The process of crossing a hybrid with an individual having the same genetic complement as one of the hybrid's parental species.

Barbel (n.): A fleshy tubular extension of the skin, usually on the head or neck, presumed to have a sensory (probably tactile) function; term derived from barb (a beard).

Basal (adj.): At or pertaining to the base.

Basal tooth replacement (n.): Tooth replacement in which the old tooth is pushed out from directly below by the new tooth.

Basibranchials (n.): Median cartilaginous or ossified elements of the hybobranchial skeleton. Synonym: copulae.

Basihyal (n.): The central large, anterior cartilaginous or ossified element of the hyoid apparatus in reptiles.

Batesian mimicry (n.): A mimetic complex where a harmless species (the mimic) shares an aposematic antipredator defense with a distasteful, toxic, or dangerous one (the model).

Batrachotoxins (n.): Toxic steroidal alkaloids known only from the skins of frogs of the genus *Phyllobates* and the feathers of one species of New Guinea bird *(Pitohui dichrous)*.

Benthic (adj.): Pertaining to the bottom of a body of water.

Benthos (n.): Organisms attached to or living on, in, or near the bottom of a body of water.

Bicondylar (adj.): Having two condyles.

Biconodont (n.): Teeth having two cusps. Synonym: bicuspid.

Bicuspid. See Biconodont

Bidder's organ (n.): Ovarian tissue retained from early development in bufonid males, associated with the functional testis.

Bifid (adj.): Partially divided into two parts; used in reference to subarticular tubercle shape in frogs (fig. 7.72).

Bilobed (adj.): Having two lobes.

Binocular vision (n.): Vision in which the visual fields of the two eyes overlap.

Biparental (adj.): Having two parents.

Bipedal (adj.): Two-legged. Usually used to describe locomotion where an animal walks or runs using only the hind limbs.

Bisexual species (n.): Species having both males and females in the population.

Blastodisc (n.): The cap or disk of cells (embryo) on the animal pole of an egg sitting on the uncleaved mass of yolk. Typically produced by meroblastic cleavage.

Body pit (n.): The depression female sea turtles make before digging the actual nest below the surface of and posterior to the body pit. Sometimes one or more are dug at other sites after the eggs are laid, presumably to deceive predators.

Bone (n.): Material produced by the deposition of a specific calcium phosphate salt, hydroxyapatite, into a fibrous or cartilaginous matrix.

Bony scales (n.): Small, thin plates of bone found in bony fishes and many caecilians. Synonym: dermal scales.

Brachial (adj.): Pertaining to or situated on the upper (humeral) part of the arm.

Bract (n.): A specialized leaf subtending a reproductive structure such as a flower.

Brille (n.): The transparent covering over the cornea derived from fusion of the eyelids. Synonym: spectacle.

Brooding. *See* Egg brooding

Brood pouch (n.): An integumentary pouch in some frogs within which eggs undergo development.

Bryozoans (n.): Various sessile colonial invertebrates commonly called "moss animals"; members of the phyla Entoprocta and Ectoprocta eaten by some marine turtles.

Buccal (adj.): Of or pertaining to the mouth cavity.

Buccopharyngeal (adj.): Of or pertaining to the combined oral (buccal) and pharyngeal cavities.

Buccopulmonary respiration (n.): Ventilation of the lungs by raising and lowering the floor of the mouth. Present in all living amphibians.

Buttons (n.): The osteoderms under the ventral scutes of some crocodilians. There may be one or two interlocking osteoderms under each scute.

Calcaneum. *See* Tibiale

Calcareous (adj.): Consisting of or containing calcium carbonate.

Calcified (adj.): Containing calcium carbonate or phosphate salts, as in calcified cartilage.

Calipash (n.): A fatty, gelatinous dull green edible substance attached to the carapace of green sea turtles *(Chelonia)*. The source of the common name green turtle.

Calipee (n.): A gelatinous, fatty light yellow edible substance attached to the plastron of green sea turtles *(Chelonia)*.

Call group (n.): One or more call notes.

Call note (n.): A single note, whether in a single- or multiple-note call group.

Calyces (n.): Cup-shaped structures.

Calyculate (adj.): Covered by cup-shaped structures, as in tripe.

Canthal (adj., n.): Of or pertaining to the canthus rostralis or scales on the canthus rostralis.

Canthus rostralis (n.): The angle of the head from the tip of the snout to the anterior corner of the eye (amphibians) or the anterior end of the eyebrow (reptiles) that separates the upper head surface from the side (lore) of the snout; synonym: canthal ridge.

Capillary action (n.): The action by which the surface of a liquid, where it is in contact with a solid (e.g., a grooved tooth), is elevated or depressed depending on the relative attractions of the involved molecules; when the surface is elevated the liquid may slowly flow by capillary movement.

Capitulum (n.): The apical area of the hemipenis in squamate reptiles that is separated from the rest of the organ by a distinct groove (fig. 11.2a).

Cardiotoxin (n.): A toxin that adversely affects the heart.

Carotid artery (n.): The vessel that carries oxygenated blood to the head in amphibians and reptiles.

Cartilage (n.): A translucent elastic tissue that forms the embryonic skeleton and is variously converted to bone in adult vertebrates.

Casque (n.): A modification of the skull and associated structures of the head to resemble a helmet. Usually involves proliferation of the dermal roofing bones of the skull or co-ossification of skin and skull.

Caudad (adv.): in the direction of the tail.

Caudal (adj.): Of or pertaining to the tail.

Centrum (pl. centra) (n.): The body of a vertebra as distinct from the neural arch above it; the shape of the articulations between vertebral centra are important systematic features.

Cephalic (adj.): Of or pertaining to the head.

Ceratohyals (n.): Paired anterolateral cartilaginous or ossified elements of the hyoid apparatus forming the anterior cornua.

Cerebral artery (n.): A branch of the internal carotid artery supplying the brain.

Cervical (adj.): Of or pertaining to the neck region.

Chemoreceptors (n.): Sense organs that sample chemicals in the environment that human beings perceive as odor and taste.

Chemosensory (adj.): Of or pertaining to taste, olfaction, and the related function of the vomeronasal (Jacobson's) organs.

Choanae (sing. choana) (n.): Internal nares that open into the mouth.

Chorioallantoic placenta (n.): A placenta found in all viviparous reptiles and most placental mammals in which the fused chorion and allantois form the embryonic contribution to the structure.

Chorion (n.): The outermost of the four extraembryonic membranes of amniotes.

Choriovitelline placenta (n.): A transitory yolk sac placenta forming a vascularized area around the embryo early in development. Present in all viviparous reptiles.

Chromatophore (n.): A cell containing pigment, especially in the integument.

Chromosome (n.): A genetic unit in the cell nucleus containing a very long molecule of DNA within a protein coating. Visible only as a morphological entity during cell division. Chromosomes usually occur in pairs (2N) in somatic cells and are unpaired (N) in the sex cells.

Ciliary muscle (n.): The eye muscle that focuses an organism's vision by contracting the lens to change its shape.

Circular backward burrowers (n.): Anurans that construct burrows by digging with the feet while rotating in a corkscrew pattern.

Cirrate (adj.): Formed of a slender spiraling coil.

Cirrus (n.): A slender, flexible appendage; e.g., the downward extension of the upper lip below the nostril containing the nasolabial groove in male plethodontid salamanders.

Clavate (adj.): Gradually thickening toward the distal end; club shaped.

Clavicles (n.): Paired dermal bones on the anterior margin of the pectoral girdle.

Cleidoic egg (n.): An egg enclosed in a shell, laid in a terrestrial situation. Typical of reptiles and birds.

Cline (n.): A continuous variation in expression of some feature along an ecological or geographic gradient; e.g., the change in the number of ventral scales from north to south in a series of snake populations of the same species.

Cloaca (n.): The common chamber that receives the contents of the digestive, excretory, and reproductive systems; it communicates to the outside through the vent.

Cloacal (adj.): Of or pertaining to the cloaca or the region of the body where the cloaca opens to the outside.

Cloacal bones (n.): Small bones associated with the cloacal sacs in some geckos.

Cloacal sacs (n.): Paired glandular organs lying posterior to the cloaca and emptying into it that produce a secretion.

Cnidarians (n.): Members of the phylum Cnidaria, which includes jellyfish, sea anemones, and corals among other groups.

Coccyx (n.): A long, rodlike endochondral bone formed by fusion of postsacral vertebral elements in frogs and toads that articulates with or is fused to the sacrum, or rarely is merely connected to it. It lies between the long shafts of the ilia and bears muscular attachments to them. Synonym: urostyle.

Cochleophagus (adj.): Of or pertaining to a predator that eats snails.

Colic (adj.): Of or pertaining to the large intestine (colon).

Columella (n.): The ear bone of amphibians and reptiles involved in transmitting pressure waves from the tympanum or through the soft head tissues to the inner ear. Synonym: stapes.

Compressed (adj.): Flattened side to side.

Conchae (**sing. concha**) (n.): Projections from the walls of the nasal sac.

Concordant (adj.): Occurring at the same time.

Condyle (n.): A rounded protruding, articulating bone surface.

Cones (n.): 1. Light-sensitive cells (photoreceptors) functioning in bright light, situated in the retina of the eye and conveying specific wavelength information interpreted as color by the brain; responsible for color vision. Structural cones may be single or double. 2. A mass of scales or bracts bearing ovules or pollen, characteristic of pines and firs and their allies.

Conical (adj.): Resembling or shaped like a cone.

Conifer (n.): A cone-bearing plant, usually a tree (e.g., a pine or cypress).

Conspecific (n., adj.): A member of the same species, or pertaining to such a member.

Convergence (n.): The taking on of similar features by distantly related organisms. Example: the beaks of turtles and birds.

Coossified (adj.): The condition in which the integument is fused to the underlying bones.

Coracoid foramen (n.): An opening in the coracoid bone of the pectoral girdle.

Cordate (adj.): Heart-shaped.

Coriolis force (n.): The deflecting force of the earth's rotation causing the water in the oceans and air masses to move clockwise in the Northern Hemisphere and counterclockwise in the Southern Hemisphere.

Cornea (n.): The transparent outermost layer of the eyeball in amphibians and reptiles having movable eyelids. In all forms having functional eyes, the cornea covers the iris and pupil and admits light to the interior.

Cornified (adj.): Covered by or impregnated with keratin.

Cornuae (**sing. cornua**) (n.): General term for various projecting parts of the hyoid apparatus.

Coronoid (n.): A dermal bone on the lower jaw that in most reptiles projects dorsally as an attachment for the temporal muscles.

Cosmopolitan (adj.): Occurring worldwide.

Costal (adj., n.): Of or pertaining to a rib; costal grooves in salamanders and primary grooves in caecilians are external indi-

cations of the position of the ribs; costal folds lie between the ribs and are external indicators of the muscle masses. The costal shields on a turtle's carapace lie above the fused ribs (fig. 12.1a).

Costiform (adj.): Riblike.

Cotyle (n.): A cuplike cavity; the concave surface of a vertebral centrum that the condyle of the neighboring vertebra fits into.

Countercurrent heat exchange (n.): The movement of heat from an artery containing warm blood to an adjacent vein containing cool blood being returned from the periphery of an organism to the body core; a mechanism to reduce heat loss and maintain a high core temperature; found in the leatherback sea turtle *Dermochelys*.

Courtship call (n.): A characteristic vocalization produced by a male anuran when a female approaches his calling site.

Cranium (n.): Skull.

Crenulate (adj.): Having a wavy outline (fig. 7.70a).

Cretaceous period (n.): A subdivision of the Mesozoic era lasting from 144 to 65 Ma.

Crista parotica (n.): A horizontal ridge on the lateral surface of the otic capsule.

Cruciform (adj.): Having the shape of a cross.

Crus. *See* Tibial segment

Crustaceans (n.): Members of the subphylum Crustacea, which includes shrimps, lobsters, crayfish, crabs, and related organisms.

Crypsis (n.): The condition of being cryptic.

Cryptic (adj.): Difficult to detect, especially visually, owing to an animal's resemblance to its surroundings.

Cryptic species (n.): Species that are essentially indistinguishable morphologically but distinct in other features.

Cuspidate (adj.): Having a cusp; used in reference to disk pad shape (fig 7.51q).

Cutaneous (adj.): Of or pertaining to the skin.

Cycoloid (adj.): Of or pertaining to scales that have a curved circular posterior margin (fig. 10.3).

Cytotypes (n.): Different karyotypes within a nominal species or population.

dbh (n.): Abbreviation for diameter at breast height (1.4 m), a measurement foresters use in evaluating tree size.

Deciduous (adj.): Of or pertaining to the act of shedding. Example: loss of leaves in the dry season.

Demersal (adj.): Living at or near the bottom of a lentic body of water but having the capacity for active swimming; also pertaining to eggs that sink to the bottom when oviposited.

Dentary (n.): Anterior tooth bearing the dermal bone of the mandible in most amphibians and reptiles.

Denticle (n.): One of a series of keratinized, toothlike structures surrounding the mouths of most tadpoles.

Denticulate (adj.): Covered with small pointed projections.

Dentigerous process (n.): That portion of a bone having teeth on it.

Depressed (adj.): Flattened (from top to bottom) in a horizontal plane.

Depressor mandibulae (n.): The muscles that open the mouth in amphibians and reptiles.

Dermal bones (n.): Bones forming from thickenings in the dermal layer of the skin; homologous to dermal scales.

Dermal fold (n.): A fold of skin.

Dermis (n.): The inner layer of the skin, generally consisting of connective tissue fibers, nerves, blood vessels, and dermal derivatives such as osteoderms.

Detritovore (n.): An organism that feeds on fragmented particulate organic matter (detritus).

Devonian period (n.): A subdivision of the Paleozoic era lasting from 408 to 360 Ma; the "age of fishes."

Dextral (adj.): Of or pertaining to the right side.

Diagonal sequence gait (n.): Locomotion where the forefoot strikes the ground after the hind foot on the opposite side.

Diapophysis (n.): A process on the neural arch of a vertebra articulating with the upper head of a rib.

Diapsid skull (n.): A skull having two openings (fossae) on each side in the temporal region, as in crocodilians and birds.

Diastema (n.): A space or gap.

Dichotomous key (n.): An identification key constructed as a sequence of paired alternatives.

Dichromatism (n.): Two colors. Species in which the sexes are of different colors are said to exhibit sexual dichromatism.

Diel (adj.): Of or pertaining to the twenty-four-hour day. A diel cycle, for example, has a period of approximately twenty-four hours.

Diencephalon (n.): The unpaired part of the forebrain that connects posteriorly to the region of the tectum.

Digit (n.): A finger or toe.

Dimorphism (n.): The existence of two morphologies, as in species where the sexes differ in form (e.g., members of one sex are larger than members of the other).

Diploid number (2N) (n.): The number of chromosomes typical of somatic cells in most organisms; double the number of chromosomes in the haploid (N) sex cells (gametes).

Direct development (n.): In salamanders and anurans, embryonic development that is completed within the egg, there being no free-living larval stage.

Distal (adj.): Situated away from the base or point of attachment of a structure. The expanded disks at the tips of the digits of hylid frogs are distal to the phalanges at the base of the digits, for example.

DNA (n.): Deoxyribose nucleic acid. Long molecules of DNA contain the genetic code that programs development to produce differences between individuals and taxa.

Dominant frequency (n.): In bioacoustics, the frequency within which the greatest amount of acoustic energy is concentrated.

Dorsad (adv.): Toward the dorsum.

Dorsal (adj.): Of or pertaining to the upper surface of the body or other structure.

Dorsolateral (adj.): A position that is intermediate between dorsal and lateral.

Dorsum (n.): The upper surface of the body or of a structure.

"Dry" bite (n.): A bite by a venomous snake during which no venom is injected into the victim.

Ecomorph (n.): A morphology shared by relatively distantly related species adapted for similar ecologic conditions.

Ectochordal vertebrae (n.): Vertebrae in which there is a persistent notochord in adults.

Ectoderm (n.): A primary embryonic germ layer giving rise to the outer covering of the body and the nervous system of vertebrates.

Ectothermic (adj.): Of or pertaining to ectothermy, the condition in which body temperature is determined largely by heat sources external to the animal. Animals that exhibit ectothermy are called ectotherms.

Edentulous (adj.): Lacking teeth.

Egg (n.): Strictly speaking, a female gamete or ovum; often used to describe the ovum plus the surrounding egg membranes in amphibians and the membranes and shell in reptiles.

Egg brooding (n.): Physical contact with the developing eggs by a parent or parents.

Egg capsule (n.): One or more gelatinous membranes around an egg, secreted by the oviduct.

Egg guarding (n.): Defense by one or both parents of the developing eggs against disturbance or predation.

Egg membranes (n.): Membranes produced around the egg as it develops in the ovary or laid down around the egg as it passes down the oviduct. The former are primary egg membranes, the latter tertiary egg membranes. The tertiary egg membrane(s) form the egg capsule in amphibians. A series of tertiary membranes including albumin layers and the shell membranes surround the ovum of reptiles. The secondary egg membrane (zona pellucida of mammals) does not occur in amphibians and reptiles.

Egg or tadpole attendance (n.): The presence of a parent or parents in the vicinity of developing eggs or larvae.

Egg tooth (n.): A true tooth or keratinized caruncle used by a hatching organism to slit the egg capsule or shell.

Electrophoresis (n.): A laboratory technique that uses the differential migration of proteins under the influence of an electrical field to establish genetic differences at the molecular level.

Electrophoretically (adv.): In relation to electrophoresis.

Electroreceptors (n.): Sensory organs stimulated by electrical pulses from the environment.

Emarginate (adj.): Of or pertaining to a margin with a single notch or indentation.

Empathic learning (n.): A response to the observed experience of another individual; e.g., avoidance of coral snakes by a predator that has observed a fatal or debilitating snakebite.

Encounter call (n.): A vocalization produced by a male frog or toad when a rival male calls or approaches closely.

Endemic (n.): A taxon restricted or peculiar to a particular area or habitat. An endemic may be allochthonous or autochthonous.

Endemic (adj.): Restricted or peculiar to a particular area or habitat.

Endochondral bones (n.): Bones preformed in cartilage that ossify during development.

Endoderm (n.): One of the embryonic germ layers that give rise to the lining of most visceral organs in vertebrates.

Endotrophic (adj.): Depending on maternal investment as the source of nutrition during development, as in lecithotrophic and viviparous forms.

Enzyme (n.): A proteinaceous substance regulating or accelerating chemical reactions at body temperatures.

Eocene epoch (n.): A subdivision of the Cenozoic era lasting from 58 to 37 Ma.

Epicardium (n.): That portion of the visceral peritoneum that covers the surface of the heart.

Epibranchials (n.): Distal cartilaginous or ossified elements of the hyobranchial skeleton.

Epicoracoids (n.): Postclavicular, midventral cartilages of anuran pectoral girdles that may be overlapping or fused on the midline.

Epidermal scales (n.): Cornified thickenings formed on the outer layer of the skin (epidermis) that cover the body or shell of reptiles. Enlarged epidermal scales on the head of reptiles and shells of turtles are usually called shields or plates. "Scute" may be used for large, flat epidermal scales, especially those on the venter of snakes. Epidermal scales also occur on the legs of birds and on some mammals.

Epidermis (n.): The superficial layer of the skin, generally consisting of an outer dead, keratinized layer and several living cell layers together with epidermal derivatives such as exocrine glands, epidermal scales, and claws.

Epigeal (adj.): Living on the surface of the substrate.

Epimere (n.): The dorsal portion of the myotome.

Epimeric muscles (adj.): Muscles derived from the epimere; epaxial muscles.

Epipelagic (adj.): Of or pertaining to the upper layer of the oceanic water column where there is some light; extending to a depth of 100 to 200 m depending on the turbidity of the water.

Epiphyllous (adj.): Used to describe organisms or eggs found on leaf surfaces.

Epithecals (n.): Polygonal osteoderms under the soft skin and connective tissue of the carapace and longitudinal ridges of the plastron in the leatherback turtle *Dermochelys coriacea*.

Escutcheon (n.): A shieldlike structure on the undersurface of the venter and legs of some geckos composed of a series of differentiated glandular scales (fig. 10.29).

Esophagus (n.): The anterior part of the digestive tube connecting the pharynx to the stomach.

Eumelanins (n.): A group of black and brown pigments found in melanophores and pterorhodophores.

Eustachian tube (n.): A tube connecting the middle ear canal to the pharynx in anurans and most reptiles; absent in all caecilians, salamanders, turtles, and snakes.

Eustatic (adj.): Pertaining to worldwide changes in sea level, such as would be produced by formation or melting of glaciers, but excluding those due to subsidence or uplift of coastlines.

Eversible (adj.): Capable of being everted (turned inside out).

Exostosed (adj.): Having the surface of the skull roughened by bony outgrowths, as in some anurans.

Exostosis (n.): Bony outgrowths that produce the pitted and roughened surface on various skull bones in some anurans (e.g., in *Osteopilus*, fig. 7.99).

Exotrophic (adj.): Having a free-living, feeding larval stage, e.g., a tadpole or polliwog.

Explosive breeders (n.): Species having reproductive periods of a few days.

Extensible (adj.): Capable of being extended (stretched out to fullest length).

Extracolumella (n.): A cartilaginous rod in the middle ear of reptiles connecting the columella to the tympanum. Synonym: extrastapedial.

Extrastapedial. *See* Extracolumella

Eye-licking behavior (n.): Licking the brille (spectacle) to clean it; found in most geckos and all xantusiid lizards.

Facultative (adj.): A behavior that may or may not occur, depending on circumstances.

Farallon plate (n.): A tectonic plate of the eastern Pacific Ocean from 50 to 25 Ma that was entirely subducted under the North American plate.

Fascia (n.): A muscle attachment consisting of a sheet of connective tissue.

Femoral (adj.): Of or pertaining to the proximal segment of the leg (thigh).

Femur (n.): The proximal long bone of the leg.

Fenestra (n.): A windowlike opening in a skeletal element.

Fenestra ovalis (n.): The opening in the otic capsule between the middle ear cavity and inner ear. The footplate of the columella and operculum (when present) abut this fenestra.

Fenestration (n.): A opening in the surface of a bone.

Fertilization (n.): The fusion of a male gamete (sperm) and a female gamete (ovum) to form a diploid (2N) zygote. In external fertilization the process takes place outside the female's body. In internal fertilization, sperm are introduced (through the cloaca in subjects of this book) into the female's body, where the zygote is formed.

Fibula (n.): The posterior (postaxial) endochondral bone that with the tibia forms the skeleton of the shank (crus) in most tetrapods.

Fibulare (n.): The posterior (postaxial) cartilage or endochondral bone articulating with the fibula or fibular portion of the tibiofibula and forming part of the greatly elongated ankle in anurans. Synonym: astragalus.

Firmisternal pectoral girdle (n.): An anuran pectoral girdle in which the epicoracoid cartilages are fused along the midline.

Flex (v.): To close a joint by contracting a muscle or muscles.

Flexor (n.): A muscle closing a joint.

Flounces (n.): Fleshy transverse ridges on squamate hemipenes.

Foramen magnum (n.): The large opening at the base of the skull through which the spinal cord passes to the brain.

Frenocular (n.): A scale lying near the lower anterior corner of the eye between the loreal and first subocular; found in some lizards.

Frequency modulation (n.): A change in the dominant frequency of a vocalization over the course of a note or call.

Frontal bones (n.): Dermal roofing bones of the skull lying between the orbits and usually bordered posteriorly by a parietal bone. They are usually paired but may fuse to form a single frontal bone or fuse with the parietal to form a single frontoparietal bone.

Frontoparietal bone (n.): A complex dermal roofing bone lying between the orbits and consisting of the fused frontal and parietal bones.

Fundamental frequency (n.): In bioacoustics, the lowest-frequency harmonic.

Gamete (n.): A sex cell of either a male (sperm) or a female (ovum).

Gastralia (n.): Large rodlike skeletal structures derived from dermal scales protecting the abdomen in crocodilians. Synonym: abdominal "ribs." Remnants of similar structures appear to contribute to the plastron in turtles.

Gastromyzophorous (adj.): Bearing a ventral sucker, as in the tadpoles of frogs of the genus *Atelopus* (fig. 7.14c).

Genioglossus muscle (n.): A muscle involved in tongue rotation originating on the mandible and inserting on the tongue pad.

Genitalia (n.): Organs of the reproductive system involved in copulation; secondary sex features produced by the effects of the different hormones secreted by the male or female gonads. Example: the hemipenes of male squamate reptiles or the spermatotheca of female salamanders.

Germ cells (n.): The cells in the gonads that produce sperm and eggs; the germ cells are usually diploid (2N) and divide to produce the gametes, which are haploid (N).

Gestation (n.): The period during which developing young are retained within the oviduct or uterus of the mother in a viviparous species.

Glans penis (n.): The apical portion of the turtle or mammalian penis.

Gliding (n.): Use of a volplaning surface to significantly reduce the angle of descent after jumping or falling from an elevated site.

Glossohyal bone (n.): A single midgular cartilage projecting anteriorly from the basihyal in support of the tongue in reptiles.

Glottis (n.): The slitlike opening leading from the pharyngeal region to the trachea.

Gonadogenesis (n.): The embryonic development of the gonads.

Gonads (n.): The primary sexual characters; organs producing sex cells (gametes); the male gonad is a testis and produces sperm; the female gonad is an ovary and produces ova (eggs).

Gondwanaland (n.): The southern supercontinent formed by the initial breakup of Pangaea about 160 Ma that later fragmented to form South America, Africa, Arabia, Madagascar, India, Australia, New Zealand, and Antarctica.

Gracilis muscle (n.): The gracilis major of anurans originates from the posterior border of the rim of the ischium, extends along the medioventral surface of the thigh, and inserts by two tendons, one (membranous) to the knee and the medial side of the head of the tibiofibula, the other to near the head of the tibiofibula. It functions to flex the knee and extend the hip joint.

Granular (adj.): Having a surface texture consisting of or resembling tiny grains (granules); used with reference to the pebbled skin in some amphibians or the small convex, nonoverlapping scales of reptiles (fig. 10.3).

Granular gland (n.): A large specialized skin gland that stores or produces noxious or toxic chemicals in amphibians; the parotoid glands of amphibians are dense concentrations of granular glands.

Gravid (adj.): Containing fully developed eggs.

Green rods (n.): Photosensitive cells in the retina of anurans and most salamanders that maximally absorb wavelengths of 432 nm.

Groin (n.): The angle formed with the body by the anterior surface of the hind limb at its insertion.

Guanine (n.): The silvery white pigment in the platelets of iridophores.

Guanophore. *See* Iridophore

Gular (adj.): Of or pertaining to the ventral surface of the throat or neck.

Habitus (n.): The characteristic form and appearance of an organism.

Hallux (n.): The first digit of the foot.

Haploid number (N) (n.): The number of each kind of chromosome typical for a species; somatic cells usually contain paired chromosomes of each kind and are diploid (2N); the sex cells (gametes) are haploid.

Harderian gland (n.): A gland at the front medial margin of the eye that secretes an oily liquid for moistening or cleaning the eye; modified in caecilians to provide fluid for transporting chemical particles from the tentacle to the vomeronasal organs.

Harmonic (n.): In bioacoustics, a frequency within which acoustic energy is concentrated; a narrow frequency band.

Hatchling (n.): A newly hatched animal.

Heliothermic (adj.): Relying on direct solar radiation to elevate the body temperature.

Hemipenis (pl. hemipenes) (n.): One of the paired copulatory organs of male snakes and lizards.

Hemipterans (n.): Members of the insect order Hemiptera, composed of bugs with sucking mouthparts.

Hertz (Hz) (n.): The frequency of a periodic phenomenon, such as "sound" waves, of which the periodic time is one second; i.e., one cycle/second.

Heterodont (adj.): Having teeth of several different morphologies.

Heterospecific (adj.): Belonging to different species.

Holarctic (adj.): Of or pertaining to the temperate and Arctic regions of the Northern Hemisphere; includes the Nearctic and Palearctic realms.

Holoblastic (adj.): Of or pertaining to the pattern of early cell divisions (cleavage) of a fertilized egg (zygote) in which the cleavage planes pass completely through the dividing cell. Example: frog and salamander eggs.

Holochordal vertebrae (n.): Vertebrae in which the notochord is entirely replaced by bone.

Homonym (n.): A name for two or more different taxa having the same spelling. Primary homonyms are species names having the same generic names when originally proposed. Secondary homonyms result when two or more species with the same specific epithets are combined into the same genus subsequent to their original descriptions in different genera.

Hybrid (n.): The offspring of or a cross between two species.

Hybridize (v.): To produce hybrids.

Hydrotroph (n.): Nutritive substance secreted by the oviduct and eaten by advanced embryos in some viviparous amphibians. *See also* Aplacental viviparity.

Hyobranchial (adj.): Of or pertaining to the gill arch muscles and skeletal components that in tetrapods and their larvae move or support the tongue, larynx, and trachea.

Hyoid apparatus (n.): The skeletal elements and muscles supporting and moving the tongue in tetrapods.

Hyoid bones (n.): Several endochondral bones supporting the tongue and involved in tongue movements.

Hyomandibula (n.): The dorsal skeletal element of the second branchial (gill) arch involved in jaw suspension in many fishes. In tetrapods modified to form one of the middle ear bones, the columella or stapes.

Hypapophysial (adj.): Of or pertaining to the process (hypapophysis) present on the ventral midline of the vertebral centrum.

Hypapophysis (n.): A median ventral ridge or process on the centrum.

Hypertrophied (adj.): Excessively developed, as an organ or part.

Hypomere (n.): The ventrolateral portion of the myotome.

Hypomeric muscles (adj.): Muscles derived from the hypomere; hypaxial muscles.

Igneous rock (n.): Rock formed by solidification of molten magma (e.g., lava) such as basalts and granites.

Ilium (n.): An endochondral bone forming the dorsal portion of the pelvic girdle that articulates with the sacrum.

Imbricate (adj.): Overlapping, as when the distal portion of one scale overlaps the proximal portion of the next scale. The scales of snakes and of many species of lizards are imbricate.

Indirect development (n.): Development where there is a free-living, feeding larval stage.

Inframarginals (n.): Epidermal shields in turtles lying on the bridge between the carapace and plastron; e.g., the axillary and inguinal shields.

Infraorbital vacuity (n.): The opening in the skull below the eyeball.

Inguinal (adj.): Of or pertaining to the groin region of the body.

Inner ear (n.): The sense organ responsible for maintaining balance and equilibrium and receiving stimuli interpreted as sound by the brain. In anurans and many reptiles pressure waves are transmitted to the inner ear by accessory structures including the tympanum and columella.

Inscriptional rib formula (n.): Notation for the number of pairs of attached and floating postxiphisternal inscriptional ribs; e.g., 3 : 1 means three attached, one floating.

Inscriptional ribs (n.): Slender calcified cartilages situated ventral to the dorsal ribs in many lizards; in the thoracic region they are tied to the dorsal ribs dorsally and the sternum and xiphisternal rods ventrally; in the abdominal region they may or may not be attached to their corresponding dorsal ribs.

Integument (n.): Skin. It consists of an outer layer, the epidermis, and an inner layer, the dermis.

Integumentary scutes (n.): Paired uncornified flaps on the upper surface of the disk covers of dendrobatid frogs.

Integumentary sense organ (ISO) (n.): A distinct porelike structure near the posterior margin of crocodilian scales, most readily seen on the ventral scutes; absent from the body and tail in all alligators and caimans; present on the body in all crocodiles and the gharial.

Intercalary element (n.): A cartilage or bone lying between the distal and penultimate phalanges in some frogs.

Intercentrum (n.): A small bone situated anteroventral to each dorsal vertebra in some reptiles; also usually present in the caudal skeleton but often in modified form as a hypapophysis.

Interclavicle (n.): A single elongate, median dermal bone in the pectoral girdle of squamates and crocodilians that attaches to a clavicle on each side at its anterior end.

Intermedium (n.): A cartilage or bone lying mediad to the ulnare and forming part of the wrist skeleton.

Internal carotid foramen (n.): An opening in the roof of the mouth by which the internal carotid artery enters the cranial cavity. Entry is by means of the carotid canal on the floor of the cranium in salamanders lacking a separate foramen.

Intromittent organ (n.): An organ inserted by a male into the female's cloaca during copulation; a penis, phallodeum, or hemipenis.

Invertebrates (n.): All animals lacking a vertebral column.

Iridophore (n.): A cell containing silvery reflecting platelets. Synonym: guanophore.

Iris (n.): The opaque contractile diaphragm perforated by the pupil that forms the colored portion of the eye.

Ischium (n.): An endochondral bone forming the posteroventral portion of the pelvic girdle.

Isotherm (n.): A line on a map connecting points of equal temperatures.

Jacobson's organs. *See* Vomeronasal organs

Jaw muscle formulas (n.): The depressor mandibulae originates on the dorsal fascia (DF) or the fasica and either the otic arm of the squamosal (SQ) or the annulus tympanicus (AT) or both of them. The adductor series consists of the posterior subexternus (s) and externus superficialis (e), one or both of which may be present. An example formula is DFSQ + e.

Jugal (n.): A dermal bone lying below the eye and forming part of the maxillary arch in reptiles.

Jurassic period (n.): A subdivision of the Mesozoic era lasting from 208 to 144 Ma.

Juxtaposed (adj.): Placed side by side; nonoverlapping.

Karyotype (n.): The number, size, and morphology characteristic of the chromosome complement of each species.

Keel (n.): A raised, ridgelike process. The term is usually applied to the scales of those reptiles in which there is a median ridge running down the long axis of the scale.

Keratins (n.): Fibrous water resistant proteins that are deposited in the epidermis of vertebrates and constitute the bulk of various horny structures including epidermal scales, claws, and beaks.

Kilohertz (kHz) (n.): One thousand cycles per second of a periodic phenomenon such as the frequency of "sound" waves.

Kinetic (adj.): Dynamic.

Lachrymal glands (n.): Tear-producing glands that furnish liquid for moistening and cleaning the eye surface; often poorly developed in amphibians and reptiles. Absent in anurans and caecilians.

Lamellae (sing. lamella) (n.): The transverse scales under the digits in lizards (fig. 10.3).

Laminate (adj.): Consisting of a series of thin parallel platelike structures.

Lanceolate (adj.): Resembling in form the sharp head of a lance or spear (fig. 7.51i).

Langley (n.): A unit of solar radiation equal to 1 calorie/cm²/ minute or 687.8 W/m²/second.

Lapse rate (n.): The rate or decline in temperature as altitude increases; the average world lapse rate is 0.64°C/100 m.

Larva (pl. larvae) (n.): A free-living developmental stage that on metamorphosis transforms into a juvenile resembling the adult.

Larynx (n.): The valvular structure preventing foreign material from entering the trachea and lungs by closing the glottis; supported by various cartilages that increase the efficiency of closure and containing the paired vocal cords in anurans.

Lateral (adj.): Of or pertaining to the side of the body or a structure; the opposite of medial. The tails of some reptiles are laterally compressed, meaning they are flattened from side to side.

Lateral line system (n.): A system of epidermal sense organs distributed over the head and body of larval and other aquatic amphibians. Most of these organs are mechanoreceptors sensitive to water currents and pressure. Others found on the head in larval caecilians and aquatic salamanders are electroreceptors.

Lateral sequence gait (n.): Locomotion where only one foot is off the ground at any one time and the forefoot strikes the ground after the hind foot on the same side followed by the hind foot on the opposite side.

Laurasia (n.): The northern supercontinent formed by the initial breakup of Pangaea about 160 Ma that later fragmented to form North America, Greenland, and Eurasia, excluding India.

Leaf stomata (**sing. stoma**) (n.): Small openings in leaf surfaces that open or close to regulate water loss.

Lecithotrophic (adj.): Depending on egg stores as the sole source of nutrition during development, as in oviparous and ovoviviparous forms.

Lens (n.): A highly transparent body in the eye that focuses light rays onto the retina.

Lentic (adj.): Of or pertaining to slow-moving or standing bodies of water (e.g., a pond).

Levator anguli oris muscle (n.): Part of the adductor mandibulae complex in snakes, variously modified in function or lost.

Levator bulbi (n.): A thin muscle under the orbit that elevates the eye during respiration in salamanders and anurans. Present but reduced in caecilians.

Lichenose (adj.): Resembling a lichen. A color pattern in which an irregular dark reticulum encloses pale patches.

Ligament (n.): The tough connective tissue attachment between two bones.

Limited-area searching (n.): Breeding period behavior where anuran males remain near one site and attempt to clasp only individuals moving into the immediate area.

Lipophilic (adj.): Of or pertaining to a substance having an affinity for lipids (fats).

Llanura (n.): Spanish for an area of little relief, a plain.

Lobules (n.): Rounded, posteriorly projecting scales bordering the anterior margin of the ear opening in skinks.

Loreal (adj.): Of or pertaining to the lore.

Lore (**pl. lores**) (n.): The side of the head between the eye and nostril; the area between the canthus rostralis and the lip; may be convex, vertical, concave, or flared.

Lotic (adj.): Of or pertaining to fast-moving bodies of water, (e.g., a stream).

Ma (abbr.): Million years ago.

Magma (n.): Molten material from within the earth that when cooled at the surface forms igneous rock.

Mandible (n.): Lower jaw.

Mantle (n.): The layer of the earth lying below the crust to a depth of 2,900 km. The uppermost rigid portion together with the crust forms the tectonic plates that "float" on the partially melted, plastic zone of the mantle that lies between 65 and 400 km below the surface.

Maxillary (n.): The third dermal bone in a complete maxillary arch, usually bearing teeth.

Maxillary arch (n.): The dermal bones forming the upper jaw; premaxillary, septomaxillary, maxillary, jugal, and quadratojugal. Not all these bones are present in every case, and fusions with other bones occur in some taxa.

Maxillopalatine bone (n.): A complex dermal bone forming the lateral tooth-bearing portion of the maxillary arch in caecilians.

Mechanoreceptors (n.): Sense organs stimulated by pressure or other physical forces; including those providing tactile information, touch, and data on equilibrium, gravity source, acceleration, and vibration (hearing), among others.

Meckel's cartilages (n.): The embryonic cartilaginous halves of the lower jaw in tetrapods, which may ossify into a small anteriorly placed Meckelian bone or the posteriorly placed angular bone.

Meckel's groove (n.): The space in the lower jaw left by the Meckel's cartilage after it partially or completely degenerates during embryonic development.

Mediad (adv.): In the direction of the midline of the body.

Medial (adj.): Of or pertaining to the midline; the opposite of lateral.

Medusae (**sing. medusa**) (n.): Jellyfish; various free-living organisms or life-history stages in members of the phylum Cnidaria.

Melanism (n.): An increased concentration of dark pigment in an individual as contrasted with others of the same species or in a species as compared with related forms.

Melanistic (adj.): Pertaining to or having predominately dark pigmentation.

Melanophore (n.): A kind of chromatophore containing melanin, a dark brown to black pigment.

Membranous labyrinth (n.): A series of fluid-filled sacs and canals forming the inner ear within the otic region on each side of the brain in vertebrates; the organ of equilibrium and receptor of pressure waves perceived as sound.

Mental (adj.): Of or pertaining to the chin or the chin region.

Meristic (adj.): Of or pertaining to discrete, discontinuously variable characters; e.g., the number of ventral scales or femoral pores in reptiles.

Meroblastic (adj.): Of or pertaining to the pattern of early cell divisions in which the cleavage planes do not readily pass through the dividing cell so that some cells are not fully separated from one another. Typical of eggs having large amounts of yolk (deutoplasm), such as reptile and bird eggs.

Mesoderm (n.): One of the primary embryonic germ layers giving rise to connective tissue, muscle, bone, and kidney tissue (among others) in vertebrates.

Mesozoic era (n.): Geologic division lasting from 245 to 65 Ma; the "age of reptiles."

Metacarpals (n.): Elongate endochondral bones forming the base of the hand, lying between the wrist bones and the phalanges.

Metachrosis (n.): Color change; the ability to change color.

Metamorph (n.): An immature anuran that has recently transformed from the larval stage.

Metamorphic rock (n.): Rock formed from preexisting rock under the effects of high temperature, pressure, or introduction of a new chemical substance.

Metamorphosis (n.): The processes by which a larva is transformed into a juvenile that resembles the adult in structure.

Metapodials (n.): Metacarpals and metatarsals, the relatively long endochondral bones lying between the phalanges and the wrist and ankle bones, respectively.

Microhabitat (n.): A small portion of the available habitat; a microenvironment.

Micro-ornamentation (n.): Structures on the ridges of the calyces or flounces on squamate hemipenes.

Middle ear (n.): The middle ear cavity and cartilages and/or bony ossicles involved in transmitting pressure waves to the inner ear from airborne stimuli.

Middle ear cavity (n.): The tube between the tympanum and inner ear connected to the pharynx by the Eustachian tube. Synonym: tympanic cavity.

Middorsal (adj.): The portion of the dorsal surface of the body that lies along the midline.

Midventral (adj.): The portion of the ventral surface of the body that lies along the midline.

Mimic (n.): The less unpalatable, toxic, or dangerous species in a mimetic complex.

Mimicry (n.): Convergences in aposematic characters between an unpalatable, toxic, or dangerous species and a less unpalatable, toxic, or dangerous one.

Model (n.): The unpalatable, toxic, or dangerous species in a mimetic complex whose aposematic antipredator defense is also present in the less distasteful, toxic, or dangerous species.

Mollusks (n.): Members of the phylum Mollusca including slugs, snails, clams, squid, and octopuses among other organisms.

Monocondylar (adj.): Having a single condyle.

Morphological (adj.): Of or pertaining to an organism's form or structure.

Morphology (n.): The form and structure of an organism or any of its parts. Also the subdiscipline of biology that involves the study of form and structure.

Mucocytes (n.): Cells that produces mucus.

Mucronate (adj.): Bearing a projecting spine on the posterior margin (fig. 10.3).

Müllerian mimicry (n.): A mimetic complex where the mimic species is somewhat distasteful, toxic, or dangerous and shares an aposematic antipredator defense with a more distasteful, toxic, or dangerous form (the model).

My (abbr.): Million years.

Myoseptum (pl. myosepta) (n.): The connective tissue septum between the main tissue masses (myotomes) that form the striated axial musculature in vertebrates; usually visible in the tail of anuran larvae as chevron-shaped divisions between the myotomes (fig. 7.13a).

Myotome (n.): The portion of a somite that develops into striated muscle.

Narial plugs (n.): Paired thickened or elevated areas on the anterior portion of the tongue in many caecilians; apparently may be inserted into the choanae when the mouth is closed.

Naris (pl. nares) (n.): A nasal opening, usually paired. Nares that open to the outside are usually termed external nares, or nostrils, whereas those that open into the mouth or pharynx are called internal nares, or choanae.

Nasal bones (n.): Dermal skull bones lying above the nasal capsule; usually paired but sometimes fused into a single element.

Nasal chamber (n.): The main portion of the nasal sac.

Nasal sac (n.): The sensory portion of the nasal apparatus lying between the external and internal nares. Synonym: olfactory sac.

Nasopremaxillary bone (n.): A complex dermal bone forming the anterior tooth-bearing portion of the maxillary arch in advanced caecilians.

Nearctic (adj.): Of or pertaining to the biogeographic realm that includes the temperate and Arctic regions of North America.

Neonate (n.): A newly born animal.

Neopalatines (n.): Paired toothless bones lying posterior to the vomers in some neobatrachian anurans and usually articulating with the maxilla and sphenethemoid; not homologous with the palatine bone of other vertebrates.

Neotony (n.): The attainment of sexual maturity in the larval stage.

Neotropics (n.): The New World tropics. Also, the biogeographic realm that includes tropical Mexico, the West Indies, and all of Central and South America.

Neural spine (n.): A dorsally directed, compressed portion of a vertebra above the neural canal.

Neuronal (adj.): Of or pertaining to a nerve cell (neuron).

Neurotoxin (n.): A toxin that adversely affects the nervous system.

Nictitating membrane (n.): A thin transparent fold of tissue attached at the inner angle of the eye, which can be drawn across the eyeball without obscuring vision. Sometimes called the third eyelid.

Nidicolous (adj.): Developing for a time in a nest, as in some species of the frog genus *Leptodacylus*.

Nomenclature (n.): The branch of systematics that is concerned with naming organisms.

Note repetition rate (n.): In anuran vocalization, the number of repetitions of a note within a call per unit time.

Notochord (n.): The axial support of all embryonic vertebrates, replaced entirely or nearly so by the vertebral column during development.

Nuchal (adj.): Of or pertaining to the dorsal surface of the neck or the neck region.

Nuchal endolymphatic sacs (n.): Sacs in the nuchal region of some lizards that are connected to the inner ear by a duct and contain calcium carbonate deposits.

Nuptial pad (n.): A thick, roughened pad on the skin of sexually active male anurans, usually on the thumb.

Obligate behavior (n.): A behavior that always occurs in a particular taxon.

Oblique septum (n.): A membrane separating the heart, pericardial cavity, lungs, and pleural cavities from the peritoneal cavity. Present in some lizards and all snakes and crocodilians.

Occipital (adj.): Of or pertaining to the posterior portion of the skull or to the occipital bones surrounding the foramen magnum.

Occipital bones (n.): Endochondral bones at the posterior end of the cranium, including unpaired basioccipital, interoccipital and supraoccipital, and paired exoccipitals, not all of which are present in a particular skull.

Occipital condyles (n.): The cranial portion of the articulation between the skull and vertebral column.

Occiput (n.): The posterior part of the head or skull.

Ocellus (pl. ocelli) (n.): An eyelike structure or pattern element.

Odontoid (n.): A bony process resembling a tooth.

Odontophores (n.): The tooth-bearing processes of the vomer and palatine bones in amphibians.

Olfaction (n.): Chemoreception by the nasal and vomeronasal organs (the sense of "smell").

Oligocene epoch (n.): A subdivision of the Cenozoic era lasting from 37 to 24 Ma.

Omosternum (n.): A single midventral skeletal element in some anurans projecting anteriorly from the pectoral girdle as a cartilaginous disk or a disk supported by an elongate style that may be ossified.

Omphalallantoic placenta (n.): A placenta in which the allantois invades the cleft between the isolated and main yolk masses seen in omphaloplacentas and fuses with the isolated yolk mass to form the embryonic portion of the placenta. Found in all viviparous snakes.

Omphaloplacenta (n.): A yolk sac placenta developed at the abembryonic pole from the outer wall of a yolk mass that is completely isolated from the main yolk mass. Unique to squamate reptiles.

Ontogeny (n.): The developmental history of an individual from fertilization until death.

Opercularis muscle (n.): A muscle of the middle ear of amphibians originating on the suprascapula and inserting on the cartilaginous or bony operculum.

Operculum (n.): The fleshy fold that grows posteriorly over the external gills in early tadpoles and at metamorphosis in other amphibians. In young tadpoles the gills become internal within the opercular chamber, which connects to the outside through a spiracle or two spiracles.

Operculum cartilage or bone (n.): An ear cartilage or bone variously present as a distinct element, fused with the columella, or absent in amphibians; absent in amniotes.

Ophiophagous (adj.): Of or pertaining to a predator that eats snakes.

Opisthocoelous vertebra (n.): A vertebra whose centrum has a convex anterior face and a concave posterior face.

Orbit (n.): The eye socket; the bony cavity that contains the eyeball and associated muscles, blood vessels, and nerves.

Orogenic (adj.): Relating to mountains.

Orogeny (n.): The process of mountain building.

Orthopterans (n.): Members of the insect order Orthoptera, including grasshoppers, katydids, and their allies.

Oscillatory tongue protrusion (n.): Up-and-down movement of a protruded tongue.

Oscillogram (n.): A record made from an oscilloscope showing variations in the electrical impulses converted from sound waves to show a plot of the waveform of an anuran cell as amplitude against time.

Oscilloscope (n.): An instrument in which sound waves from frog calls are converted to electrical impulses and appear temporarily as a visible waveform on the fluorescent screen of a cathode-ray tube.

Osmoregulation (n.): The processes involved in maintaining water balance in an organism.

Ossified (adj.): Changed into bone.

Osteoderms (n.): Superficial dermal bones underlying the epidermal scales; large and thick structures in crocodilians, variable in thickness in lizards; the epithecals of the leatherback turtle.

Osteological (adj.): Of or pertaining to bone or to the skeleton.

Ostium pharyngeum (pl. ostia pharyngea) (n.): The opening of the Eustachian tube into the pharynx.

Otic capsule (n.): The skeletal elements surrounding the inner ear.

Ovarian (adj.): Of or pertaining to the ovary.

Ovary (n.): The female gonad producing the female gametes (ova or eggs) in adults.

Oviduct (n.): The duct carrying eggs or neonates outside the female's body via the cloaca; the posterior portion of the oviduct when enlarged is called a uterus. Synonym: Müllerian duct.

Oviductal egg (n.): An egg within the oviduct.

Oviparity (n.): Reproduction in which the eggs are deposited and hatch outside the body of the mother; nutrition is lecithotrophic.

Oviposition (n.): The act of laying eggs.

Ovoviviparity (n.): Reproduction in which the eggs are retained in and hatched in the mother's oviduct with birth of live offspring; nutrition is lecithotrophic, and there is no placenta.

Ovule (n.): The structure in seed plants that becomes the seed after fertilization.

Ovum (pl. ova) (n.): A female gamete (an egg).

Palatal (adj.): Of or pertaining to the roof of the mouth.

Palatal foramen (n.): A large opening in the secondary bony palate of reptiles.

Palate (n.): The roof of the buccal cavity. In some reptiles a bony secondary palate may divide the buccal cavity into an upper respiratory part and a lower alimentary part.

Palatines (n.): Paired dermal bones in the anterior portion of the roof of the mouth that often bear teeth.

Palatoquadrate (n.): The embryonic cartilaginous upper jaw of tetrapods; the quadrate bone ossifies from the posterior part of this cartilage.

Palearctic (adj.): Of or pertaining to the biogeographic realm that includes the temperate and Arctic regions of Europe and Asia.

Paleocene epoch (n.): A subdivision of the Cenozoic era lasting from 66 to 58 Ma.

Palmate (adj.): Having the distal portion broad and flat (fig. 7.51e); having the digits united by a web (fig. 6.4c).

Palynoflora (n.): A fossil flora known from pollen and spores.

Palynology (n.): The study of pollen and spores both recent and fossil.

Pangaea (n.): The single supercontinent formed by the joining together of continental landmasses; fragmentation of this supercontinent beginning about 200 Ma and continental drift produced the present configuration of the world's continents.

Papilla amphibiorum (n.): A papilla in the wall of the sacculus of the inner ear near its junction with the utriculus.

Papillate (adj.): Covered with papillae, as on the upper surface of the tongue in many lizards; covered with tiny papillae, as on the calyces of some squamate hemipenes.

Papillae (sing. papilla) (n.): Small fleshy protuberances, as on the upper surface of the mammalian tongue.

Parachuting (n.): Flattening the body and limb extensions to reduce the rate of falling.

Parahyoid bones (n.): One or two endochondral bones associated with the central hyoid plate in anurans.

Paraphysis (n.): The dorsal brain outgrowth that forms the retina of the parietal eye and its connection to the brain.

Parasitic male. *See* Satellite male

Paravertebral (adj.): Of or pertaining to an area immediately lateral to the dorsal midline.

Parietal bones (n.): Dermal roofing bones of the skull usually bordered by frontal bones anteriorly and occipital bone(s) posteriorly. Usually these bones are paired, but they may fuse to form a single bone or fuse with the frontal to form a single frontoparietal bone.

Parietal eye (n.): The photoreceptor in the parietal region of many lizards.

Parietalectomy (n.): The surgery used to remove the parietal eye.

Parietal foramen (n.): The opening in the parietal bone providing a passage from the brain to the parietal eye in lizards.

Parotoid gland (n.): A large, swollen glandular area behind the eye in amphibians that may extend onto the neck and shoulder region (fig. 7.6a).

Parthenogen (n.): A parthenogenetic species.

Parthenogenesis (n.): Mode of reproduction involving formation of a new individual from an unfertilized egg.

Parthenogenetic (adj.): Of or pertaining to reproduction from a unfertilized egg.

Parturition (n.): The act of bringing forth young; birth.

Passerine (adj.): Of or pertaining to songbirds.

Pectoral (adj.): Of or pertaining to the chest; thus, also pertaining to the forelimb girdle.

Pedicel (n.): The area of the hemipenis between the proximal truncus and the apical region (fig. 11.2a).

Pedicellate teeth (n.): The bipartite teeth characteristic of lissamphibians. Each tooth consists of a basal pedicel and a distal crown, usually capped by enamel; the latter is connected to the former by a hinge. The pedicel is composed of dentine covered by cementum (bone); the hinge consists of uncalcified dentine or a ring of fibrous connective tissue.

Pelagic (adj.): Of or pertaining to the open ocean.

Pelvic (adj.): Of or pertaining to the hip; thus, also pertaining to the hind limb girdle.

Penis (n.): The male copulatory organ. Found in turtles and crocodilians among reptiles.

Pennsylvanian period (n.): A division of the Paleozoic era lasting from 315 to 286 Ma. The "age of amphibians." Synonym: late Carboniferous epoch.

Pentadactyl (adj.): Having five digits.

Peptide (n.): Two or more amino acids linked by a peptide bond.

Peptide bond (n.): The bond that links amino acids in a protein.

Pericapsular recess (n.): That portion of the perilymphatic system surrounding the otic capsule and proximal portion of the extracolumella in turtles.

Pericardium (n.): That part of the parietal peritoneum that forms the pericardial sac around the heart in tetrapods. *See also* Epicardium.

Perilymph (n.): The fluid contained in a series of cavities and sacs surrounding the parts of the inner ear.

Peritoneal cavity (n.): The body cavity containing the organs of digestion, excretion, and reproduction.

Peritoneal sheath (n.): The white portion of the parietal peritoneum of some frogs.

Peritoneum (n.): The thin tissue layer lining the peritoneal cavity. The portion of the peritoneum covering the organs and forming their mesenteries is the splanchnic or visceral peritoneum; the portion lining the inside of the body wall is the somatic or parietal peritoneum.

Permian period (n.): A subdivision of the Paleozoic era lasting from 286 to 245 Ma.

Petala (sing. petalum) (n.): Transverse overlapping laminate structures on some lizards' hemipenes.

Phalanges (sing. phalanx) (n.): Small bones forming the skeleton of the digits.

Phallodeum (n.): The median eversible copulatory organ derived from the cloacal wall in male caecilians.

Pharynx (n.): The region of the digestive tract just posterior to the mouth in which functional paired gill slits are present in adult and larval fishlike vertebrates, in all larval and some aquatic adult amphibians, and embryonically in other vertebrates.

Photoreceptor (n.): A light-sensitive sense organ.

Placenta (n.): A physiological connection between mother and embryo consisting of a maternal portion and an embryonic contribution to the connection. The wall of the oviduct is usually the maternal contribution in placental amphibians and reptiles. Some embryonic amphibians have pharyngeal (or gill) placentas that provide gaseous exchange between the embryo and the mother. In those reptiles having placentas the embryonic portion is provided by the extraembryonic membranes.

Placental viviparity (n.): Viviparity in which the embryo receives nutrition from the mother by means of the placental connection between them.

Plate. *See* Epidermal scales

Pleurodont (adj.): Used to describe a tooth situated on the inner side of the jawbones, not in a socket.

Plicae (sing. plica) (n.): Fleshy longitudinal folds.

Plumbeous (adj.): Resembling the color of lead.

Poison (n.): A substance that is toxic when ingested or absorbed.

Pollen, pollen grains (n.): Spheroid structures produced by male reproductive organs in seed plants that are involved in fertilization.

Pollex (n.): The first digit on the hand, the thumb.

Polychromatism (n.): The condition where a population or species is composed of individuals of several different colors or color patterns.

Polygynous (adj.): Of or pertaining to the mating system where a male mates with several females during the breeding season.

Polymorph (n.): One of the morphs in a polymorphic organism.

Polymorphic (adj.): Of or pertaining to polymorphism.

Polymorphism (n.): The existence of more than one distinct morphological type within a population.

Polyp (n.): Sessile organisms or life-history stages in members of the phylum Cnidaria, often forming large colonies. Example: sea anemones and corals.

Polypeptide (n.): A series of peptides linked by peptide bonds to form proteins.

Polyploid (adj.): Having an increased number of chromosomes over the diploid (2N) complement, such as triploid (3N), hexaploid (6N), or octoploid (8N).

Postcloacal (adj.): Of or pertaining to the region posterior to the cloacal opening.

Posterior (adj.): Pertaining to the rear or caudal end of the body or of a structure.

Posterior Vidian canal (n.): An enclosed canal on the underside of the posterior braincase containing the internal carotid artery and palatine nerve.

Postfrontal bone (n.): A dermal bone bordering the orbit posteriodorsally.

Postorbital bone (n.): A dermal bone bordering the orbit posteriorly.

Postrictal (adj.): Of or pertaining to the area just posterior to the corner of the mouth.

Precise alternation (n.): Alternation of calls leading to the formation of duets, trios, quartets, etc., within an anuran chorus.

Precloacal pore (n.): The opening of an exocrine gland situated in the precloacal region.

Prefrontal bone (n.): a dermal bone bordering the orbit anteriorly.

Prehensile (adj.): Adapted for seizing or wrapping around, like the tails of arboreal pitvipers.

Premating isolation mechanism (n.): A structure or behavior preventing mating.

Premaxillary bone (n.): One of a pair of dermal bones (sometimes fused) at the anterior end of the skull in amphibians and reptiles. The premaxillary bones, which often bear teeth, constitute the anteriormost portion of the maxillary arch. Synonym: premaxilla.

Prepollex (n.): A vestigial skeletal element at the outer base of the pollex (thumb) in some anurans.

Pretympanic (adj.): Lying just anterior to the tympanum on the side of the head.

Prevomer (n.): The name some authors prefer for a pair of dermal bones lying in the skull immediately posterior to the premaxillary bones. *See also* Vomer.

Primary note (n.): The longer of two or more notes in an anuran call sequence.

Procoelous vertebra (n.): A vertebra whose centrum has a concave anterior face and a convex posterior face.

Prolonged breeders (n.): Species having reproductive periods lasting longer than one month.

Prootic bone (n.): An endochondral bone forming part of the otic capsule.

Protractor (n.): A muscle that draws a segment forward or out from the body.

Protrusible (adj.): Capable of being protruded (thrust out and forward).

Proximal (adj.): Situated near or toward the point of attachment. The phalanges at the base of a digit lie proximal to those at the tip of the digit, for example.

Pseudautotomy (n.): Nonspontaneous loss of the tail by intervertebral breakage and usually without tail regeneration; in unspecialized pseudautotomy there is no specialized caudal morphology; in specialized pseudautotomy the tail is modified to increase its fragility.

Pseudotail (n.): The presence of a few postcloacal vertebrae in an unsegmented terminal shield in some caecilians.

Pterorhodophore (n.): A cell containing a eumelanin core surrounded by a mass of the red compound pterorhodin.

Pterygoid (n.): A dermal bone of the posterior palatal region; often bears teeth in lizards.

Pubis (n.): An endochondral bone forming the anterioventral portion of the pelvic girdle.

Pulsatile. *See* Trilled

Pulse (n.): In bioacoustics, the distinct pulsations of sound that constitute a note, often apparent as vertical marks on an audiospectrogram.

Pulse rate (n.): In bioacoustics, the number of pulses per unit time, usually expressed as pulses per second.

Pungent (adj.): Sharp but not pointed; used to describe tubercles on hands and feet of anurans (fig. 7.51s).

Pygal (adj., n.): Pertaining to the base of the tail; in turtles, the bones of the carapace above the base of the tail (fig. 12.2a).

Quadrate (n.): An endochondral bone formed on the posterior portion of the embryonic palatoquadrate cartilage that articulates with the lower jaw; fused with the pterygoid in caecilians.

Quadratojugal (n.): A dermal bone on the side of the skull lying posterior to and below the orbit. It forms the posteriormost bone in the maxillary arch.

Quaternary period (n.): A subdivision of the Cenozoic era lasting from 1.6 Ma to the present.

Radius (n.): The anterior (preaxial) endochondral bone that with the ulna forms the skeleton of the forearm.

Ramus (pl. rami) (n.): A branch.

Raphe (n.): The seamlike line of union of the halves of symmetrical anatomical parts appearing as a furrow or ridge.

Raptor (n.): A bird of prey.

Reciprocation call (n.): A call produced by female anurans in response to a male's courtship call.

Rectus abdominis lateralis (n.): A muscle of the abdominal wall that arises from the pelvic girdle, inserts on the sternum, and lies lateral to the main rectus abdominis muscle.

Red rods (n.): Photosensitive cells of amphibian retinas that maximally absorb wavelengths of 502 nm.

Release call (n.): A simple vocalization given by males or unreceptive females when amplexed by a male.

Relict (n.): A taxon formerly widespread but now restricted to a small portion of its former range.

Reniform (adj.): Kidney-shaped.

Resorption pit (n.): The basal region of the old tooth that is eroded as the new tooth replaces it.

Reticulate (adj.): Having the appearance of a net or mesh; netlike.

Retina (n.): The layer of the eye containing light-sensitive cells (photoreceptors) and several layers of sensory nerve cells.

Retractile foretongue protrusion (n.): Tongue protrusion followed by retraction of foretongue into the hindtongue.

Retractor (n.): A muscle that draws a segment back to its original position.

Rictus (n.): The corner of the mouth.

Rods (n.): Light-sensitive cells (photoreceptors) on the retina that respond to low light intensities.

Rostrum (n.): The snout.

Rugose (adj.): Usually meaning wrinkled or folded, but used in herpetology to describe any uneven or rough surface whether wrinkled, granular, or warty.

Sacculus (n.): The smaller of two saclike divisions of the membranous labyrinth of the inner ear giving information on the tilt of the head.

Sacral (adj.): Of or pertaining to the sacrum or the region of the body where the sacrum is situated.

Sacrum (n.): A vertebra or vertebrae articulating with the pelvic girdle.

Salinity (n.): A measure of the total concentration of dissolved salts in water as grams per kilogram or parts per thousand; average seawater salinity is 35.

Salps (n.): Pelagic gelatinous, filter-feeding tunicates eaten by some sea turtles.

Saltatory locomotion (n.): Locomotion by jumping (saltation).

Sargassum (n.): A genus of brown macroalgae (class Phaeophyta) forming pelagic spheroid masses (100 mm to 1 m in diameter) that support a specialized fish and crustacean fauna. High concentrations of these masses occur in the Sargasso and Arabian Seas and may form the refuge and feeding grounds for green sea turtles *(Chelonia)* during the "lost years."

Sartorius muscle (n.): A muscle in anurans that originates anterior to the acetabulum of the pelvis, extends across the ventral surface of the thigh, and inserts on a tendon that joins others at the knee. It functions with the semitendinosus muscle to abduct the femur ventrally and flex the knee. In most primitive frogs it is united with the semitendinosus into a single muscle.

Satellite males (n.): Silent male anurans that locate themselves near a calling male and attempt to intercept females attracted to the call. Synonym: parasitic male.

Savanna (n.): Lowland tropical or subtropical grasslands, often with scattered trees.

Saxicolous (adj.): Living among rocks.

Scales. *See* Bony scales; Epidermal scales

Scalloped (adj.): Covered with curved projections, as on the calyces of some squamate hemipenes.

Scansorial (adj.): Climbing.

Sclera (n.): The tough outer white supporting tunic of the eye, covering the posterior 80% of its surface and continuous anteriorly with the clear cornea.

Scramble competition (n.): Active searching for females with male-male physical interactions in some anurans during the breeding period.

Scutellation (n.): A collective term for the epidermal scales in reptiles.

Scutes. *See* Epidermal scales

Seat patch (n.): In anurans, a triangular area of skin surrounding the vent, often darkly pigmented and contrasting with the lighter coloration of the adjacent skin.

Secondary note (n.): In anuran vocalization, one or more notes that follow a primary note. Secondary notes are shorter than the primary note.

Sedimentary rocks (n.): Rocks formed by the accumulation, compaction, and cementation of layers of sediment mainly deposited under shallow parts of the sea and less frequently on stream or lake bottoms. Examples: sandstones, limestones, and shales.

Semidirect development (n.): Development in which there is a nonfeeding larval stage that receives nutrition from its yolk stores.

Semiholoblastic (adj.): Of or pertaining to a pattern of early cell divisions somewhat intermediate between holoblastic and merioblastic cleavage; characterized by producing a blastodisc and some division of the yolk, as in some fishes.

Semilunate (adj.): Shaped like a half-moon.

Seminal groove (n.): The groove in turtles carrying the sperm from the cloaca to the glans and thence into the female's cloaca.

Semitendinosus muscle (n.): A muscle in anurans having a double origin on the posterodorsal and posteroventral rims of the pelvis and extending along the ventral surface of the thigh to insert on the tibiofibula. It functions to abduct the femur and pulls it ventrally to flex the knee. In most primitive frogs it is united with the sartorius into a single muscle.

Septomaxillary bone (n.): The second dermal bone in the maxillary arch series, which is relatively small, lacks teeth, and is associated with the nasal capsule.

Septum (n.): A dividing membrane or other partition.

Serrate (adj.): Notched or toothed along the edge; resembling the toothed edge of a saw.

Sessile (adj.): Permanently attached to the substratum, not free to move about.

Setae (sing. seta) (n.): Bristlelike structures.

Sexual dichromatism (n.): Within a species, a difference in coloration between the sexes.

Sexual dimorphism (n.): Within a species, a morphological difference between the sexes.

Sexual heteromorphism (n.): Different numbers or morphology in the chromosome complement of the two sexes.

Shagreened (adj.): Rough to the touch, covered with numerous tiny close-set tubercles.

Shank. *See* Tibial segment

Shield. *See* Epidermal scales (not the same as the terminal shield in caecilians)

Simple tongue extension (n.): Protrusion of the tongue or protrusion and touching of the substrate with no oscillatory tongue movements.

Sinistral (adj.): Of or pertaining to the left side.

Siphonophores (n.): Pelagic polymorphic colonies of cnidarians composed of medusae and polyps forming the order Siphonophora; eaten by some marine turtles.

Skin retention (n.): Retention of successive layers of the dead outer layer of the epidermis to form shields of the carapace and plastron in turtles and the dorsal scutes of crocodilians.

Skin shedding (n.): Loss of the outer dead layer of the epidermis.

Sliding backward burrowers (n.): Anurans that burrow directly backward into the soil.

Somatic (adj.): Of or pertaining to the peripheral structures of a vertebrate (e.g., somatic muscles).

Somatic cells (n.): Cells of an organism's body other than the germ cells; somatic cells are usually diploid (2N).

Somites (n.): Paired, segmentally arranged mesodermal masses lying alongside the neural tube in vertebrate embryos; later in development portions of the somites will differentiate to form the dermis, striated muscles, vertebrae, and limb bones.

Sonogram (n.): A plot of an anuran vocalization showing frequency against time.

Sonograph (n.): An instrument in which sound waves are converted to electrical impulses that are graphically displayed to show frequency against time or amplitude.

Spadate (adj.): Spade-shaped, similar in form to the spade symbol on playing cards; used to describe the shape of digital disk pads in anurans (fig. 7.51h).

Species epithet (n.): The second word in a species name. Example: *marinus* in the species name *Bufo marinus*.

Spectacle. *See* Brille

Spectrogram (n.): A graphic representation of an anuran vocalization plotting amplitude against frequency.

Sphenethmoid (n.): An endochondral bone forming the floor and part of the wall of the braincase, lying posterior to the nasal capsule. It is often separated from the roof of the mouth by the various palatal bones.

Sphincter (n.): A constricting muscle encircling an opening that when contracted closes the opening.

Spicule (n.): A small, pointed tubercle, often with a keratinized or calcified tip.

Spinulate (adj.): Covered with tiny calcified spines, as on the calyces of some squamate hemipenes.

Spinulate scale organs (n.): Specialized microscopic scale organs characterized by a single median spine or filament.

Spiracle (n.): In tadpoles, an opening from the gill chamber to the outside.

Splenial bone (n.): A dermal lower jawbone usually lying below the dentary in reptiles; part of the complex anterior lower jawbone in caecilians.

Squamation (n.): The arrangements, contacts, and number of epidermal scales in reptiles.

Squamosal (n.): A dermal bone in the temporal region of the skull that articulates with the quadrate or quadratojugal.

Stapedial artery (n.): A vessel that branches from the carotid artery and carries oxygenated blood to the outer part of the head and jaws in amphibians and reptiles.

Stapes. *See* Columella

Stegochordal vertebrae (n.): Vertebrae in which only the dorsal and dorsolateral areas chondrify or ossify.

Sternal fontanelles (n.): Membrane-covered openings in the sternum.

Sternum (n.): A midventral skeletal structure supporting and protecting the body and organs in the thoracic region.

Steroids (n.): A group of alkaloids including the batrachotoxins.

Stratosphere (n.): The layer of the atmosphere in which temperature ceases to drop with increasing altitude; bordered below by the tropopause and with an upper limit at about 50 km above the earth's surface. This zone is characterized by having winds parallel to the earth's surface, has almost no water vapor, and contains the narrower ozone layer.

Structural color (n.): Color produced by differential refraction of wavelength because of the structural arrangement and shape of cells, when no pigments are involved.

Subarticular tubercles (n.): Tubercles on the lower surface of the digits under the joints between the phalanges (fig. 7.7).

Subcaudal (adj.): Of or pertaining to the inferior surface of the tail.

Subclavian artery (n.): The vessel in amphibians and reptiles supplying the forelimbs with oxygenated blood.

Submandibular gland (n.): A pair of circular glands opening on the throat near the inner posterior margins of the lower jaws in crocodilians and secreting a odoriferous mucus. The secretions are probably used in social communication.

Subspecies (n.): A geographic subdivision within a species that is afforded nomenclatural recognition. Synonym: race.

Sucking disk (n.): The abdominal suctorial disk present in some tadpoles, used to maintain position in fast-moving streams.

Suctorial (adj.): Pertaining to the large ventral facing oral disk of tadpoles that maintain position and feed in fast-moving water by adhering to rocks with the disk; also pertaining to the abdominal disk of some tadpoles used to maintain position in torrential streams.

Sulcus (n.): A shallow groove, generally open on one side. Example: the semen canal on the surface of the hemipenis of squamates, the sulcus spermaticus.

Supernumerary tubercles (n.): Tubercles on the lower surface of the digits under the phalanges and between the subarticular tubercles (fig. 7.7).

Supracaudal scutes (n.): The paired marginal epidermal shields of the carapace above the tail; also called postcentrals or postvertebrals (fig. 12.1a).

Supracloacal keels or tubercles (n.): Keels or tubercles on the dorsal scales above the anal plate in some snakes; sometimes called supra-anal keels.

Supraorbital bone (n.): A dermal bone lying above the orbit.

Suprapygal (adj.): The anteriormost one or two pygal bones in turtles; when this terminology is used, the posterior element in the series over the tail is called a pygal bone (fig. 12.2a).

Suprascapula (n.): A cartilage forming the upper portion of the scapular (shoulder) blade in amphibians and many reptiles.

Supratemporal bone (n.): A dermal bone in the temporal region of the skull lying dorsal to the squamosal.

Supratemporal foramen (n.): The upper temporal fossa.

Suturally (adv.): Relating to a sutural connection.

Suture (n.): The border line or groove between two adjacent parts, especially used with respect to abutting bones.

Sympatric (adj.): Of or pertaining to taxa that co-occur in the same geographic area.

Symphysis (n.): An articulation between bones in the median plane of the body, such as the junction between the two halves of the lower jaw.

Synchronous breeding (n.): All females arriving at the breeding site at about the same time.

Synchronous calls (n.): Calls in which male anurans rapidly answer one another so that the calls overlap and produce rhythmic bursts of sound or just a huge din.

Syncopated (adj.): Of or relating to the offbeat displacement of the regular metric ascent in music by stressing the weak beat.

Synonymize (v.): To conclude that one described taxon is a junior synonym of another having an earlier name (the senior synonym).

Synonyms (n.): Two or more names for the same taxon. A senior synonym is the earliest available name for a taxon. A junior synonym is any available name other than the senior synonym.

Syntopic (adj.): Of or pertaining to species that occupy the same microhabitat.

Systemic arteries (n.): Paired blood vessels carrying oxygenated blood from the heart to the dorsal aorta; a single pair of systemic arteries carries blood to the dorsal aorta and thence to the body in adult amphibians and nonvolant reptiles; in birds and mammals only a single systemic artery performs this function.

Tadpole (n.): The larva of frogs and toads.

Taiga (n.): The northern coniferous forest lying to the south of the Arctic tundra zone.

Tarsals (n.): Small elements of cartilage or bone forming the ankle skeleton.

Tectum (n.): One of the paired swellings of the midbrain that are the centers for the reception of visual stimuli in amphibians and reptiles.

Temporal (adj.): Of or pertaining to the temporal region, which lies behind the orbit in the skull of vertebrates.

Temporal fossae (sing. fossa) (n.): Openings in the temporal region of the skull. Reptile skulls having this region solidly roofed as in turtles are termed anapsid; those having two fossae on each side as in crocodilians are called diapsid. Turtles, lizards, and snakes are descended from ancestors having the diapsid condition. Lizards may retain the upper fossa, or it may be lost in various ways; all snakes lack temporal fossae.

Tendon (n.): A cordlike structure or membranous sheet consisting solely of tough connective tissue that connects a muscle to a bone.

Tensor fasciae latae muscle (n.): A muscle arising from the ventrolateral margin of the ilium and inserting on the fascia covering large muscles on the anterodorsal aspect of the thigh. It functions to extend the knee joint and flex the hip.

Tentacle (n.): In caecilians, a fleshy, protrusible structure situated in a groove or canal in the maxillary bone. The tentacle serves a chemosensory function.

Tentacular foramen (n.): The depression or opening on the side of the skull containing the tentacular apparatus in caecilians through which the tentacle is protruded.

Terminal (adj.): An anatomical position pertaining to the end of a structure.

Terminal shield (n.): The ventral unsegmented terminal area including the cloaca found in some caecilians.

Territory (n.): An area defended by an organism, usually against conspecifics.

Tesselated (adj.): Checkered; marked with squares or blocks of two colors to resemble a checkerboard (fig. 11.34g,h).

Testis (n.): The male gonad producing the male gametes (sperms) in adults.

Tetrapod (n.): Literally a four-footed animal; used here for members of the superclass Tetrapoda, which includes members of the classes Amphibia, Reptilia, and Mammalia regardless of the presence or absence of four feet.

Thecodont (adj.): Used to describe a tooth having a socket.

Thenar (adj.): Situated at the base of the thumb or first digit of the hand (fig. 7.7a).

Thermoconformer (n.): A species or individual whose body temperature conforms to that of its environment, showing little or no behavioral thermoregulation.

Thermoreceptors (n.): Sense organs that are stimulated by changes in temperature.

Thermoregulation (n.): The physiological and behavioral process that permits an organism to regulate its body temperature.

Thigmothermic (adj.): Term describing ectothermic organisms that rely on direct contact with the substrate to elevate body temperature.

Tibia (n.): The anterior (preaxial) endochondral bone that together with the fibula forms the shank (crus) in most tetrapods.

Tibiale (n.): The anterior (preaxial) cartilage or endochondral bone articulating with the tibia or tibiofibula and forming part of the greatly elongated ankle in anurans. Synonym: calcaneum.

Tibial epiphysis (n.): An accessory endochondral ossification center at the end of the tibia in some reptiles.

Tibial notch (n.): A notch in the distal tibial epiphysis in some lizards.

Tibial segment (n.): The portion of leg containing the tibia and fibula. Synonyms: crus, shank.

Tibial spur (n.): A bony ridge or projection on the tibia in some salamanders.

Tibiofibula (n.): A complex endochondral bone forming the shank (crus) of anurans.

Tomium (n.): The cutting edge of a turtle's beak (fig. 12.18).

Trachea (n.): The main trunk of the system of tubes by which air passes to and from the lungs and that connects the pharynx to the respiratory organs.

Tracheal lung (n.): A lung found in some snakes that lies craniad to the heart and lateral to the trachea; connected to the trachea at the caudal end; not homologous to the left lung that is lost in many serpents.

Transverse processes (n.): Short ribs that have become fused to the posterior and caudal vertebrae.

Triassic period (n.): A subdivision of the Mesozoic era lasting from 245 to 208 Ma.

Tricarinate (adj.): Having three keels; usually applied to turtles having three longitudinal keels on the carapace.

Triconodont (adj.): Having three cusps on the teeth.

Trifid (adj.): Partially divided into three parts.

Trilled (adj.): Referring to anuran calls having long, distinct pulses discernible to the human ear. Synonym: pulsatile.

Trilobate (adj.): Having three lobes.

Triploid (3N) (adj.): Having three chromosomes of each kind. (n.): An organism having 3N chromosomes.

Triramous (adj.): Having three principal branches (fig. 4.5a).

Trophosphere (n.): The lowest layer of the atmosphere, in which temperature drops rapidly with increasing altitude; characterized by active convection currents and having most of the atmosphere's water vapor and dust; the zone where most weather events occur.

Tropopause (n.): The juncture between the lowest layer of the atmosphere (the trophosphere) and the stratosphere; positioned about 10 to 15 km above the earth's surface.

Trotting locomotion (n.): Locomotion where forelimbs and hind limbs on opposite sides of body thrust backward at the same time.

Truncus (n.): The proximal portion of the hemipenis (fig. 11.2a).

Tubercle (n.): A small, rounded bump on the skin of some amphibians and reptiles.

Tuberculate (adj.): Bearing tubercles.

Tunic (n.): Generally a membrane or other tissue layer covering or lining a body part; in frogs it refers to the outer fibrous covering of the eyeball comprising the cornea and the cartilaginous sclera.

Tunicates (n.): Members of the subphylum Urochordata, including the pelagic salps eaten by some marine turtles.

Tuning (n.): The distribution of energy in respect to frequency in a vocalization; fine tuning involves several narrow bands of frequencies; poor tuning has the energy spread over the whole spectrum of frequencies in the call.

Tympanic cavity. *See* Middle ear cavity

Tympanum (n.): The membrane covering the external opening of the middle ear; the eardrum. Synonym: tympanic membrane.

Ulna (n.): The posterior (postaxial) endochondral bone that with the radius forms the skeleton of the forearm.

Ulnare (n.): An element of cartilage or bone lying distal to the ulna and forming part of the wrist skeleton.

Umbelliform (adj.): Shaped like an umbrella.

Ungual (adj.): Of or pertaining to a claw.

Unicarinate (adj.): Having a single (usually median) keel.

Uniparental (adj.): Having a single parent, as in parthenogenetic organisms.

Unisexual species (n.):An all-female species.

Urostyle. *See* Coccyx

Urotomy (n.): Tail breakage.

Utriculus (n.): The larger of two saclike divisions of the membranous labyrinth of the inner ear giving information on position and movement of the head.

Valid name (n.): The name applied under the *International Code of Zoological Nomenclature* to a particular taxon. All other available names based on that taxon are junior synonyms or rejected names.

Vegetal pole (n.): The center of the region of the egg that normally develops into the nutritive structures (e.g., digestive tract). In amphibians, usually without pigmentation; in the cleidoic egg, the region of high yolk (deutoplasm) concentration that provides nutrition to the developing embryo.

Velum (n.): A glandular organ between the buccal cavity and the pharynx in tadpoles that regulates water flow into the pharynx.

Venation (n.): A network of veinlike markings resembling, for example, the veins in a fly's wings (fig. 7.33b).

Venom (n.): A toxic substance injected by a sting or bite.

Vent (n.): The cloacal aperture. The posterior opening through which the products of the digestive, reproductive, and excretory systems pass. Synonym: cloacal slit.

Venter (n.): The lower surface of the body.

Ventrad (adv.): Toward the venter.

Ventral (adj.): Of or pertaining to the lower surface of the body or other structure.

Ventrolateral (adj.): A position intermediate between ventral and lateral.

Vermiform (adj.): Wormlike, extremely elongate.

Versants (n.): The physiographic slopes of a country.

Vertebrae (sing. vertebra) (n.): The segmentally arranged bones surrounding the spinal cord that form the backbone or vertebral column.

Vestige (n.): A poorly developed organ that earlier in ontogeny or phylogeny was more fully developed.

Vestigial (adj.): Of or pertaining to a small or imperfectly developed part or organ that was more fully developed in an earlier ontogenetic stage or ancestral species.

Viscera (sing. viscus) (n.): Collectively, all of the soft internal organs of an animal.

Visceral (adj.): Of or relating to the deep structures of a vertebrate (e.g., the intestines).

Vitelline (adj.): Of or pertaining to egg yolk (deutoplasm).

Vitelline filaments (n.): Stringlike projections from the surface membrane of the egg found in some fishes.

Vitelline membrane (n.): The cell membrane surrounding an ovum.

Vitellogenesis (n.): The provision of deutoplasm (yolk) to the egg during its development.

Viviparity (n.): Reproduction in which development occurs within the oviduct or uterus, the embryo receives nutrition from the mother, and the young are born alive.

Vocal cords (n.): Paired thickenings on the inner wall of the larynx in anurans that produce vibrations when air from the lungs passes over them.

Vocal sac (n.): An outpouching of the floor of the mouth in many male anurans that acts as a resonating chamber during vocalization.

Vocal slit (n.): A valvular opening in the floor of the mouth in male anurans leading to the vocal sac.

Volant (adj.): Flying or capable of flying.

Volplaning surface (n.): A structure providing an increased surface area that allows an animal to glide from a higher to lower site by significantly reducing the angle of descent.

Vomer (n.): Paired dermal bones immediately posterior to the premaxillary bones in the roof of the mouth of vertebrates.

Vomeronasal organs (n.): A pair of specialized chemosensory organs opening into the roof of the mouth in most tetrapods that sample fluids in the mouth. Synonym: Jacobson's organs.

Walking trot (n.): Locomotion where forelimbs and hind limbs on opposite sides of the body strike the ground at the same time and the other two feet are off the ground.

Wart (n.): A rounded, elevated bump on the skin.

Whorl (n.): A ring of scales encircling the tail of a lizard or crocodilian. Usually refers to alternating series of scales that differ in size, shape, or both.

Xanthophore (n.): A cell containing yellow, orange, or red pigments.

Xeric (adj.): Arid, lacking in moisture.

Xiphisternum (n.): The posteriormost skeletal element of the sternum in some anurans and reptiles. A cartilaginous plate in the former, and long paired rods in the latter.

Yolk sac (n.): One of four extraembryonic membranes in amniotes; encloses the yolk material and transports nutrients from the yolk through vitelline veins on its surface to the embryo.

Yolk sac placenta (n.): A placenta in which the yolk sac wall forms the embryonic contribution to the placenta. Found in most viviparous lizards.

Ypsiloid cartilage (n.): A midventrally situated Y-shaped cartilage found in many salamanders, lying anterior to the puboischium of the pelvic girdle and articulating with it.

Zygantra (n.): A pair of accessory articulating surfaces on the anterior part of the vertebrae in some lizards and all snakes; they articulate with the zygosphenes of the next anterior vertebra.

Zygosphenes (n.): A pair of posterior accessory articulating surfaces on the vertebrae of some lizards and all snakes; they articulate with the zygantra of the next posterior vertebra.

Zygote (n.): A fertilized egg.

Literature Cited

Note that for serial publications I have included in addition to the volume number the number for individual numbers or parts if it is normal practice of their publishers to include this information on separates or reprints: for example, *Copeia* 1968 (2): 203–13. Otherwise only the volume number is given: for example, *J. Morph.* 122:203–13. When the date on the publication differs from the actual date when it appeared, the former is included in quotation marks followed by the actual date of publication (e.g., "1858." 1859).

Abalos, J. W., E. C. Baez, and R. Nader. 1964. Serpientes de Santiago del Estero. *Acta Zool. Lilloana* 20:211–83.

Acuña, R. A. 1993. *Las tortugas continentales de Costa Rica.* San José: University of Costa Rica.

———. 1994. *Conservación y ecología de las tortugas terrestres, semiacuáticas y acuáticas (de agua dulce y marinas) de Costa Rica.* San José: Editorial Universidad Estatal a Distancia.

———. 1998. *Las tortugas continentales de Costa Rica.* 2d ed. San José: Editorial de la Universidad de Costa Rica.

Acuña, R. A., A. Castaing, and F. Flores. 1984. Aspectos ecológicos de la distribución de las tortugas terrestres y semiacuáticas en el Valle Central de Costa Rica. *Rev. Biol. Trop.* 31 (2): 181–92.

Acuña, R. A., and P. E. Hanson. 1990. Phorid fly larvae as predators of turtle eggs. *Herp. Rev.* 21 (1): 13–14.

Acuña, R. A., and C. A. Laurito. 1996. Nueva especie de *Rhinoclemmys* Fitzinger, 1836 (Chelonii, Cryptodira) del Cenozoico tardío de Costa Rica. *Ameghiniana* 33 (3): 271–78.

Adler, K. 1989. Herpetologists of the past. In *Contributions to the History of Herpetology*, ed. K. Adler, 4–141. Contributions to Herpetology 5. Oxford, Ohio: Society for the Study of Amphibians and Reptiles.

Agassiz, J. L. R. 1857. *Contributions to the natural history of the United States of America.* 2 vols. Boston: Little, Brown.

Alfaro, A. 1906a. Veneno de las serpientes de coral. *Gaceta Médica Costa Rica* 10 (2): 22.

———. 1906b. Catálogo de los reptiles de Costa Rica. *Gaceta Médica Costa Rica* 10 (4): 78–81.

———. 1906c. Catálogo del los batracios de Costa Rica, extractado de la biología Centrali-Americana. *Gaceta Médica Costa Rica* 10 (5): 103–5.

———. 1906d. Notas herpetológicas. *Páginas Ilustradas Rev. Semanal* 3 (108): 1734–36.

———. 1907a. Serpientes de Costa Rica. *Bol. Soc. Nac. Agrícola* 2 (6): 121–14.

———. 1907b. Notas herpetólogicas. *Páginas Ilustradas Rev. Semanal* 4 (129): 2060–62.

———. 1912. Notas herpetológicas. *Bol. Fomento Costa Rica* 2 (10): 734–41.

Allen, P. H. 1956. *The rainforests of the Golfo Dulce.* Gainesville: University of Florida Press.

Allsteadt, J. 1994. Nesting ecology of *Caiman crocodilus* in Caño Negro, Costa Rica. *J. Herp.* 28 (1): 12–19.

Allsteadt, J., and C. Vaughan. 1988. Ecological studies of the Central American caiman *(Caiman crocodilus fuscus)* in Caño Negro National Wildlife Refuge, Costa Rica. *Bull. Chicago Herp. Soc.* 23 (1): 23–26.

Almeda, F., and C. M. Pringle, eds. 1988. *Tropical rainforests: Diversity and conservation.* San Francisco: California Academy of Sciences, Pacific Division, American Association for the Advancement of Science.

Altig, R. 1970. A key to the tadpoles of the continental United States and Canada. *Herpetologica* 26 (2): 180–207.

Altig, R., and R. A. Brandon. 1971. Generic key and synopsis for free-living larvae and tadpoles of Mexican amphibians. *Tulane Stud. Zool. Botan.* 17 (1): 10–15.

Altig, R., and G. F. Johnston. 1989. Guilds of anuran larvae: Relationships among developmental modes, morphologies and habitats. *Herp. Monogr.* 3:81–109.

Altig, R., and R. W. McDiarmid. 1999. Body plan development and morphology. In *Tadpoles: The biology of anuran larvae*, ed R. W. McDiarmid and R. Altig, 24–51. Chicago: University of Chicago Press.

Alvarado, G. E. 1989. *Los volcanes de Costa Rica.* San José: Editorial Universidad Estatal a Distancia.

Alvarado, J., and J. M. Gutiérrez. 1988. Anticoagulant effect of myotoxic phospholipase A2 isolated from the venom of the snake *Bothrops asper* (Viperidae). *Rev. Biol. Trop.* 36 (2B): 563–65.

Alvarez del Toro, M. 1960. *Los reptiles de Chiapas.* Tuxtla Gutiérrez, Chiapas: Instituto Zoológico del Estado.

———. 1973. *Los reptiles de Chiapas.* 2d ed. Tuxtla Gutiérrez, Chiapas: Gobierno del Estado de Chiapas.

———. 1974. *Los Crocodylia de México: Estudio comparativo.* Mexico City: Instituto Mexicano Recursos Naturales Renovables.

———. "1982." 1983. *Los reptiles de Chiapas.* 3d ed. Tuxtla Gutiérrez, Chiapas: Instituto de Historia Natural.

Amaral, A. do. 1926. Notas de ophiologia. 3. Nota de nomeclatura ophiologia. 1. Sobre a preferência do nome genérico "Pseudoboa" Schneider, 1801, a "Clelia" Fitzinger" 1826 e "Oxyrhopus" Wagler, 1830. 2. Sobre a preferêncis do nome específico "Pseudoboa petola" (L. 1.758) a "P. petolaria" (L. 1.758). *Rev. Mus. Paulista* 14:10–16.

———. 1927. Studies of Neotropical Ophidia. 6. A new genus

of snakes from Honduras. *Bull. Antivenin Inst. Amer.* 1 (1): 22.

———. 1929a. Studies of Neotropical Ophidia. 13. A new colubrine snake in the collection of the Vienna Museum. *Bull. Antivenin Inst. Amer.* 3 (2): 40.

———. 1929b. Studies of Neotropical Ophidia. 12. On the *Bothrops lansbergii* group. *Bull. Antivenin Inst. Amer.* 3 (1): 9–27.

———. "1929c." 1930. Estudos sobre ophidos neotrópicos. 20. Revisão do genero *Phyrnonax* Cope, 1862. *Mem. Inst. Butantan* 4:301–20.

———. "1929d." 1930. Estudos sobre ophidios neotrópicos 19. Revisão do genero *Spilotes* Wagler, 1830. *Mem. Inst. Butantan* 4:275–98.

———. "1934." 1935. Estudos sobre ophidios neotrópicos. 30. Novo genero e espécie de colubrideo na fauna da Colombia. *Mem. Inst. Butantan* 8:157–82.

———. "1954." 1955. Contribuição conhecimento dos ofidios neotrópicos. 37. Sub-espécies de *Epicrates cenchria* (Linneu, 1758). *Mem. Inst. Butantan* 26:227–47.

American Society of Ichthyologists and Herpetologists. 1987. *Guidelines for the use of live amphibians and reptiles in field research.* Lawrence, Kans.: ASIH.

American Veterinary Medical Association. 2001. 2000 report of the AVMA Panel on Euthanasia. *J. Amer. Vet. Med. Assoc.* 218 (5): 670–96.

Anderson, J. 1871. A list of reptilian accessions to the Indian Museum, Calcutta, from 1865 to 1870, with description of new species. *Asiat. Soc. Bengal J.* 40 (2): 12–39.

Anderson, K. 1991. Chromosome evolution in Holarctic treefrogs. In *Amphibian cytogenetics and evolution,* ed. D. M. Green and S. K. Sessions, 299–331. San Diego: Academic Press.

Andersson, L. G. 1899. Catalogue of Linnean type-specimens of snakes in the Royal Museum in Stockholm. *Handl. Svenska Vet. Akad.,* ser. 4, 24 (6): 1–35.

———. 1900. Catalogue of Linnean type-specimens of Linnaeus's Reptilia in the Royal Museum in Stockholm. *Handl. Svenska Vet. Akad.,* ser. 4, 26 (1): 1–29.

———. 1916. Notes on reptiles and amphibians in the Zoological Museum at Gothenberg. *Medd. Göteborgs Mus. Zool.* 9:1–41.

Andrews, R. M. 1971. Structural habitat and time budget of a tropical *Anolis* lizard. *Ecology* 52 (2): 262–79.

———. 1979a. The lizard *Corytophanes cristatus:* An extreme "sit-and-wait" predator. *Biotropica* 11 (2): 136–39.

———. 1979b. Reproductive effort of female *Anolis limifrons. Copeia* 1979 (4): 620–26.

———. 1982. Spatial variation in egg mortality of the lizard *Anolis limifrons. Herpetologica* 38 (1): 165–71.

———. 1983. *Norops polylepis* (lagartija, anole, anolis lizard.) In *Costa Rican natural history,* ed. D. H. Janzen, 409–10. Chicago: University of Chicago Press.

———. 1989. Intra-population variation in age of sexual maturity of the tropical lizard *Anolis limifrons. Copeia* 1989 (3): 751–53.

Andrews, R. M., and A. S. Rand. 1974. Reproductive effort in anoline lizards. *Ecology.*

———. 1982. Seasonal breeding and long-term population fluctuations in the lizard *Anolis limifrons.* In *The ecology of a tropical forest: Seasonal rhythms and long-term changes,* ed. E. G. Leigh Jr., A. S. Rand, and D. M. Windsor, 405–12. Washington, D.C.: Smithsonian Institution Press.

Andrews, R. M., A. S. Rand, and S. Guerrero. 1983. Seasonal and spatial variation in the annual cycle of a tropical lizard. In *Advances in herpetology and evolutionary biology: Essays in honor of Ernest E. Williams,* ed. A. G. J. Rhodin and K. Miyata, 441–54. Cambridge: Museum of Comparative Zoology, Harvard University.

Antonio, F. B. 1980. Mating behavior and reproduction of the eyelash viper *(Bothrops schlegeli)* in captivity. *Herpetologica* 36 (3): 231–33.

Aragón, F. 1988. Isolation and partial characterization of *Lachesis muta melanopcephala* coagulant proteinase: Biochemical parameters of the venom. *Rev. Biol. Trop.* 36 (2B): 387–92.

Aragón, F., R. Bolaños, and O. Vargas. 1977. Carbohidratos del veneno de *Bothrops asper* de Costa Rica: Estudio cuantitativo. *Rev. Biol. Trop.* 25 (2): 171–78.

Aragón, F., and F. Gubensek. 1981. *Bothrops asper* venom from Atlantic and Pacific zones of Costa Rica. *Toxicon* 19 (6): 797–805.

Arosemena, F. A., R. Ibáñez, and F. de Sousa. "1991." 1992. Una especie nueva de *Anolis* (Squamata: Iguanidae) del grupo *latifrons* de Fortuna, Panamá. *Rev. Biol. Trop.* 39 (2): 255–62.

Ashton, R. E., Jr., and P. S. Ashton. 1985. *Handbook of reptiles and amphibians of Florida.* Part 2. *Lizards, turtles and crocodilians.* Miami: Windward.

Atwater, T. 1989. Plate tectonic history of the northeast Pacific and western North America. In *The eastern Pacific Ocean and Hawaii,* vol. N of *The geology of North America,* ed. E. L. Winterer, D. M. Hussong, and R. W. Decker, 21–72. Boulder, Colo.: Geological Society of America.

Auffenberg, W. 1974. Checklist of fossil land tortoises (Testudinidae). *Bull. Fla. State Mus.* 18 (3): 121–251.

———. 1981. *The behavioral ecology of the Komodo dragon.* Gainesville: University of Florida Press.

Auth, D. L. 1994a. Checklist and bibliography of the amphibians and reptiles of Panama. *Smithson. Herp. Info. Serv.* 98:1–59.

———. 1994b. *Corrections and additions I to the checklist and bibliography of the amphibians and reptiles of Panama.* Privately printed.

Auth, D. L., J. Mariaux, J. Clary, D. Chiszar, F. van Breuklen, and H. M. Smith. 1998. The report of the snake genus *Conophis* in South America is erroneous. *Bull. Maryland Herp. Soc.* 34 (4): 107–12.

Avery, R. A. 1982. Field studies of body temperatures and thermoregulation. In *Biology of the Reptilia,* vol. 12, *Physiological ecologyC,* ed. C. Gans and F. H. Pough, 93–166. London: Academic Press.

Avila-Pires, T. C. S. 1995. Lizards of Brazilian Amazonia (Reptilia: Squamata). *Zool. Verh. Natl. Mus. Leiden* 299:1–706.

Avise, J. C. 1994. *Molecular markers, natural history and evolution.* New York: Chapman and Hall.

Avise, J. C., B. W. Bowen, T. Lamb, A. B. Meylan, and E. Bermingham. 1992. Mitochondrial DNA evolution at a turtle's pace: Evidence for low genetic variability and reduced microevolutionary rate in the Testudines. *Mol. Biol. Evol.* 9:457–73.

Axelrod, D. I. 1958. Evolution of the Madro-Tertiary geoflora. *Botan. Rev.* 24 (7): 433–509.

———. 1975. Evolution and biogeography of Madrean-Tethyan sclerophyll vegetation. *Ann. Missouri Botan. Gard.* 62 (2): 280–334.

Axelrod, D. I., and W. S. Ting. 1960. Late Pliocene floras east of the Sierra Nevada. *Univ. Calif. Pub. Geol.* 39 (1): 1–118.

Ayala, S. C., and D. M. Harris. 1984. A new microteiid lizard *(Alopoglossus)* from the Pacific rainforest of Colombia. *Herpetologica* 40 (2): 154–57.

Bailey, J. R. "1938." 1939. A systematic revision of the snakes of the genus *Coniophanes*. *Pap. Mich. Acad. Sci. Arts Lett.* 24 (2): 1–48.

———. 1970. *Oxyrhopus*. In Catalogue of the Neotropical Squamata, part 1, Snakes, ed. J. A. Peters and B. Orejas-Miranda, 229–35. *Bull. U.S. Natl. Mus.* 297.

———. 1985. *Oxyrhopus petola* (Linnaeus). In *The snakes of Honduras*, 2d ed., ed. L. D. Wilson and J. R. Meyer. Milwaukee: Milwaukee Public Museum.

———. 1986. The *Oxyrhopus petola-petolarius* question continued. *Bull. Maryland Herp. Soc.* 22 (3): 144–45.

Baird, S. F. 1854. Descriptions of new genera and species of North American frogs. *Proc. Acad. Nat. Sci. Phila.* 7:59–62.

———. 1859a. Description of new genera and species of North American lizards in the Museum of the Smithsonian Institution. *Proc. Acad. Nat. Sci. Phila.* 10:253–56.

———. 1859b. Reptiles. In *Report on the United States and Mexican Boundary Survey*, vol. 2, part 2, *Zoology of the Boundary*, ed. W. H. Emory. Washington, D.C: U.S. Government Printing Office.

Baird, S. F., and C. Girard. 1853. *Catalogue of North American reptiles in the Museum of the Smithsonian Institution.* Part 1. *Serpents.* Washington, D.C: U.S. Government Printing Office.

Baker, R. J., G. A. Mengden, and J. J. Bull. 1972. Karyotypic studies of thirty-eight species of North American snakes. *Copeia* 1972 (2): 257–65.

Bally, A. W., and A. R. Palmer, eds. 1989. *The geology of North America: An overview.* Vol. A of *The geology of North America.* Boulder, Colo.: Geological Society of America.

Barbour, T. 1906. Vertebrata from the savanna of Panama. 4. Reptilia and Amphibia. *Bull. Mus. Comp. Zool.* 46 (12): 224–29.

———. 1914. A contribution to the zoogeography of the West Indies, with especial reference to amphibians and reptiles. *Mem. Mus. Comp. Zool.* 44 (2): 209–359.

———. 1923. Notes on reptiles and amphibians from Panama. *Occas. Pap. Mus. Zool. Univ. Mich.* 129:1–6.

———. 1924. Two noteworthy new lizards from Panama. *Proc. N. Engl. Zool. Club.* 9:7–10.

———. 1925. A new frog and a new snake from Panama. *Occas. Pap. Boston Soc. Nat. Hist.* 5:155–56.

———. 1926. New Amphibia. *Occas. Pap. Boston Soc. Nat. Hist.* 5:191–94.

———. 1928. New Central American frogs. *Proc. N. Engl. Zool. Club.* 10:25–31.

———. 1934. The anoles. 2. The mainland species from Mexico southward. *Bull. Mus. Comp. Zool.* 77 (4): 121–55.

Barbour, T., and A. do Amaral. 1924. Notes on some Central American snakes. *Occas. Pap. Boston Soc. Nat. Hist.* 5:129–32.

Barbour, T., and E. R. Dunn. 1921. Herpetological novelties. *Proc. Biol. Soc. Wash.* 34:157–62.

Barbour, T., and B. Shreve. 1934. A new race of rock iguana. *Occas. Pap. Boston Soc. Nat. Hist.* 8:197–98.

Barden, A. 1943. Food of the basilisk lizard in Panama. *Copeia* 1943 (2): 118–21.

Barrio, A. 1965. Cloricia fisiologica en batracios anuros. *Physis* 25 (69): 137–42.

Barron, E. J., C. G. A. Harrison, J. L. Sloan II, and W. W. Hay. 1981. Paleogeography, 180 million years ago to the present. *Ecol. Geol. Helvetiae* 74 (2): 443–70.

Bauer, A. M., D. A. Good, and W. R. Branch. 1997. The taxonomy of the southern African leaf-toed geckos (Squamata: Gekkonidae), with review of Old World "*Phyllodactylus*" and the description of five new genera. *Proc. Calif. Acad. Sci.* 49 (14): 447–97.

Bauer, A. M., R. Günther, and M. Klipfel. 1995. *The herpetological contributions of Wilhelm C. H. Peters (1815–1883).* Oxford, Ohio: Facsimile Report, Society for the Study of Amphibians and Reptiles.

Baur, G. H. C. L. 1888. Osteologische Notizen über Reptilien. (Fortsetzung III). *Zool. Anz.* 11:417–24.

———. 1893. Notes on the classification of the Cryptodira. *Amer. Nat.* 27:677–78.

Bayley, I. 1950. The whistling frogs of Barbados. *J. Barbados Mus. Nat. Hist.* 17:161–70.

Beauvois, A. M. F. J. P. de. 1799. Memoir on Amphibia: Serpents. *Trans. Amer. Philos. Soc.* 4:362–81.

Beçak, W. "1965." 1966. Constituição cromossômica e mecanismo de determinação do sexo em ofídios Sul-Americanos. 1. Aspectos cariotípicos. *Mem. Inst. Butantan.* 32:37–78.

Beçak, W., and M. L. Beçak. 1969. Cytotaxonomy and chromosomal evolution in Serpentes. *Cytogenetics* 8:247–62.

Beçak, W., M. L. Beçak, and H. R. S. Nazareth. 1962. Karyotypic studies on two species of South American snakes *(Boa constrictor amarali* and *Bothrops jararaca).* *Cytogenetics* 1:305–13.

Beçak, W. L., S. M. Cameiro, and M. L. Beçak. 1971. Cariologia comparada em seis espécies de colubridos (Serpentes). *Cien. Cult.*, suppl. 23:123.

Beebe, W. 1946. Field notes on the snakes of Kartabo, British Guiana, and Caripito, Venezuela. *Zoologica* 31 (1): 11–52.

Behler, J. L., and F. W. King. 1979. *The Audubon Society field guide to North American reptiles and amphibians.* New York: Knopf.

Bell, T. 1825. On Leptophina, a group of serpents comprising the genus *Dryinus* of Merrem, and a newly formed genus proposed to be named *Leptophis*. *Zool. J.* 2:322–29.

Bemis, W. E., M. H. Wake, and K. Schwenk. 1983. Morphology and function of the feeding apparatus in *Dermophis mexicanus* (Amphibia: Gymnophiona). *Zool. J. Linn. Soc. London* 77:75–96.

Berger, L., R. Speare, P. Daszak, D. E. Green, A. A. Cunningham, C. L. Goggin, R. Slocombe, M. A. Regan, A. D. Hyatt, K. R. McDonald, H. B. Hines, K. R. Lips, G. Marantelli, and H. Parkes. 1998. Chytridiomycosis causes amphibian mortality associated with population declines in the rain forests of Australia and Central America. *Proc. Natl. Acad. Sci.* 95:9031–36.

Berry, J. F. 1978. Variation and systematics in the *Kinosternon scorpioides* and *K. leucostomum* complexes (Reptilia: Testudines: Kinosternidae) of Mexico and Central America. Ph.D. diss., University of Utah.

Berthold, A. A. 1845. Ueber verschiedene neue oder seltene Reptilien aus Neu-Granada un Crustaceen aus China. *Nachr. Georg-Augustus Univ./Gesell. Wissen. Göttingen* 1845 (3): 37–48.

———. 1846. Über verschiedene neue oder seltene Reptilien aus Neu-Granada und Crustaceen aus China. *Abh. Gesell. Wissen. Göttingen* 1846:3–32.

Beuttell, K., and J. B. Losos. 1999. Ecological morphology of Caribbean anoles. *Herp. Monogr.* 13:1–28.

Bevier, C. R. 1997. Breeding activity and chorus tenure of two Neotropical hylid frogs. *Herpetologica* 53 (3): 297–311.

Bezy, R. L. 1972. Karyotypic variation and evolution of the lizards of the family Xantusiidae. *Contr. Sci. Nat. Hist. Mus. Los Angeles Co.* 227:1–29.

———. 1984. Systematics of xantusiid lizards of the genus *Lepidophyma* in northeastern Mexico. *Contr. Sci. Nat. Hist. Mus. Los Angeles Co.* 349:1–16.

———. 1989. Night lizards, the evolution of habitat specialists. *Terra* 28 (1): 29–34.

Bezy, R. L., and J. L. Camarillo. 1992. Systematics of xantusiid lizards allied to *Lepidophyma gaigeae*. *Herpetologica* 48 (1): 97–110.

Bibron, G. 1840. In J. T. Cocteau and G. Bibron, Reptiles. In *Historia física política y natural de la Isla de Cuba:. Atlas*, ed. R. de la Sagra. Paris: Libreria de Arthus Bertrand/Lib. Soc. Geogr. (Spanish ed.)

———. 1843. In J. T. Cocteau and G. Bibron, Reptiles. In *Historia física política y natural de la Isla de Cuba*, ed. R. de la Sagra, 4:121–43. Paris: Libreria de Arthus Bertrand/Lib. Soc. Geogr. (Spanish ed.)

Bickham, J. W., and R. Baker. 1976. Karyotypes of some Neotropical turtles. *Copeia* 1976 (4): 703–8.

Bickham, J. W., K. A. Bjorndal, M. W. Haiduk, and W. E. Rainey. 1980. The karyotype and chromosomal banding patterns of the green turtle *(Chelonia mydas)*. *Copeia* 1980 (3): 918–32.

Bickham, J. W., and J. L. Carr. 1983. Taxonomy and phylogeny of the higher categories of cryptodiran turtles based on a cladistic analysis of chromosome data. *Copeia* 1983 (4): 918–32.

———. 1986. Phylogenetic implications of karyotypic variation in the Batugarinae (Testudines: Emydidae). *Genetica* 70: 89–106.

Billo, R. 1986. Tentacle apparatus of caecilians. *Mem. Soc. Zool. France* 43:71–75.

Biolley, P. 1899. *Elementos de historia natural colección de textos nacionales: Segunda folleto zoología*. San José: Impresor y Libreria Española de María Va. de Lines.

———. 1902. Bibliografía extrangera acera de Costa Rica. In *Revista de Costa Rica en el siglo XIX*, vol. 1, ed. F. M. Iglesias and J. Fernández Ferraz. San José: Tipografía Nacional.

Bjorndal, K. A., ed. 1981. *Biology and conservation of sea turtles*. Washington, D.C: Smithsonian Institution Press.

———, ed. 1995. *Sea turtles*. Rev. ed. Washington, D.C: Smithsonian Institution Press.

Bjorndal, K. A., and A. B. Bolten. 1992. Spatial distribution of green turtle *(Chelonia mydas)* nests at Tortuguero, Costa Rica. *Copeia* 1992 (1): 45–53.

Bjorndal K., and A. Carr. 1989. Variation in clutch size and egg size in the green turtle nesting population at Tortuguero, Costa Rica. *Herpetologica* 45 (2): 181–89.

Blackburn, D. G. 1985. Evolutionary origins of viviparity in Reptilia. 2. Serpentes, Amphisbaenia, and Ichthyosauria. *Amph.-Reptl.* 6 (3): 259–91.

Blainville, H. M. D. de. 1816. Prodrome d'une nouvelle distribution systématique du règne animal. *Bull. Sci. Soc. Philom.* 1816:105–24.

Blair, W. F. "1958." 1959. Call structure and species groups in U.S. treefrogs *(Hyla)*. *SW Nat.* 3:77–89.

———. 1972a. *Bufo* of North and Central America. In *Evolution in the genus* Bufo, ed. W. F. Blair, 93–101. Austin: University of Texas Press.

———, ed. 1972b. *Evolution in the genus* Bufo. Austin: University of Texas Press.

Blanchard, F. N. 1921. A revision of the king snakes: Genus *Lampropeltis* Fitzinger. *Bull. U.S. Natl. Mus.* 114:1–260.

Blaney, R. M. 1973. *Lampropeltis*. *Cat. Amer. Amph. Reptl.* 150:1–2.

Blaustein, A. R. 1994. Chicken Little or Nero's fiddle? A perspective on declining amphibian populations. *Herpetologica* 50 (1): 85–97.

Blaustein, A. R., and D. B. Wake. 1990. Declining amphibian populations: A global phenomenon? *Trends Ecol. Evol.* 5 (7): 203–4.

Blaustein, A. R., D. B. Wake, and W. P. Sousa. 1994. Amphibian declines: Judging stability, persistence, and susceptibility of populations to local and global extinctions. *Cons. Biol.* 8 (1): 60–71.

Bleakney, J. S. 1965. Reports of marine turtles from New England and eastern Canada. *Canadian Field-Nat.* 19 (2): 120–28.

Blody, D. A. 1983. Notes on the reproductive biology of the eyelash viper, *Bothrops schlegelii*, in captivity. *Herp. Rev.* 14 (2): 45–46.

Blommers-Schlösser, R. M. A. 1975. Observations on the larval development of some Malagasy frogs, with notes on their ecology and biology (Anura: Dyscophinae, Scaphiophryninae and Cophylinae). *Beaufortia* 24 (1): 7–26.

Bock, B. C. 1987. *Corytophanes*. Nesting. *Herp. Rev.* 18 (2): 35.

Bocourt, M. F. 1868a. Descriptions de quelques crotaliens nouveaux appartenant au genre *Bothrops*, recueillis dans Guatemala. *Ann. Sci. Nat. Paris, Zool.*, ser 5, 10:201–2.

———. 1868b. Description de quelques cheloniens nouveaux appartenant à la faune mexicaine. *Ann. Sci. Nat. Paris, Zool.*, ser. 5, 10:121–22.

———. 1873–97. *Études sur les reptiles: Mission scientifique au Mexique et dans l'Amérique Centrale. Recherches zoologiques.* Pt. 3, sec. 1, issues 2–15. Paris: Imprimerie Impériale.

Boettger, O. 1892a. *Katalog der Batrachier-Sammlung im Museum der Seckenbergischen Naturforschenden Gesellschaft in Frankfurt am Main.* Frankfurt am Main: Knauer.

———. 1892b. Drei neue colubriforme Schlangen. *Zool. Anz.* 15:417–20.

———. 1893a. Reptilian und Batrachier aus Venezuela. *Ber. Natufr. Gesell. Frankfurt a. M.* 1893:35–42.

———. 1893b. Ein nueuer Laubforsch aus Costa Rica. *Ber. Naturf. Gesell. Frankfurt a. M.* 1892–93:251d–52.

———. 1898. *Katalog der Reptilien-Sammlung im der Museum der Senckenbergischen Naturforschenden Gesellschaft in Frankfurt am Main. Part 2. Schlangen.* Frankfurt am Main: Knauer.

Bogart, J. P. 1970a. Systematic problems in the amphibian family Leptodactylidae (Anura) as indicated by karyotype analysis. *Cytogenetics* 9:369–83.

———. "1970b." 1973. Los cromosomas de anfibios anuros del género *Eleutherodactylus. Act. IV Congr. Latinamer. Zool.* 165–78.

———. 1972. Karyotypes. In *Evolution in the genus* Bufo, ed. W. F. Blair, 171–95. Austin: University of Texas Press.

———. 1973. Evolution of anuran karyotypes. In *Evolutionary biology of the anurans,* ed. J. L. Vial, 337–49. Columbia: University of Missouri Press.

———. 1974. A karyosystematic study of frogs in the genus *Leptodactylus* (Anura: Leptodactylidae). *Copeia* 1974 (3): 728–37.

———. 1991. The influence of life history on karyotypic evolution in frogs. In *Amphibian cytogenetics and evolution,* ed. D. M. Green and S. K. Sessions, 233–58. San Diego: Academic Press.

Bogart, J. P., and C. E. Nelson. 1976. Evolutionary implications from karyotypic analysis of frogs of the families Microhylidae and Rhinophrynidae. *Herpetologica* 32 (2): 199–208.

Bogert, C. M. 1947. The status of the genus *Leptodrymus* Amaral, with comments on modifications of colubrid premaxillae. *Amer. Mus. Nat. Hist. Novitates* 1352:1–14.

———. 1968. The variation and affinities of the dwarf boas of the genus *Ungaliophis. Amer. Mus. Nat. Hist. Novitates* 2340:1–26.

Bogert, C. M., and J. A. Oliver. 1945. A preliminary analysis of the herpetofauna of Sonora. *Bull. Amer. Mus. Nat. Hist.* 83 (6): 297–426.

Böhme, W. 1988. Zur Genitalmorphologie der Sauria: Functionelle und stammesgeschichtliche Aspeckte. *Bonn. Zool. Monogr.* 27:1–176.

Bohuslavek, J. 1996. *Ninia sebae sebae* (red coffee snake): Reproduction. *Herp. Rev.* 27 (3): 146.

Boie, F. 1826. Generalubersicht der Familien und Gattung der Ophidier. *Isis von Oken* 19 (10): cols. 981–82.

———. 1827. Ueber Merrem's Versuch eines Systems der Amphbien, Marburg, 1820. 1. Ophidier. *Isis von Oken* 20 (3): cols. 508–66.

Bolaños, F., and A. C. Chaves. 1989. *Serpientes del Pacifico Sur y laderas adyacentes Costa Rica.* San José: Organization for Tropical Studies.

Bolaños, F., D. C. Robinson, and D. B. Wake. 1987. A new species of salamander (genus *Bolitoglossa*) from Costa Rica. *Rev. Biol. Trop.* 35 (1): 87–92.

Bolaños, J. R., J. Sánchez, and L. Piedra. 1997. Inventorio y estructura poblacional de crocodílidos en tres zonas de Costa Rica. *Rev. Biol. Trop.* 44 (3)/45 (1) A:283–87.

Bolaños, R. 1984. *Serpientes venenos en Centroamérica.* San José: Editorial Universidad Costa Rica.

Bolaños, R., L. O. Marin, E. Mora, and E. A. Alfaro. 1981. El accidente ofídico por cascabela *(Crotalus durissus durissus)* en Costa Rica. *Acta Med. Costarricense* 211–14.

Bolaños, R., G. Muñoz, and L. Cerdas. 1978. Toxicidad, neutralización e immunoeléctroforesis de los venenos de *Lachesis muta* de Costa Rica y Colombia. *Toxicon* 16:295–300.

Bonaparte, C. L. J. L. 1850. *C. L. Bonaparte conspectus systematum. Mastozooloiae . . . herpetologiae et amphibiologiae . . . ichthyologiae.* Rev. ed. Leiden: Brill.

Boulenger, G. A. 1882. *Catalogue of the Batrachia Salientia s. Ecaudata in the collection of the British Museum.* 2d ed. London: Taylor and Francis.

———. 1883a. Descriptions of new lizards and frogs collected by Herr A. Forrer in Mexico. *Ann. Mag. Nat. Hist.,* ser. 5, 11 (65): 42–44.

———. 1883b. Remarks on the Nyctisaura. *Ann. Mag. Nat. Hist.,* ser. 5, 12 (71): 308.

———. 1885. *Catalogue of the lizards in the British Museum (Natural History).* 2d ed. Vols. 1–2. London: Taylor and Francis.

———. 1890. First report on additions to the lizard collection in the British Musuem (Natural History). *Proc. Zool. Soc. London* 1890:77–86.

———. 1891. Notes on American batrachians. *Ann. Mag. Nat. Hist.,* ser. 6, 8 (48): 453–57.

———. 1893. *Catalogue of the snakes in the British Museum (Natural History).* Vol. 1. London: Taylor and Francis.

———. 1894. *Catalogue of the snakes in the British Museum (Natural History).* Vol. 2. London: Taylor and Francis.

———. 1896a. Descriptions of new batrachians collected by Mr. C. F. Underwood in Costa Rica. *Ann. Mag. Nat. Hist.,* ser. 6, 18 (106): 340–42.

———. 1896b. Descriptions of new batrachians in the British Musuem. *Ann. Mag. Nat. Hist.,* ser. 6, 17 (101): 401–6.

———. 1896c. *Catalogue of the snakes in the British Museum (Natural History).* Vol. 3. London: Taylor and Francis.

———. 1896d. Descriptions of new reptiles and batrachians from Colombia. *Ann. Mag. Nat. Hist.,* ser. 6, 17:16–21.

———. 1898a. A list of the reptiles and batrachians collected by the late Prof. L. Balzan in Bolivia. *Ann. Mus. Civ. Stor. Nat. Genova* 19:128–33.

———. 1898b. An account of the reptiles and batrachians collected by Mr. W. F. H. Rosenberg in estern Ecuador. *Proc. Zool. Soc. London* 1898:107–26.

———. 1898c. Fourth report on additions to the batrachian collections in the Natural History Museum. *Proc. Zool. Soc. London* 1898:473–82.

———. "1898d." 1899. Third report on additions to the lizard collection in the Natural History Museum. *Proc. Zool. Soc. London* 1898:912–23.

———. 1899. Descriptions of new batrachians in the collection of the British Museum (Natural History). *Ann. Mag. Nat. Hist.,* ser. 7, 3 (16): 273–77.

———. 1902. Descriptions of new batrachians and reptiles

from north-western Ecuador. *Ann. Mag. Nat. Hist.*, ser. 7, 9 (49): 51–57.

———. 1903. Descriptions of new snakes in the collection of the British Museum. *Ann. Mag. Nat. Hist.*, ser. 7, 12 (69): 350–54.

———. 1905. Descriptions of new snakes in the collection of the British Museum. *Ann. Mag. Nat. Hist.*, ser. 7, 15 (89): 453–56.

———. 1908. Descriptions of new batrachians and reptiles discovered by Mr. M. G. Palmer in southwestern Colombia. *Ann. Mag. Nat. Hist.*, ser. 8, 2 (12): 515–22.

———. 1913. On a collection of batrachians and reptiles made by Dr. H. G. F. Spurrell, F.Z.S., in the Choco, Colombia. *Proc. Zool. Soc. London* 1913:1019–38.

———. 1920. A monograph of the American frogs of the genus *Rana*. *Proc. Amer. Acad. Arts Sci.* 55:441–80.

Bovallius, C. 1887. *Resa i Central-Amerika, 1881–1883.* 2 vols. Upsala: Almquist och Wiksell.

Bowen, B. W., and S. A. Karl. 1999. In war, truth is the first casualty. *Cons. Biol.* 13 (5): 1013–16.

Bowland, C. L. 1993. Depositional history of the western Colombian Basin, Caribbean Sea, revealed by seismic stratigraphy. *Geol. Soc. Amer. Bull.* 105:1321–45.

Bowler, J. K. 1977. Longevity of reptiles and amphibians in North American collections. *Soc. Stud. Amph. Reptl. Herp. Circ.* 6:1–32.

Boza, M. A. 1978. *Los parques nacionales de Costa Rica.* Madrid: Instituto de la Caza Fotográfica y Ciencias Naturaleza.

———. 1986. *Parques Nacionales Costa Rica/Costa Rica National Parks.* San José: Fundación Neotrópica.

———. 1993. Conservation in action: Past, present and future of the national parks system of Costa Rica. *Cons. Biol.* 7 (2): 239–47.

Boza, M. A., and J. H. Cevo. 1998. *Los parques nacionales de Costa Rica.* San José: Instituto de la Caza Fotográfica y Ciencias Naturaleza.

Bozzolli de Wille, M. E. 1975. *Localidades indíginas costarricenses.* 2d ed. San José: Editorial Universidad Centroamerica.

Brame, A. H., Jr. II. 1963. A new Costa Rican salamander (genus *Oedipina*) with a re-examination of *O. collaris* and *O. serpens*. *Contr. Sci. Nat. Hist. Mus. Los Angeles Co.* 65: 1–12.

———. 1968. Systematics and evolution of the Mesoamerican salamander genus, *Oedipina*. *J. Herp.* 2 (1–2): 1–62.

Brame, A. H., Jr. II, and W. E. Duellman. 1970. A new salamander (genus *Oedipina*) of the *uniformis* group from western Panama. *Contr. Sci. Nat. Hist. Mus. Los Angeles Co.* 201:1–8.

Brame, A. H., Jr. II, and D. B. Wake. 1963. Redescription of the plethodontid salamander *Bolitoglossa lignicolor* (Peters), with remarks on the status of *B. palustris* Taylor. *Proc. Biol. Soc. Wash.* 76:289–96.

Branch, W. R. 1981. Hemipenes of the Madagascan boas *Acrantophis* and *Sanzinia*, with a review of hemipeneal morphology in the Boidae. *J. Herp.* 15 (1): 91–99.

———. 1986. Hemipenial morphology of African snakes: A taxonomic review. 1. Scolecophidia and Boidae. *J. Herp.* 20 (3): 285–99.

Bransford, J. F. 1881. Archaeological researches in Nicaragua. *Smithson. Contr. Knowledge.* 25 (383): 1–96.

———. 1884. Report on explorations in Central Amereica in 1881. *Smithson. Annl. Rept.* 1882:803–25.

Brattstrom, B. H. 1955. The coral snake "mimic" problem and protective coloration. *Evolution* 9 (2): 217–19.

———. 1963. A preliminary review of the thermal requirements of amphibians. *Ecology* 44 (2): 238–55.

———. 1965. Body temperatures of reptiles. *Amer. Midl. Nat.* 73 (2): 376–422.

Brattstrom, B. H., and T. R. Howell. 1954. Notes on some collections of reptiles and amphibians from Nicaragua. *Herpetologica* 10 (2): 114–23.

Brattstrom, B. H., and R. M. Yarnell. 1968. Aggressive behavior in two species of leptodactylid frogs. *Herpetologica* 24 (3): 222–29.

Breder, C. M., Jr. 1925. In Darien jungles, experiences of a student of reptile and amphibian life in a little known part of Panama. *Nat. Hist. Mag.* 25 (4): 324–37.

———. 1946. Amphibians and reptiles of the Rio Chucunaque drainage, Darien, Panama, with notes on their life histories and habitats. *Bull. Amer. Mus. Nat. Hist.* 86 (8): 375–436.

Brocchi, P. 1877. Sur quelques batraciens raniformes et bufoniformes de l'Amérique Centrale. *Bull. Soc. Philomath. Paris*, ser. 7, 1:175–97.

———. 1881–83. *Études des batraciens de l'Amérique Centrale: Mission Scientifique au Mexique et dans l'Amérique Centrale.* Paris: Imprimerie Impériale.

Brochu, C. A. 1997. Divergence timing, and the phylogenetic relationships of *Gavialis*. *Syst. Biol.* 46 (3): 479–522.

Brodie, E. D., Jr. 1983. Antipredator adaptations of salamanders: Evolution and convergence among terrestrial species. In *Plant, animal, and microbial adaptations to terrestrial environment*, ed. N. S. Margaris, S. Aríanoutsou-Faraggitaki, and R. J. Reiter, 109–33. New York: Plenum.

Brodie, E. D., Jr., P. K. Ducey, and E. A. Baness. 1991. Anipredator skin secretions of some tropical salamanders *(Bolitoglossa)* are toxic to snake predators. *Biotropica* 23 (1): 58–62.

Brodie, E. D., III. 1993. Differential avoidance of coral snake banded patterns by free-ranging avian predators in Costa Rica. *Evolution* 47 (1): 227–35.

Brodie, E. D., III, and F. J. Janzen. 1995. Experimental studies of coral snake mimicry: Generalized avoidance of ringed snake patterns by free-ranging avian predators. *Funct. Ecol.* 9:186–90.

Brodie, E. D., III, and A. J. Moore. 1995. Experimental studies of coral snake mimicry: Do snakes mimic millipedes? *Anim. Behav.* 49:534–36.

Brongersma, L. D. 1969. Miscellaneous notes on turtles, IIB. *Proc. K. Ned. Akad. Wetens.*, ser. C, 72:90–102.

Brongniart, A. 1800. Essai d'une classification naturelle des reptiles. *Bull. Soc. Philomath. Paris* 2:81–82.

Brown, L. R., C. Flavin, and H. French, eds. 2001. *State of the world 2001.* New York: Norton.

Bruce, R. C. 1997. Life history attributes of the salamander *Bolitoglossa colonnea*. *J. Herp.* 31 (4): 592–94.

———. 1998. Nesting habits, eggs, and hatchlings of the salamander *Nototriton picadoi* (Plethodontidae: Bolitoglossini). *Herpetologica* 54 (1): 13–18.

———. 1999. Life history attributes of a rare Neotropical salamander, *Nototriton picadoi* (Plethodontidae: Bolitoglossini). *Herp. Rev.* 30 (2): 76–78.

Brugger, K. E. 1989. Red-tailed hawk dies with coral snake in talons. *Copeia* 1989 (2): 508–10.

Brust, D. G. 1990. Maternal brood care by *Dendrobates pumilio:* A frog that feeds its young. Ph.D. diss., Cornell University.

———. 1993. Maternal brood care by *Dendrobates pumilio:* A frog that feeds its young. *J. Herp.* 27 (1): 96–98.

Buckley, L. J., and R. W. Axtell. 1997. Evidence of the specific status of the Honduran lizards formerly referred to *Ctenosaura palearis* (Reptilia: Squamata: Iguanidae). *Copeia* 1997 (1): 138–50.

Burger, J., and M. Gochfeld. 1991. Burrow site selection by black iguana *(Ctenosaura similis)* at Palo Verde, Costa Rica. *J. Herp.* 25 (4): 430–35.

Burger, W. L. 1971. Genera of pitvipers (Serpentes: Crotalidae). Ph.D. diss., University of Kansas.

Burger, W. L., and J. E. Werler. 1954. The subspecies of the ring-necked coffee snake *Ninia diademata*, and a short biological and taxonomic account of the genus. *Univ. Kansas Sci. Bull.* 36 (2): 643–72.

Burghardt, G. M., and M. Berkoff, eds. 1978. *The development of behavior: Comparative and evolutionary aspects.* New York: Garland STPM Press.

Burghardt, G. M., and A. S. Rand. 1982, eds. *Iguanas of the world.* Park Ridge, N.J.: Noyes.

Burnham, R. J., and A. Graham. 1999. The history of Neotropical vegetation: New developments and status. *Ann. Missouri Botan. Gard.* 86 (2): 546–89.

Burt, C. E. 1931. A study of the teiid lizards of the genus *Cnemidophorus* with special reference to their phylogenetic relationships. *Bull. U.S. Natl. Mus.* 154:1–286.

Bury, R. B., G. C. Gorman, and J. F. Lynch. 1969. Karyotypic data for five species of anguid lizards. *Experientia* 25 (3): 314–16.

Bury, R. B., F. Gress, and G. C. Gorman. 1970. Karyotypic survey of some colubrid snakes from western North America. *Herpetologica* 26 (4): 461–66.

Cabrera, J., J. R. Rojas, G. Galeano, and V. Meza. 1996. Mortalidad embrionaria y éxito de eclosión en huevos de *Trachemys scripta* (Testudines: Emydidae) incubados en un área natural protegida. *Rev. Biol. Trop.* 44 (2B): 841–46.

Cadle, J. E. 1984a. Molecular systematics of Neotropical xenodontine snakes. 1. South American xenodontines. *Herpetologica* 40 (1): 8–20.

———. 1984b. Molecular systematics of Neotropical xenodontine snakes. 2. Central American xenodontines. *Herpetologica* 40 (1): 21–30.

———. 1984c. Molecular systematics of Neotropical xenodontine snakes. 3. Overview of xenodontine phylogeny and the history of New World snakes. *Copeia* 1984 (3): 641–52.

———. 1985. The Neotropical colubrid snake fauna (Serpentes: Colubridae): Lineage components and biogeography. *Syst. Zool.* 34 (1): 1–20.

———. 1987. Geographic distribution: Problems in phylogeny and zoogeography. In *Snakes: Ecology and evolutionary biology,* ed. R. A. Siegel, J. T. Collins, and S. S. Novak, 77–105. New York: Macmillan.

———. 1988. Phylogenetic relationships among advanced snakes: A molecular perspective. *Univ. Calif. Pub. Zool.* 119:1–77.

———. 1989. A new species of *Coniophanes* (Serpentes: Colubridae) from northwestern Peru. *Herpetologica* 45 (4): 411–24.

———. 1992. On Colombian snakes. *Herpetologica* 48 (1): 134–43.

Cadle, J. E., and H. W. Greene. 1993. Phylogenetic patterns, biogeography, and the ecological structure of Neotropical snake communities. In *Species diversity in ecological communities,* ed. R. E. Ricklefs and D. Schluter, 281–93. Chicago: University of Chicago Press.

Caldwell, J. P. 1994. Natural history and survival of eggs and early larval stages of *Agalychnis calcarifer* (Anura: Hylidae). *Herp. Nat. Hist.* 2 (2): 57–66.

Caldwell, J. P., and M. C. de Araújo. 1998. Cannibalistic interactions resulting from indiscriminate predatory behavior in tadpoles of poison frogs (Anura: Dendrobatidae). *Biotropica* 30 (1): 92–103.

Caldwell, M. W., and M. S. Y. Lee. 1997. A snake with legs from the marine Cretaceous of the Middle East. *Nature* 386:705–9.

Calvert, A. S., and P. P. Calvert. 1917. *A year of Costa Rican natural history.* New York: Macmillan.

Camp, C. L. 1923. Classification of the lizards. *Bull. Amer. Mus. Nat. Hist.* 48 (11): 289–481.

Campbell, J. 1964. The masks of god: Occidental mythology. New York: Viking Press.

Campbell, J. A. 1983. The biogeography of the cloud forest herpetofauna of Middle America, with special reference to the Sierra de las Minas of Guatemala. Ph.D. diss., University of Kansas.

———. 1989. A new species of colubrid snake of the genus *Coniophanes* from the highlands of Chiapas, Mexico. *Proc. Biol. Soc. Wash.* 102 (4): 1036–44.

———. 1998. *Amphibians and reptiles of northern Guatemala, the Yucatán, and Belize.* Norman: University of Oklahoma Press.

Campbell, J. A., and E. D. Brodie Jr., eds. 1992. *Biology of the pitvipers.* Tyler, Tex.: Selva.

Campbell, J. A., and J. L. Camarillo. 1994. A new lizard of the genus *Diploglossus* (Anguidae: Diploglossinae) from Mexico, with a review of the Mexican and northern Central American species. *Herpetologica* 50 (2): 193–209.

Campbell, J. A., and W. W. Lamar. 1989. *The venomous reptiles of Latin America.* Ithaca: Cornell University Press.

———. 1992. Taxonomic status of miscellaneous Neotropical viperids, with the description of a new genus. *Occas. Pap. Mus. Texas Tech. Univ.* 153:1–31.

Campbell, J. A., and E. N. Smith. 1992. A new frog of the genus *Ptychohyla* (Hylidae) from the Sierra de Santa Cruz, Guatemala, and description of a new genus of Middle American stream-breeding treefrogs. *Herpetologica* 48 (2): 153–67.

———. 1998. A new genus and species of colubrid snake from the Sierra de las Minas of Guatemala. *Herpetologica* 54 (2): 207–20.

Campbell, J. A., and A. Solórzano. 1992. Biology of the montane pitviper, *Porthidium godmani*. In *Biology of the pitvipers*, ed. J. A. Campbell and E. D. Brodie Jr., 223–50. Tyler, Tex.: Selva.

Candeze, E. 1859. Monographie des élaterides. Vol. 2. *Mem. Soc. Sci. Liège* 1859:141–543.

Cannatella, D. C. 1980. A review of the *Phyllomedusa buckleyi* group (Anura: Hylidae). *Occas. Pap. Mus. Nat. Hist. Univ. Kansas* 87: 1–40.

Carpenter, C. C. 1965. The display of the Cocos Island anole. *Herpetologica* 21 (4): 256–60.

Carr, A. F. 1952. *Handbook of turtles*. Ithaca: Cornell University Press.

———. 1957. Notes on the zoogeography of the Atlantic sea turtles of the genus *Lepidochelys*. *Rev. Biol. Trop.* 5 (1): 45–61.

———. 1967. *So excellent a fishe*. New York: Natural History Press.

———. 1983. *Chelonia mydas* (tortuga, tortuga blanca, green turtle). In *Costa Rican natural history*, ed. D. H. Janzen, 394–96. Chicago: University of Chicago Press.

Carr, A. F., and L. Giovannoli. 1957. The ecology and migrations of sea turtles. 2. Results of field work in Costa Rica, 1955. *Amer. Mus. Nat. Hist. Novitates* 1835:1–32.

Carr, A. F., and L. Ogren. 1959. The ecology and migrations of sea turtles. 3. *Dermochelys* in Costa Rica. *Amer. Mus. Nat. Hist. Novitates* 1958:1–29.

Carriker, M. A., Jr. 1910. An annotated list of the birds of Costa Rica. *Ann. Carnegie Mus.* 6 (3): 314–970.

Carson, H. L. 1945. Delayed fertilization in a captive indigo snake with notes on feeding and shedding. *Copeia* 1945 (4): 222–25.

Carvalho, A. L. 1954. A preliminary synopsis of the genera of American microhylid frogs. *Occas. Pap. Mus. Zool. Univ. Mich.* 555:1–19.

Castillo, V. E. 1986. Factores ecológicos y de mercado de la reproducción de *Rhinoclemmys pulcherrima* y *Kinosternon scorpioides* (Testudines: Emydidae y Kinosternidae). Lic. thesis, University of Costa Rica.

Catesby, M. 1743. *The natural history of Carolina, Florida, and the Bahama Islands*. Vol. 2. London: Innys.

———. 1748. *The natural history of Carolina, Florida and the Bahama Islands; Appendix*. London: Innys.

Censky, E. J., and C. J. McCoy Jr. 1988. Female reproductive cycles of five species of snakes (Reptilia: Colubridae) from the Yucatan Peninsula, Mexico. *Biotropica* 20 (4): 326–33.

Cerdas, L., and B. Lomonte. 1982. Estudio de la capacidad ofiofaga y la resistencia de la zopilota (*Clelia clelia*, Colubridae) de Costa Rica a los venenos de serpiente. *Toxicon* 20 (15): 936–39.

Cerdas, L., and S. Martínez. 1986. Captive reproduction of the mussurana, *Clelia clelia* (Daudin) from Costa Rica. *Herp. Rev.* 17 (1): 12.

Chacón, D., W. McLarney, C. Ampie, and B. Venegas. 1996. Reproduction and conservation of the leatherback turtle *Dermochelys coriacea* (Testudines: Dermochelyidae) in Gandoca, Costa Rica. *Rev. Biol. Trop.* 44 (2B): 853–60.

Chaves, M., F. J. Alvarado, R. Aymerich, and A. Solórzano. 1993. *Aspectos básicos sobre las serpientes de Costa Rica*. San José: Instituto Clodomiro Picado, Universidad de Costa Rica.

Chippaux, J.-P. 1986. Les serpents de la Guyane française. *ORSTOM Coll. Fauna Trop.* 27:1–165.

Chiszar, D., and C. W. Radcliffe. 1989. The predatory strike of the jumping viper (*Porthidium nummifer*). *Copeia* 1989 (4): 1037–39.

Christensen, R. M. 1975. Breeding Central American wood turtles. *Chelonia* 2 (3): 8–10.

Clapperton, C. M. 1993. *Quaternary geology and geomorphology of South America*. Amsterdam: Elsevier.

Clarke, B. T. 1981. Comparative osteology and evolutionary relationships in the African Raninae (Anura, Ranidae). *Monit. Zool. Ital.*, n.s., suppl. 15 (14): 285–331.

Coates, A. G., J. B. C. Jackson, L. S. Collins, T. M. Cronin, H. J. Dowsett, L. M. Bybell, P. Jung, and J. A. Obando. 1992. Closure of the Isthmus of Panama: The near-shore marine record of Costa Rica and Panama. *Bull. Geol. Soc. Amer.* 104:814–28.

Coates, A. G., and J. A. Obando. 1996. The geologic evolution of the Central American Isthmus. In *Evolution and environment in tropical America*, ed. J. B. C. Jackson, A. F. Budd, and A. G. Coates, 21–56. Chicago: University of Chicago Press.

Cochran, D. M. 1961. Type specimens of reptiles and amphibians in the U.S. National Museum. *Bull. U.S. Natl. Mus.* 220:1–291.

Cochran, D. M., and C. J. Goin. 1970. Frogs of Colombia. *Bull. U.S. Natl. Mus.* 288:1–655.

Cocroft, R. B., R. W. McDiarmid, A. P. Jaslow, and P. M. Ruiz. 1990. Vocalizations of eight species of *Atelopus* (Anura: Bufonidae) with comments on communication in the genus. *Copeia* 1990 (3): 631–43.

Cocteau, J. T., and G. Bibron. 1841. Reptiles. In *Historia física, política y natural de la Isla de Cuba*, ed. R. de la Sagra, 4:1–143. Spanish ed. Paris: Libreria de Arthus Bertrand/Lib. Soc. Geogr. (Spanish ed.).

Coen, E. 1953. La meterológico de Costa Rica. In *Atlas estadístico de Costa Rica*, prepared for Dirección General de Estadística y Censos de Costa Rica, 34–37. San José: Casa Gráfica.

———. 1983. Climate. In *Costa Rica natural history*, ed. D. H. Janzen, 35–46. Chicago: University of Chicago Press.

Cohen, A. C., and B. C. Myers. 1970. A function of the horns (supraocular scales) in the sidewinder, *Crotalus cerastes*, with comments on horned lizards. *Copeia* 1970 (3): 574–75.

Cohen, M. M., and C. Gans. 1970. The chromosomes of the order Crocodilia. *Cytogenetics* 9:81–105.

Cohen, M. M., C. C. Huang, and H. F. Clark. 1967. The somatic chromosomes of three lizard species: *Gekko gecko*, *Iguana iguana*, and *Crotophytus collaris*. *Experientia* 23:769–71.

Cohn, J. P. 1989. Iguana conservation and economic development. *BioScience* 39 (6): 359–63.

Cole, C. J. 1971. Chromosomes of the rhinophrynid frog, *Rhinophrynus dorsalis* Duméril and Bibron. *Herp. Rev.* 3 (2): 37.

————. 1974. Chromosome evolution in selected tree-frogs, including casque-headed species *(Pternohyla, Triprion, Hyla* and *Smilisca). Amer. Mus. Nat. Hist. Novitates* 2541:1–10.

————. 1978. Karyotypes and systematics of the lizards in the *variabilis, jalapae* and *scalaris* species groups of the genus *Sceloporus. Amer. Mus. Nat. Hist. Novitates* 2653:1–13.

Cole, C. J., and H. C. Dessauer. 1993. Unisexual and bisexual whiptail lizards of the *Cnemidophorus lemniscatus* complex (Squamata: Teiidae) of the Guiana region, South America, with descriptions of new species. *Amer. Mus. Nat. Hist. Novitates* 3081:1–30.

Cole, C. J., H. C. Dessauer, and A. L. Markezich. 1993. Missing link found: The second ancestor of *Gymnophthalmus underwoodi* (Squamata: Teiidae), a South American unisexual lizard of hybrid origin. *Amer. Mus. Nat. Hist. Novitates* 3055:1–13.

Cole, C. J. H. C. Dessauer, C. R. Townsend, and M. G. Arnold. 1990. Unisexual lizards of the genus *Gymnophthalmus* (Reptilia: Teiidae) in the Neotropics: Genetics, origin, and systematics. *Amer. Mus. Nat. Hist. Novitates* 2994:1–29.

Cole, C. J., and L. M. Hardy. 1981. Systematics of North American colubrid snakes related to *Tantilla planiceps* (Blainville). *Bull. Amer. Mus. Nat. Hist.* 171 (3): 201–84.

Colinvaux, P. A. 1993. Pleistocene biogeography and diversity in tropical forests of South America. In *Biological relationships between Africa and South America,* ed. P. Goldblatt, 473–99. New Haven: Yale University Press.

————. 1996. Quaternary environmental history and forest diversity in the Neotropics. In *Evolution and environment in tropical America,* ed. J. B. C. Jackson, A. F. Budd, and A. G. Coates, 359–405. Chicago: University of Chicago Press.

Coloma, L. A. 1995. Ecuadorian frogs of the genus *Colostethus* (Anura: Dendrobatidae). *Misc. Pub. Mus. Nat. Hist. Univ. Kansas* 87:1–72.

Conant, R. 1965. Miscellaneous notes and comments on toads, lizards, and snakes from Mexico. *Amer. Mus. Nat. Hist. Novitates* 2205:1–38.

————. 1984. A new subspecies of the pit viper, *Agkistrodon bilineatus* (Reptilia: Viperidae) from Central America. *Proc. Biol. Soc. Wash.* 97 (1): 135–41.

Conant, R., and J. T. Collins. 1991. *A field guide to reptiles and amphibians: Eastern and central North America.* 3d ed. Boston: Houghton Mifflin.

Conant, R., and A. Downs Jr. 1940. Miscellaneous notes on the eggs and young of reptiles. *Zoologica* 25 (3): 33–48.

Conejo, A. 1975. *Henri Pittier.* San José: Ministerio de Cultura, Juventud y Deportes.

Connors, J. S. 1989. *Oxybelis fulgidus* (green vine snake): Reproduction. *Herp. Rev.* 20 (3): 73.

Cook, D. G. 1984. A case of envenomation by the Neotropical colubrid snake *Stenorrhina freminvillei. Toxicon* 22: 823–27.

Cope, E. D. 1860a. Catalogue of Colubridae in the Museum of the Academy of Natural Sciences of Philadelphia. 1. Calamarinae. *Proc. Acad. Nat. Sci. Phila.* 12:74–79.

————. 1860b. Catalogue of the Colubridae in the Museum of the Academy of Natural Sciences of Philadelphia, with notes and descriptions of new species. 2. *Proc. Acad. Nat. Sci. Phila.* 12:241–66.

————. "1860c." 1861. Descriptions of reptiles from tropical America and Asia. *Proc. Acad. Nat. Sci. Phila.* 12:368–79.

————. "1860d." 1861. Catalogue of the Colubridae in the Academy of Natural Sciences of Philadelphia. 3. *Proc. Acad. Nat. Sci. Phila.* 12:553–66.

————. 1861a. Remarks on reptiles (*Diphalus: Amphisbaena angustifrons: Loxocemus* Cope). *Proc. Acad. Nat. Sci. Phila.* 13:75–77.

————. 1861b. Notes and descriptions of anoles. *Proc. Acad. Nat. Sci. Phila.* 13:208–15.

————. 1861c. Contributions to the ophiology of Lower California, Mexico and Central America. *Proc. Acad. Nat. Sci. Phila.* 13:292–306.

————. 1862a. Synopsis of the species of *Holcosus* and *Ameiva* with diagnoses of new West Indian and South American Colubridae. *Proc. Acad. Nat. Sci. Phila.* 14:60–82.

————. 1862b. On some new and little known American Anura. *Proc. Acad. Nat. Sci. Phila.* 14:151–59.

————. 1862c. Contributions to Neotropical saurology. *Proc. Acad. Nat. Sci. Phila.* 14:176–88.

————. 1862d. Catalogues of the reptiles obtained during the explorations of the Parana, Paraguay, Vermejo and Uraguay Rivers, by Capt. Thos. J. Page, U.S.N.; and of those procured by Lieut. N. Mishler, U.S. Top. Eng., commander of the expedition conducting the survey of the Atrato River. *Proc. Acad. Nat. Sci. Phila:* 14:346–59.

————. 1863. Descriptions of new American Squamata in the Museum of the Smithsonian Institution, Washington. *Proc. Acad. Nat. Sci. Phila.* 15:100–106.

————. 1864a. Contributions to the herpetology of tropical America. *Proc. Acad. Nat. Sci. Phila.* 16:166–81.

————. 1864b. On the characters of the higher groups of Reptilia Squamata—especially of the Diploglossa. *Proc. Acad. Nat. Sci. Phila.* 16:224–31.

————. 1865a. Third contribution to the herpetology of tropical America. *Proc. Acad. Nat. Sci. Phila.* 17:185–98.

————. 1865b. Sketch of the primary groups of Batrachia Salientia. *Nat. Hist. Rev.,* n.s., 5:97–120.

————. 1866a. Fourth contribution to the herpetology of tropical America. *Proc. Acad. Nat. Sci. Phila.* 18:123–32.

————. "1866b." 1867. Fifth contribution to the herpetology of tropical America. *Proc. Acad. Nat. Sci. Phila.* 18:317–23.

————. 1868a. An examination of the Reptilia and Batrachia obtained by the Orton expedition to Ecuador and the upper Amazon, with notes on other species. *Proc. Acad. Nat. Sci. Phila.* 20:96–140.

————. "1868b." 1869. Sixth contribution to the herpetology of tropical America. *Proc. Acad. Nat. Sci. Phila.* 20:305–13.

————. "1868c." 1869. On the origin of genera. *Proc. Acad. Nat. Sci. Phila.* 20:242–300.

————. "1869." 1870. Seventh contribution to the herpetology of tropical America. *Proc. Amer. Philos. Soc.* 11:147–69.

————. "1870." 1871. Catalogue of Batrachia and Reptilia obtained by J. A. McNiel in Nicaragua. *Second and Third Annl. Rept. Peabody Acad. Sci.* 1869–70:82–85.

————. 1871. Ninth contribution to the herpetology of tropical America. *Proc. Acad. Nat. Sci. Phila.* 23:200–224.

————. 1874. Description of some species of reptiles obtained by Dr. John F. Bransford, assistant surgeon, U.S. Navy, while attached to the Nicaraguan surveying expedition in 1873. *Proc. Acad. Nat. Sci. Phila.* 26:64–72.

———. 1875. On the Batrachia and Reptilia of Costa Rica. *J. Acad. Nat. Sci. Phila.*, ser. 2, 8:93–157.

———. 1876. On the Batrachia and Reptilia of Costa Rica. *J. Acad. Nat. Sci. Phila.*, ser. 2, 8:93–157.

———. 1877. Tenth contribution to the herpetology of tropical America. *Proc. Amer. Philos. Soc.* 17:85–98.

———. 1879. Eleventh contribution to the herpetology of tropical America. *Proc. Amer. Philos. Soc.* 18:261–77.

———. "1884." 1885. Twelfth contribution to the herpetology of tropical America. *Proc. Amer. Philos. Soc.* 22:167–94.

———. "1885." 1886. Thirteenth contribution to the herpetology of tropical America. *Proc. Amer. Philos. Soc.* 23:271–87.

———. 1887. Catalogue of the batrachians and reptiles of Central America and Mexico. *Bull. U.S. Natl. Mus.* 32:1–98.

———. 1889. The Batrachia of North America. *Bull. U.S. Natl. Mus.* 34:1–515.

———. 1893. Second addition to the knowledge of the Batrachia and Reptilia of Costa Rica. *Proc. Acad. Nat. Sci. Phila.* 31:333–47.

———. 1894a. Third addition to a knowledge of the Batrachia and Reptilia of Costa Rica. *Proc. Acad. Nat. Sci. Phila.* 46:194–206.

———. 1894b. The classification of snakes. *Amer. Nat.* 28:831–44.

———. "1894c." 1895. The classification of the Ophidia. *Trans. Amer. Philos. Soc.* 19:185–219.

———. 1896. On the hemipenes of the Sauria. *Proc. Acad. Nat. Sci. Phila.* 48:461–67.

———. 1899. Contribution to the herpetology of New Granada and Argentina, with descriptions of new forms. *Sci. Bull. Phila. Mus.* 1:11–22.

———. 1900. The crocodilians, lizards and snakes of North America. *Annl. Rept. U.S. Nat. Mus.* 1898:153-1270.

Corn, M. J. "1974." 1975. Report of the first certain collection of *Ungaliophis panamensis* from Costa Rica. *Carib. J. Sci.* 14 (3–4): 167–75.

———. 1981. Ecological separation of *Anolis* lizards in a Costa Rican rain forest. Ph.D. diss., University of Florida.

Cornelius, S. E. 1976. Marine turtle nesting activity at Playa Naranjo, Costa Rica. *Brenesia* 8 (1): 1–27.

———. 1983. *Lepidochelys olivacea* (lora, carpintera, Pacific ridley sea turtle). In *Costa Rican natural history*, ed. D. H. Janzen, 402–5. Chicago: University of Chicago Press.

———. 1986. *The sea turtles of Santa Rosa National Park.* San José: Fundación de Parques Nacionales.

Coto, A., and R. A. Acuña. 1986. Filogenia de *Geochelone costarricensis* y la familia Testudinidae (Reptilia: Testudines) en el continent americano. *Rev. Biol. Trop.* 34 (2): 199–208.

Cott, H. B. 1926. Observations on the life-habits of some batrachians and reptiles from the lower Amazon: And a note on some mammals from Marajo Island. *Proc. Zool. Soc. London* 1926:1159–78.

———. 1940. *Adaptive coloration in animals.* London: Methuen.

Cracraft, J. 1989. Speciation and its ontology: The empirical consequences of alternative species concepts for understand-

ing patterns and processes of differentiation. In *Speciation and its consequences,* ed. D. Otte and J. A. Endler, 28–59. Sunderland, Mass.: Sinauer.

Crimmins, K. L. 1937. A case of *Oxybelis* poisoning in man. *Copeia* 1937 (4): 233.

Cronin, T. M., and H. F. Dowsett. 1996. Biotic and oceanographic responses to the Pliocene closing of the Central America Isthmus. In *Evolution and environment in tropical America,* ed. J. B. C. Jackson, A. F. Budd, and A. G. Coates, 76–104. Chicago: University of Chicago Press.

Crother, B. I. 1989. A redescription of the hemipenes of *Hydromorphus concolor* (Colubridae) with comments on its tribal affiliation. *Copeia* 1989 (1): 227–29.

Crother, B. I., J. A. Campbell, and D. M. Hillis. 1992. Phylogeny and historical biogeography of the palm-pitvipers, genus *Bothriechis*: Biochemical and morphological evidence. In *Biology of the pitvipers,* ed. J. A. Campbell and E. D. Brodie Jr., 1–19. Tyler, Tex.: Selva.

Crother, B. I., M. M. Miyamoto, and W. F. Presch. 1986. Phylogeny and biogeography of the lizard family Xantusiidae. *Syst. Zool.* 35 (1): 37–45.

Crump, M. L. 1972. Territoriality and mating behavior in *Dendrobates granuliferus* (Anura: Dendrobatidae). *Herpetologica* 28 (3): 195–98.

———. 1974. Reproductive strategies in a tropical anuran community. *Misc. Pub. Mus. Nat. Hist. Univ. Kansas* 61:1–68.

———. 1983a. Opportunistic cannibalism by amphibian larvae in temporary aquatic environments. *Amer. Nat.* 121 (2): 281–87.

———. 1983b. *Dendrobates granuliferus* and *Dendrobates pumilio* (rana venenosa, poison dart frogs). In *Costa Rican natural history,* ed. D. H. Janzen, 396–98. Chicago: University of Chicago Press.

———. 1984. Ontogenetic changes in vulnerability to predation in tadpoles of *Hyla pseudopuma. Herpetologica* 40 (3): 265–72.

———. 1986a. Homing and site fidelity in a Neotropical frog, *Atelopus varius* (Bufonidae). *Copeia* 1986 (2): 438–44.

———. 1986b. Cannibalism by younger tadpoles: Another hazard of metamorphosis. *Copeia* 1986 (4): 1007–9.

———. 1988. Aggression in harlequin frogs: Male-male competition and a possible conflict of interest between the sexes. *Anim. Behav.* 36:1064–77.

———. 1989a. Life history consequences of feeding versus nonfeeding in a facultatively non-feeding toad larva. *Oecologia* 78:486–89.

———. 1989b. Effect of habitat drying on developmental time and size at metamorphosis in *Hyla pseudopuma. Copeia* 1989 (3): 794–97.

———. 1990. Possible enhancement of growth in tadpoles through cannibalism. *Copeia* 1990 (2): 560–64.

———. 1991. Choice of oviposition site and egg load assessment by a treefrog. *Herpetologica* 47 (3): 308–15.

Crump, M. L., F. R. Hensley, and K. L. Clark. 1992. Apparent decline of the golden toad: Underground or extinct? *Copeia* 1992 (2): 413–20.

Crump, M. L., and J. A. Pounds. 1985. Lethal parasitism of an aposematic anuran *(Atelopus varius)* by *Notochaeta bufonivora* (Diptera: Sarcophagidae). *J. Parasit.* 7:1588–91.

———. 1989. Temporal variation in the dispersion of a tropical anuran. *Copeia* 1989 (1): 209–11.

Crump, M. L., and D. S. Townsend. 1990. Random mating by size in a Neotropical treefrog, *Hyla pseudopuma*. *Herpetologica* 46 (4): 383–86.

Cruz, C. A. G. "1990." 1991. Sobre as relações intergenéricas de Phyllomedusinae da Floresta Atlântia (Amphibia, Anura, Hylidae). *Rev. Brasil Biol.* 50 (3): 709–26.

Cserna, Z. de. 1989. An outline of the geology of Mexico. In *The geology of North America: An overview*, vol. A of *The geology of North America*, ed. A. W. Bally and A. R. Palmer, 233–64. Boulder, Colo.: Geological Society of America.

Cullings, K. W., D. J. Morafka, J. Hernandez, and J. W. Roberts. 1997. Reassessment of phylogenetic relationships among pitviper genera based on mitochondrial cytochrome b gene sequences. *Copeia* 1997 (2): 429–32.

Culotta, W. A., and G. V. Pickwell. 1993. *The venomous sea snakes: A comprehensive bibliography*. Malabar, Fla.: Krieger.

Cunha, O. R. da. 1970. Uma nova subspécie de quelônio, *Kinosternon scorpioides carajasensis* da Serra dos Carajás, Pará (Testudinata-Kinosternidae). *Bol. Mus. Pará "E. Goeldi" Zool.* 73:1–12.

Cunha, O. R. da, and F. P. Nascimento. "1982." 1984. Ofídios da Amazônia XV—As espécies de *Chironius* da Amazônia Oriental (Pará, Amapá e Maranhão) (Ophidia: Colubridae). *Mem. Inst. Butantan.* 46:139–72.

———. 1983. Ofídios da Amazônia. 18. O gênero *Chironius* Fitzinger, na Amazônia Oriental (Ophidia: Colubridae). *Bol. Mus. Pará "Emilio Goeldi" Zool.* 119:1–17.

Cuvier, G. L. C. F. D. 1807. Sur les différentes espèces de crocodiles vivants et sur leurs charactères distinctifs. *Ann. Mus. Natn. Hist. Nat.* 10:8–66.

———. "1817." 1816. *Le règne animal distribué d'après son organisation, pour servir de base à l'histoire naturelle des animaux et d'introduction à l'anatomie comparée.* 4 vols. Paris: Deterville.

Daly, J. W., H. M. Garraffo, C. Jaramillo, and A. S. Rand. 1994. Dietary source of skin alkaloids of poison frogs (Dendrobatidae)? *J. Chem. Ecol.* 20 (4): 943–55.

Daly, J. W., H. M. Garraffo, and C. W. Myers. 1997. The origin of frog skin alkaloids: An enigma. *Pharmaceut. News* 4 (4): 9–14.

Daly, J. W., and C. W. Myers. 1967. Toxicity of Panamanian poison frogs *(Dendrobates)*: Some biological and chemical aspects. *Sci.* 156:970–73.

Daly, J. W., C. W. Myers, and N. Whittaker. 1987. Further classification of skin alkaloids from Neotropical poison frogs (Dendrobatidae), with a general survey of toxic/noxious substances in the Amphibia. *Toxicon* 25 (10): 1023–95.

Daly, J. W., S. I. Secunda, H. M. Garraffo, T. F. Spande, Wisnieski, and J. F. Cover Jr. 1994. An uptake system for dietary alkaloids in poison frogs. *Toxicon* 32 (6): 657–63.

Daly, J. W., and T. F. Spande. 1986. Amphibian alkaloids: Chemistry, pharmacology and biology. In *Alkaloids: Chemical and biological perspectives*, ed. S. W. Pelletier, 4:1–274. New York: Wiley.

Daniel, H. 1949. Las serpientes en Colombia. *Rev. Fac. Nac. Agronom. Medellin* 9 (36): 301–33.

Darlington, P. J., Jr. 1957. *Zoogeography: The geographical distribution of animals.* New York: Wiley.

Darvesky, I. S., L. A. Kupriyanova, and V. V. Roschin. 1984. A new all-female triploid species of gecko and karyological data on the bisexual *Hemidactylus frenatus* in Vietnam. *J. Herp.* 18 (3): 277–85.

Daudin, F. M. 1801–3. *Histoire naturelle générale et particulière des reptiles.* 8 vols. Paris: Dufart.

Davenport, J. 1988. Do diving leatherbacks pursue glowing jelly? *Brit. Herp. Soc. Bull.* 24:20–21.

Davenport, J., and G. H. Balazs. 1991. "Fiery bodies": Are pyrosomas an important component of the diet of leatherback turtles? *Brit. Herp. Soc. Bull.* 37:33–38.

Davidson, E. H., and B. R. Hough. 1969. Synchronous oogenesis in *Engystomops pustulosus*, a Neotropic anuran suitable for laboratory studies: Localization in the embryo of RNA synthesized at the lampbrush stage. *J. Exp. Zool.* 172:25–48.

Davis, D. D. 1953. Behavior of the lizard *Corytophanes cristatus*. *Fieldiana Zool.* 35 (1): 1–8.

Deban, S. M., D. B. Wake, and G. Roth. 1997. Salamander with a ballistic tongue. *Nature* 389:27–28.

DeBraga, M., and O. Rieppel. 1997. Reptile phylogeny and the interrelationships of turtles. *Zool. J. Linn. Soc. London* 120:281–354.

Despax, R. 1910. Mission géodésique de l'Équateur: Collections recuellis par M. de Dr. Rivet. Liste des ophidiens et description des especès nouvelle (note préliminaire). *Bull. Mus. Natl. Hist. Nat. Paris.* 1910:368–76.

Dessauer, H. C., and C. J. Cole. 1989. Diversity between and within nominal forms of unisexual teiid lizards. *Bull. N.Y. State Mus.* 466: 49–71.

Dessauer, H. H. C., C. J. Cole, and M. S. Hafner. 1996. Collection and storage of tissues. In *Molecular systematics*, 2d ed., ed. D. M. Hillis, C. Moritz, and B. K. Mable, 29–41. Sunderland, Mass.: Sinauer.

De Vosjoli, P. 1994. Herpetoculture in a changing world. *Vivarium* 6 (3): 10–11, 36–37.

———. 1995. Herpetoculture in a changing world. 3. *Vivarium* 7 (2): 12–13, 60–61.

DeWeese, J. E. 1976. The karyotypes of Middle American frogs of the genus *Eleutherodactylus* (Anura: Leptodactylidae): A case study of the significance of the karyologic method. Ph.D. diss., University of Southern California.

Dengo, G., and J. E. Case, eds. 1990. *The Caribbean region.* Vol. H of *The geology of North America.* Boulder, Colo.: Geological Society of America.

Densmore, L. D. 1983. Biochemical and immunological systematics of the order Crocodilia. *Evol. Biol.* 16:397–465.

Densmore, L. D., and P. S. White. 1991. The systematics and evolution of the Crocodilia as suggested by restriction endonuclease analysis of mitochondrial and nuclear ribosomal DNA. *Copeia* 1991 (3): 602–15.

Dial, B. E., and L. L. Grismer. 1992. A phylogenetic analysis of physiological-ecological character evolution in the lizard genus *Coleonyx* and its implications for historical biogeographic reconstruction. *Syst. Biol.* 41 (2): 178–95.

Di-Bernardo, M. 1992. Revalidation of the genus *Echinanthera* Cope, 1894, and its conceptual amplification (Serpentes,

Colubridae). *Com. Mus. Ciências Pont. Univ. Catal. Rio Grande do Sul* 5 (13): 225–56.

Dirección General de Estadística y Censos de Costa Rica. 1953. *Atlas estadístico de Costa Rica.* San José: Casa Gráfica.

Dixon, J. R. 1960. The discovery of *Phyllodactylus tuberculosus* (Reptilia: Sauria) in Central America, the resurrection of *P. xanti*, and description of a new gecko from British Honduras. *Herpetologica* 16 (1): 1–11.

———. 1964. The systematics and distribution of lizards of the genus *Phyllodactylus* in North and Central America. *N. Mexico State Univ. Res. Cent. Bull.* 6:41–139.

———. 1968. A new species of gecko (Sauria: Gekkonidae) from the Bay Islands, Honduras. *Proc. Biol. Soc. Wash.* 81: 419–26.

———. 1973. A systematic review of the teiid lizards, genus *Bachia*, with remarks on *Heterodactylus* and *Anotosaura*. *Misc. Pub. Mus. Nat. Hist. Univ. Kansas.* 5:71–47.

———. 1980. The Neotropical colubrid snake genus *Liophis*. The generic concept. *Milwaukee Publ. Mus. Contr. Biol. Geol.* 31:1–40.

———. 1983. Systematics of the Latin American snake, *Liophis epinephelus* (Serpentes: Colubridae). In *Advances in herpetology and evolutionary biology: Essays in honor of Ernest E. Williams,* ed. A. G. J. Rhodin and K. Miyata, 132–49. Cambridge: Museum of Comparative Zoology, Harvard University.

———. 1989. A key and checklist to the Neotropical snake genus *Liophis* with country lists and maps. *Smithson. Herp. Info. Serv.* 79:1–28.

Dixon, J. R., and F. S. Hendricks. 1979. The worm snakes (family Typhlopidae) of the Neotropics, exclusive of the Antilles. *Zool. Verh. Leiden* 173:1–39.

Dixon, J. R., and C. P. Kofron. 1984. The Central and South America anomalepid snakes of the genus *Liotyphlops. Amph.-Reptl.* 4:41–64.

Dixon, J. R., and P. Soini. 1977. The reptiles of the upper Amazon basin, Iquitos region, Peru. 2. Crocodilians, turtles, and snakes. *Milwaukee Publ. Mus. Contr. Biol. Geol.* 12:1–91.

Dixon, J. R., and M. A. Stanton. 1983. *Caiman crocodilus* (caiman, lagarto, baba, babilla, cuajipalo, cayman). In *Costa Rican natural history,* ed. D. H. Janzen, 387–88. Chicago: University of Chicago Press.

Dixon, J. R., J. A. Wiest, and J. M. Cei. 1993. Revision of the Neotropical snake genus *Chironius* Fitzinger (Serpentes, Colubridae). *Monogr. Mus. Reg. Sci. Nat. Torino* 13:1–279.

Dodd, C. K., Jr. 1990a. *Caretta. Cat. Amer. Amph. Reptl.* 482: 1–2.

———. 1990b. *Caretta caretta. Cat. Amer. Amph. Reptl.* 483: 1–7.

Dodds, C. T. 1923. A note on *Bufo marinus. Copeia* 114:5–6.

Donnelly, M. A. 1987. Territoriality in the poison-dart frog *Dendrobates pumilio* (Anura: Dendrobatidae). Ph.D. diss., University of Miami.

———. 1989a. Demographic effects of reproductive resource supplementation in a terrestrial frog, *Dendrobates pumilio. Ecology* 59 (3): 207–21.

———. 1989b. Effects of reproductive resource supplementations on space-use patterns in *Dendrobates pumilio. Oecologia* 81:212–18.

———. 1989c. Reproductive phenology and age structure of *Dendrobates pumilio* in northeastern Costa Rica. *J. Herp.* 23 (4): 362–67.

———. 1991. Feeding patterns of the strawberry poison frog, *Dendrobates pumilio* (Anura: Dendrobatidae). *Copeia* 1991 (3): 723–30.

———. 1994a. Amphibian diversity and natural history. In *La Selva: Ecology and natural history of a Neotropical rainforest,* ed. L. A. McDade, K. S. Bawa, H. A. Hespenheide, and G. S. Hartshorn, 199–209. Chicago: University of Chicago Press.

———. 1994b. Amphibians: Appendix 5. In *La Selva: Ecology and natural history of a Neotropical rainforest,* ed. L. A. McDade, K. S. Bawa, H. A. Hespenheide, and G. S. Hartshorn, 380–81. Chicago: University of Chicago Press.

———. 1999. Reproductive phenology of *Eleutherodactylus bransfordii* in northeastern Costa Rica. *J. Herp.* 33 (4): 624–31.

Donnelly, M. A., and C. Guyer. 1994. Patterns of reproduction and habitat use in an assemblage of Neotropical hylid frogs. *Oecologia* 98:291–302.

Donnelly, M. A., C. Guyer, D. M. Krempels, and H. E. Braker. 1987. The tadpole of *Agalychnis calcarifer* (Anura: Hylidae). *Copeia* 1987 (1): 247–50.

Donnelly, M. A., C. Guyer, and R. O. de Sá. 1990. The tadpole of a dart-poison frog *Phyllobates lugubris* (Anura: Dendrobatidae). *Proc. Biol. Soc. Wash.* 103 (2): 427–31.

Donnelly, M. A., R. O. de Sá, and C. Guyer. 1990. Description of the tadpoles of *Gastrophryne pictiventris* and *Nelsonophryne aterrima* (Anura: Microhylidae), with a review of morphological variation in free-swimming microhylid larvae. *Amer. Mus. Nat. Hist. Novitates* 2976:1–19.

Donnelly, T. W. 1989. Geologic history of the Caribbean and Central America. In *The geology of North America: An overview,* vol. A of *The geology of North America,* ed. A. W. Bally and A. R. Palmer, 299–321. Boulder, Colo.: Geological Society of America.

Donnelly, T. W., G. S. Horne, R. C. Finch, and E. Lopez. 1990. Northern Central America: The Maya and Chortis blocks. In *The Caribbean region,* vol. H of *The geology of North America,* ed. G. Dengo and J. E. Case, 37–76. Boulder, Colo.: Geological Society of America.

Dowling, H. G. 1960. A taxonomic study of the ratsnakes, genus *Elaphe* Fitzinger. 7. The *triaspis* section. *Zoologica* 45 (2): 53–80.

———. 1975a. A provisional classification of snakes. In *Yearbook of herpetology, 1974,* ed. H. G. Dowling, 167–70. New York: Herpetological Information Search Systems, American Museum of Natural History.

———. 1975b. The Nearctic snake fauna. In *Yearbook of herpetology, 1974,* ed. H. G. Dowling, 191–201. New York: Herpetological Information Search Systems, American Museum of Natural History.

———, ed. 1975c. *Yearbook of herpetology, 1974.* New York: Herpetolotical Information Search Systems, American Museum of Natural History.

———, ed. 2002. *A new classification of snakes: A contribution to the knowledge of biodiversity.* Malabar, Fla.: Krieger. In press.

Dowling, H. G., and W. E. Duellman. "1974–78." 1978. *Systematic herpetology: A synopsis of families and higher categories.* New York: Herpetological Information Search Systems, American Museum of Natural History.

Dowling, H. G., and I. Fries. 1987. A taxonomic study of the ratsnakes. 8. A proposed new genus for *Elaphe triaspis* (Cope). *Herpetologica* 43 (2): 200–207.

Dowling, H. G., R. Highton, G. C. Maha, and L. R. Maxson. 1983. Biochemical evaluation of colubrid snake phylogeny. *J. Zool.* 201:309–29.

Dowling, H. G., and J. V. Jenner. 1987. Taxonomy of American xenodontine snakes. 2. The status and relationships of *Pseudoleptodeira. Herpetologica* 43 (2): 190–200.

———. 1988. Snakes of Burma checklist of reported species and bibliography. *Smithson. Herp. Info. Serv.* 76:1–19.

Dowling, H. G., and R. M. Price. 1988. A proposed new genus for *Elaphe subocularis* and *Elaphe rosaliae. Snake* 20:52–63.

Dowling, H. G., and J. M. Savage. 1960. A guide to the snake hemipenis: A survey of basic structure and systematic characteristics. *Zoologica* 45 (2): 17–27.

Downs, F. L. 1967. Intergeneric relations among colubrid snakes of the genus *Geophis* Wagler. *Misc. Pub. Mus. Zool. Univ. Mich.* 131:1–193.

Drewry, G. E., and K. L. Jones. 1976. A new ovoviviparous frog, *Eleutherodactylus jasperi* (Amphibia, Anura, Leptodactylidae), from Puerto Rico. *J. Herp.* 10 (3): 161–65.

Dubois, A. "1986." 1987. Miscellanea taxinomica batrachologica (1). *Alytes* 5 (4): 7–95.

———. 1992. Notes sur la classification de Ranidae (Amphibien Anoures). *Bull. Mens. Soc. Linn. Lyon* 61:305–52.

Dubois, A., and W. R. Heyer. 1992. *Leptodactylus labialis,* the valid name for the American white-lipped frog (Amphibia: Leptodactylidae). *Copeia* 1992 (2): 584–85.

Ducey, P. K., and E. D. Brodie Jr. 1991. Evolution of antipredator behavior: Individual and populational variation in a Neotropiocal salamander. *Herpetologica* 47 (1): 89–95.

Ducey, P. K., E. D. Brodie Jr., and E. A. Baness. 1993. Salamander tail autotomy and snake predation: Role of antipredator behavior and toxicity in three species of Neotropical *Bolitoglossa* (Caudata: Plethodontidae). *Biotropica* 25 (3): 344–49.

Duellman, W. E. 1956. The frogs of the hylid genus *Phrynohyas* Fitzinger, 1843. *Misc. Pub. Mus. Zool. Univ. Mich.* 96:1–47.

———. 1958. A monographic study of the colubrid snake genus *Leptodeira. Bull. Amer. Mus. Nat. Hist.* 114 (1): 1–152.

———. 1963a. A new species of tree frog, genus *Phyllomedusa,* from Costa Rica. *Rev. Biol. Trop.* 11 (1): 1–23.

———. 1963b. Amphibians and reptiles of the rainforests of southern El Petén, Guatemala. *Misc. Pub. Mus. Nat. Hist. Univ. Kansas* 15 (5): 205–49.

———. 1966a. Taxonomic notes on some Mexican and Central American hylid frogs. *Univ. Kansas Pub. Mus. Nat. Hist.* 17 (6): 263–79.

———. 1966b. The Central American herpetofauna: An ecological perspective. *Copeia* 1966 (4): 700–719.

———. 1967a. Courtship isolating mechanisms in Costa Rican hylid frogs. *Herpetologica* 23 (3): 169–83.

———. 1967b. Additional studies of chromsomes in anuran species. *Syst. Zool.* 6 (1): 38–43.

———. 1968a. The genera of phyllomedusine frogs. *Univ. Kansas Pub. Mus. Nat. Hist.* 18 (1): 1–10.

———. 1968b. Descriptions of new hylid frogs from Mexico and Central America. *Univ. Kansas Pub. Mus. Nat. Hist.* 17 (13): 559–78.

———. 1970. *The hylid frogs of Middle America.* 2 vols. Lawrence, Kans.: Museum of Natural History.

———. 1971. A taxonomic review of South American hylid frogs, genus *Phrynohyas. Occas. Pap. Mus. Nat. Hist. Univ. Kansas* 4:1–21.

———. 1972a. A review of the Neotropical frogs of the *Hyla bogotensis* group. *Occas. Pap. Mus. Nat. Hist. Univ. Kansas* 11:1–31.

———. 1972b. South American frogs of the *Hyla rostrata* group (Amphibia, Anura, Hylidae). *Zool. Meded. Natl. Mus. Leiden* 47 (14): 177–92.

———. 1974. Taxonomic reassessment of the taxonomic status of some Neotropical hylid frogs. *Occas. Pap. Mus. Nat. Hist. Univ. Kansas* 27: 1–27.

———. 1978. The biology of an equatorial herpetofauna in Amazonian Ecuador. *Misc. Pub. Mus. Nat. Hist. Univ. Kansas* 65:1–352.

———. 1983. A new species of marsupial frog (Hylidae: *Gastrotheca*) from Colombia and Ecuador. *Copeia* 1983 (4): 868–74.

———. 1988. Patterns of species diversity in anuran amphibians in the American tropics. *Ann. Missouri Botan. Gard.* 75 (1): 79–104.

———. 1989. New species of hylid frogs from the Andes of Colombia and Venezuela. *Occas. Pap. Mus. Nat. Hist. Univ. Kansas* 131:1–12.

———. 1990. Herpetofaunas in Neotropical rainforests: Comparative composition, history, and resource use. In A. H. Gentry, ed. *Four Neotropical rainforests,* ed. A. H. Gentry, 455–505. New Haven: Yale University Press.

———. 1993. Amphibian species of the world: Additions and corrections. *Spec. Pub. Mus. Nat. Hist. Univ. Kansas* 21:1–372.

———. 2001. *The hylid frogs of Middle America.* New ed. Ithaca, N.Y.: Society for the Study of Amphibians and Reptiles.

Duellman, W. E., and C. J. Cole. 1965. Studies of chromosomes of some anuran ampibians (Hylidae and Centrolenidae). *Syst. Zool.* 14 (2): 139–43.

Duellman, W. E., and R. I. Crombie. 1970. *Hyla septentrionalis. Cat. Amer. Amph. Reptl.* 92:1–4.

Duellman, W. E., and M. J. Fouquette Jr. 1968. Middle American hylid frogs of the *Hyla microcephala* group. *Univ. Kansas Pub. Mus. Nat. Hist.* 17 (12): 517–57.

Duellman, W. E., and J. D. Lynch. 1969. Descriptions of *Atelopus* tadpoles and their relevance to atelopodid classification. *Herpetologica* 25 (4): 231–40.

Duellman, W. E., L. R. Maxson, and C. A. Jeisolowski. 1988. Evolution of marsupial frogs (Hylida: Hemiphractinae): Immunological evidence. *Copeia* 1988 (3): 527–43.

Duellman, W. E., and A. Schwartz. 1958. Amphibians and reptiles of southern Florida. *Bull. Fla. State Mus.* 3 (5): 181–324.

Duellman, W. E., and L. Trueb. 1966. Neotropical hylid frogs, genus *Smilisca*. *Univ. Kansas Pub. Mus. Nat. Hist.* 17 (7): 281–375.

———. "1986." 1985. *Biology of amphibians*. New York: McGraw-Hill.

———. 1989. Two new treefrogs of the *Hyla parviceps* group from the Amazon Basin in southern Peru. *Herpetologica* 45 (1): 1–17.

———. 1994. *Biology of amphibians*. New ed. Baltimore: Johns Hopkins University Press.

Duellman, W. E., and J. B. Tulecke. 1960. The distribution, variation, and life history of the frog, *Cochranella viridissima* in Mexico. *Amer. Midl. Nat.* 63 (2): 392–97.

Duellman, W. E., and J. Wellman. 1960. A systematic study of the lizards of the *deppei* group (genus *Cnemidophorus*) in Mexico and Guatemala. *Misc. Pub. Mus. Zool. Univ. Mich.* 111:1–80.

Duellman, W. E., and J. J. Wiens. 1992. The status of the hylid frog genus *Ololygon* and the recognition of *Scinax* Wagler, 1830. *Occas. Pap. Mus. Zool. Univ. Kansas* 151: 1–23.

Dugan, B. A. 1982. The mating behavior of the green iguana, *Iguana iguana*. In *Iguanas of the world*, ed. G. M. Burghardt and A. S. Rand, 320–41. Park Ridge, N.J.: Noyes.

Duméril, A. H. A. 1853. Mémoire sur les batraciens anoures, de la famille des hylaeformes ou rainettes, comprenant la description d'un genre nouveau et de onze espèces nouvelles. *Ann. Sci. Nat. Paris, Zool.*, ser. 3, 19:135–79.

Duméril, A. H. A., M.-F. Bocourt, and M. F. Mocquard. 1870–1900. *Études sur les reptiles: Mission scientifique au Mexique et dans l'Amérique Centrale. Recherches zoologiques.* Pt. 3, sec. 1, issue 17. Paris: Imprimerie Impériale.

Duméril, A. M. C. 1853. Prodrome de la classification des reptiles ophidiens. *Mem. Acad. Sci. France* 23:399–536.

Duméril, A. M. C., and G. Bibron. 1834–44. *Erpétologie générale, ou Histoire naturelle complète des reptiles.* Vols. 1–8. Paris: Roret.

Duméril, A. M. C., G. Bibron, and A. H. A. Duméril. 1854. *Erpétologie générale, ou Histoire naturelle complète des reptiles.* Vols. 7, 9. *Atlas.* Paris: Roret.

Duméril, A. M. C., and A. H. A. Duméril. 1851. *Catalogue méthodique de la collection des reptiles (Muséum d'Histoire naturelle de Paris).* Paris: Gide et Boundry.

Dundee, H. A., and E. A. Liner. 1974. Eggs and hatchlings of the tree snake *Leptophis depressirostris* (Cope). *Brenesia* 3:11–13.

Dunn, E. R. 1921. Two new Central America salamanders. *Proc. Biol. Soc. Wash.* 34 (1): 43–46.

———. 1922. The sound-transmitting apparatus of salamanders and the phylogeny of the Caudata. *Amer. Nat.* 56: 418–87.

———. 1924a. New salamanders of the genus *Oedipus* with a synoptical key. *Field Mus. Nat. Hist. Zool. Ser.* 12 (7): 95–100.

———. 1924b. Some Panamanian frogs. *Occas. Pap. Mus. Zool. Univ. Mich.* 151:1–17.

———. 1924c. New amphibians from Panama. *Occas. Pap. Boston Soc. Nat. Hist.* 5:93–95.

———. 1926. *The salamanders of the family Plethodontidae.* Smith College Fiftieth Anniversary Publications 7. Northampton, Mass.: Smith College.

———. 1930a. A new *Geoemyda* from Costa Rica. *Proc. N. Engl. Zool. Club.* 12:32–34.

———. 1930b. Notes on Central American *Anolis*. *Proc. N. Engl. Zool. Club.* 12:15–24.

———. 1930c. New snakes from Costa Rica and Panama. *Occas. Pap. Boston Soc. Nat. Hist.* 5:329–32.

———. 1931a. New frogs from Panama and Costa Rica. *Occas. Pap. Boston Soc. Nat. Hist.* 5:385–401.

———. 1931b. The herpetological fauna of the Americas. *Copeia* 1931 (3): 106–19.

———. 1931c. The amphibians of Barro Colorado Island. *Occas. Pap. Boston Soc. Nat. Hist.* 5:403–21.

———. 1932a. Notes on blind snakes from Lower Central America. *Proc. Biol. Soc. Wash.* 45:173–76.

———. 1932b. The colubrid snakes of the Greater Antilles. *Copeia* 1932 (2): 89–92.

———. 1932c. The status of the snake genus *Rhadinaea*. *Occas. Pap. Mus. Zool. Univ. Mich.* 251:1–2.

———. 1933a. A new lizard from Nicaragua. *Proc. Biol. Soc. Wash.* 46:67–68.

———. 1933b. Amphibians and reptiles from El Valle de Anton, Panama. *Occas. Pap. Boston Soc. Nat. Hist.* 8:65–79.

———. 1933c. A new *Hyla* from the Panama Canal Zone. *Occas. Pap. Boston Soc. Nat. Hist.* 8:61–64.

———. 1934. Notes on *Iguana*. *Copeia* 1934 (1): 1–4.

———. 1935. The snakes of the genus *Ninia*. *Proc. Natl. Acad. Sci.* 21 (1): 9–12.

———. 1936a. Notes on American mabuyas. *Proc. Acad. Nat. Sci. Phila.* 87:533–57.

———. 1936b. Notes on North American *Leptodeira*. *Proc. Natl. Acad. Sci.* 22 (12): 689–98.

———. 1937a. The amphibian and reptilian fauna of bromeliads in Costa Rica and Panama. *Copeia* 1937 (3): 163–67.

———. 1937b. New or unnamed snakes from Costa Rica. *Copeia* 1937 (4): 213–15.

———. 1937c. The giant mainland anoles. *Proc. N. Engl. Zool. Club* 16:5–9.

———. 1938. A new *Rhadinaea* from Central America. *Copeia* 1938 (4): 197–98.

———. 1939a. A new pit viper from Costa Rica. *Proc. Biol. Soc. Wash.* 52:165–66.

———. 1939b. The mainland forms of the snake genus *Tretanorhinus*. *Copeia* 1939 (4): 212–17.

———. 1940a. Some aspects of herpetology in lower Central America. *Trans. New York Acad. Sci.*, ser. 2, 2 (6): 156–58.

———. 1940b. New and noteworthy herpetological material from Panama. *Proc. Acad. Nat. Sci. Phila.* 92:105–22.

———. 1940c. Notes on American lizards and snakes in the Museum of Goteberg. *Herpetologica* 1 (7): 189–94.

———. 1941a. Notes on *Dendrobates auratus*. *Copeia* 1941 (2): 88–92.

———. 1941b. Notes on the snake genus *Anomolepis*. *Bull. Mus. Comp. Zool.* 87 (7): 511–26.

———. 1942a. The American caecilians. *Bull. Mus. Comp. Zool.* 91 (6): 439–540.

———. 1942b. A new species of frog (*Eleutherodactylus*) from Costa Rica. *Not. Natl. Acad. Nat. Sci. Phila.* 104:1–2.

————. 1942c. New or noteworthy snakes from Panama. *Not. Natl. Acad. Nat. Sci. Phila.* 108:1–8.

————. 1943. An extraordinary new *Hyla* from Colombia. *Caldasia* 2 (8): 309–10.

————. 1944a. Los generos de anfibios y reptiles de Colombia. 1. Anfibios. *Caldasia* 2 (10): 497–529.

————. 1944b. Los generos de anfibios y reptiles de Colombia. 2. Reptiles, orden de los saurios. *Caldasia* 3 (11): 73–110.

————. 1944c. Los generos de anfibios y reptiles de Colombia. 3. Reptiles: Orden de las serpientes. *Caldasia* 3 (12): 156–224.

————. 1944d. A revision of the Colombian snakes of the genera *Leimadophis, Lygophis, Liophis, Rhadinaea* and *Pliocercus,* with a note on Colombian *Coniophanes. Caldasia* 2 (10): 479–95.

————. 1945. Los generos de anfibios y reptiles de Colombia. 4. Reptiles: Ordenes testudineos y cocodilinos. *Caldasia* 3 (13): 308–35.

————. 1947. Snakes of the Lerida Farm (Chiriquí Volcano, western Panama). *Copeia* 1947 (3): 153–57.

————. 1949. Notes on South American frogs of the family Microhylidae. *Amer. Mus. Nat. Hist. Novitates* 1419:1–21.

————. "1951." 1952. The status of the snake genera *Dipsas* and *Sibon,* a problem for "quantum evolution." *Evolution* 5 (4): 355–58.

————. 1954. The coral snake mimicry problem. *Evolution* 8 (2): 97–102.

————. 1957. *Contributions to the herpetology of Colombia* (1943–46). Privately printed, M. T. Dunn.

Dunn, E. R., and J. R. Bailey. 1939. Snakes from the uplands of the Canal Zone and of Darien. *Bull. Mus. Comp. Zool.* 86 (1): 1–22.

Dunn, E. R., and H. G. Dowling. 1957. The Neotropical snake genus *Nothopsis. Copeia* 1957 (4): 255–61.

Dunn, E. R., and M. T. Dunn. 1940. Generic names proposed in herpetology by E. D. Cope. *Copeia* 1940 (2): 69–76.

Dunn, E. R., and J. T. Emlen Jr. 1932. Reptiles and amphibians from Honduras. *Proc. Acad. Nat. Sci. Phila.* 84:21–32.

Dunn, E. R., and L. H. Saxe. 1950. Results of the Catherwood-Chaplin West Indies expedition of 1948. 5. Amphibians and reptiles of San Andres and Providencia. *Proc. Acad. Nat. Sci. Phila.* 102:141–65.

Dunn, E. R., and J. A. Tihen. 1944. The skeletal anatomy of *Liotyphlops albirostris. J. Morph.* 74:287–95.

Dunson, W. A., ed. 1975. *The biology of sea snakes.* Baltimore: University Park Press.

Easteal, S. 1986. *Bufo marinus. Cat. Amer. Amph. Reptl.* 395: 1–4.

Eaton, T. H., Jr. 1941. Notes on the life history of *Dendrobates auratus. Copeia* 1941 (2): 93–95.

Echelle, A. A., A. F. Echelle, and H. S. Fitch. 1971. A new anole from Costa Rica. *Herpetologica* 27 (3): 354–62.

Echternacht, A. C. 1968. Distributional and ecological notes on some reptiles from northern Honduras. *Herpetologica* 24 (2): 151–58.

————. 1971. Middle American lizards of the genus *Ameiva* (Teiidae). *Misc. Pub. Mus. Nat. Hist. Univ. Kansas.* 55:1–86.

————. 1983. *Ameiva* and *Cnemidophorus* (chisbalas, macro-

teiid lizards). In *Costa Rican natural history,* ed. D. H. Janzen, 375–79. Chicago: University of Chicago Press.

Edwards, S. R. 1974. A phenetic analysis of the genus *Colostethus* (Anura: Dendrobatidae). Ph.D. diss., University of Kansas.

Eichwald, C. E. I. von. 1831. *Zoologia specialis, quam expositi animalis tum vivis, tum fossilibus potaissimum Rossiae in Universum et Plooniae in specie, in usum lectionum publicarum in Universitate Caesarea Vilnensi habendarum. . . . Pars posterior.* Vilnius: Zawadzki.

Ely, C. A. 1944. Development of *Bufo marinus* larvae in dilute sea water. *Copeia* 1944 (4): 256.

Emerson, S. B. 1976. Burrowing in frogs. *J. Morph.* 149:437–58.

————. 1988. Convergence and morphological constraint in frogs: Variation in postcranial morphology. *Fieldiana Zool.,* n.s., 43:1–19.

Ernst, C. H. 1978. A revision of the Neotropical turtle genus *Callopsis* (Testudines: Emydidae: Batagurinae). *Herpetologica* 34 (2): 113–34.

————. 1980a. *Rhinoclemmys annulata. Cat. Amer. Amph. Reptl.* 250:1–2.

————. 1980b. *Rhinoclemmys funerea. Cat. Amer. Amph. Reptl.* 263:1–2.

————. 1981a. *Rhinoclemmys. Cat. Amer. Amph. Reptl.* 274: 1–2.

————. 1981b. *Rhinoclemmys pulcherrima. Cat. Amer. Amph. Reptl.* 275:1–2.

————. 1983a. *Rhinoclemmys annulata* (tortuga parda terrestre, jicote, jicotea, brown land turtle). In *Costa Rican natural history,* ed. D. H. Janzen, 416–17. Chicago: University of Chicago Press.

————. 1983b. *Rhinoclemmys funerea* (tortuga negra del río, jicote, black river turtle). In *Costa Rican natural history,* ed. D. H. Janzen, 417–18. Chicago: University of Chicago Press.

————. 1983c. *Rhinoclemmys pulcherrrima* (tortuga roja, red turtle). In *Costa Rican natural history,* ed. D. H. Janzen, 418–19. Chicago: University of Chicago Press.

————. 1990. Systematics, taxonomy, variation, and geographic distribution of the slider turtle. In *Life history and ecology of the slider turtle,* ed. J. W. Gibbons, 57–67. Washington, D.C: Smithsonian Institution Press.

Ernst, C. H., J. W. Gibbons, and S. S. Novak. 1988. *Chelydra. Cat. Amer. Amph. Reptl.* 419:1–4.

Ernst, C. H., J. E. Lovich, and R. W. Barbour. 1994. *Turtles of the United States and Canada.* Washington, D.C.: Smithsonian Institution Press.

Erspamer, V. 1994. Bioactive secretions of the amphibian integument. In *Amphibian biology,* vol. 1, *The integument,* ed. H. Heatwole and G. T. Barthalmus, 178–350. Chipping Norton, NSW: Surrey Beatty.

Escalante, G. 1990. The geology of southern Central America and western Colombia. In *The Caribbean region,* vol. H of *The geology of North America,* ed. G. Dengo and J. E. Case, 201–30. Boulder, Colo.: Geological Society of America.

Eschscholtz, J. F. von. 1829. *Zoologischer Atlas, enthaltens Abbildungen un Beschreibungen neuer Thierarten, wahrend des Flottcaptians von Kotzebue zweiter Reise um die Welt, auf der Russisch-Kaiserlich Kreigsshlupp Predpriaetie in den Jahren 1823–1826.* Part 1. Berlin: Reimer.

Estes, R., and G. Pregill, eds. 1988. *Phylogenetic relationships of the lizard families.* Stanford: Stanford University Press.

Estes, R., K. de Queiroz, and J. Gauthier. 1988. Phylogenetic relationships within Squamata. In *Phylogenetic relationships of the lizard families,* ed. R. Estes and G. Pregill, 119–201. Stanford: Stanford University Press.

Etheridge, R. E. 1959. Relationships of the anoles (Reptilia: Sauria: Iguanidae): An interpretation based on skeletal morphology. Ph.D. diss., University of Michigan.

———. 1967. Lizard caudal vertebrae. *Copeia* 1967 (4): 699–721.

Ewert, M. A., and C. E. Nelson. 1991. Sex determination in turtles: Diverse patterns and some possible adaptive values. *Copeia* 1991 (1): 50–69.

Feder, M. E., and W. W. Burggen, eds. 1992. *Environmental physiology of the amphibians.* Chicago: University of Chicago Press.

Ferguson, M. W. J. 1985. Reproductive biology and embryology of the crocodilians. In *Biology of the Reptilia,* vol. 14, *Development,* ed. C. Gans, F. Billett, and P. F. A. Maderson, A:329–491. New York: Wiley.

Fernandes, R. 1995. The phylogeny of dispadine snakes. Ph.D. diss., University of Texas, Arlington.

Ferrero, L. 1987. *Costa Rica precolombana: Arqueología, etnología, arte.* San José: Editorial Costa Rica.

Ferrusquía, I. 1993. Geology of Mexico: A synopsis. In *Biological diversity of Mexico: Origins and distribution,* ed. T. P. Ramamoorthy, R. Bye, A. Lot, and J. Fa, New York: Oxford University Press.

Feuer, R. C. 1966. Variation in snapping turtles, *Chelydra serpentina* Linnaeus: A study in quantitative systematics. Ph.D. diss., University of Utah.

Fisher, C. B. 1969. A systematic revision of the species *Coniophanes fissidens* (Gunther): (Serpentes, Colubridae). M.S. thesis, Northwestern Louisiana State University.

Fishman, H. K., J. Mitra, and H. G. Dowling. 1972. Chromosome characteristics of thirteen species in the order Serpentes. *Mammal Chromos. Newsl.* 13 (1): 7–9.

Fitch, H. S. 1970. Reproductive cycles in lizards and snakes. *Misc. Pub. Mus. Nat. Hist. Univ. Kansas* 52:1–247.

———. 1972. Ecology of *Anolis tropidolepis* in Costa Rican cloud forest. *Herpetologica* 28 (1): 10–21.

———. 1973a. A field study of Costa Rican lizards. *Univ. Kansas Sci. Bull.* 50 (2): 39–126.

———. 1973b. Population structure and survivorship in some Costa Rican lizards. *Occas. Pap. Mus. Nat. Hist. Univ. Kansas.* 18:1–41.

———. 1973c. Observations on the population ecology of the Central America iguanid lizard *Anolis cupreus. Carib. J. Sci.* 13(3–4):215–28.

———. 1975. Sympatry and relationships in Costa Rican anoles. *Occas. Pap. Mus. Nat. Hist. Univ. Kansas.* 40:1–60.

———. 1976. Sexual size differences in the mainland anoles. *Occas. Pap. Mus. Nat. Hist. Univ. Kansas* 50:1–21.

———. 1983. *Sphenomorphus cherriei* (escincela parda, skink). In *Costa Rican natural history,* ed. D. H. Janzen, 422–24. Chicago: University of Chicago Press.

———. 1985. Clutch and litter size in New World reptiles. *Misc. Pub. Mus. Nat. Hist. Univ. Kansas.* 76:1–76.

Fitch, H. S., A. A. Echelle, and A. F. Echelle. 1972. Variation in the Central American iguanid lizard *Anolis cupreus,* with the description of a new subspecies. *Occas. Pap. Mus. Nat. Hist. Univ. Kansas.* 8:1–20.

Fitch, H. S., A. F. Echelle, and A. A. Echelle. 1976. Field observations on rare or little known mainland anoles. *Univ. Kansas Sci. Bull.* 51 (3): 91–128.

Fitch, H. S., and J. Hackforth-Jones. 1983. *Ctenosaura similis* (garrobo, iguana negra, ctenosaura). In *Costa Rican natural history,* ed. D. H. Janzen, 394–96. Chicago: University of Chicago Press.

Fitch, H. S., and R. W. Henderson. 1978. Ecology and exploitation of *Ctenosaura similis. Univ. Kansas Sci. Bull.* 51 (5): 483–500.

Fitch, H. S., R. W. Henderson, and D. M. Hillis. 1982. Exploitation of iguanas in Central America. In *Iguanas of the world,* ed. G. M. Burghardt and A. S. Rand, 397–417. Park Ridge, N.J.: Noyes.

Fitch, H. S., and R. A. Seigel. 1984. Ecological and taxonomic notes on Nicaraguan anoles. *Milwaukee Publ. Mus. Contr. Biol. Geol.* 57:1–13.

Fitzinger, L. J. F. J. 1826. *Neue Classification der Reptilien nach ihren naturlichen Verwandschaften nebst einer Verwandschafts-Tafel und einem Verzeichnisse der Reptilien-Sammlung des K. K. zoologischen Museums zu Wien.* Wien: Huebner.

———. 1833. Text for pl. 3. In *Decriptiones et icones amphibiorum,* by J. Wagler (3 parts, 1828–33). Monaco: Cotta.

———. 1835. Entwurf einer systematischen Anordnung der Schildkröten nach den Grundsätzen der natürlichen Methode. *Ann. Naturhist. Mus. Wien.* 1:103–28.

———. 1843. *Systema reptilium.* 1. *Amblyglossae.* Vienna: Braumiller and Seidel.

Fleet, R. R., and H. S. Fitch. 1974. Food habits of *Basiliscus basiliscus* in Costa Rica. *J. Herp.* 8 (3): 260–62.

Fleischman, L. J. 1985. Cryptic movement in the vine snake *Oxybelis aeneus. Copeia* 1985 (1): 242–45.

Fleming, T., and R. S. Hooker. 1975. *Anolis cupreus:* The response of a lizard to tropical seasonality. *Ecology* 56 (6): 1243–61.

Forcart, L. 1951. Nomenclatural remarks on some generic names of the family Boidae. *Herpetologica* 7 (4): 197–99.

Ford, L. S., and D. C. Cannatella. 1993. The major clades of frogs. *Herp. Monogr.* 7:94–117.

Foster, M. S., and R. W. McDiarmid. 1983. *Rhinophrynus dorsalis* (alma de vaca, sapo borracha, Mexican burrowing toad). In *Costa Rican natural history,* ed. D. H. Janzen, 419–21. Chicago: University of Chicago Press.

Fouquette, M. J., Jr. 1960. Call structure in frogs of the family Leptodactylidae. *Texas J. Sci.* 12 (3–4): 201–15.

———. 1961. Status of the frog *Hyla albomarginata* in Central America. *Fieldiana Zool.* 39 (55): 595–601.

———. 1968. Some frogs from the Venezuelan llanos, and the status of *Hyla misera* Werner. *Herpetologica* 24 (4): 321–25.

———. 1969. Rhinophrynidae, *Rhinophrynus, R. dorsalis. Cat. Amer. Amph. Reptl.* 78:1–2.

Fouquette, M. J., Jr., and A. J. Delahoussaye. 1977. Sperm morphology in the *Hyla rubra* group (Amphibia, Anura, Hylidae), and its bearing on generic status. *J. Herp.* 11 (4): 387–96.

Fouquette, M. J., Jr., and D. A. Rossman. 1963. Noteworthy records of Mexican amphibians and reptiles in the Florida State Museum and Texas Natural History Musuem. *Herpetologica* 19 (3): 185–201.

Frair, W., R. G. Ackman, and N. Mrosovsky. 1972. Body temperature of *Dermochelys coriacea*: Warm turtle from cold water. *Sci.* 177:791–93.

Frakes, L. A., J. E. Francis, and J. I. Syktus. 1992. *Climate modes of the Phanerozoic.* Cambridge: Cambridge University Press.

Frantzius, A. von. 1861. Beträge zur Kenntniss der Vulkane Costa Rica. *Peterm. Mitt. Geogr.* 5 (9): 329–38; (10): 381–85.

———. 1862. Das rechte Ufer des San Juan-Flusses, ein bisher fastganzlich unbekaner Theil von Costa Rica. *Peterm. Mitt. Geogr.* 7 (3): 83–95.

———. 1869a. Ueber die geographische Verbreitung der Vogel Costaricas un derer Lebensweise. *J. Ornith.* 17:195–204, 289–318, 361–79.

———. 1869b. Die Säugethiere Costaricas, ein Beiträg zur Kenntnis der geographischen Verbreitung de Säugethiere Americas. *Arch. Naturgesch.* 35:245–325.

———. 1869c. Der südöstliche Theil de Republik Costa Rica. *Peterm. Mitt. Geogr.* 15 (9): 323–30.

———. 1869d. Der geographisch-kartographische Stadpunkt von Costa Rica. *Peterm. Mitt. Geogr.* 15 (3): 81–84.

———. 1873. Die warmen Mineralquellen in Costa Rica. *N. Jahr. Mines. Geol. Palon.* 1872:496–510.

Franzen, M. 1986. Herpetologische Beobachtungen im Santa Rosa–Nationalpark, Costa Rica. *Herpetofauna* 8 (41): 24–33.

———. 1988. Beobachtungen an einigen Forschlurchen aud dem Santa Rosa–Nationalpark, Costa Rica. *Herpetofauna* 10 (52): 25–28.

———. 1994. Die Herpetofauna des Maritza-Sektors im Guanacaste–Nationalpark, Costa Rica. Diplomarbeit, Universitat Bonn.

———. 1999. Notes on morphological variation and the biology of *Nototriton guanacaste* Good and Wake, 1993 (Caudata, Plethodontidae). *Alytes* 16 (3–4): 123–29.

Frazier, N. B. 1995. Preface: Herpetological research at a National Environmental Research Park. *Herpetologica* 51 (4): 383–86.

Frisch, W., M. Meschede, and M. Sick. 1992. Origin of the Central American ophiolites: Evidence from paleomagnetic results. *Geol. Soc. Amer. Bull.* 104:1301–14.

Fritts, T. H. 1969. The systematics of the parthenogenetic lizards of the *Cnemidophorus cozumela* complex. *Copeia* 1869 (3): 519–35.

Frost, D. R. 1985. *Amphibian species of the world: A taxonomic and geographical reference.* Lawrence, Kans.: Allen Press and Association of Systematics Collections.

———. 1987. A replacement name for *Glossostoma* Günther, 1900 (Anura: Microhylidae). *Copeia* 1987 (4): 1025.

Frost, D. R., and R. E. Etheridge. 1989. A phylogenetic analysis and taxonomy of iguanian lizards (Reptilia: Squamata). *Misc. Pub. Mus. Nat. Hist. Univ. Kansas* 81:1–65.

Frost, D. R., and D. M. Hillis. 1990. Species in concept and practice: Herpetological applications. *Herpetologica* 46 (1): 87–104.

Frost, J. S. 1982. Functional genetic similarity between geographically separated populations of Mexican leopard frogs (*Rana pipiens* complex). *Syst. Zool.* 31 (1): 57–67.

Funkhouser, A. 1957. A review of the Neotropical tree-frogs of the genus *Phyllomedusa*. *Occas. Pap. Nat. Hist. Mus. Stanford Univ.* 5: 1–89.

Fürbringer, M. 1900. Beiträg zur Systematik und Genealogie der Reptilien. *Jena. Zeit. Naturwissen.* 34:596–682.

Gabb, W. M. 1875. Informe sobre la exploración de Talamanca verificada durante las años de 1873–4. *An. Inst. Fisico-Geogr. Nac. Costa Rica* 5:67–92.

Gaffney, E. S., and P. A. Meylan. 1988. A phylogeny of turtles. *Syst. Assoc. Spec.* 35A:157–219.

Gaige, H. T. 1929. Three new tree-frogs from Panama and Bolivia. *Occas. Pap. Mus. Zool. Univ. Mich.* 207:1–6.

———. 1936. Some reptiles and amphibians from Yucatan and Campeche, Mexico. *Pub. Carnegie Inst. Wash.* 457:289–304.

Gaige, H. T., N. E. Hartweg, and L. C. Stuart. 1937. Notes on a collection of amphibians and reptiles from eastern Nicaragua. *Occas. Pap. Mus. Zool. Univ. Mich.* 357:1–18.

Gaige, H. T., and L. C. Stuart. 1934. A new *Hyla* from Guatemala. *Occas. Pap. Mus. Zool. Univ. Mich.* 281: 1–3.

Gans, C. 1974. *Biomechanics: An approach to vertebrate biology.* Philaelphia: Lippincott.

———. 1975. *Reptiles of the world.* Toronto: Ridge Press/Bantam Books.

———, senior ed. 1969–98. *Biology of the Reptilia.* Vols. 1–13, London: Academic Press; vols. 14–15, New York: Wiley; vol. 16, New York: Liss; vols. 17–18, Chicago: University of Chicago Press; vol. 19, Ithaca, N.Y.: Society for the Study of Amphibians and Reptiles. Continuing series.

García-París, M., D. A. Good, G. Parra-Olea, and D. B. Wake. 2000. Biodiversity of Costa Rican salamanders: Implications of high levels of genetic differentiation and phylogeographic structure for species formation. *Proc. Natl. Acad. Sci.* 97 (4): 1640–47.

García-París, M., and D. B. Wake. 2000. Molecular phylogenetic analysis of relationships of the tropical salamander genera *Oedipina* and *Nototriton*, with descriptions of a new genus and three new species. *Copeia* 2000 (1): 42–70.

Gardner, T. W., W. Buck, T. F. Bullard, P. W. Hare, R. W. Kesel, D. R. Lowe, C. M. Menges, S. C. Mora, F. J. Pazzaglia, I. D. Sasowsky, J. W. Troester, and S. G. Wells. 1987. Central America and the Caribbean. In *Geomorphic systems of North America*, ed. W. L. Graf, 343–402. Centennial Special Volume 2. Boulder, Colo.: Geological Society of America.

Garman, S. W. "1883." 1884. The reptiles and batrachians of North America. 1. Ophidia—Serpentes. *Mem. Mus. Comp. Zool.* 8 (3): 1–185.

Gaut, B. S. 1998. Molecular clocks and nucleotide substitution rates in higher plants. *Evol. Biol.* 30 (4): 93–120.

Gauthier, J., A. G. Kluge, and T. Rowe. 1988. Amniote phylogeny and the importance of fossils. *Cladistics* 4 (2): 105–209.

Gehlbach, F. R. 1971. Lyre snakes in the *Trimorphodon biscutatus* complex: A taxonomic resume. *Herpetologica* 27 (2): 200–211.

Gehlbach, F. R., J. F. Watkins II, and H. W. Reno. 1968. Blind snake defensive behavior elicited by ant attacks. *BioScience* 18 (8): 784–85.

Gehlbach, F. R., J. F. Watkins II, and J. C. Kroll. 1971. Pheromone trail-following studies of typhlopid, leptotyphlopid, and colubrid snakes. *Behaviour* 49:282–94.

Gentry, A. H., ed. 1990. *Four Neotropical rainforests.* New Haven: Yale University Press.

Gerhardt, R. P., P. M. Harris, and M. A. Vásquez. 1993. Food habits of nesting great black hawks in Tikal National Park, Guatemala. *Biotropica* 25 (3): 349–52.

Gessner, C. 1554. *Historiae animalium. Liber secundus, qui est de quadrupedibus oviparis cum appendice ad quadrupedes vivaparos.* Zurich.

Gibbons, J. W., ed. 1990. *Life history and ecology of the slider turtle.* Washington, D.C.: Smithsonian Institution Press.

Gibbons, J. W., S. S. Novak, and C. H. Ernst. 1988. *Chelydra serpentina. Cat. Amer. Amph. Reptl.* 420:1–4.

Gicca, D. F. 1983. *Enyaliosaurus quinquecarianatus. Cat. Amer. Amph. Reptl.* 329:1–2.

Gillingham, J. C., and D. L. Clark. 1981. Snake tongue-flicking: Transfer mechanics to Jacobson's organ. *Canadian J. Zool.* 59 (9): 1651–57.

Girard, C. "1854." 1855. Abstract of a report to Lieut. J. M. Gillis, U.S.N., upon the reptiles collected during the U.S.N. astronomical expedition to Chili. *Proc. Acad. Nat. Sci. Phila.* 7:226–27.

Gloyd, H. K., and R. Conant. 1990. *Snakes of the* Agkistrodon *complex: A monographic review.* Contributions to Herpetology 6. Oxford, Ohio. Society for the Study of Amphibians and Reptiles.

Gmelin, J. F. "1788." 1789. *Caroli a Linne systema naturae.* Vol. 1, pt. 3. 13th ed. Leipzig.

Goeldi, E. A. 1907. Description of *Hyla resinifictrix* Goeldi, a new Amazonian tree frog peculiar for its breeding habits. *Proc. Zool. Soc. London* 1907:135–40.

Goin, C. J., and O. B. Goin. 1968. A new green frog from Suriname. *Copeia* 1968 (3): 581–83.

Goldberg, S. R. 1995. Reproduction in the lyre snake, *Trimorphodon biscutatus* (Colubridae) from Arizona. *SW Nat.* 40 (3): 334–35.

———. 1998. Reproduction in the Mexican vine snake *Oxybelis aeneus* (Serpentes: Colubridae). *Texas J. Sci.* 50 (1): 51–56.

Goldblatt, P., ed. 1993. *Biological relationships between Africa and South America.* New Haven: Yale University Press.

Gómez, L. D. 1984. *Las plantas acuaticas y anfibias de Costa Rica y Centroamérica. 1. Liliopsida.* San José: Editorial Universidad Estatal a Distancia.

———. 1986a. Vegetación de Costa Rica. In *Vegetación y clima de Costa Rica*, ed. L. D. Gómez, 1:1–327. San José: Editorial Universidad Estatal a Distancia.

———, ed. 1986b. *Vegetación y clima de Costa Rica.* 2 vols. San José: Editorial Universidad Estatal a Distancia.

Gómez, L. D., and J. M. Savage. 1983. Searchers on that rich coast: Costa Rica field biology, 1400–1980. In *Costa Rican natural history*, ed. D. H. Janzen, 1–11. Chicago: University of Chicago Press.

———. 1991. Investigadores en aquella rica costa: Biología de campo costarricense 1400–1980. In *Historia natural de Costa Rica*, ed. D. H. Janzen, ed. San José: Editorial de la Universidad de Costa Rica.

Good, D. A. 1987. A phylogenetic analysis of cranial osteology in the gerrhonotine lizards. *J. Herp.* 21 (4): 285–97.

———. 1988. Phylogenetic relationships among gerrhonotine lizards. An analyis of external morphology. *Univ. Calif. Pub. Zool.* 121:1–139.

Good, D. A., and D. B. Wake. 1993. Systematic studies of the Costa Rican moss salamanders, genus *Nototriton*, with descriptions of three new species. *Herp. Monogr.* 7:131–59.

———. 1997. Phylogenetic and taxonomic implications of protein variation in the Mesoamerican salamander genus *Oedipina* (Caudata: Plethodontidae). *Rev. Biol. Trop.* 45 (3): 1185–1208.

Goodman, D. E. 1971. Territorial behavior in a Neotropical frog, *Dendrobates granuliferus. Copeia* 1971 (2): 365–70.

Goodrich, E. S. 1930. *Studies on the structure and development of vertebrates;* London: Macmillian.

Gorman, G. C. 1967. Studies on evolution and zoogeography of *Anolis* (Sauria: Iguanidae). Ph.D. diss., Harvard University.

———. 1970. Chromosomes and the systematics of the family Teiidae (Sauria, Reptilia). *Copeia* 1970 (2): 230–45.

———. 1973. The chromosomes of the Reptilia, a cytotaxonomic interpretation. In *Cytotaxonomy and vertebrate evolution*, ed. A. B. Chiarelli and E. Capana, 349–424. London: Academic Press.

Gorman, G. C., and L. Atkins. 1967. The relationships of the *Anolis* of the *roquet* species group (Sauria: Iguanidae). 2. Comparative chromosome cytology. *Syst. Zool.* 16 (2): 137–43.

Gorman, G. C., L. Atkins, and T. Holzinger. 1967. New karyotypic data on fifteen genera of lizards in the family Iguanidae, with discussion of taxonomic and cytological implications. *Cytogenetics* 62:86–99.

Gorman, G. C., and F. Gress. 1970. Chromosome cytology of four boid snakes and a varanid lizard, with comments on the cytosystematics of primitive snakes. *Herpetologica* 26 (3): 308–17.

Gorman, G. C., R. B. Huey, and E. E. Williams. 1969. Cytotaxonomic studies on some unusual iguanid lizards assigned to the genera *Chamaeliolis, Polychrus, Polychroides*, and *Phenacosaurus*, with behaviorial notes. *Brevoria* 316:1–17.

Gorman, G. C., R. Thomas, and L. Atkins. 1968. Intra- and interspecific chromosome variation in the lizard *Anolis cristatellus* and its closest relatives. *Brevoria* 293:1–13.

Gosner, K. L. 1960. A simplified table for staging anuran embryos and larvae with notes on identification. *Herpetologica* 16 (3): 183–90.

Graf, W. L., ed. 1987. *Geomorphic systems of North America . . .* Centennial Special Volume 2. Boulder, Colo.: Geological Society of America.

Graham, A. 1987a. Miocene communities and paleoenvironments of southern Costa Rica. *Amer. J. Botan.* 74 (10): 1501–18.

————. 1987b. Tropical American Tertiary floras and paleo-environments: Mexico, Costa Rica, and Panama. *Amer. J. Botan.* 74 (10): 1519–31.

————. 1994. Neotropical Eocene coastal floras and 18°/16°-estimated warmer and cooler equatorial waters. *Amer. J. Botan.* 81 (3): 301–6.

————. 1996. Additions and preliminary study of an Oligocene-Miocene palynoflora from Chiapas, Mexico. *Rheedea* 6 (1): 1–12.

————. 1997. Neotropical plant dynamics during the Ceno-zoic—diversification, and the ordering of evolutionary and speciation processes. *Syst. Botan.* 22 (1): 139–50.

————. 1999a. *Late Cretaceous and Cenozoic history of North American vegetation.* Oxford: Oxford University Press.

————. 1999b. Studies on Neotropical paleobotany. 13. An Oligo-Miocene palynoflora from Simojovel (Chiapas, Mexico). *Amer. J. Botan.* 86 (1): 17–31.

————. 1999c. Terrestrial Pliocene climates in northern Latin America. In *The Pliocene: Time of change*, ed. J. H. Wrenn, J.-P. Suc, and S. A. G. Leroy, 209–16. Dallas: American Association of Stratigraphic Palynologists Foundation.

Grassman, M. A., D. W. Owens, J. P. McVey, and R. M. Már-quez. 1984. Olfactory-based orientation in artificially imprinted sea turtles. *Science* 224: 83–84.

Gray, J. E. 1825. A synopsis of the genera of reptiles and Amphibia, with a description of some new species of reptiles. *Ann. Philos.*, n.s., 10: 193–217.

————. 1827. A synopsis of the genera of saurian reptiles, in which some new genera are indicated and others reviewed by actual examination. *Ann. Philos.*, ser. 2, 2: 207–9.

————. 1828. Original figures and short systematic descriptions of new and unfigured animals. *Spicilegia Zool.* 1: 1–3.

————. 1831a. A synopsis of the species of the class Reptilia. In *The animal kingdom*, vol. 9, *The class Reptilia*, ed. E. Griffith and E. Pidgeon, appendix, 1–110. London: Whittaker, Teacher.

————. 1831b. *Synopsis reptilium, or Short descriptions of the species of reptiles. 1. Cataphracta, tortoises, crocodiles and enyaliosaurs.* London: Treuttel, Wurtz.

————. 1839. A catalogue of the slender-tongued saurians, with descriptions of many new genera and species. *Ann. Mag. Nat. Hist.* 2: 287–93, 331–37.

————. 1842. Descriptions of new species of reptiles, chiefly in the collection of the British Museum. *Zool. Misc.* 4: 57–59.

————. 1844. *Catalogue of the tortoises, crocodiles, and amphisbaenians, in the collection of the British Museum.* London: Taylor and Francis.

————. 1845. *Catalogue of the specimens of lizards in the collection of the British Museum.* London: Taylor and Francis.

————. 1850. *Catalogue of the specimens of Amphibia in the collection of the British Museum. 2. Batrachia Gradienta, etc.* London: Taylor and Francis.

————. "1855." 1856. *Catalogue of the shield reptiles in the collection of the British Museum. 1. Testudinata (tortoises).* London: Taylor and Francis.

————. 1860. Description of a new species of *Geoclemmys* from Ecuador. *Proc. Zool. Soc. London.* 1860: 231–32.

————. 1868. Notice of two new salamanders from Central America. *Ann. Mag. Nat. Hist.*, ser. 4, 2: 297–98.

————. 1869. Notes on the families and genera of tortoises (Testudinata), and on the characters afforded by the study of their skulls. *Proc. Zool. Soc. London.* 1869: 165–225.

Graybeal, A. 1995. Naming species. *Syst. Biol.* 44 (2): 237–50.

Graybeal, A., and K. de Queiroiz. 1992. Inguinal amplexus in *Bufo fastidiosus*, with comments on the systematics of bufonid frogs. *J. Herp.* 26 (1): 84–87.

Greding, E. J., Jr. 1972a. An unusually large toad (Anura: Bufonidae) from the lower southeastern slope of Volcán Turrialba, with a key to the *Bufo* of Costa Rica. *Carib. J. Sci.* 12 (1–2): 91–94.

————. 1972b. Call specificity and hybrid compatibility between *Rana pipiens* and three other *Rana* species in Central America. *Copeia* 1972 (2): 383–85.

————. 1976. Call of the tropical American frog *Rana palmipes* Spix (Amphibia, Anura, Ranidae). *J. Herp.* 10 (3): 263–64.

Green, D. M. 1991. Supernumerary chromosomes in amphibians. In *Amphibian cytogenetics and evolution*, ed. D. M. Green and S. K. Sessions, 333–57. San Diego: Academic Press.

Green, D. M., and S. K. Sessions, eds. 1991a. *Amphibian cytogenetics and evolution.* San Diego: Academic Press.

————. 1991b. Nomenclature for chromosomes. In *Amphibian cytogenetics and evolution*, ed. D. M. Green and S. K. Sessions, 431–32. San Diego: Academic Press.

Greene, H. W. 1969. Unusual pattern and coloration in snakes of the genus *Pliocercus* from Veracruz. *J. Herp.* 3 (1–2): 27–33.

————. 1973. Defensive tail display by snakes and amphisbaenians. *J. Herp.* 7 (3): 143–61.

————. 1975. Ecological observations on the red coffee snake, *Ninia sebae*, in southern Veracruz, Mexico. *Amer. Midl. Nat.* 93 (2): 478–84.

————. 1983a. Dietary correlates of the origin and radiation of snakes. *Amer. Zool.* 23 (2): 431–41.

————. 1983b. *Boa constrictor* (boa, béquer). In *Costa Rican natural history*, ed. D. H. Janzen, 380–82. Chicago: University of Chicago Press.

————. 1984. Feeding behavior and diet of the eastern coral snake, *Micrurus fulvius*. *Spec. Pub. Mus. Nat. Hist. Univ. Kans.* 10: 147–62.

————. 1988. Antipredator mechanisms in reptiles. In *Biology of the Reptilia*, vol. 16, *Ecology B, Defense and life history*, ed. C. Gans and R. B. Huey, 1–152. New. York: Liss.

————. 1997. *Snakes: The evolution of mystery in nature.* Berkeley: University of California Press.

Greene, H. W., G. M. Burghardt, B. A. Dugan, and A. S. Rand. 1978. Predation and defensive behavior of green iguanas (Reptilia, Lacertilia, Iguanidae). *J. Herp.* 12 (2): 169–76.

Greene, H. W., and D. Cundall. 2000. Limbless tetrapods and snakes with legs. *Science* 287: 1939–41.

Greene, H. W., and R. W. McDiarmid. 1981. Coral snake mimicry: Does it occur? *Science* 213: 1207–12.

Greene, H. W., and R. L. Sieb. 1983. *Micrurus nigrocinctus* (corál, coral snake, coralillo). In *Costa Rican natural history*, ed. D. H. Janzen, 406–8. Chicago: University of Chicago Press.

Greer, A. E. 1967. Notes on the mode of reproduction in anguid lizards. *Herpetologica* 23 (3): 94–99.

———. 1970. A subfamily classification of scincid lizards. *Bull. Mus. Comp. Zool.* 139 (3): 151–84.

Greer, A. E., J. D. Lazell, and R. M. Wright. 1973. Anatomical evidence for a counter-current heat exchanger in the leatherback turtle *(Dermochelys coriacea)*. *Nature* 244:181.

Greer, B. J., and K. D. Wells. 1980. Territorial and reproductive behavior of the tropical American frog *Centrolenella fleischmanni. Herpetologica* 36 (4): 318–26.

Griffith, E., and E. Pidgeon, eds. 1831. *The animal kingdom arranged in conformity with its organization by the Baron Cuvier, with additional descriptions of all the species hitherto named, and of many others.* 16 vols. London: Whittaker.

Griffith, H., A. Ngo, and R. W. Murphy. 2000. A cladistic evaluation of the cosmopolitan genus *Eumeces* Wiegmann (Reptilia, Squamata, Scincidae). *Russian J. Herp.* 7 (1): 1–16.

Gronovius, L. T. 1756. *Museum ichthyologici tomus secundus sistens piscium indigenorum et exocticorum nonnullorum, quorum maxima pars in Museo Laurentii Theodorii Gronovii. . . . Accendunt nonnullorum exoticum piscium icones aeri incisae et amphibiorum animalium historia zoologica.* Leiden: Haak.

———. 1763. *Zoophylacii Gronoviani fasciculus primus exhibens Animalia Quadrupeda, Amphibia, utque Pisces quae in museo suo adservat, rite examinavit, systematice disposuit, descripsit, atque iconibus illustravit.* Leiden: Haak.

Groombridge, B. 1987. The distribution and status of world crocodilians. In *Wildlife management: Crocodiles and alligators,* ed. G. J. W. Webb, S. C. Manolis, and P. J. Whitehead, 9–21. Chipping Norton, NSW: Surrey Beatty.

Guillén, D. 1989. *Costa Rica: Textos de la historia de Centroamérica y el Caribe.* Mexico City: Instituto de Investigaciones Dr. José Maria Luis Mora; Editorial Nueva Imagen

Günther, A. C. L. G. 1858a. *Catalogue of the colubrine snakes in the collection of the British Museum.* London: Taylor and Francis.

———. "1858b." 1859. *Catalogue of the Batrachia Salientia in the collection of the British Museum.* London: Taylor and Francis.

———. 1860. Third list of cold-blooded Vertebrata collected by Mr. Fraser in Ecuador. *Proc. Zool. Soc. London.* 1860: 233–40.

———. 1863. Third account of new species of snakes in the collection of the British Museum. *Ann. Mag. Nat. Hist.,* ser. 3, 12:348–65.

———. 1865. Fourth account of new species of snakes in the collection of the British Museum. *Ann. Mag. Nat. Hist.,* ser. 3, 15:89–98.

———. 1867. Contribution to the anatomy of *Hatteria (Rhynchocephalus* Owen). *Philos. Trans. R. Soc. London.* 157: 595–629.

———. 1868. Sixth account of new species of snakes in the collection of the British Museum. *Ann. Mag. Nat. Hist.,* ser. 4, 1:413–29.

———. 1872. Seventh account of new species of snakes in the collection of the British Museum. *Ann. Mag. Nat. Hist.,* ser. 4, 9:13–37.

———. 1885–1902. *Reptilia and Batrachia: Biologia Centrali-Americana.* London: Taylor and Francis.

Gutberlet, R. L., Jr. 1998. The phylogenetic position of the Mexican black-tailed pitviper (Squamata: Viperidae: Crotalinae). *Amph.-Reptl.* 5 (4): 184–206.

Gutiérrez, J. M. 1996. Instituto Clodomiro Picado de Costa Rica: 25 aniversario. *Rev. Biol. Trop.* 44 (2A): 349–52.

Gutiérrez, J. M., and R. Bolaños. 1979. Cariotipos de la principales serpientes coral (Elapidae: *Micrurus*) de Costa Rica. *Rev. Biol. Trop.* 27 (1): 115–22.

———. 1980. Karyotype of the yellow-bellied sea snake, *Pelamis platurus. J. Herp.* 14 (2): 161–65.

———. 1981. Polimorfismo cromosómico intraespecífico en la serpiente de coral *Micrurus nigrocinctus.Rev. Biol. Trop.* 29 (1): 115–22.

Gutiérrez, J. M., A. Solórzano, and L. Cerdas. 1984. Estudios cariológicos de cinco especies de serpientes costarricenses de la familia Colubridae. *Rev. Biol. Trop.* 32 (2): 263–67.

Gutiérrez, J. M., R. T. Taylor, and R. Bolaños. 1979. Karyotypes of ten species of Costa Rican Viperidae. *Rev. Biol. Trop.* 27 (2): 309–19.

Guyer, C. 1986. The role of food in regulating population density in a tropical mainland anole, *Norops humilis.* Ph.D. diss., University of Miami.

———. 1987. Seasonal patterns of reproduction of *Norops humilis* (Sauria: Iguanidae) in Costa Rica. *Rev. Biol. Trop.* 34 (2): 247–51.

———. 1988a. Food supplementation in a tropical mainland anole, *Norops humilis:* Effects on individuals. *Ecology* 69 (2): 362–69.

———. 1988b. Food supplementation in a tropical mainland anole, *Norops humilis:* Demographic effects. *Ecology* 69 (2): 350–61.

———. 1990. The herpetofuana of La Selva, Costa Rica. In *Four Neotropical rainforests,* ed. A. H. Gentry, 371–85. New Haven: Yale University Press.

———. 1994a. The reptile fauna: Diversity and ecology. In *La Selva: Ecology and natural history of a Neotropical rainforest,* ed. L. A. McDade, K. S. Bawa, H. A. Hespenheide and G. S. Hartshorn, 210–16 Chicago: University of Chicago Press.

———. 1994b. Reptiles: Appendix 6. In *La Selva: Ecology and natural history of a Neotropical rainforest,* ed. L. A. McDade, K. S. Bawa, H. A. Hespenheide, and G. S. Hartshorn, 382–83. Chicago: University of Chicago Press.

Guyer, C., and M. A. Donnelly. 1990. Length-mass relationships among an assemblage of tropical snakes in Costa Rica. *J. Trop. Ecol.* 6:65–76.

———. 2002. A guide to the amphibians and reptiles of La Selva, Costa Rica. Manuscript.

Guyer, C., and M. S. Laska. 1996. *Coluber* (= *Masticophis) mentovarius* (tropical racer): Predation. *Herp. Rev.* 27 (4): 203.

Guyer, C., and J. M. Savage. "1986." 1987. Cladistic relationships among anoles (Sauria: Iguanidae). *Syst. Zool.* 35 (4): 509–31.

———. 1992. Anole systematics revisited. *Syst. Biol.* 41 (1): 89–110.

Haas, W., and G. Köhler. 1997. Freilandbeobachtungen, Plege un Zucht von *Bufo luetkenii* Boulenger, 1891. *Herpetofauna* 19 (109): 5–9.

Haeckel, E. 1866. *Generelle Morphologie der Organismen.* 2 vols. Berlin: Reimer.

Haffer, J. 1969. Speciation in Amazonian forest birds. *Science* 165:131–37.

Haiduk, M. W., and J. W. Bickham. 1982. Chomosomal homologies and evolution of testudinoid turtles with emphasis on the systematic placement of *Platysternon. Copeia* 1982 (1): 60–66.

Haines, T. P. 1940. Delayed fertilization in *Leptodeira annulata polysticta. Copeia* 1940 (2): 116–18.

Hall, C. A. S., C. León, and G. Leclerc, eds. 2000. *Quantifying sustainable development: The future of tropical economies.* San Diego: Academic Press.

Hall, W. P. 1973. Comparative population cytogenetics, speciation, and evolution of the crevice-using species of *Sceloporus.* Ph.D. diss., Harvard University.

Hallowell, E. 1856. Notes on the reptiles in the collection of the Academy of Natural Sciences of Philadelphia. *Proc. Acad. Nat. Sci. Phila.* 8:221–38.

———. "1860." 1861. Report upon the Reptilia of the North Pacific exploring expedition under command of Capt. John Rogers, U.S.N. *Proc. Acad. Nat. Sci. Phila.* 12:479–510.

Halpern, M., and S. Borghjid. 1997. Sublingual plicae (anterior processes) are not necessary for garter snake vomeronasal function. *J. Comp. Psych.* 111 (3): 302–6.

Hanken, J., and D. B. Wake. 1985. Genetic differentiation among plethodontid salamanders (genus *Bolitoglossa*) in Central and South America, implications for the South America invasion. *Herpetologica* 38 (2): 272–87.

Hanken, J., and R. J. Wassersug. 1981. The visible skeleton: A new double stain technique reveals the nature of the hard tissues. *Funct. Photo.* 1981 (July-August): 22–25, 44.

Hardy, D. L., Sr. 1994a. Snakebite and field biologists in Mexico and Central America: Report on ten cases with recommendations for field management. *Herp. Nat. Hist.* 2 (2): 67–82.

———. 1994b. A re-evaluation of suffocation as the cause of death during constriction by snakes. *Herp. Rev.* 25 (2): 45–46.

Harris, D. M. 1994. Review of the teiid lizard genus *Ptychoglossus. Herp. Monogr.* 8:226–75.

Harris, D. M., and A. G. Kluge. 1984. The *Sphaerodactylus* (Sauria: Gekkonidae) of Middle America. *Occas. Pap. Mus. Zool. Univ. Mich.* 706:1–59.

Harris, H. S., Jr., and R. S. Simmons. "1977." 1978. A new subspecies of *Crotalus durissus* (Serpentes: Crotalidae) from the Rupununi savanna *(sic)* of southwestern Guyana. *Mem. Inst. Butantan* 40–41:305–11.

Hartshorn, G. S. 1983. Plants: Introduction. In *Costa Rican natural history,* ed. D. H. Janzen, 118–57. Chicago: University of Chicago Press.

Hartshorn, G. S., and L. Hartshorn, eds. 1982. *Costa Rica: Country environmental profile.* San José: Trejos.

Hass, C. A. 1991. Evolution and biogeography of West Indian *Sphaerodactylus* (Sauria: Gekkonidae): A molecular approach. *J. Zool.* 225:525–61.

Hayes, M. P. 1983. Predation on the adults and prehatching stages of glass frogs (Centrolenidae). *Biotropica* 15 (1): 74–76.

———. 1985. Nest structure and attendance in the stream-dwelling frog, *Eleutherodactylus angelicus. J. Herp.* 19 (1): 168–69.

———. 1991. A study of clutch attendance in the Neotropical frog *Centrolenella fleischmanni* (Anura: Centrolenidae). Ph.D. diss., University of of Miami.

Hayes, M. P., J. A. Pounds, and D. C. Robinson. 1986. The fringe-limbed tree frog, *Hyla fimbremembra* (Anura: Hylidae): New records from Costa Rica. *Fla. Sci.* 49 (4): 193–98.

Hayes, M. P., J. A. Pounds, and W. W. Timmerman. 1989. An annotated list and guide to the amphibians and reptiles of Monteverde Costa Rica. *Soc.Stud.Amph.Reptl.Herp. Circ.* 17:1–67.

Hayes, M. P., and P. H. Starrett. 1980. Notes on a collection of centrolenid frogs from the Colombia Chocó. *Bull. South. Calif. Acad. Sci.* 79 (3): 89–96.

Heatwole, H. 1966. The effect of man on distribution of some reptiles and amphibians in eastern Panama. *Herpetologica* 22 (1): 55–59.

Heatwole, H., senior ed. 1994–2000. *Amphibian biology.* 4 vols. Chipping Norton, NSW: Surrey Beatty. Continuing series.

Heatwole, H., and G. T. Barthalmus, eds. 1994. *Amphibian biology.* Vol. 1. *The integument.* Chipping Norton, NSW: Surrey Beatty.

Hedges, S. B. 1989. Evolution and biogeography of West Indian frogs of the genus *Eleutherodactylus:* Slow-evolving loci and the major groups. In *Biogeography of the West Indies: Past, present and future,* ed. C. A. Woods, 305–69. Gainesville, Fla.: Sandhill Crane Press.

Henderson, R. W. 1976. A new insular subspecies of the colubrid snake *Leptophis mexicanus* (Reptilia, Serpentes, Colubridae) from Belize. *J. Herp.* 10 (4): 329–31.

———. 1982. Trophic relationships and foraging strategies of some New World tree snakes *(Leptophis, Oxybelis, Uromacer). Amph.-Reptl.* 3:71–80.

———. 1984. *Scaphiodontophis* (Serpentes, Colubridae): Natural history and test of a mimicry-related hypothesis. *Spec. Pub. Mus. Nat. Hist. Univ. Kans.* 10:185–97.

———. 1993a. *Corallus. Cat. Amer. Amph. Reptl.* 572:1–2.

———. 1993b. *Corallus annulatus. Cat. Amer. Amph. Reptl.* 573:1–3.

———. 1993c. *Corallus caninus. Cat. Amer. Amph. Reptl.* 574:1–4.

———. 1993d. *Corallus cropanii. Cat. Amer. Amph. Reptl.* 575:1–2.

———. 1993e. *Corallus enydris. Cat. Amer. Amph. Reptl.* 576:1–6.

———. 1993f. On the diets of some arboreal boids. *Herp. Nat. Hist.* 1 (1): 91–96.

———. 1997. A taxonomic review of the *Corallus hortulanus* complex of Neotropical tree boas. *Carib. J. Sci.* 33 (3–4): 198–221.

Henderson, R. W., and M. H. Binder. 1980. The ecology and behavior of vine snakes *(Ahaetulla, Oxybelis, Thelotornis, Uromacer):* A review. *Milwaukee Publ. Mus. Contr. Biol. Geol.* 37:1–38.

Hernández F. 1648. *Rerum medicarum novae hispaniae the-*

saurus seu plantarum animalium mineralium mexicanorum historia. Rome: Mascardi.

Herrera, W. 1986. Clima de Costa Rica. In *Vegetación y clima de Costa Rica,* ed. L. D. Gómez, 2:1–118. San José: Editorial Universidad Estatal a Distancia.

Herrera, W., and L. D. Gómez. 1993. *Mapa de unidades bióticas de Costa Rica.* San José: Instituto Geográfico Nacional de Costa Rica.

Heselhaus, R., and M. Schmidt 1994. *Harlequin frogs: A complete guide.* Neptune City, N.J.: Tropical Fish Hobbyist.

Heyer, W. R. 1967. A herpetological study of an ecological transect through the Cordillera de Tilaran, Costa Rica. *Copeia* 1967 (2): 259–87.

———. "1968." 1970. Studies on the genus *Leptodactylus* (Amphibia: Leptodactylidae). 2. Diagnosis and distribution of the *Leptodactylus* of Costa Rica. *Rev. Biol. Trop.* 16 (2): 171–205.

———. 1969. The adaptive ecology of the species groups of the genus *Leptodactylus* (Amphibia: Leptodactylidae). *Evolution* 23 (3): 421–28.

———. 1970. Studies on the frogs of the genus *Leptodactylus* (Amphibia: Leptodactylidae). 6. Biosystematics of the *melanonotus* group. *Contr. Sci. Nat. Hist. Mus. Los Angeles Co.* 191:1–48.

———. 1971. *Leptodactylus labialis. Cat. Amer. Amph. Reptl.* 104:1–3.

———. 1974. Relationships of the *marmoratus* species group (Amphibia, Leptodactylidae) within the subfamily Leptodactylinae. *Contr. Sci. Nat. Hist. Mus. Los Angeles Co.* 253:1–46.

———. 1975. A preliminary analysis of the intergeneric relationships of the frog family Leptodactylidae. *Smithson. Contr. Zool.* 199:1–55.

———. 1978. Systematics of the *fuscus* group of the frog genus *Leptodactylus* (Amphibia, Leptodactylidae). *Sci. Bull. Nat. Hist. Mus. Los Angeles Co.* 29:1–85.

———. 1979. Systematics of the *pentadactylus* species group of the frog genus *Leptodactylus* (Amphibia: Leptodactylidae). *Smithson. Contr. Zool.* 301:1–43.

———. 1994. Variation within the *Leptodactylus podicipinuswagneri* complex of frogs (Amphibia: Leptodatylidae). *Smithson. Contr. Zool.* 546:1–124.

Heyer, W. R., M. A. Donnelly, R. W. McDiarmid, L. A. C. Hayek, and M. S. Foster, eds. 1994. *Measuring and monitoring biological diversity.* Washington, D.C: Smithsonian Institution Press.

Heyer, W. R., R. W. McDiarmid, and D. L. Weigmann. 1975. Tadpoles, predation and pond habitats in the tropics. *Biotropica* 7 (2): 100–111.

Heyer, W. R., and A. S. Rand. 1977. Foam nest construction in the leptodactylid frogs *Leptodactylus pentadactylus* and *Physalaemus pustulosus* (Amphibia, Anura, Leptodactylidae). *J. Herp.* 11 (2): 225–28.

Hillis, D. M. 1985. Evolutionary genetics and systematics of New World frogs of the genus *Rana:* An analysis of ribosomal DNA, allozymes, and morphology. Ph.D. diss., University of Kansas.

———. 1988. Systematics of the *Rana pipiens* complex: Puzzle and paradigm. *Annl. Rev. Ecol. Syst.* 19:39–63.

Hillis, D. M., and S. K. Davis. 1986. Evolution of ribosomal DNA: Fifty million years of recorded history in the frog genus *Rana. Evolution* 40 (6): 1275–88.

Hillis, D. M., and C. Moritz, eds. 1990. *Molecular systematics.* Sunderland, Mass.: Sinauer.

Hillis, D. M., C. Moritz, and B. K. Mable, eds. 1996. *Molecular systematics.* 2d ed. Sunderland, Mass.: Sinauer.

Hillis, D. M., and R. de Sá. 1988. Phylogeny and taxonomy of the *Rana palmipes* group (Salientia: Ranidae). *Herp. Monogr.* 2:1–26.

Hillman, P. E. 1969. Habitat specificity in three sympatric species of *Ameiva* (Reptilia: Teiidae). *Ecology* 50 (3): 476–81.

Hirth, H. F. 1963a. Food of *Basiliscus plumifrons* on a tropical strand. *Herpetologica* 8 (4): 276–77.

———. 1963b. The ecology of two lizards on a tropical beach. *Ecol. Monogr.* 33:83–112.

———. 1964. Observations on the fer-de-lance, *Bothrops atrox,* in coastal Costa Rica. *Copeia* 1964 (2): 453–54.

———. 1971. Synopsis of biological data on the green turtle, *Chelonia mydas* (Linnaeus). *F.A.O. Fisheries Sym.* 85:1–8, 19.

———. 1980a. *Chelonia. Cat. Amer. Amph. Reptl.* 258:1–2.

———. 1980b. *Chelonia mydas. Cat. Amer. Amph. Reptl.* 249: 1–4.

Höbel, G. 1999a. Notes on the natural history and habitat use of *Eleutherodactylus fitzingeri* (Anura: Leptodactylidae). *Amph.-Reptl.* 20 (1): 65–72.

———. 1999b. Facultative nest construction in the gladiator frog *Hyla rosenbergi* (Anura: Hylidae). *Copeia* 1999 (3): 797–801.

———. 2000. Reproductive ecology of *Hyla rosenbergi* in Costa Rica. *Herpetologica* 56 (4): 446–54.

Hoffmann, C. 1856. Eine Exkursion nach dem Volcán de Cartago in Central America. *Bonplandia* 4 (6): 27–34.

———. 1857. Eine Exkursion nach dem Barba-Vulkan in Costa Rica. *Bonplandia* 6 (16–17): 301–30.

Hoffstetter, R. 1955. Squamates de type moderne. In *Traité de paléontologie,* ed. J. Piveteau, 5:606–62. Paris: Masson.

Hoge, A. R. "1953." 1954. A new genus and species of Boinae from Brazil. *Xenoboa cropanii,* gen. nov., sp. nov. *Mem. Inst. Butantan* 25:27–34.

———. "1965." 1966. Preliminary account of Neotropical Crotalinae. *Mem. Inst. Butantan* 32:109–84.

Hoge, A. R., and S. A. R. W. L. Romano. "1978–79." 1981. Poisonous snakes of the world. 1. Checklist of the pit vipers Viperoidea, Viperidae, Crotalinae. *Mem. Inst. Butantan* 42–43:179–310.

Hoge, A. R., S. L. Romano, and C. L. Cordeiro. "1976–77." 1978. Contribuição ao conhecimento das serpentes do Maranhão, Brasil (Serpentes: Boidae, Colubridae e Viperidae). *Mem. Inst. Butantan* 40–41:19–36.

Holdridge, L. R. 1967. *Life zone ecology.* Rev. ed. San José: Tropical Science Center.

———. 1982. *Life zone ecology.* Reprint. San José: Tropical Science Center.

Holdridge, L. R., W. C. Grenke, W. H. Hatheway, T. Liang, and J. A. Tosi Jr. 1971. *Forest environments in tropical life zones: A pilot study.* Oxford: Pergamon Press.

Hoogmoed, M. S. 1973. *Notes on the herpetofauna of Suri-*

nam. 4. The lizards and amphisbaenians of Surinam. The Hague: Junk.

Hoogmoed, M. S., C. J. Cole, and J. Ayarzaguena. 1992. A new cryptic species of lizard (Sauria: Teiidae) *(Gymnophthalmus) from Venezuela. Zool. Meded. Natl. Mus. Leiden* 66 (1): 1–18.

Houck, L. D. 1982. Growth rates and age at maturity for the plethodontid salamander *Bolitoglossa subpalmata. Copeia* 1982 (2): 474–78.

Houttuyn, M. 1782. Het onderscheid der salamanderen van haagdissen in't algemeen, en van de geckkos in't byzonder aangetoond. *Verh. Zeeuwsch. Gen. Wet. Vlissingen,* ser. 1, 9(2): 305–36.

Huey, R. B., E. R. Pianka, and T. W. Schoener, eds. 1983. *Lizard ecology: Studies of a model organism.* Cambridge: Harvard University Press.

Hughes, D. A., and J. D. Richard. 1974. The nesting of the Pacific ridley *Lepidochelys olivacea* on Playa Nancite, Costa Rica. *Marine Biol.* 24:97–107.

Hutchison, V. H., and R. K. Dupre. 1992. Thermoregulation. In *Environmental physiology of the amphibians,* ed. M. E. Feder and W. W. Burggren, 206–49. Chicago: University of Chicago Press.

Ibáñez, R., F. A. Arosemena, F. Solís, and C. A. Jaramillo. 1995. Anfibios y reptiles de la Serrania Piedras-Pacora, Parque Nacional Chagres. *Scientia (Panama)* 9:17–31.

Ibáñez, R., C. A. Jaramillo, and F. A. Solís. 1994. *Phyllobates lugubris* (dart-poison frog). *Herp. Rev.* 25 (4): 161.

Ibáñez, R., and A. S. Rand. 1990. *Eleutherodactylus johnstonei.* Panama. *Herp. Rev.* 21 (2): 37.

Ibáñez, R., A. S. Rand, and C. A. Jaramillo. 1999. *Los anifibios del Monumento Natural Barro Colorado, Parque Nacional Soberanía y areas adyacentes.* Panama: Editorial Mizarachi y Pujol.

Ibáñez, R., and E. M. Smith. 1995. Systematic status of *Colostethus flotator* and *C. nubicola* (Anura: Dendrobatidae) in Panama. *Copeia* 1995 (2): 446–56.

In De Bosch, H. A. J., and I. Ineich. 1994. The Typhlopidae of Sulawesi (Indonesia): A review with description of a new genus and a new species (Serpentes: Typhlopidae). *J. Herp.* 28 (2): 206–17.

Inger, R. F. 1996. Commentary on a proposed classification of the family Ranidae. *Herpetologica* 52 (2): 241–46.

Instituto C. Picado. 1989. *Tratamiento para las mordeduras por serpientes.* San José: University of Costa Rica.

International Commission on Zoological Nomenclature. 1958. Opinion 520. Suppression under the Plenary Powers of the specific name *tibiatrix* Laurenti, 1768, as published in the combination *Hyla tibiatrix,* and of the generic name, *Acrodytes* Fitzinger, 1843, and interpretation under the same powers of the nominal species *Rana venulosa* Laurenti, 1768 (Class Amphibia). *Opin. Declar. Intl. Comm. Zool. Nomen.* 19:169–200.

———. 1963. Opinion 644. *Coluber atratus* Hallowell, 1845 (Reptilia): Validation under the Plenary Powers. *Bull. Zool. Nomen.* 20 (1): 26.

———. 1999. *International Code of Zoological Nomenclature,* 4th ed. London: International Trust for Zoological Nomenclature.

Iturralde, M. A., and R. D. E. MacPhee. 1999. Paleogeography of the Caribbean region: Implications for Cenozoic biogeography. *Bull. Amer. Mus. Nat. Hist.* 238:1–96.

Iverson, J. B. 1975. Notes on courtship in *Rhinoclemmys funerea. J. Herp.* 9 (2): 249–50.

———. 1980. *Kinosternon angustipons. Cat. Amer. Amph. Reptl.* 262:1–2.

———. 1992. *A revised checklist with distribution maps of the turtles of the world.* Richmond, Ind.: Privately printed.

Jackman, T. R., A. Larson, K. de Queiroz, et al. 1999. Phylogenetic relationships and tempo of early diversification in *Anolis* lizards. *Syst. Zool.* 48 (2): 254–85.

Jackson, I. J. 1977. *Climate, water and agriculture in the tropics.* London: Longman.

Jackson, J. B. C., A. F. Budd, and A. G. Coates, eds. 1996. *Evolution and environment in tropical America.* Chicago: University of Chicago Press.

Jackson, J. F., W. Ingram III, and H. W. Campbell. 1976. The dorsal pigmentation pattern of snakes as an antipredator strategy: A multivariate approach. *Amer. Nat.* 110:1029–53.

Jacobson, S. K. 1985. Reproductive behavior and male mating success in two species of glass frogs (Centrolenidae). *Herpetologica* 41 (4): 396–404.

Jacobson, S. K., and J. J. Vandenberg. 1991. Reproductive ecology of the endangered golden toad *(Bufo periglenes). J. Herp.* 25 (3): 321–27.

James, E. 1823. *Account of an expedition from Pittsburgh to the Rocky Mountains, performed in the years 1819 and 1820.* 3 vols. Philadelphia: Carey and Lea; London: Longman, Hurst, Rees, Orne and Brown.

James, M. S. 1944. Notes on the breeding of *Hyla nigripes* in Costa Rica. *Copeia* 1944 (3): 147–48.

Jan, G. 1863. Enuerazione sistematico degli ofidi appartenenti al gruppo Coronellidae. *Arch. Zool. Nat. Fis.* 2 (2): 211–330.

Jan, G., and F. Sordelli. 1860–81. *Iconographie générale des ophidiens.* 3 vols. Milan: Jan e Sordelli.

Janzen, D. H. 1962. Injury caused by toxic secretion of *Phrynohyas spilomma* Cope. *Copeia* 1962 (3): 651.

———. 1980. Two potential coral snake mimics in a tropical deciduous forest. *Biotropica* 12 (1): 77–78.

———, ed. 1983. *Costa Rican natural history.* Chicago: University of Chicago Press.

———. 1986. *Guanacaste National Park: Tropical ecological and cultural restoration.* San José: Editorial Universidad Estatal a Distancia.

———, ed. 1992. *Historia natural de Costa Rica.* San José: Editorial de la Universidad de Costa Rica.

Janzen, F. J., and G. L. Paukstis. 1991. Environmental sex determination in reptiles: Ecology, evolution, and experimental design. *Quart. Rev. Biol.* 66 (2): 149–79.

Jaramillo, F. E., C. A. Jaramillo, and R. Ibáñez. 1997. Renacuajo de la rana de cristal *Hyalinobatrachium colymbiphyllum* (Anura: Centrolenidae). *Rev. Biol. Trop.* 45 (2): 867–70.

Jaslow, A. P. 1979. Vocalization and aggression in *Atelopus chiriquiensis* (Amphibia, Anura, Bufonidae). *J. Herp.* 13 (2): 141–45.

Jenkins, F. A., Jr., and D. M. Walsh. 1933. An early Jurassic caecilian with limbs. *Nature* 265:246–50.

Jenner, J. V. 1981. A zoogeographic study and the taxonomy of the xenodontine colubrid snakes. Ph.D. diss., New York University.

Jenner, J. V., and H. G. Dowling. 1985. Taxonomy of American xenodontine snakes: the tribe Pseudoboini. *Herpetologica* 41 (2): 161–72.

Jiménez, A., and J. M. Savage. 1962. A new blind snake (genus *Typhlops*) from Costa Rica. *Rev. Biol. Trop.* 10 (2): 199–203.

Jiménez, C. E. "1994." 1995. Utilization of *Puya dasylirioides* (Bromeliaceae: Pitcarinoidea) as foraging site by *Bolitoglossa subpalmata* (Plethodontidae: Bolitoglossini). *Rev. Biol. Trop.* 42 (3): 703–10.

Jiménez de la Espada, M. 1871. Faunae Neotropicalis species quaedum cognitae. *J. Sci. Math. Phys. Natl. Acad. Sci. Real* 3:57–65.

Jockush, E. L., and M. García-París. 1998. *Nototriton abscondens* (Cordilleran moss salamander): Reproduction. *Herp. Rev.* 29 (1): 38.

Johanbocke, M. M. 1974. Effects of a bite from *Conophis lineatus* (Squamata: Colubridae). *Bull. Phila. Herp. Soc.* 22:39.

Johnson, J. D. 1977. The taxonomy and distribution of the Neotropical whipsnake *Masticophis mentovarius* (Reptilia, Serpentes, Colubridae). *J. Herp.* 11 (3): 287–309.

———. 1982. *Maticophis mentovarius. Cat. Amer. Amph. Reptl.* 295:1–4.

Johnson, J. D., and J. R. Dixon. 1984. Taxonomic status of the Venezuelan macagua, *Bothrops colombiensis. J. Herp.* 18 (3): 329–33.

Jones, D. A. 1967. Green pigmentation in Neotropical frogs. Ph.D. diss., University of Florida.

Jungfer, K.-H. 1987. Beobachtungen an *Ololygon boulengeri* (Cope, 1887) und anderen "Knickzehen laubförschen." *Herpetofauna* 9 (46): 6–12.

———. 1996. Reproduction and parental care of the coronated treefrog, *Anotheca spinosa* (Steindachner, 1864) (Anura: Hylidae). *Herpetologica* 52 (1): 25–32.

Jungfer, K.-H., and L. C. Schiesari 1995. Description of a central Amazonian and Guianan tree frog, genus *Osteocephalus* (Anura, Hylidae), with oophagous larvae. *Alytes* 13 (1): 1–13.

Jungfer, K.-H., and P. Weygoldt. 1994. The reproductive biology of the leaf frog *Phyllomedusa lemur* Boulenger, 1882, and a comparison with other members of the Phyllomedusinae. *Rev. Fr. Aquariol.* 21 (1–2): 57–64.

Kaiser, H. 1992. The trade-mediated introduction of *Eleutherodactylus martinicensis* (Anura: Leptodactylidae) on St. Barthelemy, French Antilles, and its implications for Lesser Antillean biogeography. *J. Herp.* 26 (3): 264–73.

Kaiser, H., D. M. Green, and M. Schmid. 1994. Systematics and biogeography of eastern Caribbean frogs (Leptodactylidae: *Eleutherodactylus*), with the description of a new species from Dominica. *Canadian J. Zool.* 72:2217–37.

Kaiser, H., and J. D. Hardy Jr. 1994a. *Eleutherodactylus martinicensis. Cat. Amer. Amph. Reptl.* 582:1–4.

———. 1994b. *Eleutherodactylus johnstonei. Cat. Amer. Amph. Reptl.* 581:1–5.

Kaiser, H., C. Mais, F. Bolaños, C. Steinlein, W. Feichtinger, and M. Schmid. 1996. Chromosomal investigation of three Costa Rican frogs from the thirty-chromosome radiation of *Hyla* with description of a unique geographic variation in nucleolus organizer regions. *Genetica* 98 (1): 95–102.

Kamezaki, N., and M. Matsui. 1995. Geographic variation in skull morphology of the green turtle, *Chelonia mydas*, with a taxonomic decision. *J. Herp.* 29 (1): 51–60.

Kaplan, M. 1997. A new species of *Colostethus* from the Sierra Nevada de Santa Marta (Colombia) with comments on intergeneric relationships within the Dendrobatidae. *J. Herp.* 31 (3): 369–75.

Karl, S. A., and B. W. Bowen. 1999. Evolutionary significant units versus geopolitical taxonomy: Molecular systematics of an endangered sea turtle (genus *Chelonia*). *Cons. Biol.* 13 (5): 990–99.

Keferstein, W. 1867. Ueber einige neue oder seltene Batrachier aus Australien un dem tropischen Amerika. *Nachr. Gesell. Wissen. Göttingen* 18:341–63.

———. 1868a. Beschreibungeinigerneuen Batrachier aus Australia und Costa Rica. *Nachr. Gesell. Wissen. Göttingen* 15: 326–32.

———. 1868b. Ueber einige Batrachier aus Costa Rica. *Arch. Naturgesch.* 34 (1): 291–300.

Keiser, E. D., Jr. 1967. A monographic study of the Neotropical vine snake, *Oxybelis aeneus* (Wagler). Ph.D. diss., Louisiana State University.

———. 1969. Evidence of a dichromatic population of the vine snake *Oxybelis fulgidus* (Daudin) on the Islas de la Bahía of Honduras. *Carib. J. Sci.* 9 (1–2): 31–32.

———. 1974. A systematic study of the Neotropical vine snake *Oxybelis aeneus* (Wagler). *Bull. Texas Mem. Mus.* 22:1–51.

———. 1982. *Oxybelis aeneus. Cat. Amer. Amph. Reptl.* 305: 1–4.

———. 1989. *Oxybelis boulengeri* Proctor, a valid species of vine snake from South America. *Copeia* 1989 (3): 764–68.

Kerr, A. C., and M. A. Iturralde. 1999. New plate tectonic model of the Caribbean: Implications from a geochemical reconnaissance of Cuban Mesozoic volcanic rocks. *Geol. Soc. Amer. Bull.* 111 (11): 1581–99.

Kezer, J. 1964. Meiosis in salamander spermatocytes. In *The mechanism of inheritance*, ed. F. W. Stahl, 101–12. Englewood, N.J.: Prentice-Hall.

Kezer, J., S. K. Sessions, and P. León. 1989. The meiotic structure and behavior of the strongly heteromorphic X/Y sex chromosomes of Neotropical plethodontid salamanders of the genus *Oedipina. Chromosoma* 98:433–42.

Kiesecker, J. M., A. R. Blaustein, and L. K. Belden. 2001. Complex causes of amphibian declines. *Nature* 410:681–84.

Kim, Y. H., H. S. Mosher, F. A. Furman, and G. B. Brown. 1975. Tetrodotoxin: Occurrence in atelopoid frogs of Costa Rica. *Science* 189:151–52.

King, M. 1978. A new chromosome form of *Hemidactylus frenatus. Herpetologica* 34 (2): 216–18.

———. 1990. *Animal cytogenetics.* Vol. 4. *Chordata 2: Amphibia.* Berlin: Borntraeger.

Kingsbury, B. G., and H. D. Reed. 1909. The columella auris in Amphibia. *J. Morph.* 20:549–628.

Klauber, L. M. 1945. The geckos of the genus *Coleonyx* with descriptions of new subspecies. *Trans. San Diego Soc. Nat. Hist.* 10 (11): 133–216.

———. 1952. Taxonomic studies of the rattlesnakes of mainland Mexico. *Bull. Zool. Soc. San Diego* 26:1–143.

———. 1956. *Rattlesnakes: Their habits, life histories and influence on mankind.* 2 vols. Berkeley: University of California Press.

———. 1972. *Rattlesnakes: Their habits, life histories, and influence on mankind.* 2d ed. 2 vols. Berkeley: University of California Press.

Kley, N. J., and E. L. Brainard. 1999. Feeding by mandibular raking in a snake. *Nature* 402:369–70.

Kluge, A. G. 1967. Higher taxonomic categories of gekkonid lizards and their evolution. *Bull. Amer. Mus. Nat. Hist.* 135 (1): 1–60.

———. 1979. The gladiator frogs of Middle America: A reevaluation of their systematics (Anura: Hylidae). *Occas. Pap. Mus. Zool. Univ. Mich.* 688:1–24.

———. 1981. The life history, social organization and parental behavior of *Hyla rosenbergi* Boulenger, a nest-building gladiator frog. *Misc. Pub. Mus. Zool. Univ. Mich.* 160:1–170.

———. 1987. Cladistic relationships in the Gekkonoidea (Squamata, Sauria). *Misc. Pub. Mus. Zool. Univ. Mich.* 173:1–54.

———. 1991. Boine snake phylogeny and research cycles. *Misc. Pub. Mus. Zool. Univ. Mich.* 178:1–58.

———. 1995. Cladistic relationships of sphaerodactyl lizards. *Amer. Mus. Nat. Hist. Novitates* 3139:1–23.

Kluge, A. G., and M. J. Eckardt. 1969. *Hemidactylus garnotii* Dumeril and Bibron, a triploid all-female species of gekkonoid lizard. *Copeia* 1969 (4): 651–64.

Knight, A., L. D. Densmore, and E. P. Rael. 1992. Molecular systematics of the *Agkistrodon* complex. In *Biology of the pitvipers,* ed. J. A. Campbell and E. D. Brodie Jr., 49–70. Tyler, Tex.: Selva.

Kofron, C. P. 1985. Systematics of the Neotropical gastropod-eating snake genera, *Tropidodipsas* and *Sibon. J. Herp.* 19 (1): 84–92.

———. 1987. Systematics of Neotropical gastropod-eating snakes: The *fasciata* group of the genus *Sibon. J. Herp.* 21 (3): 210–25.

———. 1990. Systematics of Neotropical gastropod-eating snakes: The *dimidiata* group of the genus *Sibon,* with comments on the *nebulata* group. *Amph.-Reptl.* 11 (3): 207–23.

Köhler, G. 1993a. *Basilisken.* Hanau: Köhler.

———. 1993b. *Schwarze Leguane.* Hanau: Köhler.

———. 1995a. Freilanduntersuchengern zur Morphologie und Lebensweise des Funfkeil-Schwarzleguans *Ctenosaura quinquecarinata* am Isthmus von Tehuatepec, Mexiko. *Herpetofauna* 17 (97): 21–26.

———. 1995b. Freilanduntersunchungen zur Morphologie und Okologie von *Ctenosaura bakeri* und *C. oedirhina* auf den Islas de la Bahia, Honduras, mit Bemerkungen zur Schutzproblematik. *Salamandra* 31 (2): 93–106.

———. 1995c. *Ctenosaura palearis* Stejneger. *Sauria* 17 (3): 329–32.

———. 1995d. Eine neue Art der Gattung *Ctenosaura* (Sauria: Iguanidae) aus dem sudlichen Campeche, Mexico. *Salamandra* 31 (1): 1–14.

———. 1996. Freilanduntersuchungen zur Morphologie, Verbreitung und Lebensweise des Yucatán-Schwarzleguans *(Ctenosaura defensor). Salamandra* 32 (3): 153–62.

———. 1997. *Ungaliophis continentalis* (Isthmian dwarf boa). *Herp. Rev.* 28 (4): 211.

———. 1999a. The amphibians and reptiles of Nicaragua: A distributional checklist with keys. *Cour. Forsch. Senckenberg* 213:1–121.

———. 1999b. Eine neue Saumingerart der Gattung *Norops* von der Pazifikseite des nordlichen Mittelamerika. *Salamandra* 35 (1): 37–52.

Köhler, G., and K. Klemmer. 1994. Eine neue Schwarzleguanart der Gattung *Ctenosaua* aus La Paz, Honduras. *Salamandra* 30 (3): 197–208.

Köhler, G., and J. Kreutz. 1999. *Norops macrophallus* (Werner, 1917), a valid species of anole from Guatemala and El Salvador (Squamata: Sauria: Iguanidae). *Herpetozoa* 12 (1–2): 57–65.

Köhler, G., E. Lehr, and J. R. McCranie. 2000. The tadpole of the Central American toad *Bufo luetkinii* Boulenger. *J. Herp.* 34 (2): 303–6.

Köhler, G., D. Rittmann, and F. Ihringer. 1994. Pflege und Aucht des Helmsleguans *Corytophanes cristatus* (Merrem, 1821). *Iguana* 7 (13): 43–47.

Köhler, G., and B. Streit. 1996. Notes on the systematic status of the taxa *acanthura, pectinata,* and *similis* of the genus *Ctenosaura* (Reptilia: Sauria: Iguanidae). *Sencken. Biol.* 75 (1–2): 35–43.

Köhler, J., and W. Böhme. 1996. Anuran amphibians from the region of Pre-Cambrian rock outcrops (Inselbergs) in northeastern Bolivia, with a note on the gender of *Scinax* Wagler, 1830 (Hylidae). *Rev. Fr. Aquariol.* 23 (3–4): 133–40.

Köhler, J., and T. Ziegler. 1997. Amphibien und Reptilien aus Costa Rica. *DATZ* 1997 (4): 239–43.

Koller, R. M. 1996a. Herpetologische waarnemingen in Costa Rica. *Lacerta* 53 (3): 82–88.

———. 1996b. Herpetologische waarnemingen in Costa Rica (slot). *Lacerta* 54 (4): 111–19.

Krakauer, T. 1970. Tolorance limits of the toad, *Bufo marinus,* in south Florida. *Comp. Biochem. Physiol.* 33 (1): 15–16.

Kraus, F., D. G. Mink, and W. M. Brown. 1996. Crotaline intergeneric relationships based on mitochondrial DNA sequence data. *Copeia* 1996 (4): 763–73.

Krempels, D. M. 1989. "Visible light" and near infra-red reflectance of amphibians and reptiles and the visual system of avian predators (Accipiteridae: *Buteo* spp.). Ph.D. diss., University of Miami.

Kropach, C. 1972. *Pelamis platurus* as a potential colonizer of the Caribbean Sea. *Bull. Biol. Soc. Wash.* 22:67–69.

———. 1975. The yellow-bellied sea snake, *Pelamis,* in the eastern Pacific. In *The biology of sea snakes,* ed. W. A. Dunson, 185–213. Baltimore: University Park Press.

Krywicki, J. 2001. Sunlight in the rainforest. *Reptl. Mag.* 9 (3): 10–27.

Kuhl, H. 1820. *Beiträge zur Zoologie und vergleichende Anatomie.* Frankfurt am Main: Hermann.

Lacépède, B. G. E. L. 1788. *Histoire naturelle des quadrupèdes ovipares et des serpens.* Vol. 2. Paris: Imprimerie de Roi.

Lahanas, P. N. 1992. Historical biogeographic relationships of

Central America and South America: A biochemical, phylogenetic analysis of selected amphibians and reptiles. Ph.D. diss., University of Miami.

Lahanas, P. N., and J. M. Savage. 1992. A new species of caecilian from the Peninsula de Osa, Costa Rica. *Copeia* 1992 (3): 703–8.

La Marca, E. 1992. Catálogo taxonómico, biogeográfico y bibliográfico de las ranas de Venezuela. *Cuad. Geograf. Univ. Andes Venezuela* 1–197.

———. 1994a. Taxonomy of the frogs of the genus *Mannophryne* (Amphibia; Anura: Dendrobatidae). *Pub. Asoc. Amigos Donaña* 4:1–75.

———. 1994b. Descripción de un género nuevo de ranas (Amphibia: Dendrobatidae) de la Cordillera de Mérida, Venezuela. *Univ. Los Andes Inst. Geog. Conser. Recursos Nat. Anuario Invest.* 1991:39–41.

Lancini, A. R. 1979. *Serpientes de Venezuela.* Caracas: Armitano.

Landy, M. J., S. A. Langebartel, E. O. Moll, and H. M. Smith. 1966. A collection of snakes from Volcán Tacana, Chiapas, Mexico. *J. Ohio Herp. Soc.* 5 (3): 93–101.

Lang, C. 1995. Size-fecundity relationships among stream-breeding hylid frogs. *Herp. Nat. Hist.* 3 (2): 193–97.

Lang, M. 1989. Phylogenetic and biogeographic patterns of basiliscine iguanians (Reptilia: Squamata: "Iguanidae"). *Bonn. Zool. Monogr.* 28:1–172.

Langhammer, J. K. 1983. A new subspecies of boa constrictor, *Boa constrictor melanogaster,* from Ecuador. *Trop. Fish. Hob. Mag.* 32 (2): 70–79.

Lannoo, M. J., D. S. Townsend, and R. J. Wasserzug. 1987. Larval life in the leaves: Arboreal tadpole types, with special attention to the morphology, ecology, and behavior of the oophagous *Osteopilus brunneus. Fieldiana Zool.,* n.s., 38: 1–31.

Larsen, K. R., and W. W. Tanner. 1974. Numeric analysis of the lizard genus *Sceloporus* with special reference to cranial osteology. *Gt. Basin Nat.* 34 (1): 1–41.

Laurance, W. F., K. R. McDonald, and R. Speare. 1996. Catastrophic declines of Australian rainforest frogs: Is unusual weather responsible? *Biol. Cons.* 77:203–12.

Laurent, R. 1946. Mises au point dans la taxonomie de ranides. *Rev. Zool. Bot. Afr.* 39 (4): 336–38.

Laurenti, J. N. 1768. *Specimen medicum, exhibens synopsin reptilium emendatam cum experimentis circa venena et antodota reptilium austriacorum.* Vienna: Trattnern.

Lazcano, M. A., and E. Góngora. 1993. Observations and a review of the nesting and egg-laying of *Corytophanes cristatus* (Iguanidae). *Bull. Maryland Herp. Soc.* 29 (2): 67–75.

Lazell, J. D., Jr. 1964. The Lesser Antilles representatives of *Bothrops* and *Constrictor. Bull. Mus. Comp. Zool.* 132 (3): 245–73.

———. 1973. The lizard genus *Iguana* in the Lesser Antilles. *Bull. Mus. Comp. Zool.* 145 (1): 1–28.

Lee, J. C. 1980. Variation and systematics of the *Anolis sericeus* complex (Sauria: Iguanidae). *Copeia* 1980 (2): 310–20.

———. 1982. Accuracy and precision in anuran morphometrics: Artifacts of preservation. *Syst. Zool.* 31 (3): 266–81.

———. 1983. *Anolis sericeus. Cat. Amer. Amph. Reptl.* 340: 1–2.

———. 1996. *The amphibians and reptiles of the Yucatán Peninsula.* Ithaca: Cornell University Press.

Lee, M. S. Y. 1998. Convergent evolution and character correlation in burrowing reptiles: Towards a resolution of squamate relationships. *Biol. J. Linn. Soc. London.* 65:460–53.

Lee, M. S. Y., and M. W. Caldwell. 1998. Anatomy and relationships of *Pachyrhachis problematicus,* a primitive snake with hindlimbs. *Proc. R. Soc. London,* ser. B, 333:1521–52.

Leenders, T., G. Beckers, and H. Strijbosch. 1996. *Micrurus mipartitus* (NCN): Polymorphism. *Herp. Rev.* 27 (1): 25.

Legler, J. M. 1965. A new species of turtle, genus *Kinosternon,* from Central America. *Univ.Kansas Pub. Mus. Nat. Hist.* 15 (13): 615–25.

———. 1966. Notes on the natural history of a rare Central American turtle, *Kinosternon angustipons* Legler. *Herpetologica* 22 (2): 118–22.

———. 1990. The genus *Pseudemys* in Mesoamerica: Taxonomy, distribution, and origins. In *Life history and ecology of the slider turtle,* ed. J. W. Gibbons, 82–105. Washington, D.C: Smithsonian Institution Press.

Leigh, E. G., Jr., A. S. Rand, and D. M. Windsor, eds. 1982. *The ecology of a tropical forest: Seasonal rhythms and long-term changes.* Washington, D.C.: Smithsonian Institution Press.

León, J. R. 1969. The systematics of the frogs of the *Hyla rubra* group in Central America. *Univ.Kansas Pub. Mus. Nat. Hist.* 18 (6): 505–45.

Léon, P. "1969." 1970. Report of the chromosome numbers of some Costa Rican anurans. *Rev. Biol. Trop.* 17 (1): 119–24.

León, P., and J. Kezer. 1978. The localization of 5S RNA genes on chromosomes of plethodontid salamanders. *Chromosoma* 65:213–30.

Leviton, A. E., R. H. Gibbs Jr., E. Heal, and C. F. Dawson. 1985. Standards in herpetology and ichthyology. 1. Standard symbolic codes for institutional resource collections in herpetology and ichthyology. *Copeia* 1985 (3): 802–32.

Licht, L. E. 1967. Death following the possible ingestion of toad eggs. *Toxicon* 5 (2): 141–42.

———. 1968. Unpalatability and toxicity of toad eggs. *Herpetologica* 24 (2): 93–98.

Lichtenstein, H. M. C., and E. C. von Martens. 1856. *Nomenclator reptilium et amphibiorum Musei Zoologici Berolinensis.* Berlin: Akademie der Wissenschaften Berlin.

Lieb, C. S. 1981. Biochemical and karyological systematics of the Mexican lizards of the *Anolis gadovi* and *A. nebulosus* species groups (Reptilia: Iguanidae). Ph.D. diss., University of California, Los Angeles.

———. 1988. Systematic status of the Neotropical snakes *Dendrophidion dendrophis* and *D. nuchalis* (Colubridae). *Herpetologica* 44 (2): 162–75.

———. 1991a. *Dendrophidion nuchale. Cat. Amer. Amph. Reptl.* 520:1–2.

———. 1991b. *Dendrophidion vinitor. Cat. Amer. Amph. Reptl.* 522:1–2.

———. 1991c. *Dendrophidion paucicarinatum. Cat. Amer. Amph. Reptl.* 521:1–2.

———. 1996. *Dendrophidion percarinatum. Cat. Amer. Amph. Reptl.* 636:1–2.

Lieberman, S. S. 1986. Ecology of the leaf litter herpetofauna of a Neotropical rain forest La Selva, Costa Rica. *Acta Zool. Mex.*, n.s., 15:1–72.

Liem, K. F., H. Marx, and G. B. Rabb. 1971. The viperid snake *Azemiops:* Its comparative cephalic anatomy and phylogenetic position in relation to Viperinae and Crotalinae. *Fieldiana Zool.* 59 (2): 65–126.

Limbaugh, B. A., and E. P. Volpe. 1957. Early development of the Gulf Coast toad *Bufo valliceps* Wiegmann. *Amer. Mus. Nat. Hist. Novitates* 1842:1–32.

Limerick, S. 1976. Dietary differences of two sympatric Costa Rican frogs. M.S. thesis, University of Southern California.

———. 1980. Courtship behavior and oviposition of the poison-arrow frog *Dendrobates pumilio. Herpetologica* 36 (1): 69–71.

Lincoln, R., G. Boxshall, and P. Clark. 1987. *A dictionary of ecology, evolution and systematics.* Cambridge: Cambridge University Press.

———. 1998. *A dictionary of ecology, evolution and systematics.* 2d ed. Cambridge: Cambridge University Press.

Lindqkimuist, E. D. 1995. *Atelopus zeteki* (Panamanian golden frog): Pure tonal vocalization. *Herp. Rev.* 26 (4): 200–201.

Lindquist, E. D., and D. W. Swihart. 1997. *Atelopus chiriqiensis* (Chiriqui harlequin frog): Mating behavior and egg laying. *Herp. Rev.* 28 (3): 145–46.

Liner, E. A., and H. A. Dundee. 1974. Eggs and hatchlings of the tree snake *Leptophis depressirostris* (Cope). *Brenesia* 3: 11–24.

Linné, C. von. 1753. *Species plantarum.* Stockholm: Salvius.

———. 1758. *Systema naturae.* 10th ed. Vol. 1. Stockholm: Salvius.

———. 1766. *Systema naturae: Regnum animale.* 12th ed. Vol. 1. Stockholm: Salvius.

Lips, K. R. 1993a. *Bolitoglossa compacta. Herp. Rev.* 24 (3): 107.

———. 1993b. *Bolitoglossa minutula. Herp. Rev.* 24 (3): 107.

———. 1993c. *Bolitoglossa nigrescens. Herp. Rev.* 24 (3): 107.

———. 1993d. *Enulius sclateri. Herp. Rev.* 24 (3): 109.

———. 1993e. *Hyla lancasteri. Herp. Rev.* 24 (3): 108.

———. 1993f. *Oedipina grandis. Herp. Rev.* 24 (3): 107.

———. 1993g. *Urotheca myersi. Herp. Rev.* 24 (3): 111.

———. 1996a. The population biology of *Hyla calypsa,* a stream-breeding treefrog from lower Central America. Ph.D. diss., University of Miami.

———. 1996b. New treefrog from the Cordillera de Talamanca of Central America with a discussion of systematic relationships in the *Hyla lancasteri* group. *Copeia* 1996 (3): 615–26.

———. 1998. Decline of a tropical montane amphibian fauna. *Cons. Biol.* 12 (1): 106–17.

———. 1999. Mass mortality and population declines of anurans at an upland site in western Panama. *Cons. Biol.* 13 (1): 117–25.

Lips, K. R., and D. M. Krempels. 1995. Eggs and tadpoles of *Bufo fastidiosus* Cope, with comment on reproductive behavior. *Copeia* 1995 (3): 741–46.

Lips, K. R., and J. M. Savage. 1994. A new fossorial snake of the genus *Geophis* (Reptilia: Serpentes: Colubridae) from the Cordillera de Talamanca of Costa Rica. *Proc. Biol. Soc. Wash.* 107 (2): 410–16.

———. 1996a. Key to the known tadpoles (Amphibia: Anura) of Costa Rica. *Stud. Neotrop. Fauna Environ.* 31:17–26.

———. 1996b. A new species of rainfrog, *Eleutherodatylus phasma* (Anura: Leptodactylidae), from montane Costa Rica. *Proc. Biol. Soc. Wash.* 109 (4): 744–48.

Livezey, R. L. 1986. The eggs and tadpole of *Bufo coniferus* Cope in Costa Rica. *Rev. Biol. Trop.* 34 (2): 221–24.

Livezey, R. L., and A. H. Wright. 1947. A synoptic key to the salientian eggs of the United States. *Amer. Midl. Nat.* 37 (1): 179–222.

Lohmann, K. J., B. E. Witherington, C. M. F. Lohmann, and M. Salmon. 1997. Orientation, navigation, and natal beach homing in sea turtles. In *The biology of sea turtles,* ed. P. L. Lutz and J. L. Musick, 107–63. Boca Raton, Fla.: CRC Press.

Lombard, R. E., and D. B. Wake. 1987. Tongue evolution in the lungless salamanders, family Plethodontidae. 4. Phylogeny of plethodontid salamanders and the evolution of feeding dynamics. *Evolution* 41 (1): 532–51.

Lomonte, B. 1990. The serum of neonate *Clelia clelia* (Serpentes: Colubridae) neutralizes the haemorrhagic action of *Bothrops asper* venom (Serpentes: Viperidae). *Rev. Biol. Trop.* 38 (2A): 325–26.

Lomonte, B., L. Cerdas, J. A. Gene, and J. M. Gutiérrez. 1982. Neutralization of local effects of the terciopelo *(Bothrops asper)* venom by blood serum of the colubrid snake *Clelia clelia. Toxicon* 20 (3): 571–79.

Lönnberg, E. 1896. Linnean type-specimens of birds, reptiles, batrachians and fishes in the Zoological Museum of the Royal University of Uppsalà. *K. Svenska Vet.-Akad. Handl.* 22 (4): 1–45.

López, M. A., R. C. Vogt, and M. A. de la Torre. 1999. A new species of montane pitviper from Veracruz, Mexico. *Herpetologica* 55 (3): 382–89.

Losos, J. B. 1992. The evolution of convergent structure in Caribbena *Anolis* communities. *Syst. Biol.* 41 (4): 403–20.

Losos, J. B., R. M. Andrews, O. J. Sexton, and A. L. Schuler. 1991. Behavior, ecology, and locomotor performance of the giant anole, *Anolis frenatus. Carib. J. Sci.* 27 (3–4): 173–79.

Lötters. S. 1996. *The Neotropical toad genus* Atelopus. Cologne: Veneces und Glaw.

Lowe, C. H., Jr., J. W. Wright, C. J. Cole, and R. L. Bezy. 1970. Chromosomes and evolution of the species groups of *Cnemidophorus* (Reptilia: Teiidae). *Syst. Zool.* 19 (2): 128–41.

Lucas, S. G. 1986. Pyrothere systematics and a Caribbean route for land-mammal dispersal during the Paleocene. *Rev. Geol. Amer. Central.* 5:1–35.

Lutz, B. 1973. *Brazilian species of* Hyla. Austin: University of Texas Press.

Lutz, P. L., and T. B. Bentley. 1985. Respiratory physiology of diving in the sea turtle. *Copeia* 1985 (3): 671–79.

Lutz, P. L., and J. A. Musick, eds. 1997. *The biology of sea turtles.* Boca Raton, Fla.: CRC Press.

Lynch, J. D. 1970. Systematic status of the American leptodactylid frog genera *Engystomops, Eupemphix,* and *Physalaemus. Copeia* 1970 (3): 488–96.

———. 1971. Evolutionary relationships, osteology and zoogeography of leptodactylid frogs. *Misc. Pub. Mus. Nat. Hist. Univ. Kansas.* 53:1–238.

———. 1976. The species groups of the South American frogs

of the genus *Eleutherodactylus* (Leptodactylidae). *Occas. Pap. Mus. Nat. Hist. Univ. Kansas.* 61:1–24.

———. 1978. A re-assessment of the telmatobiine leptodactylid frogs of Patagonia. *Occas. Pap. Mus. Zool. Univ. Kansas.* 72:1–57.

———. 1980. Systematic status and distribution of some poorly known frogs of the genus *Eleutherodactylus* from the Chocoan lowlands of South America. *Herpetologica* 36 (2): 175–89.

———. 1985. Mimetic and non-mimetic populations of *Eleutherodactylus gaigeae* (Dunn) in lower Central America and Colombia (Amphibia: Anura, Leptodactylidae). *Stud. Neotrop. Fauna Eviron.* 20 (4): 195–202.

———. 1986. The definition of the Middle American clade of *Eleutherodactylus* based on jaw musculature (Amphibia: Leptodactylidae). *Herpetologica* 42 (2): 248–58.

———. 1993. The value of the m. depressor mandibulae in phylogenetic hypotheses for *Eleutherodactylus* and its allies (Amphibia: Leptodactylidae). *Herpetologica* 49 (1): 32–41.

———. 1996. Replacement names for three homonyms in the genus *Eleutherodactylus*. *J. Herp.* 30 (2): 278–80.

——— 2000. The relationships of an ensemble of Guatemalan and Mexican frogs (*Eleutherodactylus*: Leptodactylidae: Amphibia). *Rev. Acad. Colomb. Cien. Exact. Fisic. Nat.* 24 (90): 129–56.

Lynch, J. D., and W. E. Duellman. 1997. Frogs of the genus *Eleutherodactylus* from western Ecuador. *Spec. Pub. Mus. Nat. Hist. Univ. Kansas.* 23:1–236.

Lynch, J. D., and C. W. Myers. 1983. Frogs of the *fitzingeri* group of *Eleutherodactylus* in eastern Panama and Chocoan South America (Leptodactylidae). *Bull. Amer. Mus. Nat. Hist.* 15 (5): 481–572.

Lynch, J. D., and P. M. Ruiz. 1996. A remarkable new centrolenid frog from Colombia with a review of nuptial excrescences in the family. *Herpetologica* 52 (4): 525–35.

Lynch, J. D., and H. M. Smith. 1966. A new toad from western Mexico. *SW Nat.* 11 (1): 19–23.

Macartney, J. 1802. *Lectures on comparative anatomy, translated from the French of G. Cuvier by William Ross, under the inspection of James Macartney*. London.

McBee, K., J. W. Sites Jr., M. D. Engstrom, C. Blanco-Rivero, and J. W. Bickham. 1984. Karyotypes of four species of Neotropical gekkos. *J. Herp.* 18 (1): 83–85.

MacClean, W. P. 1974. Feeding and locomotor mechanisms of teiid lizards: Functional morphology and evolution. *Pap. Avul. Zool. Mus. Zool. Univ. São Paulo* 27 (15): 179–213.

McCoy, C. J., Jr. 1990. Additions to the herpetofauna of Belize, Central America. *Carib. J. Sci.* 26 (3–4): 164–66.

McCoy, C. J., and T. P. Maslin. 1962. A review of the teiid lizard *Cnemidophorus cozumelus* and the recognition of a new race, *Cnemidophorus cozumelus rodecki*. *Copeia* 1962 (3): 620–27.

McCranie, J. R. 1980. *Drymarchon, D. corais*. *Cat. Amer. Amph. Reptl.* 267:1–4.

———. 1993. *Crotalus durissus*. *Cat. Amer. Amph. Reptl.* 577: 1–11.

McCranie, J. R., and J. D. Villa. 1993. A new genus for the snake *Enulius sclateri* (Colubridae: Xenodontinae). *Amph.-Reptl.* 14:261–67.

McCranie, J. R., and L. D. Wilson. 1992. *Rhadinaea godmani*. *Cat. Amer. Amph. Reptl.* 546:1–3.

———. 1995. Two new species of colubrid snakes of the genus *Ninia* from Central America. *J. Herp.* 29 (2): 224–32.

———. 1996. A new arboreal lizard of the genus *Celestus* (Squamata: Anguidae) from northern Honduras. *Rev. Biol. Trop.* 44 (1): 259–64.

———. 2002. *The amphibians of Honduras*. Ithaca: Society for the Study of Amphibians and Reptiles.

McDade, L. A., K. S. Bawa, H. A. Hespenheide, and G. S. Hartshorn, eds. 1994. *La Selva: Ecology and natural history of a Neotropical rainforest*. Chicago: University of Chicago Press.

McDiarmid, R. W. 1968. Populational variation in the frog genus *Phrynohyas* Fitzinger in Middle America. *Contr. Sci. Nat. Hist. Mus. Los Angeles Co.* 134:1–25.

———. 1975. Glass frog romance along a tropical stream. *Terra* 13 (4): 14–18.

———. 1978. Evolution of parental care in frogs. In *The development of behavior: Comparative and evolutionary aspects*, ed. G. M. Burghardt and M. Bekoff, 127–47. New York: Garland STPM Press.

———. 1994. Appendix 4: Preparing amphibians as scientific specimens. In *Measuring and monitoring biological diversity: Standard methods for amphibians*, ed. W. R. Heyer, M. A. Donnelly, R. W. McDiarmid, L.-A. Hayek, and M. S. Foster, 289–97. Washington, D.C.: Smithsonian Institution Press.

McDiarmid, R. W., and K. Adler. 1974. Notes on territorial and vocal behavior of Neotropical frogs of the genus *Centrolenella*. *Herpetologica* 30 (1): 75–78.

McDiarmid, R. W., and R. Altig. 1990. Description of a bufonid and two hylid tadpoles from western Ecuador. *Alytes* 8 (2): 51–60.

———, eds. 1999a. *Tadpoles: The biology of anuran larvae*. Chicago: University of Chicago Press.

———. 1999b. Research. In *Tadpoles: The biology of anuran larvae*, ed. R. W. McDiarmid and R. Altig, 7–23. Chicago: University of Chicago Press.

McDiarmid, R. W., and R. L. Bezy. 1971. The colubrid snake *Enulius oligostichus* in western Mexico. *Copeia* 1971 (2): 350–51.

McDiarmid, R. W., J. A. Campbell, and T. Toure. 1999. *Snake species of the world: A taxonomic and geographic reference*. Vol. 1. Washington, D.C.: Herpetologists' League.

McDiarmid, R. W., and J. E. DeWeese. 1977. The systematic status of the lizard *Bachia blairi* (Dunn) 1940 (Reptilia: Teiidae) and its occurrence in Costa Rica. *Brenesia* 12–13:143–53.

McDiarmid, R. W., and M. S. Foster. 1981. Breeding habits of the toad *Bufo coccifer* in Costa Rica, with description of the tadpole. *SW Nat.* 16 (4): 353–63.

———. 1987. Cocoon formation in another hylid frog, *Smilisca baudinii*. *J. Herp.* 21 (4): 352–55.

McDiarmid, R. W., T. Toure, and J. M. Savage. 1996. The proper name of the Neotropical tree boa often referred to as *Corallus enydris* (Serpentes: Boidae). *J. Herp.* 30 (3): 320–26.

McDowell, S. B. 1964. Partition of the genus *Clemmys* and related problems in the taxonomy of the aquatic Testudinidae. *Proc. Zool. Soc. London.* 143:239–79.

———. 1967. *Aspidomorphus*, a genus of New Guinea snakes of the family Elapidae, with notes on related genera. *J. Zool.* 151:497–43.

———. 1969a. *Toxicocalamus*, a New Guinea genus of snakes of the family Elapidae. *J. Zool.* 159:443–511.

———. 1969b. Notes on the Australian sea-snake *Ephalophis greyi* M. Smith (Serpentes: Elapidae, Hydrophiinae) and the origin and classification of sea-snakes. *Zool. J. Linn. Soc. London.* 48:33–49.

———. 1970. On the status and relationships of the Solomon Island elapid snakes. *J. Zool.* 161:145–90.

———. 1972. The genera of sea-snakes of the *Hydrophis* group (Serpentes: Elapidae). *Trans. Zool. Soc. London.* 32:189–247.

———. 1987. Systematics. In *Snakes: Ecology and evolutionary biology*, ed. R. A. Seigel, J. T. Collins, and S. B. Novak, 3–50. New York: Macmillan.

Machado, S. R. 1993. A new genus of Amazonian vine snake (Xenodontinae: Alsophiini). *Acta. Biol. Lepold.* 15 (2): 99–108.

McGregor, G. R., and S. Nieuwolt. 1998. *Tropical climatology: An introducton to climates of the low latitudes.* 2d ed. Chichester, Eng.: Wiley.

McKinstry, D. M. 1983. Morphologic evidence of toxic saliva in colubrid snakes: A checklist of world genera. *Herp. Rev.* 14 (1): 12–15.

Mader, D. R., ed. 1996. *Reptile medicine and surgery.* Philadelphia: Saunders.

Maglio, V. J. 1970. West Indian xenodontine colubrid snakes: Their probable origin, phylogeny and zoogeography. *Bull. Mus. Comp. Zool.* 141 (1): 1–53.

Makino, S., and E. Momma 1949. An idiogram study of chromosomes in some species of reptiles. *Cytologia* 15:96–108.

Mann, P., ed. 1995. Geologic and tectonic development of the Caribbean plate boundary in southern Central America. *Geol. Soc. Amer. Spec. Pap.* 295:1–349.

Marcellini, D. L. 1971. Activity patterns of the gecko *Hemidactylus frenatus. Copeia* 1971 (4): 631–35.

———. 1974. Acoustic behavior of the gekkonid lizard, *Hemidactylus frenatus. Herpetologica* 30 (1): 44–52.

———. 1977. The function of the vocal display of the lizard *Hemidactylis frenatus* (Sauria: Gekkonidae). *Anim. Behav.* 25 (2): 414–17.

Marcgarf, G. 1648. *Historia naturalis brasilae.* Amsterdam.

March, D. D. H. 1929. Notes on *Bothrops nummifera*, mano de piedra or timbo. *Bull. Antivenin Inst. Amer.* 3:37–29.

Margaris, N., S. Arianoutsou-Faraggitaki, and R. J. Reiter, eds. 1983. *Plant, animal and microbial adaptations to terrestrial environment.* New York: Plenum.

Marion, K. R., and O. J. Sexton. 1971. The reproductive cycle of the lizard *Sceloporus malachiticus* in Costa Rica. *Copeia* 1971 (3): 517–26.

Marquis, R. J., M. A. Donnelly, and C. Guyer. 1986. Aggregations of calling males of *Agalychnis calcarifer* Boulenger (Anura: Hylidae) in a Costa Rican lowland wet forest. *Biotropica* 18 (2): 173–75.

Marshall, L. G., and T. Sempere. 1993. Evolution of the Neotropical Cenozoic land mammal fauna in its geochronologic, stratigraphic, and tectonic context. In *Biological relation-*

ships between Africa and South America, ed. P. Goldblatt, 329–92. New Haven: Yale University Press.

Marshall, L. G., S. D. Webb, J. J. Sepkoski, and D. M. Raup. 1982. Mammalian evolution and the great American interchange. *Science* 215:1351–57.

Martin, E. A., M. Ruse, and E. Holmes, eds. 1996. *Dictionary of biology.* 3d ed. Oxford: Oxford University Press.

Martínez, S., and L. Cerdas. 1986. Captive reproduction of the mussurana *Clelia clelia* (Daudin) from Costa Rica. *Herp. Rev.* 17 (1): 12.

Martínez, V. M. 1983. Panamá: Nuevo ámbito de distribución para la serpiente venenosa *Bothrops picadoi* (Dunn). *ConCiencia* 10 (1): 26–27.

Master, T. L. 1998. *Dendrobates auratus* (Black-and-green poison dart frog): Predation. *Herp. Rev.* 29 (3): 164–65.

Mather, C. M., and J. W. Sites Jr. 1985. *Sceloporus variabilis* Wiegmann. *Cat. Amer. Amph. Reptl.* 373:1–3.

Mather, J. R. 1974. *Climatological fundamentals and applications.* New York: McGraw-Hill.

Mattoon, A. 2001. Deciphering amphibian declines. In *State of the world 2001*, ed. L. R. Brown, C. Flavin, and H. French, 63–82. New York: Norton.

Maturana, H. R. 1962. A study of the species of the genus *Basiliscus. Bull. Mus. Comp. Zool.* 128 (1): 1–34.

Mayr, E. 1942. *Systematics and the origin of species.* New York: Columbia University Press.

Mayr, E., and P. D. Ashlock. 1991. *Principles of systematic zoology.* New York: McGraw-Hill.

Medem, F. 1962. La distribución geográfica y ecología de los Crocodylia y Testudinata en el Departamento del Chocó. *Rev. Acad. Colomb. Cien. Exact. Fisic. Nat.* 11 (44): 279–303.

———. 1977. Contribución al conocimiento sobre la taxonomía, distribución geográfica y ecología de la tortuga "bache" *(Chelydra serpentina acutirostris). Caldasia* 12 (56): 41–101.

———. 1981. *Los Crocodylia de Sur America.* Vol 1. *Los Crocodylia de Colombia.* Bogotá: Fondo Colombiano de Investigaciones Científicas y Proyectos Especiales "Francisco José de Caldas."

———. 1983. *Los Crocodylia de Sur America.* Vol. 2. *Venezuela-Trinidad-Tobago-Guyana-Suriname-Guayana Francesa-Ecuador-Peru-Bolivia-Brasil-Paraguay-Argentina-Uruguay.* Bogotá: Universidad Nacional Colombia/Fondo Colombiano Investigaciones Científicas y Proyectos Especiales "Francisco José Caldas."

Melville, R. V. 1995. *Towards stability in the names of animals: A history of the International Commission on Zoological Nomenclature, 1895–1995.* London: International Trust for Zoological Nomenclature.

Mendelson, J. R., III 1992. Frequency of tail breakage in *Coniophanes fissidens* (Serpentes: Colubridae). *Herpetologica* 48 (4): 448–55.

Mengden, G. A. 1985. Australian elapid phylogeny: A summary of the chromosomal and electrophoretic data. In *Biology of Australasian frogs and reptiles*, ed. G. Grigg, R. Shine, and H. Ehmann, 185–92. Chipping Norton, NSW: Surrey Beatty.

Merahtzakis, G. 1988. *Ungaliophis panamensis. Herp. Rev.* 19 (3): 60.

Merrem, B. 1820. *Versuch eienes Systems der Amphibien: "Tentamen systematis amphibiorum."* Marburg: Krieger.

———. "1821." 1822. Tentamen systematis amphibiorum. *Isis von Oken* 15:688–704.

Merriam, C. H. 1890. Results of a biological survey of the San Francisco Mountain region and the desert of the Little Colorado, Arizona. *N. Amer. Fauna* 3:1–113.

Mertens, R. 1942. Eine neue Schlanknatter der Gattung *Leptophis* aus Mexico. *Sencken. Biol.* 53 (5–6): 341–42.

———. 1943. Bemerkenswerts Schlanknatter de neotropischen gattung *Leptophis*. *Stud. Neotrop. Fauna Environ.* 8:141–54.

———. 1952. On snail-eating snakes. *Copeia* 1952 (4): 279.

Meyer, E. 1992. Erfolgreiche Nachzucht von *Dendrobates granuliferus* Taylor, 1958. *Herpetofauna* 14 (76): 11–21.

———. 1993. Fortpflanzung und Brutpflegeverhalten von *Dendrobates granuliferus* Taylor, 1958 aus Costa Rica (Amphibia: Anura: Dendrobatidae). *Veroff. Naturhisdt. Mus. Schleusingen* 1993 (7–8): 113–42.

———. 1996. Eine oliv-gelbe Variante con *Dendrobates granuliferus* aus dem zentral-pazifischen Tiefland Costa Rica: Erste Beobachtungen zur Fortpflanzungsbiologie. *Herpetofauna* 18 (100): 21–27.

Meyer, F. A. A. 1795. *Synopsis reptilium, novam ipsorum sistens generum methodum, nec non Gottingensium huis ordinis animalium enumerationem.* Göttingen: Vandenhoek et Ruptrecht.

Meyer, J. R. 1966. Records and observations on some amphibians and reptiles from Honduras. *Herpetologica* 22 (3): 172–81.

Meyer, J. R., and C. F. Foster. 1996. *A guide to the frogs and toads of Belize.* Malabar, Fla.: Krieger.

Meyer, J. R., and L. D. Wilson. 1971a. A distributional checklist of the amphibians of Honduras. *Contr. Sci. Nat. Hist. Mus. Los Angeles Co.* 218:1–47.

———. 1971b. Taxonomic studies and notes on some Honduran amphibians and reptiles. *Bull. South. Calif. Acad. Sci.* 70 (3): 106–14.

Meylan, A. 1984. Feeding ecology of the hawksbill turtle (*Eretmochelys imbricata*): Spongivory as a feeding niche in the coral reef community. Ph.D. diss., University of Florida.

———. 1988. Spongivory in hawksbill turtles: A diet of glass. *Science* 238:393–95.

———. 1990. Nutritional characteristics of sponges in the diet of the hawksbill turtle, *Eretmochelys imbricata*. In *New perspectives in sponge biology,* ed. K. Ruetzler, 472–77. Washington, D.C: Smithsonian Institution Press.

Michaud, E. J., and J. R. Dixon. 1989. Prey items of twenty species of the Neotropical colubrid snake genus *Liophis*. *Herp. Rev.* 20 (2): 39–41.

Milner, A. R. 1988. The relationships and origin of living amphibians. *Syst. Assoc. Spec.* 35A:59–102.

Minton, S. A., Jr., and M. R. Minton. 1991. *Masticophis mentovarius* (Neotropical whipsnake): Reproduction. *Herp. Rev.* 22 (4): 100–101.

Mittleman, M. B. 1952. A generic synopsis of lizards of the subfamily Lygosominae. *Smithson. Misc. Coll.* 117 (17): 1–35.

Miyamoto, M. M. 1983. Biochemical variation in the frog *Eleutherodactylus bransfordii*: Geographic patterns and cryptic species. *Syst. Zool.* 32 (1): 43–51.

Miyamoto, M. M., and J. H. Cane. 1980a. Notes on the reproductive behavior of a Costa Rican population of *Hyla ebraccata*. *Copeia* 1980 (4): 928–30.

———. 1980b. Behaviorial observations on noncalling males in Costa Rican *Hyla ebraccata*. *Biotropica* 12 (3): 225–27.

Mole, R. R. 1924. The Trinidad snakes. *Proc. Zool. Soc. London.* 1924:235–314.

Moll, D., and K. P. Jansen. 1995. Evidence for a role in seed dispersal by two tropical herbivorous turtles. *Biotropica* 27 (1): 121–27.

Moll, D., and E. O. Moll. 1990 The slider turtle in the Neotropics: Adaptation of a temperate species to a tropical environment. In *Life history and ecology of the slider turtle,* ed. J. W. Gibbons, 152–61. Washington, D.C: Smithsonian Institution Press.

Moll, E. O., and J. M. Legler. 1971. The life history of a Neotropical slider turtle *Pseudemys scripta* (Schoepff) in Panama. *Sci. Bull. Nat. Hist. Mus. Los Angeles Co.* 11:1–102.

Monge, J., B. Morera, and M. Chávez. 1988. Nesting behaviour of *Rhinoclemmys pulcherrima* in Costa Rica (Testudines: Emydidae). *Herp. J.* 1 (7): 308.

Moon, R. G. 1974. Heteromorphism in a kinosternid turtle. *Mammal. Chromo. Newsl.* 15: (1): 10–11.

Mora, J. M. 1987. Predation by *Loxocemus bicolor* on the eggs of *Ctenosaura similis* and *Iguana iguana*. *J. Herp.* 21 (4): 334–35.

———. 1989. Eco-behavioral aspects of two communally nesting iguanines and the structure of their shared nesting burrows. *Herpetologica* 45 (3): 293–98.

———. 1991. *Loxocemus bicolor* (burrowing python): Feeding behavior. *Herp. Rev.* 22 (2): 61.

Mora, J. M., and D. C. Robinson. 1984. Predation on sea turtle eggs (*Lepidochelys*) by the snake *Loxocemus*. *Rev. Biol. Trop.* 32 (1): 161–62.

Morafka, D. J. 1977. A biogeographical analysis of the Chihuahuan Desert through its herpetofauna. *Biogeographica* 9: 1–313.

Morales, S. A., and R. C. Vogt. 1997. Terrestrial movements in relation to aestivation and the annual reproductive cycle of *Kinosternon leucostumum*. *Copeia* 1997 (1): 123–30.

Morescalchi, A. 1968. Initial cytotaxonomic data on certain families of amphibious Anura (Diplasiocoela, after Noble). *Experientia* 24 (3): 964–66.

———. 1973. Amphibia. In *Cytotaxonomy and vertebrate evolution,* ed. A. B. Chiarelli and E. Capanna, 233–348. London: Academic Press.

———. 1975. Chromosome evolution in the caudate Amphibia. *Evol. Biol.* 8:339–87.

Morgan, E. C. 1973. Snakes of the subfamily Sibynophinae. Ph.D. diss., University of Southwestern Louisiana.

Moritz, C., and D. King. 1985. Cytogenetic perspectives on parthenogenesis in the Gekkonidae. In *Biology of Australasian frogs and reptiles,* ed. G. Grigg, R. Shine, and H. Ehmann, 327–37. Chipping Norton, NSW: Surrey Beatty.

Morreale, S. J., G. J. Ruiz, J. R. Spotila, and E. A. Standora. 1982. Temperature-dependent sex determination: Current practices threaten conservation of sea turtles. *Science* 216: 1245–47.

Mortimer, J. A. 1981. Feeding ecology of sea turtles. In *Biology and conservation of sea turtles*, ed. K. A. Bjorndal, 103–16. Washington, D.C: Smithsonian Institution Press.

Mrosovsky, N. 1983. *Conserving sea turtles*. London: British Herpetological Society.

———. 1987. Leatherback turtles off the scale? *Nature* 327:286.

Mrosovsky, N., P. H. Dutton, and C. P. Whitmore. 1984. Sex ratios of two species of sea turtle nesting in Suriname. *Canadian J. Zool.* 26:2227–39.

Mrosovsky, N., J. Lescure, J. Fretey, C. Pieau, and F. Rimblot. 1985. Sexual differentiation as a function of the incubation temperature of eggs in the sea-turtle *Dermochelys coriacea* (Vandelli, 1761). *Amph.-Rept.* 6 (1): 83–92.

Muedeking, M. H., and W. R. Heyer. 1976. Descriptions of eggs and reproductive patterns of *Leptodactylus pentadactylus* (Amphibia: Leptodactylidae). *Herpetologica* 32 (2): 137–39.

Müller, F. 1880. Erster Nachtrag zum Katalog der herpetologischen Sammlung des Basler Museums. *Verh. Naturf. Gesell. Basel.* 7:120–65 (preprint).

Muñoz, F. 1987. Autourotomy in two genera of Colubridae. *Abst. Soc. Stud Amph. Rept./Herp. League Annl. Mtg.* 124.

Murphy, J. B., and L. A. Mitchell. 1984. Miscellaneous notes on the reproductive biology of reptiles. 6. Thirteen varieties of the genus *Bothrops* (Serpentes, Crotalidae). *Acta Zool. Path. Antverpien.* 78:199–214.

Murphy, J. B., D. G. Barker, and B. W. Tyson. 1978. Miscellaneous notes on reproductive biology of reptiles. 2. Eleven species of the family Boidae, genera *Candoia, Corallus, Epicrates* and *Python. J. Herp.* 12 (3): 385–90.

Myers, C. W. 1966. A new species of colubrid snake genus *Coniophanes,* from Darien, Panama. *Copeia* 1966 (4): 665–68.

———. 1969a. Ecological geography of cloud forest in Panama. *Amer. Mus. Nat. Hist. Novitates* 2396:1–52.

———. 1969b. Snakes of the genus *Coniophanes* in Panama. *Amer. Mus. Nat. Hist. Novitates* 2372:1–28.

———. 1971a. A new species of green anole (Reptilia: Sauria) from the north coast of Veraguas, Panama. *Amer. Mus. Nat. Hist. Novitates* 2470:1–14.

———. 1971b. Central American lizards related to *Anolis pentaprion:* Two new species from the Cordillera de Talamanca. *Amer. Mus. Nat. Hist. Novitates* 2471:1–40.

———. 1973. Anguid lizards of the genus *Diploglossus* in Panama, with the description of a new species. *Amer. Mus. Nat. Hist. Novitates* 2523:1–20.

———. 1974. The systematics of *Rhadinaea* (Colubridae), a genus of New World snakes. *Bull. Amer. Mus. Nat. Hist.* 153 (1): 1–262.

———. 1982a. Spotted poison frogs: Descriptions of three new *Dendrobates* from western Amazonia, and resurrection of a lost species from "Chiriqui." *Amer. Mus. Nat. Hist. Novitates* 2721:1–23.

———. 1982b. Blunt-headed vine snakes *(Imantodes)* in Panama, including a new species and other revisionary notes. *Amer. Mus. Nat. Hist. Novitates* 2738:1–50.

———. 1986. An enigmatic new snake from the Peruvian Andes, with notes on the Xenodontini (Colubridae: Xenodontinae). *Amer. Mus. Nat. Hist. Novitates* 2853:1–12.

———. 1987. New generic names for some Neotropical poison frogs (Dendrobatidae). *Paps. Avul. Zool. Mus. Zool. Univ. São Paulo* 36 (25): 301–6.

———. 2002. Revision of *Trimetopon,* a genus of diminutive snakes endemic to Costa Rica and Panama. Manuscript.

Myers, C. W., and J. E. Cadle. 1994. A new genus for South American snakes related to *Rhadinaea obtusa* Cope (Colubridae) and resurrection of *Taeniophallus* Cope for the *"Rhadinaea brevirostris"* group. *Amer. Mus. Nat. Hist. Novitates* 3102:1–33.

Myers, C. W., and J. A. Campbell. 1981. A new genus and species of colubrid snake from the Sierra Madre del Sur of Guerrero, Mexico. *Amer. Mus. Nat. Hist. Novitates* 2708:1–20.

Myers, C. W., and J. W. Daly. 1976. Preliminary evaluation of skin toxins and vocalization in taxonomic and evolutionary studies of poison-dart frogs (Dendrobatidae). *Bull. Amer. Mus. Nat. Hist.* 157 (3): 193–262.

———. 1980. Taxonomy and ecology of *Dendrobates bombetes,* a new Andean poison frog with new skin toxins. *Amer. Mus. Nat. Hist. Novitates* 2692:1–23.

———. 1983. Dart-poison frogs. *Sci. Amer.* 248 (2): 120–33.

Myers, C. W., J. W. Daly, H. M. Garraffo, A. Wisnieski, and J. F. Cover Jr. 1995. Discovery of the Costa Rican poison frog *Dendrobates granuliferus* in sympatry with *Dendrobates pumilio,* and comments on taxonomic use of skin alkaloids. *Amer. Mus. Nat. Hist. Novitates* 3144:1–21.

Myers, C. W., J. W. Daly, and B. Malkin. 1978. A dangerously toxic new frog *(Phyllobates)* used by Embera Indians of western Colombia, with discussion of blowgun fabrication and dart poisoning. *Bull. Amer. Mus. Nat. Hist.* 161 (2): 311–65.

Myers, C. W., J. W. Daly, and S. Martínez. 1984. An arboreal poison frog *(Dendrobates)* from western Panama. *Amer. Mus. Nat. Hist. Novitates* 2783:1–20.

Myers, C. W., and M. A. Donnelly. 1991. The lizard genus *Sphenomorphus* (Scincidae) in Panama, with description of a new species. *Amer. Mus. Nat. Hist. Novitates* 3027:1–12.

Myers, C. W., and W. E. Duellman. 1982. A new species of *Hyla* from Cerro Colorado, and other tree frog records and geographical notes from western Panama. *Amer. Mus. Nat. Hist. Novitates* 2752:1–32.

Myers, C. W., A. Paolillo, and J. W. Daly. 1991. Discovery of a defensively malodorous and nocturnal frog in the family Dendrobatidae: Phylogenetic significance of a new genus and species from the Venezuelan Andes. *Amer. Mus. Nat. Hist. Novitates* 3002:1–33.

Myers, C. W., and L. Trueb. 1967. The hemipenis of an anomalepid snake. *Herpetologica* 23 (3): 235–38.

Nadkarni, N. M., and N. T. Wheelwright, eds. 2000. *Monteverde: Ecology and conservation of a tropical cloud forest.* New York: Oxford University Press.

Nakamura, K. 1949. A study in some chelonians with notes on chromosomal formula in the Chelonia. *Kromosomo* 5: 205–13.

Neill, W. T. 1960. The caudal lure of various juvenile snakes. *Quart. J. Fla. Acad. Sci.* 23:173–200.

————. 1962. The reproductive cycle of snakes in a tropical region, British Honduras. *Quart. J. Fla. Acad. Sci.* 25 (3): 234–53.

Neill, W. T., and E. R. Allen. 1959. Studies on the amphibians and reptiles of British Honduras. *Pub. Res. Div. Ross Allen's Rept. Inst.* 2 (1): 1–76.

————. 1960. Noteworthy snakes from British Honduras. *Herpetologica* 16 (3): 146–62.

Nelson, C. E. 1966a. The evolution of the family Microhylidae. Ph.D. diss., University of Texas.

————. 1966b. Systematics and distribution of snakes of the Central American genus *Hydromorphus* (Colubridae). *Texas J. Sci.* 18 (4): 365–71.

————. 1972. Systematic studies of the North American microhylid genus *Gastrophryne. J. Herp.* 6 (2): 111–37.

————. 1973. *Gastrophryne pictiventris. Cat. Amer. Amph. Reptl.* 135: 1–2.

————. 1974. Further studies on the systematics of *Hypopachus* (Anura: Microhylidae). *Herpetologica* 30 (3): 250–75.

Nelson, C. E., and J. W. Meyer. 1967. Variation and distribution of the Middle American snake genus *Loxocemus* Cope (Boidae?). *SW Nat.* 12 (4): 439–53.

Nieuwolt, S. 1977. *Tropical climatology.* New York: Wiley.

Noble, G. K. 1916. Description of a new eublepharid lizard from Costa Rica. *Proc. Biol. Soc. Wash.* 29: 87–88.

————. 1918. The amphibians collected by the American Museum expedition to Nicaragua in 1916. *Bull. Amer. Mus. Nat. Hist.* 38 (10): 311–47.

————. 1920. Two new batrachians from Colombia. *Bull. Amer. Mus. Nat. Hist.* 42 (9): 441–46.

————. 1924. Some Neotropical batrachians preserved in the United States National Museum with a note on the secondary sexual characters of these and other amphibians. *Proc. Biol. Soc. Wash.* 37: 65–71.

————. 1927. The value of life history data in the study of the evolution of the Amphibia. *Ann. N.Y. Acad. Sci.* 30 (2): 31–128.

————. 1931. *The biology of the Amphibia.* New York: McGraw-Hill.

Nopsca, F. 1923. *Eidolosaurus* und *Pacchyophis*: Zwei neue Neocom-Reptilien. *Palaentographica* 65: 97–154.

Novak, R. M., and D. C. Robinson. 1975. Observations on the reproduction and ecology of the tropical montane toad, *Bufo holdridgei* Taylor in Costa Rica. *Rev. Biol. Trop.* 23 (2): 213–37.

Nunes, V. S. 1988. Vocalizations of treefrogs *(Smilisca sila)* in response to bat predation. *Herpetologica* 44 (1): 8–10.

Nussbaum, R. A. 1988. On the status of *Copeotyphlinus syntremus, Gymnopis oligozona,* and *Minascaecilia sartoria* (Gymnophioma, Caeciilidae): A comedy of errors. *Copeia* 1988 (4): 921–28.

Nussbaum, R. A., and M. Wilkinson. 1989. On the classification and phylogeny of caecilians (Amphibia: Gymnophiona). *Herp. Monogr.* 3: 1–42.

Oftedal, O. T. 1974. A revision of the genus *Anadia* (Sauria: Teiidae). *Arq. Zool. Mus. Zool. Univ. São Paulo* 25 (4): 203–65.

Oken, L. 1817. Encyclopadische Zietung. *Isis von Oken* 21: 779–82.

Oliver, J. A. 1942. A check list of the snakes of the genus *Leptophis,* with descriptions of new forms. *Occas. Pap. Mus. Zool. Univ. Mich.* 462: 1–19.

————. 1948. The relationships and zoogeography of the genus *Thalerophis* Oliver. *Bull. Amer. Mus. Nat. Hist.* 92 (4): 157–280.

————. 1951. "Gliding" in amphibians and reptiles, with a remark on an arboreal adaptation in the lizard *Anolis carolinensis carolinensis* Voigt. *Amer. Nat.* 85: 171–76.

Olson, E. C. 1947. The family Diadectidae and its bearing on the classification of reptiles. *Fieldiana Geol.,* n.s., 11: 2–53.

Oppel, M. "1810." 1811. Mémoire sur la classification des reptiles. Ordre 2. Reptiles à écailles. 2. Ophidiens: Suite du Ier. *Ann. Mus. Hist. Nat. Paris.* 16: 376–93.

————. 1811. *Die Ordung, Familien und Gattungen der Reptilien als Prodrom einer Naturgeschichte derselben.* Munich: Lindauer.

Orcés, G., and J. C. Matheus P. 1988. Acerca de la distribución geográfica de algunas especies de la herpetofauna del Ecuador. *Rev. Mus. Ecuator. Cien. Nat.* 8 (6): 83–85.

O'Reilly, J. C., D. A. Ritter, and D. R. Carrier. 1997. Hydrostatic locomotion in a limbless tetrapod. *Nature* 386: 269–72.

Orgeix, C. A. d', and B. L. Turner. 1995. Multiple paternity in the red-eyed treefrog *Agalychnis callidryas* (Cope). *Mol. Ecol.* 4: 505–8.

Ortega, F., R. L. Sedlock, and R. C. Speed. 1994. Phanerozoic tectonic evolution of Mexico. In *Phanerozoic evolution of North American continent-ocean transitions: DNAG continent-ocean transect volume,* ed. R. C. Speed, 265–306. Boulder, Colo.: Geological Society of America.

Ortenberger, A. I. 1928. The whip snakes and racers: Genera *Masticophis* and *Coluber. Mem. Univ. Mich. Mus.* 1: 1–247.

Orton, G. L. 1943. The tadpole of *Rhinophrynus dorsalis. Occas. Pap. Mus. Zool. Univ. Mich.* 472: 1–7.

————. 1952. Key to the genera of tadpoles in the United States and Canada. *Amer. Midl. Nat.* 47 (2): 383–95.

Osborn, H. F. 1930. Biographical memoir of Edward Drinker Cope, 1840–1897. *Biograph. Monogr. Natl. Acad. Sci.* 13 (3): 127–317.

————. 1931. *Cope: Master naturalist.* Princeton: Princeton University Press.

O'Shaughnessy, A. W. E. 1874. Descriptions of new species of Scincidae in the collection of the British Museum. *Ann. Mag. Nat. Hist.,* ser. 4, 13: 298: 301.

Otte, D., and J. A. Endler, eds. 1989. *Speciation and its consequences.* Sunderland, Mass.: Sinauer.

Oviedo, G. F. de. 1535. *La historia general y natural de las Indias.* Seville.

Packard, M. J., G. C. Packard, and T. J. Boardman. 1982. Structure of eggshells and water relations of reptilian eggs. *Herpetologica* 38 (1): 136–55.

Papenfuss, T. J., and D. B. Wake. 1987. Two new species of plethodontid salamanders (genus *Nototriton*) from Mexico. *Acta Zool. Mex.,* n.s., 21: 1–16.

Parker, H. W. 1934. *A monograph of the frogs of the family Microhylidae.* London: British Museum (Natural History).

Parkinson, C. L., K. R. Zamudio, and H. W. Greene. 2000. Phylogeography of the pitviper clade *Agkistrodon:* Historical ecology, species status, and conservation of cantils. *Mol. Evol.* 9:411–20.

Paull, D., E. E. Williams, and W. P. Hall. 1976. Lizard karyotypes from the Galápagos Islands: Chromosomes in phylogeny and evolution. *Brevoria* 441:1–31.

Pavelka, L. A., Y. H. Kim, and H. S. Mosher. 1977. Tetrodotoxin and tetrodotoxin-like compounds from the eggs of the Costa Rican frog, *Atelopus chiriquiensis. Toxicon* 15 (2): 135–39.

Peccinini-Seale, D. M. 1989. Genetic studies on the bisexual and unisexual populations of *Cnemidophorus lemniscatus* (Sauria; Teiidae). *N.Y. State Mus. Bull.* 466:241–51.

Peccinini-Seale, D. M., and O. Frota-Pessoa. 1974. Structural heterozygosity in parthenogenetic populations of *Cnemidophorus lemniscatus* (Sauria, Teiidae) from the Amazonas Valley. *Chromosoma* 47:439–51.

Pechmann, J. H. K., and H. M. Wilbur. 1994. Putting declining amphibian populations in perspective: Natural fluctuations and human impacts. *Herpetologica* 50 (1): 65–84.

Pelletier, S. W., ed. 1986 *Alkaloids: Chemical and biological perspectives.* Vol. 4. New York: Wiley.

Peracca, M. G. 1897. Viaggio del Dr. Enrico Festa nell'Ecuador e regioni vicine. 4. Rettili. *Boll. Mus. Zool. Comp. Anat. Univ. Torino* 12 (300): 1–20.

Pérez-Higareda, G., H. M. Smith, and D. Chiszar. 1997. *Anolis pentaprion* (lichen anole): Frugivory and cannibalism. *Herp. Rev.* 28 (4): 201–2.

Pérez-Santos, C., and A. G. Moreno. 1988. Ofidios de Colombia. *Monogr. Mus. Reg. Sci. Nat. Torino* 6:1–517.

———. 1991. Serpientes de Ecuador. *Monogr. Mus. Reg. Sci. Nat. Torino* 11:1–538.

Pérez-Santos, C., A. G. Moreno, and A. Gearhart. 1993. Checklist of the snakes of Panama. *Rev. España Herp.* 7: 113–22.

Perfit, M. R., and E. E. Williams. 1989. Geological constraints and biological retrodictions in the evolution of the Caribbean sea and its islands. In *Biogeography of the West Indies: Past, present and future,* ed. C. A. Woods, 47–102. Gainesville, Fla.: Sandhill Crane Press.

Peters, J. A. 1957. Taxonomic notes on Ecuadorian snakes in the American Museum of Natural History. *Amer. Mus. Nat. Hist. Novitates* 1851:1–13.

———. 1960. The snakes of the subfamily Dipsadinae. *Misc. Pub. Mus. Zool. Univ. Mich.* 114:1–224.

———. 1964. *Dictionary of herpetology.* New York: Hofner.

Peters, J. A., and R. Donoso-Barros. 1970. Catalogue of the Neotropical Squamata. 2. Lizards and amphisbaenians. *Bull. U.S. Natl. Mus.* 297:1–293.

Peters, J. A., and B. Orejas-Miranda. 1970. Catalogue of the Neotropical Squamata: 1. Snakes. *Bull. U.S. Natl. Mus.* 297: 1–347.

Peters, W. C. H. 1857. Vier neue amerikanische Schlangen aus der Familie Typhlopinen vor und daruber einige vorläufige Mittheilungen. *Mber. Akad. Wissen. Berlin* 1857:402–3.

———. 1859. Über die von Hrn. Dr. Hoffmann in Costa Rica gesammelten und an das königlische zoologische Museum gesandten Schlangen. *Mber. Akad. Wissen. Berlin.* 1859: 275–78.

———. 1860. Drei neue Schlangen des königlischen zoologischen Museums aus America vor un fugte hieran Bermerkungen uber die generalle Unterscheidung von anderen bereits bekannten Arten. *Mber. Akad. Wissen. Berlin* 1860:517–21.

———. 1861. Über neue Schlangen des königlischen zoologischen Museums: *Typhlops striatus, Geophidium dubium, Streptophorus (Ninia) maculatus, Elaps hippocrepis. Mber. Akad. Wissen. Berlin* 1861:922–25.

———. 1862a. Uber die Batrachier-Gattung *Hemiphractus. Mber. Akad. Wissen. Berlin* 1862:144–52.

———. 1862b. Einen Vortrag uber die craniologischen Verschiedenheiten der Grubenotten *(Trigonocephalus)* und uber eine neue Art der Gattung *Bothriechis. Mber. Akad. Wissen. Berlin* 1862:670–74.

———. 1863a. Über einen neuen Gecko, *Brachydactylus mitratus* aus Costa Rica. *Mber. Akad. Wissen. Berlin* 1863:41–44.

———. 1863b. Über einige Arten der Saurier-Gattung *Anolis. Mber. Akad. Wissen. Berlin* 1863:135–49.

———. 1863c. Über einige neue oder weniger bekannte Schlangearten des zoologischen Musuems zu Berlin. *Mber. Akad. Wissen. Berlin* 1863:272–89.

———. 1863d. Über eine neue Schlangengattung, *Styporhynchus,* und verschiedene andere Amphibiens des zoologischen Museums. *Mber. Akad. Wissen. Berlin* 1863:399–413.

———. 1863e. Fernere Mitteilungen über neue Batrachier. *Mber. Akad. Wissen. Berlin* 1863:445–70.

———. 1868. Über eine neue Nagergattung, *Chiropodomys penicillatus,* so wie über einige oder weniger bekannte Amphibien und Fische. *Mber. Akad. Wissen. Berlin* 1868:448–60.

———. 1872. Über die von Spix in Brasilien gesammelten Batrachier des königlischen Naturalienkabinets zu München. *Mber. Akad. Wissen. Berlin* 1872:196–227.

———. 1873a. Über eine neue Schildkrötenart, *Cinsostornum Effeldtii* und einige andere neue weinger bekannte Amphibien. *Mber. Akad. Wissen. Berlin* 1873:603–18.

———. "1873b." 1874. Eine Mitteilung über neue Saurier *(Spaeriodactylus, Anolis, Phrynosoma, Tropidolepisma, Lygosoma, Ophioscincus)* aus Centralamerica, Mexico und Australien. *Mber. Akad. Wissen. Berlin* 1873:738–47.

———. 1874. Über neue amphibien *(Gymnopis, Siphonops, Polypedates, Hyla, Cyclodus, Euprepes, Clemmys). Mber. Akad. Wissen. Berlin* 1874:616–24.

———. 1876. Über neue Arten der Sauriergattung *Gerrhonotus. Mber. Akad. Wissen. Berlin* 1876:297–300.

———. "1879." 1880. Über die Eintheilung der Caecilien und insbesondere über der Gattung *Rhinatrema* und *Gymnopis. Mber. Akad. Wissen. Berlin* 1879:924–43.

———. 1881. Einige Herpetologische Mittheilungen. 1. Uebersicht der zu den Familien der Typhlopes und Stenostomi gehorigen Gattungen oder Untergattungen. *Sber. Gesell. Naturf. Freunde Berlin* 1881 (4): 69–71.

Peterson, J. A., and E. E. Williams 1981. A case history of retrograde evolution: The *onca* lineage in anoline lizards. 2. Subdigital fine structure. *Bull. Mus. Comp. Zool.* 149 (4): 215–68.

Pfennig, D. W., W. R. Harcombe, and K. S. Pfennig. 2001. Frequency-dependent Batesian mimicry. *Nature* 410:323.

Picado, C. 1913. Les broméliacées épiphytes considérées comme milieu biologique. *Bull. Sci. France Belg.* 47 (7): 215–360.

———. 1931. *Serpientes venenosas de Costa Rica.* San José: Sección Salud Pública Costa Rica.

———. 1976 *Serpientes venenosas de Costa Rica.* Reprint. San José: Editorial de la Universidad de Costa Rica.

Pickwell, G. V. 1972. The venomous sea snakes. *Fauna* 1 (4): 17–32.

Pickwell, G. V., and W. A. Culotta. 1980. *Pelamis, P. platurus. Cat. Amer. Amph. Reptl.* 255:1–4.

Pindell, J. L., and S. F. Barrett. 1990. Geological evolution of the Caribbean region: A plate-tectonic perspective. In *The Caribbean region,* vol. H of *The geology of North America,* ed. G. Dengo and J. E. Case, 405–32. Boulder, Colo.: Geological Society of America.

Pinder, A. W., K. B. Storey, and G. R. Ultsch. 1992. Estivation and hibernation. In *Environmental physiology of the amphibians,* ed. M. E. Feder and W. W. Burggren, 250–74. Chicago: University of Chicago Press.

Pino, E. M. del. 1980. Morphology of the pouch and incubatory integument in marsupial frogs. *Copeia* 1980 (1): 10–17.

Pinou, T., and H. G. Dowling. 1994. The phylogenetic relationships of the Central American snake *Tretanorhinus:* Data from morphology and karyology. *Amph.-Reptl.* 15:297–305.

Piveteau, J., ed. 1955. *Traité de paléontologie.* Paris: Masson.

Poe, S. "1996." 1997. Data incongruence and the phylogeny of crocodilians. *Syst. Biol.* 45 (4): 393–414.

———. 1998. Skull characters and the cladistic relationships of the Hispaniolian dwarf twig *Anolis. Herp. Monogr.* 12: 192–236.

Pombal, J. P., Jr., and M. Gordo. 1991. Duas novas espécies de *Hyla* da floresta Atlântica no Estado de São Paulo (Amphibia, Anura). *Mem. Inst. Butantan* 53 (1): 135–44.

Porras, L., J. R. McCranie, and L. D. Wilson. 1981. The systematics and distribution of the hog-nosed viper *Bothrops nasuta* Bocourt (Serpentes: Viperidae). *Tulane Stud. Zool. Botan.* 22 (2): 85–107.

Porter, K. R. 1962. Mating calls and noteworthy collections of some Mexican amphibians. *Herpetologica* 18 (3): 165–71.

———. 1965. Intraspecific variation in mating call in *Bufo coccifer. Amer. Midl. Nat.* 74 (2): 359–56.

———. 1966. Mating calls of six Mexican and Central American toads (genus *Bufo*). *Herpetologica* 22 (1): 60–67.

———. 1970. *Bufo valliceps. Cat. Amer. Amph. Reptl.* 94: 1–4.

Portig, W. H. 1965. Central American rainfall. *Geogr. Rev.* 55:68–90.

Pough, F. H. 1964. A coral snake "mimic" eaten by a bird. *Copeia* 1964 (1): 223.

———. 1973. Lizard energetics and diet. *Ecology* 54 (4): 837–44.

———. 1976. Multiple cryptic effects of crossbanded and ringed patterns of snakes. *Copeia* 1976 (4): 834–36.

———. 1988. Mimicry and related phenomena. In *Biology of the Reptilia,* vol. 16, *Ecology B, Defense and life history,* ed. C. Gans and R. B. Huey, 153–234. New York: Liss.

Pounds, J. A. 2000. Amphibians and reptiles. In *Monteverde: Ecology and conservation of a tropical cloud forest,* ed. N. M. Nadkarni and W. T. Wheelwright, 147–77. New York: Oxford University Press.

———. 2001. Climate and amphibian declines. *Nature* 410: 639–40.

Pounds, J. A., and M. L. Crump. 1987. Harlequin frogs along a tropical montane stream: Aggregation and the risk of predation by frog-eating flies. *Biotropica* 19 (4): 306–9.

———. 1994. Amphibian declines and climate disturbance: The case of the golden toad and the harlequin frog. *Cons. Biol.* 8 (1): 72–85.

Pounds, J. A., and M. P. L. Fogden. 2000. Appendix 8: Amphibians and reptiles of Monteverde. In *Monteverde: Ecology and conservation of a tropical cloud forest,* ed. N. M. Nadkarni and N. T. Wheelwright, 537–40. New York: Oxford University Press.

Pounds, J. A., M. P. L. Fogden, and J. H. Campbell. 1999. Biological response to climate change on a tropical mountain. *Nature* 398:611–15.

Pounds, J. A., M. P. L. Fogden, J. M. Savage, and G. C. Gorman. 1997. Tests of null models for amphibian declines on a tropical mountain. *Cons. Biol.* 11 (6): 1307–22.

Presch, W. F. 1974. Evolutionary relationships and biogeography of the macroteiid lizards (family Teiidae, subfamily Teiinae). *Bull. South. Calif. Acad. Sci.* 73 (1): 23–32.

———. 1978. Descriptions of the hemipenial morphology in eight species of microteiid lizards (family Teiidae, subfamily Gymnophthalminae). *Herpetologica* 34 (1): 108–12.

———. 1980. Evolutionary history of the South American microteiid lizards (Teiidae: Gymnophthalminae). *Copeia* 1980 (1): 36–56.

———. 1983. The lizard family Teiidae: Is it a monophyletic group? *Zool. J. Linn. Soc. London.* 77:189–97.

Price, R. M. 1991. *Sentcolis, S. triaspis. Cat. Amer. Amph. Reptl.* 525:1–4.

Price, R. M., and P. Russo. 1991. Revisionary comments on the genus *Boa* with the description of a new subspecies of *Boa constrictor* from Peru. *Snake* 23:29–35.

Pritchard, P. C. H. 1979. *Encyclopedia of turtles.* Neptune City, N.J.: Tropical Fish Hobbyist.

———. 1980. *Dermochelys coriacea. Cat. Amer. Amph. Reptl.* 238:1–4.

———. 1983. Conserving sea turtles, by N. Mrosovsky (book review). *Copeia* 1983 (4): 1108–11.

———. 1999. Status of the black turtle. *Cons. Biol.* 13 (5): 1000–1003.

Pritchard, P. C. H., and P. Trebbau. 1984. *The turtles of Venezuela.* Contributions to Herpetology 2. Oxford, Ohio: Society for the Study of Amphibians and Reptiles.

Pröhl, H. 1997a. *Los anfibios de Hitoy Cerere Costa Rica.* San José: Talamanca.

———. 1997b. Territorial behaviour of the strawberry poison-dart frog, *Dendrobates pumilio. Amph.-Reptl.* 18 (4): 437–42.

Proy, C. 1992. Zur Biologie von *Agalychnis saltator* Taylor, 1955 (Anura: Hylidae). *Herpetozoa* 5 (3–4): 99–107.

———. 1993a. Beobachtungen zur Biologie und Erfahrungen bei der Haltung und Nachzucht von *Agalychnis annae* (Duellman, 1963). *Herpetofauna* 15 (84): 27–34.

———. 1993b. Erste Terrarienbeobachtungen zur Fortpflanzung von *Hyla legleri* Taylor, 1958 aus Costa Rica (Anura: Hylidae). *Herpetozoa* 6 (3–4): 105–11.

Punzo, F. 1974. Comparative analysis of the feeding habits of two species of Arizona blind snakes, *Leptotyphlops h. humilis* and *Leptotyphlops d. dulcis. Herpetologica* 8 (2): 153–56.

Pyburn, W. F. 1963. Observations on the life history of the tree frog, *Phyllomedusa callidryas* (Cope). *Texas J. Sci.* 15 (2): 155–70.

———. 1964. Breeding behavior of the leaf-frog, *Phyllomedusa callidryas*, in southern Veracruz. *Yrbk. Amer. Phil. Soc.* 1964:291–94.

———. 1967. Breeding and larval development of the hylid frog *Phrynohyas spilomma* in southern Veracruz, Mexico. *Herpetologica* 23 (3): 184–94.

———. 1970. Breeding behavior of the leaf-frogs *Phyllomedusa callidryas* and *Phyllomedusa dacnicolor* in Mexico. *Copeia* 1970 (2): 209–18.

Quay, W. B. 1979. The parietal eye-pineal complex. In *Biology of the Reptilia*, vol. 9, *Neurology B*, ed. C. Gans, R. G. Northcutt, and P. Ulinski, 245–406. London: Academic Press.

Queiroz, K. de. 1987a. Phylogenetic systematics of iguanine lizards: A comparative osteological study. *Univ. Calif. Pub. Zool.* 118:1–203.

———. 1987b. A new spiny-tailed iguana from Honduras, with comments on the relationships within *Ctenosaura* (Squamata: Iguania). *Copeia* 1987 (4): 892–902.

Rabor, D. S. 1952. Preliminary notes on the giant toad, *Bufo marinus* (Linn.), in the Philippine Islands. *Copeia* 1952 (4): 281–82.

Raddi, G. "1822." 1823. Di alcune specie nuovi di rettilie piante brasiliane. *Atti Soc. Ital. Sci. Modena* 19:56–73.

Rafinesque, C. S. 1814. Prodrome di herpetologia siciliana. *Palermo Specch. Sci. Gio. Encicl. Siciliana* 2:65–68.

———. 1815. *Analyse de la nature, ou Tableau de l'universe et des corps organisés*. Palermo: Privately printed.

Rage, J.-C. 1978. Une connexcion continentale entre Amérique du Nord et Amérique du Sud au Crétacé supérieur? L'example des vertébrés continentaux. *C.R. Soc. Geol. France* 1978 (6): 281–85.

———. 1981. Les continents péri-atlantiques au Crétacé supérieur: Migrations des faunes continentales et problèmes paléogéographiques. *Cretaceous Res.* 2:65–84.

———. 1986. South American/North American terrestrial interchanges in the latest Cretaceous: Short comments on Brett-Surman and Paul (1985). *J. Vert. Paleont.* 63:82–83.

———. 1995. La Tethys et les dispersions transtethysiennes por voie terrestre. *Biogeographica* 71 (3): 109–26.

Rage, J.-C., and A. Richter. 1994. A snake from the Lower Cretaceous (Barremian) of Spain: The oldest known snake. *N. Jb. Geol. Palaeont. Mh.* 9:561–65.

Rameríz, J., R. C. Vogt, and J.-L. Villarreal. 1998. Population biology of a Neotropical frog *(Rana vaillanti). J. Herp.* 32 (3): 338–44.

Rand, A. S. 1969. *Leptophis ahaetulla* eggs. *Copeia* 1969 (2): 402.

———. 1983. *Physalaemus pustulosus* (rana sapito, túngra, foam frog, mud-puddle frog). In *Costa Rican natural history*, ed. D. H. Janzen, 412–15. Chicago: University of Chicago Press.

Rand, A. S., and B. A. Dugan. 1983. Structure of complex iguana nests. *Copeia* 1983 (3): 705–11.

Rand, A. S., B. A. Dugan, H. Monteza, and D. Vianda. 1990. The diet of a generalized folivore: *Iguana iguana* in Panama. *J. Herp.* 24 (4): 211–14.

Rand, A. S., and C. W. Myers. 1990. The herpetofauna of Barro Colorado Island, Panama: An ecological summary. In *Four Neotropical rainforests*, ed. A. H. Gentry, 386–409. New Haven: Yale University Press.

Rand, A. S., and M. H. Robinson. 1969. Predation on iguana nests. *Herpetologica* 25 (3): 172–74.

Rasotto, M. B., P. Cardellini, and M. Sala. 1987. Karyotypes of five species of Dendrobatidae (Anura: Amphibia). *Herpetologica* 43 (2): 177–82.

Ream, C. H. 1965. Notes on the behavior and egg laying of *Corytophanes cristatus. Herpetologica* 20 (4): 239–42.

Rebouças-Spieker, R. 1981a. Sobre uma nova espécie de *Mabuya* do nordeste do Brasil. *Pap. Avul. Zool. Mus. Zool. Univ. São Paulo* 39 (9): 121–23.

———. 1981b. Sobre uma nova espécie de *Mabuya* da Amazônia brasileira (Sauria, Scincidae. *Pap. Avul. Zool. Mus. Zool. Univ. São Paulo* 4 (16): 161–63.

Reeder, T. W. 1990. *Eumeces managuae. Cat. Amer. Amph. Reptl.* 467:1–2.

Reid, J. R., and T. E. Lott. 1963. Feeding of *Leptotyphlops dulcis dulcis* (Baird and Girard). *Herpetologica* 19 (2): 141–42.

Rendahl, H., and G. Vestergren. 1940. Notes on Colombian snakes. *Arkiv Zool.* 33A (1): 1–16.

———. 1941. On a small collection of snakes from Ecuador. *Arkiv Zool.* 33A (5): 1–16.

Rhodin, A. G. J., and K. Miyata, eds. 1983. *Advances in herpetology and evolution: Essays in honor of Ernest E. Williams*. Cambridge: Museum of Comparative Zoology, Harvard University.

Ricklefs, R. E., and D. Schluter, eds. 1993. *Species diversity in ecological communities*. Chicago: University of Chicago Press.

Ridgeway, R. 1922. Some observations on the natural history of Costa Rica. *Annl. Rept. Smithson. Inst.* 1921:303–24.

Riehl, H. 1979. *Tropical meterology*. New York: McGraw-Hill.

Rieppel, O. 1988. A review of the origin of snakes. *Evol. Biol.* 22 (2): 37–130.

Riley, J., and J. M. Winch. 1985. *Tripanurgos compressus* (Trinidad pseudofalse coral snake): Eggs. *Herp. Rev.* 16 (1): 29.

Rivero, J. A. 1961. Salientia of Venezuela. *Bull. Mus. Comp. Zool.* 126 (1): 1–207.

———. "1988." 1990. Sobre las relaciones de las especies del género *Colostethus* (Amphibia, Dendrobatidae). *Mem. Soc. Cien. Nat. La Salle* 48 (118): 9–16.

Robb, J. 1960. The internal anatomy of *Typhlops* Schneider (Reptilia). *Austral. J. Zool.* 8 (2): 181–216.

Roberts, B. K., M. G. Aronsohn, B. L Moses, R. L. Burk, J. Toll, and F. R. Weeren. 2000. *Bufo marinus* intoxication in dogs:

Ninety-four cases (1997–1998). *J. Amer. Vet. Med. Assoc.* 216 (12): 1941–44.

Roberts, W. E. 1994a. Evolution and ecology of arboreal egg-laying frogs. Ph.D. diss., University of California, Berkeley.

———. 1994b. Explosive breeding aggregations and parachuting in a Neotropical frog, *Agalychnis saltator* (Hylidae). *J. Herp.* 28 (2): 193–99.

———. 1995. *Agalychnis calcarifer* (NCN): Reproduction. *Herp. Rev.* 26 (4): 200.

———. 1997. Behaviorial observations of *Polychrus gutturosus*, a sister taxon of anoles. *Herp. Rev.* 28 (4): 184–85.

Robinson, D. C. 1961. The identity of the tadpole of *Anotheca coronata* (Stejneger). *Copeia* 1961 (4): 495.

———. 1976. A new dwarf salamander of the genus *Bolitoglossa* (Plethdontidae) from Costa Rica. *Proc. Biol. Soc. Wash.* 89 (2): 289–94.

———. 1977. Herpetofauna bromelicola costarricense y renacujaos de *Hyla picadoi*. *Hist. Nat. Costa Rica Mus. Nac. Costa Rica* 1: 31–43.

———. 1983a. *Rana palmipes* (rana, web-footed frog). In *Costa Rican natural history*, ed. D. H. Janzen, 415–16. Chicago: University of Chicago Press.

———. 1983b. *Sceloporus malachiticus* (lagartija espinosa, spiny lizard). In *Costa Rican natural history*, ed. D. H. Janzen, 421–22. Chicago: University of Chicago Press.

Rodrigues, M. T. 1991. Herpetofauna das dunas interiores do Rio São Francisco, Bahia, Brasil. 3. *Procellosaurinus:* um novo gênero de microteiídeos sem pálpebra, com redefinição do gênero *Gymnophthalmus* (Sauria, Teiidae). *Pap. Avul. Zool. Mus. Zool. Univ. São Paulo* 37 (21): 329–42.

Rokosky, E. J. 1941. Notes on new-born jumping vipers, *Bothrops nummifera*. *Copeia* 1941 (4): 267.

Roldán, J. 1985. *Caracterizacion de la herpetofauna de la Reserva Kuna Yala: Informe final.* Panama: Smithsonian Tropical Research Institute/Escuela Biología Universidad de Panamá.

Romer, A. S. 1945. *Vertebrate paleontology.* 2d ed. Chicago: University of Chicago Press.

Rose, F. L., and W. G. Weaver Jr. 1966. Two new species of *Chrysemys* (= *Pseudemys*) from the Florida Pliocene. *Tulane Stud. Geol.* 5 (1): 41–48.

Rosen, D. E. "1978." 1979. Vicariant patterns and historical explanation in biogeography. *Syst. Zool.* 27 (2): 159–88.

Rossman, D. A. 1963. The colubrid snake genus *Thamnophis:* A revision of the *sauritus* group. *Bull. Fla. State Mus.* 7 (3): 99–178.

———. 1970. *Thamnophis proximus. Cat. Amer. Amph. Reptl.* 98: 1–3.

———. 1971. Systematics of the Neotropical populations of *Thamnophis marcianus* (Serpentes: Colubridae). *Occas. Pap. Mus. Zool. Louis. State Univ.* 41: 1–13.

Rossman, D. A., N. B. Ford, and R. A. Seigel. 1996. *The garter snakes: Evolution and ecology.* Norman: University of Oklahoma Press.

Rossman, D. A., and N. J. Scott Jr. 1968. Identity of *Helicops wettsteini* Amaral (Serpentes: Colubridae). *Herpetologica* 24 (3): 262–63.

Roze, J. A. 1959. Una nueva especies del género *Drymarchon* (Serpentes: Colubridae) de la Isla de Margarita, Venezuela. *Noved. Cien. Mus. Hist. Nat. La Salle Ser. Zool.* 25: 1–4.

———. 1966. *La taxonomia y zoogeografía de los ofidios de Venezuela.* Caracas: Universidad Central de Venezuela.

———. 1967. A checklist of the New World venomous coral snakes (Elapidae) with descriptions of new forms. *Amer. Mus. Nat. Hist. Novitates* 2287: 1–60.

———. 1970. *Micrurus.* In Catalogue of the Neotopical Squamata, 1, Snakes, ed. J. A. Peters and B. Orejas-Miranda. *Bull. U.S. Natl. Mus.* 297: 196–220.

———. "1982." 1984. New World coral snakes (Elapidae): A taxonomic and biological summary. *Mem. Inst. Butantan* 46: 305–38.

———. 1996. *Coral snakes of the Americas: Biology, identification, and venoms.* Malabar, Fla.: Kreiger.

Ruetzler, K., ed. 1990. *New perspectives in sponge biology.* Washington, D.C.: Smithsonian Institution Press.

Ruibal, R. 1952. Revisionary studies of some South American teiids. *Bull. Mus. Comp. Zool.* 106 (11): 477–529.

Ruiz, P. M., M. C. Ardilla, and J. D. Lynch. 1996. Lista actualizada de la fauna de Amphibia de Colombia. *Rev. Acad. Colomb. Cien. Exact. Fisic. Nat.* 22 (77): 365–415.

Ruiz, P. M., and J. D. Lynch. 1991. Ranas Centrolenidae de Colombia. 1. Propuesta de una nueva clasificación genérica. *Lozania (Acta Zool. Colomb.)* 57: 1–30.

———. 1995. Ranas Centrolenidae de Colombia. 5. Cuatro nuevas especies de *Cochranella* de la Cordillera Central. *Lozania (Acta Zool. Colomb.)* 62: 1–23.

Rüppell, E. W. P. S. 1845. Verzeichniss der in dem Museum der seckenbergischen Gesellschaft aufgestellen Sammlung. Amphibien. *Verh. Mus. Sencken.* 3: 293–316.

Russell, F. E. 1983. *Snake venom poisoning.* Reprint. Great Neck, N.Y.: Scholium International.

Russell, F. E., F. G. Walter, T. A. Bey, and M. C. Fernández. 1997. Snakes and snakebite in Central America. *Toxicon* 35 (10): 1469–1522.

Ruthven, A., and H. T. Gaige. 1924. A new *Leposoma* from Panama. *Occas. Pap. Mus. Zool. Univ. Mich.* 147: 1–3.

Ryan, M. J. 1985. *The túngara frog: A study in sexual selection and communication.* Chicago: University of Chicago Press.

Ryan, M. J., and M. D. Tuttle. 1981. Bat predation and the evolution of frog vocalizations in the Neotropics. *Science* 214: 677–78.

Ryan, M. J., M. D. Tuttle, and A. S. Rand. 1982. Bat predation and sexual advertisement in a Neotropical anuran. *Amer. Nat.* 119 (1): 136–39.

Sagra, R. de la. 1838–43. *Historia física, politica y natural de la Isla de Cuba.* 12 vols. Paris: Libreria de Arthus Bertrand/ Lib. Soc. Geogr. (Spanish edition).

Salomão, M. da G., W. Wuster, R. S. Thorpe, and BBBSP. 1999. MtDNA phylogeny of Neotropical pitvipers of the genus *Bothrops* (Squamata: Serpentes: Viperidae). *Kaupia* 8: 127–34.

Sánchez, J., J. R. Bolaños, and C. E. Piedra. 1996. Población de *Crocodylus acutus* (Crocodylia: Crocodylidae) en dos ríos de Costa Rica. *Rev. Biol. Trop.* 44 (2B): 835–40.

Sanchiz, B., and I. de la Riva. 1993. Remarks on the tarsus of centrolenid frogs (Amphibia, Anura). *Graellsia* 49: 115–17.

San Román, L., A. L. Baez, M. Calvo, H. Gamboa, and C. Kandler. 1987. *National Museum of Costa Rica: Over one hundred years of history (May 4th 1887–May 4th 1987).* San José: Museo Nacional de Costa Rica.

Sasa, M. 1993. Distribution and reproduction of the gray earth snake *Geophis brachycephalus* (Serpentes: Colubridae) in Costa Rica. *Rev. Biol. Trop.* 41 (2): 295–97.

Sasa, M., and G. Chaves. 1992. Tamaño, estructura y distribución de un a población de *Crocodylus acutus* (Crocodylia: Crocodilidae). *Rev. Biol. Trop.* 40 (1): 131–34.

Sasa, M., and A. Solórzano. 1995. The reptiles and amphibians of Santa Rosa National Park, Costa Rica, with comments about the herpetofauna of xerophytic areas. *Herp. Nat. Hist.* 3 (2): 113–26.

Savage, J. M. 1952. Two centuries of confusion: The history of the snake name *Ahaetulla*. *Bull. Chicago Acad. Sci.* 9 (11): 203–16.

———. 1960a. Geographic variation in the tadpole of the toad *Bufo marinus*. *Copeia* 1960 (3): 233–36.

———. 1960b. A revision of the Ecuadorian snakes of the colubrid genus *Atractus*. *Misc. Pub. Mus. Zool. Univ. Mich.* 112:1–86.

———. 1965. A new bromeliad frog of the genus *Eleutherodactylus* from Costa Rica. *Bull. South. Calif. Acad. Sci.* 64 (2): 106–10.

———. 1966a. The origins and history of the Central American herpetofauna. *Copeia* 1966 (4): 719–66.

———. "1966b." 1967. An extraordinary new toad *(Bufo)* from Costa Rica. *Rev. Biol. Trop.* 14 (2): 153–67.

———. 1967. A new tree-frog (Centrolenidae) from Costa Rica. *Copeia* 1967 (2): 235–331.

———. 1968a. A new red-eyed tree-frog (family Hylidae) from Costa Rica, with a review of the *Hyla uranochroa* group. *Bull. South. Calif. Acad. Sci.* 67 (1): 1–20.

———. 1968b. The dendrobatid frogs of Central America. *Copeia* 1968 (4): 745–76.

———. 1970. On the trail of the golden frog: With Warszewicz and Gabb in Central America. *Proc. Calif. Acad. Sci.*, ser. 4, 38 (14): 273–88.

———. 1972a. The harlequin frogs, genus *Atelopus*, of Costa Rica and western Panama. *Herpetologica* 28 (2): 77–94.

———. 1972b. The systematic status of *Bufo simus* O. Schmidt with description of a new toad from western Panama. *J. Herp.* 6 (1): 25–33.

———. 1973a. Herpetological collections made by Dr. John F. Bransford, assistant surgeon, U.S.N. during the Nicaragua and Panama Canal surveys (1882–1885). *J. Herp.* 7 (1): 35–38.

———. 1973b. *A preliminary handlist of the herpetofauna of Costa Rica.* Los Angeles: University of Southern California.

———. 1973c. A revised terminology for plates in the loreal region. *Brit. J. Herp.* 5 (1): 360–62.

———. 1973d. The geographic distribution of frogs: Patterns and predictions. In *Evolutionary biology of the anurans*, ed. J. L. Vial, 351–445. Columbia: University of Missouri Press.

———. 1974a. Type localities for species of amphibians and reptiles described from Costa Rica. *Rev. Biol. Trop.* 22 (1): 71–122.

———. 1974b. On the leptodactylid frog called *Eleutherodactylus palmatus* (Boulenger) and the status of *Hylodes fitzingeri* O. Schmidt. *Herpetologica* 30 (3): 289–99.

———. 1975. Systematics and distribution of the Mexican and Central American stream frogs related to *Eleutherodactylus rugulosus*. *Copeia* 1975 (2): 254–306.

———. 1976. *A preliminary handlist of the herpetofauna of*

Costa Rica. 2d ed. San José: Editorial de la Universidad de Costa Rica.

———. 1980a. A new frog of the genus *Eleutherodactylus* (Leptodactylidae) from the Monteverde Forest Preserve, Costa Rica. *Bull. South. Calif. Acad. Sci.* 79 (1): 13–19.

———. 1980b. *A handlist with preliminary keys to the herpetofauna of Costa Rica.* Los Angeles: Hancock Foundation, University of Southern California.

———. 1981a. A synopsis of the larvae of Costa Rican frogs and toads. *Bull. South. Calif. Acad. Sci.* 79 (2): 45–54.

———. 1981b. The systematic status of Central American frogs confused with *Eleutherodactylus cruentus*. *Proc. Biol. Soc. Wash.* 94 (2): 413–20.

———. 1981c. The tadpole of the Costa Rican fringe-limbed tree-frog, *Hyla fimbrimembra*. *Proc. Biol. Soc. Wash.* 93 (4): 1177–83.

———. 1981d. A new species of the secretive colubrid snake genus *Geophis* from Costa Rica. *Copeia* 1981 (3): 549–53.

———. "1982." 1983. The enigma of the Central American herpetofauna: Dispersals or vicariance? *Ann. Missouri Botan. Gard.* 69 (3): 464–547.

———. 1987. Systematics and distribution of the Mexican rainfrogs of the *Eleutherodactylus gollmeri* group (Amphibia: Leptodactylidae). *Fieldiana Zool.*, n.s., 33:1–57.

———. 1991. Status of declining amphibian populations in lower Central America. Report of the Lower Central Amer. DAP Task Force to the National Science Foundation.

———. 1997a. On terminology for the description of the hemipenes of squamate reptiles. *Herp. J.* 7 (1): 23–25.

———. 1997b. A new species of rainfrog of the *Eleutherodactylus diastema* group from the Alta Talamanca region of Costa Rica. *Amph.-Reptl.* 18:241–47.

Savage, J. M., and B. I. Crother. 1989. The status of *Pliocercus* and *Urotheca* (Serpentes: Colubridae), with a review of included species of coral snake mimics. *Zool. J. Linn. Soc. London* 95:335–62.

Savage, J. M., and J. E. DeWeese. 1979. A new species of leptodactylid frog, genus *Eleutherodactylus*, from the Cordillera de Talamanca, Costa Rica. *Bull. South. Calif. Acad. Sci.* 78 (2): 107–15.

———. 1981. The status of the Central American leptodactylid frogs *Eleutherodactylus melanostictus* (Cope) and *Eleutherodactylus platyrhynchus* (Günther). *Proc. Biol. Soc. Wash.* 93 (4): 928–42.

Savage, J. M., and M. A. Donnelly. 1988. Variation and systematics in the colubrid snakes of the genus *Hydromorphus*. *Amph.-Reptl.* 9:289–300.

———. 1992. The second collection of, and variation in, the rare Neotropical toad *Bufo peripetetes*. *J. Herp.* 26 (1): 72–74.

Savage, J. M., and S. B. Emerson. 1970. Central American frogs allied to *Eleutherodactylus bransfordii* (Cope): A problem of polymorphism. *Copeia* 1970 (4): 623–44.

Savage, J. M., and C. Guyer. 1989. Infrageneric classification and species composition of the anole genera, *Anolis, Ctenonotus, Dactyloa, Norops* and *Semiurus* (Sauria: Iguanidae). *Amph.-Reptl.* 10:105–15.

———. "1998." 1999. A new species of anole lizard, genus *Norops* (Squamata: Polychrotidae). *Rev. Biol. Trop.* 46 (3): 805–9.

Savage, J. M., and W. R. Heyer. 1967. Variation and distribution in the tree-frog genus *Phyllomedusa* in Costa Rica, Central America. *Beitr. Neotrop. Fauna* 5 (2): 111–31.

———. "1968." 1969. The tree-frogs (family Hylidae) of Costa Rica: Diagnosis and distribution. *Rev. Biol. Trop.* 16:1–127.

———. 1997. Digital webbing formulae for anurans: A refinement. *Herp. Rev.* 28 (3): 131.

Savage, J. M., and A. G. Kluge. 1961. Rediscovery of the strange Costa Rican toad, *Crepidius eptioticus* Cope. *Rev. Biol. Trop.* 9 (1): 39–51.

Savage, J. M., and P. N. Lahanas. 1989. A new species of colubrid snake (genus *Urotheca*) from the Cordillera de Talamanca of Costa Rica. *Copeia* 1989 (4): 892–96.

———. 1991. On the species of the colubrid snake genus *Ninia* in Costa Rica and western Panama. *Herpetologica* 47 (1): 37–53.

Savage, J. M., and K. R. Lips. 1993. A review of the status and biogeography of the lizard genera *Celestus* and *Diploglossus* (Squamata: Anguidae), with description of two new species from Costa Rica. *Rev. Biol. Trop.* 41(3B):817–42.

Savage, J. M., and R. W. McDiarmid. 1992. Rediscovery of the Central American colubrid snake, *Sibon argus,* with comments on related species from the region. *Copeia* 1992 (2): 421–32.

Savage, J. M., and C. W. Myers. 2002. Frogs of the *Eleutherodactylus biporcatus* group (Leptodactylidae) of Central America and northern South America, including rediscovered, resurrected, and new taxa. *Amer. Mus. Nat. Hist. Novitates.* In press.

Savage, J. M., and N. J. Scott Jr. 1987. The *Imantodes* (Serpentes: Colubridae) of Costa Rica: One or two species? *Rev. Biol. Trop.* 33 (2): 107–32.

Savage, J. M., and J. B. Slowinski. 1990. A simple consistent terminology for the basic colour patterns of the venomous coral snakes and their mimics. *Herp. J.* 1 (11): 530–32.

———. 1992. The colouration of the venomous coral snakes (family Elapidae) and their mimics (families Aniliidae and Colubridae). *Biol. J. Linn. Soc. London* 45:235–54.

———. 1996. Evolution of colouration, urotomy and coral snake mimicry in the snake genus *Scaphiodontophis* (Serpentes: Colubridae). *Biol. J. Linn. Soc. London* 57:129–94.

Savage, J. M., and P. H. Starrett. 1967. A new fringe-limbed tree-frog (family Centrolenidae) from lower Central America. *Copeia* 1967 (3): 604–6.

Savage J. M., and J. J. Talbot. 1978. The giant anoline lizards of Costa Rica and western Panama. *Copeia* 1978 (3): 480–92.

Savage, J. M., and J. L. Vial. "1973." 1974. The venomous coral snakes (genus *Micrurus*) of Costa Rica. *Rev. Biol. Trop.* 21 (2): 295–394.

Savage, J. M., and J. D. Villa. 1986. *Introduction to the herpetofauna of Costa Rica/Introducción a la herpetofauna de Costa Rica.* Contributions to Herpetology 5. Oxford, Ohio: Society for the Study of Amphibians and Reptiles.

Savage, J. M., and M. H. Wake. 1972. Geographic variation and systematics of the Middle America caecilians, genera *Dermophis* and *Gymnopis. Copeia* 1972 (4): 680–95.

———. Wake. 2001. Reevaluation of the status of taxa of Central American caecilians (Amphibia: Gymnophiona), with

comments on their origin and evolution. *Copeia* 2001 (1): 52–64.

Savin, S. M. 1977. The history of the earth's surface temperature during the past 100 million years. *Ann. Rev. Earth Planet Sci.* 5:319–55.

Savin, S. M., and R. G. Douglas. 1985. Sea level, climate, and the Central American land bridge. In *The great American biotic interchange,* ed. F. G. Stehli and S. D. Webb, 303–424. New York: Plenum Press.

Savitsky, A. H. 1979. The origin of the New World proteroglyphous snakes and its bearing on the study of venom delivery systems in snakes. Ph.D. diss., University of Kansas.

———. 1981. Hinged teeth in snakes: An adaptation for swallowing hard-bodied prey. *Science* 212:346–49.

———. 1983. Coadapted character complexes among snakes: Fossoriality, piscivory, and durophagy. *Amer. Zool.* 23 (2): 397–409.

Say, T. 1823. In Edwin James, *Account of an expedition from Pittsburgh to the Rocky Mountains, performed in the years 1819 and 1820,* 2:339. Philadelphia: Carey and Lea; London: Longman, Hurst, Rees, Orne and Brown.

Sazima, I. 1989. Feeding behavior of the snail-eating snake, *Dipsas indica. J. Herp.* 23 (4): 464–68.

Schätti, B. 1987. The phylogenetic significance of morphological characters in the Holarctic racers of the genus *Coluber* Linnaeus, 1758 (Reptilia, Serpentes). *Amph.-Reptl.* 8:401–15.

Scherer, S. 1990. The protein molecular clock: Time for a reevaluation. *Evol. Biol.* 24 (3): 83–106.

Schlaepfer, M. A., and R. Figeroa. 1998. Female reciprocal calling in a Costa Rican leaf-litter frog, *Eleutherodactylus podiciferus. Copeia* 1998 (4): 1076–80.

Schlegel, H. 1826. Herpetologischen Nachrichten. *Isis von Oken* 20 (3): 281–94.

———. 1836. L'hemidactyle bride: *Hemidactylus frenatus.* In *Erpétologie generále, ou Histoire naturelle complète des reptiles,* by A. M. C. Duméril and G. Bibron, 3:366–68. Paris: Roret.

———. 1837. *Essai sur la physionomie des serpens: Atlas.* 2 vols. La Haye: Kips, Hz. et Van Stockum.

Schmid, K. 1819. *Naturhistorische Beschreibung der Amphibien. Systematisch bearbeitet zum gemeinnützigen Gebrauche.* Munich: Kunst-Anstalt.

Schmid, M. 1980. Chromosome banding in Amphibia. 4. Differentiation of GC- and AT-rich chromosome regions in Anura. *Chromosoma* 80:83–103.

Schmid, M., W. Feichtinger, R. Weimer, C. Mais, F. Bolaños, and P. León. 1995. Chromosome banding in Amphibia. 21. Inversion polymorphism and multiple nucleous organizer regions in *Agalychnis callidryas* (Anura, Hylidae). *Cytogenetics Cell. Genet.* 69 (1–2): 18–25.

Schmid, M., C. Steinlein, and W. Feichtinger. 1989. Chromosome banding in Amphibia. 14. The karyotype of *Centrolenella antisthenesi* (Anura, Centrolenidae). *Chromosoma* 97:434–38.

Schmidt, K. P. 1932. Stomach contents of some American coral snakes, with the description of a new species of *Geophis. Copeia* 1932 (1): 6–9.

———. 1933a. Amphibians and reptiles collected by the Smithsonian Biology Survey of the Panama Canal Zone. *Misc. Coll. Smithson. Inst.* 89:1–20.

———. 1933b. Preliminary account of the coral snakes of Central America and Mexico. *Field Mus. Nat. Hist. Zool. Ser.* 20 (6): 29–40.

———. 1936. Notes on Central American and Mexican coral snakes. *Field Mus. Nat. Hist. Zool. Ser.* 20 (20): 205–16.

Schmidt, K. P., and A. S. Rand. 1957. Geographic variation in the Central American colubrine snake, *Ninia sebae. Fieldiana Zool.* 39 (10): 73–84.

Schmidt, K. P., and T. F. Smith. 1944. Amphibians and reptiles of the Big Bend Region of Texas. *Field Mus. Nat. Hist. Zool. Ser.* 29 (5): 75–96.

Schmidt, O. 1857. Diagnosen neuer Frösche des zoologischen Cabinets zu Krakau. *Sitz. Akad. Wissen. Wien-Math.-Naturwissen. Kl.* 2410–15.

Schulte, R. 1977. Mit Geisterbeck schleicht er durchs Geäst: Der Lemurfrosch *(Phyllomedusa lemur). Aquar. Mag.* 11 (3): 98–103.

———. 1989. Nueva especies de rana venenosa del género *Epipedobates* registrada en la Cordillera Oriental, Departamento de San Martín. *Bol. Lima* 63:41–46.

Schumacher, A. H. 1996. Haltung und Zucht einer mittelamerikanische Schneckennatter—*Dipsas tenuissima* Taylor, 1954 im Terrarium. *Sauria* 18 (1): 3–10.

Schutze, H. P., and L. Trueb, eds. 1991. *Origins of the higher groups of vertebrates: Controversy and consensus.* Ithaca: Cornell University Press.

Schwalm, P. H. 1981. Ultrastructural correlates to infrared reflectance in New World tree frogs. Ph.D. diss., University of Chicago.

Schwalm, P. A., P. H. Starrett, and R. W. McDiarmid. 1977. Infrared reflectance in leaf-sitting Neotropical frogs. *Science* 196:1225–27.

Schwartz, A., and R. W. Henderson. 1988. West Indian amphibians and reptiles: A checklist. *Milwaukee Publ. Mus. Contr. Biol. Geol.* 74:1–264.

———. 1991. *Amphibians and reptiles of the West Indies.* Gainesville: University of Florida Press.

Schwartz, J. J., and K. D. Wells. 1983. An experimental study of acoustic interference between two species of Neotropical treefrogs. *Anim. Behav.* 31:181–90.

———. 1984a. Interspecific acoustic interactions of the Neotropical treefrog *Hyla ebraccata. Behav. Ecol. Sociobiol.* 14: 211–24.

———. 1984b. Vocal behavior of the Neotropical treefrog *Hyla phlebodes. Herpetologica* 40 (4): 452–64.

———. 1985. Intra- and interspecific vocal behavior of the Neotropical treefrog *Hyla microcephala. Copeia* 1985 (1): 27–38.

Schweigger, A. F. 1812. Monographiae cheloniorum. *Arch. Naturwissen. Math.* 1:271–368, 406–58.

Schwenk, K., S. K. Sessions, and D. M. Peccinini-Seale. 1982. Karyotypes of the basiliscine lizards *Corytophanes cristatus* and *Corytophanes hernandesii,* with comments on the relationship between chromosomal and morphological evolution in lizards. *Herpetologica* 38 (4): 493–501.

Scott, N. J., Jr. 1969. A zoogeographic analysis of the snakes of Costa Rica. Ph.D. diss., University of Southern California.

———. 1976. The abundance and diversity of the herpetofaunas of tropical forest litter. *Biotropica* 8 (1): 41–58.

———. 1983a. *Bolitoglossa subpalmata* (escorpiones, salamandras, mountain salamander). In *Costa Rican natural history,* ed. D. H. Janzen, 382–83. Chicago: University of Chicago Press.

———. 1983b. *Bufo haematiticus* (sapo, toad). In *Costa Rican natural history,* ed. D. H. Janzen, 385. Chicago: University of Chicago Press.

———. 1983c. *Leptodactylus pentadactylus* (rana ternero, smoky frog). In *Costa Rican natural history,* ed. D. H. Janzen, 405–6. Chicago: University of Chicago Press.

———. 1983d. *Eleutherodactylus bransfordii* (rana). In *Costa Rican natural history,* ed. D. H. Janzen, 399. Chicago: University of Chicago Press.

———. 1983e. *Eleutherodactylus diastema* (martillito, tink frog). In *Costa Rican natural history,* ed. D. H. Janzen, 399. Chicago: University of Chicago Press.

———. 1983f. *Agalychnis callidryas* (rana calzonudo, gaudy leaf frog). In *Costa Rican natural history,* ed. D. H. Janzen, 374–75. Chicago: University of Chicago Press.

———. 1983g. *Hyla boulengeri* (ranita de Boulenger, Boulenger's hyla). In *Costa Rican natural history,* ed. D. H. Janzen, 401. Chicago: University of Chicago Press.

———. 1983h. *Leptotyphlops goudotii* (culebra gusano, worm snake, blind snake). In *Costa Rican natural history,* ed. D. H. Janzen, 406. Chicago: University of Chicago Press.

———. 1983i. *Clelia clelia* (zopilota, musarana). In *Costa Rican natural history,* ed. D. H. Janzen, 392–93. Chicago: University of Chicago Press.

———. 1983j. *Conophis lineatus* (guarda camino). In *Costa Rican natural history,* ed. D. H. Janzen, 392–93. Chicago: University of Chicago Press.

———. 1983k. *Imantodes cenchoa* (bejuquilla, chunk-headed snake). In *Costa Rican natural history,* ed. D. H. Janzen, 402. Chicago: University of Chicago Press.

———. 1983l. *Rhadinaea decorata* (culebra). In *Costa Rican natural history,* ed. D. H. Janzen, 416. Chicago: University of Chicago Press.

———. 1983m. *Drymarchon corais* (zopilota, indigo). In *Costa Rican natural history,* ed. D. H. Janzen, 398–99. Chicago: University of Chicago Press.

———. 1983n. *Oxybelis aeneus* (bejuquillo). In *Costa Rican natural history,* ed. D. H. Janzen, 410–11. Chicago: University of Chicago Press.

———. 1983o. *Bothrops asper* (terciopelo, fer-de-lance). In *Costa Rican natural history,* ed. D. H. Janzen, 383–84. Chicago: University of Chicago Press.

———. 1983p. *Crotalus durissus* (cascabel, tropical rattlesnake). In *Costa Rican natural history,* ed. D. H. Janzen, 393–94. Chicago: University of Chicago Press.

Scott, N. J., Jr., and S. Limerick. 1983. Reptiles and amphibians: Introduction. In *Costa Rican natural history,* ed. D. H. Janzen, 351–67. Chicago: University of Chicago Press.

Scott, N. J., Jr., and R. W. McDiarmid. 1984a. *Trimorphodon. Cat. Amer. Amph. Reptl.* 352:1–2.

———. 1984b. *Trimorphodon biscutatus. Cat. Amer. Amph. Reptl.* 353:1–4.

Scott, N. J., Jr., J. M. Savage, and D. C. Robinson. 1983.

Checklist of amphibians and reptiles. In *Costa Rican natural history*, ed. D. H. Janzen, 367–74. Chicago: University of Chicago Press.

Scott, N. J., Jr., and A. Starrett. 1974. An unusual breeding aggregation of frogs, with notes on the ecology of *Agalychnis spurrelli* (Anura: Hylidae). *Bull. South. Calif. Acad. Sci.* 73 (2): 86–94.

Scott, N. J., Jr., D. E. Wilson, C. Jones, and R. M. Andrews. 1976. The choice of perch dimensions by lizards of the genus *Anolis* (Reptilia, Lacertilia, Iguanidae). *J. Herp.* 10 (2): 75–84.

Seba, A. 1734–65. *Locupletissimi rerum naturalium thesauri accurata descriptio, et iconibus artificiosissimis expressio, per universam physics historiam.* 4 vols. Amsterdam: Janssonio-Waesbergois and Wetstenium and Smith.

Seebach, K. von. 1865a. Reise durch Guanacaste (Costa Rica), 1864 und 1865. *Peterm. Geogr. Mitt.* 11 (7): 241–49.

———. 1865b. Besteigung des Vulkans Turrialba in Costa Rica. *Peterm. Geogr. Mitt.* 11 (9): 321–24.

Segura, A. 1944. Estudio de la primera especie nueva de tortuga fósil de Costa Rica, con algunas generalidades sobre el orden Testudines. *Escuela Farma. Guatemala* 6 (73–74): 9–29; (75–76): 16–25; (77–78): 13–14.

Seib, R. L. 1980. Human envenomation from the bite of an aglyphous false coral snake, *Pliocercus elapoides* (Serpentes; Colubridae). *Toxicon* 18:399–401.

———. 1984. Prey use in three syntopic Neotropical racers. *J. Herp.* 18 (4): 412–20.

———. 1985a. Europhagy in a tropical snake, *Coniophanes fissidens. Biotropica* 17 (1): 57–64.

———. 1985b. Feeding ecology and organization of Neotropical snake fauna. Ph.D. diss., University of California, Berkeley.

Seidel, M. E., and C. H. Ernst. 1996. *Pseudemys. Cat. Amer. Amph. Reptl.* 625:1–7.

Seidel, M. E., and D. R. Jackson. 1990. Evolution and fossil relationships of slider turtles. In *Life history and ecology of the slider turtle*, ed. J. W. Gibbons, 68–73. Washington, D.C.: Smithsonian Institution Press.

Seidel, M. E., and H. M. Smith. 1986. *Chrysemys, Pseudemys* and *Trachemys* (Testudines: Emydidae): Did Agassiz have it right? *Herpetologica* 42 (2): 242–48.

Seifert, R. P. 1983. *Bothrops schlegelii* (oropél [gold morph], bocaracá, eyelash viper, palm viper). In *Costa Rican natural history*, ed. D. H. Janzen, 384–85. Chicago: University of Chicago Press.

Seigel, R. A., and J. T. Collins, eds. 1993. *Snakes: Ecology and behavior.* New York: McGraw-Hill.

Seigel, R. A., J. T. Collins, and S. S. Novak, eds. 1987. *Snakes: Ecology and evolutionary biology.* New York: Macmillan.

Serena, M. 1985. Zoogeography of parthenogenetic whiptail lizards *(Cnemidophorus lemniscatus)* in the Guianas: Evidence from skin grafts, karyotypes and erythrocyte areas. *J. Biogeogr.* 12:49–56.

Sessions, S. K. 1984. Cytogenetics and evolution in salamanders. Ph.D. diss., University of California, Berkeley.

———. 1996. Chromosomes: Molecular cytogenetics. In *Molecular systematics*, ed. D. M. Hillis, C. Moritz, and B. K. Mable, 121–68. Sunderland, Mass.: Sinauer.

Sessions, S. K., and J. Kezer. 1991. Evolutionary cytogenetics of bolitoglossine salamanders (family Plethodontidae). In *Amphibian cytogenetics and evolution*, ed. D. M. Green and S. K. Sessions, 89–130. San Diego: Academic Press.

Sexton, O. J. 1960. Notas sobre la reproducción de una tortuga venezolana, la *Kinosternon scorpioides. Mem. Soc. Cien. Nat. La Salle* 20 (57): 189–97.

———. 1962. Apparent territorialism in *Leptodactylus insularum. Herpetologica* 18 (3): 212–14.

———. 1967. Populational changes in a tropical lizard *Anolis limifrons* on Barro Colorado Island, Panama. *Copeia* 1967 (1): 219–22.

Sexton, O. J., and H. Heatwole. 1965. Life history notes on some Panamanian snakes. *Carib. J. Sci.* 5 (1–2): 39–43.

Sexton, O. J., H. Heatwole, and E. Meseth. 1963. Seasonal population changes in the lizard *Anolis limifrons* in Panama. *Amer. Midl. Nat.* 69 (3): 482–91.

Sexton, W. J., E. P. Ortleb, L. M. Hathaway, R. C. Ballinger, and P. Licht. 1971. Reproductive cycles of three species of anoline lizards from the Isthmus of Panama. *Ecology* 52 (2): 201–15.

Shannon, F. A., and F. L. Humphrey. 1957. A new species of *Phrynohyas* from Nayarit. *Herpetologica* 13 (1): 15–18.

Shine, R. 1994. Sexual size dimorphism in snakes revisited. *Copeia* 1994 (2): 326–46.

Shreve, B. 1936. A new *Atelopus* from Panama and a new *Hemidactylus* from Colombia. *Occas. Pap. Boston Soc. Nat. Hist.* 8:269–72.

Shreve, B., and C. Gans. 1958. *Thamnophis bovallii* Dunn rediscovered (Reptilia, Serpentes). *Brevoria* 83:1–8.

Shubin, N. H., and F. A. Jenkins Jr. 1996. An Early Jurassic jumping frog. *Nature* 377:49–52.

Silverstone, P. A. 1975. A revision of the poison-arrow frogs of the genus *Dendrobates* Wagler. *Sci. Bull. Nat. Hist. Mus. Los Angeles Co.* 21:1–55.

———. 1976. A revision of the poison-arrow frogs of the genus *Phyllobates* Bibron in Sagra (Dendrobatidae). *Sci. Bull. Nat. Hist. Mus. Los Angeles Co.* 27:1–53.

Simmons, J. E. 1986. A method of preparation of anuran osteological material. In *Proceedings of the 1985 workshop on care and maintenance of natural history collections*, ed. J. Waddington and D. M. Rudkin, 37–39. Life Science Miscellaneous Publications. Toronto: Royal Ontario Museum.

———. 1987. Herpetological collection and collections management. *Soc. Stud. Amph. Reptl. Herp. Circ.* 16:1–72.

Sites, J. W., and J. R. Dixon. 1982. Geographic variation in *Sceloporus variabilis* and its relationship to *S. teapensis* (Sauria: Iguanidae). *Copeia* 1982 (1): 14–27.

Sites, J. W., Jr., J. W. Archie, C. J. Cole, and O. Flores V. 1992. A review of phylogenetic hypotheses for lizards of the genus *Sceloporus* (Phrynosomatidae): Implications for ecological and evolutionary studies. *Bull. Amer. Mus. Nat. Hist.* 213: 1–110.

Sites, J. W., Jr., D. M. Peccinini-Seale, C. Moritz, J. W. Wright, and W. M. Brown. 1990. The evolutionary history of parthenogenetic *Cnemidophorus lemniscatus* (Sauria, Teiidae). 1. Evidence for a hybrid origin. *Evolution* 44 (4): 906–21.

Slevin, J. R. 1942. Notes on a collection of reptiles from Boquete, Panama, with description of a new species of *Hydromorphus. Proc. Calif. Acad. Sci.*, ser. 4, 23 (32): 463–80.

————. 1950. A remarkable concentration of desert snakes. *Herpetologica* 6 (1): 12–13.

Slowinski, J. B. 1991. The phylogenetic relationships of the New World coral snakes (Elapidae: *Leptomicrurus*, *Micruroides*, and *Micrurus*) based on biochemical and morphological data. Ph.D. diss., University of Miami.

————. 1995. A phylogenetic analysis of the New World coral snakes (Elapidae: *Leptomicrurus*, *Micruroides*, and *Micrurus*) based on allozyme and morphological characters. *J. Herp.* 29 (3): 325–38.

Slowinski, J. B., and J. M. Savage. 1995. Urotomy in *Scaphiodontophis:* Evidence for the multiple tail break hypothesis in snakes. *Herpetologica* 51 (3): 338–41.

Smith, E. N., and M. E. Acevedo. 1997. The northernmost distribution of *Corallus annulatus* (Boidae), with comments on its natural history. *SW Nat.* 42 (3): 347–49.

Smith, H. M. 1939a. Mexican herpetological novelties. *Proc. Bio. Soc. Wash.* 52:187–96.

————. 1939b. The Mexican and Central American lizards of the genus *Sceloporus. Field Mus. Nat. Hist. Zool. Ser.* 26:1–397.

————. 1940. Descriptions of new lizards and snakes from Mexico and Guatemala. *Proc. Biol. Soc. Wash.* 53:55–64.

————. 1941a. Snakes, frogs, and bromeliads. *Chicago Nat.* 4 (2): 35–43.

————. 1941b. A new name for the Mexican snakes of the genus *Dendrophidion. Proc. Biol. Soc. Wash.* 54:73–76.

————. 1942. Mexican herpetological miscellany. *Proc. U.S. Natl. Mus.* 92:349–95.

————. 1943. Summary of the collections of snakes and crocodilians made in Mexico under the Walter Rathbone Bacon Traveling Scholarship. *Proc. U.S. Natl. Mus.* 93:393–504.

————. 1957. Curious feeding habit of a blind snake *Leptotyphlops. Herpetologica* 13 (2): 102.

————. 1959. Herpetozoa from Guatemala. 1. *Herpetologica* 15 (4): 210–16.

————. 1962. The subspecies of *Tantilla schistosa* of Middle America. *Herpetologica* 18 (1): 13–18.

————. 1969. The first herpetology of Mexico. *Bull. SW Herp. Soc.*, ser. 3, 1:1–16.

————. 1971. The snake genus *Amastridium* in Oaxaca, Mexico. *Gt. Basin Nat.* 31 (4): 254–55.

————. 1973. A tentative rearrangement of the lizards of the genus *Lepidophyma. J. Herp.* 7 (2): 109–23.

————. 1975. The blazing of a trail: The scientific career of Edward Harrison Taylor. *Monogr. Univ. Kansas Mus. Nat. Hist.* 4:115–44.

————. "1982." 1983. The gender of the nominal snake genus *Sibon. Bull. Maryland Herp. Soc.* 18 (4): 192–93.

Smith, H. M., R. G. Arndt, and W. C. Sherbrooke. 1971. A new snake of the genus *Enulius* from Mexico. *Nat. Hist. Misc. Chicago Acad. Sci.* 18:61–64.

Smith, H. M., and D. Chiszar. 1996. *Species-group taxa of the false coral snake genus* Pliocercus. Pottsville, Pa.: Ramus.

Smith, H. M., K. Fitzgerald, G. Pérez-Higareda, and D. Chiszar. 1986. A taxonomic rearrangement of the snakes of the genus *Scaphiodontophis. Bull. Maryland Herp. Soc.* 22 (4): 159–66.

Smith, H. M., and C. Grant. 1958. The proper name of some Cuban snakes: An analysis of dates of publication of Ramón de la Sagra's *Historia natural de Cuba*, and of Fitzinger's *Systema reptilium. Herpetologica* 14 (4): 215–22.

Smith, H. M., and K. R. Larsen. 1974. The nominal snake genera *Mastigodryas* Amaral, 1934, and *Dryadophis* Stuart, 1939. *Gt. Basin Nat.* 33 (4): 276.

Smith, H. M., and L. E. Laufe. 1946. A summary of Mexican lizards of the genus *Ameiva. Univ. Kansas Sci. Bull.* 31 (2): 7–73.

Smith, H. M., G. Pérez-Higareda, and D. Chiszar. 1993. A review of the members of the *Sceloporus variabilis* lizard complex. *Bull. Maryland Herp. Soc.* 29 (3): 85–125.

Smith, H. M., and R. B. Smith. 1973. *Synopsis of the herpetofauna of Mexico. Vol. 2. Analysis of the literature exclusive of the Mexican axolotl.* Augusta, W.Va.: Lundberg.

————. 1976. *Synopsis of the herpetofauna of Mexico. Vol. 3. Source analysis and index for Mexican reptiles.* North Bennington, Vt.: Johnson.

————. 1977. *Synopsis of the herpetofauna of Mexico. Vol. 5. Guide to Mexican amphisbaenians and crocodilians: Bibliographic addendum II.* North Bennington, Vt.: Johnson.

————. 1979. *Synopsis of the herpetofauna of Mexico. Vol. 6. Guide to Mexican turtles: Bibliographic addendum III.* North Bennington, Vt.: Johnson.

Smith, H. M., and E. H. Taylor. 1945. An annotated checklist and key to the snakes of Mexico. *Bull. U.S. Natl. Mus.* 187: 1–229.

————. 1948. An annotated checklist and key to the Amphibia of Mexico. *Bull. U.S. Natl. Mus.* 194:1–118.

————. 1950. An annotated checklist and key to the reptiles of Mexico exclusive of the snakes. *Bull. U.S. Natl. Mus.* 199: 1–253.

Smith, H. M., V. Wallach, and D. Chiszar. 1995. Observations of the snake genus *Pliocercus*, 1. *Bull. Maryland Herp. Soc.* 31 (4): 204–13.

Smith, H. M., K. L. Williams, and G. Pérez-Higareda. 1986. The specific name for the Linnean *Oxyrhopus* or the calico false coral snake. *Bull. Maryland Herp. Soc.* 22 (1): 10–13.

Smith, M. A. 1943. *The fauna of British India, Ceylon and Burma including the whole of the Indo-Chinese sub-region. Reptilia and Amphibia.* Vol. 3, *Serpentes.* London: Taylor and Francis.

Smith, N. G. 1969. Avian predation on coral snakes. *Copeia* 1969 (2): 402–4.

Smith, R. E. 1968a. Experimental evidence for a gonadal-fat body relationship in two teiid lizards *(Ameiva festiva, Ameiva quadrilineata). Biol. Bull.* 134:325–31.

————. 1968b. Studies on reproduction in Costa Rican *Ameiva festiva* and *Ameiva quadrilineata* (Sauria: Teiidae). *Copeia* 1968 (2): 236–39.

Smith, S. M. 1975. Innate recognition of coral snake pattern by a possible avian predator. *Science* 187:759–60.

————. 1977. Coral snake pattern recognition and stimulus generalization by naive greater kiskadee (Aves: Tyrannidae). *Nature* 265:535–36.

Smythe, R. H. 1975. *Vision in the animal world.* London: Macmillan.

Snyder, R. C. 1949. Bipedal locomotion of the lizard *Basiliscus basiliscus. Copeia* 1949 (2): 129–37.

Solís, F. "1991." 1993. Commentarios sobre dos especies de serpientes en Panamá: *Tantilla alticola* (Boulenger) y *Bothrops picadoi* (Dunn). *Scientia (Panama)* 6 (2): 107–10.

Solís, F., R. Ibáñez, and C. A. Jaramillo. 1996. Estableciendo una uniformidad sobre las especies de serpientes presentes en Panamá. *Rev. Biol. Trop.* 44 (1): 19–22.

Solórzano, A. 1989. Distribution and reproductive aspects of the snake *Bothrops nummifer* (Serpentes: Viperidae). *Rev. Biol. Trop.* 37 (2): 133–37.

———. 1990. Reproduction in the pit viper *Porthidium picadoi* Dunn (Serpentes: Viperidae) in Costa Rica. *Copeia* 1990 (4): 1154–57.

———. 1993. *Creencias populares sobre los reptiles en Costa Rica.* San José: Serpentario Tropical.

———. "1994." 1995. Una nueva especies de serpiente venenosa terrestre del género *Porthidium* (Serpentes: Viperidae), del suroeste de Costa Rica. *Rev. Biol. Trop.* 42 (3): 695–701.

———. 1995. A case of human bite by the pelagic sea snake, *Pelamis platurus* (Serpentes: Hydrophiidae). *Rev. Biol. Trop.* 43 (1–3): 321–22.

———. "1997." 1998. Reproducción de la toboba de árbol, *Bothriechis nigroviridis* (Serpentes: Viperidae), en Costa Rica. *Rev. Biol. Trop.* 45 (4): 1675–77.

Solórzano, A., and L. Cerdas. 1986. A new subspecies of the bushmaster, *Lachesis muta,* from southeastern Costa Rica. *J. Herp.* 20 (3): 462–65.

———. 1987. *Drymobius margaritiferus* (Speckled racer): Reproduction. *Herp. Rev.* 18 (4): 75–76.

———. 1988a. Ciclos reproductivos de la serpiente coral *Micrurus nigrocinctus* (Serpentes: Elapidae) en Costa Rica. *Rev. Biol. Trop.* 36 (2A): 235–39.

———. 1988b. Incubación de los huevos y nacimiento en la coral gargantilla, *Micrurus mipartitus hertwigi* (Serpentes: Elapidae) en Costa Rica. *Rev. Biol. Trop.* 36 (2B): 535–36.

———. 1988c. Biología reproductiva de la cascabel centroamericano *Crotalus durissus durissus* (Serpentes: Viperidae) en Costa Rica. *Rev. Biol. Trop.* 36 (2A): 221–26.

———. 1989. Reproductive biology and distribution of the terciopelo, *Bothrops asper* Garman (Serpentes: Viperidae) in Costa Rica. *Herpetologica* 45 (4): 444–50.

Solórzano, A., L. D. Gómez, J. Monge, and B. I. Crother. "1998." 1999. Redescription and validation of *Bothriechis supraciliaris* (Serpentes: Viperidae). *Rev. Biol. Trop.* 46 (2): 453–62.

Solórzano, A., J. M. Gutiérrez, and L. Cerdas. 1988. *Botrhops (sic) ophryomegas* Bocourt (Serpentes: Viperidae) en Costa Rica: Distribución, lepidosis, variación sexual y cariotipo. *Rev. Biol. Trop.* 36 (2A): 187–90.

———. 1989. Notes on the natural history and karyotype of the colubrid snake, *Hydromorphus concolor* Peters, from Costa Rica. *J. Herp.* 23 (3): 314–15.

Solórzano, A., W. W. Lamar, and L. Porras. "1997." 1998. The marsupial frog *(Gastrotheca cornuta)* (Amphibia: Hylidae) in Isthmian Central America. *Rev. Biol. Trop.* 45 (4): 1675–77.

Solórzano, A., M. Romero, J. M. Gutiérrez, and M. Sasa. 1999. Venom composition and diet of the cantil *Agkistrodon bilineatus howardgloydi* (Serpentes: Viperidae). *SW Nat.* 44 (4): 478–83.

Soma, M., M. L. Beçak, and W. Beçak. 1975. Estudios comparativo do contenudo de DNA em 12 espécies de lacertilios. *Cien. Cult.* 27:1322–28.

Spataro, M. 1996. Annulated boas *Corallus annulatus* captive reproduction and husbandry. *Vivarium* 8 (1): 24–25, 47.

Speed, R. C., ed. 1994. *Phanerozoic evolution of North American continent-ocean transitions.* Decade of North American Geology continent-ocean transect series. Boulder, Colo.: Geological Society of America.

Spix, J. B. von. 1824. *Animalia nova sive species novae testudinum et ranarum, quas in itinere per Brasiliam annis MDCCCXVII–MDCCCXX jussu et auspiciis Maximilliani Joséphi I. Bavariae regis suscepto collegit et descripsit Dr. J. B. de Spix.* Vol. 3. Munich: Hubschmann.

———. 1825. *Animalia nova sive species novae lacertarum quas in itinere per Brasliam annis MDCCCXVII–MDCCCXX jussu et auspicius Maximilliami Joseph I. Bavariae regis suscepto collegit et descripsit Dr. J. B. de Spix.* Leipzig: Weigel.

Spotila, J. R., and E. A. Standora. 1985. Environmental constraints on the thermal energetics of sea turtles. *Copeia* 1985 (3): 694–702.

Spotila, J. R., E. A. Standora, S. J. Morreale, and G. J. Ruiz. 1987. Temperature dependent sex determination in the green turtle *(Chelonia mydas):* Effects on the sex ratio on a natural nesting beach. *Herpetologica* 43 (1): 74–81.

Stafford, P. J. 1998. Amphibians and reptiles of the Cordillera de Guanacaste, Costa Rica: A field list with notes on colour pattern and other observations. *Brit. Herp. Soc. Bull.* 62:9–22.

Stahl, F. W., ed. 1964. *The mechanics of inheritance.* Englewood, N.J.: Prentice-Hall.

Standora, E. A., and J. R. Spotila. 1985. Temperature dependent sex determination in sea turtles. *Copeia* 1985 (3): 711–22.

Stanton, M. A., and J. R. Dixon. "1975." 1978. Studies on the dry season biology of *Caiman crocodilus crocodilus* from the Venezuelan llanos. *Mem. Soc. Cien. Nat. La Salle* 35:237–65.

———. 1977. Breeding biology of the spectacled caiman, *Caiman crocodilus crocodilus,* in the Venezuelan llanos. *U.S. Fish. Widl. Serv. Wildlife Res. Rept.* 5:1–21.

Starrett, P. H. 1960a. Descriptions of tadpoles of Middle American frogs. *Misc. Pub. Mus. Zool. Univ. Mich.* 110:1–37.

———. 1960b. A redefinition of the genus *Smilisca. Copeia* 1960 (4): 300–304.

———. 1966. Rediscovery of *Hyla pictipes* Cope, with description of a new montane stream *Hyla* from Costa Rica. *Bull. South. Calif. Acad. Sci.* 65 (1): 17–28.

———. 1967. Observations on the life history of frogs of the family Atelopodidae. *Herpetologica* 23 (3): 195–203.

———. 1973. Evolutionary patterns in larval morphology. In *Evolutionary biology of the anurans,* ed. J. L. Vial, 251–71. Columbia: University of Missouri Press.

Starrett, P. H., and J. M. Savage. 1973. The systematic status and distribution of Costa Rican glass-frogs, genus *Centrolenella* (family Centrolenidae), with description of a new species. *Bull. South. Calif. Acad. Sci.* 72 (2): 57–78.

Stehli, F. G., and S. D. Webb, eds. 1985. *The great American biotic interchange.* New York: Plenum Press.

Steindachner, F. 1864. Batrachologische Mittheilungen. *Verh. Zool. Botan. Gesell. Wien* 14:539–52.

Stejneger, L. H. "1891." 1892. Notes on some South American snakes. *Proc. U.S. Natl. Mus.* 14:501–5.

———. 1900. Description of two new lizards of the genus *Anolis* from Cocos and Malpelo Islands. *Bull. Mus. Comp. Zool.* 36 (6): 161–64.

———. 1903. Description of a new species of gecko from Cocos Island. *Proc. Biol. Soc. Wash.* 16:3–4.

———. 1904. The herpetology of Puerto Rico. *Annl. Rept. U.S. Nat. Hist. Mus.* 1902:547–724.

———. 1906. A new tree toad from Costa Rica. *Proc. U.S. Natl. Mus.* 30:817–18.

———. 1907. A new salamander from Nicargaua. *Proc. U.S. Natl. Mus.* 32:465–66.

———. 1909. Description of a new snake from Panama. *Proc. U.S. Natl. Mus.* 36:457–58.

———. 1911. Descriptions of three new batrachians from Costa Rica and Panama. *Proc. U.S. Natl. Mus.* 41:285–88.

Stewart, J. R., and D. G. Blackburn. 1988. Reptilian placentation: Structural diversity and terminology. *Copeia* 1988 (4): 839–52.

Stiles, F. G., A. F. Skutch, and D. Gardner. 1989. *A guide to the birds of Costa Rica.* Ithaca: Cornell University Press.

Stimson, A. F. 1969. Liste der rezenten Amphibian und Reptilien: Boidae (Boinae + Bolyerinae + Loxoceminae + Pythoninae). *Tierreich* 89:1–49.

Stock, A. D. 1972. Karyological relationships in turtles (Reptilia: Chelonia). *Canadian J. Genet. Cytol.* 14:859–68.

Stone, D. E. 1988. The Organization for Tropical Studies (OTS): A success story in graduate training and research. In *Tropical rainforests: Diversity and conservation,* ed. F. Almeda and C. M. Prigle, 143–87. San Francisco: California Academy of Sciences/Pacific Division, American Association for the Advancement of Science.

Straham, M. H., and A. Schwartz. 1977. Osteoderms in the anguid lizard subfamily Diploglossinae and their taxonomic importance. *Biotropica* 9 (1): 58–72.

Straughan, I. R., and W. R. Heyer. 1976. A functional analysis of the mating calls of the Neotropical frog genera of the *Leptodactylus* complex (Amphibia, Leptodactylidae). *Pap. Avul. Zool. Mus. Zool. Univ. São Paulo* 29 (23): 221–45.

Stuart, L. C. 1932. Studies on Neotropical Colubrinae. 1. The taxonomic status of the genus *Drymobius* Fitzinger. *Occas. Pap. Mus. Zool. Univ. Mich.* 236:1–16.

———. 1935. A contribution to a knowledge of the herpetology of a portion of the savanna region of central Petén, Guatemala. *Misc. Pub. Mus. Zool. Univ. Mich.* 29:1–56.

———. 1939. A new name for the genus *Eudryas* Fitzinger 1843. *Copeia* 1939 (1): 55.

———. 1940. Notes on the "*Lampropholis*" group of Middle American *Lygosoma* (Scincidae) with descriptions of two new forms. *Occas. Pap. Mus. Zool. Univ. Mich.* 421:1–16.

———. 1941. Studies on Neotropical Colubrinae. 8. A revision of the genus *Dryadophis* Stuart, 1939. *Misc. Pub. Mus. Zool. Univ. Mich.* 49:1–106.

———. 1943. Comments on the herpetofauna of the Sierra Cuchumatanes of Guatemala. *Occas. Pap. Mus. Zool. Univ. Mich.* 471:1–28.

———. 1948. The amphibians and reptiles of Alta Verapaz, Guatemala. *Misc. Pub. Mus. Zool. Univ. Mich.* 69:1–109.

———. 1953. Comments on the herpetofauna of the Sierra de los Cuchumatanes of Guatemala. *Occas. Pap. Mus. Zool. Univ. Mich.* 471:1–28.

———. 1954. A description of a subhumid corridor across northern Central America, with comments on its herpetofaunal indicators. *Contr. Lab. Vert. Biol. Univ. Mich.* 65:1–26.

———. 1955. A brief review of the Guatemalan lizards of the genus *Anolis. Misc. Pub. Mus. Zool. Univ. Mich.* 91:1–31.

———. 1963. A checklist of the herpetofauna of Guatemala. *Misc. Pub. Mus. Zool. Univ. Mich.* 122:1–150.

———. 1971. Comments on the malachite *Sceloporus* (Reptilia: Sauria: Iguanidae) of southern Mexico and Guatemala. *Herpetologica* 27 (2): 235–59.

Stull, O. G. 1932. Five new subspecies of the family Boidae. *Occas. Pap. Boston Soc. Nat. Hist.* 8:25–30.

Summers, K. 1989. Sexual selection and intra-female competition in the green poison-dart frog, *Dendrobates auratus. Anim. Behav.* 37:797–805.

———. 1990. Parental care and the cost of polygyny in the green dart-poison frog, *Dendrobates auratus. Behav. Ecol. Sociobiol.* 27:307–13.

Sundquist, E. T., and W. S. Broecker, eds. 1985. *The carbon cycle and atmospheric CO_2: Natural variations Archean to present.* Washington, D.C.: American Geophysical Union.

Talbot, J. J. 1976. Ecological and behavioral factors regulating the spatial distribution of *Norops humilis* and *N. limifrons* (Sauria: Iguanidae) at a tropical rainforest locality. Ph.D. diss., University of Southern California.

———. 1977. Habitat selection in tropical anoline lizards. *Herpetologica* 33 (4): 114–23.

———. 1979. Time budget, niche overlap, inter- and intraspecific aggression in *Anolis humilis* and *A. limifrons* from Costa Rica. *Copeia* 1979 (3): 472–81.

Tanner, W. W., and A. H. Brame Jr. II. 1961. Description of a new species of salamander from Panama. *Gt. Basin Nat.* 21 (1–2): 23–26.

Taylor, D. H., and S. I. Guttman, eds. 1977. *The reproductive biology of amphibians.* New York: Plenum Press.

Taylor, E. H. "1935." 1936. A taxonomic study of the cosmopolitan scincoid lizard genus *Eumeces. Univ. Kansas Sci. Bull.* 23 (1): 1–643.

———. 1939a. Two new species of snakes of the genus *Anomalepis* Jan, with a proposal of a new family of snakes. *Proc. N. Engl. Zool. Club.* 7:87–96.

———. "1939b." 1940. Herpetological miscellany. *Univ. Kansas Sci. Bull.* 26 (15): 489–571.

———. 1942. Tadpoles of Mexican Anura. *Univ. Kansas Sci. Bull.* 28 (3): 37–55.

———. 1944. The genera of plethodontid salamanders in Mexico. 1. *Univ. Kansas Sci. Bull.* 30 (12): 189–232.

———. 1948a. New Costa Rican salamanders. *Proc. Biol. Soc. Wash.* 61:177–80.

———. 1948b. Two new hylid frogs from Costa Rica. *Copeia* 1948 (4): 233–38.

———. 1949a. Costa Rican frogs of the genera *Centrolene* and *Centrolenella. Univ. Kansas Sci. Bull.* 33 (1): 257–70.

———. 1949b. Two new lizards from Costa Rica. *Univ. Kansas Sci. Bull.* 33 (5): 271–78.

———. 1949c. New salamanders from Costa Rica. *Univ. Kansas Sci. Bull.* 33 (6): 279–88.

———. 1951a. Two new genera and a new family of tropical American frogs. *Proc. Biol. Soc. Wash.* 64:33–40.

———. 1951b. A brief review of the snakes of Costa Rica. *Univ. Kansas Sci. Bull.* 34 (1): 1–188.

———. 1952a. A review of the frogs and toads of Costa Rica. *Univ. Kansas Sci. Bull.* 35 (5): 577–942.

———. 1952b. The salamanders and caecilians of Costa Rica. *Univ. Kansas Sci. Bull.* 34 (12): 695–791.

———. 1953. Early records of the seasnake *Pelamis platurus* in Latin America. *Copeia* 1953 (2): 124.

———. 1954a. Additions to the known herpetofauna of Costa Rica with comments on other species, no. 1. *Univ. Kansas Sci. Bull.* 36 (9): 597–639.

———. 1954b. Further studies on the serpents of Costa Rica. *Univ. Kansas Sci. Bull.* 36 (11): 673–801.

———. 1954c. Frog-eating tadpoles of *Anotheca coronata* (Stejneger) (Salientia, Hylidae). *Univ. Kansas Sci. Bull.* 36 (8): 589–96.

———. 1955. Additions to the known herpetofauna of Costa Rica with comments on other species, no. 2. *Univ. Kansas Sci. Bull.* 37 (13): 499–757.

———. 1956. A review of the lizards of Costa Rica. *Univ. Kansas Sci. Bull.* 38 (1): 3–322.

———. 1958a. Additions to the known herpetofauna of Costa Rica with comments on other species, no. 3. *Univ. Kansas Sci. Bull.* 39 (1): 3–40.

———. 1958b. Notes on Costa Rican Centrolenidae with descriptions of new forms. *Univ. Kansas Sci. Bull.* 39 (2): 41–68.

———. 1968. *The caecilians of the world: A taxonomic review.* Lawrence: University of Kansas Press.

Taylor, E. H., and H. M. Smith. 1943. A review of American sibynophine snakes, with proposal of a new genus. *Univ. Kansas Sci. Bull.* 29 (6): 301–37.

Taylor, J. D., and J. T. Bagnara. 1969. Melanosomes of the Mexican tree frog, *Agalychnis dacnicolor. J. Ultrastruct. Res.* 35:532–40.

Taylor, R. T., A. Flores, G. Flores, and R. Bolaños. "1973." 1974. Geographical distribution of Viperidae, Elapidae and Hydrophiidae in Costa Rica. *Rev. Biol. Trop.* 21 (2): 383–97.

Tchernov, E., O. Rieppel, H. Zaher, M. J. Polcyn, and L. L. Jacobs. 2000. A fossil snake with limbs. *Science* 287:2010–12.

Telford, S. R., Jr. 1971. Reproductive patterns and relative abundance of two microteiid lizard species in Panama. *Copeia* 1971 (4): 670–75.

Telford, S. R., Jr., and H. W. Campbell. 1970. Ecological observations on an all female population of the lizard *Lepidophyma flavimaculatum* (Xantusiidae) in Panama. *Copeia* 1970 (2): 379–81.

Templeton, A. R. 1989. The meaning of species and speciation: A genetic perspective. In *Speciation and its consequences*, ed. D. Otte and J. A. Endler, 3–27. Sunderland, Mass.: Sinauer.

Tennant, A. 1984. *The snakes of Texas.* Austin: Texas Monthly Press.

Teska, W. R. 1976. Terrestrial movements of the mud turtle *Kinosternon scorpioides* in Costa Rica. *Copeia* 1976 (3): 579–80.

Test, F. H., O. J. Sexton, and H. Heatwole. 1966. Reptiles of Rancho Grande and vicinity, Estado Aragua, Venezuela. *Misc. Pub. Mus. Zool. Univ. Mich.* 128:1–63.

Tihen, J. A. 1949. The genera of gerrhonotine lizards. *Amer. Midl. Nat.* 41 (3): 580–601.

Timm, R. M., D. E. Wilson, B. L. Clausen, R. K. LaVal, and C. S. Vaughan. 1990. Mammals of the La Selva–Braulio Carrillo complex, Costa Rica. *N. Amer. Fauna* 75:1–162.

Timmerman, W. W., and M. P. Hayes. 1981. *The reptiles and amphibians of Monteverde.* San José: Tropical Science Center.

Toft, C. A. 1981. Feeding ecology of Panamanian litter anurans: Patterns in diet and foraging mode. *Herpetologica* 15 (2): 139–44.

Tosi, J. A., Jr. 1969. *Ecological map of Costa Rica.* San José: Tropical Science Center.

Trueb, L. 1968. Variation in the tree frog *Hyla lancasteri. Copeia* 1968 (2): 285–99.

———. 1971. Phylogenetic relationships of certain Neotropical toads with the description of a new genus (Anura: Bufonidae). *Contr. Sci. Nat. Hist. Mus. Los Angeles Co.* 216:1–40.

Trueb, L., and R. Cloutier. 1991. A phylogenetic investigation of the inter- and intrarelationships of the Lissamphibia (Amphibia: Temnospodyli). In *Origins of the higher groups of tetrapods: Consensus*, ed. H.-P. Schultze and L. Trueb, 223–13. Ithaca: Cornell University Press.

Trueb, L., and C. Gans. 1983. Feeding specialization of the Mexican burrowing toad, *Rhinophrynus dorsalis* (Anura: Rhinophrynidae). *J. Zool.* 1991:89–208.

Trueb, L., and M. Tyler. 1974. Systematics and evolution of the Greater Antillean hylid frogs. *Occas. Pap. Mus. Nat. Hist. Univ. Kansas* 24:1–60.

Tryon, B. W.. 1985a. *Bothrops asper* (terciopelo): Caudal luring. *Herp. Rev.* 16 (1): 28.

———. 1985b. *Bothrops asper* (terciopelo): Reproduction. *Herp. Rev.* 16 (1): 28.

Tschudi, J. J. von. 1838. Classification der Batrachier, mit Berucksichtigung der fossilen Thiere dieser Abteilung der Reptilien. *Mem. Soc. Sci. Nat. Neuchatel.* 2:1–100 (preprint).

Tu, A. T. 1976. Investigation of the sea snake, *Pelamis platurus* (Reptilia, Serpentes, Hydrophiidae), on the Pacific coast of Costa Rica. *J. Herp.* 10 (1): 13–18.

Ueda, H. 1986. Reproduction of *Chirixalus eiffingeri* (Boettger). *Sci. Rept. Lab. Amph. Biol. Hiroshima Univ.* 8 (6): 109–16.

Underwood, G. L. 1954. On the classification and evolution of geckos. *Proc. Zool. Soc. London* 124:469–92.

———. 1959. A new Jamaican galliwasp (Sauria: Anguidae). *Brevoria* 102:1–13.

———. 1967. *A contribution to the classification of snakes.* London: British Museum (Natural History).

———. 1970. The eye. In *Biology of the Reptilia*, vol. 2, *Morphology B*, ed. C. Gans and T. S. Parsons, 1–97. London: Academic Press.

U.S. Soil Conservation Service. 1975. Soil taxonomy. *Agricult. Handbk.* 436:1–754.

Uzzell, T. M. 1966. Teiid lizards of the genus *Neusticurus* (Reptilia, Sauria). *Bull. Amer. Mus. Nat. Hist.* 132 (5): 277–328.

Uzzell, T. M., and J. C. Barry. 1971. *Leposoma percarinatum,* a unisexual species related to *L. guianene;* and *Leposoma ioanna,* a new spcies from Pacific coastal Colombia (Sauria, Teiidae). *Postilla* 15:41–39.

Vail, P. R., and J. Hardenbol. 1979. Sea-level changes during the Tertiary. *Oceanus* 22:71–79.

Vail, P. R., R. M. Mitchum Jr., and S. Thompson III. 1977. Seismic stratigraphy and global changes in sea level. 4. Global cycles of relative changes in sea level. *Amer. Assoc. Pet. Geol. Mem.* 26:83–97.

Vaira, M. 1997. *Leptodactylus bolivianus* (NCN): Behavior. *Herp. Rev.* 28 (4): 200.

Valerio, C. E. 1971. Ability of some tropical tadpoles to survive without water. *Copeia* 1971 (2): 364–65.

Van Andel, T. H., G. R. Heath, B. T. Malfait, D. F. Hendricks, and J. I. Ewing. 1971. Tectonics of the Panama Basin, eastern Pacific. *Geol. Soc. Amer. Bull.* 82:1489–1508.

Vandelli, D. 1761. *Epistola de holoturio, et testudine coriacea ad celeberrimum Carolum Linnaeum Equitem Naturae curiosum dioscorcidem II.* Padua: Conzatti.

Van Devender, R. W. 1975. The comparative demography of two local populations of the tropical lizard, *Basiliscus basiliscus.* Ph.D. diss., University of Michigan.

———. 1978. Growth ecology of a tropical lizard, *Basiliscus basiliscus. Ecology* 59 (5): 1031–38.

———. 1982a. Comparative demography of the lizard *Basiliscus basiliscus. Herpetologica* 38 (1): 189–208.

———. 1982b. Growth and ecology of spiny-tailed and green iguanas in Costa Rica, with comments on the evolution of herbivory and large body size. In *Iguanas of the world,* ed. G. M. Burghardt and A. S. Rand, 162–83. Park Ridge, N.J.: Noyes.

———. 1983. *Basiliscus basiliscus* (chisbala, garrobo, basilisk, Jesus Christ lizard). In *Costa Rican natural history,* ed. D. H. Janzen, 379–80. Chicago: University of Chicago Press.

Van Devender, R. W., and C. J. Cole. 1977. Notes on a colubrid snake, *Tantilla vermiformis,* from Central America. *Amer. Mus. Nat. Hist. Novitates* 2625:1–12.

Van Eijsden, E. H. T. 1977. Notities over *Phyllomedusa lemur. Lacerta* 35 (12): 175–81.

Van Wijngaarden, R., and F. Bolaños. 1992. Parental care in *Dendrobates granuliferus* (Anura: Dendrobatidae), with a description of the tadpole. *J. Herp.* 26 (1): 102–5.

Van Wijngaarden, R., and S. van Gool. 1994. Site fidelity and territoriality in the dendrobatid frog *Dendrobates granuliferus. Amph.-Reptl.* 15:171–81.

Vanzolini, P. E. 1961. *Bachia:* Espécies brasileiras e conceito genêrico (Sauria: Teiidae). *Pap. Avul. Zool. Mus. Zool. Univ. São Paulo* 14 (22): 193–209.

———. 1970. Unisexual *Cnemidophorus lemniscatus* in the Amazonas Valley: A preliminary note (Sauria, Teiidae). *Pap. Avul. Zool. Mus. Zool. Univ. São Paulo* 23 (7): 63–68.

———. 1986. Addenda and corrigenda to the catalogue of Neotropical Squamata. *Smithson. Herp. Info. Serv.* 70:1–25.

Vanzolini, P., and C. Morato de Carvalho. 1991. Two sibling and sympatric species of *Gymnophthalmus* in Roraima, Bra-

sil (Sauria, Teiidae). *Pap. Avul. Zool. Mus. Zool. Univ. São Paulo* 37 (12): 173–226.

Vanzolini, P. E., and E. E. Williams. 1962. Jamaican and Hispaniolan *Gonatodes* and allied forms (Sauria, Gekkonidae). *Bull. Mus. Comp. Zool.* 127 (10): 479–98.

Vellard, J. 1946. Morfología del hemipenis y evolución de los ofidios. *Acta Zool. Lilloana* 3:263–88.

———. 1948. Batracios del Chaco Argentino. *Acta Zool. Lilloana* 5:137–74.

Vial, J. L. 1963. A new plethodontid salamander *(Bolitoglossa sooyorum)* from Costa Rica. *Rev. Biol. Trop.* 11 (1): 89–97.

———. 1966. The taxonomic status of two Costa Rican salamanders of the genus *Bolitoglossa. Copeia* 1966 (4): 669–75.

———. "1967." 1968. The ecology of the tropical salamander, *Bolitoglossa subpalmata,* in Costa Rica. *Rev. Biol. Trop.* 15 (1): 13–115.

Vial, J. L., ed. 1973. *Evolutionary biology of the anurans: Contemporary research on major problems.* Columbia: University of Missouri Press.

———. 1975. Thermal related activity in the Mesoamerican lizard *Gerrhonotus monticolus. Brit. J. Herp.* 5 (5): 491–95.

———. 1984. Comparative field responses to diel and annual thermal regimens among sceloporine lizards, with special reference to *Sceloporus malachiticus. Rev. Biol. Trop.* 32 (1): 1–9.

Vial, J. L., and J. M. Jiménez-Porras. 1967. The ecogeography of the bushmaster, *Lachesis muta,* in Central America. *Amer. Midl. Nat.* 78 (1): 182–87.

Vial, J. L., and J. R. Stewart. 1985. The reproductive cycle of *Barisia monticola:* A unique variation among viviparous animals. *Herpetologica* 41 (1): 51–57.

———. 1989. The manifestation and significance of sexual dimorphism in anguid lizards: A case study of *Barisia monticola. Canadian J. Zool.* 67:69–72.

Vidal, N., G. Lecointre, J. C. Vié, and J. P. Gasc. 1999. What can mitochondrial gene sequences tell us about intergeneric relationships of pitvipers? *Kaupia* 8:107–12.

Villa, J. 1962. *Las serpientes venenosas de Nicaragua.* Managua: Novedades.

———. "1967." 1968. A new colubrid snake from the Corn Islands, Nicaragua. *Rev. Biol. Trop.* 15 (1): 117–21.

———. 1969a. Comportamiento defensivo de la "rana ternero," *Leptodactylus pentadactylus. Rev. Biol. Trop.* 15 (2): 323–29.

———. 1969b. Notes on *Conophis nevermanni,* an addition to the Nicaraguan herpetofauna. *J. Herp.* 3 (3–4): 269–71.

———. 1970. Notas sobre la historia natural de la serpiente de los pantanos, *Tretanorhinus nigroluteus. Rev. Biol. Trop.* 17 (1): 97–104.

———. 1971. *Crisantophis,* a new genus for *Conophis nevermanni* Dunn. *J. Herp.* 5 (3–4): 173–77.

———. 1972a. *Anfibios de Nicaragua: Introducción a su sistemática, vida y costumbres.* Managua: Instituto Geográfico Nacional/Banco Central de Nicaragua.

———. 1972b. Un coral *(Micrurus)* blanco y negro de Costa Rica. *Brenesia* 1:10–13.

———. 1973. A snake in the diet of a kinosternid turtle. *J. Herp.* 3 (4): 380–81.

———. 1977. A symbiotic relationship between frog (Amphibia. Anura, Centrolenidae) and fly larvae (Drosophilidae). *J. Herp.* 11 (3): 317–22.

———. 1978. Symbiotic relationships of the developing frog embryo, with special reference to fly larvae. Ph.D. diss., Cornell University.

———. 1983. *Peces, anfibios y reptiles Nicaraguenses: Lista y bibliografía/Nicaraguan fishes, amphibians and reptiles: A checklist and bibliography.* Managua: Escuela de Ecología y Recursos Naturales, Universidad Centroamericana.

———. 1984a. Biology of a Neotropical glass frog, *Centrolenella fleischmanni* (Boettger), with special reference to its frogfly associates. *Milwaukee Publ. Mus. Contr. Biol. Geol.* 55:1–60.

———. 1984b. The venomous snakes of Nicaragua: A synopsis. *Milwaukee Publ. Mus. Contr. Biol. Geol.* 59:1–41.

———. 1988a. *Typhlops costaricensis. Cat. Amer. Amph. Reptl.* 435:1–2.

———. 1988b. *Crisantophis, C. nevermanni. Cat. Amer. Amph. Reptl.* 429:1–2.

———. 1988c. *Rana vibicaria. Cat. Amer. Amph. Reptl.* 237:1–2.

———. 1990a. *Rana warszewitschii. Cat. Amer. Amph. Reptl.* 459:1–2.

———. 1990b. *Hydromorphus, H. concolor, H. dunni. Cat. Amer. Amph. Reptl.* 472:1–2.

———. 1990c. *Leptotyphlops nasalis. Cat. Amer. Amph. Reptl.* 473:1.

Villa, J. D., and J. R. McCranie. 1995. *Oxybelis wilsoni,* a new species of vine snake from Isla de Roatan, Honduras (Serpentes: Colubridae). *Rev. Biol. Trop.* 43 (1): 297–305.

Villa, J. D., R. W. McDiarmid, and J. M. Gallardo. 1982. Arthropod predators on leptodactylid frog foam nests. *Brenesia* 19–20:577–89.

Villa, J. D., and N. J. Scott Jr. 1967. The iguanid lizard *Enyaliosaurus* in Nicaragua. *Copeia* 1967 (2): 474–76.

Villa, J. D., and L. D. Wilson. 1988. *Celestus bivittatus. Cat. Amer. Amph. Reptl.* 423:1–2.

———. 1990. *Ungaliophis, U. continentalis, U. panamensis. Cat. Amer. Amph. Reptl.* 480:1–4.

Villa, J. D., L. D. Wilson, and J. D. Johnson. 1988. *Middle American herpetology, a bibliographic checklist.* Columbia: University of Missouri Press.

Vinton, K. W. 1951. Observations on the life history of *Leptodactylus pentadactylus. Herpetologica* 7 (2): 73–75.

Viquez, C. 1935. *Animales venenosos de Costa Rica.* San José: Imprenta Nacional.

———. 1941. *Nuestros animales venenosos.* San José: Imprenta Nacional.

———. 1942. Distribución geográfica de nuestras serpientes. *Rev. Inst. Defensa Cafe Costa Rica* 11 (87): 608–11.

Vitt, L. J. 1992. Mimicry of millipedes and centipedes by elongate terrestrial vertebrates. *Nat. Geogr. Res. Explor.* 8 (1): 76–95.

Vitt, L. J., and P. A. Zani. 1996. Ecology of the lizard *Ameiva festiva* (Teiidae) in southeastern Nicaragua. *J. Herp.* 30 (1): 110–17.

———. 1997. Ecology of the nocturnal lizard *Thecodactylus rapicauda* (Sauria: Gekkonidae) in the Amazon region. *Herpetologica* 53 (2): 165–79.

Vitt, L. J., P. A. Zani, J. P. Caldwell, and R. D. Durtsche. 1993. Ecology of the whiptail lizard *Cnemidophorus deppii* on a tropical beach. *Canadian J. Zool.* 71:2391–2400.

Vitt, L. J., P. A. Zani, and R. D. Durtsche. 1995. Ecology of the lizard *Norops oxylophus* (Polychrotidae) in lowland forest of southeastern Nicaragua. *Canadian J. Zool.* 73:1918–27.

Vivó Escoto, J. A. 1964. Weather and climate in Mexico and Central America. In *Handbook of Middle American Indians,* ed. R. C. West, 1:187–215. Austin: University of Texas Press.

Vogt, R. C., and O. Flores. 1992. Effects of incubation temperature on sex determination in a community of Neotropical freshwater turtles in southern Mexico. *Herpetologica* 48 (3): 265–70.

Vogt, R. C., and S. Guzman. 1988. Food partitioning in a Neotropical turtle community. *Copeia* 1988 (1): 37–47.

Vogt, R. C., and C. J. McCoy. 1980. Status of the emydine turtle genera *Chrysemys* and *Pseudemys. Ann. Carnegie Mus.* 49 (5): 93–102.

Volpe, E. P., and S. M. Harvey. 1958. Hybridization and larval development in *Rana palmipes* Spix. *Copeia* 1958 (3): 197–207.

Voris, H. K. 1983. *Pelamis platurus* (culebra del mar, pelagic sea snake). In *Costa Rican natural history,* ed. D. H. Janzen, 411–12. Chicago: University of Chicago Press.

Vyas, D. K., C. Moritz, D. M. Peccinini-Seale, J. W. Wright, and W. M. Brown. 1990. The evolutionary history of parthenogenetic *Cnemidophorus lemnisactus* (Sauria, Teiidae). 2. Maternal origin and age inferred from mitochondrial DNA analyses. *Evolution* 44 (4): 922–32.

Waddington, J., and D. M. Rudkin, eds. 1986. *Proceedings of the 1985 workshop on care and maintenance of natural history collections.* Life Science Miscellaneous Publications. Toronto: Royal Ontario Museum.

Wagler, J. G. 1824. *Serpentum brasiliensium species novae, ou Histoire naturelle des espècies nouvelles de serpens, recueilles et observées pendant le voyage dans l'intérieur du Brésil dans les années 1817, 1818, 1819, 1820 exécute par ordre de Sa Majesté le Roi de Bavière, publiée par J. de Spix.* Vol. 3. Munich: Hubschmann.

———. 1828–33. *Descriptiones et icones amphibiorum.* 3 parts. Munich: Cotta.

———. 1830. *Natürliches system der amphibien, mit vorangehender Classification der Säugethiere und Vogel.* Munich: Cotta.

Wake, D. B. 1966. Comparative osteology and evolution of the lungless salamander family Plethodontidae. *Mem. South. Calif. Acad. Sci.* 4:1–111.

———. 1991. Declining amphibian populations. *Science* 253:860.

Wake, D. B., and A. H. Brame Jr. II. 1963a. The status of the plethodontid salamander genera *Bolitoglossa* and *Magnadigita. Copeia* 1963 (2): 382–87.

———. 1963b. The salamanders of South America. *Contr. Sci. Nat. Hist. Mus. Los Angeles Co.* 69:1–72.

———. 1963c. A new species of Costa Rican salamander, genus *Bolitoglossa. Rev. Biol. Trop.* 11 (1): 63–73.

———. 1966. A new species of lungless salamander (genus *Bolitoglossa*) from Panama. *Fieldiana Zool.*, n.s., 51 (1): 1–10.

Wake, D. B., A. H. Brame Jr. II, and W. E. Duellman. 1973. New species of salamanders, genus *Bolitoglossa* from Panama. *Contr. Sci. Nat. Hist. Mus. Los Angeles Co.* 248:1–19.

Wake, D. B., and I. G. Dresner. 1967. Functional morphology and evolution of tail autotomy in salamanders. *J. Morph.* 122:265–306.

Wake, D. B., and P. Elias. 1983. New genera and a new species of Central American salamanders, with a review of the tropical genera (Amphibia, Caudata, Plethodontidae). *Contr. Sci. Nat. Hist. Mus. Los Angeles Co.* 345:1–119.

Wake, D. B., and A. Larson. 1987. Multidimensional analysis of an evolving lineage. *Science* 238:42–48.

Wake, D. B., and J. F. Lynch. 1976. The distribution, ecology and evolutionary history of plethodontid salamanders in tropical America. *Sci. Bull. Nat. Hist. Mus. Los Angeles Co.* 25:1–65.

Wake, M. H. 1964. The ecogeographic distribution of the lizards of Costa Rica. M.S. thesis, University of Southern California.

———. 1977. The reproductive biology of caecilians: An evolutionary perspective. In *The reproductive biology of amphibians*, ed. D. H. Taylor and S. I. Guttman, 73–101. New York: Plenum Press.

———. 1978. The reproductive biology of *Eleutherodactylus jasperi* (Amphibia, Anura, Leptodactylidae), with comments on the evolution of live-bearing systems. *J. Herp.* 12 (2): 121–33.

———. 1980. Reproduction, growth, and population structure of the Central American caecilian *Dermophis mexicanus*. *Herpetologica* 36 (3): 244–56.

———. 1983. *Gymnopis multiplicata*, *Dermophis mexicanus* and *Dermophis parviceps* (soldas, suelo con suelo, dos cabezas, caecilians). In *Costa Rican natural history*, ed. D. H. Janzen, 400–401. Chicago: University of Chicago Press.

———. 1988. *Gymnopis, G. multiplicatus*. *Cat. Amer. Amph. Reptl.* 411:1–2.

Wake, M. H., and J. A. Campbell. 1983. A new genus and species of caecilian from the Sierra de las Minas of Guatemala. *Copeia* 1983 (4): 857–63.

Wake, M. H., and S. M. Case. 1975. The chromosomes of caecilians (Amphibia: Gymnophiona). *Copeia* 1975 (3): 510–16.

Wake, T. A., and M. H. Wake. 1999. First Quaternary fossil record of caecilians from a Mexican archaeological site. *Quat. Res.* 52:138–40.

Wallach, V. 1995. Revalidation of the genus *Tropidodipsas* Günther, with notes on the Dipsadini and Nothopsini (Serpentes: Colubridae). *J. Herp.* 29 (3): 476–81.

Walls, G. L. 1942. The vertebrate eye and its adpative radiation. *Bull. Cranbrook Inst. Sci.* 19:1–785.

Walters, V. 1953. Notes on reptiles and amphibians from El Volcán de Chiriqui, Panama. *Copeia* 1953 (2): 25–127.

Ward, J. P. 1980. Comparative cranial morphology of the freshwater turtle subfamily Emydinae: An analysis of the feeeding mechanisms and the systematics. Ph.D. diss., North Carolina State University.

———. 1984. Relationships of chrysemyd turtles of North America (Testudines: Emydidae). *Spec. Pub. Mus. Texas Tech. Univ.* 21:1–50.

Warkentin, K. 1995. Adaptive plasticity in hatching age: A response to predation risk trade-offs. *Proc. Natl. Acad. Sci.* 92:3507–10.

———. 1984. The *Pseudohemisus* tadpole: A morphological link between microhylid (Orton type 2) and ranoid (Orton type 4) larvae. *Herpetologica* 40 (2): 138–49.

Wassersug, R. 1971. On the comparative palatability of some dry-season tadpoles from Costa Rica. *Amer. Midl. Nat.* 86 (1): 101–9.

Wassersug, R. J., and W. E. Duellman. 1984. Oral structures and their development in egg-brooding hylid frog embryos and larvae: Evolutionary and ecological implications. *J. Morph.* 182:1–37.

Wassersug, R. J., and W. F. Pyburn. 1987. The biology of the pe-ret' toad, *Otophryne robusta* (Microhylidae), with special consideration of its fossorial larva and systematic relationships. *Zool. J. Linn. Soc. London* 91:137–69.

Watkins, J. F., II, F. R. Gehlbach, and R. S. Baldridge. 1967. Ability of the blind snake, *Leptotyphlops dulcis*, to follow pheromone trails of army ants, *Neivamyrmex nigrescens* and *N. opacothorax*. *SW Nat.* 12:455–62.

Weaver, W. G., Jr., and F. L. Rose. 1967. Systematics, fossil history, and evolution in the genus *Chrysemys*. *Tulane Stud. Zool.* 14 (2): 63–73.

Webb, G. J. W., S. C. Manolis, and P. J. Whitehead, eds. 1987. *Wildlife management: Crocodiles and alligators*. Chipping Norton, NSW: Surrey Beatty.

Webb, S. D. 1977. A history of savanna vertebrates in the New World. 1. North America. *Ann. Rev. Ecol. Syst.* 8:355–80.

———. 1978. A history of savanna vertebrates in the New World. 2. South America and the great interchange. *Ann. Rev. Ecol. Syst.* 9:393–426.

———. 1985. Main pathways of mammalian diversification in North America. In *The great American biotic interchange*, ed. F. G. Stehi and S. D. Webb, 210–17. New York: Plenum Press.

Webb, S. D., and A. Rancy. 1996. Late Cenozoic evolution of the Neotropical mammal fauna. In *Evolution and environment in tropical America*, ed. J. B. C. Jackson, A. F. Budd, and A. G. Coates, 335–58. Chicago: University of Chicago Press.

Weber, H. "1958." 1959. Die Paramos von Costa Rica und ihre pflanzengeographische Verkettung mit den Hochanden Sudamerikas. *Abhandl. Math.-Naturwissen. Klass.* 1958 (3): 1–194.

Weimer, R., F. Bolaños, W. Feichtinger, and M. Schmid. 1994. Die Amphibien von Costa Rica herpetologische Eindrücker einer Forschungesreise. 5. Plethodontidae; Zur ökologischen Situation der Forschlurche. *Sauria* 16 (3): 11–15.

Weimer, R., W. Feichtinger, F. Bolaños, and M. Schmid. 1993a. Die Amphibien von Costa Rica herpetologische Eindrücke einer Forschungsreise.1. Einleitung, Hylidae (1). *Sauria* 15 (2): 3–8.

———. 1993b. Die Amphibien von Costa Rica herpetologische Eindrücke einer Forschungsreise. 2. Hylidae (2), Bufonidae, Centrolenidae, Dendrobatidae, Ranidae. *Sauria* 15 (3): 17–23.

———. 1993c. Die Amphibien von Costa Rica herpetologische Eindrücke einer Forschungsreise. 3. Leptodactylidae (1). *Sauria* 15 (4): 19–24.

———. 1994. Die Amphibien von Costa Rica herpetologische Eindrücke einer Forschungsreise. 4. Leptodactylidae (2), zur evolutionsbiologie der Gattung *Eleutherodactylus. Sauria* 16 (2): 15–20.

Wellman, J. 1963. A revision of the snake genus *Conophis* (family Colubridae), from Middle America. *Misc. Pub. Mus. Nat. Hist. Univ. Kansas* 15 (6): 251–95.

Wells, K. D. 1977a. The social behavior of anuran amphibians. *Anim. Behav.* 25:666–93.

———. 1977b. The courtship of frogs. In *The reproductive biology of amphibians*, ed. D. H. Taylor and S. I. Guttman, 233–62. New York: Plenum Press.

———. 1978. Courtship and parental care in a Panamanian poison-arrow frog *(Dendrobates auratus). Herpetologica* 34 (2): 148–55.

———. 1989. Vocal communication in a Neotropical treefrog, *Hyla ebraccata:* Response of males to graded aggressive calls. *Copeia* 1989 (2): 461–66.

Wells, K., and K. M. Bard. 1987. Vocal communication in a Neotropical treefrog, *Hyla ebraccata:* Responses of females to advertisement and aggressive calls. *Behaviour* 101 (1–3): 200–210.

———. 1988. Parental behavior of an aquatic-breeding tropical frog, *Leptodactylus bolivianus. J. Herp.* 22 (3): 361–64.

Wells, K. D., and B. J. Greer. 1981. Vocal responses to conspecific calls in a Neotropical hylid frog, *Hyla ebraccata. Copeia* 1981 (3): 615–24.

Wells, K. D., and J. J. Schwartz. 1984a. Vocal communication in a Neotropical frog, *Hyla ebraccata:* Advertisement calls. *Anim. Behaviour* 32:405–20.

———. 1984b. Vocal communication in a Neotropical treefrog, *Hyla ebraccata:* Aggressive calls. *Behaviour* 91 (1–3): 128–45.

Werler, J. E. 1949. Eggs and young of several Texas and Mexican snakes. *Herpetologica* 5 (2): 59–60.

———. 1951. Miscellaneous notes on the eggs and young of Texas and Mexican snakes. *Zoologica* 36 (1): 37–48.

———. 1970. Notes on young and eggs of captive reptiles. *Int. Zoo. Yearbook* 10:105–8.

Werman, S. D. 1984a. Taxonomic comments on the Costa Rican pit viper *Bothrops picadoi. J. Herp.* 18 (2): 207–10.

———. 1984b. The taxonomic status of *Bothrops supraciliaris. J. Herp.* 18 (4): 484–86.

———. 1992. Phylogenetic relationships of Central and South American pitvipers of the genus *Bothrops (sensu lato):* Cladistic analyses of biochemical and anatomical characters. In *Biology of the pitvipers*, ed. J. A. Campbell and E. D. Brodie Jr., 21–40. Tyler, Tex.: Selva.

———. 1999. Molecular phylogenetics and morphological evolution in Neotropical pitvipers: An evaluation of mitochondrial DNA sequence information and the comparative morphology of the cranium and palatomaxillary arch. *Kaupia* 8:113–26.

Werner, C., and J.-C. Rage. 1994. Mid-Cretaceous snakes from Sudan: A preliminary report on an unexpectedly diverse snake fauna. *C.R. Assoc. Franc. Avanc. Sci. Paris.* 319:247–52.

Werner, F. J. M. 1896. Beiträge zur Kenntniss der Reptilien und Batrachier von Centralamerika und Chile, sowie einiger seltenerer Schlagenarten. *Verh. Zool. Botan. Gesell. Wien.* 1896:344–64.

———. 1903a. Neue Reptilien und Batrachier aus dem Naturhistorishen Museum in Brüssel. *Zool. Anz.* 26 (693): 246–53.

———. 1903b. Ueber Reptilien und Batrachier aus Guatemala und China in der zoologischen Staats-Sammlung in München nebst einem Anhang über seltene Formen aus anderen Gebieten. *Abh. Bayer Akad. Wissen. Math.-Phys. Kl.* 22 (2): 343–84.

———. "1909." 1910. Ueber neue oder seltene Reptilien des naturhistorischen Museums in Hamburg. 1. Schlangen. *Mitt. Naturh. Mus. Hamburg.* 26 (2): 205–47.

———. 1925. Neue oder wenig bekannte Schlangen aus dem Wiener naturhistorischen Staatsmuseum. 2. *Sitz. Akad. Wissen. Wien-Math.-Naturwissen. Kl.* 134 (1): 45–66.

West, L., and W. P. Leonard. 1997. *How to photograph reptiles and amphibians.* Mechanicsburg, Pa.: Stackpole.

Wetmore, A., and J. Zetek. 1951. Report on the Canal Zone Biological Area. *Annl. Rept. Smithson. Inst.* 1950:133–44.

Wettstein, O. 1934. Ergebnisse der österreichischen biologischen Costa Rica–Expedition 1930: Die Amphibien und Reptilien. *Sitz. Akad. Wissen. Wien-Math.-Naturwissen. Kl.* 143 (1–2): 1–39.

Wever, E. G. 1978. *The reptile ear.* Princeton: Princeton University Press.

———. 1985. *The amphibian ear.* Princeton: Princeton University Press.

Weygoldt, P. 1980. Complex brood care and reproductive behavior in captive poison-arrow frogs, *Dendrobates pumilio* O. Schmidt. *Behav. Ecol. Sociobiol.* 73:29–32.

———. 1987. Evolution of parental care in dart poison frogs (Amphibia: Anura: Dendrobatidae). *Z. Zool. Syst. Evol.-Forsch.* 25 (1): 51–67.

Weyl, R. 1980. *Geology of Central America.* 2d ed. Berlin: Borntraeger.

Whistler, D. P., and J. W. Wright. 1989. A Late Miocene rear-fanged colubrid snake from California with comments on the phylogeny of North American snakes. *Herpetologica* 45 (3): 350–67.

Whitmore, F. C., Jr., and R. H. Stewart. 1965. Miocene mammals and Central American seaways. *Science* 148:180–85.

Whitmore, T. C., and G. T. Prance. 1987. *Biogeography and Quaternary history in tropical America.* Oxford: Oxford University Press.

Wicknick, J. A. 1993. *Mesaspis monticola* (NCN): Reproduction. *Herp. Rev.* 24 (1): 33–34.

Wied, Maximilian, Prinz von. 1822–31. *Abbildungen zur Naturgeschichte von Brasiliens.* 15 nos. Weimar.

Wiegmann, A. F. A. 1828. Beyträge zur Amphibienkunde. *Isis von Oken* 21 (3–4): 364–83.

———. 1833. Herpetologische Beyträge. 1. Ueber die mexicanischen Kröten nebst Bemerkungen über ihnen verwandte Arten anderer Weltgegenden. *Isis von Oken* 26:651–62.

———. 1834a. *Herpetologia Mexicana, seu Descriptio am-*

phibiorum Novae Hispaniae. 1. Saurorum species. Berlin: Lüderitz.

———. "1834b." 1835. Amphibien. In Beiträge zur Zoologie, gesammelt auf ein Reise un die Erde, by F. G. F. Meyen. Nova Acata. Acad. C. Leop.-Carol. Halle 17 (1): 183–88.

Wiens, J. J., and T. W. Reeder. 1997. Phylogeny of the spiny lizards (Sceloporus) based on molecular and morphological evidence. Herp. Monogr. 11:1–101.

Wiest, J. A. 1978. Revision of the Neotropical snake genus Chironius Fitzinger (Serpentes: Colubridae). Ph.D. diss., Texas A&M University.

Wilczynski, W., and E. A. Brenowitz. 1988. Acoustic cues mediate inter-male spacing in a Neotropical frog. Anim. Behav. 36:1054–63.

Williams, E. E. 1950. Testudo cubensis and the evolution of Western Hemisphere tortoises. Bull. Amer. Mus. Nat. Hist. 95 (1): 1–36.

———. 1966. South American anoles: Anolis biporcatus and Anolis fraseri (Sauria, Iguanidae) compared. Brevoria 239: 1–14.

———. 1972. The origin of faunas: Evolution of lizard congeners in a complex island fauna—a trial analysis. Evol. Biol. 6:47–89.

———. 1976a. West Indian anoles: A taxonomic and evolutionary summary. 1. Introduction and a series list. Brevoria 440:1–21.

———. 1976b. South American anoles: The species groups. Pap. Avul. Zool. Mus. Zool. Univ. São Paulo 29 (26): 259–68.

———. 1983. Ecomorphs, faunas, island size, and diverse end points in island radiations of Anolis. In Lizard ecology: Studies of a model organism, ed. R. B. Huey, E. R. Pianka, and T. W. Schoener, 326–70. Cambridge: Harvard University Press.

———. 1984. New or problematic Anolis from Colombia. 3. Two new semiaquatic anoles from Antioquia and Chocó, Colombia. Brevoria 478:1–22.

———. 1989. Old problems and new opportunities in West Indian biogeography. In Biogeography of the West Indies: Past, present and future, ed. C. A. Woods, 1–46. Gainesville Fla.: Sandhill Crane Press.

Williams, E. E., and W. E. Duellman. 1967. Anolis chocorum, a new punctatus-like anole from Darien, Panama (Sauria: Iguanidae). Breviora 256:1–12.

Williams, E. E., H. Rand, A. S. Rand, and R. J. O'Hara. 1995. A computer approach to the comparison and identification of species in difficult taxonomic groups. Brevoria 502:1–47.

Williams, E. E., O. A. Reig, P. Kiblisky, and C. Rivero-Blanco. 1970. Anolis jacare Boulenger, a "solitary" anole from the Andes of Venezuela. Brevoria 353:1–15.

Williams, K. L. 1978. Systematics and natural history of the American milk snake Lampropeltis triangulum. Milwaukee Publ. Mus. Spec. Pub. Biol. Geol. 2:1–258.

———. 1988. Systematics and natural history of the American milk snake, Lampropeltis triangulum. 2d ed. Milwaukee: Milwaukee Public Museum.

———. 1994. Lampropeltis triangulum. Cat. Amer. Amph. Reptl. 594:1–10.

Wilson, E. O., and W. L. Brown. 1953. The subspecies concept and its taxonomic application. Syst. Zool. 2 (3): 97–111.

Wilson, L. D. 1968. A fracture plane in the caudal vertebrae of Pliocercus elapoides (Serpentes: Colubridae). J. Herp. 1 (1–4): 93–94.

———. 1970. A review of the chloroticus group of the colubrid snake genus Drymobius, with notes on a twin-striped form of D. chloroticus (Cope) from southwestern Mexico. J. Herp. 4 (3–4): 155–64.

———. "1974." 1975. Drymobius margaritiferus. Cat. Amer. Amph. Reptl. 172:1–2.

———. 1975a. Drymobius. Cat. Amer. Amph. Reptl. 170: 1–2.

———. 1975b. Drymobius chloroticus. Cat. Amer. Amph. Reptl. 171:1–2.

———. 1982. A review of the colubrid snakes of the genus Tantilla of Central America. Milwaukee Publ. Mus. Contr. Biol. Geol. 52:1–77.

———. 1984. Additional notes on colubrid snakes of the genus Tantilla from tropical America. Herp. Rev. 15 (1): 8.

———. 1985a. Tantilla reticulata. Cat. Amer. Amph. Reptl. 370:1.

———. 1985b. Tantilla annulata. Cat. Amer. Amph. Reptl. 379:1.

———. 1986. Tantilla alticola. Cat. Amer. Amph. Reptl. 400:1.

———. 1987a. Tantilla schistosa. Cat. Amer. Amph. Reptl. 409:1–2.

———. 1987b. Tantilla vermiformis. Cat. Amer. Amph. Reptl. 410:1.

———. 1987c. A resume of the colubrid snakes of the genus Tantilla of South America. Milwaukee Publ. Mus. Contr. Biol. Geol. 68:1–35.

———. 1988. Amastridium, A. veliferum. Cat. Amer. Amph. Reptl. 449:1–3.

———. 1992. Tantilla melanocephala. Cat. Amer. Amph. Reptl. 547:1–3.

———. 1999. Checklist and key to the species of the snake genus Tantilla (Serpentes: Colubridae), with commentary on distribution. Smithson. Herp. Info. Serv. 122:1–34.

Wilson, L. D., and D. E. Hahn. 1973. The herpetofauna of the Islas de la Bahia, Honduras. Bull. Fla. State Mus. 17 (2): 93–150.

Wilson, L. D., and J. R. McCranie. 1984. Bothrops nasuta. Cat. Amer. Amph. Reptl. 349:1–2.

———. 1985. A new species of red-eyed Hyla of the uranochroa group (Anura: Hylidae) from the Sierra de Omoa of Honduras. Herpetologica 41 (2): 133–40.

———. 1986. A new species of red-eyed treefrog of the Hyla uranochroa group (Anura: Hylidae) from northern Honduras. Proc. Biol. Soc. Wash. 99 (1): 51–55.

———. 1994. Second update of the list of amphibians and reptiles known from Honduras. Herp. Rev. 25 (4): 146–50.

———. 1997. Publication in non-peer-reviewed outlets: The case of Smith and Chiszar's "Species-group taxa of the false coral snake genus Pliocercus." Herp. Rev. 28 (1): 18–21.

———. 1998. The biogeography of the herpetofauna of the subhumid forests of Middle America (Isthmus of Tehuantepec to northwestern Costa Rica). *R. Ontario Mus. Life Sci. Contr.* 163:1–50.

Wilson, L. D., J. R. McCranie, and K. L. Williams. 1985. Two new species of fringe-limbed hylid frogs from Nuclear Middle America. *Herpetologica* 41 (2): 141–50.

Wilson, L. D., and C. E. Mena. 1980. Systematics of the *melanocephala* group of the colubrid snake genus *Tantilla*. *Mem. San Diego Soc. Nat. Hist.* 11:1–58.

Wilson, L. D., and J. R. Meyer. 1969. A review of the colubrid snake genus *Amastridium*. *Bull. South. Calif. Acad. Sci.* 68 (3): 146–60.

———. 1971. A revision of the *taeniata* group of the colubrid snake genus *Tantilla*. *Herpetologica* 27 (1): 11–40.

———. 1972. The coral snake *Micrurus nigrocinctus* in Honduras (Serpentes: Elapidae). *Bull. South. Calif. Acad. Sci.* 71 (1): 139–46.

———. 1981. Systematics of the *calamarina* group of the colubrid snake genus *Tantilla*. *Milwaukee Publ. Mus. Contr. Biol. Geol.* 42:1–25.

———. 1982. *The snakes of Honduras*. Special Publications in Biology and Geology. Milwaukee: Milwaukee Public Museum.

———. 1985. *The snakes of Honduras*. 2d ed. Milwaukee: Milwaukee Public Museum.

Wilson, L. D., L. Porras, and J. R. McCranie. 1986. Distributional and taxonomic comments on some members of the Honduran herpetofauna. *Milwaukee Publ. Mus. Contr. Biol. Geol.* 66:1–18.

Wilson, L. D., and D. C. Robinson. 1971. Additional specimens of the colubrid snake *Amastridium veliferum* (Colubridae) from Costa Rica, with comments on a pseudohermaphrodite. *Bull. South. Calif. Acad. Sci.* 70 (1): 53–54.

Winterer, E. L., D. M. Hussong, and R. W. Decker, eds. 1985. *The Eastern Pacific Ocean and Hawaii*. Vol. N of *The Geology of North America*. Boulder, Colo.: Geological Society of America.

Witschi, E. 1956. *Development of vertebrates*. New York: Saunders.

Wolfe, J. A. 1985. Distribution of major vegetational types during the Tertiary. In *The carbon cycle and atmospheric CO^2: Natural variations Archean to present*, ed. E. T. Sundquist and W. S. Broecker, 357–75. Washington, D.C.: American Geophysical Union.

Wood, F., and J. Wood. 1993. Growth curve for captive-reared green sea turtles, *Chelonia mydas*. *Herp. J.* 3 (2): 49–54.

Woods, C. A., ed. 1989. *Biogeography of the West Indies: Past, present and future*. Gainesville, Fla.: Sandhill Crane Press.

Wright, J. W. 1993. Evolution of the lizards of the genus *Cnemidophorus*. In *Biology of whiptail lizards (genus* Cnemidophorus*)*, ed. J. W. Wright and L. J. Vitt, 27–77. Herpetologists' League Special Publication 3. Norman: Oklahoma Museum of Natural History and University of Oklahoma.

Wright, J. W., and L. J. Vitt, eds. 1993. *Biology of whiptail lizards (genus* Cnemidophorus*)*. Herpetologists' League Special Publication 3. Norman: Oklahoma Museum of Natural History and University of Oklahoma.

Wright, K. M., and B. R. Whitaker, eds. 2001. *Amphibian medicine and captive husbandry*. Malabar, Fla.: Kreiger.

Yntema, C. L. 1976. Effects of incubation temperatures on sexual differentiation in the turtle, *Chelydra serpentina*. *J. Morph.* 150:453–62.

Yntema, C. L., and N. Mrosovsky. 1980. Sexual differentiation in hatchling loggerheads *(Caretta caretta)* incubated at different temperatures. *Herpetologica* 36 (1): 33–36.

Young, A. M. 1982. Lizard eggs and a snake *(Imantodes)* occupy "artificial bromeliads" in Costa Rican cacao plantations. *Brenesia* 19–20:393–96.

Young, B. A. 1990. Is there a direct link between the ophidian tongue and Jacobson's organ? *Amph.-Reptl.* 11:263–76.

Zaher, H. 1994a. Les Tropidopheoidea (Serpentes: Alethinophidia) sont-ils réellement monophylétiques? Arguments en faveur de leur polyphylétisme. *C.R. Acad. Sci. Paris* 317:471–78.

———. 1994b. Phylogénie des Pseudoboini et évolution des Xenodontinae sud-américains (Serpentes, Colubridae). D.Sc. diss., Musée National d'Histoire Naturelle, Paris.

———. 1996. A new genus and species of pseudoboine snake, with a revsion of the genus *Clelia* (Serpentes, Xenodontinae). *Boll. Mus. Reg. Sci. Nat. Torino* 14 (2): 289–337.

———. 1998. The phylogenetic position of *Pachyrhachis* within snakes (Squamata, Lepidosauria). *J. Vert. Paleont.* 18 (1): 1–3.

———. 1999. Hemipenial morphology of the South American xeodontine snakes, with a proposal for a monophyletic Xenodontinae and a reappraisal of colubrid hemipenes. *Bull. Amer. Mus. Nat. Hist.* 240:1–168.

Zaher, H., and L. C. Prudente. 1999. Intraspecific variation of the hemipenis in *Siphlophis* and *Tripanurgos*. *J. Herp.* 33 (4): 698–702.

Zamudio, K. R., and H. W. Greene. 1997. Phylogeography of the bushmaster (*Lachesis muta*: Viperidae): Implications for Neotropical biogeography, systematics, and conservation. *Biol. J. Linn. Soc. London.* 62:421–42.

Zetek, J., and A. Wetmore. 1951. Report on the Canal Zone Biological Area. *Annl. Rept. Smithson. Inst. for 1950, Appendix* 10:133–43.

Zimmermann, E. 1986. *Breeding terrarium animals*. Neptune City, N.J.: Tropical Fish Hobbyist.

Zimmermann, H. 1982. Durch Nachzucht erhalten: Blattsteigerfrösche *Phyllobates vitattus* und *P. lugubris*. *Aquar. Mag.* 1982 (2): 109–12.

Zimmermann, H., and E. Zimmermann. 1988. Etho-taxonomie un zoogeographische Artengruppenbildung bei Pfeilgiftfröschen (Anura: Dendrobatidae). *Salamandra* 24 (2–3): 125–60.

Zug, G. R. 1966. The penial morphology and the relationships of cryptodiran turtles. *Occas. Pap. Mus. Zool. Univ. Mich.* 647:1–24.

———. 1971. Buoyancy, locomotion, morphology of the pelvic girdle and hindlimb, and systematics of cryptodiran turtles. *Misc. Pub. Mus. Zool. Univ. Mich.* 142:1–98.

———. 1983. *Bufo marinus* (sapo grande, sapo, giant toad, marine toad). In *Costa Rican natural history*, ed. D. H. Janzen, 386–87. Chicago: University of Chicago Press.

Zug, G. R., and C. H. Ernst. 1994. *Lepidochelys*. *Cat. Amer. Amph. Reptl.* 587:1–6.

Zug, G. R., C. H. Ernst, and R. V. Wilson. 1998. *Lepidochelys olivacea. Cat. Amer. Amph. Reptl.* 653:1–13.

Zug, G. R., S. B. Hedges, and S. Sunkel. 1979. Variation in reproductive parameters of three Neotropical snakes, *Coniophanes fissidens, Dipsas catesbyi,* and *Imantodes cenchoa. Smithson. Contr. Knowledge* 300:1–20.

Zug, G. R., E. Lindgren, and J. R. Pippett. 1975. Distribution and ecology of the marine toad, *Bufo marinus,* in Papua New Guinea. *Pacific Sci.* 29 (1): 31–50.

Zug, G. R., and P. B. Zug. 1979. The marine toad, *Bufo marinus:* A natural history resume based on native populations. *Smithson. Contr. Zool.* 284:1–58.

Zweifel, R. G. 1964a. Life history of *Phrynohyas venulosa* (Salientia: Hylidae) in Panama. *Copeia* 1964 (1): 201–8.

———. 1964b. Distribution and life history of the Central American frog *Rana vibicaria. Copeia* 1964 (2): 300–308.

———. 1965. Distribution and mating calls of the Panamanian toads, *Bufo coccifer* and *B. granulosus. Copeia* 1965 (1): 108–10.

———. 1971. Results of the Archbold expeditions. 96. Relationships and distribution of *Genyophryne thomsoni,* a microhylid frog of New Guinea. *Amer. Mus. Nat. Hist. Novitates* 2469:1–13.

Systematic Index

When more than one page reference is given, **boldface** indicates the primary discussion.

Ablabes decipiens, 645
Abronia, 532, 830
Acacia, 82, 84
Acalypha, 85
Acanthophiliinae, 703
Achras zapota, 84, 85
Aciachne pulvinata, 90
Acontinae, 501
Acrantophis, 563
Acris, 275
Acrochordidae, 535, 536, 569
Acrostichum, 81
Acutotyphlops, 556
adders. *See* Viperidae
Adelphicos, 704
Aeluroscalabotes, 481
Agalychnis, 161, 176, 276, **278–88**, 286, 610, 615
    annae, 86, 165, 180, 277, **278–79**, 284, pl. 131
    calcarifer, 277, 278, **279–81**, 285, pls. 132, 133
    callidryas, 95, 163, 164, 180, 277, **281–83**, 284, 285, 286, 300, 615, pls. 134, 135, 136
    craspedopus, 278
    helenae, 283
    litodryas, 286
    moreletti, 279
    saltator, 164, 180, 277, 279, **283–85**, 286, 300, pl. 137
    spurrelli, 164, 180, 277, 279, 282, **285–86**, 300, pls. 138, 139
Agamidae, 415, 425
agamids. *See* Agamidae
Agkistrodon, 550, 717, **718–19**, 728
    bilineatus, 35, **718–19**, 729, 733, 734, 833, 836, pl. 473
    contortrix, 718
    piscivorous, 718
    taylori, 718, 833, 836
agkistrodontines, 718–19
Aiouea costaricensis, 88
Alchornea
    costaricensis, 84, 85, 86, 87
    latifolia, 84, 87
Alethinophidia, 559–737
Alfaroa costaricensis, 85
Alligatorinae, 772, **776–78**
alligators. *See* Alligatorinae
Allobates, 376, 388

Allophrynidae, 159
allophrynid frogs. *See* Allophrynidae
Allophylus psilospermus, 85, 86
alma de vaca. See *Rhinophrynus dorsalis*
Alnus jorullensis, 87, 89
Alsophilia sp., 90
amastridines, 571
Amastridium, 550, **583–84**
    sapperi, 584
    veliferum, **583–84**, 603, 605, pl. 356
Amblyrhynchus, 433
    cristatus, 425
Ambystoma, 123
Ambystomatidae, 121
Ameiva, 420, 423, 425, 506, **508–14**, 515, 586, 652, 711
    ameiva, 507, **508–10**, 513, 516, 517, pl. 301
    festiva, 37, 507, 508, **510–11**, 512, 514, pls. 302, 303, 304
    festiva occidentalis, 511
    leptophrys, 507, 508, **511–12**, 513, 514, pls. 305, 306
    quadrilineata, 83, 507, 508, 510, **512–13**, 514, 517, pl. 307
    taylori, 719
    undulata, 36, 507, 508, 510, 512, **513–14**, 719, pl. 308
Amolops, 397
Amphibia, 839
Amphignathodon, 294
Amphisbaena fulginosa, 411
Amphisbaenidae, 415
Amphiuma, 124
amphiumas. *See* Amphiumidae
Amphiuma tridactylum, 126
Amphiumidae, 121
Anacardium, 86
    excelsum, 84, 85
Anadia, 423, **518–19**, 519
    metallica, 519
    metallica arborea, 519
    metallica metallica, 519
    ocellata, 36, **518–19**, 524, 532, pls. 311, 312
Anaxagorea costaricensis, 86
Andira inermis, 84, 85
Andiras
    davidianus, 126
    japonicus, 126
Andropogon littoralis, 83

Aneides, 127
Anguidae, 415, 417, 526, **526–34**
anguid lizards. *See* Anguidae
Anguimorpha, 525–34
anguimorph lizards. *See* Anguimorpha
Anguinae, 526
Aniliidae, 536, 559
Anisolepis, 444
Anniellinae, 526
Annona holosericea, 84
anoles, 446–80
Anolis, 444
    achilles, 469
    biscutiger, 471
    curtus, 465, 468
    frenatus, 455
    humilis marsupialis, 461, 462
    humilis uniformis, 461
    insignis, 455
    kidderi, 466
    microtus, 456
    pachypus, 465
    procellaris, 470
    pulchripes, 472
    pupurescens, 455
    tropidolepis, 465
    ustus, 466
anoloids. *See* Polychrotidae
Anomalepididae, 553–56, 703
Anomalepis, 42, 544, 553, **553–54**, 558, 711
    dentatus, 554
    mexicanus, **554**, 555, 557, pl. 329
Anomochilidae, 535, 536, 559
Anomolepis, 545
Anotheca, 175, 288, **295–97**
    coronata, 297
    spinosa, 41, 42, 175, 178, 288, **295–97**, 338, pl. 144
Anura, 840
Apeiba
    aspera, 85
    tibourbou, 84, 85
    Aracia, 86
Arctostaphylos rubescens, 90
Ardisia, 85, 87, 88, 89, 429
Aromobates, 376, 377
    nocturnus, 376
Arthroleptidae, 159
Ascaphidae, 158, 159
Ascaphus, 161

Asian pipe snake. *See* Cylindrophiidae
Aspidosperma megalocarpum, 84, 85
Asterophyrinae, 392
Astrocaryum, 82
　　alatum, 86
Atelopus, 111, 165, 166, 169, 170, 175,
　　177, 178, **186–91**, 213
　　chiriquiensis, 186, **187–88**, 189, 190, pls.
　　　63, 64
　　loomisi, 191
　　senex, 187, **188–89**, 190, 197, pls. 65, 66
　　varius, 37, 49, 177, 178, 186, 187, 188,
　　　189, 190, 192, pls. 67, 68, 69
　　varius ambulatorius, 191
　　varius varius, 191
　　zeteki, 191
Atlantic green turtle. See *Chelonia mydas*
Atractaspididae, 536, 569
Atractus, 829
Atropoides, 538, 551, 573, 717, 718, **720–
　　23**
　　nummifer, 37, 551, 659, **720–22**, 730,
　　　731, pls. 474, 475
　　olmec, 720
　　picadoi, 720, **722–23**, 731, pl. 476
Atta, 577
Australian elapids. *See* Elapidae
Australopapuan frogs. *See* Myobatrachidae
Australopapuan treefrogs. *See* Pelodryadidae
Avicennia, 81
Azemiopinae, 716

baba. See *Caiman crocodilus*
babilla. See *Caiman crocodilus*
Bachia, 423, **519–21**
　　blairi, 117, 118, 119, 120, 419, **520–21**,
　　　pl. 313
　　guianensis, 520
　　heteropa, 520
　　pallidiceps, 520, 521
　　panoplia, 520
Bactris, 82
　　gasipaes, 28
Barisia, 532
　　monticola, 534
Baryphthengus, 384
Basiliscus, 414, 424, 425, **426–32**, 438,
　　445, 829
　　basiliscus, 427, **428–30**, 430, 431, 829,
　　　pls. 234, 235
　　galerisuta, 829
　　galeritus, 429, 430
　　nausuta, 829
　　plumifrons, 427, 428, 429, **430–31**, pls.
　　　236, 237, 238
　　vittatus, 35, 83, 427, 428, 429, 430, **431–
　　　32**, pl. 239
Batagurinae, 764–69
Batrachoseps, 127
beaded lizards. *See* Helodermatidae
bejuquillas. See *Imantodes*
Bellucia costaricensis, 84
béquer. See *Boa constrictor*
big-headed turtles. *See* Platysternidae
Billia colombiana, 85, 87

Bipes, 414
black ctenosaur. See *Ctenosaura similis*
black hawk. See *Buteogallus urubitinga*
black mangrove. See *Avicennia*
Blechum, 90
　　loxense, 90
blind snakes. *See* Typhlopidae
blind worm snakes. *See* Anomalepididae
blunt-headed vine snakes. *See* Imantodes
Boa, 439, 545, **562–67**, 591, 717
　　constrictor, 35, 544, 560, **563–64**, 565,
　　　566, 568, 652, 727, 731, pls. 335,
　　　336
　　cookii, 566
　　enydris, 566
　　hortulanus, 566
boas. *See* Boidae
bocaracá. See *Bothriechis schlegelii*
Bocconia fratescens, 87
Bogertrophis, 663
Boidae, **562–69**, 703, 717
Bolitoglossa, 125, 126, **127–39**, 146, 594,
　　643, 650, 826, 829, 830
　　adespersa, 130
　　alvaradoi, **129–30**, 134, pl. 25
　　arborescandens, 130
　　cerroensis, 129, **130**, 131, 138, pl. 26
　　colonnea, 128, **131**, 134, pl. 27
　　compacta, 129, 130, **131–32**, 136, 139,
　　　pl. 28
　　diminuta, 129, **132**, 133, pl. 29
　　epimela, 129, **132–33**, 134, pl. 30
　　flaviventris, 139
　　gracilis, 129, 132, **133**, 136, 139, pls. 31,
　　　32
　　lignicolor, 129, 130, **133–34**, 137, pls. 33,
　　　34
　　marmorea, 138, **839–40**, 840
　　minutula, 128, 133, **134**, pl. 36
　　nigrescens, 129, **135**, 137, 139, pl. 37
　　palustris, 134
　　pesrubra, 45, 90, 111, 129, 132, 133,
　　　135–36, 139, 534, pls. 38, 39, 40
　　platydactyla, 130
　　robusta, 40, 126, 135, **136–37**, 146, pl.
　　　42
　　rufescens, 128
　　salvinii, 138
　　schizodactyla, 128, **137**, 138, pl. 43
　　sooyorum, 129, 130, **137–38**, 840, pl. 35
　　striatula, 128, 131, 137, **138–39**, 139, pl.
　　　44
　　subpalmata, 129, 132, 133, 135, 136,
　　　137, 138, **139**, 534, 840, pls. 45, 46
　　toresi, 136
Bolyeria, 410
Bolyeriidae, 535, 536
Bombacopsis, 86
　　quinatum, 84
Bombinatoridae, 159
Bothriechis, 538, 550, 717, 718, 719, 720,
　　723–26
　　lateralis, 675, 678, 718, **723–24**, 724, pls.
　　　477, 478
　　nigroviridis, 38, 89, 723, **724–25**, pl. 479

　　schlegelii, 37, 610, 633, 723, **725–26**,
　　　pls. 480, 481, 482, 483, 484
　　"*supraciliaris*," pl. 485
Bothriopsis, 718
Bothrops, 538, 550, 573, 718, 719, 720,
　　726–28, 829
　　asper, 14, 35, 563, 573, 583, 662, 663,
　　　716, 717, 719, **726–28**, pl. 486
　　atrox, 727, 728, 731
　　colombiensis, 728
　　godmani, 729
　　lansbergii, 733, 735
　　lateralis, 724
　　nasutus, 733
　　nummifer, 721
　　picadoi, 722
　　schlegelii supraciliaris, 726
　　supraciliaris, 726
Brachycephalidae, 159
Brachylophus, 433
Bradytriton, 127
Bravaisia integerrima, 86
Brevicipitinae, 392
Brosimum, 82, 87
　　alicastrum, 84, 85
　　costaricanum, 85
　　panamense, 86, 87
　　terrabanum, 85
　　utile, 84, 85, 87
brown forest turtle. See *Rhinoclemmys
　　annulata*
Brunellia costaricensis, 87, 89
Buddleia alpina, 89
Bufo, 82, 173, 178, **191–211**, 213, 276, 808
　　arenarum, 199
　　auritus, 207, 208
　　blombergi, 198
　　caeruleostictus, 198
　　campbelli, 203
　　canaliferus, 203
　　coccifer, 38, 164, 179, 193, 195, **203–7**,
　　　208, 209, 210, 211, 488, pl. 77
　　coniferus, 179, 195, 206, **207–8**, pl. 78
　　cristatus, 203
　　fastidiosus, 179, 193, 194, **195–96**, 197,
　　　198, 205, 212, pl. 70
　　gabbi, 208
　　gemmifer, 203
　　glaberrimus, 198
　　guttatus, 193, **198–99**
　　haematiticus, 181, 192, 193, **198–99**,
　　　205, pl. 72
　　holdridgei, 165, 177, 179, 193, 194, **196–
　　　98**, 205, 212, pl. 71
　　hypomelas, 198
　　ibarai, 203
　　ictericus, 199
　　luetkenii, 164, 181, 195, 204, 206, 207,
　　　208–9, 211, pl. 79
　　macrocristatus, 203
　　marinus, 35, 163, 165, 179, 186, 192,
　　　193, **199–202**, 205, pl. 73
　　melanochlorus, 104, 165, 193, 194, 195,
　　　206, 207, 208, **209–10**, 211, pls. 80, 81
　　occidentalis, 203

periglenes, 44, 49, 111, 170, 179, 193, 194, **202–3, 202–12**, 205, pls. 74, 75, 76
peripetates, 58, 196, 198
poeppigii, 199
rufus, 199
tacanensis, 203
turtelarius, 203
valliceps, 163, 179, 193, 195, 200, **203**, 204, 206, 208, 209, **210–11**, pl. 82
Bufonidae, 159, 160, 161, 166, 176, **185–86**
Bungarinae, 703
Bursera, 86
simaruba, 84, 85
tomentosa, 84
bushmasters. See *Lachesis*
Buteogallus urubitinga, 432, 563, 583, 659, 666, 678, 687, 721, 726, 733
Byrsonima, 85, 86
crassifolia, 84
densa, 90

Caecilia
thompsoni, 116
volcani, 58
caecilians. See Caeciliidae
Caeciliidae, **116–20**, 839
Caesalpina coriaria, 82
caiman. See *Caiman crocodilus*
Caiman, 767, **776–78**
apaporensis, 777
crocodilus, 35, 52, 220, 763, 768, 774, 775, **776–78**, 778, pl. 514
latirostris, 777
yacare, 777
caimans. See Alligatorinae
Calamagrostis, 90
Calatola costaricensis, 85
Callopistes, 506
Callopsis, 764
Calloselasma, 718
Calocarpum mammosum, 87
Calophyllum brasiliense, 85, 87
Calycophyllum candidissimum, 84
camaléon. See *Corytophanes cristatus*
Canavalis maratima, 83
cane toad. See *Bufo marinus*
Canis latrans, 748, 749, 763, 767, 768
cantil. See *Agkistrodon bilineatus*
Capparis pittieri, 86
Carapa guianensis, 85
Caretta, 52, 743, 754, **755–56**, 760, 761
caretta, 35, **755–56**, pl. 499
Cariniana pyriformis, 85
carpintera. See *Lepidochelys olivacea*
Carpotroche platypetra, 86
Carretochelyidae, 738
casabel. See *Crotalus durissus*
casabela muda. See *Lachesis melanocephala*; *Lachesis stenophrys*
Casearia, 86, 87, 88
aculeata, 84
arborea, 87
javitensis, 87

Casilla, 85
Cassia, 87
spectabilis, 84
Cassipourea guinanensis, 87
castellana. See *Agkistrodon bilineatus*
Castilla elastica, 85
Catostoma psephota, 620
Caudata, **839–40**
Cecropia, 84, 86, 90
obtusifolia, 87
peltata, 84
polyphlebia, 87
Cedrela, 86
tonduzii, 87, 88
Ceiba, 85
pentandra, 84, 85
Celestus, 424, 501, 526, **527–30**
chryoschorus, 528
cyanochloris, 40, **527–29**, 530, pl. 318
enneagrammus, 527
hylaius, 527, 528, **529**, 530, pls. 319, 320
montanus, 527
orobius, 527, 528, **529–30**
striatus, 527
Centrolene
acanthidiocephalum, 357
geckoideum, 357, 358
paezorum, 357
Centrolenella, 173, 175, 177, 178, 358, 359, 360–62, 370
albomaculata, 359
andina, 360
antioquiensis, 360
azulae, 360
ballux, 360
buckleyi, 360
euknemos, 359, 360
fernandoi, 360
grandisonae, 360
granulosa, 359, 360
guanacara, 360
heloderma, 360
hesperia, 360
huilensis, 360
ilex, 360, **360–61**, pl. 194
lemniscata, 360
mariae, 360
medemi, 360
muelleri, 360
notosticta, 360
petrophila, 360
pipilata, 360
prosoblepon, 40, 166, 179, 359, 360, **361–62**, pl. 195
puyoensis, 360
quindiana, 360
robledoi, 360
scirtetes, 360
spinosa, 359
tayrona, 360
Centrolenidae, 159, 176, 276, **357–76**
Cephaelis, 85
elata, 87
tomentosa, 87
Ceratophryinae, 166, 213

Cerrophidion, 551, 717, 718, **728–29**
barbouri, 728
godmani, 551, 718, 719, **728–29**, pl. 487
petalcalensis, 728
tzotzilorum, 728
Cespedezia macrophylla, 86
Cestrum
aurantiacum, **87–88**, 88
panamense, 87
Chalcidolepis metallica, 519
Chamaedorea, 85
Chamaeleolis, 444
Chamaeleonidae, 415, 416, 425, 426
Chamaelinorops, 444
chameleons. See Chamaeleonidae
Charina reinhardtii, 562
Chelidae, 738
Chelonia, 52, 742, 743, 754, 756
agassizii, 753, **756–57**, 759, pls. 500, 501
mydas, 35, 54, 753, 755, 756, 757, **758–59**, pl. 502
Cheloniidae, 738, **753–62**
Chelydra, 742, 743, **762–64**
serpentina, 742, 746, 747, 749, **762–64**, pls. 506, 507, 508
Chelydridae, 738, **762–64**
Chimarris latifolia, 85
Chiococca phaenostemon, 88
Chioglossa, 124
chirbala lisa. See *Sphenomorphus cherriei*
chirbalas. See Teiidae
Chirixalus eiffingeri, 296
Chironectes, 429
Chironius, 539, 545, **647–51**, 669, 717
carinatus, **648–49**, 650, 655, pl. 409
exoletus, 648, **649–50**, 662, 670, pl. 410
fuscus, 651
grandisquamis, 573, 574, 648, **650–51**, 665, 666, 686, 687, pl. 411
melas, 651
Chiropterotriton, 127
abscondens, 141
picadoi, 144
Chrisantophis nevermanni, 549
Chrysemys, 743, 764, **769–71**
adiutrix, 771
callirostris, 771
dorbigni, 771
gaigaea, 771
ornata, 36, 45, 766, 767, **769–71**, pl. 512
scripta, 769, 771
Chrysobalanus, 83
Chrysopelea, 537
Chrysophyllum panamense, 85
Chusquea, 90
subtesselata, 90
Citharexylum
donnell-smithii, 88
lankesteri, 88
Citronella costaricensis, 88
Claudius, 744
"clawed" frogs. See Pipidae
Clelia, 538, 546, **572–75**, 650, 666, 690
clelia, 35, 52, **573–79**, 697, 727, pls. 342, 343

Clelia (continued)
  clelia clelia, 574
  equatoriana, 574
  scytalina, 573, **574–75**, pls. 344, 345
Clethra gelida, 87, 89
Cleyra theaeoides, 88
Clusia, 88, 89
  alata, 87, 89
  pithecobia, 87
Cnemidophorus, 423, 425, 506, 508, 509,
  514–17, 561, 586, 652, 711, 802
  alfaronis, 516
  amivoides, 514
  angusticeps, 516
  arubensis, 516
  ceralbensis, 515
  cozumela, 516
  cryptus, 516, 517
  deppii, 36, 83, 508, **515–16**, 517, pl.
    309
  gramivagus, 516, 517
  lacertoides, 516
  leachei, 516
  lemniscatus, 58, 507, 515, **516–17**, pl.
    310
  lineatissimus, 515
  longicaudus, 516
  maslini, 516
  murinus, 516
  nigricolor, 516
  ocellifer, 516
  pseudolemniscatus, 516, 517
  rodecki, 516
  serranus, 516
  sexlineatus, 516
  vanzoi, 516
  vittatus, 516
cobras. *See* Elapidae
Coccoloba standleyana, 85
Cochlospermum, 85
  vitifolium, 84
Cochranella, 177, 178, 358, **362–67**, 373,
  pl. 196
  albomaculata, 363, **365**, 369, pl. 198
  castroviejoi, 362
  daidalea, 362
  decorata, 373
  euknemos, **363**, 366, 369
  fleischmanni, 364
  granulosa, 179, **362–64**, **363–64**, 364,
    365, 366, 369, pl. 197
  millepunctata, 373
  ocellata, 362
  prosoblepon, 366, 369
  reticulata, 375
  savagei, 362
  solitaria, 362
  spinosa, 179, 362, **365–67**, pl. 199
  talamancae, 375
Coleonyx, 423, **481–82**
  brevis, 481
  elegans, 481
  fasciatus, 481
  mitratus, 481–82, pl. 282
  reticulatus, 481

  switaki, 481
  variegatus, 481
collared lizards. *See* Crotaphytidae
Colobognathus hoffmanni, 38
Coloptychon, 424, 526, **532–33**
  rhombifer, 532–33, pl. 326
Colostethus, 165, 173, 174, 376, **377–82**
  flotator, 178, 377, **378–80**, 381, pls. 211,
    212
  nubicola, 178, 377, 378, **380–81**, pl.
    213
  pratti, 378
  talamancae, 179, 377, 378, 380, **381–82**,
    pl. 214
Coluber, 539, 541, 550, 647, **651–52**, 673
  anthonyi, 651
  constrictor, 651
  flagellum, 651
  mentovarius, 10, 543, **651–52**, 659, 674,
    pl. 412
  mentovarius mentovarius, 10
  mentovarius striolatus, 10
Colubridae, 537, 542, 569, **570–702**, 717
colubrines, 571, **647–99**
Composneura sprucei, 84, 85
cone-nosed frog. *See* Rhinophrynidae; *Rhinophrynus dorsalis*
Coniophanes, 395, 541, 548, 549, 585, **591–
  96**, 663, 704, 711
  adresemsos, 593
  alvarezi, 593
  bipunctatus, 585, 587, **592–93**, 594,
    pl. 362
  fissidens, 37, 585, 592, **593–95**, 594, 624,
    627, pl. 363
  imperialis, 587, 595
  piceivittis, 587, 592, **595–96**, pl. 364
  quinquevittatus, 592
  schmidti, 595
Conolophus, 433
Conophis, 548, 549, 571, **584–86**
  lineatus, 549, **585–86**, 587, 595, pl. 357
  vittatus, 585
Conostegia oerstediana, 88
Constrictor constrictor, 564
Copeothyphlinus, 119
Cophosaurus, 439
Cophylinae, 392
Corallus, 439, 545, 562, 563, **564–67**, 568,
  591, 717
  annulatus, 560, **564–65**, 566, pl. 337
  caninus, 564
  cookii, 567
  cropanii, 564
  grenadensis, 567
  hortulanus, 560, 564, 565, **566–67**, pl.
    338
  ruschenbergerii, 567
coral macho. *See* *Micrurus nigrocinctus*
coral pipe snake. *See* Aniliidae
coral snakes. *See* Elapidae; Elapinae
Cordia, 86, 429
  alliodora, 84, 86
  toqueve, 87
Cordylidae, 415

Cornus, 89
  disciflora, 87, 88, 89
Cornutia grandifolia, 87
Cortaderia, 90
Corucia zebrata, 500
Corytophanes, 417, 425, 426, **432–33**, 446
  cristatus, **432–33**, 474, pls. 240, 241
  hernandezi, 433
  percarinatus, 433
Corytophanidae, 426–33
Couratari panamensis, 85
Coussapoa, 85
Craugaster, 808
Crepidophryne, 174, 186, 192, 193, **211–13**
  epiotica, 186, 194, 196, 197, 206, **211–
    12**, pl. 83
Crescentia alata, 84
cribo. *See* *Drymarchon corais*
Cricosaura, 497
Crisantophis, 549, 585, **586–87**
  nevermanni, 585, **586–87**, 593, 595, 645,
    pl. 358
Croatlinae, 716
crocodiles. *See* Crocodylinae
Crocodilia, 772–79
crocodilians. *See* Crocodilia
Crocodilurus, 506
Crocodylidae, 772, **775–79**, **778–79**
Crocodylinae, 772
Crocodylus, 767, **778–79**
  acutus, 35, 36, 52, 54, 439, 768, 774,
    775, 776, **778–79**, pls. 515, 516
  crocodilus, 777
  porosus, 773
  sclerops, 777
crotalines, 718, **735–37**
Crotalus, 545, 550, 573, 717, **735–37**, 802
  catalinensis, 735
  durissus, 35, **735–37**, pl. 495
Crotaphaga, 439
Crotaphytidae, 415, 417
Croton
  glabellus, 85, 87
  panamensis, 85
Cryptobranchidae, 121
Cryptobranchoidea, 121
Cryptodira, 738, **744–71**
Cryptotis nigrescens, 729
Cryptotriton, 833
Ctenonotus, 425, 444, 446, 448, **452–53**
  bimaculatus, 452
  cristatellus, 447, **452–53**, pl. 253
  cybotes, 452
Ctenosaura, 424, 425, 433, **434–37**, 561,
  711
  acanthura, 434
  alfredschmidti, 434
  bakeri, 434
  clarki, 434
  defensor, 434
  dryinus, 736
  flavidorsalis, 434, 435
  hemilopha, 434
  melanosterna, 434
  oedirhina, 434

palearis, 434, 435

pectinata, 434

quinquecarinata, 36, **434–35**, 436, 830, 835, pl. 242

similis, 36, 421, 429, 430, 431, 434, **435–37**, 438, 439, 736, pls. 243, 244

terrificus, 736

vergrandis, 736

culebra del mar. *See Pelamis platurus*

culebrilla del café. *See Ninia sebae*

Cupania sp., 87

Cupressus, 479

Curatella americana, 84

Cyathea

divergens, 87

maxoni, 90

mexicana, 87

Cyclotyphlops, 556

Cylindrophiidae, 536, 559

Cymbopetalum costaricense, 85

Cynometra retusa, 86

Cyphomandra, 85

Cystignathus

fragilis, 222

labialis, 222

Dactyloa, 417, 425, 444, 446, 448, **453–57**, 608

aequatorialis, 453

agassizi, 453

albimaculatus, 456

apollinaris, 454

boettgeri, 456

calimae, 456

caquetae, 456

casildae, 58, 454

chloris, 456

chocorum, 452, 454, **456–57**, pl. 259

danieli, 454

deltae, 456

dissimilis, 456

fasciata, 456

festae, 456

fraseri, 454

frenata, 446, 452, **454–55**, 457, pls. 254, 255

gemmosa, 456

huilae, 456

insignis, 447, 451, 454, **455**, 474, pls. 256, 257

jacare, 456

laevis, 453

latifrons, 453

microtus, 447, 451, 454, **455–56**, 474, pl. 258

nigrolineata, 456

nigropunctata, 456

peraccae, 456

princeps, 454

propinqua, 454

punctata, 453, 456

purpurescens, 454

roquet, 453

santamartae, 456

squamatula, 454

tigrina, 453

transvalis, 456

vaupesiana, 456

Dalbergia retusa, 84

dart-poison. *See* Dendrobatidae

Darwin's frogs. *See* Rhinodermatidae

Dasypus, 763

Deinagkistrodon, 718

Deirochelys, 769

Dendrobates, 52, 170, 174, **382–88**

arboreus, 58

auratus, 37, 179, 382, **383–84**, pl. 215

azureus, 383

galactonotus, 383

granuliferus, 180, 382, **384–86**, 387, pls. 216, 217

pumilio, 37, 48, 180, 382, **384–88, 386–88**, pls. 220, 221

quinquevittatus, 382

speciosus, 58

tinctorius, 382, 383

truncatus, 383

typographicus, 387

Dendrobatidae, 159, 165, 176, 397, **376–91**

Dendropanax, 87

arboreus, 86, 87

gonatopodus, 85

Dendrophidion, 541, 547, 549, 647, **652–58**, 663, 666, 669, 670, 674, 717

bivittatum, 656

brunneum, 656

chloroticus, 662

clarkii, 655

corais, 653

dendrophis, 653, **654–56**, 655, 656

margaritiferus, 653, 654

melanotropis, 654, 662

nuchale, 543, 653, **654–55**, 656, 657, 658, 659, 660, pl. 415

paucicarinatum, 40, 543, 650, 653, 655, **656–57**, 661, 662, pl. 417

percarinatum, 40, 543, 653, 654, **656–58, 657–58**, pl. 418

rhombifer, 653

vinitor, 543, 654, **655–56**, 657, 658, pl. 416

Dendrotriton, 127

Dermatemydidae, 738, **750–53**, 802

Dermochelyidae, 738

Dermochelys, 52, 410, 739, 742, 743, **750–53**, 754, 755

coriacea, 36, 54, 738, 742, **751–53**, pl. 498

Dermophis, 116, 117–18, 554, 557, 558

balboai, 118

costaricense, 118, **839**, pls. 19, 20

glandulosus, 118, **839**

gracilior, 117–18, **839**

mexicanus, 117, 118, **839**

occidentalis, 118, **839**

parviceps, 117, 118, 839, pl. 21

Desmognathus, 127

aeneus, 123, 124

Dialanthera otoba, 85

Dialium guianense, 85

Dibamidae, 417

Dicamptodon, 124

ensatus, 126

Dicamptodontidae, 121

dicamptodontids. *See* Dicamptodontidae

Dicrodon, 506

Dicroglossinae, 397

Didymopanax

morototoni, 84

pittieri, 89, 90

Diploglossinae, 526–32

Diploglossus, 424, 501, 526, 527, **530–34**

bilobatus, 503, 505, 528, 529, **530–31**, 532, pls. 323, 324

monotropis, 530, **531–32**, pl. 325

pleii, 530

Diplolaemus, 444

Diplostephium, 90

Dipsadinae, 571

dipsadines, 571, **591–647**

Dipsas, 548, **596–99**, 608, 627, 634, 705, 717

articulata, 576, 596, **597**, 598, 629, 706

bicolor, 576, 596, **597–98**, 629, 643, 685, 706, pls. 365, 366

brevifacies, 596

costaricensis, 634

gaigeae, 596

gracilis, 596

iguieri, 596

maxillaris, 596

temporalis, 596

tenuissima, 576, 596, 597, **598–99**, 629, 706, pl. 367

viguieri, 598

Dipsosaurus, 433

Dipterodendron costaricense, 84, 85

Dipteryx panamensis, 86

Discoglossidae, 159

dos cabezas. *See* Caeciliidae

Dracaena, 506

Draco, 161, 417, 537

dragón. *See Mesaspis monticola*

Drepanoides, 572

Drimys winteri, 89, 90

Dromicus, 572, 580

Dryadophis sanguiventris, 674

Drymarchon, 548, 647, **658–59**, 673, 721

corais, 35, 543, 544, 652, 655, **658–59**, 687, pl. 419

Drymobius, 541, 546, 647, 653, **660–63**, 717

chloroticus, 657, 662

margaritiferus, 36, 543, **660–61**, pl. 420

melanotropis, 543, 655, 657, **661–62**, 678, pls. 421, 422

rhombifer, 543, 612, 614, **662–63**, pl. 423

Duellmanohyla, 176, 288, 294, 295, **297–301**

chamulae, 297

ignicolor, 297

legleri, 299

lythrodes, 291, **297–98**, 300, 318, 321, 330, pl. 145

Duellmanohyla (continued)
  rufioculis, 178, 179, 291, 297, **298–99**, 300, 318, 321, 330, pl. 146
  salvavida, 297
  schmidtorum, 297
  soralia, 297
  uranochroa, 179, 180, 276, 281, 284, 286, 291, 297, 298, **299–301**, 318, 321, pl. 147
Dussia
  cuscatlanica, 85
  macroprophyllata, 85
dwarf boa. *See* Loxocemidae
Dyscophinae, 392

Echinathera
  amoena, 622
  cynopleura, 622
  melanostigma, 622
  undulata, 622
Ehretia austen-smithii, 88
Eira barbara, 439
Elaphe, 545, 546, 572, 647, **663–65**
  flavirufa, 663, 664
  guttata, 663
  obsoleta, 663
  rosaliae, 663
  subocularis, 663
  triaspis, 543, 568, 663, **664–65**, 698, 717, 727, 729, pls. 424, 425, 426
  vulpina, 663
Elapidae, 536, 538, 569, **702–15**
Elapinae, 703, **707–14**
Eleutherodactylus, 87, 106, 107, 108, 109, 167, 169, 170, 173, 175, 177, 212, **226–75**, 275, 276, 376, 397, 584, 594, 608, 624, 648, 650, 655, 724, 808, 826, pl. 92
  altae, 235, **265–66**, 269, pl. 118
  andi, 231, 232, **237–38**, 240, 241, pl. 93
  angelicus, 232, **248–49**, 250, pl. 102
  berkenbuschii, 252
  biporcatus, 236, **253–56**, 254, 255
  bransfordii, 234, **257–58**, 259, 261, 262, 263, pl. 111
  bufoniformis, 230, 253, 255, **256–57**, pl. 110
  caryophyllaceus, 231, 235, **266–67**, 269, 270, pl. 119
  catalinae, 840, pl. 104
  cerasinus, 235, 236, **264–65**, 268, pl. 117
  crassidigitus, 232, **238–39**, 244, 246, pl. 94
  cruentus, 235, 236, 242, **265–71**, 267–68, 271, pls. 120, 121
  cuaquero, 230, 238, **239–40**, 241
  diastema, 235, 269, **272–75**, pl. 127
  dubitus, 268
  emcelae, 58
  engytympanum, 250
  escoces, 232, 248, **249–50**, pl. 103
  euryglossus, 250
  fitzingeri, 37, 231, 232, **237–45**, 240–41, 252, pl. 95

  fleischmanni, 40, 232, 248, 249, 250, **250**, 252, 840
  florulentus, 256
  gaigeae, 170, 236, **263–64**, 389, 390, pl. 116
  gollmeri, 231, 244, **245–47**, 260, pl. 99
  gulosus, 234, **253–54**, 255, 256, pl. 107
  hylaeformis, 236, 272, **273–74**, 275, pls. 128, 129
  jasperi, 169, 177, 227
  johnstonei, 236, **271–72**, pl. 126
  jota, 58
  lancasteri, 43
  lanciformis, 246
  laticeps, 246
  longirostris, 239, 241, 243
  lutosus, 268, 270
  martinicensis, 271–72
  maussi, 255
  megacephalus, 232, 234, 236, 253, **254–55**, 256, pls. 108, 109
  melanostictus, 232, 234, **241–42**, 268, pl. 96
  mimus, 231, 232, 239, 245, **246–47**, pl. 100
  monnichorum, 58
  moro, 236, **268–69**, 270, 358, pl. 122
  museosus, 58
  noblei, 236, 244, 245, 246, **247–48**, pl. 101
  nubilis, 241
  obesus, 840
  operosus, 227
  pardalis, 235, 266, **269**, pl. 123
  peraltae, 265
  persimilis, 234, **258–59**, pl. 112
  phasma, 230, **242–43**, pl. 97
  pittieri, 252
  platyrhynchus, 242
  podiciferus, 233, 234, 245, 246, 257, **259–60**, pl. 113
  polyptychus, 258
  punctariolus, 232, **250–51**, 253, 322, 840
  ranoides, 232, 241, **251–52**, pl. 105
  rayo, 230, 239, **243–44**
  rhodopis, 234, 235, **257–63**
  rhyacobatrachus, 840
  ridens, 236, 265, 267, **269–71**, pls. 124, 125
  rostralis, 246, 247
  rugosus, 232, 234, 236, 242, 253, **255–56**
  rugulosus, 165, 238, **248**, 252
  stejnegerianus, 40, 234, 257, 258, 259, **261–62**, 263, pl. 114
  talamancae, 237, 239, **244–45**, 247, pl. 98
  taurus, 232, 251, **252–53**, pl. 106
  tigrillo, 235, 272, **274**
  underwoodi, 234, 258, 261, **262–63**, pl. 115
  vocator, 235, **274–75**, pl. 130
Elgaria, 532
Emydidae, 738, **764–71**
Emydinae, 764, **769–71**
Emydoidea, 769

Engelhardtia mexicana, 87
Engystomops, 224
  pustulosus, 225
Enhydris plumbea, 58
Ensatina, 127
Enterolobium, 86
  cyclocarpum, 84
  guatemalense, 87
Enuliophis, 590
Enulius, 541, 549, 571, **587–90**, 601, 602, 605, 640, 646, 682, 683
  flavitorques, 558, **588–89**, 616, 639, pl. 359
  oligostrichus, 589
  sclateri, 547, 588, **589–90**, 616, 639, pl. 360
Enyaliosaurus, 434
Epicrates, 544, 562, 563, 564, **567–69**, 591, 717
  angulifer, 567
  cenchria, 560, 565, 566, **567–69**, pls. 339, 340
Epipedobates, 376, 382, 388
  maculatus, 58
Eretmochelys, 52, 743, 753, 754, 755, 756, **759–60**, 761
  imbricata, 54, 757, 758, **759–60**, pl. 503
Erryalius, 444
Erycinae, 562
Erythrina, 82
  cochleata, 86
Erythrolamprus, 538, 540, 547, 550, 572, **577–80**, 705
  bizona, 576, **578–79**, 580, 606, 665, 666, 696, 707, pl. 349
  mimus, 576, 578, **579–80**, 621, 706, 709, 710, 711, pls. 350, 351
Erythroxylon havanense, 84
Escallonia poasana, 89, 90
escorpión tobobo. See *Thecadactylus rapicauda*
Eublepharidae, 480–82
Eublepharus, 481
  dovii, 482
Eugenia, 86, 87, 89, 478
  storkii, 88
Eumeces, 424, **501–2**
  altamirani, 502
  managuae, 502, pl. 298
  schwartzei, 502
Eumomota, 704
Eunectes murinus, 544, 559
Eupemphix, 224
  pustulosus, 225
Eurycea, 126
Euterpe panamensis, 87
Exiliboa, 569
eyelash geckos. See Eublepharidae

falsa lora. See *Oxybelis fulgidus*
false gharial. See Crocodylinae
Farancia, 571
Felis, 439
  concolor, 748

Feyliniinae, 501
Ficus, 87, 88, 429
  glabrata, 84
  nymphaefolia, 85
  tonduzii, 86
  torresiana, 87
file snakes. *See* Acrochordidae
fire-bellied frogs. *See* Bombinatoridae
flap-foots. *See* Gekkota
Fleischmannia obscura, 584
forest turtles. *See* Emydidae
freshwater turtles. *See* Emydidae
Fuchsia
  arborescens, 88, 89, 90
  microphylla, 90

Gaiadendron poasense, 89, 90
gallina de palo. See *Ctenosaura similis*
gargantilla. See *Micrurus mipartitus*
garrobo. See *Ctenosaura similis*
Garrya laurifolia, 88, 89, 90
Gastrophryne, 173, 174, **392–94**
  carolinensis, 393
  olivacea, 393
  pictiventris, 164, 178, 184, 224, **393–94**, 395, 396, pl. 224
Gastrotheca, 166, 169, 175, 277, 288, **293–94**
  anatomia, 293
  andaquiensis, 293
  angustifrons, 293
  ceratophrys, 294
  cornuta, 106, 167, 175, 178, 288, 293, **294**, pls. 142, 143
  dendronastes, 293
  fissipes, 293
  guentheri, 158, 293
  helenae, 293
  longipes, 293
  marsupiatum, 293
  microdiscus, 293
  ovifera, 293
  testudinea, 293
  walkeri, 293
  weinlandi, 293
  williamsoni, 293
Gavialidae, 772
geckos. *See* Gekkonidae; Gekkota
Gekkonidae, 415, **482–97**
Gekkoninae, 482–89
gekkonines, 482
Gekkota, 480–97
Genipa caruto, 86
Geochelone costaricensis, 771
Geoemyda, 764
Geophis, 89, 547, 588, **599–605**, 618, 704, 711
  bakeri, 603
  brachycephalus, 599, 601, **602–3**, 604, 620, 729, pl. 370
  championi, 58, 599, **600–602**
  dolicocephala, 603
  downsi, **600**, 601, 620
  dunni, 602
  godmani, 599, **600–601**, 729

hoffmanni, 546, 584, 600, **603–4**, 605, 619, pls. 371, 372
  laticollaris, 602
  moesta, 603
  nasalis, 602
  nigroalbus, 603
  petersi, 602
  pyburni, 602
  russatus, 602
  ruthveni, 599, 600, **601–2**, 602, 604, 639
  sallaei, 602
  sieboldi, 599, 602, **602–5**
  talamancae, 599, 601, 602, **604**
  zeledoni, 584, 600, 603, **604–5**
Geothalsia meiantha, 84, 86
Gerrhonotinae, 526, **532–34**
Gerrhonotus monticola, 534
gharial. *See* Gavialidae
ghost frogs. *See* Heleophrynidae
giant salamanders. *See* Cryptobranchidae
girdle-tailed lizards. *See* Cordylidae
gladiator frogs. See *Hyla boans*
glassfrogs. *See* Centrolenidae
Gloeospermum diversipetalum, 86
Glossostoma, 397
Gloydius, 718
Godmania aesculifolia, 84
Gonatodes, 417, 422
  albogularis, **490–91**, 494, 495, 496, pls. 288, 289
  fuscus, 491
Goniurosaurus, 481
Grandsonia brevis, 116
Graptemys, 769
green iguana. See *Iguana iguana*
Grias fendleri, 85
guajipal. See *Caiman crocodilus*
guarda camino. See *Conophis lineatus*
Guarea, 85, 87, 88
  aligera, 86
Guazuma ulmifolia, 84
Gymnophiona, 839
Gymnophthalmidae, 517–25
Gymnophthalmus, 422, 518, 519, **521–22**, 711
  cryptus, 521, 522
  pleii, 521
  rubricauda, 521
  speciosus, 505, **521–22**, 614, pl. 314
  underwoodi, 521, 522
Gymnopis, 116, **118–20**, 554, 557, 558
  multiplicata, 116, 117, **119–20**, pls. 22, 23
  multiplicata multiplicata, 119
  multiplicata proxima, 119
  oligozona, 116, 119
  sartoria, 119
  syntrema, 119–20
Gynerium, 83
Gyrinophilus, 126

Haasiophis terrasanctus, 536
Haideotriton, 126
Hampea, 87
  appendiculata, 84

Haptoglossa pressicauda, 40, 154
harlequin frogs. *See* Atelopus
harmless snakes. *See* Columbridae
Hasseltia floribunda, 86, 88
Hedyosmum
  calloso-serratum, 86
  mexicanum, 89
Heisteria, 87
  densifrons, 87
Heleophrynidae, 159
Heliconia, 83, 90
Helicops wettsteini, 58
Heliocarpus appendiculatus, 87
hellbenders. *See* Cryptobranchidae
helmeted lizards. *See* Corytophanidae
Helminthophis, 544, 545, 553, **554–56**, 558, 711
  frontalis, 544, **554–55**, 557, pl. 330
Helodermatidae, 415, 417, 526, 802
Hemidactylium, 123, 126
Hemidactylus, 421, 422, **482–85**
  brooki, 483
  frenatus, 483, **484–85**
  garnotii, 483, 484, **485**, pl. 283
  giganteus, 483
  mabouia, 483
  turcicus, 483
Hemiphractinae, 275, **293–94**
Hemiphractus fasciatus, 58, 293
Hemisotidae, 159
Hemisus, 161
Hemitheconyx, 481
Hernandia, 86
  sonora, 85
Herpetotheres cachinnas, 643
Hesperomeles
  heterophylla, 89
  obovata, 90
Heterodon, 571
Hibiscus tiliaceous, 83
hidden-neck turtles. *See* Cryptodira
Hieronyma
  alchorneoides, 84, 85
  poasana, 88, 89
Hippocratea obovata, 84
Hirtella racemosa, 87
hog-nosed pitvipers. *See* Porthidium
hog-nosed viper. *See* Porthidium nasutum
Holbrookia, 439
Holodiscus fissus, 89
Hoplocercidae, 414, 415, 417, 426
hoplocercids. *See* Hoplocercidae
Hoplodactylus, 480, 481
Hura crepitans, 86
Hyalinobatrachium, 167, 175, 177, 178, 358, **367–76**
  antisthenesi, 368
  aureoguttatum, 369
  bergeri, 369
  cardiacalyptum, 369
  chirripoi, 367, 368, **369–70**, 370, pls. 201, 202
  colymbiphyllum, 179, 368, **370–71**, 375, pls. 203, 204, 205
  crurifasciatum, 369

Hyalinobatrachium (continued)
  crybetes, 369
  duranti, 369
  esmeralda, 369
  fleischmanni, 40, 49, 86, 166, 179,
    367, 368, **369–76, 370–73**, pls. 206,
    207
  fragilis, 369
  iaspidiensis, 369
  ibama, 369
  lemur, 369
  loreocarinatum, 369
  munozorum, 369
  orientale, 369
  ostracodermoides, 369
  pallidum, 369
  parvulum, 367
  pellucidum, 369
  petersi, 369
  pleurolineatum, 369
  pulveratum, 359, 363, 364, 365, 367,
    368–69, pl. 200
  ruedai, 369
  talamancae, 368, 372, **373–74**
  taylori, 369
  valerioi, 179, 367, 368, 370, **374–75**,
    pls. 208, 209
  vireovittatum, 367, 368, **375–76**, pl.
    210
Hydromantes, 127
Hydromorphus, 82, 546, 561, 569, 572,
    605–6
  clarki, 606
  concolor, 38, 579, **605–6**, 635, pl. 373
  dunni, 58, 605, 606
Hydrophiinae, 703, **714–15**
Hyla, 288, 295, **301–38**, 648, 808
  albofrenata, 301
  albomarginata, **301–3**, 307
  albopunctulata, 308
  albosignata, 301
  alleei, 301
  alvaradoi, 307
  alytolyax, 306
  angustilineata, 165, 180, 289, 306, 317,
    318, **327–28**, pls. 167, 168
  arianae, 301
  arildae, 301
  bifurca, 313
  biobeba, 303
  boans, **303**, 305, 335
  bogotensis, 306–8
  callipeza, 306
  calypsa, 176, 180, 290, **310–11**, 313,
    pl. 155
  cherrei, 317, 318
  colymba, 181, 276, 293, **306–7**, 308, 327,
    358, pl. 152
  crepitans, 303
  debilis, 180, 290, 291, **320–21**, 324, 325,
    pl. 162
  denticulata, 306
  dulcensis, 346
  ebraccata, 164, 176, 178, 276, 285, 290,

291, **313–16**, 317, 318, 319, 320, 327,
    610, 615, pls. 157, 158, 159
  echinata, 332, 334
  elaeochroa, 276, 346
  elegans, 313
  erythromma, 330
  faber, 303
  fimbrimembra, 180, 289, 304, **332–34**,
    335, pl. 173
  fuscata, 58
  gabbi, 357
  godmani, 308–9
  graceae, 58, 326
  gryllata, 313
  immensa, 335
  infucata, 327, 330
  insolita, 310
  jahni, 306
  lancasteri, 180, 290, **309–13, 311–13**,
    pl. 156
  lascinia, 306
  legleri, 180, 291, 295, 300, 318, 330,
    331–32, pl. 172
  leucophyllata, 313–20
  loquax, 180, 292, **308–9**, pl. 154
  lynchi, 306
  manisorum, 351
  mathiassoni, 313
  microcephala, 164, 178, 292, 313, 314,
    315, **316–18**, 319, 320, 327, pl. 160
  miliaria, 161, 289, 304, 333, **334–35**,
    pls. 174, 175, 176
  minera, 332, 334
  misera, 317
  moesta, 322, 328
  molitor, 354
  molitor marmorata, 354
  monticola, 322, 324, 357
  moraviensis, 312
  musica, 301
  palmeri, 181, 293, 302, 358, pl. 153
  pardalis, 303
  parviceps, 313
  pellucens, 301
  phantasmagoria, 335
  phlebodes, 41, 161, 178, 292, 309, 314,
    315, 316, 317, **318–20**, 327
  phyllognatha, 306
  picadoi, 172, 178, 181, 290, 296, 335,
    336–37, pl. 177
  piceigularis, 306
  pictipes, 165, 179, 292, **321–23**, 326,
    328, 348, 357, pl. 163
  plameri, 306, **307–8**
  platydactyla, 306
  pseudopuma, 49, 165, 180, 276, 292, 324,
    326–32, 328–30, 348, 355, pls. 169,
    170, 171
  pugnax, 303, 305
  puma, 354
  punctariola, 322
  rhodopepla, 313
  richardi, 333
  richardtaylori, 333

  rivularis, 180, 292, 321, **323–24**, 325,
    326, 328, pl. 164
  robertmertensi, 313
  rosenbergi, 167, 180, 289, 302, **303–6**,
    333, 335, pls. 150, 151
  rubra, 343
  rubracyla, 301
  rufitela, 276, 289, **302–3**, 304, 305, 306,
    308, pls. 148, 149
  salvadorensis, 330
  salvaje, 332, 334
  sartori, 313
  simmonsi, 306
  splendens, 354
  staufferi, 348
  thysanota, 334
  tica, 180, 290, 321, **324–26**, pl. 165
  torrenticola, 306
  triangulum, 313
  tuberculosa, **332–35**, 342
  underwoodi, 317, 318
  uranochroa, 299, 330
  valancifer, 334
  wavrini, 303
  wellmanorum, 354
  weyerae, 315
  xanthosticta, 292, 321, 322, **326**, pl. 166
  zeteki, 41, 180, 290, 296, **335–38, 337–
    38**, 338, 358, pl. 178
Hylactophryne, 226
Hylella fleischmanni, 338
Hylidae, 159, 161, 164, 165, 166, 176, **275–
    357**, 276
Hylinae, 288–93, **294–357**
Hylodes rugosus, 256
Hylodinae, 213
Hylomantis, 286
Hymenaea courbaril, 84, 85
Hynobiidae, 121
Hypericum
  silenoides, 90
  strictum, 90
Hyperoliidae, 159, 276
Hypnale, 718
Hypopachus, 173, 174, 392, **394–96**
  barberi, 395
  variolosus, 38, 163, 165, 178, 184, 393,
    394–96, 719, pl. 225
Hypsiglena, 615

Ideocranium russeli, 116
Iguana, 424, 433, **437–39**, 561, 711
  delicatissima, 434, 437, 439
  iguana, 52, 421, 429, 430, 431, 435, 436,
    437–39, 446, 746, pl. 246
iguana negra. See *Ctenosaura similis*
iguanas. *See* Iguanidae
iguana verde. See *Iguana iguana*
Iguanidae, 417, 425, **433–39**
Ilex, 89, 90
  pallida, 88
Imantodes, 547, 549, 598, **606–10**, 631,
    634, 717
  cenchoa, 606–8, 609, pls. 374, 375

gemmistratus, 101, 606, 607, **608–9**, pl. 376
inornatus, 282, 285, 606, **609–10**, 725, pls. 377, 378
Inga, 86, 87
coruscans, 85
marginata, 86
oerstediana, 84
punctata, 87
sapinodoides, 86
Ipomoea pes-caprae, 83
Ixalotriton, 127

Jacaranda
copaia, 84
lasiogyne, 84
Jacaratia, 86

Karwinski calderoni, 84
Kinosternidae, 738, **744–50**
Kinosternon, 81, 743, **744–50**
albogulare, 749
angustipons, 742, 745, **746**, 747, 829
dunni, 829
leucostomum, 37, 283, 315, 746, **747–48**, pl. 496
postinguinale, 748
scorpoides, 745, 747, **748–50**, pl. 497
spurrelli, 748
Klauberina, 497
kraits. *See* Elapidae

Lacertidae, 415
lachesines, 718, **719–35**
Lachesis, 538, 550, 551, 717, 718, 719, 721, 722, 727, **729–32**, 830
melanocephala, 544, **730–31**, 732
muta, 731, 732
stenophrys, 544, 730, **731–32**, pls. 489, 490
Lacistema aggregatum, 84, 85, 86
Lacmellea panamensis, 85
Ladenbergia brenesii, 84, 87
Laemanctus, 426
Laetia procera, 86
Lafoensia punicifolia, 86
lagaritja de Altura. *See Mesaspis monticola*
lagartija pintada. *See Lepidophyma reticulatum*
lagarto. *See Caiman crocodilus*
Laguncularia racemose, 81
Lampropeltis, 546, 577, 644, 647, **665–67**, 705
alterna, 665
mexicana, 665
pyromelana, 665
ruthveni, 665
triangulum, 36, 539, 573, 574, 576, 578, 643, 650, **665–67**, 682, 696, 706, 707, 709, 710, 767, pls. 427, 428
zonata, 665
land tortoises. *See* Testudinidae
Laplacea semiserrata, 87
Laticauda colubrina, 715

Leandra
costaricensis, 88
lasiopetala, 84
leatherback turtles. *See* Dermochelyidae
Lecythis costaricensis, 86
Leimadophis, 572, 580
taeniurus juvenalis, 581
Leiolopisma, 505
cherriei, 505
Leiopelma, 161, 169
Leiopelmatidae, 158, 159
Leiosaurus, 444
Leiotyphlops, **555–56**
albirostris, **555–56**, 557, pl. 331
Leiyla güntherii, 238
leopard lizards. *See* Crotaphytidae
Lepidobatrachus, 169, 212
Lepidoblepharis, 417, 422, 489, **490–92**
xanthostigma, 490, **491–92**, 494, 495, 496, 505, pl. 290
Lepidochelys, 49, 52, 106, 743, 754, 755, 756, **760–62**
kempii, 761
olivacea, 46, 48, 561, 741, **760–62**, pls. 504, 505
Lepidodactylus, 213, 414, 421, 422, **485–87**
labialis, 106
lugubris, 483, 484, 485, **486–87**, 489, pl. 284
melanonotus, 164
pentadactylus, 164, 650
poecilochilus, 106
Lepidophyma, 422, 423, 497, **498–500**, 524
anomalum, 499
flavimaculatum, 37, **498–99**, 500, pl. 296
obscurum, 499
ophiophthalmum, 499
reticulatum, 498, 499, 500, pl. 297
Lepidophyminae, 498
Leposoma, 423, 519, **522–23**
bisecta, 523
precarinatum, 522
southi, **522–23**, 525, pl. 315
southi orientalis, 523
Leptodactylidae, 159, 164, 166, 176, **212–75**, **213–26**, 276, 840
Leptodactylus, 45, 173, 175, **213–23**, 395, 397, 401, 403, 404, 648, 719
albilabris, 221, 222
bolivianus, 164, 179, 214, 215, 216, **217–19**, 220, 223, pl. 85
bufonius, 221
camaquara, 221
chaquensis, 217
colombiensis, 215
cunicularius, 221
dantasi, 215
didymus, 221
dominicensis, 221
elenae, 221
fallax, 219
fragilis, 222
furnarius, 221

fuscus, 213
geminus, 221
gracilis, 221
griseigularis, 215
insularum, 219
jolyi, 221
knudseni, 219
labialis, 164, 176, 179, 205, 214, 215, 216, **221–22**, 223, pl. 88
labrosus, 221
labyrinthicus, 219
laticeps, 219
leptodactyloides, 215
lithonaetes, 219
longirostris, 221
macrosternum, 217
maculilabris, 217, 223
marambaiae, 221
melanonotus, 164, 173, 179, 214, **215**, **215–17**, 218, 223, pl. 84
myersi, 219
mystaceus, 221
mystacinus, 221
natalensis, 215
nesiotus, 215
ocellatus, 214, **217**
pallidirostris, 215
pascoensis, 215
pentadactylus, 36, 164, 173, 213, 214, 215, 218, **219–21**, 225, 305, pls. 86, 87
petersii, 215
plaumanni, 221
podicipinus, 215
poecilochilus, 176, 179, 214, 216, 218, **222–23**, pls. 89, 90
pustulatus, 215
quadrivattatus, 223
rhodonotus, 219
riveroi, 214
rugosus, 219
sabanensis, 215
silvanimbus, 215
stenodema, 219
syphax, 219
tapiti, 221
validus, 215
ventrimaculatus, 221
viridis, 217
wagneri, 215
Leptodeira, 81, 395, 550, 608, **610–15**, 634, 705, 717
annulata, 205, 305, **611–12**, **611–13**, 613, 614, 615, 664, pl. 379
bakeri, 611
frenata, 611
nigrofasciata, 582, 610, **613–14**, pls. 381, 382
ocellata, 612
ornata, 615
polysticta, 615
rubricata, 576, 611, **612–13**, 614, 706
septentrionalis, 282, 285, 611, 612, **614–15**, 664, pls. 383, 384

Leptodrymus, 550, 647, **667**, 673
    pulcherrimus, 627, **667**, pl. 429
Leptognathus pictiventris, 630
Leptophis, 547, 647, 662, **668–73**, 678, 717
    aeruginosus, 670
    ahaetulla, 563, 662, **668–70**, 672, 673,
        pl. 430
    depressirostris, 662, 668, 669, **670–71**,
        pls. 431, 432
    mexicanus, 662, 668, **671–72**, pls. 433,
        434
    nebulosus, 662, 668, 671, **672**, pl. 435
    riveti, 543, 668, 669, **672–73**
Leptotyphlopidae, **557–59**, 703
Leptotyphlops, 544, 545, 553, **558–59**,
    704
    ater, 554, 555, 557, **558–59**, pl. 333
    dulcis, 552, 559
    goudotti, 558, 559
    humilis, 559
    nasalis, 559
Leucopternis albicollis, 432
Leurognathus, 127
lily frogs. *See* Hyperoliidae
Lineatriton, 127
Liohyla pittieri, 252
Liophis, 550, **580–82**, 705
    epinephalus, 239, 576, **580–82**, 613, 617,
        620, 657, 664, 706, pls. 352, 353
    epinephalus bimaculatus, 581
Liotyphlops, 544, 553, 558
    albirostris, 37, 554
Lioyla guentheri, 238
Lithodytes
    bransfordii, 258
    florulentus, 256
    gaigei, 264
    megacephalus, 255
Liyla rugulosa, 238
lizards. *See* Sauria
Lonchocarpus, 85
    costaricense, 84
    rugosus, 86
lora. See *Bothriechis lateralis; Lepidochelys
    olivacea*
lora falsa. See *Leptophis ahaetulla*
Lorenzochloa rectifolia, 90
Loxocemidae, **560–62**
Loxocemus, 545, 546, 548, **560–62**, 572,
    591
    bicolor, 436, 439, 560, **561–62**, 606,
        pl. 334
Luehea, 86
    seemannii, 84, 86
    speciosa, 84
lungless salamanders. *See* Plethodontidae
Lygophis, 572, 580
Lygosoma, 505
Lygosominae, 501, **502–6**
Lysiloma, 86
    seemannii, 84
Lystrophis, 572, 644

Mabuya, 420, 424, **502–4**, 711
    alliacea, 504

    brachypodia, 504
    mabouya, 504
    unimarginata, 503, 504, 505, 528, 529,
        531, 532, 719, pl. 299
Machaerium biovalatum, 84
Macroclemys temmicki, 762
macroteiid lizards. *See* Teiidae
Madagascar iguanians. *See* Opluridae
Magnolia poasana, 89
Malaclemys, 739, 769
mambas. *See* Elapidae
Manicaria, 82
Manilkara, 429
Mannophryne, 376, 377
mano de Piedra. See *Atropoides nummifer*;
    Atropoides picadoi
marsh turtles. *See* Emydidae
marsupial frogs. *See Gastrotheca*
Mastichodendron
    capiri, 88
    melanolomus, 659, 660
Masticophis, 651
Mastigodryas, 550, 647, **673–75**
    cliftoni, 673
    danieli, 673
    dorsalis, 673
    melanolomus, 543, 651, 652, 664, **673–
        75**, pls. 436, 437
matabuey. See *Lachesis melanocephala; La-
    chesis stenophrys*
Mauria, 87
    biringo, 88
    glauca, 88
Megaphyridae, 159
Megaselia scalaris, 768
Megophryidae, 159
megophryids. *See* Megaphyridae
Melanobatrachinae, 392
Meliosma
    glabrata, 88
    irazuensis, 88
    vernicosa, 87
Mertensiella
    caucasia, 123
    luschani, 123
Mesaspis, 423, 526, **533–34**
    gadovii, 532
    monticola, 90, 503, 505, 519, 531, 532,
        533–34, 729, pls. 327, 328
    moreletti, 532, 534
    viridiflava, 532, 534
Mesoamerican river turtles. *See* Dermatemy-
    didae
Mesoeumeces, 502
mica. See *Spilotes pullatus*
Miconia, 87, 89
    biperulifera, 90
Microbatrachylus
    bransfordii, 258
    persimilis, 259, 261
    polyptychus, 258, **260–61**, 262, 263
    rearki, 259
    underwoodi, 263
Microdromus virgatum, 694
Microhyla, 394, 396

Microhylidae, 159, 161, 164, 165, 166, 169,
    176, **391–97**
microhylids. *See* Microhylidae
Microhylinae, 392
microteiid lizards. *See* Gymnophthalmidae
Microtropis occidentalis, 88, 89
Micruroides, 703
    euryxanthus, 704
Micrurus, 538, 546, 563, 572, 667, 703,
    707–14, 826
    alleni, 572, 576, 667, 683, 706, 707, **709–
        10**, 711, 712, pl. 461
    alleni richardi, 709
    clarki, 576, 683, 706, 707, 709, **710**, 711,
        pl. 462
    collaris, 707
    diastema, 621
    fulvius, 707, **709–12**
    mipartitus, 576, 597, 598, 614, 629, 630,
        631, 643, 685, 706, 707, 709, **712–14**,
        pls. 467, 468, 469
    multifasciatus, 712
    multisculatus, 712
    nigrocinctus, 37, 86, 557, 576, 621, 643,
        683, 690, 707, 709, **710–12**, 768, pls.
        463, 464, 465, 466
    nigrocinctus alleni, 709
    nigrocinctus melanocephalus, 711
    nigrocinctus mosquitensis, 711
    nigrocinctus nigrocinctus, 711
    nigrocinctus yatesi, 709
    pachecoi, 711
    richardi, 709
    spixii, 707
Minascaecilia sartoria, 119
Minquartia guianensis, 85
Minyobates, 376, 377, 382
mole salamanders. *See* Ambystomatidae
Molossidae, 431
Momotus, 384
monitor lizards. *See* Varanidae
Monnia sp., 90
Monteverde golden toad. See *Bufo periglenes*
Montichardia, 81
Mora, 82
Mortoniodendron membranaceum, 86
Mosquitoxylum jamaicense, 87
Mouriri cyphocarpa, 87
mudpuppies. *See* Proteidae
mud turtles. *See* Kinosternidae
Muntingia, 429
Mus, 729
musk turtles. *See* Kinosternidae
Myobatrachidae, 159, 166
Myrica arborescens, 87, 89
Myriocarpa, 86, 87
Myrrhidendron donnell-smithii, 90

narrow-headed vine snake. See *Oxybelis
    aeneus*
Nasua narica, 439
Natator depressa, 755
natator frogs. *See* Pseudidae
Natricinae, 571
natricines, 699–702

Naultinus, 480
Nectandra, 87, 89
  panamensis, 87
Nectophrynoides, 186
Nelsonophryne, 173, 174, 392, 393, 394,
  396–97
  aequatorialis, 396
  aterrima, 178, 184, 395, **396–97**, pl.
    226
Neopareas, 598
  tricolor, 598
Neotropical frogs. *See* Leptodactylidae
Nephelobates, 377
Nerodia, 541
  sipedon, 58
Neusticurus, 423, 519, **523–24**
  apodemus, 499, 500, 519, **523–24**, pl.
    316
newts. *See* Salamandridae
New World treefrogs. *See* Hylidae
New Zealand frogs. *See* Leiopelmatidae
Nicoria, 764
night lizards. *See* Xantusiinae
Nimbaphrynoides, 169, 186
  liberiensis, 166
  occidentalis, 166
Ninia, 89, 546, 547, 588, 589, **615–22**, 704,
  711
  atrata, 618
  celata, 602, 604, 616, **617–18**, 639, pl.
    385
  cerroensis, 620
  espinali, 618
  lansbergii, 618
  maculata, 86, 616, 617, **618–19**, 729, pl.
    386
  oxynotus, 620
  pavimenta, 619
  psephota, 581, 600, 601, 602, 604, 616,
    617, 618, **619–20**, pl. 387
  sebae, 573, 574, 616, **620–22**, 690, 706,
    pl. 388
  tessellata, 619
Norops, 81, 417, 420, 425, 433, 444, 446,
  448, **457–80**, 594, 608, 829
  albii, 468
  altae, 449, 451, **468–69**, 470, 473, 474,
    pl. 268
  annectans, 446, 457
  antonii, 468
  aquaticus, 450, **458**, 472, pl. 260
  attenuatus, 480
  biporcatus, 36, 450, 455, 456, 457, **474–
    76**, pls. 274, 275, 276
  bocourtii, 468
  capito, 449, 453, 474, **476–77**, 479, pl.
    277
  carpenteri, 451, 468, **469–70**, 474, pl.
    269
  compressicauda, 461
  cupreus, 37, 101, 450, **458–60**, 466, 470,
    488, 652, pl. 261
  dariense, 460
  dolfusianus, 460
  exsul, 58, 468

fortunensis, 58, 468
fungosus, 451, 474, **477**, 478, 479
fuscoauratus, 468, 471
gibbiceps, 468
godmani, 472
humilis, 449, **460–62**, 467, pl. 262
ibague, 474
intermedius, 86, 449, 450, **462–63**, 464,
  pl. 263
kemptoni, 58, 468
laeviventris, 463
lemniscatus, 468
lemurinus, 449, 450, 462, **463–64**, 468,
  505, pl. 264
limifrons, 244, 452, 459, 468, 469, **470–
  72**, 473, 474, pl. 270
lionotus, 468, 473
loveridgei, 474
macrolepis, 468
macrophallus, 460
maculiventris, 468
mariarum, 468
medemi, 468
nannodes, 463
onca, 446, 457
ortonii, 468
oxylophus, 450, 458, **472–73**, pl. 271
pachypus, 452, 461, **464–65**, 467, 468
pandoensis, 451, 468, 469, **473**
pentaprion, 451, 474, **477–78**, 479, pl.
  276
petersii, 474
poecilopus, 468
polylepis, 452, 459, 468, 469, 470, **473–
  75**, pls. 272, 273
rivalis, 468
rodriguezii, 471
salvini, 463
scapularis, 468
sericeus, 450, 459, **465–66**, pl. 265
serranoi, 464
sulcifrons, 474
tolimensis, 468
townsendi, 41, 450, **466–67**
trachyderma, 468
tropidolepis, 449, 461, 464, 465, **467–68**,
  pls. 266, 267
tropidonotus, 461
uniformis, 461
villai, 460
vittigerus, 464, 468
vociferans, 451, 469, 474, 477, **478–79**,
  pl. 279
woodi, 449, 453, 476, **479–80**, pls. 280,
  281
woodi attenuatus, 480
Nothopsis, 545, 565, 566, 568, 570, 572,
  590–91
  affinis, 591
  rugosus, 561, **590–91**, 717, pl. 361
  torresi, 591
Nototriton, 125, 126, 127, **139–45**, 146,
  830, 833
  abscondens, 140, **141–42**, 143, 144, 145,
    pl. 47

barbouri, 140
brodei, 140
gamezi, 140, 141, **142**, pl. 48
guanacaste, 140, 141, **142–43**, pl. 49
lignicola, 140
limnospectator, 140
major, 140, 141, **143–44**
picadoi, 41, 42, 140, **141–44**, 144, pl. 50
richardi, 140, 141, 142, **144–45**, 153,
  154, 157, pl. 51
stuarti, 140
tapanti, 140, 141, 142, **145**, 153, 157,
  pl. 52
Nyctanolis, 127
Nymphaea, 81

Ochroma lagopus, 87
Ocotea
  cernua, 87
  seibertii, 88
Oedipina, 45, 50, 121, 125, 126, 127, 140,
  145–47, 146
  alfaroi, 95, 147, 148, **150–51**, 155, 156
  alleni, 146, 148, **149**, 150, 152, pl. 53
  altura, 147, 148, 150, 151, 152, 153, 154,
    155, 156, 157
  bonitaensis, 157
  carablanca, 146, 148, **149–50**
  collaris, 41, 127, 146, 147, 148, 149, 150,
    151–52
  complex, 146, 149, 150
  cyclocauda, 147, 148, 150, 151, **152–53**,
    154, 155, 156, 157, pls. 55, 56
  elongata, 149
  gephyra, 149
  gracilis, 150, 151, 152, **153**, 154, 155,
    156, 157, pls. 57, 58
  grandis, 144, 147, **153–54**, 155, pl. 59
  inusitata, 157
  longicauda, 157
  longissima, 157
  maritima, 149
  pacificensis, 148, 150, 151, 152, 153, **154**,
    155, 156, 157
  parvipes, 146, **149–50**
  paucidentata, 147, 150, 151, 153, **154–
    55**, 157
  poelzi, 148, 150, 154, **155–56**, pl. 60
  pseudouniformis, 148, 150, 151, 152,
    153, 154, **156**, 157
  savagei, 148, 149, **150**, pl. 54
  serpens, 152
  stenopodia, 150
  stuarti, 150
  syndactyla, 157
  uniformis, 10, 38, 125, 146, 147, 148,
    150–51, 152, 153, 154, 155, **156–57**,
    637, pl. 61
  vermicularis, 157
Oedipus
  gracilis, 145
  mexicanus, 130
  pacificensis, 145
  picadoi, 142, 144
  salvinii, 137–38

Oedipus (continued)
  uniformis, 145
  variegatus, 134
Oedopinola, 146
Old World treefrogs. *See* Rhacophoridae
olive ridley. *See Lepidochelys olivacea*
Ollotis coerulescens, 196
olm. *See* Proteidae
Olmedia falcifolia, 87
Ololygon
  boulengeri, 345
  elaeochroa, 346, 348
  staufferi, 348
Ophiodes, 527
Ophiophagus hannah, 544
Ophryacus, 718, 732
Opisthodelphys, 293
Opluridae, 415
Opuntia, 82
Oreopanax
  capitatum, 88
  nubigenum, 89
  xalapense, 88, 90
oropel. *See Bothriechis schlegelii*
Oryzomys, 729
Oscaecilia, 116, 120, 554, 557, 558
  bassleri, 120
  elongata, 120
  equatorialis, 120
  hypereumeces, 120
  koepckeorum, 120
  ochrocephala, 120
  osae, 117, 118, 119, 120, pl. 24
  polyzona, 120
  zweifeli, 120
Osteocephalus oophagus, 296
Osteolaemus, 778
Osteopilus, 175, 288, **338–40**
  brunneus, 296, 338
  dominicensis, 338
  septentrionalis, 180, 289, **338–40**, 350,
    352, pl. 179
Otolygon, 343
Otophryne, 169, 391
Otophryninae, 392
Ouratea
  lucens, 87
  tuerckheimii, 86
Oxybelis, 81, 546, 549, 647, 662, **675–79**,
  717
  aeneus, 35, 439, **676–77**, 679, pls. 438,
    439
  brevirostris, 669, 670, 675, **677–78**, pl.
    440
  fulgidus, 35, 36, 546, 675, 676, 677, **678–
    79**, 723, 724, pl. 441
  wildoni, 679
Oxyrhopus, 540, 546, 572, **575–76**, 705
  petola, 576
  petolarius, **575–76**, 577, 613, 682, 707,
    710, 711, pls. 346, 347

Pachira, 82
  aquatica, 85
Pachyrhachis, 536
  problematicus, 535

Palicourea, 86, 89
  angustifolia, 88
Pandanus, 486
Panthera, 746
  onca, 748
Parkia pendula, 85
parsley frogs. *See* Pelodytidae
Parvimolge, 127
pejibaye palm. *See Bactris gasipaes*
Pelamis, 538, 542, 544, 572, 703, **714–
  15**
  platurus, 9, 28, 80, 635, **714–15**, pls. 470,
    471, 472
Pelobatidae, 159, 276
Pelodryadidae, 159
Pelodryas coeruleus, 276
Pelodytidae, 159
Pelomedusidae, 738
Peltogyne purpurea, 85
Pentaclethra macroloba, 86
Peripatus, 704
Pernettya coriacea, 90
perro zompopo. *See Corytophanes cristatus*
Persea, 86, 89
  americana, 88
  pallida, 86
Petropedatinae, 397
Phaeognathus, 127
Phasmahyla, 286
Phenacosaurus, 444
Philander, 429
Phimophis, 572
Phobobates, 376, 388
Phoebe, 89
  mollicella, 88
  valeriana, 88
Phrynohyas, 176, **340–42**
  resinifictrix, 340
  venulosa, 36, 164, 180, 276, 290, 339,
    340–42, 350, 352, 355, pls. 180, 181
Phrynomerinae, 392
Phrynosoma, 439, 440
Phrynosomatidae, 417, 425, **439–44**
Phyllobates, 52, 165, 170, 174, 226, 264,
  376, 382, **388–91**
  aurotaenia, 254, 388, 829
  bicolor, 388
  lugubris, 37, 170, 179, 263, 264, 378,
    381, 388, **389–90**, 391, 829, pl. 222
  nubicola flotator, 379
  pratti, 379
  talamancae, 382
  terribilis, 170, 388
  vittatus, 40, 179, 263, 388, 389, **390–91**,
    829, pl. 223
Phyllodactylus, 421, 422, **487–88**
  apricus, 487
  bugastrolepis, 487
  eduardofischeri, 488
  homolepidurus, 487
  insularis, 487, 488
  muralis, 487
  palmeus, 487, 488
  partidus, 487
  santacruzensis, 487
  tinklei, 487

  tuberculosus, 36, 483, 484, 485, 486,
    487–88, 489, pl. 286
  xanti, 487
Phyllomedusa, 176, **286–88**
  buckleyi, 286
  danieli, 286
  hulli, 286
  lemur, 165, 180, 276, 277, **286–88**, pls.
    140, 141
  medinai, 286
  psilopygion, 286
Phyllomedusinae, 275
Physalaemus, 173, 213, **223–26**
  petersi, 224
  pustulatus, 224, 305, 393
  pustulosus, 106, 164, 165, 173, 179, **224–
    26**, pl. 91
Picramnia carpinterae, 88
Pipa, 166
pipas. *See* Pipidae
Piper, 87
  candelarianum, 86
Pipidae, 159, 166
Pitangus, 704
Pithecolobium, 86
  arboreum, 84, 85
  costaricense, 88
pitted-shell turtles. *See* Carretochelyidae
pitvipers. *See* Viperidae
Platyhyla grandis, 169
Platymiscium pinnatum, 85
Platysternidae, 738
Pleophryne, 169
Plethodon, 127
Plethodontidae, 121, **839–40**
Pleurodira, 738
Pliocercus, 640, 644
  arubricus, 643
  bicolor, 644
  dimidiatus, 643, 644
  elapoides, 644
  euryxonus, 643, 644
Podocarpus, 82
  oleifolius, 89
Polychrotidae, 444–80
Polychrus, 425, 444, **445–46**
  gutturosus, 37, 429, 430, 431, **445–46**,
    454, pl. 252
Porthidium, 551, 573, 717, 718, 720, **732–
  35**
  dunni, 733
  godmani, 729
  hespere, 733
  melanurum, 732
  nasutum, 551, 719, 720, **732–33**, 734,
    pls. 491, 492
  nummifer, 721
  ophryomegas, 732, **733–34**, pl. 493
  picadoi, 722
  volcanicum, 732, 733, **734–35**, pl. 494
  yucatanicum, 733
Pothomorphe peltata, 87
Pourouma aspera, 84, 85, 87
Pouteria neglecta, 85
primitive blind burrowing snakes. *See*
  Scolecophidia

Prionostemma, 373
Prioria, 82
Pristidactylus, 444
Prosopis juliflora, 82
Proteidae, 121
Proteus, 121
Protium, 85, 86, 87
   copal, 87
Prunus, 88
   annularis, 88, 89
Pseudemys, 769
   ornata, 771
Pseudidae, 159, 161
Pseudoboa, 572
   neuwiedii, 58
Pseudobombax, 86
   septenatum, 84
Pseudobranchus, 123
Pseudoeurycea, 127
Pseudoleptodeira latifasciata, 613
Pseudolmedia, 86
Pseudosphinx tetrio, 712
Pseudotriton, 126–27
Pseustes, 547, 647, **679–80**
   poecilonotus, 35, 37, 543, 652, 655, 659,
      679–80, 686, pl. 442
   shropshireri, 680
Psidium, 429
   friedrichsthalianum, 86
Psomophis
   genimaculatus, 622
   joberti, 622
   obtusus, 622
Psychotria, 86, 87
Pternohyla fodiens, 184
Pterocarpus, 82
Ptychadeninae, 397
Ptychoglossus, 423, 519, **524–25**
   myersi, 525
   plicatus, 523, **524–25**, pl. 317
Ptychohyla, 294, 295, 330
Puya, 136
   dasylirioides, 90
Pygopodidae, 415, 480
Pythoninae, 562
Python reticulatus, 544, 559
pythons. *See* Boidae
Pyxicephalinae, 397
Pyxicephalus, 397

Qualea paraensis, 85
Quercus, 87
   aaata, 88
   copeyensis, 89
   costaricensis, 90
   eugeniaefolia, 88
   oocarpa, 88, 89
   seemanni, 89

rabo amarillo. See *Bothrops asper*
rainbow boa. See *Epicrates cenchria*
Ramphotyphlops, 556
Rana, 220, 276, **398–405**
   berlandieri, 398, **399–402**
   bwana, 402
   esculenta, 166

   forreri, 165, 179, **399–400**, 401, pl.
      227
   juliani, 402
   maculata, 401, 402
   palmipes, 398, 401, **402–5**
   pipiens, 400, 401
   sierramadrensis, 402
   taylori, 165, 179, 398, 399, **400–402**,
      pls. 228, 229
   vaillanti, 164, 165, 180, 398, 399, **402–3**,
      pl. 230, 231
   vibicaria, 40, 165, 180, 197, **403–4**, pl.
      232
   vibocaroa, 398
   warszewitschii, 37, 180, 276, 398, **404–5**,
      pl. 233
rana ternero. See *Leptodactylus pentadacty-*
   lus
Randia, 86, 87
   karstenii, 88
Ranidae, 159, 164, 165, 166, 176, 213,
   397–405
Raninae, 398–405
Ranixalinae, 397
Rapanea
   ferruginea, 87, 89, 90
   guinanensis, 88
   pellucido-punctata, 87
   pittieri, 90
rattlesnakes. *See* crotalines
rear-fanged snakes. *See* Columbridae
red mangrove. See *Rhizophora mangle*
red turtle. See *Rhinoclemmys pulcherrima*
reed frogs. *See* Hyperoliidae
Reithrodontomys, 729
Rhachidelus, 572
Rhacodactylus, 480
Rhacoma tonduzii, 88
Rhacophoridae, 159, 161, 166, 276
Rhadinaea, 89, 549, 550, 585, 587, 592,
   621–27, 636, 640
   altamontana, 625
   anachoreta, 625
   bogertorum, 624
   brevirostris, 622
   calligaster, 585, 594, **622–23**, 636, pl.
      389
   cuneata, 624
   decipiens, 642
   decipiens rubricollis, 642
   decorata, 37, 622, **623–25**, 636, pl. 390
   flavilata, 622
   fulviceps, 642
   godmani, 585, 594, 622, **625–26**, 830,
      832, pl. 391
   guentheri, 645
   hannsteini, 625
   hempsteadae, 625
   hesperia, 624
   kanalchutchan, 625
   kinkelini, 625
   lachrymans, 625
   lateristriga, 643
   macdougalli, 624
   marcellae, 624
   montana, 624

   montecristi, 625
   myersi, 624
   pachyrua, 642
   persimilis, 645
   pilonaorum, 625
   posadai, 625
   pulveriventris, 594, 622, 623, 626, pl.
      393
   quinquelineata, 624
   rogerromani, 625
   sargenti, 626
   schistosa, 625
   serperaster, 585, 594, 622, **625**, pl. 392
   species inquirenda, 646
   stadelmani, 625
   tolpanorum, 625
   vermiculaticeps, 622, **625–27**
Rhamnus, 90
   capreaefolia, 88
   humboldtiana, 89
Rhamphastos, 439
Rhaphia, 82
Rheedia
   edulis, 86
   madruno, 85
Rheinhardtia, 86
Rhineuridae, 415
Rhinobothryum, 547, 577, 647, **680–81**,
   705
   bovallii, 576, 643, **681**, 706, 709, 710,
      711, pl. 443
   lentiginosum, 681
Rhinoclemmys, 81, 743, **764–69**, 826, 827,
   829
   annulata, **765–66**, 770, pl. 509
   areolata, 764
   funerea, 765, **766–67**, 770, pl. 510
   nicoyana, 771
   pulcherrima, 37, 712, 765, 766, **767–69**,
      770, pl. 511
Rhinoderma
   darwinii, 169
   rufum, 169
Rhinodermatidae, 159
Rhinoleptis, 557, 558
Rhinophrynidae, 159, 165, 176, **183–85**,
   802
Rhinophrynus, 160, 161, 174, **183–85**
   dorsalis, 162, 165, 174, 178, **183–85**,
      391, 393, 395, 396, pl. 62
Rhinotyphlops, 556
   braminus, 556, 557
Rhizophora mangle, 81
Rhyacotritonidae, 121
Roupala complicata, 87, 88

Saccoglottis amazonica, 84, 85
Sacroramphus papa, 439
saddleback frogs. *See* Brachycephalidae
salamanders. *See* Salamandridae
Salamandra
   atra, 123
   salamandra, 123
Salamandridae, 121
Salamandroidea, 121
Saphenophis, 572

Sapium, 85, 89
  jamaicense, 87
  sulciferum, 88
sapo grande. *See Bufo marinus*
sapos. *See Bufo*
Sapranthus palanga, 84
Sauresia, 527
Sauria, 414–534
Sauromalus, 433
Scaphiodontophis, 541, 549, 571, 587, 640,
      643, **681–85**, 705
  annulatus, 543, 576, **682–85**, 697, 707,
      709, 710, 711, pls. 444, 445, 446
  venustissimus, 575, 684
Scaphiophrynae, 392
Scaphiophryne, 391, 392
sceloporine lizards. *See* Phrynosomatidae
Sceloporus, 420, 425, 439, **440–44**
  acanthinus, 441, 442
  adleri, 441
  chryostictus, 443
  couchii, 443
  cozumelae, 443
  cryptus, 441
  formosus, 441–42
  internasalis, 441, 442
  lunnaei, 441
  malachiticus, 38, 86, 90, **441–42**, 443,
      729, pl. 247
  parvus, 443
  salvini, 441
  siniferus, 442–43
  smaragdinus, 441, 442
  smithi, 443
  squamosus, 441, **442–43**, pl. 248
  stejnegeri, 441
  subpictus, 441
  taeniocnemis, 441, 442
  tanneri, 441
  teapensis, 443
  variabilis, 36, 83, 440, 441, 442, **443–44**,
      pls. 249, 250
Scheelea, 82
  rostrata, 85
Schistometopum thomense, 113
Schizolobium parahybum, 85
Scinax, 176, 288, **342–48**
  bipunctata, 343
  boulengeri, 164, 180, 288, 312, 313, **343–
      45**, pl. 182
  elaeochroa, 164, 180, 288, 343, **345–47**,
      pl. 183
  lancasteri, 344
  rostratus, 345
  ruber, 346
  staufferi, 164, 180, 288, 343, 346, **347–
      48**, pl. 184
  staufferi altae, 348
  staufferi staufferi, 348
Scincella, 505
Scincidae, 417, **500–506**
Scincinae, 501
Scincomorpha, 497–525
scincomorph lizards. *See* Scincomorpha
Scolecophidia, 545, **552–59**
Scolecophis, 550, 647, **685–86**, 705

atrocinctus, 36, 576, 613, 630, 643, **685–
      86**, 706, pls. 447, 448
screw pine. *See Pandanus*
Scynbranchus marmoratus, 709
sea snakes. *See* Elapidae; Hydrophiinae
sea turtles. *See* Cheloniidae
Selaginella, 87
Senecio, 90
  copeyensis, 88
  Senticolis, 663
Serpentes, 535–737
Seychelles frogs. *See* Sooglossidae
sheep frogs. *See Hypopachus*
shieldtails. *See* Uropeltidae
shovel-nosed frogs. *See* Hemisotidae
Sibon, 548, 596, **627–35**, 705, 717, pl. 395
  annulatus, 576, 597, **628–32**, 634, 706,
      707, pl. 394
  anthracops, 576, 597, 598, 628, 629,
      630–31, 643, 685, 706, pl. 396
  argus, 608, 609, 627, 628, 629, **632–33**,
      pl. 398
  carri, 626
  dimidiatus, 576, 598, 628, 629, 630, **631–
      32**, 633, 707, pl. 397
  dimidiatus grandoculis, 632
  dunni, 634
  longifrens, 41, 608, 609, 628, 629, 632,
      633–34, 725, pl. 399
  nebulatus, 35, 628, 631, **634–35**, pl.
      400
  sanniolus, 627, 628
Sibymorphus
  anthracops, 631
  ruthveni, 631
Sibynomophis, 596, 683, 829
side-neck turtles. *See* Pleurodira
Simaruba
  amara, 86, 87
  glauca, 84
Siphlophis, 572, 577
Siphonops
  oligozonus, 119
  syntremus, 119
Sirenidae, 121
Sirenoidea, 121
sirens. *See* Sirenidae
Sistrurus, 717, 735
skinks. *See* Scincidae
Sloanea, 87, 429
  laurifolia, 84, 85
  medusula, 86
  megaphylla, 88
Smilisca, 176, 288, 328, **348–57**
  baudinii, 164, 180, 276, 291, 339, 341,
      348, 349, **349–51**, pl. 185
  cyanosticta, 349
  phaeota, 164, 180, 291, 297, 339, 341,
      349, 350, **351–52**, 353, pls. 186, 187,
      188
  pseudopuma, 353, 354, 356
  puma, 288, 292, 349, **352–54**, 356, pl.
      189
  sila, 180, 288, 290, 292, 325, 329, 341,
      349, **354–55**, 356, pl. 190
  sordida, 165, 180, 292, 293, 324, 329,

      348, 349, **352–57**, **355–57**, pls. 191,
      192
smoki jungle frog. *See Leptodactylus pen-
      tadactylus*
snakes. *See* Serpentes
snapping turtles. *See* Chelydridae
soft-shelled turtles. *See* Trionychidae
Solanum, 88, 89, 90
  storkii, 90
solda con soldas. *See* Caeciliidae
Sonora semiannulata, 559
Sooglossidae, 159
Sorocea
  affinis, 86
  trophoides, 87, 88
spadefoots. *See* Pelobatidae
spectacled caiman. *See Caiman crocodilus*
Sphaenorhynchus, 343
Sphaerodactylinae, 489–97
sphaerodactylines, 482
Sphaerodactylus, 417, 421, 422, 423, 489,
      492–97
  argus, 58
  continentalis, 494, 496
  dunni, 492
  glaucus, 492
  graptolaemus, 493, 494, 495, 496, pl.
      291
  heliconiae, 492
  homolepis, 492, 493, **494–95**, 496, 497,
      pls. 292, 293
  imbricatus, 495
  lineolatus, 492, 494, 495, 496
  millepunctatus, 492, 493, 494, **495–96**,
      497, pls. 294, 295
  molei, 492
  notatus, 492
  pacificus, 41, 466, 492, 493, 494, 495,
      496–97
  scapularis, 492
Sphenodon, 411
Sphenomorphus, 423, 424, **504–6**, 594,
      683, 711
  assatus, 504
  cherriei, 503, **504–6**, 522, 528, 529, 530,
      531, 532, 682, pl. 300
  cherriei cherriei, 506
  cherriei lampropholis, 506
  incertum, 504
  rarus, 58, 504
Spilotes, 545, 647, **686–87**
  pullatus, 35, 655, 659, 680, **686–87**, pl.
      449
spiny-tailed iguanas. *See Ctenosaura*
Spondias, 429
  mombin, 86
Squamata, 412–13
squeakers. *See* Arthroleptidae
Staurotypus, 744
Stenorrhina, 545, 561, 570, 586, 647, **687–
      90**, 717
  degenhardtii, 37, 543, **688–89**, 690, pl.
      450
  degenhardtii apiata, 689
  freminvillii, 573, 574, 621, 688, **689–90**,
      706, pl. 451

Sterculia
  apetala, 86
  recordiana, 84, 85
Stereochilus, 123
Sternotherus, 744
Strabomantis biporcatus, 255
Streptophorus oxynotus, 620
Stryphnodendron excelsum, 86
stumpheads. *See* Anomochilidae
Styrax polyneurus, 88, 89, 90
suelo con suelos. *See* Caeciliidae
sunbeam snakes. *See* Xenopeltidae
Swartzia simplex, 85, 86
Swietenia humilis, 84
Symphonia globulifera, 85, 87
Symplocos costaricana, 88
Syrrhophus, 226
  gaigeae, 264
  molinoi, 270

Tabebuia, 84, 86
Tabernaemontana aphlebia, 87
Tachygalia versicolor, 85
Taeniophallus, 622
"tailed" frogs. *See* Ascaphidae
Talisia nervosa, 85
tamagá. See *Porthidium nasutum*
Tantilla, 546, 547, 644, 647, **690–97**, 705
  alticola, 616, **691–92**, 696, pl. 452
  annulata, 697
  armillata, 691, **692–93**, 695, pl. 453
  costaricensis, 692, 696
  melanocephala, 693, 695
  pliolepis, 616
  reticulata, 638, 691, **693–94**, 695, pl. 454
  ruficeps, 40, 691, 693, **694–95**, pl. 455
  schistosa, 616, 639, 691, 692, **695–96**, 697, pl. 456
  supracincta, 543, 576, 578, 579, 690, 691, **696–97**, 707, pls. 457, 458
  vermiformis, 594, 640, 691, 696, **697**
  virgata, 694
Teiidae, 506–17
Teius, 506
Telmatobiinae, 213, **226–75**
Teratohyla, 366
terciopelo. See *Bothrops asper*
Terciopelo falsa. See *Xenodon rabdocephalus*
Terminalia
  amazonia, 85, 87
  bucidolies, 85
  lucida, 86
  oblonga, 86
Ternstroemia seemannii, 89
Testudinata, 738–71
Testudinidae, 738, 739
Testudinoidea, 738
Thalassia, 757
  testudinum, 758
Thamnophis, 127, 136, 541, 547, 549, 571, **699–702**
  marcianus, 700–701
  proximus, 36, 547, **699–700**, pl. 460
  radix, 700
  sauritus, 700
  sauritus chalceus, 702

  sirtalis, 700
  sirtalis chalceus, 702
Thecadactylus, 414, 421, **488–89**
  rapicauda, 36, 478, 483, 484, 485, 486, 487, **488–89**, 671, pl. 287
Thorius, 127
thread snakes. See *Leptotyphlops*
toads. See *Bufo*; Bufonidae
toboba chinga. See *Porthidium ophryomegas*
toboba de Altura. See *Cerrophidion godmani*
toboba de Pestañas. See *Bothriechis schlegelii*
toboba gato. See *Porthidium ophryomegas*
Tococa
  grandifolia, 85
  guyanensis, 87
Tomistoma, 772
  schlegelii, 778
Tomodactylus, 226
Tomopterinae, 397
torrent salamanders. *See* Rhyacotritonidae
tortuga blanca. See *Chelonia mydas*
tortuga roja. See *Rhinoclemmys pulcherrima*
Tovomita nicaraguensis, 86
Tovomitopsis psychotriaefolia, 87
Trachemys, 769
  ornata, 771
Trachops cirrhosus, 225, 355
Trachyboa, 569
Trachymenis decipiens, 645
tree boas. See *Corallus*
treefrogs. *See* Hylidae
Trema micrantha, 86, 88
Tretanorhinus, 82, 547, **635–36**
  nigroluteus, **635–36**, 747, pl. 401
  variabilis, 635
Triadobatrachus, 158
Trichilia, 86, 89
  havanensis, 88
  pittieri, 87
Trimeresurus, 718
Trimetopon, 550, 569, 588, 589, 592, 594, 618, **636–40**, 695
  barbouri, 636
  gracile, 636, **637**, 638, 639, 692, pl. 402
  pliolepis, 40, 636, **637–38**, 639
  simile, 603, 636, 637, **638–39**
  slevini, 636, 637, **639**, 640, 696
  viquezi, 636, 637, 638, **639–40**
Trimorphodon, 545, 547, 570, 647, **697–99**
  biscutatus, 436, 582, 612, 615, 664, **698–99**, 717, 727, 729, pl. 459
Trionychidae, 738
Trionychoidea, 738
Tripanurgos, 548, 572, **576–77**, 705
  compressus, 36, 575, 576, **577**, 613, 706, pl. 348
Trogonophidae, 415
Trophis, 86
Tropidodipsas, 596, 627
  fischeri, 628
Tropidodryas, 572
Tropidonotus sipedon, 58
Tropidophiidae, 535, 536
Tropiduridae, 415, 426
tropidurines. *See* Tropiduridae
tungara frog. See *Physalaemus pustulosus*

Tupinambinae, 506
Tupinambis, 506
Turpinia occidentalis, 88
turtles. *See* Testudinata
Typhlomolge, 127
Typhlophis, 553, 555
  emunctus, 555
Typhlopidae, **556–57**, 703
Typhlops, 544, 545, 552, **556–57**, 558
  costaricensis, 554, 555, **556**, pl. 332
Typhlotriton, 127

Ulmus mexicana, 87
Uma, 439
Umbrivaga, 572
Ungaliophiidae, **569–70**, 717
Ungaliophis, 546, 548, 562, **569–70**, 572, 591, 717
  continentalis, 570
  panamensis, 560, **569–70**, 582, 729, pl. 341
Urera caracasana, 88
Uropeltidae, 536, 559
Urostrophus, 444
Urotheca, 541, 549, 587, 592, 595, **640–47**, 682, 704, 705
  decipiens, **641–42**, 644, 645, 647, pls. 403, 404
  decipiens rubricollis, 647
  elapoides, 641, 643, 683
  euryzona, 576, 614, 630, **641–44**, 683, 685, 706, 709, 712, pls. 405, 406
  fulviceps, 641, 642, **644–45**, 646, 647
  guentheri, 624, 641, **645–46**, pl. 407
  myersi, 641, 644, **646**
  pachyura, 641, 642, 644, **646–47**, pl. 408
  pachyura decipiens, 647

Vaccinium, 90
  consanguineum, 89, 90
Vantanea barbouri, 85
Vanzosaura, 521
Varanidae, 417, 526
Varanus komodoensis, 421, 422
Venconcibea pleiostemona, 86
Viburnum, 87, 90
  costaricanum, 88
  stelato-tomentosum, 88
Viperidae, 536, 569, 571, 703, **715–37**
Viperinae, 716
vipers. *See* Viperidae
Virola, 85, 87
  sebifera, 84, 86, 87
Vitex, 87
Vochysia
  ferruginea, 85, 87
  hondurensis, 85, 87

Waglerophis, 572
Warscewiczia coccinea, 86
Weinamannia, 87
  pinnata, 89, 90
  wercklei, 89
Welfia georgii, 85
Wetmorena, 527

white hawk. See *Leucopternis albicollis*
white mangrove. See *Laguncularia racemose*

Xantusia, 498
Xantusiidae, **497–500**, 802
Xantusiinae, 498
Xenoboa, 564
Xenodon, 548, 570, 572, **582–83**
  bertholdi, 583
  columbrinus, 583
  rabdocephalus, **582–83**, 612, 662, 663,
    698, 717, 727, pls. 354, 355
xenodontines, 571, **572–91**

Xenopeltidae, 536, 559
Xenophidiidae, 536
Xenosauridae, 415, 417, 526, 802
Xenotyphlops, 556
Xenoxybelis, 675
  argenteus, 675
  boulengeri, 675
Xiphosoma ruschenbergerii, 567
Xiphosurus, 444, 446
Xylosma, 86
  excelsum, 86
  flexuosum, 84
  intermedium, 88

yellow-bellied sea snake. See *Pelamis platurus*

Zanthoxylum, 86, 87, 90
  chiriquinu, 88
  limoncello, 88
  melanostictum, 89
Zanthozylum panamense, 86
Zexmenia frutescens, 86
Zinowiewia integerrima, 87, 88
zopilota. See *Drymarchon corais*
zumbadora. See *Spilotes pullatus*
Zygothrica, 373

Subject Index

Abbreviations used in index: f = figure; t = table

abbreviations, 102
abundance, 97
accretionary crusts, 815t
Acuña, Rafael, 48, 50
An Addition to the Knowledge of the Batrachia and Reptilia of Costa Rica (Cope), 40
aggression, 541
Alfaro, Anastasio, 38
alkaloids, 188
allopatry, 10, 11
altitude
    amphibian distributions, 791f
    anuran distributions, 785–86t
    caecilian distributions, 784t
    crocodilian distributions, 790t
    lizard distributions, 787t
    reptile distributions, 791f
    salamander distributions, 784t
    snake distributions, 788–89t
    turtle distributions, 790t
American Society of Ichthyologists and Herpetologists, 47
Americas
    amphibian distributions, 796t
    reptile distributions, 796t
Amerind peoples, 35
amniotic membranes, reptilian, 410f
ancestry, phylogenetic, 6
Anfibios de Nicaragua (Villa), 46
Angel, Fernand, 39
Animales venenosos de Costa Rica (Viquez), 43
Animalia, 8
An Annotated List and Guide to the Amphibians and Reptiles of Monteverde, Costa Rica (Hayes, Pounds, and Timmerman), 49
annulation, caecilians, 113, 115, 115f
antipredator defenses, 98
antivenins, 22
aquariums, 26
aquatic habitats, 80–82
aquatic habits, 98
arboreal habits, 98
area, units of, 11
Arias Sánchez, Oscar, 33
ASIH. *See* American Society of Ichthyologists and Herpetologists
assemblages, 825–35
Atlantic lowland wet forest, pls. 9, 10

auditory apparatus
    anuran, 158, 170, 173
    caecilian, 114
    crocodilian, 773
    frog, 160
    lizard, 414, 415–16
    reptile, 409
    salamander, 121, 122
    snake, 536
    toad, 160
    turtle, 740
axillary webs, anuran, 171

bags, collecting, 14, 20
Baird, Spencer, F., 37, 41
bananas, 32
Barbour, Thomas, 42, 43
basking, 773, 774
batrachotoxins, 170
beaches, 83
beaks
    tadpole, 177
    turtle, 738, 745f
Berthold, Arnold, 37
Bibron, Gabriel, 36, 37
binocularity, 416
binoculars, 11–12
bioclimates
    ecological patterns, 783–84
    lower montane, 77–79
    lowland, 77
    montane, 79
    plant formation and, 69f
    premontane, 77
    subalpine, 79
biogeography, historical, 801–6
Biolley, Paul, 40
Biologia-Centrali Americana: Reptilia and Batrachia (Günther), 55
biological reserves, 52
biology, descriptions and, 97
biotemperature, 63, 75–76
blotches, definition, 96
Bocourt, Marie-Fermin, 39
body shapes
    Eleutherodactylus, 236f
    Oedipina, 146f, 148f
    ranid frogs, 399f
body temperature. *See also* thermoregulatory behaviors
    behavioral control of, 411

reptilian, 409
    turtle, 739
Boettiger, Oskar, 40
Boie, Friedrich, 36
Bolaños, Federico, 48, 49, 50
Bolaños, Róger, 46, 47, 48, 49, 50
bones
    and diet of captives, 27
    fixation, 24
    hyomandibular, 105
boots, 14
Botica de Frantzius, 38
bottles, 20
Boulenger, George A., 36, 40
Boza, Mario, 32, 52
Brame, Arden H., Jr. II, 45
Bransford, John M., 39
Braulio Carrillo National Park, 49
Breda, Treaty of, 35
Brocchi, Paul, 39
Brodie, Edmund D., III, 50
bromeliads, sampling, 18
BT. *See* biotemperature
bufodienoloids, 186, 192
bufotoxins, 192
burrowing, 161, 537

cages, 26
calcium, 27
Calderón Founier, Rafael Ángel, 33
Calderón Guardia, Rafael Ángel, 31
calls. *See* vocalizations
Calvert, Amelia, 41
Calvert, Philip, 41
cameras, 12
cannibalism, 98
captive husbandry, 25–28
carapaces. *See also* plastrons
    hatchling, 743f
    sea turtle, 751f, 753f
    turtle, 739f, 745f
Carazo, Rodrigo, 32, 52
carcinivory, 98
Carimol, Julian, 37
carnivory, 98
Carr, Archibald Fairly, Jr., 44, 45
Carrillo, Braulio, 29
Castro, José María, 29
Catalogue of Columbrine Snakes in the Collection of the British Museum (Natural History), 37

Catesby, Mark, 35
cattle production, 32
caudal morphology. *See also* tails
    crocodilian, 772, 774–75, 775f
    rattlesnake, 735
    snake, 549f, 551f
Central American Components, 809t
Central Intelligence Agency, 33
Cerdas, Luis, 46
Cerro de la Muerte, 89
Champion, George C., 39
Chaves, Anny, 48, 49
chemoreceptors
    amphibian, 111
    caecilian, 114
    lizard, 415
    reptile, 411
    tongue, 412
    turtle, 740
chemosensory organs, 113, 114
Cherrie, George K., 40
Children's Rainforest, 52
chiriquitoxin, 186, 188
Chocó block, 815, 817, 817f
Chorotega block, 815, 817, 817f, 828
Chortis block, 815, 817, 817f, 828
chromosomes, 98–99, 99f, 166
CIA. *See* Central Intelligence Agency
clades, definition, 6
cladistics, definition, 6
cladograms
    Craniata, 9f
    dispersals, 827f
    examples, 7f
    vicariances, 827f
cladogroups, 6
classes, 8
classification
    hierarchical, 9
    relationships and, 7–11
claws
    gecko, 421f, 422f
    lizard, 414, 424f
    reptilian, 410
    turtle, 742
climate
    classification, 75–76
    determinants of, 72–75
    general, 63–73
    paleoclimates and, 822–25
    sea levels and, 823f, 829
    warming, 52, 823
climatographs, 78f
cloacal probing, 417–18, 540–41
clorobutanol, 22–23
cloud forest, 83
coffee, 30, 31
collection of specimens, 13–21
    gear, 13–15, 14f, 15f
coloration
    amphibian, 110–11
    Anotheca spinosa, 288f
    anuran, 172
    anuran egg, 166
    Atelopus, 192f

caecilian, 113
    and concealment, 169–70
    crocodilian, 775
    Eleutherodactylus, 228f, 229f, 232f
    Hyla angustilineata, 289f
    infrared reflectance, 276–77
    Lepidophyma, 498f
    Leptophis, 662f
    lizard, 416, 420–21, 528f
    Oedipina, 146f, 148f
    patterns, 96f
    plastrons, 765f, 770f
    salamander, 124, 125
    snake, 543–44, 543f, 549f, 576t, 616f,
        617f, 655f, 704, 705f
    species accounts and, 96
    Sphaerodactylus, 493f
    toxins and, 170
    treefrog, 291f, 292f
    two-tone, 97
    turtle, 742
    uniform, 96
    variation in, 315f
common names, 95
Companía Bananera de Costa Rica, 30
conservation, 52–54, pl. 1
containers
    housing, 25–26
    transport, 20
continental crusts, 815t
Contra War, 33
Contributions to the History of Herpetology
        (Adler), 35
Cooper, Juan J., 38, 39
Cope, Edward Drinker, 38, 39, 40
Copeia, 42
Corn, Michael J., 46
Coronado, Juan Vásquez de, 28
cortez tree, pl. 5
Costa Rica
    bioclimates, 76–77
    cities, 4f
    climate, 63–79
    climatographs, 78f
    faunal areas, 806f, 814f
    geographic patterns, 806–10
    herpetofauna, 93f
    herpetology in, 34–52
    history of, 28–34
    hydrography, 60–63
    latitude of, 72–73
    Nicaragua and, 32, 34
    physiography, 60–63, 61f
    protected areas, 53t
    provinces, 4f
    vegetation, pl. 2
Costa Rican Natural History (Janzen), 48,
        50
courtship behaviors
    crocodilian, 773, 774
    lizard, 417, 418
    salamander, 123
    snake, 539, 541
    turtle, 740
cranium. *See* skull

creeks, 82
crests, 419
    cranial, 170, 193f, 236f, 427f
crocodile tears, 773
Crump, Martha, 49
crusts, earth, 815t

data, minimum set, 25
Daudin, François-Marie, 36
day length
    latitude and, 67f
    seasons and, 65–66
    solar radiation and, 66f
dead on road, 19, 23
deciduous forest, 82
defenses. *See particular defense mechanisms*
deforestation, 82
dehydration, 110
denticles, 177
Deppe, Ferdinand, 36
descent, phylogenetic, 6
descriptions, components of, 95–97
DeWeese, J. D., 47
dewlaps, 448f
diagnostics, 95
dichromatism, 92
diel cycles, 97
digital lamellae, 424f
digits
    anuran, 161, 171–72, 172f
    crocodilian, 772
    Eleutherodactylus, 231f, 233f, 234f, 236f
    gecko, 421f, 422f
    lamellae, 424f
    lizard, 414, 424f
    reptilian, 410
    salamander, 122–23, 124–25, 124f, 125f
    treefrog, 289f, 290f
    webbing, 176f
diseases, in captive husbandry, 28
disks, anuran, 171, 174f, 233f
dispersals, 825–35, 827f
distributions, 99, 100–101
    Abronia, 831f
    Agkistrodon, 836f
    anuran, 159, 164–66
    caecilian, 113
    crocodilian, 772
    Cryptotriton, 833f
    Ctenosaura quinquecarcinatus, 835f
    ecological patterns, 783–93
    general patterns, 814
    lizard, 415
    Nototriton, 833f
    Rhadinaea godmani, 832f
    salamander, 122
    snake, 536
    Tropical Mesoamerican genera, 795t
    turtle, 738–39
    vertical limits, 101
ditches, roadside, 81–82
diversity, 835–37, 836t
Dock, Charles F., 47
Donnelly, Maureen A., 48, 49
DOR. *See* dead on road

dorsal fields, coloration, 96
dorsolateral folds, 399f
Drosophila, 27
dry forest, 77
Duellman, William E., 45
Duméril, Auguste, 36, 37, 39
Duméril, Constant, 36, 37, 40
dunes, 83
Dunn, Emmett Reid, 41, 42
Dunn, Merle, 42
Duvernoy's gland, 537

earthworms, 27
Eastern North American Components, 810t
Echandi, Mario, 32
ecotourism, 33
ectothermism, 409
eggs. *See also* shells
    amphibian, 105, 106f, 107
    anuran, 164, 165t, 167
    cleidoic, 411
    coloration, 166
    fish, 106f
    fixation, 23–24
    key, 107–8
    lizard, 418
    photography, 13
    predation on, 167
    reptilian, 409, 410f
    salamander, 124
    snake, 541
    specimens, housing of, 27
    transport of, 20
    turtle, 741
egg teeth
    crocodilian, 772
    squamate reptile, 413
    turtle, 738
embryos. *See also* eggs; tadpoles
    anuran, 109f
    identification, 108f, 109f
    key, 107–8
    reptilian, 409
    salamander, 109f
    squamate reptile, 413
endangered taxa, 52
endemism, 810–14, 813t
endotrophs, 169t
Entisols, 79
envenomation, 538, 707
equipment, 13–15
*Erpétologie générale, ou Historie complète
    des reptiles* (Duméril), 36, 54
Eschscholtz, Johann F. von, 36
Espinoza, Gaspar de, 28
ethyl alcohol, 22
Études des batraciens de l'Amérique Centrale
    (Brocchi), 55
Études sur les reptiles (Boucourt) 54
euthanasia, 22–23
evapotranspiration, 75–76
evergreen forest, 82–83
excretion
    amphibian, 110–11
    crocodilian, 773

squamate reptile, 412
    turtle, 739–40
exotrophs, 169t
extinctions, 51, 111–12
extraembryonic membranes, 409, 410f
eyeballs
    anuran, 160
    turtle, 740
eyelids
    anuran, 160
    crocodilian, 773
    Eleutherodactylus, 231f
    Gastrotheca, 293
    lizard, 414, 416, 419
    snake, 537
    treefrog, 277
eyes
    anuran, 158, 175f
    crocodilian, 773
    licking of, 498
    lizard, 416, 419
    salamander, 121, 122
    turtle, 740
eyeshine, 19–20

family, nomenclature, 9
fangs, 537–38
faunal areas
    amphibian, 807f
    Costa Rican, 806f
    distributions, 813t
    endemism, 810–14
    genera in historical units, 810f
    reptile, 808f
    taxonomic groups by, 807t
feeding
    anuran, 161–62
    caecilian, 115
    characteristics, 98
    crocodilian, 774
    lizard, 417
    salamander, 123
    snake, 537
    tadpole, 161–62
    turtle, 740, 741
feet
    anuran, 171, 172f, 173f, 174f
    Bufo, 194f
    crocodilian, 773
    Eleutherodactylus, 231f, 233f, 234f, 236f
    Leptodactylus, 214f
    salamander, 125f
    tarsal fold, 173f
femoral pores, 417, 419–20, 420f
Fernández, Mauro, 30
Fernández, Santiago, 29
field notes, 13, 20, 25
field studies, 25
Figueres, José Maria, 31, 33–34, 52
fingers. *See* digits; hands
Fitch, Henry S., 47
fixation, 23–24, 23f
flash, photographic, 13
Fleischmann, Carl, 40
flippers, turtle, 757f

flounces, lizard, 418
folds. *See also* plicae
    dorsolateral, 399f
    glandular, 215f
    laminate, 418
    tarsal, 173f
folklore, 98, 551–52, 774
food, 27, 98
forceps, 16
forests
    broadleaf deciduous, 82
    dry, 77
    Eocene distribution, 824f
    evergreen, 82–83
    gallery, 83
    lower montane moist, 87–88
    lower montane rainforest, 89
    lower montane wet, 88–89
    lowland Atlantic moist, 85–86, 783–84
    lowland Atlantic wet, 86, 783–84
    lowland dry, 83–84, 783–84, pls. 4, 5, 6
    lowland Pacific-Atlantic moist, 84–85,
        783–84
    lowland Pacific wet, 85, 783–84, 794,
        796
    lowland wet, pls. 8, 9, 10
    moist, 77
    montane rainforest, 89–90
    montane wet, 89
    Oligocene distribution, 824f
    Pacific lowland moist, pl. 6
    Paleocene distribution, 824f
    premontane moist, 86
    premontane rainforest, 87
    premontane wet, 86–87, pl. 11
    rainforest, 77, 79
    riparian, 83
    secondary growth, 90
    semideciduous, 82, 84–85
    swamp, 82
    tropical dry, 84
    tropical moist, 85–86
    tropical premontane moist, 86
    tropical wet, 85, 86
    wet, 77, 79
formaldehyde, 23
formalin, 23
fossil records, 111–12
fossorial habits, 97
Foster, Mercedes S., 47, 48
Frantzius, Alexander von, 37, 38
freshwater marshes, 81
FSLN. *See* Sandinistas

Gabb, William More, 39
gallery forest, 83
Gámez, Rodrigo, 52
gas exchange. *See also* respiration
    amphibian, 110
    anuran ovoviviparity and, 166
    reptile, 411
genus
    definition, 8
    description, 94
    nomenclature, 9

geography
 Costa Rican patterns, 806–10
 elevational differences, 20
 paleogeographic background, 815–20
 and variation, 10–11, 92
geologic time, 816
geology
 Mesoamerican, 820–22
 North American, 816
 tectonic features, 817
Gessner, Conrad, 35
Girard, Charles, 37
glands
 development, 187f
 holocrine secretory, 417
 parotid, 170–71, 193f
 poison, 176f
 salt, 415, 773
glandular folds, 215f
global warming, 52
gloves, 15
Godman, Frederick du Cane, 39
gonadogenesis, 741
Gonzáles, Gil, 28
González, "Don Cleto," 31
Good, David A., 49, 50
Gray, John E., 36, 37
Greene, Harry W., 48, 49
Gronovius, Lorenz Theodore, 35
groups, natural, 6
gular collars, 507f
gular scalation, 548f
Günther, Albert C. L. G., 36, 40
guns, 19
Gutiérrez, José Maria, 46, 50
Guyer, Craig, 48, 49

habitats
 and declining populations, 111–12
 species descriptions, 97
 terrestrial, 82–90
 vegetation, 79–90
habits, 97
hair cells, 111
Hallowell, Edward, 37
hands
 anuran, 171, 172f
 Bufo, 194f
 crocodilian, 773
 Eleutherodactylus, 233f, 234f, 235f
 and specimen collection, 14–15
 treefrog, 289f
Hapsburg, Maximilian von, 39
Hardy, David L., 49, 50
Hayes, Marc P., 49
headlamps, 14
head plates, 545f
head shapes
 anuran, 170, 175f
 crocodilian, 774, 775f
 Eleutherodactylus, 229f, 231f
 glassfrog, 359f
 lizard, 423f, 519f
 Oedipina, 146f, 148f

snake, 544f, 545f
toad, 205f, 206f
treefrog, 290f
turtle, 745f
head shields
 snake, 542f
 turtle, 742
hearing. See auditory apparatus
heart, 105, 772
heel calcar, 277
helmets, lizard, 419
hemipenes, snake, 539–41, 539f
herbivory, 98
Hernández, Francisco, 35
herpetofauna
 development of, 815–37
 discovery by decade, 51f
 Tropical Mesoamerican unit, 794–801
Herpetologia Mexicana (Wiegmann), 36
Heyer, Miriam, 45
Heyer, W. Ronald, 45
historical source units, 825t, 837t
hobbyists, 25
Hoffmann, Carl, 37, 38
Holdridge, Leslie R., 47, 75–76
holotypes, 8
homing behaviors, 98
homoplasy, 6
hooking, 16
housing of specimens, 25–27, 26f
Houttuyn, Martin, 36
How to Photograph Reptiles and Amphibians (West and Leonard), 13
Humboldt, Alexander von, 36, 37
humeral hooks, 175f
humidity provinces
 amphibian distributions, 792f
 ecological patterns, 792–93
 reptile distributions, 792f
hybridization
 anuran, 166
 species and, 10
hydrography, 60–63
The Hylid Frogs of Middle America (Duellman), 45
hyomandibular bones, 105

Ice Ages, 830
identification of specimens, 91–93
illustrations, use of, 92, 181–82
IMF. See International Monetary Fund
INBio. See Instituto Nacional de Biodiversidad
Inceptisols, 79
incubators, 27
infrared reflectance, 276–77
insectivory, 98
Instituto Clodomiro Picado, 21, 46
Instituto Nacional de Biodiversidad, 52
integumentary pockets, 420
integumentary sense organs, 772
International Code of Zoological Nomenclature, 8, 9, 35
International Monetary Fund, 33, 34

intertropical convergence zones, 70–72,
 71f
Introduction to the Herpetofauna of Costa
 Rica (Savage and Villa), 49
isopropyl alcohol, 22
ISOs. See integumentary sense organs
ITCZ. See intertropical convergence zones

Jacobson's organs
 amphibian, 111
 caecilian, 114
 lizard, 415, 417
 snake, 536
 squamate reptile, 412
Jan, Georges, 38
Janzen, Daniel H., 48, 50
jaws
 lizard, 414, 415
 reptilian, 409
 salamander, 125f
 sea turtle, 755f
 snake, 535
 tadpole, 177
 turtle, 742, 744f, 745f
Jiménez, "Don Ricardo," 31
Journal of the American Academy of Natural
 Sciences of Philadelphia, 39
journals, 57
Juárez, Benito, 39

karyological analysis, 24
karyotypes, 98–99
Keferstein, Willhelm, 38
Keith, Minor C., 30, 39
keys, use of, 92, 94–95
Kezer, James, 45, 48
killing jars, 22–23
kingdoms, 8
Koller, O., 43
Krieger Publishing Company, 25

labeling of specimens, 20, 23, 24
Lacépède, B. G. E. L., 36
laminate folds, 418
Lankester, Charles H., 43
larvae
 amphibian, 105
 anuran, 158, 161, 169t, 173
 caecilian, 113, 114, 115
 description of, 97
 fixation, 23–24
 identification, 108f, 109f
 salamander, 124
Las Cruces Biological Station, 48, 52
La Selva Biological Station, 47, 48, 52
lateral-line organs, 111
laterosacral plicae, 236f
Laurenti, Joseph Nicholas, 36
leaves
 coral snakes in, 704
 raking of, 15–16
 sampling litter, 18
lectoparatypes, 8
lectotype, 8

Legler, John M., 45
legs. *See* limbs
Leib, Carl S., 47
Leidy, Joseph, 38
Le Lacheur, William, 29
length, units of, 11
León, Pedro, 48
Lichtenstein, Martin, 37
Lieberman, Susan S., 47
life histories, 98
    anuran, 162, 167, 172–73
    reptilian, 409
    salamander, 123
lighting, 13, 27
limbs
    amphibian, 105
    anuran, 158, 171, 214f, 232f
    caecilian, 113
    lizard, 414, 419
    reptilian, 410
    salamander, 121
    turtle, 740–41, 742
lines, coloration, 97
Linné, Carl von, 35, 36
Lips, Karen R., 50
liquid, units of, 11
listening, observing wildlife, 11–12
literature, 54–57
local residents, help from, 20
locomotion, 98
    anuran, 158, 160–61
    caecilian, 114
    crocodilian, 773–74
    lizard, 416–17
    reptilian, 411
    salamander, 122
    snake, 537
    turtle, 740–41
logs, rolling of, 15–16
Lomonte, Bruno, 46
longevity
    crocodilian, 774
    lizard, 422
    snake, 542
    turtle, 741
López, Myrna, 46
loreal pits, 714
lower montane bioclimates, 77–79
lower montane moist forest, 87–88
lower montane rainforest, 89, pls. 14, 15
lower montane wet forest, 88–89
lowland Atlantic moist forest, 85–86, 783–84
lowland Atlantic wet forest, 86, 783–84
lowland bioclimates, 77, 783–84
lowland dry forest, 831–35, pls. 4, 5, 6
    habitat, 83–84, 783–84
    herpetofauna, 831–35
lowland moist forest, pl. 6
lowland Pacific-Atlantic moist forest, 84–85, 783–84
lowland Pacific wet forest, 85, 783–84, 794, 796
lowland wet forest, pls. 8, 9, 10

lungs
    reptile, 411
    turtle, 739

magnesium carbonate, 23–24
mangrove swamps, 81, pl. 3
Marcgraf, Georg, 35
marine habitats, 80, 791t
marshes, 80–81
Martens, Carl von, 37
maternal investment, 166
Mayan block, 815, 817, 817f, 828
McDiarmid, Roy W., 47, 48
measurement, 11, 95–96
mechanoreceptors, 111, 411
Meiggs, Henry, 30
Meriam, C. Hart, 75–76
Mesoamerica. *See also* Tropical Meso-
    america
    events leading to, 827–28
    tectonic features, 818f, 819f, 820f
metacentric chromosomes, 98–99
metachrosis, 416
metamorphosis, 169. *See also* tadpoles
metatarsal tubercles, 174f
Middle American Elements, 811t
migrations, 98
*Mission scientifique au Mexique et dans
    l'Amérique Centrale* (Bocourt), 39
Mocquard, François, 39
moist forest, 77
moisture. *See also* water balance
    dehydration, 110
    ecological patterns, 792–93
    and housing of specimens, 25–26
    maintenance of, 12
    and transporting specimens, 20
molecular analysis, 24
molluscivory, 98
Monge, Carlos, 33
Monge, Julian, 48
monophyletic groups, 6
monsoons, 70
montane bioclimates, 79, 830–31, 832f
montane rainforest, 89–90, pls. 16, 18
Monteverde Forest Reserve, 48, 52
Mora Porras, Juan Rafael, 29–30, 37
morphology. *See also* particular species
    anuran, 170–73, 171f, 174f, 175f, 176f
    chromosomes, 98–99, 99f
    species accounts and, 96
    and tadpole identification, 176–83,
        177f
    toad, 194f
moss banks, 18
mountains, Mesoamerican, 820–22
mouth
    anuran, 174f, 215f
    caecilian, 115
    crocodilian, 772
    lizard, 423f
mouthparts, tadpole, 177f, 178f
Muños, Federico, 49
Museo Nacional de Costa Rica, 40

names
    common, 95
    Latin, 8
    relationships and, 7–11
    scientific, 8, 95
    Spanish, 5–6
National Liberation Party, 31
national parks, 32, 33, pl. 1
natural classification, 7, 7f
natural reserves, 32, 52
necks, lizard, 507f
Nembutal, 23
neotrophs, 169t
neotype, 8
nest building, anuran, 165t
nesting beaches, 12
nests, amphibian, 106
nets, 16–17, 17f
NF. See *nombre fundamental*
Nicaragua, 32, 34
nictitating membranes
    crocodilian, 773
    lizard, 416
    turtle, 740
nidicolous development, 169
night, observations, 12
nombre fundamental (NF), 99
nomenclature
    relationships and, 7–11
    Spanish names, 5–6
noosing, 16, 17f
nortes, 73–74
notebooks, 25
nuptial pads, anuran, 171
nutrition, 27, 169t

observations
    from captive husbandry, 25–28
    field notes, 25
    nocturnal, 12
    in wild, 11–12
oceanic crusts, 815t
oceans, heat storage, 66
ocelli, 96
odontoids, anuran, 172
Oduber, Daniel, 32, 47, 52
olfactory systems
    amphibian, 111
    crocodilian, 773
    lizard, 417
    snake, 536
    turtle, 740
omnivory, 98
On the Batrachia and Reptilia of Costa Rica
    (Cope), 39
oophagy, 98
ophiophagy, 98
oral disks, tadpole, 177, 177f, 178f
oral flaps, tadpole, 178f
orders, definition, 8
Organization for Tropical Studies, 45–49
organizations, herpetological, 57–58
Orgeix, Christopher d', 49
Orlich, Francisco J., 44

Ortega, Daniel, 32, 33
Ortega, Humberto, 32
orthotrophs, 169t
Osgood, Henry F., 38
osmoregulation, 415. *See also* water balance
osteoderms, 772
osteological preparations, 24
OTS. *See* Organization for Tropical Studies
oviparity, 541
ovoviviparity, 166

Pacific lowland moist forest, pl. 6
Pacific lowland wet forest, pl. 8
packing specimens, 24–25
paleoclimates, 822–25, 830–31
paleogeography, 815–22
palmar tubercles, 235f
Palo Verde National Park, 48
Panamanian portal, 828–30
Pangaea, 815, 817, 825
paraphasmata, 418
parasites, 28
paratype, definition, 8
parental care, anuran, 167
parotid glands, 170–71, 193f
Paso de Desengaño, 88
pastures, as habitat, 83
pectoral girdle, snake, 535
pelvic girdle, lizard, 414
pelvis, lizard, 414
penes
    crocodilian, 772
    snake, 539–40, 539f
    turtle, 738, 741
peponotrophs, 169t
permits, specimen collection, 13
PET. *See* potential evapotranspiration
petala, lizard, 418
Peters, Wilhelm, 37, 38
phalanges. *See* digits; feet; hands
photography, 12–13
photoreception
    crocodilian, 773
    lizard, 416
    reptilian, 411
    snake, 537
    turtle, 740
phylum, definition, 8
phylogenetic relationships
    anuran, 159, 159f
    batugarine turtle, 826f
    bolitoglossine salamander, 826f
    crocodilian, 773f
    lizard, 415
    Paleogene, 826f
    reptilian, 410f
    snake, 536f
    turtle, 739f
phylogenetics, 6
physiography, 60–63
Picado, Clodomiro, 41, 43
piscivory, 98
Pittier, Henri, 40
placentas
    chorioallantoic, 413f

gill, 166, 293
omphalallantoic, 413f
squamate reptile, 413f
plant formations
    anuran distribution and, 785–86t
    caecilian distribution and, 784t
    crocodilian distribution and, 790t
    lizard distribution and, 787t
    major taxa living in, 790t
    salamander distribution and, 784t
    snake distribution and, 788–89t
    turtle distribution and, 790t
plastrons. *See also* carapaces
    coloration, 765f, 770f
    turtle, 743f, 745f
plicae. *See also* folds
    laterosacral, 236f
    lizard, 418
    suprascapular, 236f
pluvial paramo, 79, 90, pl. 17
poison glands, 176f
polychromism, 227–29, 228f, 229f
polymorphic variation, 92, 227–29, 228f
polyploidy, anuran, 166
ponds, temporary, 81–82
populations
    characteristics, 98
    cohesive, 10
    intraspecies variation, 92
posture, anuran, 160
potential evapotranspiration, 75–76, 792–93
Pounds, J. Alan, 49, 50
precipitation, ecological patterns, 792–93. *See also* rainfall
precloacal pores, 417, 419–20, 420f
predation, 98, 537
predator avoidance, 98, 160–61
premontane bioclimates, 77
premontane moist forest, 86
premontane rainforest, 87, pls. 12, 13
premontane wet forest, 86–87, pl. 11
preservatives, transfer to, 24
Preston, Andrew W., 30
pupil shapes
    anuran, 170, 175f
    lizard, 423f
    snake, 544f

races, 11
rainfall. *See also* precipitation, ecological patterns
    geographic distributions, 101
    regimens of, 74
    rivers and, 63
    seasonal patterns, 72f
rainforests, 77, 79
    lower montane, 89, pls. 14, 15
    montane, 89, pls. 16, 18
    premontane, 87, pls. 12, 13
    vegetation, 83
rakes, potato, 14
Rand, A. Stanley, 48
record keeping
    captive husbandry, 27–28

field notes, 13
notebooks, 25
recording calls, 20
refuges, 52
reproductive behavior, 98
    amphibian, 105–7
    anuran, 161, 162, 163–64, 165t, 166–69
    caecilian, 113, 115
    clasping, 170
    crocodilian, 774
    lizard, 418
    salamander, 123
    snake, 539, 541
Reptiles (journal), 25
Reptilia (journal), 25
respiration. *See also* gas exchange
    amphibian, 105, 110
    caecilian, 113
    salamander, 123
    turtle, 739
reticulum, coloration, 96
Revista de Biología Tropical (journal), 45
ribs
    reptilian, 410
    turtle, 739
Ridgeway, Robert, 41
rings, coloration, 97
riparian forest, 83
rivers, 62–63, 82
road running, 18–19
Roberts, Wendy E., 49
Robinson, Douglas C., 38, 46, 47, 48, 49, 50
Rodríguez, Manuel Ángel, 34
Rodríguez, Rafael Lucas, 44, 45
Rogers, H., 39
Rojas, Gustavo, 46, 50
rubber-band guns, 19
Rüppell, Eduard, 37

Sack, Count Graf von, 36
The Salamanders of the Family Plethodontidae (Dunn), 42
salt glands, 415, 773
salt marshes, 80–81
Salvin, Osbert, 39
Sandinistas, 32–33
Santamaría, Juan, 29
Santana, Manuel, 49
Santa Rosa Park, 48
Sasa, Mahmood, 50
Savage, Jay, 49
saxicolous habits, 98
Say, Thomas, 36
scalation. *See also* scales
    anole, 451f
    blind burrowing snake, 544f
    gecko, 483f
    lizard, 418–22, 420f, 421f, 422f, 425f, 507f, 508f, 527f, 530f
    Mesapis, 423f
    postocular, 527f
    sea turtle, 743f, 755f
    skink, 424f
    snake, 542–43, 542f, 545f, 546f, 547f, 548f, 551f, 560f

Sphaerodactylus, 493f, 494f
subocular, 527f
superciliary, 425f
turtle, 739f
scale pits, 543
scales, 410, 420f. *See also* scalation
Schiede, Christian, 36
Schlegel, Hermann, 36
Schmid, Michael, 50
Schmidt, Karl Patterson, 42
Schmidt, Oscar, 37
Scott, Norman J., 46
sea levels, 823f, 829
seasonal cycles, 72–74, 72f, 97
Seba, Albertus, 35
Seebach, Karl von, 38
semiaquatic habits, 98
semifossorial habits, 97–98
sense organs, 111. *See also specific systems*
Serpientes venenosas de Costa Rica (Picado), 43
Serpientes venenosas y ofidismo en Centro- américa (Bolaños), 49
sex determination, turtle, 741
sexual dimorphism, 92
    anuran, 162, 173
    crocodilian, 774, 775
    lizard, 417–18
    salamander, 126
    snake, 539–41
    turtle, 741
shedding, 418, 541
shells. *See also* carapaces; eggs; plastrons
    egg, 411
    reptilian, 410
    snake egg, 541
    turtle, 738, 739f, 741–42
shipping, of specimens, 24–25
shooting, 19
sigatoka disease, 32
size
    anuran, 172
    crocodilian, 775
    lizard, 421–22
    salamander, 121, 125
    snake, 544
    in species accounts, 95–96
    turtle, 742
skin
    alkaloids in, 188
    amphibian, 110–11
    bufodienoloids in, 186
    bufotoxins in, 192
    chiriquitoxin in, 188
    frog, 159–60
    reptilian, 410
    salamander, 121, 124
    shedding, 418
    snake, 541
    tetrodotoxin in, 186
    toad, 159–60, 194f
    toxins in, 170
skull
    crocodilian, 772
    kinetic, 412

lizard, 414
salamander, 125f
sea turtle, 754f
turtle, 738
slingshots, 19
slugs, as food, 27
Smith, Hobart Muir, 42, 43
snakebite, 21–22, 536–37
    coral snake, 707
    and posing for photography, 13
    sea snake, 713
    viper, 713–14
Snakes (Greene), 49
snake sticks, 15, 16f
snake tongs, 15, 16f
Snake Venom Poisoning (Russell), 49
social interactions, 98
soils, 79, 80f, 84
solar radiation, 27, 65–67, 66f, 67f
Solózano, Alejandro, 48, 49, 50
Somoza Debayle, Luis, 32
Somoza García, Anastasio, 32
Soto, Bernardo, 30
South American Elements, 812t
Southwestern North American Components, 810t
Spanish personal names, 5–6
species
    definition, 6, 8
    historical source units, 801–6, 804–5t
    nomenclature, 9
    similarities, 97
    subspecies and, 10–11
    terminology used, 10
    variations, 92
species accounts, components of, 95–102
Species Plantarum (Linné), 35
specimens
    collecting, 13–21
    euthanizing, 22–23
    fixing, 23–24
    handling, 91
    housing, 25–27
    identifying, 91–93
    labeling, 24
    packing, 24–25
    posing for photography, 13
    preserving, 22–25
    shipping, 24–25
    transporting, 20–21
spines
    thumb, 215f
    turtle, 746f
spots, definition of, 96
Standard Fruit Company, 46
Starrett, Priscilla H., 47
State of the World, 2001 (Mattoon), 52
Stejneger, Leonhard H., 41
stones, turning of, 15–16
Straughan, Ian R., 47
streams, 82
stripes, coloration, 97
subalpine bioclimates, 79
subalpine pluvial paramo, 90, pl. 17
subarticular tubercles, 235f, 290f

subclavian arteries, 772
submetacentric chromosomes, 98–99
subtelocentric chromosomes, 98–99
suckers, abdominal, 178f
superciliary tubercles, 231f
supranasal scales, 493f
supraocular tubercles, 231f
suprascapular plicae, 236f
Supreme Electoral Tribunal, 31
surveys, visual, 12
swamps, 81, 82
sympatic entities, 10
syntype, 8
Systema Naturae (Linné), 35, 36
systematization, 6

tadpoles
    ecomorphological guilds, 168f, 168t
    highly derived, 169f
    identification, 108f, 176–83
    index to published illustrations, 181–82
    photographing, 13
    predation on eggs, 167
    principal types, 167–69, 167f
    Rhinophrynus, 184f
    transport of, 20
tail autonomy. *See* urotomy
tail breakage. *See* urotomy
tails. *See also* caudal morphology; urotomy
    gecko, 483f
    lizard, 419
    salamander, 124
Talbot, James J., 47
tarsal folds, 173f
tarsal tubercles, 173f
taxa
    subgroupings, 8
    systematic groups, 93–94
taxagenetics, 6
taxon, definition, 6
taxonomy, definition, 6
Taylor, Edward Harrison, 42, 43–44, 46
Taylor, Richard C., 43
teeth
    anuran, 158, 172
    caecilian, 113
    crocodilian, 772
    lizard, 414
    reptilian, 410
    salamander, 121, 125, 125f
    snake, 537
    squamate reptile, 412
telephoto lenses, 12
telocentric chromosomes, 99
temperature. *See also* altitude; body temperature
    ecological patterns and, 784, 790–92
    paleoclimactic cycles, 832f
    and specimen housing, 27
    units of, 11
terminology, systematic, 6
terrariums, 26
terrestrial habitats, 82–90, 791t
terrestrial habits, 98

territorial behaviors
    anuran, 162
    crocodilian, 773
    lizard, 418
    snake, 541
tetrodotoxin, 186, 188
thenar tubercles, 235f
thermal equator, 68f
thermal tolerance, 109–10
thermoreceptors, 542, 714
thermoregulatory behaviors, 416, 774. *See also* body temperature
throat region, 419, 507f
thumb spines, 215f
thunderstorms, 69, 70
Tinoco, Federico, 31
Tinoco, Joaquín, 31
toes. *See* digits; feet
tongues, 123, 412. *See also* chemoreceptors; olfactory systems
tourism, 33
Townsend, Charles H., 41
toxicity, infrared reflectance and, 276–77
toxins, anuran, 170. *See also* venom; *and particular toxins*
track analysis, 802–6, 803f, 804–5t
trade winds, 70, 73
trails, visual surveys, 12
transpiration, 75–76
transport, specimen, 13–21, 20–21
trapping, 17–18
Treub, Linda, 45
tribes, local, 28
Tristán, J. Fidél, 38, 41, 43
tropical dry forest, 84
Tropical Fish Hobbyist, 25
Tropical Mesoamerica, 834f. *See also* Mesoamerica
    amphibian distributions in, 795t
    anuran clades, 808t
    anuran distributions in, 798t
    caecilian distributions in, 797f, 797t
    dispersal events, 825–35
    evolution and, 828f
    herpetofauna in, 795–801, 801t
    lizard distributions in, 798–99t
    reptile distributions in, 795t
    salamander distributions in, 797f, 797t
    snake distributions in, 799–800t
    track analysis, 802–6
    turtle distributions in, 801t
tropical montane herpetofauna, 830–31
tropical premontane moist forest, 86
tropical wet forest, 85
tropics
    classification of climates, 75–76
    heat transport, 70f
    latitudinal limits, 64f
tubercles
    metatarsal 174f

palmar, 235f
    subarticular, 235f, 290f
    superciliary tubercles, 231f
    supraocular tubercles, 231f
    tarsal tubercles, 173f
    thenar tubercles, 235f
tympanum, 170, 419, 740
types, relationships and, 7–11

Ugalde, Alvaro, 32, 52
Ulate, Otilio, 31
ultraviolet radiation, 27
Underwood, Cecil F., 40
United Fruit Company, 30, 40, 42
Universidad de Santo Tomás, 30
uplands
    Mesoamerican, 820–22, 821f, 822f
    principal areas, 821
urotomy. *See also* tails
    lizard, 418
    snake, 541
    squamate reptile, 413
Utisols, 79

Vaillant, Léon-Louis, 39
Valerio, Juvenal, 43
Valerio, Manuel, 43
Valerio, Romulo, 43
Vandelli, Domenico, 36
Van Devender, R. Wayne, 47
Van Patten, Charles, 39
variation
    coloration, 315f
    continuous, 92
    geographic, 10–11, 92
    individual, 92
    intraspecific, 92–93
    polychromatic, 92
    polymorphic, 92, 227–29, 228f
    temporal, 92
vegetation
    bioclimates and, 69f
    distribution, pl. 2
    ecological patterns, 783–84
    habitats and, 79–90
    Miocene distribution of, 824f
    Neocene distribution of, 824f
    paleoclimates and, 822–25, 830–31
venom. *See also* snakebite
    lizard, 417
    snake, 537–38
The Venomous Reptiles of Latin America (Campbell and Lamar), 50
ventilation, and specimen transport, 20
ventral disks, anuran, 171
vermivory, 98
Vesco, Robert, 32
Vial, James L., 45
vicariances, 825–35, 827f
Villa, Jaime, 45, 46

Viquez, Carlos, 43
vision, 416
    crocodilian, 773
    lizard, 416
    snake, 536, 552
    turtle, 740
vitamin D, 27
viviparity, 169t, 541
vocalizations
    anuran, 161–64, 163f
    crocodilian, 773, 774
    identification using, 92–93
    learning, 20
    lizard, 416
    locating wildlife using, 12
    male, 92–93
    turtle, 740
vocal sacs, anuran, 171, 176f
voice, description of, 97
volcanoes, 820–22
vomeronasal organs
    amphibian, 111
    caecilian, 114
    lizard, 415
    snake, 536
    squamate reptile, 412

Wagler, Johann G., 36
Wake, David B., 47, 48, 49, 50
Wake, Marvalee H., 47
Walker, William, 29, 37
Warszewicz, Josef, 37, 38
water, and housing of specimens, 27
water balance. *See also* dehydration; moisture; osmoregulation
    crocodilian, 773
    snake, 536
    turtle, 739
webbing
    anuran, 172, 176f
    Bufo, 194f
    crocodilian, 772
    Eleutherodactylus, 231f
    lizard, 414
    treefrog, 289f
weight, units of, 11
Wertstein, Otto, 43
wet forest, 77, 79
Wied-Neuweid, Maximilian von (Prince), 36
Wiegmann, Arned, 36
World Bank, 34

Ximara, R., 43

A Year of Costa Rican Natural History (Calvert and Calvert), 41

Zeledón, José Castulo, 38, 39
zoological parks, 25
Zug, George R., 47